A TEXTBOOK OF
ENGINEERING PHYSICS

[For the Students of B.E., B.Tech., B.Arch., B.Sc., (Engg.)]

Dr. M.N. Avadhanulu
M.Sc., Ph.D.
Former Professor and Head, Department of Physics
Kavikulguru Institute of Technology & Science
Ramtek, Dist. Nagpur (M.S.)

Dr. P.G. Kshirsagar
M.Sc., Ph.D.
Formerly Head of the Department of Applied Physics,
Visvesvaraya National Institute of Technology
Nagpur

Dr. TVS Arun Murthy
M.Sc., Ph.D.
Professor in Physics, HoD, S&H,
Vice Principal Joginpally BR Engineering College,
Moinabad, Hyderabad

S Chand And Company Limited
(ISO 9001 Certified Company)

S Chand And Company Limited
(ISO 9001 Certified Company)
Head Office: D-92, Sector–2, Noida – 201301, U.P. (India), Ph. 91-120-4682700
Registered Office: A-27, 2nd Floor, Mohan Co-operative Industrial Estate, New Delhi – 110 044, Phone: 011-49731800
www.**schandpublishing.com**; e-mail: **info@schandpublishing.com**

Marketing Offices:

Chennai : Ph: 23632120; chennai@schandpublishing.com
Guwahati : Ph: 2738811, 2735640; guwahati@schandpublishing.com
Hyderabad : Ph: 40186018; hyderabad@schandpublishing.com
Jalandhar : Ph: 4645630; jalandhar@schandpublishing.com
Kolkata : Ph: 23357458, 23353914; kolkata@schandpublishing.com
Lucknow : Ph: 4003633; lucknow@schandpublishing.com
Mumbai : Ph: 25000297; mumbai@schandpublishing.com
Patna : Ph: 2260011; patna@schandpublishing.com

First Edition 1992
Subsequent Editions and Reprints 1993, 94, 95, 98, 2000, 2001 (Twice), 2003, 2004, 2005, 2006 (Twice), 2007, 2008, 2009 (Twice), 2010, 2011 (Twice), 2012, 2014 (Twice), Reprint 2015, 2016
Revised Edition 2019; Reprints 2019 (Thrice), 2020 (Twice), 2021 (Thrice), 2022 (Five times), 2023, 2024 (Twice)

Reprint 2025 (Twice)

ISBN: 978-93-528-3399-3 **Product Code:** H3EPH64PHYS10ENAK19R

PRINTED IN INDIA

By Shri Shyamlal Printing Press Private Limited, Plot 20/4, Site-IV, Industrial Area Sahibabad, Ghaziabad – 201 010 and Published by S. Chand And Company Limited, A-27, 2nd Floor, Mohan Co-operative Industrial Estate, New Delhi – 110 044.

Dedicated to

My Mother,
Maternal Uncles
Shri Mullapudi Suryanarayana,
Dr. Mullapudi Subba Rao and
Shri Mullapudi Satyanarayana,
and to
My Wife Suwarchala

Books are not paper and words but interaction with thinkers on a one-to-one basis, not of one generation but separated by hundreds and thousands of years

—Thomas Carlyle

Preface to the Eleventh Edition

"The value of a college education is not the learning of many facts but the training of the mind to think." – ALBERT EINSTEIN. Keeping this as our motto, we have totally revised "A Textbook of Engineering Physics". The content is strengthened at every juncture required and as many as 17 new chapters have been added to bring completeness to the subject. It may be used as a textbook or as a reference book by students and teachers for a course in Engineering Physics. The book plays the role of a source of information as well as presents the principles of physics in an easy-to-understand style. The expectations of teachers/readers and their requirements are given top-priority and the material is forged in a friendly style without sacrificing the spirit of physics at any stage. Physics and other sciences evolved by and by through the efforts of rational thinkers interested in knowing the why, what and how of natural phenomena. Engineering has emerged as an outshoot of application of their knowledge for the benefit of society at large. Thus, modern technology stands on the elaborate foundation of Physics and other sciences. The main purpose of teaching Physics to Engineering undergraduates is to leave at the end the budding engineers with a cogent appreciation of the necessary principles.

For the sake of convenience, essential material is arranged in 50 Chapters. The fundamental concepts are emphasized in each chapter and the details are developed therein. Each chapter is divided into smaller parts and sub-headings are provided to make the reading a pleasant journey from one interesting topic to another important topic.

The philosophy of presentation of the material in the book is based upon decades of classroom interaction of the authors. In each chapter, thc fundamental concepts pertinent to the topic are highlighted and the in-between continuity is emphasized. Throughout the book attention is given to the proper presentation of concepts and practical applications are cited at each opportunity to highlight the engineering aspect. A large number of problems are worked-out. Problems are included in order to expedite the learning process of students of all hues to improve their academic performance. The authors sincerely hope that this book will assist the student in learning the principles of physics much more effectively.

The authors solicit the Professors in various engineering institutions to extend unstinted support to this latest edition too as that offered to the former editions.

M.N.AVADHANULU
avadhanulu2014@gmail.com

Preface

"A Textbook of Engineering Physics" is written with two distinct objectives : to provide a single source of information for engineering undergraduates of different specializations and provide them a solid base in physics. Successive editions of the book incorporated topics as required by students pursuing their studies in various universities. In this new edition the contents are fine-tuned, modernized and updated at various stages.

Physics is not an isolated body of theories which merely serve vocational usefulness. What has been achieved in physics has sooner or later made tremendous impact on the technological growth of our society. To become active participants in the technological revolution, one has to necessarily acquaint himself with the methods of science. Mechanical memorizing of certain definitions and derivations does not belong to the method of science and as such is of little value to the student. The main purpose of teaching physics to engineering undergraduates is to equip them with an understanding of the "scientific method", so that they may use the training beneficially in their higher pursuits. An earnest attempt is made in this direction right from the first edition of this book by blending careful presentation of fundamental concepts and methods of physics.

This edition retains the original theme of emphasis on concepts with less mathematical formalism. The practical applications are discussed at each stage. The question bank given at the end of each chapter is updated. At a number of places, points for refinement are noticed and those have been incorporated. We have gladly received and carefully considered suggestions from professors and students who have used earlier editions. Further suggestions for improvement of the quality and quantity of the content are most welcome.

M.N.AVADHANULU
avadhanulu2014@gmail.com

Acknowledgements

The authors offer their special thanks to S Chand And Company Limited for all their efforts in bringing out this book nicely and in time.

M.N.AVADHANULU
avadhanulu2014@gmail.com

Contents

1 Oscillations and Waves

Learning Objectives:

By the end of this chapter, you will be able to:

- Understand basic concepts of Oscillations.
- Learn concept of Simple harmonic motion.
- Explain about different types of Oscillations – Free, Damped, Forced and Coupled.
- Describe about Waves and its different types.
- State principle of superposition specially superposition of two perpendicular SHMs.

1.1 INTRODUCTION

Oscillatory motion is a repeating motion, which occurs extensively in nature. Oscillations (vibrations) and waves pervade all sciences. We come across a number of general examples of oscillatory motion such as fluttering of tree leaves in a gentle breeze, swinging of a swing, the beating of the heart etc. Vibrations of atoms in a solid, electric and magnetic fields in light waves and radio waves, large structures swaying due to earth quake etc have an oscillatory nature. Electrical and mechanical oscillations as well as vibrations in structures are every day topics in the world of engineering. A repeating and periodic oscillation moving through a medium from one location to another give rise to a wave. Waves are also encountered extensively in sciences and technology. Waves on a string, sound waves, radio waves, microwaves, light waves, and earthquake waves are a few of the examples. Therefore, the study of oscillations and waves constitutes the core topic in engineering and technology.

1.2 OSCILLATIONS

We all know that when a *constant force* acts on a body, the body moves with a *constant acceleration*. Rectilinear motion and uniform circular motion are examples of such a motion. However, when the force acting on the body varies in time, the acceleration of the body will also change with time.

Def: Oscillatory motion is one type of motion that can occur when a body is subjected to a force that varies in time.

The first systematic observations of oscillations were made by Galileo, who had observed certain rhythm in the swinging of chandeliers in the cathedral at Pisa. He determined the time taken for the oscillations with the help of his pulse. He found to his surprise that the time taken by each oscillation was constant. Later, this property of constant time period for oscillations was exploited in making pendulum clocks. Fig. 1.1 shows the illustration of oscillatory motion.

Fig. 1.1: Illustration of Oscillatory motion.

A pendulum serves as the simplest example of an oscillating body (Fig. 1.2). When a pendulum bob is at the lowest position, the pendulum is in a *state of rest* or it is said to be in its *equilibrium position*. The forces that act on the pendulum bob are the force of gravity and the force of tension in the string. When the bob is moved from this equilibrium position, the gravitational force gives rise to a vertical component, directed along the string and a tangential component, directed at right angles to the string (see Fig. 1.8). The tangential component pulls the bob continually back to the equilibrium position and therefore it is known as the ***restoring force***. The bob will not come to the state of rest immediately. When the bob reaches the midpoint, the restoring force vanishes.

However, the ***inertia*** property causes the bob to overshoot the equilibrium position and the motion continues. Once again, the restoring force comes into action but with a change in its direction. The restoring force, which is maximal at the extreme positions of the bob, stops the bob and pulls it back toward the equilibrium position. The result is a continuing *oscillatory motion* of the bob back and forth along an arc. Thus, the constant play between the restoring force and inertia property is responsible for the oscillatory motion. The oscillatory motion is ***periodic*** and repeats itself in equal intervals of time.

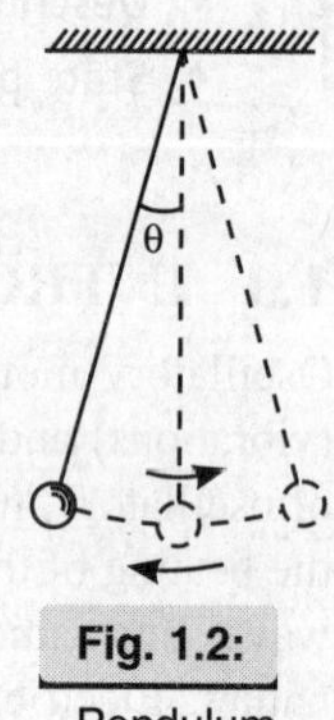

Fig. 1.2: Pendulum.

In general, an ***oscillation*** is a periodic fluctuation in the value of a physical quantity above and below some equilibrium value. In mechanical oscillations, the body undergoes linear or angular displacement whereas non-mechanical oscillations involve the variation of quantities such as voltage or current in electrical circuits or the electric and magnetic fields in TV signals, light waves, UV-rays and X-rays.

1.3 SIMPLE HARMONIC MOTION

Any motion, which repeats itself at regular intervals according to a sinusoidal law, is called a ***harmonic motion***. The oscillations of a simple pendulum or the motion of a mass m under a restoring force is an idealized model of harmonic motion (Fig. 1.3).

Def: Simple harmonic motion is an oscillatory motion in which the force is directly proportional to the displacement. S.H.M. is a *non-uniformly accelerated motion*.

Since force is proportional to the displacement, the acceleration is not constant but varies with time. Hence the equations of motion with constant acceleration are not applicable to simple harmonic motion.

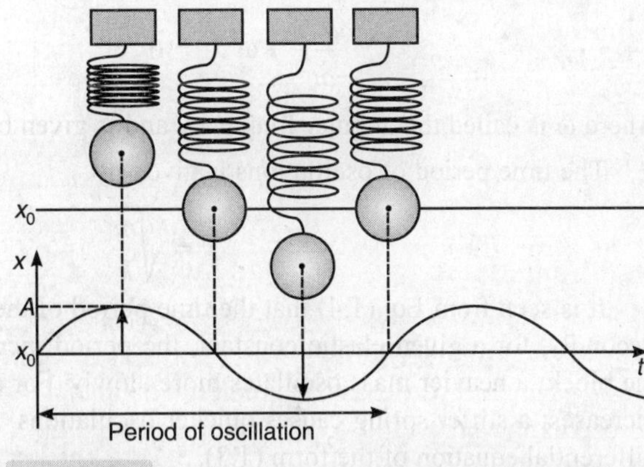

Fig. 1.3: Sinusoidal law - harmonic motion.

1.3.1 Equation of Simple Harmonic Motion

We now obtain the expressions for displacement, velocity, and acceleration of a body moving with simple harmonic motion.

Let us consider a block of mass *m* attached to a spring (see Fig. 1.4). When the mass is pulled and left to it, it oscillates about its equilibrium position. The directed distance of the mass from its equilibrium position is called its **displacement**. The restoring force *F* acting on the body is due to the *stiffness* of the spring and is given by Hooke's law.

$$F = -kx \qquad ...(1.1)$$

where *x* is the displacement from the equilibrium position and *k* is called the elastic constant.

Elastic constant represents the force required to displace the mass one unit of distance. The negative sign in Eq. (1.1) indicates that the force **F** is opposite to the displacement. When the mass is pulled to right, the spring gets stretched. Then, *x* is positive and the force is negative and is directed to the left. When *x* is negative, the spring is compressed and **F** is positive and directed to the right.

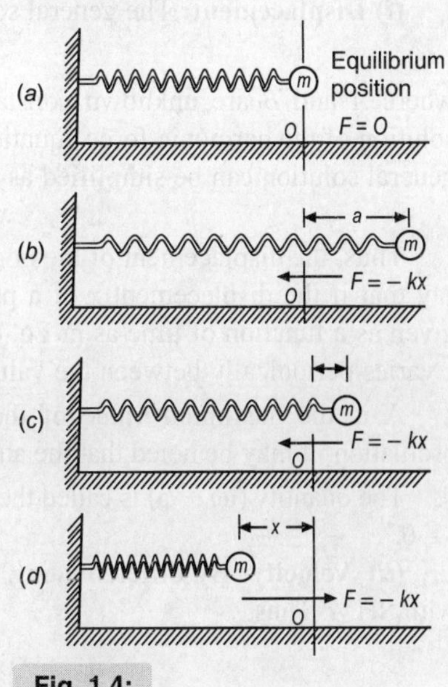

Fig. 1.4: Block of mass *m* attached to a spring.

From Newton's second law, the restoring force gives acceleration. Thus,

$$\underset{\text{(Restoring force)}}{F = -kx} = \underset{\text{(Inertial force)}}{ma}$$

$$\therefore \qquad m\frac{d^2x}{dt^2} = -kx$$

$$\text{or} \qquad \frac{d^2x}{dt^2} + \frac{k}{m}x = 0 \qquad ...(1.2)$$

Eq. (1.2) is merely another way of writing Newton's second law and it is known as the *differential equation of simple harmonic motion*.

Putting $k/m = \omega^2$ into the above equation, we get

$$\frac{d^2x}{dt^2} + \omega^2 x = 0 \quad ...(1.3)$$

where ω is called the angular frequency and is given by $\omega = \sqrt{\frac{k}{m}}$.

The time period of oscillations is given by

$$T = \sqrt{\frac{m}{k}} \quad ...(1.4)$$

It is seen from Eq. (1.4) that the time period of the mass is independent of the amplitude. Secondly, for a given elastic constant, the period increases with the increase in the mass of the block; a heavier mass oscillates more slowly. For a given mass, the period decreases as k increases; a stiffer spring causes quicker oscillations. All simple harmonic oscillators obey a differential equation of the form (1.3).

1.3.2 Characteristics of SHM

***(i)* Displacement:** The general solution of the Eq. (1.3) is given by

$$x = Ae^{i\omega t} + Be^{-i\omega t} \quad ...(1.5)$$

where A and B are unknown constants to be determined from the initial conditions. This solution of the harmonic force equation is known as the exponential form of the solution. The general solution can be simplified as

$$x = A \sin(\omega t + \varphi) \quad ...(1.6)$$

Thus, the displacement of the body at any instant is given by Eq. (1.6). Therefore, we can say that if the displacement x of a particle relative to the origin of the coordinate system is given as a function of time as in Eq. (1.6), then the motion is simple harmonic. The values of x varies periodically between the values –A and +A.

A is the maximum value of the displacement and is called as the **amplitude** of the oscillation. It may be noted that the amplitude A is constant.

The quantity $(\omega t + \varphi)$ is called the *phase angle* and φ is the initial phase, that is the phase at $t = 0$.

***(ii)* Velocity:** By differentiating Eq. (1.6) we can find the velocity of a particle moving with SHM. Thus,

$$\upsilon = \frac{dx}{dt} = \omega A \cos(\omega t + \varphi) \quad ...(1.7)$$

υ varies periodically between the values $+\omega A$ and $-\omega A$. When the amplitude is a maximum, the velocity is zero; and when the displacement is least that is at the midpoint of the motion, the velocity has its greatest magnitude.

***(iii)* Acceleration:** We can find the acceleration of the oscillating particle by differentiating the expression for υ. Thus,

$$a = \frac{d\upsilon}{dt} = -\omega^2 A \sin(\omega t + \varphi) \quad ...(1.8)$$

Acceleration varies periodically between the values $+\omega^2 A$ and $-\omega^2 A$. We can combine Eq. (1.8) and Eq. (1.6) to give

$$a = -\omega^2 x \quad ...(1.9)$$

Acceleration is proportional and opposite of the displacement in simple harmonic motion. When the displacement attains a maximum positive value, the acceleration is negative and vice versa. While the displacement is zero, acceleration is as well zero.

Basing on the Eq. (1.1) and Eq. (1.9), we can make a general statement that *whenever the force acting on a body is linearly proportional to the displacement and in the opposite direction, the body will execute simple harmonic motion.* Fig. 1.5 illustrates displacement x, velocity υ, and acceleration a as functions of time.

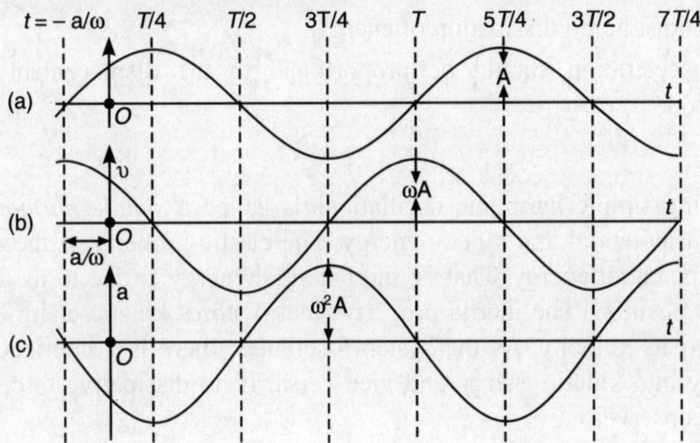

Fig. 1.5:

Displacement x, velocity v, and acceleration a as functions of time.

***(iv)* Time period:** The time period T is the time taken for one complete oscillation. The sine function Eq. (1.6) repeats itself whenever the quantity in parenthesis increases by 2π. Thus, if we start at $t = 0$, $\phi = 0$ and the time T is given by

$$\omega T = 2\pi$$

$$\therefore \qquad T = \frac{2\pi}{\omega} = 2\pi\sqrt{\frac{m}{k}} \qquad ...(1.10)$$

Note that the period of motion is determined by the mass m and elastic constant k. It *does not* depend on the amplitude of the oscillation. It means for any given values of m and k, whether the amplitude is larger or smaller, the time for one complete oscillation is the same.

***(v)* Frequency:** The frequency, ν, is the number of complete oscillations per second. It is the reciprocal of the time period.

$$\nu = \frac{1}{2\pi}\sqrt{\frac{k}{m}} \qquad ...(1.11)$$

Note that the frequency and time period are independent of the amplitude A.

***(vi)* Phase:** The angle $(\omega t + \varphi)$ called the phase of the oscillation. It represents the state of the oscillation of the body by specifying *the position and direction of motion* of the body. The constant angle ϕ is called the phase constant, which is known uniquely by the initial displacement and velocity of the particle. The constant ϕ and amplitude A gives what the displacement was at time $t = 0$. The phase of the oscillation is useful in comparing the motions of two bodies (Fig. 1.6).

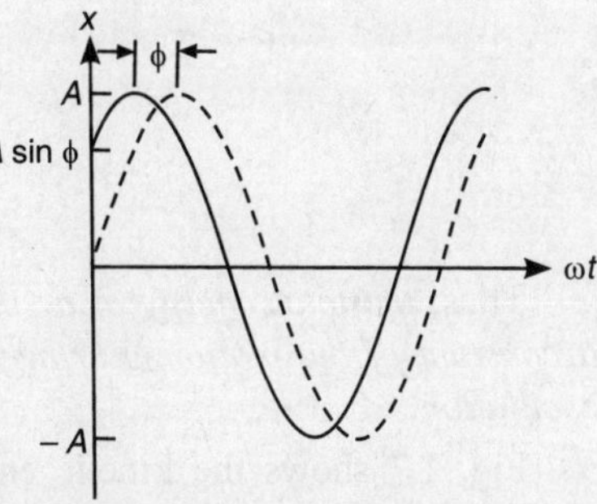

Fig. 1.6: Phase of oscillation.

1.3.3 Three Conditions for the Occurrence of Simple Harmonic Oscillations

In case of mechanical oscillators, three conditions must be satisfied for the occurrence of simple harmonic oscillations.

(*i*) There must be a position of stable equilibrium.

(*ii*) There must be no dissipation of energy.

(*iii*) The acceleration should be proportional to the displacement and opposite in direction.

1.3.4 Energy

A body executing simple harmonic oscillations is called a *simple harmonic oscillator* and it possesses both potential and kinetic energy. The elastic property of the oscillating system (spring) stores potential energy. That is, the potential energy is due to its displacement from the equilibrium position. The inertia property (mass) stores kinetic energy; thus, the kinetic energy is due to its velocity. As the system oscillates, there is a continuous conversion of potential energy into kinetic energy and vice versa. If no dissipative forces are present, the total energy is conserved

The kinetic energy of the particle is

$$E_k = \frac{1}{2}m\upsilon^2 = \frac{1}{2}m\omega^2 A^2 \cos^2(\omega t + \varphi)$$

$$= \frac{1}{2}m\omega^2 A^2[1 - \sin^2(\omega t + \varphi)]$$

$$= \frac{1}{2}m\omega^2 (A^2 - x^2) \quad \text{...(1.12)}$$

$$= \frac{1}{2}k\,(A^2 - x^2)$$

The potential energy, U, is given by

$$U = -\int_0^x F\,dx = \int_0^x kx\,dx = \frac{1}{2}k\,x^2 \quad \text{...(1.13)}$$

According to Eq. (1.13) the potential energy is proportional to the square of the displacement. It implies that the mass is in a **potential well** created by the spring. All simple harmonic oscillations are characterized by such a *parabolic potential well*.

The total energy of the simple harmonic oscillator is

$$E = K.E. + P.E.$$

$$= \frac{1}{2}k(A^2 - x^2) + \frac{1}{2}kx^2$$

or

$$E = \frac{1}{2}kA^2 = \text{constant.} \quad \text{...(1.14)}$$

Thus, *the total energy of a simple harmonic oscillator does not depend on time and is a constant of the motion. Further, it is proportional to the square of the amplitude of the oscillation.*

Fig. 1.7 shows the kinetic energy, potential energy and total energy of the oscillator plotted against the displacement x. The horizontal line represents the total energy E, which is constant and does not vary with x. The potential energy curve is parabolic with respect to x and is symmetric about the position of equilibrium, $x = 0$. The kinetic energy curve is also parabolic with respect to x and is symmetric about the position of equilibrium, $x = 0$. One curve is inverted with respect to the other, which indicates the 90° phase difference between the displacement and the velocity. The horizontal line intersects the potential energy curve at $x = -A$ and $x = A$, where the energy is entirely potential. At these points $v = 0$ and there is no kinetic energy. At the equilibrium position, $x = 0$ and $P.E. = 0$, so that the total energy is in the form of kinetic energy. At any value of x between $-A$ and $+A$, the vertical distance from the x-axis to the parabola is U; since $E = K.E. + P.E.$, the remaining vertical distance up to the horizontal line is $K.E.$ It is thus seen that energy is constantly being transferred between potential energy stored in the spring and the kinetic energy of the mass. Note that the maximum values of the potential and kinetic energies are equal.

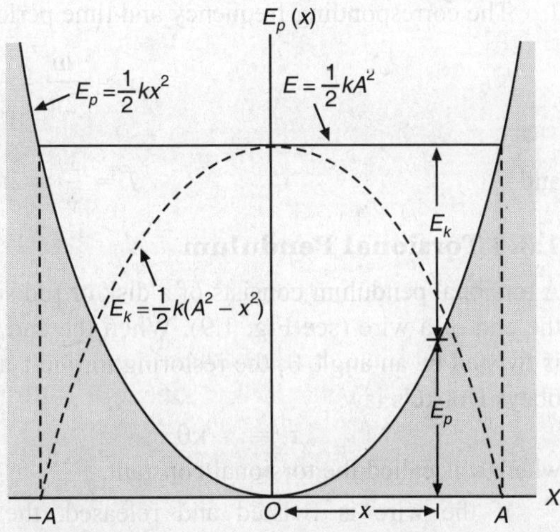

Fig. 1.7: Energy versus displacement.

We now look at two oscillating systems that exhibit simple harmonic motion, namely, a simple pendulum and a torsional pendulum.

1.3.5 The Simple Pendulum

A simple pendulum consists of a point mass, m suspended by a light string of length L. When the point mass is pulled to one side of its equilibrium position and is released, it oscillates about the equilibrium position by the force of gravity. The path of the point mass is an arc of a circle with radius L equal to the length of the string. In this case the restoring force must be proportional to x or to θ since $x = L\theta$. The forces acting on the mass are the tension, T acting along the string and the weight mg. Fig. 1.8 shows the forces on the mass in terms of tangential and radial components. Note that **F** always acts towards the equilibrium position and opposite to the displacement.

Fig. 1.8: Simple pendulum.

$$F = -mg\sin\theta \qquad ...(1.15)$$

Gravity gives the restoring force and the tension T makes the point mass move in an arc. As the restoring force is proportional to $\sin\theta$, the motion is not simple harmonic..

But, for small value of θ, $\sin\theta \approx \theta$, then

$$F = -mg\,\theta = -\frac{mg}{L}x. \qquad ...(1.16)$$

The force constant is $\quad k = mg/L. \qquad ...(1.17)$

From Eq. (1.17), the angular frequency is $\omega = \sqrt{k/m} = \sqrt{\dfrac{mg/L}{m}} = \sqrt{g/L}$

The corresponding frequency and time period relations are

$$\nu = \frac{\omega}{2\pi} = \frac{1}{2\pi}\sqrt{\frac{g}{L}} \qquad ...(1.18)$$

and

$$T = \frac{1}{\nu} = 2\pi\sqrt{\frac{L}{g}} \qquad ...(1.19)$$

1.3.6 Torsional Pendulum

A torsional pendulum consists of a disk or rod suspended at the end of a wire (see Fig. 1.9). When the end of the wire is twisted by an angle θ, the restoring torque τ arises which obeys Hooke's law.

$$\tau = -\kappa\theta \qquad ...(1.20)$$

where κ is called the torsional constant.

Suspension wire

Reference line

$+\theta_{max}$

0

$-\theta_{max}$

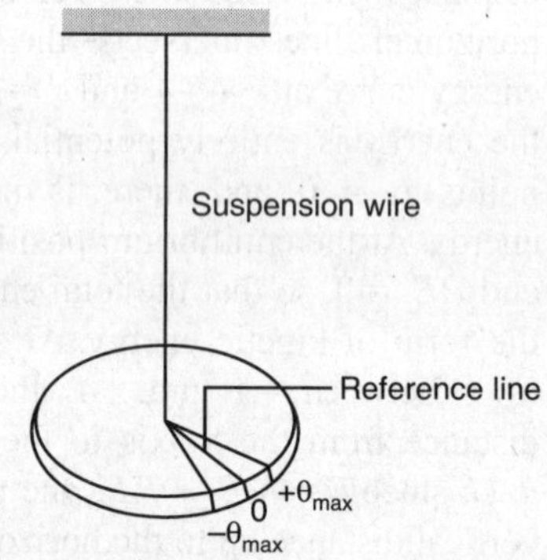

Fig. 1.9: Torsional pendulum.

If the wire is twisted and released, the oscillating system is called a torsional pendulum. The rotational form of Newton's second law is

$$\tau = I\alpha$$

or

$$-\kappa\theta = I\frac{d^2\theta}{dt^2}$$

which may be rewritten as

$$\frac{d^2\theta}{dt^2} + \frac{\kappa}{I}\theta = 0 \qquad ...(1.21)$$

Eq. (1.21) is the equation of a simple harmonic oscillator whose angular frequency is

$$\omega = \sqrt{\frac{\kappa}{I}}$$

and time period is

$$T = 2\pi\sqrt{\frac{I}{\kappa}} \qquad ...(1.22)$$

The balance wheel in a clock or wristwatch is an example of a torsional pendulum.

Example 1.1: *The displacement equation of a particle describing simple harmonic motion is x = 0.01 sin 100π (t + 0.005) meter, where x is the displacement of the particle at any instant t. Calculate the amplitude, periodic time, maximum velocity and displacement at the time of the motion.*

Solution: The displacement of the particle is,

$$x = 0.01 \sin 100\,\pi\,(t + 0.05)$$

The general equation of SHM is,

$$x = a \sin(\omega t + \varphi)$$

Comparing the two equations, we get

Amplitude, $a = 0.01$ metre

$\omega = 100\pi \quad \therefore \quad$ Periodic time $T = \frac{2\pi}{\omega} = \frac{2\pi}{100\pi} = 0.02$ seconds

The velocity of the particle at the displacement x is given by, $\upsilon = \omega\sqrt{(a^2 - x^2)}$

The maximum velocity is given by υ_{max} when $x = 0$

$$\therefore \quad \upsilon_{max} = \omega a = 100 \times 3.14 \times 0.01 = 3.14 \text{ m/s}$$

The displacement at the time of start ($t = 0$) is

$$x = 0.01 \sin 100\pi\,(0.005)$$

$$\text{or} \quad = 0.01 \sin\frac{\pi}{2} = 0.01 \text{ metre}$$

Example 1.2: *A body of mass 0.05 kg executes SHM. When the displacement from the centre of motion is 0.04 m, the force acting on the body is 18 × 10^{-3} N. If the maximum velocity is 2 m/s, find the amplitude and acceleration.*

Solution: The acceleration is given by,

$$a = -\omega^2 x$$

The force acting on the body is given by,

$$F = m \times a = m\omega^2 x$$

Accordingly, $\quad 0.05 \times \omega^2 \times 0.04 = 18 \times 10^{-3}$

$$\therefore \quad \omega^2 = \frac{18\times10^{-3}}{0.05\times0.04} = 9 \quad \text{or} \quad \omega = 3 \text{ rad./s}$$

Now, $\upsilon_{max} = \omega a$ or $\quad a = \dfrac{\upsilon_{max}}{\omega}$

$$\therefore \text{ Amplitude,} \quad a = \frac{2}{3} \text{ m} = 0.667 \text{ m}$$

The maximum acceleration $= \omega^2 a = 9 \times 0.667 = 6 \text{ m/s}^2$

Example 1.3: *A particle of mass 5 gm executes SHM and has amplitude of 8 cm. If it makes 16 vibrations per second, find its maximum velocity and energy at mean position.*

Solution: In case of SHM, $\quad x = a \sin(\omega t + \varphi)$

The maximum velocity, $\quad \upsilon_{max} = \omega a = 2\pi n a$

$$\therefore \quad \upsilon_{max} = 2 \times 3.14 \times 10 \times 8 = 803.8 \text{ cm/s}$$

$$= 8.038 \text{ m/s}$$

The energy at mean position is entirely kinetic.

$$\therefore \quad E = \frac{1}{2} m\upsilon_{max}^2 = \frac{1}{2} \times 5 \times 10^{-3} \text{ kg} \times (8.038)^2 = 0.16 \text{ J}$$

Example 1.4: *The displacement of particle executing SHM is given by,* $x = 20 \sin\left(\dfrac{2\pi t}{T} + \varphi\right)$. *The period of vibration is 60 s. At t = 0, the displacement of the particle is 1 cm. calculate (i) initial phase, (ii)phase angle when the displacement is 4 cm, (iii) phase difference between any two positions of the particle 10 s apart..*

Solution: (*i*) When $t = 0$, $x = 1$ cm

$$\therefore \quad 1 = 20 \sin(0 + \phi) \quad \text{or} \quad \sin\phi = \frac{1}{20} = 0.05$$

$\therefore \quad \phi = \sin^{-1}(0.05) = 2°52'$

(*ii*) In this case, $\quad 3 = 20 \sin\left(\frac{2\pi}{t} + \phi\right)$

$$\therefore \quad \text{Phase angle}\left(\frac{2\pi}{t} + \varphi\right) = \sin^{-1}\left(\frac{3}{20}\right)$$

$$= \sin^{-1}(0.15) = 8°38'$$

Let α_1 and α_2 be the phase angles at times t_1 and t_2 respectively such that $(t_1 - t_2) = 10$ sec.

$$\therefore \quad \alpha_1 = \left(\frac{2\pi t_1}{T} + \varphi\right) \quad \text{and}$$

$$\alpha_2 = \left(\frac{2\pi t_2}{T} + \varphi\right)$$

(*iii*) Phase difference $= \alpha_1 - \alpha_2 = \alpha_1$

$$= \frac{2\pi}{T}(t_1 - t_2) = \frac{2\pi \times 10}{60} = \frac{\pi}{3} \text{ radians}$$

Example 1.5: *A spring is stretched by 8 cm by a force of 10 N. Find the force constant. What will be the period of 4.0 kg mass suspended by it?*

Solution: $F = kx$, where k is a force constant.

$$\therefore \quad k = \frac{F}{x} = \frac{10\,\text{N}}{0.08\,\text{m}} = 125 \text{ N/m}$$

The time period $\quad T = 2\pi\sqrt{m/k} = 2\pi\sqrt{4.0\,\text{kg}/125\,\text{N/m}} = 1.12$ s

1.4 FREE OSCILLATIONS

Free oscillations are oscillations that appear in a system as a result of a single initial deviation of the system from its state of stable equilibrium. When a pendulum is displaced from its equilibrium position and left to the action of internal forces, it undergoes *free* oscillations with the frequency given by,

$$\nu = \frac{1}{2\pi}\sqrt{\frac{g}{L}}$$

The frequency of the simple pendulum depends on the length of the string and the acceleration due to gravity. It is independent of the mass. The frequency will be the same as long as the pendulum does not experience resistance to its motion. The frequency with which the pendulum oscillates freely at its own is called its ***natural frequency***. Thus, if no resistance is offered to the motion of any oscillating body by air friction or other forces, the body will keep on oscillating indefinitely at its natural frequency, as shown in Fig. 1.10.

Such an oscillator is called an *ideal* oscillator. The period of oscillations of an ideal oscillator is independent of the amplitude and a characteristic property of the oscillation. The ideal systems are frictionless and energy is not dissipated away and hence the total mechanical energy and the amplitude remain constant.

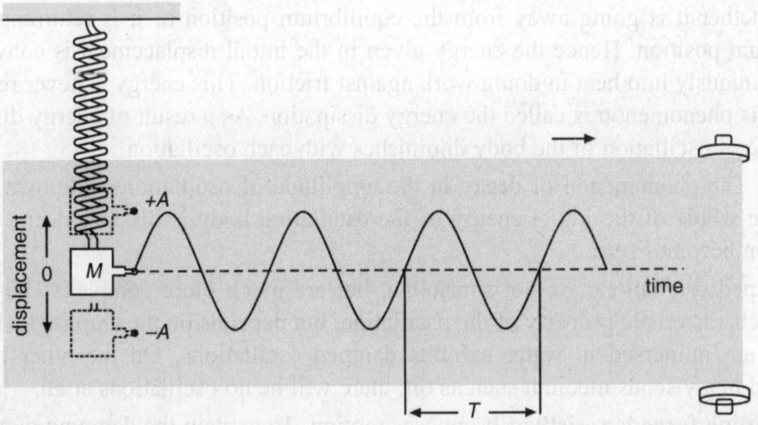

Fig. 1.10:
Natural frequency.

Example 1.6: *A body of mass 4.9 kg hangs from a spring and oscillates with a period of 0.6 seconds. How much will the spring shorten when the body is removed?*

Solution: $T = 2\pi\sqrt{m/k}$

$\therefore$ $0.6 = 2\pi\sqrt{4.9/k}$ or $k = 536.8$ N/m

Further $k = F/x$ or $x = F/k$

$$\therefore \quad x = \frac{4.9 \text{ kg} \times 9.8 \text{ m/s}^2}{536.8 \text{ N/m}} = 0.089 \text{ m} = 8.9 \text{ cm}$$

Example 1.7: *A mass M = 2 kg hangs from a vertical spring. When a mass m = 0.6 kg is gently added, the spring is further stretched by 4 cm. Now m is removed and M is set into oscillations. Calculate the period of oscillations.*

Solution: Let x be the displacement of the spring with mass M.

Now $F = kx$, $\therefore$ $Mg = kx$

or $2g = kx$

When an extra mass m (= 0.6 kg) is added, the spring is further stretched by 0.04 m.

Hence, $(M + m) \times g = k(x + 0.04)$

or $(2 + 0.06) \times g = kx + 0.04\,k = 2g + 0.04\,k$

$$\therefore \quad 2.06 \times g = 2\,g + 0.04\,k \quad \text{or} \quad k = \frac{0.6 \times 9.8}{0.04} = 14.7 \text{ N/m}$$

Now $T = 2\pi\sqrt{M/k} = 2\pi\sqrt{2/14.7} = 2.317$ sec.

1.5 DAMPED OSCILLATIONS

An *ideal* pendulum, once set into motion, continues to oscillate between two spatial positions forever without a decrease in amplitude. In actual practice, no body can oscillate for an indefinite time. If we watch an oscillating pendulum, we shall find that its amplitude of oscillation goes on decreasing due to resistance offered both at the supports and by the surrounding air; and ultimately it stops. The frictional force always opposes the motion of the body, whether it is going away from the equilibrium position or it is returning towards the equilibrium position. Hence the energy given in the initial displacement is converted slowly but continuously into heat in doing work against friction. This energy is never returned to the body. This phenomenon is called the energy dissipation. As a result of energy dissipation, the amplitude of oscillation of the body diminishes with each oscillation.

Def: The phenomenon of decay in the amplitude of oscillations is known as *damping*. When the whole of the initial energy of the oscillating body is dissipated, the amplitude of oscillation becomes zero.

Damped oscillations are not sinusoidal, but are much more complex. The period is no longer a characteristic property of the oscillation, but depends on the amplitude. For example, a pendulum immersed in water exhibits damped oscillations. On the other hand, if it is immersed in a viscous medium such as oil, there will be no oscillations at all.

Damping force is resistive: It opposes motion. To explain the damping dynamically, we may assume that in addition to the restoring force $F = -kx$, there is a *damping force* that is opposed to the velocity. Friction and viscosity are such kind of forces. A damped system is subjected to the following two forces:

(*i*) A *restoring force* proportional to displacement but oppositely directed and

(*ii*) A *frictional force* proportional to the velocity but oppositely directed.

We write the damping force as $F' = -b\upsilon$ ...(1.23)

where b is a constant that depends on the medium and the shape of the body.

The resultant force on the body is

$$F + F' = -kx - b\upsilon \qquad ...(1.24)$$

Therefore the equation of motion of the body is

$$\underset{\text{(inertial force)}}{ma} = \underset{\text{(restoring force)}}{-kx} \quad \underset{\text{(damping force)}}{-b\upsilon}$$

$$\therefore \qquad m\frac{d^2x}{dt^2} = -kx - b\frac{dx}{dt}$$

$$m\frac{d^2x}{dt^2} + b\frac{dx}{dt} + kx = 0 \qquad ...(1.25)$$

$$\frac{d^2x}{dt^2} + \frac{b}{m}\frac{dx}{dt} + \frac{k}{m}x = 0$$

$$\frac{d^2x}{dt^2} + \gamma\frac{dx}{dt} + \omega_0^2 x = 0 \qquad ...(1.26)$$

where $\gamma = b/m$ is the *damping coefficient* of the system and $\omega_0 = \sqrt{k/m}$ is the *natural frequency* of the system.

Eq. (1.26) is a differential equation. A simple solution of the differential equation is given by

$$x = A_0 e^{-\gamma t/2m} \cos(\omega t + \varphi) \quad \text{for } \omega_0{}^2 > \gamma^2 \qquad ...(1.27)$$

The angular frequency of damped oscillations is given by

$$\omega = \sqrt{\omega_0^2 - \left(\frac{\gamma}{2m}\right)^2} \qquad ...(1.28)$$

It is clear that the damped angular frequency ω is less than the natural angular frequency $\omega_0 = \sqrt{\frac{k}{m}}$.

Damping plays a beneficial role. For example, the shock absorbers in a car provide a velocity dependent damping force so that when the car goes over a bump, it does not continue bouncing forever.

1.5.1 Weak Damping

In Eq. (1.28), ω will be a real and positive quantity if the condition $\frac{\gamma}{2m} < \omega_0$ is satisfied. When ω is real, the damping force is weaker than the restoring force and the oscillations are weakly damped. It is said that the oscillations are **underdamped**. The condition represents a simple harmonic motion with amplitude $Ae^{-\gamma t/2m}$. The motion differs from the undamped motion in two ways. First, the amplitude of the oscillations is not constant but falls slowly with time over many oscillations, as shown in Fig. 1.11. We may treat the damped oscillations as nearly sinusoidal with progressively diminishing amplitude. Secondly, weakly damped systems have angular frequencies near to natural frequency of the system. The time period of damped oscillation is the time interval between two consecutive maximum displacements. It has been found that although friction affects the manner in which the amplitude decreases it has practically no effect on the period of damped oscillations.

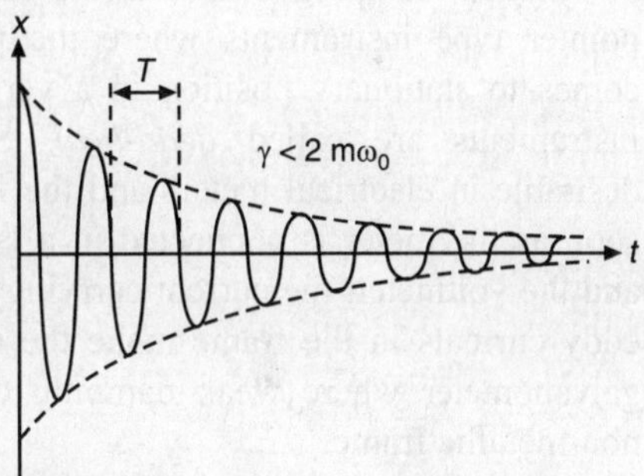

Fig. 1.11: Amplitude of the oscillations with time.

The motion of a pendulum in air and the electric oscillations of an LCR circuit belong to this category. In electric circuits containing inductance, capacitance and resistance, there is a natural frequency of oscillation and the resistance plays the role of the damping constant γ. It is usually desirable to minimize damping, but damping can never be prevented completely.

1.5.2 Heavy Damping

When the damping is very large such that $\gamma > 2\, m\omega_0$, the solution for the Eq. (1.26) assumes the form

$$x(t) = c_1 e^{-\gamma + t} + c_2 e^{-\gamma - t} \qquad ...(1.29)$$

In this case, ω is imaginary and there are no oscillations. The system once displaced, returns to its equilibrium position quite slowly without any oscillations (see Fig. 1.12). The motion is termed as ***heavy*** or ***over-damped motion***. As the damping increases, the time taken by the body to reach equilibrium also increases. For instance, if a door closing mechanism is heavily damped, when released from the open position the door gradually closes.

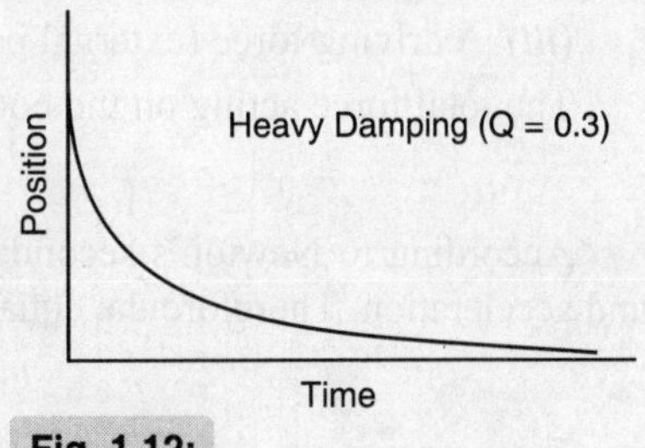

Fig. 1.12: Over-damped motion.

1.5.3 Critical Damping

When $\gamma = 2m\omega_0$, we get $\omega = 0$ and again, there is no oscillation. The motion is known as ***critical damped motion***. A critically damped system moves to the equilibrium position quickly without any oscillation (Fig. 1.13). This is a needed quality in mechanical and electrical systems.

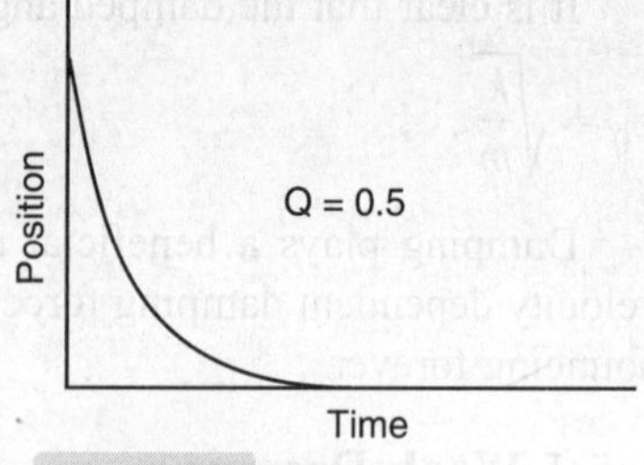

Fig. 1.13: Critical damping.

Critical damping is used in the construction of many pointer type instruments where the pointer moves and comes to stationary position in a very short time. Such instruments are called *dead-beat.* Such behaviour is desirable in electrical meters and the like where we would like to note a steady reading as soon as the meter is connected in a circuit. In the moving coil galvanometer, the ammeter and the voltmeter, the current carrying coil is wound on a metallic frame so that the induced eddy currents in the frame make the motion dead-beat. On the other hand, in the ballistic galvanometer where weak damping condition is to be fulfilled, the coil is wound on a non-metallic frame.

Shock absorbers are another example. A shock absorber is a spring with a sealed container of fluid. When a bike fitted with a shock absorber hits a bump, the force from the bump compresses the spring. The spring supplies a restoring force. The rider and bike would move forward in simple harmonic motion. The fluid inside a shock absorber supplies the *damping force* which stops oscillations and the vehicle returns to equilibrium.

1.6 FORCED OSCILLATIONS

The oscillations produced when an *external oscillatory force* is applied to a body subject to an elastic force are known as ***forced oscillations***. For instance, we would drag a swing up to a certain height, and then let it go or else we would repeatedly push the swing at any frequency. There are two frequencies in this case: the **natural frequency** ω_o of the free oscillations, and the **driving frequency** ω_f of the forced oscillations.

Def: Forced oscillations are the oscillations in which the body oscillates with a frequency other than its natural frequency under the action of an external periodic force.

The forces that act on the body undergoing forced oscillations are as follows:

(*i*) A restoring force F_1, that is proportional to the displacement and oppositely directed;

(*ii*) A damping force F_2, that is proportional to the velocity but oppositely directed; and

(*iii*) A driving force (external periodic force), $F_3 = F_o \sin \omega_f t$.

The total force acting on the body is given by

$$F = F_1 + F_2 + F_3 = -kx - \gamma\upsilon + F_o \sin \omega_f t$$

According to Newton's second law this force must be equal to the product of the mass and acceleration. Therefore the equation of the motion of the body is

$$\underset{\text{(inertial force)}}{ma} = \underset{\text{(restoring force)}}{-kx} \quad \underset{\text{(damping force)}}{-\gamma\upsilon} \quad \underset{\text{(external force)}}{+F_o \sin \omega_f t}$$

$$\therefore \quad m\frac{d^2x}{dt^2} = -kx - \gamma\frac{dx}{dt} + F_o \sin \omega_f t \qquad \text{...(1.30)}$$

Rearranging the terms in Eq. (1.30) equation, we get

$$\frac{d^2x}{dt^2}+\frac{\gamma}{m}\frac{dx}{dt}+\frac{k}{m}x=\frac{F_o}{m}\sin\omega_f t \qquad ...(1.31)$$

Eq. (1.31) is a differential equation which can be solved by standard techniques. Instead of going into the mathematical details, we confine here to the results. When the force is first applied, the motion is complex. The system wants to vibrate with a natural frequency ω_0 but is being forced by a driving frequency ω_f. Therefore, the motion is initially a superposition of free damped oscillations and forced oscillations. This initial motion is referred to as **transitory behaviour**. However, the transitory behaviour decays exponentially. After the free oscillations die out, only the forced oscillations remain. We then say that the motion has reached a **steady-state condition** and the oscillator oscillates with constant amplitude (see Fig. 1.14 b). In the steady state, the energy input per cycle equals the energy lost per cycle.

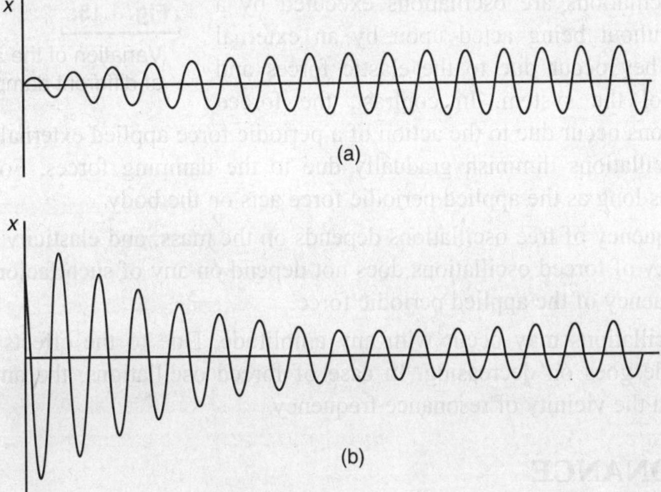

Fig. 1.14:
Steady-state condition.

In the steady-state the body is forced to oscillate with the angular frequency ω_f of the applied force. Therefore, the steady state solution to Eq. (1.31) is

$$x = A\sin(\omega_f t-\alpha) \qquad ...(1.32)$$

where A is the amplitude of the forced oscillations and α is the phase angle between the displacement x and the external force F_3. Both the amplitude A and the initial phase α are not arbitrary constants but are fixed quantities that depend on the frequency ω_f of the applied force. The larger the difference between ω_f and the natural frequency ω_0, the smaller is the amplitude of the forced oscillations. It is more difficult for the oscillator to respond to the applied force when the forcing frequency is not near the natural frequency. It can be shown that the amplitude of the forced oscillations is given by

$$A=\frac{F_0/m}{\sqrt{\left[\left(\omega_0^2-\omega_f^2\right)^2+(\gamma\omega_f/m)^2\right]}} \qquad ...(1.33)$$

At constant values of F_0, m and γ, the amplitude of free oscillations depends on the ratio of the frequencies of the driving force ω_f and of the free undamped oscillations, ω_0. The variation of the amplitude with ω_f at different damping factors γ is shown in Fig. 1.15.

Eq. (1.33) indicates that the forced oscillations are not damped but are of constant amplitude. It means that the external agent overcomes the damping forces and provides the energy necessary to maintain the oscillations.

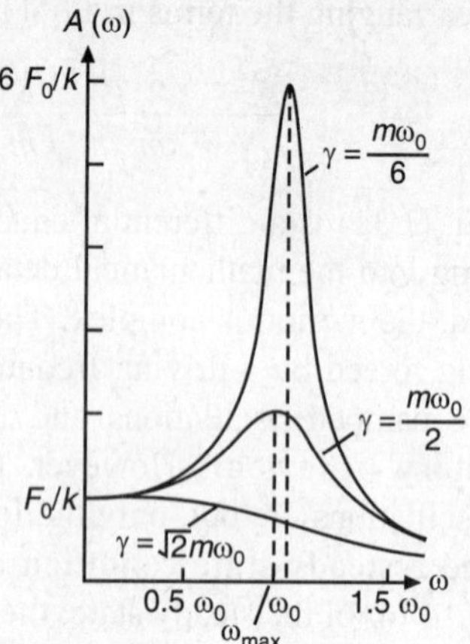

Fig. 1.15: Variation of the amplitude with ω_f at different damping factors γ.

1.6.1 Distinction between Free and Forced Oscillations

(*i*) Free oscillations are oscillations executed by a body without being acted upon by an external force. They occur due to the elastic forces and inertia of the system. In contrast, the forced oscillations occur due to the action of a periodic force applied externally.

(*ii*) Free oscillations diminish gradually due to the damping forces. Forced oscillations persist as long as the applied periodic force acts on the body.

(*iii*) The frequency of free oscillations depends on the mass, and elasticity of the body. The frequency of forced oscillations does not depend on any of such factors and is equal to the frequency of the applied periodic force.

(*iv*) Free oscillations may occur with any amplitude. Due to the effects of damping the amplitude goes on decreasing. In case of forced oscillations, the amplitude is small except in the vicinity of resonance frequency.

1.7 RESONANCE

Referring to Fig. 1.15, we observe that each driving frequency is characterized by its own amplitude. As the driving frequency ω_f is increased, the amplitude rises until it reaches a maximum at ω_{max} and at further high frequencies the amplitude again decreases.

Def: When the frequency of the driving force is near the natural frequency ω_0 of the oscillating system, the oscillation amplitude becomes very large. The raise in amplitude near the natural frequency is called *resonance* and the frequency ω_0 is called the resonance frequency of the system.

The reason for large amplitude oscillation is that the rate of energy transfer from the applied force to the forced oscillator is a maximum. Resonance occurs whenever a system is subject to an external action that varies periodically with time and with the proper frequency. At the resonance frequency, the external force and the velocity of the particle are in phase. As a result, the power transfer to the oscillator has its maximum value. At frequencies above or below the resonance value, the force and velocity are not in phase and hence the power transfer is lower.

Note the following:

(*i*) When damping is small, the amplitude of forced oscillations grows with increasing ω_f, and at $\omega_f = \omega_0$, the amplitude of the oscillations becomes equal to infinity. Further, the resonance curve is sharp, that is the amplitude falls off rapidly on either side of the resonant frequency.

(*ii*) When damping is large, the amplitude reduces very slowly on either side of the resonant frequency.

(*iii*) When the oscillator responds to a number of close by frequencies near the resonant value, the resonance is flat.

(*iv*) At resonance, the velocity is in phase with the applied force. Since the rate of work done on oscillator by the applied force is F_{υ}, this quantity is positive when F and υ are in phase and represents a favourable condition for transfer of energy to the oscillator.

(*v*) At resonance, the oscillating system absorbs energy from the external source.

The phenomenon of resonance appears in many areas of physics.

(*i*) Tuning of a radio receiver involves matching of the frequency of the tuned circuit with that of the radio waves. Only when the resonance condition is reached, we listen clearly the sounds transmitted by a particular radio station.

(*ii*) Resonance absorption of radiation by atoms takes place when the frequency of the incident light waves equals the natural frequency of the atom.

(*iii*) In a cyclotron, particles are accelerated to high energies only when the frequency of electric field accelerating the particles is equal to the frequency of revolution of the particle in magnetic field acting perpendicular to the particle path.

A dramatic example of resonance occurred when a company of soldiers was marching in step across a bridge in St. Petersburg. The bridge collapsed. The period of free oscillation of the bridge coincided with the period of an ordinary marching step and resonance took place; it caused swinging of the bridge with very high amplitude leading to the ultimate collapse. Another instance was that of the Tacoma Narrows Bridge in Washington State in 1940. The wind blowing through the Tacoma Narrows broke up into vortices, which gave wind puffs that shook the bridge at a frequency matching with one of its natural vibrational frequencies. In a couple of hours the amplitude became so large that the centre span collapsed due to resonance.

1.7.1 Sharpness of Resonance

When the driving force is increased or decreased from resonant frequency, the amplitude falls off from the maximum value.

Def: Sharpness of resonance is the rate of fall of amplitude with the change in frequency of the driving force, on either side of the resonant frequency.

The power absorbed is maximum at resonance for small values of damping. If P_r is the power absorbed at resonance, P is the power absorbed at any frequency ν. A plot between P/P_r versus ν is shown in Fig. 1.16. The frequency values, on either side of natural frequency ν_0, at which the power absorbed is half of the maximum are called half-power points.

Def: The frequency difference between these two half-power points is called the bandwidth of the oscillator.

Thus,

$$\text{Bandwidth, } \Delta\nu = \nu_2 - \nu_1 \qquad ...(1.34)$$

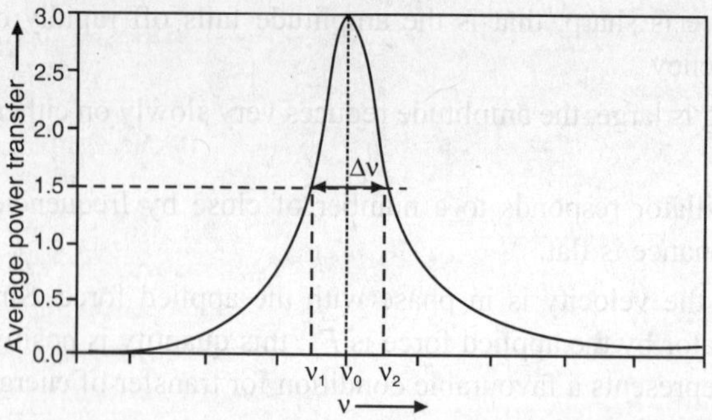

Fig. 1.16:

A plot between P/P_r versus ν.

1.8 COUPLED OSCILLATIONS

In physics we come across a variety of coupled systems that can oscillate. For example, a solid body is composed of many atoms or molecules. Every atom behaves like an oscillator, which vibrates about an equilibrium position. Each atom affects the motion of its neighbouring atom and thus the atoms of the solid are coupled together. A **coupled oscillator** is a system of two or more connected bodies that oscillates within a set of stable, predictable patterns. The number of "ways" in which a coupled system can oscillate is determined by the number of coupled bodies. Once released, a pendulum can only have one frequency that describes its motion. Two pendulums coupled together can have two oscillating frequencies. A particular motion of coupled oscillators is called a **normal mode** of oscillation. Normal modes correspond to the case where the two bodies move with the same frequency and maintain a constant phase difference. In one normal mode, the two oscillators move in phase and in the second normal mode, they move out of phase. Usually, any coupled system will exhibit a mixing of all possible modes.

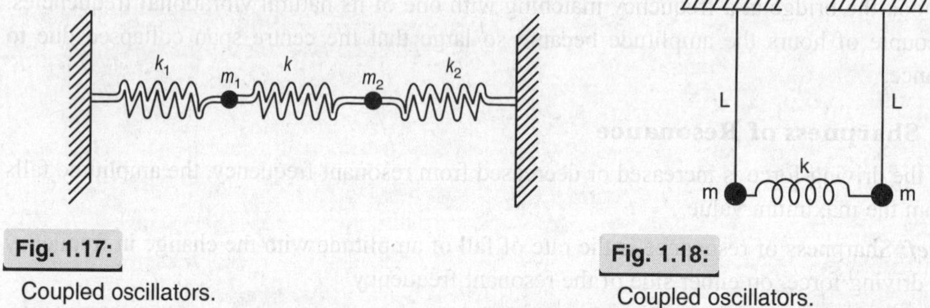

Fig. 1.17:

Coupled oscillators.

Fig. 1.18:

Coupled oscillators.

An example of coupled oscillators is two identical pendulums coupled by a spring k, as shown in Fig. 1.18. When the motion of the coupled oscillators occurs in one of the normal modes, their energies and the oscillation amplitudes remain constant. In the general case, the amplitude of each oscillator does not remain constant. The displacement of two coupled oscillators with time is shown in Fig. 1.18. It is clear form Fig. 1.18 that when one pendulum is displaced and is released, its amplitude begins to decrease while the second pendulum starts oscillating with increasing amplitude. Then the trend reverses and the amplitude of the second pendulum decreases and that of the first pendulum increases and this kind of exchange goes on.

A good example of coupled oscillators is the vibration of atoms in a molecule. The various modes of vibration of the linear tri-atomic molecule CO_2 are depicted in Fig. 1.19 and those of the non-linear molecule H_2O are shown in Fig. 1.20. Fig. 1.19 (a) depicts the oxygen atoms oscillating in phase, while the carbon atom moves in the opposite direction. Fig. 1.19 (b) depicts the oxygen atoms oscillating out of phase while the carbon atom remaining fixed. Fig. 1.19 (c) shows the oxygen atoms oscillating in a direction perpendicular to the line joining them, which leads to the bending of the molecule. Fig.1.20 shows the normal modes of vibration of the H_2O molecule.

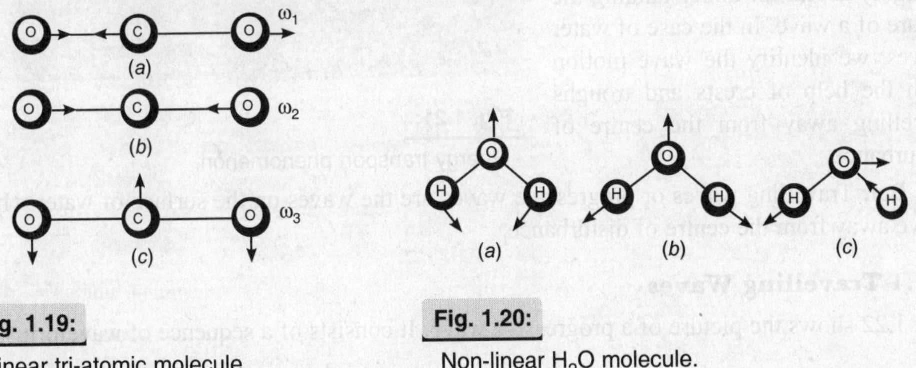

Fig. 1.19: Linear tri-atomic molecule.

Fig. 1.20: Non-linear H_2O molecule.

1.9 WAVES

The Eq. (1.3) and Eq. (1.5) describe the simple harmonic motion of a single point. Whenever a disturbance is created in a material medium, such as due to a stone thrown into still water, a local displacement of particles from equilibrium is caused. When a particle in a medium is disturbed, intermolecular forces provide the restoring force. The interactions of the particle with the next adjacent particle tend to return the former to its original position, and the latter begins to oscillate. In so doing, it affects the adjacent molecules, which are in turn set into oscillation.

In such a situation, we find that something that happens at A at t_1 causes a similar happening at B at a later time t_2. This is referred to as propagation of the disturbance. A medium is a material, which supports the propagation of the disturbance. When a continuous and repetitive disturbance passes through a medium, it gives rise to a continuous propagation of energy. In this situation, a series of particles are set into identical oscillations in succession. The disturbance may take any shape such as a finite width pulse (single disturbance) to an infinitely long sine wave.

When we observe ripples in a pond, what we see actually is a rearrangement of the surface of water. A wave transports its energy without transporting matter. The energy supplied by a stone (the agent of disturbance) is transferred from one point to another. Energy is transported through the medium, yet the water molecules are not transported. Therefore, waves are said to be an ***energy transport phenomenon***.

Def: Wave is any disturbance which travels through the medium due to the repeated periodic motion of the particles of the medium about their mean position and transporting energy from one location to another location without transporting matter.

We are familiar with waves on water surface. When a stone is thrown into still water of a pond, ripples are formed which expand in the form of circles. As shown in Fig. 1.21, one crest is followed by another and every crest is separated by a trough. This mental picture of water waves is highly useful for understanding the nature of a wave. In the case of water waves, we identify the wave motion with the help of crests and troughs travelling away from the centre of disturbance.

Fig. 1.21: Energy transport phenomenon.

Def: Travelling waves or progressive waves are the waves on the surface of water which move away from the centre of disturbance.

1.9.1 Travelling Waves

Fig. 1.22 shows the picture of a progressive wave. It consists of a sequence of waveforms.

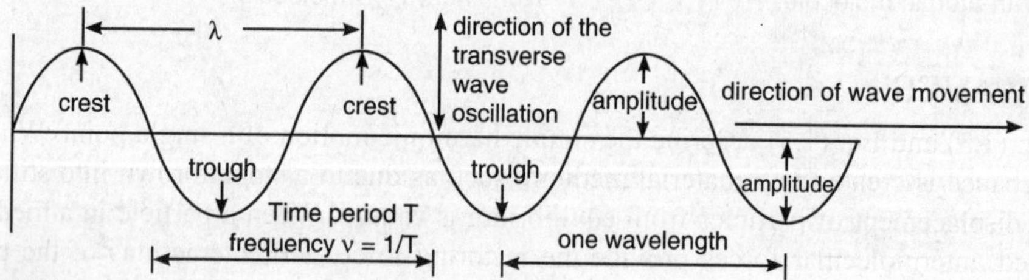

Fig. 1.22: Wave profile.

Any wave is characterized by the following parameters.

(*i*) Time Period, *T*: It is the repeat time of a point chosen on the wave profile at equal intervals of time.

(*ii*) Wavelength, λ: It is the distance between two corresponding points such as two successive crests or troughs.

(*iii*) Amplitude, A: The maximum displacement in a waveform is known as the *amplitude*.

(*iv*) Velocity, υ: The distance that the wave moves in one second is the velocity of the wave, υ.

Each time the source of disturbance vibrates once, the wave moves forward a distance λ. If there are ν vibrations in one second, the wave moves forward a distance of 'νλ.' Thus,

$$\upsilon = \nu\lambda \qquad \text{...(1.35)}$$

where

$$\nu = \frac{1}{T} \qquad \text{...(1.36)}$$

(*v*) Phase angle, ϕ: The displacement of particles in the medium and the direction of their displacement change from point to point along the wave. The quantity, which represents the displacement, is called the *phase of the vibration*, ϕ. The phase may be expressed in terms of degrees or radians; or as the ratio of time *t* to the time period *T*; or as the ratio of the distance

x to the wavelength λ. The ratios t/T and x/λ are fractional numbers and have a maximum value of 1. When expressed in terms of radians (or degrees), the maximum value that the phase can take is 2π radians (or 360°).

(*vi*) Intensity, I: The energy transferred on an average by a wave in unit time, through a unit area perpendicular to its propagation direction, is known as the *intensity* of the wave. It is established that the intensity of a wave is directly proportional to the square of the amplitude of the wave. Thus,

$$I \propto |A|^2 \quad \text{...(1.37)}$$

1.9.2 Wave Equation

Equation of motion of an object is the equation that gives the position of the object as a function of time. We obtain the entire picture of wave motion only when we consider the harmonic motion of a series of points in the medium. As the oscillations are communicated from point to point, the points in the medium will be in different states of oscillation at different times. The displacement of a particle in the medium is therefore a function of space coordinates as well as a function of time. We denote the displacement by y. Thus,

$$y = f(x, t) \quad \text{...(1.38)}$$

The displacement y is sometimes called the **wave function**.

Consider a one-dimensional wave moving along +x-axis, as in Fig. 1.23. We first consider the displacement as a function of time, at the position $x = 0$.

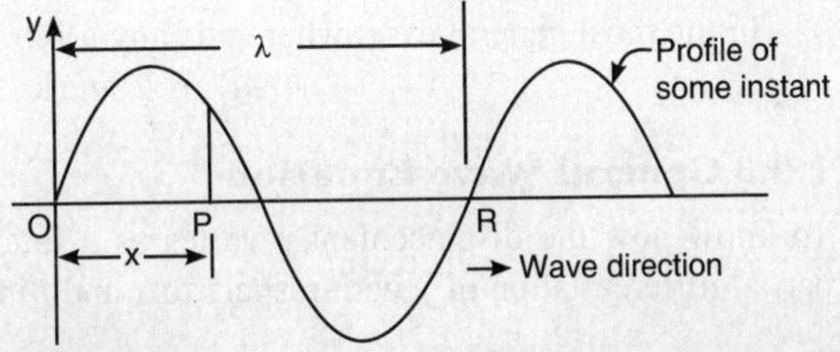

Fig. 1.23: One-dimensional wave.

Then,

$$y = f(t)$$

Since the oscillations are sinusoidal, we can describe the displacement y in terms of *time* as

$$y = A \sin \omega t$$

or

$$y = A \sin 2\pi \nu t \quad \text{...(1.39)}$$

The wave is travelling forward to the right with a velocity, say υ. Then after time t, the wave has moved through the distance $x = \upsilon t$. The displacement at x can be represented by

$$y = f(x - \upsilon t) \quad \text{...(1.40)}$$

$$\upsilon = \nu \lambda$$

Also

$$\upsilon = \frac{x}{t}. \quad \text{Therefore, } \nu\lambda = x / t.$$

or

$$\nu = \frac{x}{\lambda t} \quad \text{...(1.41)}$$

We can rewrite the Eq. (1.39) using Eq. (1.41) as

$$y = A \sin 2\pi\left(\frac{x}{\lambda}\right) \quad \text{...(1.42)}$$

This describes the displacement in terms of *space*.

Using the Eq. (1.40) and Eq. (1.42), we can describe the displacement of any point on a harmonic wave in terms of both space and time as

$$y = A \sin\left[\frac{2\pi}{\lambda}(x - \upsilon t)\right] \quad \text{...(1.43)}$$

Eq. (1.43) gives the relationship between the space and time dependence of disturbances in a medium. It is seen from the above that the wave is periodic in both space and time.

The Eq. (1.43) may be rewritten as

$$y = A \sin k\,(x - \upsilon\, t) \quad \text{...(1.44)}$$

where $$k = \frac{2\pi}{\lambda}.$$

k is known as **propagation constant or wave number**.

The Eq. (1.43) may further be rewritten as

$$y = A \sin (k\, x - \omega\, t) \quad \text{...(1.45)}$$

Eq. (1.45) can be made independent of the system of coordinates by converting it into vector form. Let vector **k** have a magnitude equal to the wave number k and a direction parallel to the positive direction of the x-axis. Such a vector is called a **wave vector**. Using **k** into Eq. (1.45), we get

$$y(x, t) = A \sin (\mathbf{k} \cdot \mathbf{x} - \omega\, t)$$

In the most general case, where r is any arbitrary direction, we replace x by r and write

$$y(r, t) = A \sin (\mathbf{k} \cdot \mathbf{r} - \omega\, t) \quad \text{...(1.45 a)}$$

1.9.3 General Wave Equation

To know how the displacement y varies as a function of space x and time t we have to do partial differentiation of y with respect to x and t in Eq. (1.45).

$$\frac{\partial y}{\partial x} = \frac{2\pi}{\lambda} A \cos\left[\frac{2\pi}{\lambda}(x - \upsilon t)\right] \quad \text{...(1.46 a)}$$

$$\frac{\partial y}{\partial t} = -\frac{2\pi v}{\lambda} A \cos\left[\frac{2\pi}{\lambda}(x - \upsilon t)\right] \quad \text{...(1.46 b)}$$

Combining both Eq. (1.46 a) and Eq. (1.46 b) and eliminating equal factors, we get

$$\frac{\partial y}{\partial x} = -\frac{1}{\upsilon}\frac{\partial y}{\partial t} \quad \text{...(1.47)}$$

If we take the second derivatives, it will hold for any sinusoidal wave, independent of the direction of travel, either $-x$ or $+x$.

$$\frac{\partial^2 y}{\partial x^2} = \frac{1}{\upsilon^2}\frac{\partial^2 y}{\partial t^2} \quad \text{or} \quad \frac{\partial^2 y}{\partial t^2} = \upsilon^2 \frac{\partial^2 y}{\partial x^2}$$

We replace y by the more general term ξ, which stands for any disturbance.

$$\frac{\partial^2 \xi}{\partial t^2} = \upsilon^2 \frac{\partial^2 \xi}{\partial x^2} \quad \text{...(1.48)}$$

This is the *one-dimensional wave equation*. It connects the variations in space and time to the velocity of propagation of the wave.

If we are to include waves propagating in any direction, we need to extend the right hand term to the y and z-axes, and replace it by

$$\frac{\partial^2 \xi}{\partial x^2} + \frac{\partial^2 \xi}{\partial y^2} + \frac{\partial^2 \xi}{\partial z^2}$$

Using the Laplacian operator $\nabla^2 = \frac{\partial^2}{\partial x^2} + \frac{\partial^2}{\partial y^2} + \frac{\partial^2}{\partial z^2}$, we can write the equation as

$$\frac{\partial^2 \xi}{\partial t^2} = \upsilon^2 \nabla^2 \xi \qquad \text{...(1.49)}$$

This is the *general three-dimensional wave equation.*

Example 1.8: *Determine the speed, wavelength and frequency of the wave described by,*

$$E(z, t) = 10^3 \sin \pi(3 \times 10^6 x - 9 \times 10^{14} t).$$

Solution: The given equation resembles the general equation,

$$E(z, t) = E_0 \sin k(x - \upsilon t) \qquad \text{(a)}$$

The given equation may be written as

$$E(z, t) = 10^3 \sin 3 \times 10^6 \pi(x - 3 \times 10^8 t) \qquad \text{(b)}$$

Comparing (b) with (a), we find that,

$\upsilon = 3 \times 10^8$ m/s and $k = 3 \times 10^6 \pi$/m.

As $k = \frac{2\pi}{\lambda}$, $\lambda = \frac{2\pi}{k} = \frac{2\pi}{3 \times 10^6 \pi} = 6666$ Å.

Frequency, $\nu = \frac{\upsilon}{\lambda} = \frac{3 \times 10^8}{6666 \times 10^{-10}} = 4.5 \times 10^{14}$ Hz.

Example 1.9: *A progressive sinusoidal wave is represented by $y(x, t) = A \sin [(0.2\ m^{-1})x - (0.4\ s^{-1})t + \pi/6]$ where x and t are in meter and second respectively. Determine the speed of propagation of the wave.*

Solution: Here, we know that,

$$y = A \sin (\omega t - kx + \Phi)$$

$\therefore$ $\omega = 0.4\ \text{s}^{-1}$, $k = 0.2\ \text{m}^{-1}$ and $\phi = \pi/6$

Now the speed of propagation of the wave is given by, $\upsilon = \frac{\omega}{k}$ $\therefore$ $\upsilon = \frac{0.4}{0.2} = 2$ m/s

1.10 TYPES OF WAVES

Waves occur in many shapes and forms. We can classify waves into two types basing on the direction of movement of the individual particles of the medium with respect to the direction in which the waves travel.

When a disturbance occurs in an elastic medium it generates an *elastic wave.* The elastic wave is a train of periodically alternating regions of compression and rarefaction; the particles of the medium being displaced in the direction of propagation of the wave (see Fig. 1.24). A wave, in which the displacement of each point of the medium is *back and forth* along the direction of propagation of the wave, is called a *longitudinal wave* or *compressional wave.*

Def: A longitudinal wave is a wave in which particles of the medium move in a direction parallel to the direction, along which the wave moves.

For example, if one end of a spring is pushed and then pulled, a longitudinal disturbance will propagate along the spring. Sound waves are longitudinal waves.

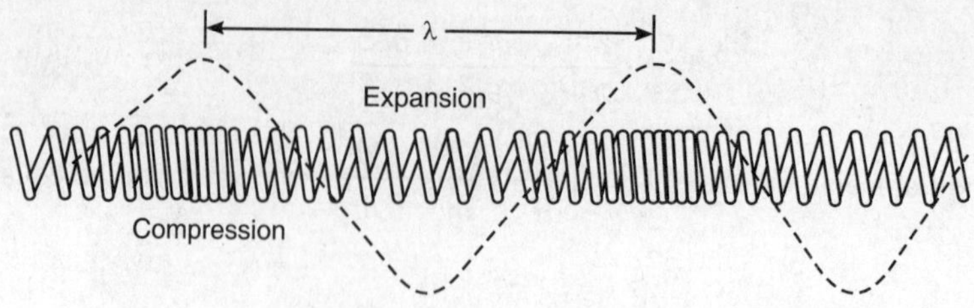

Fig. 1.24:

Elastic wave is a train with compression and rarefaction.

A wave, in which the direction of the displacement at each point of the medium is *at right angles* to the direction of wave propagation, is said to be transverse. The particles are displaced in a vertical direction, while the wave propagates in a horizontal direction.

Def: A **transverse wave** is a wave in which particles of the medium move in a direction *perpendicular* to the direction along which the wave moves.

Ripples on water surfaces are examples transverse waves. Transverse waves can exist on the surface of a liquid, the restoring force being supplied by the surface tension of the liquid.

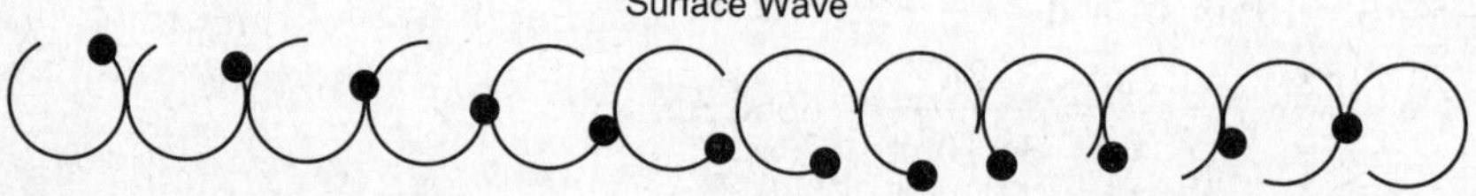

Fig. 1.25:

Surface wave.

Some waves are neither transverse nor longitudinal, but combinations of the two. The waves, which travel along the surface of the oceans, are referred to as surface waves. When a water wave travels on the surface of deep water, molecules at the surface move in nearly circular paths as shown in Fig. 1.25.

Def: A surface wave is a wave in which particles of the medium undergo a circular motion.

Rayleigh surface waves are the surface waves in solids in which the particles move in elliptical paths, with the major axis of the ellipse vertical to the surface of the solid. The width of the elliptical path reduces with the depth into the solid increases.

1.10.1 Categories of Waves

Waves can be classified as mechanical waves, electromagnetic waves, matter waves, and gravitational waves according to the source that generates them.

(*i*) **Mechanical waves:** Mechanical waves or elastic waves are governed by Newton's laws and require a material medium for their propagation. Sound waves, seismic waves, water waves in bodies of water such an ocean, river, and ponds are examples of mechanical waves.

(*ii*) **Electromagnetic waves:** Visible light, radio waves, microwaves, x-rays and γ-rays belong to this category. Electromagnetic waves consist of oscillating electric and magnetic fields and do not require material medium for their propagation. They all travel in free space with the same speed 'c.'

(*iii*) **Matter waves:** Atomic particles exhibit wave properties under certain conditions. The laws of quantum mechanics govern such *matter waves*.

(*iv*) **Gravitational waves:** It is suggested that the cosmic bodies such as galaxies, stars produce gravitational waves and interact with each other through these waves. The gravitational waves are believed to propagate with the velocity of light.

1.11 REFLECTION AND TRANSMISSION OF WAVES AT A BOUNDARY

When a wave travels through two different media, part of its energy gets reflected at the boundary. Let us consider a rope attached to a pole while the other end is held in hand. When the rope is given a jerk, a wave pulse is produced and it will travel through the rope towards the pole (Fig. 1.26 a) and it will be reflected when it reaches the boundary (See Fig. 1.26 b). The reflected pulse is inverted. When the pulse reaches the end of the rope, the rope produces an upward force on the pole. According to Newton's third law, the pole must then exert an equal and opposite reaction force on the rope. As a result, the pulse gets inverted after reflection.

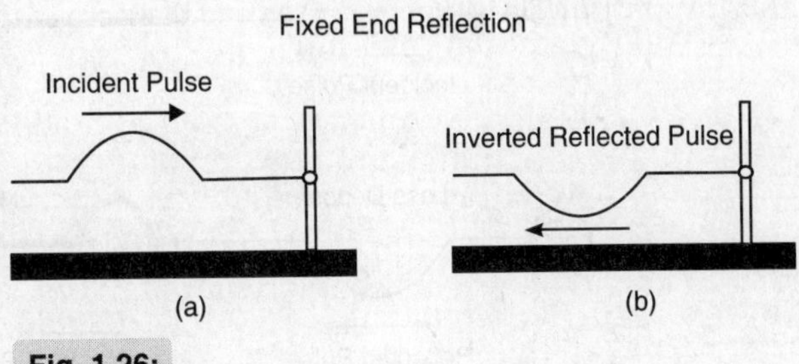

Fig. 1.26:
Fixed end reflection of waves at a boundary.

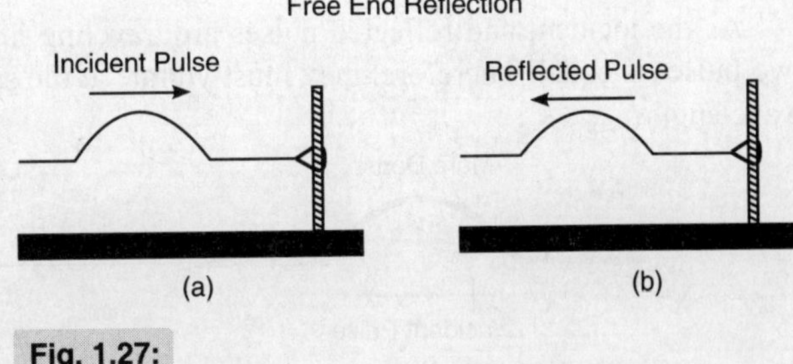

Fig. 1.27:
Free end reflection of waves at a boundary.

Let us now consider the case where the other end of the rope was free to move. Instead of being securely attached to a pole, suppose it is attached to a ring, which is loosely fit around the pole. The end of the rope is a free end. When a pulse generated at the left end of the rope reaches the free end, it will be reflected, but the reflected pulse is not inverted (see Fig. 1.27) i.e. when a crest is incident upon a free end, it returns as a crest after reflection. As the pulse reaches the pole, it exerts a force on the free end, causing the ring to accelerate upward. In the process, the ring overshoots the height of the incoming pulse and is then returned to its original position. This produces a reflected pulse that is not inverted.

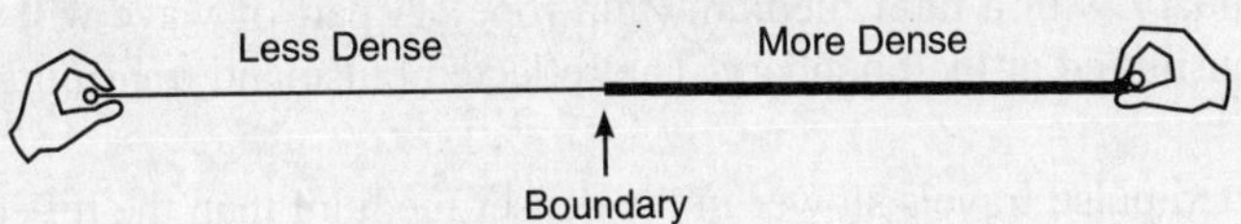

Fig. 1.28:
Reflection and transmission of waves at a neither rigid nor free Boundary.

Let us next consider a boundary, which is neither rigid nor free. For instance, let us take the case of a thin rope attached to a heavier rope, with each rope held at opposite ends (Fig. 1.28). When a pulse traveling on the thin rope reaches the boundary with a thick rope, a part of the incident pulse is reflected and returns towards the left end of the thin rope.

A part of the incident pulse is transmitted into the thick rope as the **transmitted pulse**. It will be found that the reflected pulse is inverted whereas the transmitted pulse is not inverted (see Fig. 1.29).

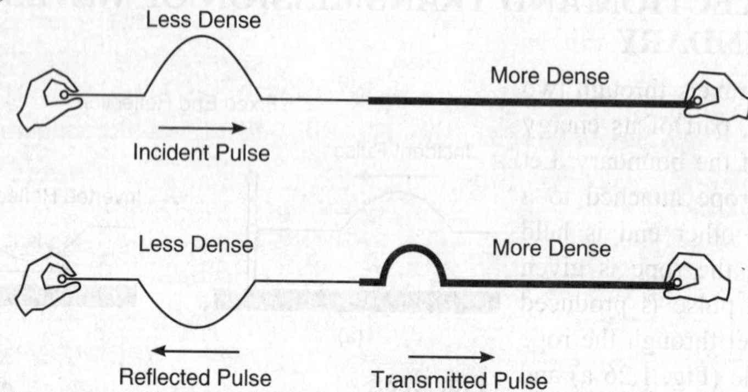

Fig. 1.29:

Reflection and transmission of waves at a neither rigid nor free Boundary.

As the incident and reflected pulses are traveling in the same medium, the speed of the two pulses is equal. Therefore, they must vibrate at the same frequency and possess the same wavelength.

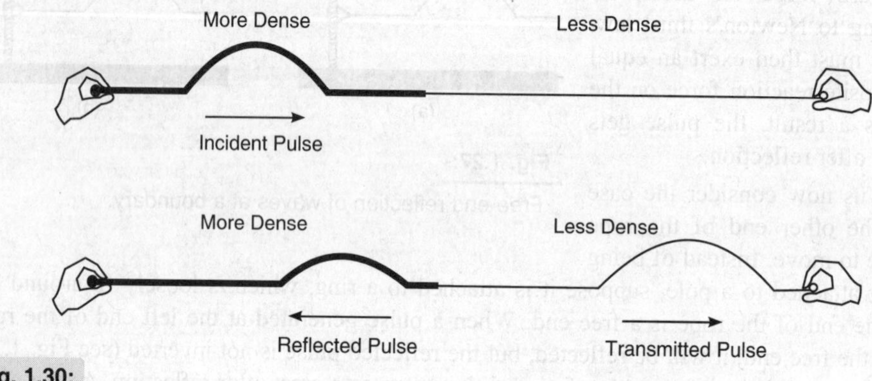

Fig. 1.30:

Reflection and transmission of waves at a neither rigid nor free Boundary.

As the last case, consider a thick rope attached to a thin rope, with the incident pulse originating in the thick rope. Here incident pulse is traveling from denser medium (thick rope) towards the boundary with a rarer medium (thin rope). A part of wave will be reflected and a part will be transmitted at the boundary. The reflected pulse and transmitted pulses are not inverted.

The transmitted pulse travels slower in the denser medium than the reflected pulse in the rarer medium. The transmitted pulse has a smaller wavelength in the denser medium than the reflected pulse has in the rarer medium. It is evident that the speed and the wavelength of the reflected pulse are the same as the speed and the wavelength of the incident pulse.

We draw the following conclusions regarding the boundary behavior of waves:

(*i*) The speed and wavelength of the wave is always a maximum in the rarer medium.

(*ii*) The frequency of a wave does not change by crossing a boundary

(*iii*) At rarer to denser medium boundary the reflected pulse becomes inverted,

The amplitudes of the reflected and transmitted waves may be expressed in terms of the amplitudes of the incident wave and the densities of the string as

$$\frac{A_r}{A_i} = \frac{\sqrt{\mu_1} - \sqrt{\mu_2}}{\sqrt{\mu_1} + \sqrt{\mu_2}} \text{ and } \frac{A_t}{A_i} = \frac{2\sqrt{\mu_1}}{\sqrt{\mu_1} + \sqrt{\mu_2}} \quad ...(1.50)$$

Eq. (1.50) are valid for *any* type of harmonic wave that meets a discontinuity in a medium.

Note the following special cases concerning the propagation of the wave.

(*i*) If $\mu_2 > \mu_1$ in the Eq. (1.50), the ratio A_r/A_i is negative and the displacement of the reflected pulse is *opposite* to that of the incident pulse. It means that the reflected pulse has a phase difference of π with respect to the incident pulse (see Fig. 1.27).

(*ii*) If $\mu_2 < \mu_1$, then the ratio A_r/A_i is always positive and the displacement of the reflected pulse is the *same* as that of the incident pulse.

(*iii*) Suppose that the string is fastened to a fixed support at $x = 0$. This is equivalent to letting μ_2 become indefinitely large. Then, $A_r/A_i = -1$ and the entire wave is reflected with a phase change of π (see Fig. 1.26).

1.12 PRINCIPLE OF SUPERPOSITION

It often happens that two or more waves propagate simultaneously through the same region in the same direction. They pass through one another as if the other wave is not present. When two pebbles are dropped at different points in a pond, the expanding water waves cross each other without either one producing any change in the other. Similarly, sound waves from different instruments in an orchestra propagate in space independent of each other and can be distinguished separately. There occur many such instances in which a number of waves meet and pass through each other without mutual effect. However, the waves act simultaneously on the particles of the medium, in the region in which they are overlapping on each other. In the region of overlap, the waves will simply add to (or subtract from) one another without permanently disrupting each other.

Once having passed through the region, each wave will move out and away without getting affected by the overlap. The resultant displacement of the medium at the location of the overlap will be different from the sum of the displacements caused by the waves individually. The resultant displacement at any point and at any instant of time can be found using the ***principle of superposition***.

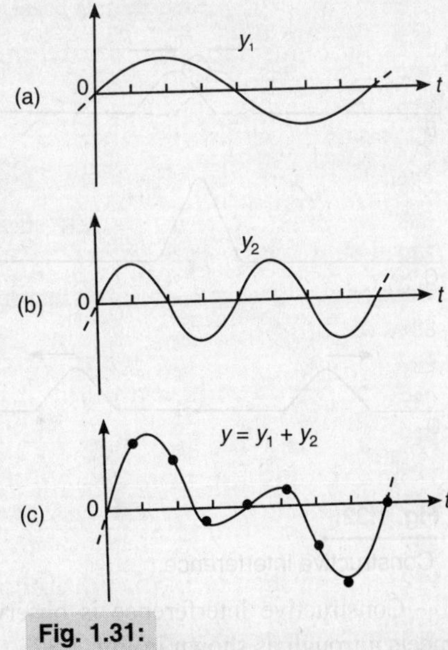

Fig. 1.31: Principle of superposition.

Statement: When a number of waves pass through a medium concurrently, the instantaneous resultant displacement of the medium at every instant is the algebraic sum of the displacements of the medium due to individual waves in the absence of others.

If $y_1, y_2, y_3, \ldots$ are the displacement vectors due to waves 1, 2, 3, ... acting separately, then the resultant displacement y is given by

$$y = y_1 + y_2 + y_3 + \ldots \qquad \ldots(1.51)$$

As an example we consider two waves traveling simultaneously along the same path. Let $y_1(x, t)$ and $y_2(x, t)$ be the displacements that the medium would experience if each wave acted alone. When both the waves act, then the displacement of the medium is

$$y(x, t) = y_1(x, t) + y_2(x, t)$$

the sum being an algebraic sum. The superposition principle holds as long as the amplitudes of the waves are not very large. This principle is applicable to all kind of waves and is very useful in the study of sound waves, electromagnetic waves and quantum physics also.

The superposition of the waves of:

(*i*) same frequency moving in the same direction leads to ***interference.***

(*ii*) slightly different frequencies moving in the same direction leads to ***beats.***

(*iii*) same frequency moving in the opposite direction leads to ***stationary waves.***

1.12.1 Interference

The superposition of two or more pulses (waves) in a given region may give rise to interference.

Def: Constructive interference occurs when the two interfering wave pulses have a displacement in the same direction; the resultant displacement is greater than the displacement of either wave.

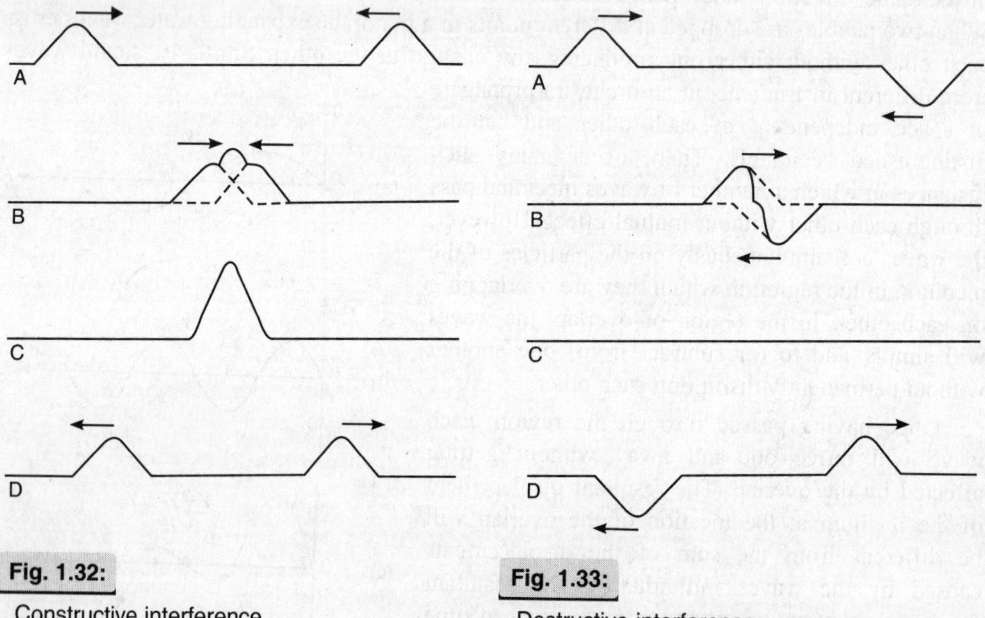

Fig. 1.32: Constructive interference.

Fig. 1.33: Destructive interference.

Constructive interference is observed when a crest meets a crest; and when a trough meets a trough as shown in Fig. 1.32.

Def: Destructive interference occurs when the pulses have displacements in opposite directions; the resultant displacement is smaller than that of either pulse.

Destructive interference is observed when a crest meets a trough; and when a trough meets a crest as shown in Fig. 1.33.

1.13 STATIONARY WAVES

Def: When two waves of equal frequency and amplitude traveling in opposite directionsinterfere with one another gives standing waves.

Let the two component waves be represented by

$$y_1(x, t) = A \sin(kx - \omega t) \quad ...(1.51a)$$

and

$$y_2(x, t) = A \sin(kx + \omega t) \quad ...(1.51b)$$

Using the principle of superposition, the resulting string displacement may be written as:

$$y(x, t) = y_1(x, t) + y_2(x, t) = A \sin(kx - \omega t) + A \sin(kx + \omega t) \quad ...(1.52)$$

By using the identity $\sin A + \sin B = 2 \sin [(A + B)/2] \cos [(A - B)/2]$, we simplify the above equation to obtain

$$y(x, t) = 2A \cos(\omega t) \sin kx \quad ...(1.53)$$

The position and time dependence have been separated in Eq. (1.59), it is not a traveling wave. The displacement of the string has an amplitude of $2A \sin kx$ and oscillates up and down according to $\cos \omega t$.

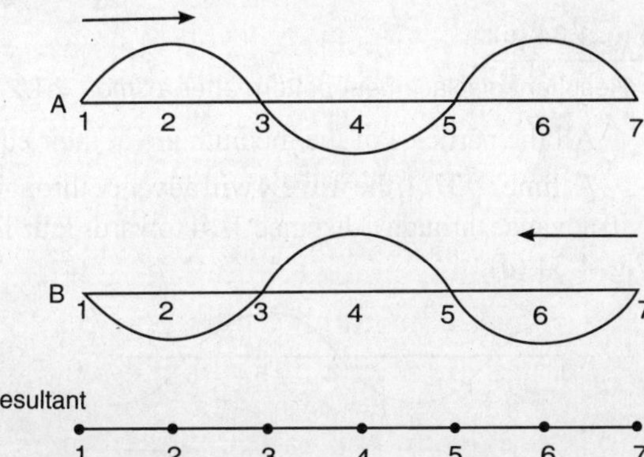

Fig. 1.34 (a): Formation of a stationary wave.

The formation of a stationary wave can be represented graphically as follows: Consider two waves A and B of the same amplitude, and frequency traveling in opposite directions. At an instant of time $t = 0$, the waves are as shown in Fig. 1.34 (a). The resultant displacement is a straight line. All the particles of the medium (depicted 1,2,3,4,5,6 and 7) are at their equilibrium position.

Consider the waves after a time $t = T/4$. During this time, the wave A will advance through a distance $\lambda/4$ towards right, and the wave B will advance through a distance $\lambda/4$ towards the left. The resultant displacement pattern is shown in Fig. 1.34 (b).

The particles at 1, 3, 5, and 7 undergo maximum displacement and the particles at 2, 4, and 6 are at their equilibrium positions.

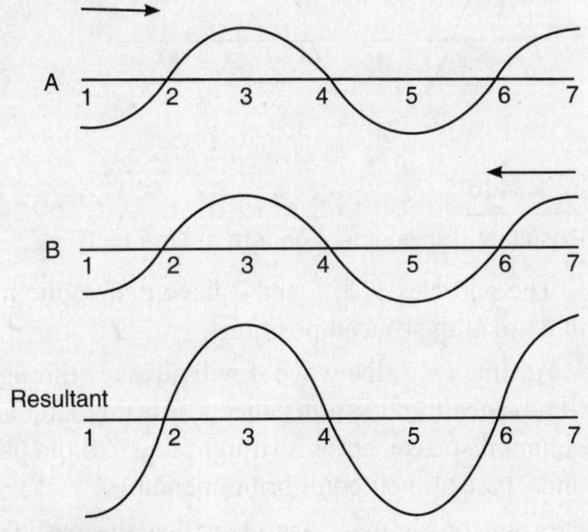

Fig. 1.34 (b): Resultant displacement pattern after a time $t = T/4$.

At time $t = T/2$, the wave A will advance through a distance $\lambda/2$ towards right while the wave B advances through a distance $\lambda/2$ towards the left. The resultant displacement pattern is shown in Fig. 1.34 (*c*).

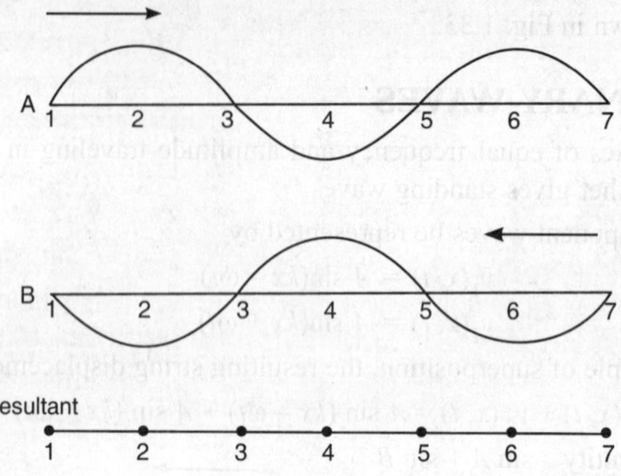

Fig. 1.34 (c):

Resultant displacement pattern after a time t = T/2.

All the particles of the medium are at their equilibrium positions.

At time $t = 3T/4$, the wave A will advance through a distance $3\lambda/4$ towards right and the wave B will advance through a distance $3\lambda/4$ towards left. The resultant displacement pattern is shown in Fig. 1.34 (*d*).

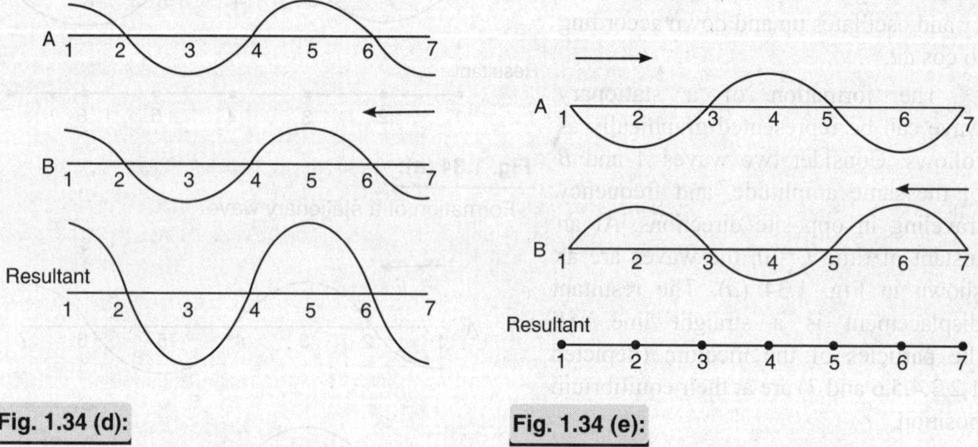

Fig. 1.34 (d):

Resultant displacement pattern at time t = 3T/4.

Fig. 1.34 (e):

Resultant displacement pattern at time t = T.

The particles 1, 3, 5 and 7 have undergone maximum displacement and particles at 2, 4, and 6 are at their mean positions.

At time $t = T$, the wave A will advance through a distance λ towards right and the wave B will advance through a distance λ towards left. The waves are as shown in Fig. 1.34 (*e*). The resultant displacement is a straight line. All the particles of the medium (depicted 1, 2, 3, 4, 5, 6 and 7) are at their equilibrium position.

From the figures, it is clear that the particles of the medium such as 2, 4, and 6 etc. always remain at their equilibrium positions. The particles such as 1, 3, 5, 7 etc. continue to vibrate simple harmonically about their equilibrium positions with double the amplitude of each wave, as shown in Fig. 1.34 (*f*). It appears as though the wave pattern is stationary in space.

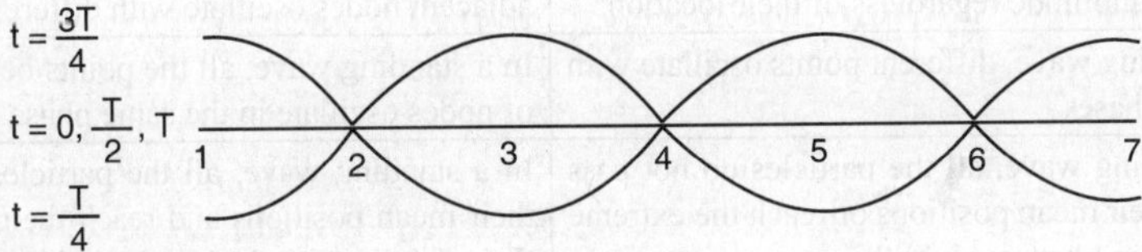

Fig. 1.34 (f):
Resultant displacement pattern.

The positions of the particles 2, 4, 6, etc. which always remain at their mean positions are called nodes. Node is a position of zero displacement and maximum strain. The positions of the particles 1, 3, 5, 7, etc. which vibrate simple harmonically with maximum amplitude are called **antinodes**. From Eq. (1.59), it is easy to see that the nodes are produced where $\sin kx = 0$, that is, where $kx = 0, \pi, 2\pi$ etc and the antinodes occur at points where $\sin kx = \pm 1$, that is, where $kx = \pi/2, 3\pi/2, 5\pi/2$, etc. The distance between any two consecutive nodes or antinodes is equal to $\lambda/2$. Between a node and an antinode, the amplitude gradually increases from zero to maximum.

Standing wave patterns are characterized by certain fixed points along the medium, which undergo no displacement, known as *nodes*. There are other points along the medium, which undergo vibrations between a large positive and large negative displacement, known as anti-nodes. A standing wave pattern consists of an alternating array of nodes and antinodes. When a standing wave pattern is formed in a medium, the nodes and the antinodes are always located at the same position along the medium; they are "standing still", so the name "standing wave."

The nodes are formed at locations where destructive interference occurs and antinodes formed at locations where constructive interference occurs. Antinodes are always vibrating up and down between the points of large positive and large negative displacement, because during a complete cycle of vibration, a crest will meet a crest; and then one-half cycle later, a trough will meet a trough. A diagram of a standing wave is usually depicted by drawing the shape of the medium at an instant in time and at an instant one-half vibrational cycle later as illustrated in Fig. 1.35. Nodes and antinodes are different from crests and troughs.

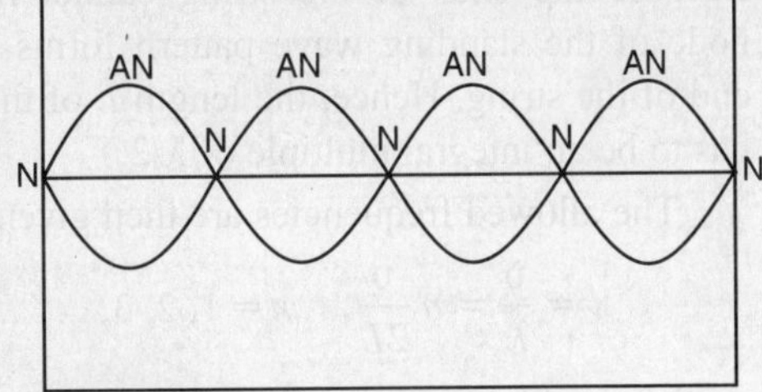

Fig. 1.35:
Nodes and anti nodes.

Table 1.1: *Distinction between Traveling Waves and Standing Waves*

	Traveling waves	Standing waves
1.	A traveling wave propagates in a medium continuously with a finite velocity.	A standing wave is stationary and does not move in the medium.
2.	A traveling wave transports energy from one location to the other. Hence there is energy flow across every plane in the direction of wave propagation.	There is no energy transfer in a standing wave. There is no flow of energy across any plane. The energy of oscillations periodically transform from kinetic energy to potential energy of the elastically deformed medium and vice versa.
3.	No particle on the wave is permanently at rest.	Nodes are permanently at rest in a standing wave.

4.	In a traveling wave all the points oscillate with the same amplitude regardless of their location.	All the points of a standing wave between two adjacent nodes oscillate with different amplitudes.
5.	In a traveling wave, different points oscillate with different phases.	In a standing wave, all the points between any pair of nodes oscillate in the same phase.
6.	In a traveling wave, all the particles do not pass through their mean positions or reach the extreme positions simultaneously.	In a standing wave, all the particles pass through their mean positions and reach their extreme positions simultaneously twice in each cycle.

Example 1.10: *Standing waves are formed by the superposition of two waves, $y_1 = 10 \sin (3\pi t - 4x)$ and $y_2 = 10 \sin (3\pi t + 4x)$. Find the amplitude of the motion, at $x = 18$.*

Solution: The resultant amplitude is given by

$$\begin{aligned} y &= y_1 + y_2 = 10 \sin (3\pi t - 4x) + 10 \sin (3\pi t + 4x) \\ &= 10[\sin 3\pi t \cos 4x - \cos 3\pi t \sin 4x + \sin 3\pi t \cos 4x + \cos 3\pi t \sin 4x] \\ &= 10[2 \sin 3\pi t \cos 4x] \\ &= 20 \cos 4x \sin 3\pi t \end{aligned}$$

The amplitude of motion is $(20 \cos 4x)$.

When $x = 18$, then $4x = 72 = [72 \times \pi/3.14]$ radians $= 22.9\,\pi$ radians

$\therefore$ Amplitude $= 20[\cos (22.9\,\pi)] = 20\,(0.9673) = 19.35$ units of length.

1.13.1 Harmonics

In general, standing waves form in a bounded medium. For instance, when a string is tied at both ends, standing waves set up, but set up only for a certain discrete set of frequencies. The standing wave patterns are called *oscillation modes.* Because the ends of the string cannot move, a node of the standing wave pattern forms at each end of the string. Hence, the length L of the string has to be an integral multiple of $\lambda/2$.

The allowed frequencies are then given by

$$\nu = \frac{\upsilon}{\lambda} = n\frac{\upsilon}{2L}, \quad n = 1, 2, 3, \ldots \quad \ldots(1.54)$$

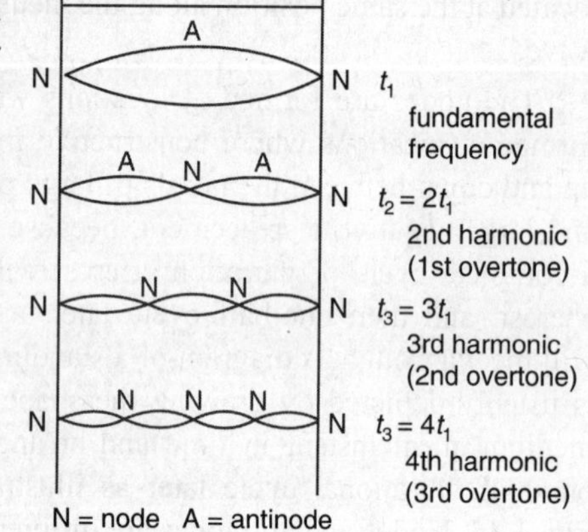

Fig. 1.36:

Harmonic series.

Def: Every allowed frequency is linked with a different standing wave pattern. These frequencies and their wave patterns are called ***harmonics***. The pattern with two nodes and one antinode is referred to as the first harmonic; that with three nodes and two antinodes is the second harmonic.

Harmonics are shown in the Fig. 1.36.

1.14 SUPERPOSITION OF TWO PERPENDICULAR SHMs

Waves that are characterized by a **scalar wave function** always interfere when they are superposed. Thus, sound waves, which are characterized by pressure fluctuations, always interfere when they are superposed. On the other hand, the displacement of a rope and the electric field in a light wave are examples of a **vector wave function**. For such waves, interference of two waves can occur only if the oscillations lie along the same line. For instance, if a rope is along the x-axis, then a wave with displacements along the y-axis would not interfere with waves having displacements along the z-axis.

In the earlier discussions on superposition, we tacitly assumed that the oscillations of the two superposing vibrations occur in the same plane. In case of two transverse waves described by vector functions, the oscillations cannot lead to interference, as the component of the oscillations in one plane onto the oscillations in the perpendicular plane is zero. However, the oscillations can mix to produce elliptical vibrations, as shown in the following discussion.

Let us consider two waves vibrating in mutually perpendicular directions with the *same* frequency and traveling along the same direction, i.e., z-direction.

Let $$x = A \cos (\omega t - kz) \qquad \text{...(1.55a)}$$

and $$y = B \cos (\omega t - kz + \delta) \qquad \text{...(1.55b)}$$

where δ is the phase difference between the x and y-oscillations and A and B are the amplitudes of oscillations.

(a) If the two waves are in phase, then $\delta = 0$, and

$$x = A \cos (\omega t - kz) \quad \text{and} \quad y = B \cos (\omega t - kz)$$

By eliminating $\cos (\omega t - kz)$ from the above two equations, we get

$$y = \frac{B}{A} x \qquad \text{...(1.56)}$$

Eq. (1.62) is the *equation of a straight line*.

(b) If the two motions are in opposite phase, then $\delta = \pi$, and it is easy to see that in this case

$$y = -\frac{B}{A} x \qquad \text{...(1.57)}$$

Eq. (1.57) also represents a straight line.

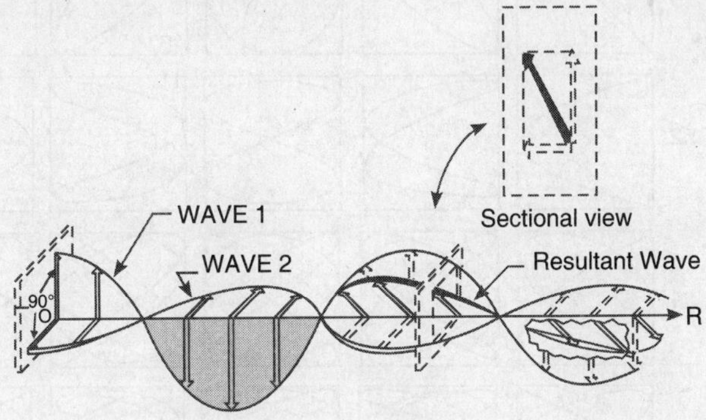

Fig. 1.37:
Resultant wave.

(c) If the phase difference δ remains constant, the tip of the resultant vector $\boldsymbol{E}$ describes a certain closed curve in the xy – plane. When $\delta = \pi/2$, the oscillations differ by a phase of 90°. The equations now become

$$x = A \cos (\omega t - kz)$$

and $$y = B \cos (\omega t - kz + \pi/2) = -B \sin (\omega t - kz)$$

By squaring and combining these relations, we get

$$\frac{x^2}{A^2} + \frac{y^2}{B^2} = 1 \quad \text{...(1.58)}$$

Eq. (1.58) is the general equation of an *ellipse*. Thus, at any particular time we find that the tip of the resultant vector traces out an ellipse. The same ellipse is obtained if $\delta = 3\pi/2$ or $-\pi/2$, but then the motion is in a counter-clockwise direction. Thus, we may say that when the phase difference δ is $\pm \pi/2$, the superposition of the two simple harmonic motions of the same frequency results in an **elliptical motion.**

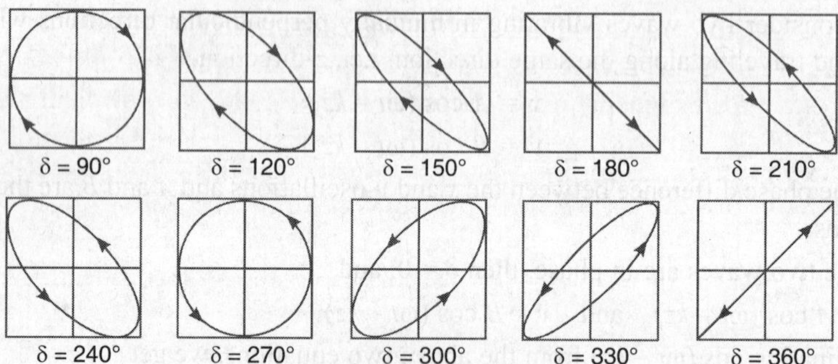

Fig. 1.38:

Phase difference of ± π/2- circular motion and Phase difference δ is ± π/2 -elliptical motion.

(*d*) In the particular case, when $\delta = \pi/2$ and $A = B$, Eq. (1.58) reduces to

$$x^2 + y^2 = A^2 \quad \text{...(1.59)}$$

φ 0 π/4 π/2 3π/4 π

frequency ratio

1:1

1:2

3:4

3:5

Fig. 1.39:

Lissajous figures.

Eq. (1.59) is the equation of a *circle*. When $A = B$, the axes of the ellipse transforms into a circle and we have **circular motion**. That is, circular motion can be generated by combining two oscillatory motions of the same frequency and amplitude along perpendicular direction but with a phase difference of $\pm \pi/2$.

If the waves are of different frequencies,

$$x = A\cos(\omega_1 t - kz) \quad \text{and} \quad y = B\cos(\omega_2 t - kz)$$

The resultant path depends on the ratio $\omega_2 : \omega_1$ and on the phase difference δ. These paths are called **Lissajous figures** and some typical figures are shown in Fig. 1.39.

1.15 DISPERSION

A truly sinusoidal wave has no beginning or end, either in space or time. In practice, waves are a mixture of waves of different frequencies, which travel with different speeds in a medium. For example, white light consists of waves of different frequencies. Similarly, a pulse is not a pure sinusoidal wave but consists of sinusoidal waves of different frequencies. A pulse traveling through a medium spreads out as it travels through a medium and undergoes a change in its shape. This phenomenon is known as **dispersion**. It is because of dispersion that sunlight spreads out into a spectrum of colours as it passes through a prism. The motion of waves in water is *dispersive*.

Waves and pulses on a stretched string travel at the same speed irrespective of their frequency as these waves are dispersionless and travel without deformation. Electromagnetic waves in unbounded free space are non-dispersive and non-dissipative and so only they can travel very long distances. Sound waves in air are also nearly non-dispersive even in the ultrasonic frequency range and so only high frequency notes and low frequency notes travel at same velocities.

1.15.1 Phase Velocity

In our discussion on wave motion, a strictly single frequency sinusoidal wave train, usually called a monochromatic wave, is used to represent the characteristics of wave propagation. A monochromatic wave train is an infinite sequence of waves in time and space of crests and troughs. Following Eq. (1.38), the equation of a harmonic wave propagating along the *x*-axis has the following form

$$y = A\sin[(kx - \omega t) + \varphi] \qquad \text{...(1.60)}$$

where ϕ is the initial phase of the wave which is determined by our choice of the beginning of counting x and t. Let us fix a value of the phase by assuming that

$$[(kx - \omega t) + \varphi] = \text{constant} \qquad \text{...(1.61)}$$

This expression determines the relation between the time t and the place x where the phase has a fixed value. The value of $\frac{dx}{dt}$ calculated from Eq. (1.61) gives the velocity with which the given value of the phase propagates.

$$k\frac{dx}{dt} - \omega = 0$$

$$\frac{dx}{dt} = \frac{\omega}{k} = \frac{2\pi v}{2\pi/\lambda} = v\lambda = \upsilon \qquad \text{...(1.62)}$$

Thus, the velocity of wave propagation υ is the velocity of phase propagation and it is therefore called the *phase velocity.* The phase velocity may be defined as the velocity of propagation of the wave front. When the waves are traveling through a non-dispersive medium, the common velocity of the waves is the phase velocity.

1.15.2 Wave Packet

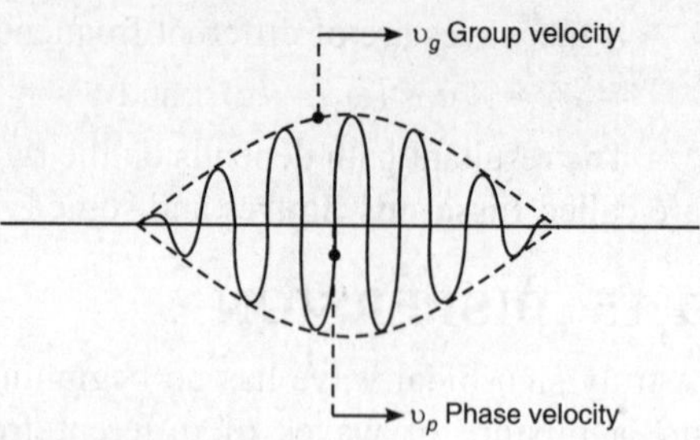

Fig. 1.40: Wave packet.

A harmonic wave is characterized by a precise wavelength λ and constant amplitude. It is non-localized and extends over a volume of space. In general real waves are of complex forms. In practice waves are far from monochromatic and can be regarded as the result of superposition of waves of a number of frequencies, each component wave having its own propagation velocity in a medium. The propagation velocity of a wave varies with frequency. The theory of Fourier analysis shows that a combination of harmonic waves with wave numbers spread over a range Δk will produce a **wave group** or **wave packet**. The superposition of a very large number of harmonic waves differing infinitesimally in frequency will produce a single wave packet. The waves cancel each other everywhere except in a small region. The wave packet is spread out in space over a length Δx (Fig. 1.40).

1.15.3 Group Velocity

If a wave packet travels through a medium without changing its shape over a long distance, then the medium is said to be a **non-dispersive** medium. Most media in nature are dispersive and no waveform can preserve its shape over a reasonable propagation distance. The wave group generally has the maximum amplitude at a particular value of x and the velocity of this maximum amplitude point is called the **group velocity**. Thus, the velocity at which a wave group (or a pulse) travels is the group velocity of the wave group. This velocity also represents the velocity with which energy of the wave group is transmitted.

1.15.4 Relation between Group Velocity and Phase Velocity

A wave packet contains several harmonic waves of differing wavelengths. Each component wave has its own phase velocity, $\upsilon = \nu\lambda$. The wave packet has amplitude that is large in a small region and very small outside it. The amplitude of the wave packet varies with x and t. Such variation of amplitude is called the *modulation* of the wave. The velocity of propagation of the modulation is known as the *group velocity*, υ_g. It is given by

$$\upsilon_g = \frac{d\omega}{dk} \qquad ...(1.63)$$

$$\upsilon_g = \frac{d(\upsilon k)}{dk} = \upsilon + k\frac{d\upsilon}{dk}$$

We further write

$$\frac{d\upsilon}{dk} = \frac{d\upsilon}{d\lambda}\frac{d\lambda}{dk}$$

But

$$\lambda = \frac{2\pi}{k}$$

Differentiating the above expression, we get

$$\frac{d\lambda}{dk} = -\frac{2\pi}{k^2} = -\frac{\lambda}{k}$$

$$\therefore \qquad k\frac{d\upsilon}{dk} = -\lambda\frac{d\upsilon}{d\lambda}$$

$$\upsilon_g = \upsilon - \lambda \frac{d\upsilon}{d\lambda} \quad ...(1.64)$$

This is the relation that connects phase velocity and group velocity.

The *phase velocity* υ of a wave is the velocity with which the wave front moves forward. It is the same as the velocity of propagation of the wave. When a number of plane waves of slightly different wavelengths travel in the same direction, they form wave groups or wave packets. The velocity with which the wave group advances in the medium is known as the *group velocity* υ_g. Phase velocity is a characteristic of an individual wave whereas group velocity characterizes a group of waves. Group velocity will be the same as phase velocity if the entire constituent waves travel with the same velocity. It means in a non-dispersive medium, $\upsilon_g = \upsilon$. However, the waves of different wavelengths travel in a medium with different velocities. Therefore, the group velocity is in general less than the phase velocity.

QUESTIONS

1. What do you understand by periodic and simple harmonic motion? What is the criterion for the motion to be simple harmonic?
2. Discuss diagrammatically the phase relationship of the velocity and acceleration of a particle executing simple harmonic motion.
3. How is the period of SHM changed when: (*a*) the mass of the particle is increased without changing the elastic constant? (*b*) When the elastic constant is increased without changing the mass? (*c*) When the mass and the elastic constant are changed by the same ratio?
4. Give some examples of simple harmonic oscillation.
5. Explain what is meant by natural frequency?
6. Assuming the damping to be proportional to the velocity, write down the differential equation for a damped harmonic oscillator.
7. Give reason for the energy dissipation in the case of a damped harmonic oscillator.
8. A damped oscillator is subjected to a damping force proportional to its velocity. Set up the differential equation of the oscillation. Discuss the under-damped, over-damped and critical damped motions of the oscillator.
9. Graphically show the displacement-time curve for oscillatory, over damped and critically damped motion of a damped oscillator. Mention the conditions of their occurrence.
10. Why are damping devices often used on machinery? Give an example.
11. What are forced vibrations? Give two examples.
12. A forced oscillator is at resonance with the external periodic force. What is the phase difference between the driving force and the velocity of the oscillator?
13. An oscillator is subjected to an external sinusoidal periodic force and a damping force proportional to its velocity. Set up a differential equation of the oscillator. Mention the condition under which velocity resonance occurs.
14. Mention any two physical phenomena where energy resonance occurs.
15. Why are the forced oscillations of a damped oscillator not damped?
16. Why must the force and the velocity be in phase at energy resonance?
17. Give some examples of common phenomena in which resonance plays an important role.
18. Give three examples of coupled oscillators.
19. What are normal modes?
20. The normal modes are independent of each other. Comment.

21. What are the main features of the normal modes of coupled oscillators?

22. Two simple pendulums of mass *m* and length *l* each, are coupled by a spring of force constant *k*. Write the expression for angular frequency of normal modes of vibration of the coupled system.

23. Give the characteristic of the in-phase mode of the motion of two coupled oscillators.

24. In the in-phase mode, the frequency of oscillation is the same as of uncoupled oscillators whereas in the out-of-phase mode, the frequency of oscillation gets raised. Comment.

25. How do the atoms in a crystal constitute coupled system?

26. Distinguish between progressive and stationary waves.

27. Define the transverse and longitudinal waves.

28. State the wave equation for one-dimensional motion and explain the terms.

29. Mention the characteristics of wave motion.

30. A wave transmits energy. Does it transfer momentum?

31. If two waves differ only in amplitude and propagate in opposite directions through a medium, will they produce standing waves?

32. In a stationary wave, the wavelength is 3.6 m. What is the distance between a node and the nearest antinode?

33. Identify the terms in the wave function given by $\Psi(r, t) = A \sin(\mathbf{k} \cdot \mathbf{r} - \omega t + \varphi)$

34. A wave packet propagates in a medium, which exhibits normal dispersion. Write the relationship between phase velocity and group velocity. Which is greater and why?

PROBLEMS

1. A particle in SHM has velocities υ_1 and υ_2 when its displacement from the mean position is x_1 and x_2 respectively. Find its period, amplitude and maximum speed.

$$\left(\textbf{Ans}: T = 2\pi\sqrt{\frac{(x_2^2 - x_1^2)}{(\upsilon_1^2 - \upsilon_2^2)}},\ a = \sqrt{\frac{(x_2^2\upsilon_1^2 - x_1^2\upsilon_2^2)}{(\upsilon_1^2 - \upsilon_2^2)}},\ \upsilon_{max} = \left[\frac{(x_2^2\upsilon_1^2 - x_1^2\upsilon_2^2)}{(x_2^2 - x_1^2)}\right]^{1/2}\right)$$

2. A particle executes SHM with a period of 0.002 s and amplitude 10 cm. Find its acceleration when it is 4 cm away from its mean position and also obtain its maximum velocity.

[Ans: $a = -3.9 \times 10^7$ cm/s^2, $\upsilon_{max} = 3.14 \times 10^4$ cm/s**]**

3. The length of a weightless spring increases by 2 cm when a weight of 1.0 kg is suspended from it. The weight is pulled down by 10 cm and released. Determine (*i*) Period of oscillation of spring and (*ii*) Kinetic energy of oscillation of the spring. **[Ans:** $T = 0.29$ s, K.E. = 2.45 J**]**

2 Mechanical Properties of Solids

Learning Objectives:

By the end of this chapter, you will be able to:

- Understand basic concepts of Solids materials.
- Explain Stress and Strain properties of elements.
- State Hooke's Law.
- Learn about plotting of Stress-Strain diagram.
- Describe about different Moduli of Elasticity and its related concepts.
- Explain about concept of Bending of Beams.

2.1 INTRODUCTION

An in-depth knowledge of the material properties under a varied range of conditions is essential for the practical use of engineering materials in manufacturing and construction engineering. The properties of material that govern its behaviour under the action of applied forces are known as *mechanical properties*. They are primarily linked to the elastic and plastic behaviour of the material. These properties are quantified in terms of stress, strain etc. A thorough knowledge of mechanical properties of materials helps us in predicting how materials behave under varied load conditions and designing the components to be made out of them for a particular end use.

2.2 SOLIDS

A solid has a definite shape and volume. It may be assumed that they are made up of particles (atoms) that are held rigidly together by interatomic forces. Ideally, the distance between particles (atoms) in a solid are fixed and remain constant. Such solids are said to be *rigid*. However, there is no perfect rigid body and diamond is nearest to rigid body.

Def: A body is assumed to be rigid if relative positions of the constituent particles of the body remains unchanged in spite of any amount of applied deforming forces.

Real solid bodies are not absolutely rigid and can be deformed by external forces. A body that is slightly deformed by an applied force, or load, will go back to its shape and dimensions when the force is removed. It means that real solids are elastic to some extent even if the deformation is not noticeable. It means that all solid materials are elastic to some extent. The property, because of which a body tends to get back its original shape and size after the taking away the load, is known as *elasticity*. On the other hand, a body which does not tend to regain its original size or shape when load is removed is said to be *plastic*. Again, there are no perfect plastic or elastic bodies. Quartz fibre is nearest to perfect elastic body and putty is nearest to perfect plastic body.

Some other useful definitions for better understanding the mechanical properties of solids are given hereunder.

Elasticity: Elasticity is the property of solid materials to return to their original shape or size after being deformed when external force applied is removed.

Elastic limit: Elastic limit is the maximum stress after the application of which the material is able to regain its original shape or size.

Plasticity: Plasticity is the characteristic of material of a body by virtue of which it does not regain its original size or shape when external force applied is removed.

Load or deforming force: Load or deforming force is the external force acting on a body which can change the form or dimensions of the body.

Restoring force: Restoring force is the force developed within the body which opposes the applied deforming force.

Fig. 2.1 shows the ball and spring model of a real solid. The balls represent the atoms and the springs the interatomic forces. Because of these forces the atoms stay in stable equilibrium positions. When a load acts, and deforms the solid the atoms are displaced from their equilibrium positions leading to a change in the interatomic distances. When the load is removed, the interatomic forces tend to restore the original positions. The body regains its original size and shape. This is how the elastic behavior of solids is visualized in microscopic nature of the solid.

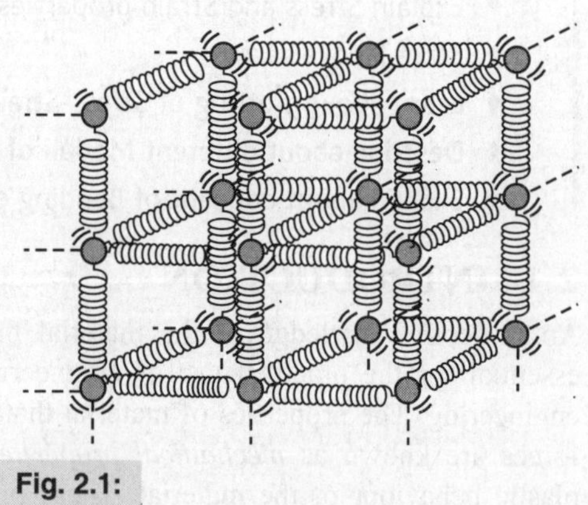

Fig. 2.1:

A Ball and Spring Model of solid.

2.3 STRESS AND STRAIN

The elastic properties of solids are generally discussed in terms of stress and strain. *Stress* is a gauge of the force causing a deformation. Strain is a relative measure of the deformation a stress causes.

When an externally applied force acts on a body, the body suffers a *deformation* which is opposed by internal forces that are set up within the particles of material due to cohesion. The sum total of all the elementary interatomic forces or internal resistances which the material exerts to counteract the applied load is called *stress*.

Let us consider a rectangular rod subjected to axial pull P (Fig. 2.2) and assume it to be cut into two halves at section XX. Each segment of rectangular bar is in equilibrium under the action of load P. Fig. 2.2 shows the internal

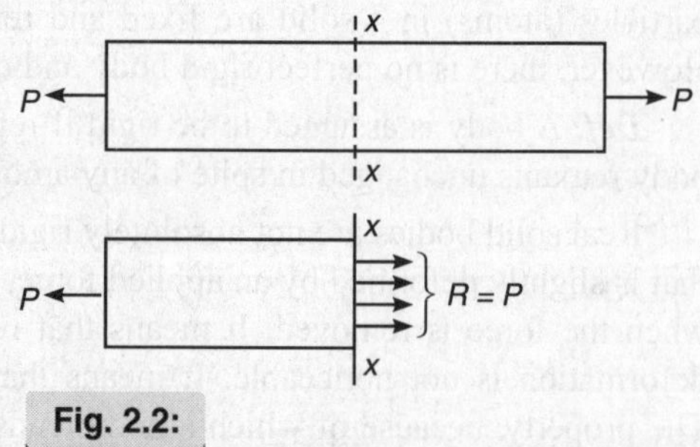

Fig. 2.2:

A rectangular rod subjected to axial pull P.

forces acting at the section XX. Internal resistance R at X-X is equal to the applied force P.

Stress is the force per unit area

$$\sigma = \frac{P}{A} \quad ...(2.1)$$

where A is the area of the X–X section.

The basic units of stress in S.I units are N/m^2 (or Pa).

Def: Stress is defined as the internal restoring force acting per unit area of cross section of the deforming body.

2.3.1 Types of Stresses:

There are only three basic stresses: (*i*) normal stress (*ii*) shear stress and (*iii*) bulk stress. Other stresses either are similar to these basic stresses or are a combination of these.

i. Normal stresses: If deforming force is applied normal to the area, then the stress is called normal stress. The normal stresses are generally denoted by a Greek letter (σ). The normal stresses can be *tensile* or *compressive* depending on the direction the stresses act out of or into the area.

(*i*) A normal stress is called *tensile* when the body is subjected to pull and there is an increase in length (Fig. 2.3). Under tensile stress the bar suffers stretching or elongation.

(*ii*) A normal stress is called *compressive* when the body is subjected to thrust and there is a decrease in length (Fig. 2.4). Under compressive stress the bar suffers shortening.

Def: When the normal stress changes the length of the body then it is called *longitudinal stress*.

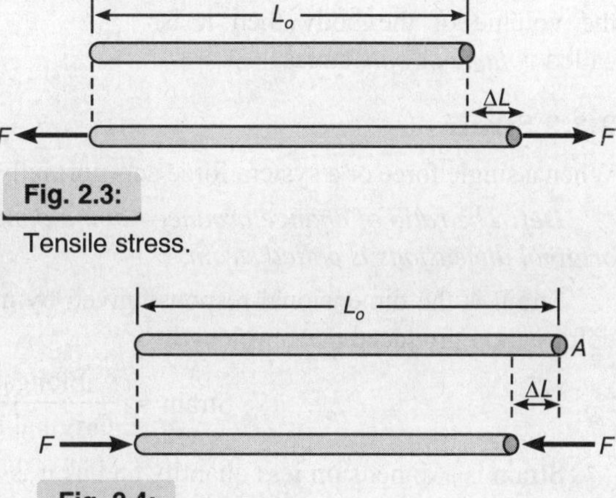

Fig. 2.3: Tensile stress.

Fig. 2.4: Compressive stress.

ii. Shear Stress: If deforming force is applied tangentially, then the stress is called tangential stress. Such forces are associated with a shearing of the material, and are known as shear forces. The resulting stress is known as *shear stress* (Fig. 2.5).

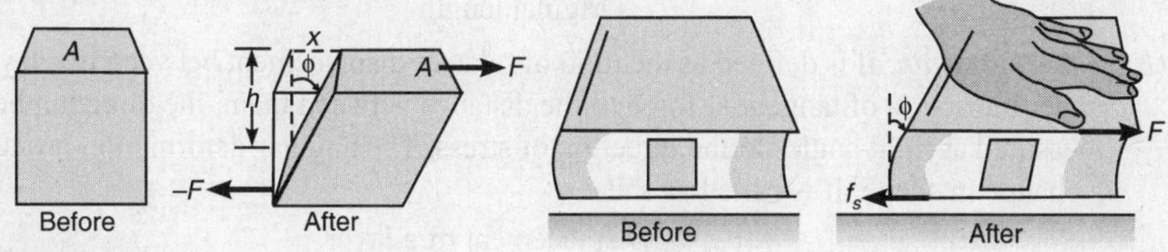

Fig. 2.5: Shear stress.

The shear stress is given by

$$\tau = \frac{F}{A} \qquad ...(2.2)$$

Where F is the total force and A is the area over which it acts.

Def: When the stress is tangential to the surface due to the application of force parallel to the surface then it is called *shearing stress*.

iii. Bulk stress: If the solid, say a rectangular block is immersed in a fluid, it is compressed uniformly on all sides. The force applied by the field acts in perpendicular direction at each point of the surface and the block is said to be under compression. This leads to decrease in its volume without any change of its geometrical shape (Fig. 2.6).

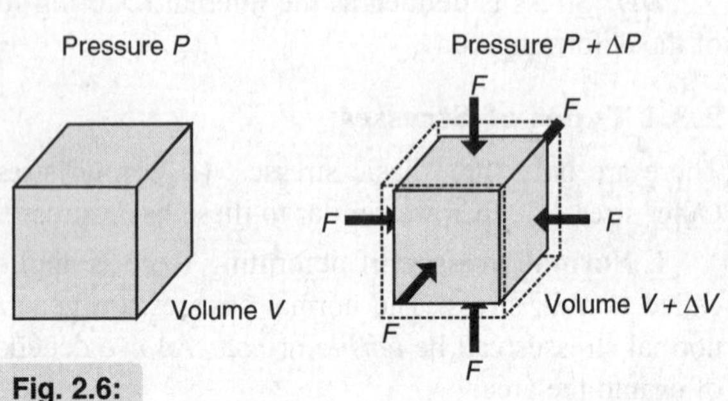

Fig. 2.6: Volume or Bulk stress.

Def: If the normal stress alters the volume of the body then it is called *volume* or *bulk* stress.

2.3.2 Strain

When a single force or a system force acts normally on a body, it undergoes some deformation.

Def: *The ratio of change produced in the dimensions of a body by system of forces to its original dimensions is called strain.*

Strain is the dimensional response given by material against mechanical loading, that is deformation produced per unit length.

$$\text{Strain} = \frac{\text{Elongation}}{\text{Original length}}$$

Strain is a dimension less quantity and as it is a ratio which has no units.

Strain is classified into three types depending upon the change produced in the body.

(*a*) *Longitudinal strain:* When the force acts along the length of the body, a change in length is produced without any change in shape. The ratio of change in length to the original length without any change in shape is called the longitudinal strain.

$$\text{Longitudinal strain} = \frac{\text{Change in the length}}{\text{Original length}} = \frac{\Delta l}{l} \qquad ...(2.3)$$

(*b*) *Shearing strain:* It is defined as the ratio of relative displacement between two layers, under the action of tangential force to the distance between them, the direction being measured at right angles to the direction of stress. The angular deformation produces a change in shape. It is called as shear.

$$\text{Shearing strain} = \frac{\text{Displacement of a layer}}{\text{Distance from fixed layer}} \qquad ...(2.4)$$

Fig. 2.7 shows the distortion produced by shear stress on an element. Let $ABCD$ be an element with side CD fixed and suppose a tangential force F is applied on the upper face AB as shown in Fig. 2.7. The shape of the block is changed to $ABC'D'$. The shear is expressed by an angle θ and is measured in radians and is dimensionless in nature.

Shearing strain is defined as the ratio of relative displacement of the faces Δx to the length of the body L.

$$\text{Shearing strain} = \frac{\Delta x}{L} = \tan\theta \qquad ...(2.5)$$

where θ is the angular displacement of the body from the vertical. Usually the angle θ is very small and $\tan\theta \approx \theta$.

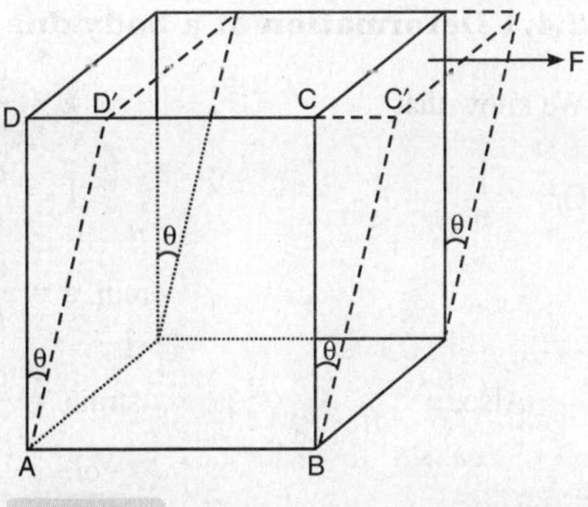

Fig. 2.7: Shifting of planes through an angle.

Thus,

$$\text{Shearing strain} = \tan\theta \approx \theta \qquad ...(2.6)$$

A shear stress cannot be effectively applied to a liquid or a gas. It is often said that fluids cannot support a shear.

(c) *Volume strain:* When the forces are applied normal to the surface of the body in all directions it undergoes a change in volume without any change in the shape. The ratio of change in volume to the original volume without any change in the shape is called the volume strain.

$$\text{Volume strain} = \frac{\text{Change in the volume}}{\text{Original volume}} = \frac{\Delta V}{V} \qquad ...(2.7)$$

2.4 HOOKE'S LAW

Robert Hooke proposed this law in the year 1679. This law gives the relationship between stress and strain.

Def: According to Hooke's law strain is proportional to stress producing it within the elastic limits.

$$k = \frac{\text{stress}}{\text{strain}} \qquad ...(2.8)$$

The constant k is coefficient of elasticity and is called *modulus of elasticity* and its unit is Pascal, N/m^2. It depends on nature of the material and conditions of the manufacturing. The elastic modulus has the same units as stress.

Hooks law holds good equally for tension and compression.

Example. 2.1: *When an iron wire of 1m length and radius 0.1 mm elongates by 0.32 mm stretched by a force 49 N. Calculate elastic constant for iron.*

Solution: Given, Force = 49 N, Radius = 0.1mm = 1×10^{-3} m, Length = 1 m, Elongation, $x = 0.32$ mm = 32×10^{-5} m

$$\text{Elastic constant} = \frac{\text{stress}}{\text{strain}} = \frac{F/\pi r^2}{x/l}$$

$$\text{Elastic constant} = \frac{49 \times 1}{\pi\,(1 \times 10^{3})\,(32 \times 10^{-5})}$$

$$\text{Elastic constant for iron} = 9.8 \times 7 \times 10^{10}\ N/m^2$$

2.4.1 Deformation of a body due to load acting on it

We know that $$k = \frac{\text{stress}}{\text{strain}}$$

Or $$k = \frac{\sigma}{\varepsilon}$$

$\therefore$ $$\text{strain, } \varepsilon = \frac{\sigma}{k} = \frac{F}{Ak}$$

Also, $$\text{strain} = \frac{\delta l}{l}$$

$\therefore$ $$\frac{\delta l}{l} = \frac{F}{Ak}$$

Or $$\text{Deformation, } \delta l = \frac{Fl}{Ak} \quad \text{...(2.9)}$$

2.5 DIFFERENT MODULI OF ELASTICITY

Def: Elastic modulus is a coefficient that relates a particular type of stress to the strain.

Three general types of elastic moduli are connected with stresses that create changes in length, shape, or volume, known as Young's modulus, shear modulus and bulk modulus.

(*i*) Linear stress - Young's Modulus or modulus of length

(*ii*) Shearing stress - Shear Modulus, Modulus of Rigidity or Shape

(*iii*) Volume stress - Bulk Modulus or Modulus of Volume

2.5.1 Young's Modulus

The quantity that describes the material response to stresses applied normal to opposite faces is called Young's modulus in honor of the British scientist Thomas Young.

Def: When a wire or rod is stretched by a longitudinal force, the ratio of longitudinal stress to longitudinal strain within elastic limit is called Young's modulus. Young's modulus is represented by *Y* and unit is as that of pressure, N/m^2.

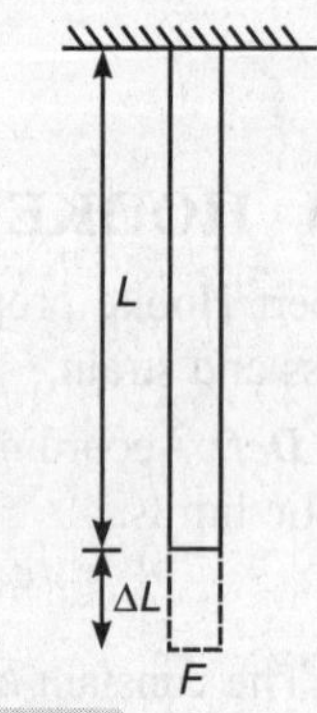

Fig. 2.8: Illustration showing the change in the length.

Young's Modulus, $$Y = \frac{\text{longitudinal stress}}{\text{longitudinal strain}} = \frac{F/A}{\Delta L/L}$$

$$Y = \frac{FL}{A\,\Delta L} \quad \text{...(2.10)}$$

where *F* is applied force, *A* is area of cross section, *l* is the length of the wire and ΔL is change in the length as shown in Fig. 2.8.

SI unit of Young's modulus is newtons per square meter (N/m^2). Some typical values of Young's modulus are shown in Table 2.1.

It is easy to see from Eq. (2.10) that

$$\Delta L = \left(\frac{FL}{A}\right)\frac{1}{Y}$$

or $$\Delta L \propto \frac{1}{Y} \qquad ...(2.11)$$

It means that the larger the Young's modulus of a material, the smaller is the relative change in length.

Young's Modulus is used to describe the elastic properties of objects like wires, rods or columns when they are stretched or compressed. It is a measure of stiffness of an elastic material and is a good overall indicator of its strength. A rigid material has a high Young's modulus. It changes its shape marginally under elastic loads (e.g. diamond). On the other hand, a *flexible* material has a low Young's modulus and changes its shape significantly under loading (e.g. rubbers).

Table 2.1: *Elastic Moduli for Various Materials (in N/m²)*

Substance	*Young's Modulus*	*Shear Modulus*	*Bulk Modulus*
Aluminium	7.0×10^{10}	2.5×10^{10}	7.0×10^{10}
Bone (limb)	1.5×10^{10}	8.0×10^{10}	
Brass	9.0×10^{10}	3.5×10^{10}	7.5×10^{10}
Copper	11×10^{10}	3.8×10^{10}	12×10^{10}
Glass	5.7×10^{10}	2.4×10^{10}	4.0×10^{10}
Iron	15×10^{10}	6.0×10^{10}	12×10^{10}
Steel	20×10^{10}	8.2×10^{10}	15×10^{10}
Alcohol, ethyl			1.0×10^{10}
Glycerin			4.5×10^{10}
Mercury			26×10^{10}
Water			2.2×10^{10}

2.5.2 Shear Modulus

The resistance to shear loads in solids is called *modulus of rigidity*.

Def: The ratio of tangential force per unit area to angular deformation produced is called rigidity modulus. Rigidity modulus is represented by G and its unit is as that of pressure, N/m².

Consider a solid cube *ABCDEFGH* as shown in the Fig. 2.9. The lower face *CDGH* is fixed and a tangential force F is applied over the upper face *ABEF*. A couple is formed by this force because an equal and opposite force comes into play on the lower fixed face. It makes the layers parallel to two faces and moves one over another. As a result that the point A shifts to A', B to B', E to E' and F to F' through an angle θ.

$$\theta = l / L$$

where $AA' = EE' = l$ = relative displacement of upper face of the cube with respect to lower fixed face. L is the length of the cube.

The angle θ measured in radians through which a line originally perpendicular to the fixed face is turned is a measure of shearing stress.

$$\text{Rigidity Modulus} \qquad G = \frac{\text{Tangential stress}}{\text{Shearing strain}}$$

$$G = \frac{F/A}{\theta} = \frac{T}{\theta} \qquad ...(2.12)$$

where A is the area of the face, L^2 and

T is the tangential stress.

From Table 2.1, it is seen that the shear modulus is generally less than the Young's modulus. G is approximately Y/3 for many materials, which implies that there is a greater response to a shear stress than a tensile stress. Liquids and gases do not possess Young's moduli or shear moduli as they do not support shear stress.

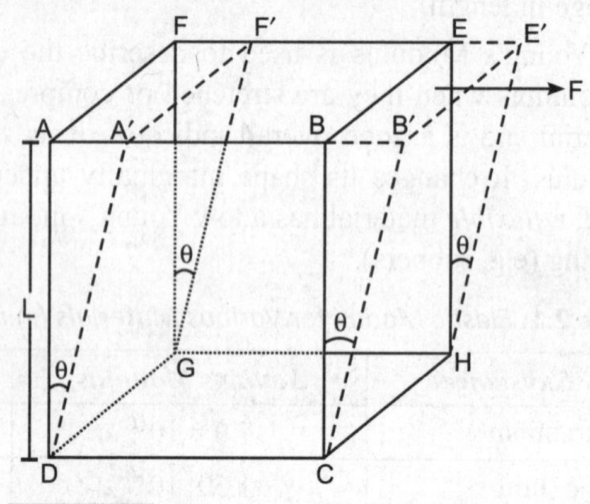

Fig. 2.9:

Shifting of planes through an angle in solid cube.

2.5.3 Bulk Modulus

When force is applied normally and uniformly to the whole surface of the body such that the shape remains the same and change occurs in volume, the body is compressed.

Def: Bulk modulus of elasticity is defined as the ratio of normal stress to the volume strain. It is represented by K and the SI unit of K is as that of pressure, N/m^2.

$$\text{Bulk Modulus, } K = \frac{\text{Bulk stress}}{\text{Volume strain}}$$

$$K = \frac{F/A}{\Delta V/V} = \frac{FV}{A\,\Delta V} = \frac{PV}{\Delta V} \qquad ...(2.13)$$

where F is applied force, A is area of cross section, V is the volume of the body, ΔV is change in the volume, and P is the bulk pressure.

2.6 POISSON'S RATIO

When a bar is subjected to a tensile loading there is longitudinal elongation (increase in length) of the bar in the direction of the applied load and also a sidewise contraction (decrease in a lateral dimension) perpendicular to the load. As a result, the bar becomes longer but also thinner. The strain produced along the direction of the force is called *longitudinal strain* and that produced in perpendicular direction is called *lateral strain.*

Def: The ratio of the lateral strain to the longitudinal strain is called the Poisson's ratio and is denoted by ν. For most steel, it lies in the range of 0.25 to 0.3, and 0.20 for concrete.

If a bar of original length L and width W are subjected to tensile stress in the direction as shown in the Fig. 2.10 and Fig 2.11, the length of the bar increases by ΔL from L to $L + \Delta L$ and the width decreases from to $W - \Delta W$.

$$\text{Poisson's ratio } \nu = \frac{\text{Lateral strain}}{\text{Longitudinal strain}} = \frac{\mu}{\lambda} = \frac{\Delta W/W}{\Delta L/L} \qquad ...(2.14)$$

For example, when an elastic band is stretched it also gets thinner. Poison ratio relates the thinning to the stretching and can be calculated by dividing the change in diameter of the elastic band by the change in length.

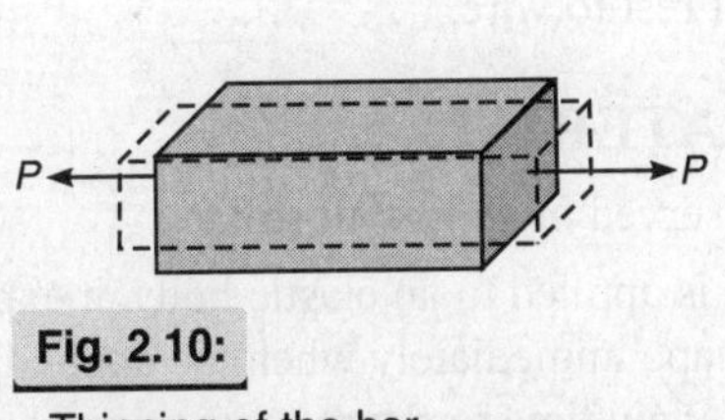

Fig. 2.10:

Thinning of the bar

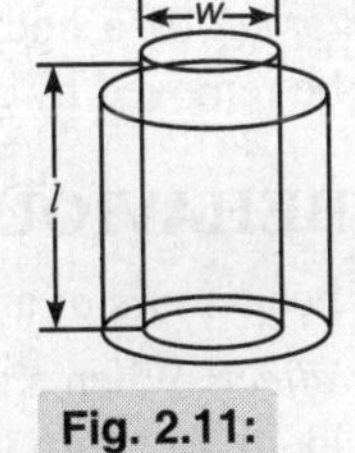

Fig. 2.11:

Thinning of a rod

Most of the materials resist a change in volume, as determined by bulk modulus K more than a change in shape, as determined by shear modulus G. So only usually Poisson's ratio is positive. At atomic level the inter-atomic bonds realign with deformation.

2.7 STRESS-STRAIN DIAGRAM

Stress-Strain diagram gives the behavior of a wire under progressive tension. We can study the connection between stress and strain by plotting a graph between the two for various values of stress and accompanying strain. The Fig. 2.12 shows the stress-strain diagram for a wire. In Fig. 2.12, a straight and sloping part *OA* of the curve shows that the strain produced is directly proportional to stress applied. *OA* obeys Hooke's law perfectly. In this region, the wire regains its original length when unloaded and so it is called the range of *perfect* elasticity. At *A* wire reaches *elastic limit.*

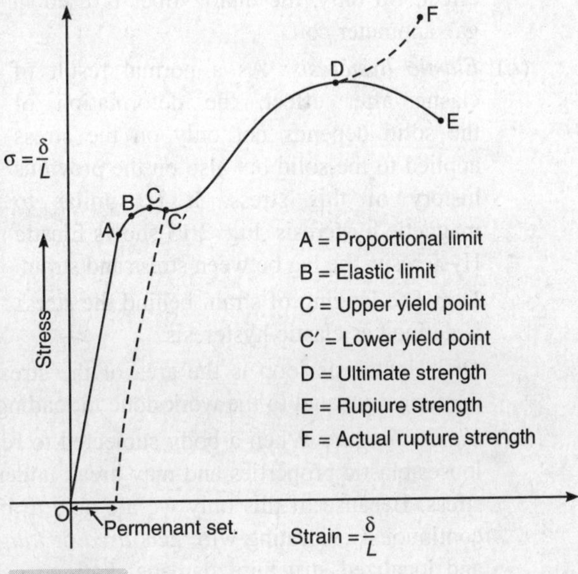

Fig. 2.12:

Stress-Strain diagram.

After the elastic limit is crossed, the strain increases more rapidly than stress. In *AB* part of the curve, the extension of the wire is partly elastic and partly plastic. At point *B* even after unloading it does not come back to original condition along *OA*. It takes dotted path *BC* so that there remains residual stress *OC* in it and *permanent set* is acquired by the wire. At point *B* even the addition of very small load causes enormous elongation. The points *B* and *D* are called super and lower yield points.

After *D* wire becomes plastic. *DF* is called tensile strength of the wire. *E* is the breaking point. The load at which the wire breaks is called the breaking weight. The maximum load to which wire can be subjected divided by the original cross sectional area is called the breaking stress. It is obvious that work has to be done in stretching the wire to destruction under increasing stress. The work is stored in the wire in the form of potential energy similar to compressed spring.

It is found that even within the elastic limit, practically all substances takes some time to reach their original length after deforming force is removed. This delay in recovering back is called *elastic after effect*. If a body is continuously subjected to stress and strain, it gets fatigued. Fatigue can be removed by giving sufficient rest to wire.

2.8 ELASTIC BEHAVIOUR OF A MATERIAL

The following elastic behaviours of a material are observed in almost all solids.

(*i*) *Elastic after effect:* When a stretching force is applied to an elastic body, it expands. The body is likely to return to its original shape immediately when the applied force is removed. However, some materials take certain time to return to its original shape. This temporary delay in attaining its original shape is called the *elastic after effect*.

Def: The delay in recovering back original condition on ending the deformation force is called elastic after effect.

Elastic after effect is more for glass. Quartz, silver, and gold have less elastic after effect. So only, the quartz fiber is used in galvanometer coil.

(*ii*) *Elastic hysteresis:* As a normal result of elastic after effect, the deformation of the solid depends not only on the stress applied to the solid but also on the previous history of this stress. It is similar to magnetic hysteresis. Fig. 2.13 shows Elastic Hysteresis- the lag between stress and strain.

Def: The lagging of strain behind the stress is defined as elastic hysteresis.

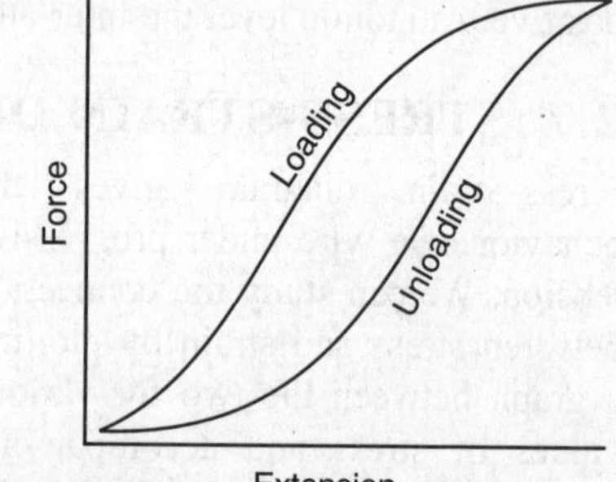

Fig. 2.13: Elastic Hysteresis- the lag between stress and strain.

Def: Hysteresis loop is the area of the stress–strain curve. The hysteresis loop is numerically equal to the work done in loading the material and then unloading it.

(*iii*) *Elastic fatigue:* When a body subjected to repeated strains beyond its elastic limits looses elastic properties and may break under a stress less than its normal breaking stress. Because of this only we are able to break a wire by repeated bending. The continuously vibrating wire gets *tired* or *fatigued* (weakening). It is the progressive and localized structural damage. Fatigue occurs when a material is subjected to repeated loading and unloading.

Def: Elastic fatigue is defined as the temporary loss of elastic properties because of the action of repeated alternating deforming force.

2.9 FACTORS AFFECTING ELASTICITY

The elastic property of materials is related to fine structure of materials. Generally the following the factors affects elasticity.

(*i*) *Effect of Hammering, Rolling and Annealing:* Effect of hammering and rolling breaks the crystal grains along the cleavage planes into smaller units. This results in an increase in elastic properties. Annealing, heating and then cooling gradually decrease elastic properties, as annealing produces larger crystal grains.

(*ii*) *Effect of Impurities:* Generally impurities are added in metals for better crystal grains without disturbing their orientation, for example, carbon in iron and potassium in gold. These impurities naturally affect the elastic properties of metal in which they are added.

(iii) *Effect of change in Temperature:* Elastic properties of materials are also affected by change in temperature. A rise in temperature usually decreases elasticity. Carbon filament which is highly elastic at ordinary temperature becomes plastic when current passes through it.

The amount of elasticity of a material is determined by modulus which is a measure of amount of force per unit area required to achieve a given amount of deformation. When a body is strained work has to be done to deform the body. This work done is stored up in the body as potential energy. When the applied forces are removed, the stress disappears and the energy of strain appears as heat.

Example 2.2: *A load of 2 kg produces an extension of 1mm in a wire of 3 m and 1mm in diameter. Calculate Young's modulus of a wire.*

Solution: Given that Load = 2 kg, $W = 2 \times 9.8 = 196$ N, Original length, $L = 3$ m, Increase in length = 1 mm = 0.001 m, Diameter of wire = 1 mm

$$\text{Young's modulus} = \frac{\text{longitudinal stress}}{\text{longitudinal strain}}$$

$$Y = \frac{FL}{Al} = \frac{FL4}{l\pi r^2} = \frac{196 \times 3 \times 4}{0.001 \times \pi \times (0.001)^2}$$

$$Y = 7.48 \times 10^{10} \text{ N/m}^2$$

Example 2.3: *What force is required to stretch a steel wire to double the length when its area of cross section is 1 cm²? Given Young's modulus of wire is 7×10^{10} N/m².*

Solution: Given that Original length = L, Increase in length = l, Area of cross section = 1 Sq cm = 10^{-4} m²

$$Y = \frac{FL}{Al}$$

$$7 \times 10^{10} = \frac{F}{10^{-4}}$$

$$\text{Force, } F = 7 \times 10^{6} \text{ N}$$

2.10 RELATION BETWEEN DIFFERENT MODULI OF ELASTICITY

Consider a cube (Fig. 2.14) of unit length of an isotropic material such that one of its vertexes coincides with the origin. Let us assume that three stress P, Q, and R are acting perpendicular to three faces $ABCD$, $BCGE$, $EBFA$. Each one of these stresses produce extension in its own direction and compression along other two perpendicular directions.

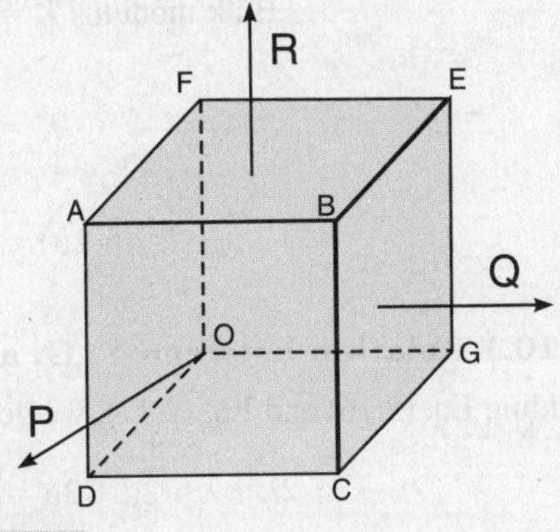

Fig. 2.14: Stresses acting on the cube faces.

If λ is the extension per unit stress, the elongation along the direction of $P = P\lambda$.

If μ is the contraction per unit length per unit stress, the contraction along the direction of P due to other two stresses will be μQ and μR.

Let us assume that all three stresses act simultaneously on the cube.

Net elongation along the direction of $P = e = \lambda P - \mu Q - \mu R$

Net elongation along the direction of $Q = f = \lambda Q - \mu P - \mu R$

Net elongation along the direction of $R = g = \lambda R - \mu P - \mu Q$

We can express three elastic constants Y, G and K in terms of λ and μ.

(*i*) Suppose only stress P acts and $Q = R = 0$: In this case we have simple longitudinal stress.

Linear strain, $e = \lambda P$

$$\text{Young's modulus } Y = \frac{\text{stress}}{\text{strain}} = \frac{P}{\lambda . P} = \frac{1}{\lambda}$$

or

$$\lambda = \frac{1}{Y} \qquad \text{...(2.15)}$$

(*ii*) Suppose stresses $R = 0$ and $P = -Q$: Elongation along the direction of $P = e = \lambda P - \mu(-P) = (\lambda + \mu) P$

The angle of shear $\theta = 2e = 2(\lambda + \mu) P$

$$\text{Rigidity modulus } G = \frac{\text{stress}}{\text{angle of shear}} = \frac{P}{\theta}$$

$$G = \frac{P}{2(\lambda + \mu)P} = \frac{1}{2(\lambda + \mu)}$$

or

$$2(\lambda + \mu) = \frac{1}{G} \qquad \text{...(2.16)}$$

(*iii*) Suppose stresses $P = Q = R$: In this case body is subjected to uniform stress in all directions.

Increase in the volume or bulk strain,

$$3e = 3(\lambda P - \mu Q - \mu R)$$

$$3e = 3(\lambda P - \mu P - \mu P)$$

$$3e = 3(\lambda P - 2\mu P)$$

$$\text{Bulk modulus } K = \frac{\text{stress}}{\text{bulk strain}}$$

$$K = \frac{P}{3(\lambda - 2\mu)P} = \frac{1}{3(\lambda - 2\mu)}$$

$$(\lambda - 2\mu) = \frac{1}{3K} \qquad \text{...(2.17)}$$

2.10.1 Relation between Y, G, and K

Adding Eq. (2.16) and Eq. (2.17), we get

$$2\lambda + \lambda - 2\mu + 2\mu = \frac{1}{G} + \frac{1}{3K}$$

$$3\lambda = \frac{1}{G} + \frac{1}{3K} = \frac{3K + G}{3GK}$$

$$\lambda = \frac{3K + G}{9\,GK} \qquad ...(2.18)$$

From Eq. (2.10), and Eq. (2.18)

$$\lambda = \frac{3K + G}{9\,GK} = \frac{1}{Y} \qquad ...(2.19)$$

$$Y = \frac{9\,GK}{3K + G} \qquad ...(2.20)$$

2.10.2 Relation between G, K and *v*

Subtracting Eq. (2.17) from Eq. (2.16)

$$(\lambda + \mu) - (\lambda - 2\mu) = \frac{1}{2G} - \frac{1}{3K}$$

$$3\mu = \frac{1}{2G} - \frac{1}{3K}$$

$$\mu = \frac{3K - 2G}{18\,GK} \qquad ...(2.21)$$

Using Eq. (2.18) and definition of Poisons ratio, $v = \dfrac{\mu}{\lambda}$

$$v = \frac{\mu}{\lambda} = \frac{3K - 2G}{18GK} \times \frac{9GK}{3K + G} = \frac{3K - 2G}{6K + 2G} \qquad ...(2.22)$$

2.10.3 Relation between Y, G and *v*

Dividing Eq. (2.16) by Eq. (2.15), we get

$$\frac{\lambda + \mu}{\lambda} = \frac{Y}{2G}$$

$$1 + \frac{\mu}{\lambda} = \frac{Y}{2G}$$

$$1 + v = \frac{Y}{2G}$$

$$v = \frac{Y}{2G} - 1 \qquad ...(2.23)$$

2.10.4 Limits to the value of Poisson's ratio

Rearranging Eq. (2.22), $v = \dfrac{3K - 2G}{6K + 2G}$, we get

$$3K\,(1 - 2v) = 2G\,(1 + v) \qquad ...(2.24)$$

where bulk modulus K and rigidity modulus G have essentially positive values. So,

(*a*) If Poisson's ratio, v be a positive quantity, the right hand side of Eq. (2.24) and hence left hand side must be positive.

For this to happen $(1 - 2v)$ must be positive or $v < \dfrac{1}{2}$

(*b*) If Poisson's ratio, v be a negative quantity, the left hand side of Eq. 2.24 and hence right hand side must be positive.

For negative v, $(1 - 2v)$ will be positive and hence $(1 + v)$ must be positive or $v < -1$

Hence, theoretically the limiting values Poisson's ratio, v are $\frac{1}{2}$ and -1.

In actual practice v is positive and lies between 0 and 0.5. Otherwise either bulk modulus or rigidity modulus would become infinite. Negative value of Poisson's ratio, v means on being extended a body should also expand internally and this condition is not possible.

2.11 DETERMINATION OF ELASTIC MODULI

The elastic moduli can be determined in a laboratory. We consider here the determination of rigidity modulus of a material. We take the material in the form of a wire, which can be treated as a thin cylinder. Modulus of rigidity value is determined by a torsion test.

2.11.1 Rigidity Modulus – Torsional Pendulum

Theory-Expression for torque per unit twist: When a body is fixed at one end and twisted about its axis by means of torque at the other end, the body is said to be under torsion. Torsion involves shearing strain and so the modulus involved is rigidity modulus.

Consider the cylinder to consist of an infinite number of hollow co-axial cylinders. Consider one such cylinder of radius x and thickness dx. A line such as AB which is initially parallel to the axis OO' of the cylinder is displaced to the position AB' through an angle ϕ due to twisting torque Eq. (2.15). The result of twisting the cylinder is a shear strain.

The angle of shear $\angle BAB' = \phi$.

Now $BB' = x\,\theta = L\varphi$

$$\therefore \quad \phi = \frac{x\theta}{L}$$

From rigidity modulus $$G = \frac{\text{shearing stress}}{\text{angle of shear } \phi}$$

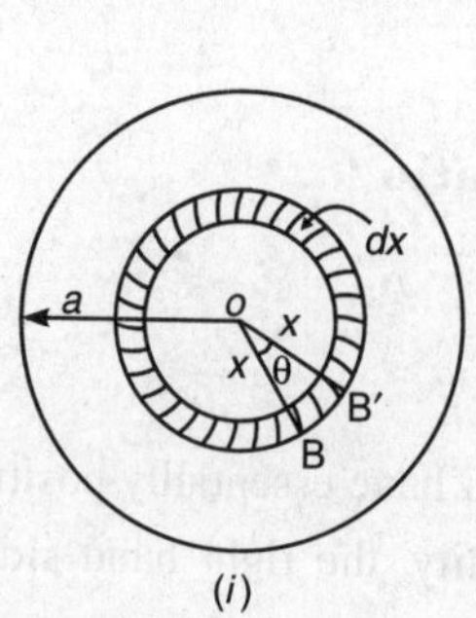

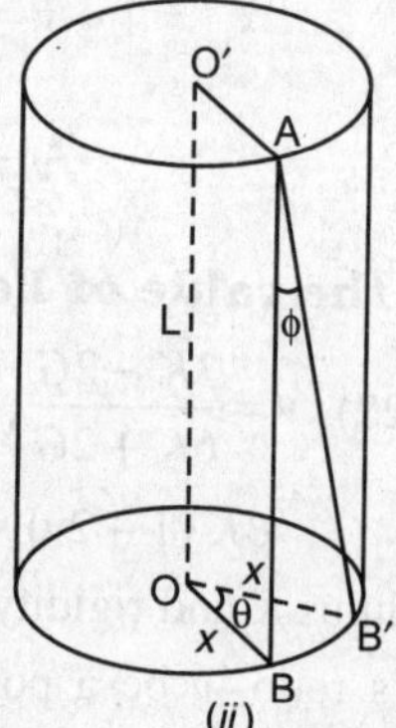

Fig. 2.15: Cylinder of radius *x*.

$$\text{Shearing stress} = G\phi = \frac{Gx\theta}{L}$$

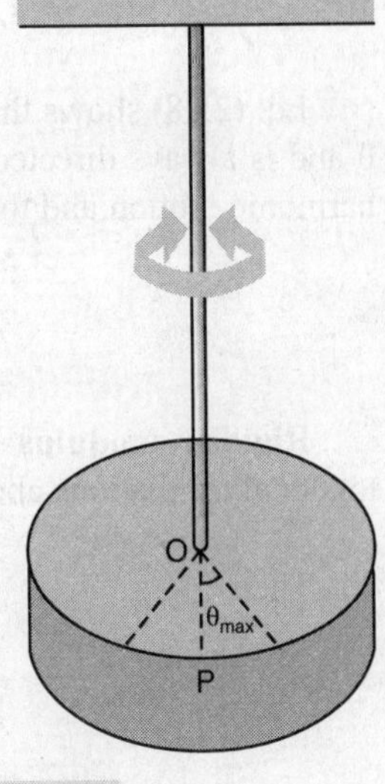

Fig. 2.16: Torsional pendulum.

As $$\text{Shearing stress} = \frac{\text{Shearing force}}{\text{Area on which force acts}}$$

So, Shearing force = shearing stress × area on which force acts.

$$\text{Area on which force acts} = 2\pi x dx$$

Hence the shearing force

$$F = \frac{Gx\,\theta}{L} 2\pi x dx$$

The moment of this force about the axis OO' of the cylinder

$$\frac{Gx\theta}{L} 2\pi x dx = \frac{2\pi\theta G}{L} x^3 dx$$

Twisting torque on the whole cylinder

$$C = \int_0^a \frac{2G\pi\theta}{L} x^3 dx$$

$$C = \frac{G\pi\theta a^4}{2L}$$

When $\theta = 1$, the torque per unit twist is given

$$C = \frac{G\pi a^4}{2L} \quad \text{...(2.25)}$$

When an external torque is applied on the cylinder to twist it, at once an internal torque due to elastic forces comes into to play and at equilibrium both are equal and opposite of each other.

Torsional oscillations of the body: Suppose a wire is clamped vertically at one end and the other end carries a disk bar or cylinder of moment of inertia I about the wire as the axis. Let the length, radius and rigidity modulus of the wire be respectively l, a, and G. When the body is given a slight rotation by applying a torque, the wire is twisted. If the body is released, the body oscillates in the horizontal plane about the wire as axis. These oscillations are called torsional oscillations and the arrangement is known as *torsion pendulum* (Fig. 2.16).

Let us consider the energy of the vibrating system when the angle of the twist is θ. Let ω be the angular velocity of the body.

$$\text{Total energy (KE + PE) of the system} = \frac{1}{2} I \frac{d^2\theta}{dt^2} + \frac{1}{2} C\theta^2 = \text{Constant} \quad \text{...(2.26)}$$

Differentiating Eq. (2.26) with respect to time, we get

$$\frac{1}{2} I^2 \frac{d\theta}{dt}\frac{d^2\theta}{dt^2} + \frac{1}{2} C\theta \frac{d\theta}{dt} = 0$$

$$I\frac{d^2\theta}{dt^2} + C\theta = 0 \quad \text{...(2.27)}$$

where I is momentum of inertia of the disk about the axis of the wire and the minus sign indicates that the couple $G\theta$ tends to decreases.

Eq. (2.27) can be rewritten as

$$\frac{d^2\theta}{dt^2} = -\frac{C}{I}\theta \qquad ...(2.28)$$

Eq. (2.28) shows that the angular acceleration is proportional to the angular displacement θ and is always directed towards the mean position. Hence the motion of the disc is simple harmonic motion and the time period of the vibration is given by

$$T = 2\pi\sqrt{\frac{\text{Displacement}}{\text{Acceleration}}} = 2\pi\sqrt{\frac{\theta}{\frac{C}{I}\theta}}$$

Rigidity modulus by torsion pendulum: The system of torsional pendulum executes torsional oscillations about the axis of the wire. The perioof oscillations T is determined.

Then
$$T = 2\pi\sqrt{\frac{I}{C}} \qquad ...(2.29)$$

$$T^2 = \frac{4\pi^2}{C}I \qquad ...(2.30)$$

$$C = \frac{4\pi^2 I}{T^2} \qquad ...(2.31)$$

where

I is the moment of inertia of the whole system about the axis of the wire and C is the torque per unit twist.

If C is known, the moment of inertia I can be determined from the measurements of T.

The torsion constant C can be written as

$$C = \frac{\eta\pi r^4}{2l} \qquad ...(2.32)$$

where, l is the length of wire, η is rigidity modulus of the material of the wire, r is the radius of the wire.

The moment of inertia of circular disc is

$$I = \frac{MR^2}{2} \qquad ...(2.33)$$

where, M is the mass of the disk and R is the radius of the disk

$$\eta = \frac{8\pi l I}{T^2 r^4} \qquad ...(2.34)$$

Using Eq. (2.33) and Eq. (2.34)

$$\eta = \frac{4\pi MR^2 l}{T^2 r^4} \qquad ...(2.35)$$

Example 2.4: *Calculate Rigidity Modulus and Poisson's ratio for a metal of $Y = 7.25 \times 10^{10}$ N/m^2 and $K = 11 \times 10^{10}$ N/m^2.*

Solution: Given $Y = 7.25 \times 10^{10}$ N/m^2 and $K = 11 \times 10^{10}$ N/m^2

Rigidity Modulus = ? Poisson's ratio = ?

From the relation between Y, G, and K we have

$$Y = \frac{9GK}{3K + G}$$

$$G = \frac{3KY}{9K - Y} = \frac{3\,(1\times10^{10})\,(7.25\times10^{10})}{9\,(11\times10^{10}) - (7.25\times10^{10})}$$

Rigidity Modulus, $G = 2.60 \times 10^{10}$ N/m^2

From the relation between Y, G and v we havePoisson's ratio $v = \frac{Y}{2G} - 1$

$$v = \frac{Y}{2G} - 1 = \frac{7.25\times10^{10}}{2\,(2.60\times10^{10})} - 1$$

$\therefore$ Poisson's ratio is 0.391

Example 2.5: *A metal disk of 0.1 m radius and mass 1kg is suspended in a horizontal plane by a vertical wire attached to its centre. If the diameter of the wire is 10^{-3} m, its length 1 m and period of torsional vibrations is 4 s find the rigidity modulus of the wire.*

Solutions: The metal disk executes torsional oscillations about the axis of the wire. According to Eq. (2.29)

$$T = 2\pi\sqrt{\frac{I}{C}}$$

where, I is the moment of inertia of the whole system about the axis of the wire and

$C = \frac{\eta\pi r^4}{2l}$ is the torque per unit twist.

Rigidity modulus of the wire, $\eta = \frac{8\pi l I}{T^2 r^4}$

As $L = 0.1$ m, $r = 0.5 \times 10^{-3}$ m, $T = 4$ s

The moment of inertia of circular disc is

$$I = \frac{MR^2}{2}$$

where, M is the mass of the disk = 1 kg, R is the radius of the disk = 0.1 m

$$I = \frac{1\times(0.1)^2}{2} = 0.005 \text{ kg m}^2$$

$$\eta = \frac{8\pi l I}{T^2 r^4}$$

$$\therefore \quad \eta = \frac{8\pi\,(0.005)\,1}{4^2(0.5\times10^{-3})^4} = 1.256\times10^{11} \text{ Nm}^{-2}$$

Example 2.6: *A body suspended symmetrically from the lower end of the wire 1m long, 1.22 × 10^{-3} m in diameter, oscillates about the wire as the axis with a period of 1.25 s. If the modulus of rigidity of the material of the wire is 8 × 10^{10} Nm^{-2}, calculate the moment of inertia of the body about the axis.*

Solution: The body executes torsional oscillations about the axis of the wire with time period

$$T = 2\pi\sqrt{\frac{I}{C}}$$

where, I is the moment of inertia of the whole system about the axis of the wire and

$C = \frac{\eta\pi r^4}{2l}$ is the torque per unit twist.

$$T^2 = \frac{4\pi^2}{C} I$$

$$\therefore \quad I = T^2 \frac{C}{4\pi^2} = T^2 \frac{\pi C r^4}{2L4\pi^2} = \frac{T^2 C r^4}{8L\pi}$$

Here $T = 1.25$s, $C = 8 \times 10^{10}$Nm^{-2}, $L = 1$m, $r = 10^{-3}$ m

$$\therefore \quad I = \frac{(1.25)^2 (8 \times 10^{10}) (0.61 \times 10^{-3})^4}{8\pi \times 1} = 6.885 \times 10^{-4} \text{kg m}^2$$

Example 2.7: *What force must be applied to a wire of 1m long, 10^{-3} m in diameter in order to twist one end of it through 90°, the other end remaining fixed? The rigidity of the material of the wire is 2.8×10^{10}Nm^{-2}.*

Solution: Here, $L = 1$m, $\eta = 2.8 \times 10^{10}$ Nm^{-2}, $r = 10^{-3}/2$, m $= 0.5 \times 10^{-3}$m, $\theta = 90° = \pi/2$ radians

$$C = \frac{\eta\pi r^4}{2l} = \frac{(2.8 \times 10^{10})\,\pi\,(0.5 \times 10^{-3})^4}{2 \times 1}$$

$$= 4.318 \times 10^{-3} \text{ Nm}$$

Example 2.8: *Explain why a hollow rod is better shaft than a solid one of the same mass, length and material.*

Solution: Consider a solid cylinder of length l, radius r, and shear modulus G.

The torque required to twist the solid cylinder through an angle θ

$$C_1 = \frac{\eta\pi r^4}{2l}$$

Let r_1 and r_2 be the inner and outer radii of hollow cylinder same mass, length and material.

The torque required to twist it through an angle θ.

$$C_2 = \frac{\eta\pi(r_2^4 - r_1^4)}{2l}$$

$$\frac{C_1}{C_2} = \frac{(r_2^4 - r_1^4)}{r^4} = \frac{(r_2^2 + r_1^2)(r_2^2 - r_1^2)}{r^4}$$

Since the two cylinders have same mass

$$\pi(r_2^2 - r_1^4)\,1\rho = \pi r^2 1\rho$$

where ρ is the density of the material of the cylinders.

$$(r_2^2 - r_1^2) = r^2$$

Adding $2r_1^2$ to the both sides,

$$\frac{C_2}{C_1} = \frac{(r^2 + 2r_1^2)\,r^2}{r^4} = \frac{(r^2 + 2r_1^2)}{r^2}$$

$$\frac{C_2}{C_1} = 1 + \frac{2r_1^2}{r^2}$$

Therefore $C_2 > C_1$

The twisting torque for a hollow cylinder is greater than that for a solid cylinder of the same mass, length and material.

Hence, a hollow cylinder is stronger and a better shaft than a solid one of the same mass, length and material.

Example 2.9: *One end of a wire of 4 mm radius and 100 cm in length is twisted through 60°. Find the angle of shear on its surface.*

Solution: If ϕ is the twist at the free end and θ is the angle of shear then

$$\theta = \frac{r l}{l}$$

As $l = 1$ m, $r = 4 \times 10^{-3}$ m, $\phi = 60°$

$$\theta = 4 \times 10^{-3} \times 60° = 0.24°$$

2.12 BENDING OF BEAMS

Following definitions are useful to understand the bending of beams and bending moments.

(*i*) *Beams:* Beams are designed to support heavy loads. Beam is defined as a rod or bar of uniform cross section whose length is much greater as compared to diameter so that shearing stresses over any section of it are negligibly small.

(*ii*) *Bending couple:* Beam bends when it is fixed at one end and loaded at other end. The load acting vertically downwards at its free end and the reaction acting vertically upward at fixed end constitute the bending couple. The couple due to this equal and opposite forces tends to bend the beam in clockwise direction and the movement of this couple is called *bending moment*. When beam is in equilibrium, external bending movement is equal to internal bending movement.

(*iii*) *Plane of bending:* Plane of bending is the plane on which bending takes place and the bending couple acts. In Fig. 2.17 the plane of the paper is plane of bending.

(*iv*) *Neutral axis:* When a beam is bent, filaments in the upper part of the beam are elongated and filaments in lower part of the beam are compressed. Some filaments in between these are neither elongated nor compressed and such filament is known as the *neutral filament* and the axis of the beam lying on the neutral filament is the *neutral axis*. The change in

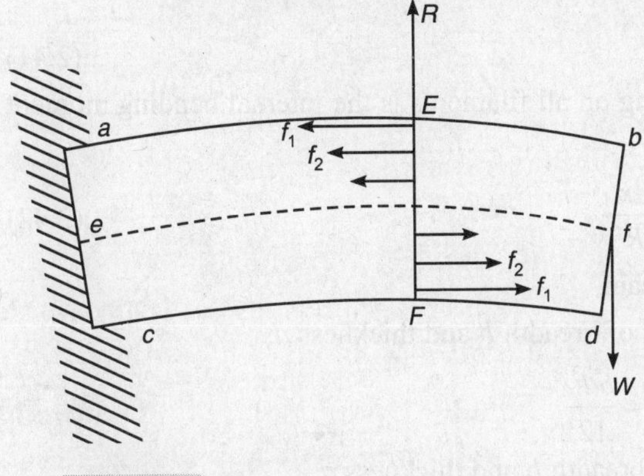

Fig. 2.17: Bending of beams.

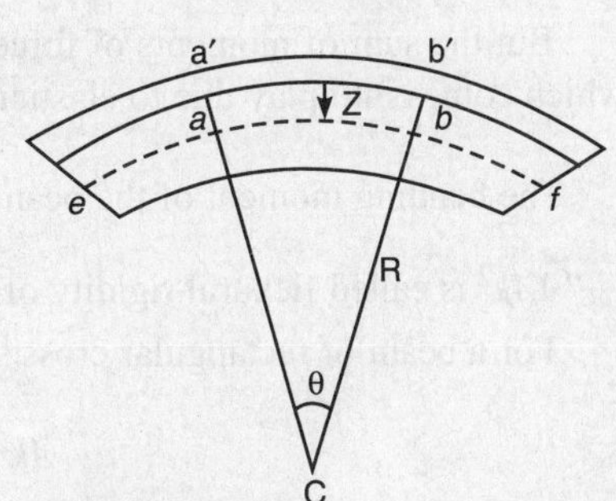

Fig. 2.18: Bending of beam into circular arc.

the length of any filament is proportional to the distance of the filament from the neutral axis. In Fig. 2.17, filament *ab* is elongated filament, filament *cd* is compressed filament and filament *ef* is neutral axis.

2.12.1 Bending Moment of Beam

Fig. 2.18 shows the bending beam with *ef* as neutral axis. Let R be the radius of the curvature of the neutral axis and θ be the angle subtended by radius at its centre of curvature C.

The filament *ef* is neutral axis and filaments above *ef* are elongated while filaments below *ef* are compressed.

Let '*ab*' be a filament at a distance z from neutral axis. The length of this filament $a'\,b'$ before bending is equal to neutral axis length ab. Let δA is the area of the cross section of the filament.

$$\text{Original length of the filament} = ab = R\,\theta$$

$$\text{Elongated length} = a'b' = (R + z)\,\theta$$

$$\text{Increase in the length of the filament} = a'b' - ab = (R + z)\theta - R\theta = z\,\theta \qquad ...(2.36)$$

$$\text{Linear strain} = \frac{\text{increase in length}}{\text{original length}} = \frac{z\theta}{R\theta} = \frac{z}{R} \qquad ...(2.37)$$

$$\text{From Young's modulus,} \quad Y = \frac{\text{longitudinal stress}}{\text{longitudinal strain}}$$

$$\text{Stress} = \text{Young's modulus, } Y \times (\text{linear strain}) = Y\frac{z}{R} \qquad ...(2.38)$$

$$\text{Tensile force on area } \delta A \text{ of the filament} = \text{stress} \times \text{area} = \frac{Yz}{R}\delta A \qquad ...(2.39)$$

$$\text{The moment of above tensile force about the filament} = \frac{Yz}{R}\delta A \times z = \frac{Y\delta A z^2}{R}$$

$$\text{Sum of the moments acting on all filaments} = \frac{Y\Sigma\delta A z^2}{R} \qquad ...(2.40)$$

$\Sigma\delta Az^2$ is called as the geometrical moment of inertia, (l_g) of the section about *ef*

$\Sigma\delta Az^2$ is written as equal to Ak^2, where A is the area of cross section of the beam and k is the radius of gyration.

$$\therefore \quad eAk^2 = l_g \qquad ...(2.41)$$

But the sum of moments of forces acting on all filaments is the internal bending moment which comes into play due to elasticity.

$$\text{The bending moment of the beam} = \frac{YAk^2}{R} \qquad ...(2.42)$$

YAk^2 is called flexural rigidity of the beam.

For a beam of rectangular cross section of breadth b and thickness d,

$$Ak^2 = l_g = \frac{bd^3}{12}$$

For a beam of circular cross section of breadth b and thickness d,

$$Ak^2 = l_g = \frac{\pi r^4}{4}$$

2.12.2 Depression of Cantilever

A beam fixed horizontally at one end and loaded at other end is known as cantilever. We can have two cases, when the weight of the beam itself produces no bending and when it does so.

Depression of cantilever loaded at the free end: Here we study the case of depression of cantilever loaded at the free end when the weight of the beam is ineffective.

Let OA be the cantilever of length l fixed at O (Fig. 2.19.) and loaded with weight W at the other end. OA' is the unstrained position of the beam. Let the depression $A'A$ of the free end be y.

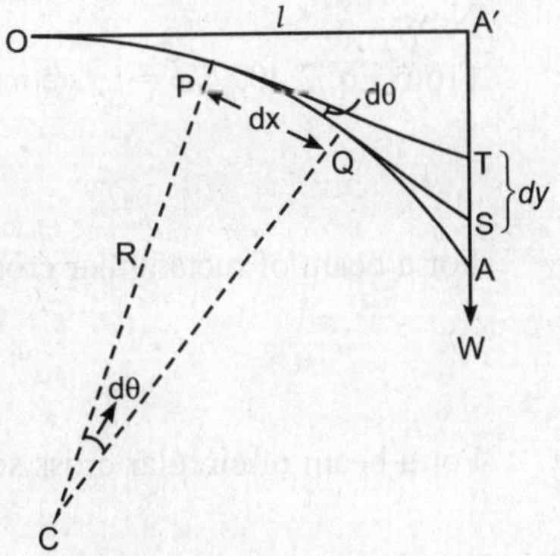

Fig. 2.19: Loaded end of cantilever.

Let us consider an element PQ of the beam of the length dx at a distance ($QA = x$) from the loaded end.

C is the centre of curvature of the element PQ and R its radius of curvature.

The load W at A and force of reaction W at Q gives external couple, which bends the beam.

So, external bending moment $= W x$

The internal bending moment $= \dfrac{YAk^2}{R}$

At equilibrium, as external bending moment is equal to internal bending moment,

$$W_x = \frac{YAk^2}{R}$$

$$R = \frac{YAk^2}{Wx} \quad \text{...(2.43)}$$

Draw tangents at P and Q meeting the vertical line at T and S respectively.

Let us assume that $TS = dy$ and angle between the tangents, $PCQ = d\theta$.

$$PQ = dx = R\,d\theta,$$

$$d\theta = \frac{dx}{R}$$

Substituting R from Eq. 2.43, $d\theta = \dfrac{dxWx}{YAk^2}$ ...(2.44)

Depression $dy = dx\theta = \dfrac{dxWx^2}{YAk^2}$ ...(2.45)

Total depression of the cantilever is obtained by integrating the expression for dy between the limits $x = 0$ and $x = l$.

$$y = \int_0^l \frac{Wx^2}{YAk^2}dx$$

$$y = \frac{Wl^3}{3YAk^2} \quad \text{...(2.46)}$$

Now,

From Eq. 2.40, $Ak^2 = I_g$, geometrical moment of inertia

$$y = \frac{Wl^3}{3YI_g} \quad ...(2.47)$$

For a beam of rectangular cross section of breadth b and thickness d, $I_g = \frac{bd^3}{12}$

$$y = \frac{4Wl^3}{Ybd^3} \quad ...(2.48)$$

For a beam of circular cross section of breadth b and thickness d, $I_g = \frac{\pi r^4}{4}$

$$y = \frac{4Wl^3}{3Y\pi r^4} \quad ...(2.49)$$

2.12.3 Depression at midpoint of cantilever loaded at the middle

Let AB represent a beam of length l supported on two knife edges at A and B and loaded with a weight W at the centre C. The reaction at each knife edge is $W/2$ acting vertically upwards. From the Fig. 2.20 of bending beam, we can observe maximum depression at the centre. The bending is non-uniform and depression is $CD = y$.

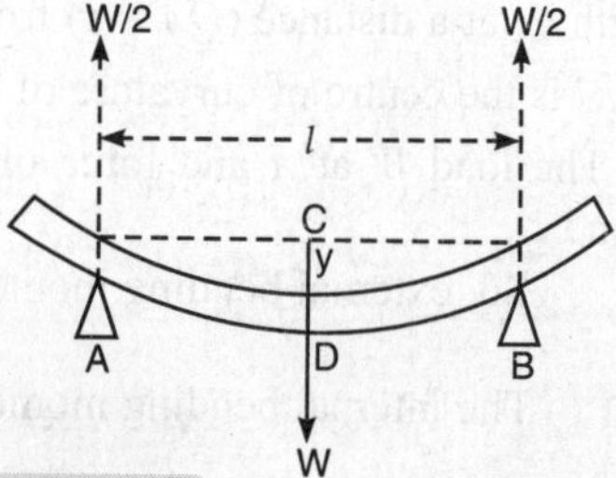

Fig. 2.20: Depression at mid point of cantilever loaded at the middle.

The portion DA of the beam may be considered as a cantilever of length $l/2$, fixed at D and bending upwards under a load $W/2$.

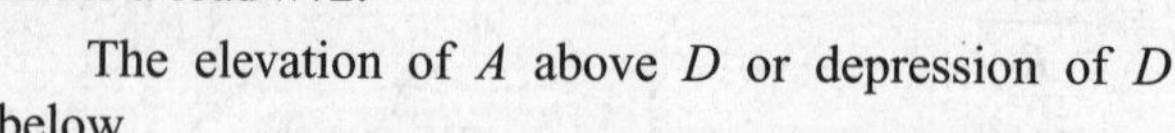

The elevation of A above D or depression of D below

$$A = y = \frac{(W/2)\,(l/2)^3}{3YAk^2}$$

$$\therefore \quad y = \frac{Wl^3}{48YAk^2} \quad ...(2.50)$$

where, Y is Young's modulus and y is the depression.

For a beam of rectangular cross section of breadth b and thickness d, $I_g = \frac{bd^3}{12}$

$$y = \frac{Wl^3}{4Ybd^3} \quad ...(2.51)$$

For a beam of circular cross section of breadth b and thickness d, $I_g = \frac{\pi r^4}{4}$

$$y = \frac{Wl^3}{12Y\pi r^4} \quad ...(2.52)$$

2.12.4 Measurement of Young's modulus of a bending Beam - Non-Uniform Bending

The Young's modulus of a beam loaded at the centre is determined by knowing its dimensions. Using Eq. 2.51.

$$y = \frac{Wl^3}{48YAk^2} \quad ...(2.53)$$

$$\therefore \quad Y = \frac{Wl^3}{48YAk^2}$$

A rectangular shaped beam is convenient for measuring moment of inertia. For a beam of rectangular cross section of breadth b and thickness d, $I_g = Ak^2 = \frac{bd^3}{12}$ and W = Mg.

$$\therefore \quad Y = \frac{Mgl^3}{4ybd^3} \quad ...(2.55)$$

where

Y = Young's Modulus of the material of the beam

y = Elevation produced for 'M' Kilogram of load

M = Mass suspended on either sides of the beam

g = Acceleration due to gravity

l = Distance between the two knife edges

b = Breadth of the beam

d = Thickness of the beam

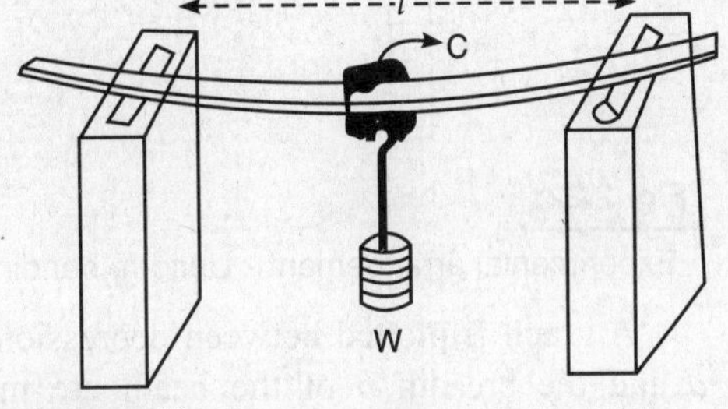

Fig. 2.21: Experimental arrangement- Non-Uniform bending.

Fig. 2.21 shows the experimental arrangement. A beam is placed symmetrically on two knife edges. A weight hanger is suspended by means of a loop of thread exactly mid-way between two knife edges. A pin is fixed vertically at C by some wax. A traveling microscope is focused on the tip of the pin such that the horizontal cross wires coincide with the tip of the pin. The reading in the vertical traverse scale of microscope is used. Weights are added in equal steps and corresponding readings are noted. Similarly, readings are taken while unloading the weights. This gives mean depression y.

A graph is plotted between depression and load W. The slope gives W/y. The distance between the knife edges, length of the beam, l is noted with the help of metre scale. Breadth b and the thickness d of beam are measured with screw gauge. By substituting the values W/y, l, b *and* d the value of Young's modulus is calculated.

2.12.5 Measurement of Young's modulus of a bending Beam - Uniform Bending

Def: When a beam is placed over the two knife edges A and B and two weight hangers are suspended, one each on either side of the knife edge at equal distance from the knife edge, the bending is uniform throughout the beam and the bending of the beam is called *Uniform Bending*.

Fig. 2.22 shows the experimental arrangement of Young's modulus of a bending Beam - Uniform Bending. A pin is fixed vertically at the middle of the beam. A traveling microscope is positioned in front of the beam. Considering the weight hangers (without load) as the dead load, the microscope is focused on the tip of the pin and the reading of the traveling microscope is recorded.

Add equal weights on both the weight hangers and note the readings using the microscope. Repeat the same procedure and reach maximum load. Repeat the same procedure by unloading the weight from both the weight hanger in same steps as that of loading and take the readings.

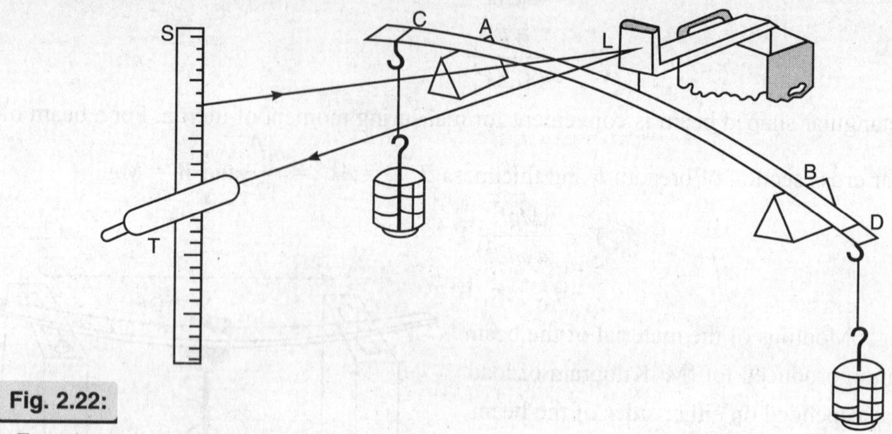

Fig. 2.22:

Experimental arrangement - Uniform bending.

A graph is plotted between depression and load W. The slope gives W/y. The thickness d and the breadth b of the beam are measured using screw gauge and vernier calipers respectively and are tabulated.

Young's modulus of the given material of the beam

$$Y = \frac{3gDMl^3}{ybd^3} \quad \text{...(2.56)}$$

where

Y = Young's modulus of the material of the beam

y = Elevation produced for 'M' Kilogram of load

M = Mass suspended on either sides of the beam

g = Acceleration due to gravity

l = Distance between the two knife edges

b = Breadth of the beam

d = Thickness of the beam

D = Distance between the weight hanger and any one of the adjacent Knife edge

Every element of the beam is bent with same radius of curvature in case of uniform bending and in non-uniform bending radius of curvature is not same for all elements in the beam.

2.12.6 Girder

Girder is an example of non-uniform bending beam. It is a supporting horizontal beam. It can be prepared from any of construction materials such as stainless steel, concrete, or a combination of these materials.

Girders shapes and designs are shown in Fig 2.23. I-beam girders so called because their shape is suggestive of a capital Roman letter I. I shaped girder is so designed that it should not bend too much or break under its own weight.

The depression (y) at the center of a beam of length l, breadth b and thickness d under a load Mg at its mid-point is given as

$$y = \frac{Mgl^3}{4bd^3Y} \quad ...(2.57)$$

where

Y = Young's modulus of the material of the beam

y = Elevation produced for 'M' Kilogram of load

M = Mass suspended on either sides of the beam

g = Acceleration due to gravity

l = Distance between the two knife edges

b = Breadth of the beam

d = Thickness of the beam

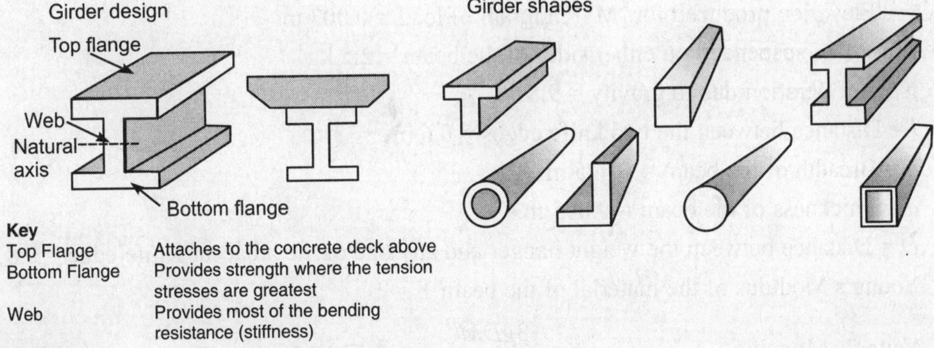

Fig. 2.23:

Girders shapes and designs.

Hence to reduce the bending for a given load, Young's modulus Y of the material of the beam should be large, b and d of the beam must also be large. The length should be as small as possible. Since depression y is inversely proportional to d^3, the depression can be reduced more effectively by increasing the thickness d rather than increasing the breadth b of the beam. But on increasing the thickness, unless the load is at the centre, the beam may bend. This is called buckling of the beam.

To prevent buckling, a large load-bearing surface is required. Hence, the beam is designed to have a large thickness to minimize bending and a large load bearing surface to prevent buckling. The shape which satisfies these conditions is *I*. So it is called the I section of the beam or girder.

Example 2.10: *The end of a cantilever depresses 10 mm under certain load. Calculate depression under same load of another cantilever of double dimensions in length and width 3 times thickness.*

Solution: Let the dimension of first cantilever be

Length = l, Width = b, Thickness = d, Depression, y = 10 mm = 10×10^{-3} m

For second cantilever, Length = $2l$, Width = $2b$, Thickness = $3d$

Depression, y' = ?

For a beam of rectangular cross section of breadth b and thickness d, $I_g = \frac{bd^3}{12}$

$$y = \frac{4Wl^3}{Ybd^3}$$

$$y' = \frac{4W(2l)^2}{Y(2b)(3d)^2} = \frac{4}{27}\left(\frac{4Wl^3}{Ybd^3}\right)$$

$$= \frac{4}{27}y = \frac{4}{27}(10 \times 10^{-3})$$

Depression $y' = 1.48$ mm

Example 2.11: *Determine the Young's modulus of a uniform bending rod by a distance of 0.6m and loads of 2.5 kg are hanging at 0.18 m away from knife edges. The breadth and thickness of the rod are 0.025 m and 0.005 m respectively elevation is 0.007 m.*

Solution: Given that

y = Elevation produced for 'M' Kilogram of load = 0.007 m.

M = Mass suspended on either sides of the beam = 2.5 kg

g = Acceleration due to gravity = 9.8

l = Distance between the two knife edges = 0.6 m

b = Breadth of the beam = 0.025 m

d = Thickness of the beam = 0.005 m

D = Distance between the weight hanger and any one of the adjacent Knife edge = 0.8 m

Young's Modulus of the material of the beam Y = ?

Young's Modulus $$Y = \frac{3gDMl^3}{ybd^3}$$

$$Y = \frac{3gDMl^3}{ybd^3} = \frac{3 \times 9.8 \times 0.18 \times 2.5 \times (0.6)^2}{2 \times 0.025 \times (0.005)^3 \times 0.007}$$

$$Y = 1.088 \times 10^{10} \text{ N/m}^2$$

QUESTIONS

1. What are the units of stress, strain and Poison's ratio?
2. Explain the factors affecting elasticity.
3. Define stress and strain and discuss the stress-strain diagram
4. Define Hooks' Law.
5. Define and discuss Poisson's ratio.
6. Give the expression and explain work done in twisting the wire.
7. Define Elasticity.
8. Define Elastic limit.
9. Define Plasticity.
10. Explain deforming force.
11. Define Restoring force.
12. Define rigid body.
13. Explain Elastic after effect.
14. Explain elastic hysteresis.
15. Discuss the types of modulus of elasticity.
16. Define the terms Y, G, K. Give the relation connecting quantities.

17. Derive the relation: $Y = \dfrac{9GK}{3K+G}$
18. Derive the relation: $\nu = \dfrac{Y}{2G} - 1$
19. Derive the relation: $\nu = \dfrac{3K-2G}{6K+2G}$
20. Derive the expression for work done in deforming a body.
21. Explain the terms: Elasticity, Elastic limit, Plasticity, deforming.
22. Derive the expression for work done in twisting the wire.
23. Derive the expression for torque per unit twist.
24. Derive the relation: $G = \dfrac{F/A}{\phi} = \dfrac{T}{\phi}$
25. Discuss: Elastic Behaviour of a material.
26. Discuss Stress-Strain diagram.
27. Derive the expression for the couple required to bend uniform metallic strip.
28. 28. What is cantilever? Derive depression at free end of thin beam clamped horizontally at one end and loaded at other end.
29. Explain an experiment to determine Young's modulus of the material of the bar by uniform bending.
30. Explain an experiment to determine Young's modulus of the material of the bar by non-uniform bending.
31. How do you differentiate uniform bending and non-uniform bending?

PROBLEMS

1. A load of 2 kg is applied to the ends of the wire 4 m long and produces an extension of 2.4×10^{-3} m calculate, stress, strain, and Y. **[Ans:** 6.24×10^{-3} N/m^2, 6×10^5, 1.04×10^{11} N/m^2]
2. A wire of length 1 m and diameter of 1 mm is fixed at one end and twisted at other end by applying a couple of 0.01 Nm. Evaluate the rigidity modulus of the wire. **[Ans:** 8.2×10^{11} N/m^2]
3. A gold wire 3.2×10^{-4} m in diameter elongates by 10^{-3} m when stretched by a force of 0.330 kg when applied equal and opposite torque of 1.45×10^{-5} Nm are applied at its end. Find the poisons ratio. **[Ans:** 0.429]
4. A brass bar of 1 sq cm in cross section is supported on two knife edges 100 cm apart. A load of 1 kg at the center of the bar depresses that point by 2.51 mm. Find Young's modulus. **[Ans:** 9.75×10^{10} N/m^2]
5. What force is required to stretch a steel wire to double the length when its area of cross section is 1 sq cm? Given Young's modulus of wire is 2×10^{10} N/m^2. **[Ans:** 2×10^7 N/m^2]
6. When an iron wire of 1 m length and radius 0.5 mm elongates by 0.32 mm is stretched by a force of 49 N. Calculate elastic constant for iron. **[Ans:** 19.5×10^{10} N/m^2]
7. A uniform rectangular bar of 1 m long 0.02 m broad and 0.003 m thick is supported on its flat surface symmetrically on two knife edges 0.7 m apart. When the loads of 0.2 kg are hung from the ends the elevation is 0.0022. Find Young's modulus. **[Ans:** 1.18×10^{10} N/m^2]
8. Calculate the load that must be suspended from steel wire of 1 mm diameter to produce an elongate of 0.02 % of its original length. Given Young's modulus of steel is 20×10^{10} N/m^2. **[Ans:** 3.027 kg]

3 Light

Learning Objectives:

By the end of this chapter, you will be able to:

- Learn about basic concepts of Light –Nature and velocity.
- Understand principles of Reflection and Refraction.
- Explain concept of Total Internal Reflection.
- Discuss various concepts and properties related to Light like – Visible range, phase change, Superposition and Coherence.

3.1 INTRODUCTION

We understand the world around us with the help of information reaching our five sensory organs namely eyes, ears, nose, tongue and skin. The sense associated with the eyes is known as *vision* or *sight*. Light is the agent that stimulates the sense of sight. The eyes convert the incoming light into electrical signals and convey them to the brain, which after processing the signals causes images or pictures to be created in our mind. Our world is mostly defined by light. We carry out daily chores with the help of sunlight and other lights. In science, we learn about the properties of atoms and their internal structure through the light emitted by them. We learn about the properties of giant stellar systems by means of the light reaching us from them after traveling for millions of years through the empty space, etc.

What is light? What is it made up of? How is it generated? How does it interact with matter? These are some of the fundamental questions that cross our mind. To tell in simple words, light is a form of energy. However, the nature of light is a difficult concept to grasp because light does not always act the same way.

3.2 NATURE OF LIGHT

(*i*) The very first thing that we observe and understand is that light travels in straight lines, in the form of **rays**. In 1675 Isaac Newton postulated **corpuscular theory**. According to this theory, the rays that are emitted by a luminous body in all directions are streams of extremely minute particles, called *corpuscles* (Fig. 3.1 *a*). This particle theory could explain formation of sharp shadows cast by objects; it could prove the laws of reflection and refraction of light (Fig. 3.1 *b*). However, the corpuscular theory could not explain the phenomenon of diffraction and the formation of Newton's rings usually observed between a glass slide and a long-focus lens placed over it.

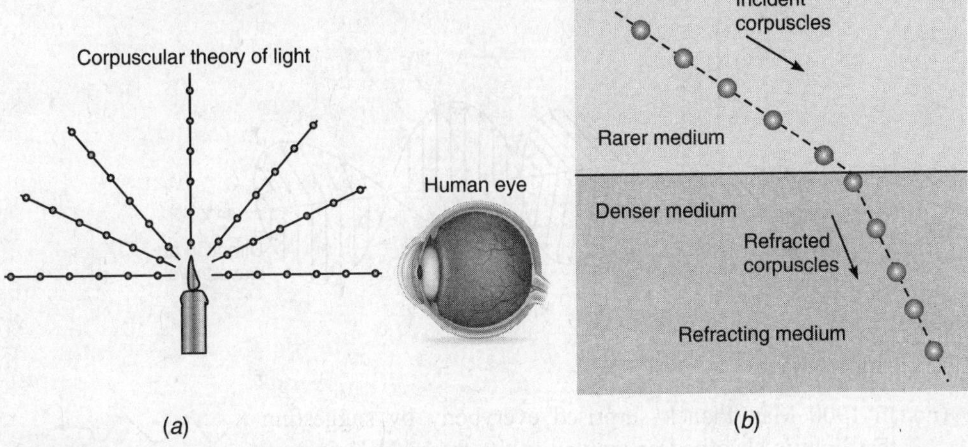

Fig. 3.1: Corpuscular theory.

(*ii*) A few years after that, i.e., in 1678, Huygens, a contemporary of Newton, proposed **wave theory** of light. According to this theory, light energy is supposed to be transferred from one point to another in the form of waves (Fig. 3.2). Huygens proved the ordinary laws of reflection and refraction on the basis of wave theory. Thomas Young in 1803 demonstrated for the first time the interference of light beams and Augustin Fresnel in 1815 provided a satisfactory explanation of the diffraction phenomenon on the basis of wave theory and both of them lent strong support to the wave theory and pushed the corpuscular theory into oblivion.

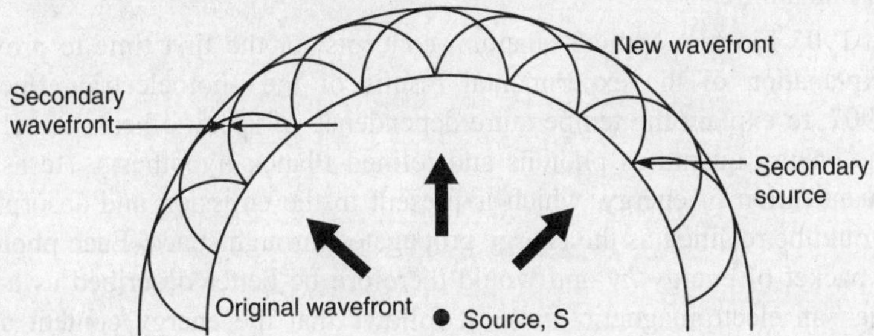

Fig. 3.2: Huygens wave theory.

(*iii*) What is waving in a light remained as puzzle till 1873, when Maxwell predicted that electric and magnetic fields are waving in light waves and concluded that light is a high frequency electromagnetic wave (Fig. 3.3). Maxwell formulated a set of four fundamental equations basing the conclusions of several other physicists and combined them using his ingenuity to bring out a unified theory of electromagnetic phenomena, known as **electromagnetic theory**. The development of electromagnetic theory suggested that in physics there is nothing else to be discovered and developed. The era up to this period in physics came to be called as **classical physics** later. However, the question about how light is emitted is left unanswered by the electromagnetic theory.

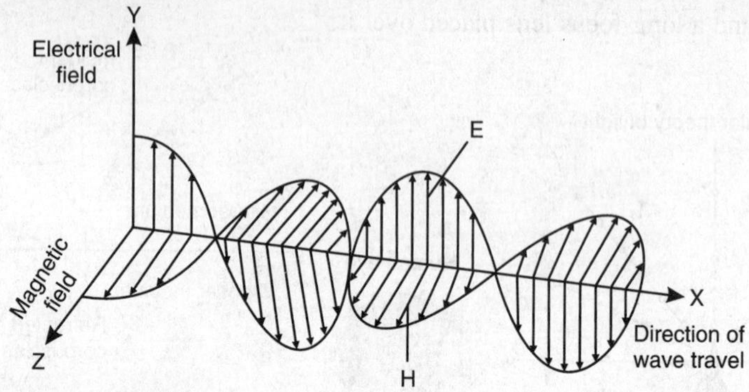

Fig. 3.3: Electromagnetic wave.

(*iv*) In 1900 Max Planck surprised everybody by suggesting that light is absorbed or emitted in the form of elementary **quanta**. In 1905, Einstein (1879-1955) made use of this yet unfamiliar quantum concept to successfully explain the photoelectric emission which defied explanation on the basis of electromagnetic theory. Einstein drastically assumed that light is a stream of photons in interpreting the photoelectric emission (Fig. 3.4). Following Einstein, in 1913, another revolutionary, namely Niels Bohr incorporated these quantum concepts and explained the emission and absorption of light by atoms. Thus, he inaugurated the age of **quantum theory** that changed the course and quality of human life.

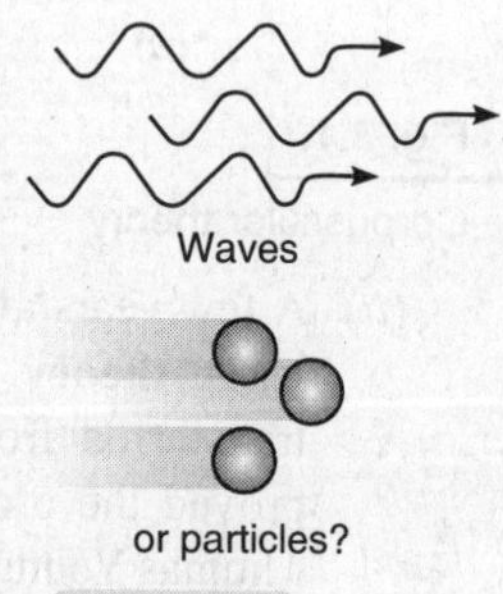

Fig. 3.4: What is the nature of light?

In 1905 Einstein applied quantum concepts for the first time to provide a brilliant explanation of the experimental results of the photoelectric effect and then in 1907, to explain the temperature dependence of specific heats of solids. He named the energy quanta as photons and refined Planck hypothesis. He asserted that the quantization of energy, which is present in the emission and absorption processes, would be retained as the energy propagates through space. Each photon represented a packet of energy $h\nu$ and would therefore be better described as a particle rather than an electromagnetic wave. It follows that the energy content of each photon gets larger as the frequency of the electromagnetic wave increases. Thus, γ-ray and x-ray photons are more energetic compared to optical photons while the photons of r.f. frequencies are the feeblest. The intensity of a monochromatic light beam, I, depends obviously on the number of photons, N, in the beam.

$$I = N h\nu \qquad ...(3.1)$$

Thus, the wave picture of radiation is replaced by particle (photon) picture. It amounted that the wave picture of radiation has to be replaced by the particle (photon) picture of radiation.

(*v*) On one hand light appears to be a continuous electromagnetic wave of frequency ν and on the other hand it appears to be a collection of photons having energy E and momentum p (Fig. 3.5). It has been found that neither of the models can separately explain all the experimental facts. A particle is precisely localized in space whereas a continuous wave cannot be attributed to a particular location in space. Thus, the corpuscular nature and wave nature appear to be mutually exclusive. However, the experimental evidence points out that light behave both as a continuous wave and a particle. Therefore, we say that light exhibits **dual nature**.

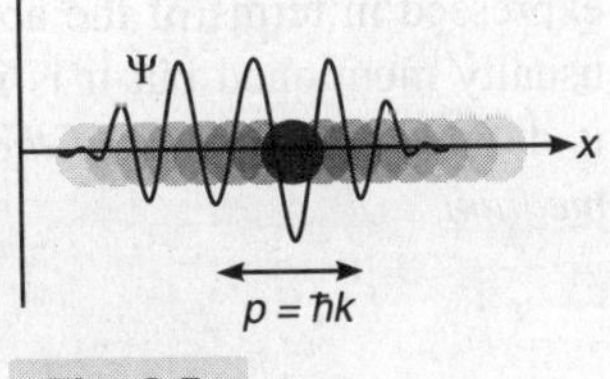

Fig. 3.5: Dual nature of light.

In photon description, the light intensity is given by $I = N\, h\nu$ and in the wave description it is given by $I = |\, E\,|^2$. It implies that the square of the amplitude of a light wave at a point in space is proportional to the number of photons arriving at that point. In other words, the amplitude of a light wave determines the probability that a photon can be found at a particular point in space. Probability to observe photons is thus proportional to $|\, E\,|^2$. Thus, the ultimate connection between the wave behaviour and the particle behaviour is interpreted.

To sum up, light *rays* consist of streams of *photons* (Fig. 3.6), which are basically *electromagnetic wave trains*. We use all the three descriptions namely rays, waves and photons appropriately to describe the behaviour of light. When light is refracted or reflected by optical elements such as lenses, we mostly use the ray description; when light propagates through space or any medium we use the electromagnetic wave description and whenever light interacts with matter we make use of the photon description.

Photon

Fig. 3.6: Light rays consist of streams of photons.

3.3 THE VELOCITY OF LIGHT

Electromagnetic waves, that are light waves, travel at an enormous speed. Measurements have shown that light travels in a vacuum at a speed of 2.998×10^8 m/s. The velocity of light is denoted by the symbol 'c'. For all practical purposes, it is taken as

$$c = 3 \times 10^8 \text{ m/s} \qquad \text{...(3.2)}$$

It is given by

$$c = \nu\lambda \qquad \text{...(3.3)}$$

where ν is the frequency and the λ is the vacuum wavelength of light.

3.4 OPTICAL MEDIUM

A material through which light propagates is called an **optical medium**. Some materials such as air, water and glass readily transmit light and are said to be **transparent**. We can see through transparent materials. Some materials transmit part of the light but scatter most of it so that objects seen through them are not clearly visible. Such materials are called **translucent**. Materials which do not transmit light at all are said to be **opaque**. We cannot see through opaque materials. Our study is related to the propagation of light through transparent media.

When light travels through any medium, its velocity reduces. The dependence of velocity of light on the medium is characterized by the quantity called **optical density**, which is expressed in terms of the absolute refractive index of the medium. The word absolute is not usually mentioned but it is implied. The **absolute refractive index** μ of an optical medium is defined as *the ratio of the velocity of light in a vacuum to the velocity of the light in the medium.*

$$\mu = \frac{\text{Velocity of light in a vacuum}}{\text{Velocity of light in the medium}} = \frac{c}{\upsilon} \quad ...(3.4)$$

The refractive index is a dimensionless number greater than unity since υ is always less than c. The refractive index is 1 for a vacuum and it is equal to 1.0003 for air. Therefore, the velocity of light in air is considered to be equal to c for all practical purposes. An optical medium with a relatively high refractive index is said to have a high optical density and one with a low refractive index has a low optical density.

According to the Eq. (3.3) the decrease in the velocity of light in a medium may be attributed to either a decrease in a wavelength or frequency or both. However, in view of the law of conservation of energy, the frequency must remain constant. The frequency equals the number of wave crests per second. The number of wave crests per second approaching the boundary in one medium must equal the number of wave crests traveling away from the boundary in the second medium, as illustrated in Fig. 3.7 (*a*). If this were not the case, either wave crests would accumulate at the boundary or would be destroyed or created at the boundary. Such a loss or gain would be contrary to the law of conservation of energy. Hence, the frequency must be constant when a wave crosses a boundary. It implies that the decrease in velocity of light in a medium must lead to a decrease in the wavelength of light, as shown in Fig. 3.7 (*b*).

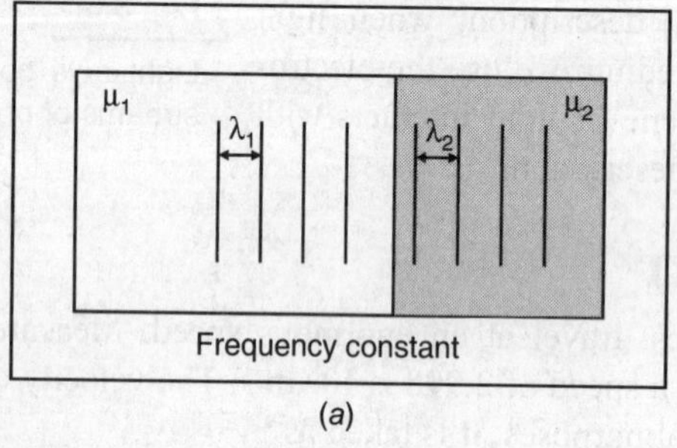

(*a*)

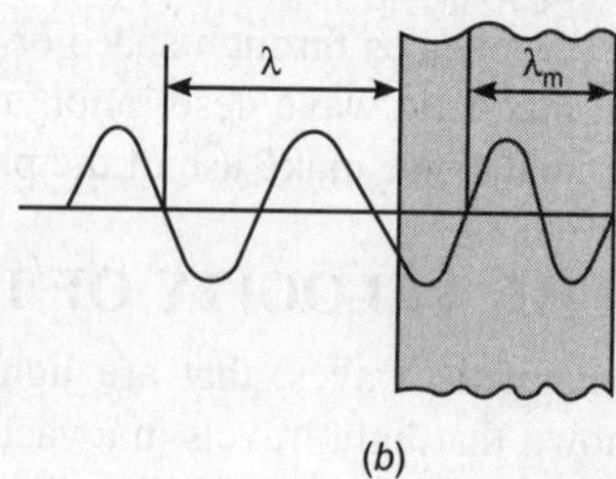

(*b*)

Fig. 3.7:

(*a*) When a light wave propagates from medium 1 to another medium 2, the frequency remains constant. (*b*) Reduction in wavelength when a light wave travels from a rarer medium into a denser medium.

In tune with the Eq. (3.3) the velocity of light in an optical medium may be expressed as

$$\upsilon = \nu\lambda_m \quad ...(3.5)$$

where λ_m is the wavelength of light in the medium. Using Eq. (3.3) and (3.4) into Eq. (3.5), we obtain

$$\mu = \frac{\lambda}{\lambda_m} \quad ...(3.6)$$

Table 3.1 shows the values of refractive index for a few materials.

Table 3.1: *Refractive indices for some materials (λ = 5893 Å)*

Material	μ	*Material*	μ
Gases (at 0° C and 1 atm)		**Solids**	
Hydrogen	1.000132	Ice	1.310
Air	1.000293	Silica Glass	1.458
Carbon dioxide	1.000541	Rock salt	1.544
Liquids		Quartz	1.55
Methylalcohol	1.329	Spinel	1.72
Water	1.333	Corundum	1.76
Ethyl alcohol	1.362	Zircon	1.923
Glycerin	1.473	Fabulite	2.409
Benzene	1.501	Diamond	2.417
Carbon disulphide	1.628	**Plastics**	
		Lucite	1.491
		Polystyrene	1.595

3.5 HOMOGENEOUS ISOTROPIC MEDIUM

An optical medium can be classified into a *homogeneous* or an *inhomogeneous* medium depending on the spatial variation of its refractive index. In a **homogeneous medium** refractive index does not vary from point to point and light propagates along straight line paths in it. On the other hand, if the refractive index varies from point to point in the medium, it is said to be optically **inhomogeneous**. In such a medium light propagates along curved paths. The Earth's atmosphere is an optically in homogeneous medium and light travels along curved paths in it. The passage of light through the atmosphere gives rise to many interesting optical effects. The distorted sun disk during the sunset, the mirages, the looming effect (Fig. 3.8), the twinkle of starts at night etc., are all due to the inhomogeneous nature of atmosphere. Such effects will not be observed on the Moon as there is no atmosphere around it.

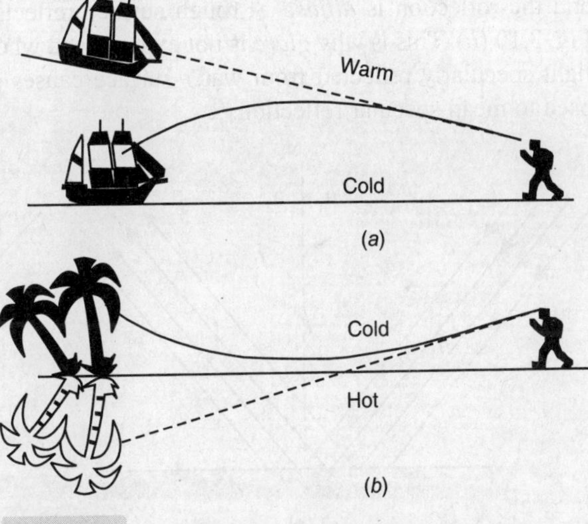

Fig. 3.8: Optical effects due to inhomogeneous nature of atmosphere (*a*) looming (*b*) mirages.

A material is said to be **isotropic**, if it exhibits the same values of physical properties in all directions. Glass, water, air, etc., are isotropic substances. They are also homogeneous. Hence we call them **homogeneous isotropic media**. Such materials have the same value for refractive index in the three principal directions. Materials showing different values of physical properties in different directions are said to be **anisotropic**. Quartz, calcite, ice etc., crystals are examples of anisotropic medium. Such materials exhibit different refractive indices in different directions as a result of which an incident light ray divides into two rays which propagate along two diverse directions and with two dissimilar velocities. This occurrence is known as *birefringence* or more commonly as *double refraction* (Fig. 3.9).

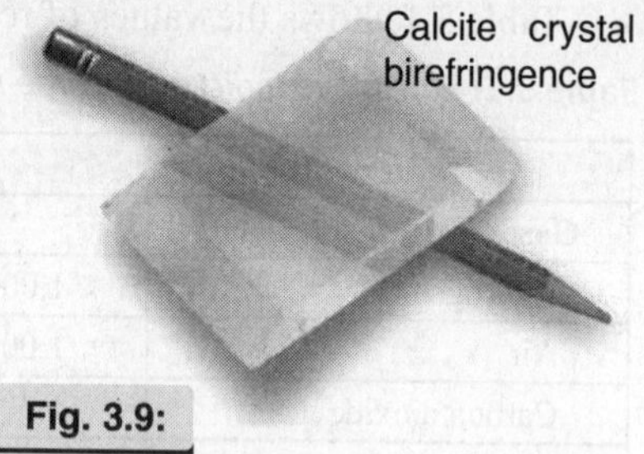

Fig. 3.9: A calcite crystal on a pencil showing two images due to double refraction.

3.6 REFLECTION AND REFRACTION

Reflection: We see most of things with the help of light reflected from them. Unless light shines on objects, we do not see them. For simplicity, we speak of the light waves in terms of rays. When a light ray is incident on the boundary separating two optical media, the ray is *partially* reflected into the first medium and *partially* transmitted into the second medium. If the irregularities of the surface are lesser than the wavelength of the incident light, the boundary is said to be *smooth*; otherwise, it is a *rough surface*. If surface is smooth, it reflects light *specularly*. **Specular reflection** means that the angle of reflection of light is equal to the angle of incidence. Polished metals, mirrors, liquid surfaces etc., smooth surfaces reflect light specularly (Fig. 3.10 *a*). However, many surfaces have microscopic irregularities and the reflection is *diffuse*. A rough surface reflects the light at many angles as shown in Fig. 3.10 (*b*). This is why *glare* is not experienced when light is reflected from a wall, whereas light specularly reflected from water surface causes glare. In physics, the term reflection is used to mean specular reflection.

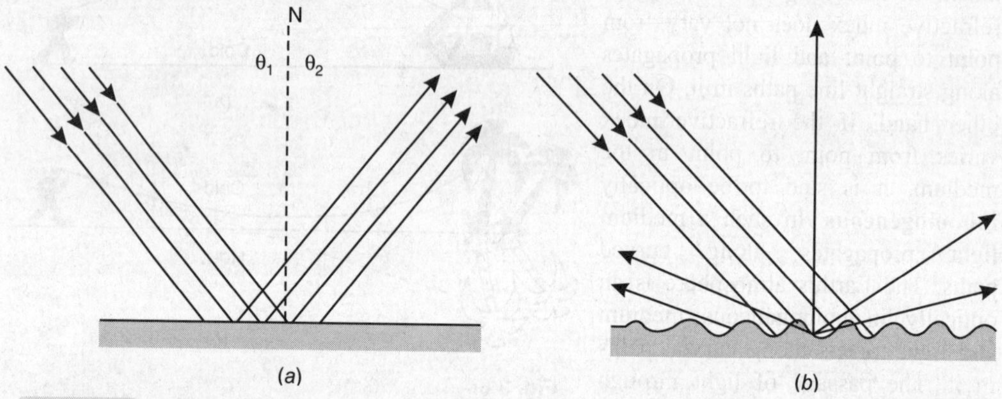

Fig. 3.10: Phenomenon of reflection (*a*) specular reflection occurs at smooth surfaces. The reflected rays are parallel; (*b*) diffuse reflection occurs at a rough surface. The rays are reflected in random directions.

Specular reflection obeys the following two laws known as **laws of reflection**:

(*i*) The incident beam, reflected beam and the normal are on the same plane, as shown in Fig. 3.10 (*a*).

(*ii*) The angle of reflection θ_2 is equal to the angle of incidence θ_1.

$$\theta_1 = \theta_2 \qquad ...(3.7)$$

Further, the angle of reflection is not a function of colour and properties of the reflecting surface.

Refraction: The light ray that enters a transparent medium from another transparent medium is *bent* at the boundary and is said to be **refracted**. The boundary is known as a **refracting surface**. The *angle of refraction r* depends on the properties of the two media and on the *angle of incidence*.

$$\frac{\sin i}{\sin r} = \frac{\upsilon_1}{\upsilon_2} = \frac{c/\mu_1}{c/\mu_2} = \frac{\mu_2}{\mu_1} = \mu_{21} \quad ...(3.8)$$

where υ_1 is the velocity of light in medium 1 having a refractive index μ_1 and υ_2 is the velocity of light in medium 2 having a refractive index μ_2.

The ratio $\frac{\upsilon_1}{\upsilon_2}$ or $\frac{\mu_2}{\mu_1}$ is called the **relative refractive index**. The Eq. (3.8) was discovered by Willebrord Snell (1591-1627), a Dutch scientist and independently by Rene Descartes (1596-1660), a French mathematician. The relation is more popularly known as **Snell's law**. It follows from Snell's law that an increase in the angle of incidence causes an increase in the angle of refraction.

Further, for a given angle of incidence, the angle of refraction depends on the wavelength of the incident light and varies from colour to colour.

Two laws of refraction: Eq. (3.8); and the reality that the incident ray, refracted ray and the normal to the refracting surface all positioned in the same plane are known as the two laws of refraction. The process of refraction is shown in Fig. 3.11.

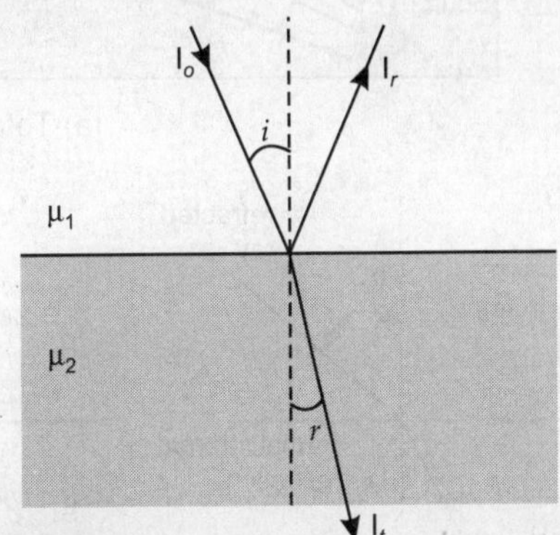

Fig. 3.11: Phenomenon of refraction – A ray obliquely incident on air-glass interference bends toward the normal in glass.

3.7 TOTAL INTERNAL REFLECTION

A medium possessing a lesser refractive index is called an optically **rarer medium** whereas a medium possessing a higher refractive index is an optically **denser medium**. While a ray of light travels through a denser medium and enters a rarer medium, it bends away from the normal in the rarer medium (see Fig. 3.12b). Snell's law for this case may be written as

$$\sin \theta_2 = \left(\frac{\mu_1}{\mu_2}\right) \sin \theta_1 \quad ...(3.9)$$

where θ_1 is the angle of incidence of light ray in the denser medium and θ_2 is the angle of refraction in the rarer medium.

When the angle of incidence, θ_1 in the denser medium is increased, the transmission angle, θ_2 increases and the refracted rays bend more and9 more away from the normal. At some particular angle θ_c the refracted ray glides along the boundary surface so that $\theta_2 = 90°$, as seen in Fig. 3.12 (*c*). At angles greater than θ_c there are no refracted rays at all. The rays are reflected back into the denser medium as though they encountered a specular reflecting surface (Fig. 3.12 *d*). Thus,

- If $\theta_1 < \theta_c$, the ray refracts into the rarer medium
- If $\theta_1 = \theta_c$, the ray just grazes the interface of rarer-to-denser media
- If $\theta_1 > \theta_c$, the ray is reflected back into the denser medium.

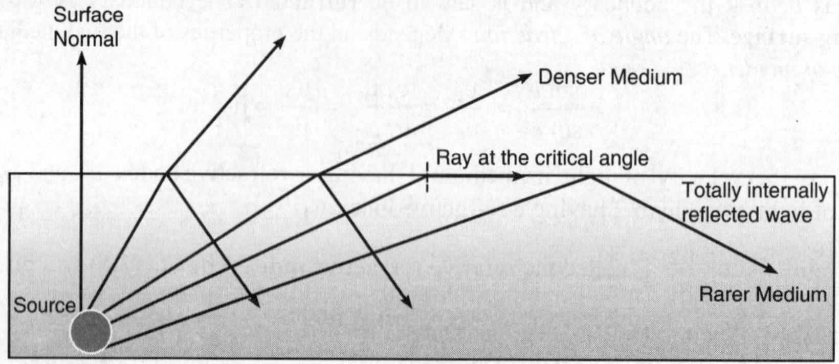

(a) Total internal reflection

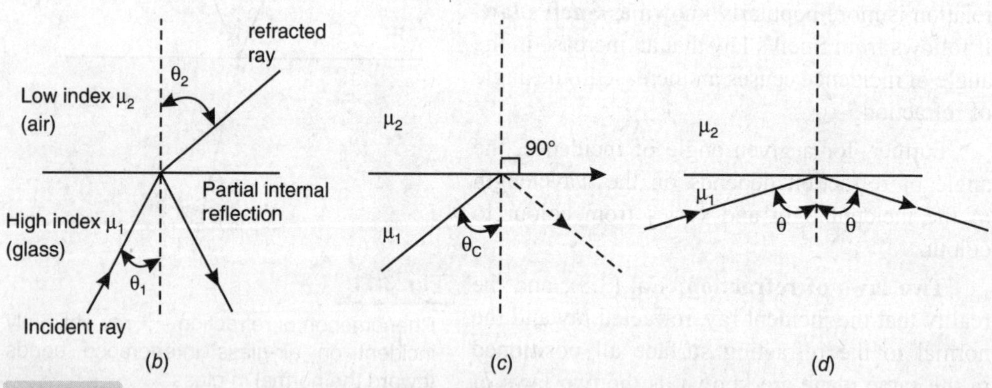

Fig. 3.12:

(*b*) when light rays travel from a denser medium into a rarer medium, they bend away from the normal in the rarer medium. As the θ_1 increases, θ_2 increases. (*c*) The angle of incidence θ_c which produces an angle of refraction 90° -the critical angle. (*d*) For larger angles of incidence, the ray does not emerge out from denser medium but gets internally reflected.

The phenomenon in which light is totally reflected from a denser-to-rarer medium boundary is known as **total internal reflection**. The rays that experience total internal reflection obey the laws of reflection. Therefore, the critical angle can be determined from Snell's law.

When $\theta_1 = \theta_c, \quad \theta_2 = 90°.$

Therefore, from Eq. (3.9), we get

$$\mu_1 \sin \theta_c = \mu_2 \sin 90° = \mu_2$$

$$\therefore \quad \sin \theta_c = \frac{\mu_2}{\mu_1} \qquad ...(3.10)$$

When the rarer medium is air, $\mu_2 - 1$ and writing $\mu_1 = \mu$, we obtain

$$\sin \theta_c = \frac{1}{\mu} \qquad \text{...(3.11)}$$

Total internal reflection does not take place when light propagates from a rarer to a denser medium. The critical angle is small for substances having high refractive index. For example, θ_c = 24° for diamond – air interface, while it is about 42° for glass-to-air interface. This phenomenon is exploited in obtaining sparkle in crystal glass and diamonds. The phenomenon of total internal reflection has made possible to guide light through optical fibres.

Example 3.1: *Determine the critical angle for a light ray traveling from water (μ = 1.333) to air.*

Solution: $\sin \theta_c = \frac{1}{\mu} = \frac{1}{1.333} = 0.75 \quad \therefore \quad \theta_c = \sin^{-1}(0.75) = \mathbf{48.61°}$.

3.8 REFLECTIVITY AND TRANSMISSIVITY

The laws of reflection and refraction only describe the relations between the directions of incident, reflected and refracted rays, but do not speak any thing about the intensities of the rays. If an energy E_i is incident on a transparent medium, it is divided into two parts – one part E_r goes with the reflected ray and the other E_t with the refracted ray. Following the law of conservation of energy,

$$E_i = E_r + E_t \qquad \text{...(3.12)}$$

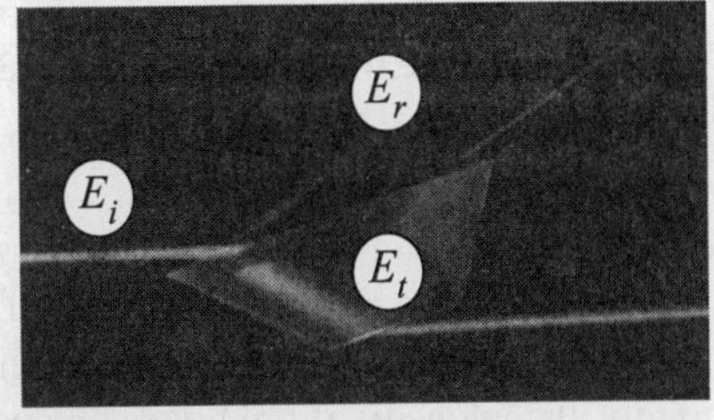

Fig. 3.13: At a planar dielectric surface showing i. Reflection, ii. Refraction, iii. Transmission.

The quantity characterizing the reflectivity of a surface is called the **reflection coefficient** ρ. It is the ratio of the reflected electric energy to the incident energy.

$$\rho = \frac{E_r}{E_i} \qquad \text{...(3.13)}$$

The reflection coefficient depends on the surface, the wavelength composition in the incident light, the angle of incidence and many other factors. However, if all the other factors remain constant, an increase in the angle of incidence i causes an increase in the reflection coefficient. For normal incidence it is shown that

$$\rho = \left[\frac{\mu_1 - \mu_2}{\mu_1 + \mu_2}\right] \qquad \text{...(3.14)}$$

It can be seen from Eq. (3.14), that the reflection coefficient ρ is positive when $\mu_2 < \mu_1$. It implies that the oscillation in the incident and reflected waves occur in the same phase. Therefore, there is no change of phase of the wave upon reflection. On the other hand, if $\mu_2 > \mu_1$, ρ is negative signifying that the oscillation in the incident and reflected waves are 180° out of phase. Thus the phase of the reflected wave changes by π rad.

The **reflectivity** r of a surface is the ratio of the reflected beam intensity to the incident beam intensity. Thus

$$r = \frac{I_r}{I_i} = \left|\frac{E_r}{E_i}\right|^2 = |\rho|^2$$

$$\therefore \quad r = \left[\frac{\mu_1 - \mu_2}{\mu_1 + \mu_2}\right]^2 \quad ...(3.15)$$

The **transmissivity** t is given by

$$t = \frac{4\mu_1\mu_2}{(\mu_1 + \mu_2)^2} \quad ...(3.16)$$

Because energy is conserved, $r + t = 1$. ...(3.17)

Example 3.2: *A solar cell is made of silicon having a refractive index 3.3. If it is exposed to sunlight, what is the percentage of light that is reflected back by it.*

Solution: $r = \left[\frac{\mu_1 - \mu_2}{\mu_1 + \mu_2}\right]^2 = \left[\frac{1-3.5}{1+3.5}\right]^2 = 0.31$

$\therefore$ **Percentage of reflected light= 31%.**

3.9 ABSORPTION

The light propagation through an optical medium is attended mainly by two effects, first the velocity of light decreases in optical medium. Secondly, the intensity of light goes on decreasing with distance in the medium. It happens even in the so called transparent media, since no material is perfectly transparent. The reduction in intensity of light with increasing length of propagation in a medium is called **absorption** or **attenuation**.

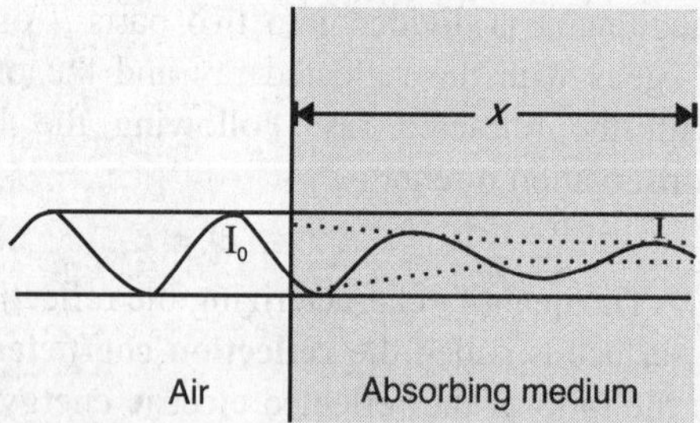

Fig. 3.14: Absorption. The intensity of light at a distance x in a medium reduces to I from I_0 as a result of absorption.

When a beam of light passes through a thin transparent material of thickness dx (Fig. 3.14), the decrease in its intensity dI is found to be proportional to the initial intensity I and to the thickness dx.

$$dI \propto I\, dx$$

or

$$dI = -\alpha I\, dx \quad ...(3.18)$$

where α is the constant of proportionality and is called the **linear absorption coefficient** of the medium. It is characteristic of the medium and also is a function of wavelength. Rearranging the terms in Eq. (3.18), we get

$$\frac{dI}{I} = -\alpha\, dx \quad ...(3.19)$$

The total loss in light intensity after passage through a slab of finite thickness x is obtained by integrating the above Eq. (3.19). Thus, if I_0 is the intensity at $x = 0$, then

$$\int_{I_0}^{I} \frac{dI}{I} = -\alpha \int_0^x dx$$

$$\ln \frac{I}{I_0} = -\alpha x$$

or $$I(x) = I_0 e^{-\alpha x} \qquad ...(3.20)$$

Eq. (3.20) shows that if a beam of light propagates in a medium, its intensity decreases exponentially with distance. α is a measure of loss of light from the direct beam. This exponential **law of absorption** was discovered independently by Pierre Bouguer (1698–1758), the French oceanographer and by J.H. Lambert (1728–1777), the German physicist. The law is, therefore, called **Lambert-Bouguer law**.

3.10 WAVE FRONT AND THE RAY

According to Huygens theory, light is considered to propagate in the form of waves. *A wave is any disturbance, from an equilibrium condition, that propagates with time from one region of space to another*.

We are all familiar with what happens when a pebble is dropped into the still water of a well or a pond. The pebble generates ripples which expand in the form of circles. If the motion of a floating object such as a leaf is watched, we observe that the leaf moves up and down but does not ravel outward with the wave. We, therefore, conclude that it is the energy supplied by the agent of disturbance which moves forward as successive waves. We identify the wave motion with the help of crests and troughs traveling away from the centre of disturbance. Floating of a leaf on water indicates that every particle of the medium oscillates about its mean position at right angles to the direction of propagation and is called as *transverse wave* motion. Water waves and light waves are examples of transverse waves.

Waves start from a source and spread out into new and new regions of space. In case of waves on a water surface, the ripples start from the point of disturbance and expand in the form of circles. The circles are in fact the crests of the waves. All the particles located at the crest will be in the same phase. Therefore, a ripple is the locus of points having the *same phase*. The propagation of ripples is carried out by circles formed one after another at a distance of λ from each other, as shown in Fig. 3.15 (*a*).

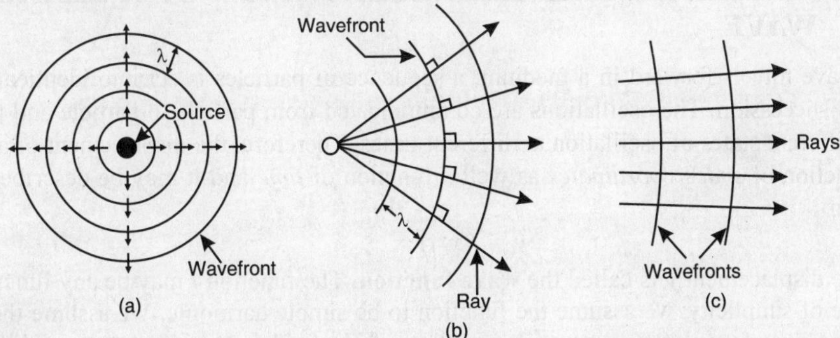

Fig. 3.15: Wavefront and rays (*a*) a point source produces spherical waves (*b*) the ray direction is perpendicular to the wavefront, (*c*) at large distance from the source, the wavefronts tend to be parallel planes.

In a similar manner, the propagation of a three dimensional wave may be visualized as being carried out by innumerable wave surfaces formed one after another at a distance of λ from each other. A **wave surface** is a spherical surface over which the phase of the wave is constant. Generally, the wave surfaces are drawn as passing through either the crest or trough of a sine wave. Therefore, the wave surfaces are separated by one wavelength. The further most wave surface from the source is called **wave front**. The wavefront is dynamic and separates that region of space into which the wave is already spread from the region into which the wave is yet to enter. The propagation of the wave is visualized by the advancing wavefront and stationary wave surfaces behind it. Usually, all the wave surfaces are referred to as wavefronts.

The shape of the wavefront is determined by the source generating the waves and by the size of the object which the waves encounter. If the source is a point source, it radiates waves in all directions uniformly and the wavefronts will be a family of concentric spheres. On the other hand, a linear source produces a family of coaxial cylindrical wavefronts (Fig. 3.16). At distances, far from the source, both the spherical and cylindrical wavefronts may be treated as plane wavefronts. When the size of the slit, through which a plane wave passes, is much larger than the wavelength, the wave front remains plane. On the other hand, if the slit is of the order of wavelength, the wavefront becomes cylindrical.

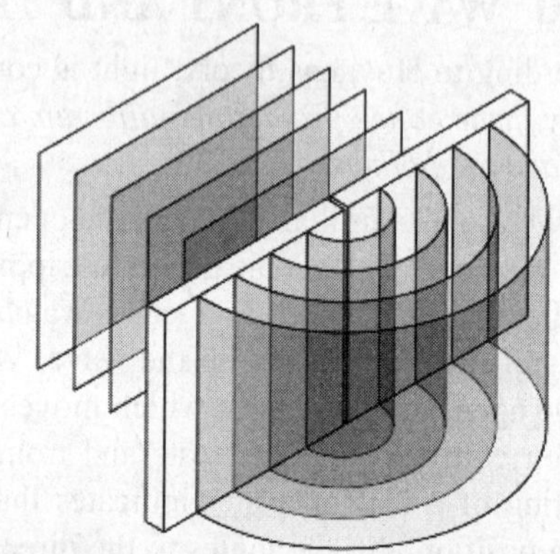

Fig. 3.16: A linear source produces a family of coaxial cylindrical wavefronts.

It is convenient sometimes to describe the wave propagation in terms of rays instead of wavefronts. Light **rays** are lines drawn normal to the wavefronts and they denote the direction of energy flow. In case of plane waves rays are parallel lines and they are radial lines in case of spherical waves. They are illustrated in Fig. 3.15.

3.11 MATHEMATICAL REPRESENTATION OF A PLANE WAVE

As a wave travels forward in a medium, a sequence of particles is set into identical oscillations in succession. The oscillations are communicated from particle to particle and they will be in different states of oscillation at different times. Therefore, the displacement of a particle is a function of *space coordinates* as well a function of *time* and it may be described by the function

$$y = f(x, t) \qquad ...(3.21)$$

The displacement *y* is called the **wave function**. The function *f* may be any function. For the sake of simplicity, we assume the function to be simple harmonic. We assume the simple harmonic waveform because an understanding of sinusoidal waves helps us understanding waves of any shape. All kinds of waveforms including pulses can be constructed by adding up sinusoidal waves of appropriately selected wavelengths and amplitudes.

We can describe the displacement of any point on a harmonic wave in terms of both space and time as

$$y = A \sin\left[\frac{2\pi}{\lambda}(x - \upsilon t)\right]$$

But $\upsilon = \nu\lambda$ and $\nu = 1/T$.

Hence, $$y = A \sin\left(\frac{2\pi x}{\lambda} - \frac{2\pi t}{T}\right)$$

or $$y = A \sin(kx - \omega t) \qquad ...(3.22)$$

where $k = \frac{2\pi}{\lambda}$. The quantity k is called the ***wave number*** or ***propagation number***. It gives the number of wavelengths contained in a distance equal to 2π metres. In Eq. (3.22) there are no limits imposed on the variables x and t. Mathematically, the equation describes a wave of infinite length, existing for all times, from the remote past to the distant future. Using slightly different arguments, the wave equation may be shown to have the form

$$y = A \sin(\omega t - kx) \qquad ...(3.23)$$

Eq. (3.22) and Eq. (3.23) stand for progressive harmonic wave traveling along the positive x-direction. As for a wave traveling in negative x-direction υ changes to $-\upsilon$, the equation for a backward traveling wave is

$$y = A \sin(kx + \omega t) \qquad ...(3.24)$$

In the above discussion, the wave equations are expressed in terms of the sine function. The wave can be represented equally well by a cosine function. The change merely implies a different choice of initial position and time. Thus,

$$y = A \cos(kx - \omega t) \qquad ...(3.25)$$

also represents a forward moving harmonic wave.

If we regard k as a vector quantity, having a magnitude equal to the wave number k and lying in a direction parallel to the positive direction of x-axis, then

$$\mathbf{i}\,k = \mathbf{k} \quad \text{and} \quad kx = \mathbf{k} \cdot \mathbf{r}.$$

Therefore, we may write Eq. (3.25) as

$$y = A \sin(\mathbf{k} \cdot \mathbf{r} - \omega t) \qquad ...(3.26)$$

Eq. (3.26) is independent of the system of coordinates. **k** is called the **wave vector**.

3.11.1 Complex Representation of a Wave

It is seen that a progressive harmonic wave may be represented by either of the following forms.

$$y = A \sin(\mathbf{k} \cdot \mathbf{r} - \omega t)$$

or $$y = A \cos(\mathbf{k} \cdot \mathbf{r} - \omega t)$$

A linear combination of the above two equations also represents a harmonic wave.

Using Euler's formula $e^{i\theta} = \cos\theta + i \sin\theta$, we can write

$$y = A[\cos(\mathbf{k} \cdot \mathbf{r} - \omega t) + i \sin(\mathbf{k} \cdot \mathbf{r} - \omega t)]$$

or $$y = Ae^{i(\mathbf{k}\cdot\mathbf{r} - \omega t)} \qquad ...(3.27)$$

Eq. (3.27) gives the complex representation of a harmonic wave.

3.11.2 Intensity

The **intensity** of a wave is equal to the energy transferred on an average by the wave in a unit time through a unit area of a surface perpendicular to the direction of propagation of the wave. It is directly proportional to the square of the amplitude of the wave. Thus,

$$I \propto |A|^2 \qquad \text{...(3.28)}$$

3.11.3 The General Wave Equation

A wave equation is an expression that shows the variation of displacement of a particle as a function of the spatial coordinates and time. The space derivative of the Eq. (3.22) is given by

$$\frac{\partial y}{\partial x} = kA \cos(kx - \omega t) \qquad \text{...(3.29)}$$

Similarly, the time derivative of the Eq. (3.22) is

$$\frac{\partial y}{\partial t} = -\omega A \cos(kx - \omega t) \qquad \text{...(3.30)}$$

Combining Eq. (3.29) and Eq. (3.30), we get

$$\frac{\partial y}{\partial x} = -\frac{1}{\upsilon}\frac{\partial y}{\partial t} \qquad \text{...(3.31)}$$

Eq. (3.31) represents differential equation for the wave moving to the right. Taking the second derivatives we can make the equation independent of the direction of x. Thus,

$$\frac{\partial^2 y}{\partial x^2} = k^2 A \sin(kx - \omega t) = k^2 y$$

and

$$\frac{\partial^2 y}{\partial t^2} = \omega^2 A \sin(kx - \omega t) = \omega^2 y$$

$$\therefore \quad \frac{\partial^2 y}{\partial x^2} = \frac{k^2}{\omega^2}\frac{\partial^2 y}{\partial t^2}$$

or

$$\frac{\partial^2 y}{\partial x^2} = \frac{1}{\upsilon^2}\frac{\partial^2 y}{\partial t^2} \qquad \text{...(3.32)}$$

Eq. (3.32) represents any wave shape, propagating towards right or left with a velocity υ. It is a one-dimensional wave equation. In three dimensions, we can write it as

$$\frac{\partial^2 \xi}{\partial x^2} + \frac{\partial^2 \xi}{\partial y^2} + \frac{\partial^2 \xi}{\partial z^2} = \frac{1}{\upsilon}\frac{\partial^2 \xi}{\partial t^2}$$

$$\left(\frac{\partial^2}{\partial x^2} + \frac{\partial^2}{\partial y^2} + \frac{\partial^2}{\partial z^2}\right)\xi = \frac{1}{\upsilon^2}\frac{\partial^2 \xi}{\partial t^2}$$

$$\nabla^2 \xi = \frac{1}{\upsilon^2}\frac{\partial^2 \xi}{\partial t^2} \qquad \text{...(3.33)}$$

where $\nabla^2 = \left(\frac{\partial^2}{\partial x^2} + \frac{\partial^2}{\partial y^2} + \frac{\partial^2}{\partial z^2}\right)$ is known as the Laplacian operator.

Eq. (3.33) is known as the *general wave equation* and it represents a three-dimensional traveling wave.

3.12 LIGHT IS AN ELECTROMAGNETIC WAVE

In contrast to the water waves etc, light waves do not require a material medium for their propagation. For example, light reaches to the earth from the sun traveling a long way in a vacuum. Maxwell showed that *light is a self-sustaining electromagnetic wave characterized by frequency and polarization.* A light wave is made up of electric and magnetic fields, which oscillate in mutually perpendicular planes and which are also at right angles to the direction of propagation of the wave (See Fig. 3.17).

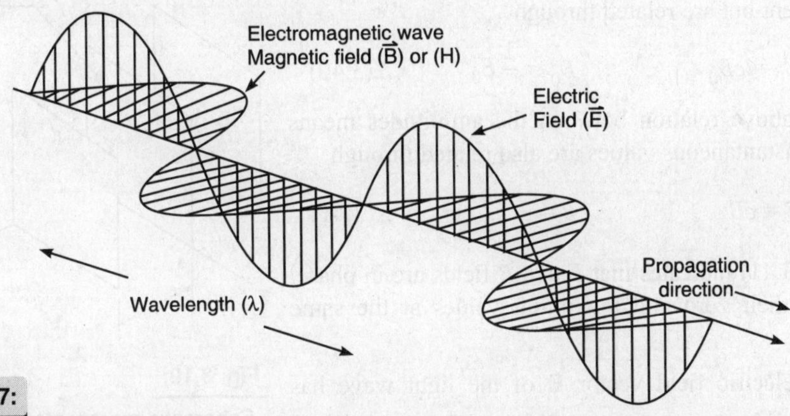

Fig. 3.17:
An electromagnetic wave.

Maxwell's equations for free space can be manipulated into the form of two concise vector equations. Assuming the direction of propagation to be *z*-axis, the **wave equation** can be written as

$$\nabla^2 \mathbf{E} = \varepsilon_0 \mu_0 \frac{\partial^2 E}{\partial t^2} \quad ...(3.34)$$

and

$$\nabla^2 \mathbf{B} = \varepsilon_0 \mu_0 \frac{\partial^2 H}{\partial t^2} \quad ...(3.35)$$

Eq. (3.34) and Eq. (3.35) are identical to the *general wave equation* by Eq. (3.33) provided that

$$\upsilon = \frac{1}{\sqrt{\varepsilon_0 \mu_0}} \quad ...(3.36)$$

Using the values of ε_0 and μ_0 into Eq. (3.36), it is found that $\upsilon \approx 3\times10^8$ m/s. Hence

$$\upsilon = c \quad \text{and} \quad c = \frac{1}{\sqrt{\varepsilon_0 \mu_0}} \quad ...(3.37)$$

The simplest solution for Eq. (3.34) is

$$E_y = E_0 \sin (kx - \omega t) \quad ...(3.38)$$

Similarly, the solution for Eq. (3.35) is

$$B_x = B_0 \sin (kx - \omega t) \quad ...(3.39)$$

E_y and B_z are the **wave functions**, which *describe the size of the electric and magnetic disturbances respectively at any time.*

The wave represented by Eq. (3.38) has a constant amplitude E_0. Therefore, it is a *plane wave*. The term *plane* means that the field vector **E** (similarly **B**) lies in a plane at each point in space, and the planes at any two different points are parallel to each other, as shown in Fig. 3.18.

It was shown that the amplitudes E_0 and B_0 are not independent but are related through

$$E_0 = cB_0 \qquad \therefore \quad B_0 = \frac{1}{c}E_0 \qquad \text{...(3.40)}$$

The above relation between the amplitudes means that the instantaneous values are also related through

$$E = cB \qquad \therefore \quad B = \frac{1}{c}E \qquad \text{...(3.41)}$$

Eq. (3.41) indicates that E and B fields are in phase, reaching their zero and maximum values at the same time.

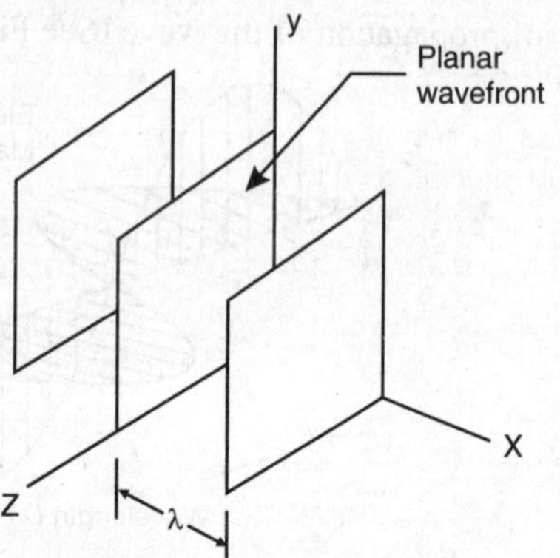

Fig. 3.18: Schematic representation of planar wave fronts.

The electric field vector **E** of the light wave has greater influence on matter than the magnetic field vector **B**. When a light wave interacts with matter, the electric and magnetic fields of light wave act upon the electrons in the material. The electric force F_E experienced by an electron due to the electric field of the light wave is given by

$$F_E = eE \qquad \text{...(3.42)}$$

The maximum magnetic force that can be experienced by an electron is given by

$$F_B = e\upsilon B \qquad \text{...(3.43)}$$

where υ is the velocity of electron. Using the relation (3.41) into (3.43), we find that

$$F_B = (\upsilon/c)eE = (\upsilon/c)F_E \qquad \text{...(3.44)}$$

As υ is always far smaller than c, F_B is insignificant in comparison to F_E. One can describe optical phenomena in terms of either the electric or magnetic field components. We prefer the electric field component since it is easier to visualize. Also, the electric field of the light wave has superior effect than the magnetic field of the wave. Further, it was established that the electric field of the light wave plays a major role in the physiological, photochemical and other actions of light radiation. Therefore, in the pictorial representation of the light wave, the vibrations of an electromagnetic wave are shown only by the direction of the electric field, **E** (Fig. 3.19) and the magnetic field vector **B** is generally omitted in order to avoid confusion.

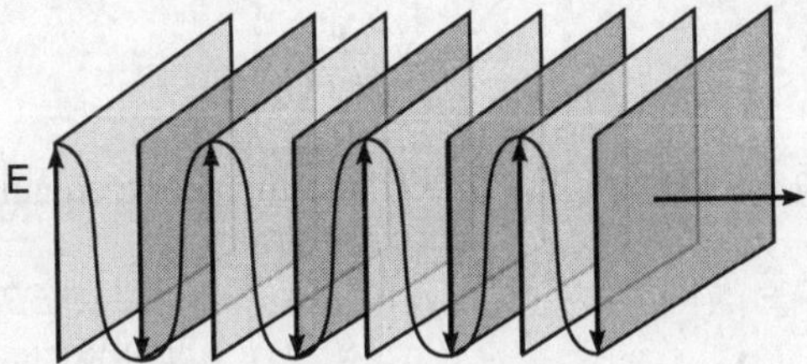

Fig. 3.19:

A planar light wave

3.12.1 Energy Density

One of the most significant properties of the electromagnetic wave is that it transports energy and momentum. Both the electric field and magnetic fields of the wave store energy. The *energy density* (energy per unit volume), u_E associated with the E-field in free space is given by

$$u_E = \frac{1}{2}\varepsilon_0 E^2 \qquad ...(3.45)$$

Similarly, the energy density u_B associated with the B-field in free space is given by

$$u_B = \frac{1}{2\mu_0} B^2 \qquad ...(3.46)$$

Using the Eq. (3.37) into Eq. (3.46), we get

$$u_B = \frac{1}{2\mu_0}\left[\frac{E}{c}\right]^2 = \left[\frac{\varepsilon_0\mu_0}{2\mu_0}\right]E^2 = u_E$$

That is $u_E = u_B$

It means that the energy of the wave is shared equally between the electric and magnetic fields of the wave. The total energy density of the electromagnetic wave is

$$u = u_E + u_B = \varepsilon_0 E^2 \qquad ...(3.47)$$

3.12.2 Poynting Vector

An electromagnetic wave transports energy from one region of space to another. We can describe the energy transfer in terms of the rate of energy flow per *unit area* or power per unit area. It is denoted by the vector **S** known as the **Poynting vector**.

Let us consider a plane wave front moving to the right (Fig. 3.20). In a time *dt*, the wave front moves a distance $dx = c.dt$.

The energy *dW* passing for the period of time *dt* through a *unit area* at right angles to the direction of the propagation of the wave is

$$dW = uc\ dt$$

where *u* is the energy density.

The rate at which the energy is transported by the wave is its power. Therefore, the power transferred per unit area, *S*, is

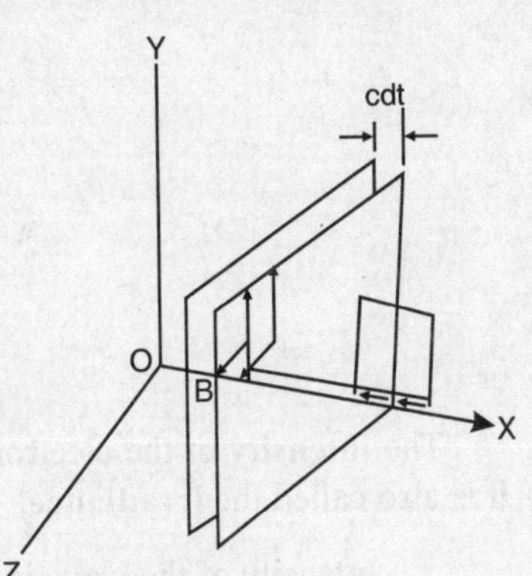

Fig. 3.20:

Wavefront.

$$S = \frac{\text{energy}}{dt} = uc$$

Making use of the Eq. (3.48) and Eq. (3.49) into the above equation, we write

$$S = \sqrt{u}\sqrt{u}\,c = \left(\sqrt{\varepsilon_0}E\right)\left(\frac{B}{\sqrt{\mu_0}}\right) = \frac{\varepsilon_0}{\sqrt{\varepsilon_0\mu_0}}EBc = \varepsilon_0 c^2 EB$$

When S is assigned the direction of propagation, it is called the **Poynting vector**. The direction of **S** is the same as that of the cross product of **E** and **B**. The vectors **E** and **B** are mutually perpendicular and form a right-handed system with the direction of propagation of the wave. Therefore,

$$\mathbf{S} = c^2\varepsilon_0\,(\mathbf{E}\times\mathbf{B}) \qquad \text{...(3.48)}$$

or

$$\mathbf{S} = \frac{1}{\mu_0}\,(\mathbf{E}\times\mathbf{B}) \qquad \text{...(3.49)}$$

3.12.3 Intensity

A plane electromagnetic wave traveling in the positive x-direction is described by the Eq. (3.38) and Eq. (3.39).

$$\therefore \quad S = \frac{1}{\mu}E_0B_0\sin^2(kx-\omega t) \qquad \text{...(3.50)}$$

$$S = c^2\varepsilon_0 E_0 B_0 \sin^2(kx-\omega t) \qquad \text{...(3.51)}$$

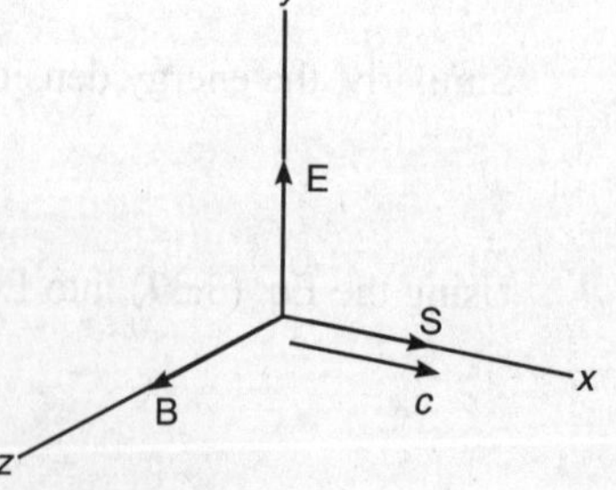

Fig. 3.21: Poynting Vector

The electromagnetic power P that the plane wave delivers to a receiver of area 'A' oriented perpendicular to x-axis is given by

$$P = SA = \frac{1}{\mu_0}E_0B_0A\sin^2(kx-\omega t)$$

or

$$P = \frac{1}{\mu_0 c}E_0^2A\sin^2(kx-\omega t) \qquad \text{...(3.52)}$$

If T is the time of observation, then the average power is given by

$$P_{av} = \frac{1}{T}\int_0^T P\,dt$$

$$= \frac{1}{\mu_0 c}\cdot E_0^2A\cdot\frac{1}{T}\int_0^T[\sin^2(kx-\omega t)]\,dt$$

$$= \frac{1}{2\mu_0 c}E_0^2A \qquad \text{...(3.53)}$$

or

$$P_{ave} = \frac{c\varepsilon_0}{2}E_0^2A \qquad \text{...(3.54)}$$

The **intensity** of the electromagnetic wave is defined as the average power per unit area. It is also called the **irradiance**.

$$\therefore \text{ Intensity of the wave is} \quad I = \frac{P_{ave}}{A} = \frac{1}{2\mu_0 c}E_0^2 \qquad \text{...(3.55)}$$

or

$$I = \frac{c\varepsilon_0}{2}E_0^2 \qquad \text{...(3.56)}$$

Thus, it is found that $I \propto |E_0|^2$ ...(3.57)

3.12.4 Wave Characteristics of Light

(*i*) Eq. (3.38) represents a plane wave propagating in the direction of $k = 2\pi / \lambda$ where λ is the *spatial period* of the wave and is known as the ***wavelength***.

(*ii*) The sine varies from +1 to –1 so that the maximum value of E_y is E_0. Therefore, electric field vector E_y of the wave oscillates parallel to the *y*-axis with values between $+E_0$ and $-E_0$. The maximum value E_0 is the ***amplitude*** of the wave.

(*iii*) The wave described by Eq. (3.38) has only one frequency, ν ($\omega = 2\pi\nu$) and hence it is a ***monochromatic wave***.

(*iv*) The wave is a simple harmonic wave extending from $-\infty$ to $+\infty$ and has no beginning and an end.

(*v*) It is a plane wave, the wave front being normal to *x*-axis.

(*vi*) The electric vector *E* of the wave always oscillates parallel to a fixed direction in space, i.e. *y* direction. In other words, the *E* vibrations are confined to *xy* plane. A wave having its vibrations confined to a single plane is called a **plane polarized** or a **linearly polarized wave**. The wave depicted in Fig. 3.17 is an ideally plane polarized wave.

(*vii*) The wave has zero divergence and does not get attenuated as it propagates through space.

(*viii*) The variation of phase along the wave extension is predictable. Therefore, if two or more such waves travel along the same direction in space, their phase variations are synchronized and the phase difference stays constant. Thus the waves will be highly coherent.

We will soon learn that actual light waves are far from ideal. It will be noted that they are wave trains of limited extension, have a certain spread in frequency around a central value, totally unpolarized and incoherent. All the natural light sources such as the sun, a lamp or a flame emit only such wave trains.

Example 3.3: *The wave function for a light wave is*

$$E(z, t) = 10^3 \sin \pi (3 \times 10^6 x - 9 \times 10^{14} t)$$

(*i*) *Determine the speed, wavelength, frequency and period of the wave.*

(*ii*) *Determine the magnetic field associated with the wave.*

Solution: (*i*)

$$E(z, t) = 10^3 \sin \pi (3 \times 10^6 x - 9 \times 10^{14} t)$$

$$= 10^3 \sin 3 \times 10^6\pi (x - 3 \times 10^8 t) \quad ...(a)$$

This equation is similar to the general wave equation

$$E(z, t) = E_0 \sin k (x - \upsilon t) \quad ...(b)$$

where E_0 is the amplitude of the wave, *k* is the wave number $k = \frac{2\pi}{\lambda}$ and υ is the velocity of the wave.

Comparing (*a*) with (*b*), we get $\upsilon = 3 \times 10^8$ m/s, $k = 3 \times 10^6\ \pi$/m

$$\therefore \quad \lambda = \frac{2\pi}{k} = \frac{2\pi}{3\times10^6 \pi / \text{m}} = 6.666 \times 10^{-7}\ \text{m} = \mathbf{6666\ Å}$$

$$\nu = \frac{\upsilon}{\lambda} = \frac{3\times10^8\ \text{m/s}}{6.666\times10^{-7}\ \text{m}} = \mathbf{4.5 \times 10^{14}\ Hz}$$

$$T = \frac{1}{\nu} = \frac{1}{4.5\times10^{14}\ \text{Hz}} = \mathbf{2.2 \times 10^{-15}\ s.}$$

(*ii*) The light wave propagates in *z*-direction while the E-vector oscillates along *x*-direction in the *xz*-plane. Since in an EM wave, magnetic field B is normal to both E and wave propagation directions, it should be in the *yz*-plane.

Thus,

$$B_x = 0; \quad B_z = 0$$

$$B = B_y\,(z, t) = B(z, t)$$

As,

$$E = cB, \quad B = \frac{E}{c}$$

$$\therefore \qquad B(z, t) = \frac{10^3 \sin \pi(3\times10^6\,y - 9\times10^{14}\,t)}{c}$$

$$B(z, t) = 3.33 \times 10^{-6} \sin \pi(3 \times 10^6\,y - 9 \times 10^{14}\,t)$$

Example 3.4: *A plane electromagnetic wave moving through free space has an E field given by,*

$$E = 100 \sin\left[8\pi\times10^{14}\left(t - \frac{z}{3\times10^8}\right)\right].$$

Calculate the corresponding intensity.

Solution:

$$E = 100 \sin\left[8\pi\times10^{14}\left(t - \frac{z}{3\times10^8}\right)\right].$$

$$\therefore \qquad E_0 = 100 \text{ V/m}$$

Intensity,

$$I = \left(\frac{c\varepsilon_0}{2}\right)E_0^2 = \frac{(3\times10^8\,\text{m/s})\left(8.85\times10^{-12}\,\dfrac{C^2}{\text{N.m}^2}\right)\left(100\,\dfrac{V}{m}\right)^2}{2}$$

$$I = 13.3\,\frac{\text{C}^2}{\text{s.N.m}}\cdot\frac{\text{V}^2}{\text{m}^2} = 13.3 \text{ W/m}^2$$

Units: $$1\,\frac{\text{C}^2\text{V}^2}{\text{s.N.m.m}^2} = 1\,\frac{\text{J}^2}{\text{s.N.m.m}^2} = 1\,\frac{\text{J.N.m}}{\text{s.N.m.m}^2} = 1\,\frac{\text{J}}{\text{s.m}^2} = 1\,\frac{\text{W}}{\text{m}^2}$$

3.13 VISIBLE RANGE

The arrangement of the various electromagnetic waves in a continuous sequence of frequencies and wavelengths, as in Fig. 3.14, is called the **electromagnetic spectrum**. The spectrum includes waves covering a broad range of wavelengths. It is bounded at one end by gigantic radio waves having wavelengths of a few hundred kilometers and by γ – rays having wavelengths of 10^{-12} m at the other end. **Visible range** is that part of the spectrum constituted by waves which can be detected by the human eye. It extends from the deepest violet to the deepest red. The limiting range of these waves depends on the individual properties of the eye and varies approximately in the interval λ = 4000 Å to λ = 7800 Å. The regions flanking the visible range are known as **infrared** on the longer wavelength side and **ultraviolet** on the shorter wavelength side. The infrared (IR) region lies in the wavelength range 7.8×10^{-7} m to 10^{-3} m and the ultraviolet (UV) region lies between 4000 Å to 10 Å. Radiation in these three regions namely IR, visible and UV together is called **optical radiation**.

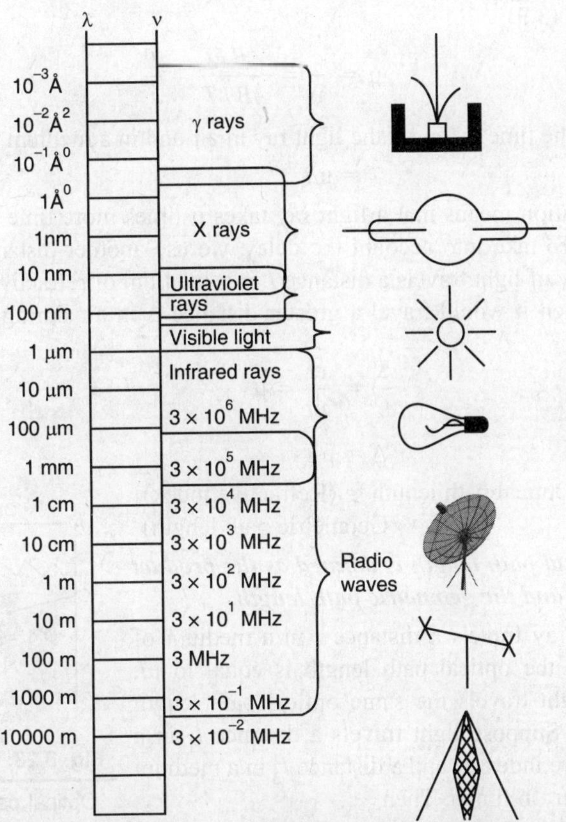

Fig. 3.22:

Electromagnetic spectrum.

Table 3.2 lists the wavelengths and the frequencies of each of the coloured region in the visible range.

Table 3.2: *Wavelengths and frequencies of different colours*

Colour	*Vacuum Wavelength (Å)*	*Frequency (Hertz)*
Red	7800 – 6220	$3.84 \times 10^{14} - 4.82 \times 10^{14}$
Orange	6220 – 5970	$4.82 \times 10^{14} - 5.03 \times 10^{14}$
Yellow	5970 – 5770	$5.03 \times 10^{14} - 5.20 \times 10^{14}$
Green	5770 – 4920	$3.84 \times 10^{14} - 6.10 \times 10^{14}$
Blue	4920 – 4550	$3.84 \times 10^{14} - 6.59 \times 10^{14}$
Violet	4550 – 3990	$3.84 \times 10^{14} - 7.69 \times 10^{14}$

3.14 OPTICAL PATH LENGTH

The shortest distance, L between two points A and B is called the *geometric path*. The length of geometric path is independent of the medium that surrounds the path AB. When a light ray travels from the point A to point B, it travels with the velocity 'c' if the medium is air and with a lesser velocity if the medium is other than air. Therefore the light ray takes more time to go from A to B located in a medium.

From equation (3.3)

$$\mu = \frac{c}{\upsilon} = \frac{AB/t}{AB/T} = \frac{T}{t}$$

where t and T are the time taken by the light ray in air and in a medium respectively.

$$\therefore \qquad T = \mu t$$

The above relation means that a light ray takes μ times more time to cover the distance AB in a medium. To take into account the delay, we use another distance called the *optical path length*. If a ray of light travels a distance L in a medium of refractive index μ in a certain interval of time, then it would travel a greater distance Δ in air during the same interval of time. Therefore,

$$\frac{\Delta}{L} = \frac{ct}{\upsilon t} = \mu$$

or
$$\Delta = \mu L \qquad \text{...(3.58)}$$

i.e., Optical path length = (Refractive index) × (Geometric path length)

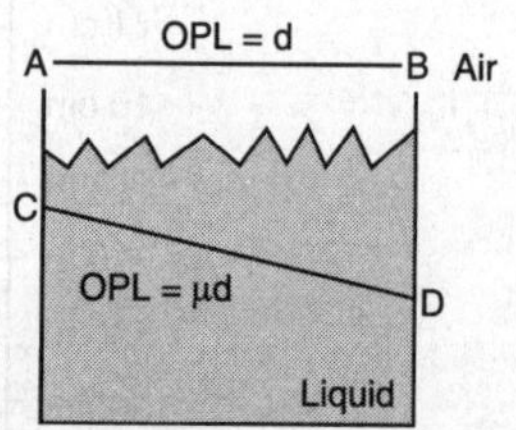

Fig. 3.23: Optical path length in medium is longer than the OPL in air.

Thus, ***the optical path length is defined as the product of refractive index and the geometric path length.***

Note that if a ray travels a distance l in a medium of refractive index μ, the optical path length is equal to μl. In a given time light travels the same optical path length in different media. Suppose light travels a distance l_1 in a medium of refractive index μ_1 and a distance l_2 in a medium of refractive index μ_2 in time t. Then

$$\mu_1 l_1 = \mu_2 l_2$$

Example 3.5: *Light of wavelength 6200 Å travels through a film of water 0.72 × 10⁻⁶ m thick. If its refractive index is 1.333, find the optical path.*

Solution: Optical path = (Refractive index) (geometric path)

$\therefore \quad \Delta = \mu L = (1.333)(0.72 \times 10^{-6}\ \text{m}) = 0.96 \times 10^{-6}\ \text{m} = \mathbf{0.96\ \mu m}$

3.15 PHASE CHANGE AND PATH DIFFERENCE

The phase of a wave changes when the wave travels in space. At any time, the phase difference between points at x_1 and x_2 is

$$\delta = \phi_2 - \phi_1$$

Taking help of Eq. (3.22), we can write the above relation as

$$\delta = (kx_2 - \omega t) - (kx_1 - \omega t) = k(x_2 - x_1) = \frac{2\pi}{\lambda}(x_2 - x_1)$$

or
$$\delta = \frac{2\pi}{\lambda} L \qquad \text{...(3.59)}$$

where $L = (x_2 - x_1)$ is the spatial separation of the points, i.e. the *geometric path*.

It follows from Eq. (3.59) that $\delta = 2\pi$ if $L = \lambda$, $\delta = \pi$ if $L = \lambda/2$ and so on. The waveform repeats over a distance of one λ. Therefore, the phase difference 2π is equivalent to zero phase difference. It implies that a displacement of wave by one complete wavelength leaves the waveform unchanged.

(*i*) Path differences between two waves may arise if two waves propagate in a medium along two different paths and overlap at some point *P* (Fig. 3.24). The result of their superposition at *P* depends on their relative disposition with respect to each other. The relationship between the waves is defined with the help of their phase difference which may be expressed in terms of the path difference. The quantity $L = (x_2 - x_1)$ in Eq. (3.59) also represents the path difference when two waves are considered. The phase difference can then be computed using Eq. (3.59) as

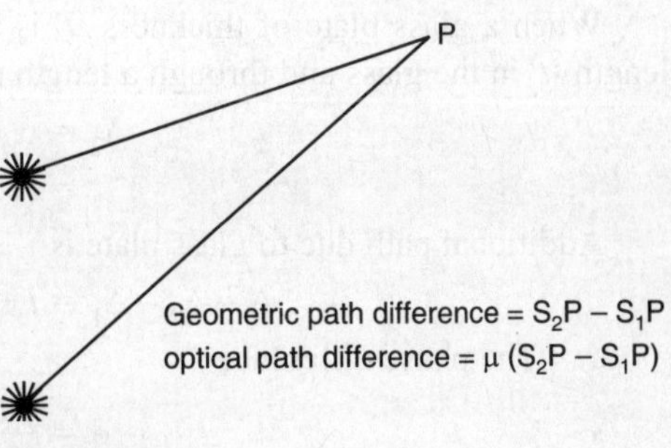

Fig. 3.24: Illustration showing Path difference between two light rays.

$$\delta = \frac{2\pi}{\lambda}\mu L$$

or

$$\delta = \frac{2\pi}{\lambda}\Delta \qquad \text{...(3.60)}$$

(*ii*) A path difference or phase difference between two light waves occurs during the course of reflection, as well. A light wave propagating in a rarer medium suffers a phase change of π rad when it gets reflected at the boundary of denser medium as shown in Fig. 3.25 (a). But phase change does not occur when it gets reflected at the boundary of denser – to – rarer medium. For example, let us consider a light wave incident on a glass block. The wave reflected from the air-to-glass boundary experiences a phase change of π radians relative to the incident wave because $\mu_g > \mu_a$, as illustrated in Fig. 3.25 (*b*). The wave transmitted through glass gets again reflected at the bottom surface which is glass-to-air boundary. The reflection at glass – air-boundary does not produce a phase change. A phase change of π radians is equivalent to a path difference of $\lambda/2$ as seen from the Eq. (3.59). Thus, a light wave gets inverted or loses a half-wave on reflection from a rarer-denser boundary.

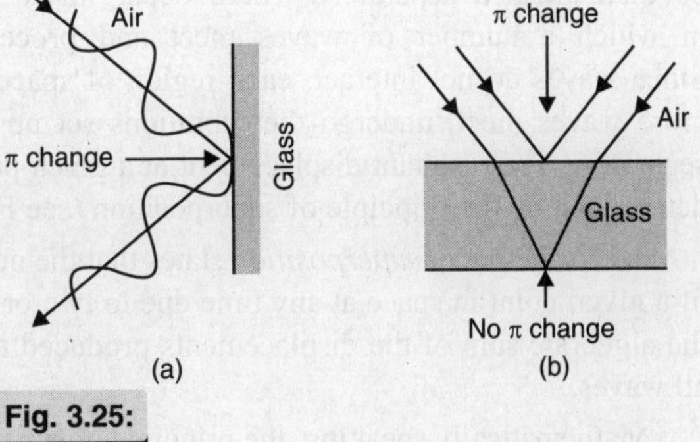

Fig. 3.25: Illustration showing phase change due to reflection.

Example 3.6: *A light wave propagates from a point A to another point B. A glass plate (μ_g = 1.5) of 1 mm thick is introduced in its path. If the wavelength of light is 5000 Å. What is the change of phase of the wave at B?*

Solution: When the light travels from *A* to *B*, a distance '*l*' (say) in air, the optical path

$$\Delta_l = \mu_a l = l$$

When a glass plate of thickness '*t*' is introduced in the path *AB*, light travels through a length '*t*' in the glass and through a length $(l - t)$ in air. Therefore the optical path is now

$$\Delta_2 = \mu_a(l - t) + \mu_g t$$
$$= (l - t) + \mu_g t = l + (\mu_g - 1)t$$

Additional path due to glass plate is

$$\Delta = \Delta_2 - \Delta_1 = l + (\mu_g - 1)t - l = (\mu_g - 1)t$$

$\therefore$ The phase difference

$$\delta = \frac{2\pi}{\lambda}\Delta = \frac{2\pi}{\lambda}(\mu_g - 1)t$$

$$= \frac{2\pi(1.5-1)\times 10^{-3}\text{ m}}{5\times 10^{-7}\text{ m}} = \mathbf{2\pi \times 10^3 \text{ radians}}$$

Phase of 1000 × 2π radians indicates that waves arriving at *B* are in phase.

3.16 THE PRINCIPLE OF SUPERPOSITION

When two pebbles are dropped at different points in a pond, the expanding ripples pass through each other without mutual effect. Likewise, sound waves from different instruments in an orchestra propagate in space independent of each other and can be distinguished separately. There occur many such instances in which a number of waves meet and proceed unaffected. Since waves do not interact, each region of space where two or more waves meet, undergo the vibrations set up by each wave separately. The resultant displacement at a given point in space is determined by the principle of superposition (see Fig. 3.26).

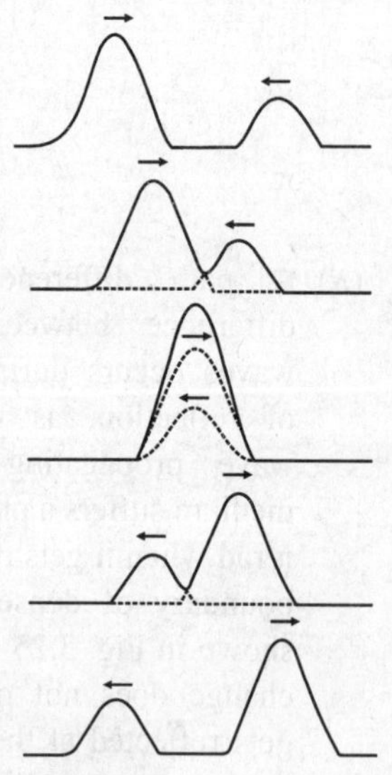

Fig. 3.26: Illustration showing the principle of superposition.

The *principle of superposition* states that the net displacement of a given point in space at any time due to two or more waves is the algebraic sum of the displacements produced at that point by all waves.

Mathematically speaking, the principle of superposition states that if two or more waves are propagating through the space, the resultant wave functions of the individual waves. Thus, if $y_1(x, t)$ and $y_2(x, t)$ are the wave functions characterizing two waves traveling in space, their resultant is given by

$$y(x, t) = y_1(x, t) + y_2(x, t) \qquad ...(3.61)$$

Wave equations that obey the superposition principle are said to be ***linear***. The superposition principle applies only to waves of small amplitude.

3.17 INTERFERENCE OF LIGHT WAVES

When harmonic waves of identical frequency propagating in a medium meet each other, they give rise to the phenomenon of interference. Let us now understand what happens when two or more light waves overlap in some region of space. Let us assume for the sake of simplicity that two sinusoidal waves of the same frequency propagate through different paths x_1 and x_2 and meet at *P* in the region of observation, as shown in Fig. 3.27. Let the waves be represented by

$$E_A = E_1 \sin \omega t \quad \text{...(3.62)}$$

and $$E_B = E_2 \sin (\omega t + \delta) \quad \text{...(3.63)}$$

where δ is the phase difference between them.

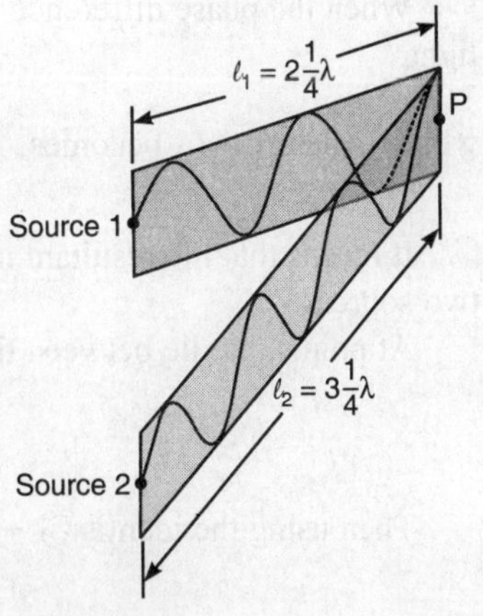

Fig. 3.27: Two light waves overlapping at point P.

As per the Young's principle of superposition, the resulting electric field at any place due to two or more harmonic waves is the algebraic total of the electric fields of the constituent waves.

Thus, the resultant electric field at the point P due to the simultaneous action of the two waves is given by

$$E_R = E_A + E_B \quad \text{...(3.64)}$$

$$= E_1 \sin \omega t + E_2 \sin (\omega t + \delta)$$

$$= E_1 \sin \omega t + E_2 (\sin \omega t \cos \delta + \cos \omega t \sin \delta)$$

$$= (E_1 + E_2 \cos \delta) \sin \omega t + E_2 \sin \delta \cos \omega t \quad \text{...(3.65)}$$

Eq. (3.65) shows that ***the superposition of two sinusoidal waves having the same frequency but with a phase difference produces a sinusoidal wave with the same frequency but with a different amplitude E.***

Let $$E_1 + E_2 \cos \delta = E \cos \phi \quad \text{...(3.66)}$$

and $$E_2 \sin \delta = E \sin \phi \quad \text{...(3.67)}$$

where E is the amplitude of the resultant wave and ϕ is the new initial phase angle. In order to solve for E and ϕ, we square the equ. (3.66) and (3.67) and add them.

$$(E_1 + E_2 \cos \delta)^2 + E_2^2 \sin^2 \delta = E^2(\cos^2\phi + \sin^2\phi)$$

or $$E^2 = E_1^2 + E_2^2 \cos^2 \delta + 2E_1E_2 \cos \delta + E_2^2 \sin^2\delta$$

or $$E^2 = E_1^2 + E_2^2 + 2E_1E_2 \cos \delta \quad \text{...(3.68)}$$

Thus, it is seen that the square of the amplitude of the resultant wave is not a simple sum of the squares of the amplitudes of the superposing waves, there is an additional term which is known as the **interference term**.

3.17.1 Intensity Distribution

The intensity of a light wave is given by the square of its amplitude.

$$I = \frac{1}{2}\varepsilon_0 c E^2 \propto E^2$$

Using this relation into Eq. (3.68), we get

$$I = I_1 + I_2 + 2\sqrt{I_1 I_2} \cos \delta \quad \text{...(3.69)}$$

We see that the resultant intensity at P on the screen is not just the sum of the intensities due to the separate waves. The term $2\sqrt{I_1 I_2} \cos \delta$ is known as the **interference term**. Whenever the phase difference between the waves is zero, i.e. $\delta = 0$, we have maximum amount of light. Thus,

$$I_{max} = I_1 + I_2 + 2\sqrt{I_1 I_2}$$

When $I_1 = I_2 = I_0$ $$I_{max} = 4I_0 \quad \text{...(3.70)}$$

It means that the resultant intensity I will be *more than the sum* of the intensities due to the two sources.

When the phase difference is $\delta = 180°$, $\cos 180° = -1$ and we have a minimum amount of light.

$$I_{min} = I_1 + I_2 - 2\sqrt{I_1 I_2}$$

which, when $I_1 = I_2$, becomes

$$I_{min} = 0 \qquad ...(3.71)$$

It means that the resultant intensity I will be *less than the sum* of the intensities due to the two sources.

At points that lie between the maxima and minima, when $I_1 = I_2 = I_0$, we get

$$I = I_0 + I_0 + 2I_0 \cos \delta$$
$$= 2I_0 (1 + \cos \delta)$$

Then using the identity, $1 + \cos \delta = 2\cos^2\left(\frac{1}{2}\delta\right)$, we get

$$I = 4I_0 \cos^2\left(\frac{1}{2}\delta\right) \qquad ...(3.72)$$

Eq. (3.72) shows that the intensity varies along the screen in accordance with the ***law of cosine square***. Fig. 3.28 shows the variation of intensity as function of phase angle δ.

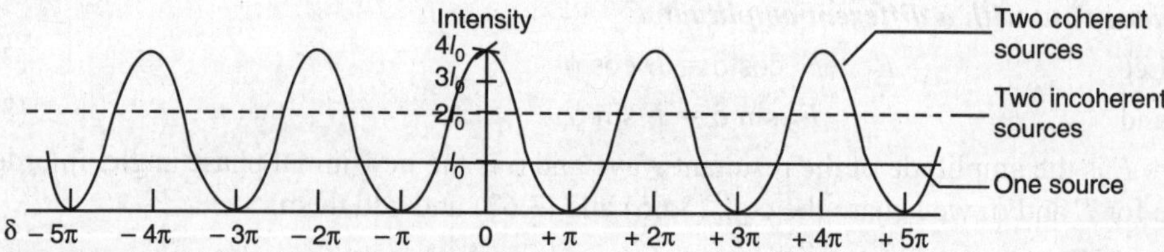

Fig. 3.28:

Intensity versus phase angle δ.

It is seen from the plot that the intensity varies from zero at the minima to $4I_0$ at the fringe maxima.

3.17.2 Superposition of Incoherent Waves

Incoherent waves are the waves that do not maintain a constant phase difference. Then the phase of the waves fluctuates irregularly with time and independently of each other. In case of light waves the phase fluctuates randomly at a rate of about 10^8 per second. Light detectors such as human eye, photographic film etc cannot respond to such rapid changes. The detected intensity is always the average intensity, averaged over a time interval which is very much larger than the time of fluctuation. Thus,

$$I_{ave} = I_1 + I_2 + 2\sqrt{I_1 I_2} < \cos \delta >$$

The average value of the cosine over a large time interval will be zero and hence the interference term becomes zero. Therefore, the average intensity of the resultant wave is

$$I_{ave} = I_1 + I_2$$

If $I_1 = I_2$, then $\qquad I_{ave} = 2I \qquad ...(3.73)$

It implies that the superposition of incoherent waves does not produce interference but gives a uniform illumination (see Fig. 3.28). The average intensity at any point is equal to the total of the intensities of the constituent waves. For instance, the collective intensity of the two headlights of a car is the total of the two separate intensities. The superposition of light from the headlights does not produce dark regions of zero intensity. Eq. (3.73) further implies that the superposition of incoherent light waves can never result in zero intensity.

3.17.3 Superposition of Many Coherent Waves

Eq. (3.70) may be written as

$$I_{max} = 2^2 I_0$$

which gives the resultant intensity when two coherent waves superpose. The resultant maximum intensity due to N coherent waves will be therefore

$$I_{max} = N^2 I_0 \qquad ...(3.74)$$

and the minimum intensity $\quad I_{min} = 0 \qquad ...(3.75)$

where N represents the number of coherent waves superposing at a point.

3.17.4 Conditions of Maximum Intensity and Zero Intensity

We may represent in a more general way the two waves meeting at P as follows:

$$E_A = E_1 \sin(\omega t - kx_1 + \phi_1)$$

and

$$E_B = E_2 \sin(\omega t - kx_2 + \phi_2)$$

where x_1 and x_2 represent the paths traveled by waves A and B. Therefore, the phase difference δ between the waves comes out to

$$\delta = (-kx_1 + \phi_1) - (-kx_2 + \phi_2) = k(x_2 - x_1) + (\phi_1 - \phi_2)$$

$$= \frac{2\pi}{\lambda}\mu L + (\phi_1 - \phi_2)$$

or

$$\delta = \frac{2\pi}{\lambda}\Delta + (\phi_1 - \phi_2) \qquad ...(3.76)$$

Thus, the phase difference δ between the waves is made up of two parts – one part arising from the difference in the paths traversed by them while the second part is on account of the difference in phases at the points of origin of the two waves. As a special case, let us assume that $\phi_1 = \phi_2$. Then $(\phi_1 - \phi_2) = 0$; which implies that the two waves under consideration are in phase initially. Eq. (3.76) reduces to

$$\delta = \frac{2\pi}{\lambda}\Delta$$

which indicates that the two waves have the same phase originally but subsequently developed the phase difference because they have traveled along different routes to reach the point P. At P they are displaced with respect to each other by a fraction of wavelength equal to Δ/λ.

The result of superposition of the waves at P is then determined solely by the path difference Δ. Eq. (3.68) becomes

$$E^2 = E_1^2 + E_2^2 + 2E_1E_2 \cos\left(\frac{2\pi}{\lambda}\Delta\right)$$

Assuming $E_1 = E_2$, the above relation reduces to

$$E^2 = 2E_1^2\left(1 + \cos\frac{2\pi}{\lambda}\Delta\right)$$

$$\therefore \qquad I = 2I_1\left(1+\cos\frac{2\pi}{\lambda}\Delta\right)$$

Using the trigonometric relation $1 + \cos 2\theta = 2\cos^2\theta$ into the above equation, we get

$$I = 2I_1\left(2\cos^2\frac{\pi}{\lambda}\Delta\right)$$

or

$$I = 4I_1\left(\cos^2\frac{\pi}{\lambda}\Delta\right) \qquad ...(3.77)$$

Eq. (3.77) helps us state the conditions of maximum intensity and zero intensity in terms of the path difference. The resultant intensity

$$I = I_{max} = 4\,I_1 \qquad \text{if} \quad \cos^2\frac{\pi}{\lambda}\Delta = 1$$

This will happen when $\Delta = 0, \lambda, 2\lambda, 3\lambda, \ldots.\ m\lambda$.

Whenever the above condition is satisfied the waves will meet in phase at P and the waves interfere constructively. Thus, *when two waves are not displaced with respect to each other or when they are displaced through an integral number of wavelengths, constructive interference takes place* Eq. (3.22). On the other hand,

$$I = I_{max} = 0 \qquad \text{if} \quad \cos^2\frac{\pi}{\lambda}\Delta = 0$$

which will happen if $\Delta = \frac{\lambda}{2}, 3\frac{\lambda}{2}, 5\frac{\lambda}{2}, \ldots\ldots\ldots(2m+1)\frac{\lambda}{2}$.

In this situation, the waves meet at P in inverted relation. The crests of one wave fall on the troughs of the second wave leading to cancellation of the wave displacements. Then the waves interfere destructively. Thus, *when two waves are displaced with respect to each other by an odd number of half-wavelengths, destructive interference results* Eq. (3.24). We summarize the conditions for the two types of interferences as follows:

$$I = I_{max}$$

if $\Delta = m\lambda$ Constructive Interference ...(3.78)

where $m = 0, 1, 2 \ldots.$

$$I = I_{min} = 0$$

if $\Delta = (2m + 1)$ Destructive Interference ...(3.79)

where $m = 0, 1, 2 \ldots.$

The condition Eq. (3.78) shows that the region, where constructive interference occurs, appears brighter than the surrounding region. Similarly, the region, where destructive interference occurs, appears dark. Therefore, the conditions Eq. (3.78) and Eq. (3.79) are as well the conditions for brightness and darkness. We restate the conditions Eq. (3.78) and Eq. (3.79) as

$$\Delta = m\lambda; \qquad (m = 0, 1, 2, \ldots.) \quad \text{Brightness} \qquad ...(3.80)$$

$$\Delta = (2m+1)\frac{\lambda}{2}; \qquad (m = 0, 1, 2, \ldots.) \quad \text{Darkness} \qquad ...(3.81)$$

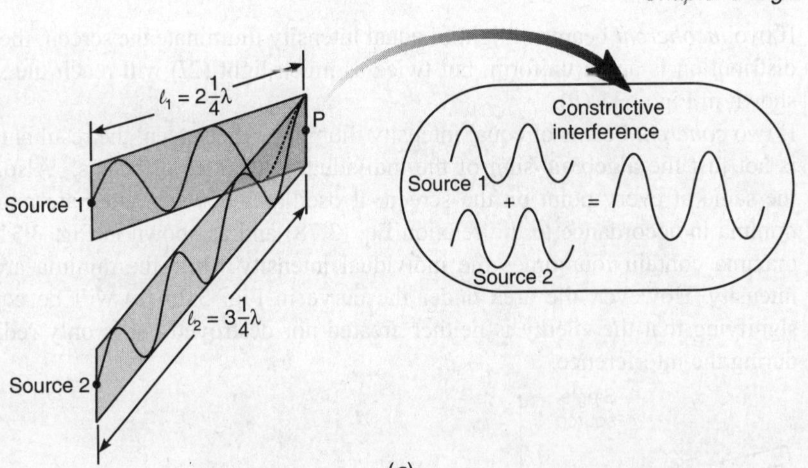

(a)

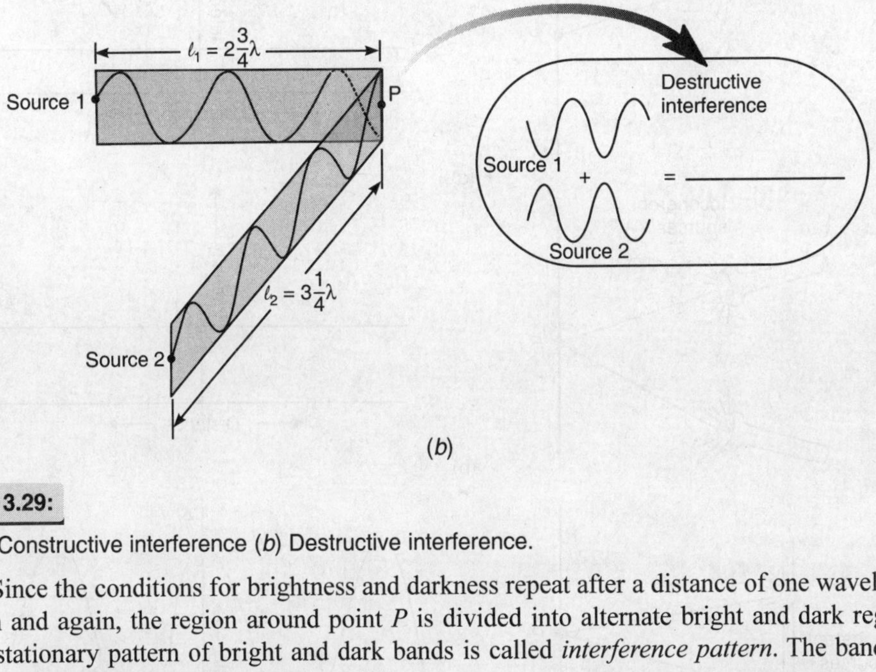

(b)

Fig. 3.29:

(a) Constructive interference (b) Destructive interference.

Since the conditions for brightness and darkness repeat after a distance of one wavelength again and again, the region around point P is divided into alternate bright and dark regions. The stationary pattern of bright and dark bands is called *interference pattern*. The bands are known as **interference fringes** (Fig. 3.30).

Interference Pattern

Fig. 3.30:

Interference pattern consisting of alternate bright and dark bands.

Fig. 3.31 shows the energy distribution on a screen under different conditions.

(a) If only one beam of light illuminates the screen the intensity distribution is uniform throughout the area of illumination, as shown in Fig. 3.31 (a).

(*b*) If two *incoherent* beams of light of equal intensity illuminate the screen, the intensity distribution is again uniform, but twice as much light ($2I$) will reach the screen, as shown in Fig. 3.31 (*b*).

(*c*) If two *coherent* beams of equal intensity illuminate the screen, the resultant intensity is not just the algebraic sum of the individual intensities of beams. Also, it is not the same at every point on the screen. It oscillates between alternate maxima and minima in accordance to the relation Eq. (3.78) and as shown in Fig. 3.31 (*c*). The maxima contain four times the individual intensity while the minima are of zero intensity. However, the area under the curve in Fig. 3.25 (*c*) will be equal to $2I$ signifying that the energy is neither created nor destroyed but is only redistributed during the interference.

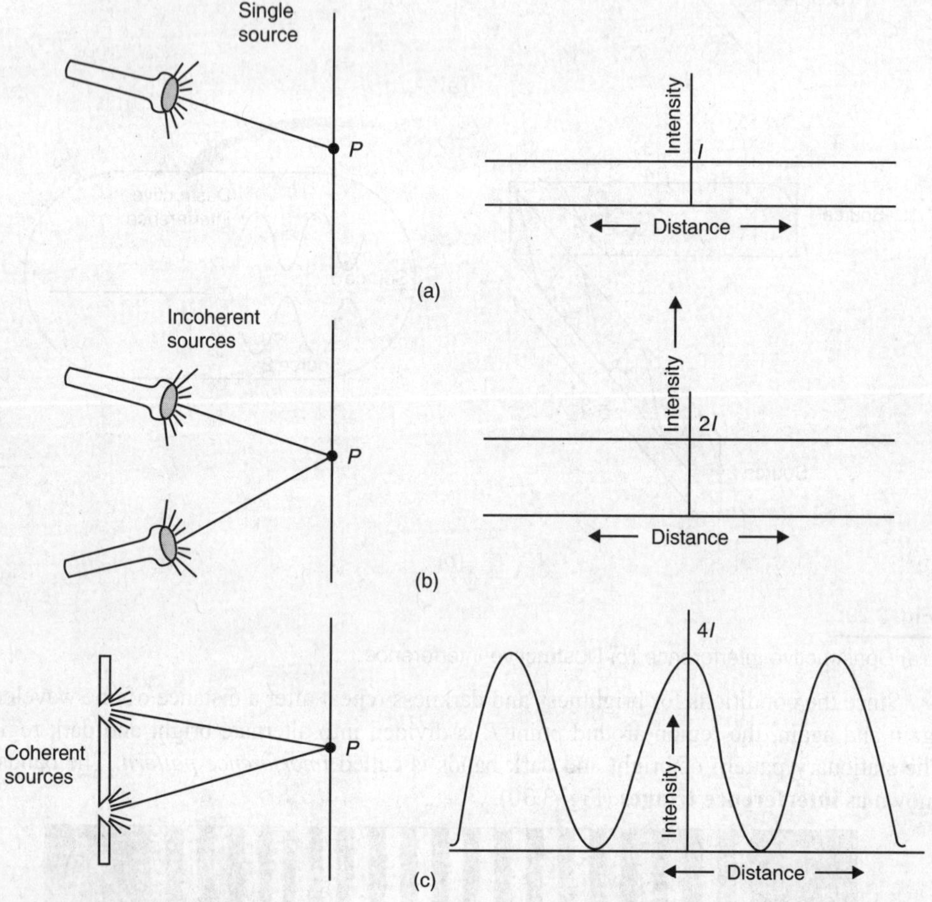

Fig. 3.31:

Intensity distribution on the screen when – (*a*) single source illuminates the screen, (*b*) two incoherent sources illuminate the screen; (*c*) two coherent sources illuminate the screen.

Example 3.7: *Two coherent light waves arrive at a particular point on a screen. The optical path difference between the waves is 3μm. Determine the nature of interference at the point if the wavelength of the wave is 3900 Å.*

Solution: Constructive interference occurs between the waves at the point, if the optical path difference contains an even number of half-waves; otherwise the nature of interference will be destructive if it contains odd number of half-waves. Thus, if

$$\Delta = m\frac{\lambda}{2} \ (m = 2, 4, 6, ...) \text{ constructive}$$

If, $$\Delta = m\frac{\lambda}{2} \ (m = 1, 3, 5, 7, ...) \text{ destructive}$$

$$\Delta = m\frac{\lambda}{2}; \ m = \frac{2\Delta}{\lambda} = \frac{2\times 3\times 10^{-6}\,\text{m}}{3.9\times 10^{-7}\,\text{m}} = 15.4 = \mathbf{15}$$

As m is an odd number, it means that the optical path difference is such that the waves arrive at the point with a phase difference of 180°. So the waves interfere destructively.

3.18 YOUNG'S DOUBLE SLIT EXPERIMENT

Young gave the first demonstration of the interference of light waves in 1801. Fig. 3.32 shows a plan view of the basic arrangement of his double slit experiment. The primary light source at S is a monochromatic source; it is generally a sodium lamp, which emits yellow light of wavelength at around 5893 Å. The expanding wave front from the primary light source S falls on two narrow closely spaced slits, S_1 and S_2 as shown in Fig. 3.32. The slits at S_1 and S_2 are very narrow and partition the incident wave front. If the slits are equidistant from S, the phase of the wave at S_1 will be the same as the phase at S_2 because parts of the same wave front emerging from S pass through S_1 and S_2. Further, waves leaving S_1 and S_2 have the same frequency as the primary source. Hence, sources S_1 and S_2 act as secondary coherent sources. The waves leaving from S_1 and S_2 interfere and produce alternate bright and dark bands on the screen at T. Thomas Young used this experiment to make the first measurement of the wavelength of the light.

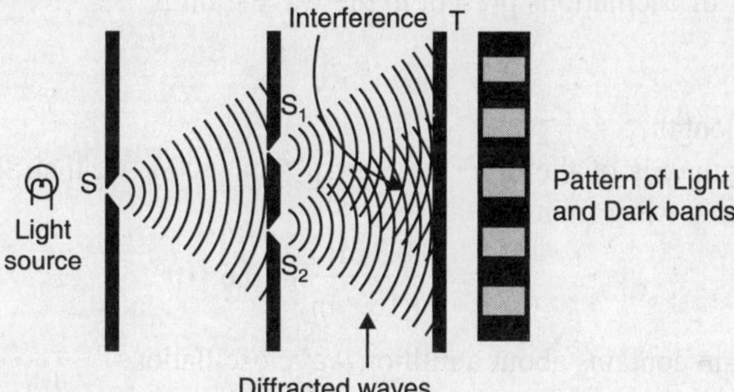

Fig. 3.32:

Young's double slit arrangement -The narrow slit S acts as a source of cylindrical waves which illuminate the slits S_1 and S_2. S_1 and S_2 behave as coherent sources and produce interference.

Now, if light is allowed to illuminate the slits S_1 and S_2 directly, instead of through slit S, interference will not be produced and the observation screen will be uniformly illuminated. We have learnt that interference is the manifestation of coherence. The absence of interference emphasizes the fact that coherence is lacking. To gain an understanding of the role of slit S in the double slit interference, let us look more intimately at the structure of real light waves.

3.19 WAVETRAINS—LIGHT FROM COMMON SOURCES

An atom leaving an excited state gives up the excess energy in the form of a burst of light, photon, and jumps to the lower normal state in an about 10^{-9}s. Therefore, the light emitted by an atom is not a continuous harmonic wave of infinite extension but is a wave *train* called a **wave packet**. After some time the atom again receives energy and gets excited and falls back to lower state by emitting another wave train. These emission events happen at random. Other atoms in the source also emit wave trains similarly but with different emission times. The wave trains generated by all atoms in the light source produces a series of wave trains with randomly distributed phases, (see Fig. 3.33).

Thus, ***the light emitted by an ordinary light source is not an infinitely long, simple harmonic wave but is a jumble of finite wave trains***. So, a real monochromatic source is a *quasi-monochromatic* source producing wave trains as shown in Fig. 3.33.

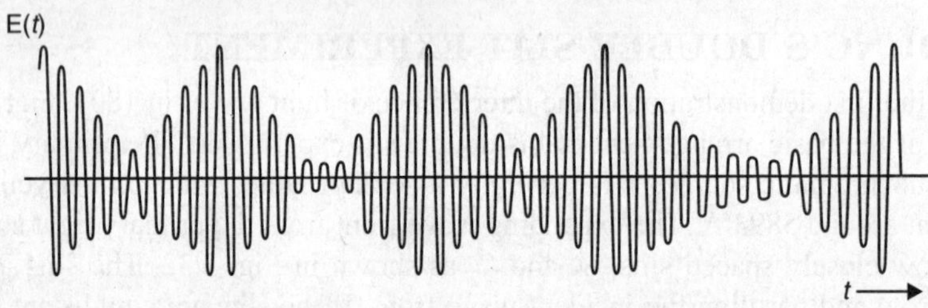

Fig. 3.33: Light is emitted in the form of wave trains.

The length of the wave train in a vacuum is

$$l = c\Delta t \qquad ...(3.82)$$

where c is the velocity of light in a vacuum and Δt is the time interval.

If $t = 10^{-9}$ s, $l = (3 \times 10^8 \text{ m/s})(10^{-9} \text{ s}) = 0.3$ m

$\therefore$ The number of oscillations present in the wave train is

$$N = \frac{l}{\lambda} \qquad ...(3.83)$$

where λ is the wavelength.

If the wave length λ is of the order of 5×10^{-7} m (5000 Å), then the number of wave oscillations present in a wave train is

$$N = \frac{0.3 \text{ m}}{5\times10^{-7} \text{ m}} = 0.6\times10^6.$$

Thus, a wave train contains about a million wave oscillations.

3.19.1 Bandwidth

A wave train is not a harmonic wave. It cannot be shown by the simple sine function such as Eq. (3.38). A wave packet is mathematically represented in terms of Fourier integrals. We sum up here some of the important features of a wave packet.

When analyzed with the help of a spectrograph, light emitted from a source is found to be made up of discrete **spectral lines**, formed by wave packets emitted by atoms. So, a spectral line and a wave packet describe the same. The wavelength of a *wave packet* or a *spectral line* is not precise but is a continuous spread of wavelengths over a finite range, $\Delta\lambda$, centered around a wavelength λ_0. The wave packet may be considered as consisting of a number of harmonic waves which differ by infinitesimal increments of wavelengths. The maximum intensity of the wave packet will occur at λ_0 and the intensity falls off quickly on either side of λ_0, as shown in Fig. 3.34. The spread of wavelength is called *line width or bandwidth*. The *band width* is the wavelength interval from $\lambda_0 - \Delta\lambda / 2$ to $\lambda_0 + \Delta\lambda / 2$ which holds the major portion of the energy of the wave packet.

According to the Fourier analysis, the bandwidth expressed in terms of frequency is governed by the relation,

$$\Delta\nu = \frac{1}{\Delta t} \quad ...(3.84)$$

where Δt is the average life time of the light emitting atom in its excited state.

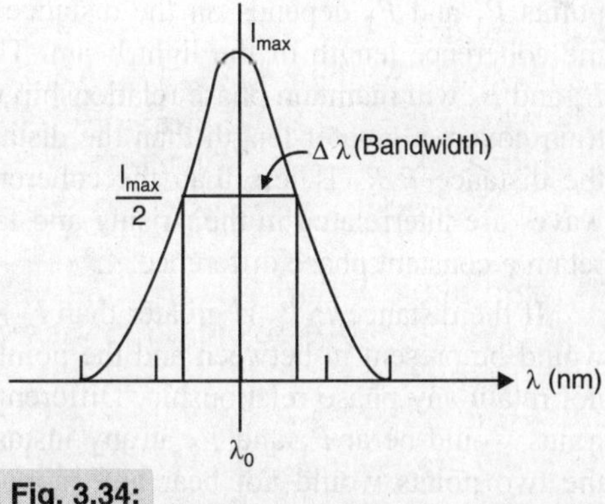

Fig. 3.34:
Band width of a spectral line or wave packet.

3.20 COHERENCE

Coherent light means that the light waves maintain constant phase difference over a period of time. A constant phase difference could be maintained easily if the waves are harmonic waves. The real light waves are not harmonic waves but a jumble of wave trains.

An abrupt change in phase occurs when we pass from one wave train to the next wave train in a light wave (see Fig. 3.35). So, relating the phase at a point in one wave train to a point in another wave train is not possible. The phase relationship fluctuates irregularly from one wave train to another wave train. As a result, interference of waves does not occur and the resultant illumination produced by conventional sources is not so high.

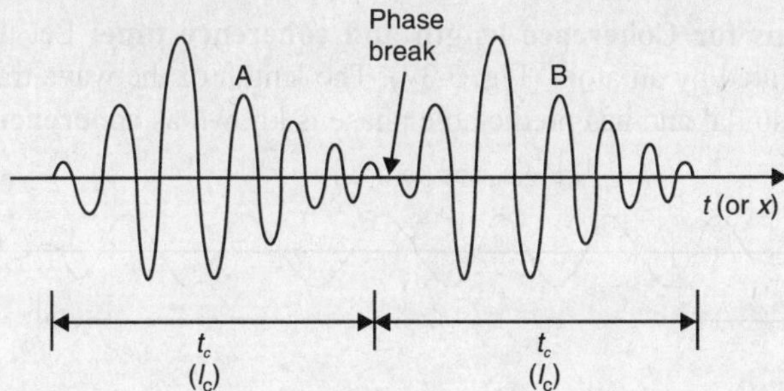

Fig. 3.35:
A light wave is made up short wave trains A, B etc. In passing from one wave train to the next wave train there is an abrupt change in phase.

In general, there are two types of coherence; *temporal* and *spatial*. The temporal coherence is connected to the finite bandwidth of the source. The spatial coherence is linked to the finite size of the source.

3.20.1 Temporal Coherence

Two waves are said to be *temporally coherent* when there is a specific relationship between their phases at a given time and at a certain time later.

Let a point source of quasi-monochromatic light S (Fig. 3.36) gives off light in all directions. For a light traveling along the line SP_1P_2, the phase relationship between the points P_1 and P_2 depends on the distance between them and the coherence length of the light beam. The electric fields at P_1 and P_2 will maintain phase relationship when a single wave train covers a greater length than the distance P_1P_2; that is if the distance P_1P_2 is less than the coherence length l_{coh}, the waves are interrelated in their rising and falling and they will retain a constant phase difference.

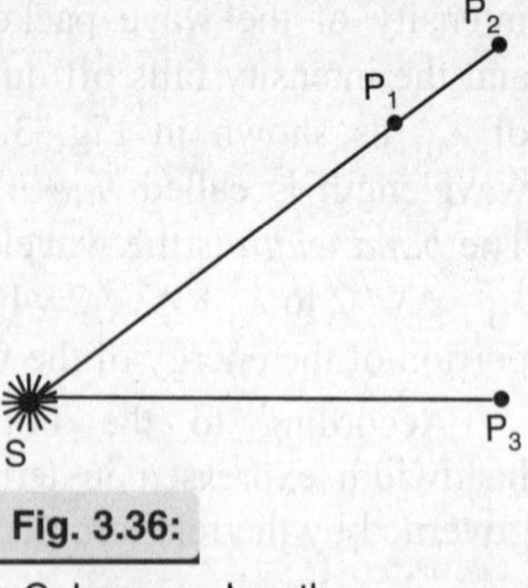

Fig. 3.36:

Coherence length.

If the distance P_1P_2 is greater than l_{coh}, many wave trains would be present in between and the points P_1 and P_2 would not retain any phase relationship. Different independent wave trains would be at P_1 and P_2 at any instant and the phase at the two points would not bear any rapport with each other. The degree to which a correlation exists is known as the amount of **longitudinal coherence**. As monochromaticity is related to the coherence length, temporal coherence is regarded as a measure of **monochromaticity**. Temporal coherence is characterized by two parameters, namely *coherence length* and *coherence time.*

(*i*) **Coherence time:** In case of conventional sources of light, light is emitted in the form of short wave trains and the phase of one wave train would remain constant with respect to the phase of another wave train from only for about 10^{-9} sec. It implies that the two wave trains are *temporally coherent* for a maximum period of about 10^{-9} sec. This time is called the **coherence time** (t_{coh}). *Coherence time is defined as the longest time interval over which the phase undergoes change in a regular way.*

(*ii*) **Coherence length:** The distance for which each wave train remains sinusoidal is called **coherence length** (l_{coh}). *Coherence length is defined as the spatial extent over which the wave train has predictable phase.*

(*iii*) **Expressions for Coherence length and coherence time:** Let us consider a wave train generated by an atom (Fig. 3.37). The length of the wave train over which it is fairly sinusoidal and has predictable phase is known as **coherence length.**

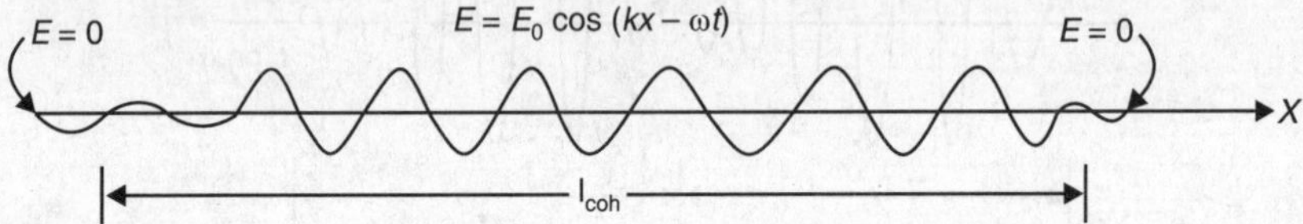

Fig. 3.37:

A typical wave train.

Coherence length may be considered as equal to the length of the wave train, $c\,\Delta t$. The time interval during which the phase of the wave train is predictable is called as **coherence time**, Δt, during which the phase of the wave train changes in a regular systematic way. So,

$$l_{\text{coh}} = c\Delta t \quad \text{...(3.85)}$$

and

$$t_{\text{coh}} = \Delta t \quad \text{...(3.86)}$$

$$\therefore \quad l_{\text{coh}} = c\, t_{\text{coh}} \quad \text{...(3.87)}$$

(*iv*) **Relation between coherence length and frequency bandwidth:** According to Fourier analysis the **frequency bandwidth** Δv is given by

$$\Delta\nu \approx \frac{1}{\Delta t} \quad ...(3.88)$$

where Δt is the average lifetime of the excited state of the atom.

Δt corresponds to the coherence time, t_{coh}, of the wave packet.

$$\therefore \quad \Delta\nu = \frac{1}{\Delta t} = \frac{1}{t_{\text{coh}}}$$

Using the relation (3.87) into the above relation, we get

$$\Delta\nu = \frac{c}{l_{\text{coh}}} \quad ...(3.89)$$

or

$$l_{\text{coh}} = \frac{c}{\Delta\nu} \quad ...(3.90)$$

(*iv*) **Relation between coherence length and wavelength bandwidth:**

$$\nu = \frac{c}{\lambda_0} \quad ...(3.91)$$

where λ_0 is the vacuum wavelength.

Differentiating Eq. (3.91) on both sides, we get

$$\Delta\nu = -\frac{c}{\lambda_0^2}\Delta\lambda \quad ...(3.92)$$

Using the Eq. (3.98) into Eq. (3.92), we obtain

$$\frac{c}{l_{\text{coh}}} = -\frac{c}{\lambda_0^2}\Delta\lambda \quad ...(3.93)$$

Rearranging the terms, we get

$$l_{\text{coh}} = \frac{\lambda_0^2}{\Delta\lambda} \quad ...(3.94)$$

The minus sign has no significance and hence is ignored. $\Delta\lambda$ is the **natural line width** or **bandwidth**.

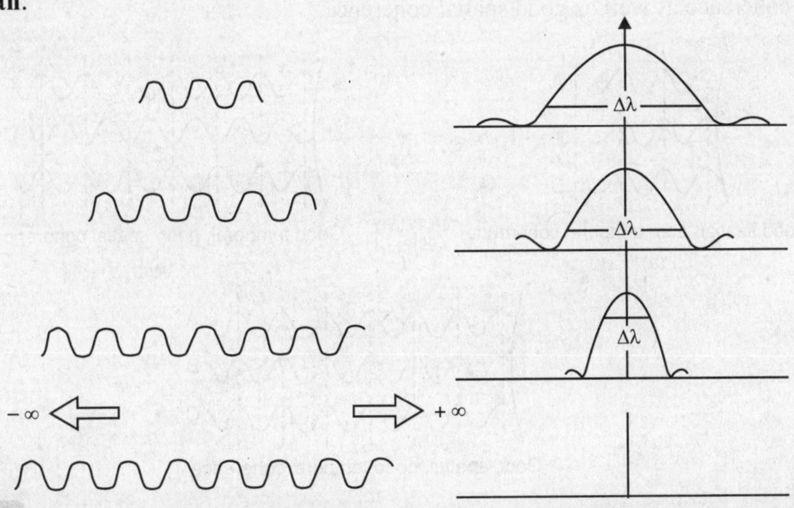

Fig. 3.38:

Dependence of bandwidth on the length of the wave train.

Eq. (3.94) implies that the longer is the length of the wave trains emitted by a light source, more monochromatic will be its light (Fig. 3.38). Thus, we conclude **that temporal coherence is a measure of the monochromaticity** of the light source.

3.20.2 Spatial Coherence

Spatial coherence denotes the continuity and uniformity of a wave in a direction at right angles to the direction of propagation. *If the phase difference for any two fixed points in a plane perpendicular to the wave propagation is constant with time, then the wave is said to possess spatial coherence*. Thus, two waves from two different points of an extended source are said to be spatially coherent if they maintain a constant phase difference over a period of time. The spatial coherence is as well recognized as **transverse coherence.**

As shown in Fig. 3.36, $SP_1 = SP_3$, the electric fields at points P_1 and P_3 would be in phase. Hence, an ideal point source displays spatial coherence.

In contrast, an extended source exhibits smaller lateral spatial coherence. Two points on the source separated by a distance greater than one wavelength will behave independently. **The degree of contrast of the interference fringes produced by a source is a measure of the degree of the spatial coherence of the waves generated by it**. The higher the contrast, the superior is the spatial coherence.

In the Young's double slit experiment, a single slit S is placed between the light source and the double slit so that the same group of wave trains is incident on slits S_1 and S_2 (Fig. 3.32). When the phase of the wave changes at S, this change is communicated simultaneously to S_1 and S_2.Therefore, the waves emerging from S_1 and S_2 will be coherent with respect to each other. Thus, the purpose of keeping the single slit in the arrangement is to obtain spatial coherence between the waves passing through slits S_1 and S_2.

Fig. 3.39 illustrates the above concepts clearly. A, B, and C are three wave trains emitted from different points on an extended source of light. Waves in Fig. 3.39 (*a*) have good spatial coherence but have poor temporal coherence (note that the wave trains are shorter and hence coherence time and coherence length are smaller). Waves in Fig. 3.39 (*b*) have good temporal coherence but have poor spatial coherence. In contrast, waves in Fig. 3.39 (*c*) possess good temporal coherence as well as good spatial coherence.

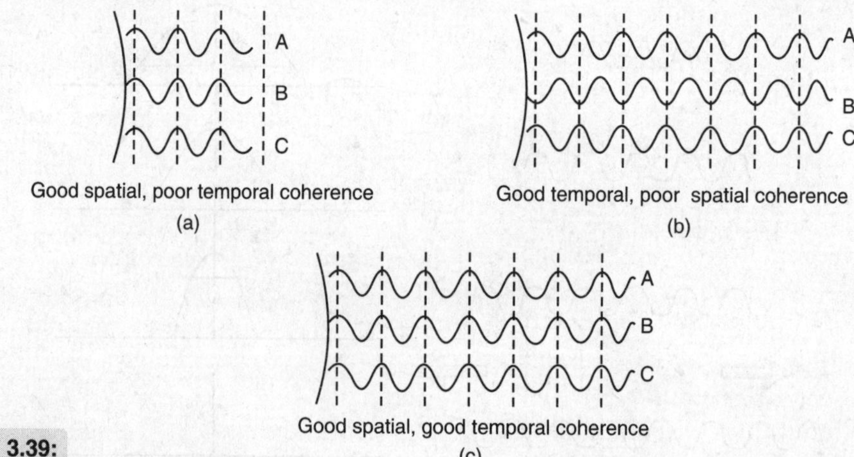

Fig. 3.39: Illustration of spatial coherence.

Example 3.8: *Calculate the coherence length for CO_2 laser whose line width is 1×10^{-5} nm at IR emission wavelength of 0.6 μm.*

Solution: $l_{coh} = \dfrac{\lambda^2}{\Delta\lambda} = \dfrac{(10.6\mu m)^2}{1\times10^{-5}\,nm} = \dfrac{(10.6\times10^{-6})^2}{10^{-5}\times10^{-9}}$ = **11.2 km**

3.21 DOUBLE SLIT EXPERIMENT AGAIN

We now illustrate the concept of coherence with the help of Young's double slit experiment (Fig. 3.32).

- In this experiment the wave train coming from the slit S is divided physically into two segments by the slits S_1 and S_2. The two sets of wave trains resulting through division of a parent wave train travel along different paths and ultimately overlap on each other. If the paths of overlapping wave trains differ by more than one coherence length, interference does not take place because the superposed waves are not parts of the same group of wave trains and are incoherent. It is therefore essential that **the optical path difference Δ must be smaller than coherence length for interference to occur**. Thus,

$$\Delta \leq l_{coh} \quad ...(3.95)$$

The number of fringes observed in the experiment is limited because of the above condition. Constructive interference takes place whenever

$$\Delta = \pm\, m\lambda \qquad (m = 0, 1, 2, 3, ...)$$

For the extreme bright fringes in the pattern, we may assume that

$$\Delta_{max} \approx l_{coh}$$

$$\therefore \qquad (m)_{max}\,\lambda \approx l_{coh} \approx \frac{\lambda^2}{\Delta\lambda}$$

$$\therefore \qquad (m)_{max} \approx \frac{\lambda}{\Delta\lambda} \quad ...(3.96)$$

The above relation indicates that the number of fringes observed is inversely proportional to the bandwidth of the incident light. The number will be large if the bandwidth is small or temporal coherence is high.

- The contrast of the fringes produced by a light beam indicates the degree of spatial coherence. The broader the source of the beam, the smaller is the degree of spatial coherence. In the double slit experiment if the slits S_1 and S_2 are directly illuminated by a light source, interference fringes are not observed and the screen is uniformly illuminated. The absence of fringes shows that the light coming out from the slits does not possess spatial coherence. If a slit S is placed between the source and the double slits, the light passing through S illuminates the slits S_1 and S_2. Fringes are not observed as long as S is wider. On gradually reducing the width of S, formation of fringes would be observed and the fringes will become distinctly clear when the slit is reduced to pin-hole size. It means that the slit S increases the spatial coherence of waves by limiting the size of the wave front from the source. It ensures that the wave trains incident on the slits S_1 and S_2 originate from a narrow region of the source. It is found that the beam will be spatially coherent when the lateral distance, d which is the slit separation, is of the order of $\lambda/2\theta$, where θ is the angle subtended by the source at the slit. Thus, the condition is that

$$d \approx \frac{\lambda}{2\theta} \quad ...(3.97)$$

The light produced by lasers is highly monochromatic and coherent. When they are used in double slit experiment, interference fringes are obtained without the necessity of slit S.

3.22 DISPERSION

Light waves of different wavelengths travel with different velocities in a transparent medium and will be bent at different angles at the refracting surface. Therefore, the refractive index of a medium varies with wavelength of incident light. Thus,

$$\mu = f(\lambda) \quad ...(3.98)$$

The **dispersion**, a relation of refractive index and wavelength is illustrated in Fig. 3.40(b). Normally, the refractive index decreases with increasing wavelength. White light passing through a medium will be separated according to wavelengths. It is referred to as **chromatic dispersion**.

When white light is incident at normal on a rectangular block of a medium, violet light would merely lag behind red light. When it is incident obliquely the colours become separated in space. When the medium is taken in a prism form, the separation of colours is much more. The angle between rays of two colours is called the **angular dispersion**.

Dispersion is expressed in terms of three refractive indices. A simple measure of the dispersions is provided by the angular separation of the red and violet rays. It is not necessary that materials of high refractive index exhibit high dispersion, as dispersion is not proportional to refractive index.

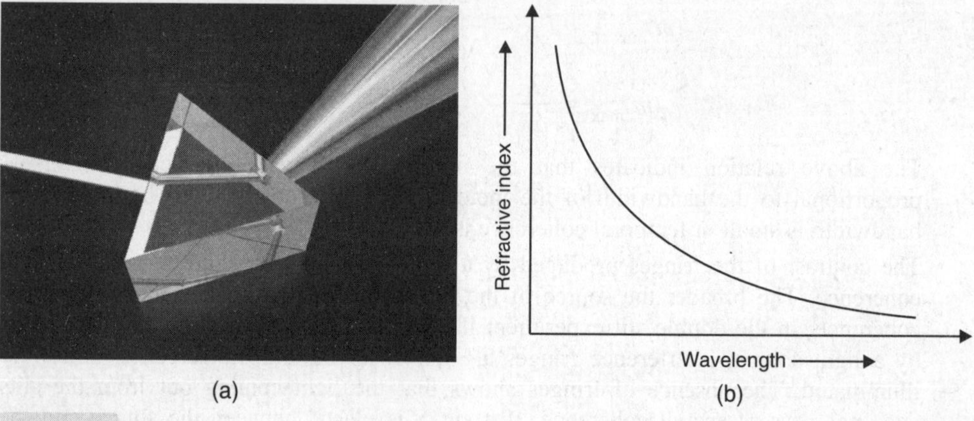

Fig. 3.40: Dispersion.

Dispersion varies as the third inverse power of wavelength as per the following relation

$$\frac{d\mu}{d\lambda} = -\frac{2B}{\lambda^3} \quad ...(3.99)$$

Dispersion of light in water droplets suspended in the atmosphere is responsible for the celestial spectacle called *rainbow*. Optical radiation emitted by bodies is studied with the spectrographs which use prisms or gratings to disperse colours.

3.23 SCATTERING

When a light beam encounters an obstacle of size comparable to its wavelength, diffraction occurs. If the size of the obstacle is much smaller than a wavelength (i.e. $d << \lambda$), light spreads in the form of spherical waves after striking the obstacle. Then light is said to have been **scattered** by the obstacle (Fig. 3.41). Scattering redirects part of the incident radiation into directions other than the direction of the incident beam. Scattering is thus a special case of diffraction. The strength of the original beam decreases during the process of scattering, causing attenuation of energy.

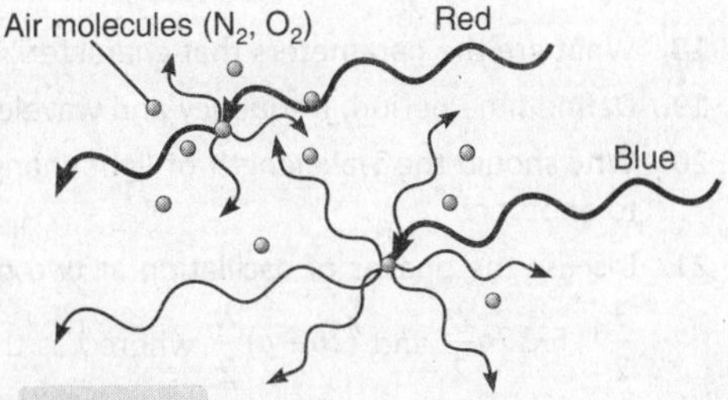

Fig. 3.41: Light scattering by air molecules.

The first quantitative study of the scattering of light was carried out by Rayleigh. In 1871, he established that the intensity of scattered light is inversely proportional to the fourth power of wavelength. Thus,

$$I_s \propto \frac{1}{\lambda^4} \qquad ...(3.100)$$

This is known as the **Rayleigh's law of scattering.** It is familiar to us that sunlight scattered from very fine particles of smoke appears bluish in colour. The Rayleigh's law of scattering explains the reason for the bluish colour of sky on a cloudless day.

QUESTIONS

1. What is meant by an optical medium?
2. Explain the terms: homogeneous optical medium, and inhomogeneous medium.
3. Explain the terms: isotropic medium and anisotropic medium.
4. Define reflectivity and transmissivity. Give the mathematical expressions.
5. What is total internal reflection? Derive the condition for total reflection.
6. Derive Lambert-Bouguer law.
7. What are the electromagnetic waves? Why light is classified in the category of electromagnetic waves?
8. What are the characteristics of a wave?
9. Define intensity of an electromagnetic wave and obtain an expression for it.
10. What is the difference between optical and geometrical paths?
11. Can the optical path between two points ever be less than the geometrical path length between these points?
12. Can two 25-watt electric bulbs with point like filament of the same material and lying close to each other produce interference?
13. What becomes to the energy of light waves in destructive interference?
14. Interference can be observed only when the two sources have some common characteristics. What are they?
15. What are coherent waves?
16. State the conditions for formation of bright and dark fringes.
17. What is a periodic wave?

18. What are the parameters that characterize a wave motion?
19. Define time period, frequency and wavelength of a wave. How are they related to each other?
20. Why should the wavelength of light change, but not its frequency, in passing from one material to another?
21. Discuss the phases of oscillation at two points on a light ray if the separation between them is $\frac{3}{2}\lambda, 5\lambda, 2n\frac{\lambda}{2}$ and $(2n+a)\frac{\lambda}{2}$ where λ is the wavelength and n is an integer.
22. What are traveling waves? When a traveling wave propagates in a medium, what is that progresses forward.
23. What do you understand by the term 'Phase Velocity'?
24. Explain the following :
    (*a*) Wave surface (*b*) Wave front and (*c*) ray
25. How are the following produced:
    (*a*) Spherical waves (*b*) Cylindrical waves and (*c*) Phase waves?
26. A plane wave of wavelength λ and amplitude A propagates in a medium with a speed v_1. Write the equation of the plane wave.
27. When are two waves said to be in-phase and our-of-phase?
28. Two sinusoidal waves of equal amplitude are ¼ of a wavelength out of phase. What is the amplitude of the resultant wave?
29. Distinguish between the geometrical path length of a light ray in a medium and its optical equivalent. How are they related? What is the difference between optical and geometrical path difference?
30. Can the optical path length between two points ever be less than the geometrical path length between those points?
31. Why light is represented by *E* vector? Explain monochromaticity and polarization of an e.m. wave?
32. What is the path change a light wave undergoes due to reflection at the interface of two media when
    i. It is traveling from a rarer medium to a denser medium and
    ii. It is traveling from denser medium to a rarer medium?
    What is the corresponding phase change?
33. Consider the superposition of two harmonic disturbances. Show that the resultant intensity due to them is not just the sum of their individual intensities?
34. Explain why two independent identical sources of light of the same frequency cannot give rise to interference fringe pattern.
35. What is meant by interference of light? State the fundamental conditions for obtaining sustained interference pattern.
36. Describe the interference pattern obtained due to superposition of the coherent waves. Write an expression for the intensity distribution over the plane of observation. Plot the resultant intensity as a function of distance on the screen?
37. Each of *N* atomic radiators is giving out radiation of same wavelength and intensity. Show under which conditions their resultant at a point can be NI or N^2I where I is intensity due to each radiator at point.
38. Explain the terms coherence length and coherence time for a light wave. Derive an expression for the coherence length of a wave train that has frequency band width $\Delta\nu$. Express the answer in terms of line width $\Delta\lambda_0$ and mean wavelength λ_0 of the wave train.

39. What do you understand by coherence? Explain temporal coherence and spatial coherence. Derive the condition for spatial coherence.
40. Explain in brief: *i.* Coherence length *ii.* Spatial coherence *iii.* Temporal coherence. Obtain an expression for coherence length.
41. Explain spatial coherence, temporal coherence and coherence and coherence length.
42. What are the processes that are responsible for emission of light from a source? What can we conclude about the character of light emitted from any real source?
43. Explain the meaning of temporal coherence and explain how duration or length of a single wave train can be a measure of temporal coherence.
44. How can you experimentally distinguish between coherent and non-coherent light?
45. What is meant by dispersion of light?
46. Explain the phenomenon of light scattering. State Rayleigh's law of scattering.

PROBLEMS

1. The frequency of green light is 5.5×10^{14} Hz. What is the wavelength in meters, microns, nanometers and angstrom units ? **[Ans:** $\lambda = 5455$ Å]
2. What is the speed of light of wavelength 5000 Å in glass whose index of refraction is 1.50? **[Ans:** $\upsilon = 2 \times 10^8$ m/s]
3. The average wavelength of sodium light is 5893 Å. Find out the number of waves in 1 cm. **[Ans:** N = 1.7×10^4 waves]
4. Monochromatic light of wavelength 4400 Å traverse from glass of refractive index 1.5 to a vacuum. Determine the change in wavelength. **[Ans:** 6600Å]
5. A soap bubble is 3×10^{-7} m thick. Red light is incident normally on the surface of the soap bubble. It gets partially reflected and partially transmitted at the top surface of the bubble. The transmitted ray travels through the film and again gets reflected partially at the bottom surface. The reflected component travels back and emerges out of the top surface. What is the geometrical path travelled by the ray in the film? What is the optical path? Refractive index is given as 1.333. **[Ans:** 0.7998 μm.]
6. A sodium atom radiates for 4×10^{-12} s. What is the coherence length of light from a sodium lamp? **[Ans:** 1.2 mm]
7. If light of 6600 Å wavelength has wave trains 20 λ long, what are its coherence length and coherence time? **[Ans:** 1.32×10^{-5}m, 4.4×10^{-14} s]
8. A He-Ne laser giving light at 6330 Å has a coherence length of 20 km. Determine (*a*) its coherence time; and (*b*) the number of waves per wave train. **[Ans:** 6.7×10^{-5}s, 3.2×10^{10} waves]
9. A mercury lamp has a band width $\Delta\nu$ = 1000 MHz. Calculate the coherence time and coherence time of its light. **[Ans:** 10^{-9}s, 30 cm]
10. Calculate the coherence length for CO_2 laser whose line width is 1×10^{-5} nm at IR emission wavelength of 10.6 μm. **[Ans:** 11.2 km]
11. White light has frequency range from 0.4×10^{15} Hz to 0.7×10^{15} Hz. Find the coherence time and coherence length. **[Ans :** 3.3×10^{-15}s, 1 μm]

4 Geometrical Optics

Learning Objectives:

By the end of this chapter, you will be able to:

- Understand basic of Thin and Co-axial Lens systems.
- Explain concepts related to Cardinal Points and Planes.
- Learn about Eyepiece instruments – Huygens and Ramsden.

4.1 INTRODUCTION

Geometrical optics will help us understand the basics of simple optical elements such as mirrors, prisms, lenses. Lenses are at the heart of many optical devices such as cameras, microscopes, binoculars, and telescopes. A lens is an image-forming device. It forms an image by refraction of light at its two bounding surfaces. It is expected that the image formed is good and free of aberrations. Ray concept is used in understanding the behaviour of light passing through lenses. As the relationship between the focal length of the lens, object distance and image distance are derived basing on ray approximation and on geometrical relations, this part of optics is known as **geometrical optics**. Geometrical optics played a very important role in the design and fabrication of optical components and instruments.

4.2 THIN LENSES

Lenses are broadly classified into *thin* and *thick* lenses. *A lens is said to be* **thin** *if the thickness of the lens can be neglected when compared to the lengths of the radii of curvature of its two refracting surfaces, and to the distances of the objects and images from it.* No lens is actually a thin lens. Yet many simple lenses commonly used can be treated as equivalent to a thin lens.

4.2.1 Lens Equation:

A lens forms an image by refraction of light at its two bounding surfaces. Each surface forms an image and contributes to the final image produced by the lens. If we know the focal length of a lens and the position of an object, the position of the image can be determined either by using graphical construction (ray diagram) or using mathematical relation. It is not always convenient to draw a ray diagram. We study here how the position of the image is determined using mathematical equation.

Now let us consider a *thin* convex lens XY with optical center at C, as shown in Fig. 4.1. Let the absolute refractive index of the lens materials be μ_2; and the lens be surrounded on all sides by air or any other rarer medium of refractive index μ_1. The centers of curvatures of the two refracting surfaces of the lens are C_1 and C_2. Let the radius of the curvature of the surfaces be R_1 and R_2 respectively. Consider a point object 'O' situated on the principal axis of the lens.

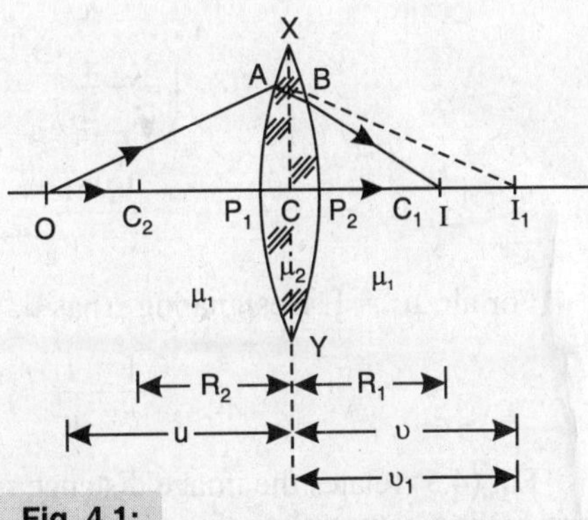

Fig. 4.1: Illustration showing thin convex lens with optical center.

Any ray of light leaving the axial point-object O and traveling along the principal axis passes without any deviation of path. Another ray OA strikes the first surface at A and is refracted in a direction BI_1. The ray is further refracted by the second surface in a direction BI and meets the ray along the principal axis at I. Therefore, I is the final point-image of the object O formed after refraction by the two surfaces of the lens.

By considering refraction at spherical surfaces we can derive an equation that describes image formation by a lens. The basic idea is that the image formed by the first refracting surface acts as a *virtual object* for the second refracting surface. We can apply **Gauss formula** for refraction at the first surface.

$$\frac{\mu_1}{u} + \frac{\mu_2}{v} = \frac{\mu_2 - \mu_1}{R} \quad ...(4.1)$$

The first surface, XAP_1Y, forms a *real* point-image, I_1. We replace υ by $υ_1$ and R by R_1 in Eq. (4.1) and apply the sign convention. We see that u is –ve. Therefore, the Eq. (4.1) is written for the first surface as

$$\frac{\mu_2}{v_1} - \frac{\mu_1}{u} = \frac{\mu_2 - \mu_1}{R_1} \quad ...(4.2)$$

In the Fig. 4.1, the light leaving the first surface would reach I_1 if the second surface XBP_2Y did not intervene. When the second surface is present, the light does not reach I_1. I_1 becomes the virtual object for the second surface and the second surface forms the image at I, which is the final image. The position of the final image is again found using Gauss formula using $u = v_1$ and $R = R_2$. Thus, we get

$$\frac{\mu_2}{v_1} + \frac{\mu_1}{v} = \frac{\mu_2 - \mu_1}{R_2}$$

Applying the sign convention, we get

$$-\frac{\mu_2}{v_1} + \frac{\mu_1}{v} = -\frac{\mu_2 - \mu_1}{R_2} \quad ...(4.3)$$

Adding Eq. (4.2) and Eq. (4.3), we obtain

$$\mu_1\left(\frac{1}{v} - \frac{1}{u}\right) = (\mu_2 - \mu_1)\left[\frac{1}{R_1} - \frac{1}{R_2}\right]$$

$$\left(\frac{1}{v}-\frac{1}{u}\right) = \left(\frac{\mu_2-\mu_1}{\mu_1}\right)\left[\frac{1}{R_1}-\frac{1}{R_2}\right]$$

$$\frac{1}{v}-\frac{1}{u} = \left(\frac{\mu_2}{\mu_1}-1\right)\left[\frac{1}{R_1}-\frac{1}{R_2}\right] \quad ...(4.4)$$

For air, $\mu_1 = 1$. Designating μ_2 as μ, we get

$$\frac{1}{v}-\frac{1}{u} = (\mu-1)\left[\frac{1}{R_1}-\frac{1}{R_2}\right] \quad ...(4.5)$$

Eq. (4.5) relates the image distance v of a thin lens to the object distance u and to the thin lens properties namely, refractive index and the radii of curvature. It is to be noted that a lens will not only focus from O to I but between any other pairs of points, as long as the points satisfy the Eq. (4.5).

4.2.2 Lens Maker's Equation:

If the object is at infinity, the image is formed at the *principal focus* of the lens. When $u = \infty$, $1/u = 0$ and $v = f$. Eq. (4.5) becomes

$$\frac{1}{f}-\frac{1}{\infty} = (\mu-1)\left[\frac{1}{R_1}-\frac{1}{R_2}\right]$$

$$\therefore \quad \frac{1}{f} = (\mu-1)\left[\frac{1}{R_1}-\frac{1}{R_2}\right] \quad ...(4.6)$$

Eq. (4.6) is known as the *lens maker's formula,* since it enables one to calculate f from the known properties of the lens. It can also be used to determine the values of R_1 and R_2 needed for a desired focal length of a lens of a given index of refraction.

If the lens is turned around, R_1 and R_2 are interchanged and the sign of each is reversed. Consequently, there is no change in f. Therefore, *for a thin lens, the focal length is independent of the order of the surfaces.*

It is clear from the lens maker's formula that to get a short focal length, the lens has to have a surface of smaller radius of curvature R and be made of a material with a high refractive index.

Comparing Eq. (4.5) and Eq. (4.6), we see that

$$\frac{1}{v}-\frac{1}{u} = \frac{1}{f} \quad ...(4.7)$$

Eq. (4.7) is known as the **Gauss' formula for a lens.** This is more useful formula than Eq. (4.5), because if we know the focal length of the lens, we can relate the position of the object and its image without necessarily knowing the index of refraction or the radii of the lens.

4.2.3 Positions of the principal foci:

Each individual surface of the lens has its own focal points and planes and the lens as a whole has its own *pair* of focal points and focal planes. The focal points and focal planes of the lens are known as *principal focal points* and *principal focal planes*. We are interested in knowing the locations of these principal focal points and principal focal planes. They are obtained as follows.

(*i*) If a point object is placed on the principal axis such that the rays refracted by the lens are parallel to the axis, then the position of the point object is called the **first principal focus** F_1 of the lens (see Fig. 4.2a). The distance at the first principal focus from the optical center C of the lens is called the **first principal focal length,** f_1. We can find f_1 as follows.

Using $u = f_1$, and $v = \infty$ into Eq. (4.6), we get

$$\frac{1}{\infty} - \left(-\frac{1}{f_1}\right) = (\mu - 1)\left[\frac{1}{R_1} - \frac{1}{R_2}\right]$$

or
$$\frac{1}{f_1} = (\mu - 1)\left[\frac{1}{R_1} - \frac{1}{R_2}\right] \quad ...(4.8)$$

The plane perpendicular to the axis and passing through the first focal point is known as the **first principal focal plane.**

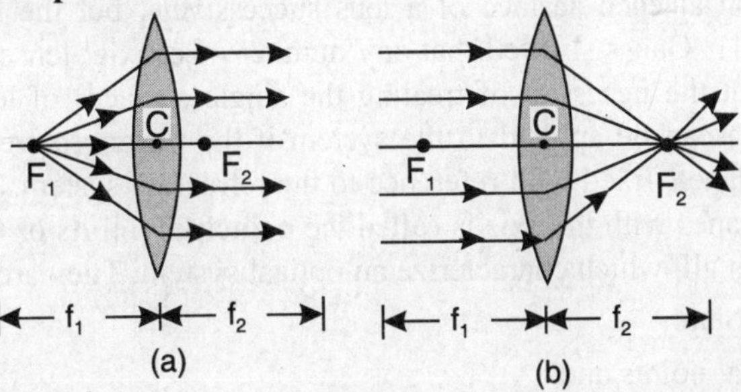

Fig. 4.2:

Positions of the principal foci.

(*ii*) If the object is situated at infinity, the position of the image on the axis is known as the **second principal focus** F_2 (see Fig. 4.2b). The distance of the second principal focus from the optical center C is called the **second principal focal length,** f_2.

Using $u = \infty$ and $v = f_2$ into Eq. (4.5), we get

$$\frac{1}{f_2} - \frac{1}{\infty} = (\mu - 1)\left[\frac{1}{R_1} - \frac{1}{R_2}\right]$$

$$\frac{1}{f_2} = (\mu - 1)\left[\frac{1}{R_1} - \frac{1}{R_2}\right] \quad ...(4.9)$$

The plane perpendicular to the axis and passing through the second focal point is known as the **second principal focal plane.**

It follows from Eq. (4.8) and Eq. (4.9) that

$$f_1 = f_2 \quad ...(4.10)$$

So, *every thin lens in air has two focal points (F_1 and F_2).They lie on each side of the lens at equal distance from the centre.*

For a converging lens the second focal length (f_2) is positive and the first (f_1) is negative and for a diverging lens the second focal length (f_2) is negative and the first (f_1) positive. *The two focal lengths of thin lens in air are numerically equal.* Frequently, one refers simply to "focal length" of a lens; it will be assumed that this always refers to the *second focal length* so that the focal length of a converging lens is positive, whilst for a diverging lens it is negative.

4.3 COAXIAL LENS SYSTEMS

Single lenses are rarely used for image formation, as they suffer from various defects. In optical instruments such as cameras, microscopes, telescopes etc, a collection of lenses are employed for forming images of objects. **An optical system consists of a number of lenses placed apart, and having a common principal axis.** The image formed by such a coaxial optical system is good and almost free of aberrations.

4.4 CARDINAL POINTS

In the case of a thin lens, the thickness of the lens has been ignored. It is not important to consider from which point of the lens the distances are measured. However, the above approximation cannot be applied for an optical system consisting of a combination of lenses. One way of calculating the position and size of the image formed by an optical system is to consider refraction at each surface of a lens successively, but the method proves to be more tedious. In 1841, Gauss showed that any number of coaxial lenses could be treated as a single unit, without the necessity of treating the single surfaces of lenses separately. The lens makers' formula can be applied to the system if the distances are measured from two imaginary parallel planes, fixed with reference to the refracting system. The meeting point of these hypothetical planes with the axis is called the **principal points** or **Gauss points**. In fact there are six points in all, which characterize an optical system. They are two

(*i*) two focal points

(*ii*) two principal points and

(*iii*) two nodal points

The above six principal points are called as **cardinal points** of an optical system. The planes passing through these cardinal points and which are normal to the principal axis are called as **cardinal planes**. The cardinal points and cardinal planes are intrinsic properties of a particular optical system and determine the image forming properties of the system. The image of an object can be found if the cardinal points and cardinal planes are known. The refraction of the rays at the various surfaces need not be considered.

4.5 DEFINITIONS AND PROPERTIES OF CARDINAL POINTS AND PLANES

(*i*) **Principal points and Principal planes:** Let us consider an optical system having its principal focal points F_1 and F_2. A ray OA traveling parallel to the principal axis and incident at A is brought to focus at F_2 in the **image space** of the optical system as shown in Fig. 4.3 (a). The actual ray is refracted at each surface of the optical system and follows the path $OABF_2$. If we extend the incident ray OA forward and the emergent ray BF_2 backward, they meet each other within the optical system at H_2. Now, we can describe the refraction of the incident ray OA in terms of a *single refraction* at a plane passing through H_2. A plane drawn through the point H_2 and perpendicular to the axis may be regarded as the surface at which refraction takes place. This plane is called the *principal plane* of the optical system.

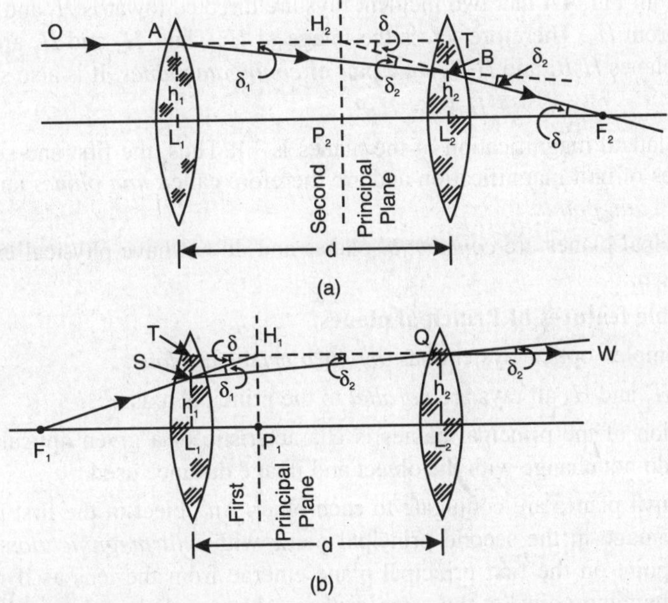

Fig. 4.3:

Principal points and Principal planes.

Thus, the four consecutive deviations of the light rays caused by the four surfaces of the optical system are equivalent to a single refraction at H_2, taking place at the *principal plane*. We now define the *principal plane* of an optical system as *the loci where we assume refraction to occur without reference as to where the refraction actually occurs*. H_2P_2 is the principal plane in the image space and is called the **second principal plane**. The point P_2, at which the second principal plane intersects the axis, is called the **second principal point**.

By adopting similar procedure, as shown in Fig. 4.3 (*b*), we can locate the *first principal plane* H_1P_1 and *first principal point* P_1 in the **object space**. Consider the ray F_1S passing through the first principal focus F_1 such that after refraction it emerges along QW parallel to the axis at the same height as that of the ray OA. The rays F_1S and QW when produced intersect at H_1. A plane perpendicular to the axis and passing through H_1 is called the **first principal plane**. The point of intersection, P_1, of the first principal plane with the axis is called the **first principal point**.

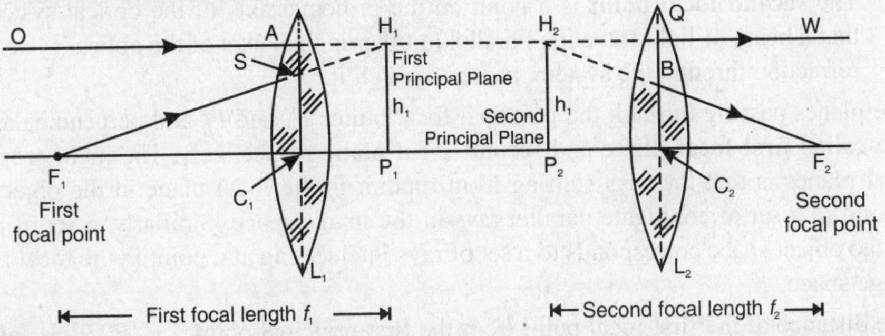

Fig. 4.4:

Conjugate points and conjugate planes.

It is seen from Fig. 4.4 that two incident rays are directed towards H_1 and after refraction seem to come from H_2. Therefore, H_2 is the image of H_1. Thus, H_1 and H_2 are the *conjugate points* and the planes H_1P_1 and H_2P_2 are a pair of *conjugate planes*. It is also seen that

$$H_2P_2 = H_1P_1.$$

Hence, the lateral magnification of the planes is +1. Thus, the first and second principal planes are planes of unit magnification and are therefore called *unit planes* and the points P_1 and P_2 are called *unit points*.

Note: The principal planes are *conceptual* planes and do not have physical existence within the optical system.

Some remarkable features of Principal planes:

(*a*) Even a complex optical system has *only two principal planes*.

(*b*) Between H_1 and H_2 all rays are *parallel* to the principal axis.

(*c*) The location of the principal planes is characteristic of a given optical system. Their positions do not change with the object and image distance used.

(*d*) The principal planes are conjugate to each other. An object in the first principal plane forms an image in the second principal plane with *unit magnification*. Any ray sent toward a point on the first principal plane emerge from the lens as if it originated at the corresponding point (at the same distance above or below the axis) on the second principal plane. Hence, the name *unit planes.*

(*e*) The principal points H_1 and H_2 provide a set of references from which several system parameters are measured.

(*ii*) **Focal points and Focal planes:** If a parallel beam of light traveling from the left to the right (in object space) is incident on the optical system, the beam comes together at a point, F_2, on the other side (in image space) of the optical system. The beam passes through the point F_2 whatever may be its path inside the system. The point, F_2, is called the **second focal point** of the optical system. A beam of light passing the point F_1 on the axis on the object side is rendered parallel to the axis after emergence through the optical system (Fig. 4.4). The point F_1 is called the **first focal point.**

We can now define the focal points as follows:

(*a*) The **first focal point** is a point on the principal axis of optical system such that a beam of light passing through it is rendered parallel to the principal axis after refraction through the optical system.

(*b*) The **second focal point** is a point on the principal axis of the optical system such that a beam of light traveling parallel to the principal axis of the optical system, after refraction through the system, passes through it.

The planes passing through the principal focal points F_1 and F_2 and perpendicular to the axis are called **first focal plane** and **second focal plane** respectively. The main property of the focal planes is that the rays starting from a point in the focal plane in the object space correspond to a set of conjugate parallel rays in the image space. Similarly, a set of parallel rays in the object space corresponds to a set of rays intersecting at a point in the focal plane in the image space.

The distance of the first focal point from the first principal point, *i.e.,* $F_1 P_1$, is called the **first focal length,** f_1 of the optical system and the distance of the second focal point from the second principal point, F_2P_2, is called the **second focal length,** f_2 . f_1 and f_2 are also known as the focal lengths in object space and image space respectively. When the medium is same on the two sides of the optical system numerically $f_1 = f_2$.

(*iii*) **Nodal Points and Nodal Planes: Nodal points** are points on the principal axis of the optical system where light rays, ***without refraction,*** intersect the optic axis. In a *thin lens* the nodal point is the centre of the lens. Light passing through the centre of a thin lens does not deviate. In an optical system the centre separates into two nodal points. The planes passing through the nodal points and perpendicular to the principal axis are called the **nodal planes**. Whereas the principal planes are planes where all refraction is assumed to occur, the nodal planes are planes where refraction does *not* take place. Fig. 4.5 represents an optical system with the help of its cardinal planes. It is seen from the Fig. 4.5 that a ray of light, AN_1, directed towards one of the nodal points, N_1, after refraction through the optical system, along N_1N_2, emerges out from the second nodal point, N_2, in a direction, N_2R, parallel to the incident ray. The distances of the nodal points are measured from the focal points.

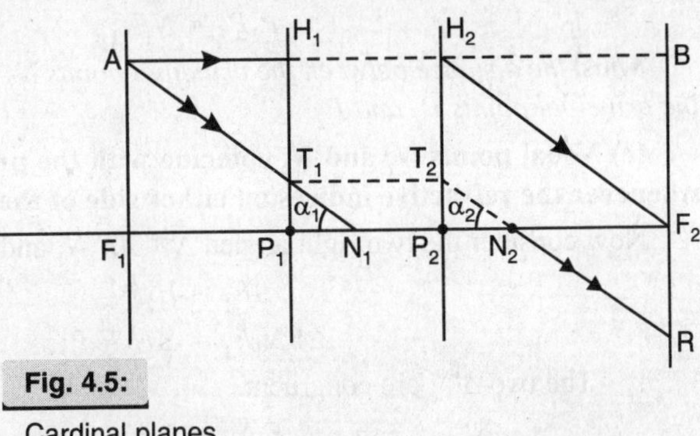

Fig. 4.5: Cardinal planes.

(*a*) **Nodal points are a pair of conjugate points on the axis having unit positive angular magnification.**

Let H_1P_1 and H_2P_2 be the first and second principal planes of an optical system. Let AF_1 and BF_2 be its first and second focal planes respectively (Refer to Fig. 4.5). Consider a point A situated on the first focal plane. From A draw a ray AH_1 parallel to the axis. The conjugate ray will proceed from H_2, a point in the second principal plane such that $H_2P_2 = H_1P_1$ and will pass through the second focus.

Take another ray AT_1 parallel to the emergent ray H_2F_2 and incident on the first principal plane at T_1. It will emerge out from T_2, a point on the second principal plane such that $T_2P_2 = T_1P_1$, and will proceed parallel to the ray H_2F_2. The points of intersection of the incident ray AT_1 and the conjugate emergent ray T_2R with the axis give the positions of the nodal points. It is clear that the two points N_1 and N_2 are a pair of conjugate points and the incident ray AN_1 is parallel to the conjugate emergent ray T_2R.

Further $\tan \alpha_1 = \tan \alpha_2$

The ratio $\dfrac{\tan \alpha_2}{\tan \alpha_1} = \gamma$ represents the angular magnification.

$$\therefore \quad \frac{\tan \alpha_2}{\tan \alpha_1} = 1 \qquad \text{...(4.11)}$$

Therefore, *the points N_1 and N_2 are a pair of conjugate points on the axis having unit positive angular magnification.*

(b) Distance between two nodal points is always equal to the distance between two principal points.

Referring to Fig. 4.5, we see that in the right angled Δ^{les} $T_1P_1N_1$ and $T_2P_2N_2$

$$T_1P_1 = T_2P_2$$

$$\angle T_1 N_1P_1 = \angle T_2 N_2P_2 = \alpha$$

Therefore, the two Δ^{les} are congruent.

$$\therefore \quad P_1N_1 = P_2N_2$$

Adding N_1P_2 to both the sides, we get

$$\therefore \quad P_1N_1 + N_1P_2 = P_2N_2 + N_1P_2$$

$$\therefore \quad P_1P_2 = N_1N_2 \qquad \text{...(4.12)}$$

Thus, *the distance between the principal points N_1 and N_2 is equal to the distance between the principal points P_1 and P_2.*

(c) Nodal points N_1 and N_2 coincide with the principal points P_1 and P_2 respectively whenever the refractive indices on either side of the lens are the same.

Now consider the two right angled Δ^{les} AF_1N_1 and $H_2P_2F_2$ in Fig. 4.5.

$$AF_1 = H_2P_2$$

$$\angle A N_1F_1 = \angle H_2F_2P_2$$

$\therefore$ The two Δ^{les} are congruent.

$$F_1N_1 = P_2F_2$$

But

$$F_1N_1 = F_1P_1 + P_1N_1$$

$$\therefore \quad F_1P_1 + P_1N_1 = P_2F_2$$

$$\therefore \quad P_1N_1 = P_2F_2 - F_1P_1$$

Also

$$P_2F_2 = +f_2 \text{ and } P_1F_1 = -f_1$$

$$\therefore \quad P_1N_1 = P_2N_2 = (f_1 + f_2)$$

As the medium is the same, say air, on both the sides of the system

$$f_2 = -f_1$$

$$\therefore \quad P_1N_1 = P_2N_2 = 0 \qquad \text{...(4.13)}$$

Thus, the principal points coincide with the nodal points when the optical system is situated in the same medium.

4.6 CONSTRUCTION OF IMAGE USING CARDINAL POINTS

From the knowledge of cardinal points of an optical system, the image of an object can be constructed. For this, it is not necessary to know the position and curvature of the refracting surfaces or the nature of the intermediate media.

Let AB be an object placed at some distance from first focal point F_1 as in Fig. 4.6. Let us consider a ray AH_1 parallel to the principal axis which meets the first principal plane at H_1. According to the property of principal planes, the ray emerges from the second principal plane from point H_2 such that $H_1P_1 = H_2P_2$ and it passes through the second focal point F_2.

Next let us consider a second ray passing through the first focal point F_1. It strikes the first focal plane at K_1 and then it emerges parallel to the principal axis. It meets the second principal plane at K_2 such that $H_1K_1 = H_2K_2$.

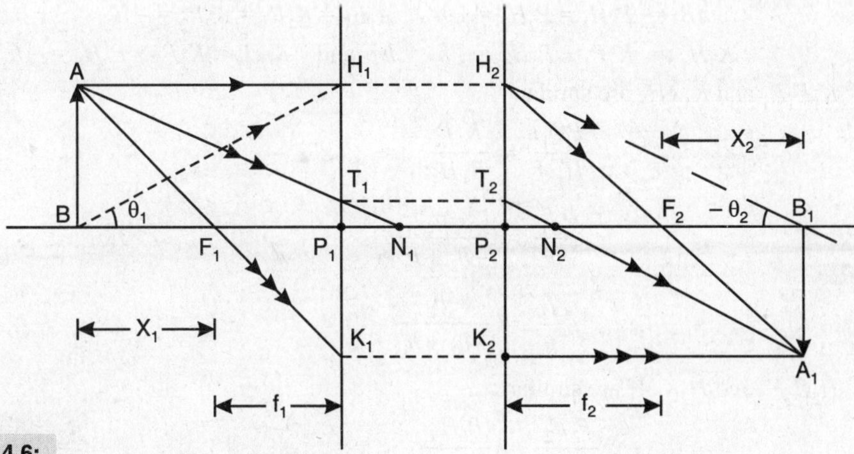

Fig. 4.6:

Image construction using Cardinal points.

Again let us consider a third ray AN_1 directed towards first nodal point N_1. This ray emerges from second nodal point N_2 along N_2A_1 parallel to its original direction.

The three emergent rays meet at A_1. Therefore, A_1 is the image of A. Similarly the images of other points of AB can be determined which give rise to image A_1B_1 of the object AB.

Newton's formula: Let the distance of the object from the first focal point F_1 be x_1 and the distance of the image from the second focal point F_2 be x_2.

Referring to the Fig. 4.6, it is seen that Δ^{les} ABF_1 and $F_1K_1P_1$ are similar.

$$\therefore \quad \frac{K_1P_1}{AB} = \frac{P_1F_1}{BF_1} \quad \text{But, } K_1P_1 = A_1B_1$$

$$\therefore \quad \frac{A_1B_1}{AB} = \frac{f_1}{x_1} \quad \text{...(4.14)}$$

Further, Δ^{les} $A_1B_1F_2$ and $H_2P_2F_2$ are similar.

$$\frac{A_1B_1}{H_2P_2} = \frac{B_1F_2}{P_2F_2} \quad \text{But, } H_2P_2 = AB$$

$$\therefore \quad \frac{A_1B_1}{AB} = \frac{x_2}{f_2} \quad \text{...(4.15)}$$

From Eq. (4.14) and Eq. (4.15), we get

$$\frac{h_2}{h_1} = \frac{f_1}{x_1} = \frac{x_2}{f_2} \quad \text{...(4.16)}$$

or
$$x_1x_2 = f_1f_2 \quad \text{...(4.17)}$$

This is the **Newton's formula**. In the foregoing discussion, the distances of the image and the object have been measured from their respective foci. But it is sometimes convenient to measure the conjugate distances from the principal points.

4.6.1 Relationship Between f_1 and f_2

Referring to Fig. 4.6 and using the sign convention,

$$P_1B = H_1A = -u, \; P_2B_1 = K_2A_1 = +\upsilon$$

Also, $AB = P_1H_1 = P_2H_2 = +h_1, \quad A_1B_1 = K_1P_1 = K_2P_2 = -h_2$

$$K_1H_1 = K_1P_1 + P_1H_1 = -h_2 + h_1 \text{ and } K_2H_2 = K_2P_2 + P_2H_2 = -h_2 + h_1$$

Δ^{les} $K_1F_1P_1$ and K_1AH_1 are similar.

$$\therefore \quad \frac{P_1F_1}{H_1A} = \frac{K_1P_1}{K_1H_1}$$

$$\frac{-f_1}{-u} = \frac{-h_2}{-h_2 + h_1}$$

$$\text{or} \quad \frac{f_1}{u} = \frac{-h_2}{-h_2 + h_1} \quad \text{...(4.18)}$$

Δ^{les} $H_2P_2F_2$ and $H_2K_2A_1$ are similar.

$$\therefore \quad \frac{P_2F_2}{K_2A_1} = \frac{P_2H_2}{K_2H_2}$$

$$\frac{f_2}{v} = \frac{h_2}{-h_2 + h_1} \quad \text{...(4.19)}$$

Adding the Eq. (4.18) and Eq. (4.19), we get

$$\frac{f_1}{u} + \frac{f_2}{v} = \frac{-h_2 + h_1}{-h_2 + h_1}$$

$$\therefore \quad \frac{f_1}{u} + \frac{f_2}{v} = 1 \quad \text{...(4.20)}$$

Eq. (4.20) can be rewritten as

$$\frac{v}{u} = \frac{(v - f_2)}{f_1} \quad \text{...(4.21)}$$

When the system is situated in air $f_2 = -f_1 = f$.

$$\therefore \quad f_1 = -f \text{ and } f_2 = f. \quad \text{...(4.22)}$$

4.7 NODAL SLIDE

A **nodal slide** is a particular type of horizontal metal support for a lens system, which is capable of rotation about a vertical axis (Fig. 4.7). The nodal slide provides a convenient method for locating the focal and nodal points and for determining the focal length of a lens system.

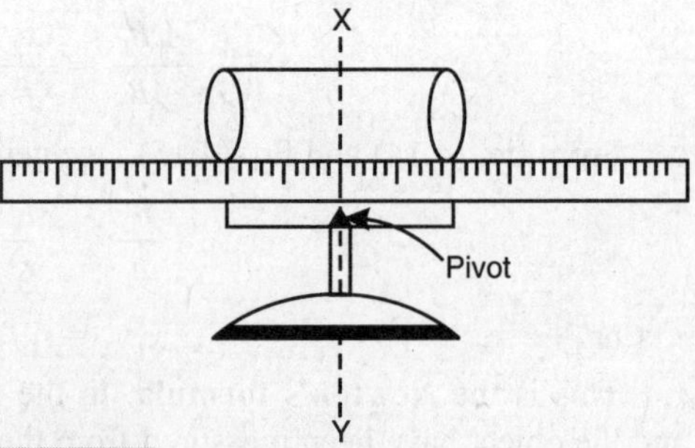

Fig. 4.7: Vertical axis of Nodal slide.

Principle: Any optical system has two nodal points N_1 and N_2. An incident light ray directed. Towards N_1, after refraction through the system, proceeds from N_2 in a direction parallel to the incident ray. The method of locating the nodal points with the help of the nodal slide involves the following principle.

"If a parallel beam of light is incident on a convergent lens system, it forms an image on a screen held at its second focal plane. When the lens system is rotated through a small angle about a vertical axis through its second nodal point, the image does not shift laterally and remains stationary."

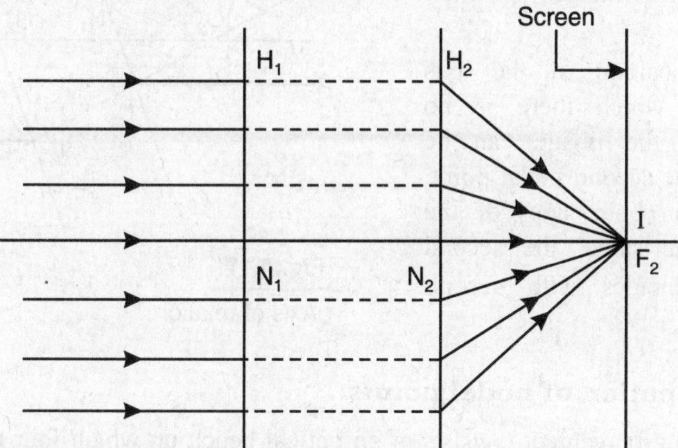

Fig. 4.8:
Beam of parallel rays incident on a coaxial lens system.

Let us suppose a beam of parallel rays is incident on a coaxial lens system (see Fig. 4.8). The beam passes through the system and converges to the second focus F_2 and a real image is formed on the screen.

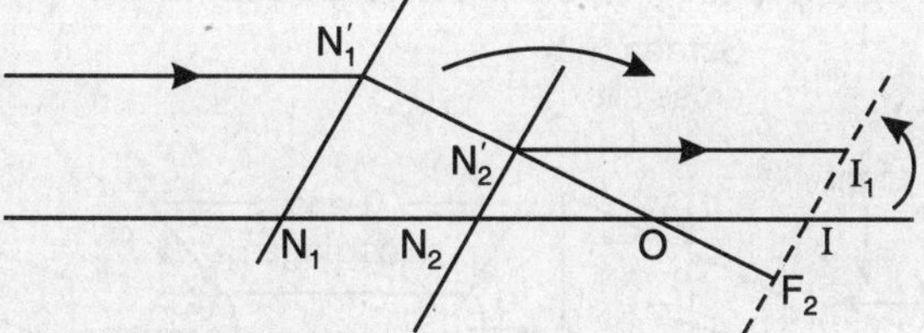

Fig. 4.9:
Parallel incident beam-image on the second focal plane.

Now let the system be rotated about a perpendicular axis through O, which lies between N_2 and F_2 (Refer to Fig. 4.9). Due to this rotation, the nodal points N_1 and N_2 shift to the positions N_1' and N_2' respectively. A ray incident at N_1' travels along $N_2'I_1$ parallel to the incident ray (see Fig. 4.9). Since the incident beam is parallel, the image lies on the second focal plane. The point of intersection of the ray $N_2'I_1$ with the focal plane gives the new position of the image. Thus, when the axis of rotation lies between N_2 and F_2, a slight rotation of the system changes the position of the image.

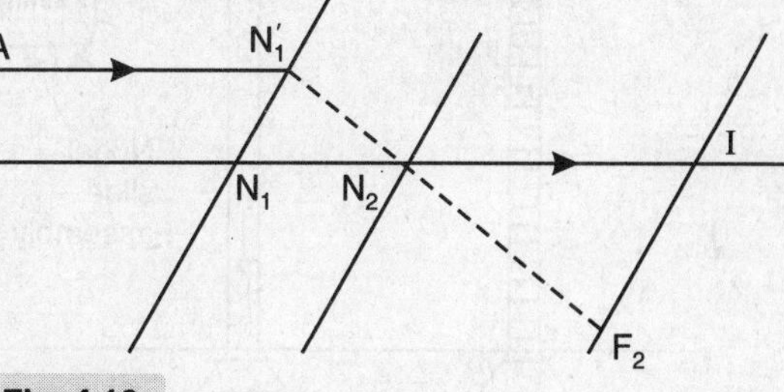

Fig. 4.10:
Stationary Image.

Next let us say the system is rotated through a small angle about an axis passing through N_2. Then N_1 shifts to N_1' while N_2 remains fixed. A parallel ray incident at N_1' on passing through N_2 follows the path N_2I. Therefore, the image remains stationary, as seen in Fig. 4.10.

If the axis of rotation lies before N_1, as in Fig. 4.11, any small amount of rotation displaces N_1 and N_2 and consequently the image position changes.

Thus, the position of the axis of rotation for which there is no displacement of the image can be found. It gives the second nodal point. As the media on both sides of the system being the same, the second nodal point coincides with second principal point.

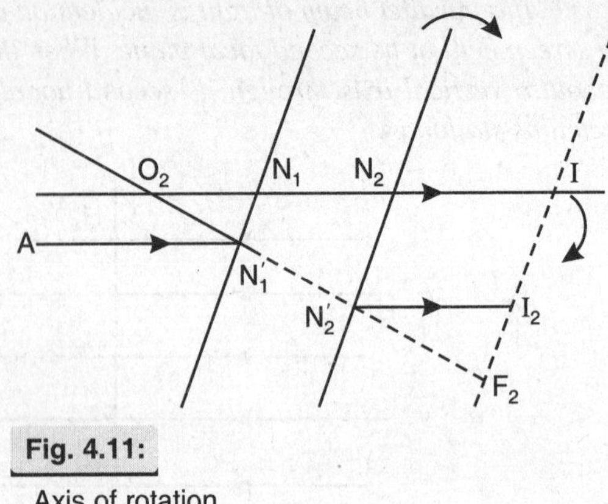

Fig. 4.11:

Axis of rotation.

4.7.1 Determination of nodal points:

The experimental arrangement consists of an optical bench on which four uprights are kept. They carry a plane mirror, the nodal slide, and a screen provided with a slit fitted with cross wires and lamp housing (see Fig. 4.12). Light from the lamp passes through the slit and is incident on the lens system. It is rendered parallel and on passing through the lens system it is reflected back by the vertical plane mirror. The reflected light once again passes through the lens system and is brought to a focus in the plane of the stilt, as shown in Fig. 4.13.

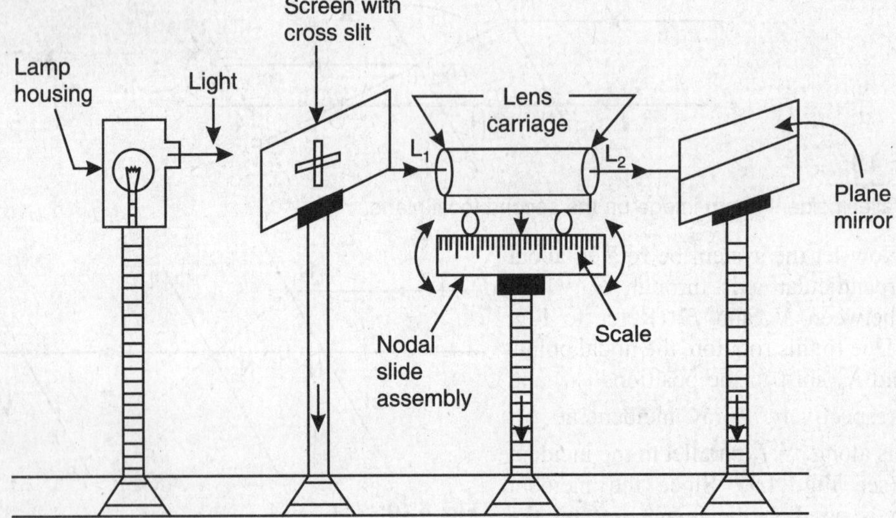

Fig. 4.12:

Experimental arrangement.

The distance between the optical system and the screen is adjusted in such a way that a well-defined image of the slit is obtained on the screen. The image of the slit is formed slightly to one side of the slit itself. It is obvious that the centre of the slit is at the first focal point of the lens. The nodal slide carrying the lens system is now rotated through a small angle and it will be found that the image shifts side ward to the right or to the left.

The nodal slide and its stand are then adjusted such that the direction of rotation of the image changes its sign and finally the image remains stationary for a slight rotation of the carriage. When this condition is reached, the axis of rotation passes through the second nodal point N_2. The other focal point nodal point can be determined by turning the nodal slide through 180° and repeating the experiment. Since the medium on both sides of the lens system is the same (air), the nodal points are also the principal points. The distance between the screen and the axis of rotation for the stationary image is an accurate measure of the first focal length of the lens system.

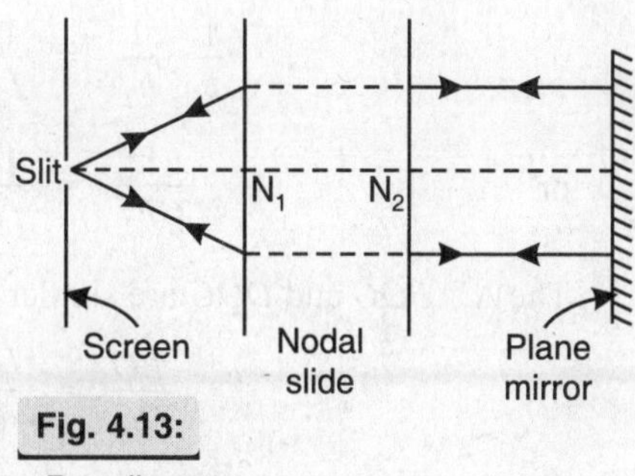

Fig. 4.13: Ray diagram.

4.8 EQUIVALENT FOCAL LENGTH OF A COAXIAL SYSTEM OF TWO THIN LENSES

We now determine the equivalent focal length of a coaxial optical system located in air medium by assuming that the object is at infinity, as shown in Fig. 4.14.

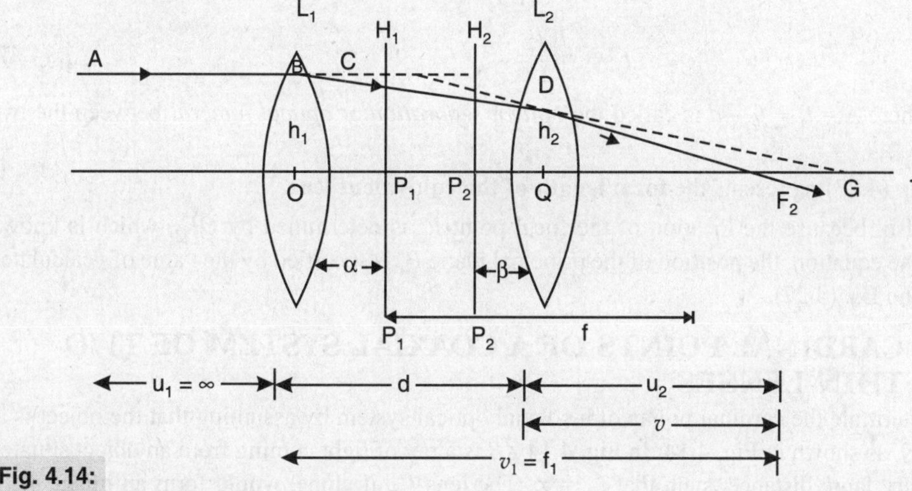

Fig. 4.14: Equivalent focal length of a coaxial optical system.

AB is a ray of light coming from an object situated at a very large distance, such that $u_1 = \infty$. The lens L_1, if alone, would form an image at G. However, because of the presence of the second lens L_2, G becomes the virtual object for L_2. The ray BD, instead of going along BDG, refracts along the path DF_2. When the ray AB is produced forward and the ray DF_2 backward, they intersect at H_1. The plane H_1P_1 normal to the axis may considered as the plane at which the refraction occurred and this plane is called **principal plane**.

Now, we can write the expression for the refraction taking place at the surface of first lens as follows.

$$\frac{1}{v_1} - \frac{1}{u_1} = \frac{1}{f_1} \quad \therefore \frac{1}{OG} - \frac{1}{u_1} = \frac{1}{f_1}$$

As $u_1 = \infty$, we obtain $OG = f_1$

The equation for the refraction at the second lens may be written as

$$\frac{1}{v}-\frac{1}{u_2} = \frac{1}{f_2} \quad \therefore \quad \frac{1}{QF_2}-\frac{1}{QG}=\frac{1}{f_2}$$

or
$$\frac{1}{QF_2} = \frac{1}{f_2}+\frac{1}{f_1-d} \quad \therefore \quad \frac{1}{QF_2} = \frac{f_1+f_2-d}{f_2(f_1-d)} \qquad ...(4.23)$$

The Δ^{les} BOG and DQG are similar and also the Δ^{les} CP_2F_2 and DQF_2 are similar.

$$\frac{BO}{OG} = \frac{DQ}{-QG} \quad \therefore \quad \frac{h_1}{f_1} = \frac{h_2}{-(f_1-d)} \qquad ...(4.24)$$

$$\frac{CP_2}{P_2F_2} = \frac{DQ}{-QF_2} \quad \therefore \quad \frac{h_1}{f} = \frac{h_2}{QF_2} \qquad ...(4.25)$$

From Eq. (4.24), Eq. (4.25) and Eq. (4.23), we get

$$\frac{h_1}{h_2} = \frac{f_1}{(f_1-d)} = \frac{f(f_1+f_2-d)}{f_2(f_1-d)}$$

or
$$\frac{1}{f} = \frac{f_1+f_2-d}{f_1 f_2} \qquad ...(4.26)$$

or
$$f = \frac{f_1 f_2}{\Delta} \qquad ...(4.27)$$

where $\Delta = f_1 + f_2 - d$ is called the *optical separation* or *optical interval* between the two lenses.

Eq. (4.27) represents the **focal length of the equivalent lens.**

Also, because the location of the focal point F_2 is determined by QF_2, which is known from the equation, the position of the principal plane P_1 is specified by the value of f calculated from the Eq. (4.27).

4.9 CARDINAL POINTS OF A COAXIAL SYSTEM OF TWO THIN LENSES

We determine the cardinal points of a coaxial optical system by assuming that the object is at infinity, as shown in Fig. 4.14. In Fig. 4.14 AB is a ray of light coming from an object situated at a very large distance, such that $u_1 = \infty$. The lens L_1, if alone, would form an image at G. However, because of the presence of the second lens L_2, G becomes the virtual object for L_2. The ray BD, instead of going along BDG, refracts along the path DF_2. When the ray AB is produced forward and the ray DF_2 backward, they intersect at H_1. The plane H_1P_1 normal to the axis may considered as the plane at which the refraction occurred and this plane is called **principal plane**.

(*i*) **Second Principal point:** Let us say the second principal plane H_2P_2 is located at a distance of $L_2P_2 = \beta$ from the second lens L_2. According to sign convention β would be *negative* as it is measured toward the left of the lens.

$$\therefore \qquad QF_2 = f-(-\beta) = f+\beta$$

We can determine β using the equation (4.23) for QF_2 into the above relation.

Thus,
$$f+\beta = \frac{f_2(f_1-d)}{f_1+f_2-d}$$

$$\therefore \qquad \beta = -f + \frac{f_2(f_1 - d)}{f_1 + f_2 - d} = \frac{-f_1 f_2}{\Delta} + \frac{f_2 f_1 - f_2 d}{\Delta}$$

where $\Delta = f_1 + f_2 - d$.

$$\therefore \qquad \beta = -f_2 \frac{d}{\Delta} \qquad ...(4.28)$$

$$\text{or} \qquad \beta = -\frac{f_2 d}{f_1 + f_2 - d} \qquad ...(4.29)$$

But from Eq. (4.26), we have $f_1 + f_2 - d = \frac{f_1 f_2}{f}$. Therefore,

$$\therefore \qquad \beta = -\frac{f\,d}{f_1} \qquad ...(4.30)$$

(*ii*) **First Principal Point:** By considering a ray of light parallel to the axis and incident on the second lens L_2 from the right side (see Fig. 4.15), we can show that the distance of first principal plane, $L_1P_1 = \alpha$ from the first lens L_1 is given by

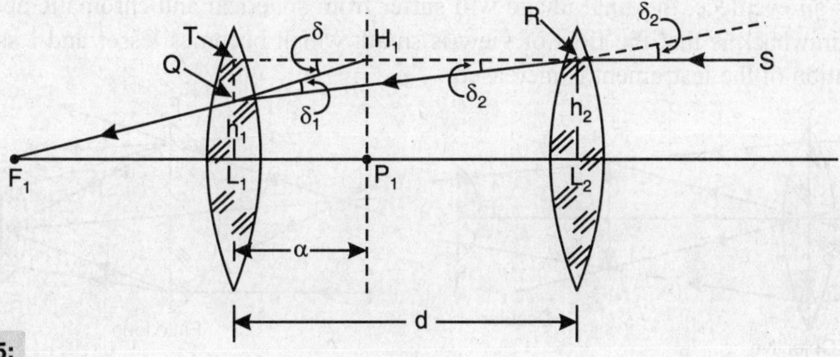

Fig. 4.15: First Principal Point.

$$\alpha = f_1 \frac{d}{\Delta} \qquad ...(4.31)$$

$$\text{Further,} \qquad \alpha = +\frac{f\,d}{f_2} \qquad ...(4.32)$$

Note: In a combination of two lenses, the sequence of the principal planes is in the reverse order- H_2P_2 is to the left of the centre and H_1P_1 is to the right.

(*iii*) **Second Focal point:** Referring to Fig. 4.14, the distance of the second focal point F_2 from the second lens L_2 is given by

$$L_2F_2 = P_2F_2 - P_2L_2$$
$$= f - (-L_2P_2) = f + \beta$$
$$= f + \left(-\frac{f\,d}{f_1}\right)$$

$$\therefore \qquad L_2F_2 = f\left(1 - \frac{d}{f_1}\right) \qquad ...(4.33)$$

(*iv*) **First Focal point:** The distance of the first focal point F_1 from the first lens L_1 (Refer to Fig. 4.15) is given by

$$L_1F_1 = P_1F_1 - P_1L_1$$

$$= -f - (-L_1P_1) = -f + \alpha$$

$$= -f + \left(\frac{f\,d}{f_2}\right)$$

$$\therefore \quad L_1F_1 = -f\left(1 - \frac{d}{f_2}\right) \quad ...(4.34)$$

(*v*) **and (*vi*) Nodal Points:** As the medium on either side of the lenses is air, the nodal points N_1 and N_2 coincide with the principal points P_1 and P_2 respectively. Then $N_1 = P_1$ and $N_2 = P_2$.

4.10 EYEPIECES

An optical instrument such as a microscope or telescope is required to produce a magnified image free from aberrations and a bright image covering a wide field of view. If a single lens is used as an eyepiece, the final image will suffer from spherical and chromatic aberrations. Another drawback is that the field of view is small, which becomes lesser and lesser as the magnification of the instrument is increased.

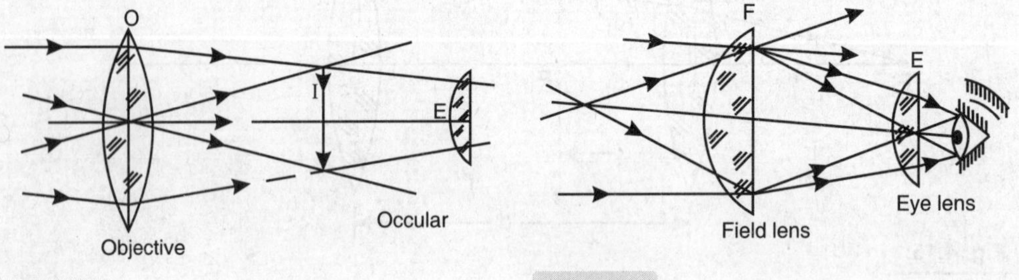

Fig. 4.16: Eyepiece with a Convex lense.

Fig. 4.17: Eyepiece with one-half of convex lens.

An **eyepiece** is a specially designed magnifier that gives a perfect and bright magnified image of an object. It consists of two convex lenses separated by a finite distance.

The first lens towards the object is called the **field lens**. It forms a real image of the object under examination. The second lens, called the **eye lens,** enlarges this image further to form a final image and which is then viewed by the eye. The rays passing through the outer portions of the image formed by the field lens are refracted through the peripheral portions of the eye lens and they cannot simultaneously enter the small aperture of the pupil of the eye placed close to the eye lens (Fig. 4.16). Hence, only that part of the image, which is nearer to the axis, will be seen. Therefore, the final image will cover a small field of view. The field of view will progressively decrease as the distance between the field lens and eye lens is increased. The distance is varied in order to increase the magnification. In other words, the greater the magnifying power of the instrument, the smaller the field of view. Therefore, field lens is used before the eye lens in the eyepiece to cause all the rays from the image to enter the eye lens. The function of the field lens is to gather in more of the rays from the objective toward the axis of the eyepiece (See Fig. 4.17). The field lens and the eye lens together constitute an **eyepiece**. The two lenses are made and kept in such a way that their combination is achromatic and free from spherical aberrations. Two common eyepieces are the Huygens and the Ramsden eyepieces.

4.11 HUYGENS EYEPIECE

Construction: Huygens' eyepiece consists of two plano-convex lenses of focal lengths $3f$ and f which are separated by a distance $2f$. The lenses are made of the same kind of glass and their curved surfaces face the incident light. The first lens acts as objective and the second lens acts as eye lens.

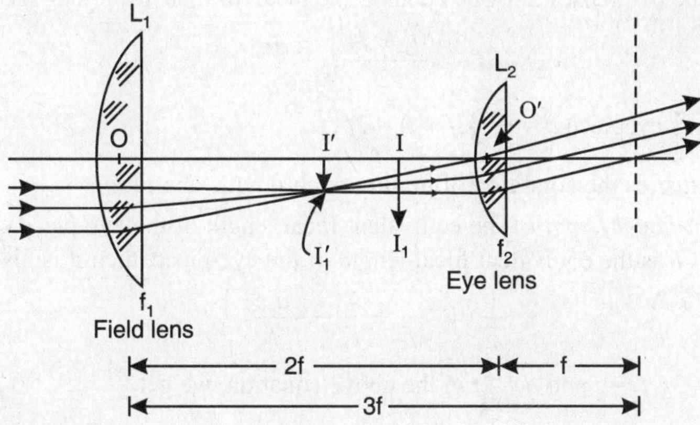

Fig. 4.18: Huygens eyepiece.

Huygen's eye-piece is a negative eye-piece: In Huygens eyepiece, the first focal plane of the eye-piece lies within the eye-piece and no real object can be placed there. Such an eye-piece is called a negative eye-piece. This eye-piece cannot be used to examine directly an object or a real image formed but it can be used only to examine a virtual image. It is used in microscopes and other optical instruments using white light only.

Position of Cross-wires: For quantitative measurements, cross-hairs or a recticle with a scale is used with the eye-piece. It is required for the recticle to be in focus with the image and so placed in the focal plane of real image. In case of Huygens eye-piece the cross-wires are to be held in between the field and eye-lenses inside the eye-piece. Therefore, they are magnified by the eye-lens only whereas the image is magnified by the field lens as well as the eye-lens. It is not a drawback in simple observations; but in case any measurement the different magnifications of the image and the cross-wire scale leads to erroneous results.

Working: The eyepiece works in conjunction with an optical instrument. Referring to the Fig. 4.18, II_1 is the image of the distant object formed by the optical instrument in the absence of the field lens L_1. With the field lens in position, the rays get refracted on passing through L_1 and the image $I'\ I_1'$ is formed. When image lies in the focal plane of the eye-lens, the magnified image is seen at infinity by the eye located at the exit pupil.

Theory:

(*i*) *Condition for minimum spherical aberration*: For spherical aberration to be a minimum the distance d between the two lenses must be equal to the difference of their focal lengths, that is

$$d = f_1 - f_2.$$

In Huygen's eyepiece $f_1 = 3f, f_2 = f.$

$$\therefore \quad d = 3f - f = 2f.$$

Thus, it satisfies the condition for minimum spherical aberration.

Also, as the convex surfaces of the field and the eye lenses face the incident ray, the total deviation due to the combination is divided into four parts which makes the combination to have minimum spherical aberration.

(*ii*) *Condition for achromatism*: For chromatic aberration to be a minimum the distance between the two lenses must be equal to the mean of their focal lengths

$$d = \frac{f_1 + f_2}{2}.$$

In Huygen's eyepiece $d = \frac{1}{2}(3f + f) = 2f.$

Thus, it satisfies the condition of minimum chromatic aberration.

(*iii*) *Equivalent Focal Length:* The equivalent focal length of the eye-piece can be found as follows. If F is the equivalent focal length of the eye-piece, then it is given by

$$\frac{1}{F} = \frac{1}{f_1} + \frac{1}{f_2} - \frac{d}{f_1 f_2}$$

Using $f_1 = 3f, f_2 = f$ and $d = 2f$ in the above equation, we get

$$\frac{1}{F} = \frac{1}{f} + \frac{1}{3f} - \frac{2f}{3f^2}$$

$$\therefore \quad \frac{1}{F} = \frac{1}{f} + \frac{1}{3f} - \frac{2}{3f} = \frac{2}{3f}$$

$$\therefore \quad F = 3f/2$$

The equivalent lens lies behind field lens at a distance of

$$x = \frac{d \times F}{f} = \frac{2f \times 3f/2}{f} = 3f$$

In other words, the equivalent lens is at a distance of $3f - 2f = f$ behind the eye lens.

(*iv*) *Cardinal points of Huygens eyepiece:*

(*a*) *Position of Principal Points*:

The distance of the first principal point H_1 from the field lens is given by

$$\alpha_1 = \frac{d \times F}{f_2} = \frac{2f \times 3f/2}{f} = 3f$$

Thus, the first principal point H_1 lies at a distance of $3f$ to the right of the field lens.

The distance of the second principal point H_2 from the field lens is given by

$$\alpha_2 = -\frac{d \times F}{f_1} = -\frac{2f \times 3f/2}{3f} = -f$$

Thus, the second principal point H_2 lies at a distance of f to the left of the eye lens.

(*b*). *Position of the Focal Points:*

The distance of the first focal point F_1 from the field lens is given by

$$\beta_1 = -F\left(1-\frac{d}{f_2}\right) = -\frac{3f}{2}\left(1-\frac{2f}{f}\right) = \frac{3f}{2}$$

Thus, the first focal point F_1 lies at a distance $\frac{3f}{2}$ to the right of the field lens.

The distance of the second focal point F_2 from the field lens is given by

$$\beta_2 = F\left(1-\frac{d}{f_1}\right) = -\frac{3f}{2}\left(1-\frac{2f}{3f}\right) = -\frac{f}{2}$$

Thus, the second focal point F_2 lies at a distance $\frac{f}{2}$ to the right of the eye lens.

(*c*). *Nodal points*: As the medium surrounding the lenses of the eye-piece is air, the nodal points N_1 and N_2 coincide with the principal points P_1 and P_2 respectively.

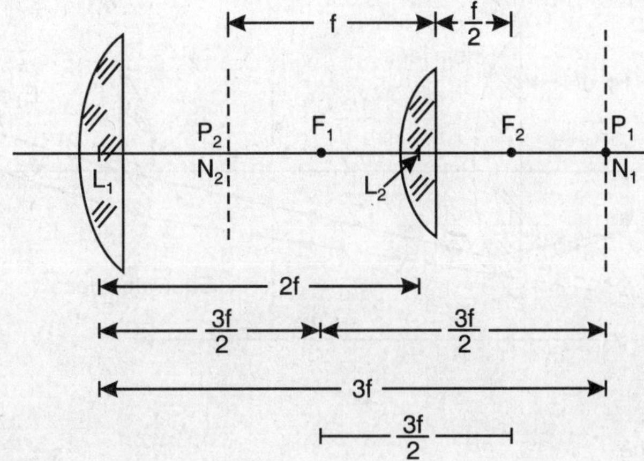

Fig. 4.19: Equivalent focal length.

Merits and Demerits

(*i*) The Huygens' eyepiece is fully free from chromatic aberration because the distance between the lenses is equal to half the sum of their focal lengths.

(*ii*) Spherical aberration is also minimum because the distance between the two lenses is equal to the difference of their focal lengths.

(*iii*) The field of view of this eyepiece is smaller than that of Ramsden's eyepiece.

4.12 RAMSDEN EYEPIECE

Construction: Ramsden's eyepiece consists of two plano-convex lenses each of focal length f separated by a distance equal to $(2/3)f$. The field lens is a little larger than the intermediate image and is placed close to this image to allow as much light as possible to pass through it. The eye lens has a smaller diameter but carries out the actual magnification.

Position of Cross-wires: The cross-wires should be placed at the position of I. Now, the position of I relative to field lens can be found as follows. If the final image is to be formed at infinity, the image I should lie in the focal plane of the equivalent lens. In other words, the distance AB should be equal to $F = 3f/4$. Since $BC = f/2$, $AC = f/4$. Therefore, the objective should produce the image at a distance of $f/4$ in front of the field lens. A fine scale may be placed here if it is desired to measure the size of the image. Since the scale and image would be magnified equally, the measurement would be trustworthy.

Ramsden eye-piece is a positive eye-piece: If the first focal plane of an eye-piece lies outside the eye-piece in the object space, then the real object can be placed on first focal plane to be in focus with the final image. Such an eye-piece is called *positive eye-piece*. In case of Ramsden eyepiece, the focal plane of the eyepiece lies to the left of the field lens and is in the object space. Hence it is a positive eyepiece and can be used to examine a real object or a real image.

Working: The eyepiece works in conjunction with an optical instrument. Referring to the Fig. 4.20, I, I_1' is the image of the distant object formed by the optical instrument which acts as an object for the field lens. The rays get refracted on passing through the field lens L_1 and the image $I\,I'$ is formed. When image lies in the focal plane of the eye-lens, the magnified image is seen at infinity by the eye located at the exit pupil.

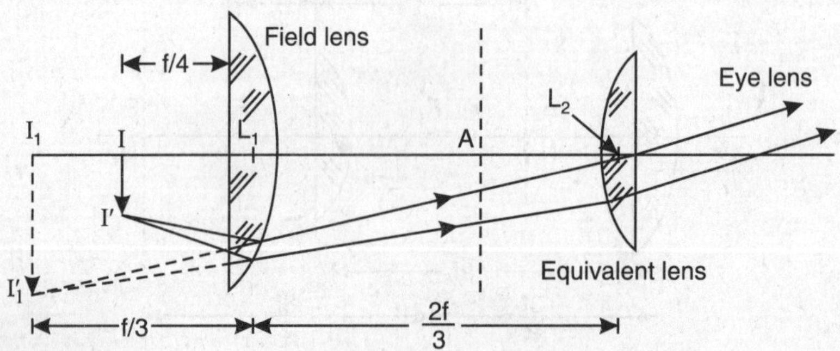

Fig. 4.20:
Ramsden's eyepiece.

(*i*) *Condition for minimum spherical aberration*: The spherical aberration is reduced by making the both the lenses plano-convex and by keeping them with their curved surfaces facing each other, as shown in Fig. 4.20.

(*ii*) *Condition for achromatism*: For chromatic aberration to be a minimum the distance between the two lenses must be equal to the mean of their focal lengths

$$d = \frac{f_1 + f_2}{2}.$$

In Ramsden's eye-piece $f_1 = f_2 = f$.

$$\therefore \qquad d = \frac{f + f}{2} = f.$$

It means that the field lens should be placed in the focal plane of the eye-lens. However, in this position any dust particle or scratch on the field lens would be magnified and the final image would be spoiled. Therefore, the distance between the two lenses is kept $\frac{2}{3}f$ (a little less than f).

(*iii*) *Equivalent focal length:* The equivalent focal length of the eyepiece can be found as follows. If F denotes the focal length of the equivalent lens, then

$$\frac{1}{F} = \frac{1}{f_1}+\frac{1}{f_2}-\frac{d}{f_1 f_2}$$

$$\therefore \quad \frac{1}{F} = \frac{1}{f}+\frac{1}{f}-\frac{(2/3)f}{f^2}$$

$$= \frac{2}{f}-\frac{2}{3f} = \frac{4}{3f}$$

$$\therefore \quad F = 3f/4$$

The final image formed by an eye-piece is at infinity. Therefore, the final image formed by an equivalent lens must be at infinity. For this, II' should be in the focal plane of the equivalent lens. Hence the equivalent lens of focal length $3f/4$ must be placed behind the field lens at a distance

$$x = \frac{F\times d}{f} = \frac{(3/4)f\times(2/3)f}{f} = \frac{f}{2}$$

Thus the equivalent lens lies between the field lens and the eye lens.

(*iv*) *Cardinal points of Ramsden eyepiece:*

(a) *Position of Principal Points*: The distance of the first principal point H_1 from the field lens is given by

$$\alpha_1 = \frac{d\times F}{f_2} = \frac{\frac{2}{3}f\times\frac{3}{4}f}{f} = +\frac{f}{2}$$

Thus, the first principal point H_1 lies at a distance of $f/2$ to the right of the field lens. The distance of the second principal point H_2 from the field lens is given by

$$\alpha_2 = -\frac{d\times F}{f_1} = -\frac{\frac{2}{3}f\times\frac{3}{4}f}{f} = -\frac{f}{2}$$

Thus, the second principal point H_2 lies at a distance of $f/2$ to the left of the eye lens.

(*b*) *Position of the Focal Points:* The distance of the first focal point F_1 from the field lens is given by

$$\beta_1 = -F\left(1-\frac{d}{f_2}\right) = -\frac{3f}{4}\left(1-\frac{\frac{2}{3}f}{f}\right) = -\frac{f}{4}$$

Thus, the first focal point F_1 lies at a distance $\frac{f}{4}$ to the left of the field lens.

The distance of the second focal point F_2 from the field lens is given by

$$\beta_2 = F\left(1-\frac{d}{f_1}\right) = \frac{3f}{4}\left(1-\frac{\frac{2}{3}f}{f}\right) = +\frac{f}{4}$$

Thus, the second focal point F_2 lies at a distance $\frac{f}{4}$ to the right of the eye lens.

(*c*) *Nodal points*: As the medium surrounding the lenses of the eye-piece is air, the nodal points N_1 and N_2 coincide with the principal points P_1 and P_2 respectively.

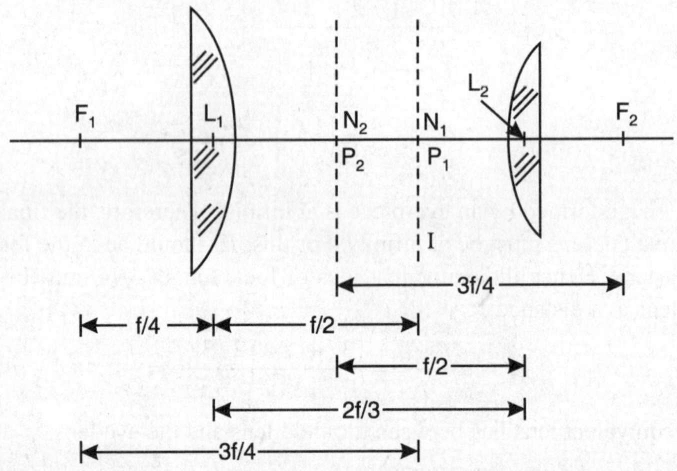

Fig. 4.21: Nodal points

Merits and Demerits

(*i*) The field of view of this eyepiece is fairly wide.

(*ii*) It is not entirely free from chromatic aberration since the distance between the two lenses is not equal to half the sum of their focal lengths. However, chromatic aberration is minimised by using an achromatic combination both for the field lens and the eye lens.

(*iii*) Spherical aberration is minimised by using two plano-convex lenses thereby spreading deviation over four surfaces.

(*iv*) Ramsden's eyepiece is used practically in all instruments where measurements of the size of the final image are to be made.

Table 4.1 shows the Comparison of Ramsden Eyepiece with Huygens Eyepiece

Table 4.1: *Comparison of Ramsden Eyepiece with Huygens Eyepiece*

	Ramsden eyepiece	Huygens Eyepiece
1.	Ramsden's eyepiece is a positive eyepiece. The image formed by the objective lies in front of the field lens. Therefore, cross-wires can be used.	Huygens' eyepiece is a negative eyepiece. The image formed by the objective lies in between the two lenses. Therefore, cross-wires cannot be used.
2.	The condition for minimum spherical aberration is not satisfied. But by spreading the deviations over four surfaces, spherical aberration is minimized.	The condition for minimum spherical aberration is satisfied.
3.	It does not satisfy the condition for achromatism but can be made achromatic by using an achromatic doublet as the eye lens.	It satisfies the condition for achromatism.
4.	It is achromatic for only two chosen colours.	It is achromatic for all colours.

5.	The other types of aberrations are better eliminated. Coma is absent and distortion is 5% less.	Other aberrations like pincushion distortion are not eliminated.
6.	The eye clearance is 5% higher.	The eye clearance is too small and less comfortable.
7.	It is used for quantitative purposes in microscopes and telescopes.	It is used for qualitative purposes in microscopes and telescopes.
8.	Its power is positive.	Its power is negative.
9.	The two principal planes are crossed.	The two principal planes are crossed.
10.	It can be used as a simple microscope because the first principal plane lies to the left of the field lens and the focal plane is real.	It cannot be used as simple microscope because the first focal plane lies to the right of the field lens and the focal plane is virtual.
11.	The nodal points coincide with the principal points.	The nodal points coincide with the principal points.

Example 4.1: *A coaxial lens system placed in air has two lenses of focal lengths 3F and F separated by a distance 2F. Find the positions of the cardinal points.*

Solution:

$$f_1 = 3F,\ f_2 = F,\ d = 2F$$

$$f = \frac{f_1 f_2}{f_1 + f_2 - d}$$

$$\therefore \quad f = \frac{3F \times F}{3F + F - 2F} = \frac{3}{2}F$$

$$\alpha = +\frac{df}{f_2} = \frac{2F \cdot 3F/2}{F} = +3F$$

Therefore, the first principal point P_1 is at a distance $3F$ to the right of the first lens, as shown in Fig. 4.22.

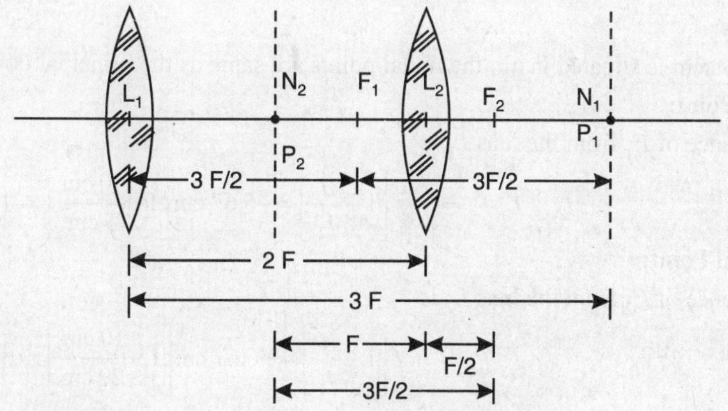

Fig. 4.22:

$$\beta = -\frac{df}{f_1} = \frac{-2F \cdot 3F/2}{3F} = -F$$

The second principal point P_2 is at a distance F to the left of the second lens.

The first focal point F_1 is at a distance $3F - 3F/2 = 3F/2$ from the first lens and F_2 is at a distance $3F/2 - F = F/2$ from the second lens. As the medium on the two sides of the lens system is the same, the nodal points N_1 and N_2 coincide with P_1 and P_2.

Example 4.2: *Two thin convex lenses of focal lengths 20 cm and 5 cm are kept coaxially separated by a distance of 10 cm. Plot the positions of the cardinal points for the combination.*

Solution: Given that $f_1 = 20$ cm, $f_2 = 5$ cm and $d = 10$ cm

$$f = \frac{f_1 f_2}{f_1 + f_2 - d} = \frac{20 \text{ cm} \times 5 \text{ cm}}{20 \text{ cm} + 5 \text{ cm} - 10 \text{ cm}} = 6.67 \text{ cm}$$

First Principal Point:

$$\alpha = \frac{fd}{f_2} = \frac{10 \text{ cm} \times 6.67 \text{ cm}}{5 \text{ cm}} = 13.33 \text{ cm}$$

Second Principal Point:

$$\beta = -\frac{fd}{f_1} = -\frac{10 \text{ cm} \times 6.67 \text{ cm}}{20 \text{ cm}} = -3.33 \text{ cm}$$

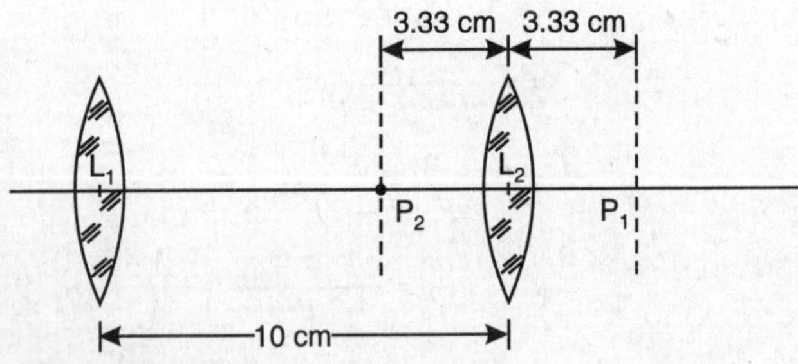

Fig. 4.23:

Nodal Points:

As the system is situated in air, the nodal points are same as the principal points.

First Focal Point:

The distance of F_1 from the lens L_1 is

$$L_1 F_1 = -f\left[1 - \frac{d}{f_2}\right] = -6.67 \text{ cm}\left[1 - \frac{10 \text{ cm}}{5 \text{ cm}}\right] = 13.33 \text{ cm}$$

Second Focal Point:

The distance of F_2 from the lens L_2 is

$$L_2 F_2 = +f\left[1 - \frac{d}{f_1}\right] = -6.67 \text{ cm}\left[1 - \frac{10 \text{ cm}}{20 \text{ cm}}\right] = +3.33 \text{ cm}$$

The first principal point P_1 is to the right of the first lens and is at a distance of 13.33 cm from it. The second principal point P_2 is to the left of the second lens and is at a distance of 3.33 cm from it. The cardinal points are plotted in Fig. 4.23.

Example 4.3: *Two thin converging lenses L_1 and L_2 of powers 5 D and 4 D are placed coaxially 10 cm apart. Find the focal length f of the combination and positions of principal points P_1 and P_2.*

Solution: The focal length of the first lens, $f_1 = \frac{1}{P} = \frac{1}{5D} = \frac{1\,m}{5} = 20$ cm

The focal length of the second lens, $f_2 = \frac{1}{P'} = \frac{1}{4D} = \frac{1\,m}{4} = 25$ cm

The equivalent focal length of the coaxial system is given by

$$f = \frac{f_1 f_2}{f_1 + f_2 - d}$$

$$\therefore \quad f = \frac{20\text{ cm} \times 25\text{ cm}}{20\text{ cm} + 25\text{ cm} - 10\text{ cm}} = 14.29\text{ cm}$$

$$\alpha = +\frac{d\,f}{f_2} = \frac{10\text{ cm} \times (14.29\text{ cm})}{25\text{ cm}} = 5.7\text{ cm}$$

The first principal point P_1 is at a distance 5.7 cm to the right of the first lens.

$$\beta = -\frac{d\,f}{f_1} = -\frac{-10\text{ cm} \times 14.29\text{ cm}}{20\text{ cm}} = -7.14\text{ cm}$$

The second principal point P_2 is at a distance of 40 cm to the left of the second lens.

Example 4.4: *Two thin convex lenses of focal lengths 30 cm and 10 cm are separated by a distance of 25 cm in air. Calculate the positions of the cardinal points.*

Solution: Given that $f_1 = +30$ cm, $f_2 = +10$ cm and $d = 25$ cm

$$f = \frac{f_1 f_2}{f_1 + f_2 - d} = \frac{30\text{ cm} \times 10\text{ cm}}{30\text{ cm} + 10\text{ cm} - 25\text{ cm}} = +20\text{ cm}$$

First Principal Point:

$$\alpha = \frac{fd}{f_2} = \frac{20\text{ cm} \times 25\text{ cm}}{10\text{ cm}} = +50\text{ cm}$$

Second Principal Point:

$$\beta = -\frac{fd}{f_1} = -\frac{20\text{ cm} \times 25\text{ cm}}{30\text{ cm}} = -16.7\text{ cm}$$

First Focal Point:

The distance of F_1 from the lens L_1 is

$$L_1 F_1 = -f\left[1 - \frac{d}{f_2}\right] = -20\text{ cm}\left[1 - \frac{25\text{ cm}}{10\text{ cm}}\right] = +30\text{ cm}$$

Second Focal Point:

The distance of F_2 from the lens L_2 is

$$L_2 F_2 = f\left[1 - \frac{d}{f_1}\right] = +20\text{ cm}\left[1 - \frac{25\text{ cm}}{30\text{ cm}}\right] = +3.3\text{ cm}$$

The first principal point P_1 is to the right of the first lens and is at a distance of 50 cm from it. The second principal point P_2 is to the left of the second lens and is at a distance of – 16.7 cm from it. The first focal point F_1 is to the right of the first lens and is at a distance of 30 cm from it. The second focal point F_2 is to the right of the second lens and is at a distance of + 3.3 cm from it.

Nodal Points: As the system is situated in air, the nodal points are same as the principal points. The cardinal points are plotted in Fig. 4.24.

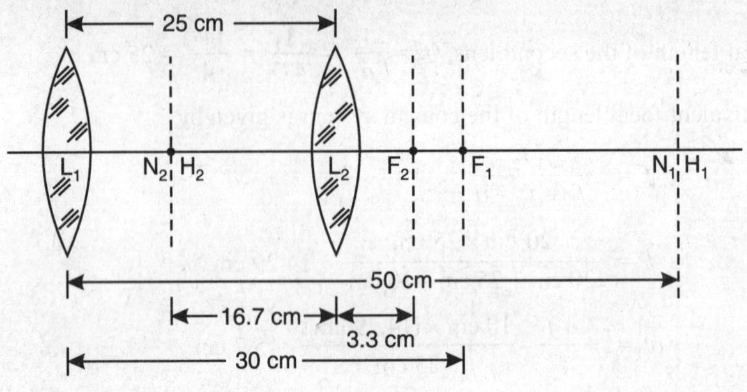

Fig. 4.24:

Example 4.5: *A convex lens of 10 cm focal length is placed in air from a concave lens of focal length 20 cm, at a distance of 5 cm. Find the distance between the two principal points of the combination.*

Solution: Given that $f_1 = +10$ cm, $f_2 = -20$ cm and $d = 5$ cm

The equivalent focal length of the coaxial system is given by

$$f = \frac{f_1 f_2}{f_1 + f_2 - d}$$

$$f = \frac{10 \text{ cm} \times (-20 \text{ cm})}{10 \text{ cm} - 20 \text{ cm} - 5 \text{ cm}} = 13.3 \text{ cm}$$

$$\alpha = +\frac{df}{f_2} = \frac{(5 \text{ cm}) \times (13.3 \text{ cm})}{-20 \text{ cm}} = -3.32 \text{ cm}$$

The first principal point P_1 is at a distance 3.32 cm to the left of the first lens.

$$\beta = -\frac{df}{f_1} = -\frac{(5 \text{ cm}) \times (13.3 \text{ cm})}{10 \text{ cm}} = -6.65 \text{ cm}$$

The second principal point P_2 is at a distance of 6.65 cm to the left of the second lens.

The distance between the two principal points of the combination

$$= \beta - \alpha - d = [-6.65 - (-3.32) - (-5)] \text{ cm} = 1.67 \text{ cm}.$$

Example 4.6: *The focal length of the more convergent lens of a Huygens' eye-piece is 1.5 cm. Calculate the focal length of the eye-piece and locate on a diagram the position of its focal points.*

Solution: The focal length of the lenses in Huygens' eye-piece are $3f$ and f respectively and the separation between them is $2f$. The focal length of the more convergent lens is given to be 1.5 cm, *i.e.*, $f_2 = 1.5$ cm.

$\therefore$ $f_1 = 3f = 4.5$ cm, $f_2 = 1.5$ cm and $d = 2f = 2 \times 1.5$ cm $= 3$ cm.

Thus in this case the focal lengths of the field and eye-lenses are 4.5 cm and 1.5 cm and the separation between them is 3 cm.

The focal length F of the eye-piece is given by

$$F = \frac{f_1 f_2}{f_1 + f_2 - d} = \frac{4.5\times1.5}{4.5+1.5-3}\text{ cm} = 2.25\text{ cm.}$$

Focal Points: The distance of the first focal point F_1 from the field lens L_1 is given by

$$\beta_1 = -F\left(1-\frac{d}{f_2}\right) = -2.25\left(1-\frac{3}{1.5}\right)\text{ cm} = +2.25\text{ cm.}$$

i.e., the first focal point F_1 lies at a distance 2.25 cm to the right of field lens L_1.

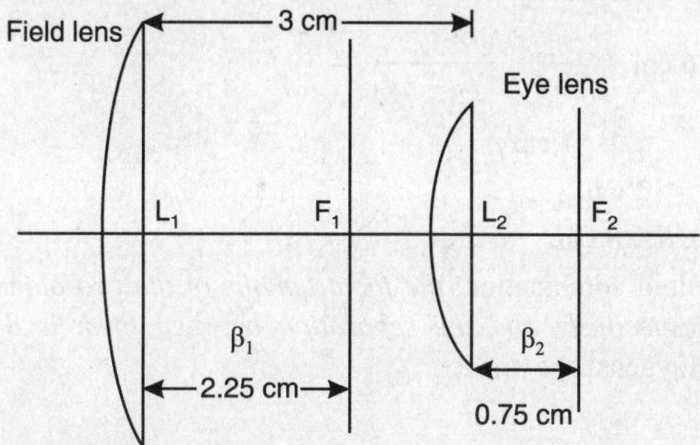

Fig. 4.25:

The distance of the second focal point F_2 from the eye-lens L_2 is given by

$$\beta_2 = F\left(1-\frac{d}{f_1}\right) = 2.25\left(1-\frac{3}{4.5}\right)\text{ cm}$$

$$= 0.75\text{ cm.}$$

i.e., the second focal point F_2 lies at a distance 0.75 cm to the right of eye-lens L_2.

The positions of the focal points F_1 and F_2 are plotted in Fig. 4.25.

Example 4.7: *From the conditions of 'no chromatic aberration' and 'minimum spherical aberration' of a combination of two separated thin plano-convex lenses, design a combination of equivalent focal length 6 cm. What type of eye-piece this combination forms? Deduce the positions of cardinal points of this eye-piece.*

Solution: If f_1 and f_2 are the focal lengths of the lenses of the combination and d is the distance between the lenses, we have for no chromatic aberration.

$$d = \frac{f_1 + f_2}{2} \quad ...(1)$$

and for minimum spherical aberration,

$$d = f_1 - f_2 \quad ...(2)$$

From Eq. (1) and Eq. (2), we have

$$f_1 - f_2 = \frac{f_1 + f_2}{2}$$

or $$f_1 = 3f_2 \quad ...(3)$$

Substituting this value of f_1 in Eq. (2), we get

$$d = 3f_2 - f_2 = 2f_2 \quad ...(4)$$

If F is the focal length of the equivalent lenses we have

$$F = \frac{f_1 f_2}{f_1 + f_2 - d}$$

Here $f_1 = 3f_2$, $d = 2f_2$ and $F = 6$ cm (given)

$$\therefore \quad 6 \text{ cm} = \frac{3f_2 \times f_2}{3f_2 + f_2 - 2f_2} = \frac{3}{2} f_2$$

$$\therefore \quad f_2 = \mathbf{4\ cm}$$

From Eq. (3), $f_1 =$ **12 cm.**

and from Eq. (4), $d =$ **8 cm.**

Thus in the required combination, *the focal lengths of the two plano-convex lenses are 12 cm and 4 cm respectively and the separation between them is 8 cm obviously this combination forms* Huygens' *eye-piece.*

Cardinal Points

***(i)* Principal Points:** The position of the first principal point H_1 from the field lens L_1 is given by

$$\alpha_1 = \frac{dF}{f_2} = \frac{8 \times 6}{4} \text{ cm} = +12 \text{ cm}$$

i.e., first principal point H_1 lies at a distance 12 cm to the right of field lens.

The distance of the second principal point H_2 from the eye-lens L_2

$$\alpha_2 = -\frac{dF}{f_1} = -\frac{8 \times 6}{12} \text{ cm} = -4 \text{ cm}$$

i.e., the second principal point H_2 lies at a distance of 4 cm to the left of eye-piece.

***(ii)* Focal Points:** The distance of the first focal point F_1 from the field lens L_1 is

$$\beta_1 = -F\left(1 - \frac{d}{f_2}\right) = -6\left(1 - \frac{8}{4}\right) \text{ cm} = +6 \text{ cm}$$

i.e., first focal point F_1 lie at a distance of 6 cm to the right of field lens.

The distance of the second rocal point F_2 from the eye-lens L_2 is

$$\beta_2 = F\left(1 - \frac{d}{f_1}\right) = 6\left(1 - \frac{8}{12}\right) \text{ cm} = +2 \text{ cm.}$$

i.e., the second focal point F_2 lies at a distance of 2 cm to the right of eye-lens L_2.

***(iii)* Nodal Point:** As the medium on either side of the combination is same, the nodal points N_1, N_2 coincide with the principal points H_1, H_2 respectively. Cardinal points H_1, H_2, F_1, F_2, N_1 and N_2 are shown in Fig. 4.26.

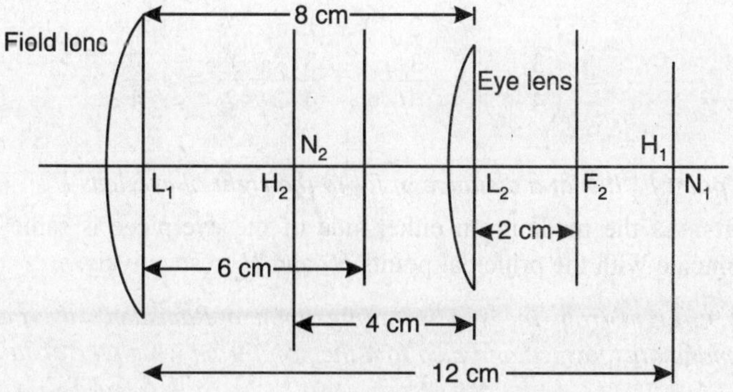

Fig. 4.26:

Example 4.8: *Focal lengths of lenses of Huygens' eye-piece are 4 cm and 12 cm. Find and plot the positions of cardinal points on a diagram. If an object is situated at 6 cm in front of the field lens, find the position of the image formed by the eye-piece.*

Solution: The focal lengths of lenses are $f_1 = 12$ cm, $f_2 = 4$ cm.

In Huygens' eye-piece, separation between lenses

$$d = f_1 - f_2 = (12 - 4) \text{ cm} = 8 \text{ cm}.$$

The cardinal points of this eye-piece have been calculated in the previous example 4.7 and plotted in Fig. 4.26.

The object is at a distance 6 cm infront of the field lens. If u and v are the distances of the object O and image I from the first and second principal points H_1 and H_2 respectively.

Then $$u = H_1O = -(H_1L_1 + L_1O) = -(12 + 6) \text{ cm} = -18 \text{ cm}, \ v = H_2I = ?$$

$$F = \frac{f_1 f_2}{f_1 + f_2 - d} = \frac{12 \times 4}{12 + 4 - 8} \text{cm} = \frac{12 \times 4}{8} \text{cm} = +6 \text{ cm}$$

Now from thin lens formula $\frac{1}{v} - \frac{1}{u} = \frac{1}{F}$, we have

$$\frac{1}{v} = \frac{1}{u} + \frac{1}{F} = -\frac{1}{18} + \frac{1}{6} = \frac{2}{18}$$

$\therefore$ $$v = H_2I = +9 \text{ cm}$$

Hence $$L_2I = H_2I - H_2L_2 = (9 - 4) \text{ cm} = +5 \text{ cm}$$

Thus the image lies at a distance of +5 cm to the right (behind) the eye-lens L_2.

$$= -\frac{3}{4}f + \frac{f}{2} = -\frac{f}{4}$$

i.e., the first focal point F_1 lies at a distance of f/4 to the left of field lens L_1.

The distance of the second focal point F_2 from the eye-lens L_2.

$$\beta_2 = F\left(1 - \frac{d}{f_1}\right)$$

$$= \left(\frac{3}{4}f\right)\left(1 - \frac{\frac{2}{3}f}{f}\right) = \frac{3}{4}f - \frac{f}{2} = +\frac{f}{4}$$

i.e., second point F_2 lies at a distance of f/4 to the right of eye-lens L_2.

Nodal Points: As the medium on either side of the eye-piece is same (air), the nodal points N_1, N_2 coincide with the principal points H_1 and H_2 respectively.

Example 4.9: *Light from the sun is falling upon a Ramsden's eye-piece. Locate the position of the image thus formed and also find the point from which the distance of the image is to be measured. The focal length of each lens of the eye-piece is $1\frac{1}{2}$ inches.*

Solution: The focal lengths of the lenses of Ramsden's eye-piece are equal and the distance between them is $\frac{2}{3}f$.

If F is the focal length of the eye-piece,

$$F = \frac{f_1 f_2}{f_1 + f_2 - d}$$

Here $$f_1 = f_2 = f = \frac{3''}{2}, \quad d = \frac{2}{3}f = \frac{2}{3} \times \frac{3''}{2} = 1''$$

$$\therefore \quad F = \frac{\frac{3''}{2} \times \frac{3''}{2}}{\frac{3''}{2} + \frac{3''}{2} - 1''} = +\frac{9''}{8}$$

As the rays coming from sun are parallel, the image will be formed by the eye-piece at second focal point F_2.

The distance of second principal point H_2 from the second lens (eye-lens)

$$\alpha_2 = -\frac{dF}{f_1} = -\frac{1'' \times \frac{9''}{8}}{\frac{3''}{2}} = -\frac{9''}{8} \times \frac{2''}{3} = -\frac{3''}{4},$$

i.e., second principal point lies at a distance $\frac{3''}{4}$ to the left of eye-piece.

The distance of the image from the eye-lens, *i.e.*, the distance of the second focal point F_2 from the eye-lens,

$$\beta_2 = F\left(1 - \frac{d}{f_1}\right) = \frac{9''}{8}\left(1 - \frac{1''}{3/2''}\right) = \frac{9''}{8}\left(1 - \frac{2''}{3}\right) = \frac{3''}{8}$$

i.e., the image is formed at a distance of $\frac{3''}{8}$ to the right of eye-lens.

QUESTIONS

1. What are cardinal points?
2. Why the principal points of a lens system are called unit points?
3. Explain the term cardinal points with reference to a coaxial system.
4. What are principal points and principal planes? Show that the principal planes are the planes of unit linear magnification.
5. What are nodal points and nodal planes? Give their properties. Show that the nodal planes are planes of unit angular magnification.
6. Derive Newton's formula for a convergent system of lenses forming a real image.
7. Describe the construction and working of a nodal slide and show how the nodal points of a system can be located with its help.
8. Show that for a co-axial lens system, $x x' = f f'$ where x and x' are the respective distances of the object and the image from the first and the second focal points and f and f' are the two focal lengths. What form does the expression take when the media on the two sides of the system are the same?
9. Define cardinal points of a system of co-axial lenses. Describe how you would determine experimentally the principal planes of a combination of two thin lenses separated by a distance.
10. What are the properties of cardinal points of a co-axial lens system? Plot the cardinal points of a Huygens eyepiece. How can they be determined experimentally?
11. Explain the term cardinal points with reference to a co-axial lens system.
12. Two thin convex lenses of focal length f_1 cm and f_2 cm are coaxial and separated by d. Show that the focal length f of the combination is given by the relation

$$f = \frac{f_1 f_2}{f_1 + f_2 - d}$$

13. Define cardinal points of a coaxial lens system. Show how the principal planes of such a system can be located using the theory of the method of deviation or otherwise. State the sign convention you use.
14. Derive an expression for the focal length of a system of two thin lenses separated by a finite distance when the space between them is filled with a medium of refractive index (μ>1).
15. Derive expressions for the focal length and the positions of principal points and focal points of a coaxial system of two thin lenses separated by a finite distance.
16. What is an eyepiece and what is its advantage over a single lens?
17. Give the construction and working of a Ramsden eyepiece. How are chromatic and spherical aberrations minimized in this eyepiece?
18. Explain the construction of a Huygens eyepiece. Why cannot a cross-wire be used with it?
19. Give the construction of Huygens eyepiece and calculate the positions of the cardinal points.
20. Give the construction and theory of Huygens eyepiece and show that it is free from spherical and chromatic aberrations.
21. Explain why it is necessary to use an eyepiece consisting of more than one lens.
22. Why is Huygens eyepiece called a negative piece and Ramsden eyepiece called a positive eyepiece?
23. Give the name and construction of the eyepiece, which satisfies the condition for achromatism.
24. Describe and point out the respective merits of Ramsden and Huygens eyepieces.
25. Compare Ramsden eyepiece with Huygens eyepiece.
26. Explain why cross-wires can be used with Ramsden eyepiece but not with Huygens eyepiece.

PROBLEMS

1. Two similar thin convex lenses of focal lengths 10 cm each are coaxial and 5 cm apart. Find the equivalent focal length and position of the principal points. Also find the position of the object for which the image is formed at infinity.

   [**Ans:** f = 6.67 cm; α = + 3.33 cm; β = – 3.33 cm; u = – 3.33 cm.]

2. Two thin converging lenses of focal lengths 15 cm and 20 cm are placed coaxially 10 cm apart. An object is placed at a distance of 24 cm from the first lens. Find (*i*) the position of the focal points and principal points and (*ii*) the position of the image.

   [**Ans:** f = 12 cm; α = + 6 cm; β = – 8 cm; υ = 12 cm.]

3. Two thin converging lenses each of 30 cm focal length are set coaxially 10 cm apart. An image of an upright pole 100 metres distant and 5 metre high is formed by the combination. Find the position of the unit and focal planes and the image. Also find the size of the image.

   [**Ans:** f = 18 cm; α = + 6 cm; β = – 6 cm; υ = 12.03 cm, h_2 = 0.9 cm.]

4. Determine the positions of the focal points, principal points and nodal points in the case of sphere of radius 12 cm. (μ = 1.5).

   [**Ans:** f = 18 cm; α = + 12 cm; β = –12 cm. The nodal points and the principal points are at the centre of the sphere.]

5. Two thin converging lenses of focal lengths 20 cm and 40 cm are placed coaxially 20 cm apart. An object is placed at a distance of 50 cm from the first lens. Calculate the positions of the principal points, focal points and the position of the image.

   [**Ans:** α = + 12 cm; β = –16 cm. P_1F_1 = –24 cm, P_2F_2 = +24 cm; Image is formed at a distance of 23.16 cm to the right of the second lens.]

5 Electrostatics

Learning Objectives:

By the end of this chapter, you will be able to:

- Understand basic concepts of Electric Charges.
- State Coulomb Law and its application.
- Explain concepts related to Electric Field.
- Describe electric potential in materials.
- State Gauss Law and its application.

5.1 INTRODUCTION

Electrons carry negative charge and protons carry a positive charge. Matter in its neutral state contains equal amounts of positive and negative charges. Accordingly, we interpret the electrification of the bodies as occurring due to transfer of charge. When two bodies are rubbed together, a redistribution of electrons takes place. The body which receives electrons becomes negatively charged while the body which loses electrons becomes positively charged. Charged bodies attract each other if they have unlike charges or repel each other if they have like charges. We say that an electric force, Coulomb's force is acting between them. The mechanical forces push and pull are contact forces whereas electrical and magnetic forces are known as field forces. A body carrying a specific charge is regarded to modify the space around it in a certain manner. We say that electric field is present around the charged body. Any other charged particle coming into the region would experience a force because of the field around the charged body.

Benjamin Franklin (1706-1790), Henry Cavendish (1731-1810), Charles Augustine Coulomb (1736-1806), Hans Christian Oersted (1777-1851), and Andre Marie Ampere (1775-1836), were some of the pioneers who contributed to the initial understanding of electrical phenomena. The study of electric forces between charged objects at rest is known as electrostatics.

5.2 ELECTRIC CHARGES

The charges either positive or negative are always built up as a collection of elementary charges, carried by fundamental particles protons and electrons. They are always an integral multiple of the smallest unit of charge, that of an electron. When a body is said to be charged, it contains either an excess of electrons or a shortage of electrons. The charge residing on a charged body is

$$Q = \pm ne \qquad ...(5.1)$$

where n is an integer taking values 1,2,3,....

The SI unit of charge is the Coulomb denoted by C. The value of e is

$$1e = 1.602 \times 10^{-19}\text{ C}$$

5.3 COULOMB'S LAW

If two charges are brought nearer, they exert forces on each other. We say that the charges are interacting. When the charges are at rest, their interaction is called *electrostatic interaction.* The electrostatic interaction between two charged particles is governed by Coulomb's law. Usually, charged particles are taken as dimensionless point charges.

Statement of Coulomb's law: The magnitude of the electrostatic force of attraction or repulsion between two point charges is directly proportional to the product of the magnitudes of charges and inversely proportional to the square of the distance between them. The force acts along the straight line joining the two point charges.

If q_1 and q_2 are two **point charges at rest**, and are separated by a distance r, they exert a force on each other which is given by

$$F = k\frac{q_1 q_2}{r^2} \qquad ...(5.2)$$

where k is a constant which has the following value in the SI system.

$$k = \frac{1}{4\pi\varepsilon_0} \qquad ...(5.3)$$

Here ε_0 is called the **permittivity of free space** and has the value

$$\varepsilon_0 = 8.85 \times 10^{-12}\text{ C}^2/\text{Nm}^2$$

$$\therefore \quad k = 9 \times 10^9\text{ Nm}^2/\text{C}^2 \qquad ...(5.4)$$

When the charges are located in a dielectric medium having a **relative permittivity** of ε_r, the force is given by

$$F = \frac{1}{4\pi\varepsilon_0}\frac{q_1 q_2}{\varepsilon_r r^2} \qquad ...(5.5)$$

The Coulomb's force is maximum when the charges in a vacuum and reduces when the charges are placed in a medium.

Vector form of the Law

The Coulomb's law (5.2) may be expressed in vector form as

$$\mathbf{F} = k\frac{q_1 q_2}{r^2}\hat{\mathbf{r}} = k\frac{q_1 q_2 \mathbf{r}}{r^3} \qquad ...(5.6)$$

In the Eq. (5.6), $\hat{\mathbf{r}}$ is the unit vector along $\mathbf{r}$. It is given by $\hat{\mathbf{r}} = \frac{\mathbf{r}}{r}$.

- In Eq. (5.6), **F** is the force produced by the charge q_1 on the charge q_2. The force produced by the charge q_2 on the charge q_1 is therefore – **F**.
- The charges are assumed to be at rest; otherwise we need to take into account the magnetic forces.

- The size of the charges must be very small compared to their separation. Hence, the charges are assumed to be point charges.

Example 5.1: *Two equal and similar charges 3 cm apart in air repel each other with a force equivalent to 1.5 kg wt. Find the magnitude of the charges.*

Solution: Now,

$$F = \frac{q_1 q_2}{4\pi\varepsilon_0 r^2}$$

$$\therefore \quad 1.5 = 9\times10^9 \times \frac{q_1^2}{\left(3\times10^{-2}\right)^2} \quad \text{as} \quad \frac{1}{4\pi\varepsilon_0} = 9\times10^9 \text{ Nm}^2\text{ C}^2$$

$$\therefore \quad q_1^2 = \frac{1.5\times9\times10^{-4}}{9\times10^9} = 1.5\times10^{-13}$$

$$\therefore \quad q_1 = \mathbf{3.87\times10^{-7}\ C}$$

Example 5.2: *Two point charges of 1 C each are separated from each other by a distance of 1 m in a vacuum.*

(a) What is the force of their interaction?

(b) What will be the force if the medium between the charges is water?

Solution: $F = \frac{q_1 q_2}{4\pi\varepsilon_0\varepsilon_r r^2}$; For air $\varepsilon_r = 1$ and for water $\varepsilon_r = 80$.

(*a*) $\therefore \quad F_{\text{air}} = 9\times10^9 \times \frac{1\times1}{1} = \mathbf{9\times10^9\ N}$

(*b*) $\quad F_{\text{water}} = 9\times10^9 \times \frac{1\times1}{80\times1} = \mathbf{1.1\times10^8\ N}$

5.4 PRINCIPLE OF SUPERPOSITION

If a number of point charges q_1, q_2, q_3, ... are present in a region, then the total force on any particular charge is the vector sum of forces it experiences due to all other charges. This is called the **principle of superposition**. For the sake of simplicity, let there be only three charges. The force on q_3 is

$$\mathbf{F} = \mathbf{F}_{13} + \mathbf{F}_{23}$$

$$= \frac{1}{4\pi\varepsilon_0}\frac{q_1 q_3}{|r_{13}|^3}\mathbf{r}_{13} + \frac{1}{4\pi\varepsilon_0}\frac{q_2 q_3}{|r_{23}|^3}\mathbf{r}_{23} \qquad ...(5.7)$$

Generalizing, one finds the force acting on a charge q_j due to a number of other charges present in the region is

$$\mathbf{F} = \frac{1}{4\pi\varepsilon_0}\sum_{i\neq j}\frac{q_i q_j}{|r_{ij}|^3}\mathbf{r}_{ij} \qquad ...(5.8)$$

5.5 ELECTRIC FIELD

Electric field is a region in which an electric charge experiences a force. The force is due to the presence of other charged body in that region. For example, a charge q is placed in a region where a charged body Q is present. According to the standard convention, we take always Q to be positive. The charge q experiences a force **F** and we say that it is in an electric field produced by the charge Q. The force on a charged particle placed in an electric field is proportional to the charge of the particle.

$$F \propto q$$

or
$$\mathbf{F} = q\mathbf{E} \quad \text{...(5.9)}$$

where E is the *electric field strength.*

The direction of E is along the direction of force, that is along the line joining Q to q.

Electric Field Intensity at a point in the electric field: It is defined as the ratio of the force to unit test charge positioned at that point where electric field intensity is desired.

Thus,

$$\mathbf{E} = \frac{\mathbf{F}}{q} \quad \text{...(5.10)}$$

Force acts on the charge in the direction of the electric field for a positive charge and acts in the opposite direction of the electric field for a negative charge. Ideally, q must be as small as possible in order to avoid possible disturbance of the original field **E**. For this reason, the following definition of **E** is more commonly used.

$$E = \underset{q \to 0}{\text{Limit}} \frac{F}{q} \quad \text{...(5.11)}$$

In SI system of units, the unit of electric field is Newton/Coulomb or N/C.

In the Eq. (5.5), let $q_1 = Q$, $q_2 = q$ and the medium be air, $\varepsilon_r = 1$. Then the force produced by the charge Q on the charge q placed at a distance r from Q is given by

$$\mathbf{F} = q\left[\frac{Q}{4\pi\varepsilon_0 r^2}\hat{r}\right] \quad \text{...(5.12)}$$

Comparing Eq. (5.12) with Eq. (5.9), we may say that the electric field E at the point where q is placed is

$$\mathbf{E} = \left[\frac{Q}{4\pi\varepsilon_0 r^2}\hat{r}\right]$$

Therefore, the field produced by the charge Q is

$$\mathbf{E} = \left[\frac{Q}{4\pi\varepsilon_0 r^2}\hat{r}\right] \quad \text{...(5.13)}$$

From Eq. (5.13) we can see that the electric field decreases as one moves away from the charged body Q. It reduces to zero at infinity.

5.5.1 Electric Field due to a Group of Point Charges

The net electric field at a point due to a group of point charges can be found by applying the superposition principle. Since the Coulomb force obeys the superposition principle, the electric field intensity (force per unit charge) obeys the superposition principle. The electric field at a point P due to n point charges $q_1, q_2, q_3, ..., q_n$ is equal to the vector sum of electric fields due to $q_1, q_2, q_3, ..., q_n$ at point P. Thus, the resultant electric field is given by

$$\mathbf{E} = \mathbf{E}_1 + \mathbf{E}_2 + \mathbf{E}_3 + ... + \mathbf{E}_n \qquad ...(5.14)$$

where $\mathbf{E}_1$ is the electric field intensity at P due to q_1,
$\mathbf{E}_2$ is the electric field intensity at P due to q_2,
$\mathbf{E}_3$ is the electric field intensity at P due to q_3, and so on.

Example 5.3: *A force of 0.015 N acts upon a charge of 2×10^{-7} C at a point in an electric field. What is the strength of electric field at that point?*

Solution: Electric field E, is given by $E = F/q$

$$\therefore \quad E = \frac{0.015\ \text{N}}{2\times10^{-7}\ \text{C}} = \mathbf{7.5 \times 10^4\ N/C}$$

5.6 COMPUTATION OF ELECTRIC FIELD IN SOME SPECIFIC CASES

1. Field due to a linear charge

Let us consider an infinitely long charged wire of negligible thickness and having a constant linear charge density λ. Let a point P be at a distance y from the wire, as shown in Fig. 5.1.

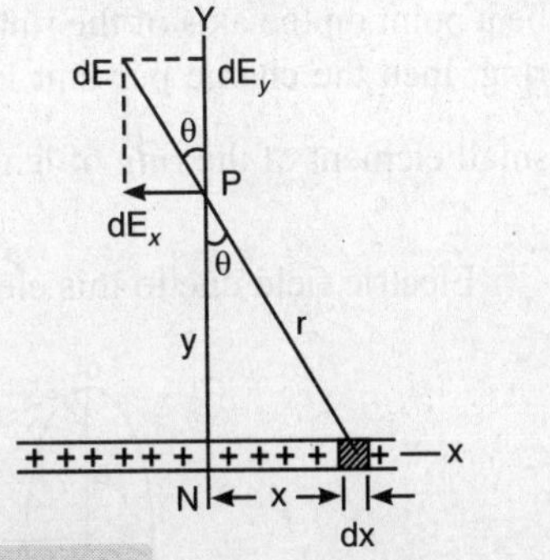

It is required to find the electric field intensity at P. Let us assume that the wire is made up of a number of infinitely small elements of length, dx. Let one of such elements be at a distance x, as shown in Fig. 5.1. Let the small charge on element be dq.

Fig. 5.1: Electric field due to a line charge.

$$dq = \lambda\, dx$$

The field due to this charge dq at point P is

$$dE = \frac{1}{4\pi\varepsilon_0} \cdot \frac{dq}{(NP)^2} = \frac{dq}{4\pi\varepsilon_0 r^2}$$

x and y components of $d\mathbf{E}$ are:

$$dE_x = -\,dE \sin\theta \quad \text{and} \quad dE_y = -\,dE \cos\theta$$

The x-components of field at P cancel out each others effect. Therefore, the net field will be due to y-components only and is directed along y-axis.

The resultant field is

$$E = \int_{x=-\infty}^{x=+\infty} dE_y = \int_{x=-\infty}^{x=+\infty} dE\cos\theta = \int_0^\infty 2\,dE\cos\theta = \int_0^\infty \frac{2\,dq}{4\pi\varepsilon_0 r^2}\cos\theta$$

$$= \int_0^\infty \frac{2\,\lambda\,dx}{4\pi\varepsilon_0 r^2}\cos\theta = \frac{\lambda}{2\pi\varepsilon_0}\int_0^\infty \frac{dx}{r^2}\cos\theta$$

From Fig. 5.1, we have $\quad \dfrac{x}{y} = \tan\theta$

$$\therefore \quad \frac{dx}{y} = \sec^2\theta\, d\theta \quad \text{or} \quad dx = y\sec^2\theta\, d\theta. \quad \text{Also, } x^2 + y^2 = r^2.$$

$$\therefore \quad E = \frac{\lambda}{2\pi\varepsilon_0}\int_0^{\theta=\pi/2}\frac{y\sec^2\theta\, d\theta}{x^2+y^2}\cos\theta = \frac{\lambda}{2\pi\varepsilon_0}\int_0^{\theta=\pi/2}\frac{y\sec^2\theta\, d\theta}{y^2\tan^2\theta+y^2}\cos\theta$$

$$= \frac{\lambda}{2\pi\varepsilon_0}\int_0^{\theta=\pi/2}\frac{y\sec^2\theta d\theta}{y^2\sec^2\theta}\cos\theta = \frac{\lambda}{2\pi\varepsilon_0}\int_0^{\theta=\pi/2}\frac{\cos\theta}{y}d\theta$$

$$= \frac{\lambda}{2\pi\varepsilon_0 y}[\sin\theta]_0^{\pi/2}$$

or $$E = \frac{\lambda}{2\pi\varepsilon_0 y} \quad ...(5.15)$$

2. Field due to a uniformly charged ring at an axial point

Let us consider a uniformly charged ring of charge q and radius a is shown in Fig. 5.2. Let P be a point on the axis of the ring at a distance x from its centre. If a positive charge q is on the ring, then the charge per unit length of the ring will be $\lambda = q / 2\pi a$. Now let us consider a small element of the ring of length dl. The charge on the element dl is $\frac{q\,dl}{2\pi a}$.

Electric field due to this element is given by $dE = \frac{1}{4\pi\varepsilon_0}\frac{\lambda dl}{r^2} = \frac{1}{4\pi\varepsilon_0}\frac{q\,dl}{2\pi a}$

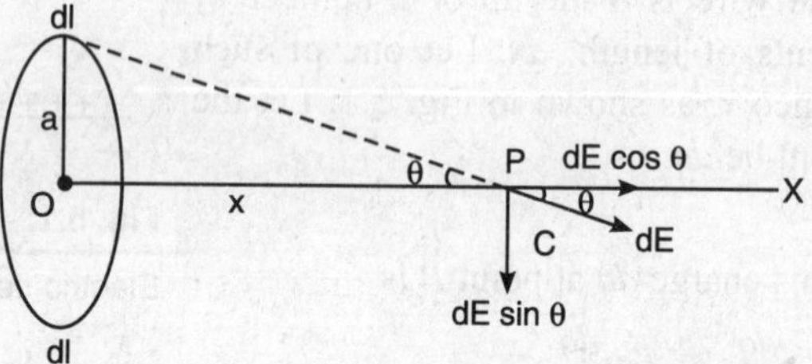

Fig. 5.2:

Field due to a uniformly charged ring at an axial point.

This field due to small element can be resolved into two components along x-axis (axis of the ring) and y-axis (perpendicular to the axis). Owing to symmetry, the perpendicular components cancel out. Hence, the resultant field will be due to the components along the axis of the ring. Therefore, the resultant field is given by

$$E = \int dE = \int dE\cos\theta$$

$$\therefore \quad E = \int\frac{1}{4\pi\varepsilon_0}\frac{q\,dl}{2\pi a}\frac{1}{r^2}\cos\theta$$

From Fig. 5.2, we have $a^2 + x^2 = r^2$ and $\cos\theta = \frac{x}{\sqrt{a^2+x^2}}$.

$$\therefore \quad E = \int\frac{1}{4\pi\varepsilon_0}\frac{q\,dl}{2\pi a}\frac{1}{(a^2+x^2)}\cdot\frac{x}{\sqrt{a^2+x^2}}$$

$$= \frac{1}{4\pi\varepsilon_0}\frac{qx}{2\pi a}\frac{1}{\left(a^2+x^2\right)^{3/2}}\int dl$$

$$= \frac{1}{4\pi\varepsilon_0}\frac{qx}{2\pi a}\frac{1}{\left(a^2+x^2\right)^{3/2}} \times 2\pi a$$

or $$E = \frac{1}{4\pi\varepsilon_0}\frac{qx}{\left(a^2+x^2\right)^{3/2}} \quad ...(5.16)$$

3. Field due to a uniformly charged disc

A disc of radius 'a' units is charged uniformly with a charge density σ C/m^2. Let P be at a distance r from the centre of the disc as shown in Fig. 5.3. The disc may be regarded as formed by several annular rings of increasing radius. Let us consider a ring of radius x and the area of the hatched ring be ds.

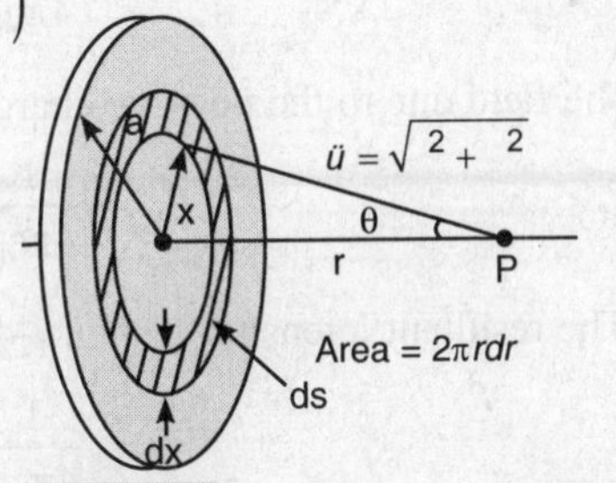

Fig. 5.3: Field due to a uniformly charged disc.

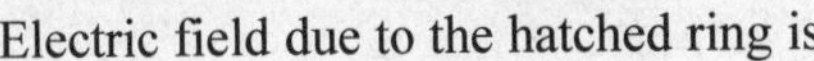

Electric field due to the hatched ring is

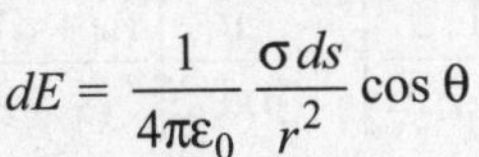

$$dE = \frac{1}{4\pi\varepsilon_0}\frac{\sigma\, ds}{r^2}\cos\theta$$

The area of the annular ring $ds = \pi\left[(x+dx)^2 - x^2\right] = \pi(2x\,dx + dx^2) \cong 2\pi x\,dx$.

The term dx^2 is negligible compared to x.

$$\therefore \quad dE = \frac{1}{4\pi\varepsilon_0}\frac{\sigma 2\pi x\, dx}{r^2}\cos\theta = \frac{\sigma x\, dx}{2\varepsilon_0 r^2}\cos\theta$$

From Fig. 5.3, it is seen that $\cos\theta = d/r \quad \therefore \quad r = d\sec\theta$

$\tan\theta = x/d \quad \therefore \quad x = d\tan\theta$ and $dx = d\sec^2\theta\, d\theta$

Using these expressions $$dE = \frac{\sigma(d\tan\theta)\cdot(d\sec^2\theta\, d\theta)}{2\varepsilon_0 d^2\sec^2\theta}\cos\theta = \frac{\sigma\sin\theta\, d\theta}{2\varepsilon_0}$$

Electric field due to the entire disc is obtained by integrating the above expression between the limits of θ from 0 to α.

$$\therefore \quad E = \frac{\sigma}{2\varepsilon_0}\int_0^{\alpha}\sin\theta\, d\theta = \frac{\sigma}{2\varepsilon_0}[-\cos\theta]_0^{\alpha} = \frac{\sigma}{2\varepsilon_0}[1-\cos\alpha]$$

From Fig. 5.3, we have $\cos\alpha = \dfrac{d}{\sqrt{a^2+d^2}}$.

$$\therefore \quad E = \frac{\sigma}{2\varepsilon_0}\left[1 - \frac{d}{\sqrt{a^2+d^2}}\right] \quad ...(5.17)$$

If $d >> a$, $$E = \frac{\sigma}{2\varepsilon_0} \quad ...(5.18)$$

4. Field due to an electric dipole

An **electric dipole** is defined as a system of two equal and opposite charges separated by a small distance as shown in Fig. 5.4.

(i) Field at an axial point of the dipole: Consider a point P on the axis of the dipole which is located at a distance x from the centre of the dipole.

The negative charge gives a field at P

$$E_1 = \frac{1}{4\pi\varepsilon_0} \cdot \frac{q}{(x-a)^2}$$

Fig. 5.4: Field at an axial point of the dipole.

The field due to the positive charge q at P is given by

$$E_2 = \frac{1}{4\pi\varepsilon_0} \cdot \frac{q}{(x+a)^2}$$

The resultant intensity $E = E_1 - E_2$

$$= \frac{1}{4\pi\varepsilon_0} \cdot \frac{q}{(x-a)^2} - \frac{1}{4\pi\varepsilon_0} \cdot \frac{q}{(x+a)^2}$$

$$= \frac{q}{4\pi\varepsilon_0}\left[\frac{1}{(x-a)^2} - \frac{1}{(x+a)^2}\right] = \frac{q}{4\pi\varepsilon_0}\left[\frac{(x+a)^2-(x-a)^2}{(x-a)^2(x+a)^2}\right]$$

$$= \frac{q}{4\pi\varepsilon_0}\left[\frac{4ax}{(x-a)^2\,(x+a)^2}\right]$$

or
$$E = \frac{qax}{\pi\varepsilon_0\left(x^2-a^2\right)^2} \qquad \text{...(5.19)}$$

If $x >> a$,
$$E = \frac{qa}{\pi\varepsilon_0 x^3} = \frac{2a\,q}{2\pi\varepsilon_0 x^3}$$

or
$$E = \frac{\mu}{2\pi\varepsilon_0 x^3}. \qquad \text{...(5.20)}$$

In the above $\mu = 2a\,q$ is known as the electric **dipole moment.**

- The resultant field **E** is along the axis of the dipole and is directed from $-q$ to q.
- The field due to the charge at P has inverse relation with the cube of the distance.

(ii) Field at a point on the perpendicular bisector of the dipole: Consider a point P on the perpendicular bisector of the dipole, Fig. 5.5. Let E_1 and E_2 be the fields due to $-q$ and q respectively.

The fields due to the charges at P has inverse relation with the cube of the distance

$$E_1 = E_2 = \frac{1}{4\pi\varepsilon_0} \cdot \frac{q}{a^2+x^2}$$

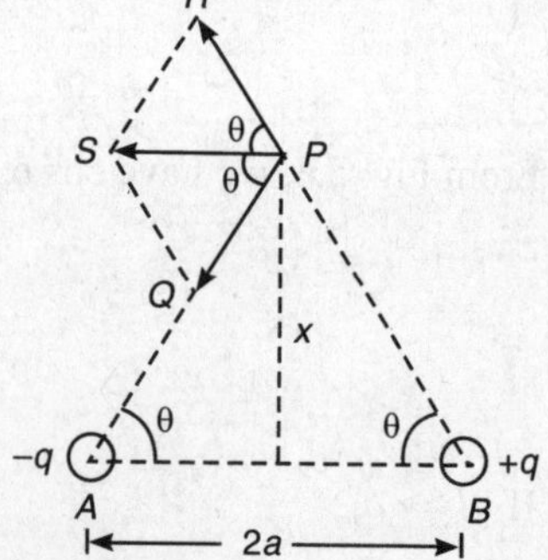

Fig. 5.5: Field at a point on the perpendicular bisector of the dipole.

The resultant intensity $E = E_1 + E_2 = 2E_1 \cos\theta$.

From the Fig. 5.5, we have $\cos\theta = \dfrac{a}{\sqrt{a^2+x^2}}$

$$\therefore \qquad E = 2 \cdot \frac{1}{4\pi\varepsilon_0} \cdot \frac{q}{a^2 + x^2} \cdot \frac{a}{\sqrt{a^2 + x^2}} = \frac{2aq}{4\pi\varepsilon_0 (a^2 + x^2)^{3/2}}$$

or
$$E = \frac{\mu}{4\pi\varepsilon_0 (a^2 + x^2)^{3/2}} \qquad \text{...(5.21)}$$

If $x >> a$,
$$E = \frac{\mu}{2\pi\varepsilon_0 x^3} \qquad \text{...(5.22)}$$

- The resultant field **E** is parallel the axis of the dipole and is directed from q to $-q$.
- The electric field due to a dipole is inversely proportional to cube of the distance.

Example 5.4: *A very large sheet of charge has density of 5 μ C/m². Determine the electric field at a distance of 25 cm. Take medium as air.*

Solution: Now,
$$\phi = \frac{q}{4\pi r^2}$$

$$\therefore \qquad q = 4\pi\phi r^2 = 4\pi \times 5 \times 10^{-6} \times (0.25)^2 = 1.57 \times 10^{-5} \text{ C}$$

But
$$E = \frac{q}{4\pi\varepsilon_0 r^2} = \frac{1.57 \times 10^{-5}}{4\pi \times 8.854 \times 10^{-12} \times (0.25)^2} = \mathbf{2.258 \times 10^6 \text{ N/C}}$$

Example 5.5: *A charged sphere of 80 μC is placed in air. Find the electric field intensity at a point 20 cm from the centre of the sphere.*

Solution: Now,
$$E = \frac{q}{4\pi\varepsilon_0\varepsilon_r r^2} = 9 \times 10^9 \times \frac{80 \times 10^{-6}}{1 \times (0.2)^2} = \mathbf{1.8 \times 10^5 \text{ N/C}}$$

Example 5.6: *The charge per unit length on a long straight filament is – 70 μC/m. Find the electric field at a distance 30 cm from the filament.*

Solution: From Gauss' law, the electric field at a distance r from a long wire, is given by,

$$E = \frac{\lambda}{2\pi\varepsilon_0 r} = 2 \times \frac{1}{4\pi\varepsilon_0} \times \frac{\lambda}{r}$$

$$\therefore \qquad E = 2 \times (9 \times 10^9) \times \frac{-70 \times 10^{-6}}{0.3} = \mathbf{-4.2 \times 10^6 \text{ N/C.}}$$

5.7 ELECTROSTATIC POTENTIAL

When an electric charge is moved towards a like charge or away from an unlike charge, work is done against the electric forces by the external agency that moves the charge. As a result, the electrical charge acquires **potential energy**. If the charge is released, work is done by the field and the charge accelerates. It means that its potential energy is converted into kinetic energy. In mechanics work done on a particle is expressed in terms of changes in potential energy. In case of electric field also, the work done on a charge can be expressed in terms of the potential energy of the charge.

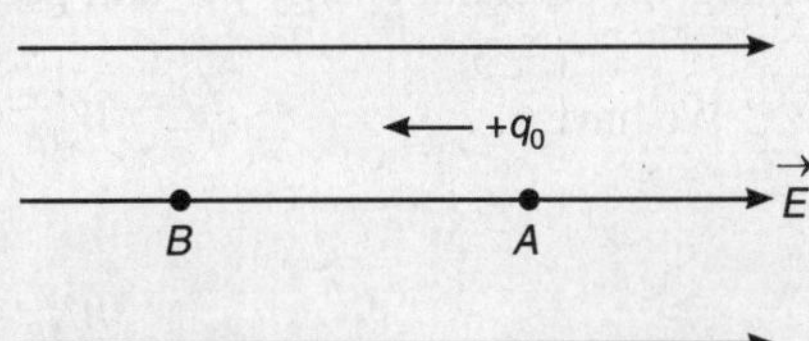

Fig. 5.6:
Accelerating charge in the field.

Work done by the electric field on a charge

Let us consider a positive point test charge q is placed at point A in a uniform electric field, **E**. The force on the charge due to electric field is $q\mathbf{E}$. If we wish to move the charge with constant velocity *against* the electric field from point A to point B, we must exert a force of qE. The work done by the external force is positive and is equal to

$$W_{ext} = F_{ext} x \cos 0°$$
$$= + F_{ext} x = + qEx.$$

The work done by the electric field on the charge is negative since force and displacement are in opposite direction. The work done by the field is

$$W_E = F_{ext} x \cos 180° = - F_{ext} x = - qEx.$$

By moving the charge from point A to point B, the external force increases the **electric potential energy**, U by an amount equal to the work done on the charge.

The change in potential energy of the charge is $\Delta U = U_B - U_A$

$$\therefore \quad \Delta U = U_B - U_A = qEx$$

where U_A and U_B are the potential energies of the charge at location A and location B respectively.

In terms of the work done *by the electric field*, the change in potential energy of the charge may be expressed as

$$\Delta U = U_B - U_A = -(-qEx) = -W_E \qquad ...(5.23)$$

Eq. (5.23) indicates that variation in the electric potential energy of the charge between the two positions B and A equals the negative of the work done by the electric field on that charge.

Line Integral

As the charge moves a distance dx along the path from A to B, the electric field does an element of work dW on it.

$$dW = -\mathbf{F} \cdot dx = -q\mathbf{E} \cdot dx$$

$\therefore$ The total work done by the field is

$$W_{AB} = -q \int_A^B \mathbf{E} \cdot dx \qquad ...(5.24)$$

The integral in the Eq. (5.24) is called the **line integral**.

Electric potential difference between any two points in an electric field is defined as the change in potential energy per unit positive test charge.

We have $$W_{AB} = -\int_A^B \mathbf{F} \cdot dx$$

$$\therefore \quad W_{AB} = -\frac{Qq}{4\pi\varepsilon_0} \int_{x_1}^{x_2} \frac{1}{x^2} dx$$

$$= -\frac{Qq}{4\pi\varepsilon_0} \left[\frac{1}{x_1} - \frac{1}{x_2}\right] \qquad ...(5.25)$$

$$= (U_B - U_A)$$

Therefore, the work done per unit charge is

$$\frac{W_{AB}}{q} = \frac{U_B}{q} - \frac{U_A}{q} = V_B - V_A \qquad \text{...(5.26)}$$

where $V_A = U_A /q$ is the potential energy per unit charge at location A and similarly $V_B = U_B /q$ is the potential energy per unit charge at location B. Then,

$$\Delta V = \frac{\Delta U}{q} = \frac{W}{q} \qquad \text{...(5.27)}$$

From now onwards, we denote $W_{AB} = W_E$ by W.

Electric potential

A charged particle positioned in an electric filed possesses potential energy. The potential energy U of a charge at any point is equal to the negative of the work done on the charge by the electric field as the charge moves from infinity to that point in the field. Usually the point A is chosen at infinity and hence $V_A = 0$. Then we denote V_B by V. Using Eq. (5.26), we define the electric potential at a point as the potential energy per unit charge placed at that point. Thus,

$$V = \frac{U}{q} \qquad \text{...(5.28)}$$

where U is the potential energy.

The electric potential is measured in unit of volts, which is equal to Joules/Coulomb or J/C.

$$1\,V = \frac{1\,\text{J}}{1\,\text{C}}$$

Electric potential is a scalar quantity and hence it is often referred to as **electrostatic scalar potential**.

We can write from Eq. (5.25) that

$$V = \frac{Q}{4\pi\varepsilon_0 r} \qquad \text{...(5.29)}$$

5.7.1 Calculating the Potential from the Field

Let a charge move from an initial point 1 to final point 2 in an electric field along the path shown. As the charge moves a distance along this path, the electric field does an element of work dW on it.

$$dW = -\mathbf{F} \cdot dx = -q\mathbf{E} \cdot dx$$

∴ The total work done by the field $W_{12} = -q \int_1^2 \mathbf{E} \cdot dx$ ...(5.30)

∴ The electric potential difference $V_2 - V_1 = -\int_1^2 \mathbf{E} \cdot dx$

If the initial point 1 is taken to be at infinity, $V_1 = 0$ and writing $V_2 = V$, the above equation yields $V = -\int_1^2 \mathbf{E} \cdot dx$

Fig. 5.7: An arbitrary path between two points.

In a more general way, we write the above expression as

$$V = -\int_1^2 \mathbf{E} \cdot dr \qquad ...(5.31)$$

5.7.2 Calculating the Field from the Potential

Let us consider a set of closely spaced equipotential surfaces perpendicular to the plane of the page and passing through it. Let the potential difference between each pair of adjacent surfaces be dV. Suppose a charge moves through a small distance dx from one equipotential surface to the adjacent equipotential surface. The work that the electric field does on the charge is given by Eq. (5.26) as

$$dW = q\, dV$$

We can also express the work done is

$$dW = -\mathbf{F} \cdot dx = -q\mathbf{E} \cdot dx = -qE \cos\theta\,(dx)$$

Equating these two equations, we obtain

$$q\, dV = -q\, E \cos\theta\,(dx)$$

or

$$E \cos\theta = -\frac{dV}{dx} \qquad ...(5.32)$$

As $E \cos\theta$ is the component of E in the x direction, we have to use partial derivative in the Eq. (5.32). Thus,

$$E_x = -\frac{\partial V}{\partial x} \qquad ...(5.33)$$

That is, the electric field is equal to the negative of the derivative of the electric potential with respect to some coordinate.

In general, the electric potential is a function of all three spatial coordinates. If V is given in terms of rectangular coordinates, and then

$$E_x = -\frac{\partial V}{\partial x},\; E_y = -\frac{\partial V}{\partial y} \text{ and } E_z = -\frac{\partial V}{\partial z} \qquad ...(5.34)$$

The electrostatic field may be expressed as the **gradient** of a scalar field. The intensity of an electrostatic field is expressed as the gradient of a scalar field called voltage. As the voltage is a number and not a vector, it is easier to handle a problem for voltage. The gradient of voltage gives a simpler way to calculate E.

$$E = -\text{grad } V = -\left[\frac{\partial V}{\partial x} i + \frac{\partial V}{\partial y} j + \frac{\partial V}{\partial z} k\right]$$

5.7.3 Potential Gradient

The rate of change of potential with distance is called the **potential gradient**. If the electric field is homogeneous and uniform, the potential gradient is given by

$$\textit{Potential gardient} = \frac{dV}{dx} \qquad ...(5.35)$$

where dV is the change in potential between two points separated by a distance x.

Example 5.7: *Two positive charges of 12×10^{-10} C and 8×10^{-10} C are placed 10 cm apart. Find the work done in bringing the charges 4 cm closer.*

Solution: The electrostatic force between the charges separated by a distance x, is given by

$$F = \frac{q_1 q_2}{4\pi\varepsilon_0 x^2}$$

$$\therefore \quad F = 9\times10^9 \times \frac{12\times10^{-10}\times8\times10^{-10}}{x^2} = \frac{8.64\times10^{-9}}{x^2}\text{ N}$$

If the charge is moved through a small distance dx, the work done dW, in bringing the charges closer by distance dx, will be,

$$dW = F \cdot dx$$

$$\therefore \quad dW = \frac{8.64\times10^{-9}}{x^2}\times dx$$

Now total work done in moving the charges from 0.10 m to 0.06 m apart will be,

$$W = -\int_{0.1}^{0.06} dW = -8.64\times10^{-9}\int_{0.1}^{0.06}\frac{dx}{x^2}$$

$$= -8.64\times10^{-9}\left[\frac{1}{x}\right]_{0.1}^{0.06}$$

$$= -8.64\times10^{-9}\left[\frac{1}{0.06}-\frac{1}{0.1}\right]$$

$$\therefore \quad \mathbf{W = 5.76\times10^{-8}\ J.}$$

5.8 EQUIPOTENTIAL SURFACES

An **equipotential surface** is a surface on which the potential has the same value at all points. In other words, the potential difference between any two points on an equipotential surface is zero.

Properties of equipotential surfaces

(*i*) *Work done in moving a charged particle over an equipotential surface is zero*: Since the potential energy of a charged particle is the same at all points of a given equipotential surface, work done in moving a charged particle over an equipotential surface is zero.

(*ii*) *Electric field is always perpendicular to an equipotential surface*: The equipotential surface through any point will be perpendicular to the direction of electric field at that point, as shown in Fig. 5.8.

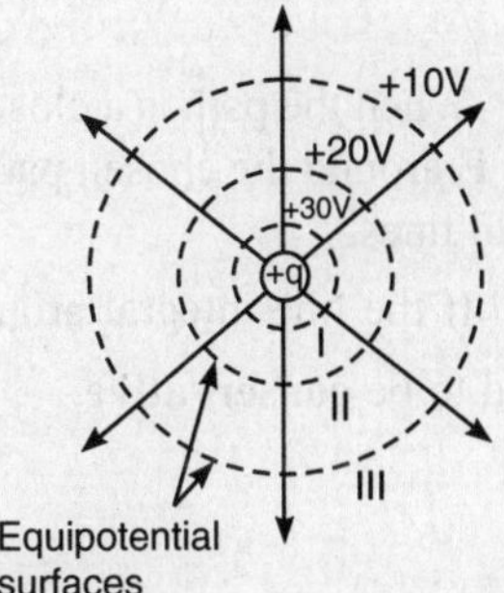

Fig. 5.8: Equipotential surfaces.

(*iii*) *The spacing between equipotential surfaces gives us indication of regions of strong and weak fields*: The region where the equipotential surfaces are crowded is the region of stronger field and the region where the surfaces are separated by larger distance is the region of weaker field.

(*iv*) *Equipotential surfaces never intersect each other*: If two equipotential surfaces could intersect, then at the point of intersection there would be two values of electric potential which is impossible.

5.9 ELECTRIC FIELD IS A CONSERVATIVE FIELD

A force is said to be **conservative**, if we can associate potential energy with it. If potential energy cannot be associated with a force, then the force is **nonconservative**. The gravitational force is conservative whereas frictional force is nonconservative.

We know that gravitational force does negative work on a body while the body is rising and an equal amount of positive work is done on its return trip. The total work done is zero for the round trip.

Definition: A field is conservative if the work done on a particle that moves through a round trip in the field is zero.

Or equivalently, a field is conservative if the work done by it on a particle between two points is the same for all paths connecting the two points.

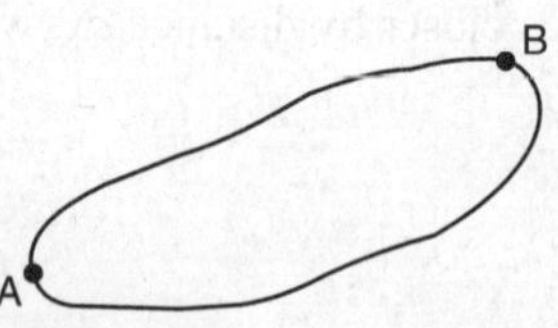

Fig. 5.9: Work done around a closed path.

Mathematically, a force field is said to be conservative if the line integral of the field along any closed path is zero. Therefore, an electric field is said to be conservative if

$$\oint \mathbf{E}\cdot \mathbf{dl} = 0 \qquad \text{...(5.35)}$$

Let a charge move from an initial point A to final point B in an electric field along the path shown in Fig. 5.9. As the charge moves a distance $d\mathbf{l}$ along this path, the electric field does an element of work dW on it.

$$dW = -\mathbf{F}\cdot d\mathbf{l} = -q\mathbf{E}\cdot d\mathbf{l}$$

∴ The total work done by the field in moving the charge from A to B is

$$W_{AB} = -q\int_A^B \mathbf{E}\cdot d\mathbf{l} \qquad \text{...(5.36)}$$

The integral in the Eq. (5.36) is called the **line integral**.

If the curve C forms a closed path, then the integral along the closed path is denoted by

$$\oint_C \mathbf{E}\cdot d\mathbf{l} \qquad \text{...(5.37)}$$

When the path is a closed curve, the line integral is referred to as "net circulation integral" for **E** around the chosen path. It is a measure of a vector field property called the curling up of field lines.

If the line integral around any closed path vanishes, that is $\oint_C \mathbf{E}\cdot d\mathbf{l} = 0$ then, the field is said to be **conservative**.

$$W_{AB} = -\frac{Qq}{4\pi\varepsilon_0}\int_{x_A}^{x_B}\frac{1}{x^2}dx = -\frac{Qq}{4\pi\varepsilon_0}\left[\frac{1}{x_A}-\frac{1}{x_B}\right]$$

It is seen from the above result that the work done depends only on the starting point x_A and the final point x_B and not on the path chosen to go from A to B. If now the charge returns from B to A through the same path or any other path the work done would be

$$W_{BA} = \frac{Qq}{4\pi\varepsilon_0}\left[\frac{1}{x_A}-\frac{1}{x_B}\right]$$

Therefore, the total work done on a point charge in the electric field over any closed path is

$$W_{\text{total}} = \oint dW = W_{AB} + W_{BA} = 0$$

$$\therefore \quad \oint_C \mathbf{E} \cdot dl = 0$$

5.10 POTENTIAL AT A POINT DUE TO A GROUP OF POINT CHARGES

The potential due to a number of charges at a point can be found by taking the algebraic sum of potentials due to all charges. If point charges $q_1, q_2, q_3, \ldots$ are at distances $r_1, r_2, r_3, \ldots$ from a point P, then the resultant potential at P is given by

$$V = \frac{q_1}{4\pi\varepsilon_0 r_1} + \frac{q_2}{4\pi\varepsilon_0 r_2} + \frac{q_3}{4\pi\varepsilon_0 r_3} + \ldots\ldots$$

$$= \frac{1}{4\pi\varepsilon_0}\left[\frac{q_1}{r_1} + \frac{q_2}{r_2} + \frac{q_3}{r_3} + \ldots\right]$$

or

$$V = \frac{1}{4\pi\varepsilon_0}\sum \frac{q_n}{r_n} \quad \ldots(5.38)$$

5.11 COMPUTATION OF ELECTRIC POTENTIAL IN SOME SPECIFIC CASES

(i) Potential due to a charged sphere

Let us consider a charged conducting sphere of radius R and carrying a charge of Q which is uniformly distributed throughout the sphere. The entire charge may be assumed to be concentrated at the centre of the sphere. Then, the charged sphere can be treated as identical to a point charge Q located at the centre of the sphere O.

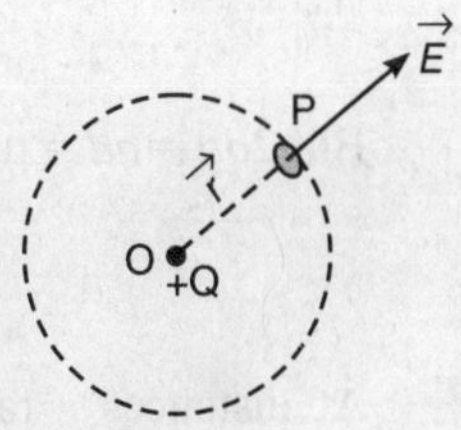

Fig. 5.10: Potential due to a charged sphere.

Case 1 : P lies outside the sphere: Let P be at a distance r from the centre of the sphere. As the entire charge is concentrated at O, the potential at P will be

$$V = \frac{Q}{4\pi\varepsilon_0 r}$$

Case 2 : P lies on the surface of the sphere: Let P be on the surface of the sphere. Therefore, $r = R$. Hence the potential at a point on the surface of sphere is

$$V = \frac{Q}{4\pi\varepsilon_0 R}$$

Case 3 : P lies inside the sphere: The potential due to a charged sphere at any point inside it is the same as on the surface of the sphere.

(ii) Potential due to an electric dipole

An electric dipole consists of two equal and opposite charges very close to each other. Let AB be an electric dipole of length $d = 2a$. Let P be a point where we would like to determine the potential due to the dipole. Let $OP = r$ and let θ be the angle between r and the dipole axis, AB (see Fig. 5.11).

From the Δ^{le} OAC, we get

$$\cos\theta = \frac{OC}{AC} = \frac{OC}{a}$$

$\therefore$ $$OC = a\cos\theta$$

Similarly, $$OD = a\cos\theta$$

If $r >> a$, we can write $$PA = PC = PO + OC$$
$$= r + a\cos\theta$$

and $$PB = PD = PO - OD$$
$$= r - a\cos\theta$$

Fig. 5.11: Potential due to an electric dipole.

The electric potential at the point P due to the electric dipole is

$$V = \frac{1}{4\pi\varepsilon_0}\left[\frac{q}{PB} - \frac{q}{PA}\right]$$

$$= \frac{q}{4\pi\varepsilon_0}\left[\frac{1}{r - a\cos\theta} - \frac{1}{r + a\cos\theta}\right]$$

$$= \frac{q}{4\pi\varepsilon_0}\left[\frac{2a\cos\theta}{r^2 - a^2\cos\theta}\right]$$

$$= \frac{1}{4\pi\varepsilon_0}\cdot\frac{2qa\cos\theta}{r^2 - a^2\cos\theta}$$

But $2qa = qd = \mu$, the electric dipole moment of the dipole.

$\therefore$ $$V = \frac{1}{4\pi\varepsilon_0}\cdot\frac{\mu\cos\theta}{r^2 - a^2\cos\theta} \qquad ...(5.39)$$

At distances r far larger than the values of a, $a^2\cos^2\theta/r^2 << 1$ and the term can be neglected. The Eq. (5.39) then reduces to

$$V = \frac{1}{4\pi\varepsilon_0}\cdot\frac{\mu\cos\theta}{r^2} \qquad ...(5.40)$$

(*a*) When the point lies on the axial line of the dipole on the side of the positive charge, then $\theta = 0$ and $\cos\theta = 1$. Therefore,

$$V = \frac{1}{4\pi\varepsilon_0}\cdot\frac{\mu}{r^2}$$

(*b*) If the point lies on the axial line of the dipole on the side of the negative charge, then $\theta = 180°$ and $\cos\theta = -1$. Therefore,

$$V = -\frac{1}{4\pi\varepsilon_0}\cdot\frac{\mu}{r^2}$$

(*c*) If the point lies on the equatorial line of the dipole, then $\theta = 90°$ and $\cos\theta = 0$. Therefore,

$$V = 0$$

Example 5.8: *A hollow sphere of radius 20 cm is charged with a charge of 30 × 10⁻⁹ C. Find the potential at a distance of 50 cm from the sphere centre.*

Solution: The potential at the surface of charged sphere of radius r is given by

$$V = \frac{q}{4\pi\varepsilon_0 r}.$$

So, the potential at a distance x will be

$$V = \frac{q}{4\pi\varepsilon_0 x} = 9\times10^9 \times \frac{30\times10^{-9}}{0.5} = \mathbf{0.539\ kV}.$$

Example 5.9: *The potential due to an isolated point charge at a point 20 cm from the charge is 400 volts. Calculate magnitude of charge.*

Solution: The potential at any point at a distance x from the point charge will be

$$V = \frac{q}{4\pi\varepsilon_0 x}$$

$$\therefore \quad 400 = 9\times10^9 \times \frac{q}{0.2}$$

$$\therefore \quad \mathbf{q = 8.9 \times 10^{-9} C}$$

Example 5.10: *Calculate electrostatic potential at a point due to charge of 50 μC at a distance of 15 cm from it.*

Solution:

$$V = \frac{q}{4\pi\varepsilon_0 r}$$

$$\therefore \quad V = 9\times10^9 \times \frac{50\times10^{-6}}{0.15} = \mathbf{3 \times 10^6\ V}$$

5.12 FLUX

Electric field is a vector field. The first important property that characterizes a vector field is *flux*. Michael Faraday made use of field lines for visualizing electric and magnetic fields. Gauss introduced the concept of flux to express the relation between a field and its source.

Fig. 5.12 gives the concept of flux and flux density. Electric flux is defined as the product of the magnitude of the electric field and surface area, A, perpendicular to the field.

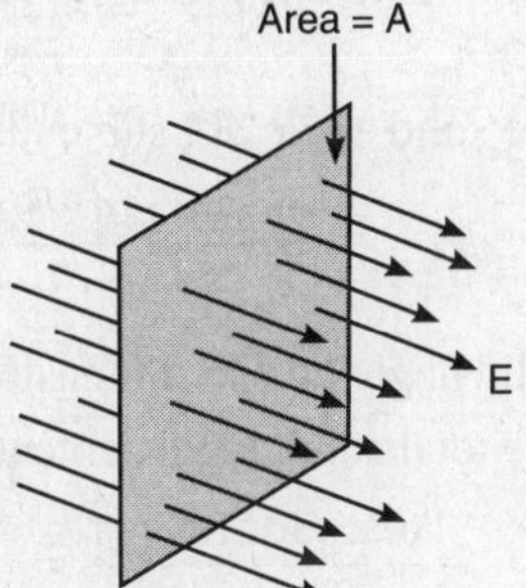

Fig. 5.12: Flux and flux density.

$$\Phi = EA$$

When the surface is not perpendicular to the field lines, then the component of E along the normal to the surface is to be multiplied by the area. Thus,

$$\Phi = (E \cos\theta)\, A$$

We may express the above relation as the scalar product of vectors **E** and **A**, as

$$\Phi = \mathbf{E} \cdot \mathbf{A} \qquad ...(5.41)$$

In more general situations, the surface is of arbitrary shape. The flux passing through any arbitrary shaped surface can be calculated by dividing the surface into a large number of small elements, each having an area dA. In the vector dA, the magnitude dA is the area of the element and the direction is given by the outward normal to the elemental surface.

The electric flux through this small element is

$$\Delta\Phi = \mathbf{E}_i \cdot \Delta\mathbf{A}_i$$

By adding the contributions of all elements, we obtain the total flux through the entire surface. Thus,

$$\Phi = \int_{\text{Surface}} \mathbf{E} \cdot d\mathbf{A}$$

Therefore, the flux of an electric field **E** through an *open surface* S is given by

$$\Phi = \int_S \mathbf{E} \cdot d\mathbf{A} \qquad ...(5.42)$$

If the surface is closed, we call it a closed surface integral and denote it by a circle on the integration symbol. Thus,

$$\Phi = \oint \mathbf{E} \cdot d\mathbf{A} \qquad ...(5.43)$$

5.13 SOLID ANGLE

The concept of solid angle is an extension of the concept of a two-dimensional angle that is measured in radians or degrees. Let us consider the small area element dS at a distance r from the point O. Let $\hat{\mathbf{n}}$ be the unit vector perpendicular to the surface element dS. When every point of the boundary of dS is joined to O, a cone is formed.

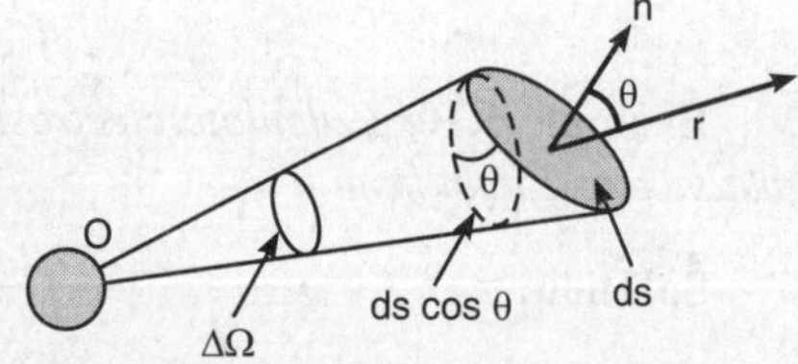

Fig. 5.13: Concept of solid angle.

The vector **OA** = **r** makes an angle θ with the unit vector $\hat{\mathbf{n}}$. The projection of the area dS on a plane normal to **r** is given by

$$\hat{\mathbf{r}} \cdot \hat{\mathbf{n}}\, dS = \hat{\mathbf{r}} \cdot dS \cos\theta$$

The solid angle $d\Omega$ subtended by the area element $d\mathbf{S}$ at a point O is defined as

$$d\Omega = \frac{\text{Projection of } d\mathbf{S} \text{ perpendicular to } \mathbf{r}}{r^2} = \frac{dS\cos\theta}{r^2} = \frac{\hat{\mathbf{r}} \cdot d\mathbf{S}}{r^2} \qquad ...(5.44)$$

Solid angle has no dimensions but is represented by the unit *steradian*.

The total solid angle subtended by the surface of a sphere at O is given by

$$\frac{\text{Total surface area}}{r^2} = \frac{4\pi r^2}{r^2} = 4\pi \qquad ...(5.45)$$

5.14 GAUSS' LAW OF ELECTROSTATICS IN FREE SPACE

Integral laws are called *global laws* since they indicate what happens over a wide range. Gauss law is one of such global laws. This law connects the electric flux through any closed surface to the net amount of charge within the surface.

Statement: Gauss law states that the total electric flux through a closed surface enclosing a charge is equal to $\frac{1}{\varepsilon_0}$ times the magnitude of the charge enclosed.

Proof: Let us consider a closed surface S surrounding a charge q, as shown in Fig. 5.14. Let $d\mathbf{S}$ be an element of area around the point P on the surface and $\hat{\mathbf{n}}$ an outward unit vector normal to it. Let θ be the angle between the electric field at P and the unit vector $\hat{\mathbf{n}}$. The electric flux through the element of area $d\mathbf{S}$ is

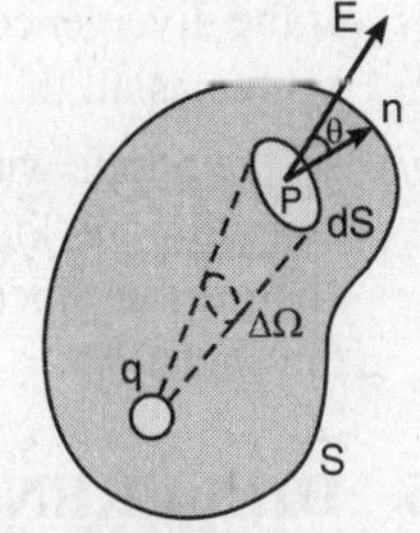

Fig. 5.14: Flux density at a point due to a charge.

$$d\phi = \mathbf{E} \cdot d\mathbf{S} = \text{E} \cos\theta \, d\text{S}$$

$$= \frac{1}{4\pi\varepsilon_0} \frac{q}{r^2} \cos\theta \, dS$$

where r is the distance of the surface element $d\mathbf{S}$ from charge q.

Now $\dfrac{d\mathbf{S}\cos\theta}{r^2} = d\Omega$ is the solid angle subtended by $d\mathbf{S}$ at q.

Therefore,

$$d\phi = \frac{1}{4\pi\varepsilon_0} q \, d\Omega$$

$\therefore$ Total flux through the entire surface S is

$$\phi = \int d\phi = \oint \mathbf{E} \cdot dS = \frac{1}{4\pi\varepsilon_0} q \oint d\Omega$$

The total solid angle subtended by S at O is 4π.

$$\therefore \qquad \phi = \oint \mathbf{E} \cdot dS = \frac{q}{\varepsilon_0} \qquad ...(5.46)$$

Gauss' law becomes very useful in calculation of electric field in cases where Coulomb's law or principle of superposition becomes tedious.

5.15 DIVERGENCE OF ELECTRIC FIELD

It is the limiting value of the ratio of the closed surface integral and the volume enclosed by the surface over which integration is carried out as the volume element reduces to zero.

$$\text{div}\,\mathbf{E} = \underset{\Delta V \to 0}{\text{Lt}} \frac{\oint_S \mathbf{E} \cdot dS}{\Delta V} \qquad ...(5.47)$$

It is common practice to denote div **E** as $\nabla \cdot \mathbf{E}$. The divergence in rectangular coordinates is found to be given by

$$\text{div}\,\mathbf{E} = \nabla \cdot \mathbf{E} = \frac{\partial E_x}{\partial x} + \frac{\partial E_y}{\partial y} + \frac{\partial E_z}{\partial z} \qquad ...(5.48)$$

$\nabla \cdot \mathbf{E}$ can be taken as a measure of the spreading out of the field **E**. If a vector function **E** spreads out from a point, then it has a *positive* divergence at that point and the point acts as a **source** of the field **E**. On the other hand, if the field converges to a point, then $\nabla \cdot \mathbf{E}$ would be negative at that point and the point acts as a **sink** for the field **E**. If the vector field neither converges nor diverges, then $\nabla \cdot \mathbf{E} = 0$. For a field at a point to have finite divergence means that an equivalent to a source must exist there. The divergence of a vector field can be considered to be a measure of scalar sources of the field.

(*i*) If the divergence of a field is zero for a small volume, the region of space is free of a source as all the flux that enters the volume also leaves it.

(*ii*) At the points where the divergence of **E** is positive, the sources of the field (positive charges) are located, and at the points where it is negative, the sinks (negative charges) are located. The field lines emerge from the field sources and terminate at the sinks.

5.16 DIFFERENTIAL FORM OF GAUSS'S LAW

The integral form of Gauss's law relates the net flux out of a finite volume to the net amount of charge enclosed in that volume. In contrast to (5.46), the differential form of the Gauss theorem establishes the relation between the volume charge density and the changes in the field intensity **E** in the vicinity of a given point in space. Thus, the differential law is a *local law*, which tells us what happens at a given point. The differential form of Gauss's law can be found by applying Gauss's integral law to an infinitesimally small volume surrounding a point. The integrals will then transform to differentials in the limit as the volume goes to zero, and we will obtain a point relationship of Gauss's law involving derivatives only. The differential form of Gauss's law is more general and will be very useful since derivatives are easier to calculate compared to integrals.

Let us represent the charge q in the volume V enclosed by a closed surface S as $q_{\text{int}} = \langle\rho\rangle V$, where $\langle\rho\rangle$ is the volume charge density, averaged over the volume V. Using this into Eq. (5.46) and dividing the equation with V, we obtain

$$\frac{1}{V}\oint E \cdot ds = \frac{\langle\rho\rangle}{\varepsilon_0} \qquad ...(5.49)$$

We now make the volume V to tend to zero by contracting it to the point we are interested in. Then, $\langle\rho\rangle$ will tend to the value of ρ at the given point of the field and hence the ratio on the R.H.S. of Eq. (5.49) will tend to ρ/ε_0. When V tends to zero, the quantity on the L.H.S. is called the **divergence of the field E**. By definition

$$\nabla \cdot \mathbf{E} = \underset{V \to 0}{\text{Lt}} \frac{1}{V}\oint_S \mathbf{E} \cdot ds$$

Consequently, the above relation transforms into

$$\nabla \cdot \mathbf{E} = \frac{\rho}{\varepsilon_0} \qquad ...(5.50)$$

Eq. (5.50) is the differential form of the Gauss theorem relating the divergence of the field and the electric charge density at a point.

5.17 DERIVATION OF COULOMB'S LAW FROM GAUSS LAW

Coulomb's law can be deduced from Gauss law. Let us consider an isolated point charge q, as shown in Fig. 5.15. Let us consider any imaginary spherical surface r centered on the charge. Such an imaginary closed surface enclosing a charge is called a *Gaussian surface*. The advantage of the spherical surface is that **E** is normal to it at all points and has the same magnitude.

In Fig. 5.15 both **E** and d**S** are directed radially outward at any point on the surface. The angle between them is zero.

$$\therefore \qquad \mathbf{E} \cdot d\mathbf{S} = \mathbf{E}d\mathbf{S}$$

Gauss law therefore reduces to

$$\varepsilon_0 \oint \mathbf{E} \cdot d\mathbf{S} = \varepsilon_0 \oint \mathbf{E}\, d\mathbf{S} = q$$

Since **E** is constant for all points on the sphere, it can be taken out of the integral.

$$\therefore \quad \varepsilon_0 \mathbf{E} \oint d\mathbf{S} = q$$

$$\oint d\mathbf{S} = \text{Area of the sphere, } 4\pi r^2$$

$$\therefore \quad \varepsilon_0 \mathbf{E}(4\pi r^2) = q$$

or $$\mathbf{E} = \frac{1}{4\pi\varepsilon_0} \frac{q}{r^2}$$

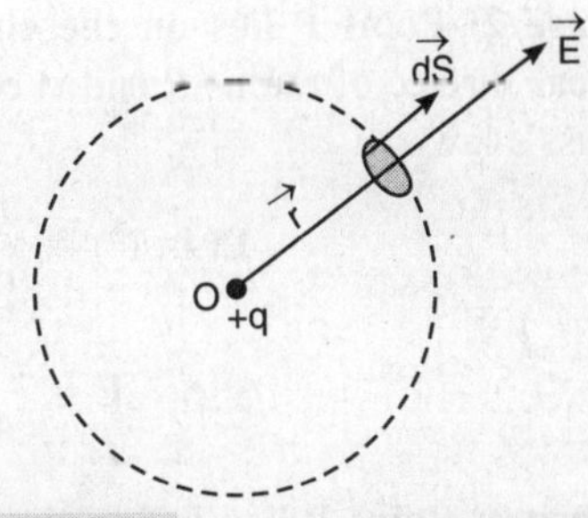

Fig. 5.15: Gaussian surface about a point charge.

The above equation gives the magnitude of the electric field strength **E** at any point at a distance r from the isolated charge q. If a second charge q_1 is kept at any point on the spherical surface, it experiences a force

$$\mathbf{F} = q_1 \mathbf{E}$$

Using the expression for **E** into the above equation, we obtain

$$F = \frac{1}{4\pi\varepsilon_0} \frac{qq_1}{r^2} \qquad ...(5.51)$$

which is the Coulomb's law.

5.18 APPLICATIONS OF GAUSS'S LAW

(i) Electric field due to a solid charged sphere

Let us consider an isolated sphere of charge Q having radius R. Let us consider a point P at a distance r from the centre O of the sphere.

Case 1: Point P lies outside the charged sphere: Let us draw the Gaussian surface through point P so that it encloses the sphere of charge. In this the Gaussian surface is a spherical surface of radius r and is centered on O, as shown in Fig. 5.16.

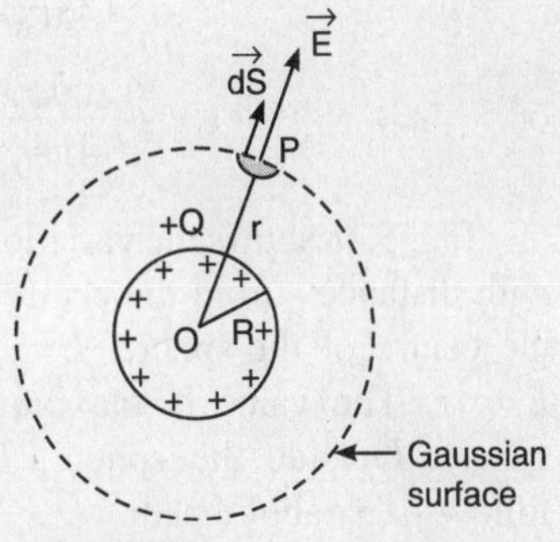

Fig. 5.16: Gaussian surface outside the charged sphere.

Let E be electric field at point P due to the sphere of charge Q. The field is spherically symmetrical. Therefore, E is along the normal to the spherical surface and has the same magnitude at all points on the surface of the sphere.

Total flux through the Gaussian surface is

$$\oint \mathbf{E} \cdot dS = \oint \mathbf{E}\, dS = E \oint dS = E(4\pi r^2)$$

According to Gauss's theorem,

$$\mathbf{E}(4\pi r^2) = \frac{Q}{\varepsilon_0}$$

or $$\mathbf{E} = \frac{1}{4\pi\varepsilon_0} \frac{Q}{r^2} \qquad ...(5.52)$$

Case 2: Point P lies on the charged sphere: In this case, the Gaussian surface is a spherical surface of radius R and is centered on O. The area of the sphere is $4\pi R^2$. According to Gauss's law

$$\mathbf{E}(4\pi R^2) = \frac{Q}{\varepsilon_0}$$

or

$$\mathbf{E} = \frac{1}{4\pi\varepsilon_0}\frac{Q}{R^2} \quad ...(5.53)$$

Case 3: Point P lies inside the charged sphere: In this case, the Gaussian surface is a spherical surface of radius $OP = r$ and is centered on O, as shown in Fig. 5.17. If Q' is the charge enclosed by the Gaussian surface, then according to Gauss's theorem

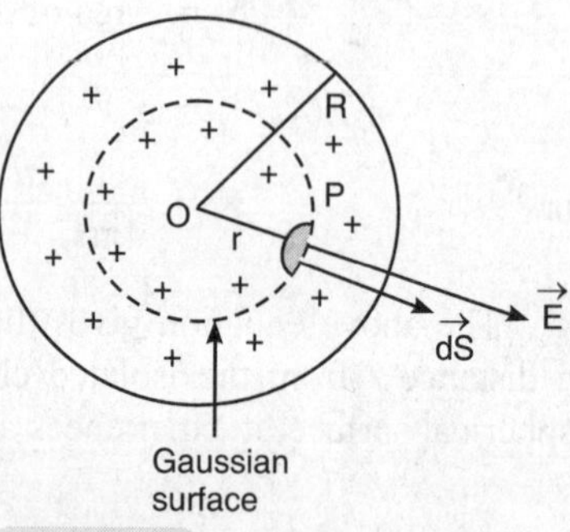

Fig. 5.17: Gaussian surface inside the charged sphere.

$$\mathbf{E}(4\pi r^2) = \frac{Q'}{\varepsilon_0} \quad (r < R)$$

or

$$\mathbf{E} = \frac{1}{4\pi\varepsilon_0}\frac{Q'}{r^2}$$

Now,

$$Q' = \frac{Q}{\frac{4}{3}\pi R^3} \times \frac{4}{3}\pi r^3 = Q\frac{r^3}{R^3}$$

$$\therefore \quad \mathbf{E} = \frac{1}{4\pi\varepsilon_0}\frac{1}{r^2} \times Q\frac{r^3}{R^3}$$

or

$$\mathbf{E} = \frac{1}{4\pi\varepsilon_0}\frac{Qr}{R^3} \quad ...(5.54)$$

Fig. 5.18 shows the variation of electric field intensity with distance r from the centre of a sphere of charge. At the centre of the sphere, $E = 0$ and within the sphere, $E \propto r$. The value is maximum at the surface of the sphere. Outside the sphere, $E \propto 1/r^2$ and the curve follows a parabolic path.

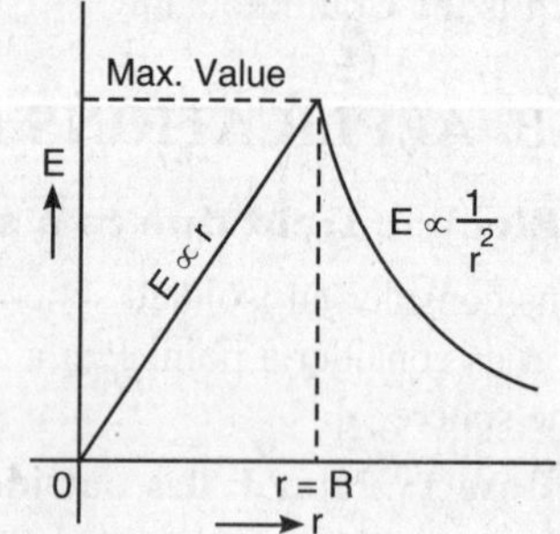

Fig. 5.18: Variation of electric field intensity with distance.

(ii) Electric field due to a uniformly charged spherical shell

Let us consider a thin spherical shell of radius R. Let O be the centre of the shell and q be the charge on the shell. The electric field due to the charged spherical shell is radial and spherically symmetric. Let us consider a point P at a distance r from the centre O of the shell.

Case 1: Point P lies outside the spherical shell: Let us draw the Gaussian surface through point P. The Gaussian surface is a spherical surface of radius r and is centered on O, as shown in Fig. 5.19.

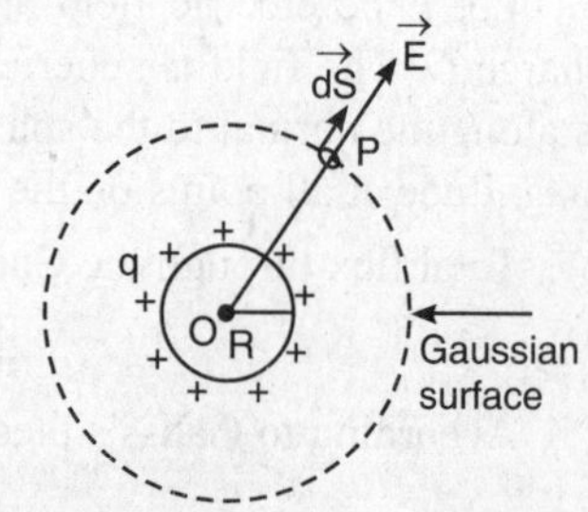

Fig. 5.19: Gaussian surface outside the spherical shell.

Let E be electric field at point P due to the sphere of charge q. The field is spherically symmetrical. Therefore, E is along the normal to the spherical surface and has the same magnitude at all points on the surface of the sphere.

Total flux through the Gaussian surface is

$$\oint \mathbf{E} \cdot dS = \oint \mathbf{E}\, dS = \mathbf{E} \oint dS = \mathbf{E}(4\pi r^2)$$

According to Gauss's theorem,

$$\mathbf{E}(4\pi r^2) = \frac{q}{\varepsilon_0} \quad (r > R)$$

or

$$\mathbf{E} = \frac{1}{4\pi\varepsilon_0}\frac{q}{r^2} \qquad \text{...(5.55)}$$

Case 2: Point P lies on the surface of spherical shell: In this case, the Gaussian surface through P just encloses the spherical shell of radius R and is centered on O. The area of the sphere is $4\pi R^2$. According to Gauss's law

$$\mathbf{E}(4\pi R^2) = \frac{q}{\varepsilon_0}$$

or

$$\mathbf{E} = \frac{1}{4\pi\varepsilon_0}\frac{q}{R^2} \qquad \text{...(5.56)}$$

Case 3: Point P lies inside the charged sphere: In this case, the Gaussian surface through the point does not enclose any charge (Fig. 5.20).

$$\mathbf{E}(4\pi r^2) = \frac{0}{\varepsilon_0} = 0$$

$\therefore$

$$\mathbf{E} = 0 \qquad (5.57)$$

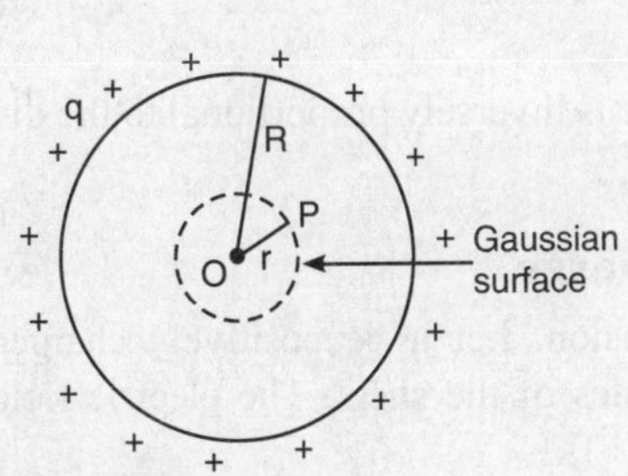

Fig. 5.20: Gaussian surface around a spherical shell.

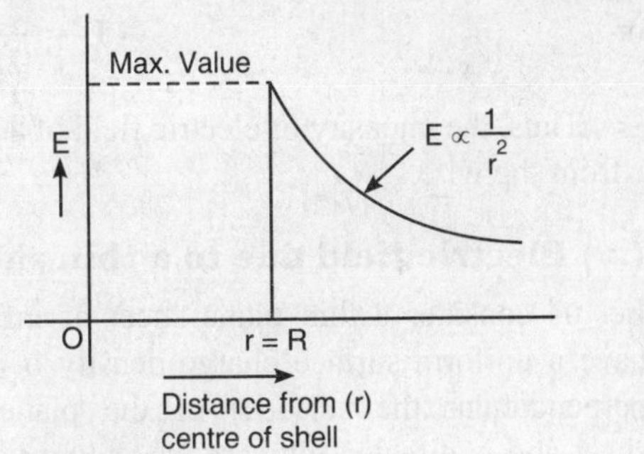

Fig. 5.21: Electric field versus distance in a shell.

Fig. 5.20 shows the variation of electric field intensity with distance from the centre O for a uniformly charged spherical shell. For r varying from 0 to R, $E = 0$. In other words, inside the shell filed is zero. The magnitude of E is maximum at the surface of the shell, i.e., at $r = R$. However, outside the shell $E \propto 1/r^2$ and the curve follows a parabolic path.

(iii) Electric field due to a line charge

Let us consider an infinitely long wire or a thin rod having a uniform linear positive charge density λ (Fig. 5.21). Let P be a point at a distance r from the wire.

Imagine a cylindrical Gaussian surface of radius r and length l around the wire. The electric flux comes from the surface, S of the cylinder and the electric flux from two ends of the cylinder is zero.

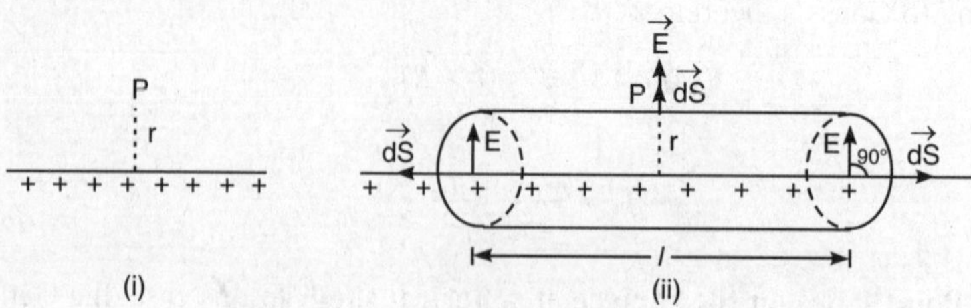

Fig. 5.22: Electric field due to a thin charged rod.

Total flux through the Gaussian surface is

$$\oint \mathbf{E} \cdot d\mathbf{S} = \oint \mathbf{E}\, dS = \mathbf{E} \oint dS = E(2\pi r l)$$

The charge enclosed inside the Gaussian surface is $= \lambda \cdot l$

According to Gauss's law

$$\mathbf{E} \oint dS = \frac{q}{\varepsilon_0}$$

$$\therefore \quad \mathbf{E}(2\pi r l) = \frac{\lambda l}{\varepsilon_0}$$

or

$$\mathbf{E} = \frac{\lambda}{2\pi\varepsilon_0 r} \qquad \text{...(5.58)}$$

Thus, the intensity of electric field of a line charge is inversely proportional to the distance r from the wire.

(iv) Electric field due to a thin sheet of charge

Let us consider a thin plane sheet of infinite extension. Let it be positively charged and have a uniform surface charge density σ on both sides of the sheet. The electric field acts perpendicular the surfaces of the plane sheet and is directed outward, as shown in Fig. 5.23.

In order to find the electric field strength due to the sheet of charge at any point P at a distance r from it, we draw a cylindrical Gaussian surface, as shown in Fig. 5.23. Since field lines are parallel to the curved surface of the cylinder, the contribution to the net electric flux from the curved surface is zero. The contribution to

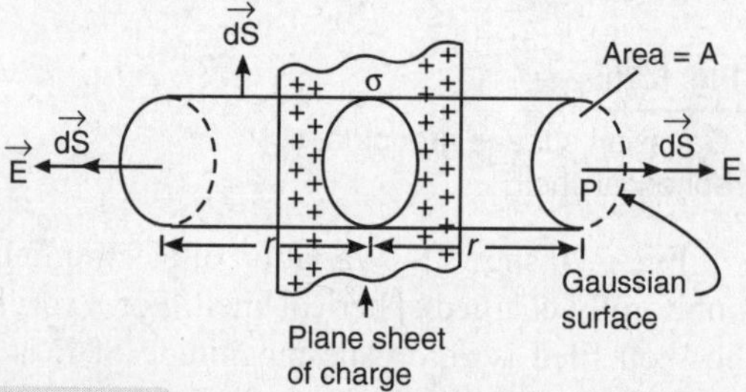

Fig. 5.23: Electric field due to a thin charged sheat.

the flux comes from the two circular faces of the cylinder. The electric flux crossing through the Gaussian surface is

$$\oint \mathbf{E} \times d\mathbf{S} = \oint_{S_1} \mathbf{E} \cdot d\mathbf{S} + \oint_{S_2} \mathbf{E} \cdot d\mathbf{S} = 2E \oint d\mathbf{S} = 2ES$$

The charge enclosed inside the Gaussian surface is $= \sigma S$

According to Gauss's law

$$\mathbf{E} \oint d\mathbf{S} = \frac{q}{\varepsilon_0}$$

$$\therefore \qquad 2\,\mathbf{E}S = \frac{\sigma S}{\varepsilon_0}$$

or

$$\mathbf{E} = \frac{\sigma}{2\varepsilon_0} \qquad \text{...(5.59)}$$

The above result indicates that the magnitude of the electric field due to a thin charged sheet is independent of the distance from the sheet.

Example 5.11: *A large planar sheet of charge has a charge per unit area of 7.5 μ C/m². Find electric field intensity just above the surface of sheet, measured from its midpoint.*

Solution: From Gauss' law, the electric field intensity due to a non conducting planar sheet with uniform charge per unit area σ, is given by,

$$E = \frac{\sigma}{2\varepsilon_0} \qquad \therefore \quad E = \frac{7.5\times10^{-6}}{2\times8.85\times10^{-12}} = \mathbf{4.24 \times 10^5\ N/C}$$

5.19 GAUSS' LAW OF ELECTROSTATICS IN A (DIELECTRIC) MEDIUM

Dielectrics do not conduct electric current as they do not have *free* electrons. But, when an external electric field is applied to the dielectric (say, held between capacitor plates, as in Fig. 5.24), the positive nuclei in each molecule are displaced along the field direction and the *bound* electrons are displaced in the opposite direction. The bound electrons are not mobile but are elastically bound to the molecule.

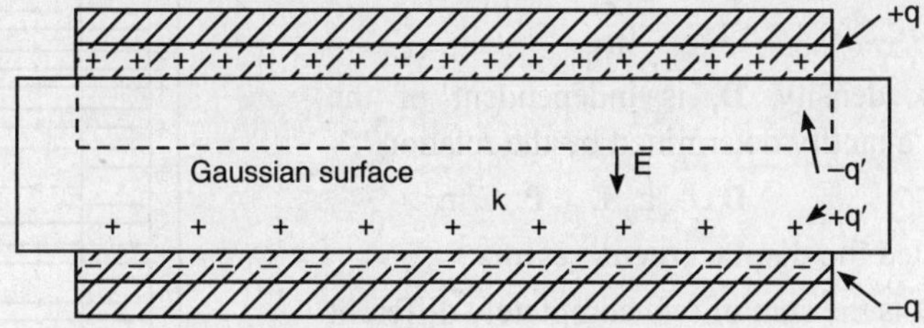

Fig. 5.24: Dielectric slab held between capacitor plates.

Therefore, they can move only within the electrically neutral molecules through a very small distance. This displacement of charges is known as **polarization** of the dielectric. As a result of polarization, uncompensated charges appear on the dielectric surface as well as in its bulk. Such charges are called **polarization charges** or **bound charges**. On the other hand, the charges that are on the plates of the capacitor are called **extraneous charges** or **free charges**. They do not belong to dielectric molecules.

The electric field **E** acting in a dielectric

$$\mathbf{E} = \mathbf{E_0} + \mathbf{E'}$$

where

$\mathbf{E_0}$ is the field due to extraneous charges

$\mathbf{E'}$ is the field due to bound charges.

As the sources of an electric field E are all electric charges-free and bound, we can express the Gauss theorem (5.46) for the field **E** as

$$\oint \varepsilon_0 \mathbf{E}\, ds = (q + q')_{\text{int}} \qquad ...(5.60)$$

where q and q' are free and bound charges respectively, enclosed by the surface S.

We now express q' in terms of P, the polarization vector, as

$$\oint \mathbf{P} ds = -q'_{\text{int}} \qquad ...(5.61)$$

Using Eq. (5.61) into Eq. (5.60), we get

$$\oint (\varepsilon_0 \mathbf{E} + \mathbf{P}) ds = q_{\text{int}} \qquad ...(5.62)$$

Let us denote the quantity in the parenthesis in the integrand of the Eq. (5.62) by **D**. We thus define an auxiliary vector **D** as

$$\mathbf{D} = \varepsilon_0\, \mathbf{E} + \mathbf{P} \qquad ...(5.63)$$

whose flux through an arbitrary enclosed surface is equal to the algebraic sum of extraneous charges enclosed by this surface.

$$\oint \mathbf{D} ds = -q_{\text{int}} \qquad ...(5.64)$$

This is the *Gauss theorem for field* **D**. The quantity D is called **dielectric displacement.**

5.20 ELECTRIC DISPLACEMENT VECTOR

Electric displacement or electric flux density D measures the influence of a charge distribution at any point in space in terms of electric field intensity, **E**. The electric field **E** depends upon the *permittivity* ε of the medium in which the charge is placed.

For example, the electric field from a point charge Q is given by

$$\mathbf{E} = \frac{Q}{4\pi\varepsilon\, r}$$

Electric flux density, **D**, is independent of the medium and is a quantity determined by the relation

$$\mathbf{D} = \varepsilon_0\, \mathbf{E} + \mathbf{P} \text{ C/m}^2$$

D is also called the **displacement vector.**

The vector **D** is the sum of two completely different quantities $\varepsilon_0\mathbf{E}$ and **P**. For this reason, **D** is regarded an *auxiliary vector* which does not have any deep physical meaning. In isotropic dielectrics vector **D** is collinear to **E** and in case of anisotropic dielectrics these vectors are generally non-collinear. Its magnitude is given by the product of the electric intensity and the dielectric constant of the medium. Its magnitude is also equal to the surface density of free charges: $D = \sigma$.

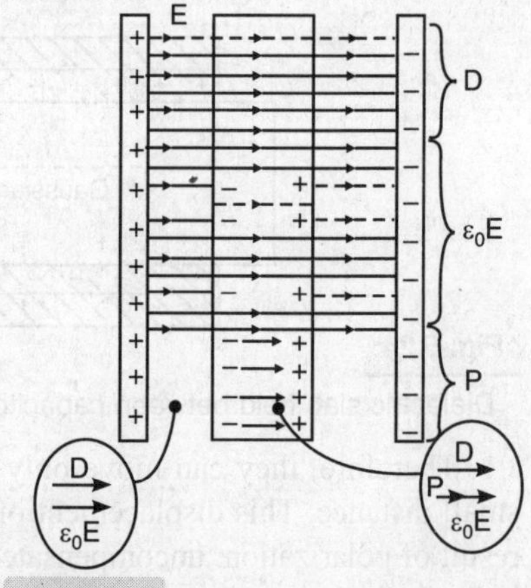

Fig. 5.25: Electric displacement vector in a dielectric.

The field **D** can be depicted with the aid of electric displacement lines. The sources of the field **D** are all charges. The displacement lines can start or stop on both *extraneous* and *bound* charges. The lines of **D** pass without interruption through the regions of the field containing bound charges.

5.21 DIFFERENTIATION OF SCALARS AND VECTORS

If $V(x, y, z)$ is a scalar function of position in space, then $\frac{\partial V}{\partial x}$ denotes the rate of change of V with respect to x when y and z remain constant, is called partial differentiation.

Similarly, $\frac{\partial V}{\partial y}$ and $\frac{\partial V}{\partial z}$ denote the rate of change of V with respect to y and z respectively.

The differentiation of a scalar function $V(x, y, z)$ is given as $dV = \frac{\partial V}{\partial x}dx + \frac{\partial V}{\partial y}dy + \frac{\partial V}{\partial z}dz$.

The differentiation of a scalar function appears in the form of gradient and that of vector function as divergence and curl.

(*i*) Gradient of a scalar field: Gradient is an operation executed on a scalar function. This operation gives a vector function representing both the magnitude and direction of maximum space rate of increase of scalar function.

A mathematical expression for the gradient can be obtained by evaluating the difference in the field between two points as

$$Grad\ V = \frac{\partial V}{\partial n}\hat{\mathbf{r}} = \nabla V \qquad \text{...(5.65)}$$

As gradient usually depends on position, so it is actually a vector field. The gradient of a scalar field is a vector whose magnitude at any point is equal to the maximum rate of increase of scalar field and directed along the direction in which this maximum change occurs.

Physical significance: The gradient of a scalar field is a vector whose magnitude at any point is equal to the maximum rate of increase of scalar field and directed along the direction in which this maximum change occurs. Gradient represent the combination of rate of change in x, y, z at a given point in three or higher dimensional curve.

For example, if a physical function describes the temperature at a point then the gradient gives the direction in which the temperature is increasing most rapidly.

(*ii*) Divergence of a vector field: Divergence operates on a vector function gives a scalar function. To illustrate the divergence concept mathematically, let us consider a closed surface S enclosing a volume V in region which has a vector function A.

It gives the net outward vector function of per unit volume at a point as the volume reduces to zero.

Since the surface integral gives a scalar, the divergence of a vector is scalar.

$$\text{div}\ \mathbf{A} = \nabla \cdot \mathbf{A} = \frac{\partial A_x}{\partial x} + \frac{\partial A_y}{\partial y} + \frac{\partial A_z}{\partial z} = \underset{\Delta V \to 0}{\text{Lt}} \frac{\oint_S \mathbf{A} \cdot d\mathbf{S}}{\Delta V} \qquad \text{...(5.66)}$$

Physical significance: The concept of divergence is valid at a point. The divergence of a vector function **A** at each point gives the rate per unit volume at which a physical quantity is emanating from that point. The divergence of a vector field is positive at source point in the field and negative at a sink point or zero where there is neither sink nor source.

For example, if the vector **A** at a particular point say *X* represents the flow of the heat and divergence is positive at that point *X*, then there must be a source of heat and heat must be leaving the point *X* so that the temperature at the point *X* is decreasing.

(iii) **Curl of a vector field:** Curl is defined as the line integral of a vector field **A** around a closed path enclosing an area ΔS

$$Curl = |\Delta \times \mathbf{A}| = \left(\lim_{\Delta S \to o} \frac{\oint_{\Delta C} \mathbf{A} \cdot dl}{\Delta S} \right) \hat{\mathbf{r}} \qquad \text{...(5.67)}$$

where $\hat{\mathbf{r}}$ is unit vector normal to the area.

The curl of a vector field is a rotational vector. Its magnitude is the maximum circulation of per unit area as the area reduces to zero and the direction is the normal direction of area.

Physical significance: Curl of a vector is associated with a point in space. Curl means circulation. Consider a pond, float a small peddle wheel and start to rotate it in the pond. The whirlpool created would be a region of large curl. The curl of a vector function gives the measure of the angular velocity at every point of the vector field. If the curl of a vector $|\nabla \times \mathbf{A}| = 0$, then the vector field is called as irrotational or non curl field.

QUESTIONS

1. State Coulomb's law in electrostatics. What do you mean by the flux of an electric field?
2. Define and explain electric intensity. What are its units? Deduce an expression for electric intensity due to a point charge.
3. State and prove Gauss's law in electrostatics.
4. State Gauss's theorem. Making use of it determine the intensity near a long charged conductor.
5. Deduce the Gauss's law in differential form $\nabla.\mathbf{E} = \rho/\varepsilon_0$.
6. Derive Coulomb's law from Gauss's law.
7. Determine the intensity of electric field due to a dipole at an equatorial and axial point.
8. Determine the intensity of electric field near a charged sphere and a plane sheet of charge.
9. Define and explain electric potential. What are its units?
10. Obtain an expression for the potential due to a uniformly charged sphere at an external and internal point.
11. Derive an expression for electric field intensity due to a uniformly charge spherical shell when point lies (*i*) outside the shell and (*ii*) inside the shell.
12. Does electric potential obey superposition principle? Explain.
13. What do you mean by line integral of electric field?
14. Show that the line integral of electric field over a closed path is zero.
15. Electrostatic force is a conservative force. Discuss.
16. What is an equipotential surface? Mention important properties of equipotential surfaces.

17. Apply Gauss's theorem to a dielectric medium and obtain the relationship between **E**, **D**, and **P** vectors.
18. Define electric displacement and explain the significance of electric displacement vector.

PROBLEMS

1. Two spheres charged with equal but opposite charges experience a force of 2.5×10^5 N when they are placed at 2 cm apart in a medium of relative permittivity 5. Determine the charge on each sphere. **[Ans:** $q = 745 \times 10^{-6}$ C]
2. A charge of 5×10^{-5} C is distributed between two small spheres which are placed with their centers 3 m apart. It is found that the spheres repel each other with a force of 0.6 N. Find the charges on two spheres. **[Ans:** $q_1 = 3 \times 10^{-5}$ C, $q_2 = 2 \times 10^{-5}$ C]
3. A charge of 0.52 μC is placed in a electric field of 4.5×10^5 N/C. What is the magnitude of the force acting on the charge? **[Ans:** F = 0.23 N]
4. Calculate the electric field intensity due to an electric dipole of length 10 cm and consisting of two charges ± 2 μC at a distance of 50 cm from each charge. **[Ans:** $E = 1.44 \times 10^4$ N/C]
5. Two positive point charges of 16×10^{-10} C and 12×10^{-10} C are placed 10 cm apart. Find the work done in bringing the two charges 4 cm closer. **[Ans:** $W = 11.52 \times 10^{-8}$ J]
6. What is the strength of electric field produced by the nucleus of a hydrogen atom at the site of electron in 1s orbit? The radius of 1s orbit is 0.53 Å. **[Ans:** $E = 5.12 \times 10^{11}$ N/C]
7. Two charges ± 1μC are placed at the corners of the base of an equilateral triangle. The length of the side of triangle is 0.5 m. Find electric field intensity at apex of the triangle. **[Ans:** E = 36 kN/C]
8. An inflated balloon in the shape of a sphere of radius 14 cm has a total charge of 8 μC uniformly distributed on its surface. Calculate the electric field intensity at 50 cm and at 13 cm from the centre of the balloon. **[Ans:** $E_{50} = 2.88 \times 10^5$ N/C, $E_{13} = 0$]
9. At what distance from a point charge of 6 μC would the potential equal to 2.7×10^4 V? **[Ans:** *r* = 2 m]
10. Find the work done in bringing a charge of 10×10^{-4} μC from infinity to a point 25 cm from charge of 3×10^{-2} μC. **[Ans:** $W = 10.79 \times 10^{-7}$ J]
11. Determine the electric field intensity and potential in air at a distance of 3 cm from a charge 6.5×10^{-2} C. **[Ans:** $E = 5 \times 10^6$ N/C, $V = 1.5 \times 10^4$ V]
12. Two parallel plates are separated by a dielectric of 4 cm thickness and relative permittivity 2.5. If the potential difference between the plates is 3000 volts. Calculate:
    (*a*) Potential gradient
    (*b*) Electric flux density **[Ans:** $V = 7.5 \times 10^5$ V, $D = 1.66 \times 10^{-5}$ C/m^2]
13. A charge of 3×10^{-9} C moves through a distance of 0.5 m in a uniform field of 200 N/C. determine,
    (*a*) The electric force on the charge
    (*b*) Work done on the charge
    (*c*) Potential difference between initial and final positions
    [Ans: $F = 600 \times 10^{-9}$ N, $W = 300 \times 10^{-9}$ J, V = 100 V]
14. Calculate the electric potential at the surface of the nucleus of the gold atom. Given – atomic number of gold = 79, charge on proton = 1.6×10^{-19} C, radius of nucleus = 6.6×10^{-15} Å. **[Ans:** $V = 1.72 \times 10^5$ V]

6 Magnetostatics and Electrodynamics

Learning Objectives:

By the end of this chapter, you will be able to:

- Learn about concepts of Magnetic Field.
- State Biot-Savart Law, Ampere Law and Gauss Law.
- Understand concepts of Magnetic Scalar and Vector Potential.
- State Faraday's Law of Equation.

6.1 INTRODUCTION

Static magnetic fields are produced by permanent magnets and steady currents flowing in conductors. Magneto-statics deals with magnetic fields produced by steady currents. Faraday discovered the method of generation of electric and magnetic fields which vary with time. Electro-dynamics deals with the study of time varying electric and magnetic fields. Maxwell unified the important laws of electricity and magnetism and formulated a unified theory in 1861. He formulated four equations that are regarded as the basis of all electric and magnetic phenomena. The consequences of Maxwell's equations are very far reaching. Maxwell predicted the existence of electromagnetic waves and that light is a form of electromagnetic radiation.

6.2 MAGNETIC FIELD

In 1820 Oersted discovered that electric currents create magnetic fields. A steady current I flowing in a straight conductor produces a magnetic field around it. The magnetic lines of force exist in the form of a series of concentric circles with conductor as the centre. The direction of the field can be found by the right hand rule.

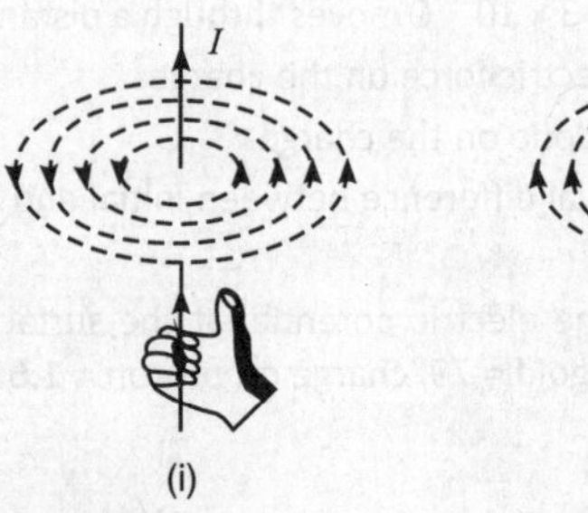

Fig. 6.1: Electric current creates a magnetic field.

Fleming Right Hand Rule: If the current carrying conductor is gripped with the right hand so that the thumb points the direction of current flow then the curled fingers shows the magnetic field direction.

Fig. 3.1 shows the method of finding the direction of magnetic field.

The region around a current carrying conductor or a permanent magnet where magnetic effects are experienced is called a *magnetic field.* A magnetic field is schematically represented by magnetic lines of force, which are also known as *field lines* or lines of *magnetic induction.* A magnetic field is described either by magnetic field strength **H** or by the *magnetic induction* (or *magnetic flux density*), **B**.

6.3 MAGNETIC FLUX DENSITY

Magnetic flux: The lines of induction are collectively called flux. The magnetic flux through a region is the number of lines of force passing normally through the region. The concentration of lines of induction is an indication of magnetic field strength. It is defined in terms of the flux density.

Magnetic induction: It is the ratio of the number of field lines (magnetic flux) per a unit area of cross-section. It is also known as magnetic flux density, and indicated by **B**.

Thus,

$$\mathbf{B} = \frac{\text{Magnetic flux}}{\text{area}} = \frac{\varphi}{A} \quad \text{...(6.1)}$$

Therefore, magnetic flux is given by $\varphi = BA$. In a more general way, let the area be inclined at an angle to the magnetic field. Let θ be the angle between the normal to the area and the direction of magnetic field. Then,

$$\varphi = \text{B A} \cos\theta = \mathbf{B} \cdot \mathbf{A} \quad \text{...(6.2)}$$

Thus, magnetic flux through an area is equal to the dot product of magnetic field **B** and area **A**.

The unit of magnetic flux is weber (Wb) and the unit of magnetic induction is weber per square metre (Wb/m^2) or tesla (T).

The magnetic flux through any surface may also be given by the surface integral of the normal component of **B**. Thus,

$$\varphi = \int_S \mathbf{B} \cdot d\mathbf{s} \quad \text{...(6.3)}$$

where $d\mathbf{S}$ is the elemental surface.

6.4 BIOT-SAVART LAW

Let a conductor of an arbitrary shape carry a steady current *I*. Let *P* be a point in the magnetic field produced by the current. Let a small element *AB* of length *dl* produce magnetic field *dB* at *P*. Let *r* be the distance of *P* from the current element *I dl* and θ be the angle between *dl* and *r*.

Statement of Biot-Savart law: The magnitude of magnetic field *dB* is directly proportional to the product *I dl* sin θ and is inversely proportional to the square of distance between current element and the point *P*.

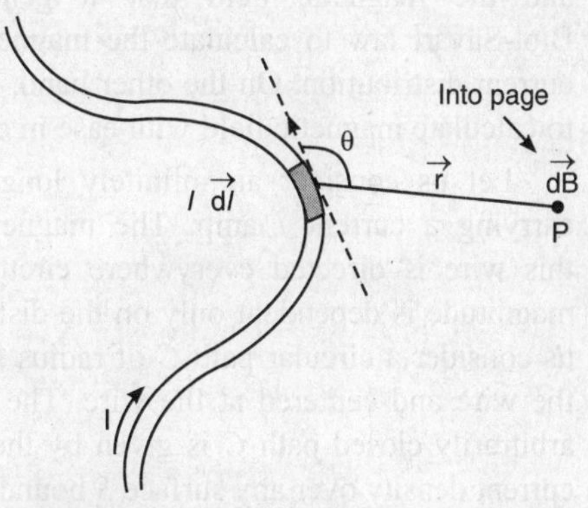

Fig. 6.2: Definition of Biot-Savart law.

Thus,

$$dB \propto I\,dl \sin\theta \quad \text{and} \quad dB \propto \frac{1}{r^2}$$

Combining these relations, $dB \propto \dfrac{I\,dl \sin\theta}{r^2}$

or

$$dB = k\frac{I\,dl \sin\theta}{r^2}$$

where k is a constant proportionality.

The value of k depends on the medium in which the conductor is situated and the system of units adopted. In SI units, its value for free space is

$$k = \frac{\mu_0}{4\pi} \quad \text{where } \mu_0 = 4\pi \times 10^{-7}\ \text{Tm/A}$$

$$\therefore \quad d\mathbf{B} = \frac{\mu_0}{4\pi}\frac{I\,dl \sin\theta}{r^2} \quad \text{Biot-Savart law} \qquad ...(6.4)$$

The Biot-Savart law is valid for steady currents only. Eq. (6.4) in the vector form

$$d\mathbf{B} = \frac{\mu_0}{4\pi}\frac{I(d\mathbf{l} \times \mathbf{r})}{r^3} \qquad ...(6.5)$$

As illustrated in Fig 6.1, the right hand thumb rule gives the direction of the magnetic field. The direction of $d\mathbf{B}$ is into the plane of the paper.

Summation of all current elements gives the total magnetic field at P.

$$\therefore \quad \mathbf{B} = \int d\mathbf{B} = \int \frac{\mu_0}{4\pi}\frac{I(d\mathbf{l} \times \mathbf{r})}{r^3} \qquad ...(6.6)$$

6.5 AMPERE'S LAW

Ampere's law statement: The magneto motive force around a closed path is equal to current enclosed by the path. The line integral of the tangential component of the magnetic field over any closed path is equivalent to the amount of the current within the loop.

$$\oint \mathbf{B} \cdot dl = \mu_0 I \qquad ...(6.7)$$

Both Ampere's law and the Biot-Savart law are relations between a current distribution and the magnetic field that it generates. We can apply Biot-Savart law to calculate the magnetic field caused by any current distribution. On the other hand, Ampere's law allows us to calculate magnetic field with ease in case of symmetry.

Let us consider an infinitely long wire along the z-axis carrying a current I amp. The magnetic flux density due to this wire is directed everywhere circular to the wire and its magnitude is dependent only on the distance from the wire. Let us consider a circular path C of radius r in the plane normal to the wire and centered at the wire. The current enclosed by an arbitrarily closed path C is given by the surface integral of the current density over any surface S bounded by the closed path C.

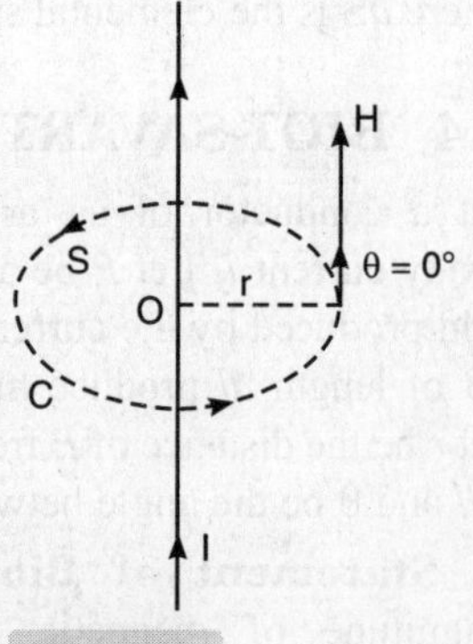

Fig. 6.3: Ampere's law.

The total current flowing through the surface area S is given by

$$I = \int \mathbf{J} \cdot ds \qquad ...(6.8)$$

Therefore, we can write

$$\oint \mathbf{H} \cdot dl = \int \mathbf{J} \cdot ds$$

$$\therefore \qquad \oint_C \mathbf{B} \cdot dl = \mu_0 \int_S \mathbf{J} \cdot ds \qquad ...(6.9)$$

This is known as *Ampere's circuital law*.

6.5.1 Ampere's Circuital Law in Differential Form

If we now shrink the path C to a very small size ΔC so that the surface area bounded by it becomes very small, ΔS, we can write Eq. (6.6) as

$$\oint_{\Delta C} \mathbf{B} \cdot dl = \mu_0 \int_{\Delta S} \mathbf{J} \cdot ds \qquad ...(6.10)$$

Since the surface area ΔS is very small we can consider the current density to be uniform over the surface so that

$$\int_{\Delta S} \mathbf{J} \cdot ds \approx \mathbf{J} \cdot \Delta S$$

The relation becomes exact in the limit $\Delta S \to 0$. Dividing both the sides of Eq. (6.10) by ΔS and letting $\Delta S \to 0$, we have

$$\lim_{\Delta S \to 0} \frac{\oint_{\Delta C} \mathbf{B} \cdot dl}{\Delta S} = \lim_{\Delta S \to 0} \frac{\mu_0 \int_{\Delta S} \mathbf{J} \cdot ds}{\Delta S}$$

$$= \mu_0 \lim_{\Delta S \to 0} \frac{\mathbf{J} \cdot \Delta \mathbf{S}}{\Delta S} \qquad ...(6.11)$$

$$= \mu_0 \, \mathbf{J} \cdot \mathbf{n}$$

Now, the curl of **B** is defined as the vector having the magnitude given by the maximum value of the quantity on the left side of Eq. (6.11). We note that this maximum value occurs for an orientation of ΔS for which the direction of its normal coincides with the direction of **J** and it is equal to μ_0 times the magnitude of **J**. Thus

$$|\nabla \times \mathbf{B}| = \text{maximum value of} \left(\lim_{\Delta S \to 0} \frac{\oint_{\Delta C} \mathbf{B} \cdot dl}{\Delta \mathbf{S}} \right) = \mu_0 \, |\mathbf{J}| \qquad ...(6.12)$$

or

$$\nabla \times \mathbf{B} = \mu_0 \, \mathbf{J} \qquad ...(6.13)$$

Eq. (6.13) is Ampere's circuital law in *differential form*.

6.6 GAUSS'S LAW FOR MAGNETISM

Just as in the case of electrostatics, the magnetic flux through an element of area d**s** is given by the dot product of **B** with *d***s**. For an arbitrary surface *ds* bounded by a closed contour *S*, total magnetic flux passing through the surface is given by

$$\phi = \oint_S \mathbf{B} \cdot ds \qquad ...(6.14)$$

As the lines of magnetic flux density are continuous loops, they have no beginning or ending. Whatever the flux enters a volume leaves the volume enclosed by a closed surface S. So, the flux of B through any closed surface is equal to zero.

Thus,

$$\oint_S \mathbf{B} \cdot ds = 0$$

Dividing both sides of the above equation by an incremental volume $\Delta \upsilon$ over which the surface is to be considered closed, we get

$$\frac{\oint \mathbf{B} \cdot ds}{\Delta \upsilon} = 0$$

The limit of the left side of the equation, as $\Delta \upsilon \to 0$, is the divergence of the vector **B**.

$$\therefore \qquad \nabla \cdot \mathbf{B} = 0 \qquad ...(6.15)$$

6.7 MAGNETIC SCALAR POTENTIAL

In electrostatics, the potential V is a **scalar**. It is related to the magnetic field **E** as

$$\mathbf{E} = -\nabla V$$

The scalar potential is related to the sources, i.e. charge distribution and can be easily calculated. In a similar way, we may define magnetic scalar potential by the following relation.

$$\mathbf{B} = -\mu_0 \nabla \phi$$

where ϕ is the **magnetic scalar potential** due to the sources.

According to Ampere's law

$$\oint \mathbf{B} \cdot dl = \mu_0 i \qquad ...(6.16)$$

In general the right hand side of the above equation is not zero and **B** is a **solenoidal** (not a conservative) **field**. In regions outside the sources of **B** (see Fig. 6.4), a closed path is not linked with a current and therefore, the line integral of **B** over such a closed path vanishes. In such cases, we may consider **B** to be a *conservative field* and therefore, express it as

$$\mathbf{B} = -\mu_0 \nabla \phi \qquad ...(6.17)$$

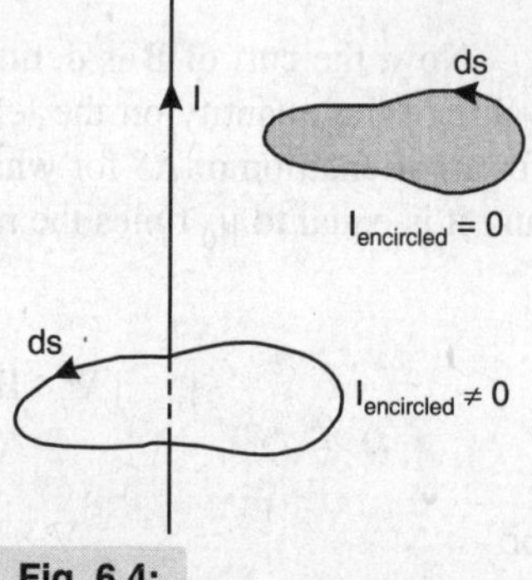

Fig. 6.4: Current through conductor and magnetic field produced.

This does not mean that **B** changes its character from a solenoidal field to a conservative field. All that we say is that in the regions outside the sources, we can simplify the mathematical formulation by expressing **B** in terms of the gradient of ϕ. Let us obtain now an expression for ϕ.

Sum of the potentials caused by the individual small loops gives the magnetic scalar potential at P due to a current loop. A current loop corresponds to a magnetic dipole. The magnetic moment of the small current loop is given by

$$dm = i\, dA$$

where dA is the area of the small loop.

The potential due to the dipole is given by an expression similar to derived in case of an electric dipole. Thus,

The scalar potential $$d\phi = \frac{1}{4\pi}\frac{i\,dA\cos\theta}{r^2}$$

where r is the distance of the elemental loop dA from P and θ is the angle between r and the vector $d\mathbf{A}$. But $dA\cos\theta = r^2\,d\Omega$, where $d\Omega$ is the solid angle subtended at P by the boundary of the given current loop.

$$\therefore \qquad \phi = \int d\phi = \frac{i}{4\pi}\int d\Omega = \frac{i\Omega}{4\pi} \qquad ...(6.18)$$

6.8 MAGNETIC VECTOR POTENTIAL

The magnetic field is a **solenoidal** field.

Two space derivatives are possible with a vector field. They are divergence and curl. We know from Eq. (6.15) that

$$\nabla \cdot \mathbf{B} = 0$$

Since the divergence of a curl is always zero, the second possibility is of expressing **B** as the curl of some vector potential function. Thus, we write

$$\mathbf{B} = \nabla \times \mathbf{A} \qquad ...(6.19)$$

A is called the **magnetic vector potential.** The potentials are a convenient way to relate a field to its scalar and vector sources. The vector sources for magnetic induction **B** are obtained by taking the curl of **B**. Thus,

$$\nabla \times \mathbf{B} = \nabla \times \nabla \times \mathbf{A} = \mu_0\,\mathbf{J} \qquad ...(6.20)$$

where the relationship between **B** and the current density **J** is given by Ampere's law. Unlike the case of the scalar potential, it is not apparent how the use of **A** will simplify the calculation of **B**.

According to vector identity

$$\nabla \times \nabla \times \mathbf{A} = \nabla(\nabla \cdot \mathbf{A}) - \nabla^2\,\mathbf{A}$$

$$\therefore \qquad \nabla(\nabla \cdot \mathbf{A}) - \nabla^2\,\mathbf{A} = \mu_0\,\mathbf{J} \qquad ...(6.21)$$

The specification of the curl of a vector does not uniquely specify the vector. It requires that the curl of vector has the same value regardless of the value of its divergence. The equations of the magnetostatic field, $\nabla \cdot \mathbf{B} = 0$ and $\nabla \times \mathbf{B} = \mu_0\,\mathbf{J}$ do not require that $\nabla \cdot \mathbf{A}$ be specified in any particular way and we can choose it at our convenience. Therefore, we choose that

$$\nabla \cdot \mathbf{A} = 0 \qquad ...(6.22)$$

This is known as a **gauge condition**, and the above Eq. (6.22) is known as the **Coulomb gauge**. Other choices of gauge (i.e. other choices of $\nabla \cdot A$) are useful in other circumstances.

Setting $\nabla \cdot \mathbf{A} = 0$ in Eq. (6.21), we get

$$\nabla^2\,\mathbf{A} = -\mu_0\,\mathbf{J} \qquad ...(6.23)$$

Thus, the **magnetic vector potential A** is defined here through the equations

$$\nabla \times \mathbf{A} = \mathbf{B}$$

and $$\nabla \cdot \mathbf{A} = 0$$

We now obtain an expression for **A**. Let d**A** be the magnetic potential due to a current element $I\,d\mathbf{l}$. By analogy with electrostatic potential, the magnetic potential may be written as

$$d\mathbf{A} \propto \frac{I\,d\mathbf{l}}{r}$$

$$d\mathbf{A} = k\frac{I\,d\mathbf{l}}{r} = \frac{\mu_0}{4\pi}\frac{I\,d\mathbf{l}}{r} \qquad ...(6.24)$$

$$\therefore \qquad \mathbf{A} = \frac{\mu_0 I}{4\pi}\int\frac{d\mathbf{l}}{r} \qquad ...(6.25)$$

For a current flowing in a circuit C, we have

$$\mathbf{A} = \frac{\mu_0 I}{4\pi}\oint_C\frac{d\mathbf{l}}{r} \qquad ...(6.26)$$

6.9 ELECTRODYNAMICS-TIME CHANGING FIELDS

Till now we have discussed the fundamentals of electrostatics and steady (d.c.) magnetic fields and derived the equations for field vectors **E, D, H** and **B**. We have observed that in static fields, the field vector pairs **E, D** are not related to field pair **H** and **B**.

James Clerk Maxwell brought together various laws of electrostatics and magnetostatics or steady magnetic fields.

While correlating various laws, Maxwell found that the result derived from Ampere's law was not consistent in time-changing fields as it was based on stationary closed circuit.

Hence in order to make it consistent, Maxwell introduced a quantity known as displacement current. This was experimentally verified but could not be explained. Thus, Maxwell gave four equations by which electromagnetic fields can be described.

The Maxwell's equations are based on fundamental laws of Faraday, Ampere and Gauss together with his concept of displacement current. These equations describe inherent properties of field vectors and the relation between them.

The significance of Maxwell's equations was the forecast of existence of electromagnetic disturbances starting off by a charged body would travel out as an electromagnetic wave with velocity of light in free space.

6.10 FARADAY'S LAWS OF INDUCTION

Michael Faraday's observations (1831) gave phenomenon known as **electromagnetic induction**. Faraday's observations on induced current in a conducting loop producing **induced emf** are

(*a*) whenever a magnet is moved toward or away from the loop

(*b*) when the magnet is held stationary and the loop is moved either toward or away from the magnet

(*c*) when both the loop and the magnet are stationary, current is not induced in the loop

The above three observations can be summarized as ***Faraday's laws of electromagnetic induction 'current is induced in the loop as long as relative motion occurs between the magnet and the loop'***.

(*i*) **First Law:** Whenever the magnetic flux linked with a circuit changes, an emf is always induced in it.

(*ii*) **Second Law:** The magnitude of the induced emf is equal to the time rate of change of the flux linkage.

Thus,

$$\mathbf{E} = -\frac{d\phi}{dt} \qquad ...(6.27)$$

where ϕ is the magnetic flux through the surface bounded by the loop.

If a coil consisting of N identical and concentric loops is used and if the field lines pass through all loops, the induced emf is

$$\mathbf{E} = -N\frac{d\phi}{dt} \qquad ...(6.28)$$

Faraday laws state that an emf is induced if the magnetic flux changes for any reason. Eq. (6.28) shows that an electric current is induced by a time-varying magnetic field.

6.10.1 Lenz's Law

An induced emf drives the current around a circuit from positive terminal to the negative terminal as in the case of conventional current.

However, the identification of positive and negative terminals in this case is not obvious. The polarity of the induced emf and direction of induced current can be determined with the help of Lenz's law. We have to note that the net magnetic field penetrating a coil results from two contributions. (*i*) The original magnetic field that produces the changing flux and hence produces induced emf and (*ii*) the **induced magnetic field** created by the induced current.

In 1834, Lenz gave the rule for determining the direction of induced current in a closed conducting loop. It is known as Lenz's law.

Lenz's law statement: *An induced current in a closed conducting loop is in a direction opposite to the original flux change.*

The Lenz's law is reflected mathematically in the minus sign on the R.H.S. of Faraday's law Eq. (6.27). Note that Lenz's law applies to induced currents and not to induced emfs.

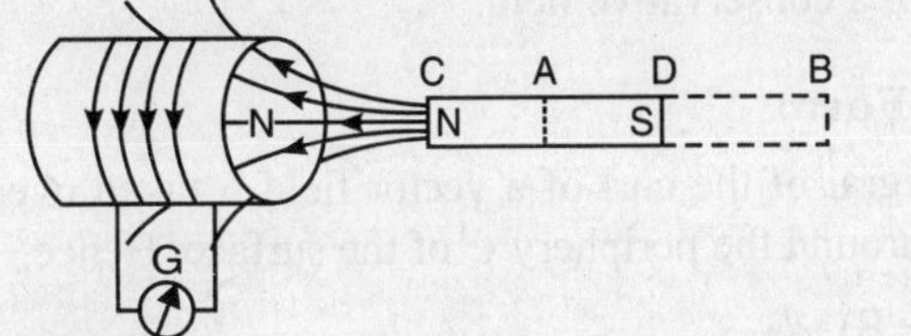

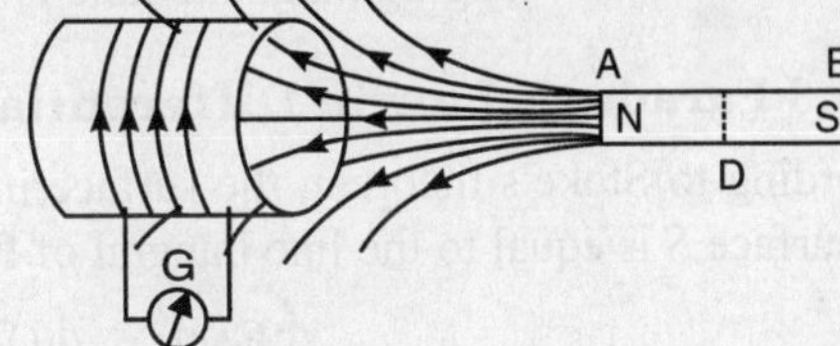

Fig. 6.5: Lenz's law.

Let us consider a coil that is approaching a magnet. As the N-pole of the magnet moves towards the coil, the magnetic flux linking the coil increases and current is induced in opposite direction of the magnetic flux in the coil. It can happen only when the right hand face of the coil becomes N-pole. The direction of the induced current can be known by the right-hand rule when we know the magnetic polarity of the coil face.

When the magnet moves away from the coil, the flux through the coil decreases. It results in an induced current flowing in a direction such that the coil's magnetic field opposes the motion. It means that the right hand face of the coil becomes S-pole. Thus, the induced current in the loop is always in such a direction to oppose the change that produces it. Lenz's law is a consequence of the law of conservation of energy.

6.10.2 Integral Form of Faraday's Law

Let us consider a closed conducting circuit (see Fig. 6.6) through which a magnetic flux of uniform flux density **B** exists. The conducting circuit is the contour of a surface, an element *ds* of which a direction specified by the unit normal, **n**. If the magnetic flux linking the circuit is decreasing, an emf **E** will be induced having the direction as shown by the arrow in the Fig. 6.6. However, an emf implies the existence of an electric field. There is thus an induced electric field set up in the circuit, which is given by

$$\mathbf{E} = \oint E_i \cdot dl \quad ...(6.29)$$

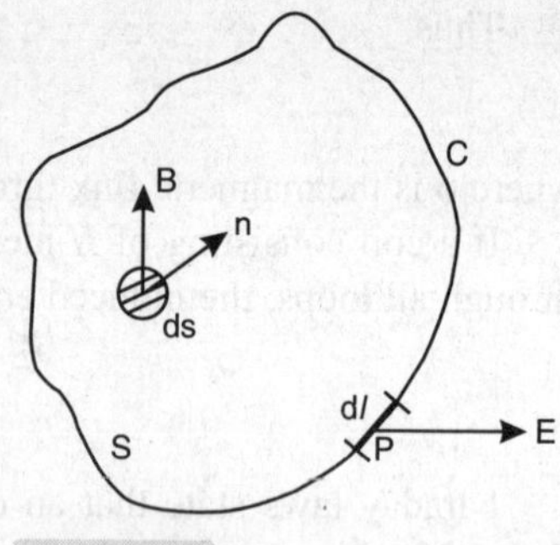

Fig. 6.6: Closed conducting circuit.

the integration being performed around the path in the direction of ε.

Combining Eq. (6.27) and Eq. (6.29), we get

$$\mathbf{E} = \oint E_i \cdot d\mathbf{l} = -\frac{d\phi}{dt}$$

where
$$\phi = \oint_S \mathbf{B} \cdot ds$$

Therefore
$$\frac{d\phi}{dt} = \frac{d}{dt}\int_S \mathbf{B} \cdot ds$$

$$\therefore \quad \oint E_i \cdot d\mathbf{l} = -\frac{d}{dt}\oint_S \mathbf{B} \cdot ds \quad ...(6.30)$$

The induced electric field exists in space irrespective of the presence of a conducting wire or not. If a conducting wire is present an induced current will flow. Eq. (6.30) shows that the electric field by a changing magnetic field is *not* a conservative field.

6.10.3 Faraday's Law in Differential Form

According to Stoke's theorem, the surface integral of the curl of a vector field **F** taken over any surface *S* is equal to the line integral of **F** around the periphery *C* of the surface. Hence,

$$\oint_C \mathbf{E} \cdot d\mathbf{l} = \oint_S (\nabla \times \mathbf{E}) \cdot ds$$

It follows from Eq. (6.30) that

$$\oint_S (\nabla \times \mathbf{E}) \cdot ds = -\frac{d}{dt}\int_S \mathbf{B} \cdot ds$$

If the surface *S* is stationary, i.e. independent of time, then

$$\frac{d}{dt}\int_S \mathbf{B} \cdot ds = \int_S \frac{\partial \mathbf{B}}{\partial t} \cdot ds$$

$$\therefore \quad \int_S (\nabla \times \mathbf{E}) \cdot ds = \int_S -\frac{\partial \mathbf{B}}{\partial t} \cdot ds$$

Comparing the integrals on the two sides, we have

$$\nabla \times \mathbf{E} = -\frac{\partial \mathbf{B}}{\partial t} \quad ...(6.31)$$

6.11 EQUATION OF CONTINUITY

The equation of continuity is based on the Principle of conservation of charge.

Principle of conservation of charge: Electric charge can neither be created nor destroyed and the net charge in an isolated system remains constant. *The time rate of change of charge within a closed volume equals the net rate of flow of charge into (out of) the volume.*

Let us consider a surface S enclosing a volume. Let $d\mathbf{S}$ be a small element of this surface. Further, let $\mathbf{J}$ be the current density at a point on the surface element. Then the current leaving the volume V bounded by the surface dS is given by

$$I = \oint_S \mathbf{J} \cdot d\mathbf{S}$$

Suppose the current is not a steady current. Then, J is a function of t as well as x, y, z and $\oint_S \mathbf{J} \cdot d\mathbf{S}$ represents the instantaneous rate at which charge is leaving the enclosed volume.

Using divergence theorem, we can write the above equation as

$$I = \oint_S \mathbf{J} \cdot d\mathbf{S} = \int_V \nabla \cdot \mathbf{J}\, d\upsilon \qquad ...(6.32)$$

As some charge is leaving the volume, correspondingly the same amount of charge diminishes with in that volume. We express this fact as

But
$$I = -\frac{dq}{dt} = -\frac{d}{dt}\int_V \rho\, d\upsilon$$

where ρ is the charge density.

Since volume V is a fixed volume, the time derivative operates only on the function ρ, when it is moved inside the integral. Further, the time derivative is a partial derivative since ρ is not only a function of time but also a function of position.

$$\therefore \qquad I = -\int_V \frac{\partial \rho}{\partial t}\, d\upsilon \qquad ...(6.33)$$

Equating Eq. (6.32) and Eq. (6.33), we obtain

$$\int_V \nabla \cdot \mathbf{J}\, d\upsilon = -\int_V \frac{\partial \rho}{\partial t}\, d\upsilon$$

or
$$\int_V \left[\nabla \cdot J + \frac{\partial \rho}{\partial t}\right] d\upsilon = 0$$

Since the volume is completely arbitrary, the integrand in the above equation must be zero. Thus,

$$\nabla \cdot \mathbf{J} + \frac{\partial \rho}{\partial t} = 0 \qquad ...(6.34)$$

Eq. (6.34) is known as the *equation of continuity*. It follows from Eq. (6.34) that

$$\nabla \cdot \mathbf{J} = -\frac{\partial \rho}{\partial t} \qquad ...(6.35)$$

It expresses the principle of conservation of charge. Charge cannot flow away from a given volume without diminishing the amount of charge existing within the volume.

6.12 DISPLACEMENT CURRENT

Ampere's law implies that a magnetic field can be produced only by a flow of charges. Ampere's law was established as a result of large number of careful experiments done on steady situations. Maxwell showed that we run into difficulty when apply the Ampere's law for time-varying situations such as charge building up on the plates of a capacitor. He showed that we have to include another current, called *displacement current*, which also can produce a time-varying magnetic field. The need for displacement current can be well understood when current flow through a capacitor is considered.

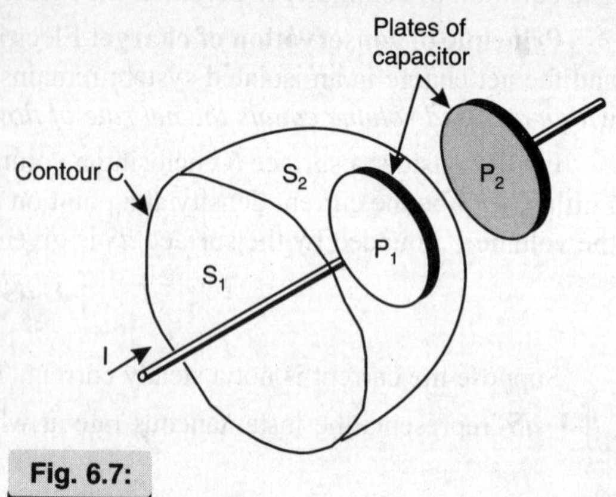

Fig. 6.7: Parallel plate capacitor and displacement current.

Let us consider the circuit shown in Fig. 6.7, which consists of a parallel plate capacitor being charged (or discharged) through a certain external resistance R. If we apply Ampere's law to the contour C and the surface S_1, we find that

$$\oint_C \mathbf{H}\cdot dl = \int_{S_1} \mathbf{J}\cdot ds = I \qquad \text{...(6.36)}$$

since the current is passing right through surface S_1.

If on the other hand, Ampere's law is applied to the contour C and the surface S_2, then J is zero at all points on S_2. As current is not flowing through the surface S_2,

$$\oint_C \mathbf{H}\cdot dl = \int_{S_2} \mathbf{J}\cdot ds = 0 \qquad \text{...(6.37)}$$

Note that the two surfaces S_1 and S_2 are bounded by the same path length l and hence the contour integrals must be equal. It is easy to see that the Eq. (6.36) and (6.37) contradict each other. Further, Eq. (6.37) cannot be wrong. Therefore, it appears that Ampere's Eq. (6.36) requires modification.

It is seen that the difficulty arises when we choose a contour between the plates of a capacitor. We know that negative charge flows from the battery up to the plate P_1 of the capacitor. Similarly, negative charge flows from plate P_2 to the battery. But there is no charge flow between the plates of the charging capacitor.

However, there is a continuous current I flowing in the circuit.

Maxwell argued that changing electric field constitutes an effective current.

From Gauss's theorem, $\oint \mathbf{D}\cdot ds = q$ and the current I is

$$I = \frac{dq}{dt} = \frac{\partial}{\partial t}\oint \mathbf{D}\cdot ds = \oint \frac{\partial \mathbf{D}}{\partial t}\cdot ds \qquad \text{...(6.38)}$$

On the other hand, according to the continuity Eq. (6.35), we have

$$\oint \mathbf{J}\cdot ds = -\frac{dq}{dt} \qquad \text{...(6.39)}$$

where $\mathbf{J}$ is the conduction current density.

Summing up the left and the right hand sides of Eq. (6.38) and Eq. (6.39) separately, we obtain

$$\oint\left(\mathbf{J}+\frac{\partial \mathbf{D}}{\partial t}\right)ds = 0 \qquad ...(6.40)$$

Eq. (6.40) is similar to the continuity equation for direct current. Maxwell called the term $\frac{\partial \mathbf{D}}{\partial t}$ as the displacement current density whose dimensions are the same as current density. Thus,

$$\mathbf{J}_d = \frac{\partial \mathbf{D}}{\partial t} \qquad ...(6.41)$$

Total current is the sum of the conduction and the displacement currents. Total current density is

$$\mathbf{J}_{\mathrm{T}} = \mathbf{J}+\mathbf{J}_d = \mathbf{J}+\frac{\partial \mathbf{D}}{\partial t} \qquad ...(6.42)$$

So, Ampere circuital law has the form for an arbitrary case

$$\oint \mathbf{H}\cdot dl = \int\left(\mathbf{J}+\frac{\partial \mathbf{D}}{\partial t}\right)\cdot ds \qquad ...(6.43)$$

Eq. (6.43) is in agreement with the results of experiments in all cases with no exception.

The equation is known as Ampere-Maxwell equation. Thus, Maxwell showed that a changing electric field causes a current, which generates a magnetic field in just the same way as an actual current. Maxwell visualized that space itself is a medium, the ether, which had dielectric properties.

When a dielectric is positioned in a rising electric field, the charges will be displaced by a continuously increasing distance giving rise to a displacement current.

This is why Maxwell called the changing electric field term as displacement current.

It is possible to convert the line integral in Eq. (6.43) into a surface integral using Stoke's theorem. Thus, we get

$$\int_S (\nabla\times\mathbf{H})\cdot ds = \int_S\left[\mathbf{J}+\frac{\partial \mathbf{D}}{\partial t}\right]\cdot ds$$

Since the above result is true for any S, the integrands can be equated and we get

$$\nabla\times \mathrm{H} = \mathbf{J}+\frac{\partial \mathbf{D}}{\partial t} \qquad ...(6.44)$$

This is the general form of Ampere's law.

Displacement currents exist only when an electric field varies with time.

As $\mathbf{D} = \varepsilon_o E + P$, (dD/dt) of displacement current is

the sum of the densities of the "true" displacement current ($\varepsilon_o d\mathbf{E}/dt$), and of the polarization current ($d\mathbf{P}/dt$).

The polarization current is due to the motion of bound charges and the displacement current is not connected with any motion of charges and is only due to the variation of the electric fields. Displacement current excites a magnetic field even in vacuum.

6.13 SUMMARY OF MAXWELL'S EQUATIONS

The field equations which are known as the **Maxwell's equations** govern the time-varying electric and magnetic fields. Maxwell's equations in differential form can be converted into integral form by integrating them over an area and applying Stoke's theorem or by integrating throughout a volume and applying Divergence theorem. Table 6.1 gives the summary of Maxwell's equations.

Table 6.1: *Summary of Maxwell's equations*

	Law	*Differential form*	*Integral form*
1.	Gauss's law	$\nabla \cdot \mathbf{D} = \rho$	$\oint_S \mathbf{D} \cdot ds = \int_V \rho dV$
2.	Gauss's law for magnetism	$\nabla \cdot \mathbf{B} = 0$	$\oint_S \mathbf{B} \cdot ds = 0$
3.	Faraday's law	$\nabla \times \mathbf{E} = -\mu \frac{\partial \mathbf{H}}{\partial t}$	$\oint_C \mathbf{E} \cdot dl = -\int_S \frac{\partial \mathbf{B}}{\partial t} \cdot ds$
4.	Ampere's law	$\nabla \times \mathbf{H} = \mathbf{J} + \varepsilon \frac{\partial \mathbf{E}}{\partial t}$	$\oint_C \mathbf{H} \cdot dl = \oint_S \left(\mathbf{J} + \frac{\partial \mathbf{D}}{\partial t} \right) \cdot ds$

The first two are the divergence equations and the last two are the curl equations. Maxwell's equations contain only the first derivatives of fields **E** and **B** with respect to time and space coordinates and the first powers of densities ρ and J of electric charges and currents. Therefore, these equations are *linear* and the fields obey superposition principle. Maxwell unified the theories of electricity and magnetism by way of deducing four very important equations, which combined the experimental observations reported by Gauss, Ampere, and Faraday with his concept of displacement current.

The equations summarize the relation between the electric field and electric charge and between the magnetic field and electric current. Maxwell's equations relate the electric and magnetic field quantities. Maxwell's equations along with some auxiliary equations outline the primary doctrine of electromagnetic theory.

In all there are four Maxwell's equations. These equations cannot be derived since they are the fundamental axioms or postulates of electrodynamics, obtained with the help of generalization of experimental results.

6.14 PHYSICAL SIGNIFICANCE OF MAXWELL'S EQUATIONS

The physical significance of Maxwell's equations can be readily interpreted from their mathematical statements.

(*i*) Maxwell's equation $\nabla.\mathbf{E} = \frac{\rho}{\varepsilon}$ shows that the total electric flux density **D** through the surface enclosing a volume is equal to the charge density ρ within the volume. It means that a charge distribution generates a steady electric field.

(*ii*) Maxwell's equation $\nabla \cdot \mathbf{H} = 0$ tells us that the net magnetic flux through a closed surface is zero. It implies that magnetic poles do not exist separately in the way as electric charges do. Thus, in other words, magnetic monopoles do not exist.

(*iii*) Maxwell's equation $\nabla \times \mathbf{E} = -\mu \dfrac{\partial \mathbf{H}}{\partial t}$ shows that the emf around a closed path is equal to the time derivative of the magnetic flux density through the surface bounded by the path. It means that an electric field can also be generated by a time-varying magnetic field.

(*iv*) Maxwell's equation $\nabla \times \mathbf{H} = \mathbf{J} + \varepsilon \dfrac{\partial \mathbf{E}}{\partial t}$ shows that the magneto motive force around a closed path is equal to the conduction current plus the time derivative of the electric flux density **D** through any surface bounded by the path. The time derivative of the electric flux density $\partial \mathbf{D}/\partial t$ is called displacement current. Thus this equation means that a magnetic field is generated by a time-varying electric field.

QUESTIONS

1. Explain the terms magnetic induction and magnetic flux.
2. State and explain Biot-Savart law.
3. State and explain Ampere's circuital law.
4. Write the integral form of the Ampere's circuital law.
5. Obtain the differential form of Ampere's circuital law.
6. Write the integral and differential forms of Gauss's law in electrostatics in vacuum.
7. Apply Gauss's law and obtain an expression for magnetic field. What is the conclusion one can draw from the expression?
8. Explain magnetic scalar potential and derive an expression for the same.
9. Express the magnetic field in terms of vector potential.
10. State and explain Faraday's laws of electromagnetic induction.
11. What is Lenz's law of electromagnetic induction? Explain how the direction of current in a circular loop can be established with the help of Lenz's law.
12. Deduce the integral form and differential forms of Faraday's law.
13. How was the concept of displacement current helpful in removing discrepancy in Ampere's law?
14. Distinguish between conduction current and displacement current.
15. What is the physical significance of equations (*i*) $\nabla \cdot \mathbf{B} = 0$ and (*ii*) $\nabla \cdot \mathbf{D} = \rho$?
16. What is displacement current? Show that the conduction current in a lead wire is identical with the displacement in gap of the capacitor.

7 Electromagnetic Waves

Learning Objectives:

By the end of this chapter, you will be able to:

- Explain basic of Electromagnetic Waves and Electromagnetic Wave Equation.
- State use of Poynting Theorem.
- Learn about Conductors and Dielectrics in Wave system.

7.1 INTRODUCTION

Maxwell unified the theories of electricity and magnetism by way of deducing four very important equations, which combined the experimental observations reported by Gauss, Ampere, and Faraday with his concept of displacement current. In all there are four Maxwell's equations. These equations cannot be derived since they are the fundamental axioms or postulates of electrodynamics, obtained with the help of generalization of experimental results.

These equations relate the electric and magnetic fields with their sources. They describe the fundamentals of electromagnetic theory and its applications

7.2 ELECTROMAGNETIC WAVES

In the preceding chapter (chapter 6) we have discussed Maxwell's equations in static fields and time changing fields. When a charged particle undergoes acceleration, it radiates electromagnetic energy. Two metal rods are connected to an a.c. generator, which causes the charge to oscillate between two rods as shown in Fig. 7.1. At $t = 0$, the upper rod is given the maximum positive charge and the bottom rod an equal negative charge. The electric field near the oscillator is as shown in Fig. 7.1 (a).

While the charges oscillate in the rod, the magnitude of field near the rod decreases and the electric field moves away from the rod.

When charges are neutralized, the electric field becomes zero and this happens at a time $t = T/4$.

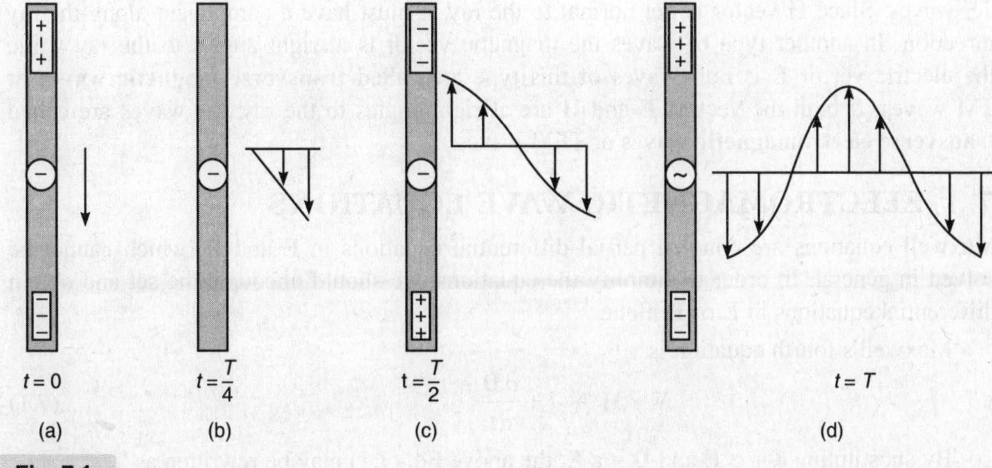

Fig. 7.1:

Production of electromagnetic wave by oscillating electric charges.

On continuing, the upper rod quickly acquires a maximum negative charge and lower rod turn into positive, resulting an electric field in upward direction and it occurs at a time $t = T/2$.

At any instant of time, the magnitude of the field depends on the quantity of charge on the rods at that instant. Oscillating charges create a current in the rods and in turn a magnetic field is also formed. The magnitude of the field has inverse relation with distance between source point and field point.

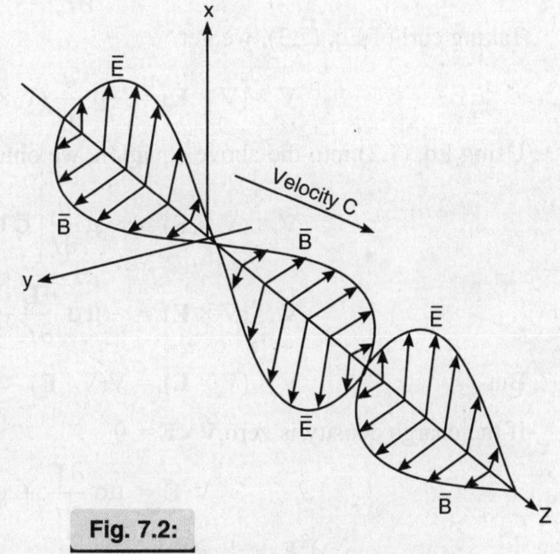

Fig. 7.2:

Electromagnetic waves propagation.

The generation of electromagnetic wave does not require any medium. **The electromagnetic waves propagate through space entirely on their own**. Maxwell's theory placed no restriction on possible wavelengths for the electromagnetic radiation. The vector cross products in Maxwell's third and fourth equations imply that the two fields, **E** and **H** are normal to each other and also normal to the direction of propagation. The electromagnetic wave travels with the velocity of light c in the vacuum.

The vectors **E** and **B** in an electromagnetic wave always oscillate in phase (see Fig. 7.2). The instantaneous values of **E** and **B** at any point are connected through the relation

$$\sqrt{\varepsilon}\, E = \sqrt{\mu}\, H$$

This means that **E** and **H** (or **B**) simultaneously attain their maximum values, vanish etc.

Note that electromagnetic waves are, in general, neither longitudinal nor transverse. In some types of electromagnetic waves, the electric vector **E** is at right angles to the ray while the magnetic vector **H** is not. Waves of this type are called **transverse electric waves** or

TE waves. Since **H** vector is not normal to the ray, it must have a component along the ray direction. In another type of waves the magnetic vector is at right angles to the ray while the electric vector **E** is not. Waves of this type are called **transverse magnetic waves** or TM waves. If both the vectors **E** and **H** are at right angles to the ray, the waves are called **transverse electromagnetic waves** or TEM waves.

7.3 ELECTROMAGNETIC WAVE EQUATIONS

Maxwell equations are coupled partial differential equations in **E** and **H**, which cannot be solved in general. In order to simplify the equations, we should uncouple the set and obtain differential equations in **E** or **H** alone.

Maxwell's fourth equation is

$$\nabla \times \mathbf{H} = \mathbf{J} + \frac{\partial \mathbf{D}}{\partial t} \qquad ...(7.1)$$

By substituting $\mathbf{J} = \sigma\,\mathbf{E}$ and $\mathbf{D} = \varepsilon\,\mathbf{E}$, the above Eq. (7.1) may be rewritten as

$$\nabla \times \mathbf{H} = \sigma\,\mathbf{E} + \varepsilon\frac{\partial \mathbf{E}}{\partial t} \qquad ...(7.2)$$

Maxwell's third equation is

$$\nabla \times \mathbf{E} = -\frac{\partial \mathbf{B}}{\partial t} = -\mu\frac{\partial \mathbf{H}}{\partial t} \qquad ...(7.3)$$

Taking curl of Eq. (7.3), we get

$$\nabla \times (\nabla \times \mathbf{E}) = -\mu\frac{\partial}{\partial t}(\nabla \times \mathbf{H})$$

Using Eq. (7.2) into the above equation, we obtain

$$\nabla \times (\nabla \times \mathbf{E}) = -\mu\frac{\partial}{\partial t}\left[\sigma\,\mathbf{E} + \varepsilon\frac{\partial \mathbf{E}}{\partial t}\right]$$

or

$$\nabla \times (\nabla \times \mathbf{E}) = -\mu\sigma\frac{\partial \mathbf{E}}{\partial t} - \mu\varepsilon\frac{\partial^2 \mathbf{E}}{\partial t^2}$$

But

$$\nabla \times (\nabla \times \mathbf{E}) = \nabla(\nabla \cdot \mathbf{E}) - \nabla^2\mathbf{E}$$

If the charge density is zero, $\nabla \cdot \mathbf{E} = 0$

$$\therefore \quad \nabla^2\mathbf{E} = \mu\sigma\frac{\partial \mathbf{E}}{\partial t} + \mu\varepsilon\frac{\partial^2 \mathbf{E}}{\partial t^2} \qquad ...(7.4)$$

or

$$\nabla^2\mathbf{E} - \mu\varepsilon\frac{\partial^2 \mathbf{E}}{\partial t^2} - \mu\sigma\frac{\partial \mathbf{E}}{\partial t} = 0 \qquad ...(7.5\text{ a})$$

This is the three-dimensional wave equation for the electric field in a conducting medium which is homogeneous and isotropic. Similar equation for magnetic field can be obtained following a similar procedure. Thus,

$$\nabla^2\mathbf{H} - \mu\varepsilon\frac{\partial^2 \mathbf{H}}{\partial t^2} - \mu\sigma\frac{\partial \mathbf{H}}{\partial t} = 0 \qquad ...(7.5\text{ b})$$

7.4 MAXWELL'S WAVE EQUATIONS FOR FREE SPACE

In order to understand the nature of waves, we consider *free space*, which is a large empty volume of space. Free space is a perfect dielectric and does not absorb ($\sigma = 0$) waves. As

$\mu = \mu_0$ and $\varepsilon = \varepsilon_0$ for free space, Eq. (7.5a) for free space (or a dielectric medium) becomes

$$\nabla^2\mathbf{E} - \mu_0\varepsilon_0 \frac{\partial^2 \mathbf{E}}{\partial t^2} = 0$$

$$\therefore \qquad \nabla^2 \mathrm{E} = \mu_0\varepsilon_0 \frac{\partial^2 \mathbf{E}}{\partial t^2} \qquad ...(7.6)$$

This is the law that **E** must obey.

A similar procedure for **H** gives us

$$\therefore \qquad \nabla^2\mathbf{H} = \mu_0\varepsilon_0 \frac{\partial^2 \mathbf{H}}{\partial t^2} \qquad ...(7.7)$$

This is the law that **H** must obey.

Eq. (7.6) and Eq. (7.7) are very much similar to the general wave equation and constitute **wave equations**. These three-dimensional vector wave equations describe the transmission of an electromagnetic wave through a uniform medium. The solutions of Eq. (7.6) and Eq. (7.7) have same form since the wave equations also have the same form.

The solutions to Eq. (7.6) and Eq. (7.7) lead to waves that can exist in free space. Even though the electric and magnetic fields of the waves start out on charges and currents, they detach themselves from them and move through free space as independent entities.

7.4.1 Velocity of the Electromagnetic Wave

The propagation characteristics of the electromagnetic wave are contained in the solution of Eq. (7.6). To bring out the characteristics, we compare it with the general wave equation

$$\nabla^2\xi = \frac{1}{\upsilon^2}\frac{\partial^2\xi}{\partial t^2} \qquad ...(7.8)$$

The comparison of Eq. (7.6) with Eq. (7.8) gives

$$\upsilon^2 = \frac{1}{\mu_0\varepsilon_0}$$

or

$$\upsilon = \frac{1}{\sqrt{\mu_0\varepsilon_0}} \qquad ...(7.9)$$

Substituting the values of μ_0 and ε_0, we find that

$$\frac{1}{\sqrt{\mu_0\varepsilon_0}} = \frac{1}{\sqrt{(4\pi\times10^{-7}\ \text{wb.A.m}^2)(8.9\times10^{-12}\text{C}^2/\text{N.m}^2)}} = 3.0\times10^8\ \text{m/s.}$$

$$\therefore \qquad \frac{1}{\sqrt{\mu_0\varepsilon_0}} = c, \text{ the velocity of light} \qquad ...(7.10)$$

Obviously, electromagnetic waves travel with the velocity of light in free space.

7.4.2 Relation between the Refractive Index and Relative Permittivity of a Medium

In case of a medium other than vacuum, we have to use $\varepsilon_0\varepsilon_r$ (= ε) and $\mu_0\,\mu_r$ (= μ), instead of ε_0 and μ_0 in Eq. (7.10). We then get

$$\upsilon = \frac{1}{\sqrt{\mu_0\mu_r\varepsilon_0\varepsilon_r}} = \frac{1}{\sqrt{\mu_0\varepsilon_0}}\cdot\frac{1}{\sqrt{\mu_r\varepsilon_r}} = \frac{c}{\sqrt{\mu_r\varepsilon_r}}$$

or $$\frac{c}{\upsilon} = \sqrt{\mu_r \varepsilon_r}$$

But $$\frac{c}{\upsilon} = n \quad \text{(refractive index of the medium)}$$

$\therefore$ $$n = \sqrt{\mu_r \varepsilon_r}$$

For a non-magnetic medium $\mu_r = 1$.

Therefore, $$n = \sqrt{\varepsilon_r} \qquad ...(7.11)$$

or $$n^2 = \varepsilon_r$$

7.5 UNIFORM PLANE WAVES

When energy is emitted by a source, it expands outwardly from the source in the form of spherical waves. The spherical wave travels at the same speed in all directions and therefore expands at the same rate. To an observer very far away from the source, the wave front of the spherical wave appears approximately planar. A plane wave is the simplest example of wave motion. In a plane wave, the electric and magnetic intensities are of constant value over any plane perpendicular to the direction of propagation. Such a plane is a surface of equal phase. **A plane wave** is thus a wave for which the phase has the same value at all points on an infinite plane. As the amplitude is also constant over the plane surface, it is called a **uniform plane wave**. Plane waves vary only in the direction of propagation and are uniform in planes normal to the direction of propagation. In this chapter we confine ourselves to the plane wave propagation in unbounded media. Plane wave propagation can be described by Cartesian coordinates, which are easier to work with mathematically than the spherical coordinates needed for describing propagation of a spherical wave.

7.5.1 The Transverse Nature of Plane Waves

Let us assume that the plane waves are traveling along the z-direction and hence **E** is constant over any given plane parallel to the xy-plane. Similarly **H** is constant over the xy-plane. We then have

$$\frac{\partial \mathbf{E}}{\partial x} = \frac{\partial \mathbf{E}}{\partial y} = 0 \text{ and } \frac{\partial \mathbf{H}}{\partial x} = \frac{\partial \mathbf{H}}{\partial y} = 0 \qquad ...(7.12)$$

These relations imply that

$$\frac{\partial E_x}{\partial x} = 0, \frac{\partial E_y}{\partial x} = 0, \frac{\partial E_z}{\partial x} = 0 \quad \text{and} \quad \frac{\partial E_x}{\partial y} = 0, \frac{\partial E_y}{\partial y} = 0, \frac{\partial E_z}{\partial y} = 0 \qquad ...(7.13\text{ a})$$

$$\frac{\partial H_x}{\partial x} = 0, \frac{\partial H_y}{\partial x} = 0, \frac{\partial H_z}{\partial x} = 0 \quad \text{and} \quad \frac{\partial H_x}{\partial y} = 0, \frac{\partial H_y}{\partial y} = 0, \frac{\partial H_z}{\partial y} = 0 \qquad ...(7.13\text{ b})$$

According to Maxwell' equation, we have

$$\nabla \cdot \mathbf{E} = 0$$

which means that

$$\frac{\partial E_x}{\partial x} + \frac{\partial E_y}{\partial y} + \frac{\partial E_z}{\partial z} = 0$$

Since $\frac{\partial E_x}{\partial x} = 0$ and $\frac{\partial E_y}{\partial y} = 0$, then $\frac{\partial E_z}{\partial z} = 0$. ...(7.14)

It means that E_z is independent of z and has the same value all along the z-axis. Wave motion consists of changing values of **E** along the direction of propagation. As E_z is constant along z-direction, it does not contribute to the wave motion and therefore E_z must be zero. $E_z = 0$ implies that **E** lies in a plane perpendicular to the direction of propagation of the wave. Hence, the electric wave is a *transverse* wave.

Again, according to Maxwell' equation, we have

$$\nabla \cdot \mathbf{B} = 0 \quad \text{and hence} \quad \nabla \cdot \mathbf{H} = 0$$

which means that

$$\frac{\partial H_x}{\partial x} + \frac{\partial H_y}{\partial y} + \frac{\partial H_z}{\partial z} = 0$$

$$\text{Since} \quad \frac{\partial H_x}{\partial x} = 0 \quad \text{and} \quad \frac{\partial H_y}{\partial y} = 0, \quad \text{then} \quad \frac{\partial H_z}{\partial z} = 0. \qquad \text{...(7.15)}$$

Thus, H_z is independent of z and has the same value all along the z-axis. As H_z is constant along z-direction, it does not contribute to the wave motion and therefore H_z must be zero. $H_z = 0$ implies that **H** lies in a plane perpendicular to the direction of propagation of the wave. Hence, the magnetic wave also is a *transverse* wave.

7.5.2 Relation between E and H in a Uniform Plane Wave

Now we assume that the variations of **E** are simple harmonic and that **E** is parallel to the y-axis. Then $E_z = 0$ and from Eq. (7.6), we may write for a wave traveling in the positive z-direction,

$$E_y = E_1\, e^{-i(\omega t - kz)} \qquad \text{...(7.16)}$$

or in a simpler form as

$$E_y = E_1 \cos[\omega(t - z/c)] \qquad \text{...(7.17)}$$

Maxwell's Eq. (7.3) can be written into the following three scalar equations.

$$\frac{\partial E_z}{\partial y} - \frac{\partial E_y}{\partial z} = -\frac{\partial B_x}{\partial t} \quad \frac{\partial E_x}{\partial z} - \frac{\partial E_z}{\partial x} = -\frac{\partial B_y}{\partial t} \quad \frac{\partial E_y}{\partial x} - \frac{\partial E_x}{\partial y} = -\frac{\partial B_z}{\partial t} \qquad \text{...(7.18)}$$

Using the results of Eq. (7.13a) into the first relation of Eq. (7.18), we get

$$\frac{\partial E_y}{\partial z} = \frac{\partial B_x}{\partial t} \qquad \text{...(7.19)}$$

The magnetic flux density B_x associated with the electric field E_y can be found by integrating Eq. (7.19). Thus,

$$B_x = \int \frac{\partial E_y}{\partial z}\, dt = -\frac{E_1 \omega}{c} \int \sin[\omega(t - z/c)]\, dt$$

$$= \frac{1}{c} E_1 \cos[\omega(t - z/c)]$$

The constant of integration is omitted since we are not interested in a constant component of the field.

$$\therefore \qquad B_x = \frac{1}{c} E_y \qquad \text{...(7.20)}$$

or

$$H_x = \frac{1}{\mu c} E_y = \frac{\sqrt{\mu_0 \varepsilon_0}}{\mu} E_y$$

$$\therefore \qquad H_x = \sqrt{\frac{\varepsilon_0}{\mu_0}} E_y \qquad \text{...(7.21)}$$

Since E_y and H_x differ only by a scalar and have the same time dependence, **E** and **H** are in phase at all points in space and are mutually perpendicular. Their cross product **E** × **H** points in the direction of propagation denoted by the vector **k**.

7.5.3 Characteristic Impedance

Eq. (7.21) may be rewritten as

$$\frac{E_y}{H_x} = \sqrt{\frac{\mu_0}{\varepsilon_0}}$$

The above equation states that the ratio of the amplitudes of the vectors **E** and **H** always equal to the square root of the ratio of μ_0 and ε_0. The ratio has the dimensions of impedance and hence it is called the **characteristic impedance** or **intrinsic impedance** of free space. It is denoted by η_0. We generalize Eq. (7.21) by writing it as

$$\frac{\mathbf{E}}{\mathbf{H}} = \eta_0 = \sqrt{\frac{\mu_0}{\varepsilon_0}} \qquad \text{...(7.22)}$$

Using the values of μ_0 and ε_0 into the above equation, we get

$$\eta_0 = \sqrt{\frac{4\pi \times 10^{-7}\ \text{H/m}}{10^{-19}/36\pi\ \text{F/m}}} = 120\ \pi = 337\ \text{ohms} \qquad \text{...(7.23)}$$

In a dielectric material, $\eta = 337 / \varepsilon_r^{1/2} = 337/n$, where n is the index of refraction of the material.

For free space and lossless dielectric media, **E** and **H** are in phase and so η is a pure resistance. For a conducting media, the ratio of E to H is a complex quantity and hence η is termed as intrinsic impedance.

An electromagnetic wave traveling in the z-direction will have no E_z component. It may have E_y or E_x component. Such a wave is called a *linearly polarized* wave. If the field is directed in x-direction, the wave is said to be x-polarized or if it is directed in y-direction, it is said to be y-polarized. The electromagnetic wave is said to be a *plane wave* since both **E** and **H** lie in a plane; E_x and H_y lie in xy-plane. The wave is also *uniform*, since neither E_x or H_y vary with distance. Since the direction of propagation of the wave is perpendicular to E_x and H_y, the wave is called **transverse electromagnetic wave** or simply **TEM wave**.

7.6 ELECTROMAGNETIC ENERGY DENSITY

One of the important characteristics of electromagnetic waves is that it transports energy from one region to another region. The *energy density*, u, of a wave is the radiant energy per unit volume. The electromagnetic wave consists of electric field and magnetic field which independently can store energy. The energy density stored in the electric field **E**, say existing between the plates of a charged capacitor, is given by

$$u_E = \frac{\varepsilon_0}{2} E^2 \qquad \text{...(7.24a)}$$

Similarly the energy density stored in a magnetic field, say produced by a current carrying loop, is given by

$$u_B = \frac{1}{2\mu_0} B^2 \qquad \text{...(7.24b)}$$

The vectors E and B are related through

$$\mathbf{E} = c\,\mathbf{B}$$

$$\therefore \qquad u_E = \frac{\varepsilon_0}{2} E^2 = \frac{\varepsilon_0}{2} c^2 B^2 = \frac{1}{2\mu_0} B^2 = u_B$$

The above expression means that the energy in the electromagnetic waves is stored equally between the electric and magnetic fields. Thus,

$$u = u_E + u_B$$

Therefore,

$$u = \varepsilon_0 E^2 \qquad \text{...(7.25a)}$$

or it can also be expressed as

$$u = \frac{1}{\mu_0} B^2. \qquad \text{...(7.25b)}$$

7.7 THE POYNTING THEOREM

An electromagnetic wave carries energy with it as it propagates through space. The energy flow of electromagnetic wave is described by the Poynting vector.

Def: Poynting vector is a vector quantity, **S** that describes the rate of energy flow per unit area or power flow per unit area. It was proposed by British physicist John H. Poynting.

Let us consider the Maxwell's curl equation

$$\nabla \times \mathbf{H} = \mathbf{J} + \frac{\partial \mathbf{D}}{\partial t}$$

$$\mathbf{E} \cdot \nabla \times \mathbf{H} = \mathbf{E} \cdot \mathbf{J} + \mathbf{E} \cdot \frac{\partial \mathbf{D}}{\partial t} \qquad \text{...(7.26)}$$

Now making use of the vector identity

$$\nabla \cdot (\mathbf{E} \times \mathbf{H}) = \mathbf{H} \cdot \nabla \times \mathbf{E} - \mathbf{E} \cdot \nabla \times \mathbf{H}$$

or

$$\mathbf{E} \cdot \nabla \times \mathbf{H} = \mathbf{H} \cdot \nabla \times \mathbf{E} - \nabla \cdot (\mathbf{E} \times \mathbf{H})$$

Putting the value of $\mathbf{E} \cdot \nabla \times \mathbf{H}$ into Eq. (7.26), we get

$$\mathbf{H} \cdot \nabla \times \mathbf{E} - \nabla \cdot (\mathbf{E} \times \mathbf{H}) = \mathbf{E} \cdot \mathbf{J} + \mathbf{E} \cdot \frac{\partial \mathbf{D}}{\partial t}$$

or

$$-\mathbf{H}\frac{\partial \mathbf{B}}{\partial t} - \nabla \cdot (\mathbf{E} \times \mathbf{H}) = \mathbf{E} \cdot \mathbf{J} + \mathbf{E} \cdot \frac{\partial \mathbf{D}}{\partial t}$$

$$-\mu\,\mathbf{H}\frac{\partial \mathbf{H}}{\partial t} - \nabla \cdot (\mathbf{E} \times \mathbf{H}) = \mathbf{E} \cdot \mathbf{J} + \mathbf{E} \cdot \frac{\partial \mathbf{D}}{\partial t}$$

$$-\nabla \cdot (\mathbf{E} \times \mathbf{H}) = \mathbf{E} \cdot \mathbf{J} + \varepsilon\,\mathbf{E} \cdot \frac{\partial \mathbf{E}}{\partial t} + \mu\,\mathbf{H}\frac{\partial \mathbf{H}}{\partial t}$$

As

$$\varepsilon\,\mathbf{E}\frac{\partial \mathbf{E}}{\partial t} = \frac{\varepsilon}{2}\frac{\partial \mathbf{E}^2}{\partial t} = \frac{\partial}{\partial t}\left(\frac{\varepsilon\,\mathbf{E}^2}{2}\right)$$

and

$$\mu\,\mathbf{H}\frac{\partial \mathbf{H}}{\partial t} = \frac{\mu}{2}\frac{\partial \mathbf{H}^2}{\partial t} = \frac{\partial}{\partial t}\left(\frac{\mu\,\mathbf{H}^2}{2}\right)$$

$$-\nabla\cdot(\mathbf{E}\times\mathbf{H}) = \mathbf{E}\cdot\mathbf{J}+\frac{\partial}{\partial t}\left(\frac{\varepsilon\mathbf{E}^2}{2}+\frac{\mu\mathbf{H}^2}{2}\right)$$

Rearranging the terms in the above equation and integrating it over a volume V, we obtain

$$\int_V(\mathbf{E}\cdot\mathbf{J})\,d\upsilon = -\frac{\partial}{\partial t}\int_V\left(\frac{\varepsilon\mathbf{E}^2}{2}+\frac{\mu\mathbf{H}^2}{2}\right)d\upsilon-\int_V\nabla\cdot(\mathbf{E}\times\mathbf{H})\cdot d\upsilon$$

Using the divergence theorem the last term can be changed from volume integral to a surface integral. Thus,

$$\int_V\nabla\cdot(\mathbf{E}\times\mathbf{H})\,d\upsilon = \int_S(\mathbf{E}\times\mathbf{H})\cdot d\mathbf{s}$$

$$\int_V(\mathbf{E}\cdot\mathbf{J})\,d\upsilon = -\frac{\partial}{\partial t}\int_V\left(\frac{\varepsilon\mathbf{E}^2}{2}+\frac{\mu\mathbf{H}^2}{2}\right)d\upsilon-\int_S(\mathbf{E}\times\mathbf{H})\cdot d\mathbf{s} \quad ...(7.27)$$

Significance of terms:

The rate of dissipation of energy in the volume V = the rate at which the stored energy in V is decreasing plus the rate at which energy is entering the volume from outside.

➢ $\int_V(\mathbf{E}\cdot\mathbf{J})\,d\upsilon$ is a generalization of Joule's law.

It represents the instantaneous power dissipated in volume V.

➢ $\frac{d}{dt}\int_V\left(\frac{\varepsilon\mathbf{E}^2}{2}+\frac{\mu\mathbf{H}^2}{2}\right)dv$ is a negative time derivative of the stored electric and magnetic energy in the volume.

Negative sign indicates the power being delivered by the field inside the volume to the region outside the volume. It represents the rate at which the stored energy in the volume is decreasing.

➢ $\int_S(\mathbf{E}\times\mathbf{H}).ds$ is from application of the Principle of conservation of energy.

A negative value represents the rate of flow of energy inward through the surface of the volume and positive value represents the rate of flow of energy outward through the surface enclosing the volume.

7.8 THE POYNTING VECTOR

The integral of $\mathbf{E}\times\mathbf{H}$ over any surface gives the rate of energy flow through that surface. The vector $\mathbf{S}=\mathbf{E}\times\mathbf{H}$ is the Poynting vector that describes the rate of energy flow per unit area at any point and has the dimensions of watts/m^2. The vector $\mathbf{S}=\mathbf{E}\times\mathbf{H}$ is perpendicular to both **E** and **H** and is in the direction of $\mathbf{E}\times\mathbf{H}$. The direction of **S** indicates the direction of the energy density flow at that point.

$$\mathbf{S}=\mathbf{E}\times\mathbf{H}$$

$$\mathbf{S}=\frac{1}{\mu_0}\mathbf{E}\times\mathbf{B}$$

or

$$\mathbf{S}=c^2\,\varepsilon_0\,(\mathbf{E}\times\mathbf{B}) \quad ...(7.28)$$

Let dU be the energy passing in a time d_t through a unit area perpendicular to the direction of propagation of the wave. In the time d_t the wave front moves a distance $dz = c.dt$.

$$dU = u\,\mathbf{c}\,\mathbf{dt}$$

where u is the energy density and is given by $u = \varepsilon_0 \mathbf{E}^2$.

$$\therefore \qquad du = \varepsilon_0\, \mathbf{E}^2\, c\, dt$$

For an electromagnetic wave

$$\varepsilon_0 \mathbf{E}^2 = \mu_0 \mathbf{H}^2 \quad \text{or} \quad \sqrt{\varepsilon_0}\, \mathbf{E} = \sqrt{\mu_0}\, \mathbf{H}$$

which implies that the electric energy density in the electromagnetic wave at any instant is equal to the magnetic energy density at the same point.

$$du = \sqrt{\varepsilon_0 \mu_0}\, EHc\, dt$$

or

$$du = EH\, dt$$

The term EH represents the magnitude of energy flux density vector **S**. The Poynting vector S indicates the *instantaneous* power density. As E and H are varying with time, the average power can be found by integrating the instantaneous Poynting vector **S** over one period and dividing by one period

7.9 WAVE PROPAGATION IN A LOSSY MEDIUM

When a medium has net conductivity σ, energy is dissipated in the medium and the medium is said to be a *lossy medium*. We assume that the medium is homogeneous, linear and isotropic. We further assume that it does not contain free volume charge. The Maxwell's equations in such a medium is given by

$$\nabla \cdot \mathbf{E} = 0 \qquad \text{...(7.29)}$$

$$\nabla \times \mathbf{E} = -\mu \frac{\partial \mathbf{H}}{\partial t} \qquad \text{...(7.30)}$$

$$\nabla \cdot \mathbf{H} = 0 \qquad \text{...(7.31)}$$

$$\nabla \times \mathbf{H} = \sigma \mathbf{E} + \varepsilon \frac{\partial \mathbf{E}}{\partial t} \qquad \text{...(7.32)}$$

Taking curl of Eq. (7.30), we get

$$\nabla \times (\nabla \times \mathbf{E}) = -\mu \frac{\partial}{\partial t} (\nabla \times \mathbf{H})$$

Using Eq. (7.32) into the above equation, we obtain

$$\nabla \times (\nabla \times \mathbf{E}) = -\mu \frac{\partial}{\partial t}\left[\sigma \mathbf{E} + \varepsilon \frac{\partial \mathbf{E}}{\partial t}\right]$$

or

$$\nabla \times (\nabla \times \mathbf{E}) = -\mu\, \sigma \frac{\partial \mathbf{E}}{\partial t} - \mu\varepsilon \frac{\partial^2 \mathbf{E}}{\partial t^2}$$

But

$$\nabla \times (\nabla \times \mathbf{E}) = \nabla(\nabla \cdot \mathbf{E}) - \nabla^2 \mathbf{E}$$

Since the charge density is zero,

$$\nabla \cdot \mathbf{E} = 0$$

$$\nabla^2 \mathbf{E} = \mu\, \sigma \frac{\partial \mathbf{E}}{\partial t} + \mu\varepsilon \frac{\partial^2 \mathbf{E}}{\partial t^2} \qquad \text{...(7.33)}$$

Similar equation for magnetic field is given by

$$\nabla^2 \mathbf{H} = \mu\varepsilon \frac{\partial^2 \mathbf{H}}{\partial t^2} - \mu\, \sigma \frac{\partial \mathbf{H}}{\partial t} \qquad \text{...(7.34)}$$

Eq. (7.33) may be written as

$$\nabla^2\mathbf{E} = j\omega\mu[\sigma + j\omega\varepsilon]\mathbf{E}$$

or

$$\nabla^2\mathbf{E} = [j\omega\mu\,\sigma - \omega^2\mu\varepsilon]\mathbf{E} \qquad ...(7.35)$$

For a plane wave traveling in x-direction and considering only the E_y component, Eq. (7.35) becomes

$$\frac{\partial^2 E_y}{\partial x^2} - [j\omega\mu\,\sigma - \omega^2\mu\varepsilon]\,E_y = 0 \qquad ...(7.36)$$

Putting $\gamma^2 = [j\omega\mu\,\sigma - \omega^2\mu\varepsilon]$

$$\frac{\partial^2 E_y}{\partial x^2} - \gamma^2 E_y = 0$$

where $\gamma = \sqrt{j\omega\mu[\sigma + j\omega\varepsilon]}$ and can be expressed as

$$\gamma = \alpha + j\beta = \sqrt{-\omega^2\mu\varepsilon}\left[1 + \frac{\sigma}{j\omega\varepsilon}\right]^{\frac{1}{2}} \qquad ...(7.37)$$

where α represents the *attenuation constant* and measures the rate at which the amplitude of the wave gets attenuated while propagating in the medium. β is the phase *shift constant* and indicates the rate of change in phase per unit length in the direction of propagation. It is also called the *phase factor*. Since γ includes α and β, which together characterize the propagation of the wave, the factor γ is called the *propagation constant*.

7.9.1 Expressions for α and β

$$\gamma = \alpha + j\beta = \sqrt{-\omega^2\mu\varepsilon}\left[1 + \frac{\sigma}{j\omega\varepsilon}\right]^{\frac{1}{2}}$$

Squaring the above equation, we get

$$\alpha^2 - \beta^2 + j2\alpha\beta = -\,\omega^2\,\mu\varepsilon + \omega\mu\sigma$$

Equating the real and imaginary parts on both sides, we get

$$\alpha^2\ \beta^2 = -\,\omega^2\mu\varepsilon$$

and

$$2\alpha\beta = \omega\mu\sigma$$

Mathematical manipulations yield the following expressions for α and β.

$$\alpha = \omega\sqrt{\frac{\mu\varepsilon}{2}}\left[\sqrt{1 + \left(\frac{\sigma}{\omega\varepsilon}\right)^2} - 1\right]^{1/2} \qquad ...(7.38)$$

$$\beta = \omega\sqrt{\frac{\mu\varepsilon}{2}}\left[\sqrt{1 + \left(\frac{\sigma}{\omega\varepsilon}\right)^2} + 1\right]^{1/2} \qquad ...(7.39)$$

7.10 CONDUCTORS AND DIELECTRICS

According to the Maxwell's fourth equation we have,

$$\nabla \times \mathbf{H} = \mathbf{J} + \frac{\partial \mathbf{D}}{dt}$$

Recalling that $\mathbf{J} = \sigma\mathbf{E}$, the above equation becomes

$$\nabla \times \mathbf{H} = \sigma\,\mathbf{E} + \frac{\partial\,\mathbf{D}}{dt}$$

For a linearly polarized plane wave traveling in the x-direction with $\mathbf{E}$ in the y-direction the above equation reduces to

$$\frac{\partial H_x}{\partial x} = \sigma E_y + \frac{\partial E_y}{dt} \qquad ...(7.40)$$

Eq. (7.40) is a scalar equation.

Assuming that E_y is a harmonic function of time we have

$$E_y = E_0 e^{j\omega t}$$

$$\frac{\partial E_y}{\partial t} = j\omega E_0 e^{j\omega t}$$

Putting these values in equation (7.40), we get

$$-\frac{\partial H_x}{\partial x} = \sigma E_y + j\omega\varepsilon E_y \qquad ...(7.41)$$

Both terms of the right hand side of equation (7.41) have the dimensions of current density. The term σE_y represents the conduction current density, while the terms $j\omega\varepsilon E_y$ represents the displacement current density. The ratio of the magnitude of the conduction current density to that of the displacement current density is given by $\frac{\sigma}{\omega\varepsilon}$.

(*i*) For a loss-less or good dielectric medium $\omega\varepsilon >> \sigma$,

the displacement current density is greatly larger than the conduction current density.

(*ii*) For a good conductor $\omega\varepsilon << \sigma$,

the conduction current density is greatly smaller than the displacement current density,

(*iii*) can be considered to be the dividing line between dielectrics and conductors.

To be more specific the media can also be classified according to the value of ratio $\sigma/\omega\varepsilon$. For dielectrics this ratio is less than 1/100; and for conductors this ratio is greater than 100.

It may be mentioned here that frequency is an important factor in determining whether a medium acts like a conductor or dielectric. For example a medium at 1kHz behaves like a conductor, while at 30 GHz its act like a dielectric. At 10MHz its behavior is that of a quasi-conductor.

7.10.1 Wave Propagation in a Good Dielectric

For a good dielectric $\sigma << \omega\varepsilon$ i.e. conduction current is very small compared to displacement current. We have

$$\gamma = \alpha + j\beta = \sqrt{-\omega^2\mu\varepsilon}\left[1 + \frac{\sigma}{j\omega\varepsilon}\right]^{\frac{1}{2}} = j\omega\sqrt{\mu\varepsilon}\left[1 + \frac{\sigma}{j\omega\varepsilon}\right]^{\frac{1}{2}}$$

We expand the factor $\left[1 + \frac{\sigma}{j\omega\varepsilon}\right]^{\frac{1}{2}}$ using the binomial expansion and retain the first two terms. Thus, we get

$$\gamma = \alpha + j\beta = j\omega\sqrt{\mu\varepsilon}\left[1 + \frac{\sigma}{2j\omega\varepsilon} + \frac{\sigma^2}{8\omega^2\varepsilon^2}\right]$$

Equating the real and imaginary parts on both sides of the above equation, we get

$$\alpha = \frac{\sigma}{2}\sqrt{\frac{\mu}{\varepsilon}}\left(1-\frac{\sigma^2}{8\omega^2\varepsilon^2}\right) \qquad ...(7.42)$$

The expression for β is given by

$$\beta = \omega\sqrt{\mu\varepsilon}\left(1+\frac{\sigma^2}{8\omega^2\varepsilon^2}\right) \qquad ...(7.43)$$

$\omega\sqrt{\mu\varepsilon}$ is the phase shift for a perfect dielectric. The effect of a small amount of loss is to add the second term as a small correction factor. The velocity of the wave in the dielectric is given by

$$\upsilon = \frac{\omega}{\beta} = \frac{1}{\sqrt{\mu\varepsilon}\left(1+\frac{\sigma^2}{8\omega^2\varepsilon^2}\right)}$$

i.e.

$$\upsilon = \frac{1}{\sqrt{\mu\varepsilon}}\left(1-\frac{\sigma^2}{8\omega^2\varepsilon^2}\right) \qquad ...(7.44)$$

where $\left(\frac{1}{\sqrt{\mu\varepsilon}}\right)$ is the velocity of the wave in the dielectric when the conductivity is zero i.e., in the perfect dielectric. The effect of a small amount of loss is to reduce slightly the velocity of propagation of the wave.

7.10.2 Wave Propagation in a Good Conductor

For a good conductor, we have $\omega\varepsilon \ll \sigma$. We write the factor γ in the following form.

$$\gamma = \alpha + j\beta = \sqrt{j\omega\mu\sigma}\left[1+\frac{j\omega\varepsilon}{\sigma}\right]^{\frac{1}{2}}$$

We can express

$$\sqrt{j} = \sqrt{\frac{2j}{2}} = \sqrt{\frac{1+2j-1}{2}} = \sqrt{\frac{(1+j)^2}{2}} = \frac{1+j}{\sqrt{2}}$$

Since $\omega\varepsilon/\sigma \ll 1$, we expand the factor $\left[1+\frac{j\omega\varepsilon}{\sigma}\right]^{\frac{1}{2}}$ as a power series and retain the first two terms. Then we obtain

$$\gamma = \alpha + j\beta = \sqrt{\frac{\omega\mu\sigma}{2}}(1+j)\left[1+\frac{j\omega\varepsilon}{2\sigma}\right]$$

Equating the real and imaginary parts in the above equation, we get

$$\alpha = \sqrt{\frac{\omega\mu\sigma}{2}}\left[1-\frac{\omega\varepsilon}{2\sigma}\right] \qquad ...(7.45)$$

and

$$\beta = \sqrt{\frac{\omega\mu\sigma}{2}}\left[1+\frac{\omega\varepsilon}{2\sigma}\right] \qquad ...(7.46)$$

It may be seen that

$$\alpha \cong \sqrt{\frac{\omega\mu\sigma}{2}} \approx \beta \qquad ...(7.47)$$

The velocity of propagation of the wave in the conductor will be

$$\upsilon = \frac{\omega}{\beta} = \sqrt{\frac{2\omega}{\mu\sigma}} \qquad ...(7.48)$$

7.10.2.1 Depth of Penetration in a Good Conductor

Using the values of α and β, the wave equation $E_y = Ae^{-(\alpha + j\beta)x}$ can be written as

$$E_y = Ae^{-\sqrt{\frac{\omega\mu\sigma}{2}}x}\, e^{-j\sqrt{\frac{\omega\mu\sigma}{2}}x} \qquad ...(7.49)$$

The above equation may be rewritten as

$$E_y = Ae^{-\frac{x}{\delta}} e^{-j\frac{x}{\delta}}$$

At $x = 0$, $E_y = A$ and at $x = \delta$, $E_y = \frac{A}{e}$.

It means that E_y decreases to $1/e$ of its initial value while the wave penetrates to a distance δ in the conducting medium. It is called the depth of penetration or **skin depth**. The skin depth is

$$\delta = \sqrt{\frac{2}{\omega\mu\sigma}} = \sqrt{\frac{1}{\pi\nu\mu\sigma}} \qquad ...(7.50)$$

The skin depth is inversely proportional to σ, μ and ν. Copper has a conductivity of $\sigma = 5.8 \times 10^7$ mhos/m and its permeability is approximately equal to that of free space. The depth of penetration of 1 MHz wave in copper is about 0.0667 mm, where as it is 8.67 mm at 60 Hz.

Example 7.1: *The wave function of a light wave is $E(z, t) = 10^3 \sin \pi(3 \times 10^6 x - 9 \times 10^{14} t)$*

(*a*) *determine the speed, wavelength, frequency and period of the wave,*
(*b*) *determine the magnetic field associated with the wave.*

Solution: (*i*)

$$E(z, t) = 10^3 \sin \pi(3 \times 10^6 x - 9 \times 10^{14} t)$$
$$= 10^3 \sin 3 \times 10^6 \pi(x - 3 \times 10^8 t) \qquad ...(1)$$

The equation is similar to the general wave equation,

$$E(z, t) = E_0 \sin k(x - vt) \qquad ...(2)$$

Comparing (1) with (2), we get,

$$\upsilon = 3 \times 10^8 \text{ m/s}, \quad \text{and} \quad k = 3 \times 10^6 \pi \text{ /m}$$

$$\therefore \quad \lambda = \frac{2\pi}{k} = \frac{2\pi}{3\times10^6\pi} = 6.666 \times 10^{-7} \text{ m} = 6666 \text{ Å}$$

$$\nu = \frac{\upsilon}{\lambda} = \frac{3\times10^8 \text{ m/s}}{6.666\times10^{-7}\text{ m}} = 4.5 \times 10^{14} \text{ Hz}$$

$$T = \frac{1}{\nu} = \frac{1}{4.5\times10^{14}\text{ Hz}} = 2.2 \times 10^{-15} \text{ s}$$

(*ii*) The light wave propagates in the *z*-direction while the *E*-vector oscillates along *x*- direction in *xz*-plane. Since in a EM wave, the magnetic field *B* is normal to both *E* and wave propagation directions, it should be in the *yz*-plane.

Thus, $B_x = 0;\ B_z = 0$

$$B = B_y(z, t) = B(z, t)$$

As $E = cB,\ B = \dfrac{E}{c}$

$$\therefore \quad B(z, t) = \frac{10^3 \sin \pi(3\times10^6 y - 9\times10^{14} t)}{c} = \frac{10^3 \sin \pi(3\times10^6 y - 9\times10^{14} t)}{3\times10^8}$$

$$= 3.33 \times 10^{-6} \sin \pi(3 \times 10^6 y - 9 \times 10^{14} t)$$

Example 7.2: *A electromagnetic wave moving through free space has an electric field given by,* $E = 100 \sin\left[8\pi \times 10^{14}\left(t - \dfrac{z}{3\times10^8}\right)\right]$. *Calculate the corresponding intensity.*

Solution: $E = 100 \sin\left[8\pi\times10^{14}\left(t - \dfrac{z}{3\times10^8}\right)\right]$; $\therefore\ E_0 = 100$ V/m

Intensity, $$I = \left(\frac{c\varepsilon_0}{2}\right)E_0^2 = \frac{(3\times10^8)\left(8.85\times10^{-12}\ \dfrac{\text{C}^2}{\text{Nm}^2}\right)(100\ \text{V/m})^2}{2} = \mathbf{13.3\ W/m^2}$$

QUESTIONS

1. State and explain Maxwell's equations for electromagnetic field. Starting from Maxwell's equations, deduce the wave equation for a plane wave in free space.
2. Show that the ratio of the electric and magnetic fields of a uniform plane wave is constant depending upon the medium.
3. What is meant by the Poynting vector? Derive Poynting vector from Maxwell's equations and explain its physical significance.
4. Solve the Maxwell's equations for perfect dielectric containing no charge and no conduction current.
5. State Maxwell's equations in their general /integral form. Derive their form for harmonically varying fields.
6. Show that the speed of propagation of electromagnetic wave in free space is given by $c = 1/\sqrt{\mu_0\varepsilon_0}$.
7. Establish Maxwell's equations for electromagnetic fields and obtain an expression for Poynting vector.
8. Write down Maxwell's field equations in differential and explain their physical meaning. Prove poynting theorem relating, the flow of energy at a point in space in an electromagnetic field.
9. Obtain Maxwells equations and deduce an expression for the velocity of propagation of plane electromagnetic wave in a medium of dielectric constant ε and permeability μ.
10. Obtain an expression for: (*i*) the electric vector, (*ii*) the poynting vector, (*iii*) the field energy density and (*iv*) the momentum density in the field for a plane wave given by $B = B_0 \cos\dfrac{2\pi}{\lambda}(z - ct)y$ moving along the z-direction in free space.
11. Write the wave equation for the electric field in an ionized medium.

12. Starting from Maxwell's electromagnetic equations in free space, in absence of charges and currents, obtain the wave equation for electric field.
13. State Poynting theorem. Explain how the Poynting vector explains the energy flow.
14. Write Maxwell's electromagnetic equations in free space in the presence of charges and currents. Name each symbol used in the equations.
15. Obtain the wave equation for electric field in a vacuum from appropriate Maxwell equations.
16. Define Poynting vector. Mention its dimension and SI unit.
17. Mention the boundary conditions satisfied by electric field and electric displacement at the boundary of two media.
18. Write the Maxwell's electromagnetic equations in differential form in a medium, in the presence of charges and currents. Identify and state the laws of electromagnetism with which these equations are associated.
19. Mention the boundary conditions satisfied by *E*, *D*, *B* and *H* at the interface between two non-conducting media.
20. Starting from Maxwell's electromagnetic equations in free space, obtain the wave equations in terms of scalar and vector potentials. Mention the gauge conditions used.

PROBLEMS

1. A plane electromagnetic wave in free space has an average poynting vector of 1 watt/m^2. Find the average energy density. **[Ans:** 3.33 × 10^{-7} J/m^3]
2. Find the velocity of a plane wave in a lossless medium having a relative permittivity of 5 and relative permeability of 1. **[Ans:** 1.34 × 10^8 m/sec]
3. A plane sinusoidal linearly polarized electromagnetic wave of wavelength $\lambda = 5 \times 10^{-7}$m travels in vacuum in the direction of the X-axis. The average intensity of the wave per unit area is 0.1 W/m^2 and plane of the vibration of the electric field is parallel to the Y-axis. Write the equations describing the electric and magnetic fields of the wave.

   [Ans: $E_y = \sqrt{24\pi}\cos\left[4\pi \times 10^6 (x - ct)\right]$ N/coulomb, $B_z = (E_y / c)$ Tesla]
4. What must be the strength of uniform electric field if it is to have the same energy density as that possessed by a 400 gauss magnetic field [1 W/m^2 = 10^4 gausses, $\varepsilon_0 = 8.85 \times 10^{12}$ F /m and $\mu_0 = 4\pi \times 10^{-7}$ H/m].
5. Find the Brewster angle for a parallel-polarized wave traveling from air into glass for which $\varepsilon_r = 5.0$ **[Ans:** 65.91°]
6. In free space, $E(z, t) = 1.0 \sin(\omega t - \beta z)a_x$ V/m. Show that the average power crossing a circular disk of radius 15.5 m in a z = const plane is 1 W.
7. In free space, $H(z, t) = 1.33 \times 10^{-1} \cos(4 \times 10^7 t - \beta z)a_x$ A/m. Obtain an expression for $E(z, t)$. Find β and λ. [E_0 = 50 V/m, 4/3 rad/m, 15 π m.]
8. A plane electromagnetic wave is traveling in an unbounded lossless dielectric medium having $\mu_r = 1$ and $\varepsilon_r = 3$ and has peak electric field intensity of 6 V/m. Find: (*a*) velocity of the wave (*b*) the intrinsic impedance of the medium (*c*) the peak value of **H**.

   [Ans: 1.73 × 10^8 m/s; 217.66 ohm; 2.76 × 10^{-2} A/m]
9. A laser beam from 100 watt source is focused on an area of 10^{-8} m^2. Evaluate the magnitude of the Poynting vector on the area.

8 Interference

Learning Objectives:

By the end of this chapter, you will be able to:

- Learn about basic concepts of an important consequence in superposition of waves - Interference.
- Understand about Plane Parallel Film and Variable Thickness Film.
- Applications of Interference.
- Use of Michelson Interferometer and its application.

8.1 INTRODUCTION

In 1678 Christian Huygens propounded the wave theory of light. He suggested that light propagates in the form of waves. Huygens did not suggest anything about the nature of light wave; whether it is a transverse wave or longitudinal wave. He had no knowledge about the speed of light or its wavelength.

In 1801 Thomas Young provided the first experimental evidence for the wave theory of light from the double slit interference experiment and determined the wavelength of light waves. Sustained interference would be obtained only when coherent sources of light are used. Special devices, such as Lloyd's mirror, Fresnel biprism are designed to create two coherent sources from a single source of light and to determine accurately the wavelength of light. The striking colour effects caused by white light on a soap bubble and on a slick of oil spreading over small puddles of water are familiar to us from our childhood. Less noticed are the colours exhibited by coatings of oxide on heated metal pans in kitchen and thin layers of mica, cellophane etc. Thomas Young explained the origin of colours in all these cases on the basis of interference. The interference of certain colours constructively and others destructively causes the exhibition of iridescent colours by the thin films in reflected light. An understanding of the thin film interference phenomenon has led to a number of practical applications such as non-reflecting coatings, interference filters, and interference mirrors.

8.2 INTERFERENCE

Interference is an important consequence of superposition of waves. Let us consider two (or more) light waves of same frequency and having a constant phase difference travel in the same region of a medium simultaneously and cross each other (see Fig. 8.1). At the point of crossing, waves overlap or superpose on each other.

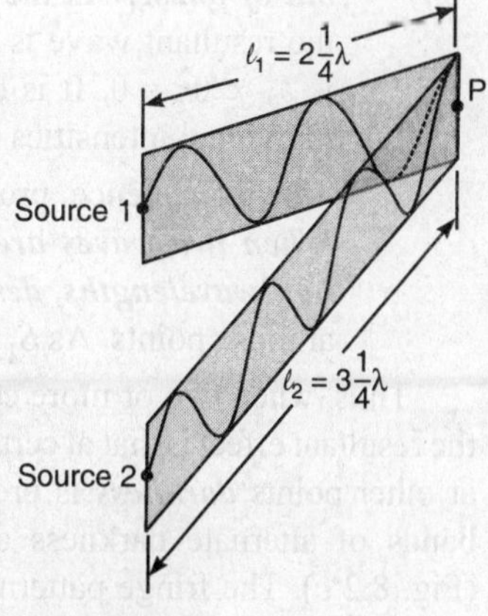

Fig. 8.1: Superposition of waves.

Def: Coherent waves are the waves having the equal frequency and a constant phase difference. Waves having fluctuating phase difference are known as incoherent waves.

According to the ***principle of superposition***, the combined effect of coherent waves at each point of the region of superposition is obtained by adding algebraically the disturbances due to individual waves. Resulting intensity at any point in the area of interference depends upon the amplitudes and the phase relationships of the component waves. Let us assume here that the component waves are of the same amplitude, say A.

We focus our attention mainly on two specific types of disposition of the waveș (see Fig. 8.2) at the point of observation, P.

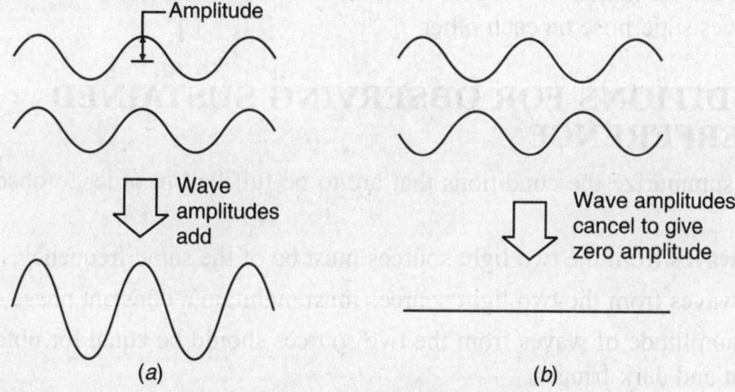

Fig. 8.2: Two types of disposition of the waves at the point of observation.

(*i*) If the two waves, meeting at P, reach their maxima and minima at the same instant of time, then their phase difference is zero or an integral multiple of 2π. Such waves will have a crest-to-crest and trough-to-trough correspondence, as shown in Fig. 8.2 (a). The amplitude of the resultant wave = $A + A = 2A$ as shown in Fig. 8.2. Hence, the intensity of the resultant wave is $I_R \propto 2^2A^2 = 2^2I$. It is obvious that the resulting intensity is greater than the sum of the individual intensities ($I + I = 2I$).

The interference produced at such points is known as *constructive interference*. ***When two waves are not displaced with respect to each other or when they are displaced through an integral number of wavelengths, constructive interference takes place***. Bright bands of light are observed at those points. As S_1 and S_2 are coherent sources, the bright bands are **stationary**.

(*ii*) When waves have a crest-to-trough correspondence then their phase difference is π radians as shown in Fig. 8.2 (b). The waves are said to be in ***opposite phase*** or ***180°***

out-of-phase. As the amplitude of one of the waves is negative, the amplitude of the resultant wave is equal to $A - A = 0$. Hence, the intensity of the resultant wave is $I_R \propto 0^2 = 0$. It is obvious that the resulting intensity is less than the sum of the individual intensities ($I + I = 2I$).

The interference produced at such points is known as *destructive interference.* ***When two waves are displaced with respect to each other by an odd number of half-wavelengths, destructive interference results.*** Dark bands of light are observed at those points. As S_1 and S_2 are coherent sources, the dark bands are **stationary**.

Thus, when two or more coherent waves of light are superposed, the resultant effect is that at certain points ***brightness*** is produced while at other points ***darkness*** is produced in the medium. The stationary bands of alternate darkness and brightness are known as ***fringes*** (Fig. 8.2 *c*). The fringe pattern is obtained only when the interfering waves are coherent.

Fig. 8.2(c): Interfrence pattern.

Def: Interference is rearrangement of light energy due to the superposition of light waves from two or more coherent sources.

When *incoherent waves* overlap on each other, the resultant intensity is a simple addition of the intensities in the region of overlap. Therefore, the resultant intensity due to two incoherent waves of equal intensity is $I + I = 2I$ and the region of overlap is uniformly bright without the formation of fringes. It means that interference does not take place when incoherent waves superpose on each other.

8.3 CONDITIONS FOR OBSERVING SUSTAINED INTERFERENCE

We may now summarize the conditions that are to be fulfilled in order to observe sustained interference.

(*i*) The waves from the two light sources must be of the same frequency.

(*ii*) The waves from the two light sources must maintain a constant phase difference.

(*iii*) The amplitude of waves from the two sources should be equal for obtaining distinct bright and dark fringes.

(*iv*) Planes of polarization of the waves must be the same. Waves polarized in perpendicular planes cannot produce interference effects.

(*v*) The two coherent sources must lie close to each other in order to distinguish the fringe pattern.

(*vi*) Background should be dark to view interference fringes.

(*vii*) Two sources must be very narrow.

(*viii*) The propagation distance over which a wave maintains a particular degree of coherence is known as *coherence length.* The path difference between the

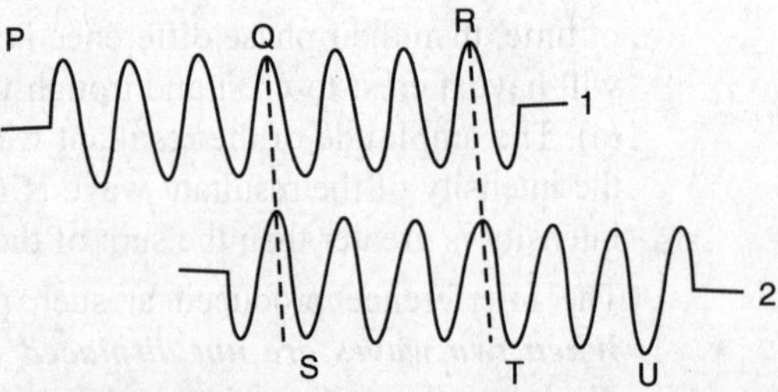

Fig. 8.3: Interfering wave trains having constant phase difference.

overlapping waves must be less than the coherence length of the waves.

$$\Delta < l_{coh} \quad ...(8.1)$$

Fig. 8.3 shows two wave trains that are having constant phase difference and about to interfere. It is easy to see that interference takes place due to parts *QR* of wave 1 and *ST* of wave 2 and that interference will not occur for the parts *PQ* and *TU*. So, the interference pattern does not become visible. When the whole wave train *PR* overlaps on the wave train *SU*, interference pattern will be distinct. In contrast, when the path difference between the waves 1 and 2 becomes very large, the wave trains arrive at different times and do not overlap on each other. Consequently, in such cases interference does not occur.

Example 8.1: *Find the coherence length for white light, the wavelength of white light range from 400 nm to 700 nm.*

Solution: $D\lambda = 700 - 400 = 300 \text{ nm} = 300 \times 10^{-9}$ m

Average wavelength $$\lambda = \frac{700 + 400}{2} = 550 \times 10^{-9} \text{ m}$$

$$l_o = \frac{\lambda^2}{D\lambda} = \frac{(550 \times 10^{-9})^2}{300 \times 10^{-9}} = 1 \times 10^{-6} \text{ m}$$

That is coherence length is of the order of one micrometer, which is very small and obtaining a path difference of this order between the interfering beams is difficult. So interference with white light is impossible.

8.4 TECHNIQUES OF OBTAINING INTERFERENCE

The phase relation between the waves emitted by two conventional light sources fluctuates rapidly and therefore they can never be coherent, though they are identical in all respects. However, two coherent sources are derived from a single source by techniques, which can be divided into two broad classes.

(*i*) **Division of wave front:** The wave front is divided using a narrow slit. In Young's double slit experiment a wave front emerging from the slit *S* is divided into two parts by the double slit S_1S_2. Fresnel's biprism, Lloyd's mirror, etc are the other examples where the ***division of wave front*** method is used.

(*ii*) **Division of amplitude:** The amplitude of a light wave is divided by partial reflection into two or more beams. Thin films (wedge, Newton's rings etc) and interferometers such as Michelson's interferometer are examples in which division of amplitude is used.

8.5 REVIEW OF IMPORTANT CONCEPTS

When studying the topic of interference, we frequently come across the terms optical path and path difference. We should clearly understand these terms and know how to calculate them.

8.5.1 Geometrical Path

Light travels along a straight line from a point *X* to another point *Y* and it is known as the *path of the light*. The shortest path between any two points *X* and *Y* is called the *geometrical path length* (GPL). GPL remains the same whether it is measured in a vacuum or in any medium.

8.5.2 Optical Path

Light travels μ times slower in a medium. Therefore, it takes μ times more time to cover the distance *AB* in the medium than it takes to cover the same distance in a vacuum. This time delay is accounted for by introducing another distance called *optical path length* (OPL). It is defined as

$$\text{O.P.L.} = \mu \times \text{G.P.L.}$$

Optical path length, $\Delta = \mu L$ ...(8.2)

The optical path length, Δ signifies the number of wavelengths that are accommodated in a given medium over the corresponding geometrical path length.

8.5.3 Path Difference

Light waves travel along different paths, which may lie in the same medium or in different media. The difference between optical paths of two rays traveling in different directions is known as the *optical path difference*.

8.5.4 Phase Difference

The phase of a wave arriving at a point depends on the optical path length it traversed. We know that if a wave covers in air a distance of one wavelength, 1λ, its phase changes by 2π radians. Therefore, we compute that if a wave travels a distance *L* in air, its phase change is given by

$$\delta = \frac{2\pi L}{\lambda} \qquad \text{...(8.3)}$$

When the wave travels the distance *L* in a medium, then

$$\delta = \frac{2\pi\Delta}{\lambda} = \frac{2\pi\mu L}{\lambda} \qquad \text{...(8.4)}$$

Comparing Eq. (8.3) and Eq. (8.4), we find that a light path of geometric length *L* in a medium of refractive index μ produces the same phase change as a light path of length μ*L* in a vacuum. Therefore, ***in the study of optics we always must calculate the optical paths traveled by light rays.***

The path difference for two

- in-phase waves will be zero or an integral multiple of a wavelength, λ
- opposite-phase waves will be λ/2 or an odd integral multiple of λ/2

Optical path difference and the consequent phase difference may arise due to two reasons. One reason is the difference in the optical paths and the other is due to reflections at optical interfaces.

(a) Phase difference due to optical path difference: Let us consider two sources of light S_1 and S_2, as shown in Fig. 8.4. Let us assume that the sources are identical and produce waves of same wavelength and that their vibrations are in the same phase at S_1 and S_2. Light from these sources travel in air along different paths, S_1P and S_2P; and meet

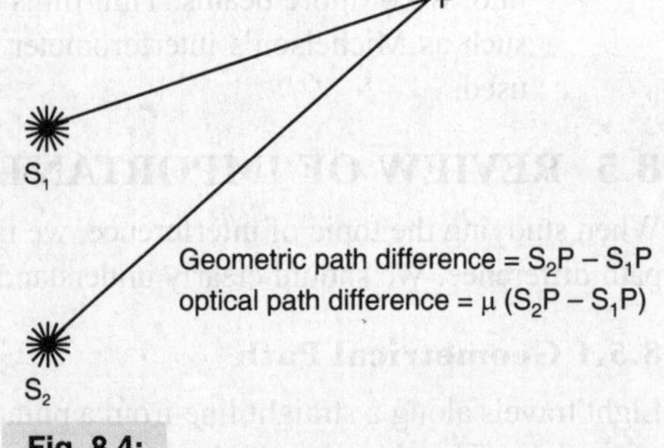

Fig. 8.4: Illustration showing Phase difference at P due to optical path difference.

at a point P. The path lengths S_1P and S_2P are different and contain different number of waves.

The geometric path difference between the waves at P is $(S_2P - S_1P)$.

The optical path difference is $\mu(S_2P - S_1P)$.

Since the paths contain different number of waves, the optical path difference will be equal either to a few integral number of waves or an integral number of wavelengths plus a fraction of one wavelength. This optical path difference leads to a phase difference between the waves meeting at P. It means that though the waves started with the same phase, they arrive at P with different phases because they traveled along different path lengths. Using Eq. (8.4), the phase difference between the waves at P may be expressed as

$$\delta = \frac{2\pi}{\lambda}\mu(S_2P - S_1P) \qquad ...(8.5)$$

(b) Phase difference due to reflection at boundaries of optical interfaces: Light waves may also undergo phase change due to reflection at some point in their path. If the waves are reflected at a rarer-to-denser medium boundary, the reflected waves suffer a phase change of π rad or 180° compared to the incident waves (See Fig. 8.5 a & b). It is seen from Eq. (8.3) that a phase change of π rad is equal to a path change of $\lambda/2$.

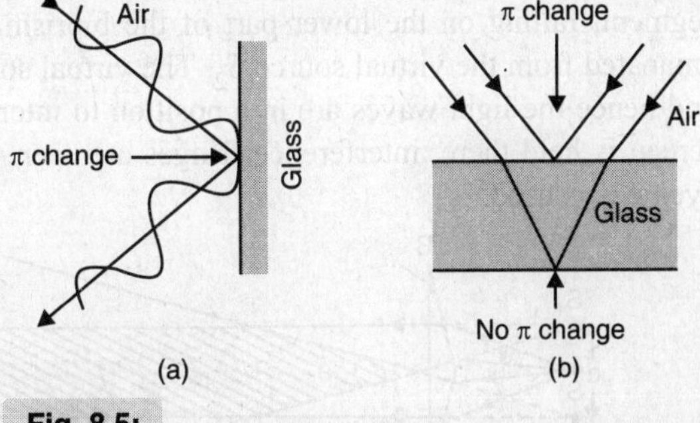

Fig. 8.5:

Illustration of Phase difference due to reflection at boundaries.

Therefore, we must add (or subtract) $\lambda/2$ in the calculation of true optical path difference whenever a reflection occurs at a denser medium. On the other hand, the waves reflected at a denser-to-rarer medium boundary do not suffer any phase change as indicated in Fig. 8.5 (b).

8.6 INTERFERENCE BY DIVISION OF WAVE FRONT-FRESNEL BIPRISM

Fresnel used a biprism to show interference phenomenon. The biprism consists of two prisms of very small refracting angles joined base to base.

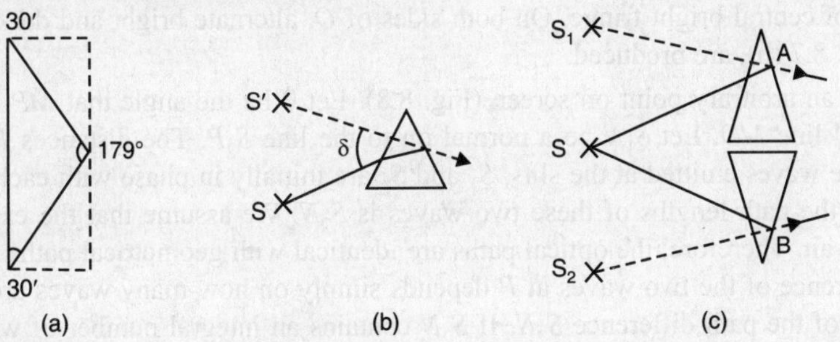

Fig. 8.6:

Fresnel biprism and formation of virtual sources.

Actually, a thin glass plate is taken and one of its faces is ground and polished till a prism (Fig. 8.6 a) is formed with an obtuse angle of about 179° and two side angles of the order of 30′. Biprism produces two virtual sources S_1 and S_2, as seen in Fig. 8.6 (b). These two virtual sources are hence *coherent as* they are produced from same source *S* by refraction.

8.6.1 Experimental Arrangement

The biprism is mounted suitably on an optical bench. An optical bench consists of two horizontal long rods, which are kept strictly parallel to each other and at the same level. The rods carry uprights on which the optical components are positioned. A monochromatic light source such as sodium vapour lamp illuminates a vertical slit *S*. Therefore, the slit *S* acts as a narrow linear monochromatic light source.

The biprism is placed in such a way that its refracting edge is parallel to the length of the slit *S*. A single cylindrical wavefront impinges on both prisms. The top portion of wavefront is refracted downward and appears to have emanated from the virtual image S_1. The lower segment, falling on the lower part of the biprism, is refracted upward and appears to have emanated from the virtual source S_2. The virtual sources S_1 and S_2 are coherent (see Fig. 8.7), and hence the light waves are in a position to interfere in the region beyond the biprism. If a screen is held there, interference fringes are seen. In order to observe fringes, a micrometer eyepiece is used.

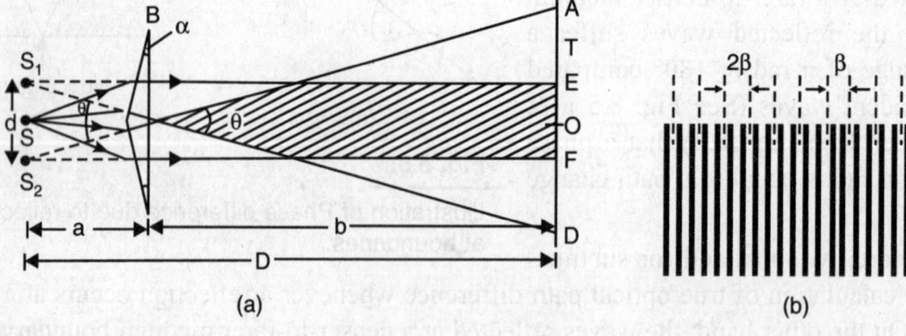

Fig. 8.7:

Fresnel biprism and fringe formation.

Theory: Let S_1 and S_2 be the two virtual images of the source *S*. Let *d* be the distance between S_1 and S_2. The fringes are formed on a screen *T* kept at a distance *D* from the biprism. The point *O* on the screen is equidistant from S_1 and S_2. Hence, the waves arrive from S_1 and S_2 arrive at *O* simultaneously and the point *O* is always bright. The point *O* corresponds to the position of central bright fringe. On both sides of *O*, alternate bright and dark fringes, as shown in Fig. 8.7 (b), are produced.

Let *P* be an arbitrary point on screen (Fig. 8.8). Let θ be the angle that *MP* makes with the horizontal line *MO*. Let S_1N be a normal on to the line S_2P. The distances PS_1 and *PN* are equal. The waves emitted at the slits, S_1 and S_2 are initially in phase with each other. The difference in the path lengths of these two waves is S_2N. We assume that the experiment is carried out in air. Therefore, the optical paths are identical with geometrical paths. The nature of the interference of the two waves at *P* depends simply on how many waves are contained in the length of the path difference S_2N. If S_2N contains an integral number of wavelengths, the two waves interfere constructively, producing a maximum in the intensity of light on the screen at *P*. If it contains an odd number of half-wavelengths, the waves interfere destructively and produce a minimum intensity at *P*.

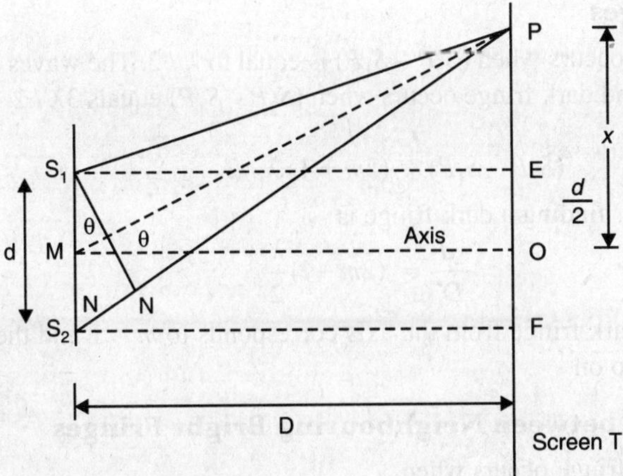

Fig. 8.8:

Interference at an arbitrary point on screen.

8.6.2 Optical Path Difference between the Waves at P

Let the point P be at a distance x from O (Fig. 8.8). Then

$$PE = x - d/2 \text{ and } PF = x + d/2.$$

$$(S_2P)^2 - (S_1P)^2 = \left[D^2 + \left(x + \frac{d}{2}\right)^2\right] - \left[D^2 + \left(x - \frac{d}{2}\right)^2\right]$$

$$\therefore \quad (S_2P)^2 - (S_1P)^2 = 2xd$$

or

$$S_2P - S_1P = \frac{2xd}{S_2P + S_1P}$$

We can approximate that $S_2P \cong S_1P \cong D$.

$$\therefore \quad \text{Path difference} = S_2P - S_1P = \frac{xd}{D} \qquad \text{...(8.6)}$$

We now find out the conditions for observing bright and dark fringes on the screen.

8.6.3 Bright Fringes

Bright fringes occur at places where the waves from S_1 and S_2 interfere constructively. The waves from S_1 and S_2 travel the same optical path length and arrive at O in phase and therefore a bright fringe occurs at axial point O. When the wave from S_2 travels one complete wavelength more than the wave from S_1 the next bright fringe occurs. Thus, constructive interference takes place at places where S_1P and S_2P arrive with a difference of a whole number of wavelengths.

The condition for finding a bright fringe at P is that

$$S_2P - S_1P = m\lambda$$

Using the Eq. (8.6), it means that

$$\frac{xd}{D} = m\lambda \qquad \text{...(8.7)}$$

where m is called the **order of the fringe**. The bright fringe at O, corresponding to $m = 0$, is called the *zero-order* fringe. The first-order bright fringe from the axis corresponds to $m = 1$ and the second order bright fringe to $m = 2$ and so on.

8.6.4 Dark Fringes

The first dark fringe occurs when $(S_2P - S_1P)$ is equal to $\lambda / 2$. The waves are now in opposite phase at P. The second dark fringe occurs when $(S_2P - S_1P)$ equals $3\lambda / 2$. The mth dark fringe occurs when

$$(S_2P - S_1P) = (2m + 1)\,\lambda / 2$$

The condition for finding a dark fringe is

$$\frac{xd}{D} = (2m+1)\frac{\lambda}{2} \qquad ...(8.8)$$

The first-order dark fringe from the axis corresponds to $m = 1$ and the second order dark fringe to $m = 2$ and so on.

8.6.5 Separation between Neighbouring Bright Fringes

The mth order bright fringe occurs when

$$x_m = \frac{m\lambda D}{d}$$

and the $(m + 1)^{th}$ order bright fringe occurs when

$$x_{m+1} = \frac{(m+1)\lambda D}{d}$$

The bright fringe separation, β is given by

$$\beta = x_{m+1} - x_m = \frac{\lambda D}{d} \qquad ...(8.9)$$

The same result will be obtained for dark fringes. Thus, neighbouring bright and dark fringes are separated by the same amount everywhere on the screen. The separation β is called the ***fringe width***.

The width of the dark or bright fringe is given by Eq. (8.9).

$$\beta = \frac{\lambda D}{d}$$

where D (= $a + b$) is the distance of the sources from the eye-piece.

Example 8.2: *A biprism is placed 5 cm from a slit illuminated by sodium light (λ = 5890Å). The fringe width obtained on the screen is 0.9424 mm. Find the d, if D = 75 cm.*

Solution: $\beta = \dfrac{\lambda D}{d}$

$$\therefore \quad 9.424 \times 10^{-2}\text{ cm} = \frac{5890\times 10^{-8}\text{ cm }(5+75)\text{cm}}{d}$$

or Distance between the two coherent sources

$$d = \frac{5890\times 10^{-8}\times 80}{9.424\times 10^{-2}}\text{ cm} = \mathbf{0.05\ cm.}$$

8.6.6 Determination of Wavelength of Light

By the Eq. (8.9) the wavelength of the light can be found. For using the relation, the values of β, D and d are to be measured. These measurements are done as follows.

(*i*) **Adjustments:** A narrow adjustable slit S, the biprism, and a micrometer eyepiece are mounted on the uprights and are adjusted to be at the same height and in a straight line. The slit is made vertical and parallel to the refracting edge of the biprism by rotating it in its own plane. It is illuminated with the light from the monochromatic source. The biprism is moved along the optical bench till, on looking through it along the axis of the optical bench, two equally bright vertical slit images are seen. Then the eyepiece is moved till the fringes appear in the focal plane of the eyepiece.

(*ii*) **Determination of fringe width β:** Once the fringes are brought into the field view of the eyepiece, the vertical cross-wire is adjusted to lie on the centre of one of the bright fringes and the position of the eyepiece is noted. The micrometer screw of the eyepiece is then turned slowly and the number of the bright fringes that cross the cross-wire is counted. The new position of the cross-wire is noted. The fringe width is then given by

$$\beta = \frac{x_{m+N} - x_m}{N}$$

(*iii*) **Determination of '*d*':** A convex lens of short focal length is placed between the slit and the eyepiece without disturbing their positions. The lens is moved back and forth near the biprism till a sharp pair of images of the slit is obtained in the field view of the eyepiece. The distance between the images is measured. Let it be denoted by d_1.

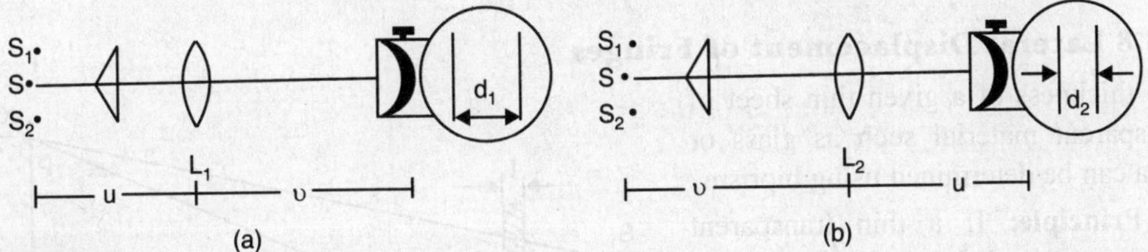

Fig. 8.9: Determination of 'd'.

If u is the distance of the slit and υ that of the eyepiece from the lens (Fig. 8.9), then the magnification is

$$\frac{\upsilon}{u} = \frac{d_1}{d} \quad ...(8.10)$$

The lens is then moved to a position nearer to the eyepiece, where again a pair of images of the slit is seen. The distance between the two sharp images is again measured. Let it be d_2. Again magnification is given by

$$\frac{u}{\upsilon} = \frac{d_2}{d} \quad ...(8.11)$$

Note that the magnification in one position is the reciprocal of the magnification in the other position.

Multiplying the Eq. (8.10) and Eq. (8.11), we obtain

$$\frac{d_1 d_2}{d^2} = 1$$

$$d = \sqrt{d_1 d_2} \quad ...(8.12)$$

Using the values of β, d and D in the Eq. (8.9), the wavelength λ can be computed.

Example 8.3: *Fresnel biprism is used to form interference fringes using sodium light of λ = 5890 Å. A convex lens interposed gives two images separated by distances 0.6 mm and 0.27 mm of the coherent sources corresponding to two positions. Calculate the fringe width. The slit to screen distance is 80 cm.*

Solution: Let d_1 and d_2 be the separation between the images in two positions. Then, the distance between the coherent sources is

$$d = \sqrt{d_1 d_2} = \sqrt{6\times10^{-4}\,\text{m}\times2.7\times10^{-4}\,\text{m}} = 4.025\times10^{-4}\ \text{m}$$

$$\therefore \quad \beta = \frac{\lambda D}{d} = \frac{5890\times10^{-10}\,\text{m}\times0.8\,\text{m}}{4.025\times10^{-4}\,\text{m}} = \mathbf{1.171\ mm.}$$

8.6.7 Interference Fringes with White Light

In the biprism experimentation, if white light is used the interference pattern has a central **white fringe** flanked on both its sides by a few coloured fringes. The central white fringe is called as the *zero-order* fringe.

While using monochromatic light all the bright fringes are of the same colour and the zero-order fringe cannot be seen.

8.6.8 Lateral Displacement of Fringes

The thickness of a given thin sheet of transparent material such as glass or mica can be determined using biprism.

Principle: If a thin transparent sheet is placed in the path of one of the two interfering beams, the fringe system gets displaced towards the beam in whose path the sheet is placed. By measuring the amount of displacement, the thickness of the sheet can be determined.

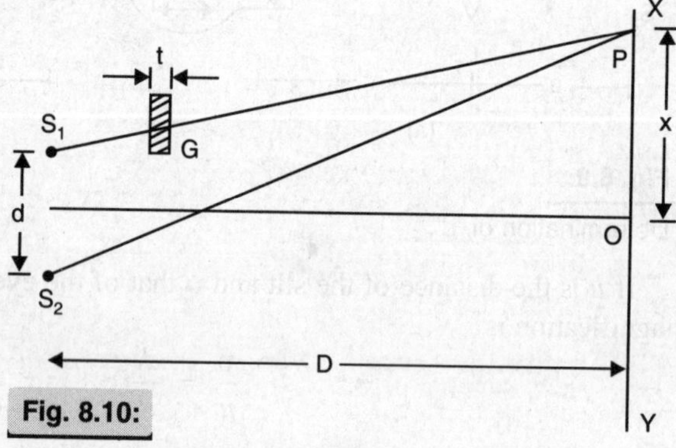

Fig. 8.10: Finding thickness of a transparent sheet.

Suppose S_1 and S_2 are the virtual coherent monochromatic sources. The point O is equidistant from S_1 and S_2, where we obtain the *central bright fringe*.

Therefore, the optical path $S_1O = S_2O$.

Let a transparent plate G of thickness t and refractive index μ be placed in the path of one of the beams (see Fig. 8.10). The optical path lengths S_1O and S_2O are now not equal and the *central bright fringe* shifts to P from O.

The light waves from S_1 to P travel partly in air and partly in the sheet G; the distance moved in air is $(S_1P - t)$ and that in the sheet is t.

The optical path $\quad \Delta_{AP} = (S_1P - t) + \mu t = S_1P + (\mu - 1)t$

The optical path $\quad \Delta_{BP} = S_2P$

The optical path difference at P is $\Delta_{AP} - \Delta_{BP} = 0$, since in the presence of the thin sheet, the optical path lengths S_1P and S_2P are *equal* and *central* zero fringe is obtained at P.

$$\therefore \quad \Delta_{S_1P} = \Delta_{S_2P}$$

$$[S_1P + (\mu - 1)t] = S_2P$$

$$\therefore \quad S_2P - S_1P = (\mu - 1)t$$

But according to the Eq. (8.7),

$$S_2P - S_1P = \frac{xd}{D}$$

where x is the **lateral shift** of the central fringe due to the introduction of the thin sheet.

$$\therefore \quad (\mu - 1)t = \frac{xd}{D}$$

Hence, the thickness of the sheet is

$$t = \frac{xd}{D(\mu - 1)} \quad \text{...(8.13)}$$

Example 8.4: *A thin sheet of glass (μ = 1.5) of thickness 6 μm is introduced in one of the paths of a biprism. It shifts the central fringe by 5 bands. Find the wavelength of light.*

Solution: The displacement of fringes is given by $(\mu - 1)t = m\lambda$. Therefore,

$$\lambda = \frac{(\mu - 1)t}{m} = \frac{(1.5 - 1) \times 6 \times 10^{-6}\ \text{m}}{5} = 600\ \text{nm}.$$

8.7 DIVISION OF WAVE AMPLITUDE- THIN FILM INTERFERENCE

An optical medium is called a *thin film* when its thickness is about the order of 1 wavelength of light in visible region. Thus, a film of thickness in the range 0.5 μm to 10 μm may be considered as a thin film. A thin film may be a thin sheet of transparent material such as glass, mica, or an air film enclosed between two transparent plates or a soap bubble.

When light is incident on a thin film, a small part of it gets reflected from the top surface and a major part is refracted into the film. Again, a small part of the refracted light is reflected back into the film by the bottom surface and the rest of it comes out of the film. A small portion of the light thus gets reflected partially several times in succession within the film (see Fig. 8.11).

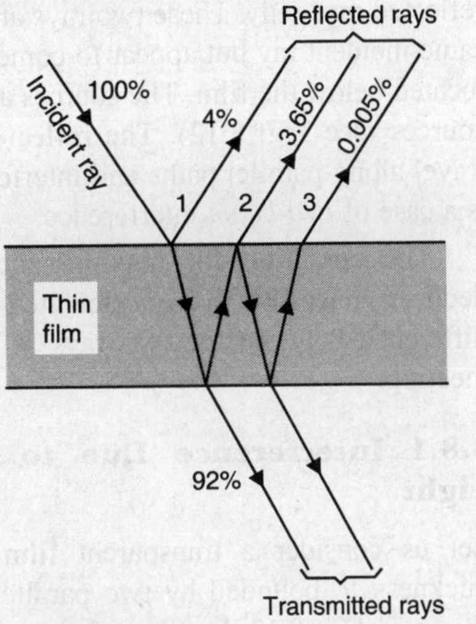

Fig. 8.11: Illustration of multiple reflections in a Thin Film.

In transparent thin films, the two bounding surfaces strongly transmit light and only weakly reflect the incident light. Therefore, only the first reflection at the top surface and the first reflection at the bottom surface will be of appreciable strength. For example, if we consider a glass plate, having a refractive index 1.52, the reflectivity of the top surface is given by

$$r = \left[\frac{1.52 - 1}{1.52 + 1}\right]^2 = 0.042 \quad \text{...(8.14)}$$

It means that about 4% of the incident light is reflected by the top surface of the glass plate, while 96% of it is transmitted into the plate. Out of the light reaching the bottom surface, again 3.8% is reflected and 92% is transmitted out of the plate. Then, again out of the 3.8% of the light 0.15% is reflected at the inner boundary of the top surface and about 3.65% is transmitted out into the air. After two reflections, the intensity will become insignificantly small. At each reflection, the intensity and hence the *amplitude of light wave is divided* into a reflected component and a refracted component. The reflected and refracted components travel along different paths and subsequently overlap to produce interference.

Therefore, the interference in thin films is referred as interference by division of amplitude. Newton and Robert Hooke first observed the thin film interference. However, Thomas Young gave the correct explanation of the phenomenon. A thin film may be uniform or non-uniform in its structure. However, as long as its thickness lies within the specified limits, interference of light occurs.

8.8 PLANE PARALLEL FILM

A transparent thin film of uniform thickness bounded by two parallel surfaces is known as a *plane parallel thin film*.

As discussed above thin films transmit incident light strongly and reflect only weakly. After two reflections, the intensities of reflected rays drop to a negligible strength. Therefore, we consider the first two reflected rays only. These two rays are derived from the same incident ray but appear to come from two sources located below the film. The sources are virtual coherent sources (see Fig. 8.12). The reflected waves 1 and 2 travel along parallel paths and interfere at infinity. This is a case of *two-beam* interference.

The condition for maxima and minima can be deduced once we have calculated the optical path difference between the two rays at the point of their meeting.

Fig. 8.12: Illustration of virtual coherent sources.

8.8.1 Interference Due to Reflected Light

Let us consider a transparent film of uniform thickness 't' bounded by two parallel surfaces as shown in Fig. 8.13. Let the refractive index of the material be μ. The film is surrounded by air on both the sides. Let us consider plane waves from a monochromatic source falling on the thin film at an angle of incidence 'i'. Part of a ray such as AB is reflected along BC, and part of it is transmitted into the film along BF. The transmitted ray BF makes an angle 'r' with the normal to the surface at the point B.

The ray BF is in turn partly reflected back into the film along FD while a major part refracts into

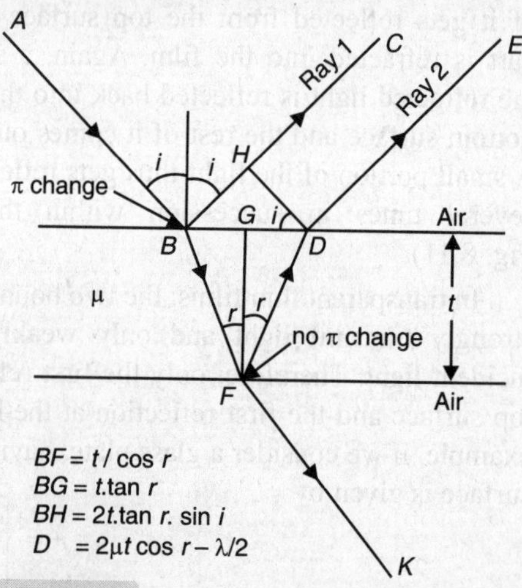

Fig. 8.13: Interference due to reflected light.

the surrounding medium along *FK*. Part of the reflected ray *FD* is transmitted at the upper surface and travels along *DE*. Since the film boundaries are parallel, the reflected rays *BC* and *DE* will be parallel to each other. The waves traveling along the paths *BC* and *BFDE* are derived from a single incident wave *AB*. Therefore they are coherent and can produce interference if they are made to overlap by a condensing lens or the eye.

(*i*) **Geometrical Path Difference:** Let *DH* be normal to *BC*. From points *H* and *D* onwards, the rays *HC* and *DE* travel equal path. The ray *BH* travels in air while the ray *BD* travels in the film of refractive index μ along the path *BF* and *FD*. The geometric path difference between the two rays is

$$BF + FD - BH.$$

(*ii*) **Optical Path Difference:**

Optical path difference $\Delta_a = \mu L$ (Δ_a is apparent O.P.D.)

$$\therefore \quad \Delta_a = \mu\,(BF + FD) - 1(BH) \qquad ...(8.15)$$

In the ΔBFD, $\angle BFG = \angle GFD = \angle r$

$$BF = FD$$

$$BF = \frac{FG}{\cos r} = \frac{t}{\cos r}$$

$$\therefore \quad BF + FD = \frac{2t}{\cos r} \qquad ...(8.16)$$

Also,

$$BG = GD$$

$$\therefore \quad BD = 2BG$$

$$BG = FG \tan r = t \tan r$$

$$\therefore \quad BD = 2t \tan r$$

In the Δ^{le} *BHD*

$$\angle HBD = (90 - i)$$

$$\angle BHD = 90^\circ$$

$$\therefore \quad \angle BDH = i$$

$$\therefore \quad BH = BD \sin i = 2\,t \tan r \sin i \qquad ...(8.17)$$

From Snell's law,

$$\sin i = \mu \sin r$$

$$\therefore \quad BH = 2t \tan r\,(\mu \sin r) = \frac{2\mu t \sin^2 r}{\cos r} \qquad ...(8.18)$$

Using the Eq. (8.17) and Eq. (8.16) into Eq. (8.15), we get

$$\Delta_a = \mu\left[\frac{2t}{\cos r}\right] - \left[\frac{2\mu t \sin^2 r}{\cos r}\right]$$

$$= \frac{2\mu t}{\cos r}\left[1 - \sin^2 r\right]$$

$$= \frac{2\mu t}{\cos r}\cos^2 r$$

$$\therefore \quad \Delta_a = 2\mu\, t \cos r \qquad ...(8.19)$$

(*iii*) **Correction on account of phase change at reflection:** When a ray is reflected at the boundary of a rarer to denser medium, a path-change of $\lambda/2$ occurs for the ray *BC* (see Fig. 8.13). There is no path difference due to transmission at *D*. Including the change in path difference due to reflection in Eq. (8.19), the *true path difference* is given by

$$\Delta_t = 2\mu\, t \cos r - \lambda/2 \qquad \text{...(8.20)}$$

8.8.2 Conditions for Maxima (Brightness) and Minima (Darkness)

Maxima occur when the optical path difference $\Delta_t = m\,\lambda$. If the difference optical path between the two rays is equal to an *integral number of full waves*, then the rays meet each other in phase.

The crests of one wave falls on the crests of the others and the waves *interfere constructively*. Thus, when the reflected rays undergo constructive interference to produce brightness or maxima at the point of their meeting.

$$2\mu\, t \cos r - \frac{\lambda}{2} = m\lambda \qquad \text{...(8.21)}$$

Condition for Brightness $2\mu\, t \cos r = m\lambda + \lambda/2$

or $$2\mu\, t \cos r = (2m + 1)\lambda/2 \qquad \text{...(8.22)}$$

Minima occur when the optical path difference is $\Delta = (2m + 1)\,\lambda/2$. If the optical path between the two rays is equal to an *odd integral number of half-waves*, then the rays meet each other in opposite phase. The crests of one wave falls on the troughs of the others and *the waves interfere destructively*. Thus, when

$$2\mu\, t \cos r - \lambda/2 = (2m + 1)\lambda/2 \qquad \text{...(8.23)}$$

the reflected rays undergo destructive interference to produce darkness. Eq. (8.23) may be rewritten as

$$2\mu\, t \cos r = (m + 1)\lambda$$

The phase relationship of the interfering waves does not change if one full wave is added to or subtracted from any of the interfering waves. Therefore $(m + 1)\lambda$ can be as well replaced by $m\lambda$ for simplicity in expression. Thus,

Condition for Darkness $2\mu\, t \cos r = m\lambda$ ...(8.24)

8.8.3 Some Important Points

(*a*) It is seen that the conditions of interference depend on three parameters, namely μt, λ and r. In the case of constant thickness (parallel) film, μt is constant. When a parallel beam of light is incident on such a film, r also remains constant. Then the interference conditions solely depend on the wavelength, λ.

(*b*) When a parallel beam of monochromatic light is incident normal to the film, the whole film will appear *uniformly* dark or *uniformly* bright. The film will appear bright in reflected light, when the film is of $\lambda/4\mu$, $3\lambda/4\mu$, $5\lambda/4\mu$, ... thick and it appears dark when its thickness is $\lambda/2\mu$, λ/μ, $3\lambda/2\mu$, ...etc. (Fig. 8.14). If the condition of constructive interference is satisfied, the film will show intense colour corresponding to the colour of the incident light.

(*c*) A change in the angle of incidence of the rays leads to a change in the path difference. Consequently, if the inclination of the film with respect to the light beam is changed gradually, we find that it will appear dark and bright (or bright and dark) in succession.

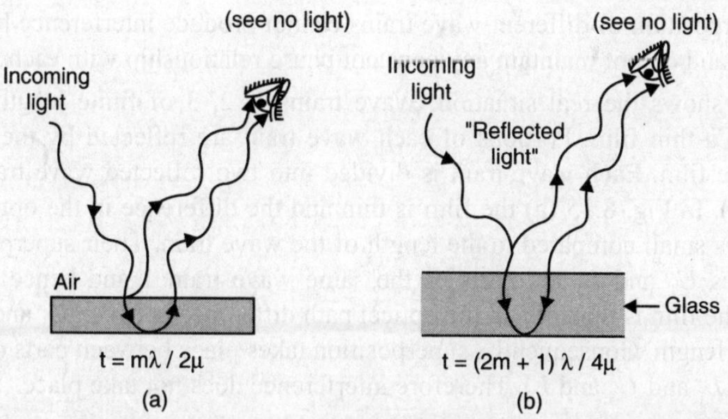

Fig. 8.14: Interference in film of constant thickness (*a*) Destructive and (*b*) Constructive Interference.

(*d*) If a parallel beam of white light falls on a parallel film, those wavelengths for which the path difference is $m\lambda$, will be absent from the reflected light. The other colours will be reflected. Therefore, the film will appear uniformly coloured with one colour being absent.

Example 8.5: *A soap film of 5 × 10⁻⁷m thick and refractive index 1.33 is seen at an angle of 35° to the normal. Find the find the missing wavelengthsthe visible spectrum.*

Solution: White light is incident on the film at an angle 30°. Let r be the angle of refraction of light into the film. r can be calculated from Snell's law $\mu = \dfrac{\sin i}{\sin r}$.

$\therefore \quad \sin r = \dfrac{\sin 30°}{1.33}$ or $r = 25.55°$ and $\cos r = 0.90$.

The absence of certain wavelengths in reflected light is due to their undergoing destructive interference and its condition is

$$2\mu\, t \cos r = m\lambda$$

To find out the missing wavelengths, we have to use different m values into the above equation and find out which of them lie in the visible region 7000 to 4000 Å.

For $m = 1$, we get $\lambda_1 = 2 \times 1.33 \times 5 \times 10^{-7}\text{m} \times 0.90 = 12 \times 10^{-7}\text{m} = 12000$ Å.

For $m = 2$, we get $\lambda_2 = (2 \times 1.33 \times 5 \times 10^{-7}\text{m} \times 0.90) \div 2 = 6 \times 10^{-7}\text{m} = 6000$ Å.

For $m = 3$, we get $\lambda_3 = (2 \times 1.33 \times 5 \times 10^{-7}\text{m} \times 0.90) \div 3 = 4 \times 10^{-7}\text{m} = 4000$ Å.

For $m = 4$, we get $\lambda_4 = (2 \times 1.33 \times 5 \times 10^{-7}\text{m} \times 0.90) \div 4 = 3 \times 10^{-7}\text{m} = 3000$ Å.

It is clear that the first wavelength lies in the infra red and the last one lies in the UV region. The middle two wavelengths lie in the visible region. Hence, missing wavelengths in the reflected light are 6000 Å and 4000 Å.

8.8.4 Restriction on Thickness of the Film

We know that interference colours are observed only in thin films but not in thick plates such as window panes or glass slabs. This is due to the fact that light waves can interfere only when both the conditions of temporal and spatial coherence are satisfied. In Fig. 8.14, we have assumed that a monochromatic wave of infinite length is incident on the film. In reality, the incident light consists of wave trains of finite length and coherence extends over the length of each wave train only. Interference can occur only when parts of the same group of wave trains overlap. Superposition of different wave trains cannot produce interference because they will be incoherent and do not maintain any constant phase relationship with each other.

Fig. 8.15 shows the real situation. Wave trains 1, 2, 3 of finite length are incident in succession on a thin film. Portions of each wave train are reflected by the top and bottom surfaces of the film. Each wave train is divided into two reflected wave trains (U_1, L_1, U_2, L_2 and U_3, L_3). In Fig. 8.15 (a) the film is thin and the difference in the optical path lengths of U_1 and L_1 is small compared to the length of the wave train. Their superposition produces interference, as U_1 and L_1 are parts of the same wave train 1 and hence are coherent. In Fig. 8.15 (b) the film is thicker and the optical path difference between U_1 and L_1 is large than the coherence length. Consequently, superposition takes place between parts of different wave trains, U_2 and L_1 and U_3 and L_2. Therefore interference does not take place.

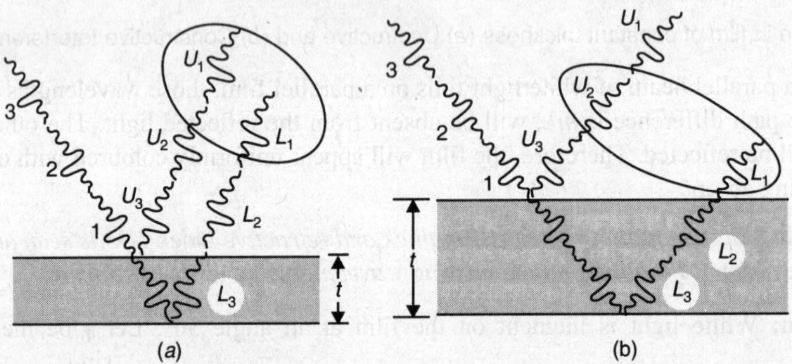

Fig. 8.15:

(a) When the film thickness is smaller than coherence length wave, superposition of reflected parts of the same wave train gives inference. (b) In thick films, incoherent wave trains superpose and no interference

It implies that interference occurs *only when* the optical path difference, Δ, between the superposing waves is less than the coherence length.

i.e., $$\Delta << l_{coh} \quad ...(8.25)$$

$$\therefore \quad (2\mu t \cos r - \lambda/2) << l_{coh} \quad ...(8.26)$$

But $$l_{coh} = \frac{\lambda^2}{\Delta\lambda}$$

$$\therefore \quad (2\mu t \cos r - \lambda/2) < \lambda^2/\Delta\lambda$$

Rearranging the terms, we obtain

$$t < \frac{\lambda\left[\frac{\lambda}{\Delta\lambda} + \frac{1}{2}\right]}{2\mu \cos r} \quad ...(8.27)$$

$\lambda/\Delta\lambda >> 1/2$ and for normal incidence $\cos r = 1$.

$$\therefore \quad t < \frac{\lambda^2}{2\mu\Delta\lambda} \quad ...(8.28)$$

Eq. (8.28) indicates that interference in thin film will be observed if the thickness of the film is less than the coherence length of the incident light waves. Normally, the coherence length of the light from ordinary sources is of the order of a fraction of a millimeter. Therefore, interference is seen with the films of thickness of the order of a few hundred microns only. It is because of this reason that thick films do not exhibit interference.

8.9 VARIABLE THICKNESS (WEDGE-SHAPED) FILM

Def: A thin film of varying thickness having a zero thickness at one end and progressively increasing to a particular thickness at the other end is known as a *wedge*.

Experimental arrangement for observing the interference pattern in a wedge shaped air-film is shown in Fig. 8.16. If a parallel beam *of monochromatic* light illuminates the wedge from above, the rays reflected from its two bounding surfaces will not be parallel. They appear to diverge from a point near the film. The path difference between the rays reflected from the upper and lower surfaces of the air film varies along its length due to variation in thickness. Therefore, alternate bright and dark fringes are observed on its top surface (Fig. 8.17). The fringes are localized at the top surface of the film.

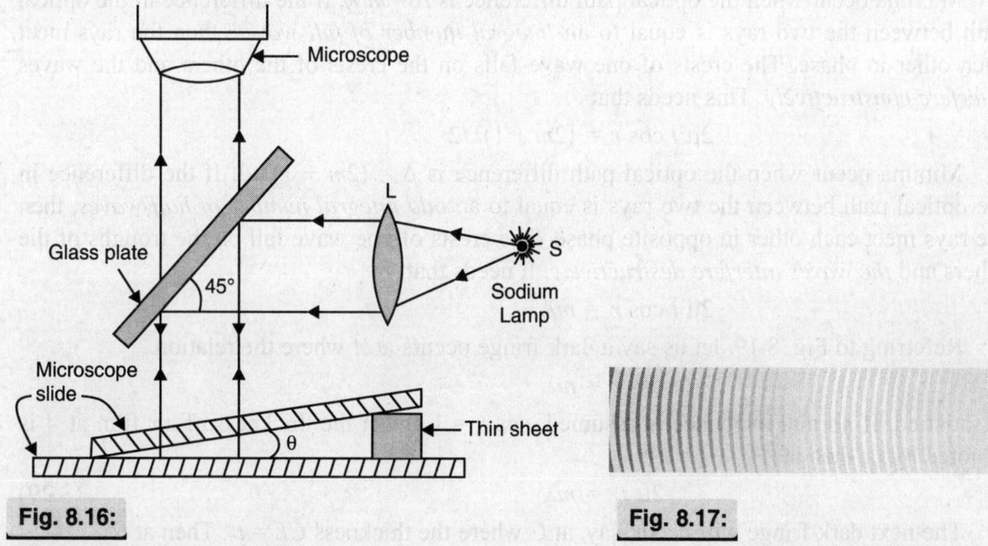

Fig. 8.16: Experimental arrangement.

Fig. 8.17: Fringes in a wedge-shaped film.

When the light is incident on the wedge from above, it gets partly reflected from the glass-to-air boundary at the top of the air film. Part of the light is transmitted through the air film and gets reflected partly at the air-to-glass boundary, as shown in Fig. 8.18. Two coherent rays *BC* and *DE* are derived by *division of amplitude* from the same ray *AB*. The rays are close enough if the thickness of the film is of the order of a wavelength of light. For small film thickness the rays interfere to produce darkness or brightness depending on the phase difference. The thickness of the glass plates is large compared with the wavelength of the incident light. Hence, the observed interference effects are entirely due to the wedge-shaped air film. The shape of the resulting fringes is shown in Fig. 8.17.

The optical difference between the two rays *BC* and *FE* is given by

$$\Delta = 2\mu\, t \cos r - \lambda/2$$

where $\lambda/2$ takes account the gain of half–wave due to the abrupt jump of π radians in the phase of the wave reflected from the bottom boundary of air – to – glass.

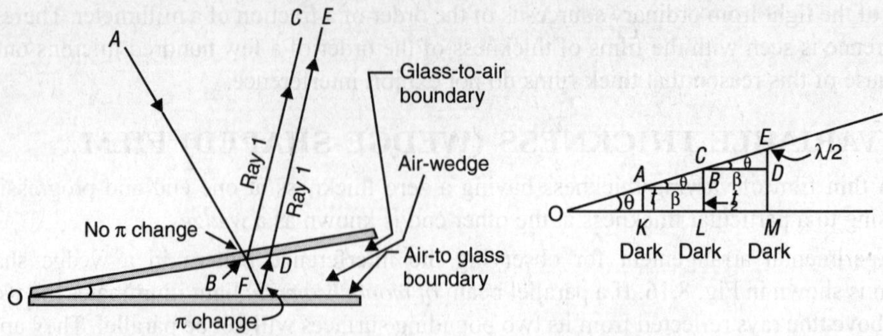

Fig. 8.18: Wedge shaped air-film.

Fig. 8.19: Fringes and Fringe width.

Maxima occur when the optical path difference is $\Delta = m\,\lambda$. If the difference in the optical path between the two rays is equal to an *integral number of full waves*, then the rays meet each other in phase. The crests of one wave falls on the crests of the others and the waves *interfere constructively*. This needs that

$$2\mu\, t \cos r = (2m + 1)\lambda/2$$

Minima occur when the optical path difference is $\Delta = (2m + 1)\lambda/2$. If the difference in the optical path between the two rays is equal to an *odd integral number of half-waves*, then the rays meet each other in opposite phase. The crests of one wave fall on the troughs of the others and *the waves interfere destructively*. It needs that

$$2\mu\, t \cos r = m\lambda.$$

Referring to Fig. 8.19, let us say a dark fringe occurs at *A* where the relation

$$2\mu\, t \cos r = m\lambda$$

is satisfied. If normal incidence is assumed, $\cos r = 1$ and if the thickness of air film at *A* is denoted by t_1, then at *A*

$$2\mu\, t_1 = m\lambda \qquad \text{...(8.29)}$$

The next dark fringe will occur, say, at *C* where the thickness $CL = t_2$. Then at C

$$2\mu\, t_2 = (m + 1)\lambda \qquad \text{...(8.30)}$$

Subtracting Eq. (8.29) from Eq. (8.30), we get

$$2\mu(t_2 - t_1) = \lambda \qquad \text{...(8.31)}$$

But $\qquad (t_2 - t_1) = BC$

$\therefore \qquad 2\mu\,(BC) = \lambda$

or
$$BC = \frac{\lambda}{2\mu} \qquad \text{...(8.32)}$$

From the $\Delta^{le} ABC$, $\angle CAB = \theta$ and $BC = AB \tan\theta$

$$\therefore \qquad (AB)\tan\theta = \frac{\lambda}{2\mu} \qquad \text{...(8.33)}$$

AB is the distance between successive dark fringes and it also equals the separation of the successive bright fringes. It is, therefore, called the **fringe width**, β. That is $AB = \beta$. We may write Eq. (8.33) as

$$\therefore \qquad \beta = \frac{\lambda}{2\mu \tan \theta} \qquad ...(8.34)$$

For small values of θ, tan θ ≈ θ.

$$\therefore \qquad \beta = \frac{\lambda}{2\mu\theta} \qquad ...(8.35)$$

Example 8.6: *Find the wedge angle for a thin glass wedge of refractive index 1.52, fringe spacing is 0.1 mm, and wavelength of light is 5893 Å.*

Solution: The fringe width $\beta = \lambda/2\mu\theta$.

$$\therefore \qquad \theta = \frac{\lambda}{2\mu\beta} = \frac{5893\times10^{-10}\,\text{m}}{2\times1.52\times10^{-4}\,\text{m}} = 1.938 \text{ rad} = \mathbf{0.11°}.$$

8.9.1 Salient Features of the Interference Pattern

(*i*) Fringe at the apex is dark

(*ii*) Fringes are straight and parallel

(*iii*) Fringes are equidistant

(*iv*) Fringes are localized

(*i*) **Fringe at the apex is dark:** At the apex, the two glass slides are in contact with each other. Therefore, the thickness of the air film at the contact edge is negligible ($t \cong 0$). The optical path difference there becomes

$$\Delta = 2\mu t - \lambda/2 = 0 - \lambda/2 = -\lambda/2 \qquad ...(8.36)$$

It implies that a path difference of λ/2 or a phase difference of π occurs between the reflected waves at the edge. The two waves interfere destructively. Therefore, the fringe at the apex is always dark (Fig. 8.20).

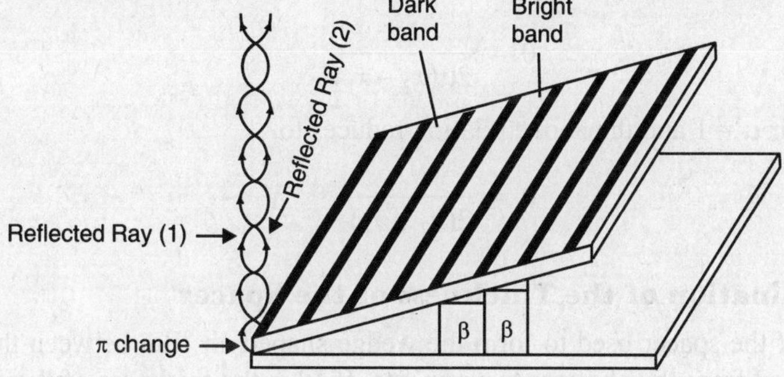

Fig. 8.20: At the contact edge, the reflected rays (1) and (2) produce a dark band

(*ii*) **Straight and parallel fringes:** Each fringe in the pattern is produced by the interference of rays reflected from sections of the wedge having the same thickness. The locus of points having the same thickness lies along lines parallel to the contact edge. Therefore, the fringes are straight. Since the fringes are equidistant, they will be parallel.

(*iii*) **Equidistant fringes:** The fringe width β is given by

$$\beta \approx \lambda/2\theta \qquad ...(8.37)$$

where λ is the wavelength of the incident light and θ is the angle of the wedge. As the quantities λ and θ are constants, β is constant for a given wedge angle. Therefore, the fringes are equidistant.

(*iv*) **Localized fringes:** The fringes are very close to the top surface of the wedge and can be seen with a microscope.

8.9.2 Determination of the Wedge Angle

The wedge angle θ can be experimentally found using a travelling microscope. Using the microscope the positions of dark fringes at two distant points *Q* and *R* are noted (Fig. 8.21). Let the distance *OQ* be x_1 and *OR* be x_2. Let the thickness of the wedge be t_1 at *Q* and t_2 at *R*.

The dark fringe at *Q* is given by

$$2\mu\, t_1 = m\lambda \qquad ...(8.38)$$

But as θ is very small, we can write

$$t_1 = x_1 \tan\theta \cong x_1\theta$$

$$\therefore \quad 2\mu\, x_1\,\theta = m\lambda \qquad ...(8.39)$$

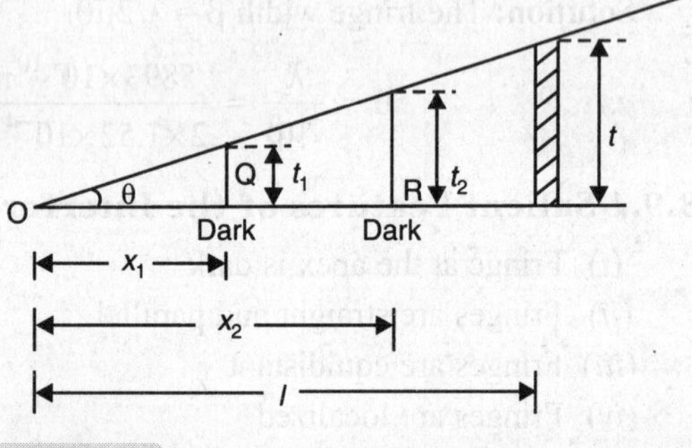

Fig. 8.21: Determination of the wedge angle

We can write similarly for the dark fringe at *R* as

$$2\mu\, x_2\,\theta = (m+N)\lambda \qquad ...(8.40)$$

where *N* is the number of dark fringes lying between the positions *Q* and *R*. Subtracting Eq. (8.39) from Eq. (8.40), we get

$$2\mu(x_2 - x_1)\theta = N\lambda$$

$$\therefore \quad \theta = \frac{N\lambda}{2\mu(x_2 - x_1)} \qquad ...(8.41)$$

In case of air μ = 1 and the above relation reduces to

$$\theta = \frac{N\lambda}{2(x_2 - x_1)} \qquad ...(8.42)$$

8.9.3 Determination of the Thickness of the Spacer

The thickness of the spacer used to form the wedge shaped air film between the glass slides can be determined from the above measurements. If '*t*' is the thickness of the spacer (foil or wire) used, we can write

$$t = l\tan\theta \cong l\,\theta \qquad ...(8.43)$$

where *l* is the length of the air wedge. Using the relation for θ (Eq. 8.42), we get

$$\therefore \quad t = \frac{l N\lambda}{2(x_2 - x_1)} \qquad ...(8.44)$$

8.9.4 Number of Dark Fringes in an Air Wedge

The fringe width in air wedge is given by $\beta = \dfrac{\lambda}{2\theta}$

The wedge angle is given by $\theta = \frac{t}{l}$

$$\therefore \quad \beta = \frac{\lambda l}{2t} \quad ...(8.45)$$

The number of dark fringes observed, N, is related to l, as

$$l = N\beta \quad ...(8.46)$$

$$\therefore \quad \beta = \frac{\lambda N \beta}{2t}$$

$$\therefore \quad N = \frac{2t}{\lambda} \quad ...(8.47)$$

8.10 COLOURS IN THIN FILMS

The colours exhibited in reflection by thin films of oil, mica, soap bubbles and coatings of oxides on heated metals etc are due to interference of light from an extended source such as sky. Thomas Young explained the origin of colours in thin films. It may be understood as follows. The reflected rays from the top and bottom surfaces of the film are very close to each other and are in a position to interfere. The optical path difference between the interfering rays is $\Delta = 2\mu t \cos r - \lambda/2$. It is seen that the path difference depends upon the thickness t of the film, the wavelength λ and the angle r, which is related to the angle of incidence of light on the film.

White light consists of a range of wavelengths and for specific values of t and r, waves of only certain wavelengths (colours) constructively interfere. Therefore, only those colours are present in the reflected light. The other wavelengths interfere destructively and hence are absent from the reflected light. Hence, the film at a particular point appears coloured. As the thickness and the angle of incidence vary from point to point, different colours are intensified at different places. The colours seen are not isolated colours, as at each place there is a mixture of colours. The composition of colours is different at different places and contours of impressive hues are observed over the entire surface of the film.

8.11 NEWTON'S RINGS

When a plano-convex lens L of a large radius of curvature is placed on a sheet of plane glass AB, a thin circular air film of variable thickness is formed between them. The locus of all points corresponding to a specific thickness of air film lies on a circle whose centre lies at O. Therefore, interference fringes are observed in the form of a series of concentric rings having their centre at C. Newton was the first to observe and study these concentric circular fringes (Fig. 8.22) and therefore, they are called Newton's rings.

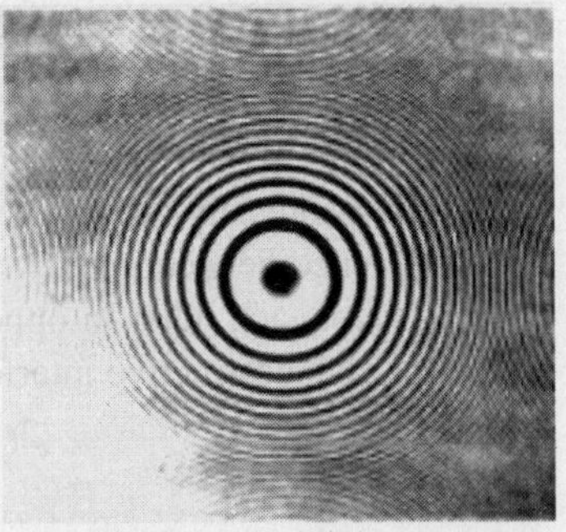

Fig. 8.22: Newton's rings as observed through a microscope.

S is the extended source and the lens L makes the light parallel (Fig. 8.23). A glass plate inclined at 45° to the horizontal is useful for reflecting light normally down onto a plano-convex lens placed on a flat glass plate. A part of the incident light is reflected from the glass-to-air boundary from point D (Fig. 8.24) and remaining light is transmitted through the air film. The transmitted light is again reflected from the air-to-glass boundary from point

J. Incident ray *CD* is divided into two coherent rays by division of amplitude. The rays 1 and 2 are close to each other and produce interference fringes. The condition of brightness or darkness depends on the path difference between the two reflected light rays, which in turn depends on the thickness of the air film at the point of incidence.

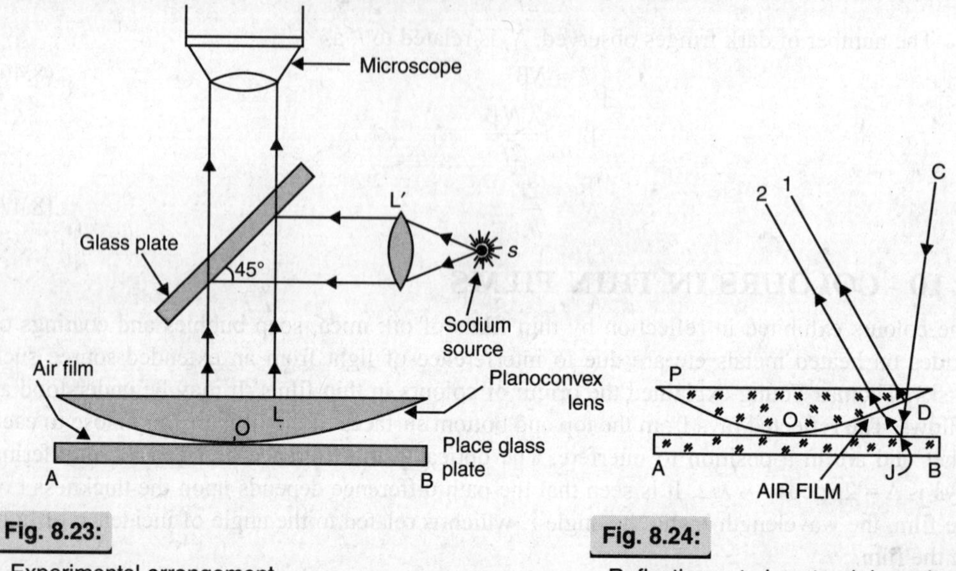

Fig. 8.23: Experimental arrangement.

Fig. 8.24: Reflections at glass-to-air boundary.

8.11.1 Condition for Bright and Dark Rings

The optical path difference between the rays is given by $\Delta = 2\mu t \cos r - \lambda/2$. Since $\mu = 1$ for air and $\cos r = 1$ for normal incidence of light,

$$\Delta = 2t - \lambda/2 \quad \text{...(8.48)}$$

Ientensity maxima occur when the optical path difference $\Delta = m\lambda$. If the optical path difference between the two rays is equal to an *integral number of full waves*, then the rays meet each other in phase. The crests of one wave falls on the crests of the others and the waves *interfere constructively*. Thus, if $2t - \lambda/2 = m\lambda$

$$2t = (2m + 1)\lambda/2 \quad \text{...(8.49)}$$

bright fringe is obtained.

Minima intensity occur when the optical path difference is $\Delta = (2m + 1)\lambda/2$. If the optical path difference between the two rays is equal to an *odd integral number of half-waves*, then the rays meet each other in opposite phase. The crests of one wave fall on the troughs of the other and give destructive interference.

Hence, if
$$2t - \lambda/2 = (2m + 1)\lambda/2$$
$$2t = m\lambda \quad \text{...(8.50)}$$

and dark fringe is produced.

8.11.2 Circular Fringes

In Newton's ring arrangement, a thin air film is between a plano-convex lens and a glass plate. The thickness of the air film at the point of contact is zero and gradually increases as we move outward. The locus of points where the air film has the same thickness then fall on a circle whose centre is the point of contact. Thus, the thickness of air film is constant at points on any circle having the point of lens-glass plate contact as the centre. The fringes are therefore circular Fig. 8.22.

8.11.3 Radii of Dark Fringes

Let R be the radius of curvature of the lens Fig. 8.25. Let a dark fringe be located at Q. Let the thickness of the air film at Q be $PQ = t$. Let the radius of the circular fringe at Q be $OQ = r_m$. By the Pythagorus theorem,

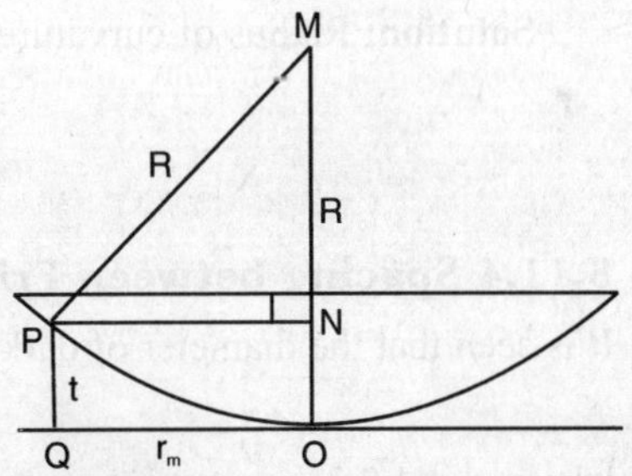

Fig. 8.25:

Finding the radius of rings.

$$PM^2 = PN^2 + MN^2$$

$$\therefore \quad R^2 = r_m^{\ 2} + (R - t)^2$$

$$\therefore \quad r_m^{\ 2} = 2Rt - t^2 \qquad \text{...(8.51)}$$

As $\quad R >> t, \quad 2Rt >> t^2$.

$$\therefore \quad r_m^{\ 2} \cong 2Rt \qquad \text{...(8.52)}$$

The condition for darkness at Q is that

$$2t = m\lambda$$

$$\therefore \quad r_m^{\ 2} \cong m\lambda R$$

$$r_m = \sqrt{m\lambda R} \qquad \text{...(8.53)}$$

The radii of dark fringes can be found by inserting values 1,2,3, ... for m.

Thus,

$$r_1 = \sqrt{1\lambda R} \quad \text{or} \quad r_1 \propto \sqrt{1}$$

$$r_2 = \sqrt{2\lambda R} \quad \text{or} \quad r_2 \propto \sqrt{2}$$

$$r_3 = \sqrt{3\lambda R} \quad \text{or} \quad r_3 \propto \sqrt{3} \text{ and so on}$$

It means that *the radii of the dark rings are proportional to square root of the natural numbers.*

The above relation also implies that $\quad r_m \propto \sqrt{\lambda}$

Thus, *the radius of the mth dark ring is proportional to square root of wavelength.*

Ring Diameter: Diameter of mth dark ring

$$D_m = 2r_m$$

$$D_m = 2\sqrt{2Rt}$$

or
$$D_m = 2\sqrt{m\lambda R} \qquad \text{...(8.54)}$$

Radius of Curvature of the Lens: By measuring diameter of the mth ring and known value of wavelength, radius of curvature of the lens can be calculated as $R = \dfrac{D^2 m}{4m\lambda}$

Example 8.7: *Calculate the radius of curvature of the lens in a Newton's rings experiment, if the diameter of 10th dark ring is 0.5 cm and wavelength 6000 Å.*

Solution: Radius of curvature,

$$R = \frac{(D/2)^2}{m\lambda} = \frac{(0.5\times10^{-2}/2)^2\,\text{m}^2}{10\times6000\times10^{-10}\,\text{m}} = \mathbf{104\ cm.}$$

8.11.4 Spacing between Fringes

It is seen that the diameter of dark rings is given by

$$D_m = 2\sqrt{m\lambda R}$$

where $m = 1,2,3, \ldots$

The diameters of dark rings are proportional to the square root of the natural numbers. Therefore, the diameter of the ring does not increase in the same proportion as the order of the ring, for example, if m increases as 1, 2, 3, 4, ... the diameters are

$$D_1 = 2\sqrt{\lambda R}$$

$$D_2 = 2(1.4)\sqrt{\lambda R}$$

$$D_3 = 2(1.7)\sqrt{\lambda R}$$

$$D_4 = 2(2)\sqrt{\lambda R} \quad \text{and so on}$$

Therefore, the rings get closer and closer, as m increases. This is why the rings are not evenly spaced.

8.11.5 Fringes of Equal Thickness

The occurrence of alternate bright and dark Newton's rings depends on the optical path difference arising between the reflected rays. If the light falls normally on the air film the optical path difference between the waves reflected from the two surfaces of the film is

$$\Delta = 2t - \lambda/2$$

It is seen that the path difference between the reflected rays arises due to the variation in the thickness 't' of the air film. Reflected light will be of minimum intensity for those thickness for which the path difference is $m\lambda$ and maximum intensity for those thickness for which the path difference is $(2m + 1)\lambda/2$. Thus, each maxima and minima is a locus of constant film thickness. Therefore, the fringes are known as fringes of equal thickness.

8.11.6 Dark Central Spot

The central spot is dark as seen by reflection. Newton's rings are produced due to superposition of light rays reflected from the top and bottom surfaces of a thin air film enclosed between a plano-convex lens and a plane glass plate. The occurrence of brightness or darkness depends on the optical path difference arising between the reflected rays. The optical path difference is given by $\Delta = 2t - \lambda/2$.

At the point of contact 'O' of the lens and glass plate (Fig. 8.26), the thickness of air film is negligibly small compared to a wavelength of light.

$$\therefore \qquad t \cong 0$$

$$\therefore \qquad \Delta \cong \lambda/2$$

The wave reflected from the lower surface of the air film suffers a phase change of π while the wave reflected from the upper surface of the film does not suffer such change.

Thus, the superposing waves are out of step by $\lambda/2$ which is equivalent to a phase difference of 180° (or π rad). Thus the two interfering waves at the centre are opposite in phase and produce a dark spot.

Newton's rings can be observed in refracted light, that is, with transmitted light. In transmitted light the centre is bright (Fig. 8.28 b). To a dark ring in reflected light, there occurs a bright ring in transmitted light. Thus, the ring system in transmitted light is complementary to that seen in reflected light. However, the rings in transmitted light are much poorer in contrast.

Dark

Reflected ray (1)

Reflected ray (2)

π change

O

Fig. 8.26: Illustration of Dark central spot.

8.11.7 Determination of Wavelength of Light

A plano-convex lens of large radius of curvature (about 100 cm) and a flat glass plate are cleaned. The lens is kept with its convex face on the glass plate and they are held in position with the help of a metal ring arrangement. The system is held under a low power traveling microscope kept before a sodium vapour lamp. It is arranged that the yellow light coming from the sodium lamp falls on a glass plate held at 45° light beam. The light is turned through 90° and is incident normally on the lens-plate system. The microscope is adjusted till the circular rings came into focus. The centre of the cross-wire is made to come into focus on the centre of the dark spot, which is at the centre of the circular ring system. Now, turning the screw the microscope is moved on the carriage slowly towards one side, say right side Fig. 8.27 (a). As the cross-wires move in the field of view, dark rings are counted. The movement is stopped when the 22nd dark ring is reached. Then the microscope is moved in the opposite side and stopped at the 20th or 19th dark ring. The vertical cross-wire is made tangential to the 19th ring and the reading is noted with the help of the scale graduated on the carriage. Thus, starting from the 19th ring, the tangential positions of the 18th, 17th, 16th, ..., 5th dark rings are noted down. Now, the microscope is moved quickly to the left side of the ring system and it is stopped at the 5th dark ring. The cross-wire is again made tangential to the 5th dark ring and its position is noted. The difference between the readings on right and left sides of the 5th dark ring gives its diameter value. The procedure is repeated till 19th

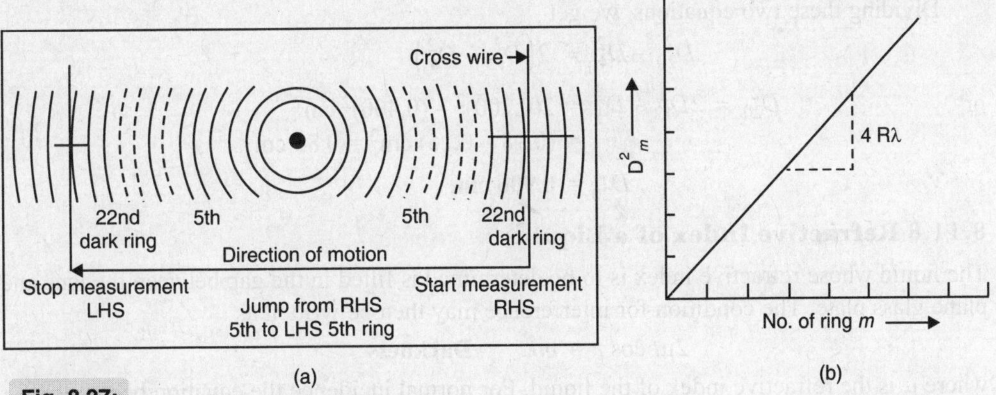

Fig. 8.27: Determination of wavelength of light: (a): Measuring ring diameters, (b): Graph between ring diameter square and its number

ring is reached and its reading is noted. From the value of the diameters the squares of the diameters are calculated. A graph is plotted between $D_m^{\ 2}$ between the ring number '*m*.' A straight line would be obtained, as shown in Fig. 8.27 (b).

We have

$$D_m^{\ 2} = 4m\lambda R \quad ...(8.55)$$

For the $(m + p)^{th}$ ring,

$$D^2_{m+p} = 4(m + p)\lambda R \quad ...(8.56)$$

$$D^2_{m+p} - D^2_{m} = 4p\lambda R$$

$$\lambda = \frac{D^2_{m+p} - D^2_m}{4pR} \quad ...(8.57)$$

The slope of the straight line (Fig. 8.27 b) gives the value of $4\lambda R$. Thus,

$$\lambda = \frac{\text{Slope}}{4R} \quad ...(8.58)$$

The radius of curvature R of the lens may be determined using a spherometer and λ is computed with the help of the above equation.

Example 8.8: *Calculate the wavelength of light used in a Newton's rings experiment if the diameter of the 15th ring is 0.59 cm and that of the 5th ring is 0.336 cm and the radius of the plano-convex lens is 100 cm.*

Solution: $\lambda = \frac{D^2_{m+p} - D^2_m}{4pR} = \frac{D^2_{15} - D^2_5}{4\times 10\times R} = \frac{(5.9 - 3.36)^2 \times 10^{-6}\,\text{m}^2}{4\times 10\times 1\text{m}}$ = **5880 Å**

Example 8.9: *In a Newton's rings experiment the diameter of the 4th and 12th dark rings are 0.400 cm and 0.700 cm respectively. Determine the diameter of 20th dark ring.*

Solution:

$$\lambda = \frac{D^2_{m+p} - D^2_m}{4pR} = \frac{D^2_{12} - D^2_4}{4\times 8\times R}$$

Considering 20th and 4th dark rings,

$$\lambda = \frac{D^2_{20} - D^2_4}{4\times 16\times R}.$$

Dividing these two equations, we get

$$D^2_{20} - D^2_4 = 2\left(D^2_{12} - D^2_4\right)$$

or

$$D^2_{20} = 2D^2_{12} - D^2_4 = 2(0.700)^2 - (0.400)^2\ \text{cm}^2$$

$$= (0.98 - 0.16)\ \text{cm}^2 = 0.82\ \text{cm}^2.$$

$$\therefore \quad D_{20} = \mathbf{0.906\ cm.}$$

8.11.8 Refractive Index of a Liquid

The liquid whose refractive index is to be determined is filled in the gap between the lens and plane glass plate. The condition for interference may then be written as

$$2\mu t \cos r = m\lambda \quad \textbf{Darkness}$$

where μ is the refractive index of the liquid. For normal incidence the equation becomes

$$2\mu t = m\lambda$$

The diameter of mth dark ring is given by

$$[D_m^2]_L = \frac{4m\lambda R}{\mu} \quad ...(8.59)$$

Similarly, the diameter of the $(m + p)^{th}$ ring is given by

$$[D_{m+p}^2]_L = \frac{4(m+p)\lambda R}{\mu} \quad ...(8.60)$$

Subtracting Eq. (8.59) from Eq. (8.60), we get

$$\left[D_{m+p}^2\right]_L - \left[D_m^2\right]_L = \frac{4p\lambda R}{\mu} \quad ...(8.61)$$

But we know that

$$(D_{m+p}^2)_{air} - (D_m^2)_{air} = 4p\lambda R \quad ...(8.62)$$

$$\therefore \quad \mu = \frac{\left(D_{m+p}^2\right)_{air} - \left(D_m^2\right)_{air}}{\left(D_{m+p}^2\right)_{liq} - \left(D_m^2\right)_{liq}} \quad ...(8.63)$$

Example 8.10: *In the example 8.8, find the diameter of 15th ring, if air film is replaced with liquid of refractive index 1.5?*

Solution:
$$\lambda = \frac{D_{m+p}^2 - D_m^2}{4pR} = \frac{D_{15}^2 - D_5^2}{4\times 10\times R}$$
$$= \frac{D_{15}^2 - D_5^2}{4\times 10\times R} = \frac{(0.59^2 - 0.336^2)\ \text{cm}^2}{4\times 10\times 100\ \text{cm}} = 5902\ \text{Å}.$$

The diameters of the Newton's rings will be reduced when the air film is replaced with a liquid film.

For example,
$$(D_{15}^2)_{liq} = \frac{(D_{15}^2)_{air}}{\mu} = \frac{0.59^2\ \text{cm}^2}{1.5} = 0.232\ \text{cm}^2$$

$$\therefore \quad D_{15} = \sqrt{0.232\ \text{cm}^2} = \mathbf{0.4732\ cm}$$

Example 8.11: *In a Newton's rings experiment the diameter of 10th ring modifies from 1.40 to 1.27 cm when the air film is replaced by a drop of liquid. Find the refractive index of the liquid.*

Solution:
$$\mu = \frac{\left(D_m^2\right)_{air}}{\left(D_m^2\right)_{liq.}} = \frac{(1.40\ \text{cm})^2}{(1.27\ \text{cm})^2} = 1.215.$$

8.11.9 Newton's Rings by Transmitted Light

Newton's rings are observed also when we view the light transmitted through the lens-plate combination. The light rays passing through the system gets partly reflected at the top and bottom surfaces of the air film enclosed between the lens and glass plate [Fig. 8.28 (a)]. The optical path difference between two rays reflected from the top of the air film and the bottom of the air film is given by

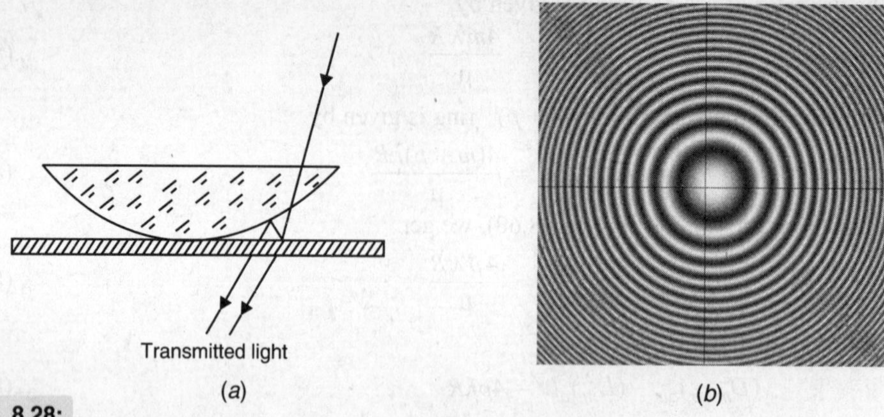

Fig. 8.28:

Transmitted light (a) Glass to air boundary, (b) Newton's rings in transmitted light.

$$\Delta = 2\ \mu t \cos r$$

Since $\mu = 1$ for air and $\cos r = 1$ for normal incidence of light,

$$\Delta = 2t$$

Intensity maxima occur when the optical path difference $\Delta = m\ \lambda$. Thus, if

$$2t = m\lambda \qquad ...(8.64)$$

bright fringe is obtained.

Intensity minima occur when the optical path difference is $\Delta = (2m + 1)\lambda/2$. Hence, if

$$2t = (2m + 1)\lambda/2 \qquad ...(8.65)$$

dark fringe is produced.

From $t = \dfrac{r_m^2}{2R}$, the radii of the bright and dark rings can be calculated. We find that for

bright rings $$r_m = \sqrt{m\lambda R} \qquad ...(8.66)$$

dark rings $$r_m = \sqrt{\frac{(2m+1)\lambda R}{2}} \qquad ...(8.67)$$

Thus, the ring system due to transmitted light is just opposite to that we observe in the reflected light. Wherever we get bright rings in the reflected light, we observe dark rings in the transmitted light and vice versa. At the center, we find bright spot in the transmitted light [Fig. 8.28 (b)].

8.12 APPLICATIONS OF INTERFERENCE

The applications of interference phenomenon are wide and varied. Interference is used for making precision measurements. For example, the wavelength of light can be measured with accuracy up to eight significant digits. Interference is used for measuring small displacements. The refractive indices of liquids and gases are measured using interference. We study here two important applications, which utilize the phenomenon of thin film interference.

8.12.1 Testing of Flatness of Surfaces

In modern technology thin film interference is widely used. One of the applications is testing of flatness of surfaces. Machine components retain surface irregularities left after machining. The extent of suitability of the component for a particular application depends on the irregularities which act as sources of stress leading to fatigue cracks. The surfaces of components which are going to be subjected to high stress and load reversals are therefore required to have a smooth surface finish.

The smoothness of a surface can be quickly inspected visually by keeping an optical flat on the component at an angle and illuminating it with a monochromatic light (Fig. 8.29). The air wedge formed between the component and optical flat produces straight and equidistant fringes if the component surface is smooth. If the fringes are curved towards the contact edge, the surface is concave and if the fringes curve away, it is convex (Fig. 8.29).

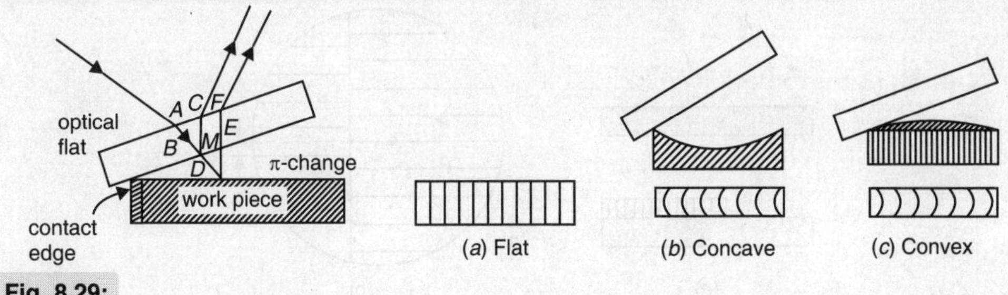

Fig. 8.29:

Testing of flatness of surfaces.

8.12.2 Testing of a Lens Surface

One of the important uses of Newton's rings is in the testing of the optical components manufactured for use in telescopes and other instruments. The grinding of a lens surface is tested by keeping it on a master. A master is an optical flat which is a cylindrical disc made of fused quartz. The two faces of the optical flat are perfectly parallel to each other. The departure from the flatness of each face is less than a light wavelength. If a lens is ground perfectly, a circular fringe pattern is observed. Otherwise variations are observed which give an indication of how the lens must be ground and polished to remove the imperfections. High quality lenses are ground with a precision of less than a light wavelength.

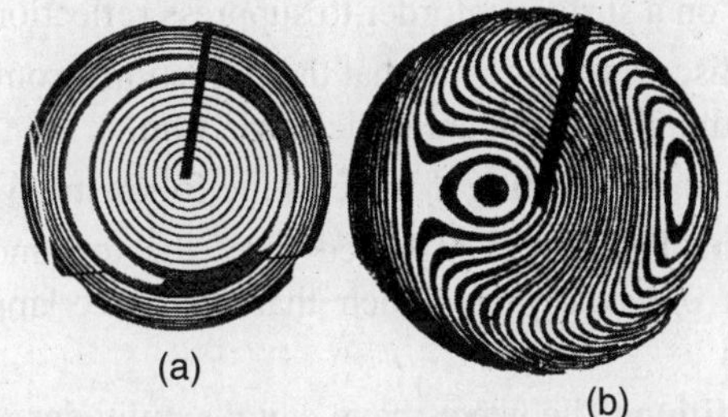

Fig. 8.30:

Lens inspection using Newton's rings (a) Circular ring pattern indicates the high quality of grinding. (b) Distorted pattern indicates irregularities.

8.12.3 Thickness of a Thin Film Coating

Dielectric and metallic thin films are often coated on optical components, solar cells etc. One of the methods of determination of thickness of such thin films is based on multiple beam interference. A partially coated substrate is used for the determination. The surfaces of the substrate and the thin film on it are coated with a transparent metallic film of uniform thickness. A glass plate is also coated on one of its surfaces with the transparent metallic film. When the substrate and the glass plate are placed in contact and examined under monochromatic light, the reflected light shows a fringe system, as shown in Fig. 8.31 (b). A shift occurs in the fringes as we pass from the region occupied by thin film to the region where thin film is absent.

The amount of displacement of one set of the fringes with respect to the second set of fringes is given by

$$s = 2t \quad \text{or} \quad t = s/2$$

where t is the thickness of the thin film. By measuring 's', t can be calculated.

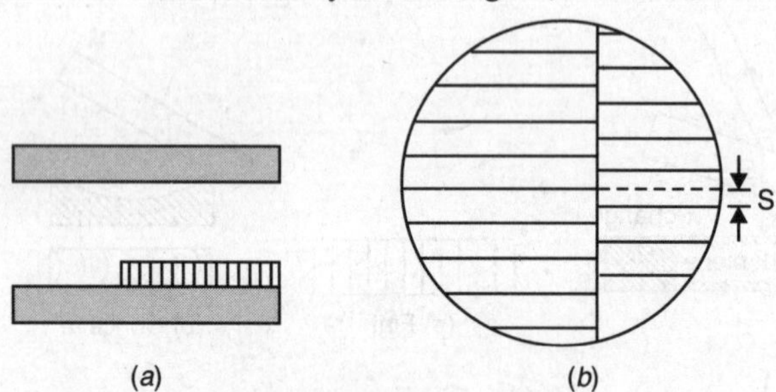

Fig. 8.31: Determination of the thickness of a thin coating.

8.12.4 Anti-Reflecting Coatings

Optical instruments such as telescopes and cameras use multicomponent glass lenses. When light is incident on the lens, part of the incident light is reflected away and that much amount of light is lost and wasted. When more surfaces are there, the number of reflections will be large and the quality of the image produced by a device will be poor. In case of solar cells, which operate on sunlight (daylight), the electrical energy produced will be less because of the loss of part of light energy due to reflection, at the cell surface. It is found that coating the surface with a thin transparent film of suitable refractive index can reduce such loss of energy due to reflections at surface. Such coatings are called **antireflection coatings**. Thus,

"Antireflection (AR) coatings are thin transparent coatings of optical thickness of one–quarter wavelength given on a surface in order to suppress reflections from the surface."

Alexander Smakula discovered in 1935 that the reflections from a surface can be reduced by coating the surface with a thin transparent dielectric film.

A thin film can act as an AR coating if it meets the following two conditions:

(*i*) **Phase condition:** The waves reflected from the top and bottom surfaces of the thin film are in *opposite phase* such that their overlapping leads to destructive interference, and

(*ii*) **Amplitude condition:** The waves have *equal* amplitudes.

The above conditions enable us determine respectively (*a*) the required thickness of the film and (*b*) the refractive index of the material to be used for forming the film.

(*i*) Phase condition and minimum thickness of the film: Let the thickness of the film be t and the refractive of the film-material be μ_f.The phase condition requires that the waves (ray 1 and ray 2) reflected from the top and bottom surfaces of thin film be 180° out of phase. It requires that the optical path difference between the two rays must equal one half-wave or an odd number of half-waves.

Referring to Fig. 8.32, the optical path difference between ray 1 and ray 2 is

$$\Delta = 2\mu_f t \cos r - \lambda/2 - \lambda/2$$

the first $\lambda/2$ corresponds to the π change at the top surface of the film (air-to-film boundary) and the second $\lambda/2$ to the π change that occurs at the film-to glass boundary. If we assume normal incidence of light, $\cos r = 1$ and the above equation reduces to

$$\Delta = 2\mu_f t - \lambda = 2\mu_f t$$

Fig. 8.32: Path difference between 2 rays.

We wrote the above equality remembering that an *addition of a full wave or subtraction of a full wave from a train of waves does not affect the original phase relation*. The ray 1 and ray 2 interfere destructively if the optical path difference satisfies the condition that

$$\Delta = (2m + 1)\lambda/2.$$

Thus, it requires that $2\mu_f t = (2m + 1)\lambda/2$

For the film to be transparent, its thickness should be a minimum, which happens when $m = 0$.

$$2\mu_f t_{\min} = \lambda/2$$

$$\therefore \quad t_{\min} = \frac{\lambda}{4\mu_f} \qquad ...(8.68)$$

It means that the optical thickness of the *AR* coating should be of one-quarter wavelength. Such quarter-wavelength coatings suppress the reflections and cause the light to pass into the transmitted component.

(*ii*) **Amplitude condition:** The amplitude condition requires that the amplitudes of reflected rays, ray 1 and ray 2 are equal. That is,

$$E_1 = E_2 \qquad ...(8.69)$$

It requires that

$$\left[\frac{\mu_f - \mu_a}{\mu_f + \mu_a}\right]^2 = \left[\frac{\mu_g - \mu_f}{\mu_g + \mu_f}\right]^2 \qquad ...(8.70)$$

where μ_a, μ_f, and μ_g are the refractive indices of air, thin film and glass substrate respectively. As $\mu_a = 1$, the above expression may be rewritten as

$$\left[\frac{\mu_f - 1}{\mu_f + 1}\right]^2 = \left[\frac{\mu_g - \mu_f}{\mu_g + \mu_f}\right]^2$$

Expanding the above equation, we get

$$\frac{\mu_f^2 - 2\mu_f + 1}{\mu_f^2 + 2\mu_f + 1} = \frac{\mu_g^2 - 2\mu_g\mu_f + \mu_f^2}{\mu_g^2 + 2\mu_g\mu_f + \mu_f^2}$$

$$4\mu_f^3\mu_g + 4\mu_f\mu_g = 4\mu_f^3 + 4\mu_f\mu_g^2$$

Dividing by $4\mu_f$ and rearranging the terms, we get

$$\mu_f^2 - \mu_g\mu_f^2 + \mu_g^2 - \mu_g = 0$$

$$\mu_f^2 = \mu_g\left(1+\mu_f^2 - \mu_g\right)$$

$$\therefore \quad \mu_f^2 \cong \mu_g \quad (\text{as } \mu_f \approx \mu_g)$$

$$\therefore \quad \mu_f = \sqrt{\mu_g} \qquad \text{...(8.71)}$$

It implies that the refractive index of thin film should be less than that of the substrate and possibly nearer to its square root.

In case of glass, if we take $\mu_g = 1.5$, $\mu_f = \sqrt{\mu_g} = 1.22$.

The materials which have refractive index nearer to this value are magnesium fluoride, MgF_2 ($\mu = 1.38$) and cryolite, $3NaF.AlF_3$ ($\mu = 1.36$). Apart from the refractive index, the material should possess some more additional properties. The film should adhere well, should be durable, scratch proof and insoluble in ordinary solvents. MgF_2 and cryolite satisfy these requirements. However, among the two, magnesium fluoride is cheaper and is hence widely used as *AR* coating.

It may be noted that the condition (8.68) is satisfied only at one particular wavelength. The wavelength normally chosen is 5500 Å for which the eye is most sensitive. This wavelength is located in the yellow-green portion of the spectrum. Consequently, the reflection of red and violet light will be larger when white light is incident on the component. Hence, the component shows *purple hue* in reflected light.

Example 8.12: *Find minimum film thickness to be deposited on the lens of $\mu = 1.5$ with magnesium fluoride ($\mu f = 1.38$) to increase the transmission of normally incident light $\lambda = 5800$ Å.*

Solution:

$$t_{min} = \frac{\lambda}{4\mu_f} = \frac{5800\times10^{-10}\text{ m}}{4\times1.38} = 1051 \text{ Å}.$$

Example 8.13: *Can a thin film of water ($\mu_f = 1.33$) formed on a glass window pane ($\mu_f = 1.52$) act as a non-reflecting film? If so, how thick should be the water film?*

Solution: A film of refractive index μ_f can act as a non-reflecting film on a substrate having refractive index μ, if $\mu_f = \sqrt{\mu}$.

Here, $\sqrt{\mu} = \sqrt{1.52} = 1.233.$

As the refractive index of water is 1.33, it is nearer to $\sqrt{\mu}$ and can act as a non-reflecting film on glass.

The minimum thickness of the film is given by

$$t_{min} = \frac{\lambda}{4\mu_f}.$$

As human eye is more sensitive to green, it may be assumed that $\lambda = 5500$ Å.

$$\therefore \quad t_{min} = \frac{5500\times10^{-10}\text{ m}}{4\times1.33} = \mathbf{1034\ Å}.$$

Multilayer AR coatings: A single layer *AR* coating is effective only at one particular wavelength. A much wider coverage across the spectrum is possible with multiple coatings, called *multilayers*. In practice three layer coatings are widely used and are highly effective over most of the visible spectrum. The central layer is half-wave (λ / 2) thick and is of high refractive index materials such as zirconium dioxide (ZrO_2, μ = 2.1). The outside layer is of magnesium fluoride having λ / 4 thickness and the layer adjacent to the substrate is again a λ /4 thick coating of cesium fluoride (CeF_3, μ =1.63) or aluminium oxide (Al_2O_3, μ = 1.76). Some of the antireflection coatings use up to 100 layers of alternating high and low refractive index materials.

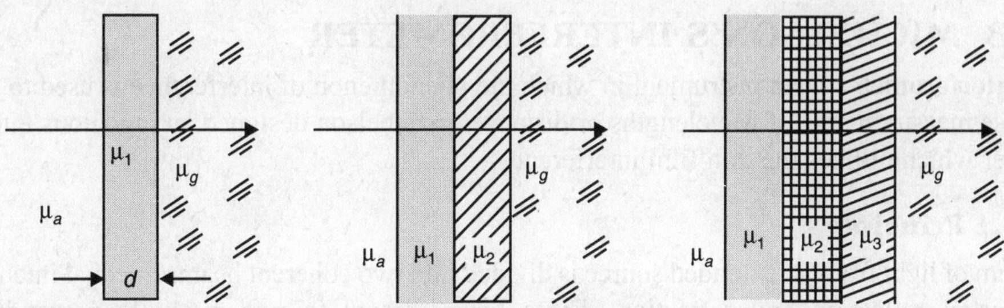

Fig. 8.33:

Multilayer AR coatings.

8.12.5 Interference Filters

An interference filter is an optical system that will transmit a very narrow range of wavelengths and thus provides a monochromatic beam of light. Interference filters are fabricated earlier as follows. A thin metallic film, usually of aluminium or silver, is deposited on a glass substrate by vacuum deposition technique. Then a thin layer of cryolite is deposited over this. The structure is again covered by another metallic film. Another plate is placed over it to protect the thin film structure. By varying the thickness of the dielectric film, any particular wavelength can be filtered out. However, the filtered light will have a narrow spectrum centered on the chosen wavelength. By increasing the reflectivity of the surfaces, the transmitted spectrum can be made narrower. But it is not possible to increase the thickness of metallic films indefinitely, as they start absorbing the light.

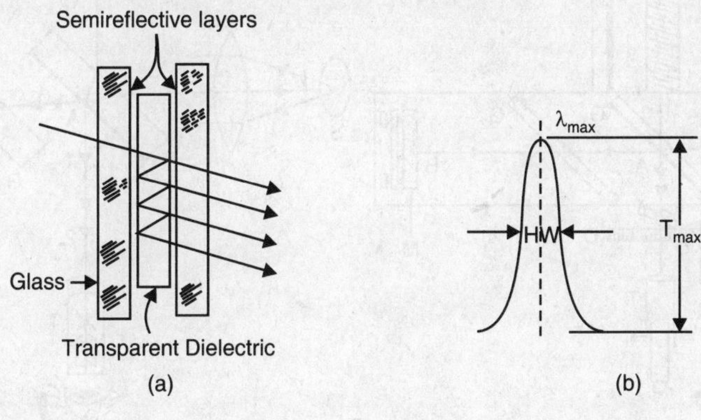

Fig. 8.34:

Interference Filters.

In modern versions metallic films are not used; instead dielectric films are used. In an all dielectric interference filter, layers of dielectric materials of appropriate refractive indices are deposited. To obtain an interference filter, a $\lambda/4$ thick film of titanium oxide is deposited and then over it a film of dielectric material with lower refractive index, such as magnesium fluoride is deposited. On this, again a $\lambda/4$ thick film of titanium oxide is deposited. In this way alternately high and low refractive index materials are deposited to obtain an interference filter. With multiple coatings, it is possible to fabricate filters, which are capable of transmitting a very narrow spectrum of a width as small as 11 Å or even less, about a chosen wavelength in the visible region. Modern filters use up to 100 layers.

8.13 MICHELSON'S INTERFEROMETER

An **interferometer** is an instrument in which the phenomenon of interference is used to make precise measurements of wavelengths or distances. Michelson designed an ingenious interferometer which utilizes the thin film interference.

8.13.1 Principle

A beam of light from an extended source is divided into two coherent beams of equal intensities by partial reflection and refraction. These beams travel in two mutually perpendicular directions and come together after reflection from plane mirrors. The beams overlap on each other and produce interference fringes. The fringes are *circular* if the reflecting mirrors are exactly perpendicular to each other or *straight* if the mirrors are inclined at a small angle.

8.13.2 Construction

The schematic of a simple Michelson interferometer is shown in Fig. 8.35 (a). It consists of a beam splitter G_1, a compensating plate G_2, and two plane mirrors M_1 and M_2. The *beam splitter* G_1 is a partially silvered plane parallel glass plate and it has the property that it transmits half the incident light and reflects the rest. The compensating plate G_2 is a simple

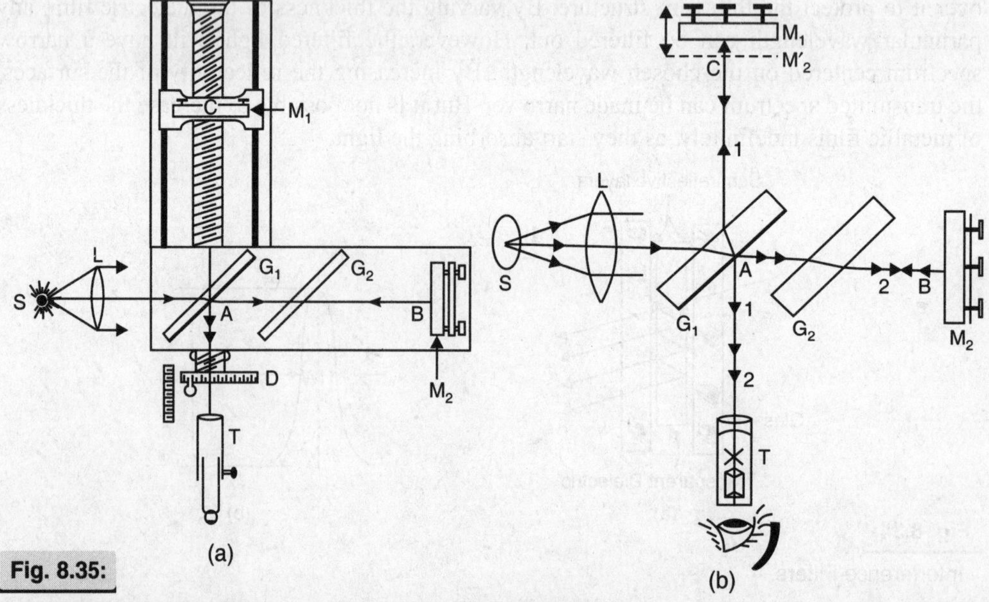

Fig. 8.35:
Michelson interferometer (a) Schematic, (b) Ray diagram.

plane parallel glass plate having the same thickness as G_1. The two plates G_1 and G_2 are held parallel to each other and are inclined at an angle of 45° with respect to the mirror M_2. The mirror is mounted on a carriage and can be moved exactly parallel to itself with the help of a micrometer screw. The distance through which the mirror M_1 is moved can be read with the help of a graduated drum D attached to the screw. Displacements of the order of 0.1 μm can be easily read. The plane mirrors M_1 and M_2 can be made perfectly perpendicular with the help of the fine screws attached to them. The interference bands are observed in the field of view of the telescope T.

8.13.3 Working

Monochromatic light from an extended source is rendered parallel by means of a collimating lens L and is made incident on the beam splitter G_1. It is partly reflected at the back surface of G_1 along AC and partly transmitted along AB. The beam AC travels normally towards the plane mirror M_1 and is reflected back along the same path and comes out along AT. The transmitted beam travels toward the mirror M_2 and is reflected along the same path. It is reflected at the back surface of G_1 and proceeds along AT. The two beams received along AT are produced from a single source through division of amplitude and are hence *coherent*. The superposition of these beams leads to interference and produces interference fringes.

From the Fig. 8.35 (b) it is clearly seen that a light ray starting from the source S and undergoing reflection at the mirror M_1 passes through the glass plate G_1 three times. On the other hand, in the absence of plate G_2, the ray reflected at M_2 travels through the glass plate G_1 only once. For compensating this path difference, a *compensating plate* G_2 of the same thickness as that of G_1 is inserted into the path AB and is held exactly parallel to G_1.

If we look into the instrument from T, we see mirror M_1 and in addition we see a virtual image, M'_2, of mirror M_2. Depending on the positions of the mirrors, image M'_2 may be in front of, or behind, or exactly coincident with mirror M_1.

Optical path difference between the two waves at T:

The optical path of the wave that travelled along

$$LG_1M_1T = 2G_1M_1 + \lambda/2 + \lambda/2 = 2G_1M_1 + \lambda.$$

The optical path of the wave that travelled along

$$LG_1M_2T = 2G_1M_2 + \lambda/2 + \lambda/2 = 2G_1M_2 + \lambda.$$

Therefore, the optical path difference between the two waves

$$\Delta = 2(G_1M_1 - G_1M_2) = 2(x_1 - x_2) = 2d.$$

The two beams interfere constructively if $\Delta = m\lambda$ or destructively if $\Delta = (2m + 1)\lambda/2$. Anything that changes the optical path difference will cause a change in the relative phase of the two waves. As an example, if mirror M_1 is moved by a distance $\frac{1}{2}\lambda$, the path difference changes by λ and the fringe pattern will shift by one fringe.

8.13.4 Circular Fringes

Circular fringes are produced with monochromatic light when the mirrors M_1 and M_2 are *exactly perpendicular to each other*. The origin of the circular fringes can be understood as follows.

If we look into the instrument from T, we see mirror M_1 directly, and in addition we will see the virtual image M'_2 of mirror M_2 formed by reflection in the glass plate G_1. It means that one of the interfering beams come from M_1 and the other beam appears to come from the virtual image M'_2. The situation is similar to an air film enclosed between mirrors M_1 and M'_2 with the difference that in case of a real film between two surfaces, multiple reflections take place, whereas in this case only two reflections take place.

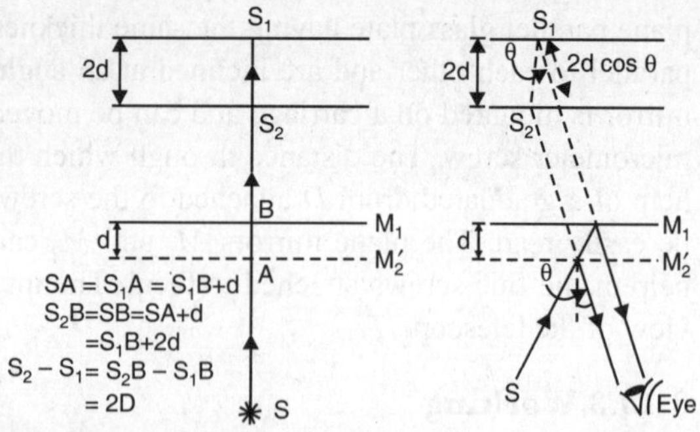

Fig. 8.36: Mirrors M_1 and M_2.

If one looks towards M_1 through T, one observes a virtual image of M_2 in G_1, parallel to M_1, say M'_2. M_1 and M'_2 act as two coherent sources formed by the thin film of thickness $M_1M'_2$.

$$M_1M'_2 = G_1M_1 - G_1M'_2 = G_1M_2 = (x_1 - x_2)$$

If the two arms of the interferometer are equal in length, $(x_1 - x_2) = 0$ and image M'_2 coincides with mirror M_1. If M'_2 and M_1 do not coincide, the distance between them is finite, $M'_2\, M_1 = (x_1 - x_2) = d$. The effect is that of light from a point source S falling on a uniformly thick film of air whose thickness is equal to d. Now, a light ray is reflected by both M'_2 and M_1 and the observer will see two virtual images- S_1 due to reflection at M'_2 and S_2 due to reflection at M_1. The virtual images are separated by a distance $2d$.

If the observer looks into the system at an angle θ, the path difference between the two beams will be $2d \cos\theta$. The light that comes from M_2 and goes to T undergoes rare-to-dense reflection and therefore a π-phase change occurs. In view of this, the total path difference between the two beams is given by

$$\Delta = 2d\cos\theta + \lambda/2. \qquad ...(8.72)$$

The condition for obtaining brightness

$$2d\cos\theta + \lambda/2 = m\lambda \qquad ...(8.73)$$

where $m = 0, 1, 2, ...$.

For a given mirror separation d, a given wavelength λ and order m, angle θ is constant. This means that the fringes are of circular shape. They are called *fringes of equal inclination.*

In case the mirror M_1 coincides with the virtual image M'_2, $d = 0$, the optical path difference between the interfering beams will be λ/2 (Refer to Eq. 8.73). Consequently, we obtain a minimum at the coincidence position and the centre of the field will be dark, as shown in Fig. 8.37 (a).

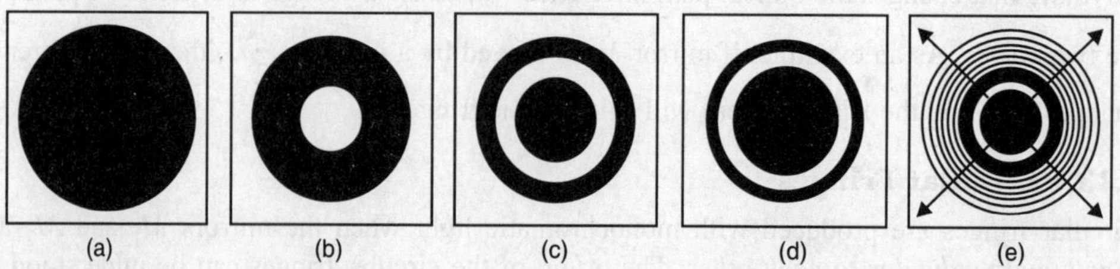

Fig. 8.37: Circular bright and dark fringes.

If one of the mirrors is now moved through a distance $\lambda/4$, the path difference changes by $\lambda/2$ and therefore a maximum is obtained. By moving the mirror through another $\lambda/4$, a minimum is obtained; moving it by another $\lambda/4$ again a maximum is obtained and so on. Therefore, a new ring appears in the centre of the field each time the mirror is moved through $\lambda/2$. As d increases new rings appear in the centre faster than rings already present disappear in the periphery; and the field becomes more crowded with thinner rings. Conversely, as d is made smaller, the rings contract and disappear in the centre.

8.13.5 Localized Fringes

When the two mirrors are tilted, they are not exactly perpendicular to each other and therefore the mirror M_1 and the virtual image M'_2 are not parallel. In this case the air path between them is wedge-shaped and the fringes appear to be straight. If one of the mirrors is moved, the fringes move across the field. The position of any particular bright fringe is taken up by the one next to it. The fringes can be counted as they pass a reference mark.

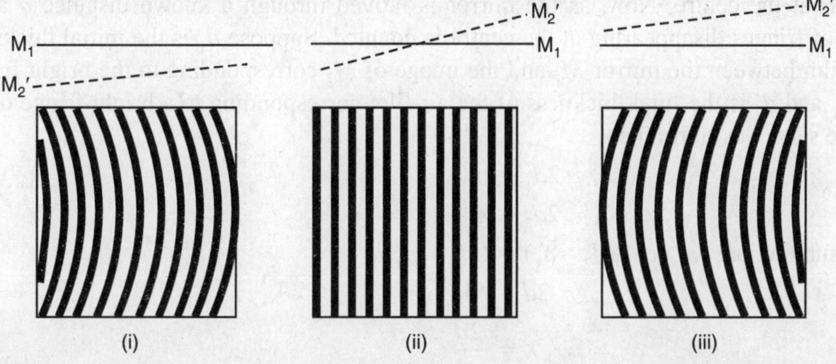

Fig. 8.38:
Localized Fringes.

If m fringes move across the field of view when M_1 moves through a distance d, then

$$d = m\,\lambda/2$$

or
$$\lambda = \frac{2d}{m} \qquad ...(8.74)$$

Example 8.14: *Find the wavelength of light using Michelson interferometer if 200 fringes cross the field of view when the movable mirror is moved a distance 0.0589 mm.*

Solution: $\lambda = \dfrac{2d}{m} = \dfrac{2\times0.0589\times10^{-3}\,\text{m}}{200} = 5890\ \text{Å}$

8.13.6 White Light Fringes

Instead of a monochromatic source, if a white light source is used; a few coloured fringes with a central dark fringe can be observed. When the position of M_1 intersects with M'_2 localised fringes are observed. Then a very slow motion of M_1 in this region using white light will bring these fringes into view, when a central dark fringe is surrounded by 8 to 10 coloured fringes on either side are observed. These fringes are useful for the determination of zero path difference.

8.14 APPLICATIONS OF MICHELSON INTERFEROMETER

Michelson interferometer can be used to determine the

(*i*) wavelength of a given monochromatic source of light

(*ii*) difference between the two neighbouring wavelengths or resolution of the spectral lines,

(*iii*) refractive index and thickness of various thin transparent materials and

(*iv*) measurement of the standard metre in terms of the wavelength of light

We study here only the first three applications.

8.14.1 Measurement of Wavelength

Wavelength of light can be determined using Michelson interferometer. The monochromatic source is kept at *S*. If the mirrors M_1 and M_2 are exactly perpendicular, circular fringes are obtained. If the mirror M_1 is moved forward or backward, the circular fringes appear or disappear at the centre. Now, as the mirror is moved through a known distance *d* and the number of fringes disappearing at the centre is counted. Suppose d_1 is the initial thickness of the air film between the mirror M_1 and the image of M_2 corresponding to the bright fringe of order m_1 and d_2 is the final thickness of the air film corresponding to a bright fringe of order m_n in the same position. Then,

$$2d_1 = m_1\lambda$$

and

$$2d_2 = m_n\lambda$$

By subtraction, we get $2(d_2 - d_1) = (m_n - m_1)\lambda$

$$\therefore \quad 2d = N\lambda$$

$$\therefore \quad \lambda = \frac{2d}{N} \qquad ...(8.75)$$

8.14.2 Determination of the Difference in the Wavelength of Two Waves

If a source of light consists of two wavelengths λ_1 and λ_2, which differ slightly, then two sets of fringes corresponding to the two wavelengths are produced in a Michelson interferometer. Adjust the position of the mirror M_1 of the interferometer to fringes which are very bright. In this location, the bright fringe due to λ_1 coincides with the bright fringes due to λ_2. When the mirror M_1 is through a certain distance, the bright fringe due to one set will coincide with the dark fringe due to the other. Here no fringes will be visible. Again by moving the mirror M_1, a position is reached when a bright fringe of one set falls on the bright fringe of the other. This can happen when the mth order of the longer wavelength agrees with the $(m + 1)^{th}$ order of the shorter wavelength.

Let m_1 and m_2 be the changes in the order and *d* be the distance between two consecutive positions of maximum.

$$\therefore \quad 2d = m_1\lambda_1 = m_2\lambda_2$$

If λ_1 is greater than λ_2

$$m_2 = m_1 + 1$$

$$\therefore \quad 2d = m_1\lambda_1 = (m_1 + 1)\lambda_2$$

$$\therefore \quad m_1 = \frac{\lambda_2}{\lambda_1 - \lambda_2}$$

$$\therefore \qquad 2d = \frac{\lambda_1\lambda_2}{\lambda_1 - \lambda_2}$$

or

$$\lambda_1 - \lambda_2 = \frac{\lambda_1\lambda_2}{2d}$$

Taking λ as the mean wavelength of the two wavelengths λ_1 and λ_2, the small difference $\Delta\lambda$ is given by

$$\Delta\lambda = \lambda_1 - \lambda_2 = \frac{\lambda^2}{2d} \qquad ...(8.76)$$

8.14.3 Thickness of a Thin Transparent Sheet

Let a transparent sheet of thickness t and refractive index μ be inserted in the path of one of the interfering beams of Michelson interferometer. The optical path of that beam increases because of the sheet. It becomes μt instead of t. The increase in the optical path is $(\mu t - t)$ or $(\mu - 1)t$. Since the beam traverses the medium twice, the extra path difference between the two interfering beams is $2(\mu - 1)t$. If m is the number of fringes by which the fringe system is displaced, then

$$2(\mu - 1)t = m\lambda$$

When monochromatic light is used, it is difficult to distinguish the sudden shift of fringes when the thin sheet is inserted. It is also not possible to count the number of fringes shifted. The difficulty is overcome by using white light first to locate the central dark fringe and it is made to coincide with the cross-wire of the telescope. The thin sheet is then introduced into the path of the beam. Position of mirror M_1 is adjusted till again a dark fringe of zero path difference coincides with the cross-wire of the telescope. The distance d through which the mirror is moved is noted. The white light is now replaced with the monochromatic light and the mirror M_1 is moved back slowly and the number of fringes contained in d is found. The thickness t is obtained from the relation

$$t = \frac{m\lambda}{2(\mu - 1)} \qquad ...(8.77)$$

8.15 MOIRE FRINGES

Lord Rayleigh first noticed the moiré effect in optics, in 1874. The term moiré is a French word and refers to the wavy finish of textiles made from wool. The moiré effect is a visual sensitivity.

It happens when seeing a set of lines or dots that is superimposed on another set of lines or dots, where the sets differ in relative size, angle, or spacing.The patterns are formed by the regions of dark and light. Moiré pattern is a pattern in the image formed in our eye.

For example, consider two window screens; one grid consists of evenly spaced straight lines sloping upwards to the left and another similar grid sloping upwards to the right. The pattern is relatively bright around the points of intersection of the two grids. The loci of these points specify the positions of the moiré fringes. Fig. 8.39 (b) shows moiré fringes.

Let us assume that the spacing of the lines in the two original grids is d and that one of the grids is turned through an angle θ with respect to the other grid (Fig. 8.39 a). Assuming that θ is small, it follows that

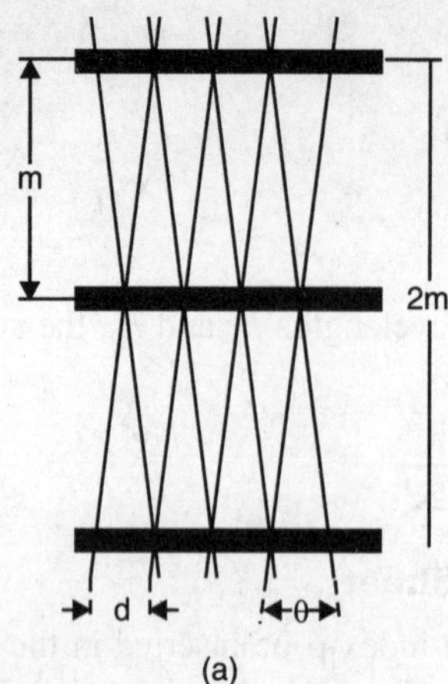

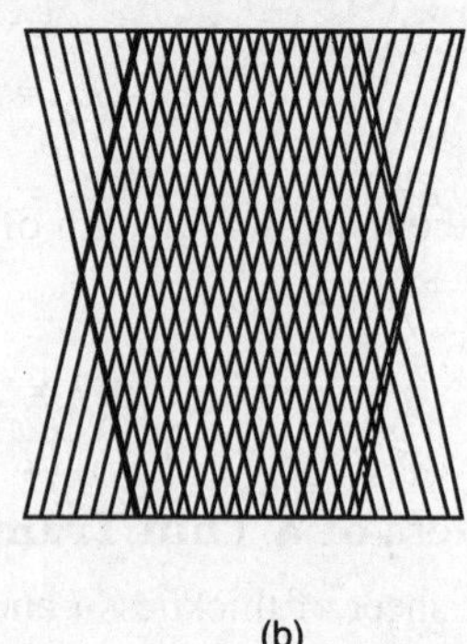

Fig. 8.39: Moire finges.

$$\theta \approx \frac{d}{m} \quad \text{or} \quad m \approx \frac{d}{\theta} \qquad ...(8.78)$$

Thus, the more parallel the grids, the wider the spacing m of the moiré lines.

QUESTIONS

1. Define interference. What do you mean by coherent sources?
2. Explain the phenomenon of interference of light. What are the necessary conditions to get clear and distinct interference fringes?
3. What are the methods used for obtaining coherent sources?
4. Write the names of two classes into which the phenomenon of interference is divided.
5. Describe Fresnel's biprism. Explain how the wavelength of light can be determined with its help.
6. Explain determination of the thickness of a given thin sheet the using Fresnel's biprism.
7. Explain the theory of Fresnelbi prism for finding the wavelength.
8. State the basic conditions for the phenomenon of interference of light.
9. Briefly discuss the effect of introducing a thin plate in the path of one of the interference beams in a biprism experiment.
10. Deduce an expression for the displacement of the fringes.
11. State the conditions necessary for obtaining sustainable interference pattern using two sources.
12. Derive conditions for path difference for interference in thin parallel film due to reflected light.
13. Derive the conditions for maxima and minima due to interference of reflected light in thin film.
14. What is thin film?
15. Explain colours of thin films.
16. Derive an expression for condition of maxima and minima for reflected light in case of thin transparent film of uniform thickness.
17. Derive the condition for path difference for interference in parallel thin film due to reflected light.
18. A thin film illuminated by white light appears coloured when observed in reflected light. Explain why.

19. Derive an equation for fringe width in the interference pattern of wedge shape film.
20. Give the theory and formation of wedge shaped films. How these can be used to find the wavelength of light used?
21. How will you find the thickness of a thin wire using air wedge method?
22. Discuss the testing of optical planeness of surfaces.
23. Explain how plane ness of a surface can be tested by air wedge.
24. Explain how interference pattern can be used for testing the optical flatness of surfaces.
25. Explain the formation of Newton's rings.
26. Explain the procedure for finding the wavelength of sodium light using Newton's rings experiment.
27. Explain Newton's rings method for determining the wavelength of monochromatic light. Why is the centre of fringes dark and how can we get a bright centre?
28. Describe Newton's rings experiment to determine the wavelength of incident monochromatic light.
29. In Newton's rings experiment show that in reflected light (*a*) the diameters of the dark rings are proportional to the square roots of natural numbers and (*b*) the diameters of the bright rings are proportional to the square roots of odd natural numbers
30. Give experimental set up to obtain Newton's rings. Explain how interference takes place.
31. Why are circular fringes obtained in Newton's rings arrangement? Why these fringes are called fringes of equal thickness? Why is the central fringe a dark spot when examined in reflected light?
32. Obtain an expression for the diameter of dark rings in reflected system. What will happen to the diameter of nth dark ring if air is replaced by water film? Explain.
33. In Newton's Ring experiment, why: (*a*) The Plano convex lens have larger radius of curvature?, (*b*) The rings get closer away from centre? (*c*) Central fringe is dark in reflected light?, (*d*) Fringes are circular and (*e*) Why the rings are not evenly spaced?
34. What will happen if the convex lens in the Newton's ring apparatus is lifted up by $\lambda/4$ where λ is the wavelength of the light used?
35. In Newton's ring experiment, if observed in reflected light, is it possible to obtain bright spot at centre? Justify your answer.
36. In a Newton's rings experiment, light of red colour is used first and then a blue light. Which set of rings would have larger diameter and therefore greater spacing between them?
37. Show that the diameter of the nth dark ring is given by $D_n = 2\sqrt{m\lambda R}$.
38. Derive an expression for wavelength of light in Newton's ring experiment.
39. Explain how the refractive index of a liquid can be found out by Newton's rings.
40. Give expressions for the total path difference in: (*a*) constant thickness film, (*b*) wedge shaped film, (*c*) film enclosed between a plano-convex lens and plane glass plate in Newton's rings apparatus.
41. Why are lenses coated with thin film to improve transmission of light?
42. Explain the use of thin film as anti-reflection coating.
43. What is anti-reflection coating? Explain its principle and application.
44. What do you understand by antireflection coating?
45. Deduce an expression for minimum thickness of antireflection coating.
46. Why does AR coated lenses look purple by reflected light?
47. Draw a neat diagram of Michelson interferometer. State the conditions for obtaining: (*a*) Circular fringes (*b*) straight line fringes (*c*) white light fringes.

48. Describe the construction and working of Michelson interferometer. How can it be used for measuring wavelength of monochromatic light?

49. Draw a neat labeled diagram of Michelson's interferometer and explain how it can be used to find the wavelength of monochromatic light.

50. Describe Michelson interferometer and explain how the fringes form in it. How can this be used for measuring the wavelength of monochromatic light. Derive the formula.

51. Explain the construction, types of fringes and applications of Michelson interferometer.

52. Explain how the wavelength of monochromatic source of light can be experimentally determined using a Michelson's interferometer.

53. Explain why circular fringes in Michelson interferometer shift in the field of view on displacing the movable mirror and deduce a relation between the wavelength and the displacement of the mirror.

54. What are the applications of Michelson interferometer?

PROBLEMS

1. A biprism forms interference fringes with monochromatic light of wavelength 546 nm. On introducing a thin glass plate of refractive index 1.5, in the path of one of the interfering beams, the central bright fringe moves to the position previously occupied by the third bright fringe. Find the thickness of the plate. **[Ans:** 3.3 μm]

2. In a biprism experiment a thin transparent sheet of refractive index 1.53 is placed in the path of one of the interfering beams. The central fringe is found to shift by a distance equal to the width of seven fringes. Calculate the thickness of the transparent sheet if the wavelength of the monochromatic light used is 5460 Å. **[Ans:** 7.2 μm]

3. In a Fresnel biprism experiment 75 fringes are obtained in the field of view with wavelength 5893 Å. What will be the number of fringes in the field of view with green light of wavelength 5200 Å. **[Ans:** 85]

4. Interference fringes are produced by monochromatic light of wavelength 5460 Å. When a thin transparent sheet of thickness 6.3×10^{-4} cm is introduced in the path of one of the interfereing beams, the central fringe shifts to a position occupied by 6th bright fringe. Compute the refractive index of the sheet. **[Ans:** 1.52]

5. Interference fringes are produced with light of wavelength 600 nm. A thin glass film of refractive index 1.5 is interposed in the path of one of the interfereing beams. The central bright band is shifted to the position previously occupied by the fifth bright band. Find the thickness of the film. **[Ans:** 6 μm]

6. A glass wedge of angle 0.01 radian is illuminated by 6000 Å incident normally. Find the distance from the edge of the wedge will the tenth fringe appears by reflected light? **[Ans:** $x = 1.5 \times 10^{-4}$ m]

7. If a parallel glass plate of $t = 4 \times 10^{-4}$ mm and $RI = 1.5$ is illuminated normally by white light, what wavelengths will be intensified in reflected beam in visible spectrum? **[Ans:** λ = 4800 Å]

8. A soap film of 5000 Å thickness is viewed at an angle of 35° to the normal. Find the wavelengths in the visible light which will be absent in the reflected light. The refractive index of the film is 1.333. **[Ans:** λ_1 = 6016 Å, λ_2 = 4011 Å]

9. In Newton's rings experiment, diameter of 10th dark ring due to wavelength 6000 Å in air is 0.5 cm. Find the radius of curvature of the lens. Yellow light of wavelength 5893 Å strikes a film of oil on water at an angle 30°. The 8th dark band is seen. Compute the thickness of the oil film if the refractive index of the oil is 1.44. **[Ans:** 1.75 μm]

10. White light is incident on a transparent film of refractive index 1.33 and thickness 1.6 μm at an angle of 45°. When the reflected light is examined, a dark band corresponding to 500 nm is seen. Find the order of the band. **[Ans:** 7]

11. Newton's rings are formed with sodium light in an experiment. What is the order of the dark ring, which has double the diameter of the 4th dark ring? **[Ans:** 16]

12. An engineer is interested in increasing the reflected and transmitted part of the incident light on a glass lens using the following thin films:

(*i*) Mg F_2 (μ = 1.38) (*ii*) ZnS (μ = 2.37)

Assume λ = 5500 Å and μ for material of lens = 1.5. Explain.

[Ans: For enhancing transmission, Mg F_2 ($\mu_f < \mu_g$) will be used and t_{min} = 996.4 Å ; for enhancing reflection, ZnS ($\mu_f > \mu_g$) should be used and t_{min} = 580.16 Å]

13. In an experiment with Michelson's interferometer the readings of two consecutive positions of the movable mirror for the maximum distinctness of fringes were found to be 1.2829 mm and 1.5774 mm. If the mean wavelength of sodium D-lines is 5893 Å, find the difference between the two wavelengths. **[Ans:** 5.896 Å]

14. Fringes of equal inclination are observed in a Michelson interferometer. As one of the mirrors is moved back by 1 mm, 3663 fringes move out from the centre of the pattern. Calculate the wavelength of light used. **[Ans:** 5460 Å]

9 Diffraction

Learning Objectives:

By the end of this chapter, you will be able to:

- Define basic concepts of Diffraction and
- Learn about different types of Diffraction - Fresnel and Fraunhoffer.
- Explain concepts of Plane Diffraction Grating.
- Resolving Power of Plane Transmission Grating, Telescope and Microscope.

9.1 INTRODUCTION

Diffraction phenomenon demonstrates wave behaviour of light. It is a familiar experience that the path of light is rectilinear. If an opaque obstacle is placed in the path of light, a sharp shadow is cast on the screen, representing rectilinear nature of light. However, when a beam of light passes through an aperture it spreads to some extent into the area of the geometrical shadow also. It is not usually noticed as in general the light sources are not point sources. Further, the obstacles used are of larger dimensions compared to the wavelength of light. We may not know the name and form, but the diffraction phenomena are encountered in our daily experience. For example, the colored spectra that one sees while viewing a distant source of light through a fine piece of cloth, the colours seen on a compact disc etc are all examples of diffraction effects. Diffraction sets a limit to the image formation ability of optical instruments.

9.2 DIFFRACTION

Diffraction phenomenon is a common characteristic of all kinds of waves. When waves pass near an obstacle, they tend to bend around the edges of the obstacle.

Def: The bending of waves around an obstacle and deviation from a rectilinear path is called diffraction.

Dependence of the Phenomenon on Wavelength

Fig. 9.1 illustrates the passing of waves through an aperture and dependence of the diffraction phenomenon on wavelength. The waves do not bend round the edges when the size of the aperture is large compared to the wavelength [Fig. 9.1 (a)]. The waves do bend round the edges when the opening is small [Fig. 9.1 (c)]. The aperture acts a source of secondary wavelets. These secondary waves propagate in all directions. This effect is observable quite *close to the opening* when the size of the opening is very small. When the opening is large, diffraction effect is observed at *greater distances from the opening*.

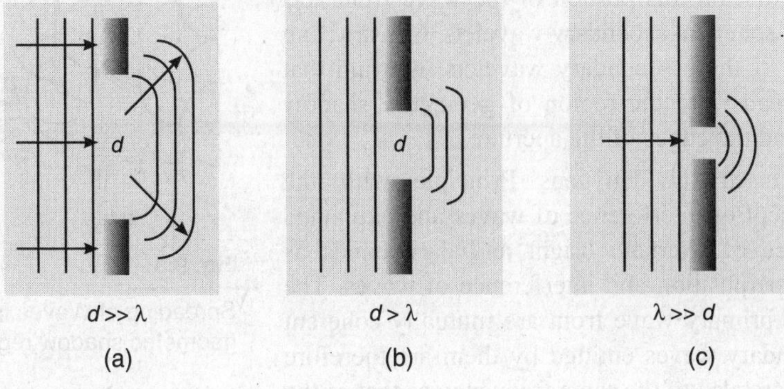

Fig. 9.1:

Illustration showing dependence of diffraction on the aperture size.

Diffraction Pattern

Diffraction of light waves is not readily apparent since light wavelength is very small compared to any physical obstacle. The rectilinear propagation of light is apparent in our day-to-day experience. However, careful experiments reveal that light passing through a tiny opening such as slit or aperture produces alternate regions of darkness and brightness beyond the region of geometric shadow.

Def: Diffraction pattern is the alternate bright and dark bands observed during diffraction phenomena. The bright central portion is known as a central maximum and it is bounded on either side by a series of secondary maxima separated by dark bands, called minima.

The successive bright band becomes less intense moving away from the central maximum. The presence of the bright region clearly indicates that light has reached there, showing thereby bending of light beam. The maxima and minima are created by interference of diffracted light waves.

A Simple Way to Observe Diffraction

A simple way to observe diffraction of light is by holding our hand in front of a strong light source and observing the light transmitted through the opening between any two fingers. When the fingers are brought very close to each other, one sees a series of dark lines parallel to the fingers. The diffraction bands produced by a razor blade are shown in Fig. 9.2.

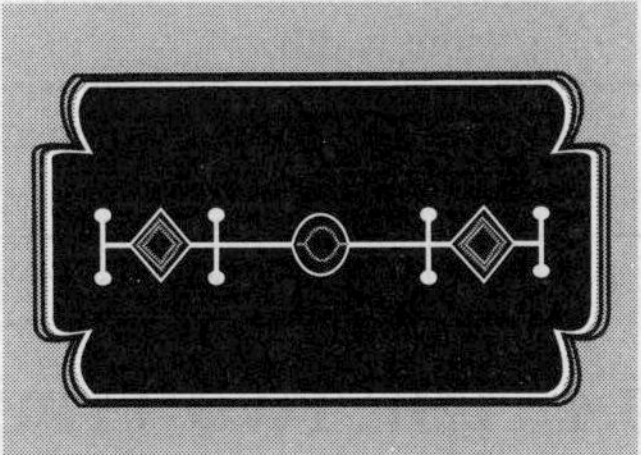

Fig. 9.2:

Diffraction bands produced by a razor blade.

Appearance of Maxima and Minima is due to Interference of Secondary Wavelets

In general, *diffraction occurs whenever a portion of a wave front is obstructed in some way.* The basic idea that explains diffraction is based on Huygens theory. The behaviour of light beyond the screen with an aperture can be qualitatively explained with the aid of Huygens Principle. The portion of the wave front that is incident on the opaque portion of the screen is obstructed while a small portion of the wave front is allowed to pass through the aperture (see Fig. 9.3). Every point on this portion of the wave front acts as a center of spherical secondary wavelets. Constructing the envelope of these secondary wavelets, we find that the wave spreads into the region of geometric shadow bending around the edges of the aperture.

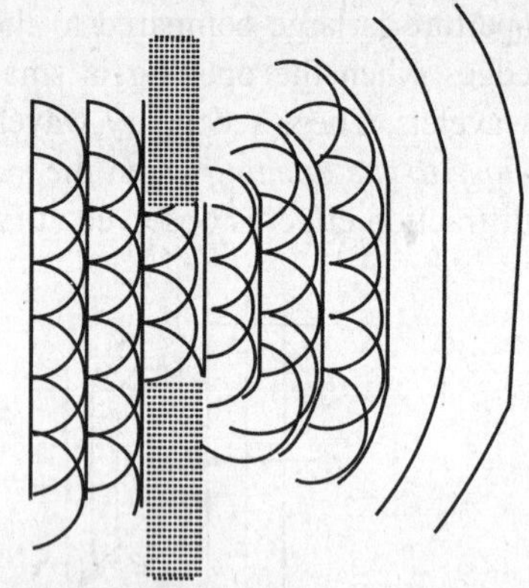

Fig. 9.3: Spreading of Waves into the geometric shadow region.

Fresnel fused the Huygens Principle with the Young's concept of interference of waves and explained the occurrence of alternate bright and dark bands as due to the superposition and interference of waves. The points on the primary wave front are mutually coherent and the secondary waves emitted by them are therefore coherent and are also of the same frequency as that of the primary wave and interfere.

9.3 DISTINCTION BETWEEN INTERFERENCE AND DIFFRACTION

The main differences between interference and diffraction are as follows:

Table 9.1: *Differences between interference and diffraction*

Interference	*Diffraction*
1. Interference is the result of interaction of light coming from different wave fronts originating from the source.	1. Diffraction is the result of interaction of light coming from different parts of the same wave front.
2. Interference fringes may or may not be of the same width.	2. Diffraction fringes are not of the same width.
3. Regions of minimum intensity are perfectly dark.	3. Points of minimum intensity are not perfectly dark.
4. All bright bands are of same intensity.	4. All bright bands are not of same intensity.

9.4 THE TWO TYPES OF DIFFRACTION

The diffraction phenomena are broadly classified into two types; Fresnel diffraction and Fraunhoffer diffraction.

(*i*) **Fresnel diffraction:** In this type of diffraction, the source of light and the screen are effectively at finite distances from the obstacle (see Fig. 9.4 a). Lenses are not used to make the rays parallel or convergent. The incident wave front is not planar. As a result, the phase of secondary wavelets is not the same at all points in the plane of the obstacle. The resultant amplitude at any point of the screen is obtained by the mutual interference of secondary wavelets from different elements of unblocked portions of wave front. It is experimentally simple but the analysis proves to be very complex.

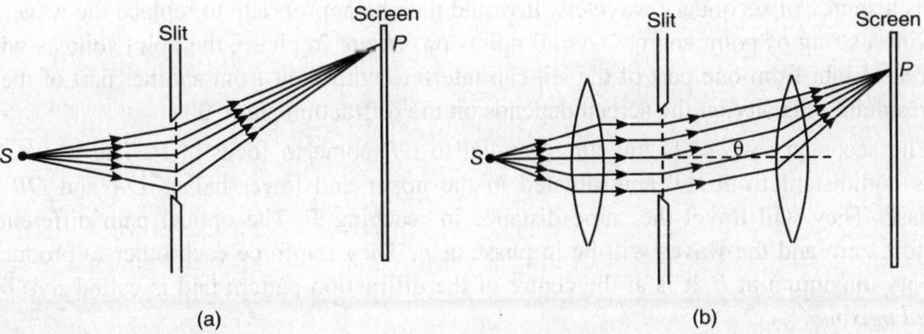

Fig. 9.4:

Conditions for (a) Fresnel diffraction and (b) Fraunhoffer diffraction.

(*ii*) **Fraunhoffer diffraction:** In this type of diffraction, the source of light and the screen are effectively at infinite distances from the obstacle (see Fig. 9.4 *b*). The conditions required for Fraunhoffer diffraction are achieved using two convex lenses, one to make the light from the source parallel and the other to focus the light after diffraction on to the screen. The incident wave front as such is plane and the secondary wavelets, which originate from the unblocked portions of the wave front, are in the same phase at every point in the plane of the obstacle. The diffraction is produced by the interference between parallel rays, which are brought into focus with the help of a convex lens. This problem is simple to handle mathematically because the rays are parallel.

9.5 FRAUNHOFFER DIFFRACTION AT A SINGLE SLIT

Let us consider the arrangement in Fig. 9.5 to obtain Fraunhoffer diffraction by a single slit. The lenses L_1 and L_2 keep the source and the screen effectively at infinity. The first lens renders the beam parallel while the second makes the screen to receive the parallel rays converging to a point, *P*.

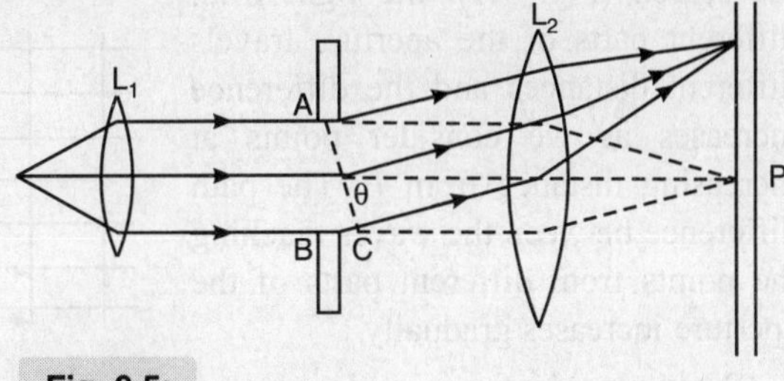

Fig. 9.5:

Fraunhoffer Diffraction at a Single Slit.

Let us consider a single narrow slit *AB* of width *d* perpendicular to the plane of the page and illuminated by a parallel beam of monochromatic light of wavelength λ. According to the considerations of geometrical optics, a sharp image of the slit is expected to form at *P* in the focal plane of the lens. However, what we obtain on the screen is a slit image of maximum brightness at the center, followed by the secondary maxima on the either side of gradually decreasing intensities, with distance. This intensity distribution on the screen is known as the Fraunhoffer diffraction pattern. Let us try to understand the reasons for this unique distribution of intensity in the image of a single slit.

9.5.1 Formation of Maxima and Minima

Fig. 9.6 shows a plane wave front (parallel rays) incident on the slit *AB*. A small part *AB* is sliced off from the incident wave front. According to Huygens Principle, each point on *AB* acts as a source of secondary wavelets. It would then be appropriate to replace the wave front *AB* with a string of point sources. As all points on *AB* are in phase, the point sources will be coherent. Light from one part of the slit can interfere with light from another part of the slit. The resultant intensity on the screen depends on the diffraction angle θ.

The secondary wavelets traveling parallel to *OP* come to focus at *P*. The waves from points equidistant from '*O*' and situated in the upper and lower halves *OA* and *OB* start in phase. They will travel the same distance in reaching *P*. The optical path difference is therefore zero and the waves will be in phase at *P*. They reinforce each other to produce an intensity maximum at *P*. It is at the centre of the diffraction pattern and is called *zero order central maxima*.

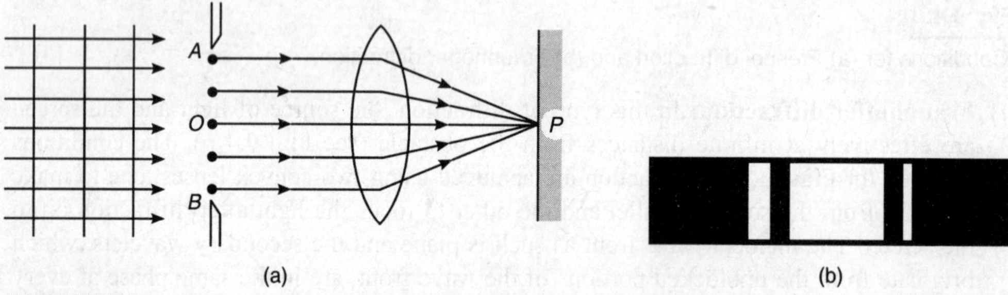

Fig. 9.6:

Fraunhoffer diffraction at a single slit (a) Conditions at the central maximum. (b) Typical diffraction pattern consisting of a central bright band flanked by weaker bright bands.

For any other point like *Q* on the screen (Fig. 9.7), the light from different parts of the aperture travels different distances and the difference increases as we consider points at increasing distance from *P*. The path difference between the waves reaching the points from different parts of the aperture increases gradually.

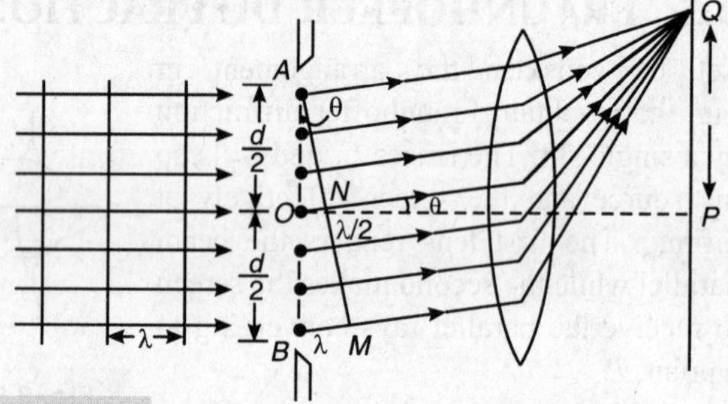

Fig. 9.7:

Conditions at the first minimum of diffraction pattern-each point on slit *AB* of the wave front acts as a point source. Any two waves that originate at points separated by *d*/2 distance are 180° out of phase and interfere destructively.

Now consider the secondary waves traveling in a direction by an angle θ with *OP*. These secondary waves are brought to focus by the lens at a point *Q*, which will have a maximum or minimum intensity, depending on the path difference between the waves arriving at *Q* from different points on the wave front *AB*.

It is convenient to divide the wavefront *AB* into two halves *AO* and *OB*. A line *AM* is drawn perpendicular to the direction of the diffracted rays. For every point in the upper half *OA*, there is a corresponding point in the lower half *OB*.

The path difference between these wavelets = $ON = (d/2) \sin \theta$.

If $ON = \lambda/2$, the two waves interfere destructively and produce darkness at *Q*. The path difference between the waves from these corresponding points will be λ/2 as any two waves that originate at points separated by ($d/2$), as the path difference between from such points will bc $\lambda/2$.

Hence, the waves from the upper half AO interfere destructively with waves from the lower half OB, if

$$\frac{d}{2}\sin\theta = \frac{\lambda}{2}$$

i.e.,
$$\sin\theta = \lambda/d \quad ...(9.1)$$

Therefore, the intensity at Q is zero and a dark band called the *first order minimum* is produced at Q.

A similar dark band occurs at Q' below P at an angular distance θ governed by the Eq. (9.1). It is also called the *first order minimum*.

We may divide the slit into four parts, six parts and so on. Arguments similar to the above show that a dark band occurs whenever

$\sin\theta = 2\lambda/d, 3\lambda/d, 4\lambda/d, 5\lambda/d$, ... etc.

They are known as second order minimum, third order minimum, etc.

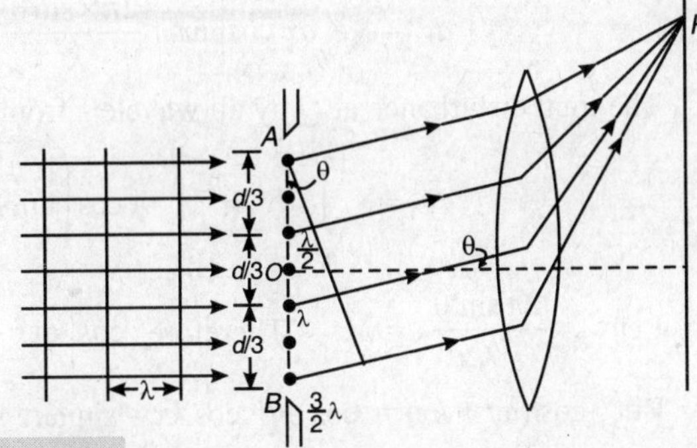

Fig. 9.8: Condition for first order maxima of diffraction pattern.

In general minima appear if the following condition is satisfied:

$$\sin\theta_m = m\lambda/d \quad \text{condition for minima} \quad ...(9.2)$$

where
$$m = 1, 2, 3, ...$$

In addition to the central maximum there are *secondary maxima*, which lie in between the secondary minima on either side of the central maximum. These are located in a direction in which the path difference ON is an odd multiple of $\lambda/2$. Hence for secondary maxima

$$ON = d\sin\theta = (2m + 1)\lambda/2$$

In general the secondary maxima are given by

$$\sin\theta_m = \left(\frac{2m+1}{d}\right)\frac{\lambda}{2} \quad \text{condition for maxima} \quad ...(9.3)$$

Thus, the diffraction pattern due to a single slit consists of a central bright maximum flanked by secondary maxima and minima on both the sides (see Fig. 9.9).

9.5.2 Intensity Distribution in Diffraction Pattern due to a Single Slit

Let a plane wave be incident normally on a long narrow slit of width d. Let us imagine that this slit width d is divided into N parallel strips of each of width Δx.

Each strip acts as a secondary source of radiation giving out wavelets leading to a characteristic distribution of intensity at a point Q. The position of the point Q in the Fig. 9.7 is fixed by the angle θ.

The wavelets from adjacent strips have a path difference of $\Delta x \sin\theta$. The resultant of all the wavelets reaching Q can be estimated by integration.

Let the disturbance caused at Q by the unit width of the slit be

$$y_0 = A\cos\omega t$$

Amplitude of the wavelet from the width dx at A, when it reaches Q is $(A\,dx)$.

Phase of this wavelet at $Q = \omega t + \frac{2\pi}{\lambda}\times OQ = \omega t + \frac{2\pi}{\lambda}\times x\sin\theta$

where x is the distance of Q from O, the mid point of the slit.

The disturbance caused at Q by the wavelet is given by

$$dy = A\,dx\cos\left[\omega t + \frac{2\pi\times\sin\theta}{\lambda}\right]$$

The total disturbance at Q by all wavelets from the slit of width d is given by

$$y = \int_{-d/2}^{+d/2} dy = \int_{-d/2}^{+d/2} A\cos\left[\omega t + \frac{2\pi x\sin\theta}{\lambda}\right]dx$$

Let $\frac{2\pi\sin\theta}{\lambda} = k.$ Therefore, $\cos\left[\omega t + \frac{2\pi x\sin\theta}{\lambda}\right] = \cos(\omega t + kx)$

But $\cos(\omega t + kx) = \cos\omega t\cdot\cos kx - \sin\omega t\cdot\sin kx$

$$\therefore\quad y = \int_{-d/2}^{+d/2} A[\cos\omega t\cdot\cos kx - \sin\omega t\cdot\sin kx]dx$$

$$y = \int_{-d/2}^{+d/2} A[\cos\omega t\cdot\cos kx]\,dx - \int_{-d/2}^{+d/2} A[\sin\omega t\cdot\sin kx]\,dx$$

$$y = A\cos\omega t\int_{-d/2}^{+d/2}\cos kx\,dx - A\sin\omega t\int_{-d/2}^{+d/2}\sin kx\,dx$$

$$y = A\cos\omega t\left[\frac{\sin kx}{k}\right]_{-d/2}^{+d/2} - A\sin\omega t\left[\frac{-\cos kx}{k}\right]_{-d/2}^{+d/2}$$

$$y = A\cos\omega t\left[\frac{\sin\left(\frac{kd}{2}\right)+\sin\left(\frac{kd}{2}\right)}{k}\right] + A\sin\omega t\left[\frac{\cos\left(\frac{kd}{2}\right)-\cos\left(\frac{kd}{2}\right)}{k}\right]$$

$$y = A\cos\omega t\left[\frac{2\sin\left(\frac{kd}{2}\right)}{k}\right] = 2A\cos\omega t\left[\frac{\sin\left(\frac{2\pi\cdot\sin\theta\cdot d}{2\lambda}\right)}{\frac{2\pi\cdot\sin\theta}{\lambda}}\right]$$

$$y = \left[A\,\frac{\sin\left(\frac{\pi d\sin\theta}{\lambda}\right)}{\frac{\pi\sin\theta}{\lambda}}\right]\cos\omega t = \left[Ad\,\frac{\sin\left(\frac{\pi d\sin\theta}{\lambda}\right)}{\frac{\pi d\sin\theta}{\lambda}}\right]\cos\omega t \quad \text{...(9.4)}$$

or $\quad y = A_\theta\cos\omega t$

The quantity in brackets gives the amplitude A_θ of the resultant disturbance at Q. For $\theta = 0$, $A_\theta = A_0 = Ad$.

So,

$$y = A_0 \left[\frac{\sin\left(\frac{\pi d \sin\theta}{\lambda}\right)}{\frac{\pi d \sin\theta}{\lambda}} \right] \cos\omega t \qquad \text{... (9.5)}$$

Let
$$\alpha = \frac{\pi d \sin\theta}{\lambda}.$$

$$\therefore \quad y = A_\theta \cos\omega t = A_0 \left[\frac{\sin\alpha}{\alpha}\right] \cos\omega t$$

$$\therefore \quad A_\theta = A_0 \left[\frac{\sin\alpha}{\alpha}\right] \qquad \text{... (9.6)}$$

Intensity distribution is given by

$$I = I_0 \left[\frac{\sin\alpha}{\alpha}\right]^2 \qquad \text{... (9.7)}$$

where I_0 is the intensity of principal maximum at $\theta = 0$.

Thus, the intensity at any point on the screen is proportional to $\left(\frac{\sin\alpha}{\alpha}\right)^2$.

A phase difference of 2π corresponds to a path difference of λ. Therefore, a phase difference of 2α is given by

$$2\alpha = \frac{2\pi}{\lambda} d \sin\theta \qquad \text{... (9.8)}$$

where $(d \sin\theta)$ is the path difference between the secondary waves from A and B.

$$\alpha = \frac{\pi}{\lambda} d \sin\theta \qquad \text{... (9.9)}$$

Thus, the value of α depends on the angle of the diffraction θ. The value of $\left(\frac{\sin^2\alpha}{\alpha^2}\right)$ for different values of θ gives the intensity at the point under consideration.

The variation of intensity with angle is shown in Fig. 9.9. It is a graph of $\left(\frac{\sin^2\alpha}{\alpha^2}\right)$ on the Y-axis and $\sin\theta$ on the X- axis. It is seen that most of the light is confined to the central band and its intensity is far greater than that of any other maxima. The intensity of the secondary maxima falls off rapidly as one

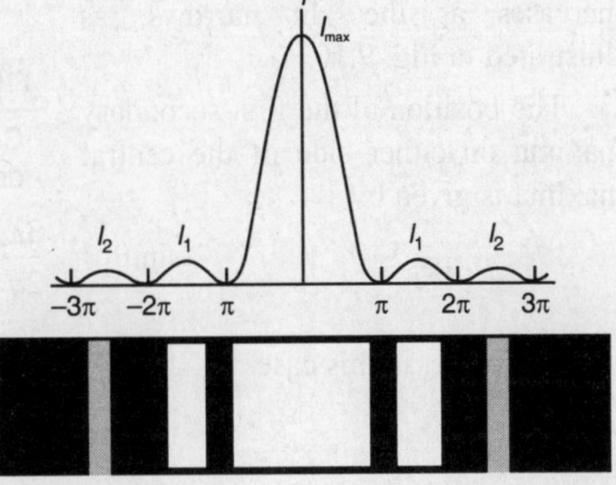

Fig. 9.9: A plot of light intensity for the single-slit Fraunhoffer diffraction pattern.

moves away from the centre. The intensity of the first secondary maximum is about 1/22 and that of the second is 1/61 of the intensity of the principal maximum. The secondary maxima are too faint to be visible ordinarily.

9.5.3 Linear Width of the Principal Maximum

If x is the distance of the first secondary minimum from the center of the principal maximum, then the width of the central maximum is

$$w = 2x.$$

If f is the focal length of the lens L, then $OP = f$ (focal length) is very large [Fig. 9.6 (a)]. As a result,

$$\sin\theta = \frac{x}{f}$$

But, for the first minimum

$$\sin\theta = \frac{\lambda}{d}$$

$$\therefore \quad \frac{x}{f} = \frac{\lambda}{d}$$

or

$$x = \frac{f\lambda}{d} \quad \text{... (9.10)}$$

Hence, the linear width of the central maximum is given by

$$W = 2x = \frac{2f\lambda}{d} \quad \text{... (9.11)}$$

Note that when the slit width $d >> \lambda$, we see on the screen uniform illumination in the shape of the slit. As the slit width is reduced, the illumination starts to spread out and dark bands become visible. Further, the width of the central maximum increases as the slit narrows, as illustrated in Fig. 9.10.

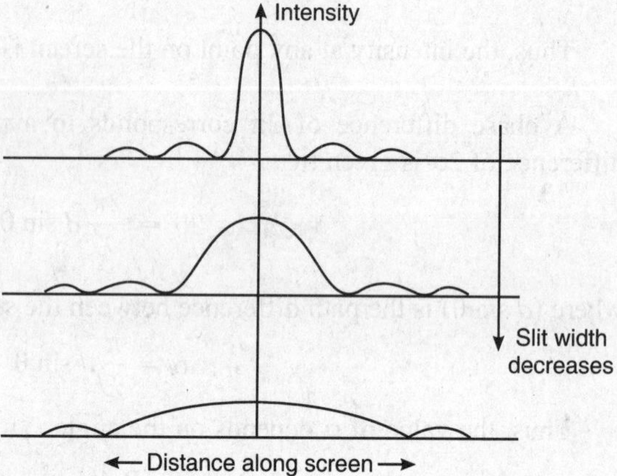

Fig. 9.10: Central maximum increases in width as the slit width decreases.

The position of the first secondary maxima on either side of the central maxima is given by

$$\sin\theta_1 = \frac{3\lambda}{2d}$$

Therefore, in this case

$$\frac{x}{f} = \frac{3\lambda}{2d}$$

and

$$x = \frac{3f\lambda}{2d} \quad \text{...(9.12)}$$

Example 9.1: *In a single slit diffraction pattern the distance between the first minima on either side of the central zero maximum is 4.4 mm. The diffraction is observed on a screen at a distance of 0.7 m. The wavelength of light used is 5890 Å. Calculate the slit width.*

Solution: The slit width, $d = \frac{f\lambda}{x} = \frac{(0.7\text{m})(5890\times10^{-10}\text{m})}{4.4\times10^{-3}\text{m}} = 0.09$ mm

Example 9.2: *Calculate the slit width for anormally incident light of wavelength 5000Å and the central maximum spreads out at (i) 30° on both sides of the direction of the incident light, (ii) 90° from the direction of the incident light?*

Solution: (*i*) In a Fraunhoffer diffraction pattern due to single slit of width *a*, the direction of minima are given by

$$a \sin\theta = \pm m\lambda \qquad m = 1, 2, 3....$$

Therefore, the angular spread of the central maximum on either side of the incident light is given by

$$\sin\theta = \lambda/d$$

Here $\theta = 30°$ so that $\sin\theta = 0.5$ and $\lambda = 5000$ Å $= 5 \times 10^{-5}$ cm.

$$\therefore \qquad d = \frac{\lambda}{\sin\theta} = \frac{5\times10^{-5}\text{ cm}}{0.5} = 1\times10^{-4}\text{ cm}$$

(*ii*) For $\theta = 90°$

$$d = \frac{\lambda}{\sin 90} = \frac{5\times10^{-5}\text{ cm}}{1}\ 5\times10^{-5}\text{ cm} = \mathbf{5000\ Å}$$

Example 9.3: *Calculate the angles at which the first dark band and the next bright band are formed in the Fraunhoffer diffraction pattern of a slit 0.3 mm wide for λ = 5890 Å.*

Solution: In a single slit Fraunhoffer's pattern, the directions θ of the minima are given by

$$d \sin\theta = \pm m\lambda \qquad m = 1,2,3....$$

First dark band: For the ***first dark band*** on either side of the central maximum, $m = 1$.

$$\therefore \qquad d\sin\theta = \lambda$$

$$\sin\theta = \frac{\lambda}{d} = \frac{5890\times10^{-8}\text{ cm}}{0.03\text{ cm}} = 0.00196$$

$$\therefore \qquad \theta = 6.7'$$

First bright band: The angle of diffraction θ′ corresponding to the ***first bright band*** on either side of the central maximum is approximately given by

$$a\sin\theta' = 3\lambda/2.$$

$$\therefore \qquad \sin\theta' = \frac{3\lambda}{2d} = \frac{3}{2}\times 0.00196 = 0.00294$$

$$\theta' = 0°\mathbf{10'}$$

9.6 FRAUNHOFFER DIFFRACTION AT DOUBLE SLIT

In Fig. 9.11 (*a*), *AB* and *CD* are two rectangular slits parallel to one another and perpendicular to the plane of the paper. The width of each *open portion* is *a* and the width of the *opaque portion* is *b*. *L* is a collecting lens and *MN* is a screen. *P* is a point on the screen such that *OP* is perpendicular to the screen. Let a plane wave front be incident on the surface of *XY*. All the secondary waves traveling in a direction parallel to *OP* come to focus at *P*. Therefore, *P* corresponds to the position of the central bright maximum.

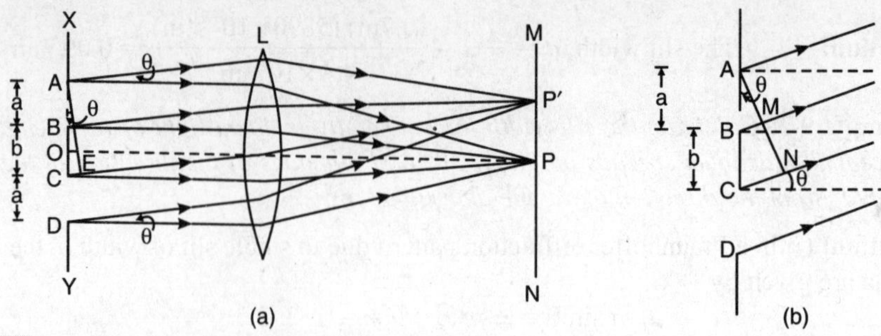

Fig. 9.11: Fraunhoffer diffraction at double slit.

In this case, the diffraction pattern has to be considered in two parts.

(*i*) The interference phenomenon due to the secondary waves emanating from the corresponding points of the two slits and

(*ii*) The diffraction pattern due to the secondary waves from the two slits individually.

The positions of interference or maxima and minima are calculated using θ, the angle between the direction of the secondary waves and the initial direction of the incident light.

(i) Interference maxima and minima

Let us consider the secondary waves traveling in a direction inclined at an angle θ with the initial direction.

In the Δ^{le} *CAN* (Fig. 9.11 *b*)

$$\sin\theta = \frac{CN}{AC} = \frac{CN}{a+b}$$

or $$CN = (a+b)\sin\theta$$

(*a*) If this path difference, *CN* is equal to ***odd multiples of λ/2***, θ gives the direction of minima due to interference of the secondary waves from the two slits.

(*b*) If the secondary waves travel in a direction θ′ such that the path difference is ***even multiples of λ/2***, then θ′ gives the direction of the maxima due to interference of light waves emanating from the two slits.

∴ For minima $$CN = (a+b)\sin\theta_n = (2n+1)\frac{\lambda}{2} \quad ...(9.13)$$

Putting $n = 1, 2, 3$, etc, the values of θ_1, θ_2, θ_3, etc, corresponding to the directions of minima can be obtained.

From equation (9.13)

$$\sin\theta_n = \frac{(2n+1)\lambda}{2(a+b)} \quad ...(9.14)$$

∴ For maxima $$CN = (a+b)\sin\theta'_n = 2n\frac{\lambda}{2}$$

or $$\sin\theta'_n = \frac{n\lambda}{(a+b)} \quad ...(9.15)$$

It may be seen that $\sin\theta'_2 - \sin\theta'_1 = \frac{\lambda}{(a+b)}$

Putting $n = 1, 2, 3$ etc, $\theta'_1, \theta'_2, \theta'_3$ etc corresponding to the directions of the maxima can be obtained.

From Eq. (9.14), we get

$$\sin\theta_1 = \frac{3\lambda}{2(a+b)}$$

and
$$\sin\theta_2 = \frac{5\lambda}{2(a+b)}$$

$$\therefore \quad \sin\theta_2 - \sin\theta_1 = \frac{\lambda}{(a+b)} \quad \text{...(9.16)}$$

Thus, the angular separation between any two consecutive minima (or maxima) is equal to $\frac{\lambda}{(a+b)}$ and inversely proportional to $(a + b)$, the distance between the two slits.

(ii) Diffraction maxima and minima

Let us consider the secondary waves traveling in a direction inclined at an angle θ with the original direction of the incident light. If the path difference BM is equal to λ the wavelength of the light used, then θ will give the direction of the diffraction minimum (Fig. 9.11 *b*).

The path difference between secondary waves emanating from the points A and B of a slit is equal to λ.

Considering the wave front on AB to be made up of the two equal parts, the path difference between the corresponding points of the upper and lower halves is equal to $\lambda/2$. The effect at P' due to the wave front incident on AB is zero. Similarly, for the same direction of the secondary waves, the effect at P' due to the wave front incident on the slit CD is also zero.

In general, $$a \sin\theta_n = n\lambda$$

Putting $n = 1, 2, 3$, etc, the values of $\theta_1, \theta_2, \theta_3$ etc, corresponding to the directions of diffraction minima can be obtained.

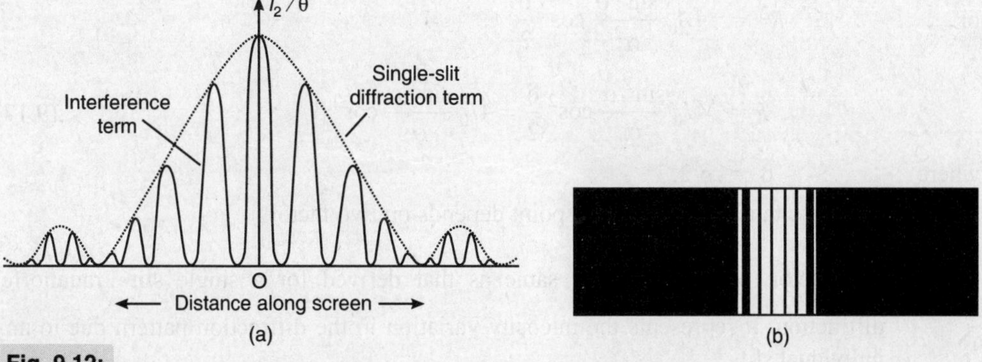

Fig. 9.12:

Fraunhoffer diffraction-Intensity distribution.

Intensity distribution due to the Fraunhoffer diffraction at two parallel slits is shown in Fig. 9.12 (*a*). The full line represents equally spaced ***interference maxima and minima*** and the dotted curve represents the ***diffraction maxima and minima***. In the region originally occupied by the central maximum of the single slit diffraction pattern, equally spaced interference maxima and minima are observed. The intensity of the central interference maximum is four times the intensity of the central maximum of the single slit diffraction pattern. The intensity of other interference maxima on the two sides of the central maximum gradually decreases. In the region of the secondary maxima due to diffraction at a single slit, equally spaced interference maxima of low intensity are observed. A typical diffraction pattern obtained on the screen is shown in Fig. 9.12 (*b*).

9.6.1 Expression for Resultant Intensity

Let us now find an expression for the resultant intensity and its variation with the angle θ. By Huygens' Principle, every point in the slits *AB* and *CD* sends out secondary wavelets in all directions. From the theory of diffraction at a single slit, the resultant amplitude due to wavelets diffracted from each slit in a direction θ is

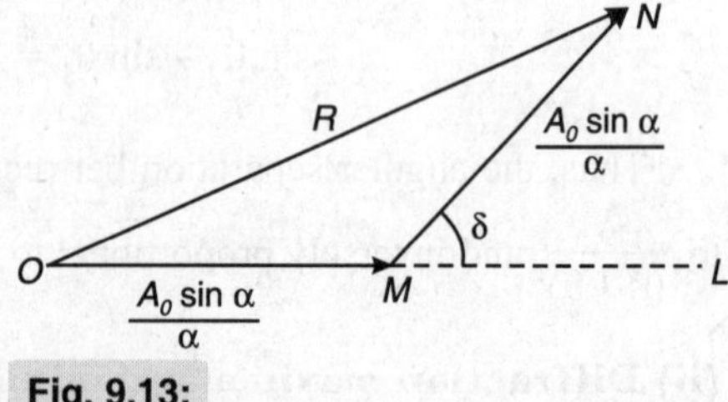

Fig. 9.13: Calculation of resultant amplitude.

$$A_\theta = A_0\left[\frac{\sin\alpha}{\alpha}\right]$$

In the case of double slit, we have interference of two waves of amplitude A_θ each, having a phase difference $\delta = \frac{2\pi}{\lambda} d\sin\theta$. The resultant amplitude *R* is given by (see Fig. 9.13).

$$ON^2 = OM^2 + MN^2 + 2(OM)(MN)\cos NML$$

$$R^2 = \left[\frac{A_0\sin\alpha}{\alpha}\right]^2 + \left[\frac{A_0\sin\alpha}{\alpha}\right]^2 + 2\left[\frac{A_0\sin\alpha}{\alpha}\right]\left[\frac{A_0\sin\alpha}{\alpha}\right]\cos\delta$$

$$= 2\left[\frac{A_0\sin\alpha}{\alpha}\right]^2(1+\cos\delta) = 2\left[\frac{A_0\sin\alpha}{\alpha}\right]^2\left(1+2\cos^2\frac{\delta}{2}-1\right)$$

or
$$R^2 = 4A_0^2\frac{\sin^2\alpha}{\alpha^2}\cos^2\frac{\delta}{2}$$

∴
$$I = 4I_0\frac{\sin^2\alpha}{\alpha}\cos^2\frac{\delta}{2} = 4I_0\frac{\sin^2\alpha}{\alpha}\cos^2\beta \qquad ...(9.17)$$

where $\beta = \delta/2$.

Thus, the resultant intensity at any point depends on two factors

(*i*) The factor $I_0\frac{\sin^2\alpha}{\alpha}$ is the same as that derived for a single slit Fraunhoffer diffraction. It represents the intensity variation in the diffraction pattern due to any individual slit.

(*ii*) The factor $\cos^2\beta$ gives the interference pattern due to waves overlapping from the two slits.

The resultant intensity at any point on the screen is the product of these two factors and will be zero when either of these factors is zero.

Maxima and minima:

(*i*) The diffraction term $I_0\left(\frac{\sin^2\alpha}{\alpha^2}\right)$ gives the central maximum in the direction $\theta = 0$ having alternate minima and secondary maxima of decreasing intensity on either side. The angular positions of minima are given by,

$$\sin\alpha = 0,$$

$$\alpha = \pm m\pi, \qquad m = 1, 2, 3, \ldots$$

i.e.,
$$\frac{\pi a \sin\theta}{\lambda} = \pm m\pi$$

or
$$\sin\theta = \pm\frac{m\lambda}{a} \qquad m = 1, 2, 3, \ldots \text{ but not zero.}$$

The angular positions of secondary maxima approach to

$$\alpha = \pm\frac{3\pi}{2}, \pm\frac{5\pi}{2}, \pm\frac{7\pi}{2}, \ldots$$

(*ii*) According to the second factor ($\cos^2\beta$), the intensity will be maximum when

$$\cos^2\beta = 1.$$

i.e.,
$$\cos^2\beta = 1, \qquad \text{where } n = 0, 1, 2, 3, \ldots$$

$$\therefore \quad \frac{\pi a \sin\theta}{\lambda} = \pm n\pi \qquad \text{when } n = 0, \theta = 0 \qquad \ldots(9.18)$$

Thus, the central maximum of interference pattern lies along the direction of incident light. This is called the ***principal maximum of zero order***. The central maximum of diffraction pattern also lies along this direction. At this point all the waves arrive with the same phase. Hence, the intensity of central maximum is the highest.

The intensity will be maximum when

$$\cos^2\beta = 0$$

i.e.,
$$\frac{\delta}{2} = \pm(2n+1)\frac{\pi}{2}$$

$$\therefore \quad \frac{\pi a \sin\theta}{\lambda} = \pm(2n+1)\frac{\pi}{2}$$

$$a\sin\theta = \pm(2n+1)\frac{\lambda}{2} \qquad \ldots(9.19)$$

It can be shown that for small values of θ, the maxima and minima are equally spaced. If a is kept constant and b is varied, the positions of maxima and minima due to diffraction remain unchanged while those due to interference undergo a change. Fig. 9.14 represents the intensity distribution determined by the factor $\cos^2\beta$ and $I_0\left(\frac{\sin^2\alpha}{\alpha^2}\right)$ respectively. The resultant of

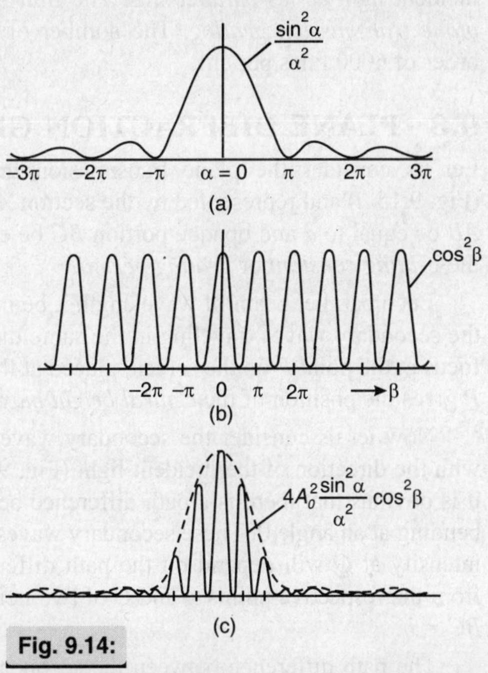

Fig. 9.14: Intensity distribution.

these curves is shown in Fig. 9.14. The resultant is obtained by multiplying the ordinates of first two curves at every point.

The entire pattern may be considered as consisting of interference fringes due to interference of light from both the slits, their intensity is being modulated by the diffraction occurring at individual slits.

9.7 DIFFRACTION DUE TO *N*-SLITS - DIFFRACTION GRATING (Normal Incidence)

Let us now consider the diffraction pattern formed by *N*-slits, each of width a.

The separation between consecutive slits is $d = a + b$,

where a is the width of the open portion and b is the width of the opaque portion.

Def: Diffraction grating is a device consisting of a large number of parallel slits of equal width and separated from one another by equal opaque spaces. The distance d between the centres of the adjacent slits is known as the *grating period.*

Rowland (1848-1901) produced ***transmission gratings*** by ruling with a diamond point extremely close, equidistant and parallel lines on an optically plane glass plate. The rulings (diamond scratch) scatter light and are effectively opaque while the parts without ruling transmit light and act as slits.

Because of the expenses and difficulty involved in fabrication, commonly used gratings are reproduced from the original ruled gratings. The ***replica gratings*** are made by pouring a thin layer of collodion solution over the surface of a ruled grating and the solution is allowed to harden. The collodion film is peeled carefully afterwards from the grating. The film retains the impression of the rulings of the original grating in the form of ridges. The ruled lines, which scatter light, act as opaque spaces whereas the spaces between them which transmit incident light act as parallel slits. The film is mounted between glass plates and it acts as a *plane transmission grating*. The number of lines on a plane transmission grating is of the order of 6000 lines per cm.

9.8 PLANE DIFFRACTION GRATING-Theory

Let us consider the plane transmission grating held normal to the plane of the page (Fig. 9.15 *a*) and represented by the section *ABC...H*. Let the width of the transparent portion *AB* be equal to a and opaque portion *BC* be equal to b. The distance $(a + b) = d$ and is called the *grating constant* or *grating period.*

Let a parallel beam of wavelength λ be incident normally on the grating surface. Then all the secondary waves traveling in the same direction as that of the incident light will come to focus at the point P on the screen, placed at the focal plane of the collecting lens, L. The point P gives the position of the *central bright maximum*.

Now let us consider the secondary waves traveling in a direction inclined at an angle θ with the direction of the incident light (Fig. 9.15 *b*). The waves travel different distances and it is obvious that there is a path difference between the waves coming out from each slit and bending at an angle θ. These secondary waves come to focus at the point Q on the screen. The intensity at Q will depend on the path difference between the secondary waves originating from the respective points A and C of two neighbouring slits. In the Fig. 9.15 (*b*), $AB = a$ and $BC = b$.

The path difference between the secondary waves starting from A and C is equal to $AC \sin\theta$.

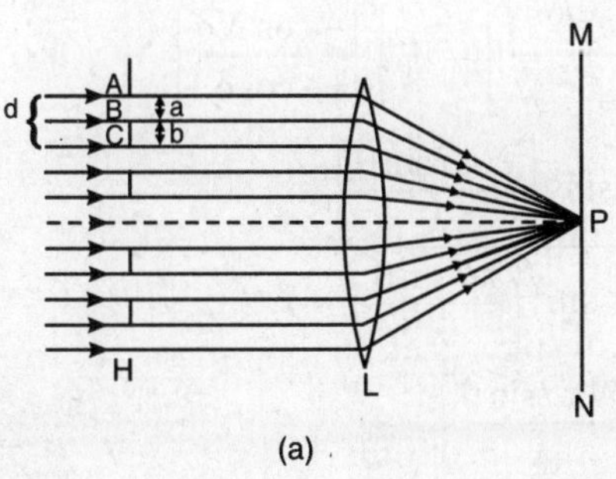

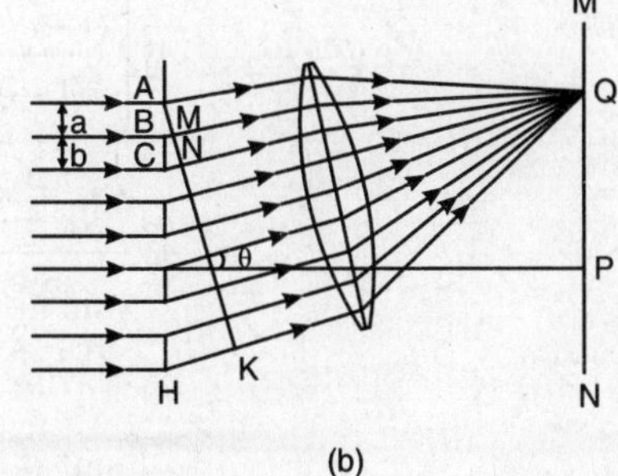

Fig. 9.15:

Diffraction pattern.

But $$AC = AB + BC = a + b$$

$$\text{Path difference} = AC \sin\theta$$

$$= (a + b) \sin\theta$$

The point Q will be of maximum intensity if this path difference is equal to integral multiples of λ.

In general $$(a + b) \sin\theta_m = m\lambda \quad \text{... (9.20)}$$

where θ_m is the direction of the mth principal maximum.

If $(a + b) \sin\theta = \lambda$, we obtain maximum intensity at Q.

When $(a + b) \sin\theta = 2\lambda$, there will be again a maximum and so on.

There will be minimum intensity in between the central maximum P and the first maximum at Q and so on.

Similar maxima and minima are obtained on the other side of central maximum. Thus, on each side of the central maximum at P, principal maxima and minimum intensity are observed due to diffracted light.

The position of mth minima is given by

$$(a + b) \sin\theta_m = (2m + 1)\lambda/2. \quad \text{... (9.21)}$$

9.8.1 Intensity Distribution

When illuminated by a beam of monochromatic radiation, the system produces N wavelets at an angle θ, each of the amplitude $A_\theta = A_0 \dfrac{\sin\alpha}{\alpha}$. The phase difference between successive wavelets is $\delta = \dfrac{2\pi d \sin\theta}{\lambda}$.

The resultant amplitude can be expressed as

$$y = A_\theta[\cos\omega t + \cos(\omega t + \delta) + \cos(\omega t + 2\delta) + \ldots + \cos(\omega t + N\delta)]$$

Expressing the amplitude terms as real parts of complex numbers, we have

$$y = A_\theta e^{j\omega t}\left[1 + e^{j\delta} + e^{2j\delta} + \ldots\ldots\right] = A_\theta e^{j\omega t}\left[\frac{1 - e^{jN\delta}}{1 - e^{j\delta}}\right]$$

We get the intensity by multiplying the amplitude with its complex conjugate.

$$I = A^2 = A_\theta^2\left[\frac{(1-e^{jN\delta})(1-e^{-jN\delta})}{(1-e^{j\delta})(1-e^{-j\delta})}\right] = A_\theta^2\left[\frac{1-\cos N\delta}{1-\cos\delta}\right]$$

$$= A_\theta^2\frac{\sin^2\frac{N}{2}\delta}{\sin^2\frac{\delta}{2}} = A_\theta^2\frac{\sin^2\left[\frac{N\pi d\sin\theta}{\lambda}\right]}{\sin^2\left[\frac{\pi d\sin\theta}{\lambda}\right]}$$

or
$$I = I_0\left(\frac{\sin^2\alpha}{\alpha^2}\right)\frac{\sin^2\left[\frac{N\pi d\sin\theta}{\lambda}\right]}{\sin^2\left[\frac{\pi d\sin\theta}{\lambda}\right]}$$

or
$$I = I_0\left(\frac{\sin^2\alpha}{\alpha^2}\right)\frac{\sin^2 N\gamma}{\sin^2\gamma} \qquad \text{... (9.22)}$$

where
$$\gamma = \frac{\pi d\sin\theta}{\lambda}$$

The expression for intensity is a product of two terms.

The term $I_0\left(\frac{\sin^2\alpha}{\alpha^2}\right)$ represents the intensity distribution due to a single slit diffraction.

The second term $\frac{\sin^2 N\gamma}{\sin^2\gamma}$ represents the distribution of intensity due to interference produced by waves from N equally spaced point sources.

Principal maxima

When $\sin\gamma = 0$, that is $\gamma = \pm n\pi \quad (n = 0, 1, 2, 3, ...)$

We have $\sin N\gamma = 0$ and hence $\frac{\sin N\gamma}{\sin\gamma}$ becomes indeterminate.

According to L'Hospital's rule,

$$\underset{\alpha\to m\pi}{\text{Lt}}\ \frac{\sin N\gamma}{\sin\gamma} = \pm N$$

Therefore,
$$\underset{\alpha\to\pm m\pi}{\text{Lt}}\left[\frac{\sin N\gamma}{\sin\gamma}\right]^2 = N^2$$

Substituting this value into Eq. (9.22), we get $I = I_0\left(\frac{\sin^2\alpha}{\alpha^2}\right)N^2$, which is maximum.

Thus the resultant intensity of maxima is

$$I = I_0\left(\frac{\sin^2\alpha}{\alpha^2}\right)N^2. \qquad \text{... (9.23)}$$

Fig. 9.16 shows the intensity distribution determined by the factor $\left(\frac{\sin^2\alpha}{\alpha^2}\right)$ and $\frac{\sin^2 N\gamma}{\sin^2\gamma}$ respectively. The resultant of these curves is obtained by multiplying the ordinates of first two curves at every point.

As the maxima are very intense, they are called ***principal maxima***. In order to find the resultant intensity of any of the principal maxima in the diffraction pattern, we have to multiply the square of the number of the slits (N^2) with the factor $I_0\left(\frac{\sin^2\alpha}{\alpha^2}\right)$, which is the intensity due to a single slit.

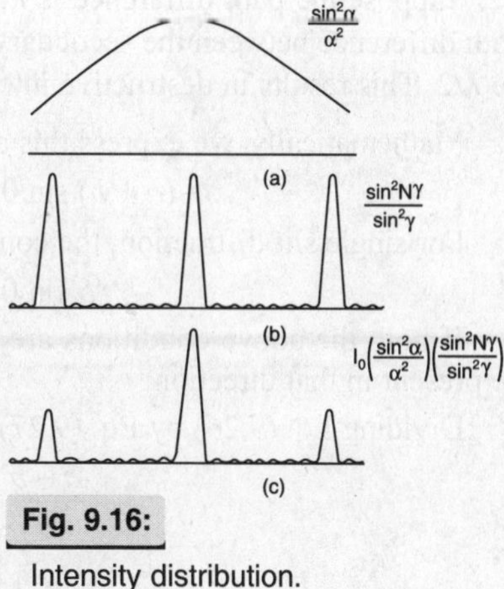

Fig. 9.16: Intensity distribution.

The ***direction of the principal maxima*** are given by $\sin\gamma = 0$, that is $\gamma = \pm\, n\,\pi$.

or $$d\sin\theta = \pm\, n\lambda \quad ...(9.24)$$

Eq. (9.24) is known as the ***grating equation***.

If we put $n = 0$, we get $\theta = 0$. This is the direction in which waves from all slits arrive in phase and produce a bright central image. This maxima is called the ***zero order principal maxima***. If we put $n = 1, 2, 3,....$ we obtain the directions of the first, second, third order principal maxima respectively. Therefore, the direction of occurrence of principal maxima is given by

$$\sin\theta_n = \frac{n\lambda}{d} = \frac{n\lambda}{a+b}$$

or $$\sin\theta_n = nN\lambda \quad ...(9.25)$$

where $N = \frac{1}{a+b}$ is the number of ruled lines per unit width of the grating.

Minima

The intensity is zero when $\sin N\gamma = 0$, or $N\gamma = \pm\, m\pi$

or $$N\gamma = \frac{N\pi d\sin\theta}{\lambda} = \pm\, m\pi$$

or $$N\cdot d\sin\theta = \pm\, m\lambda \quad ...(9.26)$$

Here, m can take all integral values except $0, N, 2N, 3N$, etc because these values give the positions of principal maxima. The positive and negative signs indicate that the minima of a given order lie symmetrically on both the sides of the central principal maxima.

It is seen from Eq. (9.26) that $m = 0$ gives principal maximum of zero order while $m = 1, 2, 3,... (N-1)$ give the minima. When $m = N$, Eq. (9.26) gives the principal maximum of first order. Thus, between zero order and first order principal maxima, we have $(N-1)$ minima.

Similarly, it can be shown that there are $(N-1)$ minima between first order and second order principal maxima and so on. Between two such consecutive minima, the intensity has to be maximum, and these maxima are known as ***secondary maxima***. The secondary maxima are not visible in the grating spectrum, as the number of slits is very large.

9.8.2 Condition for Missing of Orders

Under certain conditions, it is possible that grating forms the first and third order spectra while the second order spectrum is missing. Such a situation arises when for a given angle of diffraction θ_1 the path difference between the diffracted rays from the extreme ends of one slit is equal to an integral multiple of λ.

Suppose the path difference is λ. Then each slit can be taken as two equal parts and the path difference between the secondary waves from the respective points in the two halves will be $\lambda/2$. This results in destructive interference.

Mathematically, we express this as

$$(a + b) \sin \theta = n\lambda \quad \text{...(9.27)}$$

For single slit diffraction, the condition for minima is

$$a \sin \theta = m\lambda \quad \text{...(9.28)}$$

If both the above conditions are satisfied simultaneously, the principal maxima will not be present in that direction.

Dividing Eq. (9.26) by Eq. (9.27), we obtain

$$\frac{a+b}{a} = \frac{n}{m}$$

i.e.,

$$n = \frac{a+b}{a} m \quad \text{...(9.29)}$$

Eq. (9.29) is the condition for the nth order spectrum to be absent.

If we wish to suppress the second order spectrum, then $n = 2m = 2$ ($\because$ $m = 1$).

Then,

$$\frac{a+b}{a} = \frac{2m}{m} = 2$$

or

$$a + b = 2a$$

$$a = b$$

*Thus, if the **width of each slit a is equal to the width b** of the ruling, the second order spectrum will be absent.*

9.8.3 Maximum Number of Orders Possible

The grating equation $d \sin \theta = \pm n\lambda$ may be rewritten as

$$n = \frac{d \sin \theta}{\lambda} = \frac{\sin \theta}{N\lambda}$$

The maximum value that θ can take is 90° and hence the maximum possible value of $\sin \theta$ is 1. It implies that

$$(n)_{max} \leq \frac{1}{N\lambda} \quad \text{...(9.30)}$$

Eq. (9.30) gives the maximum number of orders that would be seen in the spectrum produced by a plane transmission grating having N lines per unit width.

9.8.4 Determination of Wavelength of Light with a Plane Transmission Grating

The wavelength of spectral lines can be determined using a diffraction grating and a spectrometer. The slit of the collimator is illuminated by an appropriate source of light (say light from sodium lamp) and the initial adjustments of the spectrometer are made.

(*i*) The collimator and the telescope are adjusted for parallel rays by Schuster's method using a prism.

(*ii*) The grating is adjusted for *normal incidence.*

(*a*) The telescope is set in line with the collimator such that direct image of the slit is obtained at the position of the vertical cross-wire. Now the axes of the collimator and the telescope are in the same line. Note the telescope position from the circular scale.

(*b*) Then the telescope is turned through 90º from this position and clamped. In this position the axis of the telescope is perpendicular to the axis of collimator.

(*c*) The given transmission grating is mounted on the prism table such that the grating surface is obtained in the centre of field of view of the telescope. This means that the parallel rays of light from the collimator are incident at an angle 45º on the grating surface because the axis of the collimator and the telescope are perpendicular to each other.

(*d*) The prism table is now rotated through 45º in the proper direction so that the grating surface is normal to the incident light faces the telescope. The prism table is clamped in this position. The above steps are shown in Fig. 9.17.

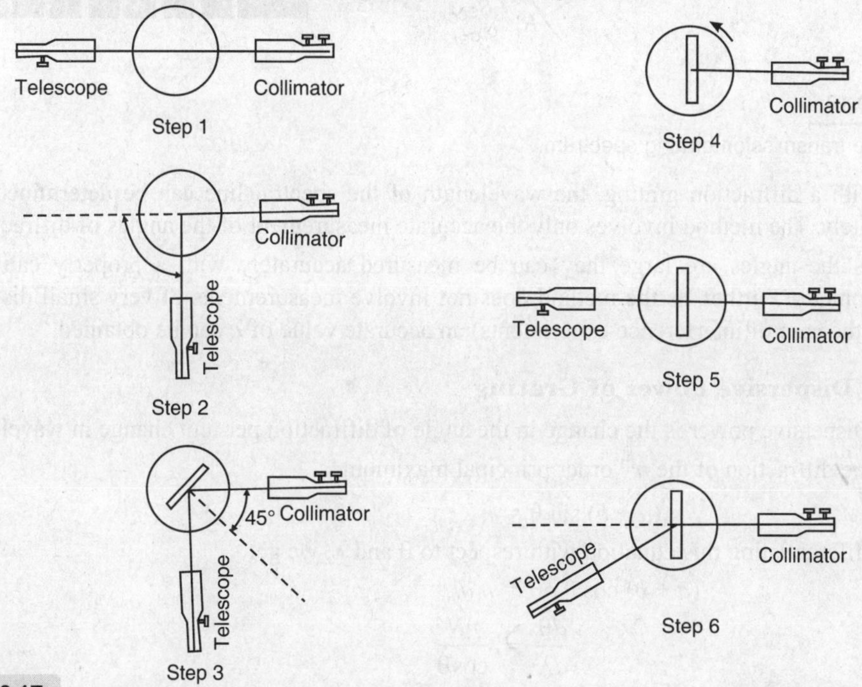

Fig. 9.17: Steps in the determination of wavelength of light with a plane transmission grating.

(*iii*) On viewing through the telescope, the grating spectrum is observed (Fig. 9.18). It consists of a direct image flanked by the different order images on its both sides. To measure θ for a spectral line of a given order, the telescope is focused on, say, first order image of that line. The position of the telescope is adjusted such that the line falls on the intersection of the cross-wires. Then the readings of both the vernier scales are noted. Then the telescope is taken to the other side of the direct image and the corresponding readings of first order image are noted. The difference between the readings of the two positions of the line gives 2θ for that line. From this the diffraction angle θ is found. The procedure is repeated for higher orders.

(*iv*) Knowing the value of d(or N) and n, the wavelength λ can be calculated from the grating equation $d \sin \theta = n\lambda$.

If the source of light emits radiations of different wavelengths, then the grating disperses the beam and in each order a spectrum of the constituent wavelengths is observed. To find the wavelength of any spectral line, the diffracting angles are noted in the first and second orders and using the grating equation, the wavelength of the spectral lines can be calculated.

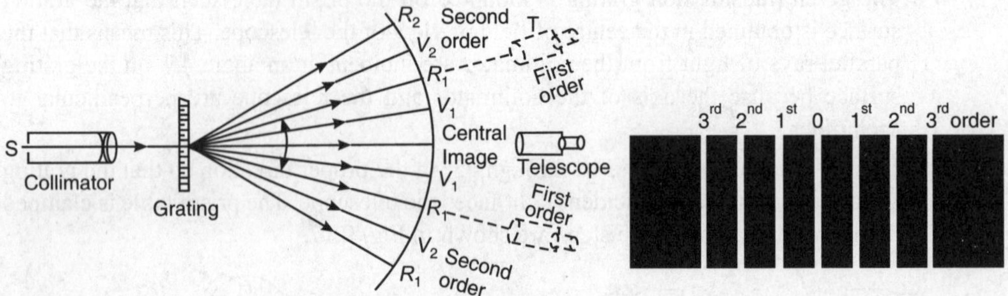

Fig. 9.18:

Plane transmission grating spectrum.

With a diffraction grating, the wavelength of the spectral line can be determined very accurately. The method involves only the accurate measurement of the angles of diffraction.

As the angles are large they can be measured accurately with a properly calibrated spectrometer. Further, as the method does not involve measurements of very small distances (as in the case of interference experiments) an accurate value of λ can be obtained.

9.8.5 Dispersive Power of Grating

Def: Dispersive power is the change in the angle of diffraction per unit change in wavelength.

The diffraction of the nth order principal maximum

$$(a + b) \sin \theta = n \lambda$$

Differentiating this equation with respect to θ and λ, we get

$$(a + b) \cos \theta \, d\theta = n \, d\lambda$$

or
$$\frac{d\theta}{d\lambda} = \frac{nN}{\cos\theta} \quad \text{... (9.31)}$$

We can note from Eq. (9.31) that the

(*i*) $d\theta/d\lambda$ is proportional

(*a*) directly to the order of the spectrum, n

(*b*) directly to the number of lines per cm, N

(*c*) inversely to $\cos \theta$.

(*ii*) angular dispersion of the lines is more with a grating having a larger number of lines per cm.

(*iii*) angular dispersion is a minimum when $\theta = 0$. If the value of θ is not large, the value of $\cos \theta$ can be taken as unity and the influence of this factor can be neglected.

Def: Normal spectrum is a spectrum in which the angular dispersion of any two spectral lines is directly proportional to the difference in wavelength of the spectral lines.

Linear dispersion: The linear dispersion is useful in studying the photographs of a spectrum and is given by

$$\frac{dx}{d\lambda}$$

where *dx* linear spacing of two spectral lines of wavelength λ and λ + *d*λ and is

$$dx = f\,d\theta$$

where *f* is the focal length of the objective.

so,
$$\frac{dx}{d\lambda} = f\frac{d\theta}{d\lambda} = \frac{fnN}{\cos\theta} \qquad \text{... (9.32)}$$

or
$$dx = \frac{fnN}{\cos\theta}\cdot d\lambda$$

Example 9.4: *Calculate the position of the third order maximum for $\lambda = 2.4 \times 10^{-4}$ cm and grating constant of 15×10^{-4} cm.*

Solution:
$$\sin\theta_n = \frac{n\lambda}{d} = \frac{3\times 2.4\times 10^{-4}\,\text{cm}}{15\times 10^{-4}\,\text{cm}} = 0.427$$

Hence,
$$\theta_3 = \sin^{-1}(0.427) = \mathbf{25.28^\circ}$$

Example 9.5: *Find the resolving power of a grating in the first order for 15 cm of the surface ruled with 6000 lines/cm.*

Solution: Total rulings on grating surface, $N = 15$ cm × 6000 lines/cm = 9×10^4

Resolving power, $R = m\,N = 1 \times 9 \times 10^4 = \mathbf{9 \times 10^4}$

Example 9.6: *In a grating spectrum, which spectral line in 4th order will overlap with 3rd order line of 5491Å?*

Solution: The grating equation is

$$(a + d)\sin\theta = n\lambda$$

If the nth order of wavelength λ_1 (say) coincides with the $(n + 1)^{th}$ order of λ_2, then

$$(a + d)\sin\theta = n\lambda_1 = (n + 1)\lambda_2.$$

Here $n = 3$, $\lambda_1 = 5461$Å, $(n + 1) = 4$, $\lambda_2 = ?$

$$\therefore \qquad \lambda_2 = \frac{n\lambda_1}{n+1} = \frac{3\times 5461\times 10^{-8}\,\text{cm}}{4}$$

$$= 4096 \times 10^{-8}\ \text{cm} = \mathbf{4096\text{Å}}$$

9.9 RESOLVING POWER

When two objects or their images are very close to each other, they appear as one and it may not be possible for the eye to see them as separate. If the objects are not seen separately, then we say that the details are *not resolved* by the eye. Optical instruments are used to assist the eye in resolving the objects or images. The method adapted to seeing the close objects as separate objects is called **resolution**.

Def: Resolving power is the capability of an optical instrument to produce clearly separate images of two objects situated very close to each other.

9.9.1 Rayleigh's Criterion

The theory of optical instruments is based on the laws of geometrical optics and rectilinear propagation of light. These laws are only approximately true. While a beam of light from a point object passes through the objective of a telescope, the lens acts like a circular aperture and produces a diffraction pattern instead of a point image. This diffraction pattern is as shown in Fig. 9.19. It is known as Airy's disc. If there are two close point objects then two diffraction patterns are produced, which may overlap on each other and it may be difficult to distinguish them as separate.

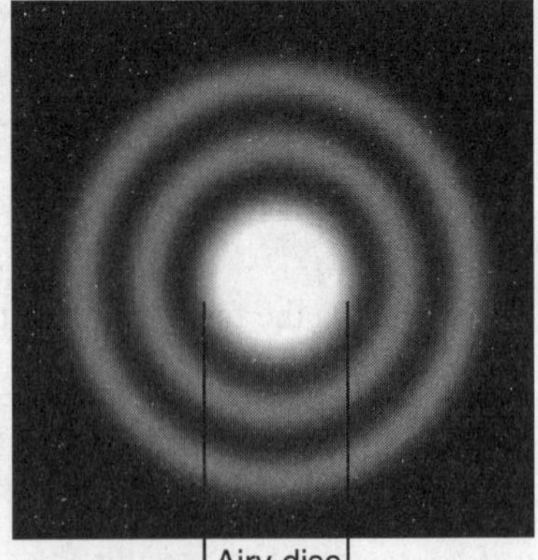

Fig. 9.19: An Airy disc is the central spot in the diffraction pattern.

To obtain the measure of the resolving power of an objective lens Rayleigh suggested a criterion known as ***Rayleigh criterion***. According to Rayleigh criterion the two images may be regarded as separated if the central maximum of one falls on the first minimum of the other (Fig. 9.20 *b*). This is equivalent to the condition that the distance between the centers of the patterns shall be equal to the radius of the central disc. This is called as ***Rayleigh's limit of resolution***.

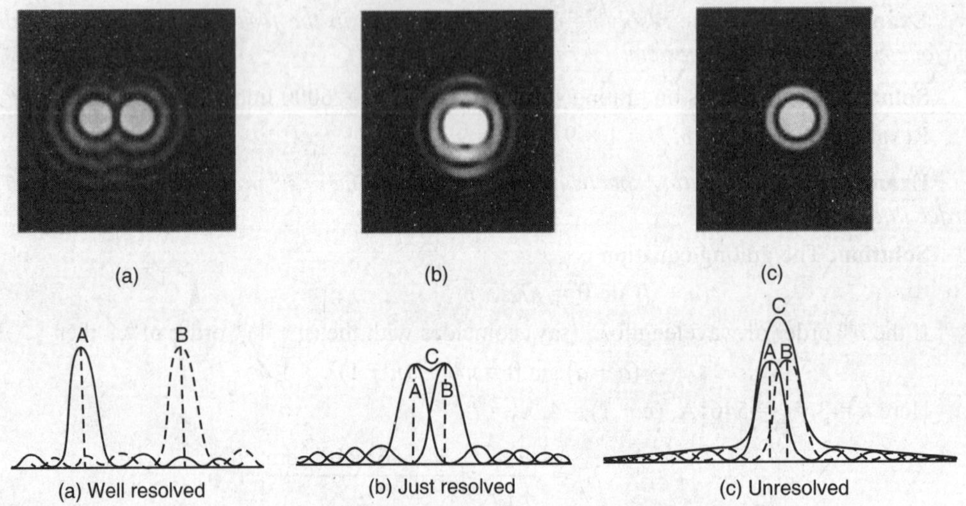

Fig. 9.20: The images of two sources are discernable when they satisfy Rayleigh criterion as seen in the second image at the top.

9.10 RESOLVING POWER OF A PLANE TRANSMISSION GRATING

One of the important properties of a diffraction grating is its ability to resolve spectral lines, which have nearly the same wavelength.

Def: The spectral resolving power of a grating is the smallest wavelength interval ($d\lambda$) that can be detected by it.

$$\text{Resolving Power} = \frac{\lambda}{d\lambda} = \frac{\lambda}{d\theta} \cdot \frac{d\theta}{d\lambda} \qquad ...(9.33)$$

Let us now find the values of $\frac{d\theta}{d\lambda}$ and $\frac{\lambda}{d\theta}$.

The diffraction grating equation is

$$(a + b) \sin \theta = n\lambda.$$

Differentiating the above equation both sides, we get

$$(a + b) \cos \theta \cdot d\theta = n\, d\lambda$$

$$\therefore \qquad \frac{d\theta}{d\lambda} = \frac{n}{(a+b)\cos\theta} \qquad ...(9.34)$$

where $d\theta$ is the angle between the two diffracted beams whose difference in wavelength is $d\lambda$.

The light diffracted from the grating enters the objective of a telescope in a grating spectrometer. If the diffracted beam completely fills the objective then width of the beam equals the diameter D of the objective lens.

The angular limit of resolution of a telescope objective

$$d\theta = \frac{\lambda}{d}$$

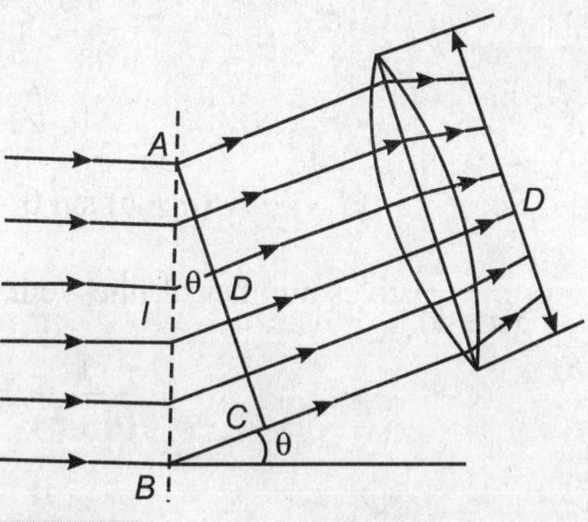

Fig. 9.21: Resolving power of a plane transmission grating.

Now $D = AB \cos \theta = l \cos \theta$ where l is the length of the grating.

$$\therefore \qquad d\theta = \frac{\lambda}{l \cos \theta}$$

$$\text{or} \qquad \frac{\lambda}{d\theta} = l \cos \theta$$

$$\text{Resolving Power} = \frac{\lambda}{d\lambda} = \frac{\lambda}{d\theta} \cdot \frac{d\theta}{d\lambda} = l\cos\theta \times \frac{n}{(a+b)\cos\theta}$$

$$= \frac{nl}{(a+b)} = n\,N \qquad ...(9.35)$$

where $N = l/(a + b)$ = number of rulings on the grating surface and n is the order of the spectrum.

Hence, the resolving power of a grating is given by the simple expression,

$$\text{R.P.} = nN$$

Example 9.7: *Find the resolving power of a grating to resolve lines of the sodium yellow doublet of wavelengths 5890 Å and 5896 Å.*

Solution: Mean wavelength $\lambda = (5890 + 5896)/2 = 5893$Å.

Wavelength difference, $\quad d\lambda = 5896\text{ Å} - 5890\text{ Å} = 6\text{Å}$

Therefore, the resolving power R required for a grating to resolve these lines is given by

$$R = \frac{\lambda}{d\lambda} = \frac{5893}{6} = \mathbf{982}$$

Example 9.8: *Find lines/cm in a grating if the angle of diffraction is 30° and for normal incidence a line of 5400 Å in a certain order superimposed on its violet line (4450 Å) of the next higher order.*

Solution: We know that $(a + b) \sin\theta = n\lambda$

Let the nth order maxima of λ_1 coincide with $(n + 1)^{th}$ order maxima λ_2.

$$(a + b) \sin\theta = n\lambda_1$$

$$(a + b) \sin\theta = (n + 1)\lambda_2$$

$$n\lambda_1 = (n + 1)\lambda_2$$

$$n = \frac{\lambda_2}{\lambda_1 - \lambda_2}$$

So, $$(a + b) \sin\theta = n\lambda_1 = \frac{\lambda_2\lambda_1}{\lambda_1 - \lambda_2}$$

$\frac{1}{(a+b)}$ gives number of lines /cm.

$$\frac{1}{(a+b)} = \sin\theta\frac{\lambda_1 - \lambda_2}{\lambda_1 - \lambda_2}$$

$$\lambda_1 = 5400 \times 10^{-8} \text{ cm}, \lambda_2 = 4050 \times 10^{-8} \text{ cm}$$

$$\text{No. of lines/cm} = \frac{\sin 30° (5400 - 4500) \times 10^{-8}}{5400 \times 4050 \times 10^{-16}} = \mathbf{3086}$$

Example 9.9: *Show that in a diffraction grating with grating element 1.5 × 10^{-6} m and light of wave length 550 nm, 3rd and higher order maxima are not visible.*

Solution: The condition for maxima $(a + b) \sin\theta = n\lambda_1$

Since maxima value of θ is $\frac{\pi}{2}$, then the highest value of n is given by $n = \frac{(a+b)}{\lambda}$

Here in this case $(a + b) = 1.5 \times 10^{-6}$ m, $\lambda = 550 \times 10^{-9}$ m

$$n = \frac{1.5 \times 10^{-6}}{550 \times 10^{-9}} = 2.7$$

Then only 2nd order principal premium can be seen and third and higher order maxima are absent.

Example 9.10: *Find the angular separation between the two sodium lines in 2nd order whose wavelength are 589 nm and 589.6 nm respectively for a grating having 8000 lines/cm.*

Solution: In plane diffraction grating

$$(a + b) \sin\theta = n\lambda$$

Given $$(a + b) = \frac{1}{8000} = 125 \times 10^{-8} \text{cm}$$

$$n = 2$$

$$\lambda_1 = 589 \times 10^{-9} \text{ m} = 5850 \times 10^{-8} \text{ cm and}$$

$$\lambda_2 = 589.6 \times 10^{-9} \text{ m} = 5896 \times 10^{-8} \text{ cm}$$

Let $\theta = \theta_1$ for $\lambda = \lambda_1$ and $\theta = \theta_2$ for $\lambda = \lambda_2$

$$(a + b) \sin \theta_2 = n\lambda_2$$

or $$\sin \theta_2 = \frac{n\lambda_2}{a+b} = \frac{2 \times 5896 \times 10^{-8}}{125 \times 10^{-6}}$$

$$\theta_2 = 70.62^{\circ}$$

Angular Spectrum of lines $\theta_2 - \theta_1 = 0.17^{\circ} = 10'$

9.11 RESOLVING POWER OF A TELESCOPE

Let us consider the rays of light from two neighboring points of a far-away object which form a Fraunhofer diffraction pattern. Let P_1 and P_2 be the positions of the central maxima of the two images. As we know, Rayleigh criterion defines the resolution.

The secondary wavelets AP_1 and BP_1 reinforce at P_1 as the path difference between them is zero and p_1 corresponds to the position of the central maximum of the first image.

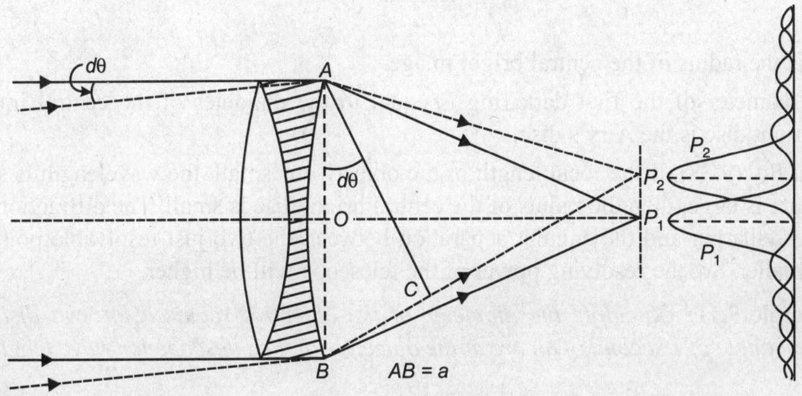

Fig. 9.22:
Resolving power of telescope.

The secondary wavelets AP_2 and BP_2 will meet at P_2 on the screen. Let the angle P_2AP_1 be $d\theta$.

From the Fig. 9.22, the path difference between the secondary waves BP_2 and $AP_2 = BC$.

From the ΔABC, $BC = AB \sin d\theta = AB.d\theta$

For small angles, $BC = a.d\theta$

where *a is* the diameter of the objective of the telescope.

If this path difference is equal to λ, P_1 represents the first minimum of the first image. But P_2 also is the position of the central maximum of the second image.

For Rayleigh's condition of resolution to be satisfied

$$a.d\theta = \lambda \quad \text{or} \quad d\theta = \frac{\lambda}{a} \qquad ...(9.36)$$

The slit (rectangular aperture) AB can be considered to be made of two equal parts AO and OB. The path difference between the secondary waves from the individual points in the two equal parts will be $\frac{\lambda}{2}$.

As the secondary waves destructively interfere with one another P_2 will be the first minimum of the first image.

For circular apertures, $d\theta$ is known as the ***limit of resolution*** of the telescope.

Resolving power of the telescope is reciprocal of $d\theta$, angle subtended by the two distant object points at the telescope

$$\frac{1}{d\theta} = \frac{a}{1.22\lambda} \quad ...(9.37)$$

Def: Resolving power of a telescope is the reciprocal of the angular separation that two distant object points must have so that their images will appear just resolved according to Rayleigh's criterion.

If f is the focal length of the telescope objective, then,

$$d\theta = \frac{r}{f} = \frac{1.22\lambda}{a}$$

or

$$r = \frac{1.22 f\lambda}{a} \quad ...(9.38)$$

where r is the radius of the central bright image.

The diameter of the first dark ring is equal to the diameter of the central image. The central bright disc is the Airy's disc.

From Eq. (9.38), if the focal length of the objective is small, the wavelength is small and the aperture is large, then the radius of the central bright disc is small. The diffraction patterns will appear sharper and the angular separation between the two just resolvable point objects will be smaller. So, the resolving power of the telescope will be higher.

Example 9.11: *Calculate the diameter of the objective required for two distant stars subtend an angle of 1 second of an arc at the objective of the telescope for wave length of light 5890 Å.*

Solution: 1 second of an arc = 4.85×10^{-6} radian.

$$\lambda = 5500 \text{ Å}.$$

Diameter of the objective $$a = \frac{1.22\lambda}{d\theta} = \frac{1.22 \times 5890 \times 10^{-8} \text{ cm}}{4.85 \times 10^{-8}} = 14.81 \text{ cm}$$

9.12 RESOLVING POWER OF A MICROSCOPE

Def: Resolving power of a microscope is the reciprocal of the minimum distance by which two points in the object are separated from each other so that their images are just seen separate.

Let A and B are two object points at a distance d apart and MN is the aperture of the objective of a microscope (Fig. 9.23). A' and B' correspond to diffraction patterns and are surrounded by alternate dark and bright diffraction rings.

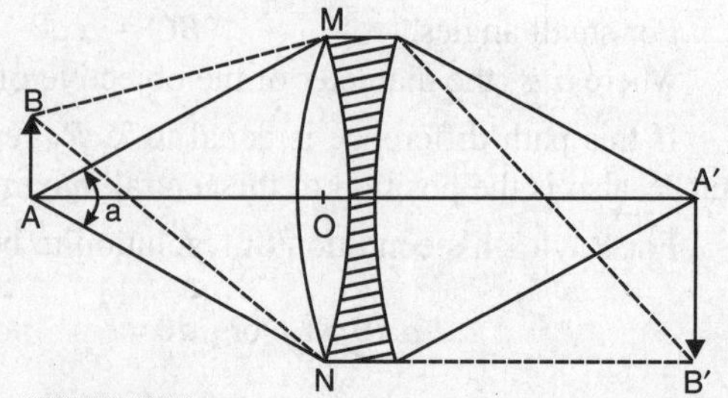

Fig. 9.23: Aperture of the objective of a microscope.

The path difference between the rays from B and reaching $A' - (BN + NA') - (BM + MA')$

As $NA' = MA'$

$\therefore$ Path difference $= BN - BM$

In Fig. 9.24, AD and AC are perpendiculars to DM and BN.

$\therefore$ $BN - BM = (BC + CN) - (DM - DB)$

But $CN = AN = AM = DM$

$\therefore$ Path difference $= BC + DB$

From the Δ^S ACB and ADB

$BC = AB \sin \alpha = d \sin \alpha$

and $DB = AB \sin \alpha = d \sin \alpha$

$\therefore$ Path difference $= 2d \sin \alpha$

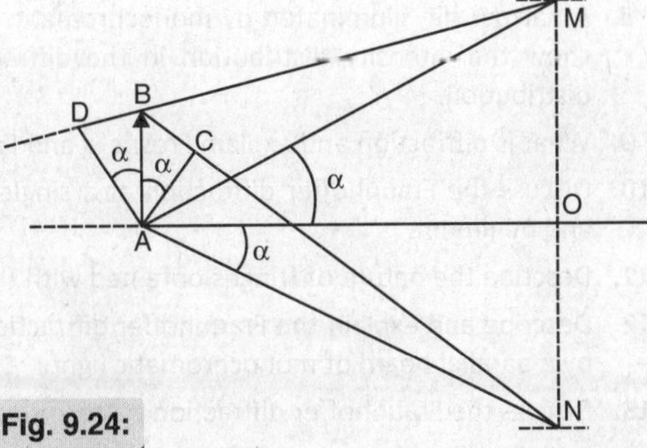

Fig. 9.24: Ray diagram for resolving power of microscope.

If this path difference $2d \sin \alpha = 1.22\ \lambda$, then, A' corresponds to the first minimum of the image B' and the two images appear just resolved.

$$\therefore \quad 2d \sin \alpha = 1.22\ \lambda$$

or
$$d = \frac{1.22\ \lambda}{2 \sin \alpha} \qquad ...(9.39)$$

The resolving power depends upon the mode of illumination.

According to Abbe,

$$d = \frac{\lambda_0}{2 \mu \sin \alpha} \qquad ...(9.40)$$

where λ_0 is the wavelength of light in vacuum and

μ is the refractive index of the medium between the object and the objective (generally cedar wood oil in microscope).

$(\mu \sin \alpha)$ is called the ***numerical aperture*** of the objective of the microscope.

Example 9.12: Find d for light of 5500 Å and $\mu \sin \alpha = 1.6$.

Solution: Linear distance between two just resolvable object points

$$d = \frac{5500 \times 10^{-8}\ \text{cm}}{2 \times 1.6} = 1.72 \times 10^{-5}\ \text{cm}$$

QUESTIONS

1. Explain with clarity what diffraction of light is.
2. Why do radio waves diffract around buildings, although light waves do not?
3. Give the difference between interference and diffraction.
4. Distinguish between Fresnel and Fraunhoffer class of diffractions.
5. What is the source of the colours seen on the surface of CDs (compact discs)?
6. Deduce the conditions for maxima and minima for diffraction at a single slit.
7. Describe and explain the nature of fringes obtained with the help of a single slit placed before a parallel beam of monochromatic light.

8. A narrow slit, illuminated by monochromatic light produces Fraunhoffer diffraction. Graphically show the intensity distribution in the diffraction pattern. Write the expression for intensity distribution.
9. What is diffraction and explain Fresnel's and Fraunhoffer diffraction?
10. Discuss the Fraunhoffer diffraction at a single slit. Obtain the condition for principal maximum and minimum.
11. Describe the nature of fringes obtained with the help of a single slit.
12. Describe and explain the Fraunhoffer diffraction pattern obtained by narrow slit and illuminated by a parallel beam of monochromatic light.
13. Discuss the Fraunhoffer diffraction pattern shown by double slit.
14. Explain diffraction at a circular aperture.
15. What is diffraction grating? How is it obtained?
16. Discuss the Fraunhoffer diffraction at a single slit. Extend this theory to the case of a plane transmission grating.
17. What is plane transmission grating? Define grating element. Explain how it can be prepared.
18. What is plane diffraction grating? How is it used to find the wavelength of light?
19. What is a plane transmission grating?
20. Obtain the expression $(a + b) \sin \theta = n\lambda$.
21. What is dispersive power of a grating? Derive equation for dispersive power.
22. Explain the theory of plane transmission grating and derive equation for maxima and minima.
23. Explain the theory of plane transmission grating with equation for maximum intensity.
24. Is it possible that no minimum is recorded in a single slit diffraction pattern? If so, under what condition?
25. Explain the theory of plane diffraction grating for getting the condition for the formation of nth order maximum.
26. State the factors on which the resolving power of grating depends.
27. Explain the construction and working of a diffraction grating.
28. Describe Rayleigh's criteria.
29. What is plane diffraction grating? Explain how it is used to determine the wavelength of a spectral line of a given source of light.
30. How do you determine the wavelength of light using diffraction grating? Explain.
31. How will you determine the wavelength of spectral line found in diffraction pattern by using plane transmission grating?
32. Compare intensity and width of principal maxima between two gratings *A* and *B*, if *A* has more number of lines than *B*.
33. Why does a diffraction grating have closely spaced rulings?
34. Why does it have a large number of rulings?
35. Why are the colours in the spectrum of a light source linear in shape?
36. What are the advantages of increasing the number of rulings in a grating?
37. In a transmission grating, how spectral lines get affected if the rulings are made closer?
38. If width of transparencies and opacities of grating are equal which spectra would be absent in a transmission grating?
39. Explain the meaning of 'missing spectra' in the diffraction pattern of a plane transmission grating.
40. What is the condition, which must be satisfied if the second order spectra are to be absent from the grating spectrum?

41. Explain the effect: if (*a*) total number of lines, (*b*) the number of lines per cm and (*c*) the width of the grating are changed?
42. Explain why increasing the number of slits in a grating sharpens the maxima.
43. Explain why decreasing the wavelength sharpens the maxima in a grating.
44. Explain why increasing slit spacing in a grating sharpens the maxima.
45. Explain what you understand by the resolving power of an optical instrument. Determine the resolving power and dispersive power of the diffraction grating.
46. Define resolving power of plane diffraction grating. Hence prove that it is independent of grating element.
47. Obtain an expression for the resolving power of grating.
48. Explain the resolving power of the plane diffraction grating.
49. Define resolving power of an optical instrument. Find the expression for resolving power of telescope and discuss in detail.
50. State and explain Rayleigh's criterion for limit of resolution.
51. Distinguish between the resolving power and dispersive power of the diffraction grating.
52. Using Rayleigh criterion of resolution, obtain an equation for the resolving power of a telescope.

PROBLEMS

1. Find the angular position of the first two minima on either side of the central maximum for given data: Normal incidence, $\lambda = 5500$ Å, slit width 22×10^{-5} cm. **[Ans:** 14°29′, 30°]
2. Calculate the total angular width of the central maximum and also the linear width for given data: Normal incidence, $\lambda = 6000$ Å, slit width 0.20 mm and screen at 2 m away. **[Ans:** 6×10^{-3} rad, 12 mm]
3. A screen is placed 200 cm away from a slit which is illuminated with light of λ 6000 Å. If the first minima lie 5 mm on either side of central maximum, find the slit width. **[Ans:** 0.24 mm]
4. Calculate the possible order of spectra with a plane transmission grating having 18,000 lines per inch when light of wavelength 4500 Å is used. **[Ans:** 3]
5. Find the normalized intensity for an angle of diffraction of 30° for $\lambda = 5000$ Å with normal incidence on a single slit of width 1 mm. **[Ans:** 0]
6. A transmission grating has 8000 rulings per cm. The first order principal maximum due to a monochromatic source of light occurs at an angle of 30°. Determine the wavelength of light.
7. A light of $\lambda = 6000$ Å is incident on a plane diffraction grating with grating element 6.0×10^{-5} cm. What is the maximum order of spectrum that can be observed?
8. The second order maximum for a wavelength of 6360 Å in a transmission grating coincides with third order maximum of an unknown light. Determine the wavelength of the unknown light.
9. A grating has 5000 lines/cm. If the slits are 10^{-4} cm wide, will there be any missing orders? If so, identify these.
10. A plane transmission grating has 40,000 lines in all with grating element 12.5×10^{-5} cm. Calculate the maximum resolving power for which it can be used in the range of wavelength 5000 Å. **[Ans:** 80,000]
11. A plane transmission grating has 16,000 lines per inch over a length of 5 inches. Find (*a*) the resolving power of the grating in the second order and (*b*) the smallest wavelength difference that can be resolved for light of wavelength 6000 Å. **[Ans:** 1,60,000; 0.0375 Å]

10 Polarization

Learning Objectives:

By the end of this chapter, you will be able to:

- Understand basic concepts of Polarization.
- Explain difference between Unpolarized and Polarized Light.
- Describe different types of Polarization.
- State Malu's Law and its applications.

10.1 INTRODUCTION

Interference and diffraction phenomena established that light travels in the form of waves. But, the two phenomena did not offer any hint about the nature of the light waves, whether they are longitudinal or transverse waves, or whether the constituent vibrations are linear or circular. All kinds of waves under suitable conditions exhibit interference and diffraction. In 1816 Arago and Fresnel confirmed that light waves vibrating in mutually perpendicular planes do not interfere.

In 1817 Thomas Young postulated that light waves are *transverse* waves and explained the absence of interference between light waves polarized in mutually perpendicular planes. Thus, the polarization property is a direct result of transverse nature of light. Light coming from common light sources is unpolarized. It can be transformed into different types of polarization using optical devices.

The state of polarization cannot be recognized by our eyes. The knowledge of polarization is essential for understanding the propagation of electromagnetic waves propagating through wave-guides and optical fibres. Polarized light has many important applications in industry and engineering. One of the most important applications is in fabrication of liquid crystal displays (LCDs), which are widely used nowadays in displays in wristwatches, calculators, TVs etc.

10.2 POLARIZATION

Waves are basically of two types: (*i*) longitudinal waves and (*ii*) transverse waves.

(*i*) **Longitudinal wave:** Waves produced on a spring and sound waves are examples of longitudinal waves. The longitudinal wave consists of alternate compressions and rarefactions, as shown in Fig. 10.1 (a & b).

Def: Longitudinal wave is a wave in which particles of the medium oscillate to and fro along the direction of propagation.

(*ii*) **Transverse wave:** Ripples on water surface and waves on a rope are examples of transverse waves. The wave propagates in the form of alternating crests and troughs, as shown in Fig. 10.1 (c & d).

Def: Transverse wave is a wave in which every particle of the medium oscillates up and down at right angles to the direction of wave propagation.

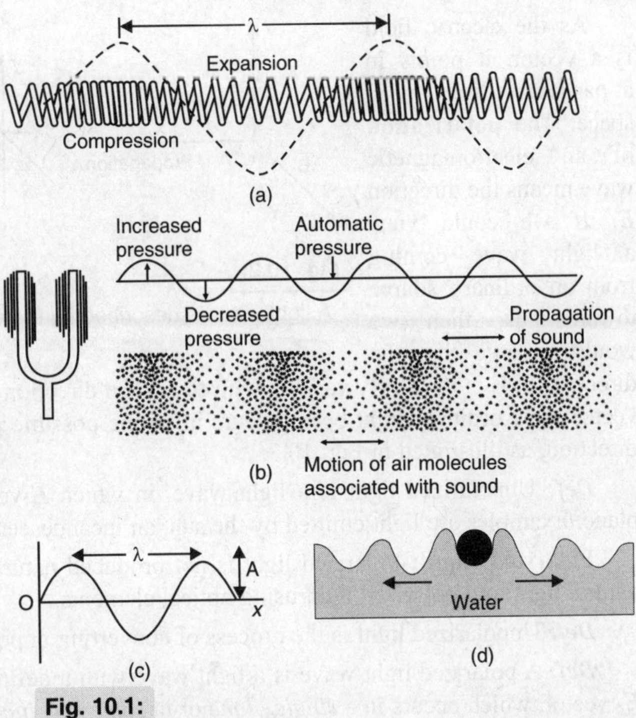

Fig. 10.1: Types of waves - (a) & (b) Longitudinal, (c) & (d) Transverse.

In a longitudinal wave, all directions perpendicular to the wave propagation are equivalent. On the other hand, a preferential direction normal to the wave propagation exists in a transverse wave. The preferential direction in a transverse wave is the direction of vibration of the particles and it differs from all other directions. The existence of a preferential direction for a transverse wave leads to the characteristic phenomenon known as *polarization.* Polarization is not found with longitudinal waves as they do not possess a directional property. Thus, polarization is specific to transverse waves.

Light waves are made up of electric and magnetic field vectors oscillate perpendicular to each other and to the direction of wave propagation. The vibrating electric field vector and the direction of propagation of the wave constitute a plane. There are an infinite number of such planes around the direction of propagation. In an ideal light wave, the vibrations of electric vector are confined to a single plane. In practice, light sources emit a mixture of light waves with random orientation of vibration planes. These planes give rise to symmetry about the wave propagation direction. As a result, the transverse nature of the wave gets concealed. The process of removing the symmetry and bringing in one-sidedness in the light wave is called ***polarization***.

10.3 UNPOLARIZED AND POLARIZED LIGHT

Fig. 10.2 (*a*) shows the electric field in the *xy*-plane, the magnetic field in the *xz*-plane and the propagation of the wave in the *x*-direction. Fig. 10.2 (*b*) shows light wave as described by the **electric field vector, E,.**

As the electric field is a vector, it points in a particular direction in space. The **polarization** of an electromagnetic wave means the direction ***E***. If we could view a light wave coming from an ordinary source towards us, then we would observe that the direction of **E** is randomly varying with time. The direction of oscillation of the electric field vector in an ordinary light beam occurs in all the possible planes perpendicular to the beam direction, as illustrated in Fig. 10.3.

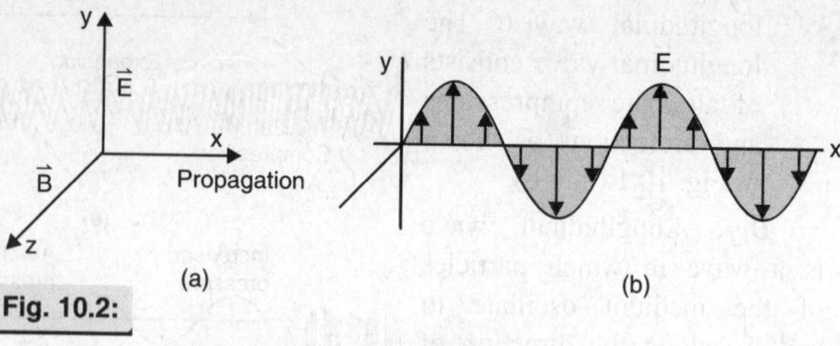

Fig. 10.2:

A light wave is described by the electric field vector, E.

Def: Unpolarized light is a light wave, in which ***E***-vector oscillates in more than one plane. Examples are light emitted by the sun, an incandescent lamp or flame.

Polarized Light: Polarized light is not produced naturally. It is obtained by converting natural light into polarized light using optical elements.

Def: Unpolarized light is the process of converting unpolarized light into polarized light.

Def: A polarized light wave is a light wave with a definite direction of oscillation of the ***E***- vector, which occurs in a *single plane* or in *some specific way*.

Def: Plane of polarization of light wave is the plane comprising the direction of oscillation of ***E*** and the direction of propagation of the beam, the *xy*-plane in Fig. 10.2 (*b*).

10.4 NATURAL LIGHT IS UNPOLARIZED LIGHT

We mentioned earlier that light from the ordinary light sources is unpolarized. Let us find out the reason. We know that atoms emit light. Any light source consists of a very large number of atomic emitters. Each atom radiates, at a specific instant, a wave packet that lasts for about 10^{-8} s. Light radiated by a source is a mixture of wave packets emitted by different atoms at different instances (see Fig. 10.3). Each wave packet is polarized in the own way with no relation to other wave packets. There is no continuity of plane of polarization and the plane of polarization varies from wave packet to wave packet in a completely random manner with no preferred direction.

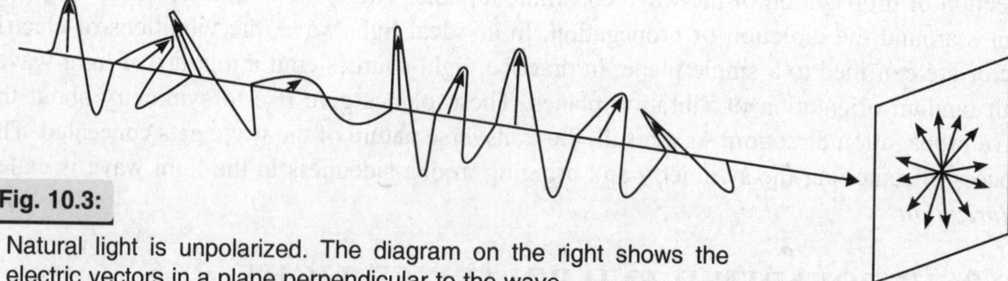

Fig. 10.3:

Natural light is unpolarized. The diagram on the right shows the electric vectors in a plane perpendicular to the wave.

Fig. 10.4 (*a*) is a fictitious diagram. It only implies that the natural light consists of electric field vectors of many possible orientations lying at different angles between 0 and 360° and hence symmetrically distributed about the direction of propagation.

(*b*) Unpolarized light is a mixture of incoherent vertically and horizontally polarized waves.

Normally, unpolarized light is looked upon as consisting of an average of half of its vibrations horizontally polarized and half of its vibrations vertically polarized. Due to the random distribution of the optical vectors, the amplitude of the vertical and horizontal component vectors is taken as equal with a randomly changing phase difference.

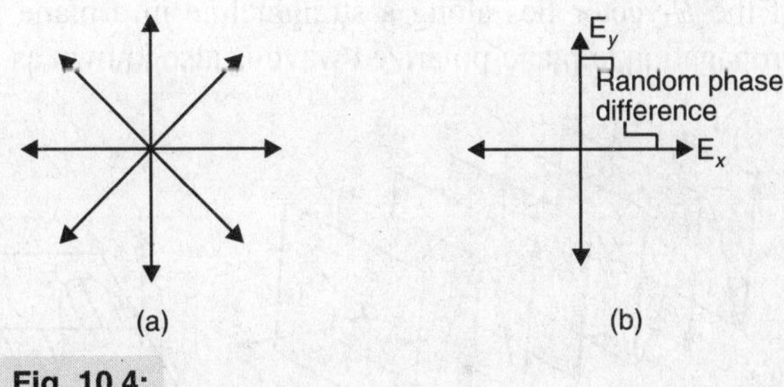

Fig. 10.4:
(a) Pictorial representation of Natural light.
(b) Unpolarized light is viewed as a combination of ***incoherent*** vertically and horizontally polarized waves

In view of this, unpolarized is pictorially represented by two electric vectors as shown in Fig. 10.4 (*b*). Table 10.1 shows the comparison of unpolarized and polarized light.

Table 10.1: *Comparison of Unpolarized and Polarized light*

S.No.	Unpolarized light	Polarized light
1.	Consists of waves with planes of vibration equally distributed in all directions about the ray direction.	Consists of waves with planes of vibration in a single plane normal to ray direction.
2.	Symmetrical about the ray direction.	Asymmetrical about the ray direction.
3.	Produced by conventional light sources.	Is to be obtained from unpolarized light with the help of polarizers.
4.	May be regarded as the resultant of two *incoherent* waves of equal intensity but polarized in mutually perpendicular planes.	May be regarded as the resultant of two mutually perpendicular *coherent* waves having zero phase difference.

10.5 TYPES OF POLARIZATION

The polarization of a light wave describes *the shape and locus of the tip of the* ***E*** in the plane perpendicular to the direction of propagation *at a given instant of time and space*. Depending upon the locus of the tip of the ***E*** vector, light may exhibit three different states of polarization. They are:

(*i*) plane or linear polarization
(*ii*) elliptical polarization
(*iii*) circular polarization.

Apart from these, the light may also be partially polarized.

An unaided human eye cannot identify the state of polarization of light. However, some insects and animals possess polarization sensitive vision.

10.5.1 Plane Polarized Light

In a plane-polarized wave, the oscillations of electric field vector ***E*** are strictly confined to a single plane *perpendicular* to the direction of wave propagation. Since the direction of the E-vector lies along a straight line in a plane perpendicular to the direction of wave propagation, a plane-polarized wave is also known as a **linearly polarized** wave.

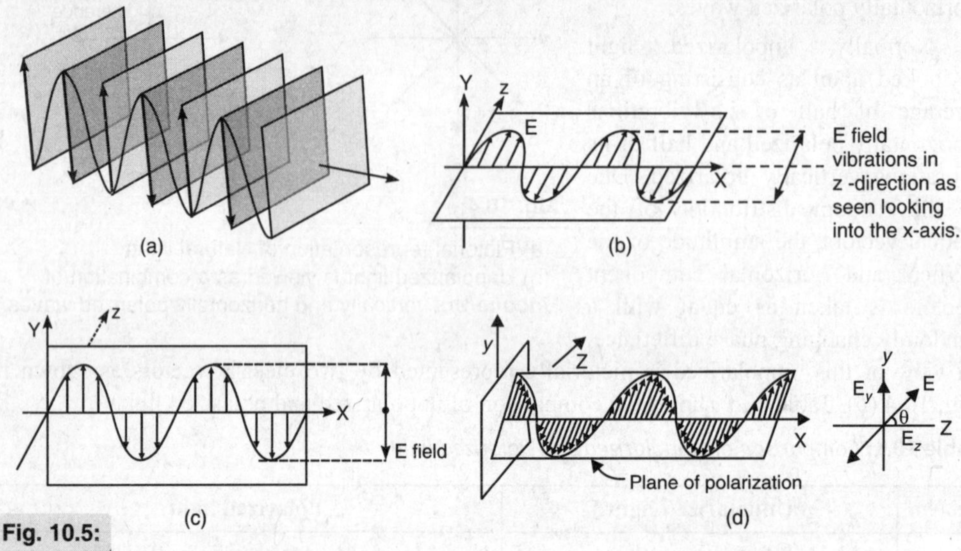

Fig. 10.5: (a) The wavefronts of a plane polarized wave traveling in space, (b) Light polarized linearly, horizontally in XZ plane, (c) Light polarized linearly, vertically in XY plane (d) A light wave linearly polarized in an arbitrary direction.

The direction of $\boldsymbol{E}$ does not vary with time, but its magnitude varies sinusoidally with time (Fig. 10.5). In vertical polarization, the field points either up or down, and in horizontal polarization it points either right or left. $\boldsymbol{E}$ is not limited to vertical or horizontal axes, but can be at any arbitrary angle to those axes. Fig. 10.5 (*d*) shows the electric field making a constant angle θ to the *z*-direction. Fig. 10.6 shows the resultant wave due to superposition of two **coherent** linearly polarized waves, oscillating in phase.

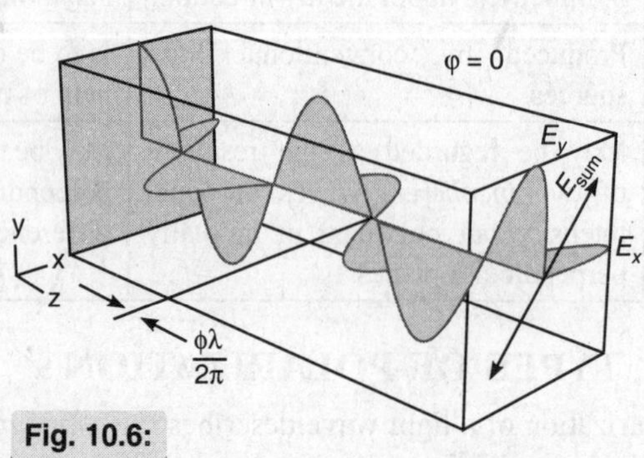

Fig. 10.6: Linear polarization - combination of coherent vertically and horizontally polarized waves.

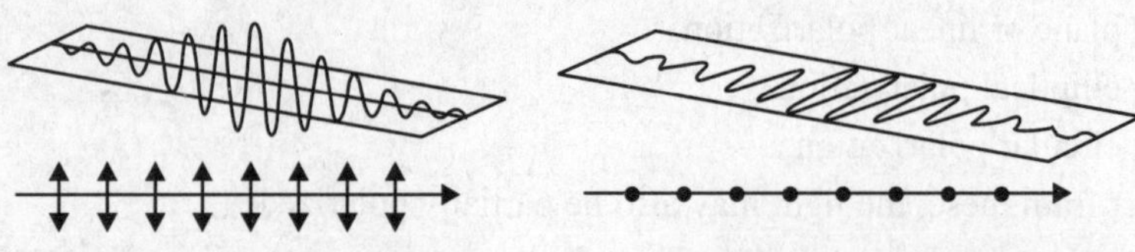

Fig. 10.7: (a) Horizontal Linear Polarization (b) Vertical Linear Polarization

Representation of linearly polarized light by (a). Arrows when $\boldsymbol{E}$ oscillates vertically in the plane of the paper (b). Dots when $\boldsymbol{E}$ oscillates perpendicular to the plane of the paper.

Representation of linearly polarized light in diagrams: Fig. 10.7 shows linearly polarized light. The short arrows symbolize an $\boldsymbol{E}$ vibrating in the plane of the page, perpendicular to the beam direction and the dots symbolize E oscillating in and out, perpendicular to the plane of the page.

10.5.2 Circularly Polarized Light

A light wave is referred as ***circularly polarized***, if in the course of wave propagation, the magnitude of the electric vector ***E*** stays constant but the vector rotates at a constant rate about the direction of propagation and sweeps a circular helix in space, as shown in Fig. 10.8. In, circularly polarized light, there is no preference to specific direction of oscillation. A circularly polarized light is the resultant of two coherent linearly polarized waves of *equal amplitude* oscillating in mutually perpendicular planes, and are out of phase by 90º (Fig. 10.9).

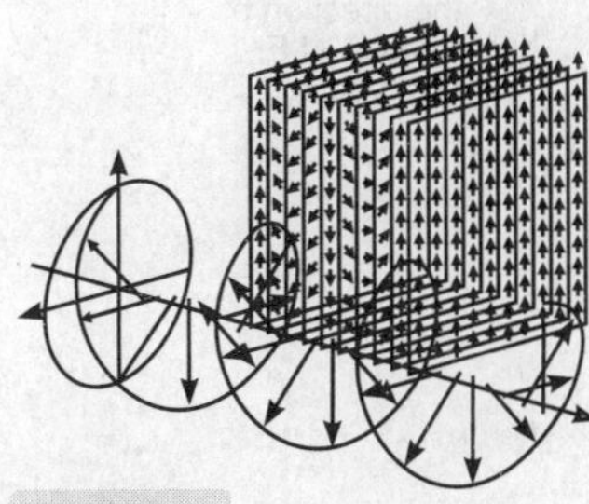

Fig. 10.8: The wavefronts of a circularly polarized wave traveling in space.

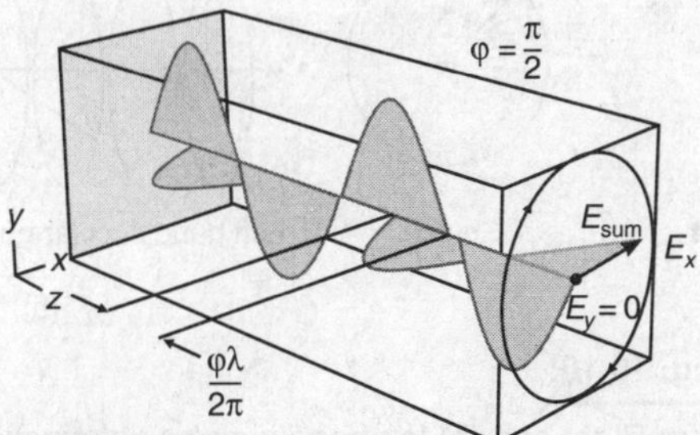

Fig. 10.9: Circular polarized - combination of horizontally polarized and vertically polarized waves that are out of phase by 90° and having equal amplitudes.

Let us consider two linearly polarized waves having equal amplitude, out of which one is polarized in *x*-direction (horizontally polarized wave) and the other in *y*-direction (vertically polarized wave). Let us further assume that they are **coherent** and out of phase by 90º (see Fig. 10.9).

At some instant, the ***E***-field vector of *x*-polarized wave will have maximum amplitude and the ***E***-field vector of *y*-polarized wave will be at zero. It is horizontal polarization.

After a small amount time, the *x*-polarized wave has decreased a little, while the *y*-polarized wave has started to increase. It is mostly horizontal but with a small vertical component.

Soon after, the *x*-component has decreased some more, and the *y*-component has increased some more and eventually becomes vertically polarized.

As the ***E***-vector sweeps a circle in space, it is called circularly polarized wave. The *oscillations of the resultant* ***E****-vector do not take place in a single plane.*

Def: If the light is viewed looking *toward the source* and the electric field rotates clockwise, the polarization is known as *right-hand circular* and in counter clockwise known as *left-hand circular polarization.*

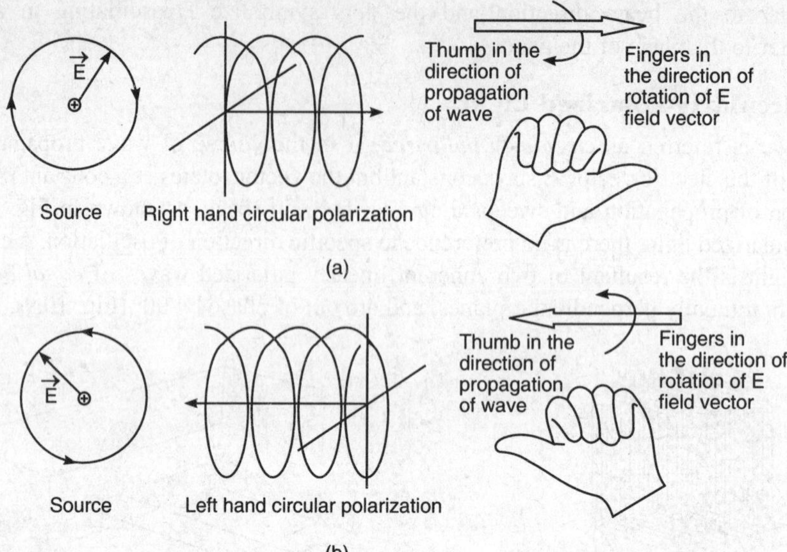

Fig. 10.10:

(a) Right and (b) left hand circular polarization.

10.5.3 Elliptically Polarized Light

A light wave is said to be ***elliptically polarized***, if the magnitude of electric vector ***E*** changes with time and the vector ***E*** rotates about the direction of propagation and sweeps a flattened helix in space, as shown in Fig. 10.11.

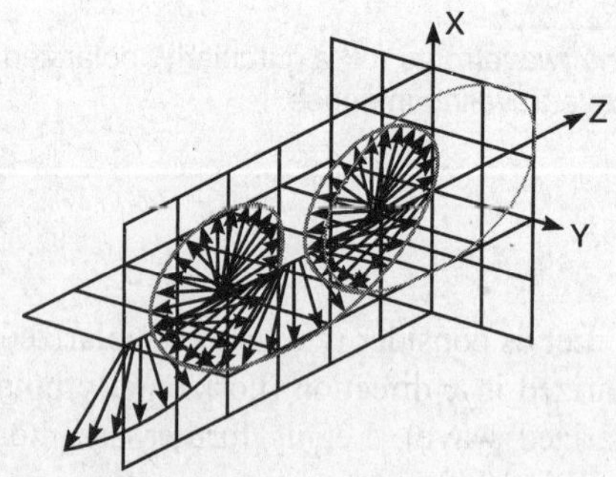

Fig. 10.11:

Space variation of **E**; **E** vector sweeps a flattened helix in space.

An elliptically polarized light wave may be regarded as the resultant wave produced due to superposition of two **coherent** linearly polarized waves of different amplitudes, oscillating in mutually perpendicular planes and are out of phase.

Let us consider two linearly polarized waves having different amplitudes, out of which one is polarized in x-direction and the other in y-direction. Let us further assume that they are **coherent** and out of phase by an arbitrary angle θ (see Fig. 10.12). The ***E***-field vector of x-polarized wave will have maximum amplitude at times when the ***E***-field vector of y-polarized wave is a minimum and vice versa (Fig. 10.12). Then the oscillations of the resultant ***E***-vector do not take place in a single plane. The magnitude of resultant ***E***-vector varies at each point in space and the overall rotation of the ***E***-vector has the appearance of a flattened helix.

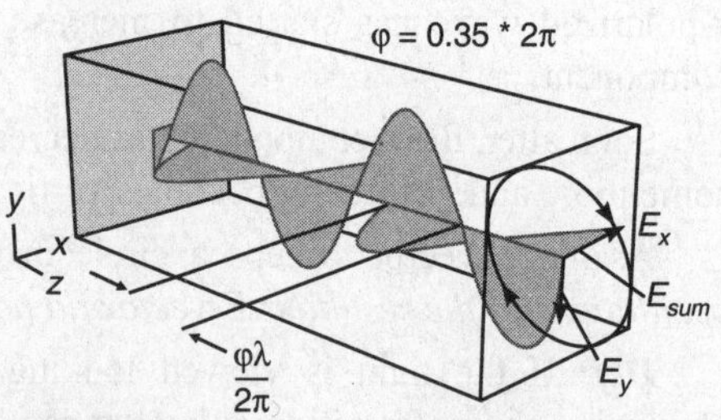

Fig. 10.12:

Elliptically polarized wave is a combination of horizontally and vertically polarized waves that are of different amplitudes and out of phase by θ.

If the rotation of ***E*** vector occurs clockwise, it is known as a *right-elliptically-polarized* wave and in counterclockwise known as *left-hand elliptically polarization*.

10.5.4 Partially Polarized Light

Usually, light is neither totally polarized nor unpolarized but a mixture of the two types. It can be seen as a combination of plane polarized light and unpolarized light. Partially polarized light is represented as shown in Fig. 10.13.

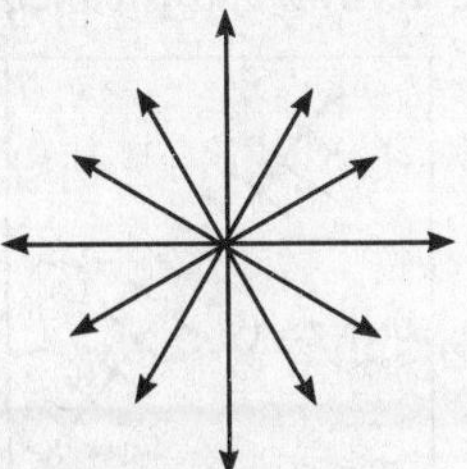

Fig. 10.13: Partially polarized light.

Partially polarized light, like natural light, can be represented in the form of a superposition of two incoherent plane-polarized waves with mutually perpendicular planes of oscillations. In case of natural light the amplitude of these waves is the same and for partially polarized light, it is different.

Degree of Polarization: If we pass partly polarized light through a polarizer, and if we rotate the polarizer about the direction of the ray, the intensity of the transmitted light will change within the limits from I_{max} to I_{min}. The transition from one of these values to the other will occur upon rotation through an angle of 90°. We define the degree of polarization with the help of the following expression.

$$P = \frac{I_{max} - I_{min}}{I_{max} + I_{min}} \quad ...(10.1)$$

% polarization

$$\% \text{ polarization} = \frac{(I_{max} - I_{min})}{(I_{max} + I_{min})} \times 100 \quad ...(10.2)$$

For plane polarized light $I_{min} = 0$, and hence $P = 1$ and the % polarization is 100%. For natural light, $I_{max} = I_{min}$, and hence $P = 0$ and the % polarization is zero. If $I_{max} = 2I_{min}$, $P = 0.33$ and % polarization = 100/3 = 33%.

The concept of the degree of polarization cannot be applied to elliptically and circularly polarized light.

10.6 PRODUCTION OF PLANE POLARIZED LIGHT

We now study the methods of producing plane-polarized light. Plane polarized light may be produced from unpolarized light using the following five optical phenomena:

(*i*) reflection, (*ii*) refraction, (*iii*) scattering, (*iv*) selective absorption (dichroism), and (*v*) double refraction. Out of these five, the phenomena of selective absorption and double refraction are helpful in practical production of plane polarized light. Plane polarized light is also known as ***linearly polarized*** light.

10.6.1 Polarization by Reflection from Dielectric Surfaces

E.L. Malus, French engineer discovered in 1808 polarization of light by reflection. He noticed that when natural light is incident on a smooth surface, at a certain angle the reflected beam is plane polarized.

When light wave is incident on a boundary between two dielectric materials, part of it is reflected, and part of it is transmitted. Fig. 10.14 (*a*) shows an unpolarized light beam *AB* incident on a glass surface.

The incident ray *AB* and the normal *NBN'* define the plane of incidence. The electric vector ***E*** of the ray *AB* can be resolved into two components, one perpendicular to the plane of incidence and the other lying in the plane of incidence. The perpendicular component is represented by **dots** and is called the ***s*-component**. The parallel component is represented by the **arrows** and is called the ***p*-component**.

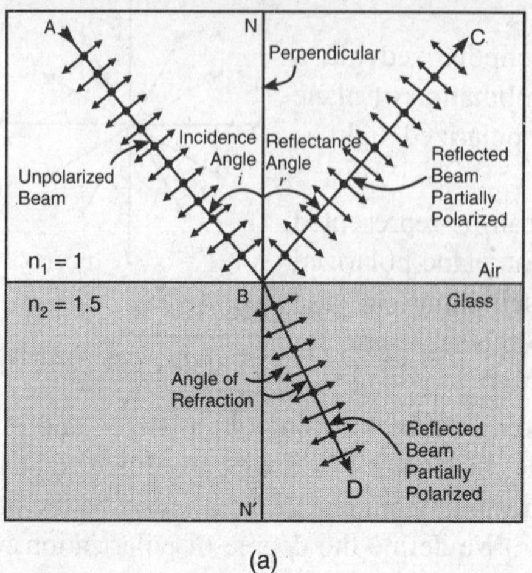

(a)

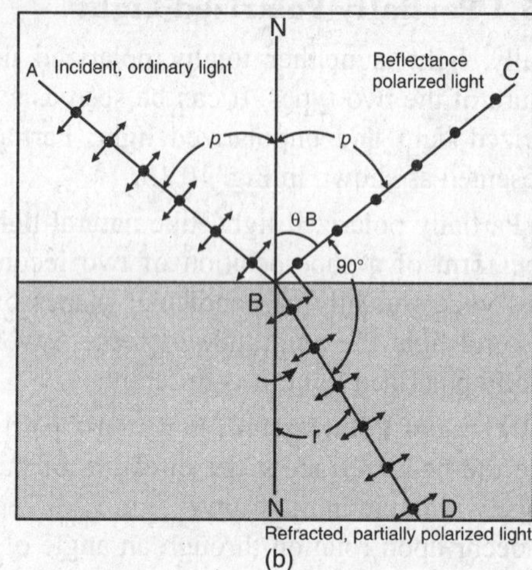

(b)

Fig. 10.14:

(a) Reflection and Refraction at the surface between two media (b) Brewster Law.

In case of completely unpolarized light the two components are of equal magnitude. At a particular angle θ_B, the reflection coefficient for *p*-component goes to zero and the reflected beam does not contain any *p*-component (see Fig. 10.14 *b*). It contains only *s*-component and is totally *plane polarized*. The angle θ_B is called the **polarizing angle** or **Brewster's angle**.

Def: Polarizing angle is the angle θ_B at which the reflected ray has no *p*-component and contains only *s*-component when a light wave is incident on a boundary between two dielectric materials.

This particular method of polarizing light is not advantageous, as the intensity of the reflected beam is very small.

10.6.1.1 Brewster's Law

Sir David Brewster (1892) performed a series of experiments on the polarization of light by reflection at a number of surfaces. He found that the polarizing angle depends upon the refractive index of the medium.

Statement: The tangent of the angle at which polarization is obtained by reflection is numerically equal to the refractive index of the medium.

If θ_B is the angle and μ is the refractive index of the medium, then

$$\mu = \tan \theta_B \qquad ...(10.3)$$

Eq. (10.3) is known as **Brewster's law**.

If natural light is incident on a smooth surface at the polarizing angle, it is reflected along *BC* and refracted along *BD*, as shown in Fig. 10.14 (*b*). Brewster found that the maximum polarization of reflected ray occurs when it is at right angles to the refracted ray. It means that $\theta_B + r = 90°$.

$$\therefore \qquad r = 90° - \theta_B \qquad ...(10.4)$$

According to Snell's law,

$$\frac{\sin\theta_B}{\sin r} = \frac{\mu_2}{\mu_1} \quad (10.5)$$

where μ_2 is the absolute refractive index of reflecting surface and μ_1 is the refractive index of the surrounding medium.

It follows from Eq. (10.4) and Eq. (10.5) that

$$\frac{\sin\theta_B}{\sin(90^\circ - \theta_B)} = \frac{\mu_2}{\mu_1}$$

or

$$\frac{\sin\theta_B}{\cos\theta_B} = \frac{\mu_2}{\mu_1}$$

As $\mu = 1$ for air, taking $\mu_2 = \mu$, we get

or

$$\tan\theta_B = \frac{\mu_2}{\mu_1} \qquad \therefore \tan\theta_B = \mu \quad ...(10.6)$$

Eq. (10.6) shows that the polarizing angle depends on the refractive index of the reflecting surface. Light reflected from any angle other than Brewster angle is partially polarized. However, this law is not applicable for metallic surfaces.

Application of Brewster's law:

(*i*) The refractive indices of opaque materials can be determined.

(*ii*) Polarizing angle necessary for total polarization of reflected light for any material can be calculated if its refractive index is known.

(*iii*) As Brewster windows in optical resonators of gas lasers. The glass windows are arranged at Brewster angle to the axis of the laser tube. The light beam traveling between the mirrors of the laser is reflected many times from these mirrors. The advantage of this arrangement is that the beam has no reflection losses, since only the transmitted polarized beam is traveling between the mirrors. The radiation out of these lasers is polarized as can be seen in Fig. 10.15.

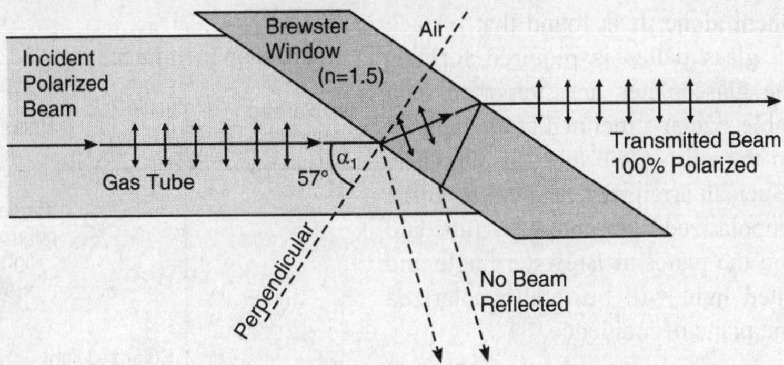

Fig. 10.15: Brewster angle window at the end of gas laser.

(*iv*) Another application utilizes the Brewster angle for transmitting a light beam into or out of an optical fibre without reflection losses.

Example 10.1: *It is desired to use a plate of glass to obtain polarized light. If the refractive index of glass is 1.5, what is the polarizing angle?*

Solution: Polarizing angle, $\theta_p = \tan^{-1}\mu = \tan^{-1}(1.5) =$ **56.31°.**

Example 10.2: *Sunlight is reflected from a calm lake. The reflected light is 100% polarized at a certain instant. What is the angle between the sun and the horizon at that instant? The refractive index of water is 1.33.*

Solution: Since the reflected is 100% polarized, the angle of incidence is equal to the Brewster angle. By Brewster's law, $\mu = \tan\theta_B$. Therefore,

$$1.33 = \tan\theta_B.$$

$$\therefore \quad \theta_B = 53.06°$$

The angle between the sun and the horizon = 90° – 53.06° = 36.94° = 36° 54′.

Example 10.3: *Find polarization angle if the critical angle of incidence for total reflection in case of water is 48°. Find the angle of refraction to the polarizing angle calculated?*

Solution: The refractive index, $\mu = \dfrac{1}{\sin\theta_C} = \dfrac{1}{\sin 48°} = 1.346.$

Now, $\mu = \tan\theta_B$ or $1.346 = \tan\theta_B$ $\therefore$ $\theta_B = 53.4° = 53°21'$

If r is the angle of refraction, $r = 90° - 53°21' = 36°39'$

10.6.2 Polarization by Refraction - Pile of Plates

When unpolarized light falls on a glass surface at Brewster angle, the reflected light is totally polarized, while the refracted light is partially polarized. If natural light is transmitted through a single plate, the transmitted beam is only partially polarized. If a stack of glass plates is used instead of a single plate, reflection from successive surfaces occur leading to the filtering of the s-component in the transmitted ray. Ultimately, the transmitted ray consists of p-component alone. It is found that a stack of about 15 glass plates is required for this purpose. The glass plates are supported in a tube of suitable size and inclined at an angle of about 33° to the axis of the tube, as shown in Fig. 10.16. Such an arrangement is called a **pile of plates**. Unpolarized light enters the tube and is incident on the plates at Brewster angle and the transmitted light will be totally polarized parallel to the plane of incidence.

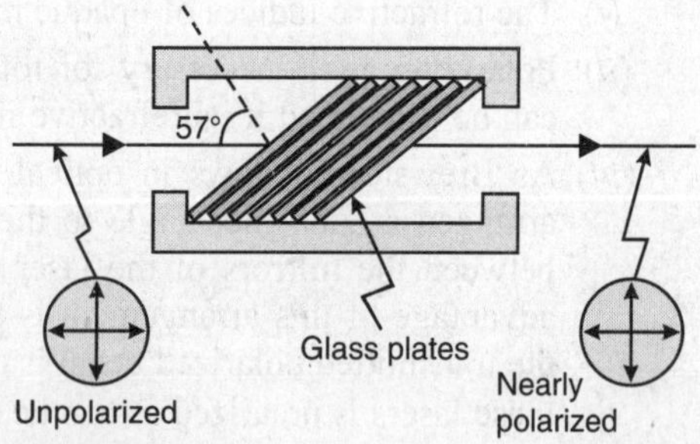

Fig. 10.16: Polarization by refraction.

10.6.3 Polarization by Scattering

If a narrow beam of natural light is incident on a transparent medium containing a suspension of ultramicroscopic particles, the light scattered is partially polarized. The alternating electric field in the incident light rays causes electrons in the scattering medium to vibrate. The vibrating electrons emit light preferably in a direction perpendicular to their vibration.

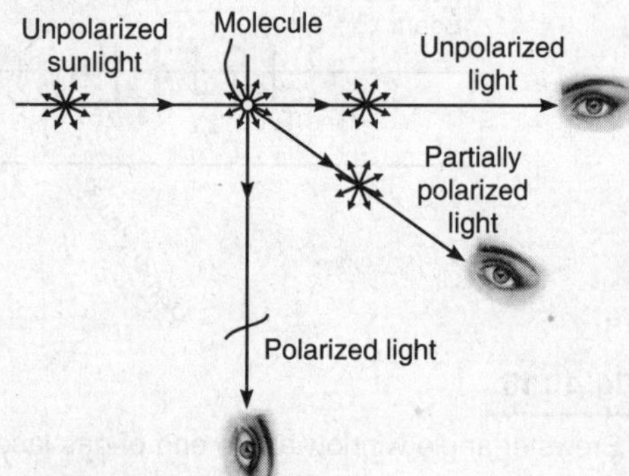

Fig. 10.17: Polarization by Scattering.

Hence light scattered through about 90° with respect to the incident direction is strongly polarized. The direction of vibration of E vector in the scattered light will be perpendicular to the plane defined by the direction of propagation and the plane of the paper, as illustrated in Fig. 10.17. The light from a blue sky is polarized, mainly at right angle from the position of the sun.

10.6.4 Polarization by Selective Absorption

Biot discovered this phenomenon in 1815. When natural light passes through a crystal for instance tourmaline; the light is split into two components. The components are polarized in planes perpendicular to each other. The crystal absorbs that light component polarized in a direction parallel to a particular plane in the crystal and transmits the light component polarized in a direction perpendicular to that particular plane. If the crystal has appropriate thickness, one of the light components is fully absorbed and the other component coming out from the crystal is linearly polarized. Selective absorption is illustrated in Fig. 10.18.

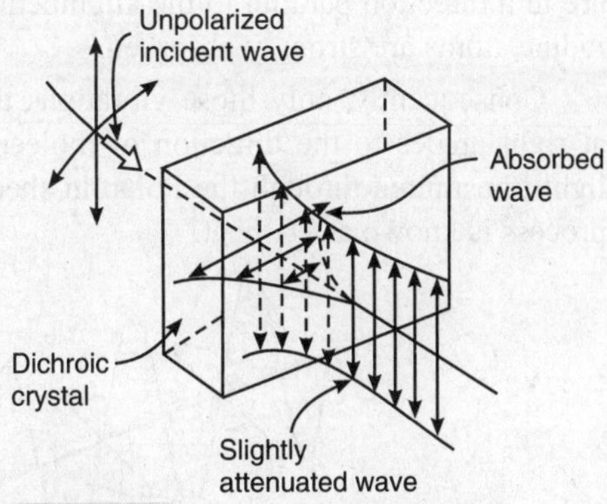

Fig. 10.18:

Polarization by Selective absorption.

10.6.5 Polarization by Double Refraction

When a beam of unpolarized light is incident on the surface of an anisotropic crystal such as calcite or quartz, it is found that it will separate into two rays (see Fig. 10.19) that travel in different directions. This phenomenon is called ***birefringence or double refraction***. The two rays are known as ***ordinary ray (o-ray)*** *and* ***extraordinary ray (e-ray)***, which are linearly polarized in mutually perpendicular directions. A single linearly polarized ray is obtained in practice through elimination of one of the two polarized rays.

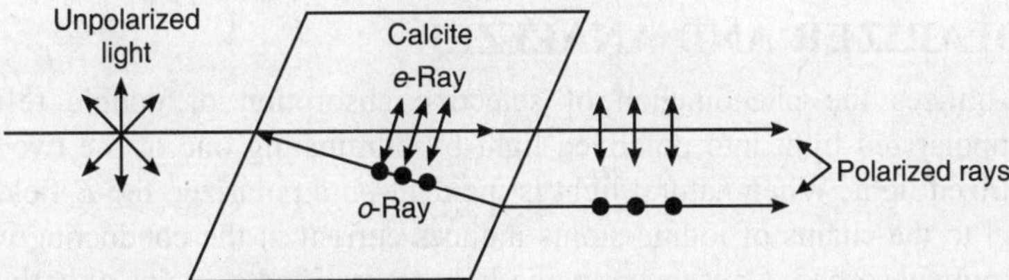

Fig. 10.19:

Polarization by double refraction.

10.7 POLAROID SHEETS

In 1928 E. H. Land invented Polaroid sheets, which utilize the phenomenon of selective absorption. The sheets are fabricated as follows. A clear plastic sheet of long chain molecules of PVA (polyvinyl alcohol) is heated and then stretched in a given direction to several times its original length. In the stretching procedure the PVA molecules become aligned along the direction of stretching (Fig. 10.20). The sheet is then laminated to a rigid sheet of plastic to stabilize its size.

It is then exposed to iodine vapor. The iodine atoms attach themselves to the straight long chain PVA molecules and as a result form long parallel conducting chains. The iodine atoms give electrons. These electrons move along the aligned chains. When natural light is incident on the sheet, the electromagnetic vibrations that are in a direction parallel to the alignment of the iodine atoms are strongly absorbed.

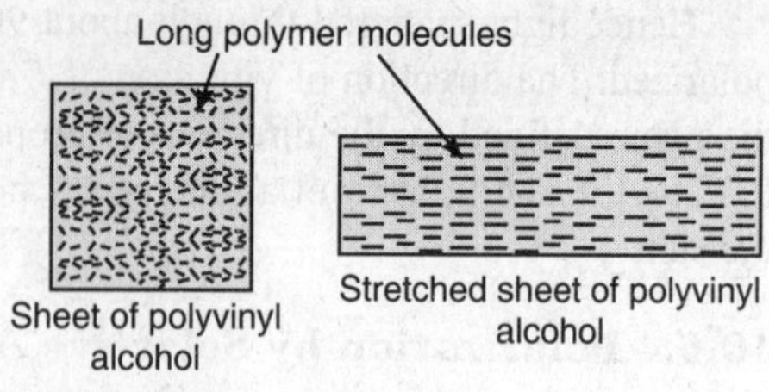

Fig. 10.20: H-polaroid preparation.

Consequently, only those vibrations that are at right angles to the direction of molecular chains are transmitted (Fig. 10.21). Thus, the light transmitted through the Polaroid sheet is polarized. A sheet fabricated according to this process is known as H-sheet.

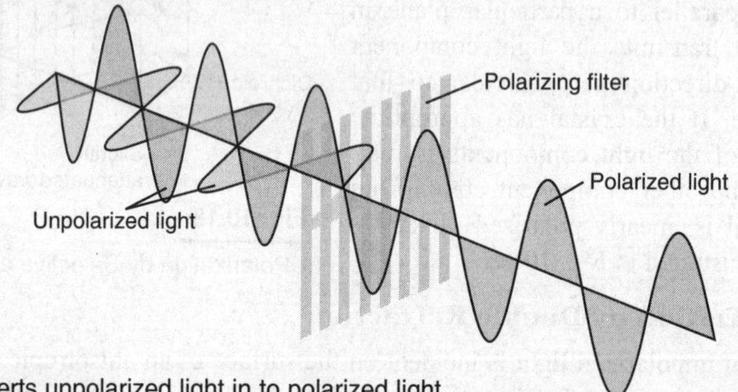

Fig. 10.21: Polaroid converts unpolarized light in to polarized light.

These sheet polarizers are inexpensive and can be made in large sizes. Polaroid sheets are widely used in sunglasses, camera filters etc to eliminate the unwanted glare from objects. Polaroid sheets are extensively used for the production and detection of linearly polarized light.

10.8 POLARIZER AND ANALYZER

A *polarizer* utilizes the phenomenon of selective absorption or double refraction, and transforms unpolarized light into polarized light by eliminating one of the two components in the unpolarized light. When natural light is incident on a polarizer, the ***E***-field component that is parallel to the chains of iodine atoms induces current in the conducting chains and is therefore strongly absorbed. Consequently, the light transmitted contains only the component that is perpendicular to the direction of molecular chains. The direction of ***E***-vector in the transmitted beam corresponds to the **transmission axis** of the Polaroid sheet.

Effect of polarizer on natural light: Let us now understand the action of polarizer on the incident unpolarized light. Consider unpolarized light falls on a polarizer with the electric vector $\boldsymbol{E}_0$ making an angle θ with reference to the *transmission axis* of the polarizer. The electric vector $\boldsymbol{E}_0$ may be resolved into its component vectors lying parallel and perpendicular to the transmission axis of the polarizer (see Fig. 10.20), that is E_y, parallel to the transmission axis and E_x, perpendicular to the transmission axis of the polarizer. The polarizer transmits the parallel component while blocking the perpendicular component. Thus, it is the parallel component E_y that is transmitted by the polarizer. But

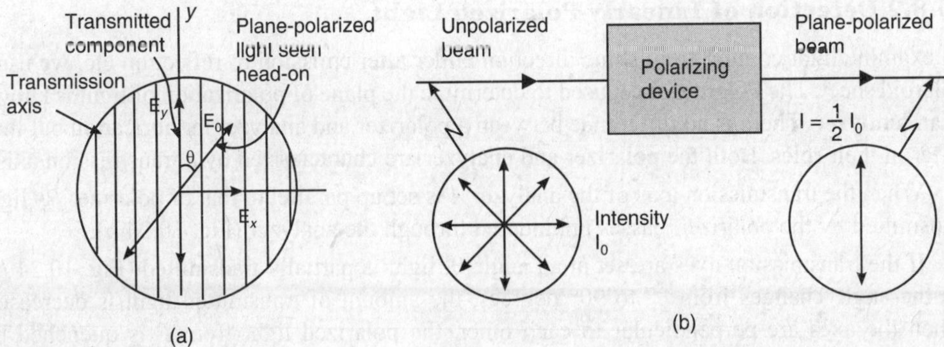

Fig. 10.22:

(a) Action of polarizer on linearly polarized wave (b) The intensity of an unpolarized beam reduces to half after passing through a polarizer.

$$E_y = E_0 \cos \theta \qquad ...(10.7)$$

and hence, the intensity of the transmitted component is given by

$$I \propto E_y^{\ 2} = E_0^{\ 2} \cos^2\theta \qquad ...(10.8)$$

In unpolarized light all the values of θ are equally probable. Therefore, the fraction of light transmitted through the polarizer equals the average value of $\cos^2\theta$, which is equal to $\frac{1}{2}$. Thus,

$$\therefore \qquad I = \frac{1}{2} E_0^2 = \frac{1}{2} I_0 \qquad ...(10.9)$$

An *analyzer* is an optical element, which is used to identify the plane of vibration of plane polarized light.

10.8.1 Production of Linearly Polarized Light Using a Polarizer

Every polarizer is associated with a specific direction of transmission called its ***transmission axis***. If natural light is incident on a polarizer, only those vibrations that are *parallel* to the transmission axis passes and other directions are totally blocked as shown in Fig. 10.23. Thus, the transmitted beam is linearly polarized. According to Eq. (10.9), when unpolarized light of intensity I_0 is incident on a polarizer, the intensity of light transmitted by the polarizer is $I_0 / 2$.

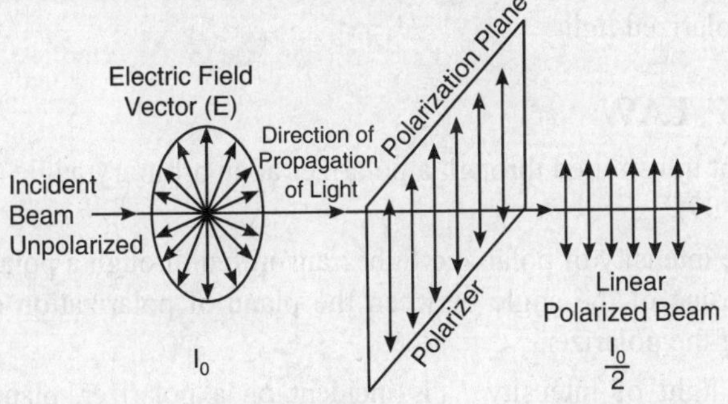

Fig. 10.23:

Production of linearly polarized light.

10.8.2 Detection of Linearly Polarized Light

To examine light coming from some direction either after emission or reflection etc, we use a Polaroid sheet. The Polaroid sheet used to determine the plane of polarization of light is known as an **analyzer**. There is no difference between a polarizer and analyzer in fabrication but they differ in their roles. Both the polarizer and analyzer are characterized by a transmission axis.

When the transmission axis of the analyzer A is set up parallel to that of polarizer P, light transmitted by the polarizer, passes unhindered through the analyzer (Fig. 10.24 *a*).

If the transmission axes are set at an angle θ, light is partially transmitted (Fig. 10.24 *b*). As the angle changes from 0° to 90° degrees, the amount of transmitted light is decreases. When the axes are perpendicular to each other, the polarized light from P is quenched by the analyzer A (Fig. 10.24 *c*). The polarizer and analyzer are known to be *crossed* in this configuration.

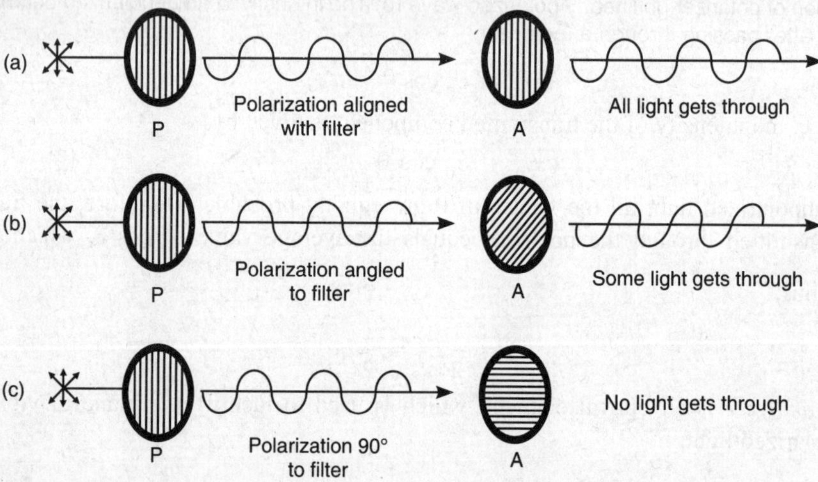

Fig. 10.24: Detection of linearly polarized light.

When we rotate the axis of the analyzer with respect to that of the polarizer, we obtain two positions of maximum intensity and two positions of zero intensity in *one full rotation*. Conversely, if we obtain two positions of maximum intensity and two positions of zero intensity in one full rotation of the analyzer, we conclude that the light incident on the analyzer is plane-polarized light.

10.9 MALUS' LAW

The quantity of light transmitted through a polarizer at an arbitrary angle (Fig. 10.25) is given by **Malus' Law** (1809).

Statement: The intensity of polarized light transmitted through a polarizer is proportional to the square of cosine of the angle between the plane of polarization of the light and the transmission axis of the polarizer.

If unpolarized light of intensity I_0 is incident on a polarizer, plane polarized light of intensity $I_0/2$ is transmitted by it. Let us denote $I_0/2$ by I_1. Let this plane polarized light pass through an analyzer. The intensity of the light transmitted (Fig. 10.25) through the analyzer is given by

$$I = E_1^2 \cos^2 \theta = I_1 \cos^2 \theta = \frac{1}{2} I_0 \cos^2 \theta \quad ...(10.10)$$

Light transmitted through the analyzer at specific settings are as follows.

Case (*i*): If $\theta = 0°$	axes parallel	$I = I_1 = \frac{1}{2} I_0$
Case (*ii*): If $\theta = 90°$	axes perpendicular	$I = 0$
Case (*i*): If $\theta = 180°$	axes parallel	$I = I_1 = \frac{1}{2} I_0$
Case (*ii*): If $\theta = 270°$	axes perpendicular	$I = 0$

Thus, we obtain two positions of maximum intensity and two positions of zero intensity when we rotate the axis of the analyzer with respect to that of the polarizer.

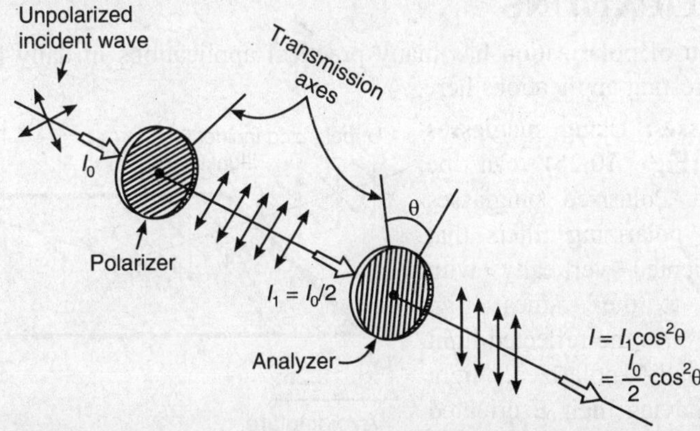

Fig. 10.25:

Illustration of Malus' law.

Example 10.4: *If the intensity of the transmitted light reduces to one-third after passing through the twp Polaroid sheets, find angle between the characteristic directions of the Polaroid sheets.*

Solution: Let I_o be the intensity of the incident unpolarized light.

Intensity of the light transmitted through the first polarizer $I_1 = I_0/2$,

Intensity of the light transmitted through the second polarizer is $I_2 = I_1 \cos^2\theta$

where θ is the angle between the characteristic directions of the polarizer sheets.

But $\quad I_2 = I_0 / 3 \quad$ (given)

$\therefore \quad I_2 = I_1 \cos^2\theta = I_0 \cos^2\theta/2 = I_0/3$

$\cos^2\theta = 2/3 \quad$ or $\quad \cos\theta = 0.8165$

$\therefore \quad \theta =$ **35.3°**

Example 10.5: *If an intensity I_o falls on a polarizer with its electric field making an angle of 30° with the transmission axis, find the intensity of the resultant if incident light is (i) unpolarized? (ii) Plane polarized.*

Solution: (*i*) If incident beam is unpolarized, then the intensity of the resultant beam will be $I_0/2$.

(*ii*) When the incident light is plane polarized, according to Malus' law

$$I = I_0 \cos^2\theta = I_0 \cos^2 30° = I_0(0.866)^2 = (3/4)I_0$$

Example 10.6: *An analyzer checks two mutually perpendicular plane polarized beams X and Y. In one position of the analyzer, beam B shows zero intensity and from that a rotation of 30° shows the two beams are matched. Deduce intensity rotation I_A/I_B.*

Solution: According to Malus law $= I\cos^2\theta$

When the beam *Y* shows zero intensity, i.e., $\theta = 90°$ and $\theta = 0$ for beam *A* for that position.

When analyzer is rotated through 30°, then $\theta = 30°$ for beam A,

$$\theta = 60° \text{ for beam } B$$

$$I = I_A \cos^2 30° = I_B \cos^2 60°$$

$$\frac{I_A}{I_B} = \frac{\cos^2 30°}{\cos^2 60°} = \frac{1}{3}$$

10.10 APPLICATIONS

The phenomenon of polarization has many practical applications in daily life. We discuss some of the interesting applications here.

(i) **Sunglasses:** Using sunglasses glare (Fig. 10.26) can be avoided. Polarized sunglasses contain polarizing filters that are oriented vertically with respect to the frames (Fig. 10.27 *a*). As the reflected light is partially polarized, light waves having their ***E*** oriented in the same direction as the polarizing lenses and perpendicular to the reflecting surface are passed through. On the other hand, light waves with ***E*** oriented parallel to the reflecting surface (and perpendicular to the filters in the lenses) are stopped. Thus, polarized sunglasses remove the glare from an illuminated surface. A polarizing filter on a camera helps reduce shiny reflections (Fig. 10.26 *b*).

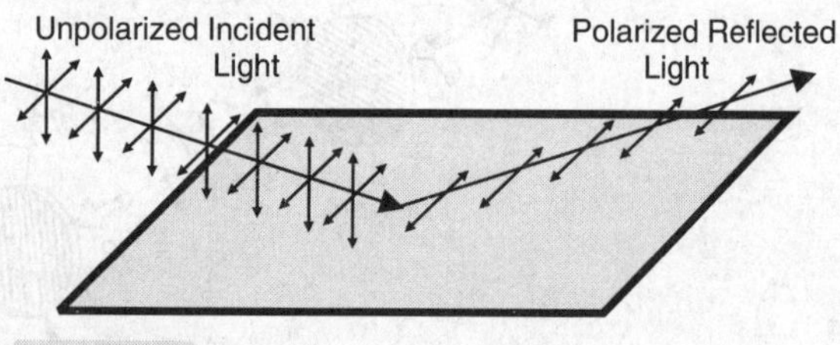

Fig. 10.26: Avoidinglare.

(ii) **Photography:** Polarization by scattering occurs as light passes through atmosphere. The scattered light gives glare. By using a polarizing filter to the camera the problem can be eliminated. As the filter is rotated, the partially polarized light is stopped and the glare is minimized. Thus, a vivid blue sky as the back drop of a beautiful fore ground is captured using polarizing filters.

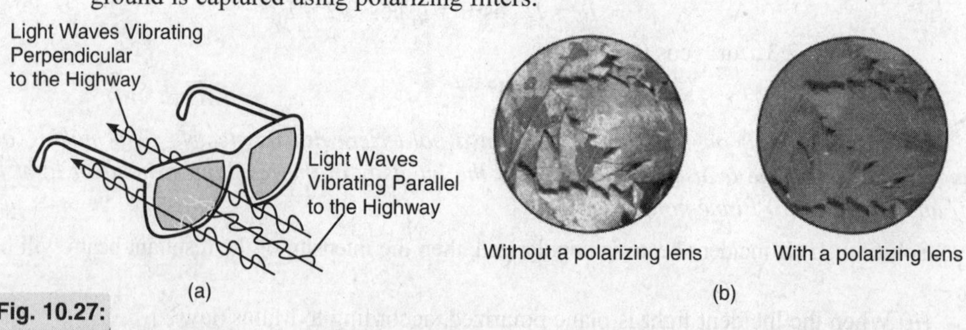

Fig. 10.27: (a) Polarized sunglasses, (b) Polarizing filter.

(iii) Stereoscopic Movies: The phenomenon of polarization is used in making and viewing stereoscopic movies. Stereoscopic movies are three dimensional movies, which give the same effect of depth as seen on a screen. The three-dimensional impression is obtained through binocular vision by projecting two movies at the same time. For making a stereoscopic movie, two views of the same scene are shot simultaneously from two slightly different camera locations. One view corresponds to that seen by the right eye and the other corresponds to the view seen by the left eye. Each of these movies are projected from different sides of the audience onto a screen through two polarizing filters, one having polarization axis aligned horizontally and the other vertically. Thus, there are two slightly different movies being projected onto the same screen by light beams which are polarized in mutually perpendicular directions. The audience wears glasses which are fitted with two Polaroid filters. Each filter has a different polarization axis - one is horizontal and the other is vertical. The consequence is that the left eye sees the movie projected from the right projector while the right eye sees the movie projected from the left projector and the audience experiences a perception of depth.

(iv) Optical Microscopy: In microscopes the polarizer is placed below the sample in the light path and the analyzer is placed above the sample, between the objective and the eyepieces. Without the sample to test, the analyzer blocks the light polarized by the polarizer. However, when a doubly refracting sample is placed on the stage between the crossed polarizers, it rotates the light. So the light successfully passes through the analyzer and various aspects of the sample can be visualized.

(v) LCDs: Light polarization concept is useful the liquid crystal displays. Fig. 10.28 shows LCD construction. LCD is simply two pieces of flat glass, over coated with a few chemical layers, and filled with liquid crystal fluid. To start with borosilicate glass plates are coated with a transparent conducting coating of indium-tin oxide (ITO). The conducting coating is etched in the form of a digit or character, as shown in Fig. 10.28 (*b*).

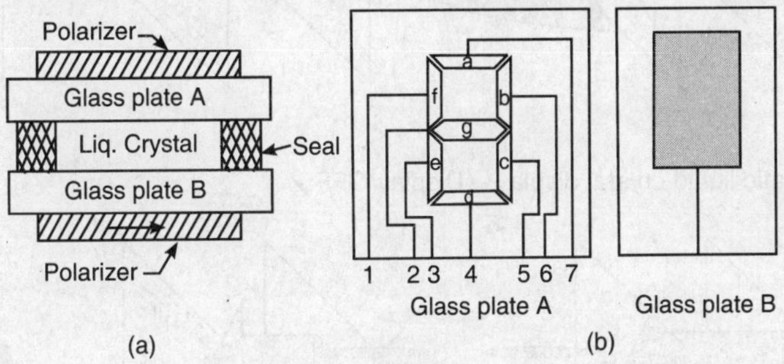

Fig. 10.28: Liquid crystal display construction.

Segments or Pixels are the smallest building blocks of a display. Segments are the smallest building block of customized displays like seven segment display which can display letters and numbers. After etching, the plates are coated with suitable chemicals which align the liquid crystal molecules in a desired direction. Then the plates are assembled and filled with a liquid crystal medium. The assembly of glass plates filled with liquid crystal material in between is then sandwiched between two crossed-polarizer sheets.

During the fabrication of LCDs, the liquid crystal molecules are aligned in such a way that their long axes undergo a 90° rotation as one move from one plate to the other plate, as illustrated in Fig. 10.29. It is called a twisted molecular arrangement. Therefore the liquid crystal display device is called a TN LCD (twisted nematic liquid crystal display).

When natural light is incident on the assembly, the front polarizer converts it into linearly polarized light. As it propagates through the LCD, the optical vector is rotated through 90° by the twisted molecular arrangement. Therefore, it passes unhindered through the rear polarizer whose transmission axis is perpendicular to that of the front polarizer. A reflecting coating at the back of the rear polarizer sends back the light, which emerges unobstructed (Fig. 10.30) by the front polarizer. Consequently, the display appears uniformly illuminated.

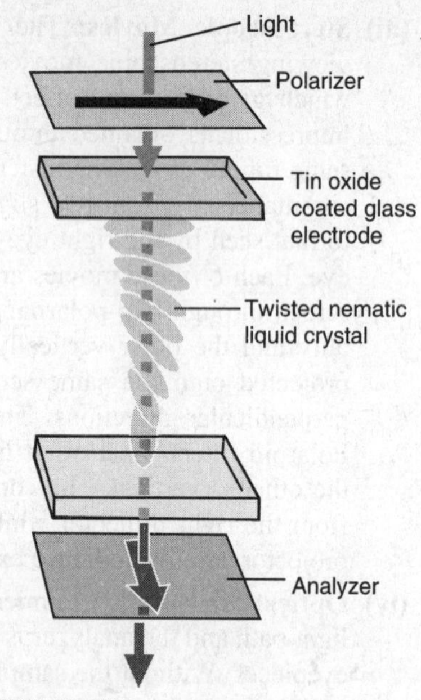

Fig. 10.29:
Twisted nematic liquid crystal display.

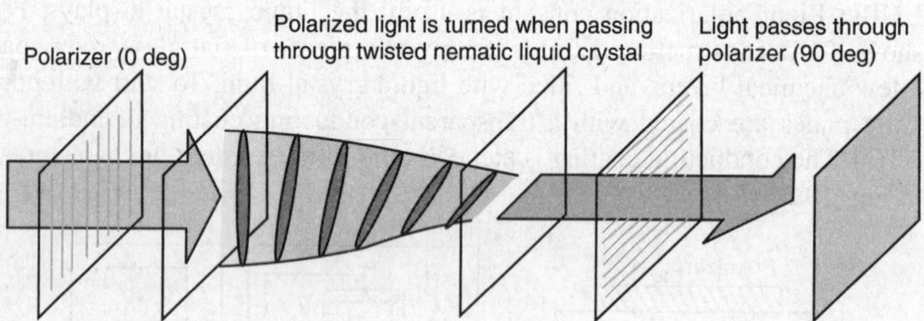

Fig. 10.30:
Twisted nematic liquid crystal display – Display OFF.

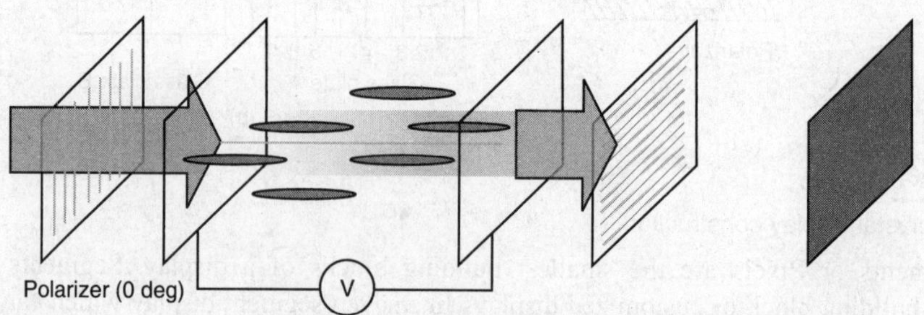

Fig. 10.31:
Liquid crystal display – Display ON.

When a voltage is applied to the device, the molecules between the electrodes untwist and align along the field direction. As a result, the optical vector does not undergo rotation as it passes through that region. The rear polarizer blocks the light (Fig. 10.31) and therefore a dark digit or character is seen in that region. The crystals in an LCD could be alternatively arranged so that light passed when there was a voltage, and not passed when there was no voltage. A table-clock using a TN LCD for displaying time is shown in Fig. 10.32.

Fig. 10.32: Alarm clock using an LCD.

QUESTIONS

1. Define and explain polarization.
2. Define the terms plane of vibration and plane of polarization.
3. Discuss the methods by which plane polarized light can be produced.
4. State Brewster's law and give its significance.
5. State and explain Brewster's law.
6. What is Brewster's law? Give any two applications of it.
7. Discuss some of the applications of Brewster's law.
8. State Brewster's law of polarization. Write mathematical expression for it.
9. How can the refractive index of a smooth opaque dielectric material be determined?
10. What is polarization? State the law of Malus.
11. State and explain the law of Malus.
12. Describe the fabrication of a Polaroid.
13. How can plane-polarized light be detected?
14. Unpolarized light falls on two polarizing sheets so oriented that no light is transmitted from the combination. If a third polarizing sheet is placed between them, can light be transmitted? Explain.
15. What is dichroism?
16. Linearly polarized light incident on a polarizer what will be the transmitted intensity?
17. Show that the transmitted intensity will be the half the incident intensity for unpolarized light.

PROBLEMS

1. Find the intensity of the resultant beam when I_o is incident on a polarizer if Incident light: (*i*) unpolarized? (*ii*) Plane polarized with E making an angle of 30° with transmission axis.
2. Unpolarized I_o is incident on a polarizer and the emergent light strikes analyser at 45° to that of first. Find the intensity of the emerging beam. **[Ans:** $I_o/4$]
3. If $\boldsymbol{E}$ of a polarized light makes 65° with the axis of a polarizer sheet. Find the fraction of the beam intensity transmitted through the sheet? **[Ans:** 0.179]
4. Find the angle between the axis of analyzer and the initial direction of $\boldsymbol{E}$ of the beam, if a beam of polarized light has one-seventh of its initial value after passing through on analyzer. **[Ans:** 67.8°]

5. Find the fraction of the I_o is transmitted, when unpolarized light is incident on a group of four polarizing sheets which are lined up so that the characteristic direction of each is rotated 30° clockwise with respect to the preceding sheet. **[Ans:** 0.211]
6. The axes of a polarizer and an analyzer are oriented at 30° to each other. (*a*). If a polarized light of intensity I_o is incident on them, what is the intensity of the transmitted light? (*b*). Polarized light of intensity I_o is incident in this polarizer, analyzer system. If the amplitude of the light makes an angle of 30° with the axis of polarizer, what is the intensity of the transmitted light? **[Ans:** $0.375I_o$, $0.56I_o$]
7. A polarizer and an analyzer are set for the maximum intensity transmission. Find the fraction of maximum intensity of the transmitted light reduced when the analyzer is rotated through
   a. 30° b. 45° c. 60°
   [Ans: $3/4\ I_1$, $½\ I_1$, $¼\ I_1$]
8. A beam of light is composed of a mixture of unpolarized and plane polarized waves. The beam is analyzed with a Polaroid sheet. As the transmission axis is rotated, the transmitted intensity varies from I_{max} to $1/3\ I_{max}$. Show that the intensities of the polarized and unpolarized beams are equal.
9. A beam of light is known to be a mixture of unpolarized light and plane polarized light. When the beam is analyzed with a polarizer sheet the transmitted intensity varies from I_{max} to $0.60\ I_{max}$. Show that $I_p = 1/3\ I_o$.
10. A plane polarized electromagnetic wave with E-field amplitude of E_o = 100 V/m strikes an ideal polarizer. The incident *E* vector makes an angle of 62° with the transmission axis of the polarizer. Determine the amplitude of the transmitted E-field. **[Ans:** *E* = 46.9 V/m.]

11 Double Refraction and Optical Activity

Learning Objectives:

By the end of this chapter, you will be able to:

- Understand concept of Light propagation in isotropic materials.
- Explain double refraction in calcite crystal.
- Learn about Phase difference between e-Ray and o-Ray.
- Identify Production of Elleptically and Circularly Light.

11.1 INTRODUCTION

E. Bartholin (1669) discovered double refraction in a crystal of Iceland spar. This phenomenon occurs in crystallographic materials that have different refractive indices with respect to different crystallographic directions. In addition to naturally occurring crystals, artificial anisotropic mediums such as glass, liquid, and other materials placed in an electric field (Kerr effect) or in a magnetic field (Cotton-Mouton effect), or under mechanical stress and under similar conditions exhibit double refraction which is known as artificial double refraction.

11.2 LIGHT PROPAGATION IN ISOTROPIC MATERIALS

A light wave consists of oscillating electric and magnetic fields perpendicular to each other. As the wave passes through the material, the oscillating electric field exerts force on movable internal charges. These moving charges may be the parts of polar molecules, vibrating ions in a crystal structure, or even simply electron clouds surrounding an atomic nucleus. Within the material, the actual electromagnetic disturbance depends not only on the applied field from the original wave, but also on the fields created by the displacement of these internal charges from their equilibrium positions.

The oscillating electric field E in the wave is given by

$$E = E\,(x, t)$$

which is assumed to be sinusoidal in time.

The effect of the displaced charges in the material can be accounted for with

$$D = \varepsilon_0 E + P$$

where P is the *polarization*, which is the field created by the displaced charges, and D is a quantity known as the *electric displacement*.

The electric field from the displaced charges adds to and interacts with the applied field from the wave. Most optical materials are linear, isotropic materials. Their internal charges respond equally to the driving electric field of the wave, regardless of its direction. Fig. 11.1 (*a*) shows a conceptual example of a negative electron cloud connected to a positive nucleus by uniform springs. Because the springs are equally strong in all directions, the displacement of the cloud due to the driving field and therefore the polarization, is proportional to E:

$$P = \varepsilon_0 \chi E$$

Further, the electric displacement is also proportional and parallel to the electric field:

$$D = \varepsilon_0 (1 + \chi) E = \varepsilon_0 \varepsilon_r E = \varepsilon E$$

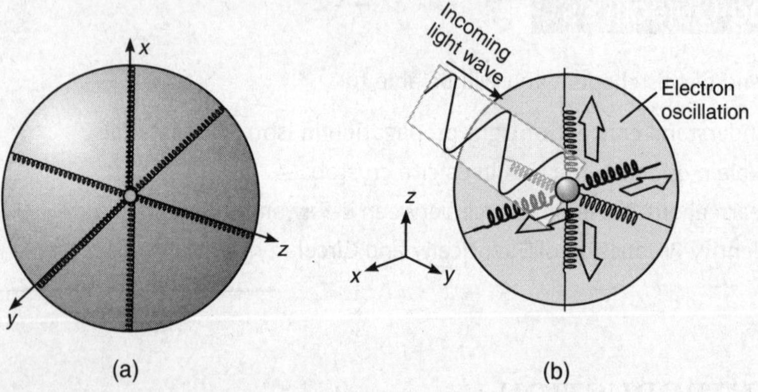

Fig. 11.1:

Conceptual model of internal charge oscillators. (a) Isotropic materials– all springs have the same stiffness. (b) Anisotropic media– springs have deferent stiffness basing on the direction.

Because these are in the same direction, the solution to the wave equation is:

$$E = E_0 e^{i(kx - \omega t)}$$

It represents a traveling plane wave moving with a velocity which is connected to the refractive index, n, as

$$\upsilon = c/n, \text{ where } n \text{ is given by } n = \sqrt{\varepsilon_r}$$

An imaginary plane connecting all the points of constant phase is referred to as a wavefront, and the wave vector k is perpendicular to the wave front everywhere.

11.3 EFFECTS OF ANISOTROPIC CRYSTALS ON LIGHT PROPAGATION

Crystal solids are asymmetric in different directions and are known as *anisotropic* crystals. As a result of asymmetry in anisotropic crystals, when an electromagnetic wave enters the material, the amount of polarization depends on the direction of the impressed electric field. Fig. 11.2 shows emission of two light rays from anisotropic solid.

An electric field in the vertical direction displaces the ions more than the same field would in the horizontal direction. The result is that the amount of polarization P now needs a different constant of proportionality (ex: $P_x = \varepsilon_0 \chi_x E_x$) for every direction of the field. In general, the electric displacement for anisotropic crystals will be related to the field through the dielectric tensor.

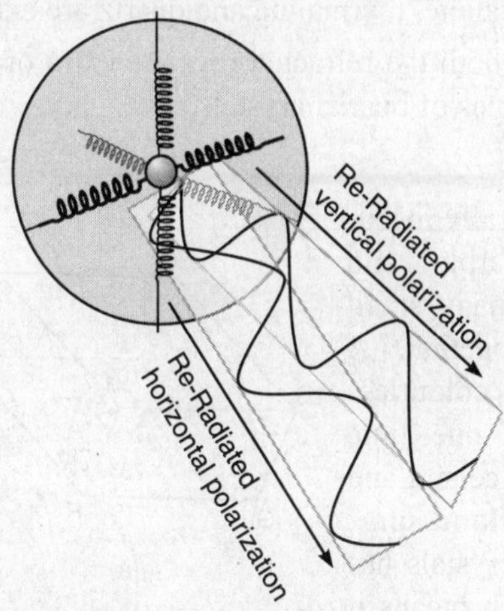

Fig. 11.2:

Emission of two light rays from anisotropic solid.

For a given light ray direction and crystal structure, if we calculate the actual direction of the field vectors, as well as the phase velocity and refractive index, using the dielectric tensor components and the two Maxwell Equations, we would find some startling results:

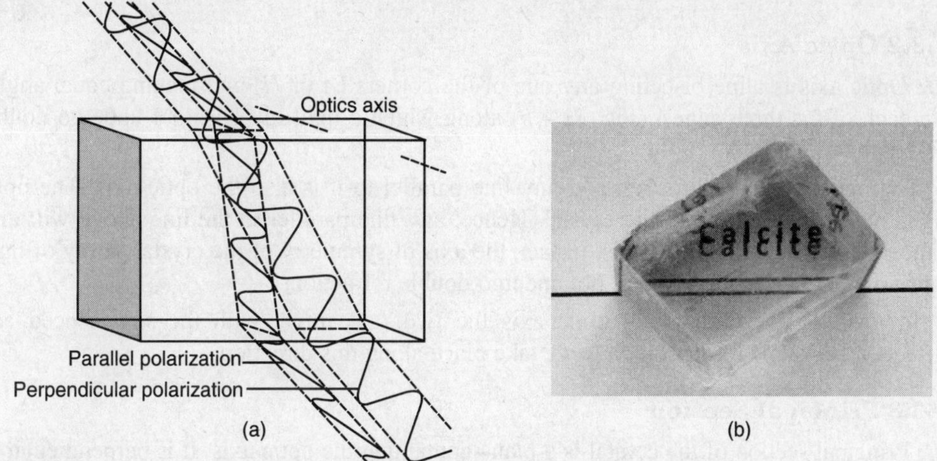

Fig. 11.3:

(a) Two polarized rays produced in anisotropic crystal. (b) Calcite splits light from the printed letters below it into two beams. One of the beams away and produces the double image.

(*i*) For a wave incident on the crystal, there arise two polarized beams, which propagate at different velocities. One of the waves is vertically polarized and the other is horizontally polarized.

(*ii*) As they travel through the crystal, one of these polarized beams moves away from the original direction of the light [Fig. 11.3 (a)].

The anisotropic crystals are divided into two classes; uniaxial and biaxial crystals.

(*i*) In **uniaxial crystals**, one of the refracted rays is an ordinary ray and the other is an extraordinary ray. Calcite, tourmaline and quartz are examples of uniaxial crystals.

(*ii*) In **biaxial crystals** both the refracted rays are extra ordinary rays. Mica, topaz and aragonite are examples of biaxial crystals.

11.3.1 Calcite Crystal

Calcite is a common naturally occur ring substance. Both marble and limestone are made up of many small calcite crystals bonded together. A large crystal of calcite is colourless and transparent. It was at one time found in great quantities in Iceland and hence it is also known as **Iceland spar**. Naturally occurring calcite crystals has rhombohedral cleavage which means it breaks into blocks with parallelogram-shaped faces.

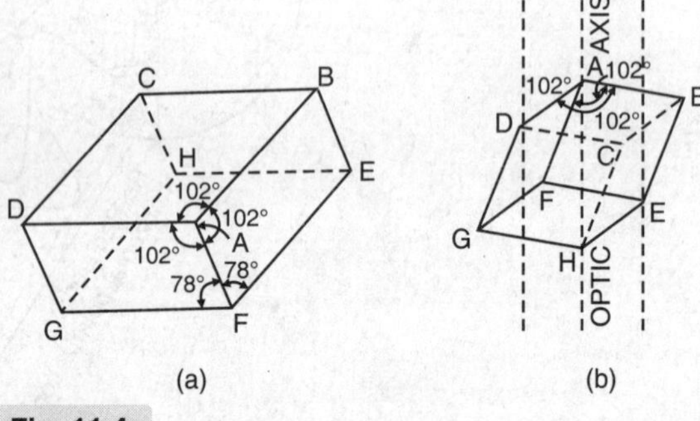

Fig. 11.4: Illustration of Optic axis.

It is bounded by six faces (Fig. 11.4 *a*), each of which is a parallelogram with angles equal to 101°55′ and 78°5′. The rhombohedron has only two corners *A* and *H* where all the face angles are obtuse (101°55′). These two corners appear as the blunt *corners* of the crystal. At the rest of six corners there is one obtuse angle and two acute angles.

11.3.2 Optic Axis

Def: Optic axis is aline bisecting any one of the corners (*A* or *H*) and making equal angles with each of the three edges, (Fig. 11.4 *b*) along which a light ray does not undergo double refraction.

The optic axis is a direction and any line parallel to it is also the optic axis. The optic axis is not a specific line in the crystal. Hence, any line parallel to the line above will also be the optic axis. The optic axis is, in fact, the axis of symmetry of the crystal. A ray of light propagating along optic axis does not undergo double refraction.

In a uniaxial crystal along optic axis the two rays travel with the *same* speed and consequently double refraction *does not* take place along this direction.

11.3.3 Principal Section

Def: Principal section of the crystal is a plane containing the optic axis. It is perpendicular to a pair of opposite faces of the crystal (Fig. 11.5 *a*).

So, there are three principal sections passing through any point within the crystal, one corresponding to each pair of opposite faces. A principal section always cuts surfaces of calcite crystal in a parallelogram having angles 71° and 109° [see Fig. 11.5 (*b*)]. Fig. 11.5 (*c*) shows a face of the crystal in which the end-view of the principal section *AB* of Fig. 11.5 (*b*) is shown by the dotted line *AB*. The lines parallel to *AB* represent the end-views of other principal sections parallel to *AB* within the crystal.

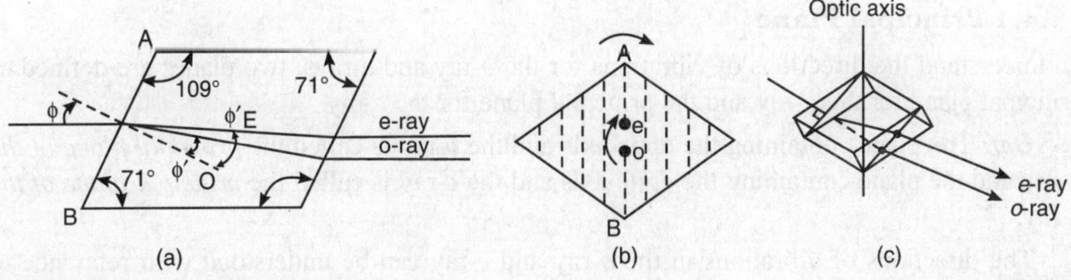

Fig. 11.5:

Principal section of the crystal.

11.4 DOUBLE REFRACTION IN CALCITE CRYSTAL

Fig. 11.5 (*a*) shows a principal section of calcite crystal. A ray of light is incident on the face *AB* of the crystal and it travels along the principal section. The ray is split into two rays, namely *o*-ray (fast ray) and *e*-ray (slow ray). The *o*-ray propagates straight through the crystal without under going refraction while the *e*-ray is bent (refracted) at some angle. The two rays emerge out parallel from the opposite faces of the crystal as the two faces are parallel. Within the crystal the *o*-ray *always* lies in the plane of incidence whereas *e*-ray does not lie in the plane of incidence. *e*-ray lies in the plane of incidence only when the plane of incidence is a principal section.

If a mark (dot or cross) is made on a paper and then the calcite crystal (*AB* face) is placed on it, two images are seen through the crystal, as illustrated in Fig. 11.5 (b). The images are produced by the *o*-ray and *e*-ray. The intensities of the images are lesser than that of the original mark. The line joining them lies in the principal section. If now the crystal is rotated slowly about an axis passing through the *o*-image, the *e*-image moves round in a circle while the *o*-image remains stationary. It shows that the velocity of propagation of *o*-ray is the same in all directions, while that of *e*-ray changes with direction. *o*-ray obeys the laws of refraction and the *e*-ray does not follow the ordinary laws of refraction.

The *e*-ray and *o*-ray are linearly polarized. The *e*-ray has its vibrations i.e., the optical vector *parallel* to the principal section whereas the vibrations in *o*-ray are *perpendicular* to the principal section, as indicated in Fig. 11.5 (*a*). The vibration directions can be established by examining the rays through a polarizer. As the polarizer is held in the path of the rays and rotated slowly, the intensity of one of the images, say the *o*-image, increases while that of the *e*-image decreases. In one position, the intensity of the *o*-image will be a maximum while the *e*-image is extinguished. Further rotation through 90° from this particular position, causes the *o*-image to disappear and *e*-image intensity to become a maximum. It proves that the *e*- and *o*-rays are linearly polarized in mutually perpendicular directions.

The *o*-ray travels with the same velocity in all directions in the crystal whereas the *e*-ray travels with different velocities in different directions. Therefore, refractive index corresponding to *o*-ray is a constant and is designated by μ_o. The refractive index corresponding to *e*-ray varies and is denoted by μ_e. The difference between the refractive indices is known as the amount of double refraction or *birefringence*. Thus,

$$\Delta\mu = \mu_e - \mu_o \qquad ...(11.1)$$

11.4.1 Principal Plane

To understand the directions of vibrations for the *o*-ray and *e*-rays, two planes are defined as principal plane for the *o*-ray and the principal plane for the *e*-ray.

Def: The plane containing the optic axis and the *o*-ray is called the *principal plane of the o-ray* and the plane containing the optic axis and the *e*-ray is called the *principal plane of the e-ray*.

The directions of vibrations in the *o*-ray and *e*-ray can be understood with reference to these planes. In general, the two principal planes do not coincide. Under the particular case, when the plane of incidence is the principal section of the crystal, then the principal planes of *o*- and *e*-rays and the principal section of the crystal coincide.

11.4.2 Huygens' Explanation of Double Refraction

In order to explain the double refraction exhibited by anisotropic crystals, Huygens postulated that the incident light breaks up into two wavelets within the crystal, one spherical wavelet connected with the ordinary waves and one ellipsoidal wavelet connected with the extraordinary waves. For example, the plane wavefront, in Fig. 11.6 (a), incident normally on the crystal surface generates spherical wavelets as well as ellipsoidal wavelets. The spherical wavelets propagate equally quickly in all directions. The wave surface corresponding to *o*-ray is therefore spherical and the plane wave propagates straight ahead with speed υ_o. The ellipsoidal wavelets [Fig. 11.6 (b)] propagate at different speeds in different directions. The wave surface corresponding to *e*-ray is therefore an ellipsoid of revolution about the optic axis.

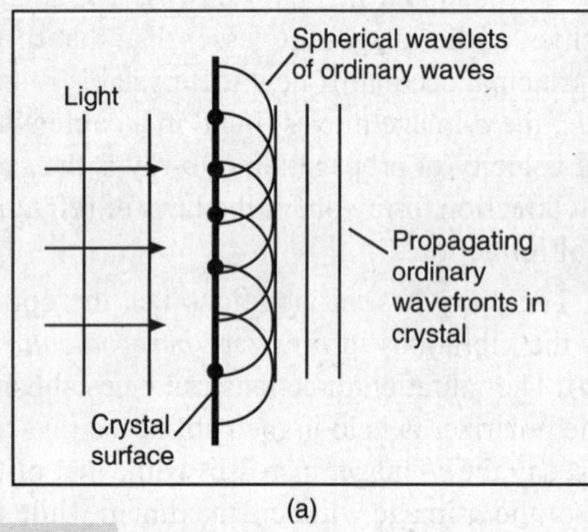

(a)

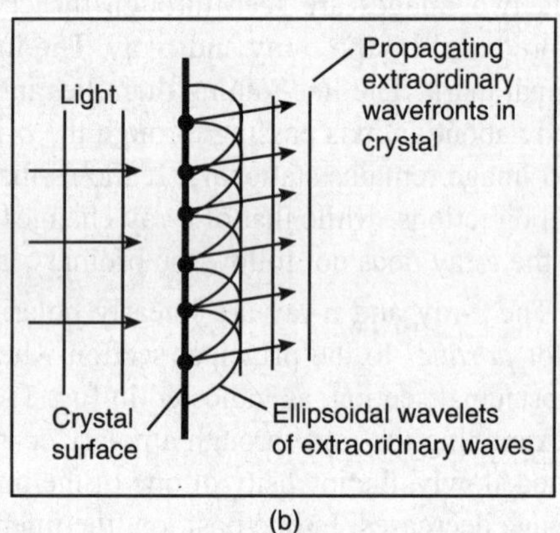

(b)

Fig. 11.6: Two wavelets propagating in a crystal.

The two wave surfaces touch each other at the two points where they are intersected by the optic axis. As light propagates through the crystal, the two wave surfaces travel in different directions in the crystal. Ultimately, two refracted rays emerge from the crystal.

11.4.3 Ordinary and Extra-ordinary Rays

Fig. 11.7 (*a*) shows the splitting of incident natural light on an anisotropic crystal at an angle to the optic axis into *o*-and *e*-rays. When natural light is incident in a direction perpendicular to the optic axis, *o*-ray and *e*-ray propagate in the *same direction* in the crystal but with *different velocities*, as shown in Fig. 11.7 (*b*). In a negative crystal *e*-ray leads *o*-ray and in case of a positive crystal *o*-ray leads *e*-ray. When natural light is incident on the crystal in a direction parallel to the optic axis, it does not split into two rays. It means that the *o*- and *e*-rays travel in the *same direction and with the same velocity* through the crystal, as shown in Fig. 11.7 (*c*). When the rays come out of the crystal, they travel with the same velocity and in the same direction.

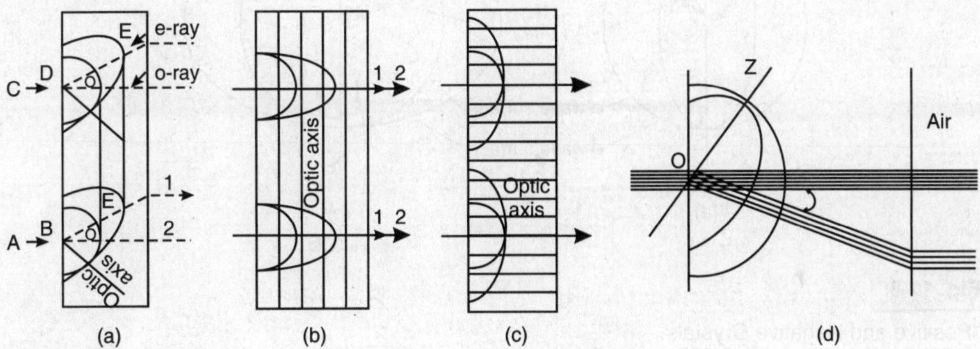

Fig. 11.7: *o*-ray and *e*-ray.

The rays outside the crystal differ only in their plane of polarization. They propagate along parallel directions and with the same velocity. They maintain the same separation as that with which they emerged outside. The designation of *o*- ray and *e*-ray has no meaning outside the crystal. Table 11.1 shows the comparison of the properties of *o*- and *e*-rays.

Table 11.1: *Comparison of the properties of o-rays and e-rays*

O-rays	E-rays
Follows the laws of ordinary refraction.	Does not follow the laws of ordinary refraction.
Plane polarized and the electric vector vibrates perpendicular to the principal section of *o*-ray.	Plane polarized and the electric vector vibrates parallel to the principal section of *e*-ray.
Travels with the same speed in all directions within the crystal.	Travels with different speeds along different directions in the crystal. The speed of *e*-ray and of *o*-ray will be equal along the optic axis direction.
Same velocity in all directions gives a constant refractive index.	The refractive index for *e*-ray varies from direction to direction.

11.4.4 Positive and Negative Crystals

Because of two different wavefronts, two different types of uniaxial crystals exist, negative crystals and positive crystals. Table 11.2 compares here the characteristics of the positive and negative crystals.

Def: Negative crystals are anisotropic crystals in which the refractive index of the *e*-ray is less than that corresponding to *o*-ray. In negative crystals the spherical wave front of *o*-ray is enclosed by the ellipsoidal wave front of *e*-ray. Ex: Calcite crystal.

Def: Positive crystals are anisotropic crystals in which the refractive index for the extraordinary ray is greater than that of *o*-ray. In positive crystals the extraordinary wavefront lies within the ordinary wavefront. Ex: Quartz crystal.

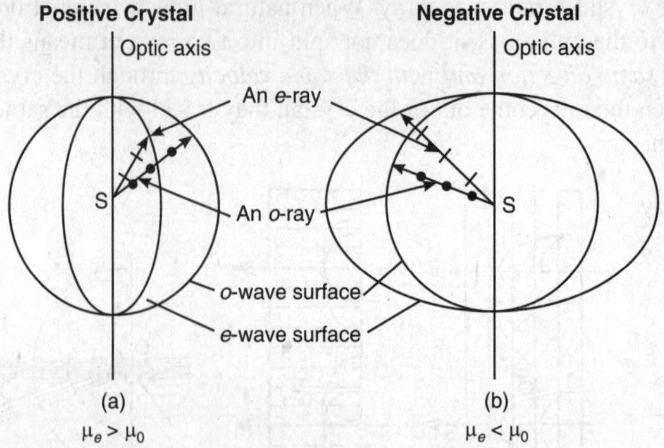

Fig. 11.8:

Positive and Negative Crystals.

The principal refractive index for *o*-ray is defined as follows:

$$\mu_0 = \frac{c}{\upsilon_0} = \frac{\text{velocity of light in a vacuum}}{\text{velocity of o-ray in the crystal}} \qquad ...(11.2)$$

The principal refractive index for *e*-ray in ***positive crystals*** is defined as follows:

$$\mu_e = \frac{c}{(\upsilon_e)_{\min}} = \frac{\text{velocity of light in a vacuum}}{\text{minimum velocity of e-ray in the crystal}} \qquad ...(11.3)$$

The principal refractive index for *e*-ray in ***negative crystals*** is defined as follows:

$$\mu_e = \frac{c}{(\upsilon_e)_{\max}} = \frac{\text{velocity of light in a vacuum}}{\text{maximum velocity of e-ray in the crystal}} \qquad ...(11.4)$$

The birefringence of a

Positive crystal $\Delta\mu = \mu_e - \mu_o$ (11.15a)

Negative crystal $\Delta\mu = \mu_o - \mu_e$ (11.15b)

Table 11.2: *Comparison of the characteristics of the positive and negative crystals*

S.N	Positive uniaxial crystals	Negative uniaxial crystals
1	The ellipsoid of revolution corresponding to the *e*-ray is totally contained within the sphere corresponding to the *o*-ray.	The ellipsoid of revolution for *e*-ray lies completely outside the sphere corresponding to *o*-ray.
2	Velocity of *e*-ray has a maximum value along the optic axis and a minimum value in a direction perpendicular to the optic axis.	velocity of *e*-ray has a minimum value parallel to the optic axis and a maximum value in a direction perpendicular to the optic axis
3	velocity of *e*-ray υ_e= velocity of *o*-ray υ_o -- parallel to optic axis $\upsilon_e < \upsilon_o$ -- other directions	$\upsilon_e = \upsilon_o$ -- parallel to optic axis $\upsilon_e > \upsilon_o$ -- other directions
4	$\mu_e > \mu_o$	$\mu_e < \mu_o$
5	As $\mu_e > \mu_o$, $\Delta\mu$ is a positive quantity.	As $\mu_e < \mu_o$, $\Delta\mu$ is a negative quantity.

11.5 NICOL PRISM

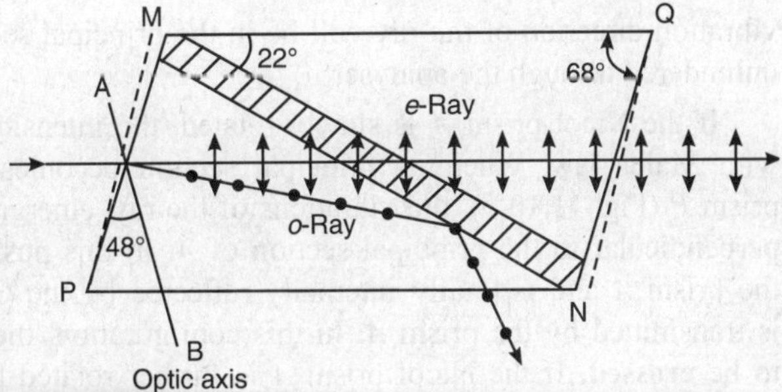

Fig. 11.9: Nicol Prism.

Nicol prism is a polarizing device fabricated from a double refracting crystal. It is similar to a Polaroid sheet in its action. A Nicol prism is made from calcite crystal. William Nicol designed it in 1820. A rhomb of calcite crystal about three times as long as it is thick, is obtained by cleavage from the original crystal. The ends of the rhombohedron are ground until they make an angle of 68° instead of 71° with the longitudinal edges. This piece is then cut into two along a plane perpendicular both to the principal axis and to the new end surfaces *AB* and *CD*. The two parts of the crystal are then cemented together with Canada balsam, whose refractive index lies between the refractive indices of calcite for the *o*-ray and *e*-ray. $\mu_o = 1.66$, $\mu_e = 1.486$ and $\mu_{\text{canada balsam}} = 1.55$. The position of the optic axis is as shown in Fig. 11.9. The refractive index for *e*-ray depends upon the direction in which *e*-ray is propagating in the crystal. The difference between the refractive index between *o*-ray and that for *e*-ray goes on increasing with the angle between the two rays in the crystal. When this angle is 90°, the difference is a maximum. Thus, for a fixed value for μ_o, the μ_e has its maximum or minimum value in perpendicular direction. In the above $\mu_e = 1.486$ represents the minimum value.

Unpolarized light is made to fall on the crystal as shown in Fig. 11.9 at an angle of about 15°. The ray after entering the crystal suffers double refraction and splits up into *o*-ray and *e*-ray. The two rays with their directions of vibrations are as shown in the Fig. 11.9. The values of the refractive indices and the angles of incidence at the Canada balsam layer are such that the *e*-ray is transmitted while the *o*-ray is internally reflected. The face where the *o*-ray is incident is blackened so that the *o*-ray is completely absorbed. Then we get only the plane-polarized *e*-ray coming out of the Nicol. Thus, the Nicol works as a polarizer.

For studying the optical properties of transparent substances, two Nicols are used – one as a polarizer and the other as an analyzer.

When two Nicol prisms *P* and *A* are placed adjacent to each other as shown in Fig. 11.10 (a), one of them acts as a *polarizer* and the other acts as an *analyser*. If unpolarized ray of light is incident on the Nicol prism *P*, a linearly polarized *e*-ray emerges from P with its vibration direction lying in the principal section of *P*. The state of the polarization of the light coming from polarizer *P* can be examined with another polarizer *A*, which for convenience is called an analyser. Let now this ray be incident on the second Nicol prism *A*, whose principal section is parallel to that to *P*. The

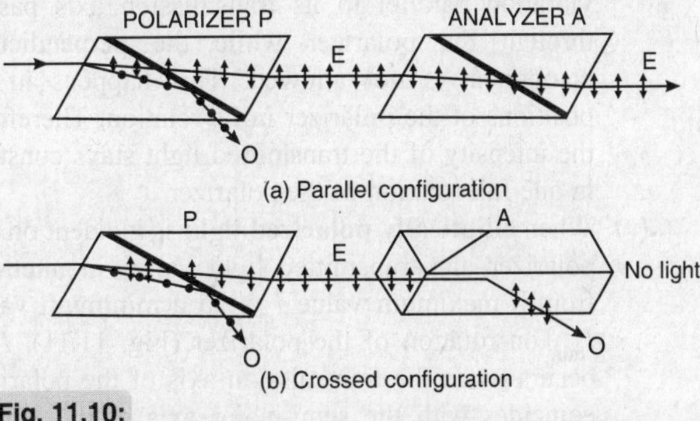

Fig. 11.10: Nicols as a polarizer and an analyzer.

vibration direction of the ray will be in the principal section of *A* and hence it is transmitted unhindered through the analyser *A*.

If the Nicol prism *A* is slowly rotated, the intensity of the *e*-ray reduces in accordance with Malus law. When its principal section becomes perpendicular to that of the Nicol prism *P* (Fig. 11.10 *b*), the vibrations of the ray, emerging from *P* and incident on *A*, will be perpendicular to the principal section of *A*. In this position the ray behaves as *o*-ray inside the prism *A* and is totally internally reflected by the Canada balsam layer. Hence no light is transmitted by the prism *A*. In this configuration, the two Nicol prisms *P* and *A* are said to be **crossed.** If the Nicol prism *A* is further rotated through another 90°, the intensity of light emerging from *A* will go on increasing. The intensity will become a maximum when its principal section is again parallel to that of the prism *P*. Thus, the prism *P* produces linearly polarised light while the prism *A* detects it. Hence the prism *P* is called a polarizer and the prism A an analyser.

11.6 EFFECT OF POLARIZER ON LIGHT OF DIFFERENT POLARIZATIONS

The action of a polarizer, whether a Nicol prism or a Polaroid sheet, on light of different types of polarization is as follows:

(*i*) If **unpolarized light** is incident on a polarizer, it transmits half the intensity of light incident on it. The transmitted light stays constant in intensity on rotation of the polarizer.

(*ii*) If **partially polarized light** is incident on the polarizer, the intensity of the transmitted light depends on the direction of the transmission axis of the polarizer. The intensity varies from a maximum value I_{max} to a minimum value I_{min}. Two positions of I_{max} and two positions of I_{min} occur in one full rotation of the polarizer.

(*iii*) If **plane polarized light** is incident on the polarizer, the intensity of the transmitted light varies from zero to a maximum twice in one full rotation of the polarizer.

(*iv*) When **circularly polarized light** is incident on the polarizer, the transmitted light remains constant in intensity on rotation of the polarizer. This is interpreted as follows: the circular vibrations may be resolved into two mutually perpendicular linear vibrations of equal amplitude. When the circularly polarized light is incident on the polarizer, the vibration parallel to its transmission axis passes through the polarizer while the perpendicular component is not allowed. This happens in all positions of the polarizer in its rotation. Therefore, the intensity of the transmitted light stays constant in one full rotation of the polarizer.

(*v*) When **elliptically polarized light** is incident on the polarizer, the transmitted light varies in intensity from a maximum value I_{max} to a minimum value I_{min} on rotation of the polarizer (Fig. 11.11). I_{max} occurs when the transmission axis of the polarizer coincides with the semi-major axis of the ellipse and I_{min} occurs when the transmission axis coincides with the semi-minor axis of the ellipse.

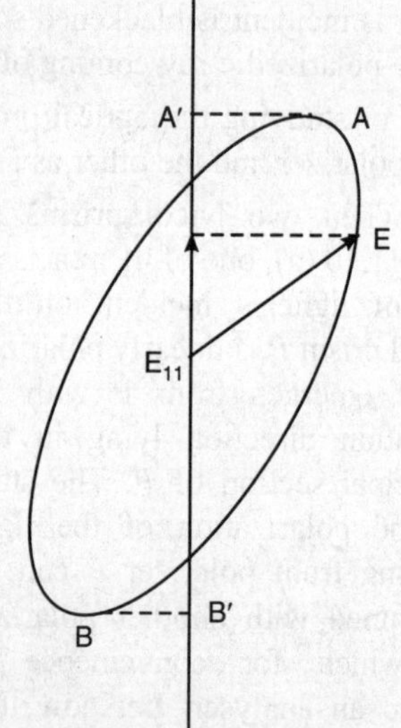

Fig. 11.11: Intensity variation from a maximum to a minimum value.

11.7 PHASE DIFFERENCE BETWEEN *e*-RAY AND *o*-RAY

We have seen that natural light incident on the surface of an anisotropic crystal undergoes double refraction and produces two plane polarized waves (Fig. 11.12 *a*). Let us consider the particular case of a slice of a negative uniaxial calcite crystal where the optic axis is parallel to refracting face of the crystal. The two waves travel along the same direction in the crystal but with different velocities.

So, when the waves come out from the back face of the crystal, an optical path difference would have developed between them (Fig. 11.12 *b*).

The optical path difference:

Let d be the thickness of the crystal.

The optical path for *o*-ray within the crystal = $\mu_o d$

The optical path for *e*-ray within the crystal = $\mu_e d$

$\therefore$ The optical path difference between *e*-ray and *o*-ray = $\Delta = (\mu_o - \mu_e)d$ ...(11.6)

So, phase difference between the two waves

$$\delta = \frac{2\pi}{\lambda}(\mu_o - \mu_e)d \qquad ...(11.7)$$

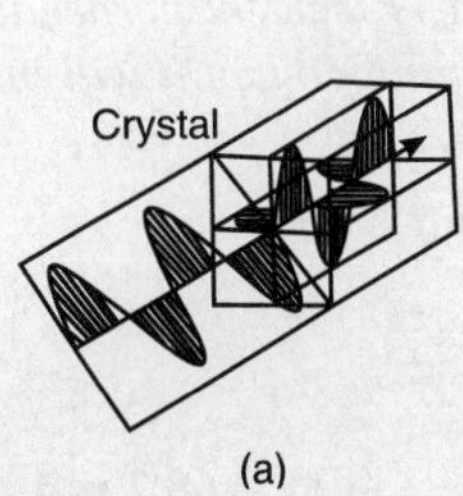

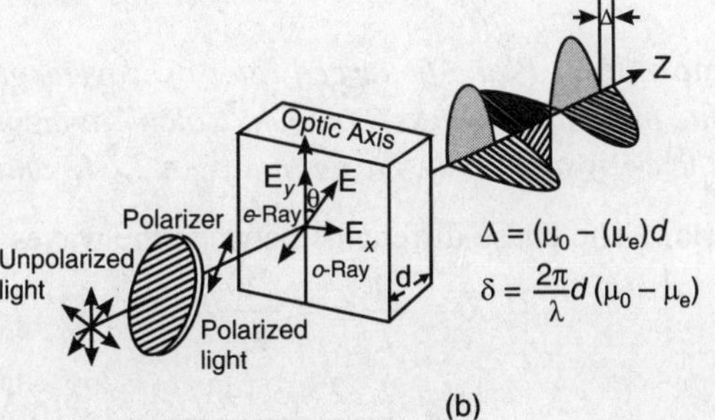

Fig. 11.12:

Phase difference between *e*-ray and *o*-ray.

As the two component waves are derived from the same incident wave, the two waves are in phase at the front face and have emerged from the crystal with a constant phase difference and hence it may be expected that the waves are in a position to interfere with each other. However, as the planes of polarization of *o*-ray and *e*-ray are perpendicular to each other, interference cannot take place between *e*- and *o*-rays. The waves instead combine with each other to give elliptically polarized wave.

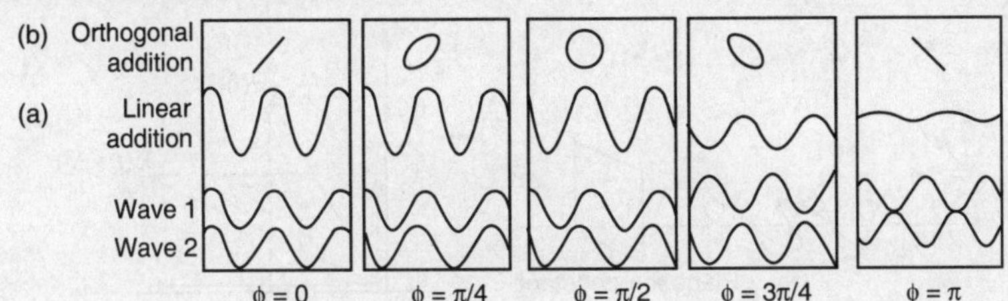

Fig. 11.13:

(a) Linear and (b) Orthogonal addition of coherent waves.

The most important point to be noted here is as follows: The superposition of two coherent waves having a *common plane* of polarization yields a wave polarized in the same plane. The superposition leads to linear addition of waves, i.e., **interference**. The resultant vibration is *linear* and occurs in the same direction as that of the two superposing vibrations [see Fig. 11.13 (a)]. On the other hand, if the two coherent waves are polarized in orthogonal planes, the resultant vibrational motion takes place in two dimensions either in the form of an ellipse, a circle or a straight line depending on the phase difference between the waves [Fig. 11.13 (b)]. This is known as *orthogonal addition*.

Example 11.1: *Calculate the thickness of quartz plate which introduces phase difference of 60° between e- and o-rays when Plane-polarized light of wavelength 5400 Å is incident perpendicular lyon it.*

Solution: The path difference between the waves is given by $\Delta = (\mu_e - \mu_o)d$

The phase difference between the waves is given by $\delta = \dfrac{2\pi}{\lambda}\Delta$

$$\Delta = \frac{60°}{360°}\lambda = \frac{\lambda}{6} \qquad \therefore \quad d = \frac{\lambda}{6(\mu_e - \mu_o)}$$

or
$$d = \frac{5400\,\text{Å}}{6(1.553 - 1.544)} = \frac{0.54}{0.054}\,\mu\text{m} = 10\ \mu\text{m}.$$

Example 11.2: *Plane-polarized light of wavelength 6000 Å is incident perpendicularly on a calcite plate of thickness 0.04 mm. Calculate the phase retardation that it will introduce between the e-ray and o-ray. Given that $\mu_o = 1.642$ and $\mu_e = 1.478$.*

Solution: The phase difference between the waves is given by

$$\delta = \frac{2\pi}{\lambda}\Delta = \frac{2\pi}{\lambda}d(\mu_e - \mu_o)$$

$$\therefore \qquad \delta = \frac{2 \times 3.143}{6000 \times 10^{-10}\,\text{m}} \times 4 \times 10^{-5}\,\text{m}(1.642 - 1.478) = \mathbf{68.7\ rad.}$$

11.8 SUPERPOSITION OF WAVES LINEARLY POLARIZED AT RIGHT ANGLES

Let us now look at the result of superposition of two waves linearly polarized at right angles to each other (see Fig. 11.14 *a*). Let us consider two light waves traveling in the same direction, *x*; one wave is polarized in the *xy* -plane and the other is polarized in *yz*-plane. We are interested to know the state of polarization of the resultant wave at the plane *x* = *constant*.

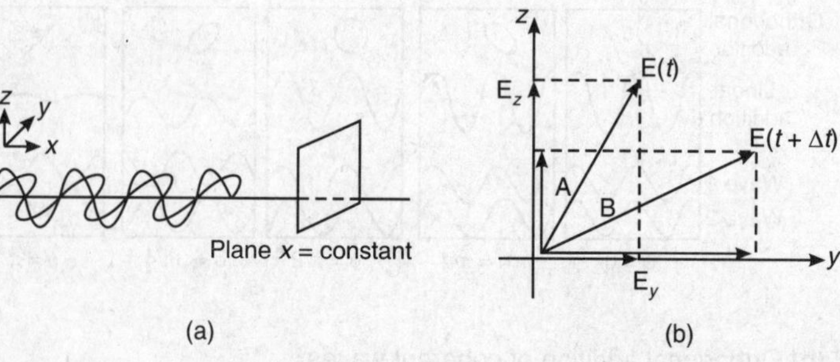

Fig. 11.14:

Superposition two linearly polarized waves.

Let the two orthogonal waves be represented by

$$E_y = E_1 \cos (k x - \omega t) \quad ...(11.8)$$

$$E_z = E_2 \cos (k x - \omega t + \delta) \quad ...(11.9)$$

The waves are of the same frequency $\nu = \omega/2\pi$. δ is the phase difference between the waves. At a given time t, the optical vectors E_y and E_z produce a resultant optical vector of magnitude, say A and at a slightly later time, $t + \Delta t$, they give rise to a resultant vector of magnitude B which points in a different direction (see Fig. 11.14 *b*). With progress of time the tip of the resultant optical vector moves along a curve in the *yz*-plane. We apply the *principle of superposition* to find the equation of the curve traced by the tip of the resultant vector of the two vectors.

According to the principle of superposition

$$E = E_y + E_z$$

$$= E_1 \cos (k x - \omega t) + E_2 \cos (k x - \omega t + \delta) \quad ...(11.10)$$

The equation of the curve may be found by eliminating 't' from the equations. We can write the expansion of Eq. (11.10) as

$$E_z = E_2 \cos (k x - \omega t) \cos \delta - E_2 \sin (k x - \omega t) \sin \delta \quad ...(11.11)$$

$$= E_2 \cos (k x - \omega t) \cos \delta \pm [1 - \cos^2 (k x - \omega t)]^{1/2} E_2 \sin \delta$$

We find from Eq. (11.8) that

$$\cos (k x - \omega t) = E_y/E_1$$

$$\therefore \quad E_z = E_2 \frac{E_y}{E_1} \cos \delta \pm \sqrt{1 - \frac{E_y^2}{E_1^2}}\, E_2 \sin \delta \quad ...(11.12)$$

Rearranging the terms, we get

$$\left[E_z - \frac{E_2}{E_1} E_y \cos \delta \right] = \pm \sqrt{1 - \left(\frac{E_y}{E_1} \right)^2}\, E_2 \sin \delta$$

On squaring both the sides, we obtain

$$E_z^2 + \frac{E_y^2 E_2^2}{E_1^2} \cos^2 \delta - \frac{2 E_y E_z E_2}{E_1} \cos \delta = E_2^2 \sin^2 \delta - \frac{E_y^2 E_2^2}{E_1^2} \sin^2 \delta$$

Rearranging the terms, we get

$$E_z^2 + \frac{E_y^2 E_2^2}{E_1^2} (\cos^2 \delta + \sin^2 \delta) - \frac{2 E_y E_z E_2}{E_1} \cos \delta = E_2^2 \sin^2 \delta \quad ...(11.13)$$

$$E_z^2 + \frac{E_y^2 E_2^2}{E_1^2} - \frac{2 E_y E_z E_2}{E_1} \cos \delta = E_2^2 \sin^2 \delta$$

Dividing both the sides by $E^2{}_2$ and rearranging the terms, we obtain

$$\frac{E_y^2}{E_1^2} + \frac{E_z^2}{E_2^2} - \frac{2 E_y E_z}{E_1 E_2} \cos \delta = E^2{}_2 \sin^2 \delta \quad ...(11.14)$$

Eq. (11.14) is the general equation of an *ellipse*. Hence, the tip of the resultant vector traces an ellipse in the *yz*-plane. The ellipse is constrained within a rectangle having sides $2E_1$

and $2E_2$. The major axis makes an angle α with the *y*-axis (see Fig. 11.15).

$$\tan 2\alpha = 2E_1E_2 \cos\delta / (E_1^2 - E_2^2)$$

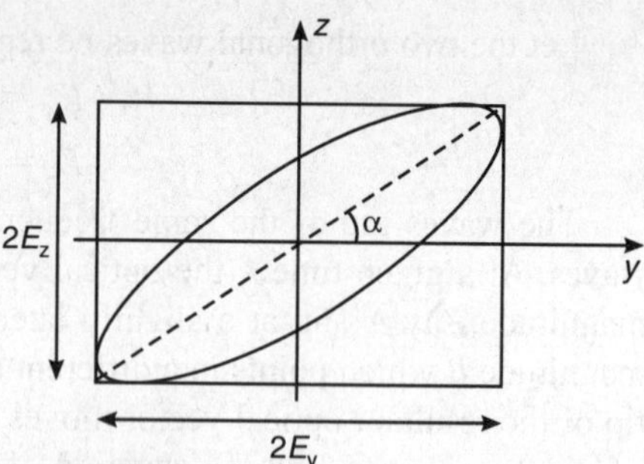

Fig. 11.15: Tip of the resultant vector traces an ellipse in *yz* - plane.

Special cases:

(*i*) When δ = 0, or ± 2 *m*π, the two waves are in phase.

cos δ = 1 and sin δ = 0 and the Eq. (11.14) reduces to

$$\frac{E_y^2}{E_1^2} + \frac{E_z^2}{E_2^2} - \frac{2E_yE_z}{E_1E_2} = 0$$

$$\left[\frac{E_y}{E_1} - \frac{E_z}{E_2}\right]^2 = 0$$

$$\frac{E_y}{E_1} - \frac{E_z}{E_2} = 0$$

$$\therefore \quad E_z = \frac{E_2}{E_1} E_y \quad \text{...(11.15)}$$

Eq. (11.15) represents a straight line, having a slope (E_2/E_1). Therefore, the equation represents a wave having its plane of polarization making an angle $\tan^{-1}(E_2/E_1)$ with respect to the y-axis. It means that **the resultant of two plane-polarized waves, which are in phase (i.e., coherent waves), is again a plane-polarised wave.**

(*ii*) When δ = π, or ± (2*m* + 1)π, the two waves are in opposite phase.

cos δ = – 1 and sin δ = 0 and the Eq. (11.14) reduces to

$$\frac{E_y^2}{E_1^2} + \frac{E_z^2}{E_2^2} + \frac{2E_yE_z}{E_1E_2} = 0$$

$$\left[\frac{E_y}{E_1} + \frac{E_z}{E_2}\right]^2 = 0$$

$$\frac{E_y}{E_1} + \frac{E_z}{E_2} = 0$$

$$\therefore \quad E_z = -\frac{E_2}{E_1} E_y \quad \text{...(11.16)}$$

Eq. (11.16) represents a straight line of a slope ($-E_2/E_1$). Therefore, the equation represents a wave having its plane of polarization making an angle $\tan^{-1}(-E_2/E_1)$ with respect to the *y*-axis. It means that **the resultant of two plane-polarized waves, which are in opposite phase (i.e., coherent waves), is again a plane-polarized wave.**

(*iii*) If δ = π/2, or ± (2*m* + 1)π/2, then cos δ = 0 and sin δ = 1. Eq. (11.14) reduces to

$$\frac{E_y^2}{E_1^2} + \frac{E_z^2}{E_2^2} = 1 \quad \text{...(11.17)}$$

This is the equation of an ellipse whose major axis and minor axis coincide with *y*- and *z*-co-ordinate axes. Therefore, when the two plane polarized waves are out of phase by 90°, their resultant is an elliptically polarized wave.

(*iv*) In the particular case, when $\delta = \pi/2$ and $E_1 = E_2 = E_o$, Eq. (11.14) reduces to

$$E_y^2 + E_z^2 = E_o^2 \qquad \text{...(11.18)}$$

This is the equation of a circle. Hence the resultant light is circularly polarized.

Fig. 11.16 shows more generally how the *E* vector changes with time for various values of δ.

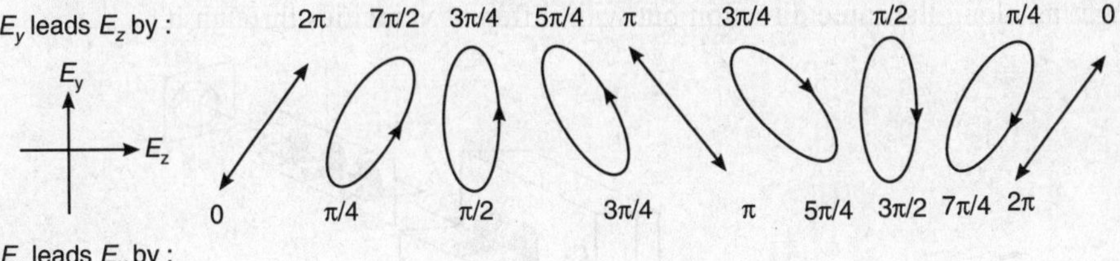

Fig. 11.16:

E vector variation with time for various values of δ.

Example 11.3: *At a given point the electric fields of component waves of a polarized wave are $E_x = 10 \cos \omega t$ and $E_y = 20 \cos(\omega t + \pi)$. Determine the type of polarization and the direction of polarization.*

Solution: At any instant, *t*, we have $E_x = 10 \cos \omega t$ and $E_y = -20 \cos \omega t$.

$$\therefore \qquad \cos \omega t = \frac{E_x}{10} = -\frac{E_y}{20}$$

or

$$E_y = -2E_x$$

Example 11.4: *Find out the state of polarization represented by the following equations. $E_x = E_0 \sin(\omega t - kz)$ and $E_y = E_0 \sin(\omega t - kz)$.*

Solution:

$$E_x = E_0 \sin(\omega t - kz)$$

$$E_y = E_0 \sin(\omega t - kz)$$

Squaring the above equations, we get

$$E_x^2 = E_0^2 \sin^2(\omega t - kz)$$

$$E_y^2 = E_0^2 \sin^2(\omega t - kz)$$

Adding the above equations, we get

$$E_x^2 + E_y^2 = E_0^2$$

The equation represents a circle. Therefore, the light is circularly polarized.

It means that the light is plane polarized in *xy*-plane with slope (–2). As the value of $\tan^{-1}(-2) = -63.4°$, the light is polarized at an angle of –63.4° with the x-axis.

11.9 RETARDERS

Def: A retarder is a uniform plate of anisotropic double refracting material having the optic axis parallel to its refracting faces that resolves a light wave into two perpendicular linear polarization components and produces a phase difference between them. They are also called as wave plate quarter-wave plates, half-wave plates and full-wave plates.

11.9.1 Quarter Wave Plate

Def: A quarter wave plate is a retarder whose thickness adjusted such that it introduces a quarter-wave ($\lambda/4$) path difference or a phase difference of 90° between the *e*-ray and *o*-ray propagating along the same direction but with different velocities through it.

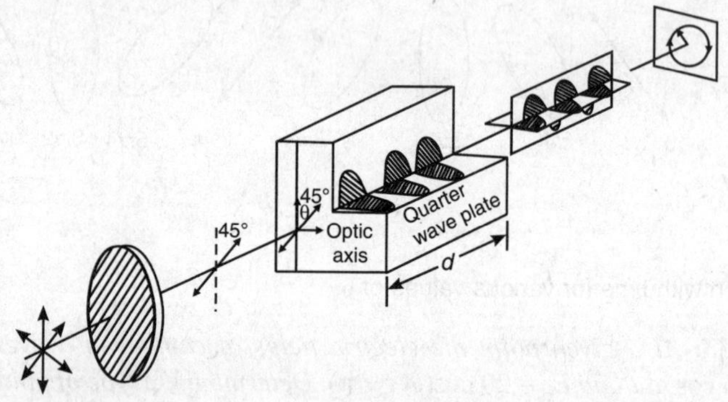

Fig. 11.17:

Quarter wave plate.

Fig. 11.17 shows the concept of working of *quarter* wave plate. The optical path difference would be developed between *e*-ray and *o*-ray is

$$\therefore \quad (\mu_o - \mu_e)d = \frac{\lambda}{4} \quad ...(11.19)$$

$$d = \frac{\lambda}{4[\mu_o - \mu_e]} \quad ...(11.20)$$

A quarter wave plate introduces a phase difference δ, between *e*-ray and *o*-ray given by $\delta = (2\pi/\lambda)\Delta = \pi/2 = 90°$.

A quarter-wave plate is used for producing elliptically or circularly polarized light. It converts plane-polarized light into elliptically or circularly polarized light depending upon the angle that the incident light vector makes with the optic axis of the quarter wave plate.

Example 11.5: *Plane-polarized light passes through a double refracting crystal of thickness 40 μm and emerges out as circularly polarized light. If the birefringence of the crystal is 0.00004, find the wavelength of the incident light.*

Solution: Quarter-wave plates alter plane-polarized light into circularly polarized light.

Its thickness is given by

$$d = \frac{\lambda}{4(\mu_e - \mu_o)}$$

$$\therefore \quad \lambda = 4d(\mu_e - \mu_o) = 4 \times 40 \times 10^{-6}\,\text{m} \times 0.00004 = 6400\ \text{Å}.$$

Example 11.6: *A plate of 0.003 cm thickness changes aplane polarized light into circularly polarized light by passing it. Find the birefringence of the crystal assuming this to be the minimum thickness that will produce the effect ($\lambda = 6 \times 10^{-5}$ cm).*

Solution: Plane polarized light is converted into circularly polarized light by a suitably oriented quarter wave plate. Its thickness is given by

$$d = \frac{\lambda}{4(\mu_e - \mu_o)} = \frac{\lambda}{4\Delta\mu}$$

$$\therefore \qquad \Delta\mu = \frac{\lambda}{4d} = \frac{6\times10^{-5}\,\text{cm}}{4\times0.003\,\text{cm}} = \mathbf{0.005}$$

11.9.2 Half Wave Plate

Def: A half wave plate is a retarder whose thickness adjusted such that it introduces a quarter-wave *(λ/2)* path difference or a phase difference of 180° between the *e*-ray and *o*-ray propagating along the same direction but with different velocities through it.

When the two waves combine, they yield a plane-polarized wave, which has its plane of polarization rotated through an angle of 2θ. The half wave plate will invert the handedness of elliptical or circular polarized light, changing right to left and vice versa. Fig.11.18 shows the concept of working of half wave plate.

The optical path difference would be developed between *e*-ray and *o*-ray is

$$\therefore \qquad (\mu_o - \mu_e)d = \frac{\lambda}{2} \qquad \text{...(11.21)}$$

$$\therefore \qquad d = \frac{\lambda}{2(\mu_o - \mu_e)} \qquad \text{...(11.22)}$$

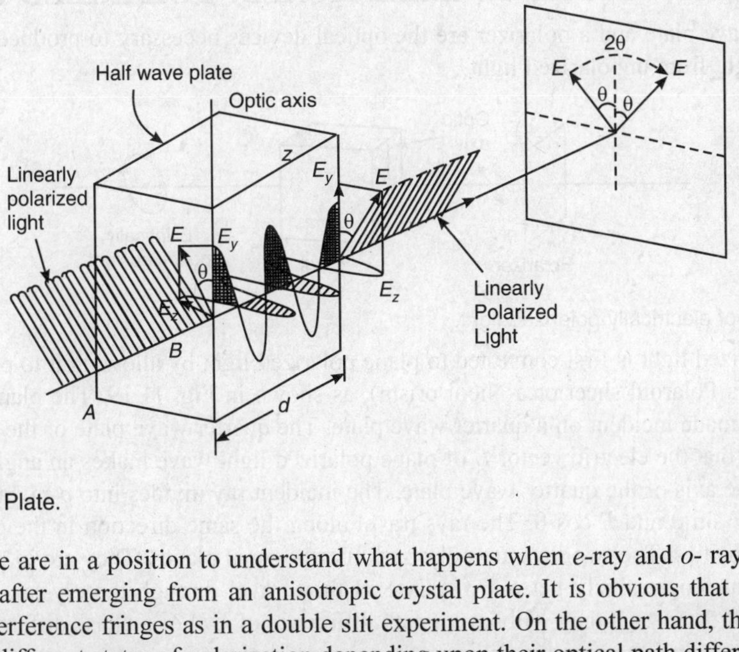

Fig. 11.18:

Half Wave Plate.

Now we are in a position to understand what happens when *e*-ray and *o*- ray overlap on each other after emerging from an anisotropic crystal plate. It is obvious that they cannot produce interference fringes as in a double slit experiment. On the other hand, they combine to produce different states of polarization depending upon their optical path difference.

(*i*) When the optical path difference is 0 or an even or odd multiple of $\lambda/2$, the resultant light wave is linearly polarized.

(*ii*) When the optical path difference is $\lambda/4$, the resultant light wave is **elliptically polarized.**

(*iii*) In the particular instance when the wave amplitudes are equal and the optical path difference is $\lambda/4$, the resultant light wave is **circularly polarized**.

Example 11.7: *Calculate the least thickness for which the o-ray and e-ray combines to form plane polarized light when plane polarized incident on a piece of quartz plate.*

Solution: When plane polarized light is incident on a half wave plate, the emergent beam will also be plane polarized. The least thickness of the quartz plate is given by

$$\therefore \qquad d = \frac{\lambda}{2(\mu_e - \mu_o)} = \frac{5 \times 10^{-5}\,\text{cm}}{2(1.5533 - 1.5542)} = \mathbf{27\ \mu m}$$

Example 11.8: *A half-wave plate is fabricated for a wavelength of 3800 Å. For what wavelength does it work as a quarter-wave plate?*

Solution: The thickness of a half-wave plate is $d = \dfrac{\lambda_1}{2(\mu_e - \mu_o)}$. The same plate is required to act as a quarter-wave plate. Therefore, we can write that $d = \dfrac{\lambda_2}{4(\mu_e - \mu_o)}$.

$$\therefore \qquad d = \frac{\lambda_1}{2(\mu_e - \mu_o)} = \frac{\lambda_2}{4(\mu_e - \mu_o)}$$

$$\therefore \qquad \lambda_2 = 2\lambda_1 = 2 \times 3800\ \text{Å} = 7600\ \text{Å}.$$

11.10 PRODUCTION OF ELLIPTICALLY POLARIZED LIGHT

A quarter wave plate and a polarizer are the optical devices necessary to produce elliptically polarized light from unpolarized light.

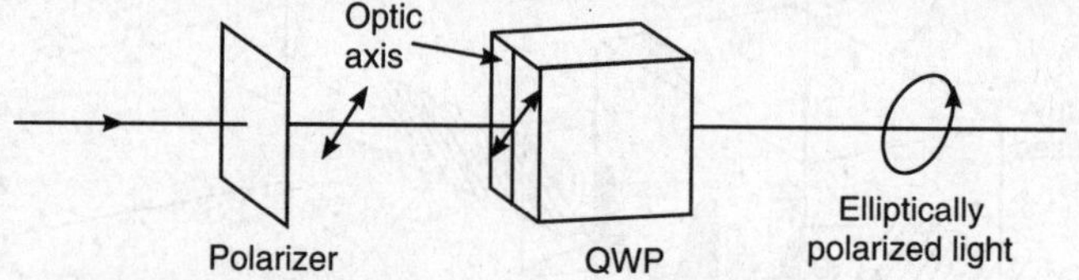

Fig. 11.19: Production of electrically polarized light.

Unpolarized light is first converted to plane polarized light by allowing it to pass through a polarizer (a Polaroid sheet or a Nicol prism), as shown in Fig. 11.19. The plane-polarized light is then made incident on a quarter wave plate. The quarter wave plate or the polarizer is rotated such that the electric vector E of plane polarized light wave makes an angle θ ($\neq 45°$) with the optic axis of the quarter wave plate. The incident ray divides into *o*-ray and *e*-ray of amplitudes $E \sin \theta$ and $E \cos \theta$. The rays travel along the same direction in the crystal with different velocities. The two rays are polarized in orthogonal planes. They are in phase at the front face but progressively get out of phase as they travel through the crystal. When they emerge out of the crystal they will have a path difference of $\lambda/4$ or a phase difference of 90°. When they combine, they produce elliptically polarized light.

11.10.1 Detection of Elliptically Polarized Light

The light beam is allowed to pass through an analyzer (a Polaroid sheet or a Nicol prism). If on rotating the analyzing Polaroid sheet or Nicol, the intensity of the emerging beam varies from a maximum to a minimum value, but never reaching zero, then the incident light is elliptically polarized. A similar result would be obtained if the incident light were partially polarized. The two cases may be distinguished by inserting a quarter-wave plate in the path of light before it falls on the analyzer.

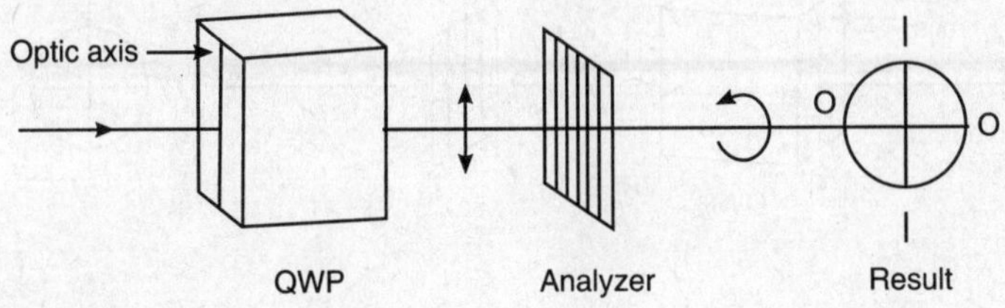

Fig. 11.20: Detection of elliptically polarized light.

If the original light is elliptically polarized, it may be considered as resultant of two coherent plane polarized waves that is *e*-ray and *o*-ray, which are out of phase by 90°. If the light passes through the quarter wave plate, an additional phase difference of 90° is introduced between the *e*-ray and *o*-ray. Therefore, the total phase difference becomes 180° between the *e*-ray and *o*-ray. On emerging from the quarter plate, the *e*-and *o*-rays combine to produce plane-polarized light. If the light coming out of quarter wave plate is examined with an analyzer, light will be extinguished twice in one full rotation of the polarizer as shown in Fig. 11.20. On the other hand, if the intensity varies between a maximum and minimum the light is partially polarized.

11.11 PRODUCTION OF CIRCULARLY POLARIZED LIGHT

A quarter wave plate and a polarizer are the optical devices required for producing circularly polarized light from unpolarized light. Unpolarized light is first allowed to pass through a polarizer which converts it into plane polarized light as shown in Fig. 11.21. This plane polarized light then falls on a quarter wave plate. The axes of the polarizer and the quarter wave plate are rotated till the electric vector E of the plane-polarized wave makes an angle of 45° with the optic axis of the quarter wave plate.

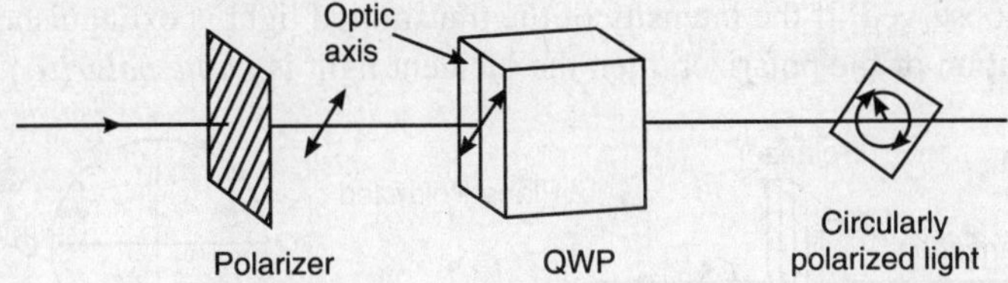

Fig. 11.21: Production of circularly polarized light.

The plane polarized wave incident on the quarter wave plate splits into two rays, *o*-ray and *e*-ray of equal amplitude ($E_1 \cos 45° = E_2 \sin 45°$). The two rays travel in the same direction inside the crystal but with different velocities. The two rays are in phase at the front face of the crystal but progressively get out of phase as they travel through the crystal. As they emerge from the rear face of the crystal, they will have a path difference of $\lambda/4$ or phase difference of 90°. The two rays are linearly polarized in mutually perpendicular directions. When they combine, they produce circularly polarized light.

11.11.1 Detection of Circularly Polarized Light

The light beam is allowed to pass through an analyzer (a Polaroid sheet or a Nicol prism). If on rotating the analyzing Polaroid sheet or Nicol, the intensity of the emerging beam remains uniform, then the incident light is circularly polarized. A similar result would be obtained if the incident light is ordinary unpolarized light. The two cases may be distinguished by inserting a quarter wave plate in the path of light before it falls on the analyzer.

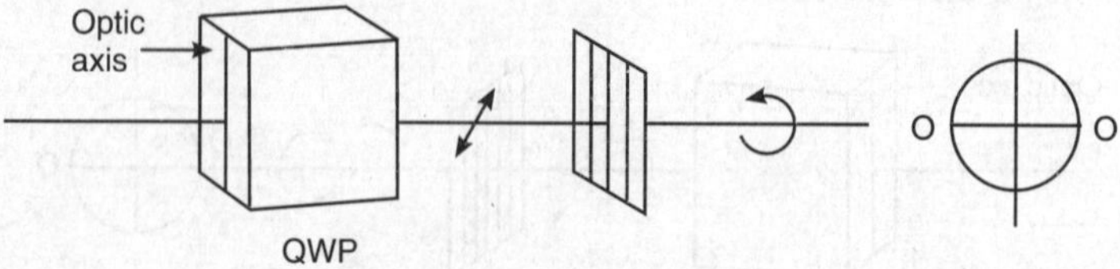

Fig. 11.22:
Detection of circularly polarized light.

If the original light is circularly polarized, it may be considered as resultant of two coherent plane-polarized waves that is *e*-ray and *o*-ray, which are out of phase by 90°. If the light passes through the quarter wave plate, an additional phase difference of 90° is introduced between the *e*-ray and *o*-ray. Therefore, the total phase difference becomes 180° between the *e*-ray and *o*-ray. On emerging from the quarter plate, the *e*- and *o*-rays combine to produce plane-polarized light. Therefore, if the light coming out of quarter wave plate is examined with an analyzer, light will be extinguished twice in one full rotation of the polarizer as shown in Fig. 11.22. If it is not extinguished,the incident light is unpolarized.

11.12 ANALYSIS OF POLARIZED LIGHT

In practice light may exhibit any one of the three types of polarization, or may be unpolarized or a mixed type. The unaided eye cannot distinguish the different types of polarization. However, using a polarizer and a quarter wave plate, the actual type of polarization of a light beam can be ascertained. The following steps are used in the analysis of the type of polarization.

(*i*) The light of unknown polarization is allowed to fall normally on a polarizer. The polarizer is slowly rotated through a full circle and the intensity of the transmitted light is observed. If the intensity of the transmitted light is extinguished twice in one full rotation of the polarizer, then the incident light is *plane polarized*.

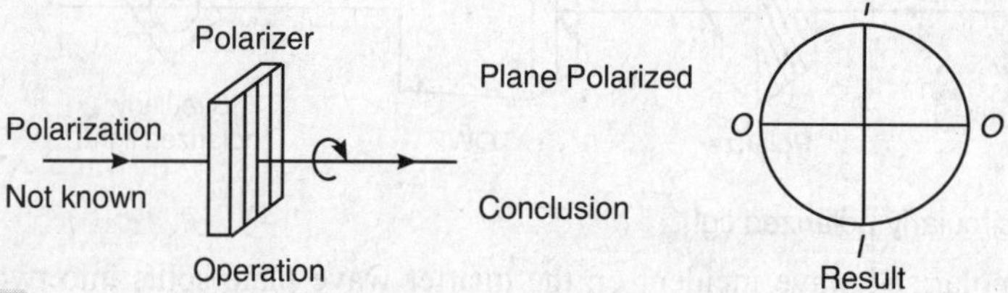

Fig. 11.23 (a):
Analysis of Polarized Light- Plane polarized.

(*ii*) If the intensity of the transmitted light varies between a maximum and a minimum value but does not become extinguished in any position of the polarizer, then the incident light is either elliptically polarized or partially polarized.

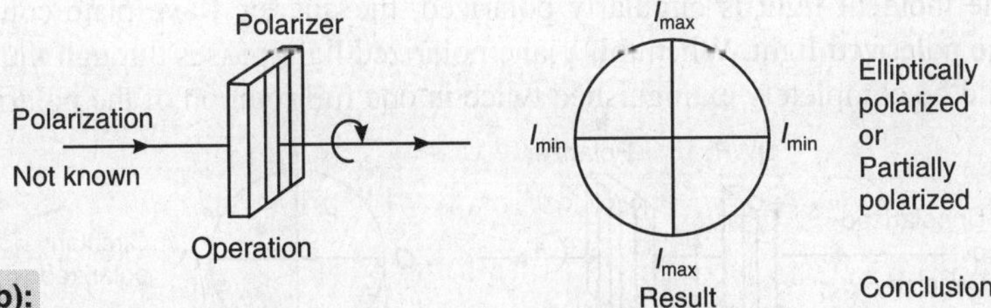

Fig. 11.23 (b): Analysis of Polarized Light- Partially polarized or elliptically polarized.

(*iii*) If the intensity of the transmitted light remains constant on rotation of the polarizer, then the incident light is either circularly polarized or unpolarized.

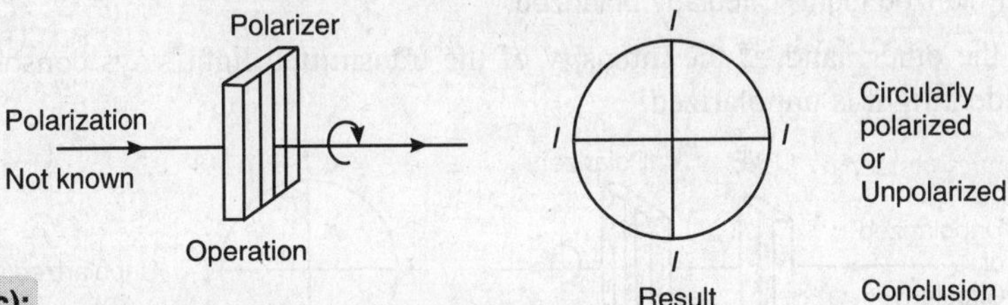

Fig. 11.23 (c): Analysis of Polarized Light- Circularly polarized.

To distinguish between elliptically polarized and partially polarized or between the circularly polarized and unpolarized light, we take the help of quarter wave plate. The light is first made to be incident on the quarter wave plate and then it is made to pass through the polarizer.

(*iv*) If the incident light is elliptically polarized, the quarter wave plate converts it into a plane polarized beam. When this polarized light passes through the polarizer, it would be extinguished twice in one full rotation of the polarizer.

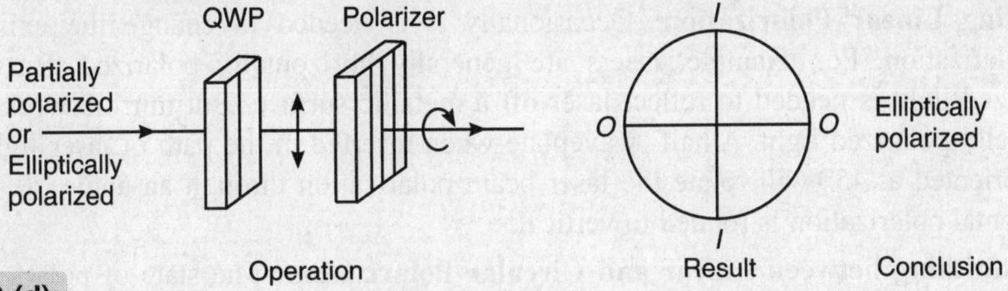

Fig. 11.23 (d): Analysis of Polarized Light- Elliptically polarized.

On the other hand, if the transmitted light intensity varies between a maximum and a minimum without becoming zero, then the incident light is partially polarized.

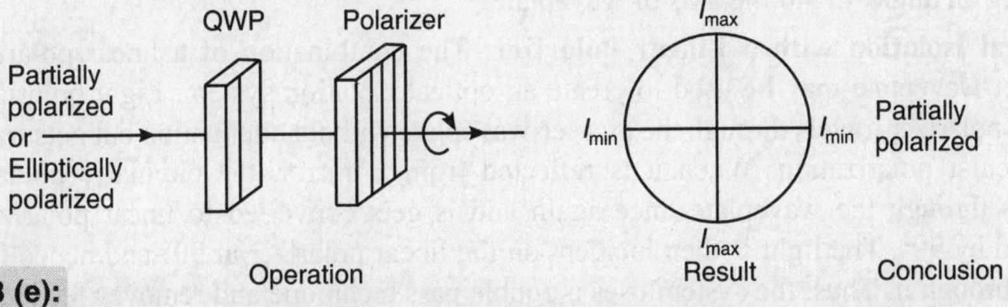

Fig. 11.23 (e): Analysis of Polarized Light- Partially polarized.

(*v*) If the incident light is circularly polarized, the quarter wave plate converts it into plane polarized light. When this plane polarized light passes through the polarizer, it would be completely extinguished twice in one full rotation of the polarizer.

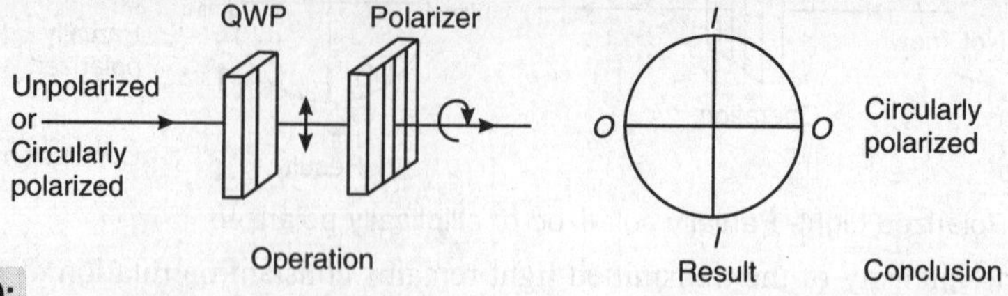

Fig. 11.23 (f):

Analysis of Polarized Light- Circularly polarized.

On the other hand, if the intensity of the transmitted light stays constant, then the incident light is unpolarized.

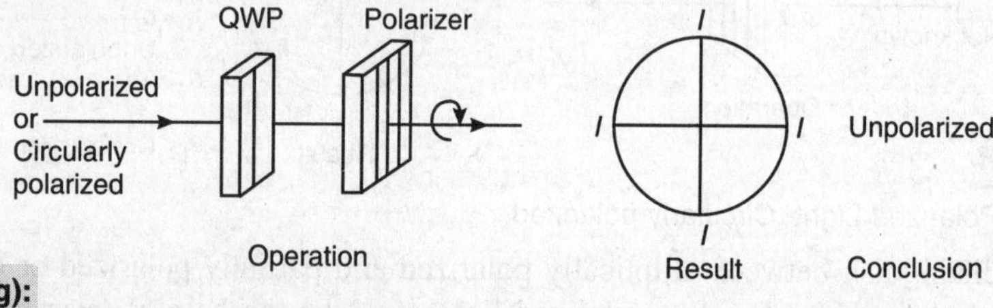

Fig. 11.23 (g):

Analysis of Polarized Light- Unpolarized.

11.13 APPLICATIONS

Waveplate applications include rotating linear polarization, or transforming linear polarization into circular polarization.

(*i*) **Rotating Linear Polarization:** Occasionally it is needed to change the existing state of polarization. For example, lasers are generally horizontally polarized. But vertically polarized light is needed to reflect laser off a metallic surface. But mirrors work best with vertically polarized light. A half- waveplate when inserted in the path of laser light with its axis oriented at 45° will rotate the laser beam polarization through an angle $2\theta = 90°$, i.e. horizontal polarization is rotated to vertical.

(*ii*) **Transforming between Linear and Circular Polarization:** The state of polarization of a linearly polarized light can be converted to circularly polarized light, and vice versa, using a linear polarizer and a quarter-waveplate. A quarter-waveplate with its axes oriented at 45° to linear polarization produces circular polarization. Conversely, circularly polarized light, going through a quarter-waveplate produces linear polarized light with its polarization making an angle 45° to the axis of waveplate.

(*iii*) **Optical Isolation with a Linear Polarizer:** The combination of a linear polarizer and a quarter-waveplate may be used to create an optical isolation system. Light polarized by the linear polarizer travels through the quarter-waveplate without attenuation but gets transformed to circular polarization. When it is reflected from a mirror, the circularly polarized light passes through the waveplate once again and is gets converted to linear polarization, but rotated by 90°. The light is then incident on the linear polarizer at 90° and hence it does not pass through it. Thus, the system uses a double pass technique and removes feedback.

(*iv*) **Enhancing visibility of digital displays:** Using circular polarizers the contrast of the digital displays can be improved. By keeping a sheet of circular polarizer placed in front of a digital display (Fig. 11.24), external light that falls on the polarizer becomes circularly polarized before it reaches the front of the display. The reflected light will have reversed state of its polarization. The polarizer blocks the reflected light. Light generated by the display passes through the polarizer and hence is seen without the background reflecting light.

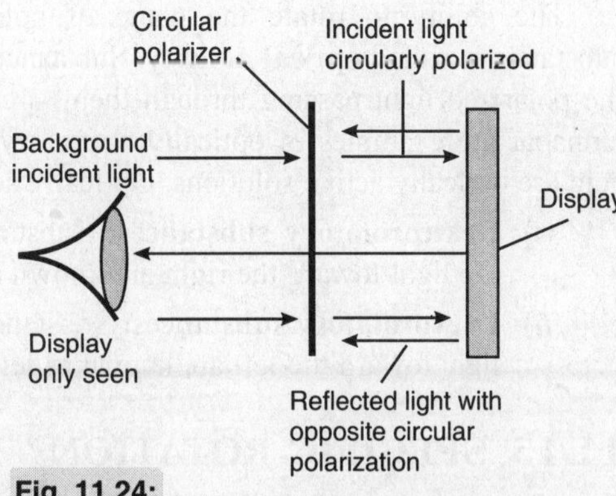

Fig. 11.24: Improving the visibility of digital display.

11.14 OPTICAL ACTIVITY

Certain crystals and solutions possess a natural ability to rotate the plane of polarization about the direction of propagation. It is known as *optical activity*. In case of crystals this ability arises due to the twisted arrangement of atomic layers with respect to one another. In liquids and solutions the optical activity is due to certain structural symmetry in their molecules. The optical activity found in bigger organic molecules provides a number of clues which help us understand biological activity. Some crystals induce birefringence with the application of a voltage or magnetic field and are useful as light-beam modulators, *Q*-switches, and deflectors.

11.14.1 Optical Rotation

When a beam of plane polarized light propagates through a quartz crystal along the optic axis, the plane of polarization steadily turns about the direction of the beam (Fig. 11.25). The optical rotation can be detected as follows. If two Polaroid sheets or Nicol prisms are held in crossed configuration and if a beam of unpolarized light is viewed through them, the field of view appears to be completely dark. Now let a quartz crystal, cut with its faces perpendicular to the optic axis, be inserted between the polarizers such that light is incident normally on the crystal. The field of view now appears lit up indicating that the light is not cut off by the analyzer. In order to cut off the transmitted light, we find that the analyzer is to be rotated through a certain angle. The experiment establishes that the plane polarized light produced by the polarizer remains plane polarized while passing through the quartz crystal but the plane of polarization is rotated through an angle. This angle is the angle through which the analyzer is rotated in order to cut off the light totally. The *optical rotation*, i.e., rotation of the plane of polarized light is shown in Fig. 11.25.

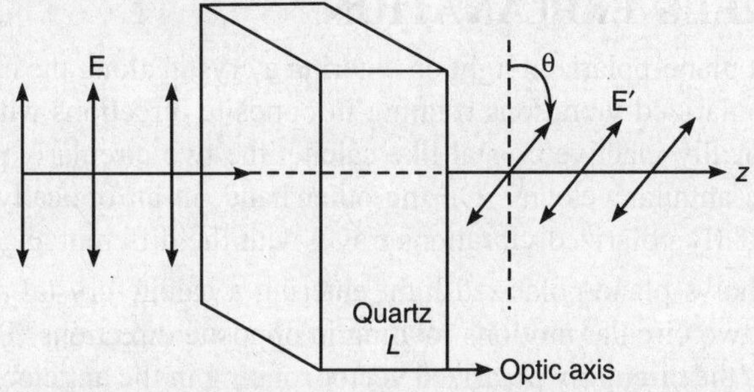

Fig. 11.25: Optical rotation.

The ability to rotate the plane of polarization of plane polarized light by certain substance is called **optical activity**. Substances, which have the ability to rotate the plane of the polarized light passing through them, are called ***optically active*** substances. Quartz and cinnabar are examples of optically active crystals while aqueous solutions of sugar, tartaric acid are optically active solutions. Optically active substances are classified into two types.

(*i*) **Dextrorotatory substances:** Substances which rotate the plane of polarization of the light toward the right are known as right-handed or dextrorotatory.

(*ii*) **Laevorotatory substances:** Substances which rotate the plane of polarization of the light toward the left are known as left-handed or laevorotatory.

11.15 SPECIFIC ROTATION

Specific rotation is a measure of the optical activity of a sample. It is the rotation produced for a 1 mm slab for a solid, or a 100 mm path length for a liquid. If an optically active material is kept between two crossed polarizers, the field of view becomes bright. In order to get darkness once again, the analyzer has to be rotated through an angle. The angle through which the analyzer is rotated equals the angle through which the plane of polarization is rotated by the optically active substance. This angle depends on

(*i*) thickness of the substance,

(*ii*) density of the material or concentration of the solution,

(*iii*) wavelength of light, and

(*iv*) temperature

The amount of rotation θ caused by crystalline materials is given by

$$\theta = \alpha l \quad ...(11.23)$$

where α is called the **rotational constant**.

In solutions the amount of rotation θ is given by

$$\theta = s\,c\,l \quad ...(11.24)$$

where *c* is the concentration and s is called the **specific rotation**.

Def: The specific rotation for a given wavelength of light at a given temperature is defined conventionally as the rotation produced by one decimetre long column of the solution containing 1 gm of optically active material per c.c .of solution.

$$[S]_{\lambda}^{t} = \frac{\theta}{l \times c} = \frac{\text{Rotation in degree}}{\text{Length in decimetres} \times \text{conc. in gm/c.c}} \quad ...(11.25)$$

11.16 FRESNEL'S EXPLANATION

Fresnel assumed that plane-polarized light on entering a crystal along the optic axis is resolved into two circularly polarized vibrations rotating in opposite directions with the same angular frequency. In an optically inactive crystal like calcite, the two circularly polarized vibrations travel with the same angular velocity. On the other hand, in an optically active crystal like quartz, the two circularly polarized vibrations travel with the different angular velocities.

Fig. 11.26 (*a*) shows plane polarized light entering a calcite crystal along the optic axis *AB* and split up into two circular motions rotating in opposite directions. They are represented by *OL* and *OR*. *OL* is the circularly polarized vector rotating in the anticlockwise direction and *OR* is the circularly polarized vector rotating in the clockwise direction. If *OL* and *OR* start simultaneously from OA and rotate with the same angular velocity, then at any subsequent time, the resultant of OL and OR will lie along OA. Hence on emerging from the crystal, the two circular waves combine to produce a linear vibration along the initial direction, OA (Fig. 11.26 *b*). Therefore, crystals like calcite do not rotate the plane of vibration.

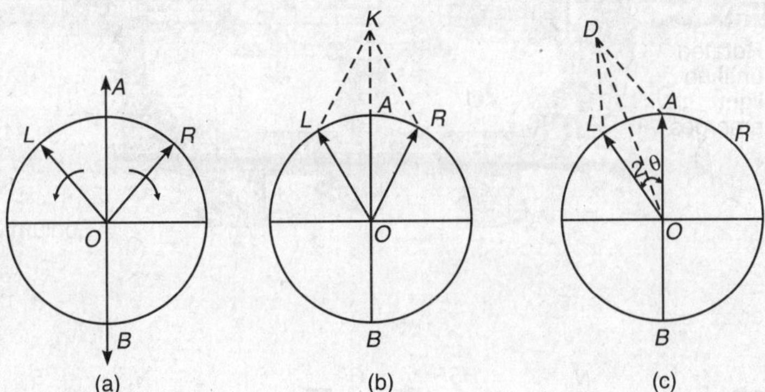

Fig. 11.26:

Fresnel's Explanation.

In case of quartz, the linearly polarized light, the component having clockwise rotation travels faster than the anticlockwise component. When the components emerge out of the crystal, they are at an angle 2θ. The resultant of these two vectors OR and OL is now along OD (Fig. 11.26 *c*). Before entering the crystal, the plane of vibration is along OA and after emerging from the crystal, it is along OD which makes an angle θ with the optic axis OA. After passing through the crystal, the circular components combine to produce a linear vibration whose direction is now along OD. Therefore, the plane of vibration has rotated through an angle θ. The angle through which the plane of vibration is rotated depends on the thickness of the crystal.

When left and right circularly polarized light travel through the crystal at different speeds, when they are combined together at any point results in a linear polarization. The angle of the polarization depends on the phase difference between the two circular polarizations. If μ_R and μ_L are the refractive indices corresponding to the right and left circularly polarized light, the angle of polarization, θ, is given by:

$$\theta = \frac{\pi}{\lambda}(\mu_R - \mu_L)d \qquad ...(11.26)$$

where d is the thickness of the material. As a result, the direction of polarisation spirals around z in space.

Helical molecules may interact with circularly polarized light differently. It depends on whether their chirality is the same handedness as the circular polarization or opposite handedness. So, there are only two refractive indices μ_R and μ_L.

11.17 POLARIMETER

A polarimeter [Fig. 11.27 (*a*)] is an instrument used for determining the optical rotation of solutions. When used for determining the optical rotation of sugar it is called a saccharimeter.

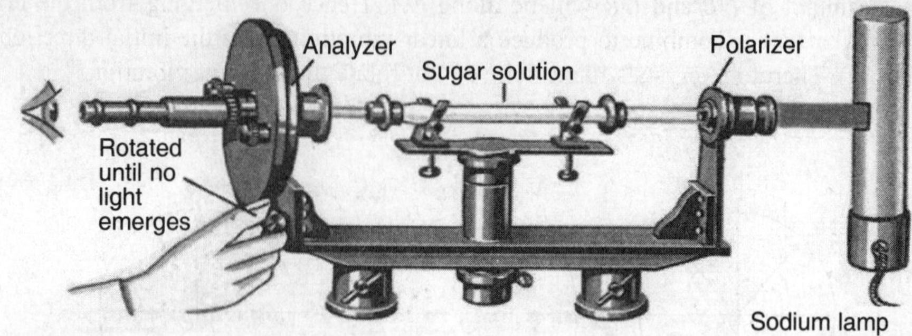

Fig. 11.27 (a): Saccharimeter.

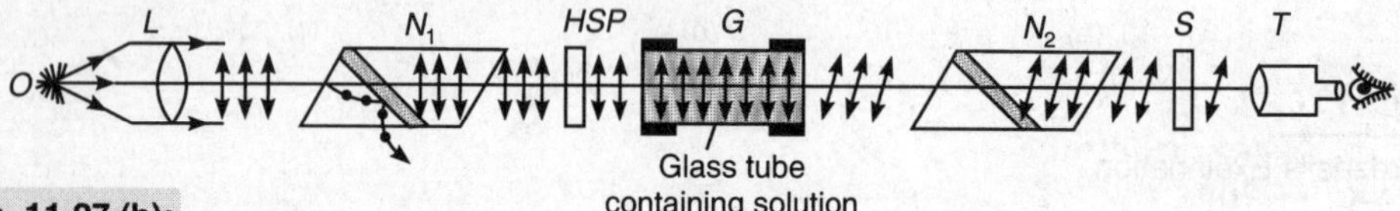

Fig. 11.27 (b): Schematic of Saccharimeter (Polarimeter).

Construction: A polarimeter consists of a glass tube for holding the solution under test held between crossed Nicol prisms. Beyond the polarizing Nicol prism a Half-Shade Plate (HSP) is located which is used for accurately adjusting the two Nicols for crossed position [Fig. 11.27 (*b*)]. Light from a monochromatic source is made parallel by the lens L and is incident on the polarizer, N_1. The light coming out of the polarizer is plane polarized. The polarized beam then passes through the half-shade plate and a glass tube G containing the solution. The light emerging from the solution will be incident on the analyzer N_2. The light is observed through a telescope T. The analyzing Nicol N_2 can be rotated about the axis of the tube and the rotation can be measured with the help of a graduated circular scale.

Working: To find the specific rotation of a solution, the analyzer is first adjusted such that field of view is completely dark. Then the glass tube is filled with the solution and is held in position. The field of view now becomes illuminated. The field of view can again be made dark by rotating the analyzer through a certain angle which gives the optical rotation of the solution. The practical difficulty in this method is in the determination of the exact position for which complete darkness is achieved. The difficulty is overcome by using what is known as a Laurent's half-shade device.

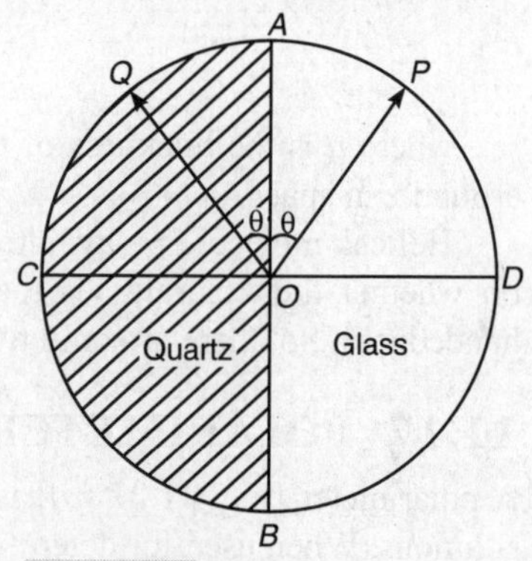

Fig. 11.28: Half shade plate.

It (Fig. 11.28) consists of a semicircular half wave plate ACB of quartz cemented to a semicircular plate ADB of glass. The optic axis of the wave plate is parallel to the line of separation AB. The half wave plate introduces a phase difference of 180º between *e*-ray and *o*-ray passing through it. The thickness of the glass plate is such that it transmits the same amount of light as done by the quartz half wave plate. One half of the incident light passes through the quartz plate *ACB* and the other half through the glass plate *ADB*. The light after passing through the polarizer is incident normally on the half shade plate and has vibrations along *OP*. On passing through the glass, half the vibrations will remain along *OP* but on passing through the quartz half, the vibrations will split into *e*- and *o*-rays. The *o*-vibrations are along *OD* and *e*-vibrations are along *OA*. The half wave plate introduces a phase difference of π rad between the two vibrations. The vibrations of *o*-ray will occur along *OC* instead of *OD* on emerging from the plate. Therefore the resultant vibration will be along *OQ* whereas the vibrations of the beam emerging from glass plate will be along *OP*. In effect, the half wave plate turns the plane of polarization of the incident light through an angle 2θ.

If the principal plane of the Nicol N_2 is aligned parallel to *OP*, the plane polarized light emerging from the glass tube will pass through the glass plate of the half shade plate and that part appears brighter. On the other hand light coming out of the quartz plate is partially obstructed and the corresponding field of view appears less bright. If the principal plane of N_2 is parallel to *OQ* the quartz half will appear brighter than the glass half. Thus, the two halves of the plate are unequally illuminated. When the principal plane of N_2 is parallel to *AB*, the two halves appear equally bright and when it is parallel to *CD*, the two halves are equally dark.

To find the specific rotation of a solution, the analyzer is first in the position for equal darkness without solution in the tube *G*. The reading on the circular scale is noted. Next, the tube is filled with the optically active solution of known concentration. The field of view is now partially illuminated. The analyzer is rotated till the field of view becomes equally dark. The reading on the circular scale is noted again. The difference between the two scale readings gives the angle of rotation of the plane of polarization caused by the solution. Knowing the values of θ, *l*, and *c*, the specific rotation is obtained using the Eq. (11.23). Or otherwise, knowing the value of the specific rotation, the concentration of the solution can be determined with the help of the Eq. (11.25).

Applications:

Polarimetry is useful as a quality parameter for raw ingredients and finished products in food, drug, and flavors industries. Fine quality standards can be maintained in process industries and formulators.

Example 11.9: *The rotation in the plane of polarization in a certain substance is 11°/cm. Calculate the difference between the refractive indices for right and left circularly polarized light in the substance. Given λ = 5893 Å.*

Solution: $$\theta = \frac{\pi}{\lambda}[\mu_R - \mu_L]d \qquad \therefore\ [\mu_R - \mu_L] = \frac{\theta\lambda}{\pi d}$$

It is given that $\frac{\theta}{d} = 10° = \frac{10\times\pi}{360°} = \frac{\pi}{36}$ radian/cm and $\lambda = 5893\ \text{Å} = 5893\times10^{-8}$ cm.

$$\therefore \qquad \mu_R - \mu_L = \frac{\pi}{36}\cdot\frac{5893\times10^{-8}\,\text{cm}}{\pi} = \mathbf{1.6\times10^{-6}}.$$

Example 11.10: *The indices of refraction of quartz for right-handed and left-handed circularly polarized waves of wavelength 7620 Å traveling in the direction of optic axis have the following values $\mu_R = 1.53914$ and $\mu_L = 1.53920$*

Calculate the rotation of the plane of polarization of light in degrees produced by a plate of 0.5 mm thick.

Solution: $\theta = \frac{\pi}{\lambda}[\mu_R - \mu_L]d = \frac{3.14 \times 0.5 \times 10^{-3}\,\text{m}}{7620 \times 10^{-10}\,\text{m}}(1.53920 - 1.53914)$

$$= 0.1236 \text{ radian} = \frac{0.1236 \times 180°}{\pi} = \frac{0.1236 \times 180°}{3.14}$$

$$= \mathbf{7°5'}.$$

Example 11.11: *A 200 mm long tube containing 48 cm³ of sugar solution produces an optical rotation of 11° when placed in a saccharimeter. If the specific rotation of sugar solution is 66°, calculate the quantity of sugar contained in the tube in the form of a solution.*

Solution: It is given that $\theta = 11°$, $l = 200$ mm = 20 cm, $S = 66°$, and $V = 48$ cm³.

$$C = \frac{10\theta}{1S} = \frac{10 \times 11°}{20 \text{ cm} \times 60°} = 0.0833 \text{ g/cm}^3$$

Mass of sugar in solution $\quad M = CV = 0.0833 \text{ g/cm}^3 \times 48 \text{ cm}^3 =$ **4 grams.**

Example 11.12: *A 20 cm long tube having sugar solution is placed between crossed Nicols and light of λ = 6000 Å. If the specific rotation is 60º and optical rotation is 12º, what is the strength of the solution?*

Solution: The specific rotation at a given temperature and for a given wavelength is given by

$$C = \frac{10\theta}{1S} = \frac{10 \times 12°}{20 \text{ cm} \times 60°} = \mathbf{0.1}$$

Therefore, it is 10% solution of sugar, i.e., 1 gm sugar is dissolved in 10 gm of water.

Example 11.13: *20 cm length of optically active solution A gives right-handed rotation of 40º and 30 cm of solution B gives left-handed rotation of 24º. Find the optical rotation produced by 30 cm length of the mixture of chemically not reacting solutions A and B in volume ratio 1:2.*

Solution: As the length of the mixtures is 30 cm and the solutions are in the volume ratio 1 : 2, we may assume that 10 cm length is of the first solution and 20 cm length is of the second solution.

The optical rotation produced by the first solution $= 40° \times \frac{10 \text{ cm}}{20 \text{ cm}}$

$$= 20° \text{ (right-handed)} = -20°.$$

The optical rotation produced by the first solution $= 24° \times \frac{20 \text{ cm}}{30 \text{ cm}}$

$$= 16° \text{ (left-handed)} = +16°$$

∴ Total optical rotation $= -20° + 16° = -4°$.

∴ The resultant optical rotation is **4° right-handed.**

QUESTIONS

1. Explain polarization by double refraction.
2. What is an optic axis?
3. Distinguish between: Ordinary ray and extraordinary ray.
4. Distinguish between: Positive and negative crystals.
5. Distinguish between: Uniaxial and biaxial crystals.
6. Describe a Nicol prism and explain how it acts as an analyzer.
7. What are the differences between positive and negative crystals?
8. Explain the phenomenon of double refraction on the basis of Huygen's wave theory of light.
9. Describe the principle, construction and working of a Nicol prism with a neat diagram.
10. Give the construction and working of Nicol prism.
11. How is Nicol prism used as an analyzer?
12. How Nicol prism can be used as polarizer and analyzer? Explain in detail with the help of diagram.
13. Explain the state of polarization when the *x* and *y* components of the electric field are given by
    (*i*) $E_x = E_0 \cos(wt + kz), E_y = \frac{E_0}{2}\cos(wt + kz + \pi)$
    (*ii*) $E_x = E_0 \sin(wt + kz)$, $E_y = E_0 \cos(wt + kz)$.
14. What is quarter plate?
15. Give the construction and theory of half wave plate.
16. Define a quarter wave plate and a half wave plate.
17. Describe the action of half wave plate on plane polarized light incident on it with it its electric vector E making an angle θ with its optic axis.
18. What do you understand by an ordinary ray and extraordinary ray? How do these differ from each other?
19. What are a quarter wave plates?
20. Derive the thickness of quarter-wave plate for a given wavelength in terms of its refractive indices.
21. What are a quarter wave plates? Explain how it is used in producing elliptically and circularly polarized light.
22. If a quarter wave plate and half wave plate are given to you, how would you proceed to distinguish them from each other?
23. What are the optical devices require producing circularly polarized light from unpolarized light? Explain how they are used to produce circular and elliptical polarized light.
24. What is circularly polarized light? How is it produced?
25. What is elliptically polarized light? How is it produced?
26. How would you produce and detect the following with the help of a Nicol prism and a quarter wave plate:
    (*i*) Plane polarized (*ii*) circularly polarized (*iii*) elliptically polarized light.
27. What is meant by circularly and elliptically polarized light? How would you produce these lights?
28. What will be the state of polarization of the transmitted light when a beam of (*a*) plane polarized light is passed through a quarter wave plate, (*b*) circularly polarized light is passed through a quarter wave plate, (*c*) elliptically polarized light is passed through a quarter wave plate, (*d*) unpolarized light is passed through a quarter wave plate.

29. Distinguish between (*i*) circularly polarized light and unpolarized light and (*ii*) elliptically polarized and partially polarized light?
30. Explain production of plane polarized and circularly polarized light.
31. Explain how you will distinguish between unpolarized light and circularly polarized light.
32. Explain how the analysis of the type of polarization of light beam can be ascertained using a polarizer and a quarter plate.
33. Explain production of elliptically and circularly polarized light.
34. Explain some of the applications of polarized light.
35. Explain the construction and working of Laurent's half shade polarimeter.
36. Discuss the Fresnel's theory of rotation of the plane of polarisation.

PROBLEMS

1. Find the minimum thickness of a positive crystal to act as quarter wave plate for light of λ = 6000 Å with μ_e = 1.31 and μ_o 1.309.
2. Calculate the thickness of a quarter wave plate for light of wavelength 5890 Å and indices of refraction for *o*-ray 1.658 and for *e*-ray 1.486.
3. Find the thickness of a half wave plate for light of wavelength 5000 Å and indices of refraction for *o*-ray 1.553 and for *e*-ray 1.544.
4. Find the thickness of plate producing a path difference of $\lambda/4$ for light of wavelength 5890 Å and indices of refraction for *o*-ray 1.530 and for *e*-ray 1.540.
5. Find the thickness of a plate that can change plane polarized into circularly polarized light of wavelength 5890 Å and indices of refraction for *o*-ray 1.658 and for *e*-ray 1.486.
6. A half wave plate is designed from a crystal for λ = 600 nm. If $(\mu_o - \mu_e)$ = 0.0057, calculate the thickness of the plate. **[Ans:** 52 μm]
7. Find the thickness of a mica sheet required for making a quarter wave plate for λ = 500 nm with μ_e = 1.586 and μ_o = 1.592? **[Ans:** 20 μm]
8. Two Nicol prisms are so arranged that the amount of light transmitted through them is a maximum. What will be the percentage reduction in intensity of the transmitted light when the analyzer is rotated through (*a*) 30° (*b*) 90°? **[Ans:** 25%, 100%]
9. Plane polarized light of λ = 500 nm is incident on a quartz crystal parallel to the optic axis. Calculate minimum thickness for which the *o*-ray and *e*-rays come together to form plane-polarized light for μ_o = 1.5442, μ_e = 1.5533. **[Ans:** 27.5 μm]
10. Find the specific rotation of a 10 cm long 10% sugar solution in a polarimeter that produces optical rotation of 13.2°. **[Ans:** 66°]
11. Calculate the specific rotation of a 20 cm long 20% sugar solution in a polarimeter that produces optical rotation of 26.4°. **[Ans:** 66°]
12. A length of 25 cm of a solution of 50 gm of solute per liter gives a rotation of the plane of polarization by 5°. Calculate the rotation of plane of polarization by a length of 75 cm of a solution containing 100 gm of solute per liter. **[Ans:** 30°]
13. For quartz the refractive indices for right-handed and left-handed vibrations are 1.55810 and 1.55821 respectively for λ = 4000 Å. Calculate the optical rotation at λ = 4000 Å by a plate of quartz 2 mm thick and with its faces perpendicular to the optic axis. **[Ans:** 98°]
14. A solution of camphor in alcohol in a tube of 20 cm long is found to rotate the plane of vibration of light by 27°. What is the mass of the camphor in unit volume of solution? The specific rotation of camphor is +54°. **[Ans:** 0.25gm/cc]

15. A plate of quartz cut with its faces perpendicular to the optic axis is found to annul exactly the rotation of plane of polarization of sodium light produced by a 30 cm length of 18% solution of lactose. Find the thickness of the quartz plate. Given that specific rotation of lactose is 52.53° and that 1 mm quartz rotates the plane of polarization of sodium light by 21.71°.

[**Ans:** 1.31 mm]

16. A sugar solution in a tube of length 20 cm gives rotation of 13°. Find the optical rotation produced by 30 cm long tube containing the solution diluted to one-third of its previous concentration.

[**Ans:** 6.5°]

17. 80 gm of impure sugar when dissolved in a litre of water gives an optical rotation of 99° when placed in a tube of length 20 cm. If the specific rotation of sugar is 66°, find the percentage purity of the sugar sample. [**Ans:** 93.75%]

18. The rotation in the plane of polarization in a certain substance, at 5893 Å, is 10°/cm. Calculate the difference between the refractive indices for right and left circularly polarized light in the substance. [**Ans:** 8.186×10^{-7}]

19. For a given wavelength 1 mm of quartz cut perpendicular to the optic axis rotates the plane of polarization by 20°. Find for what thickness will no light of this wavelength be transmitted when the quartz pieces is interposed between parallel Nicols. [**Ans:** 4.5 mm]

12 Artificial Double Refraction

Learning Objectives:

By the end of this chapter, you will be able to:

- Learn about electro-optic effects in physics.
- Understand impact of KERR effect and POCKELS effect.
- Explain the concept of magneto-optic effects.
- Describe the stress induced in birefringence.
- Learn about fundamentals of photoelasticity.

12.1 INTRODUCTION

Isotropic transparent materials such as glass do not exhibit double refraction under ordinary circumstances. However, they acquire the property of double refraction under the influence of an electrical field, magnetic field, mechanical stress and under similar conditions. It is known as ***artificial double refraction*** or ***induced birefringence***. In these cases, the medium becomes optically anisotropic and the optic axis is parallel to the direction of external field. The action of the external field is to cause distortion of the molecular arrangement within the material and thereby transform the isotropic substance into an *anisotropic* substance. The induced birefringence disappears as soon as the external force ceases to act.

The change in birefringence inducted by electric field is called ***electro-optic*** effect and the change in birefringence inducted by magnetic field is called ***magneto-optic effect***. The electro-optic effect can be exploited to electronically controling the speed of light through a material. The ability to modulate light at high speed is used in an exploding number of applications, including encoding information onto optical beams for fiber-optic communication, fast shutters for high-speed photography, Q-switching of lasers for ultra-short pulses, laser mode locking, and even for steering and deflecting optical beams.

12.2 ELECTRO-OPTIC EFFECTS

Linear optics assumes that the response of a material to an applied field, the polarizability is simply proportional to the applied field.

$$P = \varepsilon_0 \chi E \quad ...(12.1)$$

Strong electric fields applied to a material can cause higher-order changes in the polarizability. If a strong *DC* field is applied across a material, the internal charges will be displaced from their equilibrium positions. If a light wave now passes through the material, these charges will have less freedom to respond to the oscillations of the electric field of the light wave.

Def: Electro-optic effects refer to the changes in the polarizability and therefore refractive index of the material by the presence of an external electric field. These changes will be inherently anisotropic, based on the direction of the external field.

Opto-electric materials useful in preparing:

(*i*) Lens of controllable focal length- lens whose refractive index can be varied

(*ii*) Optical scanning devices- prism whose beam bending ability is controllable

(*iii*) Optical phase modulators- transparent plate whose phase shift can be controlled

The change in refractive index due to an external field can be expressed as a Taylor series expansion as

$$\Delta\mu = a_1 E + a_2 E^2 + ... \quad ...(12.2)$$

In practice, only the first two terms of the expansion are observed in materials. Therefore, considering the first two terms, the dependence of the refractive index on the applied electric field produces one of two following effects:

(*i*) ***Pockels effect:*** The refractive index changes linearly with the strength of the applied electric field. The effect is also known as the *linear electro-optic effect*.

(*ii*) ***Kerr effect:*** The refractive index changes in proportion to the square of the applied electric field. The effect is known as the *quadratic electro-optic effect*.

12.3 KERR EFFECT

Artificial double refraction induced in an isotropic liquid under the influence of an electric field is known as the Kerr effect. John Kerr discovered it in 1875. If a light beam is passed through an isotropic dielectric material placed in an electric field E perpendicular to the field direction, then the material displays induced birefringence. *O*-ray and *e*-ray propagate in the same direction perpendicular to E but have different refractive indices.

Birefringence $$\Delta\mu = K\lambda E^2 \quad ...(12.3)$$

where K is the Kerr coefficient,

λ is the wavelength of light beam,

E is the applied electric field.

As the Kerr effect is a third order effect it is also known as the quadratic electro-optic effect. The Kerr effect can be generated in all materials. A Kerr Cell (Fig. 12.1) consists of a liquid tank, made of glass, or a block of a material having a fairly high Kerr coeffcient. A liquid, such as nitrobenzene, is generally filled in the tank. The tank houses two flat electrodes, of length L, held strictly parallel to each other. By connecting a source of dc voltage to the

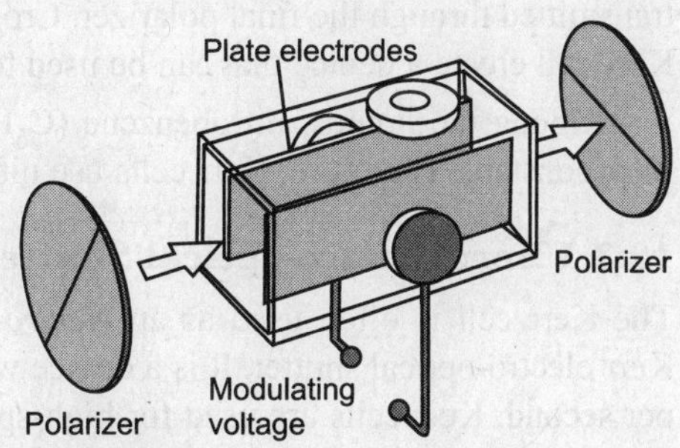

Fig. 12.1: Kerr Cell.

electrodes, an electric field V/d is applied in the vertical (z) direction. Monochromatic light beam from a source passes horizontally through the device.

The Kerr cell is filled with nitrobenzene and kept in between crossed polarizers (Fig. 12.2). The polarizers are placed at an angle of ±45 degrees to the optical axis. With no voltage applied to the cell, there is no relative phase change occurring within the cell, and the crossed polarizers block all light. When the electric potential is applied, the molecules of the liquid tend to align along the field direction. As the molecules are asymmetric, the alignment causes anisotropy and the liquid becomes double refracting.

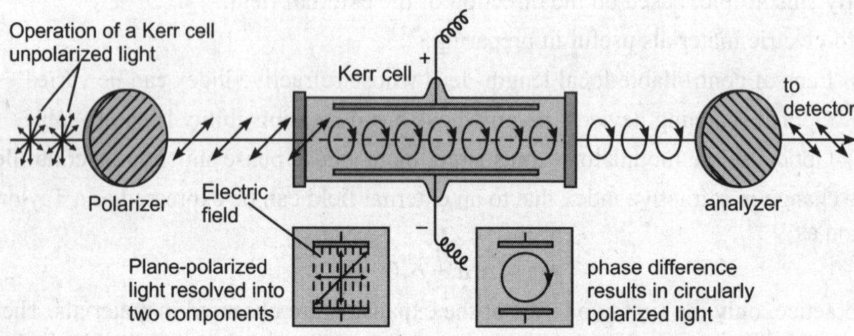

Fig. 12.2:

Operation of a Kerr cell.

Consequently, the incident plane-polarized light gets resolved into two components parallel and perpendicular to the electric field. As the components travel through the liquid with different speeds, they develop a phase difference. The relative phase change introduced between horizontal and vertical polarizations is given by

$$\Delta\varphi = (2\pi\Delta m/\ \lambda)\ L = [2\pi\ (KE^2\ \lambda)\ /\ \lambda]\ L = 2\pi LKV^2/d^2 \quad ...(12.4)$$

Thus, applying an electric field to the cell causes it to behave as a variable wave plate. To create a half-wave plate (phase change of $\pi/2$) requires an applied voltage of $V_{\lambda/2} = d/\sqrt{2LK}$ where $V_{\lambda/2}$ is characteristic quantity known as the half-wave voltage. For a typically useful device with an aperture of 2 mm, the required field strength could be in the 100 kV or MV range.

When the light components emerge from the cell, they combine to form circularly polarized light. Depending on the phase difference a variable amount of light will be transmitted through the final polarizer. Crossed linear polarizers at the entrance and exit to the Kerr cell create a device that can be used to modulate the amplitude of light.

Among the liquids, nitrobenzene ($C_6H_5NO_2$) is found to have the highest value for the Kerr constant. Therefore, Kerr cells use nitrobenzene.

12.3.1 Kerr Electro-optical Shutter

The Kerr cell is often used as an ***electro-optical shutter*** and is therefore known also as a Kerr electro-optical shutter. It is a device which can interrupt a beam of light up to 10^{10} times per second. Kerr cells are used for high-speed shutters in photography, electronic light-beam "choppers" used to replace spinning toothed wheels, and also as Q-switches for pulsed lasers. They are also used in communication studies.

Advantages and Disadvantages:

The huge advantage of Kerr cell modulators is the high modulation speed that can be attained. They can respond at speeds approaching 10 GHz.

The drawback of the Kerr cell is that the second-order effect is extremely weak in most materials. Practical devices require the generation of large electric fields (such as 100 kV or more), which is inconvenient. Often nitrobenzene ($K = 2.44 \times 10^{-12}$ m/V^2) or nitrotoluene ($K = 1.37 \times 10^{-12}$ m/V^2) are used as a Kerr medium to reduce the required field strength. However, they have the disadvantage of being toxic and explosive.

12.4 POCKELS EFFECT

If a light beam is passed through a material placed in an electric field E parallel to the field direction, then the material displays induced birefringence. O-ray and e-ray propagate in the same direction parallel to E but with different refractive indices.

Birefringence $$\Delta\mu = \mu_0 K_p E \quad \text{...(12.5)}$$

where K_p is the Pockels coefficient, and

μ_0 is the index of refraction of the material with no field applied.

As the Pockels effect is a 2nd order effect it is also known as the linear electro-optic effect (LEO). Isotropic media or media with inversion symmetry does not show Pockels effect. As electro-optic modulators use the Pockels effect, electro-optic modulators are often called Pockels cells.

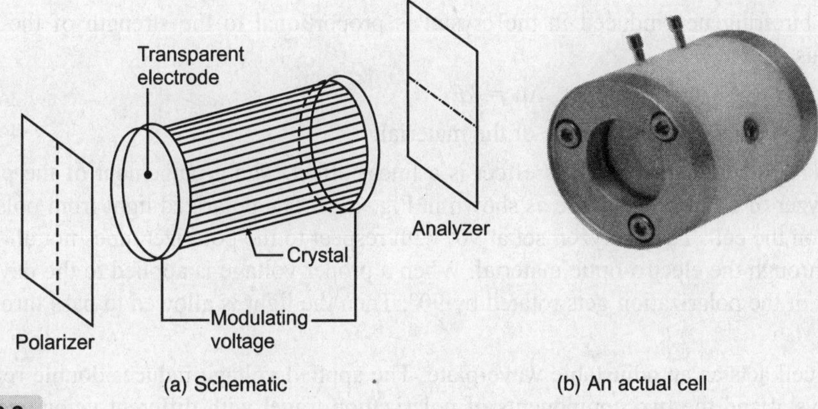

Fig. 12.3: Pockels cell.

Normally, piezoelectric crystals are birefringent but in certain directions do not exhibit double refraction. When an electric field is impressed along these directions, double refraction is induced along these directions also. Examples of piezoelectric crystals are lithium niobate ($LiNbO_3$), lithium tantalate ($LiTaO_3$), potassium di-deuterium phosphate (KD*P), β-barium borate (BBO), potassium titanium oxide phosphate (KTP), and compound semiconductors such as gallium arsenide (GaAs) and indium phosphide (InP).

A Pockels cell (Fig. 12.3) consists of a single, aligned crystal of an appropriate material such as KDP or $LiNbO_3$. Transparent electrodes (thin conducting coatings of tin oxide or indium) are deposited on opposite sides of the crystal. The crystal is oriented in such a way that its optic axis lies along the direction of the electric field applied between the electrodes.

The transparent electrodes ensure free propagation of light through the crystal. A Pockels set up is shown in Fig. 12.4.

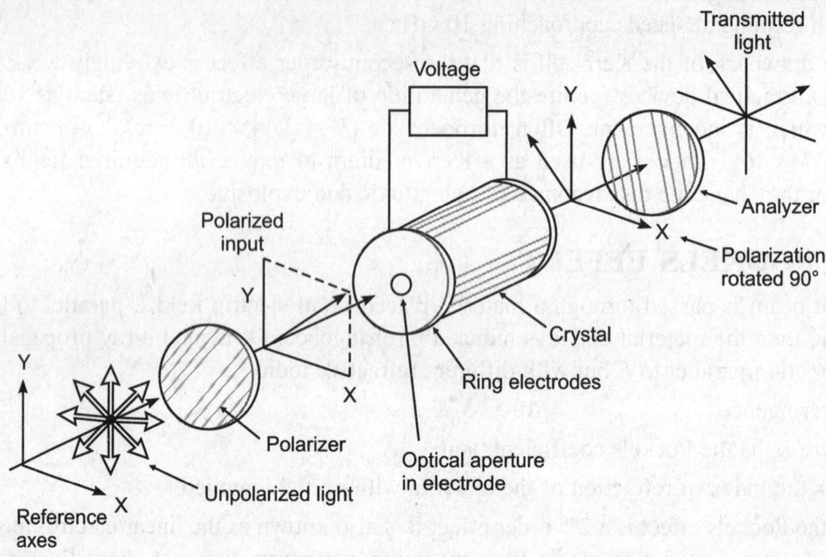

Fig. 12.4: Operation of Pockels cell.

The birefringence induced in the crystal is proportional to the strength of the applied field. Thus,

$$\Delta\mu = kE \qquad \text{...(12.6)}$$

where k is a constant characteristic of the material.

Eq. (12.6) shows that Pockels effect is a linear effect. The arrangement of the polarizer and analyzer of a Pockels cell are as shown in Fig. 12.4. The polarized light from polarizer is incident on the cell. The analyzer, set at 90º with respect to the polarizer, does not allow light to pass through the electro-optic material. When a proper voltage is applied to the device, the direction of the polarization gets rotated by 90°. Then the light is allowed to pass through the analyzer.

The cell acts as an adjustable wave-plate. The applied voltage induces double refraction in the crystal and the two components of polarization travel with different velocities within the crystal. As they travel through the crystal, they develop a phase difference progressively depending on the distance traveled and on the applied voltage. If the polarization of the emerging beam amounts to one-half wavelength (λ/2), the polarization is rotated by 90º from its original direction.

The intensity of the beam with the analyzer

$$I = I_0 \sin^2 (\pi \, \Delta\mu L/\lambda)$$

where I_0 the intrinsic intensity, $\Delta\lambda$ is the birefringence, L the length of the crystal, and λ the wavelength of the light.

The birefringence increases with the applied voltage. It is seen from the above expression that the intensity of light transmitted by the device will show an oscillatory behaviour as the applied voltage is increased. The transmission attains a maximum when

$$\Delta\mu = \lambda/2L \qquad ...(12.7)$$

It occurs at a voltage called the *half-wave voltage*, denoted $V_{1/2}$. The half-wave voltage depends on the nature of the electro-optic material and for a particular substance it is directly proportional to the wavelength.

Advantages of Pockels cells:

(*i*) Can be operated at substantially lower voltage 5 - 10 times less than a Kerr cell of the same dimensions.

(*ii*) single ionic crystal at their core is not toxic and explosive like the Kerr cell materials and can be operated at quite high speeds.

(*iii*) can be monolithically assembled into "integrated optics", particularly Lithium Niobate.

Applications:

(*i*) Pockels cells are used extensively in fiber-optics communication for modulating either the phase or amplitude of light before it enters the fiber.

(*ii*) Pockels cells are widely used as electro-optic shutters in *Q*-switching of lasers.

(*iii*) With the ability to control light beams they can be light-beam modulators, *Q*-switches, and deflectors.

12.5 MAGNETO-OPTIC EFFECTS

12.5.1 Cotton-Mouton Effect

The Cotton-Mouton effect is a magneto-optic effect. An isotropic material acquires the optical behaviour of a uniaxial crystal and exhibits double refraction under the action of an external magnetic field. The French scientists *A*. Cotton and *H*. Mouton studied the effect in detail in 1907. The set-up is shown in Fig. 12.5. To observe the Cotton-Mouton effect, monochromatic light that is linearly polarized in a plane forming a 45° angle with the direction of the magnetic field is passed through a transparent isotropic specimen placed between the poles of a strong electromagnet.

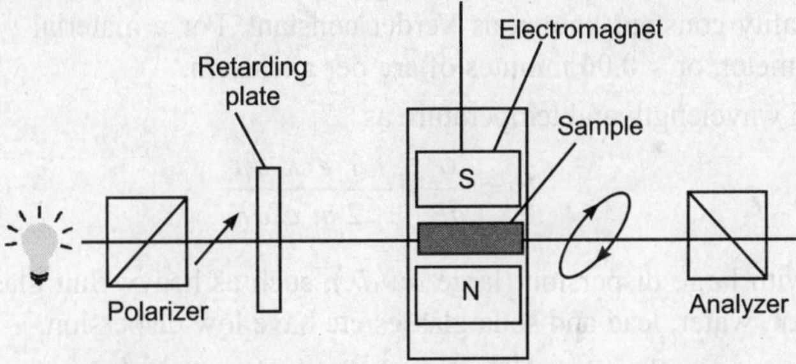

Fig. 12.5: Setup of Cotton-Mouton effect.

The material of the specimen becomes optically anisotropic in the magnetic field (its optical axis is parallel to the magnetic field *B*), and the light becomes elliptically polarized because it is propagating in the substance as two waves, the ordinary and extraordinary waves, which have different phase velocities. The difference in the indices of refraction of the ordinary beam μ_0 and the extraordinary beam μ_e, is called the induced birefringence. It is given by

$$\mu_e - \mu_0 = CB^2 \lambda \qquad ...(12.8)$$

where B is the magnetic field strength, C is a constant that depends on the substance and is called the Cotton-Mouton constant, and λ is the wavelength of the light.

The magnitude of the induced birefringence is usually very small. It may be seen that the Cotton-Mouton effect is a *quadratic effect*, the induced birefringence being proportional to the square of the magnetic field B.

12.5.2 Faraday Effect

Faraday rotation or Faraday Effect occurs in most optically transparent dielectric materials when they are subject to strong magnetic fields. It is a linear effect, the induced birefringence being proportional to the magnetic field B.

If a light beam is passed through a transparent solid or liquid dielectric placed in a uniform magnetic field parallel to the magnetic field direction, the transmitted light plane of polarization is rotated by an angle proportional to the field intensity. This optical rotation is called the ***Faraday Effect***. Fig. 12.6 shows Faraday Effect arrangement.

Becquerel's formula gives the relation between the angle of rotation of the polarization and the magnetic field as

$$\theta = VlB \qquad ...(12.9)$$

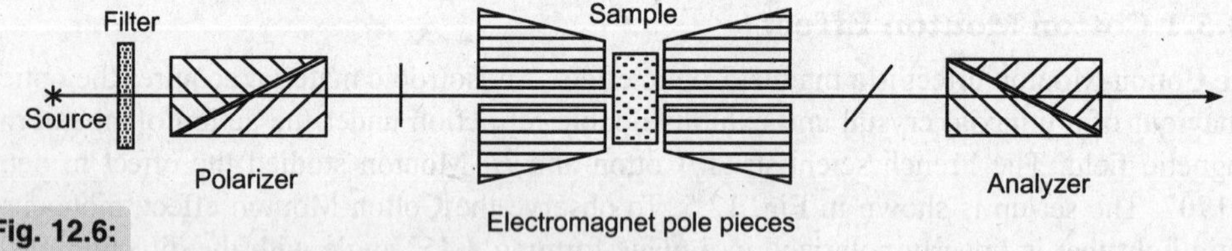

Fig. 12.6: Faraday effect arrangement.

where θ is the angle of rotation, l the length of the sample, B the manetic field and V is the proportionality constant known as Verdet constant. For a material typically, $V \approx 17.6$ radians per tesla-meter, or ≈ 0.06 minutes of arc per gauss-cm.

V varies with wavelength and temperature as

$$V = \frac{\phi}{dB} = -\frac{1}{2}\frac{e}{m}\frac{\lambda}{c}\frac{d\mu}{d\lambda} \qquad ...(12.10)$$

Substances with large dispersion (large $d\mu/d\lambda$), such as heavy flint glasses show a large Faraday Effect. Air, water, lead and soda glasses etc have low dispersion.

Difference between Faraday Effect and Optical Activity: In a sugar solution, the optical rotation occurs in the same direction, whatever may be the light direction. When a beam is reflected back through the solution it emerges with the same polarization as it entered before reflection.

In the Faraday Effect, the direction of the optical rotation is reversed when the light passes through the substance opposite to the magnetic field direction. It means that the optical rotation can be reversed by either changing the field direction or the light direction. As the reflected light passed twice through the medium, its plane of polarization gets rotated by twice the angle observed for single transmission.

One of the interesting problems encountered in satellite communications is the Faraday Effect. As radio waves pass through the ionosphere, their plane of polarization is rotated by the ionised particles in combination with the Earth's magnetic field. A horizontally polarized wave becomes vertically polarized because of the Faraday rotation in the ionosphere. This problem is solved by using an antenna with circular polarization, which ensures that the waves are received satisfactorily, no matter how they have been rotated.

12.6 STRESS INDUCED BIREFRINGENCE

In 1816 Sir David Brewster discovered that materials such as glass become double refracting when they are subjected to mechanical strain. This effect is known as ***stress induced birefringence***. Experiments show that the induced birefringence is directly proportional to the stress σ experienced at a given point of the material. Thus,

$$\Delta\mu = k\sigma \qquad ...(12.11)$$

where k is the proportionality constant characteristic of the material.

The strain birefringence is not very large: depending on the material, it can reach values of $\Delta\mu = 0.002$. When the material is held between crossed polarizers P and A, as in Fig. 12.7, the field of view appears dark as long as the external force is not applied. As soon as the force is applied, coloured contours will be seen. Dark regions indicate the absence of strain in those areas. Each coloured contour shows the areas that are identically deformed. Such contours enable us assess the distribution of the stresses in the material. Such a study is known as photoelastic study.

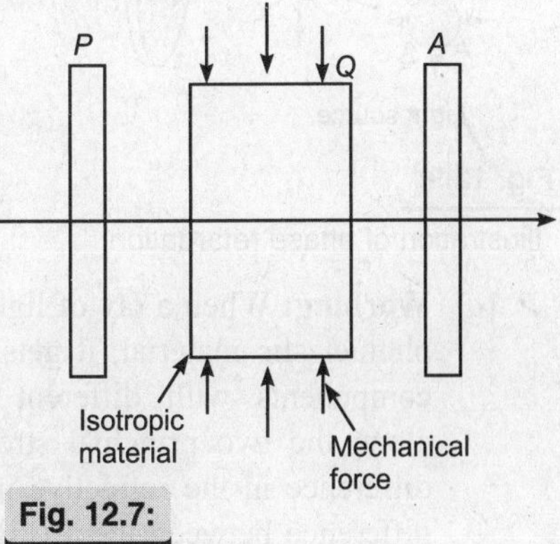

Fig. 12.7: Crossed polarizers.

12.7 PHOTOELASTICITY

Photo-elasticity is a method for analyzing stress and strain. Photo means the use of optical techniques, whereas elasticity means the study of stresses and deformations in elastic bodies. It can also be extended to inelastic bodies.

10.7.1 Photo Elastic Bench

Photo-elastic bench is an optical instrument used to analyze the stress distribution subjected to load. The instrument utilizes the properties of polarized light in its operation.

(*a*) **Construction:** A photoelastic bench is shown in Fig. 12.8. It is also called a polariscope. The arrangement consists of a polarizer and analyzer aligned at right angles to each other. A photoelastic bench mainly consists of the following parts.

Fig. 12.8: Photoelastic bench.

(*i*) Polarizers
(*ii*) Quarter wave plate
(*iii*) Loading frame
(*iv*) Light source and
(*v*) Camera.

(*b*) **Principle:** Exhibiting birefringence by photoelastic materials on the application of stress is the principle involved in photoelastic bench. The magnitude of the refractive indices at each point in the material is directly related to the state of stress at that point.

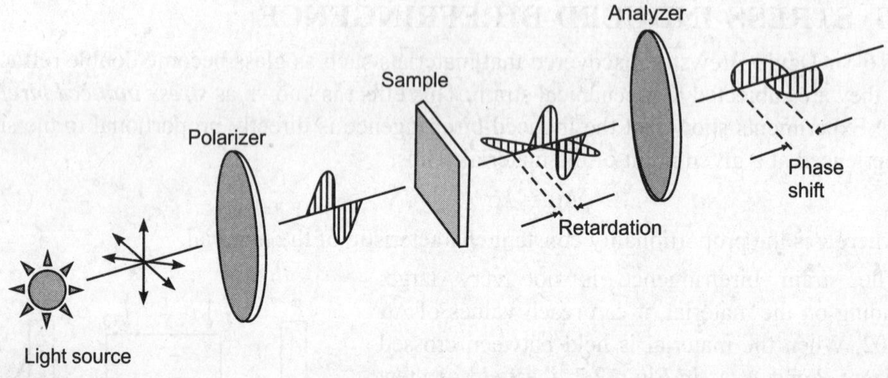

Fig. 12.9: Illustration of phase retardation.

(*c*) **Working:** When a ray of light passes through a photoelastic material, it gets resolved into two components with different refractive indices along the two principal stress directions. The difference in the refractive indices gives phase difference between them (Fig. 12.9). When the waves are brought together in the photoelastic polariscope, a photoelastic pattern of dark and light fringes, as shown in Fig. 12.10, is formed. As these fringes are related to the stresses, the magnitude and direction of stresses at any point can be found by inspection of the fringe pattern.

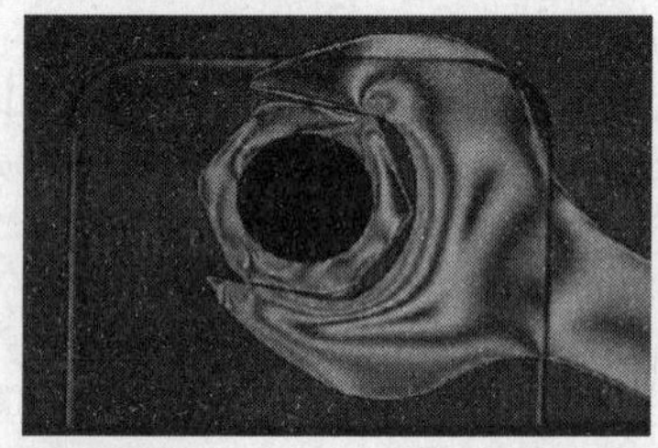

Fig. 12.10: Photoelastic fringe patterns of the nut and the spanner while the spanner is used to tighten the nut.

The magnitude of the relative retardation is given by the stress-optic law,

$$\delta = \frac{2\pi d}{\lambda} C(\sigma_1 - \sigma_2) \qquad \text{...(12.12)}$$

where $C = c_2 - c_1$ is the relative or **differential stress-optic coefficient.**

When white light is used in place of monochromatic light, coloured fringes are observed. White light is often used for demonstration, and monochromatic light is used for precise measurements. Dark lines (isoclines) occur when optic axis is parallel or perpendicular to the polarizer. Dark lines (isoclines) may be seen where the optic axis of the plastic model is aligned parallel or perpendicular to the polarizer. The isocline positions are noted for various polarizer positions, and the directions of the principle strains are built from these observations. The bright areas display a colored structure (Isochromates), as the light does not contain a component in the analyzer direction for all wavelengths.

A polariscope can convert birefringence into fringes and the number of the fringes is proportional to the stress magnitude. The experimental results obtained from the polariscope are photoelastic parameters, namely isochromatic and isoclinic, which represent the magnitude and direction of the principal stresses respectively. The photoelastic parameters are converted to maximum shear stress using an isotropic or anisotropic ***stress-optic law***.

12.7.2 Stress-Optic Law

Stress acts in every direction for any point in a loaded component. The directions in which the stresses have maximum and minimum value for the point are called as *principal directions*. The consequent stresses are called as maximum and minimum principal stresses.

Let us consider a model of uniform thickness made of a transparent high polymer material. Let the model be loaded such that it is in a plane state of stress (see Fig. 12.11). Then the state of stress can be characterized by σ_x, σ_y and τ_{xy} or by the principal stresses σ_1, σ_2 and their orientation with respect to a set of axes. Let n_o be the refractive index of the material when it was not stressed. When the model is put in a stress, the model becomes double refracting and the directions of polarization of light at the point *P* coincide with the direction of principal stress axis at that point. If n_1 and n_2 are the refractive indices for vibration corresponding to these two directions, then

$$n_1 - n_o = c_1\sigma_1 + c_2\sigma_2 \quad ...(12.13)$$

$$n_2 - n_o = c_1\sigma_2 + c_2\sigma_1 \quad ...(12.14)$$

where c_1 is called the *direct stress optic coefficient* and c_2 the *transverse stress optic coefficient*.

Since the stresses vary uniformly, σ_1, σ_2 and θ are continuously distributed functions over the model in the *xy*-plane. The direction of the polarizing axes as well as the values of n_1 and n_2 vary uniformly over the *xy*-face of the model.

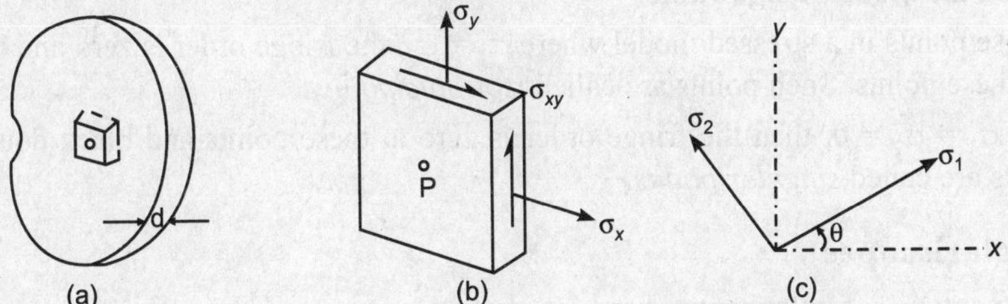

Fig. 12.11: Stress-Optic Law.

If linearly polarized light is incident normally at any point *P* of the model, the incident light gets resolved along σ_1 and σ_2; and these two vibrating components travel through the thickness of the model with different velocities. When they emerge there will be a certain amount of relative phase difference between these two components. The phase difference is given by

$$\delta = \frac{2\pi d}{\lambda}(n_1 - n_2) \quad ...(12.15)$$

$$\delta = \frac{2\pi d}{\lambda}(c_2 - c_1)(\sigma_1 - \sigma_2)$$

$$\delta = \frac{2\pi d}{\lambda}C(\sigma_1 - \sigma_2) \qquad ...(12.16)$$

where $C = c_2 - c_1$ is the ***relative or differential stress-optic coefficient*** expressed in terms of brewsters (1 brewster = 10^{-12} m^2/N).

Eq. (12.16) shows that in a transparent and isotropic model the stresses are two-dimensional. The phase difference between the two wave components traveling through the model is directly proportional to the difference of the principal stresses. When the two wave components are brought together, interference takes place. The fringe pattern depends on relative retardation and it gives the state of stress at various points in the material.

The number of wavelengths of relative path difference is given by

$$N = \frac{\delta}{2\pi} = \frac{d}{\lambda}C(\sigma_1 - \sigma_2) \qquad ...(12.17)$$

Eq. (12.16) and Eq. (12.17) are called stress-optic relations or **stress-optic law**. These equations relate the state of stress at a point to the optical behaviour of the model.

In practice one computes the values of $(\sigma_1 - \sigma_2)$ from the observed values of δ or N. Then,

$$(\sigma_1 - \sigma_2) = \frac{N\lambda}{dC} = \frac{N}{d}F \qquad ...(12.18)$$

F is called the **material fringe value**. If d = 1 cm and N = 1 wavelength, then F gives the value of $(\sigma_1 - \sigma_2)$. It produces a relative phase difference of 2π radians on a model of unit thickness. This is a property of the model material and wavelength of light used. The quantity

$$\frac{F}{d} = f \qquad ...(12.19)$$

is called the **model fringe value**.

At those points in a stressed model where $\sigma_1 = \sigma_2$, the fringe order is zero and black dots appear at these points. Such points are called *isotropic points*.

When $\sigma_1 = \sigma_2 = 0$, then the fringe order is zero at these points and black dots appears. Such points are called *singular points*.

12.7.3 Polariscopes

There are two types of polariscopes used for photoelastic analysis.

(*i*) **Plane Polariscope:** In the plane polariscope, plane-polarized light is used and the basic arrangement is as shown in Fig. 12.12.

Working: An incandescent lamp serves as a white light source. The first field lens L_1 gives a parallel beam of light in the field of view. The unpolarized light emerging from the field lens is then passed through the polarizer P and gets plane-polarized. A photoelastic material is loaded in a loading frame by which various types of loads can be applied. The plane polarized light emerging out of P passes through the model in stressed condition and splits into two beams. These beams are plane polarized in mutually perpendicular planes and are incident on the analyzer A. The analyzer combines these two beams coming from the model. The polarizer and analyzer are generally coupled together by a flexible coupling to achieve their simultaneous rotation. The second field lens L_2 makes the parallel beam of light to converge on the projection lens, which finally projects the interference fringes on to the screen or camera *C*.

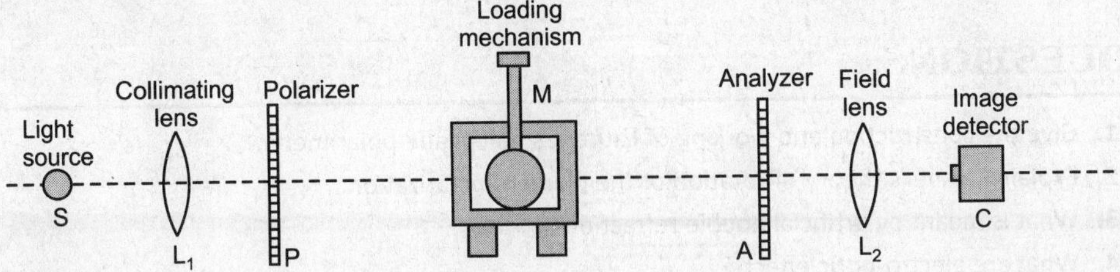

Fig. 12.12:

Plane Polariscope.

The fringe pattern consists of both the isochromatics and the isoclinics. The isoclinics fringes will vary in intensity as they pass through the extinction positions while there is no change in the isochromatic fringes.

(*ii*) **Circular Polariscope:** In the circular polariscope, circularly polarized light is used. It contains two quarter-wave plates extra in addition to all the elements of a plane polariscope. The first quarter-wave plate is kept between the polarizer *P* and the model *M*, while the second one is held between the model and the analyzer *A*. The first *QWP* converts plane polarized light into circularly polarized light and the second *QWP* converts circularly polarized light into plane-polarized light. Fig. 12.13 shows the schematic of circular polariscope.

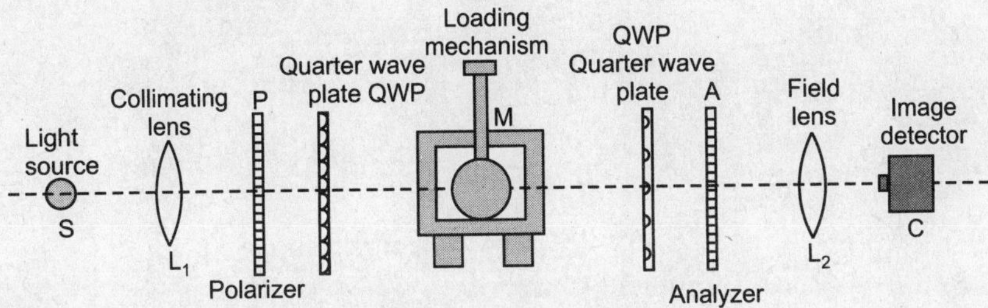

Fig. 12.13:

Circular Polariscope.

Unlike in a plane polariscope, in a circular polariscope we observe only the isochromatics and not the isoclinics. Therefore, in this setup there does not arise the necessity of differentiating between the isoclinics and the isochromatics.

Example 12.1: *In an experiment using a photo-elastic bench, the difference between principal stresses is* 8×10^9 *Nm*$^{-2}$*. The photelastic material has relative stress optic coefficient of 2 brewsters. Calculate the difference between refractive indices along the principal stresses.*

Solution:

Difference between principal stresses,

$$\sigma_1 - \sigma_2 = 8 \times 10^9 \text{ Nm}^{-2}$$

Stress optic coefficient, $C = 2$ brewsters

$$= 2 \times 10^{-12} \text{ m}^2 \text{ N}^{-1}$$

Difference between refractive indices,

$$(n_2 - n_1) = C \times (\sigma_1 - \sigma_2)$$

$$= 2 \times 10^{-12}\ \text{m}^2\ \text{N}^{-1} \times 8 \times 10^9\ \text{Nm}^{-2} = \mathbf{0.016.}$$

QUESTIONS

1. Give the construction and working of Laurent's half shade polarimeter.
2. Explain Fresnel's theory of rotation of the plane of polarization.
3. What is meant by artificial double refraction?
4. What are electro-optic effects?
5. What are magneto-optic effects?
6. Explain Kerr effect?
7. What is Pockels effect?
8. What is Cotton-Mouton effect?
9. Discuss Faraday Effect?
10. Explain photoelasticity. What is photoelastic effect?
11. State the stress-optic law and obtain an expression for the same.
12. What are isoclinic and isochromatic fringes?
13. Describe the working of a plane polariscope.
14. Explain the working of a circular polariscope.

13 Architectural Acoustics

Learning Objectives:

By the end of this chapter, you will be able to:

- Learn about basic fundamentals of sound including its types and characteristics, Intensity and pressure level and units of loudness.
- Explain about diffraction and refraction of sound waves.
- Discuss about sound resonance and sound absorption coefficient.
- Understand concept of reverberation time and Sabine's formula.
- Explain Noise effects and its related concepts.

13.1 INTRODUCTION

The study of sound plays a very important role in various branches of engineering and has developed to such a level that it has become an independent branch of engineering known as ***acoustic engineering*** or ***sound engineering***. The area of study of design of musical instruments is known as ***musical acoustics***, the technology of sound production and recording as *electro-acoustics*, use of sound in medical diagnosis and therapy as ***bio-acoustics***, the design of buildings, auditoriums, musical halls, lecture halls, recording rooms, etc as ***architectural acoustics***.

Architectural acoustics deals in general with the behaviour of sound waves in closed spaces and their design to give the best sound effects. The acoustic properties of buildings were not studied on a scientific basis till about 1900. Buildings designed to screen movies, to stage dramas or for music concerts often lacked the proper acoustic quality and were found unfit for such activities. The Fogg Art Museum hall in Harvard University, U.S.A. turned out to be highly defective when it was built. The lectures given in it were not intelligible to audience. Prof. Wallace C. Sabine, Professor of Physics in Harvard University was entrusted with the responsibility of eliminating the acoustical defects of the hall. Sabine undertook a systematic study of the problem and evolved conditions for a satisfactory acoustic quality of a hall. He found that quite often reverberation was the main cause for a defective quality of a hall. Addition of absorbent materials at appropriate surface enhances the quality of sound in the halls. Other precautions are to be taken about the shape of walls, ceiling and the hall in total so that acoustic defects do not arise. Thus, Prof. Sabine laid the foundations of acoustic engineering.

13.2 SOUND

A vibrating body excites mechanical waves in the surrounding medium. These mechanical waves propagate as a series of compressions and rarefactions of air molecules (Fig. 13.1). On reaching the ear, they cause the eardrum to vibrate, leading to the sensation of hearing. Sound cannot travel in a vacuum and requires the presence of an elastic medium for its propagation. Sound waves are longitudinal waves. The compressions and rarefactions caused by the vibrating body modulate the normal atmospheric pressure with small pressure changes occurring regularly above and below it. Thus, a sound wave is one complete cycle of pressure variation.

The wave motion of sound does not change the mean position of the vibrating particles (molecules) and the average maximum distance of a particle from its mean position is called ***amplitude***. Even a sound of 0.01 mm amplitude is enough to be audible, while a sound of 0.1 mm amplitude is quite a loud sound.

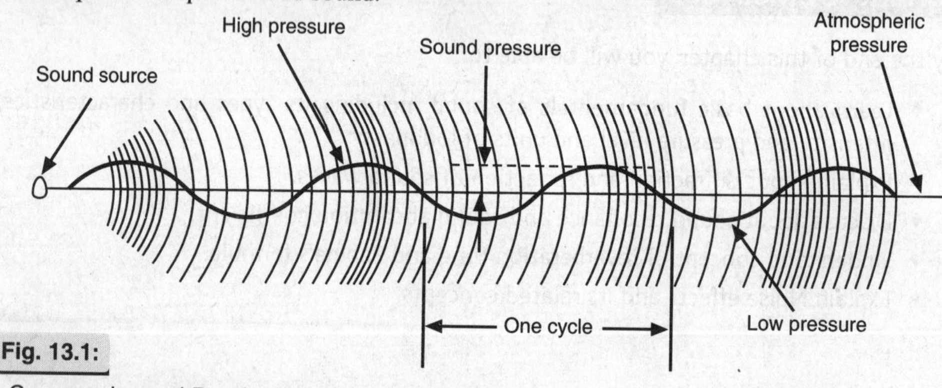

Fig. 13.1: Compression and Rarefactions of Air Molecules.

13.2.1 Sound Velocity

The velocity of sound is not a constant and depends on the nature and temperature of the medium through which it travels. In general, the velocity of sound in a gaseous medium is governed by the relation

$$\upsilon = \sqrt{\frac{B}{\rho}} \quad ...(13.1)$$

where B is the bulk modulus of the medium and ρ its density.

The speed of sound in air is commonly taken as 344 m/s for normal conditions. This is very less compared to the velocity of light. Table 13.1 lists the speed of sound in some materials. It may be noted that sound travels faster in liquid media than in gaseous media and much faster in solid media.

Table 13.1: *Speed of sound in some materials at 20°C*

Medium	Speed (m/s)	Medium	Speed (m/s)
Air	344	Brick	4300
Hydrogen	1305	Mild steel	5050
Pure water	1480	Aluminium	5150
Plexiglass	1800	Glass	5200
Soft wood	3350	Granite	6400
Concrete	3400	Gypsum Board	6800

13.2.2 Sound Wavelength

The velocity of the wave may be expressed as the product of frequency f and wavelength, λ. Thus,

$$\upsilon = f\lambda \qquad \text{...(13.2)}$$

Therefore, the wavelength may be written as

$$\lambda = \frac{\upsilon}{f} \qquad \text{...(13.3)}$$

It follows that the wavelength of sound will be larger in a medium having a higher velocity. When the medium is air, we can write

$$\lambda = \frac{344 \text{ m/s}}{f} \qquad \text{...(13.4)}$$

13.2.3 Audible Frequency Range

The limit of the audible frequency range is generally 16 Hz to 20,000 Hz. The human hearing system is sensitive to frequencies in the range of 1000 Hz-4000 Hz. Higher the audible frequency the wider is the range (Fig. 13.2). A unit ***octave*** describes the interval between two frequencies having a ratio of 2:1. Fig. 13.2 illustrates some sounds in frequency domain.

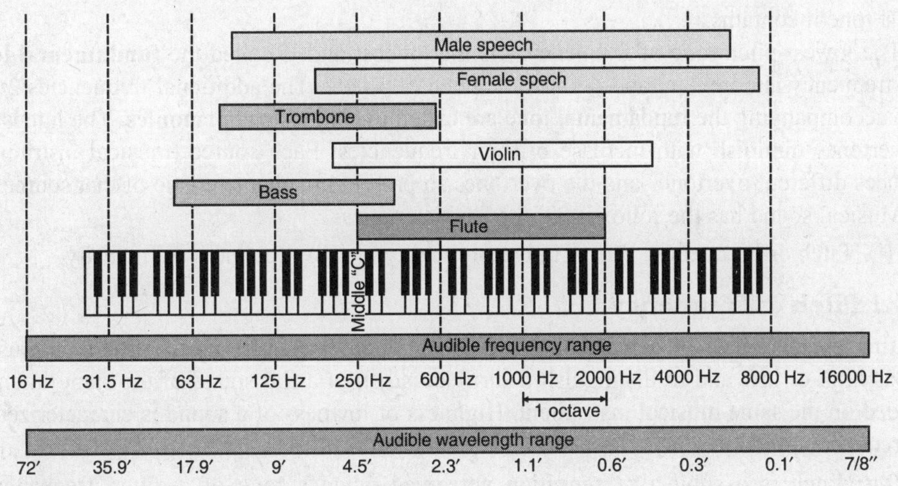

Fig. 13.2:

Audible frequency range.

13.3 CLASSIFICATION OF SOUND

Sound waves are classified based on their *frequency* into three groups, namely *sonic* (audible), *infrasonic* and *ultrasonic* waves.

(*i*) **Audible Waves:** The waves that produce a sense of sound on a human ear and lying in the range of 16 Hz to 20,000 Hz.

(*ii*) **Infrasonic Waves:** Sound waves with frequencies below 16 Hz.

(*iii*) **Ultrasonic Waves:** Sound waves with frequencies above 20 kHz.

(*iv*) **Hypersonic Waves:** Elastic waves with frequencies of 10^{10} Hz and higher and correspond to thermal waves in liquids or solids.

Audible sound waves can be further classified according to their frequency *spectrum* as **Musical sounds** and **Noise.**

(*a*) Musical sounds produce a pleasing sensation on the ear. They have the following characteristics.

(*i*) Musical sound has a line spectrum containing multiple frequencies.

(*ii*) Musical sounds are *periodic vibrations*.

(*iii*) Sudden changes in amplitude do not occur.

(*b*) Noise causes irritation and strain to our ears. If very loud, noise may cause permanent or temporary deafness. Noise has the following characteristics.

(*i*) Noise is a jumble of irregularly timed, *non-periodic vibrations*.

(*ii*) It consists of a complex spectrum of frequencies.

(*iii*) It undergoes erratic changes in amplitude and frequency.

13.4 CHARACTERISTICS OF MUSICAL SOUND

Tone is a sound wave, which has a single well-defined frequency. Sound and noise usually are not pure tones. Normally, a *musical note* consists of several tones or a series of harmonic waves of different frequencies of varying intensity. The *pitch* of that note corresponds to the lowest tone it contains.

The lowest-pitch tone of frequency, f is the loudest and is called the **fundamental** tone. This frequency is dominant and defines the pitch of a note. The additional frequencies $2f$, $3f$, $4f$ ... accompanying the fundamental tone are called **overtones** or **harmonics**. The intensities of overtones diminish with increase of their frequencies. Each source (musical instrument) produces different overtones and the overtones so present are characteristic of that source.

Musical sound has the following three characteristics:

(*i*) Pitch or Frequency (*ii*) Timbre or Quality (*iii*) Loudness or Intensity

13.4.1 Pitch or Frequency

The first characteristic of a musical sound is its pitch. It enables us to classify a musical note as high or low and to distinguish a shrill sound from a flat sound of the same intensity sounded on the same musical instrument. Highness or lowness of a sound is characterized by the frequency producing it.

Def: Pitch is a subjective sensation perceived when a tone of a given frequency is sounded. The pitch of a note depends on the frequency of the source of the sound. Greater is the frequency of a musical note, higher is the pitch and vice versa.

The frequency and pitch are two different characteristics. The frequency is a *physical quantity* and can be measured accurately, while pitch of a note is a *physiological quantity* which is merely a sensation experienced by a listener. The change in pitch with loudness is most pronounced at a frequency of about 100 Hz. In the 100 Hz range the pitch increases with increasing loudness. For frequencies between 1 kHz and 5 kHz which is the range for which the ear is most sensitive, the pitch of a tone is relatively independent of its loudness. In general, the pitch varies in a parabolic manner with frequency in the range 20 Hz to 10 kHz.

13.4.2 Timbre or Quality

The second characteristic of a musical sound is its timbre or quality. Timbre is the name given to subjective sensation, which enables us to distinguish the same note played on different instruments or sung by different singers. Memory of the timbre helps us identify different sounds such as the instrument being played or the person who is speaking or singing.

Def: Timbre gives distinguishable nature of a tone. The difference between the notes of the same pitch of two different musical instruments defines the quality of the note produced by an instrument. Quality distinguishes the note produced one instrument from that of the other instrument.

If two musical notes have the same fundamental pitch (frequency) but differ in overtones, they are said to differ in ***quality***. The frequencies and amplitudes of the overtones present in the musical notes can be analyzed with the help of Fourier analysis. The analysis showed that the quality of a note depends upon the presence or absence of particular overtones and their relative intensities.

13.4.3 Loudness or Intensity

Loudness signifies how far and to what extent, sound is audible. Loudness of sound is a subjective perception. Because of the varying sensitivity of the ear, different people perceive the same sound differently. What is loud to one person may be soft for another. An objective measure of loudness is sound *intensity*, which is applicable equally to everyone. Loudness is found to vary with frequency also. Intensity of sound is a physical quantity and does not depend on the listener.

Def: *The intensity of the sound wave is the rate of flow of sound energy through a unit area normal to the direction of propagation. Since the rate of flow of energy is power, the intensity of sound wave is measured in units of power per unit area, i.e., in W/m²*.

The sound intensity is proportional to the square of the wave amplitude.

Thus, $$I \propto P^2 \qquad ...(13.5)$$

where P is the pressure amplitude.

The intensity of the faintest sound wave that can be heard is about 10^{-12} W/m². The range of intensity of sound that an ear can hear ranges from 10^{-12} W/m² to about 1 W/m. Sounds of same intensity but of different frequencies may differ in loudness.

13.4.4 Weber-Fechner Law

Def: Weber-Fechner law states that *the degree of sensation of sound is proportional to the logarithm of the stimulus producing the sound.*

If L is the degree of loudness due to intensity I, then

$$L = k \log I \qquad ...(13.6)$$

where k is the proportionality constant. It depends on the sensitivity of the ear, quality of the sound and other factors.

Eq. (13.6) is the mathematical statement of the Weber-Fechner law. Because of the logarithmic nature, loudness is not doubled when the intensity is doubled. However, we can show that loudness increases by the same amount each time when the intensity is doubled.

Let I_1 be the initial intensity, which produces loudness L_1. Then,

$$L_1 = k \log I_1$$

Now, if the intensity is doubled, I_1 becomes $2I_1$ and the loudness due to doubled intensity will be

$$L_2 = k \log 2 I_1 = k \log 2 + k \log I_1$$

or $$L_2 = k \log 2 + L_1$$

$$\therefore \quad L_2 - L_1 = k \log 2 \qquad \text{...(13.7)}$$

As *k log 2* is a constant, it follows from Eq. (13.7) that the loudness increases by the same amount whenever the intensity is doubled, irrespective of the initial intensity.

13.5 SOUND INTENSITY LEVEL - DECIBEL

The range of variation of sound intensity is very large. The loudness of a sound as judged by the ear is proportional to the logarithm of intensity. It means that our ear is a logarithmic instrument. Therefore, the absolute intensity of sound wave is not of practical significance. Instead, the relative intensity is of more concern for us.

The lowest intensity of sound at 1 kHz to which a normal human ear can respond is $I_o = 10^{-12}$ W/m^2. This is known as the ***threshold of hearing*** and is chosen as the "zero" or "standard" intensity. Intensity of a sound is measured with reference to this standard frequency.

Def: *The ratio of the intensity of sound wave to the threshold intensity of hearing is defined as the* ***intensity level*** *of sound.*

If I and I_0 represent the intensities of two sounds of a particular frequency (chosen normally to be 1 kHz), L_1 and L_0 are their corresponding measures of loudness, then according to Weber-Fechner law

$$L_1 = k \log I$$

and

$$L_0 = k \log I_o$$

The difference in the loudness of the two sounds is given by

$$L = \log \frac{I}{I_o} \text{ bels} \qquad \text{...(13.8)}$$

L is called the *intensity level* and is expressed in bels, a unit named after Alexander Graham Bell, the inventor of telephone.

Def: One Bel is defined as the relative intensity between two sound notes if one is 10 times more intense than the other.

The unit of bel is large and in practice a smaller unit *decibel* is used.

$$1 \text{ decibel} = \frac{1}{10} \text{ bel}$$

Accordingly, the intensity level of a sound wave is defined as

$$L = 10 \log \frac{I}{I_o} \text{ decibels (dB)} \qquad \text{...(13.9)}$$

(a) **Threshold of audibility:** The intensity level corresponding to the intensity I_o will be 0 dB, since from Eq. (13.9), we get

$$L = 10 \log \frac{I_o}{I_o} = 10 \log 1 = 0$$

0 dB level represents the ***threshold of audibility***.

(b) **Physical significance of a decibel:** The smallest change in intensity level that the human ear can detect is 1 dB. Let us find the corresponding change in intensity or loudness. From Eq. (13.9), when L = 1dB,

$$1 \text{ dB} = 10 \log \frac{I}{I_o}$$

or $$\log \frac{I}{I_o} = \frac{1}{10} = 0.1$$

$$\therefore \quad \frac{I}{I_o} = 10^{0.1} = 1.26$$

or $$I = 1.26 I_o$$

It means that a change in intensity by 26% increases the intensity level by one decibel. Further,

if $I = 100\, I_o$, $L = 10 \log 10^2 = 20$ dB,
$I = 1000\, I_o$, $L = 10 \log 10^3 = 30$ dB,
$I = 10000\, I_o$, $L = 10 \log 10^4 = 40$ dB and so on.

Thus, when two sounds differ by 20 dB, the louder of them is 100 times more intense and when they differ by 40 dB, the louder one is 10,000 times more intense. The loudest sound that can be heard without pain is about 120 dB. This is known as the ***threshold of feeling*** or ***pain threshold***. Some of the typical sound levels are shown in Table 13.2

Table 13.2: *Some typical sound levels*

Source	Sound level (dB)
Rustling of tree leaves	0 - 20
Quiet living room	20 - 40
Average office	40 - 60
Average street noise	60 - 80
Train sound	80 - 100
Thunder	100 - 120
Aeroplane noise at a distance of 3 m	120-130 (painful)

13.6 SOUND PRESSURE LEVEL (SPL)

In acoustical problems, sound levels are generally dealt in terms of pressure rather than intensity. In fact, sound measuring devices respond to pressure exerted by sound. Eq. (13.5) shows that the sound intensity is proportional to the square of its pressure.

Using Eq. (13.5) into Eq. (13.9), an expression for the SPL may be obtained as follows.

$$\text{SPL} = 10 \log \frac{I}{I_o} = 10 \log \left(\frac{P}{P_o}\right)^2 = 20 \log \frac{P}{P_o} \quad \text{decibels} \qquad ...(13.10)$$

The reference pressure P_o is usually taken as $P_o = 2 \times 10^{-5}$ N/m^2.

The reference level P_o is approximately the lowest sound pressure that is audible to normal human ear. The *SPL* can be directly measured on a sound level meter.

13.7 HUMAN AUDIOGRAM

The human ear exhibits basically a nonlinear response. Human audiogram is shown in Fig. 13.3. The threshold of hearing represents the weakest sounds that can be heard while the upper curve shows the loudest sounds that can be heard without harmful effects. Further, heard the sensitivity of the ear varies with frequency. Thus, threshold of hearing of sound depends upon both the intensity and frequency of sound. A sound is audible above a certain minimum intensity and a certain minimum frequency. For a person with normal hearing, the threshold of audibility at 1 kHz is 0 dB; at 200 Hz and 15 kHz it is about 20 dB and at 50 Hz and at 18 kHz it is about 50 dB. Ordinary speech sounds lie between the frequencies of 100 and 8000 Hz, but musical sounds cover a greater range. When the intensity of sound exceeds a maximum limit, it produces a sensation of pain on the ear. The tolerable intensity limit is 120 decibels.

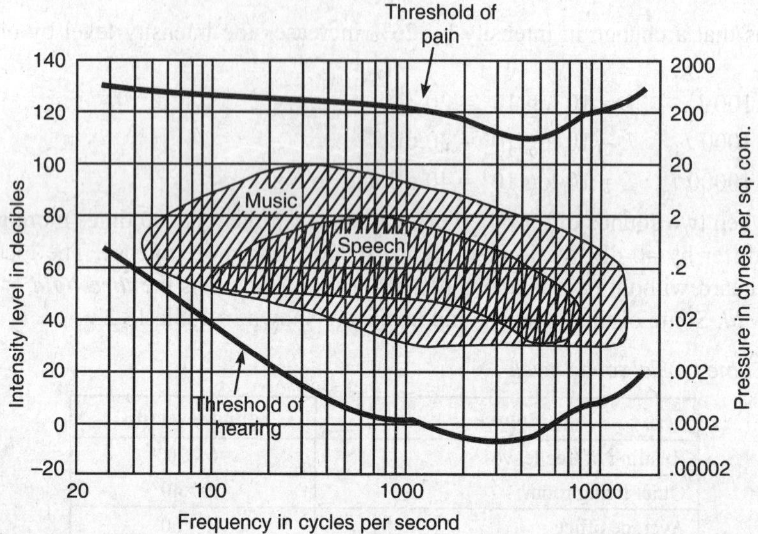

Fig. 13.3: Human Audiogram.

Similarly, there is a maximum frequency limit beyond which the sound is not heard. Thus, the maximum audible intensity of sound also depends upon both the intensity and frequency of sound. In Fig. 13.3, the two curves, the threshold of hearing and the threshold of pain enclose an area called **auditory area**. A sound, whose frequency and intensity are not within the limits set by this area, is not heard by humans.

13.8 UNITS OF LOUDNESS

Loudness is considered a subjective measurement which is influenced not only by sound pressure, but the length of the sound. ***There are two units used to measure the loudness of sound, phon and sone.***

13.8.1 Phon

The loudness of a sound depends on the frequency of the sound and the sensitivity of the ear. Generally, two different 60-decibel sounds will not have the same loudness. The loudness cannot be measured directly with a meter.

Def: The phon is a measure of loudness and it is equal to the loudness of an equally loud 1 kHz frequency note expressed in decibels.

A sound of known intensity level is produced at the frequency 1 kHz (1 kHz being taken as the standard frequency) and then a sound at some other frequency of interest is generated. The intensity of the second sound is varied in intensity till the volunteers judge it to be of the same loudness as that of 1 kHz sound. If the intensity level of both sounds is *N* decibels, then the equivalent loudness is said to be N phons. In simple terms 60 phons means "as loud as a 60 dB, 1000 Hz tone." At 1 kHz, readings in phons and dB are the same.

So, for a 1 kHz note, loudness in phons

$$= 10 \log(I/I_0) \quad \text{...(13.11)}$$

$$= 10 \log I + 10 \log (1/I_0)$$

$$= 10 \log I + 10 \log 10^{12}$$

or $$LP = 10 \log I + 120 \quad \text{...(13.12)}$$

$\therefore$ $$\log L = 0.033(10 \log I + 120 - 40)$$

$$= 0.33 \log I + 2.64$$

which reduces to $$L = 445 I^{0.33}$$

that is, $$L \propto I^{1/3} \quad \text{...(13.13)}$$

for a 1 kHz tone.

The phon unit is not an SI unit in metrology and is used by the American National Standards Institute.

13.8.2 Sone

Def: A ***sone*** is defined to be equal to 40 phons. A sone is the loudness of a 1 kHz tone of 40-db intensity level.

Experimentally it was found that a 10 dB increase in sound level corresponds approximately to a perceived doubling of loudness. So that approximation is used in the definition of the phon as

0.5 sone = 30 phon, 1 sone = 40 phon, 2 sone = 50 phon, 4 sone = 60 phon, etc.

13.9 SOUND BEHAVIOR

Sound waves are longitudinal waves and require an elastic medium for their propagation. They exhibit all the wave properties such as reflection, diffraction, interference etc. When sound encounters an obstacle it undergoes reflection as well as diffraction. The sound behavior in a room depends significantly on the ratio of the frequency of the sound to the size of the room. So, the audible spectrum can be divided into four regions illustrated in Fig. 13.4 for rectangular room.

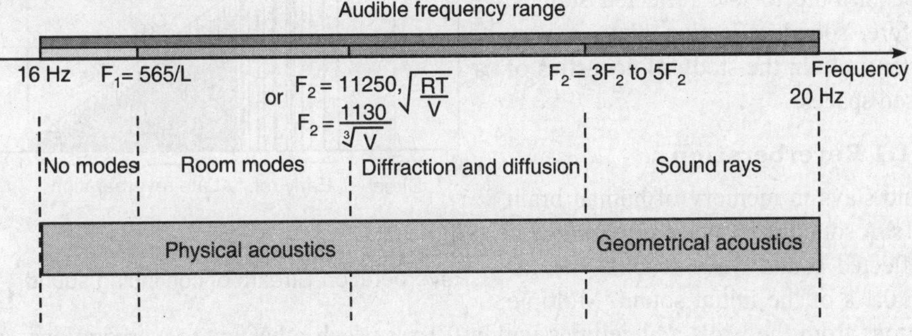

Fig. 13.4:

Four regions of audible spectrum.

13.9.1 Reflection of Sound Waves

Sound waves are reflected just as light waves and obey the similar laws. Flat surfaces reflect sound waves in such a way that the angle of incidence equals to angle of reflection (Fig. 13.5). Sound waves are reflected when the dimensions of the obstacle are large in comparison to the wavelength.

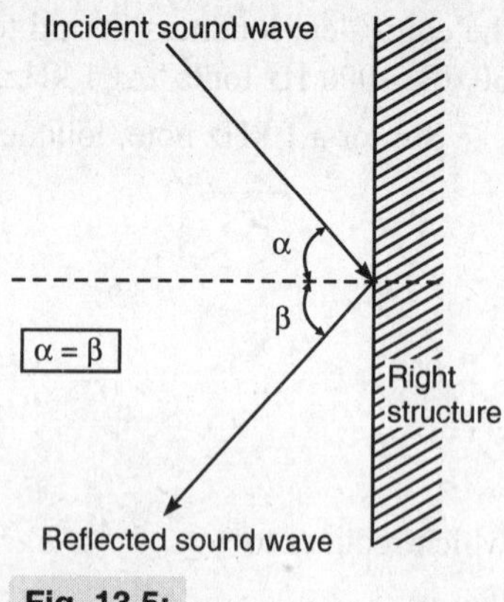

Fig. 13.5:

Reflection of Sound Waves.

Generally, in enclosed spaces such as rooms, halls etc, sound reflects from all the surfaces such as side walls, floors, ceiling etc. The sound that reaches the listener's ear directly from the sound source is called the ***direct sound***. This sound reaches the listener's ears first. The reflection of sound in the enclosed space leads to some important phenomena, namely *echo* and *reverberation*.

Sound waves that bounce off something like a wall or the ceiling and arrive at the listener's ear at almost precisely the same time as the direct sound are called "***early reflections***" (Fig. 13.6). As early reflections travel a longer path, they arrive later than the direct sound, often in a range from 5 to 100 ms.

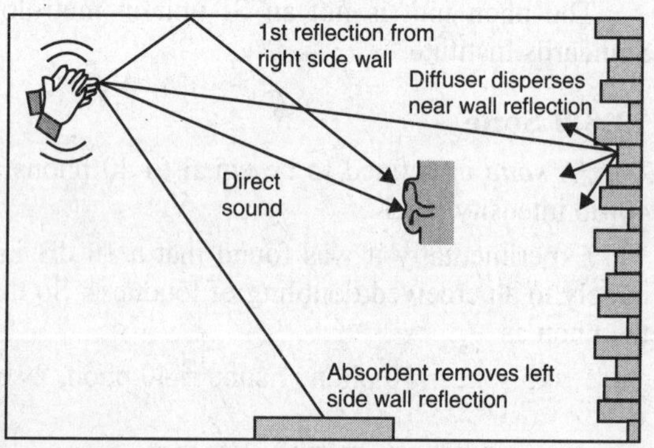

Fig. 13.6:

Early reflections-travel a longer path

When there are numerous reflections in the room, some reflections will be added together. These reflections gain in strength and produce ***echoes***. The rate of build-up of echo density is proportional to the square root of the volume of the room. The stream of continuing sound is called ***reverberation*** (Fig. 13.7). Sounds that arrive at the listener's ears after multiple reflections are termed as "***late reflections***." The speaker and room contribute to late reflected sounds. Therefore, sound reflection plays a very important role in the study of acoustics of enclosed spaces.

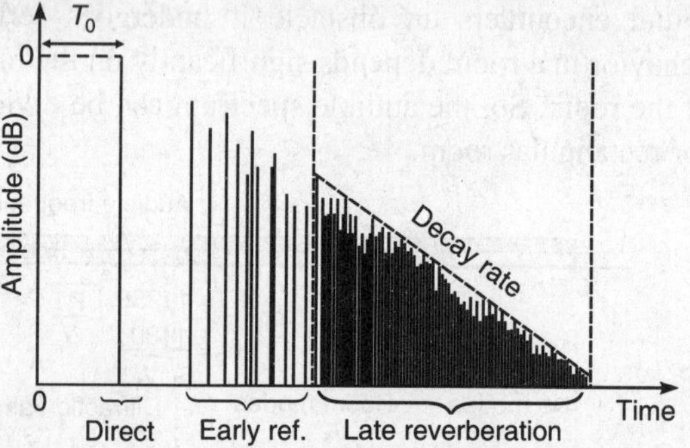

Fig. 13.7:

Reverberation-Stream of continuing sound.

13.9.1.1 Reverberation

A sound stays in memory of human brain for 0.1s. A sound is termed as *prolonged* if the reflected sound wave reaches the ear within 0.1 s of the initial sound. Multiple reflections from the walls and ceilings within 0.1 s of each other give reverberations. As the sound velocity is 340 m/s at room temperature, it will take about 0.1 s for a sound to travel the length of a 17 m room and back, thus causing a reverberation. So, only reverberations are common in rooms with dimensions of approximately 17 m or less. A person in a hall continues to receive successive reflections of sound from the walls, floor and ceiling etc whose intensity progressively diminishes (Fig. 13.8). The prolongation of sound before it falls off to a negligible intensity is called *reverberation*.

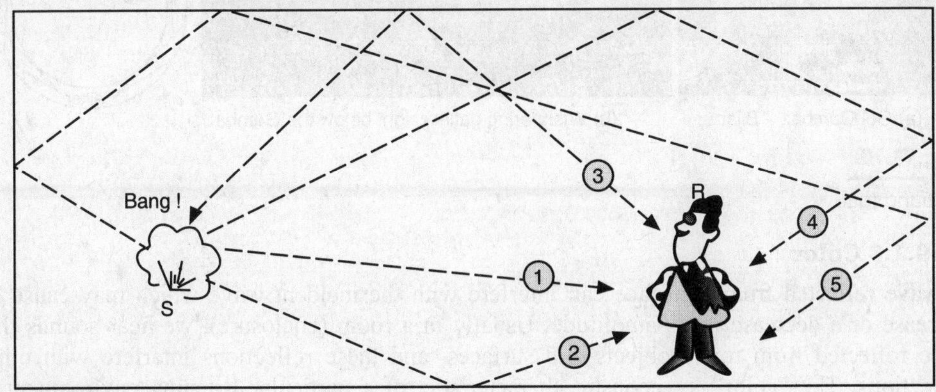

Fig. 13.8:

Successive reflections of sound waves.

In auditoriums and concert halls, reverberations occasionally occur. The auditorium would not seem to be as lively and full of sound. Reverberation in auditoriums and concert halls do not give disturbing results if the reflections are properly *designed*. Rough walls diffuse the sound by reflecting it in all directions. So that a spectator feels the sounds from every part of the room, making it seem lively and full.

Some reverberation is desirable, especially in a hall used for musical performance. A small amount of reverberation improves the original sound. However, too much reverberation causes boomy sound quality in a musical performance. Speeches given in such a hall would be unintelligible. Note that *the reverberation of sound pertains to enclosed spaces only*. In open air the sound spreads out in all directions without repeated reflections.

13.9.1.2 Sound foci

Reflection of sound waves is affected by the shape of the surface. Curved surfaces with a parabolic shape focus sound waves to a single point in space; at that point, the sound is amplified. Fig. 13.9 and Fig. 13.10 show sound reflection from concave surface -'*s*' is sound source and '*f*' is the sound focus. At the focal point even the faintest whisper can be heard.

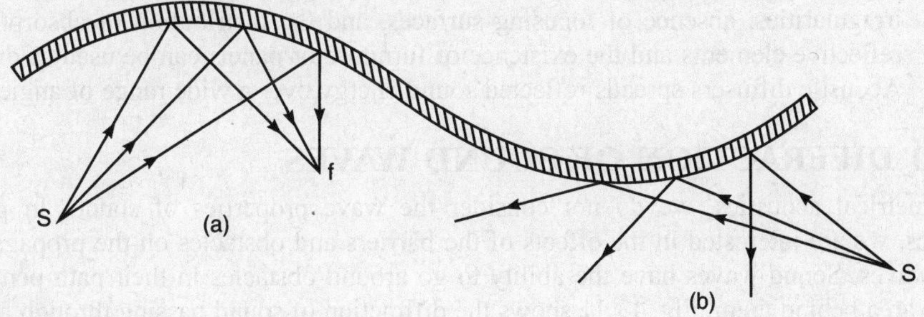

Fig. 13.9:

Sound reflection from concave surface.

(a) Gol Gumbaz – Bijapur

(b) Wishpering gallery right below the Gumbaz

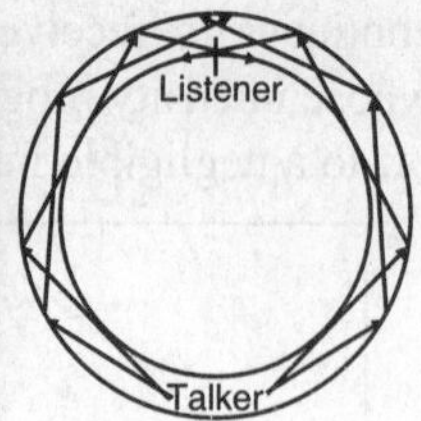

Fig. 13.10:

Sound foci.

13.9.1.3 Color

A wave reflected from a surface can interfere with the incident wave which may cause an increase or a decrease in its amplitude. Usually in a room (enclosure), we hear sounds that have reflected from many objects and surfaces, and these reflections interfere with other reflections. The "color", or acoustic characteristic, of a particular listening environment is determined by the angles and materials at which the sounds suffer reflections since different materials reflect different frequencies more efficiently than others.

13.9.2 Forms of Acoustic Treatment

There are two basic forms of acoustic treatment to deal with reflections, namely *absorption* and *diffusion*.

(*i*) **Absorption:** Sound absorbers are used to reduce reflections by dispersing the sound energy. The examples of sound absorbers are curtains, sofas, carpets and products made of mineral wool, open cell foam, or recycled fibrous material like paper-waste. The fibers of the absorbing material vibrate when a sound wave hits. This vibration causes tiny amounts of heat due to the friction. The more fibrous a material the better is its absorption. Denser materials are less absorptive. The sound absorbing characteristics of acoustical materials change with frequency. Usually low frequency sounds are absorbed less because of their long wavelength.

Def: Absorption is the property of a surface by which sound energy is converted into other form of energy.

(*ii*) **Sound Diffusion:** Acoustic diffusers are used to reduce the reflections. Sound in an enclosed space is described as diffused sound if the intensity of the sound energy is the same in each location. Diffusers scatter the sound in all, or in certain directions depending on their type and geometrical dimension. The factors such as geometrical irregularities, absence of focusing surfaces, and the distribution of absorptive and reflective elements and the existence of furniture or panels can be used as diffusers. Acoustic diffusers spreads reflected sound energy over a wide range of angles.

13.10 DIFFRACTION OF SOUND WAVES

In geometrical acoustics, we do not consider the wave properties of sound. In physical acoustics, we are interested in the effects of the barriers and obstacles on the propagation of sound waves. Sound waves have the ability to go around obstacles in their path penetrating into the area behind them. Fig. 13.11 shows the diffraction of sound passing through a slit.

Def: Diffraction of sound waves is the change in the direction of the propagation of sound waves passing the edge of an obstacle or a slit. The amount of diffraction depends on the wavelength and the size of the object.

Generally, we notice sound diffracting around corners or through door openings, allowing us to hear others who are speaking to us from adjacent rooms. Diffraction occurs whenever the wavelength of the wave is larger than the obstacle or opening. The amount of diffraction increases with increasing wavelength and decreases with decreasing wavelength.

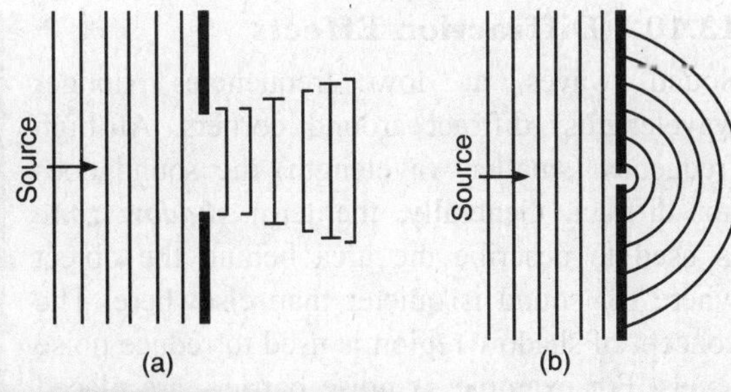

Fig. 13.11:

A. Diffraction of sound passing through a slit.

(*i*) **If the wavelength of sound is less than the size of the obstacle:** the wave diffracts around the barrier at the edges, leaving a shadow region behind it (Fig. 13.12 *a*) where the wave does not reach. In the acoustic shadow only a small portion of the sound enters behind the obstacle.

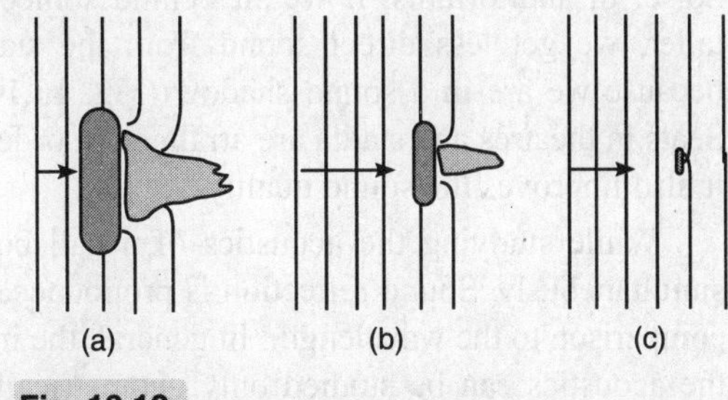

Fig. 13.12:

B. Diffraction of Sound Waves due to a hurdle.

(*ii*) **If the wavelength is comparable to the size of the object:** it surrounds partially, have an acoustic shadow just behind it (Fig. 13.12 *b*). The smaller the object is compared to the wavelength, the smaller the shadow region and the more the wave reaches behind the obstacle.

(*iii*) **If the obstacle is small compared to the wavelength:** no noticeable diffraction occurs (Fig. 13.12 *c*).

When sound encounters an obstacle, it undergoes diffraction as well as reflection. In the area of:

- high frequencies both the dissipation and reflection of the sound waves occurs in meeting an obstacle,
- lower frequencies diffraction and surrounding of the obstacle

Fig. 13.13 shows that the high frequencies can be effectively prevented by the barrier while the low frequencies suffer diffraction and bend around the barrier (Fig. 13.14). Therefore, an increase in the size of the barrier would help us stop traffic noise more efficiently from entering into residential area.

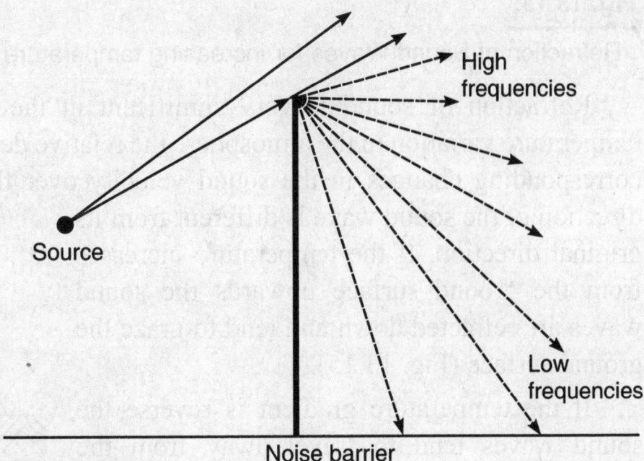

Fig. 13.13:

Effect of diffraction depending on range of frequencies.

13.10.1 Diffraction Effects

Sound waves at low frequencies (longer wavelengths) diffract around corners. At high frequency (smaller wavelength) the sound does not diffract. Generally, the term ***shadow zone*** is used to describe the area behind the object where the sound is quieter than elsewhere. The concept of shadow region is used to reduce noise levels. For example, if noise barriers are placed alongside major roads, the houses behind the barriers in the shadow zone are exposed to less noise. In auditoriums, if we sit behind someone taller, we get less direct sound from the stage because we are in a sound shadow (Fig. 13.14). Seats in theatres and stadia are arranged in order to give the audience a good line of sight, and it also improves the sound quality.

Fig. 13.14: Diffraction Effects- sound shadow.

While studying the acoustics of a hall both reflection and diffraction must be studied simultaneously. Sound reflection is pronounced only when the size of the obstacle is large in comparison to the wavelength. In general, the irregular surfaces in a hall are quite large so that the acoustics can be studied only geometrically for reflections. Whatever sound diffraction occurs in a hall because of uneven surfaces, openings, corners and sound absorbing materials, uniform diffusion of sound happens. Therefore, the acoustics of a hall may be studied by studying the sound reflections alone. It is normally done by model tests.

13.11 REFRACTION OF SOUND WAVES

Def: Refraction of sound is the phenomenon of change in the direction of waves as they pass from one medium to another and is followed by a change in speed of the waves.

Fig. 13.15: Refraction of Sound Waves for increasing temperature gradient.

Refraction of sound is very significant in the case of open-air theatres. Because of temperature variation in the atmosphere the relative densities and velocities of the air introduce corresponding changes in the sound velocity over the wave front. Therefore, the resultant direction of the sound wave is different from its original direction. If the temperature increases from the ground surface upwards the sound waves are refracted down and tend to graze the ground surface (Fig. 13.15).

If the temperature gradient is reverse the sound waves tend to travel away from the ground. It introduces loss of audibility in the open-air theatres as shown in Fig. 13.16.

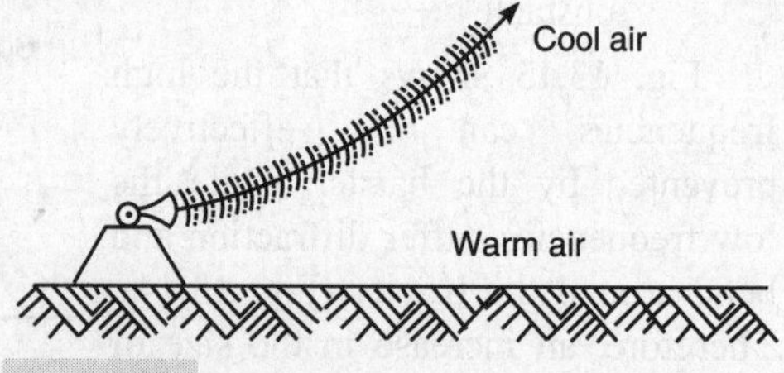

Fig. 13.16: Refraction of Sound Waves for reverse temperature gradient.

13.12 SOUND RESONANCE

Sound moving in certain directions is amplified. This result is noticeable in a bath room which has highly reflecting surfaces whereby sounds of appropriate frequency are increased in volume. This phenomenon is known as *resonance*.

Def: Resonance is a phenomenon of uniting the repetitive reflections and particular sounds becomes louder. Sound resonance is prominent in enclosed spaces.

Def: Modes are the resonant frequencies on which the waves interface and form maximums and minimums of sound pressure at different points in the room.

There are 3 types of modes.

(*a*) axial: two parallel surfaces contribute to the modes

(*b*) tangential: 4 surfaces contribute to the modes

(*c*) oblique: 6 surfaces contribute to the modes

The resonant frequencies of a rectangular room having hard plastered walls is

$$f = \frac{c}{2}\left\{\left(\frac{p}{l}\right)^2 + \left(\frac{q}{w}\right)^2 + \left(\frac{r}{h}\right)^2\right\}^{\frac{1}{2}}$$

where f = resonant frequency in Hz for a certain mode of vibration

p, q, r = numbers taken as 0, 1, 2, 3,........ for different modes.

l, w, h = length, width, and height of the room in m.

c = velocity of sound in m/s.

The lowest frequency of all modes is for the axial mode. Thus when we take $p = 1$, $q = 0$, $r = 0$, i.e., the (1, 0, 0) mode which gives vibrations parallel to the lengths, and it can be calculated from $f = c/2l$, where c is the speed of the sound and l is the room length. The (0, 1, 0) mode gives frequency of vibrations parallel to the width as $(\frac{c}{2w})$.

When the values of p, q and, r are increased gradually such that resonant frequencies of a room follow an ascending series, the vibration peaks at higher frequencies come closer or overlap each other producing uniform sound distribution. That means it is likely that the sound distribution in a hall will be erratic if only the lower modes of resonant frequencies are excited by the source frequency. In large halls, resonance may be neglected but in the resonant halls like the radio broadcasting, television or cinema studios, resonance phenomenon must be studied.

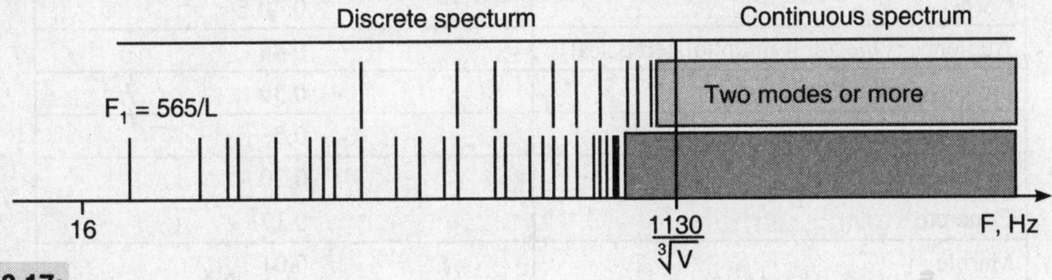

Fig. 13.17: Non-uniform of distribution of resonant frequencies over the audible spectrum.

Fig. 13.17 shows non-uniform of distribution of resonant frequencies over the audible spectrum which is discrete for low frequencies and continuous at the higher frequencies. The larger the room the lower frequency range of the continuous spectrum. Frequency distortion occurs for sounds with the discrete resonant spectrum. Resonances disturb the uniform distribution of sound in a room which is essential for satisfactory hearing. Resonance can be corrected by modifying the walls and ceiling.

13.13 SOUND ABSORPTION COEFFICIENT

A portion of the sound which is incident on a wall is reflected, a portion is absorbed (transformed into heat) and the remaining portion is transmitted through the wall to the other side as shown in Fig. 13.18. Different surfaces absorb sound to different extents. Absorption coefficient, α describes the effectiveness of a surface in absorbing sound energy.

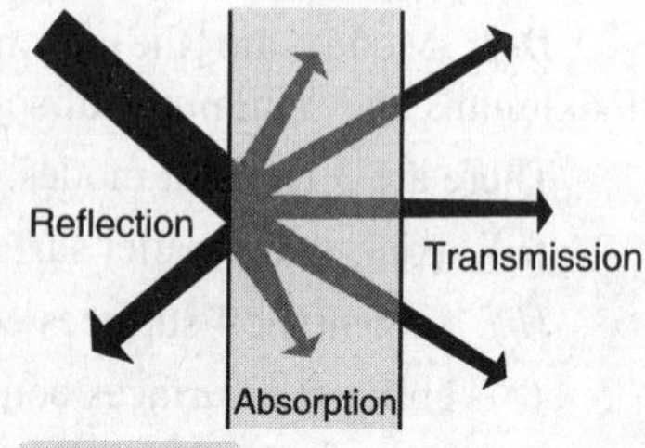

Fig. 13.18: Sound Absorption.

Def: Coefficient of absorption of a material is the ratio of sound energy absorbed by its surface to that of the total sound energy incident on the surface.

$$\text{Coefficient of absorption,} \quad \alpha = \frac{\text{Sound energy absorbed by the surface}}{\text{Total sound energy incident on the surface}} \quad \text{...(13.14)}$$

In order to compare the relative efficiency of different absorbing surfaces, it is essential to select a standard in terms of which all surfaces can be described. A unit area of open window is selected as the standard, *ideal absorber of sound*. The entire sound incident on an open window is fully transmitted and none is reflected.

The unit of absorption is the open window unit (O.W.U.), ***sabin*** termed after the scientist who established it. Table 13.3 lists the absorption coefficients of various materials.

Def: Open window unit (O.W.U.) is 1m^2 sabin amount of sound absorbed by one square metre area of fully open window.

Table 13.3: *Absorption coefficients of some materials*

Material	Absorption coefficient per m^2 at 500 Hz
Open window	1.00
Ventilators	0.10 to 0.50
Stage curtain	0.20
Curtains in heavy folds	0.40 to 0.75
Carpet	0.40
Audience (One adult in upholstered seat)	0.46
Fibrous plaster, Straw board	0.30
Perforated compressed fibre board	0.55
Wood-wool board	0.20
Concrete	0.17
Marble	0.01

The value of absorption coefficient of a material depends on its nature and the frequency of sound. The greater the frequency the larger is the value of 'α' for the same material. Therefore, the values of 'α' for a material are determined for a wide range of frequencies. Generally, value of α at 500 Hz is used in acoustic designs.

If a material has the value of "α" as 0.5, it means that 50% of the incident sound energy will be absorbed per unit area. If the material has a surface area of $S\ m^2$, then the total absorption provided by that material is

$$a = \alpha \cdot S \quad \text{sabins} \qquad ...(13.15)$$

If there are different materials in a hall, then the total sound absorption by the different materials is given by

$$A = a_1 + a_2 + a_3 + \ldots\ldots$$

$$A = \alpha_1 S_1 + \alpha_2 S_2 + \alpha_3 S_3 + \ldots\ldots \qquad ...(13.16)$$

or

$$A = \sum_{1}^{n} \alpha_n S_n \qquad ...(13.17)$$

where $\alpha_1, \alpha_2, \alpha_3$........ are absorption coefficients of materials with areas S_1, S_2, S_3,.....

13.14 DETERMINATION OF ABSORPTION COEFFICIENT

There are two methods which are used for the measurement of absorption coefficients of acoustic materials.

Method 1: One of the methods is based on the determination of reverberation times of a room without any material and with the material under test. If T_1 is the reverberation time of in the empty room, then

$$T_1 = \frac{0.161\,V}{\sum_{1}^{n} a_n S_n} = \frac{0.161\,V}{A} \quad \text{(See Eq. 13.24)}$$

where $A = \Sigma\alpha_n S_n$ denotes the absorption due to the walls, flooring and ceiling of the empty room.

Then a certain amount of absorbing material of area S and absorption coefficient α' is added in the room and again the reverberation time is measured. Let it be T_2.

$$T_2 = \frac{0.161\,V}{A + a'S}$$

Then

$$\frac{1}{T_2} - \frac{1}{T_1} = \frac{\alpha' S}{0.161\,V}$$

$$\therefore \qquad \alpha' = \frac{0.161\,V}{S}\left[\frac{1}{T_2} - \frac{1}{T_1}\right] \qquad ...(13.18)$$

Knowing the quantities on the right hand side of the Eq. (13.18) the absorption coefficient α' of the material under test can be calculated.

Method 2: The above method cannot be used if the absorbing material is already fixed to the walls and ceiling in the room. The method adopted for such cases, consists in finding the reverberation times for two sources of differing emitting powers.

Let the powers of the two sources be P_1 and P_2 respectively. The steady state energy densities of the sources will be

$$E_1 = \frac{4P_1}{vA} \text{ and } E_2 = \frac{4P_2}{vA}$$

Let T_1 and T_2 be the respective times of decay of energy density to the inaudibility level E_o. Then,

$$E_o = \frac{4P_1}{vA}e^{-\alpha T_1}$$

and

$$E_o = \frac{4P_2}{vA}e^{-\alpha T_2}$$

or

$$\frac{P_2}{P_1} = e^{\alpha(T_2 - T_1)}$$

$$\therefore \quad \alpha = \frac{\log_e P_2 - \log_e P_1}{(T_2 - T_2)}$$

But

$$\alpha = \frac{vA}{4V} \quad \therefore \quad \frac{\log_e P_2 - \log_e P_1}{(T_2 - T_1)} = \frac{vA}{4V}$$

$$A = \frac{4V\log_e(P_2/P_1)}{v(T_2 - T_1)}$$

$$\alpha S = \frac{4V\log_e(P_2/P_1)}{v(T_2 - T_1)}$$

$$\alpha = \frac{4V\log_e(P_2/P_1)}{vS(T_2 - T_1)} \qquad ...(13.19)$$

Knowing the quantities on the right hand side of the above equation, we can calculate the absorption coefficient of the material fixed in the room.

Example 13.1: *Find the total absorption in a hall with a volume of 5000 m^3 for reverberation time of 1.5 s.*

Solution:

$$T = \frac{0.161V}{\sum \alpha S}$$

$$\therefore \quad \sum \alpha S = \frac{0.161 \times 5000\ \text{m}^3}{1.5\text{s}} = 537\ \text{O.W.U.m}^2$$

Example 13.2: *If the reverberation time for an empty hall is 1.5 s and it is found to be 1 s when a curtain cloth of 20 m^2 is suspended at the center of the hall for the dimensions of the hall 10 × 8 × 6 m^3, find the coefficient of absorption of curtain cloth.*

Solution: Absorption of empty hall

$$= \frac{0.161(10 \times 8 \times 6)\text{m}^3}{1.5\text{s}} = 51.52$$

Reverberation time of the hall with curtain is

$$1\text{s} = \frac{0.161(10 \times 8 \times 6)}{51.52 + \alpha(2 \times 20)}$$

The factor 2 in the denominator takes into account the two sides of the curtain.

∴ The coefficient of absorption of curtain cloth

$$\alpha = \frac{77.28 - 51.52}{40} = 0.64$$

13.15 SOUND TRANSMISSION LOSS

Sound transmission loss means the decrease of sound intensity by any structure at a given frequency.

Def: Sound transmission loss is defined as the amount of sound energy lost in the form of heat and non-geometrical effects during transmission of sound waves.

The transmission loss for random incident sound waves is less than that of a normal incident wave. Let I_i be the intensity of incident sound waves and I_t be the intensity of transmitted sound waves. So, the transition loss per normal incident is

$$\text{TL} = 10 \log \frac{I_i}{I_t} \qquad \text{...(13.20)}$$

Let α_t be the transmission coefficient of the sound energy, which is defined as the ratio of the intensity of, transmitted to incident sound.

$$\alpha_t = \frac{I_t}{I_i} \qquad \text{...(13.21)}$$

So, substituting Eq. (13.20) in Eq. (13.21), we get transmission loss

$$\text{TL} = 10 \log \frac{1}{\alpha_t} \qquad \text{...(13.22)}$$

Consider that an enclosure consists of different kinds of surfaces with different transmission coefficients. So, the transmission loss is

$$\text{TL} = 10 \log \left[\frac{S}{S_1\alpha_{t1} + S_2\alpha_{t2} + S_3\alpha_{t3} + \text{..........}} \right] dB \qquad \text{...(13.23)}$$

where S is the total surface area of all surfaces in the enclosure

S_1, S_2, S_3, etc, are the surface areas

α_{t1}, α_{t2}, α_{t3}, etc, are the transmission coefficients of the sources 1, 2, 3 etc.

Generally, transmission loss is less for a randomly incident sound wave than that of normal incident wave.

13.16 REVERBERATION TIME

Reverberation time is the time required for the sound to disappear in a closed area. If a sound is created in a room and then the sound is stopped, the reflections from different surfaces will stay in the room afterwards for a short period. Reverberation time is defined as *the time during which the sound energy density falls from its steady state value to its one-millionth (10^{-6}) value after the source is shut off.*

$$\text{Reverbaration time} = \frac{I_{\text{final}}}{I_{\text{initial}}} = 10^{-6}$$

Reverberation time in terms of sound energy level in dB: If initial sound level is L_i and the final level is L_f, then we can write

$$L_i = 10 \log \frac{I_i}{I} \text{ and } L_f = 10 \log \frac{I_f}{I}$$

$$\therefore \quad L_i - L_f = 10\log\frac{I_i}{I_f}$$

As $$\frac{I_{\text{initial}}}{I_{\text{final}}} = 10^6, \; L_i - L_f = 10\log 10^6 = 60 \text{ dB}$$

We can define reverberation time in terms of dB.

Def: The reverberation time is the time in seconds during which sound energy diminishes by 60 dB after the sound source is stopped. RT60 is the standard reverberation time measurement.

Def: RT60, standard reverberation time is the time for the sound pressure to reduce by 60 dB. It is measured from the moment the generated test signal is abruptly ended.

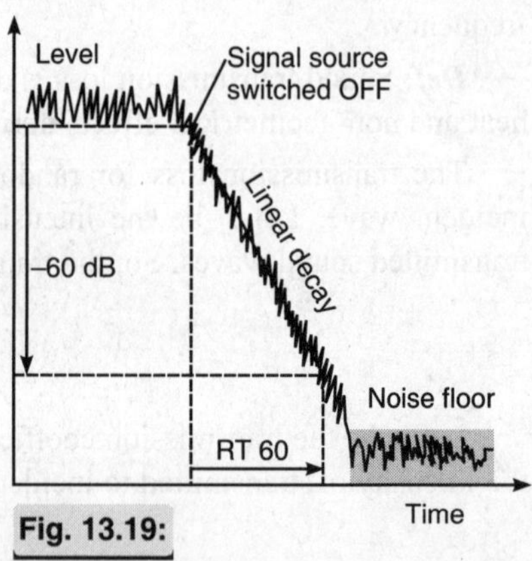

Fig. 13.19: Reverberation Time.

13.17 SABINE'S FORMULA FOR REVERBERATION TIME

Wallace C.Sabine (1868-1919) conducted some experiments and calculated the reverberation times of empty and furnished halls of different sizes.

According to his studies the reverberation time depends on:

(*i*) the reflecting properties of the surfaces such as walls, floor and ceiling of the hall. When these surfaces are good reflectors of sound, it would take longer time for the sound to die away and the reverberation time of the hall would be longer.

(*ii*) directly on the physical volume V of the enclosure.

(*iii*) the absorption coefficient of different surfaces such as carpets, cushions, curtains etc existing in the enclosure (hall, room etc).

(*iv*) the frequency of the sound waves. Absorption coefficient of the materials increases with frequency and high absorption coefficient at high frequency would lead to shorter reverberation time.

Prof. Sabine summed up his results in the following form.

Reverberation Time, $$T \propto \frac{\text{Volume of the Hall, } V}{\text{Absorption, } A}$$

or $$T = k\frac{V}{A}$$

where k is a proportionality constant. It is found to have a value of 0.161 when the dimensions are measured in metric units. Thus,

$$T = \frac{0.161\,V}{A} \quad \text{...(13.24)}$$

where A is the absorption coefficient is given by Eq. (3.16) as

$$A = \alpha_1 S_1 + \alpha_2 S_2 + \alpha_3 S_3 + \ldots = \sum_1^n \alpha_n S_n$$

Reverberation time, $$T = \frac{0.161V}{A} = \frac{0.161\,V}{\sum_1^n \alpha_n S_n} \quad \text{...(13.25)}$$

or Reverberation time, $$T = \frac{0.161V}{\alpha_1 S_1 + \alpha_2 S_2 + \alpha_3 S_3 + \ldots\ldots + \alpha_n S_n} \quad \ldots(13.26)$$

Eq. (13.26) is known as Sabine's formula and works well for large enclosures.

13.17.1 Optimum Reverberation Time

Sabine determined the time of reverberation for halls of various sizes. In these measurements, he used an organ pipe as the source, which was blown at a definite frequency and under a constant pressure. The instant of cutting off of the sound and the instant at which the observer ceased to hear the sound were recorded. And from the results, he deduced the reverberation time that is likely to be most satisfactory for the purpose for which a hall is built. Such satisfactory value is known as the ***optimum reverberation time***. The optimum reverberation time for music or speech lies between 0.6 to 0.75 s whereas it lies between 0.8 to 1.00 s for orchestra. The reverberation time of a hall can be adjusted to a desired value by arranging absorbent materials for the various surfaces of the hall.

Example 13.3: *For a hall has a volume of 1200 m^3 and total absorption of 480 m^2 of open window, find the effect on the reverberation time if audience fill the hall with additional absorption of 480 m^2 of open window?*

Solution: Reverberation time

$$T_1 = \frac{0.161V}{\sum \alpha S} = \frac{0.161 \times 1200\ \text{m}^3}{480\ \text{m}^3} = 0.40\ \text{s}$$

When the audience are present in the hall, the reverberation time is

$$T_2 = \frac{0.161V}{\sum \alpha S} = \frac{0.161 \times 1200\ \text{m}^3}{(480 + 480)\text{m}^3} = 0.20\ \text{s.}$$

Example 13.4: *A classroom has dimensions 20 × 15 × 5 m^3. The reverberation time is 3.5 sec. Calculate the total absorption of is surfaces and the average absorption coefficient.*

Solution: $$T = 3.5 = \frac{0.161V}{\sum \alpha S}$$

$$\sum \alpha S = \frac{0.161\,(20 \times 15 \times 5)\text{m}^3}{3.5\ \text{s}} = 69$$

$$\therefore \quad \alpha_{\text{average}} = \frac{69}{2(20 \times 15 + 15 \times 5 + 5 \times 20)} = \frac{69}{950} = 0.07$$

13.18 REVERBERATION THEORY

The Sabine formula Eq. (13.24) was originally an empirical formula. It was later derived from reverberation theory. The reverberation theory explains the nature of growth and decay of sound energy in an enclosure. The following assumptions are made in the theory.

(*i*) The enclosure (i.e. room) is large enough such that the sound energy is uniformly distributed in it.

(*ii*) Sound travels uniformly in all directions.

(*iii*) Absorption of sound by the air is neglected.

Let us consider a sufficiently big hall. Let a source produce sound in the hall. We assume that the sound fills the hall uniformly and hence each elemental portion of the volume of hall can be regarded as a source of sound energy. We first calculate the radiation of energy from the volume element dV in the hall and then the rate at which it is incident on the surface element ds of a plane wall, AB (see Fig. 13.20).

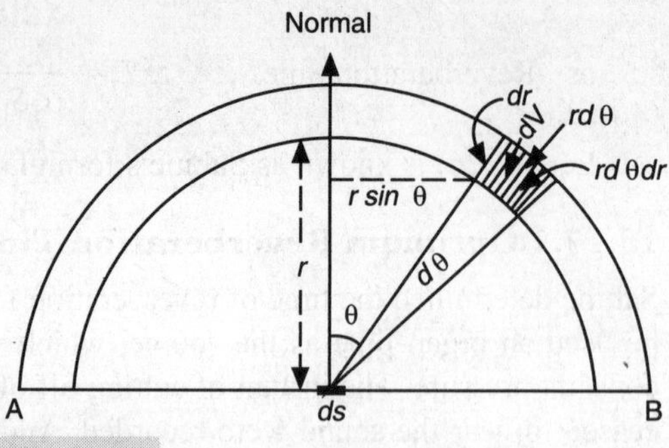

Fig. 13.20: Solid angle subtended by ds at dV.

Let us draw a normal from the center of the element ds. From the same center let us draw two circles with radii r and $r + dr$ in the plane containing the normal. Then two radii are drawn w.r.t. the normal at angles θ and $\theta + d\theta$. Consider the portion of the area (shaded area in Fig. 13.20) lying between the two radii and between the circles.

The arc length of this shaded area = $r\,d\theta$.

$$\text{Radial length} = dr$$

$\therefore$ Surface area of this element = $r\,d\theta\,dr$.

If the shaded elemental area is rotated through a small angle $d\phi$, this elemental area sweeps through a distance $r \sin\theta\,d\phi$.

Volume traced out by the elemental area, $dV = (r d\theta\, dr)(r \sin\theta\, d\phi) = r^2 \sin\theta\, d\theta\, dr\, d\phi$

The sound energy present in this volume element at any moment = $E\,r^2 \sin\theta\, d\theta\, dr\, d\phi$

where E is the sound energy per unit volume. Since the energy from the volume element travels uniformly in all directions, the energy traveling per unit solid angle along any direction is

$$dW = \frac{E\,dV}{4\pi}$$

The solid angle subtended by ds at dV is

$$d\Omega = \frac{ds \cos\theta}{r^2}$$

The amount of energy that reaches ds from dV is equal to the product of the energy traveling per unit solid angle and the solid angle subtended by ds at the volume element dV.

Thus, $$= \frac{E\,dV}{4\pi} \cdot \frac{ds \cos\theta}{r^2}$$

The amount of energy that reaches ds from dV

$$= \frac{E\,ds \sin\theta \cos\theta\, d\theta\, d\varphi\, dr}{4\pi}$$

The total energy received by ds in one second from the entire volume is obtained by integrating the above expression. Thus,

$$\text{Total energy} = \frac{E\,ds}{4\pi} \int_{\varphi=0}^{2\pi} \int_{\theta=0}^{\pi/2} \int_{r=0}^{\upsilon} \sin\theta \cos\theta\, d\theta\, d\varphi\, dr$$

$$= \frac{E\,dS}{4\pi}\upsilon \times 2\pi \times \int_{\theta=0}^{\pi/2} \sin\theta \cos\theta \, d\theta$$

$$= \frac{E\upsilon\,ds}{4} \qquad ...(13.27)$$

If α is the absorption coefficient of the surface of the wall *AB*, then the sound energy absorbed per second by the surface element *ds* is

$$dW_A = \frac{E\upsilon\alpha\,dS}{4}$$

Therefore, the total energy absorbed by all wall surfaces in the hall is

$$W_A = \frac{E\upsilon}{4}\sum \alpha\,dS$$

or

$$W_A = \frac{E\,\upsilon\,A}{4} \qquad ...(13.28)$$

where *A* is the total absorption of all surfaces.

13.18.1 Build-up of Sound in a Hall

Let *P* be the power of sound source and *V* the total volume of the hall. The total energy in the hall at a particular instant will be *EV* where *E* is the energy density at that instant.

$$\text{Rate of growth of energy in the hall} = \frac{d}{dt}(EV) = V\frac{dE}{dt}$$

At any instant,

$$\begin{pmatrix}\text{Rate of growth of}\\ \text{energy in the hall}\end{pmatrix} = \begin{pmatrix}\text{Rate of supply of}\\ \text{energy from the source}\end{pmatrix} - \begin{pmatrix}\text{Rate of absorption of}\\ \text{all surfaces in the hall}\end{pmatrix}$$

$$\therefore \quad V\frac{dE}{dt} = P - \frac{E\upsilon\,A}{4}$$

$$\frac{dE}{dt} + \frac{\upsilon\,A}{4V}E = \frac{P}{V} \qquad ...(13.29)$$

Putting $\frac{\upsilon\,A}{4V} = \alpha$, the above equation may be rewritten as

$$\frac{dE}{dt} + \alpha\,E = \frac{4P}{\upsilon A}\alpha$$

Multiplying with $e^{\alpha\,t}$ on both sides of the above equation, we get

$$\left[\frac{dE}{dt} + \alpha\,E\right]e^{\alpha t} = \frac{4P}{\upsilon\,A}\alpha\,e^{\alpha t}$$

$$\frac{d}{dt}\left[E\,e^{\alpha t}\right] = \frac{4P}{\upsilon\,A}\alpha e^{\alpha t}$$

Integrating on both the sides, we get

$$Ee^{\alpha t} = \frac{4P}{\upsilon A}e^{\alpha t} + K \qquad ...(13.30)$$

where *K* is the constant of integration.

The value of *K* may be found using the boundary conditions.

13.18.2 Growth of the Energy Density

If time t is measured from the instant the sound source emits sound, the initial condition becomes $E = 0$ at $t = 0$. Using these initial conditions into Eq. (13.30), we obtain the value of K.

$$K = -\frac{4P}{\upsilon A}$$

Using the value of K into Eq. (13.30), we get

$$E = \frac{4P}{\upsilon A}(1-e^{-\alpha t})$$

or $$E = E_m(1 - e^{-\alpha t}) \quad ...(13.31)$$

where $$E_m = 4P/\upsilon A$$

The Eq. (13.31) indicates that the sound energy density grows in an exponential manner with time, t. It increases till it attains the steady state value E_m, at $t = \infty$.

13.18.3 Decay of Sound Energy in the Hall

Let us assume that after a certain time that the energy attained the steady state value, the source of the sound is switched off. Then, P becomes zero and let that instant be taken as $t = 0$.

Then, the initial conditions would be $P = 0$ at $t = 0$ and $E = E_m$. Using these conditions into Eq. (13.30), we get

$$K = E\, e^{\alpha t}$$

As $$|K| = \frac{4P}{\upsilon A} = E_m,$$

we can write the above equation as

$$E = E_m\, e^{-\alpha t} \quad ...(13.32)$$

Eq. (13.32) indicates that the sound energy decays exponentially after the source of sound is switched off. Fig. 13.21 shows the growth and decay of sound energy in the hall.

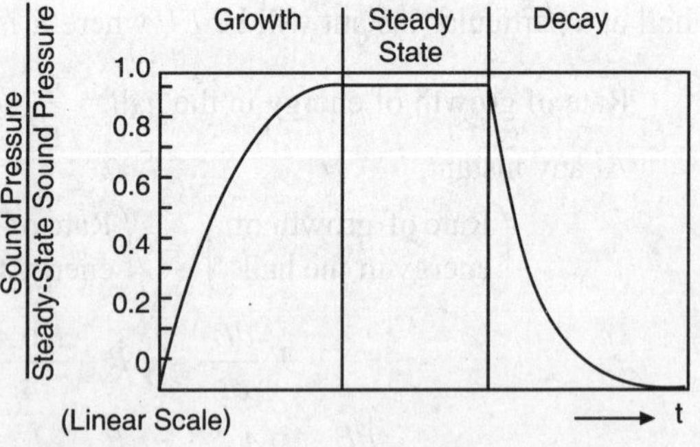

Fig. 13.21: Life cycle of sound energy in a hall.

13.18.4 Deduction of Sabine's Formula

From the definition of reverberation time,

$$\frac{E}{E_m} = 10^{-6}, \text{ at } \quad t = T.$$

It follows from Eq. (13.26) that $$e^{-\alpha t} = 10^{-6}$$

$$\therefore \quad e^{\alpha T} = 10^{6}$$

Taking logarithm on both the sides, we get

$$\alpha\, T = 6 \log_e 10 = 6 \times 2.3026$$

Using the expression for α, we get

$$\frac{\upsilon A}{4V} T = 6 \times 2.3026$$

or $$T = \frac{4 \times 6 \times 2.3026 \times V}{\upsilon A}$$

Taking $\upsilon = 344$ m/s $$T = \frac{4 \times 6 \times 2.3026 \times V}{(344)\,A} = \frac{0.161\,V}{A}$$

The above equation is identical to the empirical formula Eq. (13.18) that Sabine proposed for reverberation time.

13.19 NOISE

Generally, noise is an unwanted sound. It is undesirable. Noise, of sufficient intensity, disturbs activities such as hearing a lecture, music programme and masks the satisfactory hearing of speech and music in a hall. Noise comes as room noise, community noise, traffic noise, factory noise etc. It results in fatigue or leads to nervous breakdown. Long term exposure to high level noise gravely affects the physical and mental condition of an individual. Noise comes into a building by vibrations set up in the structure either due to air-borne or structure-borne noise

(*i*) **Air-borne noise:** The external noise coming to the building through air from distant sources is known as *air-borne noise*. It may enter a building directly through windows, doors and other openings and by transmission through walls and floors. Buildings requiring low air-borne noise must be located on quiet sites or may be shaded by introducing a buffer zone of trees, gardens. Knowing the noise level of incident air-borne noise and the desired noise criteria for a room, one can choose appropriate type of construction for walls, doors, windows etc.

(*ii*) **Structure-borne noise:** Noise travels about ten times faster in building structures than it does in air. Therefore, the structure-borne noise is more severe than the air-borne noise. Generally, in residential flats all rooms are interconnected and the structure borne-noise immediately transmits to the adjoining quarters. When such vibrations reach the partitions of a room they start oscillating and noise is radiated into the room.

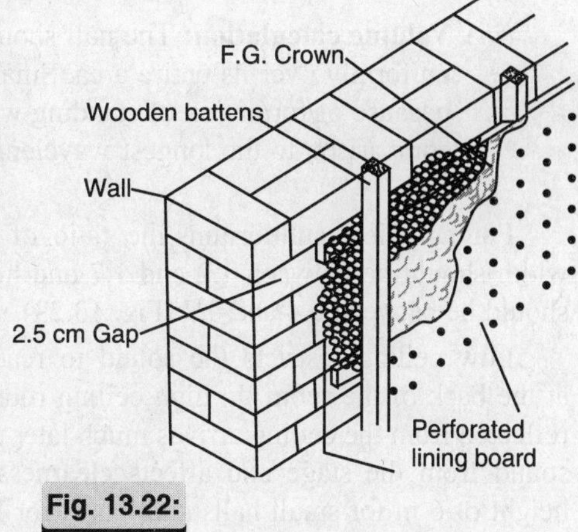

Fig. 13.22: Reducing structure-borne noise.

A noise from machine source may be reduced at the source by interposing flexible mountings (vibration isolators) between the machine and its foundation. Vibration isolators provided under machines generally consist of springs, cork, rubber felt, fiber glass or pneumatic cushions. Structure-borne noise may be prevented from passing from floor to floor by breaking the continuity of floors at their intersections. Continuity is generally broken by placing flexible materials such as bitumen. Fig. 13.22 illustrates the use of *fibre glasses* in noise-proof construction of floors.

13.20 DESIGN OF AN AUDITORIUM

We shall now discuss the general considerations for acoustical design of an auditorium aimed at using for staging dramas, for musical concerts and as public lecture halls. An auditorium is regarded to have right acoustical quality when;

(*a*) the initial sound from the source should be of adequate intensity

(*b*) the sound should be equally stretched over the whole area covered by the audience

(*c*) the successive sounds in the speech or music should be clear and distinct; and the tonal quality of music is not altered

(*d*) all undesired sound should be reduced so that they will not interfere with the normal hearing or speech

There are several factors that determine the acoustical quality of a hall. We discuss here six major factors that are to be considered in the acoustic design of a hall.

(*i*) **Selection of Site:** A proper site with quite surroundings is to be selected for an auditorium. It should be away from the busy high-way vehicular traffic, rail traffic, airport or any other noisy location. The orientation of the hall should be such that the external noise is maintained at low level. Otherwise, the vibrations produced by the traffic will be conveyed into the hall through the structures which contribute to the noise in the hall. When air-conditioning is not provided the doors and windows are to be kept open during performance. When air-conditioning is provided, care should be taken to reduce the plant noise and grill noise.

Through an appropriate orientation, layout and structural design, the background noise level in the hall should be kept at around 45 dB.

(*ii*) **Floor Area calculation:** The floor area required for the auditorium may be calculated by multiplying the proposed number of seats by area required per person and then adding the required passageway space and the desired stage area. Generally, 0.6 m^2 to 0.9 m^2 space is required per person.

(*iii*) **Volume calculation:** The hall should be big enough so that sound intensity spreads uniformly over its entire area. Smaller rooms lead to irregular distribution of sound because of formation of standing waves. When the length of the hall is very large in comparison to the longest wavelength of sound, the sound spreads uniformly in the hall.

For a typical auditorium, the ratio of the length to width should be between 1.2 and 1.7 and height to width should be between 0.4 and 0.7 (Fig. 13.23).

Low ceiling restricts the sound to reach the people at the back of the room. In high ceiling rooms, the sound reflected from the ceiling arrives much later than the direct sound from the stage and affects clearness. An average height of 6 m for small halls and 7.5 m for large halls are usually adopted. It is desirable to provide slight increase in the height of ceiling near the center of the hall.

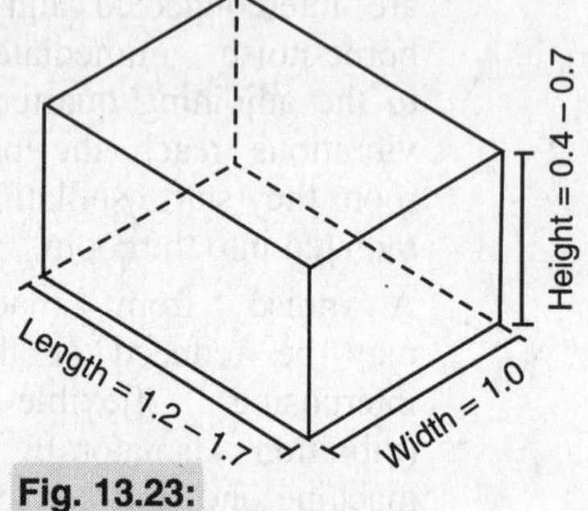

Fig. 13.23: Volume of Auditorium.

The recommended volumes for different types of auditoriums are as follows;

(*a*) Concert halls	4.0 to 5.5 m^3/person
(*b*) Theatres	4.0 to 5.0 m^3/person
(*c*) Public lecture halls	3.5 to 4.5 m^3/person

For Western symphony orchestras, high values of reverberation times are specified; therefore, higher volume per person may be required. The sound level must be around 65 dB for easy and distinct listening conditions. Sound amplification would not be required if the volume of the hall is not more than 1500 m^3.

(*iv*) **Shape of the Auditoriums Plan:** The shape of the hall plays a very important role in determining its acoustical quality. Its shape should be such that acoustical defects such as echoes flutter, sound foci are avoided and at the same time beneficial sound reflections improve the sound level at remote places. Any shapes such as traditional cross, rectangles, squares, circles, fans, pentagons, hexagons, other polygons and various irregular shapes can be used for auditorium. A rectangular hall with splayed stage walls and a pitched reflecting ceiling is quite satisfactory and easy to construct. Fig. 13.24 shows a few shapes of auditorium.

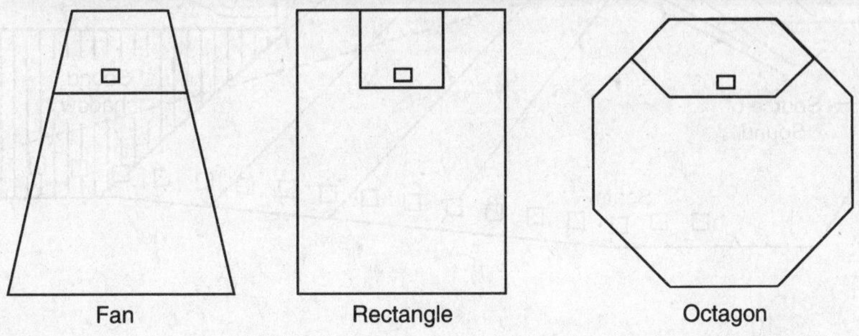

Fig. 13.24: Shapes of auditorium.

The side walls and ceiling are potentially useful reflecting surfaces and should be carefully designed to maximize their usefulness. The rear walls and floors are potential sources of useless and harmful reflections which are to be avoided. Parallel hard walls create echo problems. Use of splayed side walls greatly reduce the problem and enhance the acoustical quality of the room. In view of this a fan-shaped floor plan (Fig. 13.25) is preferred. The side walls are arranged to have an angle of not more than 100° with the curtain line. The fan shaped plan provides favorable reflection of sound from sides. A concave surface within the hall is not desirable because it focuses sound reflections. Such surface must be broken up with smaller convex surfaces so that sound is diffused in all directions.

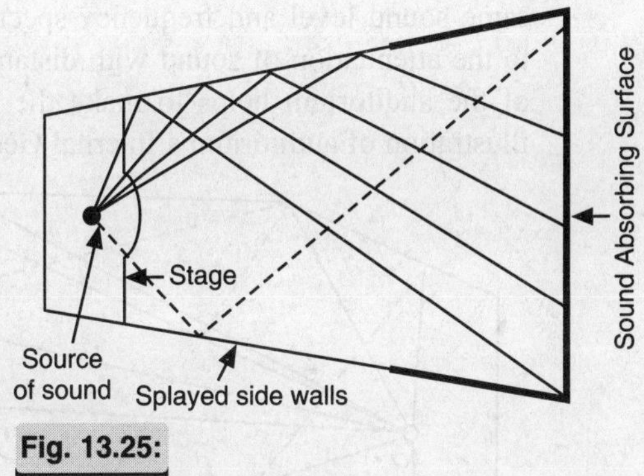

Fig. 13.25: Fan-shaped floor plan.

The length of the hall should be less than 45 m. When it exceeds 45 m, persons at the rear notice delay between the movements of the actors and the sound coming from them. To avoid excessive length of the hall, a balcony is provided at the far end for additional capacity. When balcony is provided, its projection into the hall should not be more than twice the free height of opening of balcony recess. If balconies are too deep, sound shadow forms and the persons in the seats below the balcony do not receive ceiling reflections (see Fig. 13.26). Suitable sound reflectors should be positioned at appropriate places to get rid of this defect.

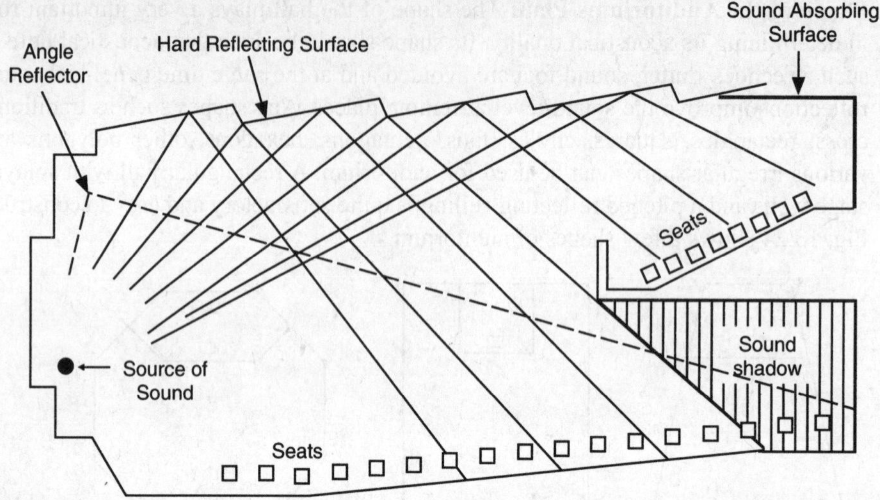

Fig. 13.26:

Ceiling reflections.

(*v*) **The Auditorium's Internal Geometry:** The angles of the walls, floors, balconies and ceilings significantly affect reflection of sound into areas that may be receiving less direct sound. The acoustically ideal room is one in which all seats receive the same sound level and frequency spectrum. But practically this is not possible due to the attenuation of sound with distance. A careful modeling of the internal angles of the auditorium helps to make the hall acoustically ideal. Fig. 13.27 shows an illustration of auditorium's Internal Geometry.

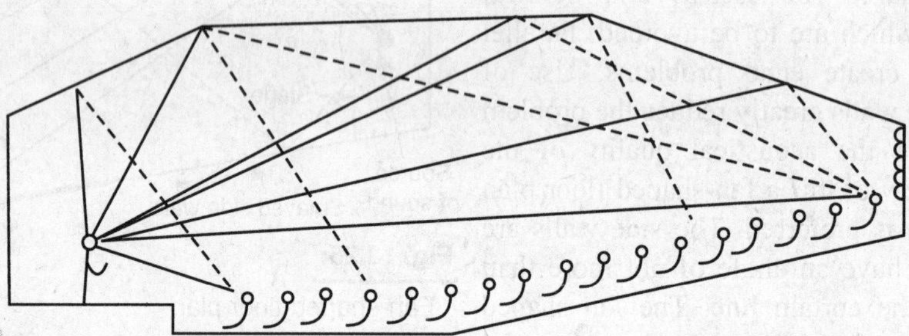

Fig. 13.27:

Auditorium's Internal Geometry.

These days, 3D computer modeling is available for showing the amount of direct and reflected sound for every seat in the auditorium. By adjusting interior wall and ceiling angles the level and concentrations of reflected sound versus direct sound can be determined, as well as the frequency spectrum of that sound.

(*vi*) **Reverberation Time:** A right reverberation time is essential for a hall to be acoustically good. The reverberation time should not be either too long or too short. A very short reverberation time makes a room 'dead' and a long reverberation time makes speech jumbled. A reverberation time of 0.5 s is acceptable for speeches and lectures, while a reverberation time of 1 to 2 s is satisfactory for concerts. In case of theatres, the optimum value varies with volume. 1.1 to 1.5 s is suitable for small theatres;

whereas for large theatres it may go up to 2.3 s. Table 13.4 shows reverberation time and acoustics; and table 13.5 shows the optimum reverberation time for halls.

Table 13.4: *Reverberation Time and Acoustics*

Reverberation time (seconds)	Acoustics
0.50 to 1.50	Excellent
1.50 to 2.00	Good
2.00 to 3.00	Satisfactory
3.00 to 5.00	Bad
> 5.00	Very bad

Table 13.5: *Optimum Reverberation Time for Halls*

Activity in Hall	Optimum Reverberation Time (in sec)	Audience Factor
Conference halls	1 to 1.5	One-third
Cinema theatre	1.3	Two-thirds
Assembly halls	1 to 1.5	Quorum
Public lecture halls	1.5 to 2	One-third
Music concert halls	1.5 to 2	Full
Churches	1.8 to 3	Two-thirds
Large halls	2 to 3	Full

Every material used in construction has an absorption coefficient and the amount of absorption varies with the frequency of sound. Carpets, drapes and curtains absorb mostly high frequencies while wood, sheetrock panels, and thin plaster on furring strips absorb lower frequencies. This must be considered in RT60 calculation. The amount of absorption in an auditorium should be fairly even throughout the frequency range. For an auditorium to be used mainly for speech needs a shorter RT60 than that mainly used for music. Contemporary music needs a very less RT60 value than orchestral music.

(*vii*) **Seating Arrangement:** The seats should be arranged in concentric arcs of the circles. Sloped floor seating is essential for a large audience to have good visibility and good acoustics. The successive rows of seats have to be raised over the preceding ones, with the result that the floor level raises towards the rear end. The rise in level may be about 8 to 13 cm per row. Further, the seats in each row should be staggered sideways in relation to those in front so that the line of sight of a person in any row is not obstructed by the person sitting in front of him. The back to back distance of chairs in successive rows should be at least 75 cm and this may be increased up to 106 cm for extra comfort.

The angle subtended with the horizontal at the front most observers, by the highest object to be seen on the stage, should not exceed 30°. On this basis, the distance of the first row should be about 4.5 m for movie watching and 3.6 m for theatres. In case of movies, synchronization of sound with lip movements is most essential; and in case of dramas, a person with normal vision should be able to discern facial expressions of the performers. In order to satisfy these conditions, it is recommended that the distance of the farthest seat from the curtain line should not exceed 23 m.

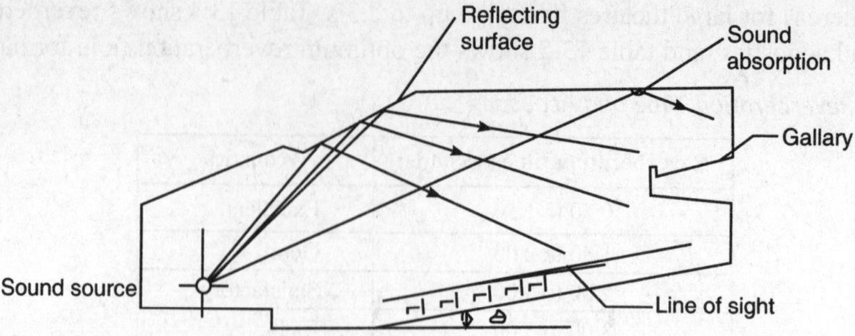

Fig. 13.28: Seating Arrangement.

The floor of the hall should be inclined at not less than about 8 degrees to the horizontal. With normal spacing of seats this inclination ensures a clearance in sight of about 12 cm from two consecutive rows. This clearance helps in auditory reception also. The seats provide better audiovisual reception if staggered, and in that case the sight clearance may be reduced to about 10 cm. The slope of the balcony floor shall be such that the line of sight is not inclined by more than 40° with the horizontal.

The seats should be so arranged that they face towards the central portion of the stage. This is achieved by arranging them in concentric arcs drawn from a point located behind the curtain at a distance equal to that between the curtain and the rear wall which carries the balcony. Back to back distance between two rows should be 85 cm to 105 cm, and the width of each seat should be 45 cm to 55 cm. The observers in the front row should not be required to tilt the line of sight by 30° form the horizontal to look at the tallest object on the stage. The front row therefore should be placed at not less than 3.5 m from the curtain line. The larger the curtain, the greater the distance from the front row to obtain easily the full horizontal and vertical field of view. Fig. 13.28 shows an illustration of seating arrangement.

(*viii*) **Acoustic Treatment of Interior Surfaces:** The interior surface of the hall should be given utmost attention to make the hall acoustically satisfactory. If the side walls are parallel, they are to be covered with absorbent materials from a length of about 7.5 m from the stage (proscenium) end. As the reflections from the near wall are of no use, the rear wall should be covered with absorbents. In large halls, a false ceiling is usually provided. The false ceiling positioned near the stage should be constructed of reflective material and inclined in a proper way to help reflections of sound from the stage to reach the rear seats in the hall. Concave shaped ceilings in the form of dome should be avoided. The rear portion of the ceiling may be treated with sound absorbing material so that build-up of audience noise is prevented. The floor should be covered with a carpet. Carpet on the floor not only covers a useless reflecting surface but also greatly reduces audience noise.

Acoustical treatment with sound absorbing materials is generally needed on the rear portions of the ceiling and side walls, the rear wall and the balcony. This will eliminate (delayed reflections) echoes. Acoustical materials must be applied in patches, preferably of irregular sizes and distribution for better sound diffusion. All seats must be fully covered (upholstered). Two-thirds of the total seating capacity may be assumed to be present at any time for calculating the sound absorption provided by the occupied seats.

(*ix*) **Sound Reinforcement System:** A sound reinforcement system is the arrangement of microphones, signal processors, amplifiers, and loudspeakers that makes sounds louder. It is also used to improve the sound of the sources on the stage. A sound reinforcement system may be very complex or as simple as a small PA system. It is to be noted that the fitting of a sound system cannot fix all acoustic problems in an auditorium. A good sound engineer has to design the reinforcement system taking building acoustics into account.

Example 13.5: *For an empty assembly hall of size 20 × 15 × 10 m³. The reverberation time is 3.5 s. Calculate the total absorption of its surfaces and the average absorption coefficient.*

Solution:

$$T = 3.5 = \frac{0.161\, V}{\Sigma \alpha S}$$

$$\Sigma \alpha S = \frac{0.161\,(20 \times 15 \times 5)\ \text{m}^3}{3.5\ s}$$

$$\therefore \quad \alpha_{\text{average}} = \frac{69}{2\,(20 \times 15 + 15 \times 5 \times 20)} = \frac{69}{950} = 0.07$$

Example 13.6: *Continuing the example 13.2, find area of the wall should be covered by the curtain so as to reduce the reverberation time to 2.5s? Given the absorption coefficient of curtain cloth is 0.5.*

Solution: Total absorption of the empty hall

$$A = \frac{0.161V}{T_1}$$

$$= \frac{0.161(20 \times 15 \times 5)}{3.5} = 69 \qquad \text{...}(i)$$

Average absorption coefficient

$$a_{av} = \frac{69}{2(20 \times 15 + 15 \times 5 + 5 \times 20)} = 0.07$$

When the walls are covered with curtain cloth of surface area S_1, the reverberation time T_2 becomes

$$T_2 = \frac{0.161\, V}{A + a_m S_1 - a_{av} S_1} \qquad \text{...}(ii)$$

where a_m is the absorption coefficient of curtain cloth.

Using (*i*) and (*ii*), we get

$$(a_m - a) = \frac{0.161\, V}{S_1}\left[\frac{1}{T_2} - \frac{1}{T_1}\right]$$

or

$$S_1 = \frac{0.161\, V}{(a_m - a)}\left[\frac{1}{T_2} - \frac{1}{T_1}\right]$$

$$= \frac{0.161(20 \times 15 \times 5)\text{m}^3}{0.5 - 0.07}\left[\frac{1}{3.5} - \frac{1}{2.5}\right]$$

The area of the wall to be covered with curtain

$$S_1 = 140 \text{ m}^2$$

Example 13.7: *A hall of volume of 2265 m³ with total absorption of 92.9 m² of open window. Calculate the effect on reverberation time if an audience fills the hall and the absorption increases by another 92.9 m² ?*

Solution: Reverberation time, $T = \dfrac{0.161V}{\sum \alpha S}$

Initial reverberation time, $T_i = \dfrac{0.161 \times 2265 \text{ m}^3}{92.9 \text{ m}^2} = 3.9 \text{ s.}$

With the audience of the total absorption

$$= 92.9 \text{ m}^2 + 92.9 \text{ m}^2 = 185.8 \text{ m}^2.$$

Final reverberation time $T_f = \dfrac{0.161 \times 2265 \text{ m}^3}{185.8 \text{ m}^2} = 1.95 \text{ s.}$

QUESTIONS

1. What is the nature of sound waves?
2. How do we classify sound?
3. What is the frequency range of sonic waves?
4. What is the difference between musical sounds and noise?
5. What are the characteristics of musical sounds?
6. Define noise.
7. How does loudness differ from intensity of a sound?
8. What are the units of measurement of loudness and of intensity?
9. Define decibel.
10. Explain phon and sone.
11. What is meant by reverberation? Is it desirable to have in a building?
12. What does sound absorption coefficient signify?
13. Why does acoustics play an important role in building designs?
14. State Weber-Fechner law.
15. Explain how the sound absorption coefficient of a material is determined.
16. Define reverberation time of a hall. Explain clearly what causes reverberation and how it can be minimized.
17. Explain the various requirements of a good auditorium.
18. Explain the terms reverberation and reverberation time. Deduce the Sabine's formula for the reverberation time.
19. Explain various factors affecting architectural acoustics and their remedies.
20. Derive Sabine's formula for reverberation time.
21. Discuss effect of following on reverberation time:
    (*a*) its size (*b*) nature of its wall surfaces and (*c*) audience
22. Define absorption coefficient of a material and describe a method for its determination.
23. Explain: Reverberation. Describe a method for determination of the sound absorption coefficient of material.

24. State and explain Sabine's formula for reverberation time and describe briefly how one can determine the sound absorption coefficient for materials like curtain cloth.
25. (*a*) A carpet is to be used to cove the floor in a music hall for absorption of sound. Describe the method to determine absorption coefficient of the carpet. (*b*) Explain reflection, reverberation and echo in case of sound in a hall.
26. Explain the terms: (*i*) reflection (*ii*) reverberation and (*iii*) echo of sound energy.
27. Define reverberation. Prove that total absorption at all the surfaces of the wall where the sound is falling is equal to *ECA* where *E* is energy density, *C* is velocity of sound and *A* is total absorption.
28. Write down the Sabine's formula. Explain the terms involved in it and describe the units of each of them.
29. What are the limitations of Sabine's formula? Discuss.
30. Explain how acoustic requirements of a good auditorium can be achieved.

PROBLEMS

1. Find the average absorption coefficient and the reverberation time for the data: volume of a room =1500 m^3, wall area of the room = 260 m^2, floor area = 1400 m^2, the ceiling area = 140 m^2 and average sound absorption coefficient for wall = 0.03 for the ceiling = 0.80 and the floor = 0.06. **[Ans:** 0.237 sabins; 1.93 s]
2. A hall has a volume of 12500 m^3 and reverberation time of 1.5 s. If 200 cushioned chairs of absorption of 1.0 O.W.U are further more added in the hall, find the new reverberation time of the hall? **[Ans:** 1.31 s]
3. For a room of volume = 600 m^3, wall area = 220 m^2, floor area =120 m^2, ceiling area = 120 m^2 and average sound absorption coefficient for the walls = 0.03, for the ceiling = 0.8 and for the floor = 0.06, find the average sound absorption coefficient and the reverberation time. **[Ans:** 0.24, 0.875 s]
4. Find the total absorption in the hall of a volume of 7500 m^3 if the reverberation time of 1.5 s. **[Ans:** 825 sabins]
5. Reverberation time is found to be 3 sec for an empty reverberation chamber and an acoustic sheet of 15 m^2 is suspended at the center of the reverberation hall. Calculate the coefficient of sound absorption of acoustic sheet if the volume of the chamber is 600 m^3.
6. Find the average absorption coefficient for a hall of volume 5500 m^3 with a sound absorbing surface of 750 m^2 and reverberation time of 2.3 s. **[Ans:** 0.513]
7. Reverberation time for a cubical chamber of 10 m width is 2.68 s. Calculate its average absorption coefficient. If one of the walls is covered with acoustic tiles the reverberation time will decrease to 2 s. Calculate the sound absorption coefficient of acoustic tiles.
8. The average reverberation time of a hall of volume 1200 m^3 is 1.5 s and area of interior surface is 3340 m^2. If the, find the absorption coefficient. **[Ans:** 0.2 sabins]
9. The volume of a hall is 475 m^3. The area of the wall is 200 m^2, area of the floor and ceiling each is 100 m^2. If the absorption coefficients of the wall, ceiling and floor are 0.025, 0.020 and 0.550 respectively, find the reverberation time for the hall. **[Ans:** 1.264s]
10. For a hall of volume 1, 20,000 m^3 and reverberation time of 1.5 s find the average absorbing power of the surface if the total absorbing surface area is 25,000 m^3? **[Ans:** 0.524 O.W.U/m^2]
11. A hall has a volume of 2265 m^3 and its total absorption of 92.9 m^2. How many persons should be accommodated in the hall so that the reverberation time becomes 2 s? Given that the absorption area of one person is 18.6 m^2 of open window. Find the reverberation time of the empty hall also. **[Ans:** 500 persons, 4s]
12. A hall of volume 1586 m^3 has reverberation time of 2 s. If the area of the sound absorbing surface is 650 m^2, find the average absorption coefficient. **[Ans:** 0.195]

14 Ultrasonics

Learning Objectives:

By the end of this chapter, you will be able to:

- **Learn about fundamentals of ultrasound including features, generation, detection and other properties of ultrasonic waves.**
- **Explain about process of determination of velocity in a liquid and solids.**
- **Describe industrial, marine and medical applications of Ultrasound.**

14.1 INTRODUCTION

The human ear is sensitive to sound waves of frequencies ranging from 16 Hz to 20 kHz. Waves of frequencies beyond the upper audible limit (f > 20 kHz) are called *ultrasonic waves* (Fig. 14.1). The range of ultrasonic frequencies lies between 20 kHz to 1 GHz. In the field of industrial material testing, ultrasonics is used at frequencies from 0.25 to 10 MHz. Frequencies, in the range 1-20 MHz, are used for medical ultrasound. Imaging ultrasound in the field of medical diagnostics ranges between 1 and 40 MHz.

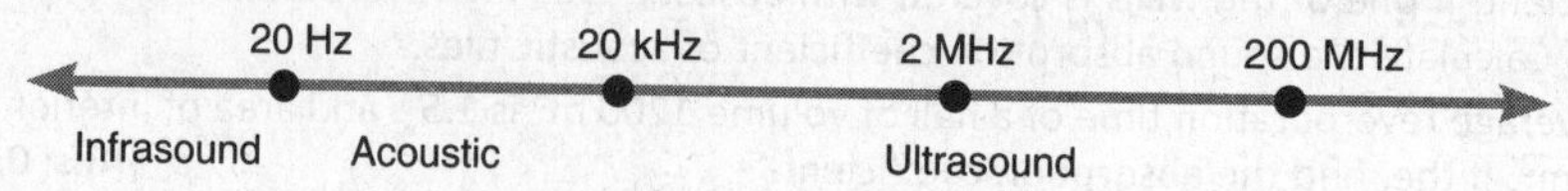

Fig. 14.1: Range of frequencies.

Human ear cannot identify the ultrasonic waves but dogs and other animals have an ability to hear the high frequency sounds. Bats and dolphins can generate ultrasonic waves and make use of the reflections of the waves to locate their way. The bat emits a series of short ultrasonic pulses at a frequency of 20 kHz to 60 kHz; the bat's large ears are specialized to detect these sounds (Fig. 14.2). The waves reflected from the surrounding objects are perceived by the bat and from the time elapsed between the generation and reflection of the pulses, the direction and distance of the objects are determined. Some of the marine animals use ultrasonic pulses to locate fish, to avoid obstacles etc. The usability of ultrasonic waves by sea animals is due to the fact that light is strongly absorbed by seawater and the radius of visibility is limited whereas ultrasonic waves are less absorbed by seawater.

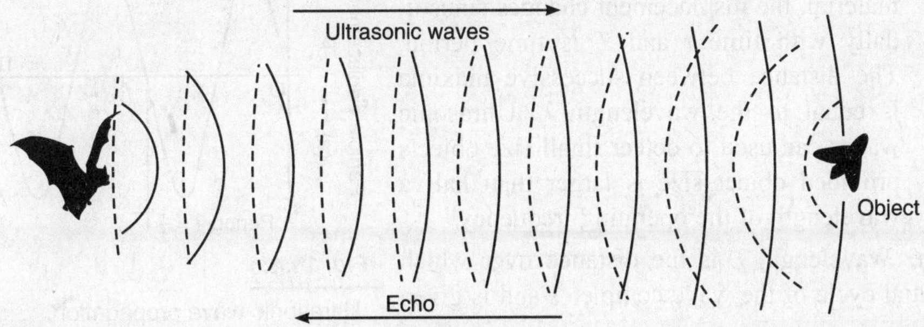

Fig. 14.2:

Prey detection by Bats, using ultrasonic waves.

In fluids ultrasonic waves are propagated as longitudinal waves whereas in solids they travel as both longitudinal as well as transverse waves. The wavelengths of ultrasonic waves are *very small and this smallness in wavelength makes them useful in many of their applications*. They travel in straight lines and can be concentrated at a given location. Ultrasonic waves are extensively used in medical diagnostics, marine applications, nondestructive testing of finished products and so on.

14.2 FUNDAMENTALS OF ULTRASOUND

(*i*) **Ultrasonic Waves in Media:** Ultrasound is sound that is above the frequency range of human hearing, 20 to 20,000 Hz. When a disturbance takes place at any point in an elastic medium, it propagates through the medium as a mechanical sound wave (elastic waves) by the vibrations of particles present. Compression (longitudinal) waves or shear (transverse) ultrasonic waves propagate through the solid. Fig. 14.3 shows schematics of ultrasonic waves in a solid.

Def: Ultrasonic waves are the elastic waves with frequency greater than 20,000 Hz and they normally exist in solids, liquids, and gases.

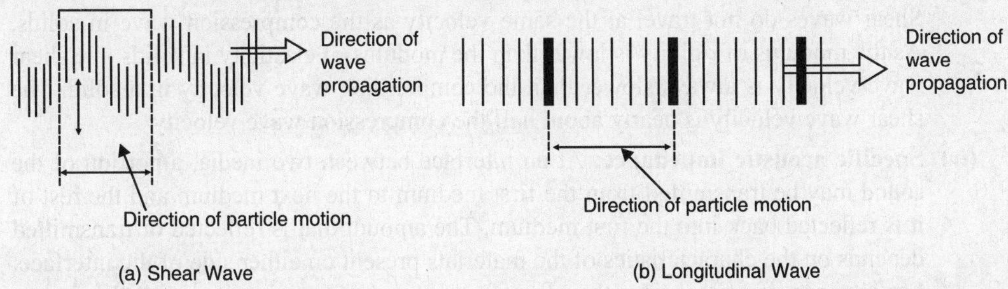

Fig. 14.3:

Schematics of ultrasonic (a) shear and (b) longitudinal in a solid waves

(*ii*) **Wavelength:** Fig. 14.4 shows ultrasonic wave propagation. At a fixed position in the material, the displacement changes sinusoidally with time t and T is time period. The distance between successive maxima is equal to the wavelength λ. Ultrasonic waves can used to detect small size objects provided object size is larger than half a wavelength of the operating frequency.

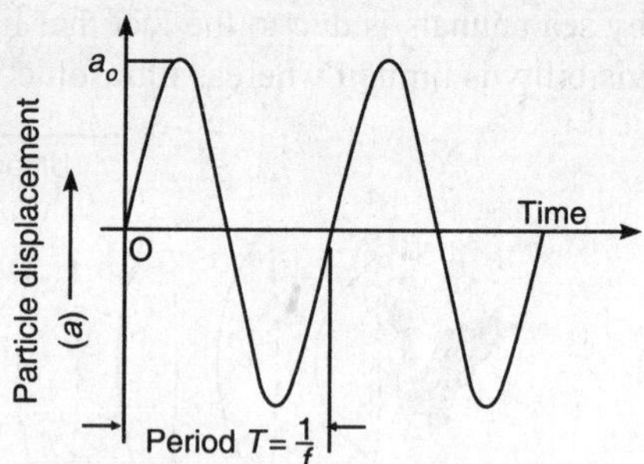

Fig. 14.4: Ultrasonic wave propagation.

Def: Wavelength λ is the distance over which one spatial cycle of the wave completes and is given as,

$$\lambda = \upsilon/f \quad \text{...(14.1)}$$

where υ is the ultrasonic velocity and f is the frequency.

(*iii*) **Velocity:** An ultrasonic wave velocity depends on the material properties and the frequency. So, ultrasonic waves do not propagate in vacuum because there are no particles. When ultrasonic wave PROPAGATES in a material medium it communicates motion to material particles which is denoted by particle velocity 'V'. Particle velocity is much smaller than wave velocity. Each medium has its own velocity which can be obtained from wave equations.

As an example for an isotropic solid.

$$\text{Compression wave velocity} \quad V_c = \sqrt{\frac{E}{\rho}\cdot\frac{1-\sigma}{(1+\sigma)(1-2\sigma)}} \quad \text{...(14.2)}$$

where Y = Young's modulus of elasticity, ρ = material density, σ = Poisson's ratio. Shear waves are also capable to exist in solids with velocity known as

$$\text{Shear wave velocity} \quad V_s = \sqrt{\frac{E}{\rho}\cdot\frac{1}{2(1+\sigma)}} = \sqrt{\frac{G}{\rho}} \quad \text{...(14.3)}$$

where G = modulus of rigidity, ρ = material density, σ = Poisson's ratio.

Shear waves do not travel at the same velocity as the compression wave in solids. As the modulus of rigidity is lower than the modulus of elasticity in solids, the shear wave velocity is always slower than the compression wave velocity in a solid. The shear wave velocity is nearly about half the compression wave velocity.

(*iv*) **Specific acoustic impedance:** At an interface between two media, a portion of the sound may be transmitted from the first medium to the next medium and the rest of it is reflected back into the first medium. The amount that is reflected or transmitted depends on the characteristics of the materials present on either side of the interface. For instance, if we consider the case of *steel to air* interface, almost all the energy gets reflected and nearly none goes into the air. On the other hand, if it is a steel to water interface, 88% of the energy is reflected and 12% is transmitted into the water.

The resistance offered to the propagation of an ultrasonic wave by a material is known as the *acoustic impedance*. It is given as

$$\text{Acoustic impedance,} \quad Z = \rho\,\upsilon \quad \text{...(14.4)}$$

where ρ is the density and υ is the wave velocity.

Def: The acoustic impedance describes the capability of a material to vibrate under an applied force. It is the resistance of the material to the passage of ultrasonic waves.

Acoustic impedance is important in knowing the acoustic transmission and reflection at a boundary of two media having different material properties. When the acoustic impedances of two media are equal the transmission of an ultrasonic wave from one medium to another becomes a maximum.

(*v*) **Acoustic pressure:** Acoustic pressure denotes the amplitude of alternating stresses on a material by a propagating ultrasonic wave. Acoustic pressure P is given in terms of acoustic impedance and the amplitude of particle vibration as

Acoustic pressure $$P = ZA \qquad \text{...(14.5)}$$

where Z is acoustic impedance and A is amplitude of particle vibration.

(*vi*) **Intensity:** Acoustic intensity describes the energy. It is defined as the energy transmitted per unit time (1s) and per unit area (1cm^2) in the direction normal to the considered area. It is measured in Wcm^{-2}. Intensity, acoustic pressure and acoustic impedance are related as:

$$I = \frac{p^2}{2Z} \qquad \text{...(14.6)}$$

(*vii*) **Reflection and Transmission:** When an ultrasonic wave is incident on an interface normally as in Fig. 14.5, a portion of the wave is reflected back into the medium 1 and the balance is transmitted to the medium 2.

For normal incidence, the reflected and transmitted waves are also normal to the interface.

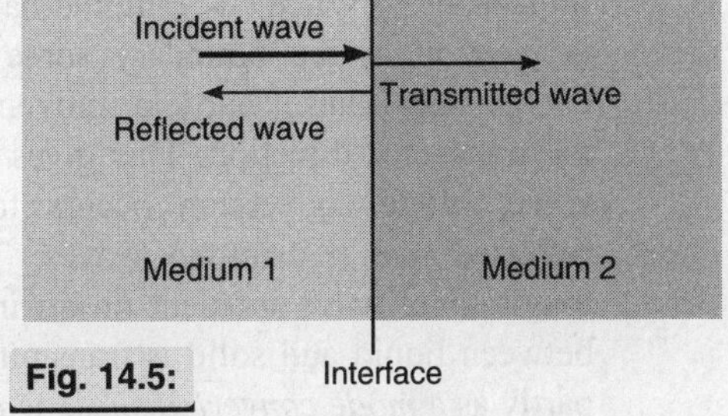

Fig. 14.5: Normal reflection and transmission at an interface.

Def: Amplitude reflection coefficient r is the ratio of the reflected to the incident acoustic pressure amplitude.

Def: Amplitude transmission coefficient t is the ratio of the transmitted to the incident acoustic amplitude.

Coefficients r and t are given by

$$r = \frac{Z_1 - Z_2}{Z_1 + Z_2}, \quad t = \frac{2Z_2}{Z_1 + Z_2} \qquad \text{...(14.7)}$$

Similarly, reflection coefficient (R) and transmission coefficient (T) are defined.

Def: Reflection coefficient R is the ratio of the amplitude of the reflected wave A_R to that of the incident wave A_I.

Def: Transmission coefficient T is the ratio of the amplitude of the transmitted wave A_T to that of the A_I.

Coefficients R and T are given by

$$R = \left(\frac{Z_1 - Z_2}{Z_1 + Z_2}\right)^2 \quad T = \frac{4Z_1Z_2}{(Z_1 + Z_2)^2} \qquad \text{...(14.8)}$$

where subscripts 1 and 2 refer to the medium 1 and 2, respectively, and

Z is the acoustic impedance.

From conservation of energy $T + R = 1$. The percentage of energy reflected at an interface

$$\text{Reflected energy} = \left(\frac{Z_1 - Z_2}{Z_1 + Z_2}\right)^2 \times 100\% \qquad ...(14.9)$$

The maximum transmission of ultrasonic wave happens when the impedances of the two media are identical and are reflected where there is a difference. This difference is known as the ***impedance mismatch***. Greater the impedance mismatch, the greater the reflection at the boundary.

(*viii*) **Refraction and Mode Conversion:** When ultrasonic wave is incident normally on an interface it transmits into second medium without any deviation. Fig. 14.6 shows oblique incidence. In oblique incidence, Snell's law gives the refraction angle if the velocities of the two media and the incidence angle are known.

In oblique incidence, a mode conversion occurs at the interface of different acoustic impedances. When a longitudinal wave is incident at the boundary some of the energy can cause particle movement in the transverse direction. This gives a shear wave. Mode conversion occurs for both reflected and transmitted waves. Thus, a longitudinal wave incident on an interface between liquid and solid is transmitted partly as a *refracted longitudinal wave* and partly as a *mode converted shear wave* in the solid.

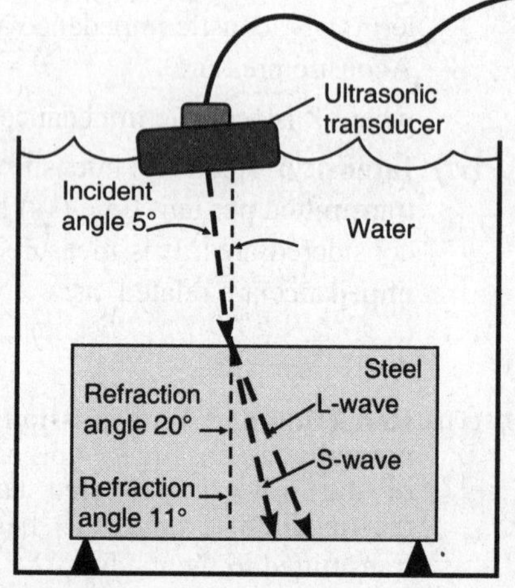

Fig. 14.6: Schematic of refraction and mode conversion at an interface.

(*ix*) ***Attenuation:*** When an ultrasonic wave propagates through a medium, there is a loss of energy, known as ultrasonic attenuation, reduction of amplitude of the wave. Some of the factors that affect the amplitude of the ultrasonic wave and cause attenuation are beam spreading, energy absorption, dispersion, scattering by inclusions and defects, Doppler Effect etc.

The ultrasonic attenuation is characterized quantitatively by *attenuation coefficient* α which is defined through the equation

$$A = A_0\, e^{-\alpha x} \qquad ...(14.10)$$

where A is the peak amplitude of the wave at propagation distance x,

A_0 is the initial peak amplitude.

Units of attenuation coefficient α are decibel per metre (dB/m) or neper per metre (Np/m). It can be determined by knowing the variation of the peak amplitude with the propagation distance. Generally it depends on frequency.

14.3 FEATURES OF ULTRASONIC WAVES

Sound travels in air as *longitudinal waves*. However, a number of different types of sound waves are possible in solids as molecules in a solid can support vibrations in other directions. There are two types of ultrasonic waves.

(*i*) **Bulk waves:** These waves propagate entirely inside an object and are independent of its boundary and shape. Bulk waves are of two types in an isotropic medium such as longitudinal and shear waves (Fig. 14.7).

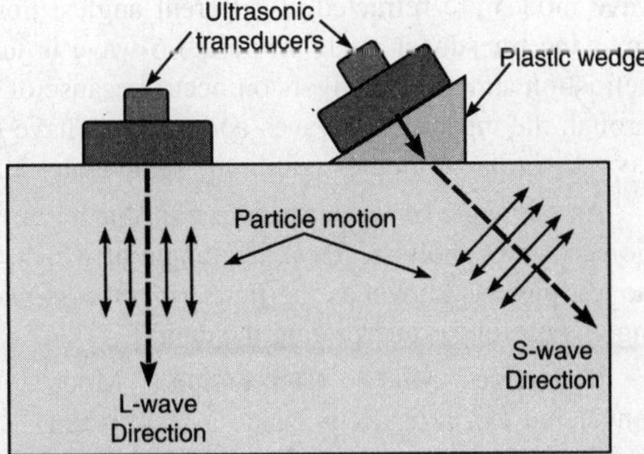

Fig. 14.7: Bulk waves-Longitudinal and shear waves.

(*a*) **Longitudinal waves:** The waves in which the particle motion occurs parallel to the direction of the wave propagation are known as ***longitudinal waves***. On the other hand, waves in which the particle motion is perpendicular to the direction of the propagation are called ***shear waves***. Both longitudinal waves and shear waves can exist in solids because they have rigidity (resistance) to shear as well as compressive loads unlike liquids and gasses. But the shear waves cannot exist in liquids and gasses because of no resistance to shear roads. Longitudinal are also called as *pressure or compression waves*. They are also occasionally known as *density waves* because material density fluctuates as the wave moves.

(b) **Shear waves:** When compared to longitudinal waves, the shear waves are weaker as shear waves are generated in materials drawing some of the energy from the longitudinal waves. The angle of incidence of ultrasonic waves determines whether longitudinal, shear, or both waves are produced. Fig. 14.8 shows an ultrasonic wave into a block of steel. For normal incidence on the steel block there is no change in path and the *L*-wave is preserved (Fig. 14.8).

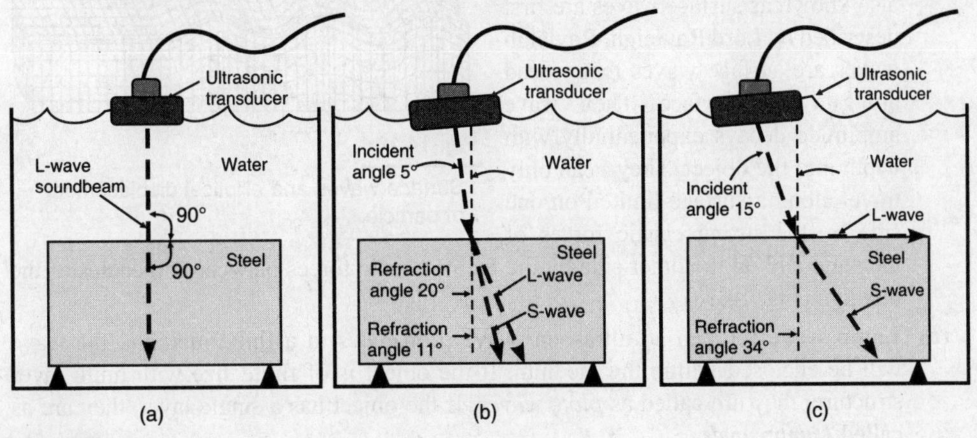

Fig. 14.8: Propagation of ultrasonic wave into a block of steel.

As the ultrasonic transducer is moved through an angle, refraction and mode conversion happens. When the transducer is at an angle of 5° [Fig. 14.8 (b)], the longitudinal wave coming from the transducer gets converted into two modes, longitudinal and shear. Both the wave modes are refracted at different angles. For example, the *L*-wave makes nearly four times the transducer angle while the *S*-wave is just at over two times the transducer angle. Refraction and mode conversion occur because of the change in *L*-wave velocity as it passes through the interface. *L*-waves and *S*-waves have different angles of refraction because they have dissimilar velocities within the same material.

As the angle of the ultrasonic transducer continues to increase, *L*-waves move closer to the surface of the test piece. The angle at which the *L*-wave is parallel with the surface of the test piece is known as the ***first critical angle and can be calculated by*** Snell's Law. This angle is useful for interpreting the data.

Reflective Mode Conversion: Mode conversion can also occur inside a solid when an ultrasonic beam gets reflected at an angle from an interface. A compression wave *C*, strikes at a steel to air interface at an angle of incidence *i*°, to the normal (Fig. 14.9). The compression wave, *C*, is reflected at an angle *r*° equal to the angle of incidence. In addition to the compression wave, there is also a mode converted shear wave, *S*, at an angle *s*°.

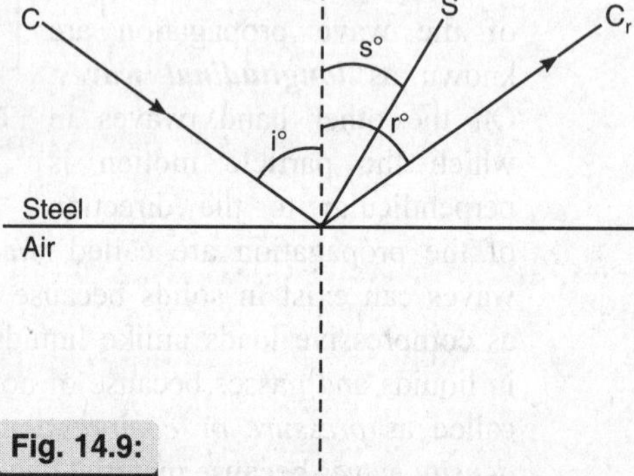

Fig. 14.9: Reflective Mode Conversion at a steel to air interface

(*ii*) Guided waves: When the influences of the boundaries or shape of an object are considered, other types of waves are produced. Guided waves are the waves that propagate near the surface or along the interface of an object. There are three types of guided waves depending on geometry of an object: Rayleigh waves, Lamb (plate) waves, and rod waves.

(*a*) Rayleigh waves: Rayleigh waves, also known as surface waves are first described by Lord Rayleigh. Rayleigh waves are surface waves that spread along a free surface. Their wave amplitude decays exponentially with depth into the object. They can only travel along a surface limited on one side by the strong elastic forces of the solid and on the other side by the feeble elastic forces between molecules of the medium (Fig. 14.10).

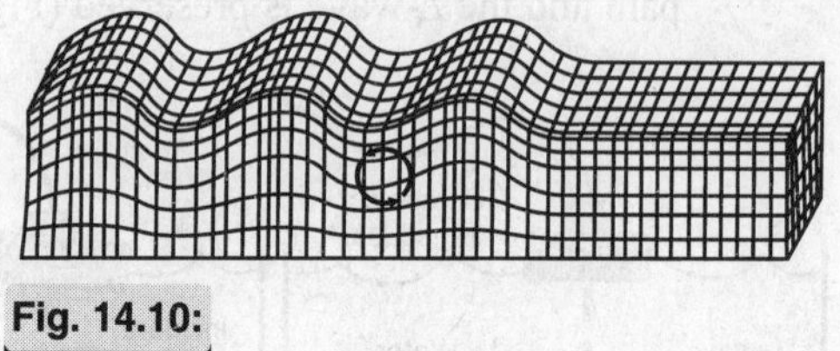

Fig. 14.10: Surface waves and elliptical displacement of particle.

(*b*) Lamb waves: When an ultrasonic wave propagates in a finite medium, the wave will be enclosed within the medium. If the object is of finite size with multi layer structure, they are called as *plate waves*. If the object has a single layer, they are as called *Lamb waves*.

Symmetrical and asymmetrical modes are the two most common modes possible for Lamb waves (Fig. 14.11). Symmetrical Lamb waves move in a symmetrical fashion about the median plane of the medium (plate). The symmetrical Lamb wave "stretches and compresses" the plate in the wave motion direction. Therefore it is also known as the extensional mode. Symmetrical mode is produced when the exciting force is parallel to the plate.

The asymmetrical Lamb wave mode called the "flexural mode", a great portion of the motion occurs in a normal direction to the plate, and a small in the direction parallel to the plate. In this mode the two surfaces move in the same direction and the body of the plate bends.

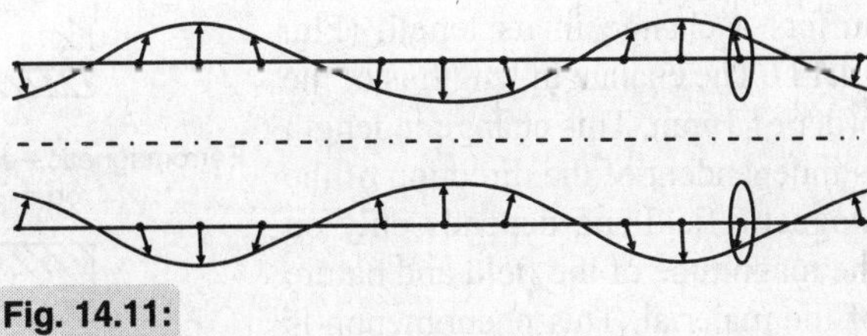

Fig. 14.11: Lamba waves bounded propagation on plate surface.

(*iii*) **Rod waves:** Rod waves are wave propagations in rod-like structures such as a thin rod and hollow cylinders. The wave propagation characteristics of guided waves strongly depend on material properties, plate thickness, rod diameter, and the frequency.

14.4 GENERATION OF ULTRASONIC WAVES

We discuss Galton method, magnetostriction method and piezoelectric effect for the generation of ultrasonic waves.

14.4.1 Galton Whistle

Galton whistle is one of method for generation of ultrasonic waves and it uses the principle of organ pipe. Fig. 14.12 shows the arrangements and comprises a closed end air column A whose length can be adjusted using a movable piston. The piston *P* can be moved to the preferred spot with the help of a screw S_2. The open end of the pipe *A* is fitted with a lip *L*. The gap between the ends of *A* and *C* can be adjusted with the help of the screw S_1.

Air is blown through the nozzle *N* located at the top. The gust of air emerging from the other end *C* strikes against the lip *L* and the column of air in the pipe is set into vibration. With the help of the piston *P* the length of the air column in *A* is adjusted and is brought to the resonant condition. The resonance frequency is governed by the length and diameter of the pipe *A*.

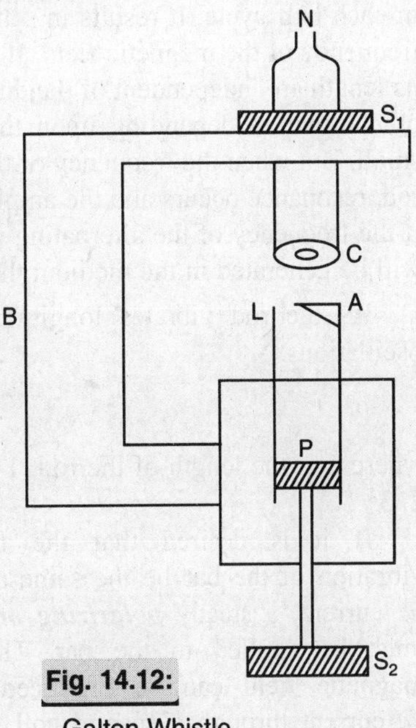

Fig. 14.12: Galton Whistle.

The wavelength is given by

$$\lambda = 4\,(l + x)$$

where l is the length of the air column in A, and x the end correction.

The frequency of sound is $$v = \frac{v}{\lambda} = \frac{v}{4\,(l + x)} \qquad \text{...(14.11)}$$

With the help of this whistle, frequencies of the order of 30,000 Hz can be produced.

14.4.2 Magnetostriction Effect

Magnetostriction method is used to produce waves in the frequency range of 20 kHz to 100 kHz. Joule discovered the phenomenon of magnetostriction in 1847. When a rod of ferromagnetic materials such as iron or nickel is kept in a magnetic field parallel to its length, the rod suffers a change in its length (Fig. 14.13). The change in length is of the order of 1ppm. This change in length is independent of the direction of the magnetic field and depends only on the magnitude of the field and nature of the material. This phenomenon is known as *magnetostriction*. Nickel exhibits a large magnetostriction effect compared to other ferromagnetic materials.

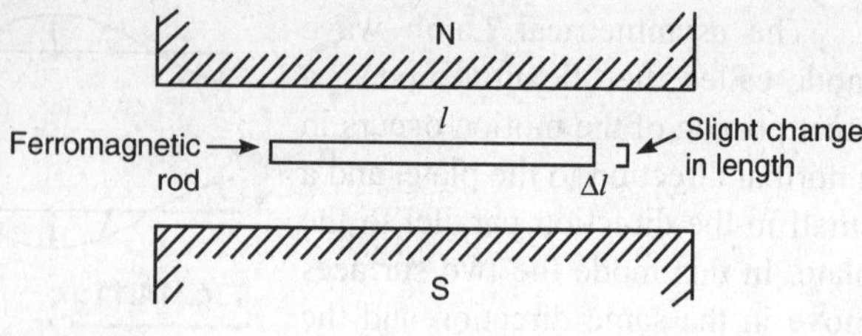

Fig. 14.13: Magnetostriction Effect.

A simple method of producing longitudinal vibrations is to apply an *ac* magnetic field parallel to the axis of the rod of a ferromagnetic material. An *ac* magnetic field is produced by wounding coil of wire around the rod and by passing an alternating current through the wire. If the alternating magnetic field oscillates at frequency *f*, the rod changes in length once in each half cycle. It results in setting up vibrations in the rod whose frequency is twice the frequency of the magnetic field. If the rod is not magnetized initially, the resulting changes in its length are independent of the direction of the field. The change may be either an elongation or contraction depending upon the material. Normally, the amplitude of the vibrations is small. But when the frequency of the alternating fields is equal to the natural frequency of the rod, resonance occurs and the amplitude of the vibrations will be considerably larger. Further, if the frequency of the alternating field lies in ultrasonic range, an ultrasound of frequency $2f$ will be generated in the medium that is in contact with the ends of the rod.

As the rod vibrates longitudinally, the following relation governs the frequency of oscillations.

$$f = \frac{m}{2L}\sqrt{\frac{Y}{\rho}} \qquad ...(14.2)$$

where L is the length of the rod, Y the Young's modulus, ρ the density of the rod and $m = 1, 2, 3, ...$.

If it is desired that the frequency of vibration of the bar be the same as that of the *ac* current, a steady ***polarizing magnetic field*** must be applied to the bar. The polarizing magnetic field can be produced by passing *dc* current through a second coil, as shown in Fig. 14.14. If the magnitude of the polarizing magnetic field is greater than that of the *ac* field, the frequency of the vibration of the bar will be equal to that of the *ac* magnetic field.

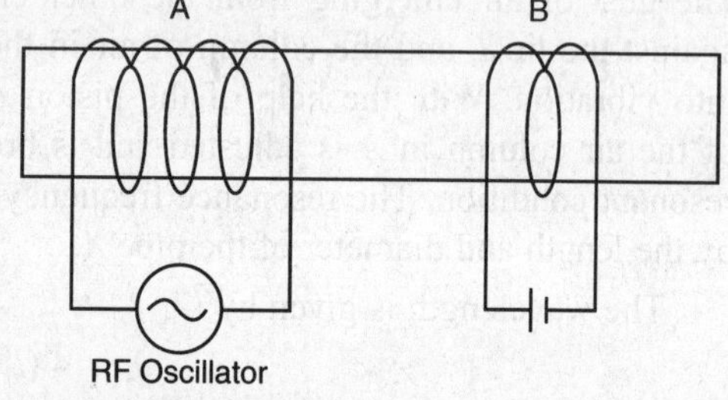

Fig. 14.14: Magnetostrictive Vibrator.

14.4.2.1 Magnetostriction Generator

The circuit diagram of a magnetostriction ultrasonic generator using transistors is shown in Fig. 14.15. It is basically a Colpitt's oscillator. The transistor T is biased with the help of the resistances R_1, R_2, R_3, and R_4. The inductance L and capacitors C_2 and C_3 constitute the tank circuit. When the circuit is switched on, oscillations build up in the tank (resonant) circuit. The oscillations are fed back to the transistor base through the feedback capacitor C_6. The appropriate frequencies are amplified and the oscillations corresponding to them are sustained. The oscillations appearing at the output terminals of the oscillator circuit are fed to a current amplifier, which raises the level of the oscillations. The output of the current amplifier is fed to the magnetostriction coil through a coupling capacitor C_5. Under the action of the high frequency electrical signal passing through the coil, the magnetization of the nickel bar within the coil varies and as a result, it produces ultrasonic waves.

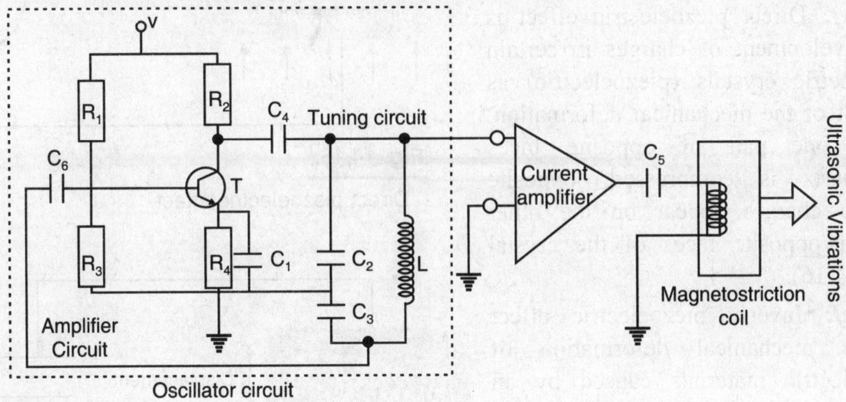

Fig. 14.15:

The circuit diagram of a magnetostriction ultrasonic generator.

Advantages

(*i*) Magnetostrictive materials are inexpensive.

(*ii*) Large output power can be produced.

Limitations

(*i*) Frequencies higher than 300 kHz cannot be generated.

(*ii*) Single frequency oscillations cannot be generated.

Example 14.1: *Calculate the natural frequency of 40 mm length of a pure iron rod. Given the density of pure iron is 7.25×10^3 kg/m^3 and its Young's modulus is 115×10^9 N/m^2. Can you use it in magnetostriction oscillator to produce ultrasonic waves?*

Solution: $f = \dfrac{n}{2L}\sqrt{\dfrac{Y}{\rho}} = \dfrac{1}{2\times40\times10^{-3}\,m}\left[\dfrac{115\times10^9\,\text{N/m}^2}{7.25\times10^3\,\text{kg/m}^3}\right]^{\frac{1}{2}}$ = **49.75 kHz.**

It can be used in magnetostriction oscillator to produce ultrasonic waves.

Example 14.2: *If Y of iron is 11.6 × 1010 N/m^2 and ρ of iron is 7.23 × 10^3 kg/m^3 find the length of an iron rod to produce ultrasonic waves of 20 kHz.*

Solution: $f = \dfrac{n}{2L}\sqrt{\dfrac{Y}{\rho}} \quad \therefore \quad L = \dfrac{n}{2f}\sqrt{\dfrac{Y}{\rho}} = \dfrac{1}{2\times20\times10^3\,\text{Hz}}\left[\dfrac{11.6\times10^{10}\,\text{N/m}^2}{7.23\times10^3\,\text{kg/m}^3}\right]^{\frac{1}{2}}$ = **1 m.**

14.4.3 Piezoelectric Effect

The French physicists Pierre Curie and Paul-Jean Curie discovered the piezoelectric effect in 1880. Crystals that exhibit piezoelectric effect are called piezoelectric crystals. Ammonium phosphate, quartz, PZT (lead zirconate titanate) etc are examples of piezoelectric materials.

Two types of piezoelectric effect, direct and inverse are possible. The inverse piezoelectric effect is used in producing ultrasonic waves.

Def: Direct piezoelectric effect is the development of charges in certain asymmetric crystals (piezoelectric) as a result of the mechanical deformation. When one pair of opposite faces of quartz is compressed, opposite electric charges appear on the other pair of opposite faces of the crystal (Fig. 14.16).

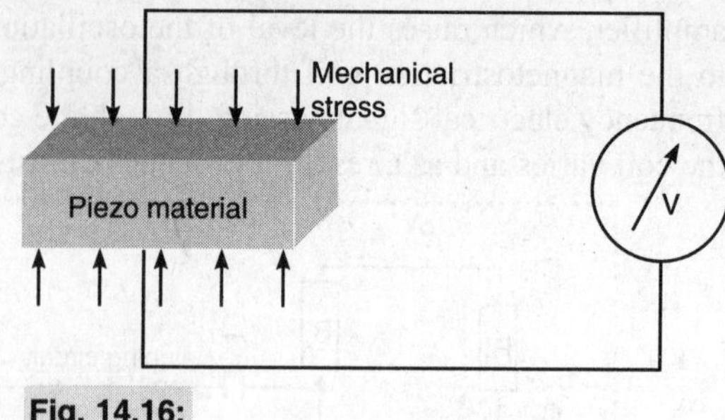

Fig. 14.16:

Direct piezoelectric effect.

Def: Inverse piezoelectric effect is the mechanical deformation of piezoelectric materials caused by an external electric. If an electric field is applied across one pair of faces of a piezoelectric crystal, it gets deformed along the direction of the other opposite pair of faces (Fig. 14.17). If an alternating voltage is applied between the two opposite faces of the crystal, it vibrates with the frequency of the field.

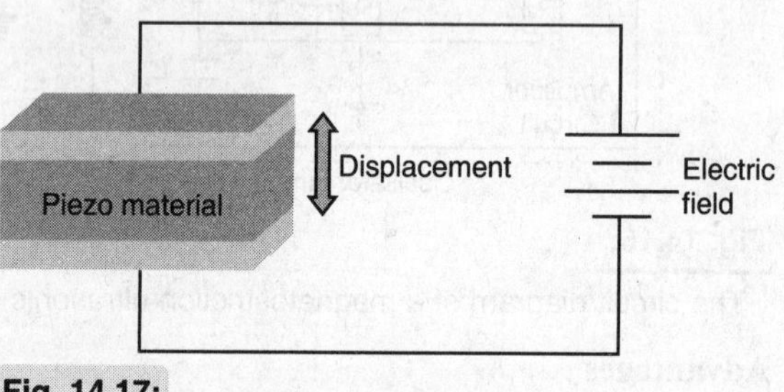

Fig. 14.17:

Inverse piezoelectric effect.

14.4.3.1 Piezoelectric Generator

The circuit diagram of a piezoelectric ultrasonic generator using transistors is shown in Fig. 14.18. It is basically a Hartley oscillator. The transistor is biased using the network of resistances R_1, R_2, R_3, and R_4. The coils L_1, L_2 and capacitor C_4 constitute the tuning (resonant) circuit. The tuning circuit is coupled to the transistor T through the coupling capacitor C_2. Capacitor C_3 provides the positive feedback to the amplifier T. The oscillations generated

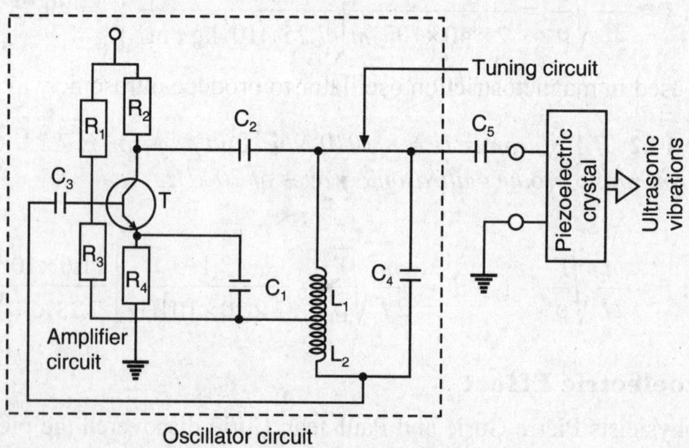

Fig. 14.18:

Piezoelectric ultrasonic generator.

by the tank circuit are sustained and the electrical signal obtained at the output is applied to the electrodes of the piezoelectric crystal through the coupling capacitor C_5. Because of high frequency electrical signal applied to it, the piezoelectric crystal produces ultrasonic waves. The frequency of these ultrasonic waves can be varied by varying the values of the components of the tuning circuit.

Piezoelectric diaphragm: It consists of a piezoelectric plate with electrodes on both sides. Applying a D.C. voltage to the electrodes causes mechanical distortion by expanding in the radial direction causing the metal plate to bend due to the piezoelectric effect. Reversing the polarity of the D.C. voltage causes the ceramic plate to shrink, bending the metal plate in the opposition direction. When A.C. voltage is applied across the electrodes (Fig. 14.19), the diaphragm alternates bending in the two directions. The repeated bending motion produces ultrasonic waves in the air.

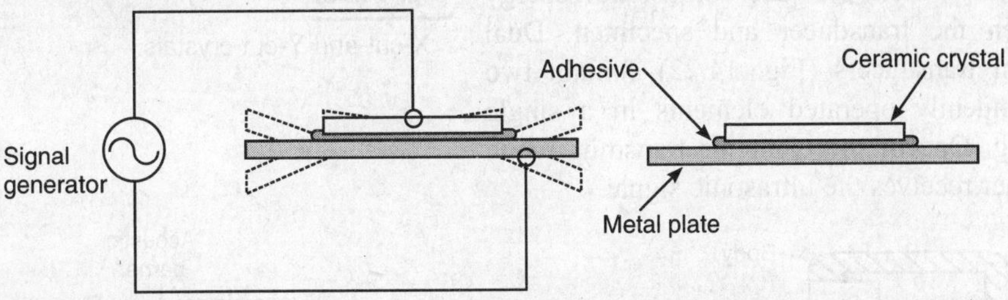

Fig. 14.19:

Piezoelectric diaphragm.

Advantages:

(*i*) High frequency waves of frequencies up to 500 MHz can be generated.

(*ii*) Single frequency output can be obtained

(*iii*) A range of frequencies can be covered using different transducers.

Example 14.3: *Calculate the natural frequency of ultrasonic waves using the following data. Thickness is 5.5 × 10^{-3} m, Young's modulus is 8.0 × 10^{10} N/m² and Density of quartz plate is 2.65 × 10^3 kg/m³.*

Solution: $$f = \frac{n}{2l}\sqrt{\frac{Y}{\rho}} = \frac{1}{2\times5.5\times10^{-3}\,m}\left[\frac{8.0\times10^{10}\,\text{N/m}^2}{2.65\times10^3\,\text{kg/m}^3}\right]^{\frac{1}{2}} = \mathbf{49.9\ kHz.}$$

14.4.3.2 Piezoelectric Transducers

Piezoelectric transducers are used to generate ultrasonic waves and are made from piezoelectric crystals such as Quartz, Lithium Niobate, Tourmaline, Gallium Orthophosphate or Lead Zirconate Titanate (PZT).

Thin plates of the quartz crystal cut perpendicular to one of its *x*-axis are known as *x*-cut plates. Similarly, thin plates cut perpendicular to one of its *y*-axis are known as *y*-cut plates (Fig. 14.20). *x*-cut plates generate longitudinal mode of ultrasonic vibrations of frequencies up to several hundred kHz, while *y*-cut plates generate transverse mode of vibrations of frequencies ranging from 1 MHz to 10 MHz.

Piezoelectric transducer construction: In a casing it consists of a thin quartz plate, electrical connections, backing materials, front layers etc. (Fig. 14.21). The front layer is to guard the piezoelectric element against outside stresses and environmental influences. The backing material is useful as a damping block. It changes the resonance frequency of the piezoelectric element and deletes redundant reflections. The electrical voltages are operated at the resonant frequency of the piezoelectric element. A suitable coupling medium effectively transmits the ultrasonic energy between the transducer and specimen. Dual element transducers (Fig. 14.22) contain two independently operated elements in a single housing. One of the elements transmits while the other receives the ultrasonic signal.

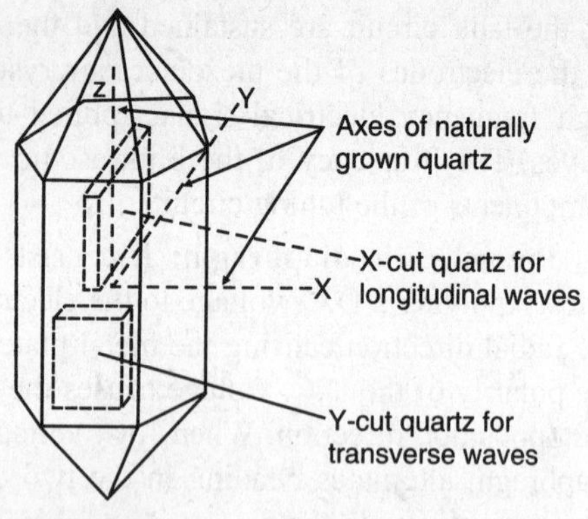

Fig. 14.20:

X-cut and Y-cut crystals.

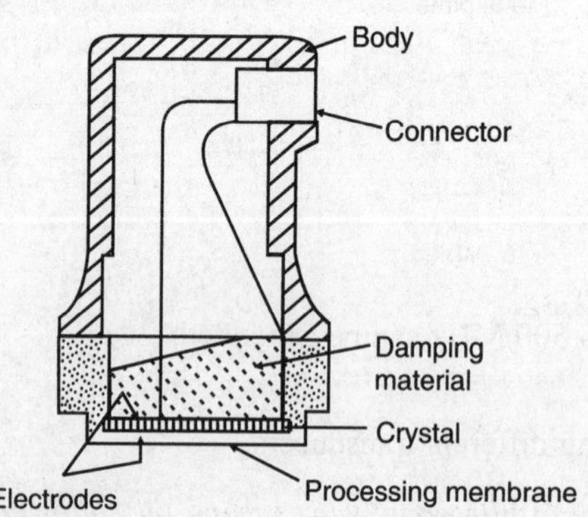

Fig. 14.21:

Typical construction of a piezoelectric transducer.

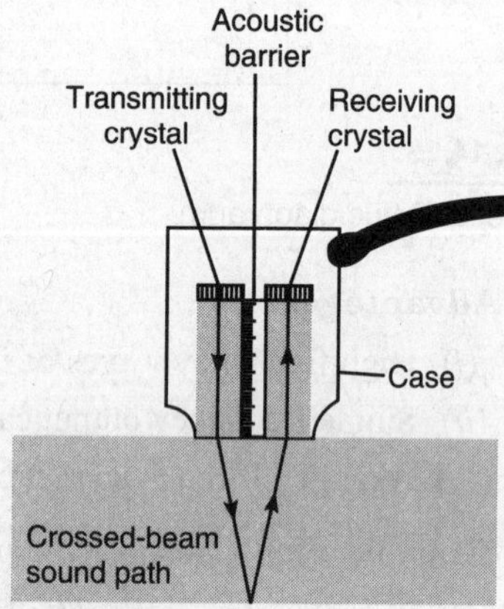

Fig. 14.22:

Dual element transducer.

14.5 DETECTION OF ULTRASONIC WAVES

(*i*) **Kundt's tube Method:** In 1866, the German scientist Adolph Kundt developed a method to detect ultrasonic waves of relatively longer wavelengths. Kundt's tube consists of a horizontal glass tube about 1 m long and 5 cm in diameter supported horizontally (Fig. 14.23). The tube has a vibrating piston at one end and is closed at the other.

A small amount of lycopodium powder is sprinkled along the inner surface of the tube. When longitudinal waves are excited at the piston *P* and pass through the tube, the lycopodium powder collects in the tube in the form of heaps at the antinodes and is blown off at the nodes. The appearance of heaps indicates the presence of waves and the mean distance between two successive heaps is equal to half the magnitude of the wavelength ($\lambda/2$) of ultrasonic waves.

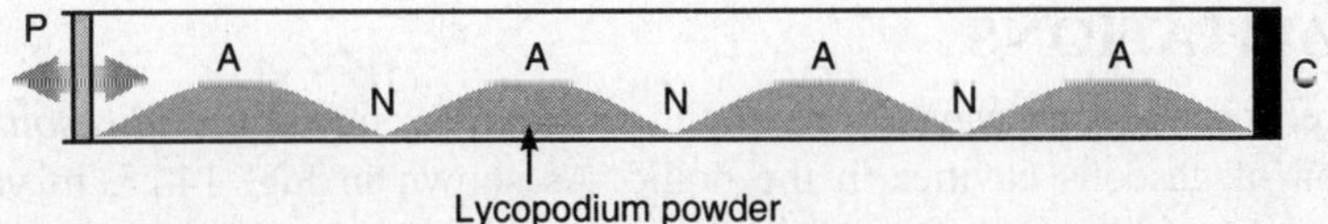

Fig. 14.23:

Kundt's tube Method.

(*ii*) **Piezoelectric transducer:** A transducer is a device that converts small amounts of energy from one kind into another. Crystals which acquire a charge when compressed are said to be piezoelectric. The Piezoelectric transducers convert mechanical energy (ultrasonic vibrations) into electrical energy and vice versa.

In case of detection of waves, ultrasonic waves are applied to one pair of faces of a quartz crystal. As a result, varying electric charges are produced on the other pair of faces of the crystal. These charges which are very small are amplified and detected.

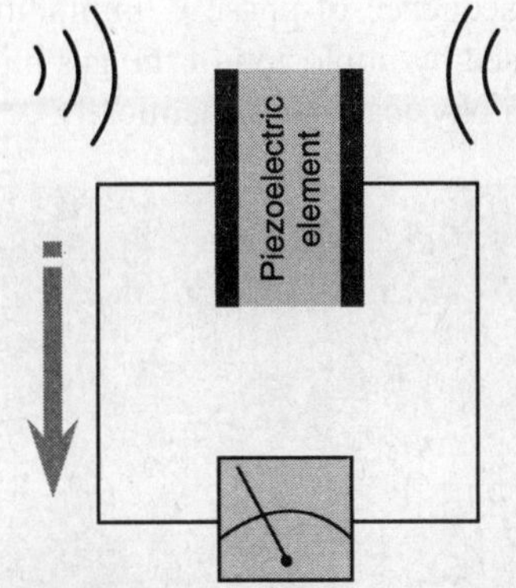

Fig. 14.24:

Piezoelectric transducer generates a voltage when an ultrasonic wave is applied.

(*iii*) **Thermal detection:** A fine platinum wire probe is used in this method of detection of ultrasonic waves. Due to alternate compressions in the medium resulting from ultrasonic waves, there occurs a change of temperature at nodes. As the platinum probe moves through the medium, its resistance changes at nodes. The change in the resistance of platinum wire is detected by using a sensitive bridge.

(*iv*) **Sensitive flame method:** As in the other methods, detection of ultrasonic waves consists of finding nodes and antinodes formed in the medium. When a narrow sensitive flame is moved along the medium, the flame remains steady at the positions of nodes, and it flickers at antinodes because there is a change in pressure. The average distance between the two adjacent antinodes is equal to half the wavelength. Given the frequency of ultrasonic wave, the velocity of ultrasonic wave through the medium can also be determined.

14.6 PROPERTIES OF ULTRASONIC WAVES

Properties of ultrasonic waves are:

(*i*) The speed of propagation of ultrasonic waves is proportional to its frequency.

(*ii*) As the wavelength of the waves is very small the diffraction effects are negligible.

(*iii*) As a highly directional beam, it can travel longer distances and without loss of energy.

(*iv*) Ultrasonic waves are highly energetic due to the high frequencies involved. Normally, 1 to 2 kW/m^2 intensities are used and can go up to 10 kW/m^2.

(*v*) They produce cavitation effects in liquids.

14.7 CAVITATIONS

The pressure and vacuum waves in a liquid medium produces the *ultrasonic cavitation*, the formation of gaseous cavities in the liquid. As shown in Fig. 14.25, in vacuum phase *A*, number of gas bubbles are formed in the liquid which expands in negative pressure. In phase *B*, the newly expanded bubble compresses by the huge negative pressure. In phase *C*, the temperature of the gas increases until the bubble collapse inward, In Phase *D*, as a consequence of phase *C* imploding gives vast release of impact energy. The impact energy caused by implosion of the gas bubble hits the surface of the object to be cleaned, interacting both physically and chemically.

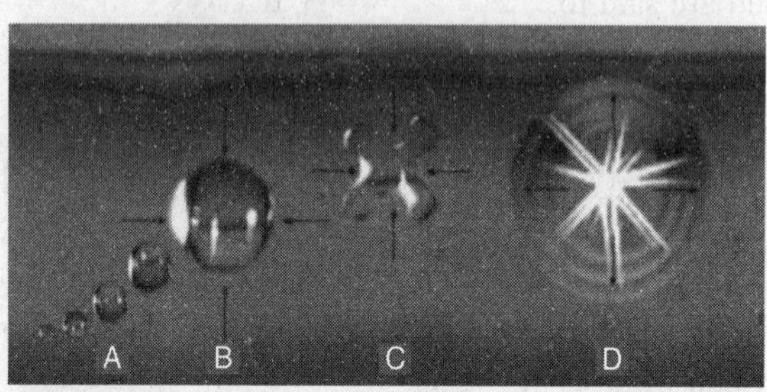

Fig. 14.25:

Ultrasonic Cavitation and different stages of a cavitation bubble.

Def: Cavitation means the formation of bubbles-empty space. These bubbles collapse inwards, implode, as fast as they are created. It happens millions of times per second. This constant implosion produces tremendous vacuum energy in the form of heat and pressure.

14.8 DETERMINATION OF VELOCITY IN A LIQUID

Determination of velocity of ultrasonic waves in a medium basically consists of determining the wavelength of the ultrasonic wave in that medium. We describe here two of the methods used for this purpose.

(*i*) Interferometer Method

(*ii*) Acoustic diffraction method

14.8.1 Interferometer Method

The velocity of ultrasonic waves is determined using an ultrasonic interferometer.

Experimental setup: The interferometer consists of an ultrasonic generator having a liquid cell connected to its tank circuit. The cell, *T*, is a vertical cylindrical tube (Fig. 14.26a) filled with the liquid medium under test. A quartz crystal is mounted at the bottom of the cell. Reflector is a metallic plate, mounted at the top end of the cell and it can be moved parallel to itself with the help of a micrometer screw. The surface of crystal and the reflector are made exactly parallel to each other.

Working: When the quartz crystal is excited, the ultrasonic waves of known frequency are produced and propagate through the liquid. These waves are reflected at the reflector and travel back to quartz crystal. Standing waves are produced when the separation among these plates is a whole multiple of the sound wavelength in the medium and acoustic resonance happens. The resonant waves cause a maximum in the anode current of the piezoelectric

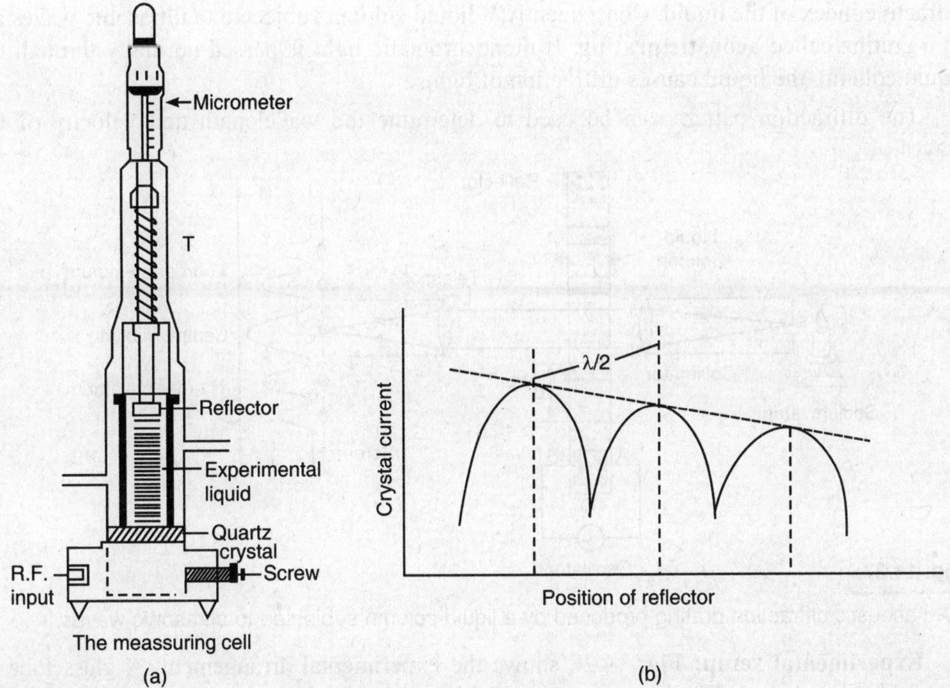

Fig. 14.26:

Ultrasonic interferometer (a) Liquid cell and (b) current variation.

generator. If the distance is now increased or decreased and the variation is exactly one-half wavelength ($l/2$) or multiple of it, anode current again becomes maximum. If d is the separation between successive adjacent maxima of anode current, then,

$$d = \frac{\lambda_u}{2}$$

Determination of velocity: The velocity (υ) of a wave is related to its wavelength (λ) by the relation, where f is the frequency of the wave.

$$\upsilon = \lambda f$$

Then, knowing the value of the frequency of the ultrasonic waves used, the velocity of the ultrasonic wave in the liquid υ_u is computed from the relation

$$\upsilon_u = f\lambda_u = 2df \quad \text{...(14.13)}$$

Determination of adiabatic compressibility of liquid: It indicates relative volume change of the fluid as a response to a pressure change. Compressibility β is the reciprocal of bulk modulus.

$$\text{Adiabatic compressibility} \qquad \beta = \frac{1}{\rho v^2} \quad \text{...(14.14)}$$

Where ρ is the density of the material of the medium and υ is the velocity of the sound wave through that medium.

14.8.2 Acoustic Diffraction Method

When ultrasonic waves travel in a liquid medium, the density of the liquid undergoes changes due to alternating compressions and rarefactions, which produces a periodic variation of refractive index of the liquid. Consequently, a liquid column subjected to ultrasonic waves act as a grating called **acoustic grating**. If monochromatic light is passed normally through the liquid column, the liquid causes diffraction of light.

The diffraction pattern can be used to determine the wavelength and velocity of the waves.

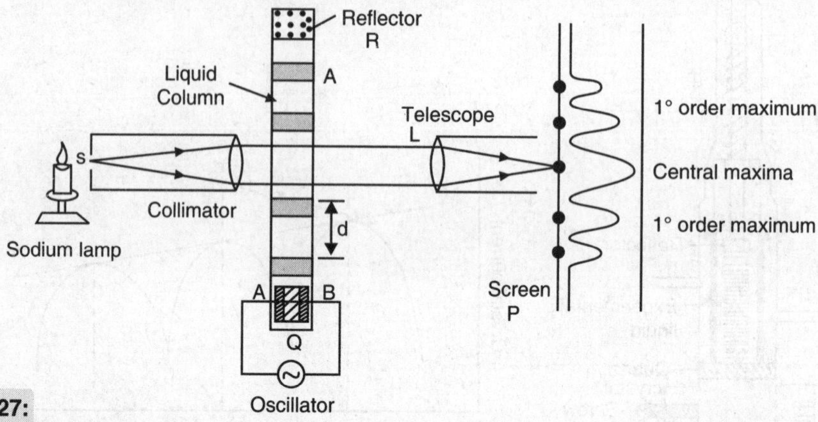

Fig. 14.27: An acoustic diffraction grating produced by a liquid column subjected to ultrasonic waves.

Experimental setup: Fig. 14.27 shows the experimental arrangement. A glass tube is filled up with a liquid. The piezoelectric transducer Q is positioned at the bottom of the glass tube and a reflector, R, is arranged at the top. The surfaces of the crystal, Q and reflector, R are held perfectly parallel to each other. S is a slit through which monochromatic light passes and gets collimated and rendered parallel. The image of the slit is focused on the screen P by the lens L.

Working: When the quartz crystal, Q is not excited and is at rest, the light forms a single image of the slit on the screen P. When the electric circuit excites the crystal, it goes into a state of vibration and produces ultrasonic waves. They propagate through the liquid column and get reflected at the reflector located at the top. The two waves combine to form stationary waves in the liquid column. The density and hence the refractive index of the liquid is maximum at nodal points and minimum at antinodal points. Therefore, the nodal points act as opaque regions while antinodal areas act as transparent regions for light. The liquid column thus resembles a ruled grating and causes diffraction of light. The image formed on the screen consists of a diffraction pattern having a central maxima flanked by first order, second order maxima and minima and so on.

Determination of wavelength: Grating equation $(a + b) \sin \theta = m\lambda$ is applicable to the acoustic grating also. The grating constant $(a + b)$ in this case equals $\lambda_u/2$ and is given by

$$\frac{\lambda_u}{2} \sin\theta = m\lambda \qquad \text{...(14.15)}$$

where λ_u is the wavelength of the ultrasonic waves,

λ is the wavelength of monochromatic light used to produce the diffraction pattern and m is the order of the maxima and takes integer values 1, 2, 3,....etc.

$\therefore$ The wavelength of the ultrasonic waves is

$$\lambda_u = \frac{2\,m\lambda}{\sin\theta} \qquad \text{...(14.16)}$$

Knowing the values of λ, m and measuring the angle θ, we can calculate the wavelength λ_u. This method of determining the wavelength of ultrasonic waves is known as **acoustic diffraction method**.

Determination of velocity: First the wavelength is determined using the values of λ, m and θ in Eq. (14.15). Then, the velocity of the ultrasonic wave in the liquid υ_u is computed from the relation

$$\upsilon_u = f\lambda_u$$

or
$$\upsilon_u = \frac{2m\lambda f}{\sin\theta} \qquad \text{...(14.17)}$$

where f is the frequency of the ultrasonic waves which is known from the frequency of the oscillator.

Example 14.4: *If the wavelength of light is 6000 Å, first order angle of diffraction is 0.046° and frequency of the ultrasonic waves produced by the transducer is 2 MHz find the velocity of ultrasonic waves in the liquid.*

Solution: Velocity of ultrasonic waves in liquid,

$$\upsilon_u = \frac{2m\lambda f}{\sin\theta} = \frac{2\times1\times6000\times10^{-10}\,\text{m}\times2\times10^{6}\,\text{Hz}}{\sin 0.046°} = \mathbf{2989\ m/s.}$$

14.9 DETERMINATION OF VELOCITY IN SOLIDS

Velocity of ultrasonic waves depend on the medium. In solids, shear and longitudinal waves generally propagate at different velocity. The longitudinal wave velocity can be determined by means of time of flight measurements using the pulse echo method.

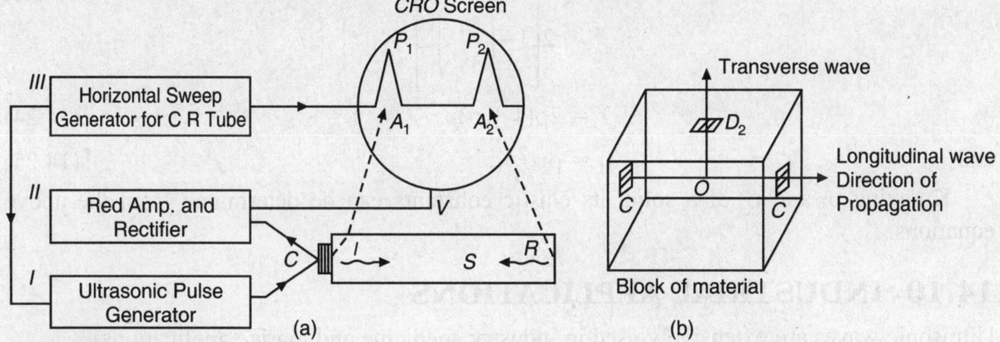

Fig. 14.28: Pulse echo method.

In this method, transducer is used which acts both as transmitter and receiver of ultrasonic waves. The ultrasonic beam is incident normally on the surface of the solid. The beam travels through the specimen and is reflected back from the rear surface of the solid. The initial pulse, the echo pulse reflected from the opposite face of the solid is displayed on a CRO screen. From the positions of the pulses on the oscilloscope screen, one can measure the time interval between two successive echoes. Knowing the thickness of the solid, d we can determine the longitudinal velocity of the ultrasonic wave in solid, which is given by

$$\upsilon_l = \frac{2d}{t} \qquad \text{...(14.18)}$$

Measurement of elastic constants in solids: It is to be noted that longitudinal as well as transverse (or shear) waves travel through solids. The velocity of propagation of longitudinal waves in solids is determined by Young's modulus and is given by

$$\upsilon_l = \sqrt{\frac{Y}{\rho}} \qquad \text{...(14.19)}$$

where Y is Young's modulus and ρ the density of the solid.

The velocity of shear waves in solids is determined by the rigidity modulus and is given by

$$\upsilon_t = \sqrt{\frac{\eta}{\rho}} \qquad \text{...(14.20)}$$

where η is the rigidity modulus of the solid.

It can be shown that the ultrasonic speeds of longitudinal and transverse waves are given by

$$\upsilon_l = \sqrt{\frac{Y(1-\sigma)}{\rho(1+\sigma)(1-2\sigma)}} \qquad \text{...(14.21)}$$

and

$$\upsilon_t = \sqrt{\frac{\eta}{\rho}} = \sqrt{\frac{Y}{2\rho(1+\sigma)}} \qquad \text{...(14.22)}$$

where σ is the Poisson's ratio of the solid.

Solving the Eq. (14.20) and Eq. (14.21), we get

$$\sigma = \frac{\left[1-2\left(\frac{\upsilon_t}{\upsilon_l}\right)^2\right]}{2\left[1-\left(\frac{\upsilon_t}{\upsilon_l}\right)^2\right]} \qquad \text{...(14.23)}$$

$$Y = 2\rho(1+\sigma)\rho_t^2 \qquad \text{...(14.24)}$$

and

$$\eta = \rho\upsilon_t^2 \qquad \text{...(14.25)}$$

Knowing υ_l and υ_t of a solid, its elastic constants can be determined from the above equations.

14.10 INDUSTRIAL APPLICATIONS

Ultrasonic waves are extensively used in industry, medicine and marine applications.

***(i)* Ultrasonic Cleaning:** It is a fast and effective cleaning method and uses the concept of cavitation. Cavitation occurs in liquids when they are subjected to ultrasonic waves. It is evenly distributed all through the volume of the liquid and is capable of reaching normally in in accessible places. Therefore, ball bearings, carburetor parts, vessels etc can be well cleaned. Ultrasonic cleaning mechanism is best to clean hard materials such as metals, glass, ceramics, and plastics etc which reflect sound. Cleaning equipment is designed to operate normally in the range of 20-50 kHz.

An ultrasonic cleaning tank (Fig. 14.29) is a stainless steel metal tank that has piezo ceramic transducers attached to the base or side. It is filled with water having an appropriate detergent dissolved in it and the parts to be cleaned are suspended in the liquid. A high frequency (20 to 50 kHz) voltage is used to excite the transducer which causes the tank

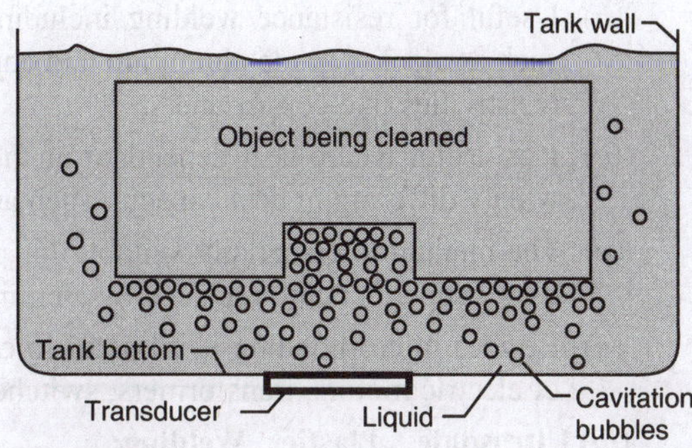

Fig. 14.29: Ultrasonic cleaning tank.

bottom to move rapidly and induces compression and rarefaction waves at that ultrasonic frequency in the liquid. When cavitation happens in water near a dirty object, creates a tiny pressure wave that reaches deep into suspended items in the liquid. This tiny pressure wave dislodges and breaks up the dirt and other contaminants and gently lifts it away.

(*ii*) Ultrasonic Metal Welding: Practically, all metals and plastics can be welded using ultrasonic waves of suitable energy. The surfaces of the work pieces are cleaned and held together. They are subjected to ultrasonic oscillations at the spot where they are to be welded. The ultrasonic energy converts to heat at the contact area as a result of friction arising between the surfaces. As the temperature of surface layers exceed the recrystallization point, the layers melt and bond together to form a strong joint. The merits of this process are that it does not cause stress at the spot of welding and that the structure of the materials remain unchanged.

Ultrasonic welding (Fig. 14.30) makes use of vibration energy at the interface of the joint to create bonding. The ultrasonic energy is delivered to the interface by a *horn* that is in contact with the top part. The horn vibrates at high frequencies. The bottom of the part is held on an "anvil." The joint forms by diffusion and there is no melting at the joint. The equipment required for ultrasonic metal welding ranges from low power microbonders (40 and 60 kHz) to machines of several kilowatt output capacity (10 and 20 kHz) for welding of larger parts.

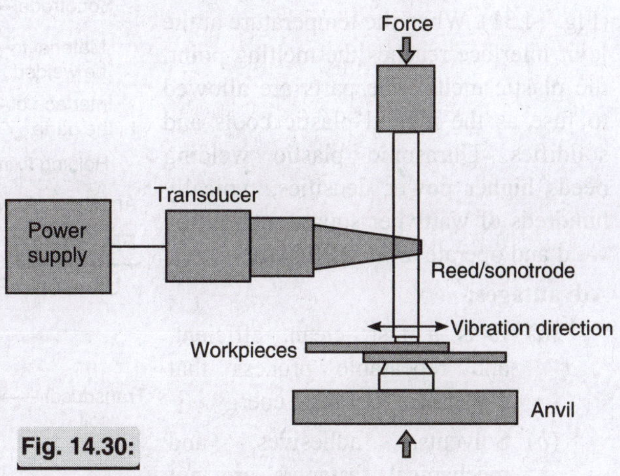

Fig. 14.30: Schematic of ultrasonic welding setup.

Drawbacks of micro-welding:

(*a*) Mechanical contact is required on either side of the joint.

(*b*) As the horn is a consumable, inspection and replacement of it is required.

(*c*) Joint geometry is limited to lap welding only.

(*d*) The speed of the welding cycle is slow.

Applications of ultrasonic metal welding:

(*a*) As the process is relatively cold and typically below the melting temperatures of the metals it is useful for high resistance inter-metallic compounds in dissimilar metal welds.

(*b*) Useful for resistance welding including the melding of high conductivity metals such as electric grade aluminum and copper also combinations of metals of different resistivities like copper and steel.

(*c*) It is useful where heat-dependent methods cannot be used such as welding of parts widely differing in heat capacity such as foil to thick sections.

(*d*) The method is essentially suitable for producing spot welds and line welds as now days copper is being replaced by aluminum.

(*e*) Besides micro-bonding it is useful in electric and electronic industries in assembly of electric motors, transformers, switches, and relays.

(*iii*) Ultrasonic Plastic Welding: This method is very useful in the automotive industry for assembly of taillights, dashboards, heater ducts, and other components. In this process welding is carried by the heat produced from high-frequency mechanical motion. High-frequency electrical energy is converted into high-frequency mechanical motion in thermoplastics (Fig. 14.31). When the temperature at the joint interface reaches the melting point, the plastic melts. The parts are allowed to fuse as the melted plastic cools and solidifies. Ultrasonic plastic welding needs higher power densities, typically hundreds of watts per square inch at the weld and operate around 20 kHz.

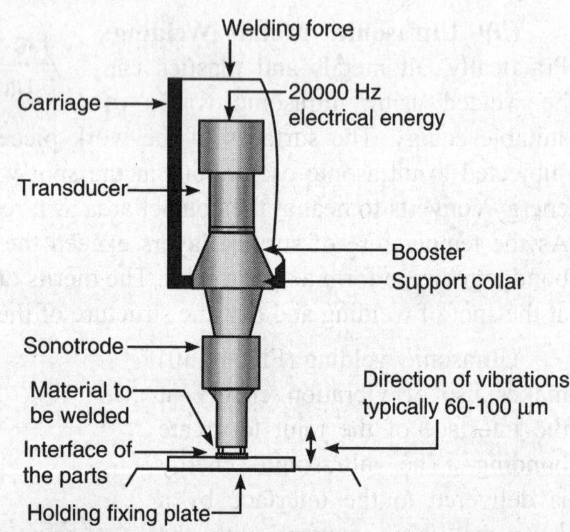

Fig. 14.31: Ultrasonic Plastic Welding.

Advantages:

(*a*) It is a fast, clean, efficient, and repeatable process that consumes very little energy.

(*b*) Solvents, adhesives and mechanical fasteners are not required.

(*c*) The finished assemblies are strong and clean.

(*d*) Process does not require a skilled operator.

(*iv*) Ultrasonic Drilling: Ultrasonic machining is a vibratory process useful for the mechanical treatment of hard and brittle solids such as ceramics, glasses, precious stones, semiconductors and hard alloys. The tool motion is produced by an acoustic concentrator to which the

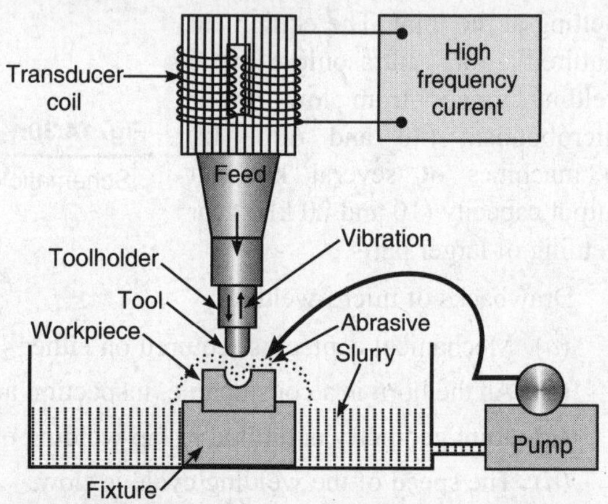

Fig. 14.32: Schematic of a typical ultrasonic drilling machine.

tool holder is threaded. The acoustic concentrator consists of a needle type magnetostriction vibrator, illustrated in Fig. 14.32. The vibrator is made of thin isolated ferromagnetic plates of high magnetostriction such as nickel. A coil is wound on the needle, through which an alternating current of frequency *f* passes. The resulting magnetic field magnetizes the core and thus changes its length. The core of the vibrator vibrates at a frequency 2*f*. By choosing the frequency to be equal to half the natural vibration frequency of the vibrator, the system is held at resonance and the vibrations of the needle will be of large amplitudes. A tapered waveguide of appropriate dimensions and rigidly attached to the vibrator concentrates the vibrational energy and communicates it to the tool.

In operation, the needle vibrator is in oscillation and the tool shank is pressed against the work piece. An aqueous suspension of a solid abrasive powder is then fed through a tube to the working zone. Abrasive particles bombard the work surface at high velocity and shear off small pieces of the material. This action rapidly chips away the work piece in a pattern controlled by the tool shape and contour.

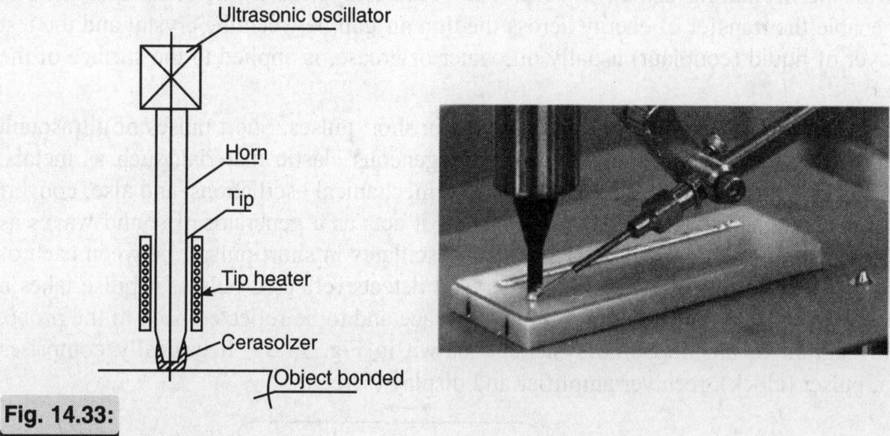

Fig. 14.33: Ultrasonic Soldering.

***(v)* Ultrasonic Soldering:** Normally, surfaces are covered with contaminants, grease and oxide films. Such films prevent formation of a good joint. Therefore, prior to soldering, the surfaces are cleaned with active fluxes. The fluxes, when heated, dissolve the oxide films and uncover the clean metal surface which readily allows the molten solder to form a firm joint. This method however is not suitable for soldering aluminium. Active metals such as aluminium can be soldered without fluxes with the help of ultrasonic waves. In this case soldering is done by a special iron that vibrates at a frequency of tens of kilo hertz.

Ultrasonic soldering is a flux free soldering method. Here surface oxide layers are cleaned by using vibrations and cavitations. During the **process**, the solder is first heated by a separate source till it melts and then vibrational energy is applied to it. The ultrasonic vibrations induce cavitation at the tip of the soldering tool which cleans the metal surfaces and subsequently the liquid solder bonds them.

Advantages:

(*a*) Solder joint is free of voids because vibrational energy pushes the liquid solder into the crevices and micro-pores in the substrate.

(*b*) Ultrasonic soldering allows for fusing dissimilar materials which is not possible with usual methods.

(c) As no flux is required, it saves time and costs of cleaning flux residues. Here corrosion is less and the durability of soldered joints is more.

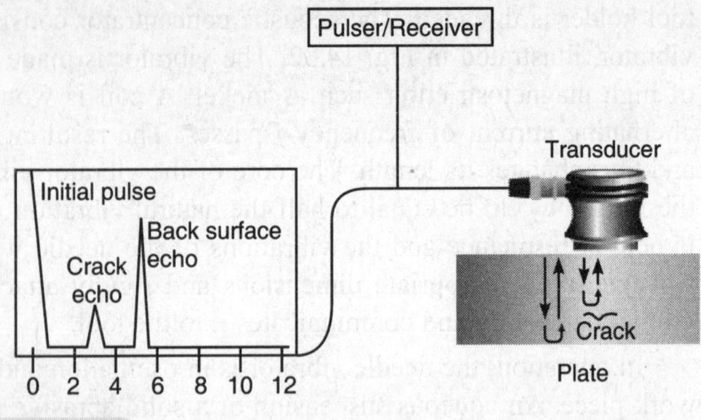

Fig. 14.34: Block diagram of an ultrasonic testing system.

(vi) Ultrasonic Testing: This method is useful for the detecting the internal and surface defects in sound conducting materials. The wave velocity is related to the Young's Modulus of the material and is characteristic of that material. For contact testing, a probe with the oscillating crystal is used. The probe is applied to the surface of the test piece. To enable the transfer of energy across the thin air gap between the crystal and the test piece, a layer of liquid (couplant) usually oil, water or grease, is applied to the surface of the test piece.

The crystal does not oscillate continuously but in short pulses. Short pulses of ultrasound (1 MHz to 6 MHz) have less attenuation in homogeneous elastic material, such as metals. Piezo electric crystals translate electrical pulses into mechanical oscillations, and also, convert mechanical oscillations into electrical pulses. Thus, it acts as a generator of sound waves as well as a detector of reflected pulses. The crystal oscillates in short pulses, between each of which it is inactive. In the inactive state the crystal detects reflected pulses. A pulse takes a short time to travel through the material to the interface and to be reflected back to the probe. The block diagram of an ultrasonic system is shown in Fig. 14.35. It basically comprises transducer, pulser (clock), receiver/amplifier and display (screen).

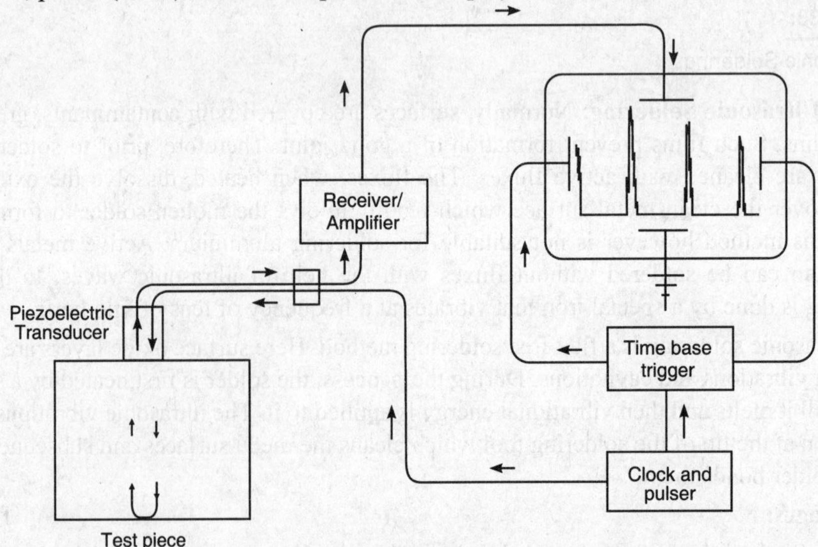

Fig. 14.35: Block diagram of an ultrasonic tester.

The sequence of operation of the system is as follows:

(*a*) The pulser provides the short, high-voltage pulse to the transducer and at the same time it supplies voltage to the time-base trigger module. It gives the "spot" on the CRT screen.

(*b*) The voltage pulse given to transducer is converted into mechanical vibrations and these vibrations pass through the test piece. All this time, the spot is moving horizontally across the CRT.

(*c*) The energy (vibration) in the test piece now reflects off the back wall towards the transducer, where it is reconverted into a voltage. This voltage is amplified by the amplifier.

(*d*) By using the delay (or zero) control, the spot on CRT is set to start at the instant that the energy goes into the test piece.

(*e*) The spot moves horizontally across the screen. The energy gets reflected at the back wall interface and returns to be received and amplified. The amplifier sends voltage to the Y- plates. The spot moves to a later position on the time base. The spot continues its journey across the screen, and the above sequence repeats repeatedly until the energy in the test piece attenuates. The display will show multiple repeat signals.

(*f*) The clock now sends a pulse second time and the next pulse is produced. It is again repeated *n* number of times per second. The *n* number of pulses per second is referred to as the pulse repetition frequency (PRF) or the pulse repetition rate (PRR).

(*vii*) Emulsification: Immiscible liquids like water and oil can mix thoroughly and form stable emulsions when their mixture is subjected to strong ultrasonic waves. The ultrasonic emulsification is used in industry to mix molten metal and form alloys of uniform composition.

14.10.1 Data Presentation

Usually data can be collected and displayed in formats such as A-scan, B-scan and C-scan.

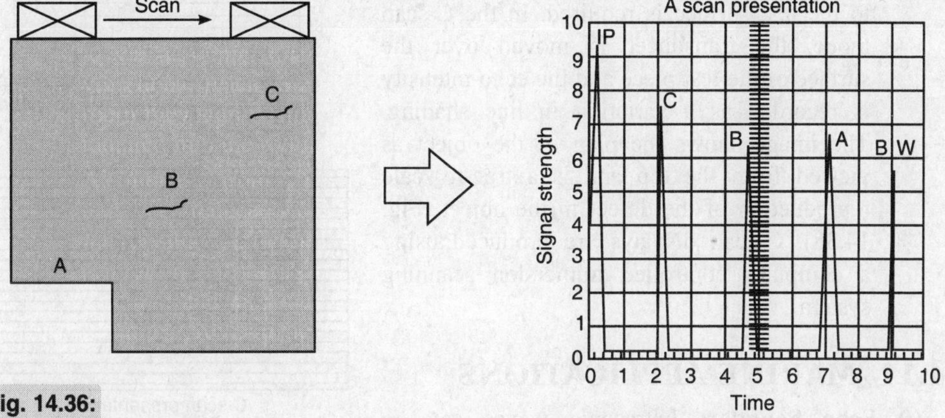

Fig. 14.36: A-Scan presentation.

(*i*) **A-scan display:** A-scan display is the most used mode of display in ultrasonic testing. In this mode of display, the relative amount of energy received is plotted along the vertical axis and the time taken by the pulse to the reflecting surface and return back to the transducer is displayed along the horizontal axis. The location of the defect is estimated by the position of the echo given by it on the horizontal axis and size of the defect from the relative amplitude of the echo. The information that is available in A-scan is one-dimensional. A typical A-scan echo pattern is shown in Fig. 14.36.

(*ii*) B-Scan Display: B-scan display gives a cross-sectional view of the test object and shows the position, orientation and depth of defects in the specimen. In this mode of display, Y-axis represents elapsed time while X-axis represents the position of the transducer along a line on the surface of the test object relative to the starting position of the transducer. Thus, the probe movement is displayed in x-direction while the distance of the defect is displayed in y-direction. Echo amplitude is indicated by the relative brightness of echo indications. If a storage oscilloscope is used, the whole picture will be displayed, which reveals the depth of the defect beneath the surface and its size in the lateral direction (see Fig. 14.37).

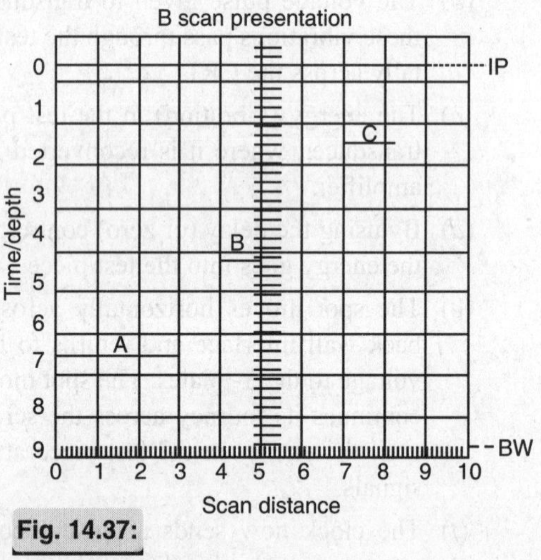

Fig. 14.37: B-scan display.

(*iii*) C-Scan Display: The depth of defects is not relevant in some testing problems, but information about their distribution parallel to the test surface is required. In the C-scan mode, the transducer is moved over the surface of the test piece and the echo intensity is recorded as a variation in line shading. The image shows the plan of the object as viewed from the top and is a true-to-scale reproduction of the defect in the object (Fig. 14.38). C-scan displays are produced using a computer controlled immersion scanning system.

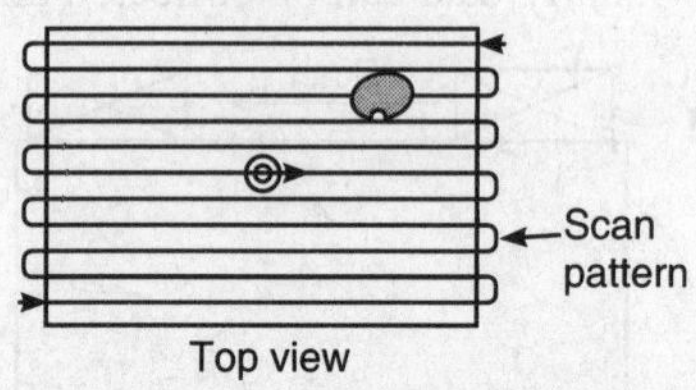

Fig. 14.38: C-Scan display.

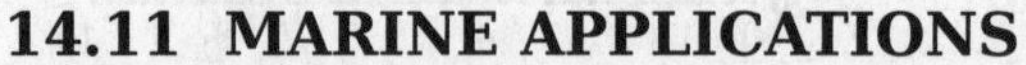

14.11 MARINE APPLICATIONS

(*i*) Echo Sounder: Ultrasonic waves can be produced in the form of directed beams like beams of light. Further, ultrasonic waves can travel long distances in water. This property is utilized in measuring the depth of ocean. A ship equipped with an echo sounder sends out short pulses of ultrasonic waves towards the bed of the ocean (Fig. 14.39). These waves are reflected back from the bed and the receiver receives the reflected pulse. The time interval between the pulse sent and the pulse received is determined. Knowing the velocity of the waves through the seawater, the depth of the ocean, can be computed with the help of the following formula.

$$l = \frac{vt}{2} \quad ...(14.26)$$

where t is the time interval between the transmitted and reflected pulses.

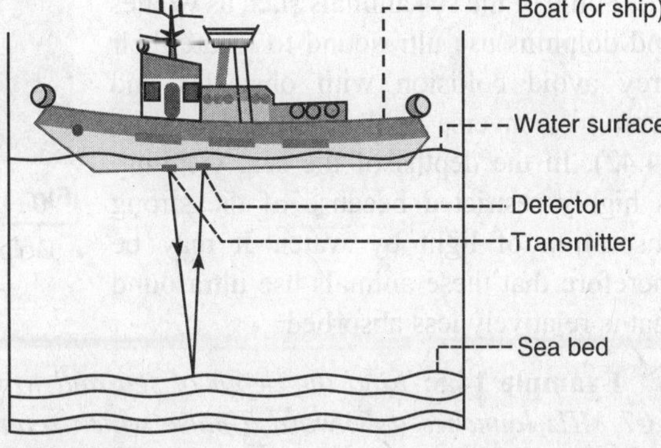

Fig. 14.39:

Depth sounding.

***(ii)* SONAR:** The word SONAR stands for Sound Navigation and Ranging. The ultrasonic waves, which are highly directional, can be used for locating objects submerged under seawater and determining their distance. The idea of ultrasonic sonar was put forward first by the French physicist Paul Langevin and was successfully used by him during the first world war for detecting enemy submarines. The sonar acts in a way much similar to an echo sounder. In sonar, an ultrasonic beam is directed in different directions into the sea. In the absence of an obstacle, the ultrasonic pulses do not return to the ship. In the presence of an obstacle, pulses are reflected from the obstacle and are picked up by the receiver. Knowing the speed of the ultrasonic waves in seawater and time elapsed between the transmitted and reflected pulses, the distance of the object is determined. Fig. 14.40 illustrates a SONAR.

Fig. 14.40:

SONAR.

(*a*) Sonar is used to guide submarines in the seas.

(*b*) It is used to detect the presence of submerged icebergs.

(*c*) It is used for direction signaling in submarines.

***(iii)* Fish-finder:** Ultra sound can be used to locate shoals of fish utilizing the fact that the swimming bladder of fish is filled with air that scatters ultrasonic waves. Ultrasonic sonar is used for this purpose. At present ultrasonic locators are mainly used for detecting icebergs, fish shoals and etc. Fig. 14.41 shows an illustration of Fish-finder.

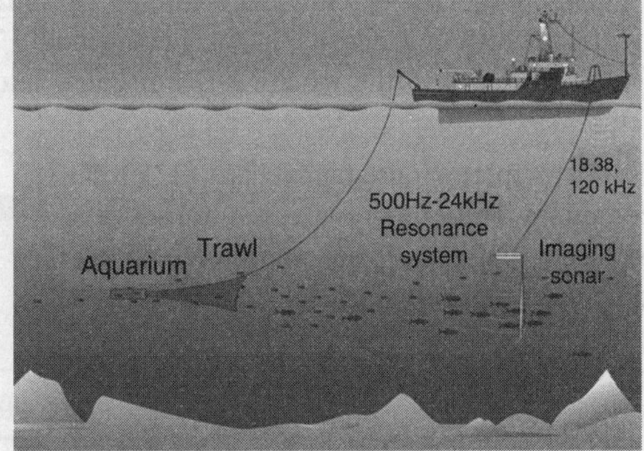

Fig. 14.41:

Illustration of Fish-finder.

Some of the sea animals such as whales and dolphins use ultrasound to locate their prey avoid collision with obstacles and even to converse with each other (Fig. 14.42). In the depths of the sea, visibility is highly restricted because of the strong absorption of light by water. It may be therefore that these animals use ultrasound that is relatively less absorbed.

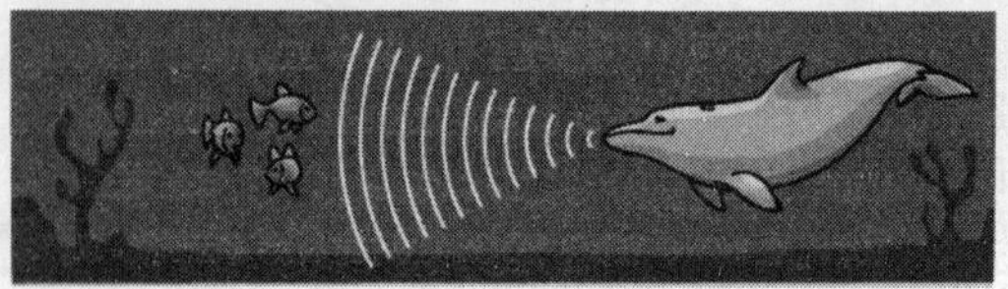

Fig. 14.42: Dolphins use ultrasound to locate their prey.

Example 14.5: *Find the depth of sea and wavelength of pulse, if ultrasonic source of 0.07 MHz launches downward a pulse which returns after 0.65 s from the seabed and the velocity of sound in seawater is 1700 m/s.*

Solution: Depth of the sea, $l = \frac{\upsilon t}{2} = \frac{(1700\text{m/s})(0.65\text{s})}{2} = \mathbf{552.5\ m.}$

Wavelength of the pulse $\lambda = \frac{\upsilon}{f} = \frac{1700\text{m/s}}{0.07 \times 10^6 \text{s}^{-1}} = \mathbf{24\ mm.}$

14.12 MEDICAL APPLICATIONS

Ultrasound scanning or Sonography, ultrasound scanner, Doppler ultrasound imaging and Ultrasonic blood flow meter are a few medical applications to mention.

14.12.1 Sonogram

Ultrasound imaging is called ultrasound scanning or Sonography and is widely used in imaging of internal organs of the human body. A sonogram is a non invasive medical procedure that helps doctors diagnose and treat medical conditions. It provides valuable information regarding the size, location, and displacement of a given structure. It can show the movement of internal organs in real time. Tumors and other regions of organ that differ in density from surrounding tissues can be detected. Interfaces between solid and fluid filled spaces can be precisely observed with the help of Ultrasound technology. Therefore, doctors use ultrasound tests to examine the fetus structure and any movement of the fetus inside a woman.

Def: Sonogram is an image formed by ultrasound waves that are bounced back from tissues of differing density in the human body. The time taken by the reflected ultrasound waves to return to the machine is translated into an image of the internal organ, or of the fetus.

Principle: Sonography uses a probe containing one or more acoustic transducers to send pulses of sound into a body. Whenever a sound pulse encounters a boundary between two tissue structures, it is partly reflected from, and partially transmitted. The sound pulse reflected back to the probe is detected as an echo (Fig. 14.43). The reflection depends on the difference in acoustic impedance of the two tissues.

P a b c

Fig. 14.43: Sonography

The acoustic impedance of a medium

$$Z = v \times \rho \qquad ...(14.27)$$

where v is the of speed of sound in the material and the ρ is density.

The time taken by the echo to travel back to the probe is used to calculate the depth of the tissue interface causing the echo. The greater the difference between acoustic impedances, the larger the echo is.

The time lagis
$$\tau = \frac{2l}{v} \qquad ...(14.28)$$

where l is the distance between tissue transducer and v is the ultrasonic velocity.

14.12.2 Ultrasound Scanner

Fig. 14.44 shows a block diagram of a simple ultrasound scanner. The **transmitter** supplies energy to the piezoelectric **transducer** which produces sharp pulses of ultrasound. The sound pulses travel through the body and get reflected from the organ or structure under investigation. The pulses with reduced strength are returned to the transducer probe which acts as a receiver as well. The returned echoes are amplified with the help of **swept-gain generator** and applied to the vertical deflection plates of CRO. Each time an echo reaches the probe, a vertical line appears on CRO screen. Corresponding to each reflecting surface in the body, one vertical line is produced on CRO screen. The **time base generator** supplies voltage to the horizontal deflection plates of CRO. The **rate generator** synchronizes the actions of the transmitter, time base generator and the swept-gain generator. The **CRO** displays the original pulse transmitted by the transducer into the body and the echoes received from different interfaces, as vertical lines on its screen. The CRO may be calibrated to give directly distances between the reflecting surfaces in the body. The display obtained using the above technique is known as a A-scan display and it is seen on CRO screen in Fig. 14.44. Generally, the frequencies in the range of 1 to 18 MHz. are used for medical imaging. But as the attenuation increases at higher frequencies, a lower frequency (3-5 MHz) is used.

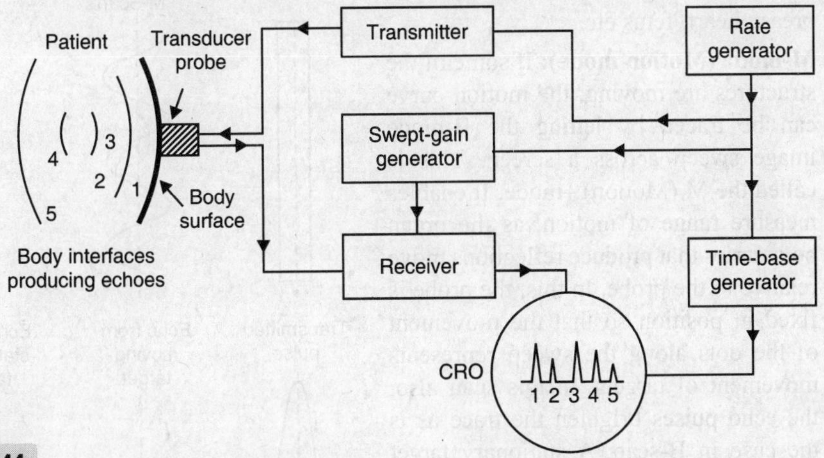

Fig. 14.44: Block diagram of a ultrasound scanner.

Display modes: The echoes are displayed as a function of time which is proportional to the distance from the source to interface. The echo information is displayed in one of several different display modes.

(*a*) **A-mode (Amplitude mode):** A-mode is the simplest type of scan mode. It is a graphic depiction of amplitude of echo versus distance into the tissue. A high energy pulse from a pulser excites the transducer. Echoes returned from the tissue are detected by the same transducer, amplified and processed for display. The returning echoes are displayed as vertical deflections on the trace (see Fig. 14.45) which represent the amplitude of the reflected energy. In most cases, the transducer is kept stationary. Hence, the echoes are static and one dimensional. Pulses are typically a few milliseconds long and are emitted at 400 to 1000 pulses/ns. It is used in ophthalmology and encephalography.

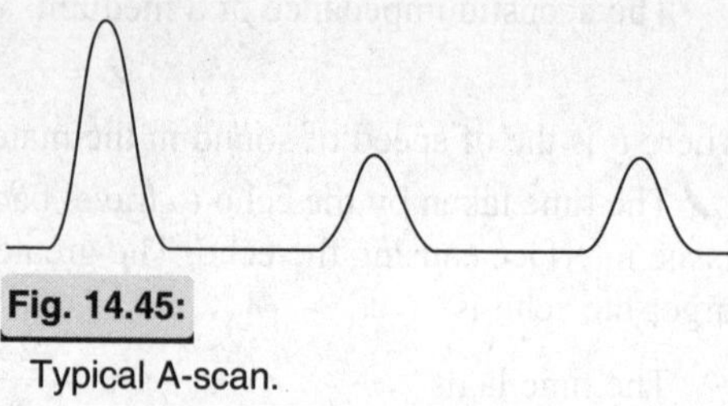

Fig. 14.45: Typical A-scan.

(*b*) **B-mode (Brightness mode):** The amplitude can also be displayed as the brightness of the certain point representing the structure, in a B- plot. In this scan, the echo signals are not applied to the horizontal deflecting plates of CRO. Instead they are used to control the brightness of the spot on the screen. Hence the reflecting surfaces appear as spots (Fig. 14.46). The brightness of the spot is proportional to the strength of returning echo. A linear array of transducers may be made to simultaneously scan a plane through the body so that we can obtain a two-dimensional image of a stationary organ or body structure on screen. It is used in diagnostic studies of liver, breast, heart, fetus etc.

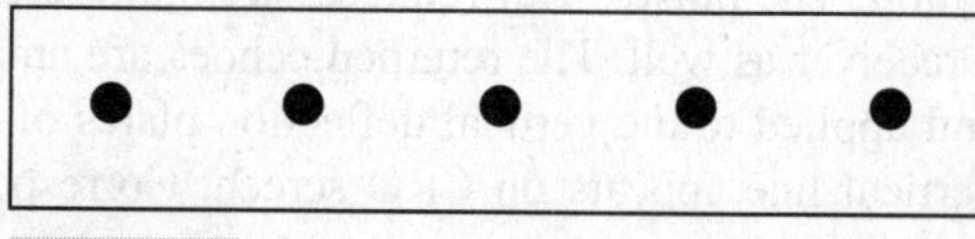

Fig. 14.46: Typical B-scan.

(*c*) **M-mode (Motion mode):** If some of the structures are moving, the motion curve can be traced by letting the B-mode image sweep across a screen. This is called the M (Motion) –mode. It enables measure range of motion, as the organ boundaries that produce reflections move relative to the probe. In this, the probe is fixed in position so that the movement of the dots along the sweep represents movement of targets. In this scan also, the echo pulses brighten the trace as is the case in B-scan. A stationary target will trace a straight line where as a moving target will trace the pattern of its movement with respect to time (see Fig. 14.47). A typical sonogram is shown in Fig. 14.48.

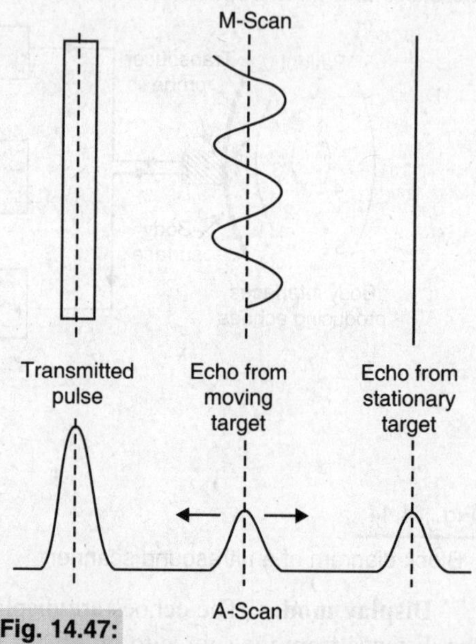

Fig. 14.47: Typical M-scan.

14.12.3 Doppler Ultrasound Imaging

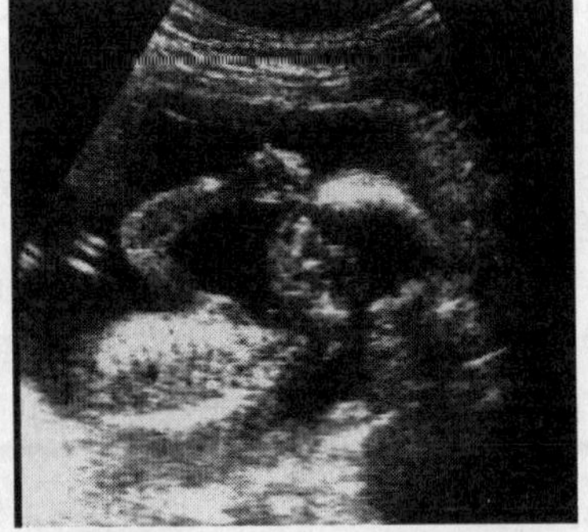

Fig. 14.48:

A primary use of ultrasound is to monitor the progress of a pregnancy. The above sonogram shows the 27 weeks stage.

The concept of Doppler Effect is used in Doppler ultrasound imaging. When a moving object reflects the ultrasound waves, the frequency of the echoes gets changed. If it is moving toward the probe a higher frequency is generated and if it is moving away from the probe a lower frequency is produced. The change in frequency depends on speed of the object. Doppler mode enables to measure the change in frequency of the echoes to calculate how fast an object is moving. Doppler ultrasound imaging is useful to measure the rate of blood flow through the heart and major arteries.

Advantages of Doppler ultrasound imaging

(*i*) It is noninvasive and uses no ionizing radiation.

(*ii*) As it causes no health problems, it can be repeatedly used.

(*iii*) It is painless, easy-to-use and less expensive.

(*iv*) It gives a better picture of soft tissues than x-ray images.

(*v*) It can be safely used for the diagnosis and monitoring of pregnant women and their unborn babies.

Limitations

(*i*) Ultrasound is not best imaging technique for the bowel (intestine) or organs covered by the bowel since ultrasound waves are disrupted by air or gas. Deeper structures such as the pancreas and aorta cannot be properly visualized because of the intestinal gas.

(*ii*) Ultrasound cannot show the internal structure of bones as it cannot penetrate the bone.

14.12.4 Ultrasonic Blood Flow Meter

The ultrasonic blood flow meter uses the concept of Doppler shift to measure the velocity of blood in veins and arteries. The speed of blood flow is measured by transmitting and receiving elements that are placed directly on the skin, as in Fig. 14.49. The transmitter emits a continuous sound of about 5 MHz. The frequency of ultrasound changes when it is reflected from the red blood cells. The electronic counter of the receiving element measures its frequency and from

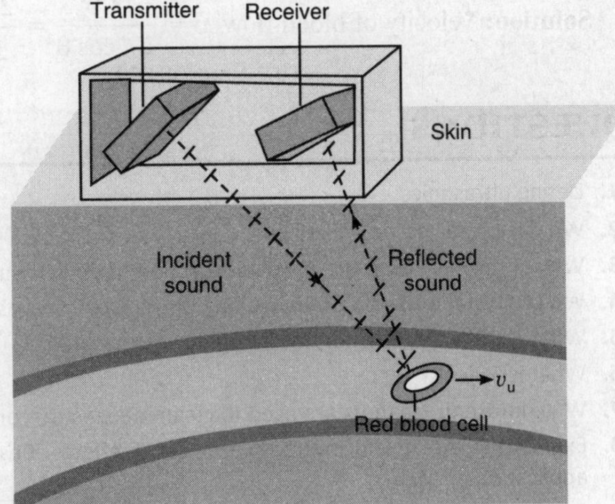

Fig. 14.49:

Doppler flow meter.

which the speed of the blood flow can be known. Usually, the frequency varies of about 600 Hz for flow speeds of about 0.1 m/s.

The Doppler flow meter can be used to find regions where blood vessels have narrowed. According to the equation of continuity, greater flow speeds takes place in the narrowed regions.

Ultrasound of frequency of about 5 to 10 MHz is directed at an angle to the blood stream. The particles of blood reflect the beam. The sound waves undergo a frequency.

The Doppler shift

$$\Delta f = \frac{2f\,v\cos\theta}{v_u}$$

where v_u is the velocity of ultrasonic waves in blood,

v is the velocity of the blood with respect to the blood vessel,

θ is the angle at which ultrasonic waves strike the blood stream and

f is the initial frequency.

$$\therefore \quad v = \frac{\Delta f\, v_u}{2f\cos\theta} \qquad \text{...(14.29)}$$

14.12.4 Other Medical Applications

(*i*) Ultrasonic therapy is used in treatment of rheumatic pains. Ultrasonic wave produce massaging action and relieves pain.

(*ii*) The waves are useful for dental cutting and they make the cutting painless.

(*iii*) Ultrasonic waves destroy bacteria and therefore they are used in sterilization of water and milk.

Example 14.6: *In an ultrasonic Doppler flow meter the direction of blood flow is along the direction of ultrasonic beam. The frequency of ultrasonic waves is 2 MHz and the velocity of waves in blood is 1500 m/s. If the Doppler shift in frequency is 267 Hz, calculate the velocity of blood flow.*

Solution: Velocity of blood flow, $\upsilon = \frac{\Delta f\, \upsilon_u}{2f\cos\theta} = \frac{267\text{Hz}\times 1500\text{m/s}}{2\times 2\times 10^6\,\text{Hz}\times 1}$ = **0.1 m/s.**

QUESTIONS

1. Define ultrasonics.
2. What is piezoelectric effect?
3. Why ultrasonic waves are used in detection of objects submerged in sea?
4. Which crystal is used more in making of ultrasonic waves?
5. What is meant by the natural frequency of a crystal?
6. What is SONAR?
7. Why ultrasonic cleaners are used to clean mechanical components and semiconductor chips etc?
8. Explain the direct and inverse piezoelectric effects. Describe the importance of each and one application of each.
9. Define piezoelectric effect and magnetostriction effect.
10. What is piezoelectric effect? With necessary circuit diagram, explain the production of ultrasonics using piezoelectric crystal.

11. Define ultrasonic waves. Describe the piezoelectric method for their production.
12. Explain the phenomenon of magnetostriction. How will you produce high frequency sound waves with its help?
13. Explain the terms magnetostriction and piezoelectric effect. Discuss any one method of production of ultrasonic waves.
14. What is inverse piezoelectric effect?
15. What is magnetostriction effect? Draw circuit diagram of magnetostriction oscillator and explain its working. What are the advantages of magnetostriction method?
16. Explain the cavitation.
17. What is piezoelectric effect? Explain how ultrasonic waves are produced by a piezoelectric transducer.
18. Explain the cavitation effect when an ultrasonic wave is passed through a liquid.
19. What are the different methods for the production of ultrasonic waves? Describe one of them in detail. How will you determine the wavelength of these waves?
20. What are ultrasonic waves? Describe a method of measuring the velocity of ultrasonic waves in solids.
21. What is an acoustic grating? Explain how an acoustic grating is used to determine the velocity of ultrasonic waves in liquids.
22. Explain the ultrasonic diffractometer with neat diagram. How will you determine the velocity of ultrasonics in a liquid?
23. Describe with theory a method of measuring velocity of ultrasonic waves in a liquid and mention how the bulk modulus of the liquid can be evaluated.
24. Determine the velocity of sound in a liquid with a neat sketch.
25. What are the applications of ultrasonics?
26. Explain echo sounding technique and cavitation with one example each.
27. What are the properties of ultrasonic waves?
28. Describe ultrasonic flaw detector. How is it used in detection of flaws in metals?
29. Explain the principle of sonar.
30. Draw a block diagram of ultrasonic flaw detector. Explain the three different scan modes used for presentation of data.
31. Discuss the use of ultrasonics for flaw detection.
32. What is the principle of ultrasonic testing?
33. What are the advantages and limitations of ultrasonic testing?
34. Explain application of ultrasound in medical field.
35. What is a sonogram and how is it obtained?
36. Explain the different scan modes used for obtaining a sonogram.
37. Explain the use of ultrasonic waves in non-destructive testing.
38. What are the applications of ultrasonics?

PROBLEMS

1. Calculate the velocity of ultrasonic waves in a certain liquid using the following data obtained in an acoustic grating experiment, if frequency of ultrasonic waves is 100 MHz, wavelength of light used is 600 nm and angle of first order diffracted beam is 5°. **[Ans:** 1475m/s]
2. An ultrasonic beam of 1 cm wavelength is sent by a ship, returns from the seabed after 2 sec. If velocity of the ultrasonic beam in seawater is 1510 m/s at 0°C, its salinity at 30°C is 29 gm/litre,

calculate the depth of the seabed at 30°C and frequency of the ultrasonic beam. (Assume that the velocity of sound in sea water is given by the equation.

$\upsilon = \upsilon_0 + 1.14s + 4.21t - 0.037t^2$ where υ is the velocity of sound at T°C in seawater, υ_0 the velocity of sound at 0°C, and s is the salinity [gm/litre]). **[Ans:** 1.6 km, 160 kHz]

3. A quartz crystal of length 0.05 cm is used in a piezoelectric oscillator. Calculate the fundamental frequency of oscillation if the velocity of the longitudinal waves in the crystal is 5.5 km/s. **[Ans:** 55 kHz]
4. An ultrasonic source of 70 kHz sends down a pulse towards the seabed which returns after 0.65s. The velocity of sound in seawater is 1700m/s. Calculate the depth of sea and the wavelength of pulse. **[Ans:** 552m, 2.4 cm]
5. Two ships are anchored at some distance away in the deep sea. An ultrasonic signal of 50 kHz is sent from one ship to another by two routes: one through water with velocity of 1472 m/s and the other through air with velocity of 342 m/s. Calculate the distance between the two ships. **[Ans:** 1472m]
6. A steel bar is tested using an ultrasonic flaw detector. The pulse arrival times were found to be 30 μs and 80 μs. If the bar is of 40 cm thick, find out the location of the defect. **[Ans:** 15 cm below the surface]
7. Calculate the depth of the sea if the time interval between the emitted signal and the echo received is 2 sec in sonar studies. Assume the velocity of sound in sea water as 1490 m/s. **[Ans:**1490 m/s]
8. Calculate the natural frequency of iron rod of 0.03 m length. The density of iron is 7.23×10^3 kg/m^3 and Young modulus is 11.6×10^{10} N/m^2.
9. A quartz crystal of thickness 0.001 m is vibrating at resonance. Calculate its fundamental frequency if Young modulus of quartz = 7.9×10^{10} N/m^2 and the density is 2650 kg/m^3. **[Ans:** 27.30 kHz]

15 Non Destructive Testing

Learning Objectives:

By the end of this chapter, you will be able to:

- Understand different types of defects in a product.
- Learn in detail about different methods of NDT - Liquid penetrant, magnetic particle, eddy current, ultrasonic inspection and radiography methods.

15.1 INTRODUCTION

An industrial product is manufactured with an aim to perform a certain function and the user expects that it performs the assigned function well. However, the product may contain certain defects and imperfections, which impair its performance level. Non - Destructive Testing (NDT) refers to the entire range of test methods that detect the harmful defects in the finished products without affecting its future usefulness.

Defects in a product might occur during the manufacturing stage, or during the assembly process, or the commissioning time or during operation. The defects might be initially present in the raw material itself or might be introduced during the subsequent processes of machining, fabrication, and assembling. A final good quality end-product can be produced by selecting suitable quality raw materials in the initial stage and then by making sure that defects are not formed during the subsequent stages of fabrication and assembly. Non-destructive testing identifies the defects and gives information about its distribution. NDT plays a critical role in modern engineering practice for accomplishing the essential standards of quality in manufacturing.

15.2 TYPES OF DEFECTS

The defects that are found in manufactured components may be broadly divided into two types, namely external defects and internal defects. Some of the commonly occurring defects are shown in Table 15.1.

Table 15.1: *Types of Defects*

External Defects	Internal Defects
Blow holes in castings	Overstressed parts
Forging defects	Fatigue blow holes
Metallic inclusions on surfaces	Porosity in casting
Surfaces finish	Micro-segregation
Micro-cracks	Axial segregation

15.3 METHODS OF NDT

There are a large number of non-destructive tests in use. We study here are some of the most common methods.

(*i*) Visual Inspection
(*ii*) Liquid Penetrant Testing
(*iii*) Magnetic Particle Testing
(*iv*) Eddy Current Testing
(*v*) Ultrasonic Testing
(*vi*) Radiography

Depending on the requirements, the above tests are used singly or in conjunction with one another.

15.4 VISUAL INSPECTION

Visual NDT is mainly used for examination and detection of surface defects. In this, the test component is first adequately cleaned. Then, it is illuminated with light and examined with unaided eye or with the help of optical devices. Visual inspection can reveal the information on

(*i*) general condition of the component
(*ii*) presence or absence of corrosive product on the surface
(*iii*) presence or absence of cracks
(*iv*) orientation and position of cracks, if present
(*v*) surface porosity etc.

In many cases, the results of visual examination will be of great assistance to other tests. Magnifying devices or lighting aids are used wherever their necessity is felt. Microscopes, borescopes, endoscopes and flexiscopes are some of the optical aids used in the visual testing. A microscope detects minute defects and details of fine structure. A borescope is an instrument used to inspect the inside of a narrow tube, bore, or chamber. The endoscope is much like a borescope with a superior optical system and high intensity light source. The flexisope is a flexible fibre-optic borescope that can be used to inspect inaccessible regions and through passages with several bends. In recent times, holography is used to inspect surfaces of precision components.

15.5 LIQUID PENETRANT TESTING

Liquid penetrant testing is also known as dye penetrant testing. This method is used for detecting minute discontinuities such as cracks and surface openings. Cracks of 150 nanometers can also be spotted. This method is useful to any non-porous clean material, metallic or non-metallic. Though it is applicable for both magnetic and non-magnetic materials, primarily it is applied for nonmagnetic materials. It is simple to perform and relatively cheaper than other testing methods.

Basic Principle: When a liquid penetrant is applied over a clean surface to be inspected, the penetrant seeps into the defect by the combined action of surface tension and capillary action gives a recognizable sign of the defect and the location, size and nature of the discontinuity.

15.5.1 Basic Processing Steps of Liquid Penetrant Inspection

The following are the basic stages on liquid penetrant method, Fig. 15.1.

(*i*) **Surface Preparation:** It is the initial cleaning and drying of the surface area of the test component. It is necessary that the surface is to be free from oil, grease, water, or etc that may avoid penetrant from entering flaws.

(*ii*) **Penetrant Application:** In the next step, the penetrant is applied by spraying, brushing and etc. The liquid should spread uniformly over the surface and seep into the crack.

(*iii*) **Penetrant Dwell:** The penetrant is left on the surface for a sufficient time (penetrant dwell time) to seep into a defect. Generally, penetrant dwell time depends on the application, materials used, the component being inspected, and the type of defect being inspected. Normally it ranges from 5 to 60 minutes.

(*iv*) **Removal of Excess Penetrant:** The excess penetrant is then carefully removed from the surface of the sample. Insufficient cleaning leaves penetrant on the surface and leads to overlooking of the defect. On the other hand over cleaning may remove penetrant from defects. Therefore, both insufficient cleaning and over cleaning must be avoided. The cleaning of the surface is carried out using a solvent and then rinsing with water, or first treating with an emulsifier and then rinsing with water.

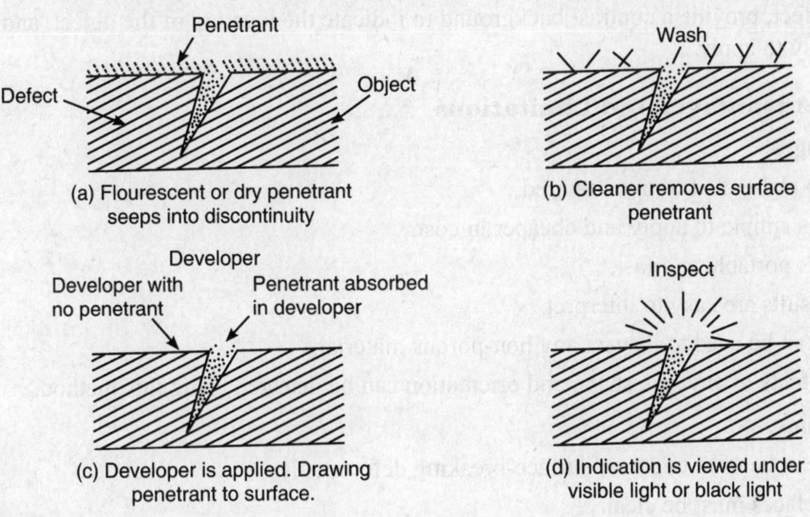

Fig. 15.1:

Sequence of steps in Liquid Penetrant Testing.

(*v*) **Developer Application:** In the next step, the surface is covered with a thin layer of developer which draws out trapped penetrant from surface imperfections and makes them visible. The developer is left on the surface for sufficient time, about 10 minutes or more so that the trapped penetrant could be removed out of surface flaws. Developers are available in different forms which may be applied by dusting, or by dipping, or by spraying.

(*vi*) **Inspection and fine cleaning:** After inspection, the surface is to be carefully cleaned.

15.5.2 Materials used in Penetrant Testing

A variety of materials are used in penetrant testing method. They are penetrant materials, cleaners, emulsifiers and developer materials.

(*i*) **Penetrants:** The penetrant materials are light, oil-like liquids, which have dissolved coloured dyes and may be classified as fluorescent, visible dye or dual-purpose materials. The role of the dye is to give colour contrast with respect to the surrounding background under white light or UV light illumination so that the defect becomes clearly visible. The penetrant materials may also be classified as (*i*) water-washable, (*ii*) solvent removable, and (*iii*) post-emulsifiable. Penetrants that are water washable can be removed from the surface by ordinary tap water. Other penetrants are removed with special solvents. For petroleum-based penetrants, an emulsifier is used that reacts with the oil-based penetrant to form a water-soluble substance, which is then washed away by water washing. The penetrant must be chemically stable, have low viscosity, chemical inertness, ease of removal, and sufficient brightness of colour.

(*ii*) **Developers:** The developer is required to enhance the conspicuity of the indication. Two types of developer materials are used in the technique: (*i*) dry developers and (*ii*) wet developers. Dry developers are dry light coloured powdery materials such as amorphous silica powder. Wet developers consist of a powdered material suspended in a suitable liquid. The developer must be highly absorptive to draw penetrant from the defect, provide a contrast background to indicate the location of the defect, and must be easy to apply.

15.5.3 Advantages and Limitations

Advantages:

(*i*) No elaborate setup is required.

(*ii*) It is simple to apply and cheaper in cost.

(*iii*) It is portable and fast.

(*iv*) Results are easy to interpret.

(*v*) It can be used to inspect any non-porous material.

(*vi*) Defects of any size shape and orientation can be detected using this method.

Limitations:

(*i*) It can be used to detect surface-breaking defects only.

(*ii*) Surfaces must be clean

(*iii*) It cannot be applied to porous materials.

15.6 MAGNETIC PARTICLE TESTING

Magnetic particle testing is useful for detecting the occurrence of cracks, inclusions, and etc at surface and subsurface in ferromagnetic materials. It is not applicable to nonmagnetic materials.

While a test sample is magnetized, the lines of force flow inside the material. Wherever there is a flaw, magnetic lines of force are interrupted and some of the lines leak out of the material. The exit and reentry points of leakage lines form the opposite magnetic poles. If minute magnetic particles are sprinkled on the material, the particles are attracted by the magnetic poles and give a image of the size and shape of the defect.

15.7 EDDY CURRENT TESTING

Eddy current testing is used to inspect electrically conducting materials for defects, irregularities in structure, and variations in composition. In eddy current testing, a varying magnetic field is produced if a source of alternating current is connected to a coil. When this field is placed near a conducting test specimen, eddy currents will be induced in the test specimen. The eddy currents, in turn will produce a magnetic field of their own. The detector measures the new magnetic field and converts the signal into a voltage that can be displayed on a CRO. This test will detect surface and sub-surface defects.

15.8 ULTRASONIC INSPECTION METHOD

Ultrasonic inspection method is a versatile and widely used NDT method. It was suggested by Sokolov in 1935 and applied by Firestone in 1940, and was further developed subsequently.

Principle: The principle of ultrasonic testing is based on the fact that solid materials are good conductors of sound waves. The waves are reflected at the interfaces and by internal faults. The interaction effect of sound waves with the material is more at the higher the frequencies of the wave and generally in the range 0.5 MHz and 25 MHz are used.

A typical UT inspection system consists of a number of parts such as the pulse transmitter, transducer, and display devices. The ultrasonic waves are generated with the help of piezoelectric devices. A pulse transmitter can generate high voltage electrical pulses. When these bursts of alternating voltage are applied to the transducer, the transducer emits high frequency ultrasonic energy. The ultrasonic beam is then transmitted into the test object which propagates through it. When there is a crack in the wave path, a part of the energy will be reflected back from the fault surface and is converted into an electrical signal by the transducer. This signal is displayed on a screen of CRT. The characteristics of the pulses produced by the transducer are used for interpretation of the nature of the defect in the specimen.

15.8.1 Transducer Probes

Piezoelectric transducers generate ultrasonic waves. They convert high frequency electrical signals into mechanical vibrations. The three common piezoelectric materials used in probes are quartz, lithium sulphate and ceramics such as barium titanate and lead zirconate titanate. Different types of transducer probes are in use. We describe here three probes, namely normal beam probe, angle beam probe and transmitter-receiver probe.

(*i*) **Normal beam probe:** The schematic of a normal beam probe is shown in Fig. 15.2 (a). Electrodes are fitted on both faces of the piezoelectric crystal and wire from these electrodes lead to the connector socket of the probe. A damping material is fitted at the back of the crystal. A wear plate is attached to the outer face of the crystal, which protects the crystal from wear and tear. The entire assembly is mounted in a casing, which protects the transducer from mechanical damage.

(*ii*) **Angle beam probe:** In an angle beam probe, the transducer is mounted on a perspex wedge, as shown in Fig. 15.2 (b). The ultrasonic beam reaches the bottom flat surface of the probe at an angle and refracts into the material to be tested. The part of the beam that is reflected from the perspex wedge is damped with a suitable damping material.

(*iii*) **Transmitter-receiver probe:** Transmitter-receiver probe consists of separate transmitter and receiver incorporated in a single housing. The transmitter and receiver transducers are of the same frequency. They are arranged such that they are inclined slightly towards each other (Fig. 15.2 c). Delay blocks of perspex are added to both the transducers and an acoustic barrier is kept between them to prevent any cross-talk.

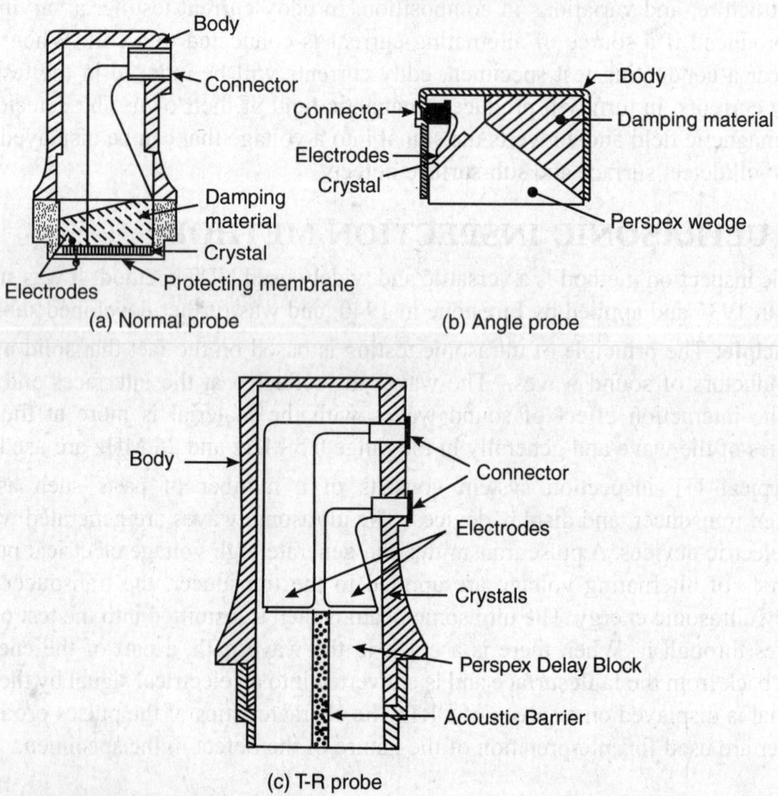

Fig. 15.2: Different types of transducer probes.

15.8.2 Coupling Materials

Acoustic impedance mismatch occurs if the transducer probe is directly placed on the test object, because of the air layer between the probe and the object. A major portion of the energy is reflected back and very less energy gets transmitted into the test specimen. Therefore, a thin film of oil or glycerin is applied to the surface of the object before placing the transducer on it. Then, the ultrasonic energy is transmitted into the test piece more effectively. Thus, oil or glycerin film acts as a couplant to couple ultrasonic energy from the probe to the test object.

15.8.3 Ultrasonic Testing Methods

Several methods have been developed for the ultrasonic testing. Among them, pulse-echo methods are most popular and widely used. These methods can be divided into two broad testing techniques, namely contact method and immersion method.

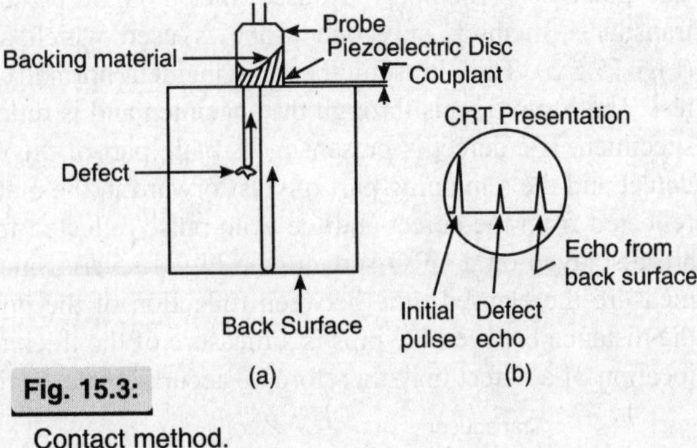

Fig. 15.3: Contact method.

In the *contact method*, the transducer is placed in direct contact with the test object (see Fig. 15.3). A very thin film of suitable coupling material is applied between the probe and the test object for maximum transfer of ultrasonic energy into the object under testing. This method can be used for bigger test objects.

In *the immersion method*, the probe head and test object are both kept immersed in water. The probes used for immersion testing are therefore made water proof. The probe head is kept at a distance from the face of the test object (see Fig. 15.4).

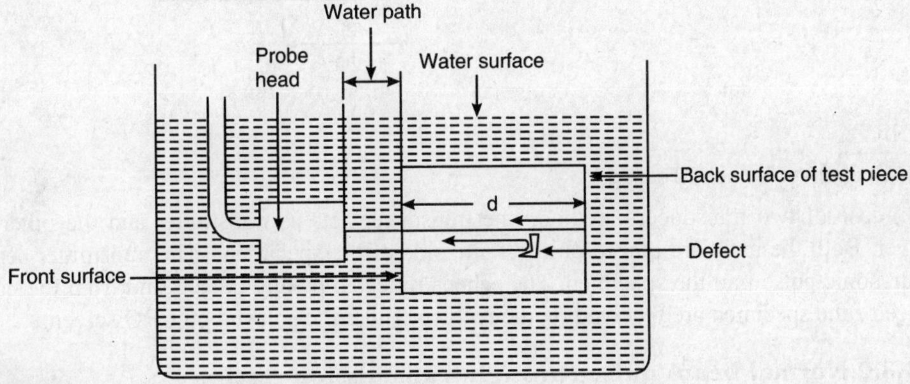

Fig. 15.4: Immersion method.

The above methods may again be grouped into three common testing methods. Two of them utilize normal beam probes and are known as (*i*) *Normal beam pulse-echo testing* and (*ii*) *Normal beam pulse through-transmission testing* respectively. In these methods, normal beam probes are used in which a transducer crystal is fixed parallel to the bottom plate of the probe. The ultrasonic beam produced by the probe propagates into the object perpendicular to the surface of contact and travel in the material in the form of longitudinal waves.

If an angle beam probe is employed for inspection, the ultrasonic beam enters the specimen at an oblique incidence. This third method is known as *angle beam pulse echo testing*. The ultrasonic beam produced by the probe propagates in the material in the form of shear waves.

15.8.3.1 Normal beam pulse-echo testing

The pulse echo method may use either a single crystal unit or two crystal units. In single transducer method, only one probe is used which acts both as transmitter and receiver (Fig. 15.5 *a*). The ultrasonic beam is incident normally on the surface of the specimen under test. The beam travels through the specimen and is reflected back from the rear surface of the specimen. If a defect is present in its path, part of the energy will be reflected back from the defect and the remaining part travels forward in the material. The initial pulse, the echo pulse reflected from the defect and the echo pulse reflected from the opposite face of the specimen are displayed on a CRO screen (see Fig. 15.5 *b*). Since the pips on the oscilloscope screen measure the elapsed time between reflection of the pulse from the front and back surfaces, the distance between the pips is a measure of the thickness of the specimen (Fig. 15.5 *b*). The location of a defect may therefore be accurately determined from the pips on the screen.

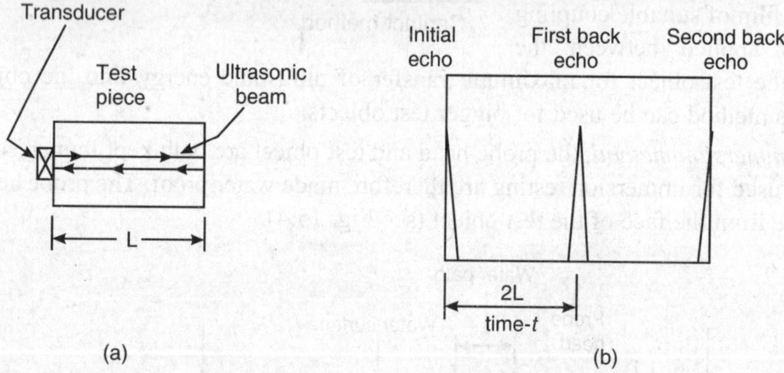

Fig. 15.5: Normal beam pulse-echo Testing.

In case of two transducers method, one transducer acts as transmitter and the other as receiver. Both the units are placed on the same side of the specimen. The transmitter sends an ultrasonic pulse into the specimen. The echoes reflected by any defects and from the back surface of the specimen are received by the receiver and are displayed on CRO screen.

15.8.3.2 Normal beam pulse through-transmission testing

In certain cases, the pulse-echo technique may not provide required information. It happens whenever a defect does not provide a suitable reflection surface or whenever its orientation is not favourable for detection. In such cases, the through-transmission method is adopted. The method uses two ultrasonic transducers on each side of the specimen being inspected. Fig. 15.6 and Fig. 15.7 show the arrangements in contact and immersion methods respectively. If an electrical pulse of the desired frequency is applied to the transmitting transducer, ultrasonic waves are produced. They travel through the specimen to the other side. The receiver transducer on the opposite side receives the vibrations and converts them into an electrical signal, which may be displayed on an oscilloscope.

If the ultrasonic pulse travels through specimen without encountering any defect, the signal received will be relatively large. If there is a defect in the path of the ultrasonic beam, part of the energy is reflected and hence the signal received at the opposite end will be reduced (see Fig. 15.6).

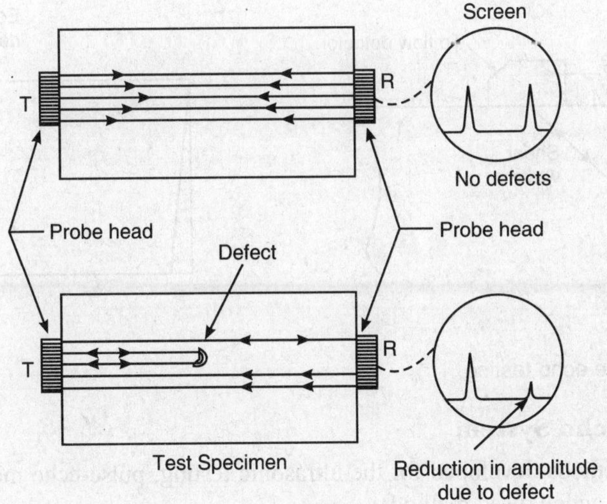

Fig. 15.6:

Normal beam pulse through-transmission Testing.

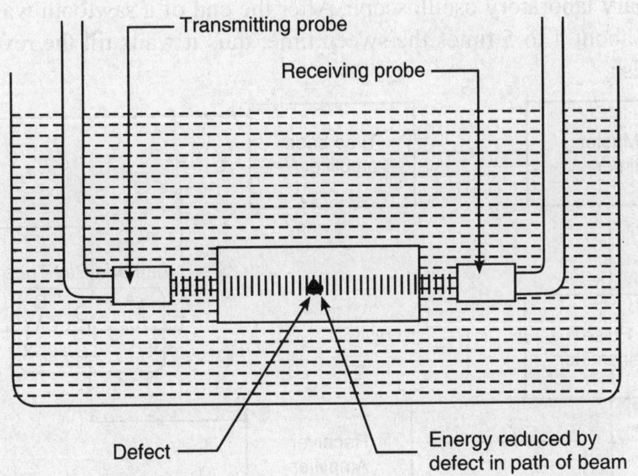

Fig. 15.7:

Normal beam pulse through-transmission Testing – Immersion.

15.8.3.3 Angle beam pulse echo testing

Fig. 15.8 shows the test arrangement. This method of testing is generally used for the inspection of sheet or plate, pipe or tubing and test pieces having shapes that prevent access for normal beam.

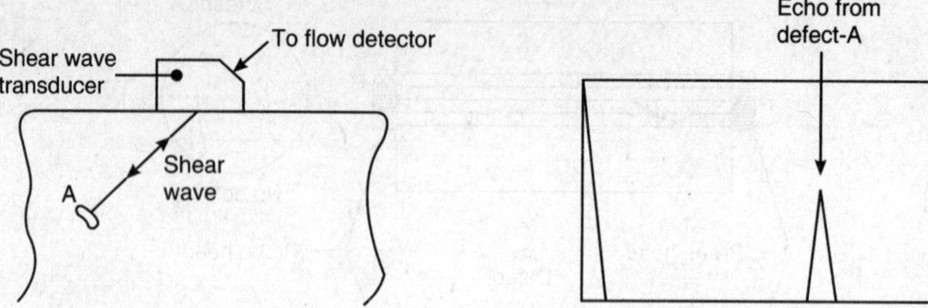

Fig. 15.8:

Angle beam pulse echo testing.

15.8.4 Pulse Echo System

Of the several methods developed for the ultrasonic testing, pulse-echo method is one of the most popular and widely used methods.

Fig. 15.9 shows the block diagram of ultrasonic pulse echo testing equipment. The essential units in the equipment are the pulse transmitter, clock or timer, receiver amplifier and cathode ray oscilloscope. The time base generator used in the CRO here differs from that used in an ordinary laboratory oscilloscope. After the end of a sawtooth wave, the next cycle starts only after about 4 to 5 times the sweep time; thus it waits till the reverberations in the specimen diminish.

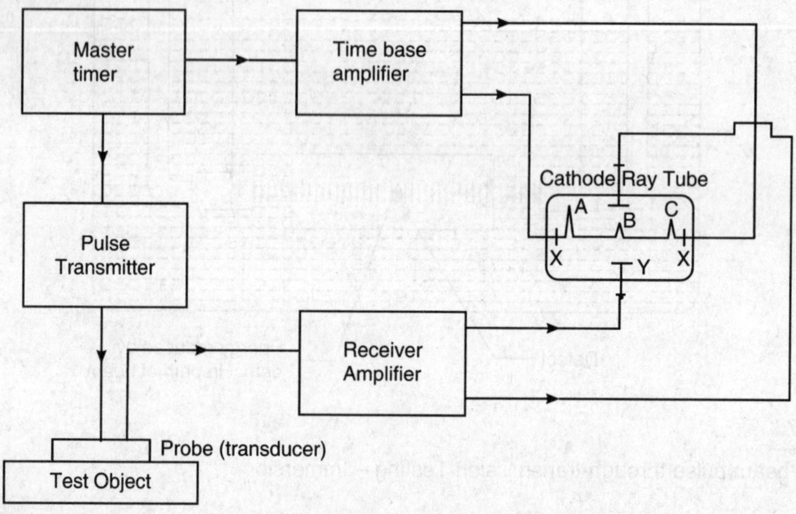

Fig. 15.9:

Ultrasonic pulse echo testing equipment.

From the transmitter, the electric pulse is fed to the piezoelectric transducer probe. Transducer is energized by the electric pulse and vibrates at its resonant frequency and produces a short ultrasonic pulse. These pulses transmit into the specimen through the couplant layer. The electric pulse also prompts the time-base generator, so that the pulse of ultrasound starts to move through the object at the same time as the luminous spot moves across the CRO screen. The output of the transducer is applied to the Y-plates of CRO through an amplifier. It produces a transmission signal (A), which represents the initial ultrasonic pulse (see Fig. 15.9). The luminous spot continues to move across the screen, as the ultrasonic pulse travels through the object. If the ultrasonic pulse encounters a defect, part of the energy is reflected back from the defect surface. The reflected part of the ultrasound returns to the transducer. Under the action of this reflected energy, the transducer vibrates and produces a small voltage pulse. This induced voltage is fed to the Y-plates of CRO through the amplifier and produces signal (B), the echo pulse from the defect. The ultrasonic energy in the transmitted pulse travels further to the bottom surface of the object and gets reflected there back to the transducer. The transducer produces an electric voltage pulse (C), which is much smaller than the transmitted pulse (A). The signal (C) constitutes the bottom surface echo.

As the Y-axis of the display on CRO screen represents time, one can determine the location of the defect in the object. The time interval between the transmitted pulse and echo pulse from the defect equals the time, t taken by the energy to travel from the transducer to the defect surface and back to the transducer. If υ is the velocity of ultrasonic waves in the material, then the distance, d of the defect from the top surface of the object is given by

$$d = \frac{\upsilon t}{2} \quad ...(15.1)$$

15.8.5 Data Presentation

Generally, **A-*scan***, **B-*scan*** and **C-*scan*** formats are used for ultrasonic data presentation.

(*i*) **A-scan presentation:** A-scan display is the most used mode of display in ultrasonic testing. In this mode of display, the *X*-axis represents time taken by the by the pulse to the reflecting surface and return back to the transducer. *Y*-axis represents the amplitude of the echoes. The location of the defect is estimated by the position of the echo given by it on the horizontal axis and size of the defect from the relative amplitude of the echo. The information that is available in A-scan is one-dimensional. A typical A-scan echo pattern is shown in Fig. 15.10.

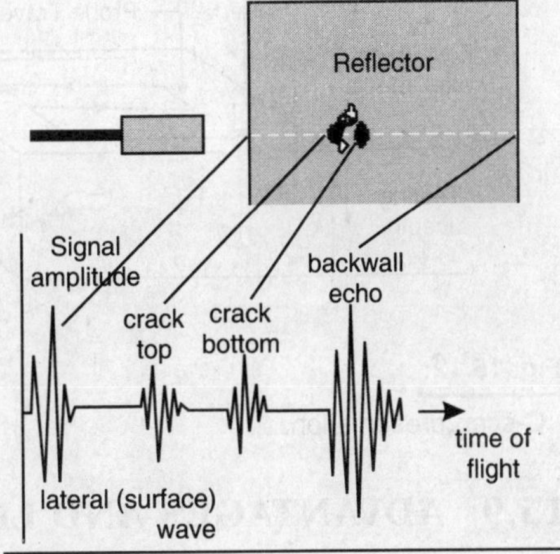

Fig. 15.10: A-scan presentation

(*ii*) **B-scan presentation:** B-scan display gives a cross-sectional view of the test object and shows the position, orientation and depth of defects in the specimen. In this mode of display, Y-axis represents elapsed time while *X*-axis gives the location of the transducer along a line on the surface of the specimen in relation to the starting position. Thus, the probe movement is displayed in *x*-direction while the distance of the defect is displayed in *y*-direction. Echo amplitude is indicated by the relative brightness of echo indications. If a storage oscilloscope is used, the whole picture will be displayed, which reveals the depth of the defect beneath the surface and its size in the lateral direction (see Fig. 15.11).

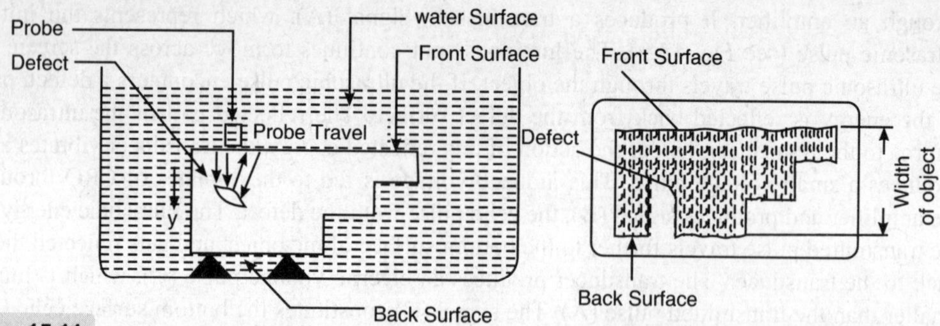

Fig. 15.11: B-scan presentation.

(*iii*) **C-scan presentation:** The depth of defects is not relevant in some testing problems, but information about their distribution parallel to the test surface is required. In the C-scan mode, the transducer is moved over the surface of the test piece and the echo intensity is recorded as a variation in line shading (Fig. 15.12). The image shows the plan of the object as viewed from the top and is a true-to-scale reproduction of the defect in the object. C-scan presentation is produced with a computer controlled immersion scanning system.

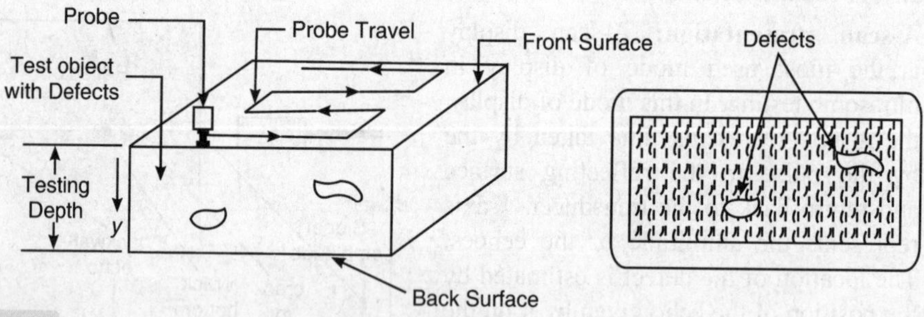

Fig. 15.12: C-scan presentation.

15.9 ADVANTAGES AND LIMITATIONS

Advantages

(*i*) The test is fast and dependable.

(*ii*) It is easy to operate and also lends itself to automation.

(*iii*) Results of tests are immediately known.

(*iv*) It can detect both surface and subsurface defects.

(*v*) Relatively portable equipment.

(*vi*) No special safety precautions are needed.

(*vii*) Can detect minute defects.

(*viii*) Objects of very large thickness can be tested.

(*ix*) High accuracy in finding the location, size and shape of the defects.

(*x*) Single side access is sufficient for conducting the test.

(*xi*) It can also be used to find out the thickness of the object.

Limitations

(*i*) It is not suited for testing of objects that are coarse, uneven in shape, very small parts, very thin and inhomogeneous objects.

(*ii*) Not suitable in coarse structure materials like castings and stainless steel.

(*iii*) Line shaped defects lying parallel to the sound beam may escape detection.

(*iv*) Interpretation of readings requires good technical knowledge.

15.10 X-RAY RADIOGRAPHY

X-ray radiography is useful for detecting the presence and location of internal cavities. X-ray examination of materials may be carried out by processes known as radiography and fluoroscopy. They differ in the manner in which the radiation emerging from the test object is detected.

Fig. 15.13: X-ray radiography.

If a beam of x-rays is directed normal to the surface of an object and if the object is of uniform thickness, the transmitted beam will be of equal intensity. However, if the object contains a cavity, groove or a similar discontinuity, a variation in X-ray transmission will result at these points. As X-rays are invisible, the effect of differential absorption caused by discontinuities has to be shown by recording the transmitted beam on a photosensitive film. The image of a cavity inside the object will appear darker than the surrounding area because of the greater X-ray transmission at this point.The essential parts are a source of radiation, the object to be radiographed and a detector that is a photographic film.

15.10.1 X-ray Source

X-rays are produced when fast moving electrons are suddenly brought to rest by colliding with matter. The source of x-rays is an X-ray tube (see Fig. 15.14). The X-ray tube consists of a glass bulb under vacuum enclosing an anode and a cathode. The cathode is a filament which, when heated by a current, emits electrons.

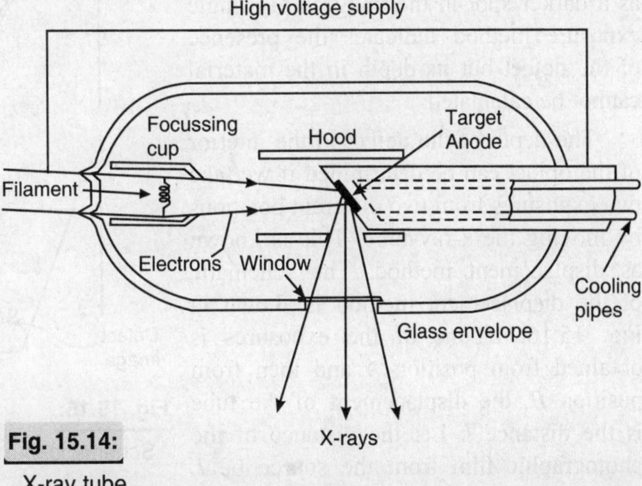

Fig. 15.14: X-ray tube.

The electrons are accelerated towards the anode under the action of the high voltage on the anode. The target is embedded in the anode and when the electron beam strikes the target, x-rays are produced. The target is usually a tungsten disc, which has a very high melting point. The anode is made of copper, which has high thermal conductivity, and a coolant such as water or oil further cools it. The anode face is angled and is hooded to spread and restrict the x-ray beam appropriately. The effective width of the source of x-rays is considerably smaller than the area of the target and is about 1 to 5 mm. This effective width is usually called *focal spot size or focus size* (see Fig. 15.15).

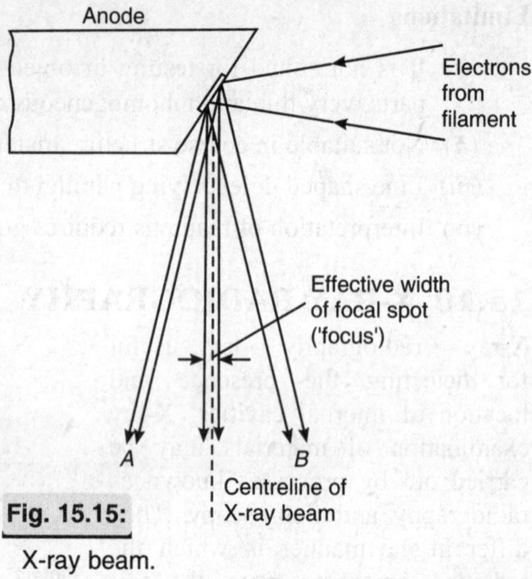

Fig. 15.15: X-ray beam.

15.10.2 Photographic Film

A radiographic film is very much similar to ordinary photographic film. It consists of a base that is made of thin transparent plastic sheet coated with a recording medium. The recording medium is an emulsion of a silver halide with gelatin binder. It is applied to both the sides of the base. A protective layer covers the emulsion. When radiation strikes the emulsion, a change takes place in the emulsion. The latent image on the film is then developed to give a visible image that is a two-dimensional shadow picture of the object.

15.10.3 Displacement Method

Fig. 15.16 is the schematic of a single exposure method. In this method, a beam of X-rays irradiates the test object and the portion of the radiation that is not absorbed by the object is transmitted and is incident on a sheet of photographic film. Absorption of radiation is less in the region of the defect as the density of the material at that place is less. When the film is developed, the defect is seen as a darker spot in the image. This single exposure method indicates the presence of the defect but its depth in the material cannot be calculated.

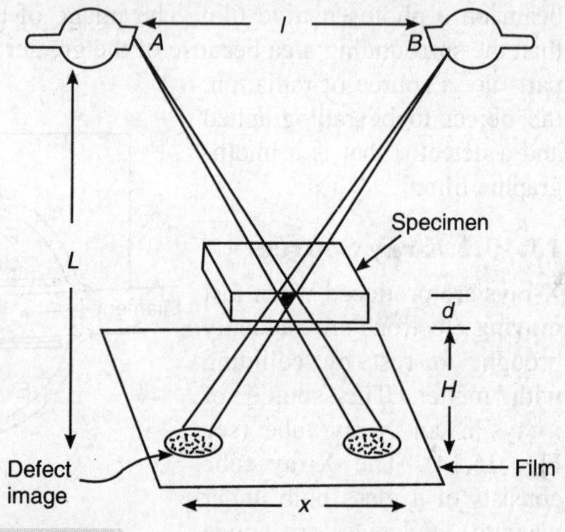

Fig. 15.16: Schematic of a single exposure method.

The depth of the defect in the interior of the object can be determined if we take two exposures from two different positions by moving the x-ray tube. This is known as displacement method. The schematic of the displacement method is shown in Fig. 15.16. If one of the exposures is obtained from position *A* and then from position *B*, the displacement of the tube is the distance *l*. Let the distance of the photographic film from the source be *L* and x be the distance between the images of the defect on the photographic film. Then, the location of the flaw from the bottom surface of the object, d, is given by

$$d = \frac{L \cdot x}{(x+1)} - H \qquad ...(15.2)$$

where H is the height of the object from the film.

15.10.4 Advantages and Limitations

Advantages

(*i*) Provides permanent record.
(*ii*) Works well on thin sections.
(*iii*) Highly sensitive.

Limitations

(*i*) Radiography is expensive.
(*ii*) Trained technicians are required to carryout the radiographic inspection.
(*iii*) It can only detect cracks that are parallel to the radiation.
(*iv*) Minute discontinuities such as inclusions, micro-porosity, micro-fissures etc cannot be detected.
(*v*) It takes more time for thick test objects.
(*vi*) It is almost impossible to detect lamination type defects in metals.
(*vii*) Safety precautions are to be taken in the use of radiographic testing.

15.11 X-RAY FLUOROSCOPY

If a fluorescent screen is held behind the test object instead of a photographic film, the x-rays are converted into visible light and an image of the object can be seen on this screen. This technique is known as **fluoroscopy**. The image obtained on the fluorescent screen is usually very faint and it is difficult to ascertain the details of defects. However, using a suitably sensitive closed-circuit television (CCTV) camera focused on the fluorescent screen, a brighter image can be obtained on the television monitor screen. Since the image is produced on a television monitor, it can be kept at a distance from the x-ray equipment, and the radiation hazards can be eliminated.

The television image can be converted into digital data and using digital image processing technique, image enhancement can be achieved. This method is known as **real time radiography**. Nowadays the term fluoroscopy is used for the real-time radiography. A basic system is shown in Fig. 15.17. It consists of a source of radiation, X-ray image intensifier, video camera, and computer with image processing arrangement and a display monitor. The x-rays from the source pass through the test object and strike the double layer screen in the image intensifier. The fluorescent layer of the screen converts the x-rays incident on it into light. The adjoining photo-cathode layer generates an electron image of the visible image produced by the fluorescent layer. The electrons from the double layer input screen are accelerated and focused on to a smaller output screen. The output screen reconverts the electron image into visible image. Thus, the dim image from the input screen is made 100 to 10,000 times more intense. The video camera views the image, which is converted into digital data by the analogue-to-digital converter. The computer processes the digital output and the real time image of the object is presented on the display monitor.

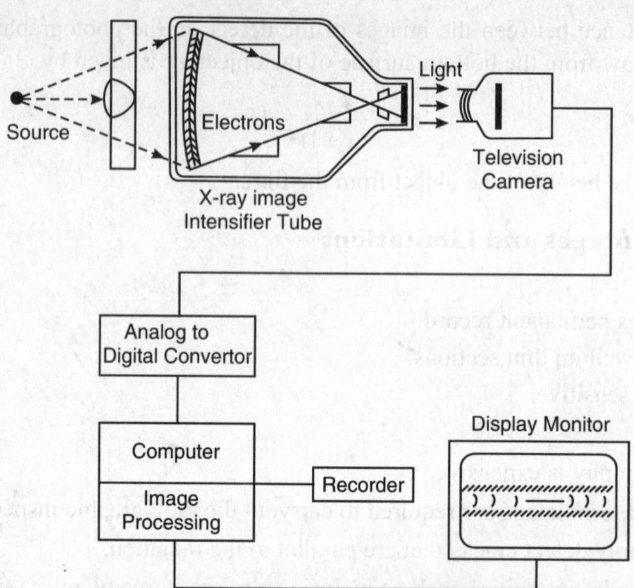

Fig. 15.17: Fluoroscopy.

15.12 COMPARISON OF CONVENTIONAL AND REAL-TIME RADIOGRAPHY

(*i*) In conventional radiography, we obtain a negative image of the object on the photographic film. In the real time radiography, we see the positive image of the object.

(*ii*) In conventional radiography the image is viewed in static mode. In real time radiography, the image is viewed in dynamic mode and is interpreted at the same time as the radiation passes through the object.

QUESTIONS

1. What is NDT? Name some important NDT methods.
2. What are the benefits nondestructive testing of products?
3. Define NDT. Discuss the liquid penetrate method of NDT in detail.
4. What is the principle of liquid penetrant method?
5. Explain in detail how liquid penetrant method is used in NDT. What are its advantages and limitations?
6. What is the basic principle of radiographic testing? Explain the displacement method for finding the depth of the defect in an object.
7. What is meant by fluoroscopy? Draw the schematic diagram of a real time radiographic testing apparatus.
8. Distinguish between radiography and fluoroscopy.
9. What is the principle of ultrasonic testing?
10. Discuss in detail the ultrasonic flaw detection.

11. What is non-destructive testing? Explain with principle how flaw in a solid can be detected by non destructive method using ultrasonics.

12. Draw a block diagram of ultrasonic testing equipment. Explain the three different scan modes used for presentation of data.

13. What are the advantages and limitations of ultrasonic testing?

14. What is the basic principle of radiographic testing? Explain the displacement method for finding the depth of the defect in an object.

15. What is meant by fluoroscopy? Draw the schematic diagram of a real time radiographic testing apparatus.

16. Distinguish between radiography and fluoroscopy.

16 Heat Transfer

Learning Objectives:

By the end of this chapter, you will be able to:

- Explain about different types of heat transfer.
- Learn about the concept of conduction and thermal conductivity.
- Describe rectilinear flow of heat-through uniform bar and cylindrical tube.
- Understand concept of heat conduction through compound media.

16.1 INTRODUCTION

All matter is made up of molecules and atoms. These molecules are always in motion. The motion of molecules creates heat or thermal energy. The more motion the molecules have the more heat or thermal energy they will have. Temperature is an average value of energy for all the molecules in a given system. Temperature is independent of how much matter there is in the system. It is only an average of the energy in the system. Unit of temperature in SI units are the Celsius and Kelvin. If two systems at different temperatures are brought together, energy is transferred from the hotter system to the colder one. We say that **heat energy** is transferred from one system to another as the result of a difference in temperature. Heat Q is energy in transit. In the SI system, common units for measuring heat are the Joule and calorie.

16.2 HEAT TRANSFER MODES

The different types of heat transfer are known as ***modes of heat transfer*** and common modes are conduction, convection and radiation.

(*i*) Conduction: Conduction heat transfer mode takes place in a solid or fluid. Heat is transmitted along temperature gradient in the medium. Conduction requires matter and does not need bulk motion of matter.

Heat can be conducted between two bodies, one hot and another cold which are in contact with each other; heat flows from one body to the other. The process of transfer of heat from a hot region to cold region is called ***thermal conduction***. Fig. 16.1 shows the transfer of heat by conduction.

Electrical conductors are good conductors of heat and electrical insulators are usually good insulators of heat.

Materials that conduct heat well are called good thermal *conductors*. Different bodies have different thermal conductivities. Free electrons are responsible for thermal conductivity.

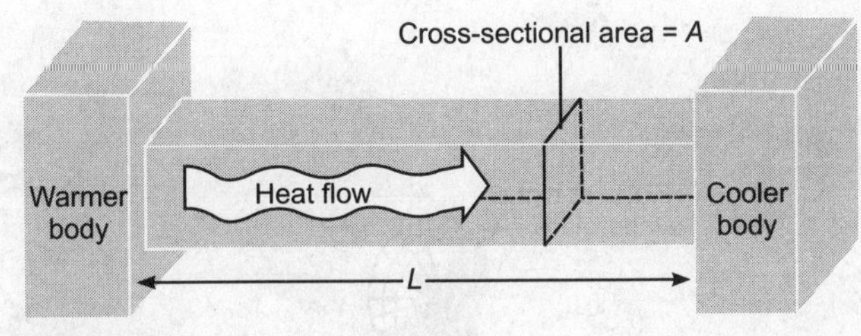

Fig. 16.1:

Transfer of Heat by Conduction.

The molecules of the body on hot portion start vibrating energetically and these vigorously vibrating molecules collide with the neighboring molecules and transfer some of the vibrating energy. Molecules which received the energy from neighboring molecules also start vibrating energetically. Thus, heat spreads through the body.

Def: *Conduction is a mode of heat transfer from the hot part of the body to its cold part without any actual motion of heated particles.*

(*ii*) Convection: Convection occurs in fluids either through random molecular motion, *diffusion* or through the bulk motion of a fluid transporting energy from place to place.

It is a different kind of conduction. Here heat transfer takes place by the bodily movement of heated particles. Convection is possible only in liquids and gases. In convection the heat is moved by bulk transfer of a non-uniform temperature fluid. The regions of the fluid which are heated expand and become less dense. When a hot fluid *mixes up* with the cold fluid heat transfer happens rapidly. Heat transfer gives the circular motion away from a source of heat. Ocean currents in water and weather patterns in air are a few examples of convection. As shown in Fig. 16.2 let us take water in a beaker and add potassium permanganate to give colour to water. On heating the beaker, the heat is conducted through the glass to water in the lower region of beaker. Hot fluids rise through surrounding and become less dense which gives the circular motion of the fluid away from a source of heat. This can be seen by the movement of coloring grains. It is clear that heat is transferred by the bodily movement of heated particles of the medium in convection process.

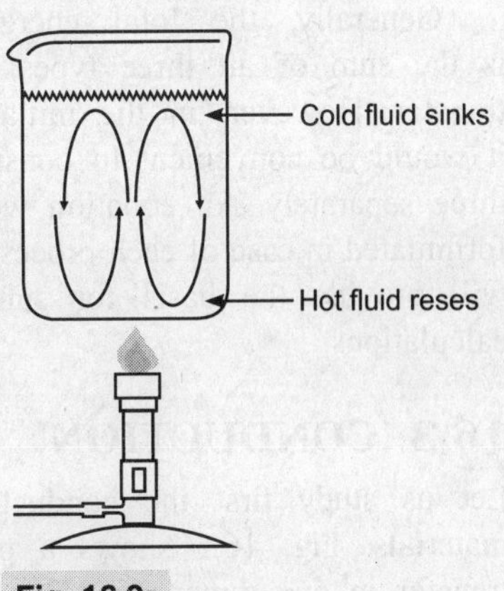

Fig. 16.2:

Transfer of Heat by Convection.

Def: *Convection is a mode of heat transfer from one part of the medium to another part by bodily motion of heated particles.*

(*iii*) Radiation: In radiation, the heat energy is transferred by electromagnetic wave phenomenon which does not require a carrying medium. Radiation is the only way that heat can move through a vacuum or space, Fig. 16.3. For example the heat loss in thermos is due to radiation. Every body radiates energy in the form of photons moving in a random direction, with random phase and frequency. When radiated photons reach another surface, they may be absorbed.

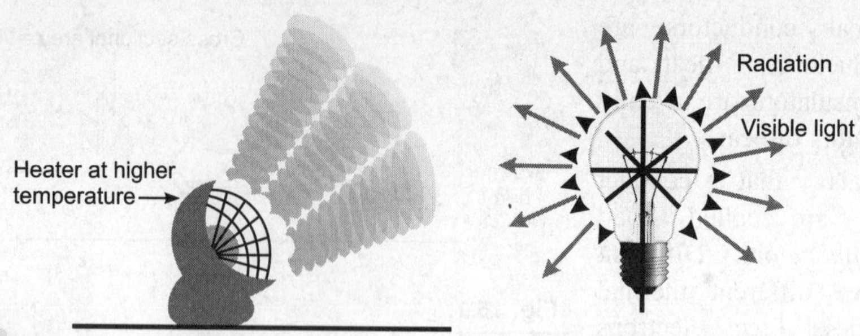

Fig. 16.3:

Transfer of Heat by Radiation.

Def: *Radiation is a mode of transmission of heat from one body to another body without heating the intervening medium.*

Convection is most efficient mode of heat transfer among all three process. Radiation is the least efficient and slowest method. All three types of modes of heat transfer are illustrated in Fig. 16.4.

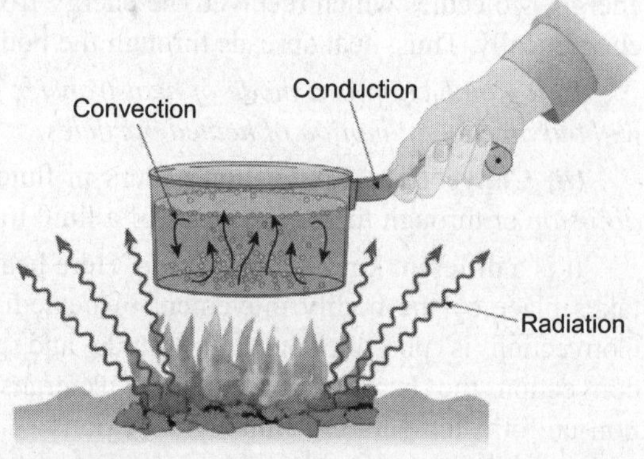

Fig. 16.4:

Three types of modes of heat transfer.

Generally, the total energy flow is the sum of all three types of heat transfer. However, in the initial stage, it would be convenient to consider the three separately. An equation would be formulated in case of each process which will provide the basis for subsequent calculations.

16.3 CONDUCTION

Let us study first the conductive transfer through solid materials. Fig. 16.5 shows a process of conductive heat transfer in one dimension. The significant quantities to be considered are identified. Heat conduction can be expressed quantitatively as the time rate of heat flow (Q/t) in a material for a given temperature difference ($T_1 – T_2$).

Let the area of cross section of solid material slab be A and thickness x and opposite parallel faces be at temperature T_1 and T_2. Heat will flow from hot face to cold face. The amount of heat Q flowing from one face to another face is

(*i*) directly proportional to the area of cross section A

(*ii*) directly proportional to difference of temperature between the two faces, $(T_1 - T_2)$ and

(*iii*) inversely proportional to length x

(*iv*) directly proportional to time t

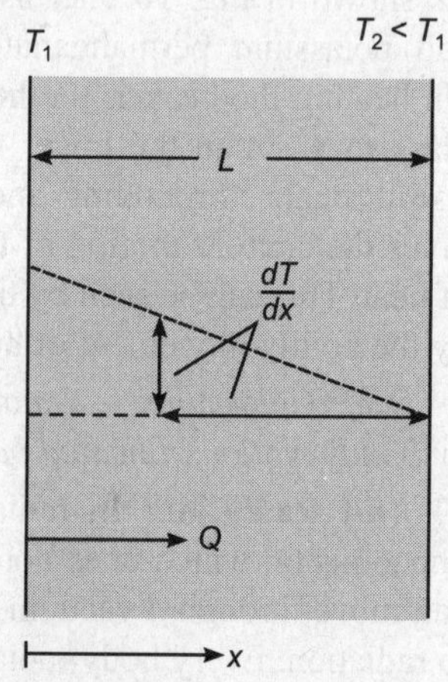

Fig. 16.5:

One dimensional conduction.

These quantities are related by Fourier's Law, a model proposed as early as 1822:

$$Q \propto \frac{A(T_1 - T_2)}{x} t \qquad ...(16.1)$$

The term $\frac{(T_1 - T_2)}{x}$ is called the thermal gradient (the change in temperature per unit of length). Using a constant of proportionality we write above equation as

$$Q = \frac{KA(T_1 - T_2)}{x} \qquad ...(16.2)$$

Where K is called the ***thermal conductivity*** and it characterizes the heat conducting ability of a material.

The greater the value of K for a material, the more rapidly it will conduct heat. The units of K are Joule per meter-sec-kelvin [J/m.s.K or W/mK].

Eq. (16.2) may be rewritten as

$$Q = -KA\frac{dT}{dx} \qquad ...(16.3)$$

where $dT = (T_2 - T_1)$.

The negative sign in the Eq. (16.3) indicates that the natural direction for the heat flow is from high temperature to low temperature and hence down the temperature gradient.

K, is the thermal conductivity (*W/m* K) of the material, through which the heat is transmitted and it depends on the temperature and pressure of the material.

$$\text{Coefficient of thermal conductivity, } K = \frac{L}{A(T_1 - T_2)} \qquad ...(16.4)$$

Coefficient of thermal conductivity of a substance is the quantity of heat which flows in one second through a slab of given material of unit thickness and unit area of cross section when its opposite faces have temperature difference of 1°C.

Considering the finite slab of material shown in Fig. 16.5, we see that for one-dimensional conduction the temperature gradient is:

$$\frac{dT}{dx} = \frac{T_2 - T_1}{L}$$

Hence for this situation the transfer law can also be written

$$Q = KA\frac{(T_1 - T_2)}{L} \qquad ...(16.5)$$

16.4 THERMAL CONDUCTIVITY:

The thermal conductivities of a few materials are listed in the following Table 16.1.

Table 16.1: *Thermal Conductivities of selected materials at room temperature*

Material	Thermal Conductivity, W/m K
Copper	401
Silver	429
Gold	317
Aluminium	237
Steel	60.5

Limestone	2.15
Bakelite	1.4
Water	0.613
Air	0.0263

Molecules/atoms of materials gain energy through various mechanisms. Gases, in which molecules are free to move, have energy in the form of kinetic energy of the molecules. Energy is gained or lost through of gas molecules.

In solids, atoms/molecules cannot transfer energy through collisions. The energy is transferred through lattice vibrations. In case of electrical conductors the electrons are weakly bound to the molecule and can drift from one molecule to another, transporting their energy in the process. Thus, flow of electrons metals is an effective transport mechanism in electrical conductors. Therefore, materials which are good electrical conductors are usually good thermal conductors.

16.4.1 Steady State

Consider a rod heated at one end and an element *PQ* of the rod at some distance from the heated end. Every cross section of the rod receives heat by conduction from the adjacent cross section towards the hot end. All the heat conducted through a layer like *P* will not leave through the layer *Q*. A portion of the total heat received is utilized in increasing the temperature of the portion *PQ* and another portion is radiated from the sides.

The rest is transmitted to next cross section. Thus for some time the temperature at various points on the rod goes on increasing. After some time a steady state will be reached and every layer will be at some steady temperature. At this stage some part of heat entering through the layer *P* is radiated from the sides and remaining is conducted through *Q*. No heat is retained by the section *PQ*. Now the temperature at each point of the rod remains constant and does not change with the time. This state is called *steady state*. The lines of heat flow along the rod are shown in the Fig. 16.6.

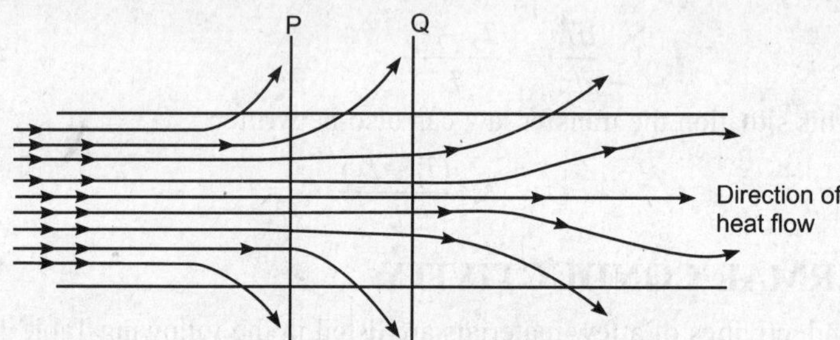

Fig. 16.6: Direction of Heat Flow.

16.4.2 Newton's Law of Cooling:

The energy flow into a system from surroundings gives heating, energy flow from a system to surroundings gives cooling of an object.

Statement: *The rate of change of the temperature of an object is proportional to the difference between its own temperature and the ambient temperature, provided the nature of radiating surface remains the same.*

Proportionality constant depends upon the shape, area and nature of exposed surface.

It is applicable under following conditions:

(*i*) Heat lost by conduction is negligible and convection and radiation are significant.

(*ii*) The hot body is cooled in a uniform streamlined air flow or forced convection.

(*iii*) The temperature of every part of the body should be the same.

Let us consider that a body with initial temperature T_1°C is allowed to cool in air which is maintained at a constant temperature T_2°C. After a time t the temperature of the body be T°C.

Newton's Law of Cooling $$\frac{dT}{dt} = -k(T - T_2) \qquad ...(16.6)$$

where k is a positive proportionality constant.

From Eq. (16.6), the term $(T - T_2)$ is positive and the rate of change of temperature is negative.

The constant k depends upon the surface properties of the material being cooled. Initial condition is given by $T = T_1$ at $t = 0$.

Solving Eq. (16.6)

$$-kt = \log(T - T_2) + \log C$$

$$(T - T_2) = Ce^{kt} \qquad ...(16.7)$$

The value of C can be found by applying initial conditions

$$C = (T_1 - T_2)$$

Substituting the value of C in Eq. (16.7) gives

$$T = T_2 + (T - T_2)e^{-kt}$$

In general, $$T(t) = T_A + (T_H - T_A)e^{-kt} \qquad ...(16.8)$$

where

$T(t)$ = Temperature at time t

T_A = Temp of surroundings

T_H = Temperature of hot object at time 0

k = positive constant and

t = time

If $k < 0$, lim $t \to \infty$, $e^{-kt} = 0$ and $T = T_2$. Eq. 16.7 gives the Newton's law of cooling. The graph between the temperature of the body and time is known as cooling curve. The slope of the tangent to the curve at any point gives the rate of fall of temperature.

(a) Derivation of Newton's law of cooling from Stefan's law:

Let T and T_0 be the absolute temperatures of hot body and surroundings. According to Stefan's law rate of loss of heat from the hot body is given by

$$\frac{dT}{dt} = \mathrm{R} = \sigma(T^4 - T_0^4) \qquad ...(16.9)$$

$$R = \sigma(T - T_0)(T^3 + T^2T_0 + TT_0^2 + T_0^3)$$

When $T \approx T_0$ $$R = \sigma(\mathrm{T}^4 - T_0^4)\,4T_0^3$$

Letting $k = \sigma 4T_0^3$ a constant $$R \propto (T - T_0) \qquad ...(16.10)$$

Thus the rate of cooling $\frac{dT}{dt}$ of the body is directly proportional to the excess of temperature of the body over that of the surroundings, if this excess temperature is small. This is nothing but Newton's law of cooling.

So, Newton's law of cooling is derived from Stefan's law.

16.5 RECTILINEAR FLOW OF HEAT- CONDUCTION ALONG A UNIFORM BAR:

Consider a rod *AB* of uniform cross section heated at the end *A* and end *B* is the cold end as shown in the Fig. 16.7. Heat flows in the direction *AB*. Consider a section of *PQ* of thickness δx at a distance *x* from the end *A*.

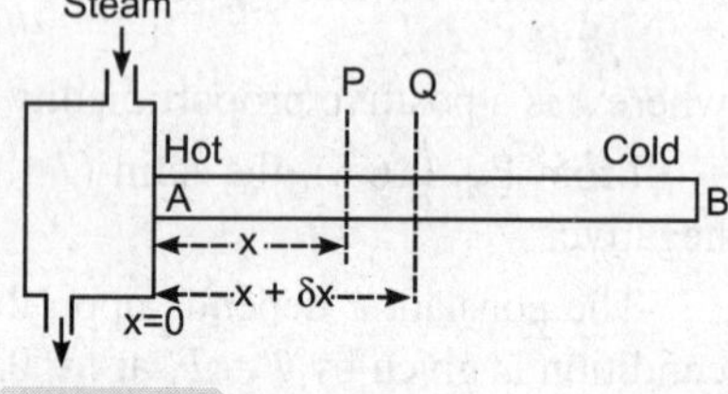

Fig. 16.7: Conduction along a Uniform Bar.

Let the temperature at *P* be *T* and the temperature gradient at *P* is $\frac{dT}{dx}$ and let *A* be the area of cross section of the rod.

Heat entering through *P* per second $= -KA\frac{dT}{dx}$

Negative sign represents the rate of fall of temperature with distance.

Temperature at $Q = T + \left(\frac{dT}{dx}\right)\delta x$.

$$\text{Heat leaving at } Q \text{ per second} = \frac{d}{dx}\left[T + \left(\frac{dT}{dx}\right)\delta x\right]$$

$$= -KA\frac{dT}{dx} - KA\frac{d^2T}{dx^2}\delta x$$

The net gain of heat by the element δx in one second $= KA\frac{d^2T}{dx^2}\delta x$ ...(16.11)

where *K* is the thermal conductivity of the material of the bar and *A* the area of cross section

Before steady state is reached a part of this heat is used in raising the temperature of the element *PQ* and remaining heat is lost by radiation from the surface of the element.

Let ρ be the density and *C* the specific heat capacity of heat material of the bar and $\frac{dT}{dt}$ is the rate of rise of temperature. Then,

Heat used per second in raising the temperature of *PQ*

$$= A\delta x\rho C\frac{dT}{dt} \qquad ...(16.12)$$

If *p* is the perimeter of the bar then the surface area of the element $PQ = p\delta x$

Let *E* be the emissivity of the surface and θ is the average excess temperature of the element over the surroundings.

The heat lost by radiation per second from

$$PQ = Ep\delta xT \qquad ...(16.13)$$

The net gain of heat by the element δx in one second = (Heat used per second in raising the temperature of *PQ* + Heat lost by radiation per second from *PQ*)

$$KA\frac{d^2T}{dx^2}\delta x = \left(A\delta x\rho C\frac{dT}{dt} + Ep\delta xT\right)$$

$$\frac{d^2T}{dx^2}\delta x = \frac{\rho C}{K}\frac{dT}{dt} + \frac{Ep}{KA}T \qquad ...(16.14)$$

Eq. (16.14) is a standard differential equation for the flow of heat flow through a bar.

After steady state is reached there is no absorption of heat by any section of the bar.

Hence $\frac{dT}{dt} = 0$

So, Eq. (16.14) becomes

$$\frac{d^2T}{dx^2}\delta x = \frac{Ep}{KA}T$$

$$\frac{d^2T}{dx^2}\delta x = \mu^2 T$$

$$\mu = \sqrt{\frac{Ep}{KA}} = \text{Constant}$$

The solution of this second degree differential equation is

$$T = ae^{\mu x} + be^{-\mu x} \qquad ...(16.15)$$

Here a and b are two constants which can be determined from the boundary conditions.

(*i*) **Suppose the bar is of infinite length:** the boundary conditions are

At $x = \infty, T = 0$, a = 0

At $x = \infty, T = T_0$, the temperature of the heated end above that of the surrounding $b = T_0$

Hence $T = T_0 e^{-\mu x}$

The temperature falls exponentially from the heated end as shown in Fig. 16.8.

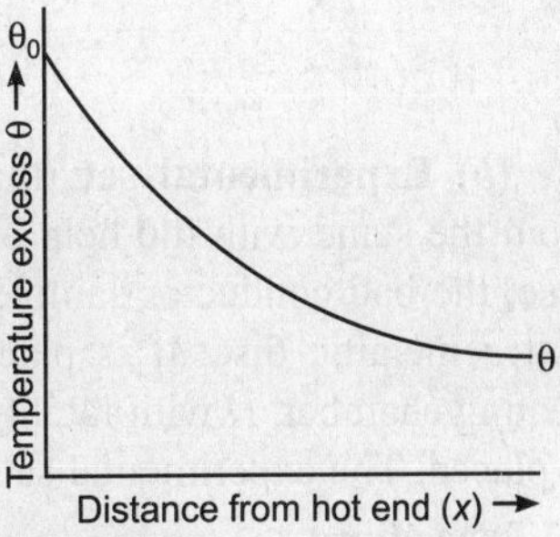

θ_0

Temperature excess θ →

θ

Distance from hot end (x) →

Fig. 16.8: Temperature Fall.

(*ii*) **Suppose the bar is sufficiently long and is of finite length L:** the boundary conditions are

At $x = \infty$, $T = T_0$

At $x = L$, $\frac{dT}{dx} = 0$

$$a = \frac{T_0}{1 + e^{2\mu L}};\ b = \frac{T_0}{1 + e^{-2\mu L}}$$

$$T = T_0\left[\frac{e^{\mu x}}{1 + e^{2\mu L}} + \frac{e^{-\mu x}}{1 + e^{-2\mu L}}\right]$$

16.5.1 Determination of Thermal Conductivity:

The most common methods of determination of thermal conductivity are Forbe's method for good conductors, Lee's method for good conductors at different temperatures, Lee's disc method for poor conductors, Lee's disc method for liquids.

16.5.2 Determination of Thermal Conductivity of Poor Conductor by Lee's disc Method:

(a) Theory: When a steady state is reached, let T_1 and T_2 be the temperatures of metallic disc M and brass disc B. So, the temperature difference between the two ends of bad conductor is $(T_1 - T_2)$.

The rate of heat conducted through the bad conductor

$$Q_1 = \frac{KA\,(T_1 - T_2)}{d} \quad ...(16.16)$$

where K is the thermal conductivity of the material of the sample, d is thickness of the bad conductor and A is the area of cross section.

The rate of heat lost by the brass disc to the surrounding under steady state is

$$Q_2 = mc\left(\frac{dT}{dt}\right)_{T_2} \quad ...(16.17)$$

where m-mass of the brass disc, c is the specific heat capacity of brass disc, JKg^{-1}K^{-1}and $\left(\frac{dT}{dt}\right)_{T_2}$ is its rate of cooling at T_2

From Eq. (16.16) and Eq. (16.17), we get

$$K = \frac{mcd\left(\frac{dT}{dt}\right)}{A\,(T_1 - T_2)} \quad ...(16.18)$$

(b) Experimental set up: The brass disc B is hung from the stand with the help of three strings. On the brass disc, the bad conductor sample disc 'S' is placed and over that a metallic disc M is placed. On the metallic disc, a heating chamber, H with facility for steam-in and steam-out is placed. The experimental set up is as shown in Fig. 16.9.

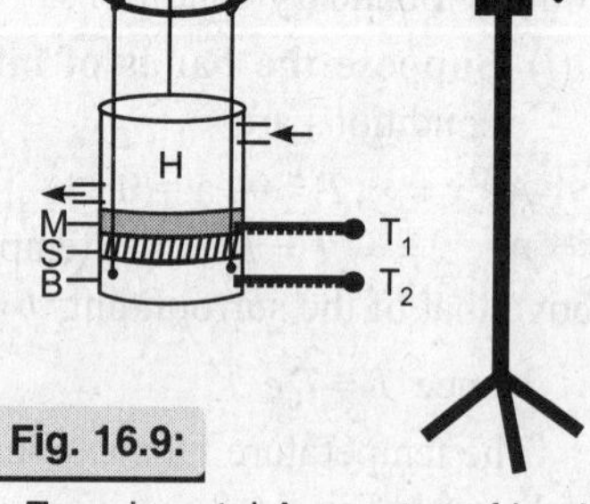

Fig. 16.9: Experimental Apparatus of Lee's disc method.

(c) Procedure:

(*i*) Measure the mass of the brass disc using a balance and take it as m kg.

(*ii*) Measure the diameter and hence the radius of the sample disc, S using vernier calipers. Calculate the circular area of the disc A, measure the thickness of the sample S using a screw gauge and take it as d.

(*iii*) Arrange the brass disc, sample disc and metallic disc as shown in the Fig. 16.9 and insert the thermometers into the grooves of metallic disc and brass disc, which measure temperatures T_1 and T_2 respectively.

(*iv*) Place the heating chamber on the above assembly and suspend it as shown in Fig. 16.9 and heat it by sending steam through the heating chamber.

(*v*) Monitor the temperatures T_1 and T_2 at a regular interval till they reach the steady state. Note the steady state temperatures T_1 and T_2.

(*vi*) To determine the rate of cooling of brass disc, lift the heating chamber and remove the metallic plate M and sample disc S, then place the heating chamber directly on the brass disc, B.

(*vii*) Allow the brass disc B to heat at least about 10°C above the steady state temperature T_2 measured in the first part of the experiment. Remove the heating chamber.

(*viii*) Switch on the stop watch and measure the temperature of brass disc at an interval of 30s as it cools down. Allow the brass disc to cool down to at least 5°C below the steady state temperature, T_2.

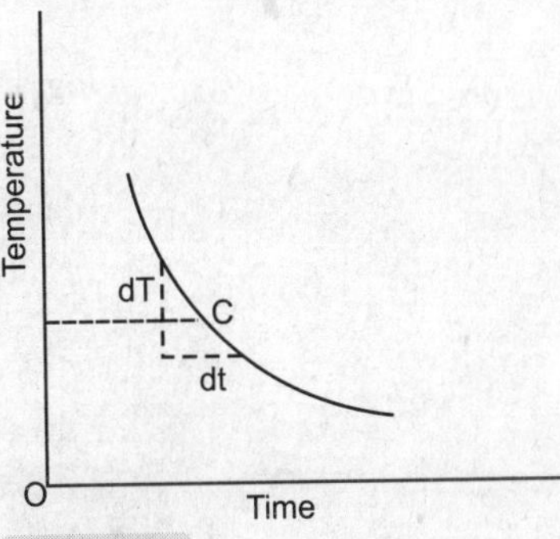

Fig. 16.10: Plot of Temperature versus Time.

(*d*) Calculations:

(*i*) Plot a graph of temperature T of brass disc as a function of time and determine its slope. The slope is equivalent to $\left(\dfrac{dT}{dt}\right)_{T_2}$ as shown in Fig. 16.10.

(*ii*) By the Lee's experimental values of $\left(\dfrac{dT}{dt}\right)_{T_2}$ and $(T_1 - T_2)$ and using Eq. 16.16 the thermal conductivity K can be calculated.

(*iii*) As cooling environment is not same as steady state condition a correction is needed and is equal to $\dfrac{r+2h}{2r+2h}$.

where h is the thickness of the brass disc and r is the radius of brass disc.

16.6 RADIAL HEAT FLOW OF HEAT THROUGH THE WALLS OF CYLINDRICAL TUBE:

Consider a cylindrical tube of internal radius r_1 and external radius r_2 and length l as shown in Fig. 16.11.

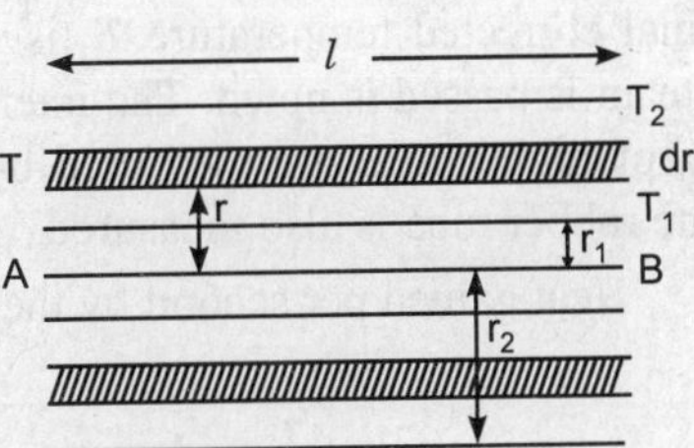

Fig. 16.11: Cylindrical tube.

Let T_1 and T_2 be the temperatures of inner and outer surfaces of the walls of cylindrical tube in the study state condition and let $T_1 > T_2$. Let the heat flow inside the tube in the direction AB. In the stady state condition there is a radial flow of heat in the cylinder.

Consider a cylinder shell of thickness dr of the tube at a distance r from its axis at temperature T, see Fig. 16.12.

Rate of flow of heat through this shell is

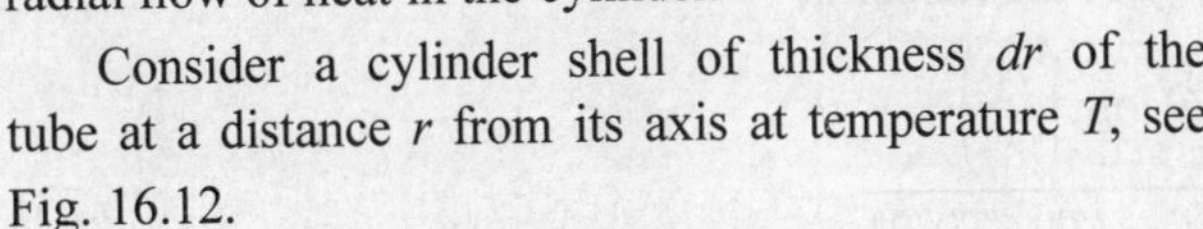

$$Q = -KA\frac{dT}{dr}\, Q = -K2\pi rl\frac{dT}{dr}$$

$$\frac{dr}{r} = -\frac{K2\pi l}{Q}dT \qquad ...(16.19)$$

Integrating Eq. 16.19

$$\int_{r_1}^{r_2}\frac{dr}{r} = -\frac{K2\pi l}{Q}\int_{T_1}^{T_2} dT \qquad ...(16.20)$$

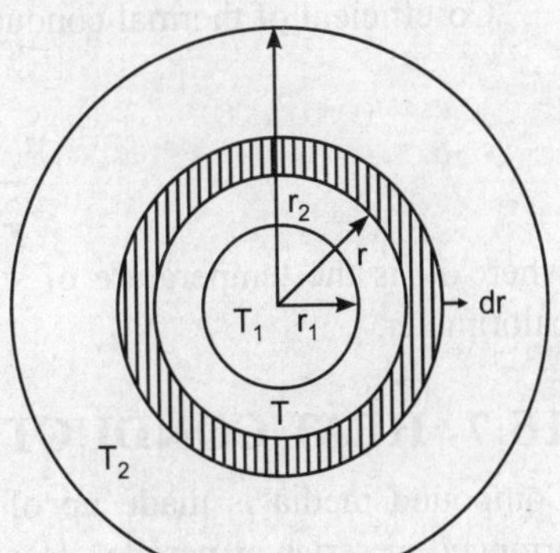

Fig. 16.12: Cylinder Shell.

$$\log_e\left(\frac{r_1}{r_2}\right) = -\frac{K2\pi l\,(T_1 - T_2)}{Q}$$

$$K = \frac{Q\log_e\left(\frac{r_1}{r_2}\right)}{2\pi l\,(T_1 - T_2)} \qquad \text{...(16.21)}$$

$$K = \frac{Q \times 2.3026 \times \log_{10}\left(\frac{r_1}{r_2}\right)}{2\pi l\,(T_1 - T_2)}$$

16.6.1 Determination of Thermal Conductivity of Rubber Tube:

A known length of rubber tube is kept immersed in a calorimeter C containing a known mass of water as shown in Fig. 16.13. The initial temperature T_1 of calorimeter and contents is noted with sensitive thermometer T; Stream is passed through one end of rubber tube and let off through the other. So, the inner side of the rubber tube attains the temperature of the stream.

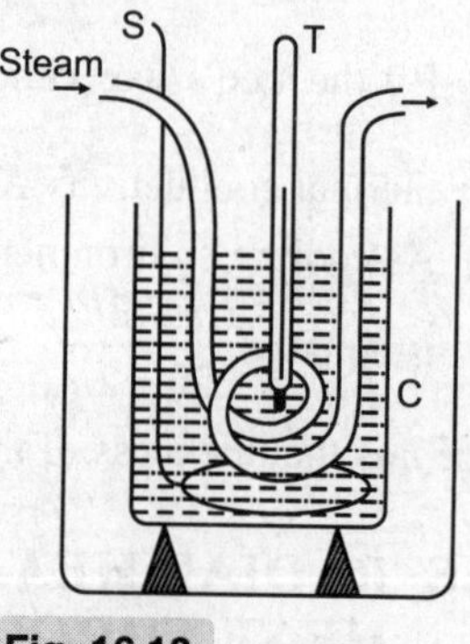

Fig. 16.13: Experimental Arrangement.

Heat flows through the wall of the rubber tube to the outer surface in contact with the water in the calorimeter. Stream is cut off when the temperature of water rises by about 10°C. The final corrected temperature T_2 is noted. The time t for which steam is passed is noted. The internal and external radii r_1 and r_2 of rubber tube are measured by a microscope. Length l of the rubber tube is also measured.

Heat gained per second by the water and the calorimeter,

$$Q = \frac{(m_1C_1 + m_2C_2)\,(T_2 - T_1)}{t} \qquad \text{...(16.22)}$$

where m_1 is mass of the calorimeter, C_1 is specific heat capacity of the material of the calorimeter, m_1 is mass of the water and C_2 is specific heat capacity of water.

Co-efficient of thermal conductivity of the rubber is

$$\frac{Q\log_e\left(\frac{r_2}{r_1}\right)}{2\pi l[T_3 - (T_2 - T_1)/2]} \qquad \text{...(16.23)}$$

where T_3 is the temperature of steam and $(T_2 - T_1)/2$ is average temperature of water and calorimeter.

16.7 HEAT CONDUCTION THROUGH COMPOUND MEDIA:

Compound media is made up of two different materials of different thermal conductivity arranged in series or parallel. Heat will flow from one end to another through the surface of contact.

(*i*) Material Bodies in Series: Let x_1 and x_2 be the thickness of two different materials A and B of thermal conductivities K_1 and K_2 of the compound slab arranged in series as shown in Fig. 16.14. Let A is the area of cross section. Let T_1 and T_2 be temperature of end faces and θ be the temperature of common surface. If $T_1 > T_2$ heat will flow in the direction AB from end face A to end face B through the surface of contact. In the steady state condition heat will flow across any cross section.

Fig. 16.14: Conduction in Material Bodies in Series.

The rate of heat conducted for the material A

$$Q = \frac{K_1 A (T_1 - T)}{x_1} \quad ...(16.24)$$

The rate of heat conducted for the material B,

$$Q = \frac{K_2 A (T - T_2)}{x_2} \quad ...(16.25)$$

From Eq. (16.24) and Eq. (16.25) $Q = \dfrac{K_1 A (T_1 - T)}{x_1} = \dfrac{K_2 A (T - T_2)}{x_2}$

$$Q = \frac{K_1 A T_1}{x_1} - \frac{K_1 A T}{x_1} = \frac{K_2 A T}{x_2} - \frac{K_2 A T_2}{x_2}$$

$$\frac{K_1 A T_1}{x_1} + \frac{K_2 A T_2}{x_2} = \frac{K_1 A T}{x_1} + \frac{K_2 A T}{x_2}$$

$$\frac{K_1 T_1}{x_1} + \frac{K_2 T_2}{x_2} = \left(\frac{K_1}{x_1} + \frac{K_2}{x_2}\right) T$$

$$T = \frac{\dfrac{K_1 T_1}{x_1} + \dfrac{K_2 T_2}{x_2}}{\dfrac{K_1}{x_1} + \dfrac{K_2}{x_2}} = \frac{\Sigma \dfrac{K_1 T_1}{x_1}}{\Sigma \dfrac{K_1}{x_1}} \quad ...(16.26)$$

Substituting Eq. (16.26) in Eq. (16.24) or Eq. (16.25) we can determine the heat flowing through the compound section.

***(ii)* Material Bodies in Parallel:** Let x_1 and x_2 be the thickness of two different materials of thermal conductivities K_1 and K_2 of the compound slab arranged in parallel as shown in Fig. 16.15. Let T_1 and T_2 be temperature of end faces. Let A_1 and A_2 be the area of cross sections of materials. If $T_1 > T_2$ heat will flow in the direction AB from end face A to end face B through the surface of contact. In the study state condition heat will flow across any cross section.

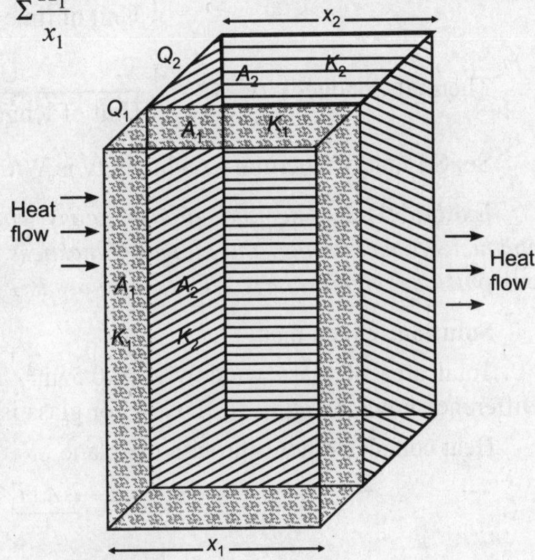

Fig. 16.15: Conduction in Material Bodies in Parallel.

The rate of heat conducted for the first material of area A_1 in t seconds

$$Q_1 = \frac{K_1A_1(T_1 - T_2)t}{x_1} \quad ...(16.27)$$

The rate of heat conducted for the second material of area A_2 in t seconds

$$Q_2 = \frac{K_2A_2(T_1 - T_2)t}{x_2} \quad ...(16.28)$$

Total heat flowing in t seconds through the compound media

$$Q = \frac{K_1A_1(T_1 - T_2)t}{x_1} + \frac{K_2A_2(T_1 - T_2)t}{x_2}$$

$$\frac{Q}{t} = (T_1 - T_2)\left(\frac{K_1A_1}{x_1} + \frac{K_2A_2}{x_2}\right)$$

Rate of heat flow through material bodies arranged in parallel is

$$\frac{Q}{t} = (T_1 - T_2)\Sigma\frac{K_1A_1}{x_1} \quad ...(16.29)$$

Example 16.1: *Show that the thermal conductivity is expressed as watt per metre Kelvin.*

Solution:

Thermal Conductivity $Q \propto \dfrac{A(T_1 - T_2)t}{x}$

Thermal Conductivity $Q = \dfrac{KA(T_1 - T_2)t}{x}$

$$= \frac{\text{Unit of heat} \times \text{unit of length}}{\text{Unit of area} \times \text{unit of temperature} \times \text{unit of time}}$$

$$= \left(\frac{\text{Unit of heat}}{\text{Unit of time}}\right)\left(\frac{1}{\text{Unit of length} \times \text{unit of temperature}}\right)$$

Thermal Conductivity $= \left(\dfrac{\text{Unit of power}}{\text{Unit of length} \times \text{unit of temperature}}\right)$

So, S.I unit of Thermal Conductivity is W/mK.

Example16.2: *The total area of glass window is 0.5m². Calculate how much heat is conducted/hour through the plane of thickness of glass 0.6 cm. Inside temperature is 23°C and outside temperature is 2°C. K for glass is 1 W/mK.*

Solution: Given that

Total area of glass window $A = 0.5$ m², Thickness of glass $x = 0.6$ cm $= 6 \times 10^{-3}$, Difference in temperature is 21°C, K for glass is 1 W/mK. $t = 1$ hr $= 3600$s, $Q = ?$

Heat conducted/hour through the plane

$$Q = \frac{KA\,(T_1 - T_2)t}{x}$$

$$Q = \frac{1 \times 0.5 \times 21 \times 3600}{6 \times 10^{-3}} = 6.3 \times 10^{-6}\,\text{J}.$$

Example 16.3: *A liquid is found to cool from 50°C to 45°C in 5 minutes and from 45°C to 41°C in rest 5 minutes. Determine the temperatures of surroundings.*

Solution: Given that for the first stage $T_1 = 50°C$, $T_2 = 45°C$, $t = 300s$

According to Newton's law of cooling

$$2.3026\log_{10}\left(\frac{T_1 - T}{T_2 - T}\right) = Kt$$

$$2.3026\log_{10}\left(\frac{50 - T}{45 - T}\right) = 300\ K \quad ...(i)$$

Given that for the second stage $T_1 = 45°C$, $T_2 = 41°C$ $t = 300s$

$$2.3026\log_{10}\left(\frac{T_1 - T}{T_2 - T}\right) = Kt$$

$$2.3026\log_{10}\left(\frac{45 - T}{41 - T}\right) = 300\ K \quad ...(ii)$$

From (*i*) and (*ii*)

$$2.3026\log_{10}\left(\frac{50 - T}{45 - T}\right) = 2.3026\log_{10}\left(\frac{45 - T}{41 - T}\right)$$

$$(45 - T)^2 = (50 - T)(41 - T)$$

$$2025 + 90T + T^2 = 2050 - 91T + T^2$$

$$T = 25$$

$\therefore$ Temperatures of surroundings is 25°C

Example16.4: *A wooden box of effective area of 1.1 m² and thickness 2 cm containing ice is covered with a cotton layer of 1 cm thickness and kept in a room of temperature 15°C. Calculate the heat flowing into the box per hour. Given that specific heat of ice = 3.3 × 10⁵J and thermal conductivity of wood and cotton are 0.16 and 0.040 W/mK.*

Solution: Given that

Specific heat of ice = 3.3×10^5J

Thermal conductivity of wood = 0.16 W/mK

Thermal conductivity of cotton = 0.040 W/mK

Box effective area = 1.1 m²

Box thickness, second layer x^2 = 2 cm = 0.02m

First layer thickness x^1 = 1 cm = 0.01m

Time of conduction = 1 hr = 3600s

Temperature of hot face T_1 = 15°C

Temperature of cold face T_2 = 0°C

As heat will flow the room into the box through the same area of cotton and wood, it is a case of compound medium in series.

Temperature at the interface face is $T = \dfrac{\dfrac{K_1T_1}{x_1} + \dfrac{K_2T_2}{x_2}}{\dfrac{K_1}{x_1} + \dfrac{K_2}{x_1}} = \dfrac{\Sigma\dfrac{K_1T_1}{x_1}}{\Sigma\dfrac{K_1}{x_1}}$

$$T = \frac{\frac{0.04 \times 15}{0.01} + \frac{0.16 \times 0}{0.02}}{\frac{0.04}{0.01} + \frac{0.16}{0.02}} = 5°C$$

Heat flowing in 3600 s is

$$Q = \frac{K_1 A(T_1 - T)t}{x_1} \text{ or } \frac{K_2 A(T_2 - T)t}{x_2}$$

$$Q = \frac{0.04 \times 1.1 \times 10 \times 3600}{1 \times 10^{-2}} = 158.4 \text{ kJ}$$

Example 16.5: *A slab of stone of area 0.5m² and a thickness of 10 cm is exposed on the lower surface to steam of 100°C and on upper surface a block of ice at 0°C is kept which is melting at 4.8 kg/hr. calculate thermal conductivity of the stone.*

Solution: Given that

Area of stone, A = 0.5 m², Thickness, x = 10 cm = 0.1 m and Difference in temperature = 100°C

Thermal conductivity, $$Q = \frac{KA(T_1 - T_2)t}{x}$$

$$Q = \frac{K \times 0.50 \times 100 \times 3600}{0.1} \text{ J} \quad ...(i)$$

Heat gained by the ice per hour = mass of the ice × specific latent heat of the ice

= 4.8 × 3.36 × 10⁵ J ...(*ii*)

Heating flowing across the slab from (*i*) and (*ii*)

$$\frac{K \times 0.50 \times 100 \times 3600}{0.1} = 4.8 \times 3.336 \times 10^5$$

Thermal conductivity, K = 1.72 W/mK

QUESTIONS

1. Define coefficient of thermal conductivity and give its units.
2. Explain the terms: temperature gradient, steady state and thermal conductivity.
3. Explain different Modes of Heats Transfer.
4. State Newton's law of cooling.
5. Explain the term Thermal Conductivity.
6. Define Steady state.
7. Derive Newton's Law of Cooling from Stefan's law.
8. Explain rectilinear flow of heat- conduction along a uniform bar.
9. Explain the experiment for Determining Thermal Conductivity.
10. Explain the Determination of Thermal Conductivity of Poor Conductor by Lee's disc method.
11. Discuss the Radial heat flow of heat through the walls of cylindrical tube.
12. Explain Determination of Thermal conductivity of rubber tube.
13. Discuss the principle of Lee's disc method.
14. Explain the method of Lee's disc to determine the coefficient of thermal conductivity of a bad conductor.
15. Describe the cylindrical shell method of determining the thermal conductivity of rubber tube.
16. Discuss the heat flow through a compound wall.

17 Elements of Thermodynamics

Learning Objectives:

By the end of this chapter, you will be able to:

- Define concepts of Thermodynamics and Zeroth Law.
- Explain basics of Heat and Work and differentiate between them.
- State Law of Conservation of Energy.
- Understand First Law of Thermodynamics and its application.
- Learn about Heat Engine and Carnot Cycle.
- Interpret Second and Third Law of Thermodynamics.

17.1 INTRODUCTION

The various efforts of converting heat to obtain mechanical work led to the development of steam engine in the eighteenth century. Initially, the steam engine was developed to pump water out of deep coal mines which were flooded with water. These steam engines were developed by practical inventors who followed trial and error methods rather than any scientific principles. In 1765, James Watt invented a superior steam engine with a separate condenser. It stimulated the development of engines that could do many kinds of jobs and marked the beginning of industrial revolution. In subsequent years, many ambitious inventors claimed to have invented perpetual motion machines which could do work without any input. The analysis of these false claims and the genuine efforts to improve the efficiency of heat engines resulted in the birth of thermodynamics as a separate branch of physics. The quintessence of experimentation and theoretical interpretation for over 200 years is summed up in the laws of thermodynamics.

The works of Julius Robert Mayer (1814-1878), James Prescott Joule (1818-1889) and Herman von Helmholtz (1821-1894) established the law of equivalence of heat and work. Rudolf Clausius (1822-1888) formulated the second law of thermodynamics mathematically.

17.2 THERMODYNAMICS

Thermodynamics is the science that deals with the rules according to which bodies exchange energy. In the early stages, thermodynamics was concerned with the relationship between mechanical and heat energy. Further developments turned thermodynamics into a science that is concerned with the relationships between heat and all other kinds of energy such as chemical, electrical etc. The entire structure of thermodynamics rests on two laws known as first and second laws of thermodynamics. The *first law of thermodynamics* embodies the law of conservation of energy which encompasses all the fields of science and engineering. The *second law of thermodynamics* expounds the idea that it is impossible to convert a given amount of heat fully into work. The second law is also known as the *law of increasing entropy*, on the basis of its interpretation at molecular level.

17.2.1 Zeroth Law of Thermodynamics

The concept of temperature originated from our subjective sense of hot and cold. We use these words to denote the degree of heating to which a body is subjected. The quantity that characterizes the degree of heating to which a body is subjected is termed the **temperature** of that body.

Experience shows that when a hot body is placed in contact with a cold body, their temperatures change. The body with the higher temperature always gets cooler and the body with the lower temperature always gets hotter. It indicates that energy exchange takes place between the two bodies. Such an exchange of energy is called **heat exchange**. Eventually, the temperature of the bodies become equal and the change in temperature stops. In technical terms, the bodies are said to be in **thermal equilibrium**. Thus, if the temperatures of two bodies are equal, no heat exchange takes place between them and the energy of each body remains constant. We may therefore define temperature as a quantity indicating the direction of heat exchange.

The concept of thermal equilibrium leads us to the **zeroth law of thermodynamics**. It may be stated formally as follows:

"*Two bodies, A and B, each in thermal equilibrium with a third body C are in thermal equilibrium with each other*."

Zeroth law of thermodynamics indicates that the necessary and sufficient condition for thermal equilibrium is equality of temperature.

According to kinetic theory of gases, temperature is a quality characterizing the average kinetic energy of translational motion of the molecules of an ideal gas. It implies that the temperature of every body is closely related to the energy of motion of its molecules. The greater the average kinetic energy per molecule of a body, the higher is its temperature. Therefore, to heat a body, energy must be supplied to it, and to cool the body energy must be taken away from it.

17.3 HEAT

Julius Robert Mayer (1814-1878), a German doctor serving as a physician on a ship, conceived on 1842 the idea of the equivalence of heat and work. He hypothesized that *heat is a form of energy*. However, the relationship between work and heat was established in 1850 experimentally by James Joule (1818-1989), the British Physicist. He conducted a long series of experiments that conclusively demonstrated the equivalence of mechanical energy and heat energy.

According to the modern view, heat is the energy that is transferred from a hot body to a cold body. For example, let us consider the case of a beaker of hot water (hot body) left on the table in an environment at room temperature (cold body). The hot water gets colder as its temperature tends to approach the room temperature. It suggests that there is some sort of exchange of energy between water (the system) and the surroundings. We now define heat in a more general way.

Heat is energy that flows from one body or system to another solely as the result of a temperature difference between them.

Thus, heat is not an intrinsic property of a body. A statement that a body contains a certain amount of heat is meaningless. On the other hand, we can say that a body can transfer a certain amount of energy as heat under certain specified conditions.

17.4 HEAT AND WORK

The conversion of mechanical energy into heat and the reverse process of obtaining mechanical work at the expense of heat are of the greatest interest in engineering. For example, in a thermal power station, thermal energy in the form of steam at high temperature and pressure drives a turbine (Fig. 17.1) and is converted into mechanical energy. The turbine in its turn drives the rotor of a generator which produces electricity.

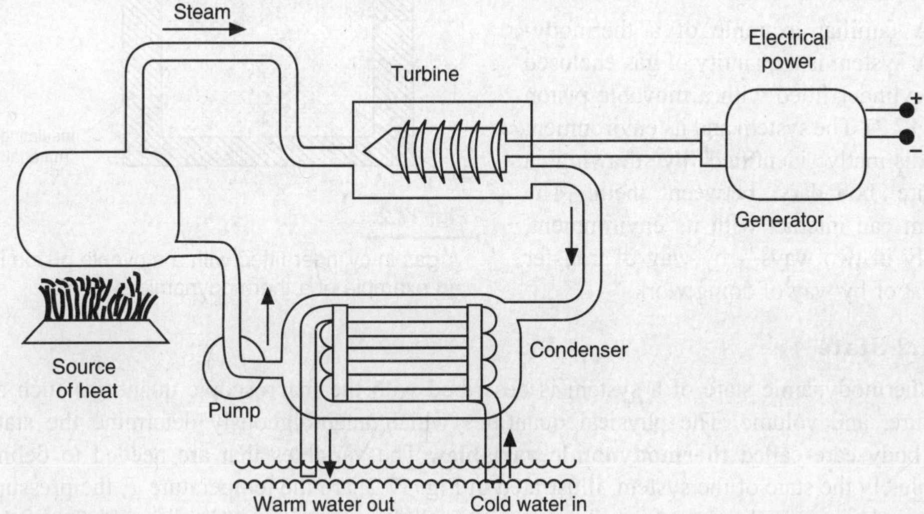

Fig. 17.1: Schematic diagram of an electrical power plant.

Thermodynamics is the science that deals with work and heat, and those properties of substances related to heat and work. Many of the properties of a substance and many phenomena can be studied either from a microscopic or macroscopic point of view. For instance, let us consider the familiar example of a monoatomic gas confined in a small container. Quantities, characteristic of the atoms, such as the energy of an atom, its velocity, its mass, etc, can be used to describe the properties of the gas. These quantities can be computed from theoretical analysis but cannot be measured. All such quantities are said to be **microscopic**. We may also describe the properties of the gas using quantities whose values can be directly measured. Such quantities are called **macroscopic** quantities. Temperature, pressure and volume are examples of macroscopic quantities. These are the quantities the gas as a whole has, but have no meaning when applied to individual atoms. Thus, we cannot speak of the pressure of temperature of one molecule. It is possible to develop the principles of thermodynamics from a microscopic point of view. Historically, the central concepts of thermodynamics were developed from a macroscopic view point without reference to microscopic models and details of the structure of matter. We study here the principles of thermodynamics from the macroscopic point of view.

17.5 TERMINOLOGY

There are certain terms in thermodynamics which are used with specific connotation. It is necessary for us to acquaint with these terms before we embark on the study of thermodynamics.

17.5.1 System

The principles of thermodynamics are usually sated with reference to a well defined system. A **system** is defined in thermodynamics as a quantity of matter of fixed mass and identity.

A familiar example of a thermodynamic system is a quantity of gas enclosed in a cylinder fitted with a movable piston (Fig. 17.2). The system and its environment are distinctly identified by drawing a definite boundary between them. The system can interact with its environment, mainly in two ways – by way of transfer of heat or by way of doing work.

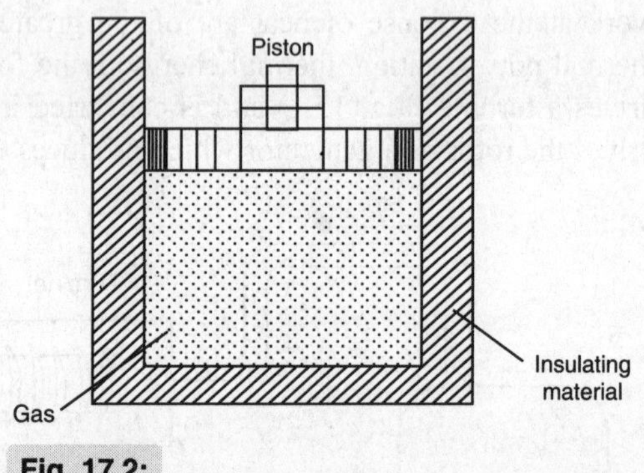

Fig. 17.2: A gas in cylinder fitted with a movable piston is an example of a thermodynamic system.

17.5.2 State

The thermodynamic state of a system is described with the macroscopic quantities such as pressure, and volume. The physical quantities which unambiguously determine the state of a body care called **thermodynamic variables**. The variables that are needed to define completely the state of the system, illustrated in Fig. 17.2, are the temperature T, the pressure P, the volume V, and the mass, m. Two states of a body are considered to be different if the value of even one of the thermodynamic variables is different for them.

17.5.3 Thermodynamic Equilibrium

In mechanics equilibrium means a state of rest. In thermodynamics the concept is somewhat broader. *A system is said to be in* ***thermodynamic equilibrium*** *if none of the thermodynamic variables determining its state changes with time.* Thermodynamic equilibrium is easily understood in the case of a monoatomic gas confined in a cylinder, as illustrated in Fig. 17.2. If the temperature of the gas in the cylinder is the same at all points in the cylinder and the temperature of the walls of the cylinder is also the same, then the gas is in thermal equilibrium with the cylinder. The heat of the gas goes not flow from one part of the cylinder to another. Further, when neither the pressure nor chemical composition of the gas changes, it is said to be in thermodynamic equilibrium. Thus, a system will be in a state of thermodynamic equilibrium if it satisfies the conditions for mechanical, thermal and chemical equilibrium.

A Cartesian coordinate system is used to plot sets of values of P, V, and T to indicate equilibrium states of a system, as in Fig. 17.3. Different points on the graph correspond to different equilibrium states of the gas.

An isolated system always reaches a state of thermodynamic equilibrium in course of time but can never depart from it spontaneously.

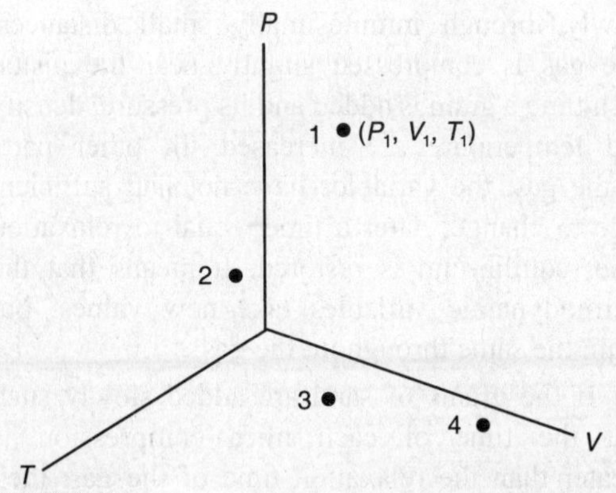

Fig. 17.3:

Different equilibrium states of a gas can be represent by points specified by values of pressure, volume different and temperature.

17.5.4 Process

We know that an increase in the pressure exerted on a gas causes a decrease in its volume. During that time the system goes through different states. Any change in the state of a system is called a **thermodynamic process**. In a thermodynamic process one or more of the thermodynamic variables change.

Different thermodynamic processes are studied and compared by depicting them graphically. Such graphs are called process charts and ***thermodynamic diagrams***. Two varying parameters of state, for example the pressure P and volume V are laid off along the axes of a two – dimensional Cartesian coordinate system. It is called a ***PV-diagram***. The dependence of P on V shows the given thermodynamic process.

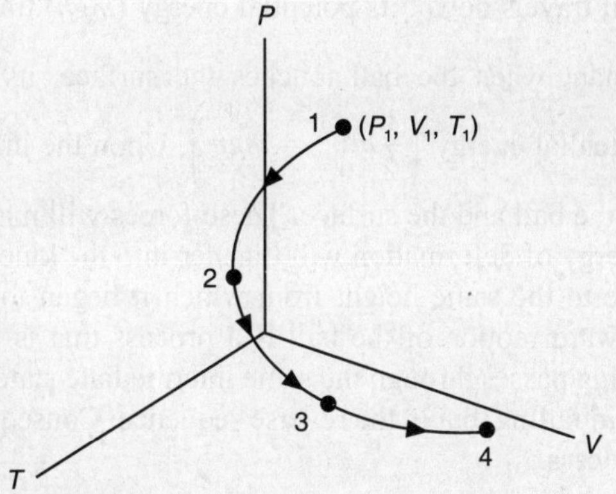

Fig. 17.4:

The curve connecting the equilibrium states of a system represents the path of the process.

17.5.5 Quasistatic Process

A **quasistatic** process is a thermodynamic process in which a system passes through a series of equilibrium states. It can be represented by a curve that joins the initial and final equilibrium states by a succession of intermediate equilibrium states, as in Fig. 17.4.

Only equilibrium states and quasistatic processes can be depicted graphically. If a system, which is originally in an equilibrium state, is momentarily disturbed and then left alone, it passes on to equilibrium state spontaneously. The process of transition to equilibrium is called **relaxation**. The time needed for the transition is known as the **relaxation time**. If a process is to be quasistatic, it must take place during a time that is longer than the relaxation time. During such a slow process the system can be considered to pass through an infinitely large number of equilibrium states.

A quasistatic process is illustrated in Fig. 17.5. A cylinder filled with a gas has piston at one end. There is no friction between the piston and the cylinder walls. Initially, the gas is in the state of equilibrium. Now, if we drop sand, grain by grain on the piston, it moves down slowly through infinitesimally small distances. The gas is compressed slightly near the piston each time a grain is added and its pressure, density and temperature are increased. In other parts of the gas, the variables have not had sufficient time to change. After a time, equal to relaxation time, equilibrium is restored. It means that the thermodynamic variables take new values, but again the same throughout the gas.

If the grains of sand are added slowly such that the time of each micro-compression is greater than the relaxation time of the gas, then the time will be sufficient for restoration of the equilibrium. Each added grain of sand takes the system into a new equilibrium state. Then, the whole process of compression can be regarded as the sum of a large number of micro-compressions and the thermodynamic process as the totality of transitions through a large number of equilibrium states.

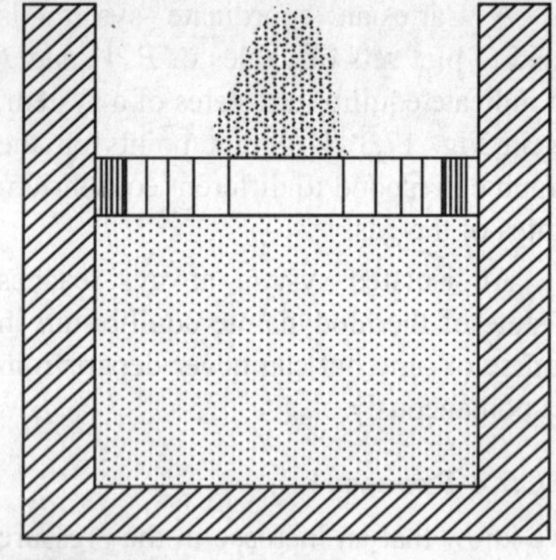

Fig. 17.5: A gas in a cylinder is compressed slowly by adding grains of sand on to the piston.

17.5.6 Reversible and Irreversible Processes

Suppose a perfectly elastic ball drops in a vacuum on to a perfectly elastic surface. As the ball travels down, its potential energy (*mgh*) transforms into kinetic energy $\left(\frac{1}{2}m\upsilon^2\right)$. At the instant when the ball touches the surface, its kinetic energy exactly equals to the initial potential energy $\left(\frac{1}{2}m\upsilon^2 = mgh\right)$. Upon the impact elastic forces appear due to deformation of the ball and the surface. These forces will make the ball begin reverse motion upward. The energy of deformation will transfer into the kinetic energy of the motion of the ball and it will rise to the same height from which it began to drop. This process repeats many times. The upward motion of the ball is a process that is reverse of downward motion. The ball when rising passes through the same intermediate states determined by its coordinates and velocities as in falling, but in the reverse sequence. Consequently, this mechanical process is a **reversible process**.

Basing on the above example, we may define a reversible process as a process that satisfies the following conditions:

(*i*) The process can be carried out with equal facility in two opposite directions.

(*ii*) In each direction the system passes through the same intermediate states; and

(*iii*) After the completion of the process, the system and its environment return to their initial state.

A process that does not satisfy even one of the above conditions is an **irreversible process**. In thermodynamics a reversible process is a quasistatic process that can be reversed by an infinitesimal change in the surroundings. Thus, a thermodynamic process is **reversible** if the system passes from the initial state to final state through a succession of equilibrium states. A reversible thermodynamics process may be understood with the help of the example given in Fig. 17.5.

In the example cited there, adding or taking out a grain of sand amounts to an infinitesimal change in the surroundings. If the sand is added grain by grain, the gas gets compressed quasistatically (infinitely slowly) and through a sequence of equilibrium states to attain the final equilibrium state.

Now, if the sand is removed by grain by grain, the gas passes through the same sequence of equilibrium states during the expansion, and reaches the initial state.

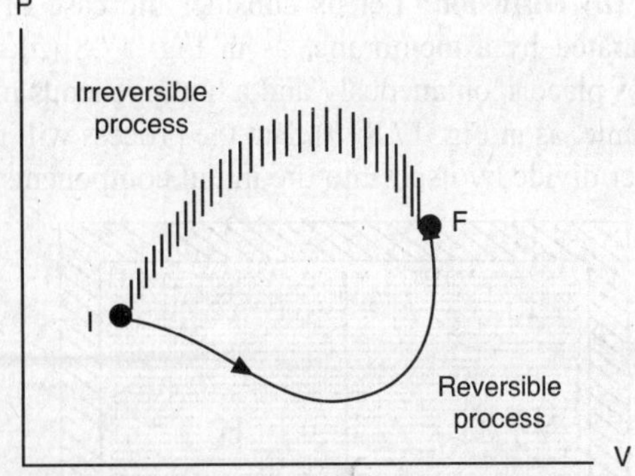

Fig. 17.6:

Reversible process between the initial state and final state can be represented by a line on PV diagram. Irreversible process passes through a series of nonequilibrium state and cannot be represented by a line.

Since a reversible process is defined by a succession of equilibrium states, it can be represented by a line on a *PV* diagram, as in Fig. 17.6. The line represents the *path* of the process. Each point ion the curve represents one of the intermediate equilibrium states. On the other hand, an irreversible process passes from the initial state to the final state through a series of nonequilibrium states. In this case, only the initial and final equilibrium states can be represented on the *PV* diagram. The intermediate nonequilibrium states are not characterized by uniquely defined sets of values for thermodynamic parameters. Hence, an irreversible process cannot be represented by a line on a *PV* diagram.

In general, all real processes in nature are irreversible processes. There are no rigorously quasistatic processes in nature because all thermal processes take place at a finite rate and not extremely slowly. Let us consider three examples that demonstrate the irreversibility of real thermal processes.

(*i*) Free expansion of a gas: Consider the example of a gas separated from a vacuum by a membrane, as in Fig. 17.7 (*a*). If the membrane breaks, the gas fills the entire vessel (Fig. 17.7 *b*). The system cannot be restored to its initial state unless the gas is compressed and the resulting heat is taken away. This, however, causes a change in the surroundings. Therefore, at the end of the process, the surroundings are not restored to their initial state. For this reason, the free expansion of a gas is an irreversible process.

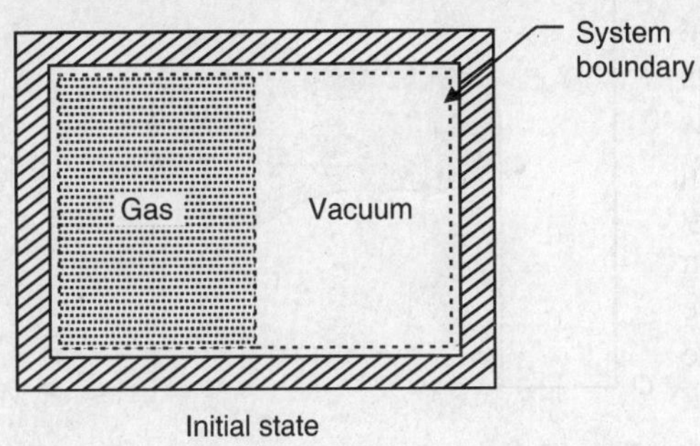

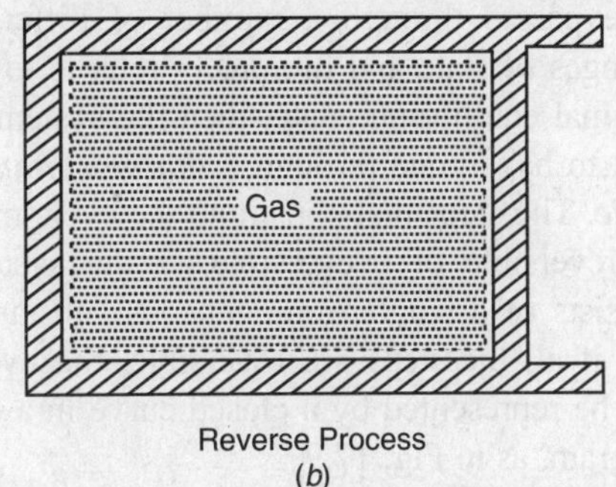

Fig. 17.7:

Free expansion of a gas. When the partition separating the gas from the evacuated space is ruptured, the gas expands freely and irreversibly.

(*ii*) Diffusion: Let us consider the case of two different gases, oxygen and hydrogen, separated by a membrane, as in Fig. 17.8 (*a*). If the membrane breaks, diffusion of gases takes place spontaneously and a homogeneous mixture of oxygen and hydrogen fill the entire volume, as in Fig. 17.8 (*b*). But the process will never reverse itself. The mixture of gases will never divide by itself into the initial components.

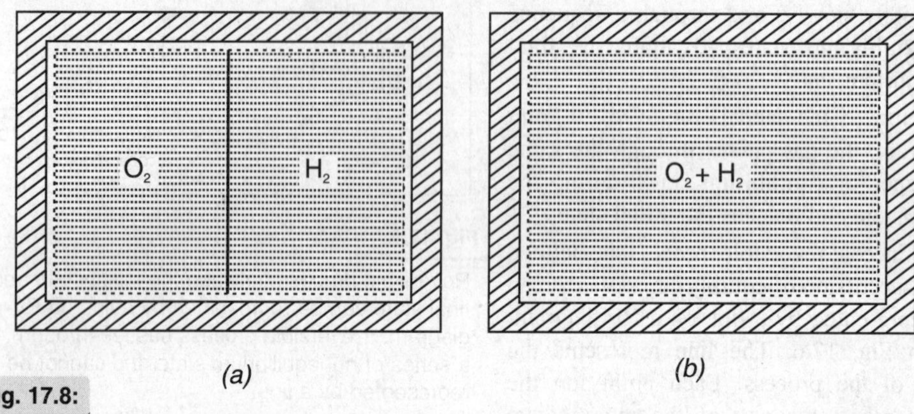

Fig. 17.8:

(*a*) Oxygen and hydrogen gases separated by a partition.

(*b*) Removal of partition leads to the irreversible mixing of the two gases.

In practice, a mixture of gases can be divided into its initial components. But first, it involves application of energy. Secondly, the system will not pass again through the same intermediate states it passed through in the diffusion process. Thirdly, the system cannot be returned to its initial state without substantially altering the properties of the surroundings.

(*iii*) Heat Exchange: Experience shows that heat exchange, like diffusion, is a one-way process. In heat exchange energy is always transmitted from a body at a higher temperature to another body at a lower temperature. Consequently, heat exchange is always accompanied by an equalization of temperatures. The reverse process of transferring energy in the form of heat from cold bodies to hot bodies never occurs by itself. For instance, a hot cup of coffee cools through heat transfer to the surroundings. However reverse will not happen.

17.5.7 Cycle

If a system in a given initial equilibrium state passes through a number of different changes of state and ultimately returns to its original equilibrium state, then the system is said to have gone through a ***thermodynamic cycle***. The steps that constitute the cycle may be reversible or irreversible. If the system consists of a single homogeneous substance and if all the steps are reversible, the cycle can be represented by a closed curve in a *PV* diagram, as in Fig. 17.9.

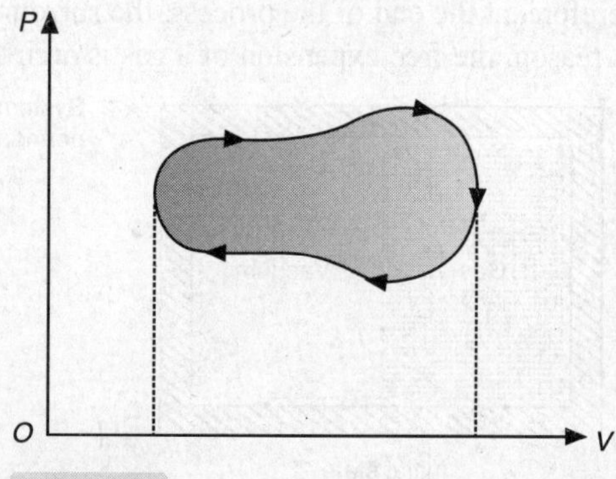

Fig. 17.9:

The PV diagram for an arbitrary cyclic process. The net work done in the process equals the area enclosed by the curve.

17.5.8 Heat Reservoir

A ***heat reservoir*** is a body from or to which heat can be transferred and it remains at constant temperature. For example, the atmosphere is a good heat reservoir. A heat reservoir from which heat is extracted by a body is called a **high-temperature reservoir**. A heat reservoir to which heat is ejected from a body is called a **low-temperature reservoir**.

17.6 WORK

The mechanical work performed by a force is the most familiar form of work. In mechanics, work is said to have been performed when a force acting on a body displaces it through a distance x, the displacement being in the direction of force. That is,

$$W = F \, . \, x \qquad ...(17.1)$$

The mechanical energy of a body changes when mechanical work is done on the body. Therefore, it serves as the measure of transfer of mechanical energy from one body to another. In general, we say work is a form of energy. Work is not stored in the body. ***Work is a form of transferring energy and is a measure of the transferred energy.***

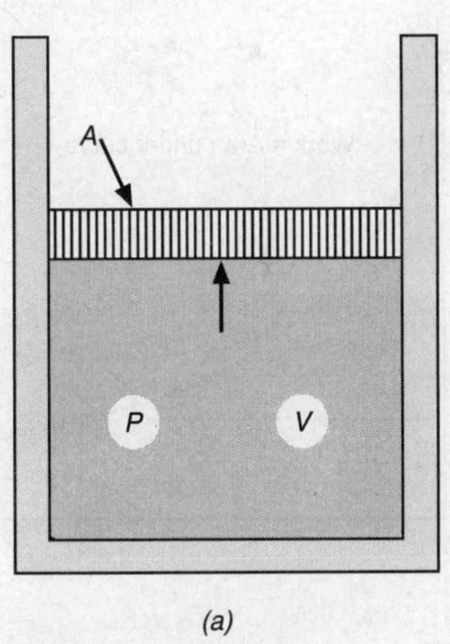

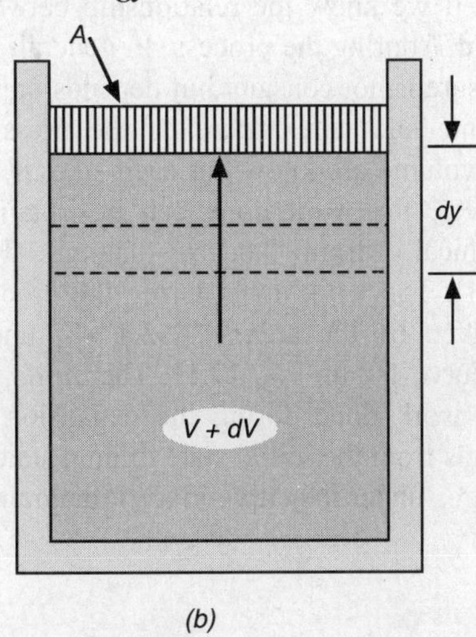

Fig. 17.10:

(a) Gas contained in a cylinder at a pressure P does work on the piston.

(b) As the piston moves the system expands from a volume V to a volume $V + dV$.

Now let us examine the work done in a thermodynamic process. Consider a thermodynamic system such as a gas contained in a cylinder fitted with a movable piston, as in Fig. 17.10 (*a*). In equilibrium the gas occupies a volume V and exerts a uniform pressure P on the piston and the walls of the cylinder. If the cross-sectional area of the piston is A, the force exerted by the gas on the piston is

$$F = PA$$

When the gas expands quasistatically as the piston goes up a distance dy (Fig. 17.10 *b*), the work done by the gas on the piston is

$$\delta W = F\,dy = PA\,dy$$

But $A\,dy = dV$, the increase in the volume of the gas. Therefore,

$$\delta W = P\,dV \qquad ...(17.2)$$

Eq. (17.2) gives the work exclusively in terms of the thermodynamic variables of the system. The nature of the outside force and other characteristics of the surroundings are not reflected in this relation. The work $\delta W = P\,dV$ is often called *thermodynamic work*. In practice, we refer to it simply as work.

The convention is that the work done by a system is regarded as positive and the work done on a system is negative. Thus, the work done by the gas expanding against a piston is positive and the work done by a piston compressing a gas is negative.

The total work done by the gas as its volume changes from V_1 to V_2 can be found by integrating Eq. (17.2). Thus,

$$_1W_2 = \int_1^2 \delta W = \int_{V_1}^{V_2} P\,dV \qquad ...(17.3)$$

The above integration can be performed only if we know the relationship between P and V during the process. In general, the pressure is not constant but depends on the volume and temperature. If the pressure and volume are known at each step of the process, the work done can be obtained graphically from the PV diagram (Fig. 17.11). The work done during the process is given by the area V_1-1-2-V_2-V_1 under the curve 1-2 on Fig. 17.11. Therefore, the total work done during the expansion of the gas from the initial state to final state is the area under the curve in a PV diagram.

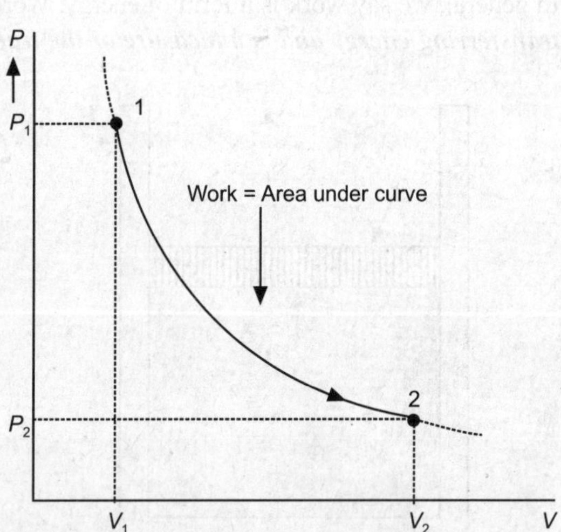

Fig. 17.11:

A gas expands from an initial state 1 to final state 2. The work done by the gas equals the area under the PV curve.

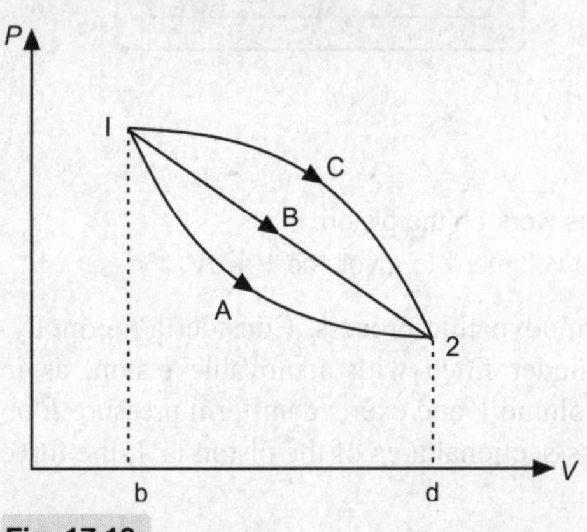

Fig. 17.12:

Three different paths connecting the same initial and final states. The work done by the gas is a maximum for the path marked C which encloses the largest area.

As illustrated in Fig. 17.12, it is possible to go from the initial state 1 to the final state 2 along any path like *A*, *B* or *C*. The area under each curve symbolizes the work for each process. The amount of work done is a function of the end states of the process and on the path that is followed. To illustrate this important point, we consider three different paths (Fig. 17.13) connecting initial and final states.

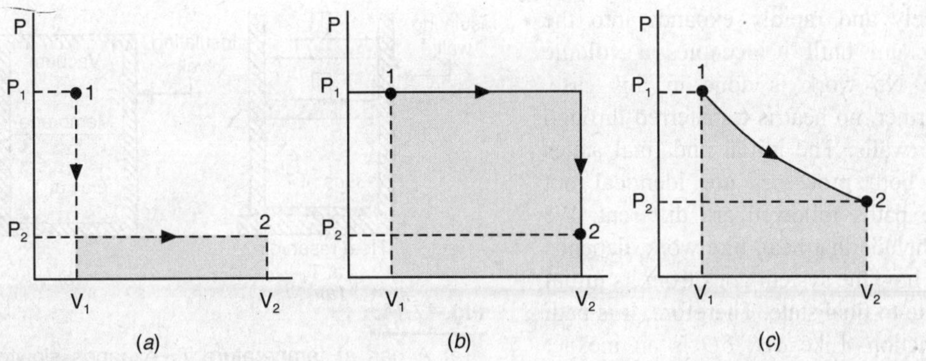

Fig. 17.13:

The work done by an ideal gas as it is taken from an initial state 1 to final state 2 depends on the path between these states 1 & 2.

In the process represented in Fig. 17.13 (*a*) the pressure of the gas is first reduced from P_1 to P_2 by cooling it at a constant volume V_1. Next, the gas is allowed to expand from V_1 to V_2 at constant pressure P_2. The work done along this path is $W_A = P_2(V_2 - V_1)$.

The gas is first permitted to expand from V_1 to V_2 at constant pressure P_1 and then the pressure is reduced to P_2 at constant volume V_2, Fig. 17.13 (*b*).

The work done along this path is $W_B = P_1(V_2 - V_1)$.

Finally in the process described in Fig. 17.13 (*c*), both P and V change continuously. To compute the work in this case, the shape of the PV diagram must be known unambiguously.

It is clear from the *PV* diagrams in Fig. 17.13 that W_A is smaller than W_B and W_C has a value intermediate between W_A and W_B. This example amply demonstrates that the work done by a system depends on how the system goes from the initial to the final state. For this reason, thermodynamic work is not a point function but is a *path function*. *dW* cannot be therefore treated as exact differential in the mathematical sense.

17.7 HEAT IN THERMODYNAMICS

A process of transfer of energy without work being done is called *heat exchange.*

Def: Heat is a form of energy that is moved from a system at a higher temperature to another system at a lower temperature.

So, heat is energy in transit. Heat transferred to a system is considered positive and from a system is negative.

Like work, heat also is a path function. The heat transferred to or from a system depends on the process. We illustrate this point with the help of the following examples.

Let us consider the case of a gas of volume V_1 which is in thermal contact with a heat reservoir (Fig. 17.14 *a*). If the pressure of the gas is infinitesimally greater than atmospheric pressure, the gas will expand and pushes up the piston. Ultimately the gas expands to a final volume V_2. The heat required to maintain the gas at a constant temperature *T*, during the process of expansion, is supplied from the reservoir to the gas. Work is done by the gas in this case.

In the second case, illustrated in Fig. 17.14 (*b*), a gas of the same volume V_1 is thermally insulated. It can neither receive nor give away heat. When the membrane is broken, the gas freely and rapidly expands into the vacuum until it occupies a volume V_2. No work is done in this case. Further, no heat is transferred through the walls. The initial and final states of both processes are identical but the paths followed are different. We conclude that heat, like work, depends on how the system goes from its initial state to final state. Therefore, it is path function. Like δW, δQ is an inexact differential.

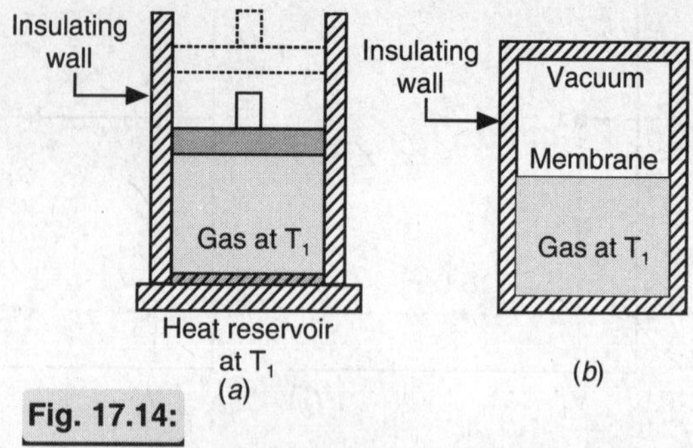

Fig. 17.14: (*a*) A gas at temperature T_1 expands slowly by absorbing heat from a reservoir at the same temperature. (*b*) A gas expands rapidly into the evacuated region due to the rupture of the membrane.

17.8 COMPARISON OF HEAT AND WORK

In thermodynamics, work includes all forms of energy except heat. When a battery is charged, electrical work was performed and stored in chemical form. When a piece of steel is magnetized, magnetic work is performed.

Any kind of energy in the course of transformations may pass through many forms but invariably ends in the form of heat energy. In the process of mechanical motion, the kinetic energy of a body decreases owing to the action of friction forces and gets transformed into heat. Similarly, the energy of an electric current, and of chemical reactions transform into heat.

Both work and heat are *transient* phenomena. Systems never contain work or heat. Heat and work exist only in a process of energy transfer, and their numerical value depends on the kind of this process. Either work or heat or both are transferred when a system undergoes a change of state. In real conditions, both ways of transferring energy to a system accompany each other. For instance, when a metal rod is heated, heat exchange takes place and at the same time thermal expansion of rod occurs. The latter implies that the work of expansion is done.

Heat and work are not equivalent forms of energy transfer from a qualitative viewpoint. The energy of ordered motion is transferred in the form of work. The energy of chaotic motion of particles constituting the body is increased when energy is transferred to a body in the form of heat.

Work can be fully converted into heat but heat cannot be entirely converted into work. Therefore work is known as *high grade energy* and heat as *low grade energy*. As both heat and work depend on the path, neither quantity is individually conserved during a thermodynamic process.

17.9 INTERNAL ENERGY

The total energy E_T of a system consists of (*i*) the kinetic energy of its macroscopic motion as a whole (*ii*) the potential energy due to the presence of external fields and (*iii*) internal energy, *U*. Thus,

$$E_T = K.E. + P.E. + U \qquad ...(17.4)$$

The internal energy *U* of the system depends on the nature of the motion and interaction of the particles in the system. It consists of (*a*) the kinetic energy of random thermal motion of molecules (*b*) the potential energy of molecules due to intermolecular interactions, (*c*) the kinctic and potcntial cncrgics of atoms and clcctrons, and (*d*) thc nuclcar cncrgy. In thermodynamics, we do not concern ourselves with the form of internal energy.

Experiments have shown that the internal energy is determined by the thermodynamic state of the system and does not depend on how the system acquired the given state. Consequently, the internal energy is not related to the process of a change in the state of the system. For instance, the internal energy of a given amount of gas depends on its pressure and temperature. All processes leading to a particular pressure and temperature leave the gas with the same internal energy.

The internal energy is not of practical interest. The change in internal energy ΔU, when a system changes from one state to another, is of actual interest. It is generally assumed that the internal energy of a system is zero at 0K.

17.10 LAW OF CONSERVATION OF ENERGY

In a closed mechanical system, the sum of kinetic energy and potential energy is constant. *The total energy can neither be created nor destroyed.* It means that energy is conserved. This is the law of conservation of mechanical energy. This law is applicable only in situation where there is no transformation of mechanical energy into heat energy. In real systems the mechanical motion is in general accompanied with heating. When the engine of a running car is switched off, it gradually slows down and ultimately stops. Apparently, its kinetic energy has disappeared. In fact, its kinetic energy is transformed into heat energy by the friction forces. Both the tire of the car and the ground are heated up. As a result, the random motion of particles constituting the interacting tires and road acquired more velocity. To sum up, the mechanical energy (K.E. of car) is transformed into the internal energy of the interacting bodies. If we take into account the internal energy, the law of conservation of energy can be extended to include thermodynamic systems also. The first law of thermodynamics generalizes the law of consideration of mechanical energy.

17.11 FIRST LAW OF THERMODYNAMICS

Thermodynamics involves the three general forms of energy, i.e. heat, work and internal energy.

- Heat is the energy transferred by virtue of temperature difference.
- Thermodynamic work is energy transferred between a system and surroundings.
- Internal energy is energy stored within a system.

If the state of a system such as the one described in Fig. 17.2 changes as a result of supplying a quantity of heat Q to it, and as a consequence the system does the work W, then the law of conservation of energy states that the quantity of heat supplied to the system will be equal to the sum of the work performed by the system and the change in the internal energy of the system. That is,

Net heat transfer = Work + Change in internal energy

Or mathematically,

$$Q = W + \Delta U \quad ...(17.5)$$

Eq. (17.5) is the **first law of thermodynamics** and is a consequence of the conservation of energy. While the quantities Q and W are path-dependent, the internal energy does not depend on the path of the process.

Suppose a thermodynamic system undergoes a change from an initial state 1 to a final state 2 in which Q units of heat are absorbed (or removed) and W is the work done by (or on) the system. Expressing both Q and W in the same units (either thermal or mechanical) the difference $(Q - W)$ can be calculated. If now we carry out this calculation for different paths between the same states 1 and 2, the quantity $(Q - W)$ will be the same for all paths connecting the states 1 and 2. It follows that the internal energy change of a system is independent of the path. If U_1 is the internal energy in state 1 and U_2 is the internal energy in state 2, then the Eq. (17.5) can be rewritten as

$$U_2 - U_1 = \Delta U = Q - W \qquad ...(17.6)$$

When a thermodynamic process proceeds smoothly, it can be treated as a continuous sequence of small changes. Mathematically, we write Eq. (17.5) in the differential form as

$$dQ = dU + dW \qquad ...(17.7)$$

It is important to note that expressing heat and work as dQ and dW does not imply the existence of properties Q and W that measure the heat and work content of a system. *dQ and dW* denote small amounts of heat and work but they are not true differentials.

There is a series limitation on the first law of thermodynamics. The law tells us whether energy considerations permit a particular process to take a system from one equilibrium state to another. But it does not tell us whether this process will actually occur or not. A certain process might be entirely consistent with the principle of energy conservation but still it will not take place.

17.12 APPLICATIONS OF THE FIRST LAW

Let us now study some special processes and the application of the first law to them.

***(i)* Isolated system:** An *isolated system* does not interact with its surroundings. Therefore, there is no heat flow and the work done is zero. That is, $Q = 0$ and $W = 0$. It follows from Eq. (17.6) that $\Delta U = 0$.

$$\therefore \quad U_2 - U_1 = 0$$

$$\text{or} \quad U_2 = U_1 \quad \text{Isolated system} \qquad ...(17.8)$$

Eq. (17.8) means that the internal energy of an isolated system remains constant.

***(ii)* Cyclic process:** In a *cyclic* process, the initial and final states of the system are the same. Thus,

$$U_2 = U_1$$

$$\therefore \quad \Delta U = 0$$

It allows from Eq. (17.5) that

$$Q - W = 0$$

$$\therefore \quad Q = W \quad \text{Cyclic process} \qquad ...(17.9)$$

Eq. (17.9) means that the net work done by the system over a cycle equals the net heat absorbed over the cycle.

***(iii)* Adiabatic process:** A process in which no heat is absorbed or ejected by the system is called an *adiabatic* process. Thus, $Q = 0$.

$$\therefore \quad \Delta U = -W \quad \text{Adiabatic process} \qquad ...(17.10)$$

Thus, the change in the internal energy of a system is equal in magnitude to the work done by the system. Heat flow into the system from the surroundings may be prevented in two ways – (i) by surrounding the system with a thick layer of insulating material or (ii) by performing the process quickly. The flow of heat requires finite time. So any process performed quickly enough will be adiabatic.

***(iv)* Isochoric process:** When a substance undergoes a process in which the volume remains unchanged, the process is called *isochoric*. If the volume of a system remains constant, it can do no work. Thus, $W = 0$ and the first law gives

$$\Delta U = Q \qquad \text{Isochoric process} \qquad ...(17.11)$$

In this case, the heat that entered the system is stored as internal energy.

***(v)* Isothermal process:** It is a process happening at constant temperature and the quantities, Q, W and ΔU are non-zero. The first law does not have any special form for an isothermal process.

***(vi)* Isobaric process:** A process taking place at constant pressure is called an *isobaric* process. As in the case of an isothermal process, Q, W and ΔU are all nonzero. The work done by a system that expands or contracts isobarically has a simple form. As the pressure is constant,

$$W = \int_1^2 P\,dV = P(V_2 - V_1) \quad \text{Isobaric Process} \qquad ...(17.12)$$

***(vii)* Isothermal expansion of an ideal gas:** Let an ideal gas on expansion in quasistatic process at constant temperature by placing the gas in good thermal contact with a heat reservoir. The ideal gas equation, $PV = nRT$ for each point on the path

$$\therefore \qquad W = \int_{V_1}^{V_2} P\,dV = \int_{V_1}^{V_2} \frac{nRT}{V}\,dV$$

At T is constant in the process, we can write

$$W = nRT \int_{V_1}^{V_2} \frac{dV}{V}$$

or
$$W = nRT \ln\left(\frac{V_2}{V_1}\right) \qquad ...(17.13)$$

***(viii)* Adiabatic expansion of an ideal gas:** Let us find the relation between P and V for an *adiabatic process* carried out on an ideal gas. According to Eq. (17.10), $\Delta U = -W$ in an adiabatic process.

For an ideal gas $\qquad \Delta U = nC_\upsilon dT \qquad$ (at constant volume)

The work done during the process is given by $W = P\,dV$.

$$\therefore \qquad PdV = -nC_\upsilon dT \qquad ...(17.14)$$

We can write the equation of the state of the gas in differential form as

$$d(PV) = d(nRT)$$

$$\therefore \qquad PdV + V\,dP = nR\,dT \qquad ...(17.15)$$

Using Eq. (17.14) into Eq. (17.15), we get

$$V\,dP = nC_\upsilon dT + nRdT = ndT(C_\upsilon + R)$$

But $\qquad C_\upsilon + R = C_p$

$$\therefore \qquad V\,dP = nC_p dT \qquad ...(17.16)$$

Taking the ratio between Eq. (17.16) and Eq. (17.14), we get

$$\frac{V\,dP}{P\,dV} = -\frac{C_p}{C_\upsilon} = -\gamma$$

$$\therefore \quad \frac{dP}{P} = -\gamma \frac{dV}{V}$$

Integrating on both sides of the above equation, we get

$$\int_{P_1}^{P_2} \frac{dP}{P} = -\gamma \int_{V_1}^{V_2} \frac{dV}{V}$$

or
$$\ln \frac{P_2}{P_1} = -\gamma \ln \frac{V_2}{V_1}$$

$$\therefore \quad P_1 V_1^{\gamma} = P_2 V_2^{\gamma} \qquad \text{...(17.17)}$$

or
$$PV^{\gamma} = \text{Constant} \qquad \text{...(17.18)}$$

17.13 HEAT ENGINE

A device that converts heat energy to work is called a *heat engine*. The automobile engines and steam turbines are examples of heat engines. Basically, they all operate on the same principle. For theoretical purposes, any heat engine can be conveniently represented by a diagram, as shown in Fig. 17.15, regardless of its particular design and features.

Any heat engine consists of three main parts: the working medium, the high-temperature reservoir and the low- temperature reservoir. The working medium is some gas (or steam) which receives a certain quantity of heat Q_H from the high-temperature reservoir, some of which is used to do useful work W and the remainder Q_L is transferred to the low-temperature reservoir.

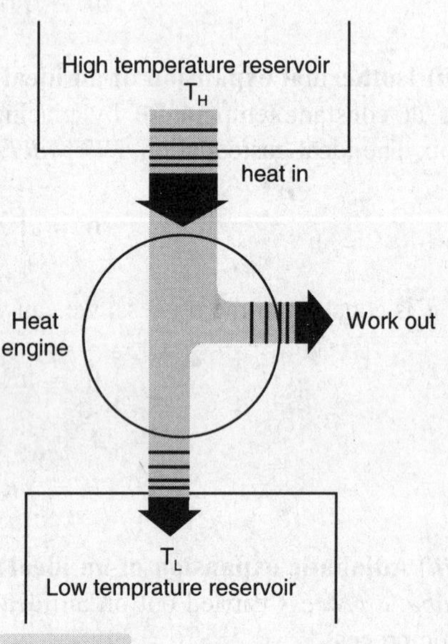

Fig. 17.15: Schematic representation of a heat engine. The engine absorbs heat Q_H from the high temperature reservoir, expels heat Q_L to low temperature reservoir and does work W

For example, in an actual engine the energy produced in the combustion of fuel is transmitted by heat exchange to some gas. As a result the gas expands and the expanding gas does work on a piston that is coupled to a crankshaft to produce useful work output. The unused heat is rejected to the surroundings through an exhaust system. However, the gas cannot continue to expand without any limit because the engine is of finite size. Secondly, in practical applications we want an engine to deliver work continuously. For these reasons, the gas must be compressed, after expansion, so that the gas and all the parts of the engine return to their initial state. Again the expansion and compression can be repeated. This implies that a heat engine must operate in cycles so that we have work output during each cycle. As mentioned above, in the real heat engine, the gas is discharged at the exhaust. A new portion of the gas is admitted and compressed instead. It is called an **open cycle** operation. However, we assume, for the sake of simplicity, a **closed cycle** in which the same portion of gas expands and is compressed as it does not influence the thermodynamics of the process.

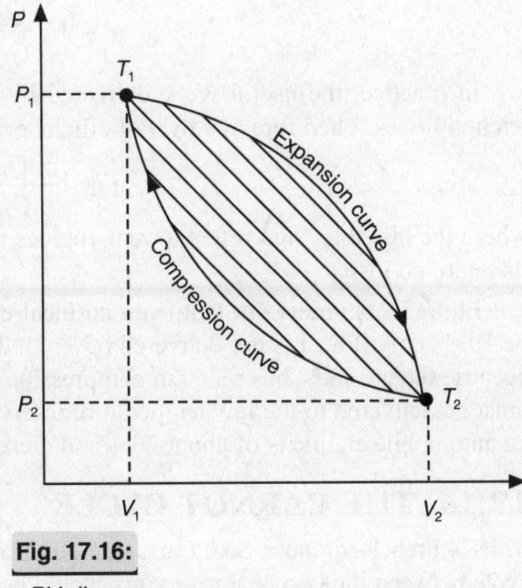

Fig. 17.16: PV diagram.

We understand the necessity of the low-temperature reservoir as follows. For an engine to do useful work during a cycle, it is necessary that the work done in expansion by the gas must be greater than the work required to compress the gas. Then only we obtain more mechanical energy. From *PV* diagram, Fig. 17.16, it is seen that the work in expansion will be larger than that in compression only if the pressure in the compression process is lower than that in expansion for all intermediate states. This is possible only when the temperature of the gas is lower in compression than in expansion at all intermediate points. Therefore, an engine will do useful work if the temperature of the gas is lower in compression than in expansion.

17.13.1 Thermal Efficiency

Efficiency is a measure of how economical an engine is. We express cost efficiency in terms of fuel economy. In the case of a mechanical device,

$$\text{Mechanical efficiency} = \frac{\text{Work output}}{\text{Work input}} \quad \text{...(17.19)}$$

It is the fraction or percent of useful work done by a machine.

In heat engines, the work input is the heat input. The work output equals the heat in minus the heat out.

Let us consider a heat engine operating in a cycle again and again.

The net heat absorbed is

$$Q = Q_H - Q_L \quad \text{...(17.20)}$$

where Q_H is the heat received

Q_L is the heat rejected by the working substance per cycle.

As the working substance comes back to its initial state after the completion of cycle its internal energy remains unchanged, $\Delta U = 0$.

From the first law

Net work $$W = Q = Q_H - Q_L \quad \text{...(17.21)}$$

Basing on the definition Eq. (17.19), we can define

Thermal efficiency, $$\eta = \frac{W}{Q_H} = \frac{Q_H - Q_L}{Q_H}$$

$$\therefore \qquad \eta = 1 - \frac{Q_L}{Q_H} \qquad \text{Ideal engine} \qquad ...(17.22)$$

In practice, the useful work delivered by an engine is less than the work W owing to friction losses. Therefore, the overall efficiency is less than the thermal efficiency. Thus,

$$\eta \leq 1 - \frac{Q_L}{Q_H} \qquad \text{Real engine} \qquad ...(17.23)$$

where the less-than sign refers to real engines and the equal sign to an ideal engine in which there are no losses.

Eq. (17.23) points out that even an ideal engine has efficiency less than 100%. It could be 100% only if heat is not delivered ($Q_L = 0$) during the compression. But this is impossible, because the gas must be cooled in compression. Therefore, a certain amount of heat ($Q_L \neq 0$) must be delivered to the low temperature reservoir. It means that $\eta < 100\%$. The efficiency of an automobile engine is of about 20% and diesel engines are from 35% to 40%.

17.14 THE CARNOT CYCLE

In 1824 French engineer, Sadi Carnot demonstrated that a heat engine operating in a reversible cycle between the two heat reservoirs would be the most efficient engine possible. The ideal engine is called the **Carnot Engine** which establishes an upper limit on the efficiencies of all engines.

The net work done by working substance taken through the Carnot cycle is the largest possible for given amount of heat supplied to the substance.

The Carnot cycle is an idealization of the cycle of a real heat engine. It is assumed that there are no losses of energy by heat exchange with the environment, that there is no friction in the machine and that processes of gas expansion and compression are quasistatic and therefore reversible.

To describe the Carnot cycle, we shall assume that the working substance is an ideal gas contained in a cylinder with a movable piston at one end. Fig. 17.17 (*a*) is a schematic diagram of a Carnot engine and Fig. 17.17 (*b*) shows the plot for a Carnot cycle.

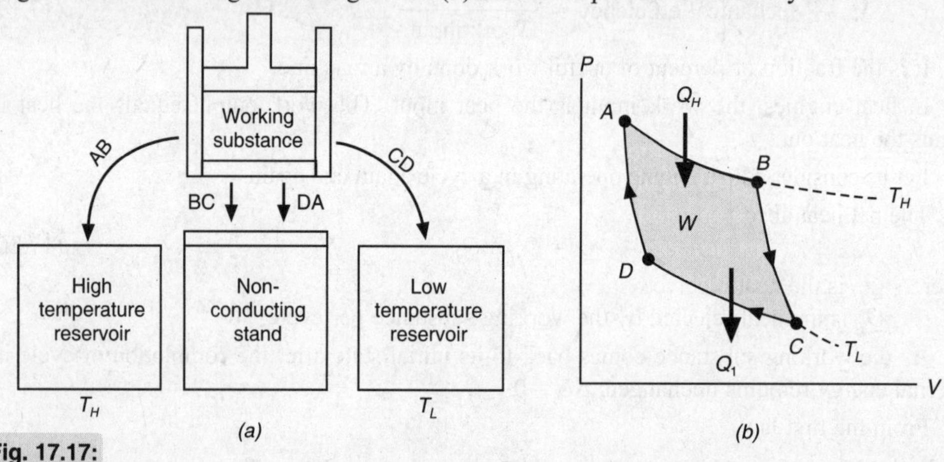

Fig. 17.17:

The Carnot cycle (*a*) In process A → B the gas expands isothermally while in contact with a heat reservoir T_H. In process B → C, the gas expands adiabatically. In process C → D, the gas is compressed isothermally while in contact with a reservoir at T_L. In process D → A, the gas is compressed adiabatically. (*b*) The *PV* diagram for the Carnot Cycle. The net work W done equals the net heat received in one cycle.

(*i*) The cycle begins at the equilibrium state A [Fig. 17.17 (*b*)] with the cylinder in contact with high-temperature reservoir at a temperature T_H (Fig. 17.17 *a*). We assume that the base of the cylinder is **diathermic** (i.e. perfectly conducting) and that the walls and the piston are ideally nonconducting. The working substance (the ideal gas) undergoes a slow quasistatic isothermal expansion to state *B*. During this part of the cycle $A \to B$, the temperature and hence the internal energy of the gas are constant. Let the total amount of heat absorbed by the gas from the reservoir during the process be Q_H. This heat energy input is converted directly to the work W_{AB} done in moving the piston.

(*ii*) With the system at *B*, the cylinder is removed from contact with the high temperature reservoir and placed on the nonconducting stand. The gas expands adiabatically along the path $B \to C$ to the state *C*. During this part of the cycle $B \to C$ no heat enters or leaves the system. However, work W_{BC} is done at the expense of reducing the internal energy. Consequently, the temperature falls from T_H to T_L.

(*iii*) Next the cylinder is placed in contact with the low temperature reservoir at a temperature T_L. The gas is now slowly compressed (by an external agency) isothermally to a predetermined volume V_D at point *D*. During this part of cycle $C \to D$, the temperature and hence the internal energy of the working substance are constant. During this process $C \to D$, the gas expels heat Q_L to the reservoir and the work done on the gas by the external agent is W_{CD}.

(*iv*) In the final part $D \to A$, the cylinder is removed from contact with the low-temperature reservoir and again placed on the non-conducting stand. The gas is compressed adiabatically and brought to the initial state *A*. The adiabatic compression is also the result of work W_{DA} done on the gas by an external energy. Consequent to the adiabatic compression, the system temperature increases from T_L to T_H.

This step completes the Carnot cycle and returns the system to its initial condition.

The net work done in this reversible cyclic process is equal to the area enclosed by the path *ABCDA* of the *PV*-diagram (Fig. 17.17 *b*). In a reversible cycle, the change in internal energy is zero. It follows from first law that the net work done in one cycle equals the net heat transferred into the system. For a system that undergoes a Carnot cycle, no heat is supplied to or rejected by the system during the adiabatic paths, BC and *DA* (Fig. 17.17b). An amount of heat Q_H is supplied to the system during the isothermal expansion *AB*, and an amount Q_L is rejected during the isothermal compression *CD*. Thus, the first law can be written as

$$W = Q_H - Q_L \qquad \text{All heat engines} \qquad ...(17.24)$$

Eq. (17.24) is applicable for a complete cycle of any heat engine. The thermal efficiency is given by Eq. (17. 22).

$$\eta = \frac{W}{Q_H} = \frac{Q_H - Q_L}{Q_H} = 1 - \frac{Q_L}{Q_H}$$

17.14.1 Efficiency

For a Carnot engine, the heat exchanges take place during the isothermal processes. During the isothermal expansion $A \to B$, the heat absorbed from the high temperature reservoir is given by

$$Q_H = W_{AB} = nRT_L \ln \frac{V_B}{V_A} \qquad ...(17.25)$$

In a similar manner, the heat rejected to the low temperature reservoir during the isothermal compression $C \to D$ is given by

$$Q_L = W_{CD} = nRT_L \ln \frac{V_C}{V_D} \quad ...(17.26)$$

Dividing these expressions, we get

$$\frac{Q_L}{Q_H} = \frac{T_L \ln (V_C / V_D)}{T_H \ln(V_B / V_A)} \quad ...(17.27)$$

For an adiabatic process, the pressure and volume are related by

$$PV^{\gamma} = \text{Constant}$$

During any reversible process, the ideal gas must also obey the equation of state,

$$PV = nRT \quad ...(17.28)$$

Eliminating pressure between the above equations, we get

$$T\,V^{\gamma-1} = \text{Constant} \quad ...(17.29)$$

Applying this result to the adiabatic processes $B \to C$ and $D \to A$, we get

$$T_H\,V_B^{\gamma-1} = T_L\,V_C^{\gamma-1} \quad ...(17.30)$$

and

$$T_H\,V_A^{\gamma-1} = T_L\,V_D^{\gamma-1} \quad ...(17.31)$$

Dividing Eq. (17.30) by Eq. (17.31) we get

$$\left(\frac{V_B}{V_A}\right)^{\gamma-1} = \left(\frac{V_C}{V_D}\right)^{\gamma-1}$$

$$\therefore \quad \left(\frac{V_B}{V_A}\right) = \left(\frac{V_C}{V_D}\right) \quad ...(17.32)$$

Substituting Eq. (17.32) into Eq. (17.27) , we obtain

$$\frac{Q_L}{Q_H} = \frac{T_L}{T_H} \qquad \text{Carnot Engine} \quad ...(17.33)$$

Using the result Eq. (17.33) into Eq. (17.23) we obtain the thermal efficiency of the Carnot engine as

$$\eta = 1 - \frac{T_L}{T_H} = \frac{T_H - T_L}{T_H} \qquad \text{Carnot Engine} \quad ...(17.34)$$

Eq. (17.34) shows that the efficiency of a Carnot engine depends only on the temperatures of the two reservoirs. When the temperature difference $(T_H - T_L)$ is large, the efficiency is nearly unity; when the difference is small the efficiency is less than unity. So, the larger the temperature difference $(T_H - T_L)$, the larger will be the efficiency of a Carnot engine. We see from Eq. (17.34) that all Carnot engines operating between the same two temperatures in a reversible manner have the same efficiency.

It brings us to the important difference in principle between heat engines and mechanical or electrical machines. In improving the design of mechanical or electrical machines we try to make their efficiency as near as possible to the theoretical limiting value of 100%. Though this limit is unattainable under real conditions, we can approach it by reducing losses. When we improve heat engines, we do not try to bring their efficiency to 100%, but to that of a Carnot engine operating in the same temperature interval. Naturally, the reduction of all kinds of losses also raises the efficiency of a heat engine but the most effective measure is to increase the temperature difference between the high temperature reservoir and low temperature reservoir.

17.14.2 Carnot's Theorem

Carnot's theorem discloses a fundamental restriction on the conversion of heat into work. Carnot engine is the most efficient heat engine operating between the two heat reservoirs.

Carnot's theorem statement: *No heat engine operating between two heat reservoirs can be more efficient than a Carnot engine operating between the same two reservoirs.*

17.15 HEAT PUMP

A device that transfers heat from a low-temperature reservoir to a high-temperature reservoir is known as a **heat pump**. This is the reverse function of a heat engine. A heat pump is represented by a diagram as shown in Fig. 17.18. It requires work input, because the heat transfer from low temperature reservoir to high-temperature reservoir will not take place spontaneously.

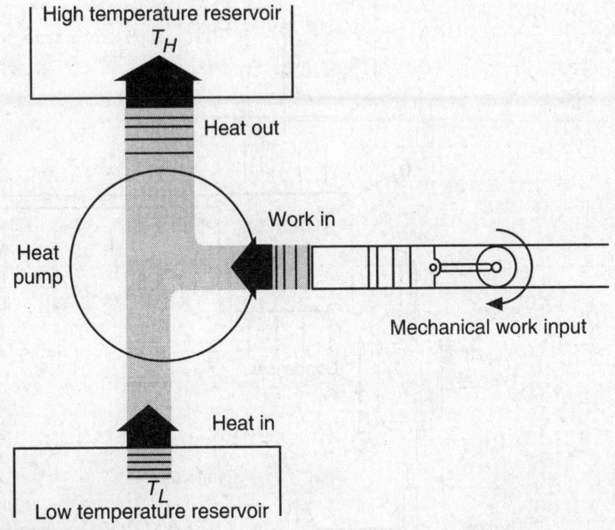

Fig. 17.18: Schematic representation of a refrigerator. It absorbs heat Q_L from the low temperature reservoir and expels and heat Q_H to the high temperature reservoir. Work W is done on the engine.

The household refrigerator and air conditioner are examples of the most common heat pumps. In an ordinary household refrigerator, the working substance is a liquid that circulates within the system. It takes heat from the low temperature reservoir, which is the cold chamber in which food articles are stored, and transfers it to the outside air in the room where the unit is kept. The work is supplied by a compressor that uses electrical energy. The working substance is a fluid that readily undergoes a liquid to gas phase change at the operating temperature. The most common refrigerants are Freon – 12 ($CC1_2F_2$) and ammonia.

A schematic diagram of a refrigerator is shown in Fig. 17.18. Starting with the gaseous refrigerant in the compressor, the gas is compressed and emerges at a high temperature and pressure. It then passes into the condenser, where it is cooled and liquefies, and the heat from the refrigerant is rejected to the surroundings. The liquid refrigerant then goes through a throttle valve. Work is done by the liquid in getting through the valve at the expense of its internal energy. It leads to the drop in the temperature of the liquid. The cooled liquid flows through the evaporation coils where it absorbs heat and boils. The gaseous refrigerant goes into the compressors and the cycle repeats.

The performance of a refrigerator is specified by a figure of merit K_r called the **coefficient of performance**.

$$K_r = \frac{Q_L}{W} = \frac{Q_L}{Q_H - Q_L} \qquad ...(17.35)$$

The quantity Q_L is the amount of heat absorbed from the cold compartment at the temperature T_L. The quantity Q_H is the heat rejected into the surroundings. Q_H is equal to the heat absorbed Q_L plus the work done on the refrigerant by the compressor. Thus, $W = Q_H - Q_L$. Assuming that the cycle is a reverse Carnot cycle, we can write.

$$K_r = \frac{T_L}{T_H - T_L} \qquad ...(17.36)$$

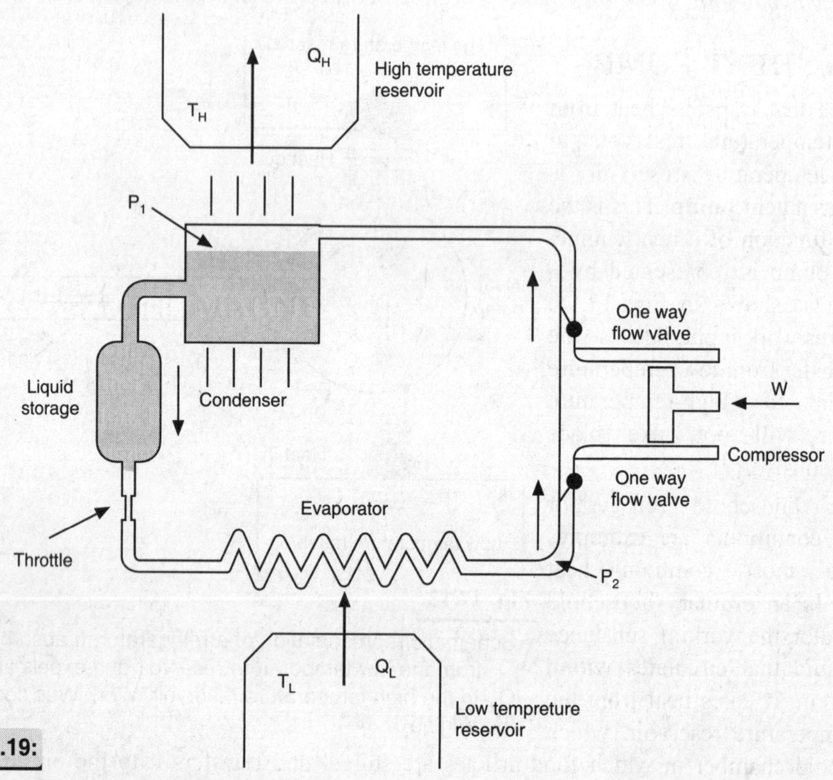

Fig. 17.19: Essential features of a common refrigerator.

For a house-hold refrigerator, typical temperatures are $T_H = 305$ K (32° C) and $T_L = 260$ K (= – 13° C) which give a typical value of 5.8 for K_r.

17.16 SECOND LAW OF THERMODYNAMICS

There are two classical statements of the second law of thermodynamics. They are known as the Kelvin-Planck statement and the Clausius statement.

An analysis of operation of heat engines led William Thomson (later Lord Kelvin) and Planck to arrive at the following conclusion:

It is impossible to construct a heat engine that will operate in a cycle and which will receive a given amount of heat from a high-temperature reservoir and does an equal amount of work. The only alternative is that some heat must be transferred from the working fluid to a low temperature reservoir. Therefore, work can be done by the transfer of heat only if there are two temperature levels involved.

The analysis of operation of heat pumps led Rudolf Clausius to conclude as follows:

It is impossible to construct a device that operates in a cycle and produces no effect other than the transfer of heat from a cold body to a hot body. In effect, it is impossible to construct a refrigerator that operates without an input of work.

The first law of thermodynamics is concerned with conservation of energy. As long as the energy is conserved in a process, the first law is satisfied. The second law underlines the fact that processes proceed in a certain direction but not in the opposite direction. For example, a hot cup of tea cools by heat transfer to the surroundings but heat will not flow from the cooler surroundings to hot the tea cup. Thus, real processes progress only in one direction.

A cycle will occur *only* when both the first and the second laws of thermodynamics are satisfied.

The first law of thermodynamics is a common declaration of the conservation of energy. It makes no distinction between the different forms of energy. The second law of thermodynamics asserts that thermal energy is different from all other forms of energy. Various forms of energy can be converted into thermal energy spontaneously and completely; where as the reverse transformation is never complete. The impossibility of converting heat completely into mechanical energy forms the basis of Kelvin-Planck statement of second law.

The basis of second law lies in the difference between the nature of mechanical energy and the nature of internal energy. Mechanical energy is the energy of ordered motion of a body. Internal energy is the energy of random motion of molecules within the body. When the body moves, the motion is due to the ordered motion of molecules within it as a whole in the direction of velocity of the body. The energy associated with this ordered motion of molecules is the kinetic energy. The kinetic and potential energies associated with the random motion is the internal energy. When a moving body comes to rest due to friction, the ordered portion of the kinetic energy becomes converted into energy of random molecular motion. It is impossible to reconvert the energy of random motion completely to the energy of ordered motion, since we cannot control the motions of individual molecules. We can convert only a portion of it. That is what a heat engine does.

17.17 ENTROPY

The differential form of the first law of thermodynamics is written as

$$dQ = dU + dW$$

The work dW done depends on the path of process and therefore it is not a function of the state. The same is the case with dQ, the quantity of heat supplied or taken away. The work dW can be expressed in terms of thermodynamic variables and their changes. For instance, we have expressed dW in Eq. (17.2) as

$$dW = P\,dV$$

It is found that dQ can also be expressed in a similar fashion, in case of reversible processes. We write

$$dQ = T\,dS \qquad ...(17.37)$$

where dS is called the change in entropy and T is the temperature. Now, we define the change in entropy as

$$dS = \frac{dQ}{T} \qquad \text{Reversible process} \qquad ...(17.38)$$

The change in entropy dS in the course of an infinitesimal change is equal to the quantity of heat dQ divided by the absolute temperature T, where dQ is the heat absorbed (or rejected) when the change is carried out in a reversible manner.

The total entropy change in a reversible process may be obtained by integrating Eq. (17.38). Thus,

$$\Delta S = S_2 - S_1 = \int_1^2 \frac{dQ}{T} \qquad ...(17.39)$$

where S_1 and S_2 are the entropies of the initial and final states of the system.

The importance of the entropy S is that it is a function of state like the internal energy U. Both these parameters depend only on the initial and final states of the system and not on the path of the process that takes the system from the initial state to final state. Eq. (17.39) assumes a simpler form when the process is an isothermal process. As T is constant in isothermal process, Eq. (17.39) may be written as

$$\Delta S = \int_1^2 \frac{dQ}{T} = \frac{1}{T}\int_1^2 dQ = \frac{Q}{T}$$

Thus, $\Delta S = \frac{Q}{T}$ Isothermal Process ...(17.40)

The units of entropy and entropy change are *J/K*.

In practice, the value of entropy S is not of much interest. We have to know the change in entropy when the system changes from one state to another.

17.17.1 Entropy, Disorder and Second Law

When processes occur, in general, they are irreversible and the degree of disorder increases as a result of these processes. As an example, let us take the case of isothermal expansion of an ideal gas (Fig. 17.10). As the gas absorbs heat, it slowly expands. At the end of the process the gas occupies a greater volume than at the beginning. The gas molecules are more disordered now. The gas will not, by its own accord, give up its thermal energy and segregate itself to confine to the initial volume. We, thus, observe that the flow of heat takes place in the direction that increases the amount of disorder. The same type of order to disorder change occurs when free expansion of gas occurs, when one gas diffuses into another, and in similar other spontaneous processes.

Rudolf Clausius (1822-1888), the German physicist, introduced the quantity entropy which is regarded as a measure of disorder in a system. An increase in disorder is equivalent to an increase in entropy. Irreversible processes are processes for which entropy increases.

These considerations led Clausius to reformulate the second law of thermodynamics in terms of entropy. According to it, the entropy of an isolate system always tends to increase. Mathematically, it is expressed as

$$\Delta S_{\text{isolated system}} \geq 0 \qquad ...(17.41)$$

17.17.2 Some Interesting Points

***(i)* The net change in entropy in any reversible cycle is zero:** Let us take the case of Carnot cycle as an example. There is no change in entropy of the working substance during the two adiabatic paths. Either during adiabatic expansion or compression, $Q = 0$. Therefore, $S = 0$. However, there is an increase in entropy during isothermal expansion, as heat Q_H is added at a constant temperature T_H. The consequent increase in entropy $\Delta S_1 = \frac{Q_H}{T_H}$. There is a decrease in entropy during the isothermal compression in which heat Q_L is rejected at a temperature T_L. Thus, $\Delta S_2 = -\frac{Q_L}{T_L}$.

∴ The net change in entropy is given by

$$\Delta S = \Delta S_1 + \Delta S_2 = \frac{Q_H}{T_H} - \frac{Q_L}{T_L}$$

But

$$\frac{Q_H}{T_H} = \frac{Q_L}{T_L}$$

∴ $$\Delta S = 0 \qquad \text{Reversible Cycle} \qquad ...(17.42)$$

(*ii*) Entropy increases in all irreversible processes: It is proved that there is a net increase in the entropy during irreversible processes. Since all real processes taking place in the universe are irreversible, there is a continuous increase in its entropy. For this reason, **entropy is not conserved**. In this respect entropy differs from energy.

We can illustrate this by taking a simple example. Suppose a small quantity of heat dQ is radiated away from a hot body *A*, at a temperature T_H, to a cold body *B* at a temperature T_C. Let dQ be so small that T_H and T_C are not altered appreciably, due to the exchange of heat. However, the entropy of A decreases by $-dQ/T_H$ where as that of *B* increases by dQ/T_C in this process.

∴ $$\Delta S = \frac{dQ}{T_C} - \frac{dQ}{T_H}$$

As $T_H > T_C$, $$\Delta S > 0 \qquad ...(17.43)$$

(*iii*) Entropy indicates the direction in which processes proceed in nature: All natural processes are irreversible. They proceed in the direction of increasing entropy.

(*iv*) Entropy represents the unavailability of energy: In the thermodynamic sense, entropy is a measure of the capability to do work or transfer heat. A system at a higher temperature will tend to do work on and/or transfer heat to its lower temperature surroundings. In the process, the entropy of the system increases and the greater the entropy the **less available** is the energy.

Let us consider the example of Carnot engine. The efficiency of Carnot engine is given by

$$\eta = 1 - \frac{T_L}{T_H}$$

As Q_H is the heat input, heat converted into work = $Q_H\left(1 - \frac{T_L}{T_H}\right)$

∴ Heat unavailable for work = $Q_H = T_L$

But Q_H / T_H represents the increase in entropy ΔS during isothermal expansion.

∴ $$\text{Energy wasted} = T_L \Delta S \qquad ...(17.44)$$

If T_L is constant, the amount of energy wasted is proportional to the increase in entropy.

17.18 THIRD LAW OF THERMODYNAMICS

With a decrease in temperature, a greater degree of order prevails in any system. If we could cool a system to 0K, the maximum conceivable order would be established in the system and the minimum entropy would correspond to this state. Now, suppose we apply a pressure on the system at 0K. What does happen to the entropy of the system? On the basis of experiments conducted at low temperatures, W.Nernst concluded that "at 0K, any change in the state of a system takes place without a change in the entropy." This is called **Nernst's theorem**. It is also called the **third law of thermodynamics**. Third law of thermodynamics is sometimes known as the principle of unattainability of absolute zero. It is stated as follows:

It is impossible to attain a temperature of 0 K.

QUESTIONS

1. Show that work is a path function, and not a property.
2. State the first law for a closed system undergoing a change of state.
3. What is thermodynamics? State the first, second and third laws of thermodynamics and discuss their significance.
4. What is a cyclic heat engine?
5. Define the thermal efficiency of a heat engine cycle. Can this be 100%?
6. Give the Kelvin-Planck statement of the second law.
7. Give the Clausius statement of the second law.
8. What is a reversible process? A reversible process should not leave any evidence to show that the process had ever occurred. Explain.
9. All spontaneous processes are irreversible. Explain.
10. Distinguish between a reversible and an irreversible process. Illustrate your answer by some examples.
11. What is a Carnot Cycle? What are the four processes which constitute the cycle?
12. Describe the different operations involved in a Carnot's cycle. Derive the efficiency of a Carnot engine in terms of source and sink temperatures.
13. What is a reversed heat engine?
14. Show that the efficiency of a reversible engine operating between two given constant temperatures is the maximum.
15. How does the efficiency of a reversible engine vary as the source and sink temperatures is varied? When does the efficiency become 100%?
16. What do you understand by the entropy principle?
17. Show that the transfer of heat through a finite temperature difference is irreversible?
18. Show that the adiabatic mixing of two fluids is irreversible.
19. What is the maximum work obtainable from two finite bodies at temperatures T_1 and T_2?
20. Why the second law is called a directional law of nature?
21. Give the Nernst statement of the third law of thermodynamics?

18 Vacuum Technology

Learning Objectives:

By the end of this chapter, you will be able to:

- Define concepts of Vacuum, its unit and ranges.
- Explain different types of Vacuum Pump – Oil Pump, Diffusion, Turbomolecular, Cryopump.
- Learn about Vaccum Gauges and application of Vacuum Technology.

18.1 INTRODUCTION

Vacuum technology was popular around 1900 in the manufacture of electric light bulbs. The various types of vacuum tubes, cathode ray tube and electron microscope followed which required a vacuum for their operation. It was found that certain processes carried out in a vacuum achieved superior results that are actually unachievable under normal atmospheric conditions. Some of such processes are increasing the light transmission of lens, the preparation of blood plasma for blood banks, and the production of reactive metals such as titanium. The nuclear energy in the 1950s gave a big impetus for development of vacuum equipments on a large scale. Further push is given by thin film deposition in semiconductor industry and other industries, space simulation and microelectronics.

18.2 VACUUM

Vacuum means a closed space with no gases or other material content inside. It is defined as a diluted gas with its pressure or density lower than that of the ambient surrounding atmosphere. In order to achieve a vacuum in a vessel it is necessary to remove from it the molecules that make up the atmosphere (mostly nitrogen, oxygen, carbon dioxide) and generate a pressure that is lower than the ambient pressure. Removal of molecules gives the desired level of vacuum is obtained and the removal process is to be continued further to maintain the desired vacuum level. Complicated and special techniques gives a vacuum of 10^{-10} - 10^{-15} Torr.

18.3 UNITS OF VACUUM

The pressure in a system is expressed in terms of the height of a column of mercury. At one standard atmosphere the pressure will support a mercury column 760 millimeter high.

Thus, one standard atmosphere equals 760 mm Hg. A smaller unit torr is introduced to express below-normal atmospheric pressures. A torr is defined as the pressure equivalent of a manometer reading of 1 mm of liquid mercury (1 torr = 1 mm Hg). Hence

$$\text{One standard atmosphere} = 760 \text{ torr.}$$

The term 'torr' was replaced in 1971 by SI unit **'pascal'** which is defined as the newton per square metre (N/m^2).

$$\text{One pascal} = 7.5 \times 10^{-3} \text{ torr.}$$

$$\therefore \quad 1 \text{ Torr} = 1/760 \text{ atm} = 133.3 \text{ Pa.}$$

18.4 VACUUM RANGES

Table 18.1 shows vacuum ranges.

Table 18.1: *Vacuum Ranges*

S.No.	Vacuum	Range
1.	Low vacuum	< 25 Torr
2.	Medium vacuum	25 to 10^{-3} Torr
3.	High vacuum	10^{-3} to 10^{-9} Torr
4.	Ultrahigh vacuum	10^{-9} to 10^{-12} Torr

18.5 PRODUCTION OF VACUUM

A vacuum is created in a chamber or vessel with the help of a vacuum pump commonly known as an exhaust pump. An **exhaust pump** is a device to remove or exhaust air, gas or vapour from a vessel. An exhaust pump consists of two sides. One side, known as **inlet** is connected to the vessel to be evacuated and the other known as **outlet** or **exhaust side** which expels the air, gas or vapour drawn from the vessel. The pressure at which the gas or air is expelled out on the exhaust side is called **exhaust pressure**. This pressure would be equal to the atmospheric pressure if the outlet is open to the atmosphere; or it would be less than that if the outlet is connected to an auxiliary pump called **backing pump**. In order to attain very low pressure or higher order vacuum in the chamber, the exhaust pressure should be very small. To achieve this, the backing pump and high vacuum pumps are arranged in series. The air or gas in the chamber are drawn in at the inlet of high vacuum pump and expelled at its outlet into the inlet of backing pump, which finally expels the gas into the atmosphere. The minimum pressure that can be produced in a given vessel is called **attainable vacuum**. This pressure depends largely on the exhaust pressure and varies from pump to pump.

18.6 CLASSIFICATION OF VACUUM PUMPS

Exhaust pumps can be classified into the following four categories:

(*i*) Oil pumps
(*ii*) Mercury pumps
(*iii*) Molecular pumps and
(*iv*) Diffusion pumps.

Oil pumps or mercury pumps can be of piston type or rotatory type. Piston type oil pumps produce low pressures of only 10^{-2} Torr and are used as backing pumps. Piston type mercury pumps are very slow and tedious in action though they produce a vacuum of the order of 10^{-5} Torr. We describe here some of the pumps that are commonly used.

18.7 ROTARY OIL PUMPS

Principle: Rotary oil pumps are based on the principle of displacement of air. A portion of the air inside a vessel is isolated by a rotating disc and is then compressed till it attains sufficiently high pressure and gets discharged to the atmosphere. As the air is expelled, more air comes to occupy the space in the vessel which in turn is also removed. The process continues and the pressure in the vessel keeps on decreasing.

Construction: Fig. 18.1 shows a typical rotary van oil-seal pump. It consists of a hollow steel chamber which is known as **stator**. Stator acts as the working chamber. A massive cylindrical shaft known as a **rotor** is installed inside the stator. The rotor is electrically driven and rotates eccentrically inside the stator such that it is always in contact with a certain peripheral point of the stator. The rotor carries two sliding rods known as the **vanes** that move radially under spring force. Rotor and **vanes** divide the working chamber into two separate spaces having variable volumes. The inlet port is connected to the vessel that is to be evacuated. A spring operated valve is fitted to the outlet port. This outlet valve is oil-sealed.

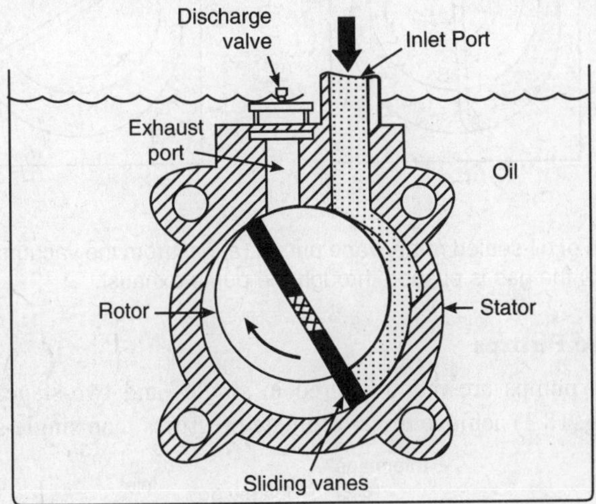

Fig. 18.1:

Chief components of a typical rotary oil-seal pump.

Working: The working principle of the pump is depicted in Fig. 18.2. As the rotor turns in the direction indicated, the space between rotor and stator on the inlet side keeps on increasing while the space between rotor and the outlet side goes on decreasing. As a result, air or gas from the vessel to be evacuated is sucked into the enlarging suction chamber until the inlet is sealed off by the vane. As the rotor moves further, the gas enclosed in the space between rotor and the outlet side is compressed.When the pressure of the compressed gas becomes high enough, it forces open the discharge valve at the outlet and escapes out. All these operations are completed in one full rotation of the rotor. The process goes on repeating till a pressure of the order of 10^{-3} torr is produced in the vessel being evacuated.

Since each revolution sweeps out a fixed volume, this pump is called a constant-displacement pump.

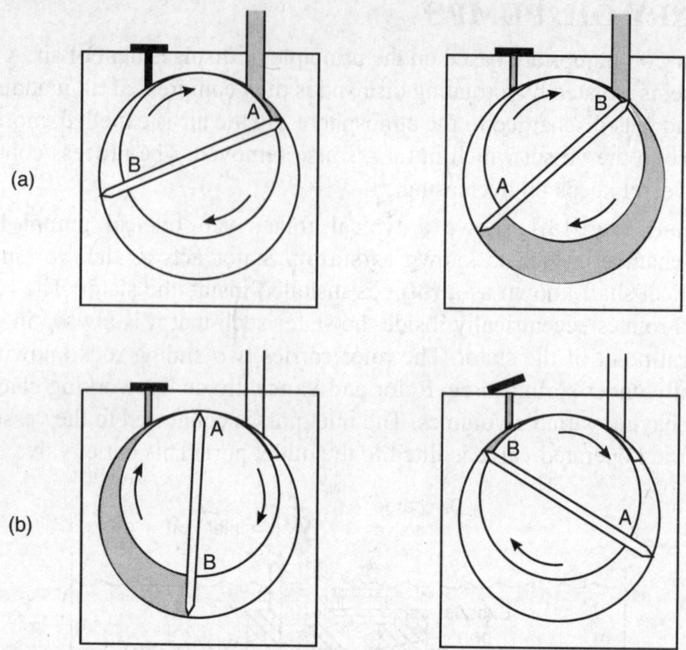

Fig. 18.2: Principle of operation of oil-sealed rotary vane pump: (*a*) gas from the vacuum system is expanded into the pump and (*b*) the gas is pushed through the pump exhaust.

18.7.1 Multi-stage Pumps

Rotary vane vacuum pumps are manufactured in single- and two-stage varieties of which two-stage pumps (Fig. 18.3) achieve lower ultimate pressures than single-stage pumps.

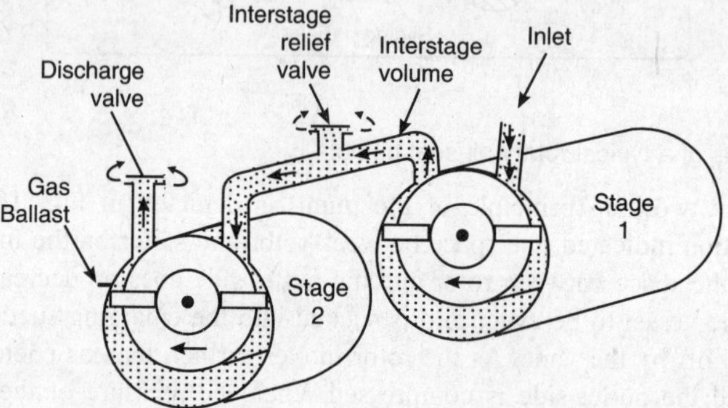

Fig. 18.3: Two stage Rotary Mechanical Pump.

Applications of Multi-stage pumps:

(*i*) in food packaging

(*ii*) high-speed centrifuges

(*iii*) ultraviolet spectrometers

18.8 DIFFUSION PUMP

In high vacuum applications, the ***diffusion pump*** (i.e., vapor jet pump) is the most commonly used pump.

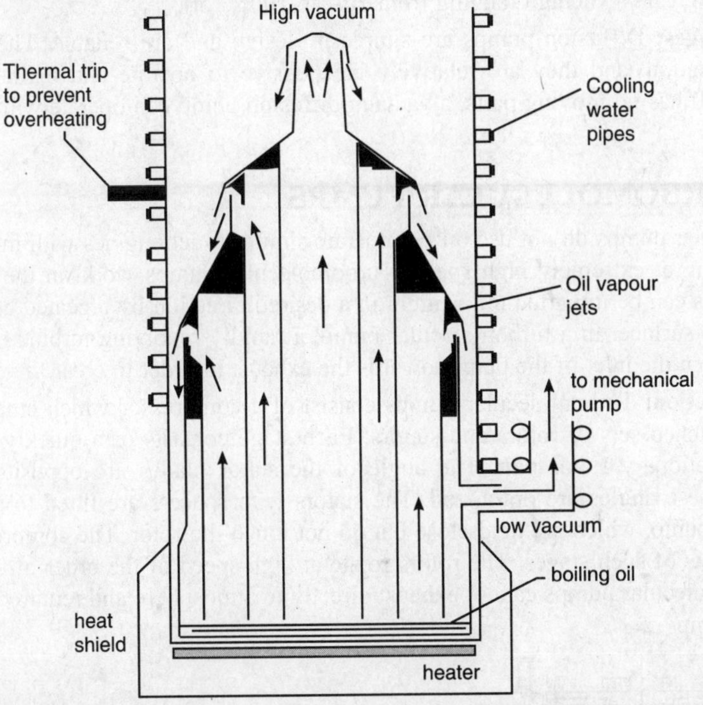

Fig. 18.4:

Schematic diagram of an oil diffusion pump. The arrows indicate the movement of boiling oil vapour.

Construction: A diffusion pump is basically a stainless steel chamber that varies in size depending on the application. It consists of three vertically stacked cone-shaped jet assemblies of decreasing sizes, with the largest one located at the bottom. At the base of the chamber is a pool of silicone-based diffusion pump oil. An electric heater is located beneath the floor of the chamber to heat the oil. The upper part of the pump wall is wrapped around with coils which circulate cold water through them. This cools the chamber.

Working: A diffusion pump functions by boiling oil at the base of the pump. Before diffusion pump started it is pumped down to an acceptable pressure. Once boiling of the fluid has begun, the oil vapour is forced up the central columns of the jet assembly. When a heavy, high speed vapour molecule collides with a gas (air) molecule, momentum transfer takes place and the gas molecule moves in a preferred direction through the pump. The vapour exits at each downward directed jet. As the pump body is externally cooled, the oil vapour striking the upper part of pump wall gets condensed and the liquid drops to the bottom of the pump carrying the adsorbed gas molecules with it. At the bottom, the heater re-boils the oil, releasing the gas molecules. The mechanical pump removes these gas molecules and produces a high vacuum in the upper portion of the chamber. This part of the chamber is connected to the sample chamber where a high vacuum is required. The diffusion pump gives a pressure of 10^{-5} to 10^{-6} Torr.

A vacuum diffusion pump cannot initiate its work with full atmospheric pressure inside the chamber. Therefore, an ancillary fore-pump is used to first bring the pressure inside the vacuum diffusion chamber down to about 10^{-3} Torr. At this point, the vacuum diffusion pump takes over to create a vacuum ranging from 10^{-3} to 10^{-10} Torr.

Advantages: Diffusion pumps are simple in design and are reliable. They run without noise or vibration, and they are relatively inexpensive to operate and maintain. Since the chamber itself has no moving parts, a vacuum diffusion pump can operate with stability over long periods.

18.9 TURBOMOLECULAR PUMPS

Turbomolecular pumps do not use oil but operate similar to jet engines with multiple, angled blades rotating at extremely high speed. Turbomolecular pumps work on the principle that gas molecules can be imparted momentum in a desired direction by repeated collision with a moving solid surface. In a turbomolecular pump, a rapidly revolving turbine rotor 'hits' gas molecules from the inlet of the pump towards the exhaust in order to create a vacuum.

Construction: Turbomolecular pumps consist of a compressor which employs multiple stages of matched sets of rotors and stators. Each stage consists of a quickly rotating rotor blade and stationary stator pair. The angle of the stator blades are opposite of the rotor blades and these angles are optimized. The stators with spacers are fitted to the outer wall of the turbo-pump, which are held close but do not touch the rotor. The compressor contains typically 9 sets of such stages. The rotors rotate at high speed of the order of 24,000-90,000 rpm. Turbomolecular pumps cannot exhaust directly to atmosphere and require the support of a backing pump.

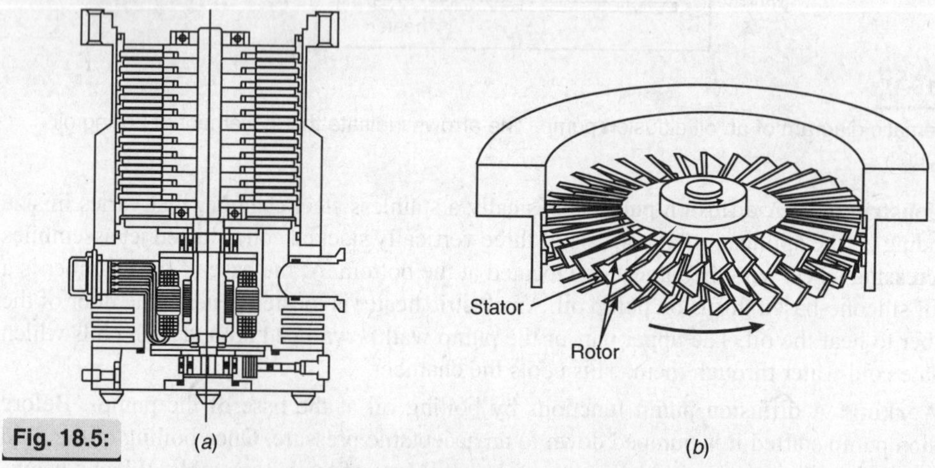

Fig. 18.5: (a) (b)

(a) Turbo molecular Pump (b) Rotors and Stators of Turbo molecular Pump.

Working: The turbomolecular pumps are momentum transfer devices. The rotor blades accelerate the molecules along the pumping axis. The moving blades of rotor hit the entering gas molecules and the mechanical energy of the blades is transferred to the molecules. The stator blades then decrease the velocity of the pumped gas. As a result, the gas is compressed. Gas captured by the first rotor is then pushed into the lower rotor/stator pairs and is again gets compressed. This leads the gas molecules to the next stage where they again collide with the rotor surface, and this process is continued until the gas reaches a pressure where it is finally led outwards through the exhaust by a mechanical backing pump. Turbo pumps can reach 10^{-7} to 10^{-10} torr vacuum.

18.10 CRYOPUMPS

Cryogenic high-vacuum pumps, or cryopumps, create high vacuum by condensing the gas molecules onto extremely cooled surfaces inside the pump. The vacuum typically ranges from 10^{-4} to 10^{-9} Torr.

Working: A schematic of a typical cryopump is shown in Fig. 18.6. A typical cryopump consists of two internal stages. Each stage in turn cools an extended surface cryopanel onto which the gases will freeze.

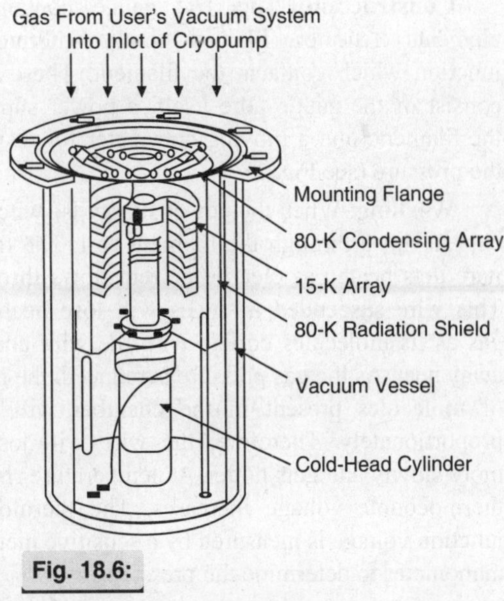

Fig. 18.6: Schematic of a cryopump.

The first stage of the pump operates in the range of 50-75 K. Its main function is to capture water vapor. It cools the outer cryopanel and louvers across the inlet opening of the cryopump. Water vapor will freeze onto these panels. The second stage, of the pump consists of a series of metal pumping surfaces. It operates between 10- 20 K and is used to cool the inner cryopanels. Gases such as nitrogen, oxygen and argon will freeze onto these panels. The metal pumping surfaces are partially covered with charcoal granules. Lastly, gases such as hydrogen and helium that have not yet frozen onto a panel will be adsorbed into charcoal and are removed from the vacuum chamber. The stages are connected to a sealed cryogenic refrigerator run by an external compressor. The refrigerator cools the stages.

Cryopumps have no moving parts that are exposed to vacuum. Hence they are naturally clean and do not contaminate products they produce.

18.11 VACUUM GAUGES

Vacuum may be defined as a space that is devoid of matter. Achieving such an empty space is impossible on earth. Instead, we describe a vacuum as a space with gaseous pressure much less than atmospheric pressure. A vacuum gauge is a device that measures the sub-atmospheric pressure in an evacuated container. Generally, depending on the measurement of range of vacuum different gauges are used.

(*i*). **Direct reading vacuum gauges:** They use the principle of gas compressibility. Bourdon tube, McLeod, and other liquid manometers and aneroid (and mercury) barometers are few examples.

(*ii*). **Indirect reading vacuum gauges:** They use the thermal conductivity or ion conduction of residual gas as a pressure indication. Thermocouple, Pirani, and ion gauges belong to this category.

We describe here the indirect reading gauges which are used to measure high vacuum.

18.11.1 Thermocouple (or T/C) Gauge

The thermocouple (or T/C) gauge is generally useful in the range 1 Torr to 1 milliTorr. T/C gauge does not measure pressure directly. Instead, it depends on changes of the thermal conductivity of the gas.

Construction: The T/C gauge contains two elements a heater (filament) and a thermocouple junction which contacts the filament. These gauges consist of the gauge tube itself, a power supply for the filament, and a moving coil meter for displaying the pressure (see Fig. 18.7).

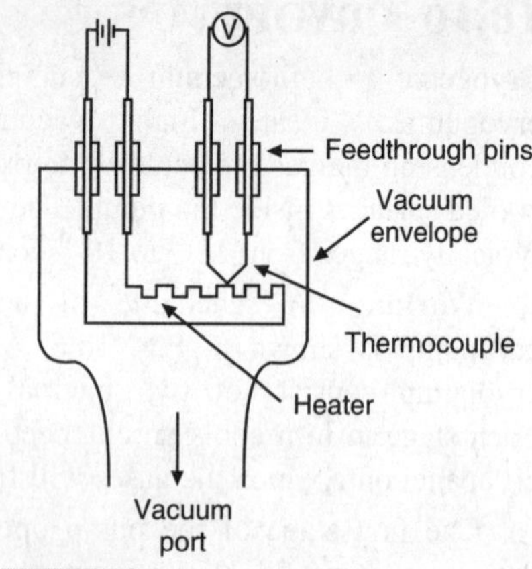

Fig. 18.7: Schematic of a thermocouple gauge.

Working: When the power supply is switched on current flows through the filament coil. The filament wire gets heated as electric current flows through it. This wire suspended in a gas will lose heat to the gas as its molecules collide with the wire and carry away heat. As the gas pressure is reduced, the number of molecules present in the chamber will reduce proportionately. Therefore, the wire will lose heat more slowly and gets hotter. As temperature rises, the thermocouple voltage increases. The thermocouple junction voltage is measured by a sensitive meter that has previously been calibrated against a manometer to determine the pressure.

18.11.2 Pirani Gauge

The Pirani gauge is a robust thermal conductivity gauge. It was invented in 1906 by Marcello Pirani. It consists of a heated filament of high temperature coefficient of electrical resistance such as platinum, tungsten hanged in a tube.

Principle: The principle of this instrument is based on the measurement of the changes in resistance of the heated wire placed into the vacuum system. The resistance depends on the temperature of the wire.

A metal filament gets heated when electric current flows through it. This wire suspended in a gas will lose heat to the gas as its molecules collide with the wire and carry away heat. As the gas pressure is reduced, the number of molecules present in the chamber will reduce proportionately. Therefore, the wire will lose heat more slowly. The electrical resistance of the wire varies with its temperature. This change in resistance of the pirani gauge filament becomes a measure of the applied pressure when calibrated.

Working: The Pirani gauge in essence is a Wheatstone bridge. A Pirani gauge chamber which encloses a platinum filament is connected in one arm of the bridge. It consists of a heated filament mounted in a bulb fitted with a connecting tube (Fig. 18.8 *a*) and is placed in the vacuum system. The filament is heated by a constant current as shown in Fig. 18.8 (*b*). A compensating cell is mounted in an adjacent arm of the bridge. The compensating cell consists of filament and is evacuated to a very low pressure. The galvanometer responds only to the changes produced by the residual gas in the gauge chamber.

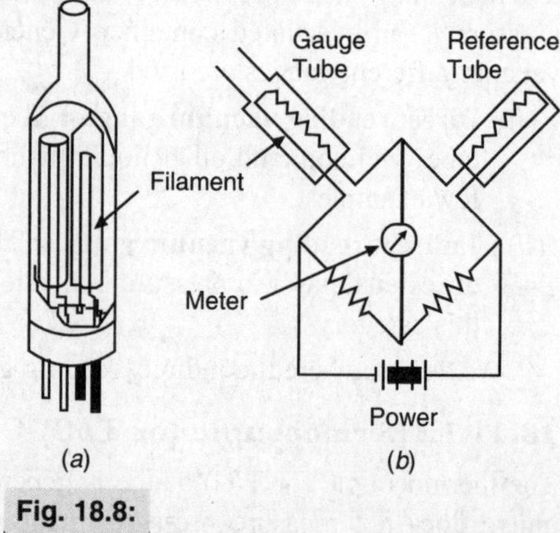

Fig. 18.8: Schematic diagram of a Pirani Gauge.

Now, a constant current is passed through the filament in the Pirani gauge chamber. Due to the current, the filament gets heated and possesses a resistance which is measured using the bridge. The vacuum to be measured is connected to the Pirani gauge chamber. Due to the decrease in pressure density of the gas surrounding the Pirani gauge filament decreases. It results in fewer collisions of gas molecules with the filament. As a result the temperature of the filament increases. Electrical resistance of a wire varies with temperature.

If the bridge is balanced at one temperature of the filament, a change in temperature causes a change in the heat conductivity of the residual gases. So, using the change in resistance of the Pirani gauge filament the applied pressure of the residual gases can be calculated.

This gauge is useful between 0.5 Torr to 10^{-4} Torr and for higher ranges instruments such as a Penning gauge are used.

18.11.3 Bayard-Alpert Gauge

In the hot cathode gauge (HCG) electrons emitted from a thermionic cathode are accelerated by suitable electrodes into an ionizing space where they cause ionization. The "Bayard-Alpert" gauge is the most common and somewhat standardized variety of hot cathode gauges. The Bayard-Alpert gauge was invented by R.T. Bayard & D. Alpert in 1950.

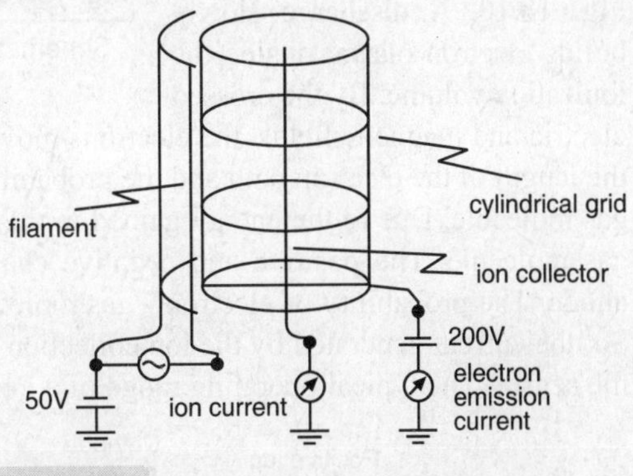

Fig. 18.9: Schematic of Bayard-Alpert ion gauge.

Construction: The Bayard-Alpert ion gauge tube (sensor) resembles a triode in construction and consists of a tungsten or coated filament located (cathode) outside of a helical grid (Fig. 18.9).

Working: Negative electrons are emitted at a controlled rate from the heated filament which is held at a positive voltage (typically +30 VDC). The electrons are accelerated toward the positively-charged helical wire grid which is maintained at a positive voltage (typically +180 VDC). Electrons pass into the space enclosed by the grid. In this space the electrons collide with the gas molecules that are in the vacuum system, and produce positive ions. The positive ions are then collected by the ion collector that is located along the axis of the cylindrical grid. The ion collector is at ground potential and is negative with respect to the grid. With a constant filament-to-grid voltage the rate that positive ions are formed is directly proportional to the density of molecules (pressure) in the gauge. The strength of the ion current is indicated on an electrometer that is calibrated in units of pressure. At constant temperature, the collector current is directly proportional to the gas pressure. The operating range is of 10^{-3} and 10^{-10} Torr.

18.11.4 Penning Gauge

"Penning" gauge is one of the varieties of **cold cathode gauges**. In this gauge, ionization is caused by circulating electron plasma trapped in crossed electric and magnetic fields.

Construction: A schematic of the Penning gauge is shown in Fig. 18.10. It consists of two electrodes, cathode and anode arranged between the pole pieces of a horse-shoe magnet that can produce a magnetic field of the order of 1-2 kG. The cathodes are two flat plates while the anode is a small cylindrical tube. The anode is held at a potential of 2000 V whereas cathode plates are held at ground potential. The magnetic field is directed along the axis of the cathode.

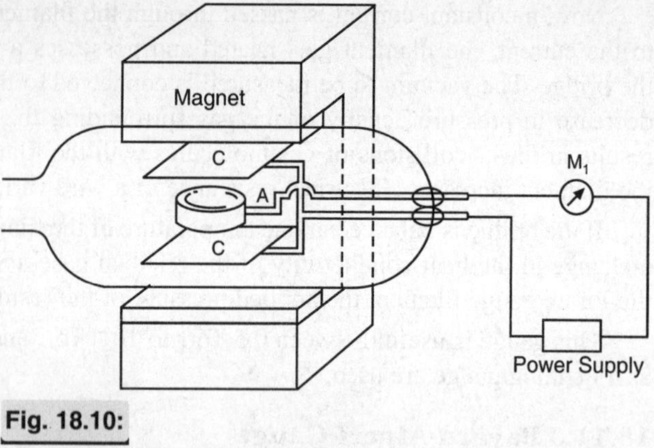

Fig. 18.10: Schematic of a Penning gauge.

Working: The working principle of this type of gauge is to generate a discharge between the anode and cathode by applying a high DC voltage between them, as shown in Fig. 18.10. A discharge slowly builds electron plasma inside the ionization volume. By the crossed electric and magnetic fields, the electrons move in spiral paths [Fig. 18.11 (*b*)]. This increases the length of the electron path and the probability that the accelerated electron strikes a neutral gas molecule. Due to the energy gained in the electric field, the energetic electron ionizes the gas molecule. The positive and negative charges produced move towards the cathode and anode. The probability of electron - gas molecule collision is proportional to the gas density. So, the current generated by the ion collection process is a measure of the existing pressure in the sensor. The typical operating range of a Penning gauge is between 10^{-2} and 10^{-7} Torr.

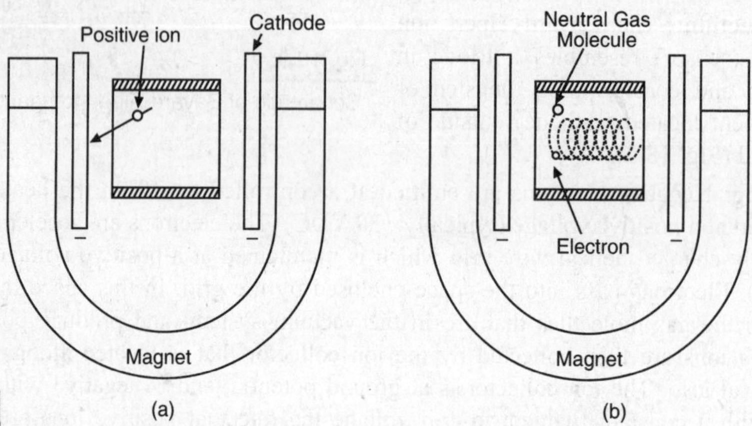

Fig. 18.11: Working concept of Penning gauge.

18.12 APPLICATIONS OF VACUUM TECHNOLOGY

Vacuum Technology is the field where processes and physical measurements are carried out under conditions of below-normal atmospheric pressure. Reasons for physical measurement under vacuum:

(*i*) It is required to avoid atmosphere where certain processes such as melting of reactive metals etc are to be conducted.

(*ii*) Removal of dissolved gas or volatile liquid from the bulk of material, such as degassing of oils, freeze-drying etc or desorption of gas from surfaces are effectively done under vacuum condition.

(*iii*) Vacuum is used to prolong in a process the distance that a particle has travel thereby helping the particles to move without collision between source and target in vacuum coating, particle accelerators, television picture tubes etc. X-ray-based, electron-probe-based equipment require that particle mean free paths be great enough to be detected by in situ detectors.

(*iv*) Vacuum is employed in clean-surface studies and preparation of pure, thin films to reduce the number of molecular impacts per second, and hence reduce chances of contamination of surfaces prepared in vacuum. The main applications of vacuum have been to deposit and etch thin films of materials used in the semiconductor industry and many other areas.

Industrial applications of vacuum are spread from mechanical handling to the deposition of integrated electronic circuits on silicon chips.

(*a*) Rough vacuum range: In the rough vacuum range from about 1torr to near atmosphere, vacuum applications are mechanical handling, vacuum packing and forming, gas sampling, filtration, degassing of oils, concentration of aqueous solutions, impregnation of electrical components, distillation, and steel stream degassing etc.

(*b*) About 10^{-4} torr: Many metallurgical processes such as melting, casting, sintering, heat treatment, and brazing require low vacuum. Chemical processes such as vacuum distillation and freeze-drying also need this range of vacuum. Freeze-drying is used extensively in the pharmaceutical industry to prepare vaccines and antibiotics and to store skin and blood plasma.

(*c*) About 10^{-6} torr: The pressure range down to about 10^{-6} torr is used for low-temperature and electrical insulation. It is used in the production of lamps; television picture tubes, X-ray tubes; decorative, optical, and electrical thin-film coatings; and mass spectrometer leak detectors.

Typical of the research equipment using vacuum down to about 10^{-6} torr are the electron microscope, analytical mass spectrometer, particle accelerator, and large space simulation equipment.

In space simulation, large units that simulate space around a complete vehicle require a vacuum of 10^{-6} torr or below.

(*d*) Below 10^{-9} torr: In the pressure region down to and below 10^{-9} torr, research applications include electrical insulation, the rmonuclear energy conversion experiments, microwave tubes, field ion microscopes, field emission microscopes, storage rings for particle accelerators, specialized space simulator experiments, and clean-surface studies.

18.13 THIN FILM DEPOSITION

The formation of a thin film requires high vacuum condition. Physical Vapor Deposition (PVD) is one of the processes used for producing thin films. PVD is the process where the material used for coating is thermally vaporized in high vacuum and the material particles move toward the object (substrate) on the ceiling of the chamber with little or no collisions with gas molecules on the way. Vacuum levels are maintained in the chamber (Fig. 18.13) to high vacuum range of 10^{-5} to 10^{-9} mbar. This produces a quality thin film of high degree of purity and adhesiveness.

In PVD techniques, resistance heating, and electron beam heating are the processes most widely used for producing thin films. The material to be deposited is usually held in a coil or in a metal boat. By means of an electrical current passing through a filament or metal boat (Evaporator) the material is heated typically to its melting point and it causes evaporation of the film material. The substrate is held facing the source at a distance from it. The evaporated material is then condensed on the substrate. A crystal monitor is mounted close to the substrate, which gives an estimate of how much and how fast the material is being deposited.

A vacuum thin-film coating system applies a thin coat on the object. The thickness of the film varies from product to product, the average being 0.1 to several dozen micrometers.

Vacuum evaporation is used in a wide variety of fields. The base material is generally plastic for decorative coatings; glass for optical coatings; and glass ceramic, or silica for electrical coatings. Thickness of the film can vary from about 1/4 wavelength of visible light to 0.001 inches or more. In the optical field, antireflection coatings are deposited on lenses for cameras, telescopes, eyeglasses, and other optical devices, considerably reducing the amount of light reflected by the lenses and thus giving a brighter transmitted image. Mirror coatings, decorative coatings, permeation barrier films on flexible packaging materials, electrically conducting films, corrosion protective coatings etc play an important role in our day to day life.

18.14 HIGH VACUUM SYSTEMS

A **high vacuum system** consists of a vacuum chamber, vacuum pumps, and valves (Fig. 18.12). In between various combinations of tubing, fittings and valves are employed. These are required for the system to operate. We do not have such a powerful pump that can exhaust air in the chamber from the level of atmosphere to high vacuum at once. Therefore, two vacuum pumps are connected to a sample chamber. One of the vacuum pumps, known as roughing pump, produces medium vacuum, and the second pump, a high vacuum pump, produces the high vacuum required for a specific process. At the start, only the roughing valve is open and the roughing pump removes air. Next, the rough valve is closed and both the roughing and diffusion pumps that are connected in series are activated by the opening the remaining two valves. This keeps the pressure inside the chamber at the required level. A vacuum gauge monitors the pressure in the sample chamber.

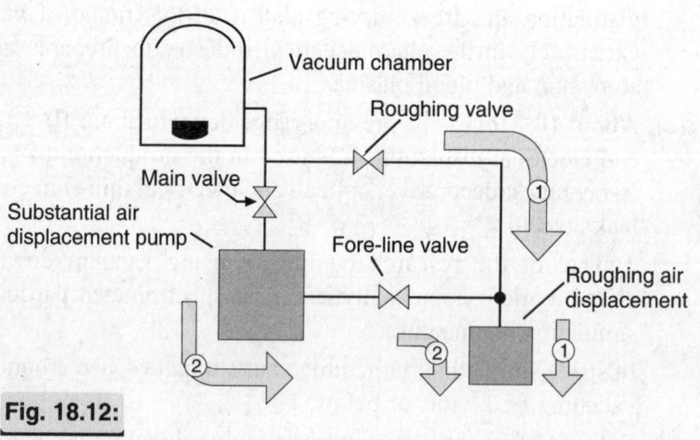

Fig. 18.12: Schematic of a typical vacuum system.

QUESTIONS

1. What is a vacuum? How it is measured? It is possible to have prefect vacuum?
2. What do you understood by vacuum technology?
3. Give the theory and working of a rotary oil pump. What is the limit of pressure that can be achieved with the pump?

4. How does diffusion pump generate vacuum? What is the order of vacuum produced by a diffusion pump?
5. Explain the construction and working of a turbo molecular pump.
6. Explain the principle and working of a cryopump.
7. Explain units of pressures, and measuring instruments used in various ranges.
8. Explain the principle and working of a thermocouple gauge.
9. Explain the principle and working of a Pirani gauge.
10. Explain the principle and working of a Bayard-Alpert gauge.
11. List the applications of vacuum technology in detail.
12. Give an outline of a high vacuum system.
13. Draw the schematic of a thin-film coating system and explain its working.

19 Electron Emission

Learning Objectives:

By the end of this chapter, you will be able to:

- Define concepts of Work Function.
- Explain fundamentals of Electron Emission.
- Learn about different types of Electron Emission – Thermionic, Photoelectric, Field and Secondary.

19.1 INTRODUCTION

Modern technology is strongly electronics-oriented. A majority of the instruments, appliances and gadgets are based on the movement of electrons in their circuits for their operation. In many of the cases, the working of a device or a circuit cannot be understood without a proper knowledge of how electrons are controlled and manipulated at various stages. The understanding of the exciting developments would be possible only when the basic physical principles regarding the generation of electrons, their response to external stimuli are amply learnt.

The methods of generation of electrons are outlined in this chapter laying stress on thermionic emission which is widely used in many of the vacuum tube devices.

19.2 WORK FUNCTION

Any metal contains a large number of free electrons. Typically, the electron concentration in a metal is $10^{29}/m^3$. Periodic potential craters are located at the sites of positive ions. Each electron moves in the common field of all positive ions. Fig. 19.1 shows the potential energy of an electron as a function of distance in the metal. It is only at distances close to an ion that the variation in potential is perceptible. At greater distances, the metal may be viewed as an equipotential region. In that case, the details in Fig. 19.1 are ignored and the potential energy curve is represented by a simple potential well (Fig. 19.2). The electrons are distributed among various energy levels in the potential well. At normal temperatures, most electrons occupy the levels up to the energy level E_F and a small number with higher energy are at levels above E_F. The level E_F is known as **Fermi level**. A conduction electron may move freely within the interior of the metal but it cannot escape from the surface. The electron has to surmount the potential barrier on the metal surface if it is to escape from the metal. The existence of the surface barrier may be understood this way. If an electron attempts to escape from a metal, it will induce a positive charge on the surface, because the metal was originally neutral. Unless the electron possesses enough energy to move away from the region of influence of the induced positive charge, it will be compelled to fall back into the potential well.

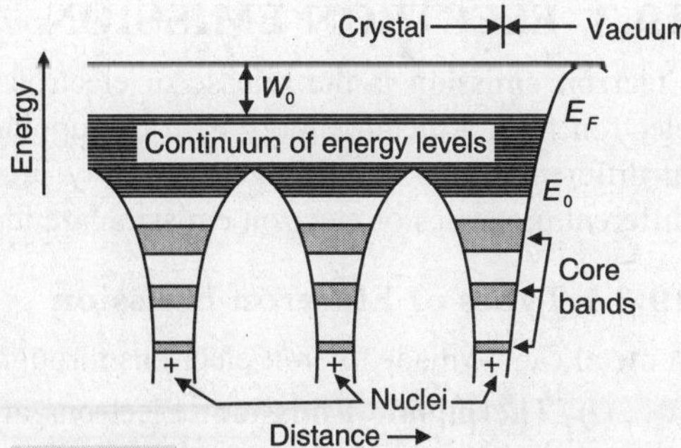

Fig. 19.1: Schematic representation of the energy barrier at the surface of a metal.

Energy is to be supplied to the electron in order to remove it from the metal. An electron at the lowest level needs an energy E_0, whereas an electron at the Fermi level needs an energy $(E_0 - E_F)$. The minimum energy required for an electron to be just emitted from a metal surface is called the **work function** of the particular metal. It is designated by W_0 and is expressed in electron – volts. Thus,

$$W_0 = E_0 - E_F \quad ...(19.1)$$

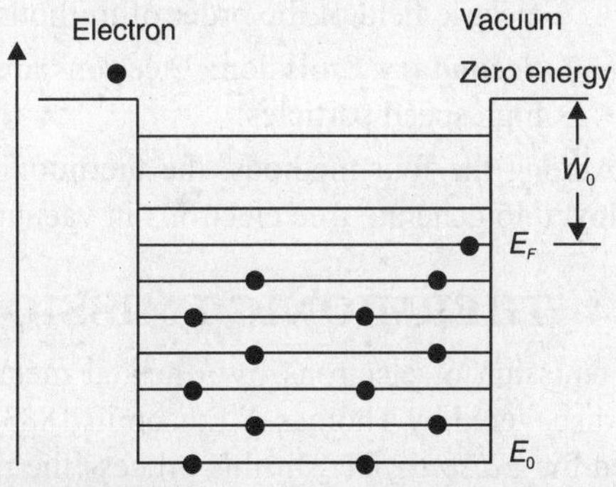

Fig. 19.2: Potential energy curve - potential well.

The magnitude of the work function depends on the energy level E_F and the height and shape of the energy barrier. Thus, the work function is a characteristic of the metal. The larger the values of work function, the harder it is to release electron emission.

The work function for some metals is shown in Table 19.1.

Table 19.1: *Work function of some metals*

Metal	Work function, eV
Caesium	1.9
Lithium	2.5
Sodium	2.3
Potassium	2.2
Silver	4.7
Platinum	5.6

The work function for a metal is approximately half the ionization energy of a free atom of that metal. For example the ionization energy of caesium atom is 3.9 eV and the work function of caesium metal is 1.9 eV. Therefore, electron emission cannot occur at low temperature.

19.3 ELECTRON EMISSION

Electron emission is the release of electrons from a metal surface. In order to liberate an electron from a metal, energy is to be supplied to the electron. The energy may be supplied in different ways. According to the way in which an electron receives the kinetic energy, different processes of electron emission are identified.

19.3.1 Types of Electron Emission

A metal can be made to emit electrons through the following four processes.

(*i*) **Thermionic Emission:** Electrons are emitted from a metal when it is heated to a high temperature.

(*ii*) **Photoelectric Emission:** Electrons are emitted from a metal when it is illuminated with a high frequency light.

(*iii*) **Field Emission:** Electrons are emitted when a metal surface is subjected to a strong electric field of the order of millions of volts/m.

(*iv*) **Secondary Emission:** Electrons are emitted from a metal when it is bombarded by high speed particles.

Among the four methods, the thermionic emission is by far the most common method employed to generate free electrons in vacuum devices.

19.4 THERMIONIC EMISSION

The emission of electrons by a heated metal is known as *thermionic emission.* This effect was discovered by Thomas A.Edison in 1883 during the development of light bulbs. It is also called the *Edison effect.* In this process the metal is heated to about 2500° C. Even at lower temperatures there are a few electrons which can surmount the energy barrier. However, the number of such electrons increases sharply with increasing temperature. Metals with lower work function emit electrons at lower temperatures. The most commonly used materials are tungsten, thoriated tungsten etc.

19.4.1 Richardson-Dushman Equation

The emission current at a particular temperature can be calculated knowing the distribution of electrons in various energy levels. Such a calculation was performed and the result is known as **Richardson – Dushman equation.** It is given as,

$$J_S = AT^2 e^{-b/T} \text{ amp/m}^2 \quad \text{...(19.2)}$$

where J_S = emission current density, i.e., current per square metre of the emitting surface,

T = Absolute temperature of emitter in K

A = Constant, depending on the type of emitter and is measured in amp/m^{2}K^2

$$= \frac{4\pi em\, k^2}{3}$$

The value of b is constant for a given metal and is given by $b = \dfrac{e\phi}{k} = 11{,}600\ \phi$ K

where $\phi = W_0$, the work function of emitter.

Using the value of b into the Eq. (19.2), we get

$$J_S = AT^2 e^{\frac{11600}{T}\phi} \quad \text{...(19.3)}$$

The theoretical value of A is $A = 120$ A/cm^{2}K^2.

The experiment values are considerably lower and differ from metal to metal.

The values of the emitter constants A and b may be obtained from the Eq. (19.2).

The Eq. (19.2) indicates that the saturation current is strongly dependent on the emitter temperature and work function. The higher the temperature the larger is the saturation current. Further, the smaller the work function the higher will be electron emission and hence the current density.

19.4.2 Thermionic Emission Materials

Thermionic emitters are operated at high temperature. They are expected to have a reasonable service life. Therefore, they have to fulfill a number of requirements.

(*i*) **Low work function:** The material should have as low a work function as possible. Then the electron emission takes place at reasonably low temperatures.

(*ii*) **High melting point:** The electron emission occurs at high temperatures (>1900° C). Therefore, the material should have a high melting point.

(*iii*) **Mechanical strength:** The emitter is held normally at zero potential or at a negative potential with respect to an anode. Hence, it is referred to as a **cathode**. The material used as cathode must be durable and the emission must be stable. Further, it should have high mechanical strength to withstand ion bombardment. Even in a high vacuum, there will be gas molecules present. The gas molecules get ionized due to the collisions with energetic electrons. The resulting positive ions will get accelerated towards the cathode. As a result, the cathode will be subjected to ion bombardment. Unless the cathode is mechanically stronger it will get damaged.

19.4.3 Thermionic Cathodes

The requirement of operation at high temperatures restricts the number of suitable emitters to be used as cathodes. The most commonly used cathodes are tungsten, thoriated tungsten and oxide coated metals.

(*i*) **Tungsten:** Tungsten is a high melting point metal (m.p. = 3300° C). It has a high work function (4.52 eV). It is operated at 2300° C. The advantages of tungsten are that it has high mechanical strength and highly consistent emission. Its use in ordinary devices such as vacuum diodes is discontinued. However, it is used in power transmitting tubes and X-ray tubes, where the anode voltages exceed 15 KV.

(*ii*) **Thoriated Tungsten:** Thoriated tungsten is an improved material. A thoriated tungsten cathode consists of a tungsten filament coated with a layer of thorium. It has a lower work function of 2.63 eV and therefore it provides emission at a lower temperature of 1700° C.

(*iii*) **Oxide Coated Cathode:** This is the most commonly used material. It consists of a nickel ribbon or cylinder coated with a thin layer of barium and strontium oxides. The oxide coated cathode has a low work function of 1.1 eV. It is operated at 700-900° C and has high emission efficiency.

The features of the various types of cathodes are summarized in Table 19.2.

Table 19.2: *Thermionic cathode materials*

Cathode	Work function eV	Working temperature °K	Js A/ cm^2	A amp/ cm^2 deg^2	b °K	Life hr.	Emission efficiency mA/watt
Tungsten	4.52	2500	0.25	60	52,400	3000	4-20
Thoriated Tungsten	2.63	1900	1.5	3	30,500	10,000	50-100
Oxide	1.1	1000-1100	1.00	0.01	12,000	20,000	1000-10,000

19.4.4 Directly and Indirectly Heated Cathodes

There are mainly two types of thermionic cathodes, namely ***directly heated cathodes** and **indirectly heated cathodes***. Some of the types are shown in Fig. 19.3.

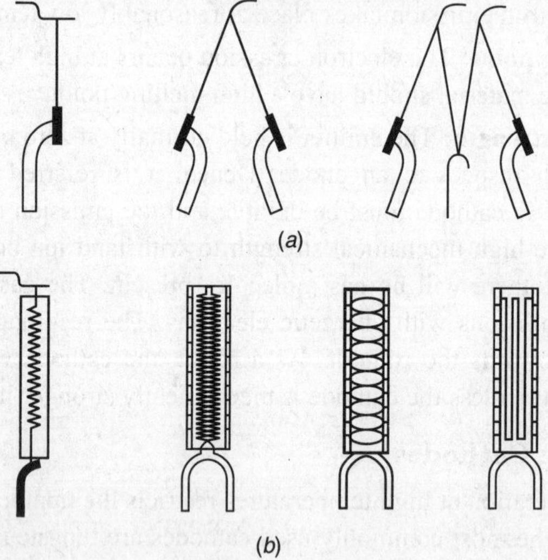

Fig. 19.3: Types of cathodes (*a*) directly heated (*b*) indirectly heated.

(*i*) **Directly heated cathode:** A directly heated cathode is a wire or ribbon filament. The filament is usually bent in zig-zag fashion. The heating current passes directly through the filament and causes thermionic emission from the filament. The advantage is that the efficiency of conversion of electric power into the thermionic emission is more. But the disadvantage is that the variations in heater voltage affect the electron emission and thus produce hum in the circuit.

(*ii*) **Indirectly heated cathode:** An indirectly heated cathode is made of a hollow nickel tube that is coated with barium and strontium oxides. A filament heater is located within the hollow of the tube and is insulated from it by means of an aluminium oxide coating. When current is passed through the filament heater, the cathode is indirectly heated through the heat emitted by the filament. The major advantages of this arrangement are that (*i*) the cathode is totally isolated from the heating circuit and it can be directly connected to any potential; (*ii*) temperature variations are practically non-existent because the cathode has a larger mass; and lastly, AC voltage can be used to heat the filament.

19.5 PHOTOELECTRIC EMISSION

Additional energy required by the electrons in a metal to surmount the surface barrier can be supplied by the application of light. When a metal is irradiated by a light of suitable frequency, electrons absorb the incident photon energy and leave the metal, as shown in Fig. 19.4. Such an emission of electrons caused by light is known as *photoelectric emission*.

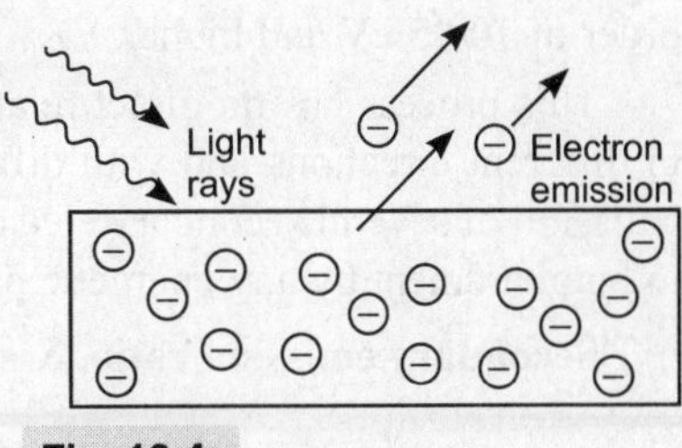

Fig. 19.4: Photoelectric emission.

In order to liberate an electron, the photon should have energy greater than the work function of the metal. Thus, if the energy of the photon is $h\nu$, $h\nu$ should be greater than the work function, W_0 for electron emission. Thus,

$$h\nu > W_0$$

or

$$\nu > \frac{W_0}{h}$$

or

$$\lambda < \frac{ch}{W_0} = \frac{12400}{W_0} \text{ Å} \quad \text{...(19.4)}$$

The electron emission in this process depends upon the frequency of the incident radiation and on its intensity.

19.6 FIELD EMISSION

When an electric field is applied to a cathode surface, the surface barrier is reduced and the work function is also reduced. When an intense electric field of the order of 10^7 to 10^8 V/m is applied, the surface barrier is reduced to such an extent that electrons can be pulled from the surface. The emission of electrons due to an intense external electric field is known as **field emission**. It is also known as **cold emission** because the cathode is not heated in this process. This phenomenon is in fact due to a quantum mechanical process known as *tunneling*. The emission current varies approximately as exp $(-\phi/E)$, where E is the intensity of the electric field.

19.7 SECONDARY EMISSION

When high energy electrons are incident on a metallic surface they cause emission of more electrons from the surface (Fig. 19.5). Such an electron emission caused by energetic electrons from a metallic surface is known as **secondary emission**. The electrons that are incident on the metal surface are called **primary electrons**. The electrons that are emitted from the metal surface are known as **secondary electrons**. The primary electrons penetrate the surface layer of the metal and give up their energy to electrons in that layer.

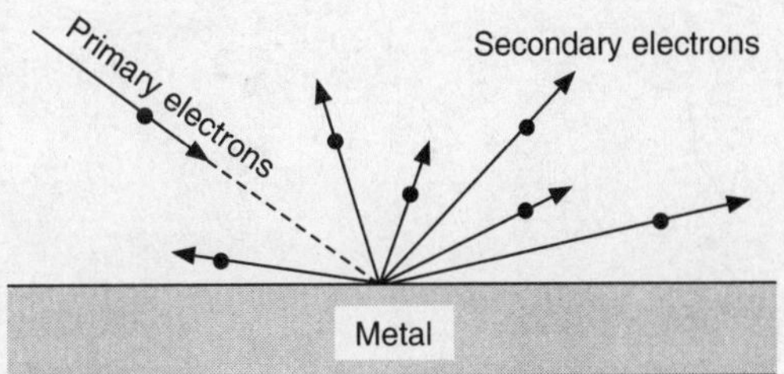

Fig. 19.5: Illustration of secondary emission.

Some of the electrons in the layer acquire sufficient energy and escape from the surface. Secondary electron emission takes place when the incident electrons have energies of the order of 10-15 eV and higher.

This process has no direct relationship to the work function. Secondary emission occurs in different directions and with different energies. Often an incident primary electron causes emission of several secondary electrons. Secondary electron emission may also be caused due to bombardment by any energetic particles.

Secondary emission ratio, δ is defined by,

$$\delta = \frac{\text{Number of secondary electrons emitted}}{\text{Number of primary electrons incident}} \quad ...(19.5)$$

δ varies between 1.5 and 2 for most metals but can be as high as 10-15 for certain surfaces.

QUESTIONS

1. Explain the concept of work function.
2. Why are electrons not able to escape from a metal under normal conditions?
3. What are the different processes that cause electron emission from a metal?
4. What is meant by thermionic emission?
5. Write down Richardson-Dushman equation and explain the various terms.
6. What are the requirements that a metal has to fulfill in order to act as a thermionic emitter?
7. Which are the materials used commonly as thermionic emitters?
8. What are directly heated and indirectly heated cathodes? Explain their functioning.
9. Explain photoelectric emission.
10. Describe field emission.
11. Describe secondary emission.

20 Electron Ballistics

Learning Objectives:

By the end of this chapter, you will be able to:

- Define concepts of Electric Field.
- Explain basics of Uniform Magnetic Field and motion of an electron.
- Understand about Magnetostatic Deflection.
- Learn the usage of Velocity Selector.
- Discuss the concepts of Charge, mass and radius of electron.

20.1 INTRODUCTION

Electric and magnetic fields affect the motion of charged particles. This was found in the experiments on discharge through gas at low pressures. We can see that if one holds a strong permanent magnet near the face of a television, the picture gets distorted. Charged particles travel in straight line path in a vacuum and behave just as any classical point mass particle, when they are in a free state. They readily respond to the commands of electric and magnetic forces. In practice, macroscopic electric and magnetic fields are employed to subject charged particles to desired forces and to obtain predictable trajectories covering measurable distances. These forces are large enough to dominate the motion so that gravitational force and mutual repulsion between the particles become negligible. Further, the particle trajectories are confined to evacuated enclosures so that the particle paths are not distorted by collisions with atmospheric gas molecules and uninterrupted travel of the particles over measurable distances is ensured. Under these conditions particles follow geometrically simpler paths which can be predicted and analyzed by the application of the laws of classical Newtonian mechanics. Various combinations and configurations of electric and magnetic fields have been used to control and manipulate the particle trajectories. A thorough appreciation of electron dynamics led to the invention of an array of vacuum tubes whose applications constituted the discipline of electronics in its early stages. The word electronics is coined by fusing the word 'mechanics' to the word 'electron.' Thus electron + mechanics = electronics.

We study in this chapter how the particle motion is affected by the electric and magnetic fields under variety of conditions, taking electron as representative of the group.

20.2 ELECTRIC FIELD

Pictorial Representation

Faraday proposed that the electric field can be easily imagined and represented in terms of **lines of force** or **electric field lines**. The field lines indicate the ***direction*** of the electric field as they are thought of to originate from positive charges and to end up on negative charges. In a given region, the number of lines per unit area through a perpendicular surface is proportional to the ***magnitude*** of the electric field in that region. When the field lines are close, the electric field strength, **E** is large; and when the lines are apart, **E** is smaller, as illustrated in Fig. 20.1. The field lines are therefore used to indicate the magnitude of the field as well as its direction.

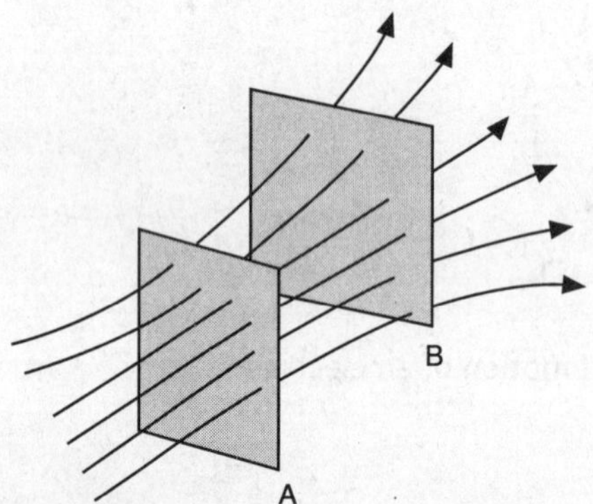

Fig. 20.1:

Field lines in a nonuniform field.
E is larger at A than at B.

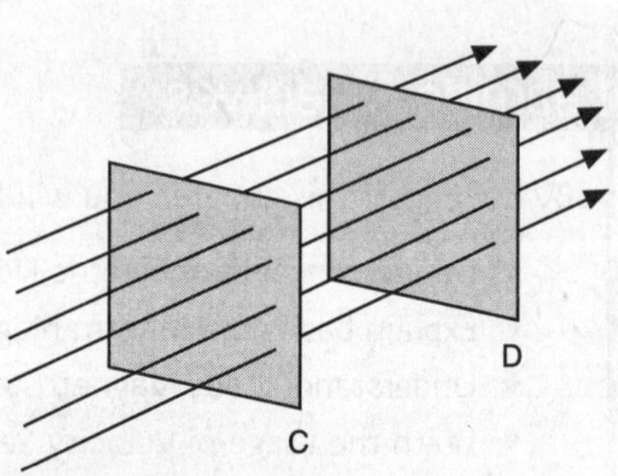

Fig. 20.2:

Field lines in a uniform field.
E is same at locations C and D.

In a non-uniform electric field, the field lines are bent and spread out non-uniformly (Fig. 20.1), whereas, the field lines are straight, parallel, and uniformly spaced in a uniform electric field (Fig. 20.2).

Uniform Electric Field

An electric field in a region is said to be **uniform**, when the field lines are straight and parallel pointing in the same direction; and the strength of the field E is constant in that region. It is also known as a **homogeneous electric field**. For a uniform electric field,

$$\mathbf{E} = \text{constant}.$$

Uniform electric fields are set up by a pair of metal plates held strictly parallel to each other and charged to opposite polarity (Fig. 20.3).

The electric field intensity **E** in the region between the plates A and B (Fig. 20.3) is given by

$$E = \frac{V}{d} \qquad ...(20.1)$$

where $V = (V_A - V_B)$ is the potential difference between the plates A and B; and d is the distance of separation of the plates.

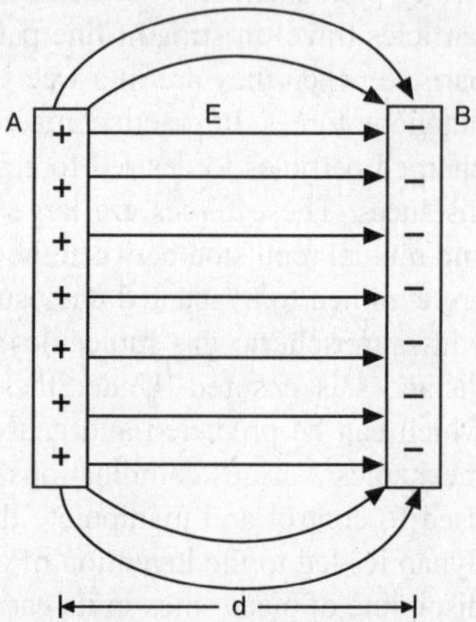

Fig. 20.3:

Uniform electric field producted by a parallel plate capacitor. Field is nonuniform at the edges.

20.3 MOTION OF AN ELECTRON IN A UNIFORM ELECTRIC FIELD

When an electron enters a uniform electric field, it experiences a force and gets accelerated. The path of electron depends upon the angle between the applied field and the initial direction of the electron velocity. We assume that electric field alone acts on the particle and the gravitational force is negligible. The region between the plates is evacuated such that the electron is not deflected away from its path due to collisions with air molecules.

20.3.1 Electric Field Parallel to Initial Velocity

Let an electron of mass m and charge $-e$ enter the region of uniform electric field. In Fig. 20.4, the uniform electric field $\boldsymbol{E}$ is directed from positive plate $\boldsymbol{A}$ to the negative plate $\boldsymbol{B}$.

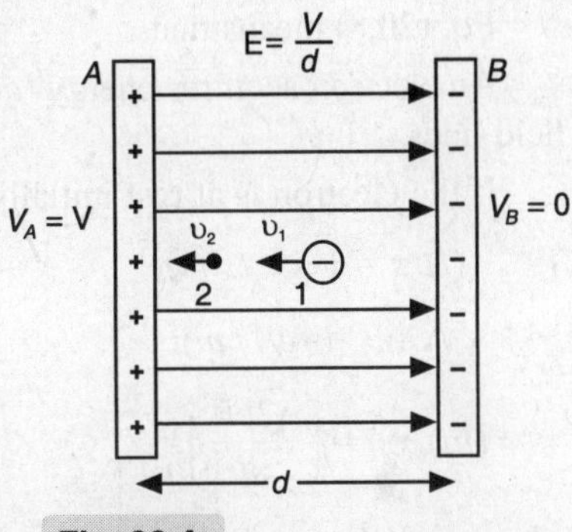

Fig. 20.4: Electron motion in a uniform electric field.

The electron experiences a force **F** due to the electric field.

$$\boldsymbol{F} = -e\boldsymbol{E} \quad ...(20.2)$$

The negative sign indicates that the force acts in a direction opposite to that of electron motion. As a result, the electron moves with acceleration a in the direction opposite to that of the electric field. According to Newton's second law, the equation of motion of the electron is given by

$$m\boldsymbol{a} = -e\boldsymbol{E} \quad ...(20.3)$$

Therefore, the acceleration $\boldsymbol{a}$ is $\boldsymbol{a} = \dfrac{-e\boldsymbol{E}}{m}$

The magnitude of the acceleration is given by

$$\boldsymbol{a} = \frac{e\boldsymbol{E}}{m} \quad ...(20.4)$$

As the parameters e and m in the Eq. 20.4 are constants and the electric field $\boldsymbol{E}$ is uniform, the acceleration a is uniform. Hence, the electron is *uniformly accelerated* in a direction opposite to that of the electric field.

- As the electric field acts along a straight line, the path of the electron is a straight line. The electron motion resembles the motion of a body freely falling in a gravitational field.
- Eq. (20.4) indicates that the acceleration of a charged particle in an electric field depends on the ratio e/m. Since the ratio e/m is different for different charged particles, their accelerations in a given electric field are different. In contrast, the acceleration due to gravity is the same for all bodies.

Equations of kinematics for rectilinear motion can now be applied to the motion of electron in an electric field. The distance traveled by the electron in the field during time t in x-direction (see Fig. 20.4) is

$$x = \upsilon_o t + (-eE/2m)t^2 \quad ...(20.5)$$

where υ_o is the initial velocity of the electron.

For convenience, we assumed that the positive x-axis is towards left in Fig. 20.4. Note that υ_o is also in x-direction.

The velocity of the electron after time t is given by

$$\upsilon = \upsilon_o + (-eE/m)t \qquad ...(20.6)$$

and

$$\upsilon^2 = \upsilon_o^2 + \left(\frac{-2eE}{m}\right)x \qquad ...(20.7)$$

The kinetic energy of the electron after moving through a distance 'x' in the field is

$$\text{K.E.} = \frac{1}{2}m\upsilon^2 = \frac{1}{2}m\upsilon_o^2 + \frac{1}{2}m\left(\frac{-2eEx}{m}\right) = (K.E.)_o + (-e)Ex \qquad ...(20.8)$$

where $(\text{K.E.})_o$ is the initial kinetic energy of the electron.

Eq. (20.8) means that

An electron acquires energy '*eEx*' when it travels in a uniform electric field parallel to the field lines.

If the electron is at rest initially, $\upsilon_o = 0$ and the above equations reduce to

$$\left.\begin{aligned} &(i)\ x = (-eE/2m)t^2 \\ &(ii)\ \upsilon = (-eE/m)t \\ &(iii)\ \upsilon^2 = \left(\frac{-2eE}{m}\right)x \\ &(iv)\ \text{K.E.} = -eEx \end{aligned}\right\} \qquad ...(20.9)$$

As $-Ex = V$, K.E. $= eV$ or $\frac{1}{2}m\upsilon^2 = eV \qquad ...(20.10)$

Eq. (20.10) shows that the potential energy is converted into kinetic energy, when the electron moves from one potential difference to another. Eq. (20.10) helps us in finding the electron velocity υ.

$$\upsilon = \sqrt{\frac{2eV}{m}} \qquad ...(20.11)$$

That is, $\upsilon \propto \sqrt{V}$

Thus, *the velocity acquired by the electron in a uniform electric field is proportional to the square root of the potential difference through which it is accelerated.*

Substituting the values of e and m of an electron into Eq. (20.11), we obtain

$$\upsilon = 5.93\sqrt{V} \times 10^5 \text{ m/s}$$

Example 20.1. *An electron starts from rest and moves freely in an electric field of intensity 1500 V/m. Determine the force on the electron and acceleration attained by the electron.*

Solution:

(*i*) $F = eE = (1.602 \times 10^{-19}\text{ C})(1500\text{V/m})$

$= \mathbf{2.4 \times 10^{-16} N}$

(*ii*) $a = \frac{eE}{m} = \frac{(1.602\times10^{-19}\text{C})(1500\text{ V/m})}{9.11\times10^{-31}\text{ kg}}$

$= \mathbf{2.6 \times 10^{14} m/s^2}$.

Example 20.2: *A proton is projected into a region of a uniform electric field of 3×10^5 N/C. The proton travels 4 cm against the field direction before it comes to rest. Determine (i) the deceleration of the proton, (ii) its initial speed and (iii) the time taken by proton to come to rest.*

Solution:

(*i*) The deceleration of the proton is given by $a = -\dfrac{eE}{m}$

$$\therefore \quad a = -\frac{(1.602\times10^{-19}\,\text{C})(3\times10^5\ \text{N/C})}{1.67\times10^{-27}\,\text{kg}} = -\frac{4.8}{1.67}\times10^{13}\ \text{N/kg}$$

$$\therefore \quad \boldsymbol{a = -2.9 \times 10^{13}\,\text{m/s}^2.}$$

(*ii*) The initial speed is given by $\upsilon_o = \sqrt{-2as}$

$$\therefore \quad \upsilon_o = \sqrt{(-2)(-2.9\times10^{13}\ \text{m/s}^2)(0.04\ \text{m})}$$

$$\therefore \quad \boldsymbol{\upsilon_o = 1.52 \times 10^6\,\text{m/s}.}$$

(*iii*) Time taken by proton to come to rest $t = -\dfrac{\upsilon_o}{a} = \dfrac{1.52\times10^6\ \text{m/s}}{2.88\times10^{13}\ \text{m/s}^2}$

$$\therefore \quad \boldsymbol{t = 5.3\times10^{-8}\,\text{s} = 53\ \text{ns}.}$$

20.3.2 Electron-Volt

An electron (or a charged particle) gains energy when it is accelerated in an electric field; the kinetic energy so gained is very small compared to a joule. Hence, in atomic physics, particle energies are expressed in terms of a smaller unit called electron-volt (eV).

An electron-volt is defined as the energy acquired by an electron when it gets accelerated through a potential difference of one volt. We see that the electron volt is related to the joule through the relation.

1 eV = Charge on the electron × 1 V = $(1.602 \times 10^{-19}$ C) (1 V) = 1.602×10^{-19} J

20.3.3 Electric Field Perpendicular to Initial Velocity

We now consider the case of a uniform electric field applied perpendicular to the initial direction of motion of the electron.

Let A and B be two plane parallel metal plates of length l oriented horizontally and separated by a small distance d. If a potential difference V is applied across the plates, a uniform electric field $E = (V/d)$ is produced in the region between the plates. The electric field is directed vertically downward from plate A to plate B, as shown in Fig. 20.5.

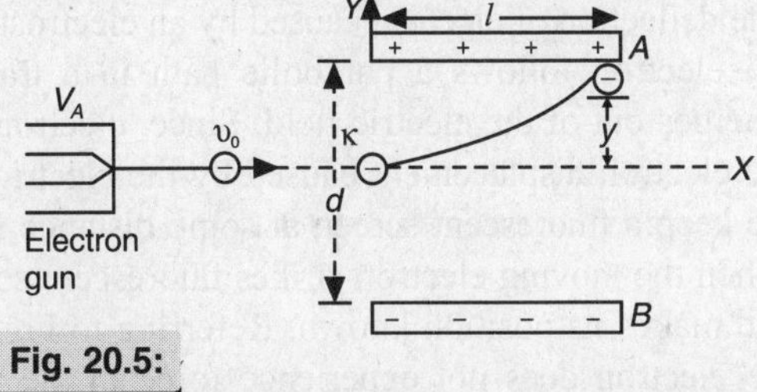

Fig. 20.5: Electron motion in a transverse uniform electric field. The electron describes a parabolic path in a transverse electric field.

Let an electron be moving in x-direction with velocity υ_o. The electron velocity in y-direction is initially zero. At point K, the electron enters the uniform electric field. As the electric field acts in y-direction, the electron experiences an upward force and gets deflected upward in y-direction. The acceleration acquired by the electron in y-direction is given by

$$\therefore \qquad a_y = \frac{eE}{m} \qquad ...(20.12)$$

The velocity attained by the electron after traveling for a time t in the electric field is

$$\upsilon_y = \frac{eE}{m}t \qquad ...(20.13)$$

If y is the **displacement of electron** in the field direction during time t, then

$$y = \frac{eE}{2m}t^2 \qquad ...(20.14)$$

Since the initial velocity υ_o is in a direction perpendicular to the electric field $\boldsymbol{E}$, it remains unchanged. Therefore, the horizontal displacement of electron in x-direction in time t is

$$x = \upsilon_o t \qquad ...(20.15)$$

Therefore, the co-ordinates of the electron after time t are $\left\{x = \upsilon_o t,\ y = \frac{eE}{2m}t^2\right\}$.

t is known as the **transit time** and is given by

$$t = \frac{x}{\upsilon_o} \qquad ...(20.16)$$

Eliminating time t from Eq. (20.14) and Eq. (20.15), we obtain the equation of the path as

$$y = \left(\frac{eE}{2m}\right)\left(\frac{x}{\upsilon_o}\right)^2$$

$$y = \left(\frac{eE}{2m\upsilon_o^2}\right)x^2 \qquad ...(20.17)$$

or

$$y = k\,x^2$$

where k is a constant.

Eq. (20.17) is an equation of a parabola. It means that ***an electron moving with uniform velocity follows a parabolic path when it enters a transverse uniform electric field.***

20.3.4 Electrostatic Deflection

The deflection of electron caused by an electrostatic field is known as **electrostatic deflection.** An electron follows a parabolic path in a transverse uniform electric field and finally, it emerges out of the electric field. Since, electron is an invisible particle, we cannot determine the electron displacement caused by the electric field, say at point M in Fig. 20.6. Therefore, we keep a fluorescent screen at some distance in the path of electron and locate its position. When the moving electron strikes fluorescent screen, it causes a luminous glow on the screen and makes its position known. Referring to Fig. 20.6, the electric field terminates at MN and the electron does not experience force to the right of the plates beyond the line MN. The electron then onwards follows a rectilinear path. It travels along the line MP with a velocity υ and strikes the fluorescent screen at point P.

If the electric field is switched off, the electron moves without deviation and strikes the screen at point Q. Therefore, QP is the linear deflection caused by the electric field. When the line MP is extended backward it cuts the axis KQ at O and the line PO represents the tangent to the parabola KM at M. The angle θ (= $\angle POQ$) represents the angular displacement of the electron path.

In Fig. 20.6, let $QP = D_E$, $OQ = L$, $AB = d$ and $AA' = l$.

From Δ^{le} OQP, $QP = OQ \tan\theta$

$\therefore$ Electrostatic deflection $\quad D_E = L \tan\theta$ ...(20.18)

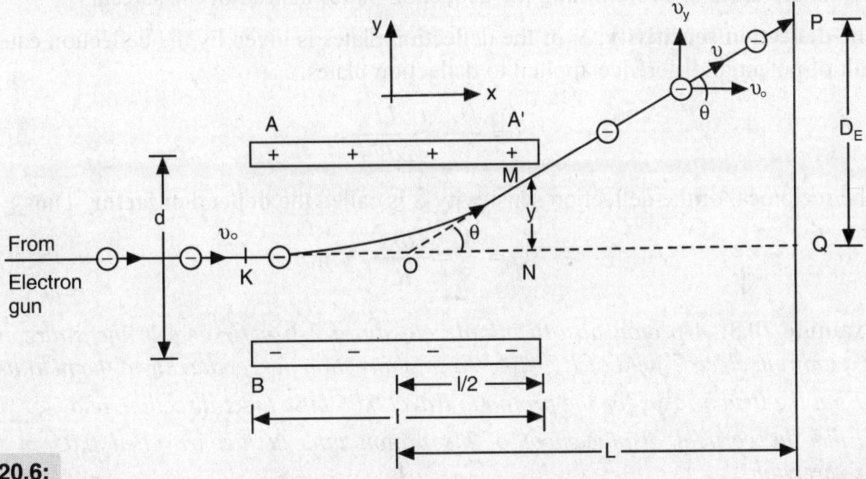

Fig. 20.6: Electrostatic deflection of an electron beam due to uniform electric field acting perpendicular to its path.

But $$\tan\theta = \left[\frac{dy}{dx}\right]_{x=l} = \frac{d}{dx}\left[\frac{eEx^2}{2m\upsilon_o^2}\right]_{x=l} = \frac{eEl}{m\upsilon_o^2} \quad ...(20.19)$$

Using Eq. (20.19) into Eq. (20.18), we get

$$D_E = \frac{LeEl}{m\upsilon_o^2} \quad ...(20.20)$$

If we have a stream of electrons, they have the same e/m value, but may have different initial velocities in the x-direction. Therefore, the electrons are deflected according to their velocities. All electrons with a given value of υ_o, will reach a point P on the screen. If the initial velocity υ_o has been obtained by passing the electron through voltage V_A, we have

$$\upsilon_o = \sqrt{\frac{2eV_A}{m}} \quad ...(20.21)$$

Using the value for υ_o (Eq. 20.21) and $E = V/d$ into the Eq. (20.20), we obtain

$$D_E = \frac{LeVl}{md} \cdot \frac{m}{2eV_A}$$

$$\therefore \quad D_E = \frac{LVl}{2dV_A} \quad ...(20.22)$$

Thus, D_E is proportional to the deflecting voltage V and inversely proportional to the accelerating voltage V_A.

- When the accelerating voltage is smaller, the electron travels through the electric field sufficiently slowly and undergoes appreciable deflection. It is required that V_A be large so that the particles will have sufficient kinetic energy for the production of a luminous spot on the fluorescent screen. But a high accelerating voltage requires a high deflection potential V for causing a sufficient amount of deflection, D_E. Therefore, the value chosen for V_A has to be a compromise to meet these conflicting requirements.

- In electrostatic deflection the angle of deflection of the electron beam is to be limited to smaller values to avoid hitting the deflection plates instead of the screen.

The **deflection sensitivity**, *S*, of the deflection plates is given by the deflection caused by one volt of potential difference applied to deflection plates.

$$S = \frac{D_E}{V} = \frac{Ll}{2dV_A} \qquad ...(20.23)$$

The reciprocal of the deflection sensitivity S is called the **deflection factor**. Thus,

$$G = \frac{1}{S} = \frac{2dV_A}{lL} \text{ V/m.} \qquad ...(20.24)$$

Example 20.3: *A proton has an initial velocity of 2.3 × 10⁵ m/s in the x-direction. It enters a unifrom electric field of 1.5×10⁴N/C in a direction perpendicular to the field lines.*

(*i*) *Find the time it takes for the proton to travel 0.05 m in the x-direction, and*

(*ii*) *Find the vertical displacement of the proton after it has travelled 0.05 m in the x-direction.*

Solution:

(*i*) Time taken by the proton $t = \frac{l}{\upsilon_o} = \frac{0.05 \text{ } m}{2.3 \times 10^5 \text{ m/s}} = \mathbf{2.2 \times 10^{-7} \text{ s}}$

(*ii*) Displacement in the vertical direction due to electric field is

$$y = \frac{eEt^2}{2m} = \frac{(1.602 \times 10^{-19} \text{ C})(1.5 \times 10^4 \text{ N/C})(2.2 \times 10^{-7} \text{ s})^2}{2 \times 1.67 \times 10^{-27} \text{ kg}}$$

$$= 3.48 \times 10^{-2} \frac{(\text{kg.m/s}^2)\text{s}^2}{\text{kg}} = \mathbf{3.5 \text{ cm.}}$$

20.3.5 Electron Projected at an Angle into a Uniform Electric Field

In general, the initial velocity of the electron may not be parallel to *x*- or *y*-axes. Suppose an electron is projected into a uniform electric field at an angle θ, and with an initial velocity υ_0 as shown in Fig. 20.7.

As the electric field direction (Fig. 20.7) is along the positive *y*-direction, the electron gets accelerated in the negative *y*-direction. The acceleration is constant and is given by

$$a = \frac{eE}{m}$$

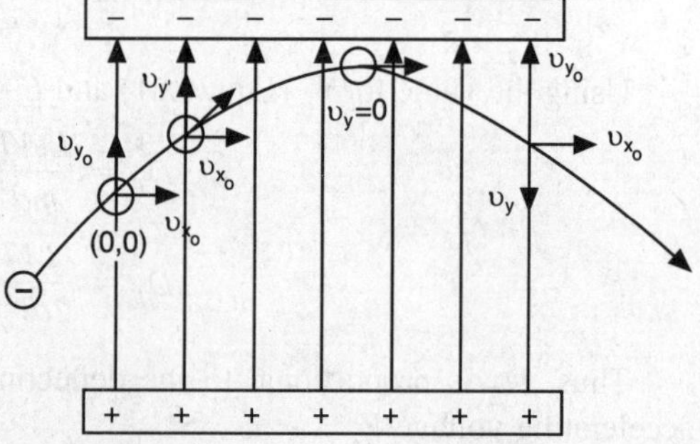

Fig. 20.7: Path of electron projected at an angle into a uniform electric field.

The acceleration is constant and the electron motion in the electric field closely resembles the motion of a projectile in the gravitational field. The component of electron velocity along *x*-axis is $\upsilon_0 \cos \theta_0$ and the component along the y-direction is $\upsilon_o \sin \theta_o$. The horizontal component $\upsilon_o \cos \theta_o$ remains constant during motion, while $\upsilon_o \sin \theta_o$ decreases initially and again increases when the electron reverses its path.

The velocity components are:

(*i*) $\upsilon_x = \upsilon_{x0} = \upsilon_o \cos\theta_o = \text{constant}$ and

(*ii*) $\upsilon_y = \upsilon_{y0} + at = \upsilon_o \sin\theta_o + at$

Therefore, the co-ordinates of the electron after time t are

$$x = \upsilon_x t = (\upsilon_o \cos\theta_o)t \text{ and}$$

$$y = \upsilon_{yo}t + \tfrac{1}{2}at^2 = (\upsilon_o \sin\theta_o)\, t + \tfrac{1}{2}at^2.$$

Eliminating t from the above equations, we get the equation for the electron motion in the electric field. Thus,

$$y = (\tan\theta_o)\, x + \left(\frac{a}{2\upsilon_o^2 \cos^2\theta_o}\right)x^2 \qquad \text{...(20.25)}$$

Eq. (20.25) is of the form $y = ax + bx^2$ which represents a parabola.

Therefore, ***the trajectory of an electron projected into a uniform electric field is a parabola.***

As is the case of a projectile in a gravitational field, the parameters of motion of electron in the uniform electric field may be calculated.

(*i*) The **maximum height** that the electron attains in the uniform electric field is

$$H = \frac{\upsilon_o^2 \sin^2\theta_o}{2a} \qquad \text{...(20.26)}$$

(*ii*) The **time taken** by the electron to reach the maximum height, H is

$$t = \frac{\upsilon_o \sin\theta_o}{a} \qquad \text{...(20.27)}$$

(*iii*) The **time of flight**, T, i.e., the time taken by electron to return to its initial position along x-direction is

$$T = \frac{2\,\upsilon_o \sin\theta_o}{a} \qquad \text{...(20.28)}$$

(*iv*) The **range**, R, i.e., the horizontal distance traveled by electron from the starting position to the point at which it returns to the initial position along x-direction is

$$R = \frac{\upsilon_o^2 \sin 2\theta_o}{a} \qquad \text{...(20.29)}$$

Example 20.4: *An electron projected at an angle of 37° to the horizontal at an initial speed of 4.5 × 10^5 ms/ in a region of uniform electric field 200 N/C.*

(*i*) *Find the time it takes the electron to return to its initial height.*

(*ii*) *Find the maximum height reached by the electron.*

(*iii*) *Find its horizontal displacement when it reaches its maximum height.*

Solution:

(*i*) The time of flight, $T = \dfrac{2m\upsilon_o \sin\theta}{eE} = \dfrac{2(9.11\times10^{-31}\text{ kg})(4.5\times10^5\text{ m/s})(\sin 37°)}{(1.602\times10^{-19}\text{ C})(200\text{ N/C})}$

$$= \frac{49.34}{320.4} \times 10^{-7} \frac{\text{kg.m/ s}}{\text{N}} = 1.54 \times 10^{-6} \frac{\text{kg.m/ s}}{\text{kg.m/ s}^2} = \mathbf{1.54\ \mu\ s}$$

(*ii*) Maximum Height, $H = \frac{m(\upsilon_0 \sin\theta)^2}{2eE} = \frac{(9.11 \times 10^{-31}\,\text{kg})(4.5 \times 10^5\,\text{m/s})^2 (\sin 37°)}{2(1.602 \times 10^{-19}\,\text{C})(200\,\text{N/C})}$

$$= 1.04 \times 10^{-3} \frac{\text{kg.m}^2/\text{s}^2}{\text{kg.m/sec}} = 104 \times 10^{-3} = \mathbf{1.04\ mm}$$

(*iii*) Horizontal displacement at the position of maximum height,

$$s = \frac{(9.11 \times 10^{-31}\,\text{kg})(4.5 \times 10^5\,\text{m/s})^2 \sin 74°}{2(1.602 \times 10^{-19}\,\text{C})(200\,\text{N/C})} = 2.77 \times 10^{-3} \frac{\text{kg.m}^2/\text{s}^2}{\text{kg.m/s}} = \mathbf{2.77\ mm}$$

20.4 UNIFORM MAGNETIC FIELD

A uniform magnetic field is a region of space where the lines of magnetic induction are parallel to each other and the magnetic induction is identical in magnitude and direction at all points in that region. It is also called a *homogeneous magnetic field.*

A uniform magnetic field is produced by a solenoid or an electromagnet.

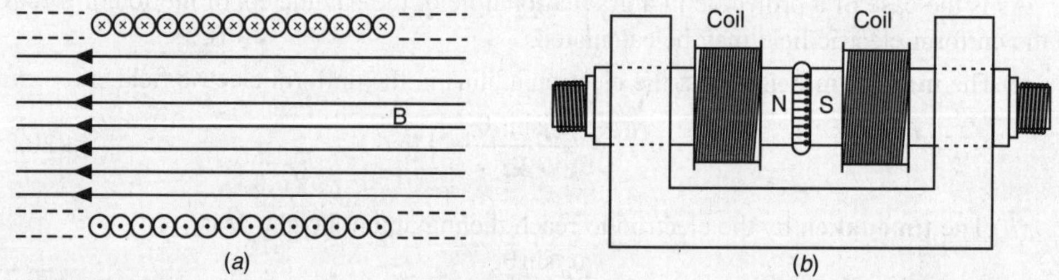

Fig. 20.8: Production of uniform magnetic field – (*a*) by a solenoid and (*b*) by an electromagnet.

20.5 MOTION OF AN ELECTRON IN A UNIFORM MAGNETIC FIELD

A uniform magnetic field is a magnetic field where the lines of magnetic induction are parallel and the magnetic induction at all points in the region is identical in magnitude and direction. A uniform electric field is produced by a solenoid. More frequently, electromagnets are used to produce uniform magnetic fields.

Let an electron enter the region where a uniform magnetic field acts. If θ is the angle between the velocity of the electron and the direction of the uniform magnetic field, then the magnetic force on the electron is given by

$$\mathbf{F} = e(\upsilon \times \mathbf{B})$$

or $$\mathbf{F} = e\upsilon\,\mathbf{B} \sin\theta \qquad ...(20.30)$$

where **B** is the magnetic induction, and υ is the velocity of the electron.

The direction of the magnetic force is determined as follows.

The vectors **υ** and **B** are drawn or visualized from a common point, as shown in Fig. 20.9. Next, a plane is visualized in which the two vectors lie. The force vector lies along a line perpendicular to this plane. To determine the direction of **F**, the vector **υ** is turned into vector **B** through the small angle between them. The direction of **F** is the direction in which a left-hand screw will advance as **υ** rotates into **B**.

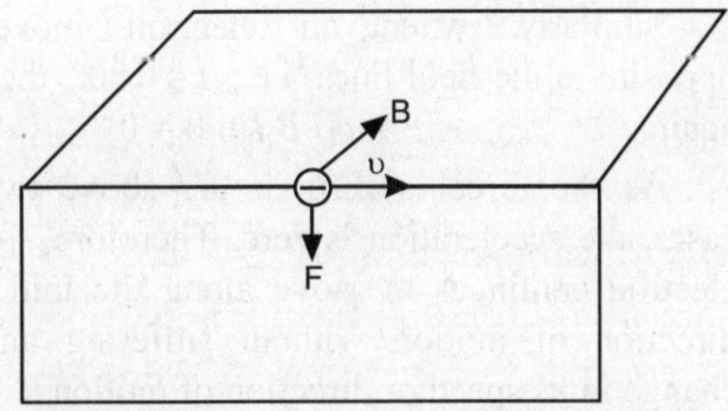

Fig. 20.9: Direction of force acting on a moving electron due to a transverse uniform magnetic field.

Example 20.5: *An electron having velocity 10^6 m/s experiences a maximum force of 1.6×10^{-14} N when it enters a uniform magnetic field. What is the magnitude of the magnetic field?*

Solution: The magnitude of the magnetic field $B = \dfrac{F}{e\,\upsilon}$

$$= \frac{1.6\times10^{-14}\,N}{\left(1.602\times10^{-19}\,C\right)\left(10^{6}\,\text{m/s}\right)} = \mathbf{0.1\ T}.$$

20.5.1 No Energy is Gained by an Electron in a Uniform Magnetic Field

The magnetic force, **F**, is ***always*** perpendicular to the velocity υ and as well as the magnetic field, *B*. Therefore, the work done by this force on the electron is zero. In other words, the magnetic field does not produce any change in either the speed or the kinetic energy of the electron.

$$\mathbf{F}.\upsilon = 0 \qquad \text{...(20.31)}$$

i.e.

$$m\,\mathbf{a}.\upsilon = 0$$

$$m\left(\frac{d\upsilon}{dt}\right)\upsilon = 0 \quad \text{or} \quad \frac{d}{dt}\left(\frac{m\upsilon^2}{2}\right) = 0$$

$$\frac{1}{2}\,m\upsilon^2 = \text{constant} \qquad \text{...(20.32)}$$

It means that ***an electron moves in the magnetic field without acquiring or losing energy.*** It follows that the magnetic field cannot change the speed *v* of the electron. The *only* other parameter that can be affected due to the magnetic field is the direction of the velocity of the electron. Therefore, the effect of the magnetic field on the electron is to change the direction of motion of the electron. The path followed by the electron in a magnetic field depends on the angle at which it enters the magnetic field.

20.5.2 Magnetic Field Parallel to Initial Velocity

When an electron enters a uniform magnetic field parallel to the magnetic field lines, the magnetic force acting on the electron is zero.

As $\theta = 0$, $\quad F = e\upsilon\, B \sin\theta = 0$

Similarly, when an electron moves opposite to the field lines, i.e., $\theta = 180^o$, then again, $F = e\upsilon B \sin\theta = 0$

As the force is zero in the above two cases, the acceleration is zero. Therefore, the electron continues to move along the initial direction of motion without suffering any change in its speed or direction of motion.

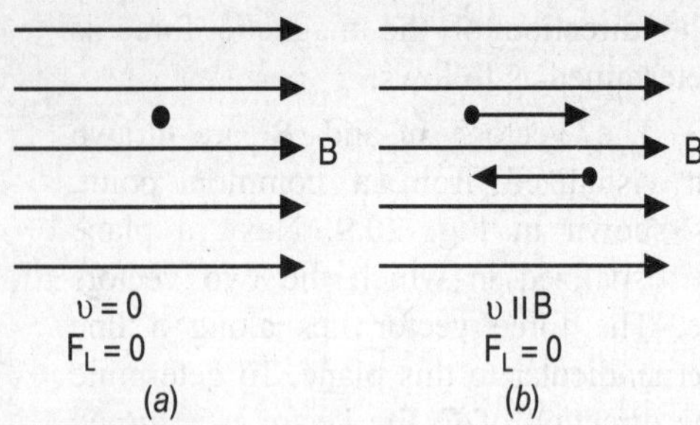

Fig. 20.10:

Electron in a uniform magnetic field. (*a*) When an electron is at rest, it is not acted upon by the magnetic field. (*b*) A longitudinal field B does not act on a charged particle.

20.5.3 Magnetic Field Perpendicular to Initial Velocity

We consider now the case of an electron entering a uniform magnetic field with its initial velocity perpendicular to the field. Then the angle θ between $\mathbf{\upsilon}$ and $\mathbf{B}$ is 90° and $\sin\theta = 1$.

$$\therefore \quad F = e\upsilon B \quad ...(20.33)$$

We assume that the magnetic field **B** is acting into the page (Fig. 20.11). This is indicated by the crosses, which represent the tail of **B** vector. The force **F** due to magnetic field acts perpendicular to the velocity $\mathbf{\upsilon}$ and **B**. This force does not perform work on the electron and hence cannot change the magnitude of electron velocity. However, the action of the force is to deflect the electron from its path and hence change the direction of velocity. After an infinitesimal change of direction, the electron has the same speed and is still moving perpendicular to **B**; but the direction of **F** changes. As the force **F** deflects the electron, the directions of $\mathbf{\upsilon}$ and **F** change continuously and the electron follows a curved path. The force that produces the curvature is **F**. Since the magnetic field is uniform everywhere, and the speed $\mathbf{\upsilon}$ is constant and is unchanged by the field, the magnitude of **F** is constant. As the direction of the force **F** needs to be always perpendicular to the velocity of the particle, it is obviously a *centripetal force* and causes the electron to move in a circular path in a plane perpendicular to the magnetic field. The sense of rotation in Fig. 20.11 is clockwise for an electron. The centripetal acceleration caused by the force due to magnetic field is υ^2/R, where R is the radius of the curvature. The centripetal force **F** is given by

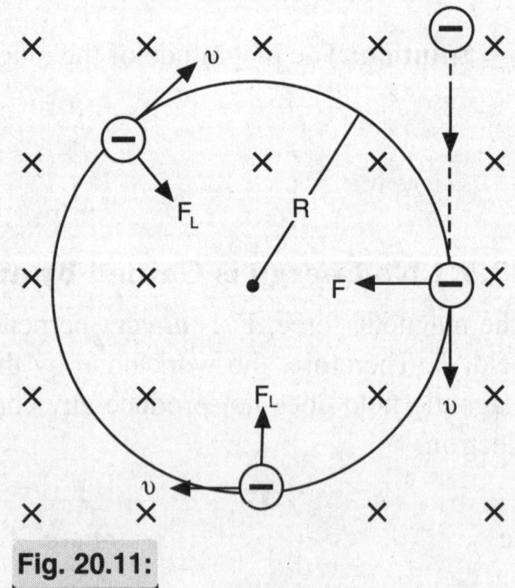

Fig. 20.11:

An electron describes a circular path in a transverse uniform magnetic field.

$$F = \frac{m\upsilon^2}{R} \quad ...(20.34)$$

Comparing Eq. (20.33) with Eq. (20.34), we get

$$e\upsilon B = \frac{m\upsilon^2}{R} \quad ...(20.35)$$

$$\therefore \qquad R - \frac{m\upsilon}{eB}$$

$$R \propto m\upsilon \qquad ...(20.36)$$

Eq. (20.36) shows that the radius of the circle is proportional to the momentum '$m\upsilon$' of the electron.

The larger the momentum, the larger is the radius of the electron path and the smaller is the curvature. It is because it becomes more difficult for the magnetic field to change the direction of motion of the electron with greater momentum.

The time taken by the electron to complete one revolution is called the **time period, T.** It is given by

$$T = \frac{\text{Distance travelled by the electron in one revolution}}{\text{Speed of the electron}}$$

$$= \frac{\text{Circumference of the path}}{\text{Speed of the electron}}$$

$$\therefore \qquad T = \frac{2\pi R}{\upsilon} = \frac{2\pi}{\upsilon} \cdot \frac{m\upsilon}{eB}$$

$$\therefore \qquad T = \frac{2\pi m}{eB} \qquad ...(20.37)$$

The frequency of revolution in the orbit is given by

$$\nu = \frac{1}{T} = \frac{eB}{2\pi m} \qquad ...(20.38)$$

Eq. (20.37) and Eq. (20.38) indicate that

- The frequency of revolution and the time period of an electron do not depend on its velocity or on the radius of circular path.

  This is because υ and R adjust themselves in such a way that for a given magnetic induction value **B**, T and υ remain constant.
- Slower electrons move in smaller circles while faster electrons move in larger circles but all of them take the same time for completion of one revolution.

Example 20.6: *An electron is accelerated through a potential difference of 5 kV and enters a uniform magnetic field of 0.02 wb/m^2 acting normal to the direction of electron motion. Determine the radius of the path.*

Solution:

The radius of the circular path described by the electron in the magnetic field is given by

$$R = \frac{m\upsilon}{eB}$$

The velocity of the electron accelerated through a potential V is given by

$$\upsilon = \sqrt{\frac{2eV}{m}}$$

$$\therefore \qquad R = \frac{1}{B}\sqrt{\frac{2mV}{e}}$$

$$= \frac{1}{0.02 wb/m^2}\left[\frac{2(9.11\times10^{-31}\text{kg})(5\times10^{3}V)}{1.602\times10^{-19}C}\right]^{1/2}$$

$$= 119.3\times10^{-4}\,\frac{\text{C.m}}{\text{N.s}}\frac{\text{N.s}}{\text{C}} = \mathbf{12\ mm.}$$

20.5.4 Relation between K.E. of the Electron and Radius of the Circular Path

Suppose an electron is moving in a circular path under the action of transverse uniform magnetic field *B*. If υ is the speed of the electron, the kinetic energy of the electron is given by

$$E_k = \frac{1}{2}m\upsilon^2 \qquad ...(20.39)$$

But,
$$\upsilon = \frac{eBR}{m} \qquad ...(20.40)$$

∴
$$E_k = \frac{e^2B^2R^2}{2m} \qquad ...(20.41)$$

Note that this is *not* the kinetic energy gained by the electron due to its motion in the magnetic field. The kinetic energy of the electron given by the Eq. (20.39) is the energy that the electron has prior to its entry into the magnetic field, and is expressed in terms of the radius of the circular path in the magnetic field in Eq. (20.41).

20.5.5 Magnetic Field Acting at an Angle to Initial Velocity

Let us consider a general case where an electron enters the uniform magnetic field *B* with its velocity at an angle θ, as shown in Fig. 20.12.

Let an electron enter a uniform magnetic field *B* with a velocity **υ** making an angle θ with the magnetic field direction. The velocity **υ** may be resolved into parallel and perpendicular components with respect to the magnetic field direction. The components are

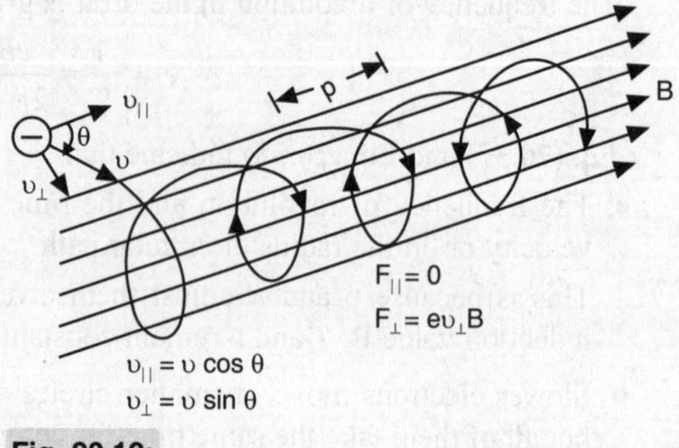

$$\upsilon_{||} = \upsilon\cos\theta$$

and
$$\upsilon_{\perp} = \upsilon\sin\theta$$

Fig. 20.12: An electron entering a uniform magnetic field at an angle describes a helical path.

(*i*) Axial velocity component causes uniform rectilinear motion: The force on the electron in the direction parallel to the magnetic field is zero.

$$F_{\perp} = e\upsilon_{||}B = 0$$

Consequently, the velocity component, $\upsilon_{||}$ which is parallel to the magnetic field, does not undergo a change due to the magnetic field. The electron continues to move along the field direction with the velocity $\upsilon_{||}$. This is rectilinear motion.

(*ii*) Perpendicular velocity component causes circular motion: The force acting on the electron due to the magnetic field in a direction perpendicular to the field is

$$F_{\perp} = e\upsilon_{\perp}B$$

Due to this normal force component, the electron travels along a circular path around the field direction with constant speed $\upsilon_{\perp} = \upsilon\sin\theta$. This is uniform circular motion.

(*iii*) **Resultant motion is helical motion around the magnetic field direction**: The actual motion of the electron is the **resultant** of the above two motions—a uniform rectilinear motion parallel to the field and a uniform circular motion perpendicular to the field. The superposition of circular motion on the rectilinear motion results in motion along a helical or spiral path. Thus, the electron describes a *helical path* in the magnetic field, the axis of the helix being parallel to the magnetic field, B.

The projection of the path onto the xy plane is a circle, whose **radius** is given by

$$R = \frac{m\upsilon_{\perp}}{eB} = \frac{m\upsilon \sin\theta}{eB} \qquad \text{...(20.42)}$$

The projections of the path onto the xz and yz planes are sinusoids.

The **time period** of the revolution is given by

$$T = \frac{2\pi R}{\upsilon_{\perp}} = \frac{2\pi m\upsilon \sin\theta}{eB\upsilon \sin\theta} = \frac{2\pi m}{eB} \qquad \text{...(20.43)}$$

The distance covered in one revolution is the **pitch of the helix**. It is given by

$$p = \upsilon_{\parallel} T = T\upsilon \cos\theta$$

$$\therefore \qquad p = \frac{2\pi m\upsilon \cos\theta}{eB} \qquad \text{...(20.44)}$$

Example 20.7: *2 keV positron is projected into a uniform magnetic field of induction 0.1 T with its velocity vector making an angle of 89° with the field. Find the period, pitch and radius of the helical path of the positron.*

Solution:

Time period is given by $T = \frac{2\pi m}{eB} = \frac{2(3.143)(9.11\times10^{-31}\,kg)}{(1.602\times10^{-19}\,C)(0.1T)}$

$$= 3.6\times10^{-10}\,\frac{kg.m}{s}\,\frac{s^2}{kg.m} = \mathbf{0.36ns}$$

Pitch of the helical path is given by $p = T\upsilon\cos\theta = T(593\times10^3\sqrt{V})\cos\theta$

$$= 3.6\times10^{-10}\,\text{s}\times593\times10^3\times\sqrt{2000}\ \text{m/s}\times\cos 89° = \mathbf{0.20\ mm}$$

Radius of the helical path is given by $R = \frac{m\upsilon}{eB} sin\,\theta$

Radius of helical path $= \frac{(9.11\times10^{-31}\ \text{kg})(593\times10^3\sqrt{2000}\ \text{m/s})\sin 89°}{(1.602\times10^{-19}C)(0.1T)}$

$$= \mathbf{1.5\ mm}$$

Example 20.8: *An electron shot into a uniform magnetic field at an angle 60° moves in a spiral of diameter of 10 cm and with a period of 6×10^{-5}s. Determine the magnetic induction and electron velocity.*

Solution:

(*i*) The period of revolution $T = \frac{2\pi m}{eB}$. $\quad\therefore\quad B = \frac{2\pi m}{eT}$

or $$B = \frac{2\times3.143\times9.11\times10^{-31}\text{kg}}{6\times10^{-5}\text{s}\times1.602\times10^{-19}\text{C}} = \mathbf{5.95\times10^{-7}\ wb/m^2}.$$

(*ii*) The radius of the spiral $R = \dfrac{m\upsilon \sin\theta}{eB}$ $\quad \therefore \upsilon = \dfrac{eBR}{m\sin\theta}$

or $$\upsilon = \frac{(1.602\times10^{-19}C)(5.95\times10^{-7}\,\text{wb/m}^2)(10\times10^{-2}\,\text{m})}{9.11\times10^{-31}\text{kg}(\sin 60°)} = \mathbf{1.2\times10^4 m/s.}$$

20.6 MAGNETOSTATIC DEFLECTION

The deflection produced in the path of electron by a magnetic field is called *magnetostatic deflection.* Fig. 20.13 shows a uniform magnetic field *B* acting in a region of space over a length *l*.

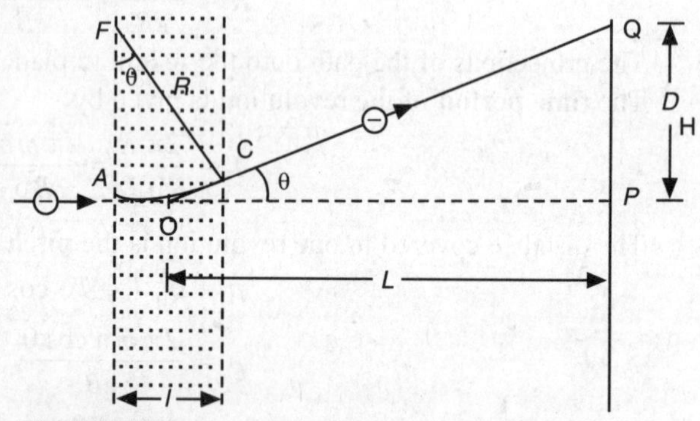

Fig. 20.13: Magnetostatic deflection. The dots represent uniform magnetic field emerging out of the plane of the page.

Let us consider an electron beam traveling in the horizontal direction with velocity υ enters a region where a uniform magnetic field acts in a transverse direction. As the beam travels through the magnetic field, it bends through an arc of radius '*R*'. Before it completes a revolution, it emerges from the field, as the magnetic field is of limited extension. After emerging from the magnetic field, the beam travels along a straight line and strikes a fluorescent screen, say at point *Q*. In the absence of magnetic field, the electron beam would have struck the screen at the point *P*. *PQ* represents the amount of deflection experienced by the electron beam due to the action of magnetic field. Let the amount of deflection be denoted by D_H.

$$\therefore \quad PQ = D_H$$

The circular arc *AC* subtends an angle θ at *F*. The radii, *FA* and *FC*, of the circle have their centre at *F*. *AO* and *OCQ* are tangents to the arc *AC* at points *A* and *C* respectively.

$$\therefore \quad \angle POQ = \theta$$

From the Fig. 20.13 we can write

$$D_H = L\tan\theta \qquad ...(20.45)$$

where *L* is the distance of the screen from the centre of the magnetic field. When θ is small, we can write

$$D_H \cong L\left(\frac{AC}{AF}\right) \cong \frac{lL}{R} \qquad ...(20.46)$$

where *l* (= *AC*) is the extent of the magnetic field. We know that

$$R = \frac{m\upsilon}{eB}$$

and $$\upsilon = \sqrt{\frac{2eV_A}{m}}$$

Using the above relations into Eq. (20.45), we obtain

$$D_H = \frac{LleB}{m\upsilon} = \left[\frac{LleB}{m}\right]\left[\sqrt{\frac{m}{2eV_A}}\right]$$

where V_A is the voltage that caused electron to attain velocity υ.

$$\therefore \quad D_H = LlB\sqrt{\frac{e}{2mV_A}} \qquad ...(20.47)$$

It is seen that *the magnetic deflection D_H is inversely proportional to the square root of the accelerating voltage V_A.*

Following the example of electrostatic deflection, we can define the *deflection sensitivity* in this case also.

$$S_H = \frac{D_H}{B} = Ll\sqrt{\frac{e}{2mV_A}} \qquad ...(20.48)$$

Similarly, the *deflection factor* is expressed as

$$G_H = \frac{1}{S_H} = \frac{1}{Ll}\sqrt{\frac{2mV_A}{e}} \qquad ...(20.49)$$

The magnetostatic deflection is inversely proportional to the square root of accelerating potential V_A.

Comparing with the electrostatic deflection Eq. (20.22), it is readily seen that

- For a given value of accelerating voltage the amount of deflection caused by a magnetic field will be more. Therefore, larger area of the fluorescent screen can be covered employing magnetic deflection than with electrostatic deflection. In view of this, magnetic deflection is employed in picture tubes of TV, radar etc.

Comparison:

(*i*) The angle of deflection of the electron beam in electrostatic deflection is to be restricted to smaller values. Otherwise, at angles greater than a certain value, the electrons hit the deflection plates instead of reaching the screen. Because of this limitation, the area that can be covered on the screen by the electron beam is smaller.
In case of magnetic deflection it is possible to cover larger area on the screen.

(*ii*) In electrostatic deflection, the deflection sensitivity decreases more rapidly with increasing accelerating voltage than in magnetic deflection.

(*iii*) The electrostatic deflection requires little power for deflection. A large power is consumed in the coils for producing magnetic field in case of magnetic deflection.

Example 20.9: *What transverse magnetic field acting over the entire length of a cathode ray tube must be applied to casue a deflection of 3 cm on the screen that is 15 cm away from the anode and if the accelerating voltage is 2000V?*

Solution: Eq. (20.47) may be rewritten in terms of B as

$$B = \frac{D_H\sqrt{2\,mV_A}}{Ll\sqrt{e}} = \frac{(0.03\text{m})\left[2\left(9.11\times10^{-31}\text{ kg}\right)(2000\text{V})\right]^{1/2}}{(0.075\text{m})(0.15\text{m})\left(1.602\times10^{-19}\text{ C}\right)^{1/2}} = 402.2\times10^{-6}\frac{(\text{kg.V})^{1/2}}{\text{C}^{1/2}.\text{m}}$$

$$= 4.02\times10^{-4}\frac{(\text{kg.J/C})}{\text{C}^{1/2}.\text{m}} = 4\times10^{-4}\frac{\text{kg.m}}{\text{C.s.m}} = \mathbf{4\times10^{-4}\ Wb/m^2}\ \left(\text{as 1kg/C.s=1Wb/m}^2\right)$$

20.7 LORENTZ EQUATION

When an electron travels in a region where both electric and magnetic fields act, it will experience both an electric force $e\mathbf{E}$ and a magnetic force $e\,(\boldsymbol{\upsilon} \times \mathbf{B})$. So, the total force is the vector sum of the electric force and the magnetic force. Thus,

$$F_L = e\,(\mathbf{E} + \boldsymbol{\upsilon} \times \mathbf{B}) \qquad ...(20.50)$$

This expression is called the **Lorentz equation** and the force **Lorentz force** because it was identified in this form by *H*. Lorentz.

20.8 CROSSED ELECTRIC AND MAGNETIC FIELD CONFIGURATION

When uniform electric and magnetic fields are perpendicular to each other and act over the same region of space, they are said to be in *crossed configuration.*

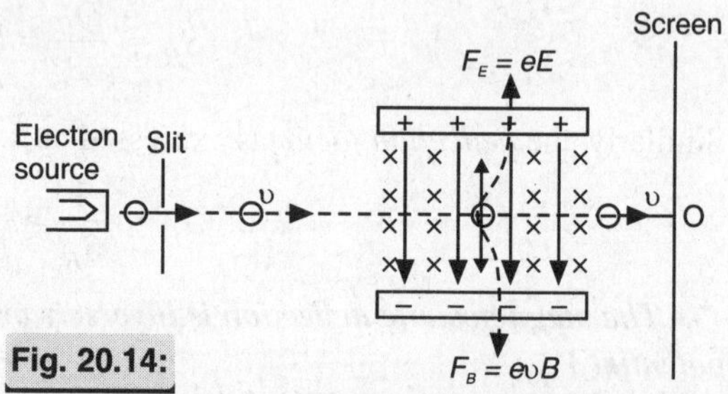

Fig. 20.14: Crossed fields acting normal to electron path. When the forces F_E and F_B balance each other the electron moves along the original straight line path.

Let us consider uniform electric field acting vertically downward, which is set up by a pair of charged parallel plates, as shown in the Fig. 20.14. A uniform magnetic field is produced by, say, a set of Helmholtz coils placed on either side of the electric field plates (not shown in Fig.) and by passing current through the coils. The magnetic field is applied perpendicular to the page and acting into the page. When an electron beam passes through the region along a direction normal to both the field directions (Fig. 20.14), it is simultaneously subjected to the action of the two fields. The electric force acts upward whereas the magnetic force acts downward.

The electron beam is deflected in a direction along which the resultant force acts. The force on the electron due to electric field is

$$F_E = eE$$

The force on the electron due to magnetic field is

$$F_B = e\upsilon B$$

The net force acting on the electron is

$$F_R = F_E - F_B$$

If $F_E > F_B$, the electron is deflected upward. On the other hand, if $F_B > F_E$, it is deflected downward. If the magnitudes of electric field **E** and that of the magnetic field **B** are adjusted such that $F_E = F_B$, then the net force on the electron beam is zero.

$$F_R = F_E - F_B = 0$$

or $$eE - e\upsilon B = 0 \qquad ...(20.51)$$

It means that the forces due to electric field and magnetic field balance each other and the electron beam passes through *E* and *B* fields without any deviation from its initial path. The forces balance each other when

$$eE = e\upsilon B \qquad ...(20.52)$$

or $$E = \upsilon B$$

The velocity of the electron may then be expressed as

$$\upsilon = \frac{E}{B} \tag{20.53}$$

This method of compensating fields was first used by J.J.Thomson in 1897 for measuring the *velocity* of an electron beam. He adjusted the field ratio E/B until the path of the investigated beam became rectilinear in the crossed fields.

Example 20.10: *An electron beam passes through a magnetic field of 2×10^{-3} wb/m² and an electric field of 4×10^4 V/m acting simultaneously. If the path of the electron remains undeviated, determine the speed of the electron.*

Solution: The speed of the undeviated electron is given by $\upsilon = \dfrac{E}{B}$

$$\therefore \quad \upsilon = \frac{4\times10^4\,\text{V/m}}{2\times10^{-3}\,\text{wb/m}^2} = 2\times10^7\,\frac{\text{V/m}}{\text{kg/C.s}} = 2\times10^7\,\frac{\text{N.s}}{\text{kg}} = \mathbf{2 \times 10^7\ m/s}$$

20.9 VELOCITY SELECTOR

A **velocity selector** or **velocity filter** is an electro-optic device which is used for picking a stream of single-velocity charged particles from a beam of particles having a range of velocities. It is based on the principle of crossed uniform electric and magnetic fields.

A narrow beam of electrons traveling in vacuum enters a region where a uniform electric field **E** and a uniform magnetic field **B** are acting at right angles to each other. Let the electrons in the beam have velocities in the range of $\upsilon_o \pm \Delta\upsilon$ spread around a central value υ_o. Let the beam enter the crossed field region in a direction normal to both the fields, as shown in Fig. 20.15.

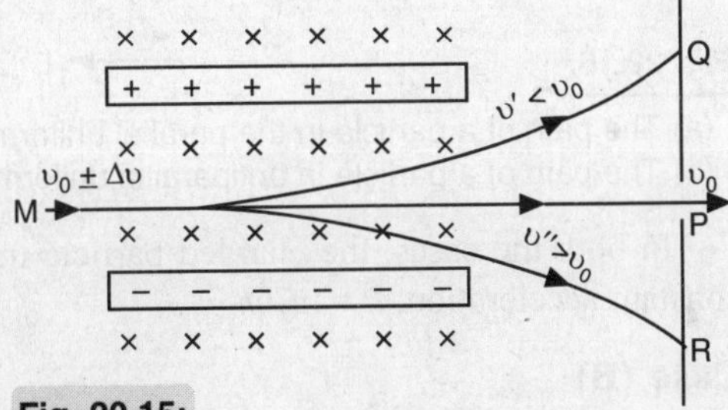

Fig. 20.15: Schematic of a velocity selector.

The uniform electric field **E** is produced in the vertical direction by a set of charged parallel plates. The uniform magnetic field **B** is applied perpendicular to **E** over the same region of space. Both **E** and **B** act normal to the direction of velocity υ_o of electrons. The electric field produces an upward force on the electrons in the beam, whereas the magnetic field produces a downward force. Now, if the field values are adjusted so that the electric force balances the magnetic force, then the electrons will not be subjected to any net force. Thus, the resultant force is

$$F_R = F_E - F_B = 0$$

or

$$eE = e\upsilon_o B$$

$$\therefore \quad \upsilon_o = \frac{E}{B}$$

Thus, there exists a unique velocity υ_o for which the electric and magnetic forces exactly cancel. Hence, the electrons traveling with the velocity $\upsilon_o = E/B$ pass through the region of fields without suffering any deflection. Subsequently, they pass through the slit *QR*. Electrons traveling with velocities less than υ_o, spend more time in the electric field region and are subject to electric force for a longer time. As a result, they get deflected upward and hit the slit walls at points such as *Q*. The force due to magnetic field is proportional to the velocity of electron. Therefore, electrons moving with velocities greater than υ_o are deflected more by the magnetic field. Hence, electrons having larger velocities are deflected downward and strike the slit at points such as *R*. The deflected electrons are absorbed by the slit walls. Electrons that have velocity υ_o are not deflected and pass through the aperture at *P*. The electron beam emerging from the aperture has a strictly single velocity.

20.10 PARALLEL ELECTRIC AND MAGNETIC FIELD CONFIGURATION

Case (A)

(*i*) If the initial velocity of the particle is zero, it is accelerated by the electric field but does not experience force due to magnetic field, because the particle has velocity parallel to magnetic field.

(*ii*) If the initial velocity of the particle is directed along the fields the magnetic field does not exert force on the particle. The resultant motion solely depends on the electric field.

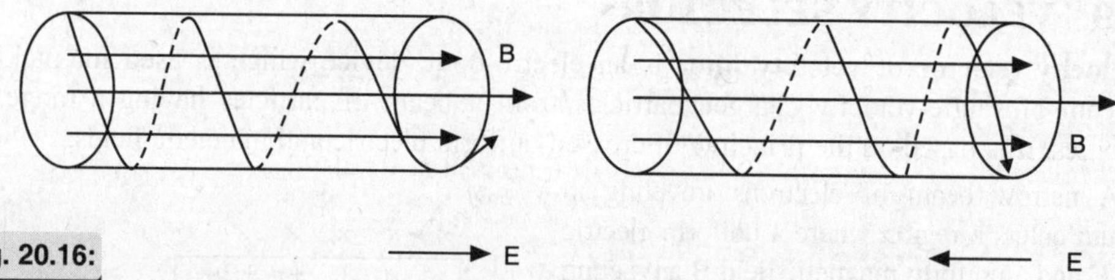

Fig. 20.16:

(a) The path of a particle in the parallel uniform electric and magnetic fields.
(b) The path of a particle in antiparallel uniform electric and magnetic fields.

In both the cases, the charged particle travels in a direction parallel to the fields with a constant acceleration, $a = eE/m$.

Case (B)

If the charged particle has an initial velocity component perpendicular to the magnetic field, then this component gives rise to a circular motion. The radius of the path is given by

$$r = \frac{m\upsilon \sin\theta}{qB}$$

which is independent of **E**.

However, the velocity component parallel to the field direction goes on changing continuously because of the acceleration due to the electric field. It may be noted that the resulting motion occurs along a helical path having a pitch $p = \upsilon_{11} T$, which changes with time.

$$\upsilon_{\parallel} = \frac{qE}{m}t \qquad \text{...(20.54)}$$

20.11 E/M OF ELECTRON

Thomson's apparatus consisted of a highly evacuated glass tube, similar to the one shown in Fig. 20.17. Electrode *C* is the cathode from which electrons emerged. Electrode *A* is the anode, which was maintained at a high positive potential. The high potential difference applied between the cathode *C* and anode *A* caused emission of electrons from the surface of the cathode. These electrons are accelerated and hit the electrode A but some of them passed through the aperture in A. These rays were further collimated by another electrode A', in which there was another aperture. The narrow electron beam passed through a region between plates P and P' and hit the fluorescent coating on the opposite side of the tube. A luminous spot is produced at point O.

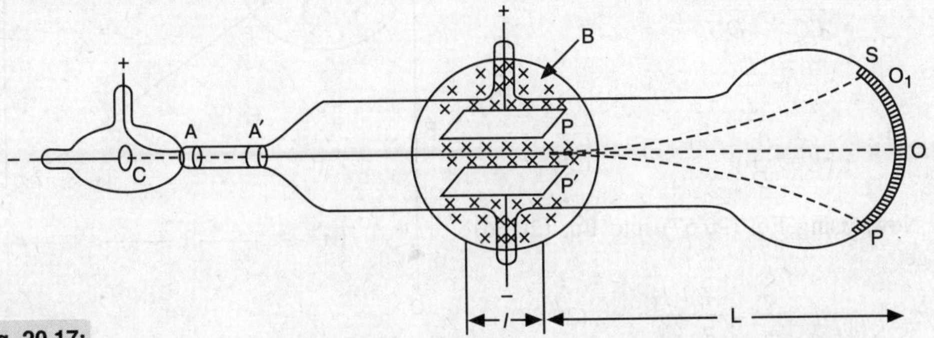

Fig. 20.17: Schematic diagram of Thomson's apparatus for the determination of e/m of electron.

A uniform electric field **E** was set up in the region between the plates P and P' when a potential difference is applied across them. A pair of Helmholtz coils was placed on either side of the tube. When a current was passed through the coils, a uniform magnetic field **B** was set up perpendicular to **E** in the same region where electric field acted. Now let us say the electric field alone is switched on and the electrons are deflected upward along OO_1. Then the magnetic field is switched on and by properly adjusting the magnetic field B, we can make the magnetic force equal to electric force. Then the two forces balance each other and the luminous spot returns from O to O_1. For the condition of balance

$$F_R = F_E - F_B = 0$$

$$\therefore \quad eE = e\upsilon B$$

$$\therefore \quad \upsilon = \frac{E}{B}$$

The electric field is now switched off. The electron beam experiences now a force $F = e\upsilon B$ due to the magnetic field alone. It gets deflected along a circular arc of radius R and strikes the screen at P. The centre of curvature of the trajectory is at C and radius, R, of the curvature of the path is given by

$$R = \frac{m\upsilon}{eB}$$

$$\therefore \quad \frac{e}{m} = \frac{\upsilon}{BR} \qquad \text{...(20.55)}$$

Using Eq. (20.54) into Eq. (20.55), we get

$$\frac{e}{m} = \frac{E}{B^2 R} \qquad \text{...(20.56)}$$

20.11.1 Determination of R

From Fig. 20.18 it is seen that CE and CF are radii of the arc of the electron path in the magnetic field. EO is tangent to the arc at E while GP is the tangent to the same arc at F. As the radius CE moved to a position CF through angle θ, the tangent EO moves through the same angle θ and goes to the position GP.

$\therefore \quad \angle ECG' = \angle OGP = \theta$

$\angle G'EC = \angle GOP = 90°$

$\therefore \quad \Delta^{le}\, ECG' \equiv \Delta^{le}\, OGP$

$$\therefore \quad \frac{EG'}{EC} = \frac{OP}{OG}$$

$$\frac{l}{R} = \frac{D}{L}$$

or
$$R = \frac{lL}{D} \qquad ...(20.57)$$

Now, using Eq. (20.57) into Eq. (20.56) we get

$$\frac{e}{m} = \frac{E}{B^2} \cdot \frac{D}{lL} \qquad ...(20.58)$$

As $E = V/d$, we may write Eq. (20.58) as

$$\frac{e}{m} = \frac{VD}{LldB^2} \qquad ...(20.59)$$

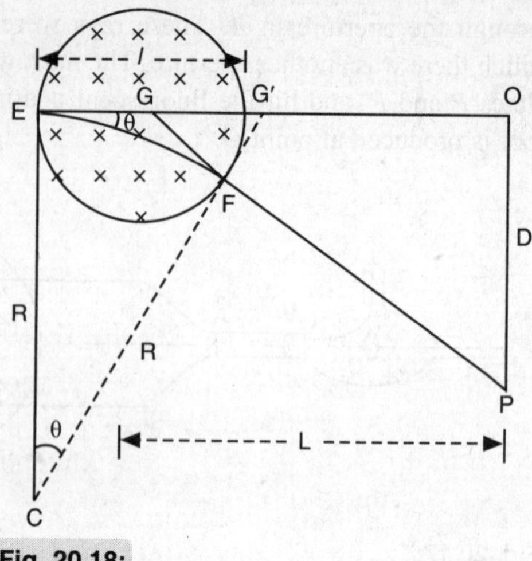

Fig. 20.18: Determination of radius of the circular path from the geometry of magnetostatic deflection.

The quantities *l*, *L*, and *d* are geometric quantities. The voltage *V*, magnetic field *B*, and deflection *D* are measurable quantities. Knowing the values of these quantities, the value of *e/m* can be computed. This experiment provided one of the first reliable experiments for measuring *e/m*. Thomson measured *e/m* for electrons and found a unique value for this quantity which was independent of the cathode material and the residual gas in the tube. The modern value of *e/m* of electrons is found to be 1.759×10^{11} C/kg.

Example 20.11: *A beam of electrons passes undeflected through two mutually perpendicular electric and magnetic fields. If the electric field is cut off and the same magnetic field is maintained, the electrons move in a circular path of radius 1.14 cm. Determine the ratio e/m, if E = 8kV/m and B = 2 × 10⁻³ T.*

Solution:

The velocity of the undeflected electrons is given by $\upsilon = E/B$.

The radius of the circular path is given by $R = \dfrac{m\upsilon}{eB} = \dfrac{mE}{eB^2}$.

$$\therefore \quad \frac{e}{m} = \frac{E}{RB^2} = \frac{8 \times 10^3\ \text{V/m}}{(0.014\,\text{m})(2 \times 10^{-3}\,\text{T})^2} = \mathbf{1.754 \times 10^{11}\ C/kg.}$$

20.12 CHARGE OF THE ELECTRON

Thomson's experiment enabled the determination of the value of *e/m* but from it either the charge *e* or mass *m* of the electron could not be determined independently. In 1920, Robert A.Millikan (1868-1953), the American physicist determined the magnitude of the elementary charge with high accuracy, using charged oil droplets supported by electric field between two parallel horizontal plates. The method is now popularly known as **Millikan oil drop method**.

Fig. 20.19 shows a schematic of the set up used by Millikan. It mainly consists of an observation chamber which contains two optically plane metal discs *A* and *B* of about 20 cm in

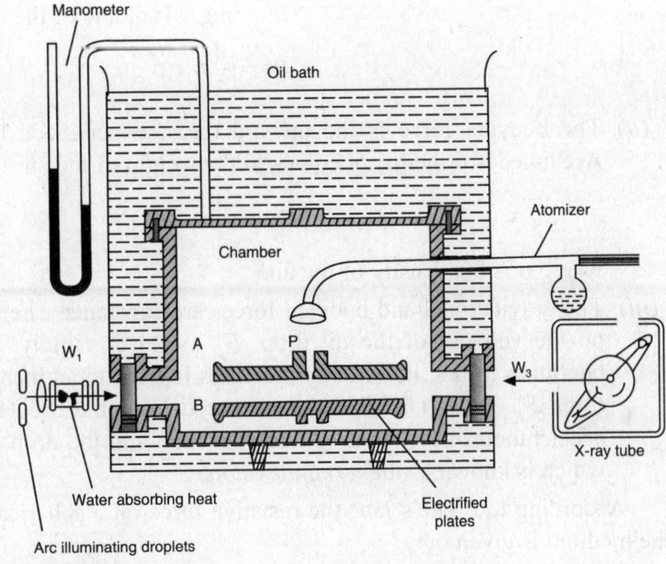

Fig. 20.19: Schematic of Millikan oil drop method set up.

diameter. The discs are held strictly parallel to each other at a distance of about 1.5 cm with the help of insulating rods of glass or ebonite. The upper disc *A* is provided with a tiny hole. The observation chamber is provided with three windows W_1, W_2 *and* W_3. One of the windows facilitates illumination of the chamber by an intense beam of light. The second window permits observation of the region between the discs. The air in this region can be ionized by X-rays which are allowed through the third window. An atomizer is used to spray minute droplets of heavy nonvolatile oil which find their way into the region between the discs through the hole *P*. The droplets of oil get charged due to the frictional effects at the nozzle of the atomizer. A high voltage power supply establishes a suitable potential difference between the discs *A* and *B*. The chamber is enclosed in a thermostat in order to avoid disturbances by thermal convection currents.

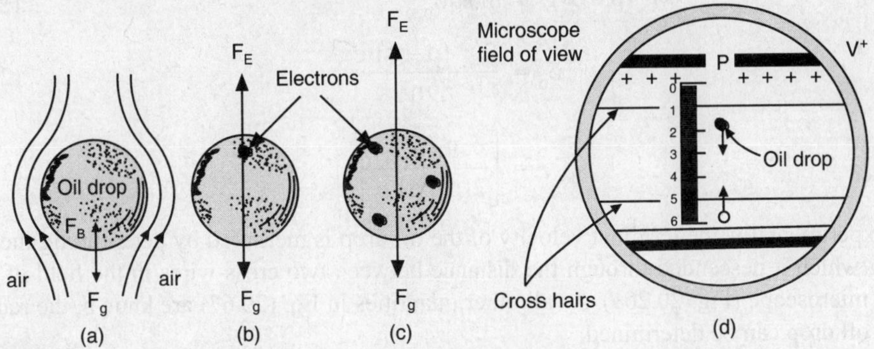

Fig. 20.20: Three forces acting on oil drop.

Now, let the atomizer spray minute oil droplets into the closed space between the horizontal discs *A* and *B*. The oil drops are illuminated by a light and one of them is selected for observation through a microscope provided with a millimeter scale in the eyepiece. As an oil droplet falls in air, three forces act on it as shown in Fig. 20.20. They are the force of gravity *W*, the buoyant force F_B and the force of resistance F_R. The net force

$$F = W - F_B - F_R \qquad ...(20.60)$$

(*i*) If ρ is the density of the oil, *r* the radius of the oil drop, then the force of gravity is given by

$$W = mg = (\text{volume of the drop} \times \text{density of oil})\, g$$

or $$W = \frac{4}{3}\pi r^3 \rho g \qquad ...(20.61)$$

(*ii*) The buoyant force is the upward thrust experienced by the oil drop. According to Archimedes principle, it equals to the weight of the air displaced by the oil drop. Thus,

$$F_B = \frac{4}{3}\pi r^3 \sigma g \qquad ...(20.62)$$

where σ is the density of the air.

(*iii*) The gravitational and buoyant forces are constants. The force of resistance F_R depends on the velocity of the oil drop. F_R increases rapidly with velocity. As the oil drop continues to fall, its velocity increases leading to an increase in F_R. When the retarding force F_R equals $(W-F_B)$, the net force F on the drop becomes zero. At this instant, acceleration of the oil drop becomes zero and the drop falls with constant velocity υ_o which is known as the *terminal velocity.*

According to Stoke's law, the resistive force on a spherical body due to the viscosity of the medium is given by

$$F_R = 6\pi r \eta \upsilon_o \qquad ...(20.63)$$

where υ_o is the terminal velocity of the body, and η is the coefficient of viscosity of air.

Under equilibrium, the net force on the oil drop is zero.

$$W - F_B - F_R = 0$$

$$\frac{4}{3}\pi r^3 \rho g - \frac{4}{3}\pi r^3 \sigma g - 6\pi r \eta \upsilon_o = 0 \qquad ...(20.64)$$

$$\frac{4}{3}\pi r^3 (\rho - \sigma) g = 6\pi r \eta \upsilon_o \qquad ...(20.65)$$

∴ $$\upsilon_o = \frac{2 r^2 (\rho - \sigma) g}{9\eta} \qquad ...(20.66)$$

or $$r = \left[\frac{9\eta \upsilon_o}{2(\rho - \sigma) g}\right]^{\frac{1}{2}} \qquad ...(20.67)$$

Experimentally, the terminal velocity of the oil drop is measured by determining the time during which it descended through the distance between two cross-wires in the field of view of the microscope (Fig. 20.20*d*). As all other quantities in Eq. (20.67) are known, the radius r of the oil drop can be determined.

Now, let the electric field be switched on. If the drop carries a charge q, it experiences an additional upward force $F_E = -qE$ along with F_B and F_R. Let the terminal velocity be reduced to υ_E.

The equation of motion of the oil drop in electric field is given by

$$\frac{4}{3}\pi r^3 (\rho - \sigma) g - 6\pi r \eta \upsilon_E - qE = 0$$

∴ $$\frac{4}{3}\pi r^3 (\rho - \sigma) g = 6\pi r \eta \upsilon_E + qE \qquad ...(20.68)$$

Using Eq. (20.65) into Eq. (20.68), we get

$$6\pi r \eta \upsilon_o = 6\pi r \eta \upsilon_E + qE$$

$$\therefore \qquad q = \frac{6\pi r\eta}{E}(\upsilon_o - \upsilon_E) \qquad ...(20.69)$$

Using Eq. (20.67) for r into Eq. (20.69) one obtains

$$q = \frac{6\,\pi\eta}{E}\left[\frac{9\eta\upsilon_o}{2(\rho-\sigma)g}\right]^{\frac{1}{2}}(\upsilon_o - \upsilon_E) \qquad ...(20.70)$$

or

$$q = \frac{9\pi}{E}\left[\frac{2\eta^3\upsilon_o}{(\rho-\sigma)g}\right]^{\frac{1}{2}}(\upsilon_o - \upsilon_E) \qquad ...(20.71)$$

When the drop begins to move upward instead of downwards, the equation of motion is

$$qE - \frac{4}{3}\pi r^3(\rho-\sigma)g = 6\,\pi\, r\,\eta\,\upsilon_E \qquad ...(20.68a)$$

which yields

$$q = \frac{9\,\pi}{E}\left[\frac{2\eta^3\upsilon_o}{(\rho-\sigma)g}\right]^{\frac{1}{2}}(\upsilon_o + \upsilon_E) \qquad ...(20.71a)$$

Thus, by measuring the speed of free fall of an oil droplet υ_o and the speed υ_E in a known electric field of strength E, one can find the charge on the droplet. In measuring the speed υ_E at a certain value of the charge q, Millikan ionized the air in the region with X-rays. New ions adhere to the droplet under observation and changed its charge, as a result of which the speed υ_E is changed. After measuring the new value of υ_E, the space between the discs was again irradiated and so on. Repeating the measurements a great number of times, Millikan established that the charge q on an oil droplet and the changes in charge Δq were some integral multiples of the same elementary charge. The value of the elementary charge obtained by Millikan was $e = 1.592 \times 10^{-19}$ C, quite close to the best modern result

$$e = 1.602 \times 10^{-19}\ \text{C} \qquad ...(20.72)$$

Millikan was awarded the Nobel prize in Physics in 1923.

20.13 MASS OF THE ELECTRON

From the values of (e/m) and e, one can compute the mass of the electron.

$$m = \frac{e}{e/m} = \frac{1.602\times10^{-19}\,\text{C}}{1.759\times10^{11}\ \text{C/kg}}$$

$$\therefore \qquad m = 9.109 \times 10^{-31}\ \text{kg} \qquad ...(20.73)$$

The mass as given by Eq. (20.73) is known as the **rest mass** of the electron. It is $\left(\frac{1}{1836}\right)^{th}$ of the mass of the hydrogen atom.

20.14 RADIUS OF THE ELECTRON

If the electron is assumed to be a charged particle having a spherical shape, its electrostatic potential energy is given by

$$E = \frac{e^2}{4\pi\,\varepsilon_o r}$$

where r is the radius of the electron.

According to Einstein's mass–energy equivalence relation, the electron rest mass energy must be equal to its electrostatic potential energy. Therefore,

$$E = m_o c^2 = \frac{e^2}{4\pi \varepsilon_o r}$$

or
$$r = \frac{e^2}{4\pi \varepsilon_o m_o c^2} \qquad ...(20.74)$$

$$= \frac{\left(1.602\times10^{-19}\text{C}\right)^2 \left(9\times10^9 \text{N.m}^2/\text{C}^2\right)}{\left(9.11\times10^{-31}\text{kg}\right)\left(3\times10^8 \text{m/s}\right)^2}$$

$$\therefore \qquad r = 2.82 \times 10^{-15} \text{ m} \qquad ...(20.75)$$

Estimates from other methods give the same order of magnitude.

20.15 POSITIVE RAYS

E. Goldstein (1850-1930) the German physicist designed a special discharge tube while investigating the properties of cathode rays and discovered positive rays in 1886. A perforated cathode is kept between an anode and a fluorescent screen in the discharge tube, as illustrated in Fig. 20.21. When a discharge is initiated in the tube, scintillations are observed on the screen. Goldstein discovered that the scintillations were due to rays passing through the canal in the cathode and hence called them **canal rays**.

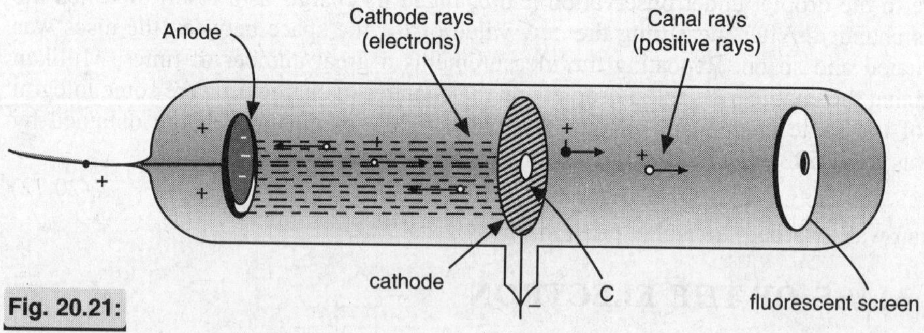

Fig. 20.21: Production of positive rays.

In 1896, W.Wien observed that the canal rays could be deflected by a magnetic field and from the direction of deflection he established that the canal rays consisted of positively charged particles. Therefore, the canal rays are commonly known as **positive rays**. The origin of the positively charged particles is as follows. In the discharge tube, the electrons released by the cathode are accelerated on the way to the anode. These electrons collide with the atoms of the gas in the tube. In the process, electrons are knocked off the neutral atoms which become positively charged. Any process by which an electron is removed from an atom is called **ionization** and the resulting positively charged particle is called a **positive ion**. The positive ions produced during electron-atom collisions are accelerated in the direction of the cathode. Some of the ions traverse the hole in the cathode. They hit the fluorescent screen causing tiny spark of light. It is hence understood that the positive rays are positively charged atoms of the gas contained in the discharge tube.

Since the mass of the electron is very small, the mass of the positive ion will be almost equal to the mass of the atom. Therefore, determination of the mass of positive ion yields the mass of the corresponding atom. The mass M and electric charge q of the positive ions can be determined by deflecting them in electric and magnetic fields.

The method of determining the charge to mass ratio using crossed field configuration is suitable only when all particles in a beam posses the same speed. In case of electrons, all electrons are emitted at cathode and all of them are accelerated by the same potential difference applied between the cathode and anode. Consequently, all the electrons in a beam possess almost the same speed. Positive ions, on the other hand, are formed as a result of ionization of atoms of a gas taking place at different regions in the discharge tube. Therefore, some of the positive ions would be accelerated through only a small part of the field, some through the entire field and others in between. The resulting positive ions consist of a stream of ions with a wide range of speeds.

20.16 THOMSON'S PARABOLA METHOD

In 1911, J.J. Thomson measured the relative masses of different atoms by using a parallel arrangement of electric and magnetic fields.

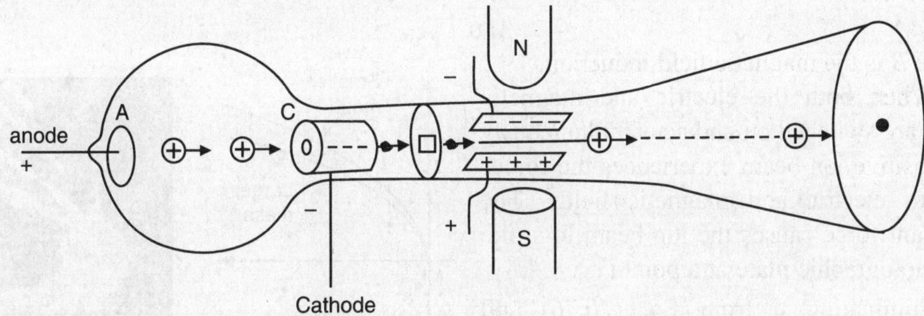

Fig. 20.22:

Schematic of the apparatus of Thomson parabola method.

Description: The apparatus used by Thomson is illustrated in Fig. 20.22. The discharge tube, in which positive rays are produced, is a large flask. A perforated cathode C is placed in the neck of the flask. An anode A is held opposite to the cathode at the head of the flask. The discharge tube is evacuated with the help of a vacuum pump. A small quantity of the gas, the mass of whose atoms is to be determined, is admitted into the tube. A pair of horizontal metal plates is arranged in the narrow tube beyond the cathode. When a potential difference is applied across these plates, a vertically acting electric field will be set up in the region. The narrow portion of the tube is held between the pole-pieces of an electromagnet in such a way that the magnetic field acts in the vertical direction over the same region where the electric field would be acting. A photographic plate is arranged at the far end of the chamber.

Working: When a large potential difference is applied between the cathode and anode, electrons emitted from the cathode ionize the gas atoms in the region between the anode and the cathode. Getting accelerated toward the cathode, many of the positive ions pass through the narrow hole in the cathode. The narrow pencil of positive rays then passes through electric and magnetic fields and strikes the photographic plate. When the electric and magnetic fields are not switched on, the positive ions will strike the photographic plate at a point in line with the initial direction of the beam.

When electric field alone is switched on the ions will be deflected upwards and hit the photographic plate at a distance y from the undeflected position. The deflection y is given by

$$y = \frac{EqLl}{M\upsilon^2} \qquad ...(20.76)$$

where E is the strength of the electric field,

L, the distance from the centre of the field to the photographic plate,

l, the length of the electric field region/plates,

υ, the velocity of the positive ion,

q, the charge of the ion,

M, the mass of the ion.

When magnetic field alone is switched on the ions will be deflected into 'the pages' and hit the photographic plate at a distance x from the undeflected spot. x is given by

$$x = \frac{BqLl}{M\upsilon} \qquad ...(20.77)$$

where B is the magnetic field induction.

When both the electric and magnetic fields are switched on and act simultaneously the positive ion beam experiences the forces due to electric and magnetic fields. The resultant force causes the ion beam to strike the photographic plate at a point (x, y).

Eliminating υ from Eq. (20.76) and Eq. (20.77), we get

$$y = \frac{E}{B^2Ll}\cdot\frac{M}{q}\cdot x^2 \qquad ...(20.78)$$

This is the equation of a parabola. Therefore, ions having identical values of q/M and different values of speed produce a trace in the form of a parabola on the photographic place. If there are ions of different masses but with the same charge, different parabolas are traced. Fig. 20.23(*b*) shows the first parabolas obtained by Thomson.

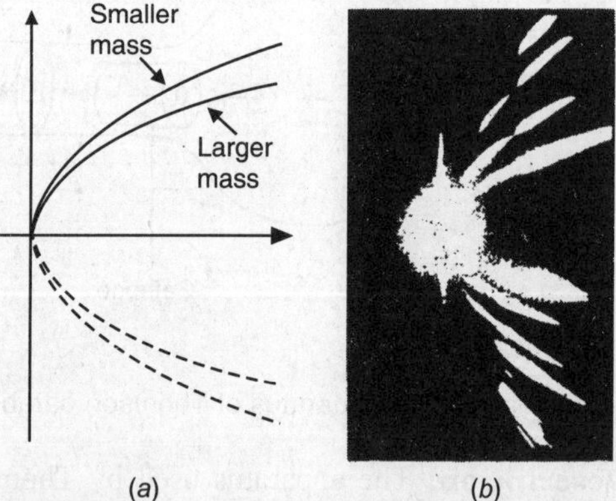

Fig. 20.23: (a) The locus of ions on the face of the positive ray tube is one of half of a parabola. Ions with different values of *e/m* fall on different parabolas. (b) Photography of parabolas with ions of xenon, krypton, argon, and neon.

We can rewrite Eq. (20.78) as

$$\frac{q}{M} = \frac{x^2}{y}\frac{E}{B^2L\,l} \qquad ...(20.79)$$

The value of q/M can be found by measuring the coordinates of the parabolic trace and using them into Eq. (20.79) along with the known values of E, B, L and l. In practice, one parabola associated with ions of known mass is identified and all the other parabolas are compared with it and their masses are deduced. Usually hydrogen is present in the discharge tube and the hydrogen parabola is taken as the standard. Then the mass of the ions of the gas under investigation is found in terms of the hydrogen ion mass.

In Fig. 20.24, let the parabola corresponding to hydrogen ions be *AB*, while the ion of unknown mass but of the same charge as the hydrogen ion produce the parabola *CD*. Since hydrogen is the lightest element, its displacement in magnetic field is more and the parabola *AB* is located at the top. The initial line *OX* cannot be identified on the plate. When the direction of magnetic field is reversed, the relevant coordinate reverses its sign and parabolas *GH* and *EF* symmetrical to the initial ones are obtained. A vertical line such as *GA* is drawn to cut the parabolas. Then,

$$\frac{e/M_H}{e/M_u} = \frac{y_H^2/x}{y_u^2/x} = \frac{y_H^2}{y_u^2} \quad ...(20.80)$$

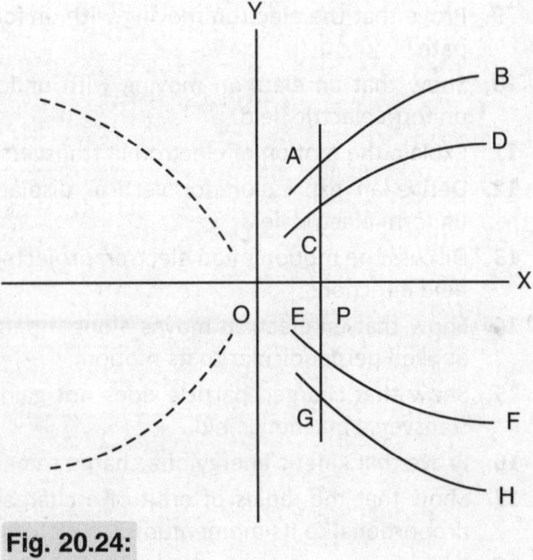

Fig. 20.24: Parabolic traces.

By measuring *GA* and *EC* on the photographic plate, and using the values into Eq. (20.80), M_u can be readily determined.

In 1912, Thomson attempted to determine the atomic weight of neon using parabola method. He discovered two parabolas, a strong one corresponding to a mass 20 and a weaker one corresponding to mass 22. Thus, for the first time, Thomson found two kinds of neon atoms, identical in chemical nature and having the same optical spectra but different in mass. In fact, it was shown later by Thomson's coworker, F.W.Aston that there are three kinds of neon atoms: 90.5% have a mass value of 20, 9.2% have a mass 22 and 0.3% have a mass 21.

Fredrick Soddy (1877-1956), the British Chemist, gave the name **isotopes** to the atoms of different weight belonging to the same element. Thus, neon has three isotopes. Thomson's method provided the first evidence to the existence of stable isotopes.

QUESTIONS

1. What is Lorentz force? Discuss the path traveled by an electron in a uniform transverse electric field.
2. Show that the velocity acquired by an electron in a uniform electrostatic field varies as the square root of potential difference through which it is accelerated.
3. When and why a charged particle entering magnetic field follows a helical path?
4. Define an electron volt. Express it in joules.
5. Describe the motion of an electron subjected to the following forces: (*i*) A uniform electric field acting normal to the electron velocity; (*ii*) A uniform magnetic field acting perpendicular to the direction of the electron motion, (*iii*) When a uniform electric and magnetic field which are perpendicular to each other act normal to the direction of the electron motion.
6. Show that the energy of a charged particle remains constant when it moves in magnetic field.
7. Describe the trajectories of electron of mass '*m*' and charge '*e*' moving with velocity v in : (*a*) Uniform magnetic field, (*b*) Uniform electric field. In both the cases field is applied perpendicular to the direction of motion of particle.
8. Explain how a charged particle fired into uniform electric field describes a parabolic motion, and when fired into a uniform magnetic field describes circular and helical motion. Find the expression for radius of the circle and pitch of the helix.

9. Prove that the electron moving with uniform velocity in transverse electric field traces parabolic path.
10. Show that an electron moving with uniform velocity follows a parabolic path in a transverse uniform electric field.
11. Explain the motion of electron in transverse electric field and obtain expression for deflection.
12. Derive an expression for vertical displacement when electron moves perpendicular to the uniform electric field.
13. Discuss the motion of an electron projected into uniform electric field at an acute angle with the field direction.
14. Show that an electron moves along a parabolic path, when it enters in a uniform electric field applied perpendicular to its motion.
15. Show that charged particle does not gain additional kinetic energy when it travels in uniform transverse magnetic field.
16. Prove that kinetic energy of a charge remains constant during its motion in magnetic field.
17. Show that the radius of orbit of a charged particle moving at right angles to magnetic field is proportional to its momentum.
18. Derive an expression for the radius and time period for an electron in a transverse magnetic field of uniform intensity.
19. State the shapes of trajectories of a charged particle of velocity $\boldsymbol{v}$ moving in an electric field **E** when: (*i*) $\boldsymbol{v} \perp$ E, (*ii*) $\boldsymbol{v}$ makes an angle with **E.** What are these shapes if **E** is replaced by a magnetic field **B**?
20. Discuss the trajectory of a charged particle moving with uniform velocity traversing uniform magnetic induction B if v_o is: (*i*) parallel to B, (*ii*) perpendicular to B, (*iii*) v_o makes an acute angle with B.
21. State conditions under which a charged article moves in a straight line in an: (*i*) electric field E, (*ii*) A magnetic field B and (*iii*) In a region having both E and B.
22. Consider a particle of mass *m* and charge *q* moving with velocity v. The particle enters a region where a perpendicular uniform magnetic field B acts. Show that in the region the kinetic energy of the particle is proportional to the square of the radius of its orbit.
23. A charged particle of charge *q* coulombs and mass *m* kg enters the field with velocity v m/s under the following conditions: (*i*) velocity perpendicular to the magnetic field, (*ii*) angle between the directions of velocity and magnetic field is at an acute angle. Indicate the trajectory of motion of charged particle and derive the concerned mathematical relationship in each case.
24. A particle of mass "*m*" and charge "*q*" enters a region of uniform magnetic induction field "B" with a velocity 'u' making an angle θ with the direction of field. Discuss its motion and derive expressions for: (*i*) radius of path (*ii*) pitch of helical path and (*iii*) frequency of revolution.
25. How is it possible for a charged particle to pass through a combination of electric and magnetic field with out any deviation? What will be its velocity? What will happen if the particle is moving slowly or faster than its velocity?
26. Obtain expressions for radius and pitch of helical path described by a charged particle when it enters a uniform magnetic field making an acute angle θ with the direction of the magnetic field.
27. Derive expression for deflection of electron beam in transverse magnetic field.
28. Describe Thomson's method for determination of e/m of the electron. Derive necessary formula.
29. Discuss the measurements of e/m for electron by Thomson's method.
30. Explain the method by which the specific charge of an electron can be determined. Is it essential in this method that the electrons have a constant speed?
31. Give the properties of positive rays.
32. Explain the method for analysis of the parabolic traces for positive rays.
33. Explain why: (*i*) No parabola passes through origin, (*ii*) Weak parabolic traces are observed on the left side of the plate, (*iii*) Thomson's parabola method for q/M., (*iv*) What is mass spectrograph? How will you determine the isotopes of Neon using parabola method?

PROBLEMS

1. An electron is liberated at rest from one of two large parallel plates separated by 2 cm. The other plate has a relative potential of 2.4 kV. How long does the electron take to reach it?

   [**Ans:** 6.9×10^{-10} s]

2. A proton, a deuteron and an alpha particle with the same kinetic energies enter a region of uniform magnetic field moving at right angles to B. Compare the radii of their circular paths.

   [**Ans:** $R_p : R_d : R_\alpha = 1 : \sqrt{2} : 1$]

3. An electron describes a helix of radius 6 cm and pitch of 2 cm in a magnetic field of intensity 30 Gauss. Obtain the component of its velocity along and perpendicular to the magnetic field.

   [**Ans:** $v_\perp$ = 3.164 × 107 m/s, $v_\parallel$ = 1.68 × 10^6 m/s]

4. Calculate the frequency of ac potential that must be applied to cyclotron dees in which protons are accelerated. Given magnetic flux density = 3wb/m^2. [**Ans:** 45.8 MHz]

5. A proton traveling at 23° with respect to magnetic field of strength 2.63 mT experiences a magnetic force of 6.48 × 10^{-17} N. Calculate (*i*) the speed and (*ii*) the kinetic energy in eV of the proton. [**Ans:** v = 3.93 × 10^5 m/s, K.E. = 0.805 keV]

6. A positive ion beam moving along X-axis enters a region of uniform electric field of 3kV/m along Y-axis and magnetic field of 1 Kilo gauss along Z-axis acting simultaneously. Calculate speed of those ions, which shall pass undeviated. What will happen if ions are moving slowly or faster than this velocity? [**Ans:** 3 × 10^4 m/s]

7. In an experiment for determination of e/m for electrons by Thompson's method, electric field between plates of 6 cm length and 2 cm separation a potential difference of 480 V produces a deflection of 10.9 cm on the screen 40.0 cm away. A magnetic field of 8.0 × 10^{-4} T restores the beam to its original position. Calculate e/m. [**Ans:** e/m = 1.75 × 10^{11} C/kg]

8. The electrons circulating in a uniform magnetic field of 4.55 ×10^{-4} wb/m^2 have a kinetic energy of 22.5 eV. What are the radius and period of revolution of the electron path?

   [**Ans:** r = 3.5 cm; T = 7.9 × 10^{-8} s]

9. The mass of an alpha particle is 7344 times that of electron and its charge is twice that of an electron. It is accelerated from rest through a potential difference of 5 × 10^6 volts. Calculate the flux density of magnetic field required to bend its path into a circle of radius 1 m if it enters the field normally and at an angle of 45° with the field. [**Ans:** 0.45 wb/m^2; 0.31 wb/m^2]

10. A proton accelerates from rest in a uniform electric field of 400V/m. After't' seconds, its speed is 3 × 10^6 m/s. (*i*) Find the acceleration of a proton, (*ii*) Find time '*t*', (*iii*) How much is the distance traveled in the time '*t*'? [**Ans:** 3.8 × 10^{10} m/s^2; 7.8 × 10^{-5}s; 117m]

11. A proton and an α-particle both move in circular paths in a uniform magnetic field with same tangential speeds. Compare the number of revolutions they make per second.

21 Electron Optics

Learning Objectives:

By the end of this chapter, you will be able to:

- Learn about application of Bethe's law.
- Define concept of electron Lens.
- Explain concept of focusing by Uniform Magnetic Field and Axially Symmetric Field.
- Understand working of Cathode Ray Tube.
- Describe Cathode Ray Oscilloscope and its seven basic functions.

21.1 INTRODUCTION

Electrons travel along straight-line path in uniform electric fields. However, in many of the practical cases the electric fields happen to be nonuniform. The analysis of electron motion in nonuniform fields is not easy to carry out and poses serious mathematical difficulties. Fortunately, there is a close resemblance between the motion of electrons in an electrostatic field and the propagation of light in an optical medium. Light rays travel along straight line path in homogeneous media and along curved paths in inhomogeneous media. Electrons travel along straight line path in an equipotential region and follow a curved path in passing through regions of varying potential. Light undergoes refraction at an optical boundary. In a similar fashion, electron path bends when electrons travel from a region of one potential into the region of another potential. The resemblance suggests that the concepts of geometrical optics can be applied to the motion of the electrons in homogeneous as well as inhomogeneous electric fields. Such an extension of the concepts of geometrical concepts to electron motion is known as ***electron optics***. This approach led to the discovery of TV tube, particle accelerators, electron microscope and perfecting of photomultiplier tube, microwave tube etc. Electron optics had its beginning in the works of H.Busch in 1926 and C.J.Davisson and C.J. Calbick in 1931.

The principles of electron lens and magnetic lens are introduced and the working of CRO and its applications are explained in this chapter.

21.2 BETHE'S LAW

Let us consider a narrow stream of electrons (or any charged particles) traveling in space. These electrons bend and follow curved path when they travel through regions of different electric field strengths. This behaviour is similar to the behaviour of light rays that change their direction of motion while passing through an optical medium of inhomogeneous refractive index. By analogy with optical refraction, the bending of electrons by electric fields is called ***electron refraction***. Bethe's law deals with electron refraction while Snell's law describes refraction of light.

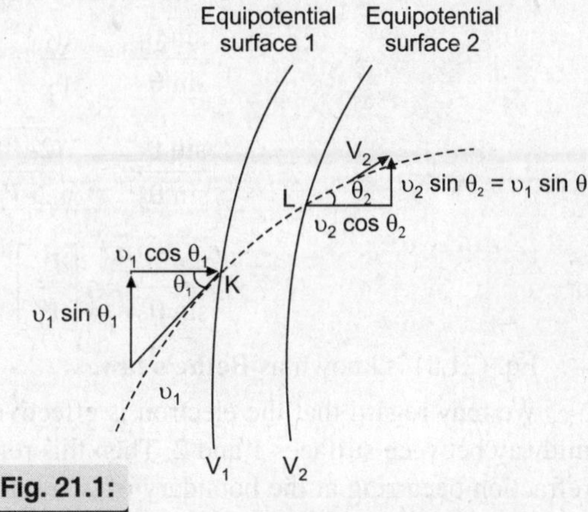

Fig. 21.1: Electron refraction through equipotential surfaces

On account of nonuniform electric fields, the equipotential surfaces shall be curved. Let us consider two equipotential surfaces 1 and 2 separated by an infinitesimal region (see Fig. 21.1). Let the potential on surface 1 be V_1 and the potential on surface 2 be V_2 and let $V_2 > V_1$.

An electron starting from rest and traveling in the region above the surface 1 acquires a velocity υ_1 when it reaches a point K on the surface V_1. This velocity is related to the potential V_1 according to the relation

$$\upsilon_1 = \sqrt{\frac{2eV_1}{m}} \qquad \text{...(21.1)}$$

where 'e' is the electronic charge and 'm' the mass of the electron.

Let the angle between the path of the electron at K and the normal to the equipotential surface 1 be θ_1. Then the velocity component of the electron perpendicular to the equipotential surface is $\upsilon_1 \cos\theta_1$ and this is also the electric field direction at K. The tangential velocity component is equal to $\upsilon_1 \sin\theta_1$. Electron velocity increases from υ_1 to υ_2, as the electron travels from the surface 1 to surface 2.When the electron reaches the point L on the second equipotential surface, it speed is υ_2.

$$\upsilon_2 = \sqrt{\frac{2eV_2}{m}} \qquad \text{...(21.2)}$$

Note that the electron is moving in nonuniform electric field since V_2 is different from V_1. Consequently, the direction of velocity shall continually change as the electron goes from K to L. If $V_2 > V_1$, the path of the electron shall be as shown in the Fig. 21.1. If the angle between the path of the electron and the normal to the equipotential surface 2 at the point L is θ_2, then the velocity component of the electron perpendicular to the equipotential surface is $\upsilon_2 \cos\theta_2$ and is along the electric field direction at L. The tangential velocity component is equal to $\upsilon_2 \sin\theta_2$.

As the electric field is perpendicular to the equipotential surface, the electron does not experience any force tangential to the surface. However, the electron is acted upon by a force due to the electric field that is acting perpendicular to the equipotential surface. Since $V_2 > V_1$,

$\upsilon_2 \cos\theta_2$ is greater than $\upsilon_1 \cos\theta_1$. Since the surfaces V_1 and V_2 are very close to each other they may be taken as plane, parallel surfaces. Hence, the components $\upsilon_1 \sin\theta_1$ and $\upsilon_2 \sin\theta_2$ are parallel to each other and equal to each other. This is due to the fact that as the electron goes from K to L, there is no force in the horizontal direction. Hence, we have

$$\upsilon_1 \sin\theta_1 = \upsilon_2 \sin\theta_2$$

$$\therefore \quad \frac{\sin\theta_1}{\sin\theta_2} = \frac{\upsilon_2}{\upsilon_1}$$

$$\therefore \quad \frac{\sin\theta_1}{\sin\theta_2} = \sqrt{\frac{2eV_2/m}{2eV_1/m}}$$

or
$$\frac{\sin\theta_1}{\sin\theta_2} = \left[\frac{V_2}{V_1}\right]^{1/2} \qquad ...(21.3)$$

Eq. (21.3) is known as **Bethe's law.**

We may regard that the electron is effectively refracted at the equipotential surface lying midway between surfaces 1 and 2. Then this refraction can be seen to be similar to the optical refraction occurring at the boundary of two refracting media. In the present case, the electron refraction is similar to the optical refraction from rarer to the denser medium, where light bends towards the normal. The light refraction is governed by the Snell's law.

$$\frac{\sin i}{\sin r} = \frac{c_1}{c_2} = \frac{\mu_2}{\mu_1} \qquad ...(21.4)$$

where c_1 and c_2 are the velocities of light in the optical media 1 and 2 respectively.

- In case of electron refraction, the rarer medium corresponds to the lower potential region and the denser medium corresponds to the higher potential region.
- When light enters a denser medium, it slows down and bends toward the normal to the surface. On the other hand, when an electron ray enters higher potential region, its velocity increases and bends toward the normal.
- The order of velocities υ_1 and υ_2 in Bethe's law is reversed from that in Snell's law.

Example 21.1: *An electron beam enters from a region of potential 75 volts into a region of potential 100 volts, making an angle of 45° with the direction of electric field. Find the angle through which the beam refracts.*

Solution:
$$\frac{\sin\theta_1}{\sin\theta_2} = \left[\frac{V_2}{V_1}\right]^{1/2}$$

$$\therefore \quad \frac{\sin 45^\circ}{\sin\theta_2} = \left[\frac{100\,V}{75\,V}\right]^{1/2}$$

$$\therefore \quad \sin\theta_2 = \frac{0.7071}{1.1547} = 0.612$$

$$\therefore \quad \theta_2 = \mathbf{37.76^\circ}.$$

Example 21.2: *Electrons accelerated under a potential of 250 V enter an electric field at an angle of incidence of 50° and gets refracted through an angle of 30°. Find the potential difference between the two equipotential surfaces.*

Solution: $\dfrac{\sin\theta_1}{\sin\theta_2} = \left[\dfrac{V_2}{V_1}\right]^{1/2}$

$\therefore$ $\dfrac{\sin 50°}{\sin 30°} = \left[\dfrac{250\,V}{V_1}\right]^{1/2}$

or $V_1 = \left[\dfrac{0.5}{0.766}\right]^2 \times 250\,V = \mathbf{106.5\ V.}$

21.3 ELECTRON LENS

An electron lens is an electrical component that focuses an electron beam to a point.

Principle: A stream of electron experiences a change in direction of motion when it passes through a non-uniform electric field. Its path is bent at each equipotential surface in the same way as a light ray is bent at an optical boundary. Hence, non-uniform electric fields can be used to focus a bundle of electron rays. In 1931 C.J. Davisson and C.J. Calbick achieved such focusing action. Just as a convex lens focuses light rays, a non-uniform electric field produced by two coaxial metal tubes maintained at different potentials can focus electron rays and therefore, such a system is called an electron lens.

Construction: An electron lens is made of two coaxial short cylindrical metal tubes T_1 and T_2 separated by some distance, as shown in Fig. 21.2. The tubes are held at different potentials V_1 and V_2 respectively. V_2 is greater than V_1. A non–uniform electric field is produced in the gap between the two tubes, as shown in Fig. 21.2.

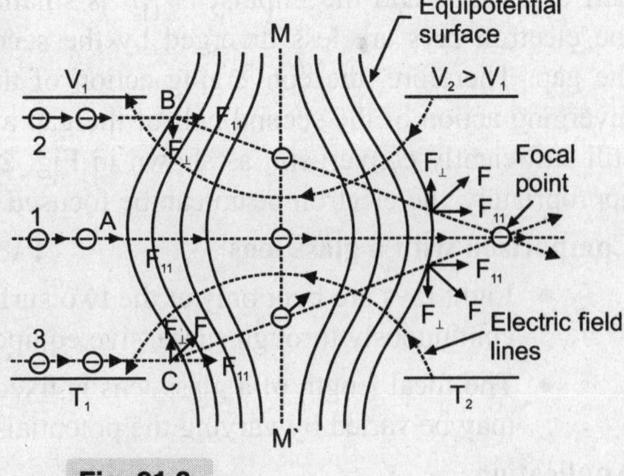

Fig. 21.2: Focusing of electron beam by an electron lens

Working: Let us consider a thin bundle of electron rays entering the lens system through the tube T_1 as shown in Fig. 21.2. The electrons are moving from a lower potential region to a higher potential region. Fig. 21.2 shows the equipotential surfaces in the region between the tubes. Electrons, labeled 1, are moving along the axis of the system. When the electrons travel through the tube T_1, they do not experience any force. As the electrons approach the gap between the cylindrical tubes, they come across first the convex shaped equipotential surfaces. The electric field is perpendicular at each point on the equipotential surface and is directed from tube T_2 to tube T_1. At point A, the electric field is along the axis and hence the electrons labeled 1 get accelerated forward along the axis. They travel forward without any deviation from their initial path. Beyond the mid-plane MM', the equipotential surfaces are concave shaped but the electric field still acts along the axis in a direction which accelerates the electrons labeled 1.

Therefore, these electrons fly forward with increased speed along the axial direction. Electrons labeled 2; reaching at B on the convex shaped equipotential surface, experience an electric force that acts at an angle to the direction of their motion. The force can be resolved into its rectangular components F_{11} and $F_\perp$. Due to the force $F_\perp$, electrons labeled

2 are deflected down toward the axis. Due to the component F_{11}, they are also accelerated toward tube T_2. Hence these electrons are continually bent downwards till the midplane MM' is reached. In the same way, the electrons, labeled 3 on reaching at C on the equipotential surface are continually deflected up toward the axis and are also accelerated forward. Because of cylindrical symmetry, all off-axis electron paths around the lens axis tend to converge toward a point on the axis, as shown in Fig. 21.2.

On crossing the mid-plane MM' of the gap, the electron rays encounter equipotential surfaces of concave shape. In the second half of the gap, the normal component of electric force, $F_{\perp}$ is directed away from the axis for all off – axis electron rays. As a result, the electron rays tend to diverge. The parallel component F_{11} is directed forward and hence the electrons are accelerated further. Before the electron rays can converge, they are diverged to some extent. However, because the potential is everywhere higher in the second half of the gap, the electrons travel faster in the second half than in first half of the gap. The electrons spend a greater time in the first half of the gap and the impulse '$F_{11}t$' is smaller for the divergence interval. Consequently, the electron rays are less diverged by the second half than converged by the first half of the gap. Therefore, the converging action of the first half of the gap will be stronger than diverging action of the second half of the gap and the electron rays emerge from the tube T_2 still sufficiently convergent, as shown in Fig. 21.3. By adjusting the potentials on the tubes appropriately, the electron beam can be focused to a suitable required point on the axis.

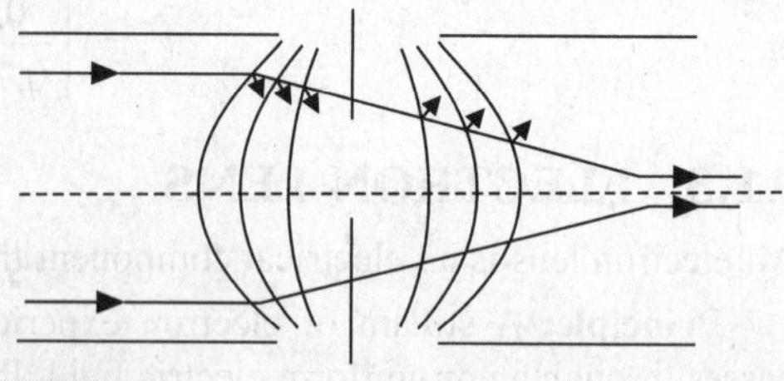

Fig. 21.3: Electron rays converging toward the lens axis

Comparison with a glass lens

- Light rays are bent only at the two surfaces of a glass lens but electron rays are bent continuously through successive equipotential surfaces.
- The focal length of a glass lens is fixed whereas the focal length of an electron lens may be varied by varying the potentials V_1 and V_2 on the cylinders.

Application

Electron lens forms the most important component of an electron gun used for producing a narrow intense electron beam. Electron lens action is utilized in particle accelerators to focus charged particles into a narrow beam.

21.4 FOCUSSING BY UNIFORM MAGNETIC FIELDS

An electron beam can be focused with the help of a magnetic field also. Depending upon the nature of the magnetic field, i.e., uniform or non-uniform, and the direction of its application, different types of focusing are possible.

21.4.1 Longitudinal Uniform Magnetic Field Focusing

In this method a uniform magnetic field acts along the direction of motion of the electron beam. When electrons enter the uniform magnetic field making an angle with the field direction, they describe a helix. The pitch of the helical path is given by

$$p = \frac{2\pi m \upsilon}{eB} \cos \phi \qquad ...(21.5)$$

Let us consider a bundle of electron rays of a given velocity 'υ' originating at a point O with the rays making different angles with the uniform magnetic field as shown in Fig. 21.4. If the divergence in the beam is assumed to be small ($\phi \leq 10°$), cos ϕ can be taken as unity. Any electron ray passing through point O on a field line, meets the field line again at a point P after a time interval T. Thus, all the electron rays passing through point O at small angles will again cross the same field line at point P after the time period T. The distance of P from O is l which is given by

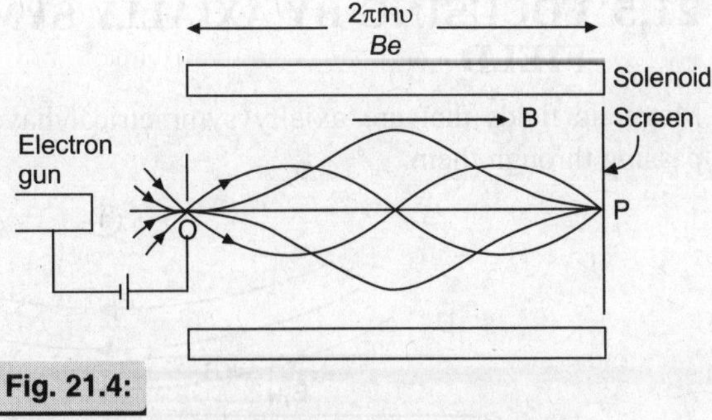

Fig. 21.4:

Focusing of electron beam in a longitudinal uniform magnetic field

$$l = \frac{2\pi m\upsilon}{eB} \qquad ...(21.6)$$

Thus a longitudinal homogeneous magnetic field focuses electrons at a distance l or at any distance which is an integral multiple of l (i.e. $n\,l$).

Application: This method is used for determining the e/m ratio of electron.

21.4.2 Transverse Uniform Magnetic Field Focusing

A transverse uniform magnetic field would also focus a beam of charged particles. The focusing would be affected after rotating the beam through 180°. The method of *semicircular focusing* was first employed by Rutherford in 1914. The principle of the method is shown in Fig. 21.5. Let us consider a beam of positively charged particles having the same velocity and very small angular divergence. If the beam enters uniform magnetic field that is acting in a perpendicular direction to its motion, the beam gets focused after completing a semicircle.

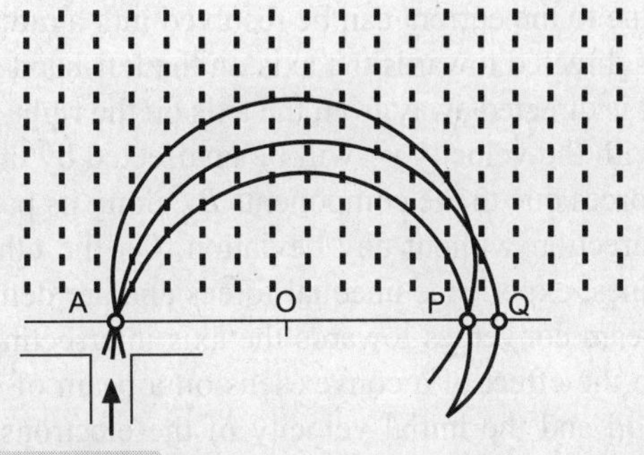

Fig. 21.5:

Focusing of positively charged particles in a transverse uniform magnetic field

This type of focusing effect is widely used in mass spectroscopy. To separate and identify the isotopes of an element, its atoms are ionized and allowed through a slit into the region of a uniform magnetic field acting transversely as shown in Fig. 21.5.

The ions belonging to different isotopes would have different masses. Therefore, even though all of them have the same velocity, they describe semicircles of different radii and come to focus at different points. The radii of the circles depend on the momenta of the different isotopes. As the separation of the isotopes is based on momentum selection by the magnetic field, such an arrangement is also known as a **momentum selector**.

Application: A momentum selector is used in Bainbridge mass spectrograph for separating positive ions of different masses.

21.5 FOCUSING BY AXIALLY SYMMETRIC MAGNETIC FIELD

Magnetic fields that are axially symmetrical have a focusing effect on an electron beam passing through them.

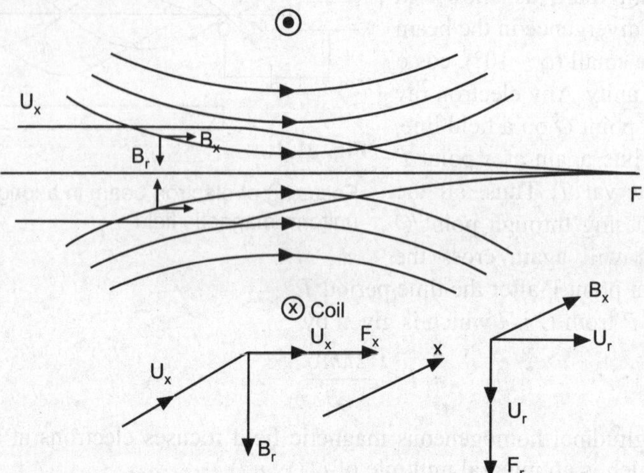

Fig. 21.6: Converging action of an axially symmetrical magnetic field

Let us consider a short coil carrying a current as shown in Fig. 21.6. The magnetic field due to the current can be resolved into a radial component B_R, and an axial component B_L. B_R is directed towards the axis on the left-hand side; it becomes zero in the plane of the coil; and it is directed away from the axis on the right-hand side. An electron moving parallel to the axis with the velocity υ_L will be unaffected by the magnetic field B_L, and experiences equal radial forces due to the components B_R along its path. As a result, the electron travels along the axial direction without any deviation. On the other hand, electrons traveling along the non-axial lines, experience unequal forces and are deflected towards the axis of the lens. Therefore, the beam converges towards the axis intersecting it at F. The effect of the coil is thus analogous to the effect of a convex lens on a beam of light. With the adjustment of current through the coil and the initial velocity of the electrons, the focal distance of the magnetic lens can be altered.

21.5.1 Magnetic Lens

Magnetic fields, which are axially symmetric, have a focusing effect on an electron beam passing through them. The axially symmetrical magnetic fields are produced by short solenoids. By encasing the coils in hollow iron shields the magnetic fields are concentrated and improved focusing action is obtained. Such solenoids are called *thin magnetic lenses*. Fig. 21.7 shows the schematic of a thin magnetic lens. *AB* is the section of the coil of an electro-magnet and the hatched portion represents

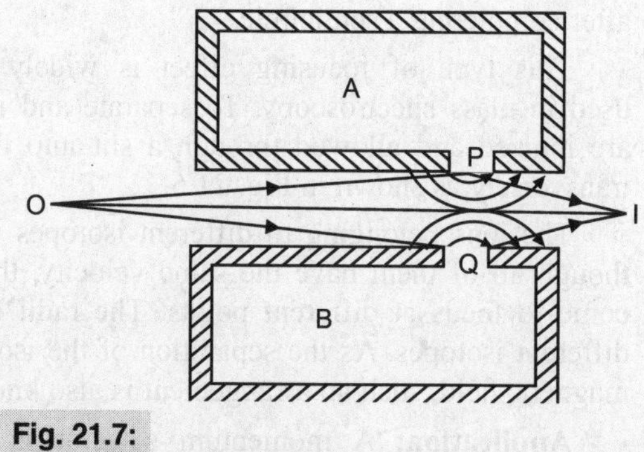

Fig. 21.7: Schematic diagram of a magnetic lens

a soft iron shield. P and Q represent the gaps at diametrically opposite positions. A strong non-uniform magnetic field is produced across the gap and it will be symmetrical about the axis OI of the coil. O is the source of electrons, which are focused at the point I due to the magnetic field. The focal length of magnetic lens is very small of the order of a few centimeters.

An electron beam can only be converged by magnetic lens. Diverging action is impossible in magnetic lenses. With the adjustment of current through the solenoid and the initial accelerating voltage of the electron, the focal distance of the magnetic lens can be adjusted.

Application: Magnetic lenses are widely employed in electron microscope and in electron beam machining equipments.

21.6 CATHODE RAY TUBE

A *cathode ray tube* is a purposefully designed vacuum tube (Fig. 21.8). In the CRT, electron beam is controlled by an electric field. The electron beam acts as a writing device, and traces on a fluorescent screen a vivid display of input signal.

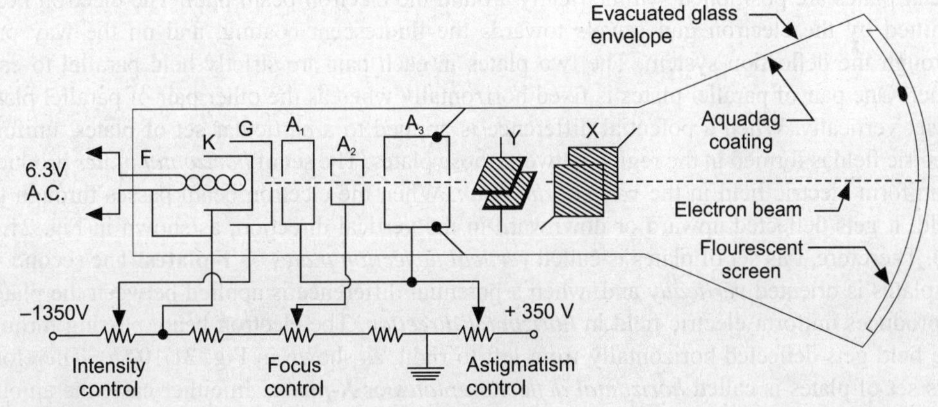

Fig. 21.8:

Schematic of Cathode ray tube (electrostatic deflection type)

It consists of the following three principal parts:

(*i*) electron gun,

(*ii*) deflection system

(*iii*) a fluorescent screen.

***(i)* Electron gun:** The electron gun consists of several electrodes (see Fig. 21.9) and is mounted at one end of the cathode ray tube. An indirectly heated cathode K releases a stream of electrons which passes through the control grid G held at a high negative potential. The active size of the aperture in the grid depends on the potential difference between the grid and cathode. The **intensity** of the sketch traced on the screen is regulated by the number of electrons striking the screen. Therefore, by varying the negative d.c. voltage on the grid, the intensity of the luminous spot on the screen may be controlled. Two anodes are positioned beyond the control grid, which are coaxial with the grid. The electron lens system formed by the combination of the grid and anodes focus the electron rays into a sharp narrow pencil of beam. The electron beam travels along the axis of the tube and strikes at the centre of a fluorescent coating located at the opposite end of the CRT. The electron beam may be brought to a sharp **focus** on the coated face by varying the relative magnitudes of voltages applied to the anodes.

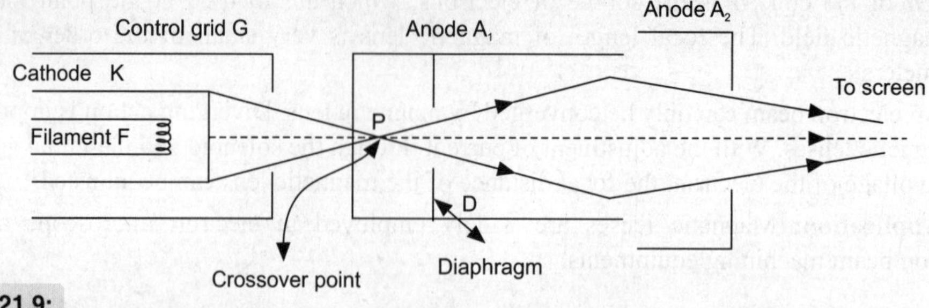

Fig. 21.9:

Schematic of an electron gun

(ii)* Deflection System:** In an electrostatic deflection type CRT, the deflection system consists of two pairs of parallel metal plates *Y* and *X* mounted in the neck of the tube. The metal plates are positioned symmetrically around the electron beam path. The electron beam emitted by the electron gun travels towards the fluorescent coating, and on the way pass through the deflection system. The two plates in each pair are strictly held parallel to each other. One pair of parallel plates is fixed horizontally whereas the other pair of parallel plates is set vertically. When a potential difference is applied to a particular set of plates, uniform electric field is formed in the region between those plates. The set of *horizontal plates* produces a uniform electric field in the *vertical direction*. When the electron beam passes through the field, it gets deflected upward or downward in the vertical direction, as shown in Fig. 21.10 (*a*). Therefore, this set of plates is called *vertical deflection plates* or ***Y*-plates**. The second set of plates is oriented *vertically* and, when a potential difference is applied between the plates, it produces uniform electric field in *horizontal direction*. The electron beam passing through the field gets deflected horizontally from left to right, as shown in Fig. 21.10 (*b*). Therefore, this set of plates is called *horizontal deflection plates* or ***X-plates. In either case, the amount of deflection of the electron beam is directly proportional to the magnitude of the voltage applied to the plates.

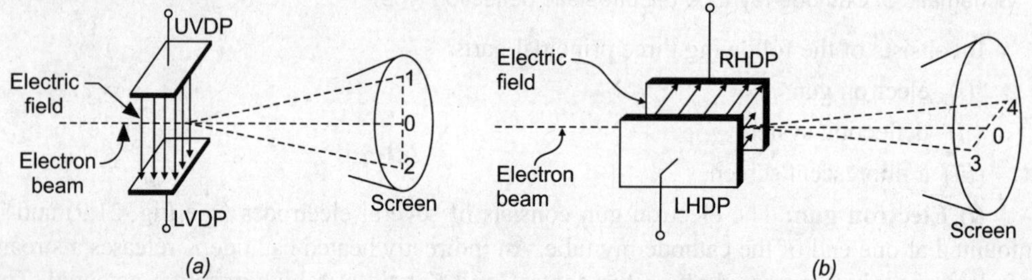

Fig. 21.10:

Electron beam deflection (*a*) Vertical deflection (*b*) Horizontal deflection

The direction of the deflection depends on the polarity of the applied voltage, as shown in Fig. 21.11. When dc voltages are applied to both the *X*- and *Y*-plates, the electron beam will be subjected to two orthogonal forces and gets deflected along the direction of their resultant, as shown in Fig. 21.11. By varying the voltages on the vertical and horizontal deflection plates, the luminous spot may be moved to any position on the screen.

***(iii)* Fluorescent Screen:** Electrons are invisible particles. However the electron beam position can be located with the help of a fluorescent screen. The interior surface of circular front face of the CRT is coated with a thin translucent layer of phosphor. The front face acts as a fluorescent screen. Phosphors have the property of emitting light when high-energy particles or high-energy radiation hits them. Therefore, when the high-energy narrow electron beam strikes the fluorescent screen at a point, it emits a glow at that point. The glow can be seen through the thin glass face and makes the position of the electron beam known.

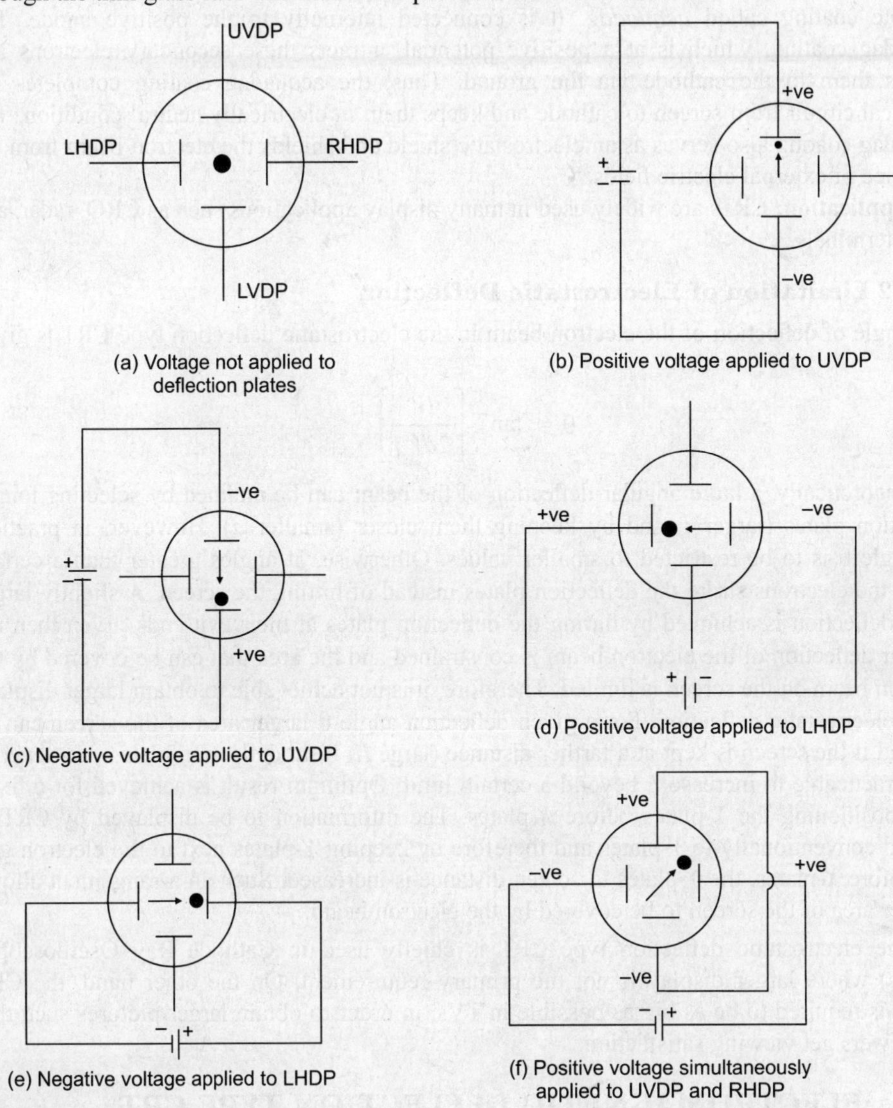

Fig. 21.10:

Electron beam deflection under the action of dc voltages given to Y and X-plates

21.6.1 Acquadag Coating

In a CRT, electrons striking the fluorescent screen emit secondary electrons, which tend to charge the screen negatively. Consequently, the screen repels the electrons that arrive afterwards. This leads to a decrease in the brightness of the trace. On the other hand, the cathode loses electrons through emission and attains positive polarity. Therefore, it drags the emitted electrons back, and this also leads to a decrease in the brightness of the trace. This problem can be eliminated if the electrons accumulating on the screen are conducted away and the shortage of electrons at the cathode is made up. This is accomplished by the acquadag coating.

The inner surface of the flare of the glass envelope of CRT is coated with conductive graphite coating called *acquadag*. It is connected internally to the positive anode. The acquadag coating, which is at a positive potential, attracts these secondary electrons and returns them to the cathode via the ground. Thus, the acquadag coating completes the electrical circuit from screen to cathode and keeps them in electrically neutral condition. The acquadag coating also serves as an electrostatic shield and shields the electron beam from the influence of external electric fields.

Application: CRTs are widely used in many display applications such as CRO, radar, and video terminals.

21.6.2 Limitation of Electrostatic Deflection

The angle of deflection of the electron beam in the electrostatic deflection type CRT is given by

$$\theta = \tan^{-1}\left[\frac{lV}{2dV_A}\right]$$

Theoretically, a large angular deflection of the beam can be attained by selecting longer deflection plates (larger l) and by keeping them closer (smaller d). However, in practice, the angle θ is to be restricted to smaller values. Otherwise, at angles greater than a certain value, the electrons strike the deflection plates instead of hitting the screen. A slightly larger beam deflection is achieved by flaring the deflection plates at their exit ends. Even then the angular deflection of the electron beam is constrained and the area that can be covered by the electron beam on the screen is limited. Therefore, it is not achievable to obtain larger displays using electrostatic deflection. For a given deflection angle θ larger area of the screen can be covered if the screen is kept at a farther distance (large L) from deflection plates. However, it is impracticable to increase L beyond a certain limit. Optimum result is achieved for a fixed L by positioning the Y-plates before X-plates. The information to be displayed by CRT is applied conventionally to Y-plates and therefore by keeping Y-plates next to the electron gun and before X-plates, the Y-plates to screen distance is increased. Such an arrangement allows a larger area of the screen to be covered by the electron beam.

The electrostatic deflection type CRT is chiefly used in Cathode Ray Oscilloscopes (CROs) where larger display is not the primary requirement. On the other hand, the CRT screen is required to be as big as possible in TVs, in order to obtain larger pictures such that the viewers get viewing satisfaction.

21.7 ELECTROMAGNETIC DEFLECTION TYPE CRT

The CRT with electromagnetic deflection is very much similar in construction to a CRT with electrostatic deflection except for the deflection system. The magnetically deflected CRT does not contain deflection plates. The electron beam deflection is caused by an external magnetic field produced by electromagnets. Two sets of coils are placed at right angles to each other and mounted on the tube neck where the electron beam leaves the electron gun. There are four coils in all, with opposite ones comprising one set. The coils in each set are connected in series such that magnetic north and south poles appear when a dc current flows through them.

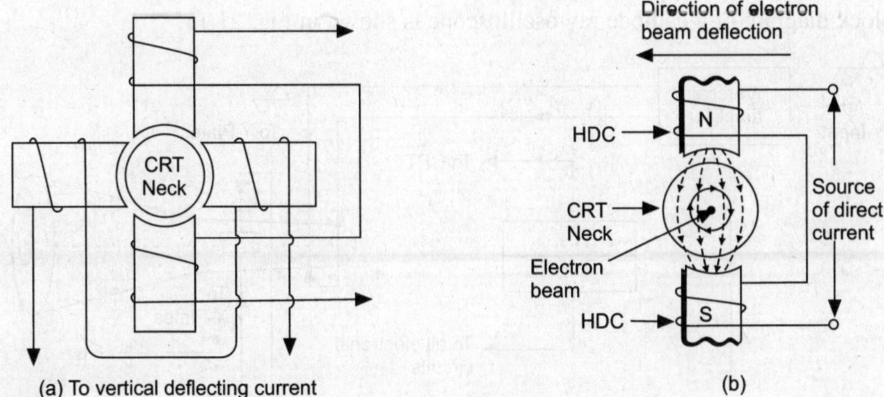

Fig. 21.12:

Deflection of luminous spot under the action of magnetic field:
a. Arrangement of deflection coils around CRT neck
b. Deflection of electron beam in horizontal direction due to X-coils.

Fig. 21.12 shows the physical placement of the coils. The horizontally mounted coils produce horizontal magnetic fields when direct current flows through them. The magnetic field acts at right angles to the electron beam. The electrons experience a Lorentz force as they pass through the magnetic field and are deflected vertically, in a direction normal to both the field direction and direction of motion. If the direction of the dc current through the coils is reversed, the magnetic field direction is reversed and the electron beam is deflected in the opposite direction. Thus, the horizontally placed coils, cause vertical deflection and are called **vertical deflection coils** or ***Y*-coils**. The coils mounted in vertical direction produce a vertically acting magnetic field which deflects the electron beam horizontally. These coils are therefore called **horizontal deflection coils** or ***X*-coils**. Both the vertical deflection coils and horizontal deflection coils are enclosed in a common former, called the **yoke**, which fits over the CRT neck. The amount of beam deflection is proportional to the strength of the magnetic field which in turn is proportional to the current in the deflection coils. By properly controlling the amount of current flowing through each set of coils, the electron beam can be made to strike any desired point on the screen.

Advantages:

(*i*) As the deflection system is mounted outside, the fabrication of CRT becomes relatively simpler.

(*ii*) The deflection system does not obstruct the deflected electron beam and as such larger angles of beam deflection can be realized leading to bigger display on the screen.

(*iii*) Electromagnetic deflection type CRTs are used in TVs, Radar receivers, VDU applications etc.

21.8 CATHODE RAY OSCILLOSCOPE

A cathode ray oscilloscope (CRO) is a very important electronic measuring instrument. It is used to measure parameters such as voltages and time intervals. The most advantageous feature of CRO is that it gives a visual display of waveform under test.

21.8.1 Block Diagram of CRO

The block diagram of a cathode ray oscilloscope is shown in Fig. 21.13.

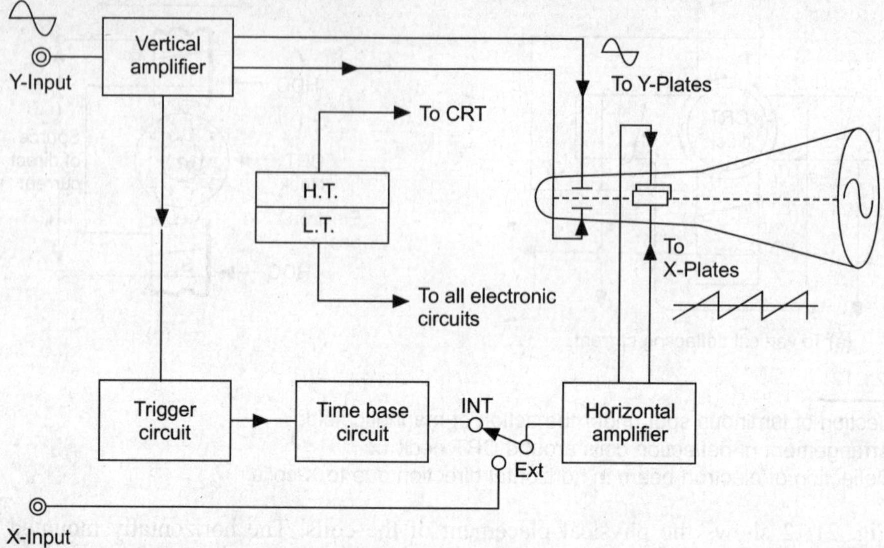

Fig. 21.13: Block diagram of a cathode ray oscilloscope

Any CRO contains the following seven basic sections:

(*i*) Cathode ray tube (CRT)
(*ii*) Time base generator
(*iii*) Trigger circuits
(*iv*) Vertical circuits
(*v*) Horizontal circuits
(*vi*) Low voltage power supply and
(*vii*) High voltage power supply

21.8.2 Cathode Ray Tube

The CRT constitutes the central part of CRO. With the help of the other sections in the CRO, the CRT produces a trace on its fluorescent screen and maintains it. An electron gun, located at one end of the CRT, produces a sharp electron beam. A small luminous spot caused on the fluorescent coating indicates the position of the electron beam. The brightness and sharpness of the luminous spot on the screen are adjusted with the help of intensity and focus controls. The vertical and horizontal controls are used to adjust the position of the luminous spot at any desired position on the screen. When voltages are not applied to *Y*-plates and *X*-plates, the luminous spot stays at the centre of the screen. When a voltage is applied to the Y-plates, an electric field is produced in the vertical direction and the electron beam passing through the field is deflected upward or downward. When a voltage is given to the *X*-plates, an electric field is produced horizontally and electron beam passing through the electric field is deflected sideways either to the left or right of the screen. The *Y*- and *X*-plates are electrically independent and different voltages can be applied to each set. A transparent graph, called a

graticule, marked in centimetre lines both vertically and horizontally, is attached to the face of the CRT for making measurements.

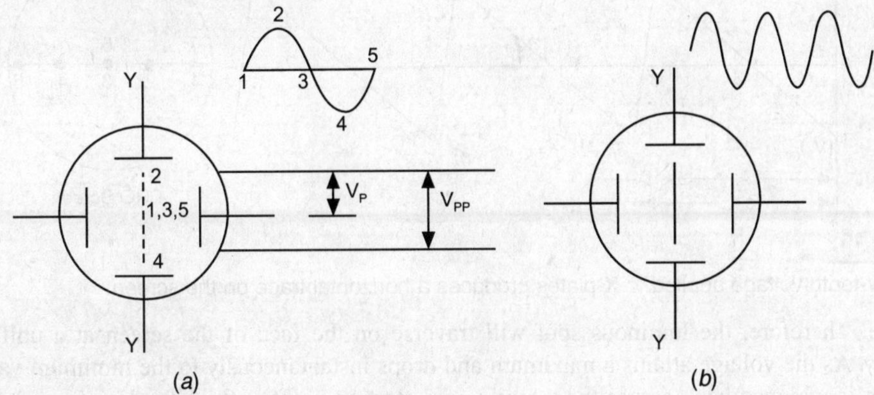

Fig. 21.14:

A vertical trace is observed when ac voltage is applied to Y-plates

Signal: When a signal is to be displayed, it is applied to the *Y*-plates of a CRO by connecting it to the *Y*-input of CRO. For example, let the signal be a simple harmonic wave. Because of the application of the signal to the Y-plates, the luminous spot moves up and down on the screen at the same frequency as that of the applied voltage provided *X*-plates are not given any voltage. The successive positions of the luminous spot can be seen (Fig. 21.14*a*) when the frequency of the signal is less than about 20 Hz. At higher frequencies the path of the beam is seen as a vertical luminous line (Fig. 21.14*b*). This is due to persistence of vision and as well as the phosphorescence of the coating. The luminous line is called a **trace**. The length of the vertical trace corresponds to the peak-to-peak value of the applied voltage, as shown in Fig. 21.14 (*a*).

21.8.3 Time Base

In a signal the voltage varies as a function of time in a specific way. It is obvious that the shape of the signal cannot be observed only through the vertical motion of the beam. It can be seen if the beam is made to move simultaneously in a horizontal direction from left to right. The signal variations can be faithfully displayed on the CRT screen when the electron beam is deflected horizontally through equal distances in equal intervals of time. It requires that the voltage applied to horizontal plates increases uniformly with time. A ramp voltage given to the *X*-plates causes uniform motion of the electron beam across the screen. The shape of the ramp voltage is shown in Fig. 21.15.

The ramp voltage is also known as sweep voltage or saw tooth voltage or time base. It is generated by an oscillator, known as time base generator.

Thus, *the time base generator is a variable frequency oscillator, which produces an output voltage of saw tooth shape.*

Ideally, the voltage increases uniformly with time from a minimum to a maximum value and from there it abruptly falls to minimum. The process is repeated again and again. When the ramp voltage is applied to *X*-plates of CRT, the electron beam is deflected horizontally from the left edge to the right edge of the screen through equal distances in equal intervals

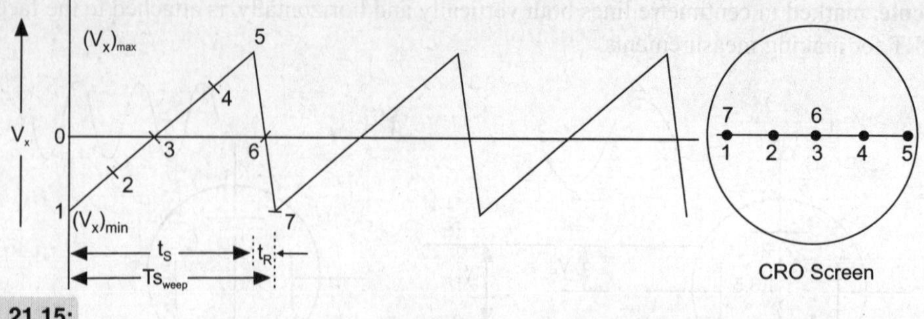

Fig. 21.15:

A saw-tooth voltage applied to X-plates produces a horizontal trace on the screen.

of time. Therefore, the luminous spot will traverse on the face of the screen at a uniform velocity. As the voltage attains a maximum and drops instantaneously to the minimum value, the spot is whipped back across the screen to the starting point at the left edge. The cycle is repeated. The sweep time may be typically from 10 ns to 5s/division.

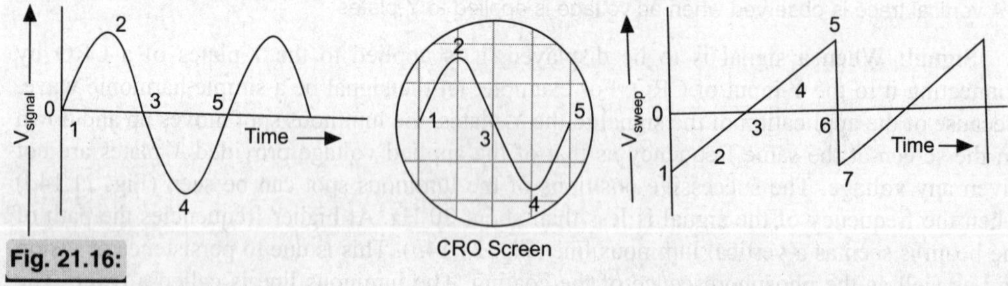

Fig. 21.16:

Display of Signal Shape.

Display of Signal Shape: For displaying the signal on the CRO screen it is necessary that the signal is fed to Y-input and the time base voltage is applied to *X*-plates. The electron beam is then subjected to two forces at the same time; one of the forces acts in the vertical direction while the second force acts in the horizontal direction. The deflection of the beam at any moment is given by the resultant of these two forces. Referring to Fig. 21.16, it is seen that at the instant 1 the input signal is zero and the sweep voltage is $(V_x)_{min}$, the resultant of the forces due to them acts along the left direction and the beam is deflected to the left extreme. At the instant 2, the signal amplitude is positive and the sweep voltage is at a lesser negative value. The beam is deflected left upward in the second quadrant of the screen. At the instant 3, both the input and sweep voltages are zero. The resultant force is zero and the beam stays at the center of the screen. At the instant 4, the signal amplitude is negative and the sweep voltage is positive. The beam is deflected to right down into the fourth quadrant of the screen. At the instant 5, the signal voltage is zero and the sweep voltage is $(V_x)_{max}$. The electron beam is deflected toward the right extreme along the horizontal direction. Then, the beam returns to position 1 and the process repeats. By joining the resultant positions of the spot, it is seen that the waveform of the input voltage is faithfully displayed.

Blanking: If the backward trace of the electron beam is displayed on the screen along with the forward trace, it gives a bad visual effect. In principle, by making the retrace time (time for the backward trace) equal to zero, the retrace path can be eliminated. But in practice, the retrace time cannot be reduced to zero; it can be only curtailed to a minimum value. To prevent the electron beam trace during retrace period, a high negative voltage pulse is applied to the control grid in the electron gun so that the electron beam is switched off momentarily. This process is known as *blanking* of the trace.

21.8.4 Trigger Circuit

Synchronization

Most of the time it is required that a stationary wave pattern to be displayed on the CRO's screen. To get this, there should be *synchronization* between the horizontal sweep voltage and the input signal. It occurs when the horizontal deflection in each sweep cycle starts at the same point of the input signal. If the sweep and signal voltages are not synchronized an stationary pattern is not obtained on the screen and the wave pattern goes on moving to the right or left of the screen.

Synchronization *is the method of locking the frequency of the time base generator to the frequency of the input signal so that a stationary display of wave pattern is seen on the CRO screen.*

The signal is synchronized when its frequency equals the sweep frequency or an integral multiple of the sweep frequency. That is

$$f_{signal} = n f_{sweep} \quad \text{...(21.7)}$$

or

$$T_{sweep} = n\, T_{signal}$$

As an example, if the sweep frequency is 50 Hz and the signal frequency is 50 Hz, one wave is displayed on the screen. On the other hand, if the sweep frequency is 50 Hz but the signal frequency is 100 Hz, the time period of sweep voltage is 20 ms and the time period of signal is 10 ms. In the sweep time which actually is horizontal trace length, the signal goes through two complete cycles. As a result two cycles of the signal voltage are displayed on the screen.

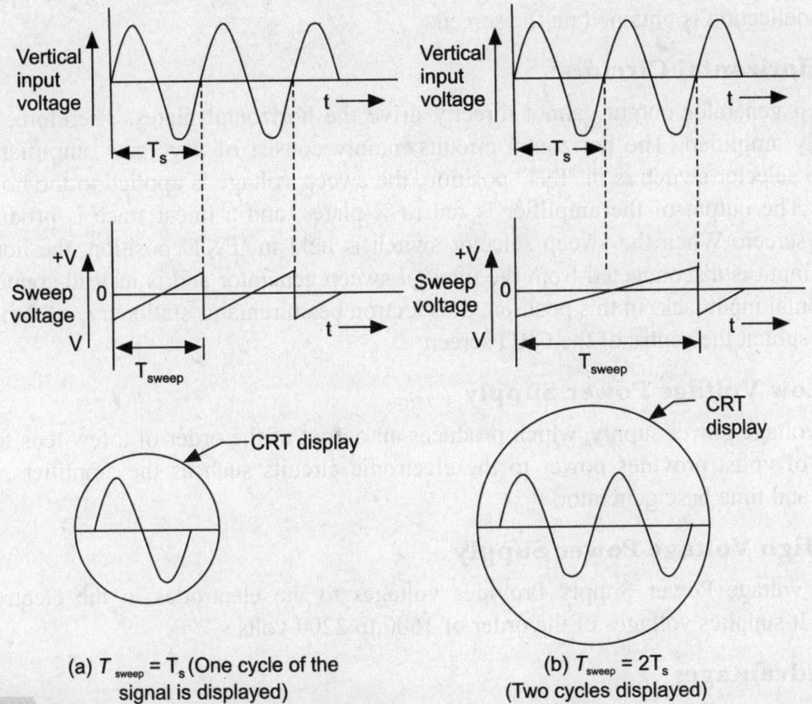

Fig. 21.17: Synchronization of signal and sweep voltages.

Achieving Synchronization

Synchronization may be achieved by using a trigger circuit. In this method, a part of the output obtained from the vertical amplifier is applied to a trigger generator. Trigger generator is sensitive to the level of the voltage applied at its input. The circuit monitors the input signal and detects the point when it reaches a selected level while moving toward a selected polarity. When the predetermined level is reached, the circuit produces a trigger pulse. This trigger pulse is fed to the time base generator, and it acts as a command signal to the time base generator and starts one sweep cycle of the time base. The sweep voltage is not developed in the trigger mode, if the input signal is not given. A portion of the trigger pulse is fed to a second circuit, which produces an un-blanking bias voltage to bring the grid of CRT to a potential, which allows electron beam to appear. Thus, a stationary display of the wave is seen only above a predetermined level of the input voltage. It happens in each cycle. Because the signal voltage is initiating the sweep cycle, both voltages will be synchronized. By proper adjustment of controls, the trigger pulse may be made to originate when the input signal is going positive on negative or at any particular voltage level.

However, in "auto" trigger mode, the trigger circuit will automatically provide a trigger pulse to the sweep generator even when the input signal is not applied to it and the horizontal trace is seen even without signal at *Y*-input.

21.8.5 Vertical Circuits

The vertical circuits consist of chiefly an attenuator, and a voltage amplifier. The signal is connected at the Y-input of CRO. It is then applied at the input of the attenuator. At this stage, the signal amplitude may be increased or decreased by changing the amount of attenuation. Subsequently, the output of the attenuator is fed to the input of the voltage amplifier and adequate deflection is obtained on the screen.

21.8.6 Horizontal Circuits

The sweep generator output cannot directly drive the horizontal plates. Therefore, it must be initially amplified. The horizontal circuits mainly consist of a voltage amplifier. When the sweep selector switch is in 'INT' position, the sweep voltage is applied to the horizontal amplifier. The output of the amplifier is fed to X-plates, and a linear trace is produced on the CRO screen. When the sweep selector switch is held in 'EXT' position, the horizontal amplifier input is disconnected from the internal sweep generator and is instead connected to the horizontal input jack. In this position, the electron beam remains stationary and produces a luminous spot at the centre of the CRO screen.

21.8.7 Low Voltage Power Supply

The low voltage power supply, which produces an output of the order of a few tens to a few hundreds of volts, provides power to the electronic circuits such as the amplifiers, trigger generator and time base generator.

21.8.8 High Voltage Power Supply

The high voltage Power Supply Provides voltages to the electrodes in the electron gun assembly. It supplies voltages of the order of 1600 to 2200 volts.

21.8.9 Advantages

(*i*) CRO is the only instrument that can give a visual display of the actual shape of the signal.

(*ii*) Electron beam, due to its negligible mass, can respond instantaneously to the variations of a signal, how rapidly they may occur. Therefore, CRO can be used to display and test signals of frequencies up to a few hundred mega hertz.

21.9 APPLICATIONS

A CRO is a universal measuring instrument capable of measuring the characteristics of a wide variety of electrical signals. The signals may be repetitive or occur only once and last a fraction of a microsecond. With the oscilloscope, we can determine the signal amplitude, time period (or duration), and also find out the phase relationship between two harmonic signals. We study here the application of CRO for measuring the frequency of a given wave and the phase difference between two harmonic waves.

21.9.1 Determination of Frequency

(*i*) Time period measurement method: The simpler method of determining the signal frequency is done by first measuring its period T and calculating the frequency using the relation

$$f = \frac{1}{T}$$

The time period of signal is measured as follows. The signal is fed to the vertical input of CRO. The scope vertical sensitivity, sweep speed and triggering controls are adjusted to obtain a stable waveform of one or two complete cycles covering a large area on the CRO screen. The number of horizontal divisions for one complete cycle is determined and the sweep speed is noted. The time period of the signal, T, is found by the formula

$$T = (\text{Number of horizontal divisions for 1 cycle}) \times (\text{sweep speed}) \quad ...(21.8)$$

Then the signal frequency is calculated.

$$f = \frac{1}{T}$$

This method of measuring frequency is accurate to approximately ± 3%, which is generally the sweep generator accuracy.

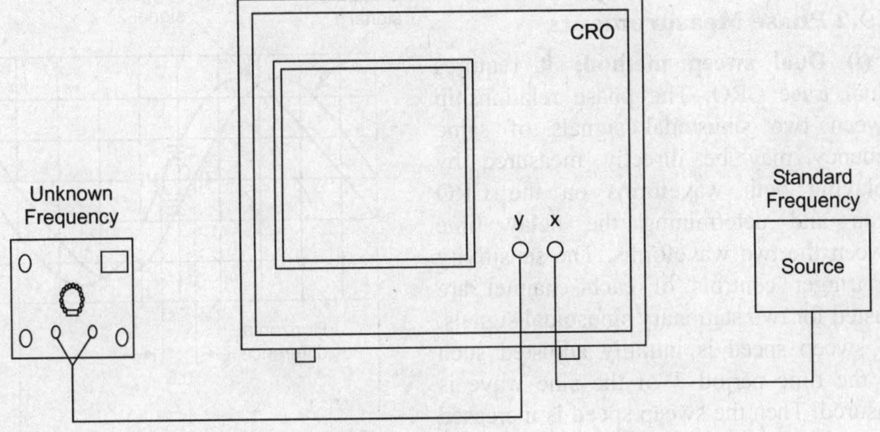

Fig. 21.18:

Experimental set up for frequency measurement

(*ii*) Lissajous Pattern method: A more accurate method is to use the CRO as a frequency comparator (see Fig. 21.18). The signal of unknown frequency is applied to the vertical input and a voltage of known frequency is given to the horizontal input. By varying the frequency of the known source, a stable loop pattern known as Lissajous pattern is obtained on the screen. The number of points at which the loops touch the horizontal and vertical tangents is noted. If L_H and L_V are the number of points touching the horizontal and vertical tangents respectively, then the unknown frequency is calculated from

$$f_y = f_x \left[\frac{L_H}{L_V}\right] \qquad ...(21.9)$$

where f_x is the known frequency. Example of measurement is shown in Fig. 21.19.

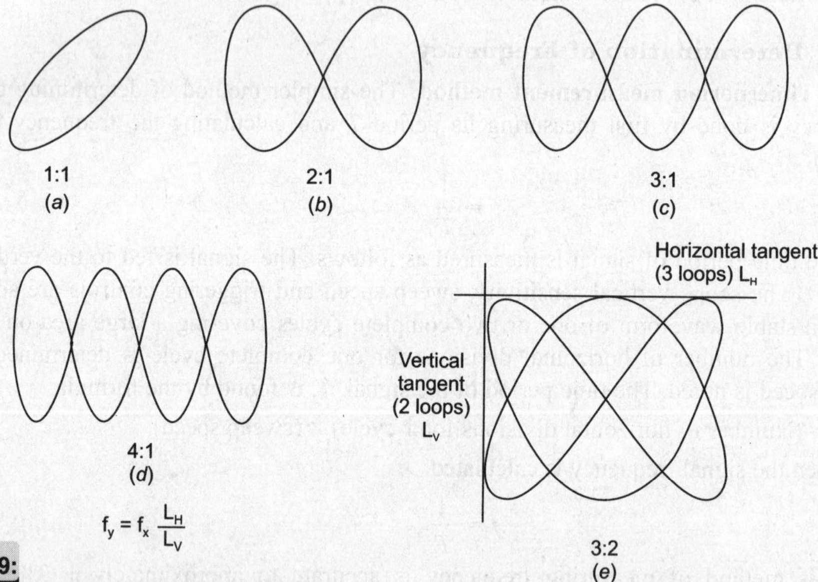

Fig. 21.19: Lissajous patterns and frequency measurement

21.9.2 Phase Measurements

(*i*) Dual sweep method: It requires a dual trace CRO. The phase relationship between two sinusoidal signals of same frequency may be directly measured by displaying both waveforms on the CRO screen and determining the delay time between the two waveforms. The sensitivity and trigger controls of each channel are adjusted for two stationary sinusoidal signals. The sweep speed is initially adjusted such that the time period T of the sine wave is measured. Then the sweep speed is increased and the delay time T_d between the two sine waves is accurately determined.

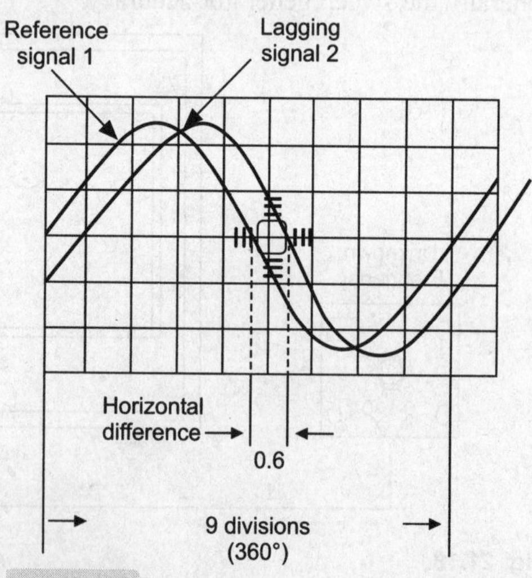

Fig. 21.20: Phase measurement using dual sweep method

The phase difference is calculated using the relation

$$\phi = \frac{T_d}{T} \times 360^\circ \qquad ...(21.10)$$

***(ii)* Lissajous Pattern method:** A second method for determining the phase difference of two sine waves of same frequency is to feed one sine wave to vertical input and the other sine wave to horizontal input (see Fig.21.21). The sweep selector switch is kept in 'EXT' position.

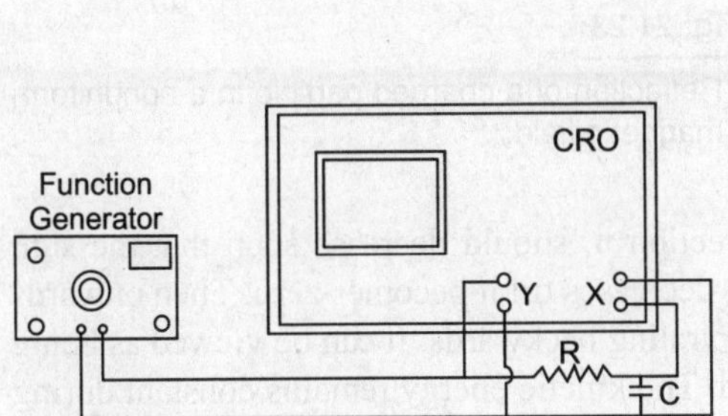

Fig. 21.21:

Experimental set up for phase measurement

A B

Vertical centre line

Fig. 21.22:

Phase measurement

A Lissajous pattern, namely ellipse (Fig.21.22), is obtained on the screen. By measuring the lengths A and B of the elliptical pattern the phase shift ϕ is calculated.

$$\phi = \sin^{-1}\left(\frac{A}{B}\right) \qquad ...(21.11)$$

21.10 OTHER APPLICATIONS OF AN ELECTRON BEAM

A focused electron beam finds important use in machining processes such as cutting, welding, etc. In the electron beam machining (EBM) a high energy electron beam is produced with the help of accelerating voltages of the order of 50 to 200 KV on the anode. The beam is focused to a sharp point of a few microns diameter using magnetic focusing. When such a high energetic beam hits the work piece, it produces very high temperatures of the order of 6000° C. Therefore, the beam can cut the material quickly and easily. In electron beam welding (EBW) the heat generated can weld materials. Materials which are difficult to weld by other means can be successfully welded with the help of EBW. It is chiefly used to weld metals such as zirconium, beryllium, tungsten, etc. In these processes, the electron gun system as well as work piece are to be kept in a vacuum.

The electron beam is also used in preparing high purity materials used in atomic, rocket and electronics engineering.

21.11 MOTION OF CHARGED PARTICLES IN A NONUNIFORM MAGNETIC FIELD

It is learnt earlier that when a charged particle enters a uniform magnetic field at an angle with respect to the field lines it spirals around the field direction. If the uniform field were of infinite extent, the charge would spiral indefinitely and the cross-section of the spiral would be uniformly the same. On the other hand, if the field is nonuniform the motion will be very complex. Let us assume that the magnetic field increases along the *X*-direction as shown in Fig. 21.23. It is seen that the field is converging in the direction of the axis of the spiral.

As the field is increasing in the X-direction the particle spirals in the X-direction with increasing cycling frequency. Also, the radius of the turns gradually decreases in the X-direction. Consequently, the particle spirals faster and faster in ever tightening loops.

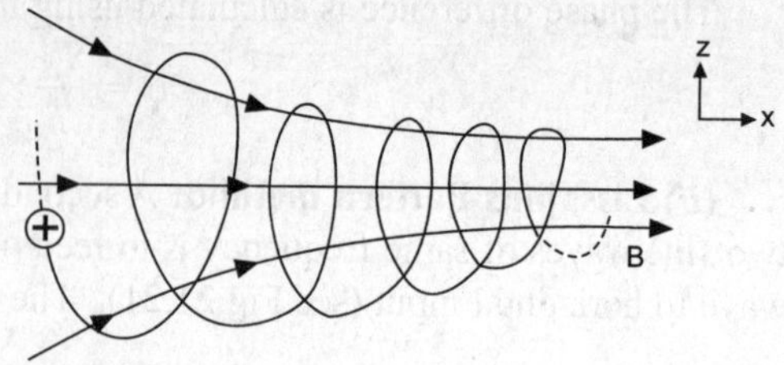

Fig. 21.23:

Reflection of a charged particle in a nonuniform magnetic field.

We have seen that a magnetic field cannot change the energy of the particle and hence it is also cannot change the velocity of the particle. Therefore, with the increasing spiraling velocity υ_x which is normal to the field B, the velocity parallel to the field direction υ_z should decrease such that the sum remains constant. The particle velocity gradually decreases till it becomes zero. Then onwards it gets reversed. As such the particle now starts spiraling backwards. It can be viewed as being reflected by the strong converging magnetic field. The kinetic energy remains constant during the process of reflection. Therefore, the backward velocity of the particle will be equal to the velocity in forward direction for the same points in space. As υ_z is increasing in magnitude, υ_x decreases in magnitude and the pitch of the spiral increases again. It amounts to state that the stronger field regions act as **mirrors** for charged particles. This effect is used to trap charged particles in a finite volume.

21.12 THE MAGNETIC BOTTLE

A **magnetic bottle** or a magnetic trap uses the above principle. It is a magnetic system with a weak field at its central region and increasingly stronger regions on both sides as illustrated in Fig. 21.24.

When a charged particle, say an electron enters the field at an angle with the field, it starts spiraling. In the region between A and B the field is diverging and the electron gets accelerated towards the right. When the particle passes into the region BC, the field lines are converging. Therefore, the electron experiences deceleration and gradually slows down and reverses its path to left. Thus the electron spirals back and forth between points M_1 and M_2, which act as mirror points. Consequently, the electron gets confined within the region AC of the nonuniform magnetic field which thus acts as a **magnetic bottle**. A cylindrical tube fitted with magnetic fields coils at its two ends will serve as a magnetic bottle. The increased fields at the ends of the cylinder serve as magnetic mirrors.

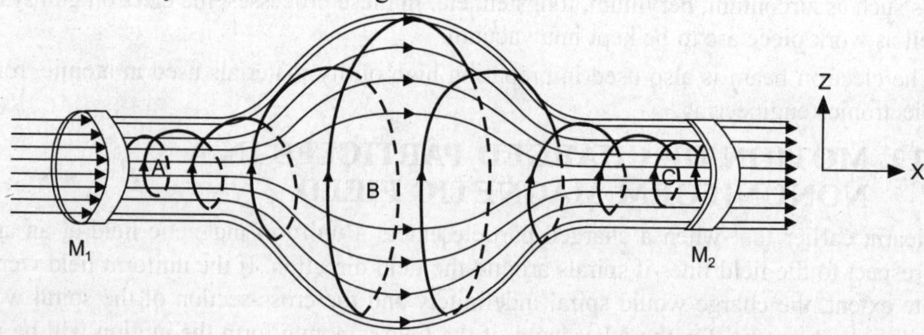

Fig. 21.24:

The magnetic bottle

It was discovered in 1958 by James A. van Allen that the earth's magnetic field forms a huge magnetic bottle in space. As charged particles of cosmic origin rush toward the earth, they get captured by the nonuniform magnetic field and spiral back and forth between the two magnetic mirrors formed by the converging field near the north and south poles. The trapped particles form radiation belts around the earth and they are called **van Allen belts**.

The magnetic traps play a very important role in confining extremely hot plasmas. Plasmas are so hot ($T = 10^6$ K) that their contact with the walls of an ordinary container would melt the material of the container. Therefore, to confine plasma, magnetic bottles are used. Such a plasma confinement scheme plays a crucial role in achieving a controlled nuclear fusion process, which is almost an endless source of energy.

QUESTIONS

1. Discuss refraction of electron beam across an equipotential surface. Hence explain the working of electrostatic lens.
2. What do you mean by a non-uniform electric field? Discuss Bethe's law. What is the refractive index of electrostatic field?
3. Explain Bethe's law of electron refraction. Compare it with Snell's law in optics.
4. State Bethe's law for electron refraction.
5. State Bethe's law. How is it analogous to Snell's law?
6. What is Bethe's law? Show how this concept helps in understanding the focusing of electron beam by a symmetrical electron lens.
7. Explain how an electron beam can be made to bend either towards or away from the normal to an equipotential surface.
8. Explain briefly the electrostatic focusing.
9. Discuss refraction of electron beam across an equipotential surface. Hence explain the working of the symmetrical and unsymmetrical electron lenses.
10. Explain working of an electrostatic electron lens.
11. Explain the working of a magnetic lens.
12. Derive an expression for electron deflection of electron beam in transverse magnetic field.
13. Draw a schematic of a cathode ray tube and show its principal parts. Briefly describe their functions.
14. What is the need of the acquadag coating inside the glass bulb in a CRT?
15. Draw a schematic of an electrostatic CRT. Describe the role of : (*i*) Electron gun, (*ii*) Deflection system, (*iii*) Fluorescent screen, (*iv*) Acquadag coating
16. Derive an expression for electrostatic deflection sensitivity of a C.R.T. in terms of d, V_A, l, and L where the terms have their usual meaning.
17. Draw the block diagram of a CRO and write in brief the working of each block.
18. Draw the block diagram of a cathode ray oscilloscope and explain the various parts.
19. Draw block diagram of CRO. How can the frequency of ac mains be determined using CRO? Explain.
20. Draw a neat block diagram of C.R.O. Explain how intensity and sharpness of the trace on the screen is controlled.
21. Draw the block diagram of CRO.
22. Draw a neat block diagram of a CRO. Explain how the actual waveform of the signal is traced on the CRO screen with the help of time base generator.
23. Draw the block diagram of CRO. Explain the function of trigger circuit.

24. Draw block diagram and explain construction, working and applications of cathode ray oscilloscope (CRO).
25. What are the important parts of CRO? Explain the function of sweep generator.
26. Explain the working of time base circuit and trigger circuit in CRO.
27. Explain how the true shape of a voltage waveform is displayed on a CRO screen.
28. In the normal operating condition of CRO, what are the voltages present on X-plates and Y-plates of its CRT?
29. How can the frequency of a given signal be determined using the CRO?
30. Discuss two applications of a CRO.
31. Draw a block diagram of CRO and explain the function of time base circuit.
32. What is CRO? Give its applications.
33. Explain the effect of a nonuniform magnetic field on the motion of a charged particle.

22 Thermoelectricity

Learning Objectives:

By the end of this chapter, you will be able to:

- Understand different Aspects of Thermoelectricty – Seebeck Effect, Peltier Effect and Thomson Effect.
- Explain basics of Thermoelectric Power.
- State Laws of Thermoelectric.
- Learn about Thermoelectric Cooling/Coolers.

22.1 INTRODUCTION

Thermoelectricity is a distinctive part of solid state technology. Thermoelectricity refers together to the Seebeck effect, Peltier effect and the Thomson effect. The Peltier-Seebeck and Thomson effects are reversible, whereas Joule heating is not. Even though the principles of thermoelectric effects are known for a long time, recently only its useful applications have become feasible. W. Thomson provided the first important theoretical basis for thermoelectricity. He related, mathematically, the Peltier and Seebeck effects and predicted the existence of the Thomson effect.

22.2 SEEBECK EFFECT

In 1821 Thomas Johann Seebeck discovered that an electromotive force appears in a circuit composed of two dissimilar metals,if the junctions between the metals are held at different temperatures (Fig. 22.1). The e.m.f. is known as thermoelectric *e.m.f.*

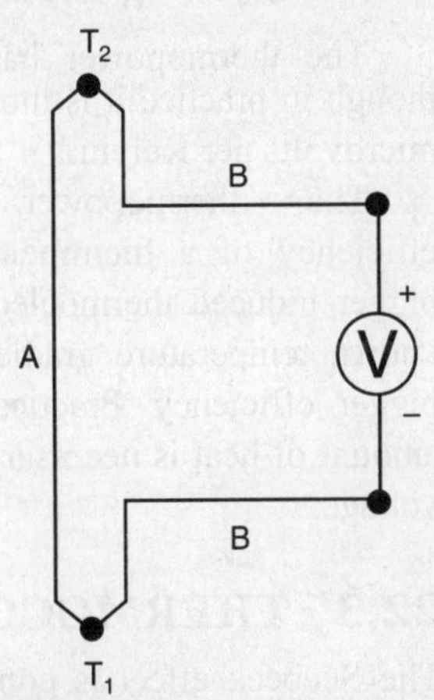

Fig. 22.1: Seebeck Effect

The thermoelectric e.m.f. causes a continuous current in the conductors if they form a complete loop and the current is known as thermoelectric current. This phenomenon is called the Seebeck effect.

The voltage (thermoelectric e.m.f.) created is of the order of several microvolts per Kelvin difference. The thermoelectric e.m.f. will exist and the current will flow in the circuit as long as the two junctions, known as the "hot" junction and "cold" junction, are at different temperatures. Thus, the Seebeck effect is the conversion of temperature differences directly into electricity. The magnitude and direction of thermoelectric current is affected by the types of metals used, and the temperature difference between the hot and cold ends; and does not depend on the temperature distribution along the conductors.

The voltage developed in the circuit, Fig. 22.1, is proportional to the temperature difference between the two junctions.

$$V = \alpha(T_2 - T_1) \quad ...(22.1)$$

where $\alpha = \alpha_B - \alpha_A$.

α_A and α_B are known as the Seebeck coefficients of the metals A and B and T_1 and T_2 are the temperatures of the two junctions.

Seebeck effect is observed not only in metals but as well in semiconductors also. It is not necessarily a junction phenomenon, but arises in a single conductor also. If a temperature gradient is caused in a conductor, electrons diffuse from the hot side to the cold side. Electrons migrating to the cold side leave behind their oppositely charged and immobile nuclei on the hot side and thus give rise to a thermoelectric voltage.

22.2.1 Seeback coefficient

Seebeck coefficient or *thermopower* of a material measures the magnitude of an induced thermoelectric voltage in response to a temperature difference across that material.

Thermopower is defined as the open circuit voltage produced between two points on a conductor, where a uniform temperature difference of 1K exists between those points.

If the temperature difference ΔT between the two ends of a material is small, then the *thermopower or Seebeck coefficient* of a material may be written as

$$\alpha = \frac{\Delta V}{\Delta T} \quad ...(22.2)$$

This can also be expressed in terms of the electric field E and the temperature gradient ∇T, as

$$\alpha = \frac{E}{\nabla T} \quad ...(22.3)$$

The thermopower has units of (V/K), though in practice it is more common to use microvolts per Kelvin.

The thermopower signifies the efficiency of a thermoelectric material. A bigger induced thermoelectric voltage for a known temperature gradient will lead to a bigger efficiency. Practically only a small amount of heat is necessary to create a large voltage.

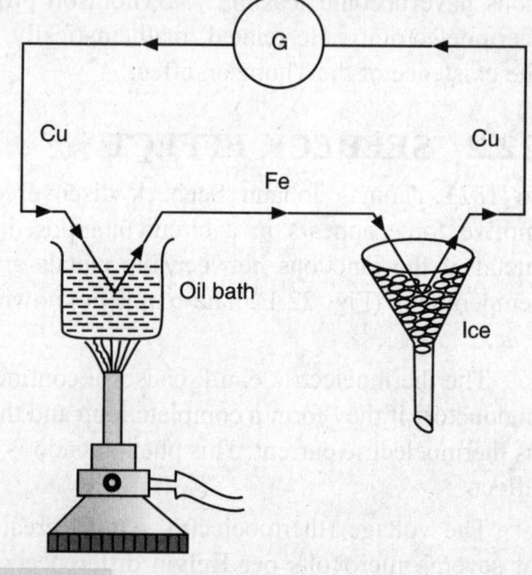

Fig. 22.2: Illustration of Thermocouple

22.3 THERMOCOUPLE

The Seebeck effect is commonly used in a device called a *thermocouple*, a widely used method of temperature measurement. A pair of junctions formed by two dissimilar metals constitutes a thermocouple. It is used to measure a temperature difference directly or to measure an absolute temperature by setting one end to a known temperature, say 0°C (Fig. 22.2).

22.4 THERMOELECTRIC SERIES

Seebeck arranged the metals into the following series, known as *Seebeck series*. Bi, Ni, Co, Pd, U, Cu, Mn, Tl, Hg, [Pb], Sn, Cr, Mo, Rh, Ir, Au, Zn, W, Cd, Fe, As, Sb, Te.

The metals to the left side of Pb are called *thermoelectrically positive metals*, while those to its right side are called *thermoelectrically negative metals*.

If a thermocouple is made with any two metals in *Seebeck* series, the thermo current flows across the hot junction from the metal found earlier in the series to the one appearing afterwards. At a given temperature difference, for the larger the separation of the two metals in the series, the higher is the e.m.f. For instance, the thermo e.m.f. for Ni-Fe couple is greater than for Cu-Fe couple. In Cu-Fe thermocouple, the current flows from Cu to Fe through the cold junction and from Fe to Cu through the hot junction.

22.5 VARIATION OF THERMOELECTRIC E.M.F. WITH TEMPERATURE

If the temperature of the cold junction of a thermocouple is kept at 0°C and the thermoelectric e.m.f. 'e' is plotted against the temperature T of the hot junction, we obtain a parabolic curve, as shown in Fig. 22.3.

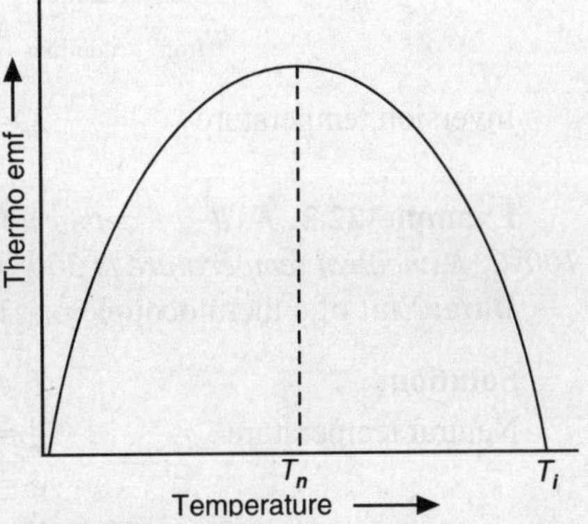

Fig. 22.3: Variation of thermoelectric e.m.f. with temperature

It is seen that the thermo e.m.f. increases with the temperature of the hot junction and becomes a maximum at a particular temperature, T_n. T_n is known as the **neutral temperature** which is a constant for the given pair of metals forming the thermocouple. *The temperature of the hot junction at which maximum thermo e.m.f. flows is a constant for a given couple and is known as neutral temperature T_n for that couple.* If the temperature of the hot junction is increased beyond the neutral temperature, the e.m.f. decreases and becomes zero at a temperature T_i, known as the **inversion temperature**.

The temperature, at which the thermo e.m.f. is zero, is known as inversion temperature.

Beyond the temperature of inversion, the e.m.f. again increases but in the reverse direction. The thermo e.m.f. varies with temperature according to the following relation.

$$e = aT + \frac{1}{2}bT^2 \qquad ...(22.4)$$

where *a* and *b* are *Seebeck constants* for the thermocouple.

Eq. (22.4) is known as *Seebeck equation*, and $t = T_2 - T_1$.

Differentiation of Eq. (22.4) gives

$$\frac{de}{dT} = a + bT \qquad ...(22.5)$$

At $T = T_n$, $\dfrac{de}{dT} = 0$ and e is a maximum. Therefore,

$$0 = a + bT_n$$

$$\therefore \quad T_n = -\frac{a}{b} \quad ...(22.6)$$

At $T = T_i$, $e = 0$. Therefore, it follows from Eq. (22.4) that

$$0 = aT_i + \frac{1}{2}bT_i^2$$

or
$$T_i(a + \frac{1}{2}bT_i) = 0$$

$$\therefore \quad T_i = -\frac{2a}{b} \quad ...(22.7)$$

It follows from Eq. (22.6) and Eq. (22.7) that

$$T_i = 2T_n \quad ...(22.8)$$

Example 22.1. *Calculate the inversion temperature for an iron-cadmium thermocouple. The value of a and b are 17.5 and –0.1 for iron and for cadmium these values are 3 and + 0.08 respectively.*

Solution:
$$a_{\text{iron - cadmium}} = a_{\text{iron}} - a_{\text{cadmium}} = 17.5 - 3 = \mathbf{14.5}$$
$$b_{\text{iron - cadmium}} = b_{\text{iron}} - b_{\text{cadmium}} = -0.1 - 0.08 = \mathbf{-0.18}$$

Inversion temperature
$$T_i = -\frac{a_{Fe-cd}}{b_{Fe-cd}} = -\frac{14.5}{-0.18} = \mathbf{80.55°C}$$

Example 22.2. *EMF of a thermocouple is 1200 μV, when working between 0°C and 100°C. Its neutral temperature is 300° C. Find the values of a and b for it.*

Data: Emf of a thermocouple $e = 1200\ \mu V = 1200 \times 10^{-6}\ V$

Solution: $T = 100°C$

Neutral temperature $T_n = 300°C$

$$e = aT + bT^2$$
$$(1200 \times 10^{-6})\ V = a\,(100) + b(100)^2$$

i.e.,
$$a + 100b = 12 \times 10^{-6} \quad ...(i)$$

$$T_n = -\frac{a}{2b}$$

$$300°C = -\frac{a}{2b}$$

$$\therefore \quad a = -600\,b \quad ...(ii)$$

From relation (*i*), $-600b + 10b = 12 \times 10^{-6}$

i.e.,
$$b = -\frac{12}{5} \times 10^{-8} = \mathbf{2.4 \times 10^{-8}}$$

From relation (*ii*) $a = (-600)\,(-2.4 \times 10^{-8}) = \mathbf{14.4 \times 10^{-6}}$

Example 22.3. *The thermo e.m.f. of a Cu-Fe thermocouple is 2160 μV when the cold junction is at 0°C and the hot junction at 250°C. Calculate the constants a and b if the neutral temperature is 330°C.*

Solution: $e = aT + bT^2$

$$2160 \times 10^{-6} \text{ volts} = a\,(250) + b\,(250)^2$$

$$\therefore \quad a + 250b = 8.64 \times 10^{-6} \quad \text{...}(i)$$

Neutral temperature $T_n = -\dfrac{a}{2b}$

$$\therefore \quad a = T_n\,(-2b) = -660\,b \quad \text{...}(ii)$$

Using (*ii*) in (*i*), we get

$$-660b + 250b = 8.64 \times 10^{-6}$$

$$\mathbf{b = -0.02107 \times 10^{-6}}$$

From (*ii*) $a = 660\,(-0.02107 \times 10^{-6})$

$$\mathbf{a = 13.91 \times 10^{-6}}$$

Example 22.4. *For Fe-Cu thermocouple it is observed that the thermo e.m.f is zero when one of the junction is at 20°C and the other one is at some higher temperature. If the neutral temperature is 285°C, calculate the higher temperature. Hence find out the temperature of inversion, if the cold junction temperature is at –20°C.*

Solution: For Fe – Cu thermocouple,

e.m.f. $e = 0$ at $T_1 = 20°C$ and $T_2 = T$ (say)

$$e = 0 = a\,(T_2 - T_1) + 2b\,(T_2 - T_1)$$

$$\therefore \quad a + 2b\,(T_2 - T_1) = 0$$

i.e., $(T_2 - T_1) = \dfrac{-a}{2b}$

We know that $-a/2b = T_n$ (neutral temperature)

$$\therefore \quad T_2 - T_1 = T_n$$

$$\therefore \quad T_2 = T_n + T_1 = 285°C + 20°C$$

$T_2 = 305°C$ is the higher temperature. If cold junction temperature $T_1 = -20°C$, then the neutral temperature $T_n = T_2 - T_1 = 305°C - (-20°C)$

or $T_n = 325°C$

$\therefore$ Inversion temperature $T_i = 2T_n = \mathbf{650°C}$

22.6 THE PELTIER EFFECT

In 1834 Peltier discovered this effect. The Peltier effect is a junction phenomenon.The Peltier effect is found to be proportional to the first power of the current.

When electric current is passed in a circuit consisting of two dissimilar metals, heat is evolved at one junction and is absorbed at the other junction. This is known as Peltier effect. It is the inverse of the Seebeck effect.

There is heat absorption or generation at the junctions depending on the direction of current flow. Heat generated by current flowing in one direction was absorbed if the current was reversed.

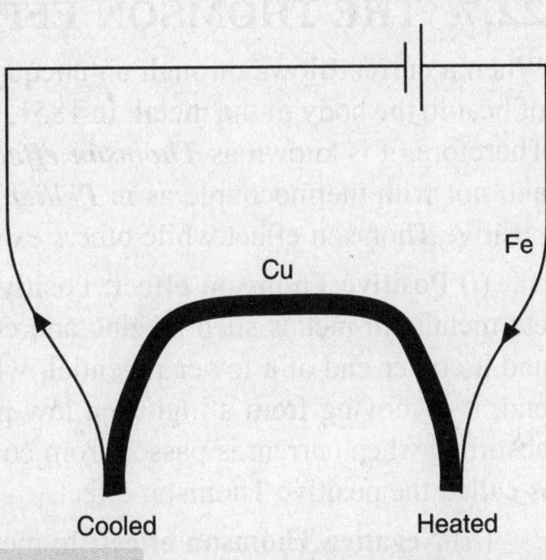

Fig. 22.4: Peltier Effect

As an example, consider the circuit Fig. 22.4. If right-hand junction is heated, the electrical energy is changed into heat energy and heat energy is changed into electrical energy at the left junction, thereby making it to be cooled. If the case is reversed; heat is absorbed at the right junction and produced at the left one.

22.6.1 The Peltier Coefficient

Peltier coefficient is defined as the amount of heat energy absorbed or evolved at the junction of two dissimilar metals when one ampere of current flows through it for one second. It is denoted by π and expressed in volts. It is a property that depends on both materials of the junction.

The heat absorbed per second at a junction carrying a current I amperes is given by

$$\text{Heat absorbed/sec} = \pi_{ab} I \qquad \text{...(22.9)}$$

$\therefore$ Heat absorbed in t seconds, $H = \pi_{ab} I\, t = \pi_{ab} q.$

where the current is from metal a to metal b. The junction emf, π_{ab}, is known as the *Peltier coefficient*.

$$\pi_{ab} = \frac{H}{It} \qquad \text{...(22.10)}$$

π_{ab} is positive if metal a is positive with respect to metal b (thus $\pi_{Cu\text{-}Fe}$ is positive). The magnitude of π_{ab} is a function of the temperature of the junction. For identical temperatures $\pi_{ab} = -\pi_{ba}$. Thus, if the direction of the current in Eq. (22.9) is reversed, the heat absorbed per second is

$$\text{Heat absorbed/sec} = \pi_{ba} I \qquad \text{...(22.11)}$$

which is opposite in sign to Eq. (22.9).

If V is the potential difference applied, then

$$\text{Heat absorbed} = V_q = VIt$$

Equating the above with Eq. (22.9), we get

$$\pi = V \qquad \text{...(22.12)}$$

Thus, the Peltier coefficient is numerically equal to the applied potential difference expressed in volts.

22.7 THE THOMSON EFFECT

When a current flows through an unequally heated metal, there is an absorption or evolution of heat in the body of the metal. In 1851, W. Thomson (later Lord Kelvin) predicted this effect. Therefore, it is known as ***Thomson effect***. In Thomson effect we deal with only metallic rod and not with thermocouple as in ***Peltier effect*** and ***Seebeck effect***. Some of the metals show positive Thomson effect while others exhibit negative Thomson effect.

(*i*) Positive Thomson effect: Positive Thomson effect is shown by Cu, Sn, Ag, Cd, Zn… etc metals. In metals such as zinc and copper, which have a hotter end at a higher potential and a cooler end at a lower potential, when current moves from the hotter end to the colder end, it is moving from a high to a low potential, so there is an evolution of heat and heat is absorbed when current is passed from colder end to hotter end, as shown in Fig. 22.5(*a*). This is called the positive Thomson effect.

(*ii*) **Negative Thomson effect:** In metals which have a cooler end at a higher potential and a hotter end at a lower potential such as cobalt, nickel, and iron, when current moves from the hotter end to the colder end, there is absorption of heat. The negative Thomson effect is the evolving of heat is when current is passed from colder end to the hotter end (Fig. 22.5 *b*). The metals which show negative Thomson effect are Fe, Co, Bi, Pt, Hg… etc.

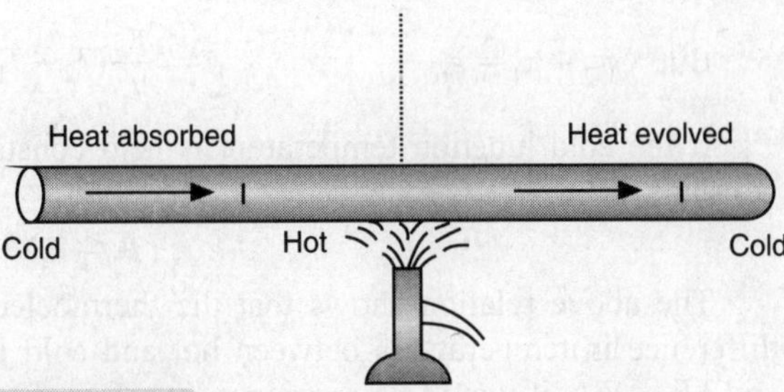

Fig. 22.5 (a): Positive Thomson effect

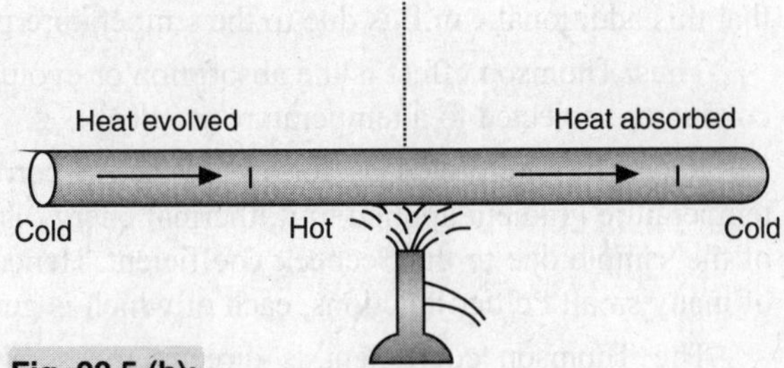

Fig. 22.5 (b): Negative Thomson effect

Now let us consider a circuit consisting of two metals A and B as shown in Fig. 22.6. The hot junction is at a temperature T_2°K and the cold junction is at a temperature T_1°K. Due to the Seebeck effect, i.e., due to temperature difference between the junctions, thermoelectric current flows through the circuit. As the current flows through the hot and cold junctions, heat is absorbed at the hot junction and evolved at the cold junction due to Peltier effect.

Let π_1 and π_2 be the Peltier coefficients at T_1 and T_2. During the passage of current an amount of heat energy equal to $\pi_2 q$ is absorbed at hot junction and heat energy $\pi_1 q$ is evolved at the cold junction. Then, the energy $(\pi_2 - \pi_1)q$ is used in driving the current through the circuit. As π_2 and π_1 are equal to the potential differences at hot and cold junctions respectively, the thermo e.m.f. developed is given by

$$e = (\pi_2 - \pi_1) \qquad ...(22.13)$$

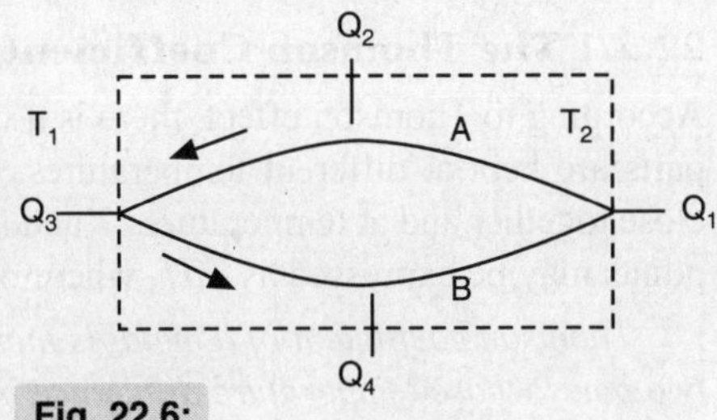

Fig. 22.6: Illustration of Peltier effect

The current in the circuit is small, and the joules heating effect is negligible. As the Peltier effect is reversible, a thermocouple may be regarded as a reversible heat engine taking heat from the source at the hot junction at temperature T_2, does work in driving the current through the circuit, and rejecting heat to the sink, the cold junction at temperature T_1. By the Carnot's engine we have

$$\frac{Q_2}{T_2} = \frac{Q_1}{T_1} \qquad ...(22.14)$$

Now during the flow of current in the thermocouple, heat absorbed at the hot junction is $Q_2 = \pi_2 q$ joules while the energy given out to the sink is $Q_1 = \pi_1 q$ joules.

$$\frac{\pi_2 q}{T_2} = \frac{\pi_1 q}{T_1} \qquad \therefore \quad \frac{\pi_2}{\pi_1} = \frac{T_2}{T_1} \qquad \therefore \quad \frac{\pi_2}{\pi_1} - 1 = \frac{T_2}{T_1} - 1$$

or
$$\frac{\pi_2 - \pi_1}{\pi_1} = \frac{T_2 - T_1}{T_1} \qquad ...(22.15)$$

or
$$\pi_2 - \pi_1 = \pi_1\left(\frac{T_2 - T_1}{T_1}\right)$$

But $\pi_2 - \pi_1 = e$. $\therefore$ $e = \frac{\pi_1}{T_1}(T_2 - T_1)$ ...(22.16)

If the cold junction temperature is held constant, the Peltier coefficient π_1 will also be constant. Then,

$$e \propto (T_2 - T_1) \quad ...(22.17)$$

The above relation shows that the thermoelectric e.m.f. is directly proportional to the difference in temperatures between hot and cold junctions. If a graph is plotted between e and $(T_2 - T_1)$, it would be a straight line. Experiments show that the graph is a parabola. It means that Peltier effect alone cannot account for thermo e.m.f. in a thermocouple. Thomson predicted that there is some additional source of e.m.f. besides the Peltier effect. He concluded that this additional e.m.f. is due to the temperature gradient in the individual metals.

Thus, Thomson effect is the absorption or evolution of energy in a single current carrying conductor subjected to a temperature gradient.

It can be interpreted as follows. When a current is flowing through a material and a temperature gradient applied to it, thermal energy is generated (or absorbed) along the length of the sample due to the Seebeck coefficient. Hence, the sample can be imagined as a series of many small Peltier junctions, each of which is generating (or absorbing) heat.

The Thomson coefficient is directly measurable for individual materials. The Peltier and Seebeck coefficients can only be determined for pairs of materials. Absolute Seebeck coefficient or absolute Peltier coefficient cannot be found for an individual material.

22.7.1 The Thomson Coefficient

According to Thomson effect, there is a variation of potential along a metal when its different parts are kept at different temperatures. Consider two points in a conductor which are very close together and at temperatures T and $(T + dT)$. The difference of potential between the two points may be expressed as σdT, where σ is the Thomson coefficient. Hence,

Thomson coefficient of a metal is numerically equal to the difference in potential between two parts whose temperature difference is 1°C.

Consider a material with a current flowing through it and a temperature gradient applied to it. In this situation, thermal energy is generated (or absorbed) all along the sample. The energy absorbed or evolved in a time 't' seconds is $(\sigma dT)\, I_t$.

Thomson coefficient may then be expressed as the ratio of the power evolved per unit volume in the sample to the applied current and temperature gradient.

$$\sigma = \frac{P}{I\, dT} \quad ...(22.18)$$

The amount of heat energy absorbed or evolved per second between two points of a conductor having a unit temperature difference, when a unit current is passed is known as ***Thomson coefficient*** for the material of a conductor.

22.8 E.M.F. IN A THERMOCOUPLE

Let us consider a thermocouple made of two dissimilar metals A and B as shown in Fig. 22.7. Let the hot junction be at absolute temperature T_2 and the cold be at T_1. Let π_1 and π_2 be the Peltier coefficients at the cold and hot junctions respectively directed from metal B to metal A through the junctions. Then, the e.m.f. due to Peltier effect is in the anticlockwise direction and is equal to $(\pi_2 - \pi_1)$.

Let σ_A and σ_B be the Thomson coefficients of the metals A and B respectively, which are assumed to be positive. Then, the Thomson e.m.f. in the metal A in the anticlockwise direction is

$$E_A = -\int_{T_1}^{T_2} \sigma_A \, dT \qquad ...(22.19)$$

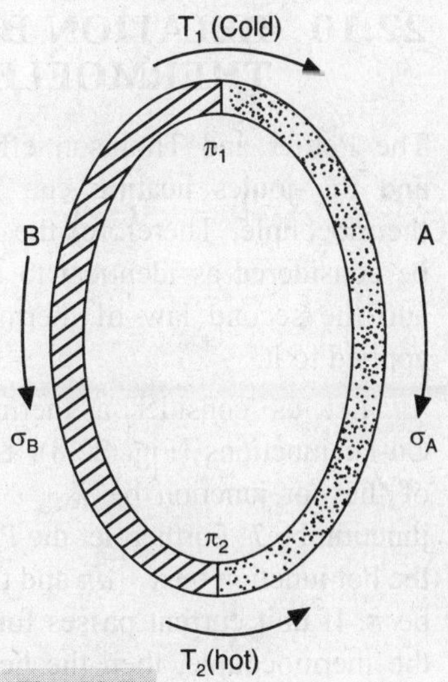

Fig. 22.7: Thermocouple made of two dissimilar metals A and B

Similarly, Thomson e.m.f. in metal B in anticlockwise direction is

$$E_B = +\int_{T_1}^{T_2} \sigma_B \, dT \qquad ...(22.20)$$

The resultant e.m.f. in the circuit is

$$E_{AB} = (\pi_2 - \pi_1) - \int_{T_1}^{T_2} \sigma_A \, dT + \int_{T_1}^{T_2} \sigma_B \, dT$$

$$\therefore \quad E_{AB} = (\pi_2 - \pi_1) - \int_{T_1}^{T_2} (\sigma_A - \sigma_B) dT \qquad ...(22.21)$$

22.9 THE THERMOELECTRIC POWER

The rate of change of e.m.f. with temperature is called *thermoelectric power* and is denoted by *P*. Thus,

$$P = \frac{de}{dT} \qquad ...(22.22)$$

Let us consider a thermocouple made of two dissimilar metals *A* and *B* with their junctions at temperatures $(T + dT)$ and T. The Peltier coefficient at temperature $(T + dT)$ is $(\pi + d\pi)$ and at temperature T, it is π. Let σ_A and σ_B be the Thomson coefficients of the metals *A* and *B* respectively, which are assumed to be positive.

Current flows in the thermocouple circuit due to Seebeck effect. Heat is absorbed at the hot junction and is evolved at the cold junction due to the Peltier effect. And as temperature gradient exists along *A* and *B*, heat is evolved in *A* and is absorbed in *B*, due to the Thomson effect.

If a unit charge passes through the circuit, the energy absorbed due to Peltier effect at the hot junction $= (\pi + d\pi)$.

Energy evolved due to Peltier effect at the cold junction $= -\pi$

Energy absorbed in metal *B* due to Thomson effect $= \sigma_B \, dT$

Energy evolved in metal *A* due to Thomson effect $= -\sigma_A \, dT$

The total gain of energy is $de = (\pi + d\pi) - \pi + \sigma_B \, dT - \sigma_A \, dT$

$$\therefore \quad de = d\pi - (\sigma_A - \sigma_B)\, dT$$

$$\therefore \quad \text{Thermoelectric power, } P = \frac{de}{dT} = \frac{d\pi}{dT} + (\sigma_B - \sigma_A) \qquad ...(22.23)$$

22.10 RELATION BETWEEN PELTIER COEFFICIENT AND THERMOELECTRIC POWER

The Peltier and Thomson effects are reversible and the joules heating can be neglected in a thermocouple. Therefore, the thermocouple may be considered as identical to a reversible engine and the second law of thermodynamics can be applied to it.

Let us consider a thermocouple made of Cu-Fe junctions (Fig. 22.8). Let the temperature of the hot junction be $T + dT$ and that of cold junction be T. Further, let the Peltier coefficient of the hot junction be $\pi + d\pi$ and that of cold junction be π. If unit current passes for unit time through the thermocouple, then the heat absorbed at hot junction will be $\frac{\pi + d\pi}{J}$ and heat given out at the cold junction will be $\frac{\pi}{J}$, where J is the mechanical equivalent of heat.

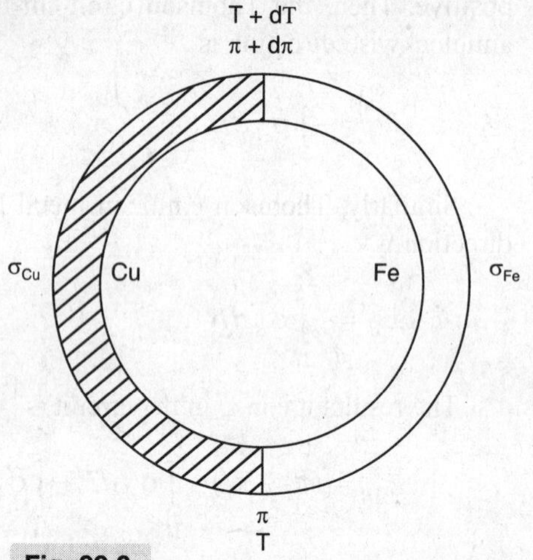

Fig. 22.8: Thermocouple made of Cu-Fe junctions

The heat energy absorbed due to the Thomson effect = $\frac{(\sigma_{Cu} - \sigma_{Fe})}{J} dT$ ...(22.24)

Using the second law of thermodynamics, $\frac{Q_1}{T_1} = \frac{Q_2}{T_2}$, we obtain

$$\frac{\pi + d\pi}{T + dT} - \frac{\pi}{T} + \frac{(\sigma_{Cu} - \sigma_{Fe})}{T} dT = 0$$

$$\frac{T d\pi}{(T + dT)T} - \frac{\pi dT}{(T + dT)T} + \frac{(\sigma_{Cu} - \sigma_{Fe})}{T^2} T\, dT = 0$$

As dT is a very small quantity, we can approximate $(T + dT)\, T \approx T^2$.

$$\therefore \quad \frac{T d\pi}{T^2} - \frac{\pi dT}{T^2} + \frac{(\sigma_{Cu} - \sigma_{Fe})}{T} dT = 0$$

or $$d\pi - \frac{\pi dT}{T} + (\sigma_{Cu} - \sigma_{Fe}) dT = 0$$

or $$d\pi + (\sigma_{Cu} - \sigma_{Fe}) dT = \frac{\pi dT}{T} \quad ...(22.25)$$

But $$d\pi + (\sigma_{Cu} - \sigma_{Fe}) dT = de$$

$$\therefore \quad de = \frac{\pi dT}{T}$$

or $$\pi = T \cdot \frac{de}{dT} \quad ...(22.26)$$

or $$\pi = TP \quad ...(22.27)$$

The above equation shows that the Peltier coefficient of a thermocouple is equal to the product of the absolute temperature of the junction and the thermoelectric power at that temperature.

Example 22.5: *The e.m.f. in micro volts (e) of a thermocouple, one junction of which is at 0°C e is given by e = 1600 T – 4 T^2 where T °C is the temperature of hot junction. Find (i) neutral temperature (ii) Peltier coefficient.*

Solution: (*i*) $e = aT + bT^2 = 1600\ T - 4\ T^2$

$$\therefore \quad a = 1600 \text{ and } b = -4$$

$$\therefore \quad \text{Neutral temperature,} \quad T_n = \frac{-a}{2b} = \frac{-1600}{2(-4)}$$

$$\therefore \quad \boldsymbol{T_n = 200°C}$$

(*ii*) Peltier coefficient $\quad \pi = T\dfrac{de}{dT} = T(a+2bT)$

$$\therefore \quad \pi = (T)\,(1600 - 8\ T) = 1600\ T - 8\ T^2$$

22.11 RELATION BETWEEN THOMSON COEFFICIENT AND THERMOELECTRIC POWER

From Eq. (22.25), we have

$$d\pi + (\sigma_{Cu} - \sigma_{Fe})\, dT = \frac{\pi dT}{T}$$

$$(\sigma_{Cu} - \sigma_{Fe}) = \frac{\pi}{T} - \frac{d\pi}{dT} \quad \text{...(22.28)}$$

Using Eq. (22.26), we may write Eq. (22.28) as

$$(\sigma_{Cu} - \sigma_{Fe}) = \frac{de}{T} - \frac{d\pi}{dT}$$

Differentiating Eq. (22.26), we obtain

$$\frac{d\pi}{dT} = \frac{de}{dT} + T\frac{d^2e}{dT^2}$$

$$\therefore \quad \frac{de}{dT} - \frac{d\pi}{dT} = -T\frac{d^2e}{dT^2} \quad \text{...(22.29)}$$

$$\therefore \quad (\sigma_{Cu} - \sigma_{Fe}) = -T\frac{d^2e}{dT^2} \quad \text{...(22.30)}$$

or

$$(\sigma_{Fe} - \sigma_{Cu}) = T\frac{d^2e}{dT^2} \quad \text{...(22.31)}$$

If we consider a thermocouple of copper and lead, the Eq. (22.31) becomes

$$(\sigma_{Cu} - \sigma_{Pb}) = T\frac{d^2e}{dT^2} \quad \text{...(22.32)}$$

But for lead, $\sigma_{Pb} = 0$.

$$\therefore \quad \sigma_{Cu} = T\frac{d^2e}{dT^2} \quad \text{...(22.33)}$$

Similarly, for a thermocouple of iron and lead

$$\sigma_{Fe} = T\frac{d^2e}{dT^2} \quad ...(22.34)$$

or

$$\sigma_{Fe} = T\frac{d}{dT}\left(\frac{de}{dT}\right) = T\frac{dP}{dT} \quad ...(22.35)$$

Thus, Thomson coefficient is the product of absolute temperature and the rate of change of thermoelectric power.

Example 22.6: *The thermoelectric power for steel is 18 μV / °C at 0°C and zero at 400 °C. That for copper is 6 μV / °C at 500 °C and zero at –50°C. Find the e.m.f. for steel-copper thermocouple with one junction at its neutral temperature and other at 0°C.*

Solution: Thermoelectric power for steel

$$\left(\frac{de}{dT}\right)_{T=0°C} = 18\ \mu V/°C = a + 2b\ (0\ °C)$$

i.e.

$$a = 18 \times 10^{-6}\ V/°C$$

$$\left(\frac{de}{dT}\right)_{T=400°C} = O = a + 2b\ (400\ °C)$$

$$\therefore \quad b = \frac{-18\times10^{-6}\ V/°C}{800°C} = -2.25\times10^{-8}\frac{V/°C}{°C}$$

Thermoelectric power for Cu

$$\left(\frac{de}{dT}\right)_{T=500°C} = 6\ \mu V/°C = a + 2b\ (500°C)$$

$$\left(\frac{de}{dT}\right)_{T=-50°C} = O = a + 2b\ (-50°C)$$

i.e

$$a = 100\ b\ °C$$

$$\therefore \quad 100\ b\ (°C) + 100\ b\ (°C) = 6 \times 10^{-6}\ V/°C$$

$$b = 0.545 \times 10^{-8}\ V/°C/°C$$

and

$$a = 0.545 \times 10^{-6}\ V/°C$$

For the Steel-Copper thermocouple

$$a = 18 \times 10^{-6}\ V/°C - 0.545 \times 10^{-6}\ V/°C$$

$$= 17.455 \times 10^{-6}\ V/°C$$

$$b = -2.25 \times 10^{-8}\ V/°C/°C - 0.545 \times 10^{-8}\ V°/C/°C$$

$$= -2.755 \times 10^{-8}\ V/°C/°C$$

Neutral temperature

$$T_n = -\frac{a}{2b} = \frac{-17.455\times10^{-6}\ V/°C}{2\left(-2.795\times10^{-8}\frac{V/°C}{°C}\right)} = 3.122\times10^{2}°C$$

$$= 312.2°C$$

The e.m.f. for the Steel – Copper thermocouple

$$e = aT_n + bT^2_n \text{ (according to data)}$$

$$\therefore \qquad e = (17.155 \times 10^{-6}\ \text{V/°C})(312.2°\text{C}) + \left(-2.795 \times 10^{-8}\ \frac{\text{V/°C}}{°\text{C}}\right)312.2^2\,(°\text{C})^2$$

$$e = 272868 \times 10^{-8}\ \text{V} = \mathbf{2.7\ mV}$$

Example 22.7. *The e.m.f. in a thermocouple, one junction of which is kept at 0°C is given by $E = a\theta + b\theta^2$. Find the neutral temperature and Peltier and Thomson coefficients.*

Solution: Let θ° be the temperature of hot junction

$$\theta°\text{C} = (T - 273)°\text{C, Where } T \text{ is in absolute degrees}$$

$$\therefore \qquad E = a\theta + b\theta^2 = a(T - 273) + b(T - 273)^2$$

$$\therefore \qquad \frac{dE}{dT} = a + 2b(T - 273)$$

and

$$\frac{d^2E}{dT^2} = 2b$$

At neutral temperature $T = T_n$

and

$$\frac{dE}{dT} = 0$$

$$\therefore \qquad 0 = a + 2b(T_n - 273)$$

or

$$T_n - 273 = -\frac{a}{2b}$$

or

$$T_n = \left[\left(273 - \frac{a}{2b}\right)\right]$$

Absolute Peltier coefficient, $\pi = T \cdot \dfrac{dE}{dT} = T\,[a + 2b(T - 273)]$

and Thomson coefficient, $\sigma = T \cdot \dfrac{d^2E}{dT^2} = \mathbf{2b.T}$

Example 22.8. *The e.m.f. in lead-iron thermocouple, one junction of which is at 0°C, is given by $E = 1784\,t - 2.4\,t^2$ (in μ volts) where t is the temperature in °C. Find the neutral temperature, π and σ.*

Solution:

$$E = at + \frac{1}{2}bt^2$$

$$a = 1784,\ b = -4.8$$

Now

$$t_n = -\frac{a}{b} = -\frac{1784}{-4.8} = 371.67°\text{C}$$

Peltier coefficient, $\pi = T \cdot \dfrac{dE}{dT} = T\,[a + b(T - 273)]$

$$= 3094.4T - 4.8T^2$$

and Thomson coefficient, $\sigma = T \cdot \dfrac{d^2E}{dT^2} = T \cdot \dfrac{d}{dT}(a + bt)$

$$= 3094.4T - 4.8T^2$$

$$\sigma = T.b = -\mathbf{4.8T}$$

22.12 THE THERMOELECTRIC LAWS

After much investigation of thermoelectric circuits several basic concepts have been conceived. These concepts can be summed up according to three fundamental "laws."

i. The Law of Homogeneous Circuits

An electric current cannot be sustained in a circuit of a single homogeneous metal, however varying in section, by the application of heat alone.

If there is a temperature circulation along the wires between the hot and cold junctions, the total thermal EMF will be unchanged. Merely the temperature at the junction between the two unlike metals will have a result on the EMF formed.

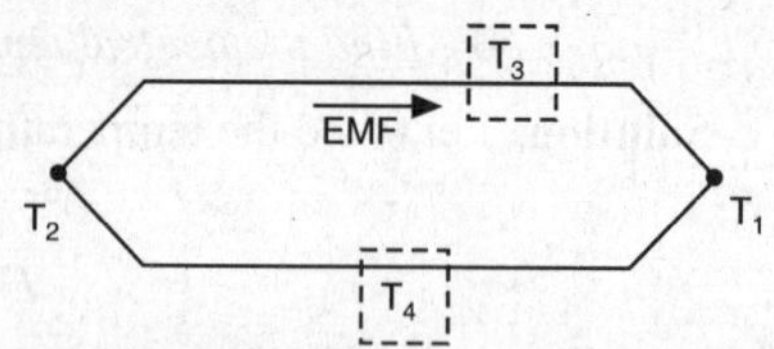

Fig. 22.9: Law of Homogeneous Circuits

ii. The Law of Intermediate Metals

If two dissimilar metals A and B with their junctions at T_1 and T_2 are joined to a third metal C at one leg, if C is kept at a uniform temperature along its entire length, the total EMF in the circuit will be unaffected.

It implies that that the addition of extra metals will not have an effect on the total EMF as long as they are kept at the same temperature as the point where they are connected.

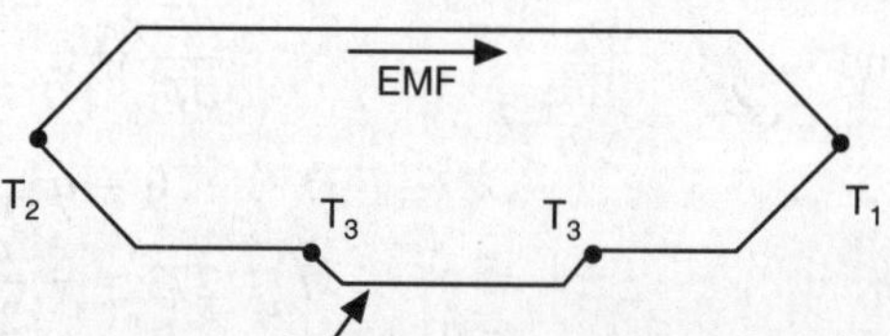

Fig. 22.10: Law of Intermediate Metals

iii. The Law of Intermediate Temperatures

Let one thermocouple be with its junction at T_1 and some reference temperature, T_2 and another thermocouple be at the same reference temperature, T_2 and the temperature, T_3 to be measured. This is equivalent to a single thermocouple with its junction at T_1 and T_3.

22.13 APPLICATIONS OF THERMOCOUPLE

The Seebeck effect is the basis for the thermocouple, which is used in measurement of temperature and detection of heat radiation.

i. **Measurement of temperature:** An unknown temperature can be measured with the help of thermocouple. First the thermocouple is calibrated. For the purpose of calibration, one of the thermocouple junctions is placed in a liquid bath while the other junction is kept in a bath of constant temperature, usually held at 0°C. As the liquid in the hot bath is heated, the thermo e.m.f. varies. The e.m.f. developed at different temperatures is recorded and a graph is plotted between the temperature and the corresponding thermo e.m.f. This graph is known as the *calibration graph*. Next, the hot junction is placed in a bath of unknown temperature and the e.m.f. developed in the thermocouple is noted. The temperature corresponding to this e.m.f. is read from the calibration graph. It is the temperature of the bath whose temperature is to be measured.

ii. Thermopile: In the detection and measurement of radiation, a thermopile is usually used. A series-connected array of thermocouples is known as a ***thermopile***. It is constructed in order to increase the output voltage since the voltage induced over each individual thermocouple is small. The thermopile was developed by Leopoldo Nobili (1784-1835) and Macedonio Melloni (1798-1854). It was initially used for measurements of temperature and infra-red radiation, but was also rapidly put to use as a stable supply of electricity for other physics experimentation.

The thermopile consists of thermocouples of metals antimony and bismuth placed end to end to enhance the thermoelectric e.m.f. The metals are widely separated in the thermo-electric series and comparatively large e.m.f. is produced for a small difference temperature between the junctions. The bismuth and antimony metals are taken in the form of thick strips which reduces the resistance of the circuit. A galvanometer of resistance equal to the resistance of the thermocouple is connected to the end of the thermocouple (Fig. 22.11). The whole assembly is enclosed in a box and the front surface of the metals is blackened. A conical piece is attached to the front side so that the incoming radiation falls on the blackened surface and increases the temperature of the hot junction. With the help of a thermopile temperature differences up to 0.001°C can be detected.

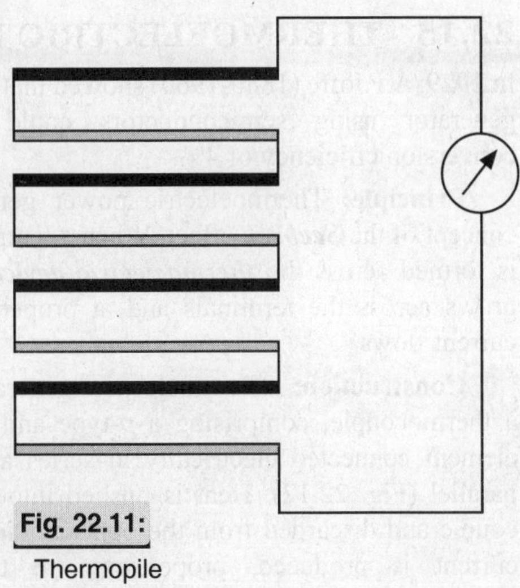

Fig. 22.11: Thermopile

22.14 FIGURE-OF-MERIT, Z

The efficiency of conversion of thermal energy to electrical energy is denoted by the parameter called the figure-of-merit of a thermoelectric material. It is defined as:

$$Z = \frac{\alpha^2 \sigma}{K} \qquad \text{...(22.36)}$$

where α is the Seebeck coefficient of the material (measured in microvolts/K), σ is the electrical conductivity of the material and K is the total thermal conductivity of the material.

Typically metals have small thermopowers. In 1909 Altenkirch proved that good thermoelectric materials should have big Seebeck coefficients, high electrical conductivity and small thermal conductivity. All these three were combined in *figure-of-merit*, Z. As Z varies with temperature, a helpful dimensionless *figure-of-merit* can be defined as ZT.

Semiconductors: Semiconductors have big positive or negative values of the thermopower basing on the charge of the excess carriers.

During the 1920's synthetic semiconductors were developed with Seebeck coefficients in excess of 100 microvolts/K.

This is expressed by *figure of merit ZT, where* $T = (T_2 + T_1) / 2$. Larger values of ZT show superior thermodynamic efficiency. ZT is used to compare the potential efficiency of devices using different materials. Values of $ZT = 1$ are considered good, and values of at least the 3–4 range are needed for thermoelectrics to compete with mechanical generation and refrigeration in efficiency.

Heat formed at a junction is a mixture of Joule heat and heat added by connections elsewhere in the circuit and likewise for heat absorbed by a junction.

Semiconductor materials put together specially for use in Peltier devices show *heat pumping* efficiency better than the majority of active metal pairings. These devices are useful in appliances ranging from lunch box coolers/warmers to laboratory instruments and communications systems. Great temperature differences can be made by cascading Peltier devices such that each level in the pyramid acts as a heat sink for the level above.

22.15 THERMOELECTRIC POWER GENERATION

In 1929, A.F.Ioffe (1880-1960) showed that a thermoelectric generator using semiconductors could accomplish a conversion efficiency of 4%.

Principle: Thermoelectric power generators use the concept of the ***Seebeck*** effect. When a temperature gradient is formed across the ***thermoelectric device***, a DC voltage grows across the terminals and at proper load, electrical current flows.

Construction: Thermoelectric generator consists of a thermocouple, comprising a *p*-type and *n*-type thermo-element connected electrically in series and thermally in parallel (Fig. 22.12). Heat is pushed into one side of the couple and discarded from the opposite side. An electrical current is produced, proportional to the temperature gradient between the hot and cold junctions.

Applications: Providing power for remote telecommunication, navigation, and petroleum installations.

Of the great number of materials studied, semiconductors based on bismuth telluride, lead telluride and silicon-germanium alloys are the best.

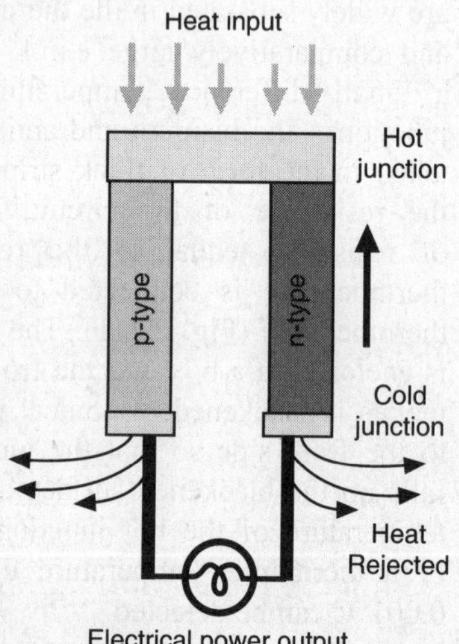

Fig. 22.12: Illustration of thermoelectric generation (Seebeck effect)

22.16 THERMOELECTRIC COOLING

The basic concept behind thermoelectric cooling is the ***Peltier effect***. Semiconductors, usually Bismuth Telluride, are the material of choice. In its simplest form, a ***Peltier device*** is a single semiconductor 'pellet' soldered to electrically-conductive material on each end to plated copper. The copper connection paths serve as the second dissimilar material required for the formation of the junctions, *i.e.*, thermocouple.

In a thermoelectric device multiple pellets are needed for better heat-pumping capacity.

By arranging *n*- and *p*-type pellets in a 'couple' and forming a junction between them with a plated copper tab, it is possible to configure a simple series circuit (Fig. 22.13).

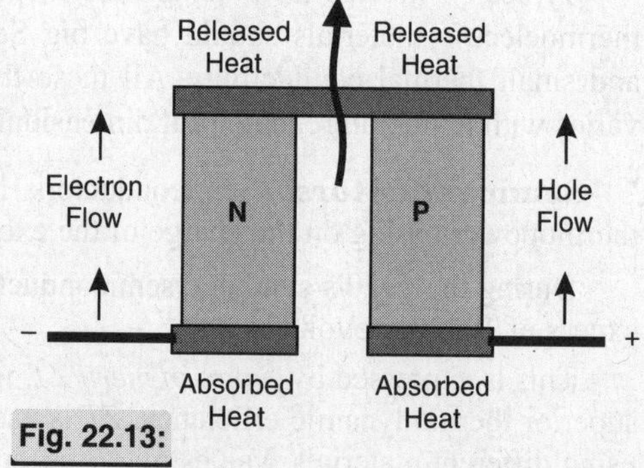

Fig. 22.13: Illustration of thermoelectric cooling (Peltier effect)

The bottom end of the p-type pellet is connected to the positive voltage potential and the bottom end of the *n*-type pellet is similarly connected to the negative side of the voltage. The holes in the *p*-material are repelled by the positive voltage potential and attracted by the negative pole; the negative charge carriers (electrons) in the n-material are likewise repelled by the negative potential and attracted by the positive pole of the voltage supply.

The electrons flow continuously from the negative pole of the voltage supply, through the *n*-pellet, the copper tab junction, the *p*-pellet, and back to the positive pole of the supply.

The heat will be moved (or 'pumped') in the direction of charge carrier movement throughout the circuit as it is actually the charge carriers that transfer the heat. Thus, heat is absorbed at the bottom junction and actively transferred to the top junction. Therefore, the bottom junction is the cold junction while the top junction forms the hot junction. The cold junction will rapidly drop below ambient temperature provided heat is removed from the hot side. The temperature gradient will vary according to the magnitude of current applied.

22.17 THE THERMOELECTRIC COOLERS

Thermoelectric coolers are solid state heat pumps used in applications where temperature stabilization, temperature cycling, or cooling below ambient are required.

A thermoelectric cooling arrangement is shown in Fig. 22.14. It consists of a thermoelectric module, a heat sink and the object to be cooled.

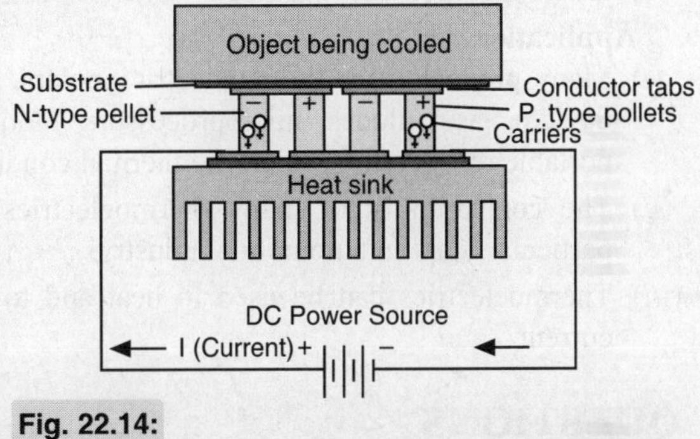

Fig. 22.14: Illustration of thermoelectric cooling arrangement

A module consists of an array of bismuth telluride semiconductor pellets. They are *doped* to make the majority charge carries. The pairs of P/N pellets are connected electrically in series and thermally in parallel. Metalized ceramic substrates give the platform for the pellets and connection them. The ceramic material on both sides of the thermoelectric gives strength and electrical insulation. The pellets, tabs and substrates thus form a layered arrangement. Module size varies from less than 0.25" × 0.25" to around 2.0" × 2.0". Thermoelectric modules can work alone or in groups. When DC voltage is applied to the module, the positive and negative charge carriers in the pellet array take up heat energy from one substrate surface and release it to the substrate at the opposite side making it as cold and other as hot.

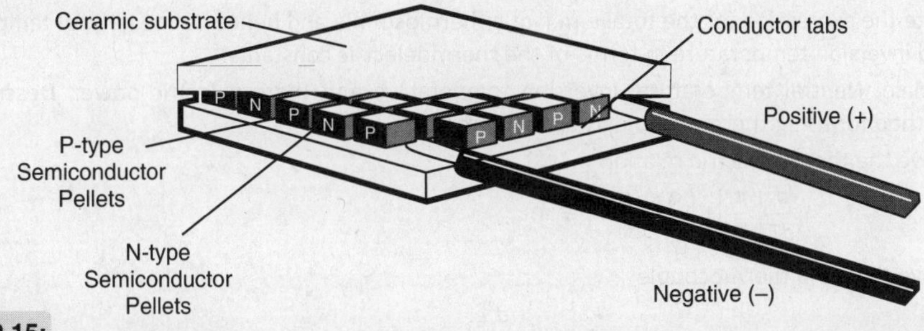

Fig. 22.15: Peltier device

These devices cannot only pump appreciable amounts of heat, but with their series electrical connection, are suitable for commonly-available DC power supplies. Thus, the most common ***thermoelectric devices*** now in use—connecting 254 alternating P and N-type pellets—can run from a 12 to 16 VDC supply and draw only 4 to 5 amps. A means to mechanically hold everything together is to mount the conductive tabs to thin ceramic substrates (Fig. 22.15); the outer faces of the ceramics are then used as the thermal interface between the ***Peltier device*** and the 'outside world.' Ceramic materials represent the best compromise between mechanical strength, electrical resistivity, and thermal conductivity.

Advantages: Thermoelectric coolers

(*i*) have no moving parts and hence less maintenance and longer life time.

(*ii*) have no materials such as chlorofluorocarbons which may require periodic replacement.

(*iii*) temperature can be controlled to the fractions of a degree with support circuitry.

(*iv*) can function in environments that are too severe, too sensitive.

(*v*) are not position-dependent.

Applications:

(*i*) Many products uses thermoelectric coolers, including CCD cameras (charge coupled device), laser diodes, microprocessors, blood analyzers and portable picnic coolers, portable refrigerators, scientific thermal conditioning, liquid coolers.

(*ii*) The compatibility of many thermoelectrics with automotive voltages makes them particularly apt in automobile industry.

(*iii*) Thermoelectrics can be used to heat and to cool, depending on the direction of the current.

QUESTIONS

1. Explain the terms Seebeck effect, Peltier effect and Thomson effect.
2. Give the theory of thermo e.m.f. Show that $\pi = T.de/dt$.
3. Explain Seebeck effect and Peltier effect. Prove that Peltier coefficient of a pair of metals is product of absolute temperature and thermoelectric power.
4. Explain characteristics of a thermocouple. Explain the terms (*i*) neutral temperature and (*ii*) inversion temperature.
5. Define thermocouple, neutral temperature, thermoelectric power and temperature of inversion.
6. What is thermoelectric emf?
7. Explain (i) the law of intermediate metals and (ii) the law of intermediate temperatures.
8. Explain the Seebeck effect. Describe how the e.m.f. of a thermocouple is measured experimentally.
9. State the expression for the total e.m.f. of a thermocouple and hence define neutral temperature and inversion temperature in terms of the thermoelectric constants.
10. Explain: Neutral temperature, inversion temperature and thermoelectric power. Describe the method for using thermocouple as a thermometer
11. Prove the thermoelectric relation

$$\frac{d}{dT}\left(\frac{\pi}{T}\right)+\left(\frac{\sigma_A-\sigma_B}{T}\right) = 0$$

12. Prove that in a thermocouple

$$\sigma = T\cdot\frac{d^2e}{dT^2}$$

13. Describe how a thermocouple is used as thermometer and discuss its advantages.

14. Obtain the law of intermediate metals for thermocouples. What is its significance?

15. What is thermoelectric effect? From the formula $e = at + \frac{1}{2}bt^2$, find out neutral temperature and thermoelectric power.

16. What is a thermocouple? Describe the Seebeck effect and the Peltier effect.

17. What are the advantages of using thermocouple as a thermometer?

18. Write a short note on thermopile.

19. Explain the laws of intermediate metals and intermediate temperatures.

20. Show that in a thermocouple composed of two metals, the total e.m.f. is given by

$$e = \pi_2 - \pi_1 + \int_{T_1}^{T_2} (\sigma_A - \sigma_B)dT$$

21. Deduce from thermodynamic consideration the expressions for Peltier and Thomson coefficients.

22. Explain thermoelectric power generation.

23. Explain the principle of thermoelectric cooling.

24. Describe the construction and working of a thermoelectric cooler.

25. Explain the advantages of thermoelectric cooling and some of the areas where thermoelectric cooling is applied profitably.

PROBLEMS

1. Calculate the neutral temperature for an iron - silver thermo couple. The values of a and bare 16. 65 and –0.096 for iron and 2.86 and 0.017 for silver respectively. [**Ans:** 61°C]

2. The neutral temperature of a thennocouple is 300 Dc. When its junctions are kept at temperatures 0°C and 100°C, the e.m.f. generated is 1300 μV. Calculate the values of the coefficients a and b. [**Ans:** 15.6×10^{-6}]

3. For Fe-Cu thermocouple it is observed that the thermo e.m.f is zero when one of the junctions is at 20°C and the other one is at some higher temperature. If the neutral temperature is 285°C, calculate the higher temperature. Hence find out the temperature of inversion, if the cold junction temperature is at –20°C. [**Ans:** 650°C]

4. A thermocouple is made of iron and constantan. Find the e.m.f. developed per °C difference of temperature between the junctions, given that then no e.m.f. of iron and constantan against platinum are +1600 and –3400 microvolts per 100°C difference of temperature. [**Ans:** –16 μV/°C (–34)μV/°C 50 μV/°C]

5. The thermoelectric power for steel is 18 μV/0 C at 0°C and zero at 400°C. That for copper is 6 μV/0 C at 500°C and zero at –50°e. Find the e.m.f. for steel-copper thermocouple with one junction at its neutral temperature and other at 0°C. [**Ans:** 2.7 mV]

7. The thermoelectric power of Fe-Pb thermocouple is 17.5 μV/ ° C at 0° C and zero at 360° C and that for Cu-Pb is 5 μV / °C at 450°C and zero at –50°C. Find the e.m.f for Fe-Cu thermocouple with its one junction at 0 0 C and other at its neutral temperature. [**Ans:** 4.97 Mev]

8. The thermoelectric power of Fe - Pb thermocouple is 17.5 μV/°C at 0°C and 5μV/°C at 125°C. The thermoelectric power of Cd-Pb thermocouple is 3 μV/°C at O°C and 15μV/°C at 150°C. Calculate the neutral temperature of the Fe – Cd junction. [**Ans:** 80.6°C]

9. The thermoelectric power of iron is 1734-4.87 *t* and that of copper is 136-0.95*t*, μV/°C where *t* is the temperature in °C, show that the e.m.f. of thermocouple of iron-copper, the junctions of which are at 0°C and 100°C is 0.14 V. [**Ans:** 130670 μV]

10. The e.m.f. of iron-lead thermocouple, where one junction is at 0°C and the other at 100°C is 1185 μV. when the second junction is at 300°C, the e.m.f. is 675 μV. Similar readings with silver-lead couple are 371 μV and 1623 μV respectively. Calculate the neutral temperature of an iron-silver thermocouple. [**Ans:** 122°C]

23 Special Theory of Relativity

Learning Objectives:

By the end of this chapter, you will be able to:

- Define concepts of space, length, time and mechanical motion.
- Learn about Frame of Reference and its different types.
- Explain the concept of Galilean transformation.
- Describe process of Michelson-Morley experiment.
- Understand Einstein's principle of relativity and Lorentz transformations.

23.1 INTRODUCTION

In 1905, at the age of only 26, Albert Einstein (1879-1955) published four scientific papers that revolutionized physics. One of the papers dealt with the photoelectric effect, which was responsible for the birth of quantum physics. Two of the papers dealt with special theory of relativity. Relativity represents the greatest intellectual achievement of twentieth century physics. Just as quantum theory showed that the classical concepts are to be revised in case of microscopic world, the relativity theory established that the classical notions are not applicable to bodies moving with velocities approximately nearer to that of light. Classical mechanics regarded space and time to be absolute and separate entities. It assumed the flow of time to be uniform in all situations. As such the moments of time and time intervals are supposed to be identical in all frames of reference. Similarly, lengths are assumed to be identical in all frames. An analysis of these concepts at high velocities revealed that they are not correct. The relativity theory leads to many unusual conclusions. It shows that the length of moving bodies contract in the direction of motion and the clocks in motion slow down. It is difficult to comprehend these conclusions because of our habit founded on routine experiences. At high velocities, space and time are no more separate but merge into space-time continuum. The results of experiments on high-energy particles provide the proof for the predictions of relativity.

23.2 SPACE, TIME AND MOTION

The concepts of space, length, time and mechanical motion appear to us as self-evident and obvious.

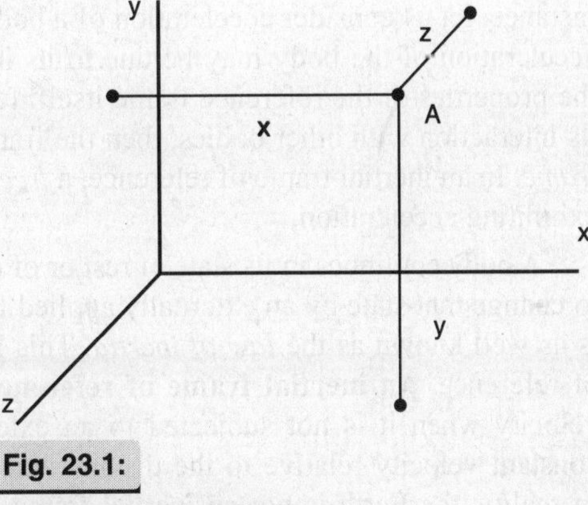

Fig. 23.1: A frame of reference

Classical mechanics presumes the space to be homogeneous in all its parts and also isotropic. It means that the properties of space are identical at all points and in all directions at each point. Classical mechanics further supposes the existence of an absolute space, which is absolutely motionless and irrelevant to the existence of any bodies. According to Newton "absolute space, in its own nature, without regard to anything external, remains always similar and immovable."

In classical mechanics, time is understood as a measure of absolute duration, which exists irrespective of physical bodies. Newton regarded that the true course of time is not liable to change; in his opinion the course of time belongs to absolute category.

If we are asked, "What is mechanical motion", we sometimes simply say that the displacement of a body in space with time is mechanical motion. In fact, the displacement of a body can be recognized only when there is a fixed body relative to which the given body changes its position. Therefore, we mean by mechanical motion the displacements of bodies relative to some other bodies. Thus, for the description of motion it is essential that there are certain real physical bodies, which can be assumed to be fixed. When we say that a train is moving, we are referring to the change of position of the train with respect to the platform, which is stationary. It is not possible to observe the change of position of a single isolated body and, therefore, it is senseless to speak of motion in such a case.

23.3 FRAME OF REFERENCE

Normally, the Cartesian coordinate system is used to specify the location of a body may it be a fixed or in motion in space. The Cartesian coordinate system attached to a fixed body is known as a "**frame of reference**."

As per the convenience any frame of reference can be taken. We choose a frame of reference rigidly connected to the earth for describing the motion of bodies on the earth. However, the Earth is not a fixed body but it is in circular motion. We attach the coordinate system to the sun, when we describe and study the Earth's motion. In studying the Sun's motion, we choose a reference frame connected to the stars. Many times, we choose the floor as the reference frame which may be at rest with respect to the Earth or which may be in motion such as in the case of a moving train or a spacecraft.

The choice of a frame of reference is subjective. A passenger in a moving train may assert that he is at rest and state that the electric poles and trees are moving backward while a person standing on the ground asserts that the train is moving forward. Both are equally right. In practice, we naturally choose a frame of reference attached to the Earth and describe the motion of other bodies. Thus, in the reference system attached to the Earth, it is the train that moves forward.

23.4 INERTIAL FRAMES OF REFERENCE

Basically, there is infinite number of reference frames available and any one of them can be used. But the laws of mechanics may take different forms in different reference frames. For instance, let us consider acceleration of a body relative to an arbitrary frame of reference. The acceleration of the body may be due to its interaction with other bodies or it may be due to the properties of the reference frame itself. If the acceleration of the body arises solely due to its interaction with other bodies, then the frame of reference is said to be an *inertial reference frame*. In an inertial frame of reference, a *free body* moves rectilinearly and uniformly, without exhibiting acceleration.

A body continues in its state of rest or of uniform rectilinear motion, unless it is compelled to change that state by an externally applied force. This is Newton's first law of motion which is as well known as the *law of inertia*. This law indicates and characterizes an inertial frame of reference. An **inertial frame of reference** is that in which a body moves with constant velocity when it is not subjected to an external force. A reference frame that moves with constant velocity relative to the distant stars is the best approximation of an *inertial frame*. In reality, the Earth is not an inertial frame because of its orbital motion about the sun and rotational motion about its own axis. However, the Earth may be assumed to be an inertial frame in many situations.

There are a vast number of inertial reference frames moving relative to one another uniformly and rectilinearly as any other reference frame moving rectilinearly and uniformly relative to an inertial frame is also inertial.

23.5 NON-INERTIAL REFERENCE FRAME

A frame of reference, which is an accelerated motion with respect to an inertial frame of reference, is known as a **non-inertial frame** of reference. The law of inertia is valid in an inertial frame, whereas it is not valid in an accelerated reference frame. A ball placed on the floor of a train will move to the rear if the train accelerates forward even though no forces act on it; likewise, a coin placed on a rotating turntable will slide to the periphery though no visible force pushes it away from the centre.

23.6 GALILEO'S PRINCIPLE OF RELATIVITY

A reference frame for which Newton's laws of dynamics hold is an inertial frame. If a body is in uniform motion (v = constant) an observer in an inertial frame, which is at rest with respect to the body, will find that the acceleration and the resultant force on the body are zero (i.e., $a = 0$ and $F = 0$). An observer in another frame of reference moving with uniform velocity in a straight line with respect to the first frame will also find that $a = 0$ and $F = 0$ for the body. Consequently, the second reference frame is inertial to the same degree as the first. It follows that any reference frame moving with uniform velocity with respect to an inertial frame is also an inertial frame.

Thus, reference frames fixed in a railway bogie or a ship that travels with a uniform velocity in a straight line, are inertial frames. Experience shows that in a railway car or on a ship traveling with uniform velocity, it is equally easy to move in any direction as it is on the earth. A body released at a height falls vertically downward. Thus, the results of an experiment performed in the stationary moving vehicle will be the same as those from the same experiment performed in the stationary laboratory. Therefore, it is not possible for us to tell by any experiment whether we are at rest or moving with uniform velocity.

It follows that the laws of mechanics have the same form in all inertial frames and none of the reference frames have any advantage over the others. It implies that there is no "exclusive" or "preferred" frame of reference and every inertial frame is as good as the other. Absolute rest or absolute motion of bodies has no sense. We can speak of their relative motion in some inertial frame reference.

The basic law of nature was recognized by Galileo and is summed up in the form of principle of relativity. **Galileo's principle of relativity states that the laws of mechanics are the same in all inertial frames of reference.**

23.7 GALILEAN TRANSFORMATIONS

The transformation from one inertial frame of reference to another is called Galilean transformation. Knowing the laws of motion of a body in a reference system *S*, one can derive the laws of motion of the same body in another inertial system *S*′.

(a) Coordinate Transformation

When a point has coordinates *x, y,* and *z* at the instant "*t*" we regard it as an event. It means that the location and time of occurrence of the event can be specified by the coordinates (*x, y, z,* and *t*), as shown in Fig. 23.2.

An event may be in reality some physical phenomenon such as the explosion of a flash-bulb. The Galilean transformation establishes the relationship between the coordinates *x, y, z* and *t* of an event in system S and the coordinates *x*′, *y*′, *z*′ and *t*′ of the same event in system *S*′.

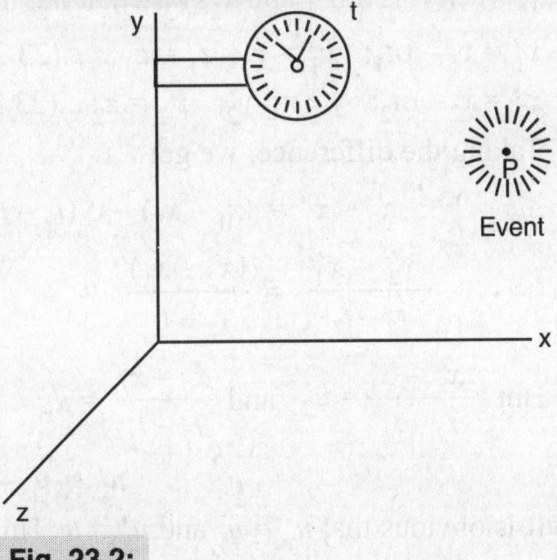

Fig. 23.2: An event is specified by the coordinates (x,y,z) at time t

Let an event occur in an inertial frame of reference *S* and let the event be located by the coordinates (*x, y, z,* and *t*) in the system *S*. Let us consider another inertial frame of reference *S*′, which moves in a straight line with respect to the frame S at a constant speed υ. For the sake of simplicity, let us assume that the reference frame *S*′ moves so that *x* and *x*′ are in one straight line and have the same positive directions. Also let the origins *O* and *O*′ coincide at the moment $t = 0$. Let the point *P* be at rest with respect to *S*. In such a case, it moves with respect to the second frame *S*′ and its coordinates change. The coordinate *x*′ depends on time, since the system *S*′ moves at a speed υ. During the time *t*, the moving system covers a distance equal to υt. Hence $x' = x - \upsilon t$. Obviously, $y' = y$ and $z = z'$. Also $t' = t$. Thus,

$$\begin{aligned} x' &= x - \upsilon t \\ y' &= y \\ z' &= z \\ t' &= t \end{aligned} \quad \textbf{Galilean Coordinate Transformation} \quad ...(23.1)$$

Galilean transformation expresses the space-time relation of an event in different inertial frames. It is seen from Eq. (23.1) that the coordinates of an event are relative and have different values in different reference frames. In the above transformation we tacitly assumed that the passage of time is the same in both the inertial systems. That is, the time of an event for an observer in *S* is the same as time for the same event in *S*′. This intuitive assumption is taken for granted in classical mechanics. It is considered natural and does not require any proof. All the phenomena in mechanics where $\upsilon << c$ do not contradict this assumption.

(b) Velocity transformation

Let us now consider the passage of a moving body. For the sake of simplicity, let us assume that the body moves parallel to the $O'x'$-axis in the positive direction at a speed u' with respect to the system S'. Let the body be at P_1 at time t_1 and at P_2 at time t_2, as shown in Fig. 23.3.

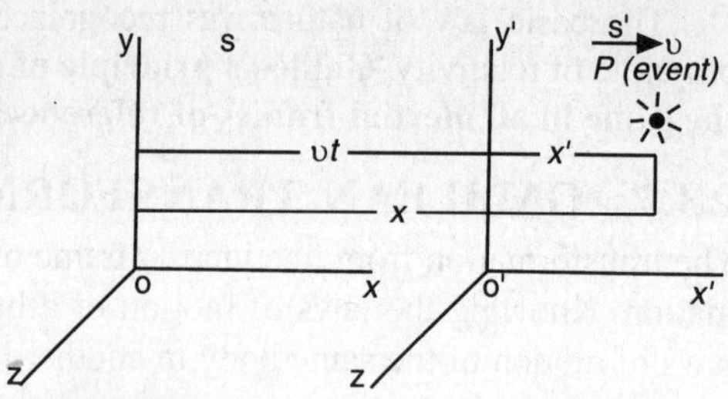

If the coordinates of the body in S are x_1, y_1, z_1, when it is at P_1 and x_2, y_2, z_2, at P_2, then

$$x_1' = x_1 - \upsilon t_1;\ \ y_1' = y_1;\ \ z_1' = z_1 \quad ...(23.2)$$

$$x_2' = x_2 - \upsilon t_2;\ \ y'_2 = y_2;\ \ z'_2 = z_2 \quad ...(23.3)$$

Fig. 23.3: An event occurring at point P is observed by an observer in two different inertial frames S and S', where S' moves with a velocity υ relative to S.

Taking the difference, we get

$$x_2' - x_1' = (x_1 - x_2) - \upsilon\,(t_2 - t_1)$$

$$\frac{x_2' - x_1'}{t_2 - t_1} = \frac{(x_2 - x_1)}{t_2 - t_1}$$

But $\dfrac{x_2' - x_1'}{t_2 - t_1} = u_2'$ and $\dfrac{x_2 - x_1}{t_2 - t_1} = u_x$

$$\therefore \qquad u_x' = u_x - \upsilon \qquad ...(23.4)$$

It is obvious that $u_y' = u_y$ and $u_z' = u_z$. Thus,

$$u_x' = u_x - \upsilon$$
$$u_y' = u_y \qquad \textbf{Galilean Velocity Transformation} \quad ...(23.5)$$
$$u_z' = u_z$$

Equation (23.5) is known as the Galilean velocity transformation. It states that the velocity of a body located at a point P measured by an observer in the moving frame equals the velocity as measured in the stationary frame minus the velocity of the frame S′.

Equation (23.4) may be rewritten as

$$u_x = u'_x + \upsilon \qquad ...(23.6)$$

Equation (23.6) implies the well-known law of addition of velocities in the classical mechanics.

The Galilean transformation equations agree with our daily experience. For example, a person in a moving railcar (Fig. 23.4) throws a ball with a speed u in the direction of motion of the railcar, where u is measured by the person in the railcar. If υ is the velocity of the moving railcar, the speed of the ball relative to a stationary observer on the ground will be $u + \upsilon$.

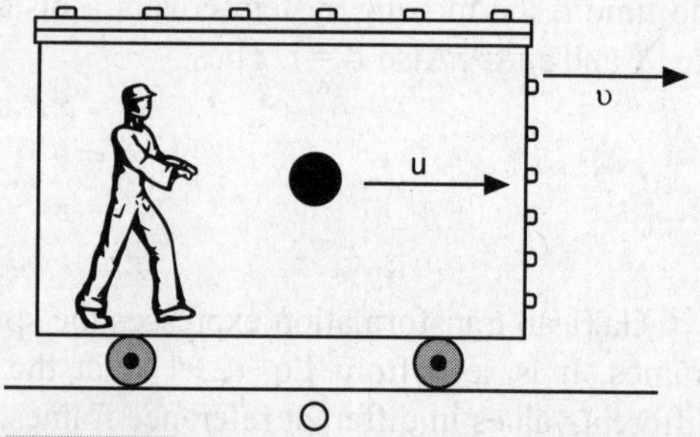

Fig. 23.4: A person in a moving railcar throws a ball with a speed u. The speed of the ball relative to a stationary observer O is u + υ

(c) Acceleration Transformation

Let us now consider the case where the body moves along x-axis of reference frame S with acceleration. At some moment of time, say

$$u = u' + \upsilon \quad ...(23.7)$$

After a time interval Δt, the velocities in both systems increase and the new velocities are given by

$$(u + \Delta u) = (u' + \Delta u') + \upsilon \quad ...(23.8)$$

Subtracting Eq. (23.7) from Eq. (23.8), we get

$$\Delta u = \Delta u'$$

Dividing both sides of the above equation with Δt, we obtain

$$\frac{\Delta u}{\Delta t} = \frac{\Delta u'}{\Delta t'}$$

or $a = a'$ **Galilean Acceleration Transformation** ...(23.9)

Multiplying Eq. (23.9) by the mass of the body, we get

$$ma = ma'$$

or

$$F = F' \quad ...(23.10)$$

It follows from equs. (23.9) and (23.10) that accelerations and forces do not change from one inertial frame to another inertial frame. Such quantities, which remain invariable under the transformation, are termed **invariants**. It is seen from the above that Newton's laws are invariant with respect to Galilean transformations.

Example 23.1. *Prove that $x^2 + y^2 + z^2 - c^2t^2$ is invariant under Lorentz transformation.*

Solution: We have $x = \dfrac{x' + \upsilon t'}{\sqrt{1-(\upsilon^2/c^2)}}$, $y = y'$, $z = z'$ and $t = \dfrac{t' + (x'\upsilon/c^2)}{\sqrt{1-(\upsilon^2/c^2)}}$. Substituting these values in the equation given, we obtain

$$x^2 + y^2 + z^2 - c^2t^2 = \frac{(x' + \upsilon t')^2}{1-(\upsilon^2/c^2)} + (y')^2 + (z')^2 - \frac{c^2\left[t' + (x'\upsilon/c^2)\right]^2}{1-(\upsilon^2/c^2)}$$

or

$$= (y')^2 + (z')^2 - \frac{1}{1-\upsilon^2/c^2}\left[c^2(t')^2 + \frac{\upsilon^2(x')^2}{c^2} - (x')^2 - \upsilon^2(t')^2\right]$$

$$= (y')^2 + (z')^2 - \frac{1}{1-(\upsilon^2/c^2)}\left[(c^2t'^2 - x'^2)\left(1-(\upsilon^2/c^2)\right)\right]$$

$$= (y')^2 + (z')^2 - (c^2t'^2 - x'^2)$$

$$= x'^2 + y'^2 + z'^2 - c^2t'^2$$

$$\therefore \quad x^2 + y^2 + z^2 - c^2t^2 = x'^2 + y'^2 + z'^2 - c^2t'^2$$

That is, $x^2 + y^2 + z^2 - c^2t^2$ is invariant under Lorentz transformation.

23.8 THE ETHER

Newton's postulate of absolute space implied the existence of an absolutely stationary reference frame. He assumed that out of the infinite inertial frames moving uniformly with respect to another, there is a preferred frame fixed in absolute space, which is really stationary. However, the mechanical experiments fail to establish the existence of an absolute reference frame. According to the Galilean principle of relativity, all inertial reference frames are equivalent as far as laws of mechanics are concerned. Mechanics was the only branch of physics advanced sufficiently at that time. With the developments in electrodynamics and optics, scientists naturally were concerned with the questions whether the principle of relativity is applicable to the phenomena in these fields and whether one can find an absolute frame of reference. As a result of the importance attached to the absolute frame, it became of considerable interest to prove its existence by experiment.

The interference and diffraction effects of light clearly established wave nature of light. As all known waves are transmitted by some medium, it was natural to assume that light waves are also transmitted by a medium, which obviously fills the space around the interplanetary and us and the interstellar space. It was supposed that this all pervasive medium would have no effect on the motion of planets and other celestial bodies. Failing to identify such an unusual medium, the scientists of 19th century concocted the medium and named it **luminiferous ether** or in short the **ether**. Whatever may be its nature, the ether cannot be at rest in all inertial frames of reference simultaneously. Therefore, one can distinguish an inertial frame that is stationary with respect to the ether. It was supposed that, in that and only in that reference frame, light propagates with the same velocity c in all directions. That frame of reference would be the absolute frame.

If a certain inertial frame moves with the velocity υ relative to the ether, the velocity of light c' in that reference frame must obey the conventional law of velocity composition. Thus, $c' = c - \upsilon$. For instance, the velocity of light on the Earth must depend on the velocity of the motion of the Earth relative to the reference system of the ether. If the velocity of the Earth in this system is υ and the speed of light relative to the medium is c, then the speed of light must be equal to $c - \upsilon$ in the direction of the Earth's motion and $c + \upsilon$ in the opposite direction.

The speed of the Earth in its orbit is about 30 km/s, which is only about 1/10,000 times the speed of light. Therefore, the observation and the measurement of the influence of the Earth's motion on the speed of light face many experimental difficulties. High precision apparatus is required to verify the addition rule for the velocities of the light and of the Earth. First experiments of this kind were carried out in 1881 by the American Physicist Albert A. Michelson (1852 – 1931) and again in 1887 by A.A. Michelson and E.W. Morley (1838 – 1923).

23.9 MICHELSON-MORLEY EXPERIMENT

A schematic diagram of the Michelson-Morley experiment is shown in Fig. 23.5. A collimated beam of light from the source S is split into two beams (A and B) by the half-silvered mirror M. Beam A travels to mirror M_1 and is reflected back to M, where it is reflected down. Beam B proceeds to mirror M_2 and is reflected back to M and is transmitted through M. The component of beam B, which is transmitted through M, combines with the reflected part of beam A.

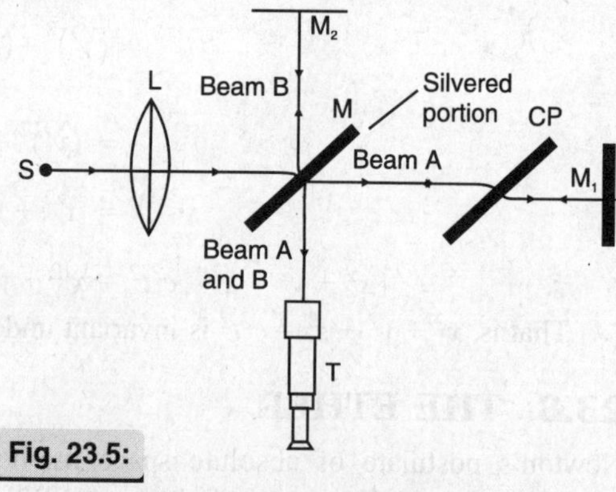

Fig. 23.5: A schematic diagram of the Michelson-Morley interferometer experiment

Since beams A and B are derived from the same beam falling on M, they are coherent and produce interference.

Let us assume that the Earth travels through stationary ether with a speed υ and that light has a speed c in the ether. We further assume that the instrument is arranged such that one of its two equal arms is parallel to the Earth's velocity υ (Fig. 23.6). The speed of beam A traveling from M to M_1 is $c - \upsilon$ relative to the instrument and the time required for light to travel from M to M_1 is $L/(c - \upsilon)$. Similarly, the speed of light beam is $(c + \upsilon)$ in the trip from M_1 to M and the time required for the return trip to M is $L/(c + \upsilon)$. The time for the round trip MM_1M is thus

$$T_A = \frac{L}{c-\upsilon} + \frac{L}{c+\upsilon} = \frac{2Lc}{c^2-\upsilon^2} = \frac{2L/c}{1-\frac{\upsilon^2}{c^2}} \quad ...(23.11)$$

The beam B that travelled toward mirror M_2 will get reflected at M_2 but only after M has moved to a new position M' (Fig. 23.6). The component of velocity of light in the direction perpendicular to the motion of the instrument is $\sqrt{c^2 - \upsilon^2}$. Therefore, the time for the round trip MM_2M is

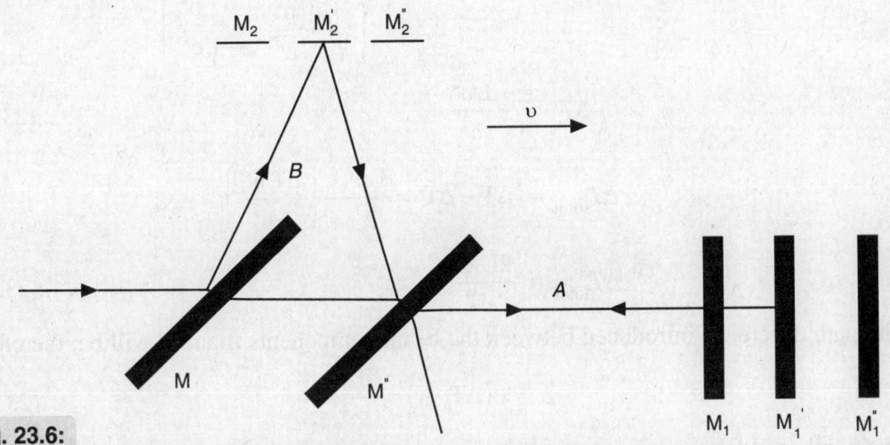

Fig. 23.6: The paths of beam A and B in reference frame at rest relative to the other. The interferometer moving is with a velocity υ relative to the other.

$$T_B = \frac{2L}{\sqrt{c^2-\upsilon^2}} = \frac{2L/c}{\sqrt{1-\frac{\upsilon^2}{c^2}}} \quad ...(23.12)$$

The time difference between the two paths due to the motion of the instrument relative to the ether is given by

$$\Delta T = T_A - T_B = \frac{2L}{c}\left[\frac{1}{1-\frac{\upsilon^2}{c^2}} - \frac{1}{\sqrt{1-\frac{\upsilon^2}{c^2}}}\right] \quad ...(23.13)$$

Since υ is very small compared to c, we may use the binomial theorem to simplify Eq. (23.13). We can approximate

$$\frac{1}{1-\frac{\upsilon^2}{c^2}}=1+\frac{\upsilon^2}{c^2} \text{ and } \frac{1}{\sqrt{1-\frac{\upsilon^2}{c^2}}}=1+\frac{\upsilon^2}{2c^2}$$

Using the above approximations into Eq. (23.13), we get

$$\Delta T = \frac{2L}{c}\left[1+\frac{\upsilon^2}{c^2}-1-\frac{\upsilon^2}{2c^2}\right]=\frac{L\upsilon^2}{c^3} \quad ...(23.14)$$

If the instrument is turned through 90°, the role of paths A and B are interchanged. In this case, the times T'_A and T'_B required to travel the paths MM'_1M' is given by

$$T'_A=\frac{2L/c}{\sqrt{1-\frac{\upsilon^2}{c^2}}} \text{ and } T'_B=\frac{2L/c}{1-\frac{\upsilon^2}{c^2}}$$

$$\Delta T' = T'_A - T'_B = \frac{2L}{c}\left[\frac{1}{\sqrt{1-\frac{\upsilon^2}{c^2}}}-\frac{1}{1-\frac{\upsilon^2}{c^2}}\right]$$

or

$$\Delta T' = -\frac{L\upsilon^2}{c^3} \quad ...(23.15)$$

$$\Delta T_{total} = \Delta T - \Delta T' = \frac{L\upsilon^2}{c^3}-\left(-\frac{L\upsilon^2}{c^3}\right)$$

$$\therefore \quad \Delta T_{total} = \frac{2L\,\upsilon^2}{c^3} \quad ...(23.16)$$

The path difference introduced between the beam components A and B will be, therefore,

$$\Delta = c(\Delta T_{total})=\frac{2L\upsilon^2}{c^2} \quad ...(23.17)$$

The number of fringes passing through a reference mark will be

$$N=\frac{\text{Path difference}}{\lambda}=\frac{2L\upsilon^2}{\lambda c^2} \quad ...(23.18)$$

Michelson and Morley mounted the entire apparatus on a large stone slab, which floated on a mercury pool (Fig. 23.7). By using multiple reflections (Fig. 23.8), the effective path L is made 11 m which improves the accuracy of the measurement. Using the light of wavelength λ = 5500 Å, and taking υ = 30 km/s, we obtain

$$N=\frac{2\times 11\text{ m}\left(3\times 10^4\right)^2\text{ m}^2/\text{s}^2}{5500\times 10^{-10}\text{ m}\left(3\times 10^8\right)^2\text{ m}^2/\text{s}^2}=0.4$$

A fringe shift of this amount is readily detected with the apparatus. It should then be possible to measure the fringe shift and from it determine the velocity of the Earth relative to the ether. Michelson and Morley failed to observe any fringe shift. Measurements were made during day and night, at various locations in Europe and U.S.A., and in various seasons of the year.

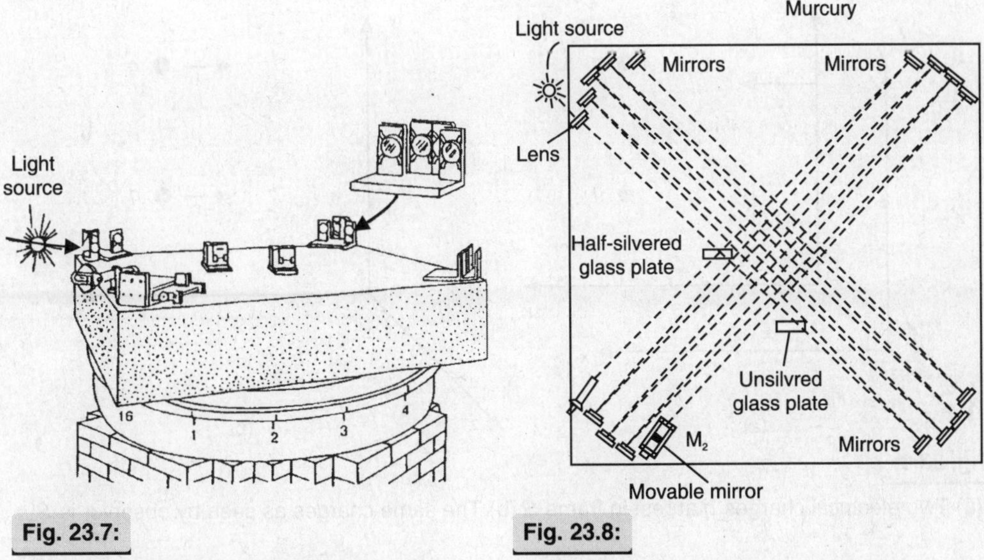

Fig. 23.7:

Drawing of the Michelson-Morley Interferometer mounted on a stone slab which floats on a pool of mercury.

Fig. 23.8:

Multiple reflections-Micheston-Morley Experiment.

The results of the experiment always proved to be negative. The negative result of Michelson-Morley experiment contradicts Galileo's relativity principle from which the ordinary addition rule for the velocities follows. It as well illustrated that motion relative to the ether cannot be detected and the velocity of light is independent of the motion of a light source.

Some astronomical observations such as double stars also indicate the fact that the velocity of light does not depend on the velocity of the source. A number of special experiments carried out later gave the same evidence.

Following the publication of the Michelson-Morley experiment in 1887, several explanations were put forward to explain the null result of the experiment. One of the ingenious explanations was offered in 1889 by G.F. Fitzgerald, an Irish Physicist. He suggested that all objects contract in the direction of their motion through the ether. It was a clever suggestion according to which the arm of the instrument moving parallel to the ether was assumed to be shortened, but similar contractions were not observed in other situations.

23.10 FAILURE OF GALILEAN TRANSFORMATIONS

We have seen that the laws of mechanics are invariant under a Galilean transformation. However, the laws of electromagnetism are found to change their form under a Galilean transformation. This fact can be demonstrated with the following simple example.

Consider two equal and like charges q which are stationary in one reference frame S. They repel each other according to Coulomb's law. Now, suppose these charges are examined by an observer in a reference frame S' moving with velocity υ along a line perpendicular to the line joining the two charges. The observer will find the two charges moving with velocity $-\upsilon$. The two moving charges constitute two currents in the same direction and two parallel currents attract each other. Therefore, the observer in the moving frame finds that the charges exert a mutual repulsive force as well as an attractive force.

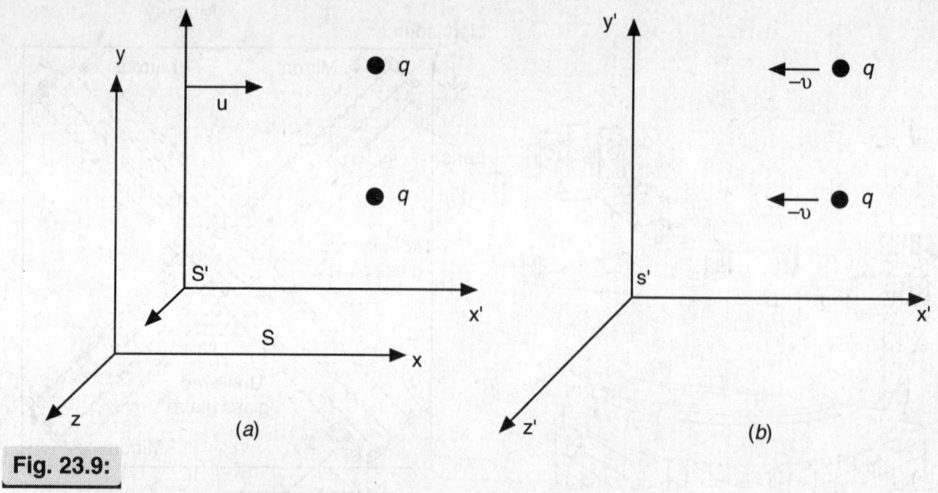

Fig. 23.9:

(a) Two electrical charges q at rest in frame S (b) The same charges as seen by observer is S'.

It means that either the Galilean transformation is wrong or Maxwell's equations are wrong. But Maxwell's equations are in total agreement with all known experiments. Albert Einstein (1879–1955) was convinced that Maxwell's equations are true and must yield the same result in all inertial frames of reference.

23.11 EINSTEIN'S PRINCIPLE OF RELATIVITY

Basing on thorough analysis of all experimental and theoretical data accumulated by the beginning of the twentieth century, Einstein concluded that the classical concept of space and time are to be revised. In 1905, at the age of 26, Einstein published a paper entitled **"On the Electrodynamics of Moving Bodies"** in the Seventeenth volume of *Annalen der Physik*, a German scientific journal. In this paper, he made the revolutionary proposal that the Newtonian concepts of space and time are to be revised; and propounded the special theory of relativity. He suggested that the experimental failure to detect uniform motion relative to an absolute reference system simply means that there is no such system at all and that all the inertial frames are completely equivalent.

The special theory of relativity is based on the following two postulates:

1. **The Principle of relativity:** All the laws of physics are the same in all inertial frames of reference.
2. **The Principle of independence of the velocity of light:** The speed of light in a vacuum is independent of the motion of the light source or receiver.

The term "special" implies that this theory considers phenomena only in inertial reference frames.

The first postulate is, in effect, a generalization of Galilean principle of relativity to cover all physical processes. All physical phenomena proceed identically in all reference frames. All physical laws are absolutely identical in all inertial systems. Basically, no experiment can distinguish one of the frames as preferable. Thus, Einstein's principle of relativity establishes the complete equality of all inertial frames and rejects the Newton's ideas of absolute space and absolute motion.

The second postulate states that the velocity of light in a vacuum has the same value for all observers and is independent of their motion or of the motion of the light source. In contrast to all other velocities, which change on transition from one reference frame to another, the velocity of light in a vacuum is invariant. This invariance of velocity of light requires that we modify some of our intuitive, everyday notions of space and time.

Let us take once again the example of an observer in a railcar moving with a velocity υ (Fig. 23.10). Suppose a flash of light is sent by him. The flash travels with a velocity c relative to the observer in the railcar. What will be the speed of the flash relative to the stationary observer on the ground? We expect the velocities to add as in classical case and the speed of the flash should be $c + \upsilon$. According to the Einstein's second postulate, both observers would measure the speed of light to be c. This conclusion appears to violate our common sense.

Fig. 23.10: Moving observer

The second postulate explains the null result of Michelson-Morley experiment clearly. It was presumed in the experiment that when light travels against the ether wind, its speed would be $c - \upsilon$ and after reflection at mirror M_1, it returns to M with a speed $c + \upsilon$. Because of the differences in speed, the beams A and B would acquire a phase shift and hence a fringe shift was predicted. According to Einstein's second postulate the value of c is the same in all directions irrespective of the direction of the motion of the Earth. Consequently, the path difference between beams A and B would be zero and as such shift in the fringe pattern does not take place.

23.12 THE LORENTZ TRANSFORMATIONS

In Newtonian mechanics the Galilean transformation equation (23.1), relate the space and time coordinates in one inertial frame to those in the other frame. The equations are not valid for cases where υ approaches the value of c. The transformation equations that apply for all speeds up to c and incorporate the invariance of speed of light were developed in 1890 by the German Physicist H.A. Lorentz (1853–1928). They are known as the **Lorentz transformations**. Their real physical significance was later established by Einstein.

(a) Coordinate transformations

Let us consider two inertial reference frames S and S′ in which the standards for measuring distances and time are the same. Let the reference system S be stationary while the system S′, moves with constant velocity υ relative to system S along the directions of the axes x and x'. The axes x and x' are in one line and the axes y and y', and z and z' are parallel. Let us assume that at the initial instant, $t = t' = 0$ and the origin O and O' of the reference systems coincide. We further assume that at the initial instant $t = t' = 0$, a light source placed at the origin emits a light pulse (flash). The light pulse (flash) travels in the form of a spherical wave at the speed c in both the reference frames. Then during the time interval t' the light pulse travels a distance of ct in the reference frame S. Similarly, in the reference frame S′, the light pulse travels a distance of ct' during the corresponding time interval t'. In both the frames the points reached by the light pulse at t and t' respectively lie on spherical surfaces of radii ct and ct'.

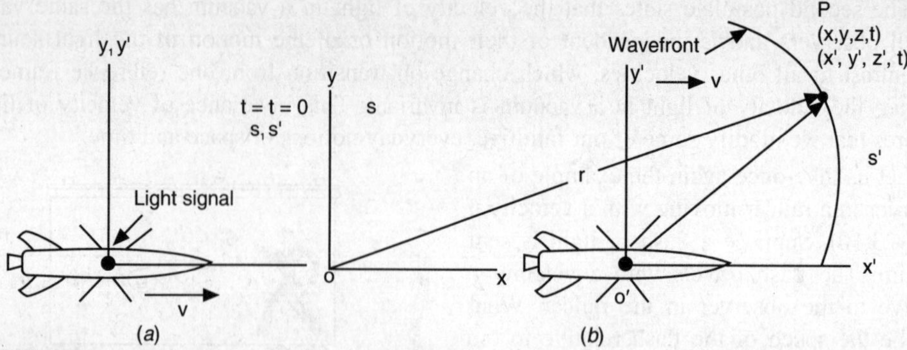

Fig. 23.11: Coordinate transformations

$$r = ct \quad ...(23.19)$$

Thus, $$r' = ct' \quad ...(23.20)$$

The wave surface in frame S is described by the equation

$$x^2 + y^2 + z^2 = r^2 = c^2t^2 \quad ...(23.21)$$

and in frame S' by the equation

$$(x')^2 + (y')^2 + (z')^2 = c^2 (t')^2 \quad ...(23.22)$$

As the space and the time are regarded as homogeneous, the relations between the coordinates and the time in different frames must be linear. Thus,

$$x' = ax + bt \quad ...(23.23)$$

$$t' = fx + gt \quad ...(23.24)$$

Equation (23.23) can be rewritten as

$$x' = a[x + (b/a)t] \quad ...(23.25)$$

At time t, the origin S' is at $x = \upsilon t$. For the position $x' = 0$, it follows from Eq. (23.25), that

$$0 = a[\upsilon t + (b/a)t]$$

or $$(b/a) = -\upsilon$$

Using the above relation into Eq. (23.25), we get,

$$x' = a\,(x - \upsilon t) \quad ...(23.26)$$

Similarly, $$x = a\,(x' + \upsilon t') \quad ...(23.27)$$

Since the systems only move in the direction of x and x' axes, the coordinates y, y' and z, z' do not change.

$$y' = y \text{ and } z' = z \quad ...(23.28)$$

From equations (23.22), (23.26) and (23.21), we may write

$$x' = ct' = a(ct - \upsilon t) = act\,(1 - \upsilon/c) \quad ...(23.29)$$

and $$x = ct = a(ct' - \upsilon t') = act'\,(1 - \upsilon/c) \quad ...(23.30)$$

From equation (23.29), we get

$$t' = at\,(1 - \upsilon/c)$$

Using the relation into Eq. (23.30), we obtain

$$ct = ac\,[at(1 - \upsilon/c)]\,(1 + \upsilon/c)$$

$$= a^2ct\,(1 - \upsilon^2/c^2)$$

$$\therefore \qquad a = \frac{1}{\sqrt{1-\frac{\upsilon^2}{c^2}}}$$

More often a is denoted by γ.

$$\therefore \qquad \gamma = \frac{1}{\sqrt{1-\frac{\upsilon^2}{c^2}}} \qquad \text{...(23.31)}$$

Using Eq. (23.31) into Eq. (23.26), we get

$$x' = \frac{x-\upsilon t}{\sqrt{1-\frac{\upsilon^2}{c^2}}} \qquad \text{...(23.32)}$$

$$t' = \frac{x'}{c} = \frac{\frac{x}{c}-\frac{\upsilon t}{c}}{\sqrt{1-\frac{\upsilon^2}{c^2}}}$$

But, $x = ct$,

$$t' = \frac{t-\frac{\upsilon x}{c^2}}{\sqrt{1-\frac{\upsilon^2}{c^2}}} \qquad \text{...(23.33)}$$

Thus, the equation for transformation of coordinates from system S to system S′ are

$$\begin{aligned} x' &= \gamma\,(x+\upsilon t) \\ y' &= y \\ z' &= z \qquad \textbf{Lorentz Transformation} \qquad \text{...(23.34)} \\ t' &= \gamma\left(t-\frac{\upsilon x}{c^2}\right) \end{aligned}$$

where

$$\gamma = \frac{1}{\sqrt{1-\frac{\upsilon^2}{c^2}}}$$

These equations were derived by Lorentz in connection with electromagnetic phenomena. Einstein pointed out that they have a universal character because they involved only spatial coordinates and time.

The inverse transformation from reference frame S′ to S is given by

$$\begin{aligned} x &= \gamma\,(x'+\upsilon t) \\ y &= y' \\ z &= z' \qquad \textbf{Lorentz Inverse Transformation} \qquad \text{...(23.35)} \\ t &= \gamma\left(t'-\frac{\upsilon x'}{c^2}\right) \end{aligned}$$

The Eq. (23.34) and Eq. (23.35) demonstrate that the space and time are intimately related to each other. What is a pure space interval or a pure time interval in one frame becomes a mixture of space and time intervals in another frame of reference. One important consequence of this mixing of space and time intervals is that the simultaneity of two events depends on the reference frame.

For the sake of comparison, the Galilean and Lorentz transformations are listed below:

Galilean Transformation	Lorentz Transformation
$x' = x - \upsilon t$	$x' = \dfrac{x - \upsilon t}{\sqrt{1 - \upsilon^2/c^2}}$
$y' = y$	$y' = y$
$z' = z$	$z' = z$
$t' = t$	$t' = \dfrac{t - \upsilon x/c^2}{\sqrt{1 - \upsilon^2/c^2}}$

It is easy to see from the table that the Lorentz transformation reduces to the Galilean transformation when speeds of motion of one inertial frame with respect to another frame are small compared to c. i.e., $\upsilon << c$.

(b) Velocity Transformation

Let us again consider two inertial frames S and S' moving with a relative velocity υ along XX' axes. Let us suppose that a body moving with a constant velocity u is observed in the reference frame S. Its positions are at x_1' at time t_1' and x_2' at time t_2' as measured by an observer in S'. Therefore, its speed u_x' measured in the system S' is given by,

$$u_x' = \frac{x_2' - x_1'}{t_2' - t_1'} = \frac{dx'}{dt'} \qquad \text{...(23.36)}$$

Differentiating Eq. (23.32), we get

$$dx' = \frac{dx - \upsilon dt}{\sqrt{1 - \upsilon^2/c^2}}$$

Differentiating Eq. (23.33), we get

$$dt' = \frac{dt - \left(\upsilon/c^2\right)dx}{\sqrt{1 - \upsilon^2/c^2}}$$

Using the expressions for dx' and dt' into Eq. (23.36), we obtain

$$u_x' = \frac{dx'}{dt'} = \frac{dx - \upsilon dt}{dt - \left(\upsilon/c^2\right)dx} = \frac{(dx/dt) - \upsilon}{1 - \left(\upsilon/c^2\right)(dx/dt)}$$

But $\dfrac{dx}{dt} = u_x$ is the velocity of the body measured in S frame,

$$\therefore \qquad u_x' = \frac{u_x - \upsilon}{1 - u_x \upsilon / c^2} \qquad \text{...(23.37)}$$

It can also be shown that

$$u_x = \frac{u_x' + \upsilon}{1 + u_x' \upsilon / c^2} \qquad \text{...(23.38)}$$

If the body has velocity components along y and z directions of S frame, we can find the corresponding components in S' as follows:

$$y' = y$$

$$dy' = dy$$

$$\frac{dy'}{dt'} = \frac{dy\sqrt{1-\upsilon^2/c^2}}{dt - \left(\upsilon/c^2\right)dx} = \frac{(dy/dt)\sqrt{1-\upsilon^2/c^2}}{1-\left(\upsilon/c^2\right)(dx/dt)}$$

or
$$u'_y = \frac{u_y\sqrt{1-\upsilon^2/c^2}}{1-u_x\upsilon/c^2} \qquad ...(23.39)$$

Similarly, we can show that,

$$u'_z = \frac{u_z\sqrt{1-\upsilon^2/c^2}}{1-u_x\upsilon/c^2} \qquad ...(23.40)$$

The following important conclusions may be drawn from the Lorentz transformation:

(*i*) The Lorentz coordinate transformation differs drastically from the Galilean transformation. However, in the limiting case $\upsilon \ll c$, both the transformation laws coincide. It means that the theory of relativity does not reject the Galilean transformation but includes it as a special case.

(*ii*) It may be seen that at $\upsilon > c$ the radicands of the Lorentz transformation become negative and the formulae lose physical meaning. It means that bodies cannot move with a velocity exceeding that of light in a vacuum. Thus no mechanical or electromagnetic agent can transport energy from one point to another with a speed exceeding c.

(*iii*) If $\upsilon = c$, the radicands become equal to zero and in this case also, the formulae do not have physical meaning. It means that there is no reference frame in which a photon is at rest.

(*iv*) The transformation law for the velocities differs radically from the ordinary addition rule known in classical mechanics. In the relativistic case $\upsilon \approx c$, the sum of the components of the velocity along the x-axis is multiplied by the quantity $\left(1+u'_x\upsilon/c^2\right)^{-1}$ which is dependent on u'_x, υ and c. The components of the velocity along y-and z- axes also change, because the time intervals change and $dt \neq dt'$.

(*v*) The Newtonian addition rule for the velocity does not apply to the speed of light. If we have $u'_x = c$ for the speed of light in system S', then in system S we also have $u_x = c$.

$$u_x = \frac{u'_x + \upsilon}{1+u'_x\upsilon/c^2} = \frac{c+\upsilon}{1+c\upsilon/c^2} = c$$

(*vi*) If a light pulse propagates along the y'-axis in S' frame ($u'_y = c$), then the components of velocity of light in system S are

$$u_y = \frac{u'_y\sqrt{1-\upsilon^2/c^2}}{1+u'_x\upsilon/c^2} = c\sqrt{1-\upsilon^2/c^2}$$

and
$$u_x = \upsilon$$

$$u_z = 0$$

Consequently, a ray of light, which is normal to the x'-axis in system S', has a different direction in system S. Of course, the magnitude of the speed of light remains c.

$$\sqrt{u_x^2 + u_y^2} = \left[\upsilon^2 + c^2\left(1 - \upsilon^2/c^2\right)\right]^{1/2} = 0$$

This change of the direction of the rays of light accounts for the aberration of starlight, which is an observed seasonal change in the position of stars. It appears due to the Earth's orbital motion.

23.12.1 Relativistic law of addition of velocities

According to Eq. (23.37), the x'-component of velocity of the body in the reference frame S' is given by $u'_x = \dfrac{u_x - \upsilon}{1 - u_x\upsilon/c^2}$ and that of the body in the reference frame S is $u_x = \dfrac{u'_x + \upsilon}{1 + u'_x\upsilon/c^2}$.

Now, if u' is along the x'-axis, $u'_x = u'$, $u'_y = 0$ and $u'_z = 0$. Therefore, the above equations may be rewritten as

$$u' = \frac{u - \upsilon}{1 - \left(u\upsilon / c^2\right)} \text{ and } u = \frac{u' + \upsilon}{1 + \left(u'\upsilon / c^2\right)} \qquad \text{...(23.41)}$$

Eq. (23.41) is known as the **relativistic law of addition of velocities**. In classical mechanics, the law of addition of velocities is simply $u = u' + \upsilon$ and $u' = u - \upsilon$.

Example 23.2. *Two particles come towards each other with speed 0.7c with respect to laboratory. What is their relative speed?*

Solution: The relative speed is given by $u = \dfrac{u' + \upsilon}{1 + \left(u'\upsilon / c^2\right)}$. Here $u' = \upsilon = 0.7c$.

$$\therefore \qquad u = \frac{2 \times 0.7c}{1 + \left(0.7c \times 0.7c / c^2\right)} = \frac{1.4c}{1.49} = 0.94c.$$

23.13 CONSEQUENCES OF SPECIAL RELATIVITY

The predictions of the special theory of relativity are very strange and startling. The strange effects, which we call *relativistic effects*, appear to conflict with our commonsense. Actually, relativistic effects require speeds that are close to the velocity of light. The relative velocities in our everyday experience are far smaller than the velocity of light. Therefore, we do not commonly observe the strange effects. The special relativity theory predicts that an observer will measure different times and length in different inertial frames. An observer finds that a clock in the moving frame appears to run more slowly. When comparing two events, he finds the events, which are simultaneous in his frame, occur at different times in a moving frame. The observer further finds moving bodies contract along the line of motion. For instance, a meter-stick is measured to be shorter in the direction of motion in a moving frame. It means that we must give up the intuitively obvious notion of absolute time and absolute length on which Newtonian mechanics is based. The new mechanics based on relativity is called **relativistic mechanics**. In relativistic mechanics, there is no such thing as absolute time and absolute length. Events that occur simultaneously at different locations in one frame are not simultaneous in another frame.

It should not be construed from the above discussion that Newtonian mechanics is incorrect. Newtonian mechanics is very much correct and valid for macroscopic bodies moving with speeds much less than c. However, in case of the motion of high-speed particles, the relativistic mechanics is to be employed. In fact, the laws of relativistic mechanics become the same as Newton's laws for cases of $\upsilon < c$. It means that Newtonian mechanics is the limiting case of relativistic mechanics.

23.14 SIMULTANEITY OF EVENTS

Before proceeding to analyze the consequence of relativity, let us first understand the concept of simultaneity of two events occurring at different places.

The site of the point at which event happens is shown in Cartesian coordinates. The corresponding moment of time can be determined by a clock placed at that point of the reference frame. This method is not useful for comparing events occurring at different points.

When two events occurs approximately at the same place, their simultaneity can be detected by means of a simple observation. When they occur at points A and B removed from each other, one has to compare the time indicated on clocks kept at A and B. There is no sense in comparing the time indicated on the two clocks unless they were synchronized before hand. For this purpose, one can bring the clocks together to some point, make them synchronous and take them apart to points A and B.

The clocks positioned at different points of a reference frame can be synchronized using light or radio signals. Let an observer located at the origin O of a given reference frame send a signal at the moment t_o as per his clock. At the moment when this signal reaches the clock located at A at a known distance r from the point O, the clock is set such that it registers $t = t_o + r/c$ which takes into account the signal delay. The same procedure is followed at different points of the reference frame. The repetition of signal after specific time intervals permits all observers to synchronize the rate of their clocks with that at the origin O. Having done this, we can claim that all the clocks of the reference frame show the same time at each moment.

In classical mechanics, temporal relationship between events is assumed to be independent of the reference frame. It implies that if two events, which occur simultaneously in one reference frame, will also be simultaneous in all other frames moving with constant velocity relative to the first frame.

According to Einstein, time and time interval measurements depend on the reference frame in which they are made.

Let us assume that two events A and B occur in the stationary reference frame S at $x_A = -X$ and $x_B = +X$ at time $t_A = t_B = t$. According to Eq. (23.33) event A occurs at

$$t'_A = \frac{t - \left(\upsilon/c^2\right)(-X)}{\sqrt{1-\upsilon^2/c^2}} = \frac{t + \upsilon X/c^2}{\sqrt{1-\upsilon^2/c^2}} \quad \text{...(23.42}a\text{)}$$

And event B occurs at

$$t'_B = \frac{t - \left(\upsilon/c^2\right)(X)}{\sqrt{1-\upsilon^2/c^2}} = \frac{t - \upsilon X/c^2}{\sqrt{1-\upsilon^2/c^2}} \quad \text{...(23.42}b\text{)}$$

It is seen that $t'_A \neq t'_B$. Event B occurs before event A in the moving frame and consequently, an observer in frame S' find that the two events did not occur simultaneously.

Two events that are simultaneous in one inertial frame of reference are in general not simultaneous in another reference frame moving with respect to the first.

Therefore, simultaneity is not an absolute concept, but is a relative concept.

Einstein devised the following thought experiment to explain this concept of relativity of events. Let us consider a railcar moving with uniform velocity. Let two lightening bolts strike at its two ends as shown in Fig. 23.12. They leave marks on the railcar as well as on the ground. Let A and B be the marks left on the ground and A' and B' be the marks on the railcar. An observer is on the ground at O midway between A and B. An observer at O midway between A' and B' is moving with the railcar. If the two light signals reach the observer at O at the same time he concludes that the events at A and B occurred simultaneously.

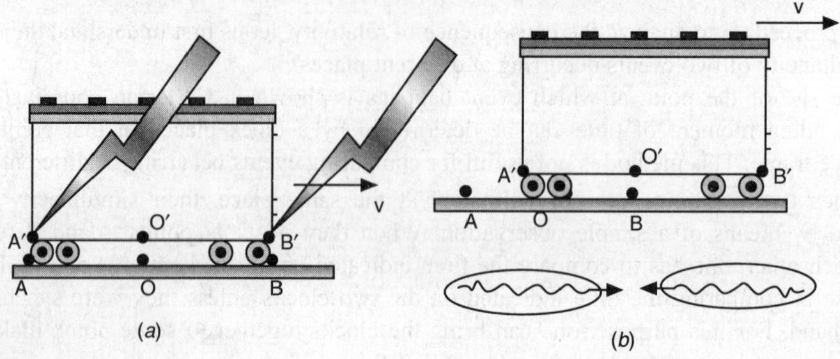

Fig. 23.12:

Einstein'sthought experiment-relativity of events

However, the observer O'observes that the light signal from B' reaches him before the light signal coming from A', because he is moving forward with the railcar. Therefore, he concludes that the front of the railcar was struck before the rear of the railcar was. Thus events, which appear simultaneous in one reference frame, appear not to be simultaneous in another inertial frame.

23.15 LENGTH CONTRACTION

Let us consider a rigid rod at rest in a moving frame S', say a spaceship or railcar moving along x-axis with a speed υ. The length of the rod is equal to the difference between the coordinates of its two end points. Thus, an observer in the spaceship measures the length of the rod to be

$$L_0 = x'_2 - x'_1 \quad ...(23.43)$$

L_o is called the **proper length** of the rod. The *proper length* of a body is defined as the length of the body measured in the reference frame in which the body is at rest.

An observer on the Earth frame S will have to determine the length of the rod by marking the forward end position and the rearward end position of the rod at the same instant of time in his stationary frame S. Thus, he will mark and measure the coordinates of rod as x_1 and x_2 in his frame. According to Lorentz coordinate transformation Eq. (23.32).

$$x'_1 = \frac{x_1 - \upsilon t}{\sqrt{1-\upsilon^2/c^2}}$$

$$x'_2 = \frac{x_2 - \upsilon t}{\sqrt{1-\upsilon^2/c^2}}$$

$$x'_2 - x'_1 = \frac{x_2 - x_1}{\sqrt{1-\upsilon^2/c^2}}$$

or

$$L_0 = \frac{L}{\sqrt{1-\upsilon^2/c^2}} \quad ...(23.44)$$

where $L = x_2 - x_1$ is the length of the rod as measured by an observer in stationary reference frame S. It follows that

$$L = L_0\sqrt{1-\upsilon^2/c^2} \quad ...(23.45)$$

Fig. 23.13: Length contraction

Since $\sqrt{1-\upsilon^2/c^2}$ is always less than 1, L is less than L_o. It means that, if an observer at rest with respect to a body measures its length to be L, an observer moving with a relative speed υ with respect to the body will find it shorter than its proper length by a factor $\sqrt{1-\upsilon^2/c^2}$ as shown in Fig. 23.13. Hence as measured from the Earth, the rod in the spaceship is contracted. This effect is symmetric; a rod at rest on Earth will undergo length contraction when measured from spaceship. This effect is known as the **length contraction**. The length contraction occurs only along the direction of motion. Thus in the case discussed above, the rod contracts only along its length, i.e., in x direction, whereas it does not suffer any change in its width and thickness along y and z directions. As compared to the shape of a body in the reference frame where it is at rest, its shape in the moving reference frame would be a oblate in the direction of motion, as shown in Fig. 23.14.

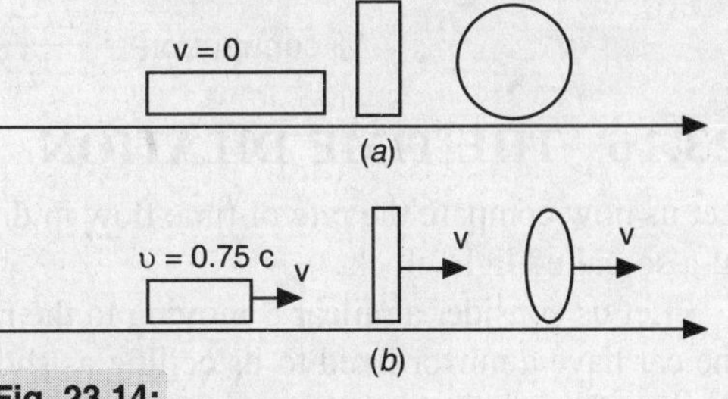

Fig. 23.14: Shape of a body in the rest reference frame and moving reference frame

Eq. (23.45) sows that the degree of contraction depends on the velocity v. As a result, in different inertial reference frames the length of the same rod turns out to be different.

Fitzgerald and Lorentz proposed independently the idea of contraction in length long before Einstein's special theory of relativity was proposed, to explain the negative result of Michelson-Morely experiment. Hence, the length contraction is also known as Lorentz – Fitzgerald contraction.

The length contraction has not been tested directly by experiments, as there is no practical method for the high precision measurement of the length of a fast moving body.

Example 23.3. *The length of a rocket ship is 100 m long on the ground. During its flight, the apparent length is found to be 99 m when measured from the ground. What is its speed?*

Solution:

$$L = L_o\sqrt{1-\upsilon^2/c^2}$$

$$\therefore \quad \upsilon = c\sqrt{1-(L/L_o)^2}$$

$$\therefore \quad \upsilon = 3\times10^8\, m/s\left[1-\left(\frac{99}{100}\right)^2\right]^{1/2} = 4.2\times10^7\, m/s$$

Example 23.4. *Calculate the percentage contraction in the length of a rod moving with a speed of 0.8 c in a direction at an angle of 60° with its own length.*

Solution: Let the rod of length L_o be at rest in the reference frame S. Let S′ be a reference frame which moves with a speed of 0.8 c in a direction making an angle 60° with x-axis. The component of L_o along the x'-axis of S' frame will be L_o cos 60°. Since S′ is moving with a speed of 0.8 c, the component L'_x is given by

$$L'_x = L_o \cos 60^\circ\sqrt{1-\upsilon^2/c^2}$$

$$= \frac{1}{2}L_o\sqrt{1-0.64} = 0.3\, L_o$$

The component of L_o along the y′-axis of S' frame will be L_o sin 60°.

Therefore, $\quad L'_y = L_o \sin 60^\circ = \sqrt{3}L_o/2$

Now, $\quad L' = \sqrt{L'^2_x + L'^2_y} = \sqrt{(0.3L_o)^2 + \left(\sqrt{3}L_o/2\right)^2} = 0.87\, L_o$

$$\%\ \text{contraction} = \frac{L_o - L'}{L_o}\times 100 = \frac{L_o - 0.87\, L_o}{L_o}\times 100 = 13.$$

23.16 THE TIME DILATION

Let us now compare the rate of time flow in different inertial reference frames, by making use of a so-called light clock.

Let us consider a railcar S moving to the right with a speed υ, as shown in Fig. 23.15. Let the car have a mirror fixed to its ceiling as shown in Fig. 23.15. An observer at rest in the car sends off a light pulse at some instant toward the mirror. The light pulse travels (upward) at the speed c gets reflected at the mirror at the top and comes back to the observer. Let us say the observer measures the time interval $\Delta t'$ for the round trip of the pulse with the help of a clock. As $\Delta t'$ is the time measured by a stationary observer in the S' frame, it is the **proper time** T_o. $\Delta t'$ is given by

$$\Delta t' = \frac{2d}{c} \qquad ...(23.46)$$

where d is the distance between the source of light pulse and the mirror at the ceiling.

With reference to an observer on the stationary frame S, the mirror and the light source are moving to the right with a speed υ. When the light pulse strikes the mirror, the mirror will have moved a distance equal to υ ($\Delta t/2$) where Δt is the time taken by the light pulse for its round trip as measured in S frame. From the geometry of the situation, we find that

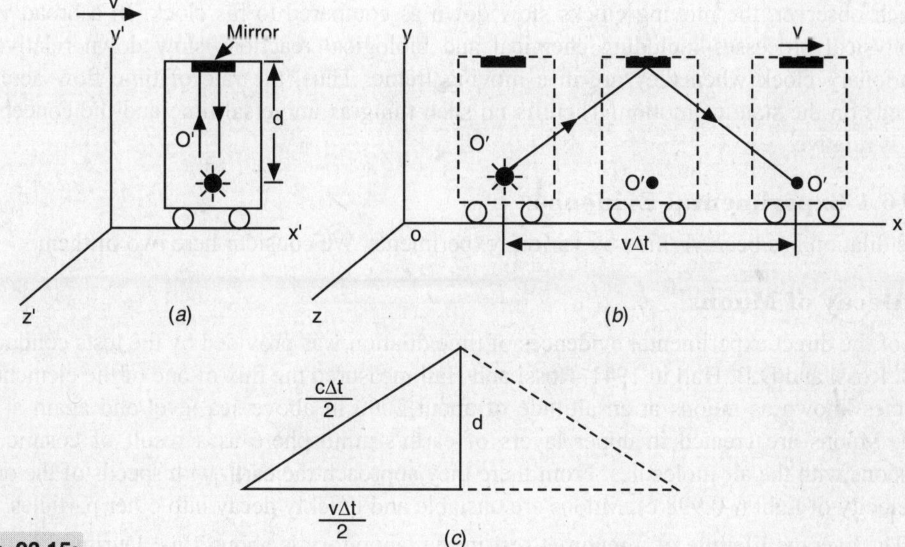

Fig. 23.15: Time Dilation

$$\left[\frac{c(\Delta t)}{2}\right]^2 = \left(\frac{\upsilon\Delta t}{2}\right)^2 + d^2$$

$$\left(c^2 - \upsilon^2\right)\left(\Delta t\right)^2 = (2d)^2$$

$$\therefore \qquad \Delta t = \frac{2d}{\sqrt{c^2 - \upsilon^2}} = \frac{2d/c}{\sqrt{1 - \upsilon^2/c^2}}$$

But $2d/c = \Delta t'$

$$\therefore \qquad \Delta t = \frac{\Delta t'}{\sqrt{1 - \upsilon^2/c^2}}$$

Designating $\Delta t = T$, we rewrite the above equation as

$$T = \frac{T_0}{\sqrt{1 - \upsilon^2/c^2}} \qquad ...(23.47)$$

It follows that a stationary clock measures a longer time interval between events occurring in a moving frame of reference than does a clock in the moving frame. Therefore, from the point of view of the observer in the stationary frame, events will be found to be happening at a slower rate in the moving frame. In other, he concludes that a moving clock runs slower compared to an identical stationary clock by a factor $\sqrt{1 - \upsilon^2/c^2}$. This relative showing of time is known as **time dilation**.

If an observer on the platform watches the events taking place in the moving railcar, he gets the impression that the clock in the car runs slower than his own. His clock measures a longer time between two events taking place in the car than the clock in the car. Conversely, if an observer is in a moving railcar and watches the events on the platform he gets the impression that the clock on the platform runs slower than clock in the railcar. With reference to each observer, the moving clocks slow down as compared to his clock. In a broad view all physical processes including chemical and biological reactions slow down relative to a stationary clock when they are in a moving frame. Thus, the rate of time flow actually depends on the state of motion. There is no such thing as universal time and the concept of time is relative.

23.16.1 Experimental Evidence

Time dilation has been verified by various experiments. We consider here two of them.

(a) Decay of Muons

One of the direct experimental evidences of time dilation was provided by the tests conducted by B. Rossi and D.B. Hall in 1941. Rossi and Hall measured the flux of one of the elementary particles known as muons at an altitude of about 2000 m above sea level and again at sea level. Muons are created in upper layers of earth's atmosphere as a result of cosmic ray collisions with the air molecules. From there they approach the earth with speeds of the order of velocity of light (~0.998 c). Muons are unstable and quickly decay into other particles.

The average lifetime of a muon at rest in the laboratory is about 2 μs. During this time, even traveling at the speed of light, a muon would travel only a distance of $d = ct = (3 \times 10^8$ m/s) $(2 \times 10^{-6}$ s) = 600 m. Therefore, muons created at high altitudes of several kilometers are expected to decay away before reaching the earth. However, measurements showed that an appreciable number of muons do reach the earth's surface (the sea level). The finding can be explained using the concept of time dilation.

The muon decays by its own clock. Relative to an observer on the Earth, muons have a lifetime equal to $\tau/\sqrt{1-\upsilon^2/c^2}$. At a speed of the order of 0.998 c, $\dfrac{1}{\sqrt{1-\upsilon^2/c^2}} = 15$. Therefore, time on earth clock is $t = \gamma\tau = 15\,(2\mu s) = 30\mu s$.

According to an observer on Earth, the average distance travelled by the muon is

$$d = \upsilon t = (0.998c)\,(30\mu s) = 8980 \text{ m}.$$

Therefore, an appreciable number of muons reach the earth before decaying.

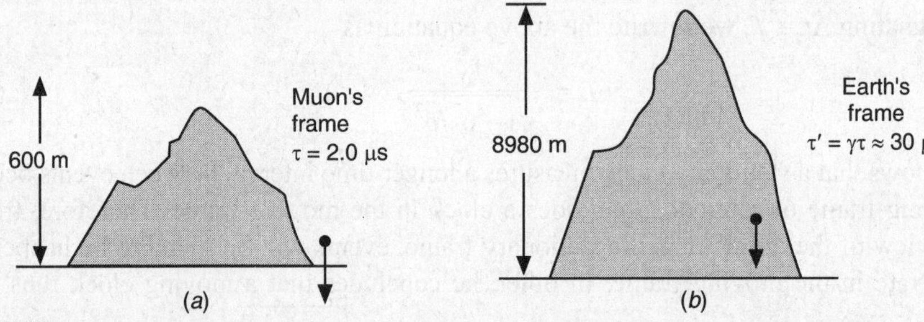

Fig. 23.16: Decay of Muons- Experimental evidences of time dilation

The above result can be interpreted from the viewpoint of muons as follows:

For muons moving at a speed close to c, the terrestrial measures of length appear greatly contracted in the direction of their motion. Because of length contractions, the muon measures the distance to be

$$L = L_0\sqrt{1-\upsilon^2/c^2} = 599\,\text{m}$$

This distance can be covered by the muon in its lifetime, i.e, in 2 μs. Therefore, the existence of appreciable number of muons at the sea level is the direct consequence of time dilation.

(b) Experiment with Clocks

Further experimental evidence for the phenomenon of time dilation was provided by the tests conducted by NASA. In 1971, J.C. Hafele, as astronomer and R.E. Keating, a physicist circled the earth twice in a jet plane, once from east to west for two days and then from west to east for two days carrying two cesium-beam atomic clocks capable of measuring time to a nanosecond. After the trip the clocks were compared with identical clocks. The clocks on the plane lost 59 ± 10 ns during their eastward trip and gained 273 ± 7 ns during the westward trip. These results are consistent with Einstein's theory of time dilation.

Example 23.5. *The proper life of π^- mesons is 2.5×10^{-8}s. What is the velocity of these mesons if the observed mean life is 2.5×10^{7}s?*

Solution.

$$\Delta t' = \frac{\Delta t}{\sqrt{1-\upsilon^2/c^2}}$$

$$\therefore \quad 2.5\times10^{-7}\text{ s} = \frac{2.5\times10^{-8}s}{\sqrt{1-\upsilon^2/c^2}}$$

or $$\sqrt{1-\upsilon^2/c^2} = \frac{1}{10} \qquad \therefore \upsilon = 0.995\,c.$$

Example 23.6. *The average lifetime of a free neutron at rest is 15 minutes. It disintegrates spontaneously into an electron, proton and neutrino. What is the average velocity with which a neutron must leave the Sun in order to reach the earth before breaking up? Given the distance of Earth from Sun is 11×10^{10}m.*

Solution. Let t_o= 15 × 60 s be the proper lifetime of the neutron. The average lifetime t of the moving neutron, as measured by an observer on the earth is $t = \frac{t_o}{\sqrt{1-\upsilon^2/c^2}}$. Also, t is the time for the neutron to reach the earth after leaving the sun.

Hence, $$t = \frac{d}{\upsilon} = \frac{11\times10^{10}\text{ m}}{\upsilon}$$

$$\therefore \quad \frac{15\times60\text{s}}{\sqrt{1-\upsilon^2/c^2}} = \frac{11\times10^{10}\text{ m}}{\upsilon}$$

or $$\frac{\upsilon}{\sqrt{1-\upsilon^2/c^2}} = \frac{11\times10^{10}\text{ m}}{900\text{s}}$$

or $$\upsilon^2 = \left[\frac{11\times10^{10}\text{ m}}{900\text{s}}\right]^2\left(1-\frac{\upsilon^2}{c^2}\right)$$

$$= 1.49\times10^{16}\text{ m}^2/\text{s}^2\left(1-\frac{\upsilon^2}{\left(3\times10^8\right)^2}\right)$$

$$= 0.16\,(9\times10^{16}-\upsilon^2)\text{ m}^2/\text{s}^2$$

$$= \frac{0.16 \times 9 \times 10^{16}}{1.16} \text{m}^2/\text{s}^2 = 1.24 \times 10^{16}\ \text{m}^2/\text{s}^2$$

$$\therefore \qquad \mathbf{\upsilon = 1.11 \times 10^8 m/s.}$$

23.17 THE TWIN PARADOX

The time dilation effect leads to the famous **"Twin Paradox"** of special relativity. Let us consider a hypothetical experiment involving twin sisters Seetha and Geetha. After celebrating their 20th Birthday, the adventurous twin Geetha sets out on a space voyage leaving behind her sister Seetha on Earth. Her spaceship travels at a speed close to the speed of light ($\upsilon = 0.998\ c$). After an year she returns back to the Earth to celebrate her 21st birthday. On return she is shocked to find her sister Seetha to be a 70 years old feeble lady while she herself is only 21 year old. It means while Geetha has aged only one year, her twin sister has aged about 50 years. This is a paradox, which challenges our common sense.

As reckoned in the reference frame of Earth, the clocks and all the physical and biological processes on the spaceship slow down due to time dilation. Hence, Geetha ages at a slower rate and at the end of her trip, when she joins her twin on Earth, she will be younger than Seetha. But this explanation appears to contradict the postulate of special theory of relativity, which asserts that all inertial frames are equivalent. The aging effect seems to provide a way of distinguishing among the frames. Further, from the point of view of the spaceship, the Earth is in motion and hence the clocks and biological processes on the Earth are slowed down due to time dilation. This leads us to the conflicting conclusion that Seetha ages at a slower rate than Geetha and when the twins are united, Seetha will be younger than Geetha.

All these contracdictions disappear when we take into account of the fact that Geetha the space traveler, was not in an inertial frame most of the time. She experienced a series of accelerations when leaving the Earth and decelerations when coming back to the Earth. Thus, she was in an acclerated frame for a greater part of her trip. Hence, the predictions based on special theory of relativity are not valid in her reference frame. On the other hand, Seetha is in an inertial frame all the time and, therefore, her predictions are reliable. Therefore, Geetha will indeed be younger than Seetha on returning to Earth.

23.18 THE RELATIVISTIC MASS

In Newtonian mechanics, the momentum of a body is defined as $p = m\upsilon$ and the mass of the body is supposed to be independent of its velocity. Further, it is supposed that the total momentum of an isolated system is conserved. Because of the immense significance of the conservation laws, the law of conservation of momentum is regarded as fundamental in the theory of relativity.

In order that the total momentum of an isolated system remains constant, it is found that the mass of a body must depend on the velocity of the body. The following relation governs the dependence of mass m on the velocity of the body:

$$m = \frac{m_0}{\sqrt{1-\upsilon^2/c^2}} \qquad \text{...(23.48)}$$

where m is called the **relativistic mass** and m_o is known as the **rest mass** of the body.

Eq. (23.48) shows that the relativistic mass is greater than the rest mass. As the relativistic mass depends on the velocity of the body, the mass of the same particle is different in different inertial reference frames. The rest mass m_0, is the same in all reference frames.

At velocities very small in comparison with c, the mass may be regarded as independent of the speed of the body. With increasing speed, the mass of the body steadily increases and requires a steadily increasing force to impart a constant speed to the body. At $\upsilon \approx c$, the mass becomes infinite and hence it is impossible to make a body move at the speed of light.

Relativistic mass increases are significant only at speed approaching that of light. At a speed $\upsilon = 0.1c$, the mass increases only by 0.5% but at speeds of $0.9c$, the increase is more than 100%. The relativistic effects are of no significance in space flight. Atomic particles such as electrons, protons etc have such high speeds which cause relativistic effects.

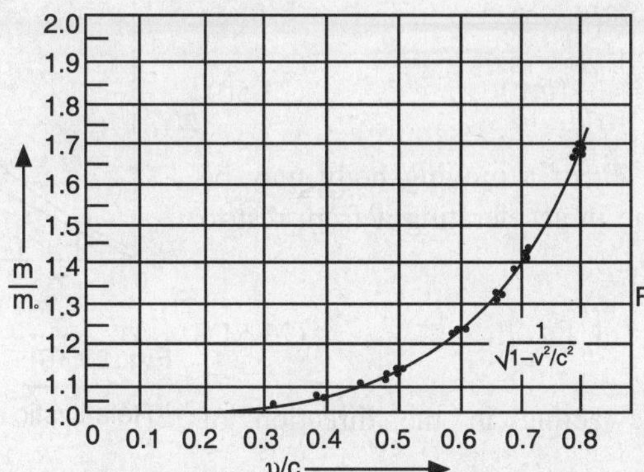

Fig. 23.17: Relativistic mass

There are experimental observations of the increase in mass of high speeds (see Fig. 23.17). It is found that a greater magnetic field is required to deflect a high-speed charged particle than that would be required if its mass is constant.

Example 23.7: *At what speed the mass of an object will be three times of its value at rest?*

Solution: Relativistic mass $m = \dfrac{m_o}{\sqrt{1-\upsilon^2/c^2}}$. Here $m = 3m_o$.

$$\therefore \qquad 3m_0 = \frac{m_o}{\sqrt{1-\upsilon^2/c^2}} \text{ or } \sqrt{1-\upsilon^2/c^2} = \frac{1}{3}$$

$$\text{or} \qquad 1-(\upsilon^2/c^2) = \frac{1}{9} \qquad \frac{\upsilon^2}{c^2} = \frac{8}{9} = 0.89$$

$$\therefore \qquad \upsilon = \upsilon = \sqrt{0.89} \times 3 \times 10^8 \, m/s = 2.67 \times 10^8 \, m/s$$

23.19 THE RELATIVISTIC MOMENTUM

The momentum of a relativistic particle is given by

$$p = m\upsilon = \frac{m_0\upsilon}{\sqrt{1-\upsilon^2/c^2}} \qquad \text{...(23.49)}$$

This is known as the **relativistic momentum** of the particle. The momentum thus defined obeys the law of conservation regardless of the inertial reference frame chosen.

For velocities $\upsilon << c$, (Eq. 23.49) yields the classical momentum $p = m_o\upsilon$. The velocity dependence of the classical and relativistic momentum of a particle is depicted in

Fig. 23.18. The difference between the momenta grows substantially as the velocity of a particle approaches that of light.

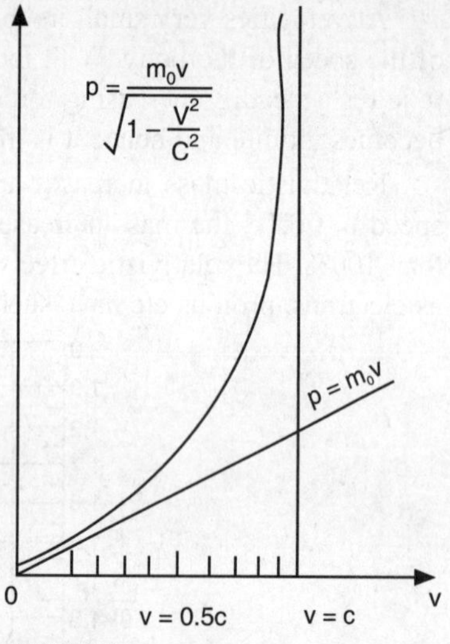

Fig. 23.18: Relativistic momentum

23.20 KINETIC ENERGY

In Newtonian mechanics force is defined as the time rate of change of momentum. This definition of force is valid in relativistic mechanics also, but the mass is a variable quantity in this case.

$$F = \frac{d}{dt}(m\upsilon) \qquad ...(23.50)$$

The kinetic energy E_k of a moving body may be defined as the work done in accelerating it from rest to its final speed υ. That is,

$$E_k = \int_0^s F ds \qquad ...(23.51)$$

where F is the force acting in the direction of displacement.

$$\therefore \qquad E_k = \int_0^s \frac{d}{dt}(m\upsilon)ds = \int_0^{m\upsilon} \upsilon d(m\upsilon) = \int_0^{\upsilon} \upsilon d\left[\frac{m_0\upsilon}{\sqrt{1-\upsilon^2/c^2}}\right]$$

Integrating the above equation by parts, we get

$$E_k = \frac{m_0\upsilon^2}{\sqrt{1-\upsilon^2/c^2}} - m_0\int_0^{\upsilon}\frac{\upsilon\, d\upsilon}{\sqrt{1-\upsilon^2/c^2}} = \left[\frac{m_0\upsilon^2}{1-\upsilon^2/c^2} + m_0c^2\sqrt{1-\upsilon^2/c^2}\right]_0^{\upsilon}$$

or

$$E_k = \frac{m_0c^2}{\sqrt{1-\upsilon^2/c^2}} - m_0c^2 \qquad ...(23.52)$$

$$\therefore \qquad E_k = mc^2 - m_0c^2 \qquad ...(23.53)$$

or

$$E_k = (m - m_0)c^2 \qquad ...(23.54)$$

It follows that the increase in kinetic energy of the body is due to the increase in mass. Eq. (23.52) may be rewritten as

$$E_k = m_0c^2\left[\frac{1}{\sqrt{1-\upsilon^2/c^2}} - 1\right]$$

This is the expression for the relativistic kinetic energy of a particle. At low velocities ($\upsilon << c$), we can write using binomial theorem that

$$\frac{1}{\sqrt{1-\upsilon^2/c^2}} = 1 + \frac{\upsilon^2}{2c^2} + \frac{3\,\upsilon^4}{8\,c^4} + - - - -$$

At $\upsilon << c$, $\upsilon^2 / c^2 << 1$ and we can ignore the higher terms in the series. Thus,

$$\frac{1}{\sqrt{1-\upsilon^2/c^2}} \approx 1 + \frac{\upsilon^2}{2c^2}$$

$$\therefore \qquad E_k = m_0c^2\left[1+\frac{\upsilon^2}{2c^2}-1\right]$$

or $$E_k = \frac{1}{2}m_0\upsilon^2$$

Thus, at velocities $\upsilon << c$, the relativistic formula reduces to the classical formula. The relativistic kinetic energy E and classical K.E. are plotted as a function of υ/c in Fig. 23.19. Their difference becomes very large at $\upsilon \approx c$.

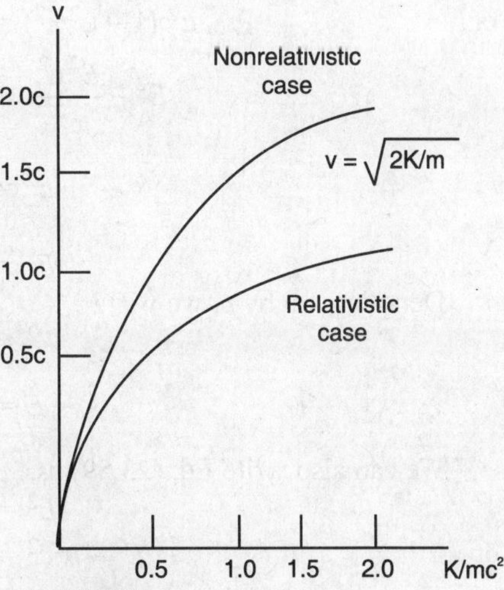

Fig. 23.19:

Relativistic KE and Classical KE as a function of v/c

23.21 MASS-ENERGY EQUIVALENCE

We rewrite equation (23.53) as $mc^2 = E_k + m_0c^2$.

If we interpret mc^2 as the total energy E of the body, Eq. (23.55) indicates that the total energy of a freely moving body consists of its rest energy plus its energy due to motion E_k. When the body is at rest, it possesses the energy m_oc^2. Therefore, m_oc^2 is called the **rest energy** E_o.

$$E = E_o + E_k$$

where $$E = mc^2 \qquad ...(23.56)$$

When $E_k = 0$, the body is motionless. However, it possesses the energy

$$E_o = m_oc^2 \qquad ...(23.57)$$

The above equations are known as **mass-energy equivalence** relations. They imply that energy manifests as mass. This is one of the most remarkable results of Einstein's theory. Till the special theory of relativity was propounded, energy and mass were believed to be independent. The theory of relativity showed that mass is a form of energy.

The validity of the Eq. (23.56) is evident in nuclear processes such as fusion and fission.

Example 23.8: *The sun radiates away energy at the rate of 4×10^{26} J/s. Calculate the rate at which its mass is decreasing.*

Solution: The change in mass $\Delta m = \frac{\Delta E}{c^2} = \frac{4\times10^{26}\text{ J/s}}{9\times10^{16}\text{ m}^2/\text{s}^2} = 4.4\times10^9\text{ kg}.$

23.22 RELATION BETWEEN MOMENTUM AND ENERGY

The energy E and the momentum of a particle have different values in different reference frames. But, the quantity $(E^2 - p^2c^2)$ is invariant and has the same value in different reference frames.

$$E = mc^2$$

or $$E = \frac{m_0c^2}{\sqrt{1-\upsilon^2/c^2}} \qquad ...(23.58)$$

Squaring the Eq. (23.58), we get

$$E^2 = \frac{m_0{}^2c^4}{1-\upsilon^2/c^2}$$

or $$E^2 (1 - \upsilon^2/c^2) = m_0^{\,2}c^4$$

$$E^2 - \frac{E^2\upsilon^2}{c^2} = m_0^{\,2}c^4$$

$$E^2 = m_0^{\,2}c^4 + \frac{E^2\upsilon^2}{c^2} = m_0^{\,2}c^4 + m^2c^4\left(\upsilon^2/c^2\right)$$

$$E^2 = m_0^{\,2}c^4 + c^2 (m\upsilon)^2$$

Denoting $m\upsilon$ by p, we write

$$E^2 = p^2c^2 + m_0^{\,2}c^4 \qquad ...(23.59)$$

$$\therefore \qquad E = c\sqrt{p^2 + m_0^{\,2}c^2} \qquad ...(23.60)$$

We can also write Eq. (23.59) as

$$E^2 = E_0^{\,2} + p^2c^2$$

or $$E^2 - p^2c^2 = E_0^{\,2}$$

The above equation does not contain υ, which implies that the quantity $(E^2 - p^2c^2)$ is independent of the velocity of the particle. It has the same value $E_0^{\,2}$ in all inertial reference frames.

Alternately, we can express momentum p in terms of energy E as

$$p = \frac{\sqrt{\mathrm{E}^2 - \mathrm{E}_0^2}}{c} \qquad ...(23.61)$$

23.23 DOPPLER EFFECT IN LIGHT

The frequency of light radiation changes when the light source or the observer moves with respect to one another. This is known as Doppler effect. Johann Christian Doppler (1803-1853) discovered the effect in case of sound waves. The response of sound waves to moving bodies is clearly observed from the sounding of the whistle of a moving train. When a train at **rest** in a station blows the whistle, **stationary** listeners who are either ahead or behind of the engine will hear the same pitch made by the whistle. But as the train **moves**, those who are **ahead** will hear the sound of the whistle at a **higher** pitch. Listeners **behind** the train, as it pulls further away from them, hear a lower pitch. The quicker the train moves the larger will be the effect of the rising and falling of the pitch.

Moreover, if the train is at rest but the listeners either move toward the sounding train whistle or away from it, the effect will be the same. Those who move toward the train will hear a greater pitch, whereas those who travel away from the train will hear a lesser pitch.

When a sound source is moving towards a stationary observer, the apparent frequency is given by

$$\nu' = \nu\left[\frac{c - \upsilon}{c}\right] \qquad ...(23.62)$$

On the other hand, when the observer is moving towards a stationary sound source, the apparent frequency is given by

$$\nu' = \nu\left[\frac{c + \upsilon}{c}\right]$$

Thus, the Doppler effect is asymmetric in case of sound waves. Sound waves require a material medium for their propagation whereas light waves do not require a medium. Therefore, the Doppler effect is symmetric in case of light and the apparent frequency is the same when either the light source moves towards the stationary observer or the observer moves towards a stationary source.

(*i*) Let us suppose that the observer is stationary and the light source is moving towards the observer with a velocity υ. As there is no medium in between, the observer will receive more waves due to the motion of the source. But the wavelength of the waves is not changed. The apparent frequency is given by

$$\nu' = \nu + \frac{\upsilon}{\lambda}$$

As, $c = \nu\lambda$, we write the above relation as

$$\nu' = \nu + \frac{\nu\upsilon}{c}$$

or $$\nu' = \nu\left[1 + \frac{\upsilon}{c}\right] \quad ...(23.63)$$

(*ii*) Let the source be stationary and the observer move towards the source with a velocity υ. Then, also

$$\nu' = \nu + \frac{\upsilon}{\lambda}$$

As, $c = \nu\lambda$, we write the above relation as

$$\nu' = \nu + \frac{\nu\upsilon}{c}$$

or $$\nu' = \nu\left[1 + \frac{\upsilon}{c}\right] \quad ...(23.64)$$

The equations (23.63) and (23.64) are similar.

(*iii*) Now, let the light source move away from the stationary observer or let the observer move away from the stationary source with a velocity υ. Then,

$$\nu' = \nu - \frac{\upsilon}{\lambda}$$

or $$\nu' = \nu - \frac{\nu\upsilon}{c}$$

$\therefore$ $$\nu' = \nu\left[1 - \frac{\upsilon}{c}\right] \quad ...(23.65)$$

(*iv*) If the source and the observer move towards each other and each moves with a velocity υ, the apparent frequency is given by

$$\nu' = \nu\left[1 + \frac{\upsilon}{c}\right]^2 \quad ...(23.66)$$

(*v*) If the source and the observer move away from each other and each moves with a velocity υ, the apparent frequency is given by

$$\nu' = \nu\left[1 - \frac{\upsilon}{c}\right]^2 \quad ...(23.67)$$

23.23.1 Applications

(*i*) The Doppler Effect in light waves can be observed by analyzing light emitted by celestial objects such as stars. When the spectrum of the Sun or any other star is obtained, a distinctive pattern of dark lines is observed in addition to red, orange and yellow through to violet colors. The wavelengths of lines can be measured extremely accurately.

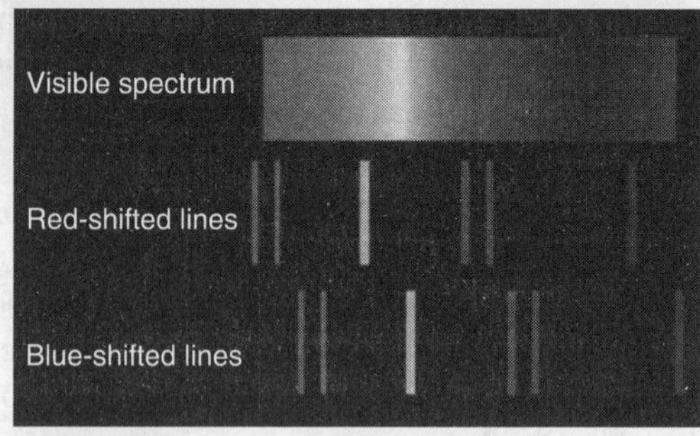

Fig. 23.20: If the star is moving towards the observer the light is blueshifted. If the star is moving away from the observer the light is redshifted.

If a star is moving towards us, the whole pattern of the spectrum gets shifted to shorter wavelengths, i.e. towards the blue end of the spectrum.

This is a **BLUESHIFT**, and we can measure it very accurately by comparing the apparent wavelengths of the spectral lines with the known laboratory wavelengths. If the star is receding, the pattern moves to longer, redder wavelengths, and this is a **REDSHIFT**.

Such an analysis of light from distant galaxies shows that the light experiences a red shift. These galaxies are moving away from the Earth. Extrapolation of this evidence, along with other findings, gives support to the "big bang" theory of the origin of the universe.

Doppler Effect is used by Police to track vehicle speed. Radio waves are transmitted out from radars, collide with a vehicle, and bounce back. The speed of the vehicle determines the change in frequency, which is detected with the radar.

(*ii*) The masses of binary stars are measured, by detecting the velocity changes as they orbit around their common center of gravity, and combining these velocities with the period of the orbit.

(*iii*) When the spectra of other galaxies are studied and their velocities are mapped, it is found that on average, all galaxies are speeding away from each other, with very high redshifts. The Doppler Shift thus indicates us that the universe as a whole is expanding.

Example 23.9. *A spectral line of wavelength 4000Å in the spectrum of light from a star is found to be displaced from its normal position towards the red end of the spectrum by 1Å. What is the velocity of the star in the line of sight?*

Solution. The star is moving away from the stationary observer. Using Eq. (23.65), we obtain

$$\nu' = \nu\left[1-\frac{\upsilon}{c}\right]$$

$$\frac{c}{\lambda'}=\frac{c}{\lambda}\left[1-\frac{\upsilon}{c}\right] \text{ or } \frac{1}{\lambda'}=\frac{1}{\lambda}\left[1-\frac{\upsilon}{c}\right] \text{ or } \frac{1}{4001\text{Å}}=\frac{1}{4000\text{ Å}}\left[1-\frac{\upsilon}{3\times10^{8}\,m/s}\right]$$

or $$\boldsymbol{\upsilon = 75 \text{ km/s}.}$$

Example 23.10. *The wavelength 5000Å of a spectral line emanating from a distant star is observed in the laboratory to be 5200Å. Is the star approaching the earth or receding from it? Calculate the velocity of approach or recession, as the case may be.*

Solution. $$\nu' = \nu\left[1-\frac{\upsilon}{c}\right]$$

$$\therefore \quad \frac{c}{\lambda'}=\frac{c}{\lambda}\left[1-\frac{\upsilon}{c}\right] \text{ or } \frac{1}{\lambda'}=\frac{1}{\lambda}\left[1-\frac{\upsilon}{c}\right] \text{ or } \frac{\lambda}{\lambda'}=1-\left(\frac{\upsilon}{c}\right)$$

$$\therefore \quad \left(\frac{\upsilon}{c}\right) = 1-\frac{\lambda}{\lambda'}=1-\frac{5000\text{Å}}{5200\text{Å}}=\frac{1}{26}$$

$$\therefore \quad \upsilon = \frac{3\times10^8 \text{ m/s}}{26} = 1154 \text{ km/s.}$$

The star is moving away from the stationary observer with a velocity of 1154 km/s.

Example 23.11. *A space ship is moving with a velocity of 10^{10} cm/s towards a star that emits radiations of wavelength 6×10^{-5}cm. Calculate the wavelength of the radiations as received by the crew in the ship.*

Solution.

$$\nu' = \nu\left[1+\frac{\upsilon}{c}\right]$$

or

$$\lambda' = \frac{\lambda c}{c+\upsilon} = \frac{(6\times10^{-5}\text{ cm})(3\times10^{10}\text{ cm/s})}{3\times10^{10}\text{ cm/s}+10^{10}\text{ cm/s}} = \mathbf{4500\ Å.}$$

QUESTIONS

1. Define an inertial frame of reference. Distinguish between special theory of relativity and general relativity.
2. Describe Michelson-Morley experiment. Show how negative results obtained from Michelson-Morley experiment were interpreted.
3. What are non-inertial frames of reference?
4. Give Galileo's principle of relativity.
5. State the fundamental postulates of special theory of relativity?
6. What is proper time?
7. What is proper length?
8. Does $E = mc^2$ apply to particles that travel with the speed of light?
9. Is mass a conserved quantity in special theory of relativity?
10. Show that the laws of mechanics are identical in any inertial reference frame.
11. Describe Michelson-Morley experiment and explain the significance of the null result of the experiment.
12. State fundamental postulates of special theory of relativity. Deduce the Lorentz transformation of space.
13. State the fundamental postulates of the special theory of relativity and deduce the Lorentz transformation equation of space and time.
14. Prove that $x^2 + y^2 + z^2 = c^2t^2$ is invariant under Lorentz transformation.
15. What is Galilean transformation? Derive Galilean transformation equations.
16. Discuss Galilean transformation for position, velocity and acceleration.
17. State the fundamental postulates of the special theory of relativity and deduce the Lorentz transformations for space and time coordinates.
18. State and explain the relativistic law of addition of velocities.
19. Discuss simultaneity of events in two inertial frames of reference. Hence explain the concept of length contraction and time dilation.

20. Show that the apparent length of a rigid body is decreased by the factor $\sqrt{1-v^2/c^2}$ in the direction of its motion.
21. Explain the concept of time dilation. Describe experimental verification of time dilation.
22. Explain the concept of time-dilation and length contraction in theory of relativity.
23. What is time dilation in special relativity? Obtain an expression for time dilation in regard to the time interval between two events measured from two different inertial frames.
24. What is time dilation? Deduce an expression for the time dilation as regards to interval between two events measured from two different inertial frames.
25. What is time dilation in special relativity? Obtain an expression for time dilation in regard to the time interval between two events measured from two different inertial frames.
26. What is the length contraction in special theory of relativity? Deduce an expression for length contraction.
27. Show how the relativistic invariance of the law of conservation of momentum leads to the concept of variation of mass with velocity.
28. Show that the addition of a velocity to the velocity of light gives the velocity of light.
29. Obtain the relativistic formula for the addition of velocities.
30. Derive the formula for the variation of mass with velocity according to special theory of relativity.
31. Deduce Einstein's mass-velocity relation $E = mc^2$, considering the variation of mass with velocity.
32. Deduce Einstein's expression for mass-energy equivalence.
33. Deduce Einstein's mass-energy relation considering the variation of mass with velocity.
34. Deduce an expression for the variation of mass with velocity.
35. State and explain the Doppler's effect of light.
36. What is Doppler's effect?

PROBLEMS

1. A person on a rocket traveling at a speed of 0.5c with respect to the earth observes a meteor coming from behind and passing him at a speed of 0.5c. How fast is the meteor moving with respect to the earth? **[Ans:** 0.8c]
2. A rocket leaves the earth at a speed of 0.6c. a second rocket leaves the first at a speed of 0.9c with respect to the first. Calculate the speed of the second rocket with respect to the earth if (*i*) it is fired in the same direction as the first one and (*ii*) if it is fixed in a direction opposite to the first. **[Ans:** 0.974c, –0.652c]
3. A particle with a mean lifetime of 10^{-6}s moves through the laboratory at a speed of 0.8c. What will be its lifetime as measured by an observer in the laboratory? **[Ans:** 1.67×10^{-6}s]
4. A meson having a mass of 2.4×10^{-28} kg travels at a speed of 0.8c. What is its kinetic energy? **[Ans:** 1.44×10^{-11}J]
5. The mean lifetime of a muon at rest is 2.4×10^{-6}s. What will its mean lifetime be as measured in the laboratory, if its velocity is 0.6c with respect to the laboratory? **[Ans:** 3×10^{-6}s]
6. The π-mesons coming out of an accelerator have a velocity of 0.99c. If they have a mean lifetime of 2.6×10^{-8}s in the rest frame, how far can they travel before decay? **[Ans:** 54.6 m]
7. A man leaves the earth in a rocket ship that makes a round trip to the nearest star, which is 4 light years away. If the speed of the rocket ship is 0.8c, how much younger will he be on his return than his twin brother who preferred to stay behind? **[Ans:** 3.4 yrs]
8. Two particles are moving in opposite directions with speeds 0.9c as observed in laboratory frame. What is the velocity of one particle relative to the other? **[Ans:** 0.994c]
9. In the laboratory the lifetime of a particle moving with speed 2.8×10^8 m/s is found to be 2.5×10^{-7}s. Calculate the proper lifetime of the particle. **[Ans:** 8.9×10^{-8}s]
10. Find the momentum and velocity of an electron having a kinetic energy of 10 MeV. The rest energy of electron is 0.51 MeV. **[Ans:** 0.998c]

24 Quantum Hypothesis

Learning Objectives:

By the end of this chapter, you will be able to:

- Define concepts of Blackbody Radiation.
- Explain basics of Planck's quantum hypothesis.
- Describe concept of Particle Picture of Radiation.
- Understand Photoelectric effect.
- Learn about X-Ray and X-Ray concepts.
- Interpret Wave-Particle Duaility principle.

24.1 INTRODUCTION

The aim of physics is *to understand the natural phenomena* around us. 19th century witnessed a rapid growth in physics. Newtonian mechanics was nearly perfected and Maxwell synthesized the fields of electricity and magnetism into a single theory which permitted inclusion of optics into the frame work of electromagnetic phenomenon. Towards the end of the 19th century it was generally believed that all that to be discovered in nature was discovered, and all the laws of nature were formulated. The Newtonian mechanics, Maxwell's electromagnetic theory and thermodynamics came to be known later as classical physics. Classical physics was developed assuming that particles are localized and we can observe them without appreciably disturbing them. The three laws of conservation, namely conservation of linear momentum, angular momentum and energy formed the basis for classical mechanics. However, discoveries that followed dispelled such illusions very soon.

At about the turn of 20th century a number of fundamental discoveries were reported which could not be explained within the framework of the above classical theories of physics. The inadequacy of classical theories was noticed first when they were applied to explain the black body radiation emitted by a body hotter than its surroundings. To explain the blackbody radiation, Max Planck put forward a revolutionary hypothesis that the molecules in a source emit energy not continuously but in small discrete packets called *quanta*. This was a radical departure from the classical theory and contrary to day-to-day experience. Einstein made use of Planck's hypothesis that a quantum of radiation carries an energy $h\nu$ and successfully explained the photoelectric effect. Other experimental results also showed that the classical concepts were entirely inadequate and one has to invoke the quantum concept to explain the behaviour of atoms and subatomic particles. A new body of ideas based on Planck's work was developed, and the new theory came to be known as **quantum physics**. *Quantum physics explains the behaviour of matter and radiation at the microscopic (atomic) level.*

24.2 WAVE-PICTURE OF RADIATION—ENERGY FLOW IS CONTINUOUS

We are familiar with several waves such as radio waves, microwaves, heat waves, light waves, UV-rays, X-rays and γ-rays etc. These waves constitute a single family called **electromagnetic waves** or simply **radiation**. Electromagnetic waves comprise varying electric and magnetic fields traveling in space at the velocity of 'c'. The propagation of electromagnetic waves and their interaction with matter is explained with the help of Maxwell's electromagnetic theory.

Maxwell's theory considered the process of emission of radiation by a source as a *continuous* process. A heated body may be supposed to be capable of emitting energy that travels in the form of waves of *all possible wavelengths*. In the same manner, the radiation incident on a body may be thought to be absorbed at all possible wavelengths. The intensity of radiation is given by

$$I = |E|^2 \qquad ...(24.1)$$

where E is the amplitude of the electromagnetic wave.

24.3 BLACKBODY RADIATION

It is well known that when a body is heated, it emits electromagnetic radiation. The radiation emitted by the source by virtue of its temperature is called **thermal radiation.** Thermal radiation is electromagnetic in nature and its energy is smoothly distributed over all wavelengths. Thus, a thermal source produces **continuous spectrum**. The intensity and the predominant wavelength of radiation vary with the temperature of the body. At low temperatures the radiation mainly lies in the infrared region. As the temperature of the body is increased, the component of maximum intensity shifts to a higher and higher frequency. For example, the filament of an incandescent bulb appears dark at room temperature and as current is passed, it gets heated up. As the current increases through the filament, it appears initially red, orange gradually and yellow and finally it emits white light. At temperatures above 1000°C, a heated body is capable of giving out energy in the form of waves of all possible wavelengths. Normally the amount of radiation emitted by a hot body depends on factors such as the properties of its surface. Apart from emitting electromagnetic radiation, a body also absorbs electromagnetic radiation incident on it.

A *body that absorbs the entire radiation incident on it is called a* ***perfect blackbody***. When a perfect blackbody is heated, it emits radiation at all frequencies and thus, it is a good radiator as well as a good absorber. *The radiation emitted by a perfect blackbody is called* ***blackbody radiation***.

In practice, there are no perfect blackbodies but an ideal black body can be made by taking a hollow sphere (cavity) and drilling a small hole in it (see Fig. 24.1). Its inner surface is coated with lampblack. Any radiation entering the cavity through O is incident on the inner surface of the cavity and is partly absorbed and partly reflected. The reflected component is incident at another point on the inner surface where again it is partly absorbed and partly reflected. Thus, light entering the cavity undergoes multiple reflections at the walls and gets trapped inside the cavity. The probability of the radiation of any wavelength escaping out of the hole is negligible. Hence, the hole acts a perfect absorber and appears perfectly dark. Conversely, when the cavity is heated, the radiation produced in the cavity comes out through the aperture and contains all the wavelengths. Therefore, the hole acts as a perfect emitter and has the characteristics of blackbody radiation. Its spectrum can be analyzed by an infrared spectrometer using a bolometer as a detector. Thus, the emissive power of the blackbody at different wavelengths can be determined. If the distribution of radiant energy as a function of wavelength at different temperatures is plotted, we obtain a set of curves as shown in Fig. 24.2.

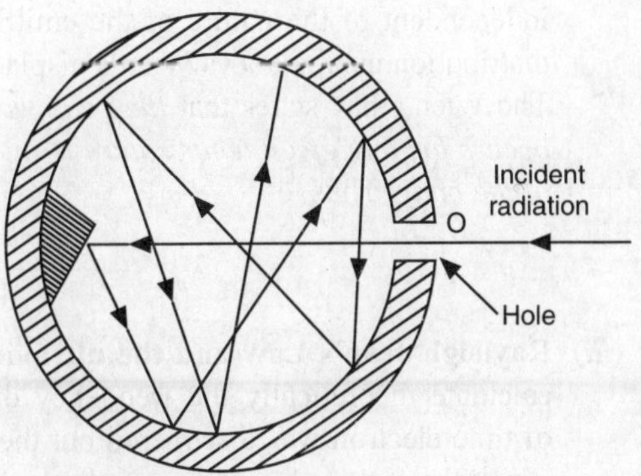

Fig. 24.1: Illustration of blackbody

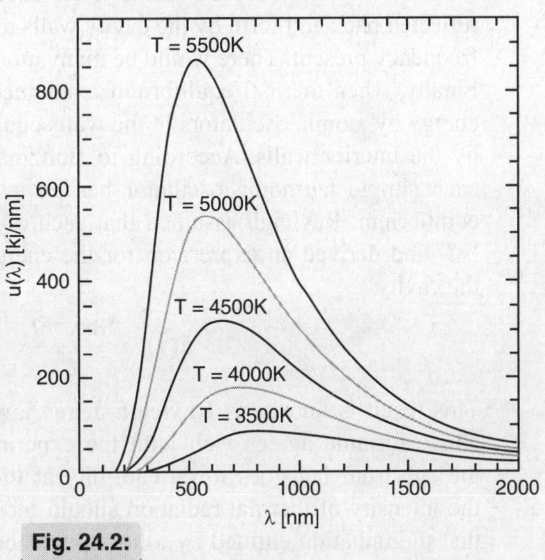

Fig. 24.2: Radiation energy density versus wavelength at a given temperature

The experimental results (Fig. 24.2) show that at a given temperature the radiation energy density initially increases with increasing wavelength then peaks at around a particular wavelength λ_m and after that decreases finally to zero at very high wavelengths. The spectral distribution of that radiation is a function of temperature alone and the material as such plays no role.

24.3.1 Laws of Blackbody Radiation

(*i*) **Stefan-Boltzmann law:** The Stefan-Boltzmann law is an empirical relationship obtained by Stefan and later derived theoretically by Boltzmann. It connects the intensity of radiation to the temperature. It states that *the total radiation emitted from a blackbody at temperature T is proportional to the fourth power of the absolute temperature of the body T^4.*

$$E = \sigma T^4 \qquad \text{...(24.2)}$$

where σ is called Stefan's constant having a numerical value of 5.67×10^{-8} W/ m^2-K^4.

(*ii*) **Wien's law:** It is seen from Fig. 24.2 that the value of λ_m depends only on the temperature T of the blackbody and decreases with increasing temperature. λ_m is independent of the nature of the emitting body. The shift in the peak of the intensity distribution curves obeys Wien's displacement law.

The Wien's law states that *the peak wavelength, λ_{max} at which the maximum emission occurs for any given temperature is inversely proportional to the absolute temperature of the body*. Thus,

$$\lambda_{max} = \frac{2.8978 \times 10^{-3}}{\mathrm{T}} m\mathrm{K} \quad ...(24.3)$$

(*iii*) **Rayleigh-Jean's Law and the ultraviolet catastrophe:** Various efforts were made to calculate theoretically the frequency distribution of thermal radiation. At that period of time electron was discovered but the inner structure of the atom was unknown. The generally accepted model was that a blackbody was made up of a huge number of atoms and the atoms are regarded as small harmonic oscillators. The random thermal motion of atoms within the walls generates electromagnetic waves, which is the thermal radiation emitted from the walls of the cavity. The radiation emitted by the atoms is reflected back and forth by the cavity walls to form a system of standing waves for each frequency present. There would be many modes of vibration present in the cavity space. Finally, when thermal equilibrium is attained, the average rate of emission of radiant energy by atomic oscillators in the walls equals the rate of absorption of radiant energy by the interior walls. According to Boltzmann's principle of equipartition of energy, each simple harmonic oscillator has an average thermal energy of 'kT' at thermal equilibrium. Rayleigh assumed that each of the standing waves ought to have energy 'kT' and derived an expression for the energy density of radiation distribution within the cavity.

$$E(\nu) = \frac{8\pi\nu^2 kT}{c^3} \quad ...(24.4)$$

This result is known as **Rayleigh-Jeans law**. The energy density calculated with the above formula agrees well with the experimental results at long wavelength end of the spectrum but goes toward infinity at the short wavelength end. It predicted that the intensity of thermal radiation should increase with square of frequency. It implied that the radiation emitted by a hot body should have a large portion of UV rays. This is contrary to our experience and violates the law of conservation of energy. This contradiction came to be known as ***ultraviolet catastrophe***. The failure of Rayleigh-Jeans formula presented a crisis and gave the first indication of inadequacy of classical physics.

(*iv*) **The Planck's Radiation Law:** When classical mechanics failed to explain the frequency distribution of the blackbody radiation, Max Planck attempted to attack the problem from an entirely novel approach. He struggled to find first a mathematical expression that could account for the blackbody radiation spectrum, and then seek to derive this new equation, which was until then unknown. This paved the path to the postulation of Planck's Radiation Law for the spectral energy density of blackbody radiation in 1900.

24.3.2 Plank's Radiation Law

After long years of struggle, Planck succeeded in 1900 to obtain the correct mathematical law for distribution of energy in the blackbody radiation. He recognized that Rayleigh's assumption about the equipartition of energy among all available modes in the cavity was leading to the absurd result that the total energy in the cavity would be infinite. Therefore, Planck put forward a revolutionary postulate that an oscillating atom can absorb or reemit energy only in quantities that are integer multiples of '$h\nu$'. All other values of energy are forbidden. The indivisible discrete unit of energy $h\nu$ is called an **energy quantum**. The assumption which allowed Planck's Radiation Law to be derived, was that the energies of the molecular oscillators must be quantized, not continuous, as it had been considered to be the case until then.

In general, the possible values of the energy of an oscillator of frequency ν are

$$E = n\,h\nu \qquad (n = 1,2,3,\ldots) \qquad \ldots(24.5)$$

where n is called the **quantum number** of the oscillator, h is a constant now known as Planck's constant.

Additionally Planck presumed that the change in energy of the oscillator due to emission or absorption of radiation possibly will occur by a discrete amount $h\nu$. As radiation is released from the oscillators, the energy possessed by the emitted radiation will be $h\nu$. Obviously, this is the energy gain of the oscillator when it absorbs the radiation. Thus, the atomic oscillators can exist in a set of discrete energy states, 0, $h\nu$, $2h\nu$, $3h\nu$, and so on. The number of oscillators in an energy state $E_n = n\,h\nu$ is determined by the Maxwell-Boltzmann distribution function.

$$N_n = N_0\,exp(-E_n/kT) = N_0\,exp(n\,h\nu/kT) \qquad \ldots(24.6)$$

where N_o is the number of oscillators in the ground state ($E_n = 0$).

Let there be N oscillators in the system in equilibrium at absolute temperature T. According to Maxwell-Boltzmann distribution law, the number of oscillators N_n with energy E_n is given by

$$N_n = Ne^{-E_n/kT}$$

Since the energies of the oscillators assume discrete values, we have to sum over all possible oscillator states to determine ⟨E⟩. The average energy ⟨E⟩ of an oscillator is then given by

$$\langle E\rangle = \frac{\sum_{n=0}^{\infty} E_n N_n}{\sum_{n=0}^{\infty} N_n} = \frac{\sum_{n=0}^{\infty} E_n e^{-E_n/kT}}{\sum_{n=0}^{\infty} e^{-E_n/kT}} \qquad \ldots(24.7)$$

$$\langle E\rangle = \frac{\sum_{n=0}^{\infty} n\,h\nu\,e^{-n\,h\nu/kT}}{\sum_{n=0}^{\infty} e^{-n\,h\nu/kT}}$$

$$= \frac{h\nu\,e^{-h\nu/kT} + 2h\nu\,e^{-2h\nu/kT} + \ldots}{1 + e^{-h\nu/kT} + e^{-2h\nu/kT} + \ldots}$$

Substituting $e^{-h\nu/kT} = x$, we get

$$\langle E\rangle = \frac{h\nu\,x\left(1 + 2x + 3x^2 + \ldots\right)}{1 + x + x^2 + \ldots}$$

$$= \frac{h\nu\,x(1-x)^{-2}}{(1-x)^{-1}} = \frac{h\nu\,x}{(1-x)}$$

$$\therefore \qquad \langle E \rangle = \frac{h\nu e^{-h\nu/kT}}{1-e^{-h\nu/kT}} = \frac{h\nu}{e^{h\nu/kT}-1} \qquad ...(24.8)$$

According to classical physics, the average energy of an oscillator is kT. The concept of quantization of energy shows that the theorem of equipartition of energy is not valid in the microscopic world. The number of oscillators in the frequency range ν and $\nu + d\nu$ are estimated to be

$$\frac{8\pi\nu^2 d\nu}{c^3}$$

Multiplying this number by the average energy of an oscillator, Planck obtained the energy density in blackbody radiation as

$$E(\nu) = \frac{8\pi h\nu^3}{c^3}\left[\frac{1}{e^{h\nu/kT}-1}\right] \qquad ...(24.9)$$

or

$$E(\lambda) = \frac{8\pi hc}{\lambda^5}\left[\frac{1}{e^{hc/\lambda kT}-1}\right] \qquad ...(24.10)$$

This is **Planck's radiation law**. It is seen from the formula that the exponential term in the denominator prevents E(ν) from rising to infinity at small wavelength values. Planck's formula gives complete fit with the experimental observations at all wavelengths and for all temperatures.

All the laws of blackbody radiation can be derived from Planck's law as follows:

***(i)* Wien's law**

When ν is large, $(h\nu/kT) >> 1$, and $(e^{h\nu/kT} - 1) \cong e^{h\nu/kT}$. Hence Eq.(24.9) reduces to

$$E(\nu) = \frac{8\pi h\nu^3}{c^3} \cdot \frac{1}{e^{h\nu/kT}} = \frac{8\pi h\nu^3}{c^3} e^{-h\nu/kT}$$

This represents Wien's law.

***(ii)* Rayleigh-Jean's law**

When ν is small, $(h\nu/kT) << 1$, and $\left(e^{h\nu/kT}-1\right) \cong \frac{h\nu}{kT}$. Hence Eq.(24.9) reduces to

$$E(\nu) = \frac{8\pi h\nu^3}{c^3} \cdot \frac{kT}{h\nu} = \frac{8\pi\nu^2 kT}{c^3}$$

This represents the Rayleigh-Jean's law.

***(iii)* Wien's displacement law**

Planck's formula (24.10) shows a maximum at $\lambda = \lambda_m$ when the denominator becomes a minimum. The denominator can be written as

$$z = \lambda^5 (e^{hc/\lambda kT} - 1)$$

Then

$$\frac{dz}{d\lambda} = 5\lambda^4\left(e^{hc/\lambda kT}-1\right) - \lambda^5 \frac{hc}{\lambda^2 kT} e^{hc/\lambda kT} = 0 \text{ for } \lambda = \lambda_m.$$

$$1 - e^{(-hc/\lambda_m kT)} = \frac{hc}{5\lambda_m kT} \qquad ...(24.11)$$

or

$$1 - e^{-x} = \frac{x}{5}$$

where

$$x = hc/\lambda_m kT. \qquad ...(24.12)$$

Eq. (24.12) is a transcendental equation which cannot be solved analytically. It has to be solved graphically. If we put $y = 1 - e^{-x}$ and $y = x/5$, then the point of intersection of the two graphs given by these two relations gives the solution. It comes out to be x = 4.9651. Using this value into Eq.(24.12), we get

$$\lambda_m T = \frac{hc}{kx} = \frac{hc}{4.9651k} = 0.0029\ m \cdot K = constant$$

This is Wien's displacement law.

24.4 PLANCK'S QUANTUM HYPOTHESIS - Energy is quantized

The empirical formula obtained by Max Planck explained the experimental observations. In formulating the formula, he presumed that the atoms of the walls of the blackbody act like small harmonic oscillators and each of them has a characteristic frequency of vibration.

Planck assumed further that:

(*i*) *An oscillating atom can absorb or reemit energy in discrete units. The discrete unit of energy hν is called an* **energy quantum**. *A quantum of energy has the magnitude given by*

$$E = h\nu \qquad ...(24.13)$$

where ν is the frequency of radiation and '*h*' is a constant now known as the **Planck's constant.**

(*ii*) *The energy of the oscillator is quantized. It can take only certain discrete amounts of energy* E_n.

$$E_n = n\,h\nu; \qquad n = 1, 2, 3\ .... \qquad ...(24.14)$$

where ***n*** *is known as the* ***quantum number***.

Planck's Quantum Hypothesis: *Radiant energy is* ***quantized*** *and as such emitted or absorbed in a discontinuous manner in the form of quanta. Planck's hypothesis implies that an atom exists in certain discrete energy states, called* ***quantum states*** *and * n* is called the *quantum number*. The atom emits or absorbs energy by jumping from one quantum state to another quantum state.

Planck's postulation of discrete energy states for an atom (Fig. 24.3) was a departure both from the point of view of classical physics as well as our day-to-day experience. We are very much familiar with a mass–spring harmonic oscillator; it can accept any amount of energy from zero to some maximum value (Fig. 24.4). Therefore, in the classical physics energy always seems to occur with continuous values. Also, energy exchange between bodies involves any random amounts of energy.

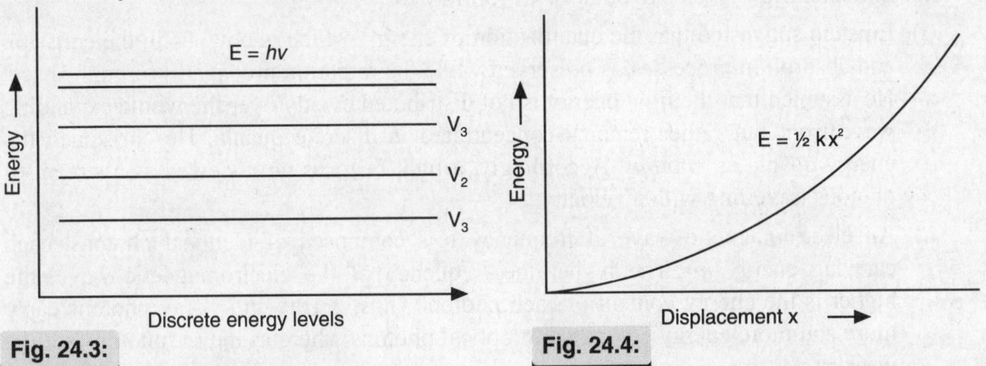

Fig. 24.3: Discrete energy states for an atomic oscillator

Fig. 24.4: Displacement versus energy of a mass–spring harmonic oscillator

The quantum concept can be understood with the help of the following example:

Consider a person walking up a hill and another person climbing a stair case, as shown in Fig. 24.5. Both of them gain in potential energy. The potential energy of the person going up the hill increases *continuously* through *arbitrary* amounts. In contrast, the potential energy of the person climbing the stair case increases through fixed doses. If all the steps are of the same height '*h*', the person acquires a quantum of energy '*mgh*' each time he climbs one step. Consequently, his potential energy increases by 1 *mgh*, 2 *mgh*,3 *mgh*, ... N *mgh*. It is not possible to acquire energy which is a fraction of *mgh* or of any intermediate value.

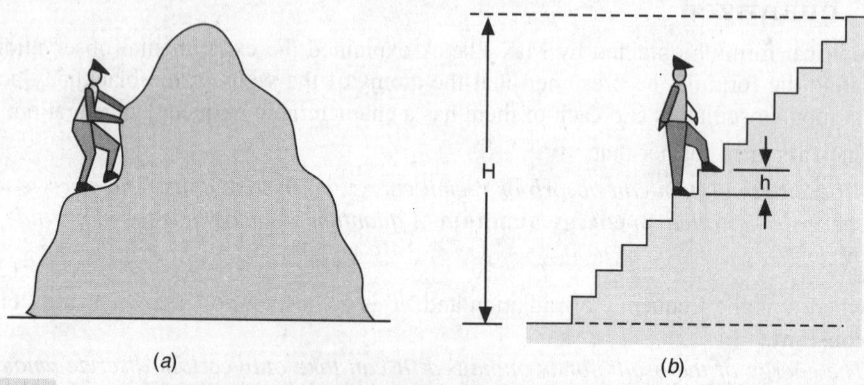

Fig. 24.5: Illustration of Quantum energy concept

24.5 PARTICLE PICTURE OF RADIATION - Radiation is a stream of photons

Though Max Planck proposed discontinuous emission and absorption of energy by bodies, but he preserved the propagation of radiation through space as taking place in the form of continuous waves. Einstein improved the Planck's postulate and endowed the quantum with a distinct identity. Using his upgraded model of energy emission and absorption, he successfully explained in 1905 the experimental results of the photoelectric effect and in 1907 the temperature dependence of specific heats of solids. The photoelectric effect convincingly established that *light behaves as a stream of particles*.

Einstein extended Planck's hypothesis as follows:

(*i*) Einstein suggested that the quantization of energy, which is obvious in the emission and absorption processes, is preserved while the radiation propagates through space. He assumed that the light energy is not distributed evenly over the whole expanding wave front but rather remains concentrated in discrete quanta. He christened the energy quanta as *photons*. Accordingly, a light beam is now viewed as a stream of photons travelling with a velocity '*c*'.

(*ii*) An electromagnetic wave of frequency v is composed of identical photons, each carrying energy hv. The higher the frequency of the electromagnetic wave, the higher is the energy content of each photon. Thus, X -ray and γ -ray photons carry more and more energy compared to optical photons where as the r.f. photons are the weakest.

(*iii*) An electromagnetic wave would have energy hv if it has only one photon, 2_{hv} if it has 2 photons and so on. Therefore, the intensity of a monochromatic light beam I, is linked to the concentration of photons, N, present in the beam. Thus,

$$I = N h\nu \qquad ...(24.15)$$

In contrast, electromagnetic theory gives the intensity of a light beam as

$$I = |E|^2$$

(*iv*) When photons meet matter, they transfer all their energy to the particles of matter and disappear. Hence, absorption of radiation is discontinuous. The number of photons radiated by even a weak light source is extremely large and the human eye cannot register the photons separately and therefore light appears as a continuous stream. Therefore, the discreteness of light is not readily perceived.

24.5.1 The Photon

As the radiant energy is viewed as made up of spatially localized photons, we may attribute particle properties to photons.

(*i*) **Energy**: The energy of a photon is determined by its frequency ν and is given by $E = h\nu$. Using the relation $\omega = 2\pi\nu$ and writing $h/2\pi = \hbar$, we may express

$$E = \hbar\omega \qquad ...(24.16)$$

(*ii*) **Velocity**: Photons always travel with the velocity of light 'c'.

(*iii*) **Rest Mass**: The rest mass of photon is zero since a photon can never be at rest. Thus,

$$m_0 = 0$$

(*iv*) **Relativistic mass**: As photon travels with the velocity of light, it has relativistic mass, given by $m = \dfrac{E}{c^2} = \dfrac{h\nu}{c^2}$. ...(24.17)

(*v*) **Linear Momentum**: The linear momentum associated with a photon may be expressed as $p = \dfrac{E}{c} = \dfrac{h\nu}{c} = \dfrac{h}{\lambda}$ ...(24.18)

As the **wave vector** $k = \dfrac{2\pi}{\lambda}$, $p = \dfrac{h}{2\pi}k = \hbar k$. ...(24.19)

(*vi*) **Angular Momentum**: Angular momentum is also known as ***spin*** which is the intrinsic property of all microparticles. Photon has a spin of one unit. Thus, $s = 1\hbar$.

(*vii*) **Electrical Charge**: Photons are electrically neutral and cannot be influenced by electric or magnetic fields. They cannot ionize matter.

Example 24.1. *The wavelength of yellow light is 5890 Å. What is the energy of the photons in the beam? Express in electron volts. Thus,*

Solution:

$$E = h\nu = \frac{hc}{\lambda} = \frac{(6.63 \times 10^{-34}\ \text{J.s})(3 \times 10^8\ \text{m/s})}{5.89 \times 10^{-7}\ \text{m}}$$

$$= \frac{19.89}{5.89} \times 10^{-19}\ \text{J} = (3.38 \times 10^{-19})(6.24 \times 10^{18}\ \text{eV})$$

$$= \mathbf{2.11\ eV}$$

Example 24.2. *The light sensitive compound on most photographic films is silver bromide, AgBr, A film is exposed when the light energy absorbed dissociates this molecule into its atoms. The energy of dissociation of AgBr is 23.9 k.cal./mol. Find the energy in electron volts,* *the wavelength and the frequency of the photon that is just able to dissociate a molecule of silver bromide.*

Solution: The number of molecules contained in one mole of any substance is equal to the Avogadro's number. N_A. The energy of dissociation of one Ag Br molecule is therefore given by $E = \frac{E_D}{N_A}$, This is equal to the energy of the photon.

(*i*) $E = \frac{E_D}{N_A} = \frac{(23.9\,\text{k.cal/mol.})(4185\,\text{J/kcal})}{6.023 \times 10^{23}\,/\text{mol}}$

$= \frac{1.0}{6.023} \times 10^{-18}\,\text{J} = (0.166 \times 10^{-18})(6.24 \times 10^{18}\,\text{eV}) = \mathbf{1.04\ eV}$

(*ii*) $\nu = \frac{E}{h} = \frac{0.166 \times 10^{-18}\,\text{J}}{6.63 \times 10^{-34}\,\text{J.s}} = 0.025 \times 10^{16}\,\frac{1}{s} = \mathbf{2.5 \times 10^{14}\ Hz}$

(*iii*) $\lambda = \frac{hc}{E} = \frac{(6.63 \times 10^{-34}\,\text{J.s})(3 \times 10^{8}\,\text{m/s})}{0.166 \times 10^{-18}\,\text{J}} = \frac{19.89}{0.166} \times 10^{-8}\,\text{m} = (119.82 \times 10^{-8})(10^{10}\,\text{Å})$

$= \mathbf{11982\ Å}$

Example 24.3. *What is the photon energy in joules corresponding to a 50 Hz wave emitted by the power line? How does it compare with the energy of yellow light?*

Solution. Photon energy, $E_{PL} = h\nu_{PL} = (6.63 \times 10^{-34}\,\text{J.s})(50\,\text{Hz}) = 3.32 \times 10^{-32}\,\text{J}$

$= (3.32 \times 10^{-32})\,(6.24 \times 10^{18}\,\text{eV}) = 2.07 \times 10^{-13}\,\text{eV}$

Yellow light photon energy, $E_y = \frac{hc}{\lambda} = \frac{12400}{5890}\,\text{eV} = 2.11\,\text{eV}$

$\therefore \quad E_y : E_{PL} = \frac{2.11\,\text{eV}}{2.07 \times 10^{-13}\,\text{eV}} = \mathbf{1.02 \times 10^{13}}$

Thus the energy of light photon is about 10^{13} times greater than the energy of the photon associated with a 50 Hz wave.

Example 24.4. *A blue lamp emits light of mean wavelength of 4500 Å. The lamp is rated 150 W and 8% of the energy appears as light. How many photons are emitted per second by the lamp?*

Solution. Total energy supplied by the lamp

$E_T = P \times 1\,\text{sec} = \text{P(J)} = 6.24 \times 10^{18}\,\text{PeV}$

Energy converted into light $E_L = \eta E_T = 6.24 \times 10^{18}\,\text{P}\eta\text{eV}$

Number of photons emitter per second $N = \frac{E_L}{E}$

Where E is the energy of a photon, $E = \frac{12400}{\lambda(\text{Å})}$

$\therefore \quad N = \frac{(6.24 \times 10^{18})\left(P\eta\lambda(\text{Å})\right)}{12{,}400} = \frac{(6.24 \times 10^{18}\,\text{eV})(150)(0.08)(4500)}{12{,}400\,\text{eV}} = 27.17 \times 10^{18}$

$\therefore \quad \mathbf{N = 2.7 \times 10^{19}}$ **photon/sec.**

24.6 PHOTOELECTRIC EFFECT

In 1887 Hertz discovered that electrons are ejected when ultraviolet radiation strikes a metal surface. Photoelectric effect is the phenomenon in which electrons are knocked off a metal surface when light is incident on it (see Fig. 24.6). The metal is said to be a *photosensitive material* and the liberated electrons are called *photoelectrons*.

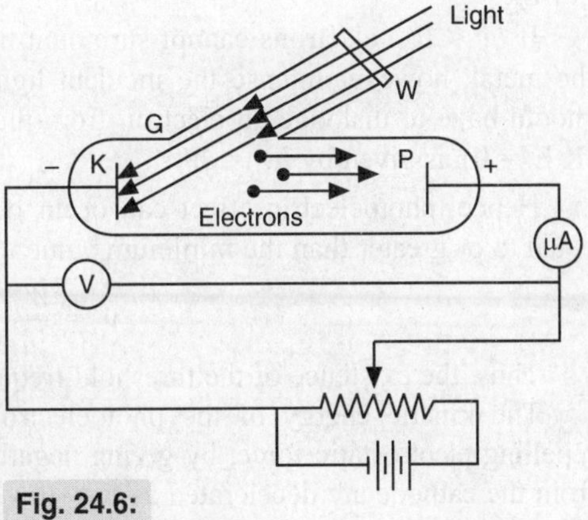

Fig. 24.6: Photoelectric effect

Electronic emission increases with the intensity of the radiation falling on the metal surface, since more energy is available to release electrons. But characteristic frequency dependence is also observed. For each substance there is a minimum frequency ν_o of light and for light of frequency less than ν_o, photoelectrons are not produced, however intense the incident light may be. The minimum frequency ν_o is called *threshold frequency*. The photoelectrons that are emitted in the process are found to have a range of energies. The maximum kinetic energy of the photoelectrons varies with the frequency of the incident light (see Fig. 24.7) and is independent of the intensity of the light.

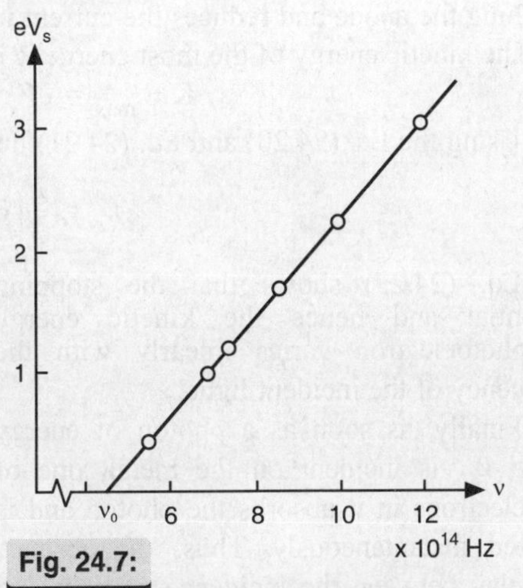

Fig. 24.7: Kinetic energy of the photoelectrons versus the frequency of the incident light

Attempts to explain the photoelectric effect on the basis of wave theory of light were unsuccessful. In 1905 Einstein successfully explained the photoelectric effect by treating the incident light as a stream of photons. In a beam of monochromic light of frequency, ν, all photons have the same energy $h\nu$. When the photons encounter electrons in a metal, they give up their energy to the electrons. However, the electrons are held within the metal by a **potential barrier** at the surface. In order to escape from the metal, an electron should have enough energy to overcome the potential barrier at the surface. It requires an input energy that is equal to the *work function*, W_o. This work function is specific for the metal. When an electron absorbs a photon, the energy gained by it is equal to $h\nu$. If $h\nu \geq W_o$, the electron will escape from the metal. Thus, out of the total energy $h\nu$, a small portion, W_o, is spent on surmounting the potential barrier and the balance energy $(h\nu - W_o)$ is given to the electron as kinetic energy. Applying the law of conservation of energy to this phenomenon, we find that

$$h\nu = W_o + \frac{1}{2}m\upsilon_{max}^2 \qquad ...(24.20)$$

Eq. (24.20) is known as ***Einstein's photoelectric equation***. It explains all the features of the photoelectric effect.

From this equation it is seen that a photoelectron is emitted from the metal only if the energy of the incident photon $h\nu \geq W_o$.

If $h\nu < W_o$, electrons cannot surmount the barrier at the metal surface and emerge from the metal, however intense the incident light may be. The minimum energy that a photon should have to dislodge an electron from the metal, without imparting it any kinetic energy (K.E. = 0), is given by $h\nu_o = W_o$.

Hence, photoelectric effect can occur only when the frequency of the incident light is equal to or greater than the minimum value ν_o, which is given by

$$\nu_o = \frac{W_o}{h} \quad ...(24.21)$$

Thus, the existence of the threshold frequency is explained.

The kinetic energy of the photoelectrons can be measured by subjecting them to a repelling electrostatic force, by giving negative potential to the anode. The electrons ejected from the cathode are decelerated as they travel toward the anode. When the repelling force is made large enough, even the most energetic electrons are stopped from reaching the anode and hence the current will drop to zero. The potential difference that stops the electrons reaching the anode and reduces the current in the circuit to zero, is called ***stopping potential***, V_s. The kinetic energy of the most energetic electrons is therefore given by

$$\text{K.E.}_{max} = eV_s \quad ...(24.22)$$

Using the Eq. (24.20) and Eq. (24.21) into Eq. (24.22), we get

$$V_s = \frac{h}{e}[\nu - \nu_o] \quad ...(24.23)$$

Eq. (24.23) shows that the stopping potential and hence the kinetic energy of photoelectron varies linearly with the frequency of the incident light.

Finally, as soon as a photon of energy $h\nu > W_o$ is incident on the metal, one of the electrons in it absorbs the photon and is ejected instantaneously. Thus, the absence of delay between the incidence of light and emission of electrons is also explained.

In 1916 Millikan verified Einstein's theory. If a plot is drawn between the potential V_s and the frequency of the incident light, the resulting graph will be a straight line (see Fig. 24.8). The slope of this graph is equal to (h/e) and is the same for all materials. Millikan showed that the value of h determined from experiments on the photoelectric effect agrees well with the value obtained from other experiments.

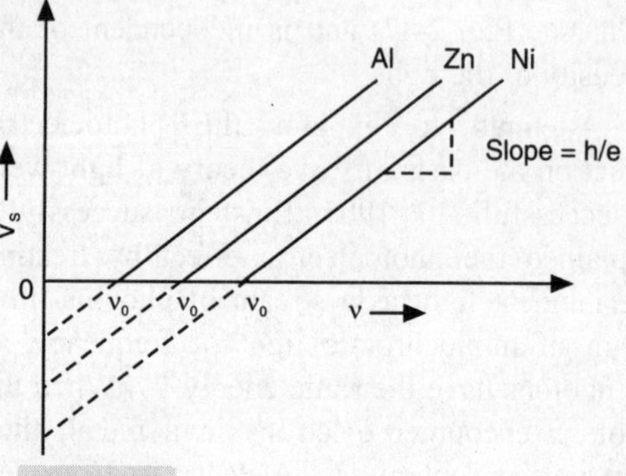

Fig. 24.8: Potential V_s versus the frequency of the incident light

Thus, the photoelectric effect confirmed the existence of energy in the form of discrete quanta.

Example 24.5. *Light of a wavelength 2000 Å falls on an aluminium surface with work function 4.2 eV. Calculate (i) threshold wavelength and (ii) stopping potential.*

Solution:

(*i*) Threshold wavelength, $\lambda_0 = \frac{hc}{W_0} = \frac{12400}{W_0(eV)}\text{Å} = \frac{12400}{4.2}\text{Å} = \mathbf{2952\,Å}$

(*ii*) If V_s is the stopping potential, then

$$eV_s = \frac{hc}{\lambda} - W_0 = \left[\frac{12400}{\lambda(Å)} - W_0\right]eV = \left[\frac{12400}{2000} - 4.2\right]eV = \mathbf{2\,eV}$$

$$\therefore \qquad V_s = \frac{2eV}{e} = \mathbf{2\,volts}$$

Example 24.6. *Work function of sodium is 2.3 eV. Obtain the maximum wavelength which will cause emission of photoelectrons from the material. What will be the maximum kinetic energy of the photoelectrons emitted by the surface exposed to the radiation of 2000 Å?*

Solution: Maximum wavelength, $\lambda = \frac{hc}{W_0} = \frac{12400}{(W_0)eV}\text{Å} = \frac{12400}{2.3eV}\text{Å} = \mathbf{5391\,Å}$

Max. kinetic energy, $K.E._{\max} = \frac{hc}{\lambda} - W_0 = \left[\frac{12400}{\lambda} - (W_0)\right]eV$

$$= \left[\frac{12400}{2000} - (2.3)\right]eV = \mathbf{3.9\,eV}$$

Example 24.7. *The work function for cadmium is 4.08 eV. What must be the wavelength of radiation incident on cadmium so that the maximum velocity of photoelectrons will be 7.2×10^5 m/s.*

Solution:

$$E = W_0 + K.E._{\max} = W_0 + \frac{1}{2}m\upsilon^2_{\max} = 4.08eV + \frac{1}{2}\left(9.11\times10^{-31}\,kg\right)\left(7.2\times10^5\,m/s\right)^2$$

$$= 4.08eV + 2.36\times10^{-19}\,kg.m^2/s^2$$

$$= 4.08eV + \left(2.36\times10^{-19}\right)\left(6.24\times10^{18}\,eV\right)$$

$$= 4.08eV + 1.47eV = 5.55eV$$

$$E = \frac{hc}{\lambda} = \frac{12400}{\lambda(Å)} = W_0 + \frac{1}{2}m\upsilon^2_{\max}$$

$$\therefore \qquad \lambda = 12400\left[\frac{1}{W_0 + \frac{1}{2}m\upsilon^2_{\max}}\right]\text{Å}$$

$$= \left[\frac{12400}{W_0 + \frac{1}{2}m\upsilon^2_{\max}}\right]\text{Å} = \frac{12400}{5.55}\text{Å} = \mathbf{2234\,Å}$$

24.7 X-RAYS

In 1895 Wilhelm Rontgen discovered X-rays while conducting experiments with a discharge tube. He observed that a screen at a distance coated with the chemical barium platinocyanide fluoresced brightly. Rontgen concluded that the invisible radiations are produced due to the bombardment of cathode rays on the walls of the discharge tube. He named them X-rays. As X-rays are electromagnetic waves with very short wavelengths, these rays are highly penetrating and can pass through many solids. X-rays wavelengths range from about 0.01 Å to about 10 Å and occur beyond the UV region in the electromagnetic spectrum.

24.8 GENERATION OF X-RAYS

Fig. 24.9 shows the schematic of the modern X-ray tube designed by Coolidge for producing X-rays. It is basically an evacuated glass bulb consisting of two electrodes, namely a cathode and an anode. The cathode is a tungsten filament that emits electrons when it is heated. The emitted electrons are concentrated into a narrow beam using a metal cup C. The anode is a tungsten or molybdenum block, embedded in a copper bar. A very high potential difference of the order of 50 kV is applied across the electrodes. The anode is cooled by water circulating through a jacket surrounding the anode. Additionally, large cooling fins conduct the heat away to the air. The face of the target is held at an angle to the approaching electron beam.

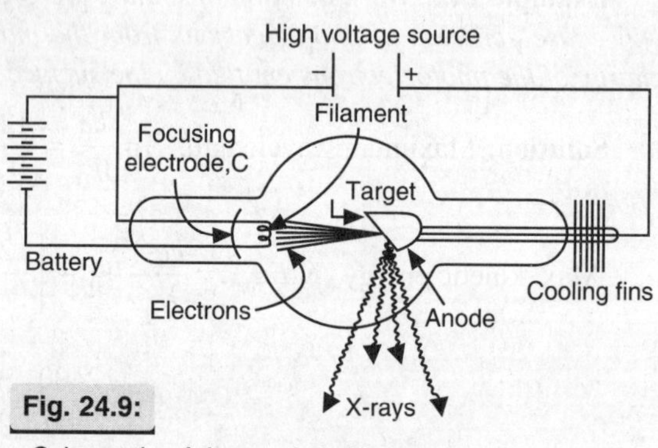

Fig. 24.9: Schematic of the modern X-ray tube

The emitted electrons are accelerated by the anode and attain high energies of about 10^5eV. X-rays are generated when an obstacle stops fast moving electrons, and that metals of high atomic weight are most effective in producing X-rays. The wavelength of the radiated X-rays depends on the speed of the electron beam. The greater the electron speed, the shorter will be the wavelength of X-rays. Only about 0.2% of the electron beam energy is transformed into X-rays and the rest of the energy gets converted into heat. The intensity of the electron beam can be adjusted by passing a suitable current through the filament. The hardness of the X-rays depends on the energy of the electron beam incident on the target. The electron beam energy can be varied by changing the potential difference between the cathode and the anode. The larger the potential difference, the *harder* are X-rays.

24.9 X-RAY SPECTRUM

When the X-rays produced by a metallic target are examined with the help of a spectrometer and their intensity is plotted against wavelength, a curve such as the one shown in Fig. 24.10 is obtained. This curve is called an X-ray spectrum. An analysis of the spectrum of radiation emitted by an X-ray tube shows the presence of two distinct contributions.

One portion of the spectrum consists of X-rays of all possible wavelengths with a lower limit. This is known as the ***continuous spectrum***. The continuous distribution of wavelengths emitted in the process is analogous to the continuous distribution of wavelengths in white light. Hence, the continuous X-ray radiation is sometimes called *white radiation*. The continuous spectrum is also called ***bremsstrahlung*** meaning breaking radiation. It is found that there is a minimum wavelength below which X-rays are not emitted. This minimum wavelength is called the *cut-off wavelength*. The other portion of the spectrum consists of several sharp lines of discrete wavelengths superimposed over the continuous spectrum. This is known as the ***characteristic spectrum***. The wavelengths of the line spectrum are characteristic of the target element.

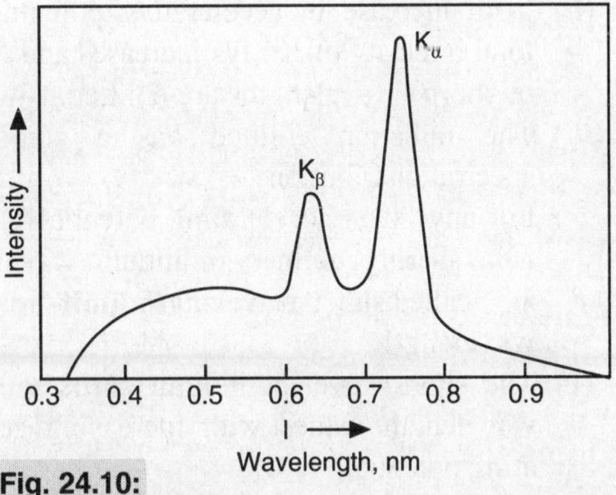

Fig. 24.10: X-Ray Spectrum-Wavelength versus intensity.

24.10 ORIGIN OF CONTINUOUS X-RAY SPECTRUM

Continuous X-rays are produced by the phenomenon of ***bremsstrahlung***, which is a German word meaning 'breaking' or 'slowing down' radiation. Electrons emitted from the cathode in the X-ray tube are accelerated toward the target. They strike the target and penetrate deep into the interior of atoms and are attracted by the nuclei due to strong Coulomb attraction. The electrons deviate from their original path (see Fig. 24.11). The deviation of an electron from its original path is equivalent to its collision with the nucleus. The electron loses some of its kinetic energy due to such collision. The energy lost appears as an X-ray photon.

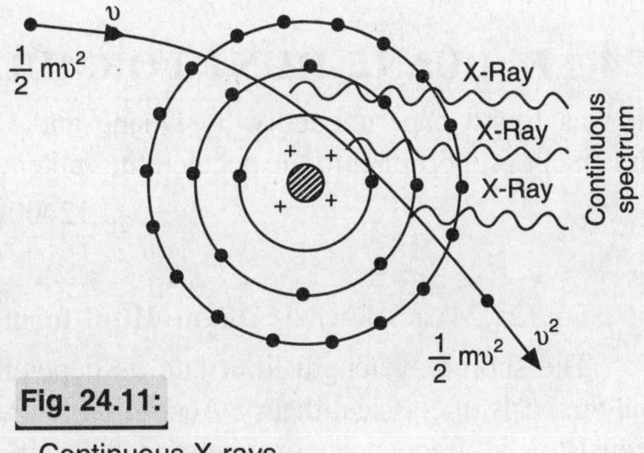

Fig. 24.11: Continuous X-rays

If E_i is the initial kinetic energy and E_f the final kinetic energy of the electron, the energy of X-ray photon is $(E_i - E_f)$.

If v is the frequency of the X-ray photon, then the energy of the photon is

$$h\nu = E_i - E_f$$

The electron may make several collisions before coming to rest. Therefore, the frequency of the emitted photon is not unique. During each collision of the electron, a photon may be given out. The energies of these photons will range from very small value up to the maximum energy E_f. The maximum energy corresponds to an electron that loses all its energy in a single encounter.

The spectral distribution of continuous X-rays and its dependence on the applied potential difference is shown in Fig. 24.12. The important features of the curves are as follows:

(*i*) The wavelengths at which the maximum intensity occurs shifts to lower values for higher accelerating potentials.

(*ii*) With increase in accelerating potential, the total intensity of X-rays increases and X-rays of shorter wavelengths are emitted.

(*iii*) The radiation emitted has a continuous spectral distribution.

(*iv*) For any given accelerating potential V, there is a sharply defined minimum wavelength λ_{min} called **short-wavelength limit** or **cut-off wavelength**.

(*v*) The short wavelength limit shifts to lower wavelength values with increasing accelerating potential.

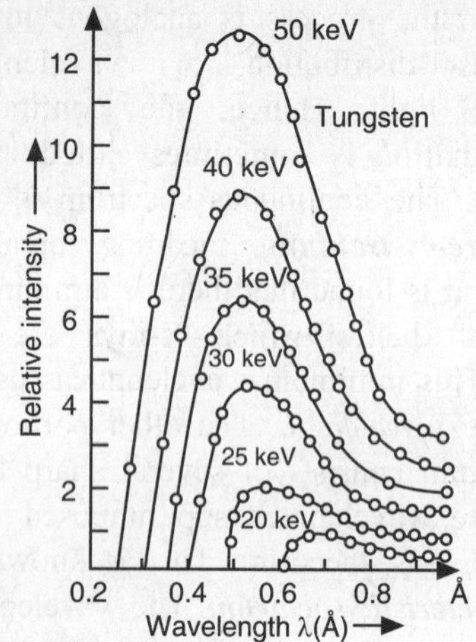

Fig. 24.12: Spectral distribution of continuous X-rays

The production of X-rays can be considered to be an *inverse photoelectric emission*. In the case of photoelectric effect, photons incident on a surface liberate electrons. In the present case, electrons cause emission of X-ray photons. Hence X-ray phenomenon is considered as **inverse photoelectric effect**.

24.11 DUANE-HUNT FORMULA

It was found experimentally by Duane and Hunt that the short wavelength limit λ_{min} is inversely proportional to the accelerating potential V and is given by

$$\lambda_{min} = \frac{12400}{V} \text{ Å} \qquad ...(24.24)$$

Eq. (24.24) is known as **Duane-Hunt formula.**

The short-wavelength limit and its dependence on applied voltage cannot be explained on the basis of classical theory. According to classical theory, the continuous spectrum should consist of all frequencies from zero to a certain maximum. The short wavelength limit can be explained only on the basis of quantum theory. In an X-ray tube, the electrons emitted at the cathode are all accelerated through the same potential difference V and they reach the target with nearly the same energy eV. When the electrons are stopped, they lose the energy eV suddenly. This energy eV goes into X-rays. According to Planck's hypothesis, the emission of X-rays occurs in the form of photons, each photon carrying energy $h\nu$. Therefore, the maximum energy that an X-ray photon can have is

$$h\nu_{max} = eV$$

$$\therefore \quad \nu_{max} = \frac{eV}{h}$$

But

$$\nu_{max} = \frac{c}{\lambda_{min}}.$$

$$\therefore \quad \lambda_{min} = \frac{hc}{eV} \qquad ...(24.25)$$

Using the values of h, c and e into Eq. (24.25), we get

$$\lambda_{min} = \frac{(6.626\times10^{-34}\text{J.s})(3\times10^{8}\text{m/s})}{(1.602\times10^{-19}\text{C})\text{V}} = \frac{1.24}{\text{V}}\times10^{-6}\text{m}$$

or
$$\lambda_{min} = \frac{12,400}{\text{V}}\text{ Å}.$$

This is the *Duane-Hunt formula.* Thus, the quantum theory explains the short-wavelength limit λ_{min} and its dependence on the accelerating voltage.

Example 24.8:

(*i*) *At what potential difference must an X-ray tube operate to produce X-rays with minimum wavelength of 0.01 Å?*

(*ii*) *What is the maximum frequency of the X-rays produced in a tube operating at 50 kV?*

Solution:

(*i*) $\lambda_{min} = \frac{hc}{eV} = \frac{12400}{V}\text{Å} \quad \therefore \quad V = \frac{12400}{\lambda_{mim}(\text{Å})} = \frac{12400}{0.01}\text{V} = \mathbf{1.24\,MV}$

(*ii*) $h\nu_{max} = \text{eV} \quad \therefore \quad \nu_{max} = \frac{\text{eV}}{h} = \frac{1.602\times10^{-19}\text{ C}\times50\times10^{3}\text{ V}}{6.63\times10^{-34}\text{ J.s}} = 1.2\times10^{19}\frac{\text{CV}}{\text{J.s}}$

$$= \mathbf{1.2\times10^{19}\ Hz.}$$

Example 24.9: *An X-ray tube operated at 40 kV emits a continuous X-ray spectrum with a short wavelength limit* λ_{min} = 0.31 Å. *Calculate Planck's constant.*

Solution:

$$\lambda_{min} = \frac{hc}{\text{eV}} = \frac{12400}{\text{V}}\text{Å}$$

$$\therefore \quad h = \frac{\lambda_{min}\,\text{eV}}{c} = \frac{(0.31\times10^{-10}\text{ m})(1.602\times10^{-19}\text{ C})(40\times10^{3}\text{ V})}{(3\times10^{8}\text{ m/s})} = \mathbf{6.61\times10^{-34}\ J.s}$$

Units: $$1\frac{\textbf{CV m}}{\textbf{m / s}} = \mathbf{1CVs = 1\,J.s}$$

24.12 COMPTON SCATTERING

Photoelectric effect provided the evidence for the existence of photons and their energy. According to the theory of relativity mass and energy are identical. If photons have energy, then they must possess mass and exhibit the property of momentum. The fact that photons have momentum was established by Compton scattering studies.

Arthur H. Compton carried out investigations on the scattering of X-rays by materials of low atomic number. Compton had observed that when a beam of X-rays was scattered by a material, the scattered radiation consisted of two components of X-rays. One of the components had the same wavelength as that of the incident beam and the other having a longer wavelength than that of the incident X-rays. **Compton effect** is the name given to this phenomenon.

Scattering of X-rays by electrons of atoms in a material without any change in their wavelength is known as **classical** or **coherent scattering**. Scattering of X-rays by electrons of atoms in a material with a change in wavelength is called **incoherent scattering**. It is now known as **Compton scattering**. In a Compton scattering event, an electron is also ejected and the electron is known as **recoil electron**.

24.12.1 Importance of Compton Effect

(*i*) Historically, Compton Effect provided the direct confirmation of the particle nature of electromagnetic radiation. It had conclusively showed that photons carry energy and momentum like any material particle. Further, it proved that the momentum as well as the energy of electromagnetic radiation is quantized.

(*ii*) Compton effect showed that the photon description applies not only to visible light but also to X-rays.

24.12.2 Experimental Arrangement

A schematic diagram of the apparatus for studying the Compton scattering is shown in Fig. 24.13(*a*). A monochromatic beam of K_α X-rays of wavelength λ_i from the molybdenum target were collimated by passing it through the slits L_1 and L_2. The collimated beam was then made incident on a graphite block, G. The graphite block scattered X-rays in different directions. The scattered X-rays were detected by a Bragg's X-ray spectrometer. Bragg's law was used in determining the wavelength of scattered X-rays. X-rays diffracted by the crystal in the spectrometer were passed through an ionisation chamber to measure their intensity. The scattering angle θ was varied and corresponding X-ray wavelengths and their intensities were determined.

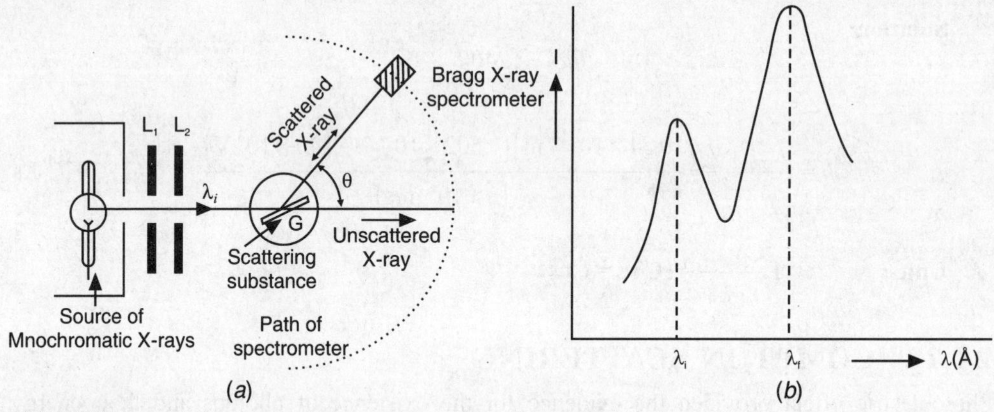

Fig. 24.13:
(*a*) Compton's apparatus for the study of scattered X-rays
(*b*) Unmodified and modified components

24.12.3 Experimental Results

(*i*) Graphs were plotted for the variation of intensity versus wavelength of the scattered X-rays. The curves were found to have only two peaks, as illustrated in Fig. 24.13(*b*). Thus, the measurements showed that though the incident X-rays are of only a single wavelength, the scattered x-rays contained two wavelengths -

the original wavelength λ_i and another wavelength λ_f which is longer than λ_i. The primary (original) wavelength is called **unmodified component** and the second wavelength the **modified component**. The peak at λ_i is caused by X-rays scattered from electrons that are tightly bound to the target atoms, and the other peak of longer wavelength, λ_f is caused by x-rays scattered from free electrons in the target.

(*ii*) The difference in the wavelength between the modified and unmodified components is called the **Compton shift**. It is denoted by

$$\Delta\lambda = \lambda_f - \lambda_i.$$

The Compton shift was found to depend solely on the scattering angle, θ and is completely independent of the primary wavelength, and the target material. The shift is also independent of the intensity of the incident radiation. It is found to vary with θ according to the following relation.

$$\Delta\lambda = \frac{h}{m_o c}(1 - \cos\theta) \qquad ...(24.26)$$

(*a*) The Compton shift $\Delta\lambda = 0$ when $\theta = 0°$ and increases to a maximum value of $2h/m_0c$ along $\theta = 180°$.

(*b*) Only the intensity of the scattered radiation depends on the target material.

24.12.4 Wave Theory Fails to Explain Compton Effect

According to the wave theory, X-rays force electrons in the atoms of the target material to execute forced oscillations. The oscillating electrons emit radiation with the same frequency as that of the incident radiation. This radiation is called *Rayleigh-scattered radiation*. Further, the electrons radiate waves uniformly in all directions. Thus, as per wave theory the

(*i*) scattered radiation should have the same wavelength as that of incident radiation.

(*ii*) wavelength of the scattered radiation should not show dependence on the scattering angle, θ.

The above conclusions are contrary to the experimental observations. It means that the wave theory fails to explain the Compton Effect.

24.12.5 Compton's Explanation

In 1923, Compton explained the effect on the basis of quantum theory of radiation. By Einstein's photon theory, the X-ray radiation consists of photons each having energy $h\nu$. The energy of a photon is directly proportional to the frequency of the light. So a change in wavelength implies a change in photon energy. An increase in wavelength of the scattered photons indicates that they had less energy than the incident ones. When a photon collides with an electron initially at rest, the photon transfers some energy and momentum to the electron. Hence, the energy of the scattered photon should decrease and its wavelength increase. Due to transfer momentum the electron gains kinetic energy and recoils with velocity υ.

Assumptions

In the above explanation Compton assumed that

(*i*) X-ray radiation is quantized and may be regarded as a stream of photons. A single photon carries an energy $h\nu$. The photons behave like true particles and have momentum like any mechanical particle.

(*ii*) The electron in the atom of the target material is loosely bound and may be considered as at rest initially.

(*iii*) The scattering event between a photon and an electron may be treated as a simple *elastic collision* between two particles. As a result of collision, the incident photon loses a part of its energy and is scattered with reduced energy at an angle θ with the direction of incident radiation. The electron gains kinetic energy equal to that of the energy lost by the photon, and recoils in a direction making an angle ϕ with the direction of incident radiation.

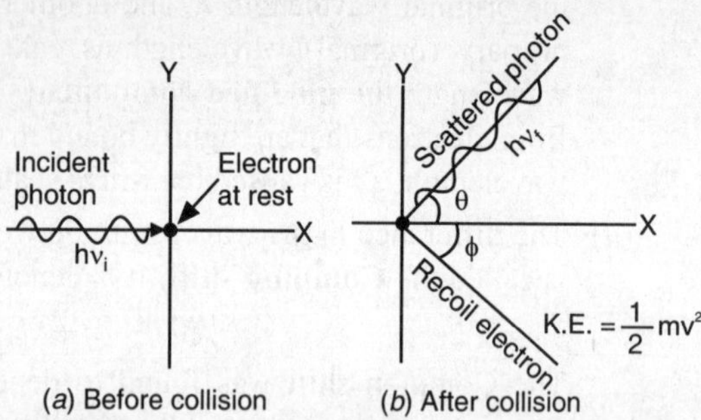

Fig. 24.14: Geometry of Compton scattering. (*a*) Before collision (*b*) After collision-The scattered photon has longer wavelength than the incident photon.

(*iv*) The geometry of Compton scattering is shown in Fig. 24.14.

24.12.6 Derivation of Expression for Compton Shift

Let an X-ray photon of energy $h\nu_i$ and momentum $h\nu_i/c$ collide with an electron at rest. As the collision is an elastic collision, the energy and momentum of the system are conserved.

The momentum and the kinetic energy of the electron before the collision is zero. However, according to relativity theory, the electron rest mass energy is m_oc^2. After the collision event, the electron acquires a momentum '*p*' and recoils with kinetic energy. Since the electron may recoil at speeds comparable to the speed of light, we must use the relativistic expression for K.E., which is given by

$$K.E. = h\nu_i - h\nu_f = mc^2 - m_oc^2 \qquad ...(24.27)$$

The *relativistic momentum* of the recoiling electron is $p = mv$. The relativistic momentum is related to the total energy through the following equation.

$$E = c\sqrt{p^2 + m_o^2c^2} \qquad ...(24.28)$$

Let the energy of the scattered photon be $h\nu_f$. We now apply the laws of conservation of energy and momentum to the *elastic collision* of photon and electron.

(a) Conservation of Energy

According to the principle of conservation of energy,

$$\left.\begin{matrix}\text{Total Energy of the}\\ \text{system before collision}\end{matrix}\right\} = \left\{\begin{matrix}\text{Total Energy of the}\\ \text{system after collision}\end{matrix}\right.$$

$$\left.\begin{matrix}\text{Initial Energy of the Photon, } h\nu_i\\ +\\ \text{Initial Energy of the electron, } m_oc^2\end{matrix}\right\} = \left\{\begin{matrix}\text{Final Energy of the photon, } h\nu_f\\ +\\ \text{Final Energy of the electron, } c\sqrt{p^2+m_o^2c^2}\end{matrix}\right.$$

$$\therefore \qquad h\nu_i + m_o c^2 = h\nu_f + c\sqrt{p^2 + m_o^2 c^2}$$

The above equation can be rewritten as

$$h(\nu_i - \nu_f) + m_o c^2 = c\sqrt{p^2 + m_o^2 c^2}$$

$$\frac{h}{c}(\nu_i - \nu_f) + m_o c = \sqrt{p^2 + m_o^2 c^2}$$

Squaring the above equation on both sides and on simplification, we get

$$P^2 = \frac{h^2}{c^2}(\nu_i - \nu_f)^2 + 2m_o h(\nu_i - \nu_f) \qquad ...(24.29)$$

(b) Conservation of Linear Momentum

Linear momentum is a vector quantity and the x- and y-components of linear momentum are conserved individually.

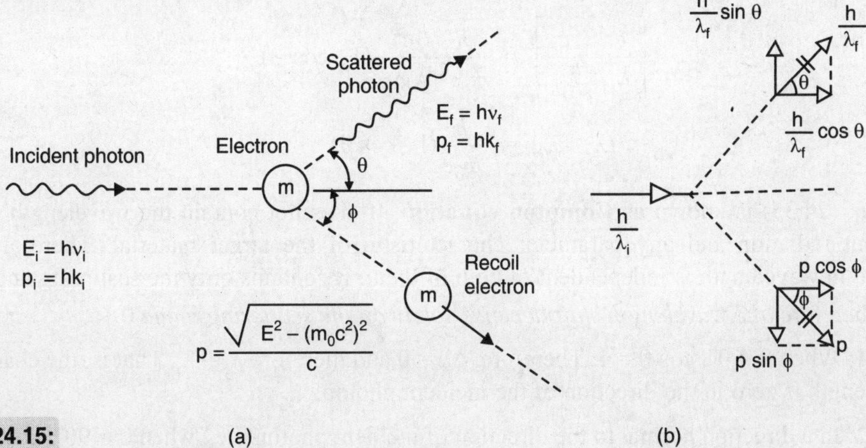

Fig. 24.15: Components of momentum before and after collision.

For the momentum in the x – direction before and after collision (Fig. 24.15), we have

$$\frac{h\nu_i}{c} + 0 = \frac{h\nu_f}{c}\cos\theta + p\cos\phi$$

$$\text{or} \qquad \frac{h\nu_i}{c} - \frac{h\nu_f}{c}\cos\theta = p\cos\phi \qquad ...(24.30)$$

For the momentum in the y – direction (Fig. 24.15),

$$0 = \frac{h\nu_f}{c}\sin\theta - p\sin\phi$$

$$\therefore \qquad \frac{h\nu_f}{c}\sin\theta = p\sin\phi \qquad ...(24.31)$$

Squaring the Eq. (24.30) and Eq. (24.31), we respectively get

$$p^2\cos^2\phi = \frac{h^2}{c^2}(\nu_i - \nu_f\cos\theta)^2 \qquad ...(24.32)$$

$$\text{and} \qquad p^2\sin^2\phi = \frac{h^2}{c^2}\nu_f^2\sin^2\theta \qquad ...(24.33)$$

Adding Eq. (24.32) to Eq. (24.33), we get

$$p^2 = \frac{h^2}{c^2}\left(\nu_i^2 + \nu_f^2 - 2\nu_i\nu_f \cos\theta\right) \qquad ...(24.34)$$

Equating Eq. (24.29) to Eq. (24.34), we obtain

$$\frac{h^2}{c^2}\left(\nu_i - \nu_f\right)^2 + 2m_o h\left(\nu_i - \nu_f\right) = \frac{h^2}{c^2}\left(\nu_i^2 + \nu_f^2 - 2\nu_i\nu_f \cos\theta\right)$$

or
$$\left(\nu_i - \nu_f\right)^2 + \frac{2m_o c^2}{h}\left(\nu_i - \nu_f\right) = \left(\nu_i^2 + \nu_f^2 - 2\nu_i\nu_f \cos\theta\right)$$

or
$$\frac{2m_o c^2}{h}\left(\nu_i - \nu_f\right) = 2\nu_i\nu_f(1 - \cos\theta)$$

or
$$(\nu_i - \nu_f) = \frac{h}{m_o c^2}\nu_i\nu_f(1 - \cos\theta)$$

or
$$\left[\frac{c}{\lambda_i} - \frac{c}{\lambda_f}\right] = \frac{h}{m_o c^2}\frac{c^2}{\lambda_i\lambda_f}(1 - \cos\theta)$$

or
$$(\lambda_f - \lambda_i) = \frac{h}{m_o c}(1 - \cos\theta) \qquad ...(24.35)$$

Eq. (24.35) is known as **Compton equation**. It does not contain the wavelength of the incident radiation and any parameter characteristic of the target material. Therefore, the change in wavelength is independent of both of them. It contains only the scattering angle θ.

Therefore, *the wavelength shift depends solely on the scattering angle* θ.

Case 1: When $\theta = 0°$, $\cos\theta = 1$. Therefore, $\Delta\lambda = 0$ and therefore $\lambda_f = \lambda_i$. That is, the change in wavelength is zero in the direction of the incident photon.

Case 2: In a direction normal to the direction of incident photon, i.e., when $\theta = 90°$, $\cos\theta = 0$.

Therefore, $\Delta\lambda = \lambda_f - \lambda_i = \dfrac{h}{m_o c} = 0.02426$ Å.

As the unit of the ratio $h/m_o c$ is angstroms, the ratio is designated as **Compton wavelength, λ_C.**

$$\therefore \qquad \lambda_C = \frac{h}{m_o c} = 0.02426 \text{ Å} \qquad ...(24.36)$$

Case 3: When $\theta = 180°$, $\cos\theta = -1$. Therefore, $\Delta\lambda = \dfrac{2h}{m_o c} = 0.04852$ Å.

The change in wavelength is maximum at $\theta = 180°$.

Thus, as θ varies from 0° to 180°, the wavelength of the scattered photon varies from λ_i to $\lambda_i + \dfrac{2h}{m_o c}$ and the Compton shift varies from 0 to a value $\dfrac{2h}{m_o c}$.

Compton effect is purely a quantum phenomenon. According to classical model, it should not occur. By letting *h* equal to zero, the quantum prediction should agree with the laws of classical physics. It is readily seen that as $h \to 0$, $\Delta\lambda \to 0$, which is in agreement with classical physics.

24.12.7 Experimental Verification

Compton measured the change in wavelength of X-rays scattered by graphite. The schematic diagram of his apparatus is shown in Fig. 24.13(*a*). The monochromatic X-rays of wavelength $\lambda_i = 0.071$ *nm* were made to be incident on a graphite block. The X-rays scattered by the graphite block were detected by a Bragg's X-ray spectrometer. Bragg's law was used in determining the wavelength of scattered X-rays. X-rays diffracted by the crystal in the spectrometer were passed through an ionisation chamber to measure their intensity. The scattering angle is θ is varied and corresponding X-ray wavelengths are determined. Thus, Compton measured how X-ray intensity depends on wavelength at various scattering angles. The wavelength of primary X-rays was measured by making them fall directly on the crystal of the spectrometer. The results obtained for four values of θ are shown in Fig. 24.16. The solid vertical line corresponds to the *primary* (*unmodified*) *wavelength*, λ_i, and the broken vertical lines correspond to the *modified wavelength* λ_f.

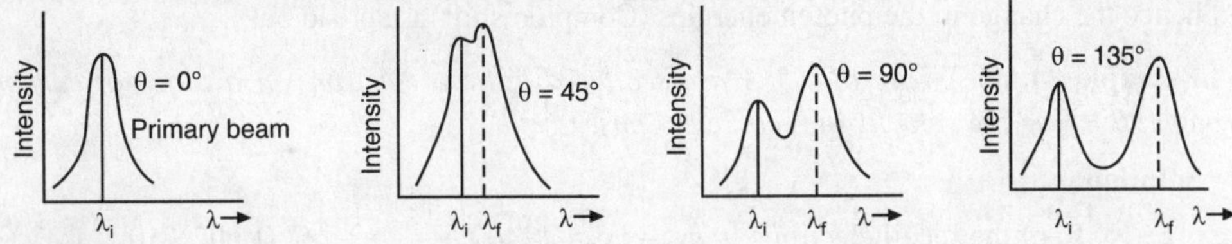

Fig. 24.16:

Wavelength versus intensity at various scattering angles

Compton found that the shift in wavelength depended solely on the scattering angle, θ and it is completely independent of the primary wavelength and the target material. Compton's experimental results were in excellent agreement with the predictions of equation (24.35).

24.12.8 Modified and Unmodified Components

In Compton scattering, the scattered radiation consists of two component wavelengths, namely, the modified and unmodified wavelengths.

The *modified wavelength* arises due to the collision of X-ray photons with either *free or loosely bound* electrons. For high-energy photons, most of the atomic electrons appear *free* and hence a large fraction of incident X-rays undergoes a wavelength shift. Therefore, in the case of low atomic number targets, the modified component will be more intense (see Fig. 24.17).

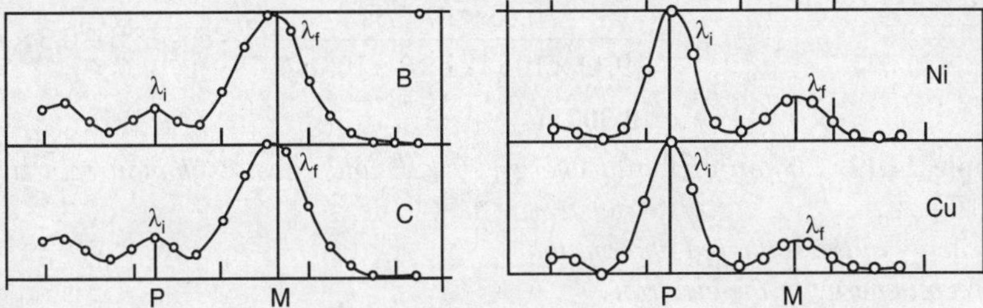

Fig. 24.17:

Modified M (λ_f) and unmodified P (λ_i) wavelengths in case of different scattering materials. Note that the modified component is more intense in lower atomic number materials whereas the unmodified component is stronger in higher atomic number materials.

The *unmodified component* arises when photons are scattered by *tightly bound* electrons. In this case, the whole atom is involved in the collision and hence the value of m_o in Eq. (24.35) is to be replaced by the mass of the atom. Since the mass of the atom is many times larger than that of an electron, the Compton shift is negligible and correspondingly there is no change in wavelength. Hence, the intensity of unmodified component is large.

In the case of scatterers of large atomic number, there are many electrons, which require high-energy X-ray photons for being scattered. Therefore, in the case of large atomic number targets, the unmodified component will be more intense (see the curves for Ni and Cu in Fig. 24.17).

For visible photons, all the atomic electrons appear bound and hence Compton shift is not observed.

Secondly, it is seen from Fig. 24.17 that the modified component is broader than the unmodified component. The broadening occurs because electrons in the target are not at rest and hence the change in the photon energies (Compton shift) is spread out.

Example 24.10. *X-rays of 0.5 Å are scattered by free electrons in a block of carbon through 90°. Find the velocity of recoil electrons.*

Solution:

$$\text{K.E. of the recoil electrons} = h\nu_i - h\nu_f = 12400\left(\frac{1}{\lambda_i}-\frac{1}{\lambda_f}\right)\text{eV}(1.602\times10^{-19}\,\text{J/eV})$$

$$\therefore \quad \text{K.E.} = 12400\left(\frac{1}{0.5}-\frac{1}{0.5243}\right)(1.602\times10^{-19})\text{J} = \mathbf{1.84\times10^{-16}J}$$

$$\therefore \quad \upsilon = \left[\frac{2(\text{K.E.})}{\text{m}}\right]^{1/2} = \left[\frac{2(1.84\times10^{-16}\,\text{J})}{9.11\times10^{-31}\,\text{Kg.}}\right]^{1/2}$$

$$= \left\{4.04\times10^{14}\,\frac{\text{kg.m}^2/\text{s}^2}{\text{kg}}\right\}^{1/2} = \mathbf{2\times10^7 m/s.}$$

Example 24.11. *X-ray photon of wavelength 0.3 Å is scattered through an angle 45° by a loosely bound electron. Find the wavelength of scattered photon.*

Solution: The wavelength of the scattered photon is given by $\lambda_f = \frac{h}{m_o c}(1-\cos\theta)+\lambda_i$

$$\therefore \quad \lambda_f = \frac{6.63\times10^{-34}\,\text{Js}}{9.11\times10^{-31}\,\text{kg}\times3\times10^8\,\text{m/s}}(1-\cos45°)+0.3\times10^{-10}\,\text{m}$$

$$= \mathbf{0.307\ Å.}$$

Example 24.12. *Photon of initial energy 90 keV undergoes Compton scattering at an angle 60°. Find:*

(*i*) *the energy of the scattered photon and*

(*ii*) *the recoil energy of the electron.*

Solution. (*i*)

$$E_i = \frac{hc}{\lambda_i} = 90\text{keV} = 90\times10^3\left(1.6\times10^{-19}\,\text{J}\right) = 1.44\times10^{-14}\,\text{J}$$

$$\therefore \quad \lambda_i = \frac{hc}{E_i} = \frac{6.63\times10^{-34}\,\text{Js}\times3\times10^8\,\text{m/s}}{1.44\times10^{-4}\,\text{J}}$$

$$= \frac{1.989\times10^{-25}\,\text{J}\cdot\text{m}}{1.44\times10^{-4}\,\text{J}} = 0.138\times10^{-10}\,\text{m}$$

$$\lambda_f = \frac{h}{m_o c}(1-\cos\theta)+\lambda_i$$

$$= 0.02426\times10^{-10}\,\text{m}(1-\cos 60°)+0.138\times10^{-10}\,\text{m}$$

$$= 0.15\times10^{-10}\,\text{m}$$

$$\therefore \quad E_f = \frac{hc}{\lambda_f} = \frac{6.63\times10^{-34}\,\text{Js}\times3\times10^{8}\,\text{m/s}}{0.15\times10^{-10}\,\text{m}}$$

$$= \frac{1.989\times10^{-25}\,\text{J}\cdot\text{m}}{0.15\times10^{-10}\,\text{m}} = 1.32\times10^{-14}\text{J} = \mathbf{82.73\ keV}.$$

(*ii*) Recoil energy of electron = $E_i - E_f = (90-82.73)\,\text{keV} = \mathbf{7.27\ keV}$.

Example 24.13. *X-rays with initial wavelength 0.5 × 10*$^{-10}$ *m undergo Compton scattering. For what scattering angle is the wavelength of the scattered X-rays greater than that of the incident X-rays by one percent?*

Solution: $\lambda_i = 0.5\times10^{-10}$ m. It is given that $\frac{\lambda_f-\lambda_i}{\lambda_i} = 1\% = 0.01$

$$\therefore \quad \Delta\lambda = \lambda_f-\lambda_i = 0.5\times10^{-10}\,\text{m}\times0.01 = 0.5\times10^{-12}\,\text{m}$$

$$\Delta\lambda = \frac{h}{m_o c}(1-\cos\theta) = 0.02426\times10^{-10}\,\text{m}\times(1-\cos\theta)$$

$$\therefore \quad 0.5\times10^{-12}\,\text{m} = 0.02426\times10^{-10}\,\text{m}\times(1-\cos\theta)$$

$$\text{or} \quad (1-\cos\theta) = 0.2061$$

$$\therefore \quad \theta = \mathbf{37.4°}.$$

Therefore, the required angle is 37.4°.

24.13 PAIR PRODUCTION

The photoelectric effect and Compton Effect provided the earliest experimental evidence in support of the quantization of electromagnetic radiation. However, there are other experiments also that can be understood only when we assume the existence of photons. Pair production is one of such processes. When a photon meets atoms, the photon loses all its energy and two particles are created: an electron and a positron. The photon energy is converted into the relativistic total energies E_+ and E_- of the positron and electron.

$$h\nu = E_+ + E_- = \left(m_e c^2 + K_+\right)+\left(m_e c^2 + K_-\right) \qquad ...(24.37)$$

As K_+ and K_- are always positive, the photon must have energy at least $2m_e c^2$ = 1.02 MeV. Such a high energy photons occur in the region of nuclear γ-rays. This process requires an atom nearby to supply the necessary recoil momentum.

24.14 WAVE-PARTICLE DUALITY

The important characteristic of waves is their ability to cause interference effects. In the Young's famous double-slit experiment, light passes through two closely spaced slits and creates a pattern of bright and dark fringes on a screen. The fringe pattern is an indication of interferences taking place between the light waves leaving from the slits. The phenomena of interference, diffraction and polarization give exclusive evidence for the wave behaviour of light. These phenomena require two waves to be present at the same position at the same time. It is impossible for two particles to occupy the same position at the same time to produce the observed effects.

On the other hand, blackbody radiation, photoelectric effect, Compton Effect etc gives exclusive evidence for the particle behaviour of light. Wave theory miserably fails to explain these effects. Thus, on one hand light resembles a continuous wave of frequency, ν and on the other hand it resembles a collection of particles having energy E and momentum p. The particle nature and wave nature are in fact contradictory.

(*i*) particle occupies a definite position in space and hence it must be very small.

(*ii*) a wave spreads out and occupies a relatively large region of space and cannot be attributed to a particular location in space.

As we need both descriptions to complete our understanding, we have to reconcile that light behaves as an advancing wave in some phenomena and it behaves as a flux of particles in some other phenomena. In short, light exhibits **wave-particle duality**.

Let us consider again the interference pattern obtained on a screen in the double-slit experiment. We may interpret the pattern as follows: at positions of bright fringes, many photons are hitting the screen and at the positions of dark fringes no photons are incident. Even if we reduce the intensity of light to the extent that only a few photons pass through the slits, we observe the usual interference pattern. Such an experiment was performed in 1909 by G.I.Taylor. We know from experiments that sending light through slits gives interference patterns and that the patterns are built by individual photons. To reconcile the wave and particle aspects, we have to regard the pattern as a *statistical distribution* that tells us how many photons, on an average go to different places. In other words, the pattern indicates the *probability* for an individual photon to arrive at a particular place on the screen. However, we cannot predict exactly where an individual photon will go.

The reason for the manifestation of wave-particle dualism can be understood if one considers the entire electromagnetic spectrum. At the lower frequency end of the spectrum (see Fig. 24.18) are radio waves whose wavelengths are of the order of a few hundred metres. The energy of these waves is therefore spreads over such a large volume of space that the energy available at any point is insignificantly small and therefore, the particle nature is not observable. On the higher frequency side of the spectrum are UV rays, *x*-rays and γ-rays. The wavelengths of these waves are so short ($<10^{-10}$m) that the wave energy is literally concentrated in a point of negligibly small dimension and the wave properties are less noticeable compared to the particle properties. Hence, the wave nature stands out at lower frequencies and at higher frequencies the particle nature dominates. The visible region represents the *transition region* where both aspects can be observed.

We know that the intensity of light is given by $I = |E|^2$

According to Eq. (24.15) $I = Nh\nu.$

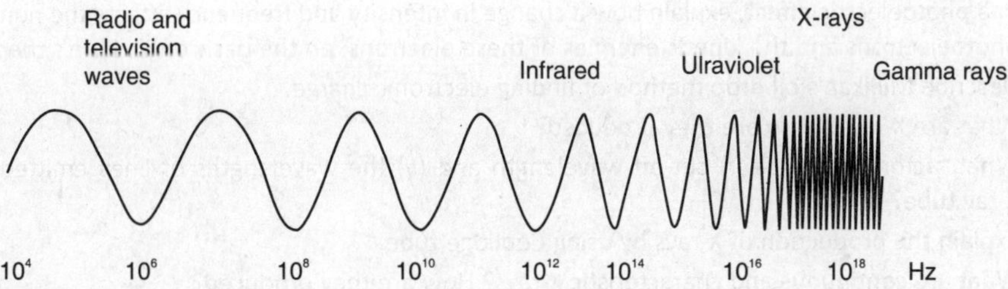

Fig. 24.18:

Electromagnetic Spectrum

A comparison of these equations indicates that the square of the amplitude of a light wave at a point in space is proportional to the number of photons arriving at that point. That is

$$N \propto |E|^2 \qquad ...(24.38)$$

Therefore, we conclude that

$$\text{Probability to observe photons} \propto |\text{Electric field amplitude}|^2 \qquad ...(24.39)$$

Eq. (24.39) provides the ultimate connection between the wave behaviour and the particle behaviour of light. The amplitude of a light wave determines the **probability** that a photon can be found at a particular point in space.

An acceptable explanation of wave-particle duality has been given by the **'quantized electromagnetic field theory'**, i.e. **'quantum electrodynamics' (QED)**. This theory extended the concept of energy levels of an atomic system and the Heisenberg uncertainty principle to electromagnetic fields. It combines particle and wave properties into a single theory which predicts as well as explains both types of behaviour.

QUESTIONS

1. What is a black body? What are the salient features of black body radiation?
2. State and explain the laws relating to the radiation and temperature of a radiating body.
3. State and explain Stefan-Boltzmann law.
4. State and explain Wien's displacement law.
5. What is a blackbody? Give Planck's hypothesis.
6. Discuss Planck's radiation law.
7. Discuss in brief Planck's quantum hypothesis.
8. State the basic postulate of Planck's theory. How did Einstein modify this theory in the light of experimental results?
9. (*a*) Explain briefly the distribution of energy in a blackbody spectrum. (*b*) What is Wien's displacement law? Explain the Wien's law of energy distribution.
10. Explain the energy distribution in the spectrum of a black body. Give an account of the attempts made through various laws to explain the spectrum.
11. Derive the Planck's law of radiation in terms of wavelength.
12. Photoelectric effect is a frequency dependent phenomenon and not intensity dependent one. Explain.
13. The photon model succeeds whereas the wave model fails in explaining the photoelectric effect. Explain.

14. In a photoelectric effect, explain how a change in intensity and frequency affects the number of photoelectrons and the kinetic energies of these electrons, on the basis of Einstein's theory.

15. Describe Millikan's oil drop method of finding electronic charge.

16. What are X-rays? How are they produced?

17. What factors influence (*i*) cut-off wavelength and (*ii*) the wavelengths of lines emitted by an X-ray tube?

18. Explain the production of X-rays by using Coolidge tube.

19. What are continuous and characteristic X-rays? How are they produced?

20. What is Duane-Hunt limit?

21. Explain the origin of the continuous spectrum.

22. What is meant by minimum wavelength limit? Derive Duane-Hunt formula for a minimum wavelength limit of continuous X-rays.

23. What are the conclusions obtained from Duan-Hunt formula?

24. What are the types of X-ray spectra? How do we get the continuous and sharp line spectrum in X-rays?

25. What is the characteristic X-ray spectrum? How do we explain it?

26. What is the origin of the cut-off wavelength? Why is it an important clue for photonature of X-rays?

27. What are characteristic X-rays? Give its origin. Explain Moseley's law of characteristic X-rays.

28. State Duane and Hunt limit and Moseley's law and its significance.

29. What are characteristic X-rays of an element? How can they be used to determine the atomic number of an element?

30. Discuss the origin and features of characteristic X-ray spectrum.

31. Discuss how Moseley explained the characteristic spectrum on the lines similar to line spectra of hydrogen-like atoms.

32. What is Compton Effect? Why is it observed in light elements?

33. Explain in brief Compton Effect on the basis of quantum hypothesis.

34. Explain how wave theory of light failed to explain Compton Effect? How the quantum theory explained this phenomenon?

35. Explain Compton Effect and derive expression for Compton shift.

36. According to the special theory of relativity, the effective mass of the particle moving with large velocity is $m = \dfrac{m_0}{\sqrt{1 - \upsilon^2/c^2}}$. Show that energy of the particle is equal to $\sqrt{m^2\upsilon^2c^2 + m_0^2c^3}$, where letters have their usual meanings.

37. How energy and momentum conservation occurs in Compton Effect?

38. Write down an expression for Compton shift and explain in brief the existence of unmodified component in Compton scattering.

39. How Planck's quantum theory can be used to explain qualitatively the existence of modified and unmodified component in Compton scattering.

40. Derive an expression for Compton shift in the wavelength of a photon after scattering from an electron.

41. In Compton Effect, considering elastic collision between a photon and a free electron, write down equations of energy and momentum conservation.

42. In Compton Effect what happens: (*i*) when photon collides with free electron in the scattering block? Write down the equations of energy and momentum conservation; (*ii*) when photon collides with bound electron?

43. Explain the occurrence of an intensity peak corresponding to an unmodified component of wavelength in a scattered X-ray beam, when a beam of X-rays is arranged to fall on a graphite target. Why it is less pronounced for scatterers of low atomic number than those belonging to high Z group?

44. Does Compton scattering expression explain unmodified scattering? Explain.

45. Explain the existence of modified and unmodified radiations in Compton Effect.

46. Derive the expression for Compton shift. Why it is not observable in the visible region of electro-magnetic radiation?

PROBLEMS

1. A black body radiator at 0°C radiates energy of 3.2×10^2 J.m^{-2}.s^{-1}. Deduce the value of Stefan's constant. [**Ans:** 5.7×10^{-8} J.m^{-2}.s^{-1}]

2. Using Wien's displacement law, estimate the temperature of sun. Given: λ_m = 4900 Å and Wien's constant = 0.292 cm.K. [**Ans:** 5963 K]

3. In an atomic explosion, the maximum temperature reached was of the order of 10^7 K. Calculate the wave lenght of maximum energy. [**Ans:** 2.93 Å]

4. In order to break a chemical bond in the molecules of human skin, photon energy of 3.5 eV is required. To what wavelength and region does this correspond? [**Ans:** 3547 Å]

5. A source is emitting 100 W of green light at a wavelength of 5000 Å. How many photons are emerging from the source per second? [**Ans:** 2.5×10^{20} Photon/Second]

6. Calculate the wavelength of which the energy of a photon becomes equal to the average thermal energy of atoms in a solid at room temperature. [**Ans:** 48 μm]

7. The work function of gold is 4.59 eV. What is the photoelectric threshold wavelength for gold? [**Ans:** 2702 Å]

8. The photoelectric threshold wavelength of tungsten is 2730 Å. Determine the maximum kinetic energy of the electrons ejected from a tungsten surface by UV light of wavelength 1800 Å. [**Ans:** 2.35 eV]

9. What is the potential difference that must be applied to stop the fastest photoelectrons emitted by a nickel surface under the action of UV light wavelength 2000 Å? The work function of nickel is 5 eV? [**Ans:** 1.2 V]

10. UV radiation of wavelength 750 Å is incident on the surface of cesium. If the work function for cesium is 1.97 eV, determine the maximum speed with which the electrons are ejected from its surface. [**Ans:** 2.3 ×106 ms]

11. Calculate the energy gap in silicon if it is given that it is transparent to radiation of wavelength greater than 11000 Å but it absorbs radiation of shorter wavelength. [**Ans:** 1.12 eV]

12. Find the energy and momentum of an X-ray photon whose wavelength is 0.2 Å. [**Ans:** 9.95×10^{15} J, 3.32×10^{-23} kgm/s]

13. X-rays of wave length 22×10^{-12} m (photon energy = 56 k eV) are scattered from a carbon target, the scattered radiation being viewed at 85° to the incident beam. (*i*) How much is Compton shift? (*ii*) What percentage of its initial energy does an incident X-ray photon lose? [**Ans:** 0.02214 Å, 9.15%]

14. In a TV tube, the operating potential is 10 kV. Find the shortest wavelength emitted by screen when struck by electrons. [**Ans:** 1.238 Å]

15. What is the maximum frequency of X-ray emitted if 12.4 kV potential is applied to a X-ray tube? [**Ans:** 3×10^{10} GHz]

16. An X-ray tube operates at 40 kV emits continuous X-ray spectrum with a short wavelength limit of 0.31 Å. Calculate Planck' constant. [**Ans:** 6.61×10^{-34} J.s]

17. A photon is Compton-scattered by a free electron at rest through an angle of 90°. What is the energy of the scattered photon if the energy of the incident photon is 10 MeV?

[**Ans:** 0.49 MeV]

18. A photon of 4Å strikes an electron at rest and is scattered at an angle of 150° to its original direction. Find the wavelength and speed of the photon after collision.

[**Ans:** 4.045 Å, 3×10^8 m/s]

19. Calculate Compton wavelength for an electron. [**Ans:** 0.0242Å]

20. X-rays of wave length 22×10^{-12} m (photon energy = 56 k eV) are scattered from a carbon target, the scattered radiation being viewed at 85° to the incident beam. (*i*) How much is Compton shift? (*ii*) What percentage of its initial energy does an incident X-ray photon lose?

[**Ans:** 0.02214 Å, 9.15%]

21. An X-ray photon of frequency 10^{19} Hz is scattered through an angle of 45° after colliding with a stationary electron. (*a*) What is its new frequency? (*b*) What is the kinetic energy of the electron after collision? [**Ans:** 9.8×10^{18} Hz, 938 eV]

25 Atomic Physics

Learning Objectives:

By the end of this chapter, you will be able to:

- Define concepts of Spectral Lines and Spectrum.
- Explain basics of electronic structure and different models - Bohr Model, Sommerfeld Atom Model and Vector Model.
- Learn about Energy Level Diagram.
- Describe Zeeman Effect on basis of Vector Atom Model.
- Understand principle of Anomalous Zeeman effect.
- Discuss about Paschen-Back and Stark Effect.

25.1 INTRODUCTION

In 1808 John Dalton postulated the atomic theory of matter. He considered the atom as the ultimate particle of matter. However, by the turn of twentieth century, scientists recognized that atoms can be further divided into subatomic particles, namely electrons, protons and neutrons. J. J. Thomson, in 1898, proposed the "plum pudding" model and in 1911 Rutherford proposed the "nuclear" model of atom. The protons and neutrons reside in the nucleus. More than 99.94% of an atom's mass is in the nucleus. The electrons of an atom are attracted to the protons in an atomic nucleus by electromagnetic force. However, these models could not explain the experimental results regarding atomic spectra. If gases are heated to a high temperature or if an electric discharge is passed through them, they emit light. They can also absorb light when light is made to pass through them. Gases are very selective in the frequencies they emit and absorb.

In 1913, Niels Bohr devised a model of the process of light emission from the atoms of elements that would explain the very particular frequencies emitted by making use of the concept of quantization New experimental results followed soon on X-rays, energy levels, and spectra. Bohr's model had been continually refined by various physicists to make it explain the successive new experimental results. Spectroscopic studies went on supplying finer and finer data on the basic line spectra, and on the effects of electric and magnetic fields on the spectral lines. These studies of atomic spectra soon guided to quantum theory of atomic structure, which marched towards a full-scale replacement for classical physics.

25.2 SPECTRAL LINES

It had been known for a long time that light is emitted by gases when they are heated to high temperature. The light emitted from a hot gas may be examined with a prism or grating. In practice, the source of light is placed behind a narrow linear slit and the narrow beam of light emerging from the slit is passed through a grating or a prism as shown in Fig. 25.1(a). The light breaks up into several beams according to wavelength. These beams produce a set of narrow lines with dark gaps between them on a photographic plate kept in their path. The lines recorded on the photographic plate are called ***spectral lines*** and the whole set of lines is known as a ***spectrum***. The term spectral line denotes light with a discrete wavelength and corresponds to the line image of the linear slit.

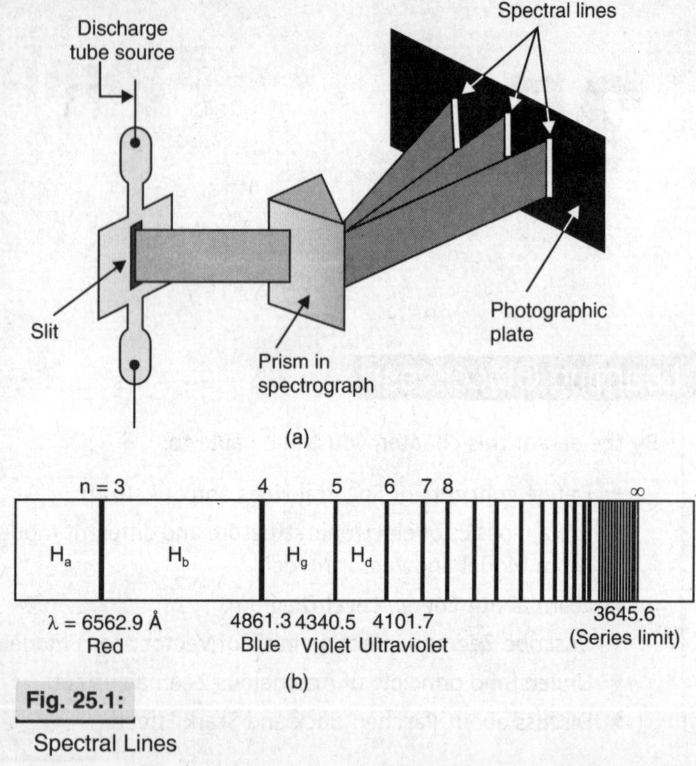

Fig. 25.1: Spectral Lines

The spectrum obtained from a hot body is called ***emission spectrum***. Each chemical element has unique spectrum lines that are characteristic of the element. Some elements, such as hydrogen and neon, have relatively few lines; others, such as iron have many thousand lines. The line spectra of elements are distinguished by their wavelength, the relative position, intensity, and number of lines. Lines characteristic of each individual element occur not only in the visible region but also in IR and UV regions. Kirchhoff and Bunsen were the first to carry out systematic investigations on line spectra during the years 1854 – 1859.

In addition to emitting light at specific wavelengths, an element can also absorb light at those specific wavelengths. The spectrum in the latter case is known as the ***absorption spectrum***. The absorption spectrum of an element is obtained by passing white light through the element in its gaseous state. It is found that each line in the absorption spectrum of a given element coincides with a line in the emission spectrum of that element. The conclusion is that an element can absorb only light of those wavelengths, which when excited, it can emit

Line spectra are produced by individual atoms of elements. The emission of light is the outcome of certain processes taking place inside the atom. The study of emission and absorption spectra of elements helped us understand the structure of the atom. In the spectrum, the wavelengths of lines form an infinite sequence following each other in an *orderly* fashion and converging to a lower limit in wavelength. A group of such lines exhibiting systematic regularities are called ***spectral series***.

Of all spectra, the emission spectrum of hydrogen is the simplest (Fig. 25.1 b). In 1885 Johann Balmer found regularity in the position of the lines of the hydrogen spectrum. He obtained the following simple empirical relation to compute the wavelengths of the lines in the visible region of the spectrum.

$$\lambda = 3646\left[\frac{n^2}{n^2-4}\right]\text{Å} \qquad n = 3, 4, 5.... \qquad ...(25.1)$$

Balmer's formula was rearranged by J.R. Rydberg in the following form.

$$\frac{1}{\lambda} = R\left[\frac{1}{2^2}-\frac{1}{n^2}\right] \qquad ...(25.2)$$

where R is known as **Rydberg constant** and has the value

$$R = 1.0973732 \times 10^7\text{m}^{-1}.$$

The reciprocal of wavelength $1/\lambda$ is called the **wave number**.

All the lines of the hydrogen spectrum satisfying the Eq. (25.2) are termed **Balmer series**. Subsequently, additional series of hydrogen spectrum were discovered. They are called Lyman, Paschen, Brackett,and Pfund series named after their discoverers. Such regularities were also observed for other atoms. An over-all formula

$$\frac{1}{\lambda} = R\left[\frac{1}{n_f^2}-\frac{1}{n_i^2}\right] \qquad ...(25.3)$$

describes all the series. The above formula which modified Balmer's formula is merely descriptive and does not provide any insight into the mechanism responsible for the emission of spectral lines. Attempts to explain the origin of discrete line spectra on the basis of classical physics met with severe failure. The Rutherford's planetary model also was not successful in explaining the atomic spectra.

25.3 ATOMIC STRUCTURE

J.J.Thomson discovered electron in 1897. He put forward the first ever model of the atom. Thomson represented it as a positively charged sphere with electrons immersed in it here and there. When unperturbed, both electrons and the positive charge were believed to be at rest. When in motion the electrons lose their energy in the form of radiation.

Basing on the results of α-scattering experiments conducted by him, Rutherford proposed 'planetary model' of the atom. The atom consists of a relatively small positive nucleus at the centre and electrons orbiting well away from it. But an electron moving in an orbit around the nucleus is in a state of accelerated motion and according to Maxwell, any accelerated charge generates electromagnetic waves. Consequently, the orbiting electron loses continually its energy until it approaches the nucleus and falls into it. This leads to the prediction that an atom exists for only about 10^{-8}s; but in fact atoms have remarkable stability. Secondly, if an electron is rotating around the nucleus, it radiates light whose frequency is the same as that of the rotation. As the electron spirals and falls onto the nucleus the light frequency continually increases and therefore the radiation given out by Rutherford atom must be continuous. However, the atomic spectra were in fact a set of discrete lines.

25.4 BOHR'S MODEL OF ATOM

In 1913, Niels Bohr removed the deficiencies of Rutherford model by incorporating some adhoc quantum hypotheses into it. Bohr explained the origin of line spectra in general terms on the basis of two central ideas. One is the concept of photon and the other is the concept of energy levels of atoms. Bohr combined these two concepts nicely to explain how light is emitted and why the spectral lines are arranged in an orderly fashion.

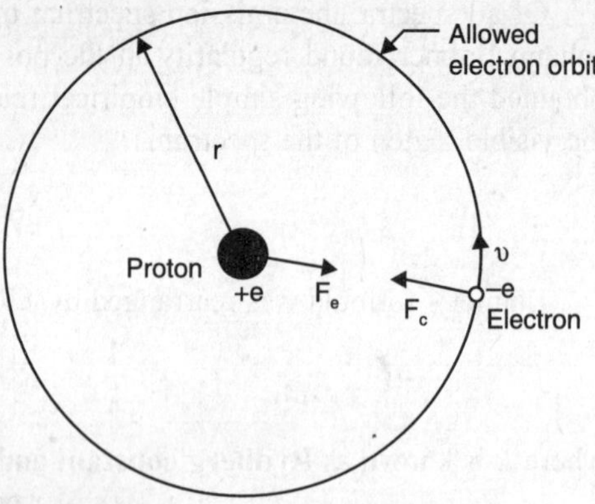

Fig. 25.2: Bohr's model of the hydrogen atom.

Bohr fused the quantum idea with the purely mechanical model of Rutherford and introduced the following three revolutionary adhoc postulates.

(*i*) Electrons revolve about a nucleus only in certain special orbits called **stationary orbits**, though an infinite number of orbits are mechanically allowed. While moving in the permitted (stationary) orbits, electrons do not emit or absorb electromagnetic radiation though they are in accelerated motion. Hence the atom is stable.

(*ii*) The allowed electron orbits are those for which the angular momentum is an integral multiple of $\hbar$, where h is Planck's constant. The angular momentum of electron is $I\omega = (mr^2)(\upsilon/r) = m\upsilon r$. Accordingly

$$m\upsilon r = n\hbar \qquad (n = 1,2,3, \ldots) \qquad \text{...(25.4)}$$

(*iii*) The atom radiates energy *only* when the electron jumps from one of the upper allowed orbits, with energy say E_2, to a lower allowed orbit, with energy say E_1. The change in energy during the transition is given by

$$E_2 - E_1 = h\nu \qquad \text{...(25.5)}$$

Note that the frequency of the emitted light is not determined by the motion of the electron but is governed by changes in its motion.

Bohr obtained values of the energies of various states of hydrogen atom by assuming that an electron with velocity 'υ' revolves in a circular orbit of radius '*r*' around the nucleus. For a dynamically stable orbit, the centripetal force experienced by the electron equals the force of electrical attraction between the nucleus and the electron. Thus,

$$\frac{m\upsilon^2}{r} = \frac{e^2}{4\pi\varepsilon_o r^2} \qquad \text{...(25.6)}$$

Use Eq. (25.4) and Eq. (25.6) to eliminate 'υ', we obtain for the radius of the orbits as

$$r = \frac{\varepsilon_o h^2}{\pi\, me^2} n^2 = r_o n^2 \qquad \text{...(25.7)}$$

where r_o is the radius of the first Bohr orbit ($n = 1$) which is given by

$$r_o = \frac{\varepsilon_o h^2}{\pi\, me^2} = 0.53\ \text{Å}. \qquad \text{...(25.8)}$$

The above value for the size of the Bohr radius agrees with the value 0.5 Å obtained from the estimates based on kinetic theory of gases.

The total energy for a specific electron orbit is obtained by adding the kinetic energy and the potential energy of the electron with respect to the nucleus. Thus,

$$E - \frac{1}{2}m\upsilon^2 - \frac{e^2}{4\pi\varepsilon_o r}$$

where the first term represents the kinetic energy of electron due to its angular motion and the second term is the potential energy due to electrical attraction.

Substituting the value of r from Eq. (25.7) and eliminating υ using Eq. (25.4), we get

$$E = -\frac{e^2}{8\pi\ \varepsilon_o r} \quad \text{...(25.9)}$$

Using the value of r into the above Eq. (25.9), we find that electron can have only certain discrete negative values of energy E_n given by

$$E_n = -\frac{me^4}{8\varepsilon_o^2 h^2}\cdot\frac{1}{n^2} = -\frac{E_o}{n^2} \quad \text{...(25.10)}$$

where $$E_o = \frac{me^4}{8\varepsilon_o^2 h^2} = 13.6\ eV \quad \text{...(25.11)}$$

Eq. (25.10) shows that the energy of an orbit varies as $-(1/n^2)$. The energy is larger for larger values of n and hence for bigger orbits. For each value of n there is a corresponding definite energy E_n. The values of E_n are called **allowed energy values** and the corresponding energy levels are called **allowed energy states**. The other values of energy or intermediate energy states are forbidden. The allowed energy states correspond to stationary orbits of electron in the atom. The energy of the system is negative. The negative energy implies that the electron is in a **bound state** and is not free to move away. It also indicates that energy is to be supplied from outside in order to separate the electron and the nucleus.

25.4.1 Rydberg Constant

According to Bohr's third postulate, an atom emits a photon of frequency ν when it jumps from a higher energy orbit to a lower energy orbit. Thus,

$$\nu = \frac{E_i - E_f}{h} \quad \text{...(25.12)}$$

Now from Eq. (25.10) we can write $E_f = -\frac{me^4}{8\varepsilon_o^2 h^2}\cdot\frac{1}{n_f^2}$ and $E_i = -\frac{me^4}{8\varepsilon_o^2 h^2}\cdot\frac{1}{n_i^2}$

$$\therefore \quad h\nu = \frac{me^4}{8\varepsilon_o^2 h^2}\left[\frac{1}{n_f^2} - \frac{1}{n_i^2}\right]$$

or $$\nu = \frac{me^4}{8\varepsilon_o^2 h^3}\left[\frac{1}{n_f^2} - \frac{1}{n_i^2}\right]$$

As $\nu = \frac{c}{\lambda}$, we write the above equation as

$$\frac{c}{\lambda} = \frac{me^4}{8\varepsilon_o^2 h^3}\left[\frac{1}{n_f^2} - \frac{1}{n_i^2}\right]$$

or $$\frac{1}{\lambda} = \frac{me^4}{8\varepsilon_o^2 ch^3}\left[\frac{1}{n_f^2} - \frac{1}{n_i^2}\right]$$

or $$\frac{1}{\lambda} = R\left[\frac{1}{n_f^2} - \frac{1}{n_i^2}\right]$$

where $$R = \frac{me^4}{8\varepsilon_o^2 ch^3} \qquad ...(25.13)$$

Substituting the known values into Eq. (25.13), we obtain $R = 1.097 \times 10^7\ m^{-1}$ which is in excellent agreement with the experimental value for Rydberg constant. It is a triumph for Bohr's theory.

25.4.2 Spectral Series of Hydrogen Atom

The origin of the spectral series can easily be understood from Bohr's theory. When hydrogen atom is excited, it returns to its ground state by emitting the energy it had absorbed earlier. The energy is emitted by the excited electron in the form of radiation of different wavelengths as it jumps down from a higher to lower orbit.

If the electron in the hydrogen atom is initially in one of the stationary orbits with $n_i = 2,3,4...$ and makes transition to the final stationary orbit with $n_f = 1$, then the wavelengths emitted due to these transitions given by Eq. (25.3) are given by $\frac{1}{\lambda} = R\left[\frac{1}{1^2} - \frac{1}{n_i^2}\right]$. These lines belong to Lyman series.

Again, if transitions take place from initial stationary orbits with $n_i = 3,4,5...$ to the final stationary orbit with $n_f = 2$, then the wavelengths emitted due to these transitions are given by $\frac{1}{\lambda} = R\left[\frac{1}{2^2} - \frac{1}{n_i^2}\right]$. These lines belong to Balmer series.

Similarly, transitions taken place from initial stationary orbits with $n_i = 4,5,6...$ to the final stationary orbit with $n_f = 3$, produce lines with wavelengths given by $\frac{1}{\lambda} = R\left[\frac{1}{3^2} - \frac{1}{n_i^2}\right]$. These lines belong to **Paschen series.**

The remaining series were not observed by the time Bohr proposed his theory of atom. He predicted in 1913 in his theory the other series in UV and far infra red regions. They were subsequently discovered establishing the validity of Bohr's theory. If transitions take place from initial stationary orbits with $n_i = 5,6,7...$ to the final stationary orbit with $n_f = 4$, then the wavelengths emitted due to these transitions are given by $\frac{1}{\lambda} = R\left[\frac{1}{4^2} - \frac{1}{n_i^2}\right]$.

These lines belong to Brackett series, and so on.

25.4.3 Limitations of Bohr's Theory

Bohr's theory employed a semi-classical model where quantum postulates are introduced into a mechanical model. But it explained with remarkable success the origin of line spectra in case of hydrogen atom, which is a one-electron atom. However the theory could not give correct predictions for two electron atoms. The Bohr hypothesis established the relation of wavelengths to energy levels, but it did not provide any general principles for predicting the energy levels of a particular atom. It did not explain why angular momentum is quantized and why atoms make transitions. However, Bohr's theory contributed immensely to the understanding of atomic structure. A more general understanding of atomic structure and energy levels is provided by the quantum mechanics.

25.5 FRANK-HERTZ EXPERIMENT

The atomic energy levels are quantized according to Bohr model. The existence of discrete energy levels of an atom was confirmed by the experiments carried out in 1914 by the German physicists James Franck and Gustav Hertz.

A schematic view of the apparatus used by Franck and Hertz is shown in Fig. 25.3. It consists of a gas tube filled with mercury vapour at low pressure of about 1 mm Hg. The glass tube contained three electrodes, namely cathode C, grid G and anode P. The cathode emits electrons through thermionic emission when it is heated by a filament. The grid and the anode are kept close and away form the cathode. The electrons emitted by the cathode are accelerated by the grid which is maintained at a positive potential V with respect to the cathode.

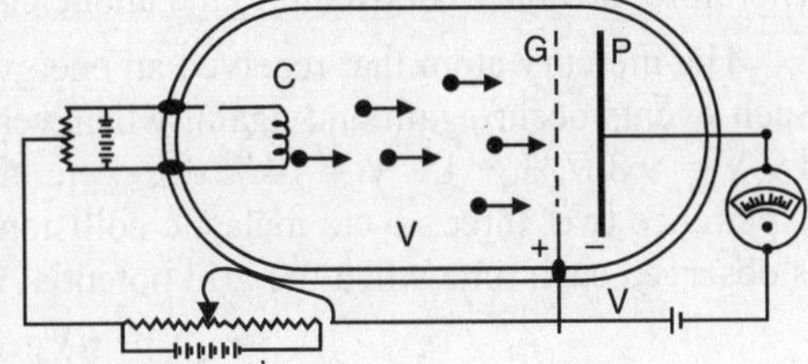

Fig. 25.3: Schematic of the apparatus used by Franck and Hertz

The potential difference between the cathode and grid can be smoothly varied with the help of potentiometer. The anode is kept at a small negative potential with respect to the grid such that the potential difference between the anode and the grid is of the order of 0.5 V. the electrons reaching the grid acquire energy eV and after passing the grid they experience a retarding electric field. Electrons that have enough kinetic energy to overcome the retarding field will reach the anode. Such electrons cause the anode current which is measured by a suitable milli-ammeter.

The variation of anode current I with the potential difference V between the cathode and grid was studied. The results obtained in the experiment are shown in Fig. 25.4. The current was found to increase first and after reaching a maximum at about 4.9 V, it decreased sharply with a further increase in voltage. It reached a minimum and then again began to increase. Maxima of current occurred at 9.8V, 14.7 V.... and so on.

The above results can be interpreted as follows. For low values of V, the collisions between an electron and a mercury atom are of elastic nature. The electron does not lose its energy in the process but only changes its direction. And also the atom is not excited. The electron on reaching G has sufficient kinetic energy to reach the anode P. As V is increased, more and more electrons are accelerated and anode current increases.

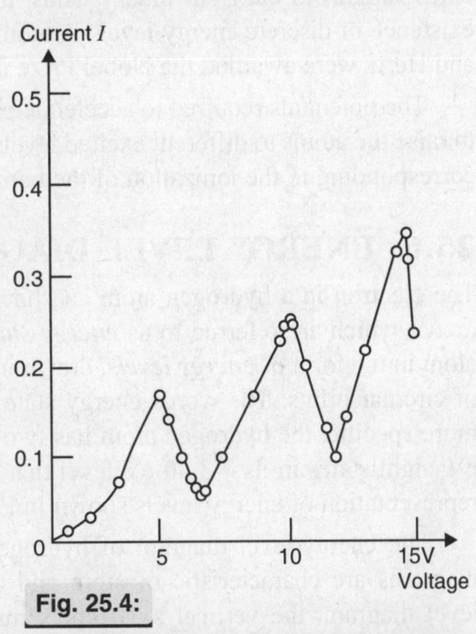

Fig. 25.4: Results of Franck and Hertz experiment

According to Bohr's third postulate, the amount of energy that a mercury atom can receive from a colliding electron would be equal to the energy difference between its ground state and one of the excited states. The excited state nearest to the ground state for a mercury atom is at about 4.9 eV above the ground state. It may therefore be expected that as long as the energy of an electron is less than 4.9 eV, it experiences only elastic collisions. When the electron accumulates energy of 4.9 eV or more, the collisions cease to be elastic. The electron has enough energy now to undergo inelastic collision with a mercury atom and transfer to the atom energy of 4.9 eV. After that the electron continues its motion with a lower velocity. Consequently, it cannot overcome the retarding field beyond the grid and returns to the grid. Therefore, it cannot contribute to the anode current and the anode current decreases.

The mercury atom that received an energy of 4.9 eV gets excited to its first excited state. Such events occur again and again, whenever the potential difference attains the values 2 × 4.9 V = 9.8 V, 3 × 4.9 V = 14.7 V,..... etc. at these values of potential difference electrons experience two, three,... etc inelastic collisions. Therefore, a sharp drop in the anode current is observed each time when the grid potential reaches the values given by

$$V = nV_1 \qquad (n = 1, 2, 3, \ldots).$$

where V_1 corresponds to the energy of the first excited state, namely 4.9 eV.

The mercury vapour in the tube was observed to emit ultraviolet light at a wavelength of 2537 Å. The UV light must have been emitted by the excited mercury atoms returning to the ground state. According to Bohr's postulate, the wavelength of light emitted by mercury atoms should be

$$\lambda = \frac{hc}{E_f - E_i} = \frac{hc}{\Delta E}$$

$$= \frac{(6.62 \times 10^{-34}\,\text{Js})\,(3 \times 10^{8}\ \text{m/s})}{(4.9\,\text{V})\,(1.602 \times 10^{-19}\,\text{C})} = 2533\ \text{Å}.$$

The good agreement between the calculated and experimental values of the wavelength lends support to the Bohr model. Thus, the Franck-Hertz experiment is a direct proof to the existence of discrete energy levels in atoms and hence to the Bohr model of the atom. Franck and Hertz were awarded the Nobel Prize in physics in 1925.

The potentials required to accelerate electrons for imparting them kinetic energy sufficient to raise the atoms to different excited levels are called **critical potentials**. The critical potential corresponding to the ionization of the bombarded atom is known as ***ionization potential***.

25.6 ENERGY LEVEL DIAGRAM

The electron in a hydrogen atom can have any one of a series of negative energies, E_2, E_3, ,....... which are referred to as ***energy states***. It is easier to represent the *energy states* of an atom in the form of ***energy levels***, drawn in the form of horizontal lines rather than in the form of circular orbits. The words energy state and energy level are used interchangeably. (To be more specific, the hydrogen atom has two states in its ground level corresponding to – 13.60 eV, eight states in its – 3.40 eV level that is the first excited level and so on). The sequential representation of energy levels shown in Fig. 25.5 is known as an ***energy level diagram***.

The energy-level diagram of hydrogen is shown in Fig. 25.5 and is the simplest. The diagrams are characteristic of atom and are complicated for complex atoms. In an energy level diagram, the vertical axis represents electron energy and the horizontal lines are the allowed energy states each identified by a specific value of quantum number *n*.

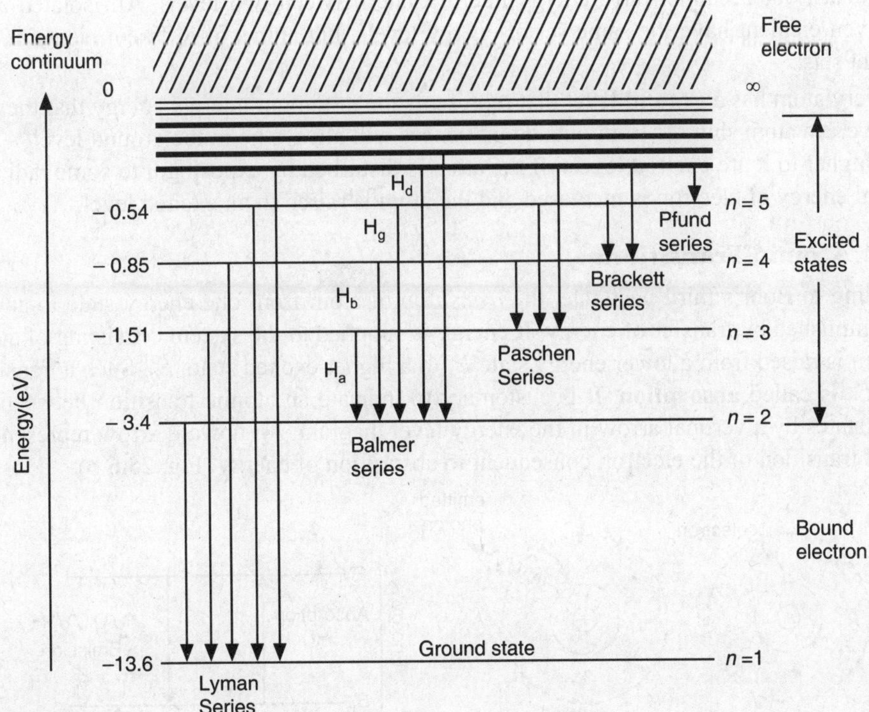

Fig. 25.5:

Energy-level diagram of Hydrogen

The lowest energy level is the level in which the electron revolves in the innermost orbit (n =1); that level is called the **ground state**. It is the normal unexcited state of the atom. The atom can reside in its ground state for an indefinite time. For hydrogen, the ground sate is at – 13.6 eV. An atom can get raised to a higher level only by absorbing energy from an external source. The energy may be supplied through inelastic collisions with other atoms, when the hydrogen gas is heated or subjected to an electric discharge. When the atoms are raised to higher levels, they are said to be **excited**. For example, the first excited level, n = 2, is at –3.4 eV; and so on. The energy difference between the energy of an excited state and the energy of the ground state is called **excitation energy**. An atom cannot stay in an excited state for more than 10^{-8} s and spontaneously returns to the ground state. It is due to Coulomb attraction exerted by the positive nucleus on the electron moved to a higher orbit. The time for which an atom stays in an excited state is known as its **lifetime** in that state.

The energy levels in the energy level diagram are not evenly spaced. They get closer and closer at the upper end of the diagram. It is because the energies of the allowed states are proportional to $1/n^2$ and as n increases the energy difference between successive levels rapidly decreases. The upper end of the diagram is bounded by zero energy level for which $n = \infty$. It is called the **ionization level** of the atom. When the electron reaches the ionization level it gets separated completely from the nucleus and the atom is said to be **ionized**. The energy needed to remove an electron from an atom in its ground state is called its **ionization** energy. In case of hydrogen it is equal to 13.6 eV.

The above picture can be extended to atoms of all elements. Each atom has a set of possible energy levels. An atom can have an amount of internal energy equal to any one of these levels, but it cannot have energy intermediate between two levels. All isolated atoms of a given element have the same set of energy levels, but atoms of different elements have different sets.

Every atom has a **ground level** that represents the minimum internal energy that the atom can have. An atom shut off from outside influences will always lie in the ground level. All the levels higher to it are **excited levels**. If the atom is disturbed by exposing it to some radiation, the total energy of electron is increased and the atom is raised to an excited level.

25.6.1 Atomic Transitions

According to Bohr's third postulate, the *transition* of atom from one energy state to another is accomplished by transfer of energy. If energy is supplied to the system consisting of atoms, the atom is raised from a lower energy state E_1 to a higher excited state, E_2. Such a transition, $E_1 \rightarrow E_2$ is called **absorption**. It is customary to indicate an atomic transition between two energy states by a vertical arrow in the energy level diagram. An upward arrow represents an upward transition of the electron consequent to absorption of energy (Fig. 25.6 b).

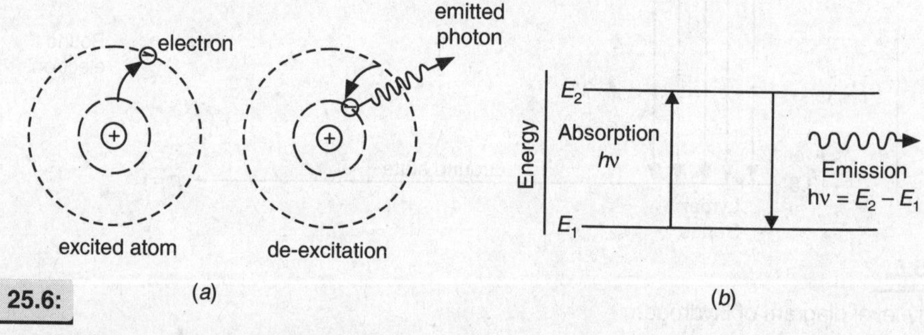

Fig. 25.6: Processes of absorption and emission by an atom.

The transition from a lower state to an excited state can occur only if the difference in energy is exactly equal to the photon energy $h\nu$. The atom does not stay in an excited state indefinitely. It usually returns on its own to the lower state after about 10 ns by emitting a photon.

The downward transition $E_2 \rightarrow E_1$ is called **emission**. A downward arrow represents a downward transition. During the emission process, as the atom returns from a higher energy state E_2 to a lower energy state E_1, it emits a quantum of energy, $h\nu$. If the photon energy is $h\nu = \frac{hc}{\lambda}$, then conservation of energy gives

$$h\nu = \frac{hc}{\lambda} = E_2 - E_1 \quad \text{...(25.14)}$$

Vertical lines in Fig. 25.5 and Fig. 25.6 show the jumps that electrons can make from one level to another. The location of these transitions, or **emission lines**, can be shown on a wavelength graph. The ensuing pattern corresponds exactly to that recorded by the spectrographic plates. Thus, **position** is one of the important properties of spectral lines.

25.6.2 Spontaneous Emission

An excited atom can lose its energy in numerous ways. The energy can be can be emitted in the form of light or it may be transferred to other atoms through collisions.

Excited electrons (atoms) will not stay at the excited level forever since nature favours a low energy level. An atom at an excited state may rapidly fall to a lower level by emitting a photon of radiation. Such emission is known as ***spontaneous emission***. Therefore, the excited electron returns to lower level by emitting a photon spontaneously after an average time of t_{sp}. t_{sp} is called the *spontaneous lifetime* of the level. The spontaneous life-time depends on the atomic species involved and can be determined for a given species. Some levels have long lifetimes measured in seconds, but others are comparatively short, of the order of nanoseconds or less.

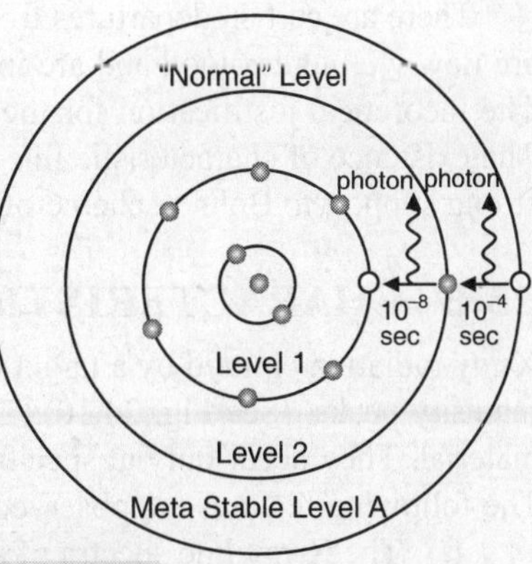

Fig. 25.7: Illustration of Metastable states

25.6.3 Metastable states

The average time spent by the atom in an excited level is called the **lifetime** of the level. The lifetime is usually of the order of 10^{-8}s. Besides the short-lived excited states, there are states with average lifetimes greater than 10 milliseconds or even as long as several seconds. They are called **metastable states**.

Example 25.1. *The energy of a particular state of an atom is 5.36 eV and the energy of another state is 3.45 eV. Find the wavelength of the light emitted when the atom makes a transition from one state to the other.*

Solution:
$$(E_1 - E_2) = h\nu = \frac{hc}{\lambda}$$

$$\therefore \quad \lambda = \frac{hc}{(E_1 - E_2)}$$

$$= \frac{(6.626 \times 10^{-34} \text{ J.s})(3 \times 10^8 \text{ m/s})}{(5.36 - 3.45)(1.602 \times 10^{-19} \text{ J})} = \mathbf{6496 \text{ Å}}$$

25.7 ELECTRON SHELLS

The simple model of hydrogen was extended to other elements by Bohr and Stoner. Each atom is composed of a positively charged nucleus and a number of electrons revolving in allowed orbits around it. The number of electrons in an atom are determined by the atomic number, Z. The orbits to which electrons are confined are the orbits characterized by different values of the quantum number. The orbits are called **electron shells**. These are named as K, L, M, N, O, etc.

When an atom is built, electrons are added one after the other filling one shell and then another starting from the shell nearest to the nucleus. A shell is filled when $2n^2$ electrons occupy it. Thus, K shell is filled when it has 2 electrons; the L shell is filled when 8 electrons occupy it and so on. Table 25.1 shows the shell, quantum number and number of electrons.

Table 25.1: *Electron Shells*

Shell	Quantum number, *n*	Capacity
K	1	2
L	2	8
M	3	18
N	4	32
O	5	50

There are certain departures from the above order in case of heavier elements. The reasons are now well understood and are indicative of the chemical behaviour of the heavy elements. The theoretical justification for the electron shells was provided by Wolfgang Pauli in 1925. The existence of characteristic line spectrum in the x-ray radiation emitted by elements lent a strong support to Bohr's scheme of organization of electrons into shells in an atom.

25.8 CHARACTERISTIC X-RAY SPECTRUM

X-ray radiation emitted by a target consists of a continuous spectrum superimposed with high intensity peaks. (See Fig. 24.10). The peaks occur at wavelengths characteristic of the target material. They occur only at specific wavelengths and hence the spectrum is a line spectrum. The following features are observed regarding the line spectrum.

(*i*) The X-ray line spectra of all elements are strikingly similar.

(*ii*) The line spectrum given off by an element remains the same irrespective of the compounds that the element forms.

(*iii*) As one proceeds from the elements of low atomic weight to elements of higher atomic weight, the line spectrum shifts progressively towards the shorter wavelengths.

The origin of X-ray line spectra could not find explanation on the basis of classical theory and is accounted only be the quantum theory.

25.8.1 Origin of Characteristic X-Rays

The origin of X-rays is explained on the basis of Bohr's theory. According to Bohr model, electrons in an atom are organized into K, L, M, N,….shells. An atom radiates energy when an electron jumps from one allowed orbit to another allowed orbit. Normally, electron transitions cannot take place from the outermost orbits to the innermost orbits because the inner orbits are occupied. However, when a high energy electron knocks off an inner core electron belonging to K, L, M, N,….shells, an electron from an outer orbit jumps to fill up the vacated inner orbit. Such a transition causes emission of high-energy photon, namely X-ray photon. For example, the binding energy of an electron in K-shell of sodium atom is 1041 eV. The binding energy of a 2s electron in L shell is 63 eV. When the 2s electron jumps into 1s level, an X-ray photon of energy 978 eV is emitted.

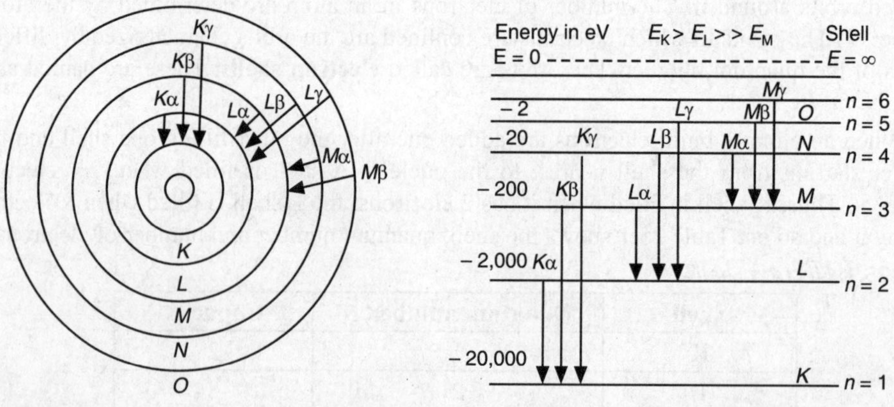

Fig. 25.8:

Characteristic spectrum - a series of discrete lines

The characteristic spectrum consists of a series of discrete lines. The group of lines having the shortest wavelength is called K series. These lines occur when an electron from the K-shell is knocked away and the resulting vacancy is filled by an electron from the next higher shells L, M, N ...etc (Fig. 25.8). If an electron from L-shell jumps into K-shell, the energy difference is emitted in the form of radiation; the transition of an electron from M-shell to K-shell produces K_β radiation.

In a similar way the L-series consisting of L_α, L_β,....lines is produced when vacancies in L-shell are filled by electrons jumping from next higher shells. K-series form the most energetic and hard X-rays whereas L- and M- series form soft X-rays.

We can summarize some more features of characteristic spectrum as follows:

(*i*) Line spectrum is produced when electrons are ejected from the innermost orbits of the atoms of a material and electrons from the outer orbits jump into those vacancies.

(*ii*) The spectrum comprises of discrete lines which form K-series, L-series and M-series. K series is formed when electron jumps end at the K-level. Similarly, L-series and M-series are produced when electron jumps end at L- and M-levels respectively.

(*iii*) K-series lines are due to the most energetic transitions and form the **hard X-rays** where as L- and M-series constitute the **soft X-rays**.

(*iv*) As the atomic number of the target is increased, a regular shift towards shorter wavelength is found in the K-spectrum. Moseley found the relationship between atomic number and wavelength as

$$\frac{\lambda_2}{\lambda_1} = \frac{\nu_1}{\nu_2} = \frac{(Z_1 - 1)^2}{(Z_2 - 1)^2}$$

where ν_1 is the frequency of the K_α line for a target material having an atomic number of Z_1 and ν_2 is the frequency of the K_α line for the target material having an atomic number of Z_2.

25.9 MOSELEY'S LAW

During the years 1913-1914, Moseley made a systematic study of characteristic X-ray spectra of various elements. He found that different elements produce similar spectra consisting of K, L, and M series. He measured the wavelengths of the characteristic X-rays of a large number of elements using Bragg's X-ray spectrometer. He found a striking similarity in the spectral lines of each series. The spectrum emitted by each element was found to be exactly similar to that of the other, the only difference being that the lines had different wavelengths. The frequency of any particular line in a series varied regularly from one element to the next one in the periodic table. On plotting a graph between the square root of the frequency of any particular line in a series and the atomic number of the element, Moseley found that it gives a straight line, as seen in Fig. 25.9. Such a graph is known as a **Moseley's plot**.

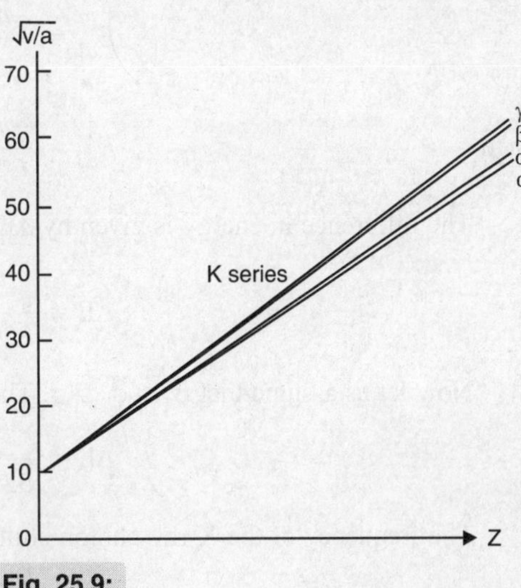

Fig. 25.9:

Moseley's Law - graph between the square root of the frequency of any particular line in a series and the atomic number of the element.

Each series is governed by the following mathematical relationship:

$$\nu = a^2 (Z - \sigma)^2 \qquad \text{...(25.15)}$$

where Z is the atomic number of the target material, and a and σ are constants. The constant σ is known as **screening constant.** The above equation is known as **Moseley's law** and is usually written in the form

$$\sqrt{\nu} = a(Z - \sigma) \qquad \text{...(25.16)}$$

Moseley's law states that *the frequency of a spectral line in the characteristic X-ray spectrum varies directly as the square of the atomic number of the element emitting it.*

25.9.1 Derivation of Moseley's Law

Moseley's law can be derived from the Bohr's theory. Let us consider an atom of fairly large atomic number Z. For an electron in an inner orbit nearer to the nucleus, the Coulomb attraction due to positive nucleus will be dominant. The energy of the electron is given by

$$E_n = \frac{-m(Ze)^2(e)^2}{32\pi^2\varepsilon_o^2\hbar^2}\left[\frac{1}{n^2}\right] \qquad \text{...(25.17)}$$

where (Ze) is the nuclear charge.

Accordingly, the energies of electron in an orbit of principal quantum number n_1 and that of n_2 can be written respectively as

$$E_{n_1} = \frac{-m(Ze)^2(e)^2}{32\pi^2\varepsilon_o^2\hbar^2}\left[\frac{1}{n_1^2}\right] \qquad \text{...(25.18)}$$

and

$$E_{n_2} = \frac{-m(Ze)^2(e)^2}{32\pi^2\varepsilon_o^2\hbar^2}\left[\frac{1}{n_2^2}\right] \qquad \text{...(25.19)}$$

Taking into account of the screening effect due to electrons, the positive charge of the nucleus seen by the electron under consideration reduces from Z to $(Z - \sigma)$.

$$\therefore \qquad E_{n_1} = \frac{-m(Z-\sigma_1)^2(e)^4}{32\pi^2\varepsilon_o^2\hbar^2}\left[\frac{1}{n_1^2}\right]$$

and

$$E_{n_2} = \frac{-m(Z-\sigma_2)^2(e)^4}{32\pi^2\varepsilon_o^2\hbar^2}\left[\frac{1}{n_2^2}\right]$$

The difference in energy is given by $\Delta E = E_{n_2} - E_{n_1}$.

$$\Delta E = \frac{me^4}{32\pi^2\varepsilon_o^2\hbar^2}\left[\frac{(Z-\sigma_1)^2}{n_1^2} - \frac{(Z-\sigma_2)^2}{n_2^2}\right] \qquad \text{...(25.20)}$$

Now let us assume that $\sigma_1 \cong \sigma_2 \cong \sigma$. Then

$$\Delta E = \frac{me^4}{32\pi^2\varepsilon_o^2\hbar^2}(Z-\sigma)^2\left[\frac{1}{n_1^2} - \frac{1}{n_2^2}\right]$$

The frequency of the X-ray photon emitted in the transition is given by

$$\nu = \frac{\Delta E}{h} = \frac{me^4}{8\varepsilon_o^2 h^3}(Z-\sigma)^2\left[\frac{1}{n_1^2} - \frac{1}{n_2^2}\right] \qquad \text{...(25.21)}$$

The K-series is due to transitions from higher states to the $n = 1$ state. Therefore, $n_1 = 1$.

Then,

$$\nu = \frac{me^4}{8\varepsilon_0^2 h^3}(Z-\sigma)^2\left[1-\frac{1}{n_2^2}\right] \qquad ...(25.22)$$

or

$$\nu = a^2 (Z-\sigma)^2 \qquad ...(25.23)$$

where $a^2 = \frac{me^4}{8\varepsilon_0^2 h^3}\left[1-\frac{1}{n_2^2}\right]$ is a constant.

Thus, Moseley's law is derived from Bohr's theory.

25.9.2 Importance of Moseley's Law

Moseley's work gave a precise method for determination of the atomic numbers of the elements. Basing on his work, Moseley concluded that the determining factor in the arrangement of elements in the periodic table should be the atomic number and not the atomic weight. Earlier, Mandeleev arranged the elements in increasing atomic weight to form the periodic table.

Moseley proposed that *the elements must be arranged in the periodic table according to their atomic numbers and not according to their atomic weights.*

Moseley found that if he arranged the wavelengths of K_α lines in the order of atomic weights, these wavelengths formed a remarkable regular progression. But, the sequence was not perfect, since some gaps were found and some of the wavelengths were out of order. Moseley attributed the gaps in the series to undiscovered elements. The atomic number is the number of positive units of electric charge on the nucleus of the atom and it is the fundamental quantity, which increases by regular steps as we pass from one element to the next. Following his conclusions, Moseley corrected the positions of the elements argon, cobalt and tellurium in the periodic table. According to the arrangement with ascending order of atomic weights, the element cobalt (Z = 27) with atomic weight 58.94 should be placed after the element nickel (Z = 28) having atomic weight 58.69.

Similarly, the element tellurium (Z = 52) with atomic weight 127.61 should be placed after the element and that of iodine (Z = 53) having atomic weight 126.92. So is the case with the elements argon and potassium. Argon (Z = 18) with atomic weight 40 should come after element potassium (Z = 19) having atomic weight 39. However, Mandeleev observed that the above elements are to be placed in reverse order to preserve the periodicity of chemical and physical properties. The discrepancy is removed if the elements are arranged as per the ascending order of atomic numbers.

Thus, Moseley had shown that the atomic number is much more important than the atomic weight in determining the properties of the elements. Moseley pointed out gaps in the group of rare earth elements at values Z = 43, 61, 72, and 75. These elements technetium (43), promethium (61), hafnium (72) and rhenium (75) were subsequently discovered.

Example 25.2. *Cobalt (Z = 27) gives strong K_α line of wavelength λ = 1.785 Å and weak lines of 2.285 Å and 1.537 Å due to impurities. Assume screening constant σ = 1 for K-series and find the atomic number Z of each of the two impurities.*

Solution:

$$\lambda_{K\alpha} = \frac{1}{R}\cdot\frac{1}{(Z-\sigma)^2}\left[\frac{1}{\left(\frac{1}{n_1^2}-\frac{1}{n_2^2}\right)}\right] = \frac{1}{R}\cdot\frac{1}{(Z-1)^2}\left[\frac{1}{\left(\frac{1}{1^2}-\frac{1}{2^2}\right)}\right]$$

$$\therefore \quad (Z-1)^2 = \frac{1}{R(1.785)}\left(\frac{4}{3}\right) = \frac{4}{3R\,(1.785)} \quad ...(1)$$

Similarly for impurity elements, we write

$$(Z_1-1)^2 = \frac{1}{R(2.285)}\left(\frac{4}{3}\right) = \frac{4}{3R\,(2.285)} \quad ...(2)$$

and

$$(Z_2-1)^2 = \frac{1}{R(1.537)}\left(\frac{4}{3}\right) = \frac{4}{3R\,(1.537)} \quad ...(3)$$

Dividing (2) by (1), we obtain

$$\frac{(Z_1-1)^2}{(Z-1)^2} = \frac{1.785}{2.285}$$

or

$$\frac{(Z_1-1)^2}{(27-1)^2} = \frac{1.785}{2.285}$$

or

$$(Z_1-1)^2 = 528.08$$

$$\therefore \quad Z_1 = 23.9 \approx 24$$

Similarly dividing (3) by (1), we obtain

$$\frac{(Z_2-1)^2}{(Z-1)^2} = \frac{1.785}{1.537}$$

or

$$\frac{(Z_2-1)^2}{(27-1)^2} = \frac{1.785}{1.537}$$

or

$$(Z_2-1)^2 = 785.08$$

$$\therefore \quad Z_2 = 28.02 \approx 28$$

25.10 THE SOMMERFELD RELATIVISTIC ATOM MODEL

The spectral lines emitted by hydrogen atom appeared to be single lines with spectroscopes of low resolving power. However, spectroscopes of high resolving power showed that these lines consist of groups of separate and distinct lines located very close to each other. It is known as **fine structure** of spectral lines. The origin of more than one spectral line can be explained only when several orbits of slightly different energies exist corresponding to a given quantum number *n*. But according to Bohr's theory, only one electron orbit is possible for each value of *n*. Obviously, Bohr's theory could not explain the fine structure of spectral lines.

In 1915, A. Sommerfeld, the German physicist extended Bohr's theory by incorporating the idea of elliptical electronic orbits and taking into consideration the relativistic variation of electron mass. The atom proposed by Sommerfeld is therefore called the Sommerfeld relativistic atom model.

25.10.1 Elliptical Orbits

In general, periodic motion under the influence of a central force leads to elliptical orbits. Therefore, it is postulated that in atom an electron moves in an elliptical orbit with the nucleus located at a focus. In the elliptical orbit the position of the electron at any instant is defined by two coordinates, namely the radius vector r and the azimuthal angle ϕ (Fig. 25.10). The electron momentum may be resolved into two components p_r along the radius vector and p_ϕ at right angles to it. The momentum component p_ϕ at right angles to the radius vector represents the angular momentum. Sommerfeld assumed that the generalized condition, postulated by Bohr in case of circular orbits, holds good for elliptical orbits also.

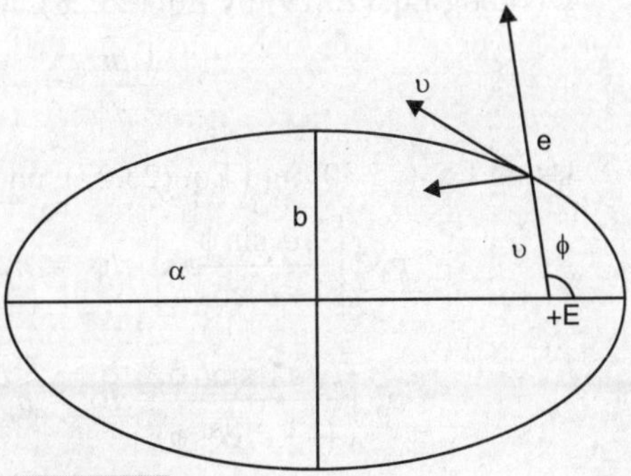

Fig. 25.10: Elliptical Orbits

Thus, $$p_r dr = n_r h \qquad ...(25.24)$$

and $$p_\phi d\phi = n_\phi h \qquad ...(25.25)$$

where n_r is called the **radial quantum number** and n_ϕ is the **azimuthal quantum number**. Further,

$$n_r + n_\phi = n$$

where n is the **principal quantum number**.

25.10.2 Allowed Elliptical Orbits

It is known form classical dynamics that the angular momentum p_ϕ remains constant during the motion. Therefore, from Eq. (25.25), we have

$$\int_0^{2\pi} p_\phi d\phi = p_\phi . 2\pi = n_\phi h$$

$$\therefore \quad p_\phi = \frac{n_\phi h}{2\pi} \qquad ...(25.26)$$

In an elliptical orbit, the radius vector is a variable quantity. The momentum may be expressed as

$$p_r = m\left(\frac{dr}{dt}\right) = m\left(\frac{dr}{d\phi}\right)\left(\frac{d\phi}{dt}\right)$$

$$= m\left(\frac{dr}{d\phi}\right)\left(\frac{p_\phi}{m\,r^2}\right) = \left(\frac{p_\phi}{r^2}\right)\left(\frac{dr}{d\phi}\right)$$

$$\therefore \quad \oint p_r dr = \oint\left(\frac{p_\phi}{r^2}\right)\left(\frac{dr}{d\phi}\right)dr = n_r h$$

or $$p_\phi \oint \left(\frac{1}{r}\frac{dr}{d\phi}\right)^2 d\phi = n_r h \qquad ...(25.27)$$

In an ellipse $$\frac{1}{r} = \frac{1+e\cos\theta}{a(1-e)^2} \qquad ...(25.28)$$

where e is the eccentricity of the ellipse and a is the semi-major axis. On differentiation of Eq. (25.28), we get

$$\frac{1}{r^2}\frac{dr}{d\phi} = \frac{e\sin\phi}{a(1-e)^2} \qquad ...(25.29)$$

Dividing Eq. (25.29) by Eq. (25.28), we obtain

$$\frac{1}{r}\frac{dr}{d\phi} = \frac{e \sin\phi}{1+e\cos\phi} \qquad ...(25.30)$$

Using Eq. (25.30) into Eq. (25.27), one obtains

$$p_\phi \oint \left[\frac{e \sin\phi}{1+e\cos\phi}\right]^2 d\phi = n_r h$$

or

$$p_\phi \int_0^{2\pi} \frac{e^2 \sin^2\phi}{(1+e\cos\phi)^2} d\phi = n_r h \qquad ...(25.31)$$

The integration of the above expression yields

$$\frac{2\pi\, p_\phi}{\sqrt{1-e^2}} - 2\pi\, p_\phi = n_r h$$

Putting $p_\phi = \frac{n_\phi h}{2\pi}$ into the above equation and simplifying it, we get

$$\sqrt{(1-e^2)} = \frac{n_\phi}{n_r + n_\phi} = \frac{n_\phi}{n}$$

$$\therefore \qquad (1-e^2) = \frac{n_\phi^2}{n^2} \qquad ...(25.32)$$

For an ellipse we have the relation

$$(1-e^2) = \frac{b^2}{a^2} \qquad ...(25.33)$$

where a and b are the semi-major and semi-minor axes of the ellipse respectively. Comparing the Eq. (25.32) and Eq. (25.33), we get

$$\frac{n_\phi}{n} = \frac{b}{a} \quad \text{Condition for allowed elliptical orbits} \qquad ...(25.34)$$

Thus, the allowed elliptical electron orbits are those for which the ratio of minor to major axes is in the ratio of n_ϕ/n. A series of elliptical orbits of different eccentricities are thus allowed for values of $n_\phi = (n-1), (n-2), (n-3), \ldots\ldots 1$. The orbit is circular for $n_\phi = n$. However, n_ϕ cannot assume the value of zero since it would imply that the electron oscillates back and forth in a straight line penetrating through the centre of nucleus, which is most unlikely. Hence, n_ϕ can have in total n possible values from n to 1.

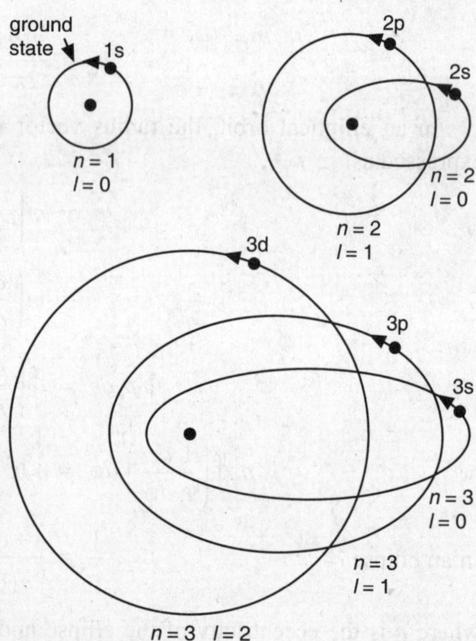

Fig. 25.11: Sommerfeld's extension elliptical orbits to Bohr's circular orbit

The net result of Sommerfeld's extension is that one elliptical orbit is added to Bohr's second orbit, two elliptical orbits are added to Bohr's third orbit; three elliptical orbits are added to Bohr's fourth orbit and so on.

25.10.3 Energy

It is shown that the total energy of the electron in an elliptical orbit is given by

$$E_n = -\frac{m e^4 Z^2}{8\varepsilon_o^2 h^2}\left(\frac{1}{n^2}\right) \quad \text{...(25.35)}$$

This energy is exactly identical to that of an electron in a Bohr' circular orbit. The introduction of elliptical orbits has increased the number of orbits but their energies being identical, the number of atomic energy levels has not increased. Therefore, additional electron transitions and spectral lines are not possible. It means that the concept of elliptical orbits could not explain the fine structure of spectral lines.

25.10.4 Relativistic Correction

Owing to the motion of an electron in an elliptical orbit, its velocity would not be uniform. When the electron comes close to the nucleus, its velocity must be much higher than its velocity in a circular orbit of the same energy. Similarly, electron velocity would be lower when it is farthest from the nucleus. Such variations in velocity cause variation of the electron mass according to the theory of relativity. Sommerfeld showed that this assumption of the relativistic variation of electron mass transforms the simple elliptical orbit into a complicated curve known as a rosette, as depicted in Fig. 25.12. As the electron describes the elliptical path, the major axis of the ellipse revolves slowly in the plane of the ellipse around the nucleus located at the focus of the ellipse.

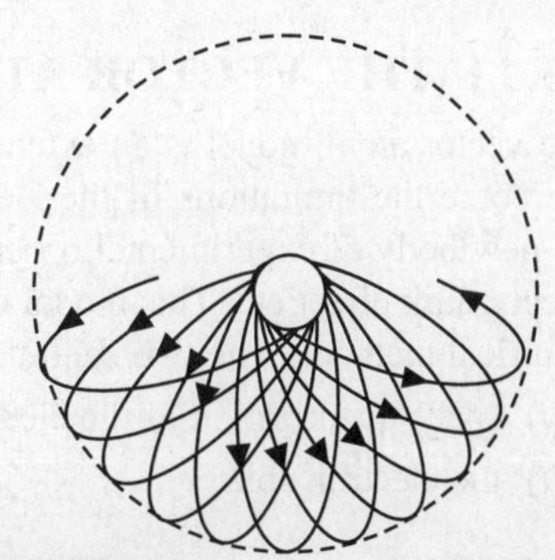

Fig. 25.12: Rosette- relativistic variation of electron mass transforms the simple elliptical orbit into a complicated curve

After relativistic correction, the energy of the electron in a particular state designated by quantum numbers n and n_ϕ is obtained as

$$E_n = -\frac{mZ^2 e^4}{8\varepsilon_o^2 n^2 h^2}\left[1+\frac{\alpha^2 Z^2}{n}\left(\frac{1}{n_\phi}-\frac{3}{4n}\right)\right] \quad \text{...(25.36)}$$

where α is called the **fine structure constant**.

The electron energy in an orbit depends not only on the principal quantum number n but also on the azimuthal quantum number n_ϕ. Consequently, transitions among different levels corresponding to the various possible n_ϕ values can occur. As the energy differences of these levels are very small, the wavelengths of the resulting spectral lines would be very close. Therefore, a group of closely associated lines is produced instead of a single line.

At first sight it would seem that the number of fine structure lines should be very large in view of the many possible values of n_ϕ. To restrict the possible number of allowed transitions, Sommerfeld proposed the following selection rule:

$$\Delta n_\phi = \pm 1.$$

Because of this **selection rule** the number of allowed electronic transititon and hence the actual spectral lines are limited in number.

25.10.5 Limitations of Sommerfeld Theory

(*i*) Sommerfeld theory gave the total number of possible azimuthal quantum numbers correctly, but the actual values are incorrect. Spectroscopic observations showed that the azimuthal quantum number can take the value of zero so that the total values are 0, 1, 2, 3, ..., $(n - 1)$. Therefore, a new azimuthal quantum number l is introduced where $l = (n_\phi - 1)$.

(*ii*) Sommerfeld theory provided a theoretical justification to the existence of fine structure of spectral lines but it failed in predicting the correct number of fine structure lines.

(*iii*) The theory does not give information of about the relative intensities of the lines.

(*iv*) Sommerfeld theory does not explain the distribution and arrangement of electrons in multielectron atoms.

25.11 THE VECTOR ATOM MODEL

The vector atom model is an extension of Bohr-Sommerfeld atom model. It was intended to overcome the limitations of the Bohr-Sommerfeld model while it also attempted to explain the new body of experimental observations such as anomalous Zeeman effect, Paschen-Back effect, Stark effect etc. This model incorporated new concepts partly by analogy and partly by empirical methods. The two central features of the vector atom model are:

(*i*) space quantization of the electron orbits, and

(*ii*) the electron spin.

25.11.1 Space Quantization

In Bohr's model of atom, the electron is assumed to move in circular orbits. The orbits are quantized as regards their magnitude, that is, their size. Thus, the electron in Bohr model has only one degree of freedom. One quantum number, namely the principal quantum number, n is enough to describe the electron motion. According to Sommerfeld theory, the electron moves in elliptical orbits, which are quantized as regards their magnitude and shape. Thus, the electron possesses two degrees of freedom. Hence, two quantum numbers, namely the principal quantum number, n and the azimuthal quantum number, n_ϕ are required to explain the motion of the electron in an atom. Description of the electron motion in terms of these two quantum numbers implies that the electron motion is confined to a single plane.

As an atom is a three-dimensional entity, the orbital plane of an electron may take different orientations in the atom. Classically speaking, the electron orbit may orient in all probable directions in space (Fig. 25.13). Sommerfeld extended the concept of quantization to the orientation of the electron orbits in space. According to Sommerfeld, out of the possible infinite orientations, only certain discrete orientations are allowed for the electron orbits.

Orientations of the orbit

The orientation of an orbit can be identified with the help of **orbital angular momentum vector,** $\boldsymbol{p_l}$, which is directed along the axis of rotation of the electron and is perpendicular to the plane of the orbit (see Fig. 25.14). We require a reference direction, with respect to which the orientation of the vector $\boldsymbol{p_l}$ could be defined. Such a reference direction may be

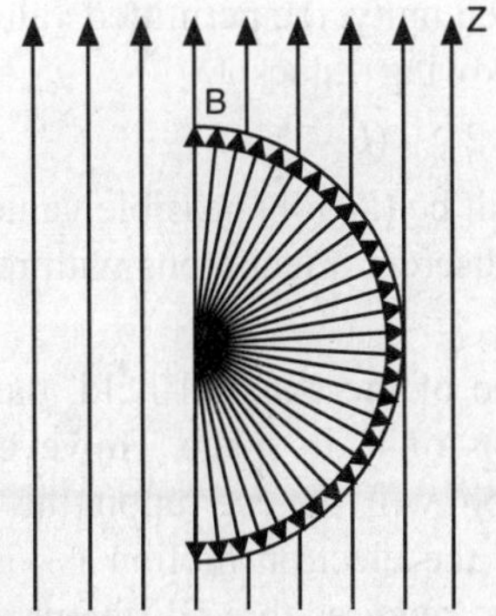

Fig. 25.13:

Probable directions of electron orbit in space

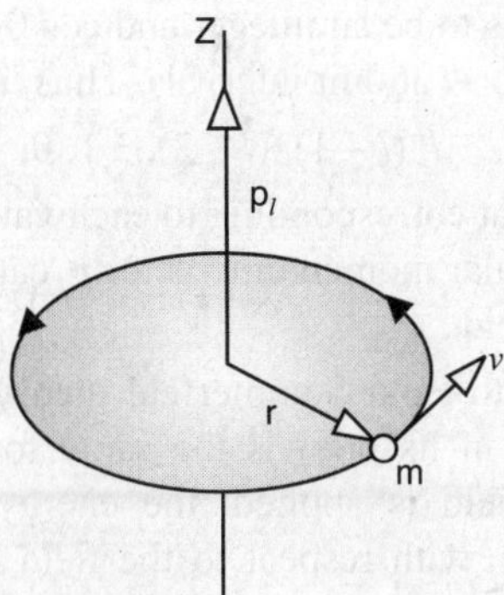

Fig. 25.14:

Orbital angular momentum vector is directed along the axis of rotation of the electron

obtained by *imagining* the atom to have been placed in an external magnetic field, **B**, whose strength tends to zero so that it may not disturb the electron orbits. Then, the magnetic field direction serves as the reference direction.

Generally, magnetic field acting along z-direction is taken as the **reference direction**. According to Sommerfeld, the vector p_l can orient itself only in certain discrete directions relative to the external magnetic field direction. This is known as **quantization of direction** or **space quantization**. Thus, in the vector atom model, the orbits are quantized in magnitude, shape and direction. The space quantization of an electron orbit is specified by the projection of its orbital angular momentum on to the reference direction; such projections being themselves quantized (see Fig. 25.15). In vector atom model, the orbital angular momentum of the electron is given by

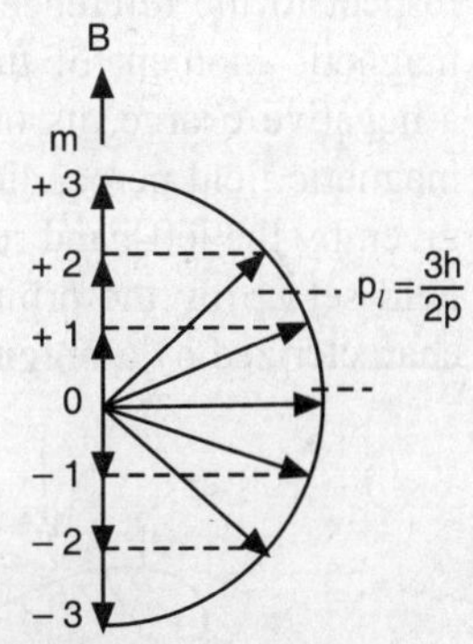

Fig. 25.15:

Allowed Orientations

$$p_l = \frac{lh}{2\pi} = l\hbar \qquad ...(25.37)$$

Allowed Orientations

According to the concept of space quantization, $\boldsymbol{p_l}$ can take only such orientations in space for which its component in the field direction **B** will take integral values of $\hbar$, see Fig. 25.15. Thus, $p_l \cos\theta$ is quantized and is given by

$$p_l \cos\theta = m_l \hbar$$

where m_l is called the **orbital magnetic quantum number** and θ is the angle between p_l and the field direction in space.

The largest component of $\boldsymbol{p_l}$ along the field direction is thus $l\hbar$ when $m_l = 1$. This value is less than the magnitude of $\boldsymbol{p_l}$ which means that the vector $\boldsymbol{p_l}$ cannot align in the direction of magnetic field. The angle between p_l and the z-axis is given by

$$\cos\theta = \frac{m_l\hbar}{l\hbar} = \frac{m_l}{l} \qquad ...(25.38)$$

Since m_l has to be an integer and cos θ can never exceed unity, the permitted values of m_l range from $+l$ to $-l$ at unit intervals. Thus, m_l takes the following values

$$l, (l-1), (l-2), \ldots 1, 0, -1, -2, \ldots -(l-2), -(l-1), -l.$$

It means that corresponding to each value of l, there will be $(2l + 1)$ possible values of m_l. Hence, the angular momentum vector $\boldsymbol{p}_l$ can take $(2l + 1)$ discrete orientations with respect to the reference field.

According to Bohr-Sommerfeld theory, in the absence of the external field, the energy of the electron in its orbit is the same for all orientations of $\boldsymbol{p}_l$ in space. However, when the reference field is applied, the energy of the electron will depend upon the relative orientation of $\boldsymbol{p}_l$ with respect to the field direction. Now the electron motion is visualized as three-dimensional motion. Space quantization, therefore, requires three different quantum numbers n, l and m_l. The energy of the electron now depends on all the three quantum numbers in an applied field.

Orbital magnetic moment: In fact, the discrete orientations of the electron orbit with respect to the reference magnetic field direction occur due to the interaction of the orbital magnetic moment of the electron and the reference magnetic field. Since an electron has a negative charge, its orbital motion is equivalent to a tiny current loop, which produces a magnetic field perpendicular to the plane of the orbit. The direction of the magnetic field is given by the left-hand rule and is pointed opposite to the direction of vector $\boldsymbol{p}_l$. The magnetic field set up by the orbital electron is quite similar to the field around a bar magnet and is characterized by a **magnetic dipole moment**.

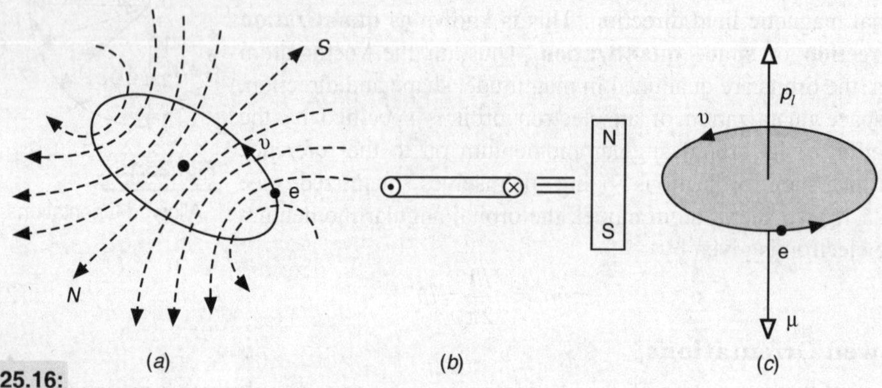

Fig. 25.16: Magnetic dipole moment

The magnetic moment of a current loop is given by

$$\mu = IA$$

The magnetic moment of the electron orbit can be expressed as

$$\mu_l = IA = \frac{e}{T}\pi r^2$$

where r is the radius of the electron orbit and T is the time period for electron revolution.

But $$\frac{2\pi r}{T} = \upsilon$$

Therefore, $$\mu_l = e\pi r^2\left(\frac{\upsilon}{2\pi r}\right) = \frac{e\upsilon r}{2}$$

Using $p_l = m\upsilon r$ into the above equation, we get

$$\mu_l = \left(\frac{e}{2m}\right)p_l = \left(\frac{e}{2m}\right)l\hbar \qquad ...(25.39)$$

All the symbols on the right side of the above equation are atomic constants. Hence, we write

$$\mu_l = \mu_B l$$

Bohr Magneton: is the basic unit of atomic magnetic moments.

$$\mu_B = \left(\frac{e\hbar}{2m}\right) \qquad ...(25.40)$$

The numerical value of Bohr magneton is given by

$$\mu_B = 9.273\times10^{-24}\,Am^2 \qquad ...(25.41)$$

Gyromagnetic Ratio: The ratio

$$G = \frac{\mu_l}{p_l} = \frac{e}{2m} \qquad ...(25.42)$$

is called the *gyromagnetic* ratio. It is written as

$$G = g\cdot\frac{e}{2m} \qquad ...(25.43)$$

where g is known as the **Lande splitting factor**. $g = 1$ for orbital motion of electron.

25.11.2 Electron Spin

In 1925, Uhlenbeck and Goudsmith put forward the hypothesis of the spinning electron, in order to explain some of the spectral phenomena such as fine structure, Zeeman effect etc. According to it, the electron revolves about its own axis while orbiting round the nucleus. The concept is very much akin to the spinning of the planets in the solar system. The spinning motion, like the orbital motion, is quantized not only in magnitude but also in direction according to the rules of space quantization.

The intrinsic **spin angular momentum** is given by

$$p_s = s\hbar \qquad ...(25.44)$$

where $s = \frac{1}{2}$ is known as the *spin quantum number*. The spin angular momentum is a vector.

As both the orbital angular momentum $\mathbf{p}_l$ and the intrinsic spin angular momentum $\mathbf{p}_s$ that determine the state of the electron are quantized vectors and are added as per the rules of vectors, the atom model is called **vector atom model**.

Allowed Orientations: Classically, a body can have spin in clockwise or anti-clockwise. Correspondingly, the spin angular momentum can take only two orientations with respect to the reference magnetic field direction, as illustrated in Fig. 25.17.

In the presence of magnetic field, the projection of p_s on to the reference direction is given by

$$p_{sz} = m_s\hbar \qquad ...(25.45)$$

By analogy with the orbital angular momentum vector, the spin angular momentum vector can take $(2s + 1)$ orientations in space relative to the reference magnetic field direction. As $s = \frac{1}{2}$, p_s can take only two orientations. It implies that the spin angular momentum $\boldsymbol{p_s}$ can align either parallel or antiparallel to the applied field direction. The spin magnetic quantum number m_s can have only two values, namely $+\frac{1}{2}$ or $-\frac{1}{2}$. They correspond to "spin-up" (parallel) and "spin-down" (antiparallel) orientations.

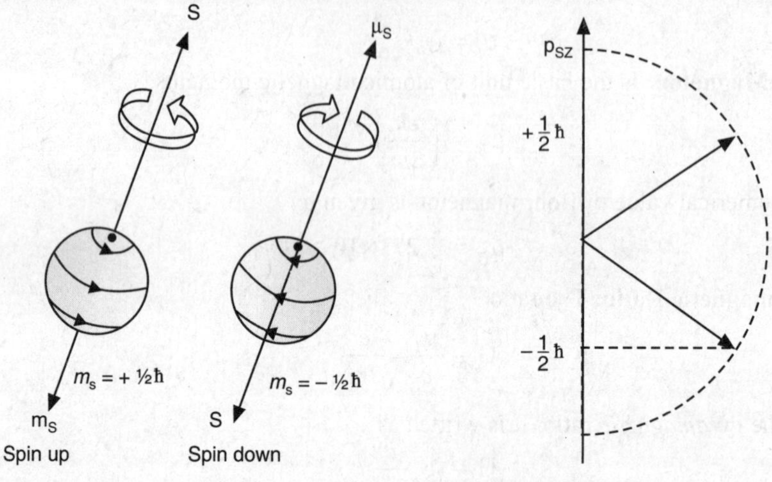

Fig. 25.17: Allowed Orientations with respect to the reference magnetic field direction

Intrinsic magnetic moment: The spinning of an electron around its own axis generates a magnetic field. This field is similar to the magnetic field around a bar magnet and is characterized by a magnetic moment. The magnetic moment μ_s of a spinning electron is given by

$$\mu_s = 2\frac{e}{2m}p_s$$

The Lande splitting factor $g = 2$ in case of spin motion.

$$\mu_s = 2\frac{e}{2m}\cdot\frac{\hbar}{2} = \frac{e\hbar}{2m} \qquad ...(25.46)$$

or $$\mu_s = 1\mu_B$$

Resultant angular momentum: Since an electron in an atom has two different angular momenta $\boldsymbol{p}_l$ and $\boldsymbol{p}_s$. The resultant angular momentum is obtained by the vector addition of $\boldsymbol{p}_l$ and $\boldsymbol{p}_s$. Thus,

$$\boldsymbol{p}_j = \boldsymbol{p}_l + \boldsymbol{p}_s \qquad ...(25.47)$$

where $\boldsymbol{p}_l = j\hbar$.. j is called the **total angular momentum quantum number.**

Thus, $$p_j = \frac{jh}{2\pi} = j\hbar \qquad ...(25.48)$$

For each value of l, there are two possibilities.

$$j\hbar = l\hbar + s\hbar \qquad ...(25.49a)$$

and $$j\hbar = l\hbar - s\hbar \qquad ...(25.49b)$$

As all the vectors have the common factor $\hbar$, the quantum numbers themselves are used as vectors. Thus, Eq. (25.49) may be written in simple terms as

$$j = l + s \text{ and } j = l - s \qquad ...(25.50)$$

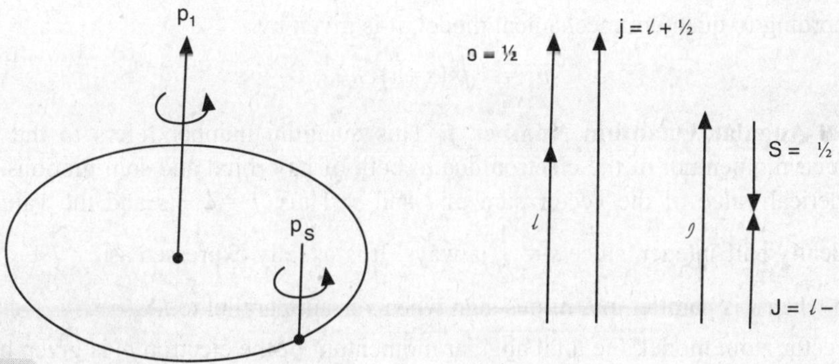

Fig. 25.18:
Resultant angular momentum of p_l and p_s.

For a single electron, $s = \frac{1}{2}$ and hence j can take one of the two values, namely $j = l + \frac{1}{2}$ or $j = l - \frac{1}{2}$. For example, if $l = 0$, j can have only one value, i.e., $j = \frac{1}{2}$. If $l = 1$, then j can be either 3/2 or 1/2.

25.11.3 Quantum Numbers Associated With the Vector Atom Model

Let us now sum up the various quantum numbers used in the vector atom model at one place. In this model, a quantum number is assigned to each of the component parts. The numerical value of the quantum number may be thought of as the length of the vector. The additions of the components are carried out as per the rules of vector analysis. A straight line, whose direction is parallel to the axis of rotation and whose length is proportional to the magnitude of the momentum, represents angular momentum. The quantum numbers associated with an electron in a given atom (in the absence of reference field) are as follows:

(*i*) **Principal Quantum Number, *n*:** An electron is characterized by a principal quantum number, n which describes the quantization of the electron energy. It can take only integral values, 1,2,3,……n.

(*ii*) **Orbital Quantum Number, *l*:** The shape of the orbit is characterized by l which may take any integral value between 0 and $(n - l)$. Thus, if $n = 4$, l can take the values 1,2,3.

In vector atom model, the orbital angular momentum of the electron is given by

$$p_l = \frac{lh}{2\pi} = l\hbar.$$

According to quantum mechanical model, it is given by

$$p_l = \sqrt{l(l+1)}\,\hbar \qquad ...(25.51)$$

It is a common practice to designate the states having different l values as follows.

Orbital quantum number	0	1	2	3	4	5
Designation of the state	*s*	*p*	*d*	*f*	*g*	*h*

(*iii*) **Spin Quantum Number, *s*:** The spin quantum number has a magnitude $\frac{1}{2}$ always. In the vector atom model, the electron spin angular momentum is given by

$$p_s = \frac{sh}{2\pi} = s\hbar.$$

According to quantum mechanical model, it is given by

$$p_s = \sqrt{s(s+1)}\ \hbar \qquad ...(25.52)$$

(*iv*) **Total Angular Quantum Number, *j*:** This quantum number refers to the resultant angular momentum of the electron due to both of its orbital and spin motions. It is the numerical value of the vector sum of l and s. Thus, $j = l + s$ and the value of j is evidently half-integer, since $s = \frac{1}{2}$ always. It is usually expressed as $j = l \pm s$; the plus sign when s is parallel to l, minus sign when s is antiparallel to l.

In vector atom model, the total angular momentum of the electron p_j is given by

$$p_j = \frac{jh}{2\pi} = j\hbar.$$

According to quantum mechanical model, it is given by

$$p_j = \sqrt{j(j+1)}\ \hbar \qquad ...(25.53)$$

When the atom is placed *in a reference magnetic field*, three more quantum numbers are associated with the electron due to space quantization.

(*v*) **Orbital magnetic quantum number, m_l:** m_l is the numerical value of projection of the orbital vector l in the reference field direction. It is an integer and takes a total of $(2l + 1)$ values ranging from $-l$ to $+l$ including zero. According to space quantization, the projection of l in the field direction must itself be quantized. Hence l can orient only at such discrete angles that its projection m_l may also be an integer. For example, if the vector l is inclined to the field direction at an angle θ, the projection is $m_l = l \cos \theta$. Now since m_l has to be an integer and $\cos \theta$ can never exceed unity, the permitted values of m_l are from $-l$ to $+l$ including zero.

The total number of permitted orientations of the l vector relative to the field direction is also $(2l + 1)$. Thus, if $l = 1$, the permitted orientations of l are three for which m_l has the values $+1, 0, -1$.

(*vi*) **Spin magnetic quantum number, m_s:** m_s is the numerical value of the projection of the spin vector **s** on the reference field direction. By analogy with the orbital angular momentum vector l, the spin vector s can have only $(2l + 1)$ values from $-s$ to $+s$ at unit intervals. Therefore, the value zero is excluded and m_s can take only two values $+\frac{1}{2}$ and $-\frac{1}{2}$.

(*vii*) **Total magnetic quantum number, m_j :** m_j is the numerical value of the projection of the total angular momentum vector ***j*** on the reference field direction. In case of a single electron, ***j*** can have only half-integral value, i.e., $l \pm \frac{1}{2}$. Consequently, m_j has to take only half-integral values. The permitted orientations of ***j*** with respect to the field direction are $(2j + 1)$ and hence m_j can have only $(2j + 1)$ values from $-j$ to $+j$, excluding zero.

25.11.4 Multielectron Atom

When there is more than one electron in an atom, the ***l*** vectors for the different electrons must be added up vectorially to obtain the resultant orbital angular momentum of the atom.

$$L = \sum_i l_i = l_1 + l_2 + l_3 + ...(25.54)$$

where l_1, l_2, l_3,... are the orbital angular momenta of the individual electrons in the atom. From now onwards, we use the symbol l in place of $\mathbf{p}_l$.

The value of the total orbital quantum number L of the atom must be an integer. For example, if $l_1 = 2$ and $l_2 = 1$, then L can have one of the values 3, 2, or 1 (see Fig. 25.19).

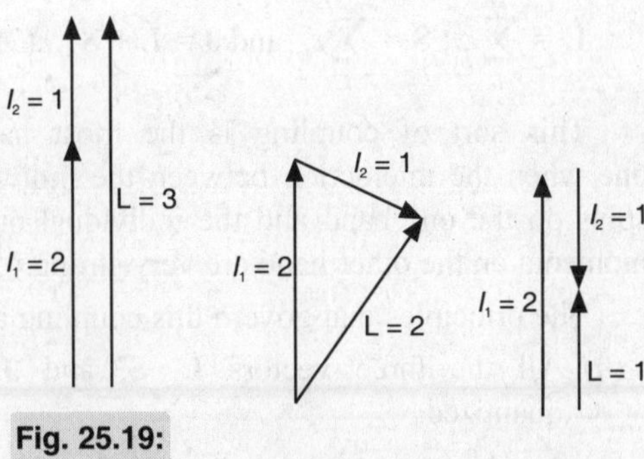

Fig. 25.19:

Total orbital quantum number

The resultant spin angular momentum is given by

$$S = \sum_i s_i = s_1 + s_2 + s_3 + ... \quad ...(25.55)$$

where s_1, s_2, s_3,........are the spins of the individual electrons in the atom. Each of these is equal to $\frac{1}{2}$.

The rule for the vector addition of the spins of the different electrons requires that these can be either parallel or antiparallel to one another. Therefore, the value of S can be either zero or an integer, if the atom has an even number of electrons. It will be half-odd integer, if the atom has an odd number of electrons. These two cases are illustrated in Fig. 25.20.

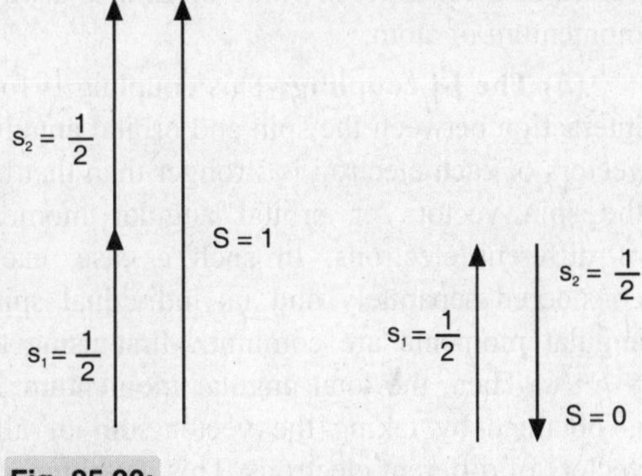

Fig. 25.20:

Vector addition of the spins

25.11.5 Coupling Schemes

There are several ways in which the different vectors of the electrons may combine to give the vectors representing the atom as a whole. The method of combination depends on the interaction or "coupling" between the component vectors, since the orbital and spin motions of the electron produce magnetic fields and thereby result in mutual perturbation. It is usual to distinguish two types of coupling known as the *L-S coupling* and *j-j coupling*.

(*i*) L-S Coupling: It is also known as Russell-Saunders coupling. It is the most frequently occurring coupling and hence is known as the *normal coupling*. In this, all the spin vectors of the electrons combine to form a resultant vector **S** and all the orbital angular momentum vectors of the electrons likewise combine to form a resultant vector **L**. Then the resultant vectors **S** and **L** combine to produce the total angular momentum **J** of the atom. The scheme may be represented as

$$\mathbf{L} = \sum_i l_i \,;\ \mathbf{S} = \sum_i s_i \text{ and } \mathbf{J} = \mathbf{L} + \mathbf{S} \qquad ...(25.56)$$

This sort of coupling is the most natural one when the interaction between the individual spins on the one hand and the individual orbital momenta on the other hand are very strong.

The principles that govern this coupling are:

(*a*) All the three vectors **L**, **S**, and **J** are quantized.

(*b*) **L** is always an integer. It may take values 0, 1, 2, 3,….etc

(*c*) **S** is an integer or half-integer depending on the number of electrons involved and the direction of the spin vectors. Thus, for an atom containing a single electron **S** can have only one value $\frac{1}{2}$; for a two electron system **S** may be either 1 or 0 depending on whether the spin vectors are parallel or antiparallel and so on.

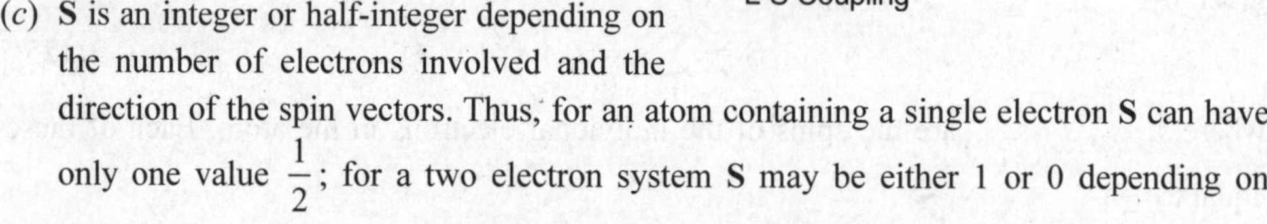

Fig. 25.21: L-S Coupling

(*d*) **J**, the vector sum of **L** and **S** must be an integer (0,1, 2, 3,..) for odd electron systems and it is an half-integer ($\frac{1}{2}, \frac{3}{2}, \frac{5}{2}$, ….) for even electron systems. In general, the possible number of values which J can assume are

$$(2\mathrm{S} + 1) \text{ if } \mathrm{L} \geq \mathrm{S}$$

and $$(2\mathbf{L} + 1) \text{ if } \mathrm{L} < \mathrm{S}.$$

The quantity (2**S** + 1) is known as the multiplicity of the **L** state. It gives the permitted values of **J** for a given value of **L**. **J** is always positive since it represents the total angular momentum of atom.

(*ii*) The j-j coupling: This coupling is found when the interaction between the spin and orbital angular momentum vectors of each electron is stronger than that between either the spin vectors or orbital angular momentum vectors of different electrons. In such a case, each electron is considered separately and its individual spin and orbital angular momenta are combined first using the relation, $j = l + s$. Then, the total angular momentum **J** of the atom is obtained by taking the vector sum of all individual j vectors of different electrons. This coupling scheme may be symbolically represented as

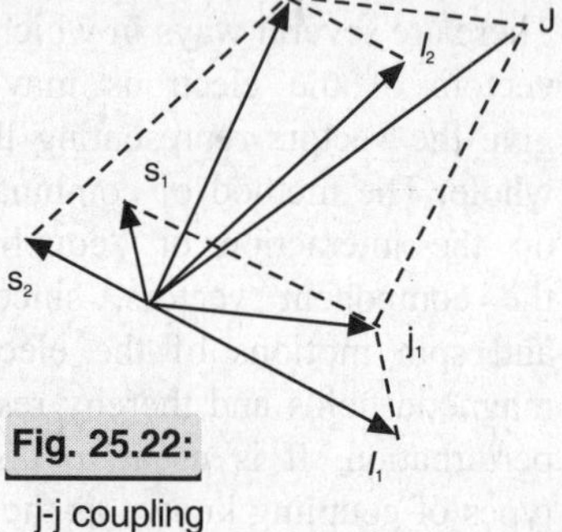

Fig. 25.22: j-j coupling

$$j_i = l_i + s_i \text{ and } \mathrm{J} = \sum_i j_i \qquad ...(25.57)$$

This type of coupling exists mainly in heavy atoms.

25.11.6 Pauli's Exclusion Principle

Pauli's exclusion principle is an empirical rule, which is used in conjunction with the vector atom model to explain the electronic structure in an atom. Among the several quantum numbers associated with the electron, four quantum numbers, namely n, l, m_l and m_s are needed in order to specify completely its state. According to exclusion principle, every completely defined quantum state in an atom can be occupied by only one electron. In other words, it is impossible for two electrons in an atom to be identical as regards all their quantum numbers. It is variously stated as "no two electrons in an atom can occupy the same quantum state" or "no two electrons in an atom can have the same four quantum numbers."

The exclusion principle explains certain experimental observations such as the occurrence or non-occurrence of spectra lines and the scheme of arrangement of electrons in atoms etc.

25.11.7 Selection Rules and Intensity Rules

It has been deduced from spectroscopic observations that electronic transitions from each energy state to all other states do not occur. In the vector model of atom, three selection rules have been devised and they are supplemented by the intensity rules in order to predict the intensity of the lines that occur.

(*i*) Selection Rules: The rules that govern the electronic transitions are known as *selection rules*. They are devised one for L, another for J, and a third for S.

(*a*) *The selection rule for L*: Most of the observed spectral lines are due to transitions between states, in which a single electron jumps from one orbit to another. The rule that governs the transitions is that $\Delta L = \pm 1$. Only those transitions are observed for which the value of L changes by ±1.

(*b*) *The selection rule for J*: Spectral lines are observed only when transitions take place between states for which $\Delta J = \pm 1$ or 0. The transition $0 \rightarrow 0$ is not permitted.

(*c*) *The selection rule for S*: Transitions take place between states for which $\Delta S = 0$.

(*ii*) Intensity Rules: Whether a spectral line due to an allowed transition is weak or intense is determined by the intensity rules. They are as follows:

(*a*) A spectral line will be *intense* for which electron transition takes place in such a way that L and J change in the same sense.

(*b*) A spectral line will be *weak* for which electron transition L and J change in the opposite sense.

(*c*) A spectral line will be *intense* for which electron transition L and J decrease, that is, $L \rightarrow (L-1)$ and $J \rightarrow (J-1)$.

(*d*) A spectral line will be *less intense* for which electron transition L and J increase, that is, $L \rightarrow (L+1)$ and $J \rightarrow (J+1)$.

(*e*) The oppositely directed transitions do not occur.

The above rules may be summarized as follows:

$\Delta L = -1$,	$\Delta J = -1$	Very intense line
$\Delta L = -1$,	$\Delta J = 0$	Less intense line
$\Delta L = +1$,	$\Delta J = +1$	Weak line
$\Delta L = +1$,	$\Delta J = 0$	Weak line
$\Delta L = -1$,	$\Delta J = +1$	No transition
$\Delta L = +1$,	$\Delta J = -1$	No transition

25.12 APPLICATIONS OF THE VECTOR ATOM MODEL

The vector atom model provided a rational explanation of several complex atomic and spectral phenomena. Out of them, the most important are (*i*) the electronic structure in atoms

(*ii*) the Zeeman effect

(*iii*) anomalous Zeeman effect

(*iv*) the Stark effect

We now study these phenomena and understand how vector atom model succeeded in explaining them.

25.13 THE ELECTRONIC STRUCTURE OF ATOMS

An atom of a certain kind cannot be said to have a definite size but from a practical point of view a definite size is attributed to it. Thus, hydrogen atom is supposed to have a diameter of about 1Å.

The electrons in an atom occupy energy or orbitals characterized by the quantum numbers n, l and m_l. A set of these numbers defines the **state** of the electron. The states (or orbitals) are organized into subshells and shells. The possible orbitals corresponding to a particular value of n are said to constitute a **shell.** All the electrons in an atom which have the same principal quantum number n belong to the electron shell. The electrons having $n = 1$ are said to form the K shell of the atom; electrons in the $n = 2$ state form the L shell and so on. The designation of the electron shells is as follows,

Principal quantum number, n	1	2	3	4	5	6
Designation of the electron shell	K	L	M	N	O	P

A shell characterized by the principal quantum number n accommodates a maximum of $2n^2$ electrons. The value of n increases as the distance of a shell from the nucleus increases. The number of electrons that a shell can accommodate increases in accordance with the value of n. Thus, a K shell with $n = 1$ accommodates 2 electrons, an L shell with $n = 2$ accommodates 8 electrons and so on. The maximum capacities of the shells are given below.

Electron shell	K $n = 1$	L $n = 2$	M $n = 3$	N $n = 4$	O $n = 5$	P $n = 6$
Maximum capacity, $2n^2$	2	8	18	32	50	72

Shells are built from subshells which accommodate electrons of the same value of the orbital quantum number l. Thus, electrons that share a certain value of l in a shell said to occupy the same subshell. The number of subshells in a shell is equal to the value of n. Thus a K shell has only one subshell corresponding to $l = 0$, an l shell has two subshells and so on. In atomic physics, states with a particular value of l have a particular name. A state with $l = 0$ is called an s state; a state with $l = 1$ is called a p state and so on. The designations are listed in the table below;

Orbital quantum number, l	0	1	2	3	4	5
Designation of the state or subshell	s	p	d	f	g	h

For a given set of values of n and l, there are $(2l + 1)$ possible values of m_l. It means that for a given value of n and l, there are $(2l + 1)$ orbitals or electron **states.** Thus, if $l = 0$, there is only one electron state; if $l = 1$, three electronic states etc. The electronic states correspond to the different orientations of the orbitals.

An energy state described by the three quantum numbers n, l and m_l can be occupied by only two electrons which have opposite spin directions. This is known as **Pauli's exclusion**

principle. This principle indicates why electrons do not crowd into the lowest energy state. The electron spin is characterized by quantum number m_s. The state of the electron described by the four quantum numbers n, l, m_l and m_s is called the ***quantum state.*** Thus each energy state consists of two quantum states.

The terms energy level and energy state are used often in the same meaning. It may be noted that an energy level is not equivalent with an electronic state. An energy level is determined by the value of the principal quantum number n, and such a level corresponds to n^2 electronic states. Thus, for $n = 2$, there are $2^2 = 4$ different electronic states. The energy of an electron is mainly determined by the value of n, but to some extent by the quantum numbers l and m_l also.

The occupancy of various subshells and shells in a complex atom is governed by three basic rules;

(*i*) **Minimum energy condition.** Electrons tend to occupy the lowest available energy state such that their total energy is a minimum. They go to higher energy states only when the lower energy states are not vacant.

(*ii*) **Pauli's exclusion principle.** An orbital in an atom described by the quantum numbers n, l and m_l can be occupied by only two electrons having opposite spin directions.

(*iii*) **Hund's rule.** The order of filling of the orbitals of subshells obeys Hund's rule. According to this rule the *total spin number of the electrons of a shell must be maximum.* It means that the orbitals of a subshell are filled first with one electron each and then with the second electron respectively.

For instance, in nitrogen atom, there are three electrons in $2p$ subshell. Three electrons occupy the three orbitals p_x, p_y and p_z, instead of occupying only p_x and p_y orbitals.

The sequence of the energy states in order of increasing energy of the orbitals of a multi electron atom is in the following order.

$1s < 2s < 2p < 3s < 3p < 4s < 3d < 4p < 5s < 4d < 5p < 6s < 4f < 5d < 6p.$

As each orbital characterized by the set of quantum numbers n, l, m_l can accommodate two electrons having m_s values $+\frac{1}{2}$ and $-\frac{1}{2}$, the electron capacity of a subshell is a maximum of $2(2l + 1)$ electrons.

Therefore, the maximum capacity N of a shell is given by,

$$N = \sum_{l=0}^{n-1} 2(2l+1) = 2\,[1 + 3 + 5 + \ldots\ldots\ldots + 2(n-1) + 1]$$

$$= 2[1 + 3 + 5 + \ldots\ldots\ldots + 2(n-1) + 1]$$

$$= 2 \times \frac{n}{2}\,[1 + 2(n-1)]$$

$$= 2n^2 \qquad \ldots(25.58)$$

The Table 25.2 shows at a glance the capacities of individual orbitals and the shell in total.

Table 25.2: *Capacities of individual orbitals and the shell in total*

n	*shell*	*l=0*	*Electron capacity*	*l=1*	*Electron capacity*	*l=2*	*Electron capacity*	*l=3*	*Electron capacity*	*l=4*	*Electron capacity*	*l=5*	*Electron capacity*	*Maximum capacity of shell* $2n^2$
1	K	1s	2	-	-		-		-	-	-		-	2
2	L	2s	2	2p	6	-	-	-	-	-	-		-	8
3	M	3s	2	3p	6	3d	10		-	-	-		-	18
4	N	4s	2	4p	6	4d	10	4f	14				-	32
5	O	5s	2	5p	6	5d	10	5f	14	5g	18	-	-	50
6	P	6s	2	6p	6	6d	10	6f	14	6g	18	6h	22	72

The electron configuration of an atom is described using nl^x notation, where x denotes the number of electrons occupying the subshell. For example, we write the electron configuration of the sodium atom. $1s^2$, $2s^2$, $2p^6$, $3s^1$.

A shell or subshell that contains its full quota of electrons is said to be **closed** or **completely filled.** Thus, the K shell and L shell of sodium atom are closed while the M shell is partially filled. The 3s subshell of M shell contains only one electron and is half filled.

The completely filled shells and subshells of atoms are stable and are not readily disturbed. The electrons in the outer shell will be in a position to interact with similar electrons in adjacent atoms. The number of such electrons determines the **valency** of the atom and the chemical behaviour of the element.

25.14 ZEEMAN EFFECT

One of the basic concepts on which vector atom model rests is the concept of space quantization. Sommerfeld proposed the concept of space quantization and the Zeeman effect is considered as one of the experiments that confirmed the existence of space quantization.

In 1896, Zeeman found that if a source of light producing line spectrum is placed in a magnetic field, the spectral lines are split up into a number of components distributed symmetrically about the original lines. The splitting of spectral lines under the action of a magnetic field is called as **Zeeman Effect**. Splitting of the spectral lines takes place because the energy levels are split when a magnetic field acts on atoms.

Zeeman observed that in the presence of a strong magnetic field of the order of 1 tesla (104 gauss), each spectral line is split into two components when viewed along the field direction and into three components when viewed in a direction perpendicular to the magnetic field direction. This is known as **normal Zeeman Effect.**

Experimental arrangement: The normal Zeeman Effect can be readily observed. The experimental arrangement is shown in Fig. 25.23. It consists of an electromagnet capable of producing strong magnetic field. A source of light such as a sodium flame or a mercury arc is placed between the pole pieces of the electromagnet. The pole pieces are conical in shape. A spectroscope of high resolving power is used to study the light coming from the source. To enable viewing the light parallel to the field (longitudinal direction) a hole is drilled lengthwise in one of the pole pieces.

First, the spectroscope is focused on one of the spectral lines emitted by the source of light without switching on the magnetic field. When the magnetic field is switched on, the spectral line is seen to be split into three components with the original line. One of the lines is in the same position as the original line and the other two lines are located symmetrically about the original line. The splitting is illustrated in Fig. 25.24 (*b*). When the outer two lines are viewed through a Nicol prism, it was found that they are polarized at right angles to the undisplaced line. This effect is sometimes called *transverse Zeeman Effect*.

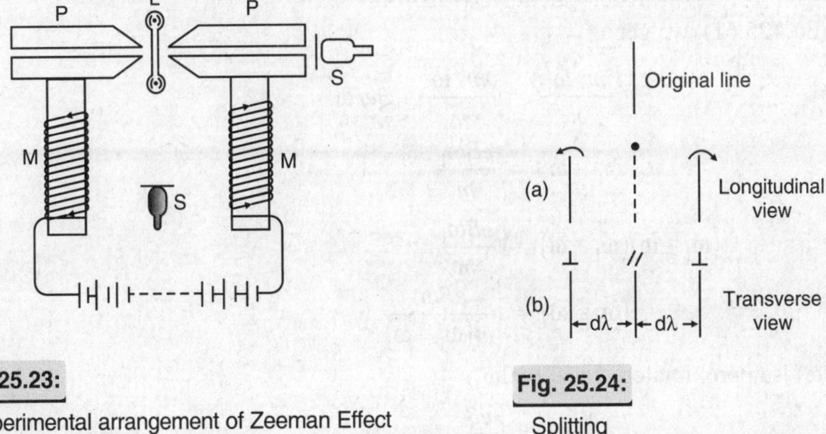

Fig. 25.23: Experimental arrangement of Zeeman Effect

Fig. 25.24: Splitting

When the source of light is viewed in a direction parallel to the magnetic field direction, spectral line is not found in the position of the original line. However, two outer lines are present indicating that the original line is split into two lines (Fig. 25.24 a). When the outer two lines are examined with a Nicol prism, it is revealed that they are circularly polarized in opposite directions. This result is called as *longitudinal Zeeman Effect*.

25.14.1 Classical Explanation of Zeeman effect by Lorentz on the Basis of Electron Theory

A satisfactory explanation of the normal Zeeman effect was given by H.A. Lorentz on the basis of the electron theory.

Let us consider an electron moving in a circular orbit of radius r with a velocity υ. The centripetal force acting on the electron is

$$F = \frac{m\upsilon^2}{r}. \qquad ...(25.59)$$

Fig. 25.25: Lorentz explanation of the normal Zeeman Effect

When an external magnetic field is applied, an additional force acts on electron perpendicular to the direction of the magnetic field and along the radius of the orbit. If the force acts outward along the radius, the velocity of the electron decreases and it increases if it acts inward.

Let the force due to magnetic field be and let the velocity increase to due to this force. Then,

$$F_1 = Be\upsilon_1 \qquad ...(25.60)$$

As this force is in a direction to add to the initial force, the total force on the electron now is

$$F_R = F + F_1$$

or $$\frac{m\upsilon_1^2}{r} = \frac{m\upsilon^2}{r} + Be\upsilon_1 \qquad ...(25.61)$$

But $$\upsilon = \omega r \text{ and } \upsilon_1 = \omega_1 r$$

where ω and ω_1 are the respective angular velocities.

From Eq. (25.61), we get

$$\frac{mr^2\omega_1^2}{r} = \frac{mr^2\omega^2}{r} + Ber\omega_1$$

or $$\omega_1^2 - \omega^2 = \frac{eB\omega_1}{m} \qquad ...(25.62)$$

or $$(\omega_1 - \omega)(\omega_1 + \omega) = \frac{eB\omega_1}{m}$$

or $$(\omega_1 - \omega) = \frac{eB\omega_1}{m(\omega_1 + \omega)}$$

$(\omega_1 + \omega)$ is approximately equal to $2\omega_1$.

$$\therefore \quad (\omega_1 - \omega) = \frac{eB}{2m}$$

$$\therefore \quad \omega_1 = \omega + \frac{eB}{2m} \qquad ...(25.63)$$

If ν_1 and ν are the corresponding frequencies, then $\omega_1 = 2\pi\nu_1$ and $\omega = 2\pi\nu$. Then,

$$2\pi\nu_1 = 2\pi\nu + \frac{eB}{2m}$$

or $$\nu_1 = \nu + \frac{eB}{4\pi\, m} \qquad ...(25.64)$$

When the electron revolves in an opposite direction, the magnetic field produces a force in the opposite direction as a result of which its velocity decreases to υ_2. In that case, the total force on the electron is

$$F_R = F - F_2$$

$$\therefore \quad \frac{m\upsilon_2^2}{r} = \frac{m\upsilon^2}{r} - Be\upsilon_2$$

or $$\frac{mr^2\omega_2^2}{r} = \frac{mr^2\omega^2}{r} - Ber\omega_2$$

$$\therefore \quad \omega_2^2 - \omega^2 = -\frac{eB\omega_2}{m}$$

$$\therefore \quad (\omega_2 - \omega) = -\frac{eB}{2m}$$

$$\therefore \quad \nu_2 = \nu - \frac{eB}{4\pi\, m} \qquad ...(25.65)$$

Subtracting Eq. (25.65) from Eq. (25.64), we obtain

$$\nu_1 - \nu_2 = \frac{eB}{4\pi\, m}$$

or $$\Delta\nu = \frac{eB}{4\pi m} \qquad ...(25.66)$$

The quantity $\frac{eB}{4\pi m}$ is known as normal Zeeman separation. Knowing the values of ν and B, the ratio e/m can be calculated. The value of e/m calculated from these measurements is in satisfactory agreement with the value 1.75×10^{11} C/kg obtained from Thomson's experiment.

This effect established that electrons emit the spectral lines.

25.14.2 Explanation of Zeeman Effect by Debye on the Basis of Vector Atom Model

P. Debye explained the Zeeman effect on the basis of vector atom model. He, however, did not take into account the concept of electron spin.

Let us consider an electron revolving in an orbit in an atom. Its orbital angular momentum is given by $p_l = l\hbar$. The magnetic moment of the orbit μ_l is directed opposite to p_l and is related to it through the expression

$$\mu_l = -\left(\frac{e}{2m}\right)p_l$$

$$\therefore \qquad \mu_l = -\left(\frac{eh}{4\pi m}\right)l$$

Let the energy of the electron in its orbit be E. When a magnetic field of induction B is applied, the orbital angular momentum vector l precesses around the magnetic field direction with the Larmor frequency. The change in the energy of the electron is now given by

$$dE = -\mu_l \cdot B = \left(\frac{eh}{4\pi m}\right) lB\cos\theta \qquad ...(25.67)$$

where θ is the angle between l and B. θ is given by

$$\cos\theta = \frac{m_l}{l}$$

where m_l is the projection of l on to B.

$$\therefore \qquad dE = \left(\frac{eh}{4\pi m}\right) m_l B$$

or $$dE = m_l \mu_B B \qquad ...(25.68)$$

The Eq. (25.68) suggests that, in the absence of the magnetic field, the states differing in the values of the quantum number m_l have identical energy determined by the quantum numbers n and l. Consequently the electron orbit does not prefer any particular orientation. In the presence of a magnetic field, the energy depends on the value of m_l as well as on that of l. A state characterized by orbital quantum number l breaks up into $(2l + 1)$ substates due to the magnetic field, since m_l can take $(2l + 1)$ values. The substates will have slightly more or slightly less energy than the energy of the state in the absence of the field. All this means that the electron orbits prefer only certain orientations with respect to magnetic field and the energy of the orbit varies from orientation to orientation by a discrete amount $\mu_l B$.

If E_1 represents the energy of a state of the atom before the application of the magnetic field and E_A its energy after applying the magnetic field, then

$$E_A = E_1 + (m_l)_1 \mu_B B \quad ...(25.69)$$

If E_2 represents the energy of a second state which changes to E_B after applying the magnetic field, then

$$E_B = E_2 + (m_l)_2 \mu_B B \quad (25.70)$$

Both E_A and E_B are multiple states, as $(m_l)_1$ and $(m_l)_2$ can take $(2l_1 + 1)$ and $(2l_2 + 1)$ values respectively. Several components are likely to arise due to transitions between these various states. The frequency ν of any of these components may be computed from the Eq. (25.69) and Eq. (25.70). Thus,

$$\nu = \frac{(E_B - E_A)}{h} \quad (25.71)$$

$$= \frac{E_2 - E_1}{h} + \frac{(m_l)_2 - (m_l)_1}{h} \mu_B B$$

$$= \nu_o + \Delta m_l \left(\frac{eB}{4\pi m} \right) \quad (25.72)$$

where $\Delta m_l = (m_l)_2 - (m_l)_1$. $\nu_o = \frac{E_2 - E_1}{h}$ represents the frequency of the line in the absence of the magnetic field.

The allowed transitions are determined by the selection rule $\Delta m_l = 0, \pm 1$. Using this selection rule, we find that the allowed transitions are only the following three.

$$\left.\begin{aligned} \nu_1 &= \nu_o - \frac{eB}{4\pi m} && \text{corresponding to } \Delta m_1 = -1 \\ \nu_1 &= \nu_0 && \text{corresponding to } \Delta m_1 = 0 \\ \nu_1 &= \nu_o + \frac{eB}{4\pi m} && \text{corresponding to } \Delta m_1 = +1 \end{aligned}\right\} \quad (25.73)$$

The frequency shift due to the magnetic field is

$$\Delta \nu = \frac{eB}{4\pi m} \quad (25.74)$$

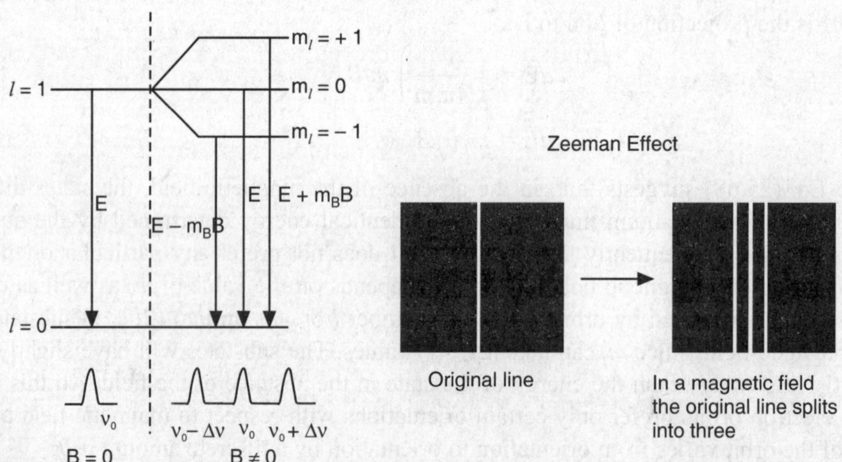

Fig. 25.26:

Splitting of the levels and spectral lines- Zeeman Effect

Fig. 25.26 shows the splitting of the levels and spectral lines for the transition between the states $l - 1$ and $l - 0$. In the absence of the magnetic field, one spectral line is observed whose frequency is denoted by ν_o. When the magnetic field is switched on, two lines appear in addition to the line ν_o. The two lines having frequencies $\nu_0 + \Delta\nu$ and $\nu_0 - \Delta\nu$ are symmetrically located relative to the initial line ν_0. The satisfactory agreement between the measured values of frequency shift and the values calculated from the Eq. (25.74) provided supporting evidence to the concept of space quantization. Thus, it confirmed that the electrons in an atom take on certain specified orientations in space and that these orientations are determined by the angular momentum projections on to the field direction.

Example 25.3. *Determine the normal Zeeman splitting of the cadmium red line of 6438Å when the atoms are placed in a magnetic field of 0.009 T.*

Solution: We have $E = \dfrac{hc}{\lambda}$. $\therefore\ dE = -hc\dfrac{d\lambda}{\lambda^2}$ or $|dE| = hc\dfrac{|d\lambda|}{\lambda^2}$ $\therefore\ |d\lambda| = \dfrac{\lambda^2 |dE|}{hc}$

$$\therefore \quad |dE| = \Delta E_{Zeeman} = \frac{eh}{4\pi\, m}B = \left(5.79\times10^{-5}\frac{\text{eV}}{T}\right)(0.009\ \text{T}) = 5.21\times10^{-7}\text{eV}$$

$$|d\lambda| = \frac{\lambda^2|dE|}{hc} = \frac{\left(6438\times10^{-10}\,\text{m}\right)^2\left(5.21\times10^{-7}\,\text{eV}\right)}{12.4\times10^3\,\text{eVÅ}} = \mathbf{1.74 \times 10^{-3}\ Å.}$$

25.15 THE STERN-GERLACH EXPERIMENT

In 1921 O. Stern and W. Gerlach directly measured the magnetic moments of atoms. These celebrated experiments provided direct proof for the space quantization of the angular momentum and for the spin of the electron.

Experimental Arrangement: Fig. 25.27 shows a schematic of their experimental set up. A beam of silver atoms is produced by evaporating silver in a small electric oven. The beam is collimated by the slit S and passed through a very inhomogeneous magnetic field produced between two pole pieces and strikes a photographic plate. The pole pieces are 6 cm long and arranged 2 mm apart. One of the pole pieces is shaped in the form of a knife-edge and the other has a cylindrical groove, the lines of force converge very strongly towards the sharp edge. The magnetic field is thus highly non-uniform, having a gradient $\dfrac{dB}{dz}$ parallel to the flux density B acting in the z-direction.

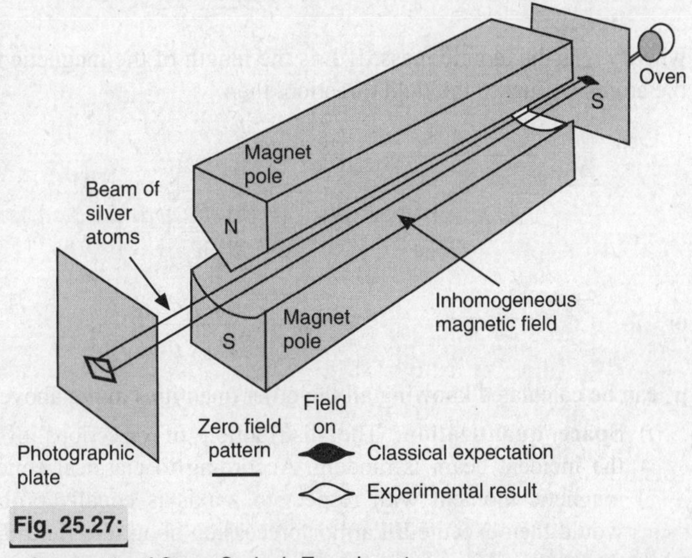

Fig. 25.27:

Schematic of Stern-Gerlach Experiment

Principle: The experiment is based on the behaviour of a magnetic dipole in a non-uniform magnetic field. In a uniform magnetic field, the magnetic dipole merely experiences a torque that tends to align the dipole in the field direction. If a beam of atomic magnets pass through a homogeneous magnetic field in a direction normal to the field, the beam will travel along a straight line path without any deviation. In an inhomogeneous magnetic field, the dipoles rotate into the field direction and in addition experience a translatory force in the direction of the field. Therefore, the atomic beam gets deflected along the field direction.

Let the magnetic field vary along the z-direction so that the field gradient is $\frac{dB}{dz}$ and is positive. Let MN be the atomic magnet having a pole strength m, length l and dipole moment $\mu = ml$. Let MN be inclined at an angle θ to the field direction. If the field strength at the pole M is B, then the field strength at the other pole N will be $B+\frac{dB}{dz}l \cos\theta$. Hence, the forces acting on the two poles of the magnet are now different and are given by mB and $m\left(B+\frac{dB}{dz}l \cos\theta\right)$. The translatory force on the magnet is

$$F_z = \frac{dB}{dz} ml \cos \theta$$

or

$$F_z = \mu_z \frac{dB}{dz} \qquad ...(25.75)$$

where $\mu_z = ml \cos \theta$ is the projection of the magnetic moment on to the field direction. This force deflects the atoms up or down depending on whether μ_z is positive or negative.

Expression for displacement in the field direction: The acceleration, a, acquired by the atomic magnet along the field direction = $\frac{F_z}{m_A}$.

The displacement of the atom along the field direction, $D = \frac{1}{2}at^2 = \frac{1}{2}\left(\frac{F_z}{m_A}\right)t^2$

where m_A is the atomic mass. If L is the length of the magnetic field and υ is the velocity of the atoms normal to the field direction, then

$$t = \frac{L}{\upsilon}$$

$$\therefore \quad D = \frac{1}{2}\left(\frac{F_z}{m_A}\right)\left(\frac{L}{\upsilon}\right)^2$$

or

$$D = \frac{\mu_z}{2m_A}\left(\frac{dB}{dz}\right)\left(\frac{L}{\upsilon}\right)^2 \qquad ...(25.76)$$

μ_z can be calculated knowing all the other quantities in the above equation.

(*i*) **Space quantization**: The distribution of directions of the magnetic moments in the incident beam is random. According to classical concepts, any orientation of the magnetic moment with respect to z-axis is equally probable. Therefore, each atom would then execute a Larmor precession about the field direction at some tilt angle θ. Since all values of θ occur among the atoms, the projection of the magnetic moment may have any value from $-\mu_z$ to $+\mu_z$, including zero value. Since there are several atoms in the beam, their deflections would be distributed in a continuous fashion and the trace formed by them on the photographic plate should be a *continuous band*.

But in the Stern-Gerlach experiment, two discrete traces are found for silver or hydrogen atomic beam. That is, the atomic beam is divided into two, each being deflected by the same amount but in opposite directions. It suggests that the z-projection of the magnetic moment can have only two values, equal in magnitude and opposite in direction. It proves that the classical theory, which expects equally probable orientation of the magnetic moment in random directions, is wrong. Discrete orientations are expected according to vector atom model. Owing to space quantization restriction, each atomic magnet can take up only certain specific orientations in the magnetic field. It appears in case of silver or hydrogen atoms, only two orientations are allowed. They correspond to μ_z parallel to B and μ_z antiparallel to B. This result establishes the existence of space quantization of angular momentum.

(*ii*) **Electron spin**: By the Stern-Gerlach experiment, the magnetic moment of the silver atom was found to be equal to one Bohr magneton.

$$\mu = \frac{eh}{4\pi m} = \mu_B \qquad ...(25.77)$$

The magnetic moment μ was initially attributed to the orbital magnetic moment of the electron in a fundamental Bohr orbit. At the time of Stern-Gerlach experiments, the electron configuration of silver atom was not clearly known. It came to be known later that monovalent atoms like silver have all their electrons, except one, arranged in shells, which are completed and therefore cannot contribute to the magnetic moment. The lone valence electron occupies a *S*– subshell which has no orbital angular momentum. The absence of angular momentum in the atom leads to the conclusion that the quantized magnetic moment must be associated with some kind of non-orbital angular momentum. Obviously, it must be due to spinning of the electron. Thus, the Stern-Gerlach experiment provided direct experimental evidence of electron spin.

25.16 ANOMALOUS ZEEMAN EFFECT

The normal Zeeman effect is observed at high magnetic field strengths. In weak fields, a spectral line is found to split into more than three components. The splitting of spectral lines into even number of components in the presence of weak fields is known as *anomalous Zeeman effect*. The anomalous Zeeman effect can be explained only when the existence of electron spin is taken into account in addition to its orbital motion.

If **L** and **S** represent the orbital and spin angular momentum vectors, then their resultant **J** will precess about the direction of the magnetic field. There are (2J + 1) possible orientations of J with respect to the field direction. Its components in the field direction are given by $m_j\hbar$ where m_j = J, (J – 1), (J – 2), ……., –J. In a given magnetic field, an atomic energy level with a definite set of values for L and J splits up into (2J + 1) closely lying component levels. The splitting between the adjacent levels depends on the value of J.

We know that the vectors **L** and **S** precess about the direction of **J**. Since the magnetic moments $\boldsymbol{\mu}_L$ and $\boldsymbol{\mu}_S$ due to the orbital and spin motions are aligned antiparallel to the vectors **L** and **S** respectively, these two vectors also precess about the direction of **J**. However, the magnitudes of the vectors $\boldsymbol{\mu}_L$ and $\boldsymbol{\mu}_S$ are different and hence, their resultant $\boldsymbol{\mu} = (\boldsymbol{\mu}_L + \boldsymbol{\mu}_S)$ has a direction different from that of **J**. As $\boldsymbol{\mu}_L$ and $\boldsymbol{\mu}_S$ precess about the direction of **J**, their resultant also precesses about **J** (see Fig. 25.28). The resultant magnetic moment μ can be resolved into its rectangular components $\boldsymbol{\mu}_\perp$ and $\boldsymbol{\mu}_J$ with respect to **J**. The component $\boldsymbol{\mu}_\perp$ averages out to zero whereas $\boldsymbol{\mu}_J$ is equal to the sum of the components of $\boldsymbol{\mu}_L$ and $\boldsymbol{\mu}_S$ in the direction of **J**. Thus,

$$\mu_J = \mu_L \cos(L, J) + \mu_S \cos(S, J) \qquad ...(25.78)$$

where cos (L, J) and cos (S, J) are the cosines of the angles between **J** ands the vectors **L** and **S** respectively. Using cosines law, we get

$$\cos(L, J) = \frac{L(L+1)+J(J+1)-S(S+1)}{2\sqrt{L(L+1)}\cdot\sqrt{J(J+1)}} \qquad ...(25.79)$$

$$\cos(S, J) = \frac{S(S+1)+J(J+1)-L(L+1)}{2\sqrt{S(S+1)}\cdot\sqrt{J(J+1)}} \qquad ...(25.80)$$

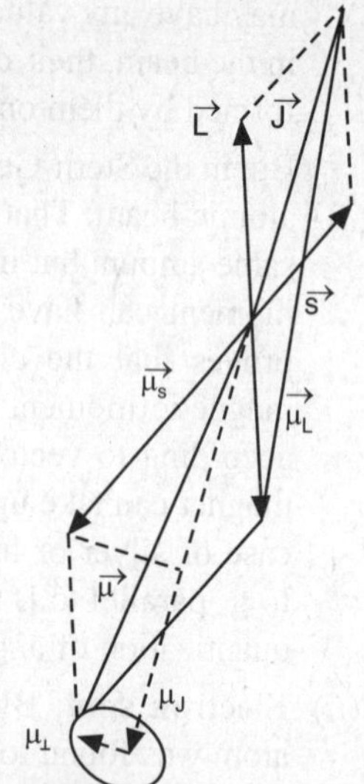

Fig. 25.28: Anomalous Zeeman Effect

Further, the quantum mechanical values of $\boldsymbol{\mu}_L$ and $\boldsymbol{\mu}_S$ are given by

$$\mu_L = \frac{e}{2m}p_L = \frac{e\hbar}{2m}\sqrt{L(L+1)} \qquad ...(25.81)$$

and
$$\mu_S = \frac{e}{m}p_S = \frac{e\hbar}{2m}\cdot 2\sqrt{S(S+1)} \qquad ...(25.82)$$

Using the above expressions into Eq. (25.72), we get

$$\mu_j = \frac{e\hbar}{2m}\left[\sqrt{L(L+1)}\frac{L(L+1)+J(J+1)-S(S+1)}{2\sqrt{L(L+1)}\sqrt{J(J+1)}} + 2\sqrt{S(S+1)}\frac{S(S+1)+J(J+1)-L(L+1)}{2\sqrt{S(S+1)}\sqrt{J(J+1)}}\right]$$

or
$$\mu_J = \frac{e\hbar}{2m}\sqrt{J(J+1)}\left[1+\frac{J(J+1)+S(S+1)-L(L+1)}{2J(J+1)}\right] \qquad ...(25.83)$$

or
$$\mu_J = g\mu_B\sqrt{J(J+1)} \qquad ...(25.84)$$

where g is the Lande splitting factor given by

$$g = 1+\frac{J(J+1)+S(S+1)-L(L+1)}{2J(J+1)} \qquad ...(25.85)$$

The additional energy ΔE due to the action of the magnetic field on the atom is

$$\Delta E = \mu_J B\cos(\boldsymbol{J},\boldsymbol{B}) = g\mu_B B\sqrt{J(J+1)}\cos(\boldsymbol{J},\boldsymbol{B}) \qquad ...(25.86)$$

The factor $\sqrt{J(J+1)}\cos(\boldsymbol{J},\boldsymbol{B})$ is the component of the vector **J** along the magnetic field **B** and has the value m_J. Therefore,

$$\Delta E = \mu_B B g m_J \qquad ...(25.87)$$

As the values of L, S, and J are different for different energy levels, g is different for different energy levels. Hence the magnetic splitting is different for different energy levels. The upper and lower levels split up by different amounts in a magnetic field. The transitions between the levels are governed by the selection rule

$$\Delta m_J = 0, \pm 1$$

For example, the transitions for the sodium lines are shown in Fig. 25.29.

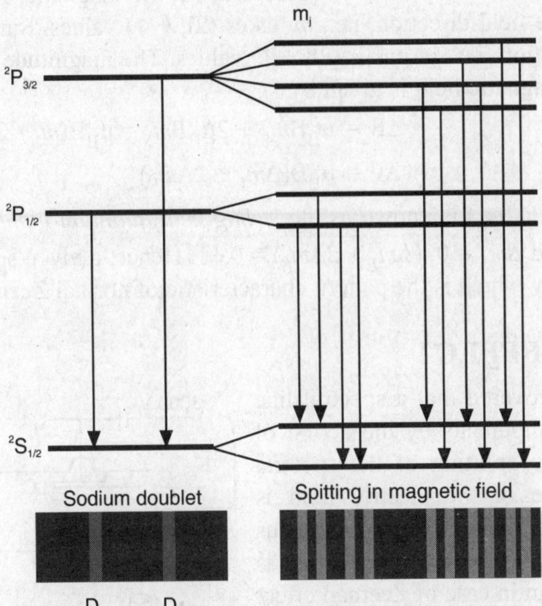

Fig. 25.29: Transitions for the sodium lines

The ground state $^2S_{1/2}$ splits into two sublevels. This state is characterized by L = 0, and $S = \frac{1}{2}$. Therefore, $J = \frac{1}{2}$ and

$$g = 1 + \frac{\frac{1}{2}\left(\frac{1}{2}+1\right)+\frac{1}{2}\left(\frac{1}{2}+1\right)-0}{2\times\frac{1}{2}\left(\frac{1}{2}+1\right)} = 2.$$

Since m_J can have the values $\frac{1}{2}$ and $-\frac{1}{2}$, gm_J can have the values +1 and –1. Thus, the ground state splits into two levels. Similar calculations are carried out for gm_J corresponding to the other states. Applying the selection rule.

$$\Delta m_J = 0, \pm 1$$

It is predicted that the spectral line D_1 corresponding to the transition $3^2P_{1/2} \rightarrow 3^2S_{1/2}$ splits into four levels and the spectral line D_2 corresponding to the transition $3^2P_{3/2} \rightarrow 3^2S_{1/2}$ splits into six levels. These conclusions agree with the experimentally observed splitting.

25.17 PASCHEN-BACK EFFECT

The anomalous Zeeman effect is observed when the applied magnetic field is relatively weak. In this effect, a spectral line splits into several lines. As the magnetic field is progressively increased, the splitting pattern becomes similar to normal Zeeman splitting where the spectral line splits into two or three components only. The reduction of the number of component lines takes place either through the coalescence of the lines or through the disappearance of certain lines. This transition is called Paschen-Back effect.

As the magnetic field becomes stronger, the coupling between the orbital and spin angular momentum vectors **L** and **S** breaks down. The vector **J** loses it significance and the two vectors **L** and **S** now independently precess about the magnetic field direction, **B**. The projection of **L** in the field direction, i.e., m_l takes (2L + 1) values. Similarly, the projection of **S** in the field direction, i.e., m_s takes (2S + 1) values. The magnitude of the splitting of an energy level in the magnetic field is given by

$$\Delta E = \mu_B B m_L + 2\mu_B B m_S = \mu_B B(m_l + 2m_s)$$

$$\therefore \quad \Delta\nu = \mu_B B(\Delta m_L + 2\Delta m_S) \quad ...(25.88)$$

The quantity $(m_l + 2m_S)$ is known as the *strong field quantum number* and is an integer. Since $\Delta m_L = 0,\pm1$, and $\Delta m_S = 0$, $(\Delta m_L + 2\Delta m_S) = 0,\pm1$. Hence, a given spectral line splits into three components only which is the pattern characteristic of normal Zeeman effect.

25.18 STARK EFFECT

In 1913, J. Stark discovered that a spectral line splits into several components by the action of the electric field. The splitting of the spectral lines in the presence of the electric field is known as Stark effect. This effect is analogous to the Zeeman effect. However, the splitting is very much smaller than in case of Zeeman effect and can be observed only with the help of high resolving power spectroscopes.

The experimental arrangement for the study of Stark effect is shown in Fig. 25.30. It is an ordinary glass discharge tube provided with a perforated cathode, C. An auxiliary electrode F is placed close to the cathode. A very strong electric field of several thousand volts per metre is maintained between F and C. The discharge tube is filled with hydrogen gas. When the pressure in the tube is not very low, discharge takes place between the anode and cathode maintained at a suitable potential difference. The canal rays stream through the perforations in the cathode. The effect of the electric field can be studied both in transverse and longitudinal directions. With the help of a spectrometer, Stark noticed that the spectral lines emitted by the canal rays were split into numerous components under the action of the electric field. Observation perpendicular to the direction of the electric field showed that the components were polarized; some of them were polarized parallel to the direction of the field and others perpendicular to it.

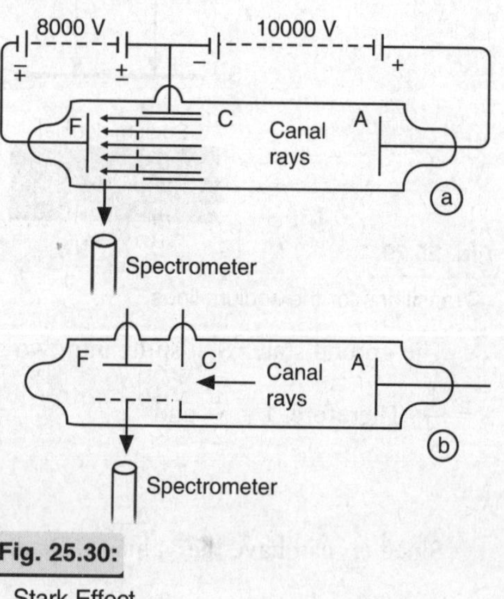

Fig. 25.30: Stark Effect

QUESTIONS

1. Write short notes on spectral series of atomic hydrogen.
2. Derive expression for radius of electron orbit and velocity of electron revolving in the orbit of hydrogen atom.

3. Calculate radius of Bohr's orbit in ground state in a hydrogen atom. Derive an expression for allowed energy of an electron in hydrogen atom. y
4. Obtain an expression for energy of an electron in nth orbit of Bohr's atom.
5. Give an account of Bohr's theory of hydrogen atom and deduce the expression for the wavelength of the first two lines of Balmer series.
6. Describe the various series of lines in the spectrum of atomic hydrogen and show how they have been explained in Bohr's theory.
7. Explain excitation and ionization of atom.
8. On the basis of Bohr's theory deduce expression for (*i*) the ground state radius of the hydrogen atom and (*ii*) the ionization energy of the hydrogen atom.
9. Determine shortest wavelength in Bracket series of spectral lines.
10. Describe Frank-Hertz experiment and explain how the results of the experiment confirm Bohr's postulates.
11. Prove that energy of radiation is absorbed by atom in quanta.
12. Describe Frank-Hertz experiment to prove the existence of discrete stationary states for electrons in atoms.
13. Draw energy level diagram for hydrogen atom and show different spectral series of hydrogen atom.
14. What do you mean by the fine structure of spectral lines? Give an account of Bohr-Sommerfeld model of elliptical orbits of hydrogen atom. How does it account for the fine structure of hydrogen?
15. What are the different suggestions given by Sommerfeld in order to explain fine structure of Hα line of Balmer series and why?
16. Explain the concepts of vector atom model.
17. Describe the vector model of the atom and explain the different quantum numbers associated with it.
18. Explain space quantization and spin of electron in vector atom model.
19. What do you mean space quantization and explain spinning of electron?
20. Show that magnetic moment due to spin motion of electron is always 1 Bohr magneton.
21. Derive an expression for magnetic moment of orbital electron.
22. State Pauli's exclusion principle. Explain how it helped in fixing up the electronic configuration of the elements in the periodic table.
23. Justify Pauli's exclusion principle for $n = 2$ when spin orbit interaction is considered.
24. Describe the Stern and Gerlach experiment and explain the importance of the results obtained.
25. Show that total magnetic moment offered by S electron is always 1 Bohr magneton.
26. What are the four quantum numbers required to specify the energy state of an electron in an atom having strong spin-orbit interaction? Explain their significance.
27. Prove that maximum number of electrons in a shell is $2n^2$ where is the number of shell.
28. Write n, l, m_l and m_s for $3d^{10}$ electrons.
29. What is Zeeman Effect? Explain normal Zeeman Effect on the basis of classical ideas.
30. Outline the quantum theory of Zeeman Effect and illustrate it with specific reference to sodium D-line.
31. Explain clearly the phenomenon of anomalous Zeeman Effect. Describe the spectral patterns expected for the yellow lines of sodium.
32. Write short notes on: (*i*) Paschen-Back effect and (*ii*) Stark effect.

PROBLEMS

1. Calculate the radii of the first, second and third permitted electron Bohr-orbits in a hydrogen atom. **[Ans:** 0.0527 Å, 2.108 Å, 4.743 Å]
2. If the Rydberg constant is $1.097 \times 107\ m^{-1}$, what are the wavelengths of the first three lines of the Paschen series? **[Ans:** 18750 Å, 12810 Å, 10930 Å]
3. Calculate the frequency of electron in the first Bohr orbit in hydrogen atom. **[Ans:** 6.5×10^{15}Hz]
4. When hydrogen was bombarded in Frank-Hertz experiment by 10.21 eV and 12.10 eV electrons, emission of three spectral lines was observed. Calculate their wavelengths. **[Ans:** 1026 Å, 1216 Å, 5672 Å]
5. If the average life time of an excited state of hydrogen is 10^{-8} s, estimate how many orbits an electron makes when it is in the state n = 2 and before it undergoes a transition to state n =1. Given Bohr radius = 0.53 Å. **[Ans:** 80,000]
6. In a normal Zeeman experiment, the calcium 4226 Å splits into 3 lines separated by 0.25 Å in a magnetic field of 3 T. Find e/m for the electron. **[Ans:** 1.76×10^{11} C/kg]
7. Determine normal Zeeman splitting in mercury 4916 Å line when in a magnetic field of 0.3 T.
8. A 5000 Å line exhibits a normal Zeeman splitting of 1.1×10^{-3} Å. Find the magnetic field.
9. What magnetic flux density is required to observe the normal Zeeman affect if a spectrometer can resolve spectral lines separated by 0.5 Å at 500 Å. **[Ans:** 4.28 T]
10. A sample of certain element is placed in magnetic field of density 0.3 T. How far apart are the Zeeman components of wavelength 4500 Å, if e/m = 1.76×10^{11} C/kg? **[Ans:** 0.0283Å]
11. The experimental value of Bohr magneton is 9.21×10^{-25} SI units. Calculate the value of e/m for the electron.
12. Find the processional frequency of an electron orbit when placed in a magnetic field of 5T. **[Ans:** 7×10^{10}Hz]
13. A beam of silver atoms in a Stern-Gerlach experiment obtained from an oven heated to a temperature of 1500 K, passes through an inhomogeneous magnetic field having a field gradient of 2 Wb/m^2/cm perpendicular to the beam. The pole pieces are 10 cm long. What is the separation between the two components of the beam on a photographic plate placed at a distance of 50 cm? **[Ans:** 0.15 mm]

26 Matter Waves

Learning Objectives:

By the end of this chapter, you will be able to:

- Understand De-Broglie hypothesis and justification of Bohr's postulate.
- Learn about different properties of matter waves.
- Explain Davisson– Germer experiment and G.P. Thomson's experiment.
- Discuss about applications of de broglie waves.

26.1 INTRODUCTION

In 1924 Louis de Broglie proposed that matter has dual character just like radiation. He suggested that the idea of wave-particle duality should be extended to all microparticles. His notion, while as yet ungrounded in experimental evidence, did provide surprising insight into the quantum conditions for Bohr's orbits. In 1927, Davisson and Germer observed the diffraction of an electron beam incident on a nickel crystal. In 1927 the wave behaviour of microparticles was confirmed by experiments on electron diffraction conducted simultaneously in several different laboratories. In 1929, soon after the discovery of wave properties of electrons, the German physicist Otto Stern and his coworkers detected diffraction phenomena with neutral atomic and molecular beams.

These experiments indicated that the wave properties such as interference, diffraction etc are exhibited by the matter when it is moving and in the state of rest it shows particle properties. Consequently the matter displays dual nature. Erwin Schrödinger took the hint from the matter waves and formulated of quantum mechanics in 1926.

26.2 DE BROGLIE HYPOTHESIS

In 1924, Louis de Broglie extended the wave–particle dualism of light to the material particles. He reasoned out that nature exhibits a great amount of symmetry. Therefore, if a light wave can act as a wave sometimes and as a particle at other times, then particles such as electrons should also act as waves at times. This is known as *de Broglie hypothesis*.

According to de Broglie hypothesis any moving particle is associated with a wave. The waves associated with particles are known as **de Broglie waves** or **matter waves**. The wavelength λ of matter waves associated with a particle moving with velocity v is inversely proportional to the magnitude of the momentum of the particle. Thus,

$$\lambda = \frac{h}{m\upsilon} \quad ...(26.1)$$

De Broglie deduced the connection between the particle and wave properties as follows.

De Broglie wavelength of Matter waves

From Planck's relation and Einstein mass-energy relation, the wavelength and momentum p of a photon are related to each other through the expression

$$\lambda = \frac{h}{p} \quad ...(26.2)$$

De Broglie proposed that the relation (26.2) between the momentum and the wavelength of a photon is a universal one and must be applicable to photons and material particles as well.

The quantities ν and λ are wave properties and the quantities E and p are particle properties. They are tied to each other through the relations

$$E = h\nu \text{ and } p = h/\lambda$$

which demonstrate that the wave and particle natures of a photon are intimately tied up to each other. These equations reflect the wave-particle dualism of light.

Now let us consider a moving particle. A particle of mass 'm' moving with a velocity υ carries a momentum $p = m\upsilon$ and it must be associated with a wave of wavelength

$$\lambda = \frac{h}{p} = \frac{h}{m\upsilon} \quad ...(26.3)$$

The waves associated with moving particles are called **matter waves** or **de Broglie waves.**

The relation $\lambda = h/m\upsilon$ is known as **de Broglie equation** and the wavelength λ is called the **de Broglie wavelength**.

From Eq. (26.3), we may draw the following conclusions.

(*i*) $\lambda \rightarrow \infty$ when the velocity of the particle is zero. It means that matter waves are detectable only for moving particles.

(*ii*) Lighter the particle, smaller the value of mass m and hence the longer is the wavelength of the matter wave associated with it. Therefore, wave behaviour of micro-particles will be significant whereas waves associated with macro-bodies can never be detected.

(*iii*) The smaller the velocity of the micro-particle, the longer is the wavelength of the matter wave associated with it.

If a photon is considered to be a particle, then the corresponding electromagnetic wave is the de Broglie wave for the photon. Similarly, atomic particles also can be viewed as associated with matter waves, which do not have any similarity to any known waves. It is later understood that the waves associated with particles are not real three dimensional waves in the way sound waves are, but are **probability waves** related to the probabilities of finding the particles in various places and with various properties.

De-Broglie wavelength associated with an accelerated charged particle

If a charged particle, say an electron is accelerated by a potential difference of V volts, then its kinetic energy is given by $K.E. = eV$.

or $$\frac{1}{2}m\upsilon^2 = eV$$

∴ $$\upsilon = \sqrt{\frac{2eV}{m}}$$

Then the electron wavelength is given by

$$\lambda = \frac{h}{m\upsilon} = \frac{h}{m}\cdot\sqrt{\frac{m}{2eV}}.$$

∴ $$\lambda = \frac{h}{\sqrt{2emV}} \quad ...(26.4)$$

De Broglie Wavelength expressed in terms of K.E.

If a particle has kinetic energy K.E., then $\text{K.E.} = \frac{1}{2}m\upsilon^2 = \frac{m^2\upsilon^2}{2m} = \frac{p^2}{2m}$

or $$p = \sqrt{2m(\text{K.E.})}$$

∴ $$\lambda = \frac{h}{\sqrt{2m(\text{K.E.})}} \quad ...(26.5)$$

De Broglie wavelength associated with particles in thermal equilibrium

If particles are in thermal equilibrium at temperature T, then their kinetic energy is given by

$$\text{K.E.} = \frac{3}{2}kT$$

∴ $$\lambda = \frac{h}{\sqrt{2m(\text{K.E.})}} = \frac{h}{\sqrt{3mkT}} \quad ...(26.6)$$

26.3 DE BROGLIE'S JUSTIFICATION OF BOHR'S POSTULATE

Bohr's theory of atomic structure was successful in explaining a large body of experimental observations concerning atomic behaviour but the three adhoc postulates underlying his theory remained without a valid theoretical justification for a long time. In support of his hypothesis of matter waves, de Broglie demonstrated that it could provide an explanation for the postulate regarding quantization of angular momentum of electron in Bohr's model of atom.

One of the postulates that Bohr used in formulating a model of atom is that the angular momentum L of the electron revolving in a stationary orbit is quantized. Thus,

$$L = n\hbar \quad ...(26.7)$$

The above postulate of Bohr follows directly from the concept of matter waves.

As the electron travels round in one of its circular orbits, the associated matter waves propagate along the circumference again and again. A wave must meet itself after going round one full circumference. If it did not meet, the wave would be out of phase with itself after going round one orbit. After a large number of orbits, all possible phases would be obtained and the wave would be annihilated by destructive interference. It implies that the wave should produce a standing wave profile in the orbit to preclude the electron energy from radiating away.

If a stretched string is fastened at both ends and is made to vibrate, standing waves are formed provided the length of the string is an integral number of half–wavelengths of the disturbance. If the string is formed into a circular loop, the condition for standing waves is that the circumference of the loop should be an integral number of whole wavelengths of the disturbance. Thus, if r is the radius of the circular loop,

$$2\pi r = n\lambda \qquad n = (1,2,3,....) \qquad ...(26.8)$$

We may regard the stationary electron orbits in an atom to be analogous to the circular loop of string. We conclude that stationary electron wave pattern can form in the orbit if only an integral number of electron wavelengths fit into the orbit, as shown in Fig. 26.1 (*b*).

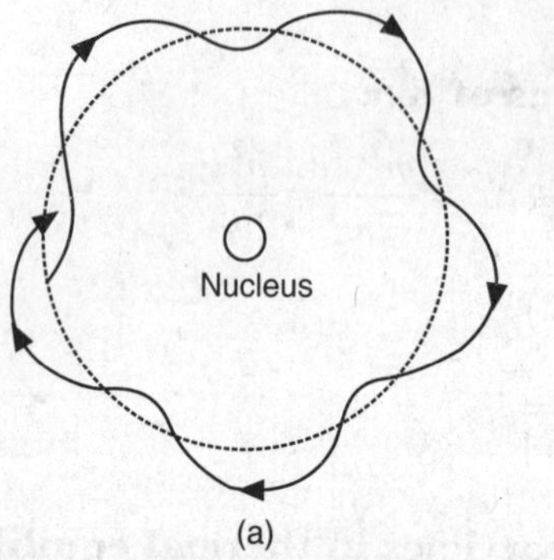

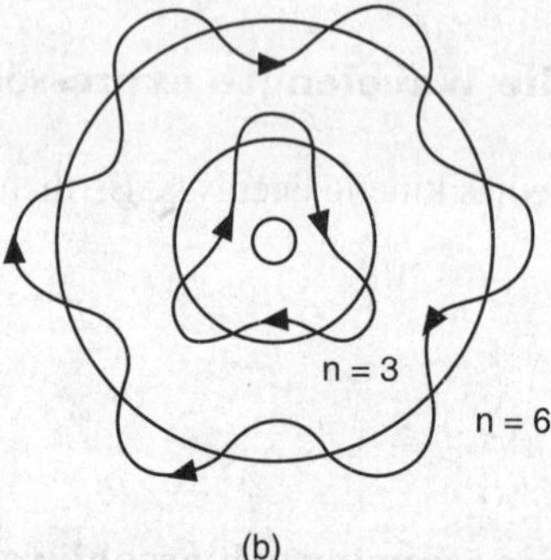

Fig. 26.1:

Waves in atomic orbits.

Eq. (26.8) can be applied for electron waves, taking λ as de Broglie wavelength of electron waves.

The de Broglie wavelength of electron wave is given by

$$\lambda = \frac{h}{m\upsilon}$$

where υ is the speed of the electron in the orbit.

Using the de Broglie wavelength into Eq. (26.8), we obtain

$$2\pi r = \frac{nh}{m\upsilon}$$

$$\therefore \qquad m\upsilon r = \frac{nh}{2\pi} \qquad ...(26.9)$$

But the quantity '$m\upsilon r$' is the angular momentum, L, of electron in the orbit of radius r.

Thus, $$L = m\upsilon r$$

It, therefore, follows that

$$L = n\hbar$$

which is precisely Bohr's postulate. De Broglie thus demonstrated that the quantization of angular momentum is a direct consequence of wave nature of electron.

Let us calculate the wavelength of the electron in the first orbit of hydrogen atom. The electron speed υ in the orbit is given by

$$\upsilon = \frac{e}{\sqrt{4\pi\varepsilon_o mr}}$$

$$\therefore \qquad \lambda = \frac{h}{e}\sqrt{\frac{4\pi\varepsilon_o r}{m}} \qquad ...(21.10)$$

Taking $r = 5.3 \times 10^{-11}$ m, we get $\lambda = 3.3$ Å. The circumference of the orbit is $2\pi r = 3.3$ Å.

Hence, the first orbit of the electron in a hydrogen atom corresponds to one complete electron wave joined on itself. The de Broglie hypothesis thus offered a new meaning to the quantum number 'n'. n is the number of de Broglie wavelengths that fit into the circumference of Bohr allowed orbits.

Now the electron can be visualized in the atom as a

(*i*) particle moving in a fixed orbit with a certain quantized angular momentum, $m\upsilon r$

(*ii*) standing wave spread in a certain region around the nucleus

26.4 DE BROGLIE WAVES ARE INSIGNIFICANT IN CASE OF MACRO-BODIES

According to de Broglie hypothesis a moving body is associated with matter waves and the wavelength of the waves is given by

$$\lambda = \frac{h}{m\upsilon}$$

where υ is the velocity with which the body moves.

As the mass m of the body increases, the wavelength tends to be insignificant. Therefore, the wavelength associated with macroscopic bodies become insignificant in comparison to the size of the bodies themselves even at very low velocities. Because of the smaller magnitude of Planck's constant h, the wavelength λ will be significant only in case of micro-particles.

For example, if we consider a cricket ball of mass 500 gm flying with a velocity of 50 km/hr, its wavelength comes to

$$\lambda = \frac{6.62 \times 10^{-34}\ \text{J.s}}{0.5\text{kg} \times 13.9\text{m}/s} = 10^{-34}\ \text{m} = 10^{-24}\ \text{Å}.$$

It is easy to see that this wavelength is insignificant in comparison to the size of the ball.

On the other hand, if we consider the case of an electron, having energy 100 eV, the de Broglie wavelength of the electron is given by

$$\lambda = \frac{h}{\sqrt{2\text{meV}}}$$

$$= \frac{6.62 \times 10^{-34}\ \text{J.s}}{\sqrt{2 \times 9.11 \times 10^{-31}\ \text{kg} \times 1.602 \times 10^{-19}\ C \times 100\text{V}}} = 1.33\ \text{Å}.$$

The size of an electron is about 10^{-5}Å, which is far smaller than the wavelength of 1.33Å. It means that the electron behaves more as a wave than a particle under the circumstances.

26.5 PROPERTIES OF MATTER WAVES

(*i*) Matter waves are produced by the motion of the particles and are independent of the charge. Therefore, they are neither electromagnetic nor acoustic waves but are new kind of waves.

(*ii*) They can travel through vacuum and do not require any material medium for their propagation.

(*iii*) The smaller the velocity of the particle, the longer is the wavelength of the matter wave associated with it.

(*iv*) The lighter the particle, the longer is the wavelength of the matter wave associated with it.

(*v*) The velocity of matter waves depends on the velocity of the material particle and is not a constant quantity.

(*vi*) The velocity of matter waves is greater than the velocity of light.

(*vii*) They exhibit diffraction phenomenon as any other waves.

26.6 DAVISSON-GERMER EXPERIMENT

Waves exhibit diffraction. If the de Broglie hypothesis is valid, then the matter waves should exhibit diffraction effects. Diffraction is observed when the wavelength is comparable to the size of the object causing diffraction. The wavelength of 100 eV electrons is of the order of 1 Å and the interatomic spacing in a crystal is of the order of 2 to 3 Å. Therefore, we expect that the wave behaviour of a micro-particle such as an electron becomes noticeable when a beam of particles interact with crystals. In 1927, Davisson and Germer observed the diffraction of an electron beam incident on a nickel crystal. The experiment provided a convincing proof of the wave nature of matter.

Apparatus: The experimental arrangement of Davisson and Germer is shown in Fig. 26.2. The apparatus consisted of an electron gun, which produced collimated beam of electrons. An anode, A, connected to a variable voltage source accelerated the electrons. The energy of the electrons can be computed from the accelerating potential. These electrons were scattered by a nickel crystal located at C. The crystal can be rotated on the axis. The number of electrons scattered by the crystal in different directions was measured with the help of a detector D, which can be moved on a scale.

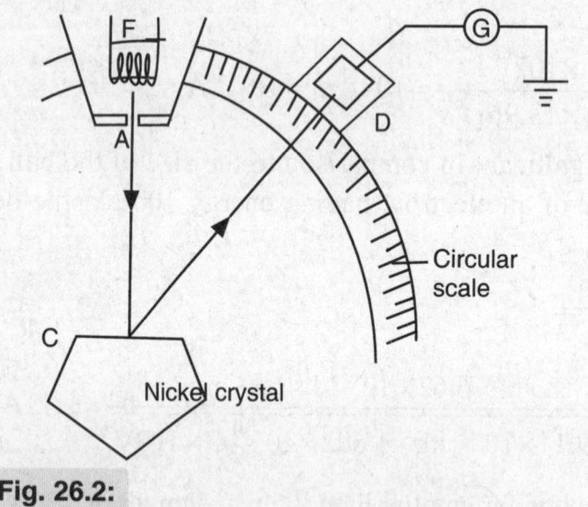

Fig. 26.2:

Schematic of Davisson and Germer experiment

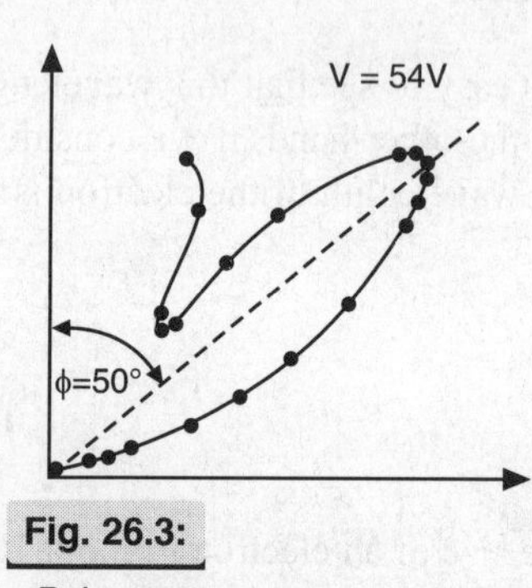

Fig. 26.3:

Polar curve

The experimental arrangement of Davisson and Germer is shown in Fig. 26.2. An electron beam is generated from a hot tungsten filament F and an anode A connected to a variable voltage source accelerated the electrons. The energy of the electrons can be computed from the accelerating potential, V applied between the filament F and the anode A. The electrons emerge through an opening in the anode and fall normally on the surface of a nickel crystal, C. These electrons are scattered by a nickel crystal located at C. The crystal can be rotated on the axis. The detector D measured the number of electrons scattered by the crystal in different directions. The detector could be moved on a graduated semicircular scale. Thus, the intensity of the scattered electron beam was determined as a function of the scattering angle, ϕ.

Investigations: During their experiments Davisson and Germer moved the detector on the circular scale to various positions and the current was measured. The detector current is a measure of the intensity of the diffracted beam. A polar graph was then plotted between the detector current and the angle between the incident beam and the diffracted beam. Such polar curves were obtained for electrons accelerated through different voltages. It was found that a hump appears in the polar curve when 44 eV electrons were incident on the crystal. The hump grew in size as the accelerating voltage is increased and became most pronounced at 54 volts. The polar curve corresponding to 54 V is shown in Fig. 26.3. It is found that for the accelerating voltage of 54 volts, the electrons are scattered more pronouncedly at an angle of 50° with the direction of the incident beam. The maximum is an indication that electrons are being diffracted.

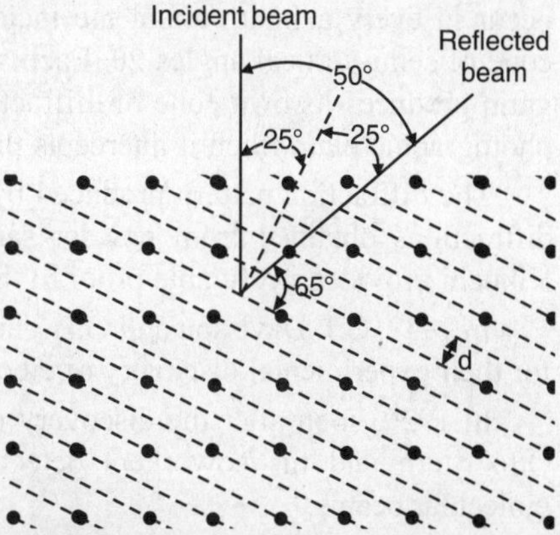

Analysis: It may be interpreted that the rows of atoms at the surface of the nickel crystal act like rulings of a natural diffraction grating and the de Broglie waves associated with the electrons underwent diffraction when they were incident on the crystal. The hump produced at 50° in Fig. 26.3 then corresponds to the first order diffraction maxima. Braggs' law, applicable for X-ray diffraction by crystals, would be valid for electron wave diffraction also. Fig. 26.4 shows atomic planes and the incident and scattered beams. The interplanar spacing is obtained from X-ray analysis to be $d = 0.91$ Å.

Fig. 26.4: Atomic Planes and the Incident and Scattered Beams

From the Fig. 26.4, it is seen that the glancing angle $\theta = 65°$. Applying Braggs' equation,

$$\lambda = 2d \sin \theta = 2 \times 0.91 \text{ Å} \times \sin 65° = \mathbf{1.65 \text{ Å}}.$$

The wavelength of the electron wave is thus determined to be 1.65 Å. The wavelength of electron wave can be computed from the accelerating potential V using de Broglie equation.

$$\lambda = \frac{h}{\sqrt{2\text{me V}}}$$

$$= \frac{6.63 \times 10^{-34} \text{ Js}}{\left[2 \times 9.1 \times 10^{-31} \text{ kg} \times 1.602 \times 10^{-19} \text{ C} \times 54 \text{V}\right]} = \mathbf{1.66 \text{Å}}.$$

It is seen that the values obtained experimentally using Braggs' equation and de Broglie equation agreed well. Therefore, Davisson-Germer experiment gave conclusive evidence that electrons exhibit diffraction property.

26.7 G.P. THOMSONS'S EXPERIMENT

The de Broglie hypothesis was further confirmed in 1927 by the experiments conducted independently by G.P.Thomson in England and by Kikuchi in Japan.

Thomson's experimental arrangement is shown in Fig. 26.5. Electrons are produced from a heated filament F and accelerated through a high positive potential given to the anode A.

The whole apparatus is kept highly evacuated. The electron beam passes through a fine hole in a metal block B and falls on a gold foil of thickness 0.1 μm. The electrons passing through the foil are received on a photographic plate P.

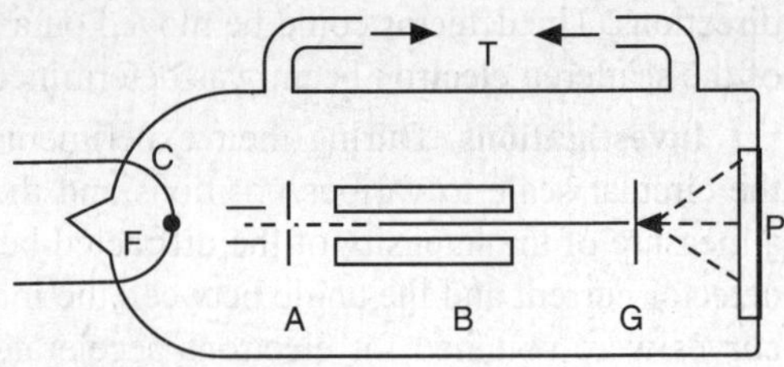

Fig. 26.5: Experimental arrangement

Metals are polycrystalline in which the grains are oriented completely at random. Therefore, some grains have always the right inclination θ towards the incident beam in order to produce a Bragg reflection. Owing to the random orientation, the reflections from a given set of lattice planes at the glancing angle occur in every azimuth about the incident beam. Consequently, the reflected beams form a cone of semi-vertical angles 2θ. Each set of lattice planes with its particular spacing *d* in the grain produces its own cone of diffracted rays. A concentric rings pattern is produced on the photographic plate when it intercepts the coaxial cones of diffracted rays.

The diffraction pattern produced by the electron beam was strikingly similar to the x-ray diffractions obtained from powder samples. Thus, the experiments of G.P. Thomson and Kikuchi provided irrefutable proof to the existence of de Broglie waves.

In 1937, C.J. Davisson and G.P. Thomson were jointly awarded the Noble Prize in physics for their experimental discovery of electron diffraction.

In 1929, soon after the discovery of wave properties of electrons, the German physicist Otto Stern and his coworkers detected diffraction phenomena with neutral atomic and molecular beams.

26.8 VELOCITY OF DE BROGLIE WAVES

Any harmonic wave is characterized by a precise wavelength λ and constant amplitude. It is non-localized and has no beginning and end. It means that such a wave extends over a very large volume of space.

26.8.1 Phase Velocity

If we consider a harmonic wave, the wave has a single wavelength and a single frequency. The velocity of propagation of the wave is given by

$$\upsilon_p = \nu\lambda$$

Using, $\nu = \omega/2\pi$ and $\lambda = 2\pi/k$ into the above equation, we get

$$\upsilon_p = \frac{\omega}{2\pi}\cdot\frac{2\pi}{k} = \frac{\omega}{k} \quad ...(26.11)$$

υ_p is called the *phase velocity*.

The velocity with which the plane of equal phase travels through a medium is known as the phase velocity. It thus represents the velocity of propagation of the wave front.

As $E = h\nu$ and $p = h/\lambda$, we get

$$\upsilon_p = \frac{E}{h}\cdot\frac{h}{p} = \frac{E}{p} \quad ...(26.12)$$

(*i*) When the atomic particle velocity is *non-relativistic*, the total energy $E = mc^2$ and momentum $p = m\upsilon$.

Therefore, the phase velocity of the de Broglie wave associated with the particle is

$$\upsilon_p = \frac{E}{p} = \frac{mc^2}{m\upsilon} = \frac{c^2}{\upsilon} \qquad ...(26.13)$$

As $\upsilon < c$, the phase velocity of the de Broglie wave associated with the atomic particle is always greater than c.

(*ii*) When the atomic particle velocity is *relativistic*, the total energy $E = \sqrt{m_o^2c^4 + p^2c^2}$, where m_o is the rest mass of the particle.

Therefore, the phase velocity of the de Broglie wave associated with the particle is

$$\upsilon_p = \frac{E}{p} = \left[\frac{m_o^2c^4 + p^2c^2}{p^2}\right]^{1/2}$$

$$= c\left[\frac{m_o^2c^2}{p^2} + 1\right]^{1/2} = c\left[\frac{m_o^2c^2\lambda^2}{h^2} + 1\right]^{1/2} \qquad ...(26.14)$$

As the term $\frac{m_o^2c^2\lambda^2}{h^2}$ is always a positive quantity, the phase velocity of the de Broglie wave associated with the atomic particle is always greater than c.

According to the theory of relativity, it is not possible that the velocity of the particle wave be greater than or equal to the velocity of light. Hence, a harmonic wave of wavelength λ cannot represent a moving atomic particle. Thus, ***de Broglie waves cannot be harmonic waves***.

26.9 WAVE PACKET - REPRESENTS A MICROPARTICLE

We have so far assumed that a particle may be represented by a monochromatic de Broglie wave. However, a wave spreads over a large region of space and cannot represent a highly localized particle. Schrödinger postulated that a **wave packet** rather than a single harmonic wave represents a particle. A wave packet consists of a group of harmonic waves. Each wave has slightly different wavelength. The superposition of a very large number of harmonic waves differing infinitesimally in frequency will produce a single wave packet (see Fig. 26.6 c). The waves interfere constructively over only a small region of space and cancel each other everywhere except in that small region. The position of the particle would then be approximately determined by the position of the wave packet.

The velocity with which the wave packet propagates is called the **group velocity** υ_g. The individual waves forming the wave packet propagate at a velocity known as the **phase velocity** υ_p.

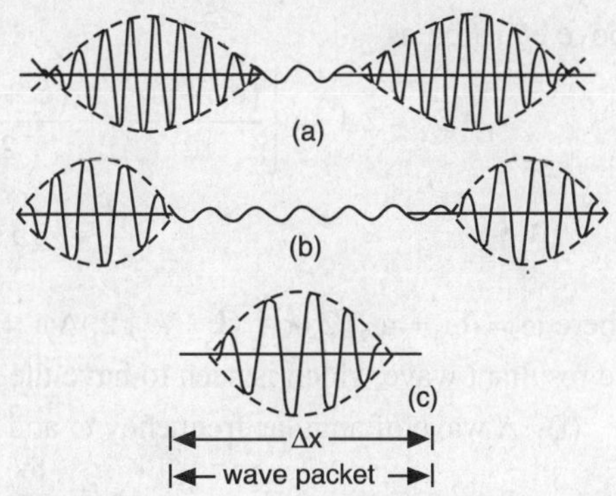

Fig. 26.6: Formation of a Wave packet. (a) two waves of slightly different frequencies produce constructive interference. (b) three waves produce interference maxima of larger size separated by larger distance. (c) A large number of waves-having slightly different frequencies produces only one maxima and it is called a wave packet.

26.9.1 Group Velocity

When a number of plane waves of slightly different wavelengths travel in the same direction, they form wave groups or wave packets. The velocity with which the wave group advances in the medium is known as the *group velocity* υ_g. Each component wave has its own phase velocity, $\upsilon_p = \nu\lambda$. The wave packet has amplitude that is large in a small region and very small outside it. The amplitude of the wave packet varies with x and t. Such a variation of amplitude is called the *modulation* of the wave. The velocity of propagation of the modulation is known as the *group velocity*, υ_g.

Here, we should note that *wave packets are only theoretical artifices to aid our visualization of various phenomena in the micro-world.*

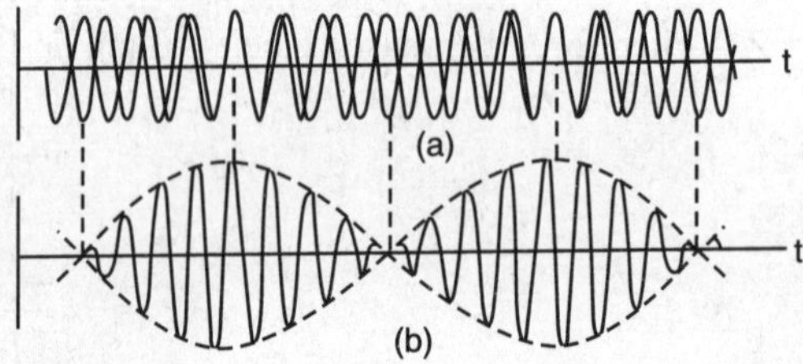

Fig. 26.7: Beats are formed when two waves of slightly different frequencies combine (a) the individual waves (b) the resultant wave.

Expression for the Group Velocity

We derive now an expression for group velocity considering a group of waves consisting of two components of equal amplitude and slightly differing angular velocities ω_1 and ω_2.

Let the waves in Fig. 26.7 (a) be represented by the equations

$$y_1 = \text{A} \sin(\omega_1 t - k_1 x)$$

$$y_2 = \text{A} \sin(\omega_2 t - k_2 x)$$

The superposition of these two waves is given by

$$y_1 + y_2 = \text{A} \sin(\omega_1 t - k_1 x) + \text{A} \sin(\omega_2 t - k_2 x)$$

Using the trigonometric relation $\sin\alpha + \sin\beta = 2\sin\left(\frac{\alpha+\beta}{2}\right)\cos\left(\frac{\alpha-\beta}{2}\right)$, we write the above equation as

$$y_1 + y_2 = 2A \sin\left[\frac{(\omega_1+\omega_2)}{2}t - \frac{(k_1+k_2)}{2}x\right]\cos\left[\frac{(\omega_1-\omega_2)}{2}t - \frac{(k_1-k_2)}{2}x\right]$$

$$= 2A\sin(\omega t - kx)\cos\left(\frac{\Delta\omega t}{2} - \frac{\Delta kx}{2}\right) \quad \text{...(26.15)}$$

where $\omega = (\omega_1 + \omega_2)/2$, $k = (k_1 + k_2)/2$, $\Delta\omega = \omega_1 - \omega_2$ and $\Delta k = k_1 - k_2$. Eq. (26.15) represents the resultant wave which is seen to have the following two parts.

(i) A wave of angular frequency ω and propagation constant k, moving with a velocity

$$\upsilon_p = \frac{\omega}{k} = \nu\lambda \text{ and}$$

(ii) A second wave of angular frequency $\Delta\omega/2$ and propagation constant $\Delta k/2$, moving with a velocity $\upsilon_g = \dfrac{\Delta\omega}{\Delta k}$

When $\Delta\omega$ and Δk are very small, we can write the above equation as

$$\upsilon_g = \frac{d\omega}{dk} \quad \text{...(26.16)}$$

or $$\upsilon_g = \frac{2\pi\, d\nu}{2\pi\, d(1/\lambda)} = -\lambda^2 \frac{d\nu}{d\lambda}$$

26.9.2 Relation between Phase Velocity and Group Velocity

The velocity of the individual component wave of the wave packet is given by

$$\upsilon_p = \nu\lambda$$

Using, $\nu = \omega / 2\pi$ and $\lambda = 2\pi / k$ into the above equation, we get

$$\upsilon_p = \frac{\omega}{2\pi}\cdot\frac{2\pi}{k} = \frac{\omega}{k} \qquad ...(26.17)$$

$$\therefore \qquad \omega = k\upsilon_p$$

The group velocity is given by the relation (26.16) as

$$\upsilon_g = \frac{d\omega}{dk} = \frac{d}{dk}\left(k\upsilon_p\right) = \upsilon_p + k\frac{d\upsilon_p}{dk}$$

But $$k = \frac{2\pi}{\lambda}.$$

Therefore, $$dk = -\frac{2\pi}{\lambda^2}d\lambda$$

and $$\frac{k}{dk} = -\frac{\lambda}{d\lambda}.$$

$$\therefore \qquad \upsilon_g = \upsilon_p - \lambda\frac{d\upsilon_p}{d\lambda} \qquad ...(26.18)$$

Group velocity will be the same as phase velocity if the entire constituent waves travel with the same velocity. It means that in a nondispersive medium, $\upsilon_g = \upsilon_p$. However, the waves of different wavelengths travel in a medium with different velocities. Therefore, the group velocity is in general less than the phase velocity.

26.9.3 The Velocity of a Particle Equals the Group Velocity of the Associated Matter Waves

A particle moving with a velocity υ is supposed to consist of a group of de Broglie waves. The group velocity of a wave packet is given by

$$\upsilon_g = \frac{d\omega}{dk}$$

which we can write as $$\upsilon_g = \left(\frac{d\omega}{dE}\right)\left(\frac{dE}{dp}\right)\left(\frac{dp}{dk}\right) \qquad ...(26.19)$$

As $$E = h\nu = h\cdot\frac{\omega}{2\pi} = \frac{h}{2\pi}\cdot\omega = \hbar\omega, \quad \frac{d\omega}{dE} = \frac{1}{\hbar}$$

and $$p = \frac{h}{\lambda} = h\cdot\frac{1}{\lambda} = h\cdot\frac{k}{2\pi} = \frac{h}{2\pi}\cdot k = \hbar k, \quad \frac{dp}{dk} = \hbar$$

$$\therefore \qquad \upsilon_g = \left(\frac{d\omega}{dE}\right)\left(\frac{dE}{dp}\right)\left(\frac{dp}{dk}\right) = \frac{1}{\hbar}\left(\frac{dE}{dp}\right)\hbar = \left(\frac{dE}{dp}\right) \qquad ...(26.20)$$

For a particle, $E = \frac{1}{2}m\upsilon^2 = \frac{1}{2}\frac{(m\upsilon)^2}{m} = \frac{p^2}{2m}$.

$$\therefore \qquad \upsilon_g = \frac{dE}{dp} = \frac{p}{m} = \upsilon. \qquad ...(26.21)$$

Thus, the de Broglie wave group associated with an atomic particle travels with the same velocity as that of the particle itself.

26.9.4 Relation Between the Group Velocity and Particle Velocity (in a Non-dispersive Medium)

A particle moving with a velocity υ is supposed to consist of a group of de Broglie waves. For an atomic particle of rest mass m_o moving with a velocity υ, the total energy and momentum are given by

$$E = mc^2 = \frac{m_o c^2}{\sqrt{1-\upsilon^2/c^2}} \text{ and } p = m\upsilon = \frac{m_o \upsilon}{\sqrt{1-\upsilon^2/c^2}} \text{ respectively.}$$

The frequency of the associated de Broglie wave is

$$\nu = \frac{E}{h} = \frac{m_o c^2}{h\sqrt{1-\upsilon^2/c^2}} \text{ and } \omega = 2\pi\nu = \frac{2\pi\, m_o c^2}{h\sqrt{1-\upsilon^2/c^2}}.$$

Therefore,
$$d\omega = \frac{2\pi m_o}{h\left(1-\upsilon^2/c^2\right)^{3/2}}\upsilon \cdot d\upsilon. \qquad ...(26.22)$$

The wavelength of the de Broglie wave is

$$\lambda = \frac{h}{p} = \frac{h\left(1-\upsilon^2/c^2\right)^{1/2}}{m_o \upsilon} \text{ and } k = \frac{2\pi}{\lambda} = \frac{2\pi\, m_o \upsilon}{h\left(1-\upsilon^2/c^2\right)^{1/2}}$$

$$\therefore \quad dk = \frac{2\pi m_o}{h}\left[\left(1-\upsilon^2/c^2\right)^{-1/2} d\upsilon + \upsilon \cdot \frac{\upsilon}{c^2}\left(1-\upsilon^2/c^2\right)^{-3/2} d\upsilon\right]$$

or
$$dk = \frac{2\pi m_o d\upsilon}{h\left(1-\upsilon^2/c^2\right)^{3/2}} \qquad ...(26.23)$$

Dividing Eq. (26.22) by Eq.(26.23), we get

$$\upsilon_g = \frac{d\omega}{dk} = \upsilon \qquad ...(26.24)$$

Thus, the de Broglie wave group associated with an atomic particle travels with the same velocity as that of the particle itself in a non-dispersive medium.

26.10 APPLICATIONS OF DE BROGLIE WAVES

We discuss here some of the applications of de Broglie waves.

***(i)* Possible energy states of a microparticle trapped in a box:** Let us consider a microparticle trapped in a one-dimensional box of length L. According to de Broglie hypothesis, the particle is associated with a wave having a wavelength λ. The particle cannot move beyond the walls of the box. Hence, the amplitude of the de Broglie wave drops to zero at the walls. It implies that the de Broglie wave of the particle forms a standing wave pattern with nodes at the walls. The formation of standing wave pattern requires that the distance L must be an integral multiple of half-wavelength. Thus,

$$L = n\frac{\lambda}{2} \text{ or } \lambda = \frac{2L}{n} \quad ...(26.25)$$

where $n = 1,2,3,......$

The possible values of linear momentum are given by

$$p = \frac{h}{\lambda} = n\frac{h}{2L} \quad ...(26.26)$$

The possible values of the kinetic energy of the microparticle are given by

$$\text{K.E.} = \frac{p^2}{2m} = \frac{n^2h^2}{8mL^2} \quad ...(26.27)$$

The above expression indicates that a microparticle trapped in a box can take only certain discrete energy states. Secondly, the particle cannot have zero energy and the minimum kinetic energy that it can take is $\frac{h^2}{8mL^2}$ when $n = 1$.

***(ii)* Neutron diffraction:** Experiments have showed that neutrons exhibit diffraction. The diffraction of neutrons is used to study atomic structures of solids containing hydrogen atoms. Normally, x-ray and electron diffraction analysis are used to determine the structures of solids. In solids containing hydrogen atoms, the scattering of x-rays and electrons by hydrogen is not sufficient. Therefore, it is difficult to locate the position of hydrogen atoms using these techniques. In contrast, neutrons interact largely with atomic nuclei, notably hydrogen nuclei. Therefore, the scattering of neutrons by hydrogen-containing atoms in solids reveals the presence and location of hydrogen atoms in the lattice.

Matter waves are less penetrating than x-rays and so are useful in studying surface features of materials.

***(iii)* Transmission Electron microscope:** A microscope is an optical instrument used to magnify small objects in order to study their structural details. It consists of two high power lenses called the *objective* and the *eyepiece*. The objective forms a real image of the object kept in front of it and the image is viewed by the eye through the eyepiece. The magnification of the object by the microscope is given by the product of the magnifications of the objective and eyepiece. It may appear at the first instance that one may get an image magnified to any desired extent by increasing the magnifications of the objective and eyepiece. It does not happen so in practice. The light reflected from each point of the object has to pass through the objective, which is a circular aperture. Consequently, circular diffraction is produced by the objective corresponding to each point. The higher the objective power, the smaller the aperture and the larger is the diffraction pattern. When the image is viewed further through the eyepiece of higher power, the points appear as blurred patches overlapping on each other and the details are not discernible. Thus, the diffraction effects restrict the ultimate useful magnification achievable by a microscope. The maximum useful magnification of an optical microscope is about 1000×. Any further magnification does not show more details though a larger image is obtained. The diffraction effects depend on the wavelength of light. The useful magnification can be increased by making use of shorter wavelength radiation. Thus, UV radiation can give higher magnification of around 2000×.

Very large magnification is obtained by exploiting the wave character of electrons. The de Broglie wavelengths of electrons are extremely small, of the order of 0.001 to 1 Å as against the light wavelength of 5000 Å. As a result, magnifications of the order of $10^6 \times$ can be easily attained using electron waves. The Transmission Electron Microscope (TEM) was modeled on the lines of a Light Transmission Microscope where a focused beam of electrons is used. The instrument was developed by Max Knoll and Ernst Ruska in Germany in 1931.

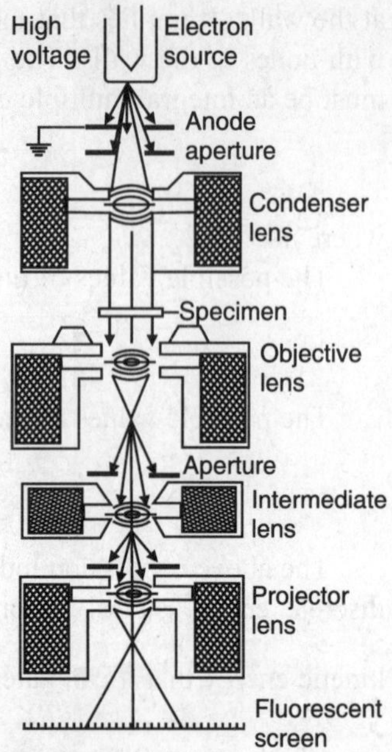

High voltage
Electron source
Anode aperture
Condenser lens
Specimen
Objective lens
Aperture
Intermediate lens
Projector lens
Fluorescent screen

Fig. 26.8: Transmission Electron Microscope (TEM)

Construction: A schematic diagram of the electron microscope is shown in Fig. 26.8. It is essentially a very large modified cathode ray tube. It consists of an electron gun at one end of the tube, a number of magnetic lenses in the path of the electron beam and a fluorescent screen at the other end of the tube. Each magnetic lens is a solenoid encased in soft iron and has a soft iron pole piece to concentrate the magnetic field lines. The focal lengths are of the order of a few millimeters and can be varied by varying current through the solenoid. The tube is mounted vertically as illustrated in Fig. 26.8.

Working: Electrons are emitted by a hot cathode and are accelerated to high velocities with the help of an anode held at about 50 to 100 kV. The electrons pass through a magnetic lens that acts as a condenser lens and is formed into a parallel beam. The electron beam then passes through the specimen to be viewed. The specimen is prepared in the form of a very thin slice of thickness of about 100 to 1000 Å so that electrons are not scattered and blur the image. Different number of electrons passes through different portions of the specimen depending on its structure. After passing through the specimen, the electron beam goes through a second lens, which acts as the objective lens and forms an intermediate image of the object. After that the beam passes through a third lens that acts as the eyepiece and forms the final image of the object on a fluorescent screen. The fluorescent screen converts the image into an optical image. The magnified image of the object is viewed through a side window. The resolving power of an electron microscope is of the order of 10 to 100 Å and the magnification of the order of 10^6 to 10^7 is easily attainable.

In view of high magnification power the electron microscope proved to be a very valuable tool for studying the microstructures of a variety of materials and also for studying micro-organisms. The structural details of virus, proteins etc could be understood only with the help of an electron microscope.

***(iv)* Scanning electron microscope (SEM):** An important variation is the scanning electron microscope. Fig. 26.9 shows a schematic diagram of the scanning electron microscope. Scanning electron microscope (SEM) is a microscope that uses electrons to form an image. An electron gun situated at the top of the SEM column generates a beam of electrons. The electron beam passes through the anode, is condensed by a condenser lens, and gets focused by the objective lens as a very fine point on the sample. The scan coils produce a magnetic field which deflects the electron beam back and forth in a controlled way. The varying voltage is simultaneously applied to the coils around the neck of the Cathode-ray tube (CRT) which produces a pattern of light deflected back and forth on the CRT screen. The pattern of deflection of the electron beam is the same as the pattern of deflection of the spot of light on the CRT screen. When the electron beam hits the sample, it produces secondary electrons which are collected by a collecting anode that is held at a positive potential with respect to the specimen. The current in the electron collector is converted to a voltage, and amplified. The amplified voltage is applied to the grid of the CRT and causes the intensity of the spot of light to change. The image consists of thousands of spots of varying intensity on the CRT screen that corresponds to the topography of the sample.

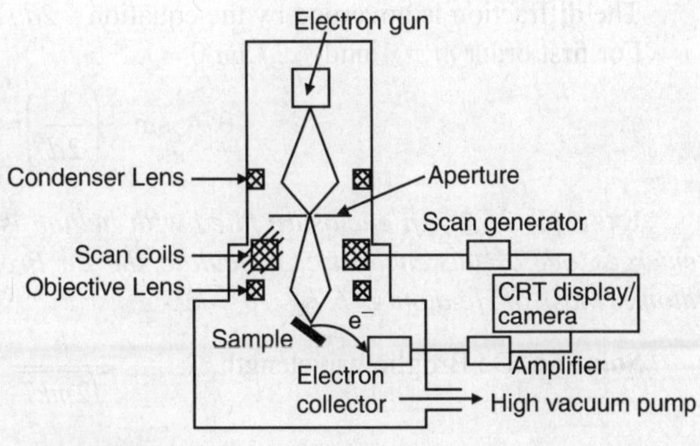

Fig. 26.9: Scanning electron microscope (SEM)

The combination of higher magnification, larger depth of field, greater resolution and compositional and crystallographic information makes the SEM a useful instrument.

Principle: In a scanning electron microscope, the specimen is exposed to a narrow electron beam from an electron gun, which scans the surface of the specimen. This causes the release of a shower of secondary electrons and other types (backscattered electrons, diffracted backscattered electrons, photons) of radiations from the specimen surface. Due to the very narrow electron beam, SEM micrographs have a large depth of field giving structure of a sample.

Components: Electron Source, Electron Lenses, Sample Stage, Detectors for all signals of interest, Display / Data output devices.

Advantages: Generally, the TEM resolution is about an order of magnitude greater than the SEM resolution. However, the advantages of SEM are as follows. (*i*) As the process of forming image involves surface processes, bulk samples can be used. (*ii*) It provides a much greater depth of view, and so can produce images that are a good representation of the 3D structure of the sample.

Example 26.1: *An electron beam is accelerated from rest through a potential difference of 200 V.*

(*i*) *Calculate the associated wavelength.*

(*ii*) *This beam is passed through a diffraction grating of spacing 3 Å. At what angle of deviation from the incident direction will be the first maximum observed?*

Solution: The wavelength of the waves associated with the electron beam is given by

$$\lambda = \frac{h}{\sqrt{2meV}}$$

$$= \frac{6.63 \times 10^{-34}\,\text{Js}}{\sqrt{2 \times 9.11 \times 10^{-31}\,\text{kg} \times 1.602 \times 10^{-19}\,\text{C} \times 200\,\text{V}}}$$

$$= 0.86\ \text{Å}.$$

The diffraction is governed by the equation $2d \sin \theta = m\lambda$

For first order $m = 1$ and $2d \sin \theta = \lambda$.

$$\therefore \quad \theta = \sin^{-1}\left(\frac{\lambda}{2d}\right) = \sin^{-1}\left(\frac{0.86 \times 10^{-10}\text{ m}}{2 \times 3 \times 10^{-10}\text{ m}}\right) = \mathbf{8.31°}.$$

Example 26.2. *An enclosure filled with helium is heated to 400K. A beam of He-atoms emerges out of the enclosure. Calculate the de Broglie wavelength corresponding to He atoms. Mass of He atom is 6.7×10^{-27} kg.*

Solution. De Broglie wavelength $\lambda = \dfrac{h}{\sqrt{2mkT}}$

$$= \frac{6.63 \times 10^{-34}\text{ Js}}{\sqrt{2 \times 6.7 \times 10^{-27}\text{ kg} \times 1.376 \times 10^{-21}\text{ J/deg} \times 400}}$$

$$= \mathbf{0.769\ Å}$$

Example 26.3: *Find the de Broglie wavelength of*

(*i*) *an electron accelerated through a potential difference of 182 volts, and*

(*ii*) *a 1 kg object moving with a speed 1 m/s. Comparing the results explain why the wave nature of matter is not more apparent in daily observations.*

Solution:

(*i*) $$\lambda_e = \frac{h}{\sqrt{2\text{emV}}} = \frac{6.626 \times 10^{-34}\text{ J.s}}{\sqrt{2(1.602 \times 10^{-19}\text{ C})(9.11 \times 10^{-31}\text{ kg})182\text{ V}}}$$

$$= \frac{6.626 \times 10^{-34}\text{ J.s.}}{7.29 \times 10^{-24}\text{ kg.m/s}} = 0.91 \times 10^{-10}\,\frac{\text{kg.m}^2\text{/s}}{\text{kg.m/s}} = 9.1 \times 10^{-11}\text{ m} = \mathbf{0.91\ Å}$$

(*ii*) $$\lambda_m = \frac{h}{M\upsilon} = \frac{6.626 \times 10^{-34}\text{ J.s}}{1\text{ kg} \times 1\text{ m/s}} = 6.6 \times 10^{-34}\,\frac{\text{kg.m}^2\text{/s}}{\text{kg.m/s}} = \mathbf{6.6 \times 10^{-34}}\ \mathbf{m}.$$

It is seen from the above that the wavelength of the accelerated electron is about 10^5 times larger then its own size ($\approx 10^{-15}$ m) and is therefore significant. On the other hand, the wavelength associated with the macroscopic object is negligibly small and is thus not apparent in its interactions with other objects.

Example 26.4. *Find the energy of neutron in eV whose de-Broglie wave length is 1Å.*

Solution: We have $$\lambda = \frac{h}{mv} = \frac{h}{\sqrt{2mE}}$$

So, $$E = \frac{h^2}{2m\lambda^2}$$

Given $m = 1.674 \times 10^{-27}$ kg, $\lambda = 1$Å, $h = 6.62 \times 10^{-2}$ eV.

$$E = \frac{(6.62 \times 10^{-34})^2}{2 \times 1.674 \times 10^{-27} \times (10^{-10})} = 13.01 \times 10^{-31}\text{ J}$$

$$= 8.13 \times 10^{-12}\text{ eV}$$

QUESTIONS

1. What is de Broglie hypothesis?
2. What is a matter wave? Explain its significance with change in mass of particle.
3. Show that the de Broglie wavelength for electron is found to be equal to $(12.26/\sqrt{V})$ Å.
4. What are matter waves? Describe the experiment that supports the existence of matter waves.
5. Why is the wave nature of matter not apparent to our daily observation? Give a suitable example to illustrate the point.
6. Describe an experiment which proves the validity of de-Broglie hypothesis regarding wave nature of matter.
7. Describe Davisson and Germer experiment. What does it confirm?
8. Calculate the wavelengths of electron using Bragg law and de Broglie hypothesis using the data as obtained in the Davisson and Germer experiment. Explain the importance of results obtained.
9. What are matter waves? Why does a single monochromatic wave not represent a localized particle? Explain synthesis of a wave packet.
10. Explain in detail the experiment which establishes the relationship between momentum and wavelength of a moving particle.
11. Explain the duality of matter waves from the inferences drawn from photoelectric effect and Davisson-Germer effect.
12. Explain the de Broglie concept of matter waves. How is the wave nature of electron demonstrated experimentally? Explain with the help of an experiment.
13. Discuss de-Broglie's concept of matter waves in light of wave-particle dualism of radiation.
14. Distinguish between phase and group velocities. Show that the de Broglie wave group (wave packet) associated with a moving particle travels with same velocity as that of particle.
15. Explain the terms phase velocity, group velocity and wave packet.
16. Define phase velocity and group velocity. Derive an expression for de-Broglie wavelength from group velocity.
17. Derive a relationship between group and phase velocity of matter waves.
18. Show that the phase velocity of deBroglie wave is greater than the velocity of light, but the group velocity is equa;l to the velocity of the particle with which the waves are associated.
19. Define group velocity and obtain an expression for the same.
20. How could Davison and Germer be sure that the peak obtained for 54 volts electron was a first order diffraction peak?
21. What do you understand by a wave packet? Using the concept of matte waves, obtain the Bohr's condition for quantization of angular momentum.
22. Show how the quantization of angular momentum follows from the concept of matter waves.
23. Describe an experiment which supports the existence of matter waves.
24. Explain de Broglie's concept of matter waves. Give an account of Davisson and Germer experiment to show the wavelike character of a beam of electrons.
25. How Bohr's condition of stationary orbits of an atom can be obtained from concept of matter wave?
26. Discuss similarities and differences between a mater wave and an electromagnetic wave. Why is the wave nature of matter less apparent in our daily observations?
27. Explain how de Broglie hypothesis regarding wave nature of matter leads to Bohr's quantization condition for angular momentum of an electron.

28. Describe the construction, and working of an electron microscope.
29. With a neat sketch, explain the working of an electron microscope. Which are the features that differentiate it from an optical microscope?
30. Describe the construction, principle, and working of a scanning electron microscope.
31. An electron microscope is made by using short wavelength electrons to provide high resolution. Then is it possible to make a proton microscope and a neutron microscope? Explain your answer in each case.
32. If the particles listed below all have the same energy, which has the shortest wavelength: electron, α-particle, neutron, and proton?
33. "If an electron is localized in space, its momentum becomes uncertain. If it is localized in time, its energy becomes uncertain." Explain this statement.

PROBLEMS

1. Calculate the wavelength associated with 1MeV electron. **[Ans:** 1.23×10^{-12} m**]**
2. Calculate the energy of the neutron in eV if its de Broglie wavelength is 3×10^{-10} m. **[Ans:** 9×10^{-3} eV**]**
3. Calculate the velocity and kinetic energy of an electron of wavelength 1.66 Å. **[Ans:** 4386km/s, 54.7 eV**]**
4. Calculate de Broglie wavelength of an electron moving with velocity 10^7 m/s. **[Ans:** 0.72 Å**]**
5. Calculate the de Broglie wavelength of an α-particle accelerated through a potential difference of 400 volts. **[Ans:** 1.6×10^{-13} m**]**
6. Calculate the de Broglie wavelength of neutron of energy 12.8 MeV. **[Ans:** 8×10^{-5} Å**]**
7. Determine the velocity and kinetic energy of a neutron having de Broglie wavelength of 1Å. **[Ans:** 3.97×10^{3} m/s, 0.0825 eV**]**

27 Quantum Mechanics

Learning Objectives:

By the end of this chapter, you will be able to:

- Explain the principle of Heisenberg uncertainty and its applications as well.
- Understand the concept of Schrödinger wave equation.
- Learn about free particle.
- Describe about Harmonic Oscillator and its working.
- State Superposition principle and its use.
- Learn about fundamental postulates of quantum mechanics.

27.1 INTRODUCTION

Werner Heisenberg had formulated the first rational mathematical system for quantum theory in 1925. This was based upon connections between the quantities that are experimentally measurable, either directly or indirectly. Heisenberg took atomic transitions arising from interactions of the atoms with light quanta as the basic relevant observables. Basing on the studies on these observables Heisenberg formulated a coherent mathematical structure, later known as ***matrix mechanics*** for the quantum theory of atoms.

Following an entirely different approach Schrödinger developed in 1926 a differential equation for a wave function. The problem of calculating the energy levels of a bound microparticle was reduced by Schrödinger to the problem of finding eigenvalues. Schrödinger assumed that the discrete allowed states of any system of particles are the stable forms of continuous matter waves. He used the techniques of continuum mechanics and formulated an equation for the matter waves.

Schrodinger viewed the quantum as the apparent atomicity of matter disguised under the underlying continuum. Schrödinger theory came to be known as ***wave mechanics***. Max Born advocated that Schrödinger's waves are only pure densities of probability for finding discrete particles. Heisenberg postulated the uncertainty principle which restricted the very possibility of measurement. The uncertainty principle implied the departure of quantum mechanics from classical determinism and confirmed that quantum mechanics is a statistical theory. In

1930, P.A.M. Dirac proposed a general formalism which is a unifying concept of the matrix mechanics and wave mechanics. The finally unified theory is known as **quantum mechanics.**

27.2 HEISENBERG UNCERTAINTY PRINCIPLE

The uncertainty principle is the unique aspect of which quantum mechanics. The uncertainty principle states that we cannot determine *precise simultaneous values* of the position and momentum of a physical system. Instead, we can determine these quantities only with some typical 'uncertainties' which cannot become arbitrarily small *simultaneously*.

In 1927 Heisenberg showed that the product of uncertainty Δx in the x-coordinate of a quantum particle and the uncertainty Δp_x in the x-component of the momentum would always be of the order of Planck's constant h. Thus,

$$\Delta x \cdot \Delta p_x \approx h$$

or more precisely

$$\Delta x \cdot \Delta p_x \geq \frac{\hbar}{2} \qquad ...(27.1)$$

This is known as Heisenberg's uncertainty principle for position and momentum, which may be stated as follows:

"*It is not possible to know simultaneously and with exactness both the position and the momentum of a microparticle*".

The Uncertainty Principle indicates a built-in, inevitable *limit to the accuracy with which we can make measurements*. Classical physics considers that the precision of any measurement was limited only by the accuracy of the instruments used. Heisenberg showed that whatever may be the accuracy of the instruments used, quantum mechanics limits the precision in microworld when two properties of a particle or a system of particles are measured at the same time. These are *not just any* two properties but pairs of measurable quantities whose product has dimensions of energy × time. Such quantities are called *conjugate quantities* in quantum mechanics, and have a special relation to each other. Position–linear momentum, energy-time, time-frequency and angular momentum-angular displacement are conjugate pairs of variables.

The uncertainty principle asserts that it is physically impossible to know simultaneously the exact position ($\Delta x = 0$) and exact momentum ($\Delta p_x = 0$) of a micro-particle. According to it, the more precisely we know the position of the particle, the less precise is our information about its momentum. To localize a wave packet, we have to add more wavelengths to form the wave packet. More wavelengths mean larger $\Delta\lambda$ and more uncertainty in momentum (note that $\Delta p \propto \Delta\lambda$). Conversely, in order to have more precise value of momentum, the wave packet should contain less number of waves. Less number of waves produces a longer wave packet. Thus, the momentum of a particle cannot be precisely specified without our loss of knowledge of the position of the particle at that time.

Similarly, a particle cannot be precisely localized in a particular direction without our loss of knowledge of momentum in that particular direction. We can at best specify that certain momentum of the particle is more probable than the other or that the particle is more likely to be here than there. We cannot use classical notions like coordinates and momentum to describe the motion of quantum particles. Thus, the uncertainty principle implies that we can never define the path of an atomic particle with the absolute precision indicated in classical mechanics. Therefore, concepts such as velocity, position, and acceleration are of limited use in quantum world. Relations similar to Eq. (27.2) hold good for other components of position and linear momentum. Thus,

$$\Delta y \cdot \Delta p_y \geq \frac{\hbar}{2} \qquad ...(27.2a)$$

$$\Delta z \cdot \Delta p_z \geq \frac{\hbar}{2} \qquad ...(27.2b)$$

We may interpret the situation as follows: Classically, the state of a particle can be defined by specifying its position and momentum at any given time t. If a body is moving along x-direction with a velocity v, its position is given by $x = vt$ and its momentum is given by $p = mv$. From this,

$$x = \frac{p}{m}t \qquad ...(27.3)$$

At each instant, the position and momentum can be measured to a very high accuracy. When an atomic particle is conceptualized as a de Broglie wave packet such a precision becomes restricted. A wave packet spreads over a region of space. Therefore, it is difficult to locate the exact position of the microparticle. Although the particle is somewhere within the wave packet, it is impossible to know where exactly the particle is at a given instant. If the linear spread of the wave packet is Δx, the particle would be located somewhere within the region Δx. The probability of finding the particle is a maximum at the centre of the wave packet and falls off to zero at its ends. Therefore, there is an **uncertainty** Δx in the position of the particle. As a result, the momentum of the particle at that instant cannot be determined precisely. It means that the location and momentum of a microparticle cannot be **simultaneously** determined with certainty. Any attempt to determine these variables will lead to uncertainties in each of the variables.

27.2.1 Energy - Momentum Uncertainty

The uncertainty relation for the simultaneous measurement of energy E and time t is expressed as

$$\Delta E \cdot \Delta t \geq \frac{\hbar}{2} \qquad ...(27.4)$$

The physical significance of the energy-time uncertainty relation is different from that of the position –momentum uncertainty. If ΔE is the maximum uncertainty in the determination of the energy of a particle, then the minimum time interval for which the particle remains in that state is given by

$$\Delta t = \frac{\hbar/2}{\Delta E}$$

And, if a particle remains in a particular energy state for a maximum time Δt, then the minimum uncertainty in the particle energy is given by

$$\Delta E = \frac{\hbar/2}{\Delta t}$$

Derivation: We can obtain the Eq. (27.4) as follows. Let us consider a microparticle of mass m moving with a velocity υ. Its kinetic energy will be

$$E = \frac{1}{2}m\upsilon^2$$

If the uncertainty in the energy is ΔE, then $\Delta E = \Delta\left[\frac{1}{2}m\upsilon^2\right] = m\upsilon\,\Delta\upsilon = \upsilon\,\Delta p$

As the velocity $\upsilon = \frac{\Delta x}{\Delta t}$, the uncertainty in energy may be written as $\Delta E = \frac{\Delta x}{\Delta t}\Delta p$

Thus, $$\Delta E \cdot \Delta t = \Delta x \cdot \Delta p$$

But $$\Delta x \cdot \Delta p \geq \frac{\hbar}{2}$$

Therefore, $$\Delta E \cdot \Delta t \geq \frac{\hbar}{2}$$

The above relations are to be supplemented by the following uncertainty relation

$$\Delta M_x \cdot \Delta \varphi_x \geq \frac{\hbar}{2} \qquad ...(27.5)$$

where ΔM_x is the uncertainty in the projection of the angular momentum on the x-axis and $\Delta \Phi_x$ is the uncertainty in the angular coordinates of the microparticle.

By analogy with Eq. (27.2a) and Eq. (27.2b), we may write down relations for other projections of momentum and angular momentum as follows:

$$\Delta M_x \cdot \Delta \varphi_y \geq \hbar/2 \text{ and } \Delta M_z \cdot \Delta \varphi_z \geq \hbar/2 \qquad ...(27.5a)$$

In general if q and p denote two canonically conjugate variables, the uncertainty relation is given by

$$\Delta q \cdot \Delta p \geq \hbar/2 \qquad ...(27.6)$$

The above relations do not mean that the uncertainty principle creates certain obstacles to the understanding of the atomic phenomena; it only reflects certain peculiarities of the objective properties of a quantum particle.

27.3 ELEMENTARY PROOF OF UNCERTAINTY PRINCIPLE USING DE BROGLIE WAVE CONCEPT

A wave packet produced by a superposition of large number of harmonic waves is shown in Fig. 27.1. Since a wave packet is not an infinite harmonic wave, it has a range of wave numbers Δk instead of one definite wave number. Δk is thus the uncertainty in wave number. Further, the position of the particle cannot be given with certainty. It will lie somewhere between the two consecutive nodes (see Fig. 27.1). Thus, the uncertainty in the position of the particle is equal to the distance between two consecutive nodes.

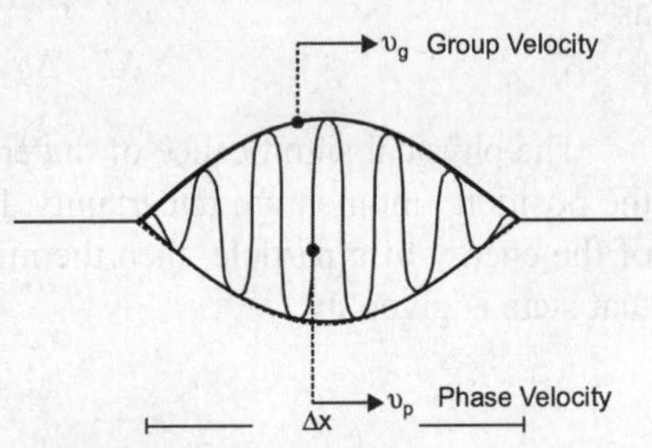

Fig. 27.1: Wave packet

Referring to Eq. (26.15), the condition for formation of a node is

$$\cos\left(\frac{\Delta\omega t}{2} - \frac{\Delta k x}{2}\right) = 0$$

i.e., $$\frac{\Delta\omega}{2}t - \frac{\Delta k}{2}x = \frac{\pi}{2}, \frac{3\pi}{2}, \frac{5\pi}{2}, ..., (2n+1)\frac{\pi}{2}$$

Thus, if x_1 and x_2 are the positions of two consecutive nodes, then

$$\frac{\Delta\omega}{2}t - \frac{\Delta k}{2}x_1 = (2n+1)\frac{\pi}{2}$$

and $$\frac{\Delta\omega}{2}t - \frac{\Delta k}{2}x_2 = (2n+3)\frac{\pi}{2}$$

Subtracting the above upper equation from the lower one, we find that

$$\frac{\Delta k}{2}(x_2 - x_1) = \pi \quad \text{or} \quad \frac{\Delta k}{2}\Delta x = \pi$$

where $\Delta x = (x_2 - x_1)$ is the uncertainty in the position of the particle. Thus,

$$\Delta x = \frac{2\pi}{\Delta k}.$$

Since $$k = \frac{2\pi}{\lambda} = \frac{2\pi p}{h},$$

$$\Delta k = \frac{2\pi}{h}\Delta p$$

Therefore, $$\Delta x = \frac{2\pi}{\Delta k} = \frac{2\pi h}{2\pi\,\Delta p} = \frac{h}{\Delta p}$$

or $$\Delta x \cdot \Delta p = h$$

When we consider a group consisting of very large number of harmonic waves of continuously varying frequencies, the product of the uncertainties comes to

$$\Delta x \cdot \Delta p \geq \frac{1}{2}\hbar$$

27.4 IMPLICATION OF UNCERTAINTY PRINCIPLE

The uncertainty principle expresses a fundamental limitation in nature that also limits the precision of our measurements. According to classical mechanics the position and momentum of a macro-particle can be determined exactly. But the uncertainty principle asserts that it is physically impossible to know simultaneously the exact position ($\Delta x = 0$) and exact momentum ($\Delta p_x = 0$) of a microparticle. According to it, the more precisely we know the posistion of the particle, the less precise is our information about its momentum. Thus, the momentum of a particle cannot be precisely specified without our loss of knowledge of the position of the particle at that time. Similarly, a particle cannot be precisely localized in a particular direction without our loss of knowledge of momentum in that particular direction. We can at best specify that certain momentum of the particle is more probable than the other or that the particle is more likely to be here than there. It means that our classical notions like coordinates and momentum derived from ordinary macroscopic experiences are inadequate to describe the atomic world.

The uncertainty principle points out that in the microscopic world,

(*i*) the dynamical variables of a particle are combined in sets of *simultaneously determined* quantities which are known as *complete sets* of quantities;

(*ii*) the coordinate and momentum components of a particle etc are **pairs of concepts** which are interrelated and fall in different complete sets of quantities. They cannot be defined simultaneously in a precise way.

Thus, the uncertainty principle implies that we can never define the path of an atomic particle with the absolute precision indicated in classical mechanics. Therefore, concepts such as velocity, position, and acceleration are of limited use in quantum world. To describe the quantum particle the concept of *energy* becomes important since it is related to the *state* of the system rather than to its path.

27.5 UNCERTAINTY PRINCIPLE IS NOT SIGNIFICANT IN CASE OF MACRO-BODIES

The Heisenberg Principle is of no practical importance for heavy bodies where the de Broglie wavelength is negligibly small.

For example, let us take the case of a cricket ball in flight. The indeterminacy in the position of the ball is, say, 1 mm. We can determine the indeterminacy of velocity of the ball ($m = 0.5$ kg) from uncertainty principle.

$$\Delta x \cdot \Delta p \approx h$$

$$\therefore \qquad \Delta x \cdot m\Delta \upsilon \approx h$$

$$\Delta \upsilon \approx \frac{h}{m\Delta x} = \frac{6.62 \times 10^{-34}\,\text{J.s}}{0.5\text{kg} \times 10^{-3}\,\text{m}} \cong 10^{-30}\,\text{m/s}.$$

The above inaccuracy is negligible and not detectable. It implies that the uncertainties are of no importance in case of macro bodies; and the position and velocity of a macro body can be simultaneously determined with a high degree of accuracy. As a result, macroscopic body follows a well defined trajectory.

In contrast if we take the example of an electron orbiting in a hydrogen atom, the inaccuracy in its position is ± 1Å. The uncertainty in its speed is

$$\Delta \upsilon = \frac{h}{m\Delta x} = \frac{6.62 \times 10^{-34}\,\text{J.s}}{9.11 \times 10^{-31}\,\text{kg} \times 2 \times 10^{-10}\,\text{m}} \approx 2 \times 10^{5}\,\text{m/s}$$

which is of the same order as the velocity of the electron in the orbit. It means that it is not possible to determine the velocity and the position of a microparticle with certainty and as such we cannot talk of a specific trajectory. Instead we have to be content knowing only the probable values.

27.6 THOUGHT EXPERIMENTS

It is not possible to verify Heisenberg uncertainty principle in the laboratory. Therefore, we illustrate it with the help of two *thought experiments*.

(i) Treating the electron as a wave: We assume that an electron has wave character. Suppose that we want to determine the *y*-coordinate of an electron moving along the *x*-axis. We place a slit of width '*d*' perpendicular to the direction of motion of the electron (Fig. 27.2). The precision of the position of electron, Δy, in *y*-direction is limited by the size of the slit. That is, $\Delta y \approx d$.

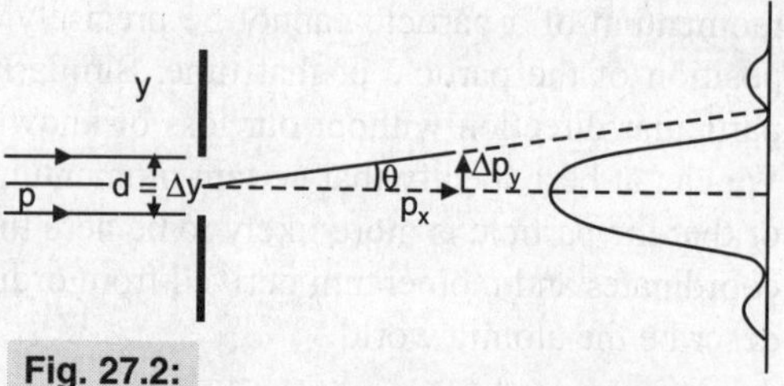

Fig. 27.2: Thought experiment -Treating the electron as a wave

If the slit is narrow enough, it causes a change in the motion of the electron after going through the slit and brings out the wave character of electron as evidenced from the diffraction pattern observed on the screen. The uncertainty in the electron momentum parallel to *y*-axis depends on the diffraction angle θ. According to the theory of diffraction at a single slit, the angle θ is given by

$$\sin\theta = \lambda/d \qquad ...(27.7)$$

The uncertainty in the momentum of the electron parallel to y-axis is given by

$$\Delta p_y = p \sin\theta = \frac{h\lambda}{\lambda d} = \frac{h}{d} \approx \frac{h}{\Delta y}$$

$$\therefore \qquad \Delta y\, \Delta p_y \approx h \qquad \text{...(27.8)}$$

If we wish to determine the exact position of the electron along the y-axis, we have to use a very narrow slit. However, a very narrow slit produces a wider diffraction pattern, which leads to a larger uncertainty in our knowledge of the y-component of the momentum. Conversely, if we attempt to reduce the uncertainty in our knowledge of y-component of momentum, the diffraction pattern should be very narrow. Therefore, we have to use a very wide slit, which in turn results in a large uncertainty in the y-coordinate of the electron. Thus, our efforts to simultaneously reduce the uncertainties Δy and Δp will get frustrated.

(ii) Treating the Electron as a Particle: We now consider that electron is a particle and attempt to measure its position. For the sake of convenience, we assume that the electron is at rest and use a microscope to locate it. We cannot use an optical microscope for this purpose. The act of observing an object requires that light from a source be reflected by the object and enter a recording instrument such as eye. An object reflects a wave when the size of the object is sufficiently larger than the wavelength of the wave. In the present case, if we use light waves, they would pass on without getting reflected by the electron, since the size of the electron is about 10^5 times smaller than the wavelength of light. Therefore, we use a γ-ray microscope to detect and locate an electron.

Let a free electron be directly beneath the center of the γ-ray microscope's lens. The circular lens forms a cone of angle 2α from the electron. The electron is illuminated from the left by γ-rays. The microscope can resolve objects to a size of Δx. Δx is given by the expression

$$\Delta x = \frac{\lambda}{2\sin\alpha} \qquad \text{...(27.9)}$$

To be observed by the microscope, the γ-ray must be scattered into any angle within the cone of angle 2α. A γ-photon carries a very large momentum. When the γ-photon strikes the electron, part of the momentum and energy are transferred to the electron due to Compton scattering. Consequently, as the scattered photon enters the microscope, the electron has already moved away in a certain direction (Fig. 27.3). The total momentum p is related to the wavelength by the formula $p = \dfrac{h}{\lambda}$.

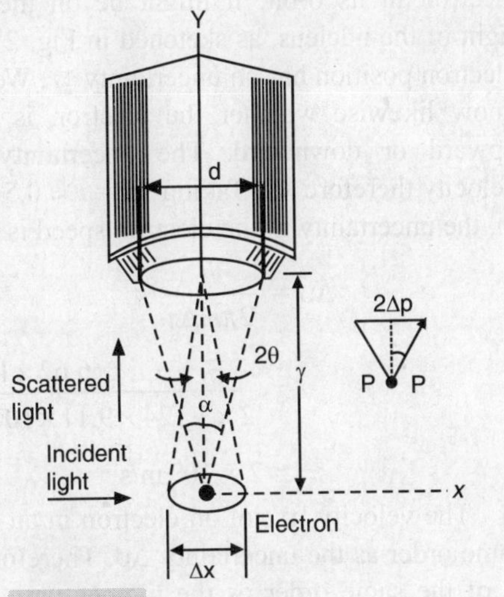

Fig. 27.3: Measurement of position and momentum of an electron by means of a γ-microscope.

The total momentum in the x direction would be the sum of the electron's momentum p'_x in the x-direction in the extreme case of diffraction of the gamma ray to the right edge of the lens.

Gamma ray's momentum in the x-direction is

$$p'_x + \left(\frac{h \sin\alpha}{\lambda'}\right) \qquad \text{...(27.10)}$$

where λ' is the wavelength of the deflected gamma ray. In the other extreme, the observed gamma ray recoils backward, just hitting the left edge of the lens. In this case, the total momentum in the x-direction is:

$$p_x'' - \left(\frac{h \sin\alpha}{\lambda''}\right) \qquad (27.11)$$

As the momentum is conserved, in each case the final x-momentum must equal the initial x-momentum.

So, $$p_x' + \left(\frac{h \sin\alpha}{\lambda'}\right) = p_x'' - \left(\frac{h \sin\alpha}{\lambda''}\right)$$

If α is small, then the wavelengths are approximately the same, $\lambda' \approx \lambda'' \approx \lambda$. And we have

$$p_x'' - p_x' = \frac{2h \sin\alpha}{\lambda}$$

or $$\Delta p_x = \frac{2h \sin\alpha}{\lambda} \qquad ...(27.12)$$

Using Eq. (27.9) into the above equation, we obtain

$$\Delta p_x = \frac{h}{\Delta x}$$

$$\therefore \qquad \Delta x \cdot \Delta p_x = h$$

27.7 APPLICATIONS OF UNCERTAINTY PRINCIPLE

We deal here with three simple examples to illustrate the application of uncertainty principle.

***(a)* Bohr's Orbit and Energy:** Let us consider the electron in a hydrogen atom. We cannot know at any instant the position of the electron in its orbit. It might be on the left or right of the nucleus, as sketched in Fig. 27.4. The electron position has an uncertainty $\pm r$. We cannot know likewise whether the electron is moving upward or downward. The uncertainty in its velocity therefore $\pm\upsilon$. Taking $\Delta x = r \approx 0.5 \times 10^{-10}$ m, the uncertainty in the electron speed is

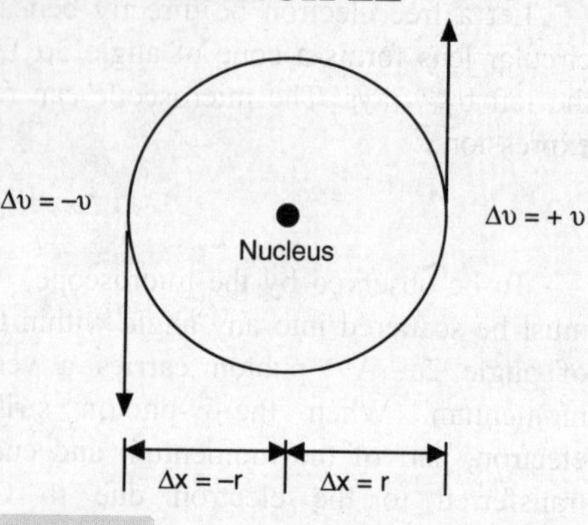

Fig. 27.4: The uncertainties in the position and velocity of electron in an atom.

$$\Delta\upsilon = \frac{h}{2\pi m\Delta x}$$

$$= \frac{6.62\times10^{-34}\text{ J.s}}{2\times3.124\times9.11\times10^{-31}\text{ kg}\times0.5\times10^{-10}\text{ m}}$$

$$\cong 2 \times 10^6 \text{ m/s}$$

The velocity 'υ', of an electron in an atom is of the order of 1.0×10^6 m/s and is of the same order as the uncertainty $\Delta\upsilon$. Therefore, we conclude that the uncertainty in momentum is of the same order as the momentum. That is $\Delta p \approx p$. It means that sharp position and momentum do not exist simultaneously for the electron in an atom. Hence it is not possible to ascribe any specific trajectory to an electron in an atom. It can only be said that atomic electrons traverse the whole of the space about the nucleus, but however, they move most of the time at a distance corresponding to a permitted Bohr radius.

Now let us calculate the energy of the Bohr's first orbit. The total energy of the electron in the first orbit is given by

$$E = \text{K.E.} + \text{P.E.} = \left[\frac{1}{2}m\upsilon^2\right] - \frac{e^2}{4\pi\varepsilon_o r} = \left[\frac{p^2}{2m}\right] - \frac{e^2}{4\pi\varepsilon_o r} \qquad \text{...(27.13)}$$

where p is the momentum of the electron.

As $\Delta p \approx p$, we can write $p \approx \dfrac{h}{2\pi\Delta x} = \dfrac{h}{2\pi r}$. ...(27.14)

$$\therefore \qquad E = \frac{h^2}{8\pi^2 m r^2} - \frac{e^2}{4\pi\varepsilon_o r} \qquad \text{...(27.15)}$$

But r is given by $r = \dfrac{\varepsilon_o h^2}{\pi\, me^2}$.

$$\therefore \qquad E = -\frac{me^4}{8\varepsilon_o^2 h^2} \qquad \text{...(27.16)}$$

Eq. (27.16) is the same as that is given by Bohr theory.

***(b)* Particle in a Box:** Let us consider a particle confined to a box of length l. The uncertainty Δx in the position is l.

$$\Delta x \cdot \Delta p \cong \hbar$$

$$\therefore \qquad \Delta p = \frac{\hbar}{\Delta x} = \frac{\hbar}{l} \qquad \text{...(27.17)}$$

Energy is given by $$E = \frac{p^2}{2m} \approx \frac{(\Delta p)^2}{2m} = \frac{(\hbar / l)^2}{2m} = \frac{\hbar^2}{2ml^2} \qquad \text{...(27.18)}$$

This result agrees with the result obtained from Schrödinger equation.

***(c)* Electrons cannot be present in the nucleus:** The radiation emitted by radioactive nuclei consists of α, β and γ-rays, out of which β-rays are identified to be electrons. We apply uncertainty principle to find whether electrons are coming out of the nucleus. The radius of the nucleus is of the order of 10^{-14} m. Therefore, if electrons were to be in the nucleus, the maximum uncertainty Δx in the position of the electron is equal to the diameter of the nucleus. Thus,

$$\Delta x = 2 \times 10^{-14}\ \text{m}.$$

The minimum uncertainty in its momentum is then given by

$$\Delta p = \frac{\hbar}{\Delta x} = \frac{1.04\times10^{-34}\ \text{J.s}}{2\times10^{-14}\ \text{m}} = 5.2 \times 10^{-21}\ \text{kg-m/s}.$$

The minimum uncertainty in momentum can be taken as the momentum of the electron. Thus,

$$p = 5.2 \times 10^{-21}\ \text{kg-m/s}.$$

The minimum energy of the electron in the nucleus is then given by

$$E_{\text{min}} = p_{\text{min}}\, c = (5.2 \times 10^{-21}\ \text{kg-m/s})(3 \times 10^8\ \text{m/s}) = 1.56 \times 10^{-12}\ \text{J} = 9.7\ \text{MeV}.$$

It implies that if an electron exists within the nucleus, it must have a minimum energy of about 10 MeV. But the experimental measurements showed that the maximum kinetic energies of β-particles were of the order of 4 MeV only. Hence electrons are not present in the nucleus. It is subsequently established that emission of β-particles occurs due to transformations in the nucleus. The transformation of a neutron into a proton produces an electron.

Example 27.1: *Uncertainty in time of an excited atom is about 10^{-8}s. What are the uncertainties in energy and in frequency of the radiation?*

Solution:
$$\Delta E\,\Delta t \approx \frac{h}{2\pi}$$

$$\therefore \quad \Delta E = \frac{1.054\times10^{-34}\,\text{J.s}}{10^{-8}\,\text{s}} = \frac{1.054\times10^{-26}}{1.602\times10^{-19}}\text{eV} = 6.58\times10^{-8}\,\text{eV}.$$

$$\Delta\nu = \frac{\Delta E}{h} = \frac{1.054\times10^{-26}\,\text{J}}{6.626\times10^{-34}\,\text{J.s}} = \mathbf{15.9\ MHz.}$$

Example 27.2. *An electron is confined to a potential well of width 10 nm. Calculate the minimum uncertainty in its velocity.*

Solution:
$$\Delta x\,\Delta p \approx \frac{h}{2\pi} \quad \text{or } \Delta x\,\Delta p \approx \frac{h}{2\pi}$$

$$\therefore \quad \Delta \upsilon = \frac{h}{2\pi\, m\cdot\Delta x}$$

$$\therefore \quad \Delta \upsilon = \frac{6.63\times10^{-34}\,\text{Js}}{2\times3.143\times9.11\times10^{-31}\,\text{kg}\times10\times10^{-9}\,\text{m}} = \mathbf{12.1\ km/s.}$$

Example 27.3: *If the kinetic energy of an electron known to be about 1 eV, must be measured to within 0.0001 eV, what accuracy can its position be measured simultaneously?*

Solution:
$$E = \frac{p^2}{2m} \quad \therefore \quad \Delta\text{E} = \frac{2p\,\Delta p}{2m} \quad \therefore \quad \Delta\text{p} = \frac{m}{p}\Delta E$$

$$\Delta x\Delta p = \frac{h}{2\pi}$$

$$\therefore \quad \Delta x = \frac{h}{2\pi\cdot\Delta p} = \frac{h}{2\pi}\cdot\frac{p}{m\Delta E} = \frac{h\sqrt{2mE}}{2\pi\, m\Delta E} = \frac{h}{\pi\,\Delta E}\sqrt{\frac{E}{2m}}$$

$$\therefore \quad \Delta x = \frac{6.63\times10^{-34}\,\text{Js}}{3.143\times0.0001\times1.602\times10^{-19}\,\text{J}}\cdot\sqrt{\frac{1.602\times10^{-19}\,\text{J}}{2\times9.11\times10^{-31}\,\text{kg}}}$$

$$= \mathbf{1.95\ \mu m.}$$

Example 27.4: *An electron and a 150 gm base ball are traveling at a velocity of 220 m/s, measured to an accuracy of 0.005 %. Calculate and compare uncertainty in position of each.*

Solution: The uncertainty in the velocity is $\Delta\upsilon = \upsilon\times0.065\% = (220\text{m/s})\times\frac{0.065}{100} =$ 0.143 m/s.

(*i*) The uncertainty in the position of electron is

$$\Delta x_e = \frac{\hbar}{2\,m\Delta\upsilon} = \frac{1.05\times10^{-34}\,\text{J.s}}{2\times9.11\times10^{-31}\,\text{kg}\times0.143\,\text{m/s}} = \mathbf{0.4\ mm.}$$

(*ii*) The uncertainty in the position of baseball is

$$\Delta x_B = \frac{\hbar}{2\,M\Delta\upsilon} = \frac{1.05\times10^{-34}\ \text{J.s}}{2\times0.15\ \text{kg}\times0.143\ \text{m/s}} = \mathbf{2.5\times10^{-33}m.}$$

27.8 WAVE FUNCTION AND PROBABILITY INTERPRETATION

A propagating disturbance in a medium is waves and light waves, sound waves, and water waves are typical examples. These waves are described by a certain quantity that varies with position and time. Light waves consist of variations of electric and magnetic fields in space, and sound waves consist of pressure variations. We cannot specify in a similar way what is actually varying in de Broglie waves. Since microparticles exhibit wave properties, it is assumed that a quantity ψ represents a de Broglie wave. This quantity ψ is called a **wave function**. ψ describes the wave as a function of position and time. However, it has no direct physical significance, as it is not an *observable* quantity. In general, ψ is a complex-valued function. According to Heisenberg uncertainty principle, we can only know the probable value in a measurement. The probability cannot be negative. Hence ψ cannot be a measure of the presence of the particle at the location (x, y, z). But it is certain that it is in someway an index of the presence of the particle at around (x, y, z, t).

Probability Interpretation of Wave Function given by Max Born

A probability interpretation of the wave function was given by Max Born in 1926. He suggested that *the square of the magnitude of the wave function* $|\psi|^2$ *evaluated in a particular region represents the probability of finding the particle in that region.*

In other words, probability, P, of finding the particle in an infinitesimal volume dV $(= dx\,dy\,dz)$ is proportional to $|\psi(x, y, z)|^2\,dx\,dy\,dz$ at time t.

or $$P \propto |\psi(x,y,z)|^2\,dV \qquad ...(27.19)$$

$|\psi|^2$ is called the **probability density** and ψ is the **probability amplitude.**

Since the particle is certainly somewhere in the space, the probability $P = 1$ and the integral of $|\psi|^2 dV$ over the entire space must be equal to unity. That is

$$\int_{-\infty}^{+\infty} |\psi|^2\,dV = 1 \qquad ...(27.20)$$

The wave function ψ is in general a complex function. But the probability must be real. Therefore to make probability a real quantity, ψ is to be multiplied by its complex conjugate ψ^*.

$$\int_{-\infty}^{+\infty} \psi\psi^*\,dV = 1$$

Thus, ψ has no physical significance but $|\psi|^2$ gives the probability of finding the atomic particle in a particular region.

27.8.1 Normalization Condition

If at all the particle exists, it must be found somewhere in the universe. Since we are sure that the particle must be somewhere in space, the sum of the probabilities over all values of x,y,z must be unity. If $\Psi(x, y, z, t)$ is multiplied by a constant C such that $\Psi_N(x,y,z,t) = C\,\Psi(x,y,z,t)$, where $\Psi_N(x, y, z, t)$ satisfies the relation

$$\int_{-\infty}^{\infty} |\Psi_N(x,y,z,t)|^2 \, dxdydz = |C|^2 \int_{-\infty}^{\infty} |\Psi(x,y,z,t)|^2 \, dxdydz = 1 \qquad ...(27.21)$$

then Ψ_N (x,y,z,t) is said to be **normalized** wave function and C the ***normalization constant.***

This condition Eq. (27.21) is known as the *normalization condition*. From Eq. (27.21), we have

$$|C|^2 = \frac{1}{\int_{-\infty}^{\infty} |\Psi(x,y,z,t)|^2 \, dxdydz} \qquad ...(27.22)$$

$|\psi_N(x, y, z, t)|^2 \, dxdydz$ is called the ***probability density***.

Whenever wave functions are normalised, $|\psi|^2 dV$ equals the probability that a particle will be found in an elemental volume dV.

Thus, Probability

$$P = |\psi(x, y, z)|^2 dV \qquad ...(27.23)$$

27.8.2 Well-behaved Wave Functions - Conditions to be satisfied by ψ-function

An acceptable wave function ψ must be normalized and fulfill the following requirements:

(*i*) **ψ function must be finite:** The wave function must be finite everywhere. Even if $x \to \infty$ or $-\infty$, $y \to \infty$ or $-\infty$, $z \to \infty$ or $-\infty$, the wave function should not tend to infinity. It must remain finite for all values of x, y, z. If ψ is infinite, it would imply an infinitely large probability of finding the particle at that point. This would violate the uncertainty principle.

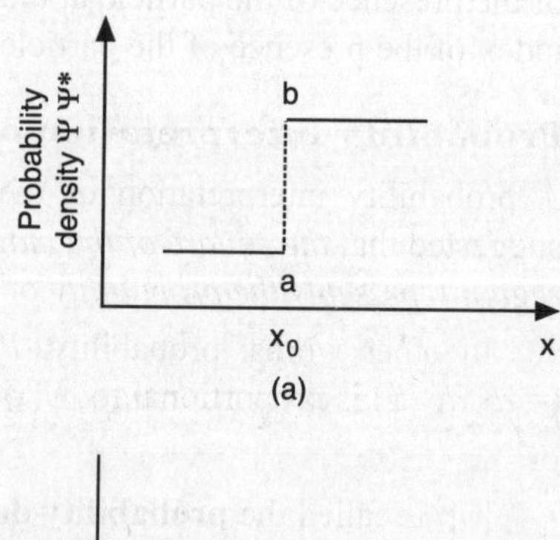

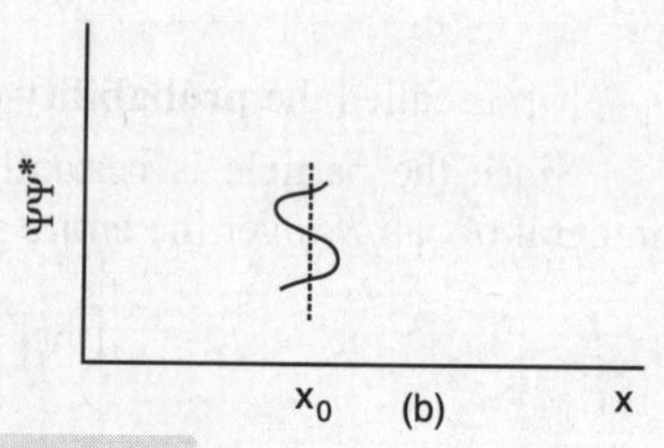

Fig. 27.5: Probability of finding the particle

(*ii*) **ψ function must be single–valued:** Any physical quantity can have only one value at a point. For this reason, the function related to a physical quantity cannot have more than one value at that point. If it has more than one value at a point, it means that there is more than one value of probability of finding the particle at that point (Fig. 27.5).

(*iii*) **ψ function must be continuous:** ψ function should be continuous across any boundary. Since ψ is related to a physical quantity, it cannot have a discontinuity at any point. Therefore, the wave function ψ and its space derivatives $\frac{\partial \psi}{\partial x}$, $\frac{\partial \psi}{\partial y}$ and $\frac{\partial \psi}{\partial z}$ should be continuous across any boundary. Since ψ is related to a real particle, it cannot have a discontinuity at any boundary where potential changes.

Wave functions satisfying the above mathematical conditions are called ***well-behaved wave functions.***

27.9 SCHRÖDINGER WAVE EQUATION

Schrödinger assumed that the de Broglie waves associated with electrons are similar to the light waves and formulated wave equation that describes the behaviour of matter waves. It is known as Schrodinger wave equation. This equation defines the wave properties of electrons and also predicts their particle-like behaviour.

Time dependent Schrödinger Wave Equation

Since the concept of de Broglie waves is not a result of previous physical theories, it is not possible to derive a wave equation for the wave associated with a particle. However, a wave equation can be developed. Let us assume that a particle of mass 'm' is in motion along the x-direction. Let the wave function ψ be the dependent variable of the de Broglie wave which is a function of the coordinates x and t. Analogous to the classical wave, we may expect that, ψ will be a function of $(x - \upsilon t)$. As $\upsilon = \omega / k$, the wave function may be written as a function of $(kx - \omega t)$. Using the relation $p = \hbar k$ and $E = \hbar\omega$, we can write

$$\psi = f\left(\frac{px - Et}{\hbar}\right) \qquad ...(27.24)$$

The more general wave would be a sum of a sine and cosine waves. Taking help of Euler's identity, we write the Eq. (27.24) in an exponential form as follows:

$$\psi = A \exp\left[\frac{i}{\hbar}(px - Et)\right] \qquad ...(27.25)$$

We assume that the energy and momentum of the particle are constant. Differentiating the Eq. (27.25) with respect to x, we get

$$\frac{\partial\psi}{\partial x} = A\frac{ip}{\hbar} \exp\left[\frac{i}{\hbar}(px - Et)\right]$$

$$\therefore \quad \frac{\partial\psi}{\partial x} = \frac{ip}{\hbar}\psi$$

Rearranging the terms in the above equation, we get

$$p\psi = \frac{\hbar}{i}\frac{\partial\psi}{\partial x} = -i\hbar\frac{\partial\psi}{\partial x} \qquad ...(27.26)$$

Differentiating the Eq. (27.25) with respect to t gives

$$\frac{\partial\psi}{\partial t} = -\frac{iE}{\hbar}A \exp\left[\frac{i}{\hbar}(px - Et)\right] = -\frac{iE}{\hbar}\psi$$

Rearranging the terms in the above equation, we get

$$E\psi = -\frac{\hbar}{i}\frac{\partial\psi}{\partial t} = i\hbar\frac{\partial\psi}{\partial t} \qquad ...(27.27)$$

The partial derivatives with respect to x and t are connected by means of the relation between the energy and momentum. The classical expression for the kinetic energy in terms of the momentum is

$$E_k = \frac{m\upsilon^2}{2} = \frac{(m\upsilon)^2}{2m} = \frac{p^2}{2m} \qquad ...(27.28)$$

The total energy and momentum are related by the expression

$$\frac{p^2}{2m} + V = E \qquad ...(27.29)$$

where V is the potential energy of the particle. Multiplying the Eq. (27.29) with ψ, we obtain

$$\frac{p^2}{2m}\psi + V\psi = E\psi$$

Using the Eq. (27.26) and Eq. (27.27) into Eq. (27.29), we get

$$-\frac{\hbar^2}{2m}\frac{\partial^2\psi}{\partial x^2} + V\psi = i\hbar\frac{\partial\psi}{\partial t}$$

The above equation may be rewritten as

$$i\hbar\frac{\partial\Psi}{\partial t} = \left[-\frac{\hbar^2}{2m}\frac{\partial^2}{\partial x^2} + V\right]\Psi \qquad ...(27.30)$$

Eq. (27.30) is known as the **time-dependent Schrödinger wave equation.**

When extended to the three-dimensional case, we find that

$$-\frac{\hbar^2}{2m}\left(\frac{\partial^2\psi}{\partial x^2} + \frac{\partial^2\psi}{\partial y^2} + \frac{\partial^2\psi}{\partial z^2}\right) + V\psi = i\hbar\frac{\partial\psi}{\partial t}$$

$$-\frac{\hbar^2}{2m}\left(\frac{\partial^2}{\partial x^2} + \frac{\partial^2}{\partial y^2} + \frac{\partial^2}{\partial z^2}\right)\psi + V\psi = i\hbar\frac{\partial\psi}{\partial t}$$

or

$$-\frac{\hbar^2}{2m}\nabla^2\psi + V\psi = i\hbar\frac{\partial\psi}{\partial t}$$

where

$$\nabla^2 = \frac{\partial^2}{\partial x^2} + \frac{\partial^2}{\partial y^2} + \frac{\partial^2}{\partial z^2}.$$

Therefore, the Schrödinger equation for three dimensional motion may be written as

$$i\hbar\frac{\partial}{\partial t}\Psi(r,t) = \left[-\frac{\hbar^2}{2m}\nabla^2 + V(r)\right]\Psi(r,t) \qquad ...(27.31)$$

Time independent Schrödinger Wave Equation

Knowing the form of V, Eq. (27.30) can be solved for the wave function ψ. In a number of cases the potential energy V of a particle does not depend on time; it varies with the position of the particle only and the field is said to be **stationary**. In the stationary problems Schrödinger equation can be simplified by separating out time and position-dependent parts. Accordingly, we can write the wave function as a product of x, $\psi(x)$ and a function of t, $\varphi(t)$.

We, therefore, write that

$$\psi(x,t) = \psi(x)\varphi(t) \qquad ...(27.32)$$

Eq. (27.30) may be written as

$$-\frac{\hbar^2}{2m}\frac{\partial^2}{\partial x^2}(\psi,\phi) + V\psi\phi = i\hbar\frac{\partial}{\partial t}(\psi,\phi)$$

$$-\frac{\hbar^2}{2m}\phi\frac{\partial^2}{\partial x^2}(\psi,\phi) + V\psi\phi = i\hbar\frac{\partial}{\partial t}(\psi,\phi)$$

Dividing the above equation with $\psi\phi$, we get

$$-\frac{\hbar^2}{2m}\frac{1}{\psi}\frac{d^2\psi}{dx^2} + V = i\hbar\frac{1}{\phi}\frac{d\phi}{dt} \qquad ...(27.33)$$

If we assume that the potential energy V is a function of x only, the entire left hand side of Eq. (27.33) is a function of x only while the right hand side is a function of t only. Since x and t are independent variables, both the function of x and t must be equal to a constant. The constant that each side must equal is called the separation constant E.

Thus,
$$-\frac{\hbar^2}{2m}\frac{1}{\psi}\frac{d^2\psi}{dx^2}+V = E \qquad ...(27.34)$$

and
$$i\hbar\frac{1}{\phi}\frac{d\phi}{dt} = E$$

Eq. (27.34) may be rewritten as
$$-\frac{\hbar^2}{2m}\frac{d^2\psi}{dx^2}+V\psi = E\psi \qquad ...(27.35)$$

Eq. (27.35) involves only the space coordinates and is called the **time-independent Schrödinger wave equation.**

Eq. (27.35) may be rewritten as
$$\frac{d^2\psi}{dx^2}+\frac{2m}{\hbar^2}(E-V)\psi = 0$$

$$\therefore \quad \frac{d^2\psi}{dx^2}+\frac{8\pi^2 m}{h^2}(E-V)\psi = 0 \qquad ...(27.35a)$$

In three dimensions, Eq. (27.35) may be written as
$$-\frac{\hbar^2}{2m}\nabla^2\psi+V\psi = 0 \qquad ...(27.35b)$$

27.9.1 Allowed Wave Functions and Energies

The time-independent wave equation (27.35) is the pertinent equation for studying properties of atomic systems in stationary conditions. Wave mechanical methods of solving the problem of particle motion are essentially based on ψ functions. Appropriate wave equation is formulated by incorporating the particle mass 'm' and potential energy function V for the region in which the particle is located. The next step consists of solving the differential equation for solutions, namely for the ψ functions which will satisfy the differential equation. By solving the Schrödinger equation, we obtain the possible set of ψ functions. In case of bound particles the acceptable solutions for the differential equation are possible only for certain specified values of energy. These energy values will be the only possible results of precise measurements of the total energy of the particle. These discrete values of energy E_1, E_2,.....E_n are called **eigenvalues** or **allowed values** of the energy of the particle. The solutions ψ_1, ψ_2, ...ψ_n corresponding to the eigen energy values E_n are called the *eigen functions.* The quantization of energy thus appears as a natural element of the wave equation.

In the course of formulation, Schrödinger did not consider the existence of the stationary states. The quantization of energy for the different stationary states followed directly from the mathematical development of the wave equation. The wave mechanics reproduced all the results of the Bohr's theory without resorting to any inconsistent hypotheses. It also accounts for the experimental information such as the probability of an electron changing from one state to another, which the Bohr's theory failed to explain.

We now apply the Schrödinger wave equation to different cases of particles confined to move in different enclosures and obtain information regarding the energy values it can take and its probability density etc.

27.10 THE FREE PARTICLE

A particle is said to be a ***free particle*** when it is moving in space without being subjected to any external force in any region of space, and its potential energy is constant [V = constant] everywhere. We study here the one dimensional motion of a particle under no forces.

Let us assume that the particle is moving in space along the positive x-direction. As the particle is not acted upon by a force, the potential energy of the particle is constant. For convenience, we take the constant potential to be zero. Then, the time-independent Schrödinger wave equation for the free particle is written as

$$\frac{d^2\psi}{dx^2}+\frac{8\pi^2 m}{h^2}E\psi = 0 \qquad ...(27.36)$$

As the particle is moving freely with zero potential energy, its total energy will be kinetic energy which is given by

$$E = \frac{p_x^2}{2m} \qquad ...(27.37)$$

where p_x is the momentum of the particle along the x-direction.

We may rewrite the Eq. (27.36) as

$$\frac{d^2\psi}{dx^2}+k^2\psi = 0 \qquad ...(27.38)$$

where $$\frac{8\pi^2 mE}{h^2} = k^2 \qquad ...(27.39)$$

(a) Wave function: The general solution of the Eq. (27.38) is of the form

$$\psi(x) = \mathrm{A}e^{ikx} + \mathrm{B}e^{-ikx} \qquad ...(27.40)$$

where A and B are constants of integration. Eq.(27.66) gives the time-independent part of the wave function. The complete wave function is given by

$$\begin{aligned}\psi(x, t) &= \psi(x)e^{-i\omega t}\\ &= (Ae^{ikx} + Be^{-ikx})\, e^{-i\omega t}\\ &= Ae^{-i(\omega t-kx)} + Be^{-i(\omega t+kx)}\end{aligned} \qquad ...(27.41)$$

Eq. (27.41) represents a continuous plane simple harmonic wave. The first term represents the wave traveling along the positive x-direction while the second term represents the wave traveling along the negative x-direction. The particle traveling in the positive x-direction is represented by

$$\psi(x,t) = Ae^{-i(\omega t-kx)} \qquad ...(27.42)$$

(b) Energy: It is seen from Eq. (27.39) that k has the following value.

$$k = \sqrt{\frac{8\pi^2 mE}{h^2}} = \sqrt{\frac{2mE}{\hbar^2}} \qquad ...(27.43)$$

Therefore, the particle energy is given by

$$E = \frac{\hbar^2 k^2}{2m} \qquad ...(27.44)$$

There are no boundary conditions to be applied to the particle motion and hence there are no restrictions placed on k. It follows from Eq. (27.44) that the particle is permitted to have any value of energy. In other words, the energy is not quantized. It means that a freely moving particle possesses a continuous energy spectrum as shown in Fig. 27.6.

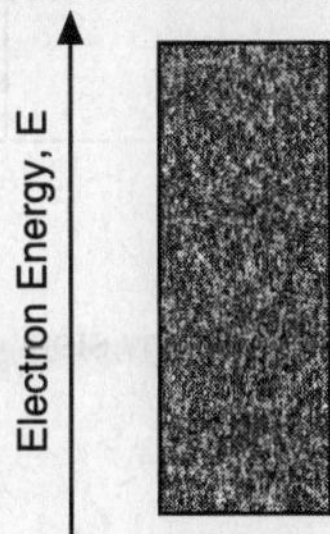

Fig. 27.6:

Continuous energy spectrum for freely moving particle

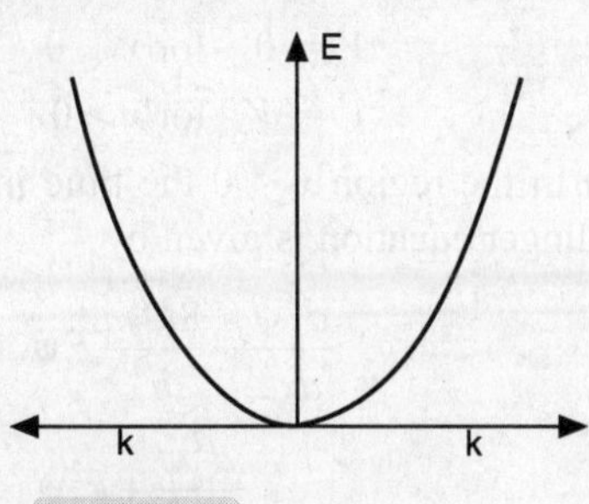

Fig. 27.7:

E-k curve

The k-vector describes the wave properties of the particle. It may be seen from Eq.(27.44) that $E \propto k^2$. The plot of E as a function of k gives a parabola, as illustrated in Fig. 27.7.

(c) Momentum: The momentum operator, operating on the wave function $\psi(x, t)$ gives

$$\frac{\hbar}{i}\frac{\partial\psi}{\partial x} = \frac{\hbar}{i}\left[A(ik)e^{-i(\omega t - kx)}\right]$$

$$= \hbar kAe^{-i(\omega t-kx)} = \hbar k\psi$$

As $$p_x = \hbar k$$

$\therefore$ $$\frac{\hbar}{i}\frac{\partial\psi}{\partial x} = p_x\psi \qquad \text{...(27.45)}$$

It follows from Eq.(27.45) that the wave function $\psi(x, t)$ for the particle is an eigen function of the linear momentum operator, and the momentum p_x is the eigen value of the operator. Therefore, the particle has a sharp value for momentum.

(d) Position of the paticle: The probability of finding the particle between x and $x + dx$ is given by

$$P\,dx = \psi^*(x,t)\psi(x,t)dx$$

or $$P\,dx = A^2 dx \qquad \text{...(27.47)}$$

The probability density P for the position of the particle with the definite value of momentum is constant over the x-axis. It means that the particle is not located at a definite point but all positions are equally probable. This conclusion is in agreement with the Heisenberg uncertainty principle. As the momentum is well defined in this case, it is difficult to assign a position to the particle. The uncertainty in position will be infinity which means that the particle position is indeterminate.

27.11 POTENTIAL ENERGY STEP

A particle moving in a region of constant potential energy region may suddenly encounter a region of higher potential energy, which is also constant. Thus, the potential energy step is a square potential step as shown in Fig. 27.18. We will not discuss in detail the solutions but only give the outline of the salient steps and conclusions.

(a) Potential energy step, $E > V_0$:

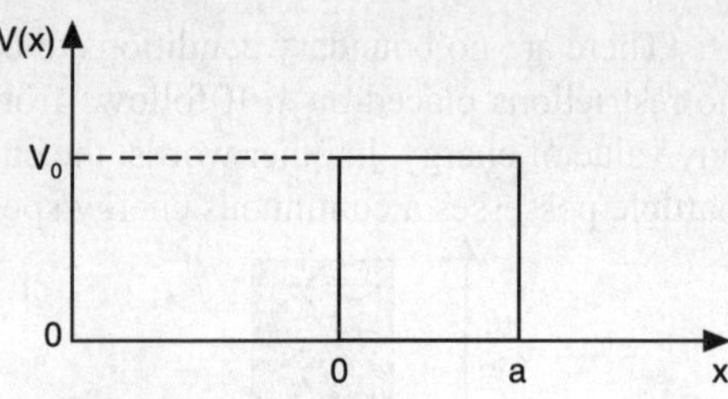

Fig. 27.8: Square Potential energy step

Let E be the fixed total energy of the particle and V be the value of the constant potential energy.

It is seen from Fig. 27.8 that

$$V = 0 \quad \text{for } x < 0$$

and

$$V = V_0 \quad \text{for } x > 0$$

(*i*) In the region $x < 0$ the time independent Schrödinger equation is given by

$$\frac{d^2\psi}{dx^2} + \frac{8\pi^2 m}{h^2} E\psi = 0$$

or

$$\frac{d^2\psi}{dx^2} + k_o^2\psi = 0 \qquad \text{...(27.47)}$$

where

$$\frac{8\pi^2 mE}{h^2} = k_o^2.$$

(*ii*) In the region $x > 0$ the time independent Schrödinger equation is given by

$$\frac{d^2\psi}{dx^2} + \frac{8\pi^2 m}{h^2}(E - V_o)\psi = 0$$

or

$$\frac{d^2\psi}{dx^2} + k^2\psi = 0 \qquad \text{...(27.48)}$$

where

$$k^2 = \frac{8\pi^2 m}{h^2}(E - V_o).$$

The solutions of Eq. (27.47) and Eq. (27.48) are

$$\psi(x) = Ae^{ik_o x} + Be^{-ik_o x} \quad \text{for } x < 0 \qquad \text{...(27.49)}$$

$$\psi(x) = Ce^{ikx} + De^{-ikx} \quad \text{for } x > 0 \qquad \text{...(27.50)}$$

where A, B, C and D are constants of integration.

The Schrödinger equation changes discontinuously at $x = 0$, as there is a discontinuity in V at $x = 0$. Relationships among the four coefficients A, B, C and D may be found by applying the condition that $\psi(x)$ and its derivative $\frac{d\psi}{dx}$ must be continuous at the boundary. Thus,

when $x = 0$,

$$\psi(x) = A + B \text{ and } \psi(x) = C + D$$

$\therefore$

$$A + B = C + D \qquad \text{...(27.51 a)}$$

At $x = 0$,

$$\frac{d\psi}{dx} = k_o(A - B)$$

and

$$\frac{d\psi}{dx} = k(C - D)$$

$\therefore$

$$k_o(A - B) = k(C - D)$$

or

$$(A - B) = \left(\frac{k}{k_o}\right)(C - D) \qquad \text{(27.51 b)}$$

Solving the Eq. (27.51 a) and Eq. (27.51 b), we get

$$B = \left[\frac{k_o - k}{k_o + k}\right] A + \left[\frac{2k}{k_o + k}\right] D \qquad ...(27.52\text{ a})$$

and

$$C = \left[\frac{2k_o}{k_o + k}\right] A - \left[\frac{k_o - k}{k_o + k}\right] D \qquad ...(27.52\text{ b})$$

Substituting the value of B in Eq.(27.49), we get

$$\psi(x) = Ae^{ik_o x} + \left[\left(\frac{k_o - k}{k_o + k}\right) A + \left(\frac{2k}{k_o + k}\right) D\right] e^{-ik_o x}$$

$$\psi(x) = A\left[e^{ik_o x} + \left(\frac{k_o - k}{k_o + k}\right) e^{-ik_o x}\right] + D\left[\left(\frac{2k}{k_o + k}\right) e^{-ik_o x}\right]$$

or

$$\psi(x) = A\psi_1 + D\psi_2 \qquad ...(27.53\text{ a})$$

Substituting the value for C in Eq.(27.50 b), we get

$$\psi(x) = \left[\left(\frac{2k_o}{k_o + k}\right) e^{ikx}\right] A + \left[e^{-ik\,x} - \left(\frac{k_o - k}{k_o + k}\right) e^{ikx}\right] D$$

or

$$\psi(x) = A\psi_1 + D\psi_2 \qquad ...(27.53\text{ b})$$

Eq. (27.53 a) and Eq. (27.53 b) are the same and conditions of continuity at $x = 0$ are satisfied by both ψ_1 and ψ_2.

The complete solution of Eq.(27.47) is obtained when the Eq.(27.49 a) is multiplied by the time dependent term $e^{-i\omega t}$. Thus,

$$\psi(x, t) = Ae^{-i(\omega t - k_o x)} + Be^{-i(\omega t + k_o x)} \qquad ...(27.54)$$

The first term in the above equation represents a wave moving along the positive x-direction while the second term represents a wave moving along the negative x-direction. Therefore, Eq.(27.54) describes the superposition of a wave of intensity $|A|^2$ moving along the positive x-direction (from $-\infty$ to zero) and a wave of intensity $|B|^2$ moving along the negative x-direction. Therefore, when the particles are incident on the step from left on the potential energy step, $|A|^2$ gives the intensity of the **incident** wave and $|B|^2$ gives that of the **refelcted** wave. The ratio $\frac{|B|^2}{|A|^2}$ gives us the reflected fraction of the incident wave intensity.

(*b*) Potential energy step, *E* < V$_o$:

Whenever E becomes less than V, a different solution results.

$$\psi(x) = Ae^{kx} + Be^{-kx} \qquad ...(27.55\text{ a})$$

If E were less than V_o, then the solution for ψ_o(for $x < 0$) would still be given by Eq.(27.50 a), but the solution for ψ (for $x > 0$) becomes

$$\psi_1(x) = Ce^{k_1 x} + De^{-k_1 x} \qquad ...(27.56\text{ b})$$

where

$$k_1 = \sqrt{\frac{8\pi^2 m}{h^2}(V_o - E)}.$$

We set, $C = 0$ to keep $\psi_1(x)$ from becoming infinite as $x \to \infty$, and apply the boundary conditions on $\psi_1(x)$ at $x = 0$. The resulting solution is shown in Fig. 27.18 (b).

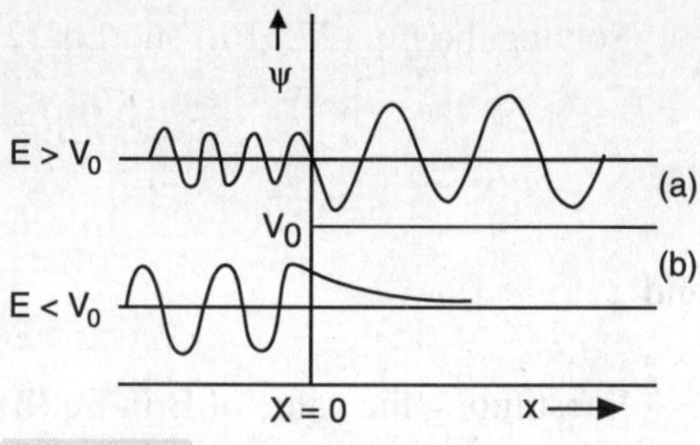

Fig. 27.9:

Solution of boundary condition

This illustrates an important difference between classical and quantum mechanics. Classically, the particle can never be found in the region $x > 0$, since its total energy is not sufficient to overcome the potential energy step. However, quantum mechanics allows the wave function, and therefore the particle, to penetrate into the classically forbidden region. The particle can never be observed in the forbidden region, but a particle can pass through a classically forbidden region and emerge into an allowed region where it can be observed.

Penetration into the forbidden region is associated with the wave nature of the particle. The penetration distance is consistent with the uncertainty in defining the location of the particle.

27.12 RECTANGULAR POTENTIAL BARRIER

According to classical ideas, a particle striking a hard wall has no chance of leaking through it. But the behaviour of a quantum particle is different owing to the wave nature associated with it. We know that when an electromagnetic wave strikes at the interface of two media, it is partly reflected and partly transmitted through the interface and enters the second medium. In a similar way the de Broglie wave also has a possibility of getting partly reflected from the boundary of the potential well and partly penetrating through the barrier. The penetration of a barrier by a quantum particle is called **tunneling**.

Fig. 27.10 represents a potential barrier of height V_o and thickness L. The region around the barrier can be divided into three regions as shown in Fig. 27.11. The potential energy is zero in regions I and III, i.e., for $x < 0$ and $x > L$ and has a constant value V_o in the region II, i.e., for $0 < x < L$. Let us consider a particle of total energy E approaching the barrier from the left. In regions I and III the total energy of the particle is kinetic energy. In region II, the total energy is partly kinetic and partly potential energy. From the view-point of classical physics, the electron would be reflected from the barrier because its energy E is less than V_o. For the particle to overcome the potential barrier, it must have energy equal to or greater than V_o. Quantum mechanics leads to an entirely new result.

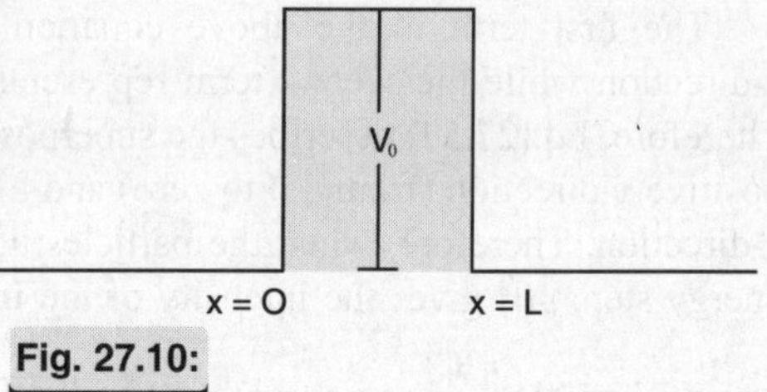

Fig. 27.10:

Potential barrier of height V_o and thickness L

If $E < V_o$, according to quantum mechanics, there is a finite chance for the electron to leak to the other side of the barrier. We say that the electron tunneled through the potential barrier and hence in quantum mechanics, the phenomenon is called **tunneling.**

We write down the Schrödinger wave equation for the electron wave in the three regions and solve them. The Schrödinger wave equation for a particle constrained to move along x-axis is given bye

$$\frac{d^2\psi}{dx^2}+\frac{8\pi^2 m}{h^2}(E-V)\psi = 0 \qquad ...(27.56)$$

In the region I, $V = 0$ and the Eq. (27.56) takes the form

$$\frac{d^2\psi_1}{dx^2}+\frac{8\pi^2 mE}{h^2}\psi_1 = 0$$

or

$$\frac{d^2\psi_1}{dx^2}+k_o^2\psi_1 = 0 \qquad ...(27.57)$$

$\therefore$ In the region II, $E < V_o$, $V = V_o$ and the Eq. (27.56) takes the form

$$\frac{d^2\psi_2}{dx^2}+\frac{8\pi^2 m}{h^2}(E-V_o)\psi_2 = 0$$

or

$$\frac{d^2\psi_2}{dx^2}-\frac{8\pi^2 m}{h^2}(V_o-E)\psi_2 = 0 \qquad ...(27.58)$$

or

$$\frac{d^2\psi_2}{dx^2}-k^2\psi_2 = 0 \qquad ...(27.59)$$

where

$$k = \sqrt{\frac{8\pi^2 m}{h^2}(V_o-E)}.$$

$\therefore$ In the region III, $V = 0$. Therefore,

$$\frac{d^2\psi_3}{dx^2}+k_o^2\psi_3 = 0$$

The solutions of Eq.s (27.57), Eq. (27.58), and Eq.(27.59) are

$$\psi_1 = Ae^{ik_o x}+Be^{-ik_o x} \qquad ...(27.60\text{ a})$$

$$\psi_2 = Ce^{k x}+De^{-k x} \qquad ...(27.60\text{ b})$$

$$\psi_3 = Le^{ik_o x}+Me^{-ik_o x} \qquad ...(27.60\text{ c})$$

In the region III, no particle comes from right and hence $M = 0$.

$\therefore$

$$\psi_3 = Le^{ik_o x} \qquad ...(27.60\text{ d})$$

In Eq. (27.60 a), $Ae^{ik_o x}$ represents the de Broglie wave traveling along the *x*-direction in the region I, with amplitude A and $Be^{ik_o x}$ represents the wave reflected along the negative x-direction with amplitude B. Since the probability that the particle is present in a region of the space is proportional to the square of the de Broglie wave amplitude, the ratio

$$R = \frac{|B|^2}{|A|^2} \qquad ...(27.61)$$

is the **coefficient of reflection** of the particle from the barrier.

The wave function Ψ_2 is *not* zero inside the barrier (the region forbidden by classical mechanics), but decreases exponentially. De^{-kx} represents the exponentially decreasing wave in the barrier. The square of the amplitude of this wave defines the probability of penetration of the particle into the region II. The ratio

$$T = \frac{|D|^2}{|A|^2} \qquad ...(27.62)$$

is the **coefficient of penetration** of the interface. Ce^{kx} is the reflected wave within the barrier. Since such a wave is nonexistent, C should be equal to zero.

In Eq. (27.60 d) $Le^{ik_o x}$ represents the transmitted wave moving along the x-direction in the region III.

The form of the wave function in the region (I), (II) and (III) is also shown in the Fig. 27.11. The wave function Ψ_1 corresponds to the free electron with momentum $p = \sqrt{2mE}$. Since Ψ_3 is not equal to zero at $x = L$, there is a finite probability of finding the electron in the region III. That means the particle that is initially to the *left* of the barrier has some probability of being found to the *right* of the barrier. The wave function Ψ_3 represents the wave transmitted through the barrier and the free particle on the right side of the barrier. The particle has the same momentum as the incident particle but has smaller amplitude. Thus, it is possible for a particle to penetrate through the potential barrier even if its kinetic energy is less than the height of the potential barrier.

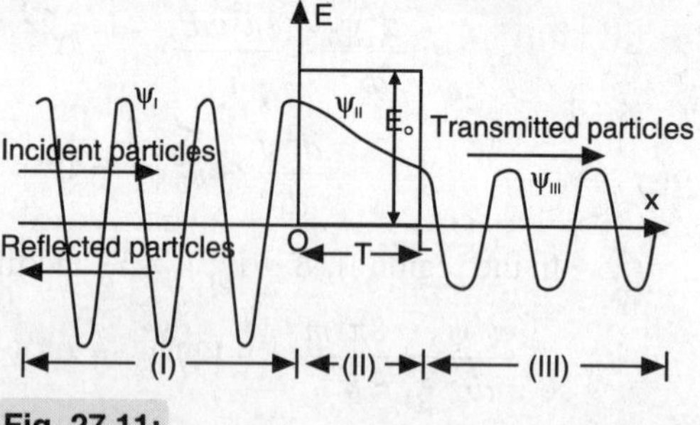

Fig. 27.11: Wave function in three regions

This probability is proportional to the square of the modulus of the wave function Ψ_2.

$$P \propto |\psi_2|^2 = |C|^2 \exp\left(-\frac{2x}{\hbar}\sqrt{2m(V_o - E)}\right) \qquad ...(27.63)$$

This probability is indicative of the fact that the particle is able to penetrate the potential barrier of finite width L. Such a penetration is called the **tunnel effect**. The probability that the particle gets through the barrier is called the **transmission coefficient**. It is defined as

$$T = \frac{\text{Probability density of the transmitted wave}}{\text{Probability density of the incident wave}}$$

It is shown that the transmission coefficient is given approximately by

$$T = Ge^{-2kL} = \left[\frac{16E}{V}\left(1 - \frac{E}{V}\right)\right] e^{-2L\sqrt{8\pi^2 m(V-E)}/h} \qquad ...(27.64)$$

where G is a constant close to unity.

Eq. (27.64) shows that the probability of particle penetration through a potential barrier depends on the height, V and width, L of the barrier. The probability decreases rapidly with increasing barrier width, L. Tunneling is significant in many areas of physics.

27.12.1 Application of Tunnelling

(*i*) The **tunnel diode** is a semiconductor diode. The current in this device is largely due to tunneling of electrons through a potential barrier. The rate of tunneling or current can be controlled over a wide range by varying the height of the barrier, which is done by varying the applied voltage.

(*ii*) **α-decay**: The quantum possibility of penetrating potential barriers is the basis of the explanation of the *α-decay* of radioactive nuclei. In *α-decay*, an unstable nucleus disintegrates into a lighter nucleus and an α-particle. For example, a uranium nucleus

^{238}U undergoes α-*decay* and forms thorium nucleus ^{234}Th.

In 1928 George Gamow explained α-decay of unstable nuclei on the basis of quantum tunneling. The forces in the nucleus set up a potential barrier of height of the order of 30 MeV against α-particle emission. Classically, the α-particle would be trapped unless its energy exceeds 30 MeV. The α-particles have energies in the range of 4 to 9 MeV only. Therefore, it is impossible for a α-particle to jump the barrier. According to quantum mechanics, the α-particle tunnels through the potential barrier (Fig. 27.12). The experimental studies of Geiger-Nuttal on α- *decay* confirmed the predictions of Gamow's theory.

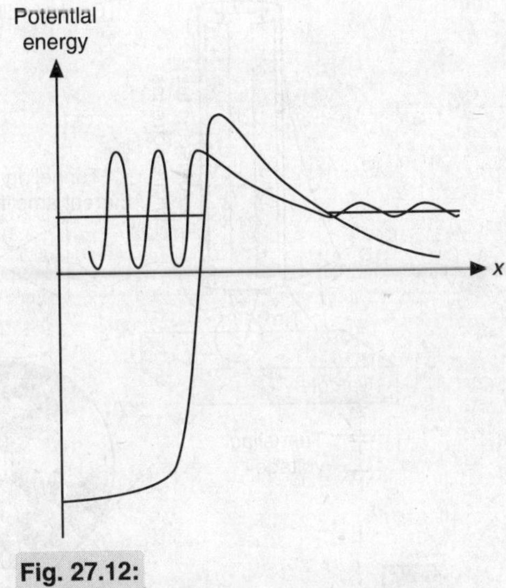

Fig. 27.12:
Alpha particle having less energy than the height of the potential barrier tunnels through the barrier

(*iii*) **Scanning tunnelling electron microscope (STM):** The instrument was invented in the early 1979 by Gerd Binnig and Heinrich Rohrer, who were awarded the 1986 Nobel prize in physics for their work. The scanning tunnelling electron microscope uses electron tunneling to produce images of surfaces down to the scale of individual atoms. When two conducting materials are brought in juxtaposition to each other, electron wave from one material propagate into the other. The electrons "tunnel" through the barrier into the adjacent sample. This happens when their separation is of the order of 1 nm i.e. of the order of several atomic spacings. The probability of an electron to tunnel through barrier decreases exponentially with the barrier width. Therefore, the tunneling current is a sensitive measure of the distance between two conducting samples. The STM makes use of this sensitivity.

Working: The schematic diagram of a scanning tunneling electron microscope is shown in Fig. 27.13 (*a*). In STM a sharp metallic tip scans the sample, which is connected to an XYZ positioning device of the scanner. The sharp metal needle is brought near, of the order of a few angstroms, to the surface to be imaged. A bias voltage is applied between the sample and the tip. When the needle is at a positive potential with respect to the surface, electrons can tunnel across the gap and cause a small "tunnelling current" to flow in the needle. This feeble tunnelling current is amplified and measured. The tunnelling current keeps the feedback electronics the distance between tip and sample constant. The sensitivity of the STM is so large that electronic corrugation of surface atoms and the electron distribution around them can be detected. Fig. 27.13 (*b*) shows the tip of the sharp needle and Fig. 27.13 (*c*) shows the image of silicon crystal surface.

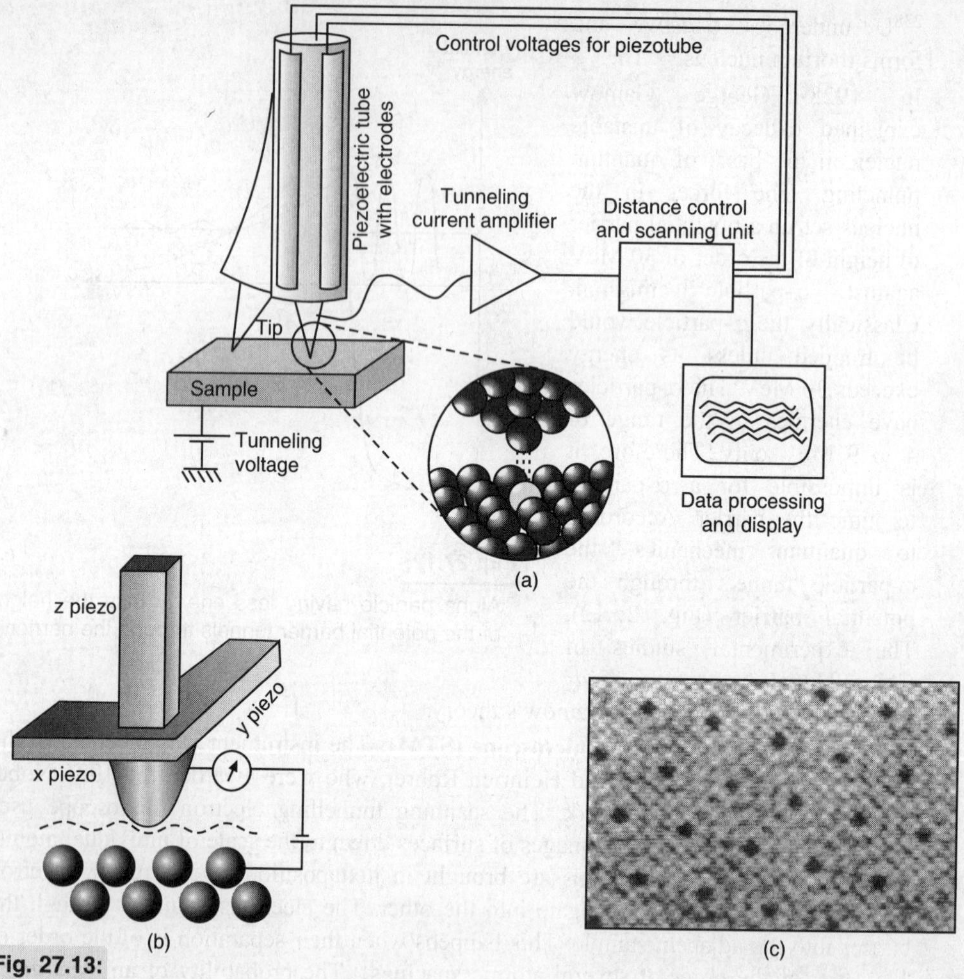

Fig. 27.13:

(a) The schematic diagram of STM (b) The tunneling tip is scanned over the specimen, producing an image of the tunneling current (c) An image of a silicon crystal surface produced by a STM.

Example 27.5. *A stream of electrons, each of energy E = 3eV, is incident on a potential barrier of height V = 4eV. The width of the barrier is 20 Å. Calculate the percentage transmission of the beam through this barrier.*

Solution: The probability of transmission through a potential barrier is given by

$$T = Ge^{-2kl} = \left[\frac{16E}{V}\left(1-\frac{E}{V}\right)\right]e^{-2L\sqrt{8\pi^2 m(V-E)}/h}$$

$$\frac{2L}{h}\sqrt{8\pi^2 m\,(V-E)} = \frac{2\times 20\times 10^{-10}\,\text{m}}{6.63\times 10^{-34}\,\text{Js}}\sqrt{8\times(3.143)^2\times 9.11\times 10^{-31}\,kg\,(4-3)\times 1.602\times 10^{-19}\,\text{J}}$$

$$= 20.49$$

$$\therefore \qquad T = \left[\frac{16E}{V}\left(1-\frac{E}{V}\right)\right]e^{-20.49}$$

$$= \frac{16 \times 3\text{eV}}{4\text{eV}}\left(1 - \frac{3\text{eV}}{4\text{eV}}\right) e^{-20.49}$$

$$= 3 \times e^{-20.49} = 3.8 \times 10^{-9}$$

Therefore, the percentage of transmission = $\mathbf{3.8 \times 10^{-7}}$**%.**

27.13 INFINITE POTENTIAL WELL

A **potential well** is a potential energy function $V(x)$ that has a minimum (see Fig. 27.14). A potential well is the opposite of a potential barrier; it is a potential-energy function with a minimum. If a particle is left in the well and the total energy of the particle is less than the height of the potential well, we say that the particle is *trapped* in the well. In classical mechanics a particle trapped in a potential well can vibrate back and forth with periodic motion but cannot leave the well. In quantum mechanics, such a trapped state is called a **bound state.**

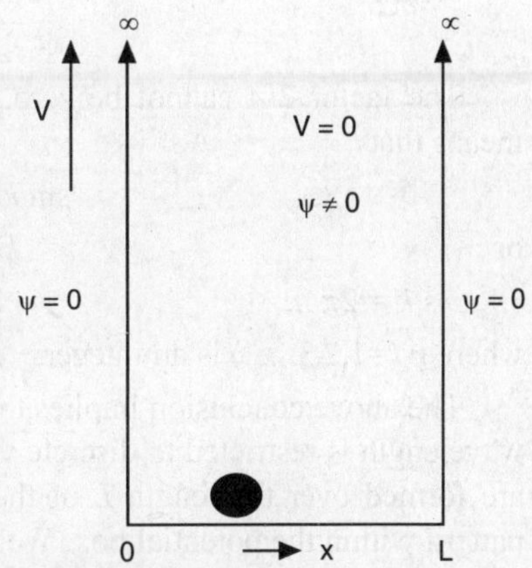

Fig. 27.14: Potential energy function-Infinite Potential Well

Let us consider a particle confined to the region $0 < x < L$. It can move freely within the region $0 < x < L$ but subject to strong forces at $x = 0$ and $x = L$. Therefore, it can never cross to the right to the region $x > L$ or to the left of 0. It means that $V = 0$ in the region $0 < x < L$ and rises to infinity ($V = \infty$) at $x = 0$ and $x = L$. This situation is called a **one-dimensional potential box.**

For a particle trapped in a one-dimensional potential box, $V = 0$ and the Schrodinger equation takes the following form

$$\frac{d^2\psi}{dx^2} + \frac{8\pi^2 mE}{h^2}\psi = 0 \qquad \text{...(27.65)}$$

Putting $\frac{8\pi^2 mE}{h^2} = k^2$, we rewrite Eq. (27.65)

$$\frac{d^2\psi}{dx^2} + k^2\psi = 0 \qquad \text{...(27.66)}$$

Because the particle can move back and forth freely between $x = 0$ and $x = \text{L}$, the solution of the Eq. (27.66) is of the form

$$\psi(x) = Ae^{ikx} + Be^{-ikx} \qquad \text{...(27.67)}$$

which contains motion in both directions. We can evaluate the constants A and B with the help of boundary conditions. The boundary conditions are as follows. The particle cannot jump over the walls and therefore the function does not exist outside the box. The particle is located within the box and therefore it exists within the box. It means that

$$\psi(x) = 0 \text{ at } x = 0$$

$$\psi(x) = 0 \text{ at } x = L$$

$\therefore$ $$\psi_{x=o} = A+B = 0$$

$\therefore$ $$B = -A$$

Using this result into Eq.(27.67), we get

$$\psi(x) = A(e^{ikx} - e^{-ikx})$$

or $$\psi(x) = 2iA \sin kx$$

Again $$\psi(x) = 0 \text{ at } x = L$$

$\therefore$ $$\psi_{x=L} = 2iA \sin kL = 0$$

The factor $2iA$ cannot be zero, because we would not then have the wave function. It means that

$$\sin kL = 0$$

or $$kL = n\pi \qquad ...(27.68)$$

As $k = 2\pi/n$, $$\lambda = 2L/n$$

where n (=1,2,3, ...) is an integer.

The above conclusion implies that the wave equation has solutions only when the particle wavelength is restricted to discrete values such that only a whole number of half-wavelengths are formed over the length L of the box. It means that particle waves form standing wave pattern within the potential box. We write Eq. (27.68) as

$$k = n\pi / L$$

(*i*) The possible values of the momentum of the particle are then given by

$$p = \hbar k = \frac{n\pi\hbar}{L} = \frac{nh}{2L} \qquad ...(27.69)$$

(*ii*) The possible values of the energy are

$$E = \frac{p^2}{2m} = \frac{n^2h^2}{8mL^2} \qquad ...(27.70)$$

The above equation indicates that a particle confined in a certain region can have only certain values of energy. Other energy values are not allowed. In other words, ***energy quantization*** is a consequence of restricting a microparticle to a certain region.

The allowed wave functions which are the solutions of the Schrodinger equation are given by

$$\psi_n = 2iA \sin kx = C \sin kx$$

As $k = n\pi / L$, we write the above as

$$\psi_n = C \sin\left(\frac{n\pi x}{L}\right) \qquad ...(27.71)$$

Applying normalization condition, we can determine the value of the constant C in the above equation. The normalisation condition is that

$$\int_0^L \psi_n \psi_n^* dx = 1$$

$\therefore$ $$C^2 \int_0^L \sin^2 \frac{n\pi x}{L} dx = 1$$

$$\frac{C^2}{2}\int_0^L\left[1 - \cos\frac{2n\pi x}{L}\right]dx = 1$$

$$\frac{C^2}{2}\left[x - \frac{\sin\frac{2n\pi x}{L}}{\frac{2n\pi}{L}}\right]_0^L = 1$$

$$\therefore \qquad C^2\frac{L}{2} = 1$$

i.e., $$C^2 = \frac{2}{L}$$

or $$C = \sqrt{2/L} \quad ...(27.72)$$

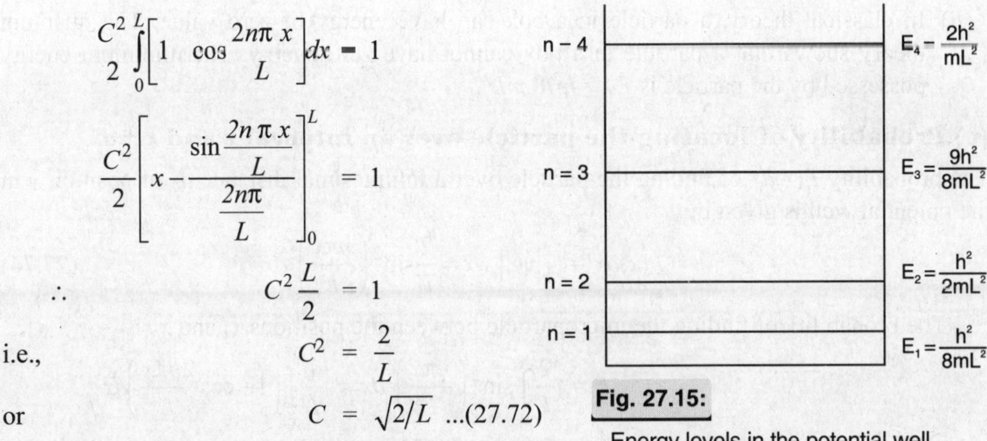

Fig. 27.15: Energy levels in the potential well

Using the value of C into Eq. (27.71), we obtain the wave function as

$$\psi_n = \sqrt{\frac{2}{L}}\sin\left(\frac{n\pi x}{L}\right) \quad ...(27.73)$$

(a) Energy Levels:

The allowed energy states are given by

$$E_n = \frac{h^2}{8mL^2}n^2$$

where n is the quantum number given by $n = 1,2,3, ...$

The energy states are obtained by inserting the value of n in the above equation. Thus

First energy level $$E_1 = \frac{h^2}{8mL^2}$$

Second energy level $$E_2 = \frac{h^2}{2mL^2}$$

Third energy level $$E_3 = \frac{9h^2}{8mL^2}$$ and so on.

The energy levels are shown in Fig. 27.15.

It is important to note that the particle cannot have zero energy. The lowest energy allowed for the particle is $E_1 = h^2/8\,mL^2$. It cannot possess energy less than this in the one–dimensional potential box.

The energy value E_1 is called **zero–point energy**. The zero point energy is a consequence of the uncertainty principle. If the energy of the particle is zero, its momentum also would be zero, and the uncertainty principle requires that the wavelength λ be infinite. In such case the particle cannot be confined to the box. Therefore, the particle must have a certain minimum amount of kinetic energy.

(b) Comparison with Classical Results:

(*i*) Classically, a particle in a box could have any positive energy. Quantum theory shows that the particle is restricted to take only certain discrete values.

(*ii*) In classical theory, a particle in a box can have energy of zero value. The quantum theory shows that a particle in a box cannot have zero energy. The minimum energy possessed by the particle is $E_1 = h^2/8\,mL^2$.

(c) Probability of locating the particle over an interval x and $x + dx$

The probability $P(x)dx$ of finding the particle over a infinitesimal distance dx at position x in the potential well is given by

$$P(x)dx = \int |\psi_n|^2 dx = \frac{2}{L}\sin^2\left(\frac{n\pi x}{L}\right) dx \qquad ...(27.74)$$

The probability of finding the microparticle between the positions x_1 and x_2 is

$$P = \frac{2}{L}\int_{x_1}^{x_2} \sin^2\left(\frac{n\pi x}{L}\right) dx = \frac{1}{L}\int_{x_1}^{x_2}\left(1-\cos\frac{2n\pi x}{L}\right)dx$$

$$= \frac{1}{L}[x]_{x_1}^{x_2} - \frac{1}{2\pi}\left[\sin\frac{2n\pi x}{L}\right]_{x_1}^{x_2} \qquad ...(27.75)$$

The probability density is given by

$$P(x) = \frac{2}{L}\sin^2\left(\frac{n\pi x}{L}\right) dx \qquad ...(27.76)$$

The probability density is a maximum when $\frac{n\pi x}{L} = \frac{\pi}{2}, \frac{3\pi}{2}, \frac{5\pi}{2}, ...$ or $x = \frac{L}{2n}, \frac{3L}{2n}, \frac{5L}{2n}, ...$

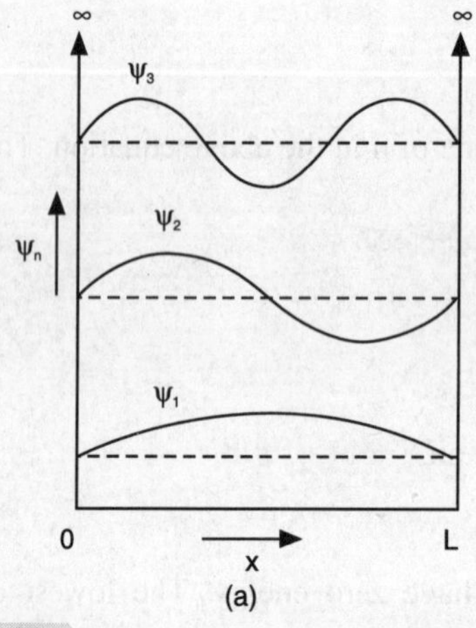

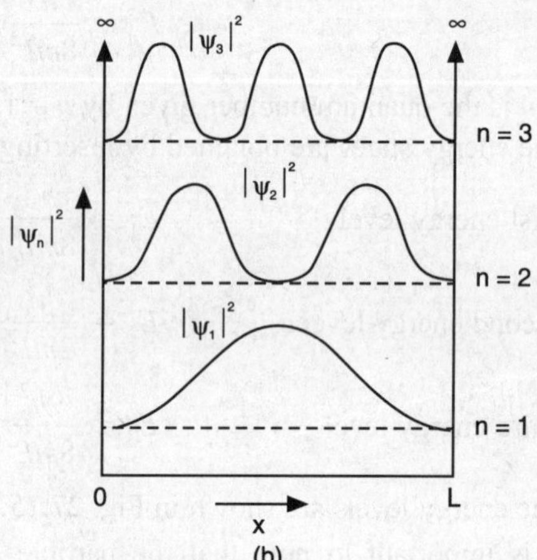

Fig. 27.16: Variation of the probability densities with x

The variation of the probability densities with x for $n = 1$, $n = 2$, and $n = 3$ are shown in Fig. 27.16.

It is easy to see that in the state ψ_1 the probability is seen to be largest at $x = L/2$, i.e., in the middle of the box and it decreases towards the walls. It means that the particle will stay more in the centre of the box and avoids the region near walls. Similarly, in the state ψ_2 the most probable positions are at $x = L/4$ and $x = 3L/4$. Therefore, the particle will be found either in the right–half or in the left–half of the box but never in the middle. In classical theory, all positions within the box are equally probable for the particle. Classically, the particle passes back and forth between the walls and it has an equal probability of being found any where between $x = 0$ and $x = L$. The classical probability function is a constant with a value of $1/L$. The quantum results show peaks of magnitude $2/L$ and valleys where the probability is very small. The number of peaks is equal to the quantum number n. As n increases, the number of peaks increase and when n is very large, the distribution approaches the classical distribution.

(d) Dependence of quantization on the width of the box:

Eq. (27.70) suggests that the quantization of energy is dependent on the width of the potential box, L and on the quantum number n. Let us now study the effect of L on the quantization of particle energy. If we designate two adjacent energy levels by E_{n+1} and E_n, the separation between these levels is given by

$$\Delta E = E_{n+1} - E_n$$

$$= \frac{h^2}{8mL^2}\left[(n+1)^2 - n^2\right]$$

$$= \frac{h^2}{8mL^2}(2n+1) \qquad ...(27.77)$$

As $2n > 1$, $2n+1 \approx 2n$

$$\therefore \qquad \Delta E = \frac{h^2}{4mL^2}n \qquad ...(27.78)$$

Case (i): Let us take the case of an electron moving in a box of side $L = 1$ cm $= 10^{-2}$ m.

From Eq. (27.28) we get

$$\Delta E \cong n \times 10^{-15} \text{ eV}$$

The thermal energy possessed by the electron is

$$kT \cong 10^{-3}\text{eV at } T = 1°\text{K}$$

$$\therefore \qquad \boldsymbol{\Delta E << kT}$$

It means that the electron can move form a lower energy level to an upper energy due to its own thermal energy and without the need of any energy input from external agency. As the levels are nearly continuous, the electron behaves as a classical particle without revealing the quantum aspects [Fig. 27.17 (a).

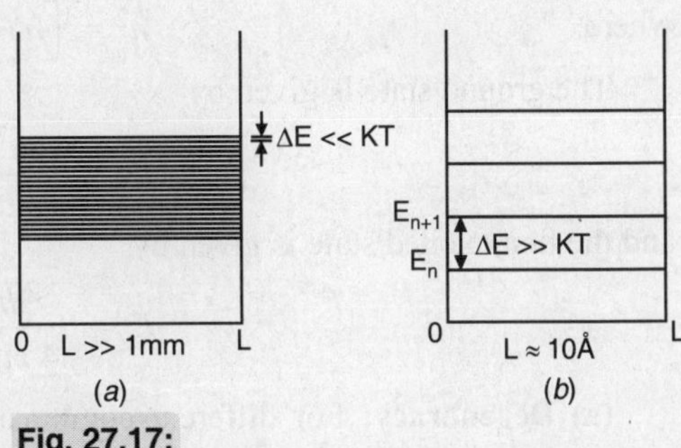

Fig. 27.17: Variation of quantisation on the width of the box

Case (ii) Let us next consider the case of an electron moving in a box of side $L = 10$ Å.

Then

$$\Delta E \cong 0.75\,(n)\,eV.$$

$$\therefore \qquad \Delta E >> kT.$$

This energy difference between the adjacent energy levels is very large and the electron cannot go to higher levels on its own, as the energy levels are separated by large energy values (Fig. 27.17b). Therefore, it becomes necessary to supply energy from an external agency.

The above example amply demonstrates that the quantization of energy assumes a great importance in the atomic world.

27.14 EXTENSION TO THREE-DIMENSIONAL CASE

The case can be generalized to the three-dimensional box with sides *a, b* and *c*. As the coordinates are orthogonal, the general time-independent Schrodinger equation can be separated into three equations and can be solved separately for the eigen values and eigen functions. The complete eigen function is

$$\psi(x, y, z) = \psi_{nx}(x)\,\psi_{ny}(y)\,\psi_{nz}(z)$$

$$= \sqrt{\frac{8}{abc}}\sin\frac{n_x \pi x}{a}\sin\frac{n_y \pi y}{b}\sin\frac{n_z \pi z}{c} \qquad ...(27.79)$$

The total energy of the particle is

$$E_n = E_{nx} + E_{ny} + E_{nz}$$

$$= \frac{h^2}{8m}\left[\frac{n_x^2}{a^2}+\frac{n_y^2}{b^2}+\frac{n_z^2}{c^2}\right] \qquad ...(27.80)$$

where n_x, n_y, and n_z are the three quantum numbers for this case.

In case of a cubical box, $a = b = c$ and we get

$$\psi(n_x, n_y, n_z) = \left[\frac{8}{a}\right]^{3/2}\sin\frac{n_x \pi x}{a}\sin\frac{n_y \pi y}{b}\sin\frac{n_z \pi z}{c} \qquad ...(27.81)$$

and

$$E_n = \frac{h^2}{8m\,a^2}\left[n_x^2 + n_y^2 + n_z^2\right] \qquad (27.82)$$

or

$$E_n = \frac{n^2 h^2}{8ma^2} \qquad (27.83)$$

where

$$n^2 = \left[n_x^2 + n_y^2 + n_z^2\right]$$

The ground state is given by

$$E_1 = \frac{3h^2}{8ma^2}$$

and the first excited state is given by

$$E_2 = \frac{3h^2}{4\,m\,a^2} = 2E_1.$$

(a) Degeneracy: For different combinations of quantum numbers, we may obtain the same energy value but the wave functions are different. Such quantum states having the same energy are called ***degenerate***. Thus, for example, the wave functions

$$\psi_{112} = \sqrt{\frac{8}{a^3}}\sin\frac{\pi x}{a}\sin\frac{\pi y}{b}\sin\frac{2\pi z}{c}$$

$$\psi_{121} = \sqrt{\frac{8}{a^3}}\sin\frac{\pi x}{a}\sin\frac{2\pi y}{b}\sin\frac{\pi z}{c}$$

$$\psi_{211} = \sqrt{\frac{8}{a^3}}\sin\frac{2\pi x}{a}\sin\frac{\pi y}{b}\sin\frac{\pi z}{c}$$

are different. But the corresponding energies are the same. The quantum numbers corresponding to the first wave function are

$$n_x = 1, n_y = 1, \text{ and } n_z = 2$$

and

$$n^2 = \left[n_x^2 + n_y^2 + n_z^2\right] = 6$$

$$\therefore \qquad E_{112} = \frac{6h^2}{8ma^2}$$

Similarly, for the other two wave functions, we get $E_{121} = E_{211} = \dfrac{6h^2}{8ma^2}$

$$\therefore \qquad E_{112} = E_{121} = E_{211} = \frac{6h^2}{8ma^2}$$

Thus, the first excited state is degenerate, since the same n value is given by three sets (1,1,2), (1,2,1), and (2,1,1). The number of different states with a certain value of the energy is known as the ***degree of degeneracy***. Thus, we say that the first excited state is three-fold degenerate.

Example 27.6. *Find the probability that a particle trapped in a box L wide can be found between 0.45 L and 0.55 L for the ground state and first excited state.*

Solution:

$$P = \frac{2}{L}\int_{x_1}^{x_2} \sin^2\left(\frac{n\pi x}{L}\right)dx$$

$$= \frac{1}{L}\int_{x_1}^{x_2}\left(1-\cos\frac{2n\pi x}{L}\right)dx = \frac{1}{L}[x]_{x_1}^{x_2} - \frac{1}{2n\pi}\left[\sin\frac{2n\pi x}{L}\right]_{x_1}^{x_2}$$

For the ground state, $n = 1$ and

$$P = \frac{1}{L}[x]_{x_1}^{x_2} - \frac{1}{2\pi}\left[\sin\frac{2\pi x}{L}\right]_{x_1}^{x_2}$$

$$= \frac{1}{L}[x]_{0.45L}^{0.55L} - \frac{1}{2\pi}\left[\sin\frac{2\pi x}{L}\right]_{0.45L}^{0.55L}$$

or

$$P = \frac{1}{L}[x_2 - x_1] - \frac{1}{2\pi}\left\{\left[\sin\frac{2\pi x_2}{L}\right] - \left[\sin\frac{2\pi x_1}{L}\right]\right\}$$

$$P = \frac{1}{L}[0.55L - 0.45L] - \frac{1}{2\pi}\left\{\left[\sin\frac{2\times 180^o \times 0.55L}{L}\right] - \left[\sin\frac{2\times 180^o \times 0.45L}{L}\right]\right\}$$

$$0.1 - \frac{1}{2\times 3.143}(-0.309 - 0.309) = \mathbf{0.198}$$

For the 1st excited state, $n = 2$ and

$$P = \frac{1}{L}[0.55L - 0.45L] - \frac{1}{4\pi}\left\{\left[\sin\frac{4\times 180^o \times 0.55L}{L}\right] - \left[\sin\frac{4\times 180^o \times 0.45L}{L}\right]\right\}$$

$$= 0.1 - \frac{1}{4\times 3.143}(0.5878 + 0.5878) = \mathbf{0.0065}.$$

Example 27.7. *An electron is confined to move in a one dimensional potential well of length 5Å. Find the quantized energy values for the three lowest energy states.*

Solution: The quantized energy values of an electron confined in a potential well are given by

$$E_n = \frac{h^2}{8mL^2}n^2$$

Energy of the ground state $E_1 = \frac{h^2}{8mL^2}1^2 = \frac{h^2}{8mL^2} = \frac{(6.63\times10^{-34}\,\text{J.s})^2}{8\times9.11\times10^{-31}\,\text{kg}\times(5\times10^{-10}\,\text{m})^2}$

$= 2.41\times10^{-19}$ J = **1.5 eV.**

Energy of the 1st excited state $E_2 = \frac{h^2}{8mL^2}2^2 = \frac{4h^2}{8mL^2} = 4\times1.5$ eV = **6 eV.**

Energy of the 2nd excited state $E_3 = \frac{h^2}{8mL^2}3^2 = \frac{9h^2}{8mL^2} = 9\times1.5$ eV = **13.5 eV.**

Example 27.8. *Calculate the energy required for an electron to jump from ground state to the second excited state in a potential well of width L.*

Solution: $$E_n = \frac{h^2}{8mL^2}n^2$$

Energy of the ground state $E_1 = \frac{h^2}{8mL^2}1^2 = \frac{h^2}{8mL^2}$

Energy of the 2nd excited state $E_3 = \frac{h^2}{8mL^2}3^2 = \frac{9h^2}{8mL^2}$

Energy required for the transition $E = E_3 - E_1 = \frac{9h^2}{8mL^2} - \frac{h^2}{8mL^2} = \boldsymbol{\frac{h^2}{mL^2}}$.

Example 27.9. *An electron is trapped in a one-dimensional box of length 0.1 nm. Calculate the energy required to excite the electron from its ground state to the fourth excited state.*

Solution: $$E_n = \frac{h^2}{8mL^2}n^2$$

∴ $$E_1 = \frac{1^2\times(6.63\times10^{-34}\,\text{J.s})^2}{8\times9.11\times10^{-31}\,\text{kg}\times(0.1\times10^{-9}\,\text{m})^2}$$

$= 60.3\times10^{-19}$ J = 37.6 eV.

and $$E_4 = \frac{4^2\times(6.63\times10^{-34}\,\text{J.s})^2}{8\times9.11\times10^{-31}\,\text{kg}\times(0.1\times10^{-9}\,\text{m})^2} = 601.6\text{ eV}$$

The energy required to excite the electron from its ground state to the fourth excited state is

$E_4 - E_1$ = 601.6eV – 37.6eV = **564 eV.**

Example 27.10. *An electron is confined to move between two rigid walls separated by 1 nm. Find the de Broglie wavelength representing the first two allowed energy states of the electron and the corresponding energies.*

Solution: The eigen values of electron energy confined in a potential well are given by

$$E_n = \frac{h^2}{8mL^2}n^2$$

Energy of the 1st allowed state $E_1 = \frac{h^2}{8mL^2}1^2 = \frac{h^2}{8mL^2}$

$$\therefore \quad E_1 = \frac{1^2 \times \left(6.63\times10^{-34}\,\text{J.s}\right)^2}{8\times9.11\times10^{-31}\,\text{kg}\times\left(1\times10^{-9}\,\text{m}\right)^2} = 6.03\times10^{-20}\text{J}$$

Energy of the 2nd allowed state $E_1 = \frac{h^2}{8mL^2}2^2 = 4\times\frac{h^2}{8mL^2} = 4\times6.03\times10^{-20}\text{J} = 2.4\times10^{-19}\text{J}$

De Broglie wavelength of the electron at the 1st energy level, $\lambda_1 = \frac{h}{\sqrt{2mE_2}}$

$$= \frac{6.63\times10^{-34}\,\text{J.s}}{\sqrt{2\times9.11\times10^{-31}\,\text{kg}\times6.03\times10^{-20}\,\text{J}}} = \mathbf{2\ nm.}$$

De Broglie wavelength of the electron at the 2nd energy level, $\lambda_2 = \frac{h}{\sqrt{2mE_2}}$

$$= \frac{6.63\times10^{-34}\,\text{J.s}}{\sqrt{2\times9.11\times10^{-31}\,\text{kg}\times2.4\times10^{-19}\,\text{J}}} = \mathbf{1\ nm.}$$

Example 27.11. *Find the lowest energy of an electron confined to move in a one-dimensional box of length 1Å. Express the result in electron volts.*

Solution: The energy values of electron in a one-dimensional box are given by

$$E_n = \frac{n^2h^2}{8mL^2}$$

For the lowest energy level $n = 1$.

$$\therefore \quad E_1 = \frac{(6.626\times10^{-34}\,\text{J.s})^2}{8\times9.11\times10^{-31}\,\text{kg}\times(10^{-10}\,\text{m})^2}$$

$$= 6\times10^{-18}\text{J} = (6\times10^{-18})(6.242\times10^{18}\,\text{eV}) = \mathbf{37.4\ eV.}$$

Example 27.12: *Compare the lowest three energy states for (i) an electron confined in the infinite potential well of width 10 Å and (ii) a grain of dust ($m = 10^{-6}$ gm) moving with a speed of 10^6 m/s in an angle potential well pf width 0.1 mm. What can you conclude from this comparison?*

Solution: $E_n = \frac{n^2h^2}{8mL^2}; \qquad n = 1,2,3, ...$

(*i*) Electron

$$E_1 = \frac{(1)^2\left(6.626\times10^{-34}\text{ J.s.}\right)^2}{8\left(9.11\times10^{-31}\text{ kg}\right)\left(10^{-9}\text{ m}\right)^2}$$

$$E_1 = 6.03\times10^{-20}\text{ J} = 0.37\text{ eV}$$

$$E_2 = 4\times6.03\times10^{-20}\text{ J} = 24\times10^{-20}\text{ J} = 1.48\text{eV}$$

$$E_3 = 9\times6.03\times10^{-20}\text{ J} = 54\times10^{-20}\text{ J} = 3.33\text{eV}$$

(*ii*) Grain of dust,

$$E_1 = \frac{(1)^2\left(6.626\times10^{-34}\text{ J.s.}\right)^2}{8\left(10^{-9}\text{ kg}\right)\left(10^{-4}\text{ m}\right)^2} = 5.49\times10^{-51}\text{ J}$$

$$E_1 = 5.49\times10^{-51}\text{ J}$$

$$E_2 = 4\times5.49\times10^{-51}\text{ J} = 2\times10^{-51}\text{ J}$$

$$E_3 = 9\times5.49\times10^{-51}\text{ J} = 48.7\times10^{-51}\text{ J}$$

$$K.E. = \frac{1}{2}m\upsilon^2 = \frac{1}{2}\times10^{-9}\text{ kg}\times\left(10^6\text{ m/s}\right)^2 = 500\text{ J}$$

It may be inferred from the above calculations that

(*i*) Energy levels of an electron in an infinite potential well are quantized and the energy difference between the successive levels is quite large. The electron cannot jump from one level to another level on the strength of the thermal energy. Hence quantization of energy plays a significant role in case of an electron.

(*ii*) The energy levels of a grain of dust are so near to each other that they constitute a continuum. These energy values are far smaller than the kinetic energy (=500J) possessed by the grain of dust. It can move through all these energy levels without an external supply of energy. Thus quantization of energy levels is not at all significant in case of macroscopic bodies.

27.15 HARMONIC OSCILLATOR

Atoms in solids execute harmonic vibrations about their equilibrium positions and hence can be treated as simple harmonic oscillators. As the atom displaces through a distance x, from the equilibrium position ($x = 0$), a restoring force proportional to x appears. It is given by

$$F = -kx$$

where k is the force constant. The potential energy of a harmonic oscillator is given by

$$V = \tfrac{1}{2}kx^2$$

The Schrödinger equation for this case may be written as

$$\frac{d^2\psi}{dx^2} + \frac{8\pi^2 m}{h^2}(E - \tfrac{1}{2}kx^2)\psi = 0 \qquad ...(27.84)$$

This equation has a solution only at the following values of energy

$$E_n = \left(n+\tfrac{1}{2}\right)\hbar\omega \tag{27.85}$$

where $n = 0, 1, 2, 3,\ldots$.

Thus the energy levels of a harmonic oscillator are quantized. The lowest value of the energy which the oscillator can take, corresponds to $n = 0$.

$$\therefore \qquad E_o = \tfrac{1}{2}\hbar\omega \tag{27.86}$$

This level is the ground state. The energy E_o is called **zero-point energy**, as it does not vanish even at absolute zero temperature. The energy levels of the harmonic oscillator are shown in Fig. 27.18.

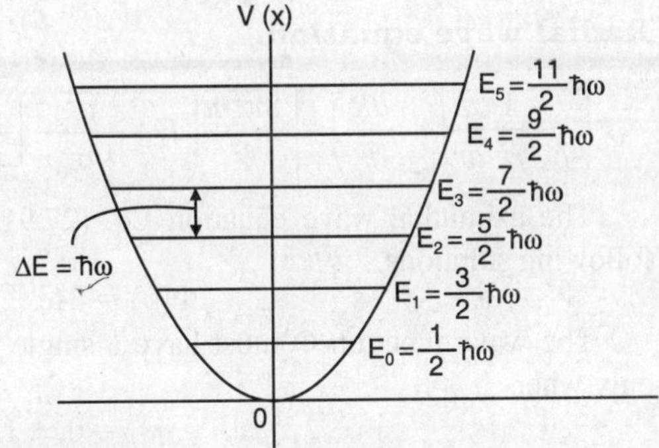

Fig. 27.18: Energy levels of the harmonic oscillator

The energy levels are equidistant and are separated by an energy value

$$\Delta E = \hbar\omega \qquad \ldots(27.87)$$

The existence of zero-point energy is the consequence of uncertainty principle. At $T = 0K$, if the atomic vibrations come to a halt, it would imply that the position and momentum of the atoms can be determined precisely. But such a determination is not possible in the microworld, since the behaviour of micro particles obeys uncertainty principle.

27.16 THE WAVE MECHANICAL MODEL OF ATOM

We now apply the time-independent Schrodinger equation to the hydrogen atom and obtain the eigen values of energy and eigen functions related to the electronic states of the hydrogen atom. The Schrodinger equation for the electron in three dimensions is

$$\nabla^2\psi + \frac{8\pi^2 m}{h^2}(E - V)\,\psi = 0$$

The potential V is that of a positive charge and negative charge separated by a distance 'r' which is given by

$$V = -\frac{e^2}{4\pi\varepsilon_o r} \qquad \ldots(27.88)$$

$$\therefore \qquad \nabla^2\psi + \frac{8\pi^2 m}{h^2}\left(E + \frac{e^2}{4\pi\varepsilon_o r}\right)\psi = 0 \qquad \ldots(27.89)$$

It is difficult to solve this partial differential equation. It can be easily solved if it is expressed in spherical polar coordinates. The equation takes the following form in spherical polar coordinates.

$$\frac{1}{r^2}\left[\frac{\partial}{\partial r}\left(r^2\frac{\partial}{\partial r}\right) + \frac{1}{\sin\theta}\frac{\partial}{\partial\theta}\left(\sin\theta\frac{\partial}{\partial\theta}\right) + \frac{1}{\sin^2\theta}\frac{\partial^2}{\partial\varphi^2}\right]\psi + \frac{8\pi^2 m}{h^2}\left(E + \frac{e^2}{4\pi\varepsilon_o r}\right)\psi = 0 \qquad \ldots(27.90)$$

Eq. (27.90) can be easily separated into three independent equations, each involving only in one variable. After carrying out appropriate substitutions and separating the variables, we obtain the following three total differential equations.

Azimuthal wave equation:

$$\frac{d^2\Phi}{d\varphi^2} + m_l^2\Phi = 0 \qquad ...(27.91)$$

Polar wave equation:

$$\frac{1}{\sin\theta}\frac{d}{d\theta}\left(\sin\theta\frac{d\Theta}{d\theta}\right) + \left[l(l+1) - \frac{m_l^2}{\sin^2\theta}\right]\Theta = 0 \qquad ...(27.92)$$

Radial wave equation:

$$\frac{1}{r^2}\frac{d}{dr}\left(r^2\frac{dR}{dr}\right) + \left[\frac{8\pi^2 m}{h^2}\left(E + \frac{e^2}{4\pi\varepsilon_o r}\right) - \frac{l(l+1)}{r^2}\right]R = 0 \qquad ...(27.93)$$

The azimuthal wave equation Eq. (27.91) is a simple differential equation having the following solutions.

$$\Phi(\phi) = Ae^{i\, m_l\varphi}$$

The wave function Φ must have a single value at a given point in space. It can happen only when

$$m_l = 0, \pm 1, \pm 2, \pm 3, \ldots\ldots \qquad ...(27.94)$$

The polar equation Eq. (27.92) has a complicated solution in terms of associated Legender functions. The solution exists only when the constant l is an integer equal to or greater than $|m|_l$. This condition implies that for a given value of l, m_l can have values

$$m_l = 0, \pm 1, \pm 2, \pm 3, \ldots \pm l. \qquad ...(27.95)$$

The radial wave equation Eq. (27.93) also has a complicated solution in terms of polynomials called the associated Laguerre functions. In case of the electron bound to atom, the radial wave equation can be solved only under the following conditions.

(*i*) E has one of the negative values E_n given by

$$E_n = -\frac{me^4}{8\varepsilon_o^2 h^2}\cdot\frac{1}{n^2} \qquad ...(27.96)$$

(*ii*) n is an integer equal to or greater than $(l + 1)$. This condition means that

$$l = 0,1,2,3,\ldots, (n-1). \qquad ...(27.97)$$

The energy eigen values specified by eq.(27.121) are precisely the same as those obtained by Bohr on the basis of the semi-classical model. The three quantum numbers needed to describe the three-dimensional motion of the electron in the hydrogen atom are n, l, and m_l. They must satisfy the following conditions:

(*i*) Principal quantum number, $n \geq 1$; $n = 1, 2, 3, \ldots\ldots$

(*ii*) Orbital quantum number, $l \leq (n-1)$; $l = 0,1,2,3,\ldots\ldots, (n-1)$

(*iii*) Orbital magnetic quantum number, $|m_l| \leq l$; $m_l = 0, \pm 1, \pm 2, \pm 3, \ldots \pm l$.

Using the above conditions, we can write down all the allowed eigen functions describing the various possible quantum states for the hydrogen atom by using the notation $\psi(n, l, m)$ for each state.

27.16.1 Wave Functions

The wave function ψ_1 of the ground state ($n = 1$) of the hydrogen atom is relatively simple to find out. As ψ_1 has to be completely spherically symmetric, we assume the solution for it.

$$\psi_1 = C_1 e^{-\alpha r} \qquad ...(27.98)$$

α is a constant which turns out to be the reciprocal of the value of the first Bohr radius. For a spherically symmetric volume element $4\pi r^2 dr$, the normalization constant C_1 has a value

$$C_1 = \frac{1}{\pi^{1/2} r_o^{3/2}}$$

Therefore, the wave function for the ground state of hydrogen is

$$\psi_1 = \left(\frac{1}{\sqrt{\pi} r_o^{3/2}}\right) e^{-r/r_o} \qquad ...(27.99)$$

The wave function does not tell us where we actually would find the electron. The probability $P(r)dr$ of finding an electron at any radial distance within a spherically symmetric shell of thickness dr and volume $4\pi r^2\ dr$ is given by

$$\psi_1^* \psi_1 4\pi r^2 dr = 4\pi r^2 C_1^2 e^{-2r/r_o} dr$$

or

$$P(r)dr = \left[\frac{1}{\pi r_o^3} \exp\left(-\frac{2r}{r_o}\right)\right] 4\pi r^2 dr \qquad ...(27.100)$$

The probability $P(r)dr$ of finding the electron at a given distance from the nucleus is plotted in Fig. 27.19. The curve for $n = 1$ shows a maximum at a distance r_o. The total probability of finding the electron at all points of distance r from the nucleus is greatest when $r = r_o$. This value is the same as the value determined by the Bohr theory for the radius of the first orbit. In wave mechanics, r_o is the distance at which the electron is likely to be found most often. It is not possible to draw a shape that bounds a region in which the probability of finding the electron is 100%. The electron probability is distributed over the entire volume of the atom. However, a surface can be drawn that connects points of equal probability and that encloses a volume in which the probability of finding the electron is high. An electron in an atom may therefore be visualized as an electron cloud. There is a region where the electron probability density falls steeply to a low value and it corresponds to the boundary of the atom. The concept of orbit is replaced with that of an orbital. An **orbital** is the region around the nucleus in which the probability of finding the electron is the highest. Depending upon the configuration and size of the boundary surface, the shape and size of the orbital is determined.

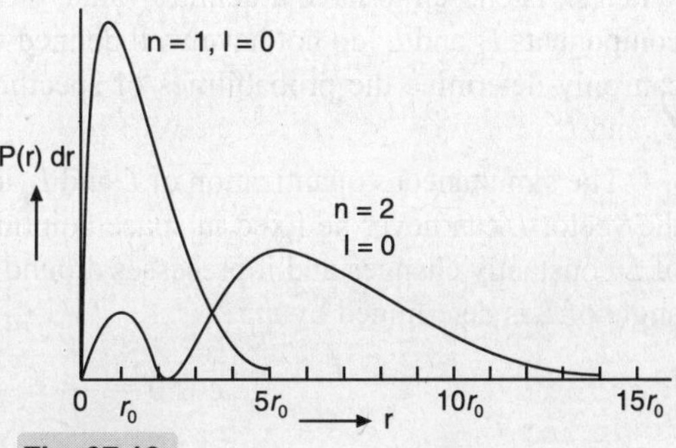

Fig. 27.19: The probability distribution as function of r

27.16.2 Orbital Angular Momentum

It is found that like total energy E, the angular momentum of electron is both conserved and quantized. Quantum theory shows that the magnitude L of the electron orbital angular momentum L is given by

$$L = \sqrt{l(l+1)}\hbar \qquad ...(27.101)$$

The law of conservation angular momentum implies that an electron in a definite stationary state can have a definite angular momentum. However ($L = m\upsilon r$) contains simultaneously both $p(= m\upsilon)$ and r. The uncertainty principle forbids a particle to have a definite momentum p and a definite coordinate r at the same time. The quantum theory shows that the conserved angular momentum of the electron can be characterized by the magnitude L and one of its rectangular components, L_z. The component L_z is also quantized which implies the quantization of direction of L. Thus, space quantization emerges as a natural element of quantum mechanical solution of electron motion in the atom.

$$L_z = m_l \hbar \qquad ...(27.102)$$

The choice of L_z among the three components L_x, L_y, L_z is simply a matter of convention where a magnetic field is assumed parallel to z-direction for the sake of reference. When L_z is chosen to have a definite value, then the other components L_x and L_y do not have well defined values; one can only determine the probabilities of specific values of L_x and L_y.

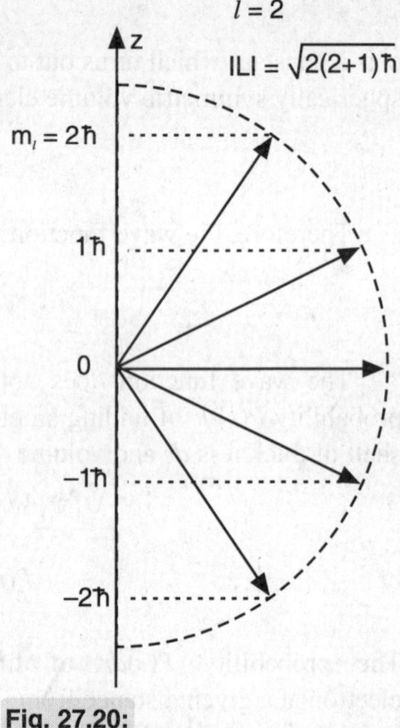

Fig. 27.20: Orbital angular momentum

The simultaneous quantization of L and L_z implies that the vector L can never be fixed in space pointing any specific direction. In fact, the direction of L constantly changes and it precesses around the z-axis tracing out a cone in space. The tilt angle of L is determined by m_l.

$$\cos\theta = \frac{L_z}{L} = \frac{m_l}{\sqrt{l(l+1)}} \qquad ...(27.103)$$

27.16.3 Intrinsic Angular Momentum

Electron is known to possess spin angular momentum. Schrödinger equation does not give any hint regarding electron spin. When relativity is included into the quantum mechanical treatment of electron motion in an atom, the intrinsic non-orbital angular momentum of the electron emerges in a natural way. The relativistic wave equation was first formulated in 1928 by P.A.M.Dirac (1902–1984). According to Dirac's theory, the only value that the spin quantum numbers can have is $\frac{1}{2}$. The magnitude S of the intrinsic angular momentum of the electron is given by

$$S = \sqrt{s(s+1)}\hbar \qquad ...(27.104)$$

The direction of this intrinsic angular momentum is also quantized. The space quantization of intrinsic angular momentum is described by the spin magnetic quantum number m_s. The components S_z along a magnetic field in the z-direction is given by

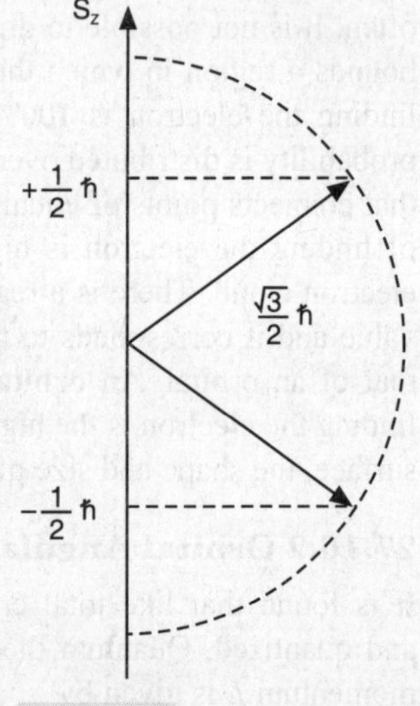

Fig. 27.21: Intrinsic Angular Momentum

$$S_z = m_s\hbar \quad ...(27.105)$$

27.16.4 Selection Rules

The fact that each level in a hydrogen-like atom is composed of several angular momentum states is important from the point of view of light emission. The electron transitions are restricted by **selection rules**. The selection rules for electric dipole transitions are

$$\Delta l = \pm 1 \text{ and } \Delta m_l = 0, \pm 1$$

These selection rules are imposed by the law of conservation of angular momentum since the emitted or absorbed photon carries a spin angular momentum of 1. Therefore, the angular momentum of the atom must change by one unit to compensate for the angular momentum carried by the emitted or absorbed photon.

27.17 THE TRANSITION FROM DETERMINISTIC TO PROBABILISTIC NATURE

The probabilistic interpretation is a fundamental feature of quantum mechanics. Every particle has a wave function associated with it. The wave function determines the corpuscular characteristics like position, momentum etc in a statistical sense. For instance a particle with a wave function $\psi(x, t)$ has a probability proportional to $|\psi(x, t)|^2$ for being found in the neighbourhood of the point x.

Nature does not permit any more precise specification of the sate of a particle than what is provided by the wave function and the probability distributions obtainable from it.

Quantum mechanics yields probability functions, not precise trajectories.

It should be noted that due to uncertainty principle, our idea regarding causality needs revision. The classical concept of cause and effect applies to systems which are left undisturbed. Since a small system cannot be observed without producing serious disturbances in it, we cannot expect to find casual connection between the results of our observations.

There is an inevitable indeterminacy in the estimation of the observational results. Quantum mechanics allows only calculating the probability of getting a particular result when observations are made.

In classical mechanics the motion of a body under the action of a force is described by Newton's second law. If the initial position and the initial velocity are specified, the position and velocity of the body at any next instant can be found by Newton's second law. Thus, giving the position and velocity of a body at any instant chosen as reference ($t = 0$) will fully specify the state of the body. If one knows the state of the system at some instant, one can accurately predict the state of the system at any other instant. This is known as **mechanical determinism**. It is assumed that the accuracy with which the initial position and velocity of bodies can be specified is solely dependent on the quality of the instruments measuring position and velocity.

In contrast, since the position and velocity of quantum mechanical particles can be determined simultaneously only with uncertainties given by Heisenberg relation, one is prompted to conclude that the state of a system cannot be accurately specified at the initial instant, t_0, or at any other succeeding instants. The probabilistic interpretation of the wave function makes quantum mechanics an inherently statistical theory.

27.18 SUPERPOSITION PRINCIPLE

Quantum mechanics equations are linear with respect to the wave function ψ. The sum of two solutions of a linear equation again satisfies the same equation. Therefore, any solution of a wave equation can be represented in the form of a certain set of standard solutions. The opportunity of indicating a single wave function in terms of the sum of other wave functions is known as the **superposition principle**. The *principle of superposition of states* is one of the important principles of quantum mechanics.

According to this principle, if a certain quantum-mechanical system has an allowed state ψ' and also state ψ'', then the system can also exist in a state described by the function

$$\psi = c'\,\psi' + c''\,\psi'' \qquad \text{...(27.106)}$$

where c' and c'' are arbitrary complex numbers.

Given several solutions of the time-independent Schrödinger equation with different values E_n of the energy, any linear superposition of the form

$$\psi = \sum_n c_n \psi_n$$

will also be a solution of the Schrödinger wave equation.

Let ψ_1 and ψ_2 be two wave functions representing two physically allowed states of a given object. Then any linear superposition of the form

$$\psi = a_1\psi_1 + a_2\psi_2 \qquad \text{...(27.107)}$$

also describes the state of the object. a_1 and a_2 are arbitrary complex coefficients. Repeated application of the superposition principle shows that if the principle holds for the superposition of two wave functions, it also holds for the superposition of an arbitrary number of wave functions.

27.19 OBSERVABLES AND OPERATORS

An **observable** is a quantity obtained through the process of measurement on a physical system. The measurement of an observable in a system is expressed as a number. An observable is always a *real quantity* as it is the result of an actual measurement.

An **operator** is an entity which when applied to a function, transforms it into a new function.

An operator is denoted by the symbol $\hat{L}$. The equation

$$\hat{L}\psi(x) = \phi(x) \qquad \text{...(27.108)}$$

means that the operator denoted by $\hat{L}$ acts on the function $\psi(x)$ and as a result we get the function $\phi(x)$. The action could be a *mathematical operation* in algebra such as to multiply the input function by a constant, add something else to it, or an operation in calculus such as to differentiate it, integrate it, or so on. A quantum mechanical operator $\hat{L}$ does not work on an algebraic function but on a state vector like $|\psi\rangle$, which is an abstract description of a physical situation. An operator performs an operation on a state $|\psi\rangle$ and produces a new state $|\phi\rangle$.

The operator $\hat{L}$ is called a ***linear operator*** if it satisfies the following conditions:

$$\hat{L}(\psi_1 + \psi_2) = \hat{L}\psi_1 + \hat{L}\psi_2 \text{ and } \hat{L}(a\psi) = a\hat{L}\psi \qquad \text{...(27.109)}$$

where a is some number. We are concerned here with linear operators only.

The effect of an operator acting on a function may be represented as a definite or an improper integral.

$$\hat{L}\psi(x) = \int L(x,y)\psi(y)dy$$

If the variable is discrete, we will have

$$\hat{L}\psi_n = \sum_m L_{nm}\psi_m \qquad ...(27.110)$$

The totality of coefficients L_{nm} forms the *matrix of the operator* $\hat{L}$ and we speak of the matrix representation of the operator.

Suppose $\hat{L}\,\psi = \varphi$; the operator $\hat{L}^*$ is called the *complex conjugate* of the operator $\hat{L}$, if by the action of this operation on the function ψ^* we get the function φ^*:

$$\hat{L}^*\psi^*(x) = \varphi^*(x)$$

The expectation value of a dynamical variable gives the expected average of the result of measurement of the dynamical variable. So it must be a real number since it is a physically measurable quantity. The operators representing dynamical variables whose expectation values are real are said to be ***Hermitian operators***.

If $\hat{A}$ is the operator, then the expectation value of A is given by

$$< A > = \int \psi^* \hat{A}\psi \; dx \qquad ...(27.110)$$

where ψ is a normalised wave function specifying the state of the system. The complex conjugate of the expectation value of A is

$$< A >^* = \left(\int \psi^* \hat{A}\psi \; dx\right)^* = \int \psi \, \hat{A}^* \psi^* \, dx$$

For the expectation value to be real, we must have $<A> = <A>^*$.

or

$$\int \psi^* \hat{A}\psi \; dx = \int \psi \, \hat{A}^* \psi^* \, dx \qquad (27.111)$$

An operator is said to be Hermitian if it satisfies the above condition.

A linear operator commutes with constants and obeys distributive and associative laws. The sum of two linear operators is commutative, but their product may or may not be commutative. Thus, if $\hat{A}$ and $\hat{B}$ are two linear operators, then these may or may not commute. It means that the product operator $\hat{A}\hat{B}$ may or may not be equal to the operator $\hat{B}\hat{A}$.The ***commutator*** of two operators $\hat{A}$ and $\hat{B}$ corresponding to two physical quantities A and B is defined as

$$[\hat{A}, \hat{B}] = \hat{A}\hat{B} - \hat{B}\hat{A} \qquad ...(27.112)$$

When $\hat{A}$ and $\hat{B}$ commute, the commutator vanishes. That is,

$$[\hat{A}, \hat{B}] = 0$$

If $[A, B]\psi = 0$, then the eigen values of A and B can be measured simultaneously.

If $\hat{A}$ and $\hat{B}$ do not commute, then the commutator $[\hat{A}, \hat{B}] \neq 0$.

If [A, B] when operated on some wave function ψ yields a non-zero result, then there will be an uncertainty relation limiting A and B of the form

$$(\Delta A)\,(\Delta B) \geq \frac{1}{2}\sqrt{-[A,B]^2}$$

Pairs of physical variables that are linked by an uncertainty relation are known as ***conjugate*** or ***complementary*** variables.

27.20 IMPORTANT OPERATORS OF QUANTUM MECHANICS

In quantum mechanics an operator when applied to a wave function gives the corresponding observable quantity of the system multiplied by the wave function. The most important operators of quantum mechanics are position operator, momentum operator, energy operator and Hamiltonian operator.

Let us consider the wave function $\psi = A \exp\left[\frac{i}{\hbar}(p_x x - Et)\right]$. ...(27.114)

Differentiating the above expression with respect to x, we obtain

$$\frac{\partial \psi}{\partial x} = A\frac{ip_x}{\hbar} \exp\left[\frac{i}{\hbar}(p_x x - Et)\right]$$

or

$$\frac{\hbar}{i}\frac{\partial}{\partial x}\Psi = p_x \Psi$$

From the above we see that the operator corresponding to the observable quantity momentum p_x can be represented by

$$\hat{p}_x = \frac{\hbar}{i}\frac{\partial}{\partial x} = -i\hbar\frac{\partial}{\partial x} \quad ...(27.115)$$

In three dimensions the operator is given by $\hat{p} \to -i\hbar\nabla$.

Now, the linear momentum operator $\hat{p}_x$ when applied to the wave function ψ gives the corresponding observable quantity, linear momentum p_x multiplied by the wave function.

Similarly, when the Eq. (27.114) is differentiated with respect to t, we get

$$\frac{\partial \psi}{\partial t} = -\frac{iE}{\hbar}A \exp\left[\frac{i}{\hbar}(p_x x - Et)\right] = -\frac{iE}{\hbar}\psi$$

or

$$i\hbar\frac{\partial}{\partial t}\Psi = \mathrm{E}\Psi$$

We denote the operator E by $\hat{E}$. When $\hat{E}$ operates on the wave function ψ, we get the total observable property 'total energy' E multiplied by the wave function.

$$\hat{E}_\psi = i\hbar\frac{\partial}{\partial t}\Psi = E\Psi \quad ...(27.116)$$

The operator corresponding to energy is thus $\hat{E} \to i\hbar\frac{\partial}{\partial t}$. Thus, the dynamical variables momentum and energy of a quantum particle can be represented by the two mathematical operators $-i\hbar\frac{\partial}{\partial x}$ and $i\hbar\frac{\partial}{\partial t}$ respectively. In addition since the coordinate x is a multiplying operator, we get $x \to \hat{x} = x$.

Some useful physical operators are given in the following table.

Table 27.1: *Useful Physical Operators*

Dynamical variable	Quantum Mechanical Operator	Result (Multiplication of ψ by)
Position, x	$\hat{x}$	x
Linear momentum, p	$-i\hbar\nabla$	p
Kinetic Energy, E	$-\frac{\hbar^2}{2m}\nabla^2$	E
Potential Energy	$V(r)$	V

In classical mechanics, Hamiltonian is defined as $H = \frac{p^2}{2m} + V$ and it gives the total energy E of the system. Hence, the total energy operator for a non-relativistic particle is represented by Hamiltonian operator $\hat{H}$. As $\hat{p} = i\hbar\nabla$, the Hamiltonian operator is

$$\hat{H} = -\frac{\hbar^2}{2m}\nabla^2 + V \qquad ...(27.117)$$

When the Hamiltonian operator operates on the function ψ, the result is the function multiplied by total energy E. Thus,

$$\hat{H}\Psi = \left[-\frac{\hbar^2}{2m}\nabla^2 + V(r)\right]\Psi = E\,\Psi$$

It may be noted here that the quantum mechanical operators for x and p_x satisfy the following commutation relation.

$$[\hat{x}, p_x] = \left[\hat{x}, \frac{\hbar}{i}\frac{\partial}{\partial x}\right] = -i\hbar\left(x\frac{\partial}{\partial x} - \frac{\partial}{\partial x}x\right) = i\hbar$$

Thus, the commutator $[\hat{x}, \hat{p}_x] \neq 0$. Therefore, the precise simultaneous measurement of these variables is not possible. If the operators corresponding to variables are commutative, that is, $[\hat{A}, \hat{B}] = 0$, then the variables can be determined to a high precision simultaneously.

27.21 EXPECTATION VALUES

Some questions in quantum contexts will yield exact answers whereas others can only be answered in terms of a probability distribution. For example, if we ask about the energy of an electron in 1s orbit of the hydrogen atom, the answer is definite and is –13.6 eV. But if we ask what its position is, our answer is not definite. If we take a large number of hydrogen atoms and somehow measure the positions of electrons in them simultaneously, and average the results, we get the *average* position of the electron from the nucleus. The theoretically predicted value of this average is known as the ***expectation value*** of the position. It is designated by $<x>$.

Average value of x is $<x> = \dfrac{x_1 + x_2 + x_3 + ... + x_i + ...x_n}{n} = \dfrac{\sum_{i=1}^{n} x_i}{n}$

Thus, if a variable takes n discrete values, we add all these values of the variable and divide the sum by the total number of values to obtain the average value of that variable. On the other hand if the variable takes the value of $x_1 m_1$ times, the value $x_2 m_2$, the value $x_3 m_3$ and so on, then

Average value of x is $<x> = \dfrac{m_1x_1 + m_2x_2 + m_3x_3 + ... + m_ix_i + ... + m_nx_n}{m_1 + m_2 + ... + m_i + ...m_n}$

$$= \frac{\sum_{i=1}^{n} m_i x_i}{\sum_{i}^{n} m_i} = \frac{\sum_{i=1}^{n} m_i x_i}{n}$$

or

$$<x> = \sum_{i=1}^{n} x_i \frac{m_i}{n}$$

But $\frac{m_i}{n} = P_i$, the probability of occurrence of the value x_i.

$$\therefore \quad \langle x \rangle = \sum_{i=1}^{n} x_i P_i$$

If the variable is continuous instead of discrete, then in place of summation we use the integral and get

$$\langle x \rangle = \int x\, P\, dx \qquad \text{...(27.118)}$$

Given a probability distribution for any measurable physical quantity A, the simplest and most important property of that distribution is the average value of A, designated as <A> and understood as an average taken over many individual measurements under identical conditions on the same initial state.

27.21.1 Expectation Value in Operator Notation

When a given system is in a state ψ, the average of a large number of independent measurements of an observable corresponding to the operator $\hat{L}$ is given by

$$\langle L \rangle = \frac{\int\limits_{\text{Available region}} \psi^* \hat{L} \psi\, dr}{\int\limits_{\text{Available region}} \psi^* \psi\, dr} \qquad \text{...(27.119)}$$

<L> is called the expectation value of the observable. For a normalised function, $\int \psi^* \psi\, dr = 1$.

$$\therefore \quad \langle L \rangle = \int\limits_{\text{Available region}} \psi^* \hat{L} \psi\, dr \qquad \text{...(27.120)}$$

The general rule for calculating expectation values is

$$\langle \text{Observable} \rangle = \int \psi^*(x) \{(operator)\, \psi(x)\} dx \qquad \text{...(27.121)}$$

where operator denotes the operator corresponding to the observable of interest. The order of computation is important here. First {(operator)ψ(*x*)} is computed. Then, the result is multiplied by ψ* and finally the integral is carried out.

Position: The simplest example is the expectation value of the position **r** of an object. From Eq. (27.120), it follows that

$$\langle r \rangle = \int_{-\infty}^{+\infty} \psi^* \hat{r} \psi\, dr = \int_{-\infty}^{+\infty} \psi^* r \psi\, dr = \int_{-\infty}^{+\infty} \mathrm{r} |\psi|^2\, dr \qquad \text{...(27.122)}$$

The operator $\hat{r} \equiv r$ operates on the wave function ψ to its right, the operation being ordinary multiplication with the vector **r.** Thus, $\hat{r}$ is the *position operator*. It is an **algebraic operator**.

Momentum: The expectation value for the momentum **p** is given by

$$\langle \mathrm{p} \rangle = \int \psi^* \hat{p} \psi\, dr = \int \psi^* (-i\hbar \nabla) \psi\, dr$$

In one-dimensional case $p_x = -i\hbar \frac{\partial}{\partial x}$.

$$\therefore \qquad \langle p_x \rangle = -i\hbar \int \psi^*(x) \frac{\partial}{\partial x} \psi(x) \qquad \text{...(27.123)}$$

Integrals of the general form of equation (20.148) arise frequently in quantum mechanics. To reduce the tedium of writing these out, it is convenient to adopt a short hand notation introduced by Dirac and known as Dirac "**bra-ket**" notation. In this notation, the expectation value of *f(x)* is written simply as

$$\langle \Psi^*(x) | f(x) | \Psi(x) \rangle$$

Thus, $$\langle \Psi^*(x) | f(x) | \Psi(x) \rangle = \int \psi^*(x) \{Op[f(x)]\} \psi(x) dx$$

where $|f(x)|$ denotes Op[*f(x)*] which is the operator corresponding to *f(x)*.

Example 27.13: *Find the expectation value ⟨x⟩ of the position of a particle trapped in a box of width L.*

Solution:

$$\langle x \rangle = \int_{-\infty}^{\infty} x \, |\psi|^2 \, dx$$

$$= \frac{2}{L} \int_0^L x \sin^2 \frac{n\pi x}{L} dx$$

$$= \frac{2}{L} \left[\frac{x^2}{4} - \frac{x \sin(2n\pi x/L)}{4n\pi/L} - \frac{\cos(2n\pi x/L)}{8(n\pi/L)^2} \right]_0^L$$

or $$\langle x \rangle = \frac{2}{L} \left[\frac{L^2}{4} \right] = \frac{L}{2}.$$

Thus, the average position of the particle is the middle of the box in all quantum states.

Example 27.14: *A Particle is moving in 1D potential box of infinite height and width 25Å. Calculate the probability of finding the particle within an interval of 5Å at the centers of the box when it is in the state of least energy.*

Solution: The wave function of a particle enclosed in an infinite potential wall is

$$\psi(x) = \sqrt{\frac{2}{a}} \sin \frac{n\pi x}{a}$$

For least energy state $n = 1$, $\psi(x) = \sqrt{\frac{2}{a}} \sin \frac{\pi x}{a}$

At the center of the box $x = \frac{a}{2}$

The probability of finding the particle in the interval at the center of the box is

$$[\psi(x)]^2 = \left[\sqrt{\frac{2}{a}} \sin \frac{n\pi a/2}{a} \right]^2 = \frac{2}{a}$$

$\therefore$ The probability P in the interval Δx is

$$P = [\psi(x)]^2 \Delta x = \frac{2}{a}\Delta x$$

Given: $a = 25Å$, $\Delta x = 5Å = 5 \times 10^{-10}$ m

$$P = \frac{2 \times 5 \times 10^{-10}}{25 \times 10^{-10}} = 0.4$$

Example 27.15: *The wave function of a particle confined to a box of length L is* $\psi(x) = \sqrt{\frac{2}{L}}\sin\frac{\pi x}{2}$ *in the region 0<x<L and zero elsewhere. Calculate the probability of finding the particle in the region 0 < x < (L/2).*

Solution: The probability of finding a particle in Δx at a distance x is $P(x)\,\Delta x = |\psi(x)|^2 \Delta x$

$$\psi(x) = \sqrt{\frac{2}{L}}\sin\frac{\pi x}{L}$$

$$P(x)\Delta x = \frac{2}{L}\sin^2\left(\frac{\pi x}{L}\right)\Delta x$$

The probability of finding the particle in the region $0 < x < (L/2)$ is

$$P = \int_0^{L/2} P(x)dx = \frac{2}{L}\int_0^{L/2} \sin^2\frac{\pi x}{L}dx$$

$$= \frac{1}{L}\int_0^{L/2}\left(1 - \cos\frac{\pi x}{L}\right)dx = \frac{1}{L}\times\frac{L}{2} = \frac{1}{2}$$

$\therefore$ The probability of finding the particle in $0 < x < (L/2)$ is $\frac{1}{2}$.

Example 27.16: *Calculate the expectation value <Ps> of the momentum of a particle trapped in a one-dimensional box.*

Solution: The normalized wave functions of the particle are

$$\psi_n^* = \psi_n = \sqrt{\frac{2}{L}}\sin\frac{n\pi x}{L}$$

$$\frac{d\psi}{dx} = \sqrt{\frac{2}{L}}\left(\frac{n\pi x}{L}\right)\cos\frac{n\pi x}{L}$$

Now,
$$\langle p_x \rangle = \int_{\infty}^{-\infty} \psi^*\left(-ih\frac{d}{dx}\right)\psi dx$$

$$= -ih\frac{2}{L}\frac{n\pi}{L}\int_0^L \sin\frac{n\pi x}{L}\cos\frac{n\pi x}{L}dx$$

$$= 0$$

∴ The expectation value $<p_x>$ of the particle's momentum is 0.

Example 27.17: *Find the expectation value <x> of the position of a particle trapped in a box 1 wide.*

Solution:

$$<x> = \int_{\infty}^{-\infty} x\,|\psi|^2\,dx = \frac{2}{L}\int_0^L x\sin^2\frac{n\pi x}{L}dx$$

$$= \frac{2}{L}\left[\frac{x^2}{4} - \frac{x\sin(2n\pi x/L)}{4n\pi x/L} - \frac{\cos(2n\pi x/L)}{8\,(n\pi/L)^2}\right]_0^L$$

$$<x> = \frac{2}{L}\left[\frac{L^2}{4}\right] = \frac{L}{2}$$

This result means that the average position of the particle is the middle of the box in all quarter states.

Example 27.18: *Verify the orthogonal property of wave functions of a particle in a one dimensional box.*

Solution: The wave functions of a particle in one dimensional box are

$$\psi_n(x) = \sqrt{\frac{2}{L}}\sin\frac{n\pi x}{L}$$

We want to verify

$$\int_0^L \psi_n^*(x)\psi_m(x)dx = 0 \quad \text{For } n \neq m$$

$$\text{L.H.S.} = \int_0^L \sqrt{\frac{2}{L}}\sin\frac{n\pi x}{L}\sqrt{\frac{2}{L}}\sin\frac{m\pi x}{L}dx$$

$$= \frac{1}{L}\int_0^L\left(2\sin\frac{n\pi x}{L}\sin\frac{m\pi x}{L}\right)$$

$$= \frac{1}{L}\int_0^L\left(\cos(n-m)\frac{\pi x}{L} - \cos(n+m)\frac{\pi x}{L}\right)dx$$

L.H.S. = 0, Hence verified.

27.22 FUNDAMENTAL POSTULATES OF QUANTUM MECHANICS

We summarize here the fundamental postulates which we have introduced in the course of our above study.

Postulate I

(*a*) The state of a quantum mechanical system is described or represented by a wave function $\psi(x, t)$. All constant multiples of a given ψ describe one and the same state.

(*b*) The state ψ of the system can be built up by applying the principle of superposition. Thus,

$$\psi = \sum_n c_n\psi_n$$

where c_n are complex numbers.

Postulate II

Each dynamical variable A(x, p) is represented by a linear operator in quantum mechanics. Dynamical variables in general do not commute with each other.

Postulate III

If large number of measurements of a dynamical variable A are made on a system which is prepared to be in one and the same state before each measurement, the results of the measurements are distributed over an average value known as expectation value <A>. The expectation value of the dynamical variable is given by

$$\langle A\rangle = \frac{\int \psi^* \hat{A}\psi \, dt}{\int \psi\psi^* dt}$$

Postulate IV

The only possible results of measurement of a dynamical variable A are the eigen values of the operator Â satisfying the eigen value equation

$$\hat{A}\, u_n = a_n u_n$$

where u_n is the eigenfunction of the operator Â belonging to the eigen value a_n.

The eigen value equation satisfied by the Hamiltonian operator associated with the Hamiltonian function H can be written as

$$\hat{H}\psi_n = E_n\psi_n$$

where E_n is the energy of the system. The above equation is the Schrödinger equation.

QUESTIONS

1. Write the statement of Heisenberg uncertainty principle.
2. Show that electrons cannot exist within the nucleus on the basis of the above principle.
3. Arrive at Heisenberg's uncertainty principle with the help of a thought experiment.
4. Show that the uncertainty in measurement of electron momentum is equal to its momentum itself; if the uncertainty in measurement of location of electron is equal to its de Broglie wavelength.
5. State uncertainty principle. Write its mathematical form for the following pairs of variables: (i) Position and momentum, (ii) Energy and time and (iii) Angular position and angular momentum.
6. What is Heisenberg Uncertainty principle? Explain how it is the outcome of the wave description of a particle.
7. Explain Heisenberg uncertainty principle. Based on this, show the non-existence of electrons inside the nucleus.
8. Can a wave given by an equation $Y = A \sin (wt\text{-}kx)$ represents a particle? Explain the concept of a wave packet. How does this concept lead to Heisenberg's uncertainty principle?
9. Explain why a single monochromatic wave cannot represent a particle. Explain the physical significance of wave function ψ.
10. Define a wave function. Show that it represents the probability density of finding a particle at a given position and given time.
11. What is a wave function? What are the necessary conditions of physically acceptable wave function?

12. What is meant by normalized wave function?
13. Explain concept of electron wave. Give physical significance of wave function.
14. Write down Schrödinger's time dependent and time independent wave equations of matter waves. Explain, why: (i) the wave function ψ must be single valued and continuous function of position. (ii) The integral of $|\psi|$ overall space must be equal to unity.
15. What is meant by expectation value in quantum mechanics?
16. State time independent Schrödinger equation. Find out the wave function associated with free electron and discuss the relationship between energy and wave vector in case of free electrons.
17. Write down: (i) time independent and (ii) time dependent Schrödinger equations.
18. Derive the time dependent Schrodinger equation for a free particle.
19. Distinguish between Newtonian and quantum mechanics.
20. Derive time independent Schrödinger's wave equation. Apply the above equation for a particle confined to a rigid box and discuss its wave functions and energy levels.
21. Show that the energy of an electron confined in a 1-D potential well of length L and infinite depth is quantized. Is the electron trapped in potential well allowed to take zero energy? If not, why?
22. Show that the solution of Schrödinger's equation for a particle in an infinite potential well leads to the concept of quantization of energy.
23. Assuming the time independent Schrödinger wave equation, discuss the solution for a particle in one dimensional potential well of infinite height. Hence obtain the normalized wave function.
24. Show that the wave function for a particle confined in an infinite one-dimensional potential well of length '*l*' is given by $\psi_n(x) = \sqrt{\frac{2}{l}}\sin\left(\frac{n\pi x}{l}\right)$. Hence discuss the energy levels and their discreteness.
25. Show that the energy of a microparticle confined in an infinite one-dimensional potential well of length '*l*' is given by $E_n = \frac{n^2h^2}{8ml^2}$ where the symbols have their usual meaning. In the above situation the particle cannot have zero energy. Explain, why?
26. Using normalization condition on ψ show that A is given by $\sqrt{2/l}$.
27. Derive the Eigen values and Eigen functions for a particle in a one dimensional box.
28. Explain the terms potential well and potential barrier. How does a particle with energy lower than the barrier height, tunnel through it? Give one example.
29. Explain in short the phenomenon of tunnelling that occurs when a beam of particles are incident on a potential barrier of finite width.
30. Explain the barrier tunnelling of electron on the basis of quantum mechanics.
31. Formulate Schrödinger wave equation for a linear harmonic oscillator.
32. Write the Schrödinger wave equation of the hydrogen atom in polar coordinates. Explain the origin and significance of the quantum numbers n, l and m_l.

PROBLEMS

1. Calculate the wavelength associated with 1 MeV electron. **[Ans:** 1.23×10^{-12} m]
2. Calculate the energy of the neutron in eV if its de Broglie wavelength is 3×10^{-10} m.
   [Ans: 9×10^{-3} eV]
3. Calculate the velocity and kinetic energy of an electron of wavelength 1.66 Å.
   [Ans: 4386km/s, 54.7 eV]

4. Calculate de Broglie wavelength of an electron moving with velocity 10^7 m/s. **[Ans:** 0.72 Å]
5. Calculate the de Broglie wavelength of an α-particle accelerated through a potential difference of 400 volts. **[Ans:** 1.6×10^{-13} m]
6. Calculate the de Broglie wavelength of neutron of energy 12.8 MeV. **[Ans:** 8×10^{-5} Å]
7. Determine the velocity and kinetic energy of a neutron having de Broglie wavelength of 1Å. **[Ans:** 3.97×10^3m/s, 0.0825 eV]
8. An electron is confined to a spherical box of diameter 10^{-8}m. Calculate the minimum uncertainty in its velocity. **[Ans:** 1.16×10^4 m/s]
9. If the uncertainty in the position of an electron is 4×10^{-10} m, calculate the uncertainty in its momentum. **[Ans:** 26.4×10^{-26} kg.m/s]
10. An electron has a speed of 800 m/s with an accuracy of 0.004%. Calculate the certainty with which we can locate the position of the electron. **[Ans:** 3.6×10^{-3} m]
11. Compute the minimum uncertainty in the location of a mass of 2.0 gm moving with a speed of 1.5 m/s and the minimum uncertainty in the location of an electron moving with a speed of 0.5×10^8 m/s given that the uncertainty in the momentum p for both is $\Delta_p = 10^{-3}$ p. **[Ans:** 23.2 Å]
12. A meson has a life time of 10^{-23}s. To what accuracy can its mass be known? **[Ans:** 1.17×10^{-28} kg]
13. An electron is trapped in an infinite potential well of length 2 Å. In the ground state, evaluate the probability of finding the electron in the region **[Ans:** 0.0125]
14. An electron is moving in a one-dimensional infinite well of width 2×10^{-10} m. What is the probability of finding the electron between $x = 0$ and $x = 10^{-10}$ m in the first excited state? **[Ans:** 0.5]
15. A particle is moving in one dimension infinite potential box of width 25Å. Calculate the probability of finding the particle with in a small interval of 0.05 Å at the centre of the box when it is in its state of least energy. **[Ans:** 0.022]
16. The position and momentum of a 1 keV electron are simultaneously determined. If its position is located to within 1 Å, what is the percentage of uncertainty in its momentum? **[Ans:** 6.17%]
17. An electron is confined to a box of length 100 Å. Calculate the minimum uncertainty in its velocity. **[Ans:** 72.8 km/s]
18. Compare the uncertainties in the velocities of an electron and a proton confined in a 10 Å box. **[Ans:** 1.16×10^5 m/s; 63 m/s]
19. Calculate the probability of transmission for a proton of energy 1 MeV through a 4 MeV high rectangular potential energy barrier of width 10^{-2} cm. **[Ans:** 0.0015]
20. Calculate the probability of transmission for an electron of energy 2eV incident upon a rectangular potential energy barrier of height 4 eV and width 10^{-9} m. **[Ans:** 2×10^{-6}]

28 Atomic Nucleus and Nuclear Energy

Learning Objectives:

By the end of this chapter, you will be able to:

- Understand basic concepts of atomic nucleus.
- Learn about different static properties of nucleus.
- Identify different types of Nuclear models.
- Explain concept of radioactivity, radioactive decay and radioactive series.
- Learn about concept of Activity, Half-life, average life time and units of activity.
- Describe about neutrons, neutron induced reactions and nuclear fission and chain reaction.
- Understand basics of nuclear reactors, nuclear power plant, nuclear waste disposal and nuclear fusion.

28.1 INTRODUCTION

The nucleus of an atom occupies an astonishingly small volume at the center of the atom and is most densely packed with protons and neutrons. Almost all the entire mass of the atom is accounted by the nucleus alone. The particles in the nucleus are held together by nuclear forces which are a fundamental force like gravitational force. The heavier nuclei exhibit instability and transform into stable nuclei through radioactive disintegration. The natural radioactivity of nuclei is effectively used in the determination of ages of organic as well as organic substances. Radioactivity can be induced in nuclei through nuclear reactions, which assist mankind in different fields such as medicine, industry and agriculture. The neutron induced nuclear reactions occupy special place. The neutrons have the ability of causing fission of uranium and plutonium nuclei which liberates tremendous energy. The nucleus is thus a storehouse of enormous power, which can be utilized in wars as well as in production of electrical power.

For the past century fossil fuels namely, coal, oil and natural gas have supplied the major portion of our energy requirements. These sources will be nearly exhausted in the near future and it is imminent that alternative sources of power are to be searched for. Nuclear power is one of the alternative sources. The energy reserve in the form of uranium is many times greater than that of fossil fuels. The generation of power through nuclear fission, however, poses its own threats and problems. Nuclear fusion power is believed to be an inexhaustible source of energy and is without any attendant hazards. Efforts are in progress for producing electrical power through nuclear fusion on commercial scale.

28.2 THE ATOMIC NUCLEUS

The British physicist Ernest Rutherford proposed the nuclear model of atom in 1911 in an attempt to explain the large angle scattering of α-particles which were directed at a thin gold foil. He postulated that all the positive charges of an atom were concentrated in a central massive core and named it *nucleus*. A nucleus is the centre of the atom where most of the mass of the atom and all of the positive charge are concentrated.

There is a vast difference between the size of an atom and its nucleus. The nucleus is a very small part of an atom. The radius of nucleus is of the order of 10^{-14}m while that of an atom is of the order of a few angstroms (10^{-10}m). Therefore, an atom occupies about a million times more space than does a nucleus. In between the atomic electrons and the nucleus, there is a lot of void of space.

An atomic nucleus is not a single indivisible point mass. It is composed of smaller particles. It is obvious that nucleus contains protons. However, all of the nuclear mass could not be accounted for by protons. In 1920, Rutherford suggested the existence of a neutral particle in the nucleus which he called a **neutron**. Neutron was discovered in 1932 by James Chadwick. All nuclei, with the exception of hydrogen nucleus, contain neutrons. Protons and neutrons are about 2000 times more massive than electrons. They are collectively known as **nucleons**. As the number of protons in the nucleus is always equal to the number of electrons around the nucleus, a stable atom is electrically neutral. **Atomic number**, Z is the number of protons or electrons in an atom and **mass number**, A is the total number of protons and neutrons in an atomic nucleus. The nucleus of an atom is denoted by the chemical symbol of the atom subscripted and superscripted respectively by atomic number Z and mass number A. Thus, if the chemical of an atom is X, its nucleus is denoted by ${}_{Z}^{A}X$. For example, ${}_{6}^{12}C$ (read as carbon-six-twelve) denotes the nucleus of carbon atom, containing 6 protons and 12 – 6 = 6 neutrons. *There are 92 stable species of atom available in nature.*

28.3 ISOTOPES

A particular kind of atom of any element is called a *nuclide*. A nuclide is distinguished from other nuclides by the number of protons and neutrons it contains. The atomic number Z determines the chemical nature of an element. Although for a particular element the number of electrons and protons is fixed, the number of neutrons in the nucleus may vary. It implies that the mass number A may differ though the atomic number Z remains the same. Such atoms will be chemically identical but their nuclei show marked differences in stability. The *isotopes* are atoms of a particular element that have different masses. To cite an example, hydrogen has three isotopes. ${}_{1}^{1}H$ is the most common isotope. It is just a proton. The other isotopes ${}_{1}^{2}H$ and ${}_{1}^{3}H$ are called **deuteron** and **triton** respectively. In atomic form they are known as deuterium and tritium respectively. The deuteron denoted by D is made of one proton and one neutron. For about every 6500 atoms of ordinary hydrogen in water, there is one atom of deuterium. Triton denoted by T is made of one proton and two neutrons. It is radioactive with a half-life period of 12.26 years.

The uranium element which plays a very important role in the production of nuclear energy exists in three isotopic forms.

Isotope	Relative abundance	Half-life
$^{238}_{92}U$	99.28%	4.5×10^9 years
$^{235}_{92}U$	0.714%	7.1×10^8 years
$^{234}_{92}U$	0.006%	2.5×10^5 years

28.4 THE NUCLEAR FORCE

The nucleons are clustered together within the small volume of the nucleus. Large repulsive electrical forces operate between the positively charged protons, which tend to tear the nucleus apart. Yet the nuclei of most atoms are stable. It means that there must be some strong attractive force which more than balances the electrostatic repulsion and holds the nucleus together. This strong attractive force is called the **nuclear force**. The nuclear interaction between nucleons is, therefore, a **strong interaction**. The nuclear force is a fundamental force like the gravitational and electrical forces, but is more complicated and not completely understood. The nuclear forces have the following properties.

Salient Features of the Nuclear Forces:

(*i*) The nuclear force is strongly attractive and much larger in magnitude than either the electrostatic force or gravitational force. They act between any pair of nucleons – proton-proton, proton-neutron, and neutron-neutron.

(*ii*) Nuclear forces are very **short-ranged** forces. A nucleon interacts only with its nearest neighbours, over distances of the order of 10^{-15} m.

(*iii*) Nuclear forces are **charge-independent**. They do not depend on the charge of nucleons. The nuclear forces acting between two protons, between a proton and a neutron or between two neutrons have the same magnitude. That is *n-n* force is no different from the *p-p* and *n-p* force.

(*iv*) Nuclear forces are **spin-dependent** and depend on the mutual orientation of the spins of the nucleons. For example, a neutron and a proton are kept together forming a nucleus of deuteron, only if their spins are parallel to each other.

(*v*) Nuclear forces are **not central** forces. They cannot be represented as directed along the straight line connecting the centers of the interacting nucleons.

(*vi*) Nuclear forces have the property of **saturation**. It means that each nucleon in a nucleus interacts with a limited number of neighbours. Saturation manifests itself in that the binding energy per nucleon does not grow with an increase in the number of nucleons, but remains approximately constant. Further, the saturation of the nuclear forces is indicated by the volume of a nucleus being proportional to the number of nucleons forming it.

(*vii*) The magnitude of nuclear force is so high that the work required to divide a nucleus into its constituents is about 8 MeV in contrast to a few eV required to separate the extra-nuclear electrons from its atom.

28.4.1 Proton-Neutron Theory

For elements of low mass numbers, the atomic number Z is nearly half the mass number A. Thus the number of protons is nearly equal to the number of neutrons. With increasing mass number, the value of Z becomes less than half of A which means that the number of neutrons exceeds that of protons. In case of $^{238}_{92}U$, $Z = 92$ and $A = 238$. Therefore, $(A–Z) = 146$. Thus, uranium nucleus consists of 92 protons and 146 neutrons. For elements of low mass numbers, the neutron-to-proton ratio is close to unity. But as the number of protons in a nucleus exceeds 20, the neutron to proton ratio is greater than unity in stable nuclides. It appears that in order to maintain nuclear stability, the neutrons must exceed protons in number (Fig. 28.1). Further, the neutron excess increases with the increasing atomic number. For the heaviest stable nuclides such as $^{208}_{82}Pb$ the neutron-to-proton ratio exceeds 1.5. It is inferred from the above facts that neutrons act as a nuclear "glue" to hold the nucleus together. As the number of protons increases for larger stable nucleus, the number of neutrons also increases, so that the short range nuclear forces are greater than the long range repulsive electrical forces. Neutrons provide attractive nuclear forces between both protons and other neutrons. The proton-neutron theory was put forward by Heisenberg in 1932.

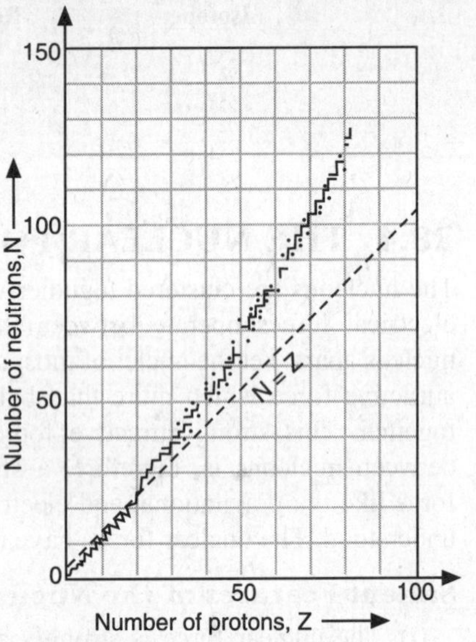

Fig. 28.1: A Plot of neutron number N versus proton number Z in nuclides. Nuclides with proton numbers greater than 20 have more neutrons than protons. The dashed line corresponding to the condition N = Z is called the line of stability.

Origin of the strong interaction: In 1935, H. Yukawa suggested a successful theoretical model to explain the origin of nuclear force. He proposed that the nuclear force is an ***exchange force*** acting through virtual particles. The virtual particles responsible for nuclear binding forces are called π mesons (or pions). The existence of mesons was discovered experimentally in 1947. Three different types of pions, namely positive, negative and neutral pions exist. The mass of a π meson is about 273 times the mass of an electron. It is postulated that nucleons consist of some sort of common core surrounded by a pulsating cloud of pions. The pions are supposed to rapidly jump back and forth between the nucleons, thereby changing their identity equally fast and at the same time keeping them bound together. When neutral pions are exchanged, the nucleons do not undergo any change. Thus, neutral π° mesons are associated with the forces that exist between neutrons and neutrons. When a negative pion is exchanged between a neutron and proton, the neutron emits a negative pion and turns into a proton and the proton absorbing the negative pion turns into a neutron.

Thus, negative π^- mesons are associated with the forces that exist between neutrons and protons. When a positive pion is exchanged between a neutron and proton, the proton emits a positive pion and turns into a neutron and the neutron absorbing the positive pion turns into a proton. No particular proton will remain a proton in the nucleus and no particular neutron remains a neutron in the nucleus. If a proton is in the field of the negative meson, it is converted into a neutron. When a neutron is in the field of a positive meson, it is converted into a proton. Thus, the nucleus is seen to be an ever-changing complex structure. Thus, according to Yukawa's theory, the protons and neutrons in the nucleus are held together by the continual exchange of π mesons.

The interactions are summarized as follows:

$$n - \pi^- \rightarrow p$$
$$p + \pi^- \rightarrow n$$
$$p - \pi^+ \rightarrow n$$
$$n + \pi^+ \rightarrow p$$
$$n + \pi^{o} \rightarrow n$$
$$p + \pi^{o} \rightarrow p$$

28.5 STATIC PROPERTIES OF NUCLEUS

28.5.1 Nuclear Mass

Mass of an atom, a nucleus or an elementary particle is extremely small. Expressing their masses in conventional units involves cumbersome negative powers of ten. Therefore, a separate unit called **atomic mass unit** has been devised for expressing the masses of atomic particles. Nuclear mass means the mass or weight of the nucleus alone. Atomic mass unit (a.m.u.) is defined as 1/12th the mass of the ${}^{12}_{6}C$ atom.

$$1 \text{ a.m.u.} = \frac{1}{12} \times \text{Mass of one carbon atom} = \frac{1}{12} \times \frac{12 \text{ kg}}{6.02 \times 10^{26}} = 1.66 \times 10^{-27} \text{ kg.}$$

The uranium nucleus U-238 has a mass of 238 a.m.u. As the mass of the electron is negligibly small in comparison to that of protons and neutrons, the mass of the nucleus is taken as the mass of the corresponding atom also. Thus, the mass of the uranium atom is 238 a.m.u.

The a.m.u. is now called ***unified atomic mass*** and is denoted by the letter "u".

28.5.2 Nuclear Radius

Nuclear radii are estimated from the measurements on the maximum scattering angles of α–particles when they approached the target nuclei. The nucleus is assumed to be spherical and the following empirical relation gives its radius.

$$R = R_o A^{1/3} \qquad \text{...(28.1)}$$

where R_o has an average value of 1.4×10^{-15} m.

∴ The radius of uranium nucleus = $1.4 \times 1^{-15} \text{ m} \times 238^{1/3} = 8.68 \times 10^{-15}$ m.

28.5.3 Nuclear Density

The density of nuclear matter is tremendous. All nuclei have nearly the same density. The density is calculated as follows.

$$\rho = \frac{\text{Mass of nucleus}}{\text{Volume of the nucleus}} = \frac{M_N}{\frac{4}{3}\pi R^3} = \frac{m_p A}{\frac{4}{3}\pi R^3} \qquad \text{...(28.2)}$$

$$= \frac{1.673 \times 10^{-27} \text{kg}}{11.5 \times 10^{-45} \text{m}^3} = 1.45 \times 10^{17} \text{ kg/m}^3.$$

28.5.4 Nuclear Charge

Nucleus is electrically positive and the magnitude of its charge is equal to the number of protons in the nucleus. Thus, the charge on a nucleus is equal to its atomic number, Z. Thus, a uranium nucleus carries a charge of 92 units.

28.5.5 Nuclear Quantum States

The studies of α- and γ-ray spectra show that every nucleus possesses a set of quantum states. Transitions between different nuclear states cause emission of γ-rays.

28.5.6 Spin and Magnetic Moment

The hyperfine structures observed in atomic spectra indicated that the nucleus has spin motion. It is concluded on the basis of experimental evidence that the protons and neutrons are in continuous motion in discrete quantized orbits. Because of this motion, the nucleus possesses angular momentum and magnetic moment.

The magnetic moment of nuclei is given by

$$\mu_1 = g\frac{h}{2\pi}\cdot\frac{e}{2M} \qquad ...(28.3)$$

The product $\frac{h}{2\pi}\cdot\frac{e}{2M}$, often written as $\frac{e\hbar}{2M}$ is known as ***nuclear magneton***.

28.6 MASS DEFECT

The nucleus contains protons and neutrons. But the mass of nucleus is less than the sum of the masses of the constituent protons and neutrons. This mass variation is termed **mass defect**, Δm.

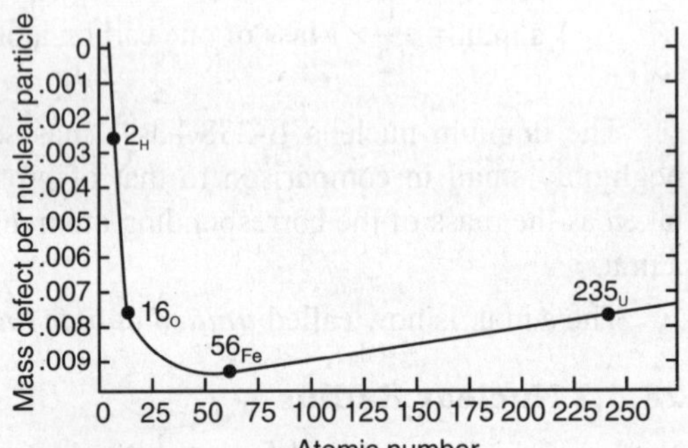

Fig. 28.2: Mass defect for different nuclides. The most stable nuclei center around Fe-56 which has the largest mass defect per nucleon.

The sum of the masses of the protons and neutrons = $Z\, m_p + (A - Z)\, m_n$

where Z is the number of protons in the nucleus

$(A–Z)$ is number of neutrons in the nucleus

m_p is the mass of the proton and

m_n is that of the neutron

If M is the actual mass of the nucleus, then mass defect is

$$\Delta m = Zm_p + (A - Z)m_n - M \qquad ...(28.4)$$

Because the nucleus is more stable than the separated neutrons and protons, the nucleus is in a lower energy state. It implies that the mass, which disappears, is released in the form of energy when nucleons are bound together in a nucleus.

For example, a $_2He^4$ nucleus is formed from two neutrons and two protons.

Mass of two protons	= 2 × 1.007826 = 2.015652 amu
Mass of two neutrons	= 2 × 1.008665 = 2.017330 amu
Total mass	= 4.032982 amu

Measured mass of $_2He^4$ = 4.002604 amu

∴ Mass defect Δm = (4.032982 – 4.002604) amu = 0.030378 amu

28.7 BINDING ENERGY

The binding energy of the nucleon is the energy requisite to take out any nucleon from the nucleus. It is equivalent to the work to be done so as to remove the nucleon from the nucleus without imparting it any kinetic energy.

The total binding energy of a nucleus is defined as the energy required breaking up the nucleus into its constituent protons and neutrons and placing them at rest at infinite distances from one another.

It means that the binding energy is the energy equivalent of the mass defect. Thus,

$$\Delta E_b = (\Delta m)c^2$$

$$\therefore \quad \Delta E_b = [Zm_p + (A-Z)m_n - M]c^2$$

The binding energy of the neucleus in MeV is expressed as

$$\Delta E_b = 931.4\ [Zm_p + (A-Z)m_n - M]\ MeV/u \quad ...(28.5)$$

The **average binding energy per nucleon** is given by

$$\Delta \xi = \Delta E_b / A \quad ...(28.6)$$

The binding energy per nucleon is around 8 MeV. Fig. 28.3 illustrates the average binding energy per nucleon as a function of mass number, A.

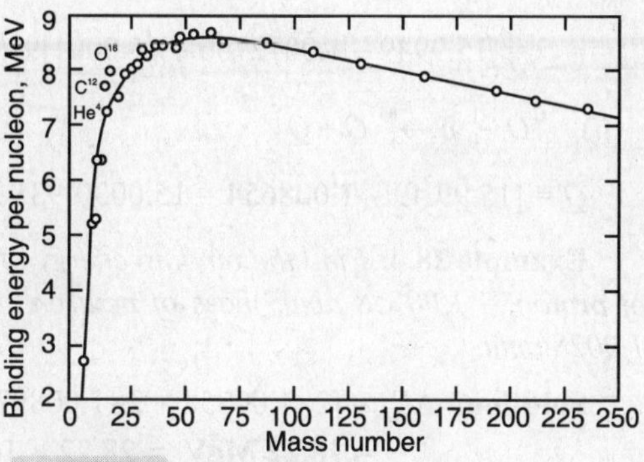

The following important features are seen from the binding energy curve.

(*i*) The binding energies of very light nuclei such as $_1H^2$ are very small. In case of light nuclei, most nucleons will be in the nuclear surface and as such they will not be in a position to use all their bonds. Consequently, $\Delta E_b/A$ is lower in these cases.

Fig. 28.3: Average binding energy per nucleon as a function of mass number

(*ii*) The binding energy is high in the middle of the periodic table, for elements whose A is in the range 28 < A < 138, i.e. from $_{14}Si^{28}$ to $_{56}Ba^{138}$. For these nuclei, the binding energy per nucleon is nearly 8.7 MeV.

(*iii*) The binding energy per nucleon decreases in case of elements having A>138. It is found to decrease to 7.6 MeV for uranium. The decrease may be attributed to the repulsive forces between protons, whose number increases in heavy nuclei.

Example 28.1: *Calculate the binding energy of a nitrogen nucleus in MeV from the following data:*

$m_H = 1.00783$ u, ***and*** $m_n = 1.00867$ u ***and*** $m\left({}^{14}_{7}N\right) = 14.00307$ u

Solution: Mass defect $\Delta M = 7 \times 1.00783 + 7 \times 1.00867 - 14.00307$ u = **0.11243 u**

$\therefore$ Binding energy = 0.11243 u × 931.4 MeV/u = **104.7 MeV.**

Example 28.2: *What is the binding energy per nucleon in ${}^7_3Li$ nuclide?*

Proton mass, m_p = 1.00814 amu

Neutron mass, m_n = 1.008665 amu

Mass of lithium nucleus. M = 7.01822 amu

and *1 amu = 931 MeV*

Solution: Number of neutrons in lithium nucleus = 7 – 3 = 4

Total binding energy ΔE_b = [3 × 1.00814 + 4 × 1.008665 – 7.01822]931 *MeV* = **38.041 MeV**

Binding energy per nucleon $\Delta\xi = \dfrac{\Delta E_b}{A} = \dfrac{38.041}{7}\text{MeV} = \mathbf{5.43\,MeV}$

Example 28.3: *The atomic masses of* ${}^{15}_{7}N, {}^{16}_{8}O$ *and O are 15.0001u, 15.0030u and 15.9949u respectively. How much energy is needed to remove one proton from* ${}^{16}_{8}O$*? How much energy is needed to remove one neutron from* ${}^{16}_{8}O$*? Why are these figures different from each other? (Proton rest mass = 1.0072766 u, Neutron rest mass = 1.0086654 u)*

Solution:

(*i*) ${}^{16}_{8}O - {}^{1}_{1}H \rightarrow {}^{15}_{7}N + Q$

$$Q = [15.9949 - 1.0072766 - 15.0001]931 \text{ MeV} = [-0.0124766]931 \text{ MeV} = \mathbf{-11.62\ MeV}$$

(*ii*) ${}^{16}_{8}O - {}^{1}_{0}n \rightarrow {}^{15}_{8}O + Q$

$$Q = [15.99.49 - 1.008654 - 15.0030]931 \text{ MeV} = [-0.0167654]931 \text{ MeV} = \mathbf{-15.62\ MeV}$$

Example 28.4: *Find the binding energy of an* α*-particle in MeV and Joule. Given : Mass of proton = 1.00758 amu, mass of neutron = 1.00897 amu and mass of helium nucleus = 1.0028 amu.*

Solution: $\Delta E_b = [2\times1.00758 - 2\times1.00897 - 4.0028]931$ MeV

$= \mathbf{28.22\ MeV} = 28.22 \times 10^{6} \times 1.602 \times 10^{-19}\text{ J} = \mathbf{4.5\times10^{-12}\ J}$

28.8 NUCLEAR MODELS

Models are devised to account for the properties and behaviour of a system. In case of an atom, Thomson's model, Rutherford model, Bohr model, de Broglie model, vector atom model and quantum mechanical model are devised which successively incorporated refinements one over the other. Ultimately, the quantum-mechanical model succeeded in interpreting all the properties and behaviour of an atom. In the same way efforts are made to develop a model, which can successfully explain the properties of the nucleus such as stability, spin, magnetic moment, etc. Various models have been proposed for nucleus. However, each of the models can explain only a limited number of properties of the nucleus. One of the earliest models was α - particle model proposed by Gamow.

28.8.1 Gamow Model

It is also known as α-particle model. According to this model, nucleus is assumed to have sub-groups in the form of α-particles. Each sub group has two protons and two neutrons. Hydrogen and deuterium nuclei are exceptions. The model was successful in explaining the emission of α-particles by the radioactive nuclei. The model is discarded subsequently.

Out of the other different models, two models namely *shell model* and *liquid drop model* are of importance and we study them here.

28.8.2 Nuclear Shell Model

Nuclear shell model is similar to the Bohr's model for the atom. By analogy with the closed sub-shells and shells in the case of atoms, it is assumed that nucleons also form similar closed sub-shells and shells within the nucleus. The electrons in an atom are supposed to revolve in the Coulomb electrostatic field of the nucleus in allowed orbits. In a similar manner, it is assumed in the shell model that each nucleon moves inside the nucleus in a fixed orbit under the influence of a central field of force produced by the average interaction between all the remaining nucleons.

The nuclear properties vary periodically with Z and N in a way like the periodic deviation of atomic properties with Z. A nucleus is stable if it has a definite number of either protons or neutrons.

Nuclei containing the following numbers of protons and neutrons exhibit high stability.

Z	2	8	20	50	82	
N	2	8	20	50	82	126

These values of Z and N are commonly called *magic numbers*. These are some features of the nucleus which indicate that the nucleus exhibits a shell structure. To interpret the existence of magic number and periodic variation of nuclear properties, one may assume that the protons and neutrons are arranged in shells in the nucleus just as extra nuclear electrons are distributed in various shells outside the nucleus. It is assumed that nucleons are arranged in shells, each shell containing a restricted number of nucleons in accordance with the Pauli principle. With increase of the number of nucleons, the first shell is filled, then the next nucleon shell etc. Since there are two classes of particles in a nucleus, there is a double shell arrangement, one for protons and another for neutrons.

We may characterize the nuclear energy states by quantum number n and l, specifying the energy level and the orbital angular momentum. The shells are regarded as 'filled' when they have their full quota of nucleons. The magic properties are observed when the outer nucleon shell is completely filled. The first shell is filled in $_2He^4$ which constists of two protons and two neutrons; the next shell is filled in $_8O^{16}$ etc. The enhanced stability of magic nuclei resembles the chemical inertness of helium, neon, etc.

Even Z-even N nucleus has zero total nuclear spin and even–odd and odd–even nuclei have half integral nuclear spin. It means that the protons and neutrons fill their levels in pairs with each pair having antiparallel spins. When the number of both the nucleons is even, they yield zero net spin; if one of them is even and the other is odd, the net spin is obviously half-integral spin. In case of odd–odd nucleus the proton, left over after the other protons are paired off, has its half spin added to the half spin of the left over neutron resulting in integral nuclear spin.

On excitation of the nucleus, a nucleon or several nucleons move into excited levels. Transition to the ground state is accompanied by the emission of γ-quanta.

28.8.3 Liquid Drop Model

The liquid drop model was proposed in 1936 by Niels Bohr basing on the external analogy between the nucleus and a liquid drop. The nucleus of an atom is similar to a drop of a liquid in many ways.

The following are the similarities between a small drop of liquid and the nucleus.

(*i*) Molecules at the surface of liquid are attracted to the inner molecules and the result is surface tension because of which the liquid drop takes spherical shape. Nucleus is also assumed to be spherical in shape.

(*ii*) The density of the nucleus is independent of its volume and the density of liquid is also independent of the volume.

(*iii*) In a liquid the forces are short-range forces; each molecule interacts only with its immediate neighbours. The nuclear force is also a short-range force and would primarily act between a nucleon and its nearest neighbours.

(*iv*) The molecules in the liquid drop are in random motion and frequently collide with each other. Similarly, the nucleons in a nucleus are in random motion and undergo frequent collisions.

(*v*) Each collision between nucleons involves an exchange of energy and momentum between them. If a certain amount of energy is imparted to the nucleus, it gets excited. The energy gets redistributed rapidly between the nucleons due to their mutual collisions. Some of the energy may be transferred to a surface nucleon. If the energy of the nucleon exceeds its binding energy in the nucleus, the nucleon overcomes the nuclear forces and leaves the nucleus. The expulsion of a nuclear particle from the nucleus is similar to the evaporation of a molecule from a liquid drop.

The liquid drop model successfully explained the process of nuclear fission and α-decay.

28.9 NATURAL RADIOACTIVITY

In 1896, Henry Becquerel discovered that uranium salts emit invisible radiation. *The phenomenon of emission of radiation by elements is called* **radioactivity**. The elements, which emit radiations, are called **radioactive** elements. It was found that radioactivity is the property of certain nuclei and hence such nuclei are called **radioactive nuclides**. Investigations on radioactive materials (Fig. 28.4) showed that the emitted radiations consist of three types of radiations, namely, α-rays, β-rays and γ-rays. Soon it was established that α-rays are streams of ${}^{4}_{2}He$ nuclei; β-rays are streams of very fast electrons and γ-rays are electromagnetic radiations which are much more penetrating than X-rays. During the emission of α-rays and β- rays, the composition of the nucleus changes, whereas in γ-rays emission the nuclear composition remains unaltered. γ-rays are emitted whenever the nucleus goes over from a higher energy state to a lower energy state.

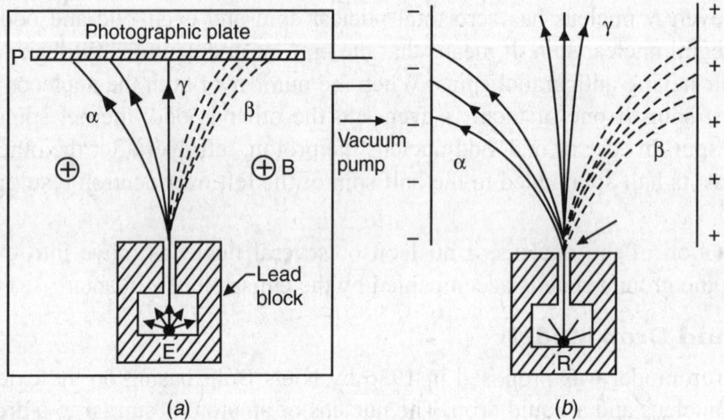

Fig. 28.4: Passing the radiation from a radioactive source through (a) a magnetic field or (b) an electric field shows that there are three components

Radioactivity may be natural or artificial. The **natural radioactivity** is the radioactivity found in nature and is exhibited by only a very small number of nuclei. Radioactivity exhibited by certain nuclei produced in the laboratory through nuclear reactions is called **induced radioactivity** or **artificial radioactivity**.

28.10 RADIOACTIVE DECAY

The spontaneous transformation occurring in the constitution of radioactive nuclides due to their radioactive property is called **radioactive decay** or **radioactive disintegration**. The transformation accompanied by the emission of α-rays is called **α-decay** and that accompanied by the emission of β-rays is called **β-decay**. The nucleus that goes through radioactive decay is termed as the **parent**, the intermediate products as **daughters** and the final stable nonradioactive nucleus as the **end product**.

α-decay: An α-particle is the same as a helium nucleus and consists of two protons and two neutrons. It carries a positive charge of two units. In α-decay, the mass number of the nucleus decreases by four units and the charge on the nucleus by two units. Therefore, the original element is transformed into an element two steps down in the periodic table. For example, α-decay of uranium produces thorium. Thus,

$$^{238}_{92}\text{U} \rightarrow {}^{4}_{2}\text{He} + {}^{234}_{90}\text{Th}$$

β-decay: A beta particle is simply an electron. β-decay of a nucleus leads to a decrease of one negative charge and no change in mass. It produces an element one step higher in the periodic table. For example, Th-234 emits β-particles and transforms into protactinium (Pa).

$$^{234}_{90}\text{Th} \rightarrow {}^{0}_{-1}\text{e} + {}^{234}_{91}\text{Pa}$$

γ-decay: A γ-particle is a quantum of energy. γ-radiation may or may not be emitted simultaneously with α- or β-rays. The nucleus has energy levels similar to the atomic energy levels. A γ-ray is emitted when an excited nucleus jumps to a lower energy level. For example,

$$^{209}_{82}\text{Pb}^* \rightarrow {}^{209}_{82}\text{Pb} + \gamma$$

Since γ-rays do not involve mass or charge, the emission of a gamma photon does not cause a transformation of species of nucleus.

28.11 RADIOACTIVE SERIES

The nuclei of all the naturally occurring radioactive elements (Z = 81 to Z = 92) are unstable and disintegrate by emitting either α or β particles and occasionally γ-rays. Radioactive nuclei frequently decay to another radioactive nucleus. This daughter nucleus decays to a third nucleus, which is also radioactive. The chain of radioactive decays continues until finally a stable nucleus forms. It is known that all natural disintegration processes end with the formation of stable lead atoms. The chain of successive radioactive decays is said to form a **radioactive series**.

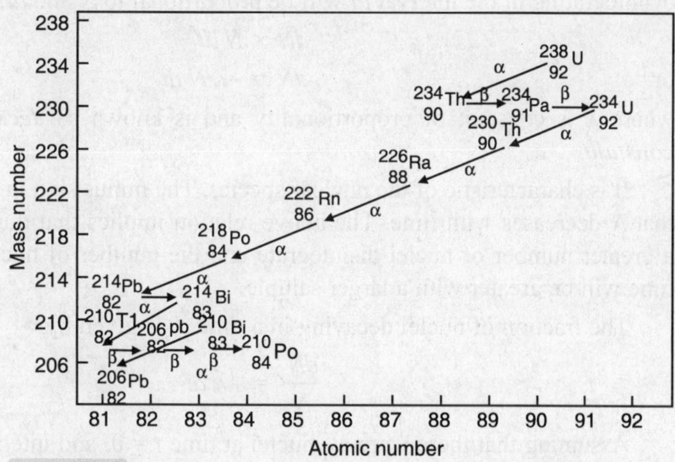

Fig. 28.5: Uranium radioactive decay series

There are three series of naturally occurring radioactive elements and a fourth one produced artificially. The three series of naturally occurring radioactive elements are (*i*) Uranium series

(*ii*) Actinium series

(*iii*) Thorium series

The series of artificially produced radio-active elements is Neptunium series.

Each series is named after the name of the parent nucleus from which the series starts. These series follow a decay pattern similar to that of the uranium series. In each radioactive series, each nuclide transforms into the next through a chain of α- and β-decays and ultimately a stable nucleus forms at the end. Thus, the uranium series terminates in the $^{206}_{82}Pb$ nucleus, the actinium series in $^{207}_{82}Pb$ nucleus and the thorium series in $^{208}_{82}Pb$ nucleus. The uranium radioactive decay series is shown in Fig. 28.5.

28.12 LAW OF RADIOACTIVE DECAY

The number of parent nuclei in a radioactive material decreases with time because of radioactive disintegration. The disintegration is a statistical and random process. Which nucleus in the material disintegrates first is only a matter of chance. The number of decay of nuclei over a specified time period can be determined as each nucleus has the same probability of decaying in one second.

The number of nuclei that disintegrate per second is called *rate of radioactive decay*. The rate of decay is proportional to the number of nuclei that have not yet disintegrated at any instant.

Let there be N untransformed nuclei present in a radioactive sample at time t and let dN be the number of decays in a short duration between $t+dt$. Then, dN, the number of nuclei disintegrating in the interval dt will be proportional to N and dt. That is,

$$dN \propto N\, dt$$

or

$$dN = -\lambda N\, dt \qquad ...(28.7)$$

where λ is constant of proportionality and is known as ***decay constant*** or ***disintegration constant***.

It is characteristic of the nuclear species. The minus sign in the above Eq. (28.7) indicates that N decreases with time. The above relation implies that during a longer interval of time, a greater number of nuclei disintegrate and the number of nuclei undergoing decay per unit time will be greater with a larger sample.

The fraction of nuclei decaying in a time dt is given by

$$\frac{dN}{N} = -\lambda\, dt$$

Assuming that there were N_o nuclei at time $t = 0$, and integrating the above equation, we can find out the nature of the decay.

$$\int_{N_o}^{N} \frac{dN}{N} = -\lambda \int_{0}^{t} dt$$

$$ln[N/N_o] = -\lambda\, t$$

Taking exponential on both the sides of the above equation, we obtain

$$\frac{N}{N_o} = e^{-\lambda t}$$

or $$N = N_o e^{-\lambda t} \qquad ...(28.8)$$

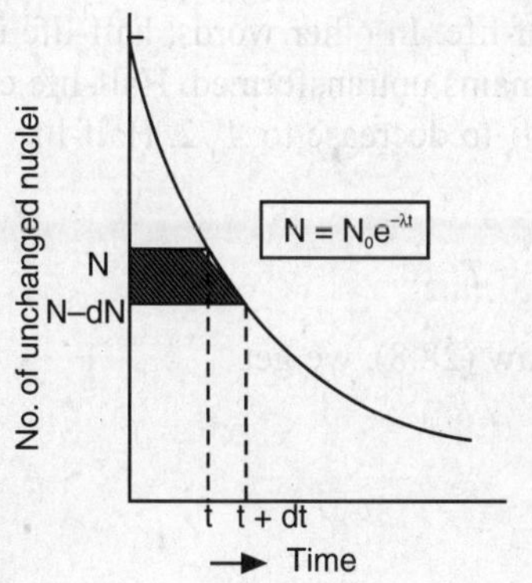

Fig. 28.6:

ln (N/N_o) versus time

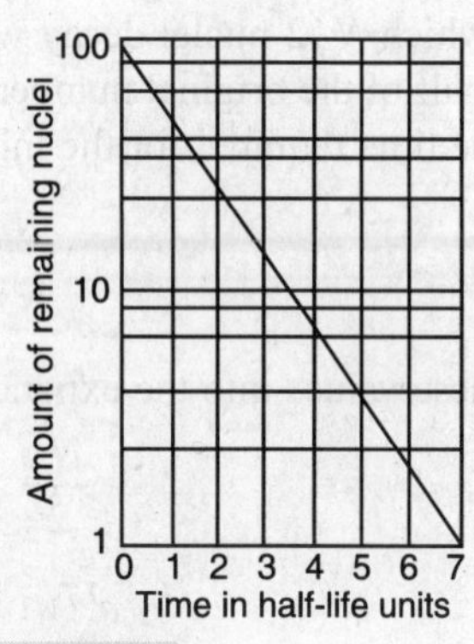

Fig. 28.7:

Slope gives λ

Eq. (28.8) is known as the ***law of radioactive decay***.

Law of radioactive decay shows that the number of nuclei of a given species decreases exponentially with time provided no new nuclei are added. The decay occurs rapidly initially and then becomes slower and slower as depicted in Fig. 28.6.

A plot of *ln* (N/N_o) versus time is plotted in Fig. 28.7. It is a straight line and the slope of the line gives the value of λ. Eq. (28.7) may be rearranged as

$$\lambda = -\frac{(dN/N)}{dt} \qquad ...(28.9)$$

λ may be defined as the fractional decrease in the number of nuclei decaying per unit time.

28.13 ACTIVITY

In the decay process, we are generally interested in the number of disintegrations per second. This is called the **activity**, A of the sample. Thus,

$$A = \left|\frac{dN}{dt}\right| = \lambda N_o e^{-\lambda t} = \lambda N \qquad ...(28.10)$$

It is obvious from Eq. (28.10) that the activity at $t = 0$ is given by $A_o = \lambda N_o$. Therefore,

$$A = A_o e^{-\lambda t} \qquad ...(28.11)$$

The plot of activity as a function of time is shown in Fig. 28.8.

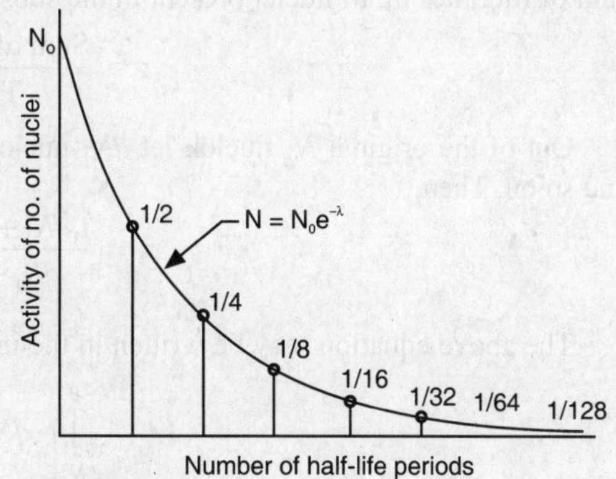

Fig. 28.8:

Activity versus Time

28.14 HALF-LIFE

The rate of decay or activity of a radioactive element is measured in terms of a characteristic time called the half-life. The **half-life** of an element is defined as the time taken by half of the original quantity to undergo decay. If N_0 are the number of nuclei present at $t = 0$, then the time in which $N_o/2$ nuclei decay will be the half-life. In other words, half-life is the time after which half of the original number of nuclei remains untransformed. Half-life can also be defined as the time required for the initial activity A_0 to decrease to $A_0/2$. Half-life is denoted by $T_{1/2}$.

$$N = \frac{N_o}{2} \quad \text{at } t = T_{1/2}$$

Using these values into the exponential decay law (28.8), we get

$$\frac{N_o}{2} = N_o e^{-\lambda T_{1/2}}$$

or

$$e^{\lambda T_{1/2}} = 2$$

$$\therefore \quad \lambda T_{1/2} = ln\, 2 = 0.693$$

$$\therefore \quad T_{1/2} = \frac{0.693}{\lambda} \qquad ...(28.12)$$

Eq. (28.12) suggests that the longer the half-life of an element, the slower it decays.

The half-life of uranium is 4500 million years, for radium it is 1620 years and for radon it is 3.8 days.

28.15 AVERAGE LIFE TIME

The atoms of radioactive substances are continuously disintegrating. Therefore, the life of each nucleus is different. The nuclei, which disintegrate earlier, have shorter life whereas those that disintegrate at the end have a longer life. It is difficult to specify why some atoms have shorter existence while others remain untransformed for a long time. Therefore, it becomes necessary to specify the **mean lifetime** or **average lifetime** of a radioactive species. The average lifetime, τ of a nuclide of a radioactive substance is defined as the ratio of the sum of lifetimes of all nuclei present in the substance to the total number of nuclei. That is,

$$\tau = \frac{\text{Sum of the life times of all nuclei}}{\text{Total number of nuclei}}$$

Out of the original N_o nuclei, let dN_1 nuclei live for time t_1, dN_2 nuclei live for time t_2, and so on. Then,

$$\tau = \frac{t_1 dN_1 + t_2 dN_2 + ...}{dN_1 + dN_2 + ...}$$

The above equation may be written in the integral form as

$$\tau = \frac{\int_{N_o}^{0} t \cdot dN}{\int_{N_o}^{0} dN} = \frac{\int_{0}^{N_o} t \cdot dN}{\int_{0}^{N_o} dN}$$

As, $N = N_o e^{-\lambda t}$, $dN = -\lambda N_o e^{-\lambda t}\, dt$.

$$\therefore \qquad \tau = \frac{\int_{\infty}^{0} (-\lambda) t N_o e^{-\lambda t} dt}{N_o} \qquad \because \int_{0}^{N_o} dN = N_o$$

$$= \lambda \int_{0}^{\infty} t e^{-\lambda t} dt$$

$$= \lambda \left[\frac{e^{-\lambda t}}{-\lambda} t - \frac{e^{-\lambda t}}{-\lambda^2} \right]_0^{\infty} = -\frac{1}{\lambda}\left[(\lambda t + 1) e^{-\lambda t} \right]_0^{\infty} = \frac{1}{\lambda}$$

$$\therefore \qquad \tau = \frac{1}{\lambda} \qquad \text{...(28.13)}$$

Thus, the mean lifetime of a radioactive species is equal to the reciprocal of the decay constant.

Using Eq. (28.13) into Eq. (28.12), we obtain

$$T_{1/2} = 0.693\ \tau \qquad \text{...(28.14)}$$

28.16 UNITS OF ACTIVITY

The activity of a radioactive source is defined as the number of disintegrations that occur per second. The traditional unit of activity is **curie** (Ci). It is defined as

1 Ci = 3.7×10^{10} disintegrations per second.

This definition is based on the activity of 1 gm of radium. Curie is rather a large unit, so millicuries (*m Ci*) and microcuries (μ Ci) are commonly used.

The *SI* unit of radioactivity is the Becquerel (*Bq*). It is defined as

1 Bq = 1 decay per second

Example 28.5: *A certain radioactive substance has a disintegration constant $\lambda = 1.44 \times 10^{-3}$ per hour. In what time will 75% of the initial number of atoms disintegrate?*

Solution: $N = N_o - \frac{3}{4} N_o = \frac{1}{4} N_o$

But $N = N_o e^{-\lambda t}$ $\quad \therefore \quad \frac{1}{4} = e^{-\lambda t}$ or $e^{\lambda t} = 4$

$$\therefore \qquad t = \frac{\log_e 4}{\lambda} = \frac{2.3026 \times 0.6021}{1.44 \times 10^{-3}} \text{hours} = \mathbf{962.9\ hrs.}$$

Example 28.6: *Calculate the activity of 1 mg sample of ${}^{90}_{38}Sr$ whose half life is 28 years.*

Solution: $\lambda = \frac{0.693}{T_{1/2}} = \frac{0.693}{28 \text{ yrs}} = \frac{0.693}{28 \times 365 \times 24 \times 3600 \text{s}}$

$$= 7.85 \times 10^{-10}/\text{s}.$$

Number of nuclei in one mg sample,

$$N = \frac{mN_A}{M} = \frac{10^{-3} \times 6.02 \times 10^{26}}{90} = 6.69 \times 10^{21}$$

Activity of the sample $A = \lambda N = 7.85 \times 10^{-10}/\text{s} \times 6.69 \times 10^{21}$

$= \mathbf{5.25 \times 10^{12}}$ **disintegrations/s.**

Example 28.7: *One gm of a radioactive material having a half-life of 2 years is kept in store for a duration of 4 years. Calculate how much of the material remains unchanged.*

Solution: $$T_{1/2} = \frac{0.6931}{\lambda} = 2 \text{ yrs}$$

$\therefore$ $$\lambda = \frac{0.6931}{2 \text{ yrs.}}$$

As $$N = N_0 e^{-\lambda t}, \; t = \frac{1}{\lambda} \ln \frac{N_0}{N}$$

$\therefore$ $$4 \text{ yrs.} = \left(\frac{2 \text{ yrs.}}{0.6931}\right) \ln \frac{N_0}{N}$$

$\therefore$ $$\ln \frac{N_0}{N} = \frac{4 \times 0.6931}{2} = 1.386$$

$\therefore$ $$\frac{N_0}{N} = 4$$

or $$N = \frac{N_0}{4} = \frac{1\,gm}{4} = \mathbf{0.25}\ \boldsymbol{gm}$$

The material that remains unchanged after 4 years is 0.25 gm.

Example 28.8: *5 gm of radium is reduced by 10.5 mg in 5 years. Calculate the half-life of radium.*

Solution: The initial quantity of radium $N_o = 5$ gm.

Quantity of radium present $N = 5 - 10.5 \times 10^3$ gm $= 4.9895$ gm

$$\frac{N}{N_0} = e^{-\lambda t}$$

$\therefore$ $$\frac{4.9895\,\text{gm}}{5\,\text{gm}} = e^{-\lambda t}$$

$\therefore$ $$\lambda t = \ln\left[\frac{5}{4.9895}\right]$$

As $t = 5$ years, $$\lambda = \frac{\ln\left(\frac{5}{4.9895}\right)}{5 \text{ years}} = 4.2 \times 10^{-4} \text{ dis./year}$$

$$T_{\frac{1}{2}} = \frac{0.693}{\lambda} = \frac{0.693}{4.2 \times 10^{-4}} \text{ yrs.} = 1650 \text{ years}$$

28.17 INDUCED RADIOACTIVITY

Joliot-Curie and Frederic discovered in 1934 the phenomenon of **induced** or **artificial radioactivity**. They discovered that aluminium, boron and magnesium become radioactive when irradiated with α-particles. The irradiation of aluminium induces the nuclear reaction

$$^{27}_{13}\text{Al} \quad + \quad ^{4}_{2}\text{He} \quad \rightarrow \quad ^{30}_{15}\text{P}^{*} \quad + \quad ^{1}_{0}\text{n}$$

The phosphorous isotope $^{30}_{15}P*$ is radioactive. It turns into a stable silicon isotope after emitting a positron.

$$^{30}_{15}\text{P}^{*} \quad \rightarrow \quad ^{30}_{14}\text{Si} \quad + \quad ^{0}_{+1}\text{e}$$

The discovery of induced radioactivity is important on two counts. It was for the first time that radioactive materials were synthesized and secondly, it proved that not only heavy nuclei have radioactive isotopes but light nuclei also have radioactive isotopes. Subsequent works demonstrated that radioactive isotopes of all elements could be synthesized. Radioactive isotopes are produced by irradiating the nuclei with α-particles, protons, deuterons or high energy γ-rays.

Fermi studied radioactivity induced by neutrons. At present the most widely used method for producing radioactive isotopes is by neutron irradiation. All nuclei except $^{4}_{2}\text{He}$ absorb neutrons and transform into β-active isotopes.

28.18 APPLICATIONS OF RADIOACTIVITY

Radioactivity is widely used in many areas of science and technology.

(*i*) Wide use is made of high penetration power of γ-rays. Since the absorption of radiation increases with the distance it travels through an object, the variations in intensity of radiation can be used to measure the thickness of an object or to detect internal defects. Metal castings in industry are tested using this method. The radioisotope ^{60}Co which emits γ-rays is housed in an aluminium thimble and is placed inside the casting under test. A photographic film is positioned outside the object. The gamma rays penetrate the metal part and make structural flaws such as cavities in the metal observable on the photographic film. The structural members of big boilers in aircrafts are tested by this method. The thickness measurement and monitoring is carried out in metal sheet and foil manufacturing with the help of this method.

(*ii*) Radioactive isotopes behave chemically just like nonradioactive isotopes of the same element. Therefore, artificial radioactive isotopes are widely used as **tracers**.

(*iii*) The wearing out of piston rings etc is studied by introducing radioactive iron isotopes into the parts. During operation of an engine, the radioactive atoms appear in the lubricating oil. By periodically determining the activity of the oil, the wear of a particular component can be determined.

(*iv*) Similarly the location of an underground pipe or a clog can be determined by using a liquid radioactive tracer.

(*v*) In medicine radioactive tracers are used for diagnostic purpose and treatment of sickness. For example, the extent of affliction of thyroid gland can be determined using radioactive iodine. The minimum daily requirement of iodine is 150 micrograms. If a patient is given radioactive iodine-131, the thyroid gland does not know the difference. By measuring the counting rate near the gland, we can know how the radioactive iodine is distributed in the gland. The count rate is affected if there is some abnormality in the functioning of the gland.

(*vi*) Radioactive ^{131}I is also used for therapy, in the treatment of thyroid cancer. A patient is given a solution of ^{131}I as potassium iodide. The iodine makes its way to the thyroid gland and the beta rays emitted by iodine-131 destroy the cancerous thyroid cells.

(*vii*) The ionizing effect of nuclear radiation is used in medicine to destroy malignant tumours. The heating effect caused by nuclear radiation is used to generate electricity which in turn powers a cadiac pacemaker. The electricity generated by plutonium is sent as a pulse directly to the ventricles of the heart at a predetermined rate.

(*viii*) Tracer atoms are used to study photosynthesis in plants. Tracer atoms help determine the effect of fertilizers on plants. This type of research leads to better yields per acre, and more food at less expense.

(*ix*) Nuclear radiation is used to preserve foods. Foods are pasteurized by irradiation to retard the growth of organisms such as bacteria, molds and yeasts. The irradiation prolongs shelf life under refrigeration.

28.18.1 Radioactive Dating

One of the most important applications of radioactivity lies in the determination of the age of rocks, planets and solar system, and the age of fossils and objects discovered in excavations. The exponential character of disintegration of radioactive nuclei provides time scale for these determinations. From Eq. (28.8) and Eq. (28.12), the time interval between the instants when the number of radioactive nuclei is N_o and N may be given by

$$t = \frac{1}{\lambda}\ln\frac{N_o}{N} = 1.44\ T_{1/2}\ln\frac{N_o}{N} \qquad \text{...(28.15)}$$

If N is the number of unchanged nuclei at the present time, the above equation gives the age of the given species of radioactive nuclei.

Each application requires a different time scale in practice. In geology a sufficiently slow radioactive time scale is required and uranium is the most suitable scale for this purpose. The age of uranium ore can be determined from the ratio of uranium to the final product of lead contained in it. Age determinations for various rocks taken from different parts of the world indicate their ages to be in the neighbourhood of 3 to 4 billion years. It suggests that the crust of the earth was formed about 4 billion years ago. The age of some meteorites is found to be 4.5 billion years. Therefore, the age of the planets in the solar system may be thought to be 4.5 billion years. In archaeology, a time scale with a half-life of only a few centuries is required. Uranium time is very long and also many of the excavated objects do not contain uranium, making the uranium dating unsuitable. Carbon-14 dating is employed in these studies.

The interaction of cosmic rays with the nuclei of nitrogen in the atmosphere turns them into radioactive carbon isotope $^{14}_{6}C$. The half-life of carbon-14 is 5730 years. The radioactive carbon produces radioactive carbon dioxide which mixes with the ordinary carbon dioxide in the atmosphere and, in turn, is assimilated by all living matter through natural food chains. Upon the death of the organism, the intake of food ceases and the natural level of radioactive carbon present within the structure begins to decrease. Therefore, radioactive carbon-14 can be used to determine the date of ancient artifacts derived from the living matter.

Example 28.9: *In a uranium mineral, lead -206 is found predominantly. The mineral contains 0.093 of lead in 1 gm of uranium. Calulcate the age of the mineral if half-life of uranium is* 4.3×10^9 *years.*

Solution: *As t =0, let N_0 be the uranium nuclei. Number of uranium nuclei at t be N.*

$$N - N_0\, e^{-\lambda t}$$

Number of uranium nuclei disintegrated in time *t* = *N* = Number of lead nuclei at time *t*.

$$N' = N_0 - N_0 e^{-\lambda t} = N_0\,(1 - e^{-\lambda t})$$

$$= \frac{\text{Number of lead nuclei at } t}{\text{Number of uranium nuclei at } t} = \frac{N_0\left(1-e^{-\lambda t}\right)}{N_0\, e^{-\lambda t}} = \frac{N'}{N}$$

$$\therefore \quad e^{\lambda t} - 1 = \frac{N'}{N} \quad \text{or } e^{\lambda t} = 1 + \frac{N'}{N}$$

$$\lambda t = ln\left(1 + \frac{N'}{N}\right)$$

$$t = \frac{1}{\lambda} ln\left(1 + \frac{N'}{N}\right)$$

Number of nuclei in 1 gm of uranium, $N = \frac{N_A}{238}$

Number of nuclei in 0.093 gm of lead, $N' = \frac{0.093\, N_A}{206}$

$$\frac{N'}{N} = \frac{0.093 \times 238}{206} = \frac{22.134}{206}$$

$$1 + \frac{N'}{N} = 1 + \frac{22.134}{206} = \frac{228.134}{206} = 1.107$$

$$t = \frac{1}{\lambda} \ln(1.107) = \frac{4.5 \times 10^9 \text{ yrs.}}{0.693} \ln(1.107)$$

$$= 6.6 \times 10^8 \; years = \textbf{660 Million Years}$$

Example 28.10: *A wooden piece of antiquity weighs 50 gm and shows ^{14}C activity of 320 disintegrations per minute. Estimate its age, assuming that the living tree, of which the wooden piece was a part, shows ^{14}C activity of 12 disintegrations per minute per gm. The half-life of ^{14}C is 5730 years.*

Solution: Let thre be N_o radioactive C-14 atoms in the tree just before it died. Its activity is

$$A_0 = \lambda N_0$$

After its death, the activity decreases exponentially

Thus
$$A = -\frac{dN}{dt} = \lambda N = \lambda\, N_0\, e^{-\lambda t}$$

$$\therefore \quad \frac{A}{A_0} = e^{-\lambda t}$$

$$A_0 = 12 \text{ disintegrations/min/gm}$$

$$A = \frac{320}{50} \text{ disintegrations/min/gm}$$

$$\lambda = \frac{0.693}{5730}(\text{year})^{-1}$$

$$\ln\left(\frac{A}{A_0}\right) = -\lambda t$$

$$t = \frac{1}{\lambda}\ln\left(\frac{A_0}{A}\right) = \left(\frac{5730\text{ Yrs.}}{0.693}\right)\ln\left(\frac{50\times 12}{320}\right) = \mathbf{5197.5\ yrs.}$$

28.19 NUCLEAR REACTIONS

Natural radioactivity is a nuclear reaction in which a radioactive nucleus of one element *spontaneously* changes into the nucleus of another element. It is also possible to induce a nuclear reaction artificially, by bombarding a nucleus with high-energy particles. *The reaction, in which an external bombarding particle successfully changes the identity of a target nucleus, is known as an* **artificial nuclear reaction** *or simply* **nuclear reaction**.

The first artificial nuclear reaction was produced by Rutherford in 1919 by bombarding nitrogen gas with α-particles emitted by a natural bismuth-214 source.

$${}^{14}_{7}\text{N} + {}^{4}_{2}\text{He} \rightarrow {}^{17}_{8}\text{O} + {}^{1}_{1}\text{H}$$

The bombardment resulted in conversion of nitrogen nucleus into oxygen nucleus. Such transformation of one nucleus into another nucleus is called **transmutation**. The important implication of this reaction is that one element can be changed into another by use of bombardment process. Subsequent to Rutherford's experiment, many isotopes were subjected to beams of high energy particles and numerous reactions were found. Not all nuclear reactions produce stable isotopes. If ${}^{25}_{12}\text{Mg}$ is bombarded with an alpha source, a radioactive isotope of aluminium ${}^{28}_{13}\text{Al}$ is produced which does not exist in nature.

$${}^{25}_{12}\text{Mg} + {}^{4}_{2}\text{He} \rightarrow {}^{28}_{13}\text{Al}^* + {}^{1}_{1}\text{H}$$

The asterisk (*) is used to denote a **radioactive isotope**. Artificial radioactive isotopes have characteristic half-life as do the naturally occurring ones.

In a typical nuclear reaction experiment, a target material is bombarded by high-energy particles such as protons, neutrons, deuterons, electrons or α-particles. The interaction between the target material and the incident particle depends on the nature and the energy of the particles and the target material.

Most of the alpha particles emitted from natural radioactive decay do not have enough energy to penetrate a heavy positively charged nucleus to induce a nuclear reaction. If the artificial nuclear reactions are to be studied for heavier elements, the kinetic energy of projectile particles must be increased.

A nuclear reaction is represented in a form similar to that of a chemical reaction. Thus, the reaction equation is shown as

$${}^{A}_{Z}X + a \rightarrow {}^{A_1}_{Z_1}Y + b$$

where a is the projectile particle that bombards the target nucleus X resulting in the product nucleus Y and an outgoing particle b. The reaction is represented in an abbreviated form as

$${}^{A}_{Z}X\ (a, b)\ {}^{A_1}_{Z_1}Y$$

Thus, the first artificial nuclear reaction is written as ${}^{14}_{7}\text{N}\,(\alpha, p)\,{}^{17}_{8}\text{O}$. The type of (a, b) nuclear reaction is determined by the nature of the projectile and the emitted particles.

In any nuclear reaction, mass and charge are conserved. Conservation of mass requires that the sums of the superscript mass numbers on each side of the equation are equal. Conservation of charge requires that the sums of the subscript protons on each side of the equation are equal.

28.20 Q-VALUE

Nuclear reactions resemble ordinary chemical reactions and are accompanied by energy changes. The energy liberated or absorbed during a nuclear reaction is called **nuclear reaction energy**. This energy is denoted by Q in the reaction equation and is called the *energy balance* of the reaction or more commonly, its **Q-value**. The Q-value of a reaction can be positive or negative depending on the nature of the reaction. According to Einstein's mass-energy equivalence principle, Q-value must be balanced by the changes in mass associated with the nuclear reaction.

Let us consider a general nuclear reaction

$$X + a \rightarrow Y + b.$$

Let M_0 be the mass of the stationary target nucleus X,

M_1, E_1 be the mass and energy of the projectile a,

M_2, E_2 be the mass and energy of the product nucleus Y, and

M_3, E_3 be the mass and energy of the emitted particle b.

The nuclear reaction may be written as

$$M_0 + (M_1 + E_1) = (M_2 + E_2) + (M_3 + E_3)$$

or $$(M_0 + M_1) - (M_2 + M_3) = (E_2 + E_3) - E_1$$

As Q-value is the energy balance, it may be expressed as

$$Q = (E_2 + E_3) - E_1 \qquad \text{...(28.16 a)}$$

or $$Q = (M_0 + M_1) - (M_2 + M_3) \qquad \text{...(28.16 b)}$$

Eq. (28.16 a & b) shows that the Q-value of a nuclear reaction can be calculated from the known kinetic energies of the particles involved or from the known masses of the reactants and the product nuclei.

But Q-value generally defined as the difference in masses of the reactants and the products.

(*i*) If $Q > 0$, the reaction is said to be **exothermic** or **exoergic**. That is, energy is released in the reaction. In this case, $(M_0 + M_1) > (M_2 + M_3)$. Hence, the total mass of the products is less than that of the reactants. The difference in masses (mass defect) is converted into energy.

(*ii*) If $Q < 0$, the reaction is said to be **endothermic** or **endoergic**. In this case $(M_0 + M_1) < (M_2 + M_3)$. That is, the total mass of the reactants is less than that of the products. It means that there is a gain of mass in the reaction, which could happen when there is absorption of energy. Therefore, energy is to be supplied in the form of kinetic energy of the projectile a. The kinetic energy of the projectile particle must have some minimum value, below which the reaction cannot occur. The minimum energy necessary for an endothermic reaction to occur is called the **threshold energy**.

1. Example of Exothermic Nuclear Reaction:

${}_{3}^{7}\text{Li}\ (p, \alpha)\ {}_{2}^{4}\text{He}$ is an example of exothermic reaction. Let us calculate the Q-value of this reaction.

Reactants total mass	**Products total mass**
$M_0 = 7.01822$ amu	$M_2 = 4.00387$ amu
$M_1 = 1.00814$ amu	$M_3 = 4.00387$ amu
$M_0 + M_1 = 8.02636$ amu	$M_2 + M_3 = 8.00744$ amu

$$\Delta M = (M_0 + M_1) - (M_2 + M_3) = 0.01862 \text{ amu}$$

$$Q = c^2 \Delta M = 931.4\ (\Delta M)\ MeV = 931.4\ (0.01862) \text{ MeV}$$

or $$\mathbf{Q = 17.34\ MeV.}$$

2. Example of Endothermic Nuclear Reaction:

${}_{7}^{14}\text{N}\ (\alpha, p)\ {}_{8}^{17}\text{O}$ is an example of endothermic reaction. Let us calculate the Q-value of this reaction.

Reactants total mass	**Products total mass**
$M_0 = 14.00753$ amu	$M_2 = 17.00450$ amu
$M_1 = 4.00387$ amu	$M_3 = 1.00814$ amu
$M_0 + M_1 = 18.01140$ amu	$M_2 + M_3 = 18.01264$ amu

$$\Delta M = (M_0 + M_1) - (M_2 + M_3) = -\ 0.00124 \text{ amu}$$

$$\text{Q} = c^2 \Delta M = -931.4\ (\Delta M) \text{ MeV} = -931.4\ (0.00124) \text{ MeV}$$

or $$\mathbf{Q = -1.15\ MeV.}$$

Example 28.11: *A nuclear reaction is given by* ${}_{5}^{10}B + {}_{2}^{4}He \rightarrow {}_{6}^{13}C + {}_{1}^{1}H$

Given that: ${}_{5}^{10}B = 10.016125$ amu, ${}_{2}^{4}He = 4.003874$ amu

${}_{6}^{13}C = 13.007440$ amu and ${}_{1}^{1}H = 1.008146$ amu

Compute the energy released.

Solution: $Q = [10.016125 + 4.003874 - (13.007440 + 1.008146)]\ 931.4$ MeV

$= \mathbf{4.11\ MeV}$

28.21 NUCLEAR REACTION CROSS-SECTION

All particles that are incident on a target material do not produce nuclear reactions; only a small fraction of them interact with the nuclei of the target material. Through the concept of nuclear cross-section, σ the probability of a bombarding particle to interact with a target particle can be expressed. Greater the cross-section, the greater is the probability of interaction.

The nuclear reaction cross-section may be defined as

(*i*) *the probability that an event may occur when a single nucleus is exposed to a beam of particles of total flux of one particle per unit area, or*

(*ii*) *the probability that an event may occur when a single particle is shot perpendicularly at a target consisting of one nucleus per unit area.*

28.21.1 Calculation of Microscopic Cross-section

Let us suppose that we have a slab of some material whose area is A and thickness is dx. We assume here that the target is thin and there is no overlapping of nuclei.

The volume of the slab = $A\,dx$.

Let the material contain n atoms per unit volume,

then the total number of nuclei in the slab = $n\,A\,dx$.

If each nucleus has a cross-section σ for some particular interaction,

then the aggregate cross-section for all the nuclei in the slab = $\sigma\,n\,A\,dx$.

If N is the number of incident particles in a beam, and dN is the number of particles that interact out of them with the nuclei in the slab, then

$$\frac{\text{Number of interacting particles}}{\text{Number of incident particles}} = \frac{\text{Aggregate corss-section}}{\text{Target area}}$$

$$\frac{dN}{N} = \frac{\sigma n\,Adx}{A} = \sigma\,n\,dx$$

or
$$\sigma = \frac{dN/N}{n\,dx} \qquad ...(28.17)$$

The above equation gives the cross-section per nucleus. Here, σ is known as the *microscopic cross-section.*

28.21.2 Calculation of Macroscopic Cross-section

Let a beam of particles is incident on a slab of finite thickness (Fig. 28.9). If each incident particle can interact with nuclei only once, then

$$\frac{-dN}{N} = \sigma\,ndx$$

where dN is the number of particles removed from the beam in passing through the slab

dx is thickness of the slab

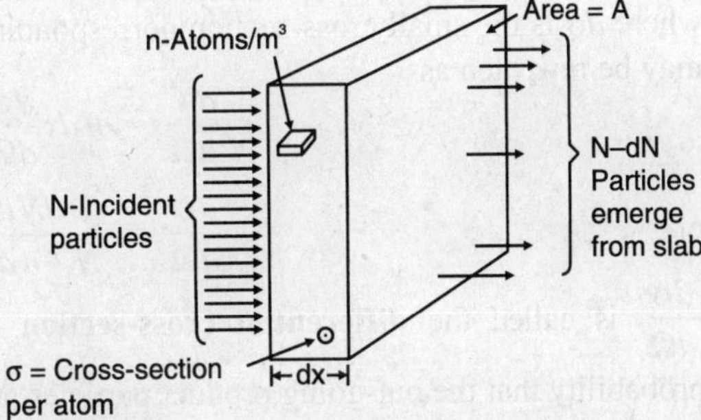

Fig. 28.9: Calculation of Macroscopic Cross-section

The negative sign in the above equation indicates that the number of particles decrease as the beam passes through the slab. Let N_0 be the initial number of incident particles. Then

$$\int_{N_0}^{N} \frac{dN}{N} = -\sigma\,n\int_0^x dx$$

$$ln\frac{N}{N_0} = -\,\sigma nx$$

$$\therefore \qquad N = N_0 e^{-\sigma nx}$$

Thus, N the number of surviving particles decreases exponentially with increasing thickness of the slab.

The product of microscopic cross-section and the number of nuclei per unit volume, i.e. σn is known as the macrosopic cross-section. It is denoted by Σ.

σ has the dimensions of area and Σ has the dimension of L^{-1}. It is important to note that the effective cross-section of a nucleus is not equal to its geometric cross-section. It is only a convenient artifice to describe the probability of occurrence of nuclear reactions. In fact, the cross-sections of target nuclei depend upon the type of interaction, type of incident particle and its energy.

The unit used for nuclear cross-section is barn.

$$1 \text{ barn} = 10^{-28} \text{ m}^2$$

28.21.3 Differential Cross-section

In many nuclear reactions, the particles are not produced in an isotropic manner. In such a case, if dN is the number of product nuclei emitted per unit time in a small solid angle $d\Omega$ at some angle θ with the direction of the incident beam (Fig. 28.10), then from Eq. (28.17) we can write

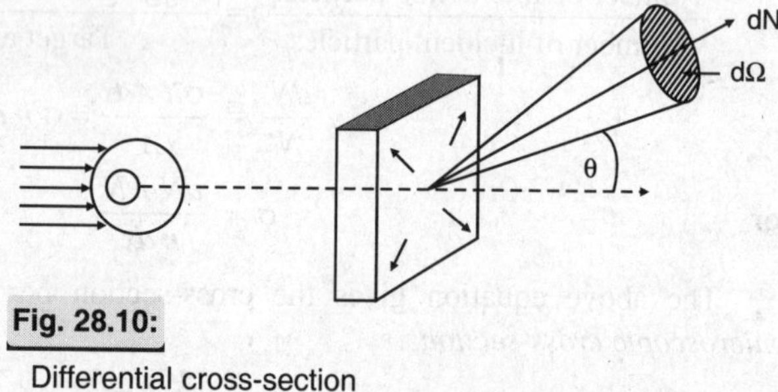

Fig. 28.10: Differential cross-section

$$\frac{dN}{N} = n\,dx\,d\sigma$$

where $d\sigma$ is the small cross-section corresponding to the solid angle $d\Omega$. the above equation may be rewritten as

$$\frac{1}{N}\frac{dN}{d\Omega} = n\,dx\frac{d\sigma}{d\Omega}$$

or

$$\frac{d\sigma}{d\Omega} = \frac{1}{N}\frac{dN/d\Omega}{n\,dx} \qquad ...(28.18)$$

$\dfrac{d\sigma}{d\Omega}$ is called the **differential cross-section** of the nuclear reaction and represents the probability that the out-going product particles are confined to an element of solid angle $d\Omega$.

Example 28.12. *A 0.01 mm thick* $^{7}_{3}Li$ *target is bombarded with* 10^{13} *protons/s. As a result* 10^8 *neutrons/s are produced. What would be the cross-section for this reaction? The density of lithium is 500 kg/m³.*

Solution.

$$\sigma = \frac{N}{N_0 N_t}$$

The number of target nuclei per unit volume $= \dfrac{\rho N_A}{M} = \dfrac{(500\text{ kg/m}^3)(6.02\times10^{26})}{7\text{ kg}} = 4.3 \times 10^{28}/\text{m}^3$.

The number of target nuclei/unit area $N_t = \left(\dfrac{\rho N_A}{M}\right)t = (4.3 \times 10^{28}/\text{m}^3)\,(0.01 \times 10^{-3}\text{ m})$

$$= 4.3 \times 10^{23}/\text{m}^2.$$

Number of nuclei undergoing interaction per second = Number of neutrons produced /s

$$= 10^8.$$

$$\therefore \qquad \sigma = \frac{10^8}{10^{13} \times 4.3 \times 10^{23}/\text{m}^2} = \mathbf{2.3 \times 10^{-29} m^2}.$$

28.22 NEUTRONS AND NEUTRON INDUCED REACTIONS

The neutron was discovered by the English physicist James Chadwick in 1932. He found that when beryllium is bombarded with α-particles, it emits neutral particles with a mass close to that of the proton.

$$^{9}_{4}\text{Be} \quad + \quad ^{4}_{2}\text{He} \quad \rightarrow \quad ^{12}_{6}\text{C} \quad + \quad ^{1}_{o}\text{n}$$

The neutron produced in this reaction carries sufficient energy to cause additional nuclear reactions in nuclei with which the neutron collides. Chadwick received Nobel Prize in physics in 1935 for his discovery of neutron.

Neutrons are very effective in initiating nuclear reactions because they do not possess electric charge. As such they penetrate nuclei more deeply than any other particle. Neutrons interact with nucleus differently, according to whether they are fast or slow. **Fast neutrons** are those having energies in the range of 100 keV to 50 MeV. **Slow neutrons** are those which have energies not exceeding 100 keV.

The heaviest naturally available element is uranium having an atomic number 92. Enrico Fermi, an Italian physicist, who undertook a systematic study of neutron-induced nuclear reactions, speculated that neutron bombardment of uranium would yield new elements that would be more massive than uranium. In 1934 Fermi found that uranium bombarded with neutrons yield radioactive products which were assumed to be trans-uranium elements. However, in 1938 Otto Hahn and Fritz Strassmann, the German chemists, using precise radiochemical identification established beyond doubt that the neutron bombardment of uranium produced an isotope of barium $\left(^{139}_{56}\text{Ba}\right)$. More similar identifications followed. In 1939, Lise Meitner and Otto R. Frisch suggested a process called **nuclear fission** where the uranium nucleus divided into 'two nuclei of roughly equal size". Thus, neutron bombardment of uranium leads sometimes to fission and sometimes to formation of transuranium elements neptunium $\left(^{239}_{93}\text{Np}\right)$ and plutonium $\left(^{239}_{94}\text{Pu}\right)$.

28.23 NUCLEAR FISSION

Nuclear fission is a neutron-induced nuclear reaction in which a heavy nucleus such as uranium splits into two intermediate lighter nuclei. In a nucleus there is a competition between the nuclear force, which holds the nucleus together, and the electrostatic repulsion of the protons which tires to tear the nucleus apart. In case of heavy nuclei there is a delicate balance between the nuclear and electric forces. This balance can be easily upset. The energy necessary to cause fission is about 6 MeV.

Only certain heavy nuclei can undergo fission and the probability of fission reaction for a particular nucleus depends on the energy of the incident neutrons. Nuclei with odd number of neutrons, $^{233}_{92}\text{U}$, $^{235}_{92}\text{U}$ and $^{239}_{94}\text{Pu}$ undergo fission with slow neutrons. On the other hand, the nucleus $^{232}_{90}\text{Th}$ with an even number of neutrons requires fast neutrons with energies of 1 MeV or more.

Nuclear fission *is the phenomenon of breaking up the nucleus of a heavy atom into two more or less equal segments with the release of a large amount of energy.*

Hahn discovered in 1938 that when uranium was bombarded with thermal neutrons, the uranium nucleus broke up into barium and krypton nuclei of atomic numbers 56 and 36 and liberated 3 neutrons accompanied by a tremendous amount of energy (Fig. 28.11). The nuclear reaction is represented as follows:

$$ {}^{235}_{92}\text{U} + {}^{1}_{0}\text{n} \rightarrow {}^{141}_{56}\text{Ba} + {}^{92}_{36}\text{Kr} + 3\,{}^{1}_{0}\text{n} + \text{Q} \qquad ...(28.19)$$

Fig. 28.11: Nuclear Fission

The U^{235} nuclei do not all split up into those of Ba and Kr. The fission reaction can proceed in about 40 different ways, yielding different final products. The heavy nuclei may split up into nuclei of several pairs of elements lying in the central region of the periodic table with slightly unequal masses. These are known as *fission fragments*.

The fission products such as ${}^{141}_{56}\text{Ba}$ and ${}^{92}_{36}\text{Kr}$ are radioactive and undergo disintegrations until stable isotopes are formed. After several steps they form the stable isotopes ${}^{141}_{59}\text{Pr}$ and ${}^{90}_{40}\text{Zr}$ respectively. Radio-chemical analysis showed that nuclides resulting from fission have atomic numbers between 30 and 63 and mass numbers between 72 and 158. Fig. 28.12 shows the mass distribution of the fission fragments from the fission of U-235 nuclei. It is most likely that one fragment will have a mass number of about 95 and the other about 140. Most of the energy (about 80%) released in fission goes into kinetic energy of the two fission fragments, and the remaining (20%) appears as decay products (β and γ rays) and kinetic energy of neutrons emitted in the fission process. The neutrons typically have energies of one to several MeV.

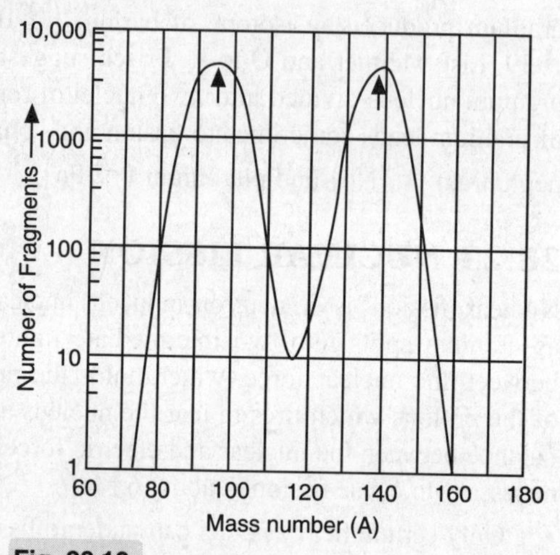

Fig. 28.12: Mass distribution of the fission fragments from the fission of U-235 nuclei

The most important features of nuclear fission is that the process is accompanied by the

(*i*) release of a large amount of energy and

(*ii*) emission of two or more energetic neutrons which under appropriate conditions cause fission in neighbouring nuclei.

28.23.1 Energy Released during Nuclear Fission

The energy released during the process of fission is known as *nuclear energy or atomic energy*. The amount of energy released during fission may be estimated using mass defect method. We illustrate the method by taking the fission reaction (28.19) as an example.

$$\therefore \quad {}^{235}_{92}\text{U} + {}^{1}_{0}\text{n} \rightarrow {}^{141}_{56}\text{Ba} + {}^{92}_{36}\text{Kr} + 3\,{}^{1}_{0}\text{n} + \text{Q}$$

Actual mass before the fission reaction

Mass of U-235 nucleus = 235.125 amu
Mass of the neutron = 1.009 amu
Total mass = 236.134 amu

Actual mass after the fission reaction

Mass of Ba-141 nucleus = 140.958 amu
Mass of Kr-92 nucleus = 91.926 amu
Mass of three neutrons = 3.027 amu
Total mass = 235.911 amu

Mass decrease during the reaction = (236.134 – 235.911) amu = 0.223 amu
= 0.223 amu × 931.4 MeV/amu = **207 MeV**

Therefore, each fission event produces about 200 MeV of energy. The energy converted per atom is roughly 10^8 times greater in nuclear reactions than in chemical reactions. The energy, 200 MeV = $200 \times 10^6 \times 1.6 \times 10^{-19}$ J $= 3.2 \times 10^{-11}$ J, is in fact very small. We spend about 5J when we do a simple job, say picking up an object from a table. However, as there are billions of nuclei even in a small sample of uranium, the total energy that could be produced will be enormous.

As an example, les us compute the energy produced when nuclei contained in 1 gram of U-235 undergo fission.

Atomic weight of U-235 = 235
There are 6.023×10^{26} atoms in 235 kg of Uranium.

$$\therefore \quad \text{Number of nuclei in one gram of U} - 235 = \frac{6.023\times10^{6}}{235\times1000} = 2.56\times10^{21}$$

$$\begin{aligned}\text{Energy produced by 1 gm of U–235} &= (2.56 \times 10^{21}) \times 200 \text{ MeV}\\ &= 2.56 \times 10^{21} \times 3.2 \times 10^{-11} \text{ J}\\ &= 8.2 \times 10^{10} \text{ J.}\\ 1 \text{ kWh} &= (1 \times 10^3 \text{ J/s})\,(3600\text{s}) = 3.6 \times 10^6 \text{ J}\end{aligned}$$

$$\therefore \textit{ Energy produced by 1 gm of U–235} = \frac{8.2\times10^{10} \text{ J}}{3.6\times10^{6} \text{ J/kWh}} = 22.8\times10^{3} \text{ kWh} = 22.8 \text{ MWh.}$$

We can as well express the energy in calories.

$$1 \text{ cal} = 4.187 \text{ J}$$

$$\text{Energy produced by 1gm of U-235} = \frac{8.2\times10^{10}}{4.187} \text{cal} = 2 \times 10^{10} \text{ cal}$$

The quantity of coal required to produce an energy equivalent to the above figure may be calculated assuming that the particular grade of coal used in thermal power plant yields 7×10^3 kcal energy per kg.

$$\text{Quantity of coal required} = \frac{2\times10^{10}\,\text{cal}}{7\times10^{3}\,\text{kcal/kg}} = \frac{2\times10^{10}}{7\times10^{6}}\,\text{kg} \cong 3000\ \text{kg}.$$

It means that about 3 tonnes of coal is required to produce as much energy as 1 gm of U-235 produces.

Enrico Fermi, the Italian physicist was honoured in 1938 with the Nobel prize in physics for his discovery of nuclear reactions brought about by slow neutrons.

28.23.2 Theory of Nuclear Fission

Niels Bohr and John A. Wheeler explained in 1939 the process of nuclear fission using the liquid drop model. According to this model the stable nucleus may be compared to a spherical liquid drop. The shape of the drop depends on the balance between the short range forces and coulomb repulsion forces. When a nucleus captures a neutron, a compound nucleus is formed. The nucleus is excited by being given mechanical energy and is in a state of higher energy. Just as an excited liquid drop oscillates, the nucleus will be in a state of oscillations. The oscillations tend to distort the spherical shape so that the nucleus assumes an ellipsoid shape. The restoring forces arising from short-range nuclear forces tend to make the nucleus return to is original spherical shape. If the excitation energy is sufficiently large, the ellipsoid deforms further into a dumb bell shape. The short-range attractive forces of the deformed nucleus are balanced by the coulomb force of repulsion. Each of the fragments will then quickly take spherical form because within it the attractive nuclear forces predominate again. A possible sequence of stages of fission is shown in Fig. 28.13. Generally, the two fission fragments are not of the same size. As neutron / proton ratio for heavy nuclei is higher than for lighter nuclei, the fragments contain excess neutrons. To reduce the excess, the fragments release two or three neutrons as soon as they are formed.

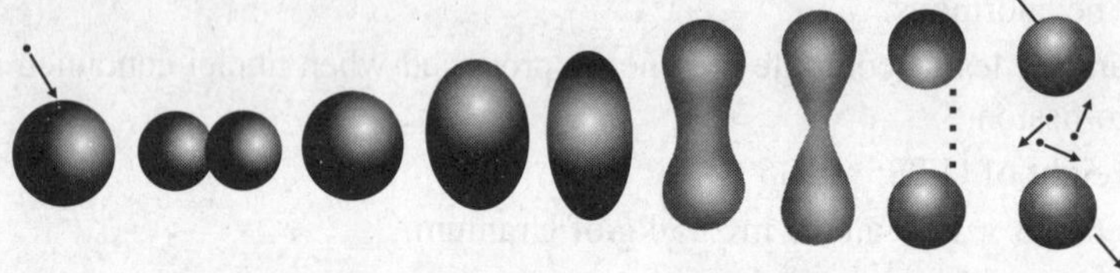

Fig. 28.13: Nuclear Fission

Obviously, a certain minimum amount of energy must be available to the nucleus to deform it enough and cause fission. The energy is called the **threshold** or **critical energy**, can be calculated. Basing on such calculations, Bohr and Wheeler predicted that U-235 nucleus undergoes fission with slow neutrons. They also showed that U-238 does not undergo fission unless it is bombarded with fast neutrons having energies of 0.9 MeV or more.

28.24 NUCLEAR CHAIN REACTION

*A **nuclear chain reaction** is a self-propagating process in which the number of neutrons goes on multiplying rapidly almost in geometrical progression till the total fissionable nuclei in the material are fissioned.*

A neutron striking a large fissionable nucleus initiates the fission process. The neutrons released in the process strike other fissionable nuclei in turn and induce fission in them. The average number of neutrons emitted per one thermal neutron absorbed is about 2.5. This abundant emission of neutrons enhances the possibility of fission occurring in more nuclei. If each of the newly emitted neutrons induces fission in a neighbour nucleus, the number of fission events grows cumulatively at a rapid rate. For example, let us say that three neutrons are emitted in a uranium fission act. The three neutrons could produce fission in three more uranium nuclei, the 3^2 neutrons emitted by the nuclei could produce the nine fissions, the 3^3 neutrons produce 27 fissions, the resulting 3^4 (= 81) neutrons produce 81 fissions and 3^5 (= 243) neutrons could produce 243 fissions and 3^6 (= 729) neutrons and so on. This process is called a ***chain reaction.*** It is shown in Fig. 28.14.

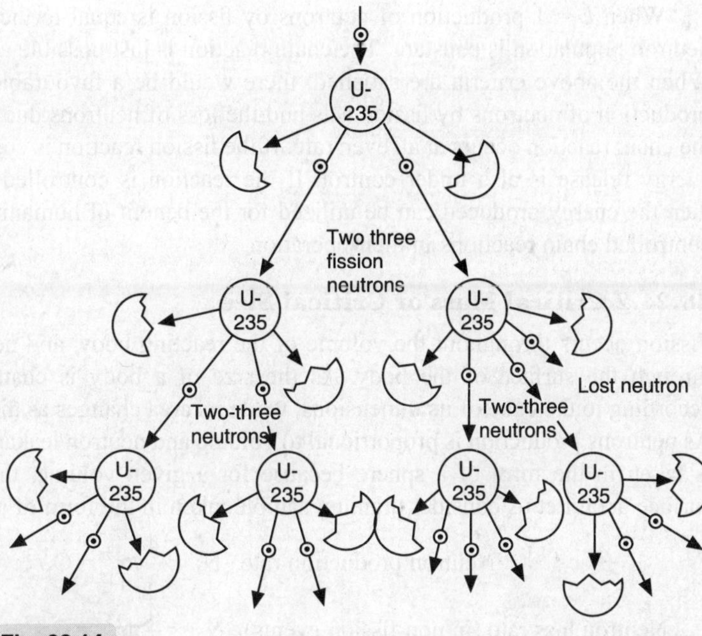

Fig. 28.14: Chain Reaction

A nuclear chain reaction is defined as the process in which the number of neutrons goes on multiplying rapidly in a geometrical progression till the total fissionable nuclei are fissioned.

The neutrons released in the process strike other fissionable nuclei in turn and induce fission in them. The average number of neutrons emitted per one thermal neutron absorbed is about 2.5. This abundant emission of neutrons enhances the possibility of fission occurring in more nuclei. If each of the new neutrons induces fission in neighbouring nuclei, the number of fissions will grow cumulatively, at each successive stage at a rapid rate.

In practice, not all neutrons produced in fission participate in further fission acts. Some of the neutrons are lost as a result of the following processes.

1. capture of neutrons by uranium, which may not lead to fission.
2. capture of neutrons by other nuclei in the sample and
3. escape out of the sample without being captured.

These losses affect the course of chain reaction.

28.24.1 Controlled Chain Reaction

The course of a chain reaction is determined by the availability of neutrons to continue fission events in the sample. The condition required is conveniently expressed in terms of a **multiplication factor,** *k,* of the system, which is defined as

$$k = \frac{\text{Number of neutrons in a particular generation}}{\text{Number of neutrons in the preceeding generation}} \quad ...(28.20)$$

When $k = 1$ production of neutrons by fission is equal to their loss by leakage and the neutron population is constant. The chain reaction is just possible and will be **self-sustaining**. When the above criteria are satisfied, there would be a favourable balance between the net production of neutrons by fission acts and the loss of neutrons due to the various reasons and the chain reaction occurs at an even rate. If the fission reaction is controlled, the corresponding energy release is also under control. If the reaction is controlled over an extended period, then the energy produced can be utilized for the benefit of humanity. Nuclear reactors utilize controlled chain reactions in their operation.

28.24.2 Critical Mass or Critical Size

Fission occurs throughout the volume of the reacting body, and neutron leakage takes place through the surface of the body. As the size of a body is changed, the volume changes according to the cube of its dimensions, while its area changes as the square of its dimensions. As neutrons production is proportional to volume and neutron leakage to surface area, uranium is taken in the form of a sphere because for a given volume, the sphere has the smallest surface area. Let us consider uranium sample taken in the form of a sphere of radius, r. Then

$$\text{Neutron production rate, } N_1 \propto \frac{4}{3}\pi r^3 = C_1 r^3$$

$$\text{Neutron loss rate (in non-fission events), } N_2 \propto \frac{4}{3}\pi r^3 = C_2 r^3$$

and

$$\text{Neutron leakage rate, } N_3 \propto \frac{4}{3}\pi r^2 = C_3 r^2$$

where C_1, C_2, and C_3 are proportionality constants.

When $k = 1$

$$N_1 > N_2 + N_3,$$

i.e.,

$$C_1 r^3 > C_2 r^3 + C_3 r^2$$

or

$$(C_1 - C_2) r > C_3$$

$$\therefore \quad r > \frac{C_3}{(C_1 - C_2)} = C \text{ (say)} \qquad ...(28.21)$$

The above equation implies that the larger the size of the body, lesser the escape rate as compared to the production rate. Therefore, there is always a certain size called **critical size,** C of the material. For a self-sustained reaction to take place, the sample size must be greater than the critical size.

If the size of the body is less than the critical size, i.e. it is **subcritical** so that $k < 1$, then loss of neutrons is greater than their production. Each new generation of neutrons produce a decreasing number of fission acts and the reaction terminates soon for want of neutrons. Consequently, chain reaction is not possible under these conditions.

28.24.3 Uncontrolled Chain Reaction

If $k > 1$, the size is greater than the critical size and is called **supercritical** size. Each new generation of neutrons produces an increasing number of fission acts and chain reaction becomes uncontrollable. The neutron population increases exponentially with time and the chain reaction builds up at an alarming rate leading to explosion, accompanied by the liberation of enormous energy and a rise of temperature of the surrounding medium to several million degrees. Atomic bombs utilize uncontrolled chain reactions for their operation.

Atom Bomb

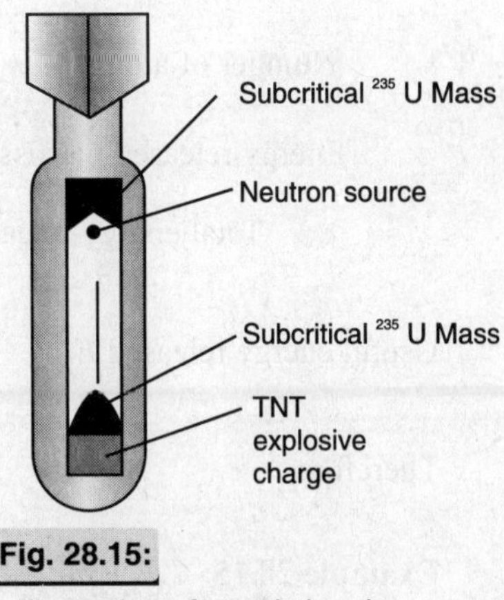

Fig. 28.15: Illustration of atomic bomb

An atom bomb is an example where uncontrolled chain reaction takes place leading to an explosion. Natural uranium is composed of 99.3% *U*-238 and only 0.7% of the fissionable *U*-235. The material used in weapons is enriched to 99% or more *U*-235. The critical mass of such highly enriched uranium is kept separated into several subcritical masses, so that chain reaction is not initiated (Fig. 28.15). The masses arc brought together by suddenly firing explosives like TNT. The segments form one large mass which becomes supercritical. Neutrons entering this supercritical mass from a Ra-Be source initiate a chain reaction which develops into an uncontrollable reaction. It results in a violent explosion with the liberation of tremendous energy. Everything in the immediate vicinity will be heated to temperatures of 5 to 10 million degrees. The atmosphere gets heated up and expands suddenly exploding every thing near by.

There is no danger of atomic explosion in the uranium mineral deposits in the earth. First, only 0.7% of natural uranium contains fissionable *U*-235. Secondly, uranium occurs only in compound form and is not found in pure form in nature.

28.25 NUCLEAR ENERGY

Energy requirements of mankind are at present met by fossil fuels, namely coal, oil and natural gas. These sources are being fast depleted and will be nearly exhausted in the near future. It is necessary that alternative sources of power are to be searched for. Nuclear power is one of the alternative sources. The energy reserve in the form of uranium is many times greater than that of fossil fuels.

The nucleus is a store-house of enormous power. Heavier nuclei, uranium and plutonium, can be split into fragments by neutrons and such a fission process liberates tremendous energy. Nuclear fusion power is believed to be an inexhaustible source of energy and is without the radioactive hazards and pollution threats.

Example 28.13: *A reactor is developing energy at the rate of 32 MW. How many atoms of U-235 undergo fission per sec.? Assume that on the average an energy of 200 MeV is released per fission.*

Solution: Energy released per second $E_1 = 32 \times 10^6$ J

Energy released per fission $E_2 = 200 \times 10^6$ eV $= 200 \times 10^6 \times 1.6 \times 10^{-19}$ J

$$\text{Number of atoms undergoing fission /second } N = \frac{E_1}{E_2} = \frac{32\times10^6\text{J}}{200\times10^6\times1.6\times10^{-19}\text{ J}} = \mathbf{10^{18}}.$$

Example 28.14: *A city requires 100 MW of electrical power on the average. This is to be supplied by a nuclear reactor of efficiency 20% using U-235 as nuclear fuel. Calculate of the fuel required for one-day operation. Given that the energy released per fission of U-235 nuclide is 200 MeV.*

Solution: Energy required per day $E_1 = 100 \times 10^6 \times 86400$ J $= 8.64 \times 10^{12}$ J

Let the fuel required be *m* kg.

$$\text{Number of atoms in } m \text{ kg.} = \frac{m \times 6.023 \times 10^{26}}{235}$$

$$\text{Energy released per fission} = 200 \text{ MeV} = 3.2 \times 10^{-11} \text{ J}$$

$$\text{Total energy released} = \frac{m \times 6.023 \times 10^{26} \times 3.2 \times 10^{-11}}{235} \text{ J}$$

$$\text{Useful energy released } E_2 = \frac{m \times 6.023 \times 10^{26} \times 3.2 \times 10^{11}}{235} \text{ J} \times \frac{20}{100} = m(16.403 \times 10^{12} \text{ J})$$

Therefore,
$$m = \frac{8.64 \times 10^{12} \text{ J}}{16.403 \times 10^{12} \text{ J}} = \mathbf{0.5267 \text{ kg/day}}.$$

Example 28.15: *Calculate the power output of a nuclear reactor which consumes 25 gm of U-235 per day. Assume 5% reactor efficiency. Energy released per fission is 200 MeV.*

Solution: Number of nuclei in 235 kg of *U-235* = N_A

$$\text{Number of nuclei in 25 gm} = \frac{6.02 \times 10^{26}}{235 \times 40} = 6.4 \times 10^{22} \left(\because 25 \text{ gm} = \frac{1}{40} \text{kg} \right)$$

Energy released per fission = 200 MeV = 3.2×10^{-11} J

Energy produced by 6.4×10^{22} nuclei = $6.4 \times 10^{22} \times 3.2 \times 10^{-11}$ J = 2.05×10^{12} J

Efficiency = 5%

$$\text{Energy converted to electrical power} = \frac{2.05 \times 10^{12} \times 5}{100} \text{ J} = 1.025 \times 10^{11} \text{ J}$$

Time taken to consume 25 gm of *U-235* = 1 day = 24 × 3600 sec = 8.64×10^4 secs.

$$\text{Power output} = \frac{\text{Energy}}{\text{time}} = \frac{1.025 \times 10^{11}}{8.64 \times 10^4} \text{ J/s} = \mathbf{1.19 \text{ MW}}$$

Example 28.16: *A nuclear reactor consumes 20.4 kg of U-235 in 1000 hrs. of operation. Assuming that on an average 200 MeV of energy is released per fission of a single U-235 nucleus, determine the power developed by the reactor.*

Solution: Number of nuclei in 20.4 kg of *U-235* $= \frac{N_A m}{M} = \frac{6.02 \times 10^{26} \times 20.4 \text{ kg}}{235 \text{ kg}}$

$$= 523 \times 10^{25}$$

Energy released per fission = 200 MeV = 3.2×10^{-11} J

Total energy released = $5.3 \times 10^{25} \times 3.2 \times 10^{-11}$ J = 1.67×10^{15} J

$$\text{Power} = \frac{\text{Energy}}{\text{time}} = \frac{E}{t} = \frac{1.67 \times 10^{15} \text{ J}}{1000 \text{ hrs.}} = \frac{1.67 \times 10^{15} \text{ J}}{3.6 \times 10^{6} \text{ s}} 4.65 \times 10^{8} \text{ W} = \mathbf{465 \text{ MW}}$$

28.26 NUCLEAR REACTORS

A system in which nuclear fission is produced in a controlled, self-sustaining chain reaction is known as nuclear reactor.

We can sum up the three important features of nuclear fission reaction as follows:

(*a*) **Energy Emission:** Each fission reaction produces about 200 MeV of energy and the total energy produced by a small sample of uranium, say 1 gm, is as enormous as 8.2×10^{10} J = 22.8 MWh. Most of the energy is carried by fission fragments in the form of kinetic energy.

(*b*) **Neutron Multiplicity:** Each fission event is accompanied by the emission of a number of neutrons which in turn cause fission of other nuclei.

(*c*) **Delayed Neutrons:** Many of the neutrons are emitted at the instant of fission. They are known as *prompt neutrons.* About 1% of the neutrons are emitted due to decays of fission fragments and they are called *delayed neutrons.*

The last feature enables mechanical control of the reaction rate and keeps the reaction from proceeding too rapidly. The second feature can be used to build a self-sustaining chain of nuclear fissions. The energy produced in the reactions can be extracted as heat and used to boil water; the resulting steam can then be used to drive a turbine to generate electrical power. Enrico Fermi built the first nuclear reactor in 1942. A very large number of nuclear reactors are in operation in different countries as on today. They differ widely in design and construction depending on their end use.

28.26.1 Components of nuclear reactor:

A nuclear reactor essentially consists of the following seven components arranged in different zones, each of which serves a definite purpose.

(*i*) **Fuel:** Nuclear reactors can use pure fissile materials but it is easier and cheaper to use mixtures of isotopes. Naturally occurring uranium consists of 99.3% U-238 and only 0.7% U-235. U-238 is for al practical purposes not fissionable. Often natural uranium in which there is one U-235 atom per 140 U-238 atoms is used as nuclear fuel. In order to be useful as a fuel, the concentration of U-235 must be substantially increased (up to 3%) in naturally available uranium. This process is called *enrichment*. In many reactors, uranium enriched in U-235 is used as fuel. Another fissionable material is Pu-239. This does not occur in nature. It is produced through neutron capture by the non-fissionable U-238. The process of plutonium fuel production from uranium is known as *breeding*.

(*ii*) **Moderators:** The kinetic energy of neutrons emitted fission process is of the order of a few MeV. Such high energy neutrons are known as *fast neutrons*. The fast neutrons have relatively low probability of inducing fission. Nuclear fission is much more probable with *slow neutrons*. The fast neutrons must be therefore, slowed down in order to increase their chances of initiating fission events. If a neutron is scattered from a heavy nucleus like uranium, the energy of the neutron is not changed. On the other hand in a collision with a very light nucleus, the neutron can lose substantial energy. Materials consisting of atoms of lower atomic mass are used for slowing down neutrons. Such materials are called *moderators*.

The commonly used moderators are water, heavy water, graphite and beryllium. Heavy water is considered to be the best moderator, since it does not absorb neutrons. When a solid moderator is required, graphite is commonly used. Ordinary water is a good moderator, but has high neutron absorption cross-section. Beryllium and its oxide are also good moderators, but they are expensive, toxic and have poor mechanical property.

(*iii*) **Reflectors:** Some of the neutrons generated in the fission process may leak away without being absorbed. It is necessary to conserve neutrons so that we minimize the consumption of fissile material and keep the size of the reactors small. To reduce the neutron loss due to leakage the inner surface of the reactor is surrounded by a material which reflects the neutrons back into the core. Such materials are called *reflectors*. Reflectors are made of nickel, thorium or other suitable materials.

(*iv*) **Coolants:** Intense heat is generated within the reactor core due to nuclear fission reactions. This heat must be removed using coolants for the safe operation of reactor. Ordinary water, heavy water and liquid metals are used as coolants. The *coolant* keeps the fuel assembly at a safe temperature; at the same time the heat carried away by the coolant is used in the heat exchange for utilization in power generation.

(*v*) **Control Rods:** The rate of reactions in a nuclear reactor is controlled by control rods. Since the neutrons are responsible for the progress of the chain reactions, suitable neutron absorbers are employed to achieve control of reaction rate. Cadmium and boron are the most frequently used materials. The *control* procedure involves the insertion or withdrawal of these materials, taken in the form of rods, into or from the reactor core. With the control rod fully inserted, enough neutrons are absorbed so that the average number of neutrons available to cause new fissions is less than one per fission reaction. As the rod is slowly withdrawn, the average number of available neutrons increases until it is just equal to one per reaction. At that time the reactor is said to be *critical*. During the operation, the position of the control rod is continually adjusted so that energy is released at a steady rate.

(*vi*) **Structural and Cladding materials:** In a reactor, structural materials have to be used in the forms of mechanical support for the various components. They are also used for holding the fuel, coolants, control rods and measuring instruments. Uranium readily reacts with air, water and other fluids. Hence *cladding* is required to pack fuel elements to isolate them. The cladding also prevents escape of the fission fragments. Zirconium has been found to be an excellent structural and cladding material. Titanium is also a good material but it is very expensive.

(*vii*) **Reactor Shielding:** A nuclear reactor is a powerful source of highly penetrating neutrons and γ–radiation, which are very harmful to life. Therefore, a thermal shield and a biological shield are used to minimize the effects of the harmful radiation. The thermal shield, usually a wall of steel, is placed between the reactor and the biological shield to protect the later from excessive heating and damage. The biological shield is usually made of concrete. Concrete wall of about 2 m thick surrounding the reactor serves as an adequate shield.

28.26.2 Construction of Nuclear Reactor

A schematic of a nuclear reactor is shown in Fig. 28.16.

It consists of several zones each of which serves a definite purpose. The central part is called the ***active zone*** or ***core***. The core is made up of moderator blocks with slots in which fuel channels are housed. A ***fuel channel*** is a metal tube

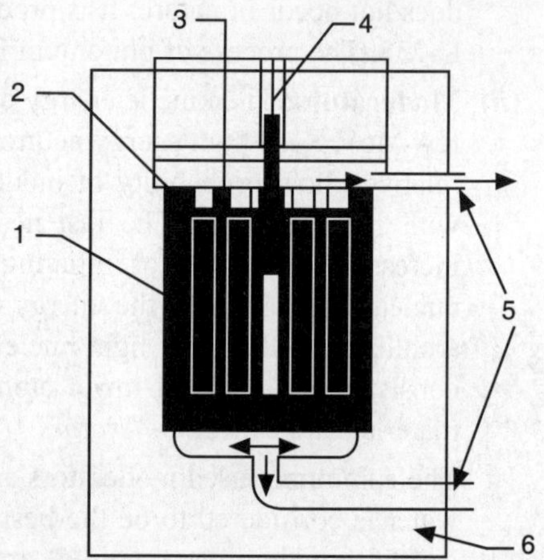

Fig. 28.16: Nuclear reactor indicating
1. Nuclear fuel, 2. Nuclear reflector,
3. Moderator, 4. Control rod,
5. Coolant, 6. Shield

containing the fuel elements. The fuel element consists of uranium containing slug clad in a stainless steel or zirconium metal casing. The fuel elements are grouped lengthwise in the fuel channel and form a common construction called the ***fuel assembly***. Inlet and outlet pipes are incorporated in the fuel channel for circulation of coolant. The active zone is filled with a moderator. The fuel assembly and moderator are cooled by the coolant passing through the fuel channels. The heated coolant then passes through an outlet pipe to the collecting tank. The heat can be used to generated steam and drive the turbine.

28.26.3 Types of Reactors

Nuclear reactors are used as

(*i*) ***Civilian reactors*** - generate energy for electricity

(*ii*) ***Military reactors*** - produce materials for nuclear weapons

(*iii*) ***Research reactors*** - develop weapons or energy production technology, for training purposes, for nuclear physics experimentation, and for producing radio-isotopes for medicine and research.

(*i*) **Thermal reactor:** In a power reactor, heat produced in the nuclear reactor fuel is extracted to drive a turbine connected to an electrical generator. In practice two different types of reactors are used for producing power. In a thermal reactor, neutrons produced as a result of neutron absorption in fuel possess high kinetic energy of the order of MeV. Neutrons formed by fission have high energies and therefore travel extremely fast. These *fast neutrons* do not cause fission as efficiently as *thermal neutrons*. So they are slowed down by elastic collisions resulting in neutrons with energies of the order of 0.025 eV. These slow neutrons are called thermal neutrons as they are cooled to the temperature of the neighboring coolant.

(*a*) **Boiling – Water Reactor:** In this type of reactors, a stream of water circulates through the core. The heat turns the water into steam, which is then directly fed to the turbine. Therefore, an external steam generator is not required. There is a considerable financial saving due to this. The disadvantage of this type of reactor is that the water can become radioactive; and a rupture of the pipes near the turbines could result in a serious accident with the spread of radioactive materials.

(*b*) **Pressurized – Water Reactor:** In this type of reactor, heat is extracted in two steps. Water is circulated through the core under high pressure, which prevents the water turning into steam. This hot water in turn heats a secondary water system, which delivers steam to the turbine. Since steam is not taken from the reactor core, it is not radioactive. The main disadvantages of these reactors are the necessity of using highly enriched fuel and expensive structural materials.

(*c*) **Heavy water reactor:** Heavy water reactor for the purpose of electricity; plutonium production. Coolant used is heavy water (deuterium oxide, D_2O). Natural uranium is used as fuel.

(*ii*) **Breeder reactors:** Some reactors produce more fuel than they consume. These reactors are called breeder reactors.

(*iii*) **Fast breeder reactor (FBR):** The *fast breeder reactor* (FBR) is a fast neutron reactor. It breeds fuel by generating more fissile material than it consumes. They can generate up to 30 percent more fuel than they use. Layers of blankets containing fertile material are used to ensure that more fuel is breed than that burnt, qualifying the definition of a breeder. As moderation is undesirable in a fast reactor, they do not contain moderating materials like *water, heavy water and graphite* in the core. Non-moderating materials like Helium and liquid metals like sodium, lead, lead-bismuth eutectic are used as coolants owing to their non-moderating nature. The lack of a moderator to slow down neutrons gives the name 'fast' to the reactor.

A fast reactor employs a coolant which is not an efficient moderator. Therefore its neutrons maintain high-energy. These fast neutrons are readily captured by an isotope of uranium (U_{238}), which then converts it into plutonium (Pu_{239}). Nuclear reactors normally use uranium fuel that has had its U_{235} content enriched to somewhere between 3 and 8 percent by weight. Even if the U_{235} does most of the fissioning, more than 90 percent of the atoms in the fuel are U_{238}. U_{238} atoms are potential neutron capture targets and prospective plutonium atoms.

Pu_{239} is formed when U_{238} captures a neutron. It forms U_{239} and then goes through two beta decays. It is better than U_{235} at getting fissioned. Pu_{239} is formed in every reactor and also fissions as the reactor operate. A nuclear reactor can derive a significant amount of energy from such plutonium fission. Because the plutonium fissions, it reduces the amount of plutonium that is available in the fuel. As the plutonium is depleted it breeds more plutonium from the uranium around it. Fig. 28.17 shows Pu-239 breeding. Fig. 28.18 shows the illustration of fast breed reactor.

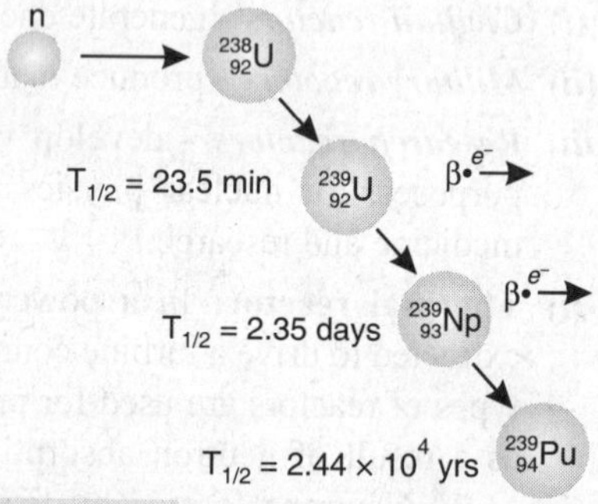

Fig. 28.17:

The uranium nucleus absorbs neutron and leads to Pu-239 breeding.

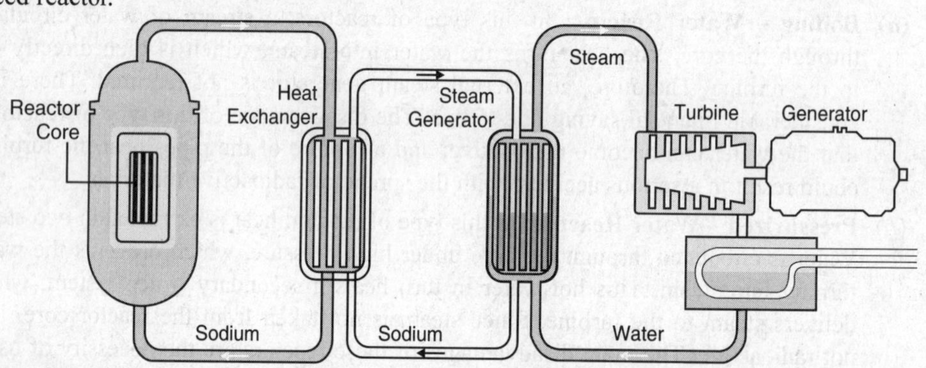

Fig. 28.18:

Diagram of fast breed reactor

One of the advantages of using sodium as coolant is the possibility of achieving a ***high coolant (sodium) outlet temperature***, while maintaining a pressure much lower than those maintained for light water and heavy water reactors. This is due to the high boiling point of sodium even at atmospheric pressure. Hence problems associated with high pressures are circumvented to a large extent.

Sodium cooled fast breeder reactors use two cycles of coolant flows. The primary circuit involves the circulation of sodium through the core. Relatively low temperature sodium enters the core at the bottom and leaves at the top at higher temperature. This sodium, called primary sodium is radioactive due to exposure to neutrons while passing through the core. Another circuit involves heat transfer between the radioactive primary sodium and secondary sodium in separate heat exchangers. The secondary sodium in turn transfers heat to water in steam generator, thus producing steam. The use of secondary coolant between primary coolant and steam is aimed at preventing contact of radioactive sodium with water in case of leakage. The steam drives a turbine and generates electricity.

Advantages:

(*i*) The neutrons from the uranium + plutonium are used to produce more fissile material.

(*ii*) No moderator is needed.

(*iii*) Small core is enough.

Disadvantages:

(*i*) It has to be cooled with liquid sodium.

(*ii*) It is more difficult and costly than a normal reactor.

(*iii*) Specific power of the reactor is low.

(*iv*) Sodium is radioactive.

Research Reactors: Research reactors produce high neutron fluxes for research. These neutrons are used as projectiles in nuclear reactions to produce new nuclides. These nuclides are used as tracers and for medical purposes.

28.27 NUCLEAR POWER PLANT

A schematic diagram of nuclear power plant is shown in Fig. 28.19. A nuclear power plant consists of a nuclear reactor, heat exchanger (i.e., steam generator), turbine, electric generator and condenser.

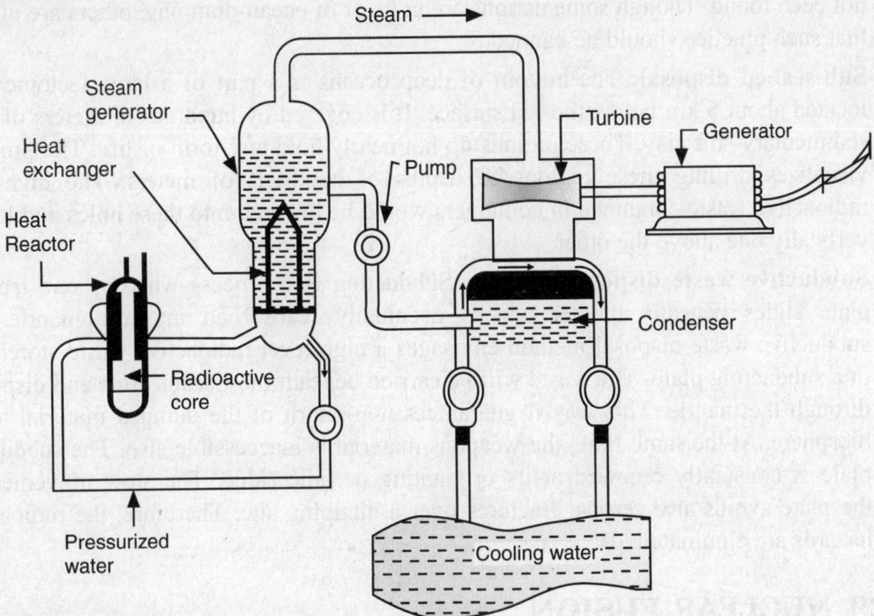

Fig. 28.19:

A Schematic of Nuclear Power Plant.

The heat liberated in the nuclear fission is conducted to the coolant circulating through the reactor core. Hot coolant leaves the reactor at top end and then flows out into tubes of boiler and passes on its heat to the feed water. The steam produced drives the turbine of special design. The turbine in turn drives an electric generator developing electrical power. From the turbine, after work is performed, steam flows to condenser. Pumps are provided to maintain the flow of coolant and feed water.

28.28 NUCLEAR WASTE DISPOSAL

Waste is any material that has been or will be discarded *as being of no further use*. Waste that emits nuclear radiation is radioactive waste. Radioactive waste can lead to radioactive exposure such as direct exposure to radioactive materials, breathing of airborne radioactive materials or consuming food that is contaminated with radioactive material. The harmful effects depend on the duration of exposure.

Waste disposal is throwing away of the waste with no intention of recovery. As in other industries, unusable and unwanted *waste* products accumulate in the nuclear industry also and the residues are harmful. Radioactive waste poses a tough problem for handling and protecting the environment.

The following options and each one of the options demands serious studies and technical assessments:

(i) **Geologic disposal:** The subterranean geological sites offer a natural isolation to contain long-lived radioactive waste. Low-level radioactive waste is generally disposed in old mines and near-surface facilities. High-level radioactive waste is disposed in host rocks.

(ii) **Ocean-dumping:** For many years the industrialized countries of the world used to dump the wastes into the oceans. Proof of bad effects on the ocean and marine life has not been found. Though some nations are in favor of ocean-dumping, others are of view that such practice should be banned.

(iii) **Sub-seabed disposal:** The bottom of deep oceans is a part of a large tectonic plate located about 5 km below the sea surface. It is covered by hundreds of meters of thick sedimentary soft clay. These regions do not nearly hold any form of life. The proposal visualizes drilling these regions to depths of hundreds of meters. The high-level radioactive waste contained in containers would be lowered into these holes and loaded vertically one above the other.

(iv) **Subductive waste disposal method:** Subduction is a process whereby one tectonic plate slides beneath another and is eventually reabsorbed into the mantle. The subductive waste disposal method envisages a high-level radioactive waste storehouse in a subducting plate. The waste will be carried beneath the Earth's crust and dispersed through the mantle. This way it guarantees non-return of the dumped material to the biosphere. At the same time, the weapons material is inaccessible also. The subducting plate is constantly renewed at its originating oceanic ridge. The slow movement of the plate avoids any vertical fractures over a dumping site. Therefore, the radioactive hazards are eliminated.

28.29 NUCLEAR FUSION

Nuclear fusion is *the process of combining two lighter nuclei into a heavier nucleus*. Since the mass of final nucleus is less than the rest masses of the original nuclei, there is a loss of mass, which gets converted into large amounts of energy. Fusion is the source of energy for the sun and the stars.

28.30 FUSION REACTION IN STARS

Sun is an average sized star and has been emitting enormous energy at the rate of 4×10^{26} J/s for the past 4×10^9 years. Such huge amount of energy cannot be liberated through ordinary sources of energy such as chemical reactions. Therefore, a series of nuclear fusion reactions are the only possible source of energy being produced by stars and the sun. The sun is mostly made up of hydrogen and helium. It was suggested that four hydrogen nuclei combine, through suitable series of nuclear reactions, and produce one helium nucleus releasing energy in these processes. The series of nuclear reactions that convert four protons into a helium nucleus are known as *thermonuclear reaction cycles*. Though many nuclear reactions are possible in sun, the two most common ones are the proton-proton cycle and the carbon cycle.

Proton-proton cycle: The simplest process of fusion is known as *proton-proton cycle.* It consists of fusing two hydrogen nuclei, which finally give a helium nucleus. The sequence of the fusion reactions is

(*i*) ${}_1^1\text{H} + {}_1^1\text{H} \rightarrow {}_1^2\text{H} + e^+ + \nu + 0.42\ \text{MeV}$

(*ii*) ${}_1^1\text{H} + {}_1^2\text{H} \rightarrow {}_2^3\text{He} + \gamma + 5.5\ \text{MeV}$

(*iii*) ${}_2^3\text{He} + {}_2^3\text{He} \rightarrow {}_2^4\text{He} + {}_1^1\text{H} + {}_1^1\text{H} + 12.8\ \text{MeV}$

In the first stage, two hydrogen nuclei combine to form a deuteron with release of 0.42 Mev of energy, a positron and a neutrino. In the second stage, the deuteron nucleus combines with another hydrogen nucleus to form a ${}_2^3\text{He}$ nucleus with release of 5.5 MeV energy and γ-radiation. In the final state, two ${}_2^3\text{He}$ nuclei combine to give a helium nucleus and two hydrogen nuclei. The net effect of the sequence is

$$4\,{}_1^1\text{H} \rightarrow {}_2^4\text{He} + 2e^+ + 2\nu + 2\gamma + 26.72\ \text{MeV}$$

Thus, the result is the gradual depletion of protons and the building up of helium concentration. The production of energy in sun is mainly due to proton-proton cycle.

Carbon cycle: Another self-sustaining process is known as *carbon-nitrogen* cycle. In this process, carbon acts as a catalyst and four hydrogen nuclei fuse to yield a helium nucleus, positrons, neutrinos and energy of 26.72 MeV. The steps in the process are as follows:

(*i*) ${}_1^1\text{H} + {}_6^{12}\text{C} \rightarrow {}_7^{13}\text{N} + \gamma$

(*ii*) ${}_7^{13}\text{N} \rightarrow {}_6^{13}\text{C} + e^+ + \nu$

(*iii*) ${}_1^1\text{H} + {}_6^{13}\text{C} \rightarrow {}_7^{14}\text{N} + \gamma$

(*iv*) ${}_1^1\text{H} + {}_7^{14}\text{N} \rightarrow {}_8^{15}\text{O} + \gamma$

(*v*) ${}_8^{15}\text{O} \rightarrow {}_7^{15}\text{N} + e^+ + \nu$

(*vi*) ${}_1^1\text{H} + {}_7^{15}\text{N} \rightarrow {}_6^{12}\text{C} + {}_2^4\text{He}$

The net result is

$${}_6^{12}\text{C} + 4\,{}_1^1\text{H} \rightarrow {}_6^{12}\text{C} + {}_2^4\text{He} + 2e^+ + 2\nu + \gamma + 26.72\ \text{MeV}$$

In case of medium and heavy stars the main part of the energy production is due to the carbon cycle. On the other hand, the main part of the energy production in lighter stars is due to the proton cycle.

When all the available hydrogen has been converted to helium, the star (or sun) will contract and its temperature will increase till helium starts burning. Helium fusion requires more thermal energy than that required for hydrogen fusion. When helium is exhausted, the star will again contract and its temperature will increase, till carbon burning occurs. Such processes will continue until ^{56}Fe is reached. The star then collapses under its own gravitational forces to become a white dwarf, or a neutron star or a black hole. It may even undergo supernova explosion.

28.31 CONTROLLED THERMONUCLEAR REACTIONS

Fusion can take place at a very high pressure and temperature of the order of 10^6 K because it is only under these conditions that nuclei are able to overcome their mutual coulomb repulsion. The energy released in nuclear fusion is known as **thermal nuclear energy** and it is possible to produce nuclear fusion under controlled conditions.

For achieving the fusion the two nuclei should have enough kinetic energies to overcome the mutual Coulomb repulsion. The force of repulsion will go on increasing as the nuclei approach each other. It becomes therefore necessary that the nuclei approach each other at velocities high enough to overcome the Coulomb repulsion. It is calculated that a deuterium atom should have about 0.25 MeV energy to overcome Coulomb repulsion. We can estimate the temperature to which the deuterium gas is to be heated to attain the required energy. Using $(3/2)\ kT = 0.25$ MeV, we find that it corresponds to a temperature of the order of 10^9 K. Therefore, fusion reaction can be accomplished by heating the fuel to extremely high temperature. Because of the requirement of such high temperatures, the fusion reactions are called *thermonuclear reactions.* If a gas is heated to such a very high temperature, the electrons are stripped off the atoms; and the bare positive nuclei and the negative electrons move about freely. This mixture is known as **plasma**, which is regarded as the fourth state of matter. The plasma as a whole is electrically neutral. In the state of plasma and at temperatures of the order of 10^8 K, the nuclei will have energies of the order of a few keV and the bare nuclei have a probability to overcome Coulomb repulsion and produce fusion reactions.

Hydrogen and its isotopes are more suitable for controlled thermonuclear reactions as their nuclei carry smaller charge and hence the smaller is the repulsive force. The following are the more promising fusion reactions.

D-D reactions: $${}_1^2\text{H} + {}_1^2\text{H} \rightarrow {}_2^3\text{H} + {}_0^1\text{n} + 3.27\ \text{MeV}$$

$${}_1^2\text{H} + {}_1^2\text{H} \rightarrow {}_1^3\text{H} + {}_1^1\text{H} + 4.03\ \text{MeV}$$

D-T reaction: $${}_1^2\text{H} + {}_1^3\text{H} \rightarrow {}_2^4\text{He} + {}_0^1\text{n} + 17.59\ \text{MeV}$$

D-He reaction: $${}_2^3\text{He} + {}_1^2\text{H} \rightarrow {}_2^4\text{He} + {}_1^1\text{H} + 18.30\ \text{MeV}$$

Out of the above four reactions, the D-T (deuterium-tritium) reaction has large energy release and perhaps the best one. This reaction liberates energy about 5.9 MeV/nucleon which is about six times as that produced in a fission reaction. Such enormous amount of energy release in fusion reaction makes it an attractive process for generating electrical power.

28.31.1 Ignition Temperature

Plasma emits energy as radiation due to the electrical interactions between the nuclei and electrons. This constitutes a loss of energy. At moderately high temperatures, more energy is radiated away from the plasma than is produced by fusion. As the temperature is further increased, production of fusion energy increases more rapidly than emission of radiation. At a sufficiently high temperature, the fusion becomes sustained. The temperature at which the rate of energy production becomes greater than the rate of energy loss is called the *critical ignition temperature*. It has been estimated that for *D-T* reaction $T_c \approx 4$ KeV while it is 10 times higher for *D-D* reaction. (In fusion studies, temperatures are expressed in KeV; 1 KeV = 11×10^6 K). This is why the *D-T* reaction is considered to be the most promising reaction for power generation.

28.31.2 Lawson Criterion

Apart from high temperature requirement, there are two other essential requirements to increase the probability of collisions, which would result in fusion of nuclei.

(*i*) high density of nuclei, n and

(*ii*) a long *confinement time,* τ during which the high temperature and density must be maintained.

In 1957 J.D. Lawson showed that for the condition of self-sustained fusion reaction the product $n\tau$ must exceed a certain a minimum value. This is known as *Lawson criterion.*

In case of *D-T* reaction, calculations showed that

$$n\tau \geq 10^{20}\ s/m^3 \quad \text{...(28.22)}$$

Two techniques are under development for heating the fusion fuel to the required high temperature and to confine the heated plasma long enough for the fusion to produce power. They are known as magnetic confinement and inertial confinement.

28.31.3 Magnetic Confinement

The plasma can be heated by passing electric current through the plasma. We may regard the plasma column as made up of a large number of plasma threads, each carrying an electric current in the same direction. They attract one another as a result of which the whole plasma column is compresses perpendicular to its axis. This is known as the *pinch effect*. The sudden adiabatic compression results in the heating of the plasma column.

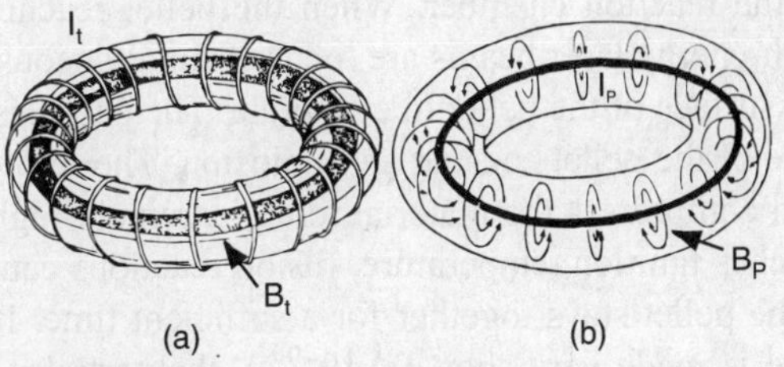

Fig. 28.20: Toroidal magnetic confinement geometry-Tokomak

The hot plasma may be confined by the application of a magnetic field. The particles in the plasma must be prevented from reaching the container wall for adequately long time, since contact with the wall will immediately cool them. Fig. 28.20 shows toroidal magnetic confinement geometry. The device is called a *tokomak.*

The tokomak is a doughnut-shaped reaction chamber. The chamber contains plasma with a circular cross-section where a toroidal magnetic field is provided for confinement and a toroidal electric field to maintain current through the plasma. The electric current generates a circular magnetic field in the cross-section of the toroid. The combination of this field and the toroidal field produces helical magnetic field lines which confine the plasma.

28.31.4 Inertial Confinement

An alternative approach to achieve fusion in a self-sustained manner is known as the inertial confinement.

In case of plasma, particle density is of the order of 10^{20} per m^3. Therefore, Lawson criterion can be satisfied if the confinement time is of the order of 1 sec or more. If the fusion fuel could be used in the condensed state where the density is of the order of 10^{28} per m^3, then the confinement time would be of the order of 10^{-8} s. Therefore in inertial confinement the fuel is compressed to high densities for very short confinement times.

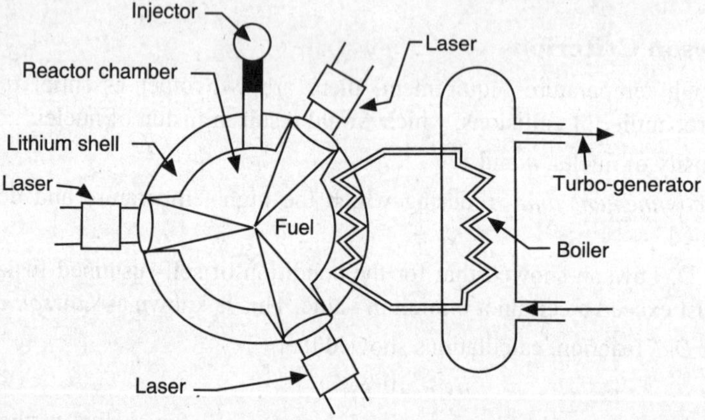

Fig. 28.21:
Inertial confinement

A small D-T fuel pellet of about 1 mm diameter is injected into the spherical cavity of the reaction chamber. When the pellet reaches the center, several symmetrically located high-energy laser beams are focused simultaneously on the pellet for about 10^{-9} s. This causes the surface of the pellet to evaporate. The particles coming out produce a reaction force on the core of the pellet causing an implosion. The implosion causes an increase in temperature and also compresses the material to an extremely high density. When the temperature of the core reaches ignition temperature, fusion reactions cause the pellet to explode. Fusion could occur if the pellet stays together for a sufficient time. It depends on the inertia. If the confinement time is made very short ($\leq 10^{-9}$ s), the particles do not move appreciably from their initial positions because of their own inertia. Hence this technique is called *inertial confinement.* After an interval of a second or more, another pellet is injected and the process is repeated.

Example 28.17: *The fusion of* $^2_1\text{H} + ^2_1\text{H} \rightarrow ^4_2\text{He} + \text{Q}$ *is proposed to be used for the production of industrial power. Assuming the efficiency of the process to be 30%, find how many kg of deuterium will be consumed in a day for an output of 50 MW. Given: mass of* ^2_1H *= 2.01478 u and mass of* $^4_2He = 4.00388$ *u.*

Solution: Energy released per D-D reaction

$$Q = [2(2.01478) - 4.00388]931.4 \text{ MeV} = 23.92 \text{ MeV}$$

Actual output per two atoms = 30% Q = (0.3)(23.92) MeV = 7.175 MeV

Actual output per one atom = 3.5875 MeV.

$$\text{Number of deuterium atoms required} = \frac{50\times10^6 \text{ J}}{3.5875\times10^6\times1.602\times10^{-19}\text{ J}} = 8.7\times10^{19}.$$

$$\text{Equivalent mass of deuterium required per second} = \frac{8.7\times10^{19}\times2.01478}{6.02\times10^{26}}\text{kg/s} = 2.9\times10^{-7}\text{kg/s}$$

∴ Deuterium requirement per day = $(2.9 \times 10^{-7}$ kg/s$)(8.64 \times 10^4$ s$)$ = **0.025 kg.**

28.32 FUSION REACTOR

A fusion power reactor is shown in Fig. 28.22.

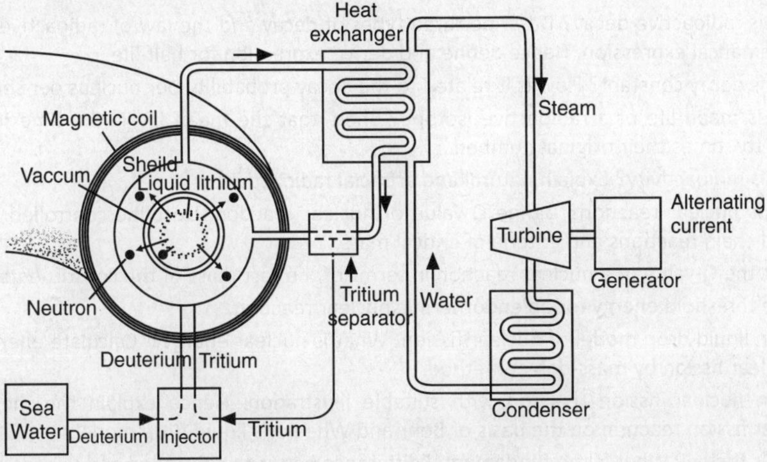

Fig. 28.22:

Thermal conversion for nuclear reaction

The fusion energy of the nuclei is conveyed to the neutrons that are released in the reaction. The energy is extracted by a lithium blanket surrounding the reactor core. Lithium transfers the energy to the turbogenerator through a heat exchanger.

The chief advantage of fusion over fission and other processes is that it offers a low cost and pollution-free energy.

Fusion is believed to be the ultimate and inexhaustible energy source because the fuel is abundantly available and at a cheaper price. One atom of deuterium for every 6500 atoms of ordinary hydrogen is present in water. This would be sufficient to meet the energy needs of the world for very long time.

QUESTIONS

1. Define binding energy of a nucleus.
2. Why is $Z < N$ for the heavier nuclei?
3. What do you understand by mass defect of a nuclide? Write an expression for the mass defect.
4. What do you understand by binding energy of a nucleus?

5. Draw a curve showing the variation of binding energy per nucleon against the mass number. What information do you derive from such a curve?
6. What is meant by binding energy of a nucleus?
7. How does the binding energy per nucleon vary with mass numbers for different elements in the periodic table? Explain its significance.
8. What are the properties of nuclear forces?
9. Explain briefly the shell model of the nucleus.
10. Explain salient features of nuclear shell model. What are magic numbers given for nucleus of an atom?
11. Explain briefly the liquid drop model of the nucleus.
12. Write a short note on the structure of the nucleus.
13. Sketch the plot of binding energy per nucleon versus A and mention its salient features.
14. Define activity of a radioactive substance.
15. What is radioactive decay? Describe three types of decay and the law of radioactive decay with mathematical expression. Hence define and derive expression for half-life.
16. What is decay constant? How is it related to the decay probability per nucleus per second?
17. What is mean life of a radioactive isotope? Show that the mean life is the time for nuclei to decay to times their original number.
18. What is radioactivity? Explain natural and artificial radioactivity.
19. What is nuclear reaction? Define Q-value of nuclear reaction. Describe controlled and uncontrolled chain reactions and criteria of critical mass.
20. Define the Q-value of a nuclear reaction in terms of the rest mass of the constituents.
21. Define threshold energy for an endothermic nuclear reaction.
22. Explain liquid drop model of nuclear fission. What is nuclear energy? Calculate energy released in nuclear fission by mass-defect method.
23. Explain nuclear fission reaction with suitable illustration. Hence explain the mechanism of nuclear fission reaction on the basis of Bohr and Wheeler's liquid drop model.
24. What is fission? What is the fundamental difference between controlled and uncontrolled fission chain reaction? How do you use controlled fission energy in a constructive way?
25. What do you understand by the multiplication factor of a nuclear chain reaction? Discuss the factors on which it depends.
26. Classify nuclear reactions.
27. What is fission? Explain with neat diagram the chain fission reaction. What are the difficulties encountered in the reaction? Suggest suitable remedy for their removal.
28. When do you say a nuclear fission reaction is (*i*) critical (*ii*) supercritical and (*iii*) subcritical?
29. What is chain reaction in nuclear fission? Discuss the factors on which the chain reaction depends.
30. Explain nuclear chain reaction and construction and working of a nuclear reactor.
31. Define multiplication factor. How is this related with size of fissionable material?
32. Explain the terms: (*i*) Explosive chain reaction (*ii*) Binding energy curve.
33. Explain the main features of the design and working of a nuclear fission reactor.
34. Explain the function of moderator, reflector, shield, coolant and control rod for a nuclear reactor.
35. Discuss how nuclear chain reaction is used in a constructive way in a nuclear reactor. Give details of different parts of the reactor along with a neat diagram.
36. Explain chain reaction. How can the chain reaction be controlled?
37. What are the essential components of a nuclear reactor? Describe the function of each of these components.
38. What do you mean by nuclear reactor? Give the sketch of it. What is the function of control rod? Write two factors which should be considered while selecting the site for nuclear reactor.

39. Explain the function of moderator in a fission reactor.

40. Explain the following: (*i*) Q value, (*ii*) Critical size, (*iii*) Nuclear cross-section, (*iv*) Chain reaction

41. Explain the function of control rods in a fission reactor.

42. Why is enriched U-235 used as fuel in a nuclear reactor?

43. Explain nuclear fusion. What is ignition temperature?

44. Define nuclear fission.

45. Define thermonuclear reaction.

46. What do you mean by nuclear fusion? Give an account of proton-proton cycle as the cause of stellar energy. State the conditions required to initiate a self sustaining fusion reaction.

47. Explain what do you understand by nuclear fusion? Describe the necessary conditions to bring about fusion process.

48. Explain the proton-proton cycle, which is the main source of energy from sun.

49. Show that in the carbon cycle, carbon is not consumed and the net effect is the same as for the proton-proton cycle.

50. What do you mean by thermonuclear reaction?

51. Describe proton-proton cycle in nuclear fusion.

52. Define critical ignition temperature, confinement time and Lawson criterion.

53. What do you mean by inertial confinement in fusion?

54. Distinguish between the nuclear fission and fusion reactions. Explain the release of energy during fission and fusion reactions with example.

PROBLEMS

1. Calculate the binding energy and average binding energy per nucleon of ${}^{12}_{6}C$. Given: Mass of proton = 1.007825 amu, Mass of neutron = 1.008665 amu [**Ans:** 92.16 MeV; 7.68 MeV]

2. The disintegration constant λ of a radioactive element is 0.00231 per day. Calculate its half-life and average life. [**Ans:** 300 days; 432.9 days]

3. The activity of certain radionuclide decreases to 15% of its original value in 10 days. Find its half-life. [**Ans:** 3.65 days]

4. Compute the Q-value of the reaction Given: Mass of Be^9 = 9.012182 u, Mass of B^{10} = 10.012938 u, m_d = 2.014102 u and m_n = 1.008665u. [**Ans:** 4.36 MeV]

5. Determine the quantity of coal required to produce an energy equivalent to 1 gram of U-235 if the heat of combustion of coal is 7000 k.cal/kg. [**Ans:** 2.7 tons]

6. Assuming the energy released per fission of U-235 is 200 MeV, calculate the rate in kg per year at which fission should occur in a nuclear reactor in order to operate at a power level of 1 watt. [**Ans:** 1.64×10^{-9} kg]

7. If Fission of one atom of U-235 releases 200 M eV energy, calculate the amount of energy released by 10^{-3} kg of uranium. [**Ans:** 5.12×10^{23} MeV]

8. A nuclear reactor consumes 20.4 kg of U-235 in 1000 hrs operation. Determine the power developed by the reactor, assuming that on an average 200 MeV of energy is released per fission of a single U-235 nucleus. [**Ans:** 465 MW]

9. Calculate the energy release per gram of fuel for the reaction ${}^{2}_{1}H + {}^{2}_{1}H \rightarrow {}^{3}_{1}H + {}^{2}_{1}H$. Given: Mass of ${}^{2}_{1}H$ = 1.007825 u and mass of ${}^{3}_{1}H$ = 3.016090 u. [**Ans:** 6.05 MeV]

10. Calculate the energy released if 1 kg deuterium undergoes fusion. Assume the energy released per deuterium-tritium event is 17.6 MeV. [**Ans:** 2.4×10^{8} kWh]

29 Nuclear Instruments

Learning Objectives:

By the end of this chapter, you will be able to:

- Understand principle, working and application of Geiger-muller counter, Wilson cloud chamber, Bubble chamber, Spark chamber and Scintillation counter.
- Learn about mass spectrographs and its different types.
- Explain principle and working of Particle accelerators and Drift tube accelerator.
- Describe about different kinds of Cyclotron.

29.1 INTRODUCTION

Radioactive decay and nuclear reactions are accompanied by the emission of charged particles like α-particles, β-particles, protons and γ-rays. Our senses cannot detect these products directly, and detection must be done by indirect means. More immediate and quantitative methods of detection are desirable and a variety of instruments have been developed for this purpose. As a result of studies of radioactive transmutations, it was found that there are atomic nuclei with equal atomic numbers but with different mass numbers. Such nuclei are called isotopes. Isotopes frequently appear in radioactive transmutations. Radium, radon, uranium, thorium have several isotopes each, and so on. Mass spectrometers are utilized to analyze the isotopes. The mass spectrometer helped in the discovery of isotopes of uranium. Investigations on nuclear reactions require charged particles accelerated to very high energies. The high-energy particles are used in producing radio-active isotopes required in therapy and agriculture. They are also used in discovering new elementary particles. Particle accelerators are used to accelerate the charged particles. Some of the common radiation detectors, mass spectrometers and particle accelerators are described in this chapter.

29.2 GEIGER-MULLER COUNTER

Nuclear radiations emitted by disintegrating nuclei cannot be sensed directly. Indirect methods are to be employed to detect them. Alpha, beta and gamma rays have the ability to ionize neutral atoms. This property is used in radiation detecting instruments.

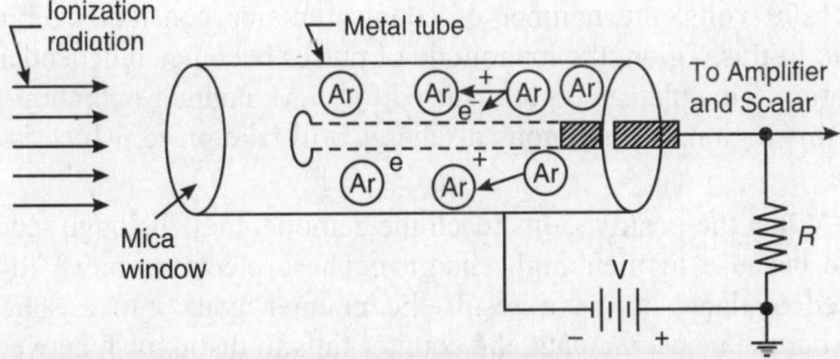

Fig. 29.1:

A Schematic of Geiger -Muller counter. Radiation ionizes the argon gas molecules in the tube giving rise to electrical pulse, which is counted.

Construction: The Geiger-Muller counter is a radiation detector. It is a modified cathode ray tube with electrical circuits needed to amplify the current and detect it. The Geiger-Muller (G.M.) tube (Fig. 29.1) consists of a sturdy metal case surrounded by a thin glass tube. The hollow metal case constitutes the cathode. A fine tungsten wire, which acts as anode, is fixed at center of the tube and is insulated from the metal case. The glass tube is evacuated and then partially filed with a mixture of 90% argon at 10 cm pressure and 10% ethyl alcohol vapour at 1 cm pressure. At one end of the tube a thin mica window is fixed to permit radiation into the tube. A dc potential of about 1200 volts is applied between the cathode and the wire. The value of the voltage is adjusted to be somewhat below the breakdown voltage of the gaseous mixture. A high resistance R is connected in series with battery.

A high energy particle entering through the mica window will ionize one or more of the argon atoms. These electrons and ions of argon ionize other argon atoms. Consequently, massive electrical discharge takes place which causes a current pulse. The current through R produces a voltage pulse of the order of 10 μV. An electronic pulse amplifier accepts the small pulse voltages and amplifies them to about 5 to 50 volts. The amplified output is then applied to a counter. As each incoming particle produces a pulse, the number of incoming particles can be counted.

The number of secondary electrons is independent of the number of the primary ions produced by incoming particle. The incoming particle acts as a trigger to release an avalanche of secondary electrons. The electrons reach the anode and cause ionization current in the circuit, whereas the positive ions move slowly and form a sheath around the anode for a short time. They reduce the potential difference to such a low value that the current in the circuit is stopped. Therefore, a brief pulse of current is produced by each incoming particle.

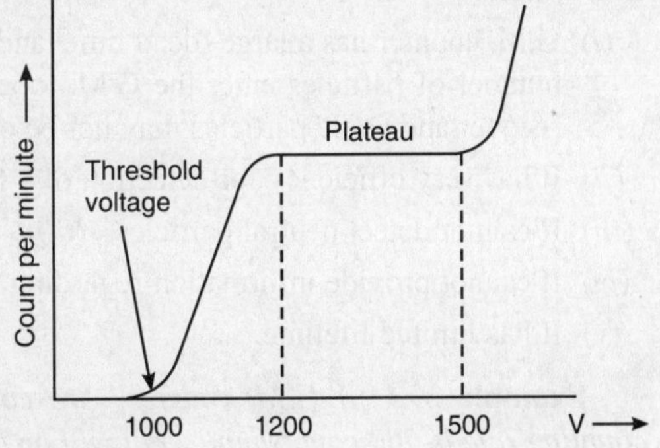

Fig. 29.2:

A plot of potential difference applied across the electrodes in a G.M. counter versus count rate. G.M. counter is operated in the plateau region.

Working: Fig. 29.2 shows a plot of counts per minute as a function of voltage. For voltages less than 1000 volts there is no discharge and hence no counts. Between 1000 to 1200 volts the number of pulses increases with the applied voltage almost linearly. Above 1200 volts, the number of counts remains constant over a certain region known as *plateau*. In this region, the magnitude of pulses becomes independent of the amount of original ionization. This plateau region is used for G.M. counter operation. If the voltage is increased above this region, a continuous discharge will take place, which is undesirable and is hence avoided.

Quenching: When the positive ions reach the cathode, they dislodge secondary electrons from the cathode because of their high energies. These electrons move toward anode and produce unwanted avalanches. As a result the counter goes into a state of continuous avalanching. During the measurements, the counter fails to distinguish between the two types of pulses, one that is due to an incoming particle and the other due to unwanted avalanching. The process of preventing the undesirable continuous avalanching is known as *quenching*. In other words, quenching is the elimination of sheath of positive ions around the cathode.

Self-quenching: To cause internal automatic quenching, a small percentage of ethyl alcohol vapour is added to the argon gas in the tube which prevents undesirable continuous avalanching.

Counting rate: The G.M. counter can count about 5000 particles per second. The counting rate depends upon the dead time and recovery time of the GM counter.

Dead time: *Dead time* refers to the time taken by the tube to recover between counts. In the counter, the slowly moving positive ions take about 100μs to reach the cathode. If a second particle enters the tube during this time, it will not be registered, as the potential difference across the electrodes is very low. Hence, the time interval is known as the dead time.

Recovery time: After dead time, the tube takes approximately 100μs before it regains the original working conditions. This time interval is known as recovery time. Thus, *recovery time* is the time after which the original pulse levels are restored.

Paralysis time: The sum of dead time and recovery time is known as paralysis time, which is 200 μs. The tube can respond to the second incoming particle only after 200 μs.

Applications:

(*i*) The G.M. counter is very useful for detecting nuclear radiations and charged particles.

(*ii*) It is largely used for recording cosmic ray events and measuring cosmic ray intensities.

Limitations

(*i*) G.M. counter has a large 'dead time' and recovery time of the order of 200 μs. If a large number of particles enter the G.M. tube at a rapid rate, the tube will not have time to recover and some particles may not be counted.

(*ii*) It has very efficiency for detection of γ-radiation.

(*iii*) It cannot detect neutral particles.

(*iv*) It cannot provide information regarding the nature of the ionizing particle.

(*v*) It has limited lifetime.

Example 29.1. *A G.M. counter wire collects 10^8 electrons per discharge. When the counting rate is 500 counts/min, what will be the average current in the circuit?*

Solution: Number of electrons collected in one minute, $n = 10^8 \times 500$

Charge per minute, $Q = n\,e = (10^8 \times 500)\,(1.602 \times 10^{-19}\text{C})\,/\text{min}.$

Average current, $$I = \frac{Q}{60\,s} = \frac{(10^8 \times 500)\,(1.602 \times 10^{-19}\,\text{C})}{60\,s} = 1.3 \times 10^{-10}\text{A}.$$

29.3 THE WILSON CLOUD CHAMBER

In 1911, C.T.R.Wilson devised an instrument known as cloud chamber using which charged particles can be detected and their paths can be photographed. It was extensively used in the study of cosmic rays and played an important role in the discovery of new elementary particles such as positron, meson etc.

Principle: The working of the cloud chamber is based on the fact that supercooled vapour in a chamber condenses when dust particles or ions are present in the chamber. In case of completely dust-free and ion-free super-cooled vapour, condensation does not take place. But if ions are available in the chamber, they serve as nuclei for condensation. If saturated vapour in the chamber is suddenly subjected to an adiabatic expansion, there will be an increase in volume which produces cooling rendering the saturated vapour to a supersaturated unstable state. If an ionizing particle, such as an α-particle or β-particle passes through the chamber, ions are produced along its path and droplets condense on these ions. Consequently, the track of the particle becomes visible.

Construction: The apparatus (Fig. 29.3) consists of an air-tight cylindrical chamber *A*, with walls and ceiling made of glass. It is provided with a movable piston *P*. The chamber contains a dust-free mixture of alcohol vapour and air. A small amount of water and alcohol is kept in a trough at the bottom of the chamber to ensure that the vapour is in saturated state. When the piston moves down rapidly, adiabatic expansion of the vapour inside the chamber takes place. The wooden blocks WW reduce the air space inside the piston. The piston is connected to a large evacuated vessel *F* through a valve. The chamber is illuminated by a mercury vapour lamp *L*. Two cameras *CC* are used to photograph the tracks.

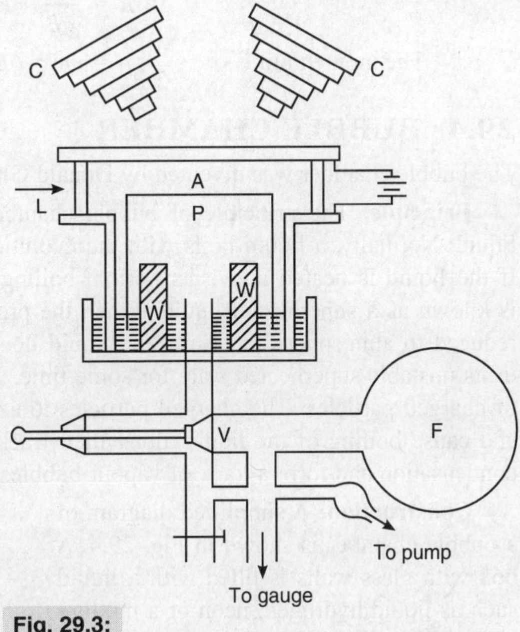

Fig. 29.3: Schematic of apparatus of Wilson cloud chamber

Working: Initially, the chamber *A* is filled with the saturated vapour. When the valve is opened, the air under the piston rushes into the evacuated vessel *F*, and causes the piston to drop suddenly. As a result, the gaseous mixture in the expansion chamber is subjected to adiabatic expansion and is supercooled. When ionizing particles pass through, they ionize the gas. Negative and positive ions are formed all along the path of the ionizing particle. A large number of extremely fine droplets are formed on the newly produced ions. These droplets form a **track** of the moving ionizing particles. The droplets are clearly visible when the chamber is illuminated by light. The tracks can be photographed with the help of cameras.

In fact, the operation of the cloud chamber is automatic. The process of expansion, entry of the ionizing particles into the expansion chamber, illuminating the chamber and clicking the camera are all be carried out in rapid succession automatically. Immediately after photographing of the track, the particle track is removed by applying an electric field across the air gap. Then, the chamber gets ready for taking the next photograph.

Different particles produce different types of tracks. The ionizing particle can be easily identified from its path in the cloud chamber. α-particles, being highly ionizing, produce short, broad, thick, and straight line tacks. β-particles being very light and less ionizing, produce thin, beaded and crooked tracks. As the ionizing power of γ-rays is very low, they are never observed in a cloud chamber.

If the cloud chamber is placed between the pole pieces of a strong electromagnet, the paths of the charged particles are deflected. From the direction of curvature, the positive and negative charged particles can be distinguished. The momentum of the particle can be estimated from the measurement of the radius of the curved track as follows.

The radius R, of the curved path of a charged particle moving in a transverse uniform magnetic field is given by

$$R = \frac{m\upsilon}{qB}$$

∴ The momentum $$p = m\upsilon = qBR. \qquad ...(29.1)$$

29.4 BUBBLE CHAMBER

The bubble chamber was invented by Donald Glaser in 1952.

Principle: The principle of bubble chamber is based on the property of superheated liquid. Normally, a liquid boils with the evolution of bubbles of vapour at its boiling point. If the liquid is heated above its normal boiling point by increasing the pressure over it, it is known as a superheated liquid. When the pressure on the superheated liquid is suddenly reduced to atmospheric pressure, the liquid does not start boiling immediately but remains in its unstable superheated state for some time. This liquid is highly sensitive to the passage of charged particles. The charged particles ionize the atoms as they pass through the liquid and cause boiling of the liquid along their tracks. The newly created ions act as centers of condensation and form a track of vapour bubbles.

Construction: A simplified diagram of a bubble chamber is shown in Fig. 29.4. A box with glass walls is filled with a liquid such as liquid hydrogen, neon or a mixture of these. The box is connected to a pressure system. The ionizing rays enter the liquid through a thin window. The box can be illuminated by a flash of flood lights. A camera is attached to photograph the track of the particles.

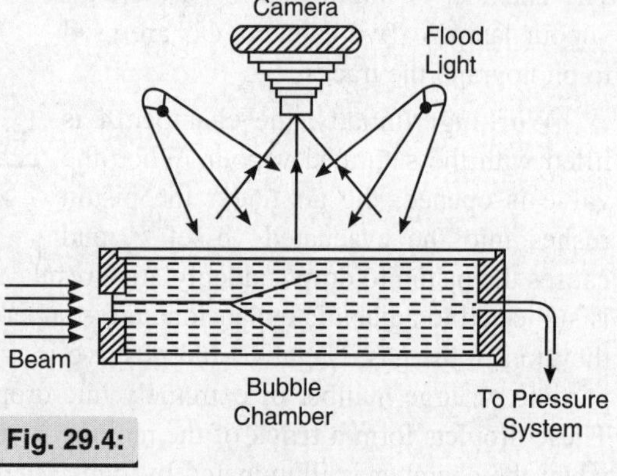

Fig. 29.4: Schematic of bubble chamber

Working: The liquid of 2 : 1 neon-hydrogen mix is held under a pressure of about 5 atmospheres and the pressure is reduced to about 2 atmospheres just before the ionizing particles arrive. As charged beam particles pass through this superheated liquid they ionize atoms and cause the liquid to boil along their paths. The bubbles formed are allowed to grow up to a diameter of about

1 mm and a flash photograph is then taken. The pressure is then increased once again to clear the bubbles and wait for the arrival of the next surge of particles.

Advantages:

The bubble chamber has the following advantages over the cloud chamber.

(*i*) The tracks obtained in a bubble chamber are sharper.

(*ii*) Highly energetic particles can be stopped within the liquid and hence their ranges can be measured.

(*iii*) Particles of low ionizing ability produce good tracks in bubble chamber.

29.5 SPARK CHAMBER

Spark chamber is one of the important detectors developed for use in the study of high-energy nuclear reactions.

Construction: A spark chamber consists of a stack of equally spaced conducting plates (Fig. 29.5). Typically, there are about 25 to 100 plates, each about 1 mm thick and 1 m square. They are accurately spaced about 6 mm apart. Alternate plates are connected together and the two sets are connected to a 10 kV to 15 kV d.c. voltage source. The chamber is filled with pure helium of a mixture of 90% neon and 10% helium at a pressure of 1 atmosphere.

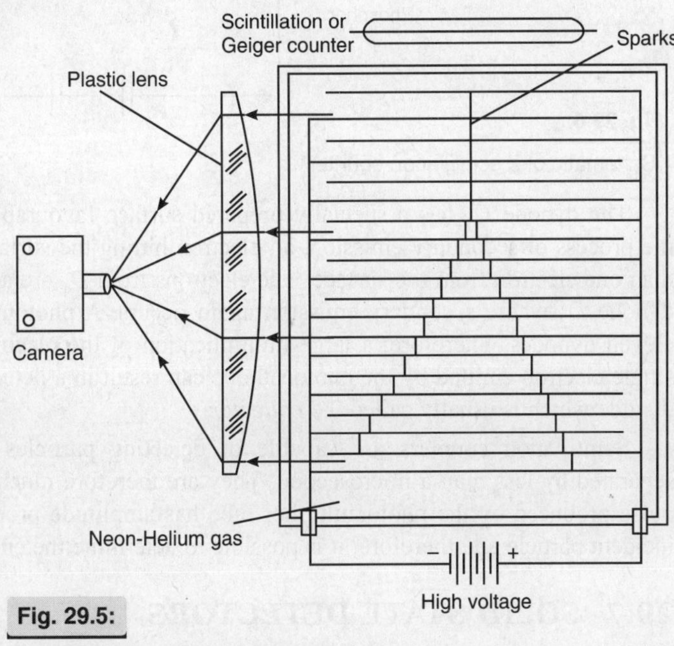

Fig. 29.5: Schematic of spark chamber

Working: When a high-energy charged particle enters the spark chamber, a detector triggers the high voltage to the plates and opens the camera shutter. Traversing the chamber the charged particle produces many ion pairs along its path and sparks jump between the pairs of oppositely charged plates. A large plastic lens makes it possible to photograph the light from sparks between the plates. A second large lens and camera looking into an adjacent side of the chamber permits the simultaneous recording of a second photograph, thereby recording stereoscopic photos of each event. A series of sparks marks the track of the particle.

29.6 SCINTILLATION COUNTER

Fig. 29.6 shows a schematic of a scintillation counter. It is a combination of a scintillator and a photomultiplier tube. The passage of a charged particle through the scintillating material causes some of the scintillator atoms to get excited. As the electrons in the excited atoms return to their ground states, photons are emitted. The photons strike the semitransparent photocathode in the photomultiplier tube and cause emission of electrons through photoelectric effect. Ideally each photon would produce one electron at the photocathode. In practice the efficiency of electron production is about 10%. Electrons produced at the photocathode are accelerated towards the first *dynode* D_1 which is held positive with respect to the cathode. The number of electrons emitted at the photocathode will be proportional to the number of photons incident on it and therefore, to the energy of the original incoming particle.

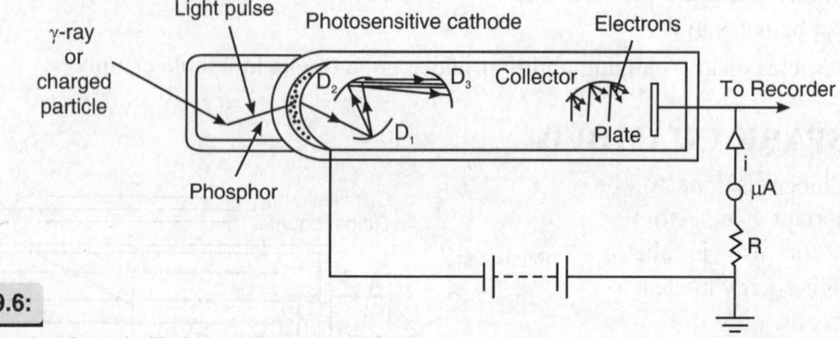

Fig. 29.6: Schematic of a scintillation counter

The dynode D_1 has a specially prepared surface favourable for secondary emission. In the process of secondary emission, an electron hitting the surface results in ejection of more than one electron from the surface. The electrons from D_1 are accelerated towards the second dynode D_2 where secondary emission again occurs. A photomultiplier tube contains ten to eleven dynodes. Therefore, a large multiplication of the electrons takes place. Therefore, a single electron emitted by the photocathode can result in a detectable current pulse at the last dynode which is usually called the *collector*.

Scintillation counters are capable of detecting particles whose times of arrival are separated by less than a microsecond. They are therefore much faster. Secondly, the voltage pulse produced by the photomultiplier tube has amplitude proportional to the energy of the incident particle and therefore, it is possible to determine the energies of incoming particles.

29.7 SOLID STATE DETECTORS

Fig. 29.7 shows a semiconductor diode used as a particle detector. The detector consists of a *p-n* junction formed between *p*-type and *n*-type silicon. When a *p-n* junction is made, a depletion region arises between the *p*- and *n*-type semiconductor regions. The depletion region does not contain any mobile charges. The *p-n* junction is reverse biased with the help of battery *B*. The reverse bias causes a widening of the depletion region. It also minimizes the current flowing in the detector when no radiation is incident on it.

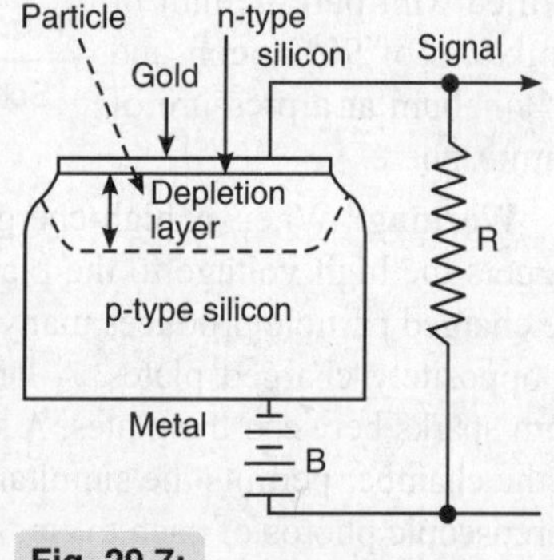

Fig. 29.7: Schematic of semiconductor diode as a particle detector

When a charged particle travels through the depletion region, it interacts with the atoms in the depletion region and produces electron-hole pairs. The electric field acting across the depletion region separates the electrons and holes which move through the external circuit. As the electrons and holes are oppositely charged, they cause current in the same direction in the external circuit. As they pass through the resistor *R*, they produce a voltage pulse. The output pulse is then amplified. The strength of the output pulse depends on the number of carriers collected.

29.8 CERENKOV DETECTOR

When a charged particle moves through a transparent dielectric medium with a velocity greater than the velocity of light in that medium, a cone of light waves is emitted. The light waves are known as Cerenkov radiation. In a dielectric medium of refractive index μ, photons move with a velocity c / μ. Let us consider a charged particle moving with a velocity υ through the dielectric medium. Fig. 29.8 (a) shows the interactions when $\upsilon > c / \upsilon$. The envelope of the radiation is a cone of half-angle θ with the particle at its apex. θ is given by the following relation.

$$\sin\theta = \frac{(c/\mu)t}{\upsilon t} = \frac{c}{\mu\upsilon} \quad \text{...(29.2)}$$

The angle θ of the cone of radiation depends on the speed υ of the particle. When a beam of fast charged particles moves through a medium such as glass or transparent plastic, the radiation can be detected and the angle θ can be measured. Thus, the speed of the particle can be determined.

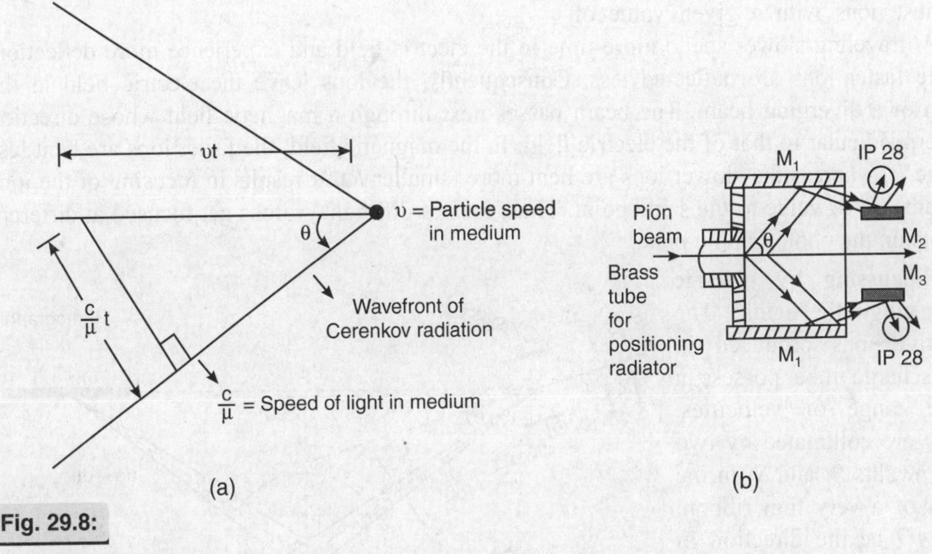

Fig. 29.8: Cerenkov Detector

Fig. 29.8 (b) shows a Cerenkov detector designed by Marshall. A collimated beam of pions is allowed to pass along the axis of a large hemispherical Perspex lens. The Cerenkov radiation radiated in this lens is reflected by the cylindrical mirror M_1 on to one of the plane mirrors M_2, M_3 and then on to the cathode of photomultiplier tubes (IP 28). The two photomultipliers operate in coincidence. The position of the radiator lens is adjusted so that Cerenkov light cone is correctly focused on to the phototubes to give maximum counting rate.

29.9 MASS SPECTROGRAPHS

Atoms that are chemically identical but differ in masses are called *isotopes*. **Mass spectrograph** is an instrument that separates different isotopes from a stream of positive ions of an element by using electric and magnetic fields and measures their individual masses as well as their relative abundance. By analogy with the instrument that separates light of different wavelengths, the instrument that separates ions of different masses is called a mass *spectrograph.* In the mass spectrograph positive ions having the same q / M value are brought to a common focus and produce a line on a photographic plate. Ions having different q / M values produce different lines of different intensity on the photographic plate. Owing to its similarity to an optical line spectrum, the image obtained on the plate is called a **mass spectrum**.

29.10 ASTON MASS SPECTROGRAPH

Principle: A schematic of Aston's mass spectrograph is shown in Fig. 29.9. A beam of positive ions issuing out of a discharge tube is collimated by a system of slits and is allowed to pass through a uniform electric field first. The electric field produces a dispersion of the ions with respect to the velocity. It is because ions with a given value of q / M traveling slower spend more time in the electric field and experience more deflection, while faster ions are deflected less. Consequently, the ions leave the electric field in the form of a diverging beam. The beam passes next through a magnetic field whose direction is perpendicular to that of the electric field. In the magnetic field, the faster ions are bent less (large R) whereas the slower ions are bent more (smaller R). It results in focusing of the ions of same q / M value to the same point. Thus, ions of different values get focused at different points on the photographic plate.

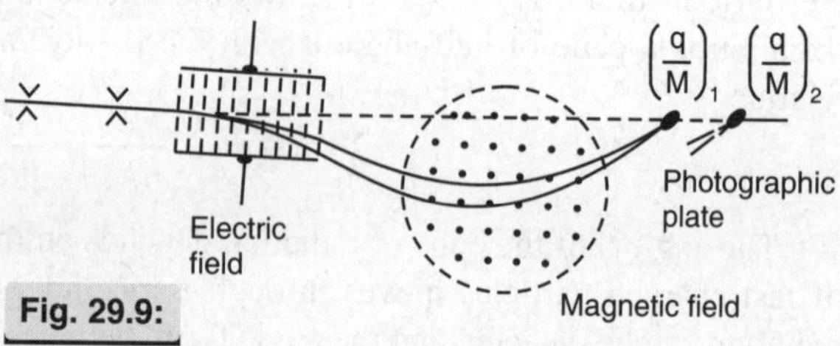

Fig. 29.9: Schematic of Aston's mass spectrograph

Focussing by electric and magnetic fields: The positive ions produced in a discharge tube possess a wide range of velocities. They are collimated by two narrow slits S_1 and S_2 in the form of a very thin ribbon. Let AO be the direction in which the well-collimated stream of positive ions enters the electric field region (see Fig. 29.10). The electric field, E, acts vertically downward from plate P to plate Q and hence the positive ions are deflected downward. As the positive ion beam contains ions having a range of velocities spread between υ and $\upsilon + d\upsilon$, they will be dispersed through an angle $d\theta$.

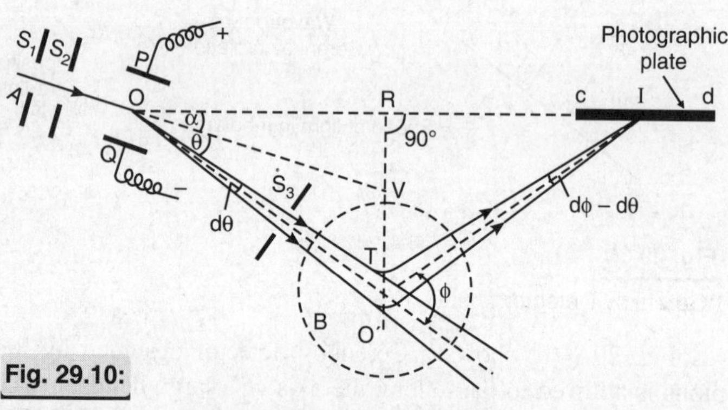

Fig. 29.10: Stream of positive ions in electric field region

Thus, let θ be the angle of deflection of the beam with respect to the axis of the electric field and $d\theta$ be the angle of dispersion of the beam. Let a be the distance between the centers of the electric and magnetic fields and b be the distance from the center of the magnetic field to the photographic plate. The magnetic field, B, acts into the plane of the page and under its action the positive ion beam deflects upward. Because of the velocity focusing action of the magnetic field the ion beam converges on the plane of the photographic plate. Let ϕ be the angle of deflection due to magnetic field and $d\phi$ the angle of refocusing.

The deflection of an ion of mass M and charge e in the electric field is inversely proportional to the square of the ion velocity and is given by

$$x = k_1 \frac{eE}{m\upsilon^2}$$

But $$x \propto \theta$$

$\therefore$ $$\theta \propto k_1 \frac{eE}{m\upsilon^2}$$

or $$\theta = L \frac{e}{m\upsilon^2} \qquad \text{...(29.3)}$$

where L is a constant depending on the values of E and the distance traveled in the field.

Differentiating Eq. (29.3), we get

$$d\theta = -\frac{2Le}{m\upsilon^3} d\upsilon$$

$\therefore$ $$\frac{d\theta}{\theta} = -\frac{2Le\,d\upsilon}{m\upsilon^3} \cdot \frac{m\upsilon^2}{Le} = -\frac{2\,d\upsilon}{\upsilon} \qquad \text{...(29.4)}$$

After emerging from the electric field, the ion beam enters a uniform magnetic field at a mean distance 'a' from the center of the electric field. The magnetic field causes deflection of the ion beam in the same plane as that of electric deflection. The magnetic deflection is given by

$$y = k_2 \frac{eB}{m\upsilon}$$

But $$y \propto \phi$$

Therefore, $$\phi \propto k_2 \frac{eB}{m\upsilon}$$

or $$\phi = M \frac{e}{m\upsilon} \qquad \text{...(29.5)}$$

where M is a constant depending on the values of B and the distance traveled in the field.

Differentiating Eq. (29.5), we get

$$d\phi = -\frac{Me}{m\upsilon^2} d\upsilon$$

$\therefore$ $$\frac{d\phi}{\phi} = \frac{Me\,d\upsilon}{m\upsilon^2} \cdot \frac{m\upsilon}{Me} = -\frac{d\upsilon}{\upsilon} \qquad \text{...(29.6)}$$

It follows from Eq. (29.4) and Eq. (29.6) that

$$\frac{d\theta}{\theta} = \frac{2\,d\phi}{\phi} \quad \text{or} \quad \frac{d\phi}{d\theta} = \frac{\phi}{2\theta} \qquad \text{...(29.7)}$$

In the absence of the magnetic field, the dispersion produced in the beam for a distance $(a + b)$ is equal to $(a + b)d\theta$. The magnetic field acting in a direction perpendicular to the electric field compensates the electric dispersion $d\theta$ completely and refocuses the ion beam at some distance b. Dispersion produced by the magnetic field in a distance b equals $b\ d\phi$. The ions are all focused to the same position, under the condition that the linear electric dispersion $(a + b)d\theta$ is equal and opposite to the linear magnetic dispersion $b\ d\phi$. Thus,

$$(a+b)d\theta = b\,d\phi$$

or $$\frac{d\phi}{d\theta} = \frac{a+b}{b} \quad ...(29.8)$$

Comparing Eq. (29.7) and Eq. (29.8), we get

$$\frac{a+b}{b} = \frac{\phi}{2\theta}$$

or $$2a\theta = b(\phi - 2\theta)$$

or $$\frac{b}{a} = \frac{2\theta}{\phi - 2\theta} \quad ...(29.9)$$

The above equation gives the ion-focusing distance b from the magnetic field.

If a photographic plate is placed along the locus of the ion focusing points, the ions register at different points on the plate.

Now, let $O'R$ be perpendicular to the line cd produced backward. Then in the $\Delta^{le}ROO'$

$$RO' = OO' \sin(\alpha + \theta) = a\sin(\alpha + \theta) \quad ...(29.10)$$

In the $\Delta^{le}RIO'$, $RO' = IO' \sin RIO' = b\sin[180 - (\phi - \alpha - \theta)] = b\sin(\phi - \alpha - \theta)$ ...(29.11)

Equating Eq. (29.10) and Eq. (29.11), we obtain

$$a\sin(\alpha + \theta) = b\sin(\phi - \alpha - \theta)$$

For small angles, we can write the above equation as

$$a(\alpha + \theta) = b(\phi - \alpha - \theta) \quad ...(29.12)$$

Comparing Eq. (29.9) and Eq. (29.12), it is clear that the two equations are identical if $\alpha = \theta$. Thus, the condition for focusing is that the photographic plate must be placed at an angle θ with the direction of the incident positive ion beam such that $\alpha = \theta$, where θ is the deviation produced by the electric field.

29.11 DEMPSTER MASS SPECTROGRAPH

In 1918 A.J.Dempster, the American physicist built a mass spectrograph which was further improved in 1922.

Principle: When all the positive ions produced are accelerated by the same high potential difference, it may be assumed that all of them have the same velocity and a homogeneous magnetic field acting normal to the path of the ions will separate the ions of different masses.

Construction: A schematic diagram of one of the early mass spectrometers developed by Dempster is shown in Fig. 29.11. The material under investigation

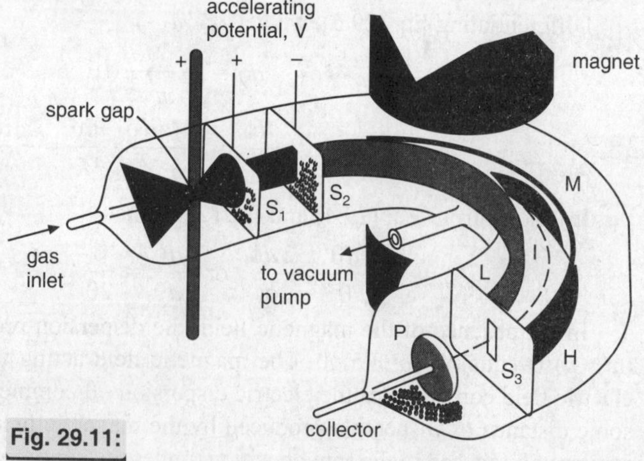

Fig. 29.11: Schematic of Dempster mass spectrometers

is fed into the spectrometer in the form of gas. It is ionized with the help of high voltage applied across the spark gap. The positively charged ions of the element are collimated by slits S_1 and are accelerated toward the slit S_2 by a high negative potential of about 1000 V. A stream of positive ions which may be regarded as emerging with uniform energy, spread at high velocity through the slit S_2 into the semicircular space M. A magnetic field of sufficient strength is applied normal to the path of the ions. The magnetic field causes the stream of ions to follow semicircular path. The ions pass through a third slit S_3 and impinge on a plate P connected to an electrometer or a similar device for measuring the ion current. The instrument is called a mass spectrometer because the ion current is measured in the instrument rather than recording the ions on a photographic plate.

Working: If V is the accelerating potential through which the positive ions are accelerated in the region between the slits S_1 and S_2, the velocity of the ions which may be assumed to be the same for all ions, is given by

$$\upsilon = \left[\frac{2qV}{M}\right]^{\frac{1}{2}} \qquad \text{...(29.13)}$$

In the magnetic field the ions describe a semicircular path whose radius is given by

$$R = \frac{M\upsilon}{qB} \qquad \text{...(29.14)}$$

Using Eq. (29.13) into Eq. (29.14), we get

$$R = \frac{M}{qB}\left[\frac{2qV}{M}\right]^{\frac{1}{2}}$$

or

$$R^2 = \frac{2MV}{qB^2}$$

$$\therefore \quad \frac{q}{M} = \frac{2V}{B^2R^2} \qquad \text{...(29.15)}$$

In the mass spectrometer only positive ions describing a definite value of R can pass through the third slit S_3 and reach the plate P. The radius R is defined by slits S_2 and S_3; and the q/M values of ions that describe a path of this radius are determined by the accelerating potential V and the magnetic field induction B. In practice, the magnetic field B is kept constant and the potential V is varied steadily. The corresponding ion current is measured for each value of V. Since each accelerating potential corresponds to a definite mass of ions reaching the electrometer, the current can be plotted against the mass of the ions. The resulting graph shows a series of maxima. Each maximum corresponds to a set of ions having a definite q/M value. Knowing the value of the potential V at which a maximum occurs, it is possible to calculate the value of q/M using Eq. (29.15).

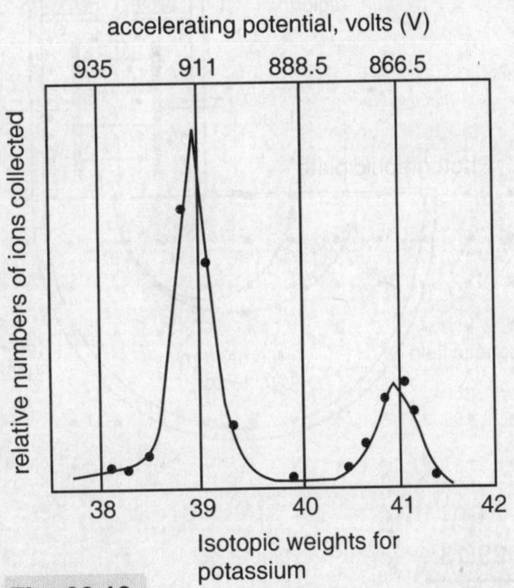

Fig. 29.12: Isotopic weight versus relative number of ions collected at various accelerating voltages

With the element potassium, Dempster obtained the results that are shown in Fig. 29.12. Ions of potassium passed through the slit S_3 at $V_1 = 866.5$ V and $V_2 = 911$ V. From Eq. (29.15) we obtain

$$M_1V_1 = \text{constant} = M_2V_2$$

$$\therefore \quad M_1 = \frac{M_2V_2}{V_1} \qquad \text{...(29.16)}$$

Dempster knew from other measurements that most potassium atoms have a mass of about 39 amu. Hence he assigned the maximum at 911 V to ^{39}K. Then using the values into Eq. (29.16), he found that

$$M_1 = 39 \text{ amu} \frac{911\,V}{866.5\,V} = 41 \text{ amu}$$

The chief advantage of the mass spectrometer is that the relative amounts of the various particles can be estimated from the magnitude of the ion currents at the corresponding maxima. It is easy to see from Fig. 29.12 that the two isotopes of potassium have relative abundances in the ratio 18 : 1.

29.12 BAINBRIDGE MASS SPECTROGRAPH

Principle: The Bainbridge mass spectrograph is based on the principle that uniform magnetic field acting perpendicular to the path of ions deflects them along circular paths. Ions having the same velocity but different masses are deflected along circular paths of different radii. It is necessary that all the ions in the beam entering the transverse magnetic field must have a single velocity. However, the beam of positive ions produced in a discharge tube and entering the spectrograph will have a wide range of velocities. Therefore, a velocity selector is used to form a single-velocity ion beam.

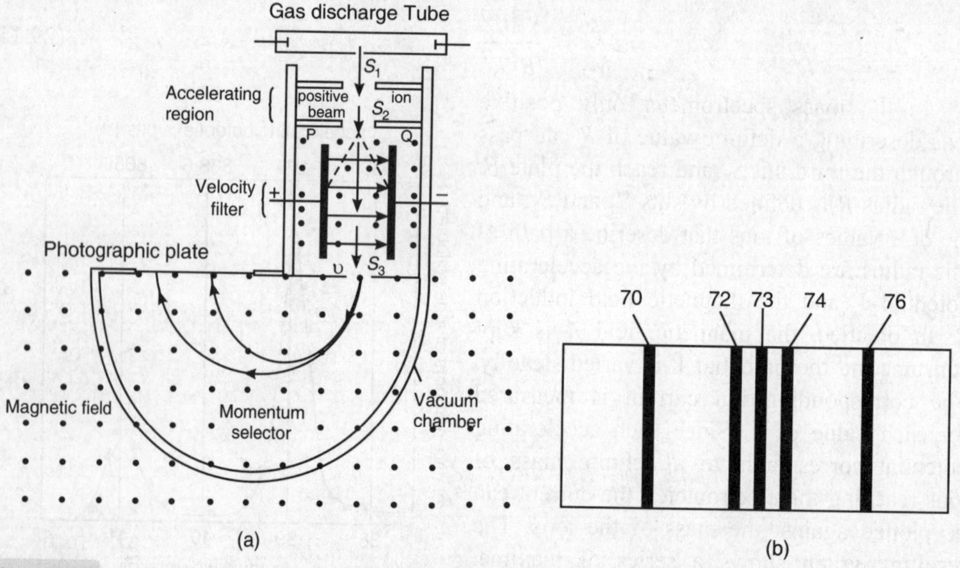

Fig. 29.13: (a) Schematic of Bainbridge Mass Spectrograph (b) Mass spectrum of Germanium

Construction: Fig. 29.13 (a) shows the schematic of Bainbridge mass spectrograph. It is essentially a vacuum chamber placed in uniform magnetic field. A discharge tube produces positive ions of the element under investigation. The slits S_1 and S_2 accelerate and also collimate the incoming ion beam. The ion beam passes through a velocity selector. Beyond the velocity filter, another slit S_3 is arranged which further collimate the mono-velocity beam. A photographic plate is mounted in the analyzing chamber in line with the slit S_3. A photographic plate is mounted in the analyzing chamber in line with the slit S_3.

Working: The element under study is taken in the form of gas and introduced into a discharge tube. The gas is ionized under the action of the applied voltage and the positive ions, which are formed in the discharge tube, are accelerated and conducted into the spectrograph through slits S_1 and S_2. Most of the positive ions formed will carry a charge of +1 because it is much more difficult to remove further electrons from an already positive ion. The ion beam, thus consisting of ions carrying a positive charge of +1, possesses a wide range of velocities $\upsilon \pm \Delta\upsilon$. It is then passed through the velocity selector.

A velocity selector is an electro-optic device, which selects a stream of single velocity ions from a beam of ions having a wide range of velocities. It consists of a combination of uniform electric and magnetic fields in crossed configuration.

Referring to Fig. 29.13 (a), the electric field E is produced in the horizontal direction by a set of charged parallel plates PQ and the uniform magnetic field B is applied normal to E in the same region. Both E and B act normal to the ion path. The electric field produces a dispersion of the ions with respect to the velocity. The ions travelling with a velocity υ (=E/B) do not experience any net force and they continue to travel along the initial direction. Ions travelling with velocities differing from υ are deflected sideways and are absorbed by the metal plates. Therefore, a strictly mono-velocity ion stream, having a velocity $\upsilon = E / B$, issues out of velocity selector. This beam, of ribbon shape, then passes through the slit S_3 and enters the analyzing chamber. The uniform magnetic field B acting on the chamber plays the role of a *momentum selector*. Ions of different masses are deflected along circular paths of different radii depending on their momentum and strike the photographic plate after completion of a semicircular movement. They leave vertical lines on the plate, which resemble spectral lines formed by light. Hence, it is called **mass spectrum**. The number of lines on the plate corresponds to the number of isotopes of the element. Fig. 29.13 (b) shows the mass spectrum of germanium.

Considering the isotope with a mass M_1, the circular path described by it has the radius R.

Now $$R = \frac{M\upsilon}{qB}$$

But as $\upsilon = \dfrac{E}{B}$, $$R = \frac{ME}{qB^2} \qquad ...(29.17)$$

The distance of the line formed by the isotope on the photographic plate can be measured from a reference line. If x is the distance of the line from the slit, then

$$x = 2R$$

$$\therefore \quad M = \frac{qB^2}{2E}x \qquad ...(29.18)$$

or $$M = kx$$

where k is a constant because q, B and E are constant parameters.

$$\therefore \quad M \propto x$$

The above relation is linear in x and hence the mass scale is *linear*.

As the relation between M and x is linear, the mass of an ion may be obtained from the measurements of E, B and x by proper calibration of the photographic plate. The relative masses of two isotopes involve only measurement of x and therefore, it can be obtained with high precision. If M_1 and M_2 are the masses of two isotopes and if x_1 and x_2 are the distances of the corresponding lines respectively from S_3, the line separation is given by

$$\Delta x = (x_2 - x_1) = \frac{2E}{qB^2}(M_2 - M_1) \qquad \text{...(29.19)}$$

In practice, an ion with known mass is used as a standard and the mass of the ion under study is compared with it by means of Eq. (29.19).

Example 29.2: *In a Bainbridge mass spectrograph, the magnetic field in the velocity selector is 1.0 T and the ions having a speed of 4×10^6 m/s pass through it undeflected.*

(i) What should be the electric field between the plates?

(ii) If the separation of the plates is 0.5 cm, what is the potential difference between the plates?

Solution: The velocity of the undeflected ions is given by $\upsilon = E / B$.

$$\mathbf{E} = \upsilon B = (4 \times 10^6 \text{ m/s})(1.0T) = \mathbf{4 \times 10^6 \text{ V/m.}}$$

The potential difference between the plates is given by $V = Ed$.

$$\therefore \quad V = (4 \times 10^6 \text{ V/m})(5 \times 10^{-3}\text{m}) = \mathbf{20 \text{ kV.}}$$

Example 29.3. *The electric field between the plates of velocity selector is 150 V/cm and the magnetic field is 0.5 T. If the source contains the three isotopes of magnesium Mg^{24}, Mg^{25} and Mg^{26} and the ions are singly charged, find the distance between the lines formed by the these isotopes on the photographic plate.*

Solution: The separation between the successive lines is given by $\Delta x = \dfrac{2E}{qB^2}(M_2 - M_1)$.

$$\therefore \quad \Delta x = \frac{2(1500 \text{ V / m})}{(1.602 \times 10^{-19} \text{C})\,(0.5 \text{ T})^2}(25\text{u} - 24\text{u}) \frac{3000\text{V / m}}{(1.602 \times 10^{-19}\text{ C})(0.5\text{T})^2}(1.66 \times 10^{-27}\text{ kg})$$

$$= \mathbf{1.2 \text{ mm}}$$

Example 29.4: *In a Bainbridge mass spectrograph, the electric field used is 8×10^4 V/m and the magnetic field common to both places is 0.55 wb/m^2. If the ion source consists of singly ionized neon isotopes of atomic masses 20 and 22, calculate linear separation of lines formed on photographic plate.*

Solution: The separation between the successive lines is given by $\Delta x = \dfrac{2E}{qB^2}(M_2 - M_1)$.

$$\therefore \quad \Delta x = \frac{2(8 \times 10^4 \text{ V / m})}{(1.602 \times 10^{-19}\text{ C})(0.55 \text{ wb / m}^2)^2}(22\text{u} - 20\text{u})$$

$$= \frac{16 \times 10^4 \text{ V / m}}{(1.602 \times 10^{-19}\text{ C})(0.55\text{T})^2}(2 \times 1.66 \times 10^{-27}\text{ kg}) = \mathbf{11 \text{ mm}}$$

29.13 PARTICLE ACCELERATORS

Particles are accelerated to very high energies using electric and magnetic fields. A charged particle acquires energy 'qV' when it is accelerated by a potential difference V. Choosing V to be very large, it should be possible to achieve any desired energy. In practice, the maximum voltage that one can produce limits the maximum energy of the particles. The production of maximum voltage in fact depends on the problem of providing adequate insulation. The Cockcroft-Walton and the Van de Graaff accelerators are examples of single-step high voltage accelerators. We can bypass the problem of insulation if we use small voltages and provide acceleration through small steps. By making the number of steps large, we can produce very high-energy charged particle beam. Two different types of small voltage accelerators are built which differ in their principle of operation. They are known as linear accelerators and cyclic accelerators.

***(i)* Linear accelerator:** A linear accelerator (LINAC) accelerates charged particles to high energies without the need for very high voltages. In a linear accelerator, there is a succession of electrodes to which an alternating voltage is applied. Successive tubes have opposite voltages, but the voltage alternates with the frequency of the applied voltage. If the lengths of the tubes are correctly chosen, the motion of the charged particles is synchronised with the alternation of the voltage so that they cross the gap between the successive tubes at the right time to receive a push that increases their energy.

The largest proton linear accelerator is at the University of Minnesota, U.S.A., which can accelerate protons to 68 MeV. The largest electron linear accelerator is at Stanford University, U.S.A. It is 3.2 km long and can accelerate electrons to 25 GeV.

***(ii)* Cyclic accelerator:** In cyclic accelerators, the particles are forced by a magnetic field to describe a curved path along which they receive increase in energy from electric fields at certain points on its path. Cyclotron is the first cyclic accelerator. The maximum energy that can be reached with a cyclotron is limited to several MeV. As demand arose for more energetic particles, cyclotrons were replaced by synchrotrons. The most powerful of these machines is the super proton synchrotron (SPS) at the European Centre for Nuclear Research (CERN), Switzerland, which can accelerate protons to energy of 26 GeV.

Examples: Van de Graff generator and **LINAC** are examples of linear accelerators, while cyclotron and betatron are examples of cyclic accelerators.

29.14 DRIFT TUBE ACCELERATOR

A linear accelerator (LINAC) accelerates charged particles to high energies without the need for very high voltages. In linear accelerators, there is a series of coaxial hollow cylindrical electrodes, known as *drift tubes* to which an alternating voltage is applied. Successive tubes have opposite voltages, but the voltage alternates with the frequency of the applied voltage. If the lengths of the tubes are correctly chosen, the motion of the charges particles is synchronized with the alteration of the voltage so that they cross the gap between the successive tubes at the right time to receive a push that increases their energy.

Construction and working: The LINAC consists of a set of drift tubes of increasing lengths arranged linearly in an evacuated glass chamber (see Fig. 29.14). Alternate tubes are connected together. The odd numbered tubes are joined to one terminal and the even numbered tubes are joined to the other terminal of a high frequency power supply, which is in fact a radio-frequency oscillator. Thus, when tubes C_1, C_3, C_5 are positive, the other tubes C_2, C_4 and C_6 are negative and reversal of potential takes place in step with the frequency of the power supply. The positive ions produced by the source S travel along the axis of the tubes and are accelerated on crossing the gaps between the drift tubes. The ions do not receive acceleration while traveling inside the tubes because electric field does not exist there.

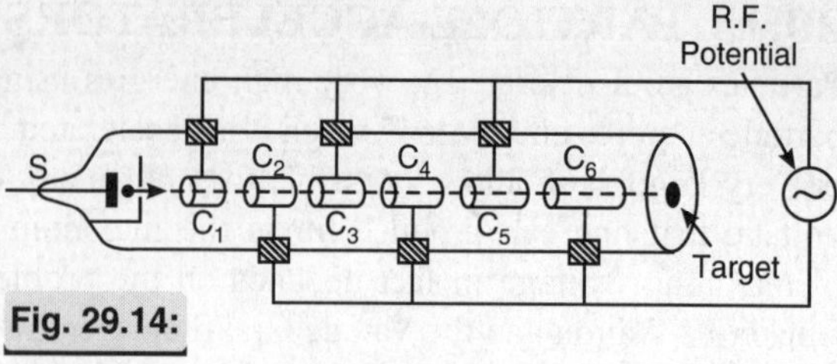

Fig. 29.14: Schematic of linear accelerator

Initially, the positive ions move towards C_1 when it is negative. Then they travel through it with constant velocity. Length of C_1 is so selected in relation to the frequency of the oscillator that these ions arrive at the gap between C_1 and C_2 at the instant when C_2 has just become negative. It will happen only when the time taken by ions to travel through C_1 is exactly equal to half the time period of the oscillator. The ions are further accelerated across the gap and travel through C_2 at a higher uniform velocity. The length of C_2 is such that the ions arrive in the gap at the instant when C_3 has just become negative and are further accelerated. Thus, the ions are accelerated in the successive gaps and emerge from the final drift tube with extremely high velocities.

If N is the number of tubes, V is the maximum voltage of the oscillator then the energy acquired by an ion of charge q is given by

$$E = \frac{1}{2} m\upsilon_N^2 = NqV \qquad \text{...(29.20)}$$

The time required to travel through any tube is equal to half time period of the applied radio-frequency voltage. Hence,

$$t = \frac{T}{2} = \frac{1}{2\nu}$$

If l_N is the length of the Nth tube and υ_N the velocity of the ion while traveling through it, then

$$t_N = \frac{l_N}{\upsilon_N}$$

$$\therefore \quad \frac{l_N}{\upsilon_N} = \frac{1}{2\nu}$$

or

$$l_N = \frac{\upsilon_N}{2\nu} \qquad \text{...(29.21)}$$

Substituting the value of υ_N from Eq. (29.20) into Eq. (29.21), we get

$$l_N = \sqrt{\frac{NqV}{2m\nu^2}} \qquad \text{...(29.22)}$$

The largest proton linear accelerator is at the University of Minnesota, USA, which can accerlate protons to 68 MeV. The largest electron linear accelerator is at Stanford University, USA. It is 3.2 km long and can accelerate electrons to 25 GeV.

Example 29.5: *In a linear accelerator, proton accelerated thrice by a potential of 40 kV leaves a tube and enters an accelerating space of length 30 cm before entering the next tube. Calculate the frequency of the r.f.voltage and the length of the tube entered by the proton.*

Solution: Let υ_1 and υ_2 be the velocities of the proton entering and leaving the accelerating space. Let q and m be the mass and charge of the proton respectively. Then

$$\frac{1}{2}m\upsilon_1^2 = 3 \times 1.602 \times 10^{-19}\text{ C} \times 4 \times 10^4\text{ V}$$

$$\therefore \quad \upsilon_1 = \left[\frac{2\times3\times1.602\times10^{-19}\text{C}\times4\times10^4\text{V}}{1.67\times10^{-27}\text{kg}}\right]^{\frac{1}{2}} = 4.8 \times 10^6 \text{ m/s}$$

$$\text{Similarly} \quad \upsilon_2 = \left[\frac{2\times4\times1.602\times10^{-19}\text{C}\times4\times10^4\text{V}}{1.67\times10^{-27}\text{kg}}\right]^{\frac{1}{2}} = 5.5 \times 10^6 \text{ m/s}$$

Mean velocity $\upsilon = 5.165 \times 10^6$ m/s.

The time taken to travel 30 cm = 0.3 m equals the half-period of the r.f. voltage.

$$\therefore \quad \frac{T}{2} = \frac{0.3\text{ m}}{5.165\times10^6\text{ m/s}}$$

$$\therefore \text{ Frequency of the r.f.voltage } \nu = \frac{1}{T} = \frac{5.165\times10^6\text{ m/s}}{2\times0.3\text{m}} = 8.6 \text{ MHz.}$$

$$\text{Length of the next tube entered by the protons } l = \frac{\upsilon_2}{2\nu} = \frac{5.5\times10^6\text{ m/s}}{2(8.6\times10^6\text{ Hz})} = 0.32 \text{ m.}$$

29.15 CYCLOTRON

Cyclotron is the first **cyclic accelerator** built by E.O. Lawrence and M.S. Livingston in 1932.

Principle: A moving charged particle describes a circular path in the presence of a transverse uniform magnetic field. The frequency of revolution of the charged particle is given by

$$\nu = \frac{qB}{2\pi m} \qquad ...(29.23)$$

which is independent of the particle velocity. It means that a faster particle moving in a circle of bigger radius and a slower particle moving in a smaller circle take the same time for completing one revolution in a given uniform magnetic field. Hence, charged particles having different initial velocities can be uniformly accelerated to produce high-energy particle beam using a combination of crossed electric and magnetic fields.

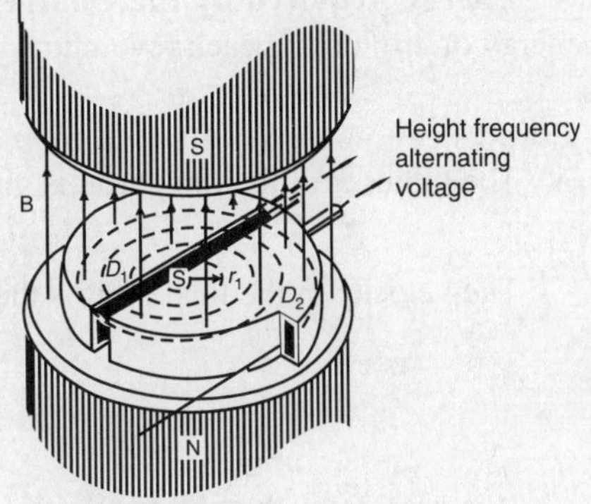

Fig. 29.15: Schematic of Cyclotron

Construction: The schematic of a cyclotron is shown in Fig. 29.15. A cyclotron consists of two hollow metal dees formed by cutting a short, cylindrical box along its diameter. The dees are separated by a few centimetres from each other. They are insulated from each other and are placed in a vacuum chamber located between the pole pieces of an electromagnet. The magnet produces a uniform magnetic field in a direction perpendicular to the semicircular faces of dees. A high frequency oscillator is connected to the dees, which produces an r.f. electric field in the gap between the dees. A source of charged particles is located at the centre of the gap between the dees.

Working: Charged particles, say protons, are injected by the ion source into the gap between the dees. The protons are accelerated by the radio-frequency electric field existing in the gap toward the dee, which is at a negative potential at that instant. However, the magnetic field that is acting perpendicular to the protons deflects them along a circular path. The protons travel in the hollow region of the dee and come back into the gap after completion of a half-revolution. The time taken for a half-revolution is given by

$$\frac{T}{2} = \frac{\pi m}{qB} \quad \text{...(29.24)}$$

where T is the time period for the circular path in the magnetic field. The protons will be further accelerated if the dees reverse their polarity at the instant when the protons emerge into the gap. In such a case, the protons travel further in a semicircular path in the other dee and reach the gap in a time $T/2$. If, again at the same instant, the dees reverse their polarity, the protons receive another dose of acceleration. The process is repeated over and again many times. As the velocity increases with each dose of acceleration, the protons describe a spiral path in the dees, as shown in Fig. 29.15. During each revolution, each proton receives energy of $2qV$ electron volts and after about a hundred or more revolutions, the protons acquire energies of the order of several million electron volts. At the end of the journey, the proton beam is pulled out of its circular path by a negatively charged deflector plate and emerges out of the chamber through a narrow aperture.

Condition of Resonance: In a cyclotron the protons are progressively accelerated provided that the time period T_0 of the r.f. electric field equals the time period T of revolution of protons in the magnetic field, B. Thus, it is required that the condition $T_0 = T$ is fulfilled. It means that

$$T_0 = \frac{2\pi m}{qB} \quad \text{...(29.25)}$$

or

$$\nu_0 = \frac{qB}{2\pi m} \quad \text{...(29.26)}$$

The above relation is known as the ***condition of resonance***.

Energy Acquired by the Charged Particles: A proton makes N revolutions, receiving energy of $2qV$ during each revolution. The total kinetic energy acquired is

$$E = 2NqV \quad \text{...(29.27)}$$

The radius R of the final orbit is given by $R = \dfrac{m\upsilon_{max}}{qB}$

The velocity in the final orbit is therefore $\upsilon_{max} = \dfrac{qBR}{m}$

$$E = \frac{1}{2} m\upsilon_{max}^2 = \frac{B^2 q^2 R^2}{2m} \quad \text{...(29.28)}$$

It is seen from the above expression that ***the final energy acquired by the particles in a cyclotron does not depend on the magnitude of the voltage applied across the dees***. A comparison of the two equations for final energy of the particles suggests that *the particles will have to execute a larger number of revolutions if the applied voltage is low or to make* *smaller number revolutions if the applied voltage is higher, to gain the same amount of kinetic energy.*

Role of Electric and Magnetic Fields: The primary function of the r.f. electric field consists in imparting high kinetic energy to the charged particles. The primary function of magnetic field is to deflect the charged particles along circular path so that they repeatedly pass through the r.f. electric field region, and acquire more and more energy.

The second function of r.f. electric field is to focus the charged particles into a sharp beam. The non-uniform electric field in the gap between the dees plays the role of electron lens (Fig. 29.16 a.) and causes focussing of charged particles. The second function of magnetic field is to correct the paths of charged particles revolving nearer to periphery. Particles, which tend to stray from the median plane, are brought back to the median plane due to the Lorentz force component produced by non-uniform field at the outer edges of the magnet poles (Fig. 29.16b).

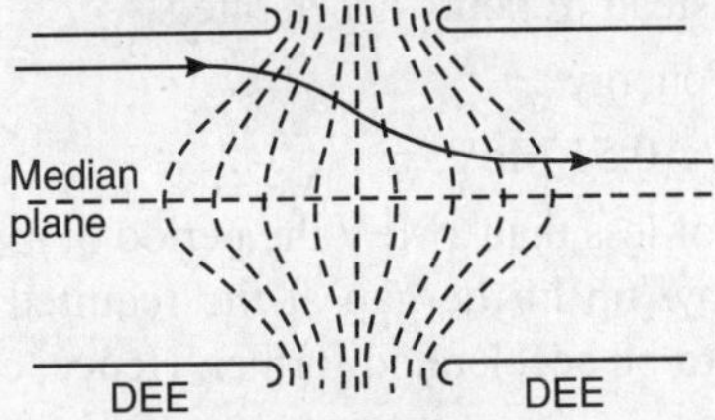

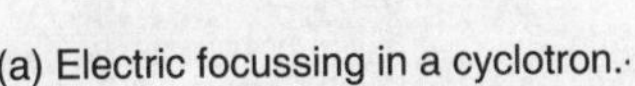
(a) Electric focussing in a cyclotron.

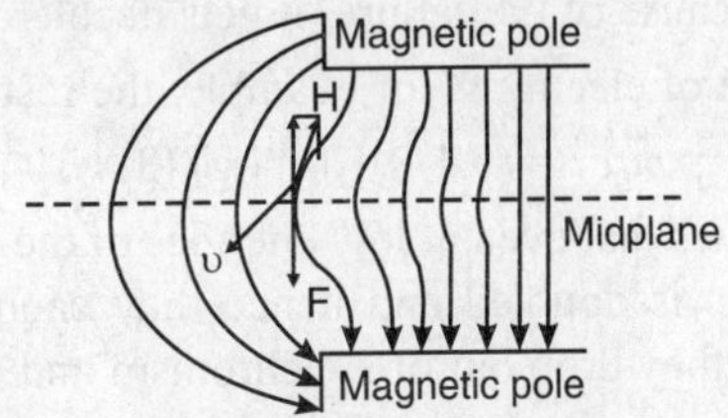

(b) Magnetic focussing in a cyclotron.

Fig. 29.16:

Focussing by Electric and Magnetic Fields

Limitation of Cyclotron: The kinetic energy acquired by charged particles in a cyclotron is given by

$$E = \frac{B^2 R^2 q^2}{2m}$$

According to this relation, it appears that the maximum energy of the particle beam is limited by the magnetic field *B*. Increasing the size of the magnet can increase B. However, there exists an ultimate limit for the size of magnet and the electric current to drive the electromagnet.

There arises a more basic limitation due to the relativistic variation of particle mass at velocity $\upsilon \approx c$. It is known that the mass '*m*' of a particle increases with velocity according to the relation

$$m = \frac{m_0}{\sqrt{1 - \upsilon^2 / c^2}} \qquad \text{...(29.29)}$$

where m_0 is known as the rest mass of the particle.

Therefore, the time taken by the particle to complete a revolution increases from

$$T_0 = \frac{2\pi m_0}{qB}$$

to $$T = \frac{2\pi m}{qB}, \text{ at velocities comparable to } c.$$

$$= \frac{2\pi m_0}{qB}\left[\frac{1}{\sqrt{1 - \upsilon^2 / c^2}}\right] \qquad \text{...(29.30)}$$

We can express T as

$$T = T_0\left(\frac{m}{m_0}\right) = T_0\left(\frac{mc^2}{m_0c^2}\right) = T_0\left(\frac{E}{E_0}\right)$$

where E_0 is the energy of the particle at $\upsilon << c$ and E is the energy of the particle at $\upsilon \approx c$.

Writing $$E = E_0 + K,$$

$$T = \left(1 + \frac{K}{E_0}\right) \qquad ...(29.31)$$

As K increases with increasing particle velocity, $T > T_0$, the above inequality shows that the period of particle revolution significantly increases with the increasing velocity. The result is that the particle fails to reach the gap at the right moment when the electric field reversal occurs. Because of the failure, it gets decelerated instead of being accelerated.

In case of electrons, for example, the rest mass energy

$$E_0 = m_0c^2 = (9.1 \times 10^{-31}\text{kg})\,(3 \times 10^8\ \text{m/s})^2 = 0.51\ \text{MeV}$$

It implies that even at low energies of the order of less than 1MeV the period of revolution of electrons is doubled and hence they cannot show up in the gap at the required instant. Therefore, they drop out of synchronism and cease to be accelerated further. Hence, electrons cannot be accelerated to high energies in a cyclotron.

Example 29.6: *A cyclotron with its dees of radius 2 m has a magnetic field of 0.75 wb/m². Calculate the maximum energies to which (i) protons and (ii) deuterons can be accelerated.*

Solution:

The maximum energy to which particles are accelerated in a cyclotron is given by

$$E_{max} = \frac{B^2q^2R^2}{2m}$$

(*i*) In case of protons
$$E_{max} = \frac{(0.75\ \text{wb/m}^2)^2(1.602\times10^{-19}\text{C})^2(2\text{m})^2}{2(1.67\times10^{-27}\text{kg})}$$

$$= 1.73\times10^{-11}\frac{wb^2\cdot c^2\cdot m^2}{m^4\cdot kg} = 1.73\times10^{-11}\frac{\text{V}^2\cdot s^2\cdot \text{C}^2}{m^2\cdot kg}$$

$$= 1.73 \times 10^{-11}\text{J} = (1.73 \times 10^{-11})(6.24 \times 10^{18}\text{eV})$$

$$= 107.9 \times 10^6\text{eV}$$

$$\therefore \qquad \mathbf{E_{max} = 108\ MeV}$$

(*ii*) In case of deuterons
$$E_{max} = \frac{(0.75\text{wb/m}^2)^2(1.602\times10^{-19}\text{C})^2(2\text{m})^2}{2(3.34\times10^{-27}\text{kg})}$$

$$= 8.64 \times 10^{-12}\text{J} = (8.64 \times 10^{-12})(6.24 \times 10^{18}\text{eV})$$

$$= \mathbf{53.91\ MeV.}$$

Example 29.7: *Protons are accelerated in a cyclotron. The magnetic field strength is 1.3 wb/m² and the radius of the last semicircle is 0.5 m.*

(*i*) *What must be the frequency of the oscillator supplying power to the dees?*

(*ii*) *What is the final energy acquired by the proton beam?*

(iii) If the total transit time of a proton is 3.3 μs, how much energy is imparted to protons in each passage from one dee to the other?

Solution: (*i*) Frequency $\nu = \dfrac{Bq}{2\pi m} = \dfrac{1.3\ \text{wb/m}^2 \times 1.602\times10^{-19}\,\text{C}}{2\times3.143\times1.67\times10^{-27}\,\text{kg}}$ = **19.85 MHz.**

(*ii*) Final energy $E_{max} = \dfrac{B^2q^2R^2}{2m} = \dfrac{(1.3\ \text{wb/m}^2)^2(1.602\times10^{-19}\,\text{C})^2(0.5\text{m})^2}{2(1.67\times10^{-27}\,\text{kg})}$

= **20.28 MeV.**

(*iii*) Number of revolutions $N = 2\nu T = 2 \times 19.85 \times 10^6\ \text{Hz} \times 3.3 \times 10^{-6}\ \text{s} = 131$

Energy gained by proton during one transit = $\dfrac{2NeV}{N} = \dfrac{20.28\ MeV}{131}$ = **155 keV.**

Example 29.8: *The magnetic field strength in a certain cyclotron is 0.9 Wb/m2. If light hydrogen ions (protons) are accelerated in the cyclotron*

(i) What must be the frequency of the oscillator supplying power to the dees?

(ii) If each passage of ions across the accelerating gap increases the energy of the ion by 60,000 eV, how long does it take for the ion introduced at the centre of the dees to emerge at the rim of the dee with energy of 6 MeV?

(iii) Calculate the radius of the last semicircle before the ion emerges from the cyclotron.

Solution: (*i*) Frequency $\nu = \dfrac{Bq}{2\pi m} = \dfrac{0.9\,\text{wb/m}^2 \times 1.602\times10^{-19}\,\text{C}}{2\times3.143\times1.67\times10^{-27}\,\text{kg}}$ = **13.72 MHz.**

(*ii*) Transit time $T = \dfrac{N}{2\nu} = \dfrac{E_{max}/E_1}{2\nu} = \dfrac{6\times10^6(1.602\times10^{-19})\text{J}}{6\times10^4(1.602\times10^{-19})\text{J}\times2\times13.72\times10^6\,\text{Hz}}$

= **3.64 μs.**

(*iii*) Maximum Energy $E_{max} = \dfrac{B^2q^2R^2}{2m}$ $\therefore$ $R = \dfrac{\sqrt{2mE_{max}}}{bq}$

$$\therefore\ R = \frac{(2\times1.67\times10^{-27}\,\text{kg}\times6\times10^6\,\text{eV}\times1.602\times10^{-19}\,\text{J/eV})^{1/2}}{(0.9\ \text{wb/m}^2)(1.602\times10^{-19}\,\text{C})} = \mathbf{0.39\ m.}$$

Example 29.9. *Deuterons are accelerated in a fixed frequency cyclotron to a maximum dee orbit radius of 0.88 m. The magnetic field is 1.4 T. Calculate the energy of the emerging deuteron beam and the frequency of the dee voltage. What change in magnetic flux density is necessary if doubly charged helium ions are accelerated?*

Given atomic masses: $H^2 = 2.014102$ amu *and* $He^4 = 4.002603$ amu

Solution: (*i*) Frequency $\nu = \dfrac{Bq}{2\pi m} = \dfrac{1.4\text{wb/m}^2 \times 1.602\times10^{-19}\,\text{C}}{2\times3.143\times2.014102\times1.67\times10^{-27}\,\text{kg}}$ = **10.61 MHz.**

(*ii*) Energy $E_{max} = \dfrac{B^2q^2R^2}{2m}$

$$= \frac{(1.4\text{wb/m}^2)^2(1.602\times10^{-19}\text{C})^2(0.88\text{m})^2}{2\times2.014102\times1.67\times10^{-27}\text{kg}}$$

$$= \mathbf{36.19\ MeV.}$$

(*iii*) Magnetic flux density $B_{\text{He}} = \frac{2\pi m\nu}{q}$

$$= \frac{2\times3.143\times4.002603\times1.67\times10^{-27}\text{kg}\times10.61\times10^6\text{Hz}}{2\times1.602\times10^{-19}\text{C}}$$

$$= \mathbf{1.39\ T}$$

Example 29.10: *A particle cyclotron is designed with dees of radius 75 cm and with magnets that can provide a field of 1.5 T.*

(*i*) *To what frequency should the oscillator be set if deuterons are to be accelerated?*

(*ii*) *What is the maximum energy of deuterons that can be obtained?*

Solution: (*i*) Frequency $\nu = \frac{Bq}{2\pi m} = \frac{1.5\text{wb/m}^2\times1.602\times10^{-19}\text{C}}{2\times3.143\times2\times1.67\times10^{-27}\text{kg}} = \mathbf{11.45\ MHz.}$

(*ii*) Energy $E_{\max} = \frac{B^2q^2R^2}{2m} = \frac{(1.5\text{wb/m}^2)^2(1.602\times10^{-19}\text{C})^2(0.75\text{m})^2}{2\times2\times(1.67\times10^{-27}\text{kg})}$

$$= \mathbf{30.26\ MeV.}$$

29.16 SYNCHROCYCLOTRON

Synchrocyclotron is a modified form of cyclotron. In a cyclotron the loss of resonance between the applied r.f. voltage and the moving ion is caused by the relativistic increase in mass of the ion. With growth of mass, the frequency of revolution of the ions in the cyclotron decreases and the ion gets out of phase with the h.f. voltage of fixed frequency applied across the dees and as a result cannot be accelerated. The decrease in ion frequency may be compensated and the ion may be kept in phase with the h.f. voltage in two ways. One of the ways is to keep the value of the magnetic field B constant and decreasing the frequency of h.f. generator in step with the decrease in the frequency of revolution of the ion. The frequency of the applied alternating electric field is gradually changed at such a rate that as the ion lags a little due to increase in mass, the electric field frequency also automatically decreases and the ion always enters the dee at the right moment when it experiences maximum acceleration. Accelerators based on this principle are known as **synchrocyclotrons.**

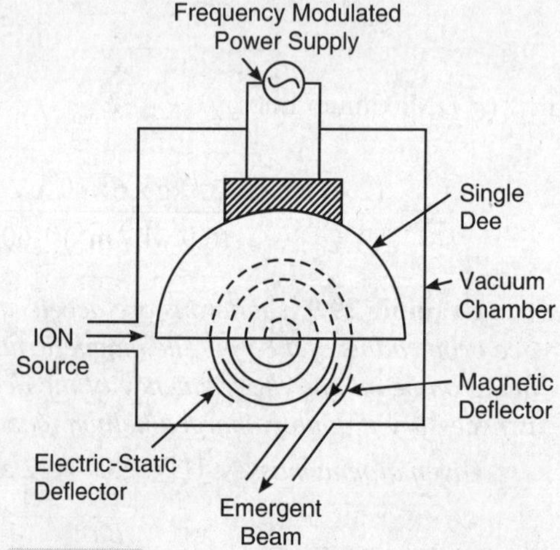

Fig. 29.17: A Schematic diagram of a synchrocyclotron. The spiral illustrates the typical path of a charged particle accelerated in the synchrocyclotron

A synchrocyclotron (Fig. 29.17) consists of only one dee enclosed in a vacuum chamber.

The entire unit is kept between the pole pieces of a huge electromagnet. Instead of the second dee, a metal plate (ejector) is held opposite the opening of the dee. The output of a h.f. generator is connected between the single dee and the earthed chamber. The frequency of the h.f. generator is modulated with a low frequency of 120 Hz. The high-energy ions are extracted from the chamber with the help of a high positive pulse applied to the ejector plate. A synchrocyclotron is also known as a *frequency modulated cyclotron*.

The following are the basic differences between a cyclotron and a synchrocyclotron.

(*i*) A cyclotron uses two dees, whereas a synchrocyclotron uses only one dee.

(*ii*) The frequency of the h.f. generator is fixed in a cyclotron operation, whereas the h.f. generator is modulated and hence is variable in a synchrocyclotron.

(*iii*) A continuous beam of high-energy ions is produced in a cyclotron. In case of synchrocyclotron ions come out in bursts of a few hundred per sec. Each burst lasts for about 100 μs.

29.17 BETATRON

Beta-particles are fast moving electrons. The accelerator that produces fast moving electrons is called a **betatron**. D.W.Kerst constructed the first betatron in 1941. With a betatron it is possible to obtain electrons of 300 MeV energy.

Principle of Betatron: The principle of betatron is very much similar to that of transformer. In the transformer, if an alternating electric current is passed through the primary coil an alternating field will appear in the core. This field induces an e.m.f. in the secondary coil.

In a betatron, the electrons are accelerated with the help of an electric field produced by a changing magnetic field. The electrons are held in an orbit of fixed radius by varying transverse magnetic field acting on the orbit and simultaneously these electrons are imparted energy by an induced e.m.f. resulting from an increase in the magnetic field.

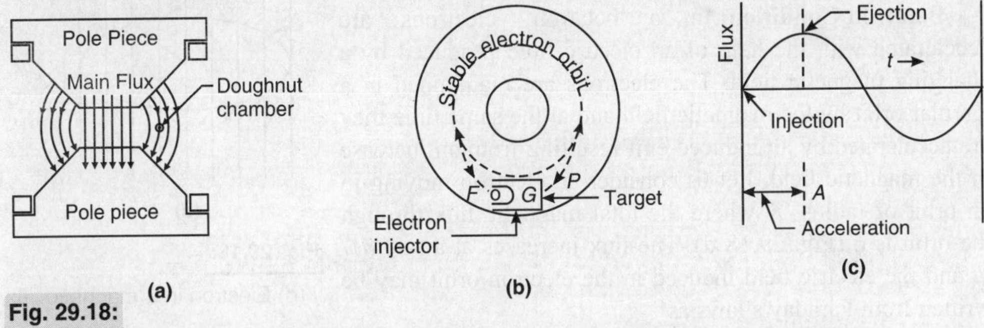

Fig. 29.18: (a), (b) Schematic of a betatron, (c) Ejection of electrons versus flux

Construction: A schematic of a betatron is shown in Fig. 29.18 (a). The betatron consists of a highly evacuated chamber made of glass or porcelain and has the shape of a doughnut. The chamber is mounted between the pole pieces of an electromagnet fed by an alternating current of 50 Hz. The poles of electromagnet are tapered (Fig. 29.18 a) so that the doughnut shaped chamber lies in the region of the magnetic field where flux density is half the main flux density. It maintains electrons in an orbit of constant radius in the annular space of doughnut chamber. To introduce electrons into the stable orbit, an electron gun is used. It consists of a filament F that gives out thermionic electrons, a focusing grid G and the positive plate P (Fig. 29.18 b).

The electrons are accelerated for a time 1/200 second at intervals of 1/50 seconds.

Working: The electron gun injects electrons into annular tube at the instant when the magnetic flux just started increasing. When current increases from zero to its maximum value, the magnetic flux increases in a given direction. This happens during the quarter cycle as shown in the Fig. 29.18 c. The electrons injected in a plane close to the stable orbit position will be accelerated by the magnetic field and execute a damped oscillatory motion in the beginning but will finally settle in the orbit. In 1/200 s, the electron will make several hundred thousand revolutions in the stable orbit gaining energy constantly. The electrons must be ejected from the betatron when the magnetic field reaches its maximum value; otherwise the electrons would slow down as the magnetic flux decreases and finally will reverse their direction. The electrons are deflected from their stable orbit by sending a pulse of current through an auxiliary coil. The high energy electron beam can be made to strike a target T within the tube, thus producing an intense x-ray beam (more correctly γ-rays) or the electron beam can be guided out through a window

Betatron is like a transformer: The action of the betatron depends upon the same principle as that of the transformer in which an ac current applied to a primary coil induces a similar current in the secondary windings. The primary current produces an oscillating magnetic field which in turn induces an oscillating emf in the secondary coil. The betatron is also like a transformer in which a cloud of electrons located inside a dough-nut shaped vacuum chamber takes the place of the secondary windings. The chamber is placed within the pole pieces of an electromagnet energized by an alternating pulsed current and the magnet produces a strong varying field in the central place or hole of the dough-nut. The electrons move in a circular orbit of constant radius within the vacuum chamber and gain energy by induction because of the change with time of the magnetic flux linking the orbit. Thus the electromagnet plays the role of the primary coil.

Betatron Condition: In a betatron, electrons are accelerated with the help of an electric field produced by a changing magnetic field. The electrons are maintained in a circular orbit by the magnetic field and at the same time they are accelerated by an induced emf resulting from an increase in the magnetic field. Let us consider an electron moving in an orbit of radius 'r' where the total magnetic flux through the orbit is ϕ (Fig. 29.18 d). The flux increases at a rate $d\phi/dt$ and the electric field induced in the electron orbit may be written from Faraday's laws as

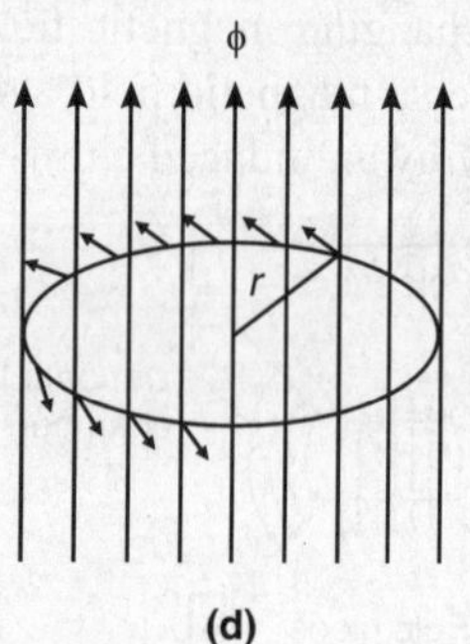

Fig. 29.18:

(d) Electron in a magnetic flux

$$V = -\frac{d\phi}{dt} \quad \text{...(29.32)}$$

The work done on the electron in one revolution = $V e$.

The force, F, on the electron acts along the tangent to the circular path at any point.

$$\text{Work done} = \text{Force} \times \text{distance} = F \cdot 2\pi r$$

$$\therefore \quad Ve = F \cdot 2\pi r$$

$$\therefore \quad F = \frac{Ve}{2\pi r}$$

or

$$F = -\frac{e}{2\pi r} \cdot \frac{d\phi}{dt} \quad \text{...(29.33)}$$

Under the influence of the induced electric field, the energy of each electron increases as it goes round the orbit. Consequently, its radius will tend to increase causing the electron to spiral outward. The condition for confining the electron to a circular orbit of a constant radius can be obtained as follows.

The momentum of the electron in the circular orbit under the action of magnetic field is given by

$$m\upsilon = Ber$$

$$\therefore \quad F = \frac{d}{dt}(m\upsilon) = e\frac{d}{dt}(Br)$$

When r is constant, the tangential force becomes

$$\frac{d}{dt}(m\upsilon) = er\frac{dB}{dt} \quad \text{...(29.34)}$$

To maintain a constant radius of the orbit, the net tangential force on the electron must be zero, that is, the values of F given by relations Eq. (29.33) and Eq. (29.34) must be numerically equal.

Eq. (29.33) and Eq. (29.34), we get

$$\frac{e}{2\pi r}\frac{d\phi}{dt} = er\frac{dB}{dt}$$

or

$$d\phi = 2\pi r^2 \times dB$$

On integration of the above equation, we obtain

$$\phi = 2\pi r^2 B \quad \text{...(29.35)}$$

The above condition is known as the **betatron condition.**

If a uniform magnetic field B acts over a ring of area πr^2, the magnetic flux

$$\phi = \pi r^2 B \quad \text{...(29.36)}$$

Eq. (29.35) and Eq. (29.36) imply that *the electron moves in an orbit of constant radius in an increasing magnetic field if the field in the orbit itself is half of the field over the whole area of the orbit.*

Number of revolutions in the fixed orbits: Electrons are accelerated in betatron for a time interval equal to a quarter of time period. i.e. $T/4$ s. Assuming the electron velocity in the orbit to be equal to the velocity of light 'c', the total distance traveled by the electron is

$$L = c\times\frac{T}{4} = \frac{c}{4\nu} = \frac{\pi c}{2\omega} \quad \text{...(29.37)}$$

$$\therefore \text{ Number of revolutions,} \quad N = \frac{L}{2\pi r} = \frac{c}{4\omega r} \quad \text{...(29.38)}$$

Energy acquired by the electrons: The momentum of the electron,

$$m\upsilon = \frac{E}{c}$$

Also,

$$m\upsilon = Ber$$

$$\therefore \quad \frac{E}{c} = Ber$$

The energy acquired by the electrons, $E = Berc$ ...(29.39)

Alternative Derivation: Let us suppose that magnetic flux in the betatron is given by the relation

$$\phi = \phi_0 \sin \omega t$$

As an increasing magnetic field in a given direction is only obtained during the quarter cycle in which the current in the electromagnet increases from zero to maximum value.

$\therefore$ The time of acceleration $= \dfrac{T}{4} = \dfrac{1}{4} \times \dfrac{2\pi}{\omega} = \dfrac{\pi}{2\omega}$

where T is time period of the changing magnetic flux and ω the corresponding angular frequency.

Energy gained by the electron per turn = $V e$

$$= e\frac{d\phi}{dt} = e\frac{d}{dt}(\phi_0 \sin \omega t) = e\phi_0 \frac{d}{dt}(\sin \omega t)$$

This energy is gained in a time $\dfrac{T}{4} = \dfrac{\pi}{2\omega}$.

$$\text{Average value of energy per turn} = \frac{e\phi_0}{\pi/2\omega}\int_0^{\pi/2}\frac{d}{dt}(\sin \omega t)dt = \frac{2e\omega\phi_0}{\pi}$$

Substituting the value of $\phi_0 = 2\pi r^2 B$ from Eq. (29.35), we get

$$\text{Average energy per turn} = \frac{2e\omega}{\pi} \times 2\pi r^2 B = 4e\omega r^2 B$$

Total (final) energy, E = Number of revolutions made × Average energy per revolution

$$= \frac{c}{4\omega r} \times 4e\omega r^2 B = cer\,B$$

Example 29.11: *In a certain betatron the maximum magnetic field at the electron orbit is 0.5 wb/m². The diameter of the stable orbit is 1.5 m. If the frequency of the alternating current through electromagnet coils is 59 Hz, what is the (i) final energy gained by the electrons and (ii) the number of revolutions taken by the electrons?*

Solution: Final energy $E = B\,eR\,c = \dfrac{0.5\,\text{wb}/\text{m}^2 \times 1.6\times10^{-19}\,\text{C}\times 0.75\,\text{m}\times 3\times10^8\,\text{m/s}}{1.6\times10^{-19}\,\text{J/eV}}$

$=$ **112.5 MeV.**

Number of revolutions $N = \dfrac{c}{4\omega r} = \dfrac{3\times10^8\,\text{m/s}}{4(2\pi\times 59\,\text{Hz})\times 0.75\,\text{m}} = \mathbf{2.7\times10^5}$.

Example 29.12: *A betatron working on an operating frequency of 60 Hz has a stable orbit of diameter 1.6 m. Find the energy gained per turn as also the final energy if the magnetic field at the orbit is 0.5 T.*

Solution: Average energy per turn = $4\,e\,\omega\,r^2\,B$

$$= \frac{4(1.602\times10^{-19}\,\text{C})\times 2\pi\times 60\times(0.8\,\text{m})^2\times 0.5\,\text{T}}{1.602\times10^{-19}\,\text{J/eV}} = \mathbf{482.6\ eV.}$$

Final energy $= E = B\,eR\,c = \dfrac{0.5\,\text{T}\times 1.602\times10^{-19}\,\text{C}\times 0.8\,\text{m}\times 3\times10^8\,\text{m/s}}{1.602\times10^{-19}\,\text{J/MeV}} = \mathbf{120\ MeV.}$

Example 29.13: *In a betatron, the magnetic flux at a stable orbit changes at a constant rate of 15 Wb/s. What would be the energy of an electron which undergoes 10^6 revolutions?*

Solution: The increase in electron energy per revolution $= e\dfrac{d\phi}{dt} = 15$ eV

Final energy of electron after 10^6 revolutions = 15 eV × 10^6 = **15 M eV**.

29.18 ELECTRON SYNCHROTRON

The electron synchrotron is based on the principle of the combined working of betatron and cyclotron. Here, first electrons are accelerated by the betatron to energy of about 2 MeV and gain a velocity of 0.98 c. Consequently, the electrons travel at almost constant speed, but increase in mass as energy is imparted to them.

For an electron traveling with an angular velocity ω in a circular orbit of radius r

$$m\omega^2 r = Be\,\omega\, r$$

or

$$\omega = \frac{Be}{m} \qquad ...(29.40)$$

where B is the magnetic flux density at the orbit.

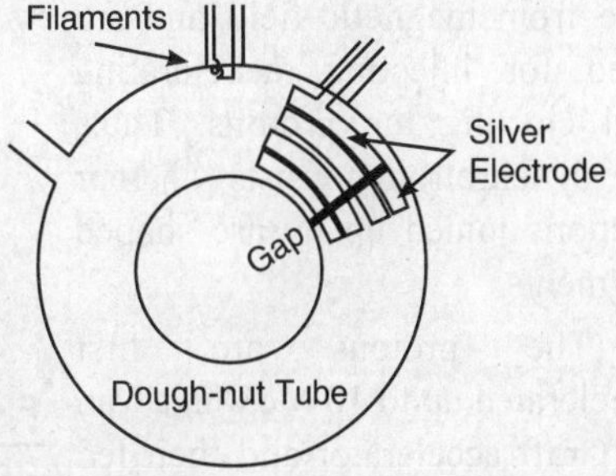

Fig. 29.19: Dough-nut tube -to keep the electrons in a stable orbit

If ω is to remain constant, B must increase in the same ratio as 'm'. In order to keep the electrons in a stable orbit, a small magnet is used inside the dough-nut tube. The magnet is less massive as the acceleration of the electrons beyond energy 2 MeV is achieved by radio-frequency electric field. This r.f. electric field is obtained between silver electrodes deposited over a short length along the arc of the tube (see Fig. 29.19). The silver coating has a short gap in it across which the output of an r.f. oscillator is connected. The frequency of the radio-frequency supply is adjusted to be equal to the time of one revolution of the electron in the circular orbit. Thus, the electrons are accelerated each time they cross the gap and gain additional energy. The radio-frequency supply is kept on while the magnetic flux is increasing and is automatically cut off when the electrons attain the required energy.

Maximum energy: In an electron synchrotron the maximum energy of electrons depends upon the radius r of the orbit and on maximum magnetic field strength B and is given by

$$E = Berc \qquad ...(29.41)$$

or

$$E = \frac{(3\times10^8\,\text{m/s})(1.602\times10^{-19}\,\text{C})}{1.602\times10^{-13}}\,rB\,MeV = 300\,rB\text{ MeV} \qquad ...(29.42)$$

Frequency of the radio-frequency electric field: As the synchrotron acceleration starts when the velocity of electrons are very close to c, the frequency is given by

$$\nu = \frac{c}{2\pi r} = \frac{4.7}{r}\text{ MHz} \qquad ...(29.43)$$

The radio-frequency accelerating electric field must have a frequency equal to the above frequency.

Electrons can be accelerated up to 1 BeV using the electron-synchrotron.

29.19 PROTON SYNCHROTRON

The principle of proton synchrotron is essentially the same as that of the electron synchrotron. A proton synchrotron consists of a dough-nut shaped vacuum chamber which has the form of a race-track (Fig. 29.20). It is made of stainless steel, porcelain or plastic supported in the gap of an annular magnet. The annular magnet is made of four quadrants separated by straight gaps. The straight sections are free from magnetic field and are used for injecting, accelerating and ejecting the protons. Thus, the synchrotron consists of four sections joined up by arc shaped segments.

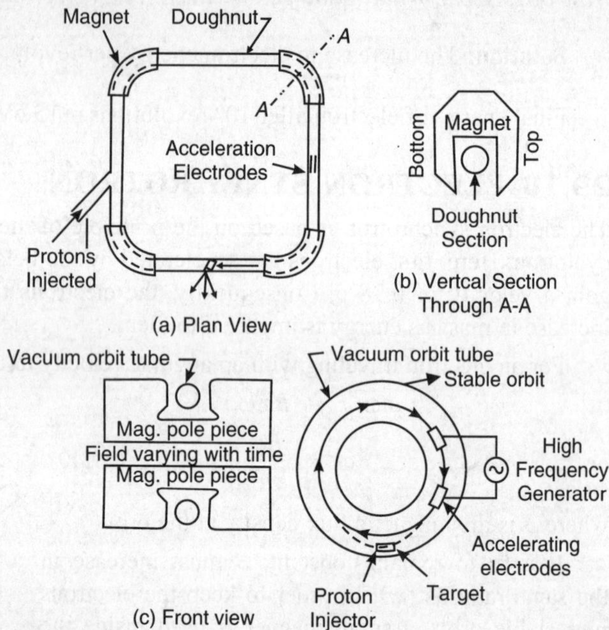

Fig. 29.20: Schematic of proton synchrotron

The protons are first accelerated up to 10 MeV in a Van de Graff accelerator and then fed into the synchrotron. Protons at low energy are injected in periodic pulses into the orbit. The protons are made to go in circular orbit by the magnetic field. They are accelerated once in each revolution when they pass between the electrodes connected to a radio-frequency oscillator. The magnet is excited periodically from 300 gauss to 15,000 gauss and the protons are accelerated during the time the magnetic field is increasing. Simultaneous control over the variation of magnetic field strength and the frequency of the radio-frequency oscillator is maintained in such a way that the protons travel in an orbit of constant radius and arrive at the electrodes when the applied radio-frequency voltage is in phase of acceleration.

When the protons have attained maximum energy, the frequency is distorted so that the orbit does not remain stable. By a suitable adjustment, the high energy protons are allowed to strike the target.

QUESTIONS

1. Describe the construction and working of Geiger-Muller counter. What are its limitations?
2. Explain the construction and working of Wilson's cloud chamber. What are the limitations of this apparatus?
3. Describe the principle and working of a bubble chamber.
4. Describe the principle, construction and working of a Wilson Cloud Chamber.
5. Describe the construction and working of a scintillation counter.
6. Write a short note on: (*i*) solid state detector (*ii*) Cerenkov detector

7. Give the construction and theory of Aston's mass spectrograph. Discuss the experimental results obtained.
8. What is mass spectrograph? Explain construction and working of Bainbridge mass spectrograph. How the relative proportions of isotopes are determined?
9. Explain construction and working of Aston's Mass Spectrograph. Discuss the condition of focusing of ion beam.
10. Explain the construction and theory of Dumpster's mass spectrograph.
11. What are isotopes? Explain the principle of their detection, in case of Bainbridge mass spectrograph.
12. Explain with a neat diagram the principle, construction and working of Bainbridge's Mass spectrograph.
13. Explain the function of velocity selector in Bainbridge mass spectrograph. Obtain an expression for linear separation in it.
14. Describe the construction and working of Bainbridge mass spectrograph and show that it has linear mass scale.
15. Explain the working of velocity selector.
16. What is a mass spectrograph? Explain the working of velocity selector arrangement in Bainbridge mass spectrograph. Show that the mass scale is linear.
17. Describe construction, principle and working of Bainbridge mass spectrograph using a well labeled diagram.
18. What are particle accelerators? Explain construction and working of Linac and cyclotron and the difference between them.
19. Describe the theory, construction and working of cyclotron. What are the limitations of a cyclotron?
20. In a cyclotron, show that the time spent by the particle in a dee is independent of its speed and radius of its circular path. Discuss whether a cyclotron can be used to accelerate electrons.
21. Explain the principle and working of cyclotron. Show that all the particles of the same mass and charge take the same time '*t*' to describe a semicircle and '*t*' is independent of the radius of the path and velocity of the particles.
22. With the help of neat diagram explain the principle, construction and working of cyclotron. Obtain the expression for the cyclotron frequency and maximum energy of the particle.
23. What is the primary function of electric and magnetic fields in a cyclotron?
24. Describe construction and working of a cyclotron with a neat diagram. Derive and explain the resonance condition. Find the energy acquired by a charged particle after n complete rounds.
25. How do electric and magnetic fields operate in a cyclotron? Why electrons cannot be accelerated to high energies in a cyclotron?
26. Describe the construction and working of a cyclotron. What are its limitations?
27. Give the construction and working of Bainbridge mass spectrograph.
28. What is a synchrocyclotron? How does it work?
29. Discuss betatron condition. How does it help in maintaining circular orbit?
30. Explain construction, principle and working of betatron and what is betatron condition?
31. Describe the construction of a betatron. How is energy gained in it? Obtain betatron condition. How is it achieved?
32. Give the construction and working of betatron with neat diagram. Obtain the betatron condition.
33. Explain the working of Betatron. Obtain the betatron condition.
34. Give the principle and working of a Betatron. What is Betatron condition?
35. Give the principle and working of a synchrotron. Explain how the magnetic field is modulated in an electron-synchrotron.
36. Give the principle and working of proton synchrotron.

PROBLEMS

1. Calculate the frequency of ac potential that must be applied to cyclotron dees in which protons are accelerated. Given magnetic flux density = 3wb/m^2. **[Ans:** f = 45.8 MHz]
2. The voltage across the dees of a cyclotron is 50 kV. How many revolutions do protons make to reach a kinetic energy of 20 MeV? **[Ans:** 200]
3. What strength of magnetic field is used in a cyclotron in which protons make 1.9×10^7 revolutions per second? **[Ans:** 1.2 T]
4. A cyclotron with dee radius 40 cm is to be used to accelerate protons using an oscillator of frequency 8 MHz. Calculate the magnetic field needed to maintain resonance and the final energy of the protons. **[Ans:** 0.52 T, 2.07 MeV]
5. Deuterons are accelerated in a cyclotron with dees of radius 0.3 m and magnetic field of flux density 0.7 Wb/m^2. Calculate (*i*) the velocity of deuterons (*ii*) the energy and (*iii*) the frequency of the field between the dees. **[Ans:** 10^7m/s, 1.08 MeV, 3.3 MHz]
6. A cyclotron accelerating protons employs a magnetic flux density 0.6 wb/m^2. How rapidly should the electric field between the metal dees be reversed? Also obtain the energy of the emerging proton. **[Ans:** 1.1×10^{-7}s, 1.08 MeV]
7. A proton is accelerated by cyclotron and acquires energy of 1.0 MeV. The voltage between dees is 1000 volts. How many rotations the proton shall have to make to achieve this energy? **[Ans:** 500]
8. Singly ionized Ne^{20} and Ne^{22} ions enter a Bainbridge mass spectrograph with a velocity of 10^5 m/s. If they are deflected by a magnetic filed of flux density 0.08 Wb/m^2, calculate the radii of their paths. **[Ans:** 0.26 m, 0.29 m]
9. Singly charged neon ions with mass numbers 20 and 22 and having kinetic energy 6.2×10^{-16}J come into a uniform magnetic field of induction 0.24 T in vacuum at right angles to the induction lines and traveled in a semicircle separate into two beams. What is the separation between the beams? **[Ans:** 33 mm]
10. The element tin is being analyzed in a Bainbridge mass spectrometer. Among the isotopes present are those of masses 116, 117, 118, 119 and 120 u. The electric and magnetic fields are E = 20kV/m and B = 0.25 T. What is the spacing between the marks produced on the photographic plate by the ions of tin-116 and tin-120? **[Ans:** 29 mm]
11. In a betatron, the maximum magnetic field 0.4 wb/m^2 is operating at 50 Hz. The diameter of the stable orbit is 1.524 m. Determine the final energy. **[Ans:** 91.1 MeV]
12. A betatron of energy 100 MeV has a stable radius of 0.84 m. Calculate (*i*) the value of magnetic field intensity at the orbit for this energy and (*ii*) the frequency of the applied electric field if average energy gain per turn is 420 eV. **[Ans:** 0.4 T, 60 Hz]

30 Cosmic Rays and Elementary Particles

Learning Objectives:

By the end of this chapter, you will be able to:

- Understand the concept of Primary and Secondary cosmic rays.
- Learn about elementary particles and classify its different types.
- Explain about Leptons and Hadrons.
- Describe key concepts of the Quark model.

30.1 INTRODUCTION

Cosmic rays are highly penetrating radiations which are continually entering the earth's atmosphere in all directions from outer space. Cosmic rays consist of high energy particles. Most of the particles have energy of the order of 15 GeV.

Cosmic rays were first detected by Elster and Geital and by C.T.R. Wilson. They found that the charge of a well insulated electroscope leaked away slowly even in the absence of any ionizing agent. At first, it was thought that the loss of charge of electroscope is due to ionizing radiations coming from radioactive substances present in the earth. In that case, the rate of discharge must decrease when the electroscope is taken to higher altitudes. Hess in 1911, recorded ionization of air at different altitudes by sending electroscope in balloons and found that the intensity of ionizing radiation increases as we go to higher altitudes. On the other hand, the intensity of ionization decreases as we go down into mines or the sea. This means that the source of the radiation is not on the earth but somewhere in space. Hess suggested that some kind of penetrating rays were entering the earth's atmosphere from outer space in all directions. These rays were named as cosmic rays by Millikan. Hess is credited with the discovery of cosmic rays and was awarded the Nobel Prize in physics for the year 1936.

The fundamental building blocks of matter are known as the elementary particles. Various types of elementary particles were discovered in the study of cosmic rays. With advent of sophisticated particle accelerators and detectors, more and more elementary particles were subsequently discovered. The study of elementary particles helps us understand about matter and the basic forces that hold matter together.

30.2 PRIMARY COSMIC RAYS

The cosmic rays which are just entering the earth's atmosphere are known as **primary cosmic rays**. The primary cosmic rays consist of stable high-energy particles flying in space in all possible directions. The intensity of cosmic radiation in the solar system is on the average 2 to 4 particles per square centimeter per second. It was established that the intensity of the cosmic rays near the poles of the earth is about 1.5 times greater than at its equator. Primary cosmic radiation, on account of its deflection by the earth's magnetic field, was identified as *consisting of positively charged particles*. The radiation mainly consists of protons (about 91%), α-particles (about 6.6%) and small percentages of nuclei of other elements (under 1%) and electrons (about 1.5%). The energies of primary cosmic rays range from 1 MeV to 10^{12} MeV.

30.3 SECONDARY COSMIC RAYS

When primary cosmic rays coming out of space interact with the nuclei of atoms in the upper layers of the atmosphere, **secondary cosmic rays** are produced. Below an altitude of 20 km, all cosmic radiation is secondary. On entering the atmosphere, the primary cosmic rays collide with air nuclei and produce mostly π-mesons and some hyperons. The π-mesons carry sufficient energy and decay into lighter particles, namely μ-mesons, electrons, positrons, neutrinos and photons. All such particles constitute the secondary cosmic rays. The μ-mesons do not take part in nuclear interactions and spend their energy only on ionization. Because of this they have great penetration power. At sea level the secondary cosmic rays contain nearly 70% μ-mesons, 29% electron-positron pairs and 1% heavy particles. μ-mesons penetrate the atmosphere and are observed even deep below the surface of the earth. Mesons constitute the ***hard component*** of cosmic radiation and the electrons, positrons and photons constitute the ***soft component***.

The π-mesons and μ-mesons in cosmic rays fly at speeds close to the speed of light and because of the relativistic time dilatation they are able to cover large distances before decaying.

30.4 ORIGIN OF COSMIC RAYS

It is, as yet, a mystery where the cosmic ray particles originate and where they are accelerated to such enormous energies. Different theories have been put forward to explain the possible sources of cosmic rays.

(*i*) There is enough matter in the interstellar spaces in the galaxy in the form of hydrogen clouds out of which stars evolve in the course of time due to condensation. In the process, lighter elements are converted into heavier elements. The temperature of the star rises due to more and more gravitational contraction. At the end, the star exhausts most of its hydrogen and develops so much compression that it suffers violent explosion. Such exploding stars are called Nova and Supernova. It is presumed that cosmic ray particles are ejected in the outbursts of nova and supernova and are accelerated in the non-homogeneous magnetic fields of interstellar space.

(*ii*) Another view is that the sun may be the source of at least some of the cosmic rays. At the time of solar activity violent eruptions occur and ionized gases shoot out from the sun. Thus, some of the protons and α-particles in the sun are thrown out into interplanetary space. The points in favour of this view are that the cosmic ray intensity increases during solar flares. But since the cosmic ray intensity remains uniform at all hours of the day and night, the sun cannot be the source for the majority of the primary cosmic rays. It may be the source of a small fraction of the low energy primary particles.

30.5 ALTITUDE EFFECT

The intensity of cosmic rays was determined by measuring the ionization produced by cosmic rays at different altitudes. The variation of cosmic ray intensity with altitude is shown in Fig. 30.1.

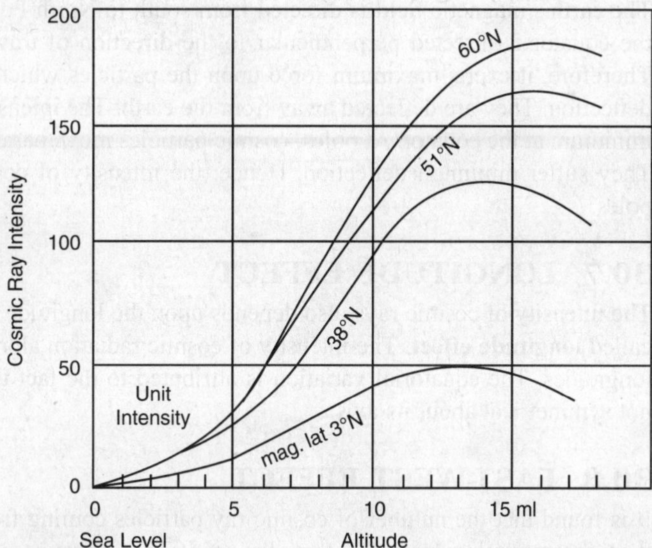

Fig. 30.1: Variation of cosmic ray intensity with altitude

The intensity of cosmic rays rises slowly up to a height of about 8 km, after which the rise becomes faster up to about 19 to 24 km. At heights above 24 km the intensity decreases gradually. The experiments were conducted at 3°, 38° 51° and 60° N and the results are similar.

The maximum of intensity is not at the top of the atmosphere but well below it. The reason is that at this height the primary cosmic rays produce quite a good number of secondary particles due to interaction with nuclei of atmospheric gases. Thus, both primary and secondary rays are present in abundance at that height. With decrease of altitude the absorption increases and therefore, intensity falls.

30.6 LATITUDE EFFECT

The variation of cosmic ray intensity with latitude is shown in Fig. 30.2. The curve shows that as one proceeds from magnetic north to magnetic south, at sea level, the cosmic ray intensity remains constant until magnetic latitude of about 42° is reached. In this region the intensity begins to drop appreciably, reaches a minimum at the equator and rises again to symmetrical intensities in the southern Hemisphere. Thus, the cosmic ray intensity is maximum at the geomagnetic poles ($\lambda = 90°$) and minimum at

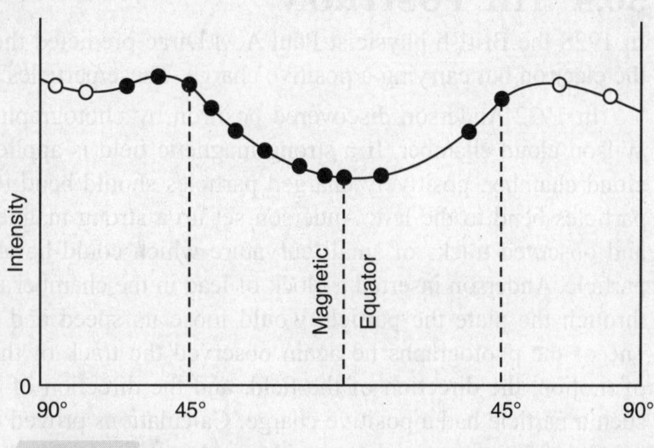

Fig. 30.2: Variation of cosmic ray intensity with latitude

the geomagnetic equator ($\lambda = 0°$). The intensity remains constant between 42° and 90°. This variation of cosmic ray intensity with geomagnetic latitude is called **latitude effect**.

The latitude effect may be explained on the basis of effect of earth's magnetic field on cosmic rays. The presence of such an effect indicates that the cosmic rays are charged particles. The earth's magnetic field is directed from south to North Pole. The earth's magnetic field at the equator is directed perpendicular to the direction of travel of charged cosmic particles. Therefore, it exerts maximum force upon the particles which consequently suffer maximum deflection. They are deflected away from the earth. The intensity of cosmic rays is therefore a minimum at the equator. At poles, cosmic particles move parallel to the earth's magnetic field. They suffer minimum deflection. Hence, the intensity of cosmic rays is a maximum at the poles.

30.7 LONGITUDE EFFECT

The intensity of cosmic rays also depends upon the longitude of the point of observation. It is called **longitude effect**. The intensity of cosmic radiation along the equator varies at different longitudes. The equatorial variation is attributed to the fact that the earth's magnetic field is not symmetrical about its axis.

30.8 EAST-WEST EFFECT

It is found that the number of cosmic ray particles coming from the west direction is greater than those coming from the east direction. This effect is called **east-west effect** and is a maximum at the equator. At the equator, the number of particles coming from west is 14% more than the particles coming from east. This phenomenon gave evidence to the fact that cosmic rays are composed predominantly of positively charged particles.

The charged particles approaching the earth's atmosphere are deflected by the earth's magnetic field in a direction perpendicular to the magnetic field and to the direction of their motion. The positively charged particles are deflected towards the east by earth's magnetic field and thus appear to come from the west of the vertical. Thus at any azimuth, more particles approach the earth from the west than from the east.

30.9 THE POSITRON

In 1928 the British physicist Paul A.M.Dirac predicted the existence of a particle similar to the electron but carrying a positive charge. These particles are named as *positrons*.

In 1932 Anderson discovered positron by photographing the tracks of cosmic rays in a Wilson cloud chamber. If a strong magnetic field is applied perpendicular to the face of the cloud chamber, positively charged particles should bend to the right and negatively charged particles bend to the left. Anderson set up a strong magnetic field in Wilson cloud chamber and observed tracks of small curvature which could be attributed to a high-energy charged particle. Anderson inserted a block of lead in the chamber to slow down the particles. Passing through the plate the particle would loose its speed and move in a more curved path. On one of the photographs he again observed the track of this particle. Knowing the direction of motion, the direction of the field, and the direction of bending, Anderson concluded that such a particle had a positive charge. Calculations proved the mass and the magnitude of the charge of the new particle to be identical to those of the electron.

30.10 PAIR PRODUCTION

An extension of the quantum theory of the electron proposed by Dirac led scientists to the prediction that if a high energy photon (γ-ray) were to come close enough to the nucleus of an atom, the electric field of the nucleus would annihilate the γ-ray and create in its place a pair of particles, an electron and a positron. According to the theory, these two particles should have the same mass and equal but opposite charges. Blackett, Anderson and others discovered such pairs in a cloud chamber.

The reaction for pair production is written as

$$\gamma \rightarrow e^- + e^+$$

If $h\nu$ is the energy of the γ-ray, then

$$h\nu = 2m_0c^2 + E^- + E^+$$

where m_0c^2 is the rest energy of electron (or positron), E^- is the kinetic energy of electron and E^+ is the kinetic energy of positron.

Pair production is materialization of radiant energy. From the above equation, it is clear that the *threshold energy* for electron-positron pair production is $2m_0c^2$.

The converse of materialization of energy is the annihilation of matter. When a positron combines with an electron, both the particles disappear and in their place two γ-rays are produced. Thus,

$$e^- + e^+ \rightarrow \gamma + \gamma$$

This process is called ***annihilation of matter***.

Positron is known as an ***antiparticle*** of electron.

30.11 COSMIC RAY SHOWERS

During the study of cosmic rays with the help of Wilson cloud chamber, the experimenter occasionally finds a picture of a group of tracks, instead of the usual one or two tracks. The group of tracks is termed as *comic ray showers*. An extensive study of showers has led to the conclusion that each shower is produced by a single, high-energy cosmic ray. Rossi investigated the phenomenon extensively and the results were explained by Homi J. Bhabha and Heitler on the basis of cascade theory.

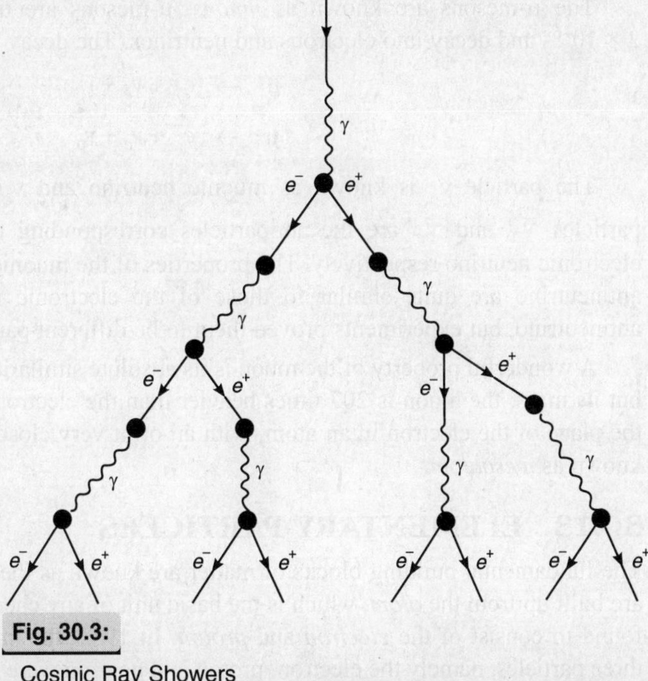

Fig. 30.3: Cosmic Ray Showers

According to the cascade theory, a high energy electron (or positron) in the cosmic rays, entering in the earth's atmosphere, loses its energy when it meets the atomic nuclei. The energy transforms into high energy photon. The photon in turn interacts with an atomic nucleus, gets wholly absorbed and produces an electron-positron pair. The electron and positron possess enough energy to produce more photons on their subsequent interaction with nuclei. The resulting photons cause further pair production. This results in the generation of a large number of photons, electrons and positrons having a common starting point (Fig. 30.3). The multiplication continues until the initial energy turn into a large number of pairs and the individual energies of the particles fall below the critical energy when photon emission and pair production halts.

30.12 THE MESONS

In 1938 Anderson and Nedermeyer discovered the presence, in cosmic rays, of charged particles having a mass greater than that of an electron but considerably smaller than that of proton. These particles are called mesons. They are produced high in the atmosphere by the collisions of cosmic rays with atomic nuclei. When a high energy proton collides with a nucleus it knocks out protons, neutrons and ***π-mesons*** or ***pions***. Some pions are positively charged (π^+), some are negatively charged (π^+) while others are neutral (π°). The π^+ and π^- are antiparticles to each other. They have a rest mass of 273 m_e. The rest mass of π° meson is 264 m_e. All charged π-mesons are unstable and have a half-life of 2×10^{-8} s and decay into a charged μ particle and a lightweight neutral particle called a *neutrino*.

The decay of pions may be written as interactions as follows.

$$\pi^+ \rightarrow \mu^+ + \nu_\mu$$

$$\pi^- \rightarrow \mu^- + \nu_\mu$$

The neutral pion has a half-life of 8.7×10^{-17}s and decays into γ-rays.

$$\pi^\circ \rightarrow \gamma + \gamma$$

The μ-mesons are known as *muons*. μ-mesons are unstable and have a half-life of 2×10^{-6}s and decay into electrons and neutrinos. The decay of the muons may be written as

$$\mu^+ \rightarrow e^+ + \nu_e + \bar{\nu}_\mu$$

$$\mu^- \rightarrow e^- + \nu_e + \nu_\mu$$

The particle ν_μ is known as muonic neutrino and ν_e is the electronic neutrino. The particles $\bar{\nu}_\mu$ and $\bar{\nu}_e$ are the antiparticles corresponding to the muonic neutrino and the electronic neutrino respectively. The properties of the muonic neutrino and the corresponding antineutrino are quite similar to those of the electronic neutrino and the corresponding antineutrino, but experiments proved them to be different particles.

A wonderful property of the muon is its absolute similarity with the electron in everything but its mass; the muon is 207 times heavier than the electron. It can also temporarily occupy the place of the electron in an atom, with an orbit very close to the nucleus. Such an atom is known as *mesoatom*.

30.13 ELEMENTARY PARTICLES

The fundamental building blocks of matter are known as the elementary particles. Molecules are built up from the *atom*, which is the basic unit of any chemical. Early in 1930, atoms were found to consist of the *electron* and *proton*. In 1932, the *neutron* was discovered and these three particles, namely the electron, proton and neutron were recognized to be the constituents of the atom. Hence, they were thought to be the elementary particles. Subsequently, the positron and neutrino were discovered. Various types of mesons were discovered in the study of cosmic rays. With the advent of sophisticated particle accelerators and detectors, more and more elementary particles were discovered.

30.14 CLASSIFICATION OF ELEMENTARY PARTICLES

Several systems classify the elementary particles on the basis of their various properties. Particles having their half-life of the order of 10^{-16} s are known as *stable particles* and those having half-life of the order of 10^{-22} s are known as *resonances*. One of the classifications of stable elementary particles is based on the basic nuclear force interactions.

30.15 BASIC FORCES IN NATURE

Four basic forces are recognized which are responsible for interaction of elementary particles. They are

(*i*) Weak nuclear interaction

(*ii*) Strong nuclear interaction

(*iii*) Electromagnetic interaction and

(*iv*) Gravitational interaction

Each of the above forces is carried by an elementary particle. The electromagnetic force is facilitated by the photon. The strong force is facilitated by the ***gluon***. The weak force is mediated by the W and Z particles while gravity force is supposed to be facilitated by the graviton.

Gravitational interaction is the weakest nuclear force. The particle that mediates the gravitational interaction is called the ***graviton***. It has not been detected so far but is assumed to be massless and possesses a spin of 2. It travels at the speed of light and interacts with all particles that have mass.

Electromagnetic interaction operates between charged particles and particles having electric and magnetic moments. It is responsible for the binding of electrons in the atoms, for formation of molecules, and for all chemical and biological processes. Almost all non-gravitational interactions are due to electromagnetic interaction. The particle that mediates the electromagnetic interaction is the ***photon***. Photon is mass-less and has a spin of 1.

Strong nuclear interaction operates between the nucleons in a nucleus. It is the strongest of all the forces. It keeps a number of protons together in a nucleus against the strong Coulomb force of repulsion. It is a short-range force. The nucleons and mesons are part of a general group of particles called ***quarks***. The particle that mediates the strong interaction between quarks is called a ***gluon***. Its mass is zero and has a spin 1. A free quark or a gluon is not observed since they remain hidden within particles confined by the strong force.

Weak nuclear interaction is also a short-range force operating up to a distance of 0.001 fm. It is responsible for the decay of some heavier particles.

The electromagnetic and weak interactions are unified in the electroweak theory. According to this theory, the particle responsible for the weak interaction is called the ***W particle***. There are three types of the W particle: W^+, W^- and Z°. All these particles have spin one.

30.16 CLASSIFICATION OF ELEMENTARY PARTICLES BASING ON THE BASIC FORCES

The stable elementary particles are divided into four groups basing on the above basic forces.

(*i*) **Gravitons:** They mediate the gravitational interaction and are yet to be discovered. They have spin 2 and therefore, they are ***bosons***.

(*ii*) **Photons:** They mediate the electromagnetic interaction and obey Bose-Einstein statistics. Therefore, they are also ***bosons***.

(*iii*) **Leptons:** Particles that interact via the weaker nuclear force but not the strong force are called ***leptons*** ("light ones"). The lepton family includes the electron, muon, and the neutrinos. They obey Fermi-Dirac statistics. Hence they are called ***fermions***.

(*iv*) **Hadrons:** Particles that interact via the strong nuclear force are called hadrons. These include the proton, neutron, and pion. These are also bosons. Hadrons are further subdivided into two groups, namely ***mesons*** and ***baryons***.

The classification may be depicted as follows in a chart.

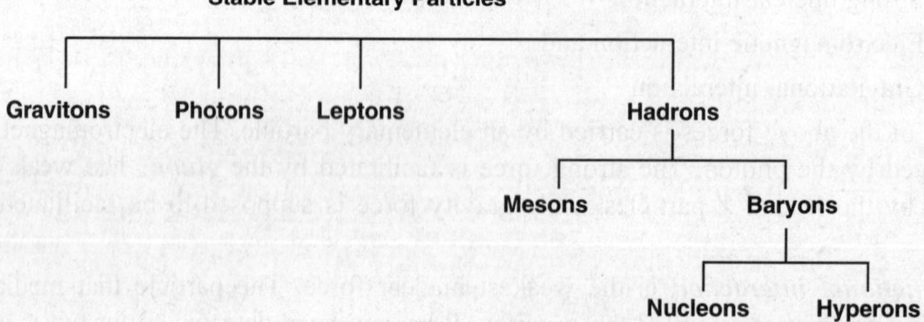

30.17 ANTIPARTICLES

An *antiparticle* is a particle having the same mass, spin and lifetime, but its charge having opposite sign to that of the particle. For example, positron e^+ is an antiparticle to electron e^-.

30.18 LEPTONS

There are only six leptons and six of their antiparticles. They are

(*i*) electron e^- and its antiparticle positron e^+

(*ii*) negative muon μ^- and its antiparticle positive muon μ^+

(*iii*) tau τ^- and its antiparticle τ^+.

(*iv*) electron-neutrino and its antiparticle

(*v*) muon neutrino and its antiparticle

(*vi*) tau neutrino and its antiparticle

The electron, muon and tau are negatively charged particles and each of them has an associated neutrino.

(*i*) **Electron and positron:** The most familiar lepton is the **electron**. It is a stable particle and the only lepton that exists naturally in atoms. It has spin $\frac{1}{2}$ and obeys Fermi-Dirac statistics. Therefore, it is a **fermion**. There is no evidence of any internal structure, at least down to the measurement limit of 10^{-17} m. Thus, at present the electron is considered to be a point particle.

Positron is the antiparticle of the electron. It is a positively charged particle having the charge, mass, and spin equal to that of electron and it is a fermion.

(*ii*) **Muons:** Muons were first observed in cosmic rays. Muon is electrically charged and 207 times more massive than an electron. Its spin is $\frac{1}{2}$ and it is a fermion. Muons are unstable and emit electrons. Since muons also appear not to have any internal structure, they are sometimes called *heavy electrons*. Muons are unstable and decay in about 2.22×10^{-6} s, according to the following schemes:

$$\mu^- \rightarrow e^- + \bar{\nu}_e + \nu_\mu$$
$$\mu^+ \rightarrow e^+ + \nu_e + \bar{\nu}_\mu$$

Thus, for each muon decay, the neutrino and antineutrino particles are of two different types: one is electron-neutrino and the other is muon-neutrino.

(*iii*) **Tauon:** A third charged lepton is known as a tau (τ^-) particle, or **tauon**. It is 3000 times heavier than the electron.

The electron, muon, and tauon are negatively charged. They have no apparent internal structure, and have positively charged antiparticles. The remaining leptons are neutrinos, which are electrically neutral.

(*iv*) **Neutrinos and antineutrinos:** Neutrinos are present in cosmic rays and are emitted by the Sun and by some radioactive decays. Neutrinos have very little or no mass and thus travel at the speed of light. They are not influenced either by the electromagnetic force or the strong force and so pass through matter as if it were not there. They are electrically neutral and stable particles with spin $\frac{1}{2}$.

There are three types of neutrinos, each associated with a different charged lepton $\begin{pmatrix} + + + \\ - - - \\ e, \mu, \tau \end{pmatrix}$. They are known as the electron neutrino (ν_e), the muon neutrino (ν_μ) and the tau neutrino (ν_τ). There is an antineutrino for each of these, for a total six different neutrinos. Thus, there are twelve different leptons in all including six leptons and their antiparticles. The neutrino and antineutrino pair annihilates to give 2 γ-ray photons. For example,

$$\nu_e + \bar{\nu}_e \rightarrow 2\gamma \quad \text{or} \quad \nu_\mu + \bar{\nu}_\mu \rightarrow 2\gamma$$

30.19 HADRONS

Hadrons are said to be strongly interacting particles. They interact via the strong force, the weak force, and gravity. The electrically charged members can also interact by the electromagnetic force. The hadrons are divided into two subgroups - *baryons* and *mesons*.

Baryons: Baryons have masses at least as large as the proton mass. The proton and neutron also belong to this group. They have half- integer spins $\left(\frac{1}{2} \text{ or } \frac{3}{2}\right)$. Except for the stable proton, baryons decay into products that eventually include a proton.

Mesons: Mesons, which include pions, have integer spin values (0 or 1) and eventually decay into leptons and photons. For example the neutral pion decays into two gamma rays. Their masses are greater than the muon mass and have fairly short lifetime. They strongly interact with matter.

Hyperons: Hyperons are unstable particles having masses greater than that of nucleon. They were found in cosmic rays. Three groups of hyperons, called lambda (Λ), sigma (Σ) and Xi (Ξ) are now known. The hyperons have a half-life of the order of 10^{-10} s and generally decay by several alternative modes to nucleons and pions.

30.20 RESONANCES

The term stable particle is applied to those particles which have half-lives far greater than the time required for light to travel a distance equal to the diameter of the elementary particle. The diameter is taken to be around 10^{-15} m and the time required to travel this distance is of the order of 10^{-23} s. Decays caused by weak forces have lifetimes of the order of 10^{-10} s. A number of particles in the meson and baryon families cannot be noticed directly. Their lifetimes are very short and they cannot be observed prior to decaying. These particles are identified as *resonances* or *resonance states*, as the manner of their creation is very much similar to the resonance occurring in an electrical circuit.

The first such particle was observed by Fermi when he bombarded protons with a beam of π^+ pions. The graph drawn between the number of interactions and the kinetic energy of pions showed a large peak around 200 MeV. Fermi concluded that the proton and pion combined momentarily to form a short-lived particle before separating apart again. Subsequently, a large number of resonance particles were discovered.

30.21 THE QUARK MODEL

The large numbers of hadrons suggests that they may be composites of other truly elementary particles. In 1963 Murray Gell-Mann and George Zweig put forth the *quark theory*. They proposed that hadrons are not elementary particles in the sense that they are not fundamental building blocks, but are composite particles composed of truly elementary (fundamental) particles. They named these particles quarks. Originally quark model consisted of three different quarks (with fractional charges) and their antiparticles (antiquarks). These quarks were named the up quark (u), the down quark (d), and the strange quark (s). By combining three quarks, the relatively heavy hadrons, the baryons could be built. Quark-antiquark pairs could account for the lighter hadrons, the mesons.

Quarks are the fundamental particles of the hadron family. Since several quarks have to combine to give the charge on the hadron, it was clear that they have fractions of an electron charge, e. Thus the theory proposed that u, d, and s quarks have charges of $+\frac{2}{3}e, -\frac{1}{3}e$ and $-\frac{1}{3}e$, respectively. The antiquarks, designated by overbars, such as $\bar{u}$, have opposite charges (for example, the charge on the $\bar{u}$ is $-\frac{2}{3}e$). Thus three quark combinations could produce any baryons. A proton, for instance, consists of two up quarks and a down quark, and a neutron consists of two down quarks and an up quark. Thus, the quark composition of the proton and neutron would be *uud* and *udd*, respectively (see Fig. 30.4).

A meson is made up of one quark and one antiquark. For example, the π^+ meson is the combination of u quark and u antiquark $\left(u\bar{d}\right)$.

The discovery of new elementary particles in the 1970s led to the addition of three more quark types: charm (c); top, or truth (t); and bottom or beauty (b). Today there is firm experimental evidence of the existence of all six quarks and their antiparticles.

According to the quark model, the truly elementary particles may be grouped as: leptons, quarks, photons and gravitons.

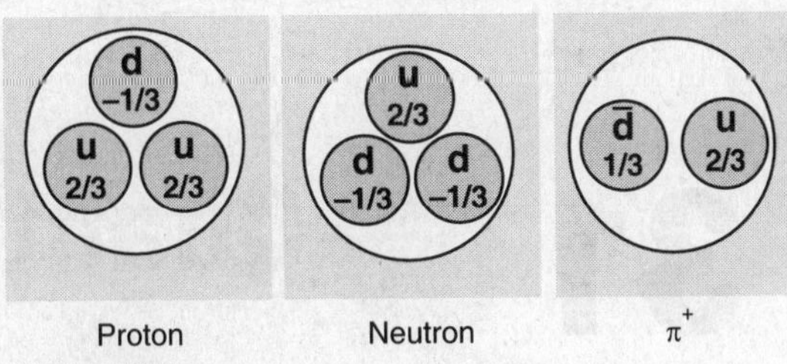

Fig. 30.4: Quark composition

30.22 OTHER MODELS

A model known as the Standard Model, describes the behavior of all known subatomic particles within a single theoretical framework. In this model each quark is supposed to carry a **colour charge** of the strong nuclear force, which is similar to the electric charge; antiquarks similarly carry anticolor. There are three colors assumed for each quark- red (R), green (G) and blue (B). Just as charged particles interact through photon exchange, color charged particles interact through **gluon** exchange. However, gluons are themselves color charged. Therefore, the strong force gets intensified as color charged particles are separated. Consequently, color charged particles feel increasing force and can never separate from one another. The Standard Model incorporates the quarks and leptons as well as their interactions through the strong, weak and electromagnetic forces, whereas gravity remains outside the Standard Model.

The ***string theory*** assumes that each type of fundamental particle matches to a different design of fundamental string. All strings are basically the same, although they may be open or closed. Different particles vary in the coordination of their strings. One particular forecast of string theory is the existence of huge counterparts of ordinary particles due to vibrational excitations of the fundamental string. One more significant prediction of string theory is the existence of a mass-less spin-2 particle behaving like the ***graviton***.

QUESTIONS

1. Why is a free quark not observed?
2. What are cosmic rays? How are they affected with latitude and altitude?
3. What are primary and secondary cosmic rays?
4. Given an account of the production and properties of π and μ mesons.
5. What are cosmic showers?
6. What are fundamental interactions in nature? Explain their relative range and strength.
7. Briefly explain the mediating particles in the different fundamental interactions.
8. Write a brief note on the elementary particles as identified today.
9. What are leptons? Name them.
10. Discuss the properties of leptons in brief.
11. Compare the properties of leptons and baryons.
12. What is a resonance particle?
13. Briefly explain the quark model.

31 Quantum Statistics

Learning Objectives:

By the end of this chapter, you will be able to:

- Understand concept of Statistical mechanics.
- Learn about Classical -Maxwell- Boltzmann statistics and its assumptions.
- Explain Quantum statistics concepts and its different types -fermi-dirac and bose-einstein statistics.

31.1 INTRODUCTION

Till seventeenth century the physical phenomenon were explained by the use of the mathematical analysis of the motion of bodies as a result of impressed forces (dynamical principle). But when system contains a large number of particles, dynamical principle is not useful due to mathematical complications or due to their insufficiency for interpretation. In general it is impossible to calculate the exact average value of the properties of systems from dynamics which is useful for making interpretations.

Such problems were solved by statistical methods. The statistical methods are applied only to physical systems containing large number of identical particles in some sense but distinguishable by means of some property. These methods are used to study the law governing the motion of large assemblies of electrons, atoms, molecules and etc. In statistics average values are found without going into the details of the characteristics of individual particles of the system. It should be noted that the statistical methods do not give information of actual behaviour of system, but that of greatest possibility. The larger is the number of particles in the physical system considered, the more nearly correct are statistical predictions.

Before advent of quantum theory Maxwell, Boltzmann, Gibbs etc applied statistical methods to classical physics. Classical, Maxwell-Boltzmann statistics successfully explained many observed physical phenomenon like temperature, pressure, energy etc, but could not explain phenomenon like blackbody radiation, specific heat at low temperatures etc. For explaining this entire phenomenon, a new approach was introduced by Bose, Einstein, Fermi and Dirac which made use of new quantum idea of discrete exchange of energy between the systems. The new statistics was subdivided into two categories; Bose-Einstein statistics and Fermi-Dirac statistics.

31.2 STATISTICAL MECHANICS

Statistical mechanics intends to explain the behaviour of macroscopic systems in terms of the dynamical laws operating among the microscopic constituents. The term 'statistical' is used due to the fact because a mechanical explanation can be given only if we also introduce probabilistic elements into the theory. Statistical mechanics deal with the connection between the largely behavior of the systems and the properties of the particles of the system.

We apply statistical mechanics to solve a system for many particles. We can easily solve the Schrodinger's equation for one particle/atom/molecule to know energy and its distribution. For many particles, for example in an ideal gas, assuming that the molecules are non-interacting, the levels and each particle contain certain energy. At any temperature, $T > 0$ the system possesses a total energy, E. It is difficult to know how total E is distributed among the particles and how the particles are distributed over the energy levels. Statistical mechanics offers a basis for linking the microscopic properties of individual atoms and molecules to the macroscopic bulk properties of materials.

Statistical mechanics can be applied both classical and quantum concepts.

- *Fundamental particles and composite particles are treated as distinguishable in Classical Mechanics.*
  - *One can tag and track all individual particles in a system. As a result varying the position of any two particles in the system leads to a totally different configuration of the entire system. There is no restriction on assigning more than one particle to any given state accessible to the system.*
  - *Classical statistics is known as Maxwell-Boltzmann Statistics.*
- *All particles of a particular type are indistinguishable from one another in Quantum Mechanics*
  - *In an assemble consisting of similar particles, interchanging any two particles does not give a new configuration of the system. The wave function of the system is invariant with respect to the interchange of the constituent particles.*
  - *In an assemble consisting of particles belonging to different nature for example electrons and protons, the wave function of the system is invariant independently for the assembly of the two particles.*
  - *Basing on symmetry of the system Quantum statistics is divided in to two classes - (i) Bose-Einstein statistics and (ii) Fermi –Dirac statistics.*

31.2.1 Important Concepts of Statistical Mechanics

The following concepts are useful in our understanding of statistical mechanics.

(*a*) **Micro and Macro states of a system:** Consider a specific example of three distinguishable particles *a*, *b* and *c* which are distributed in two exactly similar compartments in an open box.

Each distinct arrangement is known as *microstate* of the system. Each compartment distribution of a system of particles is known as a *macrostate*.

Macrostate	Possible Arrangements		Number of Microstates
	Compartment 1	Compartment 2	
0, 3	0	*a, b, c*	1
1, 2	*a* *b* *c*	*b, c* *a, c* *a, b*	3
2, 1	*a, b* *b, c* *c, a*	*C* *a* *b*	3
3, 0	*a, b, c*	0	1

We assume that the macrostate is the most stable and contains an enormous number of microstates. The stability of the macrostate depends on the stability of microstates. This provides the probabilistic approach in physics.

A ***microstate*** *is a state of the system where all the parameters of the particles are specified. Many microstates exist for each state of the system specified in macroscopic variables (E, V, N ...) and there are many parameters for each state.*

In classical mechanics to specify the parameters of a microstate, the position (x, y, z) and momentum (p_x, p_y, p_z) with 6 N degrees of freedom in a phase space are used.

According to quantum mechanics the energy levels and the state of particles in terms of quantum numbers are used to specify the parameters of a microstate.

A ***macrostate*** *is any state of a system that describes the actual observations of its macroscopic statistical properties.*

From quantum mechanical view, a macrostate is defined as a state of the system where the distribution of particles over the energy levels is specified. The macrostate consists of the different energy levels and the number of particles having specific energies.

In thermodynamically equilibrium state three macroscopic variables (P, V, T) or (P, V, N) or (E, V, N), where P is pressure, V is Volume, T is Temperature, N is Number of particles and E is Energy are to be specified. The equation of state for the system gives the relationship of the three variables to a fourth, for e.g. for an ideal gas $PV = NkT$.

(*b*) **Phase space:** Phase space is a concept of classical statistical mechanics and it unifies classical (Hamiltonian) and quantum mechanics. It is often convenient in statistics to imagine a six-dimensional space. Each *phase space* dimension corresponds to a particle degree of freedom. Phase space is a combination of position and momentum space. A point phase space is completely specified by six

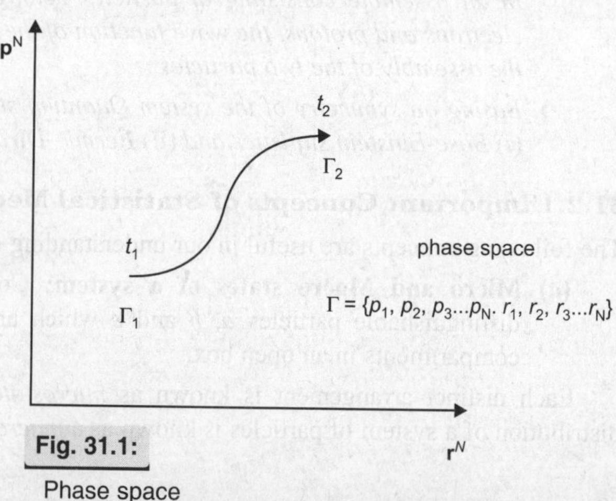

Fig. 31.1: Phase space

coordinates. A quantum mechanical state is not a point in phase space, but is a wave function, which is a complex-valued function.

It can then be visualized in a geometrical framework where each point in phase space specifies a particular position and momentum. Each point in phase space symbolizes a unique state of the particle. The state of a system of particles corresponds to a certain distribution of points in phase space. Fig. 31.1 shows the phase space.

Position space: Position space is the three dimensional space in which the *location* of a particle is completely specified by three position coordinates along three coordinates.

Small volume element in position space is given by $dV = dxdydz$

If the system is a dynamic, the particles move with various velocities and gains momentum and momentum space.

Momentum space: Momentum space is a three dimensional space in which momentum of the particle is completely specified by the momentum coordinates, p_x, p_y, p_z.

The counting of the number of states available to a particle gives the available volume in phase space. One can anticipate that for a continuous phase space, any finite volume would contain an infinite number of states. However from the uncertainty principle we cannot really say that a particle is at a mathematical point in phase space. So we consider an element of "volume" in phase space.

Small volume element in momentum space is given by $d\Gamma = dp_x dp_y dp_z$

Small volume element in phase space is given by

$$d\tau = dxdydzdpxdpydpz$$

$$d\tau = dVd\,\Gamma$$

Total phase space volume

$$\tau = \int d\tau = \int dV \int d\Gamma$$

$$\tau = V\Gamma$$

where V is the physical volume the particle of the system and Γ is the momentum space occupied by all the particles.

If the momentum space is divided into large number of spherical shells such that width of each shell is same, each shell is called energy or momentum compartment. A compartment in phase space is also known as molecular space or μ space.

(*c*) **Cells in phase space:** From the above equations

$$d\tau = dx\, dydzdp_x dp_y dp_z$$

The molecules specified by the above phase space has its position in the range

x and $x + dx$ y and $y + dy$ z and $z + dz$

Momentum in the range

p_x and $p_x + dp_x$ p_y and $p_y + dp_y$ p_z and $p_x + dp_z$

The small volume $d\tau$ in the phase space is ***called cell or phase point.***

A cell is a sub-compartment. Each compartment is divided into large number of cells in such a way that each cell is of same size. Hence all the cells have equal probability of a particle going into the cell.

In quantum mechanics, in view of Heisenberg uncertainty principle

$$d\tau = dxdydzdp_x dp_y dp_z \geq h^3$$

$$d\tau = dVd\Gamma = dxdydzdp_x dp_y dp_z = h_0^{\ 3}$$

where $h_0^{\ 3}$ represents the minimum size of the cell.

(*d*) **Ensemble:** In the year 1902 J.W Gibbs introduced the concept of ensemble. An ensemble is a collection of a large number of replicas of the microstates of the system under the same macroscopic condition or having the same macrostate. However, the microstates of the members of an ensemble can be arbitrarily different. Thus, for a given macroscopic condition, a system of an ensemble is represented by a point in the phase space. The ensemble of a macroscopic system of given macrostate then corresponds to a large number of points in the phase space. During time evolution of a macroscopic system in a fixed macrostate, the microstate is supposed to pass through all these phase points.

***Ensembles** can be defined as the collection of a number macroscopically identical but essentially independent microscopic systems.*

Statistical ensembles are also classified into three different types. In an isolated system, neither energy nor matter is exchanged and the corresponding ensemble is known as *microcanonical* ensemble. A closed system exchanging only energy (not matter) with its surroundings is described by *canonical* ensemble. Both energy and matter are exchanged between the system and the surroundings in an open system and the corresponding ensemble is called a *grand canonical* ensemble. Table 31.1 shows the comparisons of ensembles.

Table 31.1: *Comparisons of ensembles*

Micro-canonical	Canonical	Grand canonical
Neither energy nor matter is exchanged between systems	Energy is exchanged but not matter	Both energy nor matter is exchanged between systems
No contact with environment	Contact with heat bath	Contact with heat and particles
N, V and T remain constant	N and V constant but E varies	V constant and N varies
ρ = constant if $E < E + dE$	$\rho(E) = Ae^{E/kT}$	$\rho(E) = e^{(\Omega + n\mu)/kT}$
$\rho = 0$ outside the range E and $E + dE$		

31.3 CLASSICAL -MAXWELL-BOLTZMANN STATISTICS

As discussed earlier in classical mechanics all the particles in a system are considered to be distinguishable. Classical statistics is called Maxwell - Boltzmann Statistics. It is mostly used in kinetic theory of gases to find various aspects like distribution of molecular velocities, probable speed of molecules of gas and internal energy of a gas.

Assumptions of Maxwell - Boltzmann Statistics:

(*i*) All particles which obey M-B statistics are known Maxons and are identical and distinguishable and there are no symmetry restrictions.

(*ii*) The chosen volume has large number of cells as the volume of each phase space is extremely small.

(*iii*) In a phase space cell, there is no limit on the number of particles. Each cell can have one or more particles.

(*iv*) The total number of particles and their total energy remains constant in the system, i.e. the system is isolated.

(*v*) The state of each particle is specified by instantaneous position and momentum coordinates.

(*vi*) Energy is continuous.

The probability $F(E)$ that a boson occupy a state of energy E, Maxwell – Boltzmann distribution function is given by

$$F(E) = A \exp^{-\frac{E}{kT}} \qquad ...(31.1)$$

Number of molecules with energies between E and $E + dE$

$$n(E)\, dE = \frac{2\pi N}{(\pi kT)^{3/2}} \sqrt{E} \exp^{-\frac{E}{kT}} dE \qquad ...(31.2)$$

Fig. 31.2 shows Maxwell-Boltzmann energy distribution of molecules of an ideal gas.

Limitations of Classical Statistics: It fails to explain the

(*i*) observed energy distribution of electrons in metals.

(*ii*) observed energy distribution of photons inside an enclosure.

This failure of classical statistics forced the issue in favour of the new quantum idea of discrete exchange of energy between the systems. Thus, we have quantum statistical mechanics.

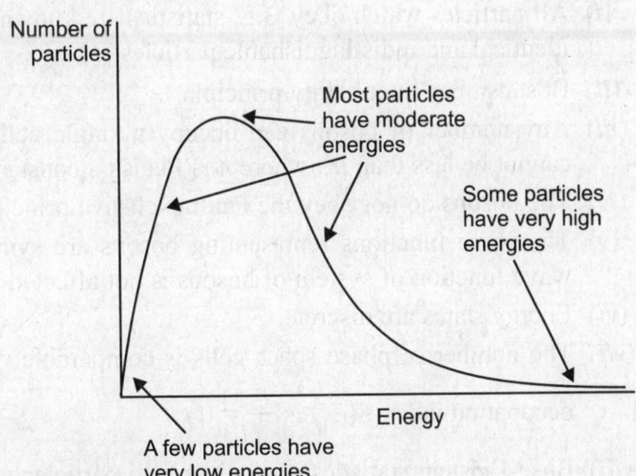

Fig. 31.2: Maxwell-Boltzmann energy distribution of molecules of an ideal gas

31.4 QUANTUM STATISTICS

Quantum statistical mechanics is statistical mechanics applied to quantum mechanical systems. The quantum statistics was first formulated in 1924 by Bose in the deduction of Planck's radiation law by purely statistical reasoning on the basis of quantum concepts. In the same year Einstein utilized practically same principle as a substitute for classical Boltzmann statistics. Thus new statistics known as Bose-Einstein statistics came into existence. In quantum mechanics a probability distribution over possible quantum states is described. According to quantum statistics the particles of the system are indistinguishable.

In 1926 Fermi and Dirac quite independently modified Bose-Einstein statistics in certain cases on the basis of Pauli exclusive principle. Addition of Pauli principle leads to the recognition of second kind of quantum statistics known as Fermi-Dirac statistics. So, we have two Quantum statistics categories; Fermi-Dirac statistics and Bose-Einstein statistics.

31.4.1 Bose-Einstein statistics

Statistics displayed by particles that show quantum mechanical behavior, but do not obey the Pauli's exclusion principle is known as Bose-Einstein statistics. Particles that have integer spin such as photons, several atoms, and even pairs of electrons under some conditions, are examples of this category. The statistical behavior of these particles was described by Satyendra Nath Bose, and Albert Einstein, and goes by their names as Bose- Einstein statistics. Particles obeying these statistics are called Bosons. **Bosons** are those particles which are *identical, indistinguishable* and have *integral spin.*

Exemption from Pauli's exclusion principle implies that *any number of particles can occupy the same state at the same time.* In the derivation that follows, we are examining particles that are identical, indistinguishable, and exempt from the Pauli's exclusion principle.

Bose used Planck's hypothesis, according to which radiation in a temperature encloser are composed of light quanta known as photons each of energy, $E = h\nu$. These photons in the encloser are indistinguishable.

Basic Postulates:

(*i*) All particles which obey B.E. statistics are known as Bosons. Bosons of the system are identical and indistinguishable particles.

(*ii*) Bosons obey uncertainty principle.

(*iii*) Any number of bosons can occupy a single cell in phase space.The size of the cell cannot be less than h^3, where h is Planck's constant (6.63×10^{-32} J S).

(*iv*) The bosons do not obey the Pauli exclusive principle.

(*v*) The wave functions representing bosons are symmetric and do overlap slightly. The wave function of system of bosons is not affected by the exchange of any pair of them.

(*vi*) Energy states are discrete.

(*vii*) The number of phase space cells is comparable with the number of particles, i.e., the occupation index $f(E_1)$ is $\frac{n_i}{g_i} = 1$.

(*viii*) Bose-Einstein statistics is applicable to particles with integral spin angular momentum in units of $\frac{h}{2\pi}$.

Let us consider a system containing of n independent and indistinguishable particles. The particles have definite energies (or momenta) and occupy the definite positions. Therefore, they can be represented by phase points in the phase space. In order to determine the energy (or momenta) distribution of these particles in the most probable state known as equilibrium state, we divide the available volume in the phase space into large number (say k) of compartments (quantum group or energy levels), each compartment representing a small interval of energy (or momentum) as shown in Fig. 31.2(a). Further we divide each compartment into

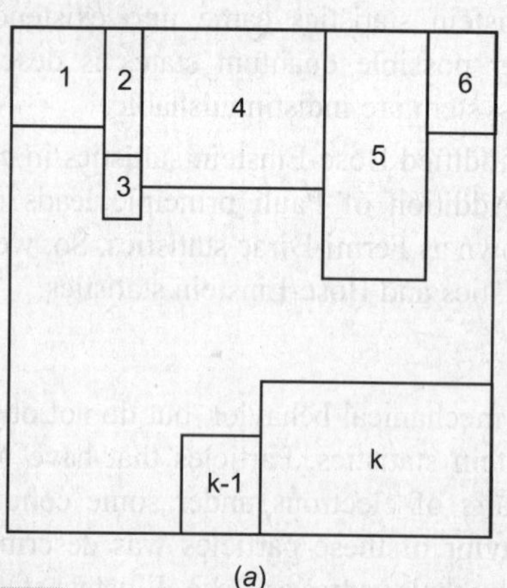

(*a*)

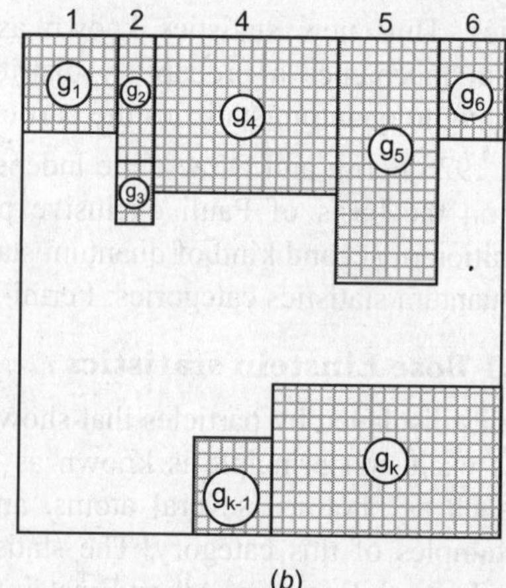

(*b*)

Fig. 31.3: Compartments in Phase space

elementary cells, each of size h^3 where h is Planck's constant Fig. 31.2 (b). We suppose that the size of the compartment is very large as compared to the size of the cell so that each compartment contains a very large number of elementary cells.

Let there be $n_1, n_2, n_3,..n_i,...n_k$ particles having mean energy values $E_1, E_2, E_3,..E_i... E_k$ respectively in compartments numbered as 1, 2....i...k containing $g_1, g_2,....g_i,....g_k$ cells respectively in them.

The total number of particles in the system is $N = n_1 + n_2 + ... + n_i + ...n_k$

Now consider i^{th} compartment. It has n_i distinguishable particles among its g_i cells. Suppose, that n_i particles are arranged in a row and distributed among g_i quantum states with $(g_i - 1)$ partitions in between. The total number of possible arrangements of particles and partitions is equal to the total number of permutations of $(n_i + g_i - 1)$ objects in a row.

Therefore, the total possible ways of arranging ni particles with $g_i - 1$ partitions

$$= (n_i + g_i - 1)!$$

As the particles are indistinguishable, the rearrangement of particles among themselves will not give rise to any new distribution. The number of such meaningless permutations is $n_i!$. Hence above result would be divided by $n_i!$. Moreover, this also includes permutations of $(g_i - 1)$ partitions among themselves. These permutations also do not produce different states and hence are meaningless. The above result is further divided by $(g_i - 1)!$.

Hence, the actual number of ways in which n_i particles are to be distributed in g_i cells in the ith compartment is given by

$$W_i = \frac{(n_i + g_i - 1)!}{n_i!\,(g_i - 1)!} \qquad ...(31.3)$$

Similar expressions will be obtained for other compartments. Therefore, the total number of different arrangements for all n particles of the system gives thermodynamic probability.

$$W_{(n_1, n_2, ...n_i....n_k)} = \frac{(n_1 + g_1 - 1)!}{n_1!(g_i - 1)!} \times \frac{(n_2 + g_2 - 1)!}{n_2!(g_2 - 1)!} \times ... \times \frac{(n_1 + g_1 - 1)!}{n_1!(g_1 - 1)!} \times ... \times \frac{(n_k + g_k - 1)!}{n_k!(g_k - 1)!}$$

$$= \prod_{i=1}^{k} \frac{(n_i + g_i - 1)!}{n_i!(g_1 - 1)!} \qquad ...(31.4)$$

where Π denotes multiplication of terms stated above for various values of i from $i = 1$ to k.

Most Probable Microstate: The most probable microstate corresponds to the state of maximum thermodynamic probability. In Eq. (31.4), n_i and g_i both are very large numbers. Hence we may neglect 1 in the above expression

$$W_{(n_1, n_2, ...n_i....n_k)} = \prod_{i=1}^{k} \frac{(n_i + g_i)!}{n_i!\,g_i!} \qquad ...(31.5)$$

Taking natural logarithm on both sides, we have

$$\ln W = \sum_{i=1}^{k} [\ln (n_i + g_i)! - \ln n_i! - \ln g_i!] \qquad ...(31.6)$$

As n_i and g_i are very large numbers, we can use Sterling approximation.

$$\ln n! = (n \ln n) - n$$

Applying Sterling approximations, we get

$$\ln W = \sum_{i=1}^{k}[(n_i - g_i)\ln(n_i + g_i) - (n_i + g_i) - n_i \ln n_i + n_i - g_i \ln g_i + g_i]$$

$$= \sum_{i=1}^{k}[(n_i - g_i)\ln(n_i + g_i) - n_i \ln n_i - g_i \ln g_i] \qquad ...(31.7)$$

Here g_i is not subject to variation whereas n_i varies continuously.

To get the state of maximum thermodynamic probability, we differentiate Eq. (31.7) and equate it to zero; i.e.

$$\delta(\ln W) = 0$$

$$\delta(\ln W) = \sum_{i=1}^{k}\left[\delta n_i \ln(n_i + g_i) + (n_i + g_i)\frac{1}{(n_i + g_i)}\delta n_i - \delta n_i \ln n_i - n_i \frac{1}{n_i}\delta n_i\right]$$

$$= \sum_{i=1}^{k}\delta n_i \ \text{In}(n_i + g_i)_i - \delta n_i \ \text{In}\ n_i \qquad ...(31.8)$$

Since g_i is number only, $\delta g_i = 0$

or $$\delta(\ln W) = \sum_{i=1}^{k}\left[\ln\frac{(n_i + g_i)}{n_i}\right]\delta n_i = -\sum_{i=1}^{k}\left[\ln\frac{n_i}{(n_i + g_i)}\right]\delta n_i = 0$$

or $$\sum_{i=1}^{k}\left[\ln\frac{n_i}{(n_i + g_i)}\right]\delta n_i = 0 \qquad ...(31.9)$$

In addition, our system must satisfy two subsidiary or auxiliary conditions:

(*i*) Conservation of total number of particles, i.e., $n = a$ constant.

$$\therefore \qquad n = \sum_i n_i = \text{constant}$$

i.e., $$\delta n = \sum_i \delta n_i = 0 \qquad ...(31.10)$$

(*ii*) Conservation of total energy of the system, i.e., $E = a$ constant.

$$\therefore \qquad E = \sum_i n_i E_i = \text{constant}$$

i.e., $$\delta E = \sum_i E_i \delta n_i = 0 \qquad ...(31.11)$$

Now we shall apply the Langrangian method of undetermined multipliers. For this let us multiply Eq. (31.10) by α and Eq. (31.11) by β, and add the resulting expression to Eq. (31.9) so that we get

$$= \sum_{i=1}^{k}\left[\ln\left(\frac{n_i}{(n_i + g_i)}\right)\alpha + \beta E_i\right]\delta n_i = 0 \qquad ...(31.12)$$

The variations δn_i are independent of each other. Hence we get

$$\ln\left(\frac{n_i}{(n_i + g_i)}\right) + \alpha + \beta E_i = 0$$

or $$\ln\left(\frac{n_i}{(n_i + g_i)}\right) = -\alpha - \beta E_i$$

or $$\frac{n_i}{(n_i + g_i)} = e^{-\alpha - \beta E_i} = e^{-(\alpha + \beta E_i)} \qquad ...(31.13)$$

or $$\frac{n_i + g_i}{n_i} = e^{\alpha + \beta E_i}$$

or $$1 + \frac{g_i}{n_i} = e^{\alpha + \beta E_i} - 1$$

or $$\frac{g_i}{n_i} = e^{\alpha + \beta E_i} - 1$$

or $$n_i = \frac{g_i}{e^{\alpha + \beta E_i} - 1} \qquad ...(31.14)$$

Eq. (31.14) represents the most probable distribution of the particles among various energy levels for a system obeying Bose-Einstein statistics and is therefore, known as Bose-Einstein's Distribution Law for an assembly of bosons.

Substituting $\beta = \frac{1}{kT}$, the Eq. (31.14) becomes

$$n_i = \frac{g_i}{e^{\alpha}.e^{\beta E_i} - 1} = \frac{g_i}{e^{\alpha}.e^{E_i/kT} - 1} \qquad ...(31.15)$$

B-E Energy distribution function.

The energy distribution function $f(E_i)$ is the average number of particles per quantum state in the energy levels E_i. It is given by

$$f(E_i) = \frac{n_i}{g_i}$$

Substituting the value of (n_i / g_i) from Eq. (31.15), we get

$$f(E_i) = \frac{1}{e^{\alpha}.e^{E_i/kT} - 1} \qquad ...(31.16)$$

For continuous distribution of energy, the distribution function is written as

$$f(E) = \frac{1}{e^{\alpha}.e^{E/kT} - 1} \qquad ...(31.17)$$

B-E Energy distribution law for continuous variation of energy

When the energy levels of the system are very closely packed, they form a quasi-continuous spectrum. In such a case, if $g(E)\,dE$ is the number of energy states between the energy range E and $(E + dE)$, then

$$g_i = g(E)\,dE$$

If $n(E)\,dE$ is the number of particles whose energy lies between E and $(E + dE)$, then $n_i = n(E)\,dE$

Substituting in Eq. (31.15), we get

$$n(E)\,dE = \frac{g(E)dE}{e^{\alpha}.e^{E_i/kT} - 1} \qquad ...(31.18)$$

The quantity $g(E)$ denotes the density of the states.

For the system containing of free particles with no spin $g(E)\, dE$ is given by

$$g(E)\, dE = 2\pi V \left(\frac{2m}{h^2}\right)^{3/2} E^{\frac{1}{2}} dE \qquad ...(31.19)$$

Substituting this value in Eq. (31.16), we get

$$n(E)\, dE = 2\pi V \left(\frac{2m}{h^2}\right)^{3/2} \frac{E^{\frac{1}{2}} dE}{e^{\alpha}.e^{E_i/kT} - 1} \qquad ...(31.20)$$

This is ***Bose-Einstein energy distribution law*** for continuous distribution of energy among the particles with no spin.

The probability $F(E)$ that a boson occupy a state of energy E, Bose-Einstein distribution function is given by

$$F(E) = \frac{1}{\exp[\alpha + \frac{E}{kT}] - 1}$$

where α is a constant is depends on the property of the system and temperature T.

If n is the number of bosons with total energy E in equilibrium at temperature T, the most probable distribution of bosons among the various energy levels

$$n_i = \frac{g_i}{\exp[\alpha + \frac{E_i}{kT}] - 1} \qquad ...(31.21)$$

where g_i is the number of quantum states available for the same energy E_i, called as degeneracy.

α is a constant

n_1, n_2, ...n_k are particles at various energy states E_1, E_2, E_3,....

The Bose-Einstein distribution predicts that at very low temperatures, nearly absolute zero, all particles in the system will condense into a single state, called the Bose-Einstein condensate. This was a prediction made around the year 1924. At that time, the experimental facilities available were not capable of attaining the low temperatures required to test this prediction. To attain temperatures of the order of 4 K using liquid He was a significant experimental challenge. It took almost 70 years from the time of the prediction, for experimental facilities to evolve to the stage where it was possible to demonstrate the formation of this condensate.

In 1995 Eric Cornell and Carl Weiman, demonstrated the formation of the condensate using a small collection of Rb atoms, which they cooled to a few hundred nano Kelvin. For demonstrating the formation of the Bose-Einstein condensate, they were awarded the Nobel Prize in Physics for the year 2001 along with Wolfgang Ketterle.

31.4.2 Fermi –Dirac statistics

Fermi-Dirac statistics is followed by Fermions. There is a restriction on the number of particles which can occupy a particular energy level. Examples of Fermions are electrons in conductors, protons, neutrons and He^3.

Assumptions of Fermi –Dirac statistics:

(*i*) All particles which obey F.D. statistics are known as Fermions. Fermions of the system are identical and indistinguishable.

(*ii*) The bosons have half integral spin.

(*iii*) Fermions obey uncertainty principle.

(*iv*) Fermions do obey the Pauli exclusive principle.

(*v*) The wave functions representing bosons are anti-symmetric.

(*vi*) Week interaction exists between the particles.

(*vii*) Energy states are discrete.

Let us consider a system consisting of n independent and indistinguishable particles. The particles have definite energies (or momenta) and occupy definite positions. Therefore, they can be represented by phase points in the phase space. In order to determine the energy (or momenta) distribution of these particles in the most probable state known as equilibrium state; we divide the available volume in the phase space into large numbers (say k) of compartments (quantum groups or energy levels), each compartment representing a small interval of energy (or momentum). Further each compartment is divided into elementary cells; each size h^3, where h is Planck's constant.

Further, we suppose that the size of the compartment is very large as compared to the size of the cell so that each compartment contains a very large number of elementary cells. Let the compartments be marked 1, 2,……, i,……,k and their mean energy values be represented by E_1, E_2,……,E_i,….E_k containing g_1, g_2,……, g_i,…..g_k cells respectively in them. The total number of particles in the system is $n = n_1 + n_2 + \ldots\ldots + n_i + \ldots\ldots + n_k$.

Our aim is find the Fermi-Dirac distribution law i.e., the distribution of n_i Fermions out of the total n Fermions in g_i cells of the ith compartment.

For this, the basic postulates are:

Basic Postulates

(*i*) The particles of indistinguishable so that there is no distinction between the various ways in which n_i particle are chosen.

(*ii*) The particles obey Pauli's exclusion principle according to which there can be either no particle or only one particle in a given cell. Therefore, the number of cells must be much greater than the number of particles, i.e., $g_i >> n_i$.

Out of n_1 number of particles in the compartment with mean value E_i, the first particle can be placed in any one of the available g_i state, i.e., this particle can be assigned to any of the g_i sets of quantum numbers. Thus, the first particle can be distributed in g_i different ways, in accordance with Pauli's exclusion principle and the remaining $(g_i - 1)$ will remain vacant. The second particle can be arranged in $(g_i - 1)$ different ways and the process continues. Thus,

Total number of different ways of arranging g_i states with energy level E_i is given by

$$= g_i(g_i - 1)(g_i - 2)(g_i - 3)\ldots\ldots[g_i - (n_i - 1)]$$

$$= \frac{g_i!}{(g_i - n_i)!} \qquad \ldots(31.22)$$

Further, since the particles are indistinguishable, it would not be possible to detect any difference when n_i particles are reshuffled into different ways occupied by them in the energy level E_i. Thus, out of these n_i permutations of the indistinguishable particles among themselves will be meaningless. Therefore,

Total number of different and distinguishable ways is

$$= \frac{g_i!}{n_i!(g_i - n_i)!} \qquad \ldots(31.23)$$

Therefore, the thermodynamic probability for the microstate $(n_1, n_2, \ldots\ldots n_i, \ldots\ldots n_k)$ of the system is given by

$$W_{(n_1, n_2, \ldots\ldots n_i \ldots n_k)} = \frac{g_1!}{n_1!(g_i - n_1)!} \times \frac{g_2!}{n_2!(g_2 - n_2)!} \times \ldots \times \frac{g_i!}{n_1!(g_1 - n_i)!} \times \ldots \times \frac{g_k!}{n_k!(g_i - n_k)!}$$

$$= \prod_{i=1}^{k} \frac{g_i!}{n_i!(g_i - n_i)!} \qquad \ldots(31.24)$$

where Π denotes the multiplication of terms for various values of i from 1 to k.

Most Probable Microstate

Taking natural logarithm on both sides of Eq. (31.24),

$$\ln W = \sum_{i=1}^{k} [\ln g_i! - \ln n_i! - \ln (g_i - n_i)!]$$

As n_i and g_i are very large numbers, we can use Sterling's approximation

$$\ln n_i! = (n \ln n) - n$$

Applying Sterling's approximation, we have

$$\ln W = \sum_{i=1}^{k} [g_i \ln g_i - g_i - n_i \ln n_i + n_i - (g_i - n_i) \ln (g_i - n_i) + (g_i - n_i)]$$

$$= \sum_{i=1}^{k} [g_i \ln g_i - n_i \ln n_i - (g_i - n_i) \ln g_i (g_i - n_i)]$$

Here g_i is not subject to variation whereas n_i varies continuously. Differentiating both sides, we have

$$\delta (\ln W) = \sum_{i=1}^{k} \left[-n_i \frac{1}{n_i} \delta n_i - \ln n_i \, \delta n_i + (g_i - n_i) \frac{1}{(g_i - n_i)} \delta n_i + \ln (g_i - n_i) \, \delta n_i \right]$$

$$= \sum_{i=1}^{k} [\ln (g_i - n_i) - \ln n_i] \, \delta n_i$$

To get the state of maximum thermodynamic probability

$$\delta (\ln W) = 0$$

$$\therefore \quad \sum_{i=1}^{k} [\ln (g_i - n_i) - \ln n_i] \, \delta n_i = 0$$

or $$\sum_{i=1}^{k} \left[\ln \frac{(g_i - n_i)}{n_i} \right] \delta n_i = 0$$

or $$-\sum_{i=1}^{k} \left[\ln \frac{n_i}{(g_i - n_i)} \right] \delta n_i = 0$$

or $$\sum_{i=1}^{k} \left[\ln \frac{n_i}{(g_i - n_i)} \right] \delta n_i = 0 \qquad \ldots(31.25)$$

In addition, our system must satisfy the two auxiliary conditions:

(*i*) Conservation of total number of particles, i.e., $n = a$ constant

$$n = \sum_{i=1}^{k} n_i = \text{constant}$$

i.e.,
$$\delta E = \sum_i \delta n_i = 0 \qquad \text{...(31.26)}$$

(*ii*) Conservation of total energy of the system, i.e., $E = a$ constant.

$$\therefore \qquad E = \sum_i n_i E_i = \text{constant}$$

i.e.,
$$\delta E = \sum_i E_i \, \delta n_i = 0 \qquad \text{...(31.27)}$$

We shall apply now the Langrangian method of undetermined multipliers. For this, let us multiply Eq. (31.26) by α and Eq. (31.27) by β, and add the resulting expression to Eq. (31.23).

$$= \sum_{i=1}^{k} \left[\ln\left(\frac{n_i}{(g_i - n_i)} \right) + \alpha + \beta E_i \right] \delta n_i = 0$$

The variations δn_i are independent of each other. Hence we get

$$\ln\left(\frac{n_i}{(g_i - n_i)} \right) + \alpha + \beta E_i = 0$$

or
$$\frac{n_i}{(g_i - n_i)} = e^{-\alpha - \beta E_i} = e^{-(\alpha + \beta E_i)}$$

or
$$\frac{g_i - n_i}{n_i} = e^{\alpha + \beta E_i}$$

or
$$\frac{g_i}{n_i} - 1 = e^{\alpha + \beta E_i}$$

or
$$\frac{g_i}{n_i} = e^{\alpha + \beta E_i} + 1$$

or
$$n_i = \frac{g_i}{e^{\alpha + \beta E_i} + 1} \qquad \text{...(31.28)}$$

Eq. (31.28) represents the most probable distribution of the particles among various energy levels for a system obeying Fermi-Dirac statistics and is therefore, known as Fermi-Dirac Distribution Law for an assembly of fermions.

Substituting $\beta = \dfrac{1}{kT}$ in Eq. (31.28),

where *k* is Boltzmann's constant and has the same role as in case of M-E distribution law.

We have
$$n_i = \frac{g_i}{e^{\alpha} . e^{\beta E_i} + 1} = \frac{g_i}{e^{\alpha} . e^{E_i / kT} + 1} \qquad \text{...(31.29)}$$

Fermi-Dirac energy distribution function:

Since there can be a maximum of one particle per quantum state, the function $f(E_i)$ is the ratio of the number of quantum states of energy E_i occupied by electrons to the total number of quantum states available in the energy level E_i. Therefore, the value of $f(E_i)$ for the Fermi-Dirac distribution at a particular energy E_i is the probability that under equilibrium a quantum state of that energy is occupied by a particle.

From Eq. (31.29), $f(E_i)$ is given by

$$f(E_i) = \frac{n_i}{g_i}$$

$$= \frac{1}{e^{\alpha}.e^{E_i/kT}+1} \quad ...(31.30)$$

For continuous distribution of energy at energy E, the distribution function is written as

$$f(E_i) = \frac{1}{e^{\alpha}.e^{E_i/kT}+1} \quad (31.31)$$

F-D energy distribution law for continuous variation of energy:

If the energy levels are very close together, then the distribution of energy of the particles may be considered continuous. For this distribution the number of particles $n(E)dE$ whose energies lie between E and $E + dE$, then $n = n\,(E)dE$ and is given by

$$n\,(E)dE = f(E)\,g(E)dE \quad ...(31.32)$$

whereg (E) dE is the number of quantum states of energy between E and $E + dE$.

Substituting the value of $f(E)$ in Eq. (31.32), we have

$$n\,(E)dE = \frac{g\,(E)dE}{e^{\alpha}.e^{E_i/kT}+1} \quad ...(31.33)$$

For particles like electrons of spin angular momentum $\pm\frac{1}{2}h$, there are two possible spin orientations.

For a system of such particles $g(E)$ dE is given by Eq. (31.19) and substituting it we get

$$n(E)dE = 2\times 2\pi V\left(\frac{2m}{h^2}\right)^{3/2}\frac{E^{\frac{1}{2}}dE}{e^{\alpha}.e^{E_i/kT}+1} \quad ...(31.34)$$

This is the Fermi-Dirac energy distribution law for continuous variation of energy among free particles with spin $\frac{1}{2}$.

The probability $F(E)$ that a Fermion occupy a state of energy E, Fermi-Dirac distribution function is given by

$$F(E) = \frac{1}{\exp[\alpha+\frac{E}{kT}]+1}$$

where α is a constant is depends on the property of the system and temperature T.

If n is the number of Fermions with total energy E in equilibrium at temperature T, the most probable distribution of Fermions among the various energy levels

$$n_i = \frac{g_i}{\exp[\alpha+\frac{E_i}{kT}]+1}$$

where g_i is the number of quantum states available for the same energy E_i, called as degeneracy. α is a constant

$n_1, n_2, \ldots n_k$ are particles at various energy states $E_1, E_2, E_3, \ldots$.

If we assume continuous distribution of energy among electrons, the number of electrons within the energy interval E and $E + dE$ is given by

$$n(E)\,dE = \frac{g(E)dE}{\exp[\alpha + \frac{E}{kT}] + 1}$$

where $g(E)\,dE$ is the number of phase space cells within the energy interval E and $E + dE$.

When there is large number of cells and comparatively a few particles are to be arranged the probability that a cell containing more than one particle is very small. Therefore, the three statistics give similar results. However in case when the number of particles is equivalent with number of cells, the three statistics give different results. A comparison between the three statistical distribution functions is shown in Table 31.2.

Table 31.2: *Comparison between the three statistical distribution functions*

Maxwell-Boltzmann Statistics	**Bose-Einstein Statistics**	**Fermi-Dirac Statistics**
Classical statistics	Quantum statistics	Quantum statistics
Applicable to identical and distinguishable particles. Each particle has recognizable identity.	Applicable to identical and indistinguishable particles. Each particle loses their individuality.	Applicable to identical and indistinguishable particles. Each particle loses their individuality.
Molecules of a gas is an example	Photons in a cavity, photons in a solid, liquid, helium at low temperature are examples	Free electrons in conductors are examples.
Applicable to particles of any spin and wave functions do not overlap	Applicable to particles of integral spin (0,1,2,3..) and wave functions are symmetric	Applicable to particles of half integral spin (1/2, 3/2, 5/2…) and wave functions anti symmetric.
$F(E) = A\exp^{-\frac{E}{kT}}$	$F(E) = \frac{1}{\exp[\alpha + \frac{E}{kT}] - 1}$	$F(E) = \frac{1}{\exp[\alpha + \frac{E}{kT}] + 1}$
Average occupancy number $n_i = \frac{g_i}{\exp[\alpha + \beta E_i]}$	Average occupancy number $n_i = \frac{g_i}{\exp[\alpha + \frac{E_i}{kT}] - 1}$	Average occupancy number $n_i = \frac{g_i}{\exp[\alpha + \frac{E_i}{kT}] + 1}$
Total number of independent ways of realizing a distribution $W = \frac{N!}{n_1!n_2!n_3!...}$.	Total number of independent ways of realizing a distribution $W = \underset{i=1}{\overset{n}{\pi}} \frac{(n_i + g_i - 1)!}{n_i!(g_i - 1)!}$.	Total number of independent ways of realizing a distribution $W = \underset{i=1}{\overset{n}{\pi}} \frac{g_i!}{n_i!(g_i - 1)!}$.
Particle does not obey uncertainty principle.	Particle does obey uncertainty principle.	Particle does obey uncertainty principle.
No limit on number of particles per state. Each particle is as likely to be in one cell as in another. [p | q] [q | p] [pq |] [| qp]	Do not obey Pauli's principle. No limit on number of particles per state. [• | •] [•• |] [| ••]	Obey Pauli's principle. Never more than one in a state. [| ••]

APPLICATION OF F-D DISTRIBUTION FUNCTION

(a) Electrons in Metals

Let us consider an example of electron gas which obeys Fermi-Dirac statistics. Electrons are fermions as they possess spin 1/2. The free electrons in a metal may be treated as gaseous particles, the Coulomb force between them and their interactions with the ionic cores may be ignored.

The Fermi-Dirac distribution function is given by

$$\frac{n_i}{g_i} = \frac{1}{e^{-\alpha} e^{\frac{E_i}{kT}} + 1} \quad ...(31.35)$$

Again, if we assume that we have a lot of closely spaced energy levels,we can write

$$n\,(E)dE = \frac{g\,(E)dE}{e^{-\alpha} e^{\frac{E_i}{kT}} + 1} \quad ...(31.36)$$

Now, the number of electrons in our electron gas is fixed, and so we cannot just set $e^{-\alpha} = 1$. Furthermore, it should be obvious that ± must be some sort of function of temperature. This can be seen by considering the equation:

$$N = \int_0^\infty \frac{g(E)dE}{e^{-\alpha} e^{\frac{E_i}{kT}} + 1} \quad ...(31.37)$$

For the integral always to have the same value (as the L.H.S. is a constant), α must be some function of temperature. We let

$$e^{-\alpha} = e^{\frac{-\mu}{kT}} \quad ...(31.38)$$

where μ is called the chemical potential and so

$$f(E) = \frac{n_i}{g_i} = \frac{1}{e^{\frac{(E_i - \mu)}{kT}} + 1} \quad ...(31.39)$$

Let us consider what happens as we tend towards absolute zero, ($T \to 0$). Careful examination of Eq. (31.39) shows that under these conditions μ must be positive, otherwise n_i/g_i would be zero at all energies.Taking this into account we see that the $e^{\frac{(E-\mu)}{kT}}$ factor will look like $e^{-large\ number}$ (i.e. 0) for $E < \mu$, and $e^{+largenumber}$, (i.e. ∞), for $E > \mu$. Therefore,

$$f(E) = \frac{n_i}{g_i} = 1 \quad (E < \mu) \quad ...(31.40)$$

$$f(E) = \frac{n_i}{g_i} = 0 \quad (E < \mu) \quad ...(31.41)$$

Thus, the number of particles in a particular state is unity up to the energy μ, and is zero afterwards. At the lowest of temperatures the particles are going to have the lowest energy they possibly can. But, because of the Pauli Exclusion Principle, only two one of each spin can be in the lowest energy level. Each spin orientation corresponds to a different wave function. (Fig. 31.4).

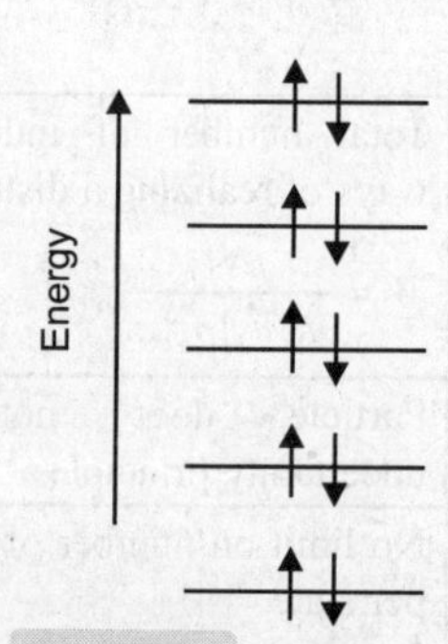

Fig. 31.4: Illustration of Pauli Exclusion Principle

The next pair of electrons must go into the next energy level, and so on - the particles have to stack up in energy in order to obey the condition that no two are allowed with the same wave function. We call the value of μ at absolute zero the Fermi Energy, denoted by E_F. It should be clear that:-

$$N = \int_0^{E_F} g(E)dE = \int_0^{E_F} 4\pi m V \frac{(2m)^{1/2}}{h^3} E^{1/2} dE \times 2 = \left(\frac{2m\, E_F}{h^2}\right)^{3/2} \frac{V}{3\pi^2} \qquad ...(31.42)$$

where we have used the normal density of states for free particles, and multiplied by an extra factor of two to take into account that each energy level has two wave functions associated with it, as the electrons can have two spin states. Therefore

$$E_F = \frac{h^2}{2m}\left(\frac{3\pi^2 N}{V}\right)^{2/3} \qquad ...(31.43)$$

Thus, the Fermi Energy depends only on the density of the system. It is of the order of 10^{29} m^{-3} and $E_F \sim 8 \times 10^{-19}$ J. When expressed in electron volts, the Fermi Energy for a typical metal is of order 5 eV. In eV units, kT at room temperature is about 1/40 eV. Thus, for typical metals, it is about 200 times greater than the energy of a molecule of air at room temperature. As no two electrons are allowed in the same energy state, they "stack up", two by two, until the topmost ones reach the Fermi Energy level.

However, the electrons that have an energy of 5-eV, do not have a temperature of 5-eV. It is not allowable to equate energy with kT in FD statistics as we do in the classical case which obey Boltzmann statistics.

(b) Heat capacity of metals

Fig. 31.5 shows the occupation number (i.e. the number of particles per state $-n_i/g_i$) for different temperatures. As the system heats up, most of the electrons which possess the lower energies are not affected. The high energy electrons start to occupy higher states (Fig. 31.6) and the occupation number gets 'rounded' at the edges (Fig. 31.5). Ultimately, if we heat up the Fermi gas to a high enough temperature, the distribution changes once again to a decaying exponential, that is back to the Boltzmann distribution. At this high temperature the particles are spread out in energy and the occupation number is everywhere less than 1/2. The electrons are no longer contending to be in the same quantum state.

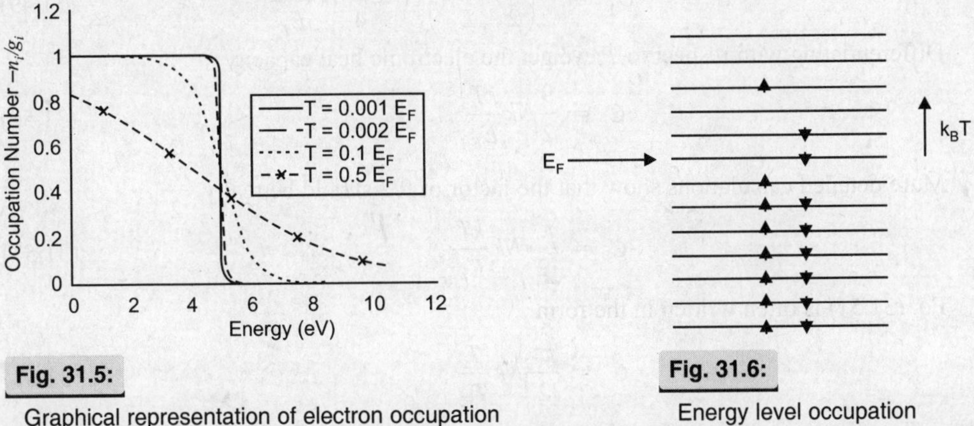

Fig. 31.5: Graphical representation of electron occupation

Fig. 31.6: Energy level occupation

So, for a small increase in temperature, only electrons close to the Fermi energy are excited. Other electrons are below the Fermi energy level and are not excited at all; so these electrons would not contribute to the heat capacity. Thus, it is chiefly the electrons close to the Fermi energy that would contribute to the heat capacity.

These electrons close to the Fermi energy behave like the ideal gas. We know that the energy of an ideal gas is

$$U_1 = 3/2\, N_1 kT \qquad \text{...(31.44)}$$

where N_1 is the number of particles in the ideal gas.

In the case of the electrons, N_1 ought to indicate the number of electrons above the Fermi energy, and not the total number of electrons.

As the electrons do get excited in the small energy range of order kT from the Fermi energy, we can estimate N_1 with the number of electrons that are within the energy interval of $dE = kT$ from the Fermi energy.

We know that the number of particles in a given energy interval is

$$n(E)\, dE = 2g(E) f(E)\, dE \qquad \text{...(31.45)}$$

where the factor of 2 again comes from the spin states of the electrons.

At 0 K, the energy states below E_F are fully occupied, i.e. $f(E) = 1$. So the number would be given by

$$N_1 \approx 2g(E_F)\, kT \qquad \text{...(31.46)}$$

From the definitions of the density of state and the Fermi energy, it can be shown that

$$g(E_F) = \frac{3N}{4E_F} \qquad \text{...(31.47)}$$

where N is the total number of electrons.

This gives

$$N_1 \approx 2\frac{3N}{4E_F}kT = \frac{3}{2}N\frac{kT}{E_F} \qquad \text{...(31.48)}$$

Substituting into the energy for ideal gas $U_1 = \frac{3}{2}N_1 kT$ we get the energy for the excited electrons

$$U_1 = \frac{3}{2}\left(\frac{3}{2}N\frac{kT}{E_F}\right)kT = \frac{9}{4}Nk\frac{kT^2}{E_F} \qquad \text{...(31.49)}$$

Differentiating with respect to T, we get the electronic heat capacity

$$C = \frac{9}{2}Nk\frac{kT}{E_F} \qquad \text{...(31.50)}$$

More detailed calculations show that the factor of 9/2 should be really $\frac{\pi^2}{2}$.

$$C = \frac{\pi^2}{2}Nk\frac{kT}{E_F} \qquad \text{...(31.51)}$$

Eq. (31.51) is often written in the form

$$C = \frac{\pi^2}{2}Nk\frac{T}{T_F} \qquad \text{...(31.52)}$$

where $T_F = E_F / k$ is called the Fermi temperature.

Notice that the heat capacity is directly proportional to T. This is often written in the form

$$C = \gamma T \qquad ...(31.53)$$

There is another contribution to the heat capacity. This comes from the vibrations of the atoms, and it is proportional to T^3.

We could imagine writing the total heat capacity in the form:

$$C_v = \gamma T + AT^3 \qquad ...(31.54)$$

We can measure this heat capacity to check if the formula is correct. Suppose that we have obtained a table of values for T and C_v. To check if the formula is correct, we can rewrite it in this form:

$$(Cv/T) = \gamma + AT^2 \qquad ...(31.55)$$

From Eq. (31.55) the plot of (C_V/T) against T^2 should be a straight line. In 1955, William Corak et al working on the heat capacities of copper, gold and silver from 1 K to 5 K plotted (C_V/T) against T^2 and found that it was a straight line, Fig. 31.7. This proves the predictions of the Fermi-Dirac statistics are correct.

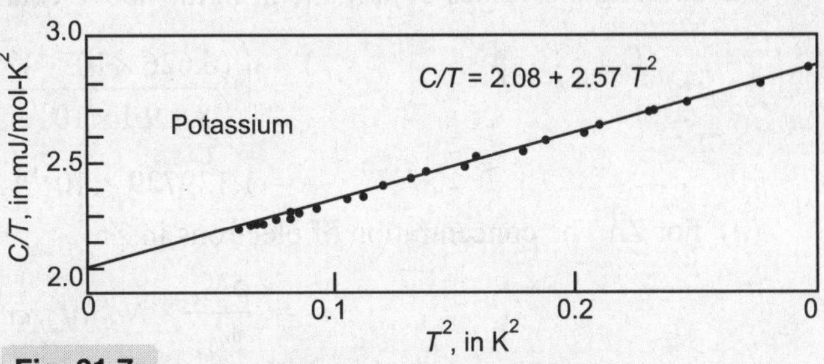

Fig. 31.7: Plot of C_V/T against T^2

Example: 31.1: *Fermi temperature of a material is 24600 K. Calculate Fermi Velocity.*

Solution: We know

$$\frac{1}{2}mv_F^2 = KT$$

$$v_F^2 = \frac{2KT}{m} = \frac{2 \times 1.38 \times 10^{-23} \times 246000}{9.108 \times 10^{-31}} = 0.74 \times 10^{13}$$

$$v_F = 0.86 \times 10^6 \text{ m/s.}$$

Example: 31.2: *Calculate the Fermi energy of the following metals:*

	Cu	Zn	Al
ρ(g cm^{-3})	8.96	7.14	2.70
M_{at}	63.55	65.38	27.0

Solution: Given data:

For Cu

$$\text{Density of Cu} = 8.96 \times 10^3 \text{ kg m}^{-3}$$
$$\text{Atomic weight of Cu} = 63.55$$

For Zn

$$\text{Density of Zn} = 7.14 \times 10^3 \text{ kg m}^{-3}$$
$$\text{Atomic weight of Cu} = 65.38$$

For Al

$$\text{Density of Al} = 2.7 \times 10^3 \text{ kg m}^{-3}$$
$$\text{Atomic weight of Al} = 27.0$$

(*i*) For Cu The concentration of electrons in Cu

$$n = \frac{\text{Density} \times \text{Avogadro's constant} \times \text{No. of free electrons per atom}}{\text{Atomic weight}}$$

$$= \frac{8.96 \times 10^3 \times 6.022 \times 10^{26}}{63.55} = 8.4905 \times 10^{28} \text{ m}^{-3}$$

For energy at 0 K,

$$E_{FO} = \left(\frac{h^2}{8m_e}\right)\left(\frac{3n}{\pi}\right)^{2/3}$$

Substituting the values of *h*, *n* and *m* in the above equation, we get

$$= \left(\frac{(6.626 \times 10^{-34})^2}{8 \times 9.1 \times 10^{31}}\right)\left(\frac{3 \times 8.4905 \times 10^{28}}{\pi}\right)^{2/3}$$

$$= 1.129729 \times 10^{-18} \text{ J} = 7.0608 \text{ eV}$$

(*ii*) For Zn The concentration of electrons in Zn

$$n = \frac{\rho N_A}{M_{at}} \times \textit{No. of free electrons/atom}$$

$$= \frac{7.14 \times 10^3 \times 6.022 \times 10^{26} \times 2}{65.38} = 13.152976 \times 10^{28} \text{ m}^{-3}$$

For energy at 0 K,

$$E_{FO} = \left(\frac{h^2}{8m_e}\right)\left(\frac{3n}{\pi}\right)^{2/3}$$

Substituting the values of *h*, *n* and *m* in the above equation, we get

$$= \left(\frac{(6.626 \times 10^{-34})^2}{8 \times 9.1 \times 10^{31}}\right)\left(\frac{3 \times 13.153 \times 10^{28}}{\pi}\right)^{2/3}$$

$$= 1.5152397 \times 10^{-18} \text{ J} = 9.45 \text{ eV}$$

(iii) For Al The concentration of electrons in Al

$$n = \frac{\rho N_A}{M_{at}} \times \textit{No. of free electrons/atom}$$

$$= \frac{2.7 \times 10^3 \times 6.022 \times 10^{26} \times 2}{27} = 18.066 \times 10^{28} \text{ m}^{-3}$$

For energy at 0 K,

$$E_{FO} = \left(\frac{h^2}{8m_e}\right)\left(\frac{3n}{\pi}\right)^{2/3}$$

Substituting the values of *h*, *n* and *m* in the above equation, we get

$$= \left(\frac{(6.626 \times 10^{-34})^2}{8 \times 9.1 \times 10^{31}}\right)\left(\frac{3 \times 18.066 \times 10^{28}}{\pi}\right)^{2/3}$$

$$= 1.8689 \times 10^{-18} \text{ J} = 11.68 \text{ eV}$$

The Fermi energy of Cu is 7.0608 eV.

The Fermi energy of Zn is 9.45eV.

The Fermi energy of Al is 11.68eV.

Example: 31.3: *Calculate the density of states for Cu at the Fermi level for T = 0 K, if (m*/m) = 1.01.*

Solution: Given data: Temperature = 0 K, The effective mass of electron in Cu = 1.01

The density of the states is given by $Z(E) = \gamma E^{1/2}/2$

Where $\gamma = 6.82 \times 10^{27}\ \text{m}^{-3}\ \text{eV}^{-3/2}$

For energy at 0 K,

$$E_{FO} = \left(\frac{h^2}{8m_e}\right)\left(\frac{3n}{\pi}\right)^{2/3}$$

Substituting the values of *h*, *n* and *m* in the above equation, we get

$$= \left(\frac{(6.626\times10^{-34})^2}{8\times9.1\times10^{31}}\right)\left(\frac{3\times8.4905\times10^{28}}{\pi}\right)^{2/3}$$

$$= 1.129729 \times 10^{-18}\ \text{J} = 7.0608\ \text{eV}$$

Therefore, $$Z(E) = \gamma E_F^{1/2}/2 = 3.39\times10^{27}\times\sqrt{7.0608} = 9\times10^{27}\ \text{m}^{-3}$$

The density of states for Cu at the Fermi level for $T = 0$ is $9 \times 10^{27}\ \text{m}^{-3}$.

Example: 31.4: *Use Fermi distribution function to obtain the value of F(E) for $E - E_F$ = 0.01 eV at 200K.*

Solution: Given data: The difference between energy level to Fermi level $E - E_F$ = 0.01 eV

$$= 0.01 \times 1.6 \times 10^{-19}\ \text{J} = 1.6 \times 10^{-21}\ \text{J}$$

Temperature $T = 200$ K

We know that the Fermi distribution function

$$F(E) = \frac{1}{1+e^{(E-E_F)ikT}}$$

Substitution the value of $E - E_F$ and *T*, we get

$$= \frac{1}{1+e^{(1.6\times10^{-21})/(1.38\times10^{-23}\times200)}} = \frac{1}{1+e^{0.5797}} = 0.3589$$

The Fermi distribution for energy *E* is 0.3589.

Example: 31.5: *Calculate Fermi energy and Fermi temperature in a metal. Fermi velocity of electron in the metal is $0.86 \times 10^6\ ms^{-1}$.*

Solution: Given data:

Velocity of electron $v = 0.86 \times 10^6\ \text{ms}^{-1}$

Mass of electron $m = 9.1\times10^{-31}$ kg

Electronic change $e = 1.6 \times 10^{-19}$ C

Boltzmann's constant $k = 1.38 \times 10^{-23}\ \text{JK}^{-1}$

We know that the Fermi energy

$$E_F = \frac{1}{2}mv^2$$

Substituting the values, we get

$$= \frac{1}{2} \times 9.11 \times 10^{-31} \times 0.86 \times 10^6 = 3.368 \times 10^{-19}\,\text{J}$$

We also know that the Fermi temperature

$$T_F = \frac{E_F}{k}$$

Substituting the value, we get

$$= \frac{3.368 \times 10^{-19}}{1.38 \times 10^{-23}} = 2.43 \times 10^4\,\text{K}$$

Therefore, the Fermi temperature T_F is 2.43×10^4 K.

Example: 31.6: *Calculate the number of states lying in an energy interval of 0.01 eV above the Fermi level for a crystal of unit volume with Fermi energy EF= 3.0 eV.*

Solution: Given data:

Mass of electron	m	$= 9.1 \times 10^{-31}$ kg
Energy interval	ΔE	$= 0.01$ eV
Plank's constant	h	$= 6.63 \times 10^{-34}$ Js
Fermi energy	E_F	$= 3.0$ ev

Let us take E_1 is ground state,

i.e., $E_1 = 3.0\text{ eV} = 3.0 \times 1.6 \times 10^{-19}$

$E_1 = 4.8 \times 10^{-19}$ J

E_2 is given by $E_2 = E_1 + \Delta E = (3.0 + 0.01)\text{ eV} = 3.01 \times 1.6 \times 10^{-19}$ J

$E_2 = 4.816 \times 10^{-19}$ J

Number of status per unit volume lying between E_1 and E_2 is given by

$$n = \int_{E_1}^{E_2} \frac{4\pi}{h^3}(2m)^{3/2} E^{1/2} dE$$

Here, density of status $Z(E)$ is unity

$$= \frac{4\pi}{h^3}(2m)^{3/2} \int_{E_1}^{E_2} E^{1/2} dE$$

Integrating the above equation, we get

$$= \frac{4\pi}{h^3}(2m)^{3/2} \left[\frac{2}{3}(E)^{3/2}\right]_{E_1}^{E_2}$$

Simplifying the above equation, we get

$$= \frac{4\pi}{h^3}(2m)^{3/2} \left[\frac{2}{3}(E)^{3/2}\right]_{E_1}^{E_2}$$

$$= \frac{4\pi}{h^3}(2m)^{3/2} \cdot \frac{2}{3}\left[E_2^{3/2} - E_1^{3/2}\right]$$

Substituting the value, we get

$$= \frac{4\pi \times (2 \times 9.1 \times 10^{-31})^{3/2}}{(6.63 \times 10^{-34})^3} \times \frac{2}{3}\left[\left\{(4.816)^{3/2} - (4.8)^{3/2}\right\} \times 10^{-19}\right]^{3/2}$$

$$= 3.74 \times 10^{55} \times (1.108 \times 10^{-30}) = 4.14 \times 10^{25}$$

The number of states lying between the energy level is 4.14×10^{25}.

Example: 31.7: *Fermi temperature of a metal is 24600 K. Calculate the Fermi velocity.*

Solution: Given data:

Fermi temperature of the metal $T_F = 24600$ K

Mass of an electron $m = 9.11 \times 10^{-31}$ kg

We know that the Fermi velocity is

$$v_F = \sqrt{\frac{2kT_F}{m}}$$

Substituting the values, we get $= \sqrt{\frac{2 \times 1.38 \times 10^{-23} \times 24600}{9.11 \times 10^{-31}}} = 0.8633 \times 10^6 \text{ ms}^{-1}$

The Fermi velocity v_F is 0.8633×10^6 ms^{-1}.

Example: 31.8: *Free electron density of aluminium is 18.10×10^{28} m^{-3}. Calculate its Fermi energy at 0 K. Plank's constant and mass of free electron are 6.62×10^{-34} J s and 9.1×10^{-31} kg.*

Solution: Given data:

Electron density of aluminium $n = 18.10 \times 10^{28}$ m^{-3}

Plank's constant $h = 6.62 \times 10^{-34}$ J s

Mass of electron $m = 9.1 \times 10^{-31}$ kg

We know that the Fermi energy at 0 K

$$E_{F0} = \left(\frac{3n_c}{8\pi}\right)^{2/3} \frac{h^2}{2m}$$

Substituting the given values, we get

$$= \left(\frac{3 \times 18.10 \times 10^{28}}{8 \times 3.14}\right)^{2/3} \times \frac{(6.62 \times 10^{-34})^2}{2 \times 9.1 \times 10^{-31}}$$

$$= 1.8689 \times 10^{-18} \text{ J} = \frac{1.8689 \times 10^{-18}}{1.6 \times 10^{-19}} = 11.68 \text{ eV}$$

The Fermi energy at 0 k is 11.68 eV

QUESTIONS

1. Explain the need for introducing quantum statistics.
2. What is indistinguishability of particles? What role it plays in quantum statistics.
3. What is B-E statistics and derive the expression for n_i.
4. What are bosons?
5. Explain the postulates of B-E statistics.
6. What are fermions?
7. Explain the postulates of F-D statistics.
8. Explain the postulates of M-B statistics.
9. What is F-D statistics and derive the expression for n_i?
10. Compare M-B, B-E and F-D statistics.
11. Explain position and momentum space.
12. Explain phase space.
13. Compare different types of ensembles.
14. What is Fermi Energy?
15. Give the difference between bosons and Fermions.
16. What is M-B statistics and derive the expression for n_i?
17. What are bosons and fermions? How they are distributed among the enegy levels.

32 Bonding in Solids

Learning Objectives:

By the end of this chapter, you will be able to:

- Learn about attractive and repulsive forces.
- Classify bonds into different types.
- Estimate cohesive energy for the materials specially ionic solids.

32.1 INTRODUCTION

Elements often interact with each other. Such interactions create a variety of structures and compounds. In the simplest form these interactions occur as *bond formation*. Every pure substance has an exclusive set of properties which discriminates it from other substances. Bond formation between the atoms, ions, or molecules in a sample of matter largely determines its behavior; and so, bonding can be correlated with the properties of the material sample.

32.2 ATTRACTIVE AND REPULSIVE FORCES

Two different forces are necessary to establish bonding in a solid or in a molecule.

(*i*) an *attractive force* is necessary for bonding and

(*ii*) a *repulsive force* is required to keep the atoms from getting extremely close to each other.

An expression for an interatomic potential can be written as

$$F(r) = \frac{A}{r^m} - \frac{B}{r^n} \quad ...(32.1)$$

where r is the distance between atoms; m, n, A and B are constants depends on type of the bond, $\frac{A}{r^m}$ represents attractive force, $\frac{B}{r^n}$ represents repulsive force and $n > m$, and it means the repulsive part has to prevail for short distances.

The potential and the resulting force are shown in Fig. 32.1. Pauli Exclusion Principle is the reason for the strong repulsion of electron clouds of two atoms at short distances.

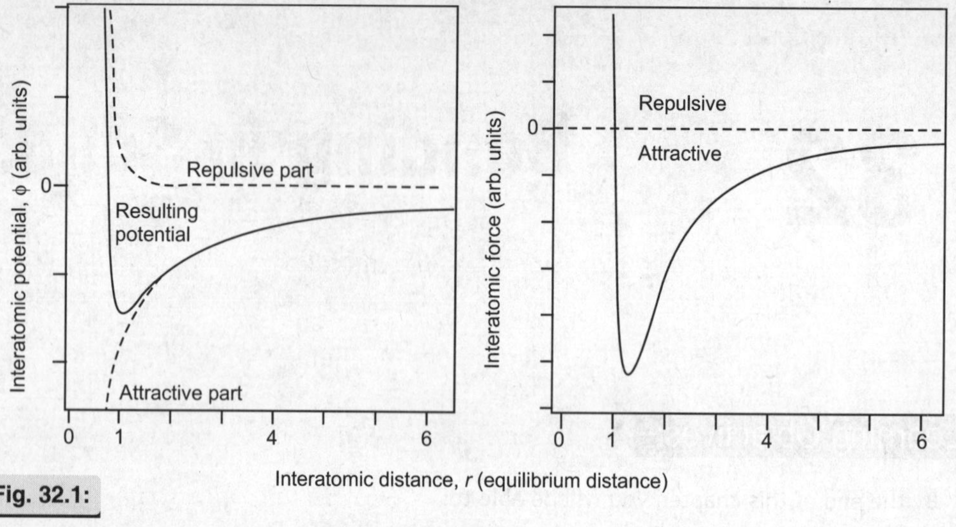

Fig. 32.1: Interatomic distance versus interatomic potential

32.3 CLASSIFICATION OF BONDS

The following are the forces responsible for the cohesion of solids.

(*i*) *Coulomb force:* This is the major contribution to the cohesion of a solid; it is the attractive force between electrons and the nuclei.

(*ii*) *Magnetic force:* This has a weak contribution to cohesion.

(*iii*) *Gravitational force:* It has a negligible contribution to cohesion.

*The **binding** or **cohesive energy** of a substance is the energy required to break all the bonds associated with one of its constituent molecules. It is measure of the **inter-molecular energy** of a substance.*

Types of bonds in solids: Five types of chemical bonding - covalent bonding in nonmetals, metallic bonding, ionic bonding, van der Waals attraction, and hydrogen bonding- are sufficient to explain atomic and molecular interactions in solids, liquids, and gases. Each type contributes a different kind of stability to the atoms it binds.

The bonds, based on bonds strength and directionality, are classified into the following two types.

(*i*) Primary bonds: These bonds have energies in the range of 0.1-10 eV. Primary bonds differ by the virtue of their nature and inter atomic forces. Primary bonds are interatomic bonds. The attractive forces in primary bonds are associated with valence electrons. The bonds are of three types based on borrowing, sharing the valence electrons.

(*a*) Ionic bond

(*b*) Covalent bond

(*c*) Metallic bond

(*ii*) Secondary bonds: These bonds have energies in the range of 0.01-0.5 eV. The secondary bonds are intermolecular bonds.

(*a*) Hydrogen bond

(*b*) Van der Waal's bond

(*iii*) **Other Types of Crystalline Solids:** Ceramics, Zeolites, Organic molecules are briefly discussed.

(i) Primary Bonds

(*a*) **Ionic Bonds and Ionic Compounds:** Ionic bonds require an electron donor, often a metal, and an electron acceptor, a nonmetal. When a metal atom and a nonmetal atom interact to form a bond, the metal atom loses one or more of its electrons and forms a positively charged ion called a ***cation***. The nonmetal atom acquires one or more electrons and forms a negatively charged ion called an ***union***. The two ions are held together by the Coulomb attraction and the bond is called the *ionic bond*.

Ionic bonding involves the complete transfer of valence electrons between atoms. NaCl, KCl, Br, MgO, $MgCl_2$ KOH are a few examples. The cations and anions form a rigid crystal lattice wherein each positive ion is bounded by a number of negative ions and each negative ion is surrounded by a number of positive ions. A compound made up of ions linked by ionic bonds - is called an *ionic compound*.

Consider the case of sodium chloride, see Fig. 32.2. Here a single valance electron is transferred from sodium atom to the chlorine atom to get stable configuration. There exists an electrostatic attraction between positive and negative ions. The Na^+ and Cl^- ions so formed are arranged in FFC structure.

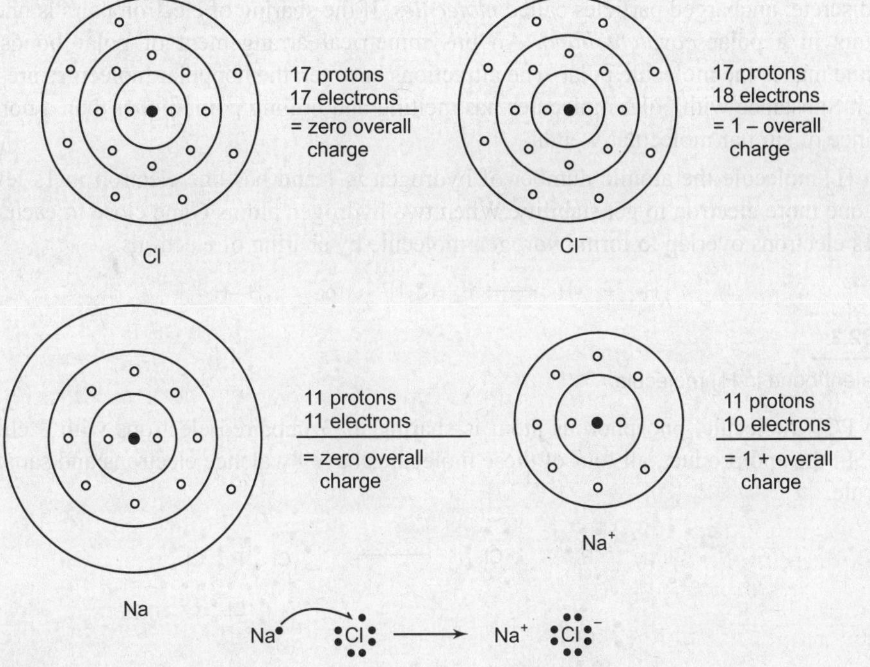

Fig. 32.2: Formation of Ionic bond

Binding energy of NaCl molecule:

The ionization energy of Na atom is 5.1 eV and electron affinity of chlorine is 3.6 eV.

So, net energy spent in ionizing both Na and Cl is

$$Na + Cl + (5.2 - 3.6) \rightarrow Na^+ + Cl^-$$

The bonding energy

$$V = \frac{-e^2}{4\pi\varepsilon_0 r_o} = 6 \text{ eV}$$

Hence net energy released when NaCl molecule is formed from Na and Cl atoms, is

$$(5.1 - 3.6 - 6) = -4.5 \text{ eV}$$

Properties of ionic solids:

(*i*) Most ionic salts are hard, brittle and have high melting and boiling points.

(*ii*) They are soluble in polar liquids like water.

(*iii*) Their electrical conductivity is less at room temperature. But conductivity increases with temperature due to ionic conductivity.

(*iv*) Ionic solids crystallize in close packed structures.

(*v*) Exists in the compounds of metallic and non-metallic elements.

(*vi*) Ionic bonds are strongest and equal in all directions around the atom.

***(b)* Covalent Bonds and Covalent Compounds:** When two nonmetal atoms share a pair or several pairs of electrons, the shared pairs of electrons form *covalent bonds*. A compound having only covalent bonds is called a *covalent compound*. The atoms which share electrons form discrete, uncharged particles called *molecules*. If the sharing of electron pairs is unequal, it results in a polar *covalent bond*. An unsymmetrical arrangement of polar bonds in a molecule makes the molecule polar. The attractions between the nonpolar molecules are much weaker. Substance with polar molecules has melting and boiling point higher than a nonpolar substance of similar molecular weight.

In H_2 molecule the atomic number of hydrogen is 1 and has one electron in 1s level. It needs one more electron to get stability. When two hydrogen atoms come close to each other, their 1s electrons overlap to form hydrogen molecule by sharing of electrons.

H• + •H ⟶ H : H or H—H

Fig. 32.3:

Covalent bond in H_2 molecule

In PCl_3 molecule, phosphorous atom is sharing its 3 unpaired electrons with 3 chlorine atoms. In the end product, all four of these molecules have 8 valence electrons and satisfy the octet rule.

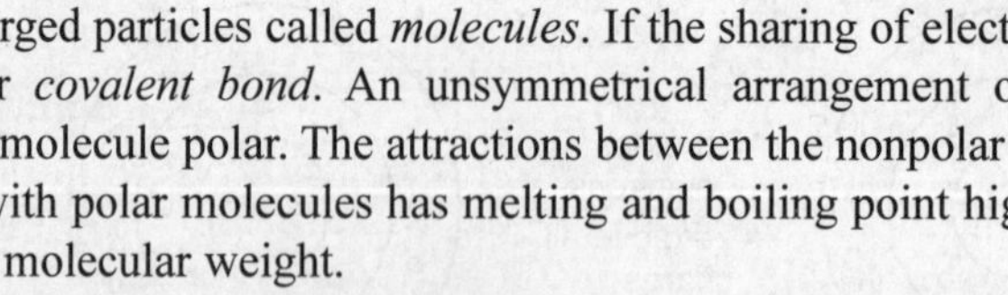

Cl P Cl ⟶ Cl P Cl

Cl Cl

Fig. 32.4:

Covalent bond in PCl_3 molecule

Covalent bonds include interactions of the sigma and pi orbitals; therefore, covalent bonds lead to formation of single, double and triple bonds.

(*a*) A ***single bond*** is formed when two electrons (one pair of electrons) are shared between two atoms and is shown by a single line between the two atoms. Single bond is a stable bond as it has a lower level of reactivity (losing electrons is less). Example is HCl

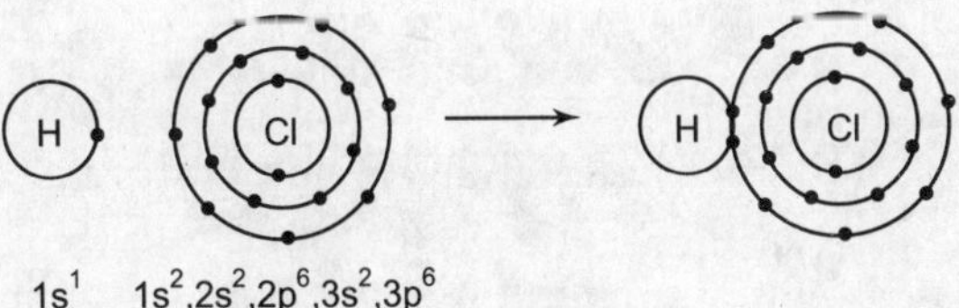

Fig. 31.5:

Single bond in HCl

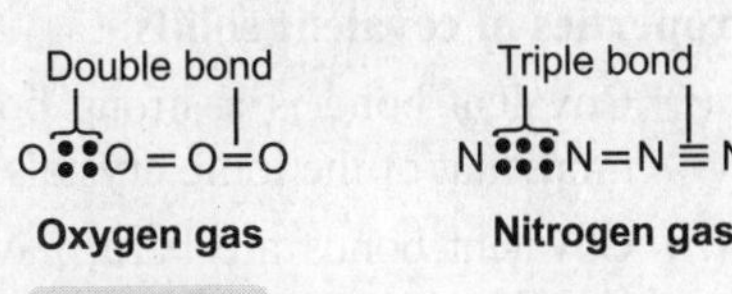

Fig. 32.6:

Bonds in Methane, O_2, N_2

(*b*) A ***double bond*** is formed when two atoms share two pairs of electrons with each other and is shown by two horizontal lines between two atoms in a molecule. This bond is less stable as it has greater amount of reactivity compared to single bond. Example is CO_2.

(*c*) A ***triple bond*** is formed when three pairs of electrons are shared between two atoms in a molecule. This bond is the least stable. Example is Acetylene.

(*d*) A ***polar covalent bond*** is formed between atoms having different electronegativities. So, the electrons are not equally shared and shared electrons will be closer to the atom with the higher electronegativity (Fig. 32.7 b). Examples are Water, Sulfide, Ozone, etc.

(*e*) A ***nonpolar covalent bond*** is formed between atoms having equal electronegativity. So, electrons are equally shared. Examples are Hydrogen gas atom, Nitrogen gas atoms, etc.

Fig. 32.7:

Polar and (*a*) noon-polar covalent bonds

Covalent Network Solids

In some of the covalent solids, networks or chains of atoms or molecules are held together by covalent bonds. An ideal single crystal of a covalent solid is a single giant molecule. For example, diamond consists of sp^3 hybridized carbon atoms; each bonded to four other carbon atoms in a tetrahedral array and creates a giant network. In graphite, the two-dimensional planes of carbon atoms are stacked to form a three-dimensional solid. Therefore, graphite exhibits properties typical of both covalent and molecular solids.

Silicon dioxide, SiO_2 (quartz, sand) and silicon carbide, SiC (Carborundum) are examples of covalent network solids. Such macromolecular solids are very hard, brittle, and very high-melting (above 1000°C). They are insoluble in all solvents because they have no loosely bound electrons. They are insulators to electricity.

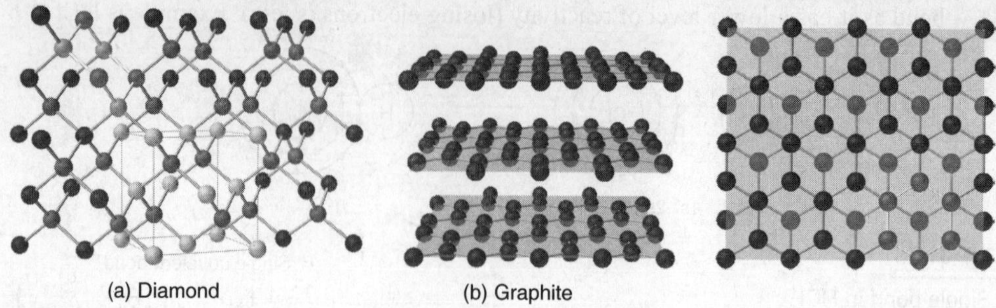

Fig. 32.8: Covalent network solids

Properties of covalent solids:

(*i*) Covalent bond is a strong bond. Cohesive energies are of 6 to 12 eV/atom which is more that of the ionic crystals.

(*ii*) Covalent bonds are hard, have high melting points and are insoluble in all ordinary liquids.

(*iii*) Covalent bonds are strongly directional. The change in the direction of the bond gives the formation of different substance.

(*iv*) Covalent solids are good insulators of electricity at low temperatures. The conductivity increases with increase in temperatures. Covalent crystals doped with certain impurities become semiconductors.

(*v*) Covalent crystals are transparent to long wavelength radiation but opaque to short wavelength radiation.

(*c*) Metallic Bonds and Metallic Solids: Metal atoms form metallic bonds through their interaction with other metal atoms. These bonds are found in pure metals, alloys, and certain inter metallic compounds. The metal cations form a lattice which is immersed in a *sea of valence electrons* which are mobile and delocalized. Particular valence electrons have no associations with particular cations. It accounts for the lack of directional preference in the bonding.

Free electrons from outer shells of metal atoms

Metal ions

Fig. 32.9: Atomic arrangements in metal crystals

Metals are good conductors of electricity because of the mobility of the valence electrons. Metals are malleable and ductile because of the delocalization of the electrons. Metals have characteristic shiny or lustrous appearance as the energy states of the electrons in the sea of electrons are virtually continuous and all the visible light frequencies are both absorbed and reradiated.

The cohesion of the metallic crystals is from the attractive force of electrons cloud and ion cloud, mutual repulsion of electrons and mutual repulsion of ion cores. Fig. 32.10 shows the potential energy contribution as a function of nuclear spacing from each of these charge interactions.

Metallic bond arises from the electrostatic attraction between cation cores and electrons that are not confined to any cation. The electrons are free to move between the cation cores.

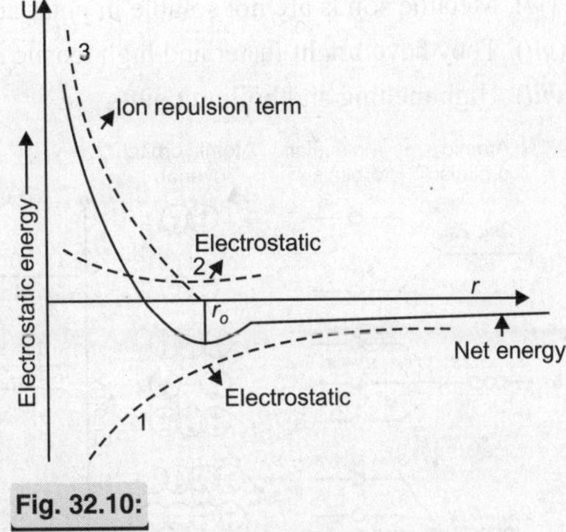

Fig. 32.10: Potential Energy versus Nuclear Spacing

Band Theory

According to the bonding model, the metal atom cores are assumed to be held together by delocalized valence electrons that are free to move about the crystal to conduct electricity. According to a formal approach, which is similar to the molecular orbital theory, the valence molecular orbitals in crystals involve all of the atoms in the crystal and so extend over the whole crystal.

According to molecular orbital theory, when two identical orbitals from two identical atoms combine, two *molecular orbitals*, one of lower energy and one of higher energy, are produced. In a crystal that contains a mole of atoms, a mole of such orbitals will combine to produce a mole of molecular orbitals that extend over the crystal and are very closely spaced in energy. A group of such orbitals is called a *band* because the orbitals are nearly continuous in energy.

The orbitals in the valence shells of the atoms combine to give a sequence of bands which are separated by small energy gaps. In the isolated metal crystal, valence electrons occupy the lowest energy orbitals possible in what is called the *valence band* to account for bonding. The next lowest energy orbitals available act as the *conduction band*.

In order for a material to be a conductor, electrons in the top of the valence band must be able to jump into the conduction band when an electric field is applied. It happens when the difference in energy between the valence band and the conduction band is sufficiently small.

In metals, either both the valence band and the conduction band are contiguous, or the valence band and the conduction band overlap. Therefore, there is no energy gap between the valence and conduction bands. Metals fall into this category and are therefore conductors.

Fig. 32.11 shows the delocalized molecular-orbital bands in lithium. The original 2*s* and 2*p* atomic orbitals are so close in the energy that the molecular orbital bands overlap. Lithium has one electron in every 2*s* atomic orbital, hence only half as many as electrons can be accommodated in the 2*s* atomic orbitals or in delocalized –molecular-orbital band. There are unfilled energy states an infinitesimal distance above highest –energy filled state. So, an infinitesimal energy is required to excite an electron and send it moving through the metals. Thus lithium is a conductor.

Properties of metallic solids:

(*i*) High electrical and thermal conductivity due to number of free electrons.

(*ii*) Metallic bonds are relatively week

(*iii*) High optical reflection and absorption coefficients.

(*iv*) Metallic solids are malleable and ductile.

(*v*) Metallic bonds are non directional.

(*vi*) Metallic solids are not soluble in polar and nonpolar solvents.

(*vii*) They have bright luster and high tensile strength

(*viii*) High melting and boiling points.

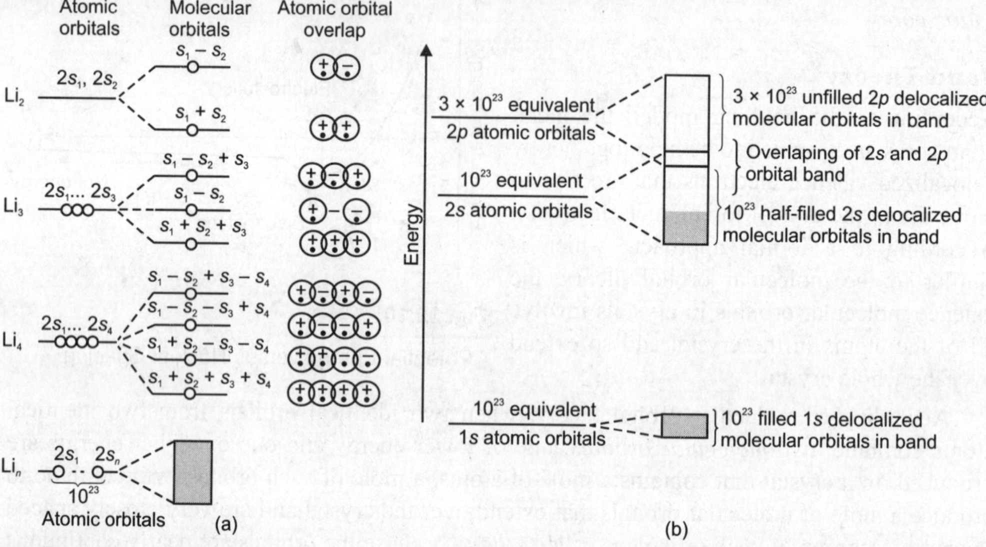

Fig. 32.11:

(a) Delocalized-molecular-orbital bands in lithium, (b) The original 2s and 2p atomic orbitals are so close in energy that the molecular orbital bands overlap

(ii) Secondary Bonds

(a) Hydrogen Bonding: Hydrogen bonds are formed between hydrogen atom and atoms of the most electronegative elements (N, O, and F). Hydrogen being the smallest atom offers very little repulsion when it interacts with highly electronegative atoms in other molecules and forms partial bonds with them. Hydrogen bonds are strong but weaker compared to primary bonds. Hydrogen bonds are of two types- intermolecular and intramolecular. In intermolecular hydrogen bonds, the bonds are between hydrogen atom of one molecule and electronegative atom of another, for example, o-nitrophenol. In intramolecular hydrogen bonds, the bonds are between hydrogen atom and electronegative atom of the same molecule. However they do not have any covalent interactions, for example, p-nitrophenol.

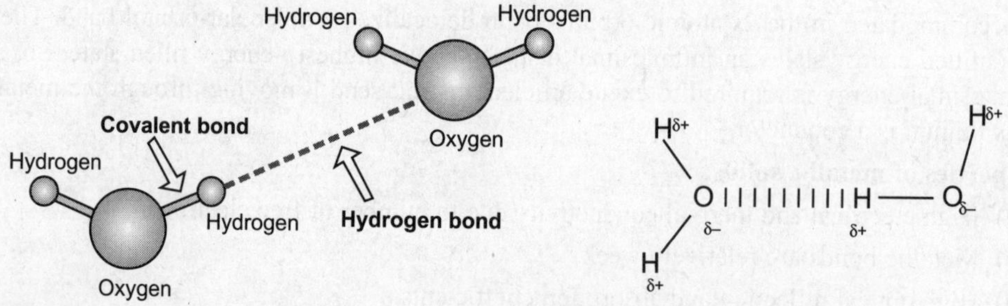

Fig. 32.12:

Hydrogen bonding

Hydrogen bond is formed when a charged part of a molecule having polar covalent bonds forms electrostatic interaction with a substance of opposite charge. Nonpolar molecules do not form hydrogen bond and Hydrogen bonds are directional.

(b) Van der Waal's Forces: Permanent dipoles are produced in asymmetric molecules in which there are permanent positive and negative regions. For example, water molecule is made of one oxygen and two hydrogen atoms. When these atoms approach each other they share a pair of electrons between hydrogen and the oxygen atom. As oxygen is a highly electronegative atom, the shared electron cloud is attracted more towards oxygen than the hydrogen atoms, giving rise to a permanent dipole. When this water molecule approaches another water molecule, a partial bond is formed between the partially positive hydrogen atom of one molecule and the partially negative oxygen of another. This partial bond is called a Van der Waal's bond. This type of bond which forms as a result of electrostatic attraction between two permanent dipoles is known as ***first type*** of Van der Waal's bond.

The ***second type*** of Van der Waal's bond is formed due to temporary dipoles. A temporary dipole is formed in a symmetric molecule but which has fluctuations of charges giving rise to partial dipole moments for only a few moments. Methane is a symmetric molecule. In the compound, the intermolecular bonds between the molecules are of weak Van der Waal's forces. When it is solidified, the solid cannot exist for a long time because of the weak nature of intermolecular bonding.

(iii) Other Types of Crystalline Solids

(a) Ceramics: Ceramics, in general insulators are solid inorganic materials that are prepared by heating at raised temperatures. But high temperature superconductor materials such as lanthanum-barium-copper oxide ceramic, yttrium-barium-copper oxide ceramic etc are discovered in ceramics. These ceramics consist of layers of copper oxide separated by layers containing other atoms.

(b) Zeolites: Zeolites are solid inorganic materials that have an aluminosilicate framework. They are network-covalent structures. These substances can act as *molecular sieves* because they have pore structures with molecular dimensions of variable sizes that can hold some molecules or ions. For example, zeolites are extensively used for ion-exchange in water purification and softening.

(c) Organic molecules: Organic molecules also form crystals. Acetamide, CH_3CONH_2, is a small organic molecule which occurs as a crystalline solid at room temperature.

Table 32.1 shows a comparison of various types of bonds.

Table 32.1: *Comparison of various types of bonds*

Type	Ionic	Covalent	Molecular	Metallic
Bond	Electron attraction - transfer	Sharing	Van der Waals	Electron gas
Properties	Hard, high mp, soluble in polar liquids, electrical insulators, Non-directional, Low conductivity.	Very hard, High mp Insoluble, Semiconductors, Directional, Low conductivity.	Soft Low mp & bp, Soluble in covalent liquids, Electrical insulators, Directional, Low conductivity.	Ductile, Metallic luster, High electrical and thermal conductivity, Non-directional, Good conductivity.

Example	NaCl, MgO Cohesive energy 3.28 eV/atom	Diamond Cohesive energy 7.4 eV/atom	Methane, water Cohesive energy 0.1 eV/atom	Na, Al, Fe Cohesive energy 1.1 eV/atom

32.4 ESTIMATION OF COHESIVE ENERGY

Let us consider a general situation of two identical atoms. Potential between a pair of atoms is called the interatomic potential. From the existence of solids we know

(*i*) There must be attractive forces between the atoms or molecules which keep them together.

(*ii*) There must be repulsive forces between the atoms or molecules as well, since large external pressures are required to compress a solid to an appreciable extent.

The forces between atoms can be attractive and repulsive. As the atoms approach, the potential energy decreases due to work done by atoms during the attractions. When separation decreases between the atoms, the repulsive force begins to act. The repulsive force is positive and hence potential energy increases.

In solids the two forces exist between the atoms, attractive forces to prevent atoms from moving away from each other and repulsive force to prevent atoms from collapsing. At equilibrium position these forces are equal. The potential energy U of the either atoms is

U = Decrease in potential energy due to attraction + Increase in potential energy due to repulsion.

The potential energy, the work done on the system can be calculated by integrating the binding force. The work done in moving through a small distance dr is

$$du(r) = F(r)dr$$

$$F(r) = \frac{A}{r^m} - \frac{B}{r^n}(n < m) \qquad \text{...(32.1)}$$

where r is the distance between atoms; m, n, A and B are constants which depend on type of the bond; $\frac{A}{r^m}$ represents attractive force; and $\frac{B}{r^n}$ represents repulsive force.

From Coulomb's inverse square law m is 2 and n takes usually 7 to 10 for metallic bonds and 10 to 12 for covalent bonds.

Potential energy of the atom

$$U(r) = \int du(r) = \int F(r)dr$$

$$U(r) = \int\left(\frac{A}{r^M} - \frac{B}{r^N}\right)dr = \int (A\bar{r}^{M} - B\bar{r}^{N})dr$$

$$U(r) = \left(\frac{Ar^{1-M}}{1-M} - \frac{Br^{1-N}}{1-N}\right) + C$$

where C is constant of integration.

$$\therefore \qquad U(r) = -\left[\frac{A}{M-1}\right]\frac{1}{r^{M-1}} + \left[\frac{B}{N-1}\right]\frac{1}{r^{N-1}} + C$$

Lct $a = \left[\dfrac{A}{M-1}\right]$ and $b - \left[\dfrac{B}{N-1}\right]$

$$m = M - 1 \text{ and } n = N - 1$$

$$\therefore \qquad U(r) = -\frac{a}{r^m} + \frac{b}{r^n} + C$$

Applying boundary condition, $r = \infty$, $U = 0$,

We get $C = 0$

$$\therefore \qquad U(r) = -\frac{a}{r^m} + \frac{b}{r^n} \qquad ...(32.2)$$

At equilibrium distance r_0, the potential energy should be a minimum. Thus,

$$\left[\frac{dU}{dr}\right]_{r=r_0} = 0$$

From Eq. (32. 2)

$$\left[\frac{dU}{dr}\right]_{r=r_0} = \frac{am}{r_0^{m+1}} - \frac{bn}{r_0^{n+1}} = 0$$

$$\frac{am}{r_0^{m+1}} = \frac{bn}{r_0^{n+1}} \qquad ...(32.3)$$

$$r_0^{n-m} = \left(\frac{b}{a}\right)\left(\frac{n}{m}\right)$$

$$r_0 = \left(\frac{b}{a}\frac{n}{m}\right)^{\frac{1}{n-m}} \qquad ...(32.4)$$

Fig. 32.13 shows the relation of inter atomic force versus interatomic spacing. Eq. (32.4) can be written as

$$\therefore \qquad U(r_0) = -\frac{a}{r_0^m} + \frac{b}{r_0^n} \qquad ...(32.5)$$

This energy corresponding to the equilibrium position, $r = r_0$, is known as *binding energy* or *energy of cohesion* of the molecule. This is also called as energy of dissociation as this much amount of energy is needed to dissociate atoms.

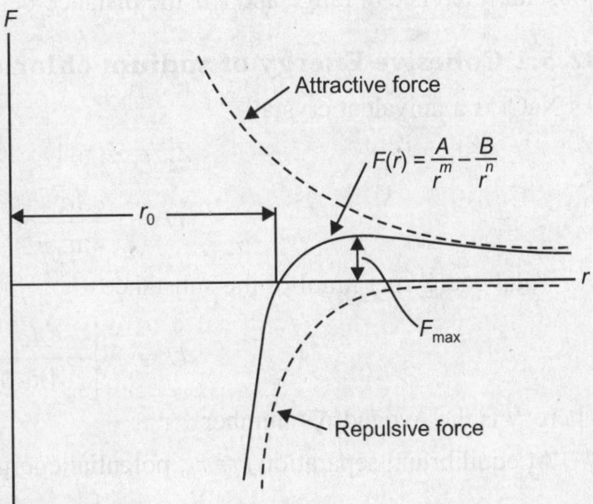

Fig. 32.13: Inter atomic force versus interatomic spacing

Dissociation energy

The positive quantity $-U(r_0)$ is the dissociation energy of the molecule, energy required to separate two atoms. Dissociation energy may occur at high temperatures or as a result of other process in which the molecule can absorb sufficient energy. A minimum in the energy curve is possible only if $n > m$.

Using minimum energy condition

$$\left[\frac{d^2U}{dr^2}\right]_{r=r_0} > 0 \qquad ...(32.6)$$

The second derivative gives

$$\left[\frac{d^2U}{dr^2}\right]_{r=r_0} = \frac{-am\,(m+1)}{r_0^{m+2}} - \frac{bn\,(n+1)}{r_0^{n+2}} > 0$$

$$-am(m+1)r_0^{n+2} + bn\,(n+1)r_0^{m+2} > 0$$

$$bn\,(n+1)r_0^{m+2} > am\,(m+1)r_0^{n+2}$$

$$bn\,(n+1) > am\,(m+1)r_0^{n-m}$$

Substituting r_0 from Eq. (32.5)

$$bn\,(n+1) > am\,(m+1)\frac{bn}{am}$$

$$(n+1) > (m+1)\; n > m \qquad ...(32.7)$$

Eq. 32.7 shows that the force acting between the atoms is mostly electrostatic in nature.

32.5 COHESIVE ENERGY OF IONIC SOLIDS

Ions can be treated as point charges. The cohesive energy of a crystal containing oppositely charged ions with charges Z_1 and Z_2

$$U = -\frac{AZ_1Z_2e^2}{4\pi\varepsilon_0 r} + B\exp\left(-\frac{r}{\rho}\right) \qquad ...(32.8)$$

where A is Madelung constant, B is repulsion constant reference to strength, ρ is the repulsion constant reference to range and r is the distance between oppositely charged ions.

32.5.1 Cohesive Energy of sodium chloride

As NaCl is a univalent crystal

$$Z_1 = Z_2 = 1$$

$$U = -\frac{Ae^2}{4\pi\varepsilon_0 r} + B\exp\left(-\frac{r}{\rho}\right) \qquad ...(32.9)$$

Total energy per kmol of the substance

$$U = N\left[-\frac{Ae^2}{4\pi\varepsilon_0 r} + B\exp\left(\frac{-r}{\rho}\right)\right] \qquad ...(32.10)$$

where N is the Avogadro's number

At equilibrium separation $r = r_0$, potential energy U is minimum.

$$\left(\frac{dU}{dr}\right)_{r=r_0} = N\left[-\frac{Ae^2}{4\pi\varepsilon_0 r} + \frac{B}{\rho}\exp\left(-\frac{r_0}{\rho}\right)\right] = 0$$

$$\left[\frac{Ae^2}{4\pi\varepsilon_0 r_0^2} - \frac{B}{\rho}\exp\left(-\frac{r_0}{\rho}\right)\right] = 0$$

$$\frac{B}{\rho}\exp\left(-\frac{r_0}{\rho}\right) = \frac{Ae^2}{4\pi\varepsilon_0 r_0^2} \qquad ...(32.11)$$

Putting Eq. (32.11) in Eq. (32.10), the potential energy at equilibrium separation

$$U_{r_0} = -\frac{Ae^2 N}{4\pi\varepsilon_0 r_0}\left(1 - \frac{\rho}{r_0}\right) \qquad ...(32.12)$$

where $-\frac{Ae^2 N}{4\pi\varepsilon_0 r_0}$ is known as Madelung constant or electrostatic energy.

Eq. (32.12) gives the cohesive energy of ionic crystal such as NaCl. This is the energy released during formation of NaCl or energy spent to separate ionic crystal into constituent ions.

The equilibrium distance r_0 is determined empirically, the range ρ is determined from the knowledge of the bulk modulus of the crystal, the Madelung constant; A is named after Erwin Madelung, a German physicist and can be calculated from the geometrical arrangement of ions in the crystal. If A and ρ are known,the lattice energy of the crystal can be determined.

For NaCl

$$\frac{r_0}{\rho} = \frac{18 r_0^4 \beta}{Ae^2} + 2 \qquad ...(32.13)$$

where β is the bulk modulus

On simplification

$$U_{r_0} = \frac{Ae^2 N}{4\pi\varepsilon_0 r_0}\left(\frac{n-1}{n}\right) = -\frac{Ae^2 N}{4\pi\varepsilon_0 r_0}\left(1 - \frac{1}{n}\right) \qquad ...(32.14)$$

where n is the repulsive constant, A is Madelung constant is a function of crystal structure (the geometrical arrangement of ions in the crystal) and its value is 1.74756; $r_0 = 0.281$ and n, Born repulsive constant = 9.

Estimation of the lattice energy for NaCl:

$$U_{r_0} = \frac{Ae^2 N}{4\pi\varepsilon_0 r_0}\left(1 - \frac{1}{n}\right)$$

$$U_{r_0} = \frac{(1.74756)\,(1.6022 \times 10^{-19} C)^2\, 6.022 \times 10^{23}/\text{mol}}{4\pi(8.854 \times 10^{-12})\,(282 \times 10^{-12}\,\text{m})}$$

$$U_{r_0} = 756 \text{ kJ/mol}$$

$$U_{r_0} = 7.96 \text{ eV}$$

The cohesive energy of NaCl is 7.95 eV.

Evaluation of theMadelung constant for NaCl:

The definition of Madelung constant is

$$\frac{A}{R} = \sum_j \frac{\pm}{r_j} \qquad ...(32.15)$$

wherer j is the distance of the jth ion from the reference ion and R is the nearest neighbour distance.

Consider a one dimensional crystal consisting of alternative positive and negative ions as shown in Fig. 32.14 and R is inter-ionic spacing.

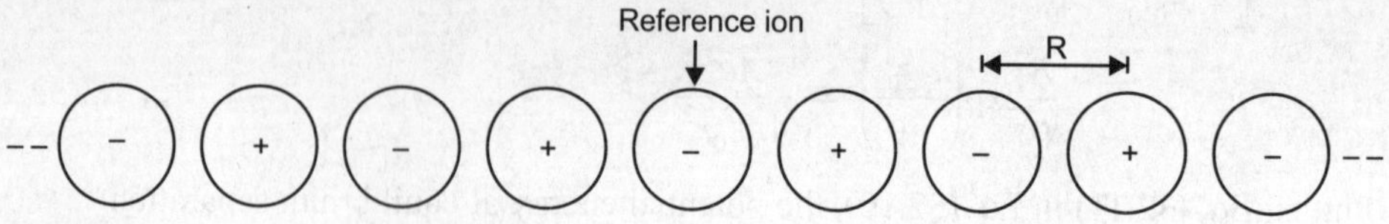

Fig. 32.14:

One dimensional NaCl crystal

Consider negative ion as the reference ion. Then,

$$\frac{A}{R} = \sum_j \frac{\pm}{r_j} = 2\left[\frac{1}{R} - \frac{1}{2R} + \frac{1}{3R} - \frac{1}{4R} + \ldots\right] \quad \ldots(32.16)$$

$$A = 2\left[1 - \frac{1}{2} + \frac{1}{3} - \frac{1}{4} + \ldots\right] \quad \ldots(32.17)$$

The factor 2 occurs because of two equivalent ions one to the right and one to left each at distance $r_j = nR$.

Using the series expansion

$$\ln(1 + x) = x - \frac{x^2}{2} + \frac{x^3}{3} - \frac{x^4}{4} + \ldots$$

Putting $x = 1$, we get $\ln 2 = 1 - \frac{1}{2} + \frac{1}{3} - \frac{1}{4} + \ldots$

Eq. (32.17) gives $A = 2 \ln 2 = 1.38$ ...(32.18)

For NaCl, the central Na^+ is surrounded by 6 Cl^- ions as first neighbours. Let us consider them at unit distance as shown in Fig. 32.15.

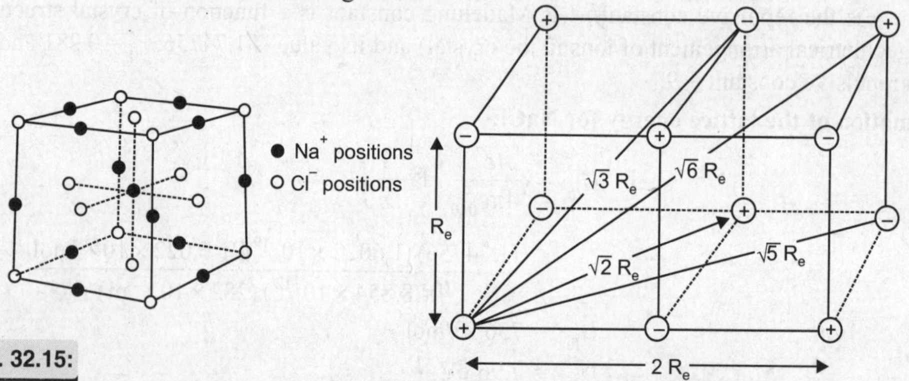

Fig. 32.15:

Geometry of NaCl, Central Na^+ is surrounded by 6 Cl^- ions as first neighbours

Twelve Na^+ ions are the second nearest neighbours at a distance $\sqrt{2}$

Eight Cl^- ions the third nearest neighbours at a distance $\sqrt{3}$ and so on

Madelung constant A for NaCl structure can be written as a summation serious.

$$A = \frac{6}{1} - \frac{12}{\sqrt{2}} + \frac{8}{\sqrt{3}} - \frac{6}{\sqrt{4}} + \frac{24}{\sqrt{5}} - \frac{24}{\sqrt{6}}$$

This converges to a value 1.74756.

QUESTIONS

1. Discuss about the forces between two interacting atoms.
2. What is meant by bonding?
3. Define the binding energy.
4. Explain the variation of (i) attractive potential energy, (ii) repulsive potential energy, (iii) resultant potential energy with inter atomic distance.
5. Derive the expression for cohesive energy.
6. Derive the expression for equilibrium spacing of two atoms for which potential energy is a minimum.
7. Explain the formation of ionic bonds with suitable examples.
8. Explain the formation of covalent bonds with suitable examples.
9. Explain the formation of metallic bonds with suitable examples.
10. Explain the formation of hydrogen bonds with suitable examples.
11. Explain the formation of Van der waals bonds with suitable examples.
12. Compare the properties of ionic, covalent and metallic bonds.
13. Explain the formation of secondary bonds.
14. What is Madelung's constant. Give its value for NaCl.
15. Give the properties of Ionic bonds.

33 Crystal Growth and Crystal Defects

Learning Objectives:

By the end of this chapter, you will be able to:

- Understand basic concepts of crystal growth - from solution, melt and vapour phase.
- Learn in detail about Crystal defects and its different types - Point defects (vacancies and schottky, interstitials and frenkel, impurities and electronic), Planar defects and Volume defects.

33.1 INTRODUCTION

Crystals are the pillars on which the base of modern technology is founded. In fact there is no field of technology that does not utilize crystals. When we say crystals, we are referring to single crystals. Single crystals are more difficult to be prepared than polycrystalline materials and extra efforts are required to be put in to produce single crystals. There are several techniques for preparation of single crystals but we dwell here only on some selected methods.

33.2 CRYSTAL GROWTH FROM SOLUTION

In this method the material from which a crystal is to be grown is dissolved in a solution. In a solution, a solvent (say, water) can only hold a certain amount of solute. This is called the *solubility* of a solution. When no more of the solid substance can be dissolved, the solution is said to be saturated. A *saturated solution* is a solution that contains the maximum concentration of a solute dissolved without forming a precipitate.

Crystals grow by a process termed *nucleation*. Fig. 33.1 shows the illustration of nucleation of unsaturated solution to saturated solution.

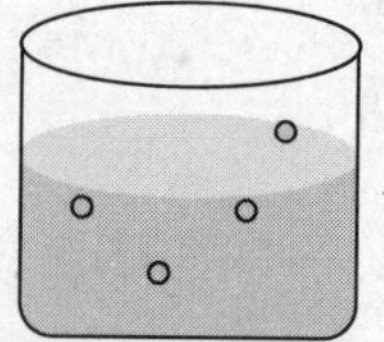

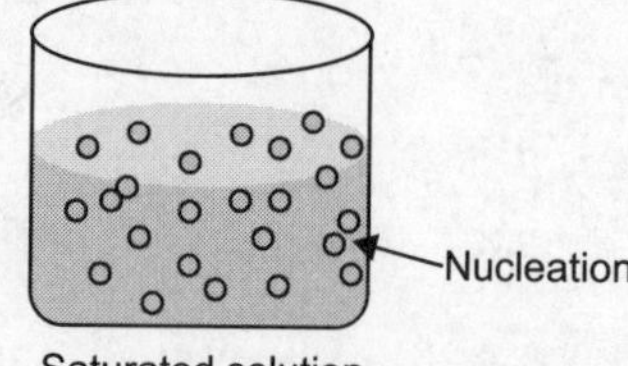

Fig. 33.1: Illustration of Nucleation of unsaturated to saturated solution

In nucleation, the solute molecules are separated into their specific units in the solvent. As the solvent evaporates, the solute particles come into contact and connect with each other into a subunit. As this subunit is bigger than a separate particle more particles will attach to it. The crystal nucleus crystallizes. Other solute molecules also slowly start to attach to the surface of the crystal. The crystal grows until equilibrium is reached between the solute molecules in the crystal and solute molecules that stay in the solution.

A seed crystal may be obtained by taking saturated solution in a glass jar and hang a small string into the liquid. Small crystals will start to grow on the string, which can be used as seed crystals. A good seed crystal is selected and is tied to a nylon thread. It is suspended in a saturated or supersaturated crystal solution so that it is completely covered as shown in Fig. 33.2. In a few days large crystals grow around the seed crystal.

Small crystals

Large crystals

Fig. 33.2: Growing of large crystal

Solution method is very useful for growing materials that have superior solubility at room temperature and have variation in solubility with temperature. However solution method is not useful to grow crystals that have low solubility.

33.3 CRYSTAL GROWTH FROM MELT

Czochralski and Bridgman techniques are prominent methods for growing single crystals from molten material.

33.3.1 Czochralski Method

Materials that are sparingly soluble in solvents may be grown into crystals from their melts. Polish scientist, Jan Czochralski developed a technique of growing crystals from molten material. Czochralski method is widely used for growing large-size single crystals of semiconductors (Silicon etc), metals (Palladium, Platinum, Silver and Gold etc) and many oxide crystals ($LaAlO_2$, YAG etc).

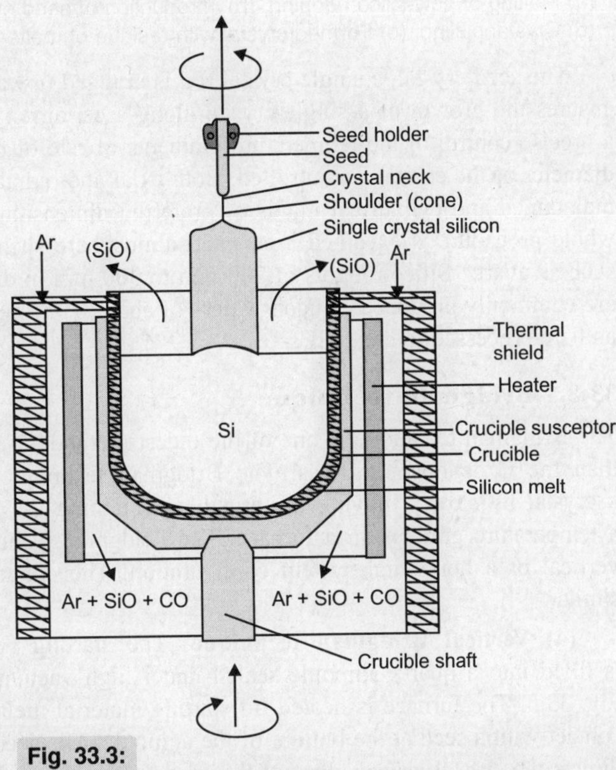

Fig. 33.3: Schematic of apparatus of the Czochralski method

In the beginning the polycrystalline material (say, silicon) to be grown into a crystal is melted under a controlled atmosphere in a non-reacting quartz crucible by induction heating or resistance heating. The melt is kept for a certain time at a temperature above the melting point and the temperature is then reduced to a value slightly above the freezing point.

A seed crystal mounted on a rod is then immersed into the melt. Predetermined amounts of suitable impurity atoms can be added into the molten intrinsic silicon to transform it into extrinsic silicon.

The apparatus of the Czochralski method is shown in Fig. 33.3 and the principle is shown in Fig. 33.4.

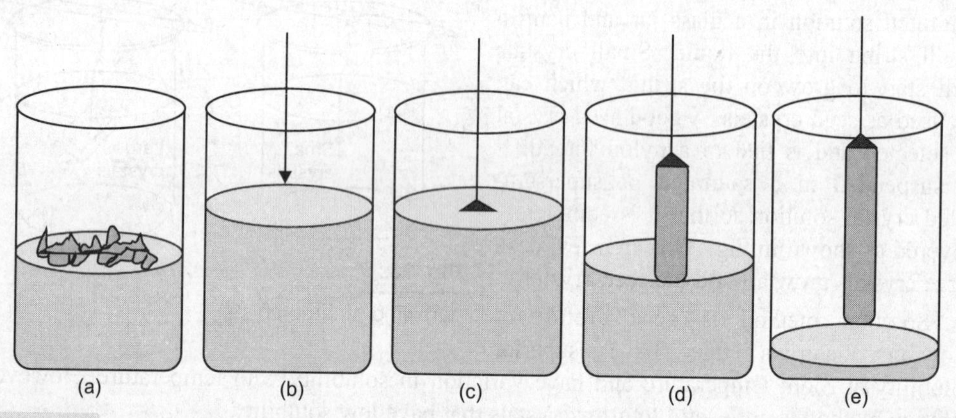

Fig. 33.4:

Schematic of principle of the Czochralski method
(a) Melting of poly-silicon doping, (b) introduction of seed crystal, (c) Beginning of crystal growth, (d) Crystal pulling, (e) Formed crystal with residue of melted silicon

The seed crystal, a single crystal rod is dragged upwards and rotated at the same time. It ensures the growth of a single-crystal along a certain crystallographic orientation. Through a precise control of the temperature gradients of rate of pulling and of speed of rotation the diameter of the crystal is controlled. Both axial and radial temperature gradients are properly maintained and cylindrical ingots of preferred dimensions are obtained from the melt. This whole procedure is executed in an inert atmosphere such as argon and in an inert chamber such as quartz. Silicon ingots of 200 mm to 300 mm in diameter and 1 to 2 meters in length are commonly produced. Silicon wafers of about 0.75 mm thick are sliced from these ingots and are processed to create ICs.

33.3.2 Bridgman technique

The Bridgman technique is one of the oldest methods used for growing crystals. It is easier than the Czochralski process. The Bridgman technique uses directional solidification of a crystal by slowly moving the crucible containing the molten material along the axis of a temperature gradient in a furnace. The Bridgman technique can be employed in either a vertical or a horizontal system configuration. The concept of these two configurations is similar.

(a) Vertical Bridgman technique: The starting material, a polycrystalline powder is filled into a quartz ampoule sealed under high vacuum, put into the Bridgman furnace, Fig. 33.5. The furnace is heated till starting material melted completely and is brought into contact with a seed at the bottom of the ampoule. The seed is a piece of single crystal which directs the crystal growth. Part of the seed gets melted when it comes into contact with the melt. This provides a fresh interface for the crystal growth. The ampoule is then moved slowly into the cooler part of the furnace. When the temperature at the bottom of the ampoule drops below the solidification temperature the crystal growth is commenced at the melt-seed interface. As the ampoule gets out through the cold zone, the melt forms in to a single-crystalline ingot. It takes generally 10-14 days. In the vertical Bridgman technique crystals grow in circular shape. This method is useful for producing semiconductor crystals, such as GaAs, ZnSe, CdTe, BGO etc.

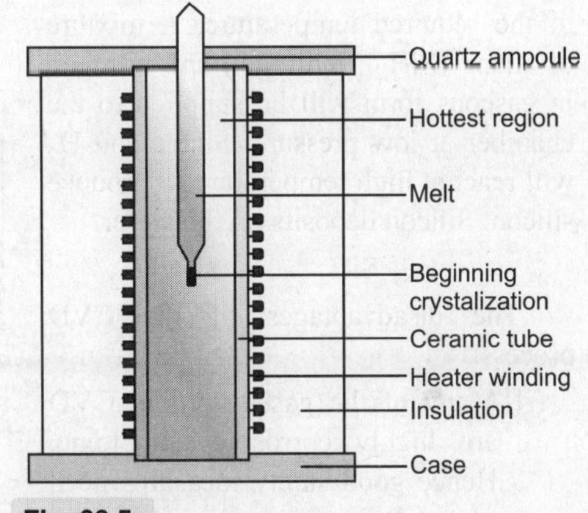

Fig. 33.5: Cross section of Bridgman furnace

(b) Horizontal Bridgman technique: It uses a two-zone furnace, Fig. 33.6. The left hand zone is maintained at a low temperature. The right hand zone of the furnace contains the polycrystalline raw material held in a crucible at a temperature just above its melting point. As the furnace moves from left to right, the melt cools and solidifies. A seed crystal is placed at the left hand side of the melt at a point where the temperature gradient is such that only the end melts. A specific orientation of single crystal propagates at the liquid-solid interface eventually to produce a single crystal. The concept of these two configurations is similar.

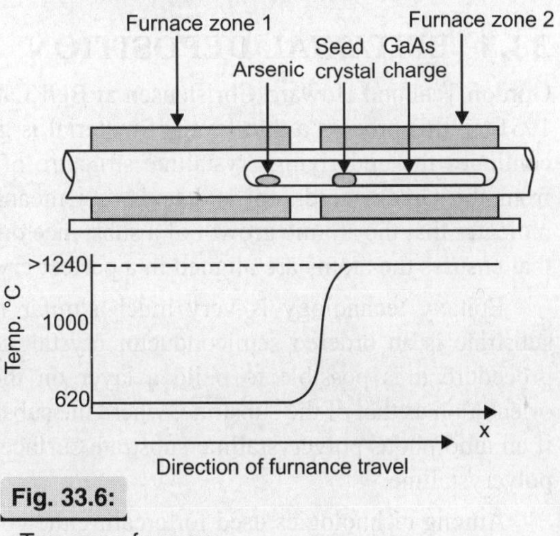

Fig. 33.6: Two-zone furnace

33.4 CRYSTAL GROWTH FROM VAPOUR PHASE

Single crystals of materials having high melting point can be obtained from vapour phase by sublimation, condensation and sputtering of elemental materials. In frequently used chemical transport reaction method, the source material is transported as a volatile species to the crystallization region. Transport of the material occurs between two zones having different temperatures. The starting reaction occurs at higher temperature and is reversed at the lower temperature to deposit molecules of compound at the favourable crystalline sites. The commercial importance of vapour growth is the production of thin layers by chemical vapor deposition.

33.4.1 Chemical Vapor Deposition

A large CVD chamber may be of 6 ft long and 3 ft diameter. The CVD chamber will also have pipes and automatic valves to create vacuum and to supply desired gases at controlled rates. The chamber is housed in an RF power furnace. Wafers on which crystals are to be grown are loaded in a tray and are placed in the reaction chamber (Fig. 33.7). Then the wafers are heated to the required temperature. A mixture of silane (SiH_4), hydrogen and nitrogen in gaseous form will be supplied to the chamber at low pressure. Silane and H_2 will react at high temperature to produce silicon. Silicon deposits on the wafer.

$$SiH_4 \rightarrow Si + 2H_2$$

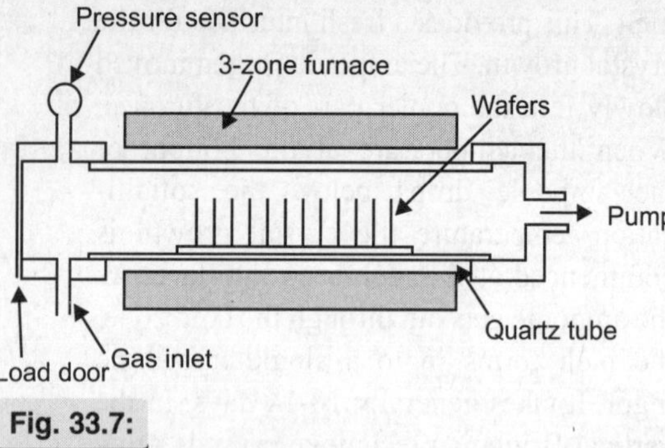

Fig. 33.7: Schematic of a CVD chamber

The disadvantages of the CVD process:

(*i*) Many of the gases used in CVD are highly corrosive and toxic. Hence good safety measures need to be taken.

(*ii*) The cost of the high purity chemicals is high.

33.5 EPITAXIAL DEPOSITION

Gordon Teal and Howard Christensen at Bell Labs developed epitaxial deposition process in 1951. In this process a thin layer of material is grown on a substrate and the deposited layer continues the underlying crystalline structure of the substrate. The word epitaxy is derived from the Greek words epi and taxis. epi means ''on'' and taxis means "arrangement". It indicates that the atomic growth of a substance on to the surface of a substrate occurs in a way that ensures the atoms are aligned in a perfect crystal lattice.

Epitaxy technology is very much similar to CVD process. However, in epitaxy, the substrate is an ordered semiconductor crystal such as silicon, gallium arsenide. Using this procedure it is possible to build a layer on the substrate with the same crystallographic orientation as that of the substrate, where the substrate acts as a seed for the deposition. Hence, if an amorphous/polycrystalline substrate surface is used, the film will also be amorphous or polycrystalline.

Among technologies used for creating the conditions needed to support epitaxial growth, Vapor Phase Epitaxy (VPE) is a prominent one.

33.5.1 Vapor Phase Epitaxy

Epitaxy technology is used for making silicon on insulator (SOI) substrates. When the deposited layer is the same material as the substrate for example silicon on silicon it is called *homoepitaxy*. If a material different from the substrate is deposited for example silicon on silicon oxide it is called *heteroepitaxy*. The epitaxial layer deposition is carried out in a chamber called an epitaxial reactor. Fig. 33.8 shows the schematic of a horizontal vapor phase epitaxial reactor.

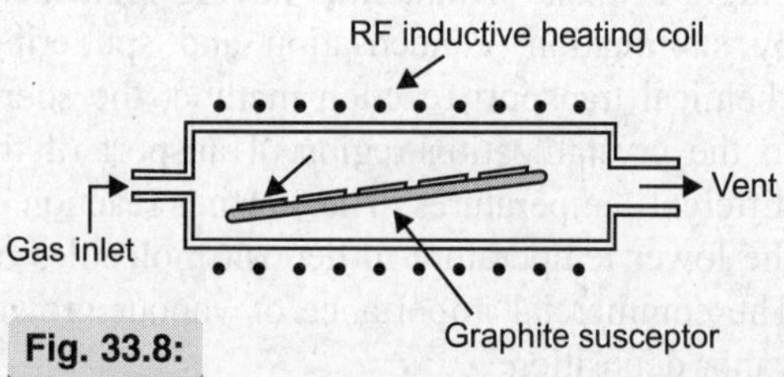

Fig. 33.8: A schematic diagram of a typical VPE reactor

A hydrogen carrier gas is employed first to drive away air in the reactor. The reactor is then heated to a high temperature. Anhydrous HCl gas is fed into the reactor. At about 1150°C the HCl gas reacts with the silicon at the surface of wafers and etches the silicon surface. The etching is necessary to remove a small amount of Si and other contaminants from the wafer surfaces to ensure that a clear freshly etched silicon surface will be available for epitaxial layer deposition. The wafers are now heated to a temperature 1150°C -1250°C. $SiCl_4$ and hydrogen as a carrier gas are fed into the chamber.

The reaction proceeds as follows:

$$SiCl_4 + 2H_2 \rightarrow Si + 4HCl$$

With reduction of $SiCl_4$ the reaction gives rise to free silicon atoms. Atoms from the gas phase skid about on the surface of the growing epitaxial film until they find correct position in the lattice before becoming fastened into the growing structure. Once growth is complete, the temperature reduced and the reactor is allowed to cool to ambient temperature. The hydrogen flow is replaced by a nitrogen flow so that the reactor may be opened safely. Diborane (B_2H_6) or phosphine (PH_3) is added to the process gases, to create doped layers. Fig. 33.9 shows illustration of a barrel reactor for epitaxial processes.

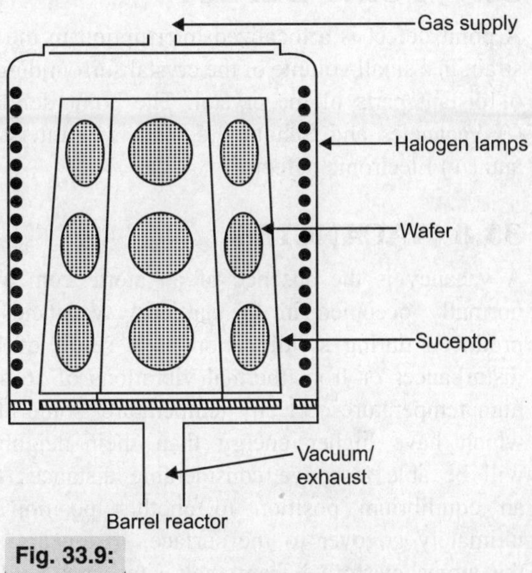

Fig. 33.9: Barrel reactor for epitaxial process

33.6 CRYSTAL DEFECTS

Solids are characterized by regular arrangement of atoms over the entire volume of the material. This is the description of an *ideal* solid. Some of the properties of solids such as specific heat, elasticity, magnetic properties etc can be explained on the basis of the above consideration. However, we cannot explain other properties such as colour of crystals, luminescence etc. In the real solids, on account of thermodynamic considerations, and also on account of conditions of crystal growth, some atoms are not found at places where they should be. The atoms may occupy sites where they are not expected. These are known as *defects* in crystals. Almost all the crystals contain defects, and these defects change the physical properties of crystals to some extent.

The ideal crystal has an infinite 3D repetition of identical units, which may be atoms or molecules. Therefore, it does not contain lattice imperfections. Therefore, in an ideally perfect crystal long range order would be observed over unlimited distances. However, ideal crystals neither occur in nature nor can be produced by artificial methods. The real crystals are limited in size and repeated violations of long range order would be found in them.

A defect or an imperfection is any deviation from the perfect periodic arrangement of atoms observed in real crystals.

Classification of defects: The defects found in real crystals are classified into four main categories. They are:

(*i*) Point defects (Zero dimensional defects)

(*ii*) Line defects or dislocations (One dimensional defects)

(*iii*) Planar defects (Two dimensional defects) and

(*iv*) Volume defects (Three dimensional defects).

33.7 POINT DEFECTS

A point defect is a localized interruption in the regularity of the crystal lattice. It produces strain in a small volume of the crystal surrounding the defect, but does not affect the perfection of distant parts of the crystal. The point defects can be divided into four types, namely (*i*) Vacancies and Schottky defects, (*ii*) Interstitials and Frenkel defects, (*iii*) Impurities, and (*iv*) Electronic defects.

33.8 VACANCIES

A vacancy is the absence of an atom from a site normally occupied in the lattice. Vacancies are produced during solidification as a result of local disturbances or from thermal vibrations of atoms at high temperatures. At any temperature, some atoms which have higher energy than their neighbours will be able to move considerable distances from an equilibrium position to another position and ultimately go over to the surface. It appears as if the atoms evaporate from their sites and condense elsewhere. The site left by the atom is a vacancy in the lattice, as shown in Fig. 33.10. Such vacancies arise in close packed structures, that is, in metallic structures. The number of vacancies per unit volume of the crystal depends upon the temperature.

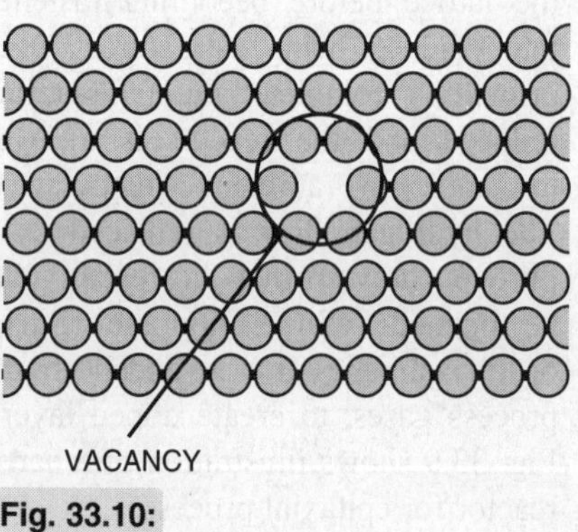

Fig. 33.10: A Vacancy

Summary:

(*i*) A vacancy is an atomic site from which the atom is missing.

(*ii*) Vacancies form in close packed metallic structures.

(*iii*) The concentration of vacancies in a crystal depends on the temperature.

33.9 ENERGY OF FORMATION OF VACANCY IN A METALLIC CRYSTAL

Point defects are produced in a crystal due to thermal vibrations. At any given temperature, some of the lattice points normally occupied by metal ions are vacant, giving rise to vacancies. The energy of formation of a vacancy may be defined as the energy difference between the energy needed to remove an atom from the interior of the crystal and take it to infinity, and the energy gained when the atom is placed on the surface of the crystal.

When a vacancy is produced, the internal energy of the crystal increases. At the same time, it increases the entropy of the crystal also. The *free energy* of the system, denoted by G is defined as

$$G = H - TS$$

where H is the *enthalpy* of the system, and S is *entropy* of the system.

At thermal equilibrium at any temperature, the free energy of the crystal is a minimum. That is, $\partial G = 0$.

Let N be the total atomic sites in the crystal. Let n be the number of vacancies at temperature T. The remaining $(N - n)$ sites are occupied by atoms. The vacancies are indistinguishable particles and can be arranged among themselves in $n!$ ways. Similarly, the atoms are indistinguishable and can be arranged among themselves in $(N - n)!$ ways. The number of ways in which we can arrange n vacancies and $(N - n)$ atoms among N sites will be equal to ${}^{N}C_n$. The probability of a distribution in which n vacancies can be created is given by

$$W = {}^{N}C_n = \frac{N!}{(N-n)!n!} \qquad ...(33.1)$$

The change in entropy as a result of creation of vacancies, is given by Boltzmann equation as

$$\Delta S = k \log_e W \qquad ...(33.2)$$

where k is the Boltzmann constant.

$$\therefore \qquad \Delta S = k \log_e \frac{N!}{(N-n)!n!}$$

$$= k[\log_e N! - \log_e (N-n)! - \log_e n!] \qquad ...(33.3)$$

Using Stirling's approximation $\log_e x! \approx x \log_e x - x$ for $x >> 1$ into the above equation, we get

$$\Delta S = k[(N \log_e N - N) - (N-n)\log_e (N-n) + (N-n) - n \log_e n + n]$$

or

$$\Delta S = k[N \log_e N - (N-n)\log_e (N-n) - n \log_e n] \qquad ...(33.4)$$

In formation of vacancies, a change in free energy occurs due to a change in enthalpy and a change in entropy. It is expressed as

$$\Delta G = \Delta H - T\Delta S$$

where ΔH is enthalpy of vacancy formation. It is the change in enthalpy resulting from the addition of vacancies and arises from the increase in internal energy caused by breaking interatomic bonds (i.e. removing atoms).

As the process takes place at constant volume,

$$\Delta H = \Delta E,$$

where ΔE is the change in the internal energy of the crystal. Hence,

$$\Delta G = \Delta E - T\Delta S \qquad ...(33.5)$$

If E_υ is the average energy required to create a vacancy, then ΔE is the energy change due to creation of n vacancies. Thus,

$$\Delta E = n E_\upsilon \qquad ...(33.6)$$

Substituting for ΔE and ΔS into the Eq. (33.5), we get

$$\Delta G = n E_\upsilon - kT[N \log_e N - (N-n)\log_e (N-n) - n \log_e n] \qquad ...(33.7)$$

The equilibrium state of the crystal at any temperature requires that its free energy is a minimum. As the process takes place at constant volume, the change of free energy occurs due to a change in the number of vacancies.

$$\therefore \qquad \frac{\partial \Delta G}{\partial n} = \frac{\partial}{\partial n}\{n E_\upsilon - kT[N \log_e N - (N-n)\log_e (N-n) - n \log_e n]\}$$

$$= E_\upsilon - kT \frac{\partial}{\partial n}[-(N-n)\log_e (N-n) - n \log_e n] \qquad (\because N \text{ is a constant})$$

$$= E_\upsilon - kT\left\{-\left[\left(\frac{N-n}{N-n}\right)(-1) + \log_e(N-n)(-1) + \frac{n}{n} + \log_e n\right]\right\}$$

$$= E_\upsilon - kT\{-(-1-\log_e(N-n)+1+\log_e n)\}$$

$$= E_\upsilon - kT\log_e\left(\frac{N-n}{n}\right) \qquad ...(33.8)$$

Therefore, the equilibrium condition is that $\frac{\partial \Delta G}{\partial n} = 0$ ...(33.9)

$$\therefore \qquad E_\upsilon - kT\log_e\left(\frac{N-n}{n}\right) = 0$$

or

$$E_\upsilon = kT\log_e\left(\frac{N-n}{n}\right) \qquad ...(33.10)$$

Eq.(33.10) indicates the average energy of formation of a vacancy in a crystal.

33.9.1 Equilibrium Concentration of Vacancies in a Metallic Solid

The energy of formation of a vacancy in a crystal is given by

$$E_\upsilon = kT\log_e\left(\frac{N-n}{n}\right)$$

$$\therefore \qquad \left(\frac{N-n}{n}\right) = \exp\left(\frac{E_\upsilon}{kT}\right)$$

The number of vacancies is much smaller than the total number of atoms. That is, $n \ll N$ and therefore we can approximate that $(N-n) \approx N$.

$$\therefore \qquad \left(\frac{N}{n}\right) = \exp\left(\frac{E_\upsilon}{kT}\right) \quad \text{or} \quad \left(\frac{n}{N}\right) = \exp\left(\frac{-E_\upsilon}{kT}\right)$$

$$\therefore \qquad n = N\exp\left(\frac{-E_\upsilon}{kT}\right) \qquad ...(33.11)$$

Eq. (33.11) shows that the concentration of vacancies in a metal depends on temperature. It increases rapidly as the temperature of the metal is increased. The formation energy of vacancy is characteristic of the metal; it depends on the nature of the metal.

Example 33.1: *In the germanium crystal the equilibrium vacancy concentration decreased by six orders of magnitude when the temperature was reduced from 600 to 300°C. Calculate the energy of formation of the vacancy.*

Solution: The equilibrium vacancy concentration is given by $n = N\exp\left(\frac{-E_\upsilon}{kT}\right)$.

$\therefore$ Equilibrium vacancy concentration at 873 K is $n_{873} = N\exp\left(\frac{-E_\upsilon}{873k}\right)$

and the equilibrium vacancy concentration at 573 K is $n_{573} = N\exp\left(\frac{-E_\upsilon}{573k}\right)$

$$\therefore \quad \frac{n_{873}}{n_{573}} = \exp\left[-\frac{E_\upsilon}{k}\left(\frac{1}{873}-\frac{1}{573}\right)\right] = \exp\left[\frac{E_\upsilon}{k}\left(\frac{1}{573}-\frac{1}{873}\right)\right]$$

$$= \exp\left[\frac{E_\upsilon \times 6\times10^{-4}}{8.625\times10^{-5}\,\text{eV}}\right]$$

$$\therefore \quad 10^6 = e^{0.696\,E_\upsilon}$$

$$ln\ 10^6 = 0.696\ E_\upsilon$$

$$\therefore \quad E_\upsilon = \mathbf{1.99\ eV}$$

33.10 SCHOTTKY DEFECT

In case of ionic crystals, the formation of a vacancy requires a local readjustment of charge such that charge neutrality is maintained in the crystal as a whole. Therefore, a pair of vacancies arises due to missing of one cation and one anion from the structure. Such a pair of vacant sites is called a **Schottky defect.** A Schottky defect is shown in Fig. 33.11. The vacancies disturb the arrangement of atoms around them. The atoms surrounding a vacancy collapse towards the vacancy producing tensile stress.

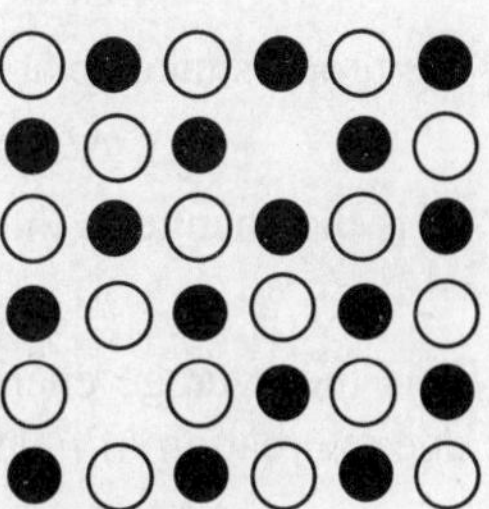

Fig. 33.11: A missing cation (vacancy) and also a missing anion nearby is called a Schottky defect

Summary:

(*i*) Schottky defects are ion vacancies. This type of point defect forms in ionic crystals.

(*ii*) A Schottky defect is the combination of one cation vacancy and one anion vacancy.

(*iii*) The concentration of Schottky defects decreases the density of the crystal.

33.11 EQUILIBRIUM CONCENTRATION OF SCHOTTKY DEFECTS IN AN IONIC CRYSTAL

In case of ionic crystals, point defects occur in pairs. A Schottky defect occurs when a cation-anion pair goes missing from the respective lattice positions.

Let us consider that the crystal contains equal number of cations and anions and let N be the total number of ions. Let n be the number of Schottky defects, i.e., n cation vacancies and n anion vacancies in the crystal. The number of different ways in which cation defects can be produced is given by ${}^NC_n$. Similarly, the number of different ways in which anion defects can be produced is given by ${}^NC_n$. The probability of a distribution in which n Schottky defects (defect pairs) can be produced is given by the compound probability, which is the product of the two probabilities. Thus,

$$W = \left[\frac{N!}{(N-n)!n!}\right]^2 \qquad ...(33.12)$$

The change in entropy as a result of creation of vacancies, is given by Boltzmann equation as

$$\Delta S = k \log_e W \qquad ...(33.13)$$

where k is the Boltzmann constant.

$$\therefore \qquad \Delta S = 2k \log_e \frac{N!}{(N-n)!\,n!}$$

$$= 2k[\log_e N! - \log_e(N-n)! - \log_e n!] \qquad ...(33.14)$$

Using Stirling's approximation $\log_e x! \approx x \log_e x - x$ for $x >> 1$ into the above equation, we get

$$\Delta S = 2k\,[(N \log_e N - N) - (N-n)\log_e(N-n) + (N-n) - n \log_e n + n]$$

or
$$\Delta S = 2k\,[N \log_e N - (N-n)\log_e(N-n) - n \log_e n] \qquad ...(33.15)$$

In formation of defects, a change in free energy occurs due to a change in internal energy and a change in entropy. It is expressed as

$$\Delta G = \Delta H - T\Delta S$$

As the process takes place at constant volume,

$$\Delta H = \Delta E,$$

where ΔE is the change in the internal energy of the crystal. Hence,

$$\Delta G = \Delta E - T\Delta S \qquad ...(33.16)$$

If E_S is the average energy required to create a Schottky defect, then ΔE is the energy change due to creation of n defects. Thus,

$$\Delta E = nE_S \qquad ...(33.17)$$

Substituting for ΔE and ΔS into the Eq. (33.16), we get

$$\Delta G = nE_S - 2kT[N \log_e N - (N-n)\log_e(N-n) - n \log_e n] \qquad ...(33.18)$$

The equilibrium state of the crystal at any temperature requires that its free energy is a minimum. As the process takes place at constant volume, the change of free energy occurs due to a change in the number of vacancies.

$$\therefore \qquad \frac{\partial \Delta G}{\partial n} = \frac{\partial}{\partial n}\{nE_S - 2kT\,[N \log_e N - (N-n)\log_e(N-n) - n \log_e n]\}$$

$$= E_S - 2kT\frac{\partial}{\partial n}[-(N-n)\log_e(N-n) - n\log_e n] \qquad (\because N \text{ is a constant})$$

$$= E_S - 2kT\left\{-\left[\left(\frac{N-n}{N-n}\right)(-1) + \log_e(N-n)(-1) + \frac{n}{n} + \log_e n\right]\right\}$$

$$= E_S - 2kT\{-(-1 - \log_e(N-n) + 1 + \log_e n)\}$$

$$= E_S - 2kT\log_e\left(\frac{N-n}{n}\right) \qquad ...(33.19)$$

Therefore, the equilibrium condition is that $\dfrac{\partial \Delta G}{\partial n} = 0$ ...(33.20)

$$\therefore \qquad E_S - 2kT\log_e\left(\frac{N-n}{n}\right) = 0$$

or
$$E_S = 2kT\log_e\left(\frac{N-n}{n}\right) \qquad ...(33.21)$$

$$\therefore \qquad \left(\frac{N-n}{n}\right) = \exp\left(\frac{E_S}{2\,kT}\right)$$

The number of defects is much smaller than the total number of ion pairs. That is, $n << N$ and therefore we can approximate that $(N - n) \approx N$.

$$\therefore \quad \left(\frac{N}{n}\right) = \exp\left(\frac{E_S}{2kT}\right) \quad \text{or} \quad \left(\frac{n}{N}\right) = \exp\left(-\frac{E_S}{2kT}\right)$$

$$\therefore \quad n = N \exp\left(-\frac{E_S}{2kT}\right) \qquad \text{...(33.22)}$$

Eq. (33.22) shows that the concentration of defects in an ionic crystal depends on temperature. It increases exponentially as the temperature of the crystal is increased. The average energy required to produce Schottky defects is characteristic of the solid; it depends on the nature of the solid.

Example 33.2: *If the average energy required for producing a Schottky defect is 1.97 eV in the ionic crystal NaCl, calculate the density of Schottky defects at 27°C. Given that the interionic distance is 2.82 Å.*

Solution: A unit cell of NaCl crystal contains **4** molecules of NaCl.

Volume of unit cell = $(2 \times 2.82 \times 10^{-3}\text{ m}) = 1.794 \times 10^{-28}\text{ m}^3$.

$$\text{Concentration of molecules} = \frac{4}{1.794 \times 10^{-28}} = 2.23 \times 10^{28}.$$

$$E_S = 1.97 \times 1.602 \times 10^{-19}\text{ J}$$

The concentration of Schottky defects is given by $n = N \exp\left(-\frac{E_S}{2kT}\right)$.

$$\therefore \quad n = 2.23 \times 10^{28} \times \exp\left[-\frac{1.97 \times 1.602 \times 10^{-19}\text{ J}}{2 \times 1.38 \times 10^{-23}\text{ J/K} \times 300\text{K}}\right]$$

$$= 2.23 \times 10^{28} \times e^{-38}$$

$$= 2.23 \times 10^{28} \times 3.14 \times 10^{-17} = \mathbf{7 \times 10^{11}}.$$

33.12 INTERSTITIALS

An Interstitial is an atom on a non-lattice site, as shown in Fig. 33.12.

An interstitial defect arises when an atom occupies a position in the lattice that is not normally occupied in the perfect crystal. Interstitial atom produces lattice strain because it tends to push the surrounding atoms further apart.

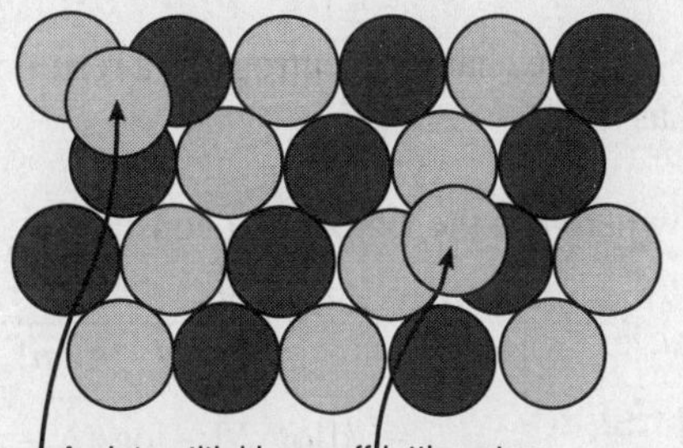

An interstitial is an off-lattice atom which may be a foreign atom or a regular atom

Fig. 33.12: Interstitial

33.13 FRENKEL DEFECT

In ionic crystals, a cation goes into an interstitial position and the vacancy-interstitialcy pair is called a ***Frenkel defect***. Frenkel defect is illustrated in Fig. 33.13.

Summary:

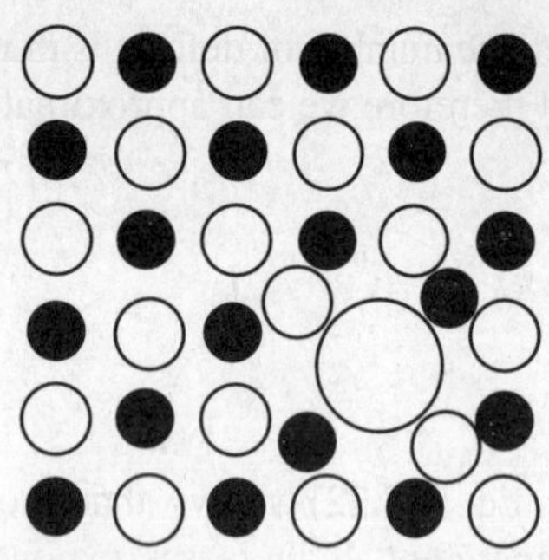

Fig. 33.13: A cation displaced from its structurally correct site becomes interstitial - Frenkel defect

(*i*) In case of ionic crystals, an ion displaced from the lattice site into an interstitial site is called a Frenkel defect.

(*ii*) As cations are generally smaller in size, they may get displaced into the void space present between in the crystal. Anions are bigger in size and they do not go into void spaces.

(*iii*) A Frenkel defect is the combination of one cation vacancy and one cation interstitial defect.

(*iv*) A Frenkel defect does not change the overall electrical neutrality of the crystal.

(*v*) The concentration of Frenkel defects does not change the density of the crystal.

33.14 EQUILIBRIUM CONCENTRATION OF FRENKEL DEFECTS IN AN IONIC CRYSTAL

Frenkel defects are produced when the crystal is heated or a high energy radiation is used to irradiate the crystal. Some of the atoms are knocked out of their normal positions in the lattice and enter into the interstices of the structure. Thus, vacancies and interstitials are simultaneously produced.

Let us consider that the crystal contains equal number of cations and anions and let N be the total number of ions. Let n be the number of Frenkel defects, i.e., n cation or anion vacancies and n interstitial ions in the crystal. The number of different ways in which vacancy defects can be produced is given by ${}^{N}C_n$. If N_i is the number of interstitial positions, the number of different ways in which interstitial defects can be produced is given by ${}^{N_i}C_n$. The probability of a distribution in which n Frenkel defects can be produced is given by the compound probability, which is the product of the two probabilities. Thus,

$$W = \frac{N!}{(N-n)!n!} \cdot \frac{N_i!}{(N_i-n)!n!} \qquad \text{...(33.23)}$$

The change in entropy as a result of creation of vacancies, is given by Boltzmann equation as

$$\Delta S = k \log_e W \qquad \text{...(33.24)}$$

where k is the Boltzmann constant.

$$\therefore \quad \Delta S = k \log_e \left[\frac{N!}{(N-n)!\,n!} \cdot \frac{N_i!}{(N_i-n)!\,n!} \right]$$

$$= k[\log_e N! - \log_e (N-n)! - \log_e n! + \log_e N_i! - \log_e (N_i - n)! - \log_e n!] \qquad \text{...(33.25)}$$

Using Stirling's approximation $\log_e x! \approx x \log_e x - x$ for $x >> 1$ into the above equation, we get

$$\Delta S = k \begin{bmatrix} (N \log_e N - N) - (N-n)\log_e (N-n) + (N-n) - n \log_e n + n \\ + (N_i \log_e N_i - N_i) - (N_i - n)\log_e (N_i - n) + (N_i - n) - n \log_e n + n \end{bmatrix}$$

or

$$\Delta S = k[N \log_e N + N_i \log_e N_i - (N-n)\log_e (N-n) - (N_i - n)\log_e (N_i - n) - 2n \log_e n] \qquad \text{...(33.26)}$$

In formation of defects, a change in free energy occurs due to a change in internal energy and a change in entropy. It is expressed as

$$\Delta G = \Delta H - T\Delta S$$

As the process takes place at constant volume,

$$\Delta H = \Delta E,$$

where ΔE is the change in the internal energy of the crystal. Hence,

$$\Delta G = \Delta E - T\Delta S \qquad ...(33.27)$$

If E_f is the average energy required to create a Frenkel defect, then ΔE is the energy change due to creation of n defects. Thus,

$$\Delta E = nE_f \qquad ...(33.28)$$

Substituting for ΔE and ΔS into the Eq. (33.27), we get

$$\Delta G = nE_f - kT[N \log_e N + N_i \log_e N_i - (N-n)\log_e(N-n) - (N_i - n)\log_e(N_i - n) - 2n \log_e n] \qquad ...(33.29)$$

The equilibrium state of the crystal at any temperature requires that its free energy is a minimum. As the process takes place at constant volume, the change of free energy occurs due to a change in the number of vacancies.

$$\therefore \quad \frac{\partial \Delta G}{\partial n} = \frac{\partial}{\partial n}\{nE_f - kT[N\log_e N + N_i \log_e N_i - (N-n)\log_e(N-n) - (N_i - n)\log_e(N_i - n) - 2n \log_e n]\}$$

$$= E_f - kT\frac{\partial}{\partial n}[-(N-n)\log_e(N-n) - (N_i - n)\log_e(N_i - n) - 2n\log_e n]$$

$$(\because N \text{ is a constant})$$

$$= E_f - kT\left\{-\left[\left(\frac{N-n}{N-n}\right)(-1) + \log_e(N-n)(-1) + \left(\frac{N_i - n}{N_i - n}\right)(-1) + \log_e(N_i - n)(-1) + \frac{2n}{n} + \log_e n\right]\right\}$$

$$= E_f - kT\{-(-1 - \log_e(N-n) - 1 - \log_e(N_i - n) + 2 + \log_e n)\}$$

$$= E_f - kT\log_e\left(\frac{N-n}{n}\right)\left(\frac{N_i - n}{n}\right)$$

$$= E_f - kT\log_e\frac{(N-n)(N_i-n)}{n^2} \qquad ...(33.30)$$

Therefore, the equilibrium condition is that $\dfrac{\partial \Delta G}{\partial n} = 0$ ...(33.31)

$$\therefore \quad E_f - kT\log_e\frac{(N-n)(N_i-n)}{n^2} = 0$$

or

$$E_f = kT\log_e\frac{(N-n)(N_i-n)}{n^2} \qquad ...(33.32)$$

$$\therefore \quad \frac{(N-n)(N_i-n)}{n^2} = \exp\left(\frac{E_f}{kT}\right)$$

The number of defects is much smaller than the total number of ion pairs. That is, $n << N$ and $n << N_i$, therefore we can approximate that $(N-n) \approx N$ and $(N_i - n) \approx N_i$.

$$\therefore \quad \left(\frac{NN_i}{n^2}\right) = \exp\left(\frac{E_f}{kT}\right)$$

or

$$\left(\frac{n^2}{NN_i}\right) = \exp\left(-\frac{E_f}{kT}\right)$$

$$\therefore \quad n^2 = NN_i \exp\left(-\frac{E_f}{kT}\right)$$

or

$$n = \sqrt{NN_i} \exp\left(-\frac{E_f}{2kT}\right) \qquad ...(33.33)$$

Eq. (33.33) shows that the concentration of defects in an ionic crystal depends on temperature. It increases exponentially as the temperature of the crystal is increased. The average energy required to produce Frenkel defects is characteristic of the solid and depends on the nature of the solid.

Example 33.3: *If the average energy required to produce a Frenkel defect in an ionic crystal is 1.4 eV, find out the ratio of the number of Frenkel defects at 20°C and 300°C in one gram of the crystal.*

Solution: The concentration of Frenkel defects is given by

$$n = \sqrt{NN_i} \exp\left(-\frac{E_f}{2kT}\right).$$

$$\therefore \quad n_{293} = (NN_i)^{1/2} \exp\left(-\frac{1.4}{586k}\right)$$

and

$$n_{573} = (NN_i)^{1/2} \exp\left(-\frac{1.4}{1146k}\right)$$

$$\therefore \quad \frac{n_{293}}{n_{573}} = \exp\left[\left(-\frac{1.4}{k}\right)\left(\frac{1}{586} - \frac{1}{1146}\right)\right] = \exp\left(-\frac{1.4}{8.625 \times 10^{-5}}\right) \times 8.3 \times 10^{-4} = e^{-13.47}$$

or

$$\frac{n_{293}}{n_{573}} = \mathbf{1.4 \times 10^{-6}}$$

33.15 IMPURITIES

An *impurity* is the substitution of a regular lattice atom with an atom that does not normally occupy that site.The impurities can occupy two different sites in the lattice. An impurity atom may replace an atom of the parent lattice site. Such impurities are known as *substitutional impurity* or *substitutional defect*. Or an impurity may occupy an interstitial position which is a non-atomic site. These impurities are known as *interstitial impurities*.

(a) Substitutional impurity: A substitutional impurity arises when a host atom in the lattice is replaced by a foreign atom. The substitutional atom remains in at the regular lattice site. The substitutional defects are illustrated in Fig. 33.14.

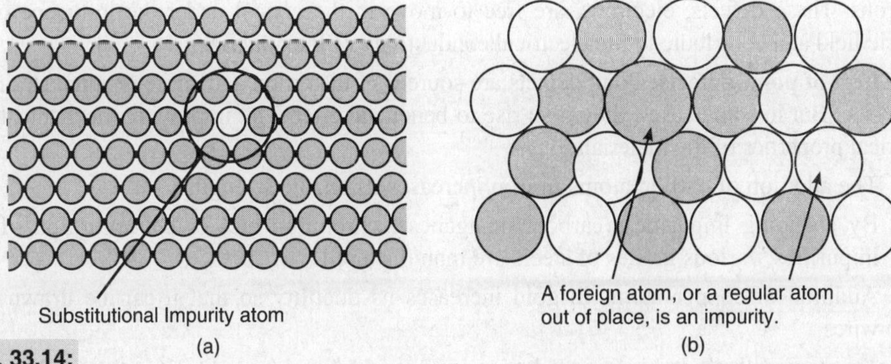

Fig. 33.14: Impurities

A small substitutional atom causes a tensile stress in the lattice (Fig. 33.14a), while a large substitutional atom produces a compressive stress (Fig. 33.14b).

A controlled addition of impurity to an intrinsic semiconductor crystal is the basis of manufacture of many semiconductor devices. Here, impurity atoms of nearly the same size substitute in place of some of the host semiconductor atoms. During the production of brass alloy, zinc atoms are doped in copper lattice. Such alloys are called substitutional solid solutions.

(b) Interstitial impurity: An interstitial impurity is a small sized atom occupying the void space in the parent crystal, without dislodging any of the host atoms from its site (Fig. 33.15). A foreign atom can occupy an interstitial space when it is substantially smaller than the host atom. Carbon steel is made by introducing carbon atoms in the void spaces formed in iron lattice. Such addition of carbon to iron increases the mechanical strength of iron.

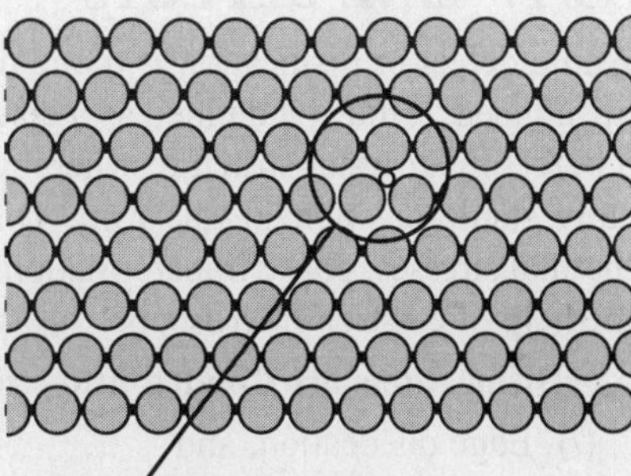

Fig. 33.15: Interstitial impurity

33.16 ELECTRONIC DEFECTS

Electronic defects are the defects in charge distribution in solids. When the composition of an ionic crystal does not correspond to the exact stoichiometric formula these defects arise. For instance, NaCl is stoichiometric if the ratio of Na^+ ions to Cl^- ions is exactly 1:1.

The non-stoichiometric composition, having deviation from the exact ratio is caused by an excess of metal ions. When a stoichiometric compound ZnO is heated in the presence of zinc vapour, a nonstoichiometric compound Zn_yO is produced. Some of ionized zinc atoms enter the lattice of ZnO and stay around as interstitial cations, as shown in Fig. 33.16. Since the crystal must be electrically neutral, the excess of cations must be balanced by an equal number of

Zn^{2+} O^{2-} Zn^{2+} O^{2-} Zn^{2+} O^{2-}
O^{2-} Zn^{2+} O^{2-} Zn^{2+} O^{2-} Zn^{2+}
⊖ Zn^{2+} ⊖
Zn^{2+} O^{2-} Zn^{2+} O^{2-} Zn^{2+} O^{2-}
⊖ Zn^{2+}
O^{2-} Zn^{2+} O^{2-} Zn^{2+} O^{2-} ⊖ Zn^{2+}

Fig. 33.16: Electronic Defects

electrons. These defects, electrons, are free to move in the crystal under the influence of an electric field and contribute to the electrical conductivity of the material.

Effect of point defects: Point defects are sources of distortion and cause residual strains in the lattice. But in some cases, they give rise to beneficial effects by improving mechanical and electrical properties of the materials.

(*i*) The addition of carbon atoms in iron increases its tensile strength.

(*ii*) By changing impurities (carbon, manganese, chromium etc) or concentration of the impurities, various grades of steels are manufactured.

(*iii*) Addition of copper atoms in gold increases its ductility so that it can be drawn into wires.

(*iv*) Tin atoms substituted in copper lattice increase the bearing properties of copper.

(*v*) Pure germanium and silicon semiconductor materials are doped with pentavalent and trivalent impurities and as a result their electrical conductivity increases considerably.

(*vi*) Copper atoms in silver increases the electrical resistivity of silver.

(*vii*) Vacancies help the transport of atoms through the lattice for annealing etc purposes.

33.17 LINE DEFECTS

A line defect or dislocation is a one-dimensional defect around which some of the atoms are misaligned. It may be defined as a disturbed region between two perfect parts of a crystal. It is a defect in a crystal structure whereby a part-plane of atoms is dislocated from its symmetrically stable position in the array. Dislocations appear in crystals due to growth accidents, thermal stresses, phase transformations etc. They can be observed in crystalline materials with the help of electron microscope.

Dislocations are divided into two basic types:

(*i*) Edge dislocation, and

(*ii*) Screw dislocation.

Most of the dislocations found in crystals are neither pure edge nor pure screw dislocations, but contain components of both these types. They are called mixed dislocations.

33.17.1 Edge dislocation

In a perfect crystal, atoms are arranged in both vertical and horizontal planes. The atoms are in perfect equilibrium in their positions and all bond lengths are in equilibrium state. If one of the vertical planes does not extend to the full length, but abruptly terminates in between, within the crystal, then it may be viewed as an extra half-plane inserted between a set of parallel planes (Fig. 33.17). The edge of such a plane constitutes a linear defect. It is a linear defect centered on the line along the end of the extra

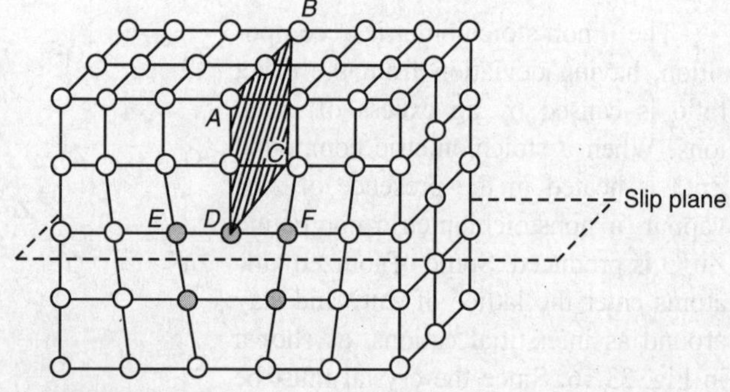

Fig. 33.17: Edge dislocation

half-plane of atoms. Within the region around the edge, there is some localized distortion. It may be seen that the atoms above the edge are squeezed and are in a state of compression while the atoms below the edge are pulled apart and are in a state of tension. The region of distortion spreads a few lattice distances around the edge. This distortion is known as ***edge dislocation***. The edge is in the direction perpendicular to the plane of the Fig. 33.17 and constitutes the dislocation line.

As the locus of defective points formed in the lattice by the dislocation lie along a line, the dislocation is known as a *line defect*. Dislocation runs along the bottom of the extra half-plane. The inter-atomic bonds are considerably distorted only in the *immediate vicinity* of the dislocation line. The plane in which the dislocation line is present is called a *slip plane*. The dislocation line forms a boundary between slipped and unslipped portions.

An edge dislocation is said to be positive when the half-plane appears to have been inserted from the top of the crystal, as in Fig. 33.17. It is denoted by $\perp$. An edge dislocation is said to be negative if the extra half-plane lies in the lower part of the crystal. It is denoted by *T*.

33.17.2 Screw dislocation

The screw dislocation is also called Burger's dislocation. The geometry of screw dislocation is shown in Fig. 33.18. Let us visualize a perfect crystal cut by a plane *P*. Then let the crystal on the right side of the cut be shifted (sheared) relative to that on the left side by an amount *b*. The resulting arrangement is shown in Fig. 33.18. From this figure it is seen that the shift has occurred for the length of one-atomic spacing in the predominant part of the crystal and reached the line *OO′*. Away from this line, the crystal structure is undisturbed and shows an ordered atomic arrangement. Near the line *OO′*, atoms of atomic planes lying to the left and right of the shear plane *P* have undergone displacement with respect to each other. This displacement causes a local distortion called a screw dislocation. The line *OO′* is the *line of dislocation*.

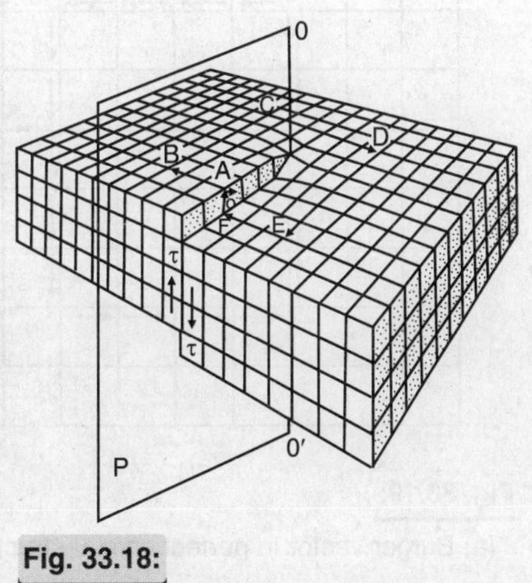

Fig. 33.18: Screw dislocation

It may be viewed now that the atoms in the dislocation region lie on a surface which spirals from one end of the crystal to the other with the line of dislocation as the axis of the spiral. The displacement of the atoms from their original positions in the perfect crystal is described by the following equation.

$$r = \frac{b}{2\pi}\theta \qquad ...(33.34)$$

where *r* is the displacement along the dislocation line and the angle θ is measured from a line perpendicular to the dislocation line.

33.18 BURGERS VECTOR

Dislocations are quantitatively described by the Burgers vector, and is denoted by **b**. It tells us the direction and the magnitude of the lattice distortion associated with a dislocation in a crystal. It describes the slip which one part of the crystal undergoes relative to the rest of the crystal. The nature of the dislocation is defined by the relative orientations of dislocation line and Burgers vector. The Burger vector of an edge dislocation is perpendicular to the dislocation line and that of a screw dislocation is parallel to the dislocation line. In case of dislocations of mixed nature, the vector will be neither perpendicular nor parallel to the dislocation line.

The magnitude of Burgers vector is found by drawing a closed circuit, called the Burgers circuit surrounding the dislocation line. First, let us consider the perfect crystal as given in Fig. 33.19 (a). In this perfect crystal structure, a rectangle whose lengths and widths are integer multiples of lattice vector is drawn encompassing a centre. The number of lattice vectors travelled along each side of the rectangle is noted. As an example, starting from the point *P*, let us go to right by n steps ($n = 4$), then take m steps ($m = 5$) down, then *n* steps to the left and finally m steps up. We reach the point where we started, having described a close path. This movement around a centre is called a ***Burgers circuit***. In a perfect crystal we reach the starting point after completion of the circuit.

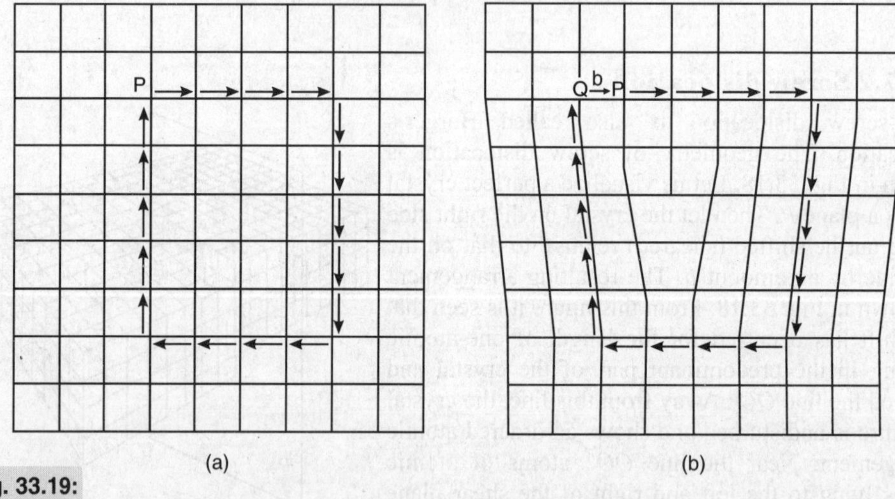

Fig. 33.19:

(a) Burger vector in perfect crystal structure, (b) Burger vector in edge dislocation

(a) Edge dislocation: Now we consider the Burgers circuit in a crystal that contains edge location (Fig. 33.19 b). Burgers circuit in this case is drawn by proceeding through the undistorted region surrounding the dislocation, moving the same number of lattice vectors along each direction as in Fig. 33.19 (a).

Starting from *P*, if we trace the Burgers circuit in a plane normal to the dislocation line, the circuit would not be completed since the end point *Q* and the starting point *P* do not coincide. From *Q*, we need an extra step "*b*" to arrive at the starting point *P*. This extra step *QP* = *b* is called *Burgers vector* (BV). The Burgers vector gives both the magnitude and direction of the displacement.

$$BV = \mathbf{QP} = \mathbf{b}$$

It is seen from Fig. 33.19 (b) that the Burgers vector of an edge dislocation is perpendicular to the dislocation.

(b) Screw dislocation: Now we consider the Burgers circuit in a crystal that contains screw dislocation. In the perfect crystal, starting from *P*, if we trace the Burgers circuit, the circuit is a closed path *PMNOP* (Fig. 33.20 a). In case of a crystal with a screw dislocation, the circuit would not be completed and requires an extra step ***b*** = ***FA*** (Fig. 33.20 b), parallel to the dislocation axis, to close the circuit. This additional vector b is *Burgers vector*. The Burgers vector gives both the magnitude and direction of the displacement. The Burgers vector of a screw dislocation is parallel to the dislocation.

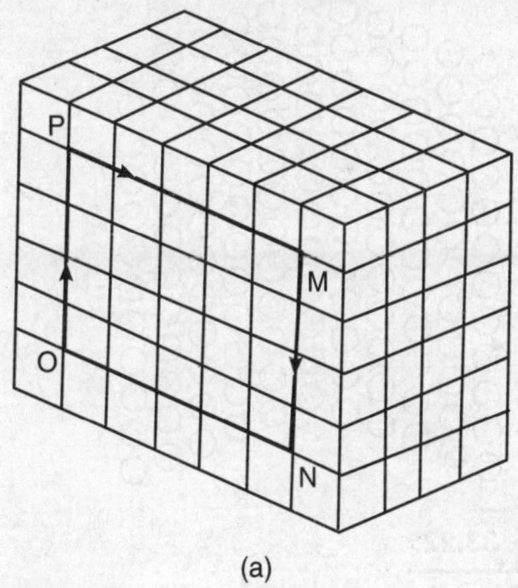

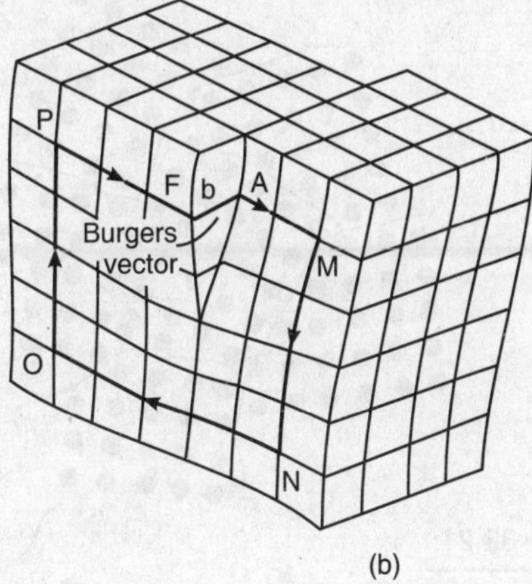

Fig. 33.20:

(a) Burger vector in perfect crystal structure, (b) Burger vector in screw dislocation

33.19 PLANAR DEFECTS OR SURFACE DEFECTS

Planar defects are two dimensional defects that separate two regions of the crystal. These defects are: the external surfaces, grain boundaries, twin boundaries, and stacking faults.

(i) External surfaces: The surface of a crystal itself is a defect. The surface atoms have formed bonds with neighbours only on one side, whereas the atoms inside the crystal formed bonds with neighbours on either side of them. Since the surface atoms are not bonded to the maximum number of nearest neighbours, they are in a higher energy state than the atoms in the interior of the crystal. This gives rise to a surface energy. The crystal tends to minimize the total surface area in order to reduce the surface energy, but it is not possible because solids are mechanically rigid.

(ii) Grain boundaries: The most important defect in crystalline samples is the grain boundary. A grain boundary is a general planar defect that separates regions of different crystalline orientation (i.e. *grains*) within a polycrystalline solid. Solids are formed when the melt of materials are cooled. Depending upon the rate of cooling, either single crystals or polycrystalline solids form. During solidification, infinitesimally small crystallites are formed first at some places in the melt. They are called nuclei. As the solidification proceeds, these nuclei grow in size. They are randomly oriented. Crystallites with particular orientation tend to join those with the same orientation. All the crystallites that have the same orientation form a bigger block called grain. When the solidification is completed, the solid consists of a number of grains attached to each other. The result is a polycrystalline sample. In between the grains, there exists a boundary called grain boundary. The boundary region is of a few atomic diameters wide. There occurs some atomic mismatch in the region of transition from one grain to its adjacent grain. The atoms are bonded less regularly along a grain boundary and there is grain boundary energy. The magnitude of this energy is a function of the degree of misalignment. Grain boundaries are therefore more chemically reactive than the grains themselves. Fig. 33.21 shows Grain boundaries.

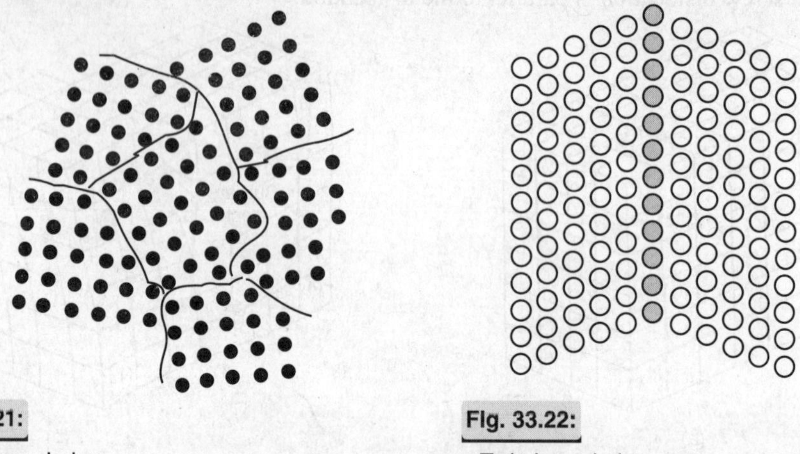

Fig. 33.21: Grain boundaries

Fig. 33.22: Twin boundaries

(iii) Twin boundaries: A twin boundary is a particular kind of grain boundary where atoms on one side of a twin boundary are located in mirror-image positions of the atoms on the other side. The boundary between the twinned crystals will be a single plane of atoms. The crystals on either side of the plane are mirror images of each other. The boundary atoms look as if they belong to the crystal structures of both twins. Fig. 33.22 shows Twin boundaries.

(iv) Stacking faults: An *FCC* (face centered cubic) crystal is regarded as formed by the stacking of hexagonal close packed atomic layers, *A*, *B* and *C* in the following sequence:

ABC ABCABC ....

If one of the layers go missing, then stacking fault occurs in the crystal. For example, if the A layer in the third set of sequence fails to appear, the sequence changes to *ABC ABC BCA* .... , then the sequence is interrupted and a stacking fault is produced. The stacking faults are usually produced during the growth of the crystals. Stacking faults are observed in *FCC* metals. Fig. 33.23 shows stacking faults.

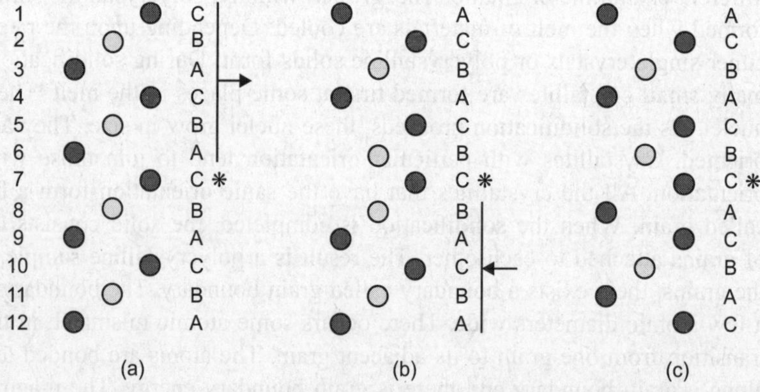

Fig. 33.23: (a) Perfect crystal, (b). With intrinsic stacking fault, (c) with extrinsic stacking fault

33.20 VOLUME DEFECTS

Volume defects such as cracks may arise in crystals during the process of crystal growth. A crack may result during the growth due to a possible small electrostatic dissimilarity between the stacking layers. A large vacancy may arise due to missing of clusters of atoms which is a volume defect. Inclusion of foreign particles or non-crystalline regions of the size of 20 Å also belong to the category of volume defects.

QUESTIONS

1. What are crystal imperfections and how they affect properties of crystals? Explain point defect.
2. What are crystal defects? Mention the different kinds of crystal imperfections.
3. What is crystal defect? How does it arise? Explain effect of point defect and line defect on properties of crystals.
4. Explain the various point defects in a crystal.
5. How are vacancies created in a lattice?
6. Derive the expression for the energy required to produce a vacancy in a crystal.
7. What are point defects in crystals? Obtain an expression for the equilibrium concentration of vacancies in a metal at a given temperature.
8. Derive an expression for the concentration of Schottky defects in a crystal.
9. Distinguish between Schottky and Frenkel defects in ionic crystals.
10. Discuss the Schottky defect in ionic crystals.
11. Obtain an expression for the equilibrium concentration of Frenkel defects at a given temperature in an ionic crystal.
12. Explain interstitial defects.
13. (*a*) Explain edge and screw dislocations with neat diagrams.

    (*b*) What is Burger circuit? Draw Burger's circuit for an edge dislocation and screw dislocation. What is the significance of the Burger's vector?
14. What are grain boundaries? Explain.
15. Write notes on: Stacking faults and twin boundary.

PROBLEMS

1. If the average energy required producing a vacancy in a metal is 1 eV, calculate the ration of vacancies in a metal at 1000K and 500K. **[Ans:** 1.08×10^5**]**
2. The fraction of vacancy sites in a metal is 1×10^{-10} at 500°C. What will be the fraction of vacancy sites at 1000°C? **[Ans:** 8.4×10^{-7}**]**
3. Calculate the equilibrium number of vacancies per unit volume at a temperature of 1000°C. The energy of formation of vacancy in copper is 0.90 eV. What is the vacancy fraction at 500°C? Given that density of copper metal is 8960 kg/m^3 and the atomic weight is 63.5. **[Ans:** 1.6×10^{-6}**]**
4. Calculate the equilibrium number of vacancies at 300K in aluminium to that produced by rapid quenching at 800k. Enthalpy of formation of vacancies in aluminium is 68 kJ/mol. **[Ans:** 3.75×10^{-8}**]**
5. The density of Schottky defects in a certain sample of NaCl is 5×10^{11}/m^3 at 25°C. If the interionic distance is 2.82 Å, what is the energy required to create one Schottky defect? **[Ans:** 1.97 eV**]**
6. The concentration of Schottky defects in an ionic crystal is 1 in 10^{10} at a temperature of 300K. Estimate the average separation in terms of the lattice spacings between the defects at 300K and calculate the value of concentration to be expected at 1000K.

34 Crystal Structures

Learning Objectives:

By the end of this chapter, you will be able to:

- Learn about different types of solids and their properties.
- Understand difference between BCC and FCC cells.
- Explain about Lattice planes, miller indices and diamond cubic and ZNS structure.
- State Braggs' law and talk about Braggs' spectrometer.
- Learn about powder crystal method and rotating crystal method.

34.1 INTRODUCTION

A solid consists of atoms or clusters of atoms arranged in close proximity. The physical structure of a solid and its properties are closely related to the scheme of arrangement of atoms within the solid. In amorphous solids the arrangement of atoms is random while in crystals there is a regular arrangement of atoms. Simple geometrical concepts of a lattice and unit cell are used to describe the atomic arrangement in crystals. The advantage of such a description is that all possible crystal structures are represented by a limited number of basic unit cell geometries. Majority of metals are found to belong to three simple structures. The crystal structures are analyzed using x-ray diffraction technique invented by Max von Laue and extensively employed by Bragg and Bragg. The study of crystal geometry helps us understand the diverse behaviour of solids in their mechanical, metallurgical, electrical, magnetic and optical properties.

34.2 CLASSIFICATION OF SOLIDS

Solids are classified into three categories basing on the atomic arrangement within the solid, single crystals, polycrystalline solids and amorphous solids. Single crystals follow law of constancy of angles. The same substance may crystallize under different conditions of crystal growth to form different geometrical shapes but the angles between the faces are always constant for different shapes.

(*i*) **Single crystals:** X-ray diffraction studies have shown that the atoms in crystalline solids are arranged in a regular periodic pattern in three dimensions, as shown in Fig. 34.1 (a).

Def: Single crystals are polyhedrons that have a distinctive shape for each material and are bounded by smooth shiny faces and straight edges. When a crystal is broken, it cleaves along certain preferred directions.

The arrangement of atoms in specific relation to each other is called ***order.*** In crystals, the order exists in the immediate neighbourhood of a given atom as well as over large distances corresponding to several layers of atoms. Therefore, crystals possess both ***short-range order*** and ***long-range order***.

In single crystals, since the periodic arrangement of atoms differs in the three directions, the physical properties vary with direction and therefore, they are called ***anisotropic*** substances. Quartz, alum, diamond and rock salt are examples of solids that occur as large size single crystals.

(*ii*) **Polycrystalline solids:** Majority of the natural solids has polycrystalline structure. Metals are examples of polycrystalline solids. Polycrystalline solids consist of fine grains, having a size of 10^3 to 10^4 Å, separated by well-defined boundaries but oriented in different directions (Fig. 34.1b). Each such grain is a single crystal of an irregular shape. Since the grains are oriented randomly, a polycrystalline material is ***isotropic*** and the physical properties do not vary with direction.

Def: Polycrystalline solids or polycrystals are solids that are composed of many crystallites of varying size and orientation. Crystallites are also referred to as grains.

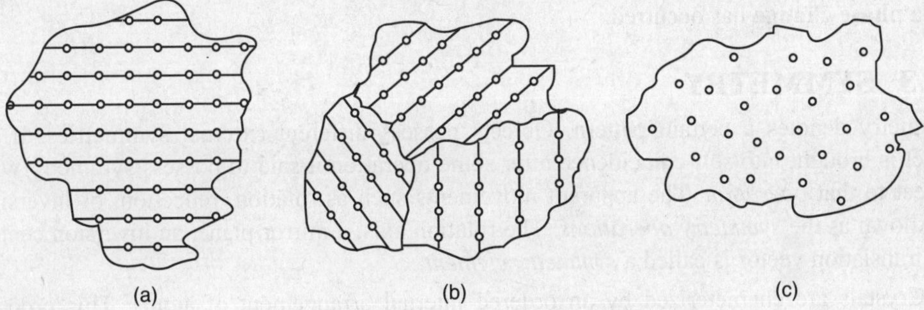

Fig. 34.1:

Atomic arrangement in (*a*) Single crystal (*b*) Polycrystalline solid and (*c*) Amorphous substance

Types of Polycrystalline solids: The atoms are bound to each other by electrostatic forces. Depending on the nature of the bond between the neighbouring atoms, the crystals can be classified as metallic crystals, covalent crystals, and ionic crystals.

(*a*) In ionic crystals, the electron is transferred from the cation to anion and the atoms exist in the crystal as ions and hence the name ***ionic crystals***. Inter atomic forces are non-directional.

(*b*) Inter-atomic forces are directional in ***covalent crystals*** where electron is shared by the neighbouring atoms.

(*c*) In ***metallic crystals***, the electrons are very loosely bound to the corresponding nuclei.

(*iii*) **Amorphous solids:** The amorphous materials are generally referred to as glass because they have features similar to glass. The physical properties of amorphous solids are not dependent on the direction of measurement and therefore they are ***isotropic*** substances (Fig. 34.1c).

The atomic arrangement is random in these materials. They are in fact considered as super-cooled liquids having a very high viscosity. Glass, rubber and many polymers are amorphous solids. Though the word *solid* has been traditionally used for crystalline, polycrystalline and amorphous materials, nowadays the word *solid* is meant to convey only crystalline materials.

Def: An amorphous solid is any non-crystalline solid in which the atoms and molecules are not organized in a definite lattice pattern. They posses only short-range order and maintain a fixed volume and shape and resemble solids in their external features, but internally they do not have the ordered crystalline arrangement.

34.2.1 Phase

When the crystals are grown there are two important requirements to be satisfied. (*i*) The atoms must occupy minimum space and (*ii*) the bond angle requirements must be fulfilled. For metallic crystals, the second condition does not exist and therefore, the atomic arrangement in metallic crystals is usually much simpler than the atomic arrangement in nonmetallic crystals.

Since the atomic arrangement everywhere in the crystal is the same, the crystal growth occurs at a fixed temperature. The atoms in the crystal arrange and disperse like soldiers in the army, each soldier going through identical movement. A given atomic arrangement in the material is called a **phase**. Thus, a given chemical like water may exist in vapour, water or ice phase. If the atomic arrangement in the crystal changes, we say that a new crystal has formed and a **phase change** has occurred.

34.3 SYMMETRY

Symmetry denotes a certain pattern. Objects possess different external symmetries. If an object is brought into self-coincidence after some operation it said to possess symmetry with respect to that *operation*. The apparent movements such as rotation, reflection, or inversion are known as the *symmetry operations*. The rotation axis, a mirror plane, an inversion center, or a translation vector is called a *symmetry element*.

Crystals are characterized by an ordered internal arrangement of atoms. This ordered arrangement consists of the atoms being arranged in a symmetrical fashion on a three dimensional *lattice*. The internal symmetry is mirrored in the external shape of perfect crystals. We therefore study the external symmetry.

34.3.1 Symmetry Operations

Let us now briefly look at symmetry operations that can be used to describe crystals.

(*i*) **Inversion:** Generally, the centre of the crystal is taken as the centre of symmetry. The line drawn through centre of symmetry intersects the surface of the crystal at equal distance on either side. This means that if there is an atom at some distance at one end, there is also an atom at the other end at the same distance. Centre of symmetry is also called the ***center of inversion***. The inversion operation can be carried out through center of inversion. In this operation, every part of the object is reflected through an inversion center which is denoted as *i*. The object is reproduced inverted from its original position. Fig. 34.2 shows the inversion axis.

(*ii*) **Rotation:** A crystal is said to possess an ***axis of symmetry*** if when the crystal is rotated about the axis, the atomic arrangement looks exactly the same more than once during one complete revolution. If the atomic arrangement looks the same twice in one revolution, the axis is said to be a *diad* axis. Similarly, the axis may be a triad axis, tetrad axis, and *hexad* axis. The diad axis is represented by 2, triad axis by 3, tetrad axis by 4, and the *hexad* axis by 6, corresponding to 180°, 120°, 90°, 60°, rotations respectively. Note that only these types of axes are possible. The other types of axes are not consistent with the condition of periodicity.

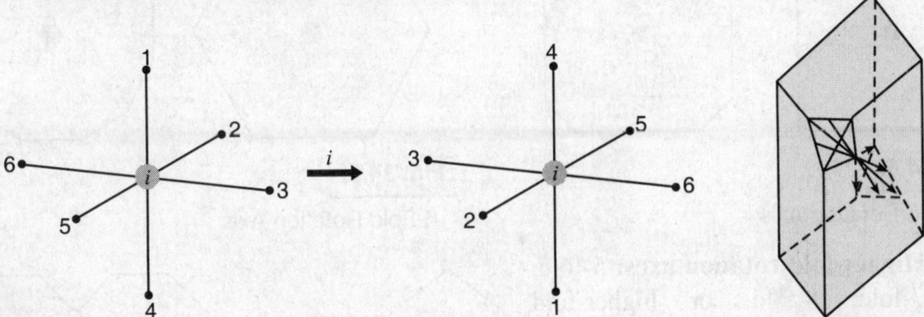

Fig. 34.2:

Inversion axis

(*a*) **1-Fold Rotation Axis:** By the rotation of one full 360° an object repeats itself and hence it is supposed to have a 1-fold axis of rotational symmetry. Fig. 34.3 shows 1-fold rotation axis.

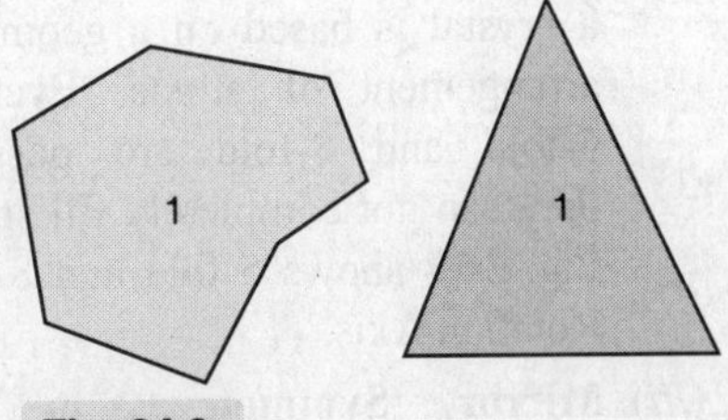

Fig. 34.3:

1-Fold Rotation Axis

(*b*) **2-Fold Rotation Axis:** Suppose an object shows identical figure after a rotation of 180° then it is known to have a 2-fold rotation axis. As illustrated in Fig. 34.4, the dark oval shape corresponds to the 2-fold rotation axis. It is perpendicular to the page.

(*c*) **3-Fold Rotation Axis:** Suppose an object shows identical figure after a rotation of 120° then it is known to have a 3-fold rotation axis. As illustrated in Fig. 34.5, the dark triangle shape corresponds to the 3-fold rotation axis.

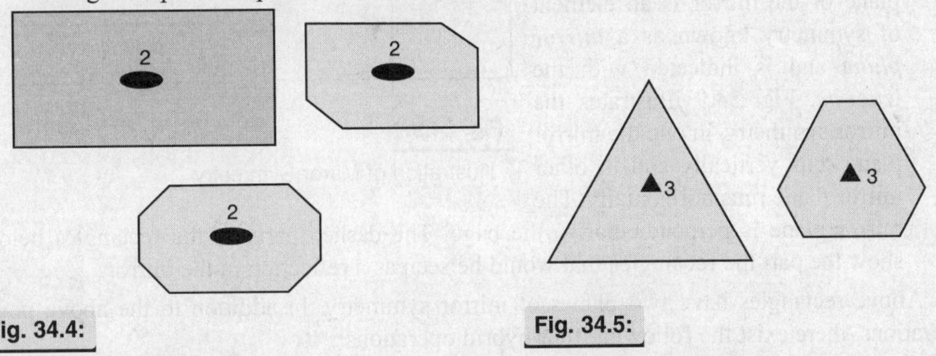

Fig. 34.4:

2-Fold Rotation Axis

Fig. 34.5:

3-Fold Rotation Axis

(*d*) **4-Fold Rotation Axis:** Suppose an object shows identical figure after a rotation of 90° then it is known to have a 4-fold rotation axis. As illustrated in Fig. 34.6, the dark square shape corresponds to the 4-fold rotation axis.

(*e*) **6-Fold Rotation Axis:** Suppose an object shows identical figure after a rotation of 60° then it is known to have a 6-fold rotation axis. As illustrated in Fig. 34.7, the dark square shape corresponds to the 6-fold rotation axis.

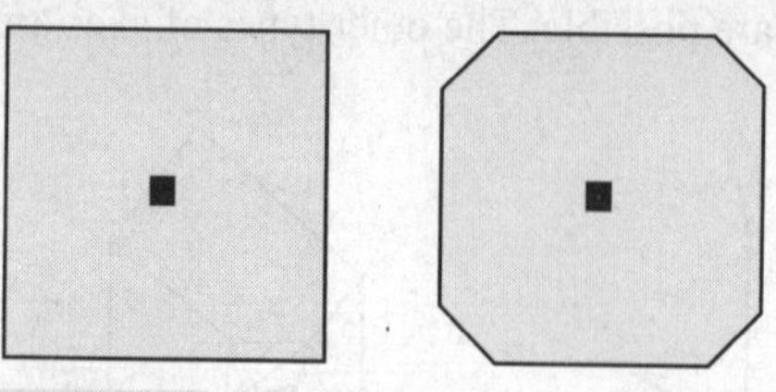

Fig. 34.6:
4-Fold Rotation Axis

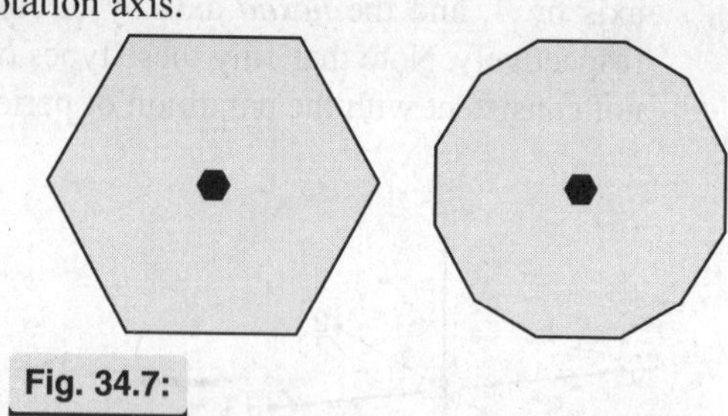

Fig. 34.7:
6-Fold Rotation Axis

(*f*) **Higher-fold rotation axes:** 5-fold, 7-fold, 8-fold, or higher-fold rotation axes are not possible in crystals since the external shape of a crystal is based on a geometric arrangement of atoms. Even if 5-fold and 8-fold are possible they can not completely fill space. Fig. 34.8 shows 5-fold and 8-fold Rotation Axis.

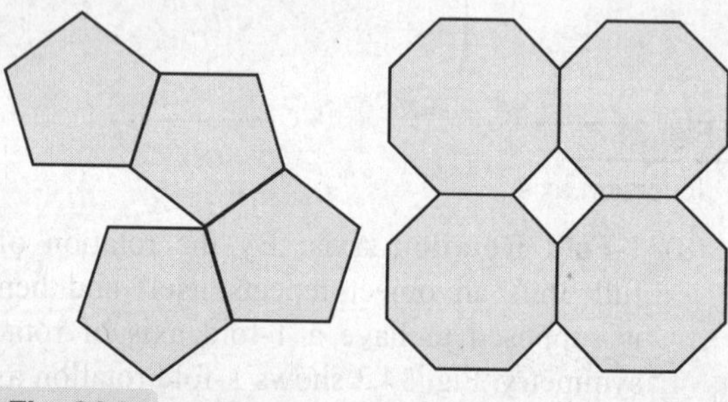

Fig. 34.8:
5-fold and 8-fold Rotation Axis

(*iii*) **Mirror Symmetry:** In the imaginary mirror symmetry operation, it is assumed that an object possessing mirror symmetry is cut into two halves and placing one of the halves on a mirror, the reflection in the mirror reproduces the other half of the object. The plane of the mirror is an element of symmetry known as a ***mirror plane*** and is indicated with the letter *m*. Fig. 34.9 illustrates the mirror symmetry, in one the mirror plane runs vertically and in other mirror plane runs horizontally. The mirror plane is perpendicular to the page. The dashed parts of the rectangles below show the part the rectangles that would be seen as a reflection in the mirror.

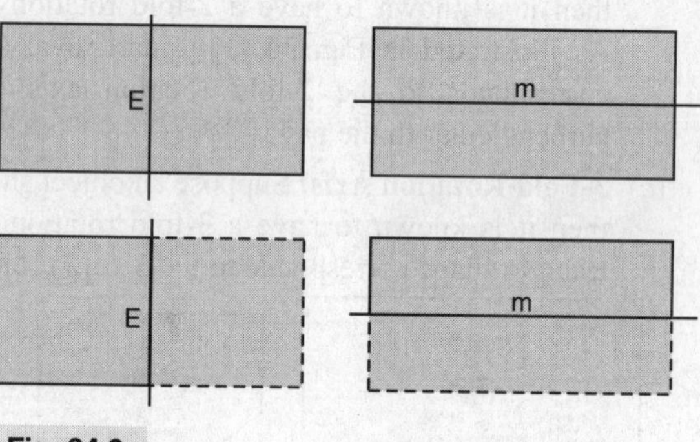

Fig. 34.9:
Illustration of Mirror Symmetry

Above rectangles have two planes of mirror symmetry. In addition to the above point operations, there exist the following four hybrid operations.

(*a*) **Rotoreflection:** This is the combination of an *n*-fold rotation followed by a reflection in a plane perpendicular to the rotation axis.

(*b*) **Rotoinversion:** This is the combination of an *n*-fold rotation followed by an inversion.

(*c*) **Screw translation:** In this the *n*-fold rotation axis is coupled with the translation parallel to rotation axis.

(*d*) **Glide reflection:** In this a mirror plane is coupled with a translation parallel to the reflecting plane.

34.3.2 Nomenclature

Symmetry operations are indicated by two different sets of nomenclature. Schönflies nomenclature is used in spectroscopy, while Hermann-Mauguin nomenclature is used in crystallography. The notations are different, but the operations are the same, Table 34.1 shows different sets of nomenclature of symmetry.

Table 34.1: *Different Sets of Nomenclature of Symmetry*

Symmetry operation	Hermann-Mauguin symbol	Schönflies symbol
Identity	*E* or *I*	*I*
Inversion (centro symmetric)	*i*	$\bar{1}$
Rotation axis	Cn (*n* = fold)	*n*
Mirror plane (reflection)	σ_h, σ_v horizontal and vertical mirror planes	*m*

34.3.3 Point groups and Space groups:

Point group: As all the symmetry elements, points, lines, and planes will intersect at a single point, they form a ***point group*** called point group. Only *32 point groups* are possible in the crystalline state of matter in view of crystalline-translational periodicity and are also known as the 32 crystal classes. Of the *32 crystal classes*, only 11 contain the operator *center of symmetry*, and these *11 centro-symmetric crystal classes* are known as *Laue groups*.

Space groups: The complete group of all symmetry elements including translations of a crystal is called its ***space group***. When linear translation, glide plane (translation + reflection) and screw axis (translation + rotation) are applied to the point groups, a large number of 3D structures can be generated. 230 unique shapes called space groups can be generated. These are grouped into the *seven crystal systems*.

34.4 SPACE LATTICE

A regular and periodic arrangement of atoms is the most important feature of crystals. The actual arrangement of atoms is called the **structure**. Suppose the atoms or clusters of atoms in a crystal are represented by points that correspond to their mean positions. Then we obtain a regular distribution of points in space. These points are called **lattice points** or **lattice sites**. Although lattice points represent atom locations, the lattice points and atom centres need not be coincident. The three-dimensional network of regularly arranged points is known as a **space lattice**. A point is a dimensionless and shapeless entity; therefore, a lattice is merely an imaginary geometrical framework. The study of crystal structure becomes simpler when it is represented by a space lattice.

Def: A space lattice is as an array of points in three dimensions in which every point has surroundings identical to that of every other point.

t_1 t_1 t_1

Fig. 34.10 a:
A sequence of periodically spaced points

A space lattice can be generated by successive translations of an initial point. A very simple operation of repetition consists of repeating the unit without change after translating it a distance '*t*'. Repeated application of a translation of given length and direction, t_1, to an initial point generates a sequence of periodically spaced points or a row of points.

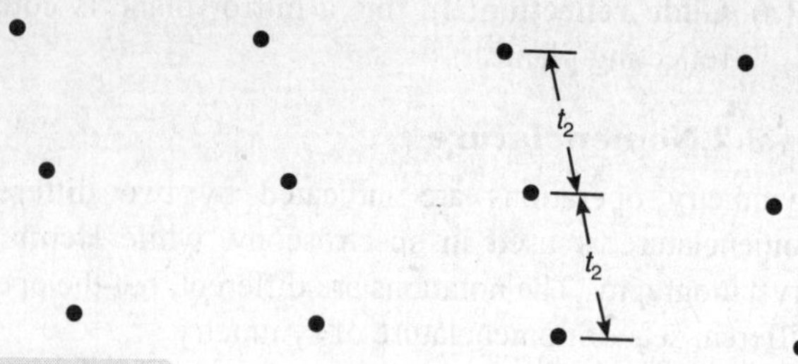

Fig. 34.10 b:

A planar array of points

The repeated application of some other translation which are non-collinear (not along the same line), t_2, to the above row generates a planar array of points.

A third translation is non-coplanar (not in the same plane), t_3, applied to a two-dimensional lattice generates a three-dimensional array of points called a ***space lattice*** or simply ***lattice***.

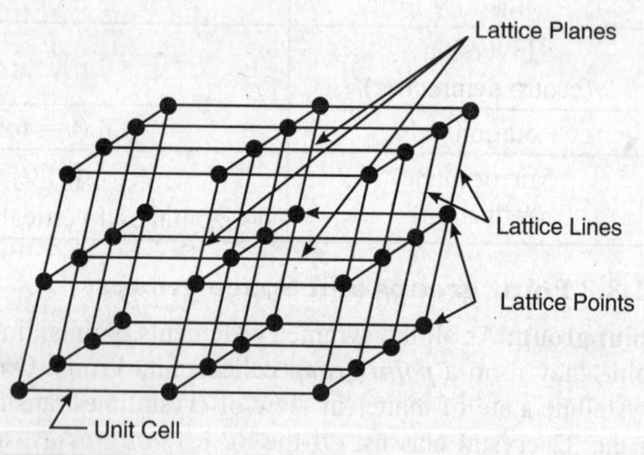

Fig. 34.10 c:

Three dimensional lattice or space lattice

Thus, lattice points satisfy the condition of periodicity. Hence, the crystal structure, *i.e.,* the arrangement of atoms in the crystal can be understood with reference to a lattice. It is also observed that a given point in the lattice (Fig. 34.10c) is surrounded by a specific number of lattice points located at equal distances. The arrangement of points around a given lattice point is called its ***environment***. The environment continues to change as we traverse in a particular direction from one lattice point to the next lattice point, but at the next lattice point it is exactly the same as that at the initial point. Thus, the same environment would be found after equal intervals of distance in the same direction. The environment would be different in different directions. The space lattice is the skeleton upon which crystal structure is built by placing atoms on or near the lattice points.

34.5 CRYSTAL STRUCTURE

A space lattice is a mathematical abstraction. The crystal structure is formed only when a group of atoms is identically attached to each lattice point, as shown in Fig. 34.11. The group of atoms that is associated with every lattice point is called a **basis**. The basis must be identical in composition, arrangement and orientation such that the crystal appears exactly the same at one point as it does at other equivalent point.

A crystal structure is thus the result of two quantities; namely a lattice and a basis. Thus,

$$\textbf{lattice + basis} \rightarrow \textbf{crystal structure}$$

In the simplest crystals such as copper, silver, gold, iron and the alkali metals, the basis is a single atom. Often the basis contains a few atoms.

Fig. 34.11:

Crystal structure is obtained when a basis is added to each lattice site

34.6 UNIT CELL

It is the smallest volume that carries a full description of the entire lattice. Because of the inherent periodicity, the space lattice can be represented by a unit cell.

Def: Unit cell is the smallest geometrical unit, which when repeated in space indefinitely, generates the space lattice. It is described by lattice parameters.

The lines drawn parallel to the lines of intersection of any three faces of the unit cell which do not lie in the same plane are called ***crystallographic axes***. Three translational vectors a, b, and c lie along the crystallographic axes. The intercepts ***a, b,*** and ***c*** (Fig. 34.12 a) define the dimensions of the unit cell and are known as **primitives**. The angle α represents the angle between *a* and *b* axes, the angle γ represents the angle between *b* and *c* axes and the angle α represents the angle between the *c* and *a* axes. The volume of the unit cell is given by $[(a \times b).c]$

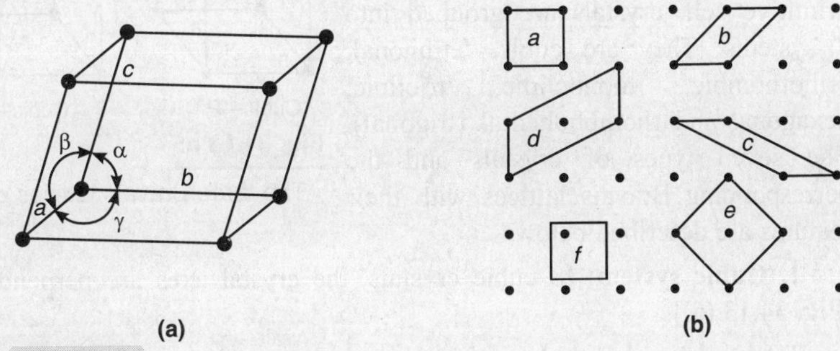

Fig. 34.12:

(a) A primitive unit cell, (b) Primitive and non-primitive unit cells in a two-dimensional lattice

Def: Basic lattice parameters are the axial lengths *a, b, c* and the three inter-axial angles α, β, γ of the unit cell.

Def: Primitive unit cell is the unit cell whose volume in space contains only one lattice point.

It is not necessary that every unit cell should be a primitive cell (Fig. 34.12 b). Depending on the requirement and symmetry of the lattice we can choose larger cell containing more than one lattice point as a unit cell.

Def: If there are two or more lattice points per unit cell, then it is called a ***non-primitive unit cell***. Most of the unit cells of various crystal lattices contain two or more lattice points and are non-primitive cells.

Example 34.1: *The unit cell of copper is a cube. The side of the cube is 3.6 Å. If the unit cell in copper solid are lined up side by side, how many unit cells will be there along 10 mm length of the solid?*

Solution: The number of unit cells in a length l of solid is given by

$$N = \frac{\text{Length of the solid}}{\text{side of the unit cell}} = \frac{l}{a} = \frac{10^{-2}\,\text{m}}{3.6 \times 10^{-10}\,\text{m}} = \mathbf{28 \times 10^6}$$

34.7 BRAVAIS LATTICES

Bravais introduced the concept of space lattice in the study of crystal structures. A three-dimensional space lattice is generated by repeated translations of three non-coplanar vectors. One would expect that many lattices can be generated in three dimensions with different primitive and non-primitive cells. When the discrete points are atoms, ions, or polymer strings of solid matter, the Bravais lattice concept is used to formally define a crystalline arrangement.

Def: Bravais showed that there are only fourteen different ways of arranging identical points in three-dimensional space, satisfying the condition of periodicity, so that they are in every way equivalent in their surroundings. These fourteen types of arrangements are called the ***space lattices*** or ***Bravais lattices***.

There are seven primitive cells and seven non-primitive cells. With the 14 types of lattices and on the basis of primitive cell, crystals are grouped into 7 systems. They are cubic, tetragonal, orthorhombic, monoclinic, triclinic, hexagonal and rhombhohedral (trigonal). The seven types of crystals and the corresponding Bravais lattices with their features are described below.

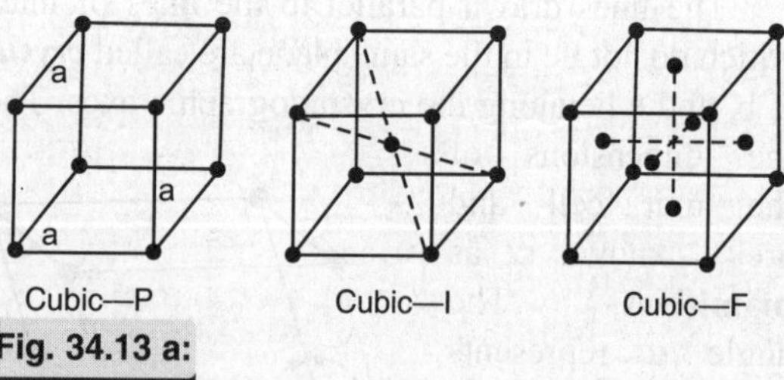

Fig. 34.13 a: The three Bravais lattices of a cubic system

1. Cubic system: In cubic crystals, the crystal axes are perpendicular to one another [Fig. 34.13 (a)].

Thus, relation between inter axial angles is $\alpha = \beta = \gamma = 90°$.

The length of the primitives (edges of unit cell) is the same along the three axes.

Thus, relation between axial lengths is $a = b = c$.

Cubic lattice has three possible types of arrangements, simple (SC) or cubic-P, body centered (BCC) or cubic-I and face centered (FCC) or cubic-F.

(*i*) **Simple cubic lattice:** It has lattice points at all 8 corners of the unit cell.

(*ii*) **Body-centered cubic lattice:** It has lattice points at all 8 corners of the unit cell and one lattice point at the centre of the body.

(*iii*) **Face-centered cubic lattice:** It has lattice points at all 8 corners of the unit cell and one lattice point each at the centre of six faces of the cube.

2. Tetragonal system: In tetragonal crystals, the crystal axes are perpendicular to one another [Fig. 34.13 (b)].

Thus, relation between inter axial angles is $\alpha = \beta = \gamma = 90°$.

The lengths of the edges of unit cell along two axes are the same but the edge along the third axis is different.

Thus relation between axial lengths is $a = b \neq c$.

Tetragonal lattice has two possible types of arrangements, simple (P) and body centered (I).

(*i*) ***Simple tetragonal lattice:*** It has lattice points at all 8 corners of the unit cell.

(*ii*) ***Body-centered tetragonal lattice:*** It has lattice points at all 8 corners of the unit cell and one lattice point at the centre of the body.

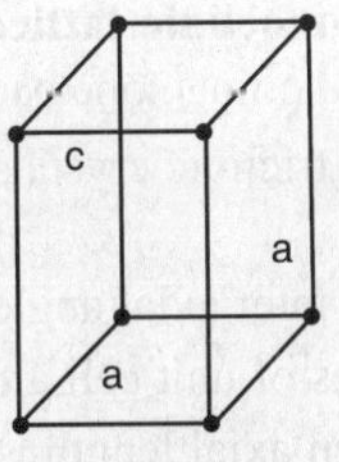

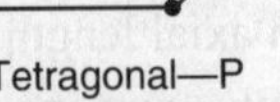

Tetragonal—P

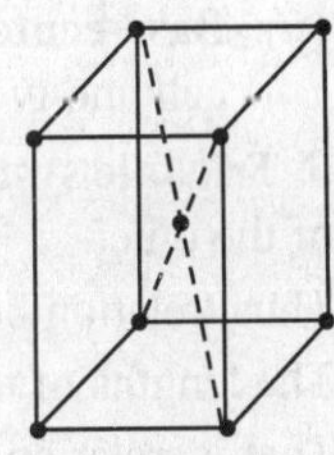

Tetragonal—I

Fig. 34.13 b: The two Bravais lattices of a tetragonal system

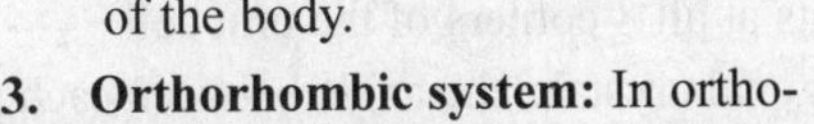

3. Orthorhombic system: In orthorhombic crystals, the crystal axes are perpendicular to one another.

Thus, relation between inter axial angles is $\alpha = \beta = \gamma = 90°$.

But the lengths of the edges of unit cell along the three axes are different.

Thus, relation between axial lengths is $a = b \neq c$.

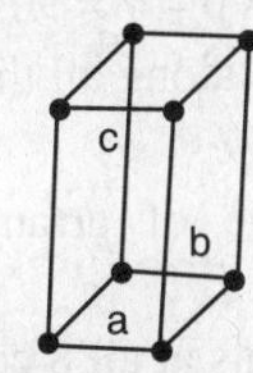

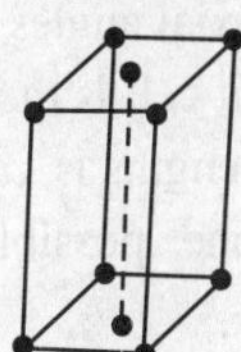

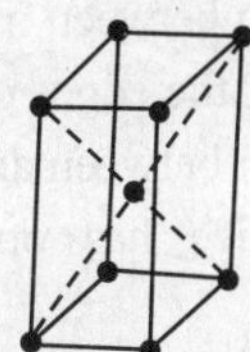

Orthorhombic—P Orthorhombic—C Orthorhombic—I Orthorhombic—F

Fig. 34.13 c: Four Bravais lattices of an orthorhombic system

Orthorhombic lattice has four possible types of arrangements, simple (P), base centered (C), body centered (I) and face centered (F) [Fig. 34.13 (c)].

(*i*) **Simple orthorhombic lattice:** It has lattice points at all 8 corners of the unit cell.

(*ii*) **Base-centered orthorhombic lattice:** It has lattice points at all 8 corners of the unit cell and two lattice points, one each at the base and top face of the body.

(*iii*) **Body-centered orthorhombic lattice:** It has lattice points at all 8 corners of the unit cell and one lattice point at the centre of the body.

(*iv*) **Face-centered orthorhombic lattice:** It has lattice points at all 8 corners of the unit cell and six lattice points, one each at the centre of six faces of the unit cell.

4. Monoclinic system: In monoclinic crystals, out of the three crystal axes, two are not perpendicular to each other but the third is perpendicular to both of them.

Thus, relation between inter axial angles is $\alpha = \gamma = 90° \neq \beta$.

The lengths of the edges of unit cell along the three axes are different.

Thus, relation between axial lengths is $a \neq b \neq c$.

Monoclinic lattice has two possible types of arrangements, simple (P) and base centered (C) [Fig. 34.13 (d)].

(*i*) **Simple monoclinic lattice:** It has lattice points at all 8 corners of the unit cell.

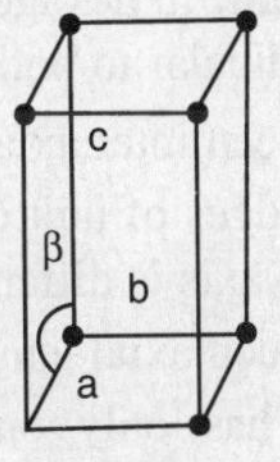

Monoclinic—P

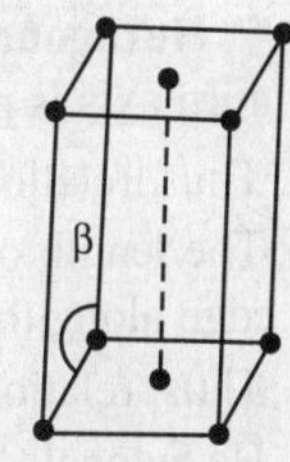

Monoclinic—C

Fig. 34.13 d: The two Bravais lattices of a monoclinic system

(*ii*) **Base-centered monoclinic lattice:** It has lattice points at all 8 corners of the unit cell and two lattice points, one each at the base and top face of the unit cell.

5. Triclinic system: In triclinic crystals, none of the three crystal axes is perpendicular to any of the other.

Thus, relation between inter axial angles is $\alpha \neq \gamma \neq 90° \neq \beta$.

The lengths of the edges of unit cell along the three axes are different.

That is, relation between axial length is $a \neq b \neq c$.

Triclinic lattice has only one possible type of arrangement, namely, simple (*P*) [Fig. 34.13 (e)].

(*i*) **Simple triclinic lattice:** It has lattice points at all 8 corners of the unit cell.

6. Trigonal or Rhombhohedral system: In trigonal crystals, the angle between each pair of crystal axes is the same but is not equal to 90°.

Thus, relation between inter axial angles is $\alpha = \beta = \gamma = 90° < 120°$

The length of the edges of unit cell is the same along all the three axes.

Thus, relation between axial lengths is, $a \neq b \neq c$.

Trigonal lattice has only one possible type of arrangement, namely, simple (R) [Fig. 34.13 (f)].

(*i*) **Simple trigonal lattice:** It has lattice points at all 8 corners of the unit cell.

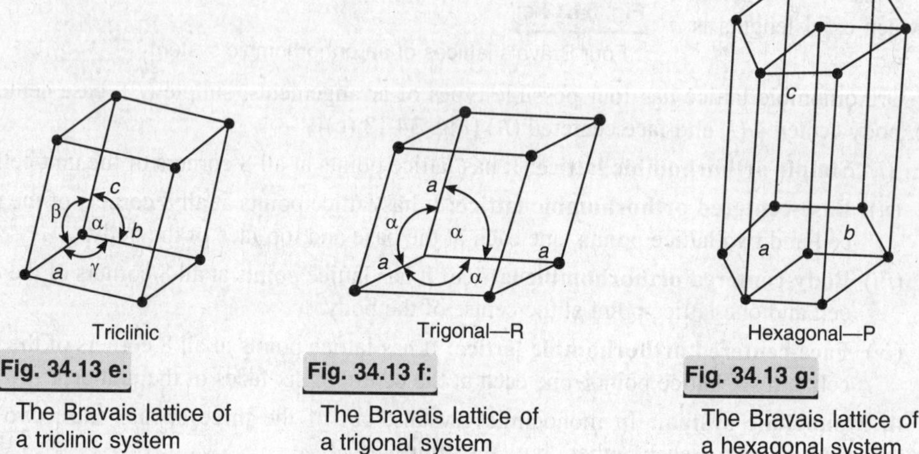

Fig. 34.13 e: The Bravais lattice of a triclinic system

Fig. 34.13 f: The Bravais lattice of a trigonal system

Fig. 34.13 g: The Bravais lattice of a hexagonal system

7. Hexagonal system: In hexagonal crystals, two of the crystal axes are 120° apart, while the third axis is perpendicular to both of them.

Thus, relation between inter axial angles is $\alpha = \beta = 90°$ and $\lambda = 120°$.

The length of the edges of unit cell is the same along all the axes that are 120° apart but the edge along the third axis is different.

Thus, relation between axial lengths is $a = b \neq c$.

Hexagonal lattice has only one possible type of arrangement, namely, simple (P) [Fig. 34.13 (g)].

(*i*) **Simple hexagonal lattice:** It has lattice points at all 12 corners of the hexagonal unit cell and two lattice points, one each at the base and top face of the hexagonal prism.

The seven crystal systems and the parameters of the corresponding unit cells are summarized in Table 34.2.

Table 34.2: *The seven crystal systems*

Sr. No.	Crystal System	Type and number of Bravais lattices	Unit cell axes	Angle between axes
1.	Triclinic	P,1	$a \neq b \neq c$	$\alpha \neq \beta \neq \gamma \neq 90°$
2.	Monoclinic	P, C } 2	$a \neq b \neq c$	$\alpha = \gamma = 90° \neq \beta$
3.	Orthorhombic	P, C, I, F } 4	$a \neq b \neq c$	$\alpha = \beta = \gamma = 90°$
4.	Tetragonal	P, I } 2	$a = b \neq c$	$\alpha = \beta = \gamma = 90°$
5.	Cubic	P, I, F } 3	$a = b = c$	$\alpha = \beta = \gamma = 90°$
6.	Hexagonal	P, I	$a = b \neq c$	$\alpha = \beta = 90°, \gamma = 120°$
7.	Trigonal (Rhombohedral)	R, I	$a = b = c$	$\alpha = \beta = \gamma \neq 90° < 120°$

Note: P-Primitive (simple), C-Base Centred, I-Body Centred, F-Face Centred, R- Rhombhohedral

34.8 SYMMETRY ELEMENTS IN A CUBIC CRYSTAL

We now take the example of a cubic crystal and illustrate the elements of symmetry it exhibits.

(*i*) Let us consider the cube shown in Fig. 34.14. If the body center point is *I*, and body diagonals are drawn through it, each diagonal connects identical lattice points located at equal distances and in opposite directions from *I*, The point *I* acts as a point mirror which generates the second lattice point at an equal distance in opposite direction. Therefore, the point *I* is the center of symmetry or inversion point. Thus, for a cubic crystal inversion point *I* is located at the body center.

(*ii*) Let us consider a normal MN through mid-points of the pair of opposite parallel faces of a cube. If the cube is rotated through 90° rotation, it goes into an indistinguishable configuration. In one complete rotation of 360°, the cube becomes indistinguishable four times. Therefore, the face-normal is

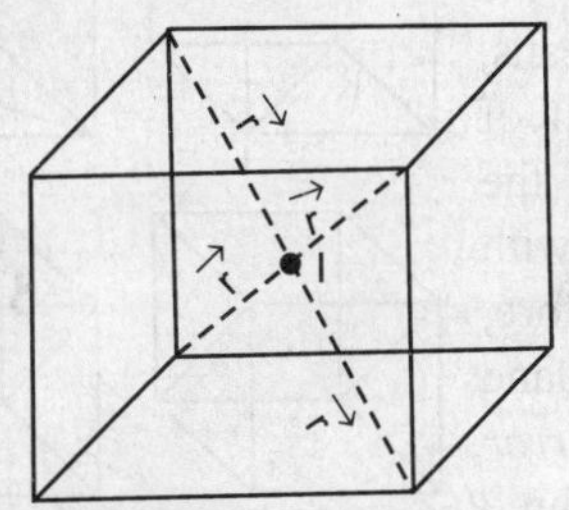

Fig. 34.14: The center of symmetry or inversion point

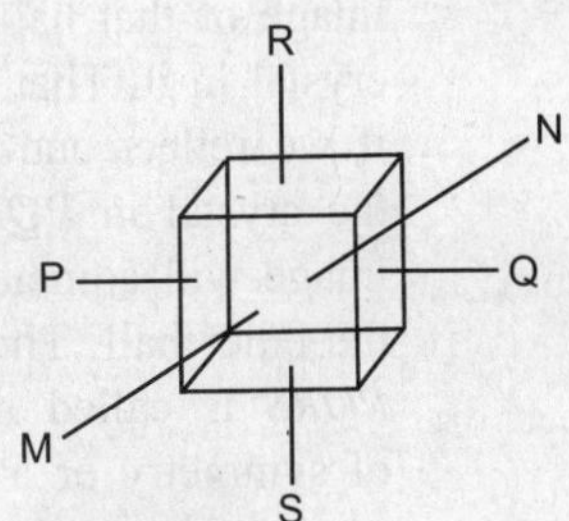

Fig. 34.15: 4 -Fold Rotation Axis

a 4-fold axis of symmetry. There are three such 4-fold axes of symmetry, *MN*, *PQ* and *RS*, one normal to each of the three pairs of parallel faces (see Fig. 34.15).

Now consider the axis CD passing through the body diagonal of the cube (Fig. 34.16). If the cube is rotated around the body diagonal through 120°, it comes into a congruent position. In one full rotation, the cube becomes indistinguishable three times. Therefore, the body diagonal is a 3-fold axis of symmetry. A cube has four body diagonals and therefore possesses four 3-fold axes of symmetry.

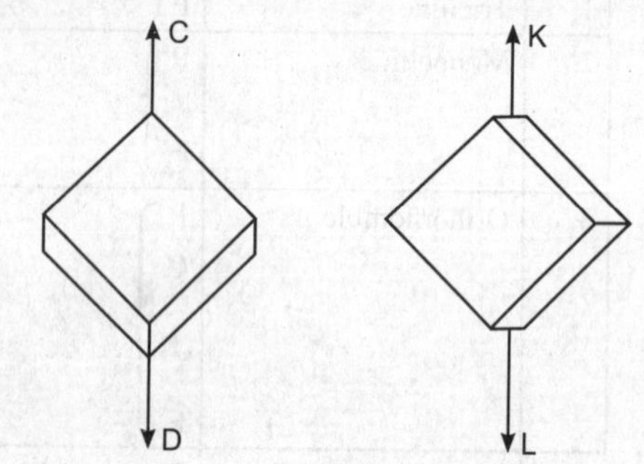

Fig. 34.16: 3-Fold Rotation Axis

Fig. 34.17: 2-Fold Rotation Axis

Next, consider a normal KL at mid points of parallel edges of Fig. 34.17. If the cube is rotated through 180°, it becomes congruent. In one rotation of 360°, the cube assumes indistinguishable configuration twice. Therefore, *KL* is a 2-fold axis of symmetry. As there are 12 edges in a cube, the number of 2-fold axes of symmetry is six.

The axes of symmetry in a cube are thus

2 - fold axes (diad axes)	6
3 - fold axes (triad axes)	4
4 - fold axes (tetrad axes)	3
	13

There are in total 13 axes of symmetry in a cube.

(*iii*) Let us consider a plane such as *PQRS* in the middle of the cube and parallel to one pair of the faces, Fig. 34.18 (*a*). If it is a plane mirror and one half of the crystal is cut and removed, the plane *PQRS* forms the image of that half of the crystal in it. That means, if we reflect one half of the crystal in *PQRS*, the image will coincide with the other half. Therefore, *PQRS* is called a plane of symmetry or a *mirror plane*. There are three such planes of symmetry parallel to the faces of the cube.

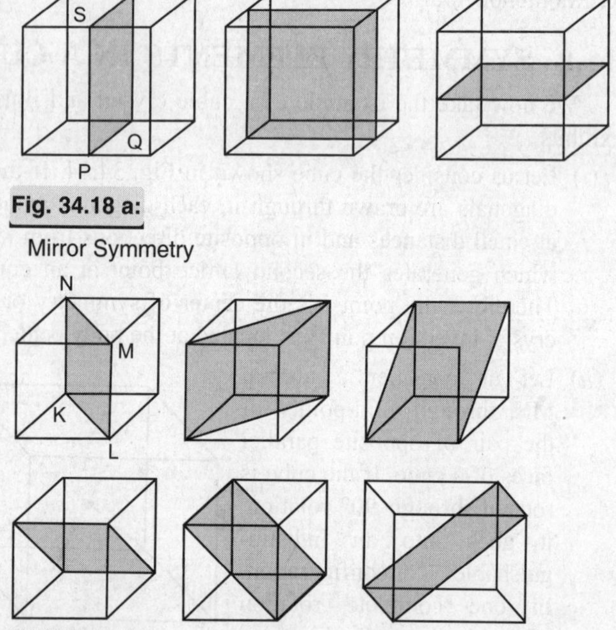

Fig. 34.18 a: Mirror Symmetry

Fig. 34.18 b: Mirror Symmetry

Further, consider the diagonal plane *KLMN* in the cube, Fig. 34.18 (*b*). If *KLMN* is imagined to be a mirror, it is readily seen that the prism behind it is the reflection of the prism in contact with it in its front. Thus it is also a mirror plane. There are six such diagonal mirror planes in a cube.

The total elements of symmetry of a cubic crystal are 23 comprising of one center of inversion, 13 axes of rotation and 9 mirror planes. Table 34.3 shows the summary of symmetry.

Table 34.3: *Summary of Symmetry*

Symmetry	
Translation symmetry	Point symmetry
Unit cell, lattice, crystallographic coordination system	Reflection, rotation, inversion, roto-inversion
14 Bravais lattices	32 Crystal classes (point groups)
6 Crystal systems	
Screw axis, glide planes	
230 Space groups	

34.9 CALCULATION OF PARAMETERS OF A CUBIC LATTICE

A unit cell is characterized by a set of characteristics, namely, volume, effective number of atoms, coordination number, atomic packing fraction, and density. These parameters can be computed with relative ease in case of cubic unit cells. There are three varieties of the cubic lattice as shown in Fig. 34.19. They are simple cubic (SC), body centered cubic (BCC), and face centered cubic (FCC) unit cells.

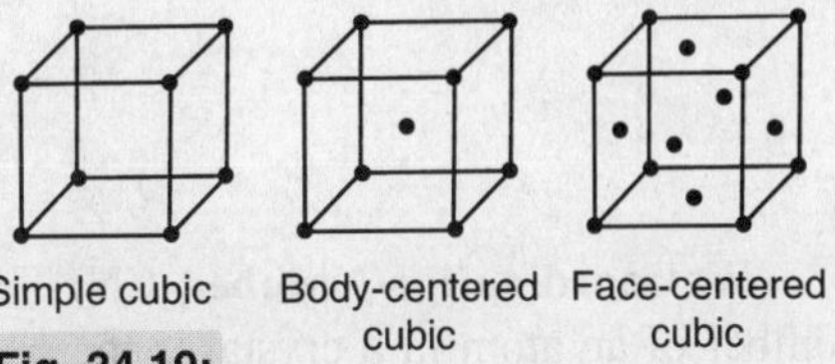

Fig. 34.19: The three space lattices of a cubic system

34.9.1 Simple Cubic (SC) Cell

A unit cell is said to be *primitive* when the cell has lattice points only at its corners. The primitive cubic unit cell is also called a simple cubic (SC) unit cell. Fig. 34.20 shows a SC cell.

(*i*) Unit cell volume, V: In case of cubic cell all the edges of the cube are of equal length, '*a*'. Therefore, the volume is given by

$$V = a^3 \qquad ...(34.1)$$

(*ii*) Effective number of atoms per unit cell, Z: A unit cell is a part of an infinite scheme, and is not an isolated entity. Therefore, several adjacent cells share each lattice point. As a result, the basis attached to a lattice site contributes only a fraction of its mass and volume to one unit cell. Let only one atom, having a radius *R*, be attached to one lattice point. The effective

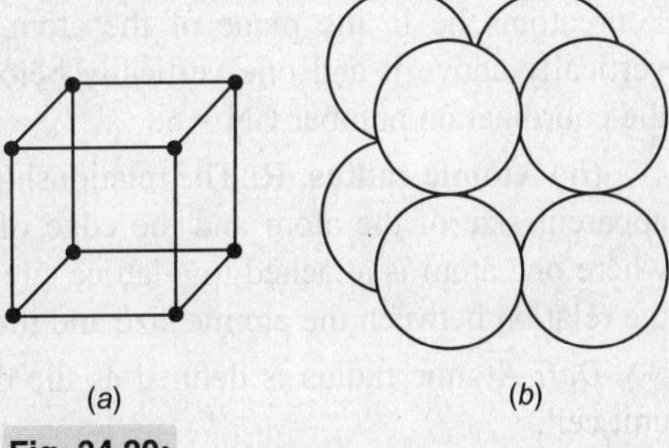

Fig. 34.20: Simple cubic cell (a) atomic site model (b) hard sphere model

number of atoms per unit cell is given by

$$Z = Z_B + \frac{Z_F}{2} + \frac{Z_C}{8} \quad ...(34.2)$$

where Z_B = Number of body centered atoms,

Z_F = Number of face centered atoms and

Z_C = Number of corner atoms

In the three-dimensional array, each corner atom is linked to eight surrounding cells, as shown in Fig. 34.21 (a). Hence, in effect, the atom contributes 1/8th of its content to a unit cell, as shown in Fig. 34.21 (b). SC cell being a primitive cell does not contain lattice points within the body volume or in the centre of faces. Therefore, the total contribution to Z comes from the corner atoms of the unit cell and it is given by

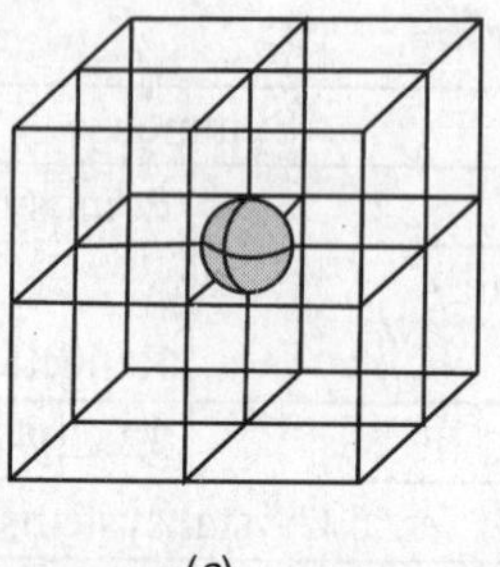

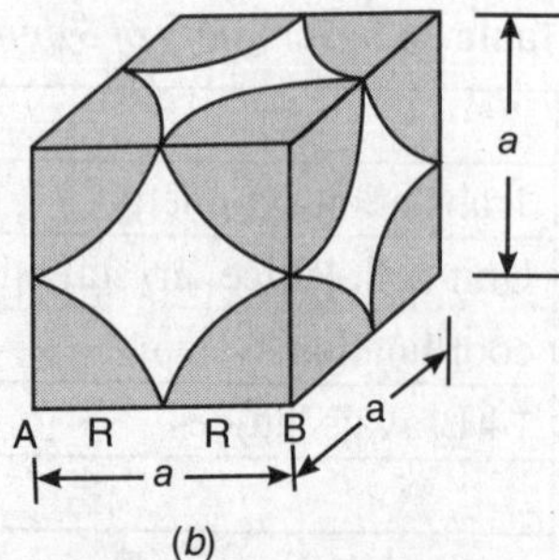

Fig. 34.21: (a) An atom contributes one eighth to a unit cell when it is located at the corner of the cell. (b) Isolated SC cell

$$Z = Z_B + \frac{Z_F}{2} + \frac{Z_C}{8} = 0 + 0 + \frac{8}{8} = 0 + 0 + 1 \text{ atom / cell}$$

$$\therefore \quad Z = 1 \text{ atom / cell} \quad ...(34.3)$$

***(iii)* Coordination Number, CN:** The coordination number of an atom in a crystal is the number of nearest neighbour atoms. It signifies the tightness of packing of atoms in the crystal.

Def: The coordination number of an atom or ion is the number of nearest neighbours of a point in space lattice of a unit cell.

Around an atom in a SC cell, there would be six equally spaced nearest neighbour atoms each at a distance '*a*' from that atom, as shown in Fig. 34.22. Four atoms lie in the plane of the atom while one is vertically above it and one vertically below. Therefore, the coordination number CN = 6.

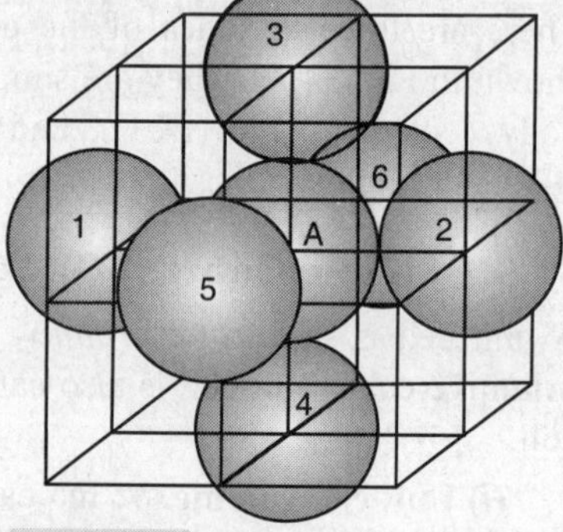

Fig. 34.22: There are six close neighbours for any selected atom in a SC cell.

***(iv)* Atomic radius, R:** The relationship between the apparent size of the atom and the edge of the unit cell where one atom is attached to a lattice site can be determined. By applying simple geometry the relation between the atomic size and the unit cell edge can be computed.

Def: Atomic radius is defined as the distance between nearest neighbouring atoms in a unit cell.

In a SC cell the atoms would be in contact along the edges of the cube, as seen from Fig. 34.21 (b). If 'a' is the edge of the cubic cell and R is the radius of the atom.

$$a = 2R \quad \text{or} \quad R = a/2 \qquad ...(34.4)$$

(v) Packing Fraction, APF: The fraction of space occupied by atoms in a unit cell is known as atomic packing fraction.

Def: Packing fraction. It is defined as the ratio of volume of effective number of atoms in the unit cell to the total volume of the unit cell. Thus,

$$APF = \frac{\text{(Number of atoms/unit cell) (Volume of each atom)}}{\text{Volume of the unit cell}} \qquad ...(34.5)$$

$$\therefore \quad APF = \frac{Z\upsilon}{V}$$

In case of SC cell, $Z = 1$

Volume of unit cell, $V = a^3 = (2R)^3 = 8R^3$.

Volume of spherical atom, $\upsilon = \frac{4}{3}\pi R^3$

$$\therefore \quad APF = \frac{1 \times \frac{4}{3}\pi R^3}{8R^3} = \frac{\pi}{6} = 0.52 \qquad ...(34.6)$$

(vi) Percentage void space: *The void space in the unit cell is the vacant space left unutilised in the cell.* It is often expressed as percentage.

$$\begin{aligned} \text{\% void space} &= (1 - APF) \times 100 \\ &= (1 - 0.52) \times 100 \\ &= 48\% \end{aligned} \qquad ...(34.7)$$

(vii) Density of SC crystal, ρ: As a unit cell possesses all the structural properties of a bulk crystal, the density of a unit cell must be the same as that of the bulk crystal. Thus,

$$\text{Density, } \rho = \frac{\text{Mass}}{\text{Volume}} = \frac{ZW}{V} \qquad ...(34.8)$$

where W is the mass of each atom, which is given by

$$W = \frac{M}{N_A} \qquad ...(34.9)$$

M is the molecular weight of the material and N_A is the Avogadro number.

Therefore, the density of a cubic crystal is given by

$$\rho = \frac{ZM}{N_A V} = \frac{ZM}{N_A a^3} \qquad ...(34.10)$$

In case of SC crystal $Z = 1$.

Hence, $$\rho = \frac{M}{N_A a^3} \qquad ...(34.11)$$

34.10 BODY CENTRED CUBIC (BCC) CELL

The body centered cubic cell has a lattice point within the cell in addition to the eight corner lattice points, as shown in Fig. 34.23 (a).

(i) Unit cell volume, V: In case of cubic cell all the edges of the cube are of equal length 'a'. Therefore, the volume is given by

$$V = a^3$$

***(ii)* Effective number of atoms per unit cell, Z:** A body centered cubic cell has eight lattice points at the eight corners of the cube and one lattice point at centre within the cell, as shown in Fig. 34.23 (a). There are no points in the face. An atom at the lattice point within the cell belongs completely to the cell and the atoms at the corners of the cell contribute 1/8th each, as shown in Fig. 34.23 (b).

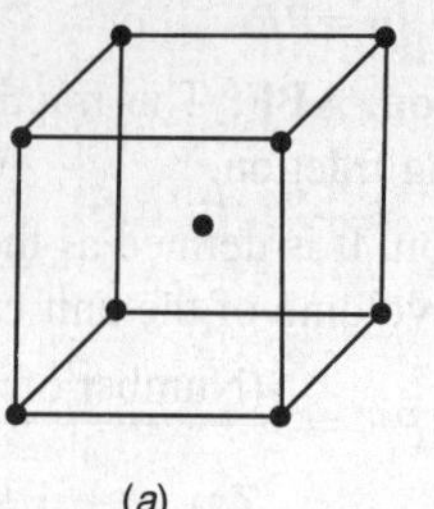

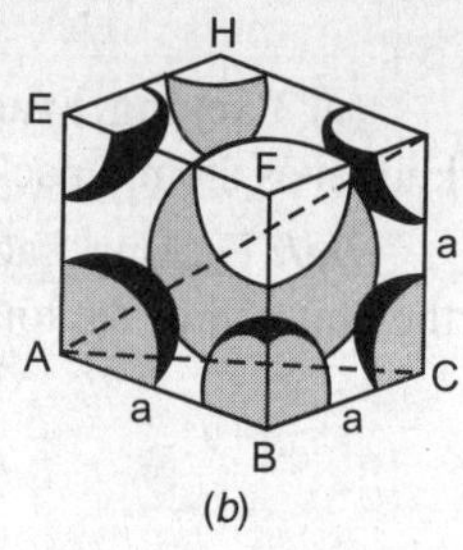

Fig. 34.23:

(a) BCC cell, (b) Isolated BCC unit cell

The effective number of atoms per BCC cell is then

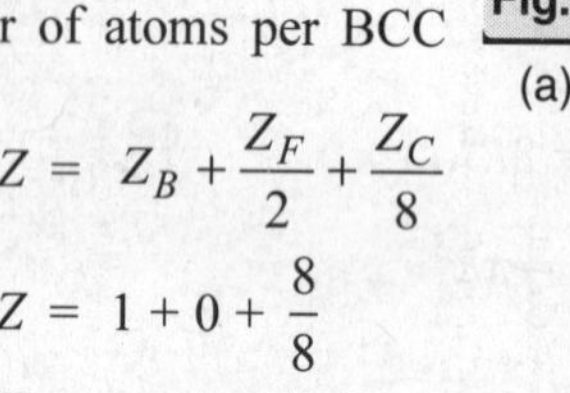

$$Z = Z_B + \frac{Z_F}{2} + \frac{Z_C}{8}$$

$$Z = 1 + 0 + \frac{8}{8}$$

$$\therefore \quad Z = 2 \text{ atoms/cell} \qquad ...(34.12)$$

***(iii)* Coordination Number, CN:** In the BCC cell, atoms occupying the corner lattice points do not touch each other. However, each corner atom is in contact with the atom at the body center (Fig. 34.24). As there are eight unit cells around each corner of the cell, the atom located at a corner would be simultaneously touched by the eight body centered atoms of the eight surrounding cells.

$$\text{CN} = 8$$

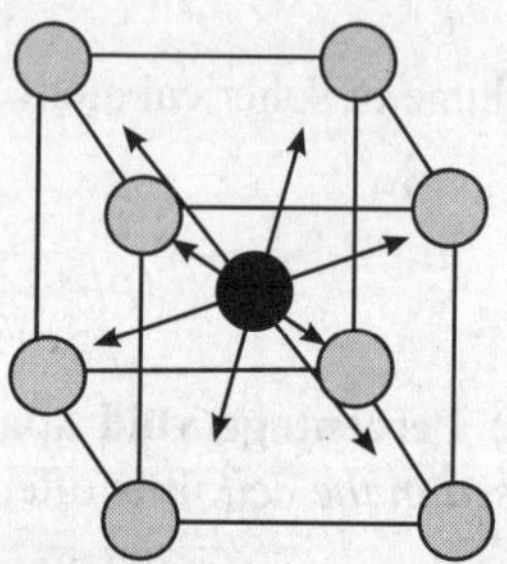

Fig. 34.24:

A given atom in a BCC cell is surrounded by 8 closed neighbours

***(iv)* Atomic radius, R:** In a BCC cell atoms would be in contact along the body diagonal. From the Fig. 34.25,

$$(AG)^2 = (AC)^2 + (CG)^2 = (AB)^2 + (BC)^2 + (CG)^2 \qquad ...(34.13)$$

$$AG = 4R \quad \text{and} \quad AB = BC = CG = a$$

$$\therefore \quad (4R)^2 = 3(a)^2$$

$$\therefore \quad 4R = a\sqrt{3}$$

$$\therefore \quad R = \frac{a\sqrt{3}}{4} \qquad ...(34.14)$$

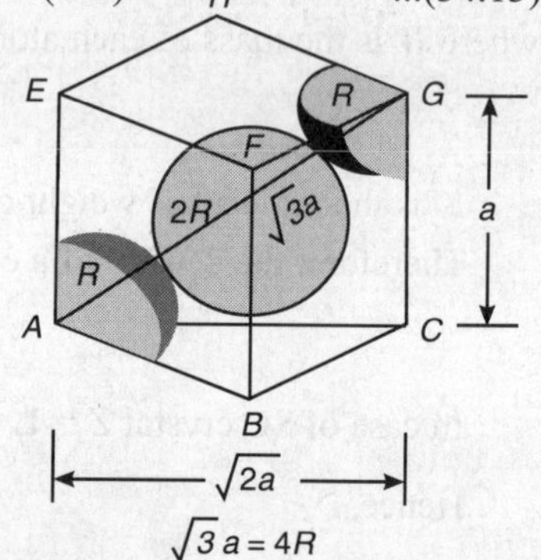

Fig. 34.25:

Determination of the relation between atomic radius and lattice edge in a BCC cell

***(v)* Atomic packing fraction, APF:**

For BCC cell, $Z = 2$ and $a = \frac{4R}{\sqrt{3}}$

Volume of the spherical atom $\upsilon = \frac{4}{3}\pi R^3$

Volume of the unit cell $V = a^3 = \frac{64R^3}{3\sqrt{3}}$

$$APF = \frac{Z\upsilon}{V} = \frac{2 \times \frac{4}{3}\pi R^3}{\frac{64}{3\sqrt{3}}R^3} = \frac{\pi\sqrt{3}}{8} = 0.68 \qquad ...(34.15)$$

(vi) Percentage void space:

$$\text{Percentage void space} = (1 - \text{APF}) \times 100 = (1 - 0.68) \times 100 = 32\% \quad ...(34.16)$$

(vii) Density of BCC crystal, ρ:

In case of BCC crystal $Z = 2$.

Hence
$$\rho = \frac{2M}{N_A a^3} \quad ...(34.17)$$

Example 34.2. *Molybdenum has a BCC structure. Its density is 10.2 × 10³ kg/m³ and its atomic weight is 95.94. Determine the radius of molybdenum atom.*

Solution: Density of a cubic unit cell is given by $\rho = \dfrac{ZM}{N_A a^3}$

$$\therefore \quad a^3 = \frac{ZM}{\rho N_A} = \frac{2 \times 95.94\,\text{kg / k.mol}}{10.2 \times 10^3\,\text{kg / m}^3 \times 6.02 \times 10^{26}\ \text{/ k.mol}} = 31.25 \times 10^{-30}\text{m}^3$$

$$\therefore \quad a = \mathbf{3.15\ Å.}$$

The atomic radius in BCC structure is given by $r = \dfrac{a\sqrt{3}}{4} = \dfrac{3.15\,\text{Å} \times 1.732}{4} = \mathbf{1.364\ Å.}$

Example 34.3. *Sodium crystallizes in a cubic lattice. The edge of the unit cell is 4.3Å. The density of sodium is 963 kg/m³ and its atomic weight is 23. What type of unit cell does sodium form?*

Solution:

Density of a cubic unit cell is given by $\rho = \dfrac{ZM}{N_A a^3}$

The effective number atoms per unit cell is given by $Z = \dfrac{\rho N_A a^3}{M}$

$$\therefore \quad Z = \frac{963\,\text{kg/m}^3 \times 6.02 \times 10^{26}\,\text{/k.mol} \times (4.3 \times 10^{-10}\,\text{m})^3}{23\,\text{kg/k.mol}} = \frac{46.09}{23} = 2\ \text{atoms/unit cell}$$

∴ Sodium forms BCC structure.

34.11 FACE CENTRED CUBIC (FCC) CELL

The face centred cubic cell is a non-primitive cell having six lattice points at the centres of its six faces and eight atoms at the eight corners of the cube as shown in Fig. 34.26.

(*i*) **Unit cell volume, V:** In case of cubic cell all the edges of the cube are of equal length 'a'. Therefore, the volume is given by

$$V = a^3$$

(*ii*) **Effective number of atoms per unit cell, Z:** In a three dimensional array, a unit cell is surrounded by other unit cells, while each corner of the cell is shared by eight adjacent cells, and each face is shared by two adjacent cells, as shown in Fig. 34.27. Therefore, each face-centred atom contributes half of its mass and volume to one cell, while each corner atom contributes 1/8th, as shown in Fig. 34.28. There are 6 faces and 8 corners in a cubic cell.

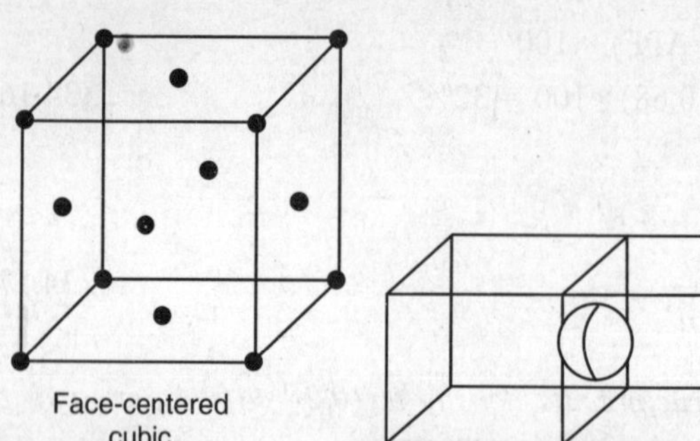

Fig. 34.26: Face Centered Cubic cell

Fig. 34.27: An atom contributes half of its volume to the unit cell when it is located in the face

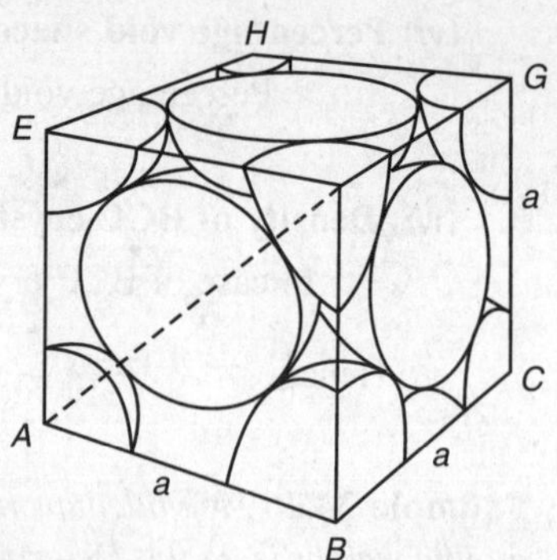

Fig. 34.28: Determination of effective number of atoms per unit cell

Thus, the number of atoms effectvely contributing to the FCC cell is,

$$Z = Z_B + \frac{Z_F}{2} + \frac{Z_C}{8} = 0 + \frac{6}{2} + \frac{8}{8} \qquad ...(34.18)$$

$$\therefore \qquad Z = 4 \text{ atoms/cell.}$$

(iii) Coordination Number, CN: In a FCC cell, each corner atom is in contact with the face-centered atom. It would therefore be in contact with 4 atoms in the *xy* plane, 4 atoms in the *yz*-plane and 4 atoms in the *zx*-plane, as shown in Fig. 34.29.

Therefore, Coordination number = 4 + 4 + 4 = 12.

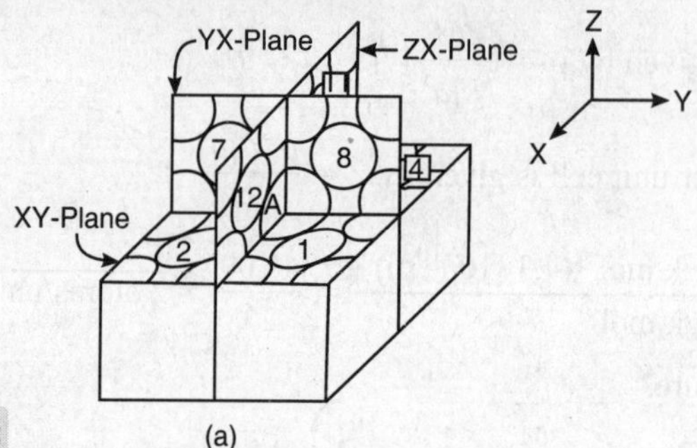

Fig. 34.29: Determination of nearest neighbours in a FCC cell

(iv) Atomic radius, R: In a FCC cell, atoms are in contact along the face diagonal of the cube, as shown in Fig. 34.30. It is seen from the figure that

$$AF^2 = AB^2 + BF^2 \qquad ...(34.19)$$

But $AF = 4R$ and $AB = a = BF$

$$\therefore \qquad (4R)^2 = 2a^2$$

$$\therefore \qquad 4R = a\sqrt{2}$$

$$\therefore \qquad R = \frac{a\sqrt{2}}{4} = \frac{a}{2\sqrt{2}} \qquad ...(34.20)$$

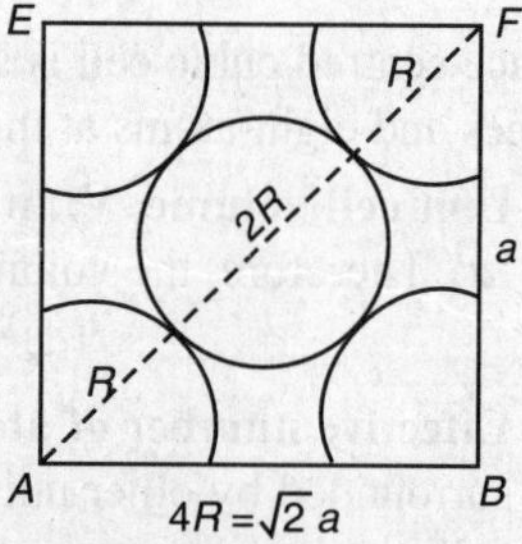

Fig. 34.30: Determination of atomic radius in a FCC cell

(v) Atomic packing fraction, APF:

$$APF = \frac{Z\,\upsilon}{V} = \frac{4\upsilon}{a^3}$$

$$a^3 = \left[2\sqrt{2}\,R\right]^3 = 16\sqrt{2}R^3$$

$$\therefore \qquad APF = \frac{4\times\frac{4}{3}\pi R^3}{16\sqrt{2}R^3} = \frac{\pi}{3\sqrt{2}} = 0.74 \qquad ...(34.21)$$

(vi) Percentage void space:

Percentage void space = (1 – *APF*) × 100 = (1 – 0.74) × 100 = 26 % ...(34.22)

(vii) Density of FCC crystal, ρ:

The density of a cubic crystal is given by

$$\rho = \frac{ZM}{N_A V} = \frac{ZM}{N_A a^3}$$

In case of FCC crystal $Z = 4$.

Hence

$$\rho = \frac{4M}{N_A a^3} \qquad ...(34.23)$$

Example 34.4. *Lead exhibits FCC structure. Each side of the unit cell is of 4.95 Å. Calculate the radius of a lead atom.*

Solution: The radius of the lead atom is given by $R = \frac{a}{2\sqrt{2}} = \frac{4.95\,\text{Å}}{2(1.414)}$ = **1.75 Å.**

34.12 HCP STRUCTURE

Hexagonal close packed (HCP) structure is one of the most common metallic structures. About 25 metals exhibit this structure. Metals do not crystallize into the simple hexagonal crystal structure. The atoms attain a lower energy and a more stable condition only by forming the HCP structure.

***(i)* Effective number of atoms per unit cell:** The isolated HCP unit cell is shown in Fig. 34.31. Each corner atom of the hexagonal face is shared by six unit cells. Consequently it contributes 1/6th of its volume and mass to one unit cell. There are six corner atoms in the base. Further, there is an atom at the center of the each hexagonal face, which is shared by two adjacent unit cells. Three atoms forming a triangle in the middle layer are within the body of the cell and cannot be shared by adjacent cells.

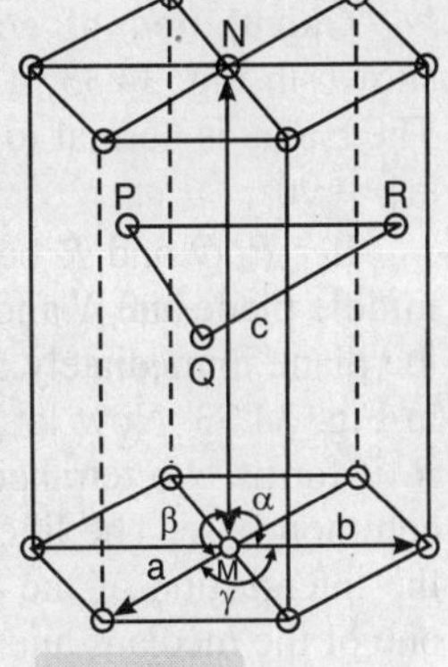

Fig. 34.31: HCP Unit Cell

The total contribution to the effective number of atoms in the cell is thus

$$Z = \left(2\,\frac{\text{base}}{\text{cell}}\right)\times\left(\frac{1}{6}\,\frac{\text{atom}}{\text{corner}}\right)\times\left(6\,\frac{\text{corners}}{\text{base}}\right)+\left(2\,\frac{\text{bases}}{\text{cell}}\right)\times\left(\frac{1}{2}\,\frac{\text{atoms}}{\text{base}}\right)+\left(3\,\frac{\text{atoms}}{\text{cell}}\right)$$

or $\quad Z = 6$ atoms/unit cell.

***(ii)* Atomic radius, r:** The atoms are in contact along the edges of the hexagon as seen from Fig. 34.31. Therefore,

$$2r = a \quad \text{or} \quad r = a/2$$

***(iii)* Coordination number, CN:** Each atom in the structure is positioned in valleys formed by three adjacent atoms of the top layer and by three adjacent atoms in the bottom layer; and is surrounded by six neighbour atoms. Thus, twelve atoms are in contact with the given atom under consideration. Therefore

$$\text{CN} = 12$$

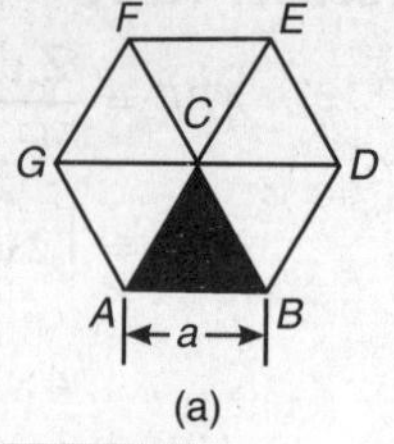

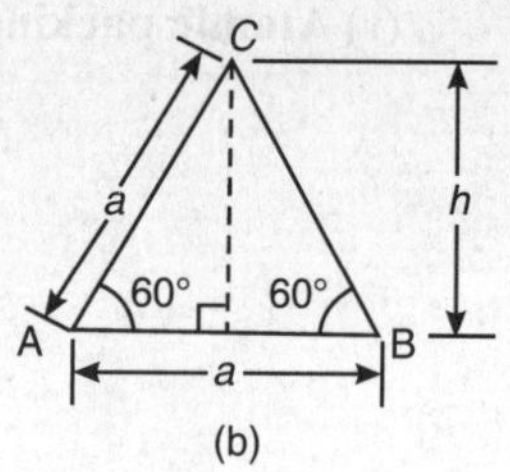

Fig. 34.32: Area of the base of the HCP unit cell

***(iv)* Volume of the unit cell, V:** The volume of the unit cell may be determined by computing the area of the base of the unit cell and then by multiplying it by the cell height. The area of the base of the unit cell is the area of the hexagon *ABDEFG*, shown in Fig. 34.32 (a).

It is equal to the sum of the areas of the six equal-sized triangles.

$$\text{Area of the hexagon } ABDEFG = 6 \text{ (area of } \Delta \text{ ABC)} = 6(1/2a)(a \sin 60°)$$

$$= 3a^2 \sin 60° = 3a^2\left(\frac{\sqrt{3}}{2}\right) = \frac{3\sqrt{3}}{2}a^2$$

$\therefore$ Volume of the unit cell = (Area of the hexagon) × (height of the unit cell)

$$= \left(\frac{3\sqrt{3}}{2}a^2\right)c$$

or

$$V = \frac{3\sqrt{3}}{2}a^2c \qquad \text{...(34.24)}$$

Calculation of c/a ratio: Referring to the HCP unit cell shown in Fig. 34.33, it is seen that $a = b$ and the angle $\gamma = 120°$. The c axis is normal to the plane containing a and b. Therefore $\alpha = \beta = 90°$,

Let *P, Q* and *R* be the centers of the adjacent atoms in the middle plane and *N* and *M* be the centers of the adjacent atoms in the plane immediately above and below the plane *PQR*, as shown in Fig. 34.33. Now let us join *M* to *N* and *M* and *N* to *P, Q* and *R*. it forms two tetrahedrons *MPQR* and *NPQR* with *PQR* as the common base. The line *MN* passes through the point *T*, which is the intersection of the three medians of the triangle *PQR*. *RS* is one of the medians and therefore,

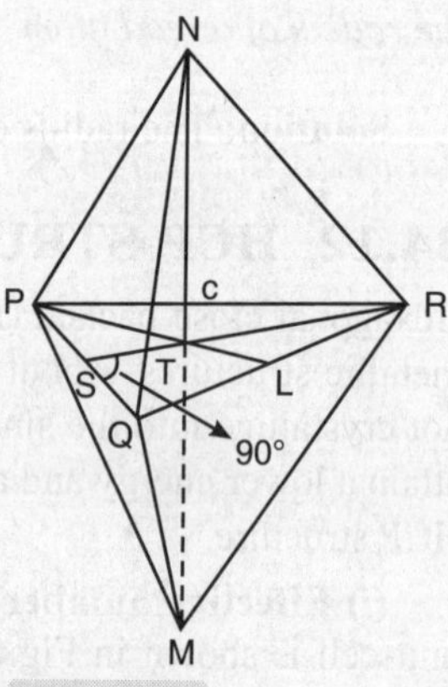

Fig. 34.33: Centre of adjacent atoms

$$TR = 2\ TS$$

Further,

$$MN = c \quad \text{and} \quad PQ = QR = a$$

As the median of an equilateral triangle is also perpendicular to the opposite side, $\angle QSR = 90°$.

$$\therefore \quad \text{RS} = \sqrt{a^2 - \left(\frac{a}{2}\right)^2} = \frac{\sqrt{3}}{2}a$$

Also,

$$RT = 2TS = \frac{2}{3}(TS + TR) = \frac{2}{3}RS = \frac{2}{3}\cdot\frac{\sqrt{3}}{2}a = \frac{a}{\sqrt{3}}$$

And

$$NT = \sqrt{NR^2 - RT^2} = \sqrt{a^2 - \frac{a^2}{3}} = \sqrt{\frac{2}{3}}a \qquad (\because NR = NQ = NP = a)$$

$$c - 2NT - 2\sqrt{\frac{2}{3}}a = \sqrt{\frac{8}{3}}a$$

$$\therefore \quad \frac{c}{a} = \sqrt{\frac{8}{3}} \qquad ...(34.25)$$

$\therefore$ The volume of the hexagonal cell, $V = \frac{3\sqrt{3}}{2}a^2 c = \frac{3\sqrt{3}}{2}a^2 \cdot \sqrt{\frac{8}{3}}a = 3\sqrt{2}\,a^3$ ...(34.26)

(*v*) Atomic Packing Fraction (APF):

$$\text{APF} = \frac{Z\upsilon}{V} = \frac{6\times\frac{4}{3}\pi r^3}{3\sqrt{2}(2r)^3} = \frac{\pi}{3\sqrt{2}} = 0.74$$

(*vi*) Void space

Percentage void space = (1 – APF) × 100 = (1 – 0.74) × 100 = 26%

(*vii*) Density, ρ:

The theoretical density of HCP cell is given by

$$\rho = \frac{ZM}{N_A V} = \frac{6M}{N_A(3\sqrt{2}a^3)} = \frac{\sqrt{2}M}{N_A a^3} \qquad ...(34.27)$$

The characteristics of the three types of cubic unit cells and the HCP cell are summarized in Table 34.4.

Table 34.4: *Summary of Unit cell characteristics*

Sr. No.	Characteristics	Unit Cell			
		SC	BCC	FCC	HCP
1.	Unit cell volume, V	a^3	a^3	a^3	$3\sqrt{2}\,a^3$
2.	Atoms per unit cell, Z	1	2	4	6
3.	Atomic radius, *r*	*a*/2	$a\sqrt{3}/4$	$a/2\sqrt{2}$	*a*/2
4.	Coordination number, CN	6	8	12	12
5.	Atomic packing fraction, APF	π/6 = 0.52	$\pi\sqrt{3}/8$ = 0.68	$\pi/3\sqrt{2}$ = 0.74	$\pi/3\sqrt{2}$ = 0.74
6.	Void space	48%	32%	26%	26%
7.	Density, ρ	$\frac{M}{N_A a^3}$	$\frac{2M}{N_A a^3}$	$\frac{4M}{N_A a^3}$	$\frac{\sqrt{2}M}{N_A a^3}$

It is seen from the above table that the SC cell is loosely packed and the FCC and HCP cells are the close packed cells among the four types of unit cells.

34.13 ATOM POSITIONS IN CUBIC UNIT CELLS

Let a rectangular right handed coordinate system be attached to the cubic lattice, as shown in Fig. 34.34. The position of a lattice site is described by three coordinates (*x*, *y*, *z*) which are expressed as

$$x = pa, \quad y = qb, \quad z = rc \qquad ...(34.28)$$

where a, b and c are lattice constants in x, y and z directions

p, q and r are integers.

If the lattice constants a, b and c are taken as **unit** axial lengths, $a = 1$, $b = 1$, and $c = 1$, and the co-ordinates of the lattice site will be (p, q, r). These are the indices of the lattice site and are written with commas separating the numbers and enclosed in parenthesis.

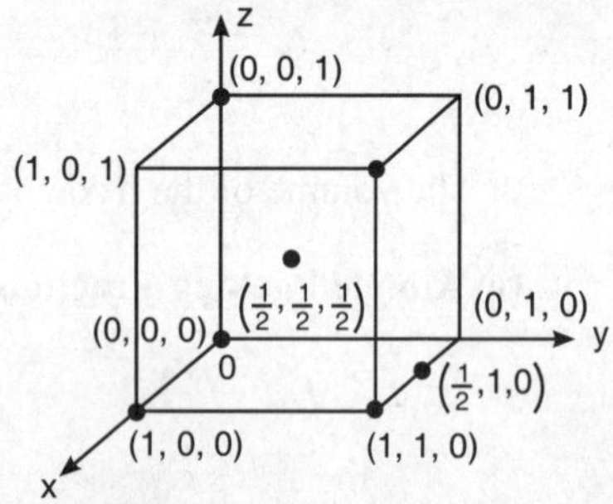

As an example, the indices of atomic sites in the BCC cell are shown in Fig. 34.34. The atom positions for the eight corner atoms are (0,0,0), (1,0,0), (0,1,0), (0,0,1), (1,1,1), (1,1,0), (1,0,1), and (0,1,1). The body cnetre atom in the cell has the coordinates $\left(\frac{1}{2}, \frac{1}{2}, \frac{1}{2}\right)$. By convention, the set of coordinates (0,0,0) stands for the locations of all eight corners of cubic unit cell. Therefore, it is sufficient if the atom positions (0,0,0) and $\left(\frac{1}{2}, \frac{1}{2}, \frac{1}{2}\right)$ are specified for the BCC cell.

Fig. 34.34: Rectangular right handed coordinate BCC lattice

The minimum number of coordinate sets necessary to specify the locations of all atoms in a unit cell are equal to the value of Z, the effective number atoms/cell.

34.14 INDICES OF CRYSTALLOGRAPHIC DIRECTION

Many crystals exhibit properties that are dependent on the direction in which they are measured and on the direction of the external stimuli, such as an electric field, magnetic field or mechanical stress, acting on the crystal. Therefore, it is essential that we know reference directions in crystals. To describe a direction in a crystal lattice, a straight line passing through the origin is chosen. The co-ordinates of the first lattice point lying on this line are utilized to denote the direction of the line. Thus, the indices of direction are the vector components of direction resolved along each of the axes and reduced to the smallest integers. The vector components are multiples of lattice constants. If the direction passes through the origin, then the indices of the lattice will also be the indices of direction.

The indices of direction are calculated by the following procedure.

(*i*) First, find the coordinates of the lattice site nearest to the origin in a given direction.

(*ii*) The coordinates are then divided by appropriate unit translations.

(*iii*) If fractions are obtained, each of the fractions is multiplied by smallest common divisor.

(*iv*) The integers obtained are the indices of direction written in square brackets, as [p q r].

Thus, *the indices of direction in a crystal are the set of smallest integers, which have the same ratios as the components of a vector in the desired direction, referred to the axes.*

As an example let us consider a cubic cell shown in Fig. 34.35. Let the point 'O' be the origin. The point P is at the position (0, b, 0). 'b' represents one unit distance. Therefore, the indices of the point P are (0,1,0). The direction OP is given as [0 1 0]. On the other hand, the direction of PO is specified as $\left[0\,\bar{1}\,0\right]$. A bar on the number indicates a negative direction. Similarly, the directions OQ and OR are indicated as [1 0 0] and [0 0 1] respectively.

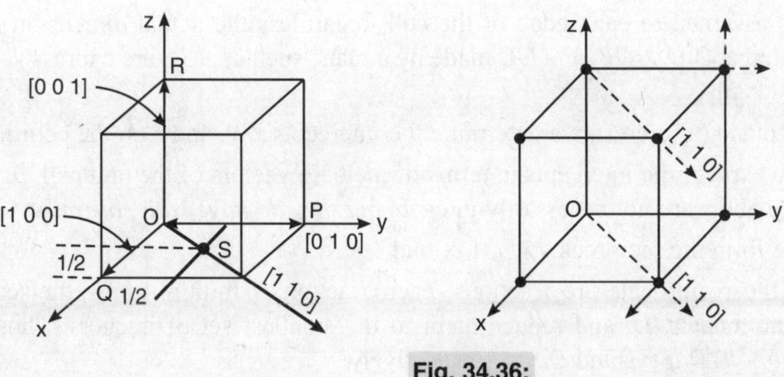

Fig. 34.35: Indices of direction

Fig. 34.36: Indices of parallel directions

The direction of *OS* may be determined as follows. The position indices of the point *S* are $\left(\frac{1}{2}, \frac{1}{2}, 0\right)$. Then, using the instruction (*iii*), the direction of the line *OS* is obtained as $\left[2 \times \frac{1}{2}, 2 \times \frac{1}{2}, 0\right] = [110]$.

All parallel directions have the same direction indices (Fig. 34.36) and the parallel directions are equivalent. <*pqr*> represents the family of directions [0 1 0], [0 0 1], [1 0 0], $[0\,\bar{1}\,0]$, $[0\,0\,\bar{1}]$, $[\bar{1}\,0\,0]$. All of them are grouped as <1 0 0>.

34.15 LATTICE PLANES AND MILLER INDICES

An aggregate of a set of parallel, equally spaced planes passing through the lattice points gives a crystal lattice. These planes are known as ***lattice planes*** and the perpendicular distance between neighboring planes is termed as called ***interplanar spacing***. Each of infinite set of lattice planes of a given space lattice will have their own characteristic interplanar spacing. In a crystal the geometrical location of a plane is not important. All the planes that are parallel to each other play a similar role. Three smallest whole numbers known as Miller indices represents the position and orientation of a lattice plane in a crystal.

Def: *Miller Indices* of a plane are the three smallest possible integers which have the same ratios as the reciprocals of intercepts of the plane concerned in three axes.

We find the Miller indices of a given plane as follows. We choose a set of coordinate axes parallel to unit cell edges at a convenient point. Let *OX*, *OY* and *OZ* be positive directions of the axes along the three edges of the unit cell, as shown in Fig. 34.37. We consider a plane *ABC* oriented with respect to the coordinate axes and which intercepts the *x*-, *y*-, and *z*-coordinates at distances *OA*, *OB* and *OC* respectively. It will be seen that Miller indices for all the planes parallel to the plane *ABC* and containing similar type of atomic array are the same. The intercepts are not measured in metrical distances of cm or Å, but are measured in terms of respective

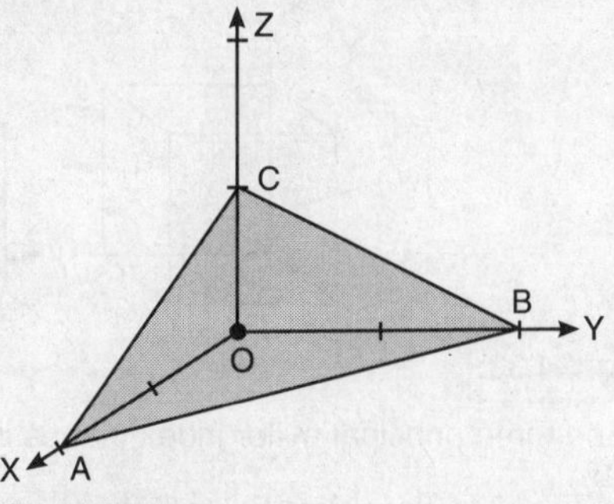

Fig. 34.37: A lattice plane intercepting the three crystallographic axes.

unit lengths assigned to each edge of the cell, regardless the actual dimension of the edge. Thus, the intercepts *OA*, *OB*, and *OC* made by a plane such as *ABC* are expressed as multiples of the axial lengths.

For the plane of interest, we determine the intercepts *x, y*, and *z* on the coordinate axes.

(*i*) We express the intercepts in terms of the base vectors of the unit cell, *pa*, *qb*, and *rc*; *p, q* and *r* are not necessarily integers *but they do have rational ratios.*

(*ii*) We form the reciprocals $1/p$, $1/q$, and $1/r$.

(*iii*) If the reciprocals are fractions, each fraction is multiplied by the least common denominator, *D*, and reduce them to the smallest set of integers. Thus, we obtain $D/p = h$, $D/q = k$ and $D/r = l$ respectively.

(*iv*) We put the integers in parenthesis (*hkl*). These are the **Miller indices** of the given plane.

All the planes, having the same structure of atoms, which are parallel to each other and therefore have the same orientation with respect to the three edges of the unit cell belong to the same family of planes and are represented by the same Miller indices. Consider a plane, of a family of similar planes, lying within the unit cell and nearest to the origin. This plane cuts off intercepts equal to $(1/h)a$, $(1/k)b$,$(1/l)c$. This implies that the (*hkl*) plane divides "*a*" into *h* parts, "*b*" into *k* parts and "*c*" into *l* parts. Other members of the family cut off intercepts $2a/h$, $2b/k$, $2c/l$,……. $n\,a/h$, $n\,b/k$, $n\,c/l$ so that the ratio between the intercepts is the same.

Some important features of Miller indices

(*i*) Parallel planes spaced equally have the same Miller indices. Thus, all the planes in a set parallel to a particular plane (*hkl*) are denoted by (*hkl*).

(*ii*) A plane parallel to one coordinate axis has Miller index 0 for that direction.

(*iii*) A plane passing through origin is denoted by Miller indices of a parallel plane having non-zero intercepts.

(*iv*) When the intercept of a plane is negative on an axis, a bar is placed on the corresponding Miller index.

34.15.1 Miller indices of principal planes in a cubic cell

Any plane in the crystal containing some atoms is a crystal plane. However, the effect of the planes with small atomic density on the physical properties of the crystal is negligible. Hence, the planes with appreciable atomic density are called principal planes.

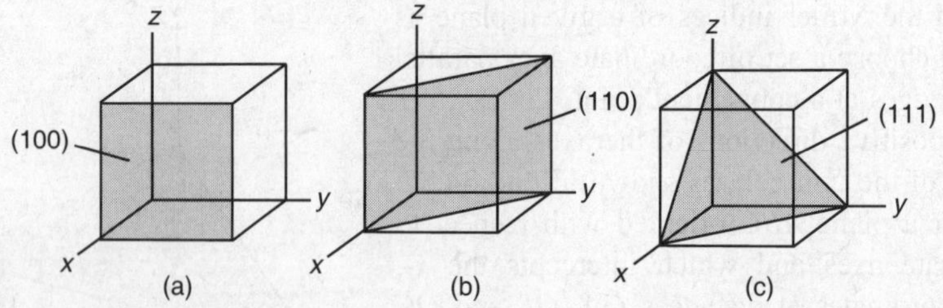

Fig. 34.38: The three principal miller index planes in a cubic lattice.

Fig. 34.38 depicts the principal crystallographic planes of cubic crystal structure. One of the faces of the cube is shown shaded in Fig. 34.38 (a). The shaded plane intercepts the axes *x, y, z* at 1, ∞, ∞ respectively. The reciprocals of the intercepts are 1, 0, 0. The Miller indices for this plane are (100). The plane is called one-zero-zero plane. The plane in Fig. 34.38 (b) has the intercepts at 1, 1, ∞. Therefore, the Miller indices are (110) and the plane is called one-one-zero plane. Finally, the plane in Fig. 34.38 (c) has intercepts at 1, 1, 1 and hence the Miller indices are (111). This plane is called one-one-one plane.

The three (100), (110) and (111) planes are known as the ***principal planes of cubic crystal system***.

34.15.2 Sketching a lattice plane (hkl)

The intercepts made by a plane can be found if the Miller indices of the plane are known.

(*i*) Let the Miller indices of a plane be (*hkl*).

(*ii*) A unit cell (for e.g. a cube) is drawn with three coordinate axes.

(*iii*) From the Miller indices, the reciprocals are obtained → $\frac{1}{h}, \frac{1}{k}, \frac{1}{l}$

(*iv*) The intercepts are marked on the coordinate axes.

(*v*) The three non-planar points are joined and the plane is obtained.

Example: Draw a plane (321)

Step (*i*): A cube having *unit axial lengths* ($a = 1$) along the three coordinate axes is drawn as shown in Fig. 34.39.

Step (*ii*): The reciprocals of Miller indices are $p = \frac{1}{3}, q = \frac{1}{2}$ and $r = 1$.

Step (*iii*): The intercepts, $\frac{1}{3}, \frac{1}{2}$ and 1, are marked within the cube on *x*, *y* and *z* axes respectively.

Step (*iv*): A plane is drawn through the points and is shaded.

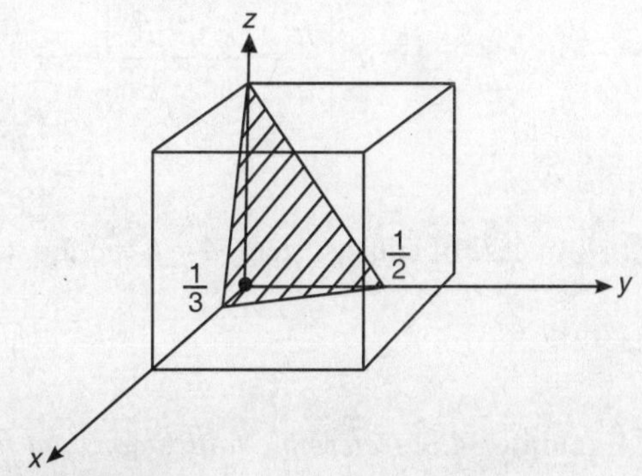

Fig. 34.39: Sketching a lattice plane **(321) plane**

34.16 INTERPLANAR SPACING IN A CUBIC LATTICE

Def: The distance '*d*' between successive lattice planes is known as the *interplanar distance*.

The interplanar distance '*d*' involves the axial lengths of the unit cell and the Miller indices of the planes. We shall derive here an expression for d in the case of a cubic system only. We know that the three axes of a cubic crystal are mutually perpendicular.

Let *ABC* be one of the family of parallel lattice planes in the crystal (Fig. 34.40). Let its Miller indices be (*hkl*) and the intercepts on the crystallographic axes be $OA = a/h$, $OB = b/k$ and $OC = c/l$. The next plane of the set parallel to *ABC* passes through the origin of the coordinates *O* (which is not shown in the Fig. 34.40). Therefore, *ON*, the length of the normal from the origin to the plane is equal to '*d*'.

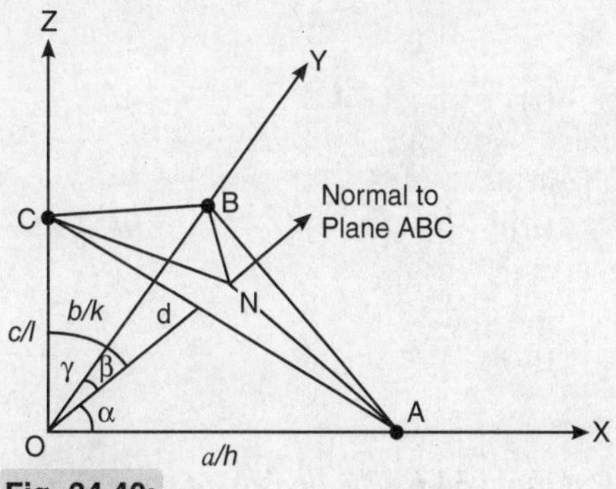

Fig. 34.40: Determination of interplanar distance in a cubic crystal

Let α, β, and γ be the angles *ON* makes with the three crystallographic axes respectively. Then the direction cosines of *ON* are

$$\cos\alpha = \frac{\text{ON}}{\text{OA}} = \frac{d}{a/h}$$

$$\cos\beta = \frac{\text{ON}}{\text{OB}} = \frac{d}{b/k}$$

$$\cos\gamma = \frac{\text{ON}}{\text{OC}} = \frac{d}{c/l} \qquad \text{...(34.29)}$$

The sum of the squares of the direction cosines of a line equals unity. Thus,

$$\cos^2\alpha + \cos^2\beta + \cos^2\gamma = 1 \qquad \text{...(34.30)}$$

Substituting the values of cos α, cos β, and cos γ in Eq. (34.30), we get

$$\frac{d^2}{a^2/h^2} + \frac{d^2}{b^2/k^2} + \frac{d^2}{c^2/l^2} = 1$$

$$d^2\left[\frac{h^2}{a^2} + \frac{k^2}{b^2} + \frac{l^2}{c^2}\right] = 1$$

$$\therefore \qquad d = \left[\frac{h^2}{a^2} + \frac{k^2}{b^2} + \frac{l^2}{c^2}\right]^{-1/2} \qquad \text{...(34.31)}$$

In case of cubic system, $a = b = c$. Eq. (34.31), then reduces to

$$d_{hkl} = \frac{a}{\sqrt{(h^2 + k^2 + l^2)}} \qquad \text{...(34.32)}$$

Example 34.5. *Determine lattice constant for FCC lead crystal of radius 1.746 Å. Also find the spacing of (i) (111) planes, (ii) (200) planes and (iii) (220) planes.*

Solution: The lattice constant is related to the atomic radius in FCC structure through

$a = 2\sqrt{2}\,r = 2 \times 1.414 \times 1.746$ Å = **4.938 Å.**

The interplanar spacing is given by $d = \dfrac{a}{\sqrt{h^2 + k^2 + l^2}}$

(*i*) $$d_{111} = \frac{4.938}{\sqrt{1^2 + 1^2 + 1^2}}\text{ Å} = \mathbf{2.851\ Å}$$

(*ii*) $$d_{200} = \frac{4.938}{\sqrt{2^2 + 0^2 + 0^2}}\text{ Å} = \mathbf{2.469\ Å}$$

(*iii*) $$d_{220} = \frac{4.938}{\sqrt{2^2 + 2^2 + 0^2}}\text{ Å} = \mathbf{1.746\ Å}$$

Example 34.6. *The density of copper is 8980 kg/m³ and unit cell dimension is 3.61Å. Atomic wt. of copper is 63.54. Determine crystal structure. Calculate atomic radius and interplanar spacing of (1 1 0) plane.*

Solution: (*i*) The effective number atoms per unit cell is given by $Z = \dfrac{\rho N_A a^3}{M}$

$$\therefore \quad Z = \frac{8980\,\text{kg/m}^3 \times 6.02\times10^{26}\ /\ \text{k.mol} \times (3.61\times10^{-10}\ \text{m})^3}{63.54\,\text{kg/k.mol}} = \frac{254.32}{63.54} = \textbf{4 atoms/unit cell.}$$

$\therefore$ **Copper exhibits FCC structure.**

(*ii*) The lattice constant is related to the atomic radius in FCC structure through $a = 2\sqrt{2}\,r$

$$r = \frac{a}{2\sqrt{2}} = \frac{3.61\text{Å}}{2\times1.4142} = \textbf{1.276 Å}$$

(*iii*) The interplanar spacing is given by $d = \dfrac{a}{\sqrt{h^2+k^2+l^2}}$

$$\therefore \quad d = \frac{3.61\text{Å}}{\sqrt{1^2+1^2+0^2}} = \textbf{2.553 Å.}$$

Example 34.7. *A crystal with primitives 1.2 Å, 1.8 Å and 2 Å has a plane (231) which cuts an intercept 1.2 Å along x-axis. Calculate the intercepts along y- and z-axes.*

Solution: Intercept on X-axis, $pa = 1.2$ Å $\quad \therefore p = \dfrac{1.2\text{Å}}{1.2\text{Å}} = 1.$

L.C.M, $D = hp = 2 \times 1 = 2.$ $\quad \therefore \quad q = \dfrac{D}{k} = \dfrac{2}{3}$ and $r = \dfrac{D}{l} = \dfrac{2}{1}.$

$\therefore$ Y-intercept, $qb = \dfrac{2}{3} \times 1.8\text{Å} = 1.2$ Å.

and Z-intercept, $rc = 2 \times 2\text{Å} = 4$ Å.

34.17 ATOMIC PACKING

The unit cells are geometric models connecting points which are purported to be the atomic centres. Atoms are space-filling entities. Therefore, the crystal structure may be viewed, alternatively, as resulting from the packing of hard spheres in three dimensions. The most efficient packing of spheres is called **closest packing**, or simply **close packing**.

Let us consider the packing of hard spheres of identical size. They can be close packed in a row by arranging them in contact with each other. A two dimensional layer of spheres can be built in two ways as shown in Fig. 34.41. Fig. 34.41 (a) shows square packing and triangular packing of two rows. A comparison of the two types suggests that the spheres fit more compactly in triangular packing rather than in square packing. When more rows are added to build the layer, a close packing such as the one shown in Fig. 34.41 (b) is obtained.

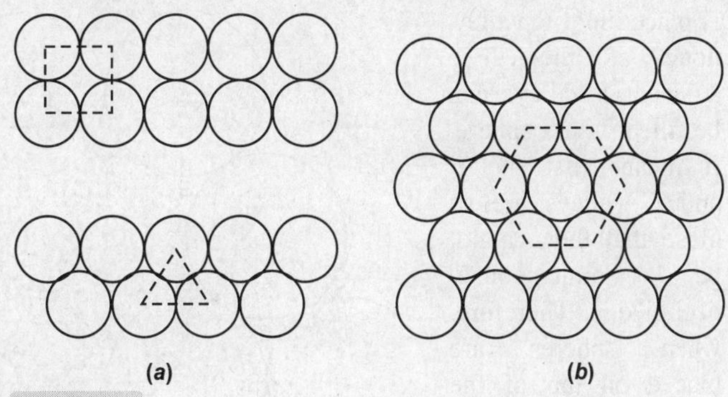

Fig. 34.41: Two dimensional layers of spheres (a) Square packing, and Triangular packing (b) Hexagonal packing.

In a layer of close packed spheres, each sphere is in contact with six neighbours. The lines joining the centres of the six spheres form a hexagon. Therefore, such a packing is commonly called **hexagonal packing**. Hexagonal packing is the most compact packing possible within a layer. Such a layer is known as a hexagonal close packed layer. Three dimensional close packed structures are built by keeping hexagonal close packed layers on top of one another in a regular sequence.

In practice, identical spheres can be packed in different ways to produce three dimensional structures. Different kinds of close packing are theoretically possible but all are combinations of two different basic structures, namely HCP structure and CCP structure. They are built starting with a hexagonal packed layer.

34.17.1 HCP Structure

Fig. 34.42 (a) shows the top view of a hexagonal close packed layer of identical spheres. Let us designate the spheres as *A* and hence the layer as *A*-layer. Successive layers of spheres are added on top of the *A*-layer to form three dimensional structures. As spheres have curved surfaces, a regular array of valleys are formed between the spheres in the *A* layer. Each valley is formed where three spheres are in contact with each other. Each sphere in the layer is surrounded by six neighbour spheres and six valleys (Fig. 34.42 a). The valleys may be divided into two sets of differing shapes and are labeled *B* and *C*.

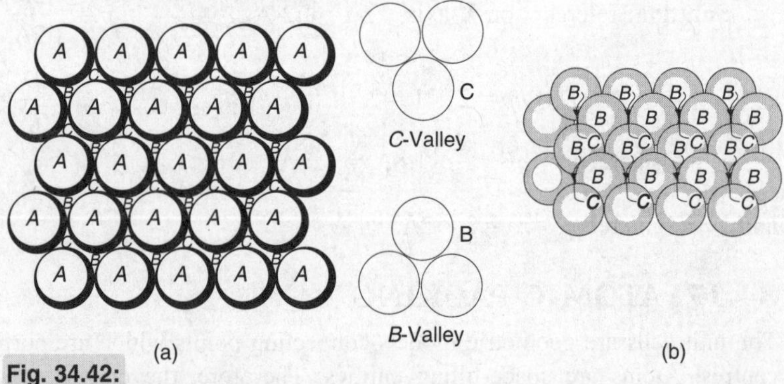

Fig. 34.42: Top view of a hexagonal close packed layer and valleys between spheres.

Thus, there are three *B*-type valleys and three type *C*-valleys. If a sphere is placed in a *B* valley, none of the three adjacent *C*-valleys can be filled with a sphere. Similarly, if a sphere is placed in a *C* valley, none of the three adjacent *B*-valleys can be filled with a sphere. It means that the *B* and *C* valleys are so close that they cannot be simultaneously occupied. Therefore, when spheres are placed on top of the *A*-layer, all of them must roll down into one kind of valley, either *B* or *C*.

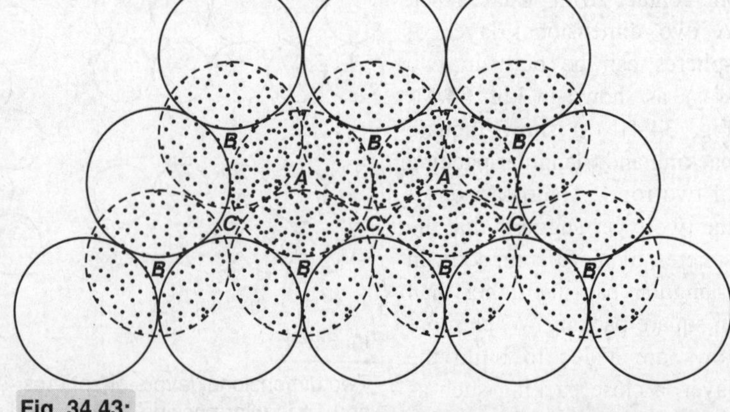

Fig. 34.43: Close packing of identical spheres in two layers with B&C valleys

Thus, the second layer spheres occupy only one kind of valley leaving the other kind vacant. It means that there is only one way of arranging the second layer. Fig. 34.43 shows the close packing of identical spheres in two layers. The centres of the spheres in the upper layer lie above the *B*-valleys. Let us designate this layer as *B*-layer.

Again in the *B*-layer, each sphere is surrounded by two sets of valleys marked *CCC* and *AAA*. Now there are two options. The spheres can be positioned in *A* valleys or *C* valleys to form the third layer. If spheres are placed in *A* valleys, the third layer becomes equivalent to the first layer at the bottom. It means that it is again an *A*-layer. Similarly, the fourth layer may be arranged such that its spheres sit in *B* valleys of third layer, forming a *B*-layer. Spheres may be arranged to form the fifth layer as layer, sixth layer as *B*-layer and so on. This kind of stacking of spheres is essentially a repeating two-layer-close-packing. The packing is known as *AB AB AB* ……layer sequence. In this structure, the *C* valleys are left vacant from the bottom layer to the topmost layer. Fig. 34.44 shows the schematic of *AB AB AB*….layer sequence.

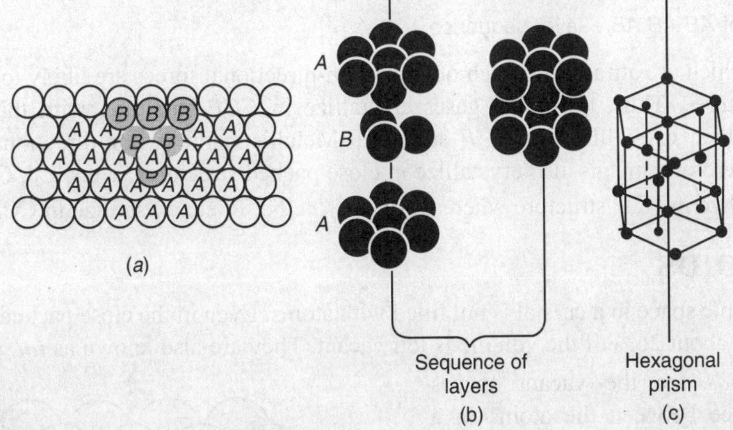

Fig. 34.44:

Schematic of *AB AB AB*….layer sequence.

If the centres of six spheres surrounding a particular sphere in the first layer spheres and those of the third layer spheres are joined, a right prism having a regular hexagon for its base is obtained. Therefore, the unit cell of this structure is a hexagon. In view of this, the *AB AB AB*….layer packing is called hexagonal close packing and the resulting three dimensional structures is known as **hexagonal close packed structure** or **HCP structure**.

34.17.2 CCP Structure

If the spheres of the third layer are positioned in *C* valleys instead of *A* valleys of second layer, a different sequence of layers is produced. The third layer built on *C* valleys is denoted as *C*-layer. In the *C*-layer, each sphere is surrounded by *AAA* and *BBB* valleys. If the spheres of fourth layer are stacked on *A*-valleys, it becomes a repetition of the first layer. Similarly, the fifth layer may be arranged as a repetition of *B*-layer, the sixth that of *C*-layer, the seventh that of *A*-layer and so on. The structure is symbolically represented as *ABC ABC ABC*…….. layer sequence. Fig. 34.45 shows the sequence of layers.

The fundamental block of *ABC ABC ABC*…..structure is a cube. Therefore, the packing is known as **cubic close packed structure or *CCP* structure**. The atoms are located at the corners of the cube and at the centres of the faces of the cube. The *CCP* structure is in fact *FCC* structure only.

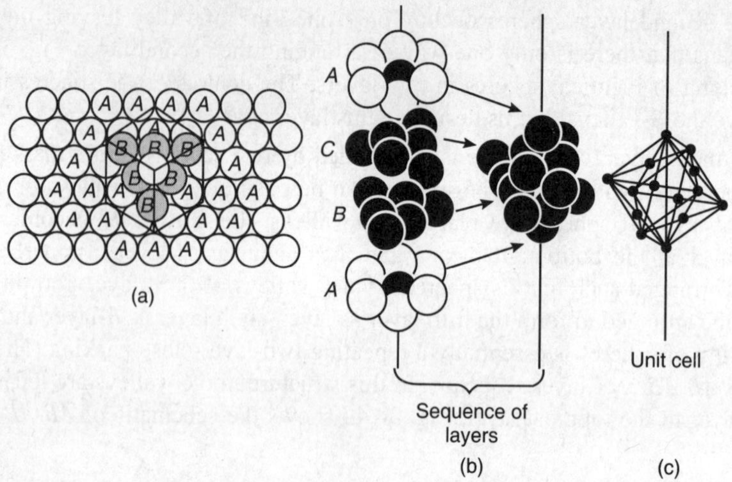

Fig. 34.45: Schematic of *AB AB AB*....layer sequence.

Atoms which are attracted to each other by non-directional forces are likely to form close packed structures. Thus, the noble gases crystallize in *CCP* structure, with the exception of helium which crystallizes in *HCP* structure. Metallic bond is non-directional and non saturable. Therefore, metals also crystallize in close packed structures. Thus, Cd, Co, Mg, Zn, Ti etc crystallize in HCP structure whereas Ag, Al, Cu, Ni, Pt etc crystallize in CCP structure.

34.18 VOIDS

All the available space in a crystal is not filled with atoms. Even in the close packed structures, it is seen that about 26% of the volume is left vacant. They are also known as ***interstices***.

Def: *Voids* are the vacant spaces available space between the atoms in a crystal.

Two kinds of voids form in close packed structures, such as FCC close packed structures. They are tetrahedral *voids* and *octahedral voids* (see Fig. 34.46). A **tetrahedral void** is formed by a sphere fitting into the valley formed between three adjacent spheres of a close packed layer, as shown in Fig. 34.47 (a). If the centres of the four spheres are joined, a regular tetrahedron is produced. The vacant site enclosed by the tetrahedron is known as the tetrahedral void. If the fourth sphere belongs to the upper layer and is on top of three spheres, an upright tetrahedral void forms.

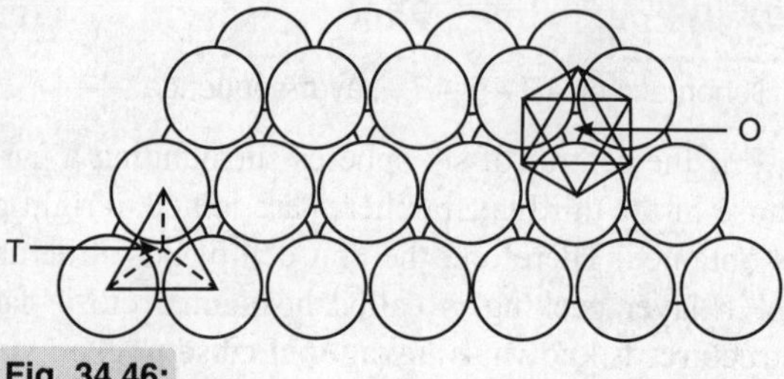

Fig. 34.46: Formation of tetrahedral and octahedral voids in a three dimensional close packed structure

On the other hand, if the fourth sphere belongs to the bottom layer an inverted tetrahedral void forms. Thus, in a three-dimensional structure two tetrahedral voids form for every sphere.

An **octahedral void** is produced when a void formed by three spheres in one layer comes on the top of the void formed by three spheres in the top layer, as depicted in Fig. 34.47 (c). When the centres of the six spheres are joined, an octahedron is formed. The space enclosed is an octahedral void. An octahedral void is surrounded by six spheres. There is no sphere in the central valley. Octahedral voids are bigger in size and smaller in number compared to tetrahedral voids. In a three-dimensional close packed structure, one octahedral void forms per one sphere.

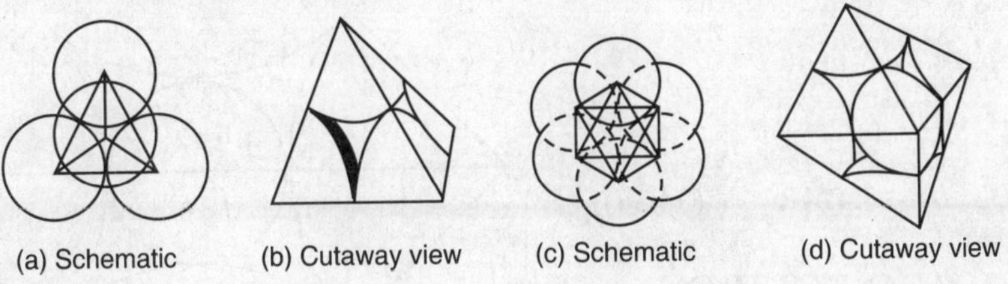

Fig. 34.47:

Tetrahedral Void (*a*) Schematic (*b*) cutaway view; octahedral void (*c*) Schematic (*d*) cutaway view

34.18.1 Sizes of the Spheres that fit into Voids

(*i*) The tetrahedron in Fig. 34.47 (b) may be used to determine the radius of the sphere that can fit into the tetrahedral void. A cube *ABCDEFGH* is formed using the tetrahedron *CAEH* (Fig. 34.48 a). The point *C* is the apex of the tetrahedron and is one of the corners of the cube. The edges *AE, AH* and *EH* of the tetrahedron are the face diagonals of the faces *ABEF, ADFH*, and *EHGF* of the cube. The body diagonal *AEGD* of the cube is shown separately in Fig. 34.48 (b).

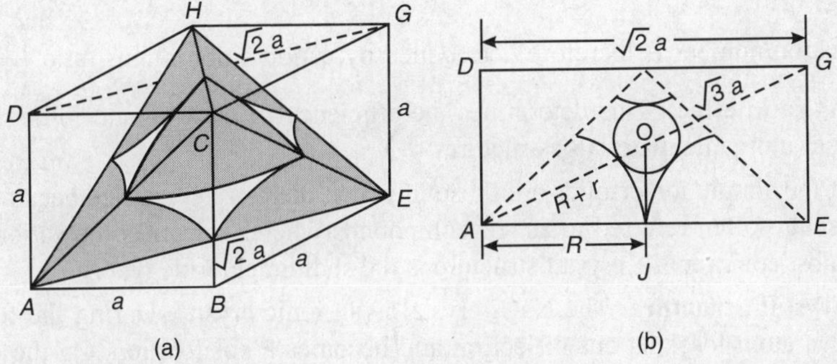

Fig. 34.48:

Body diagonal *AEGD* of the cube

From the Fig. 34.48 (*b*) it is seen that the Δ^{les} *AJO* and *AEG* are similar.

$$\therefore \quad \frac{AO}{AJ} = \frac{AG}{AE} \qquad \therefore \quad \frac{R+r}{R} = \frac{\sqrt{3}\,a}{\sqrt{2}\,a}$$

$$\therefore \quad 1+\frac{r}{R} = \frac{\sqrt{3}}{\sqrt{2}} \qquad \therefore \quad r = \left[\frac{\sqrt{3}-\sqrt{2}}{\sqrt{2}}\right]R = 0.225\text{ R} \qquad ...(34.33)$$

(*ii*) For finding the size of the sphere that can fit tightly into an octahedral void, let us consider the octahedron. In Fig. 34.48(*d*). The plane *ABCD* passing through four spheres surrounding the central valley is shown in Fig. 34.49.

The triangles *ABC* and *ADC* are right isosceles triangles. If each side of the square is *a*, in the Δ^{le} *ABC*.

$$\frac{AC}{BC} = \frac{\sqrt{2}\,a}{a}$$

$$\therefore \quad \frac{2R+2r}{2R} = \sqrt{2}$$

or $$R + r = \sqrt{2}R$$

$$\therefore \quad r = (\sqrt{2}-1)R$$

or $$\mathbf{r = 0.414R} \quad ...(34.34)$$

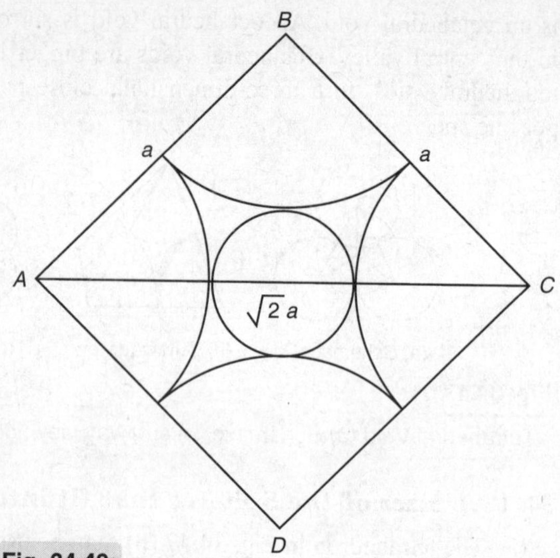

Fig. 34.49: Illustration showing four spheres surrounding the central valley

34.19 IONIC SOLIDS

Ionic solids are made up of ions of different sizes. The cations are ordinarily smaller than anions as cations have given up their valence electrons to the anions. The anions that are relatively larger in size than the cations, form close packed structures and the cations are accommodated in the voids. The packing depends upon the ionic radii of cations, r_C and anions, r_A.

Each cation prefers to have as many nearest neighbours as possible. So does each anion. The coordination number is therefore determined by cation-anion radius ratio $\left(\frac{r_C}{r_A}\right)$. This is called **radius ratio effect**. This determines the efficiency of packing and forms the basis of many ceramic (inorganic materials) structures.

Some of the simple ionic solids are those in which there are equal number of cations and anions. These are often referred to as *AX* compounds where *A* denotes the cation and *X* the anion. The most common *AX* crystal structure is the sodium chloride type.

NaCl Crystal Structure: The NaCl crystal is an ionic crystal. During the formation of the crystal, Na atom loses its outer electron and becomes a positive ion. On the other hand, the chlorine ion acquires the electron lost by the Na atom and becomes a negative ion. When the two ions approach each other, at a certain distance the forces of attraction are balanced by forces of repulsion and a stable system is produced. Na^+ and Cl^- ions arranged alternately in a cubic pattern in space so that the electrostatic attraction between the nearest neighbours is a maximum. The resultant structure has a coordination number 6. Six anions surround each cation and each anion is surrounded by six cations. Thus, the cations occupy the octahedral voids formed by the anions (Fig. 34.50 a). Since the ratio of anions to cations is 1:1 and there is one octahedral void per anion, all octahedral voids in the structure are occupied.

NaCl is an example of face centered cubic (FCC) Bravais lattice. Each cell has 8 corners and 8 cells meet at each corner. Thus, an ion at a corner of the cell is shared by 8 cells, i.e. only 1/8 ion belongs to any one cell. Similarly, an ion at the center of a face of the cell is shared by 2 cells, i.e. only 1/2 ion belongs to any one cell. Since a cell has 8 corners and, 6 faces, it has $[(8 \times 1/8) + (6 \times 1/2)] = 4$ ions of one kind, and similarly 4 ions other kind.

Thus, *there are 4 Na^+ - Cl^- ion pairs (molecules) per unit cell*. The position coordinates of Na^+ and Cl^- are as follows.

Na	0 0 0	$\frac{1}{2}\ \frac{1}{2}\ 0$	$\frac{1}{2}\ 0\ \frac{1}{2}$	$0\ \frac{1}{2}\ \frac{1}{2}$
Cl	$\frac{1}{2}\ \frac{1}{2}\ \frac{1}{2}$	$0\ 0\ \frac{1}{2}$	$0\ \frac{1}{2}\ 0$	$\frac{1}{2}\ 0\ 0$

Fig. 34.43 (b) shows a unit cell of NaCl lattice. NaCl lattice can be regarded to be made up of two fcc sub-lattices, one of Na^+ ions having origin at (0,0,0) and the other of Cl^- ions having its origin midway along the cube edge.

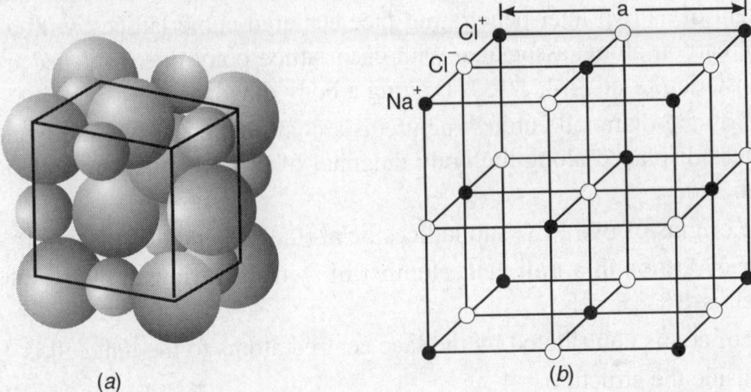

Fig. 34.50: NaCl Crystal Structure

Each Na^+ ion has 6 Cl^- ions as nearest neighbours and similarly, each Cl^- ion has 6 Na^+ ions. Hence, the coordination number of NaCl is 6, the same as that for, simple cubic lattice.

34.20 DIAMOND CUBIC STRUCTURE

In diamond crystal structure each carbon atom forms directional covalent bonds with four other carbon atoms that occupy four corners of a cube in a tetrahedral structure (Fig. 34.51a). Diamond lattice is formed with tetrahedral units with bond length of 1.53 Å and the bond angle of 109.5°.

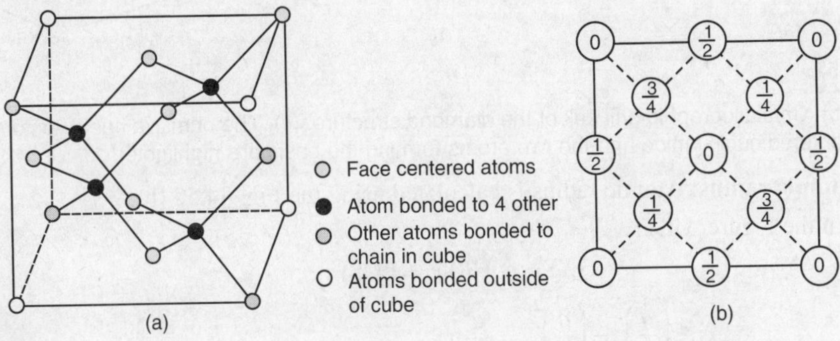

Fig. 34.51: Diamond Unit Cell

Diamond exhibits both cubic and hexagonal type structure. The diamond cubic (DC) structure is more common. The space lattice of diamond is FCC with a basis of two carbon atoms. Fig. 34.51 (b) shows the position of atoms. The points at 0 and 1/2 are on the FCC lattice, those at 1/4 and 3/4 are on a similar lattice displaced along the body diagonal by one fourth of its length. The fractions denote the height above the base in units of a cube edge.

It can be considered that the diamond lattice is composed of two interpenetrating FCC sub lattices. One FCC structure is shifted relative to the other by one fourth of a body diagonal. Each atom has four nearest neighbours with which it forms covalent bonds. Accordingly, the coordination number of diamond crystal is 4 and the number of atoms per unit cell is 8. Silicon, germanium and gray tin crystallize in the diamond structure.

(*i*) **Lattice Vector and Basis Atoms:** The structure consists of two basis atoms and may be thought of as two inter-penetrating face centered cubic lattices, with a basis of two identical carbon atoms associated with each lattice point one displaced from the other by a translation of ao (1/4, 1/4, 1/4) along a body diagonal.

We can say the diamond cubic structure as a combination of two interpenetrating FCC sub lattices displaced along the body diagonal of the cubic cell by 1/4th length of that diagonal.

Thus the origins of two FCC sub lattices lie at (0, 0, 0) and (1/4, 1/4, 1/4).

(*ii*) **Number of Atoms in a unit cell:** Number of atoms contributed by the corner atoms to a unit cell is $1/8 \times 8 = 1$

Number of atoms contributed by the face centred atoms to the unit cell is $1/2 \times 6 = 3$

Atoms inside the structure = 4

So, total number of atoms present in a diamond cubic unit cell is $1 + 3 + 4 = 8$.

(*iii*) **Coordination number:** Since each carbon atom is surrounded by four more carbon atoms, the co-ordination number is 4.

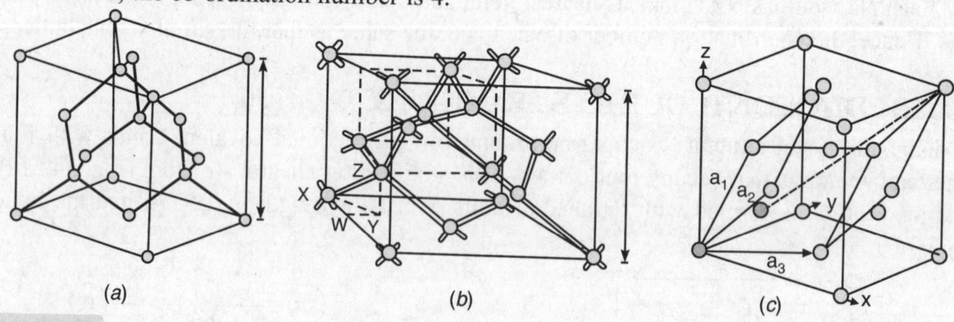

Fig. 34.52:

(a) & (b). Crystallographic unit cell of the diamond structure, (c). The primitive basis vectors of the face centered cubic lattice and the two atoms forming the basis are highlighted

(*iv*) **Atomic radius:** Atomic radius is calculated using the Fig. 34.52 (b).

From the Figure ΔWXY

$$(XY)^2 = (WY)^2 + (XW)^2$$

$$\left(\frac{a}{4}\right)^2 + \left(\frac{a}{4}\right)^2 = \frac{a^2}{8}$$

From the Figure ΔXYZ

$$(XZ)^2 = (XY)^2 + (YZ)^2$$

$$\left(\frac{a}{8}\right)^2 + \left(\frac{a}{4}\right)^2 = \frac{3a^2}{16}$$

But
$$XZ = 2r$$
$$(2r)^2 = \frac{3a^2}{16}$$
$$a = \frac{8r}{\sqrt{3}}$$
$$r = \frac{\sqrt{3}a}{8}$$

(*v*) **Packing fraction:** Packing fraction is calculated using the Fig. 34.53.

From the Fig. 34.53,
$$AD = DC = CE = \frac{a}{4}$$

In ΔADC, $\angle D = 90°$
$$(AC)^2 = (AD)^2 + (DC)^2$$
$$\left(\frac{a}{4}\right)^2 + \left(\frac{a}{4}\right)^2 = \frac{a^2}{8}$$
$$AC = \frac{a}{2\sqrt{2}}$$

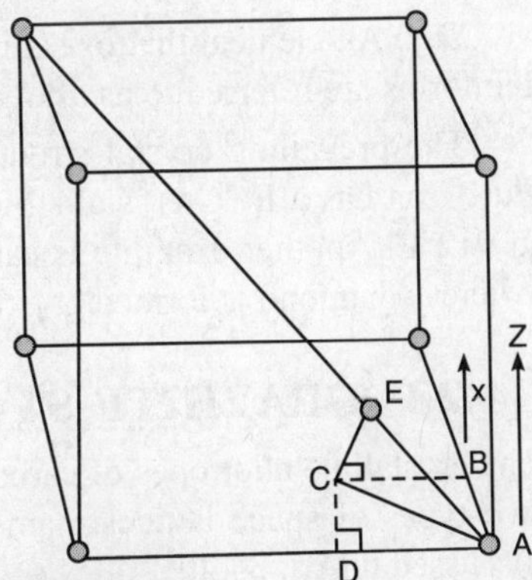

Fig. 34.53:
Calculation of Packing fraction

In ΔACE, $\angle C = 90°$
$$(AE)^2 = (AC)^2 + (CE)^2$$
$$\left(\frac{a}{8}\right)^2 + \left(\frac{a}{16}\right)^2 = \frac{3a^2}{16}$$
$$AE = \frac{\sqrt{3}a}{8} = d = 2r$$
$$r = \frac{\sqrt{3}a}{8}; \quad a = \frac{8r}{\sqrt{3}}$$
$$PF = \frac{\text{Volume of 8 carbon atoms}}{\text{Volume of unit cell}} = \frac{8 \times \frac{4\pi r^3}{3}}{\left(\frac{8r}{\sqrt{3}}\right)^2} = 0.34$$

Atomic Packing Fraction = 0.34 = 34%,
Thus, diamond is a loosely packed structure.

34.21 ZnS STRUCTURE

The structure of zinc sulphide is identical to the diamond cubic structure. It consists of two interpenetrating FCC sub lattices which are occupied by two different elements and are displaced from each other by one quarter of the body diagonal. Zinc sulphide structure results when Zn atoms are placed on one FCC lattice and S atoms on the other FCC lattice, as shown in Fig. 34.54. The conventional pattern of this structure is a cube. There are four molecules per conventional cell. For each atom there are four

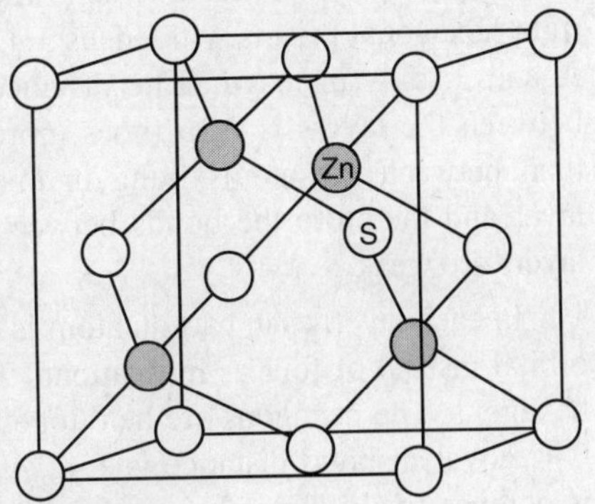

Fig. 34.54:
ZnS Structure

equidistant atoms of other kind arranged at the vertices of a regular tetrahedron. Some of the important compounds which possess this structure are semiconductors such as InSb, GaAs, and CdS.

34.22 POLYMORPHISM AND ALLOTROPY

Polymorphism is found among pure elements as well as among chemical compounds. Polymorphism of an element is frequently called allotropy.

Def: The phenomenon by which a single substance crystallizes in two or more different forms under different conditions is called **polymorphism** and the different crystalline forms are said to be polymorphous with each other.

Def: An element that exists in more than one crystalline form is said to be **allotropic** and the forms are referred to as **allotropes** and the phenomenon is called **allotropy**.

The prevailing crystal structure depends on the temperature and the external pressure. Pure iron has a BCC crystal structure at room temperature, which changes to FCC structure at 912°C. Another example is carbon. Graphite is the stable allotrope at ambient conditions, whereas diamond is formed at extremely high pressures.

34.23 GRAPHITE STRUCTURE

The crystalline allotropes of carbon are diamond and graphite. The atoms of carbon can form two types of space lattices, namely diamond and graphite structures. Diamond structure is discussed in Art. 34.20.

The second type of packing of carbon atoms occurs in graphite. In graphite each atom is bound to four neighbours but the forces and the directions of the bonds differ (Fig. 34.55). An atom has a strong bond with three other atoms that lie together with it in the same layer at angles of 120° and a weaker bond with a fourth atom lying in the adjacent layer. This last bond makes an angle of 90° with the plane of the layer. The unit cells are right hexagonal prisms with edges 1.42 Å and 3.35 Å in length. The distance between the layers is 2.36 times more than between the atoms lying in one layer and therefore the bonds between layers are weak.

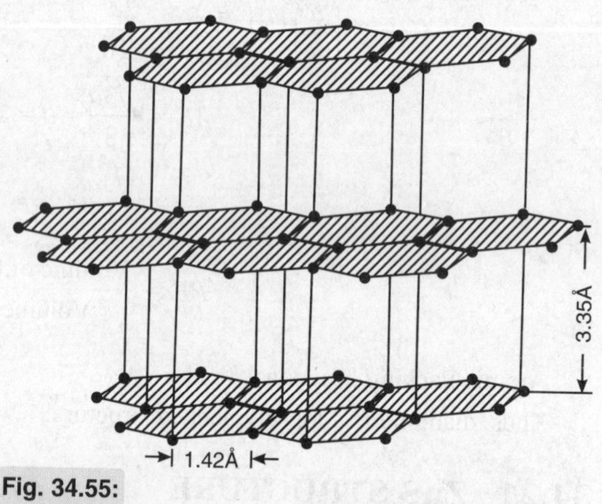

Fig. 34.55: Graphite structure

In each layer, each carbon atom is covalently bonded to three others involving sp^2 hybrid orbital instead of four as in diamond. Thus, all atoms in a single plane are linked to give flat hexagons. The hexagons are held together in sheet-like structure, parallel to one another. The C-C covalent bond distance is 1.42 Å. The distance between the sheets of layers, however is comparatively larger being about 3.35 Å. So, there is no possibility of covalent bonding between the layers. Thus graphite has a **layer structure**.

The difference in structure of the diamond and graphite lattices is the cause of the striking differences in their physical properties. Diamond is one of the hardest substances in nature; it can cut glass, as well as any hard rock like granite. Graphite is very soft; it writes on the paper because its layers slide easily with respect to each other. The diamond is an insulator as it has no free electrons; but graphite is a conductor. Diamond is transparent whereas graphite is opaque. The density of diamond is 3500 kg/m^3; that of graphite is 2100 kg/m^3.

34.24 CRYSTAL STRUCTURE ANALYSIS

Much of our knowledge of the internal structure of crystals has been obtained from x-ray diffraction experiments. Diffraction of waves occurs when the waves are scattered by a periodic array of scattering centres separated by distances of the order of a wavelength. A plane transmission grating is a familiar device used to study diffraction of light. It is made by ruling thousands of parallel lines on a piece of glass. The lines are opaque and the spaces between the lines form openings for light to pass through. A grating normally used in laboratories has 15000 lines per inch. The slit width or grating period in the grating comes to about 17,000 Å, which is about three times the wavelength of light. When light passes through these narrow slits, the waves spread in all directions.

The diffracted waves interfere constructively in a few specific directions and destructively in all other directions, producing a spatial pattern of alternate regions of brightness and darkness. The dimension of atoms and the interatomic spacing in a crystal are of the order of 2 to 5 Å which is of the order of the wavelength of x-rays. In view of this Max von Laue, the German physicist predicted in 1912 that a crystal acts as a natural three dimensional grating for x-rays, where the regular periodic lines of atoms serve the role of parallel ruled lines. His associates, W.Friedrich and P.Knipping duly tested and confirmed the prediction. Laue was awarded the Noble prize in Physics in 1914.

34.24.1 Laue Method

X-rays produced by an x-ray tube are made into a narrow beam by a set of lead slits S_1 and S_2. A thin crystal C is mounted in the path of the x-ray beam and a photographic film is positioned beyond it, as shown in Fig. 34.56.

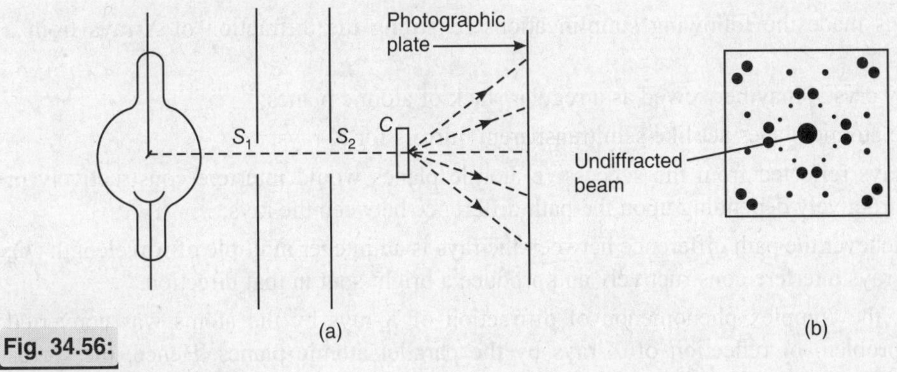

Fig. 34.56: Laue Method

When the x-ray beam goes through the crystal C, some of the rays are scattered by atoms from their initial direction. The x-rays coming out from the crystal in specific directions as highly narrow beams and they are captured by the photographic film. On developing the exposed film, a pattern of bright spots corresponding to maximum intensity are observed and are called as **Laue spots**. The pattern of Laue spots distinctively features the crystal C. The central bright spot on the film matches the main beam. A hole is often cut in the film so that the central spot is not recorded. In the typical experiment of Friedrich and Knipping the wavelength of x-rays was calculated from the measurement of angular positions of Laue spots and from inters atomic distance.

34.25 BRAGGS' LAW

In 1912 W.H.Bragg and W.L.Bragg discovered that X-rays can be regularly reflected by the cleavage planes of the crystals. The cleavage planes are successive atomic planes in the crystal (Fig. 34.57). They are also known as **Braggs' planes**. Thus, a crystal may be regarded as consisting of a stack of several parallel planes of atoms. Each plane in a given set has the same distribution of atoms. When an x-ray beam is incident on the crystal, each atom in the crystal scatters a portion of the incident beam. The scattered waves travel in all directions. It is convenient to consider the net scattering by atoms in terms of the diffraction by the crystal planes. Each family of planes gives rise to scattering but only a certain family scatters constructively in a given direction. W.H.Bragg and W.L Bragg derived a relation between the wavelength of x-rays and the angular positions of the scattered beam and the separation of atomic planes in the crystal.

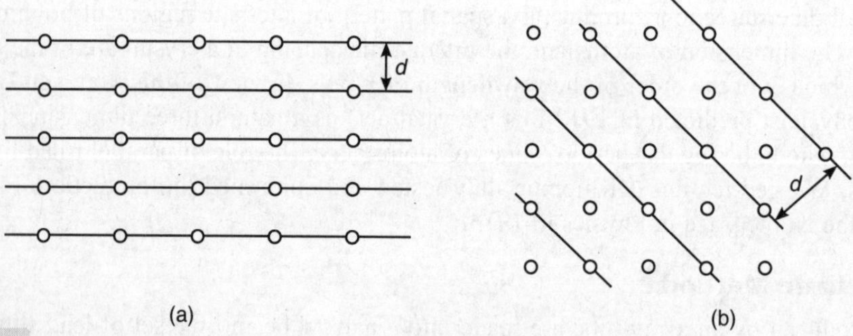

Fig. 34.57: Two different sets of Bragg planes

Braggs made the following simplifications regarding the diffraction of X-rays from a crystal:

(*i*) Any crystal may be viewed as a regular stack of atomic planes;

(*ii*) The atomic planes act like semitransparent mirrors for X-rays;

(*iii*) X-rays reflected from the successive atomic planes would interfere constructively or destructively depending upon the path difference between the rays;

(*iv*) Whenever the path difference between the rays is an integer multiple of wavelength (λ), the rays interfere constructively and produce a bright spot in that direction.

Thus, the complex phenomenon of diffraction of X-rays by the atoms was converted into the problem of reflection of X-rays by the parallel atomic planes. Hence, the words 'diffraction' and 'reflection' are mutually interchangeable in Braggs' treatment. Based on these considerations, Braggs derived a simple mathematical relationship that serves as a condition for the Bragg reflection to occur. This condition is known as Braggs' law.

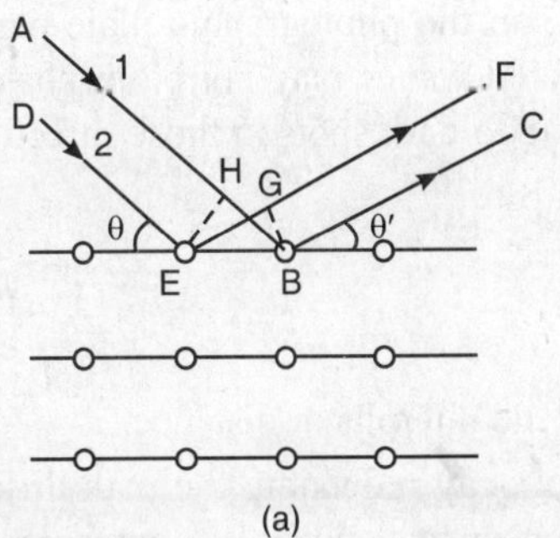

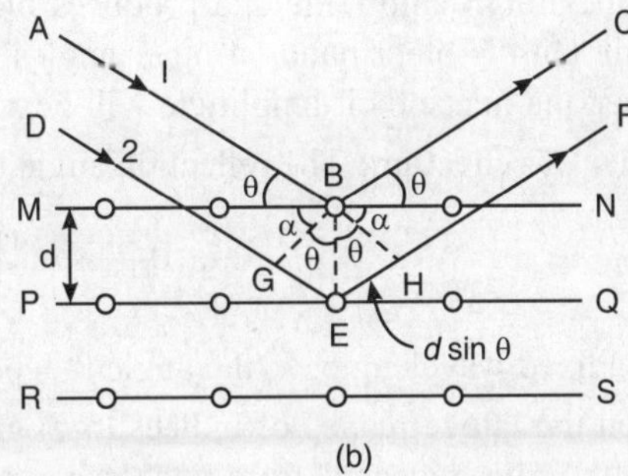

Fig. 34.58:

Determination of path difference between x-rays diffracted by consecutive lattice planes

Let us consider a set of parallel atomic planes with interplanar spacing *d*. We represent the row of atoms as a single horizontal line, as shown in Fig. 34.58. Let a parallel beam of monochromatic X-rays of wavelength λ, represented by the parallel lines *AB* and *DE*, be incident on these planes at a glancing angle (Fig. 34.58a). The scattered beam emerges along *BC* and *EF*. The contributions of two adjacent atoms in the same plane are considered in the Fig. 34.58 (a).

(*a*) The rays *BC* and *EF* are coherent and reinforce each other, if they are in phase. It requires that the path lengths *BH* and *EG* are equal. They will be equal when $\theta = \theta'$. This is the *condition of reflection*.

(*b*) We consider next the contributions of two adjacent atoms to the reflection from **successive planes** *MN* and *PQ* are considered (see Fig. 34.58 *b*). The path difference Δ between the reflected rays *BC* and *EF* is

$$\Delta = GE + EH \quad ...(34.35)$$

$$\angle ABG = \theta + \alpha = 90° \quad ...(34.36)$$

BE is normal to the plane *MN*. Therefore,

$$\angle MBE = \alpha + \angle GBE = 90° \quad ...(34.37)$$

From Eq. (34.36) and (34.37) we get

$$\angle GBE = \theta$$

Similarly, $\angle EBH = \theta$

In the Δ^{le} *BGE*, $BE = d$

$\therefore$ $GE = d \sin\theta$

Similarly, in the Δ^{le} *EBH*, $EH = d \sin\theta$

$\therefore$ The path difference $\Delta = GE + EH = d\sin\theta + d\sin\theta = 2d\sin\theta$ ...(34.38)

The rays *BC* and *EF* will constructively interfere only when $\Delta = m\lambda$, where $m = 1,2,3\ldots$

$\therefore$ The condition for reinforcement of scattered waves is

$$2d\sin\theta = m\lambda \quad ...(34.39)$$

Eq. 34.39 is called **Bragg's equation** or **Braggs' law**.

Eq. 34.39 means that the reinforcement of reflected waves will take place only at certain values of θ, corresponding to specific values of λ and *d*. In these directions the resultant wave has the maximum amplitude and produces bright spots on the photographic plate placed in their path. On the other hand, at other angles, the scattered waves may not be in phase with each other and hence their amplitude will be zero, leading to dark spots in those directions.

Order of reflection: The reflection angle θ is given by

$$\theta = \sin^{-1}\left(\frac{m\lambda}{2d}\right) \qquad ...(34.40)$$

For a fixed wavelength, λ, the angle θ depends d_{hkl}, the interplanar distance.

When the lattice planes of indices (*hkl*) give rise to x-ray reflection, the path difference between the rays reflected from successive planes is one wavelength. For example, if the reflection from (100) planes of the lattice occurs, say, at an angle θ_1, then the path difference between the reflections will be λ and the order of reflection $m = 1$. When the path difference between the reflected rays is 2λ, the reflection will occur at an angle and the Bragg equation is satisfied with $m = 2$. Thus, as θ is increased gradually, a number of positions will be found at which the reflections are intense. These positions correspond to $m = 1,2,3...$ values. The diffraction lines appearing for m = 1, 2, and 3 are called first, second, third order diffraction lines respectively. Thus, *m* denotes the **order of reflection**. The highest possible order is determined by the condition that sin θ cannot exceed unity.

If the value of θ is determined experimentally, and knowing the wavelength λ, the interplanar spacing 'd' can be determined with the help of Bragg law.

Example 34.8. *X-rays of unknown wavelength give first order Bragg reflection at glancing angle 20° with (212) planes of copper having FCC structure. Find the wavelength of X-rays, if the lattice constant for copper is 3.615 Å.*

Solution: The interplanar spacing is given by $d = \dfrac{a}{\sqrt{h^2+k^2+l^2}}$

$$\therefore \qquad d = \frac{3.615\text{Å}}{\sqrt{2^2+1^2+2^2}} = 1.205\ \text{Å}.$$

According to Braggs' law

$$2d \sin\theta = m\lambda.$$

$$\therefore \qquad \lambda = \frac{2d\sin\theta}{m} = \frac{2\times 1.205\,\text{Å}\times \sin 20°}{1} = \mathbf{0.824\ Å}.$$

34.26 BRAGGS' SPECTROMETER

W.H.Bragg and W.L:Bragg devised an X-ray spectrometer in which a crystal is used as a reflection grating. It is used to measure glancing angle θ.

Construction: Braggs' spectrometer is very much similar in construction to the optical spectrometer. The schematic diagram of the Braggs' spectrometer is shown in Fig. 34.59. X-rays from an X-ray tube are collimated into a narrow beam by two slits S_1 and S_2 cut in lead plates. The beam is then allowed to be incident at a glancing angle on the face of the crystal *D*, which is mounted in wax on the turntable of the spectrometer. The turntable is capable of rotation about a vertical axis passing through its center and the position of the crystal can be read from the circular scale, *C*. Most of the incident beam passes straight through the crystal. Some of the X-rays are however scattered by the regularly arranged atoms in different crystal planes. The scattered X-rays can be regarded as having been reflected from the crystal planes rich in atoms. The reflected X-ray beam enters an ionization chamber carried by the spectrometer arm, which is capable of rotation about the same axis as the turntable. The turntable and the arm are so linked together that when the turntable rotates through an angle θ, the arm turns through an angle 2θ. In this way, the X-ray beam is always reflected into the ionization chamber whatever its incidence angle is at the crystal face. The ionization current produced by the reflected beam is measured by a sensitive electrometer *E* or recorded on a photographic plate.

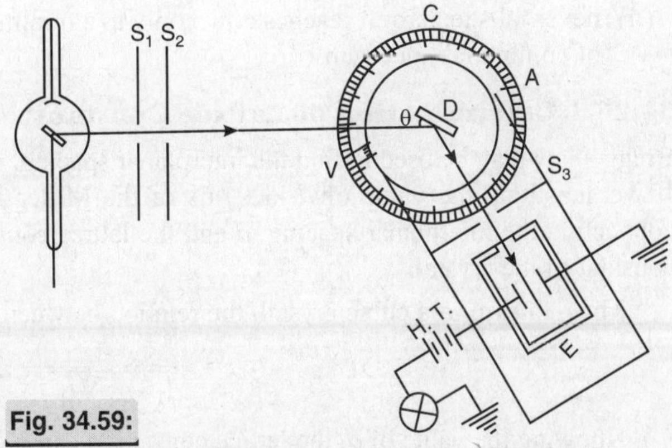

Fig. 34.59: The schematic diagram of the Bragg's spectrometer

Working: Initially, the single crystal under investigation is mounted on the turntable such that the glancing angle $\theta = 0°$ and the ionization chamber is adjusted to receive X-rays. Then the crystal is moved in small angles and the corresponding deflection obtained in the electrometer is noted down. A graph is plotted for glancing angle versus intensity of diffracted x-rays.

X-ray spectrum: Fig. 34.60 shows a typical diffraction spectrum produced by a monochromatic X-rays scattered from a crystal. The diffraction spectrum is a graph plotted between the intensity of ionization and the glancing angle. It is seen from the graph that the intensity of ionization current increases abruptly for certain values of glancing angle. The peaks A_1, A_2, A_3 etc in intensity occur whenever the Braggs' equation is satisfied. If θ_1, θ_2 and θ_3 are the angles of incidence where intense peaks occur, then they correspond to the different orders of reflection with n = 1, 2 and 3 respectively for a given wavelength.

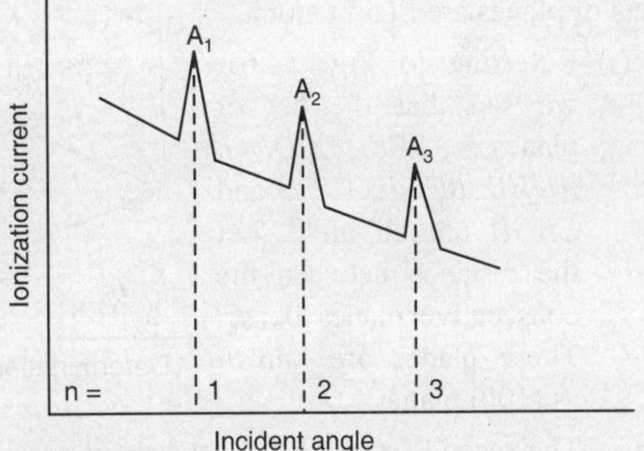

Fig. 34.60: A typical diffraction spectrum produced by a monochromatic X-rays scattered from a crystal

Using Braggs' law

$$2d \sin \theta_1 = \lambda, \quad 2d \sin \theta_2 = 2\lambda \quad \text{and} \quad 2d \sin \theta_3 = 3\lambda$$

That is, $2d \sin \theta_1 : 2d \sin \theta_2 : 2d \sin \theta_3 = \lambda : 2\lambda : 3\lambda$

$$\therefore \quad \sin \theta_1 : \sin \theta_2 : \sin \theta_3 = 1 : 2 : 3 \qquad ...(34.41)$$

The ratio in Eq. (34.41) indicates that the peaks correspond to first, second and third order reflections respectively.

It may be seen from Fig. 34.60 that the intensity of the reflected beam

(*i*) decreases as the order of the spectrum increases.

(*ii*) never falls to zero. It reaches down only to a minimum value. This is due to the presence of continuous spectrum.

34.26.1 Determination of Lattice Constant

Bragg's law may be used to find the interplanar spacing, if the wavelength of *x*-rays is known. If we substitute the value of *d* in terms of the Miller indices of the planes in the relation connecting the interplanar spacing, *d* and the lattice constant, *a*, we can determine the lattice constant of the crystal.

Thus, in case of a cubic crystal, the relation between d and a is given by

$$d_{hkl} = \frac{a}{\sqrt{h^2 + k^2 + l^2}} \qquad ...(34.42)$$

Knowing the value of *d*, the lattice constant *a* can be calculated from the above equation.

34.26.2 Determination of Crystal Structure

The Braggs' method of determining crystal structure consists in finding out the interplanar spacing, d_{hkl} of the various sets of planes rich in atoms. From the knowledge of d_{hkl} and from the relative intensities of diffracted beams of different orders, the crystal structure can be deduced.

For example, let us consider a crystal of a simple cubic system. In the unit cell the atoms are located at the corners of the cell. It may be seen that three sets of planes are rich in atom.

(*i*) Referring to Fig. 34.61, we see that the set of planes *ABFE, CDHG, ADHE, BCGF, ABCD* and *EFGH* are all alike. Let the distance between the consecutive planes be d_1. These planes are known as (100) planes.

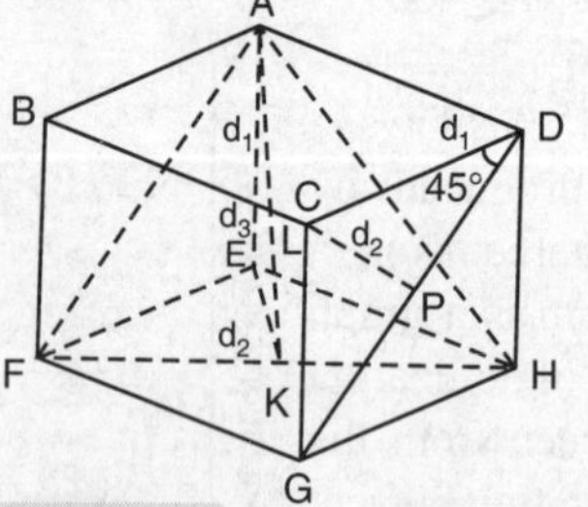

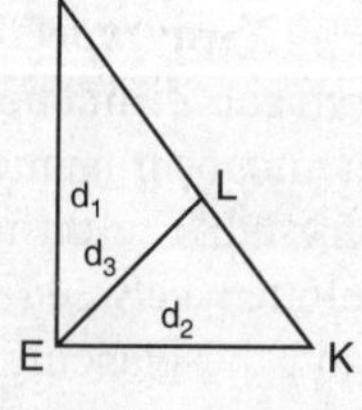

Fig. 34.61: Determination of crystal structure

(*ii*) The second set of planes consists of parallel planes like *ADGF*, inclined at an angle of 45° to the planes mentioned in (*i*). Let be the spacing between the second set of planes, as shown in Fig. 34.61 by *CP*. Note that a plane parallel to *ADGF* contains the line *BC*. Then, from the Δ*CDP*, we find that

$$\frac{d_2}{d_1} = \sin 45° = \frac{1}{\sqrt{2}}$$

or

$$\frac{d_2}{d_1} = \frac{1}{\sqrt{2}} \qquad ...(34.43)$$

These planes are known as (110) planes.

(*iii*) The third set of planes consists of planes like *AFH*. Let us draw *EK* perpendicular to *FH* and join *AK* to complete the triangle *AEK*. The perpendicular *EL* denoted by d_3 represents the distance between the plane *AFH* and a parallel plane passing through *E*.

In Δ*AEK*, $AE = d_1$, $EK = d_2$ and $AK = \sqrt{d_1^2 + d_2^2}$. Further,

$$\sin A = \frac{d_3}{d_1} \quad \text{and also} \quad \sin A = \frac{d_2}{\sqrt{d_1^2 + d_2^2}}$$

$$\therefore \qquad d_3 = \frac{d_1 d_2}{\sqrt{d_1^2 + d_2^2}} = \frac{d_1 d_2}{\sqrt{2d_2^2 + d_2^2}} = \frac{d_1}{\sqrt{3}} \qquad \text{...(34.44)}$$

These planes are known as the (111) planes. The interplanar distances are in the ratio

$$d_1 : d_2 : d_3 = 1 : 1/\sqrt{2} : 1/\sqrt{3} \qquad \text{...(34.45)}$$

Using a KCl crystal, Bragg found the values of θ for reflections from the faces *ABCD*, *ADGF* and *AHF* respectively as

$$\theta_1 = 5°22', \theta_2 = 7°36' \quad \text{and} \quad \theta_3 = 9°25'.$$

Hence for the first order spectrum,

$$\begin{aligned} d_1 : d_2 : d_3 &= 1/\sin 5°22' : 1/\sin 7°36' : 1/\sin 9°25' \\ &= 1/0.0936 : 1/0.1323 : 1/0.1636 \\ &= 1/1 : 1/1.410 : 1/1.748 \end{aligned}$$

These ratios, within the limits of experimental error, are

$$d_1 : d_2 : d_3 = 1 : 1/\sqrt{2} : 1/\sqrt{3}$$

Hence it is concluded that KCl crystal has a simple cubic structure. On the other hand, NaCl crystal which is very much similar to KCl is found to have the ratio of interplanar distances as

$$d_1 : d_2 : d_3 = 1 : 1/\sqrt{2} : 2/\sqrt{3}$$

which agree with the theoretical values for a face centred cubic system. The observed dissimilarity between the conclusions on these similar crystals can be resolved once we determine the relative intensities of the diffracted beams. From such a study, it has been confirmed that both the KCL and NaCl crystals have face centred cubic crystal structure.

Example 34.9. *A beam of X-rays λ = 0.842 Å is incident on a crystal at a glancing angle of 8°35′ when the first order Braggs' diffraction occurs. Calculate the glancing angle for 3rd order diffraction.*

Solution: According to Braggs' law, $2d \sin \theta = m\lambda$.

$$d = \frac{m\lambda}{2\sin\theta} = \frac{1 \times 0.842\,\text{Å}}{2 \times \sin 8°35'} = 2.826\ \text{Å}$$

$$\theta = \sin^{-1}\left(\frac{m\lambda}{2d}\right) = \sin^{-1}\left(\frac{3 \times 0.842}{2 \times 2.826}\right) = \mathbf{26.55°}.$$

Example 34.10. *X-rays of wavelength 1.5 Å are incident on Na Cl crystal having a grating spacing of 2.8 Å. What is the highest order that the crystal can diffract?*

Solution: According to Braggs' law, $2d \sin \theta = m\lambda$. m will be maximum if $\sin \theta = 1$

$$\therefore \qquad m = \frac{2d\sin\theta}{\lambda} = \frac{2 \times 2.8\text{Å} \times 1}{1.5\text{Å}} = 3.73.$$

It means that the highest order of diffraction is **3**.

34.27 POWDER CRYSTAL METHOD

Powder photography is highly suitable for the materials for which it is impossible to obtain single crystals of required size. One form of the powder photography is known as Debye-Scherrer method invented by P. Debye and Scherrer.

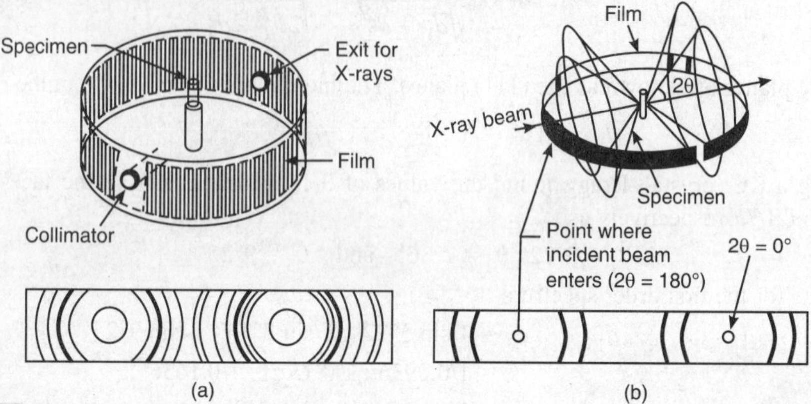

Fig. 34.62: Experimental arrangement

Debye-Scherrer camera is a cylindrical camera whose width is smaller as compared to its diameter. The material to be studied is taken in form of fine powder packed into a small capillary tube and placed vertically at the centre of the drum. A photographic film strip is mounted round the inside of the cylindrical drum of the camera. A narrow beam of monochromatic x-rays enter and leave the drum through the apertures on the opposite sides of the drum. Each crystallite has the same system of atomic planes. Some of the crystallites are bound to lie with their planes at glancing angle θ to the incident ray such that Braggs' condition is satisfied. Each such crystal produces a spot on the photographic plate. Fig. 34.62 (a) shows the experimental arrangement.

As the specimen contains a large number of crystallites oriented in all directions, almost all the possible values of θ and d are available. Also, for a particular value of the angle of incidence θ, numerous orientations of a particular set of planes are possible. The diffracted rays corresponding to fixed values of θ and d lie on the surface of a cone with its apex at the sample and the semi-vertical angle 2θ [Fig. 34.62 (b)].

Different cones are observed for different sets of θ and d for a particular value of m and also for different combinations of θ and m for a particular value of d. Each cone of the diffracted beam leaves two impressions on the film in the form of arcs on either side of the exit aperture and their centres coinciding with the aperture.

If x is the distance at which a diffracted beam strikes the film from the centre O, then

$$2\theta = \frac{180^\circ x}{\pi R}$$

where R is the radius of the drum.

$$\therefore \qquad \theta = \frac{90^\circ x}{\pi R} \qquad \text{...(34.46)}$$

Let x_1, x_2, x_3 ...... be the distances between symmetrical arcs on the stretched photographic film. Then,

$$\theta_1 = \frac{90° x_1}{\pi R}, \quad \theta_2 = \frac{90° x_2}{\pi R}, \quad \theta_3 = \frac{90° x_3}{\pi R} \quad \text{and so on.}$$

Using the value of θ into Braggs' equation, the interplanar spacing *d* can be calculated.

The diffraction pattern obtained in this method helps us distinguish amorphous materials from crystalline materials. Amorphous materials do not have reflecting planes. Therefore, diffraction rings are not produced on the film. However, they may produce a smeared ring as there is some sort of short range order in the arrangement of its molecules.

34.28 ROTATING CRYSTAL METHOD

The rotating crystal method is used when single crystals of moderate size are available. The experimental set up of the method is shown in Fig. 34.63 (a). The crystal is mounted on a rotating spindle which is rotated by a rotator. The crystal is mounted in such a way that one of the crystal axes is along the axis of the spindle. A photographic film is mounted inside the drum which is concentric with the rotating spindle.

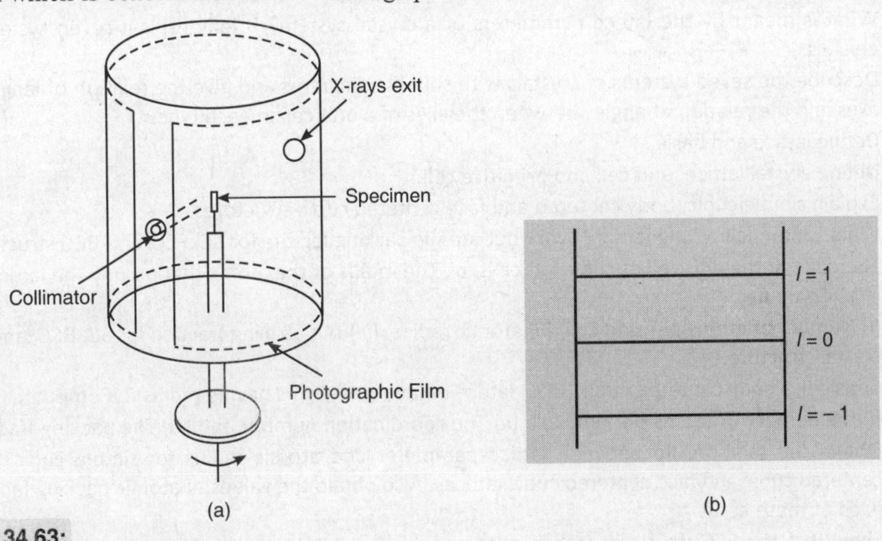

Fig. 34.63: Experimental arrangement

A monochromatic beam of X-rays is allowed through the aperture in the drum to impinge on the crystal at right angles. As the crystal rotates, various planes come successively into positions for which Bragg's condition is satisfied. The diffracted beams produce spots on the photographic film. The planes parallel to the axis of rotation diffract the incident rays in a horizontal plane. The planes tending at angle to the rotation axis makes reflections above or below the horizontal plane basing on the angle of inclination. The horizontal lines formed by diffraction spots on the film are known as *layer lines*.

If the crystal is placed such that its c-axis coincides with the axis of rotation, all the planes with Miller indices (*h*, *k*, 0) will produce the central layer line (Fig. 34.54 b). The planes having Miller indices (*h*, *k*, *l*) and $\left(h, k, \bar{l}\right)$ will produce layer lines above and below the central line respectively, and so on. The vertical spacing between the layer lines depends on the distance between the lattice points along the c-axis. Similarly, we can determine the translational vectors a and b on mounting the crystal along a and b axes, respectively. Thus, the dimensions of the unit cell of the crystal are determined.

QUESTIONS

1. Distinguish between crystalline solids and amorphous solids.
2. Differentiate the following: Isotropic solids and anisotropic solids.
3. What are crystalline and amorphous solids?
4. What are elements of crystal symmetry? Explain them with reference to simple cubic structure.
5. Elaborate the statement, lattice + basis = crystal structure.
6. What do you mean by space lattice and unit cell of crystal?
7. Define the following terms:
   (*i*) Unit cell, (*ii*) Crystallography, (*iii*) Axis of symmetry, (*iv*) Lattice parameter
8. What are Bravais lattices?
9. Find out the total number of axes of symmetry and planes of symmetry for a cubic crystal. Illustrate your answer with the help of proper diagrams.
10. What is meant by the lattice parameters of a crystal system? Briefly explain seven systems of crystals.
11. Describe the seven systems of crystals with suitable diagrams and give the relation of lengths of axes and the relation of angles between the axes of a unit cell in each type.
12. Define lattice and basis.
13. Define crystal lattice, unit cell and primitive cell.
14. Explain simple cubic, body centered and face centered cubic structures.
15. What is atomic packing factor? Work out atomic packing factors for SC, FCC, and BCC structures.
16. The edge of the unit cell of cubic lattice is 'a'. The radius of the atoms that occupy the lattice site is 'r'. Compute:
    (*i*) Number of atoms per unit cell, (*ii*) atomic radius, (*iii*) the packing fraction for SC, BCC, and FCC crystal structure.
17. Consider a body centered cubic (BCC) lattice of identical atoms having radius R. Compute:
    (*i*) the number of atoms per unit cell, (*ii*) The coordination number and (*iii*) The packing fraction.
18. Obtain the relationship between lattice parameter and atomic radius for simple cubic, body centered cubic and face centered cubic lattices. Also obtain the values of atomic packing factor in each of these cases.
19. Show that the FCC structure possesses maximum packing density and minimum percentage of void space among the three crystal structures SC, BCC and FCC.
20. Show that the FCC structure possesses least percentage of void space among SC, BCC and FCC cubic structures.
21. Explain how to find number of atoms per unit cell for simple cubic, BCC and FCC structure in a crystal.
22. Define atomic packing factor. Show that atomic packing factor for BC structure is 0.68 and for FCC structure it is 0.74.
23. Verify that simple cubic structure possesses maximum void space amongst all cubic structures.
24. Tabulate the characteristics of SC, BCC and FCC unit cells.
25. Show that FCC structure possesses least percentage void among SC, BCC and FCC cubic structure.
26. What are Miller indices? Find the Miller indices for a crystal plane. Derive the expression for interplanar distance between consecutive planes described by Miller indices (hkl).
27. Explain the term Miller indices. What is their role in crystal structure? Give the important features of Miller indices.
28. Explain with an example, the different steps to be followed, to identify the Miller indices of a crystallographic plane.

29. Draw separately the principal planes (100), (110) and (111) in a simple cubic crystal.
30. Obtain an expression for interplanar spacing between two adjacent planes of Miller indices (*hkl*) in a cubic crystal.
31. Show that the interplanar spacing between (*hkl*) planes in a cubic lattice is given by

$$d_{hkl} = \frac{a}{\sqrt{h^2 + k^2 + l^2}}$$ where, a is the lattice constant.

32. What are tetrahedral and octahedral voids?
33. What do you understand by Miller indices of a crystal plane? Show that the spacing between consecutive parallel planes defined by Miller indices (*hkl*) is given by

$$d_{hkl} = \left[\frac{h^2}{a^2} + \frac{k^2}{b^2} + \frac{l^2}{c^2}\right]^{-\frac{1}{2}}$$

where a, b and c are primitive vectors along three mutually perpendicular axes.
34. Describe crystal structure of NaCl. Derive the expression for lattice constant.
35. Describe the structure of zinc Sulphide.
36. What is meant by polymorphism and allotropy?
37. Discuss the structures of diamond and graphite.
38. Explain with neat sketch the diamond crystal.
39. Deduce Braggs' law of x-ray diffraction in crystals. Describe and explain how Braggs' spectrometer can be used in the study of crystal structure analysis.
40. Explain crystal state of matter and obtain the expression for Braggs' law of X-ray diffraction in crystals.
41. Deduce Braggs' law of X-ray diffraction.
42. State and derive Braggs' law for diffraction in crystals.
43. Derive and explain Bragg's relation. How is it useful in crystal structure determination? Why is this law not useful for amorphous solids? Why X-rays are preferred to visible light in crystal structure determination?
44. Explain how Braggs' x-ray spectrometer can be used for verification of Braggs' relation.
45. Describe how Braggs' spectrometer is used for determination of crystal structure.
46. Explain how Braggs' x-ray spectrometer can be used to determine the interplanar spacing.
47. Give the theory of Braggs' x-ray diffraction. Describe Braggs' x-ray spectrometer. How is it used to verify Braggs' law?
48. Explain and deduce Braggs' law in x-ray diffraction. Describe a Braggs' spectrometer. Explain how it is used to determine the wavelength of x-rays.
49. Explain how the structure of KCl and NaCl is determined using Braggs' x-ray technique.
50. Describe with suitable diagram the powder method for determination of crystal structure.
51. Describe the rotating crystal method for crystal structure analysis. What is the advantage of this method?
52. Write a note on rotating crystal method of x-ray diffraction.

PROBLEMS

1. The density of copper is 8980 Kg/m^3 and unit cell dimension is 3.6Å. Atomic wt. of copper is 63.54. Determine crystal structure. Calculate atomic radius and interplanar spacing of (110) plane.
2. The lattice constant for BCC iron at 20° C is 2.87 Å. The density of iron is 7870 kg/m^3. Determine its atomic mass. (Avogadro's number = 6.02 × 10^{26} atoms/kmol). **[Ans:** 56]
3. Draw crystal planes having Miller indices (111), (110) and (211).

4. A certain crystal has axial units $x:y:z$ of 0.424:1:0.367. Find the Miller indices of crystal faces whose intercepts are 0.212:1:0.183. **[Ans:** (212)]
5. Silver has fcc structure and its atomic radius is 1.441 Å. Find the spacing of (220) planes. **[Ans:** 1.441 Å]
6. Calculate the interplanar spacing for a (321) plane in a simple cubic lattice whose lattice constant is 4.2 Å. **[Ans:** 1.12 Å]
7. The lattice constant of BCC iron at 20° C is 2.87Å. The density of iron is 7870 kg/m^3. Determine its atomic mass?
8. Nickel is characterized by FCC lattice. The edge of the unit cell is 3.52Å. The atomic weight of nickel is 56.71kg/k-mol. Determine the density of the metal.
9. Draw crystal planes having Miller indices (111), (110) and (211).
10. Find the Miller indices of a set of planes with intercepts a, $2a$ and $3a$ on X, Y and Z axes respectively for a cubic crystal.
11. Draw separately the planes (100), (110) and (111) in a simple cubic crystal.
12. What are Miller indices? Draw crystal planes having Miller indices:(2,1,0), (1,0,1) and (0,1,0) for simple cubic structure.
13. Draw separately the principle planes (100), (110) and (111) in a simple cubic crystal.
14. Find the Miller indices of a set of planes with intercepts a, 2a and 3a on X, Y and Z axes respectively for a cubic crystal.
15. Draw crystal planes having Miller indices (111), (110) and (211). Lead is face centered cubic with an atomic radius r = 1.746Å. Find the spacing of:(200), (220) and (111) planes.
16. Calculate the longest wavelength that can be analyzed by rock salt crystal of spacing 2.82Å in the first order.
17. If the average energy required producing a vacancy in a metal is 1 eV, calculate the ration of vacancies in a metal at 1000K and 500K. **[Ans:** 1.08×10^5]

35 Conductors

Learning Objectives:

By the end of this chapter, you will be able to:

- Understand resistivities and conductivities of some solids materials.
- Learn concept of electrical conductivity, mobility, relaxation time and thermal conductivity.
- Explain classical free electron theory and its drawbacks.
- Describe Quantum free electron theory in detail.

35.1 INTRODUCTION

Materials having low electrical resistivity are known as conductors. Metals and their alloys belong to this group of materials. In metals the valence electrons are loosely bound to their individual atoms. They become free and are responsible for the conduction of electricity and heat in metals. Free electron theory was proposed by Drude by assuming that the valence electrons become free in metals and move about randomly within the metal much the same way as molecules in a gas confined to a container. H.A. Lorentz later refined Drude's theory by assuming that the velocity distribution of the electrons obeyed the classical Maxwell-Boltzmann law. The Drude-Lorentz theory could successfully explain the Ohm's law and the high electrical conductivity of metals. However, the theory suffered from a number of setbacks, because of the various assumptions made. Sommerfeld later extended the free electron model by incorporating Fermi-Dirac statistics. It too could not succeed to explain all the experimental observations found on solids. Ultimately, the band theory of solids successfully explained the electrical behaviour of solids.

35.2 ELECTRICAL CONDUCTION

If a potential difference V is applied across a solid, as shown in Fig. 35.1, it establishes an electric field E in the solid. E is given by

$$E = \frac{V}{L} \qquad ...(35.1)$$

where L is the length of the solid along which charge carriers move. The electric field accelerates the charge carriers and causes a flow of electric current through the solid. The current I passing across an area A is defined as the net charge Q transported through the area per unit time. Thus,

$$I = \frac{Q}{t} \qquad ...(35.2)$$

When electric current flows through a material, it is said to be **conducting** electricity. Any material can conduct electricity if it contains mobile charge carriers. A free electron is an example of charge carrier. Other carriers include mobile positive or negative ions, holes etc.

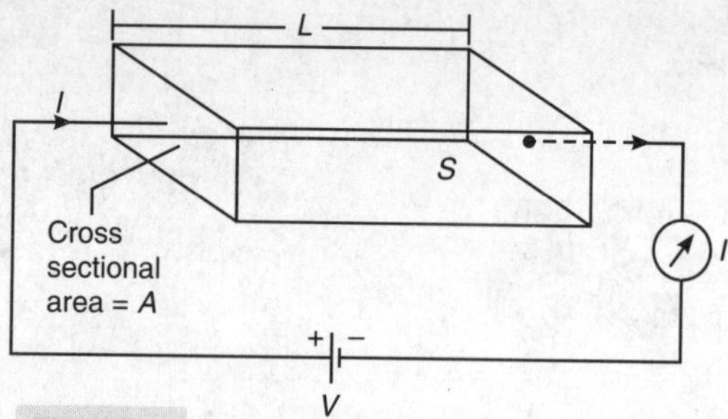

Fig. 35.1: Schematic of the circuit for determining the conductivity of a solid

The magnitude of the electrical current, I, passing through a solid (Fig. 35.1) at a constant temperature is directly proportional to the potential difference V applied across the solid.

$$I = \frac{V}{R} \qquad \textbf{Ohm's law} \qquad ...(35.3)$$

where R is the electrical **resistance** offered by the solid to the current flow.

When electrons travel in a vacuum, they are not impeded in their motion. They travel along straight line paths and acquire kinetic energy which is equal to the work done by the accelerating electric field. On the contrary, electrons encounter opposition while moving through a solid. The opposition to the electron motion in materials is manifest as electrical resistance. The electrical resistance offered by a solid is found to depend on the physical dimensions of the solid.

$$R \propto \frac{L}{A}$$

where L is the length and A the area of cross-section of the solid. Thus,

$$R = \rho \frac{L}{A} \qquad ...(35.4)$$

ρ is called **electrical resistivity**. It is a material constant and does not depend on the dimensions of the solid.

$$\rho = \frac{RA}{L} \quad \text{Ohm.m} \qquad ...(35.5)$$

The reciprocal of the electrical resistivity is known as **electrical conductivity**, ρ. Thus,

$$\sigma = \frac{1}{\rho} = \frac{L}{RA} \qquad ...(35.5a)$$

Using the Eq. (35.3) into Eq. (35.5a), we get

$$\sigma = \frac{IL}{VA} \quad \text{S/m} \qquad ...(35.6)$$

Electrical conductivity, σ, characterizes the ability of a material to conduct electricity.

35.3 CLASSIFICATION OF MATERIALS

The resistivity of a solid can be determined using the circuit shown in Fig. 35.1. The electrical resistivity and conductivity of some materials are given in Table. 35.1

Table 35.1: *Resistivities and conductivities of some solids*

Material	Resistivity Ohm.m	Conductivity S/m
Silver	1.47×10^{-8}	68×10^{6}
Copper	1.78×10^{-8}	58×10^{6}
Aluminium	2.63×10^{-8}	38×10^{6}
Steel	20.00×10^{-8}	5×10^{6}
Lead	22.00×10^{-8}	4×10^{6}
Carbon	3500×10^{-8}	0.03×10^{6}
Germanium	6.00×10^{-1}	1.67
Silicon	2300.00	4.35×10^{-4}
Aluminium glass	10^{10} to 10^{12}	10^{-10} to 10^{-12}
Borosilicate glass	10^{13}	10^{-13}
Polyethylene	10^{13} to 10^{13}	10^{-13} to 10^{-15}

It is seen from Table 35.1 that materials differ greatly in their conductivity. However, they may be broadly classified into three groups.

(*i*) Metals and alloys exhibit large conductivity of the order of 10^8 S/m and are therefore, called **conductors**.

(*ii*) Materials such as metal oxides, glasses, plastics are found to possess very low conductivity of the order less than 10^{-12} S/m. They are called **insulators**.

(*iii*) Materials such as silicon and germanium have values of conductivity, of the order of 10^4 to 10^{-4} S/m, intermediate to those of conductors and insulators. They are hence called semiconductors.

35.4 FREE ELECTRON MODEL OF SOLIDS

Electrical conduction is one of the important properties of solids. It was generally accepted that the valence electrons are involved in electrical conduction in metals and alloys. The evidence was provided at a later stage by the experiments of Stewart and Tolman. Much earlier to the experimental confirmation, the free electron model of solids was proposed by Paul Drude. He assumed that the valence electrons become **free** in solids and move about randomly within the solids much the same way as molecules in a gas confined to a container. This is the **free electron model** which is applicable to all the three categories of solids. This model is used in explaining not only the electrical properties but also the thermal, optical and magnetic properties of solids.

The free electron theory underwent successive modifications in an attempt to explain the electrical behaviour and the distinction between the three types of solids.

(i) Classical free electron theory: This theory was proposed by Paul Drude in 1900 and later was extended by Lorentz. Therefore this theory is also known as the **Drude-Lorentz theory**. In this theory it was assumed that valence electrons become free in metals and move about randomly within the metal. Further, it was assumed that the free electrons move in a region of constant potential. Just as the velocities of molecules in a container, the velocities of electrons in a solid obey the classical Maxwell-Boltzmann distribution. This theory successfully explained the Ohm's law and the high electrical conductivity of metals. However, the theory failed to explain other features and the distinction between conductors, insulators and semiconductors.

(ii) Quantum free electron theory: This theory was developed by Sommerfeld in 1928. Unlike neutral molecules, electrons are charged particles and obey Pauli Exclusion Principle. An assembly of free electrons obeys Fermi-Dirac statistics. Sommerfeld took into account these facts and modified the Drude's classical free electron theory. However, in this theory also it was assumed that the free electrons move in a region of constant potential. This theory is based on the particle character of electron and did not take into account of its wave character. The theory failed to explain other features and the distinction between conductors, insulators and semiconductors.

(iii) Band theory of solids: This theory was formulated by Felix Bloch in 1928. This theory takes into account that electrons exhibit wave character as they move between atoms in a solid. It further assumed that the potential varies in a periodic manner in the solid. This theory successfully explained the classification of solids into three groups, namely conductors, insulators and semiconductors.

35.5 CLASSICAL FREE ELECTRON THEORY OF METALS

Free electron gas

After the discovery of electron by J.J.Thomson in 1897, Drude developed the free electron theory of metals. According to this theory, metals consist of positive ion cores and valence electrons. The ions cores are immobile and consist of positive nucleus and the bound electrons. The valence electrons get detached from the parent atoms during the process of formation of the metal and move randomly among these cores. Hence they are known as **free electrons**.

In Drude's model, the potential field of the ion cores is considered to be constant all over the metal and the mutual repulsion among the electrons is neglected. The free electrons moving within the metal are supposed to be similar to the freely moving atoms in the perfect gas. These free electrons are collectively referred to as **free electron gas**. The potential energy of a stationary electron within the metal is less than the potential energy of an identical electron just outside the metal. This energy difference acts as a potential barrier and prevents the free electrons from escaping from the surface of the metal. Therefore, free electrons are constrained to stay within the metal. In other words, the free electron gas is held in a potential energy box. As the free electrons are responsible for conduction of electricity in the metals, they are also called **conduction electrons**.

Thermal motion of free electrons

The free electrons keep moving randomly in all directions through the lattice structure of the metal due to thermal energy. The average speed is very high and is of the order of 10^6m/s. As free electrons in a metal move through a maze of positive ion cores, they suffer repeated collisions with the ion cores or other electrons and get deflected. The direction of motion of each free electron changes on every collision and the electron moves along a zigzag path. On the average for every electron moving in a particular direction, there is another electron moving in the opposite direction. Therefore, the thermal motion of free electrons does not cause flow of current through the metal. The electron motion between two collisions is linear and uniform.

Drift motion of free electrons

When the ends of a piece of metal are connected to the terminals of a battery, an electric field is impressed across the metal and the equilibrium condition is disturbed. The electric field accelerates the electrons. The electrons acquire velocity and move in a direction opposite to that of electric field. The directional motion of electrons due to the action of electric field is called **drift**. The drift velocity gained by an electron due to acceleration is lost completely whenever a collision occurs. After that, the electron gets accelerated once again and loses its velocity at the next collision. The process goes on repeating and the electron moves on an average with a **mean drift velocity** υ_d. The magnitude of the drift velocity is limited by the decelerations caused by collisions. Thus, the collisions play the role of a frictional force.

The drift speed is typically of the order of 10^{-2}m/s, which is very small compared to the thermal speed, 10^6 m/s. Thus, the motion of a free electron in the presence of electric field consists of a much slower directional drift motion superposed over random zigzag motion. The drift motion is directional and causes current flow in a conductor called ***drift current*** or ***conduction current***.

Mean Collision time, τ

An electron moving in a particular direction inside a metal encounters another electron or fixed positive ion and deviates from its direction through some angle with the initial direction. This is the process of **scattering** of electrons. For simplicity we say that electron underwent a collision during its motion. After some time another collision occurs and the electron deviates in another direction and this goes on. The time elapsed between two successive collisions is not a constant but varies. The number of collisions per second that an electron makes with the ion cores is proportional to its speed. The average duration of time that elapses between two successive collisions is called **mean collision time**, τ of the electron. It means that the electron on the average travels for a time τ before its next collision, and has traveled for a time τ since its last collision. The collision time is given by

$$\tau = \frac{\lambda}{\bar{\upsilon}} \qquad ...(35.7)$$

where λ is known as mean free path and $\bar{\upsilon}$ the *rms* (root mean square) velocity.

Mean free path, λ

The average distance traveled by an electron between any two consecutive collisions is known as **mean free path**, λ. The mean free path is given by the product of rms velocity of electrons and mean collision time.

$$\therefore \qquad \lambda = \bar{\upsilon}\tau \qquad ...(35.8)$$

35.6 DRIFT VELOCITY

When an electric field E is impressed on a metal the electrons in it experience a force $-eE$.

If υ is the velocity of free electron and τ is the average time between two consecutive collisions, the frictional force opposing the continuous acceleration of the electron may be written as $-m\upsilon/\tau$. Using Newton's second law, the equation of motion of free electron may be written as

$$m\frac{d\upsilon}{dt} = -eE - m\frac{\upsilon}{\tau} \qquad ...(35.9)$$

where m is the mass of the electron and E is the intensity of the applied electric field.

Under the steady state condition, $\frac{d\upsilon}{dt} = 0$, and the electron attains a steady value of velocity υ_d in dynamic equilibrium. Therefore, under steady state condition, Eq. (35.9) reduces to

$$0 = -eE - m\frac{\upsilon_d}{\tau}$$

$$\therefore \qquad \upsilon_d = -\frac{eE\tau}{m} \qquad ...(35.10)$$

where υ_d is the steady state velocity of the electron and is known as **drift velocity**. The drift motion of electrons causes current flow in a conductor.

35.7 ELECTRICAL CONDUCTIVITY

Let n be the number of free electrons per unit volume of the conductor, S (Fig. 35.1). n is called the free **electron density** or free **electron concentration** in the solid.

The total number of electrons in the metal specimen is given by

$$N = \text{(electrons per unit Volume) (Total volume)}$$

$$\therefore \qquad N = nAL \qquad ...(35.11)$$

The total charge present in the solid block may be written as

$$Q = Ne = nAle \qquad ...(35.12)$$

Current flowing through the solid is given by

$$I = \frac{Q}{t} = \frac{nALe}{t} \qquad ...(35.13)$$

The term L/t represents velocity and gives the average *drift velocity*, υ_d of electrons in the solid.

$$I = neA\upsilon_d \qquad ...(35.14)$$

The current density is defined as $J = \frac{I}{A}$. ...(35.15)

$$\therefore \qquad J = ne\upsilon_d \qquad ...(35.16)$$

Also, $$J = ne\left(\frac{eE\tau}{m}\right) = \frac{ne^2\tau}{m}E \qquad ...(35.17)$$

According to Eq. (35.6), $$\sigma = \frac{IL}{VA}$$

As $\frac{V}{L} = E$, the above equation may be rewritten as $\sigma = \frac{J}{E}$

or **Point form of Ohm's law** $J = \sigma E$ ...(35.18)

Equating R.H.S. of Eq. (35.17) and Eq. (35.18), we obtain

$$\sigma = \frac{ne^2\tau}{m} \qquad ...(35.19)$$

Similarly, equating R.H.S. of Eq. (35.16) and Eq. (35.18), we obtain

$$\sigma = \frac{ne\upsilon_d}{E} \quad ...(35.20)$$

or $$\sigma = ne\mu \quad ...(35.21)$$

where μ is called electron **mobility**.

Eq. (35.19) or Eq. (35.21) indicates that electrical conductivity of a material depends mainly on the free electron concentration in it. The large electrical conductivity of metals is explained as due to the presence of large number of free electrons in them. Similarly, one may explain the very low conductivity of insulators as due to non-availability of free electrons in them and the moderate conductivity in semiconductors as due to the presence of modest number of free electrons in them. However, the equation cannot explain the variation in conductivity due to external influences such as the temperature, radiation and impurities in case of semiconductors and insulators.

Example 35.1: *Find the drift velocity of free electrons in a copper wire of cross-sectional area 10 mm³ when the wire carries a current of 100 A. Assume that each copper atom contributes one electron to the free electron gas. Density of copper is 8969 kg/m³ and its atomic weight is 63.54.*

Solution: The electron concentration in copper, n = *Atomic density, N × contribution from each atom, x*

$$n = \frac{\text{Density} \times N_A}{M} \times x = \frac{8969\,\text{kg/m}^3 \times 6.02 \times 10^{26}\ /\ \text{k.mol}}{63.54\,\text{kg/k/mol}} \times 1 = 8.49 \times 10^{28}/\text{m}^3.$$

Drift velocity of free electrons in a metal is given by

$$\upsilon_d = \frac{I}{nAe} = \frac{100\ \text{A}}{8.49 \times 10^{28}\ /\ \text{m}^3 \times (10^{-5}\ \text{m}^2)(1.602 \times 10^{-19}\ \text{C})} = \mathbf{7.4 \times 10^{-4}\,m/s.}$$

35.8 MOBILITY

When an electric field is applied across a solid, it accelerates the electrons in the direction of electric field. As electrons moving through a solid undergo repeated collisions with the atoms in the solid and therefore, move with a steady velocity known as drift velocity, υ_d. The drift velocity is proportional to the electric field applied. Thus,

$$\upsilon_d \propto E$$

or $$\upsilon_d = \mu E \quad ...(35.22)$$

where μ is the proportionality constant and is called electron **mobility**.

We define *electron mobility* as the drift velocity of electrons per unit electric field.

$$\therefore \quad \mu = \frac{\upsilon_d}{E} \quad ...(35.23)$$

Mobility indicates the ease with which electrons move in a solid. We find that electron mobility in metals is of the order of 10^{-3} m²/V.s and in semiconductors it is of the order of 10^{-1} m²/V.s.

Example 35.2: *Find the mobility of electrons in copper if there are 9 × 10²⁸ valence electrons/m³ and the conductivity of copper is 6 × 10⁷ mho/m.*

Solution: Electrical conductivity is given by $\sigma = n e\mu$.

$$\therefore \quad \mu = \frac{\sigma}{ne} = \frac{6 \times 10^7\,\Omega^{-1}/\text{m}}{9 \times 10^{28}\ /\ \text{m}^3 \times 1.602 \times 10^{-19}\,\text{C}} = \mathbf{4.16 \times 10^{-3}\ m^2/V.s}$$

35.9 RELAXATION TIME

When electric field is applied to a metal, the free electrons in it move with an average velocity υ_d in a direction opposite to that of the applied field. If the external field is switched off, the velocity of the free electrons decays exponentially as a result of collisions with positive ions. The decay is given by the expression

$$\upsilon_d(t) = \upsilon_d(0)e^{-t/\tau}$$

where υ_0 is the initial velocity of the electrons before application of electric field.

If $t = \tau$, $$\upsilon_d(t) = \upsilon_d(0)e^{-\tau/\tau} = \frac{\upsilon_d(0)}{e} \qquad ...(35.24)$$

The duration of time in which the drift velocity of an electron decays to $\frac{1}{e}$ times of its initial velocity is known as **relaxation time**. It gives the time taken by electrons in a conductor to return from non-equilibrium condition to the equilibrium condition, after the applied electric field is turned off. Its value is of the order of 10^{-14}s.

Example 35.3: *Find the relaxation time of conduction electrons in a metal if its resistivity is 1.54 × 10^{-8}Ωm and it has 5.8 × 10^{28} conduction electrons/m^3.*

Solution: $$\sigma = \frac{ne^2\tau}{m}$$

$$\therefore \quad \tau = \frac{m}{\rho ne^2} = \frac{9.11\times10^{-31}\,\text{kg}}{1.54\times10^{-8}\,\Omega\text{m}\times5.8\times10^{28}\,/\,\text{m}^3\times(1.602\times10^{-19}\,\text{C})^2} = \mathbf{3.9 \times 10^{-14}s.}$$

35.10 THERMAL CONDUCTIVITY

Heat conduction is the transfer of thermal energy from the hotter to the colder part of a material. All solids conduct heat. The rate at which heat is conducted depends on the temperature gradient and the material of the solid. Let us consider a metal bar of length x whose ends are held at different temperatures. Heat that flows through a cross-section of the bar divided by time and area is proportional to the temperature gradient $\frac{dT}{dx}$. Thus,

$$\frac{\Delta Q}{A \cdot \Delta t} \propto \frac{dT}{dx}$$

where ΔQ is the thermal energy conducted through a cross-sectional area A in time Δt between two planes with a temperature gradient of $\frac{dT}{dx}$. The above relation may be written in equation form as

$$J = -K\frac{dT}{dx} \qquad ...(35.25)$$

where J is the thermal flux and K is the proportionality constant, called the **thermal conductivity**.

The negative sign in the above equation indicates that the heat flows from the hot end to the cold end of the bar. Using the above expressions, we can define the thermal conductivity K of a solid.

Thermal conductivity *is defined as the quantity of heat crossing per unit time through a unit area and maintaining a unit temperature difference across the body*. Thus,

$$K = Q\left[\frac{dT}{dx}\right]^{-1} \qquad ...(35.26)$$

35.10.1 Expression for Thermal Conductivity

Good electrical conductors are also good conductors of thermal energy. It may therefore be presumed that the free electron gas is also responsible for transporting thermal energy through the metal. They carry out this via random collision processes. Treating free electron gas as a perfect gas, we obtain an expression for thermal conductivity making use of the kinetic theory of gases.

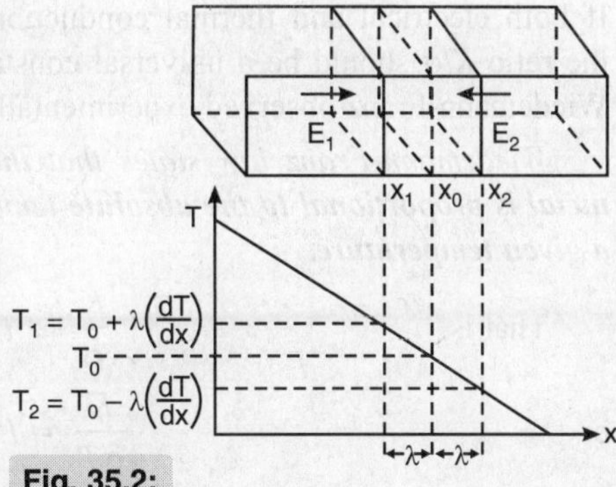

Fig. 35.2: Uniform metal bar

Let us now consider a uniform metal bar, which is heated at one of its ends (Fig. 35.2). Let the left side of the metal bar be hot and the right side be cold. Thus, a temperature gradient $\dfrac{dT}{dx}$ exists in the x – direction. Let us select a volume at the center of the bar whose faces have the size of a unit area and whose length is 2λ, where λ is the mean free path of electrons. Let us assume that at the distance λ from the centre x_0 the average electron had its last collision and gained energy of that place. Now we calculate the energy E_1 carried by electrons that drift from the left into the selected volume.

$$E_1 = \text{Number of electrons} \times \text{average energy of an electron} = z \times \frac{3}{2}kT_1.$$

Let n be the electron concentration in the material and υ be the velocity of electrons. From the kinetic theory of gases, z is given by

$$z = \frac{1}{6}n\upsilon$$

$$\therefore \quad E_1 = \frac{n\upsilon}{6}\cdot\frac{3}{2}k\left[T_0 + \lambda\left(-\frac{dT}{dx}\right)\right] = \frac{n\upsilon}{6}\cdot\frac{3}{2}k\left[T_0 - \lambda\frac{dT}{dx}\right] \qquad ...(35.27)$$

The same number of electrons drifts from right to left through the volume under consideration. These electrons carry a lower energy E_2 because of the lower temperature of the particles at the site of interaction. Thus,

$$E_2 = \frac{n\upsilon}{6}\cdot\frac{3}{2}k\left[T_0 + \lambda\frac{dT}{dx}\right] \qquad ...(35.28)$$

The excess thermal energy transferred per unit time into the unit volume is

$$J = E_1 - E_2 = \frac{n\upsilon}{6}\cdot\frac{3}{2}k\left[2\lambda\frac{dT}{dx}\right] \qquad ...(35.29)$$

Comparing Eq. (35.29) with Eq. (35.25), we obtain the expression for thermal conductivity as

$$K = \frac{n\upsilon k\lambda}{2} \qquad ...(35.30)$$

As $\lambda = \upsilon\tau$, the above equation may be expressed as

$$K = \frac{n\upsilon^2 k\tau}{2} \qquad ...(35.31)$$

35.11 WIEDEMANN-FRANZ LAW

If both electrical and thermal conduction in metals are conducted by free electrons, then the ratio K/σ should be a universal constant, having the same value for all metals. In 1853, Wiedemann-Franz observed experimentally such a simple connection between K and σ.

Wiedemann-Franz law states that ***the ratio of thermal to electrical conductivity of a metal is proportional to the absolute temperature and the ratio is constant for all metals at a given temperature.***

That is, $$\frac{K}{\sigma} \propto T \qquad ...(35.32)$$

or $$\frac{K}{\sigma T} = L \qquad ...(35.33)$$

where L is a constant called the **Lorentz number**.

Derivation

Let us consider a metal. The thermal conductivity of the metal is given by

$$K = \frac{n\upsilon^2 k\tau}{2}$$

And its electrical conductivity is given by

$$\sigma = \frac{ne^2\tau}{m}$$

The ratio of thermal conductivity to electrical conductivity is given by

$$\frac{K}{\sigma} = \frac{m\upsilon^2 k}{2e^2} \qquad ...(35.34)$$

But $\frac{1}{2}m\upsilon^2 = \textit{kinetic energy} = \frac{3}{2}kT$.

$$\therefore \qquad \frac{K}{\sigma} = \frac{3}{2}kT \cdot \frac{k}{e^2}$$

or $$\frac{K}{\sigma} = \frac{3}{2}\left(\frac{k}{e}\right)^2 T$$

or $$\frac{K}{\sigma} = LT \qquad ...(35.35)$$

where $$L = \frac{3}{2}\left(\frac{k}{e}\right)^2 \qquad ...(35.36)$$

Thus, it is seen from Eq. (35.35) that the ratio of thermal to electrical conductivity of a metal is proportional to the absolute temperature and the ratio is a constant.

Wiedemann-Franz law confirmed that the motion of the free electrons is mainly responsible for both electrical and thermal conductivity.

35.12 LORENTZ NUMBER

The Lorentz number is the ratio of thermal conductivity of a metal to the product of its electrical conductivity and its absolute temperature.

Thus, $$L = \frac{K}{\sigma T}$$

Using the classical theory, we find that

$$L = \frac{3}{2}\left(\frac{k}{e}\right)^2$$

Now substituting the values of k and e in the above expression, we get the value of L.

$$L = \frac{3}{2}\left(\frac{1.38\times10^{-23}\ \text{J/K}}{1.602\times10^{-19}\ \text{C}}\right)^2 = 1.12\times10^{-8}\ \text{watt.ohm/deg}^2$$

This classical value is only half of the experimental value. Using the expressions derived from quantum theory, we find that L is given by

$$L = \frac{\pi^2}{3}\left(\frac{k}{e}\right)^2 \qquad \text{...(35.37)}$$

Substituting the values of k and e into the above expression, we get

$$L = 2.45\times10^{-8}\ \text{watt.ohm/deg}^2$$

The above value agrees well with the experimental value.

Example 35.4: *The electrical resistivity of copper at 27°C is $1.72\times10^{-8}\Omega$m. Compute its thermal conductivity if the Lorentz number is 2.26×10^{-8} WΩK^{-2}.*

Solution: $\frac{K}{\sigma T} = L \quad \therefore \quad K = \frac{LT}{\rho} = \frac{2.26\times10^{-8}\ \text{W}\Omega\Omega^{-2}\times300\,\text{K}}{1.72\times10^{-8}\ \Omega\text{m}}$ = **394 Wm^{-1}K^{-1}.**

Example 35.5. *The thermal and electrical conductivities of copper at 200°C are 390 Wm^{-1}K^{-1} and $5.87\times10^7\ \Omega^{-1}\ m^{-1}$. Calculate Lorentz number.*

Solution: $L = \frac{K}{\sigma T} = \frac{390\ \text{Wm}^{-1}\text{K}^{-1}}{5.87\times10^7\ \Omega^{-1}\text{m}^{-1}\times293\,\text{K}}$ = **2.267 × 10^{-8} WΩK^{-2}.**

35.13 RESISTANCE

Origin of Resistance

The origin of electrical resistance is assumed to be due to repeated collisions of electrons with obstacles in the material. In the classical free electron model, these obstacles were thought to be the stationary positive ions in the lattice. If it *were* true, the value of electron mean free path should be around a few angstroms, which is the order of distance between the ions. Experimentally determined values for mean free path are of the order of a few hundred angstroms. Secondly, the mean free path of electrons should not depend on the temperature of the material and should depend only on the concentration of the ions in it. However, resistivity of a metal is inversely proportional to the electron mean free path and a decrease in resistivity with temperature indicates an increase in the value of the mean free path. At extremely low temperatures, it is then expected that the mean free path values reach macroscopic values. It is obvious that the classical picture was incorrect because we have not taken into account the wave nature of electrons.

According to new interpretation, electrical resistance arises due to the scattering of electrons at the imperfections in the lattice. The imperfections are the deviations from perfect periodicity of the lattice. Vibrations of the ions in the lattice, vacant lattice sites, different lattice defects etc are the imperfections that exist in a real solid and all these cause electron scattering and contribute to the resistivity of the material.

Temperature dependence of Resistivity

Electrical resistivity of metals has been observed to increase linearly with increase of temperature. However, at very low temperatures, the resistivity varies as T^5 and remains practically constant and independent of temperature below 10 K down to nearly 0 K, as shown in Fig. 35.3(a).

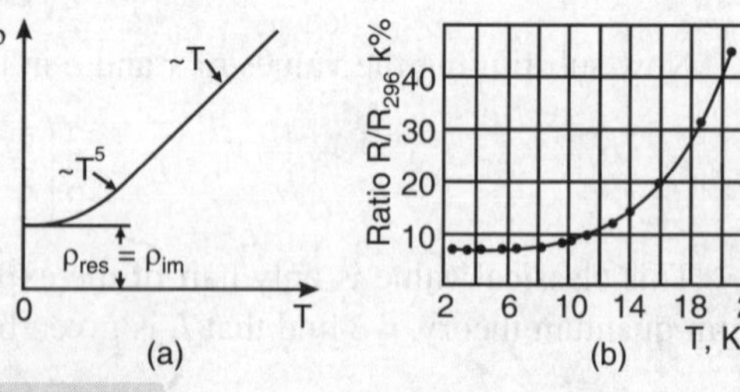

Fig. 35.3: Temperature dependence of Resistivity

(*i*) The constant value of resistivity is attributed to the presence of impurity atoms in small concentrations and also to geometrical imperfections like grain boundaries, point defects etc in the metal. This resistance is called **residual resistivity**. The residual resistivity, ρ_i, arises due to the electron scattering at the imperfections in the metal. A relaxation time, τ_i, can be assigned to the impurity and imperfection scattering.

(*ii*) At temperatures well above 10 K, the positive ions, which are arranged in a periodic array in a metal, starts vibrating like simple harmonic oscillators. Since, the ions are bound to each other through elastic and electric forces, they vibrate with different frequencies. These vibrations are similar to standing waves with fixed energies. Therefore, they are considered to be packets of energy called **phonons**. The phonon spectrum and phonon concentration are strongly dependent on temperature. At low temperatures, phonon number reduces considerably. At low temperatures, the specific heat of metallic lattice is proportional to T^3 and therefore, the density of phonons and the probability of scattering are proportional to T^3. In addition to this, the relaxation time is dependent on the scattering angle that would give rise to resistivity variation with temperature as T^2. Consequently, $\rho_{ph} \propto T^3 \cdot T^2 = T^5$. At much higher temperatures, the electrical resistivity increases linearly with temperature. The relaxation time, τ_{ph} can be assigned to phonon scattering.

The total resistivity of a metal at any given temperature is a sum of resistivity due to impurity scattering, ρ_i, and phonon scattering, ρ_{ph}. Thus,

$$\rho(T) = \rho_i + \rho_{ph} \qquad \text{...(35.38)}$$

Eq. (35.38) known as ***"Matthiessen's rule"***.

As $\rho = \dfrac{m}{ne^2\tau}$, we can write Eq. (35.38) as

$$\frac{m}{ne^2\tau} = \frac{m}{ne^2\tau_i} + \frac{m}{ne^2\tau_{ph}}$$

or

$$\frac{1}{\tau} = \frac{1}{\tau_i} + \frac{1}{\tau_{ph}} \qquad \text{...(35.39)}$$

35.14 DRAWBACKS OF CLASSICAL FREE ELECTRON THEORY

The free electron model is highly successful in explaining many physical properties of metals such as their high electrical and thermal conductivities, their high luster etc. However, it failed to account for some of the other properties. We cite some important failures of the model here.

(*i*) Monovalent metals (copper, silver etc) have been found to have higher electrical conductivity than divalent (cadmium, zinc etc) and trivalent (aluminum, indium etc) metals. If the conductivity is proportional to electron concentration, then monovalent metals should have lesser electrical conductivity compared to the divalent and trivalent metals.

(*ii*) The model cannot explain the classification of materials into conductors, semiconductors and insulators.

(*iii*) Metals are expected to exhibit negative Hall coefficient since the current carriers in them are electrons. However, some of the metals such as zinc have positive values for Hall coefficient. The free electron model cannot explain why zinc and other metals have positive Hall coefficient.

(*iv*) According to the classical theory, the conductivity of a metal is given by the Eq. (35.19). Thus, the resistivity is given by

$$\rho = \frac{m}{ne^2\tau} \qquad \text{...(35.40)}$$

When the above expression is used to calculate the mean free path of electron in metals, wefind it to be the order of about 3 Å and is consistent with the assumption of classical theory that the origin of resistivity is due to frequent collisions of electrons with the lattice ions. However, the experimentally measured values are very high and are of the order of 50 Å. The experimental results suggest that electrons pass a long distance through the lattice without collisions.

(*v*) According to classical free electron theory, all the valence electrons can absorb thermal energy. According to the law of equipartition energy, each free electron possesses an average kinetic energy of (3/2)kT. If we consider a monovalent crystal, each atom contributes one valence electron to the electron gas and there will be N free electrons per unit volume of the crystal. Then the total energy of electrons is given by

$$E = \frac{3}{2}NkT$$

When the metal is heated, the free electrons also absorb part of the heat energy and the electronic specific heat is given by

$$[C_\upsilon]_{el} = \left(\frac{dE}{dT}\right) = \frac{3}{2}Nk \qquad (35.41)$$

Substituting the values of N and k, we get $[C_\upsilon]_{el}$ = 12.5kJ/kmol/K. This value is about hundred times greater than the experimentally measured value. This result implies that free electrons do not contribute significantly to the heat capacity of the metal. We conclude that the law of equipartition and hence Maxwell-Boltzmann statistics is not applicable to the free electrons in a metal.

35.15 QUANTUM FREE ELECTRON THEORY

The quantum theory for the assembly of free electrons in a metal was first advanced by Sommerfeld in 1928. He had retained the vital features of classical free electron theory and included (*i*) the Pauli Exclusion Principle and (*ii*) the Fermi-Dirac statistics to formulate his quantum free electron theory.

The main assumptions of this theory are:

It is assumed that the eigen values of conduction electrons are quantized and are expressed in terms of a set of energy levels. Further, the distribution of electrons in various allowed energy levels takes place according to Pauli's exclusion principle.

35.16 DENSITY OF ENERGY STATES

Let us consider a specimen of a metal. For the sake of simplicity, we consider it to have the shape of a cube with the side L. We assume that the free electrons travel absolutely freely within the volume of the specimen. The sea of electrons obeying the Pauli Exclusion Principle is called ***Fermi gas***. Since the electrons are confined inside the specimen, their wave properties limit the energy values that they may take.

The application of Schrödinger equation to the electron motion in the three directions reveals that the electron energy is quantized. The quantized value of energy is given by

$$E = \frac{h^2}{8mL^2}\left(n_x^2 + n_y^2 + n_z^2\right) \qquad ...(35.42)$$

The state of a free electron is determined by the four quantum numbers n_x, n_y, n_z and by the spin quantum number $m_s = \pm\frac{1}{2}$. Eq. (35.42) indicates that the energy of an electron is determined by the sum of the squares of the quantum numbers n_x, n_y, and n_z.

We get the same value for energy for several different combinations of these three quantum numbers. Corresponding to each set we can find a specific energy state (or energy level) E.

We can therefore represent a *quantum state* by a point in quantum number space (Fig. 35.4 *a*). Quantum number space is an imaginary space where the values of quantum numbers are denoted along the three axes respectively.

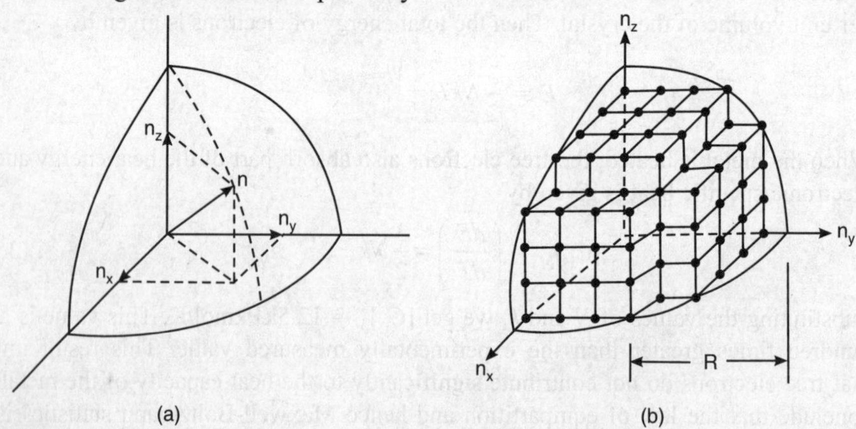

Fig. 35.4:

Representation of a *quantum state* by a point in quantum number space

In this space, a radius vector n may be drawn from the origin of the coordinate system to a point (n_x, n_y, n_z) where

$$n^2 = n_x^2 + n_y^2 + n_z^2 \quad ...(35.43)$$

It is easy to see that all points on the surface of a sphere of radius n will correspond to the same energy. All points within the sphere represent quantum states with energies smaller than E.

Number of energy states within a sphere of radius n = Volume of the sphere = $\frac{4\pi}{3}n^3$.

Since the quantum numbers can have only positive integer values, the n-values can only be defined in the positive octant of the sphere (Fig. 35.4 *b*).

Therefore, the number of quantum states with energy equal to or smaller than E is proportional to the first octant of the sphere.

Number of energy states within one octant of the sphere of radius $n = \frac{1}{8} \times \frac{4\pi}{3} n^3$ (35.44)

Similarly, the number of energy states within one octant of a sphere of radius $(n + dn)$ corresponding to energy

$$(E + dE) = \frac{1}{8}\left[\frac{4\pi}{3}(n+dn)^3\right]$$

The number of energy states having energy values between E and $(E + dE)$ is given by

$$N(E)\, dE = \frac{1}{8}\left[\frac{4\pi}{3}(n+dn)^3\right] - \frac{1}{8}\left(\frac{4\pi}{3}\right)n^3 \cong \frac{\pi}{6}(3n^2 dn) = \frac{\pi}{2}n^2 dn$$

Terms corresponding to higher powers of dn are negligibly small and are hence neglected.

$$\therefore \quad N(E)dE = \frac{\pi}{2}n^2 dn \quad ...(35.45)$$

We know that

$$E = \frac{h^2}{8mL^2}\left(n_x^2 + n_y^2 + n_z^2\right) = \frac{n^2 h^2}{8mL^2}$$

$$\therefore \quad n^2 = \frac{8mL^2}{h^2}E \quad ...(35.46)$$

or

$$n = \left[\frac{8mL^2}{h^2}\right]^{1/2} E^{1/2}$$

Differentiating the Eq. (35.46), we get

$$n\, dn = \frac{4mL^2}{h^2}dE$$

$$\therefore \quad N(E)dE = \frac{\pi}{2}n^2 dn = \frac{\pi}{2}n(n\, dn)$$

$$= \frac{\pi}{2}\left[\frac{8mL^2}{h^2}\right]^{1/2} E^{1/2} \times \frac{4mL^2}{h^2}dE$$

$$= \frac{\pi}{4}\left[\frac{8mL^2}{h^2}\right]^{3/2} E^{1/2} dE \quad ...(35.47)$$

There are two spin states $m_s = \pm\frac{1}{2}$ for an electron. According to Pauli Exclusion Principle, two electrons of opposite spin can occupy each state.

Hence, the number of *energy states* available for electron occupancy is double the above value Eq. (35.47) and equals to

$$N(E)dE = \frac{\pi}{2}\left[\frac{8mL^2}{h^2}\right]^{3/2} E^{1/2}dE$$

or $$N(E)dE = \frac{4\pi}{h^3}(2m)^{3/2} L^3 E^{1/2} dE \quad ...(35.48)$$

Density of Energy States: The *density of states* is given by the number of available electron states per unit volume per unit energy range at a certain energy level, E.

$\therefore$ $$Z(E)dE = \frac{N(E)\,dE}{L^3} = \frac{4\pi}{h^3}(2m)^{3/2} E^{1/2} dE \quad ...(35.49)$$

Where $Z(E)$ is called the **density of states function.**

It may noted that $Z(E)$ is independent of the dimensions (L) of the potential box and hence is applicable for any case.

We define $Z(E)$ *as the number of available states per unit energy interval centered around E.*

35.16.1 Energy Distribution of Electrons

Density of energy states in the energy interval E and $E + dE$ is given by

$$Z(E)\,dE = \frac{4\pi}{h^3}(2m)^{3/2} E^{1/2} dE$$

The density of states plotted against the energy gives a parabola (Fig. 35.5). The area under the curve represents the number of electrons in the metal. Note that *the number of available energy levels at the lower end of the parabola is considerably less than at higher energies.*

The plot of $Z(E)$ versus E at $T = 0$ drops abruptly to zero at $E = E_F$. This distribution of electron energies is understandable. The electrons cannot crowd in the lower energy levels since they obey Pauli exclusion principle. They start with the lowest energy level and go on occupying higher and higher levels in pairs until all of them are accommodated. The highest energy occupied is E_F.

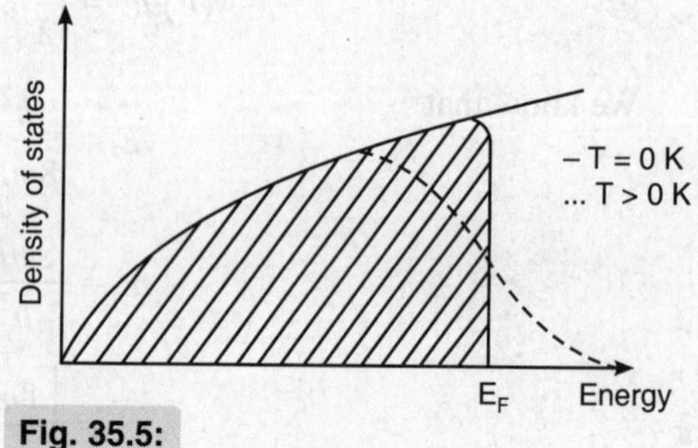

Fig. 35.5: Density of states versus the energy

The plot of $Z(E)$ versus E at a much higher temperature is also shown in the Fig. 35.5. Since at any temperature the area under the corresponding curve gives the number of electrons in the metal, the areas under the two curves, shown in Fig. 35.5, must be equal. It is seen that even for a large increase in temperature, the distribution curve changes only very slightly.

On heating the conductor, electrons are excited to higher energy levels. The curve shows that for most of the electrons lying deep in the conduction band, the thermal energy is not sufficient to cause transitions to upper levels and electrons in lower levels are left undisturbed. Only those electrons occupying the energy levels near the Fermi level are thermally excited. These levels make up a narrow band of width kT directly adjacent to the Fermi level.

Therefore, electrons having energy a little below E_F jump into levels with energy somewhat above E_F and a new energy distribution of electrons is obtained.

35.17 CARRIER CONCENTRATION IN METALS

The density of states represents the number of states that could be occupied by charge carriers. However, all the available energy states are not filled in a metal. A particular energy level E is occupied or not, is determined by the probability $f(E)$ that a charge carrier can have the energy E. Hence, the number of carriers per unit volume within a given energy range depends both on the number of available states lying in that range and on the probability that carriers acquire sufficient energy to occupy the states. Hence, the carrier concentration in energy range dE is obtained by multiplying the density of states in that range with the probability of their being occupied. Thus, at thermal equilibrium, the concentration of electrons, dn, having energy between E and $E + dE$ is given by

$$dn = f(E) \cdot Z(E)dE \qquad \text{...(35.50)}$$

The probability that an electron occupies an energy level E at thermal equilibrium is given by

$$f(E) = \frac{1}{1+\exp[(E-E_F)/kT]} \qquad \text{...(35.51)}$$

where E_F is known as **Fermi level**.

By substituting for $f(E)$ and for $N(E)$ in Eq. (35.41), we get

$$dn = \frac{4\pi}{h^3}(2m)^{3/2} \cdot \frac{E^{1/2}dE}{e^{(E-E_F)/kT}+1} \qquad \text{...(35.52)}$$

Therefore, the electron concentration, i.e., number of electrons per unit volume of the conductor is given by

$$n_C = \frac{4\pi}{h^3}(2m)^{3/2}\int_0^\infty \frac{E^{1/2}dE}{e^{(E-E_F)/kT}+1} \qquad \text{...(35.53)}$$

35.17.1 Fermi Energy, E_F

Let the specimen of metal contain N free electrons. In a conductor at absolute zero temperature, the electrons fill the available states starting from the lowest energy level. Therefore, all the levels with an energy E less than a certain value $E_F(0)$ will be filled with electrons, whereas the levels with $E > E_F(0)$ will remain vacant. The energy $E_F(0)$ is known as the **Fermi energy** and the corresponding energy level is known as **Fermi level**. The total number of free electrons is equal to the number of quantum states up to the energy E_F.

Note the distinction between energy level and energy state as used here; a *quantum state* accommodates only *one electron* whereas an *energy level* accommodates *two electrons.* If we set the number of states equal to the number of electrons in Eq. (35.48), then we find that

$$N = \int_0^{E_F} N(E)\,dE = \frac{4\pi}{h^3}(2m)^{3/2}L^3\int_0^{E_F} E^{1/2}dE$$

or

$$N = \frac{8\pi V}{3}\left(\frac{2m}{h^2}\right)^{3/2}[E_F(0)]^{3/2} \qquad \text{...(35.54)}$$

where we have denoted $L^3 = V$ and E_F at absolute zero temperature as $E_F(0)$.

$$\therefore \qquad E_F(0) = \frac{h^2}{2m}\left[\frac{3N}{8\pi V}\right]^{2/3} \qquad ...(35.55)$$

In terms of electron concentration, the above equation may be written as

$$E_F(0) = \frac{h^2}{2m}\left[\frac{3n_C}{8\pi}\right]^{2/3} \qquad ...(35.56)$$

where $n_C = \frac{N}{V}$. Eq. (35.56) shows that Fermi energy of a metal depends only on the electron concentration in the metal.

Let us estimate the value of $E_F(0)$. The concentration of conduction electrons, n_C in metals is of the order of $5 \times 10^{28}/\text{m}^3$. Therefore, $E_F(0)$ is of the order of 5 eV.

Significance of Fermi energy: Fermi energy levels are used to explain the flow of electrons when two metals are in contact, *p*-type and *n*-type semiconductors, and Seebeck Effect.

35.17.2 Variation of Fermi Energy with Temperature

Fermi energy decreases when the temperature of the metal is increased. It can be shown that when $k_T << E_F$, we obtain the following expression for the Fermi level E_F.

$$E_F \approx E_F(0)\left[1 - \frac{\pi^2}{12}\left(\frac{kT}{E_F(0)}\right)^2\right] \qquad ...(35.57)$$

From the above expression it follows that the temperature dependence of the Fermi level is very slight and for all practical purposes we assume that $E_F = E_F(0)$.

Sommerfeld theory was successful to a good extent but failed to give any basis for the classification of solids into conductors, insulators and semiconductors.

Example 35.6: *Calculate the Fermi energy of sodium at 0K assuming that it has one free electron per atom and density of sodium is 970 kg/m³ and atomic weight 23.*

Solution: $n_C = \frac{N\rho}{M} = \frac{6.02 \times 10^{26}\,/\,\text{k.mol} \times 970\text{kg/m}^3}{23} = 2.5 \times 10^{28}/\text{m}^3$

$$\therefore \qquad E_F(0) = \frac{h^2}{2m}\left[\frac{3n_C}{8\pi}\right]^{2/3} = \frac{(6.625 \times 10^{-34}\,\text{Js})^2}{2 \times 9.11 \times 10^{-31}\,\text{kg}}\left[\frac{3 \times 2.5 \times 10^{28}\,/\,\text{m}^3}{8 \times 3.142}\right]^{2/3} = 5.11 \times 10^{-19}\text{J}$$

$$= \frac{5.11 \times 10^{-19}}{1.602}\,\text{eV} = \mathbf{3.1\ eV.}$$

35.18 QUANTUM FREE ELECTRON THEORY OF ELECTRICAL CONDUCTION

It is assumed in the classical free electron theory that electrons follow Maxwell-Boltzmann distribution of velocities and energies. According to Maxwell-Boltzmann distribution many electrons can simultaneously possess the same energy (or velocity). According to quantum theory, electrons obey Pauli Exclusion Principle and hence follow Fermi-Dirac distribution. According to quantum theory, only two electrons can occupy the same energy level. Hence even at 0K, conduction (or free) electrons occupy different discrete energy levels. Electrons occupying the higher energy levels would possess higher energies. The highest energy level is called the *Fermi energy* level, E_F. Therefore, the largest energy which the electrons can have in a metal at 0K is the Fermi energy, E_F. A large number of electrons possess this energy since the density of states is highest around E_F.

Each conduction electron occupies a particular energy state and possess a particular momentum, $p = \hbar k$. If the momentum of the electron is plotted in *k*-space, we obtain a point (Fig. 35.6). Similarly, if we plot the momenta of conduction electrons in *k*-space, we obtain points. The points correspond to the tips of the *k*-vectors. In the absence of the electric field, the electron momenta are random and the momentum surface will have a spherical symmetry at any given temperature. At 0K, all the levels from the lowest level to the highest level would be occupied, with the highest level being the Fermi level, E_F. The outermost boundary containing all possible momenta will be a sphere, of radius, k_F (Fig. 35.6). Such a sphere is called a **Fermi surface**. In two dimensions, the Fermi surface is a circle (Fig. 35.6). All points inside the Fermi sphere are occupied. Thus, Fermi surface separates occupied states from unoccupied states. For every occupied state *k* there is an occupied state (–*k*). As a consequence, the net momentum of the electrons is zero. Therefore, there is no current flow across the material.

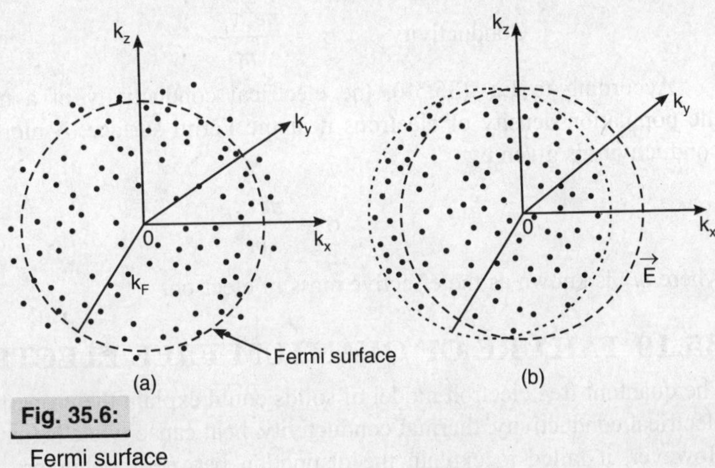

Fig. 35.6: Fermi surface

When an electric field E is applied, electrons experience a force F, which is given by

$$F = \frac{dp}{dt} = -eE$$

or

$$\hbar \frac{dk}{dt} = -eE \qquad ...(35.58)$$

Eq. (35.58) represents the equation of motion of electron in the presence of applied field. Under the action of the electric field, electrons near the Fermi surface move into vacant levels adjacent to them. However, the electrons deep in the lower levels cannot move as all the energy levels lying above them are occupied. Since the electric field acts on all of the electrons, the equilibrium is disturbed for a moment. Some of the electrons undergo collisions while others are accelerated by the field. Soon, the system returns to equilibrium. In this process, the Fermi sphere gets displaced slightly opposite to the field direction. The majority of the electron velocities cancel each other pair wise. However, some electrons remain uncompensated and cause the observed current.

Integrating Eq. (35.58), we get

$$k(t) - k(0) = -\frac{eEt}{\hbar}$$

It means that the center of the Fermi sphere moves in a time $t = \tau_F$ to a center at

$$\Delta k = -\frac{eE\tau_F}{\hbar}$$

If n is the electron density per unit volume near the Fermi surface, the steady state current density would be given by

$$J = n(-e)\upsilon = -\frac{ne\hbar\Delta k}{m} = \sigma E \qquad \left[\because \upsilon = \frac{p}{m}\right]$$

$$\frac{ne\hbar}{m}\left[\frac{eE\,\tau_F}{\hbar}\right] = \sigma E$$

$$\frac{ne^2\tau_F}{m}E = \sigma E$$

$$\therefore \quad \text{Conductivity} \quad \sigma = \frac{ne^2\tau_F}{m} \qquad ...(35.59)$$

According to Eq. (35.59), the electrical conductivity of a metal depends largely on the population density of electrons near the Fermi surface. A more rigorous expression for conductivity is given by

$$\sigma = \frac{ne^2\tau_F}{m^*} \qquad ...(35.60)$$

where m^* is known as the effective mass of electron.

35.19 FAILURE OF QUANTUM FREE ELECTRON THEORY

The quantum free electron model of solids could explain the properties of conductors such as electrical conductivity, thermal conductivity, heat capacity etc better than the classical theory. However, it failed to explain the distinction between conductors, insulators and semiconductors. Occurrence of positive Hall coefficients in case of some metals like zinc could not be accounted on the basis of this model.

Quantum mechanical calculations are to be applied to determine the behaviour of an electron in solid. The solution of the Schrodinger equation in a periodic lattice shows that an electron moving in a crystal have a large number of possible energy values, and these values are distributed into *energy bands*, each band consisting of a large number of closely spaced energy levels. The band theory successfully explained the classification of solids into conductors, insulators and semiconductors.

QUESTIONS

1. A rectangular block of a solid is connected to a dc voltage source. Obtain the expression for the: (*i*) current density flowing through the block and (*ii*) conductivity of the material in terms of the concentration of carriers in it.
2. Define mean free path of electrons.
3. Explain the terms: (*i*) Drift velocity and (*ii*) Carrier mobility.
4. What are bound and free electrons?
5. Discuss the important postulates of free electron theory of metals.
6. Define the terms mobility and relaxation time of free electrons in a metal.

7. Using the free electron model derive an expression for electrical conductivity in metals.
8. How does classical free electron theory lead to Ohm's law?
9. Based on free electron theory, derive an expression for electrical conductivity of metals. How does electrical resistance change with impurity and temperature?
10. State Mathiessien's rule and give an account of the nature of total resistivity both at high and low temperatures.
11. What are the special features of classical free electron theory of metals?
12. Derive an expression for the electrical conductivity of a metal.
13. Based on Drude-Lorentz theory, derive the expression for electrical conductivity and assuming the classical expression for thermal conductivity derive Wiedemann-Franz law.
14. State and derive Wiedemann-Franz law.
15. What is meant by Lorentz number?
16. Write any two drawbacks of the classical free electron theory of metals.

PROBLEMS

1. The density of silver is 10.5×10^3 kg/m^3. The atomic weight of silver is 107.9. Assuming that each silver atom provides one conduction electron (*i*) calculate the density of free electrons. The conductivity of silver at 20°C is 6.8×10^7 ohm^{-1}m^{-1}. (*ii*) Calculate the electron mobility in silver.

   [Ans: 5.86×10^{28}; 7.25×10^{-3} m^2/V.s]

36 Band Theory of Solids

Learning Objectives:

By the end of this chapter, you will be able to:

- Discuss about Band Theory of Solids.
- Learn about Energy Band structure of Solid.
- Understand concept of Energy Band Diagram for different Solids.
- Explain Energy Band Structure of Constructor, Insulator and Semi-Conductor.

36.1 INTRODUCTION

X- ray diffraction studies show that a solid is an ordered structure. In the solid atoms occupy the lattice sites and the spacing between the atoms is of the same order as that of the linear dimensions of atoms. Therefore, atoms in a solid interact strongly and set up an internal electric field, which is *periodic* in nature. The periodic electric field affects the motion of free electrons. Electron is not an ordinary particle but possesses a wave character also. The application of quantum mechanics to the motion of electrons in solid shows that the allowed values of electron energy are distributed into bands, each band consisting of a sequence of closely spaced energy levels arranged in a manner akin to the steps of a ladder. In 1928 Felix Bloch developed **zone theory** for the electrons moving in a periodic field provided by a crystal lattice. This theory is popularly known as the **band theory of solids**. Knowledge of the formation of energy bands and the consequent restrictions imposed on electron motion in a solid are obtained from the band theory. Such considerations led to the invention of a gamut of solid state devices, which has revolutionized the field of electronics leading to miniaturization, micro-miniaturization and mass production of devices and systems.

36.2 THE BAND THEORY OF SOLIDS-A qualitative explanation

A solid may be imagined as formed by allowing initially free atoms to gradually approach one another. As long as the atoms are widely separated their interactions are negligible. Every atom has the *same* energy-level diagram. The energy-level diagram for the entire system of N atoms resembles the energy-level diagram for a single atom; now each state of the system can be occupied by N electrons instead of just one. As the atoms come together to create a close packed periodic structure, they interact strongly due to their proximity to each other. By **interaction** we mean that the positive nucleus of one atom attracts the electrons and repels the nucleus of the adjacent atom and vice versa. As a result, instead of one energy level the same for all N isolated atoms, there arise N closely spaced separate levels, which fall into groups. The energy levels are so closely spaced in the group that they form a virtual continuum, which is called an **energy band**. The formation of energy bands can be qualitatively understood using molecular orbital concept.

Let us consider an imaginary situation where N hydrogen atoms approach each other to form solid hydrogen. Each hydrogen atom is characterized by one electron residing at 1s energy level corresponding to an atomic orbital ψ. As long as the separation r of two atoms A and B is much larger than the size d of the atoms ($r >> d$), the atoms do not interact and the atomic orbitals ψ_A and ψ_B are not affected (see Fig. 36.1 *a*). The two atoms have identical energy levels marked E in Fig. 36.2. When the two atoms come into contact ($r = d$), the orbitals ψ_A and ψ_B overlap and form molecular orbitals, where each electron may be said to orbit around both nuclei. A linear combination of these two atomic orbitals yields two types molecular orbitals, namely

$$\psi_+ = C_1\psi_A + C_2\psi_B \qquad \text{and} \qquad \psi_- = C_3\psi_A - C_4\psi_B$$

where ψ_+ and ψ_- denote molecular orbitals and C_1, C_2, C_3 and C_4 are constants.

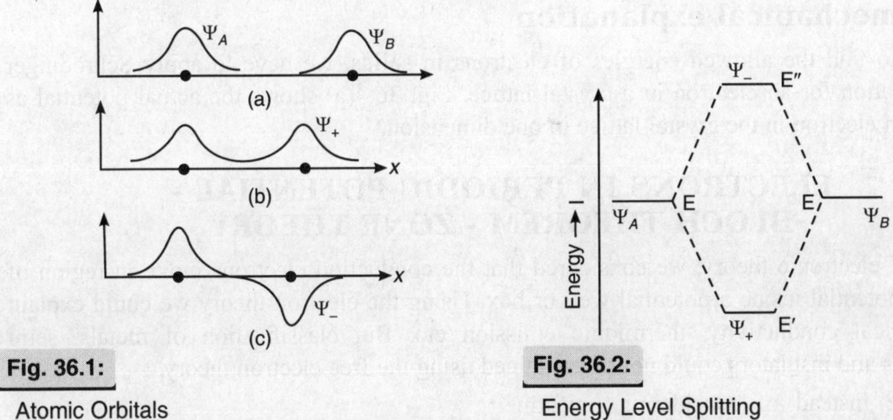

Fig. 36.1: Atomic Orbitals

Fig. 36.2: Energy Level Splitting

The molecular orbital ψ_+ is called the **bonding orbital** and has a lower energy E' ($E' < E$). The molecular orbital ψ_- is called the **antibonding orbital** and has a higher energy E''($E'' > E$). Thus, the combination of two atomic orbitals results in two molecular orbitals, which extend over both the atoms. The first important consequence of the molecular orbital formation is that the individual valence electrons are not localized to their original atomic positions and they belong to both the atoms. The second important result is that one of the molecular orbitals (bonding orbital) is of lower energy than either of the individual atomic orbitals and the other molecular orbital (antibonding orbital) is of higher energy. It means that the original energy level E of each electron is split into two energy levels (see Fig. 36.2). *The transformation of a single energy level into two or more separate energy levels is known as* **energy level splitting**.

When two atoms come close, one energy level splits into two energy levels (Fig. 36.3 b). When three atoms approach each other closely, the original level splits into three levels; four atoms produce four levels and so on (Fig. 36.3 c). In general, N interacting atoms cause a particular energy level to split into N levels. The group of energy levels resulting from splitting is so closely spaced that it is called an *energy band* (see Fig. 36.3 d).

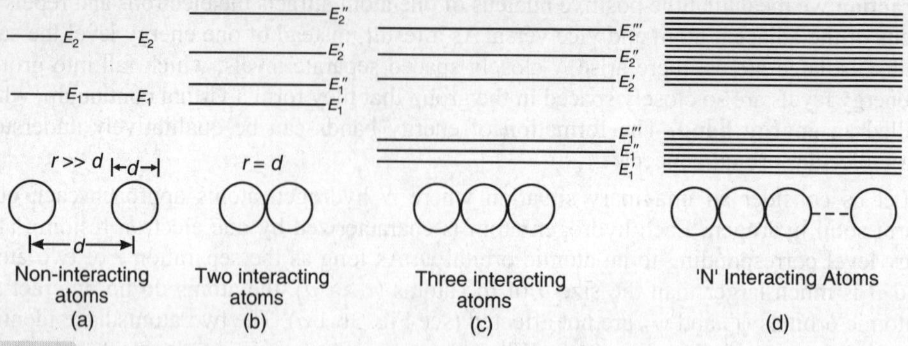

Fig. 36.3: Energy level splitting and band formation

The individual valence electrons no longer belong to individual atoms; but they now belong to all nuclei in the solid.

36.3 THE BAND THEORY OF SOLIDS—Quantum mechanical explanation

In order to find the allowed energies of electrons in solids, we have to apply Schrödinger wave equation for an electron in a crystal lattice. Fig. 36.4(a) shows the actual potential as seen by an electron in the crystal lattice in one dimension.

ELECTRONS IN PERIODIC POTENTIAL - BLOCH THEOREM - ZONE THEORY

In case of electronic theory, we considered that the conduction electrons move in region of constant potential inside a potential well or box. Using the electron theory we could explain the electrical conductivity, thermionic emission etc. But classification of metals, semi conductors and insulators could not be explained using the free electron theory.

Hence instead of considering uniform constant potential as we have done in free electron theory, we have to consider the variation of potential inside the metallic crystal with the periodicity of the lattice.

Bloch stated the Zone theory in 1928. According to this theory, the free electrons move in a periodic field provided by the lattice. This theory is also called Band theory of solids.

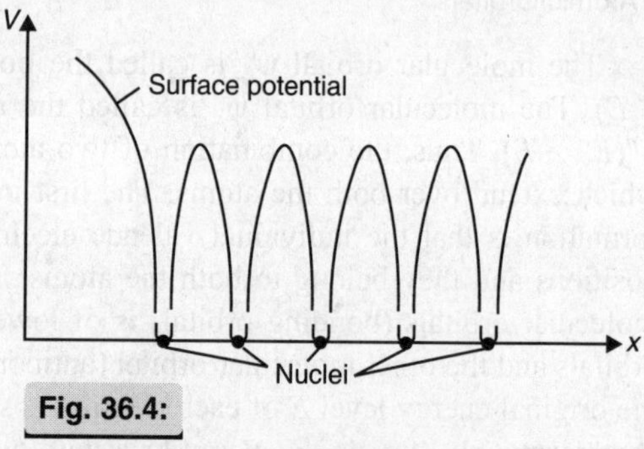

Fig. 36.4: Density of states versus the energy

The potential is minimum at the positive ions sites and maximum between the two ions as shown in the Fig. 36.4.

Three dimensional Schrödinger's equation corresponding to this can be

$$\frac{d^2\psi}{dx^2} + \frac{8\pi^2}{h^2}\left(E - V(x)\right) = 0$$

The periodic potential $V(x)$ may be defined by means of lattice constant a as

$$V(x) - V(x + a) \quad ...(36.1)$$

Bloch has shown that one dimensional solution of Schrödinger equation is of the form

$$\psi(x) = e^{(ikx)} U_k(x) \quad ...(36.2)$$

In the three dimensional form

$$\psi(x) = e^{(ikr)} U_k(r) \quad ...(36.3)$$

Eq. (36.2) and Eq. (36.3) are Bloch functions in 1D and 3D. $U_k(x)$ and $U_k(r)$ are periodic with the periodicity of the crystal lattice. The free electron wave is modulated by periodic functions $U_k(x)$ and $U_k(r)$.

Let us now consider a linear chain of atoms of length L in one dimensional case with N number of atoms in the chain.

Then

$$U_k(x) = U_k(x + Na) \quad ...(36.4)$$

where a is a lattice constant.

Eq. (36.2) and Eq. (36.4) we have

$$\psi_k(x + Na) = U_k(x + Na) \exp [ik(x + Na)]$$

$$\exp [ik(xNa)]\ U_k(x) \exp (ikx)$$

$$\psi_k(x + Na) = \psi_k \exp (ikNa) \quad ...(36.5)$$

Eq. (36.5) is referenced to as Bloch condition or Bloch Theorem.

By taking complex conjugate of Eq. (36.5) we understand that electron is not localized around any particular atoms and the probability of finding the electron is same throughout the crystal.

Hence in Eq. (36.5), $\exp (ikNa) = 1$

$$k = (2n\pi)/L$$

where L is the length of the chain of the atoms and $n = 0, \pm 1, \pm 2, \pm 3, \ldots\ldots$

When $n = N/2$ we have $k = \pi/2$. This is the edge of the first Billoullin zone.

Quantum mechanical explanation of Band theory of solids-Kronig-Penny model

In order to find the allowed energies of the electrons in solids, we have to apply Schrodinger wave equation for an electron in crystal lattice. According to the Zone theory, the potential of the solid varies periodically with the *periodicity* of *space lattice a*, which is lattice constant-inter atomic spacing. It is assumed that the potential energy of the electron is zero near the nucleus of the positive ion in the lattice and maximum when it is halfway between the adjacent nuclei.

The Kronig-Penny Model: Kronig and Penny suggested a simplified model consisting of an infinite row of rectangular potential wells separated by barriers of width b. This one-dimensional representation of periodic lattice is known as *Kronig-Penny model* and is shown in Fig. 36.5. Each well has a width b and a depth V_o.

The period of the potential is $(a + b)$. In regions where $0 < x < a$, the potential energy is assumed to be zero and in regions such as $-b < x < 0$, it is V_o.

Through this model, Schrodinger equation can be solved explicitly in terms of elementary functions.

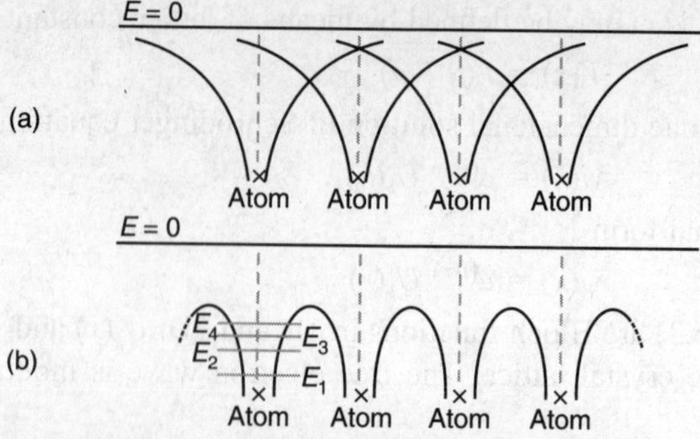

(a) Overlapping Energy functions of adjacent atoms (b) Net energy function of a one-dimensional single crystal.

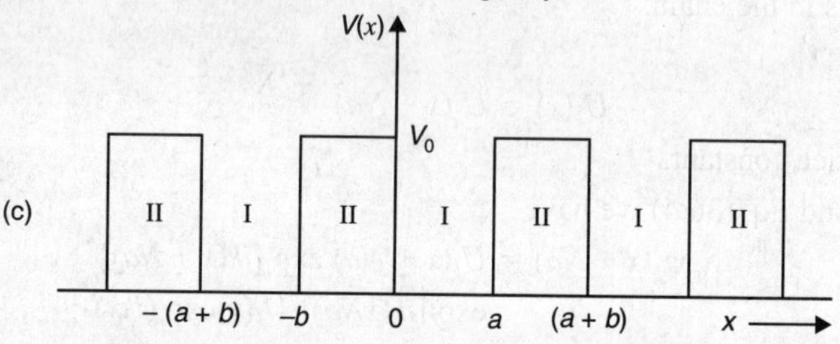

Fig. 36.5: (c) 1D periodic potential function of Kronig-Penny Model

Kronig-Penny Model

The Schrodinger equation for the two regions can be written as

$$\frac{d^2\psi}{dx^2} + \frac{8\pi^2 m}{h^2} E\psi = 0 \qquad \text{for } 0 < x < a \qquad \text{...(36.6)}$$

and

$$\frac{d^2\psi}{dx^2} + \frac{8\pi^2 m}{h^2} (E - V_0)\psi = 0 \qquad \text{for } -b < x < 0 \qquad \text{...(36.7)}$$

We rewrite the above equation as

$$\frac{d^2\psi}{dx^2} + \alpha^2 \psi = 0 \qquad \text{for } 0 < x < a \qquad \text{...(36.8)}$$

and

$$\frac{d^2\psi}{dx^2} + -\beta^2 \psi = 0 \qquad \text{for } -b < x < 0 \qquad \text{...(36.9)}$$

where

$$\alpha^2 = \frac{8\pi^2 m}{h^2}$$

$$\beta^2 = \frac{8\pi^2 m}{h^2} (V_0 - E)$$

Eq. (36.8) and Eq. (36.9) is known as Bloch theorem and according to which the solution of the Schrodinger equation for a periodic potential would be of the form of a plane wave modulated with the periodicity of the lattice.

It means that the solution can be represented as the product of two functions: a free particle wave function and a periodic function $u(x)$ that has the same period as the lattice. Thus

$$\psi(x) = u(x)e^{ikx} \quad \text{...(36.10)}$$

with $$u(x) = u(x + a) \quad \text{...(36.11)}$$

The wave functions of the above type are known as **Bloch functions** which change periodically with increasing x.

36.3.1 Energy Bands

We substitute the above wave functions into the Schrodinger equation and solve it in the usual way. When we apply the periodic boundary condition, we get the following expression.

$$\frac{maV_0b}{\hbar^2} \cdot \frac{\sin \alpha a}{a} + \cos \alpha a = \cos k a \quad \text{...(36.12)}$$

where $$\alpha = \frac{\sqrt{2mE}}{\hbar}$$

Eq. (36.12) provides the allowed solutions to the Schrodinger equation. As the relation involves trigonometric functions, only certain values of α are possible. The right hand side of equation (36.12) is cosine function and can take values only between –1 and +1. Therefore, the left-hand side of the equation is restricted to vary between those two limits. Hence, only certain values of α are allowed. It means that energy E is restricted to lie within certain ranges.

This concept is best understood by drawing the plot of energy as a function of the wave number, k. The plot is shown in Fig. 36.6. The parabolic relation between E and k obtained in case of free electron is interrupted at certain values of k, as shown by the broken curve.

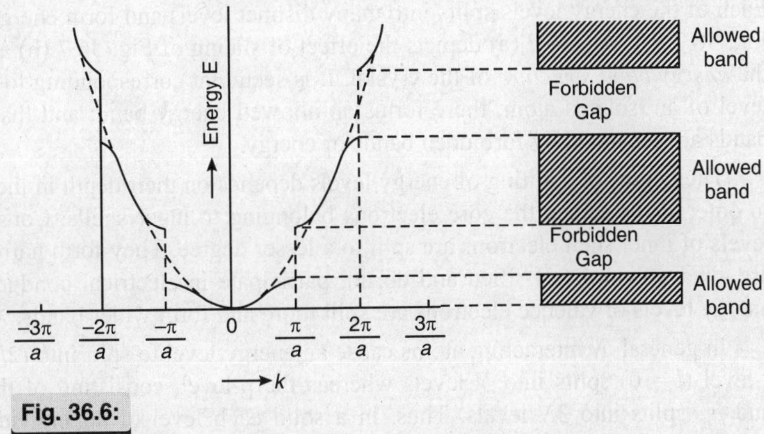

Fig. 36.6: Electron energy, E versus wave number, k plot for a solid

Fig. 36.6 shows discontinuities in E. The discontinuities occur at $ka = \pm n\pi$ i.e., at

$$k = \pm\frac{\pi}{a}, \pm\frac{2\pi}{a}, \pm\frac{3\pi}{a}, ... \quad \text{...(36.13)}$$

The origin of the allowed energy bands and forbidden gaps is clear from Fig. 36.6.

36.4 ENERGY BAND STRUCTURE OF A SOLID

A crystal (i.e., solid) consists of an enormous number of atoms arranged in a regular periodic structure. The extent of energy level splitting in the solid depends on the nearness of atoms in it. Let us assume that N identical atoms form the crystal. The energy levels of the isolated atoms are shown in Fig. 36.7 (c). All the N atoms have identical sets of energy levels. The electrons fill the energy levels in each atom independently. Fig. 36.7 (b) shows an atom sitting at the origin of the coordinate system. Now let us imagine that other atoms approach this atom along the three axial directions and halt at the distance a_0, which is the lattice constant of the crystal.

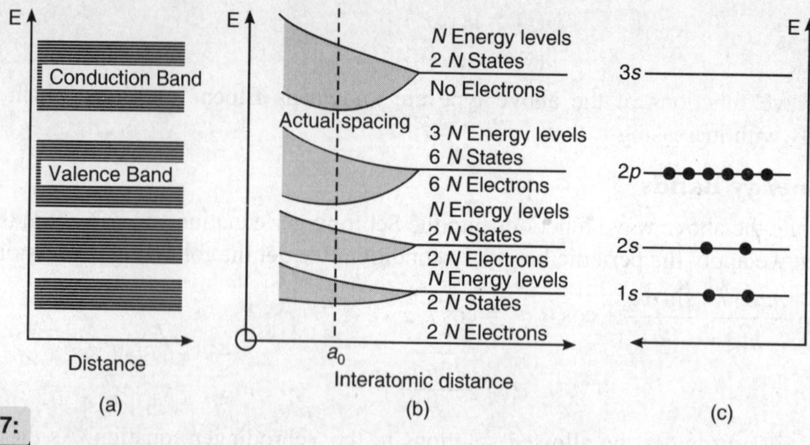

Fig. 36.7: Energy level splitting in a solid as a function of interatomic distance (a) Energy band structure of the solid corresponding to the actual spacing of atoms in the solid. (b) Energy level splitting as a function of distance. (c) Discrete energy levels in an isolated atom

As the atoms approach, a continuously increasing interaction occurs between the atoms. Each of the energy levels splits into many distinct levels and form energy bands, as shown in Fig. 36.7 (b). Fig. 36.7 (a) depicts the effect of slicing of Fig. 36.7 (b) at a_0 and it represents the *energy band structure* of the crystal. It is seen that corresponding to each allowed energy level of an isolated atom, there forms an allowed energy band; and that the allowed energy bands are separated by forbidden bands of energy.

The degree of splitting of energy levels depends on their depth in the atom. The electrons in outer shells screen the core electrons belonging to inner shells. Consequently, the energy levels of inner shell electrons are split to a lesser degree. They form narrow core bands. They are always completely filled and do not participate in electrical conduction. In contrast the energy levels of valence electrons are split more and form wider bands.

In general, N interacting atoms cause an energy level to split into $(2l + 1)\,N$ levels. Thus, s-level ($l = 0$) splits into N levels whereas the p-level, consisting of three sublevels p_x, p_y and p_z, splits into $3N$ levels. Thus, in a solid each level of an individual atom splits into $(2l + 1)\,N$ number of levels where N is the number of atoms in the system. Consequently, the maximum electron capacity of an s-band is $2N$ electrons whereas the capacity of a p-band is $6N$ electrons.

While occupying an energy band, electrons start from the lowest energy level in the band and fill the levels one after the other in the ascending order of energy. When $2N$ electrons occupy the N levels available in the band, the band is said to be **completely filled**. In case of non-availability of $2N$ electrons, the energy band gets **partially filled**. When there are no electrons to occupy the levels, the energy band remains **vacant**.

The width of an allowed or forbidden energy band is generally of the order of a few electron-volts. As N is very large, the energy separation ΔE between successive energy levels in an allowed band is very small and is of the order of 10^{-27}eV. At room temperature, the kinetic energy of the electronis of the order of kT (≈ 0.026 eV) which is very large compared to the energy level separation in an allowed band. Consequently, electrons can easily move into higher vacant levels within the allowed energy band either due to thermal energy or due to a small externally applied electric field. On the other hand, electrons cannot jump across a forbidden band under normal thermal energy possessed by them or due to an applied electric field. High temperatures are required to cause inter-band electron transitions.

36.5 ELECTRICAL CONDUCTION FROM THE VIEW POINT OF BAND THEORY

In a solid, the allowed values of electron energy are distributed into bands [Fig. 36.7 (a)], each band consisting of a sequence of closely spaced discrete energy levels arranged in a manner akin to the rungs of a ladder. The electrons are distributed in the energy levels according to the Pauli Exclusion Principle. The motion of an electron corresponds to its transition from a lower energy level to an upper vacant energy level. This implies that the following two conditions are to be fulfilled for electrical conduction to take place in a solid:

(*i*) There should be free electrons available in the solid.

(*ii*) Vacant energy levels should be available immediately above the levels occupied by free electrons.

If a band has vacant energy levels but is devoid of electrons, there would be no carriers to move through the vacant levels when energy is supplied to the solid from a source such as a battery. Hence current does not flow through the solid.

On the other hand, if all the energy levels within a band were completely occupied by electrons, there would be no energy level to which an electron can jump. Therefore, even though the energy is supplied to the solid from a source such as a battery, the electrons cannot acquire energy and electrical conduction cannot occur in the solid.

If an energy band is partially filled, then the electrons will have vacant upper energy levels into which they can jump. On acquiring energy from the electric field applied across the solid, the electrons move into successive upper energy levels and cause electrical conduction. Thus, *partially filled energy band is required for electrical conduction* in a solid.

36.6 ENERGY BAND DIAGRAM

An energy band diagram is a graphic representation of the energy levels associated with top energy band and the next lower energy band in a solid. The energy band diagram shows two bands with a gap in between (see Fig. 36.8). The upper band is called the **conduction band** and the lower energy band is called the **valence band**. These two bands are separated by a forbidden gap. This energy gap is more popularly called **band gap** and is denoted by the symbol E_g. The conduction band corresponds to the energy values of *free electrons* that have broken their valence bonds, and hence have become free to move in the crystal. The bottom of the conduction band represents the smallest energy that the electron must possess to become free. Only the free electrons can move in the crystal under the influence of the externally applied electric field. Hence, these electrons are called *conduction electrons* and the energies of such electrons constitute the **conduction band**. The band showing the energy values of *valence electrons* that are engaged in covalent bonding is called the **valence band**.

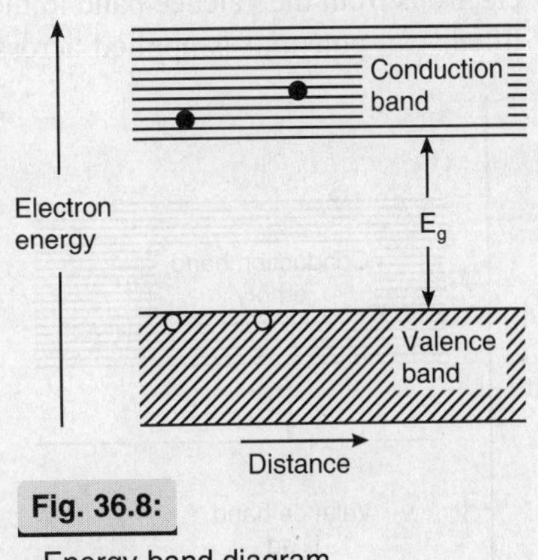

Fig. 36.8: Energy band diagram

36.7 CLASSIFICATION OF SOLIDS

The concept of energy bands helps us in understanding the division of solids into three groups. The nature of the energy bands determines whether the solid is an electrical conductor or insulator. According to the band theory, the electrical conductivity of a solid is characterized by the energy gap E_g separating the outermost energy bands namely, the valence band and the conduction band. The electrical conductivity of a material depends on magnitude of the energy gap E_g.

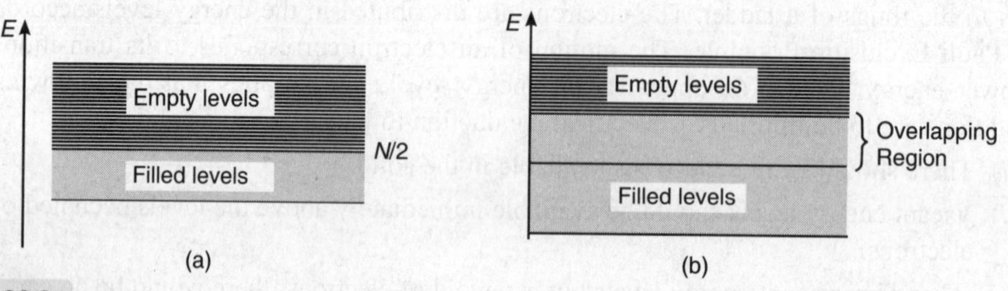

Fig. 36.9: Energy band formation in a conductor (a) half filled conduction band, (b) empty upper band overlaps on a totally filled lower band

In some solids, an upper vacant band overlaps the valence band or the valence band itself is half-filled, as shown in Fig. 36.9. It means that electrons in the valence band have easy access to levels in the upper vacant band. For this reason, very large numbers of electrons are available for conduction, even at extremely low temperatures. When electric field is impressed across the solid, electrons readily jump into upper unoccupied energy levels of the vacant band and current flows in a large measure in the solid. Therefore, these solids exhibit good electrical conductivity and are called **conductors**.

In some solids the band gap is narrow and of the order of 2 eV or less, as shown in Fig. 36.10. Acquisition of small amounts of energy from the vibrations of atom can raise electrons from the valence band to the conduction band. The conduction band is then partially filled. If a potential is applied across the material, it causes the electrons in the conduction

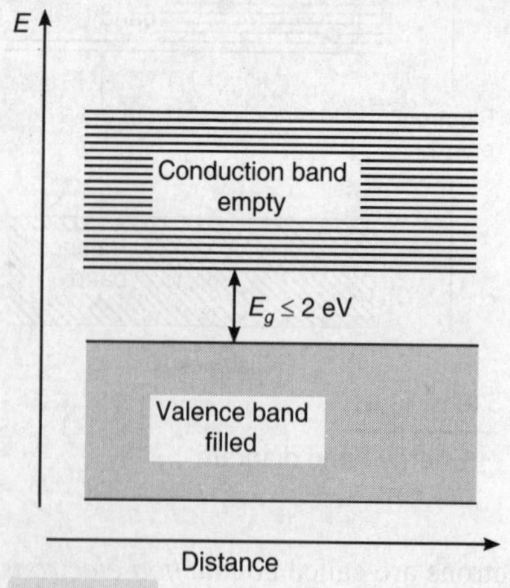

Fig. 36.10: Energy band structure of a semiconductor

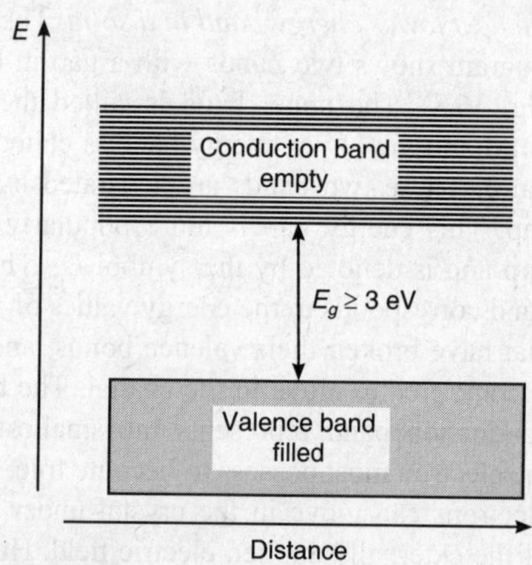

Fig. 36.11: Energy band structure of an insulator

band to move to upper levels. As a result, current flows in a modest measure in the solid. Such solids are called ***semiconductors***.

Some solids (Fig. 36.11) have band gaps that are very wide (Eg> 3 eV). It would require the acquisition of very large amounts of energy to cause an electron to jump from the valence band to the conduction band. A small number of electrons can acquire such large amount of energy to jump from valence band to conduction band. Therefore, very few electrons would be present in the conduction band. On application of external voltage across the solid, negligible current flows through it. These solids are called **insulators**. (Insulators show very low electrical conductivity.)

36.8 ENERGY BAND DIAGRAMS FOR SOME TYPICAL SOLIDS

(a) Lithium: Let us consider the element lithium belonging to Group I in the periodic table. The electron configuration of lithium atom is $1s^2$ $2s^1$. The $1s$ shell is closed and there is only one electron at the $2s$ level. In solid lithium, $1s$ and $2s$ bands form corresponding to the $1s$ and $2s$ levels, as illustrated in Fig. 36.12.

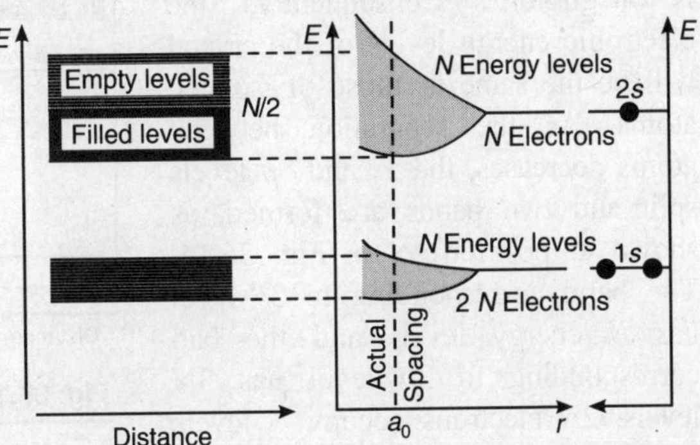

Fig. 36.12: Energy level splitting and energy band configuration in lithium solid showing half-filled 2s band

Both $1s$ and $2s$ bands have N levels each. The $1s$ band is completely filled as 2N electrons occupy N energy levels whereas the $2s$ band is half-filled because the N available electrons fill N/2 lower levels in the band leaving the upper N/2 levels vacant. In general, the solids of Group-I elements form such half-filled energy bands at the top and therefore belong to the group of conductors.

(b) Beryllium: Let us next consider the case of alkaline earth elements of Group-II. The first element in this group is beryllium. Its electron configuration is $1s^2$ $2s^2$. From the electron configuration, it is expected that beryllium solid would be an insulator. However, it is known to be a conductor. The reason is that the upper vacant $2p$ band overlaps the lower completely filled $2s$ band leading to the formation of a partially filled hybrid band, as shown in Fig. 36.13.

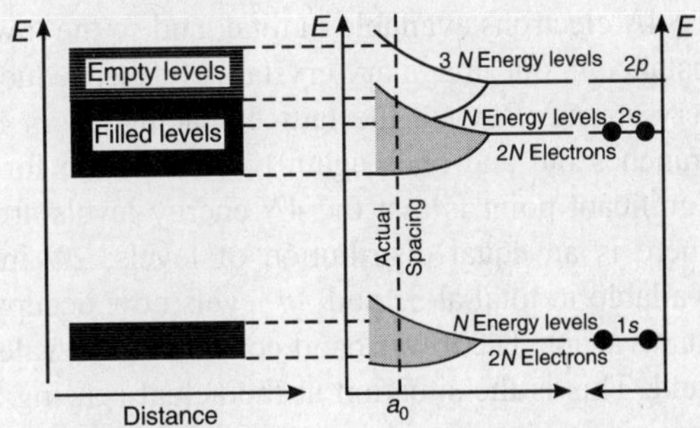

Fig. 36.13: Energy level splitting and energy band configuration in beryllium solid showing overlapping of completely filled $2s$ band and vacant $2p$ band

In general, solids of Group-II elements exhibit such partially filled bands, and therefore belong to the conductors group.

(c) Silicon: Silicon belongs to Group IV elements in the periodic table. The electron configuration of silicon atom is $1s^2\ 2s^2\ 2p^6\ 3s^2\ 3p^2$. It is seen that the inner K and L shells are closed and the corresponding bands would be completely filled. In the outer subshells 3*s* and 3*p*, 3*s*-subshell is closed. The 3*p* sub-shell is partially filled. Hence it is expected to behave as a good conductor. But because of formation of a hybrid band, which later branches out, the Si solid behaves as a semiconductor.

In the crystal formation process, when the atoms are very far apart, as at position '*d*' in Fig. 36.14, the electrostatic interaction among them is negligible. Consequently, the electronic energy levels of the crystal will be the same as those of isolated atoms. As the separation between atoms decreases, the 3*s* and 3*p* levels split and two bands are formed, as shown at position '*c*' in Fig. 36.14. The band corresponding to 3*s* level has *N* energy levels and the bad corresponding to 3*p* level has 3N levels. 2*N* electrons occupy *N* levels in 3s-band and 2*N* electrons occupy *N* levels in 3*p*-band. It may now be noted that there is an energy gap between the two bands. The energy gap is seen to decrease with the decrease in atomic spacing. At position '*b*' in Fig. 36.12 the two bands merge and form a composite band. The 3N upper levels merge with *N* lower levels, giving rise to a total of 4*N* levels. These levels have to be occupied by the 4*N* electrons available in total, and so the lowermost 2*N* levels are filled. When the atomic distance in our imaginary crystal is further reduced, the interaction among the atoms becomes very strong. Beyond the lattice spacing '*b*' in Fig. 36.14, we find that the composite band branches out and once again two bands are formed, separated by a forbidden gap, E_g. The significant point is that the 4*N* energy levels are equally divided between the two branches. There is an equal distribution of levels, 2*N* in each, in the two bands. The 4*N* electrons available in total at 3*s* and 3*p* levels now occupy the lower energy band and leave the upper band vacant. The lower band constitutes the valence band and the upper band the conduction band. This is the situation at the actual spacing 'a_o' in the silicon crystal. At position ao the two bands are not widely separated from each other. The value of E_g at 0 K is 1.12 eV. At normal temperatures, a significant number of electrons will be thermally excited from valence band to conduction band. The electrons excited to conduction band respond to the external voltage and produce a modest flow of current. Thus, Si behaves as a semiconductor.

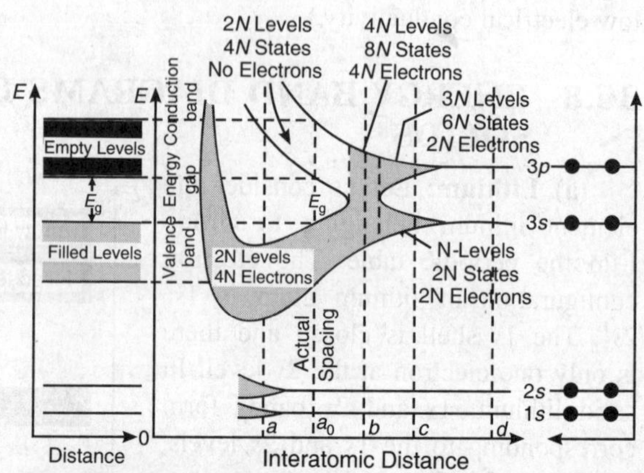

Fig. 36.14: Energy level splitting and band formation in crystals of Group IVA elements

(d) Diamond: It is evident from Fig.36.14 that the energy gap between the two branches goes on increasing with decreasing atomic distance. At the interatomic distance corresponding to line at '*a*' in Fig. 36.14, the distance between the two bands becomes considerably large. In case of diamond the two bands are separated by 5.47 eV. Even at high temperatures, the thermal energy would be insufficient to excite enough number of electrons to the conduction band. Because of the non-availability of electrons in the conduction band electrical conduction cannot take place in the material and hence diamond behaves as insulator.

36.9 ENERGY BAND STRUCTURE OF A CONDUCTOR

Conductors are characterized by a single energy band, namely conduction band which contains more energy levels than free electrons. At the temperature 0K, electrons occupy the lower energy levels in the conduction band up to a certain energy level called the Fermi level E_F.

36.9.1 Fermi-Dirac Distribution Function

We are next interested in knowing how electrons are distributed among the various energy levels in the conduction band at a given temperature. We cannot apply Maxwell-Boltzmann distribution to electrons because (*i*) they obey *exclusion principle* and (*ii*) they are *indistinguishable* particles. The statistical distribution function applicable to quantum particles is the *Fermi–Dirac distribution* function.

The probability that an electron occupies an energy level E at thermal equilibrium is given by

$$f(E) = \frac{1}{1 + \exp[E - E_F)/kT]}$$

where E_F is known as **Fermi level.**

In general E_F may or may not correspond to an energy level but it provides a reference with which other energies can be compared.The function $f(E)$ is known as ***Fermi factor***.

The above equation is known as *Fermi-Dirac equation* or *Fermi-Dirac distribution function*. Note that the probability of the electron to occupy the energy level E increases with temperature. We first discuss about the distribution function and the related topics with reference to conductors. We shall find later that these concepts are equally applicable to other cases.

Example 36.1: *Evaluate the Fermi function for energy kT above the Fermi energy.*

Solution: The Fermi function is given by $f(E) = \dfrac{1}{1 + e^{(E-E_F)/kT}}$

If $(E - E_F) = kT$, then $f(E) = \dfrac{1}{1 + e^{(E-E_F)/kT}} = \dfrac{1}{1 + e^1} = \dfrac{1}{1 + 2.78} = \dfrac{1}{3.78} = \mathbf{0.269.}$

Example 36.2: *In a solid, consider the energy level lying 0.01 eV below Fermi level. What is the probability of this level not being occupied by an electron?*

Solution: $(E_F - E) = [E_F - (E_F - 0.01)] = 0.01$ eV and $kT = 0.026$ eV at T = 300 K

The probability of an energy level E not being occupied by an electron is given by $[1 - f(E)]$.

$$[1 - f(E)] = 1 - \frac{1}{1 + e^{(E-E_F)/kT}} = \frac{1}{1 + e^{(E_F-E)/kT}} = \frac{1}{1 + e^{0.01eV/0.026eV}} = \frac{1}{1 + e^{0.385}}$$

$$= \frac{1}{1 + 1.47} = \mathbf{0.405}$$

36.9.2 Fermi Level

The occupancy of the energy levels by electrons in conductors is described by the Fermi-Dirac distribution function.

$$f(E) = \frac{1}{1 + \exp[(E - E_F)/kT]} \quad (36.14)$$

We distinguish two situations - one at absolute zero and the other at higher temperatures.

Case 1: At T = 0K

At absolute zero, electrons occupy energy levels in pairs starting from the bottom of the band up to an upper level designated as E_F. **Fermi level** can be, therefore, defined as the uppermost filled energy level in a conductor at 0 K. Correspondingly, **Fermi energy** is defined as maximum energy that a free electron can have in a conductor at 0 K. The energy band and the Fermi function at 0 K are shown in Fig. 36.15.

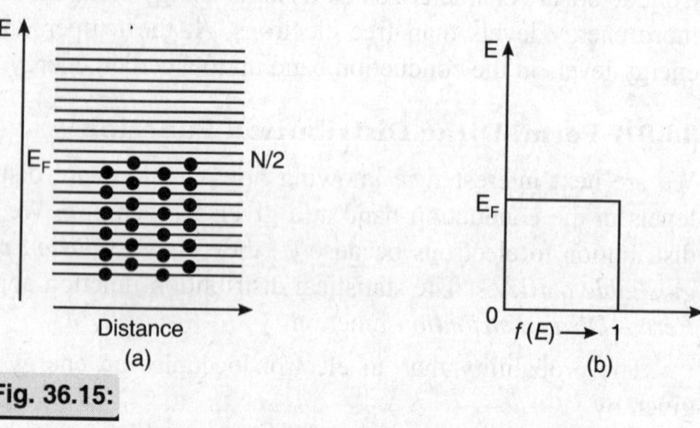

Fig. 36.15: (a) Energy band and (b) Fermi function in a conductor

Let us now apply Eq. (36.14) to the solid taking the value of T as 0K.

(*i*) For energy levels E lying below E_F, $E < E_F$, $(E - E_F)$ is a negative quantity.

$$\therefore \qquad f(E) = \frac{1}{1 + e^{-(E-E_F)/0}} = \frac{1}{1 + e^{-\infty}} = \frac{1}{1+0} = 1$$

$f(E) = 1$ indicates that all the energy levels lying below the level E_F are occupied.

(*ii*) For energy levels located above E_F, $E > E_F$, $(E - E_F)$ is a positive quantity.

$$\therefore \qquad f(E) = \frac{1}{1 + e^{(E-E_F)/0}} = \frac{1}{1 + e^{\infty}} = \frac{1}{1+\infty} = \frac{1}{\infty} = 0$$

The result $f(E) = 0$ implies that all the levels above E_F are vacant at T = 0K.

(*iii*) For $E = E_F$, the quantity $(E - E_F) = 0$.

$$\therefore \qquad f(E) = \frac{1}{1 + e^{0/0}} = \text{indeterminate}$$

The above result implies that the occupancy of Fermi level at 0K ranges from zero to one [Fig. 36.15 (b)].

Case 2: At T > 0K

On heating the conductor, electrons are excited to higher energy levels. In general, $E_F >> kT$. Therefore, for most of the electrons lying deep in the conduction band, the thermal energy is not sufficient to cause a transition to an upper unoccupied level. At normal temperatures, only those electrons occupying the energy levels near the Fermi level can be thermally excited. These levels make up a narrow band of width kT directly adjacent to the Fermi level. Therefore, upon heating the solid, electrons having energy a little below E_F, jump into levels with energy somewhat above E_F and a new energy distribution of electrons is obtained.

Thus, as a result of thermal excitation, the probability of finding electrons in the levels immediately below E_F will decrease. On the same hand, the probability of finding electrons in the levels immediately above E_F increases. This fact is reflected in the graph (Fig. 36.16 b) as a blurring of the step plot.

At $T > 0$K, if we consider an electron at Fermi level, then $E = E_F$

$$\therefore \quad f(E) = \frac{1}{1+e^{0/kT}} = \frac{1}{1+1} = \frac{1}{2}$$

This implies that the probability of occupancy of Fermi level at any temperature above 0K is 0.5 or 50%. Now we can define Fermi level as the energy level, which has a probability of occupancy of 0.5. An operational definition of Fermi energy can be given now. *Fermi energy is the average energy possessed by electrons participating in conduction in metals at temperatures above 0* K.

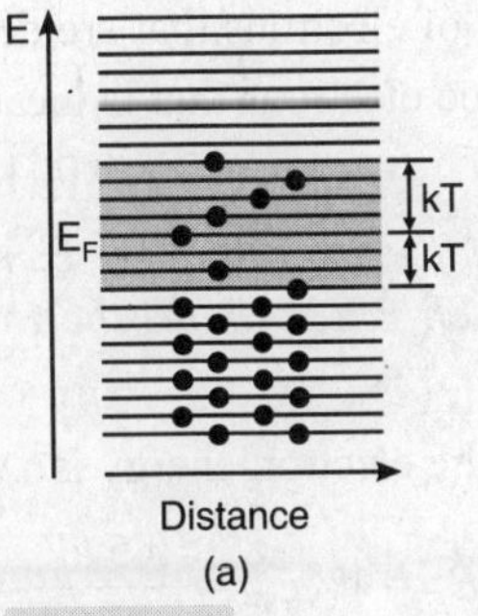

(a)

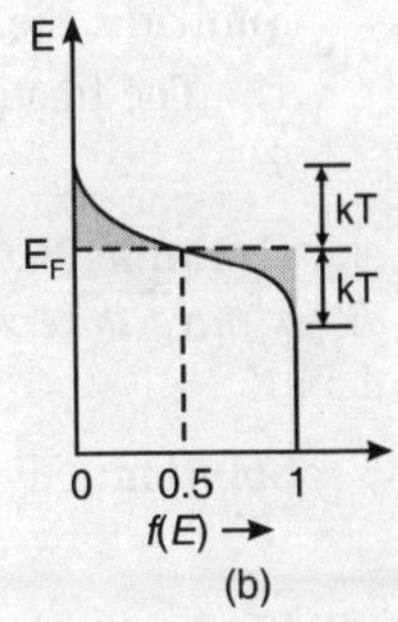

(b)

Fig. 36.16:

36.9.3 Effect of Temperature on Fermi Function

The Fermi-Dirac distribution curves for different temperatures are shown in Fig. 36.17. At T = 0 K, there is an abrupt jump in the value of $f(E)$ from 1 to zero at E_F. At $T > 0$ K the change is gradual. The higher the temperature, more gradual is the change.

It is seen from the curves for different temperatures in Fig. 36.17 that they all pass through a crossover point C, at which the probability of occupancy is 0.5. This is due to the fact that $f(E)$ has a value of 0.5 for any temperature greater than 0 K.

We may deduce from the curves that *Fermi energy E_F is the average energy possessed by electrons that participate in conduction process in a conductor at temperatures above absolute zero*.

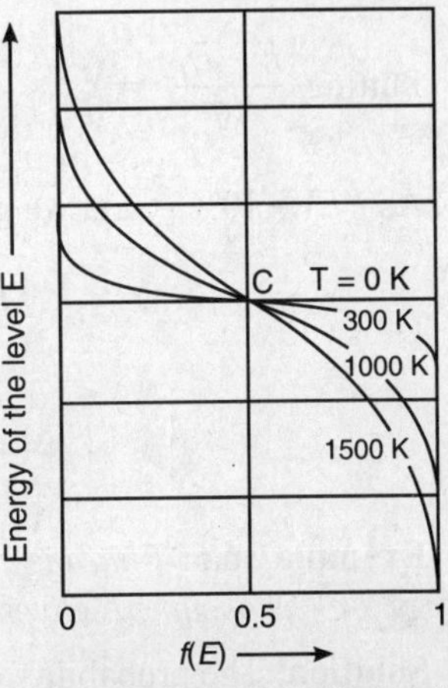

Fig. 36.17: Fermi-Dirac distribution curves for different temperatures

Example 36.3: *The Fermi level for potassium is 2.1 eV. Calculate the velocity of the electrons at the Fermi level.*

Solution: $E_F = \frac{1}{2} m\upsilon_F^2$

$$\therefore \quad \upsilon_F^2 = \frac{2E_F}{m} = \frac{2\times2.1\times1.602\times10^{-19}\,\text{C}\cdot\text{V}}{9.10\times10^{-31}\,\text{kg}} = 0.74 \times 10^{12}\ \text{m}^2/\text{s}^2$$

$$\upsilon_F = \mathbf{8.6 \times 10^5\ m/s.}$$

Example 36.4: *The Fermi level of silver is 5.5 eV. Calculate the fraction of free electrons at room temperature located up to a width of kT otn either side of the Fermi level.*

Solution: The number of electrons that occupy levels above E_F at a temperature T is proportional to kT. Therefore, the fraction of electrons that occupies levels higher than E_F is given by

$$\frac{kT}{E_F} = \frac{0.026\ \text{eV}}{5.5\ \text{eV}} = 0.0047$$

Similarly, the fraction of electrons that are deprived of partners ≅ 0.0047

∴ The fraction of free electrons that is located up to a width kT on either side of E_F

$$= 2 \times 0.0047 = 0.0094 \approx \mathbf{0.01}$$

Example 36.5: *At what temperature we can expect a 10% probability that electrons in silver have an energy which is 1% above the Fermi energy? The Fermi energy of silver is 5.5 eV.*

Solution: Given that the electron energy is $E = E_F + 1\%E_F$.

$$\therefore \quad E - E_F = 1\%E_F = \frac{5.5\,eV}{100} = 0.055 \text{ eV}$$

Also

$$f(E) = \frac{1}{1+e^{(E-E_F)/kT}}$$

Putting $\dfrac{E-E_F}{kT} = x$,

$$f(E) = \frac{1}{1+e^x}$$

As $f(E) = 10\% = 0.1$, we get $\dfrac{1}{1+e^x} = 0.1$ or $x = 2.197$

$$\therefore \quad \frac{E-E_F}{kT} = 2.197$$

or

$$T = \frac{E-E_F}{2.197 \times k} = \frac{0.055 \text{ eV}}{2.197 \times 8.61 \times 10^{-5} \text{ eV/K}} = \mathbf{290\ K.}$$

Example 36.6: *Find the temperature at which there is 1% probability that a state with energy 2 eV is occupied. Given that Fermi energy is 1.5 eV.*

Solution: The probability of an energy state E being occupied by an electron is given by

$$f(E) = \frac{1}{1+e^{(E-E_F)/kT}}$$

$$E - E_F = 2 \text{ eV} - 1.5 \text{ eV} = 0.5 \text{ eV} \quad \text{and} \quad f(E) = 1\%$$

$$\therefore \quad \frac{1}{100} = \frac{1}{1+e^{0.5/kT}} \quad \text{or} \quad e^{0.5/kT} = \frac{0.99}{0.01} = 99$$

Taking logarithm on both the sides, we get

$$\frac{0.5 \text{ eV}}{kT} = 2.303 \log 99$$

or

$$\frac{0.5 \text{ eV}}{kT} = T = \frac{0.5 \text{ eV}}{2.303 \log 99 \times 8.61 \times 10^{-5} \text{eV}} = 1262 \text{ K.}$$

Example 36.7: *Show that the probability of finding an electron of energy ΔE above the Fermi level is same as probability of not finding an electron at energy ΔE below the Fermi level.*

OR

Show that the probability that a state ΔE above the Fermi level E_F is filled equals the probability that a state ΔE below E_F is empty.

Solution: Let us consider an energy level E_2 that is above the Fermi level by an amount of energy ΔE. The probability that the energy level E_2 is occupied is given by

$$f(E_2) = f(E_F + \Delta E) = \frac{1}{1+\exp[(E_2 - E_F)/kT}$$

$$= \frac{1}{1+\exp[(E_F + \Delta E - E_F)/kT]}$$

$$\therefore \quad f(E_2) = \frac{1}{1+\exp(\Delta E/kT)} \qquad ...(1)$$

Next, let us consider the energy level E_1 that is below the Fermi level by energy ΔE. $[1 - f(E_1)]$ gives the probability that the level E_1 is not occupied.

$$[1 - f(E_1)] = [1 - f(E_F - \Delta E)]$$

$$= 1 - \frac{1}{1+\exp[(E_F - \Delta E - E_F)/kT]}$$

$$= 1 - \frac{1}{1+\exp[(-\Delta E)/kT]}$$

$$= \frac{\exp[(-\Delta E)/kT]}{1+\exp[(-\Delta E)/kT]}$$

$$\text{or} \quad [1 - f(E_1)] = \frac{1}{1+\exp[(\Delta E)/kT]} \qquad ...(2)$$

The R.H.S of equation (1) and (2) are the same.

$$\therefore \quad f(E_2) = [1 - f(E_1)]$$

It means that the probability of an energy level $[E_F + \Delta E]$ (ΔE above the Fermi level) being occupied is the same as the probability of an energy level $[E_F - \Delta E]$ (ΔE below E_F), being vacant.

Example 36.8: *Show that the occupancy probabilities of two states whose energies are equally spaced above and below the Fermi energy add up to one.*

Solution: Let us consider two energy levels E_2 and E_1, which are equally spaced above and below the Fermi energy E_F.

Let $E_2 = E_F + \Delta E$ and

$E_1 = E_F - \Delta E$

The probability of occupancy of the level E_2 is given by

$$f(E_2) = f(E_F + \Delta E) = \frac{1}{1+\exp[(E_2 - E_F)/kT]}$$

$$= \frac{1}{1+\exp[(E_F + \Delta E - E_F)/kT]}$$

$$\therefore \quad f(E_2) = \frac{1}{1+\exp(\Delta E/kT)}$$

The probability of occupancy of the level E_1 is given by

$$f(E_1) = f(E_F - \Delta E) = \frac{1}{1+\exp[(E_F - \Delta E - E_F)/kT]}$$

$$\text{or} \quad f(E_1) = \frac{1}{1+\exp[-\Delta E/kT]}$$

$$\therefore \quad f(E_1) + f(E_2) = \frac{1}{1+\exp(\Delta E / kT)} + \frac{1}{1+\exp(-\Delta E / kT)}$$

$$= \frac{1}{1+\exp(\Delta E / kT)} + \frac{\exp(\Delta E / kT)}{1+\exp(\Delta E / kT)}$$

$$= \frac{1+\exp(\Delta E / kT)}{1+\exp(\Delta E / kT)}$$

$$\therefore \quad f(E_1) + f(E_2) = 1$$

Thus, the occupancy probabilities of two states whose energies are equally spaced above and below the Fermi energy add up to one.

36.10 ENERGY BAND STRUCTURE OF AN INSULATOR

Insulators are characterized by two energy bands – conduction band and valence band, separated by a large energy gap. At 0K all valence electrons are engaged in covalent bonds, the valence band is full. The absence of any mobile charge carriers keeps the conduction band vacant. The situation is same even at higher temperatures (300K), as the valence band and conduction bands are separated by a large gap (> 3 eV) and it is not possible to excite electrons from valence band to conduction band by thermal energy (kT ≅ 0.026 eV). Consequently, insulators do not allow flow of current even at temperatures higher than room temperature.

The concept of Fermi level can be extended to insulators also. As the energy levels in valence band are filled, $f(E)$ is equal to unity throughout the valence band. As there are no electrons in the conduction band, $f(E)$ is equal to zero throughout the conduction band. Since the Fermi function is symmetrical about E_F, the Fermi level may be expected to be situated in the middle of the energy gap. Even though there are no energy levels and no electrons in the band gap, the meaning of Fermi level remains the same. It is a *reference energy position*. The energy band diagram for an insulator is shown in Fig. 36.18 along with probability function.

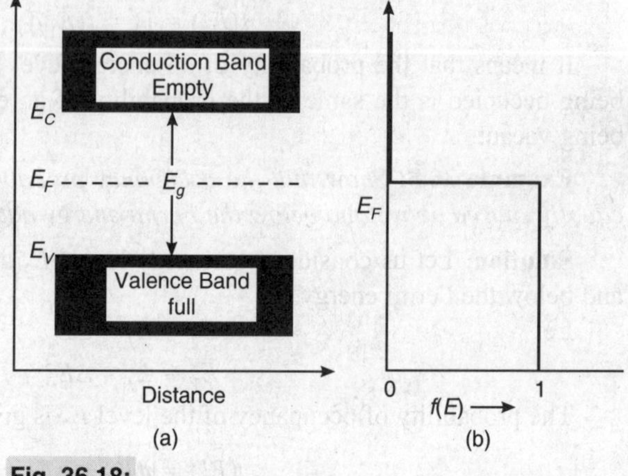

Fig. 36.18: Energy band diagram for an Insulator

36.11 ENERGY BAND STRUCTURE OF A SEMICONDUCTOR

In a semiconductor the conduction band and valence band are separated by a smaller energy gap. At room temperatures, a significant number of electrons are thermally excited from valence band to conduction band and an equal number of vacancies are produced in the valence band. These vacancies are treated as particles having a mass equal to that of electron and carry positive charge. They are called **holes**.

The Fermi-Dirac distribution function is applicable to a semiconductor. Fig. 36.19 depicts the probability function plotted alongside the band diagram for a semiconductor. Because the probability of electron occupancy of the conduction band increases at temperatures greater than 0K, the probability function is blurred and tapers off towards higher energy in the conduction band. Similarly, the probability of hole occupancy of the valence band increases, and the probability curve is blurred near the top edge of the valence band. The extent of blurring of probability curve in both the bands is equal indicating that the concentration of electrons in the conduction band and that of holes in the valence band are equal. Secondly, the probability function rapidly approaches zero value with increasing *E*.

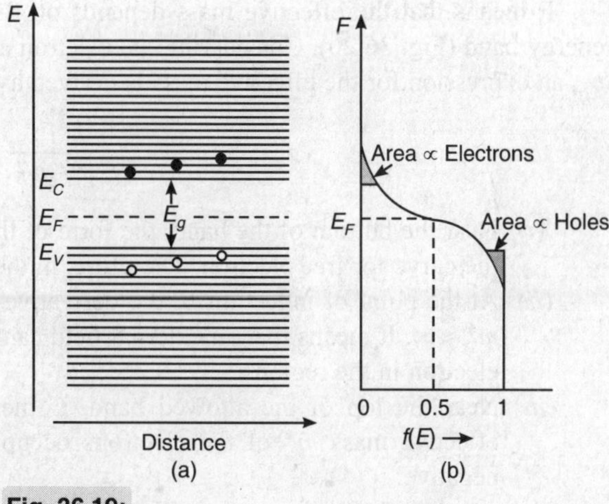

Fig. 36.19:
(a) Band diagram for a semiconductor,
(b) Probability function

It indicates that the electrons in the conduction band are crowded close to the bottom edge of the band while the holes are clustered very close to the top edge of the valence band. As a result, we approximate that electrons are located at the bottom edge of the conduction band whereas the holes are located at the top edge of the valence band.

The Fermi level represents the average energy of charge carriers participating in conduction. Both electrons and holes take part in conduction in semiconductor; and they lie in two different bands separated by a forbidden gap. We can understand that the Fermi level lies in the middle of the forbidden gap. If the Fermi level is located elsewhere, it would mean that the number of electrons in the conduction band would be different from the number of holes in the valence band. It, in turn, would imply that the material does not exhibit overall neutrality which is not at all true.

36.12 EFFECTIVE MASS

We generally assume that the mass of an electron in a solid is the same as the mass of a free electron. However, experimentally measured values show that for some solids the mass is larger while in case of other solids it is slightly smaller than the free electron mass. This experimentally determined electron mass is known as the **effective mass** *m**. It has been found that the effective mass is inversely proportional to the curvature of an allowed energy band.

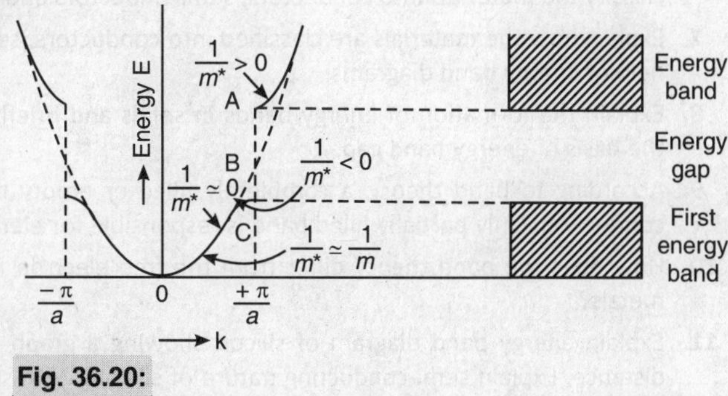

Fig. 36.20:
The effective mass of electron in a solid depends on electron's location in the band.

It means that the effective mass depends on the location of an electron in the allowed energy band (Fig. 36.20). Considering the electron as a wave packet having a group velocity, v_g, an expression for the effective mass is derived as

$$m^* = \frac{\hbar^2}{d^2E / dk^2} \qquad ...(36.15)$$

(*i*) Near the bottom of the band, the form of the $E - k$ curve does not differ much from the curve for free electron. Therefore, in these regions $m^* \approx m$.

(*ii*) At the point of inflection B, the derivative $d^2E/dk^2 = 0$. Therefore, in these regions $m^* \approx \infty$. It means that an external field cannot exert any action on the motion of the electron in the region.

(*iii*) Near the top of the allowed band, C the derivative $d^2E/dk^2 < 0$. Therefore, the effective mass m^* of the electrons occupying levels near the top of the band is negative.

The concept of effective mass explains suitable account of the charge carriers in crystals. In alkali metals that have partially filled energy band, the conduction takes place chiefly through electrons. But in crystals for which the energy band is nearly full, the negative charge and negative mass vacancies may be considered as acting as positive charge and positive mass particles called *holes*. It explains the origin of positive Hall coefficient in certain metals such as zinc.

QUESTIONS

1. Define energy level and energy band. Explain with proper diagrams, how on the basis of band theory, solids are classified as conductors, insulators and semiconductors.
2. Describe the formation of energy bands in a crystalline solid.
   Define valence band, conduction band and forbidden gap in the energy band structure.
   Hence classify solids into conductors, semiconductors and insulators.
3. Explain formation of energy bands in solids on the basis of band theory of solids.
4. Explain quantitatively band theory of solids. Explain energy band diagram and distinguish metal, semiconductor and insulator on the basis of above theory.
5. Explain how solids are classified on the basis of energy band gap.
6. Describe in short the formation of energy bands in solids and hence explain how it helps to classify the materials into conductors, semiconductors and insulators (with an example in each).
7. Explain how the materials are classified into conductors, semiconductors and insulators with the help of energy band diagrams.
8. Explain the formation of energy bands in solids and briefly explain how solids are classified on the basis of energy band gap.
9. According to band theory, a completely filled or empty band is not associated with electrical conduction. Only partially filled band is responsible for electrical conduction. Explain, why.
10. How does the band theory differ from the free electron model in explaining the properties of metals?
11. Explain energy band diagram of silicon showing a graph of variation of potential energy with distance. Explain semi-conducting nature of silicon. With similar band structure why is diamond insulator?
12. Draw a graph showing variation of electron energy in germanium crystals as a function of interatomic distance and explain why it shows semiconducting behaviour.

13. Draw a graph showing variation of electron energy levels of germanium as a function of its interatomic distance. Explain from it why germanium is an insulator at 0°K and semiconductor at 7°K.
14. What is Fermi level and Fermi energy?
15. Explain Fermi energy function. How does it vary with temperature?
16. Write and explain Fermi function. Explain with the help of a diagram how it varies with change of temperature.
17. Write down the Fermi-Dirac equation for the probability of occupation of an energy level E by an electron. Show that the probability of its occupancy by an electron is zero if $E > E_F$ and unity if $E < E_F$ at temperature 0 K.
18. Define Fermi distribution function. Show that at all temperatures (T > 0 K) probability of occupancy of Fermi level is 50%.
19. What is Fermi function? Draw a graph showing its variation with energy at different temperatures and discuss it.
20. Write down Fermi distribution function $f(E)$. Show graphically and analytically that $f(E)$ as function of E always passes through a point $\left(E_F, \frac{1}{2}\right)$ at different temperatures.
21. Why is that only the electrons near the Fermi level contribute to electrical conductivity?
22. Explain the concept of hole.
23. What is meant by effective mass of electron?
24. Explain the concept of negative mass on the basis of band theory.
25. Write down an expression for the probability of occupancy of a particular energy state of an electron in an intrinsic semiconductor. Represent it graphically at 0° K and at room temperature.
26. Explain in brief the concept of Fermi level. Show diagrammatically the Fermi level in metals, intrinsic semiconductors and insulators at 0°K and at higher temperature.
27. How are the band structures of insulators and semiconductors similar? How are they different?

PROBLEMS

1. The density of silver is 10.5×10^3 kg/m^3. the atomic weight of silver is 107.9. Assuming that each silver atom provides one conduction electron (*i*) calculate the density of free electrons. The conductivity of silver at 20°C is 6.8×10^7 ohm^{-1}m^{-1}. (*ii*) Calculate the electron mobility in silver. **[Ans:** 5.86×10^{28}; 7.25×10^{-3}m^2/V.s]
2. In a solid, consider the energy level lying 0.01 eV above Fermi level. What is the probability of this level being *occupied* by an electron at 200 K? **[Ans:** 0.359]
3. In a solid, consider the energy level lying 0.01 eV below Fermi level. What is the probability of this level being occupied by an electron at 300 K? **[Ans:** 0.595]
4. In a solid, consider the energy level lying 0.01 eV above Fermi level. What is the probability of this level being *occupied* by an electron at 300 K? **[Ans:** 0.405]
5. In a solid, consider the energy level lying 0.01 eV above Fermi level. What is the probability of this level being not *occupied* by an electron at 300 K? **[Ans:** 0.595]

37 Semiconductors

Learning Objectives:

By the end of this chapter, you will be able to:

- Explain concepts related to Intrinsic Semi-Conductor including its conductivity, band diagram and limitations.
- Understand concepts related to Extrinsic Semi-Conductor (n-type and p-type) including its conductivity, band diagram and limitations.
- Learn about Drift and Diffusion Current.
- Describe about Compound Semi-Conductors.

37.1 INTRODUCTION

Semiconductors are materials having electrical conductivity considerably greater than that of an insulator but significantly lower than that of a conductor. Of all the elements in the periodic table, eleven elements are semiconductors. Germanium and silicon are the most widely used semiconductors in device manufacturing applications. They are known as ***elemental semiconductors***. Besides these, there are certain ***compound semiconductors*** such as Gallium Arsenide, Indium Phosphide etc which are formed from the combinations of the elements of groups III and V or groups II and VI. The unique and interesting feature of semiconductors is that they are bipolar and two charge carriers, namely electrons and holes, transport current in these materials. The electrical conductivity of a *pure semiconductor*, known as intrinsic conductivity, is significantly low and is drastically influenced by temperature. As such pure semiconductors cannot be used in device fabrication. Through the technique of doping, the conductivity of a semiconductor can be increased in magnitude to a desired value and can be made independent of temperature in a certain temperature interval. Doped semiconductors are known as *extrinsic semiconductors*. The remarkable feature of extrinsic semiconductors is that current is transported in them by two different charge carriers, *electrons* and *holes*; and through two different processes, *drift* and *diffusion*. Extrinsic semiconductors are widely used in fabrication of solid-state devices. An understanding of the mechanism of conduction in intrinsic and extrinsic semiconductors helps us understand the working of solid-state devices.

37.2 CRYSTAL STRUCTURE

Silicon and germanium are elemental semiconductors. They belong to Group IVA in the periodic table. Silicon atom has 14 electrons and germanium atom 32 electrons. Each of them has four valence electrons which are distributed among the outermost s and p orbitals. In case of silicon atom, there are four energy levels, $3s$, $3p_x$, $3p_y$, and $3p_z$, out of which the lower two levels are filled by four valence electrons. In an actual piece of silicon crystal, the $3s$ and $3p$ orbitals of the atom combine to yield $3sp^3$ *hybrid molecular orbitals*. The $3sp^3$ orbitals form four covalent bonds of equal angular separation, leading to a tetrahedral arrangement of atoms in space (Fig. 37.1 a). Atoms occupy the corners of a regular tetrahedron and each atom in turn bonds with four neighbours and so on. The resulting crystal structure is known as ***diamond cubic (DC) crystal structure***, shown in Fig. 37.1 (b).

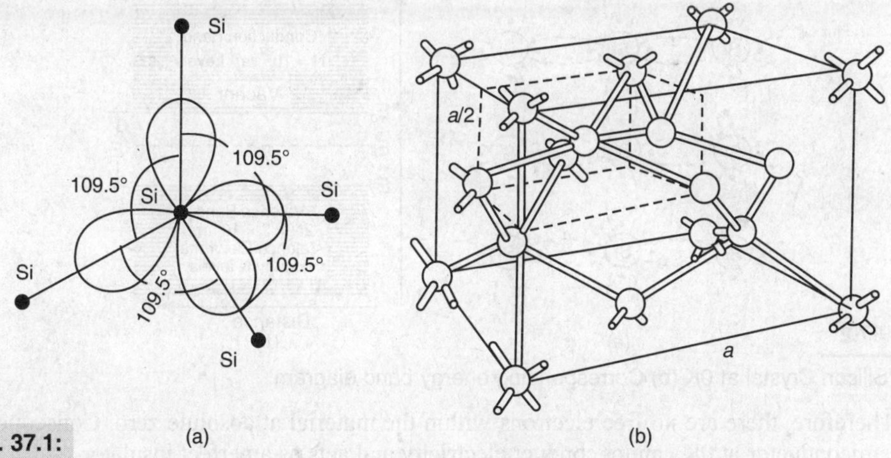

Fig. 37.1: (a) Tetrahedral arrangement of atoms in space (b) diamond cubic crystal structure

37.3 INTRINSIC SEMICONDUCTOR

Chemically *pure* semiconductors are known as **intrinsic semiconductors**. A semiconductor is considered to be pure when there is less than one impurity atom in a billion host atoms.

A two-dimensional representation of silicon crystal is shown in Fig. 37.2 (a). Each silicon atom forms covalent bonds with four surrounding atoms. The hatched circles in Fig. 37.2 (a) represent the cores of the silicon atoms. The four valence electrons are shown by the small black dots surrounding each hatched circle. The probability of valence electrons being in any place between the bonding atoms is indicated by the dashed curves. In terms of energy band diagram, a conduction band and a valence band separated by a smaller energy gap characterize a semiconductor (Fig. 37.2b).

In a real crystal, the concentration of atoms N is given by

$$N = \frac{N_A \rho}{M} \qquad \text{...(37.1)}$$

where N_A is the Avogadro number, ρ the density and M the atomic weight of the material.

Using the data for silicon into Eq. (37.1), we obtain

$$N = \frac{\left(6.02\times10^{26}\ \text{atoms/k.mol}\right)\left(2330\ \text{kg/m}^3\right)}{28.09\ \text{kg / k.mol}}$$

or

$$N = 5 \times 10^{28}\ \text{atoms/m}^3 \qquad \text{...(37.2)}$$

The valence and conduction bands of silicon crystal contain $2N$ energy levels each. Therefore, the number of energy levels in each band is 10^{29} levels/m^3. In other words, there are

2×10^{29} states/m^3. The number of valence electrons available in the silicon crystal is $4N = 2 \times 10^{29}$ electrons/m^3. These electrons occupy the valence band and leave the conduction band vacant.

(a) At 0K an Intrinsic Semiconductor Behaves as a Perfect Insulator: At 0K and temperatures close to 0K, all valence electrons are locked in covalent bonds (Fig. 37.2 a) spend most of the time between neighbouring atoms. Since all the valence electrons are engaged in covalent bonds, the bonds are complete. The energy available at 0K is not sufficient to break the covalent bonds.

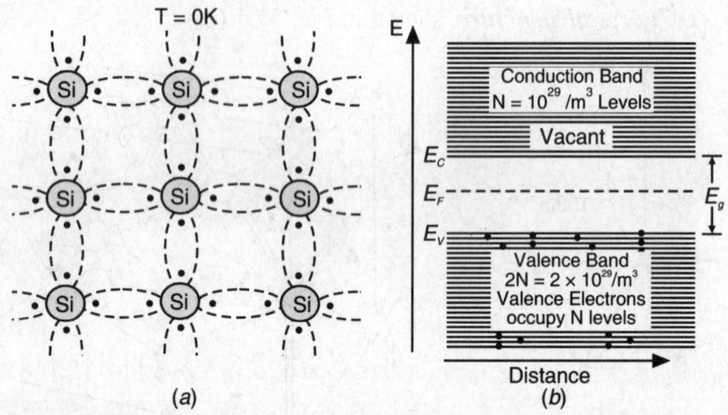

Fig. 37.2: (a) Silicon Crystal at 0K (b) Corresponding energy band diagram.

Therefore, there are no free electrons within the material at absolute zero. Consequently, the semiconductor at 0K cannot conduct electricity and acts as a perfect insulator.

In terms of energy band diagram, the situation is as follows. There are $2N = 10^{29}$ energy levels/m^3 in the valence band and $2N = 10^{29}$ energy levels/m^3 in the conduction band. The total number of valence band electrons available in the crystal is $4N = 2 \times 10^{29}$ electrons/m^3. At 0 K these 2×10^{29} electrons/m^3 completely occupy the 10^{29} energy levels/m^3 available in the valence band. There are no electrons left to go into the conduction band (Fig. 37.2b). At 0 K, electrons in the valence band do not possess enough energy to jump into the conduction band. As free electrons do not exist in the conduction band, an externally applied electric field cannot cause flow of current through the crystal. Hence, the intrinsic semiconductor behaves as a *perfect insulator* at 0 K.

(b) Mechanism of Conduction in an Intrinsic Semiconductor: At temperatures above absolute zero, the finite thermal energy causes each atom in the crystal to vibrate about its mean position. When the vibrations become violent, some of the electrons acquire sufficient energy and break away from covalent bonds (Fig. 37.3a). Whenever a covalent bond is ruptured by thermal energy, a valence electron becomes free. The higher the temperature, the more covalent bonds are broken. The electrons liberated from bonds move randomly in the void spaces between the atoms in the crystal. If an electric field is applied, these free electrons cause electrical conduction.

From the energy band point of view, it means that some of the electrons in the valence band convert part of their thermal energy into potential energy. Those electrons which acquire energy equal to or in excess of the band gap energy E_g are excited to the conduction band (Fig. 37.3b). Thus, the band gap energy E_g is the minimum amount of energy required to excite an electron from valence band to conduction band. E_g is characteristic of the material. The number of electrons excited to the conduction band depends on the amount of thermal energy received by the crystal. For example, the concentration of broken bonds in a silicon crystal at 300K is 1.5×10^{16} /m^3. Thus, there are 1.5×10^{16} electrons/m^3 in the conduction band at 300 K. In fact, the conduction band can accommodate 2×10^{29} electrons/m^3 and hence it is partially filled.

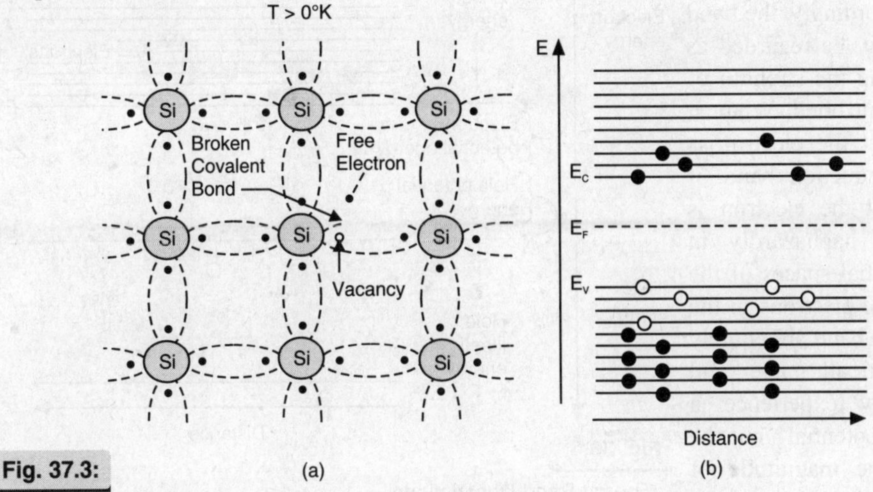

Fig. 37.3: (a) Silicon Crystal at T > 0 K (b) Energy band diagram.

When an electron from the valence band jumps to the conduction band, an ***empty state*** (quantum vacancy) arises in the valence band. In silicon crystal at 300K, 1.5×10^{16} vacant states/m^3 appear in the valence band, which are very small in number compared to the number of electrons remained in the band. Thus, now both the bands are partially filled. The electrons in the conduction band and the electrons in the valence band can be excited to upper vacant levels within the respective bands. Therefore, if an electric field is applied, these electrons can move into higher vacant levels and current flows in the crystal at ordinary temperatures. The motion of valence electrons in the valence band is customarily described in terms of a fictitious particle called **hole** which is bequeathed with a positive charge $+e$ and a mass m_h equal to that of an electron.

In pure semiconductors all available charge carriers, electrons and holes, arise due to thermally ruptured bonds and these thermally generated electron-hole pairs cause electrical conduction. Thermal generation is an *intrinsic process*. Therefore, we may define an intrinsic semiconductor as follows.

An intrinsic semiconductor is a semiconductor crystal in which electrical conduction arises due to thermally excited electrons and holes.

(c) Significance of Band gap E_g: The band gap energy E_g is the minimum amount of energy required for breaking a covalent bond. It is the minimum amount of energy required to excite an electron from valence band to conduction band. Also, it is the minimum amount of energy required to convert a bound electron into a free electron. The energy required to break a covalent bond in a germanium crystal is about 0.72 eV at 300 K and that in silicon is 1.12 eV.

37.4 CORRELATION BETWEEN CRYSTAL LATTICE AND ENERGY BAND DESCRIPTIONS

The correlation between the crystal lattice description and the energy band description of a semiconductor can be summed up as follows:

The statement that an electron is in the valence band means that the electron is participating in a covalent band. The excitation of the electron from the valence band to the conduction band implies rupturing of a covalent bond. Accordingly, the band gap E_g may be regarded as representing the strength of the covalent bond. When it is said that the electron is in the conduction band, it means that the electron is wandering haphazardly in the interstitial spaces of the crystal. The electrons in the conduction band are equally attracted to all nuclei and hence they experience a constant potential in the crystal. The magnitude of the potential energy may be taken as E_g. It is the energy corresponding to the lowest energy level in the conduction band.

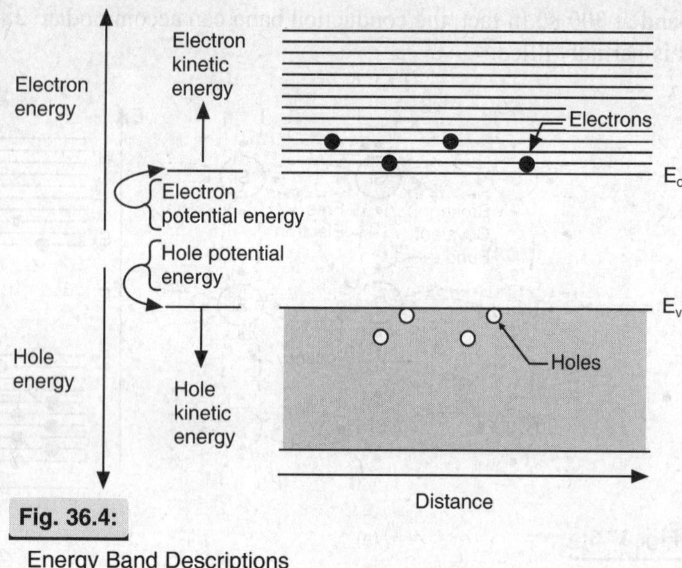

Fig. 36.4: Energy Band Descriptions

Out of the thermal energy E supplied to a valence electron, an amount E_g is translated into the potential energy and the electron reaches the bottom edge E_C, of the conduction band (Fig. 37.4). Any additional energy goes into the form of kinetic energy. Thus,

$$E = E_g + \tfrac{1}{2}m\upsilon^2 \qquad ...(37.3)$$

Consequently, the electrons in the conduction band are characterized by their velocity only and act as **free electrons**. It is seen from the relation (37.3) that an electron with an energy E_g just reaches the bottom edge of the conduction band and will be at rest since. With more energy, it is positioned higher and higher above the conduction band edge E_C. The situation is very much akin to the billiard balls on a billiard table. The balls are confined to the table but can have wide ranging velocities.

The situation is similar for holes. The top of the valence band, E_V represents the potential energy of the hole (Fig. 37.4). The energies below E_V indicate the increasing kinetic energy of a hole. The directions of increasing energy for electrons and holes are opposite to each other because they have charges of opposite sign.

37.5 HOLES

In a silicon crystal, when a covalent bond is broken and an electron is set free, there arises a ***quantum vacancy*** at the site of the broken bond. The removal of a negative charge from an otherwise neutral atmosphere accords an effective positive charge to the resulting vacancy. Obviously, the actual seat of the positive charge is the nucleus of the silicon atom which is deprived of its valence electron. The positively charged vacancy, say, created at the site *A* can attract an electron from the adjacent bonds. An electron from site, say, *B* jumps into the vacancy at A and bridges the bond but leaves a vacancy at *B* and so on. In this way, a new type of electron motion is created in the valence band. In quantum mechanical terms, the electron at a lower energy level makes an upward transition to a vacant level. As the electron moved up, the vacancy moves down. Note that the vacancy and the electron move in opposite directions in the valence band. As both motions are related, the hopping of a large number of valence electrons can be equivalently described in terms of normal motion of a small number of holes. The concept of hole helps in differentiating the two different kinds of electron motion, namely the motion of high energy electrons in the conduction band and the motion of low energy electrons in the valence band.

The concept of holes simplifies the understanding of electrical conduction in semiconductors. It is seen as follows:

The instantaneous current produced by one electron moving at a speed υ_k is given by

$$i = -e\upsilon_k$$

The current set up by the motion of many electrons in a valence band is therefore

$$I = -e\sum_i \upsilon_i$$

For a completely filled band, the resultant current is zero. That is,

$$I = -e\sum_i \upsilon_i = 0 \qquad \text{...(37.4)}$$

The above equation may be rewritten as

$$I = -e\left[\sum_{i\neq k} \upsilon_i + \upsilon_k\right] = 0$$

$$\therefore \qquad -e\sum_{i\neq k} \upsilon_i = -e(-\upsilon_k) = e\upsilon_k \qquad \text{...(37.5)}$$

$$\therefore \qquad \sum_{i\neq k} \upsilon_i = -\upsilon_k \qquad \text{...(37.6)}$$

Eq. (37.6) implies that if the kth electron is absent in the valence band, then the sum of the velocities of the remaining electrons is $-\upsilon_k$. Eq. (37.5) shows that these electrons will set up a current equal to $+e\upsilon_k$. It means that the current set up by all the electrons in the valence band having one vacancy is equivalent to the current resulting from the motion of a single particle with a positive charge $+e$ that occupies the vacancy.

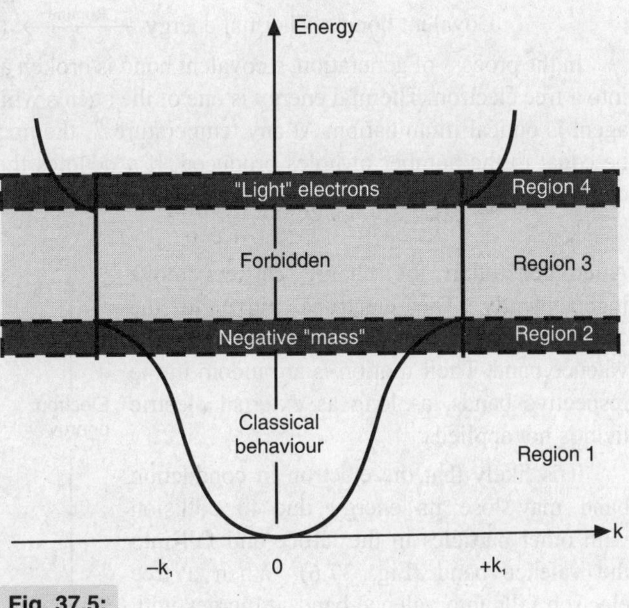

Fig. 37.5: *E-k* Curve

If a vacancy is to be given the status of a particle, a mass is also to be ascribed to it. We can arrive at the mass from the band theory. In solids, electrons experience forces due to atoms and other electrons. The motion of electron in a solid subjected to an electric field may be written in the form

$$F = m_e^* a$$

where m_e^* is the effective mass of the electron.

The effective mass of an electron heavily depends on its location in the energy band. Electrons near the bottom of the conduction band have an effective mass which is nearly identical to the mass of a free electron. Electrons near the top of valence band have negative effective mass. The concept of negative effective mass is not easy to grasp. The removal of an electron with a negative effective mass is identical to creating in its place a particle of positive effective mass. Thus, a hole is assigned an effective mass, m_h^*.

A hole in the sea of valence electrons in the valence band is analogous to a bubble in a liquid. A bubble has shape, size and also a velocity. Its behaviour is the same as that of a particle. In reality, it is the absence of liquid. One can in principle describe the motion of a bubble by describing the motion of the surrounding liquid; but it will be a tedious task. In a similar way, the motion of the valence electrons is difficult to be described. On the other hand, the description of their motion in terms of holes is a far simpler task.

Direct evidence for the existence of holes furnished by the results of Hall Effect measurements justifies the use of parlance of holes.

37.6 GENERATION AND RECOMBINATION

In semiconductors a single event of covalent bond breaking leads to the generation of two charge carriers, an electron in the conduction band and a hole in the valence band (Fig. 37.3). The electron and hole are produced simultaneously as a pair and the process is called **electron-hole pair generation**. The process may be represented as

$$\text{Covalent bond + Thermal energy} \xrightarrow{\text{Rupture}} \text{(Electron + Hole) pair}$$

In the process of generation, a covalent bond is broken and a bound electron is transformed into a free electron. Thermal energy is one of the agents which causes pair generation. Another agent is optical illumination. At any temperature T, the number of electrons generated would be equal to the number of holes produced. If n denotes the concentration of electrons in the conduction band and p is the concentration of holes in the valence band, then

$$n = p \qquad ...(37.7)$$

After generation the charge carriers move independently. The electrons move in the conduction band and the holes move in the valence band. Their motion is at random in the respective bands, as long as external electric field is not applied.

It is likely that the electron in conduction band may lose its energy due to collision with other particles in the lattice and fall into the valence band (Fig. 37.6). When a free electron falls into valence band, it merges with a hole. This process is called *recombination*. When a recombination event occurs, the free electron enters a ruptured covalent bond and re-bridges it.

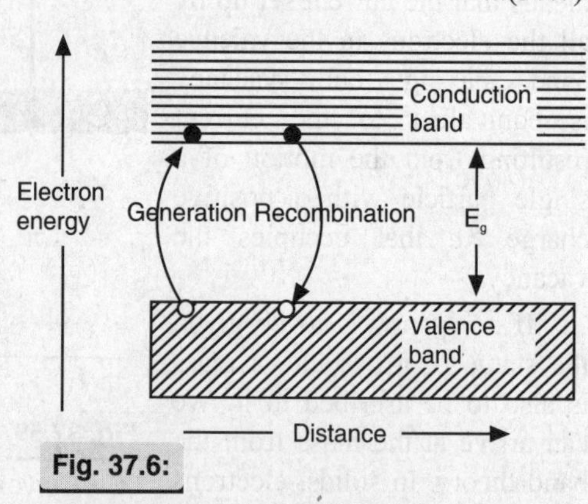

Fig. 37.6: Generation and recombination of electron-hole pairs

$$\text{Electron} + \text{Hole} \rightarrow \text{Covalent bond} + \text{Energy}\uparrow$$

Therefore, recombination means that a free electron transforms into a valence electron and that a ruptured covalent bond is re-bridged. In the process the electron-hole pair disappears and energy is released. The released energy is mainly in the form of thermal energy.

At a steady temperature a dynamic equilibrium exists which balances the two processes of electron-hole pair generation and electron-hole recombination (Fig. 37.6).

Just as thermal energy generates electron-hole pairs, light radiation can produce electron-hole pairs in a semiconductor. When a semiconductor material is irradiated with optical radiation, electron-hole pairs are generated if the frequency ν of the radiation satisfies the condition

$$h\nu \geq E_g \qquad ...(37.8)$$

When the radiation is switched off the excess electron-hole pairs recombine and the material returns to thermal equilibrium. Optically generated electron-hole pairs are responsible for the working of LDRs, photodiodes, etc optical detectors.

37.7 INTRINSIC CONDUCTIVITY

A single event of bond breaking in a pure semiconductor leads to generation of an ***electron-hole*** pair. At any temperature T, the number of electrons generated will be equal to the number of holes generated per unit volume. As the two charge carrier concentrations are equal, they are denoted by a common symbol n_i, which is called *intrinsic density* or *intrinsic concentration*. Thus,

$$n = p = n_i \qquad ...(37.9)$$

In a semiconductor under thermal equilibrium condition, free electrons move in the conduction band and holes in the valence band, which are in a state of random motion. At a temperature T, they possess an average kinetic energy given by

$$\frac{1}{2}m\upsilon_{th}^2 = \frac{3}{2}kT$$

where υ_{th} is the mean ***thermal velocity***.

When a potential difference is applied across the solid, the equilibrium condition is disturbed. The electric field accelerates the electrons and holes but their motion is hindered due to interactions with the lattice vibrations. In the steady state condition there arises a net movement of electrons in a direction opposite to that of the electric field and movement of holes in the direction of the electric field. This net movement of electrons and holes is called *drift*, and the corresponding mean velocity is known as ***drift velocity***, υ_d. The drift motion is superposed on the random thermal motion of the charge carriers. The drift motion is directional and causes ***drift current*** flow, which is more often called ***conduction current***.

The drift velocity is given by

$$\upsilon_d = \mu E \qquad ...(37.10)$$

Since the electrons move in relatively less populated conduction band the properties such as mobility, conductivity etc of electrons are larger compared to those of holes as the latter move in nearly full valence band. We designate the drift velocity of electron with υ_{de} and that of hole with υ_{dh}. Similarly, we denote the mobility of electron and hole with μ_e and μ_h respectively. Then, the current density due to electrons is given by

$$J_e = ne\upsilon_{de} = ne\mu_e E \qquad ...(37.11)$$

and the current density due to holes is

$$J_h = pe\upsilon_{dh} = pe\mu_h E \quad ...(37.12)$$

Comparing the above expressions with Ohm's law $J = \sigma E$, we obtain the expressions for electronic and hole conductivities as follows.

$$\sigma_e = ne\mu_e \quad ...(37.13)$$

and
$$\sigma_h = pe\mu_h \quad ...(37.14)$$

Let us now consider a sample of semiconductor across which a potential difference V is applied. The potential difference V establishes an electric field E in the semiconductor. It causes a current I_e due to electrons drifting in the conduction band and a current I_h due to holes drifting in the valence band (see Fig. 37.7). The total current through the semiconductor is

$$I = I_e + I_h \quad ...(37.15)$$

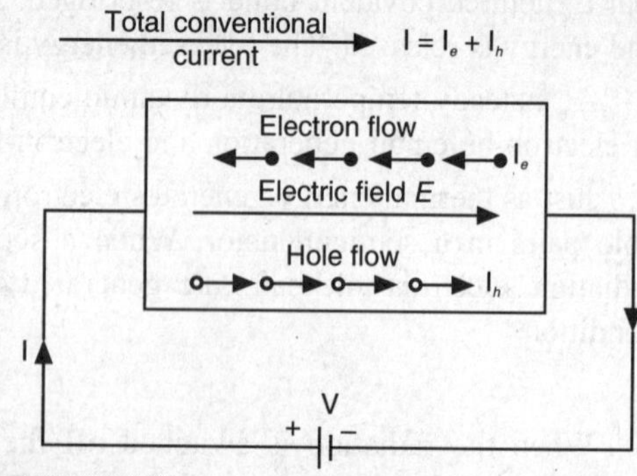

Fig. 37.7: Components of drift current due to electron and hole motion in an intrinsic semiconductor

∴ Current density $J = \frac{I_e}{A} + \frac{I_h}{A}$ where A is cross-sectional area of the crystal.

or
$$J = J_e + J_h$$
$$= (ne\mu_e + pe\mu_h)E \quad ...(37.16)$$

Therefore, the intrinsic conductivity is given by

$$\sigma = (ne\mu_e + pe\mu_h) \quad ...(37.17)$$

As $n = p = n_i$ in an intrinsic semiconductor, Eq. (37.17) may be rewritten as

$$\sigma = en_i\,(\mu_e + \mu_h) \quad ...(37.18)$$

Eq. (37.18) does not explain the temperature dependence of electrical conductivity in semiconductors. In general the variation of mobility with temperature is too small and the large variation in electrical conductivity in semiconductors is linked to the variation of electron concentration with temperature.

Thus, as a first approximation, taking e, μ_e and μ_h as constants, we find that

$$\sigma(T) \propto n_i\,(T) \quad ...(37.19)$$

The expression for the intrinsic carrier concentration is derived in the band theory. Here, we take the final results from the band theory and use them to obtain an expression for $n(T)$, the variation of intrinsic carrier concentration with temperature.

37.8 CARRIER CONCENTRATIONS

With an increase in temperature covalent bonds are broken in an intrinsic semiconductor and electron-hole pairs are generated. Electrons and holes are charged particles they are collectively called as ***charge carriers***. The number of electrons in the conduction band per unit volume (n) and the number of holes in the valence band per unit volume (p) of the material is known as ***carrier concentration***. Carrier concentration is also known as the ***density of charge carriers***. We now intend to calculate the electron concentration, n, in the conduction band and the hole concentration, p, in the valence band.

37.8.1 Calculation of Electron Density

Let dn be the number of electrons whose energy lies in the energy interval E and $E + dE$ in the conduction band. Then,

$$dn = Z(E)f(E)dE \qquad ...(37.20)$$

where $Z(E)\, dE$ is the density of states in the energy interval E and $E + dE$ and $f(E)$ probability that a state of energy is occupied by an electron.

The electron density in the conduction band is given by integrating the above equation between the limits E_C and ∞. E_C is the energy corresponding to the bottom edge of the conduction band and ∞ the energy corresponding to the top edge of the conduction band. As the probability of electrons occupying upper levels of conduction band $f(E)$ readily approaches zero for higher energies, the upper limit, namely the top of conduction band is taken as ∞. Thus,

$$n = \int_{E_C}^{\infty} Z(E) f(E) dE \qquad ...(37.21)$$

The density of states in the conduction band is given by

$$Z(E)dE = \frac{4\pi}{h^3}\left(2m_e^*\right)^{3/2} E^{1/2} dE \quad \text{for } E > E_C. \qquad ...(37.22)$$

The bottom edge of the conduction band E_C corresponds to the potential energy of an electron at rest. Therefore, $(E - E_C)$ will be the kinetic energy of the conduction electron at higher energy levels. Hence, Eq. (37.22) is to be modified as follows.

$$Z(E)dE = \frac{4\pi}{h^3}\left(2m_e^*\right)^{3/2} \left(E - E_C\right)^{1/2} dE \qquad ...(37.23)$$

The probability of an electron occupying an energy level is given by

$$f(E) = \frac{1}{1+\exp\left[(E - E_F)/kT\right]}$$

When the number of particles is very small compared to the available energy levels, the probability of an energy state being occupied by more than one electron is very small. Such a situation is valid when $(E - E_F) >> 3kT$. Under this circumstance, the number of available states in the conduction band is far larger than the number of electrons in the band. Then, Fermi-Dirac function can be approximated to Boltzmann function,

$$f(E) = \exp[-(E - E_F)/kT]. \qquad ...(37.24)$$

Using the Eq. (37.23) and Eq. (37.24) into Eq. (37.21), we obtain

$$\therefore \qquad n = \frac{4\pi}{h^3}\left(2m_e^*\right)^{3/2} \int_{E_C}^{\infty} \left(E - E_C\right)^{1/2} e^{-(E-E_F)/kT} dE \qquad ...(37.25)$$

or

$$n = \frac{4\pi}{h^3}\left(2m_e^*\right)^{3/2} e^{(E_F-E_C)/kT} \int_{E_C}^{\infty} \left(E - E_C\right)^{1/2} e^{-(E-E_C)/kT} dE \qquad ...(37.26)$$

The integral in Eq. (37.26) is of the standard form which has a solution of the following form.

$$\int_0^{\infty} x^{1/2} e^{-ax} dx = \frac{\sqrt{\pi}}{2a\sqrt{a}}$$

where $a = 1/kT$ and $x = (E - E_C)$.

$$\therefore \quad n = \frac{4\pi}{h^3}\left(2m_e^*\right)^{3/2} e^{(E_F-E_C)/kT}\left[\frac{\sqrt{\pi}}{2}(kT)^{3/2}\right] \quad ...(37.27)$$

Rearranging the terms, we get

$$n = 2\left[\frac{2\pi m_e^* kT}{h^2}\right]^{3/2} e^{-(E_C-E_F)/kT} \quad ...(37.28)$$

Eq. (37.28) is the expression for the ***electron concentration*** in the conduction band of an intrinsic semiconductor.

In Eq. (37.28) designating $N_C = 2\left[\frac{2\pi m_e^* kT}{h^2}\right]^{3/2}$ ...(37.29)

we obtain

$$n = N_C e^{-(E_C - E_F)/kT} \quad ...(37.30)$$

where N_C is a temperature dependent material constant known as the ***effective density of states*** in the conduction band. In silicon at 300 K, $N_C = 2.8 \times 10^{25}/\text{m}^3$.

The importance of the Eq. (37.30) is that it relates the equilibrium electron concentration to a single variable, namely the Fermi level E_F. Therefore, electron concentration is specified, if E_F is specified.

37.8.2 Calculation of Hole Density

Let dp be the number of holes whose energy lies in the energy interval E and $E + dE$ in the valence band. Then,

$$dp = Z(E)[1 - f(E)]dE \quad ...(37.31)$$

where $Z(E)dE$ is the density of states in the energy interval E and $E + dE$ and $[1 - f(E)]$ the probability that a state of energy is vacant and not occupied by an electron.

If $f(E)$ is the probability for occupancy of an energy state at E by an electron, then the probability that the energy state is vacant is given by $[1 - f(E)]$. Since a hole represents a vacant state in valence band, the probability for occupancy of a state at E by a hole is equal to the probability of absence of electron at that state.

We can write

$$[1 - f(E)] = 1 - \frac{1}{1+e^{(E-E_F)/kT}} = \frac{1}{1+e^{(E_F-E)/kT}} \approx e^{-(E_F-E)/kT} \quad ...(37.32)$$

The density of states in the valence band is given by

$$Z(E)dE = \frac{4\pi}{h^3}\left(2m_h^*\right)^{3/2} E^{1/2} dE \quad ...(37.33)$$

The top edge of the valence band E_V corresponds to the potential energy of a hole at rest. Therefore, $(E_V - E)$ will be the kinetic energy of the hole at lower energy levels. Hence, equ. (37.23) is to be modified as follows.

$$Z(E)dE = \frac{4\pi}{h^3}\left(2m_h^*\right)^{3/2}\left(E_V - E\right)^{1/2} dE \quad ...(37.34)$$

The number of holes in the energy interval E and $E + dE$ is

$$dp = \frac{4\pi}{h^3}\left(2m_h^*\right)^{3/2}\left(E_V - E\right)^{1/2} e^{-(E_F-E)/kT} dE \quad ...(37.35)$$

In order to calculate the number of holes in the valence band Eq. (37.35) is to be integrated between the limits ∞ and E_V. The hole density in the valence band is therefore given by

$$p = \frac{4\pi}{h^3}\left(2m_h^*\right)^{3/2}\int_{-\infty}^{E_V}\left(E_V - E\right)^{1/2} e^{-(E_F-E)/kT}\,dE \quad ...(37.36)$$

$$= \frac{4\pi}{h^3}\left(2m_h^*\right)^{3/2} e^{-(E_F-E_V)/kT}\int_{-\infty}^{E_V}\left(E_V - E\right)^{1/2} e^{-(E_V-E)/kT}\,dE \quad ...(37.37)$$

The integral in Eq. (37.37) is of the standard form which has a solution of the following form.

$$\int_0^{\infty} x^{1/2} e^{-ax}\,dx = \frac{\sqrt{\pi}}{2a\sqrt{a}}$$

where $a = 1/kT$ and $x = (E_V - E)$.

$$\therefore \quad p = \frac{4\pi}{h^3}\left(2m_h^*\right)^{3/2} e^{-(E_F-E_V)/kT}\left[\frac{\sqrt{\pi}}{2}(kT)^{3/2}\right] \quad ...(37.38)$$

Rearranging the terms we get

or
$$p = 2\left[\frac{2\pi\, m_h^* kT}{h^2}\right]^{3/2} e^{-(E_F-E_V)/kT} \quad ...(37.39)$$

The above equation is the ***expression for the hole concentration*** in the valence band of an intrinsic semiconductor.

Denoting
$$N_V = 2\left[\frac{2\pi\, m_h^* kT}{h^2}\right]^{3/2} \quad ...(37.40)$$

$$\therefore \quad p = N_V e^{-(E_F - E_V)/kT} \quad ...(37.41)$$

where N_V is called the ***effective density of states*** in the valence band. For silicon at 300 K, $N_V = 10^{25}/\text{m}^3$. It is seen that $N_C/N_V = 2.8$.

To a first approximation, N_C is taken to be equal to N_V. The electron and hole concentrations in an intrinsic semiconductor are shown in Fig. 37.8.

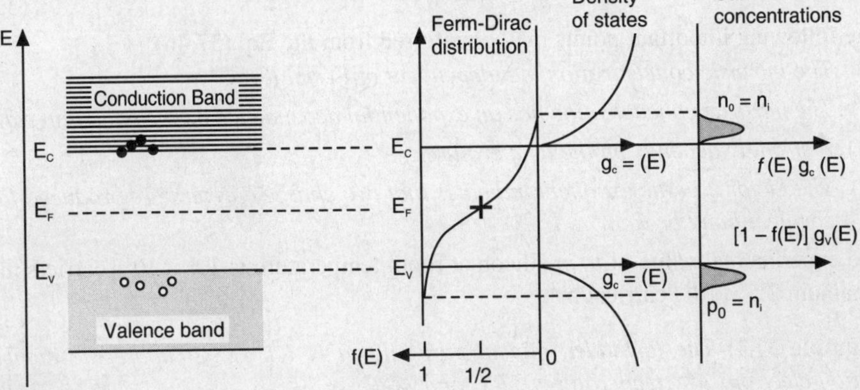

Fig. 37.8:

Electron and hole concentrations in an intrinsic semiconductor

37.9 INTRINSIC CARRIER CONCENTRATION

A single event of bond breaking in a pure semiconductor leads to generation of an ***electron-hole*** pair. At any temperature T, the number of electrons generated will be equal to the number of holes generated. As the two charge carrier concentrations are equal, they are denoted by a common symbol n_i, which is called *intrinsic density* or *intrinsic concentration*. Thus,

$$n = p = n_i$$

We can write that

$$n^2{}_i = np \quad \text{...(37.42)}$$

$$= (N_C e^{-(E_C - E_F)/kT})(N_V e^{-(E_F - E_V)/kT})$$

$$= (N_C N_V)^{-(E_C - E_V)/kT} \quad \text{...(37.43)}$$

The term $(E_C - E_V)$ stands for the difference in energy between the top level of valence band and the bottom level of conduction band. Thus, it represents the separation between the valence and conduction bands, which is the band gap E_g.

$$(E_C - E_V) = E_g \quad (37.44)$$

$$\therefore \quad n_i^2 = (N_C N_V) e^{-E_g/kT}$$

Substituting the values of N_C and N_V into the above equation, we obtain

$$= 4\left[\frac{2\pi\, kT}{h^2}\right]^3 \left(m_e^* m_h^*\right)^{3/2} e^{-E_g/kT}$$

$$\therefore \quad n_i = 2\left[\frac{2\pi kT}{h^2}\right]^{3/2} \left(m_e^* m_h^*\right)^{3/4} e^{-E_g/2kT} \quad \text{...(37.45)}$$

This is the expression for intrinsic carrier concentration.

37.9.1 Variation of Intrinsic Carrier Concentration with Temperature

Eq. (37.45) may be written as

$$n_i = 2\left[\frac{2\pi\, k}{h^2}\right]^{3/2} \left(m_e^* m_h^*\right)^{3/4} T^{3/2} e^{-E_g/2kT} \quad \text{...(37.46)}$$

Eq. (37.45) shows that the free charge carrier concentration varies with temperature and it may be approximated to

$$n_i = 10^{21.7} T^{3/2} \times 10^{-2500\, E_g/T} \quad \text{...(37.46a)}$$

The following important points may be inferred from the Eq. (37.46):

(*i*) *The intrinsic concentration is independent of Fermi level.*

(*ii*) *The intrinsic concentration has an exponential dependence on the band gap value* E_g.

(*iii*) *It strongly depends on the temperature.*

(*iv*) *The factor 2 in the exponent indicates that two charge carriers are produced for one covalent bond broken.*

The experimental value of n_i *in* silicon at room temperature is 1.5×10^{16} carriers/m^3 and in germanium 2.5×10^{19} carriers/m^3.

Example 37.1: *The forbidden gap in pure silicon is 1.1 eV. Compare the number of conduction electrons at temeperatures 37°C and 27°C.*

Solution: Let n_1 be the number of conduction electrons at 27°C and n_2 at 37°C. Using Eq. (37.46 a), we get

$$n_1 = 10^{21.7}\,(300\text{ K})^{3/2}\,(10^{-2500\times 1.1\text{eV}/300\text{ K}})$$

and

$$n_2 = 10^{21.7}\,(310\text{ K})^{3/2}\,(10^{-2500\times 1.1\text{eV}/310\text{ K}})$$

$$\frac{n_2}{n_1} = \frac{(310\text{K})^{3/2}\times 10^{-8.87}}{(300\text{K})^{3/2}\times 10^{-9.2}} = \mathbf{2.96}$$

Thus the number of electrons in the conduction band increased nearly three-fold as the temperature of the material increased from 27°C to 37°C.

Example. 37.2: *Compute the concentration of intrinsic charge carriers in a germanium crystal at 300 K. Given that $E_g = 0.72$ eV and assume $m_e^* = m_e$.*

Solution: Intrinsic charge carrier concentration, $n_i = 2\left[\dfrac{2\pi\, m_e^* kT}{h^2}\right]^{3/2} \exp\left(-\dfrac{E_g}{2kT}\right)$

$$2\left[\frac{2\pi m_e^* k}{h^2}\right]^{3/2} = 2\left[\frac{2\times 3.143\times 9.11\times 10^{-31}\text{kg}\times 1.38\times 10^{-23}\text{ J/K}}{\left(6.626\times 10^{-34}\text{ J.s}\right)^2}\right]^{3/2}$$

$$= 4.83\times 10^{21}\,,\ T^{3/2} = 300^{3/2} = 5196$$

and

$$\exp\left(-\frac{E_g}{2kT}\right) = \exp\left(-\frac{0.72\text{ eV}}{2\times 8.61\times 10^{-5}\text{eV/K}\times 300\text{K}}\right)$$

$$= \exp(-13.846) = 9.7\times 10^{-7}$$

$$\therefore \quad n_i = 4.83\times 10^{21}\times 5196\times 9.7\times 10^{-7} = \mathbf{3.4\times 10^{19}/m^3}.$$

37.10 THE FRACTION OF ELECTRONS IN THE CONDUCTION BAND

The Fermi-Dirac probability function gives the fractional occupancy of the energy states. Thus, $f(E) = \dfrac{1}{1+\exp\left[(E-E_F)/kT\right]}$ gives the probability that an electron occupies the energy state E.

The probability that an electron occupies the energy state E_C may then be found from

$$f(E_C) = \frac{1}{1+\exp\left[(E_C-E_F)/kT\right]} \qquad \text{...(37.47)}$$

But, according to the definition of probability, $f(E_C) = \dfrac{n}{N}$

where n is the number of electrons excited to conduction band levels and n is the total number of electrons available in the valence band initially.

$$\therefore \quad \frac{n}{N} = \frac{1}{1+\exp\left[(E_C-E_F)/kT\right]} \qquad \text{...(37.48)}$$

As $(E_C - E_F) = E_g/2$, we write

$$\frac{n}{N} = \frac{1}{1+\exp\left[E_g/2kT\right]}$$

Since $E_g > 2kT$, the factor unity may be neglected in comparison to the exponential term.

$$\therefore \quad \frac{n}{N} = \frac{1}{\exp\left[E_g/2kT\right]} = e^{-E_g/2kT} \qquad ...(37.49)$$

Eq. (37.49) gives the fraction of electrons in the conduction band.

Example 37.3: *What is the probability of an electron being thermally promoted to the conduction band in diamond at 27° C, if the bandgap is 5.6 eV wide?*

Solution: The probability that an electron being thermally promoted to the conduction band is given by

$$f(E_C) = \frac{1}{1+\exp\left(E_g/2kT\right)} = \frac{1}{1+\exp\left[\frac{5.6\,\text{eV}}{2(0.026)}\text{eV}\right]}$$

$$= \frac{1}{1+e^{107.69}} = \mathbf{1.7\times10^{47}}$$

Example 37.4: *Estimate the fraction of electrons in the conduction band at 300°K of (i) Germanium (E_g = 0.72 eV) (ii) Silicon (E_g = 1.1 eV) and (iii) Diamond (E_g = 5.6 eV) What is the significance of these results?*

Solution:

(*i*) **Germanium:** $f(E_C) = e^{-E_g/2kT} = e^{\frac{0.72\,\text{eV}}{2\times0.026\,\text{eV}}} = e^{-13.85} = \mathbf{9.66\times10^{-7}}$

(*ii*) **Silicon:** $f(E_C) = e^{-E_g/2\,kT} = e^{\frac{1.1\text{eV}}{2\times0.026\,\text{eV}}} = \mathbf{6.5\times10^{-10}}$

(*iii*) **Diamond:** $f(E_C) = e^{-E_g/2kT} = e^{\frac{5.6\text{eV}}{5\times0.026\,\text{eV}}} = e^{-107.7} = \mathbf{1.7\times10^{-47}}$

The above results show that the larger the band gap the smaller the electrons that can go into the conduction band, at a given temperature.

Example 37.5: *Assuming that the number of electrons near the top of the valence band available for thermal excitation is 5 × 10^25/m^3 and the intrinsic carrier density is 2.5 × 10^19/m^3, calculate the energy gap of germanium at room temperature.*

Solution: Fraction of electrons in conduction band $\frac{n}{N} = \exp\left[-\frac{E_g}{2kT}\right]$

$$\therefore \quad E_g = -2kT\ \ln\left(\frac{n}{N}\right)$$

$$= -2(8.61\times10^{-5}\text{eV/K})(300\,\text{K})\ \ln\left[\frac{2.5\times10^{19}/\text{m}^3}{5\times10^{25}/\text{m}^3}\right]$$

$$= 0.052\times14.509\ \text{eV} = \mathbf{0.75\ eV}.$$

Example 37.6: *Estimate the fraction of electrons in conduction band at room temperature in Ge with E_g = 0.72 eV and in diamond with E_g = 5.6 eV.*

Solution: Fraction of electrons in conduction band $\frac{n}{N} = \exp\left[-\frac{E_g}{2kT}\right]$

(*i*) Fraction of electrons in conduction band in Ge,

$$\frac{n}{N} = \exp\left[-\frac{0.72 \text{ eV}}{2\times 0.026 \text{ eV}}\right] = \exp(-13.846) = \mathbf{9.7 \times 10^{-7}}.$$

(*ii*) Fraction of electrons in conduction band in diamod,

$$\frac{n}{N} = \exp\left[-\frac{5.6 \text{ eV}}{2\times 0.026 \text{ eV}}\right]$$

$$= \exp(-107.692) = \mathbf{1.7 \times 10^{-47}}.$$

37.11 FERMI LEVEL IN INTRINSIC SEMICONDUCTOR

In a pure semiconductor, the electrons in the conduction band gather close to the bottom edge of the band as shown in Fig. 37.9.

Similarly, we assume that the holes are at the top edge of the valence band. The electron concentration in the conduction band is given by

$$n = N_C e^{-(E_C - E_F)/kT}$$

The hole concentration in the valence band is given by

$$p = N_V e^{-(E_F - E_V)/kT}$$

In an intrinsic semiconductor, the electron and hole concentrations are equal. Thus, $n = p$

$$N_C e^{-(E_C - E_F)/kT} = N_V e^{-(E_F - E_V)/kT} \quad ...(37.50)$$

Fig. 37.9: Fermi level in intrinsic semiconductor

Taking logarithm on both sides, we get

$$-\frac{(E_C - E_F)}{kT} = \ln\frac{N_V}{N_C} - \frac{(E_F - E_V)}{kT}$$

$$-E_C + E_F = kT \ln\frac{N_V}{N_C} - E_F + E_V$$

$$2E_F = (E_C + E_V) + kT \ln\frac{N_V}{N_C}$$

$$\therefore \quad E_F = \frac{E_C + E_V}{2} + \frac{1}{2}kT \ln\frac{N_V}{N_C} \quad ...(37.51)$$

But $N_C = 2\left[\dfrac{2\pi\, m_e^* kT}{h^2}\right]^{3/2}$ and $N_V = 2\left[\dfrac{2\pi\, m_h^* kT}{h^2}\right]^{3/2}$

$$\therefore \quad \frac{N_V}{N_C} = \left(\frac{m_h^*}{m_e^*}\right)^{3/2}$$

$$\therefore \quad \ln\left(\frac{N_V}{N_C}\right) = \frac{3}{2}\ln\left(\frac{m_h^*}{m_e^*}\right)$$

$$\therefore \qquad E_F = \frac{E_C + E_V}{2} + \frac{3}{4}kT \ \ln\left(\frac{m_h^*}{m_e^*}\right) \qquad ...(37.52)$$

We can also write the above equation as

$$E_F = \frac{E_C + E_V}{2} - \frac{3}{4}kT \ \ln\left(\frac{m_e^*}{m_h^*}\right) \qquad ...(37.53)$$

If the effective mass of a free electron is assumed to be equal to the effective mass of a hole, i.e.,

$$m_h^* = m_e^*$$

$$\ln\left(\frac{m_h^*}{m_e^*}\right) = 0$$

$$\therefore \qquad E_F = \frac{E_C + E_V}{2} \qquad ...(37.54)$$

To make the meaning of the Eq. (37.54) more explicit, we write

$$E_F = \frac{E_C - E_V}{2} + E_V$$

$$E_F = \frac{E_g}{2} + E_V$$

If we denote the top of the valence band E_V as zero level, $E_V = 0$.

$$\therefore \qquad E_F = \frac{E_g}{2} \qquad ...(37.55)$$

Eq. (37.55) shows that in an intrinsic semiconductor the Fermi level lies in the middle of the forbidden gap. We made the following assumptions in obtaining the Eq. (37.55).

(*i*) It was assumed that the electrons undergo transitions from the top edge level, E_V, of the valence band to the bottom edge level, E_C, of the conduction band. In reality, transitions are possible between the other levels also. However, the above result does not appreciably differ if other transitions are also taken into account.

(*ii*) It was assumed that the effective mass of electrons in the conduction band, m_e^*, is exactly equal to the effective mass of the holes in the valence band, m_h^*. In practice, the effective masses differ from each other. However, the difference does not alter the above result significantly.

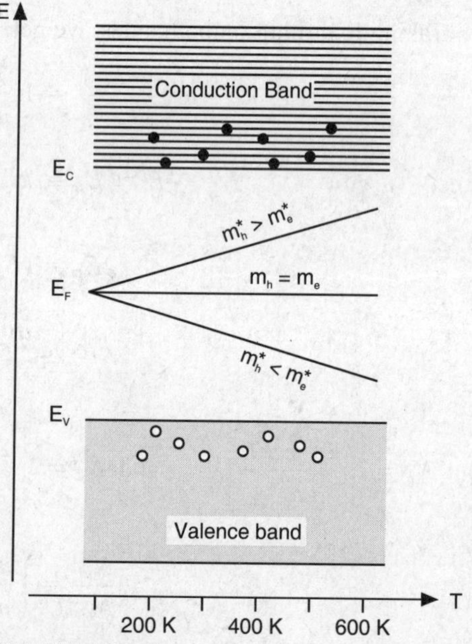

Fig. 37.10: Variation of Fermi level with temperature in an intrinsic semiconductor

An important point to be noted here is that the ***Fermi level is not an allowed energy level*** in semiconductors. It only serves as a reference energy with reference to which we specify the energies of electrons and holes in a semiconductor.

37.11.1 Variation of Fermi Level with Temperature in an Intrinsic Semiconductor

With an increase in temperature, the Fermi level gets displaced upward to the bottom edge of the conduction band if $m_h^* > m_e^*$ or downward to the top edge of the valence band if $m_h^* < m_e^*$, as indicated in Fig. 37.10.

In most of the materials, the shift of Fermi level on account of $m_h^* \neq m_e^*$ is insignificant. The Fermi level in an intrinsic semiconductor may be considered as independent of temperature and as staying in the middle of the band gap.

Example 37.7: *Determine the position of Fermi level in silicon semiconductor at 300 K. Given that the band gap is 1.12 eV, and $m_e^* = 0.12\ m$ and $m_h^* = 0.28m$.*

Solution:

$$E_F = \frac{E_g}{2} + \frac{3kT}{4} \ln\left(\frac{m_h^*}{m_e^*}\right)$$

$$= \frac{1.12 \text{ eV}}{2} + \frac{3 \times 8.61 \times 10^{-5} \text{ eV/K} \times 300 \text{ K}}{4} \ln\left(\frac{0.28 \text{m}}{0.12 \text{m}}\right)$$

$$= 0.56 \text{ eV} + (0.0194) \ln 2.333 \text{ eV}$$

$$= 0.56 \text{ eV} + (0.0194)(0.8473) \text{ eV}$$

$$= 0.56 \text{ eV} + 0.016 \text{ eV} = \mathbf{0.576\ eV.}$$

The Fermi level is 0.016 eV above the centre of the forbidden gap. In other words, it is at 0.576 eV from the top of the valence band.

Example 37.8: *If the effective mass of an electron is equal to twice the effective mass of hole, determine the position of the Fermi level in an intrinsic semiconductor from the center of forbidden gap at room temperature.*

Solution:

$$E_F = \frac{E_g}{2} + \frac{3}{4} kT \ln\left(\frac{m_h^*}{m_e^*}\right)$$

$$\frac{E_g}{2} - \frac{3}{4} kT \ln 2 = \frac{E_g}{2} - \frac{3}{4}(0.026 \text{ eV})(0.69)$$

or

$$E_F = \frac{E_g}{2} - 0.0135 \text{ eV}$$

The Fermi level is below the centre of the forbidden gap by 0.014 eV.

37.12 VARIATION OF INTRINSIC CONDUCTIVITY WITH TEMPERATURE

Incorporating the expression for n_i from Eq. (37.43) into the Eq. (37.18), we get

$$\sigma = \sqrt{N_C N_V}\, e(\mu_e + \mu_h)\, e^{-E_g/2kT}$$

or

$$\sigma = \sigma_o e^{-E_g/2kT} \quad ...(37.56)$$

where σ_o is a constant.

Eq. (37.56) gives the temperature dependence of conductivity of an intrinsic semiconductor, which is dominated by the exponential term $e^{-E_g/2kT}$.

Taking logarithm on both sides of Eq. (37.56), we get

$$\ln \sigma_i = \ln \sigma_o - \frac{E_g}{2kT} \qquad ...(37.57)$$

Fig. 37.11 shows a plot of ln σ_i versus 1/T. It shows that the conductivity increases with temperature.

Fig. 37.11: Variation of intrinsic conductivity with temperature

Example 37.9: *Find the resistivity of intrinsic germanium at 300°K. Given that the intrinsic density carriers is $2.5 \times 10^{19}/m^3$.*

Solution:

$$\sigma_i = en_i(\mu_e + \mu_h)$$

$$= 1.602\times10^{-19}\,\text{C}\times2.5\times10^{19}\times(0.39+0.19)\,\text{m}^2\text{V}^{-1}\text{s}^{-1}$$

$$= 2.32 \text{ mho/m}$$

$$\therefore \qquad \rho_i = \frac{1}{\rho_i} = \mathbf{0.43\,ohm - m.}$$

37.13 DETERMINATION OF BAND GAP

Eq. (37.56) may be rewritten in terms of resistivity of the intrinsic semiconductor as follows:

$$\rho_i = \rho_o e^{E_g/2kT} \qquad ...(37.58)$$

or

$$\frac{R_i A}{L} = \rho_o e^{E_g/2kT}$$

$$\therefore \qquad R_i = Ce^{E_g/2kT} \qquad ...(37.59)$$

Taking logarithms on both sides, we obtain

$$\ln R_i = \ln C + \frac{E_g}{2kT} \qquad ...(37.60)$$

A plot of *ln* R_i versus $1/T$ gives a straight line, as shown in Fig. 37.12. The slope of the straight line gives the value of E_g. In practice we measure the resistance of intrinsic semiconductor at different temperatures with the help of four-point probe technique and draw a plot between *ln* R_i and $10^3/T$. The slope of the straight line thus obtained gives the value of $E_g/(2k)$. Hence,

$$E_g = m(2k) \qquad ...(37.61)$$

where m is the slope of the straight line.

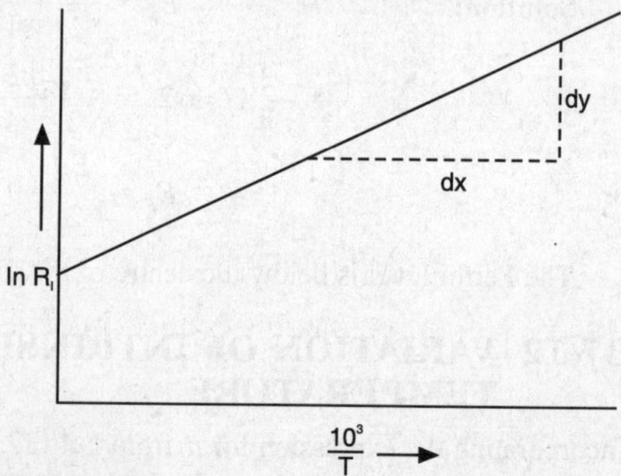

Fig. 37.12: Variation of resistance with temperature in intrinsic semiconductor

Example 37.10: *The resistivity of an intrinsic semiconductor is 4.5 Ω.m at 20°C and 2.0 Ω.m at 32°C. Find the energy gap.*

Solution:

$$E_g = 2k\left(\frac{dy}{dx}\right) = 2k\left(\frac{\log\rho_2 - \log\rho_1}{(1/T_1 - 1/T_2)}\right)$$

$$= 2\times 8.61\times 10^{-5}\,\text{eV/K}\left(\frac{\log 2 - \log 4.5}{(1/293)-(1/305)}\right)\text{K}$$

$$= 2\times 8.61\times 10^{-5}\,\text{eV}\left(\frac{0.6532 - 0.3010}{0.134\times 10^{-3}}\right)$$

$$= 2 \times 8.61 \times 10^{-5}\ \text{eV} \times 2.63 \times 10^{3} = \mathbf{0.45\ eV}$$

Example 37.11: *The resistivity of intrinsic silicon is 2.3 × 10³ Ω-m at 300K. Calculate its resistivity at 100°C. Assume E_g = 1 eV and k = 1.38 × 10⁻²³ J/K.*

Solution: The resistivity of an intrinsic semiconductor is given by $\rho_i = \rho_o e^{E_g/2kT}$.

∴ The resistivity of the intrinsic semiconductor at 300 K is given by $\rho_1 = \rho_o e^{E_g/2k(300\text{ K})}$

Resistivity of the semiconductor at 373 K is given by $\rho_2 = \rho_o e^{E_g/2k(373\text{K})}$.

$$\therefore \quad \frac{\rho_2}{\rho_1} = \frac{\exp\left[\dfrac{1\text{eV}}{2\times 8.61\times 10^{-5}\text{eV}\times 373\text{K}}\right]}{\exp\left[\dfrac{1\text{eV}}{2\times 8.61\times 10^{-5}\text{eV}\times 300\text{K}}\right]}$$

$$\therefore \quad \rho_2 = \rho_1\frac{\exp\left[\dfrac{1\text{eV}}{2\times 8.61\times 10^{-5}\text{eV}\times 373\text{K}}\right]}{\exp\left[\dfrac{1\text{eV}}{2\times 8.61\times 10^{-5}\text{eV}\times 300\text{K}}\right]}$$

$$= 2.3\times 10^{3}\,\frac{\exp\left[\dfrac{1\text{eV}}{2\times 8.61\times 10^{-5}\text{eV}\times 373\text{K}}\right]}{\exp\left[\dfrac{1\text{eV}}{2\times 8.61\times 10^{-5}\text{eV}\times 300\text{K}}\right]}\,\Omega.\text{m}$$

$$= \mathbf{52\ \Omega.m.}$$

37.14 LIMITATIONS OF INTRINSIC SEMICONDUCTOR

We cannot use intrinsic semiconductors, as they are, in device manufacture. They possess low conductivity which strongly depends on temperature. If we take a crystal of pure silicon or germanium and connect it in the circuit, the current in the circuit will increase with voltage. *Increasing the voltage increases the current only proportionally*, obeying Ohm's law. We will also find that the current in the circuit will gradually increase as the temperature of the crystal is increased. *Increasing the temperature increases the current at an exponential rate.*

We summarize the limitations as follows:

(*i*) Conductivity is low. Germanium has a conductivity of 1.67 S/m, which is nearly 10^7 times smaller than that of copper.

(*ii*) Conductivity is a function of temperature and increases exponentially as the temperature increases.

(*iii*) Conductivity cannot be controlled from outside.

37.15 EXTRINSIC SEMICONDUCTORS

A judicious introduction of impurity atoms in an otherwise perfect semiconductor crystal produces useful modifications of its electrical conductivity. It makes the current more voltage dependent than temperature dependent. A controlled amount of impurity added into an intrinsic semiconductor is known as ***doping***. The impurity which is introduced is known as a ***dopant***. A semiconductor doped with impurity atoms is called an ***extrinsic semiconductor***.

The impurity-produced electrons are not temperature-dependent but are voltage-dependent and they will be under our control.

One of the important methods of doping is to add precisely determined quantities of impurity to the melt from which the semiconductor crystal is grown. This way, a crystal having a constant impurity density is obtained. Typical doping levels range from 10^{20} to 10^{27} impurity atoms/m^3. Pentavalent elements from Group V or trivalent elements from Group III are used as dopants. The atoms belonging to these two groups are nearly of the same size as silicon or germanium atoms and easily substitute themselves in place of some of the host atoms in the semiconductor crystal. Thus, they are substitutional impurities and do not cause any distortion in the original crystal structure. Depending on the two different types of doping, two types of extrinsic semiconductors are possible. They are *n-type and p-type* semiconductors.

Common Dopant Elements for Silicon and Germanium	
***n*-type**	***p*-type**
Phosphorous	Aluminium
Arsenic	Boron
Antimony	Gallium
Indium	

Advantages of Extrinsic Semiconductors

(*i*) Conductivity is high.

(*ii*) Conductivity can be tailored to the desired value through the control of doping concentration.

(*iii*) Conductivity is not a function of temperature

37.16 *n*-TYPE SEMICONDUCTOR

An *n*-type semiconductor is produced when a pure semiconductor is doped with a pentavalent impurity such as phosphorous. A phosphorous atom has five valence electrons. Out of the five electrons, only four participate in bonding with four host silicon atoms while the fifth electron remains loosely bound. The host silicon lattice is a dielectric medium having a dielectric constant of 12. As a result, the Coulomb force between the phosphorous nucleus and the fifth electron is smaller than that it would be in free space. Therefore, the ionization energy of the fifth electron is very small. It is found to be 0.045 eV. The ionization energy is so small that the thermal energy can easily liberate the fifth electron from the nucleus.

It means that the energy levels corresponding to phosphorous atoms are nearer to the bottom edge of the conduction band. At normal temperatures, the fifth electron becomes free to move about in the crystal and act as a charge carrier. That is the electron jumps into the conduction band leaving behind the ***positive phosphorous ion*** that is fixed in the crystal lattice. As the phosphorous atom is donating an electron for the purpose of electrical conduction, it is called a ***donor atom***.

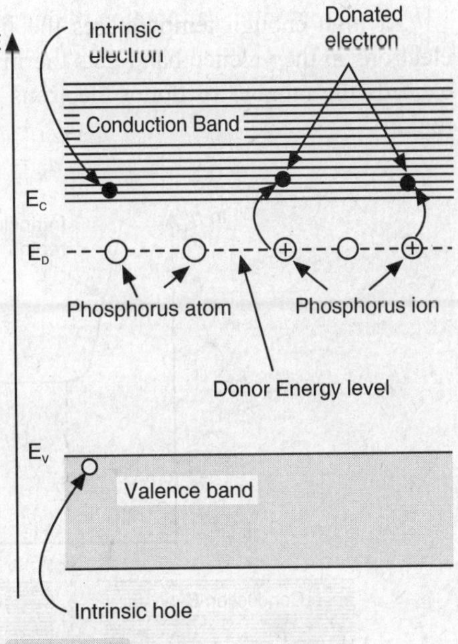

Fig. 37.13: Energy band diagram of *n*-type semiconductor

Energy Band Diagram

The energy band diagram of n-type semiconductor is shown in Fig. 37.13. If the donor atom density is low, the donor atoms are distantly spaced from one another, approximately, by 100 atom spacings. In such a situation, the donor atoms cannot interact with each other and their energy levels are discrete levels, E_D. They are called ***donor levels*** and represent the ground state of the fifth electron of impurity atom. As even small amount of thermal energy can readily liberate the fifth electron from the atom and send it into the conduction band, the donor levels are expected to be located very near to the bottom edge of the conduction band.

37.16.1 Temperature Variation of Carrier Concentration

The general dependence of electron concentration on temperature is shown in Fig. 37.14. At 0 K, the donor atoms are not ionized which means that all the donor electrons are bound to the donor atoms. The conduction band is empty, while the valence band is full. The material behaves essentially as an insulator.

At slightly elevated temperatures, the donor atoms are ionized and the donor electrons go into the conduction band (Fig. 37.14, ionization region) by getting energy from lattice vibrations. In this process holes are not produced in the valence band. At about 100K, the donor levels are all ionized. Once all the donor atoms are ionized, further increase in temperature does not produce electrons and the curve levels off. The plateau region is called the *depletion region*. In the depletion region, the electron concentration in the conduction band is nearly identical to the concentration of the dopant atoms.

If N_D is the concentration of donor atoms, then

$$n_n \cong N_D \quad ...(37.62)$$

where n_n is the electron concentration in the conduction band of *n*-type material.

As temperature increases further, some electrons from the valence band are also excited into the conduction band (Fig. 37.14, depletion region). The conduction band, therefore, contains electrons that have come through two different processes-namely (*i*) donor atom ionization and (*ii*) intrinsic process. The intrinsic process produces holes in the valence band.

At high enough temperatures the intrinsic behaviour takes over since large number of electrons in the valence band gets thermally excited to conduction band and their number far exceeds the number of donor electrons (Fig. 37.14, intrinsic region). Therefore, in intrinsic region

$$n_n = n_i \quad \text{...(37.63)}$$

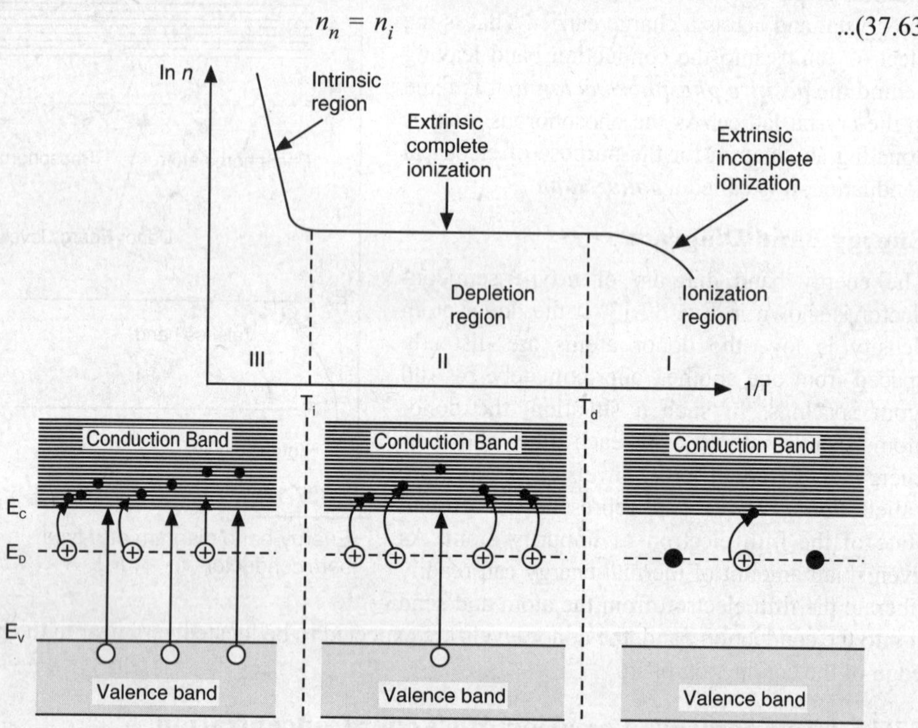

Fig. 37.14: Variation of electron concentration in an *n*-type semiconductor as a function of the inverse of temperature

In an *n* type semiconductor at moderate temperatures, there are some 10^{10} electrons and 10^{10} holes which have been generated due to heat-ruptured bonds in addition to the 10^{16} free electrons received from the donors. Donor electrons are in the majority and their number is essentially constant at all usual temperatures; this is because they are not part of covalent bonds and only 0.05 eV is required to free them. Thus, by doping we produce a less temperature-sensitive semiconductor with an abundance of electrons.

In *n*-type material the electrons outnumber the holes and constitute the *majority carriers* (in region II). Holes are *minority carriers*. The number of carriers is independent of temperature in the depletion region. The current in this type of crystal is mainly due to the negatively charged electrons and hence the material is called ***n-type semiconductor***.

37.16.2 Carrier Concentration in n-type Semiconductor at Low Temperatures: (In the Ionization Region)

The energy band diagram of a *n*-type semiconductor is shown in Fig. 37.13. Let N_D be the concentration of donors in the material. At 0K, the donor atoms are not ionized and are at the level E_D which is very near to E_C. When the temperature is raised above 0°K, the donor atoms get ionized and free electrons appear in the conduction band. With increase in temperature more and more donor atoms get ionized and the electron concentration in the conduction band increases. Electrons require an energy E_D, for their transition to the conduction band from the donor levels. Therefore, we may assume that the electron concentration, n, in the conduction band is

$$n = N_D^+$$

or

$$n = N_D - N_D^0$$

where N_D^+ is the number of donor atoms that are ionized and N_D^0 is the number of atoms left unionized at the energy level E_D.

The concentration of ionized donors $N_D^+ = \left(N_D - N_D^0\right) = N_D[1 - f(E_D)] = \dfrac{N_D}{1+e^{-(E_D-E_F)/kT}}$

$$\therefore \quad n = \frac{N_D}{1+e^{-(E_D-E_F)/kT}}$$

From the operational definition of Fermi level it is expected that the Fermi level in n-type semiconductor lies a few kT above E_D. Therefore, the above equation may be simplified as

$$n = N_D e^{(E_D-E_F)/kT} \quad \text{...(37.64)}$$

But the electron concentration, n, in the conduction band is given by

$$n = N_C e^{-(E_C-E_F)/kT} \quad \text{...(37.65)}$$

$$\therefore \quad N_D e^{(E_D-E_F)/kT} = N_C e^{-(E_C-E_F)/kT}$$

Taking logarithm and rearranging the terms we get

$$\left(\frac{E_D - E_F}{kT}\right) + \left(\frac{E_C - E_F}{kT}\right) = \ln \frac{N_C}{N_D} \quad \text{...(37.66)}$$

$$(E_D + E_C) - 2E_F = (kT) \ln \frac{N_C}{N_D}$$

or

$$E_F = \frac{E_D + E_C}{2} - \left(\frac{kT}{2}\right) \ln \frac{N_C}{N_D}$$

or

$$E_F = \frac{E_D + E_C}{2} + \left(\frac{kT}{2}\right) \ln \frac{N_D}{N_C}$$

or

$$E_F = \frac{E_D + E_C}{2} + \left(\frac{kT}{2}\right) \ln \frac{N_D}{2\left(2\pi\, m_e^* kT/h^2\right)^{3/2}} \quad \text{...(37.67)}$$

It is seen from Eq. (37.67) that at T = 0K,

$$E_F = \frac{E_D + E_C}{2} \quad \text{...(37.68)}$$

That is, the equilibrium Fermi level lies midway between the bottom of the conduction band and donor levels. Now,

$$\exp\left[\frac{E_F - E_C}{kT}\right] = \exp\left[\frac{E_D + E_C}{2kT} + \left(\frac{1}{2}\right) \ln \frac{N_D}{2\left(2\pi\, m_e^* kT/\text{h}^2\right)^{3/2}} - \frac{E_C}{kT}\right]$$

$$= \exp\left[\frac{E_D - E_C}{2kT} + \left(\frac{1}{2}\right) \ln \frac{N_D}{2\left(2\pi\, m_e^* kT/\text{h}^2\right)^{3/2}}\right]$$

$$= \exp\left[\frac{E_D - E_C}{2kT} + \ln\sqrt{\frac{N_D}{2\left(2\pi\, m_e^* kT/\text{h}^2\right)^{3/2}}}\right] \qquad \left[\because \frac{1}{2}\ln x = \ln\sqrt{x}\right]$$

$$= \exp\left[\left(\frac{E_D - E_C}{2kT}\right)\right]\cdot \exp\left[\ln\sqrt{\frac{N_D}{2\left(2\pi\, m_e^* kT/\text{h}^2\right)^{3/2}}}\right] \left[\begin{matrix}\because \exp\,(a+b) = \\ \exp(a)+\exp(b)\end{matrix}\right]$$

$$= \exp\left[\left(\frac{E_D - E_C}{2kT}\right)\right]\cdot\left[\sqrt{\frac{N_D}{2\left(2\pi\, m_e^* kT/\text{h}^2\right)^{3/2}}}\right] \qquad [\exp\,(\ln x) = x]$$

$$\therefore\ n = N_C \exp\left[\frac{E_F - E_C}{kT}\right] = 2\left[\frac{2\pi\, m_e^* kT}{\text{h}^2}\right]^{3/2} \exp\left[\left(\frac{E_D - E_C}{2kT}\right)\right]\cdot\left[\sqrt{\frac{N_D}{2\left(2\pi\, m_e^* kT/\text{h}^2\right)^{3/2}}}\right]$$

or

$$n = \left(2N_D\right)^{\frac{1}{2}}\left[\frac{2\pi\, m_e^* kT}{\text{h}^2}\right]^{3/4} \exp\left[\left(\frac{E_D - E_C}{2kT}\right)\right] \qquad ...(37.69)$$

From Eq. (37.69) gives the majority carrier concentration. The electron concentration in the conduction band of an *n*-type semiconductor is proportional to the square root of the donor concentration at fairly low temperatures.

37.17 p-TYPE SEMICONDUCTOR

A *p*-type semiconductor is produced when a pure semiconductor is doped with a trivalent impurity such as boron. Boron atom has three valence electrons. Therefore, it falls short of one electron for completing the four covalent bonds with its neighbours. When an electron from a neighbouring atom acquires energy and jumps into the vacancy to form the fourth bond, it leaves behind a hole. The boron atom having acquired an additional electron becomes a negative ion. The hole can move freely in the valence band whereas the impurity ion is fixed in position by the covalent bonds. As the boron atom accepted an electron from the valence band, it is called an *acceptor atom*. The acceptor impurity atoms produce holes without the simultaneous generation of the electrons in the conduction band.

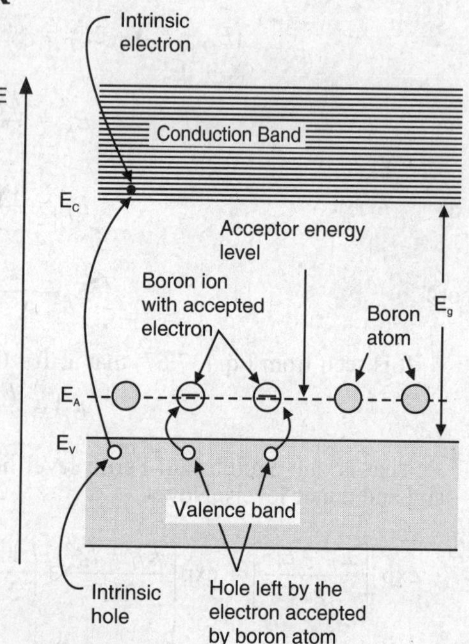

Fig. 37.15: Energy band diagram for a *p*-type semiconductor

Energy Band Diagram

The energy band diagram of p-type semiconductor is shown in Fig. 37.15. If the acceptor atom density is low, the acceptor atoms are distantly spaced from one another. As such, the acceptor atoms cannot interact with each other and their energy levels are discrete levels, E_A. They are called ***acceptor levels*** and represent the ground state of the hole. As even small amount of thermal energy can make an electron in the valence band jump into the acceptor level, the acceptor levels are expected to be located very near to the top edge of the valence band. They are at about 0.01 eV above the valence band. A hole may be said to have moved from the acceptor atom to the valence band.

At 0 K the acceptor levels are vacant and the valence band is full. The conduction band is also vacant. The material behaves essentially as an insulator.

At slightly elevated temperature, electrons from the valence band jump into the acceptor levels and holes are generated in the valence band (Fig. 37.15). In this process holes are generated without the simultaneous generation of electrons. At normal temperatures the acceptor levels are saturated and a few electrons are excited to the conduction band also. At about 100K, the acceptor atoms are all ionized. Once all the acceptor atoms are ionized, further increase in temperature does not produce holes and we say the acceptor levels are saturated. The region is called the *saturation region*. In the saturation region, the hole concentration in the valence band is nearly identical to the concentration of the acceptor atoms.

If N_A is the concentration of acceptor atoms, then

$$p_p \cong N_A \qquad ...(37.70)$$

where p_p is the hole concentration in the valence band of *p*-type material.

As temperature is increased further, some electrons from the valence band are excited into the conduction band. The valence band now contains holes that have been generated by two different processes-namely, (*i*) acceptor atom ionization and (*ii*) intrinsic process. The intrinsic process causes electrons to appear in the conduction band.

At high enough temperatures, a large number of electron-hole pairs are generated and the number of holes generated thermally far exceeds the number of holes due to acceptor impurity. The material behaves as an intrinsic semiconductor. In the intrinsic region

$$p_p = n_i \qquad ...(37.71)$$

In *p*-type material the holes outnumber the electrons and constitute the *majority carriers*. Electrons are *minority carriers*. The number of majority carriers is independent of temperature in the depletion region. The current in this type of crystal is mainly due to the positively charged holes and hence the material is called ***p-type semiconductor***.

37.17.1 Carrier Concentration in p-type Semiconductor

The energy band diagram of a *p*-type semiconductor is shown in Fig. 37.15. Let N_A be the concentration of acceptors in the material. At 0K, the acceptor atoms are not ionized and are at the level E_A which is very near to E_V. When the temperature is raised above 0°K, the acceptor atoms get ionized and holes appear in the valence band. With increase in temperature more and more acceptor atoms get ionized and the hole concentration in the valence band increases. Since transition of electrons to the acceptor levels from the valence band requires an energy E_A, they can go to acceptor levels and ionize acceptor atoms. Therefore, we may assume that the hole concentration, *p*, in the conduction band is

$$p = N_A^-$$

where N_A^- is the number of acceptor atoms that are ionized.

The concentration of ionized acceptors $N_A^- = N_A f(E_A) = N_A \exp\left(\frac{E_F - E_A}{kT}\right)$

$$\therefore \qquad p = N_A \exp\left(\frac{E_F - E_A}{kT}\right) \qquad ...(37.72)$$

But the hole concentration, p, in the valence band is given by

$$p = N_V e^{-(E_F - E_V)/kT}$$

$$= N_V \exp\left(\frac{E_V - E_F}{kT}\right) \qquad ...(37.73)$$

$$\therefore \qquad N_A \exp\left(\frac{E_F - E_A}{kT}\right) = N_V \exp\left(\frac{E_V - E_F}{kT}\right)$$

Taking logarithm and rearranging the terms we get

$$\left(\frac{E_F - E_A}{kT}\right) - \left(\frac{E_V - E_F}{kT}\right) = \ln \frac{N_V}{N_A}$$

$$-(E_V + E_A) + 2E_F = (kT) \ln \frac{N_V}{N_A}$$

or
$$E_F = \frac{E_V + E_A}{2} + \left(\frac{kT}{2}\right) \ln \frac{N_V}{N_A}$$

or
$$E_F = \frac{E_V + E_A}{2} - \left(\frac{kT}{2}\right) \ln \frac{N_A}{N_V}$$

or
$$E_F = \frac{E_V + E_A}{2} - \left(\frac{kT}{2}\right) \ln \frac{N_A}{2\left(2\pi\, m_h^* kT/h^2\right)^{3/2}} \qquad ...(37.74)$$

It is seen from Eq. (37.74) that at $T = 0$K,

$$E_F = \frac{E_V + E_A}{2} \qquad ...(37.75)$$

That is, the equilibrium Fermi level lies midway between the bottom of the valence band and acceptor levels. Now,

$$\exp\left[\frac{E_V - E_F}{kT}\right] = \exp\left[\frac{E_V}{kT} - \frac{E_V + E_A}{2kT} + \left(\frac{1}{2}\right) \ln \frac{N_A}{2\left(2\pi m_h^* kT/h^2\right)^{3/2}}\right]$$

$$= \exp\left[\frac{E_V - E_A}{2kT} + \left(\frac{1}{2}\right) \ln \frac{N_A}{2\left(2\pi m_h^* kT/h^2\right)^{3/2}}\right]$$

$$= \exp\left[\frac{E_V - E_A}{2kT} + \ln \sqrt{\frac{N_A}{2\left(2\pi m_h^* kT/h^2\right)^{3/2}}}\right] \qquad \left[\because \frac{1}{2} \ln x = \ln \sqrt{x}\right]$$

$$= \exp\left[\left(\frac{E_V - E_A}{2kT}\right)\right] \cdot \exp\left[\ln\sqrt{\frac{N_A}{2\left(2\pi\, m_h^* kT/h^2\right)^{3/2}}}\right] \left[\because \exp\,(a+b) = \exp(a) + \exp(b)\right]$$

$$= \exp\left[\left(\frac{E_V - E_A}{2kT}\right)\right] \cdot \left[\sqrt{\frac{N_A}{2\left(2\pi\, m_h^* kT/h^2\right)^{3/2}}}\right] \qquad [\because \exp\,(\ln x) = x]$$

$$\therefore p = N_V \exp\left[\frac{E_V - E_A}{kT}\right] = 2\left[\frac{2\pi\, m_h^* kT}{h^2}\right]^{3/2} \exp\left[\left(\frac{E_V - E_A}{2kT}\right)\right] \cdot \left[\sqrt{\frac{N_A}{2\left(2\pi\, m_h^* kT/h^2\right)^{3/2}}}\right]$$

or
$$p = \left(2N_A\right)^{\frac{1}{2}} \left[\frac{2\pi\, m_h^* kT}{h^2}\right]^{3/4} \exp\left[\left(\frac{E_V - E_A}{2kT}\right)\right] \qquad ...(37.76)$$

From Eq. (37.76) gives the majority carrier concentration of a *p*-type semiconductor. The hole concentration in the valance band is proportional to the square root of the acceptor concentration at fairly low temperatures.

37.18 BAND DIAGRAMS OF EXTRINSIC SEMICONDUCTORS AT 0K AND 300K

The energy band diagrams of extrinsic semiconductors at 0K and 300K showing the position of Fermi level in each case is shown in Fig. 37.16 and Fig. 37.17 respectively.

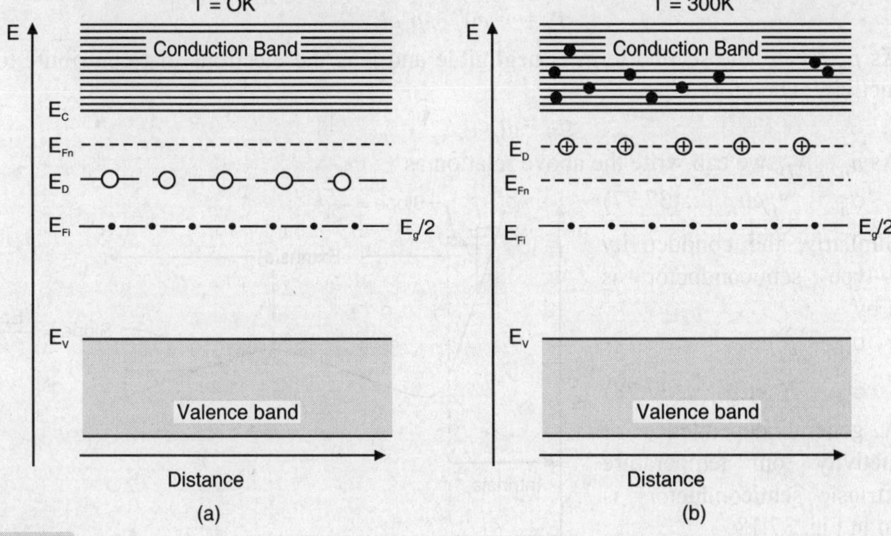

Fig. 37.16:

Energy band diagram of an *n*-type semiconductor (a) at 0K (b) at 300 K

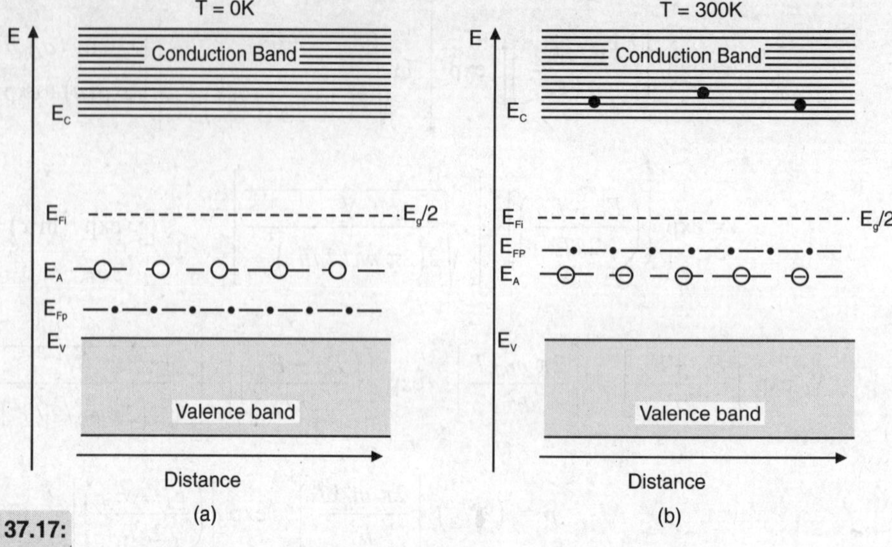

Fig. 37.17: Energy band diagram of an p-type semiconductor (a) at 0K (b) at 300 K

37.19 EXTRINSIC CONDUCTIVITY

The temperature range in which a doped semiconductor used is the extrinsic region where all the impurity atoms are ionized. This region is of practical interest since the carrier concentration is essentially independent of temperature in this region and any desired conductivity can be achieved by controlling the amount of impurities added from outside. Note that whenever extrinsic semiconductor is discussed, we invariably imply its characteristics in the extrinsic region. The conductivity of n-type semiconductor is given by

$$\sigma_n = n_n e\mu_e + p_n e\mu_h$$

As $p_n << n_n$, the second term is negligible and it is the electrons that contribute to the conductivity. Therefore,

$$\sigma_n = n_n e\mu_e$$

As $n_n = N_D$, we can write the above relation as

$$\sigma_n = N_D e\mu_e \quad ...(37.77)$$

Similarly, the conductivity of *p*-type semiconductor is given by

$$\sigma_p = p_p e\mu_h$$

or $$\sigma_p = N_A e\mu_h \quad ...(37.78)$$

A general dependence of conductivity on temperature in extrinsic semiconductors is shown in Fig. 37.18.

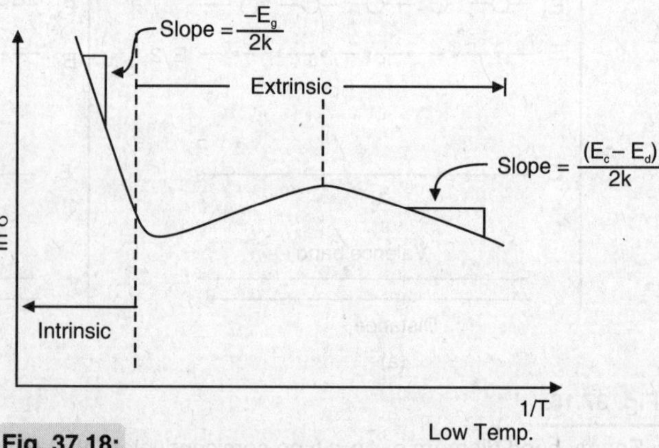

Fig. 37.18: Dependence of conductivity on temperature in extrinsic semiconductors

Example 37.12: *In a doped semiconductor, there are 4.52 × 10^{24} holes and 1.25 × 10^{14} electrons per cubic metre. What will be the carrier density in undoped specimen? Electron and hole mobilities are 0.38 m^2/V.s and 0.18 m^2/V.s respectively. Calculate the conductivity of intrinsic and the doped semiconductors.*

Solution. Intrinsic carrier density, $n_i = \sqrt{p_p n_p}$

$$\therefore \quad n_i = \sqrt{(4.52\times10^{24}\ /\ \text{m}^3)(1.25\times10^{14}\ /\ \text{m}^3)}$$

$$= (5.65 \times 10^{38}\ /\ \text{m}^6)^{1/2} = 2.38 \times 10^{19}\ /\text{m}^3$$

Conductivity of the intrinsic semiconductor

$$\sigma_i = en_i\,(\mu_e + \mu_h)$$

$$= (1.602\times10^{-19}\,\text{C})(2.38\times10^{19}\ /\ \text{m}^3)(0.38+0.18)\frac{\text{m}^2}{\text{V.s}}$$

$$= 2.14\frac{\text{C}}{\text{m.V.s}} = \mathbf{2.14\ S/m}$$

Conductivity of the doped semiconductor

$$\sigma_p = ep_p\mu_h$$

$$= (1.602\times10^{-19}\,\text{C})(4.52\times10^{24}\ /\ \text{m}^3)(0.18)\frac{\text{m}^2}{\text{V.s}}$$

$$= 1.30\times10^5\frac{\text{C}}{\text{m.V.s}} = \mathbf{130\ kS/m}$$

Unit: $1\dfrac{\text{C}}{\text{m.V.s}} = 1\dfrac{\text{A}}{\text{mV}} = 1\Omega^{-1}\,\text{m}^{-1} = 1\ \text{S /m}$

Example 37.13: *Silicon has a conductivity of only 5 × 10^{-4} $\Omega^{-1}m^{-1}$ in its pure form. An engineer wanted it to have conductivity of 200 $\Omega^{-1}m^{-1}$ and doped it with aluminium to produce p-type semiconductor. Calculate the impurity concentration. Assume μ_h=0.05 m^2/V.s.*

Solution: $\sigma = pe\mu_h = N_A e\mu_h.$

$$\therefore \quad N_A = \frac{\sigma}{e\mu_h} = \frac{200\ \Omega^{-1}\text{m}^{-1}}{1.602\times10^{-19}\,\text{C}\times0.05\ \text{m}^2/\text{V.s}}$$

$$= \mathbf{2.5 \times 10^{22}\ atoms/m^3}.$$

37.20 LAW OF MASS ACTION

In case of intrinsic semiconductors the product of *n* and *p* is a constant for a certain semiconductor at a certain temperature and is given by

$$np = n_i^2 = N_C N_V e^{-E_g/kT}$$

There is no condition in the expression that restricts it to intrinsic semiconductors because E_g does not change with impurity concentration and N_C and N_V are constants. This product is therefore a constant equally valid for intrinsic as well as for extrinsic semiconductors.

The electron and hole concentrations in extrinsic semiconductors may be given by expressions similar to Eq. (37.30) and Eq. (37.41). Denoting the electron concentration in *n*-type semiconductor by n_n and the hole concentration by p_n, we can write

$$n_n = N_C e^{-(E_C - E_F)/kT} \qquad ...(37.79)$$

and
$$p_n = N_V e^{-(E_F - E_V)/kT} \qquad ...(37.80)$$

$\therefore$
$$n_n p_n = N_C N_V e^{-E_g/kT} \qquad ...(37.81)$$

$\therefore$
$$n_n p_n = n_i^2 \qquad ...(37.82)$$

Similarly, if we denote the hole concentration in a *p*-type semiconductor by p_p and electron concentration by n_p, we can express

$$p_p = N_V e^{-(E_F - E_V)/kT} \qquad ...(37.83)$$

and
$$n_p = N_C e^{-(E_C - E_F)/kT} \qquad ...(37.84)$$

$\therefore$
$$p_p n_p = N_C N_V e^{-E_g/kT} \qquad ...(37.85)$$

$\therefore$
$$p_p n_p = n_i^2 \qquad ...(37.86)$$

Eq. (37.82) and Eq. (37.86) show that *the product of majority and minority carrier concentrations in an extrinsic semiconductor at a particular temperature is a constant and is equal to the square of intrinsic carrier concentration at that temperature.*

The law of mass action is very important relation because it in conjunction with charge neutrality condition enables us calculate minority carrier concentration. The law suggests that the addition of impurities to an intrinsic semiconductor increases the concentration of one type of carrier, which consequently becomes majority carrier and simultaneously decreases the concentration of the other carrier, which as a result becomes minority carrier. The minority carriers decrease in number below the intrinsic value because the majority carriers increase the rate of recombinations. The law of mass action states that the product of majority and minority carriers remains constant in an extrinsic semiconductor and it is independent of the amount of donor and acceptor impurity concentrations. Note that when the doping is heavy, the minority carrier concentration will be low and if doping is lighter, the minority carrier concentration will be larger.

37.21 CHARGE NEUTRALITY CONDITION

An extrinsic semiconductor is an electrically neutral body in its equilibrium condition. In *n*-type semiconductor, the total number of electrons in the conduction band must be equal to the sum of electrons originated from the donor atoms and electrons excited from the valence band. Electrons coming from donor levels leave behind positive donor ions while electrons excited from the valence band leave behind holes. These processes have not created any additional charges, and the equality between positive and negative charges remain undisturbed. The ***charge neutrality condition*** applied to the *n*-type semiconductor implies that the total negative charge of mobile electrons is equal to the total positive charge created in the crystal. It means that

$$n_n = N_D + p_n$$

where N_D is the donor impurity concentration and all the donor atoms are assumed to have got ionized.

But
$$n_n >> p_n$$

$\therefore$
$$n_n = N_D \qquad ...(37.87)$$

Eq. (37.87) indicates that *the* ***majority carrier concentration****, n_n, in an n-type semiconductor is equal to the donor impurity concentration, N_D.*

The charge neutrality for a *p*-type semiconductor requires that

$$p_p = N_A + n_p$$

As $p_p >> n_p$

$$p_p = N_A \quad ...(37.88)$$

37.22 CALCULATION OF MINORITY CARRIER CONCENTRATION

In the case of an *n*-type semiconductor, the majority carrier concentration n_n is given by

$$n_n = N_D$$

where N_D is the donor impurity concentration and all the donor atoms are assumed to have got ionized. The above relation indicates that *the **majority carrier concentration**, n_n, in an n-type semiconductor is equal to the donor impurity concentration, N_D*. The ***minority carrier concentration***, p_n, is then given by Eq. (37.82), as

$$p_n = \frac{n_i^2}{n_n}$$

or

$$p_n = \frac{n_i^2}{N_D} \quad ...(37.89)$$

Similarly, in case of a *p*-type semiconductor, the minority carrier concentration, n_p is given by

$$n_p = \frac{n_i^2}{p_p}$$

or

$$n_p = \frac{n_i^2}{N_A} \quad ...(37.90)$$

Example 37.14: *A sample of intrinsic germanium at room temperature has a carrier concentration of* $2.4 \times 10^{19} / m^3$. *It is doped with antimony at a rate of one antimony atom per million atoms of germanium. If the concentration of the germanium atoms is* $4 \times 10^{28} /m^3$, *determine the hole concentration.*

Solution:

$$N_D = \frac{4\times10^{28}\,\text{atoms/m}^3}{10^6\,\text{atoms/m}^3} = 4\times10^{22}\ \text{donors/m}^3$$

$$n_n = N_D = 4\times10^{22}\ \text{electrons/m}^3$$

$$\therefore \quad p_n = \frac{n_i^2}{n_n} = \frac{(2.4\times10^{19}\ \text{carriers/m}^3)^2}{(4\times10^{22}\ \text{electrons/m}^3)}$$

or $\mathbf{p_n = 1.4 \times 10^{16}}$ **holes/m³.**

Example 37.15: *A sample of intrinsic silicon at room temperature has a carrier concentration of* $1.5 \times 10^{16}/m^3$. *A donor impurity is added to the extent of 1 donor atom per* 10^8 *atoms of silicon. If the concentration of silicon atoms is* 5×10^{28}*atoms/m³, determine the resistivity of the material. (Given* μ_e*=0.135 m²/V.s and* μ_h*=0.048 m²/V.s)*

Solution:

Donor atom density is given as $N_D = \frac{N}{10^8} = \frac{5\times10^{28}\,/\text{m}^3}{10^8} = 5\times10^{20}\,/\text{m}^3$

Free electron concentration is given by $n_n = N_D = 5 \times 10^{20}\,/\text{m}^3$

Hole concentration is given by $p_n = \dfrac{n_i^2}{n_n} = \dfrac{(1.5\times10^{16}/\text{m}^3)^2}{5\times10^{20}/\text{m}^3}$ = $\mathbf{4.5{\times}10^{11}/m^3}$.

Resistivity of the material $\rho = \dfrac{1}{\sigma} = \dfrac{1}{e(n_n\mu_e + p_n\mu_h)}$

$$= \frac{1}{(1.602\times10^{-19}\,\text{C})(5\times10^{20}\times0.135\,\text{m}^2/\text{V.s}+4.5\times10^{11}\times0.048\,\text{m}^2/\text{V.s})} = \mathbf{0.092\ \Omega m}.$$

Example 37.16: *A sample of intrinsic germanium at room temperature has a carrier concentration of 2.4 × 10*^{19}*/m*^{3}*. It is doped with antimony at a rate of one antimony atom per million atoms of germanium. If the concentration of the germanium atoms is 4 × 10*^{28}*/m*^{3}*, determine the hole concentration.*

Solution:

Donor atom density is given as $N_D = \dfrac{N}{10^6} = \dfrac{4\times10^{28}/\text{m}^3}{10^6} = 4 \times 10^{22}\,/\text{m}^3$

Free electron concentration is given by $n_n = N_D = 4 \times 10^{22}\,/\text{m}^3$

Hole concentration is given by $p_n = \dfrac{n_i^2}{n_n} = \dfrac{(2.4\times10^{19}/\text{m}^3)^2}{4\times10^{22}/\text{m}^3}$ = $\mathbf{1.4 \times 10^{16}/m^3}$.

37.23 FERMI LEVEL IN EXTRINSIC SEMICONDUCTORS

In intrinsic semiconductors, the Fermi level, E_F lies in the middle of the band gap. However, the situation is different with extrinsic semiconductors. In an n-type semiconductor, the Fermi level lies in the upper half of the gap, as the majority carriers reside in the conduction band and their average energy is more than E_{Fn}. In a *p*-type semiconductor, the Fermi level lies in the lower half of the gap, as the majority carriers reside in the valence band and their average energy is less than E_{Fp}. The carrier concentrations in extrinsic semiconductors vary with temperature and impurity concentration. It means that the probability of occupancy of respective bands varies and consequently the position of Fermi level changes with temperature and impurity concentration.

(a) Variation of Fermi Level with Temperature in an *n*-type Semiconductor: The donor atoms in the *n-type semiconductor* at low temperatures are ionized and supply electrons to the conduction band. Electrons in the conduction band are only from the donor levels. Therefore, at T = 0 K, E_{Fn} lies midway between the donor levels and the bottom of the conduction band. It is thus,

$$E_{Fn} = \frac{E_C + E_D}{2} \text{ at } T = 0\,K \qquad ...(37.91)$$

As the temperature increases the donor levels gradually get depleted and the Fermi level moves downward. At the temperature of complete depletion of donor levels, T_d, the Fermi level coincides with the donor level E_D. Thus

$$E_{Fn} = E_D \text{ at } T = T_d \qquad ...(37.92)$$

As the temperature grows further above T_d, the Fermi level shifts downward in an approximately linear fashion. At a temperature T_i, the intrinsic process contributes to electron concentration significantly. At higher temperatures, the *n*-type semiconductor loses its extrinsic character and behaves as an intrinsic semiconductor. In the intrinsic region, the electron concentration in conduction band increases exponentially and the Fermi level approaches the intrinsic value. Thus,

$$E_{Fn} = E_{Fi} = \frac{E_g}{2} \quad \text{at } T \geq T_i \qquad ...(37.93)$$

The variation of Fermi level E_{Fn} in an *n*-type semiconductor with temperature is illustrated in Fig. 37.19.

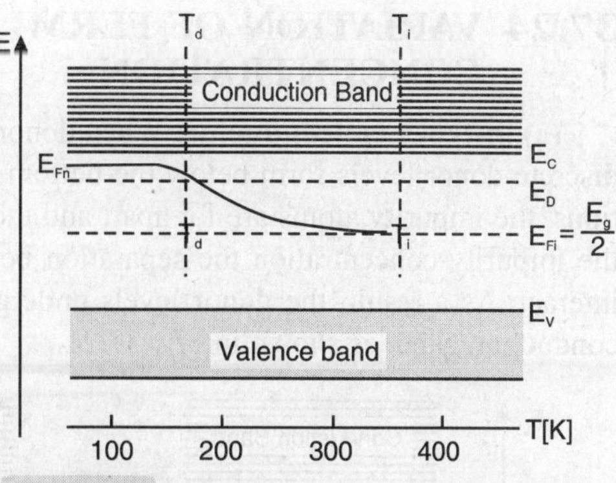

Fig. 37.19: Qualitative dependence of Fermi level on temperature in an *n*-type semiconductor

(b) Variation of Fermi Level with Temperature in a *p*-type Semiconductor: In case of *p*-type semiconductor, in the low temperature region, holes in the valence band are only due to the transitions of electrons from the valence band to the acceptor levels. As the valence band is the source of electrons and the acceptor levels are the recipients for them, the Fermi level must lie between the top of the valence band and the impurity acceptor levels. When $T = 0$, Fermi level lies midway between the acceptor levels and the top of the valence band. Thus,

$$E_{FP} = \frac{E_V + E_A}{2} \qquad ...(37.94)$$

As the temperature increases the acceptor levels gradually get filled and the Fermi level moves upward. At the temperature of saturation T_s, the Fermi level coincides with the acceptor level E_A. Thus,

$$E_{FP} = E_A \quad \text{at } T = T_s \qquad ...(37.95)$$

As the temperature grows above T_s, the Fermi level shifts upward in an approximately linear fashion.

At a temperature T_i intrinsic behaviour sets in. At higher temperatures, the *p*-type semiconductor loses its extrinsic character and behaves as an intrinsic semiconductor. In the intrinsic region, the hole concentration in the valence band increases exponentially and the Fermi level approaches the intrinsic value. Thus

$$E_{FP} = E_i = \frac{E_g}{2} \quad \text{at } T = T_i \; ...(37.96)$$

The variation of Fermi level E_{Fn} in an *p*-type semiconductor with temperature is illustrated in Fig. 37.20.

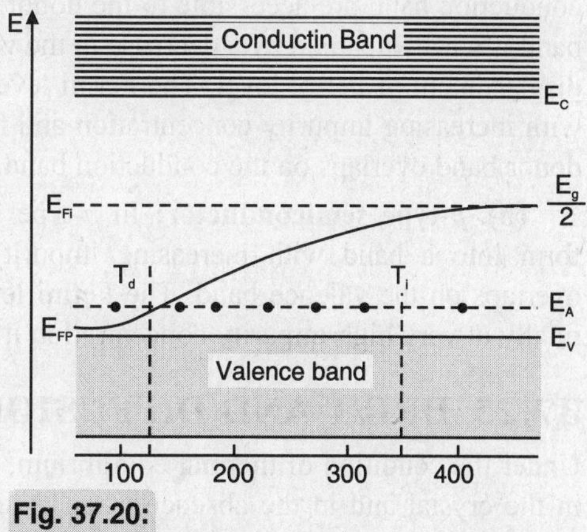

Fig. 37.20: Qualitative dependence of Fermi level on temperature in a *p*-type semiconductor

37.24 VARIATION OF FERMI LEVEL WITH IMPURITY CONCENTRATION

(a). *n*-type Semiconductor: When donor impurity is added to an intrinsic semiconductor, discrete donor levels form below the bottom edge of the conduction band. At low concentrations, the impurity atoms are far apart and they do not interact with each other. On increasing the impurity concentration the separation between impurity atoms reduces and they tend to interact. As a result, the donor levels undergo splitting and form an energy band below the conduction band, as shown in Fig. 37.21.

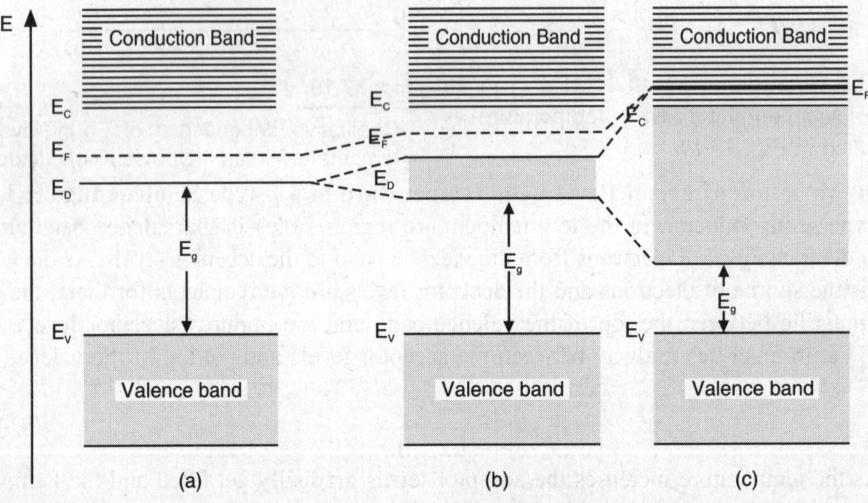

Fig. 37.21:

Energy band diagrams of an *n*-type semiconductor at three different levels of doping; (a) lowlevel doping; (b) medium doping; (c) heavy doping

The larger the doping concentration, the broader is the impurity band; and at one stage the impurity band overlaps on the conduction band. Then the upper vacant levels in the conduction band are accessible to the donor electrons. The broadening of donor levels into a band is a accompanied by a decrease in the width of the forbidden gap and also by the upward displacement of Fermi level. The Fermi level shifts closer and closer to the conduction band with increasing impurity concentration and finally moves into the conduction band when the donor band overlaps on the conduction band.

(b). *p*-type semiconductor: In *p*-type semiconductor, the acceptor levels broaden and form into a band with increasing impurity concentration. The acceptor band ultimately overlaps on the valence band. The Fermi level moves down closer to the valence band and finally at very high impurity concentration it will shift into the valence band.

37.25 DRIFT AND DIFFUSION CURRENTS

Under the condition of thermal equilibrium, the electrons and holes are uniformly distributed in the crystal and in the absence of an external stimulus their average velocity is zero and no current flows through the crystal. This is equally true for an intrinsic or an extrinsic semiconductor.

The thermal equilibrium may be disturbed by an external agent and the chaotic motion of charge carriers acquire a directional movement leading to a flow of current in the material. Electric field and concentration gradients are examples of such disturbing agents.

***(i)* Drift Current:** When an electric field E is applied across a semiconductor, the charge carriers acquire a directional motion over and above their thermal motion and produce drift current.

The electrons drifting in the conduction band produce a current component J_e given by

$$J_e\,(drift) = ne\mu_e E$$

The holes drifting in the valence band cause a current component J_h given by

$$J_h\,(drift) = pe\mu_h E$$

Therefore, the total drift current density is,

$$J_{dr} = J_e\,(\text{drift}) + J_h\,(\text{drift})$$

Drift current occurs only when external electric field is present across the solid. Although electrons and holes move in opposite directions, the direction of conventional current flow due to both the carriers is in the same direction.

$$\therefore \quad J_{dr} = e(n\mu_e + p\mu_h)E \qquad \text{...(37.97)}$$

(ii)* Diffusion Current:** In case of semiconductors, current can also flow without the application of an external electric field. If a spatial variation of carrier density is created in the semiconductor, current flows in it. If we consider an arbitrary surface in the volume of the solid and if there are more charge carriers on its one side than on the other side, we say there is a ***concentration gradient. This concentration gradient causes a directional movement of charge carriers, which continues until all the carriers are evenly distributed throughout the material. Any movement of charge carriers constitutes an electric current, and this type of movement produces a current component known as ***diffusion current***.

Concentration gradient may be produced in an extrinsic semiconductor by applying heat or light locally at one region. Suppose an external agent such as light or heat acts momentarily at one end of a *p*-type semiconductor, as shown in Fig. 37.22. The external agent generates additional electron-hole pairs leading to a sudden increase in the concentration of charge carriers at that end. In the rest of the volume, the concentration of carriers is at equilibrium value. The difference in the concentration of charge carriers initiates the carriers to diffuse from the region of higher concentration to the region of lower concentration in order to restore the equilibrium condition. As the carriers are charged particles, their migration produces a current flow, which is the diffusion current. The diffusion current strength is proportional to the concentration gradient, i.e. the rate of change of carrier concentration per unit length. In case of electrons moving left to right (see Fig. 37.22), current flows from right to left in the negative *x*-direction.

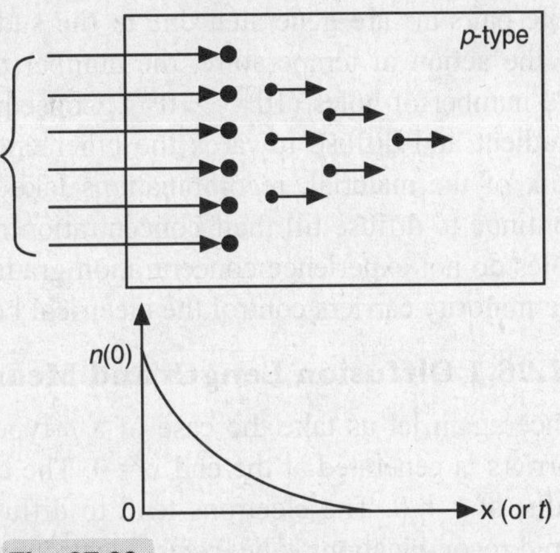

Fig. 37.22: Illustration of Diffusion current and its variation

The current component due to electron diffusion is given by

$$J_e\,(\text{diff}) = eD_e\frac{dn}{dx} \qquad ...(37.98)$$

The current component due to hole diffusion is given by

$$J_h\,(\text{diff}) = -eD_h\frac{dp}{dx} \qquad ...(37.99)$$

where D_e and D_h are *diffusion coefficients* for electrons and holes respectively.

Drift and diffusion currents coexist in semiconductors. The total current density due to drift and diffusion of electrons may be written as

$$J_e = e\left(n\mu_e E + D_e\frac{dn}{dx}\right) \qquad ...(37.100)$$

Similarly, for holes we can write

$$J_h = e\left(p\mu_h E - D_h\frac{dp}{dx}\right) \qquad ...(37.101)$$

37.26 MINORITY CARRIER DIFFUSION

In extrinsic semiconductors the external agent disturbs the equilibrium concentration of minority carriers only. For instance, let us again consider *p*-type semiconductor with an acceptor impurity concentration of 2.25×10^{22} *atoms / m*^{3}. Under the equilibrium condition, the majority carrier concentration is 2.25×10^{22} *holes / m*^{3} whereas the minority carrier concentration is

$$n_p = \frac{n_i^2}{N_A} = \frac{(1.5\times10^{16}\,\text{carriers/m}^3)^2}{2.25\times10^{22}\,\text{atoms/m}^3} = 10^{10}\ \text{electrons/m}^3$$

Let us assume that one end of the semiconductor is heated momentarily and 10^{10} electron-hole pairs/m^3 are generated due to the sudden local rise in temperature. It is seen that due to the action of temperature, the number of electrons is doubled but there is no change in the number of holes ($10^{22} >> 10^{10}$). Consequently, electrons experience a large concentration gradient and diffuse towards the other end of the semiconductor. As they diffuse into the bulk of the material, recombinations take place along the length of their travel. Electrons continue to diffuse till their concentration reaches the equilibrium value. On the other hand, holes do not experience concentration gradient and therefore do not diffuse. It all means that the minority carriers control the electrical behaviour of a semiconductor device.

37.26.1 Diffusion Length and Mean Lifetime

Once again let us take the case of a *p*-type semiconductor (Fig. 37.22) where an excess of carriers is generated at the end $x = 0$. The concentration of electrons rises above equilibrium value at $x = 0$. The electrons tend to diffuse to the right and as they diffuse, they undergo rapid recombinations. The excess electron concentration gradually falls of with distance in an exponential manner and the decrease with distance is governed by the relation

$$\Delta n(x) = n(0)e^{-x/L_n} \qquad ...(37.102)$$

where $n(0)$ is the electron concentration at $x = 0$ and L_n is the *diffusion length* for electrons.

The ***diffusion length*** is the average distance covered by an excess carrier during its lifetime τ. It may also be defined as the distance covered by an excess carrier between its generation and recombination.

The recombination of carriers occurs with distance and in time. The decrease in the excess of electrons with time is also governed by an exponential relation.

$$\Delta n(t) = n(0)e^{-t/\tau_n} \quad ...(37.103)$$

where τ_n is called the ***mean lifetime*** of electrons.

The diffusion length L_n and the mean lifetime τ_n are intimately related to each other.

$$L_n = \sqrt{D_n \tau_n} \quad ...(37.104)$$

Similar equations are valid for excess holes generated in an *n*-type semiconductor.

$$\Delta p(x) = p(0)\, e^{-x/L_h} \quad ...(37.105)$$

$$\Delta p(t) = p(0)e^{-t/\tau_h} \quad ...(37.106)$$

$$L_h = \sqrt{D_h \tau_h} \quad ...(37.107)$$

37.26.2 Einstein Relations

Although drift and diffusion are two seemingly different processes, the parameters μ, the mobility and *D*, the diffusion length are not independent. There exists a close relationship between them, since both these parameters are determined by the thermal motion and scattering of the free carriers. They are related as follows:

$$\frac{D_h}{\mu_h} = \frac{kT}{e} \quad ...(37.108)$$

Similar relation holds good for electrons also.

$$\frac{D_n}{\mu_e} = \frac{kT}{e} \quad ...(37.109)$$

Eq. (37.108) and Eq. (37.109) are known as **Einstein relations**. From these relations we get

$$\frac{D_n}{D_h} = \frac{\mu_e}{\mu_h} \quad ...(37.110)$$

Example 37.17: *Find the diffusion coefficient of electrons in silicon at 300K if μ_e is 0.19 m² /V.s.*

Solution:

$$\frac{D_n}{\mu_e} = \frac{kT}{e}.$$

Therefore,

$$D_n = \frac{kT}{e}\mu_e = \frac{1.38\times10^{-23}\,\text{J/K}\times300\text{K}}{1.602\times10^{-19}\,\text{C}}\times0.19\ \text{m}^2/\text{V.s}$$

$$= \mathbf{0.0045\ m^2\ /V.s.}$$

37.27 COMPOUND SEMICONDUCTORS

Silicon and germanium are the commonly used semiconductors. They are elements belonging to the IV group. They are used either in pure form or doped form. These intrinsic and extrinsic semiconductors are known as ***elemental semiconductors***.

Compound semiconductors are semiconductor compounds composed of elements from two or more different groups of the periodic table, such as III-IV, II-VI, IV-VI semiconductors. They are formed by combining equal atomic fractions of the above group elements. III-V semiconductors are composed of elements from group III (B, Al, Ga, In) and Group V (N, P, As, Sb, Bi). Thus, GaP, GaAs, InP, InAs, and AlSb are examples of the III-IV compound semiconductors. Similarly, ZnO, ZnS, ZnSe, CdS, and HgS are examples of II-VI compounds. The range of possible compounds is very broad. Binary compounds are formed by combining two elements as in GaAs. Ternary compounds are obtained when three elements are combined as in InGaAs. Quaternary compounds are formed when four elements are combined, as in AlInGaP. It is possible to adjust the bandgap and change the optical properties by forming their ternary and quaternary derivatives by doping with impurities. GaAs is of great technical interest because it has large band gap and electron mobility is large in the material.

The elemental semiconductors Si, and Ge are indirect band gap semiconductors whereas III-IV compound semiconductors are direct gap semiconductors. The elemental semiconductors are not suitable for producing light, while III-IV compound semiconductors have high optoelectronic conversion efficiency and are widely used in fabrication of optoelectronic devices, such as lasers, LEDs etc.

37.28 HALL EFFECT

In 1879 E. H. Hall discovered that "if a metal or a semiconductor carrying a current *I* is subjected to a transverse magnetic field *B*, a potential difference V_H is produced in a direction normal to both the magnetic field and current directions." This is called *Hall Effect*. This effect established that it is negatively charged particles that carry current in metals.

37.28.1 Importance of Hall Effect

The importance of Hall Effect in the field of semiconductors is that it helps to determine the

(*i*) type of semiconductor,
(*ii*) sign of majority charge carriers,
(*iii*) majority charge carrier concentration,
(*iv*) mobility of majority charge carriers, and
(*v*) mean drift velocity of majority charge carriers.

37.28.2 Experimental Arrangement

The experimental set up for the measurement of Hall voltage and determination of Hall coefficient is shown in Fig. 37.23 (*a*). A thin rectangular semiconductor wafer is mounted on

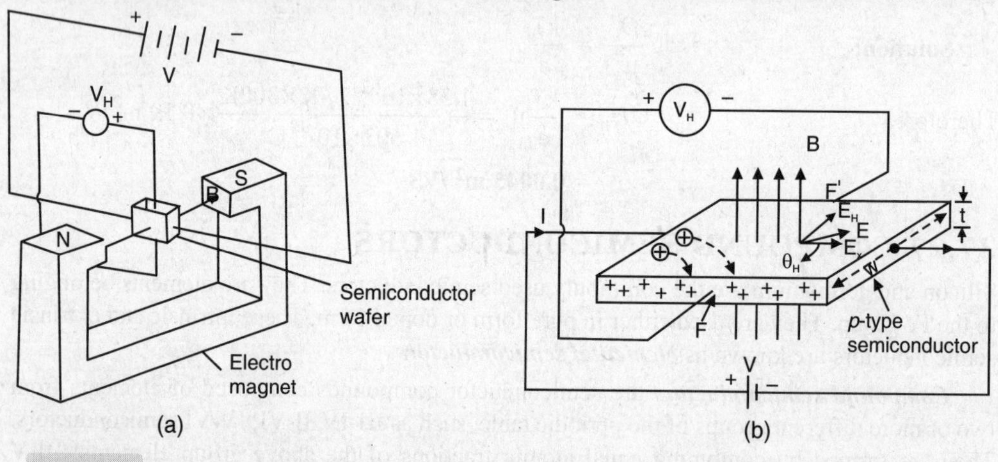

Fig. 37.23: (a) Basic experimental arrangement to study Hall effect (b) Generation of Hall voltage

an insulating strip and two pairs of electrical contacts are provided on opposite sides of the wafer. One pair of contacts is connected to a constant current source. And the other pair is connected to a sensitive voltmeter. This arrangement is mounted in between two pole pieces of an electromagnet such that the magnetic field acts perpendicular to the lateral faces of the semiconductor wafer.

37.28.3 Hall Voltage

Let us assume that the semiconductor is a *p*-type semiconductor. Let a potential difference *V* be applied across its ends. A current of strength *I* flows through it along the *x*-direction (Fig. 37.23 *b*). Holes are the majority charge carriers in the *p*-type semiconductor. The current through the wafer is given by

$$I = peA\upsilon_d \quad ...(37.111)$$

where *p* is the hole concentration
A is the area of cross-section of the end face of semiconductor wafer,
e is the electrical charge associated with a hole, and
υ_d is the average drift velocity of holes.

The current density $J_x = \frac{I}{A} = pe\upsilon_d$ ...(37.112)

Any plane perpendicular to the current flow direction is an equipotential surface. Therefore, the potential difference between the front and rear faces *F* and *F′* is zero (see Fig. 37.24).

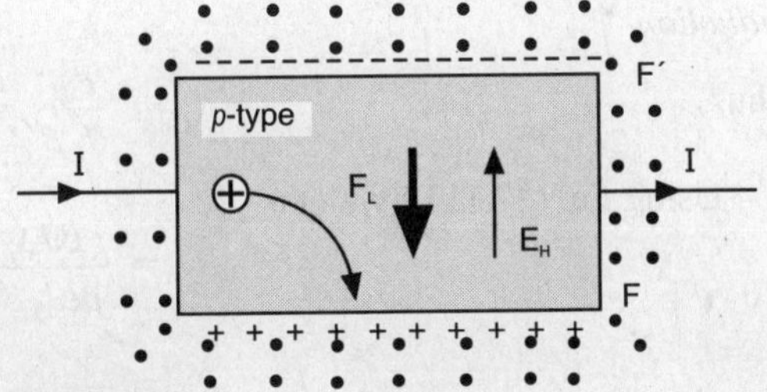

Fig. 37.24:
Top view of the wafer - The directions of magnetic force and Hall field in p-type semiconductor

Now, if a magnetic field *B* is applied normal to the wafer surface and hence to the direction of current flow in it, then a transverse potential difference is produced between faces *F* and *F ′*. It is known as ***Hall Voltage*** V_H.

Without the application of magnetic field, holes move parallel to faces *F* and *F ′*. On application of magnetic field *B*, the holes experience a sideways deflection due to the magnetic force F_L, which is given by

$$F_L = eB\upsilon_d \quad ...(37.113)$$

Holes are deflected toward the front face *F* and pile up there. Due to this, a corresponding equivalent negative charge is left on the rear face *F ′*. These opposite charges produce a transverse electric field, $E_H{}'$ whose direction is from the front to the rear face. Due to the action of E_H, holes experience an electric force in addition to the Lorentz force. When the force F_E due to this transverse electric field balances the magnetic force F_L, equilibrium condition is attained and the holes once again flow along *x*-direction parallel to the faces *F* and *F ′*.

In the equilibrium condition

$$F_E = F_L$$

$$\therefore \quad eE_H = e\upsilon_d B$$

If 'w' is the width of the semiconductor wafer, $E_H = \dfrac{V_H}{w}$

$$\therefore \quad \left(\frac{V_H}{w}\right) = B\upsilon_d \qquad ...(37.114)$$

From Eq. (37.112), we have $\upsilon_d = \dfrac{J_x}{pe}$

Therefore, we can write Eq. (37.114) as

$$\therefore \quad \frac{V_H}{w} = \frac{BJ_x}{pe} \qquad ...(37.115)$$

$$V_H = \frac{wBJ_x}{pe} = \frac{wBI}{peA}$$

If 't' is the thickness of the semiconductor plate, $A = w\,t$.

$$\therefore \quad V_H = \frac{BI}{pet} \qquad ...(37.116)$$

37.28.4 Hall Coefficient

Hall coefficient, R_H is defined as Hall field per unit current density per unit magnetic induction.

Thus,
$$R_H = \frac{E_H}{J_x B} = \frac{V_H / w}{J_x B}$$

Using Eq. (37.114), we get

$$R_H = \frac{BJ_x}{peJ_x B}$$

or
$$R_H = \frac{1}{pe} \qquad ...(37.117)$$

Hall voltage, V_H can now be written as

$$V_H = R_H \frac{BI}{t} \qquad ...(37.118)$$

$$\therefore \quad R_H = \frac{V_H t}{BI} \qquad ...(37.119)$$

Eq. (37.119) is derived assuming that the p-semiconductor contains only holes. When the two types of charge carriers are taken into account, it is shown that the Hall coefficient is given by

$$R_H = \frac{(p\mu_h^2 - n\mu_e^2)}{e(p\mu_h + n\mu_e)^2} \qquad ...(37.120)$$

From the Eq. (37.120), we see that the Hall coefficient and Hall voltage are smaller for intrinsic materials than for extrinsic materials.

37.28.5 Drift Velocity

According to the equilibrium condition, namely $F_E = F_L$, we have

$$e\left(\frac{V_H}{w}\right) = eB\upsilon_d$$

$$\upsilon_d = \frac{V_H}{Bw} \qquad ...(37.121)$$

As it is possible to experimentally measure the Hall voltage, magnetic field B and width of the wafer w, the mean drift velocity of the charge carriers can be obtained from the above equation.

37.28.6 Carrier Concentration

From the measurement of Hall voltage and current through the wafer and knowing the strength of magnetic induction and thickness of the wafer, R_H can computed using Eq. (37.119). Once R_H is known, the carrier concentration can be determined with the help of Eq. (37.108), which may be rewritten as follows:

$$p = \frac{1}{R_H e}$$

In case of conductors and n-type semiconductors, the expression for Hall coefficient is

$$R_H = -\frac{1}{ne} \qquad ...(37.122)$$

where n represents the concentration of electrons in the conductor or n-type semiconductor.

$$\therefore \qquad n = -\frac{1}{R_H e} \qquad ...(37.123)$$

37.28.7 Doping Level

It is possible to estimate the doping concentration, N_A, from the value of p using the relation

$$p_p \cong N_A.$$

37.28.8 Carrier Sign and Type of Semiconductor

With the direction of the magnetic field and the current as depicted in Fig. 37.23 (*b*), the sign of the Hall voltage is positive. For an n-type semiconductor, the Hall voltage will be negative when the directions of B and I are kept same. Therefore, knowing the sign of Hall voltage, the type of the semiconductor and hence the sign of the majority carrier can be determined.

37.28.9 Hall Mobility

Mobility is defined as the drift velocity acquired in unit electric field. We know that the current density J is given by the following expressions.

$$J = pe\upsilon_d \text{ and } J = \sigma E$$

$$\therefore \qquad pe\upsilon_d = \sigma E$$

or

$$\frac{\upsilon_d}{E} = \frac{\sigma}{pe}$$

$$\therefore \qquad \mu_h = \sigma R_H \qquad ...(37.124)$$

37.28.10 Hall Angle

The net electric field E in the semiconductor is a vector sum of E_x and E_H. It acts at an angle θ_H to the x-axis, where θ_H is called the ***Hall angle***. From the Fig. 37.23 (*b*), we have

$$\tan\theta_H = \frac{E_H}{E_x} \qquad ...(37.125)$$

But $$E_H = \frac{V_H}{w} = \frac{BJ_x}{pe}$$

Also $$E_x = \frac{J_x}{\sigma} = \rho J_x$$

Thus, $$\tan\theta_H = \frac{BJ_x}{pe\rho J_x}$$

$\therefore$ $$\tan\theta_H = \frac{B}{pe\rho}$$

$$\frac{E_H}{E_x} = \frac{B}{pe\rho} \qquad \text{...(37.126)}$$

Using $\frac{1}{pe} = R_H$ and $\frac{1}{\rho} = \sigma$, Eq. (37.125) can be rewritten as

$$\tan\theta_H = \sigma R_H B$$

or $$\tan\theta_H = \mu_h B \qquad \text{...(37.127)}$$

37.28.11 Factors Affecting Hall Voltage

Hall voltage is given the following expression.

$$V_H = \frac{BI}{pet}$$

It is seen from the above equation that Hall voltage is directly proportional to the magnetic field strength and the current passing through the wafer. Therefore, the Hall voltage will be larger, the larger the magnetic field or the current. Further, the Hall voltage is larger, the smaller is the carrier concentration, and the thinner is the wafer. It is easy to select thin wafers of the material, so making t small. The fact that the concentration of carriers must be small implies that large Hall voltage is obtained with poor conductors, i.e. those with relatively few free electrons or holes per unit volume. Thus, the best voltage is obtained with semiconductors like silicon and germanium that have been lightly doped with impurities to make them conducting slightly.

For the semiconductors the number of charge carriers per unit volume is about $10^{23}/m^3$ while in case of metals it is about $10^{28}/m^3$. Therefore, the Hall voltage is about 10^5 greater in semiconductors than in metals.

37.28.12 Variation of Hall Coefficient with Temperature

(*i*) In metals it was found that the Hall coefficient does not depend on temperature. Hence, it may be concluded that the concentration of free electrons does not vary with temperature in metals.

(*ii*) In semiconductors, the Hall coefficient decreases sharply with increase in temperature. It indicates that the concentration of charge carriers in semiconductors increases with increasing temperature.

37.28.13 Applications

Hall effect is widely used in various fields for a variety of applications. In almost all cases a Hall effect sensor is employed. A Hall effect sensor is a transducer that produces its output voltage in response to changes in magnetic fields.

(*i*) **Determination of semiconductor type:** The Hall coefficient is negative for a *n*-type semiconductor and positive for a *p*-type semiconductor. Therefore, the sign of the Hall coefficient can be used to determine whether a given semiconductor is *n*- or *p*- type.

(*ii*) **Determination of carrier concentration:** By measuring the Hall coefficient, the carrier concentration in a semiconductor can be determined making use of the relations

$$n = \left|\frac{1}{R_H e}\right| \text{ or } p = \frac{1}{R_H e}.$$

(*iii*) **Determination of carrier mobility:** By measuring the Hall coefficient and conductivity of the semiconductor, the carrier mobility can be determined using the relation

$$\mu_h = \sigma|R_H|$$

(*iv*) **Measurement of magnetic fields:** Hall voltage is proportional to the magnetic field intensity, for a given current through the sample. Therefore, one of the important applications of Hall Effect consists in measuring magnetic fields. Knowing the parameters of the Hall probe, and applied current, we can determine the intensity of the magnetic field. Hall probes can be used for static as well as high-frequency magnetic fields. Hall probes measure variable magnetic fields up to a frequency of 10^{12} Hz.

(*v*) **Measurement of power in an electromagnetic wave**: In an electromagnetic wave the electric and magnetic fields are mutually perpendicular to each other. If a semiconductor is held parallel to electric field *E*, it will cause a flow of current *I* in the semiconductor. As the semiconductor is subjected to a transverse magnetic field at the same time, Hall voltage is produced in the sample. This Hall voltage is proportional to the product *EH* which signifies the power of the wave. So, Hall Effect can be employed to measure the power of an electromagnetic wave.

(*vi*) **Miscellaneous applications:** Hall sensors are useful for proximity switching, positioning, speed detection, and current sensing applications.

(*a*) Current passing through a conductor generates a magnetic field which changes with current. A Hall sensor can be used to calculate the current without breaking the circuit. They are especially used in measuring extremely heavy currents, where conventional ammeters cannot be used.

(*b*) A Hall sensor is combined with circuitry to act as digital switches that are useful to detect missing paper in computer printers and open covers.

Example 37.18: *An n-type germnanium sample has a donor density of $10^{21}/m^3$. It is arranged in a Hall experiment having magnetic field of 0.5 T and the current density is 500 A/m^2. Find the Hall voltage if the sample is 3 mm wide.*

Solution: Hall voltage $V_H = \dfrac{BI}{net} = \dfrac{BJA}{net} = \dfrac{BJwt}{net}$

As $n = N_D$

$\therefore$
$$V_H = \frac{BJw}{N_D e}$$

$$= \frac{(0.5\text{ T})(500\text{ A/m}^2)(3\times10^{-3}\text{ m})}{(10^{21}\text{/m}^3)(1.602\times10^{-19}\text{ C})}$$

$$= 4.7\times10^{-3}\text{ V} = \mathbf{4.7\text{ mV}}$$

Units: $$1\frac{TAm^2}{C} = 1.\frac{N}{A.m}.\frac{A.m^2}{C} = 1.\frac{N.m}{C} = 1\frac{J}{C} = 1\text{ V}$$

Example 37.19: *A copper strip 2.0cm wide and 1.0mm thick is placed in a magnetic field with B=1.5 wb/m². If a current of 200 A is set up in the strip, calculate Hall voltage that appears across the strip. Assume* $R_H = 6 \times 10^{-7} m^3/C$.

Solution: Hall voltage $$V_H = R_H \frac{IB}{t}$$

$$= 6\times10^{-7}\text{ m}^3/\text{C}\cdot\frac{200A\times1.5\text{ wb/m}^2}{10^{-3}\text{m}}$$

$$= \mathbf{0.18\ V}$$

Example 37.20: *An electric field of 100 V/m is applied to a sample of n-type semiconductor whose Hall coefficient is −0.0125 m³/C. Determine the current density in the sample, assuming* $\mu_e = 0.6\ m^2/V.s$.

Solution: Current density $$J = \frac{\mu_e E}{R_H}$$

$$= \frac{(0.36\text{ m}^2/\text{V.s})(100\text{ V/m})}{-0.0125\text{ m}^3/\text{C}}$$

$$= -2880\text{ C/s.m}^2 = -\mathbf{2880\ A/m^2}.$$

Example 37.21: *In a Hall coefficient experiment, a current of 0.25 A is sent through a metal strip having thickness 0.2 mm and width 5 mm. The Hall voltage is found to be 0.15 mV when a magnetic field of 2000 gauss is used.*

(*a*) *What is the carrier concentration?*

(*b*) *What is the drift velocity of the carriers?*

Solution: (*a*) The carrier concentration,

$$n = \frac{IB}{V_H e t}$$

$$= \frac{(0.25\text{ A})(0.2\text{ T})}{(0.15\times10^{-3}\text{V})(1.602\times10^{-19}\text{ C})(0.2\times10^{-3}\text{m})}$$

$$= \mathbf{1.04 \times 10^{25}\ carriers/m^3}.$$

(*b*) Drift velocity of the carriers,

$$\upsilon_d = \frac{V_H}{wB}$$

$$= \frac{0.15\times10^{-3}\text{V}}{5\times10^{-3}\text{m}\times0.2\text{ T}}$$

$$= 0.15\text{ V/m.T} = \mathbf{0.15\ m/s}.$$

QUESTIONS

1. Show that the intrinsic concentration n_i for a semiconductor is given by $n_i = (N_C N_V)^{1/2} e^{-E_g/2kT}$ where the symbols have their usual meaning.
2. Write the expressions for electron and hole concentrations in an intrinsic semiconductors and hence derive the expression: $E_F = \frac{E_C + E_V}{2} + \frac{3}{4} kT \ln\left(\frac{m_h^*}{m_e^*}\right)$ for Fermi level in the intrinsic semiconductor. Assume the symbols to have their usual meanings.
3. Draw a neat sketch of a band diagram of intrinsic semiconductor at room temperature and show that the Fermi level in an intrinsic semiconductor lies in the middle of the energy gap.
4. Derive an expression for Fermi energy in intrinsic semiconductor. What is the effect of temperature on Fermi level in an intrinsic semiconductor?
5. Describe a method of determining the band gap of a semiconductor. How does electrical conductivity vary with temperature for an intrinsic semiconductor?
6. Using the expressions of electron concentration and hole concentration for an intrinsic semiconductor show that the intrinsic carrier density is independent of Fermi level.
7. Obtain an equation for the conductivity of an intrinsic semiconductor in terms of carrier concentration and carrier mobility.
8. Why are holes not generated in metals? Can the electrical conduction of semiconductors be improved to the extent possible in metals? Discuss.
9. Differentiate the *n*-type and *p*-type semiconductors with their Fermi level diagram.
10. *P* and *N*-type semiconductors have more holes and electrons respectively. Explain why they are electrically neutral.
11. What is the effect of temperature on the working of *n*- and *p*-type semiconductors? Draw diagrams to show the variation in the position of Fermi level with rise in temperature.
12. Draw energy band diagrams for *n*-type semiconductor at 0 K and at room temperature. Show that at 0K Fermi level lies midway between valence band and conduction band,
13. Distinguish between extrinsic and intrinsic semiconductors.
14. Discuss the effect of increasing amounts of dopants on the Fermi level in extrinsic semiconductors.
15. Explain the effect of impurity concentration on the Fermi level in an extrinsic semiconductor.
16. Obtain an expression for Hall coefficient for an extrinsic semiconductor.
17. What is Hall Effect? Obtain an expression for Hall coefficient. Does RH depend on the doping concentration?
18. Derive expressions for Hall voltage and Hall coefficient. Mention important applications of Hall Effect.
19. Explain in brief the concept of Fermi level. Show diagrammatically the Fermi level in metals, intrinsic semiconductors and insulators at OoK and at higher temperature.
20. Explain Hall Effect and its significance. Give its applications.
21. How does the Fermi level change with temperature in extrinsic semiconductors? Discuss the effect of increasing amounts of dopants in extrinsic semiconductors.
22. What do you understand by intrinsic and extrinsic semi-conductors?
23. Write short notes on (*i*) Elemental and compound semiconductors, (*ii*) Intrinsic and extrinsic semiconductors.
24. Explain in brief the concept of Fermi level. Derive an expression for Fermi energy in intrinsic semiconductor. What is the effect of temperature on Fermi level in an intrinsic semiconductor?

25. Differentiate between an intrinsic and an *n*-type semiconductor on the basis of (*i*) Crystal representation and (*ii*) band representation and (*iii*) relation between *n* and *p*.
26. Draw energy band diagrams at 0 K and room temperature for (*i*) Intrinsic semiconductor, (*ii*) *n*-type semiconductor, (*iii*) *p*-type semiconductor.
27. An n-type extrinsic semiconductor is in equilibrium at room temperature. What is the net charge on the impurity atom? Discuss.
28. How does the Fermi level change with temperature in extrinsic semiconductors? Discuss the effect of increasing amounts of dopants in extrinsic semiconductors.
29. Derive the expression: $E_F = \frac{E_g}{2} + \frac{3}{4}kT \ln\left(\frac{m_h^*}{m_e^*}\right)$, for Fermi level in an intrinsic semiconductor.
30. For a *n*-type semiconductor material with high doping concentration and low temperature show that: $n = (N_c N_d)^{\frac{1}{2}} \exp\left(\frac{E_D - E_g}{2kT}\right)$ Where the symbols have their usual meaning.
31. Obtain an expression for density of electrons in the conduction band of an *n*-type extrinsic semiconductor by assuming Fermi-Dirac distribution.
32. Obtain an expression for density of holes in the valence band of *p*-type extrinsic semiconductor.
33. What is law of mass action? How does this lead to the relation $n_p = n_i^2$, where symbols have their usual meaning.
34. Derive the expression for the Fermi energy at any temperature for an n-type semiconductor.
35. Write short note on: Drift and Diffusion currents.
36. What is Hall Effect? Derive a formula for density of charge carriers in a *p*-type semiconductor give two of its applications.
37. Explain Hall effect and its importance. Show that the ratio of Hall electric field E_H to the electric field *E* which is responsible for the current in a *n*-type semiconductor wafer kept in a uniform magnetic field *B* is given by
$$\frac{E_H}{E} = \frac{B}{ne\rho}$$
38. Show that the minimum conductivity of a semiconductor sample occurs when $n_o = n_i \sqrt{\mu_p / \mu_n}$.

PROBLEMS

1. If effective mass of an electron is equal to twice the effective mass of hole, determine the position of the Fermi level in an intrinsic semiconductor from the centre of forbidden gap at room temperature. [**Ans:** The Fermi level is 0.014 eV below the centre of forbidden gap]
2. Determine the fraction of electrons in conduction band in silicon at 27°C and 227°C. Given: $E_g = 1.1$ eV and $k = 1.38 \times 10^{-23}$ J/K. [**Ans:** 5.7×10^{-10}; 2.8×10^{-6}]
3. A silicon wafer is doped with 10^{21} phosphorus atoms/m^3. Calculate: (*i*) The majority carrier concentration, (*ii*) The minority concentration and, (*iii*) The electrical resistivity of the doped silicon at room temperature. Assume complete ionization of the dopant atoms; $n_i = 1.5 \times 10^{16}$ / m^3, $\mu_e = 0.135$ m^2/V.s. and $\mu_h = 0.048$ m^2/V.s.
4. The resistivity of intrinsic silicon at 270° C is 3000 Ωcm. Calculate the intrinsic carrier density. Assume: μe = 0.17 m^2/V.s. and μh = 0.035 m^2/V.s. [**Ans:** 1.0×10^{16}/m^3]
5. Calculate the temperature at which Si ($E_g = 1.14$ eV) will have the same fraction of electrons in the conduction band as germanium ($E_g = 0.72$ eV) has at 300 K. [**Ans:** 475K]
6. Compute the current produced in a Ge plate of 1 cm^2 area, 0.03 mm thickness across its faces. Assume a free electron concentration of 2×10^{19} m^{-3}, the electron and hole mobility's being 0.39 m^2/V.s and 0.19 m^2/V.s respectively. [**Ans:** 12.4 A]

7. Find the fraction of electrons in the valence band of intrinsic geranium which can be thermally excited across the forbidden energy gap of 0.7 eV into the conduction band at: (*i*) 50 K, (*ii*) 300 K, (*iii*) 1000 K [**Ans:** 5.1×10^{-36}; 1.3×10^{-6}; 0.017]

8. What fraction of the conductivity of intrinsic silicon at room temperature is due to Electrons and Holes? [**Ans:** 73.8% due to electrons and 26.2% due to holes]

9. The electrons and the hole mobility's are, $\mu_e = 0.135$ m^2/V.s. & $\mu_h = 0.048$ m2/V.s. The energy gap in silicon crystal is 1.12 eV. Its electron and hole mobilities at room temperature are 0.48 m^2/V.s and 0.013 m^2/V.s respectively. Find its conductivity. [**Ans:** $1.29 \times 10^{-3}\Omega^{-1}m^{-1}$]

10. The forbidden band gap in pure silicon is 1.1 eV. Compare the number of conduction electrons at temperature 27°C and 37°C.

11. A silicon wafer is doped with 10^{21} phosphorus atoms/m^3. Calculate the (*i*) majority carrier concentration, (*ii*) minority concentration and (*iii*) electrical resistivity of the doped silicon at room temperature. Assume complete ionization of the dopant atoms; $n_i = 1.5 \times 10^{16}$ /m^3, $\mu_e = 0.135$ m^2/V.s. $\mu_h = 0.048$ m^2/V.s. [**Ans:** 10^{21}/m^3; 2.25×10^{11}/m^3; 0.046 Ω.m]

12. A sample of intrinsic silicon at room temperature has a carrier concentration of 1.5×10^{16} m^3. If a donor impurity is added to the extent of 1 donor atom per 10^8 atom/m^3, determine the resistivity of the material. Given: $\mu_e = 0.135$ m^2/V.s.; $\mu_h = 0.048$ m^2/V.s. [**Ans:** 0.09 Ω.m]

13. A current density of 10^3 A/m^2 flows through an *n*-type germanium crystal which has resistivity 0.05 ohm-m. Calculate the time taken for electrons in material to drift a 50 μm distance. The mobility of electron is 0.38m^2/V.s. [**Ans:** 2.6×10^{-6}s]

14. A sample of Ge is doped to the extent of 10^{20} donor atoms/m^3 and 7×10^{19} acceptor atoms/m^3. At the temperature of the sample the resistivity of intrinsic germanium is 0.6 ohm-m. If the applied electric field is 200 V/m, find the total conduction current density. Assume: $\mu_e = 0.38$ m^2/V.s., $\mu_h = 0.18$ m^2/V.s [**Ans:** 431.6 A/m^2]

15. An *n*-type Ge sample has a donor density of 1021 /m^3. It is arranged in Hall Effect experiment having magnetic field of 0.5 Tesla and current density 500A/m^2. Find the hall voltage, if the sample is 3 mm wide. [**Ans:** 4.7 mV]

38 Solid State Diodes

Learning Objectives:

By the end of this chapter, you will be able to:

- Explain concepts related to P-N Junction diode – forward and reverse bias and its applications as well.
- Learn about principles of Zener Diode and Varactor Diode.
- Describe about other special diodes like LED, Photo-detectors and Solar Cells.

38.1 INTRODUCTION

Pure semiconductors are of very limited use. Semiconductors that are doped with impurities form the basis of the practical devices. A semiconductor that is doped with acceptor impurities forms a *p*-type semiconductor and then into it donor atoms are diffused so that a thin layer of *n*-type region is formed. This combination constitutes a *p-n* junction diode. A *p-n* junction diode is also known as a semiconductor diode. The *p-n* junction allows the current to flow in one direction and resists it in the opposite direction. This property is known as *rectifying action*. Semiconductor diodes are widely used as rectifiers, which convert input ac voltage to dc voltage. Practically, all semiconductor devices contain at least one *p-n* junction. The production techniques enable the fabrication of *p-n* junction to suit specific purposes. Thus, a varicap that acts as a variable capacitor, a tunnel diode and a Gunn diode as oscillators, a zener diode as a voltage stabilizer, a photodiode as a light detector, a solar cell as a voltage source, an LED and a laser as light sources are all *p-n* junctions. A junction transistor is fabricated with two *p-n* junctions in close proximity. Therefore, *p-n* junction constitutes the most basic component of solid state devices and a thorough understanding of its electrical behaviour is essential for appreciation of the operation of many semiconductor devices.

38.2 p-n JUNCTION DIODE

By doping an intrinsic semiconductor crystal, it can be converted into an *n*-type or *p*- type extrinsic semiconductor crystal. When taken individually, *p*-type and *n*-type semiconductors conduct with equal facility in both the directions just like a resistor (Fig. 38.1) and exhibit linear conduction characteristic. Such behaviour is known as ***Ohmic behaviour***. It is indeed possible to grow a semiconductor crystal with part of the crystal *n*-type and the other part *p*-type. ***p-n junction*** is the boundary between one region of a semiconductor with *p*-type impurities and another region containing *n*-type impurities. The technology of junction fabrication is beyond the scope of our discussion. The most remarkable property of the *p-n* junction is that it has a *non-linear* conduction characteristic (see Fig. 38.2) and allows current in one direction and opposes it in the opposite direction. This is known as *rectifying property*.

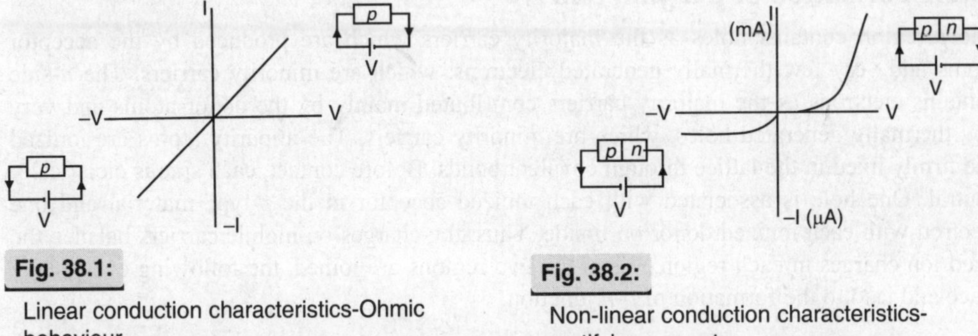

Fig. 38.1:

Linear conduction characteristics-Ohmic behaviour

Fig. 38.2:

Non-linear conduction characteristics-rectifying property

p-n junction diodes are widely used in rectifiers, which convert ac voltages to dc voltages. A *p-n* junction is the basic component of solid state devices and is used as a varicap, a tunnel diode, a Gun diode, a Zener diode, a photodiode, a solar cell, an LED, a laser, etc.

38.2.1 A Physical Description of the p-n Junction

A *p-n* junction is produced when a *p*-type and an *n*-type region are formed side by side within a thin semiconductor chip such that a rapid transition from *p* type to *n* type material takes place. This transition is called a ***junction***. In the ideal case, the transition from one type to the other would take place abruptly. Such junctions are known as ***abrupt junctions***. The distance over which the change from *p*-type to *n*-type occurs takes place in a very short length much less than 1 μm.

We assume here uniform *p*-doping on one side of the junction and uniform *n*-doping on the other side; and that doping concentrations at either side of the junction are identical, i.e., $N_A = N_D$.

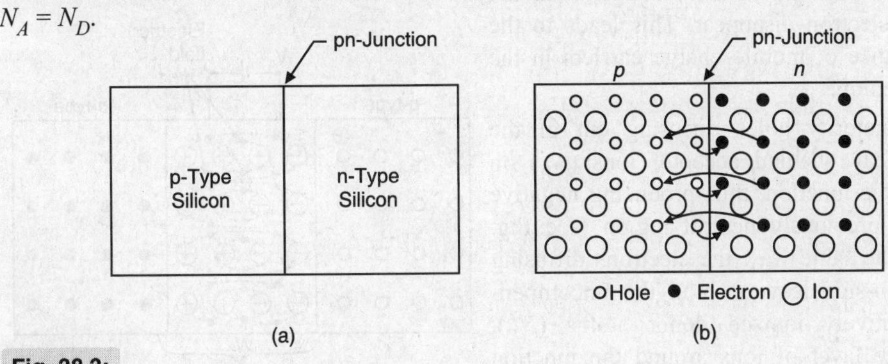

Fig. 38.3:

p-n junction (a) a *p-n* junction is formed when a *p*-type and *n*-type semiconductors are metallurgically joined (b) majority carriers diffuse across the junction at the instant of formation of the junction.

Our primary interest lies in understanding the formation of the junction and the properties of junction. The line in Fig. 38.3 (a) implies some sort of barrier between the two regions. Without such a barrier all the free electrons in the *n* type region gradually diffuse into the holes in the *p* region, leaving neither free electrons nor holes there. In the beginning, when the crystal was formed, it was without a barrier and electrons began to diffuse into holes and formed a transition zone, which had neither free electrons nor holes. The transition zone is the barrier between the regions. The question now concerns the nature of this barrier and its genesis.

38.2.2 Formation of p-n Junction

The *p*-region contains holes as the majority carriers which are produced by the acceptor atoms and very few thermally generated electrons, which are minority carriers. The *n*-side contains electrons as the majority carriers contributed mainly by the donor atoms and very few thermally generated holes, which are minority carriers. The impurity atoms are ionized and firmly fixed in the lattice through covalent bonds. Before contact, each side is electrically neutral. One hole is associated with each ionized acceptor in the *p*-type material and one electron with each ionized donor on *n*-side. Thus, the charges on mobile carriers balance the fixed ion charges in each region. When the two regions are joined, the following events take place and lead to the formation of *p-n* junction.

38.2.3 Diffusion of Majority Carriers and Formation of Space Charge

When the two sections are brought together, the holes in the *p*-region are more than the holes in the *n*-region. Because of the concentration difference, holes nearer to the junction begin to diffuse from *p*-side to the *n*-side (see Fig. 38.3 *b*). Similarly, the electrons in the *n*-region are more than the electrons in the *p*-region. Due to the concentration gradient, electrons nearer to the junction begin to diffuse from *n*-side to the *p*-side. Thus, majority carriers start moving into opposite regions. As electrons and holes are charged particles, their motion produces *electron diffusion current*, J_{en} and *hole diffusion current*, J_{hp} respectively. The current components due to holes and electrons add up although the carriers are moving in opposite directions (the electron current is opposite to the direction of electron flow). The first letters '*h*' and '*e*' of the subscripts in the designation denote the carrier and the second letters '*p*' and '*n*' indicate the region of their origin.

At the junction the holes and electrons meet each other and undergo recombination. As a hole recombines with an electron, both the hole and electron disappear. This leads to the disappearance of mobile charge carriers in the junction region.

The majority holes diffusing out of the *p*-region leave behind acceptor ions (N_A) on *p*-side of the junction, thus producing negative ions in a previously neutral region (see Fig. 38.4). In the same way, the electrons diffusing from the *n*-side leave behind the uncompensated positively ionized donor atoms (N_D). The double layer of ions around the junction is known as the ***space charge region***. This narrow space-charge region is depleted of mobile charges and contains only the immobile

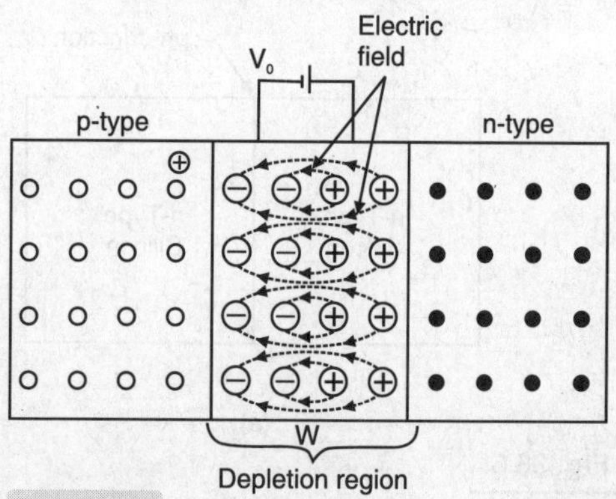

Fig. 38.4: Formation of space charge

uncompensated ions. Therefore, this region is also called the ***depletion region***, which is about 1 μm.

The array of fixed ions produces ***electric field, E***, which is directed from the donor ions on *n*-side toward the acceptor ions on *p*-side. This electric field is in a direction that opposes the diffusion of majority carriers into opposite sides.

The space charge layers tend to reduce the diffusion of holes and electrons. The holes leaving *p*-region must overcome the repulsion of the positive ion layer on *n*-side of the junction in order to continue their forward movement. Similarly, the electrons diffusing from the *n*-side must overcome the repulsion of the negative ion layer on *p*-side of the junction (Fig. 38.5 a). Thus, the ***internal electric field*** stops the majority charge carriers movement.

As the barrier builds up, diffusion of majority carriers is halted. The process is self-arresting. Ultimately, the electric field establishes an equilibrium potential difference V_o across the depletion region. V_o is known as the ***internal potential barrier***. Thus, the very electric field which is produced due to the diffusion of the majority carriers across the junction, inhibits the diffusion.

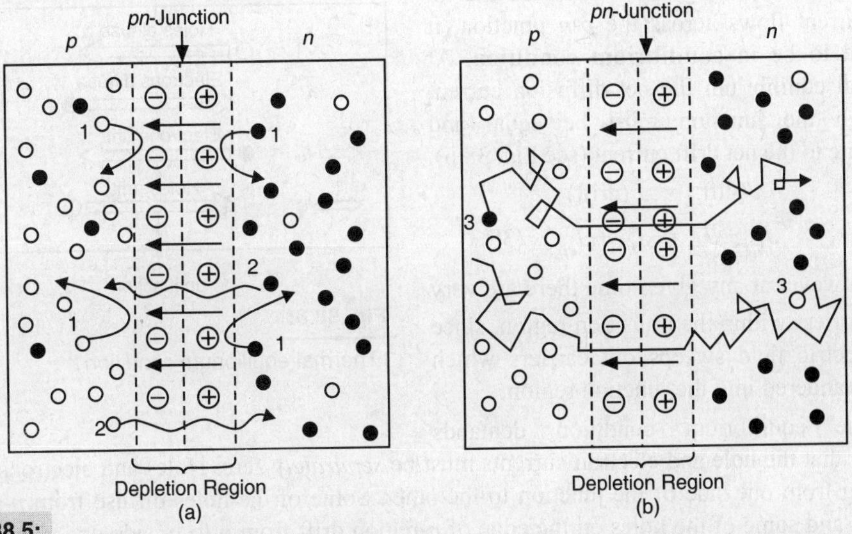

Fig. 38.5: *P-N* junction formation (a) majority carriers movement is halted (b) occasional drift of minority carriers.

However, occasionally a majority hole in *p*-region diffuses into *n*-region whenever it has energy equal to or greater than 'eV_o'. Similarly, a majority electron in *n*-region diffuses into the *p*-region when it has a minimum energy 'eV_o'.

The initial diffusion of charge carriers and the creation of the resultant barrier potential occur when the junction is formed during the crystal growth process. Without such a barrier all the free electrons in the *n*-type region gradually diffuse and recombine with the holes in the *p*-region, leaving neither free electrons nor holes.

The diffusion of majority carriers causes *diffusion current* to flow across the junction. It is easy to see that the current components due to holes and electrons add up although the carriers are moving in opposite directions (the electron current is opposite to the direction of electron flow). The net diffusion current density flowing across the junction is given by

$$J\,(\text{diff}) = J_{hp} + J_{en} \quad ...(38.1)$$

38.2.4 Space Charge and Drift of Minority Carriers

It may however be noted that the field due to space charge has the right direction to cause the flow of *minority carriers* across the junction. Electrons reaching the edge of the junction on *p*-side are accelerated by the electric field into *n*-region and similarly, the holes reaching the edge of the junction on *n*-side are accelerated into *p*-region.

As a consequence, an electric current flows across the junction. This current, which is caused by electric field, is called *drift current*. We may readily see that the current components due to drift motion of holes and electrons are in the same direction and add to each other. If we designate the hole drift current density as J_{hn} and the electron drift density as J_{ep}, then the net drift current through the junction is

$$J(\text{drift}) = J_{hn} + J_{ep} \quad ...(38.2)$$

38.2.5 Thermal Equilibrium Condition

When external voltage is not applied and no net current flows across the *p-n* junction, it is said to be in **equilibrium condition.** At thermal equilibrium the net diffusion current through the junction must be equal and opposite to the net drift current (see Fig. 38.6).

$$J(\text{diff.}) = J\,(\text{drift})$$

i.e., $$J_{hp} + J_{en} = J_{hn} + J_{ep} \quad ...(38.3)$$

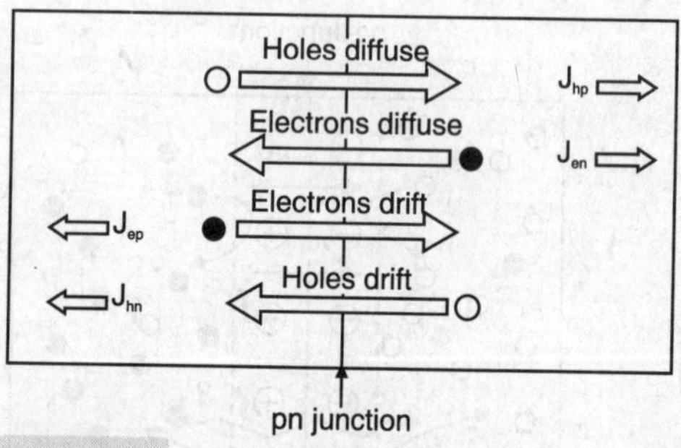

Fig. 38.6: Thermal equilibrium condition

However at any given time there are very few carriers within the transition region, since the electric field sweeps out carriers which have wandered into the junction region.

The equilibrium condition demands further that the hole and electron currents must be *separately* zero. Holes and electrons go on moving from one side of the junction to the other. Some of the holes diffuse from *p*-side to *n*-side, and some of the holes on the edge of *n*-region drift from *n* to *p*- side under the action of the junction field.

Conversely, some electrons diffuse from *n*-side to *p*-side and some drift from *p* to *n*-side. In the equilibrium condition, the neutrality of *p*- and *n*-regions is preserved. Since there can be no net build up of holes or electrons on either side as a function of time, the drift and diffusion currents must cancel for *each* type of carrier. In other words the hole drift current must exactly cancel the hole diffusion current and the electron drift current must cancel the electron diffusion current. Thus,

$$J_{hp} - J_{hn} = 0$$

$$J_{en} - J_{ep} = 0$$

or

$$\left.\begin{aligned} J_{hp} &= J_{hn} \\ J_{en} &= J_{ep} \end{aligned}\right\} \quad ...(38.4)$$

38.2.6 Energy Band Diagram of p-n Junction at Equilibrium

Let us now understand the formation of *p-n* junction from the point of view of energy band structure. When the two semiconductors are in contact, equilibrium is attained only when there is no net current flow across the junction region. Current can be zero when the probability of occupancy of a given energy level is the same in both the semiconductor regions. It implies that the reference energy level, namely Fermi level E_F must be at the same level in the two semiconductor regions. If it is not so, the different probability of occupancy in neighbouring regions would lead to carrier migration and current flow. Fermi level is similar to the liquid level in a container, temperature of a body etc. Their equality throughout a medium ensures no-flow-condition and equilibrium state. In the same way, an energy band structure having equalized Fermi level in both the regions is characteristic of thermal equilibrium of a *p-n* junction.

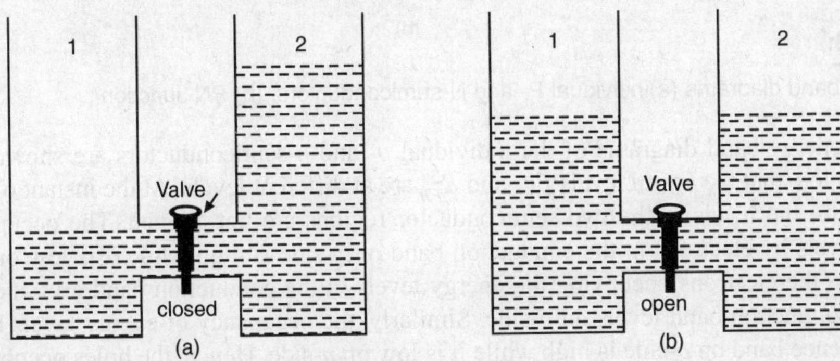

Fig. 38.7:
Illustration of Fermi Levels

The equalization of Fermi levels in two different regions can be understood with the help of the following analogy. Let us consider two containers (1) and (2) which are interconnected through a valve (Fig. 38.7a). Initially, the valve is closed and the containers are filled with water to different levels. When the valve is opened, water flows from container (2) into container (1). As a result, the water level in container (1) rises and water level in container (2) falls down. The process continues till the levels in both the containers are equalized as shown in Fig. 38.7 (b).

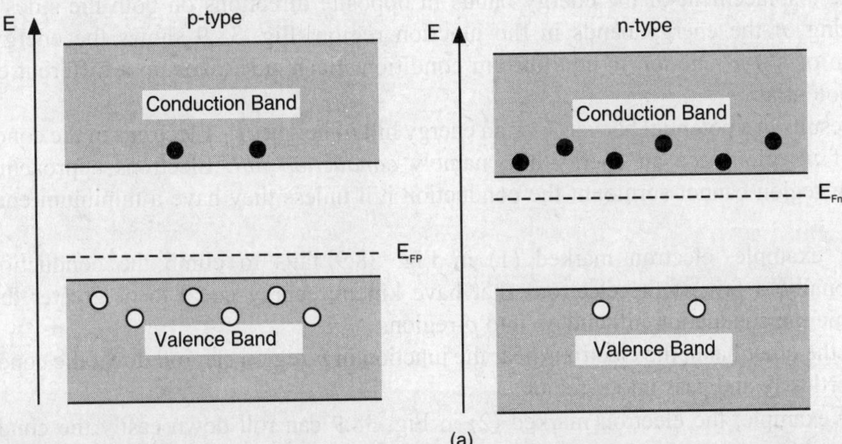

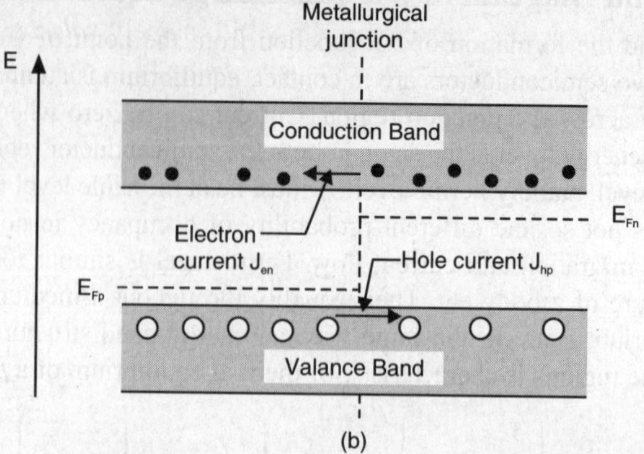

Fig. 38.8:

Energy band diagrams (a)individual P- and N-semiconductors, (b) PN Junction

The energy band diagrams of the individual *p*- and *n*-semiconductors are shown in Fig. 38.8 (a). Note that the Fermi levels E_{Fp} and E_{Fn} are at different levels. At the instant of joining (Fig. 38.8b) the levels in the two semiconductor regions are not aligned. The occupancy of energy levels by electrons in the conduction band on *n*-side is high while it is low on *p*-side. Therefore, the electrons occupying the energy levels in the conduction band on *n*-side move into the conduction band levels on *p*-side. Similarly, the occupancy of energy levels by holes in the valence band on *p*-side is high while it is low on *n*-side. Hence, the holes occupying the energy levels in the valence band on *p*-side move into the valence band levels on *n*-side. As high energy electrons leave *n*-region, the Fermi level E_{Fn} which represents the average energy of electrons moves downwards. Since the Fermi level is fixed relative to the band structure of the region, its movement causes downward shift of the entire band structure in the *n*-region. On the *p*-side, holes having higher energy leave the valence band in that region. The direction of decrease in hole energy is upward and hence the Fermi level E_{Fp} moves upward. Along with E_{Fp}, the entire band structure in the *p*-region shifts upward. The shifting of energy bands continues till the energy levels E_{Fp} and E_{Fn} attain the same level in both the regions. When the two levels are equalized, the carrier migration comes to a halt and equilibrium condition is established.

The displacement of the energy bands in opposite directions on both the sides causes a bending of the energy bands in the junction region. Fig. 38.9 shows the energy band diagram of a *p-n* junction in equilibrium condition. Each side takes up a different electrostatic potential.

It results in a potential barrier V_o or an energy hill of height eV_o. Electrons in the conduction band of *n*-region face an energy hill, namely *conduction hill*. Electrons approaching the junction region cannot surmount the conduction hill unless they have a minimum energy of eV_o.

For example, electron marked (1) in Fig. 38.9 fails to climb the conduction hill. Occasionally, a few of the electrons that have kinetic energy equal to or greater than eV_o overcome the conduction hill and go into *p*-region.

On the other hand, the electrons near the junction in *p*-region can roll down the conduction hill effortlessly and pass into *n*-region.

For example, the electron marked (2) in Fig. 38.9 can roll down easily the conduction hill. The components of currents due to such migration of electrons are in opposite directions and balance each other. As the direction of increasing energy is downward for holes, the holes in the valence band of p-region encounter an energy hill, namely *valence hill*. The holes on the p-side cannot go into the n-region unless they have a minimum energy of eV_o.

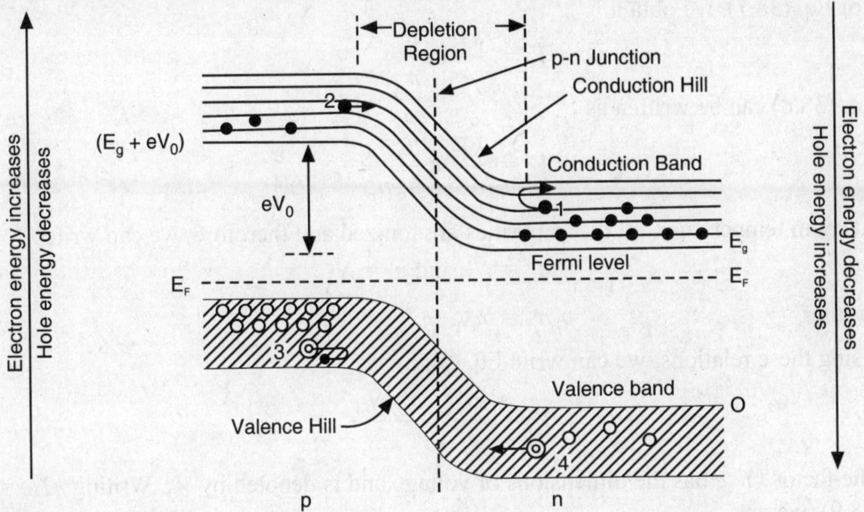

Fig. 38.9:

Energy band diagram of p-njunction at equilibrium

The hole marked (3) in Fig. 38.9 fails to surmount the valence hill whereas a few holes having kinetic energy equal to or greater than eV_o succeed in going into n-region.

On the other hand, holes near the junction on the n-side can readily float up the hill irrespective of their energy.

For example, hole marked (4) in Fig. 38.9 floats up the valence hill. The two components of current due to the opposite flow of holes balance each other.

Thus, the current due to occasional diffusion of majority carriers is balanced by the occasional drift of minority carriers and the net current across the p-n junction is zero.

38.2.6.1 Calculation of the internal potential barrier, V_o

The magnitude of the potential barrier V_o can be estimated from the knowledge of the electron concentrations in p- and n-region of the diode. Referring to Fig. 38.9, E_g is the edge of the conduction band on the n-side. The electron concentration in the conduction band on the n-side can be written as

$$n_n = N_C \exp[-(E_g - E_F)/kT \quad ...(38.5)$$

The edge of the conduction band on the p-side is given by $(E_g + eV_o)$. The electron concentration on p-side can be expressed as

$$n_p = N \exp[-\{(E_g + eV_o) - E_F\}/kT \quad ...(38.6)$$

Dividing Eq. (38.5) by Eq. (38.6) we get

$$\frac{n_n}{n_p} = \exp\left[\frac{eV_o}{kT}\right] \quad ...(38.7)$$

Eq. (38.7) shows that at thermal equilibrium the concentrations of electrons on both sides of the junction are related through the Boltzmann factor e^{eV_o/E_T}. The concentrations of holes on both the sides are related by an equation similar to Eq. (38.7). Taking logarithms on both sides of Eq. (38.7), we obtain

$$V_o = \frac{kT}{e} \ln \frac{n_n}{n_p} \quad \text{...(38.8)}$$

Eq. (38.8) can be written as

$$V_o = \frac{kT}{e} \ln \frac{n_n p_p}{n_p p_p}$$

At room temperature, all the impurities are ionized and therefore, we can write

$$n_n = N_D \text{ and } p_p = N_A$$

Further, $$p_p n_p = n_i^2$$

Using these relations, we can write Eq. (38.8) as

$$V_o = \frac{kT}{e} \ln \frac{N_D N_A}{n_i^2} \quad \text{...(38.9)}$$

The factor kT/e has the dimensions of voltage and is denoted by V_T. Writing $kT/e = V_T$ in Eq. (38.9), we get

$$V_o = V_T \ln \frac{N_D N_A}{n_i^2} \quad \text{...(38.10)}$$

Equ. (38.10) indicates that the barrier potential in a junction diode depends on the equilibrium concentrations of the impurities in *p*- and *n*-regions and does not depend on the charge density in the depletion region.

Example 38.1: *Calculate the potential barrier for a germanium pn junction at room temperature, if both the p- and n-regions are doped equally and to the extent of one atom per 10^6 germanium atoms.*

Solution: $$V_o = \frac{kT}{e} \ln \frac{N_A N_D}{n_i^2}$$

$$= \frac{1.38 \times 10^{-23} \text{ J/°K} \times 300\text{K}}{1.602 \times 10^{-19} C} \ln \frac{(4.4 \times 10^{22})^2}{(2.4 \times 10^{19})^2}$$

$$= 0.0258 \times 16.12\text{V} = \mathbf{0.42 \text{ Volts}}$$

38.2.6.2 Circuit symbol and bias

Fig. 38.10 shows a schematic of *p-n* junction diode. The arrowhead signifies the conventional direction of current flow in the diode when it is forward biased. The *p*-side of the diode is positive and the *n*-side is negative.

A device is said to be ***biased*** when a dc voltage is applied to it. A *p-n* junction can be biased in two ways:

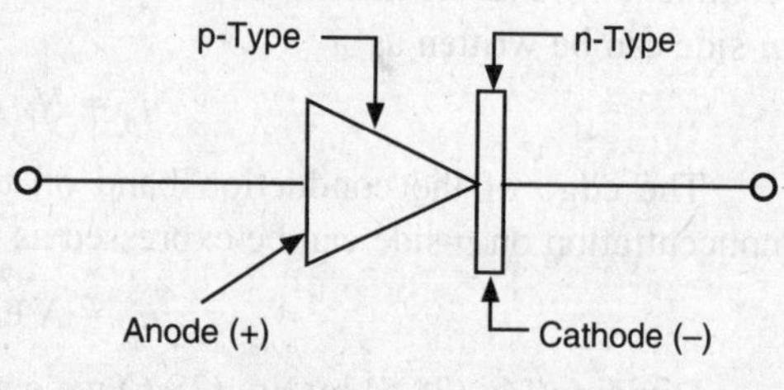

Fig. 38.10: Circuit symbol of *p-n* diode

(*i*) ***forward biasd***: the positive terminal of the dc voltage source is connected to the *p*-region and the negative terminal to the *n*-region.

(*ii*) ***reverse biasd:*** the positive terminal of the dc voltage source is connected to the *n*-region and the negative terminal to the *p*-region.

38.3 p-n JUNCTION UNDER FORWARD BIAS

When a dc voltage, V_F is connected to the diode in such a way that the positive terminal of the source is connected to the *p*-region and the negative terminal to the n-region, then the junction is forward biased (Fig. 38.11 a). Since the junction region is depleted of mobile charges, its resistance is very high. The resistance of the remaining parts of the semiconductor, where mobile charges are in plenty, is very low. Therefore, the forward bias voltage V_F appears across the junction and acts opposite to the internal potential V_o. The effective voltage across the junction is $(V_o - V_F)$ now.

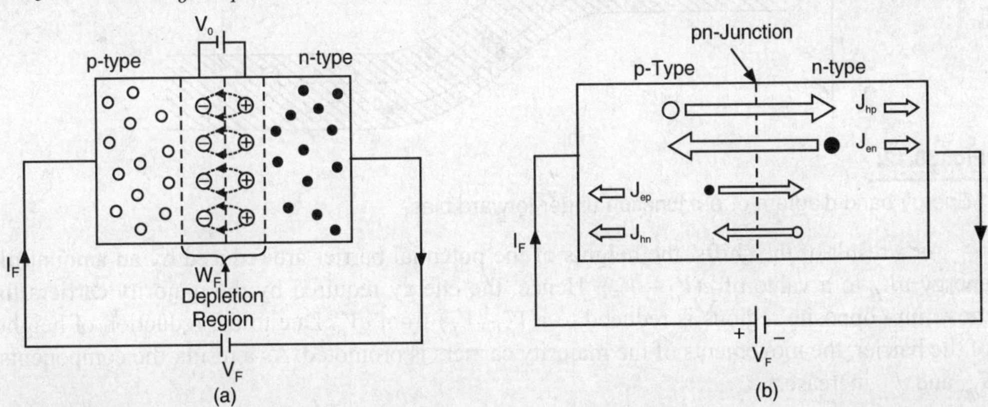

Fig. 38.11:
(a) *p-n* junction under forward bias (b) Current components across forward bias junction

Therefore, the potential barrier is reduced from V_o to $(V_o - V_F)$. The majority carriers push into the depletion region and the width of the depletion region is reduced from the equilibrium value of W to W_F. As a result, the diffusion of majority carriers across the junction increases. The diffusion current density increases from J_{drift} to a value J^*_{diff}.

However, the reduction in the potential does not influence the minority carriers. As usual, they are in a favourable position to move into opposite regions. Hence the drift current density J_{drift} remains unaffected by the forward bias. Consequently, $J^*_{diff} > J_{drift}$. Therefore, a net current density equal to $J^*_{diff} - J_{drift}$ flows through the junction (Fig. 38.11 b).

38.3.1 Energy Band Diagram

The steady state situation of the energy band structure under forward bias condition is shown in Fig. 38.12. The negative terminal of the external voltage source causes an increase in electron energy and an upward shift of all energy levels on the *n*-side. Similarly, the positive terminal connected to *p*-side causes an increase in hole energy and hence a lowering of all levels on *p*-side. As the displacements of the energy levels occur in opposite directions, the Fermi levels E_{Fn} and E_{Fp} get separated by an amount of energy eV_F added by the voltage source.

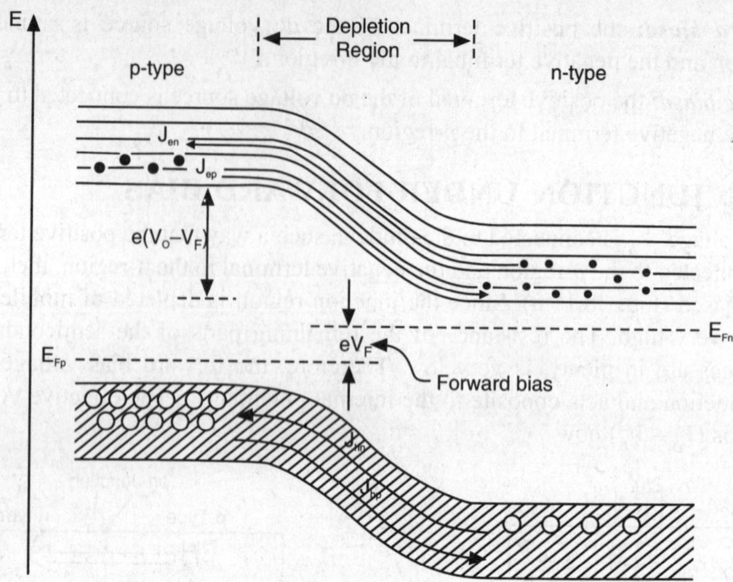

Fig. 38.12:

Energy band diagram of *p-n* junction under forward bias.

As a result of the shifts, the heights of the potential barrier are reduced by an amount of energy eV_F to a value of $e(V_o - V_F)$. Hence, the energy required by the majority carriers to move into opposite regions is reduced to $e(V_o - V_F)$ from eV_o. Due to the reduction of height of the barrier, the movements of the majority carriers is promoted. As a result, the components J_{hp} and J_{en} increase.

The reduction in the height of the potential barrier does not affect the minority carrier movement. As usual those carriers which are near the depletion region are in a favourable position to cross the junction. Electrons slide down the conduction hill from *p*-side to *n*-side, while holes float up the valence hill from *n*-side to *p*-side. The small drift current density components J_{ep} and J_{hn} due to minority carriers do not change due to forward bias.

Therefore, a net current flows through the junction, which is determined by the diffusion of majority carriers.

38.3.2 The Law of the Junction

Note that the majority carriers that cross the junction and arrive in the opposite region become minority carriers in that region. Therefore, due to forward bias the number of minority carriers in that region increases to a very large value. The increase in minority carrier concentration can be estimated as follows.

Under equilibrium condition, the electron concentration on *p*-side of the junction is given by Eq. (38.7). It may be rewritten as

$$n_p = n_n \exp\left[-\frac{eV_o}{kT}\right] \qquad ...(38.11)$$

Due to the application of forward bias, V_o reduces to $(V_o - V_F)$. It causes an increase in the value of n_p. Let n_p^* be the enhanced concentration of electrons at the junction on *p*-side.

Then, Eq. (38.11) may be written as

$$n_p^* = n_n \exp\left[-\frac{e(V_o - V_F)}{kT}\right] \quad ...(38.12)$$

$$= n_n \exp\left[-\frac{eV_o}{kT}\right] \cdot \exp\left[\frac{eV_F}{kT}\right]$$

or

$$n_p^* = n_p \exp\left[\frac{eV_F}{kT}\right] \quad ...(38.13)$$

Eq. (38.13) indicates that upon applying a small forward bias, an exponential increase in the number of electrons occurs on *p*-side of the junction. To get an idea of increase in electron concentration on *p*-side, let us assume that $V_F = 0.1$ *volt*. The new value of the electron concentration on *p*-side is

$$n_p^* = n_p \exp\left[\frac{0.1}{0.025}\right] \approx 50\, n_p.$$

38.3.3 Injection of Minority Carriers

The penetration of majority carriers through the *p-n* junction under forward bias and the consequent increase in the number of minority carriers on the opposite side of the junction is called the ***injection of minority carriers***.

When the doping levels on the two sides are different, the junction is said to be asymmetric and the number of carriers injected from the region with higher doping level exceeds the number of carriers injected from the weakly doped region. The region from which larger number of carriers is injected is called the ***emitter*** and the region into which the injection is directed is called the ***base***.

Note that the majority carrier concentrations are not affected in each region by the application of forward bias. Each region acts as a reservoir of majority carriers and does not suffer in strength when some of the carriers leave the region, since the number of carriers leaving the region is only a small fraction.

38.3.4 Diffusion of Minority Carriers

It is seen from the Eq. (38.13) that under forward bias, a large concentration of minority carriers are found at the depletion layer edge of both the sides. For example, the hole concentration is large at the depletion layer edge on *n*-side and the electron concentration suddenly rises on *p*-side edge. The concentration of minority carriers is low in the rest of the bulk of the material. Thus, the forward bias produces a concentration gradient for minority carriers in each region.

Consequently, to reach equilibrium state diffusion of excess minority carriers occurs away from the junction and into the bulk of the semiconductor on both sides of the depletion layer. The concentration of minority carriers decreases quickly with distance from the junction. This is due to the reason that they are surrounded by majority carriers and a swift recombination of carriers takes place. If the supply were limited, the concentration of these carriers would be that at thermal equilibrium. However, as the majority carriers diffuse across the junction, new carriers are supplied due to the forward bias. Therefore, thermal equilibrium is never reached and a continuous diffusion current is maintained. The external voltage source replenishes the carriers lost through recombinations. The processes of diffusion and recombination transport charge through the device and cause current flow. The current continues to flow in the device as long as the external voltage source is connected in the circuit.

It should be recognized at this stage that the flow of minority carriers in the two regions is due to the concentration gradient rather than due to the electric field established by the battery. There will be only an insignificantly small voltage drop across the bulk of the semiconductor due to its high conductivity, while the major portion of the voltage appears across the depletion layer which is a high resistance region. As a result, the concentration gradient for minority carriers is more effective in causing diffusion of minority carriers.

38.3.5 Carrier Movement in the Forward Bias Circuit

The movement of charge carriers through the forward biased *p-n* junction may be traced as follows:

An electron leaves the negative terminal of the battery and moves toward the *n*-region. It enters the *n*-region and becomes a majority carrier. Gradually, it drifts towards the junction. As it approaches the junction, the electron slows down under the repelling influence of electric field acting across the junction. While slowing down it gives its energy to the neighbouring electrons through collisions. The electron that absorbs the extra energy overcomes the barrier and goes into *p*-region, where it becomes a minority carrier. In the *p*-region the electron diffuses due to concentration gradient. As it diffuses, it is surrounded by a large number of holes and the electron recombines with one of the holes. Recombination means that a free electron becomes a valence electron.

Thus, the moving electron suddenly stops and its kinetic energy is communicated to other valence electrons in the lattice. As a result, a covalent bond is broken elsewhere and an electron-hole pair is generated. The electron of the pair diffuses forward and it in turn undergoes recombination after some time. Another electron-hole pair is generated. This process goes on till the electron reaches the terminal end of the *p*-region. At the end, out of the electron-hole pair, the hole drifts towards the junction while the electron flows into the connecting wire and returns to the positive terminal of the battery. This completes the circuit. The current density in the circuit is composed of two components J_e and J_h, as shown in Fig. 38.13.

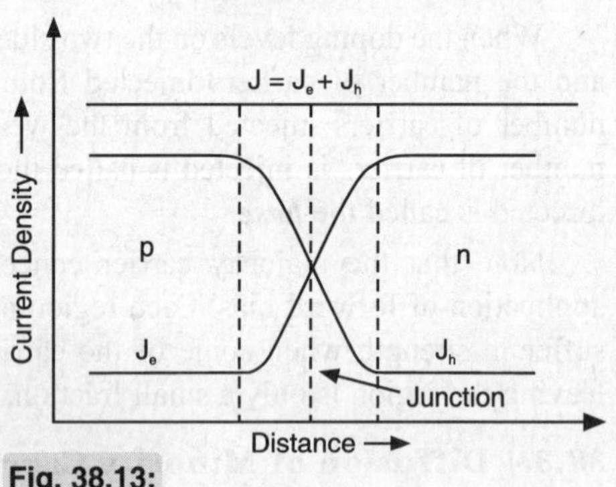

Fig. 38.13: Current density in the circuit - components J_e and J_h

38.4 p-n JUNCTION UNDER REVERSE BIAS

Fig. 38.14 shows a reverse biased *p-n* junction. It is seen that the external voltage V_R adds to the barrier voltage V_o and the potential barrier increases to $e(V_o + V_R)$. As a result, the electric field in the depletion region increases and the majority carriers are pushed farther from the depletion region. This action leads to an increase in the width of the depletion region from W_f to W_{rb} (see Fig. 38.14). The increase in potential barrier inhibits the diffusion of majority carriers and the diffusion current falls to zero at higher bias values (Fig. 38.15).

However, the increase in the potential barrier does not influence minority carriers. The increase in the electric field increases the acceleration of minority carriers without causing a change in their number. The drift current density remains the same as in unbiased and forward bias conditions. The current through the junction is only due to drift current caused by the minority carriers. The components of drift current J_{hn} and J_{ep} are shown in Fig. 38.15. The current is of the order of a nanoampere or less. Thus, in a reverse biased *p-n* junction

$$J_{diff} = 0$$

$$J_{drift} = -(J_{hn} + J_{ep}) = -J_o \quad \text{...(38.14)}$$

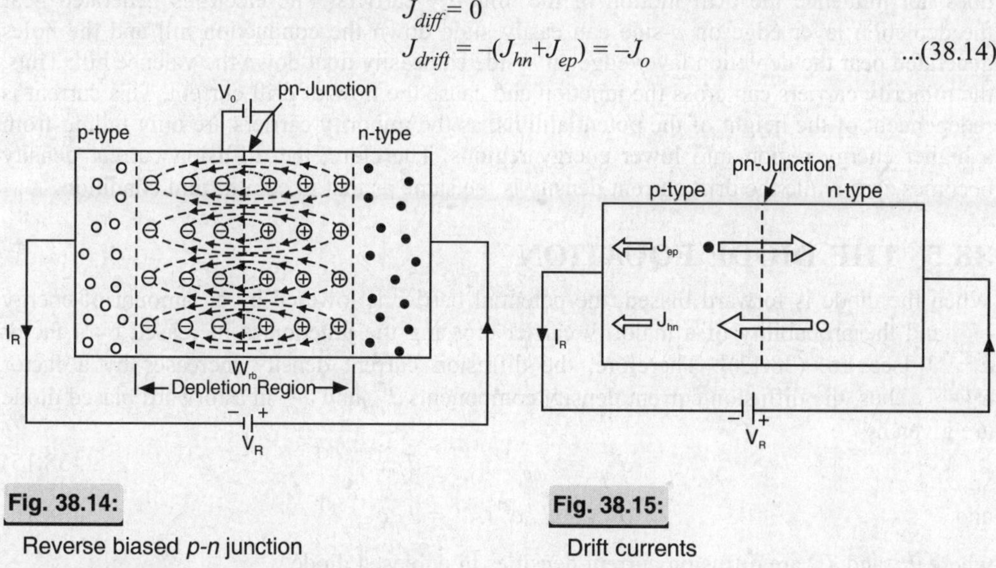

Fig. 38.14: Reverse biased *p-n* junction

Fig. 38.15: Drift currents

38.4.1 Energy Band Diagram

The energy band diagram for a *p-n* junction under reverse bias condition is shown in Fig. 38.16. The positive terminal of the voltage source, V_R connected to the *n*-side reduces the energy of the electrons. Therefore, the energy levels on *n*-side are displaced downwards. Similarly, the negative terminal connected to *p*-side reduces the energy of holes on *p*-side. Therefore, the energy levels on *p*-side are displaced upwards. Due to such displacement of energy levels on both sides, the Fermi levels E_{Fn} and E_{Fp} get separated by an amount of energy eV_R. Now, the barrier height increase to a value of $e(V_o + V_R)$. The majority carriers in *n*-region, electrons cannot climb the conduction hill to go into *p*-region.

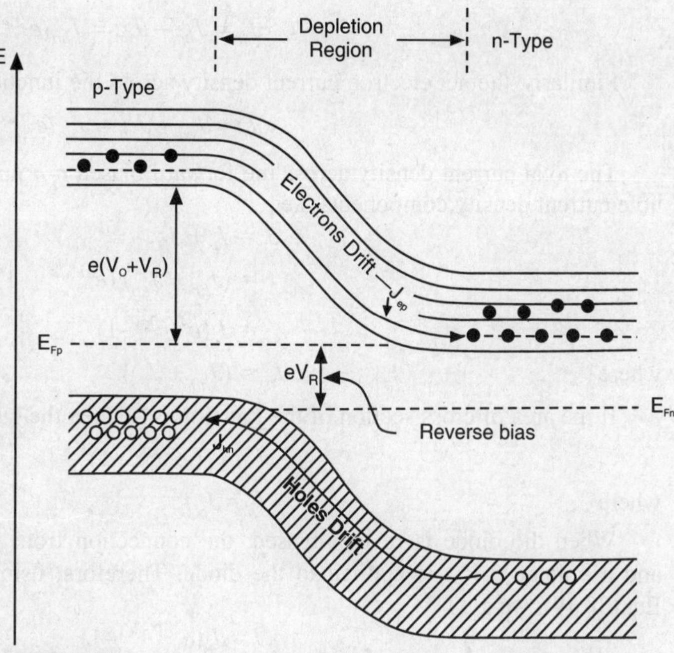

Fig. 38.16: The energy band diagram for a *PN* junction under reverse bias condition

Similarly, the majority carriers in the *p*-region, holes cannot float up the valence hill to go into *n*-region. The diffusion of majority carriers is totally stopped. However, the barrier does not influence the drift motion of the minority carriers. The electrons generated near the depletion layer edge on *p*-side can easily slide down the conduction hill and the holes generated near the depletion layer edge on *n*-side can easily float down the valence hill. Thus, the minority carriers can cross the junction and cause the flow of drift current. This current is independent of the height of the potential hills, as the minority carriers are only falling from a higher energy region into lower energy regions. Therefore, the diffusion current density becomes zero while the drift current density is the same as that in equilibrium condition.

38.5 THE DIODE EQUATION

When the diode is forward biased, the potential barrier is lowered by an amount of energy eV_F and the probability of a majority carrier crossing the junction is increased by a factor $e^{eV_F/kT}$ [see Eq. (38.13)]. Therefore, the diffusion current density increases by a factor $e^{eV_F/kT}$. Thus, the diffusion current density components J^*_{hp} and J^*_{en} in a forward biased diode are given by

$$J^*_{hp} = J_{hp}e^{eV_F/kT} = J_{hn}e^{eV_F/kT} \qquad ...(38.15)$$

and

$$J^*_{en} = J_{en}e^{eV_F/kT} = J_{ep}e^{eV_F/kT} \qquad ...(38.16)$$

where J_{hp} and J_{en} are diffusion current densities in unbiased diode.

The drift current density components have not changed and have the same magnitude as in equilibrium case.

Therefore, the net hole current density across the forward biased junction is

$$J_h = J^*_{hp} - J_{hn} = J_{hn}(e^{eV_F/kT} - 1) \qquad ...(38.17)$$

Similarly, the net electron current density across the junction is

$$J_e = J^*_{en} - J_{ep} = J_{ep}(e^{eV_F/kT} - 1) \qquad ...(38.18)$$

The total current density across the forward biased *p-n* junction is a sum of electron and hole current density components, i.e.,

$$J = J_e + J_h$$

$$\therefore \quad = (J_{hn} + J_{ep})\,(e^{eV_F/kT} - 1)$$

$$= J_o(e^{eV_F/kT} - 1) \qquad ...(38.19)$$

where $J_o = (J_{hn} + J_{ep})$.

If the area of cross section of the junction is A, then the current is $I = JA$. Therefore,

$$I = I_o\,(e^{eV_F/kT} - 1) \qquad ...(38.20)$$

where $I_o = J_oA$.

When the diode is reverse-biased, the connection from the voltage source is reversed and a voltage $(-V_R)$ is applied to the diode. Therefore, using $(-V_R)$ in place of V_F in the Eq. (38.19), we get

$$J = J_o(e^{-eV_R/kT} - 1) \qquad ...(38.21)$$

For larger values of V_R, $e^{-eV_R/kT} \to 0$.

$$\therefore \quad J = -J_0 \qquad ...(38.22)$$

$$\therefore \quad I = -I_0 \qquad ...(38.23)$$

Eq. (38.19) and Eq. (38.21) can be combined into a single equation by denoting the forward and reverse voltages by a single symbol V. Thus, we write

$$J = J_o\,(e^{eV/kT} - 1) \qquad ...(38.24)$$

In terms of current, $$I = I_o(e^{eV/kT} - 1) \qquad ...(38.25)$$

where $V = V_F$ for forward bias and $V = -V_R$ in case of reverse bias.

Equ. (38.25) is known as the ***rectifier equation*** or ***diode equation***.

The term I_o is known as ***reverse saturation current***.

Example 38.2: *Current flowing in a p-n junction is 0.2 μA at room temperature when a large reverse bias voltage is applied. Calculate the current when a forward bias of 0.1 V is applied.*

Solution: The current flowing through a junction is given by $I = I_o\,[\exp(eV/kT) - 1]$

$\therefore$ $$I = 2 \times 10^{-7} A\,[\exp(0.1\text{ V}/0.026\text{ V}) - 1] = \mathbf{9.2\ \mu A}.$$

38.6 VOLTAGE-AMPERE CHARACTERISTIC

A graph that shows the variation in current in a device with the variation of voltage applied across it is called **volt-ampere characteristic**. Fig. 38.17 shows the volt-ampere characteristic of a *p-n* junction diode. It is seen that the characteristic is not linear and hence a *p-n* junction is a **non-linear** device. The *p-n* junction acts as a *closed switch* in forward bias condition allowing large current to flow through it; and acts as an *open switch* in reverse bias condition causing an insignificantly low current through it. This unidirectional conduction is called ***rectifying*** action.

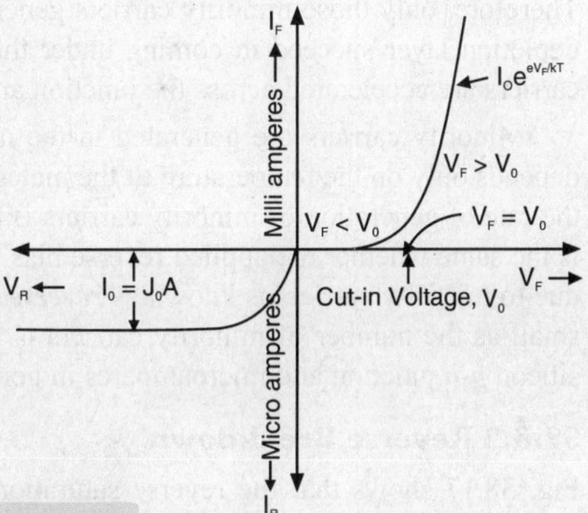

Fig. 38.17: PN junction Volt-Ampere characteristics

For a forward bias voltage V_F, the current through the junction is negligibly small as long as V_F is less than V_o. The voltage V_o at which the current raises sharply is called the **cut-in voltage**. It is in fact the same as barrier voltage. As V_F increases beyond V_o, the barrier disappears and an exponential increase in forward current I_F occurs. The forward current is equal to the difference between the diffusion current and drift current.

$$I_F = I_o e^{eV_F/kT} - I_o = I_{\text{diffusion}} - I_{\text{drift}} \qquad ...(38.26)$$

When V_F exceeds V_o, the barrier collapses and the drift current becomes nearly zero. However, the barrier does not totally vanish. The bulk resistance of the diode causes the bias voltage to be distributed over the entire length of the semiconductor regions. Therefore, at higher values of V_F the characteristic becomes practically linear (i.e., ohmic).

The reverse current through the junction is very much smaller than the forward current. Therefore, a different scale is used for plotting the reverse bias characteristic. The reverse current I_R is given by

$$I_R = I_{\text{drift}} - I_{\text{diffusion}} = (-I_o) - I_o e^{-eV_R/kT} \qquad ...(38.27)$$

In the initial stage the reverse current increases rapidly due to an exponential decrease of diffusion current with increasing voltage. At higher V_R values, the diffusion current drops to zero and the reverse current becomes equal to the drift current due to the minority carriers. As the number of minority carriers at a given temperature is constant and does not change with the bias voltage, the reverse current stays constant with increase in the reverse bias.

38.6.1 Reverse Saturation Current

The reverse saturation current I_o is constant and is caused by the minority carriers. Minority carriers are generated in the material due to rupture of covalent bonds and have succeeded in arriving into the depletion region. Diffusion length decides the number of minority carriers that can arrive into the depletion region. Electrons generated in p-region at distances greater than the diffusion length L_e from the junction cannot reach the depletion region since they undergo recombinations before reaching the depletion layer. So is the case with holes in n-region. Holes generated in n-region at distances greater than the diffusion length L_h from the junction cannot reach the depletion region since they undergo recombinations before reaching the depletion layer. The diffusion length is of the order of 10^{-4} m in case of germanium. Therefore, only those minority carriers generated in the neighbourhood of the outer edges of depletion layer succeed in coming under the influence of internal electric field. Only those carriers are accelerated across the junction and cause the drift current.

Minority carriers are generated in the material due to rupture of covalent bonds which depends only on the temperature of the material. As long as the temperature remains constant, the rate of generation of minority carriers is constant. Therefore, the current due to their flow is the same whether the applied reverse bias voltage is small or large. Hence, the drift current due to minority carriers is known as ***reverse saturation current.*** It is denoted by I_o. It is very small as the number of minority carriers is very small. It is of the order of nanoamperes in silicon *p-n* junction and microamperes in germanium *p-n* junction.

38.6.2 Reverse Breakdown

Fig. 38.17 shows that the reverse saturation current is very small. Nevertheless, when the reverse bias voltage is increased steadily, the junction breaks down at a particular stage and starts conducting heavily. This critical value of voltage is called the **breakdown voltage**. Once the breakdown occurs, even a very small increase in voltage causes large change in the reverse current. In normal operation the condition of breakdown should be avoided as it permanently damages the crystal structure of the two regions and renders the junction useless. As the breakdown phenomenon is irreversible and damages the diode permanently, ordinary diodes are never operated in this region.

The breakdown occurs in reverse biased diode mainly due to two different mechanisms.

(*i*) **Avalanche breakdown**: In lightly doped junctions on large reverse voltage across the junction **avalanche** breakdown occurs. When large reverse bias is applied to the diode, the electric field in the depletion region will be high. Therefore, minority electrons entering the depletion region from the *p*-side acquire high kinetic energy. The high energy electrons on a collision with host atoms knock off valence electrons from some of the covalent bonds. These new carriers produce additional charge carriers and the process continues. A flood of carriers is produced in a very short time and causes large reverse current. This is known as ***avalanche breakdown***. The avalanche breakdown is self-sustaining as long as the reverse voltage is present across the depletion region.

The breakdown voltage is found to increase with temperature.

(*ii*) **Zener breakdown:** The Zener breakdown mechanism occurs in thin, abrupt and heavily doped *p-n* junctions and requires relatively low reverse voltage for its operation.

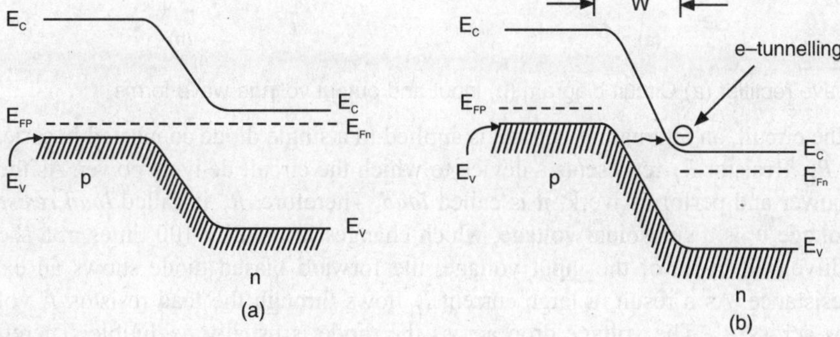

Fig. 38.18:

Illustration of Zener breakdown mechanism

When the *p* and *n* regions are heavily doped, the depletion region becomes very thin, of the order of 10 nm. Application of reverse bias voltage causes the conduction and valence bands to bend to the extent that the *n*-side conduction band appears opposite to the *p*-side valence band. As a result, a number of vacant states in the *n*-side conduction band are brought directly opposite to large number of filled states in the *p*-side valence band (Fig. 38.18). As the barrier separating the two bands is very narrow, the electrons tunnel through it under the effect of a small reverse bias. The electrons from the valence band on *p*-side reach the conduction band on *n*-side, thus causing a sudden large reverse current from *n* to *p* side. This is the *zener effect*. Zener breakdown takes place usually at low reverse voltages of the order of 4 volts or less in silicon diodes.

38.7 APPLICATIONS

A *p-n* junction diode is used as a rectifier, a switch, a clamper etc. We shall study here the application of diode as a rectifier. Government supplies electrical energy in the form of ac voltage because it is more economic and efficient to distribute ac power. However, dc voltage is required for operating most of the electronic equipments and gadgets. It is therefore necessary to convert the incoming mains supply to dc voltage. The process of converting ac voltage into dc voltage is called **rectification** and the device or circuit that converts ac voltage into dc voltage is called **rectifier**. The unidirectional current conduction property of diode is used in rectifiers. A single diode or more diodes can be connected into a circuit to form different types of rectifier circuits.

38.7.1 Half-Wave Rectifier

Half-wave rectifier is a circuit with a single diode that conducts in only one half-cycle of the input ac cycle. The basic half wave rectifier circuit is shown in Fig. 38.19(a).

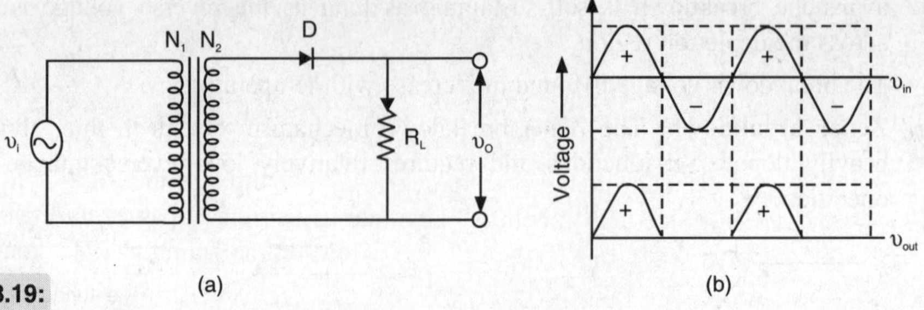

Fig. 38.19: Half-wave rectifier (a) Circuit diagram (b) input and output voltage wave forms

In the circuit, an alternating voltage is applied to a single diode connected in series with a resistor R_L. Resistor R_L represents a device to which the circuit delivers power. As the device draws power and performs work, it is called ***load***. Therefore, R_L is called ***load resistor***. The input voltage v_i is a sinusoidal voltage, which changes in polarity 100 times in a second. In the positive half cycle of the input voltage, the forward biased diode shows an extremely small resistance. As a result, a large current I_L flows through the load resistor. A voltage v_0 develops across R_L. The voltage drop across the diode is usually negligible. Therefore, the entire power appears across R_L. In the negative half cycle, the reverse-biased diode shows a very high resistance. Therefore, only a negligible reverse saturation current flows through the circuit. As such $I_L = 0$ and hence $v_0 = 0$. The negative half-cycle of input voltage waveform is not useful for carrying power to the load and all the input voltage appears across the diode itself.

The input and output waveforms are shown in Fig. 38.19 (b). It is seen that the output voltage is no longer an ac voltage. It is a ***unidirectional fluctuating voltage.*** It has an average dc value over which a number of ac components are superimposed. The undesired ac components are called the ***ripple***. In case of the half-wave rectifier, the lowest ripple frequency is the same as the frequency of the input voltage.

38.7.2 Full Wave Rectifier

A full wave rectifier uses two diodes and rectifies both the half cycles of the input voltage. The input is supplied with a center-tapped secondary winding, as shown in Fig. 38.20 (a).

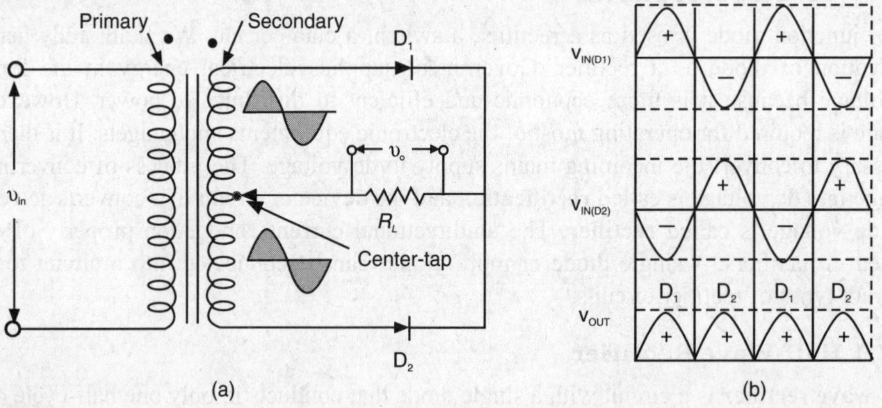

Fig. 38.20: Full-wave rectifier (a) Circuit diagram (b) input and output voltage wave forms

In the positive half-cycles of the secondary voltage, the diode D_1 is forward biased and the diode D_2 is reverse biased. The current i_L flows through the diode D_1, the load resistor R_L and the upper half-winding of the secondary. The output voltage v_0 develops across R_L. During the negative half cycles, the diode D_1 is reverse biased and the diode D_2 is forward biased. The current now flows through the diode D_2, the load resistor R_L and the lower half winding of the secondary. The direction of the current flow through the resistor R_L is the same in both the cases. Hence during both the alternations, current i_L passes through R_L and produces an output voltage v_0. Thus, in the full wave rectifier, the output has of a continuous series of positive half cycles of alternating voltage. The input and output waveforms are shown in Fig. 38.20 (b).

The pulsating dc voltage obtained from a rectifier is smoothened out with the help of filter circuits and a stable dc voltage is obtained.

38.8 ZENER DIODE

Zener diode is a semiconductor diode specially designed to operate in the breakdown region of the reverse bias. Zener diodes are always operated in reverse bias condition. By varying the impurity concentration and other parameters, it is possible to design the breakdown voltage to suit specific applications. In Zener diodes, the breakdown phenomenon is reversible and harmless.

Fig. 38.21 (a) shows the symbol of a zener diode. It may be seen that it is like that of an ordinary diode except the bar is turned into Z-shape. The volt-ampere characteristic of the Zener diode is shown in Fig. 38.21 (b). Zener diode acts very much similar to ordinary diode under forward bias condition. However, the zener diode is generally not used in the forward bias condition. The reverse bias characteristic is very much different from that of an ordinary diode. As the reverse voltage is increased, the reverse current remains constant till a certain value is reached. At that value, the reverse current increases abruptly. The voltage at which such sudden increase in reverse current occurs is called *zener breakdown voltage* or *zener voltage,* V_Z.

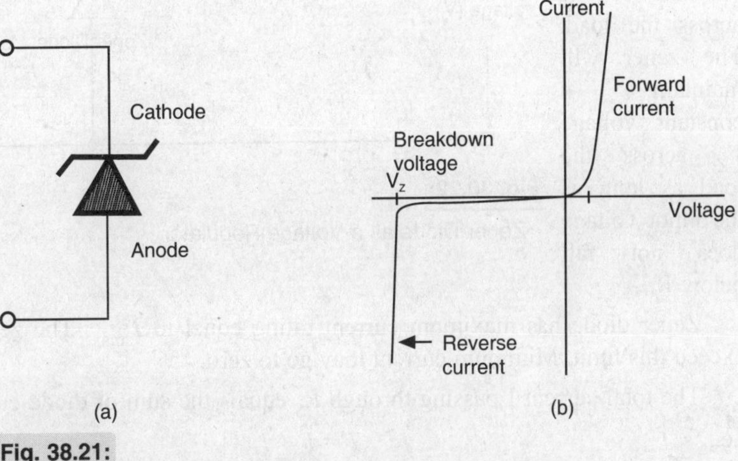

Fig. 38.21: Zener diode (a) symbol, (b) VI characteristics

If a reverse current above breakdown value is allowed the diode is permanently damaged. ***Zener diodes*** are designed to lower zener voltage such as 2.4 V. In the zener region the voltage across zener diode remains constant but the current changes depending on the supply voltage. The voltage drop across the zener diode is equal to the zener voltage of that diode. Doping levels decides the position of zener region, an increase in doping will decrease the zener potential. Zener diodes are available in the range from 2V to 200V.

Applications: Zener diodes are widely used in electronic circuits. Some of the important uses are in

1. voltage regulators
2. reference element
3. meter protection etc.

Zener Diode as a Voltage Regulator: The capacity of an electronic circuit to keep a constant output voltage when either input voltage or load current varies is known as *regulation* or ***voltage regulation***. A zener diode can serve as a voltage regulator. It can be used to provide a constant voltage from a source whose voltage may vary over a considerable range. Fig. 38.22 shows the use of zener diode as a voltage regulator. An unregulated power supply provides the input voltage V_i, R_S is the current limiting resistor. A load resistance R_L is connected parallel with the diode.

The series resistance R_S absorbs the output voltage fluctuations in order to keep constant voltage across the load. The zener will maintain a constant voltage V_Z across the load as long as the input voltage does not fall below V_i.

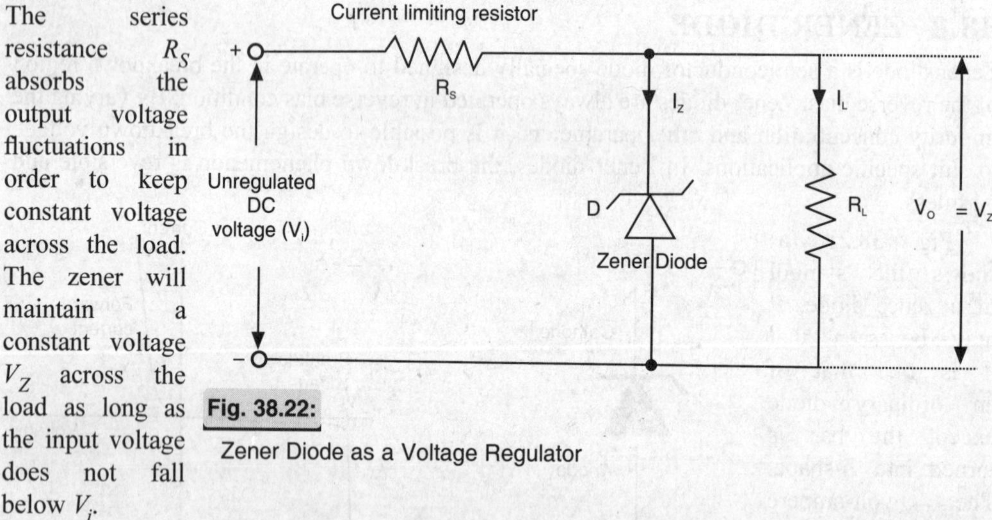

Fig. 38.22: Zener Diode as a Voltage Regulator

Zener diode has maximum current rating equal to I_{zmax}. The zener current should not exceed this limit. Minimum current may go to zero.

The total current I passing through R_S equals the sum of diode current and load current, i.e.,

$$I = I_z + I_L.$$

This circuit makes use of the fact that under reverse bias breakdown voltage, the voltage across the zener diode remains constant even if larger current is drawn. Since the load resistance, R_L is parallel to the zener diode, the voltage across the load resistance does not vary even though current through the load changes. Hence, the voltage across the load is regulated against the variations in the load current.

Operation:

Case 1: We assume that the load resistance R_L is constant and V_i is varying.

As R_L is constant, I_L is also constant because $I_L = \dfrac{V_z}{R_L}$. But supply current keeps changing due to change in V_i. The current through R_S is $I = \dfrac{V_i - V_z}{R_S}$ and is the sum of $I = I_z + I_L$.

If V_i increases, then the current I will increase. But as V_z and R_L are constant, the load current I_L will remain constant. Naturally, the increase in current will increase the zener current I_z. Thus, the increase in I will be absorbed by the zener diode without affecting I_L. The increase in V_i results in a larger voltage drop across R thereby keeping V_o as constant.

If V_i decreases, I will decrease causing I_z to decrease. The diode takes a smaller current and voltage drop across R is reduced. As a result, the output V_o remains constant. Thus, whenever V_i changes, I and IR drop in such a way as to keep V_o constant.

Case 2: We assume that V_i is constant and the load resistance R_L *is varying.*

If R_L decreases then I_L will increase. But as I is constant, the zener current will decrease, thereby keeping I and IR drop constant. The output voltage remains constant.

If R_L increases then I_L will decrease. With decrease in I_L, the zener current I_z will increase in order to keep I and IR drop constant. Again, the output voltage remains constant.

38.9 VARACTOR DIODE

The varactor is a *p-n* junction diode which acts as a voltage-dependent capacitor. It makes use of dependency of the depletion-area capacitance on the voltage impressed across the diode. The depletion region of a junction diode has a large concentration of positive charges stored on the n-side of the junction and a large concentration of negative charges on the *p*-side of the junction, separated by a distance '*w*', the width of the depletion region. The situation is similar to a charged capacitor where one of the parallel plates is positively charged and the other is negatively charged. The depletion region gives a junction capacitance known as junction *capacitance* or *transition capacitance*, C_T. It is given by

$$C_T = \frac{\varepsilon A}{w} \qquad ...(38.28)$$

where A is the area of cross-section of the junction,

ε is the permittivity of semiconductor and

w is the width of the depletion region.

The junction capacitance of a normal diode is about 20 pf with no external bias. Junction capacitance depends on reverse bias and doping concentration.

(*i*) Reverse bias causes an increase in the depletion region width and hence the junction capacitance decreases (Fig. 38.23 a). The change in the width of the depletion region in case of an abrupt junction is given by

$$w \propto \sqrt{V_R}$$

$$\therefore \qquad C_T \propto \frac{1}{\sqrt{V_R}} \qquad ...(38.29)$$

Thus, junction capacitance C_T decreases with an increase in reverse bias.

We have seen that the deplection region acts as a capacitor.

(*ii*) If doping level is increased, the width of the depletion layer reduces. Therefore, C_T increases with increase in doping level.

Varactor diodes are also known as ***varicaps***. The circuit symbol of the varactor diode is shown in Fig. 38.23 (b). Varactor diode shows normal volt-ampere characteristics as the junction diode in both forward and reverse bias conditions. Fig. 38.23 (c) shows the variation of transition capacitance with applied reverse bias in a typical silicon diode.

The semiconductor material used for the fabrication of varactor diodes should have high carrier mobility, low dielectric constant, large band gap, low ionization potential, and high thermal conductivity etc. The material having most of the required properties is GaAs and therefore, GaAs is used for the fabrication of varactor diodes.

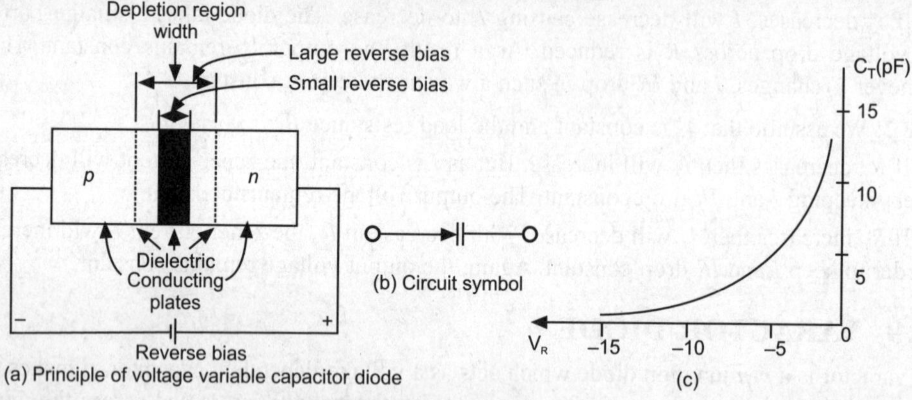

Fig. 38.23: (a) Junction capacitance, (b) Circuit symbol, (c) Varactor diode volt-ampere characteristics

Applications:

(*i*) TV receivers,

(*ii*) F.M. receivers,

(*iii*) Tuned devices,

(*iv*) Communication equipments.

If a diode is used in the reverse-biased mode as a capacitor in an *LC* resonant circuit, the value of *C* and hence the resonant frequency can be varied by changing the reverse bias voltage applied to the diode; no moving parts are required.

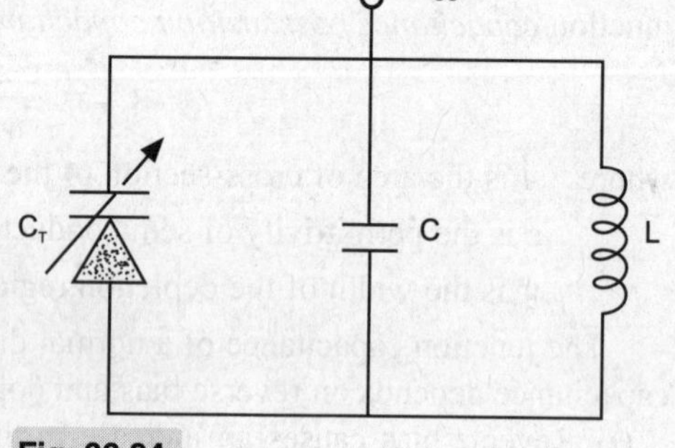

Fig. 38.24: *LC* resonant circuit

Fig. 38.24 shows the use of a varactor diode in an *LC* resonant circuit. The resonant frequency is given by

$$f = \frac{1}{2\pi\sqrt{L(C+C_T)}}$$

By changing the C_T by changing the applied voltage, the resonant frequency of the resonant circuit can be changed.

Example 38.3: *Calculate the junction capacitance of a germanium diode whose area is 1 mm × 1 mm and depletion region width is 2 μm. The relative permittivity of germanium is 16.*

Solution: $$C_T = \frac{\varepsilon A}{w}$$

$$= \frac{8.854\times10^{-12}\,F/m\times16\times10^{-3}\,\text{m}\times10^{-3}\,\text{m}}{2\times10^{-6}\,\text{m}} = \mathbf{70.8\ pF.}$$

Example 38.4: *Junction capacitance of an abrupt junction diode is 20 pF at 5 V. Compute the decrease in capacitance for 1 V increase in bias.*

Solution:

$$\frac{C_{T2}}{C_{T1}} = \left(\frac{V_{R1}}{V_{R2}}\right)^{\frac{1}{2}}$$

$$\therefore \quad C_{T2} = C_{T1}\left(\frac{V_{R1}}{V_{R2}}\right)^{\frac{1}{2}} = 20 \times 10^{-12}\,\text{F} \times \left(\frac{5\text{ V}}{6\text{ V}}\right)^{\frac{1}{2}} = 18.25\text{ pF}.$$

Decrease in capacitance = (20 – 18.25) pF = **1.75 pF.**

38.10 LIGHT EMITTING DIODE (LED)

A light emitting diode (LED) is fabricated using III-IV compound semiconductors, such as GaAs, which have a direct band gap. LEDs emit light in forward bias.

Principle: When a *p-n* junction is forward biased, minority carriers flow in large numbers into regions where they can recombine with majority carriers producing light in the visible or infra red region. The wavelength of light is given by

$$\lambda = \frac{hc}{E_g} = \frac{1.24}{E_g(\text{eV})}\,\mu\text{m}\,. \qquad \text{...(38.30)}$$

This effect is known as ***injection electroluminescence.*** A significant light output is obtained only when there is large number of electro-hole recombinations occurring per second. To ensure this, the *p* and *n* regions are heavily doped.

Theory: Fig. 38.25 (a) shows the energy band diagram of a heavily doped *p-n* junction. The *p-n* junction has a large concentration of electrons in the conduction band of *n*-region and a large concentration of holes in the valence band of *p*-region. Under forward bias conditions the electrons move forward into the depletion region and occupy energy levels in the conduction band.

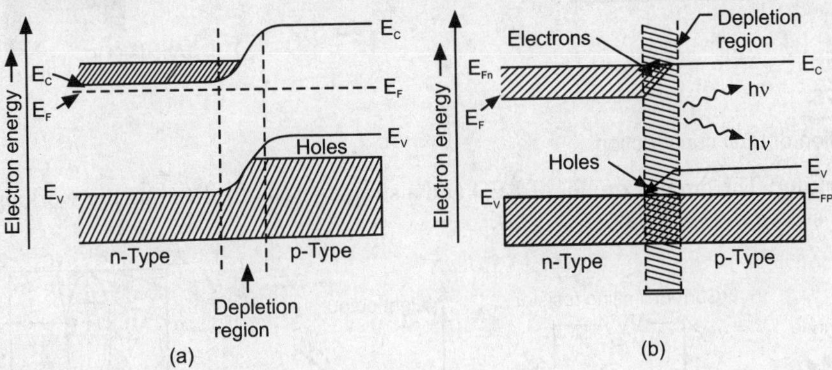

Fig. 38.25:

Energy band diagram of an LED- (a) without bias (b) under forward bias

Similarly, holes push forward into the depletion region and occupy energy levels in the valence band. The electrons in the conduction band are directly above the holes at the edge of the valence band (Fig. 38.25 b). The situation is highly conducive for direct recombination of electrons and holes. When an electron from the conduction band jumps into the hole in the valence band, recombination occurs and the excess energy is emitted in the form of a light photon.

The colour of the emitted light depends on the type of material used.

SN	Material used	Colour of the emitted light
1.	Gallium Arsenide, GaAs	Infrared
2.	Gallium Arsenide-phosphide, GaAsP	Red or Yellow
3.	Gallium Phosphide, GaP	Red or Green

Construction: We describe here the structure of a surface emitting LEDs. These LEDs emit light in a direction perpendicular to the *p-n* junction plane. The construction of a surface emitting LED is shown in Fig. 38.26 (a). An n-type layer is grown on a substrate and a *p*-type layer is grown on it by the process of diffusion. The *p*-layer is made very thin to prevent loss of photons due to absorption in the layer. Metal connections are made at the edges of the *p*-layer in order to allow more central surface for the light to escape. A metal film is deposited at the bottom of the substrate for reflecting as much light as possible towards the surface of the device and also to provide electrode connection. The light generated at the junction may not emerge from the surface of the device as it is likely to suffer total internal reflection at the semiconductor-air boundary. Therefore, the device is encapsulated in a clear epoxy resin of suitable refractive index (Fig. 38.26 b).

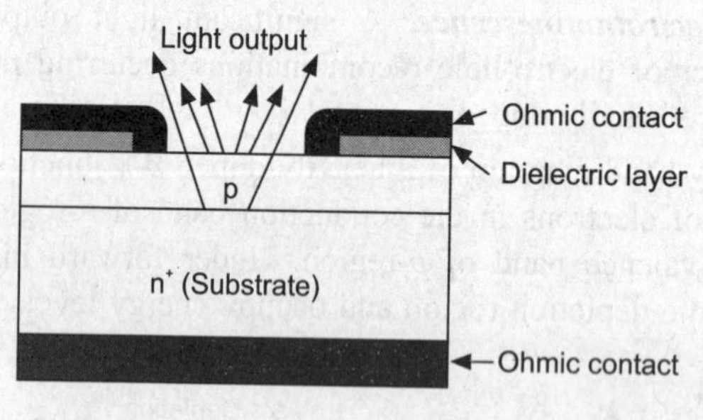

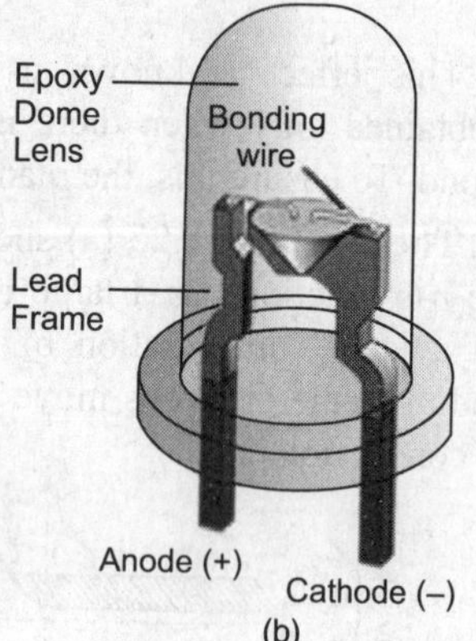

Fig. 38.26: Illustration of LED construction

Working: The circuit symbol of LED and a simple circuit to illustrate the working of an LED are shown in Fig. 38.27.

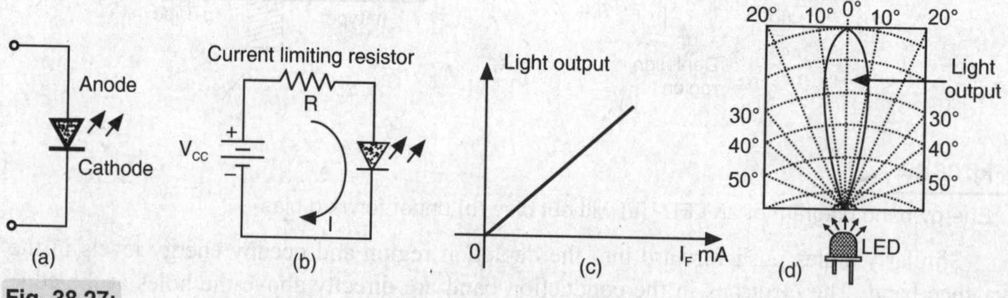

Fig. 38.27: LED (a) symbol, (b) LED in FB bias, (c) amount of light emitted versus forward current, (d) Light illumination

LED is always forward biased. The forward voltage across an LED is significantly greater than an ordinary diode in between 1.2 V and 3.2 V. The amount of light emitted is directly proportional to the forward current, Fig. 38.27 (c). The reverse breakdown voltage of LED is of the order of 3V and an LED is never reverse biased.

Applications: LEDs are used in many applications. Discrete LEDs are used as indicators and as light sources in fibre-optic communications. A number of LEDs may be grouped to form a display. The LEDs may be arranged in the form of a seven-segment display where by energizing a proper combination of segments the decimal numbers 0 to 9 may be displayed. Or they may be arranged in the form of a 5 × 7 matrix which may be used to generate a decimal number or alphabetical character.

Example 38.5: *A light emitting diode is made of GaAsP having a band gap of 1.9 eV. Determine the wavelength and colour of radiation emitted.*

Solution: $\lambda = \frac{hc}{E_g} = \frac{1.24}{E_g(\text{eV})}\ \mu\text{m} = \frac{1.24}{1.9}\ \mu\text{m} = 652.6\ \text{nm}.$

∴ Colour of the radiation is red.

38.11 PHOTODETECTORS

Photoconductive Effect: When light is incident on an intrinsic semiconductor, electrons are excited from the valence band to the conduction band. Such electrons leave behind holes in the valence band. Thus, free electrons and holes are generated in the material; but they do not leave the material. Therefore, an increase of free charge carrier concentration occurs within the semiconductor. This is known as ***internal photoelectric effect***. An electron gets excited to the conduction band from the valence band by a light photon provided the photon energy, $h\nu$ is greater than the band gap energy, E_g. That is, $h\nu \geq E_g$. It means that the frequency of the photon should satisfy the following condition.

$$\nu \geq \frac{E_g}{h}$$

We can express the above condition in terms of wavelength as

$$\lambda \leq \frac{hc}{E_g}.$$

The largest wavelength that can cause the electron transition is therefore given by

$$\lambda_g = \frac{1.24}{E_g(eV)}\mu\text{m} \qquad ...(38.31)$$

A raise in the availability of free charge carriers raises the conductivity of the semiconductor and is called as ***photoconductivity***. The application of an electric field to the photoconductor causes the drifting of electrons and holes through the material which results in an electric current in the circuit.

Photodetectors are devices that convert optical energy to electrical energy. Internal photoelectric effect is the principle of photoelectric detectors. When light is incident on an intrinsic semiconductor, electrons are excited from the valence band to the conduction band.

Such electrons leave behind holes in the valence band. Thus, free electrons and holes are generated in the material; but they do not leave the material. Therefore, an increase of free charge carrier concentration occurs within the semiconductor. This is known as ***internal photoelectric effect***.

There are two types of photodiode structures used as photodetectors:

(*i*) Positive Intrinsic Negative (PIN) and

(*ii*) Avalanche Photo Diode (APD).

PIN is the favored component in most of the systems. It can be operated normally between 5 and 15 V. APD devices have much better sensitivity and better bandwidth. But, they cannot be used on a 5V printed circuit board. They also require a stable power supply. APD devices are usually found in longer communications links.

Photodetectors are devices that absorb optical energy and convert it to electrical energy. The operation of photoelectric detectors is based on the internal photoelectric effect. There are three main types of photodetectors, namely, photodiodes, pin diodes and avalanche photodiodes, which are widely used in optical communication systems.

38.11.1 Photodiode

Photodiodes are essentially the same as the *p-n* junction diodes. During the fabrication of the *p-n* diode, a depletion layer forms at the junction region by immobile negatively charged acceptor atoms in the *p* type material and immobile positively charged donor ions in the *n* type material. The electric field due to these ions stops the motion of majority carriers but accelerates minority carriers across the junction. When a photon is incident on the device, an electron-hole pairs are generated. In case of electron-hole pairs generated within the depletion region, the electric field acting across the region causes the pair to separate as shown in Fig. 38.28. This charge separation can be utilized in two ways. If the diode is short-circuited externally, a current flows between *p* and *n* regions. It is known as the photoconductive mode of operation. The diode is reverse biased for photoconductive operation. If the diode is left on open-circuit, a voltage appears between *p* and *n* regions and is known as **photovoltaic mode** of operation. Solar cells use this mode of operation.

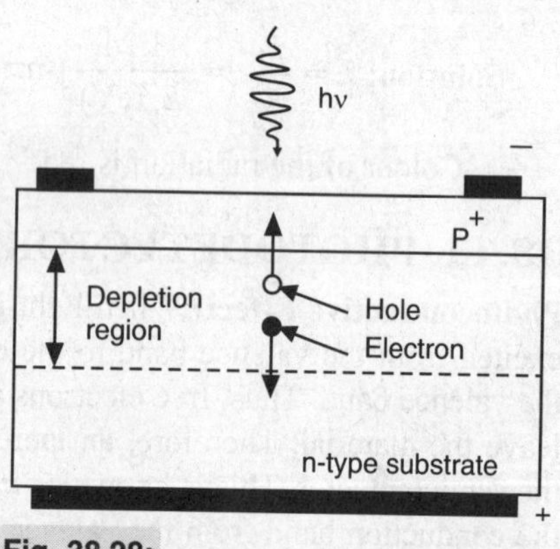

Fig. 38.28: Motion of photo-generated carriers in a *p-n* photodiode

A ***semiconductor photodiode*** is a reverse biased *p-n* junction. The structure of a photodiode is shown in Fig. 38.29 (a). When a reverse bias is applied across the junction (Fig. 38.29 b), the depletion layer widens as mobile carriers are swept to their respective majority sides. The motion of minority carriers causes the *reverse leakage current* of the diode. Thus, even when no light radiation is present (zero light), a small leakage current exists. This leakage current is called ***dark current***. The amount of dark current depends on the reverse bias voltage, the series resistance and the ambient temperature.

When light shines on the diode photons are absorbed in the depletion layer and neutral regions. A photon of energy $h\nu \geq E_g$ incident in or near the depletion layer will excite an electron from the valence band to the conduction band. This process produces a hole in the valence band. The electron and hole generated by the optical photon are known as ***photocarriers***. The electrons and holes in the depletion layer drift in opposite directions under the action of the electric field and induce an electric current in the external circuit over and above the already existing dark current (reverse leakage current). The photocurrent flowing in the external circuit is always in the reverse direction.

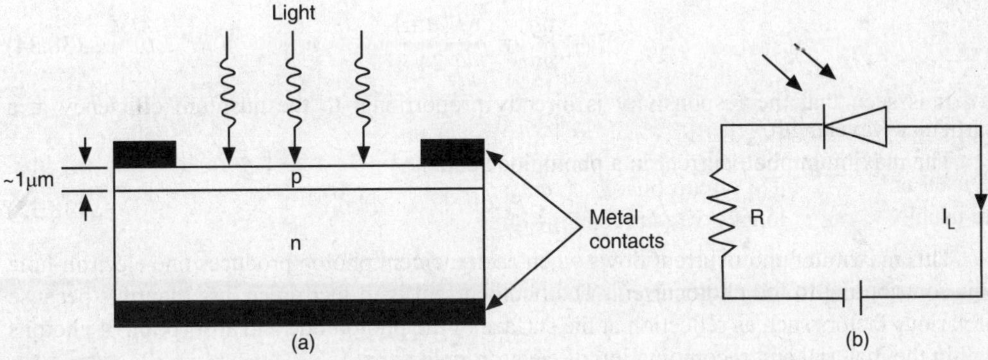

Fig. 38.29:

A photo diode (a) Side view (b) A reverse biased *pn*-junction

Increasing the level of illumination increases the reverse current flowing. The light incident in the neutral region, on either side of the depletion layer, also produces electron-hole pairs. Electrons and holes generated within a diffusion length of the depletion layer will move randomly and slowly diffuse into the depletion region and are accelerated by the bias, thereby contributing to the photocurrent. Thus, optical excitation leads to an increase in the reverse-biased current. It is required that the depletion region be suitably wide to enable a large fraction of incident light to be absorbed. The photocurrent is proportional to the incident light intensity. Consequently, the diode under a reverse bias voltage can be used as a photodetector.

The illumination (I-V) characteristic of a photodiode is very much similar to that of a p-n junction diode and is shown in Fig. 38.30. It may be noted that the I-V characteristic passes through the first, third and fourth quadrants. When the diode is used as a photodetector, it is operated in the third quadrant.

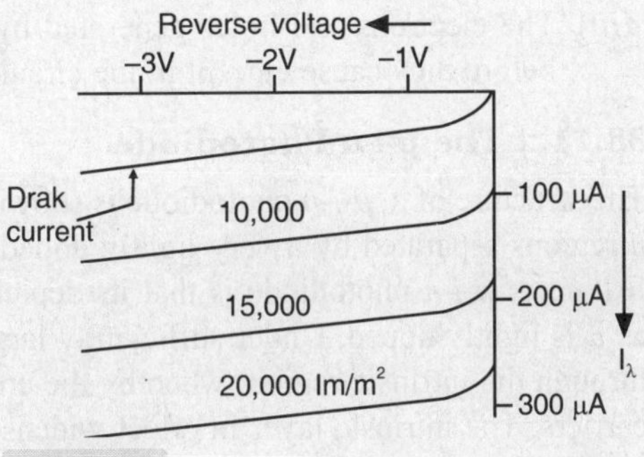

Fig. 38.30:

I-V characteristic of a photodiode

Characteristics of a Photodetector: Responsivity, dark current and bandwidth are the major characteristics of a photodetector.

$$\text{Quantum efficiency,} \qquad \eta = \frac{\text{number of electron-hole pairs generated}}{\text{number of incident photons}}$$

$$= \frac{I_L / q}{P_o} \qquad \text{...(38.32)}$$

The performance of a photodiode is characterized by the responsivity $\Re$, which specifies the photocurrent generated per unit optical power. This is given by

$$\Re = \frac{I_L}{P_o} \qquad \text{...(38.33)}$$

where I_p is the output photocurrent in amperes and P_o is the incident optical power in watts.

It can be shown that the responsivity $\Re$ is related to quantum efficiency through the following relation.

$$\Re = \frac{\eta q}{h\nu} = \frac{\eta\lambda(\mu m)}{1.24} \quad ...(38.34)$$

It is seen that the responsivity is directly proportional to the quantum efficiency at a particular wavelength.

The maximum photocurrent in a photodiode equals

$$I_{L(max)} = \frac{q}{h\nu} P_o \quad ...(38.35)$$

This maximum photocurrent flows when each incident photon produces one electron-hole pair contributing to the photocurrent. The photocurrent is in fact much less than this because of various factors such as reflection at the surface of the photodiode and absorption of photons within the material and recombination of electron-hole pairs.

Disadvantages of a p-n Photodiode

(*i*) The depletion region is a relatively small portion of the total volume of the diode. Hence, only those few photons, which are absorbed in the depletion region, cause current in the circuit whereas many of the photons absorbed in the bulk of the diode do not result in current.

(*ii*) To increase the width of the depletion region, the reverse bias applied across the junction has to be increased; which may not be possible beyond a certain value.

(*iii*) The electrons and holes generated by the photons in the bulk of the diode recombine before they cause current in the circuit.

38.11.2 The *p-i-n* Photodiode

The structure of a *p-i-n* photodiode is shown in Fig. 38.31. It is a device that consists of *p* and *n* regions separated by a very lightly doped intrinsic region (*i*). The first and most important feature of *p-i-n* photodiode is that its depletion region extends well into the intrinsic region, as it is lightly doped. Under sufficiently large reverse bias, the depletion region could extend through the intrinsic region, whereby the entire intrinsic region could be made free of charge carriers. The intrinsic layer in effect widens the depletion region and therefore increases area available for capturing light. Thus, the *p-i-n* structure of the photodiode enables us to increase the width of the depletion region to a value far greater than what it could be in a simple *pn*-junction.

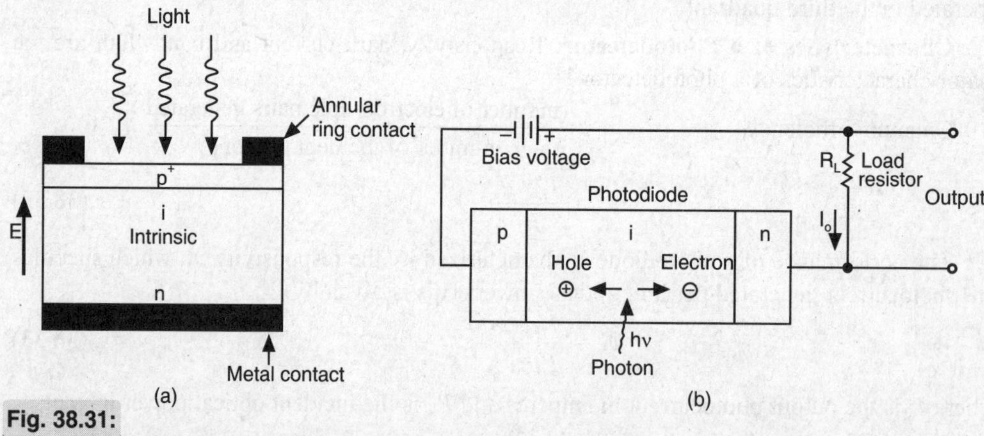

Fig. 38.31: *p-i-n* photodiode (a) sides view of the structure (b) circuit diagram

The *p-i-n* diode has a broad intrinsic semiconductor layer between *p-* and *n*-regions. As the intrinsic layer has no free charges its resistance is high and the largest part of the bias voltage drops across it. When an incident photon has energy greater than or equal to the band gap energy of the semiconductor material, it can excite an electron from the valence band to the conduction band. This produces free electron-hole pairs in the depleted intrinsic region where most of the incident light is absorbed. The high electric field in the depletion region causes the free carriers to separate and move across the reverse biased junction. This gives rise to a current flow in the external circuit. As the intrinsic layer is wide enough, most of the photons are absorbed and larger photocurrent is produced. Hence, *p-i-n* photodiode is more sensitive than *pn-* photodiode.

Advantages of p-i-n-photodiode

(*i*) The depletion region is very wide and extends throughout the intrinsic region and hence the reverse bias need not be varied to widen the depletion region.

(*ii*) The reverse bias applied is small, of the order of 5V.

(*iii*) As the depletion area is wider, most of the incident photons are absorbed in this region and hence the efficiency of this device is high.

(*iv*) The dark current in this device is smaller.

38.11.3 Avalanche Photodiode

An avalanche photodiode (APD) is more sophisticated than a *p-i-n* diode and incorporates internal gain mechanism so that the photoelectric current is amplified within the detector. It will be very much useful when very low levels of light are to be detected. The structure of a typical APD is shown in Fig. 38.32. This configuration is known as $p^+\pi\, p\, n^+$ reach-through structure.

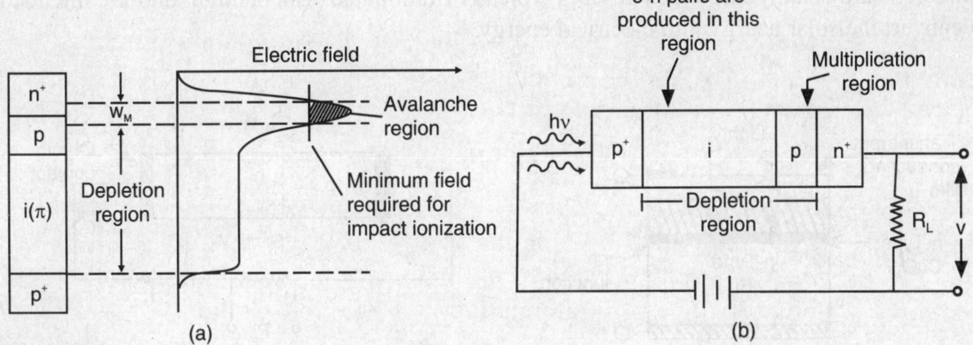

Fig. 38.32:

APD photodiode (a) side view of the structure (b) circuit diagram

The device is essentially a reverse-biased *p-n* junction. The n^+ and p^+ are heavily doped semiconductors and have very low resistance. The π region is very lightly doped and hence is nearly intrinsic. Most of the incident light passes into the intrinsic region through the thin p^+ region and electron-hole pairs are generated in the intrinsic region. Under reverse bias most of the applied voltage drops across the p-n^+ junction. With increase in the reverse bias voltage, the depletion region across this junction widens. Under sufficient reverse bias, the depletion region widens enough to reach through to the intrinsic layer.

In this condition, the internal field intensity near the junction becomes very high and the junction approaches the breakdown condition. Therefore, the electrons and holes photogenerated in the depletion layer acquire sufficient energy from the field to liberate secondary electrons and holes within the layer by a process of *impact ionization*. The newly generated carriers are also accelerated by the high electric field, thus gaining enough energy to cause further impact ionization. The number of carriers multiplies in geometrical progression and this phenomenon is called ***avalanche effect***. In this device multiplication is initiated by electrons. Holes generated in the π region drift to p^+ electrode and hence do not take part in the multiplication process.

A photon that enters through the p^+ region is absorbed in the intrinsic region and the resulting electron-hole pair is separated by the electric field in the π region. The hole drifts towards the p^+ and do not take part in the multiplication process. The electron drifts through the π region to the pn^+ junction. There, the electric field due to high reverse bias accelerates the electron. The electron acquires enough kinetic energy to ionize neutral atoms in its path. The electrons thus produced get in turn accelerated and ionize atoms lying in their paths. The effect is cumulative and builds up into an avalanche. As a result, one electron-hole pair will on an average produce M electron-hole pairs in the process, where M is the multiplication factor. Thus there occurs a carrier multiplication and internal amplification. This internal amplification process enhances the responsivity of the detector.

Advantages of APD:

(*i*) Internal current gain due to carrier multiplication.

(*ii*) High frequency response.

38.12 SOLAR CELL

A solar cell is basically a *p-n* junction that can generate electrical power, when illuminated. Solar cells are usually large area devices typically illuminated with sunlight and are intended to convert the solar energy into electrical energy.

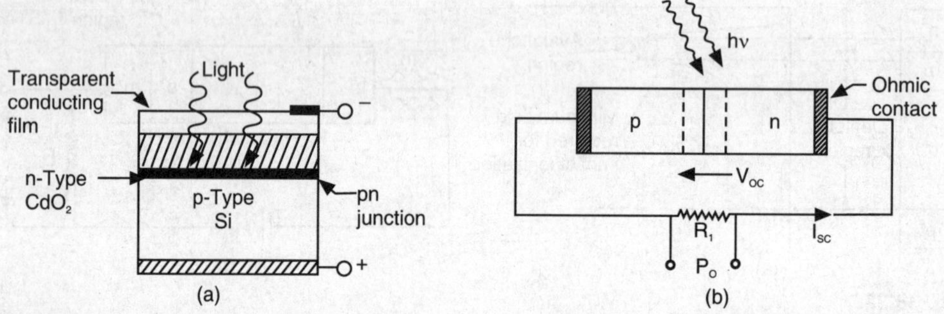

Fig. 38.33:

Schematic of a solar cell

The schematic of a solar cell is shown in Fig. 38.33 (a). It consists of a *p*-type chip on which a thin layer of *n*-type material is grown. When the solar radiation is incident on the cell, electron-hole pairs are generated in the *n* and *p* regions. The majority of them cannot recombine in the regions. They reach the depletion region at the junction where an electric field due to the space charge separates them. Electrons in the *p*-region are drawn into the *n*-region and holes in the *n*-region are drawn into the *p*-region. This gives charge accumulation on the two sides of the junction and produces a potential difference called

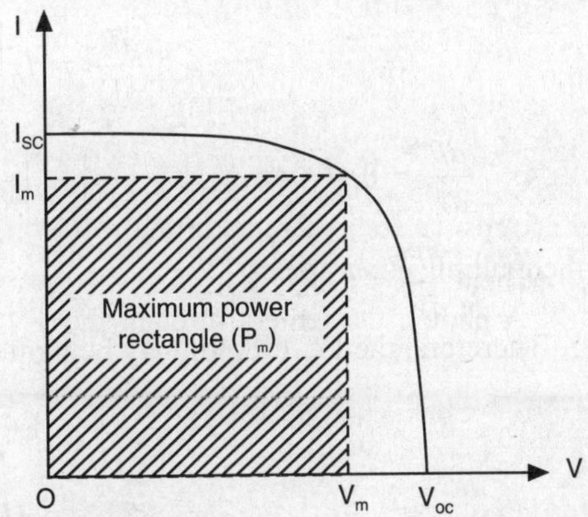

Fig. 38.34: I-V characteristic of a solar cell

photo emf of magnitude of order of 0.5 V. Generally power-conversion efficiency of single-crystalline solar cells is in the ranges of 10 to 30 %. Fig. 38.33 (b) shows the sign convention of the current and voltage. Current coming out of the cell is taken as positive as it leads to electrical power generation. The power generated depends on the solar cell itself and the load connected to it.

The *I-V* characteristic of a solar cell is shown in Fig. 38.34. The open-circuit voltage, V_{oc}, is the voltage across the illuminated cell at zero current and the short-circuit current, I_{sc}, is the current through the illuminated cell if the voltage across the cell is zero. The short-circuit current is close to the photocurrent while the open-circuit voltage is close to the turn-on voltage of the diode. The power equals the product of the diode voltage and current and at first increases linearly with the diode voltage but then rapidly goes to zero around the turn-on voltage of the diode. The maximum power is attained at V_m with I_m.

Solar cells can be connected in parallel or series into solar panels, which can deliver power output of several kilowatts. Solar panels are used in numerous applications in remote locations and in space. Solar cells of all kinds are used in different consumer products – from watches and calculators to power supplies for laptop computers.

38.12.1 Expressions for V_m and I_m

The expression for the total current in a solar cell is given by (38.36)

$$I = I_o\,(e^{eV/kT} - 1) - I_L \qquad ...(38.36)$$

We obtain the expression for short circuit current if we put $V = 0$ in Eq. (38.36). Thus,

$$I_{sc} = -I_L \qquad ...(38.37)$$

The open circuit voltage is obtained if we put $I = 0$ in the Eq. (38.36). Then

$$I_o\,(e^{eV/kT} - 1) = I_L$$

or

$$e^{eV/kT} = 1 + \frac{I_L}{I_o}$$

Taking logarithms on both the sides and designating $V = V_{oc}$ and $kT/e = V_T$, we get

$$V_{oc} = V_T \ln\left(1 + \frac{I_L}{I_o}\right) \approx V_T \ln\left(\frac{I_L}{I_o}\right) \qquad ...(38.38)$$

However, no power is delivered when the solar cell operates at either of these points. It is necessary that the operating point should be selected such that the output power becomes maximum. The output power is given by

$$P = VI = V[I_o\,(e^{V/V_T} - 1) - I_L] \qquad ...(38.39)$$

The condition for maximum power is given by $\frac{dP}{dV} = 0$.

Therefore, by differentiating the Eq. (38.39) and equating it to zero, we get the expressions for V_m and I_m at the point of maximum power. Thus,

$$\frac{dP}{dV} = I_o\left[e^{eV/kT}\left(1+\frac{eV}{kT}\right)-1\right]-I_L$$

As $\frac{dP}{dV} = 0,\ I_o\left[e^{eV/kT}\left(1+\frac{eV}{kT}\right)-1\right]-I_L = 0$ ...(38.40)

When $\frac{dP}{dV} = 0$, $V = V_m$.

Therefore, the Eq. (38.40) may be written as

$$e^{eV_m/kT}\left(1+\frac{eV_m}{kT}\right) = 1+\frac{I_L}{I_o}$$

or

$$e^{eV_m/kT} = \frac{(1+I_L/I_o)}{(1+eV_m/kT)}$$

Taking logarithm on both sides of the above equation, we obtain

$$\frac{eV_m}{kT} = \ln\left[\frac{(1+I_L/I_o)}{(1+eV_m/\text{kT})}\right]$$

or

$$V_m = V_T \ln\frac{1+\text{I}_\text{L}/I_o}{1+V_m/V_T} = V_{oc} - V_T \ln\left(1+\frac{V_m}{V_T}\right) \quad ...(38.41)$$

where $kT/e = V_T$.

Now, using the Eq. (38.41) into the Eq. (38.36) we obtain an expression for I_m.

$$I_m = I_o\left[\exp\left\{\frac{e}{kT}\cdot\frac{kT}{e}\ln\left[\frac{(1+I_L/I_o)}{(1+eV_m/kT)}\right]\right\}\right]-(I_o+I_L)$$

$$= I_o\left[\frac{(1+I_L/I_o)}{(1+eV_m/\text{kT})}\right]-(I_o+I_L)$$

$$= \left(\frac{I_o+I_L}{1+eV_m/\text{kT}}\right)-(I_o+I_L)$$

$$= (I_o+I_L)\left[\frac{1}{1+V_m/V_T}-1\right]$$

or

$$I_m = -(I_o+I_L)\left[\frac{V_m}{V_m+V_T}\right] \quad ...(38.42)$$

The maximum power is given by

$$P_m = V_m I_m \cong I_o\left[V_{oc} - V_T\ln\left(1+\frac{V_m}{V_T}\right)-V_T\right] \quad ...(38.43)$$

Conversion efficiency of solar cell is defined as

$$\eta = \frac{\text{Electrical Power delivered}}{\text{Solar Power incident}} = \frac{V_m I_m}{P_i}$$

$$= \frac{I_L\left[V_{oc} \quad V_T \ln\left(1+\frac{V_m}{V_T}\right) - V_T\right]}{P_i}$$

$$= \frac{FF \cdot I_L V_{oc}}{P_i} \qquad \text{...(38.44)}$$

where FF is known as the **fill factor**.

38.13 LIGHT SOURCES FOR FIBER OPTIC SYSTEMS

Light sources are a key element in any fiber optic system which converts the electrical signal into a corresponding light signal to be injected into the fiber.

Requirements of light sources to work as transmitters:

(*i*) Source must emit light in a cone with cross sectional diameter 8-100 microns to couple the light into the fiber optic cable.

(*ii*) It must be able to produce adequate optical power so that transmission over longer distances is possible.

(*iii*) High efficiency in coupling the light into the fiber optic cable is needed.

(*iv*) The optical source must be capable of easily modulated with an electrical signal.

(*v*) Other requirements are small size, low weight, low cost and high reliability.

There are two types of light sources that satisfy the above requirements. They are

(*a*) light emitting diode (LED) and

(*b*) laser diode (LD).

LEDs are simpler and produce incoherent, lower power, light. It can be modulated at the desired speeds. LDs are more complex and produce coherent, higher power light. LDs have greater coupling efficiency as output beam from them have much less spatial width than an LED.

38.13.1 Light Emitting Diodes (LEDs)

The basic LED types used for fiber optic communication systems are the surface-emitting LED (SLED) and the edge-emitting LED (ELED).

(a) Surface-Emitting LEDs: C.A. Burrus developed the surface-emitting LED known as the Burrus LED, Fig. 38.35. In SLEDs, the primary active region is a small circular area of 20 μm to 50 μm diameter. The active region is below the surface of the semiconductor substrate at right angles to the axis of the fiber and from which photons are emitted.

A **well** is carved into the substrate for direct coupling of the emitted light to the optical fiber and by getting a close contact with the emitting surface. The epoxy resin that binds the optical fiber to the SLED reduces the refractive index mismatch, increasing coupling efficiency.

(b) Edge-Emitting LEDs: A typical ELED structure is as shown in Fig. 38.36. The main active region of the ELED is a narrow stripe which lies below the surface of the semiconductor substrate. The semiconductor substrate is cut so that the stripe runs between the front and back of the device. The cut surfaces at each end of the stripe are called *facets*. In an ELED the rear facet is highly reflective and the front facet is coated with antireflection material. The rear facet reflects the light propagating toward the rear end-face back toward the

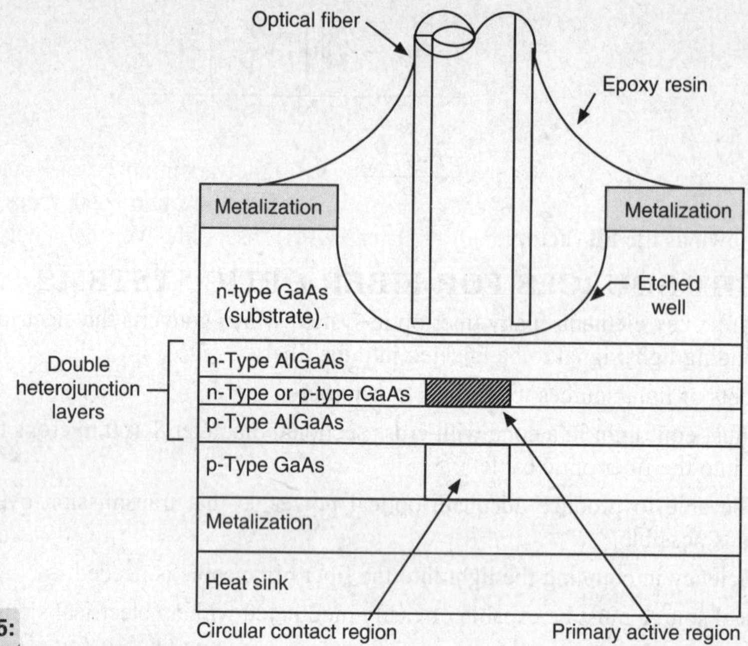

Fig. 38.35:

Illustration of surface-emitting LED

front facet. By coating the front facet with antireflection material, the front facet allows light emission through it. ELEDs emit light in a narrow emission angle. It allows better source-to-fiber coupling.

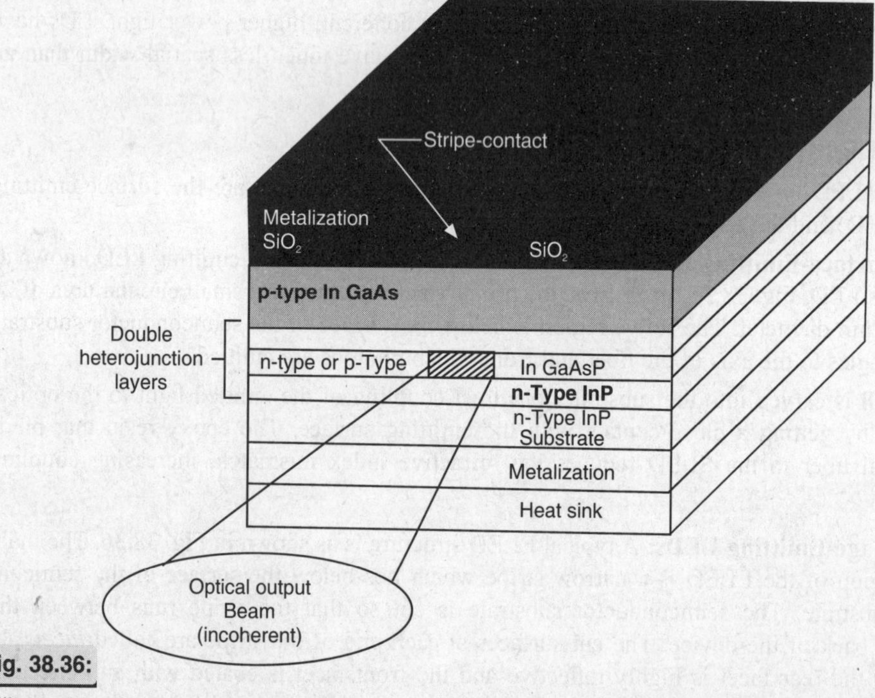

Fig. 38.36:

Illustration of typical ELED structure

GaAs and AlGaAs are used to fabricate LEDs in the 850-nm region and InGaAsP and InP for the 1300 nm and 1550-nm regions.

38.13.2 Laser Diodes (LD)

Ordinary lasers, having Fabry-Perot structure are multimode lasers and emit light at a number of discrete wavelengths. However, in high-speed, long-distance communications, single-mode lasers that emit light at a single frequency are required. There are two basic types of single mode laser diode structures: Distributed feedback (DFB) lasers and Vertical cavity surface-emitting lasers (VCSELs).

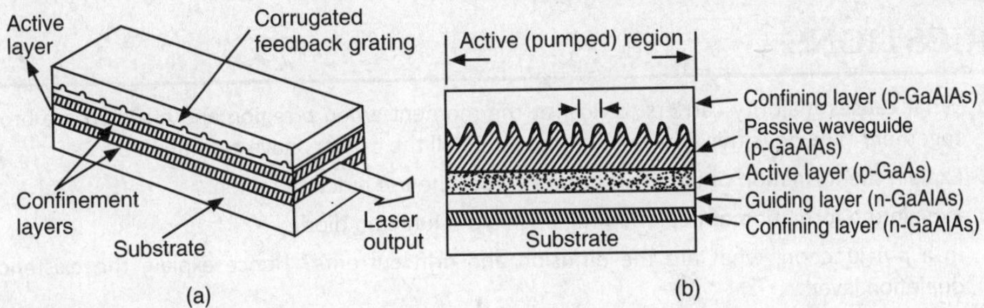

Fig. 38.37:

Iillustration of the structure of a DFB laser

(a) Distributed feedback (DFB) lasers: The structure of a DFB laser is shown in Fig. 38.37 (a). It has a corrugated layer etched internally just above the active region. The corrugation functions as an optical grating. The wave propagates parallel to the grating. At each discontinuity in the side of the wave guide, some reflection occurs. If the period of the corrugations is equal to ½ of the internal wavelength, then the reflections reinforce. A laser having one side of the active region corrugated in this way can oscillate at a frequency defined by the corrugations. There is no need for end mirrors in this laser.

The operating wavelength is given by Braggs' law

$$\lambda_o = \frac{2n_e\Lambda}{m} \qquad ...(38.45)$$

where Λ is the period of the corrugations, m is the order of reflection, and n_e is the effective refractive index of the mode.

(b) Vertical Cavity Surface-Emitting Lasers (VCSELs): VCSEL emits laser light vertically from its surface and has vertical laser cavity. There are many designs of VCSEL structure. Fig. 38.38 illustrates one of the structures of VCSEL. The main part of the VCSEL is the light emitting an active region. The cavity length is very short, about 1-3 wavelengths of the emitted light. Hence,

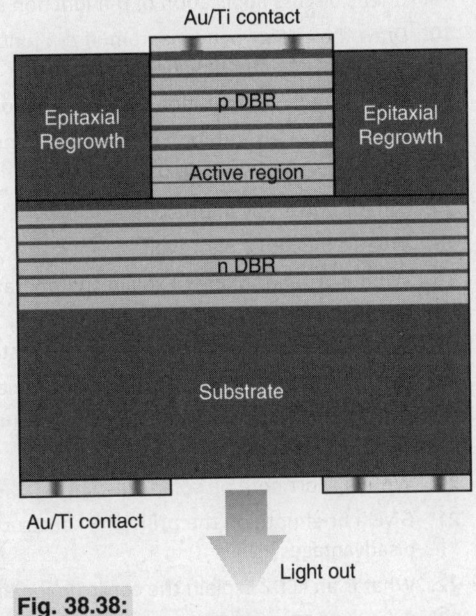

Fig. 38.38:

A typical VCSEL

a photon has negligible chance of triggering stimulated emission at low carrier densities in a single pass of the cavity. So, VCSELs require highly reflective mirrors. High reflectivity cannot be realized by the use of metallic mirrors.

In VCSELs, the reflectors above and below the active region are made from a stack of dielectric materials. The refractive index of the layers alternate between high and low values, resulting in high reflection. They are known as Distributed Bragg Reflectors (DBRs). Bragg-reflectors with as many as 120 mirror layers form the laser reflectors. Each mirror reflects a narrow range of wavelengths back into the cavity initiating light emission at a single wavelength.

QUESTIONS

1. What causes majority carriers to flow at the moment when *p*-region and *n*-region are brought together? Why does this flow not continue until all the carriers have recombined?
2. Explain the formation of potential barrier across the *p-n* junction region.
3. Explain the formation of depletion region in a *p-n* junction diode.
4. In a *p-n* junction, what are the diffusion and drift currents? Hence explain the existence of depletion layer.
5. Draw neat energy band diagrams for symmetrically doped *p-n* junction when it is (*i*) Unbiased, (*ii*) Forwardbiased and (*iii*) Reverse biased.
6. Draw neat energy bands diagram for *p-n* junction when unbiased, forward biased and reverse biased. Obtain the condition of equilibrium of current when unbiased. How does biasing change this equilibrium and hence explain forward and reverse bias characteristics.
7. Discuss (draw) the energy diagram for (*i*) forward biased *p-n* junction and (*ii*) reverse biased *p-n* junction.
8. Differentiate between drift and diffusion currents.
9. Briefly discuss application of *p-n* junction as half wave rectifier.
10. Draw the energy band diagram of *p-n* junction diode in equilibrium. Hence, obtain an expression for height of potential energy barrier.
11. What is reverse saturation current in a diode?
12. Why is reverse saturation current independent of reverse bias?
13. What is reverse break down?
14. What is meant by photoconductivity?
15. What is meant by photovoltaic emf?
16. What is a zener diode? Explain the operation of a zener diode in the forward and reverse bias condition.
17. Explain how a zener diode maintains constant voltage across the load.
18. Discuss the avalanche breakdown and zener breakdown.
19. Discuss the avalanche and breakdown voltage of zener diode and also discuss the zener diode as a voltage regulator.
20. Write a short note on solar cells.
21. Give a brief note on the principle, construction and working of LED. What are its advantages and disadvantages?
22. What is an LED? Explain the construction and working of LED.
23. Compare LED and LCD.
24. What is an LDR? Explain its construction and working.

25. Draw the I-V characteristic of a simple photodiode. What is meant by dark current?

26. Explain the principle of a photoconductive cell.

27. Explain construction and discuss I-V characteristics of the following semiconductor devices: (*i*) Photodiode, (*ii*) Solar cell.

28. Sketch typical illumination characteristics for a photodiode and explain the theory of the device.

29. Explain the phenomenon of photovoltaic effect.

30. Explain the principle of photovoltaic cell.

31. Explain the working of a solar cell.

32. Draw and explain the V-I characteristics of a solar cell.

33. What are the parameters of solar cell?

34. Write short notes on (*i*) solar cell, (*ii*) *p-n* junction diode.

35. Write short notes on: (*i*) Solar cell, (*ii*) Phototransistor, (*iii*) Photo-resistor

39 Bipolar Junction Transistor

Learning Objectives:

By the end of this chapter, you will be able to:

- Learn about structure of Transistor.
- Interpret about Energy band diagram of Unbiased transistor.
- Describe about different ways of Biasing the Transistor.
- Explain the concepts of Common-Base Configuration and Common-Emitter Configuration in detail.

39.1 INTRODUCTION

The transistor was invented in 1947 by the American physicists John Bardeen, Walter Brattain, and William Schockley at Bell telephone Laboratories. The transistor heralded the modern era of solid-state electronics. The electronic circuits grew smaller and smaller, became lighter and inexpensive. Profound changes are brought in all fields of human endeavour because of the advancements in the field of electronics. Exploration of outer space and deep sea, advancements in atomic power and arsenal, communications and computers, entertainment and medicine, automatization of production processes – all these and other fields are enriched. The inventors of transistor were honoured with Nobel Prize in physics in 1956.

The term transistor is mainly associated with the bipolar junction transistor (BJT). In BJT the action of both holes and electrons is important and therefore, it is called bipolar junction transistor. There are other types of transistors such as FET, MOSFET, and UJT which operate on the action of only one type of charge carrier. We study about BJT in this chapter.

39.2 TRANSISTOR STRUCTURE

A ***transistor*** is a semiconductor device consisting of three regions separated by two distinct *p-n* junctions. The central region is called the ***base***. It may be a *p*-type or *n*-type semiconductor. The two outer regions are called ***emitter*** and ***collector***. They are of the same type extrinsic semiconductor but different from that of base (Fig. 39.1). Thus, if the base is *p*-type the emitter and collector are *n*-type and if the base is *n*-type the emitter and collector are *p*-type.

Thus, two types of transistors are available, as shown in Fig. 39.2. They are called *npn* and *pnp* transistors. The ***npn transistor*** is constructed using *n* type material as the emitter and collector while the base is made of *p* type material. The ***pnp transistor*** is constructed using *p* type material as the emitter and collector while the base is made of *n* type.

The *n*-region contains free electrons which are negative charge carriers and *p*-region contains mobile holes which are positive carriers. Thus, two types of charge carriers, namely electrons and holes, are involved in current flow through an *npn* or *pnp* transistor. Therefore, these transistors are known as bipolar junction transistors.

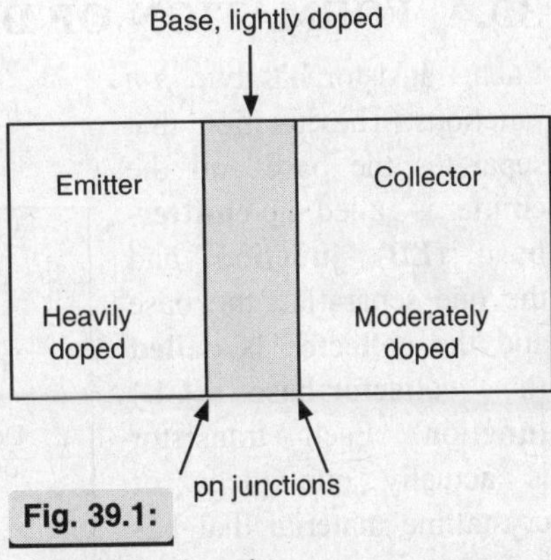

Fig. 39.1: Transistor junctions

The function of each element is as follows:

(*i*) The emitter provides the majority carriers necessary to support current flow.

(*ii*) The base controls the flow of the majority carriers within all elements of the transistor.

(*iii*) The collector supports the majority of the current flow in the transistor. In most cases the current that flows through the collector accomplishes the work done by a transistor.

39.3 SCHEMATIC REPRESENTATION

The schematic symbols of npn and pnp transistors are shown in Fig. 39.2.

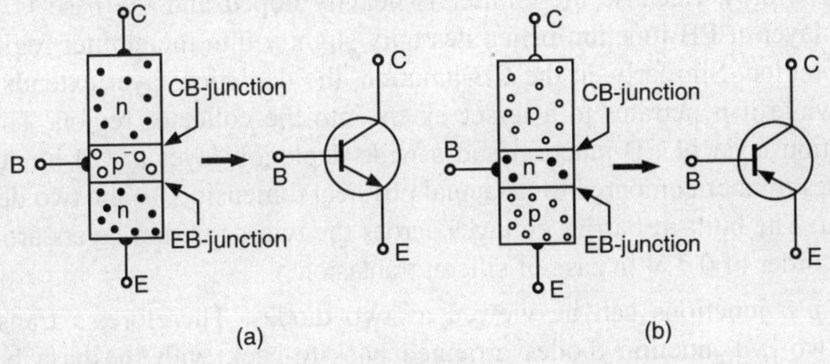

Fig. 39.2: Circuit symbol of transistor (a) *n-p-n* transistor (n) *p-n-p* transistor

In the symbols, the emitter is always indicated by an arrowhead. This arrowhead serves to tell us three things:

(*i*) location of the emitter;

(*ii*) type of the transistor that is being represented; and

(*iii*) direction in which the conventional current flows.

In the *npn* transistor, the arrow points away from the base. In this device electrons flow from the emitter into the base and hence the current flows from the base to the emitter. In the pnp transistor, the arrow points toward the base. The holes flow from the emitter into the base and the current flows from the emitter into the base.

A simple way of remembering the direction of arrow is as follows. In an npn transistor, *n*-regions are outside *p*-region and the arrow points outward and in a pnp-transistor, *n*-region is in between *p*-regions and the arrow points inward.

npn ….. *n* outside, arrow outward

pnp ….. *n* inside, arrow inward

39.4 FORMATION OF DEPLETION REGIONS

Each transistor has two *p-n* junctions. The junction that separates the base and the emitter is called the **emitter-base (EB) junction** and the one separating the base and the **collector is called the collector-base (CB) junction**. Each transistor is actually one piece of crystalline material that has been doped to create the three elements.

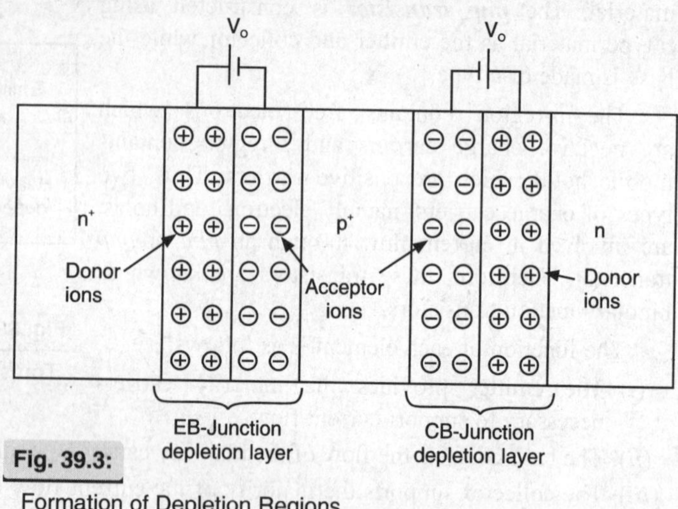

Fig. 39.3: Formation of Depletion Regions

During the process of formation of junctions, diffusion of majority carriers takes place and depletion layers form (Fig. 39.3).

As the doping levels in the three regions are different, the two depletion layers form with different widths. Because the emitter is heavily doped and the base is lightly doped, the depletion layer at EB-junction penetrates only slightly into the emitter region and deeply into the base region. Similarly, at the CB-junction, the depletion layer extends deep into the base region while it penetrates to a lesser extent into the collector region. This results in a narrow depletion layer at EB-junction and a wide depletion layer at CB-junction. The base region becomes thinner compared to its actual physical dimension, as the two depletion layers encroach on it. The built-in barrier voltages across the two depletion layers are the same and will be of the order of 0.7 V in case of silicon transistor.

The two *p-n* junctions can be viewed as two diodes. Therefore, a transistor may be regarded as two *p-n* junction diodes arranged back-to-back with the base being common to both the diodes. As both the diodes have the base in common, they influence each other strongly.

39.5 ENERGY BAND DIAGRAM OF UNBIASED TRANSISTOR

(a) *n-p-n* transistor: The energy band diagram of an unbiased *n-p-n* transistor is shown in Fig. 39.4. At the instant of formation of the junctions, diffusion of majority carriers occurs and the energy bands in each region get shifted. The energy bands in *n*-region move down while those in *p*-region are shifted upward. The displacement of energy bands and diffusion of majority charge carriers come to a halt as soon as the Fermi levels in the three regions are equalized. The process of relative displacement of energy bands causes bending of bands in the junction regions and creation of potential barriers of height eV_o. The potential barriers inhibit the majority carrier diffusion but promote minority carrier drifting. In the equilibrium condition, the current components due to diffusion and drift of carriers balance each other. Hence, the net current through each junction is zero.

(b) *p-n-p* transistor: The energy band diagram of an unbiased *p-n*-p transistor is shown in Fig. 39.5. At the instant of formation of the junctions, diffusion of majority carriers occurs and the energy bands in each region get shifted. The energy bands in *n*-region move down while those in *p*-region are shifted upward. The displacement of energy bands and diffusion of majority charge carriers come to a halt as soon as the Fermi levels in the three regions are equalized. The process of relative displacement of energy bands causes bending of bands in the junction regions and creation of potential barriers of height eV_o. The potential barriers inhibit the majority carrier diffusion but promote minority carrier drifting. In the equilibrium condition, the current components due to diffusion and drift of carriers balance each other. Hence, the net current through each junction is zero.

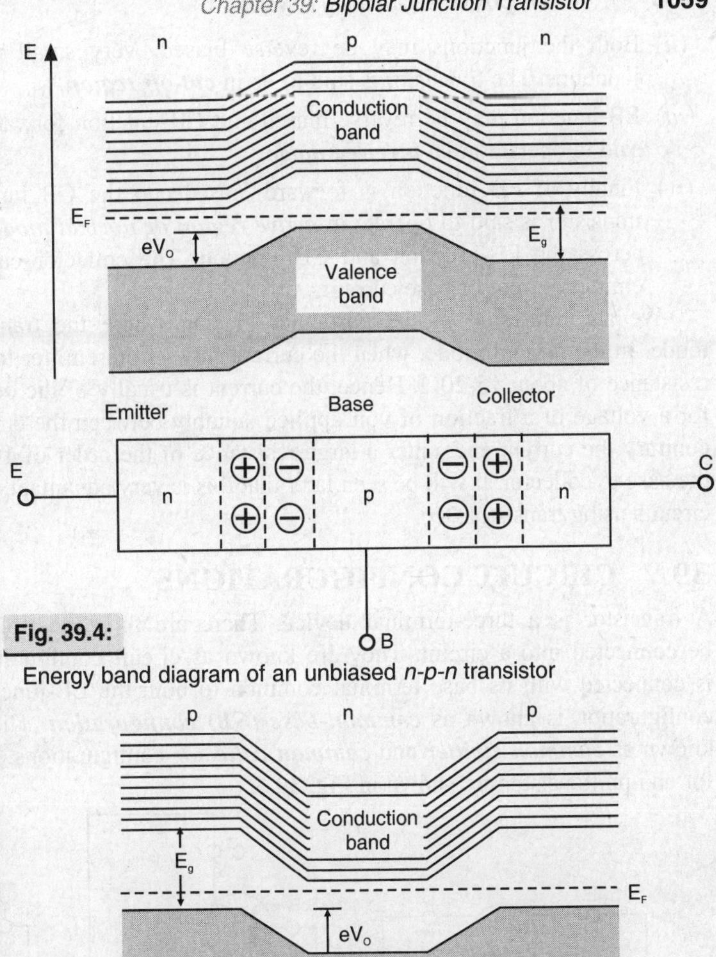

Fig. 39.4: Energy band diagram of an unbiased *n-p-n* transistor

Fig. 39.5: Energy band diagrams of an unbiased *p-n-p* transistor

39.6 BIASING THE TRANSISTOR

The two junctions of a transistor can be biased in four different ways.

(*i*) Both the junctions may be forward biased. It causes large currents to flow across the junctions. The currents join together in the base and flow down the common lead. Then the transistor is said to be operating in ***saturation region***.

(*ii*) Both the junctions may be reverse biased. Very small currents flow through the junctions. The transistor is said to be in ***cut-off region***.

(*iii*) EB-junction may be reverse biased and CB-junction forward biased. The transistor is said to operate in an ***inverted mode***.

(*iv*) Finally, if EB junction is forward biased and the CB junction in reverse bias, the transistor is said to operate in ***active region or normal mode*** and a large current flows across the EB-junction and CB-junction. The collector current is controlled by the emitter current or base current.

We are interested in the particular biasing where the transistor operates in normal mode. In the normal mode, when the current flows from emitter to base, it meets with a low resistance of about 15-20Ω. Hence, the current is usually of the order of a few milliamperes for a voltage of a fraction of volt applied suitably between the emitter and the base. On the contrary the current encounter a large resistance of the order of 100 kΩ when it flows from the base to collector. It will be seen later that this is very advantageous for setting up amplifier circuits using transistors.

39.7 CIRCUIT CONFIGURATIONS

A transistor is a three-terminal device. There are three possible ways in which it may be connected into a circuit. They are known as circuit configurations. When the transistor is connected with its base terminal common to both the EB-junction and CB-junction, the configuration is known as ***common-base (CB) configuration***. The other configurations are known as ***common emitter*** and ***common collector*** configurations. The circuit configurations for an npn transistor are shown in Fig. 39.6.

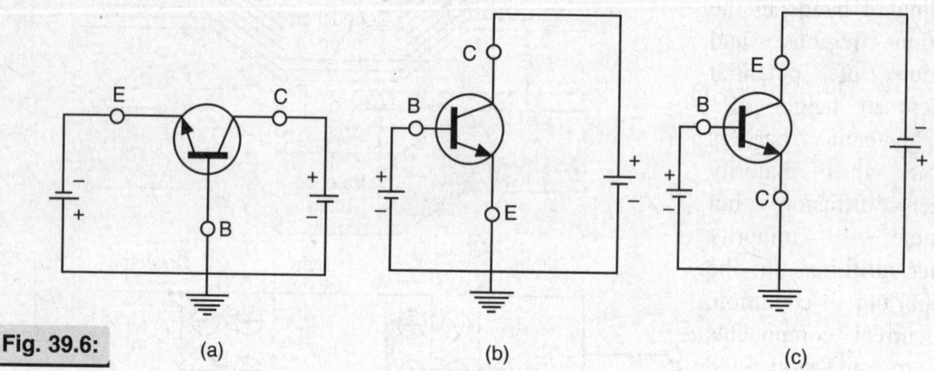

Fig. 39.6: The three different ways of connecting a transistor - (a) Common-base configuration. (b) Common-emitter configuration. (c) Common-collector configuration.

39.8 ACTION OF THE BIAS

Let an *npn* transistor be connected into the circuit as shown in Fig. 39.7. The battery V_{EE} forward biases the EB-junction when the switch S_1 is closed. The source V_{CC} reverse biases CB-junction when the switch S_2 is closed.

In the first step, let the switch S_1 be closed and switch S_2 be kept open. The closing of the switch S_1 forward biases the EB-junction. As a result, the potential barrier at EB-junction gets reduced and majority carriers diffuse across the junction. Electrons diffuse from emitter to base and holes from base to emitter. In a transistor the base region is intentionally doped lightly and holes are smaller in number in that region. Therefore, hole diffusion current will be very much small compared to the electron diffusion current. The total current across the junction practically consists of electron current flowing from emitter to base. It constitutes the emitter current I_E. In other words, the forward bias causes intense injection of electrons across EB-junction. As a result the electron concentration in the base region in the vicinity of EB-junction will rise to high value. The electrons entering the base become minority carriers and diffuse in the base region. They move towards the terminal end B and flow out into the connecting wire and proceed towards the positive terminal. The emitter current I_E flowing through the base becomes the base current I_B. Hence I_B is large and is equal to I_E. As the collector is open, current does not flow into collector and the collector current I_C is zero.

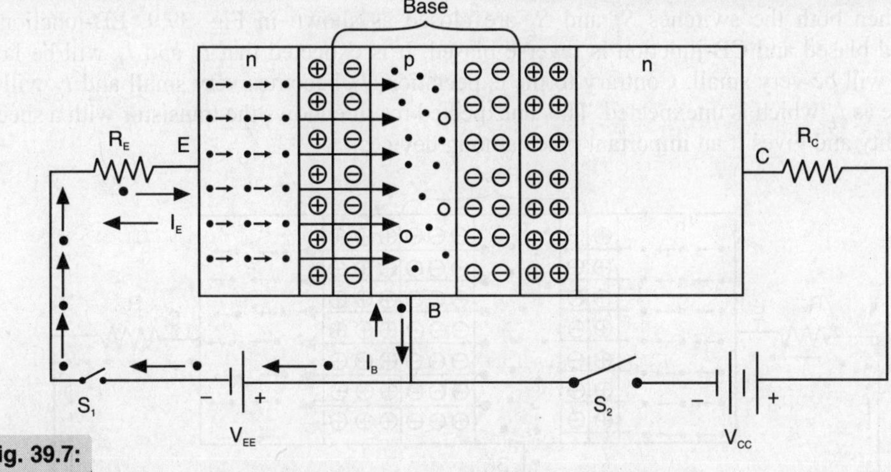

Fig. 39.7:

Action of the Bias on NPN transistor– S_1 is closed and S_2 is kept open.

In the next step, let us consider the situation where the switch S_1 is kept open and switch S_2 is closed (Fig. 39.8). The closing of switch S_2 reverse biases the CB-junction. The reverse bias causes drift of minority carriers. Holes from the collector region move into base region and electrons from base region flow into collector region. Because of the large difference in doping levels of base and collector regions, there are more electrons in the base region than are holes in the collector region. The electric field due to reverse bias acting across the collector depletion layer accelerates electrons and sweeps them into the collector region.

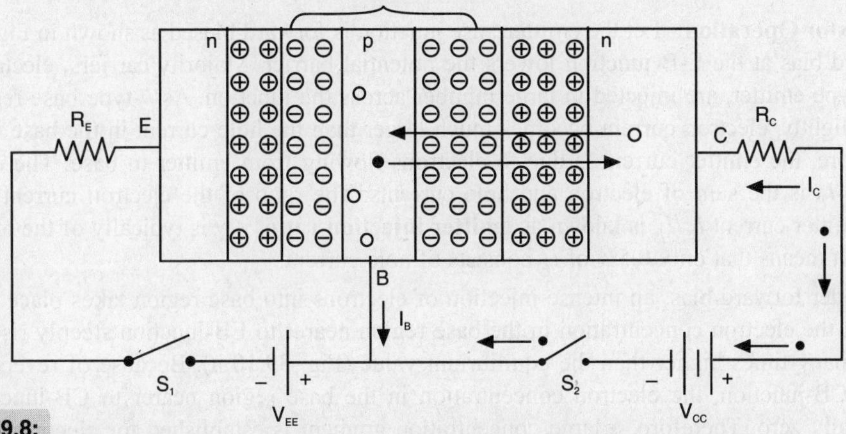

Fig. 39.8:

Action of the Bias on NPN transistor-switch S_1 is kept open and switch S_2 is closed

Therefore, electron concentration in the base region in the vicinity of CB-junction becomes practically equal to zero. The current through the CB-junction is mainly due to electrons moving from base to collector. This current is known as the reverse saturation current. As the minority carrier concentration is small, the magnitude of the reverse saturation current is very small. This current is not at all dependent on the magnitude of the reverse bias. It is often called the **collector leakage current** I_{CBO}, where the subscript implies that it is a current through CB-junction when the emitter lead is open. Thus, in this case the collector current is I_{CBO} and as the same current flows through the base lead, the base current I_B is equal to I_{CBO} and is thus very small.

When both the switches S_1 and S_2 are closed as shown in Fig. 39.9, EB-junction is forward biased and CB-junction is reverse biased. It is expected that I_E and I_B will be large and I_C will be very small. Contrary to the expectation, I_B becomes very small and I_C will be as large as I_E which is unexpected. This unexpected result endows the transistor with a special capability and gives it an important place among devices.

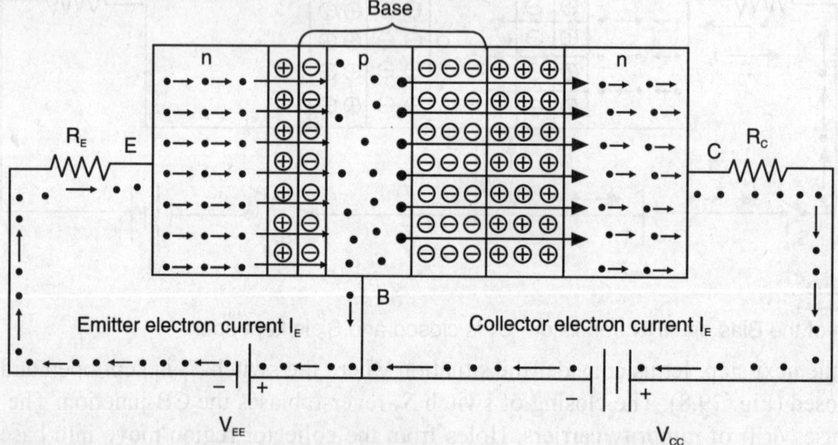

Fig. 39.9:

Action of the Bias on NPN transistor- switches S_1 and S_2 are closed

39.9 TRANSISTOR ACTION

Transistor Operation: Let the emitter-base junction is forward biased as shown in Fig. 39.9. Forward bias at the E-B junction lowers the potential barrier. Majority carriers, electrons in the *n*-type emitter, are injected in large number across the junction. As *p*-type base region is doped lightly, electron current becomes much larger than the hole current in the base region. Therefore, the emitter current is due to electrons flowing from emitter to base. The emitter current I_E is the sum of electron and hole currents. The ratio of the electron current to the total emitter current I_e/I_E is known as **emitter injection ratio**, γ. γ is typically of the order of 0.995. It means that only 0.5% of I_E consists of hole current.

Under forward bias, an intense injection of electrons into base region takes place and as a result the electron concentration in the base region nearer to EB-junction steeply rises to a value many times higher than the equilibrium value (Fig. 39.10 a). Because of reverse bias at the CB-junction, the electron concentration in the base region nearer to CB-junction is practically zero. Therefore, a large concentration gradient is established for electrons in the base region.

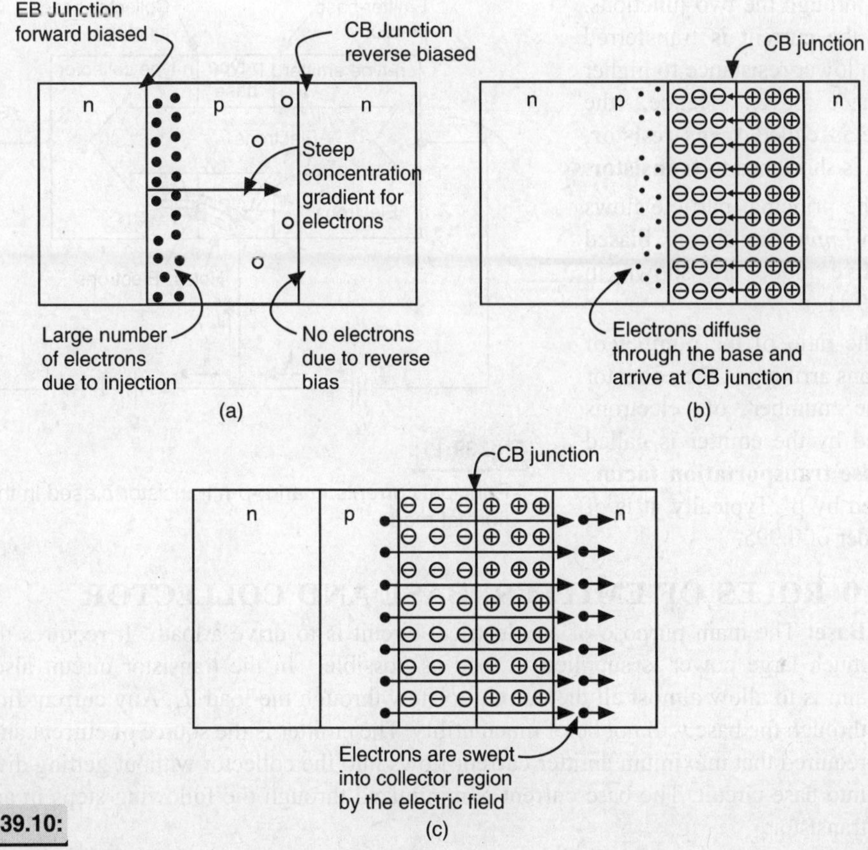

Fig. 39.10:

Transistor Action

Now the electrons in the base region have two options.

(*i*) One is that they may recombine with the holes in the base causing large base current. But due to light doping of the base region, sufficient number of holes is not available; recombination cannot take place in large way. It is necessary that recombinations should take place for the electrons to flow into the base-emitter circuit. Since recombinations are precluded in the base region, the base current I_B is very small.

(*ii*) As electron concentration is very high on the emitter side and zero on the collector side of the base region, the possibility is that electrons swiftly diffuse towards the collector-base junction under the influence of the concentration gradient across the base. The base region is narrow originally and is made further narrower due to the encroachment of depletion layers into the base and due to the action of the biases applied. Owing to this electrons quickly reach the CB-junction. Once they arrive in the vicinity of the junction they will be acted upon by the strong electric field due to reverse bias and get swept into the collector region, as shown in Fig. 39.10 (c). Consequently, a great majority of electrons emitted by the emitter flow into the collector. It causes a large reverse current I_C which is nearly equal to I_E to flow across CB-junction.

A small base current I_B is caused by the few electrons that undergo recombination in the base. The emitter-base junction is forward biased and therefore, it has a low resistance. The collector-base junction is reverse biased and has a high resistance. Almost the same current flows through the two junctions. Thus, the current is transferred from a lower resistance to higher resistance level. Hence, the device is called **trans-resistor**, which is shortened to **transistor**.

The principal particle flows in an *npn* transistor biased in active mode are shown in Fig. 39.11.

The ratio of the number of electrons arriving at the collector to the number of electrons emitted by the emitter is called the **base transportation factor**, denoted by β′. Typically, it is of the order of 0.995.

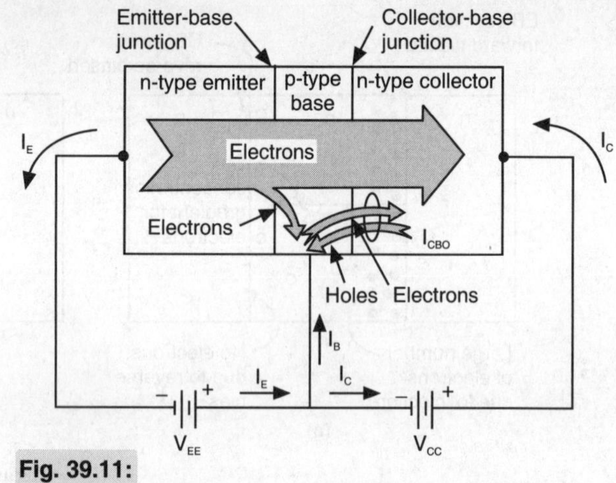

Fig. 39.11: Principal currents in an *n-p-n* transistor biased in the normal mode

39.10 ROLES OF EMITTER, BASE AND COLLECTOR

(*i*) **Base:** The main purpose of an electrical circuit is to drive a load. It requires that as much large power is supplied to load as possible. In the transistor circuit also, the aim is to allow almost all the current to flow through the load R_L. Any current flowing through the base will not be of much utility. The emitter is the source of current and it is required that maximum emitter current flows into the collector without getting diverted into base circuit. The base current is minimized through the following steps in an *npn* transistor:

(*a*) **Base region is lightly doped:** If base region is heavily doped, more holes would be present in the base and the incoming electrons would have more chance of undergoing recombination. The number of electrons flowing into collector would have thus decreased. It leads to lesser collector current and more base current. To reduce this possibility, base region is lightly doped.

(*b*) **Base region is made narrow:** It enables the electrons injected into base to quickly diffuse and come under the action of electric field due to reverse bias across junction, which sweeps them into the collector. Thus, the chance of electrons recombining with holes and causing a base current is precluded.

(*ii*) **Emitter is heavily doped:** In a transistor, the emitter is the source of current. It is required that a maximum of the majority carriers is injected into the base so that the emitter current I_E will be large. The function of the emitter is to provide charge carriers in large number. Hence emitter is heavily doped compared to base and collector.

(*iii*) **Collector is wider:** Collector current is produced by minority carriers. Current by minority carriers is a drift current and requires only the presence of electric field acting in a favourable direction. Whatever may be the strength of the electric field, minority carriers are accelerated into the collector region. The minority carriers are in fact rolling down the barrier. Whether the barrier is high or low it does not matter for rolling down it. Therefore whether the reverse bias (collector voltage) is large or small it does not influence the strength of the collector current. The minority carriers rolling down the high potential barrier acquire large kinetic energy. They produce large amount of heat while transferring part of their energy to the lattice through collisions. In order to dissipate away the heat, the collector region is made larger.

39.11 RELATION BETWEEN CURRENTS IN CB CONFIGURATION

It is seen from Fig. 39.11 that in a transistor connected in CB configuration the emitter current divides itself into the base current and collector current. Thus,

$$I_E = I_B + I_C \qquad \text{...(39.1)}$$

Assuming that the emitter current I_E remains constant, it is seen from the Eq. (39.1) that the smaller the base current, the larger the collector current. In order to make the collector current as large as possible, the base current must be minimized.

The collector current I_C consists of two components:

(*i*) a fraction of the emitter current $\alpha_{dc} I_E$ and (ii) the reverse leakage current I_{CBO}. Thus,

$$I_C = \alpha_{dc} I_E + I_{CBO} \qquad \text{...(39.2)}$$

where α_{dc} represents the fraction.

$$\therefore \qquad \alpha_{dc} = \frac{I_C - I_{CBO}}{I_E}$$

The reverse leakage current is negligible compared to the total collector current I_C. Therefore,

$$\alpha_{dc} = \frac{I_C}{I_E} \qquad \text{...(39.3)}$$

where α_{dc} is called the ***common base current gain factor.***

As I_C is always less than I_E, α_{dc} is less than unity. α_{dc} is called ***current gain*** though there is no gain in current. However, as the ratio of the output current to the input current is termed gain, α_{dc} representing the ratio of I_C to I_E is also called current gain. Its value ranges from 0.95 to 0.99.

Example 39.1: *In an npn transistor circuit, the collector current is 15 mA. If 95% of the electrons emitted by the emitter reach the collector, what is the base current?*

Solution: Base current is given by $I_B = I_E - I_C$.

$$\therefore \qquad I_B = (I_C/0.95) - I_C = 15 \text{ mA } (1/0.95 - 1)$$

$$= \mathbf{0.79\ mA}$$

Example 39.2: *A transistor is connected in common-base configuration. If the emitter current is 2 mA and base current is 20 μA, what are the values of I_C and α?*

Solution: $$I_C = I_E - I_B = (2 - 0.02) \text{ mA} = \mathbf{1.98\ mA}$$

$$\alpha = \frac{I_C}{I_E} = \frac{1.98 \text{ mA}}{2 \text{ mA}} = 0.99$$

39.12 ENERGY BAND DIAGRAM OF A TRANSISTOR BIASED IN NORMAL MODE

(a) *n-p-n* transistor: The energy band diagram for an *n-p-n* transistor biased in normal mode is shown in Fig. 39.12. The transistor is connected in common base configuration. As the EB-junction is forward biased, the energy levels in the emitter (*n*-region) and base (*p*-region) undergo relative displacement. The energy bands in the *n*-region are pushed up and those in the *p*-region are pulled down due to addition of energy. It causes a reduction in the height of the potential barrier by an amount eV_{EE}. Hence, a large number of electrons from emitter can move forward into the base region causing a current flow. Let J_{en} be the electron current density. Similarly, holes from the base region can move into the emitter region causing a hole current density, J_{hp}. As the emitter is heavily doped and larger than the base, $J_{en} >> J_{hp}$.

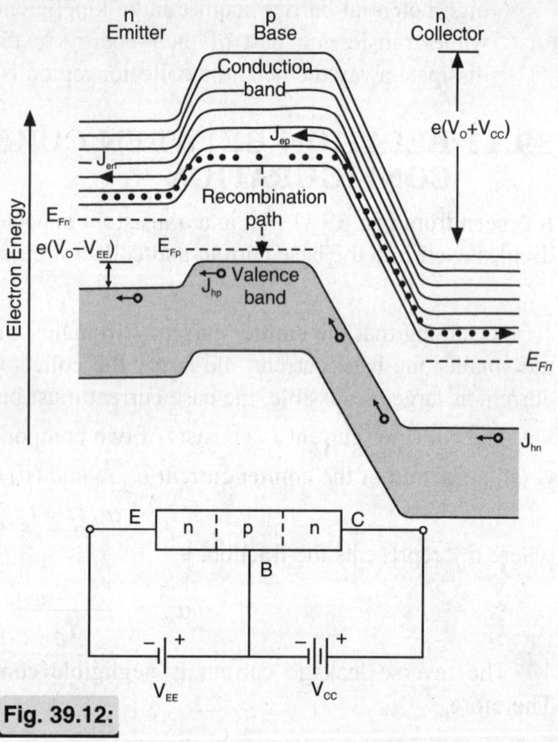

Fig. 39.12: Energy band diagram of an *n-p-n* transistor biased in normal mode

The reverse bias at the collector-base junction results in an increase in the barrier height to $e(V_o + V_{CC})$ from eV_o. Majority carrier diffusion cannot occur across CB-junction as it is reverse biased. The current that flows across the CB-junction is due to minority carriers. Electrons roll down from the base region into collector region and holes from collector region float up into the base region. They produce the current density components J_{ep} and J_{hn}. The current density J_{ep} will be larger than the current density J_{hn}, as the base is lightly doped. The electrons moving into the base due to forward bias become minority carriers in the base region and are in a position to roll down into the collector region. As a result, the total current density that flows across CB junction is nearly equal to the current density flowing through EB junction. It would be nearly equal to $(J_{en} + J_{ep})$. Out of the electrons injected into the base, about 2% undergo recombination by falling into the valence band from the conduction band. Such recombinations produce a small base current.

(b) *p-n-p* Transistor: The energy band diagram for a *p-n-*p transistor biased in normal mode is shown in Fig. 39.13. The transistor is connected in common base configuration. As the EB-junction is forward biased, the energy levels in the emitter (*p*-region) and base (*n*-region) undergo relative displacement. The energy bands in the *p*-region are pushed down and those in the *n*-region are pulled up due to addition of energy. It causes a reduction in the height of the potential barrier by an amount eV_{EE}. Hence, a large number of holes from emitter can move forward into the base region causing a current flow. Let J_{hp} be the hole current density. Similarly, electrons from the base region can move into the emitter region causing an electron current density, J_{en}. As the emitter is heavily doped and larger than the base, $J_{hp} >> J_{en}$.

The reverse bias at the collector-base junction results in an increase in the barrier height to $e(V_o + V_{CC})$ from eV_o. Majority carrier diffusion cannot occur across CB-junction as it is reverse biased. The current that flows across the CB-junction is due to minority carriers. Holes float up from the base region into collector region and electrons from collector region roll down into the base region. They produce the current density components J_{ep} and J_{hn}. The current density J_{hn} will be larger than the current density J_{ep}, as the base is lightly doped. The holes moving into the base due to forward bias become minority carriers in the base region and are in a position to float up into the collector region. As a result, the total current density that flows across CB junction is nearly equal to the current density flowing through EB junction. It would be nearly equal to ($J_{hp} + J_{hn}$). Out of the holes injected into the base, about 2% undergo recombination because of electrons falling into the valence band from the conduction band. Such recombinations produce a small base current.

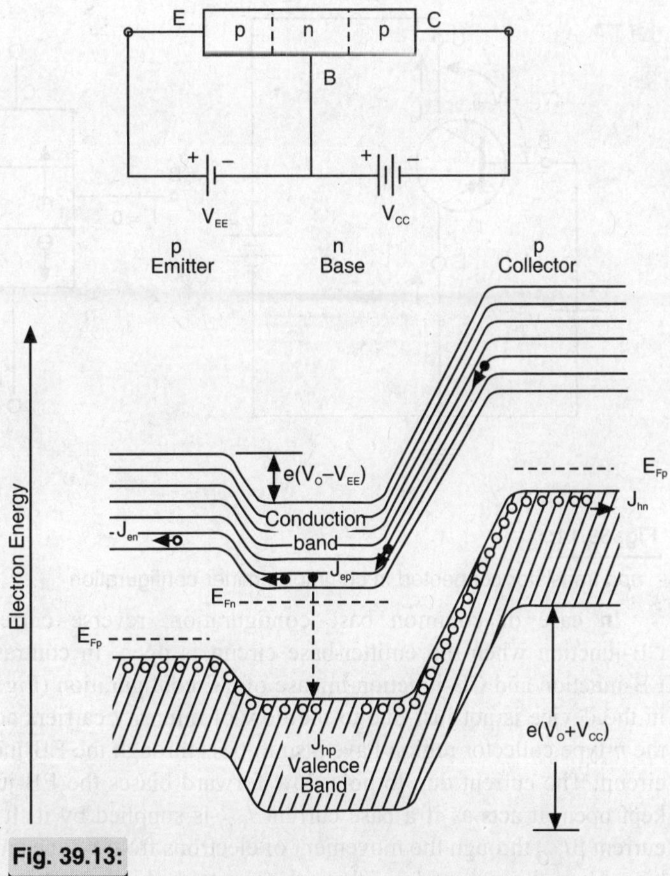

Fig. 39.13: Energy band diagram of a *p-n-p* transistor biased in normal mode

39.13 COMMON EMITTER CONFIGURATION

The emitter is made common to the input and the output in common emitter configuration. Fig. 39.14 shows an *npn* transistor in common emitter configuration. The transistor is biased in active region. The battery V_{BB} forward-biases the EB-junction. The battery V_{CC} is connected between the emitter and collector. The base is at $+V_{BB}$ potential with respect to the emitter and the collector is at $+V_{CC}$ potential with respect to the emitter. Therefore, the net potential of the collector with respect to the base is ($V_{CC} - V_{BB}$). This potential reverse biases the CB-junction. As V_{CC} is much larger than V_{BB}, the reverse bias voltage at the CB-junction is nearly V_{CC}.

The voltage source V_{BB} forward biases the EB-junction and electrons are injected into the base from the emitter. The electrons pass through the thin weakly doped base and reach the collector. About 98% of the injected electrons are swept into the collector region and a large reverse current flows through the reverse biased CB-junction. It constitutes the current I_C. About 2% electrons recombine with holes in the base region and then onwards travel as valence electrons and flow into the base lead causing a base current I_B. A small change in forward bias of EB-junction causes a change in the value of I_B and hence of I_C.

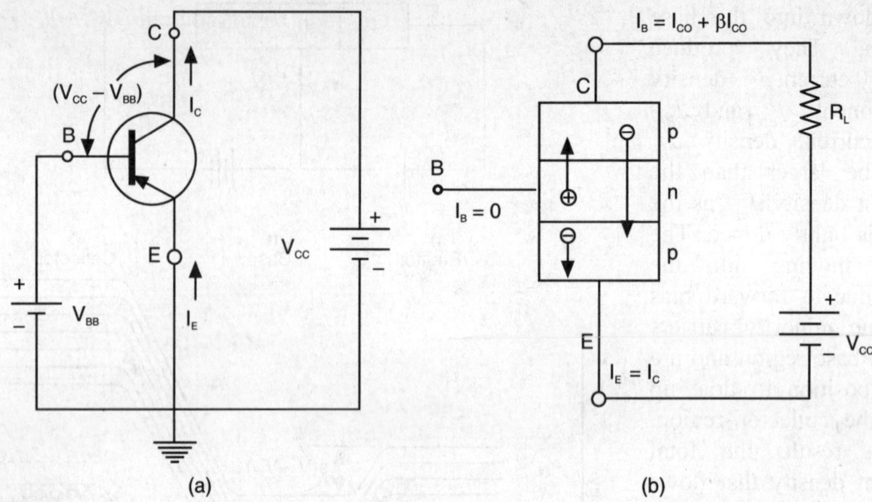

Fig. 39.14:

npn transistor connected in common emitter configuration

In case of common base configuration, reverse current I_{CBO} flows only through CB-junction when the emitter-base circuit is open. In contrast, current flows through both EB-junction and CB-junction in case of CE configuration (Fig. 39.14 b). The leakage current in the device is not just due to the flow of minority carriers across CB-junction. Holes from the *n*-type collector region have also to pass through the EB-junction in order to complete the circuit. The current due to hole flow forward biases the EB-junction, though the junction is kept open. It acts as if a base current I_{CO} is supplied by it. It causes an additional collector current βI_{CO} through the movement of electrons from *n*-type emitter to the base under forward bias. Thus, the current I_{CEO} through the reverse biased CB-junction will be

$$I_{CEO} = I_{CO} + \beta I_{CO}$$

or $$I_{CEO} = (1 + \beta)\, I_{CO} \quad ...(39.4)$$

39.14 CURRENT RELATIONS IN CE CONFIGURATION

In CB configuration, I_E is the input current and I_C is the output current. They are related through the equations

$$I_E = I_C + I_B$$

and $$I_C = \alpha_{dc} I_E + I_{CBO}$$

$\therefore$ $$I_C = \alpha_{dc}(I_C + I_B) + I_{CBO}$$

or $$(1 - \alpha_{dc}) I_C = \alpha_{dc} I_B + I_{CBO}$$

or $$I_C = \frac{\alpha_{dc}}{1 - \alpha_{dc}} I_B + \frac{1}{1 - \alpha_{dc}} I_{CBO}$$

or $$I_C = \beta_{dc} I_B + I_{CEO} \quad ...(39.5)$$

where $$\beta_{dc} = \frac{\alpha_{dc}}{1 - \alpha_{dc}} \quad ...(39.6)$$

and $$I_{CEO} = \frac{I_{CBO}}{1 - \alpha_{dc}} \quad ...(39.7)$$

In *CE* configuration, I_B becomes the input current and I_C is the output current. As α_{dc} is less than unity, β_{dc} will be much larger than α_{dc} and I_{CEO} much larger than I_{CBO}.

The factor β_{dc} is called the **common-emitter dc current gain**. Typically, β_{dc} has values in the range from 20 to 300. From the Eq. (39.5) we can write

$$\beta_{dc} = \frac{I_C - I_{CEO}}{I_B}$$

As $I_C >> I_{CEO}$,

$$\beta_{dc} = \frac{I_C}{I_B} \quad ...(39.8)$$

Eq. (39.8) is a linear relation. A small change in I_B causes a large variation in I_C. Thus, I_C is controlled by I_B. In view of this transistor is said to be a ***current controlled device***.

39.15 TRANSISTOR AS AN AMPLIFIER

An ***amplifier*** is an electronic circuit that causes an increase in the voltage or power level of a given signal. Fig. 39.15 shows an *npn* transistor connected in common base configuration. The transistor is biased to operate in the active region. The battery V_{EE} forward biases the *EB* junction and the battery V_{CC} reverse biases the *CB* junction. A signal source υ_i is connected in the input circuit. A load resistance R_L is connected in the output circuit. An output voltage υ_o is developed across R_L.

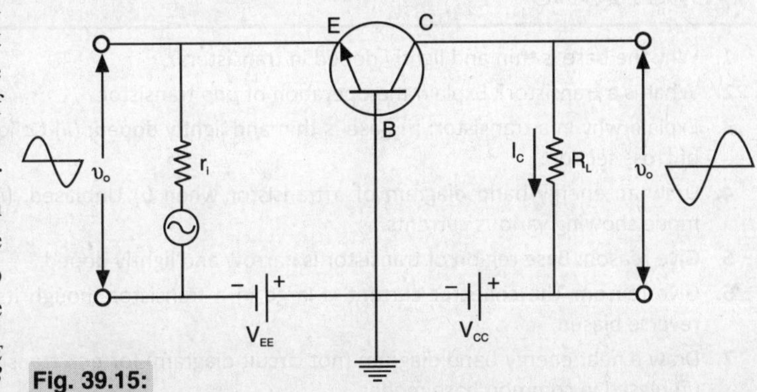

Fig. 39.15: Transistor amplifier

The dc voltage V_{EE} is a fixed voltage and causes a dc current I_E to flow through *EB* junction. When the ac voltage υ_i is superimposed on V_{EE}, the emitter-base voltage varies with time. For example, if V_{EE} is10 volts and the peak voltage of the signal υ_i is 1 V, the emitter-base voltage swings from 9 volts to 11 volts. This variation causes corresponding variations in emitter current I_E and the collector current I_C. The varying current flows through the load resistor R_L and develops a varying voltage across it. It is the output voltage υ_o.

Since the *EB* junction is forward biased, it has a low resistance r_i of the order of 100 ohms. The emitter current variation ΔI_E due to emitter-base voltage variation can be expressed as

$$\Delta I_E = \upsilon_i / r_i \quad ...(39.9)$$

The collector current I_C changes by ΔI_C due to the variation ΔI_E caused in I_E.

$$\Delta I_C = \alpha_{dc} \Delta I_E \quad ...(39.10)$$

This current ΔI_C flows through R_L causing a voltage drop $(\Delta I_C)\, R_L$. Therefore,

$$\upsilon_o = (\Delta I_C)\, R_L \quad ...(39.11)$$

$$= \alpha\, (\Delta I_E)\, R_L$$

$$= \alpha\, (\upsilon_i / r_i)\, R_L \quad ...(39.12)$$

The voltage gain of an amplifier is defined as the ratio of output signal voltage υ_o to the input signal voltage υ_i. Thus,

$$\text{Gain} = \frac{\text{Output Voltage, } \upsilon_o}{\text{Input Voltage, } \upsilon_i} \quad \text{...(39.13)}$$

$$= \frac{\alpha R_L}{r_i}$$

$$\cong \frac{R_L}{r_i} \qquad (\because \alpha \approx 1) \quad \text{...(39.14)}$$

R_L is of the order of kilo ohms and is far larger than r_i. Consequently, υ_o is larger than υ_i and the gain of the circuit is larger than unity. It means that the transistor amplifies a small input voltage to give a larger output voltage.

QUESTIONS

1. Why the base is thin and lightly doped in transistors?
2. What is a transistor? Explain the operation of *pnp* transistor.
3. Explain why in a transistor: (*i*) Base is thin and lightly doped; (*ii*) Collector region has larger area of cross section.
4. Draw an energy band diagram of a transistor when (*i*) Unbiased, (*ii*) biased in common base mode showing various currents.
5. Give reason: Base region of transistor is narrow and lightly doped.
6. Give reason: The collector current is large in a transistor though its collector-base junction is reverse biased.
7. Draw a neat energy band diagram (not circuit diagram) for *pnp* transistor when (*i*) unbiased and (*ii*) biased in common base mode.
8. Draw the energy band diagram for *npn* transistor when biased in common base mode.
9. With a neat labeled diagram, explain transistor action, when it is biased to operate in the active region.
10. Draw energy band diagram of npn transistor in common-base mode.
11. What will happen to the various currents in a transistor if its base is made thick?
12. Why transistor doeshave low input resistance and high output resistance?
13. Why the collector is current large in a transistor though its collector-base junction is reverse biased?
14. Explain the working of transistor as an amplifier?

40 Dielectric Materials

Learning Objectives:

By the end of this chapter, you will be able to:

- Learn about fundamentals of Dielectric.
- Explain about Electric Polarization and its different types.
- Describe about temperature and frequency dependence of Total Polarization.
- Understand about Dielectric Loss and Breakdown.
- Learn about Ferro-electricity, Piezoelectricity and Pyro-electricity.

40.1 INTRODUCTION

Electrical circuits use insulator materials in various forms. Capacitors use dielectrics as medium. The dielectric materials, which are also known as insulators, constitute very important group of electrical and electronic engineering materials. They are characterized by dielectric constant, dielectric loss, dielectric strength and high resistivity. The two properties namely, dielectric constant and dielectric loss are strongly frequency dependent. The use of a dielectric in a specific application is dictated by the frequency of the applied voltage. The dielectric loss is required to be a minimum so that the performance of the dielectric does not deteriorate with time. Ferroelectric materials exhibit very high dielectric constant and low dielectric loss. They are used in making miniature capacitors. Piezoelectric materials are dielectric materials which have very interesting properties.

Mechanical deformation is produced in them in response to electrical force and electrical effects are produced in response to mechanical forces. The two piezoelectric effects are vastly exploited in all areas of technology. The dielectric loss manifests in the form of heat. The dielectric heating can be employed in cases where heating by other means is difficult or is not possible. It finds application in the fields such as food processing. Material such as glass, ceramics, polymers and paper are nonconducting materials. They prevent flow of current through them.

40.2 DIELECTRICS

As dielectric materials are non-conducting materials, there are no free charge carriers in them. When they are placed in an electric field, they undergo considerable changes which make them act as stores of electrical charges. A good dielectric is always an insulator. So any insulator is a dielectric.

An insulator is a nonconducting material and its main function is to provide electrical insulation.

A dielectric material is an electrical insulator that can be polarized by an applied electric field and its main function is charge storage.

A dielectric may be described in terms of the energy band structure. The forbidden gap E_g (Fig. 40.1) is very large in dielectrics and excitation of electrons from the normally full valence band to the empty conduction band is not possible under ordinary conditions. Therefore, conduction cannot occur in a dielectric. Even if a dielectric contains impurities, extrinsic conduction such as the one observed in doped semiconductors is not possible. The resistivity of an ideal dielectric should be therefore infinitely high. However, in practice, dielectrics conduct electric current to a negligible extent and their resistivities range from 10^{10} to 10^{20} ohm-m.

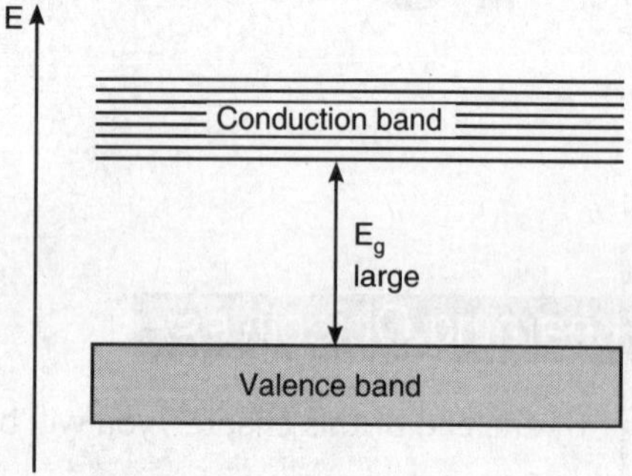

Fig. 40.1: Energy band diagram of an insulator

40.3 DIELECTRIC CONSTANT

A dielectric is chiefly characterized by its dielectric constant. *Dielectric constant of a dielectric is defined and measured as the ratio of capacitance of a capacitor containing the dielectric medium to the capacitance of the same capacitor with air as the medium.*

$$\varepsilon_r = \frac{C}{C_0} \qquad ...(40.1)$$

where C_0 is the capacitance with air as the medium between the plates and

C is the capacitance with dielectric as medium.

ε_r is called ***dielectric constant*** or ***relative permittivity***.

Dielectric constant is a dimensionless quantity, which is always greater than unity in case of dielectrics, and it is independent of the size or shape of the dielectric. In fact, ε_r *describes the ability of the dielectric material to store electric charges*. At times another quantity known as the *permittivity* of the medium, ε, is used. It is given by

$$\varepsilon = \varepsilon_0 \, \varepsilon_r \qquad ...(40.2)$$

where $\varepsilon_0 = 8.854 \times 10^{-12}$ F/m represents the permittivity of free space.

40.4 DIELECTRIC POLARIZATION

Let us consider an electrically neutral slab of an isotropic dielectric inserted between the plates of a charged parallel plate capacitor, as shown in Fig. 40.2.

Very low conductivity of the dielectric rules out the presence of free charges and their possible motion in the electric field. Hence current does not flow in the material. However, the electric field can act on the bound charges in the dielectric. These bound charges are not free to migrate through the dielectric. The action of the field E_0 on the bound charge consists in

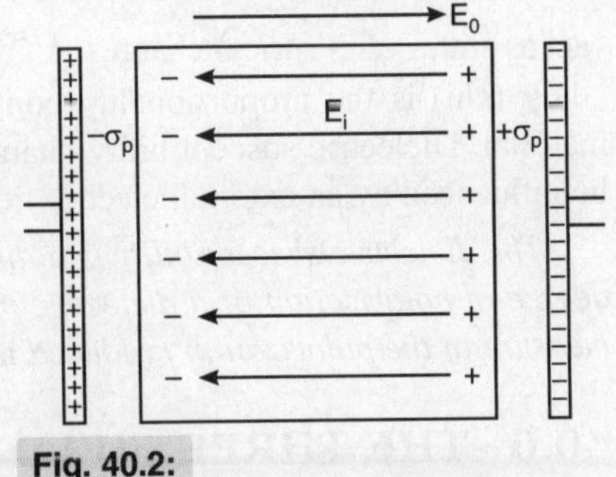

Fig. 40.2:
Induced charges appear on the oppoisite sides of the dielectric

displacing the bound charges relative to one another. The negative charges (electrons) are displaced in a direction opposite to that of the electric field, while the positive charges (nuclei)) are displaced in the same direction as that of the electric field.

Each atom or molecule then acts as an elementary dipole and acquires an electric dipole moment in the direction of the field. The cumulative effect of formation of such dipoles is that negative charge is induced by the electric field on the dielectric surface adjacent to the positive capacitor plate while a positive charge of equal magnitude is induced on the dielectric surface adjacent to negative capacitor plate.

Thus, the action of the electric field on a dielectric is to induce charges on its surfaces. When charges of opposite polarity are induced on the surfaces of a dielectric, the dielectric is said to be ***polarized***. The effect is known as dielectric ***polarization***. The polarized dielectric is equivalent to a big dipole consisting of polarization charges separated by a distance d, which is the thickness of the slab.

When a dielectric is placed in an electric field, electric charges slightly shift from their average equilibrium positions and induce charges on its surfaces is known as ***dielectric polarization***.

The intensity of polarization P is defined as the total dipole moment per unit volume of the material.

The intensity of polarization is therefore given by

$$P = \frac{\sum d\mu}{V} \qquad ...(40.3)$$

where $d\mu$ is the dipole moment of an elemental volume and V is the total volume of the dielectric. In fact, we can consider the polarized dielectric as a big dipole consisting of induced charges separated by distance d.

Thus, $$\mu = (A\sigma_p)d = \sigma_p V \qquad ...(40.4)$$

where A is the area of the slab and σ_p is the surface charge density due to polarization. Comparing the Eq. (40.3) and Eq. (40.4), we find that

$$P = \sigma_p \qquad ...(40.5)$$

It follows that polarization is equal to the surface density of the induced charges in a dielectric.

The effect of polarization is to reduce the magnitude of the external field E_0. The induced surface charges on the dielectric give rise to an induced electric field E_i which opposes the external field E_0.

Therefore, the net electric field E in the dielectric has a magnitude given by

$$E = E_0 - E_i \qquad ...(40.6)$$

40.5 DIELECTRIC SUSCEPTIBILITY

The magnitude of polarization is directly proportional to the intensity of the electric field. Thus,

$$P = \chi\varepsilon_0 \mathbf{E} \qquad ...(40.7)$$

χ (chi) is the proportionality constant and is called the ***dielectric susceptibility*** of the material. Dielectric susceptibility characterizes the ease with which a dielectric material can be influenced by an external electric field.

The dielectric susceptibility is a dimensionless proportionality constant that indicates the degree of polarization of a dielectric material in response to an applied electric field. It is a measure of the polarization produced in the material per unit electric field.

40.6 THE THREE FIELD VECTORS

The three field vectors **E** electric field, **D** electric displacement, and **P** polarization are related by

$$\mathbf{D} = \varepsilon_0\mathbf{E} + \mathbf{P} \qquad ...(40.8)$$

The three electric vectors, **E, D** and **P** are shown in Fig. 40.3. Eq. 40.8 for **D** shows that this vector is related to the free charges only; they are the charges stored on the capacitor plates. Eq. 40.5 for **P** shows that it is related to the bound charges only. The expression for the vector **E** shows that it is connected with both types of charges present.

It may be seen from Fig. 40.3 that the lines of **D** begin and end on free charges; the lines of **P** begin and terminate on induced charge and the lines of **E** change from one medium to the other.

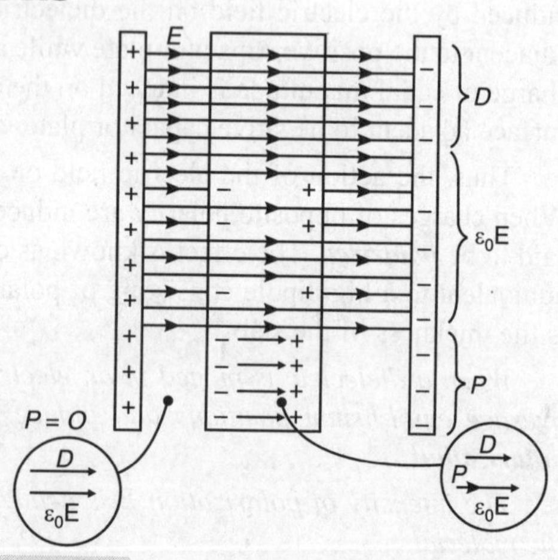

Fig. 40.3: The electric field **E** is connected to the free charges, the vector **P** is related to polarization charges and the vector **D** to all charges

40.7 RELATION BETWEEN ε_r AND χ

In order to describe the combined effects of the applied electric field **E** and electric polarization **P**, we have introduced the auxiliary vector **D**, called the ***displacement vector***.

$$\mathbf{D} = \varepsilon_0\,\mathbf{E} + \mathbf{P}$$

Substituting the expression for P from Eq. (40.7), we get

$$\mathbf{D} = (1 + \chi)\varepsilon_0\mathbf{E}$$

which we can write as

$$\mathrm{D} = \varepsilon_0\,\varepsilon_r E = \varepsilon E$$

$$\varepsilon_r = 1 + \chi \qquad ...(40.9)$$

40.8 RELATION BETWEEN P AND E

When the Gauss law is applied to capacitor (Fig. 40.4), the relation between induced charges q' on the surface of the dielectric and enclosed charge q on one of the plates and dielectric constant ε_r is obtained as

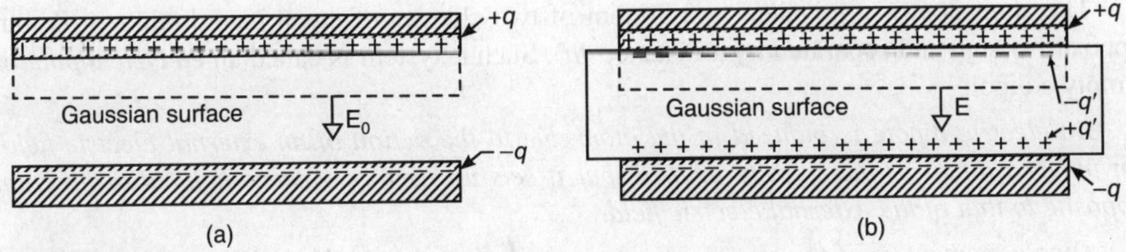

Fig. 40.4:

A parallel plate capacitor (a) without and (b) with a dielectric

$$\mathbf{q'} = \text{q}\left[1 - \frac{1}{\varepsilon_r}\right] \qquad \text{...(40.10)}$$

But from Eq. 40.5 $\qquad \mathbf{P} = \sigma_\text{p} = \dfrac{\mathbf{q'}}{\mathbf{A}}$

Therefore, $\qquad \mathbf{P} = \dfrac{q'}{A} = \dfrac{q}{\varepsilon_r A}(\varepsilon_r - 1)$

Writing $\qquad \dfrac{q}{A} = \mathbf{D}$

$$\mathbf{P} = \frac{\mathbf{D}}{\varepsilon_r}(\varepsilon_r - 1) \qquad \text{...(40.11)}$$

$$\mathbf{P} = \varepsilon_0\,(\varepsilon_r - 1)\mathbf{E} \qquad \text{...(40.12)}$$

Example 40.1. *When NaCl crystal is subjected to an electric field of 50 V/cm, the resulting polarization is 2.215 × 10^{-7} C/m^2. Calculate relative permittivity of NaCl.*

Solution: $\qquad \mathbf{P} = \varepsilon_0(\varepsilon_r - 1)\mathbf{E}$

$$\therefore \qquad \varepsilon_r = 1 + \frac{P}{\varepsilon_0 E} = 1 + \frac{2.215\times10^{-7}\,\mathbf{C/m^2}}{8.85\times10^{-12}\,\mathbf{F/m}\times 50\mathbf{V/cm}\times100\mathbf{cm/m}} = \mathbf{6.006}$$

40.9 INDUCED DIPOLES

In order to understand the action of electric field on a dielectric, it is necessary to understand its action on an atom. In an atom the nucleus is about 10^{-15} m in diameter and it can be regarded as a point. The electron cloud is about 10^{-10} m in diameter and it may be assumed that its negative charge is concentrated at its centre. Therefore, the centres of gravity of positive and negative charges in an atom coincide (Fig. 40.5 a).

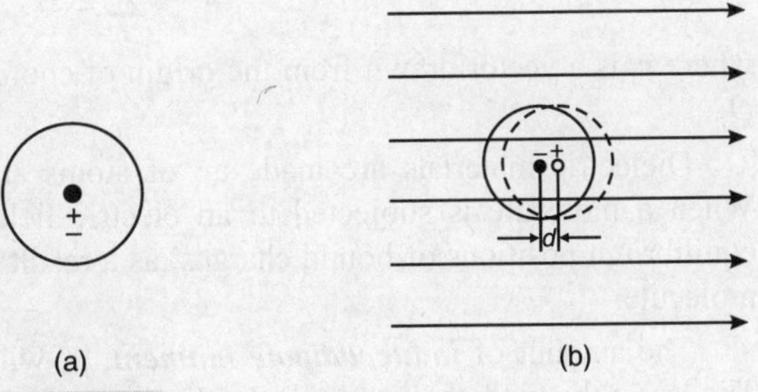

Fig. 40.5:

(a) Centres of gravity of positive and negative charges coincide in an atom, (b) Atom in an electric field of strength *E* with the charges separated by 'd'.

Consequently, such an atom neither produces any electric field of its own nor is acted upon by an external field. If now the atom is placed in an electric field of strength *E*, the electron cloud will be displaced in the direction opposite to that of *E* by a distance "*d*" with respect to the nucleus (Fig. 40.5 b). The centres of gravity of positive and negative charges in the atom no more coincide.

The atom is now equivalent to a system of two charges of equal magnitude $q = Ze$ but opposite in sign and separated by a distance "d". Such a system is called an ***electric dipole*** or simply a dipole.

An electric dipole is induced in the atom due to the action of an external electric field. An induced dipole is electrically neutral, but it sets up its own electric field in a direction opposite to that of the external electric field.

Dipole moment, μ of the electric dipole is defined as the product of the magnitude of the charges and the distance of their separation.

Thus, $$\mu = q\,d \qquad ...(40.13)$$

Fig. 40.6 shows diploe in electric field. The dipole moment is a vector directed along the axis of the dipole from the negative charge to the positive charge Fig. 40.7.

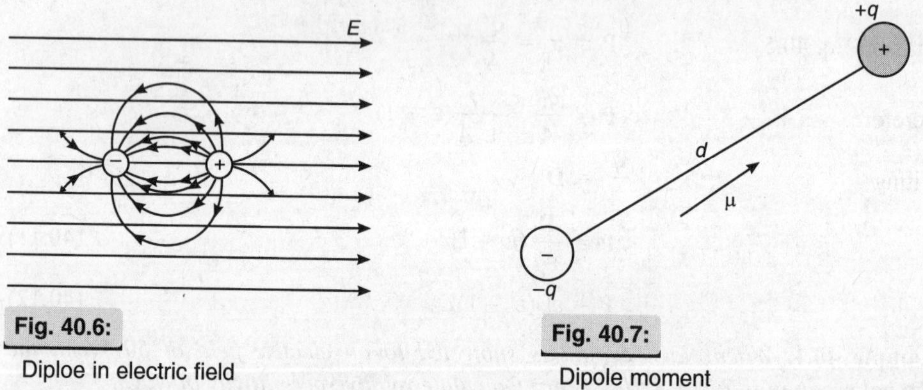

Fig. 40.6: Diploe in electric field

Fig. 40.7: Dipole moment

In general any neutral system of N point charges $Q_1, Q_2, Q_3, \ldots\ldots Q_N$ occupying a volume having linear dimensions "d" acts as a dipole. The sum of charges $\sum Q_i$ in the volume should be equal to zero to ensure the neutrality of the system.

The dipole moment of such a neutral system of point charges is given by

$$\mu = \sum_i Q_i \mathbf{r}_i \qquad ...(40.14)$$

where $\mathbf{r}_i$ is a vector drawn from the origin of coordinate system to the position of the charge Q_i.

Dielectric materials are made up of atoms and molecules, which are neutral systems. When a molecule is subjected to an electric field, the electric field tends to displace the equilibrium positions of bound charges, as a result of which dipole moment is induced in the molecule.

The amount of ***induceddipole moment***, μ, will be proportional to the field strength, **E**. The larger the field, the greater is the displacement of charges and hence the larger the induced dipole moment. As the charges are displaced along the field direction, the dipole moment is induced in the same direction. The molecule is then said to be ***polarized*** by the field.

When a molecule becomes polarized, restoring forces due to coulomb attraction come into play, which tend to pull the displaced charges together. The charges are separated until the restoring force balances the force due to the electric field.

Restoring forces vary in magnitude from one kind of molecule to another and therefore, the extent of dipole moment induced differs. As the amount of induced dipole moment is proportional to the field strength, we write

$$\mu_{ind} \propto E$$

or $$\mu_{ind} = \alpha E \quad ...(40.15)$$

where α is the proportionality constant and is known as the **polarizability** of the molecule.

Polarizability characterizes the capacity of the electric charges in the molecule to suffer displacement in an external field. It has the dimensions of volume. Induced dipole moment vanishes as soon as the electric field is switched off.

40.10 PERMANENT DIPOLES

A polar molecule possesses a ***permanent dipole*** moment. Permanent dipole occurs in a molecule when its atoms have largely different electronegativity. One atom attracts electrons more than another, becoming more negative; while the other atom becomes more positive and the centers of gravity of the charges of opposite sign are separated even in the absence of an external electric field.

When a molecule having intrinsic dipole moment is placed in a uniform electric field E, the field exerts a force $+ qE$ on charge and $- qE$ on charge $-q$. The net force on the dipole is zero since the two forces acting on it are equal and opposite of each other. Therefore, there is no translational force on the dipole in a uniform electric field. However, the forces are antiparallel and constitute a couple which tends to rotate the dipole Fig. 40.8.

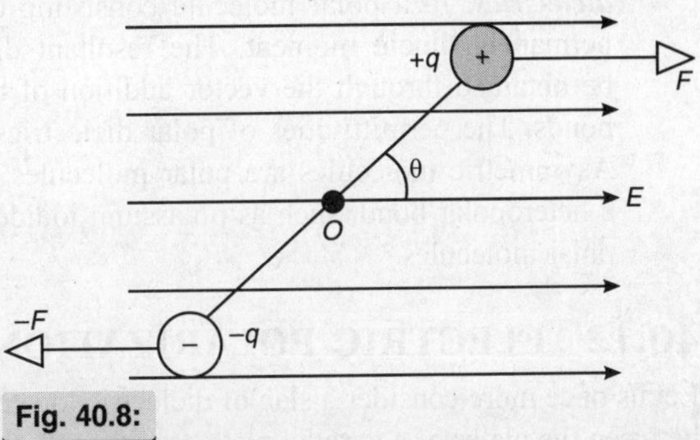

Fig. 40.8: Torqueacting on the dipole

The torque acting on the dipole is given by

$$\tau = q\,E\,d \sin\theta = \mu\,E \sin\theta$$

or $$\tau = \mu \times \mathbf{E} \quad ...(40.16)$$

Thus, a dipole experiences a torque in a uniform electric field and rotates in an attempt to align with the field direction. In fact, a free dipole aligns its axis with the field direction.

Further, the electric field can also induce a dipole moment in the molecule. Therefore, the total dipole moment of the molecule is a sum of the induced and permanent dipole moments. Thus,

$$\mu_T = \mu_{ind} + \mu_{per} \quad ...(40.17)$$

However, in case of polar molecules, $\mu_{ind} << \mu_{per}$ and therefore, $\mu_T \approx \mu_{per}$

40.11 NONPOLAR AND POLAR DIELECTRICS

Dielectrics are broadly divided into two major groups, namely, *nonpolar* and *polar* dielectrics basing on dipole moment. A molecule is a neutral system in which the algebraic sum of all the charges is equal to zero. However, the spatial arrangement of charges in a molecule may differ from material to material. All positive charges of a molecule may be replaced by one equivalent positive charge located at the center of gravity of positive charges. Similarly, all negative charges in it may be replaced with a single equivalent negative charge located at the center of gravity of all negative charges. The two resultant charges are equal in magnitude. Their points of action in space may coincide or may not coincide. Whether a molecule is a polar or nonpolar can be judged from its structure.

(*i*) **Nonpolar dielectrics:** When the points of action of two resultant charges coincide, the molecule will not possess a ***permanent dipole moment***. Such molecules are called nonpolar molecules and the material is known as a *nonpolar dielectric*. Their permittivities are low and range from 1 to 2.2. It is obvious that symmetric molecules are nonpolar since the centres of gravity of positive and negative charges coincide with each other. Thus, monoatomic molecules He, Ne, Ar, and Xe are nonpolar. Molecules consisting of two identical atoms linked by a homopolar bonds such as H_2, N_2, Cl_2 are nonpolar.

(*ii*) **Polar dielectrics:** If the points of the resultant charges of a molecule do not coincide in space, the molecule will possess an ***intrinsic dipole moment***. Such molecules are called polar molecules and the materials made up polar molecules are called ***polar dielectrics***. In a polar molecule consisting of several bonds, each bond may carry a permanent dipole moment. The resultant dipole moment of the molecule may then be obtained through the vector addition of the moments associated with the different bonds. The permittivities of polar dielectrics are high ranging from 3 to 8 and more. Asymmetric molecules are polar molecules. The molecules of ionic compounds with a heteropolar bond, such as potassium iodide (KI) have a high dipole moment and are polar molecules.

40.12 ELECTRIC POLARIZATION-AN ATOMIC VIEW

Let us once more consider a slab of dielectric located between the plates of a parallel plate capacitor. In the absence of an external electric field each elementary volume of the dielectric has no dipole moment. If the dielectric is a nonpolar material, the constituent molecules do not possess intrinsic dipole moments and in case the dielectric is a polar material, the individual molecular dipoles are randomly oriented so that in an elementary volume $\sum \mu = 0$. Hence, the polarization is zero. When the electric field is switched on dipoles are induced in nonpolar molecules, which form chains along the field lines, as shown in Fig. 40.9.

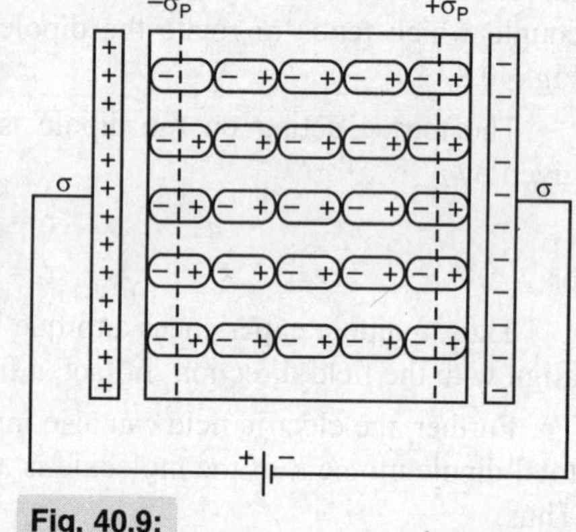

Fig. 40.9: Aatomic point of view of Polarization of a dielectric

The induced charge density σ_p on the dielectric is less than the free charge density σ on the plates.

The polarization for ***nonpolar dielectric*** is given by

$$P = N\mu = N\alpha E \qquad ...(40.18)$$

where N is the number of molecules per unit volume.

In a polar dielectric, the molecular dipoles experience a torque that tends to align them with the field direction. Total alignment is not achieved because of the disordering effects of thermal agitation. An average alignment $\langle\mu\rangle$ is achieved in the direction of the field.

The polarization for ***polar dielectric*** is therefore given by

$$P = N \langle \mu \rangle \qquad ...(40.19)$$

Thus, the action of electric field brings the dipoles into a certain ordered arrangement in space. It is seen that the ends of adjacent dipoles carrying opposite charges neutralize each other. Only the charges of the dipole ends terminating on the opposite faces of the slab remain uncompensated.

Thus, the application of an electric field to a dielectric produces a displacement of charge within the material through a progressive orientation of intrinsic or induced dipoles. This is known as ***dielectric polarization***.

40.13 TYPES OF POLARIZATION

Dielectric polarization is classified into four basic types.

(*i*) Electronic polarization,

(*ii*) Ionic polarization,

(*iii*) Orientation polarization, and

(*iv*) Space charge polarization.

One or two of the polarizations are always present and at a particular temperature or frequency of applied field, one or another may contribute in a large measure to the total polarization.

40.13.1 Electronic Polarization

Electronic Polarization results from the displacement of the electron clouds of atoms, molecules and ions with respect to heavy 'fixed' nuclei to a distance that is less than the dimensions of the atoms, molecules or ions. It occurs in all dielectrics for any state of aggregation. The electronic polarization sets in over a very short period of time, of the order of 10^{-14} to 10^{-15} s.

It is independent of temperature. The phenomenon is illustrated in Fig. 40.10.

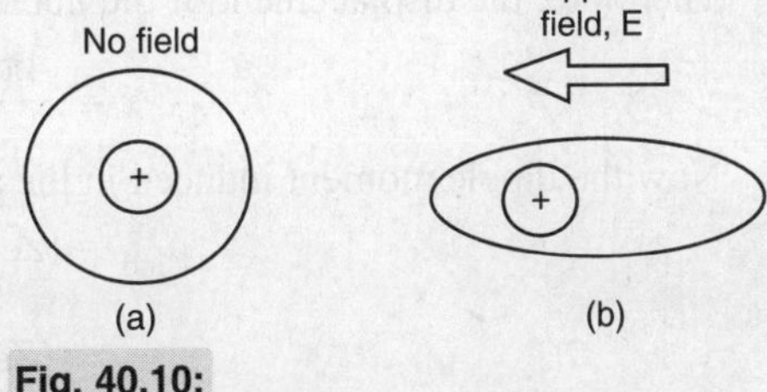

Fig. 40.10: Electronic polarization - (a) Atoms are not polarized in the absence of the electric field, (b) Electronic polarization results from the distortion of electron cloud by an applied electric field.

Expression for Electronic Polarization

Let us consider a single atom with atomic number *Z*. The charge on its nucleus is +Ze and *Z* electrons move around the nucleus. Let us assume that the nucleus is a point charge and the total negative charge is homogeneously distributed throughout a sphere of radius, *R*.

When this atom is subjected to an electric field **E**, the nucleus and the electron cloud will move in

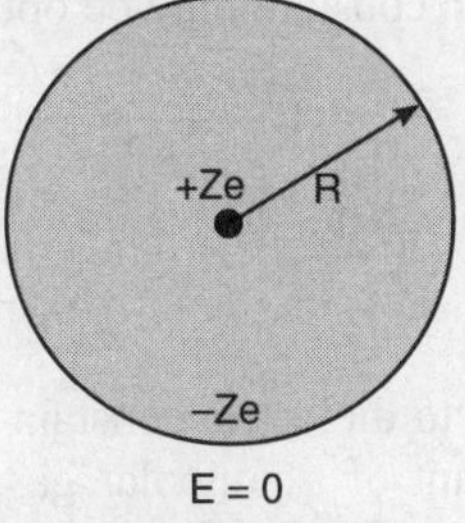

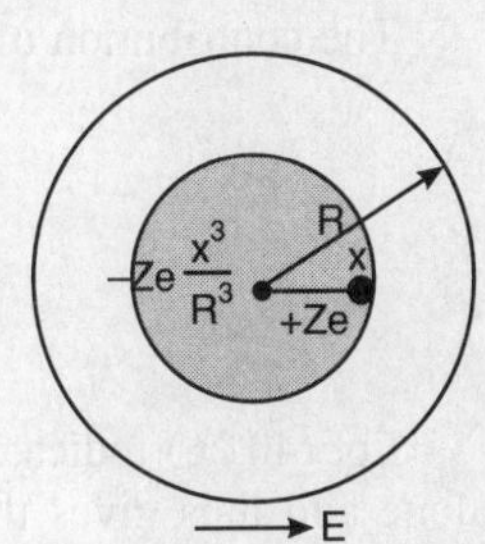

Fig. 40.11: An atom in an electric field

opposite directions. The coulomb attractive force opposes the movement, which acts as the restoring force here. Equilibrium condition will be attained in which the nucleus is displaced relative to the center of the electron cloud by the amount x.

The force on the nucleus along the field direction is

$$F = ZeE \qquad ...(40.20)$$

To determine the coulomb attraction on the nucleus, we divide the electron cloud into two regions. One region is the one that is inside the sphere of radius x and the other is the annular region lying between the two spherical surfaces of radii x and R.

By applying Gauss theorem, we find that the force experienced by the nucleus arises due to the negative charge lying within the spherical region of radius x. The charge inside this region is given by $-\frac{Zex^3}{R^3}$.

The force exerted by this charge on the nucleus is given by

$$F = \frac{1}{4\pi\varepsilon_0} \cdot \frac{(Ze)(Zex^3 / R^3)}{x^2} \qquad ...(40.21)$$

The equilibrium condition is that the above two forces balance each other. Thus,

$$ZeE = \frac{1}{4\pi\varepsilon_0} \cdot \frac{(Ze)(Zex^3 / R^3)}{x^2}$$

Therefore, the displacement of the nucleus is

$$x = \frac{4\pi\varepsilon_0 R^3}{Ze} E \qquad ...(40.22)$$

Now the dipole moment induced in the atom due to the displacement is

$$\mu_{ind} = (Ze)x$$

$$= (Ze)\frac{4\pi\varepsilon_0 R^3}{Ze} E$$

or $$\mu_{ind} = 4\pi\varepsilon_0 R^3 E \qquad ...(40.23)$$

$\therefore$ $$\mu_{ind} = \alpha_e E$$

where $$\alpha_e = 4\pi\varepsilon_0 R^3 \qquad ...(40.24)$$

It follows that the electronic polarization in a unit volume of the dielectric is given by

$$P_e = N\alpha_e E \qquad ...(40.25)$$

where α_e is the ***electronic polarizability***.

The contribution of P_e to the dielectric constant may be obtained as follows

$$\varepsilon = 1 + \chi = 1 + \frac{P_e}{\varepsilon_0 E} = 1 + \frac{N\alpha_E}{\varepsilon_0 E}$$

$$\varepsilon_r = 1 + \frac{N\alpha_e}{\varepsilon_0} \qquad ...(40.26)$$

Eq. (40.26) indicates the contribution to dielectric constant due to electronic polarization alone and thus gives the dielectric constant of a nonpolar gas. Note that it depends on the polarizability of a molecule and the number of molecules in a unit volume of the dielectric. In case of monoatomic gas $\alpha_e = 4\pi\varepsilon_0 R^3$.

$\therefore$ $$\varepsilon_r = 1 + 4\pi N R^3 \qquad ...(40.27)$$

Example 40.2. *Calculate the electronic polarizability of argon atom. Given $\varepsilon_r = 1.0024$ at NTP and $N = 2.7 \times 10^{25}$ atoms/m^3.*

Solution: $P = \varepsilon_r (\varepsilon_r - 1)\, E$ and $P = N\alpha_e E$

$$\therefore \quad \alpha_e = \frac{\varepsilon_0(\varepsilon_r - 1)}{N} = \frac{(8.85\times10^{-12}\,\text{F/m})(1.0024-1)}{2.7\times10^{25}\,/\,\text{m}^3} = \mathbf{7.9 \times 10^{-40}\ F.m^2}.$$

Example 40.3. *The number of atoms in hydrogen gas is 9.8×10^{20} atoms/cc. The radius of hydrogen atom is 0.053 nm. Calculate its electronic polarizability and relative permittivity.*

Solution: $\alpha_e = 4\pi\varepsilon_0 R^3 = 4 \times 3.14 \times 8.85 \times 10^{-12}\text{F.m} \times (0.053\ \text{nm})^3 = \mathbf{1.657 \times 10^{-41} F.m^2}$.

$\varepsilon_r = 1 + 4\pi N R^3 = 1 + 4 \times 3.14 \times 9.8 \times 10^{26}/\text{m}^3 \times (0.053\ \text{nm})^3 = \mathbf{1.0018}$.

40.13.2 Ionic Polarization

Ionic polarization occurs in ionic crystals. It occurs due to the elastic displacement of positive and negative ions from their equilibrium positions.

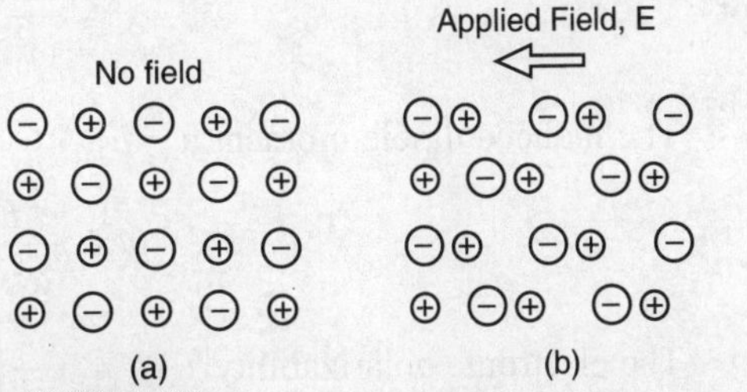

Let us take the example of sodium chloride crystal. A sodium chloride molecule consists of Na^+ ions bound to Cl^- ions through ionic bond. If the interatomic distance is d, the molecule exhibits an intrinsic dipole moment equal to 'ed'. When a dc electric field is applied to the molecule, the sodium and chlorine ions are displaced in opposite directions (Fig. 40.12) until ionic bonding forces stop the process. The dipole moment of the molecule increases consequently. When the field direction is reversed the ions move closer and again the dipole moment undergoes a change.

No field

Applied Field, E

(a) (b)

Fig. 40.12: Ionic Polarization - (a) unpolarized dielectric. (b) Ionic polarization results from the relative displacement of electrically charged ions in response to an applied electric field

Thus, dipoles are induced. The induced dipole moment is proportional to the applied field and is expressed as

$$\mu_i = \alpha_i E \qquad \text{...(40.28)}$$

where α_i is known as ***ionic polarizability***.

Expression for ionic Polarization

Let us consider the sodium crystal being subjected to an external electric field E. Due to the action of electric field, the positive ions (Na^+) displace in the direction of electric field through a distance of, say, x_1 units and the negative ions (Cl^-) by units in a direction opposite to that of the field.

The net displacement of the ions $x = x_1 + x_2$ ...(40.29)

The force on the Na^+ ion due to electric field $= +eE$

The force on the Cl^- ion due to electric field $= -\,eE$

The restoring force acting on the Na^+ ion $= -\,k_1x_1$

The restoring force acting on the Cl^- ion $= +k_2x_2$

where k_1 and k_2 are the force constants. They are given by

M is the mass of the positive ion (Na^+) and

m is the mass of the positive ion (Cl^-) and

ω_0 is the natural angular frequency of the molecule.

$$k_1 = M\omega_0^2 \quad \text{and} \quad k_2 = m\omega_0^2.$$

The shifting of ions halts and equilibrium is attained when the electric force and restoring force are equal and opposite to each other.

Thus, the equilibrium condition requires that

$$eE = k_1x_1 \quad \text{and} \quad eE = k_2x_2$$

$$x_1 = \frac{eE}{k_1} = \frac{eE}{M\omega_0^2} \quad \text{and} \quad x_2 = \frac{eE}{k_2} = \frac{eE}{m\omega_0^2}$$

Using the above expressions into Eq. (40.29), we obtain the net displacement of ions as

$$x = \frac{eE}{M\omega_0^2} + \frac{eE}{m\omega_0^2}$$

or

$$x = \frac{eE}{\omega_0^2}\left[\frac{1}{M} + \frac{1}{m}\right] \qquad \text{...(40.30)}$$

The induced dipole moment $\mu = e\,x$.

$$\mu = \frac{e^2E}{\omega_0^2}\left[\frac{1}{M} + \frac{1}{m}\right] \qquad \text{...(40.31)}$$

∴ The electronic polarizability $\alpha_i = \dfrac{e^2}{\omega_0^2}\left[\dfrac{1}{M} + \dfrac{1}{m}\right]$ ...(40.32)

It is seen from the above expression for ionic polarizability that

(*i*) It is inversely proportional to the square of the natural frequency of the molecule.

(*ii*) It is directly proportional to the reduced mass of the molecule, $\left[\dfrac{1}{M} + \dfrac{1}{m}\right]$.

(*iii*) It does not depend on temperature.

The ions experience electronic polarization in addition. For most materials, the ionic polarizability is less than the electronic polarizability. Typically

$$\alpha_i = \frac{1}{10}\alpha_e$$

The ionic polarization is given by

$$P_i = N\,\alpha_i\,E$$

or

$$P_i = \frac{Ne^2}{\omega_0^2}\left[\frac{1}{M} + \frac{1}{m}\right]E \qquad \text{...(40.33)}$$

Ionic polarization takes 10^{-11} to 10^{-14} s to build up, and is not influenced by temperature.

40.13.3 Orientation Polarization

The orientation polarization is characteristic of polar dielectrics, which consist of molecules having permanent dipole moment. It is also known as dipole polarization. In the absence of external electric field, the orientation of dipoles is random resulting in a complete cancellation of each other's effect, as illustrated in Fig. 40.13 (a).

Orientation Polarization is the polarization arising from the orientationof molecules which have permanent dipoles occuring from an asymmetric charge distribution.

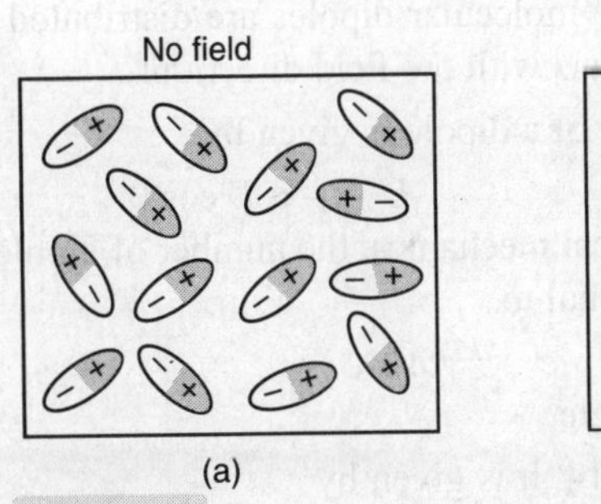

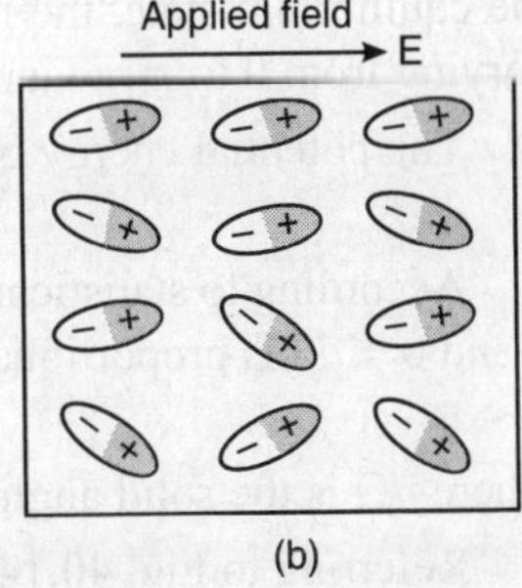

Fig. 40.13:

Orientaiton polarization-(a) molecular dipoles are randomly oreinted when E = 0.
(b) When the field is applied the dipoles are partially aligned

When the electric field is impressed the molecular dipoles rotate about their axis of symmetry to align with the applied field. In case of electronic and ionic polarizations, the force due to the external field is balanced by a restoring force due to coulomb attraction, but for orientation polarization, restoring forces do not exist. However, the dipole alignment is counteracted by thermal agitation. The higher the temperature, the greater is the thermal agitation.

The dipoles can turn only through a small angle, as illustrated in Fig. 40.13 (b). Even in case of liquids or gases, where molecules are free to rotate, a complete alignment cannot be achieved due to the randomizing effect of the temperature. However, it is estimated that it is enough if one molecular dipole in 10^5 completely aligns with the field to produce orientation polarization of the order of electronic polarization.

Thus, orientation polarization is strongly temperature dependent. This type of polarization occurs in gases, liquids and amorphous viscous substances. In case of solids, the molecules are fixed in their positions and their rotation is highly restricted by the lattice forces, leading to a great reduction in their contribution to orientation polarization. Because of this reason, while the dielectric constant of water is about 80, that of solid ice is about only 10.

As the process of orientation polarization involves rotation of molecules, it takes relatively longer time than the electronic and ionic polarizations. The build up time is of the order of 10^{-10} s or more.

Expression for Orientation Polarizability - Langevin-Debye theory

Let us consider a polar gas dielectric. If it is subjected to an electric field, E of appropriate low frequency, the individual dipoles experience torque and tend to align themselves with the electric field. If the electric field is sufficiently strong, all the dipoles in the gas would be completely aligned into the field direction and the orientation polarization would reach the saturation value

$$P_0 = N\mu$$

where N is the number of molecular dipoles per unit volume of the gas.

However, at the modest electric field strengths usually employed, the dipole alignment is not complete and is much less than the saturation value, given above. This is due to the effect of temperature, which counteracts the ordering effect of the electric field. The thermal agitation of molecules tends to randomize the orientation of dipoles. A modest orientation of dipoles is achieved in thermal equilibrium. In

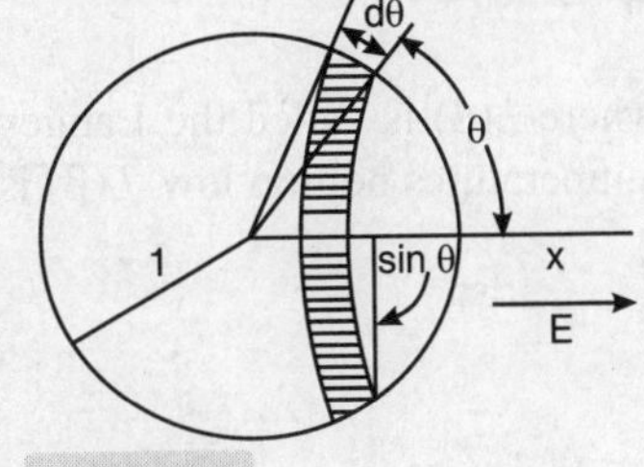

Fig. 40.14:

Distribution of molecular dipoles over all directions with the field direction the equilibrium state, the molecular dipoles are distributed over all directions making angles varying from 0 to π radians with the field direction.

The potential energy of a dipole is given by

$$U = -\mu E \cos\theta \qquad \text{...(40.34)}$$

According to statistical mechanics, the number of dipoles, dN, having orientation between θ and θ + $d\theta$ is proportional to

$$e^{(-U/kT)} d\Omega$$

where $d\Omega$ is the solid angle.

Referring to Fig. 40.14, it is given by

$$d\Omega = 2\pi \sin\theta\, d\theta$$

A dipole making an angle θ with the field direction contributes a component of dipole moment μ cos θ parallel to the field.

Hence the contribution of the dN dipoles to the orientation polarization is

$$dP_o = \mu\, d\Omega\, dN \cos\theta$$

$$\therefore \quad dP_o = 2\pi\mu\, e^{\mu E \cos\theta/kT} \cos\theta \sin\theta\, d\theta \qquad \text{...(40.35)}$$

Therefore, the average contribution to polarization is given by

$$P_{\text{ave}} = \frac{\text{Total polarization due to all dipoles}}{\text{Total number of dipoles}} \qquad \text{...(40.36)}$$

$$= \frac{\mu \int_0^{\pi} \exp\left[\frac{\mu E \cos\theta}{kT}\right] \cos\theta \sin\theta\, d\theta}{\int_0^{\pi} \exp\left[\frac{\mu E \cos\theta}{kT}\right] \sin\theta\, d\theta} \qquad \text{...(40.37)}$$

Putting $\mu E/kT = \beta$, and $\cos\theta = y$ into the above equation, we rewrite it as follows.

$$P_{\text{ave}} = \frac{\mu \int_{+1}^{-1} y e^{\beta y}\, dy}{\int_{+1}^{-1} e^{\beta y}\, dy}$$

or

$$\frac{P_{ave}}{\mu} = \frac{(e^{\beta} + e^{-\beta})}{(e^{\beta} - e^{-\beta})} - \frac{1}{\beta} = \coth\beta - \frac{1}{\beta} \equiv L(\beta) \qquad \text{...(40.38)}$$

where $L(\beta)$ is called the Langevin function. For electric fields that are not too high and for temperatures not too low, $L(\beta)$ is given by

$$L(\beta) \cong \frac{1}{3}\beta = \frac{\mu E}{3kT} \qquad \text{...(40.39)}$$

$$\therefore \quad P_{\text{ave}} = \frac{\mu^2 E}{3kT} \qquad \text{...(40.40)}$$

The total orientation polarization of the dielectric is

$$P_0 = NP_{\text{ave}} = \frac{N\mu^2 E}{3kT} \qquad \text{...(40.41)}$$

which states that the orientation polarization is directly proportional to the square of the permanent dipole moment and inversely proportional to the temperature.

As $$P_0 = \frac{N\mu^2 E}{2kT} = N\alpha_0 E,$$

The orientation polarizability α_0 is given by

$$\alpha_0 = \frac{\mu^2}{3kT} \quad ...(40.42)$$

We have from Eq. (40.12) that $\mathbf{P} = \varepsilon_0 (\varepsilon_r - 1)\mathbf{E}$.

Considering the contribution only from orientational polarization, we write the above expression as

$$\mathbf{P}_0 = \varepsilon_0 (\varepsilon_r - 1)\mathbf{E} \quad ...(40.43)$$

Equating the Eq. (40.41) and Eq. (40.43), we obtain

$$\varepsilon_0(\varepsilon_r - 1) = \frac{N\mu^2}{3kT} \quad ...(40.44)$$

40.13.4 Space Charge Polarization

Space charge polarization occurs in heterogeneous dielectric materials in which there is a change of electrical properties between different phases and in the homogeneous dielectrics that contain impurities, pores filled with air, inclusions of hygroscopic water etc. In particular, the properties of plastics, ceramics etc materials in their outer layers are likely to differ from those in depth because of the environmental effects and as a result they may behave as heterogeneous materials.

When an electric field is applied, the electric charges those migrate within the impurity regions store up at the interfaces. The accumulation of charges takes place with opposite polarity on the interfaces, as shown in Fig. 40.15. The space charge polarization takes generally a longer time and this polarization therefore occurs at low frequencies. Space charge polarization is also known as **interfacial polarization** or **migrational polarization**.

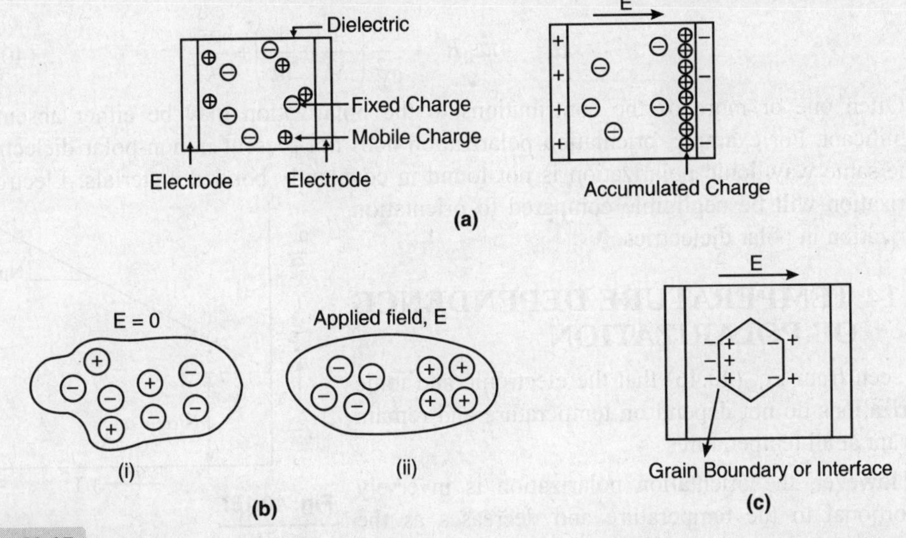

Fig. 40.15:

Space charge polarization

Space-charge polarization is the polarizationof a dielectric which occurs when charge carriers are present which can migrate an appreciable distance through the dielectric but which become trapped or cannot discharge at an electrode.

One of the typical examples of interfacial polarization is the grain boundaries that frequently lead to interfacial polarization as they can trap charges migrating under the influence of an applied field. Normally, interfacial polarization exists in materials, however perfect they may be, as they cotain crystal defects, impurities, and various mobile charge carriers such as electrons, holes, or ionized impurity ions. H^+, Li^+ ions cause this type of polarization in ceramics and glasses.

In case of interfacial polarization there is no general way to compute the charges on either interfaces or their contribution to the total polarization of a dielectric. Many dielectrics in real capacitors depend on interfacial polarization.

40.13.5 Total Polarization

In a material, which can experience all forms of polarization, the total polarization is equal to the sum of the electronic, ionic, orientation and migrational polarizations. The total polarization is given by

$$P_{\text{total}} = P_e + P_i + P_0 + P_m$$

In general, the migrational polarization is very small and negligible. Therefore, total polarization in a material may be taken as due to the other three contributions only. Thus,

$$P_{\text{total}} = P_e + P_i + P_0 \qquad ...(40.45)$$

The total polarization of a polar dielectric is therefore given by

$$P = N[\alpha_e + \alpha_i + \alpha_0]E$$

or

$$= N\left[4\pi\varepsilon_0 R^3 + \frac{e^2}{\omega_0^2}\left(\frac{1}{M} + \frac{1}{m}\right) + \frac{\mu^2}{3kT}\right]E \qquad ...(40.46)$$

The total polarizability is given by

$$\alpha = \alpha_e + \alpha_i + \alpha_0$$

or

$$\alpha = 4\pi\varepsilon_0 R^3 + \frac{e^2}{\omega_0^2}\left(\frac{1}{M} + \frac{1}{m}\right) + \frac{\mu^2}{3kT} \qquad ...(40.47)$$

Often one or more of the contributions to the polarization may be either absent or insignificant. For example, orientation polarization does not present in non-polar dielectrics. In the same way, ionic polarization is not found in covalently bonded materials. Electronic polarization will be negligible compared to orientation polarization in polar dielectrics.

40.14 TEMPERATURE DEPENDENCE OF POLARIZATION

It is seen from Eq. (40.46) that the electronic and ionic polarizations do not depend on temperature and remain constant at all temperatures.

However, the orientation polarization is inversely proportional to the temperature and decreases as the temperature increases.

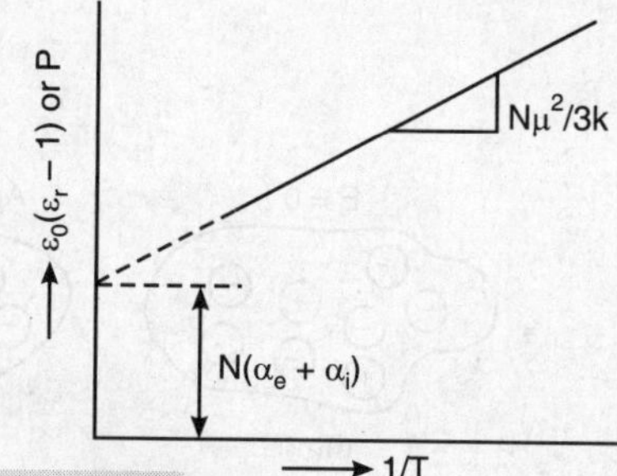

Fig. 40.16: The variation of total polarization as a function of 1/T.

If polarization P is plotted as function of $1/T$, a straight line will be obtained, as shown in Fig. 40.16. The intercept of the line with y-axis at $(1/T) = 0$ gives the value of $N(\alpha_e + \alpha_i)$ from which $(\alpha_e + \alpha_i)$ can be evaluated. The dipole moment μ can be computed from the slope of the straight line, knowing the value of N, and the number of molecules per m^3.

40.15 FREQUENCY DEPENDENCE OF TOTAL POLARIZATION

In many practical situations, a dielectric is subjected to an alternating electric field. With each direction reversal of ac field, the polarization component also reverses. The total polarization depends on the ability of dipoles to orient themselves in the direction of the field during each alternation of the field. Fig. 40.17 illustrates the dependence of P on frequency of the electric field for a polar dielectric.

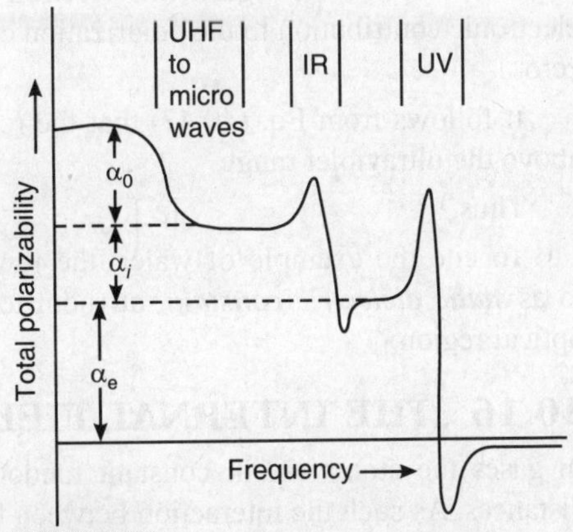

Fig. 40.17: Dependence of total polarization on the frequency of alternating electric field

In ***audio frequency region***, all types of polarization are possible and the dielectric is characterized by a polarizability $\alpha = \alpha_e + \alpha_i + \alpha_o$ and the polarization $P = P_e + P_i + P_o$.

At ***low frequencies***, the dipoles will get sufficient time to orient themselves completely along the instantaneous direction of the field. This orientation occurs first in one direction and then in the other, following the changes in the direction of the field (Fig. 40.18). The average time taken by the dipoles to reorient in the field direction is known as the ***relaxation time*** τ. The reciprocal of the relaxation time is called the ***relaxation frequency*** ν. If the frequency of the applied electric field is much higher than the relaxation frequency of the dipoles, the dipoles cannot reverse fast enough. If the dipole relaxation time τ is less than half the period of the electric field T ($<< T/2$), the dipole can easily follow electric field alternations and contribute to orientation polarization. Consequently, the orientation polarization, which is effective at low frequencies, is damped out for higher frequencies, ($f_{\text{field}} > f_{\text{relax}}$).

In the ***radio frequency or microwave band region***, the permanent dipoles fail to follow the field reversals and the polarization falls to a value corresponding to $(Pi + Pe)$. As a result, ε_r decreases considerably.

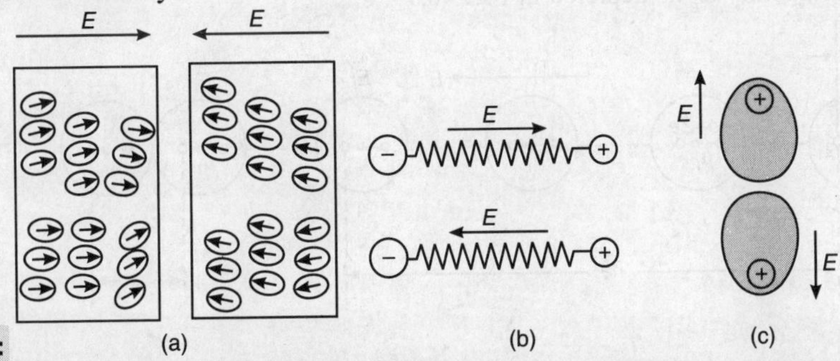

Fig. 40.18: The behaviour of dipoles in alternating electric field (a) permanent and (b) and (c) induced dipoles

In the **infrared region** the ionic polarization fails to follow the field reversals due to the inertia of the system and the contribution of ionic polarizability ceases. In this region, only electronic polarization contributes to the total polarization. Therefore $P = Pe$.

In the ***optical region***, the electron cloud follows the field variations and the material exhibits an electronic polarizability α_e. The relative permittivity in the optical region will be equal to the square of the refractive index 'n' of the dielectric.

$$[\varepsilon_r]_{\text{optical region}} = n^2 \qquad ...(40.48)$$

In the ***ultraviolet region***, the electron cloud too fails to follow the field alternations and electronic contribution to the polarization ceases. Consequently, the total polarization becomes zero.

It follows from Eq. (40.47) that the relative permittivity approaches unity at frequencies above the ultraviolet range.

Thus, $$[\varepsilon_r]_{\text{X-ray}} = 1$$

To cite the example of water, the low frequency dielectric constant, generally referred to as ***static dielectric constant***, at room temperature is about 80. It falls to about 1.9 in the optical region.

40.16 THE INTERNAL FIELD IN SOLIDS

In gases the atoms are in constant random motion and are separated by sufficiently large distances. As such the interaction between the atoms can be neglected. When an external field E is applied, the intensity of the electric field experienced by a given atom in the gas will be equal to the applied field E.

In case of solids and liquids subjected to external electric field, the atoms are surrounded on all sides by other polarized atoms, and the internal intensity of the electric field at a given point of the material is, in general, not equal to the intensity of the applied field E.

The ***internal field E_i***, which is defined as the electric field acting at the location of a given atom, is given by the sum of the electric fields created by the neighbouring atoms and the applied field. In evaluation of the bulk polarization, the additional effects of the surrounding polarized atoms are to be taken into account.

The effective field intensity E_i in the dielectric is given by

$$\mathbf{E_i} = \mathbf{E} + \mathbf{E'}$$

where E' is the field due to neighbouring atoms. The value of E' can be evaluated by the summation of all the effects of the surrounding atoms. To illustrate the method of evaluation, let us consider a one-dimensional solid consisting of a string of equidistant identical atoms, each of polarizability α_e as depicted in Fig. 40.19.

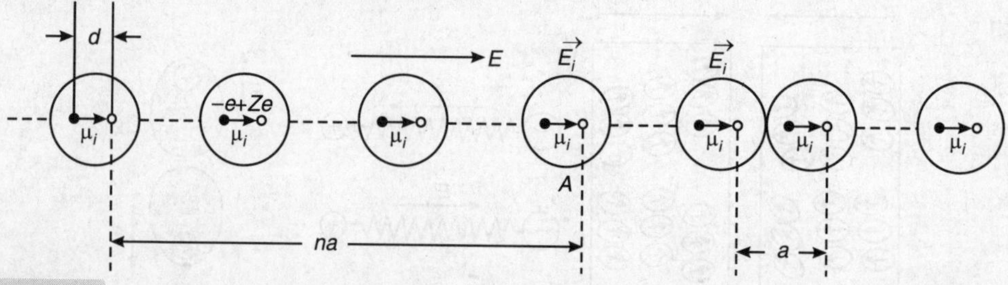

Fig. 40.19: The electric field E_i seen by the atoms is different from the external fields E which is, say given by $E = V/L$. A one-dimensional solid is considered for computation of local field

Let us consider an external field **E** applied in a direction parallel to the string. We shall determine the net internal field $\boldsymbol{E}_i$ experienced by one of the atoms, say A. The field seen by all other atoms will also be the same, and from the consideration of symmetry $\boldsymbol{E}_i$ will be parallel to **E**.

The dipole moment induced in each of the atoms of the string is therefore

$$\mu_{\text{ind}} = \alpha_e E_i$$

The field at A due to the dipole induced in an atom located at a distance 'na' from it is given

$$E_n = \frac{Ze}{4\pi\varepsilon_0}\left[\frac{1}{(na)^2} - \frac{1}{(na+d)^2}\right] = \frac{Ze}{4\pi\varepsilon_0}\left[\frac{(na+d)^2-(na)^2}{(na)^2(na+d)^2}\right]$$

$$= \frac{Ze}{4\pi\varepsilon_0}\left[\frac{2nad+d^2}{(na)^2(na+d)^2}\right] \cong \frac{2Zed}{4\pi\varepsilon_0(na)^3} \qquad \text{(since } d \ll na\text{)}$$

$$= \frac{\mu_i}{2\pi\varepsilon_0(na)^3} \qquad \text{(since } \mu_i = Zed\text{)}$$

The total field E_i at A is given by

$$E_i = E + \frac{\mu_i}{2\pi\varepsilon_0}\left[2\sum_{n=1}^{\infty}\frac{1}{(na)^3}\right] \qquad \text{...(40.49)}$$

The factor 2 in the parenthesis takes into account the atoms to the left and to the right of atom A.

$$E_i = E + \frac{\mu_i}{\pi\varepsilon_0 a^3}\sum_{n=1}^{\infty}\frac{1}{n^3}$$

or

$$E_i = E + \frac{1.2\mu_i}{\pi\varepsilon_0 a^3} \qquad \text{...(40.50)}$$

Thus, the combined effect of induced dipoles of neighbouring atoms is to produce a net field at the location of a given atom, which is larger than the applied field. It is seen from the Eq. (40.50) that the greater the polarizability α_e or the smaller the intermolecular spacing 'a', the larger is the internal field.

40.17 LORENTZ FIELD

The local field in a three dimensional solid is determined by the structure of the solid. An accurate calculation of the internal field in solids and liquids is in general very complicated.

Let us consider a dielectric slab kept in a uniform electric field, **E** (Fig. 40.20). Let a molecule be at the point O and be surrounded by a spherical cavity of radius r. Let r be arbitrary but sufficiently large compared to molecular dimensions and sufficiently small compared to the dimensions of the dielectric slab. The spherical cavity contains many molecules within it.

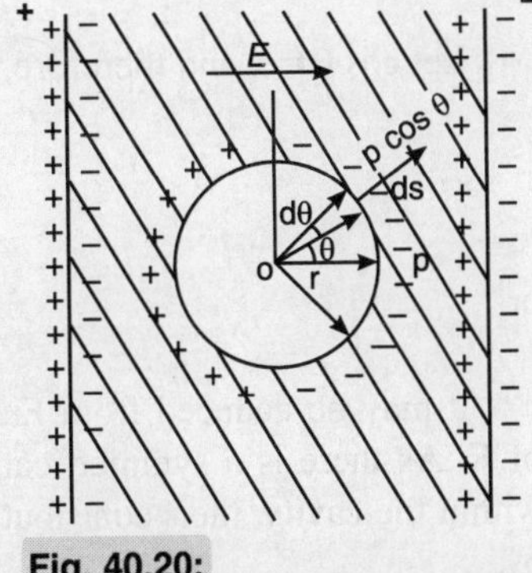

Fig. 40.20: Computation of Lorentz field

The molecule at O experiences three electric fields acting on it.

(*i*) The external electric field **E**

(*ii*) The field $\mathbf{E}_1$ due to induced charges on the surface of the spherical cavity

(*iii*) The field $\mathbf{E}_2$ due to the molecular dipoles present in the spherical cavity.

Therefore, the total internal field intensity, E_i is given by

$$\mathbf{E}_i = \mathbf{E} + \mathbf{E}_1 + \mathbf{E}_2 \qquad ...(40.51)$$

To calculate E_1, let us imagine that the dielectric is removed from the sphere. For the actual pattern of the electric field not to be distorted, a surface electric charge should be placed on the spherical surface. At each point of the sphere, the surface charge density is given by

$$\sigma = P \cos \theta$$

where θ is the angle between radius vector **r** and the direction of **E**.

The charge on element dS of the surface of the sphere will be

$$dq = \sigma\, dS = P \cos \theta\, dS \qquad ...(40.52)$$

This charge will produce electric field intensity $d\mathbf{E}_1$ at the center of the sphere.

$$dE_1 = \frac{dq}{4\pi\varepsilon_0 r^2} = \frac{P}{4\pi\varepsilon_0 r^2} \cos \theta\, dS \qquad ...(40.53)$$

This electric field can be resolved into two components: one component $dE_1 \cos \theta$ parallel to the direction of **E** and the other $dE_1 \sin \theta$ perpendicular to the direction of **E**.

$$dE_1 \cos \theta = \frac{P}{4\pi\varepsilon_0 r^2} \cos^2 \theta\, dS \qquad ...(40.54)$$

and
$$dE_1 \sin \theta = \frac{P}{4\pi\varepsilon_0 r^2} \cos \theta \sin \theta\, dS \qquad ...(40.55)$$

It is obvious that the perpendicular components of the upper and lower half of the sphere cancel each other and only the parallel components contribute to the total intensity E_1. E_1 is obtained by integrating dE_1 over the whole surface area of the sphere.

Thus,
$$E_1 = \int_0^{\pi} dE_1 \cos\theta\, dS = \frac{P}{4\pi\varepsilon_0 r^2} \int_0^{\pi} \cos^2 \theta\, dS$$

But $dS = 2\pi\, r^2 \sin \theta\, d\theta$.

Therefore,
$$E_1 = \frac{P}{2\varepsilon_0} \int_0^{\pi} \cos^2 \theta \sin \theta\, d\theta$$

Let $\cos \theta = x$ and therefore, $- \sin \theta\, d\theta = dx$.

∴
$$E_1 = -\frac{P}{2\varepsilon_0} \int_1^{-1} x^2 dx = -\frac{P}{2\varepsilon_0} \left[\frac{x^3}{3}\right]_1^{-1} = \frac{2P}{6\varepsilon_0}$$

or
$$E_1 = \frac{P}{3\varepsilon_0} \qquad ...(40.56)$$

It may be deduced from Fig. 40.20 that the direction of $\mathbf{E}_1$ coincides with the direction of **E**. As there is a symmetrical distribution of molecular dipoles around the molecule at O within the cavity, their contributions cancel each other. Hence the total internal field is given by.

$$\mathbf{E}_i = \mathbf{E} + \mathbf{E}_1$$

$$\therefore \quad \mathbf{F_l} = \mathbf{E} + \frac{\mathbf{P}}{3\varepsilon_0} \qquad ...(40.57)$$

The field given by the Eq. (40.57) is called ***Lorentz field*** or ***local field.***

40.18 CLAUSIUS-MOSOTTI EQUATION

Let us consider now the simple case of an elemental solid dielectric, which exhibits only electronic polarizability. Solids such as diamond, silicon and germanium crystals are made up of single type of atoms. If α_e is the electronic polarizability per atom, it is related to the bulk polarization P through the relation

$$\alpha_e = \frac{P}{NE_i} \qquad ...(40.58)$$

where N is the number of atoms per m^3 and E_i is the local field.

Using the Eq. (40.61) for E_i, we get

$$\alpha_e = \frac{P}{N\left[E + \frac{\gamma P}{\varepsilon_0}\right]} \qquad ...(40.59)$$

where γ is known as the internal field constant.

According to Eq. (40.12)

$$E = \frac{P}{\varepsilon_0(\varepsilon_r - 1)}$$

If the internal field is assumed to be Lorentz field, $\gamma = 1/3$ and Eq. (40.59) becomes

$$\alpha_e = \frac{P}{N\left[E + \frac{P}{3\varepsilon_0}\right]}$$

Using the Eq. (40.12) into the above equation, we obtain

$$\alpha_e = \frac{P}{N\left[\frac{P}{\varepsilon_0(\varepsilon_r - 1)} + \frac{P}{3\varepsilon_0}\right]}$$

or

$$\frac{N\alpha_e}{\varepsilon_0} = \frac{1}{\left[\frac{1}{\varepsilon_r - 1} + \frac{1}{3}\right]} = \frac{1}{\left[\frac{\varepsilon_r + 2}{3(\varepsilon_r - 1)}\right]}$$

or

$$\frac{3(\varepsilon_r - 1)}{\varepsilon_r + 2} = \frac{N\alpha_e}{\varepsilon_0}$$

$$\therefore \quad \frac{\varepsilon_r - 1}{\varepsilon_r + 2} = \frac{N\,\alpha_e}{3\varepsilon_0} \qquad ...(40.60)$$

Eq. (40.60) is known as ***Clausius-Mosotti equation*** which is valid for ***nonpolar solids having cubic crystal structure.***

The measured values of ε_r for the three elements of IV group of the periodic table are: Diamond is 5.68, Silicon is 12 and Germanium is 16.

Example 40.4: *The atomic weight and density of sulphur are 32 and 2.08 gm/cm*^{3} *respectively. The electronic polarizability of the atom is* 3.28×10^{-40} *F.m*^{2}*. If sulphur solid has cubic symmetry, what will be its relative permittivity?*

Solution: $$\frac{\varepsilon_r - 1}{\varepsilon_r + 2} = \frac{N\alpha_e}{3\varepsilon_0} = \frac{N_A \rho \alpha_e}{3M\varepsilon_0}$$

$$\frac{\varepsilon_r - 1}{\varepsilon_r + 2} = \frac{(6.023\times10^{26})(2.08\times10^3 \text{ kg/m}^3)(3.28\times10^{-40} \text{ F.m}^2)}{3\times32\times8.85\times10^{-12} \text{ F/m}}$$

$$= 0.483$$

$\therefore$ $$\varepsilon_r = \frac{1.966}{0.517} = \mathbf{3.8}$$

Example 40.5. *A dielectric material has* $\varepsilon_r = 4.94$ *and* $n^2 = 2.69$. *Calculate the ratio between electronic and ionic polarizability of this material.*

Solution: $$\frac{\varepsilon_r - 1}{\varepsilon_r + 2} = \frac{N\alpha}{3\varepsilon_0} = \frac{N(\alpha_e + \alpha_i)}{3\varepsilon_0}$$ $\because \alpha_0$ is negligibly small.

$\therefore$ $$\frac{N(\alpha_e + \alpha_i)}{3\varepsilon_0} = \frac{4.94 - 1}{4.94 + 2} = 0.568 \qquad (i)$$

At optical frequencies, $$\varepsilon_r = n^2$$

$\therefore$ $$\frac{n^2 - 1}{n^2 + 2} = \frac{N\alpha}{3\varepsilon_0}$$

or $$\frac{N\alpha}{3\varepsilon_0} = \frac{2.69 - 1}{2.69 + 2} = 0.360 \qquad (ii)$$

Dividing equ. (*i*) by (*ii*), we get

$$\frac{N(\alpha_e + \alpha_i)}{N\alpha_e} = \frac{0.568}{0.360} = 1.578$$

or $$1 + \frac{\alpha_i}{\alpha_e} = 1.578$$

or $$\frac{\alpha_i}{\alpha_e} = 0.578$$

$\therefore$ $$\frac{\alpha_e}{\alpha_i} = \mathbf{1.73}$$

40.19 DIELECTRIC LOSS

When a conductor is subjected to an electric field, it dissipates part of the electrical energy, which gets converted to heat energy. That part of energy is lost or wasted. The term ***power loss*** denotes the average electrical power dissipated in a material during a certain interval of time. The power loss in conductors is also called ***I^2R loss*** or ***Joules heat***. As distinct from conductors, the power (I^2R) loss in dielectrics subjected to dc voltages will be very small due to the high resistance of the dielectric materials, whereas the power loss in ac fields will be quite large.

Def: The absorption of electrical energy by a dielectric subjected to an alternating electric field is known as the *dielectric loss*.

The dielectric loss caused by an ac field also results in dissipation of the electrical energy as heat in the materials. An ideal dielectric does not absorb electrical energy. However, a real dielectric always causes some loss of electrical energy.

The origin of dielectric loss: An ac field changes its direction with time. With each direction reversal, the molecules are required to follow the field reversals in order to contribute to the polarization of the dielectric. When a capacitor is charged in one half-cycle, the molecules of the dielectric medium are polarized. When the capacitor is discharged in the second half-cycle, the molecules should revert to their initial condition. When it happens, the energy spent in charging the capacitor is completely returned. However, in the process of returning to their initial state, the molecules push each other and lose energy due to friction. The energy lost due to friction takes the form of heat. This energy loss will increase with increase in frequency.

40.19.1 Loss Angle and Loss Tangent

Let us consider a parallel plate capacitor C, constituted by plates of area A and separated by a distance d (Fig. 40.21). Let a dielectric having permittivity ε_r fill the space between the plates. Let a sinusoidal voltage V of angular frequency ω be applied to the capacitor. The current through the capacitor is given by

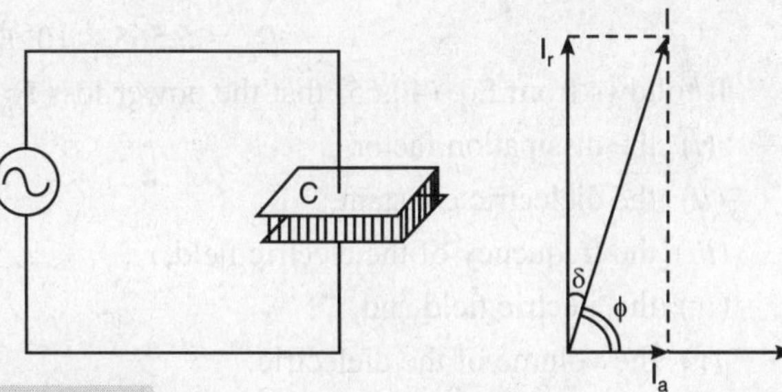

Fig. 40.21: A dielectric in an alternating field. The current through the capacitor is sum of reactive and absorption current components

$$I = j\omega\, CV + \frac{V}{R} \qquad \text{...(40.61)}$$

or

$$I = j\, I_r + I_a \qquad \text{...(40.62)}$$

Eq. (40.62) indicates that two kinds of current flow through the dielectric the conduction current $I_a = V/R$ and the displacement current I_r, given by

$$I_r = \omega\, CV$$

The resultant current $I = \sqrt{I_r^2 + I_a^2}$ lags behind the displacement current by an angle δ.

In case of an ideal dielectric, $R \approx \infty$ and $I_a = 0$; and it would not absorb electric energy. In such a case, the resultant current I would be ahead of the voltage V precisely by a phase angle $\phi = 90°$ and the current would have been purely reactive current I_r.

$$I = I_r = \frac{\omega\varepsilon_0\varepsilon_r AV}{d}$$

However, for a lossy dielectric, the total current is

$$I = I_a + j\,|\,I_r\,|$$

The phase angle ϕ between V and I is now slightly less than 90°. The angle $\delta = (90 - \phi)$ is called the **loss angle**. It is given by

$$\tan\delta = \frac{I_a}{I_r} \qquad \text{...(40.63)}$$

or

$$\tan\delta = \frac{V/R}{\omega CV} = \frac{1}{\omega CR} \qquad \text{...(40.64)}$$

$\tan\delta$ is known as the ***loss tangent***. It represents the electrical power lost which is often in the form of heat. Hence it is also called ***dissipation factor***.

The real power loss in the dielectric is given by

$$P_L = V I_a = V I_r \tan\delta = \omega C V^2 \tan\delta$$

$$= \frac{\omega\varepsilon_0\varepsilon_r A V^2}{d}\tan\delta = \frac{2\pi f\varepsilon_0\,\varepsilon_r\,(Ad)V^2}{d^2}\tan\delta$$

or $$P_L = 2\pi\,\upsilon\,\varepsilon_0\,E^2 f\,\varepsilon_r\,\tan\delta \qquad ...(40.65)$$

where $\upsilon = Ad$ is the volume of the dielectric and $E = V/d$.

Substituting the values of 2π and ε_0 into the above equation, we obtain

$$P_L = 5.565 \times 10^{-11}\upsilon\,E^2 f\varepsilon_r \tan\delta$$

It follows from Eq. (40.65) that the power loss P_L in a dielectric is related to the

(*i*) the dissipation factor
(*ii*) the dielectric constant,
(*iii*) the frequency of the electric field,
(*iv*) the electric field and
(*v*) the volume of the dielectric.

40.19.2 Complex Relative Permittivity

In most of the materials, the dielectric behaviour is more complex indicating the presence of other sources of dielectric loss. To include losses from all sources, we rewrite the Eq. (40.62) as

$$I = I_a + j|I_r| = j|I_r|\left[1 - j\frac{I_a}{|I_r|}\right] = \frac{j\omega A V\varepsilon_0\varepsilon_r}{d}(1 - j\tan\delta) \qquad ...(40.66)$$

Eq. (40.66) suggests that the lossy dielectric can be described with the aid of a complex relative permittivity ε^*_r given by

$$\varepsilon^*_r = \varepsilon_r(1 - j\tan\delta)$$

or $$\varepsilon^*_r = \varepsilon'_r - j\,\varepsilon''_r \qquad ...(40.67)$$

where $$\tan\delta = \frac{\varepsilon''_r}{\varepsilon'_r} \qquad ...(40.68)$$

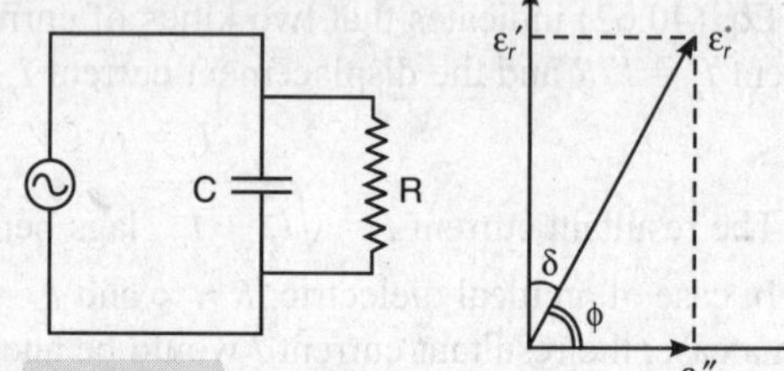

Fig. 40.22: Representaion of a lossy dielectric. The complex dielectric constant is at angle to the real part of dielectric constant

Therefore the product (ε'_r $\tan\delta$) is known as the ***loss factor***. A lossy dielectric is represented by a resistance parallel to the capacitor, as shown in Fig. 40.22.

Using Eq. (40.68) into Eq. (40.65), we can write

Power dissipation per unit volume $$\frac{P_L}{v} = \omega\varepsilon_0\varepsilon''_r E^2 \qquad ...(40.69)$$

At any given frequency, ε''_r produces the same type of macroscopic effect as the conductivity σ and for all practical purposes the two are indistinguishable.

40.19.3 Dielectric Loss Spectrum

The typical variation of ε'_r and ε''_r of a polar dielectric with frequency are shown in Fig. 40.23.

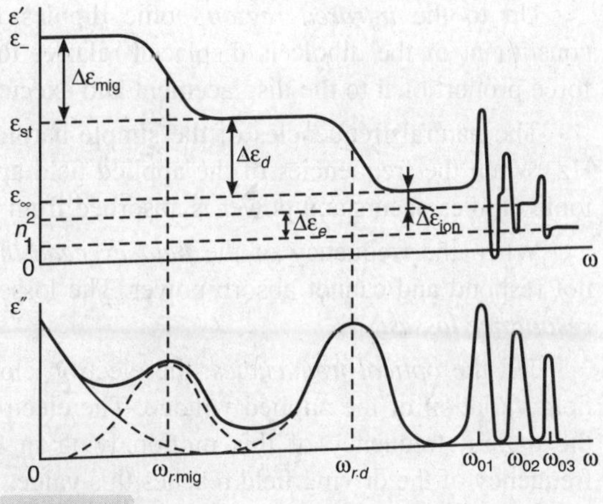

Fig. 40.23: Schematic representation of the frequency dependence of the real and imaginary parts of relative permittivity of a dielectric

At *audio frequencies*, the electric field reverses slowly so that the molecular dipoles can keep shifting their orientation directions in step with the field alternations (Fig. 40.24). There is no power loss in this region.

In the *radio frequency region*, the rotation of the dipoles in order to align with the electric field direction is opposed by the internal friction of the material and the thermal agitation of the molecules. Consequently, the dipoles lag behind the forces that cause the motion. A phase difference develops between polarization and the electric field resulting in a fall of ε'_r as the frequency increases. It is accompanied by heating of the dielectric and therefore by a loss of energy. Energy supplied to maintain the rotation of the dipoles accounts for power loss.

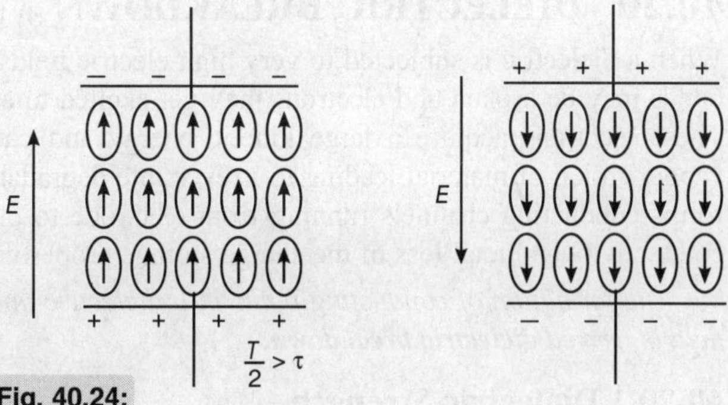

Fig. 40.24: The shifting of molecular dipoles in step with the field alternations

At frequencies of the field near the *relaxation frequency* of dipoles, the rotation becomes more rapid and the energy loss approaches a maximum at the relaxation frequency.

At frequencies *above the relaxation frequency*, the electric field reverses so rapidly that the dipoles cannot follow the field reversals due to inertia. By the time the dipole attempts to align along the particular direction, the field direction changes. Therefore, the dipoles fail to respond and maintain a random orientation and never become aligned with the field. Consequently, the dielectric constant is reduced and the power loss decreases after going through a maximum.

At *much higher frequencies*, the losses become negligible. As a result the ε''_r versus frequency variation exhibits a bell shaped profile with a maximum at the molecular dipole relaxation frequency and reaching zero on either side. It is seen thus that the greatest loss occurs at frequencies at which the dipoles can almost but not completely be reoriented.

At lower frequencies, losses are low because the dipoles have time to rotate. At higher frequencies losses are low because the dipoles do not rotate at all. The losses in this process are known as ***relaxation losses***.

Up to the *infrared region*, ionic dipoles follow electric field variations. When one constituent of the dipole is displaced relative to the other, it again experiences a restoring force proportional to the displacement and executes simple harmonic motion.

The natural frequencies of the simple harmonic motion lie in the range of 10^{13} to 10^{14} Hz. When the frequencies of the applied field approaches *natural frequencies*, ***resonance*** of ionic dipoles occurs and power is absorbed from the field.

When the frequency of the *field exceeds their natural frequency*, the ionic dipoles do not respond and cannot absorb power. The losses encountered in this process are known as ***resonance losses***.

Till the *optical frequencies*, the electron clouds in the molecule respond to the electric field variation of the applied voltage. The electron cloud executes simple harmonic motion, the natural frequency of this motion lying in the range of 10^{17} and 10^{18} Hz. When the frequency of the driving field reaches this value, resonance absorption takes place because of which electrons get excited to higher energy levels. Subsequently, they reemit this energy in a random fashion, in the form of optical photons. For frequencies greater than these, electrons fail to follow the excursions of the field and cannot absorb power any more. The losses encountered in this process are also known as ***resonance losses***.

40.20 DIELECTRIC BREAKDOWN

When a dielectric is subjected to very high electric fields, a considerable number of covalent bonds may be broken and electrons may get excited to energies within the conduction bond. These electrons acquire a large kinetic energy and cause localized melting, burning and vapourization of material leading to irreversible degradation and failure of the material. This forms conducting channels running from electrode to electrode. It results in high electrical conductivity and total loss of the charge storage property of the dielectric.

The formation of conducting paths in a dielectric under the action of an applied electric field is termed ***dielectric breakdown***.

40.20.1 Dielectric Strength

The dielectric strengthof a material is a measure of the ability of that material to withstand high electric fields.

Dielectric strength is defined as the maximum electric field that the dielectric can withstand without suffering electrical breakdown.

Thus, Dielectric strength, $E_{max} = \left(\frac{V_{max}}{d}\right)$ ...(40.70)

In other words, the dielectric strength is the limiting electric field intensity above which a breakdown occurs and the charge storage property of the dielectric disappears.

The dielectric strength depends on the thickness of the insulating material and on the length of time for which the dielectric is subjected to electric field. Moisture, contamination, elevated temperature, ageing and mechanical stress usually tend to decrease the dielectric strength of the material.

40.20.2 Breakdown Mechanisms in Solid Dielectrics

The physical pattern of breakdown of solid dielectrics may differ in various cases. However, the fundamental breakdown mechanisms are as follows.

(*i*) Intrinsic breakdown
(*ii*) Thermal breakdown
(*iii*) Discharge breakdown
(*iv*) Electrochemical breakdown and
(*v*) Defect breakdown

(*i*) Intrinsic breakdown: When a dielectric is subjected to very high electric fields, a considerable number of covalent bonds may be torn away and electrons may get excited to energies within the conduction band. The high energy electrons collide with other atoms and molecules and release more electrons, which in turn collide with more atoms thereby liberating more electrons. The number of electrons increases very rapidly with time. Conducting channels form running from electrode to electrode and it results in high electrical conductivity. Ultimately, dielectric breakdown occurs. Localized melting, burning and vaporization of material take place at this stage causing irreversible degradation and failure of the material. This type of breakdown is called ***avalanche breakdown***.

The characteristics of intrinsic breakdown are as follows:

(*a*) It occurs at larger electric fields.
(*b*) It occurs at ordinary temperatures.
(*c*) It occurs in thin samples.
(*d*) The breakdown time is of the order of microseconds.

(*ii*) Thermal breakdown: In dielectric materials, energy due to the dielectric loss appears as heat. This heat must be dissipated away to the surroundings. If the rate of heat generation is larger than the rate of heat dissipation, the temperature of the dielectric increases which results in local melting. Eventually the dielectric breaks down. Thus, dielectric breakdown occurs when the rate of heat generation is larger than the rate of heat dissipation.

The characteristics of thermal breakdown are as follows:

(*a*) It occurs at high temperatures.
(*b*) The breakdown time is of the order of milliseconds.
(*c*) In ac fields, the breakdown strength is lower.
(*d*) The breakdown strength depends on the size and shape of the material sample.

(*iii*) Discharge breakdown: Breakdown of dielectrics by gas discharge is classified as external breakdown and internal breakdown.

Corona discharge produces ***external breakdown*** and is observed at sharp edges of electrodes. The solid dielectric placed between the electrodes weekens by this discharge and forms carbon. So, the damaged areas become conducting.

Eventually, conducting paths are formed leading to a powerful arc and total failure of the dielectric. Such breakdown is caused mainly due to the contamination of the dielectric surface by conducting impurities such as dust, moisture etc.

Internal breakdown occurs due to the presence of gas or liquid filled cavities within the solid dielectric. The inherent strength to electric stress of such cavities is low relative to the solid portion. As a result partial discharges may occur in such cavities and cause gradual deterioration of the adjacent solid dielectric.

The characteristics of thermal breakdown are as follows:

(*a*) It occurs at low electric fields.

(*b*) It depends on the frequency of applied voltage.

(*iv*) Electrochemical breakdown: Electrochemical breakdown is very much related to thermal breakdown. Many materials have free ions which cause leakage current in the presence of electric field. When temperature increases, mobility of ions increases and also increases leakage current. Electrochemical reaction takes place in the material. Field induced chemical reactions reduces the resistance of the dielectric and finally results in breakdown.

The characteristics of electrochemical breakdown are as follows:

(*a*) It depends on the concentration of ions and magnitude of leakage current.

(*b*) It occurs at ordinary temperatures.

(*v*) Defect breakdown: In surface defects of the dielectric material such as cracks and porosity, impurities for instance dust or moisture may deposit and that leads to breakdown.

40.21 APPLICATIONS OF DIELECTRIC MATERIALS

Two most important applications of dielectric materials are: (*i*) as insulating materials and (*ii*) as medium in capacitors. For insulating materials application the dielectric is required to have low dielectric constant, low dielectric loss, high resistance and high dielectric strength. Further, they should possess adequate chemical stability, high moisture resistance, and suitable mechanical properties for particular service condition.

(*i*) Solid Insulating Materials: Polymers and ceramics are the widely used solid insulators. A variety of plastics, rubbers, waxes, paper, synthetic fibres and fabrics are applied in the form of films, sheets, slabs, tapes, sleeving, tubing, rods and moulding. Plastics such as polyethylene, polytetrafluroethylene (PTFE) and polysterene have low ε_r and practically no dielectric loss. Porcelain towers are used in high voltage power lines because of their high dielectric strength. The dielectric strength of porcelain bodies is enhanced by glazing their surfaces. Porcelain, glass, mica, alumina and asbestos are widely used ceramics.

Capacitors: A capacitor is an electronic component that stores energy in the form of electric field. Basically, it consists of two conducting plates separated by a dielectric. Capacitors are widely used in electrical and electronic equipments.

(*a*) **Paper Capacitors:** In this type of capacitors, one or more layers of extremely thin kraft or linen paper are used as the dielectric medium. The paper is kept between aluminium foils which act as the metal plates. The whole assembly is rolled into a cylindrical element. The dielectric is impregnated with mineral oil or waxes to prevent absorption of moisture.

(*b*) **Plastic Capacitors:** Plastics can be formed in thin, uniform and non-porous films. Such thin plastic films are used as dielectric medium in these capacitors. Some of the materials used are polyester, polycarbonate, polyethylene, polystyrene, polypropylene, poly tetrafluoroethylene (PTFE) and polythene Terephthalate films.

(*c*) **Ceramic Capacitors:** These capacitors use ceramic as the dielectric medium. Low loss low permittivity capacitors are made from steatite which formed in the form of a thin plate or foil. High permittivity capacitors use barium titanate as the dielectric material.

(*d*) **Mica Capacitors:** Muscovite mica is a naturally occurring material and can be laminated into very thin sheets. This material has good mechanical strength and can be used up to high temperatures of the order of 500°C. Impregnants like polystyrene improve the properties of mica.

(*e*) **Glass Capacitors:** Very thin plates of glass are used as dielectric in these capacitors. The plates are interleaved with aluminium foil and fused together to form a solid block.

(*f*) **Electrolytic Capacitors:** In electrolytic capacitors, a metallic anode has oxide film grown over it and this oxide layer acts as a dielectric. The anode is surrounded by an electrolytic solution of ammonium borate or sodium phosphate which acts as cathode. Etched aluminum foil is used as anode in aluminum electrolytic capacitors. Aluminum oxide film is grown over the aluminum foil and it acts as a dielectric film. A liquid electrolyte is held in contact with dielectric film. Another etched aluminum foil is used as the cathode. The assembly is sealed in an aluminum can.

(*ii*) Liquid Insulating materials: Liquid insulating materials are mainly mineral oils and synthetic oils, which are used for the *purpose of insulation* as well as *cooling* in transformers.
Transformers: A transformer is a device used for transmitting power from one circuit to another or from one place to another place. It consists of two windings, primary and secondary windings, linked by a common magnetic flux. During the construction of transformers, the windings are impregnated by varnishes. In case of H.V. transformers used in distribution of power where very high voltages are present, proper provisions are to be provided to distribute away the heat produced and to provide high dielectric strength. These transformers are usually immersed in liquid dielectrics.

(*a*) **Mineral insulating oil:** Mineral oil has very high dielectric strength and is highly viscous. It transfers heat from the transformer windings and core to the outer shield and enables dissipation of the heat generated. The oil should be perfectly free from moisture to maintain its high dielectric strength. Even small traces of water significantly reduce the dielectric strength. Therefore, the oil is periodically dehydrated. Secondly, sludge formation takes place in the oil due to constant heating of the oil during its working and it also should be removed periodically to maintain its initial quality.

(*b*) **Synthetic insulating oil:** Nowadays, synthetic oils are being used in place of mineral oils because synthetic oils are much more resistant to oxidation and fire hazards. Sovol, sovotol etc are some of the synthetic oils widely used in H.V. transformers.

(*c*) **Miscellaneous insulating oils:** Petroleum oils, silicone oils, and vegetable oils belong to this category. They have high thermal stability. They are mainly used as filling medium for transformers, circuit breakers etc and as impregnants for high voltage cables.

(*iii*) Dielectric heating: Insulating materials can be efficiently heated up by subjecting them to a high voltage of suitable frequency, namely the frequency at which dielectric loss is maximum. The dielectric loss manifests in the form of heat. Adequate heating may be obtained at high voltages of the order of 20 kV having a frequency of about 30 MHz. The chief advantage of this method is that the material is heated up quickly as the heat is produced in the insulating material itself.

Cooking in microwave oven is one of the popular examples of dielectric heating. Water invariably exists in all articles of food, which exhibits dielectric loss in microwave region. In a oven, microwaves produced by a source are distributed by reflection from the metal walls. They pass through the glass-cooking dish and are absorbed by water molecules. The food is cooked due to the heat produced in the absorption process.

Dielectric heating is widely employed in dehydration of food, tobacco etc. Wooden sheets are preferred to be glued by this method. The heat produced in the glue due to the dielectric absorption leads to binding of the wooden sheets. The advantage of this method is that the moisture content of wooden sheets remains unaltered.

40.22 CLASSIFICATION OF DIELECTRIC MATERIALS

Dielectric materials may be divided into the two following categories; linear dielectrics and nonlinear dielectrics.

(*i*) **Linear dielectrics** are those materials in which the polarization **P** and displacement **D** are directly proportional to the intensity of the electric field **E**; and relative permittivity ε_r and susceptibility χ do not depend on the intensity of electric field. They are also known as ***passive dielectrics***.

(*ii*) **Nonlinear dielectrics** are those materials in which relative permittivity ε_r and susceptibility χ depend on the intensity of electric field. These materials are known as ***active dielectrics***. Piezoelectrics, pyroelectrics, ferroelectrics and some of the optical media belong to the category of nonlinear dielectrics. Piezoelectric crystals provide a coupling between electrical and mechanical forces and hence serve as transducers which produce or detect electrical or mechanical signals. Hence they are used to detect very small mechanical displacements and small amounts of electric charge. All commercial piezoelectric materials used today are ferroelectrics.

40.23 FERROELECTRICITY

Ferroelectric materials constitute a very important group of dielectrics. They are anisotropic crystals that exhibit spontaneous polarization. In the absence of an electric field, if the centers of gravity of the positive and negative charges do not coincide, it results in a resultant dipole moment, which is the cause of spontaneous polarization.

Spontaneous polarization is the dielectric polarization, which occurs under the action of the internal processes and without the application of an electric field.

The materials, which possess special structure that permits spontaneous polarization, are called ***ferroelectrics*** and the phenomenon of spontaneous polarization is called ***ferroelectricity***. Ferroelectricity was first discovered in Rochelle salt (Na KC_4 H_4 O_6. $4H_2O$). It exhibits spontaneous polarization over a range of temperature –18°C to 22°C. Barium titanate, potassium phosphate, and potassium niobate are other examples of ferroelectrics.

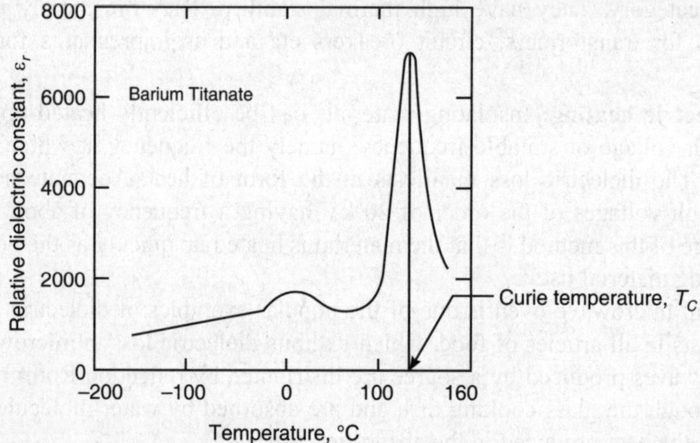

Fig. 40.25: The variation of ε_r with temperature in barium titanate

The main characteristics of ferroelectric substances are as follows:

(*i*) They possess very high values of permittivity ε_r of the order of 1000 to 10,000

(*ii*) The static dielectric constant of ferroelectric materials change with temperature according to the following relation.

$$\varepsilon = \frac{C}{T - T_C} \quad (T > T_C) \qquad ...(40.71)$$

Eq. (40.71) is known as ***Curie-Weiss law***. C is called the ***Curie constant*** and T_C the ***Curie temperature***. The variation of ε_r with temperature in barium titanate is shown in Fig. 40.25.

(*iii*) They possess spontaneous electric polarization, that is, polarization without the help of an external electric field. However, the spontaneous polarization occurs only within a definite temperature range and up to the Curie temperature T_C. The variation of spontaneous polarization with temperature is shown in Fig. 40.26.

(*iv*) In a ferroelectric, the dielectric polarization depends nonlinearly on the applied electric field. In ordinary dielectrics, the polarization varies linearly with the applied electric field. Because of this, ferroelectrics are known as ***nonlinear dielectrics***. They exhibit hysteresis under the action of an alternating voltage. The polarization versus electric field curve is known as a *ferroelectric hysteresis loop* (Fig. 40.27).

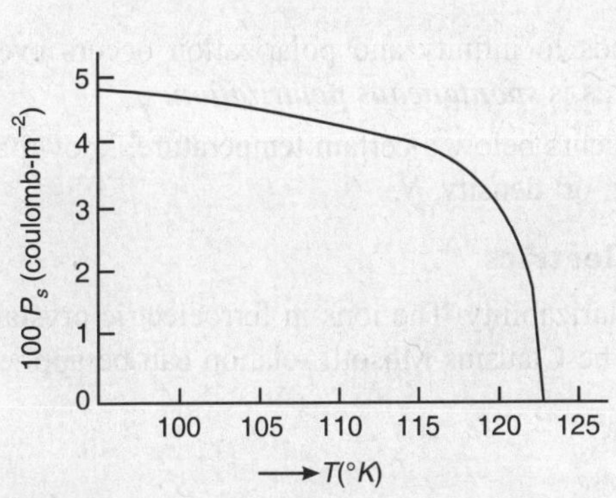

Fig. 40.26:

The variation of spontaneous polarization with temperature

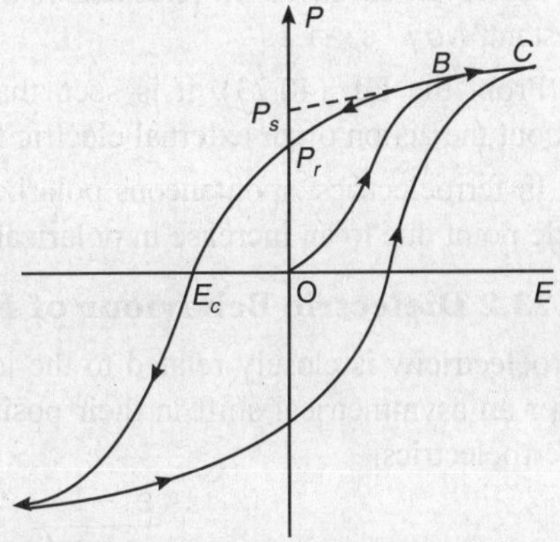

Fig. 40.27:

A Ferroelectric Hysteresis loop

40.23.1 Polarization Catastrophe

We know that

$$P = \varepsilon_0 (\varepsilon_r - 1)E$$

$$\therefore \qquad E = \frac{P}{\varepsilon_0(\varepsilon_r - 1)}$$

We have earlier obtained expression for the internal field in non-polar dielectrics. Now let us assume that it could as well be applied to polar dielectrics. If α is the polarizability per atom, it is related to the bulk polarization P through the following relation

$$P = N\alpha E_i$$

where N is the number of atoms per m^3 and E_i is the local field. The internal field in a solid is given by

$$E_i = E + \frac{\gamma P}{\varepsilon_0}$$

Then,

$$P = N\alpha\left[E + \frac{\gamma P}{\varepsilon_0}\right]$$

$$P\left[1 - N\alpha\frac{\gamma}{\varepsilon_0}\right] = N\alpha E$$

$$\therefore \quad P = \frac{N\alpha E}{I - N\alpha\gamma/\varepsilon_0} \quad ...(40.72)$$

and the dielectric susceptibility

$$\chi = \frac{N\alpha/\varepsilon_0}{1 - N\alpha\gamma/\varepsilon_0} \quad ...(40.73)$$

In common solid dielectrics of average permittivity, $N\alpha$ is approximately equal to ε_0. For cubic and isotropic materials, the internal field constant $\gamma = 1/3$. Therefore, $N\alpha\gamma / \varepsilon_0 \approx 1/3$.

In these cases, the dipole interaction does not exert a substantial influence on the dielectric properties.

However, in cases of certain crystals of higher polarizability and larger internal field constant, $N\alpha\gamma / \varepsilon_0 \to 1$.

From the Eq. (40.73), it is seen that χ tends to infinity and polarization occurs even without the action of an external electric field. This is ***spontaneous polarization***.

In ferroelectrics, spontaneous polarization occurs below a certain temperature, known as Curie point due to an increase in polarizability, α, or density, N.

40.23.2 Dielectric Behaviour of Ferroelectrics

Ferroelectricity is closely related to the ionic polarizability. The ions in ferroelectric crystals suffer an asymmetrical shift in their positions. The Clausius-Mosotti relation can be applied to ferroelectrics.

$$\frac{\varepsilon_r - 1}{\varepsilon_r + 2} = \frac{N\alpha}{3\varepsilon_0}$$

where α is the polarizability of an ion and N the number of ions per unit volume.

Now, we assume that α is independent of temperature and the temperature dependence of ε_r is solely due to the variation in N which changes due to thermal expansion of the crystal. We rewrite the above equation as

$$\frac{\varepsilon_r - 1}{\varepsilon_r + 2} = \beta N \quad ...(40.74)$$

where β is a constant. Differentiating Eq. (40.74) w.r.t. T, we get

$$\frac{1}{(\varepsilon_r + 2)}\left[1 - \frac{(\varepsilon_r - 1)}{(\varepsilon_r + 2)}\right]\frac{d\varepsilon_r}{dT} = \beta\frac{dN}{dT}$$

Dividing the above equation by Eq. (40.74), we obtain

$$\frac{1}{(\varepsilon_r - 1)}\left[1 - \frac{(\varepsilon_r - 1)}{(\varepsilon_r + 2)}\right]\frac{d\varepsilon_r}{dT} = \frac{1}{N}\frac{dN}{dT}$$

or $$\left[\frac{3}{(\varepsilon_r - 1)(\varepsilon_r + 2)}\right]\frac{d\varepsilon_r}{dT} - \frac{1}{N}\frac{dN}{dT} = -\gamma'$$

where γ' is the volume expansion constant of the material.

As $\varepsilon_r >> 1$, we can approximate $(\varepsilon_r - 1)(\varepsilon_r + 2) \approx \varepsilon_r^2$.

Therefore, $$\frac{3}{\varepsilon_r^2} d\varepsilon_r = -\gamma' dT$$

Integrating the above equation,

$$\int_{\varepsilon_r}^{\infty} \frac{3}{\varepsilon_r^2} d\varepsilon_r = -\int_{T}^{T_C} \gamma' \, dT$$

we obtain $$\varepsilon_r = \frac{3/\gamma'}{T - T_C} \quad ...(40.75)$$

This relation is similar in form to the Curie-Wiess law. It implies that the temperature dependence of the dielectric constant of the ferroelectric crystal is associated with the expansion of the lattice.

40.23.3 Spontaneous Polarization in a Ferroelectric

Let us consider the case of barium titanate crystal. A unit cell of the barium titanate is shown in Fig. 40.28 (a). It exhibits tetragonal symmetry at temperatures below Curie temperature. The Ba^{2+} ions are located at the corners of the unit cell, the O^{2-} ions in the centers of faces and the Ti^{4+} ion nearly at the center of the unit cell body. In effect one ion of barium and three ions of oxygen belong to the unit cell and correspond to the empirical formula $BaTiO_3$. The titanium ion does not occupy the exact body center of the unit cell. The relative displacement of the O^{2-} ions from their symmetrical positions is shown in the side view of the unit cell (Fig. 40.28 b). It is seen that the Ti^{4+} ion is displaced upward from the center of the unit cell while the O^{2-} ions are located slightly below the centers of each of the six faces.

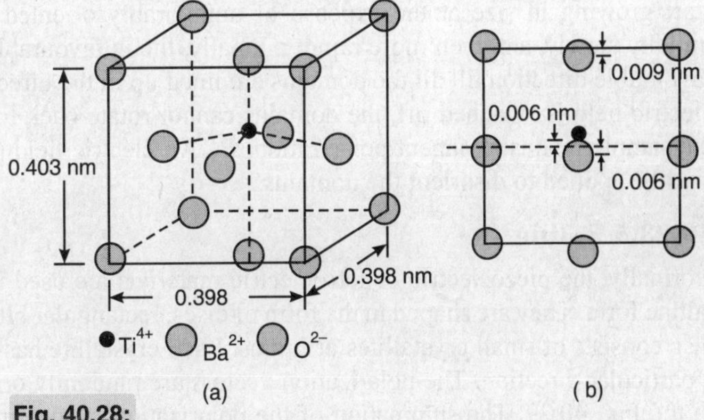

Fig. 40.28:

(a) The strcuture of $BaTiO_3$ below its Curie temperature 120° C which is slightly tetragonal

(b) Due to a slight shift of the central ion Ti^{4+} with respect to the surrounding O^{2-} ionsof the unit cell, an electric dipole moment arises in the unit cell.

Consequently, a permanent ionic dipole moment arises in each unit ell. Strong interactions between the adjacent permanent dipoles cause all the dipoles to mutually align in the same direction within some volume of the solid. Such regions of spontaneous polarization are known as *ferroelectric domains*. In an unpolarized ferroelectric solid, the polarization vectors of different domains orient in different directions, so that the net polarization of the solid is zero.

Above the Curie temperature, the thermal energy causes transformation of tetragonal unit cell into a cubic unit cell. It leads to a change in the relative positions of Ti^{4+} and O^{2-} ions so that the center of action of negative charges is coincident with of positive charges. Therefore, the net dipole moment becomes zero and the spontaneous polarization vanishes

40.23.4 Ferroelectric Hysteresis Loop

When a virgin ferroelectric crystal is subjected to an alternating electric field, the polarization P versus electric field E describes a closed loop called a ferroelectric hysteresis loop (Fig. 40.27). The polarization increases nonlinearly and reaches saturation at a certain value, P_s. The polarization will not increase further even if the electric field is increased. When the electric field is switched off, the value of the polarization does not return to zero and the crystal retains a *residual polarization* P_r. In order to bring back the polarization to zero, an electric field E_c must be applied in the opposite direction. E_c is known as the *coercive field.*

There is a close similarity between the electric properties of nonlinear dielectrics and the magnetic properties of ferromagnetic materials. It is for this reason that the *nonlinear dielectrics are called ferroelectrics.*

The hysteresis in ferroelectrics is explained on the basis of ferroelectric domains. In the absence of an external electric field, the domains are oriented randomly and the net polarization is zero. When electric field is applied domains oriented in a favorable direction start growing in size at the expense of unfavorably oriented domains. The growth occurs initially slowly, and then more rapidly. Finally, the unfavourable domains are rotated into the favourable direction till all the domains are lined up in the direction of the applied field. If the electric field is switched off, the domains cannot rotate back to their original orientation and the sample retains remanent polarization, P_r. An electric field ($-E_c$) in the opposite direction is to be applied to disorient the domains.

40.23.5 Poling

Normally, the piezoelectric or ferroelectric materials are used in ***ceramic*** form *i.e.* polycrystalline form. They are shaped in the form of discs, rectangular blocks or rings. A polycrystalline disc consists of small crystallites or grains. Each crystallite has a polarization (P_s) oriented in a particular direction. The polarization vectors are randomly oriented in the sample as shown in the Fig. 40.29. The summation of the polarization in the entire sample comes out to be a small value or zero compared to that of a single crystal.

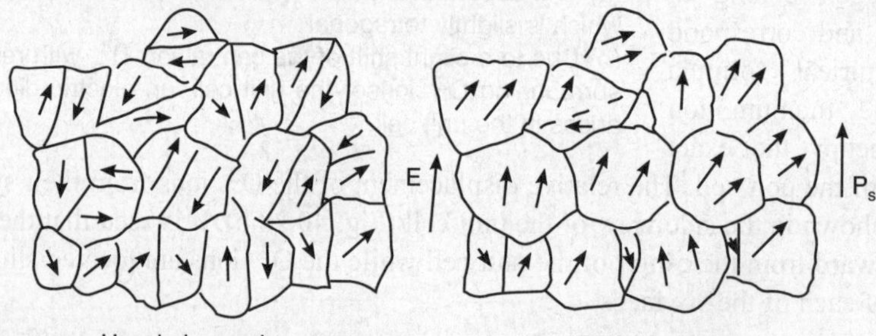

Fig. 40.29: Unpoled ceramic Poled ceramic

The polarization vectors in the poled ceramic are oriented in the field direction

Such a sample doe not exhibit piezoelectric effect. To impart piezoelectric properties to a ceramic, it is necessary to subject the material to polarization. For this purpose, the sample is held in a strong electric field of the order of 2 to 4 MV/m at temperature of 100° to 150°C for about an hour.

After such a treatment, the polarization vectors in the grains become oriented in a direction close to the field direction and such a sample is called ***electrically poled*** ceramic sample. After field removal, the sample stays polarized owing to stable remnant polarization. The ceramic sample thus transforms into an anisotropic body and has a preferred axis of polarization in the direction of the remnant polarization.

Def: Poling is the keeping ceramic sample in a strong electric field so that the polarization vectors in the grains become oriented in a direction close to the field direction and imparts piezoelectric properties to the sample.

Hence, poled ceramics exhibit the piezoelectric or ferroelectrics effects nearly as single crystals even though they cannot replace a single crystal. Their physical properties related to piezoelectricity and ferroelectricity depends on the amount of poling. The electrical poling of a ceramic is analogous to the magnetizing of a permanent magnet.

40.23.6 Ferroelectric Materials

All ferroelectric materials are ***pyroelectric***. All pyroelectric materials are ***piezoelectric***. Ferroelectric materials are naturally piezoelectric. Thus, ferroelectric materials display pyroelectric and piezoelectric properties. A material is said to be **piezoelectric**, if in response to an applied mechanical load, the material produces an electric charge proportional to the load.

Def: Pyroelectric effect *is the change in spontaneous polarization when the temperature of the specimen is changed.*

The extent of the spontaneous polarization is a maximum at temperatures well below the Curie temperature and reaches zero as the Curie temperature is approached. Likewise, the material produces a mechanical deformation in response to an applied voltage.

Ammonium phosphate, quartz and PZT (lead zirconatetitanate) are examples of piezoelectric materials. Quartz crystals are widely used in filter and resonator applications. Rochelle salt is used as transducer in ear phones, microphones and hearing aids.

Barium titanate, lead zirconate and lead titanate are ceramic piezoelectric materials used in gas lighters, accelerometers and transducers.

GaS, ZnO, CdS etc are piezoelectric semiconductors which are used in making ultrasonic wave amplifiers. Ferroelectric materials such as quartz, lithium niobate, barium titanate, calcium barium titanate and lead barium niobate are used in making pressure transducers, ultrasonic transducers etc.

Pyroelectric materials such as barium titanate and lithium niobate are used in making infrared detectors. TGS, $NaNO_2$ and PZT ceramics are used in fabrication of pyroelectric image tubes.

Piezoelectrics, pyroelectrics, ferroelectrics and some of the optical media belong to the category of nonlinear dielectrics. Piezoelectric crystals provide a coupling between electrical and mechanical forces and hence serve as *transducers* which produce or detect electrical or mechanical signals. Hence they are used to detect very small mechanical displacements and small amounts of electric charge. All commercial piezoelectric materials used today are ferroelectrics.

40.23.7 Application of Ferroelectric Materials

Application of the ferroelectric materials utilizes the pyroelectric, piezoelectric or ferroelectric properties of the materials.

(i) **Capacitors:** A capacitor consists of a dielectric material packed in between two electrodes. The total capacitance for this device

$$C = \frac{\varepsilon_0 \varepsilon_r A}{d} \quad ...(40.76)$$

where 'C' is the capacitance, ε_0 is the permittivity of free space, ε_r is the relative dielectric permittivity, 'd' is the distance between the electrodes, and 'A' is the area of the electrodes.

From Eq. (40.76) it is seen that in order to obtain a high capacitance per unit volume the dielectric material placed between the electrodes should have a large dielectric constant, a large area and a small thickness. $BaTiO_3$ based ceramics show dielectric constant values as high as 15,000 whereas common ceramic and polymer materials have 5 or 10. The use of a ceramic like $BaTiO_3$, yields large capacitance values in relatively small volume capacitors.

(ii) **Generation of ultrasonic waves:** Piezoelectric crystals provide a coupling between electrical and mechanical forces and hence serve as transducers which produce or detect electrical or mechanical signals. Hence they are used to detect very small mechanical displacements and small amounts of electric charge. All commercial piezoelectric materials used today are ferroelectrics.

High frequency oscillations are transformed into mechanical oscillations in the production of ultrasonic waves. Inverse piezoelectric effect is used in this. Especially cut quartz crystal discs are generally used in this application, frequencies as high as 50 MHz can be achieved.

(iii) **Vibrators:** Another important application is their use as vibrators. When an alternating voltage is applied across a piezoelectric element, it vibrates and at a particular frequency of the field, it resonates. The resonant frequency of the element depends on the thickness. The mechanical vibrations can be transferred into solids or liquids as ultrasonic waves. This frequency range lies from 1000 Hz to 10 MHz. Hard type piezoelectric ceramics are used in this application. The speakers and buzzers utilize the piezoelectric vibrations.

(iv) **Detectors:** Piezoelectric ceramics are used in the generation and reception of sound waves in water. They are used in ultrasonic cleaners, and under water detectors of sounds. The ultrasound is also useful for fault detection i.e. for finding internal cracks and other hidden defects in solid bodies like bars, rods, plates etc.

(v) **Pyroelectric Detectors:** Pyroelectricity is the polarization obtained due to a small change in temperature. Single crystals of triglycine sulfate (TGS), $LiTaO_3$, and (Sr, Ba) Nb_2O_6 are used in applications where heat is to be sensed. $PbTiO_3$, (Pb, La) TiO_3 and PZT are extensively used for thin film pyroelectric sensing applications.

(vi) **Gas Ignitors:** In the gas igniter, a very high voltage is generated spontaneously across a piezoelectric ceramic element when a strong mechanical stress pulse is given to the element. This voltage gives rise to a spark. A typical design for a voltage generator for gas ignitors is shown in Fig. 40.30.

Gas igniter consists of two oppositely poled ceramic cylinders attached end to end. It doubles the charge available for the spark.

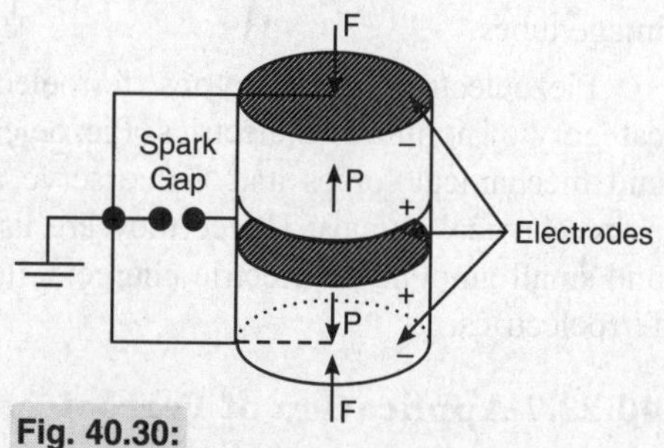

Fig. 40.30: Voltage generator in a gas ignitor

The compressive force is applied fast to see that the leakage of charge across the surfaces of the piezoelectric ceramic does not occur. Usually PZT ceramic disks are used for this application.

(*vii*) **Accelerometers:** An accelerometer is a device that measures acceleration. Fig. 40.31 depicts a typical accelerometer. The transducer is a piezoelectric cylinder which is poled along its axis. Sensing electrodes are attached to its inner and outer surfaces. The cylinder is connected to the fixed central pole on the inside and a cylindrical mass on the outside. When an axial acceleration takes place the cylinder is subjected to a shear force between the outer mass and inner pole. Motion in the radial direction does not produce output. Thus, the device is highly directional.

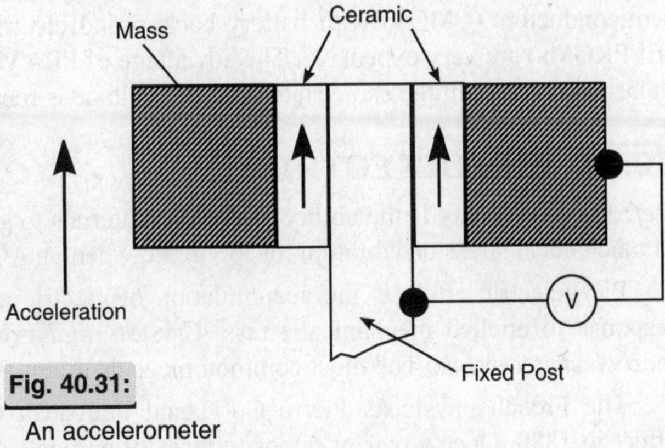

Fig. 40.31: An accelerometer

(*viii*) **Piezoelectric Transformers:** A piezoelectric plate can transform low voltage to high voltage. Fig. 40.32 illustrates a flat plate containing electrodes on half of it larger face and on an edge. The region between the larger face electrodes and the edge electrode are poled separately. A length mode resonance is excited by applying a low AC voltage source between the larger face electrodes. The step up voltage ratio would be proportional to the ratio of the input to output capacitance and the efficiency of the device. This principle is useful in making EHT transformers for miniature television receivers.

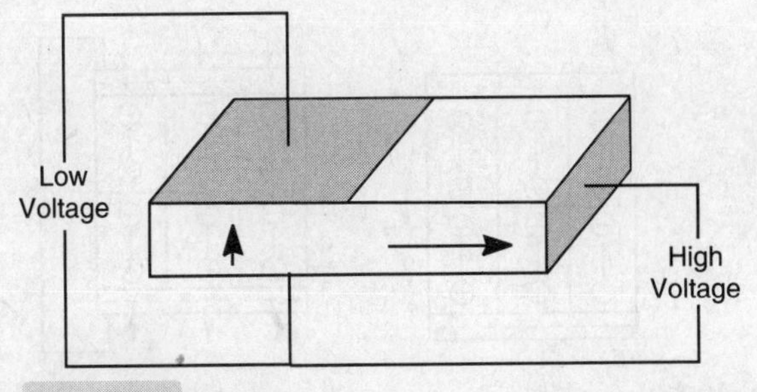

Fig. 40.32: A piezoelectric transformer with the arrows indicating the poling directions

(*ix*) **Impact Printer Head:** Dot matrix impact printer heads have piezoelectric displacement sensors. The advantages of the dot matrix printer over the conventional electromagnetic drive printers are high printing speeds, low energy consumption and noiseless printing. Inkjet printer heads are also based on displacement piezoelectric elements.

(*x*) **Ferroelectric Memories:** Ferroelectric materials suddenly polarize on cooling below the T_c. The magnitude and direction of polarization can be inverted by the use of an external electric field. The ferroelectric RAMs (FRAMs) made from ferroelectric thin films uses of this phenomenon to store data. Data is stored by localized polarization switching in the microscopic regions of ferroelectric thin films.

The ferroelectrics, which exhibit a square loop in polarization versus electric field characteristics, are more useful in memory technology. Barium- strontium titanate, strontium bismuth titanates are used in this application.

The disadvantage of semiconductor memories such as dynamic random access memories (DRAMs) and static random access memories (SRAMs) is that the information stored in them is lost when the power fails. The non-volatile memories such as complementary metal oxide semiconductors (CMOS) with battery backup and electrically erasable read only memories (EEPROMs) are very expensive. The advantage of FRAMs is that they are non-volatile as the polarization stays in the same state after the voltage is removed.

40.24 PIEZOELECTRICITY

Def: Piezoelectricity is the ability of certain materials to generate AC voltage when subjected to mechanical stress or vibration, or to vibrate when an AC voltage is applied.

Piezoelectric effect is the accumulation of electric charge in certain solid materials in response to applied mechanical stress. *Crystals that exhibit piezoelectric effect are called piezoelectric crystals*. The most common piezoelectric material is quartz.

The French physicists Pierre Curie and Paul-Jean Curie discovered the piezoelectric effect in 1880. Once a pair of opposite faces of certain asymmetric crystals such as quartz is compressed, opposite electric charges emerge on the other pair of opposite faces of the crystal (Fig. 40.33). If the crystals are subjected to tension, the polarities of the charges are reversed.

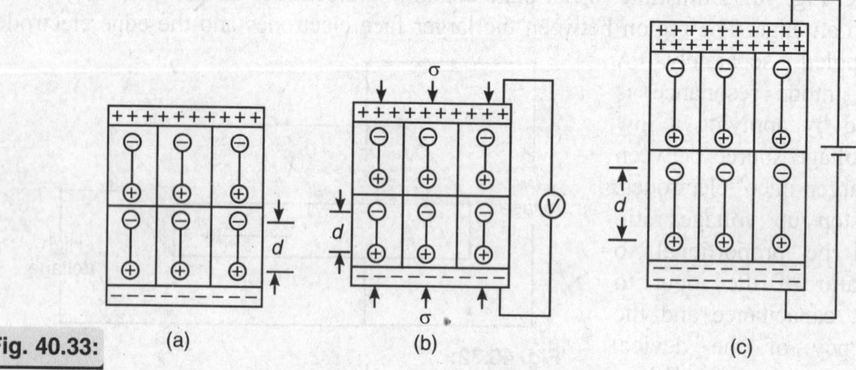

Fig. 40.33:
(a) Electric dipoles in a piezoelectric crystal.
(b) Mechanical forces cause appearance of polarization charges.
(c) An external voltage applied across ends of the crystal causes dimensional changes.

Def: *The development of charges as a result of the mechanical deformation is known as the **direct piezoelectric effect**.*

Def: The mechanical deformation of piezoelectric materials caused by an external electric field is known as the ***inverse piezoelectric effect***.

Piezoelectric effect is exhibited by a crystal only if the crystalline symmetry is non-centrosymmetric. The crystals which do not have center of symmetry are piezoelectric.

Ammonium phosphate, quartz and PZT (lead zirconate titanate) are examples of piezoelectric materials.

In ordinary solids, a stress causes a proportional strain '*s*' related by an elastic modulus. Piezoelectricity is the additional creation of the electric charge by an applied stress. The induced polarization P, in *direct piezoelectric effect*, is directly proportional to the applied mechanical stress, σ. Thus,

$$P = d\sigma \qquad \text{...(40.77)}$$

where d is the proportionality constant and is known as **piezoelectric coefficient** and is expressed in Coulombs/Newton.

Piezoelectric Coefficient may be defined as the charge developed per unit force. A change in sign of σ, reverses the sign of polarization. The value of d should be high for practical applications.

In the *inverse piezoelectric effect*, an electric field **E** produces a proportional strain, s.

Thus,

$$s = d\mathbf{E} \qquad \text{...(40.78)}$$

In practice, ***electromechanical coupling factor*** k is used to describe the piezoelectric effect in actual piezoelectric elements. Energy can be given to a piezoelectric element either mechanically by stressing it or electrically by charging it. All the energy given to it is not converted in producing the effect. Therefore, the piezoelectrics are characterized by *strength of piezoelectric effect*.

This strength is measured by the electromechanical coupling factor, k. In case of direct piezoelectric effect, the external force is expended not only on the deformation of the element but also on its polarization. The square of the piezoelectric coupling factor is defined as the ratio of the electrical energy generated by the piezoelectric element to the total energy expended on the deformation. Thus,

$$k^2 = \frac{\text{Mechanical energy converted to electrical energy}}{\text{Total input mechanical energy}} \qquad \text{...(40.79)}$$

In case of inverse piezoelectric effect, the external voltage is expended not only on charging the element but also on its deformation. The square of the piezoelectric coupling factor is defined as the ratio of the electrical energy generated by the piezoelectric element to the total energy expended on the deformation. Thus,

$$k^2 = \frac{\text{Electrical energy converted to mechanical energy}}{\text{Total input electrical energy}} \qquad \text{...(40.80)}$$

The numerical values of k obtained from the direct and inverse piezoelectric effects are found to coincide.

Quartz Crystal: Among piezoelectric crystals, quartz crystal is the most popular piezoelectric. It is the most common variety of silica, SiO_2. The natural quartz crystal has the shape of a hexagonal prism with a pyramid attached to each end. The axis along the longest dimension of the natural crystal is called **optic axis** or **z-axis**, see Fig. 40.34 (a). The three lines, which pass through the opposite corners of the crystal, constitute its three **x-axes** or *electrical axes*. Similarly, the three lines, which are perpendicular to the sides of the hexagon, form the three **y-axes**, which are known as **mechanical axes**. The arrangement of atoms in the crystal is shown in Fig. 40.34 (b). In the absence of the external stress, all the charges in a unit cell are balanced and the net polarization is zero. When, an external stress is applied to the crystal the balance is disturbed and the crystal gets polarized. The charge developed per unit force is the piezoelectric coefficient d. The d-coefficient for quartz is 2.3×10^{12} C/N at 550°C.

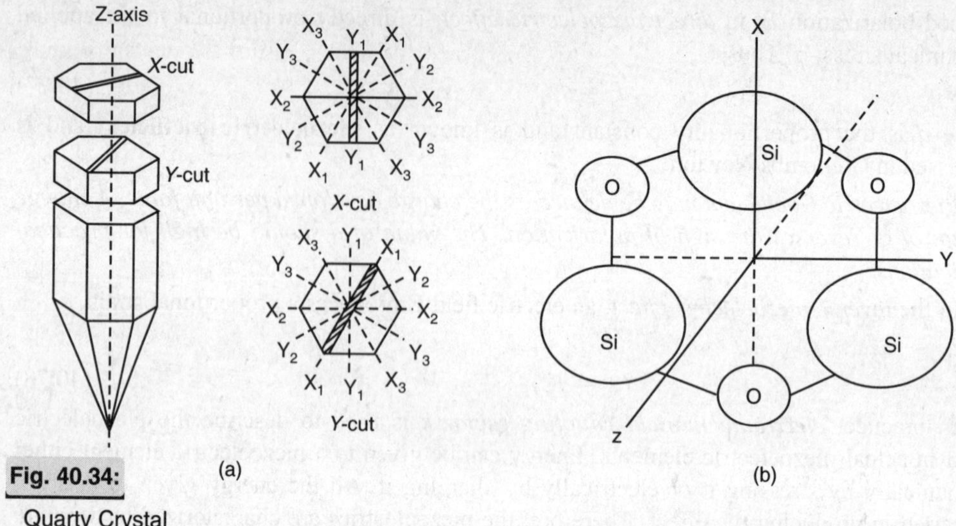

Fig. 40.34: Quarty Crystal

40.25 PYROELECTRICITY

The term Pyroelectricity was originated from Greek word Pyro- meaning fire and the term electricity. Thus, pyroelectricity means heat-generated electricity. ***Pyroelectricity*** or the ***pyroelectric effect*** is the polarization of a dielectric subjected to uniform heating or cooling. Fig. 40.35 shows pyroelectric effect in pyroelectric crystals.

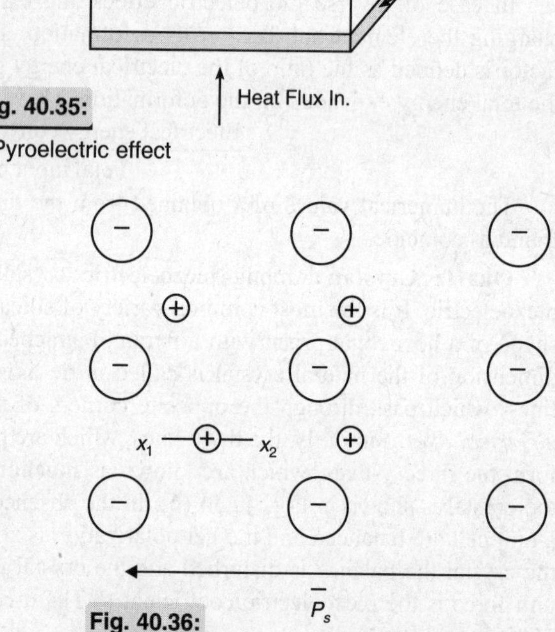

Fig. 40.35: Pyroelectric effect

Fig. 40.36: Schematic of two-dimensional array of ions in the lattice

Minor changes occur in the position of atoms within the crystals due to the variation in the temperature and the polarization of the crystal changes. This polarization change produces a voltage across the crystal. The voltage that develops across the crystal is not stable and gradually disappears due to leakage of current.

Fig. 40.36 shows a two-dimensional lattice of cations and anions.

It is seen that the cations are displaced relative to the 'centre of gravity' of unit cells and produce spontaneous polarization P_s along the line (x_1-x_2). The potential energy of any cation along this line will be asymmetric. Any excitation caused by an increase in lattice temperature changes its quantized energy level and leads to a change in its mean equilibrium position in the lattice. This gives a change in the overall electrical dipole moment, which appears as the *macroscopic pyroelectric effect*.

Quantitatively, the pyroelectric effect could be described by the equation

$$\Delta P_s = p_g \Delta T$$

where ΔT denotes the change in temperature,

ΔP_s is the change in the polarization and

p_g is the pyroelectric coefficient.

Spontaneous polarization *is the dielectric polarization, which occurs under the action of the internal processes and without the application of an electric field.*

Pyroelectric effect is the change in spontaneous polarization in certain anisotropic solids when the temperature of the material is changed. The pyroelectric properties of a material are described by a pyroelectric coefficient.

Pyroelectric coefficient *is defined as the change in polarization per unit temperature change of the material.*

Thus,

$$p_g = \frac{dp}{dT} \quad \text{...(40.81)}$$

The largest pyroelectric effects are observed in ferroelectrics.

Applications of Pyroelectrics: The most important pyroelectric materials are ceramics, synthetic polymers, and ceramic-polymer composites. Lead zirconate (PZ)-based ceramics, such as PZT and PLZT, have a high value of pyroelectric coefficient and also of dielectric constant. Pyroelectric materials are used as thermal detectors and calorimeters. Some applications of pyroelectric devices are discussed here.

(*i*) **Intruder alarms:** The thermal scene of the interior of an unoccupied building is fairly constant.When a moving intruder enters the scene he would produce variable infrared radiation of about 10 μm in wavelength which on reaching a pyroelectric detector in the premises, will trigger alarm system. The detector is insensitive to visible radiation.

(*ii*) **Fire alarms:** It operates at shorter wavelengths than for intruder alarms, typically around **4** pm. The systems usually sense the flicker frequency of the flames, at **5-40** Hz to avoid false alarms.

(*iii*) **Radiometers:** A pyroelectric detector can be used for radiometry by equipping it with a lens to define the field of view and a chopper to modulate the radiation. The magnitude of the pyroelectric signal can be used to measure the temperature of a target.

QUESTIONS

1. Explain the behaviour of dielectrics under static electric fields. Derive a relation between polarization *P*, the external electric field *E* and displacement vector *D*.
2. What do you understand by dielectric constant? Define dielectric susceptibility. Derive a relation between dielectric constant ε_r and dielectric susceptibility.
3. With usual notations show that $P = \varepsilon_0 (\varepsilon_r - 1)E$.
4. Explain dielectric polarization.

5. What is meant by dipole moment?
6. What are non-polar dielectrics?
7. What are polar dielectrics?
8. Distinguish between electronic, ionic and orientation polarization and discuss the effect of temperature on each of them.
9. Explain the behaviour of dielectrics under static electric fields. Derive the relation between the polarization P and the external field E.
10. Explain Gauss's law for dielectrics and derive relationship between **D**, **E** and **P** vectors.
11. Define dielectric susceptibility and polarizability of a dielectric. Derive the relation connecting the two.
12. Discuss the dependence of electronic polarizability on the frequency of the applied field.
13. Explain electronic polarizability and show that electronic polarizability for a monoatomic gas increases as the size of the atoms becomes larger.
14. Explain ionic polarizability and derive an expression for ionic polarizability.
15. Define dipole moment and classify dielectric materials on its basis.
16. Explain briefly the various types of polarization in dielectrics.
17. Explain what is meant by a permanent dipole moment. Obtain an expression for orientation polarization.
18. Write one difference between polar and non-polar dielectrics.
19. Discuss the frequency dependence of various polarization processes in dielectric materials.
20. Obtain an expression for the internal field seen by an atom in an infinite array of atoms subjected to an external field.
21. Explain the meaning of internal field in solids. Incorporating internal field in the expression for polarization, derive Clausius-Mosotti relation for elemental solid dielectrics.
22. What is meant by local field in a dielectric and how it is calculated for a cubic structure? Deduce the Clausius-Mosotti relation.
23. Explain the concept of internal field in solids and hence obtain an expression for the static dielectric constant of elemental solid dielectric.
24. Derive an expression for internal field in case of liquids and solids.
25. What do you mean by internal field? Derive Clausius-Mosotti relationship for cubic solids.
26. Write short notes on dielectric loss.
27. Explain loss tangent and loss factor.
28. What are the causes behind the dielectric losses occurring in r.f., infrared and visible regions of the electromagnetic spectrum?
29. Discuss in detail the various dielectric breakdown mechanisms.
30. What is meant by polarization catastrophe?
31. What is meant by spontaneous polarization?
32. Explain the electromechanical coupling factor.
33. Explain the phenomenon of ferroelectricity with particular reference to barium titanate.
34. What are the important characteristics of ferroelectric materials?
35. How does the dielectric constant of a ferroelectric vary with temperature? Mention some of the uses of ferroelectric materials.
36. Describe the ionic displacement theory and show how it explains the ferroelectric nature of barium titanate.
37. Explain the important requirements of insulators.
38. Explain the characteristics and function of transformer oil in transformers.
39. What is electrical poling? Why is it done?
40. Discuss the applications of piezoelectric and ferroelectric materials.

PROBLEMS

1. A parallel plate capacitor, having an area 6.45×10^{-4} m^2 and plate separation of 2×10^{-3} m, and across the plates a potential of 12 V is applied. If a material having a dielectric constant 5 is placed within the region between the plates, calculate the polarization. **[Ans:** 5.8×10^{-11} F.m^2]
2. Carbon tetra chloride contains 74 electrons in its molecule. Its relative permittivity is 2.26 when its density is 1.68×10^3 kg/m^3. If the field acting on the liquid is 5×10^6 V/m, what is its electronic polarizability and average electron displacement. **[Ans:** 8.57×10^{-33}C.m; 0.72×10^{-15}m]
3. A monoatomic gas contains 3×10^{25} atoms/m^3 at a certain temperature at one atmosphere pressure; the radius of the atom is 0.19 nm. What is the relative permittivity of the gas at the g_i.
4. Ven pressure and temperature? What is the polarizability of the atom? **[Ans:** 1.0026; 7.66×10^{-40} F.m]
5. A water molecule has a dipole moment of 6.2×10^{-30}C.m. What is the polarization of a water drop of 0.1 cm radius polarized in the same direction? **[Ans:** 26.4×10^{-10} C/m^2]
6. The centers of two identical atoms of polarizability $\alpha = 2 \times 10^{-40}$ F.m^2 are separated by a distance of 5×10^{-10}m. A uniform electric field is applied in a direction parallel to the line joining the centres of the two atoms. Calculate the ratio between the internal field, E_i and E.
7. An elemental dielectric material has $\varepsilon_r = 12$ and it contains 5×10^{28} atoms/m^3. Calculate its electronic polarizability assuming Lorentz field.
8. Find the total polarizability of CO_2, if its susceptibility is 0.985×10^{-3} and density is 1.977 kg/m^3. **[Ans:** 3.24×10^{-40} F.m^2]
9. A solid elemental dielectric having density of 3×10^{28} atoms/m^3 shows an electronic polarizability of 10^{-40} F.m^2. Assuming the internal electric field to be a Lorentz field, find the dielectric constant of the material. **[Ans:** 1.339]
10. A parallel plate capacitor of area 650 mm^2 and a plate separation of 4 mm has a charge of 2×10^{-10} C on it. When a material of dielectric constant 3.5 is introduced between the plates, what is the resultant voltage across the capacitor? **[Ans:** 13.9V]
11. The relative permittivity of sulphur is 4. Calculate its atomic polarizability. Given that sulphur is in cubic form and has a density of 2.08×10^3 kg/m^3 and atomic weight of 32. **[Ans:** 3.39×10^{-40} F.m^2]
12. Three identical atoms in a string are subjected to a uniform electric field E V/m along the line joining their centers. If the polarizability is 2.5×10^{-40} F.m^2 and the center to center spacing is 0.3 nm, find (E_i/E) at the position of the center of the atom.
13. Carbon tetrachloride contains 74 electrons in its molecule. Its relative permittivity is 2.26 when its density is 1.68×10^3 kg/m^3. If the field acting on the liquid is 5×10^6 V/m, what is its electronic polarizability and average electronic displacement? **[Ans:** 8.57×10^{-33} F.m^2, 7.2×10^{-16} m]
14. A water molecule has a dipole moment of 6.2×10^{-30} C.m. What is the polarization of a water drop of 0.1 cm radius polarized in the same direction? **[Ans:** 2.64×10^{-9} C/m^2]

41 Magnetic Materials

Learning Objectives:

By the end of this chapter, you will be able to:

- Understand basic concepts of magnetic moment.
- Learn about different types of magnetic materials.
- Describe basic principle of Soft and Hard Magnetic materials.
- Learn about different magnetic devices.

41.1 INTRODUCTION

The development of quantum physics helped us understand the phenomenon of magnetism to a great extent. Magnetic materials play a prominent role in modern technology. They are widely used in industrial electronics, entertainment electronics and computer industry. The methods of information storage and retrieval are based on magnetic storage techniques. Substances which when placed into an external magnetic field becomes sources of an additional magnetic field are known as magnetic materials. Michael Faraday classified the magnetic materials into three groups, namely diamagnetic, paramagnetic and ferromagnetic materials. The diamagnetic and paramagnetic materials are weakly magnetic and at any temperature they interact weakly with a magnetic field. Ferromagnetic materials interact strongly with a field at definite temperatures.

The development of quantum physics helped us understand the phenomenon of magnetism to a great extent. A large number of devices utilize mainly two magnetic phenomena, ferromagnetism and ferrimagnetism. The magnetic materials are classified into soft and hard materials. Soft magnetic materials are easily magnetized and demagnetized and are therefore used in ac applications. Hard magnetic materials retain magnetism on a permanent basis and are used in producing permanent magnets. A basic understanding of the magnetic phenomena is essential to appreciate the operating principles of the various magnetic devices.

41.2 TERMS AND DEFINITIONS

When a solid is placed in magnetic field, it gets magnetized. The following terms are of great assistance in understanding the behaviour of materials in the presence of magnetic field.

(i) Magnetic field, H: The magnetic field in which a material is kept is called *magnetizing field*. The strength (or intensity) of the magnetic field is denoted by H. The units of **H** are ampere-turns per metre (A/m) in SI system. Magnetic field is produced by permanent magnets such as a horse shoe magnet and temporarily by electromagnets or superconductor-magnets.

(ii) Magnetization, M: *Magnetization* is defined as the magnetic moment per unit volume developed inside a solid and is denoted by **M**. In SI system,magnetization is measured in amperes per metre (A/m).

Since the magnetization is induced by the field, we may assume that **M** is proportional to **H**.

Thus, $$\mathbf{M} \propto \mathbf{H}$$

or $$M = \chi H \qquad \text{...(41.1)}$$

where χ is the proportionality constant and is known as *magnetic susceptibility*.

(iii) Magnetic Susceptibility, χ: The *magnetic susceptibility* of a material is a measure of the ease with which the material can be magnetized.

Thus, $$\chi = \frac{\mathbf{M}}{\mathbf{H}} \qquad \text{...(41.2)}$$

In general, the vectors **M** and **H** can have different directions and χ is a tensor. However, in isotropic media, **M** and **H** point in the same direction and χ is a scalar quantity. Materials having high susceptibility are easily magnetized.

(iv) Magnetic Induction, B:

A magnetic field is schematically represented by lines of magnetic induction. It is described either by magnetic field strength **H** or by the *magnetic induction* (or *magnetic flux density*), **B**.

The lines of induction are collectively called ***flux***. The number of field lines passing through a unit area of cross-section is called the ***magnetic flux density***.

$$\mathbf{B} = \frac{\text{Magnetic flux}}{\text{area}} = \frac{\varphi}{\mathbf{A}} \qquad \text{...(41.3)}$$

The quantity B is measured in weber per square metre (Wb/m^2) or tesla (T). The cgs unit for magnetic induction is the gauss (G).

$$1\ G = 10^{-4}\ T$$

Relationship between B and H: When a material is kept in a magnetic field, two types of induction arise; one due to the magnetizing field, **H** and the other as a consequence of the magnetization, **M** of the material itself.

The magnetic induction, **B**, produced inside the material is given by

$$\mathbf{B} = \mu_0 (\mathbf{H} + \mathbf{M}) \qquad \text{...(41.4)}$$

where μ_0 is known as the *permeability of the free space*. It is equal to $4\pi \times 10^{-7}$ H/m.

Using Eq. (41.1) in Eq. (41.4), we get

$$\mathbf{B} = \mu_0 (1 + \chi)\mathbf{H} \qquad \text{...(41.5)}$$

or $$\mathbf{B} = \mu\, \mathbf{H} \qquad \text{...(41.6)}$$

where μ is called the *absolute permeability* of the medium.

Like χ, μ is in general a tensor. In isotropic medium, it is a scalar quantity. In case of free space, M = 0 and Eq. (41.4) reduces to

$$\mathbf{B} = \mu_0\, \mathbf{H} \qquad \text{...(41.7)}$$

(v) Absolute Permeability, μ: When a magnetic material is placed in a magnetic field, the magnetic field lines are redistributed and tend to pass more (or less in some cases) through the material. The *absolute permeability* of the material is a measure of the degree of which the field lines penetrate or permeate the material.

Absolute permeability is defined as the ratio of the magnetic induction, **B**, in the medium to the magnetizing field, H.

$$\mu = \frac{\mathbf{B}}{\mathbf{H}} \qquad ...(41.8)$$

The unit of absolute permeability is henry per metre (H/m).

(vi) Relative Permeability, μ_r: *The relative permeability* of a material is defined as the ratio of the absolute permeability of that material to the permeability of free space. That is,

$$\mu_r = \frac{\mu}{\mu_0} \qquad ...(41.9)$$

μ_r is only a number and has no units. Its value for air or vacuum is equal to unity.

Thus, $\mu_r = 1$

41.3 RELATION BETWEEN μ_r AND χ

Comparing Eq. (41.5) and Eq. (41.6), we find that

$$\mu = \mu_0(1 + \chi) \qquad ...(41.10)$$

Using Eq. (41.9) into the above equation, we obtain

$$\mu_r = (1 + \chi) \qquad ...(41.11)$$

Eq. (41.11) relates the permeability to susceptibility of the material.

Example 41.1: *A magnetic material has a magnetization of 2300 A/m and produces a flux density of 0.00314 Wb/m^2. Calculate magnetizing force and relative permeability of the material.*

Solution. Magnetizing force, $H = \dfrac{B}{\mu_0} - M = \dfrac{0.00314 \text{ Wb/m}^2}{12.57 \times 10^{-7} \text{ H/m}} - 2300 \text{ A/m} = \mathbf{198 \text{ A/m}}$.

Relative Permittivity, $\mu_r = \dfrac{B}{\mu_0 H} = \dfrac{0.00314 \text{ Wb/m}^2}{(12.57 \times 10^{-7} \text{ H/m})(198 \text{A/m})} = \mathbf{12.56}.$

41.4 ORIGIN OF MAGNETIZATION-MAGNETIC MOMENT

The magnetic properties of solids arise due to electrons undergoing different motions in the atoms, which give rise to magnetic dipole moments. These magnetic dipole moments are responsible for the magnetic properties of materials. In general, the magnetic dipole moment of the atom arises from three sources:

(i) The orbital motion of electrons: The atom of any material consists of a central nucleus and the electrons move around the nucleus in specific orbits. Each electron orbit is equivalent to a tiny current loop and behaves as an elementary magnet having a magnetic dipole moment. The total orbital magnetic moment of an atom is the sum of orbital magnetic moments of individual electrons.

(ii) The electron spin: Each electron is spinning about an axis through itself and this spin also gives rise to a magnetic dipole moment.

(iii) The nuclear spin: In addition to electronic contribution, nuclear spin also contributes to magnetic moment of atoms. The magnetic moment due to nuclear spin is not taken into account in understanding the magnetic properties of solids as the magnetic moment of the nucleus is very small, nearly 1/2000 of the magnetic moment of electron. We consider that the resulting magnetic moment of an atom is the sum of the orbital and spin magnetic moments of its electrons. The main contribution to atomic magnetic moment arises from the spin of unpaired valence electrons. Normally, the magnetic moments carried by atoms align in different directions to produce a net non-zero magnetic moment. When the substance is placed in a magnetic field, the atoms are aligned with their directions of magnetic moment along the direction of the magnetic field. Thus, the material is magnetized.

41.4.1 Bohr Magneton

***Bohr magneton** is defined as the elementary electron magnetic moment. It is elementary in the sense that no electron can have a magnetic moment below it. It is the natural unit for the measurement of atomic magnetic moments.*

It is denoted by μ_B. It has a value

$$\mu_B = \frac{eh}{4\pi m} = 9.28 \times 10^{-24} \text{A.m}^2.$$

41.5 CLASSIFICATION OF MAGNETIC MATERIALS

Solids are classified into three groups basing on the magnitude and sign of relative permeability, μ_r exhibited by them, Diamagnetic materials ($\mu_r < 1$), Paramagnetic materials ($\mu_r > 1$), Ferromagnetic materials ($\mu_r >> 1$). However, the solids can be divided into two broad groups,on the basis of *magnetic dipole moments*. The atoms of one group do not possess permanent magnetic dipole moment whereas atoms of the other group possess permanent magnetic dipole moment.

The materials consisting of atoms with zero magnetic dipole moment are called as ***Diamagnetic materials***. Materials composed of atoms or molecules having permanent magnetic moment are classified into four categories depending on the interaction between the atomic magnetic dipoles. If the interaction between the atomic magnetic dipoles is negligible, the material is ***paramagnetic***. If the magnetic dipoles interact in such a way that they tend to orient in the same direction, the material is ***ferromagnetic***. If neighbouring dipoles orient in opposite directions and if the dipoles are of equal magnitude, the material is ***antiferromagnetic***. If the neighbouring dipoles are of different magnitude and orient antiparallel, the material is ***ferrimagnetic***. The above four kinds of orientations of atomic dipoles are illustrated in Fig. 41.1.

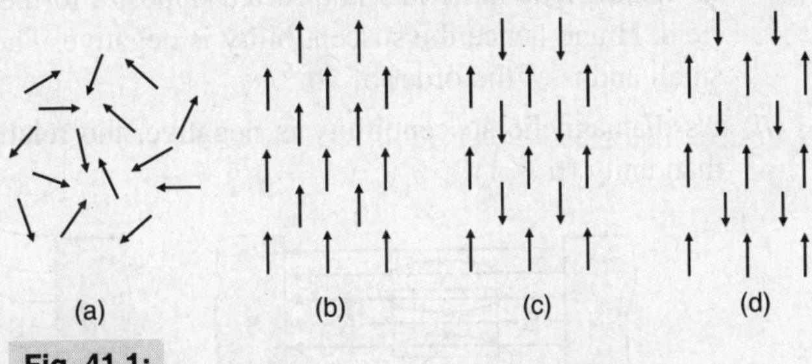

Fig. 41.1: Schematic illustration of the orientation of spins in (a) paramagnetic , (b) ferromagnetic, (c) antiferromagnetic and (d) ferrimagnetic materials

41.6 DIAMAGNETISM

Let us consider materials whose atoms consist of an even number of electrons. In such atoms, the electrons get paired up. The electrons in each pair have orbital motions as well as spin motions occurring in opposite sense. The resulting magnetic moments cancel each other and hence the total magnetic moment of the atom is zero. Such materials exhibit ***diamagnetism***. When a diamagnetic material is placed in an external magnetic field, alignment of magnetic dipoles does not occur, as there are no dipoles in the material.

However, the external magnetic field modifies the motion of electrons in the orbits. Following Lenz's law one of the electrons in each pair is accelerated while the other is decelerated. Hence, now each electron pair gives rise to a resultant magnetic moment. Consequently, an effective magnetic moment is *induced* in the atom as a whole in a direction opposite to that of the external magnetic field (Fig. 41.2). As a result the material becomes magnetized.

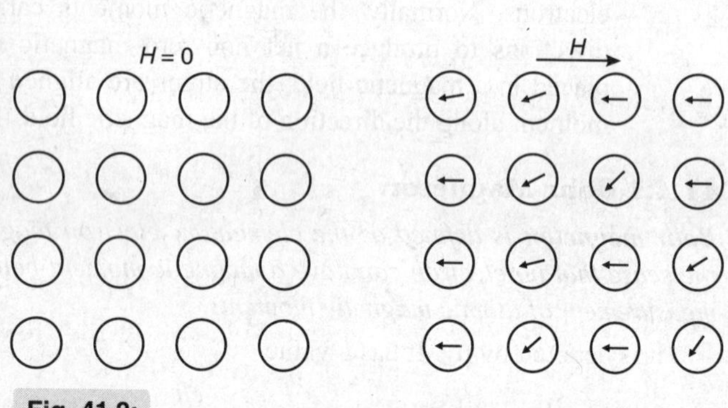

Fig. 41.2:

(a) In the absence of an external field, dipoles do not exist.
(b) In the presence of external field, dipoles are induced opposite to the field direction.

The strength of the induced magnetic moment is proportional to the applied field and hence the magnetization of the material varies linearly with the strength of the magnetic field. The induced dipoles and magnetization vanish as soon as the applied field is switched off. The induced dipoles and their orientation in external field are not at all influenced by the temperature, with the result that the diamagnetic susceptibility does not depend on the temperature.

Characteristics: The salient features of diamagnetic materials are as follows:

(*i*) Diamagnetic materials exhibit *negative* magnetic susceptibility. The magnetization **M** in diamagnetic materials is directed opposite to the direction of the applied magnetic field, **H** and hence the susceptibility is negative. The absolute value of susceptibility is small and is of the order of 10^{-6}.

(*ii*) As diamagnetic susceptibility is negative, the relative permeability μ_r is slightly less than unity ($\mu_r < 1$).

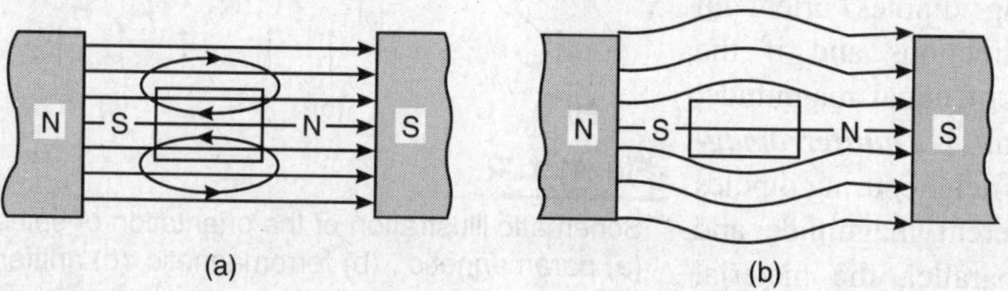

Fig. 41.3:

(a) Magnetization is induced in a direction opposite to that of external field in a diamagnetics.
(b) Magnetic field lines are pushed away from the diamagnetic material

(*iii*) Diamagnetic materials are substances that are repelled by a magnetic field. If a diamagnetic body is placed in an inhomogeneous magnetic field, it tends to be pushed into the regions of weaker field. When the material is taken in the form of a small rod and is freely suspended in the magnetic field, it turns to a position perpendicular to the field lines. Diamagnetic materials push aside the field lines, as illustrated in Fig. 41.3 (b).

(*iv*) The magnetic susceptibility of diamagnetic materials is practically independent of temperature. Pierre Currie discovered the independence of diamagnetic susceptibility on temperature in 1895.

(*v*) The magnetization, M is a linear function of magnetic field H, as illustrated in Fig. 41.4.

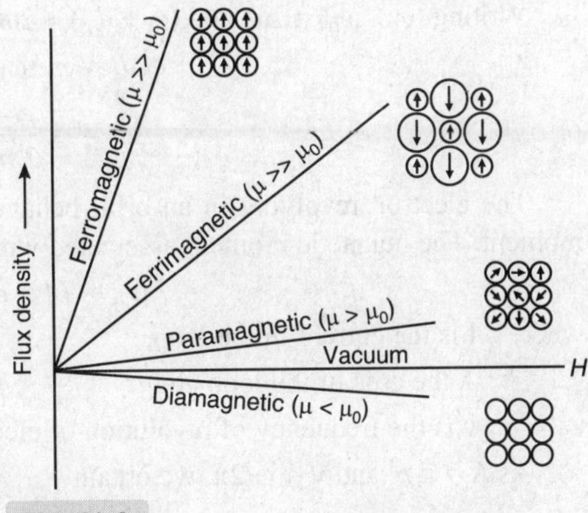

Fig. 41.4: *M* is a linear function of magnetic field *H*

(*vi*) In general, diamagnetic materials do not have significant engineering applications. There is one special group known as *superconducting materials*, which are perfect diamagnetic materials. Their susceptibility is given by

$$\chi_{dia} = \frac{M}{H} = -1 \qquad ...(41.12)$$

which is very large. This strong diamagnetism finds application in making frictionless bearings leading to a levitation effect.

(*vii*) Hydrogen, air, water, gold, silver, and bismuth are examples. Inert gases, a majority of metals, and many organic compounds are diamagnetic substances. As induced dipole effect is present in all materials, it is *a universal property of matter*.

41.6.1 Langevin's Theory of Diamagnetism

In 1905 Langevin explained diamagnetism considering orbital motion of electrons in the atoms. Let us consider an electron revolving in a stationary orbit around the nucleus in an atom. The electrostatic force F_C exerted by the nucleus is the centripetal force acting on the electron. Thus,

$$F_C = m\omega_0^2 r$$

where ω_0 is the angular velocity of the electron in its orbit.

Now, let an external magnetic field **B** be applied such that it acts perpendicular to the plane of the orbit. It produces an additional force F_m, which will also be centripetal force acting on the electron. The net centripetal force on the electron is

$$F_C \pm F_m$$

depending upon whether the force F_m acts radially inward or outward. Then,

$$F_C \pm F_m = m\omega^2 r \qquad ...(41.13)$$

where ω is the new angular velocity.

Here, we assume that the magnetic field will change only the velocity of the electron but not radius of its orbit. We know that

$$F_m = e\upsilon B = e\omega r B$$

Therefore, we can rewrite Eq. (41.13) as

$$m\omega_0{}^2 r \pm e\omega r B = m\omega^2 r$$

or

$$m\omega_0{}^2 \pm e\omega B = m\omega^2$$

or

$$\pm e\omega B = m(\omega^2 - \omega_0{}^2) = m(\omega - \omega_0)(\omega + \omega_0)$$

Writing $(\omega - \omega_0) = \Delta\omega$ and $(\omega + \omega_0) \approx 2\omega$ into the above equation, we get

$$\pm e\omega B = m\Delta\omega(2\omega)$$

or

$$\Delta\omega = \pm \frac{eB}{2m} \qquad \text{...(41.14)}$$

The electron revolving in an orbit behaves as a current loop and possesses a magnetic moment. The magnetic moment associated with a current loop is given by

$$\mu_m = IA = e\nu A$$

where I is the current in the loop,
A the area of the loop and
ν is the frequency of revolution of electron in the orbit.

As A = πr^2 and $\nu = \omega/2\pi$, we obtain

$$\mu_m = e\left(\frac{\omega}{2\pi}\right)\pi r^2 = \frac{1}{2} e\omega r^2$$

A change in angular velocity produces a change in magnetic moment and using Eq. (41.14), the magnitude of ***induced magnetic moment*** is given by

$$\Delta\mu_m = \frac{1}{2} e\, r^2\, \Delta\omega = \frac{1}{2} er^2\left(\frac{eB}{2m}\right)$$

or

$$\Delta\mu_m = \frac{e^2 r^2 B}{4m} \qquad \text{...(41.15)}$$

If there are i electrons in an atom, then the total induced magnetic moment is

$$\sum_i \Delta\mu_m = \frac{e^2 \sum_i r_i^2}{4m} B$$

If there are N atoms per unit volume of the material, then the magnitude of the magnetization, is given by

$$M = N \sum_i \Delta\mu_m = \frac{Ne^2 \sum_i r_i^2}{4m} B = \frac{Ne^2 \sum_i r_i^2}{4m} \mu_0 H$$

The vectors μ_m and **H** are opposite in direction (due to Lenz's law). Therefore, the above equation in vector form is written as

$$\mathbf{M} = -\frac{\mu_0 Ne^2 \sum_i r_i^2}{4m} \mathbf{H}$$

The diamagnetic susceptibility is thus,

$$\chi_{\text{dia}} = \frac{\mathbf{M}}{\mathbf{H}} = -\frac{\mu_0 N e^2 \sum_i r_i^2}{4m}$$

As all the i electron orbits may not be perpendicular to the applied magnetic field, the right hand side of the above equation is multiplied by a factor of (2/3). Thus, the diamagnetic susceptibility is given by

$$\chi_{dia} = -\frac{\mu_0 N e^2 \sum_i r_i^2}{6m} \quad ...(41.16)$$

It is seen from the above Eq. (41.16) that diamagnetic susceptibility is

(*i*) negative

(*ii*) independent of temperature

(*iii*) directly proportional to the atomic number

(*iv*) larger, if the atom is bigger.

In fact, the induced dipole effect is present in all materials. Therefore, *diamagnetism is a universal property of matter* and all substances have a diamagnetic contribution to their susceptibility. However, in a majority of the materials, diamagnetism is overshadowed by other magnetic phenomenon.

Example 41.2: *Diamagnetic Al_2O_3 is subjected an external magnetic field of 10^5 A/m. Evaluate magnetization and magnetic flux density in Al_2O_3. (Susceptibility of Al_2O_3 = -5×10^{-5}).*

Solution: Magnetization $M = \chi H = (-5 \times 10^{-5})(10^5 \text{ A/m}) = \mathbf{-5 \text{ A/m}}$

Magnetic flux density, $B = \mu_0(H + M) = (12.57 \times 10^{-7} \text{H/m})(10^5 \text{A/m} - 5 \text{ A/m})$

$= \mathbf{0.126 \text{ wb/m}^2}$.

41.7 PARAMAGNETISM

Molecules of paramagnetic materials possess a net permanent magnetic moment even in the absence of an external magnetic field. The magnetic moments are randomly oriented in the absence of an external magnetic field. Therefore, the net magnetization of the material is equal to zero (Fig. 41.5 a). When the material is subjected to the influence of magnetic field, the magnetic dipoles tend to align in the direction of the field (Fig. 41.5 b) and the material becomes magnetized.

H = 0

H

Fig. 41.5: (a) Atomic dipoles are distributed randomly in the absence of an external field. (b) In the presence of external field, dipoles tend to align along the field direction.

The thermal agitation tends to counteract the orientation of dipoles. When the temperature is increased, the thermal agitation increases and the alignment of dipoles become more and more difficult. Therefore, the magnetization and hence susceptibility of paramagnetic materials decrease with an increase in the temperature.

Characteristics: The salient features of paramagnetic materials are as follows:

(*i*) Paramagnetic materials exhibit positive magnetic susceptibility as the magnetization coincides in direction with the magnetic field **H**. The susceptibility is of the order of 10^{-6}.

(*ii*) The relative permeability μ_r is slightly more than unity ($\mu_r > 1$) for paramagnetic materials. Field lines are pulled towards the material and permeate through it when it is placed in a magnetic field, as illustrated in Fig. 41.6.

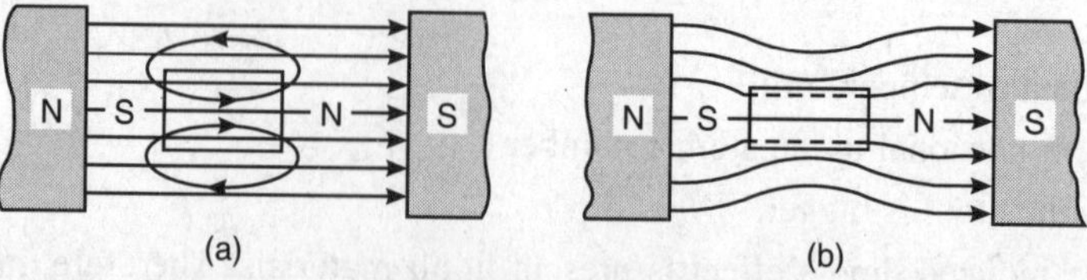

Fig. 41.6:
(a) Paramagnetic material gets magnetized in the direction of the field.
(b) Magnetic field lines are pulled towards the paramagnetic material.

(*iii*) When a rod of paramagnetic material is freely suspended in a magnetic field, it aligns itself along the lines of induction. In a non-uniform field, the paramagnetic substances are attracted towards stronger region of magnetic field. If a paramagnetic liquid is taken in a U-shaped glass tube and one leg is placed in a magnetic field, the level of the liquid rises in this leg.

(*iv*) The magnetization, **M** is a linear function of the magnetic field **H**, when the field is not too strong, as shown in Fig. 41.4.

(*v*) The paramagnetic susceptibility is strongly dependent on temperature. Pierre Curie discovered in 1895 that the susceptibility of a paramagnetic substance varies inversely with the temperature. The magnetization and susceptibility of paramagnetic materials decrease with an increase in the temperature.

Thus,

$$\chi_{\text{para}} = \frac{C}{T} \quad \text{Curie Law} \qquad \text{...(41.17)}$$

where C is a constant known as Curie constant.

Eq. (41.17) is called **Curie's Law**. The temperature dependence of the paramagnetic susceptibility is conveniently illustrated by $1/\chi_{para}$ plotted as a function of T, as in Fig. 41.7.

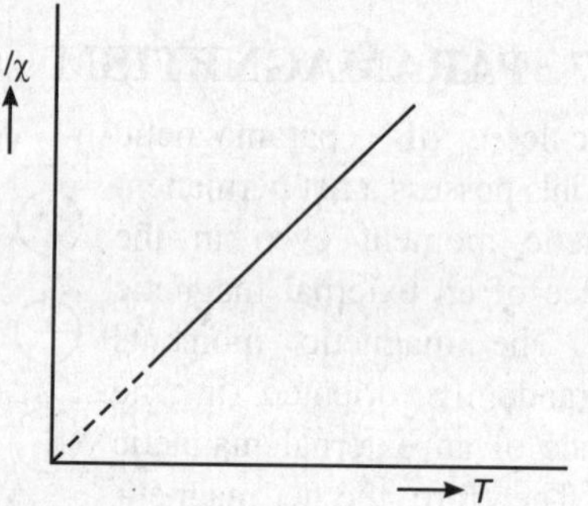

Fig. 41.7: Illustration ofthe temperature dependence of the paramagnetic susceptibility

(*vi*) Oxygen, solutions of iron salts, copper chloride, chromium and platinum are examples of paramagnetic materials.

41.7.1 Langevin's Theory of Paramagnetism

The French physicist, Paul Langevin in 1905, developed the classical theory of paramagnetism. The Langevin theory attempts to explain the experimental observations of Curie and Weiss. It is considered in the theory that the magnetic moments arising from the orbital motion of electrons causes paramagnetic behaviour and that a paramagnetic material is conveniently represented by "a gas of magnetic needles". The word gas denotes the fact that the interaction between magnetic moments is neglected.

In the absence of an external magnetic field, the magnetic axes of molecules are uniformly distributed in all directions, and the sum of projections of magnetic moments in any chosen direction is zero. Therefore, there is no net magnetization at the macroscopic level.

The potential energy corresponding to a magnetic dipole in a magnetic field of induction **B** is given by

$$W = -\mu_m \cdot \mathbf{B} = \mu_m \, \mathrm{B} \cos\theta \qquad \text{...(41.18)}$$

where θ is the angle between the magnetic field and the magnetic dipole.

The potential energy attains its minimum value when the direction of dipole orientation coincides with the direction of the magnetic field. Since a system is stable when its potential energy is minimum, the magnetic dipole moments tend to orient along the external magnetic field direction.

Therefore, when a paramagnetic material is subjected to a magnetic field, the action of the field turns the dipoles so that they are aligned with the field. Thermal agitation however counteracts the ordering action of the magnetic field. The degree of orientation of magnetic dipoles in a magnetic field is therefore determined both by the field strength and the temperature of the substance.

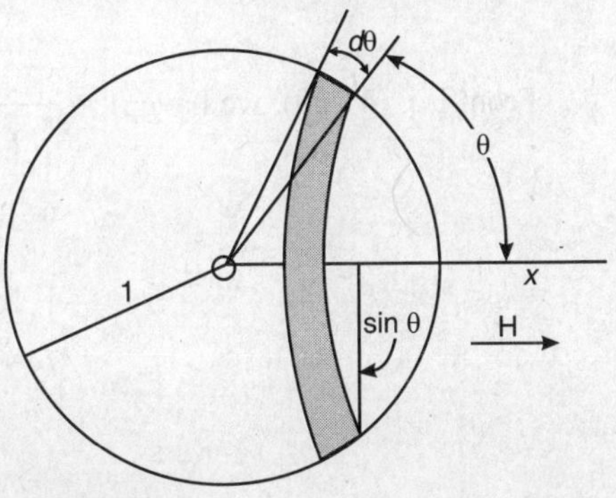

Let us consider a system of N dipoles each carrying a dipole moment μ_m. They are randomly oriented in all directions in the absence of an external magnetic field. Therefore, all directions of orientation are equally probable as a result of which the net magnetic moment of a unit volume of material is zero.

Fig. 41.8: Schematic representation of a unit sphere along whose radii atomic dipoles are imagnied to be distributed.

Let us assume that the system is in thermal equilibrium at a temperature T. At any instant, the number of magnetic moments $N(\theta)\, d\theta$ that lie within a solid angle dΩ is proportional to the solid angle, which is equal to $2\pi \sin\theta . d\theta$. (Fig. 41.8).

Thus, $\quad N(\theta)d\theta \propto 2\,\pi \sin\theta . d\theta.$

Let us now suppose that the gas of magnetic moments is subjected to a magnetic field of induction **B**. The magnetic field causes orientation of dipoles along its direction.

According to Maxwell-Boltzmann statistics, the number of dipoles whose potential energy is W, is proportional to the Boltzmann factor $e^{-W/kT}$. Therefore,

$$\left.\begin{array}{l}\text{The number of dipoles } dN \\ \text{which have potential energy} \\ W \text{ and lying within the solid} \\ \text{angle } d\Omega\end{array}\right\} = A'\, d\Omega\, e^{-W/kT}$$

$$\therefore \qquad dN = A'(2\pi \sin\theta \, d\theta)\, e^{\mu_m B \cos\theta / kT}$$

or

$$dN = A\, e^{\mu_m B \cos\theta / kT} \sin\theta \, d\theta, \quad \text{where } A = 2\pi A'.$$

Denoting $\beta = \dfrac{\mu_m B}{k\,T}$, we write $dN = A\, e^{\beta \cos\theta} \sin\theta \, d\theta \qquad$ ...(41.19)

The total number of dipoles N lying within a unit volume and having potential energy W is given by

$$N = \int_0^{\pi} dN = A \int_0^{\pi} e^{\beta \cos\theta} \sin\theta \, d\theta \qquad ...(41.20)$$

If a magnetic dipole μ_m inclined at an angle θ to the applied magnetic field, the component of its moment in the field direction is $\mu_m \cos\theta$. The magnetic moment contributed by dN dipoles is given by $dN \cdot \mu_m \cos\theta$.

The total magnetic moment due to N molecules per m^3 is given by

$$p_m = \int_0^{\pi} dN \, \mu_m \cos\theta$$

But p_m is the same as magnetization. Therefore, $M = \int_0^{\pi} dN \, \mu_m \cos\theta$.

or
$$M = A\mu_m \int_0^{\pi} e^{\beta\cos\theta} \cos\theta \sin\theta \, d\theta$$

From Eq. (41.20), we have $A = \dfrac{N}{\int_0^{\pi} e^{\beta\cos\theta} \sin\theta \, d\theta}$

$\therefore$
$$M = N\mu_m \frac{\int_0^{\pi} e^{\beta\cos\theta} \cos\theta \sin\theta \, d\theta}{\int_0^{\pi} e^{\beta\cos\theta} \sin\theta \, d\theta}$$

Setting $\cos\theta = y$, we get $dy = -\sin\theta \, d\theta$. We can write the above equation as

$$M = N\mu_m \frac{\int_{+1}^{-1} e^{\beta y} y(-dy)}{\int_{+1}^{-1} e^{\beta y}(-dy)} = N\mu_m \frac{\left[\frac{e^{\beta y}}{\beta} \cdot y\right]_{+1}^{-1} - \left[\frac{e^{\beta y}}{\beta^2}\right]_{+1}^{-1}}{\left[\frac{e^{\beta y}}{\beta}\right]_{+1}^{-1}}$$

$$= N\mu_m \frac{\left[-\frac{e^{-\beta}}{\beta} - \frac{e^{\beta}}{\beta}\right] - \left[\frac{e^{-\beta}}{\beta^2} - \frac{e^{\beta}}{\beta^2}\right]}{\left[\frac{e^{-\beta}}{\beta} - \frac{e^{\beta}}{\beta}\right]} = N\mu_m \frac{\frac{2}{\beta}\left[\left(\frac{e^{\beta} + e^{-\beta}}{2}\right) - \left(\frac{e^{\beta} - e^{-\beta}}{2\beta}\right)\right]}{-\frac{2}{\beta}\left(\frac{e^{\beta} - e^{-\beta}}{2}\right)}$$

$$= N\mu_m \frac{\cosh\beta - \frac{1}{\beta}\sinh\beta}{\sinh\beta} = N\mu_m \left(\coth\beta - \frac{1}{\beta}\right) \qquad ...(41.21)$$

or
$$M = N\mu_m L(\beta) \qquad ...(41.22)$$

The function $L(\beta) = \left(\coth\beta - \dfrac{1}{\beta}\right)$ is known as the **Langevin function.**

Case(i): $\mu_m\beta >> kT$ and $\beta >> 1$.

Then
$$|L(\beta)|_{\beta\to\infty} = \left(\cos h\beta - \frac{1}{\beta}\right)_{\beta\to\infty} = \left|\frac{e^{\beta}+e^{-\beta}}{e^{\beta}-e^{-\beta}} - \frac{1}{\beta}\right|_{\beta\to\infty}$$

$$= \left|\frac{1+e^{-2\beta}}{1-e^{-2\beta}} - \frac{1}{\beta}\right|_{\beta\to\infty} = 1$$

$$\therefore \qquad M_s = N\mu_m$$

where M_s denotes the saturation value of magnetization.

It will be attained either when the applied field **B** is very large or the temperature is very low. It corresponds to the condition where all the dipoles are completely aligned in the field direction.

Case(ii): $\mu_m\beta << kT$ and $\beta << 1$.

The magnetic moment μ_m of an atom has a magnitude of the order of 10^{-23} J/T. At moderate fields of the order of 1Tesla, $\mu_m B = 10^{-23}$ J. The factor kT is about 4×10^{-21} J at room temperature. It follows that $\mu_m\beta << kT$ for not very strong magnetic fields.

For values of $\beta << 1$, the Langevin function is given by

$$\therefore \qquad L(\beta) = \left[\frac{1}{\beta} + \frac{\beta}{3} - \frac{\beta^3}{45} + \frac{2\beta^5}{945} + \ldots\ldots\right] - \frac{1}{\beta} \approx \frac{1}{\beta} + \frac{\beta}{3} - \frac{1}{\beta} = \frac{\beta}{3}.$$

$$\therefore \qquad L(\beta) = \frac{\mu_m\beta}{3kT}$$

Substituting the above value of $L(\beta)$ into Eq. (41.22), we obtain

$$M = \frac{N\mu_m\beta}{3} = \frac{N\mu_m^2 B}{3kT}$$

or
$$M = \frac{N\mu_0\mu_m^2 H}{3kT} \qquad \text{...(41.23)}$$

$$\therefore \qquad \chi_{\text{para}}^{\text{orb}} = \frac{M}{H} = \frac{N\mu_0\mu_m^2}{3kT} = \frac{C}{T} \qquad \text{...(41.24)}$$

where the Curie constant C is given by

$$C = \frac{N\mu_0\mu_m^2}{3k} \qquad \text{...(41.25)}$$

Eq. (41.24) is the Curie's law.

Example 41.3: *The susceptibility of paramagnetic $FeCl_3$ is 3.7×10^{-3} at 27°C. What will be the value of its relative permeability at 200°K and 500° K?*

Solution: Curie constant $C = \chi T = (3.7 \times 10^{-3})$ (300 K) = 1.11 K

$$\chi_{200\text{K}} = \frac{C}{T} = \frac{1.11\text{K}}{200\text{K}} = \mathbf{5.55 \times 10^{-3}}$$

$$\chi_{500\text{K}} = \frac{C}{T} = \frac{1.11\text{K}}{500\text{K}} = \mathbf{2.22 \times 10^{-3}}.$$

41.8 FERROMAGNETISM

Ferromagnetic materials are metallic crystals which when placed in a magnetic field become strongly magnetized in the direction of the field. Iron, nickel and some steels are examples of ferromagnetic materials. The experiments of Einstein and de Haas showed that the spin magnetic moments of electrons are responsible for ferromagnetism. In definite conditions, the magnetic moments of adjacent atoms become aligned parallel to one another even in the absence of external magnetic field.

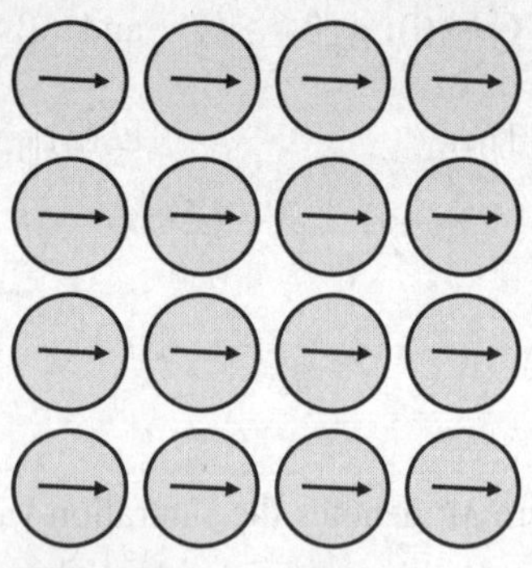

Fig. 41.9: In the absence of applied magnetic field magnetic moments are aligned in a common direction

Consequently, ferromagnetic materials exhibit spontaneous magnetic moment even in the absence of applied magnetic field (Fig. 41.9). The internal magnetic field has an induction that is hundreds and thousands of times greater than the induction of external magnetic field producing the magnetization.

It means that $M >> H$ and we may write Eq. (41.4) as

$$B \cong \mu_0 M$$

Characteristics: The salient features of ferromagnetic materials are as follows:

(*i*) Ferromagnetic materials exhibit very high values of magnetic susceptibility and relative permeability. Susceptibilities as large as 10^6 and relative permeability of the order of a few thousands are common.

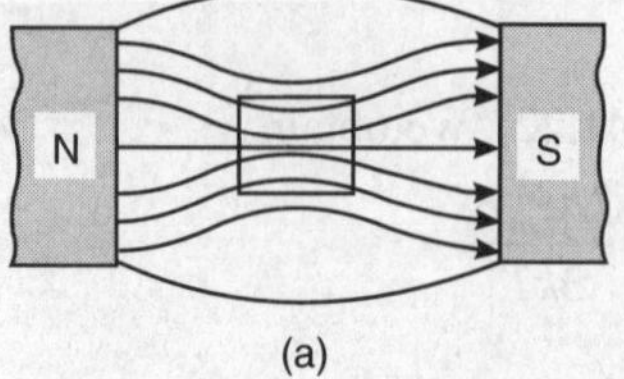

Fig. 41.10: (a) (b)

(a) Magnetic fields are strongly pulled towards a ferromagnetic material.
(b) Magnetic shielding affected by a ferromagnetic shell.

(*ii*) When a ferromagnetic material is kept in a magnetic field, the field lines crowd into the material, as shown in Fig. 41.10 (a). Thus, ferromagnetic materials **conduct magnetic flux** much as metals conduct electric current.

If the material is taken in a shell form, the magnetic lines are concentrated mainly in the shell without penetrating into the cavity, as shown in Fig. 41.10 (b). It means that a shell made of a ferromagnetic material acts as a ***magnetic shield*** which does not allow the magnetic field to penetrate into the space enclosed by the shell. Thus, ferromagnetic materials **conduct magnetic flux** much as metals conduct electric current.

(*iii*) Magnetization of a ferromagnetic material does not vary linearly with the applied field **H**. It is a very complex nonlinear function of the field strength. It is illustrated in Fig. 41.11 (a).

Initially, **M** (therefore **B**) increases very fast with a small increase in **H**. Then its increase slows down and at large enough values of **H**, magnetic saturation is reached in which magnetization becomes practically independent of the strength of the magnetizing field. The magnetization proceeds along the curve *OA* (Fig. 41.11 a) in case of an initially non-magnetized ferromagnetic substance. The curve *OA* is known as the ***initial magnetizing curve***.

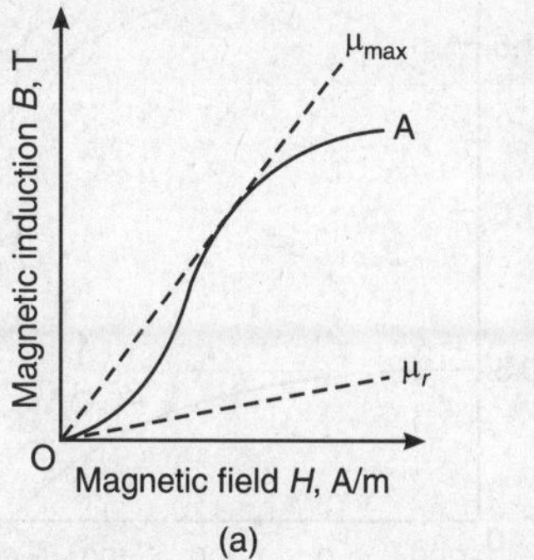

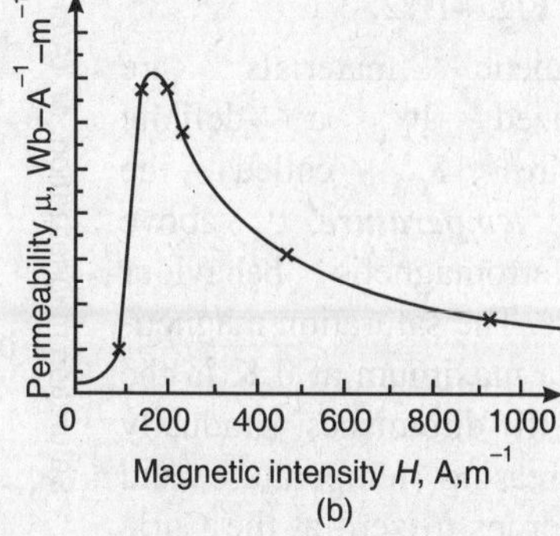

Fig. 41.11: (a) Initial magnetization curve for a virgin ferromagnetic material. (b) The permeability of a ferromagnetic material as a function of magnetic field H.

(*iv*) Because of the nonlinear relationship between **B** and **H**, the permeability of a ferromagnetic material does not have a constant value. Eq. (41.6) is not valid. Therefore, in practice, the magnetic permeability of a ferromagnetic is measured by either its initial permeability μ_i or its maximum permeability μ_{max}.

The permeability calculated from the curve *OA* is plotted as a function of **H** in Fig. 41.11(b). It is seen that μ rises with increasing **H**, reaches its maximum value and then rapidly decreases as the magnetic saturation is attained.

(*v*) The specific properties of a ferromagnetic material manifest only in the *crystalline state* at temperatures lower than a certain temperature. In contrast, diamagnetism and paramagnetism are displayed in any state of aggregation. In the liquid and gaseous states, ferromagnetic materials behave like ordinary paramagnetic substances.

(*vi*) The ferromagnetic properties of crystals are found to be dependent on the direction of magnetization. The magnetization is more readily obtained along certain crystallographic axes than along others.

The direction in which the magnetization is the strongest for a given value of the field is called the ***direction of easy magnetization*** while the direction corresponding to the lowest magnetization for a given field is called the direction of ***hard magnetization***.

For example iron belongs to BCC lattice. If the magnetic field is applied along <111>, <110> and <111> directions, the crystal is easily magnetized along the cube edge <100> and it is not easy to magnetize iron crystal along the body diagonal direction <111>.

It requires a much higher field to obtain the same magneti-

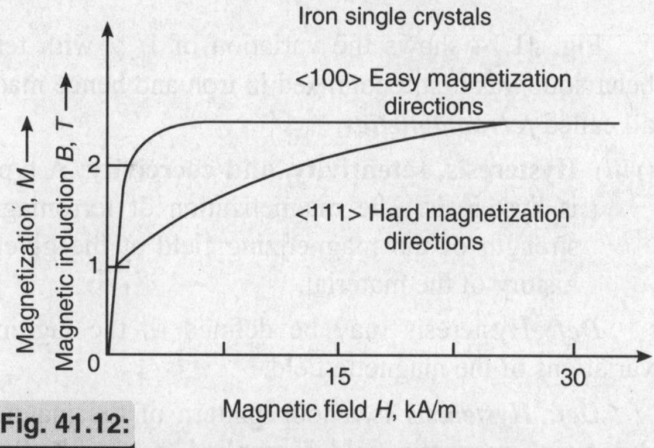

Fig. 41.12: Magnetization versus applied field for different crystallographic directions in an iron crystal.

zation. The magnetization along the different directions for BCC iron is shown in Fig. 41.12.

(*vii*) Ferromagnetic materials are characterized by a definite temperature T_C, called the ***Curie temperature***, above which ferromagnetic behaviour disappears. The saturation magnetization is a maximum at 0 K in the material. It diminishes gradually with increasing temperature and abruptly drops to zero at the Curie temperature, as shown in Fig. 41.13.

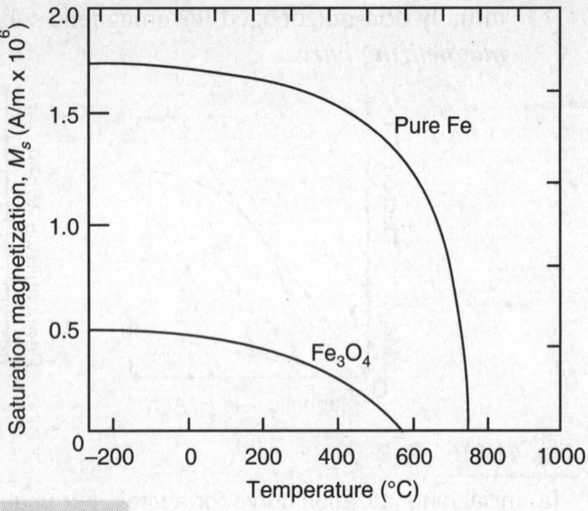

Fig. 41.13: Plot of saturation magnetization of pure iron Fe_3O_4 as a function of temperature.

The ferromagnetic behaviour vanishes beyond T_C and the material transforms into a paramagnetic material. It is a reversible transition. When the material is cooled back through T_C, the ferromagnetic character reappears. The disappearance of ferromagnetic character at T_C is not accompanied by changes in the crystal lattice. The lattice is not broken up but changes its type of symmetry.

Well above the Curie temperature T_C, the magnetic susceptibility of the paramagnetic state obeys Curie –Weiss law.

$$\chi = \frac{C}{T - \theta};\ T > T_C \qquad ...(41.26)$$

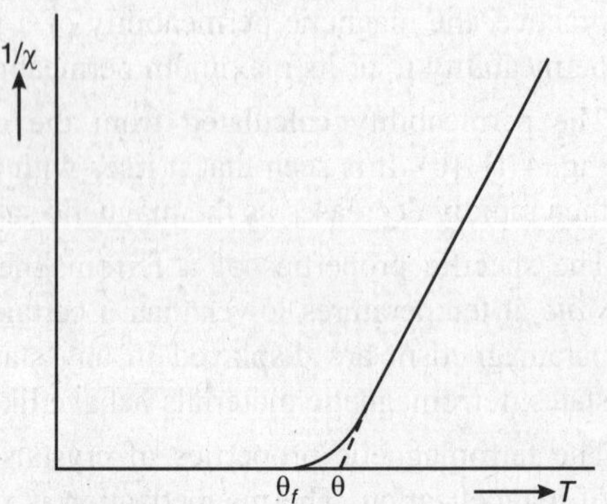

Fig. 41.14: A plot of reciprocal of the susceptibility versus temperature for a ferromagnetic material

Fig. 41.14 shows the variation of $1/\chi$ with temperature above T_C. The ferromagnetic behaviour was first recognized in iron and hence materials, which show similar behaviour are all called ***ferromagnetics***.

(*viii*) **Hysteresis, retentivity, and coercivity:** A typical property of ferromagnetic materials is hysteresis. The magnetization of ferromagnetic material depends not only on the strength of the magnetizing field at the given instant but also on the magnetization history of the material.

Def: Hysteresis may be defined as the lag in the changes of magnetization behind variations of the magnetic field.

Def: Hysteresis is a loop pattern of the magnetic field induction **B** and increasing or decreasing magnetic field **H** applied to an initially unmagnetized ferromagnetic specimen (Fig. 41.15).

The curve starts at the origin **O**. As **H** is increased, the field **B** begins to increase slowly, then more rapidly. It finally reaches a saturation value and becomes independent of **H**. The maximum value of **B** is known as the saturation flux density B_s and the corresponding magnetization is the saturation magnetization M_s.

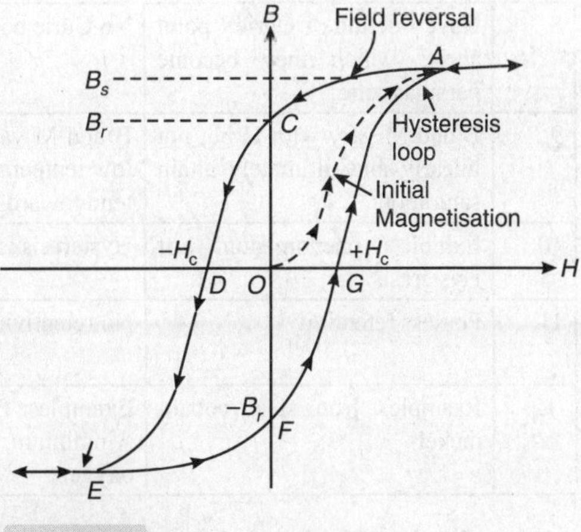

Fig. 41.15: Hysteresis loop

If H is now decreased, B also decreases but following a path AC instead of the original path AO. Thus, B lags behind H. When H becomes zero, B does not become zero but has a value equal to OC (B_r).

It indicates that the material remains magnetized even in the absence of an external applied field H. The power of retaining the magnetism is called the ***retentivity*** or ***remanence*** of the material.

Def: The retentivity of a material is a measure of the magnetic flux density remaining in the material when the magnetizing field is removed.

When the magnetic field H is applied in the reverse direction, a field of magnitude $-H_c$ must be applied to reduce the flux density to zero. H_c is called the ***coercivity*** or the ***coercive force***.

Def: Coercivity is a measure of the magnetic field strength required to destroy the residual magnetism in the material.

As the applied field H is increased further in the negative direction, saturation is ultimately reached in the reverse direction (point E).

The closed curve ACDEFGA represents a ***cycle of magnetization*** of the specimen and is known as the ***hysteresis loop*** of the specimen.

Table 41.1 presents a brief comparison of the three magnetic materials.

Table 41.1: *Comparison of the Three Magnetic Materials.*

S.N.	Ferromagnetics	Paramagnetics	Diamagnetics
1.	Solids and possess crystalline structure	Solid, liquid or gas	Solid, liquid or gas
2.	Strongly attracted towards magnetic field	Feeble attraction towards magnetic field	Feeble repulsion from the magnetic field
3.	Field lines are concentrated in the material	More number of field lines pass through the material than outside	Less number of field lines pass through the material than outside
4.	Set along the direction of the magnetic field	Tend to align along the magnetic field direction	Tend to align perpendicular to the magnetic field direction
5.	Susceptibility, χ is large and positive	Susceptibility, χ is <1 but positive	Susceptibility, χ is small but negative
6.	Relative permeability, μ_r is greater than unity	Relative permeability, μ_r is slightly greater than unity	Relative permeability, μ_r is less than unity
7.	Susceptibility, χ decreases with temperature in a complex manner	Obeys Curie law, i.e. $\chi = \frac{1}{T}$	χ is independent of temperature

8.	Have definite Curie point above which they become paramagnetic	No Curie point	No Curie point
9.	B and M vary with H but not linearly and ultimately attain saturation	B and M vary with H linearly at low temperature and at high field tend towards saturation	B and M vary with H linearly but no saturation is reached
10.	Exhibit phenomenon of Hysteresis	Hysteresis is not exhibited	Hysteresis is not exhibited
11.	Possess retentivity	No retentivity	No retentivity
12.	Examples: Iron, steel, cobalt, nickel	Examples: Platinum, Chromium, Aluminium, salts of Fe & Ni, oxygen	Examples: Bismuth, Mercury, Silver, Copper, water, air

41.8.1 Weiss Theory of Ferromagnetism

In 1907, P.Weiss proposed the molecular field theory to explain qualitatively the salient features of ferromagnetism. He postulated that the internal molecular field sets up spontaneous magnetization and that a macroscopic ferromagnetic specimen is divided up into small regions called *domains*.

Each domain is at all times spontaneously magnetized to saturation and has a definite magnetic moment. In a non-magnetized specimen, the directions of the magnetic moments of the domains are distributed randomly so that the material as a whole will have a zero magnetization.

(*i*) **Spontaneous Magnetization:** Weiss postulated that an *internal molecular field* causes a parallel alignment of magnetic dipoles and sets up spontaneous magnetization in a ferromagnetic material. In 1928, Heisenberg explained the origin of internal molecular field as due to quantum ***exchange interactions*** between the electrons. The spontaneous magnetization exists in the material only below the Curie temperature, T_C.

The magnetization in the material is a maximum temperature at 0K and decreases with increase in temperature. It falls off rapidly to zero value at T_C. If the material is cooled from above the Curie temperature, the spontaneous magnetization reappears at T_C. At temperatures above Curie point, a ferromagnetic material goes into paramagnetic state (Fig. 41.16 a).

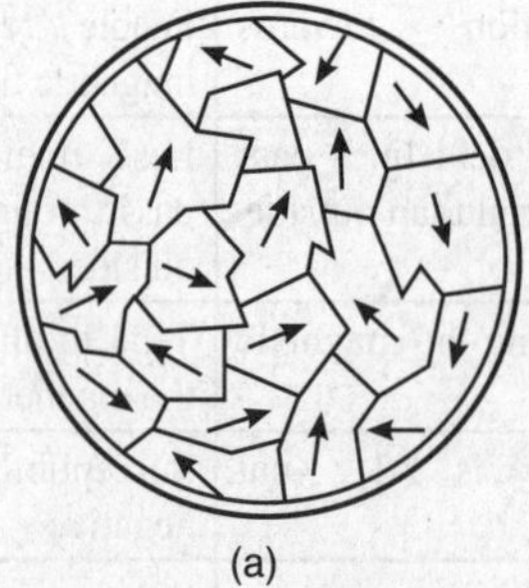
(a)

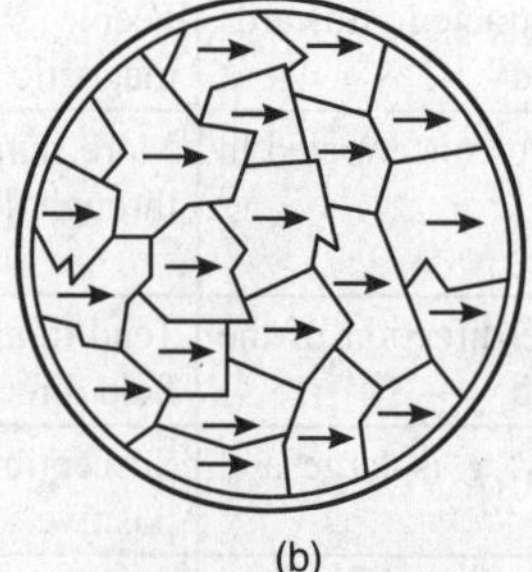
(b)

Fig. 41.16:

(a) Schematic illustration of magnetic domains in a ferromagnetic material.
(b) Domain configuration in a magnetized body.

(*ii*) **Ferromagnetic Domains:** Even though ferromagnetic materials are spontaneously magnetized, virgin samples do not show any noticeable magnetization. A virgin sample does not possess net magnetic moment.

According to Weiss hypothesis, the entire ferromagnetic volume splits into a large number of small regions of spontaneous magnetization (Fig. 41.16 a). The regions are called the domains. Each domain is spontaneously magnetized to saturation and carries a definite magnetic moment. In the absence of an external magnetic field, the magnetic moment vectors of the domains are oriented in all probable directions such that the net magnetic moment of the entire body equals to zero and therefore, the sample does not exhibit magnetization.

Weiss postulated that in ferromagnetic materials the internal molecular field is responsible for lining up of dipoles in the same direction. The internal field H_i is given by

$$H_i = H + \gamma M \qquad ...(41.27)$$

where H is the applied field, and γ M is a measure for the tendency of the environment to align a given dipole parallel to the magnetization already existing.

The factor γ is the ***molecular field constant*** which is also known as ***Weiss constant***. It determines the intensity of interaction between the dipoles.

The Eq. (41.27) explains the Curie-Weiss law and occurrence of spontaneous magnetization by assuming that ferromagnetics are essentially paramagnetic materials having a very large molecular field.

For paramagnetic materials, the susceptibility is given by the expression

$$\chi_m = \frac{C}{T}$$

In case of the paramagnetic state of a ferromagnetic material,

$$\chi_{\text{para}} = \frac{M}{H_i} = \frac{C}{T}$$

Using Eq. (41.27) into the above equation, we find that

$$\frac{M}{H + \gamma M} = \frac{C}{T}$$

or

$$M = \frac{HC}{T - \gamma C} \qquad ...(41.28)$$

$$\therefore \qquad \chi = \frac{C}{T - \gamma C} = \frac{C}{T - \theta}$$

It is obvious from the Eq. (41.28) that the value of magnetization tends to infinity at $T = \theta$. It means that the interactions of the individual magnetic moments reinforce each other causing them to align parallel at $T = \theta$.

From measurements of the susceptibility as a function of temperature, the internal field constant γ is of the order of 10^3. This value is thousand times larger than the value theoretically calculated basing on the assumption that the internal field arises due to the interaction of atomic dipoles. This discrepancy arises because of the fact that the interaction between dipoles is considered to be the classical magnetic interaction. In fact, the interaction is of a quantum mechanical origin and the forces operating between the magnetic dipoles are ***exchange forces***.

Eq. (41.28) suggests nonvanishing magnetization for $T = \theta$ even in the absence of magnetic field ($H = 0$). It means that the magnetization at temperature $T \leq \theta$, is spontaneous.

We confirm it by the following considerations. When $H = 0$, the magnetic dipoles are subjected only to the molecular field γM. The Langevin variable b may be written now as

$$\beta = \frac{\mu_m H}{kT} = \frac{\mu_m \gamma M}{kT}$$

$$M = \frac{kT}{\gamma \mu_m} \beta \qquad ...(41.29)$$

The magnetization M is thus a linear function of β.

Magnetization M is also given by the Eq. (41.21).

$$M = N \mu_m \left[\coth \beta - \frac{1}{\beta} \right]$$

Therefore, the Eq. (41.21) and Eq. (41.29) simultaneously determine the condition of spontaneous magnetization.

Fig. 41.17 shows the plots drawn between M and β. The straight lines depict the linear relationship represented by Eq. (41.29) at different temperatures whereas the curve corresponds to the Langevin curve corresponding to the Eq. (41.21). The intersection point A of a given temperature with the Langevin curve represents the finite spontaneous magnetization at that temperature. If the dipoles assume state C, then the local magnetization is less than the corresponding equilibrium state A.

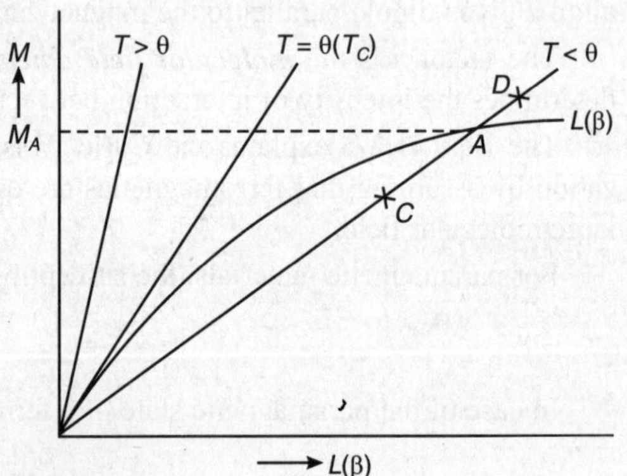

Fig. 41.17: Langevin function L (ß) and plot of Eq. 3.29 for three temperatures.

The magnetization and Langevin variable β will increase until the state A is reached. On the other hand, if the dipoles assume state D, then the local magnetization is more than the equilibrium value. Now, the magnetization M and the value of β decrease, until the state A is reached.

With increasing temperature, the straight lines increase in slope according to Eq. (41.29), thus bringing down the point of intercept A and the value of spontaneous magnetization M. Finally at the Curie temperature T_C, there is no intercept and there is no more spontaneous magnetization. At T_C, the slope of the line $(kT_C / \gamma\mu_m)$ is identical to the slope of the Langevin curve near the origin. It is equal to

$$\therefore \qquad \frac{k\theta}{\gamma\mu_m} = \frac{N\mu_m}{3} = \frac{M_s}{3}$$

$$\therefore \qquad \gamma = \frac{3k\theta}{\mu_m M_s}$$

$$\therefore \qquad \gamma M_s = \frac{3k\theta}{\mu_m} \qquad ...(41.30)$$

which is the internal molecular field or Weiss field.

For iron, $\mu_m = 2.2\mu\beta = 9.273 \times 10^{-24}$ Am^2 and $\theta = 1043$ K

$$\therefore \qquad \gamma M_s = \frac{3\times 1.38\times 10^{-23}\ J/K\times 1043\,\text{K}}{2.2\times 9.273\times 10^{-24}\ \text{Am}^2} = 2117\ T$$

The condition for stable spontaneous magnetization is given by

$$T < \frac{\mu_m M_s \gamma}{3k}$$

$$\therefore \qquad T < \theta$$

Hence below the Curie point θ (*i.e.* T_C), the material is ***spontaneously*** magnetized to a degree depending upon temperature. The magnetization approaches saturation value as the temperature approaches 0 K. Above the Curie point θ, spontaneous magnetization does not occur and the ferromagnetic material transforms into paramagnetic material.

Explanation of Hysteresis Effect on the basis of domains

When a magnetic field is applied, the magnetization of the material takes place through two processes.

(*i*) **By motion of domain walls:** In weak fields, domains, which are favourably oriented with respect to the applied field, increase their volume (grow in size) at the expense of unfavourably oriented domains.

(*ii*) **By the rotation of domains:** In strong fields, domains rotate in an attempt to align their magnetic moments with the field direction.

Hysteresis effect observed in ferromagnetic materials can be explained on the basis of ferromagnetic domains. In the absence of a magnetic field, the domains in the material are randomly directed and the resultant magnetic moment is equal to zero. When the material is placed in a magnetic field H, the orientations of the magnetization vectors of various domains with respect to the magnetic field direction are different. The situation is illustrated Fig. 41.18.

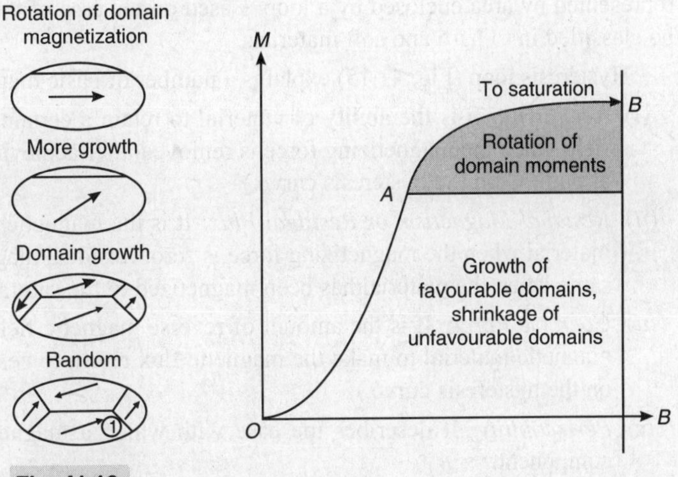

Fig. 41.18: Magnetic domain growth and rotation as a virgin ferromagnetic specimen is magnetized to saturation by an applied magnetic field.

It is seen that the magnetization vector of the first domain forms the smallest angle *H*. With an increase in *H*, the growth of the most favorably oriented domain 1 becomes energetically advantageous. Therefore, domain 1 grows at the expense of its neighbour domains by sidewise motion of domain walls. Domain 1 grows until the whole crystal becomes a single domain.

Fig. 41.18 shows the magnetization curve for a single crystal. As the applied field is increased, magnetization proceeds along the curve *OAB*. The portion *OA* of the curve corresponds to the stage of wall motion. A further increase in the field strength causes the onset of rotation of domain moments into the field direction. The process of magnetization goes on more slowly in this direction. The magnetization at this stage reaches technical saturation. It occurs along the *AB* portion of the curve.

Beyond *B*, the growth of magnetization with *H* is very slow. In the single domain formed under the action of the magnetic field, the individual magnetic moment alignment would not be complete due to the disturbing effect of temperature. An increasing *H* prevails over the thermal disturbance and tends to orient atomic dipoles.

When the magnetic field is reduced back to zero, the magnetization curve does not retrace the initial curve *BAO*. The failure to retrace the original path is due to irreversible processes occurring in the course of demagnetization. Various defects such as impurity atoms, dislocations and nonmagnetic inclusions present in the material have a strong effect on the domain wall motion. They may aid or hinder the domain wall motion. During the course of demagnetization some walls may get stuck up.

Consequently, the domains magnetized along the field direction retain their former orientation even after the applied field is made zero. It leads to remanent induction B_r. Elimination of the remanent magnetization requires the application of H $(= H_C)$ in opposite direction.

The amount of energy spent in this region is a loss. Hysteresis loss is the loss of energy in taking a ferromagnetic body through a complete cycle of magnetization and the loss is represented by area enclosed by a loop. Based on the area of the loop magnetic materials can be classified in to hard and soft materials.

Hysteresis loop (Fig. 41.15) explains a number of basic magnetic properties of a material.

(*i*) *Retentivity:* It is the ability of material to retain a certain amount of residual magnetic field when the magnetizing force is removed after achieving saturation. (The value of B at point C on the hysteresis curve.)

(*ii*) *Residual Magnetism or Residual Flux:* It is the magnetic flux density that remains in a material when the magnetizing force is zero. Residual magnetism and retentivity are the same when the material has been magnetized to the saturation point.

(*iii*) *Coercive Force:* It is the amount of reverse magnetic field which must be applied to a magnetic material to make the magnetic flux return to zero. (The value of H at point D on the hysteresis curve.)

(*iv*) *Permeability:* It describes the ease with which a magnetic flux is established in the component.

(*v*) *Reluctance:* It is the opposition that a ferromagnetic material shows to the establishment of a magnetic field. It is analogous to the resistance in an electrical circuit.

Hysteresis property of ferromagnetic materials is useful as a magnetic "memory". Some compositions of ferromagnetic materials will retain an imposed magnetization indefinitely and are useful as "permanent magnets".

A few evidences of domain theory are listed hereunder.

(*i*) Cutting a magnet gives two magnets.

(*ii*) Magnetizing by rubbing directionally lines domains up.

(*iii*) Saturation means all domains lined up.

(*iv*) Hard steel difficult to magnetize and demagnetize.

(*v*) Larger magnets are stronger because of more domains.

(*vi*) Heating and AC demagnetizes because they scramble domains

(*vii*) Hysteresis, a magnetic inertia as domains resist being moved.

Example 41.4: *Find the relative permeability of the ferromagnetic material if a magnetic field of strength 220 A/m produces magnetization of 3300 A/m in it.*

Solution: $\mu_r = (1+\chi) = 1 + \frac{M}{H} = 1 + \frac{3300 \text{ A/m}}{220 \text{ A/m}} = \mathbf{16.}$

Example 41.5: *The saturation magnetic induction of nickel is 0.65 T. If the density of nickel is 8906 kg/m³ and atomic weight is 58.7, find out the magnetic moment of the nickel atom in Bohr magneton.*

Solution: Number of atoms /m³ $N = \frac{\rho N_A}{M} = \frac{8906 \text{ kg/m}^3 \times 6.023 \times 10^{26} \text{ atoms/k.mol}}{58.7}$

$= 9.14 \times 10^{28}/\text{m}^3.$

Magnetic moment $\mu_m = \frac{B}{N\mu_0} = \frac{0.65T}{9.14 \times 10^{28} / \text{m}^3 \times 4\pi \times 10^{-7}} = 5.66 \times 10^{-24} \text{A.m}^2.$

$= \frac{5.66 \times 10^{-24}}{9.27 \times 10^{-24}} = \mathbf{0.61}\mu_B.$

41.9 MAGNETOSTRICTION

Once a ferromagnetic material is magnetized, a small change occurs in its dimensions. It may either expand or contract in the direction of magnetization. The material reverts to the original dimensions on removal of the magnetic field. Since the process is reversible, the change in the magnetization can be produced when the dimensions are changed as the effect of an external force.

Magnetically induced reversible elastic strain ($\Delta l/l$) is called magnetostriction. Magnetostriction is caused by the rotation of domains of a ferromagnetic material under the action of a magnetic field. The rotation of domains gives rise to internal strains in the material causing its contraction or expansion.

The magnitude of magnetostriction for a given material depends on the initial orientation of domains which can be controlled by heat treatment and cold work. Ferromagnetic materials may show both positive and negative magnetostriction, varying in magnitude according to the intensity of the magnetic field. Materials showing negative magnetostriction contract when the magnetic field increases in strength and expand when it decreases. The converse is true for materials showing positive values of magnetostriction. Iron shows positive magnetostriction in weak magnetic fields and negative magnetostriction in strong magnetic fields. Nickel and cobalt show negative magnetostriction increasing in magnitude with the strength of the magnetic field.

41.10 ANTIFERROMAGNETISM

Def: When the magnetic moments of sublattices in a crystal unit cell are equal in magnitude but opposite in direction, they cancel each other giving rise to antiferromagnetism.

Antiferromagnetic materials are crystalline materials, which exhibit a small positive susceptibility of the order of 10^{-3} and 10^{-5}. The variation of susceptibility with temperature follows a peculiar pattern in these materials. The susceptibility increases with increasing temperature and reaches a maximum at a certain temperature called ***Neel temperature***, T_N (Fig. 41.19 a). With a further increase in temperature, the material goes into paramagnetic state. The material is antiferromagnetic below T_N. The transition temperature T_N lies far below room temperature for most of the materials.

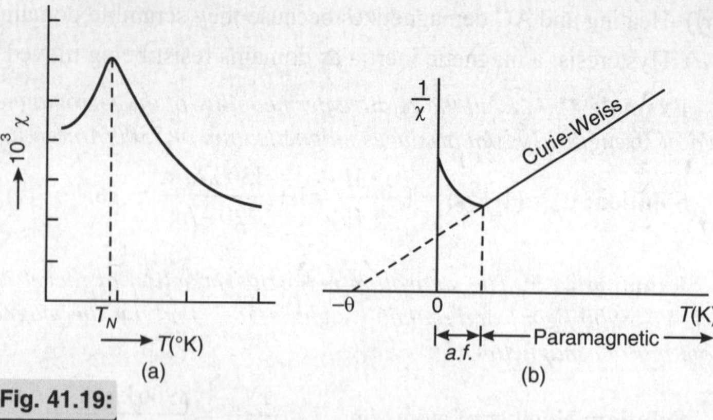

Fig. 41.19: (a) Variation of susceptibility of an antiferromagnetic crystal as a function of temperature. (b) Reciprocal of susceptibility of an antiferromagnetic crystal versus temperature

In the paramagnetic state, the variation of inverse susceptibility ($1/\chi$) with temperature is linear, as shown in Fig. 41.19 (b). The extrapolation of the paramagnetic line, in Fig. 41.19 (b) to $1/\chi = 0$ yields a negative θ. The variation of susceptibility with temperature obeys therefore a modified Curie- Weiss law.

$$\chi_{\text{a.f.}} = \frac{C}{T-(\theta)} = \frac{C}{T+(\theta)} \qquad T > T_N \qquad \text{...(41.31)}$$

where θ is called the paramagnetic Curie temperature and C the Curie constant.

The elements, manganese and chromium exhibit antiferromagnetism at room temperature. Most of the antiferromagnetic materials are ionic compounds. MnO, MnS, Cr_2O_3, NiCr are some of the compounds which exhibit antiferromagnetism. Antiferromagnetic materials are of little practical interest.

Antiferromagnetism arises in materials in which the magnetic moments are equal, but adjacent magnetic moments point in opposite directions (Fig. 41.20 a). An antiferromagnetic unit cell can be considered as composed of two interpenetrating ferromagnetic sublattices, say A and B. The atoms of one of these sublattices have their spins oriented in one direction and the atoms of the other sub-lattice have their spins oriented in opposite direction, as shown in Fig. 41.20 (b).

When there is no external magnetic field acting on them, the magnetization of the unit cell is zero since the magnetic moments of the sublattices are mutually compensated giving a zero net magnetization. When the material is placed in an external magnetic field

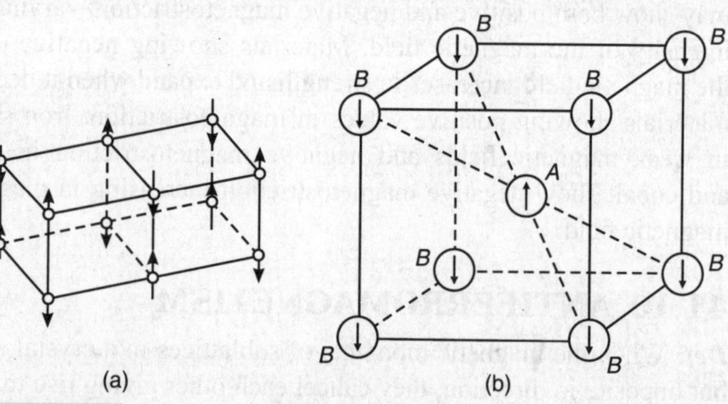

Fig. 41.20: An antiferromagnetic unit cell

and the temperature is 0K, the magnetization is still zero. As the temperature is increased in the presence of an external field, the antiparallel arrangement of spins is slowly disturbed and magnetization (and hence the susceptibility) increases (Fig. 41.19 a). The susceptibility reaches a maximum at Neel temperature, T_N. At the Neel temperature, the spin ordering is lost completely. The susceptibility decreases with further rise in temperature and the material transforms into a paramagnetic above T_N.

41.11 FERRIMAGNETISM

Def: Certain crystals in which the magnetic moments of the two sublattices are not exactly equal in magnitude and are oriented in opposite directions results a net magnetic moment. Such crystals possess spontaneous magnetization and exhibit most of the properties of ferromagnets. This uncompensated antiferromagnetism is known as ***ferrimagnetism***, shown in Fig. 41.1 (d).

Ferrimagnetic materials are very much similar to ferromagnetic materials in their macroscopic magnetic characteristics. The saturation magnetization in these materials is not as high as for ferromagnetic materials. The saturation magnetization decreases with increasing temperature until it vanishes at a Curie temperature T_c. Above T_c, they exhibit paramagnetic state, the Curie-Weiss linear relationship is not valid, and the susceptibility is a nonlinear function of temperature.

The materials exhibit hysteresis character. They have a residual magnetization, are characterized by a coercive force and so on. They possess small domains in which electron spins are spontaneously aligned. However, the spins are antiparallel. Ferrimagnetism is therefore referred to as uncompensated antiferromagnetism. Ferrimagnetic materials are widely used in high frequency applications where ferromagnetic materials cannot be utilized.

41.12 FERRITES

Ferrimagnetic materials are ceramic materials and are therefore good electrical insulators. The most important group of these materials is ferries. ***Ferrites*** consist chiefly of ferric oxide Fe_2O_3, combined with one or more oxides of divalent metals. They are represented by a general formula written as $M\,Fe_2O_4$ or $M.O.\,Fe_2\,O_3$ in which M represents any one of several metallic elements.

The name of the divalent or monovalent metal, whose oxide is in the composition of ferrite, is given to the ferrite. Thus a ferrite is called a zinc ferrite or nickel ferrite depending on whether the composition contains ZnO or NiO.

Composite ferrites are solid solutions of one simple ferrite in another. They show the best magnetic properties. Depending upon the composition and treatment, ferrites with a wide variety of magnetic properties can be produced. Ni – Zn ferrite is an example of a composite ferrite.

Garnets constitute another important group of ferrimagnetic materials. The crystal structure of garnets may be corresponding to a general formula $M_3Fe_5O_{12}$ where *M* represents a rare earth ion such as samarium, europium, gadolinium or yttrium. Yttrium iron garnet ($Y_3Fe_5O_{12}$) denoted as Y I G is the most common material of this type.

Characteristics: The salient characteristics of ferrimagnetic materials are as follows.

(*i*) They exhibit nonlinear magnetization curve, that is, hysteresis loop. There is a characteristic retentivity and coercivity for each ferrite. Ferrites are polycrystalline samples and the coercivity, retentivity and permeability depend on the grain size.

(*ii*) Ferrites possess high permeability values normally. But some ferrites can be made to have low permeability values. The range of permeabilities extends from 15 to 20 for nickel ferrite to several thousand for some manganese-zinc ferrites.

(*iii*) Ferrites have high resistivity ranging from 10^2 to 10^{10} ohm-m.

41.13 HYSTERESIS LOSS

When a ferromagnetic specimen is magnetized by keeping it in an external magnetic field, those domains in the specimen that are favourably oriented with respect to the external field grow in size, and also each domain rotates as a single unit such that its direction of magnetization becomes aligned with the field direction. Energy is consumed in this process of domain growth and rotation. When the external field is removed, the domain boundaries do not move back to their original positions and the specimen is left with some magnetization. It means that the energy supplied to the specimen during magnetization is not fully recovered. The difference of energy is lost as heat in the material. This is known as *hysteresis loss*.

It is calculated that energy dissipated per cycle of magnetization

$$= \oint H\, dB\,(\text{J/m}^3) = \text{area of the B-H curve.} \qquad ...(41.32)$$

A ferromagnetic or ferrimagnetic material that exhibits a narrow B-H curve produces smaller hysteresis loss. Such a material will be better suited in fabrication of the cores of the transformers and armatures of motors.

41.14 SOFT AND HARD MAGNETIC MATERIALS

Ferromagnetic materials are known as ***metallic magnets*** and ferrimagnetic materials as ***ceramic magnets***. The magnetic flux produced by an electric current is proportional to the relative permeability of the material through which the flux passes. As ferromagnetic and ferrimagnetic materials possess high permeabilities, they produce required flux densities with relatively small currents.

Therefore, both these materials are widely used as flux paths and flux multipliers in electrical machines. Transformers, motors and generators etc rotating electrical machines use these magnetic materials as core materials. Basing on the area of the hysteresis loop, these magnetic materials are broadly categorized into two types, namely, ***soft magnetic materials*** and ***hard magnetic materials***.

41.14.1 Soft Magnetic Materials

The soft magnetic material has a narrow hysteresis loop which makes it easy to magnetize or demagnetize (Fig. 41.21). It is imperative that a soft magnetic material should have a high initial permeability and a low coercivity. In view of these properties, the material reaches its saturation magnetization with a relatively low applied field, and exhibits low hysteresis energy losses. This becomes an important consideration when the material is used in alternating current applications, since the area of hysteresis loop represents the energy lost as heat during a cycle.

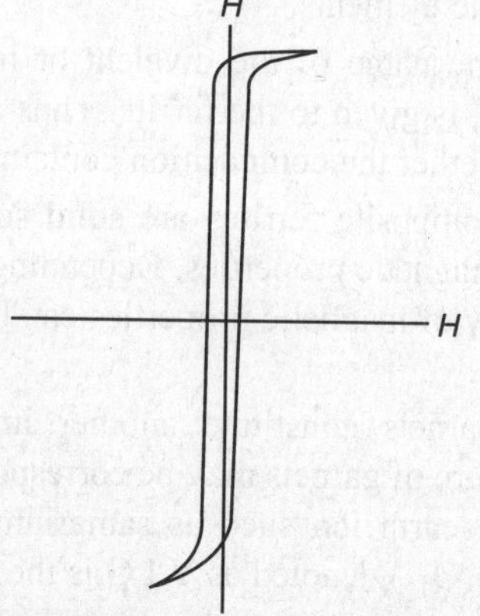

Fig. 41.21: Hysteresis loop for a soft magnetic material.

The smaller the area, the lower are the power losses and the greater the possibility of using the material at higher frequencies. A low value of coercivity means the easy movement of domain walls with the applied magnetic field. Structural defects hinder the domain wall motion and increase coercivity. Hence, a soft magnetic material must be free of such defects.

Magnetic materials, which are easily magnetized and demagnetized, are known as ***soft magnetic materials***.

When the magnetic flux through a conducting material changes, voltages are induced which cause the flow of eddy currents. Eddy currents constitute the second major contributor to energy loss. It is, therefore, necessary to minimize eddy current losses in soft magnetic materials. Increasing the electrical resistivity of the material can reduce eddy current losses.

The resistivity of ferromagnetic materials is very low in their pure form. Forming solid solutions increases their resistivity. For example, iron containing a few percent of silicon has higher resistivity instead of pure iron. Using thin laminations of ferromagnetic material, which are insulated from each other, can further increase the resistivity of the material. Further, eddy current losses increase with the square of the frequency.

Therefore, the laminations are made extremely thin of the order of 1 to 0.025 mm thickness. In very high frequency applications, even laminated sheets cause excessive power losses. Therefore powdered materials are used. Very fine metallic or ceramic magnet powders of a 2 to 10 mm particle size are mixed with a binder that insulates the particles from one another. The material is fabricated into the required shape and size.

The soft magnetic material can be broadly divided into four groups:

(*i*) **Heavy duty flux multipliers:** These are the cores of transformers, motors and generators. Electrical steel is used for manufacturing cores.

(*ii*) **Light duty flux multipliers:** These are cores of small special purpose transformers, inductors etc used in communication equipment. Ni- Fe alloys and soft ferrites are used in these applications.

(*iii*) **Square loop materials:** They are used in magnetic amplifiers, saturable core devices, computers etc. For these applications also, nickel-iron alloys and soft ferrites are used.

(*iv*) **Microwave system components:** Soft ferrites and garnets are used in these applications.

41.14.2 Hard Magnetic Materials

Hard magnetic materials are those, which have a high resistance to demagnetization. A high remanence, high permeability, a high coercive field and a large hysteresis loop characterize the hard magnetic materials (Fig. 41.22). A desirable requirement is protection of the material to loss of magnetization by ac fields, mechanical ill-treatment and changes in temperature.

The hard magnetic materials are magnetized in a magnetic field strong enough to orient the magnetic moments of their domains in the direction of the applied field. Part of the energy of the applied field is converted into potential energy, which is stored in the permanent magnet produced.

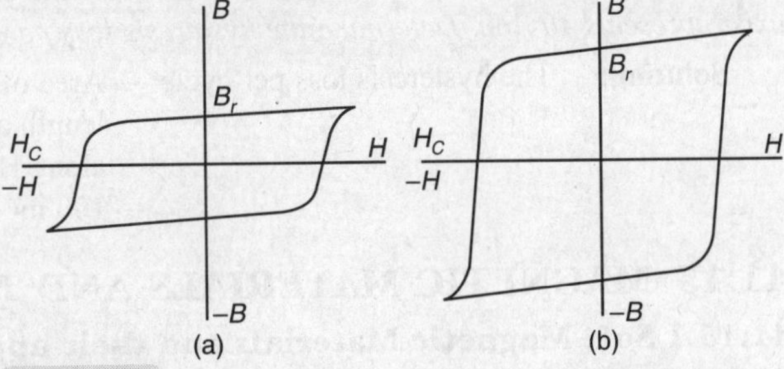

Fig. 41.22: Two types of wide hysteresis loops for hard magnetic materials.

A permanent magnet in the fully magnetized condition is thus in a relatively high-energy state as compared to a demagnetized state. The power or external energy of a hard magnetic material is directly related to the size of its hysteresis loop. The magnetic potential energy of a hard material is measured by its maximum energy product $(BH)_{max}$.

Energy product: The second and fourth quadrants of the B-H curve represent the magnetizing curves of the material, and are related to the energy required to demagnetize the magnet. It is customary to choose the second quadrant as the demagnetizing curve. The energy product versus the induction B curve is plotted either separately or along with the demagnetizing curve. Fig. 41.23 shows the demagnetizing curve and the energy product (BH) curve for a hypothetical hard material.

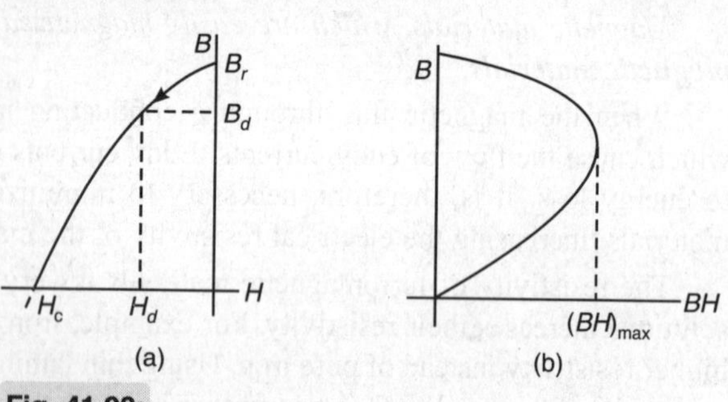

Fig. 41.23:

(a) A Schematic diagram of the B-H curve in the second quadrant. (b) The energy product versus magnetic induction for a hard magnetic material.

It is seen that the energy product is zero at point H_c and at point B_r and it attains a maximum at an intermediate point, known as the maximum energy product $[BH]_{max}$. Basically, the maximum energy product of hard material is the area occupied by the largest rectangle that can be inscribed in the demagnetizing curve.

The hysteresis behavior is related to the ease with which the magnetic domain boundaries move. Through changes caused in the microstructure of the material, domain wall motion can be hindered and the energy product value may be enhanced.

Usually, the alloys used in making permanent magnets are subjected to precipitation hardening in such a way that microscopically heterogeneous structures composed of acicular particles oriented in a particular direction are produced. The size of the particles is of the order of 1mm. Each particle constitutes a magnetically saturated domain. The reversal of magnetization in this case can take place only by a complete rotation of the domain. It requires a very high field. In view of each particle being a domain, the displacement of domain walls is suppressed and consequently the coercive force becomes large.

Example 41.6: *The area of a hysteresis loop drawn between B and H is 100 m^2. Each unit space along the vertical axis represent 0.01 wb/m^2 and each unit space along the horizontal axis represents 40 A/m. Determine the hysteresis loss per cycle.*

Solution: The hysteresis loss per cycle = Area of the hysteresis loop × Value of unit length along B-axis × Value of unit length along H-axis

$= 100\ m^2 \times 0.01\ wb/m^2 \times 40\ A/m = \mathbf{40\ J/m^3}$.

41.15 MAGNETIC MATERIALS AND THEIR APPLICATIONS

41.15.1 Soft Magnetic Materials and their applications

Soft magnetic materials are useful for electrical machines such as power transformers, output transformers, motors, generators, electromagnets etc. Electrical steel is used as core materials of these machines.

(i) Low carbon steel: Pure iron has a higher permeability but its higher electrical conductivity causes more eddy current losses. Low carbon steel (Fe-0.05% C) has a relatively small permeability and a higher resistivity. It is the least expensive core material and is used where low cost is more important than other factors.

(ii) Iron-silicon alloys: The addition of about 3-4% silicon to iron produces iron-silicon alloys with improved characteristics. Silicon increases the electrical resistivity of low carbon steel and thus reduces the eddy current losses. It increases the magnetic permeability and lowers hysteresis losses. It decreases magnetostriction and therefore reduces transformer noise known as hum. However, silicon addition tends to decrease the ductility of iron and makes the material brittle. The maximum limit is only about 4 wt%.

(iii) Grain orientation: The permeability can be substantially increased and hysteresis losses can be decreased by making use of favourable grain orientation in the material. In case of iron crystal <100> direction is the easy direction and the spin moments in a virgin crystal are aligned along <100> directions. Further, when sheets of iron or iron-alloys are manufactured by rolling and annealing, the <100> direction is parallel to the rolling direction. Therefore, cold rolled grain oriented (CRGO) steels possess better magnetic properties in a direction parallel to the direction of rolling. Consequently, the cores require less material.

Often, aluminium and manganese of less than 1% are added in iron-silicon alloys to improve the grain orientation and reduce the hysteresis losses.

(iv) Nickel-iron alloys: A pure nickel-iron alloy with nickel content of about 25% is practically non-magnetic. With the increase in nickel content, a wide range of magnetic properties are obtained. The magnetic permeability of iron-silicon alloys is relatively low at low fields. Low initial permeability is not important for power applications such as transformer cores. For communications applications much higher permeabilities at low fields are required. Nickel-iron alloys are used for these applications.

These alloys can be broadly divided into three groups: 36% nickel, 50% nickel and 77% nickel.

(*a*) 36% nickel alloys have high resistivity and low permeabilities. They are used for high frequency devices such as high-speed relays, wideband transformers and inductors. They are comparatively cheaper.

(*b*) The 50% nickel alloy have moderate permeability (μ_{max} = 25,000) and high saturation induction (B_{sat} = 1.6 T).They are used where low loss and small size are required, such as in relays, small motors and synchros etc.

(*c*) The 79% nickel alloy has high permeability (μ_{max} = 10^6) but lower saturation induction (B_{sat} = 0.8T). They are used in recording heads, pulse transformers, sensitive relays etc.

(v) Mumetal: The highest permeabilities of the order of 105 to 106 are found in multicomponent nickel-iron alloys, such as Permalloy, Supermalloy etc. Mumetal of 77% nickel, 16% iron, 5% copper and 2% chromium which can be rolled into thin sheets are useful to shield electronic equipment from stray magnetic fields.

(vi) Soft ferrites: A remarkable feature of ferrites is that they have a high electrical resistance, 10^5 to 10^{15} times the resistance of metallic ferromagnets. Ferrites possess the electrical properties of dielectrics combined with the magnetic properties of ferromagnetic materials. Therefore, they can be used for applications at high frequencies, without the eddy current losses. Soft ferrites are used for low signal memory core, audiovisual and recording head appliances. At low signal levels, soft ferrite cores are used for transformers and low energy inductors. A large use of soft ferrites is in making deflection yoke cores, flyback transformers and convergence coils for television receivers.

Mn-Zn and Ni-Zn ferrites are widely used for these applications. Mn-Zn ferrites are used for operation upto 500 kHz whereas Ni-Zn ferrites can be used for high frequency operation upto 100 MHz.

Materials having nearly square-shaped hysteresis loop are earlier used as memory or logic operation devices in computers, as switching devices, and in information storage. Magnesium-manganese ferrite, manganese-copper ferrite and lithium-nickel ferrite are used for these applications. These ferrites can exist in two stable but different magnetic states capable of storing a 'bit' of information. Usually they are made in the form of tiny rings called 'cores' which are assembled using external connections into a large matrix of wires containing cores at each junction. Microwave devices in the frequency range 1 GHz to 100 GHz rely for their operation on the Faraday rotation. The Faraday effect consists in rotation of the plane of polarization of a plane-polarized electromagnetic wave as it travels through a medium in the direction of an applied magnetic field.

41.15.2 Hard Magnetic Materials- Magnetic Storage Materials

Hard magnetic materials are used in fabrication of permanent magnets, which are required to retain their magnetic field indefinitely. Magnetic materials are widely used for storage of information. The entertainment electronics and computer industries heavily rely on magnetic tapes for the storage and reproduction of audio, video and digital sequences.

The magnetic storage medium consists of magnetic oxide particles distributed in a binder, which strongly bonds to a plastic film. The oxide particles are needle shaped and each particle is single domain magnetized to saturation along its major axis. The most commonly used magnetic particles are γ-Fe2O_3 crystallites. Videotapes use cobalt-doped γ-Fe_2O_3, which has a higher coercivity.

(i) Hard magnetic materials: Hard magnetic materials have a high resistance to demagnetization. A high remanence, high permeability, a high coercive field and a large hysteresis loop characterize the hard magnetic materials.

(ii) Alnico alloys: Permanent magnets are made mainly using alnico alloys. Alnico alloys contain various amounts of aluminium, nickel, cobalt and iron along with some minor constituents such as copper and titanium. Alnico alloys are mechanically hard and brittle. Their magnetic properties are highly stable against variations in temperature.

(iii) Hard Ferrites: Hard ferrites are used in making permanent magnets. The most important hard ferrites are the barium ferrite ($BaFe_2O_3$) and strontium ferrite ($SrFe_2O_3$). They are widely used in generators, relays, loud speakers, telephone ringers, toys etc. Hard ferrite powders are often mixed with plastic materials to form flexible magnets for door closer and other holding devices.

41.16 MAGNETIC DEVICES

(i) Transformers Cores: Transformers consist of a number of parts. The core makes up the bulk of a transformer. It is a piece of high permeability magnetic material used to confine and guide magnetic fields in electrical devices. The composition of a transformer core depends on voltage, current, and frequency. It is usually made of ferromagnetic metal such as iron, or ferrites. The high permeability, relative to the surrounding air, causes the magnetic field lines to be concentrated in the core material. The presence of the core can increase the magnetic field of a coil by a factor of several thousand over what it would be without the core. A range of cores exist, such as steel laminated, solid, toroidal, pot and planar cores, etc. (Fig. 41.24).

Steel Laminated Cores: Steel laminated cores are used for transmitting voltage at the audio frequency level. Laminated core with several steel laminations, protected by an insulator material between layers, confines the eddy currents and lessens magnetizing effects. Although thin laminations are harder to manufacture and are more expensive, they are effective in high frequency transformers.

Several designs of steel laminated transformers are available, each offering its own advantages. An E shaped core is affordable to manufacture, but tends to exhibit more energy loss. A C type core offers reduced resistance because the metal grains run parallel to the energy flux.

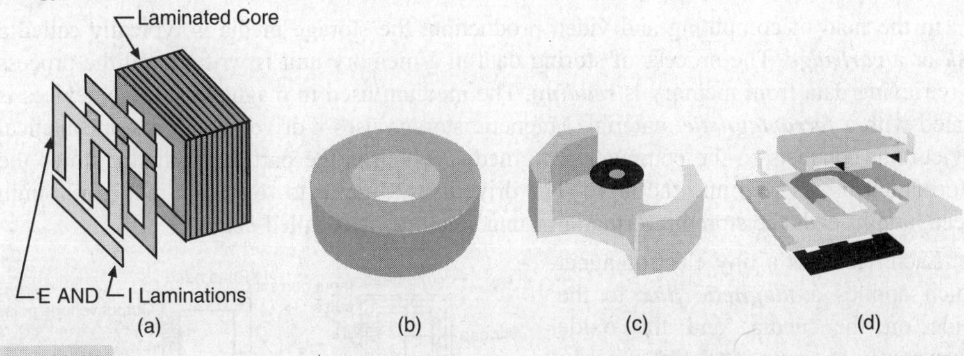

Fig. 41.24: Different types of transformer cores- (a) Steel laminated core (b) toroidal core (c) Pot core (d) planar core

Solid Cores: Solid cores, particularly the powdered iron cores used in circuits, have high magnetic permeability as well as electrical resistance. When used in circuits, they tend to work best for transmission levels above main frequencies. For frequencies that tend to range even higher, such as those beyond the VHF (very high frequency) band, powdered iron is replaced by ferrites.

Toroidal Cores: A range of materials are available for use in toroidal cores, including steel, coiled permalloys, powdered iron, or ferrites. These cores can be circular in structure, with the rest of the transformer built around the core ring—the lack of an opening in the core ring means no air gaps—or they can be a long strip of material. The advantage of using a strip is reduced resistance as a result of properly aligned grain boundaries.

Toroidal cores are more efficient at handling the same kind of energy load than steel laminated E shape cores, and can be made smaller, lighter, and with a lower magnetic field. However, windings tend to be more expensive for toroidal cores.

Pot core: The shape of a pot core is round with an internal hollow that almost completely encloses the coil. Usually a pot core is made in two halves which fit together around a coil former. This design of core has a shielding effect which reduces electromagnetic interference.

Planar core: A planar core consists of two flat pieces of magnetic material, one above and one below the coil. It is typically used with a flat coil that is part of a printed circuit board. This design is excellent for **mass production** and allows a high **power**, small **volume transformer** to be constructed for low cost. It is not as ideal as either a pot core or toroidal core but costs less to produce.

(ii) Magnetic Storage: Data storage devices are the essential requiements of entertainment electronics and computer systems. There are two types of storage devices available, namely semiconductor memories in which flip-flops are the storage elements and magnetic memories in which the magnetic domains are the storage elements. The semiconductor memories are *volatile* whereas the magnetic memories are permanent and *non-volatile*.

The most common form of storage technology is magnetic storage. The data and programs required for computer operations are stored on magnetic bulk storage devices like tapes and disks. Magnetic storage and magnetic recording means storage of information on a magnetized medium. Magnetic storage was first suggested by Oberlin Smith in 1888. The first working magnetic recorder was invented by Valdemar Poulsen in 1898 and in 1928; Fritz Pfleumer developed the first magnetic tape recorder.

In the field of computing and video production, the storage media is typically called a ***disk*** or a ***cartridge***. The process of storing data in a memory unit is writing and the process of retrieving data from memory is ***reading***. The medium used in magnetic-storage devices is coated with a ***ferromagnetic*** material. Magnetic storage uses a drive, which is a mechanical device that connects to the computer. The media, which is the part that actually stores the information, is inserted into the drive. The drive uses a motor to rotate the media at a high speed, and it reads the stored information using small devices called ***heads***.

Each head has a tiny electromagnet which applies a ***magnetic flux*** to the oxide on the media, and the oxide permanently "remembers" the flux. In writing, the data signal is sent through the coil of wire to create a magnetic field in the core (Fig. 41.25). At the gap, the magnetic flux forms a fringe pattern and the flux magnetizes the oxide on the media. When the data is read by the drive, the ***read head*** pulls a varying magnetic field across the gap, creating a varying magnetic field in the core and therefore a signal in the coil. This signal is then sent to the computer as binary data.

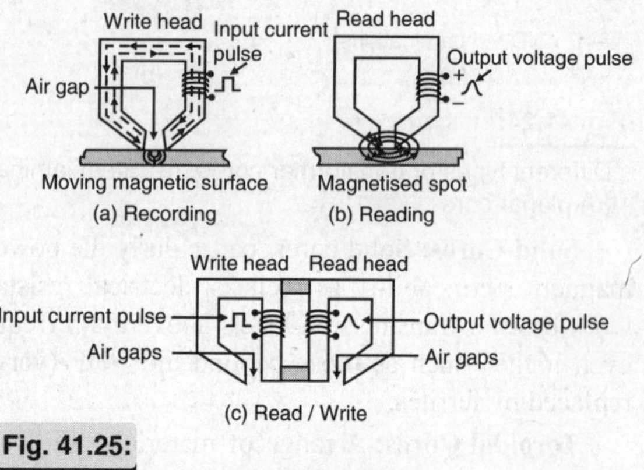

Fig. 41.25: Magnetic recording and reading

(a) Hard disks: Magnetic disk memories provide large storage capabilities with moderate operating speed. A magnetic disk is a flat, circular plate called ***platter*** which has a surface that is coated with magnetic iron oxide particles. It also has a read-write head that hovers over the surface to read data.

A *hard disk* is one or more platters and their associated read-write heads. The platters rotate at very high speed of the order of 3600 rpm. The information is stored on their surface by magnetic heads mounted on access arms. Information is recorded in the form of bands.

Each band of information on a particular disk is called a ***track***. The tracks are divided into **sectors** (see Fig. 41.26). Tracks and sectors are created when the hard disc is first ***formatted*** and this must take place before the disc can be used. The tracks are arranged in ***concentric rings*** so the software can jump from "file 1" to "file 19" without having to fast forward through files 2 through 18. There will be several thousand data tracks on one side of a disk in which the bits are recorded in a track at a density of the order of 20,000 to 1, 00,000 bits/inch.

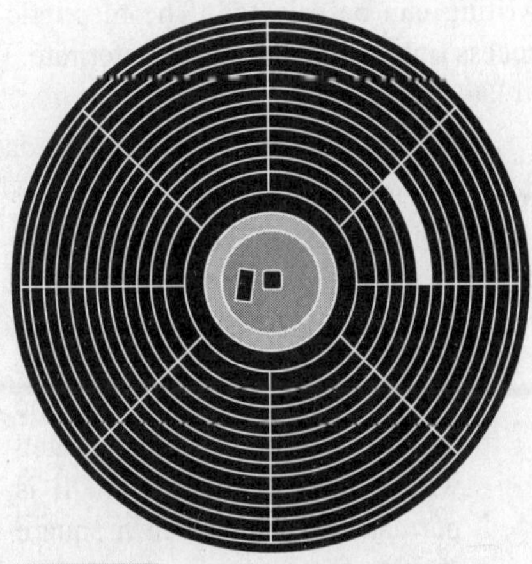

Fig. 41.26:

Tracks and sectors on a hard disk

A typical outer track contains more bits than the inner tracks since the circumference of an outer track is greater. The disk or cartridge spins like a record and the heads move to the correct track, providing what is known as ***direct-access storage***.

The total time taken by the head to begin reading or to begin writing on a selected track is called ***access time***. The time taken to position a head on the selected track is called ***seek time*** which is of the order of milliseconds. The time required for the desired data to reach the magnetic head after the positioning of the head is called ***rotational delay***.

The time requirement will be of the order of milliseconds. The total access time for a disk is the sum of the seek time and rotational delay. The number of bits transferred per second once reading or writing begins is called transfer rate of disk. A schematic diagram of a flying head is shown in Fig. 41.27.

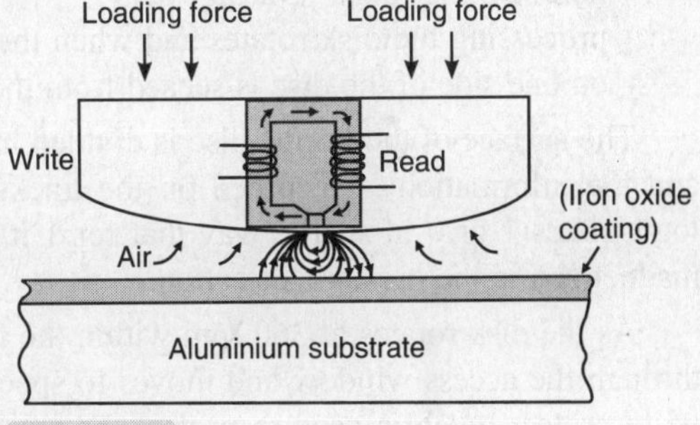

Fig. 41.27:

Schematic diagram of a flying head

When the magnetic disk rotates, a thin but resilient layer of air rotates along with the disk. The shape of the head is designed in such a way that it rides on this layer of air causing the disk to maintain separation from the head. It floats on a cushion of air a fraction of a millimetre above the surface of the disc. The drive is inside a sealed unit because even a speck of dust could cause the heads to crash.

The storage capacity in the modern disk memories is increased by mounting several disks on a common drive unit known as disk pack as shown in Fig. 41.28. During the operation of the memory, the disks are rotated at a uniform speed by a disk drive unit. Each recording surface consists of one read-write head. Using the movable arm connected with the disks, a particular set of track for reading/

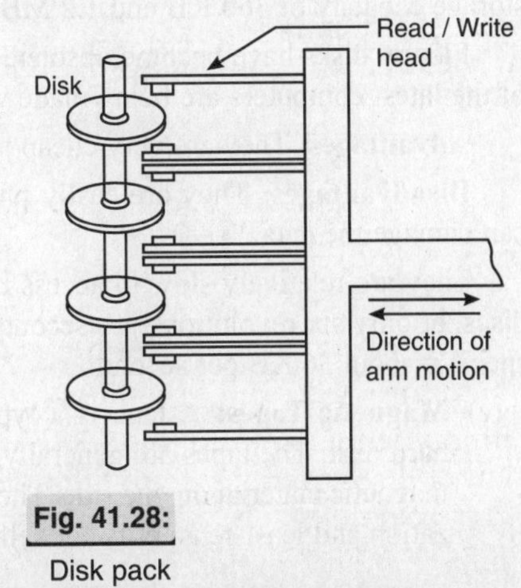

Fig. 41.28:

Disk pack

writing can be selected. The magnetic disk memories are relatively inexpensive, have low access time and high disk transfer rate. Hard disks of size 80 to 250 GB have become common in latest personal computers.

Advantages: Very fast access to data. Data can be read directly from any part of the hard disc ***(random access)***. The access speed is about ***1000 KB per second.***

(b) Floppy disks: The floppy disk is a small flexible direct access magnetic storage device. It is made of very thin and flexible Mylar (plastic) material. The surface of the disk is coated with a thin magnetic film. It is permanently housed in a square jacket for protective purpose as illustrated in Fig. 41.29. The index hole establishes a reference point for all the tracks on the disc. The outer circle shows a hole in the jacket while the inner circle shows the index hole in the disk. During data processing the disk rotates and when the two holes are aligned a beam of light shining on one side of the disc is sensed from the other side and is used for timing functions.

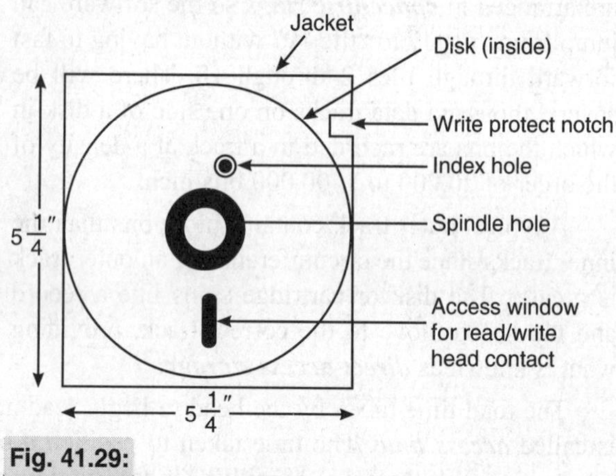

Fig. 41.29: Illustration of floppy disk

The surface of the floppy disc is divided into a number of concentric circles called tracks and the information is recorded on the tracks. Tiny magnetic spots are used to record the logic states 1 or 0 in such a way that for 1 it is magnetized in one direction and for 0, it is magnetized in the opposite direction.

As the disk rotates at 360 rpm within the fixed jacket, the read/write head makes contact through the access window and moves to specified position along the length of the slot. The write protect notch is used to protect the stored information. A typical 5¼ inch floppy disk is organized into 77 tracks and is divided into 26 sectors. Thus, each track is divided into 26 equal-sized sectors. The 5¼ inch disk is available with double density and high density with storage capacity of 360 KB and 1.2 MB.

Floppy disks have become obsolete because of their low capacity range and so now most of the latest computers are being made without a floppy disk drive.

Advantages: They are very cheap to buy and floppy disc drives are very common.

Disadvantages: They are easily physically damaged if unprotected and magnetic fields can damage the data.

They are relatively slow to access because floppy discs rotate far more slowly than hard discs, at only six revolutions per second, and only start spinning when requested. The access speed is about 36 KB per second.

(c) Magnetic Tapes: A tape is a type of backup storage device used to copy data on a hard disk. The tapes are generally thin flexible plastic tapes with a thin coating of hard magnetic material on one side. The magnetic material must have high remanent magnetization and least sensitivity for self-demagnetization. As such materials with rectangular hysteresis loop are used in making the coating. Just like the tape in a tape-recorder, the data is written to or read from the tape as it passes the magnetic heads.

Data on a tape is arranged as a long sequence, beginning at one end of the tape and stretching to the other end. Therefore a tape is a sequential storage device whereas hard disks and floppies are random access storage devices. Access time on a tape is measured in seconds, unlike hard disk drives which are measured in milliseconds. Tape drives have good storage capacity, being able to hold as much information as a hard disk. Most tape drives can back up 1GB of data in 15-20 minutes. Tape storage is not suitable for daily tasks because it is too slow to be the computers main storage device.

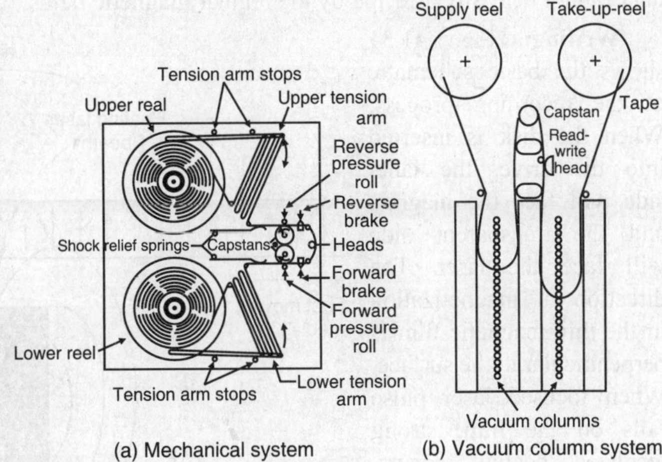

Fig. 41.30: Illustration of tape mechanism

Fig. 41.30 shows a high speed start-stop tape mechanism which uses a set of tension arms around which the tape is laced. The tension arms are movable and when the tape is suddenly driven past the heads by the capstan, the mechanism provides a buffering supply of tape. A servo mechanism is used to drive the upper and lower reels to maintain sufficient tape between capstan and tape reels.

Output signals from read heads are generally in the range of 0.1 to 0.5 V. the recording density varies from few hundred bits/inch to several thousand bits/inch. Data are recorded on magnetic tape by using some coding system. Usually one character is stored per row along the tape.

Advantages: Magnetic tape is relatively cheap and tape cassettes can store very large quantities of data (*typically 26 GB*).

Disadvantages: Accessing data is very slow and we cannot go directly to an item of data on the tape as we can with a disc. It is necessary to start at the beginning of the tape and search for the data as the tape goes past the heads (serial access).

(iii) Magneto-optical Recording: Compact discs are classified based on the storage techniques and capabilities as CD-ROM (Read only Memory), CD-WORM (Write Once Read Many), CD-R/W (Read/Write) etc. They use different materials but write and read the data using a laser beam. In case of CD(R/W), we can write the data, read and rewrite after erasure. Erasability denotes that the recording media can undergo a very large number of write/erase operations without any loss. Magneto-optical (MO) systems write magnetically and read optically.

Principle: A laser beam is used for writing, reading and erasing of data on a magneto-optical medium. A weak laser also generates high local temperatures when focused at a small spot on the medium. Generally the Curie temperature is of the order of 200°C for magnetic materials. As the coercivity of material drops at higher temperatures, the magnetization can be changed by applying a weak magnetic field at higher temperatures. When the material is cooled to room temperature, its coercivity rises back to such a high level that the magnetic data cannot be easily affected by the regular magnetic fields.

Writing: Fig. 41.31 shows the basic schematic of the recording process. When the disk is inserted into the drive, the label side will face the magnet, and the transparent side will face the laser. The direction of magnetization in the thin magnetic film is perpendicular to the surface. When focused laser pulse falls on the film, strong localized heating occurs. At the high temperature, domain reversal takes place due to the action of the downward magnetic field and the upward atomic magnets turn downward.

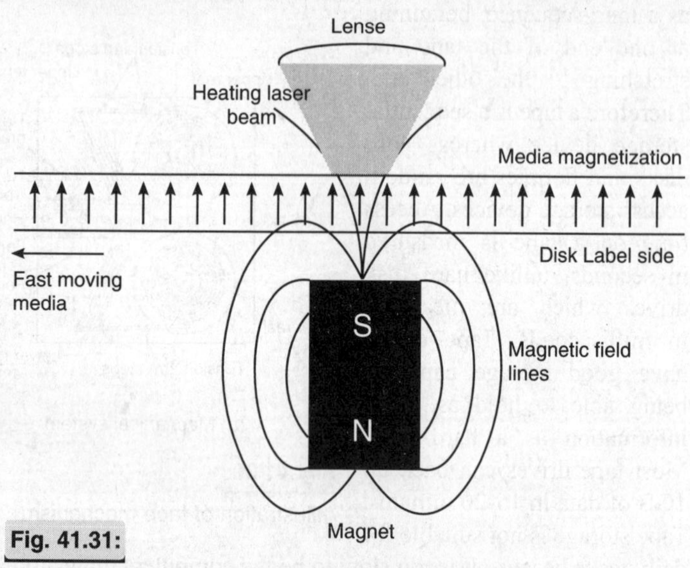

Fig. 41.31: Schematic of the magneto-optic recording process

Reading: Magneto-optic systems use polarized light to read the data from the disk. The changes in light polarization occur due to the presence of a magnetic field on the surface of the disk. If a beam of polarized light is reflected from a magnetized surface, the polarization of the reflected beam will change a little. If the magnetization is reversed, the change in polarization is also reversed. The change in direction of magnetization could be associated with numbers 0 or 1, making this technique useful for binary data storage.

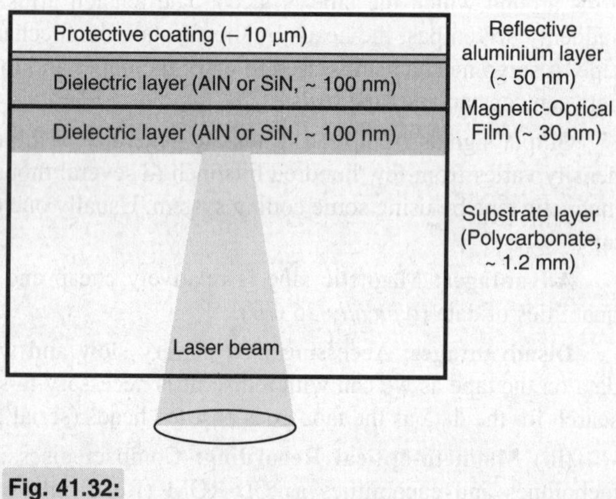

Fig. 41.32: Design of a quadrilayer magneto-optical disk

Magneto-optic Erasing: To erase the data, magnetic field is applied in the upward direction. The focused laser pulse causes local heating which assists the atomic magnets to turn in the upward direction.Thus, data is written, read, erased and rewritten on the MO disc.

Design of MO disk: A thin film of amorphous Terbium iron cobalt (TbFeCo) magnetic film is coated on a substrate. It is the active medium. The active magneto-optical layer is in fact enclosed between thin dielectric layers. The layers act as 'anti-reflection layers' and increase light absorption by the active layer. On top of these layers, a thin aluminium layer is coated. The aluminium layer acts as a light reflector and a heat sink to minimize lateral heating of the active layer.

QUESTIONS

1. Explain magnetic flux density, B, magnetic field strength, H, and magnetization, M. How are they related to each other?
2. Define magnetic susceptibility and permeability. Obtain the relation between them.
3. Why diamagnetic materials have negative susceptibility?
4. What happens to paramagnetic susceptibility when the temperature is increased?
5. Explain how substances are classified according to their magnetic behaviour.
6. Describe diamagnetic, paramagnetic and ferromagnetic materials.
7. Explain their classification of magnetic materials on the basis of permanent magnetic moment.
8. How do you distinguish between diamagnetic, paramagnetic and ferromagnetic materials?
9. What are ferromagnetic materials? Discuss the importance of hysteresis curve. How would you use the hysteresis curve for selecting the material for the construction of a permanent magnet?
10. Define magnetic susceptibility.
11. Explain the significance of Curie temperature for a ferromagnetic material.
12. State the Curie law of paramagnetism. What is the Curie temperature?
13. Explain important magnetic properties of ferromagnetic materials.
14. Explain the terms diamagnetism, paramagnetism, ferromagnetism, antiferromagnetism and ferrimagnetism and ferrimagnetism on the basis of magnetic dipoles of the atoms.
15. What is spontaneous magnetization?
16. What are domains?
17. What is the hysteresis loop? What does it represent? What is the significance?
18. Explain ferromagnetic hysteresis on the basis of domains.
19. What are the losses in ferromagnetic materials? How are they reduced?
20. What is meant by hysteresis? Explain hysteresis loss. How would you use the hysteresis curves to select material for the construction of permanent magnets?
21. Discuss the variation of spontaneous magnetization with temperature for ferromagnetic materials.
22. Compare and contrast ferromagnetism with ferrimagnetism.
23. What are ferrites? What is their importance?
24. Why are ferrites preferred over ferromagnetic materials at high frequency as core material?
25. What are soft and hard magnetic materials? Give their characteristic properties and applications.
26. Show that $B = H + 4\pi I$ and give an expression for energy loss in a complete hysteresis cycle.
27. What are hard and soft magnetic materials? Indicate the properties sought in each case. Give their applications.
28. Laminated iron sheet is commonly used for transformer cores. Why are not castings used?
29. How can the permeability of a ferrite be improved?
30. Explain the difference between the terms 'Curie temperature' and 'Neel temperature'.
31. A polycrystalline iron rod does not show any magnetization. Explain.
32. Write a short note on materials used for magnetic tape.

33. Distinguish between ferro-, ferri-, and antiferromagnetic materials. Out of these, which materials are used as permanent magnets? Why?

34. A ferromagnetic material is placed in a magnetic field of intensity H. What is the energy stored in the material?

35. What should be the characteristics of permanent magnetic materials?

36. Discuss the applications of soft ferrites?

37. What are hard ferrites?

38. What are different types of compact discs? How are data stored, read and rewritten?

PROBLEMS

1. A magnetic material has a magnetization of 2300 A/m and produces a flux density of 31.4 gauss. Calculate magnetizing force and relative permeability of the material.

2. A magnetic field of 1800 A/m produces a magnetic flux of 3×10^{-5} Wb in an iron bar of cross-sectional area 0.2 cm^2. Determine the permeability of iron. **[Ans:** 8.3×10^{-3} H/m]

3. A magnetic material has a magnetization of 3000 A/m and a flux density of 0.005 wb/m^2. Calculate the magnetic force and the relative permeability of the material.
[Ans: 980.9 A/m, 4.06]

4. Find the relative permeability of a ferromagnetic material if field of strength 220 A/m produces a magnetization of 3300 A/m in it. **[Ans:** 16]

5. The molecular weight of a paramagnetic salt is 168.5 and its density is 4370 kg/m^3 at 27°C. Calculate its susceptibility and the magnetization produced in it in a field of 2×10^5 A/m. Assume that the contribution to paramagnetism is two Bohr magnetons per molecule.
[Ans: 5.4×10^{-4}, 108.5 A/m]

6. Assume that iron atoms have magnetic moment of two Bohr magnetons. Calculate the Curie constant if its density is 7150 kg/m^3 and atomic weight is 55.84. **[Ans:** 0.209]

7. Assuming a bcc lattice for iron with lattice constant a = 3 Å, calculate the Curie constant and Weiss constant. Given that the Curie temperature of iron is 1050 K. **[Ans:** 0.193, 5440]

8. A paramagnetic system of electron spin magnetic dipole moment is placed in an external field of 10^5A/m. Calculate the average magnetic moment per dipole at 300 K. **[Ans:** 2.6×10^{-27} A.m^2]

9. A paramagnetic salt contains 10^{28} ions/m^3 with magnetic moment of one Bohr magneton. Calculate the paramagnetic susceptibility and the magnetization produced in a uniform magnetic field of 10^6A/m when the temperature is 27°C.

10. An electron of hydrogen revolves in an orbit of radius 0.51 Å. Calculate the change in magnetic moment for this electron if a magnetic field of induction 2 T acts at right angles to the plane of the orbit. **[Ans:** 3.66×10^{-29} A.m^2]

42 Superconductors

Learning Objectives:

By the end of this chapter, you will be able to:

- Define concepts of Superconductivity and its properties.
- Learn about different types of Superconductors.
- Explain different applications of Superconductivity.

42.1 INTRODUCTION

Superconductivity is a state of matter exhibited usually at very low temperatures where the resistivity of the material drops to zero. The superconducting state is a state in which quantum mechanics operates on a macroscopic scale of the order of many atomic distances rather than the usual atomic and subatomic scale. The superconducting state is influenced by temperature, current and magnetic field. There exist critical values for these three parameters, above which the material passes into normal state. Besides being of immense theoretical interest, the phenomenon has many possible practical applications. Superconducting magnets are designed for use in the fields of medicine and particle physics. It is being vigorously tried for use in transportation and transmission of power. Until recent times the main hurdle in the way of their use is the requirement of extremely low temperatures of the order of 20 K and less. Recent discoveries of high temperature ceramic superconductors raised the hopes of using superconductors in making more efficient and smaller electrical and electronic devices at normal temperatures. Vigorous research is being carried out all over the world to improve the properties of ceramic superconductors to make them suitable for various practical applications.

42.2 PHENOMENON OF SUPERCONDUCTIVITY

Metals are good conductors of electricity as they contain a tremendous number of free electrons. The low resistance offered by them to the flow of current is attributed to the scattering of the free electrons by vibrating ions of the lattice. When temperature increases the amplitude of the lattice vibrations increases and cause more scattering of electrons leading to more resistance. Even at 0 K, metals offer finite resistance, called *residual resistance*, which is attributed to the scattering of electrons by impurities and crystal defects present in the material. The variation of resistivity of normal metals with temperature is shown in Fig. 42.1, which indicates the existence of residual resistance, ρ_r.

H.K. Onnes was verifying the behaviour of metals at very low temperatures and in 1911, he discovered that the electrical resistance of highly purified mercury dropped abruptly to zero at a temperature of 4.15 K (see Fig. 42.2). The sudden drop in resistivity was quite unexpected and Onnes recognized it to be an entirely new phenomenon. Onnes also found that the transition was reversible. When heated above the transition temperature 4.15 K, mercury regained its resistivity. Onnes named the phenomenon as *superconductivity*. Subsequently, superconductivity was discovered in lead, zinc, aluminium, and other metals as well as in a number of alloys. Superconductivity was strictly a low temperature phenomenon till 1980's, when certain ceramic materials were found to exhibit superconductivity at higher temperatures of about 120 K.

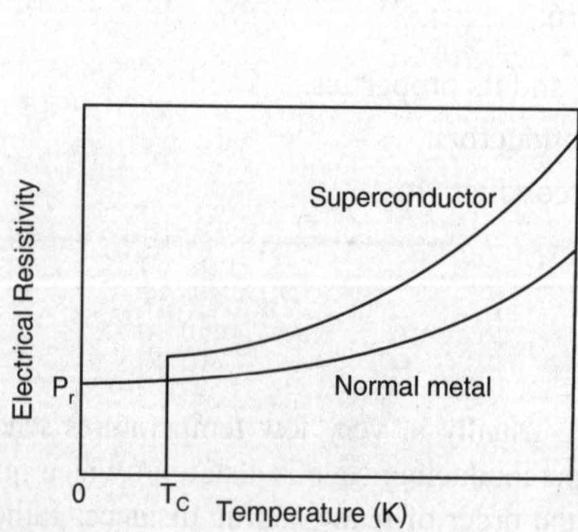

Fig. 42.1: Plot of electrical resistivity of a normally conducting metal and superconducting metal as a function of temperature

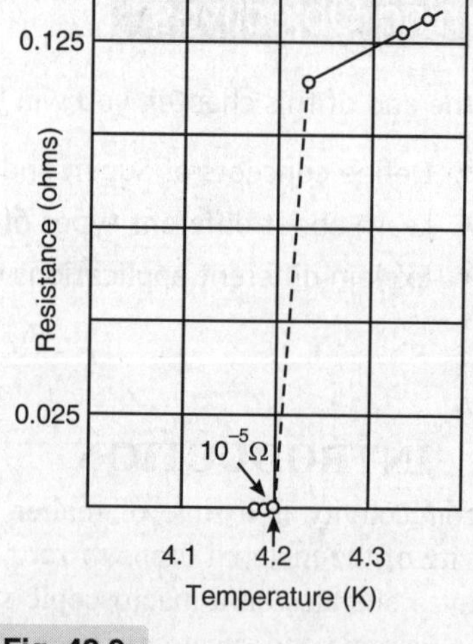

Fig. 42.2: Abrupt change in resistance of mercury at liquid helium temperature

***Superconductivity** is the phenomenon in which electrical resistance of materials suddenly disappears below a certain temperature. Superconductivity is a state of matter exhibited usually at very low temperatures where the resistivity of the material drops to zero and becomes perfect diamagnetic.*

The superconducting state is influenced by temperature, current and magnetic field. The low resistance for flow of current in metals is due to the scattering of the free electrons by vibrating ions of the lattice.

*The temperature at which a normal material abruptly changes into a superconductor is called **transition temperature**, T_C. It is also known as the critical temperature.*

42.3 MATERIALS (LOW T_C MATERIALS)

Since the discovery of superconductivity in the metal mercury, many metals have been found to exhibit superconductivity at liquid helium temperatures. About 26 elements from the periodic table are found to exhibit superconductivity (Fig. 42.3). Apart from these, thousands of alloys display superconductivity.

Good conductors of electric current such as silver, copper and gold are not superconductors. In fact, the superconductors are relatively poor conductors at room temperatures. Non- transition metals such as Nb, Mo and Zn exhibit superconductivity.

H																	He
Li	Be 0.03											B	C	N	O	F	Ne
Na	Mg											Al 1.19	Si 6.7	P 4.6-6.1	S	Cl	Ar
K	Ca	Se	Ti 0.39	V 5.3	Cr	Mn	Fe	Co	Ni	Cu	Zn 0.9	Ga 1.09	Ge 5.4	As 0.5	Se 6.9	Br	Kr
Rb	Sr	Y 0.5;2.7	Zr 0.55	Nb 9.2	Mo 0.92	Tc 7.8	Ru 0.5	Rh 325 μ	Pb	Ag	Cd 0.55	In 3.4	Sn 3.7: 5.3	Sb 3.6	Te 4.5	I	Xe
Ca 1.5	Ba 1.8: 5.1	La 4.8;5.9	Hf	Ta 4.4	W 0.01	Re 1.7	Os 0.65	Ir 0.14	Pt	Au	Hg 4.15 3.95	Ti 2.39 1.45	Pb 7.2	Bi 3.9 7.2;8.5	Po	At	Rn
Fr	Ra	Ac															
			Ce 1.7	Pr	Nd	Pm	Sm	Eu	Gd	Tb	Dy	Ho	Er	Tm	Yb	Lu 0.1-0.7	
			Th 1.37	Pa 1.3	U 0.2	Np	Pu	Am	Cm	Bk	Cf	Es	Fm	Md	No	Lw	

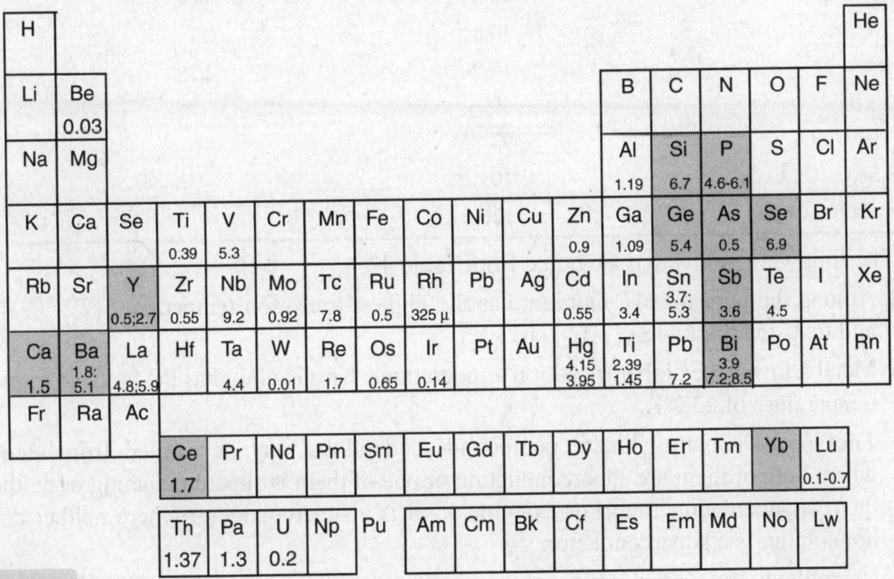

Fig. 42.1:

Elements found to exhibit superconductivity

Semiconductors like Si, Ge, Se and Te change to a metallic phase when subjected to high pressure and then become superconducting at low temperatures. The ferromagnetic materials like Fe, Co, Ni do not exhibit superconductivity phase. Table 42.1 lists some of the superconducting materials and their transition temperatures T_C.

Table 42.1: *Transition Temperatures and Critical Field of Some Superconducting Materials*

Material		*Transition Temperature (K)*	*Critical Field at 0K (Wb/m²)*
		Type I. Superconductors	
Tungsten	(W)	0.015	0.0001
Titanium	(Ti)	0.390	0.0100
Cadmium	(Cd)	0.560	0.0030
Zinc	(Zn)	0.850	0.0054
Molybdenum	(Mo)	0.920	0.0095
Indium	(In)	3.408	0.0281
Tin	(Sn)	3.722	0.0305
Mercury	(Hg)	4.153	0.0411
Vanadium	(V)	5.380	0.1420
Lead	(Pb)	7.193	0.0803
Niobium	(Nb)	9.460	0.1980

	Type II. Superconductors	
Nb Ti	10.0	15.7
Nb NN	15.7	1.5
B_3Al	18.7	32.4
Nb_3Ge	23.2	38.4
V_3G	14.8	2.1
V_3Si	16.9	2.35
La – Ba – Cu – O	40	–
$Y\,Ba_2\,Cu_3O_7$	92	–
$Bi_2Sr_2Ca_2Cu_3O_{10}+\delta$	105	–
$Th_2Ca\,Ba_2\,Cu_2O_{10}+\delta$	125	–

The following points may be noted from Table 42.1.

(*i*) Among the pure metals, niobium has the highest transition temperature (T_C = 9.46 K) and tungsten the lowest (T_C = 0.015 K).

(*ii*) Metal alloys have higher transition temperatures. Nb_3Ge alloy has the highest transition temperature of 23.2 K.

(*iii*) The superconducting binary compounds and alloys may be formed from elements where both of them are superconducting or one of them is superconducting or neither of them is superconducting. For example, CuS is a superconductor where neither copper nor sulphur is a superconductor.

(*iv*) Crystalline structure is not a necessary condition for super conductivity. Amorphous and polycrystalline samples also display superconductivity.

42.4 PROPERTIES OF SUPERCONDUCTORS

Superconductors exhibit many unusual and interesting properties.

42.4.1 Zero Electrical Resistance

A super conductor is characterized by zero electrical resistivity. It is not fundamentally possible to test experimentally whether the resistance is zero. A method devised by Onnes consists of measuring the decrease of the current in a closed ring of superconducting wire. The superconducting ring is kept in a magnetic field and it is cooled to below the critical temperature so that it goes into the superconducting state. When the external magnetic field is switched off, a current is induced in the ring. If the ring had a finite resistance, *R*, the current circulating in the ring would decrease according to the equation

$$I(t) = I(0)e^{-Rt/L} \qquad ...(42.1)$$

where *L* is the inductance of the ring. The decay current is monitored by a change in the magnetic flux through a test coil held close to the superconducting ring.

Any change in the magnetic flux of the superconducting ring will induce an emf in the test coil. Careful measurements established that the resistivity of superconductors could be taken as zero.

42.4.2 Persistent Current

Once a current is started in a closed loop of superconducting material, it will continue to keep flowing, of its own accord, around the loop as long as the loop is held below the critical temperature (see Fig. 42.4). Such a steady current, which flows without diminishing in strength, is called a ***persistent current***.

*A steady current, which flows without diminishing in strength, is called a **persistent current.***

The persistent current does not need external power to maintain it because there do not exist I^2R losses. Persistent current is one of the most important properties of a superconductor. Super conductor coils with persistent current flowing through them produce magnetic fields and can therefore act as magnets. Such a superconducting magnet does not require power supply to maintain its magnetic field.

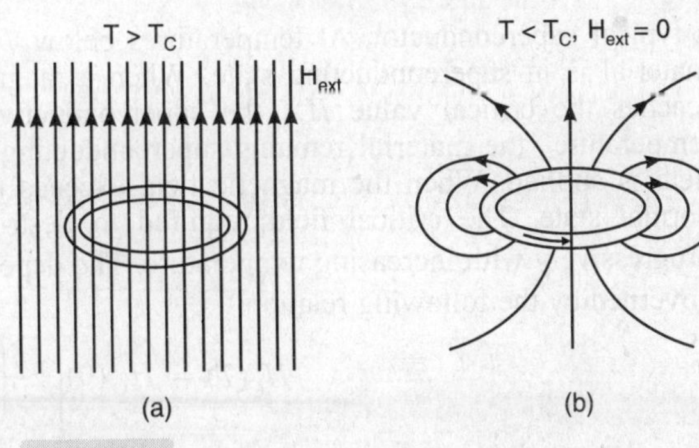

Fig. 42.4:

(a) Superconducting ring is cooled in the presence of a magnetic field. (b) At temperature T less than T_c the magnetic field is switched off. Persistent current is induced in the ring which supports the magnetic flux

42.4.3 Critical Temperature

When a superconducting material is cooled below a certain temperature, it goes into the superconducting state from normal state.

*The temperature at which a material in normal state goes into superconducting state is known as the **critical temperature**, T_C.*

Different materials have different critical temperatures. The transition is reversible.

When the temperature of the material is increased above the critical temperature, it passes into the normal state. The transition is a thermodynamic phase transition. Just as the order in the arrangement of atoms increases at the transition of a material from liquid to solid state, a rearrangement of conduction electrons takes place leading to an increase in the order at the transition from normal to superconducting state. The superconducting transition is sharp for a chemically pure and structurally perfect specimen while the transition range is broad for specimens which are structurally imperfect or which contain impurities.

42.4.4 Critical Magnetic Field

Superconducting state depends on the strength of the magnetic field in which the material is placed. Superconductivity vanishes if a sufficiently strong magnetic field is applied.

*The minimum magnetic field, which is necessary to regain the normal resistivity, is called the **critical magnetic field**, H_C.*

When the applied magnetic field exceeds the critical value H_C, the superconducting state is destroyed and the material goes into normal state.

The value of H_C varies with temperature. Fig. 42.5 shows the dependence of H_C on temperature in

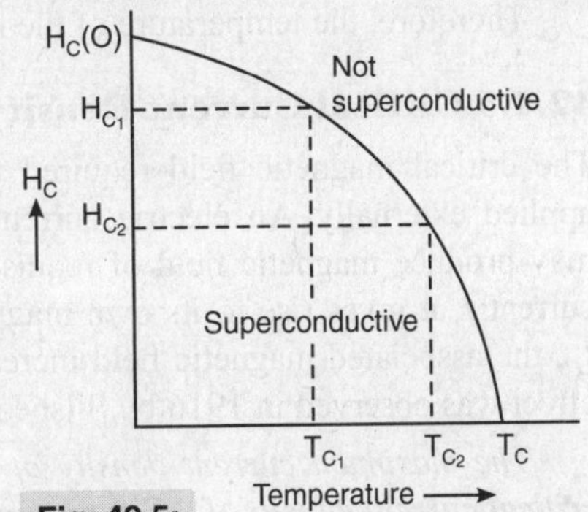

Fig. 42.5:

Schematic representation of the critical magnetic field as a function of temperature a typical superconductor. At temperatures below T_C, in the absence of magnetic field, the material is in superconducting state. When a magnetic field is applied and as its strength reaches the critical value H_C, the superconductivity in the material disappears. At any temperature the material remains superconducting until a corresponding critical magnetic field is applied. When the magnetic field exceeds the critical value, the material goes into normal state. The critical field required to destroy the superconducting state decreases progressively with increasing temperature. The dependence of critical field on temperature is governed by the following relation.

$$H_C(T) = H_C(0)\left[1-\left(\frac{T}{T_C}\right)^2\right] \quad ...(42.2)$$

where $H_C(0)$ is the critical field at 0 K.

Example 42.1: *The transition temperature for Pb is 7.2 K. However, at 5 K it loses the superconducting property if subjected to magnetic field of 3.3 × 10⁴ A/m. Find the maximum value of H which will allow the metal to retain its superconductivity at 0 K.*

Solution: $H_C(0) = \dfrac{H_C(T)}{1-(T^2/T_C^2)} = \dfrac{3.3\times10^4\ \text{A/m}}{1-(25/51.28)} = \mathbf{6.37\times10^4 A/m.}$

Example 42.2: *The critical field of niobium is 1 × 10⁵ A/m at 8 K and 2 × 10⁵ A/m at 0 K. Calculate the transition temperature of the element.*

Solution: $T_C = \dfrac{T}{\left[1-\dfrac{H_C(T)}{H_C(0)}\right]^{\frac{1}{2}}} = \dfrac{8K}{\left[1-\dfrac{1\times10^5\ \text{A/m}}{2\times10^5\ \text{A/m}}\right]^{\frac{1}{2}}} = \mathbf{11.3\ K}$

Example 42.3: *The transition temperature for lead is 7.26 K. The maximum critical field for the material is 8 × 10⁵ A/m. Lead has to be used as a superconductor subjected to a magnetic field of 4 × 10⁴ A/m. What precaution will have to be taken?*

Solution: $T = T_C\left[1-\dfrac{H_C(T)}{H_C(0)}\right]^{\frac{1}{2}} = 7.26\text{K}\left[1-\dfrac{4\times10^4\ \text{A/m}}{8\times10^5\ \text{A/m}}\right]^{\frac{1}{2}} = \mathbf{7.08\ K.}$

Therefore, the temperature of the metal should be held below 7.08 K.

42.4.5 Critical Current Density

The critical magnetic field required to destroy superconductivity need not be necessarily applied externally. An electric current flowing through the superconducting material itself may produce magnetic field of requisite strength. Thus, if a superconducting ring carries a current I, it gives rise to its own magnetic field. As the current increases to a critical value, I_C, the associated magnetic field increases to H_C and the superconductivity disappears. This effect was observed in 1916 by Silsbee and is known as Silsbee effect.

*The maximum current density at which the superconductivity disappears is called the **critical current density**, J_C. For any value of $J < J_C$, the current can sustain itself whereas for values $J > J_C$, the current cannot sustain itself.*

A superconducting ring of radius R ceases to be a superconductor when the current is

$$I_C = 2\pi R H_C \quad ...(42.3)$$

Thus, the existence of a critical current sets a definite limit to the size of the current that can flow through a superconducting coil without disturbing its superconducting state. As the temperature of superconductor is increased, the maximum current that it can carry decreases and falls to zero at the transition temperature of the material. As the critical current falls with temperature, the critical magnetic field will also decrease as the transition temperature is approached. The variation of critical current density J_C and critical magnetic field H_C with temperature is shown in Fig. 42.6.

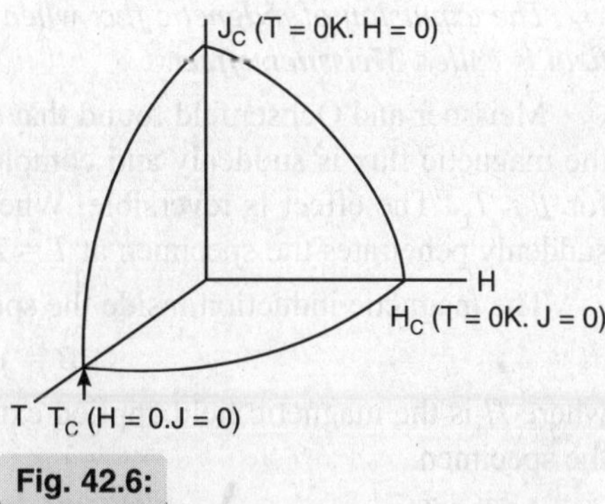

Fig. 42.6:

Critical parameters

Example 42.4: *The critical magnetic field at 5 K is 2 × 10³ A/m in a superconductor ring of radius 0.02 m. Find the value of critical current.*

Solution: $I_C = 2\pi R H_C = 2 \times 3.143 \times 0.02\text{m} \times 2 \times 10^3$ A/m = **251.4 A**

Example 42.5: *Calculate the critical current for a wire of lead having a diameter of 1 mm at 4.2 K. The critical temperature for lead is 7.18 K and $H_C(0) = 6.5 \times 10^4$ A/m.*

Solution: $H_C(T) = H_C(0)\left[1-\left(\frac{T}{T_C}\right)^2\right] = 6.5\times10^4 \text{ A/m}\left[1-\left(\frac{4.2K}{7.18K}\right)^2\right] = 4.28 \times 10^4 \text{A/m}$

The critical current $I_C = 2\pi r H_C = \pi d H_C = 1 \times 3.14 \times 10^{-3}\text{m} \times 4.28 \times 10^4$ A/m = **134.5 A.**

42.4.6 Perfect Diamagnetism – Meissner Effect

In 1933 W.Hans Meissner and Robert Ochsenfeld found that when superconductors are cooled below their critical temperature in the presence of a magnetic field, the magnetic flux is expelled from the interior of the specimen and the superconductor becomes a perfect diamagnetic. This phenomenon is known as ***Meissner effect*** (see Fig. 42.7).

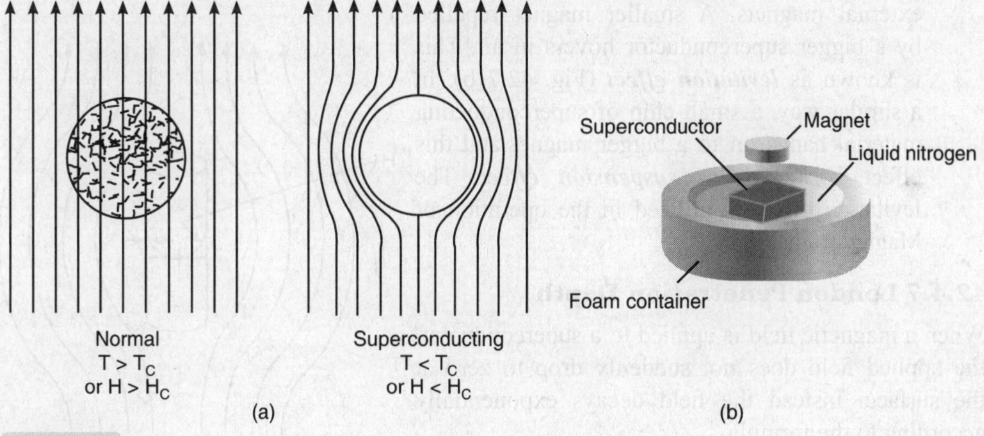

Fig. 42.7:

The Meissner effect- (a) When the specimen becomes superconducting, all the magnetic field is expelled from it (b) A magnet gets repelled by the superconductor, and hovers in air.

*The expulsion of magnetic flux when a material becomes superconducting in a magnetic field is called **Meissner effect**.*

Meissner and Ochsenfeld found that as the temperature of the specimen is lowered to T_C, the magnetic flux is suddenly and completely expelled from it. The flux expulsion continues for $T < T_C$. The effect is reversible. When the temperature is raised from below T_C, the flux suddenly penetrates the specimen at $T = T_C$ and the material returns to the normal state.

The magnetic induction inside the specimen is given by

$$B = \mu_0(H + M) = \mu_0(1 + \chi)H$$

where H is the magnetic field applied externally and M is the magnetization produced within the specimen.

At $T < T_C$, $\quad B = 0$ and therefore $\mu_0(H + M) = 0$

It follows that $\quad M = -H$

The susceptibility of the material is

$$\chi = \frac{M}{H} = -1 \qquad ...(42.4)$$

The specimen is therefore diamagnetic and the state in which magnetization cancels the external magnetic field completely is referred to as ***perfect diamagnetism***.

The Meissner effect contradicts the fundamental principles of electromagnetism. The condition of perfect diamagnetism cannot be explained from the simple definition that superconductivity is a state of zero resistivity. Meissner effect shows that in the superconductor not only $\frac{dB}{dt} = 0$ but also $B = 0$. Thus, two mutually independent properties, namely zero resistivity and perfect diamagnetism are the essential properties that characterize the superconducting state.

Applications of Meissner effect

(*i*) The Meissner effect is the standard test used to conclusively prove whether a particular-material is a superconductor or not.

(*ii*) A material in superconducting state is a perfect diamagnetic and hence strongly repels external magnets. A smaller magnet repelled by a bigger superconductor hovers in air. This is known as ***levitation effect*** (Fig. 42.7 b). In a similar way, a small chip of superconducting material hangs on to a bigger magnet and this effect is known as ***suspension effect***. The levitation effect is utilized in the operation of Maglev trains.

42.4.7 London Penetration Depth

When a magnetic field is applied to a superconductor, the applied field does not suddenly drop to zero at the surface. Instead the field decays exponentially according to the formula

$$H(x) = H(0)e^{-x/\lambda} \qquad ...(42.5)$$

where $H(0)$ is the field applied at the surface at $x = 0$ and x is the distance from the surface.

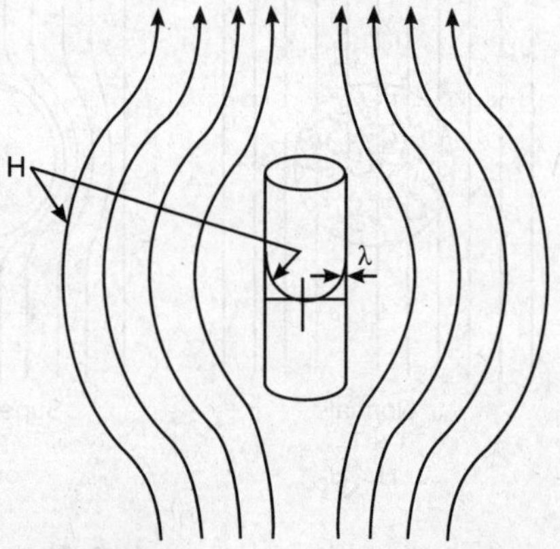

Fig. 42.8: Field penetration in a cylindrical superconducting body

The length λ is called the ***London penetration depth***. It may be defined as the effective depth to which a magnetic field penetrates a superconductor. The penetration depth λ ranges from 300 to about 5000 Å depending on the material. It is independent of frequency of the magnetic field but it strongly depends on temperature.

The temperature dependence of λ is given by the relation

$$\lambda(T) = \frac{\lambda(0)}{\sqrt{\left[1-\left(\frac{T}{T_C}\right)^4\right]}} \qquad ...(42.6)$$

where λ(*T*) and λ(0) are the penetration depths at *T* and 0 K.

It follows from Eq. (42.6) that λ increases with the increase in temperature and at the critical temperature, it becomes infinite. At $T = T_C$ the material goes into the normal state and hence the magnetic field penetrates the whole specimen.

Example 42.6: *Calculate the penetration depth of lead at 5.2 K if the London penetration depth at 0 K is 37 nm. The critical temperature of lead is 7.193 K.*

Solution: $\lambda(T) = \lambda(0)\left[1-\left(\frac{T}{T_C}\right)^4\right]^{-1/2} = 37\text{nm}\left[1-\left(\frac{5.2K}{7.193K}\right)^4\right]^{-1/2}$ = **43.4 nm.**

42.4.8 Flux Quantization

In 1957 A.A. Abrikosov predicted the existence of magnetic flux quanta. According to him, a closed superconducting loop can enclose magnetic flux only in integral multiples of a fundamental quantum of flux. The magnetic flux enclosed by a superconducting ring is given by

$$\varphi = n\frac{h}{2e} = n\varphi_0 \qquad n = 1, 2, 3,....$$

where

$$\varphi_0 = \frac{h}{2e} \qquad ...(42.7)$$

is the flux quantum and is called a **fluxon**. Its value is

$$\varphi_0 = 2.07 \times 10^{-15} \text{ weber}$$

The quantization of magnetic flux has been confirmed experimentally in 1961 by Deaver and Fairbank.

42.4.9 Entropy

Entropy is a measure of the disorder of a system. Generally, entropy decreases with decreasing temperature. In superconducting materials, the entropy decreases linearly up to critical temperature and more remarkably below the critical temperature (see Fig. 42.9). It indicates that a superconducting state is more ordered than the normal state.

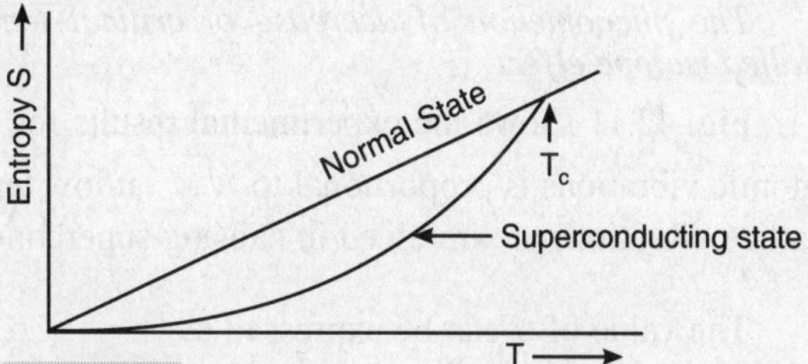

Fig. 42.9: Entropy of Aluminium in the normal and superconducting states as a function of temperature

42.4.10 Heat Capacity

The transition of a metal from its normal state to superconducting state does not involve a change of crystallographic structure. It is in fact a thermodynamic phase transition in which the specific heat changes intermittently at the transition temperature T_C. In case of a normally conducting metal, the specific heat is composed of a lattice part and an electronic part. The electronic contribution varies smoothly at low temperatures. In superconductors it is found that the electronic part decreases exponentially, as shown in Fig. 42.10, and it is given by

$$C_\upsilon = e^{-b\,T_C/T} \qquad \text{...(42.8)}$$

where b is a constant.

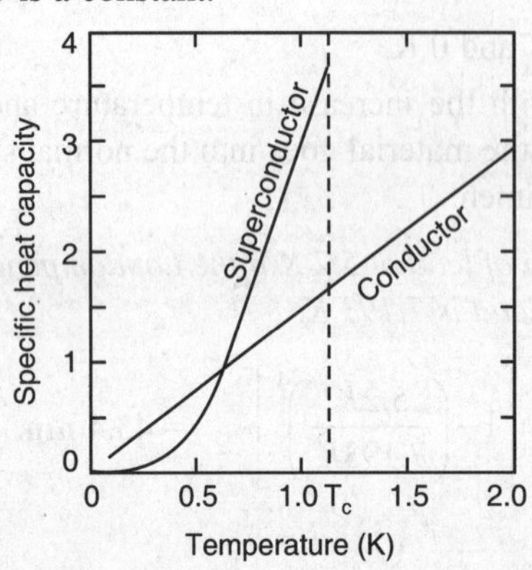

Fig. 42.10: A plot of specific heat as function of temperature in a superconductor and normal conductor.

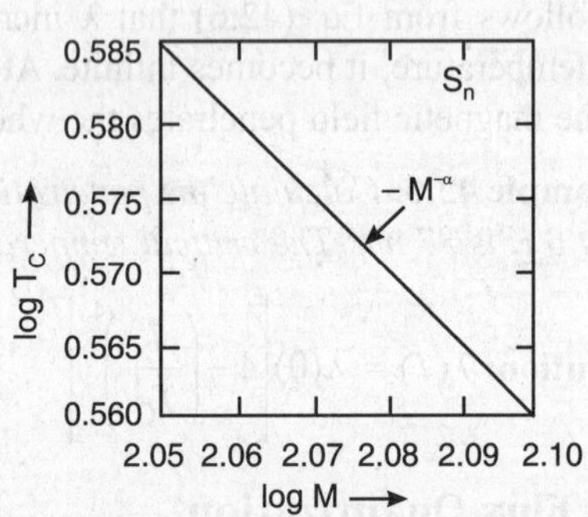

Fig. 42.11: Isotope effect for tin metal

42.4.11 Isotope Effect

In 1950, C.A. Reynolds and E. Maxwell found that the critical temperature decreases with increasing isotopic mass M. The variation is given by the relation

$$T_C \propto M^{-\alpha}$$

or

$$M^{\alpha} T_C = \text{constant} \qquad \text{...(42.9)}$$

where α is a constant and is approximately equal to $\frac{1}{2}$.

The phenomenon of decrease of critical temperature with increasing atomic mass is called ***isotope effect.***

Fig. 42.11 shows the experimental results for tin. Since the mean square of amplitude of atomic vibrations is proportional to $\sqrt{M}$ at low temperatures, the Eq. (42.9) suggests that the lattice vibrations are involved in causing superconductivity.

The value of α can be expressed as

$$\alpha = -\frac{\partial(\ln T_C)}{\partial(\ln M)} \qquad \text{...(42.10)}$$

Change in the isotopic mass does not change the electronic structure. Dependence of the transition temperature on the isotopic mass suggests involvement of electron-lattice interaction.

Example 42.7: *In a superconducting material isotopic mass is 199.5 amu and critical temperature is 5K. Calculate isotopic mass at 5.1 K.*

Solution. $T_C \propto \frac{1}{\sqrt{M}}$. Therefore, $T_1\sqrt{M_1} = T_2\sqrt{M_2}$

or $$M_2 = M_1\left[\frac{T_1}{T_2}\right]^2 = 199.5 \text{ amu}\left[\frac{5\text{K}}{5.1\text{K}}\right]^2 = \mathbf{191.68 \text{ amu.}}$$

42.5 OTHER EXTERNAL FACTORS THAT AFFECT SUPERCONDUCTIVITY

(*i*) **Stress:** The transition temperature can be changed by application of stress. Usually, stress which increases the dimensions increases the transition temperature. Highly sensitive methods are required to detect the change.

(*ii*) **Impurities:** Addition of chemical impurities modifies nearly all the superconductivity properties and in particular the magnetic properties.

(*iii*) **Size:** If the size of the specimen is reduced below about 100 μm,its superconducting properties are modified in many respects. The magnetic permeability of the small specimen is no longer zero and varies with temperature while the critical magnetic field becomes greater than that of bulk material.

(*iv*) **Frequency:** The zero resistance of superconductors is modified at very high frequencies of alternating current. Up to 10 MHz the resistance is still zero but for frequencies of the order of GHz, it is found that considerable resistance is observed even below T_C.

42.6 TYPE-I AND TYPE-II SUPERCONDUCTORS

Superconductors are of two types such as Type I and Type II.

(a) Type-I Superconductors: Superconductors are divided into two categories depending on the way in which the transition from superconducting to normal state proceeds when the externally applied magnetic field exceeds H_C.

In type I superconductors, the transition from superconducting state to normal state in the presence of magnetic field occurs sharply at the critical value of H_C, as shown in Fig. 42.12 (a). Type I superconductors are perfectly diamagnetic below H_C and completely expel the magnetic field from the interior of the superconducting phase. The magnetization of the material grows in proportion to the external field up to the critical field strength Hc and then it suddenly drops to zero at the transition to the normal state, Fig. 42.12 (b). The magnetic field can penetrate only the surface layer and current can flow only in this layer. Consequently, type I superconductors are poor carriers of electrical current.

Aluminium, lead and indium are examples of Type I superconductors.

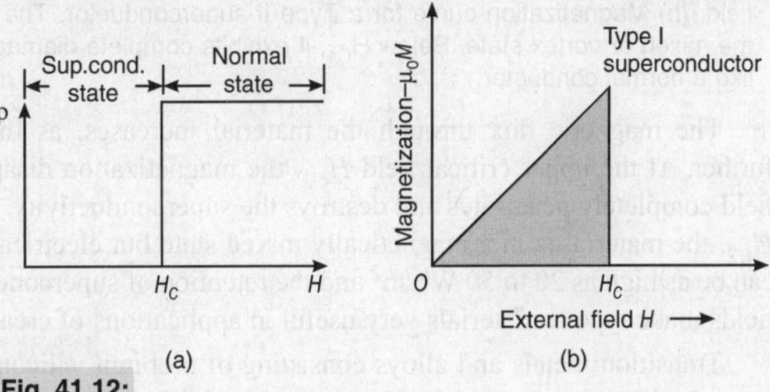

Fig. 41.12: Typical variation of resistivity of a Type I superconductor as a function of applied magnetic field strength. The resistivity abruptly jumps from zero to a high value at H_C. (b) Magnetization curve for a Type I.

The critical field is relatively low for type I superconductors. They would generate magnetic fields of about 100 to 2000 G only. Hence, they are not of much use in production of high magnetic fields. Type I superconductors are also called ***soft superconductors***. We summarize here the characteristics of Type I superconductors.

Characteristics of Type-I superconductors:

(*i*) They are perfectly diamagnetic and exhibit complete Meissner effect.

(*ii*) They have only one critical field. At the critical field the magnetization drops to zero.

(*iii*) The maximum critical field for type I superconductor is of the order of 0.1 Wb/m^2.

(*iv*) The transition at H_C is reversible. Below H_C the material behaves as a superconductor, and above H_C it behaves as a normal conductor.

Disadvantages: Type I superconductors cannot carry large currents and hence are not of much use in producing high magnetic fields.

(b) Type-II Superconductors: Type II superconductivity was discovered by Schubnikov in 1930s and was explained by Abrikosov in 1957. Type II superconductors are characterized by two critical fields H_{C1} and H_{C2}. The transition from superconducting state to normal state occurs gradually as the magnetic field is increased from H_{C1} to H_{C2}, as shown in Fig. 42.13. The magnetization of the material grows in proportion to the external field up to the lower critical field H_{C1}. The external magnetic flux is expelled from the interior of the material till then. At H_{C1} the magnetic field lines begin penetrating the material.

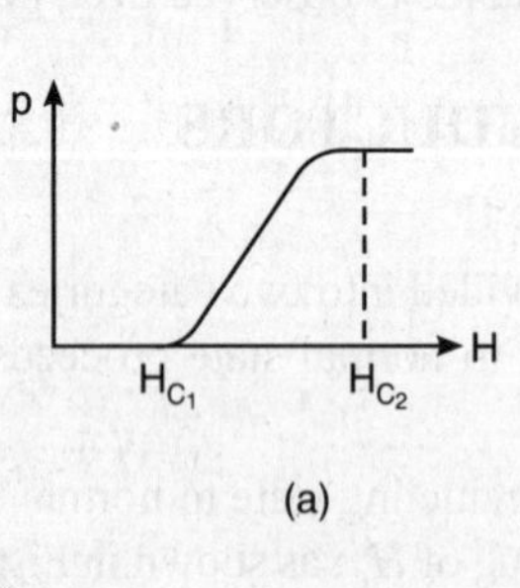

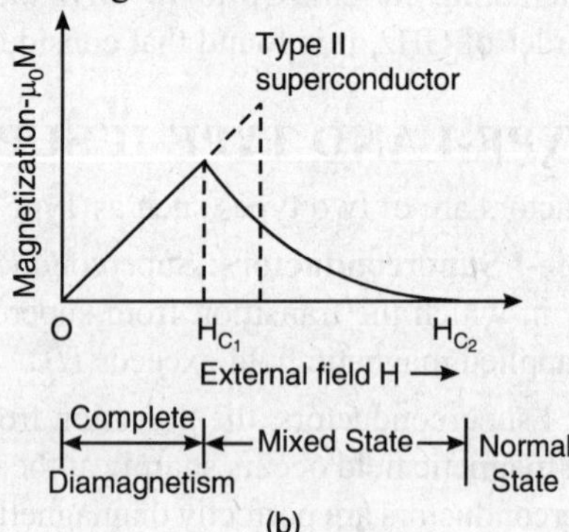

Fig. 42.13: (a) Typical variation of the resistivity of a Type II superconductor as a function of applied magnetic field. (b) Magnetization curve for a Type II superconductor. The region between H_{C1} and H_{C2} is the mixed or vortex state. Below H_{C1}, it exhibits complete diamagnetism and above H_{C2} behaves like a normal conductor.

The magnetic flux through the material increases, as the magnetic field is increased further. At the upper critical field H_{C2}, the magnetization disappears totally and the external field completely penetrates and destroys the superconductivity. In the region between H_{C1} and H_{C2}, the material is in a magnetically mixed state but electrically it is a superconductor. H_{C2} can be as high as 20 to 50 Wb/m^2 and the retention of superconductivity in such high magnetic fields make type II materials very useful in applications of creating very high magnetic fields.

Transition metals and alloys consisting of niobium, silicon and vanadium exhibit type II superconductivity. Ceramic superconductors also belong to this category.

A distinguishing feature of type II superconductors is that super currents arising in an external magnetic field can flow not only over the surface of a conductor but also in the bulk. Above the lower critical field H_{C1}, it becomes energetically more favourable to admit a single flux quantum rather than have the superconductor exclude H_{C1}. The superconductor passes into a mixed state where the bulk of the material is superconducting but is threaded by very thin filaments of normal material (see Fig. 42.14). The thin filaments of normal material serve as the paths along which the magnetic field penetrates. In the center of the filament superconductivity is absent. These normal regions are surrounded by vortices of super currents. At a flux line together with its current vortex called fluxoids appear in the material. Its number increases as the magnetic field is increased.

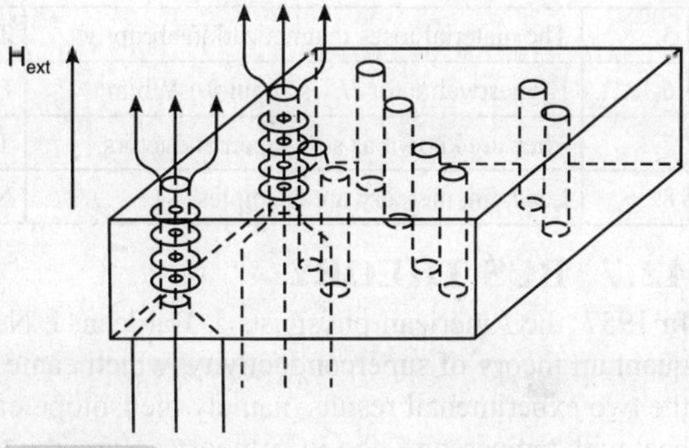

Fig. 42.14: At H_{C2} the fluxoids fill the entire specimen and superconductivity disappears

An increase in the magnetic field will not cause an increase of the flux in each vortex line but will cause an increase in the number of fluxoids threading the superconductor. At H_{C2} the fluxoids fill the entire specimen and superconductivity disappears.

Type II superconductors can carry larger currents when the magnetic field lies between and H_{C2}. Type II superconductors are called ***hard superconductors***. We summarize here the characteristics of Type II superconductors.

Characteristics of Type-II superconductors

(*i*) They have two critical magnetic fields, H_{C1} and H_{C2}.

(*ii*) The material is perfect diamagnetic below the *lower critical field*, H_{C1}. Meissner effect is complete in this region. Above the *upper critical field*, H_{C2}, magnetic flux enters the specimen.

(*iii*) Above H_{C1} they do not show complete Meissner effect and therefore do not behave as perfect diamagnetic materials.

(*iv*) They exist in an *intermediate state* in between the critical fields, H_{C1} and H_{C2}. The intermediate state is a mixture of the normal and superconducting states, magnetically but electrically the material is a superconductor.

(*v*) At H_{C2} the magnetization vanishes and the specimen returns to normal conducting state.

(*vi*) The upper critical field is very high and is of the order of 30 Wb/m^2.

Applications

They are used in applications of generating very high magnetic fields.

Table 42.2: *Comparison between Type I And Type II Superconductors*

Sl.No.	Type-I Superconductors	Type-II Superconductors
1.	They exhibit complete Meissner effect	They do not exhibit complete Meissner effect
2.	They show perfect diamagnetic behaviour	They do not show perfect diamagnetic behaviour
3.	They have only one critical magnetic field, H_C	They have two critical magnetic fields, lower critical magnetic field, H_{C1} and upper critical magnetic field, H_{C2}
4.	There is no mixed state or intermediate state in case of these materials	Mixed state or intermediate state is present in these materials

5.	The material loses magnetization abruptly	The material loses magnetization gradually
6.	Highest value for H_C is about 0.1 Wb/m^2	Upper critical field is of the order of 30 Wb/m^2
7.	They are known as soft superconductors	They are known as hard superconductors
8.	Lead, tin, mercury are examples	Nb-Sn, Nb-Ti, Nb-Zr, Va-Ga are examples

42.7 BCS THEORY

In 1957, the American physicists J. Bardeen, L.N. Cooper and J.R. Schrieffer developed the quantum theory of superconductivity, which came to be known as BCS theory. Starting from the two experimental results, namely the isotope effect and the variation of electronic specific heat with temperature, the BCS theory assumed interaction of two electrons through quanta of lattice vibrations. It successfully explained the effects like zero resistivity, Meissner effect etc. The two principal features of BCS theory are

(*i*) Cooper pairs, electrons pairs are formed and propagate throughout the lattice.

(*ii*) Cooper pairs propagate without resistance as the electrons move in resonance with phonons.

To appreciate the formation of Cooper pair, let us consider the model in Fig.42.15, in which two electrons propagate along a single lattice row. Each electron experiences an attraction towards its nearest positive ion. When the electrons get very close to each other in the region between ions, they repel each other due to Coulomb force. In an equilibrium condition, a balance between attraction and repulsion is established and the two electrons combine to form a Cooper pair. At normal temperatures, the attractive force is too small and pairing of electrons does not take place. However, at lower temperatures, such pairing is energetically advantageous. In a typical superconductor, the dense cloud of cooper pairs forms a collective state and the motion of all the cooper pairs is correlated. As such the pairs drift cooperatively through the material. Thus, the superconducting state is an *ordered state of the conduction* electrons. Since the density of cooper pairs is very high, even large currents require only a small velocity. The small velocity of ordered cooper pairs minimizes collision processes and leads to zero resistivity.

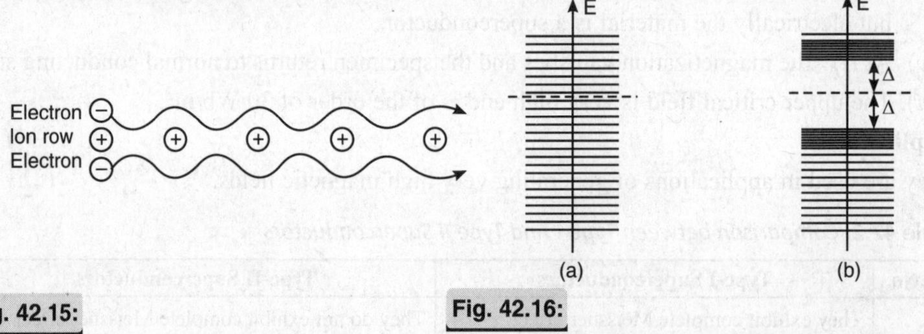

Fig. 42.15: Illustration of cooper pair

Fig. 42.16: Energy band diagram of superconducting state

The electrons of a cooper pair have a lower energy than two unpaired electrons. The theory predicted the existence of an ***energy gap*** between the ground state (superconducting state) and first excited state (see Fig. 42.16). The energy gap represents the energy required to break up a cooper pair. Hence, larger energy gaps correspond to more stable superconductors. According to BCS theory, the energy gap at 0K is given by

$$E_g(0) = 2\Delta \cong 3.52kT_C \quad ...(42.11)$$

The energy gap is generally of the order of 10^{-3} eV. The existence of energy gap can be proved experimentally. One such experiment involves studying the absorption of microwaves by a superconductor. At temperatures close to absolute zero, a superconductor does not absorb energy until the energy quanta of incident radiation is equal to greater than 2Δ. The absorption then grows fast to a value typical for the normal metal, because electrons can now absorb photons and go to higher energy states that lie above the energy gap.

The BCS theory also predicted flux quantization.

42.8 JOSEPHSON EFFECT

In 1962, Brian Josephson predicted that cooper pairs could tunnel through an insulating layer, which separates two super conductors. The superconductor-insulator-superconductor layer constitutes the ***Josephson junction***, as shown in Fig. 42.17 (b). The insulating layer is of the order of 1 nm thickness. Josephson predicted that the tunneling can occur without any resistance, giving rise to a direct current when the voltage applied across the junction is zero and an alternating current when the applied voltage is a dc voltage.

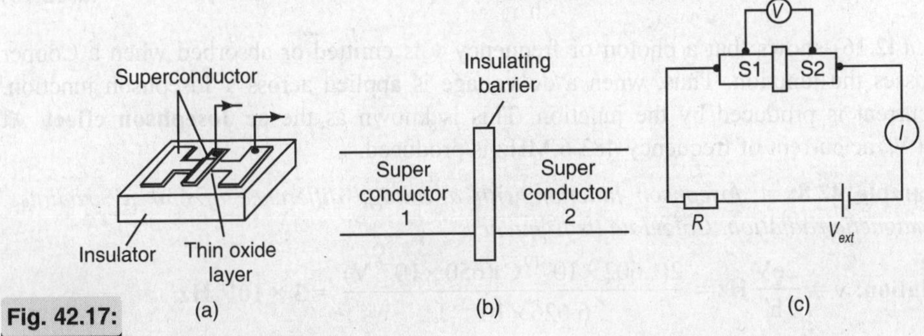

Fig. 42.17: Illustration of Josephson Effect

42.8.1 The dc Josephson Effect

Two superconductors separated by a thick insulating layer, say of 10 nm thickness, are essentially two independent superconductors without any joint properties. When the insulator layer is thin, say 1nm thick, they become a system of coupled conductors. The cooper pairs tunnel through the barrier (insulating layer) as a single unit.

Let us consider a Josephson junction consisting of two superconducting metal films separated by a thin oxide barrier of 10 to 20 Å thick. Let it be connected in a circuit as shown in Fig. 42.17 (c). The cooper pairs in a superconductor can be represented by a wave function, which is the same for all pairs. The cooper pairs tunnel from one side of the junction to the other side easily. The effect of the insulating layer is that it introduces a phase difference between the wave function of cooper pairs on one side of the insulating layer and the wave function of the pairs on the other side. Because of this phase difference, a super current appears across the junction even though the applied voltage is zero. This is known as the **dc Josephson Effect**. Josephson showed that the super current through the junction is given by

$$I_S = I_C \sin \varphi_0 \quad ...(42.12)$$

where φ_0 is the phase difference between the wave functions describing Cooper pairs on both sides of the barrier, and I_C is the critical current at zero voltage condition.

I_C depends on the thickness and width of the insulating layer and the temperature.

42.8.2 The ac Josephson Effect

If we apply a dc voltage across the Josephson junction, it introduces an additional phase on cooper pairs during tunneling. As a result a strikingly new phenomenon will be observed. The dc voltage generates an alternating current I given by

$$I = I_C \sin(\varphi_0 + \Delta\varphi) \quad ...(42.13)$$

Because of the dc voltage V applied across the barrier, the energies of cooper pairs on both sides of the barrier differ in energy by 2eV. Using the quantum mechanical calculations, it can be shown that

$$\Delta\varphi = 2\pi t\left(\frac{2eV}{h}\right) \quad ...(42.14)$$

$$\therefore \quad I = I_C \sin\left[\varphi_0 + 2\pi t\left(\frac{2eV}{h}\right)\right] \quad ...(42.15)$$

The current given by Eq. (42.15) represents an alternating current of frequency

$$\nu = \frac{2eV}{h} \quad ...(42.16)$$

Eq. (42.16) shows that a photon of frequency ν is emitted or absorbed when a Cooper pair crosses the junction. Thus, when a dc voltage is applied across a Josephson junction, an ac current is produced by the junction. This is known as the **ac Josephson effect**. At V = 1 μ V, ac current of frequency 483.6 MHz is produced.

Example 42.8: *A Josephson junction with a voltage difference of 650 μV radiates electromagnetic radiation. Calculate its frequency.*

Solution: $\nu = \frac{2eV}{h}\text{Hz} = \frac{2(1.602\times10^{-19}\text{C})(650\times10^{-6}\text{V})}{6.626\times10^{-34}\text{J}}$ = **3 × 10^{11} Hz.**

42.9 HIGH T_C SUPERCONDUCTORS

Super conductors are divided into low T_C and high T_C superconductors based on their transition temperature. Broadly, materials having T_C below 24 K are regarded as low T_C superconductors and those having T_C above 27 K are regarded as high T_C superconductors. However, in practice, materials for which liquid nitrogen cooling can cause transition to superconducting state may be regarded as high T_C superconductors while those that require liquid helium coolant are considered as low T_C superconductors. The maximum transition temperature that could be achieved before 1980's was 23.2 K in Nb_3Ge, which is a metallic alloy. Therefore, it was hoped that metallic alloy systems could be made to have higher transition temperatures but such systems were not discovered. Therefore, the focus has shifted to ceramic oxides, which are insulating materials at normal temperatures.

In 1986 Bednortz and Muller discovered superconductivity in ceramic materials. They found that the mixed metallic oxide of lanthanum-barium-copper ($La_1Ba_2Cu_3O_7$) exhibited superconductivity at about 30 K. The superconductivity of the oxide was linked with the deficiency of oxygen ions in the oxide compound. When this deficiency of oxygen was carefully controlled, by keeping the samples in oxygen atmosphere, it was found that the material would exhibit superconductivity in the temperature range 30 to 42 K. In 1987 Chu and coworkers replaced lanthanum with yttrium and prepared $YBa_2Cu_3O_7$ with transition temperature of about 95 K. This was a major breakthrough, as this can be maintained in the superconducting state with far less expensive liquid nitrogen coolant (nitrogen exists in liquid state below 77 K) and marks the beginning of preparation of high superconductors. This oxide also contains a deficiency of oxygen with the chemical formula $YBa_2Cu_3O_{7-\delta}$ where δ indicates the deficiency of oxygen and is in the range 0 to 0.1. $YBa_2Cu_3O_7$ is the most extensively studied high superconductor and exhibits a defective perovskite structure with three perovskite cubic unit cells stacked on top of each other (Fig. 42.18). For an ideal stack of three perovskite cubic unit cells, the $YBa_2Cu_3O_x$ compound should have the composition $YBa_2Cu_3O_9$, in which x would be equal to 9. However, analysis shows that x ranges from 6.65 to 6.90 for this material to exhibit superconductivity. At $x = 6.90$, its T_C is highest (≈ 90 K) and at $x = 6.65$, superconductivity disappears.

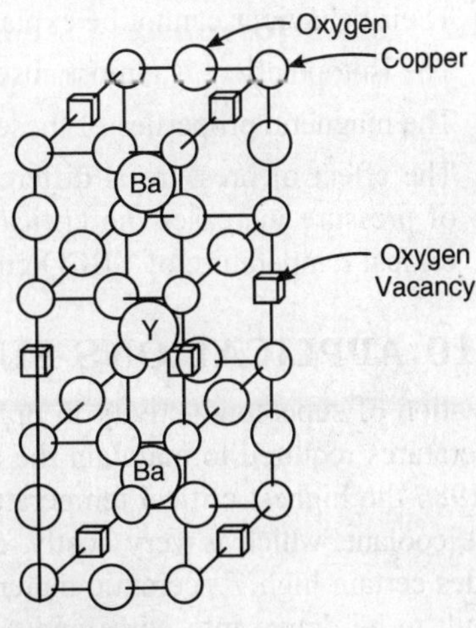

Fig. 42.18: Superconductivity in ceramic materials- yittrium-barium-copper oxide

Thus, oxygen vacancies are found to play a key role in the superconducting behaviour of ceramic oxides. When the cell contains one atom of rare earth metal, two barium atoms, three copper atoms and seven oxygen atoms, then such compounds are called 1-2-3 superconductors.

These high T_C copper oxide superconductors belong to type II and their upper critical field is very high of the order 150 to 200 tesla. Apart from $YBa_2Cu_3O7_{-\delta}$ compounds, the compounds based on bismuth like Bi-Sr-Cu-O or Ba-Ca-Cu-O systems are also superconducting. For example, the compound $Bi_2CaSr_2Cu_2O_{8+x}$ exhibits superconductivity and have the transition temperature 85 K and $Bi_2Ca_2Sr_2Cu_3O_{10+x}$ has the transition temperature 110 K. In 1993, a still higher of 133 K was achieved in mercury based copper oxide $HgBa_2Ca_2Cu_3O_{1+x}$.

The above discussed mixed oxide materials may be arranged into four major families.

(*i*) LBCO (mixed oxide of lanthanum, barium and copper)

(*ii*) YBCO (mixed oxides of yttrium, barium and copper)

(*iii*) BSCCO (mixed oxides of bismuth, strontium, calcium and copper)

(*iv*) TBCCO (mixed oxides of thallium, barium, calcium and copper)

Continuous search is going on to discover materials that may exhibit superconducting state around room temperature. Interesting practical applications of superconductors are visualized in many fields, but they could not be used on large scale because of the requirement of liquid helium, which is highly expensive. Therefore, they remained as a laboratory curiosity for a long time. The discovery of high temperature materials opened up possibilities of putting superconductors to large-scale use. They require only liquid nitrogen, which is easily available and also cheaper.

Properties of high T_C superconductors: Some of the properties of high T_C superconductors are as follows:

(*i*) The high T_C superconductors are brittle in nature.

(*ii*) The properties of the normal state of these materials are highly anisotropic.

(*iii*) The Hall coefficient is positive indicating that the charge carriers are holes.

(*iv*) Their behaviour cannot be explained by BCS theory.

(*v*) The isotope effect is almost absent in these materials.

(*vi*) The magnetic properties of these materials are highly anisotropic.

(*vii*) The effect of pressure is different on different materials. For example the application of pressure increases the critical temperature of LBCO compounds but decreases the critical temperature of YBCO compounds.

42.10 APPLICATIONS SUPERCONDUCTIVITY

Utilization of superconductivity in practical applications is severely limited by the very low temperatures required to maintain the superconducting material in the superconducting state. Till 1986 the highest critical temperature known was about 27 K. Only using liquid helium as the coolant, which is very costly, one can attain such low temperatures. In the last two decades certain high T_C ceramic materials have been discovered. These materials are brittle, difficult to be drawn into wires and cannot carry large currents. Vigorous research is going on around the world to overcome the drawbacks of high T_C materials and to gainfully utilize them in different applications. We discuss here some of the interesting applications.

(*i*) **Power transmission:** The most obvious application of superconductors is in power transmission. In a country, if the electrical grid is made of superconductor wires rather than aluminium, there would be no need to transform the electricity to a higher voltage and then back down again. The resulting savings in cost and materials would be enormous.

(*ii*) **Transformers and electrical machines:** The transformers and electrical machines with superconducting coils generate greatly stronger magnetic fields than magnetic circuits that use ferromagnetic materials produce.

The normal eddy current losses and hysteresis losses will not be present in superconducting devices and hence the size of motors and generators will be drastically reduced. Thus, superconductors are likely to revolutionize the whole range of rotating electrical machines, making them smaller, lighter and highly efficient. For example, a superconducting generator about half the size of a copper wire generator is about 99% efficient; typical generators are around 50% efficient.

(*iii*) **Diagnostic equipments:** High magnetic fields are essential in many areas of research and diagnostic equipments in medicine. They are at present produced using big electromagnets. The electromagnets are unwieldy being very big; require large electrical power and continuous cooling. On the other hand, superconducting solenoids produce very strong magnetic fields.

They are small in size and are less cumbersome. They do not need either large power supplies or the means of removing heat. The only power required is to bring the solenoid into the superconducting state and maintain it in that state. The low power requirement and simple cooling technique leads to a large saving in cost. The superconductor magnets have improved the MRI, as they can be smaller and more efficient than an equivalent conventional magnet.

(*iv*) **Electronic switches:** Type II superconductors can be used as very fast electronic switches. This has allowed building a 4-bit computer microchip operating at about 500 times the speed of earlier processors.

(*v*) **Cryotrons:** The application of a magnetic field greater than its critical magnetic field changes the superconducting state of a superconducting material to normal state and removal of the field brings the material back from normal state to the superconducting state. This fact is used in developing ***cryotron switches***.

Fig. 42.19 (a) shows a schematic diagram of a simple cryotron. It consists of a superconducting material (control) coil, *B*, wound around another superconducting (gate) wire, *A*. The current in the central wire *A* can be controlled by the current in the coil *B*. Thus, whenever the current passing through the coil *B* exceeds the critical current value, the wire *A* will become a normal conductor exhibiting a finite resistance. This closes the gate for the flow of current through the core wire *A*. Removal of the current reopens the gate. Thus the system acts as a fast acting relay or switching element and is highly suitable as fast acting memory element in computers.

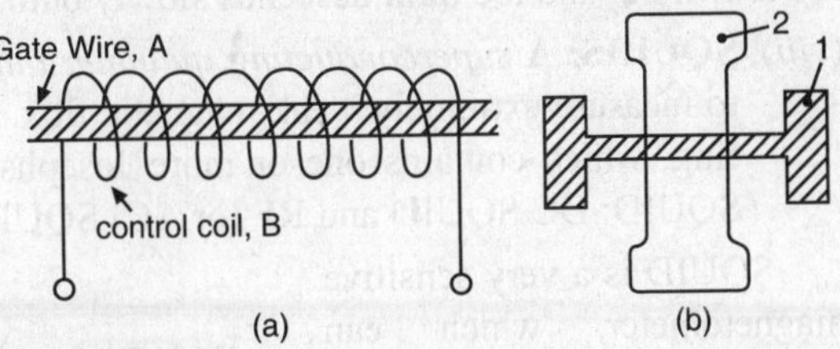

Fig. 42.19: Schematic diagram of a cryotron

The speed of the cryotron switch is dependent on its time constant $\tau = L/R$, where *L* is the inductance of the control coil *B* and *R* is the resistance of the gate coil *A* in the normal state. Wire wound cryotrons have τ of the order of 10^{-3} s. To reduce the value of τ, *R* is to be increased and *L* is to be decreased as far as possible. This objective is achieved by depositing two crosses strips on a substrate and separating them by a thin dielectric layer (Fig. 42.19 b). Strip 1 acts as the gate and is usually made of tin (T_C = 3.7K) and strip 2 acts as the control element usually made of lead (T_C = 7.2 K). Varying the current the strip 2 enables switching the strip 1 from the superconducting state to the normal state and vice versa. Thus opening and closing of the circuit is achieved. Through an appropriate design, the switching can be made faster as the value of τ can be decreased up to 10^{-7} s.

(*vii*) **MagLev Trains:** The most spectacular application would be the so-called 'MagLev' trains. The coaches of the train do not slide over steel rails but float on a four inch air cushion above the track using superconducting magnets; this eliminates friction and energy loss as heat, allowing the train to reach high speeds of the order of 500 km/hr. Such magnetic levitation trains would make train travel much faster, smoother, and more efficient due to the lack of friction between the tracks and train.

Operation: A typical plan of maglev train is shown in Fig. 42.20. The train has superconducting magnets built into the base of its carriages. An aluminium guide-way is laid on the ground and carries electric current. The repulsion between the two powerful magnetic fields, namely the field produced by the superconductor magnet and the field produced by the electric current in the aluminium guide-way causes magnetic levitation of the train. A levitation of about

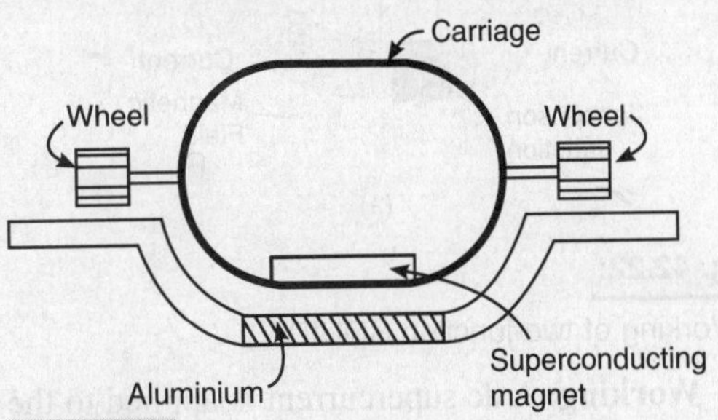

Fig. 42.20: Schematic cross-sectional view of Maglev Vehicle

10 to 15 cm is achieved. The train is fitted with retractable wheels, which act in a way similar to the wheels of an airplane. Once the train is levitated in air, the wheels are retracted into the body and the train glides forward on the air cushion. When the train is to be halted, the wheels are drawn out and the train descends slowly onto the guide-way and runs forward till it stops.

(*viii*) **SQUIDS:** A *superconducting quantum interference device* (SQUID) is a device used to measure extremely weak magnetic flux. The heart of a SQUID is a superconducting ring, which contains one or more Josephson junctions. There are two main types of SQUID: DC SQUID and RF (or AC) SQUID.

SQUID is a very sensitive magnetometer, which can measure very weak magnetic fields of the order of 10^{-13} Wb/m^2. The sensitivity of a SQUID to magnetic fields can be enhanced by using a flux transformer (Fig. 42.21). A flux transformer consists of a loop of superconducting material, which is coupled to the SQUID. An external magnetic field produces a persistent super-current in the loop and this current induces a flux in the SQUID. As the loop encloses a much larger area than can a SQUID, the sensitivity of the device gets enhanced.

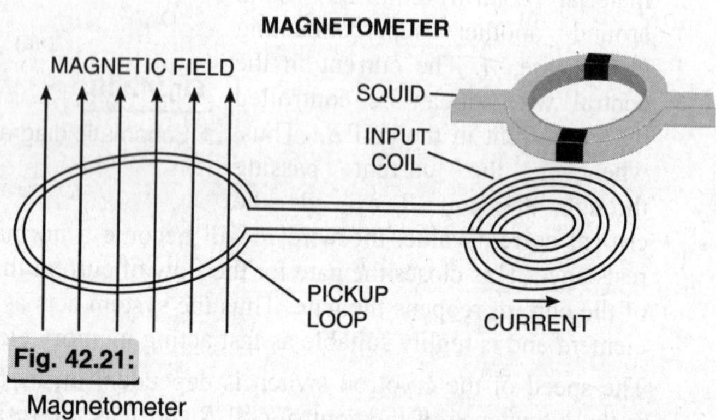

Fig. 42.21: Magnetometer

Fabrication: SQUIDs are fabricated by depositing a thin niobium layer on an alloy having 10% gold or indium. It acts as the base electrode of the SQUID and the tunnel barrier is oxidized onto this niobium surface. The top electrode is a layer of lead alloy deposited on top of the other two. The entire device is cooled to nearly absolute zero with liquid helium.

The schematic of a two-junction SQUID [direct current (DC) SQUID] is shown in Fig. 42.22 (a). It consists of two Joseph junctions arranged in parallel so that electrons tunneling through the junctions demonstrate quantum interference.

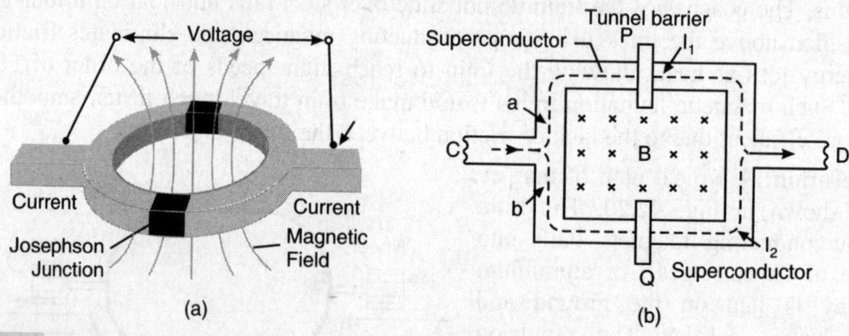

Fig. 42.22: Working of two-junction SQUID

Working: A dc supercurrent is applied to the SQUID (Fig. 42.22 b). This current, known as *bias* current, enters the device through the arm *A*. It is divided along two paths *a* and *b* and again merge into one and leaves through the arm *B*. *P* and *Q* are the Josephson junctions and the insulating layers at P and Q are of different thickness. I_1 and I_2 are the currents tunneling through the junctions P and Q respectively. In a superconductor, a single wave function describes all the Cooper pairs. The wave function experiences a phase shift at the junctions P and Q. Let the phase difference between points A and B taken on a path through junction P be δ_a and the phase difference between points A and B taken on a path through junction Q be δ_b. In the absence of magnetic field these two phases are equal. That is, $\delta_b - \delta_a$.

When a magnetic field is applied perpendicular to the loop, the flux passes through the loop, and changes the quantum mechanical phase difference across each of the two junctions. The wave functions at the two Josephson junctions interfere with each other. In other words, the supercurrents flowing along the paths a and b interfere. Hence, the device is named SQUID. The interference closely resembles the optical interference observed with Young's double slit. In the case of light, the phase difference between light waves is due to the difference in optical path lengths. In case of super-current interference, the waves are the de Broglie waves of Cooper pairs, and the phase difference is caused by the applied magnetic field. According to Josephson's theory, the phase difference between the reunited currents is directly proportional to the magnetic flux, Φ, through the ring. It can be shown that the total current through two parallel Josephson junctions is given by

$$I_T = 2(I_0 \sin \delta_0) \cos \frac{e\Phi}{\hbar c} \quad \text{...(42.17)}$$

The above relation indicates that a progressive increase or decrease of the magnetic flux, causes the current to oscillate between a maximum and a minimum value. Maxima in the current occur whenever the magnetic flux increases by one flux quantum. Thus, the period of these oscillations is one flux quantum Φ_0.

$$\Phi_0 = \frac{h}{2e} = 2.06 \times 10^{-15} \text{ webers}$$

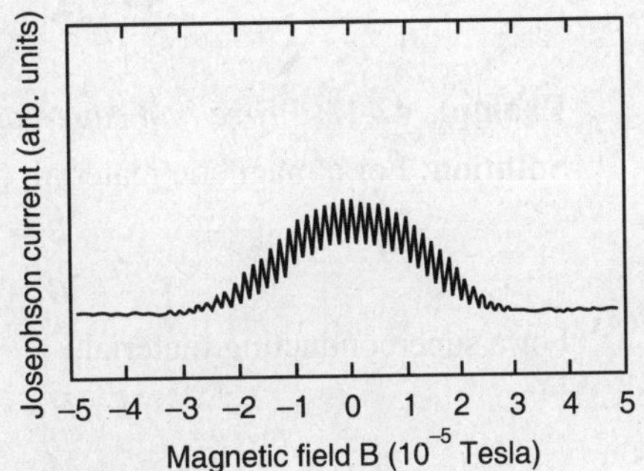

Fig. 42.23: Variation of the current as a function of the magnetic flux applied.

Fig. 42.23 shows the variation of the current through a pair of Josephson junctions as a function of the magnetic flux applied.

In practice, instead of the current, we measure the voltage across the SQUID, which also oscillates with the changing magnetic field. Thus, the SQUID is a flux-to-voltage transducer, converting a tiny change in magnetic flux into voltage. The flux Φ is related to the magnetic field B through the relation

$$\Phi = BA$$

where A is the area of the ring. With the help of this relation, flux measurements are converted into magnetic field measurements.

Applications of Squids: SQUIDs are used to measure very small magnetic fields. Since the current is sensitive to very small changes in the magnetic field, the SQUID acts as a very sensitive magnetometer. As ordinary magnetometers, SQUIDs are capable of measuring magnetic fluctuations of the order of 10^{-13} T. Because of their extreme sensitivity, SQUIDs find applications in many fields, engineering, medicine and many other fields. For example, geologists use them for measuring rock magnetism and continental drift. Physical processes, such as muscular or neural activity, in humans create magnetic fields as small as a thousand billionth of a tesla. Human heart generates magnetic fields of about 10^{-14} Wb/m^2 and the human brain generates magnetic fields of about 10^{-15} Wb/m^2. SQUIDs can detect these feeble fields and an array of SQUIDs is used in *magnetoencephalography* (MEG), the process of brain imaging. The SQUIDs are also used in nondestructive testing. In testing for corrosion of aluminum sheets riveted together in aircraft, the SQUID measures the influence of the aircraft skin on an applied oscillating magnetic field; a change in electrical conductivity reveals the defects.

Example 42.9: *For a certain metal the critical magnetic field is 5 × 10³ A/m at 6K and 2 × 10⁴ A/m at 0K. Determine its transition temperature.*

Solution: $$T_C = \frac{T}{\left[1-\frac{H_C(T)}{H_C(0)}\right]^{\frac{1}{2}}} = \frac{6K}{\left[1-\frac{5\times10^3 \text{ A/m}}{2\times10^4 \text{ A/m}}\right]^{\frac{1}{2}}} = \mathbf{6.93\ K.}$$

Example 42.10: *Calculate the critical current which can flow through a long thin superconducting wire of diameter 10³ m. Given H_c =7.9 × 10³ A/m.*

Solution: From Silsbee rule $I_c = 2\pi r H_c$

$$I_c = \frac{2\pi\times10^{-3}\times7.9\times10^9}{3} = \mathbf{24.81\ A}$$

Example 42.11: *Prove that superconductors are perfectly diamagnetic.*

Solution: For a magnetic material, the flux density is given by

$$B = \mu_0 (M+H)$$

$$B = 0, \text{ i.e., } 0 = \mu_0 (M+H)$$

For a superconducting material,

$$M+H=0$$

Or

$$M = -H$$

$$\frac{M}{H} = \chi = -1$$

For a diamagnetic material, the susceptibility is negative. This equation proves that a superconductor behaves as a perfect diamagnetic material.

Example 42.12: *Superconducting tin has a critical temperature of 3.7K. at magnetic field and a critical field of 0.0306 Tesla at Find the critical field at 2K.*

Solution: Given data:Critical temperature $T_c = 3.7$K

Critical temperature $H_0 = 0.0306$ T

We know that the critical field

$$H_c = H_0\left[1+\frac{T^2}{T_c^2}\right]$$

Substituting the values, we get $$= 0.0306\left[1-\frac{(2)^2}{(3.7)^2}\right] = 0.02166 \text{ T}$$

Therefore, the critical field at 2K is 0.02166T.

Example 42.13: *Calculate the critical current for a wire of lead having a diameter pf 1mm at 4.2K. Critical temperature for lead is 7.18K and $H_c(0) - 6.5 \times 10^4$ Am^{-1}.*

Solution: Given data:

Critical temperature $T_c = 7.18K$

Critical magnetic field $H_c(0) = 6.5 \times 10^4\ Am^{-1}$

We know that, the critical field

$$H_c(0) = H_0\left[1 - \frac{T^2}{T_c^2}\right]$$

Substituting the values, we get $= 6.5 \times 10^4 \times \left[1 - \left(\frac{4.2}{7.18}\right)^2\right] = 42.758 \times 10^3\ Am^{-1}$

Therefore, the critical current through the wire is

$$I_c = 2\pi r H_c$$

Substituting the values, we get $I_c = 2 \times 3.14 \times 0.5 \times 10^{-3} \times 42.758 \times 10^3\ I_c = 134.26$ A

The critical current through a wire of lead is 134.26 A.

Example 42.14: *The critical temperature for a metal with isotopic mass 199.5 is 4.185K. Calculate the isotopic mass if the critical temperature falls to 4.133K.*

Solution: Given data: Critical temperature of lead $T_{c1} = 4.133$ K

Critical magnetic field at 0°K $T_{c2} = 4.133$ K

Isotopic mass $M_1 = 199.5$

We know that, $$\frac{T_{c_2}}{T_{c_1}} = \left[\frac{M_1}{M_2}\right]^{\frac{1}{2}}$$

$$\frac{T_{c_2}^{\ 2}}{T_{c_1}^{\ 2}} = \frac{M_1}{M_2}$$

$$\therefore \quad M_2 = \frac{M_1 \times T_{c_1}^{\ 2}}{T_{c_1}^{\ 2}} = \frac{199.5 \times (4.185)^2}{(4.133)^2}$$

Therefore, the isotopic mass = 2.04.

QUESTIONS

1. The dc resistance of a superconductor is practically zero. What about its ac resistance?
2. How can you change a superconductor from type I to type II?
3. What is the importance of isotope effect in superconductivity?
4. What are Cooper pairs?
5. What is coherence length?
6. How is Josephson tunneling different from single particle tunneling?
7. A superconducting wire and a copper wire are connected in parallel. Does the copper wire carry current when a potential difference is applied?
8. Give an account of the phenomenon of superconductivity.

9. What is the significance of critical temperature, critical magnetic field and critical current density for superconductors?
10. Explain the Silsbee effect.
11. Explain Meissner effect and isotope effect.
12. Compare the dependence of resistance on temperature of a superconductor with that of a normal conductor.
13. Define superconductivity. Explain the effect of isotopes on superconductors.
14. Discuss the relation between isotopic mass and transition temperature.
15. What is flux quantization?
16. What do you mean by "perfect diamagnetism" of a superconductor?
17. Describe how Cooper pairs are formed and explain the salient features of superconductivity.
18. Give a short account of high temperature superconductivity.
19. Explain the term high temperature superconductivity. Give the various applications of superconductors.
20. Explain high temperature superconductivity. Explain its advantage.
21. Explain B.C.S. with key note of Cooper pairs.
22. Explain the BCS theory with key note of Cooper pairs. Distinguish between type I and II superconductors.
23. Write short notes on BCS theory of superconductivity.
24. Why are type I superconductors poor current carrying conductors?
25. Explain the various properties and important applications of superconducting materials.
26. Explain ac and dc Josephson's effect.
27. Explain the theory of d.c. Josephson's effect.
28. Explain the Josephson's effect in superconductivity. Describe the principle of SQUID and mention its applications.
29. Explain what is meant by Quantum interference?
30. Discuss the principle and working of SQUID.
31. What is meant by SQUIDs? Mention anyone of its applications.
32. Describe the following practical applications of superconducting materials: (*i*) Superconducting magnet (*ii*) MAGLEV vehicles (*iii*) SQUIDs.

PROBLEMS

1. For a specimen of superconductor, the critical fields are respectively 1.4×10^5 A/m and 4.2×10^5 A/m for temperatures 14 K and 13 K respectively. Calculate the transition temperature and the critical fields at 0 K and 4.2 K.
2. Estimate the current through a tin wire of diameter 2mm at 2K, if $T_C = 3.722$ K and critical field at 0 K is 0.0305 T.
3. A Josephson junction has a voltage of 8.5 μV across its terminals. Calculate the frequency of the e.m. waves generated by it. **[Ans:** 410 GHz]
4. Calculate the critical current for a wire of Pb having a diameter of 3mm at 5K. The critical temperature for Pb is 8 K and critical magnetic field is 5×10^4 A/m at 0K. **[Ans:** 286.9 A]
5. The transition temperature of mercury with an average atomic mass of 200.59 amu is 4.153 K. Determine the transition temperature of one of its isotopes $_{80}Hg^{204}$. **[Ans:** 4.118 K]

6. The critical temperature of a superconductor at zero magnetic field is T_C. Determine the temperature at which the critical field becomes half of its value at 0K. **[Ans:** 0.707 T_C]

7. A long thin superconducting wire of a metal produces a magnetic field of 105 × 10^3 A/m on its surface due to the current through it at a certain temperature T. The critical magnetic field of the metal is 150 × 10^3 A/m at 0K. The critical temperature T_C of the metal is 9.2 K. What is the value of T? **[Ans:** 4.3 K]

8. A lead wire has a critical magnetic field of 6.5 × 10^3 A/m at 0K. The critical temperature is 7.18 K. At what temperature the critical field would drop to 4.5 × 10^3 A/m. The diameter of the wire is 2mm. What is the critical current density at that temperature?

   [Ans: 5.19 K, 9.6 × 10^6 A/m^2]

9. The London penetration depth of mercury at 3.5 K is 75 nm. Estimate the penetration depth at 0K. **[Ans:** 51.9 nm]

10. The penetration depths for lead at 3 K and 7.1 K are 39.6 nm and 173 nm respectively. Calculate the critical temperature for lead. **[Ans:** 7.193 K]

43 Masers

Learning Objectives:

By the end of this chapter, you will be able to:

- Define basic principles of Masers.
- Discuss about Ammonia Masers.
- Learn about Solid State Masers and its different types.
- Explain about Celestial Masers.

43.1 INTRODUCTION

In 1906 Lee de Forest invented the triode tube. Since then it was known how to build an oscillator which could generate electromagnetic waves. This knowledge was extended into the millimeter wave region with new and new amplifier tubes. Cavity magnetron and klystron were invented by 1938 and could amplify signals in VHF (30 MHz to 300 MHz) and UHF(300 MHz to 3000 MHz) range *i.e.* up to up to 1 cm wavelength. These were used in radar signals generation and detection during the World War II. Microwaves, which have wavelengths between 1 millimeter and 1 meter, are another small segment of the invisible part of the electromagnetic spectrum.

In 1954 Gordon, Townes and Zeiger working from a totally different direction unveiled a new device, a MASER, which generates electromagnetic radiation at mm wavelengths. The word MASER is an acronym for *Microwave Amplification by Stimulated Emission of Radiation*. It is the first oscillator (quantum amplifier) which produced sub-centimeter wavelength radiation based on discrete energy levels in molecules. It was a pulsed maser. Two Soviet scientists, Nikolai G. Basov and Alexander M. Prokhorov developed the first continuous-wave maser during the same period.

Masers are difficult to build and expensive to operate. They either needed special vacuum chambers to work or temperatures close to absolute zero (–273.15 C, –459.67 F). Further, they need intense magnetic fields, too. With these difficulties, masers remained confined to mainly research laboratories. Masers have been developed to operate at many different wavelengths, so that the original designation "microwave" is no longer strictly accurate.

43.2 PRINCIPLE OF MASERS

When atoms or molecules of a *medium* are bombarded with photons of a suitable frequency, the atoms or molecules go to a higher energy state. This is known as *pumping*. Pumping leads to a state of population inversion where the number of molecules in an excited state is far greater than the number of molecules in a lower state. When the excited atoms are hit by photons of appropriate frequency they emit photons of the same frequency and trigger an avalanche of photons from other excited molecules. This is known as "*stimulated emission*" of radiation. By keeping the excited medium in a resonant cavity (or cavity resonator), feedback action occurs and "*coherent*" radiation is produced.

Electromagnetic waves are said to be coherent when they propagate at the same frequency, bear the same phase, and travel in the same direction.

The basic *requirements* for realizing maser action are:

(*i*) **Amplifying medium:** A medium having an appropriate set of energy levels.

(*ii*) **Pumping:** Any process that supplies energy to atoms or molecules and causes them to jump to a higher energy state is called as an energy pump, or simply as a pump.

A suitable pumping mechanism is to be selected to excite atoms or molecules from a lower energy level to a chosen higher energy level till the population of atoms at higher level far exceeds the population at the lower level. This state of population excess at a higher level over the population at the lower level is known as ***population inversion*** (See Fig. 43.12).

In 1950 Edward Purcell and Robert Pound, at Harvard University, demonstrated population inversion of molecules in a crystal of lithium fluoride placed in a magnetic field.

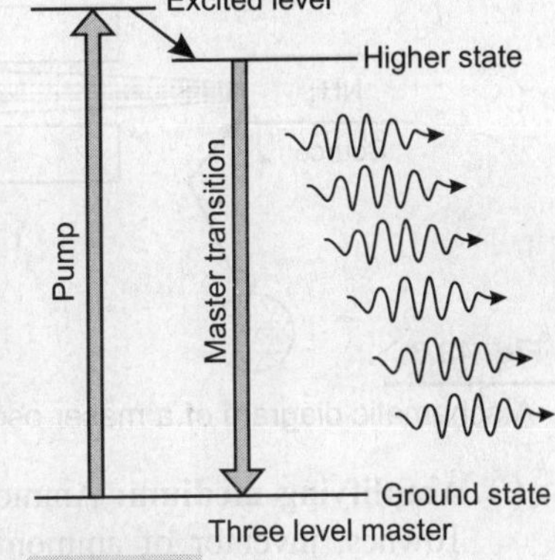

Fig. 43.1: Three level Maser

(*iii*) **Stimulated emission:** All excitations occur *instantaneously*. The time that an atom spends in an excited level is known as the ***lifetime*** of that state. Excited state lifetimes are typically a few nanoseconds. On the other hand, the de-excitation may be delayed by a measurable time interval depending on the properties of the excited state. In certain cases, the lifetime of an excited state may be as long as a few milliseconds (10^{-3} s). Therefore in such instances, an atom in the excited state may not instantaneously de-excite. It may take a long time make a transition by itself. Another alternative is that a passer-by photon which has energy exactly equal to the transition energy may stimulate the atom to emit a photon instantly (Fig. 43.1). In the process there arise two identical photons travelling in the same direction. As, they travel, they trigger a cascade of photons (Fig. 43.1).

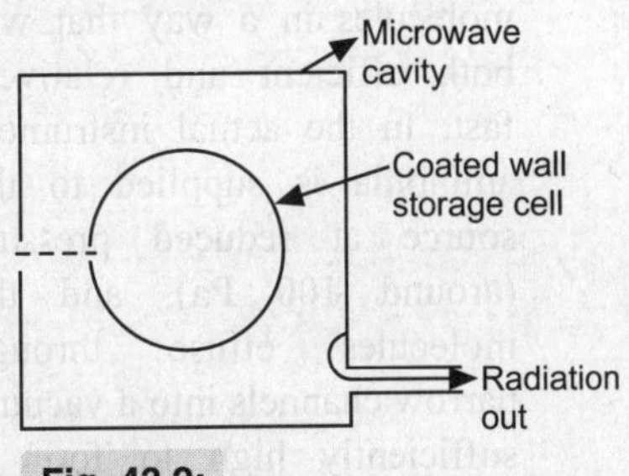

Fig. 43.2: Cavity resonator

(*iv*) **Cavity Resonator:** The resonant cavity (Fig. 43.2) is an enclosure where the excited medium is held and photons that bounce back and forth between the walls lead to amplification of the stimulated emission. This amplification grows exponentially.

The required condition is that the cavity wall to wall distance, L, must be equal to a multiple of half a wavelength of e.m. wave. In symbols, $L = n[\lambda/2]$, where λ is the wavelength of the e.m. wave and n is some integer.

In the case of microwaves, n would be a small number. Therefore, there are a very small number of resonant frequencies. However, only those resonant frequencies that are amplified by the maser medium will have large amplitudes. So there are only a few *maser modes* or maser resonant frequencies present in the output from a maser.

(*v*) **Outlet:** An outlet guides the radiation to an outside agency.

43.3 AMMONIA MASER

Ammonia maser is the first maser and first quantum amplifier. However, it is a *two-level maser*. Now we shall see how the above requirements are met and maser action is realized in practice. Fig. 43.3 shows schematically the essential parts of an ammonia maser.

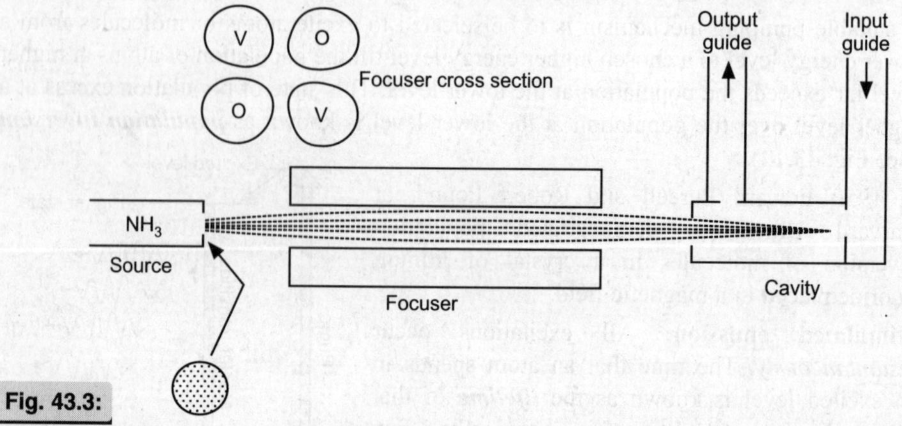

Fig. 43.3: A schematic diagram of a maser oscillator

(*i*) **Amplifying medium:** Ammonia gas was chosen as the amplifying medium by C.H. Townes, inventor of ammonia maser. The gas was chosen because it enabled him to physically separate individual molecules in a way that was both efficient and relatively fast. In the actual instrument ammonia is supplied to the source at reduced pressure (around 100 Pa), and the molecules effuse through narrow channels into a vacuum sufficiently high to form a beam.

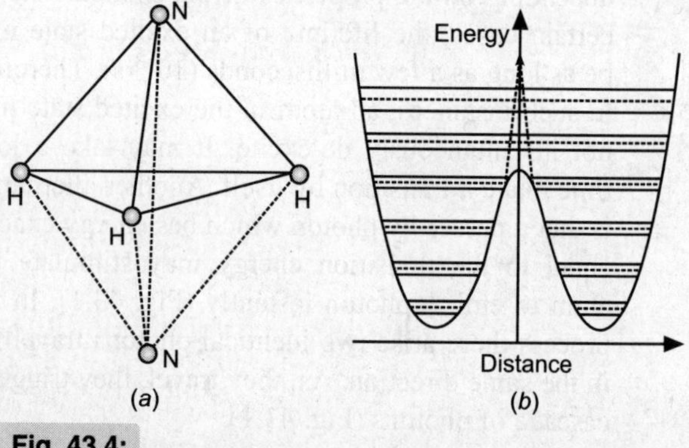

Fig. 43.4: (a) Pyramidal structure of the ammonia molecule. (b) Potential energy of the nitrogen atom as a function of its distance from the hydrogen atom.

The ammonia molecule has the shape of a pyramid. At the apex of the pyramid is a nitrogen atom; at the base, three hydrogen atoms. The nitrogen atom is able to move through the plane of the hydrogen atoms thus turning the pyramid inside out.

The ammonia molecule has a large number of energy states (Fig. 43.4). Particularly, there is a pair of quantum states between which transitions occur in the microwave region (Fig. 43.5) and they can be exploited in building a maser.

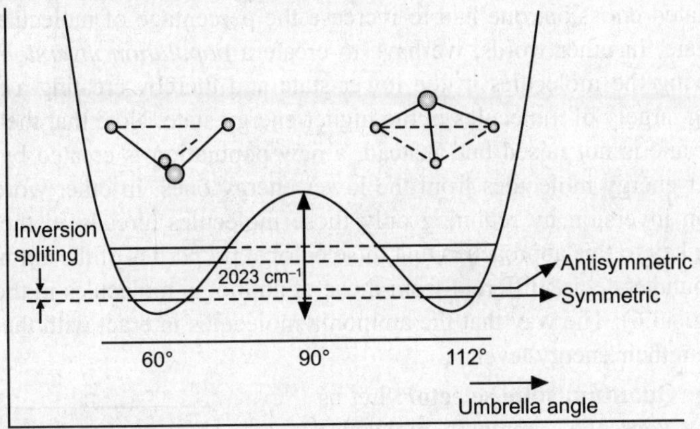

Fig. 43.5:

NH_3 levels (splitting ~10^{-4} eV, ν = 23800 MHz)

The nitrogen atom flips back and forth at a characteristic frequency of about 24,000 megacycles per second. Looking from the quantum point of view, at a given time the nitrogen has a certain probability of being on either side of the plane containing hydrogens. Consequently, the molecule as a whole possesses two distinct energy states.

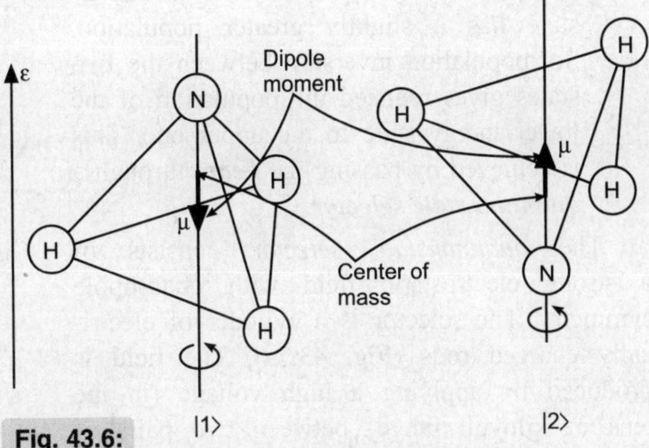

Fig. 43.6:

Electricdipole moment in NH_3 molecule

The nitrogen atom (N) in the ammonia molecule (NH_3) can tunnel through the plane of the three hydrogen (H) atoms, so it can be'up' or 'down'. This 'up' or 'down' *state* has nothing to do with the classical or quantum-mechanical notion of spin, which is related to the *magnetic* moment. The nitrogen atom could be either beneath or, else, above the plane of the hydrogens, with 'beneath' and 'above' being defined in regard to the molecule's direction of rotation around its axis of symmetry. We call the two states state '1' and state '2' respectively. The NH_3 molecule has *electric* dipole moment, which is denoted by μ in Fig. 43.6 and which, depending on the *state* of the molecule (*i.e.* the nitrogen atom being above or beneath the plane of the hydrogens), changes the total energy of the molecule by an amount that is equal to +με or −με, with ε some *external* electric field, as illustrated by the ε arrow on the left-hand side of the diagram. An electric field will change the total energy of the molecule by an amount that is equal to +με or −με.

The difference in energy between the states equals the energy of a photon with a frequency of 24,000 megacycles per second.

(*ii*) **Population Inversion:** In any population of ammonia molecules at least a few of the molecules are already in the higher energy state. Under normal conditions, however, the higher energy molecules simply are not dense enough in the gas to enable the phenomenon of stimulated emission to occur. In order to facilitate the phenomenon of stimulated emission, one had to increase the percentage of molecules in the higher energy state. In other words, we have to create a *population inversion*. It is achieved by removing the molecules in the lower state and thereby creating a new population consisting largely of molecules in the higher energy state. Note that the energy level of any molecule is not raised but, instead, a new population is created by separating out the higher energy molecules from the lower energy ones. In other words, we obtain a population inversion by retaining only those molecules already in the higher energy state. The key to this approach is that the electrical properties of the ammonia molecules in the ground state are different from those of ammonia molecules in the higher energy level (Fig. 43.6). The way that the ammonia molecules interact with the electrical field depends on their energy level.

(*iii*) **Pumping: Quantum state selector:** Let us consider a source of ammonia molecules in equilibrium at ordinary temperatures. The populations of the selected two states are very nearly equal. The lower-energy state has a slightly greater population. So, population inversion between the two states gives reduced the population of the lower-energy state to the upper one. This is achieved by passing the ***beam*** through a ***quantum state selector***.

The *quantumstate selector* consists of a strong electrostatic field with quadrupole symmetry. The selector is a cylinder of electrically charged rods (Fig. 43.7a). The field is produced by applying a high voltage (in the tens of kilovolt range) between two pairs of equally spaced parallel cylinders, as shown in Fig. 43.7(b). In the vicinity of the rods the field is strong; along the axis of the cylinder the field is weak. When a beam of ammonia molecules is sent through the selector, molecules in the lower energy state are pulled toward the rods whereas the molecules in the higher state are attracted to the axis of the cylinder. Out of the far end of the selector emerges a stream of molecules, virtually all of which are in the upper energy state. The molecules in the higher state then enter a resonant cavity (right), where they may be used to amplify a microwave signal.

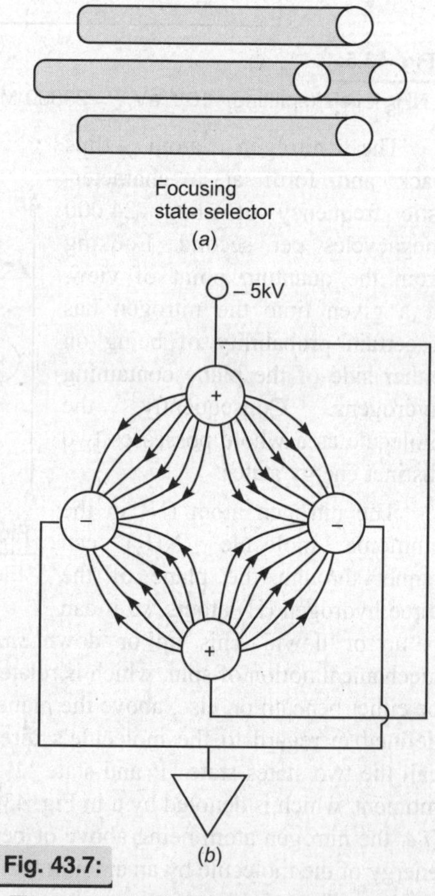

Fig. 43.7: The electrostatic quadrupole state selector of the ammonia maser

The probability for the *spontaneous* transition of molecules to the lower state is small. As a result, most of the molecules inside the cavity will be in the upper state.

If these molecules are irradiated with 24,000-megacycle microwaves, only downward transitions will be induced. Energy will be given up by the molecules to the microwave field, and the incoming wave will be amplified.

(*iv*) **Cavity Resonator:** On passing the electric field, the ammonia molecular beam is admitted into a microwave cavity tuned to resonance at the frequency of around 23,870 MHz. The cavity is cylindrical; having either a circular cross section, with ends closed (Fig. 43.8). Each incoming photon can bounce back and forth across the chamber thousands of times before it escapes again, greatly increasing its chance of interacting with a molecule in the beam. Whenever there is a collision, a new photon is born. It too is trapped in the chamber for a time and may collide with another molecule, producing a second new photon, and so on. If there are enough molecules in the cavity, this chain reaction becomes self-sustaining; the amplifier turns into an oscillator, generating its own - wave without any input signal.

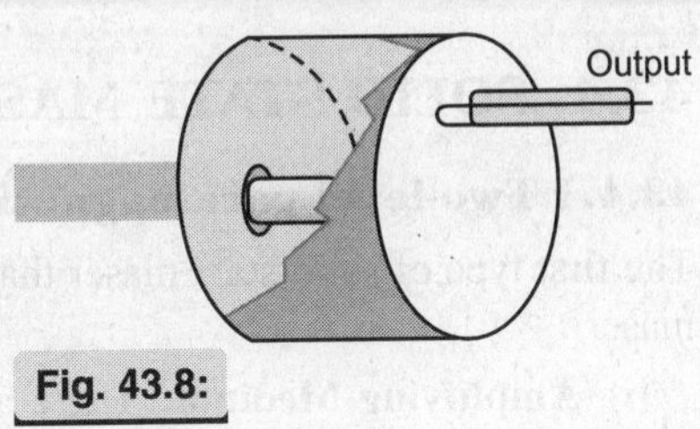

Fig. 43.8: Resonant cavity

(*v*) **Output:** An output waveguide is coupled to the cavity (Fig. 43.9) to draw out a part of the microwave power generated by the maser action.

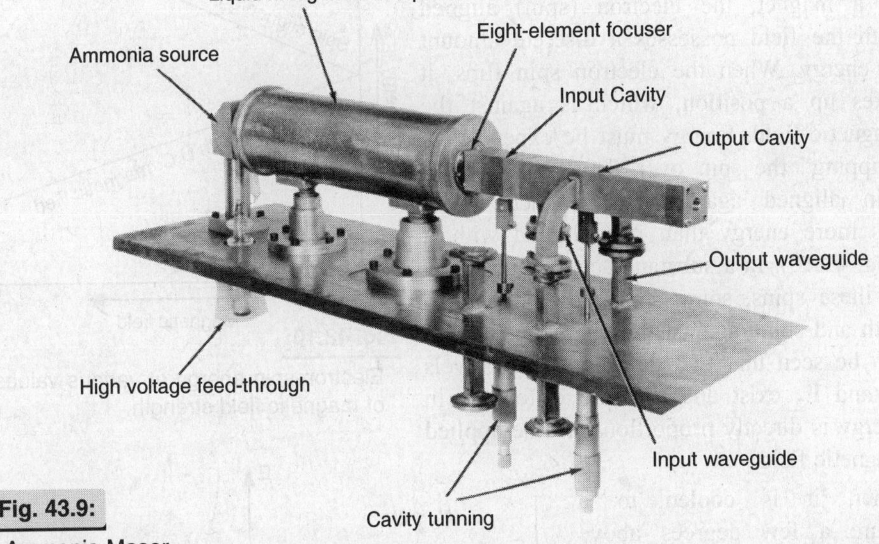

Fig. 43.9: Ammonia Maser

(*vi*) **Features:** As an amplifier the ammonia maser has a remarkably **narrow band-width:** it will not amplify waves which depart from its central frequency by more than 3,000 to 5,000 cycles. The ammonia maser is **not readily tunable**; the central frequency cannot easily be changed. This means that it is **not really a practical amplifier**. If it were used in a communications channel, it could transmit only one voice at a time; it could not come close to receiving a television station. The ammonia maser was, however, the instrument which first demonstrated the great potentialities of maser amplifiers.

However, the ammonia maser is an extraordinarily stable oscillator. If in a cavity tuned to a specific transition, atoms are retained such that there is more number of excited atoms than at the lower level, they can be prompted to deliver their excess energy to the electromagnetic radiation present in the cavity. This radiation, in turn, stimulates more atoms in the excited state to emit radiation. Thus an oscillator is formed that resonates at the atomic frequency.

(*vi*) **Advantages:** Ammonia maser is an excellent frequency standard and is an excellent candidate to make very precise standards of frequency or to build atomic clocks.

(*vii*) **Disadvantage:** It emitted a very narrow line and the frequency of this line could not be changed to tune the device to other frequencies.

43.4 SOLID STATE MASERS

43.4.1 Two-level paramagnetic maser

The first type of solid-state maser that was developed is known as the two-level paramagnetic maser.

(*i*) **Amplifying Medium:** There are a few substances known as *paramagnetic* in which *some electrons are unpaired and the material as a whole is magnetic*. A silicon crystal containing some impurity atoms, such as those of phosphorus, which have one more electronsis an example of a paramagnetic material and serves as amplifying medium.

(*ii*) **Quantum Selector:** When the paramagnetic crystal is placed between the pole pieces of a magnet, the electron (spin) aligned with the field possesses a discrete amount of energy. When the electron spin flips, it takes up a position, which is against the magnetic field. Energy must be expended in "flipping" the spin over. Thus, an electron spin aligned against the magnetic field has more energy than one aligned with it (Fig. 43.10). In a substance containing many of these spins, some spins will be aligned with and some against the magnetic field. It can be seen that two definite energy levels E_1 and E_2 exist and that the difference in energy is directly proportional to the applied magnetic field.

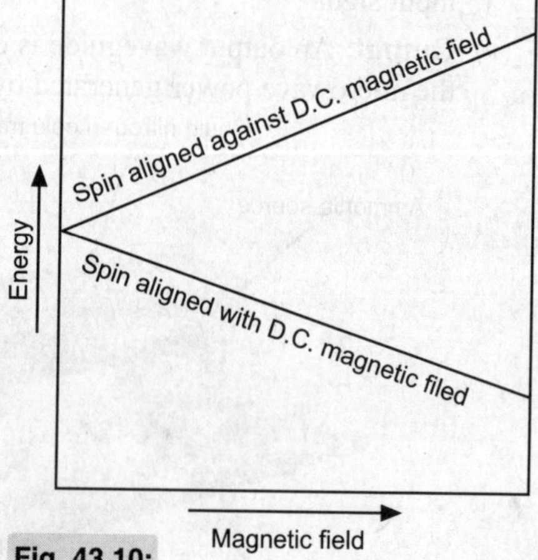

Fig. 43.10: Electron-spin energy at various values of magnetic field strength.

Further, it is cooled to a temperature a few degrees above absolute zero in a bath of liquid helium, so that most of its unpaired electrons fall into the lower of their two possible energy states (Fig. 43.11).

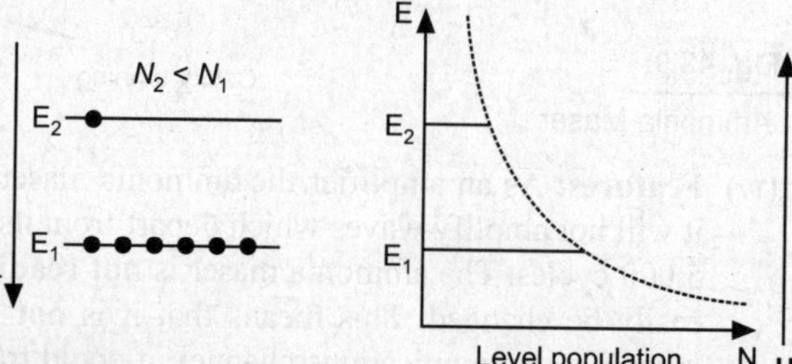

Fig. 43.11: Energy states of electron spins

(*iii*) **Pumping:** When a fairly high-powered microwave pulse is applied to the assembly of electrons, it pumps electrons in the crystal to a higher state E_2 (Fig. 43.12).

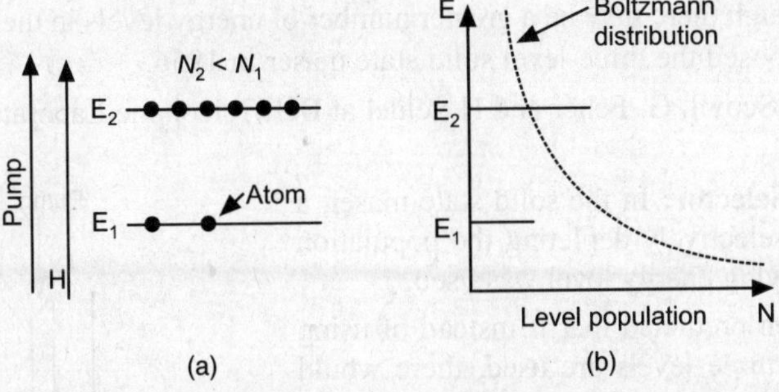

Fig. 43.12:
Pumping to higher state (a) More spins in the higher energy state E_2 (b) the distribution

(*iv*) **Population Inversion:** Thus, for a short time the majority of the electrons are raised from the lower state E_1 to the higher state E_2. Consequently, the number of electrons in the upper state E_2 exceeds the population in the lower state E_1. Thus, the condition of 'population inversion' is attained. This is contrary to the normal Boltzmann distribution of population.

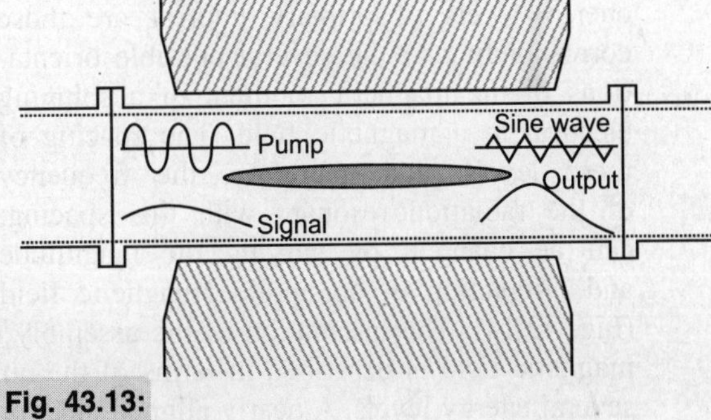

Fig. 43.13:
Signal amplified due to stimulated emissions

(*v*) **Amplification:** Now if an input signal of lower frequency having the appropriate energy (E_2-E_1) is applied, it promotes an avalanche of stimulated emissions and produces an amplified output (Fig. 43.13).

(*vi*) **Advantages:** The difference between the energy of the upper and lower levels depends upon the strength of the magnetic field. Hence by adjusting the strength of the magnet, the maser can be tuned over a wide range of frequencies.

In a solid crystal the external magnetic field is not the only one to act on unpaired electrons. The electrons are also influenced by the magnetism of neighboring atoms. The internal magnetic effect varies from point to point in the crystal, so that not all the electrons are subjected to exactly the same field. Thus they respond to slightly different frequencies, and this is the reason for the wider band-width of the solid-state devices. Using solids instead of gases, the power is also enormously increased because now the paramagnetic ion concentrations in a solid could easily be one hundred thousand times greater than the number of ammonia molecules in a beam.

(*vi*) **Disadvantages:** The chief disadvantage of the two-level maser is that it can be operated only in bursts; its amplifying action stops each time its electrons drop down again to the lower level.

This problem has been overcome with the development of the newest member of the maser family: the three-level paramagnetic maser.

43.4.2 Three-level paramagnetic maser

The term "3-level maser" means that only three energy levels are used in attaining maser operation eventhough there may be a greater number of energy levels in the material. Nicolaas Bloembergen proposed the three-level solid state maser in 1956.

In 1957 H.E. Scovil, G. Feher and H.Seidel at Bell Telephone Laboratories built the first three-level maser.

(*i*) **Quantum Selector:** In the solid state maser, a method of selectively depleting the population of a lower lying energy level was used.

Bloembergen predicted that if instead of using two levels, three levels are used, there would be no longer a necessity to separate physically the molecules in the upper energy state, but could select the population of the different levels by a clever use of interactions.

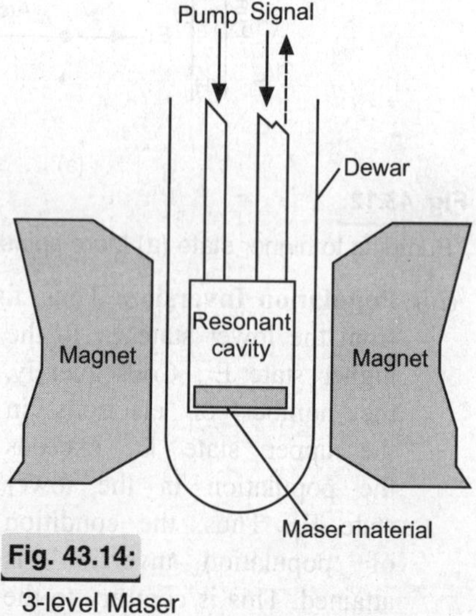

Fig. 43.14: 3-level Maser

(*ii*) **Pump:** For a paramagnetic solid, the energy levels of primary interest are those corresponding to the several possible orientations of the magnetic moment of a spinning electron in a magnetic field. The spacing of these levels, and, therefore, the frequency of the radiation resonant with this spacing, can be changed by varying the magnitude and orientation of an applied magnetic field (Fig. 43.14). The orientation of the assembly of electron spins relative to the applied magnetic field is described in terms of the number of spins (particles) occupying the several energy levels. A nearly aligned electron spin with the applied magnetic field is represented by one occupying the lowest energy level. Perfect alignment (all spins in the lowest level) is prevented by the effect of thermal vibrations in the solid. High temperatures prevent spin alignment even in the presence of fairly large magnetic fields. The distribution, at equilibrium, of spins over the energy levels is given by Boltzmann distribution. For positive temperatures T the population of higher energy levels will always be less than that of lower levels.

(*iii*) **Population Inversion:** If microwave energy is supplied to a solid having a suitable set of energy levels, say E_1 and E_3, then energy will be absorbed. The absorption causes the elevation of particles in the lower level E_1 to the upper level E_3 (Fig. 43.15). If the microwave energy has adequate power, the rate at which the particles are excited from the lower level E_1 to the upper level E_3 will exceed the rate at

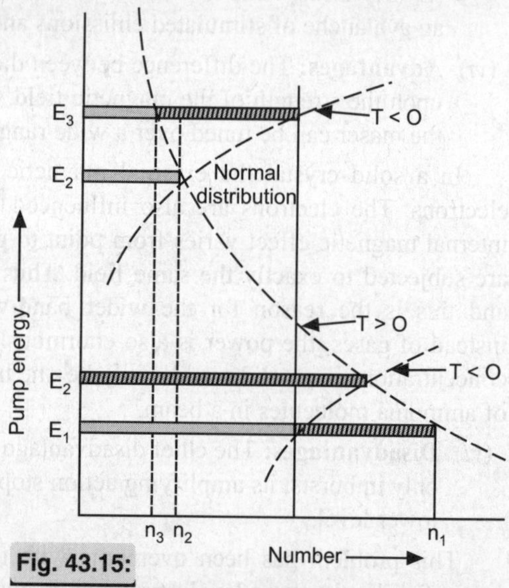

Fig. 43.15: Population inversion shown by T < O curves

which they come down to a lower level. By raising the incident power, it is possible to make the population of the two energy levels equal. When the populations are equalized, the incident radiation stimulates absorption and emission by an equal amount and no further change can be made by increasing power. This is called as ***saturation***.

However, it is seen that levels lying in-between the saturated levels exhibit *population inversion*. If, the intermediate level is close to the upper level E_3, then the upper level is populated in excess of the intermediate level. If, on the other hand, the intermediate level is located close to the lower level E_1, then intermediate level is populated in excess of the lower level. Thus by the selective depletion of a low-lying energy level (E_1), an inversion of the population is achieved.

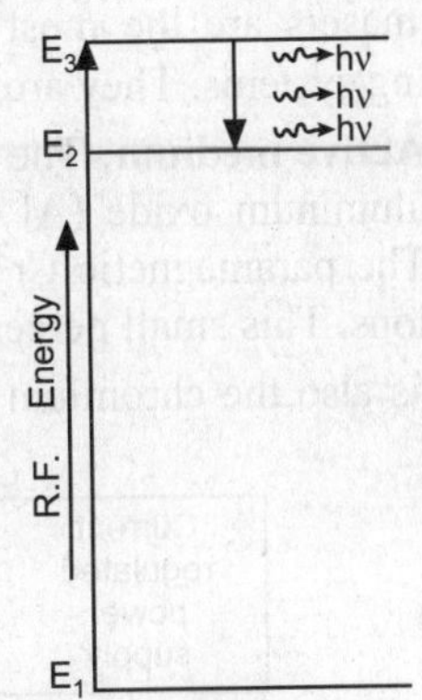

Fig. 43.16: Three-levels configuration of a paramagnetic material

(*iv*) **Amplification:** When we inject into the crystal a radio frequency signal, equal in frequency to the difference between energy levels E_1 and E_3, some of the electron spins in the lower level absorb energy from the radio frequency field and jump to level E_3 (Fig. 43.16). If the radio frequency signal (pump signal), is strong enough, the "pump" transition (level E_1 to level E_3) can be saturated, i.e., the number of spins at E_3 can be made equal to that at E_1. The total number of spins in the two levels has not changed, but now they are equally populated.

E_3 now possesses more electron spins than E_2. This is contrary to the normal equilibrium condition. It means population inversion is established between these two levels, namely E_3 and E_2. Weak incident radiation resonant with the separation of inverted levels will stimulate a net excess of coherent radiation, hence produce amplification.

By varying the magnetic field, thereby changing the spacing of the energy levels, "maser action" can be obtained at other frequencies.

(*v*) **Applications:** The ammonia maser was useful as a frequency standard due to the stability of its emission frequency. The solid state maser being tunable could be used for communications and radar. It could be tuned in a continuous way on a band of frequencies, maintaining the low noise characteristic of the maser.

Fig. 43.17 shows a complete three-level maser. In center, between the poles of a large electromagnet, is a silvered flask which is filled with liquid helium. The maser cell is inside the flask between the two magnet poles.

The three-level maser is small and robust but it requires cooling to liquid helium temperature. Further, it is to be kept in a powerful magnetic field, and the cooling system and the magnet are very cumbersome and heavy (Fig. 43.17). So, the device is not suitable for the battlefield or for setting up within an aircraft.

Fig. 43.17: Complete three-level maser

Also for the radio astronomical application, its weight and encumbrance were unfavourable, considering that to fully exploit its low-noise characteristics it was necessary to mount it at the centre of a gigantic receiving antenna, because otherwise the components that would have to be employed to transfer the signal to a more suitable place would introduce their own noise and effectively cancel the advantages of the device.

43.4.2.1 Ruby Maser

Ruby masers are the most sensitive and low-noise pre-amplifiers used in the microwave receiving systems. They are rugged and are not susceptible to the microscopic failures.

(*i*) **Active medium:** The pink ruby crystal used in maser amplifiers is about 99.95 percent aluminum oxide (Al_2O_3) with approximately 0.05 percent chromium oxide (Cr_2O_3). The paramagnetic Cr^{3+} ions occupy some of the sites normally occupied by the Al^{3+} ions. This small percentage of chromium is what gives ruby its characteristic red color.

It is also the chromium which provides the three net electron spins necessary for maser action.

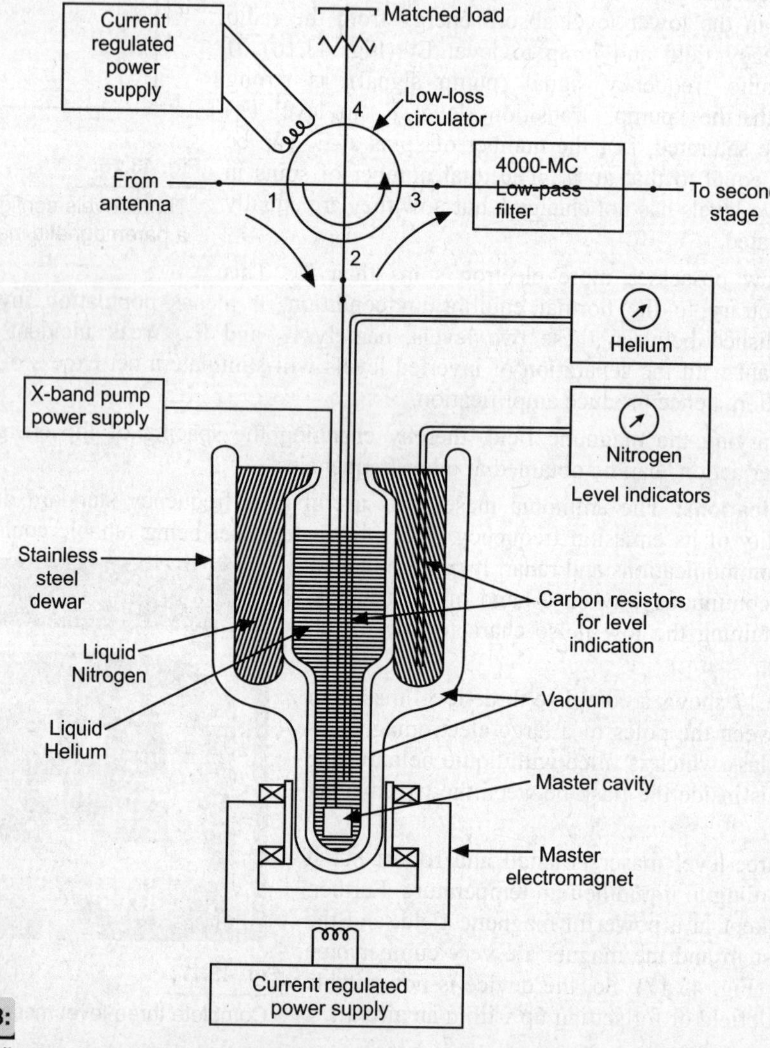

Fig. 43.18:

System diagram of Ruby maser operation.

Each chromium ion has three unpaired electrons. Due to the intrinsic spin angular momentum each of these ions possesses magnetic moment. These three magnetic moments add together and form a larger effective permanent magnetic dipole moment. This permanent magnetic dipole moment is called a "spin."

An important parameter in the construction of a maser is the angle that the external magnetic field makes with the crystal axis. The energy separation between the levels, and therefore the operation frequency of the maser, depend on it.

(*ii*) **Population inversion:** If the substance possesses two net electron spins per atom, it will have three energy levels - one for both spins aligned with the d.c. field, one for both spins aligned against, and one for one spin aligned with and one spin aligned against the d.c. field. In the case of three net electron spins, there are four energy levels and so on. Quantum physics states that the difference in energy (ΔE) between two levels can be equated to frequency (f) by the simple formula: $\Delta E = hf$, where h is Planck's constant. Now we can change the graph of energy levels and write frequency in place of energy.

Fig. 43.19 shows the Energy-level diagram for ruby with its axis making an angle 54.44° with respect to the magnetic field.

There are approximately 7×10^{10} electron spins in a cubic centimeter of ruby and, of these, a certain percentage of spins are in each of the four energy levels. The number of spins in each level is a function of the temperature of the crystal. At room temperature the four levels are almost equally populated but at liquid helium temperature, 4.2° above absolute zero, the populations of the four levels are substantially unbalanced.

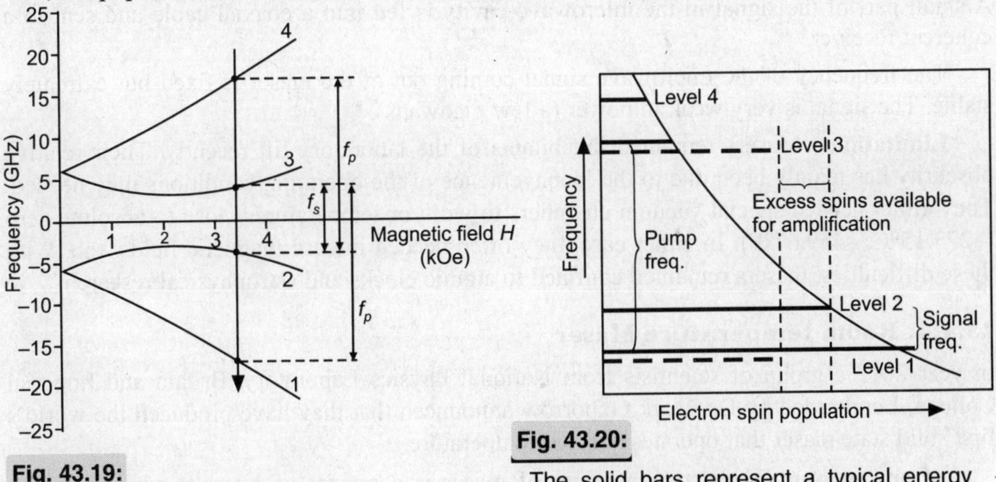

Fig. 43.19: Energy-level diagram for ruby

Fig. 43.20: The solid bars represent a typical energy population distribution for ruby.

(*iii*) **Amplification:** Spins in particular energy levels are excited by *pumping*, by applying microwave energy to the ruby. Incoming signal photons stimulate the excited spins. The spins drop to a lower energy level by emitting photons in-phase with the stimulating signal photons. This amplification process allows linear low-noise amplification. Ruby masers can withstand input signal power levels of many watts and are not damaged by voltage transients.

43.4.3. HYDROGEN MASER

Hydrogen maser is used as an atomic frequency standard. Norman Ramsey and his colleagues were first to develop hydrogen maser. It provides a sharp and constant oscillating signal. Here transitions in atomic hydrogen that occurs at a frequency of 1421 megahertz.

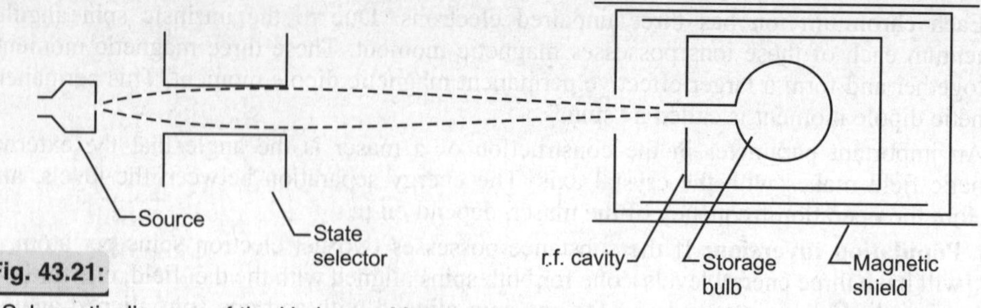

Fig. 43.21: Schematic diagram of the Hydrogen Maser

To start with, a beam of atomic hydrogen is formed by exposing hydrogen gas at low pressure to a radio-frequency discharge (Fig. 43.21). The beam is allowed to pass through an aperture and a magnetic field. After passing through the magnetic field, many of the atoms in the beam are in the upper energy level of the lasing transition. Most of the lower energy level atoms are deflected away. Thus, the population inversion of the atoms is achieved through "state selection". Now, the atoms in the upper energy level are ready for stimulated emission. The atoms in the upper energy level can on their own decay to the lower energy state and emit microwave radiation. The microwave cavity confines the microwaves and injects them repeatedly into the atom beam. The stimulated emissions that followed amplify the microwaves on each pass through the beam. The resonant frequency of the microwave cavity is 1420 405 751.768 Hz, which is exactly tuned to the hyperfine structure of hydrogen. A small part of the signal in the microwave cavity is fed into a coaxial cable and sent to a coherent receiver.

The frequency of the microwave signal coming out of the maser is fixed but extremely stable. The signal is very weak in power (a few picowatts).

Limitations: Masers remained inhabitants of the laboratory till recently. Their relative obscurity has mainly been due to the inconvenience of the operating conditions they needed. They either needed special vacuum chambers to work or temperatures close to absolute zero (–273.15°C, –459.67°F). In either case, they often needed intense magnetic fields, too. With these difficulties, masers remained confined to atomic clocks and astrophysical research.

43.4.4. Room temperature Maser

In year 2012 a group of scientists from National Physics Laboratory, Britain and Imperial College, London led by Dr. Mark Oxborrow announced that they have produced the world's first solid state maser that operates at room temperature.

(*i*) **Construction:** The active medium of maser is a crystal of *p*-terphenyl doped with pentacene. The pentacene: *p*-terphenyl crystal is held in a dielectric sapphire ring. The sapphire ring is housed in a cylindrical box (Fig. 43.22).The box acts as the microwave cavity and supports a high-Q electromagnetic mode. A pulsed dye laser sends strong pulses of yellow light onto the crystal through a side hole in the ring.

(*ii*) **Working:** Working is understood by three concepts.

(*a*) **Pumping (Jablonski diagram):** The *p*-terphenyl crystal hosts pentacene guest molecules. Pulsed dye laser sends strong pulses of pump light onto the pentacene: p-terphenyl maser crystal at its optimal absorption wavelength (585 nm). Pentacene molecules are pumped by photons of yellow light from their singlet ground states, S_0, to their first-excited single states, S_1. From S_1, the molecules undergo intersystem crossing into their triplet ground states, T_1. This process is spin selective; the relative

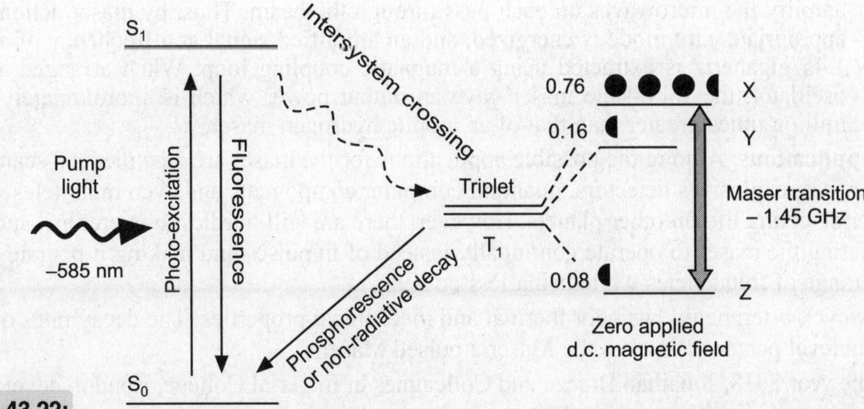

Fig. 43.22:

Pumping scheme

population in the three spin sublevels of T_1 is as shown by solid black circles in Fig. 43.23.

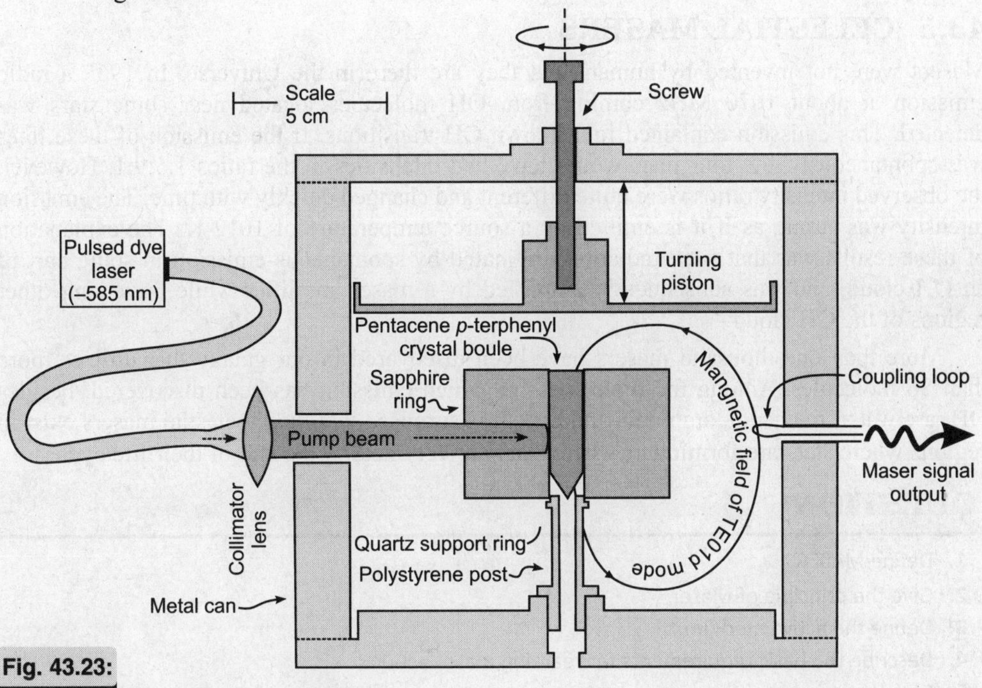

Fig. 43.23:

The first room temperature Maser

(*b*) **Population Inversion:** As shown in Fig. 43.23, the uppermost sublevel, *X*, is better populated, resulting in a strong initial population inversion between it and the lowest sublevel, *Z*. It offers the conditions for masing through the $X \rightarrow Z$ transition. The relatively high zero-field splitting makes it possible to use a microwave resonator with no applied d.c. magnetic field.

(*c*) **Output:** Now, the molecules in the uppermost sublevel, X are ready for stimulated emission. They can on their own decay to the lower energy state and emit microwave radiation. The microwave cavity confines the microwaves and introduces them repeatedly into the atom beam. The stimulated emissions that followed amplify the microwaves on each pass through the beam. Thus, by maser action, the appropriate e.m. mode is energized, and an amplified signal at a frequency of about 1.45 gigahertz is extracted using a magnetic coupling loop. When arranged as an oscillator, the solid-state maser gave an output power which is approximately 100 million times greater than that of an atomic hydrogen maser.

(*iii*) **Applications:** Among the possible applications for the maser are new medical scanning devices, explosives detectors, quantum computer components and even radio telescopes for detecting life on other planets. However, there are still hurdles to overcome, such as getting the maser to operate continually instead of in pulses and making it operate over a range of frequencies while using less power.

However, *p*-terphenyl has poor thermal and mechanical properties. The decay rates of the triplet sublevel pentacene makes the Maser a pulsed Maser.

In the year 2018, Jonathan Breeze and Colleagues in Imperial College, London developed a continuous working Maser using a synthetic diamond crystal grown in nitrogen atmosphere, and having vacancies created in it. The crystal was placed in a sapphire ring to concentrate microwave energy and was illuminated with green laser. The Maser worked at room temperature and also continuously.

43.5 CELESTIAL MASERS

Masers were not invented by humans but they are there in the Universe. In 1965 a radio emission at about 1670 MHz coming from OH molecules located near some stars was detected. This emission contained four known OH transitions. If the emission of these lines was spontaneously, the four lines would have had intensities in the ratios 1:5:9:1. However, the observed intensity ratios were quite different and changed quickly with time. The emission intensity was strong as if it is emitted by a source temperature of 1012 K. The explanation of these results was that such radiation originated by spontaneous emission in some part of an OH cloud, and was subsequently amplified by a maser amplifier while traversing other regions of the OH cloud.

More than one thousand masers have been discovered in our galaxy that utilizes more than 36 molecules. Among the molecules for which emission has been discovered, besides OH, are water, methanol, ammonia and SiO. It is surmised that these celestial masers exist in regions where stars are forming or where stars are very near to the end of their life cycle.

QUESTIONS

1. Define MASERS.
2. Give the principle of Maser.
3. Define the active medium.
4. Describe the basic requirements for realizing maser action.
5. Give the working of a cavity resonator.
6. Explain the principle of population inversion.
7. Explain the principle, construction of working of Ammonia Maser.
8. Explain the mechanism of pumping.
9. Enlist Advantages and disadvantages of ammonia maser.
10. Explain energy level diagram of the two level solid state masers.
11. Explain energy level diagram of the three level solid state masers.
12. Explain the principle, construction of working of ruby maser.
13. Explain the principle, construction of working of hydrogen maser.
14. Explain the principle, construction of working of room temperature.

44 Lasers

Learning Objectives:

By the end of this chapter, you will be able to:

- Learn about basic concepts of Laser.
- Identify different components of Laser.
- Describe concept of Threshold Conditioning.
- Understand Pumping Method and its types – Electrical and Optical.
- Explain different types of Laser.

44.1 INTRODUCTION

Laser is one of the outstanding inventions of the 20th century. The word 'LASER' is the acronym for *Light Amplification through Stimulated Emission of Radiation*. However, laser is not a simple amplifier of light but is a generator of light. It is an artificial light source that differs vastly from the traditional light sources. Laser is more akin to radio and microwave transmitters and produces a highly directional coherent monochromatic polarized light beam. Einstein gave the theoretical basis for the development of laser in 1916, when he predicted the possibility of stimulated emission. In 1954, C.H.Townes and his co-workers put Einstein's prediction for practical realization. They developed a microwave amplifier based on stimulated emission of radiation. It was called a **MASER**.

Shortly thereafter, in 1958, A. Schawlow and C.H. Townes extended the principle of masers to light and T.H. Maiman built the first laser device in 1960. In 1961, A. Javan and associates developed the first gas laser, the helium-neon laser. Laser is a high technology device and is the most sought after tool in a wide variety of fields such as metalworking, entertainment, communications, surgery, and ophthalmology and weapon guidance in wars.

44.2 INTERACTION OF LIGHT WITH MATTER AND THE THREE QUANTUM PROCESSES

It is familiar to us that when light travels through a medium, it undergoes absorption and scattering processes. Light absorption means the transfer of energy from light to atoms and light scattering involves change in the direction of travel of waves. As a result of these two processes, light intensity decreases with distance in the medium. Transfer of energy from atom to light is not conceivable from classical point of view. However, it is found to be possible when the interaction of light with medium is considered from the point of view of quantum mechanics. The transfer of energy from atom to light results in ***light amplification***. A light amplifier can be further converted into a source of light having superior characteristics compared to traditional light sources. A ***laser is a monochromatic coherent light source*** that depends on quantum processes for its operation. It is therefore necessary to first understand the quantum processes in order to understand the operation of a laser.

Let us consider a material medium, which is composed of *identical atoms*. Atoms are characterized by many energy levels but for the sake of simplicity to understand, let us assume that the atoms of the material medium under consideration be characterized by only two energy levels, namely energy level E_1 and energy level E_2. E_1 is the ground state while E_2 is the excited state. As the atoms of the material are identical, majority of them occupy the energy level E_1 and the others the energy level E_2.

The number of atoms per unit volume at an energy level is called the ***population density****. Let the populations at the two energy levels E_1 and E_2 be N_1 and N_2 respectively.*

Under normal conditions higher the energy, lesser is its population. Hence,

$$N_1 > N_2$$

Now, let light radiation be incident on the material and we assume that the radiation and the medium are in thermal equilibrium. The incident radiation may be viewed as a stream of photons, and let the photon density be $\rho(v)$. Let each photon carry an energy E, where $E = E_2 - E_1 = hv$. When photons travel through the medium, they are likely to cause three different processes. They are absorption, spontaneous emission and stimulated emission.

44.2.1 Absorption

Suppose an atom is in the lower energy level E_1. If a photon of energy $(E_2 - E_1)$ is incident on the atom, it imparts its energy to the atom and disappears. Then we say that the atom absorbed the incident photon. As a result of absorption of adequate energy, the atom jumps to the excited state E_2 (Fig. 44.1). This is known as an **absorption transition**. In each absorption transition, one atom is excited and one photon is lost from the incident light beam. We may express the process as

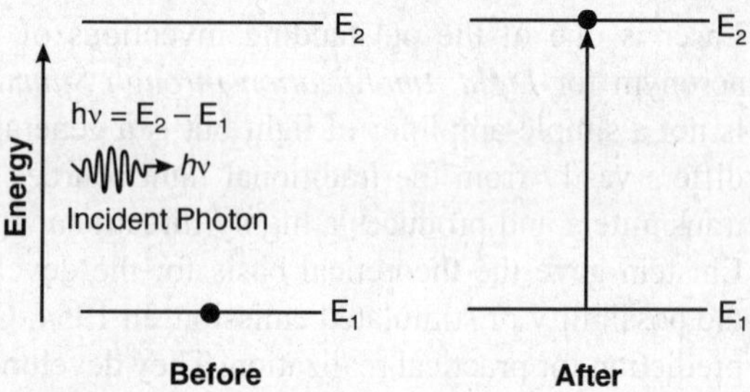

Fig. 44.1: Process of Absorption of photons

$$A + h\nu = A^*$$

where A is an atom in the lower state and A^* is an excited atom.

The probability that an absorption transition occurs is proportional to the photon density $\rho(\nu)$.

$$P_{12} \propto \rho(\nu)$$

or $$P_{12} = B_{12}\rho(\nu) \qquad ...(44.1)$$

where B_{12} is the constant of proportionality.

B_{12} is known as the *Einstein coefficient for induced absorption*. It is a constant characteristic of the atom and represents the properties of the energy states E_1 and E_2.

The number of absorption transitions occurring in the material at any instant will be equal to the product of the number of atoms at the energy level E_1 and the probability P_{12} for the absorption transition. Thus, the number of atoms, N_{ab}, excited during the time Δt is

$$N_{ab} = B_{12}N_1\rho(\nu)\Delta t \qquad \text{...(44.2)}$$

where N_1 is the population of atoms at E_1.

When the atoms are more at the lower energy level, then more atoms can jump into the excited state. Similarly, when more photons are incident on the assembly of atoms, then more atoms can get excited to the higher energy level. Induced absorption involves the excitation of the atom to the fixed higher level only. As a result of this absorption, N_1 decreases while N_2 increases. But under normal conditions N_2 cannot be greater than N_1.

44.2.2 Spontaneous Emission

When an atom at lower energy level is excited to a higher energy level, it cannot stay in the excited state for a relatively longer time. In a time of about 10^{-8}s, the atom reverts to the lower energy state by releasing a photon of energy $h\nu$ where $h\nu = E_2 - E_1$. The emission of photon occurs on its own and without any external impetus given to the excited atom (Fig. 44.2).

*Emission of a photon by an atom without any external impetus is called **spontaneous emission**.*

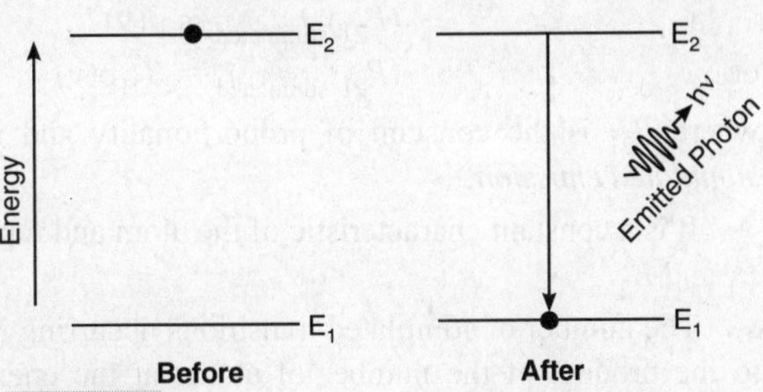

Fig. 44.2: Process of Spontaneous emission of photons

We may write the process as

$$A^* \rightarrow A + h\nu$$

The probability that a spontaneous transition occurs depends only on the properties of energy states E_2 and E_1 and is independent of the photon density. It is equal to the lifetime of level E_2. Thus,

$$(P_{21})_{\text{Spont.}} = A_{21} \qquad \text{...(44.3)}$$

where A_{21} is a constant and known as the *Einstein coefficient for spontaneous emission.*

A_{21} is a constant characteristic of the atom. *$1/A_{21}$ is a measure of the lifetime of the upper state against spontaneous transition to the lower state.*

The number of spontaneous transitions, N_{sp}, taking place during the time Δt depends *only* on the number of atoms N_2 staying at the excited state E_2. Thus,

$$N_{sp} = A_{21}N_2\Delta t \qquad \text{...(44.4)}$$

It is the process of spontaneous emission that dominates in conventional light sources.

44.2.3 Stimulated Emission

In 1916, Einstein showed the existence of equilibrium between matter and radiation required a new radiation process called stimulated radiation. It requires the presence of external radiation. If an atom in the excited state interacts with a photon with energy $h\nu = E_2 - E_1$, the photon induces the excited atom to make a downward transition well before the atom can make a spontaneous transition. The atom emits the excess energy in the form of a photon,

$h\nu = E_2 - E_1$, as it drops to the lower energy state. The passing photon is not affected while the excited atom emits a photon (Fig. 44.3).

*The phenomenon of forced photon emission by an excited atom due to the action of an external agency is called **stimulated emission**.*

It is also known as induced emission. The process may be expressed as

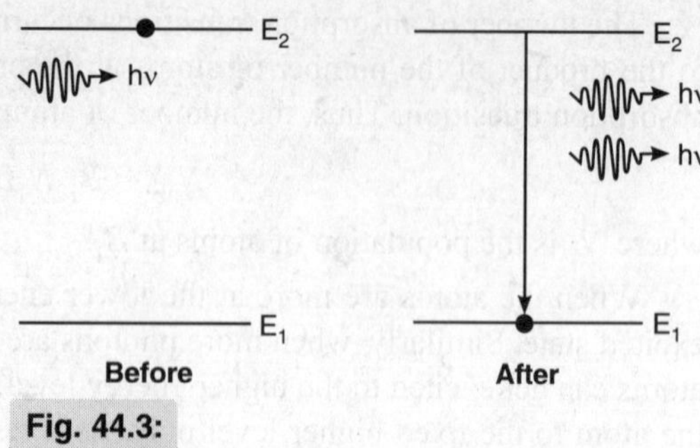

Fig. 44.3: Process of Stimulated Emission of photons

$$A^* + h\nu \rightarrow A + 2h\nu$$

The probability that a stimulated transition occurs is given by

$$(P_{21})_{\text{stimulated}} \propto \rho(\nu)$$

or

$$(P_{21})_{\text{stimulated}} = B_{21}\rho(\nu) \qquad ...(44.5)$$

where B_{21} is the constant of proportionality and is known as the *Einstein coefficient for stimulated emission.*

It is a constant characteristic of the atom and represents the properties of the energy states E_1 and E_2.

The number of stimulated transitions occurring in the material at any instant will be equal to the product of the number of atoms at the energy level E_2 and the probability P_{21} for the stimulated transition. Thus, the number of atoms, N_{st}, that undergo downward transition during the time Δt is

$$N_{st} = B_{21}N_2\rho(\nu)\Delta t \qquad ...(44.6)$$

44.2.4 Multiplication of Stimulated Photons

The photon induced in this process propagates in the *same direction* as that of stimulating photon. The induced photon has features identical to that of the inducing photon. It has the *same frequency, phase and plane of polarization* as that of the stimulating photon. The outstanding feature of this process is the *multiplication of photons*.

For one photon interacting with an excited atom, there are two photons emerging. The two photons travelling in the same direction interact with two more excited atoms and generate two more photons and produce a total of four photons. These four photons in turn stimulate four excited atoms and generate eight photons, and so on. The number of photons builds up in an avalanche like manner, as shown in Fig. 44.4.

All the light waves generated in the medium are due to one initial wave and all of the waves are in phase.

Fig. 44.4: Multiplication of Stimulated photons into an avalanche

Thus, the waves are coherent and interfere constructively. The net intensity of light will be proportional to the square of the number of atoms radiating light. Thus,

$$I_{total} = N^2 I$$

Since the number of atoms in the material medium is very large, coherent emission leads to an enormously high intense light and we say that the incident *light is amplified*. Therefore, the process of stimulated emission is the key to the operation of a laser.

Table 44.1: *Distinction between Spontaneous and Stimulated Emission*

Sl. No.	Spontaneous emission	Stimulated emission
1.	Spontaneous emission is a random and probabilistic process.	Not a random process.
2.	Not amenable for control from outside.	Amenable for control from outside.
3.	The photons are emitted haphazardly. The instant of emission, direction of emission, phase, polarization state of photon are all random quantities and cannot be controlled.	The stimulating photon imposes its characteristics on the photon emitted.
4.	Photons are emitted uniformly in all directions from an assembly of atoms. As a result, the light is ***non-directional***.	The photons emitted in the process travel in the same direction as that of stimulating photon. The light produced by the process is essentially ***directional***.
5.	Photons of slightly different frequencies are generated. As a result, the light is ***not monochromatic***.	The spread of photon frequencies is relatively very narrow. As such the light is nearly ***monochromatic***.
6.	Photons do not have any correlation in their phases, which fluctuate randomly. Therefore, the light produced by this process is ***incoherent***.	The photons emitted by this process are all in phase and therefore, the light is ***coherent***.
7.	In this process multiplication of photons does not take place. Hence there is **no amplification** of light due to the process.	One stimulating photon causes emission of two more photons. These two produce four photons, which in turn generate eight photons and so on. Thus, if there are N excited atoms, 2^N photons will be produced. ***Light amplification*** occurs due to such multiplication of photons.
8.	The net intensity of the generated light is given by $I_T = N I$ where N is the number of atoms emitting photons and I is the intensity of each photon.	As all the photons are in phase, they constructively interfere and produce an intensity $I_T = N^2 I$
9.	The planes of polarization of the photons are oriented randomly. Hence, light from the source is ***unpolarized***.	The planes of polarization are identical for all photons. Consequently, light is ***polarized***.

44.2.5 Steady State

The three processes described above occur simultaneously (Fig. 44.5). Under steady state condition the absorption and emission balance each other. Thus,

$$N_{absorption} = N_{spont.emission} + N_{stim.emission} \quad ...(44.7)$$

$$B_{12} N_1 \rho(\nu) \Delta t = A_{21} N_2 \Delta t + B_{21} N_2 \rho(\nu) \Delta t$$

$$B_{12}\, N_1\, \rho(\nu) = A_{21}\, N_2 + B_{21}\, N_2\, \rho(\nu) \quad ...(44.8)$$

If we consider a medium in thermal equilibrium, there would be more atoms in the lower level than at higher level. That is $N_1 >> N_2$. As the probability for absorption transition is equal to the probability for stimulated transition, a photon traveling through the medium is *more likely to get absorbed* than to stimulate an excited atom to emit a photon. Therefore, usually the process of absorption dominates the process of stimulated emission. Similarly, an atom that is at the excited state is more likely to jump to the lower level on its own than being stimulated by a photon. It is due to the fact that the photon density in the incident beam is not sufficient to interact with the excited atoms; and the photons interact with atoms at lower level because of the large population available at that level. Owing to this, the spontaneous emission dominates the stimulated emission.

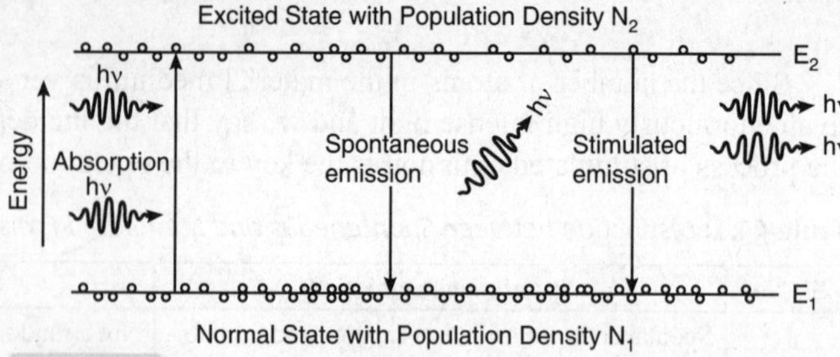

Fig. 44.5: Absorption, spontaneous and stimulated emission processes

44.3 EINSTEIN COEFFICIENTS AND THEIR RELATIONS

44.3.1 Einstein Coefficients

We summarize here the three Einstein coefficients, which are the proportionality constants introduced in the above discussions.

(*i*) The probability that an absorption transition occurs is given by

$$P_{12} = B_{12}\rho(\nu)$$

where B_{12} is the constant of proportionality known as the *Einstein coefficient for induced absorption*. It is a constant characteristic of the atom and represents the properties of the energy states E_1 and E_2.

(*ii*) The probability that a spontaneous transition occurs is given by

$$(P_{21})_{\text{Spontaneous}} = A_{21}$$

where A_{21} is a constant known as the *Einstein coefficient for spontaneous emission*. A_{21} is a constant characteristic of the atom and is known as the *radiative rate* measured in units of s^{-1}. $1/A_{21}$ *is the lifetime of the upper state against spontaneous decay to the lower state*.

(*iii*) The probability that a stimulated transition occurs is given by

$$(P_{21})_{\text{stimulated}} = B_{21}\rho(\nu)$$

where B_{21} is the constant of proportionality known as the *Einstein coefficient for stimulated emission*. It is a constant characteristic of the atom and represents the properties of the energy states E_1 and E_2.

We may note the following points here regarding the Einstein coefficients.

(*a*) The coefficients indicated by B are related to the induced transitions, i.e., transitions induced by external photons. Thus, B_{12} represents the transition induced by a photon from lower energy level E_1 to the higher energy level E_2, whereas B_{21} denotes the transition induced by a photon from higher energy level E_2 to the lower energy level E_1. It turns out that B_{12} and B_{21} are equal under the special condition that the quantum states E_1 and E_2 are single energy levels (i.e., nondegenerate levels).

(*b*) The coefficient indicated by A is related to the spontaneous transition, i.e., transition occurred on its own without the assistance of external agent. Since a spontaneous transition cannot take place from lower energy state E_1 to the higher energy state E_2, we do not have the coefficient A_{12}. In other words, $A_{12} = 0$.

44.3.2 Relation between the Einstein Coefficients

The Einstein coefficients A_{21}, B_{12} and B_{21} are interrelated. To find out the relation, we assume that

(*i*) The atoms and the radiation are in thermal equilibrium.

(*ii*) The radiation is identical with black body radiation and consistent with Planck's radiation law for any value of T.

(*iii*) The population densities N_1 and N_2 at the lower and upper energy levels respectively are constant in time and are distributed according to Boltzmann law in the energy levels.

The above conditions require that the rate of change of atoms at the level E_2 must equal to zero. It means that the number of transitions from E_2 to E_1 must be equal to the number of transitions from E_1 to E_2 (see Fig. 44.5).

Thus,

$$\left.\begin{array}{l}\text{The number of atoms absorbing}\\ \text{photons per second per unit volume}\end{array}\right\} = \left\{\begin{array}{l}\text{The number of atoms emitting}\\ \text{photons per second per unit volume}\end{array}\right.$$

The number of atoms absorbing photons per second per unit volume = $B_{12}\,\rho(\nu)\,N_1$

The number of atoms emitting photons per second per unit volume = $A_{21}N_2 + B_{21}\,\rho(\nu)\,N_2$

As the number of transitions from E_1 to E_2 must equal the number of transitions from E_2 to E_1, we have

$$B_{12}\,\rho(\nu)\,N_1 = A_{21}N_2 + B_{21}\,\rho(\nu)\,N_2 \qquad ...(44.9)$$

$$\rho(\nu)\,[B_{12}\,N_1 - B_{21}\,N_2] = A_{21}N_2$$

$$\therefore \qquad \rho(\nu) = \frac{A_{21}N_2}{[B_{12}N_1 - B_{21}N_2]} \qquad ...(44.10)$$

By dividing both the numerator and denominator on the right hand side of the above equation with $B_{12}N_2$, we obtain

$$\rho(\nu) = \frac{A_{21}/B_{12}}{\left[\dfrac{N_1}{N_2} - \dfrac{B_{21}}{B_{12}}\right]} \qquad ...(44.11)$$

But
$$\frac{N_2}{N_1} = e^{-(E_2-E_1)/kT}$$

As $E_2 - E_1 = h\nu$,
$$\frac{N_2}{N_1} = e^{-h\nu/kT} \quad \text{or} \quad \frac{N_1}{N_2} = e^{h\nu/kT}$$

$$\therefore \qquad \rho(\nu) = \frac{A_{21}}{B_{12}}\left[\frac{1}{e^{h\nu/kT} - B_{21}/B_{12}}\right] \qquad ...(44.12)$$

To maintain thermal equilibrium, the system must release energy in the form of electromagnetic radiation. It is required that the radiation be identical with black body radiation and be consistent with Planck's radiation law for any value of T. According to Planck's law

$$\rho(\nu) = \left(\frac{8\pi h\nu^3\mu^3}{c^3}\right)\left[\frac{1}{e^{h\nu/kT}-1}\right] \quad ...(44.13)$$

where μ is the refractive index of the medium and c is the velocity of light in free space.

Energy density $\rho(\nu)$ given by Eq. (44.12) will be consistent with Planck's law Eq. (44.13), only if

$$\frac{A_{21}}{B_{12}} = \frac{8\pi h\nu^3\mu^3}{c^3} \quad ...(44.14)$$

and $$\frac{B_{21}}{B_{12}} = 1 \quad \text{or} \quad B_{12} = B_{21} \quad ...(44.15)$$

Eq. (44.14) and Eq. (44.15) are known as the **Einstein relations**. It follows that the coefficients are related through

$$B_{12} = B_{21} = \frac{c^3}{8\pi h\nu^3\mu^3}A_{21} \quad ...(44.16)$$

Eq. (44.15) shows that the coefficients for both absorption and stimulated emission are numerically equal. The equality implies that when an atom with two energy levels is placed in the radiation field, the probability for an upward (absorption) transition is equal to the probability for a downward (stimulated) transition.

Eq. (44.14) shows that the ratio of coefficients of spontaneous versus stimulated emission is proportional to the third power of frequency of the radiation. This is why it is difficult to achieve laser action in higher frequency ranges such as *x*-rays.

44.4 LIGHT AMPLIFICATION

Light amplification requires that stimulated emission occur almost exclusively. In practice, absorption and spontaneous emission always occur together with stimulated emission. The laser operation is achieved when stimulated emission exceeds in a large way the other two processes. Let us now look at the conditions under which the number of stimulated transitions can be made larger than the other two transitions.

44.4.1 Condition for Stimulated Emission to Dominate Spontaneous Emission

The ratio of Eq. (44.6) to Eq. (44.4) gives

$$\frac{\text{Stimulated transitions}}{\text{Spontaneous transitions}} = \frac{B_{21}N_2\rho(\nu)}{A_{21}N_2} = \frac{B_{21}}{A_{21}}\rho(\nu) \quad ...(44.17)$$

Eq. (44.17) indicates that stimulated transitions will dominate the spontaneous transitions if the radiation density $\rho(\nu)$ is very large. Thus, the presence of a large number of photons in the active medium is required. However, it will lead to more absorption transitions. Hence, large photon density alone will not guarantee more stimulated emissions.

44.4.2 Requirement of States of Larger Lifetimes

Eq. (44.17) further indicates that stimulated transitions will dominate the spontaneous transitions if the value of the ratio B_{21}/A_{21} is also large. To increase the probability of stimulated emissions, the lifetime of atoms at the excited state should be larger. In other words, it is necessary that the excited state has a longer lifetime (remember that $1/A_{21}$ represents the lifetime of the excited state).

44.4.3 Condition for Stimulated Emission to Dominate Absorption Transitions

The ratio of Eq. (44.6) to Eq. (44.2) yields

$$\frac{\text{Stimulated transition}}{\text{Absorption transition}} = \frac{B_{21}N_2\rho(\nu)}{B_{12}N_1\rho(\nu)} = \frac{N_2}{N_1} \qquad ...(44.18)$$

We used here the fact that $B_{12} = B_{21}$.

The above condition indicates that the stimulated transitions will overwhelm the absorption process if N_2 is greater than N_1. It means that there should be more atoms present in the higher energy level than in the lower energy level for stimulated emissions to dominate over the spontaneous emissions.

A medium amplifies light only when the above ***three conditions*** are fulfilled. Therefore to achieve high percentage of stimulated emissions, an artificial situation known as *population inversion* is to be created in the medium.

44.5 MEETING THE THREE REQUIREMENTS

44.5.1 Population Inversion

When the material is in thermal equilibrium condition, the population ratio is governed by the Boltzmann factor according to the following equation:

$$\frac{N_2}{N_1} = e^{-[E_2 - E_1]/kT} \qquad ...(44.19)$$

It means that the population N_2 at the excited level E_2 will be far smaller than the population N_1 at the level E_1. For example, if we take typical values for E_1 and E_2, the population N_1 would be 10^{30} times of N_2. The condition in which there are more atoms in the lower energy level and relatively lesser number of atoms in the higher energy level is called *normal state or equilibrium state* (Fig. 44.6 a).

E_2 — N_2 E_2 — N_2

E_1 — N_1 E_1 — N_1

(a) Normal state, Thermal equilibrium condition $N_1 >> N_2$

(b) Inverted state, Non-equilibrium condition $N_1 << N_2$

Fig. 44.6: Illustration of Population Inversion (a) Normal state (b) Inverted state

Thus, under thermal equilibrium condition, $N_1 >> N_2$.

Population inversion *is a non-equilibrium state and exists only for a short time. This is the condition of the material in which population of the upper energy level N_2 far exceeds the population of the lower energy level, N_1* (Fig. 44.6 b).

That is,

$$N_2 >> N_1 \quad ...(44.20)$$

In this condition the population distribution between the levels E_1 and E_2 is inverted and hence it is known as the *inverted state*. This is a *non-equilibrium state* and exists only for a short time. Population inversion is obtained by employing *pumping techniques*, which transfer large number of atoms from lower energy level to higher energy level.

Example 44.1: *A 10 mW He-Ne laser has efficiency of 1%. Assume that all input energy is utilized in pumping the atoms from the ground state to the excited state, which is 20 eV above the ground state. Find how many atoms are promoted to the excited state in one second.*

Solution: Efficiency of laser = 1% = 0.01

$$\text{Power input} = \frac{\text{Power output}}{\text{Efficiency}} = \frac{10\,mw}{0.01} = 1\text{ W.}$$

Therefore, energy input in one second = 1 J.

$$\text{Number of atoms excited in one second} = \frac{1\text{J}}{20\text{ eV}} = \frac{1\text{J}}{20\times1.602\times10^{-19}\text{ J}} = \mathbf{3.12\times10^{17}}.$$

Example 44.2: *Find the ratio of populations of the two states in a He-Ne laser that produces light of wavelength 6328Å at 27°C.*

Solution: The ratio of population is given by $\frac{N_2}{N_1} = e^{-(E_2-E_1)/kT}$

$$E_2 - E_1 = \frac{12400}{6328}\text{eV} = 1.96\text{ eV}$$

$$\therefore \quad \frac{N_2}{N_1} = \exp\left[\frac{-1.96eV}{(8.61\times10^{-5}\,eV)\,(300\text{K})}\right] = e^{-75.88} = \mathbf{1.1\times10^{-33}}$$

44.5.2 Metastable States

An atom can be excited to a higher level by supplying energy to it. Generally, excited atoms have very short lifetimes and spontaneously give out their energy in a few nanoseconds (10^{-9}s). Thus, atoms do not stay for long enough time at the excited state. Even though the pumping agent continuously raises the atoms to the excited level, they undergo spontaneous transition and quickly return to the lower energy level. Population inversion cannot be attained under such circumstances. In order to achieve the condition of population inversion, the excited atoms are to be made to 'wait' at the upper energy level without undergoing spontaneous transition till a huge number of atoms build up at that level.

Such an opportunity would be provided by metastable states.

Metastable state *is an excited state of atoms where atoms remain excited for an appreciable time, which is of the order of 10^{-6} to 10^{-3}s. This is 10^3 to 10^6 times the lifetimes of the ordinary excited energy levels. Therefore, the metastable state allows accumulation of a large number of excited atoms at that level.*

The metastable state population can exceed the population at a lower level and establish the condition of population inversion in the lasing medium. It would be impossible to create the state of population inversion without a metastable state.

Metastable state can be readily obtained in a crystal system containing impurity atoms. These levels lie in the forbidden band gap of the host crystal. Population inversion readily takes place as the lifetimes of these levels are large, and secondly, there is no competition in filling these levels, as they are localized levels. For example, **phosphorescent** materials are made up of atoms with metastable states.

There could be no population inversion and hence no laser action, if metastable states do not exist.

44.5.3 Confining Radiation within the Medium

According to Eq. (42.17) a high radiation density $\rho(\nu)$ is required to be present in the active medium so that stimulated emission dominates spontaneous emission. If laser medium is enclosed in between a pair of optically plane parallel mirrors, photon density builds up to a very high value through repeated reflections of photons which remain within the medium. Such an arrangement is known as an ***optical resonant cavity*** or ***optical resonator***.

44.6 COMPONENTS OF LASER

The essential components of a laser are (*i*) an active medium; (*ii*) a pumping agent and (*iii*) an optical resonator (see Fig. 44.7).

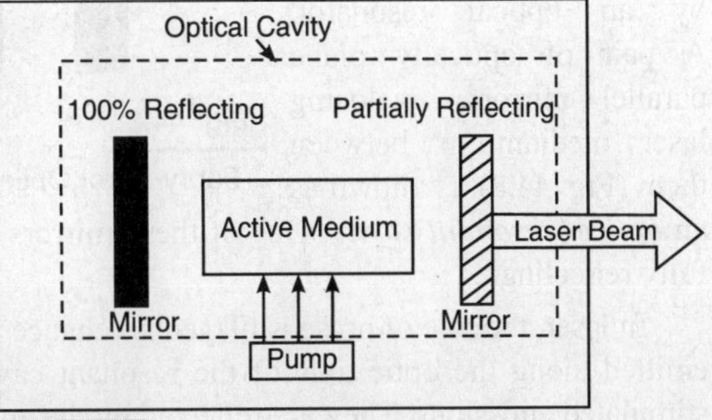

Fig. 44.7: Components of a Laser

44.6.1 Active Medium

Active medium *is the material in which the laser action takes place. The most important requirement for the laser medium is that we should be able to obtain population inversion in it.*

Atoms are in general characterized by a large number of energy levels. However, all types of atoms are not suitable for laser operation. Even in a medium consisting of different species of atoms, only a small fraction of atoms of a particular type have energy level system suitable for achieving population inversion. Such atoms can produce more stimulated emission than spontaneous emission and cause amplification of light. Those atoms, which cause laser action, are called *active centers*. The rest of the medium acts as host and supports active centers. The medium hosting the active centers is called the ***active medium***. An active medium is a medium which when excited reaches the state of population inversion and promotes stimulated emissions leading to light amplification.

44.6.2 The Pump

For achieving and maintaining the condition of population inversion, we have to raise continuously the atoms in the lower energy level to the upper energy level. It requires energy to be supplied to the system. Pumping is the process of supplying energy.

The pump is an external source that supplies energy needed to transfer the laser medium into the state of population inversion.

There are a number of techniques for pumping a collection of atoms to an inverted state. Optical pumping, electrical discharge and direct conversion are some of the methods of pumping. In optical pumping, a light source such as a flash discharge tube is used. This method is adopted in solid-state lasers. In electrical discharge method, the electric field causes ionization of the medium and raises it to the excited state. In semiconductor diode lasers, a direct conversion of electrical energy into light energy takes place.

44.6.3 Optical Resonator

Laser is a light source and it is analogous to an electronic oscillator. An electronic oscillator is essentially an amplifier supplied with a positive feed back. A part of the output of the amplifier is taken and fed back at its input. Then, electrical noise signal of appropriate frequency present at the input gets amplified; the output is again fed back to the input and amplified. It goes on till a stable output is reached. At this stage the oscillator acts as a source of a particular frequency.

In laser the active medium is the amplifier, which is converted into an oscillator through the feed back mechanism established by an optical resonator. A pair of optically plane parallel mirrors, enclosing laser medium in between them (Fig. 44.8), is known as an ***optical resonant cavity***. One of these mirrors is partially reflecting and the other is made fully reflecting.

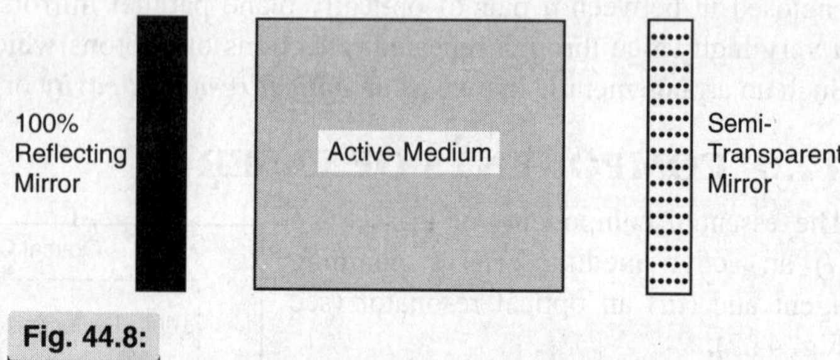

Fig. 44.8: Fabry-Perot Optical resonator

In laser, the role of noise is played by chance photons emitted spontaneously. The photons emitted along the optic axis of the resonant cavity travel through the medium and trigger stimulated emissions. They are reflected by the end mirror and reverse their path. The photons are thus fed back into the medium and travel toward the opposite end mirror causing more stimulated emissions. The photons are once more reflected at the mirror and travel toward the opposite mirror. Substantial light amplification takes place because the light beam is reflected several times at the mirrors and gains strength in each passage. Ultimately, when the amplification balances the losses in the cavity, the laser beam emerges out from the front – end mirror. *In the absence of resonator cavity, there would be no amplification of light.*

Role of the Optical Resonator

(*i*) The primary function of the optical resonator is to provide positive feed back of photons into the medium so that stimulated emission is sustained and the laser acts as a generator of light.

(*ii*) The laser oscillation is initiated by photons spontaneously emitted by some of the excited atoms. Each spontaneous photon can trigger many stimulated transitions along the path of its travel. As the initial spontaneous photons are emitted in different directions, the stimulated photons would travel in different directions. The optical resonator selects the direction in which the light is to be amplified; the direction being the optical axis of the pair of mirrors. Thus, optical cavity makes the laser beam directional.

(*iii*) In order to make the stimulated emission dominate spontaneous emission, a high radiation density $\rho(\nu)$ is required to be present in the active medium. The optical cavity builds up the photon density to a very high value through repeated reflections of photons and confines them within the medium.

(*iv*) Optical cavity selects and amplifies only certain frequencies causing the laser output to be highly monochromatic.

44.7 LASING ACTION

Fig.44.8 shows the active medium enclosed in optical resonator and being excited by a pumping agent. The resulting laser action, which is shown in Fig. 44.9, consists of the following steps:

Step-1: Pumping: The atoms (active centers) in the medium are in the ground state initially, as shown in Fig. 44.9 (a). By supplying energy from an external source, the atoms are excited from the ground level to an excited state.

Step-2: Population inversion: The lifetime of atoms at the excited state is extremely small, of the order of 10^{-8} sec. Therefore, the atoms drop spontaneously from the excited state to the metastable state. As the lifetime of atoms at the metastable state is comparatively longer (10^{-3} sec), the atoms go on accumulating at the metastable state. As soon as the number of atoms at the metastable state exceeds that of the ground state, the medium goes into the state of population inversion (Fig. 44.9 b).

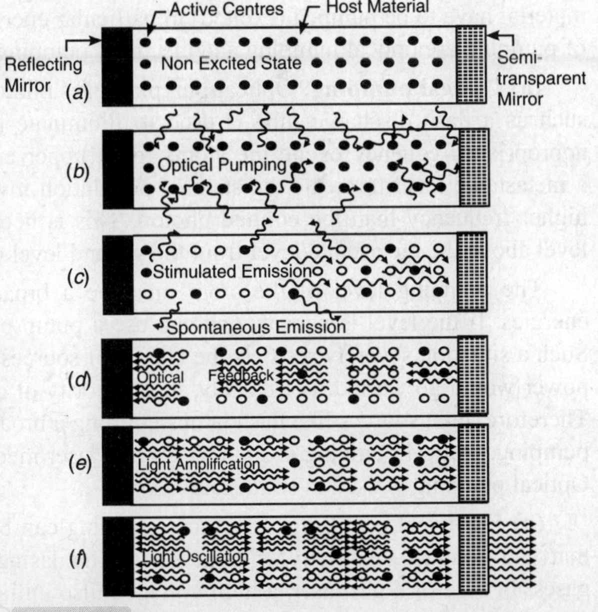

Fig. 44.9: Light amplification and oscillations due to the action of optical resonator

Step-3: Spontaneous emissions: Some of the excited atoms at the metastable state may emit photons spontaneously in various directions (Fig. 44.9 c). Each spontaneous photon can trigger many stimulated transitions along the direction of its propagation. As the initial spontaneous photons are moving in different directions, the photons stimulated by them also travel in different directions. Many of such photons leave the medium without reinforcing their strength. The photons emitted in a direction other than the axial direction will pass through the sides of the medium and are lost forever.

Step-4: Amplification: A majority of photons traveling along the axis cause stimulated emission and on reaching the end mirror they are reflected back into the medium. On their way to opposite mirror, stimulates more and more atoms and boost up the photon strength, Fig. 44.9 (d). The photons that reach the opposite mirror are reflected once again into the medium, as shown in Fig. 44.9 (e). The photons travel once more through the medium generating more photons and more amplification. The photons are then reflected again at the mirror and travel through the medium. As the photons are reflected back and forth between the mirrors, stimulated emission sharply increases and the amplification of light takes place. The mirrors thus provide *positive feed* back of light into the medium so that stimulated emission acts are sustained and the medium operates as an oscillator.

Step-5: Oscillations: On every reflection at the front-end mirror, light is partly transmitted through it. It is a *loss of energy* from the resonator. When the losses occurring at the mirrors and within the medium together balance the gain a stable and strong laser beam will come out from the front-end mirror as shown in Fig. 44.9 (*f*).

44.8 PUMPING METHODS

In order to create the state of population inversion in an active medium, the atoms in the material have to be pumped (excited) to particular energy levels. The most common methods of pumping are optical pumping and electrical pumping.

(a) Optical pumping: Optical pumping uses photons to excite the atoms. A light source such as a flash discharge tube is used to illuminate the laser medium and the photons of appropriate frequency excite the atoms to an upper energy level. From there, they drop to a metastable level to create the state of population inversion. The pump photon must have higher frequency than the emitted photon. This is because the atoms are to be excited to a level above the metastable level from the ground level or a lower energy level.

The pumping level of the atom must be a broader level. It should span a range of energies. If the level is narrow, one can use a pump photon of only one specific frequency. Such a situation severely restricts the choice of sources and also a large portion of the source power would go wasted. Fortunately, in a majority of cases the pump levels are wide bands. Therefore, light sources like flash lamps emitting a broad range of frequencies can be used for pumping. Optical pumping is suitable for any laser medium that is transparent to pump light. Optical pumping is used in solid state lasers.

(b) Electrical pumping: Electrical pumping can be used only in case of laser materials that can conduct electricity without destroying lasing activity. This method is limited to gases. In case of a gas laser, a high voltage pulse initially ionizes the gas so that it conducts electricity. An electric current flowing through the gas excites atoms to the excited level from where they drop to the metastable level leading to population inversion.

44.8.1 Principal Pumping Schemes

Atoms in general are characterized by a large number of energy levels. Among them only three or four levels will be pertinent to the pumping process. Therefore, only those levels are depicted in the pumping scheme diagrams. Two important pumping schemes are widely employed. They are known as three-level and four-level pumping schemes.

Two levels scheme is not generally feasible for laser action. The main reason is that the energy being used to pump the atoms into the upper laser state has an equal probability of stimulating them back down. Therefore, it is not possible in general to pump more than half of atoms into the excited state.

1. Three-Level Pumping Scheme: In the *three levels scheme* the atoms are first pumped to a state higher in energy than the upper laser state. The pumped state should have a shorter lifetime for spontaneous emission so that the atoms can spontaneously come down to the upper laser state. The upper laser state is selected such that it has as long a lifetime as possible. Then the atoms stay in the upper laser state long enough to be stimulated.

A typical *three-level pumping scheme* is shown in Fig. 44.10. The state E_1 is the ground state; E_3 is the pump state and E_2 is upper lasing level, which is a metastable state. When the medium is exposed to pump frequency radiation, a large number of atoms will be excited to E_3 level. However, they do not stay at that level but rapidly undergo downward transitions to the metastable state E_2 through non-radiative transitions. The atoms are trapped at this state as spontaneous transition from the level E_2 to the level E_1 is forbidden. The pumping continues and after a short time there will be a large accumulation of atoms at the level E_2. When more than half of the ground state atoms accumulate at E_2, the population inversion condition is achieved between the two states E_1 and E_2. Now a chance photon can trigger stimulated emission.

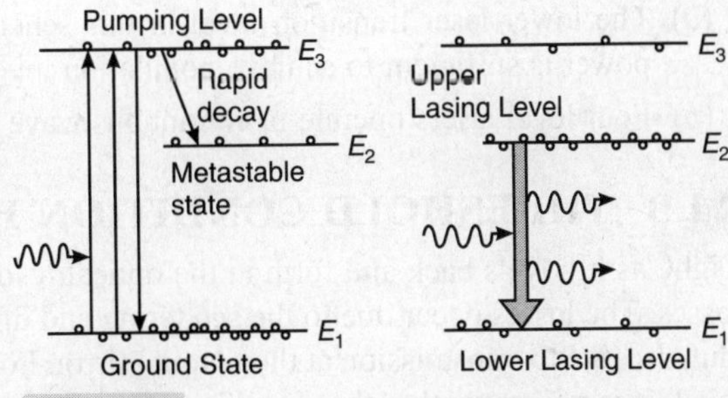

Fig. 44.10: A typical three level pumping scheme (a) Pumping (b) Lasing action

(*a*) In this scheme, the terminal state of the laser transition is simultaneously the ground state. Therefore, population inversion is achieved only when more than half of the ground state atoms are pumped to the upper state. Thus, the scheme requires very high pump power.

(*b*) The three levels scheme produces light only in pulses. Once stimulated emission commences, the metastable state is quickly emptied and the population of the ground state increases rapidly. As a result, the population inversion ends. One has to wait till the population inversion is re-established. Thus, the three-level lasers operate in pulsed mode.

2. Four-Level Pumping Scheme: A typical *four level pumping scheme* is shown in Fig. 44.11. The state E_1 is the ground state, E_4 the pumping level, E_3 the metastable upper lasing level and E_2 the lower lasing level. E_2, E_3 and E_4 are the excited states. When light of pump frequency v_p is incident on the lasing medium, the active centers are readily excited from the ground state to the pumping level E_4. The atoms stay at the E_4 level for only about 10^{-8}s, and quickly drop down to the metastable state E_3. As spontaneous transitions from the level E_3 to level E_2 cannot take place, the atoms get trapped in the state E_3. The population at the state E_3 grows rapidly. The level E_2 is well above the ground state such that $(E_2 - E_1) > kT$.

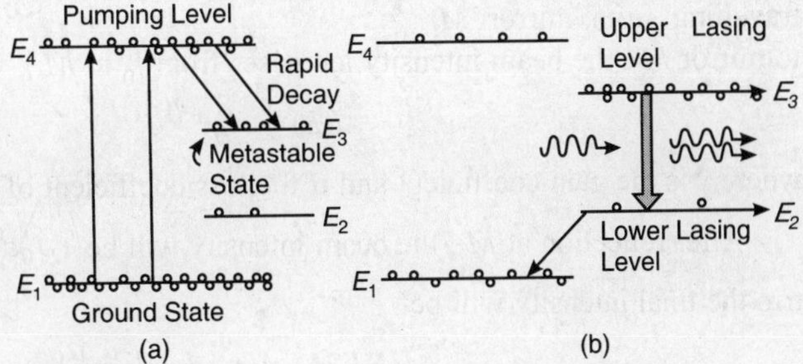

Fig. 44.11: Illustration of fourlevels pumping scheme (a) Pumping (b) Lasing Action.

Therefore, at normal temperature atoms cannot jump to level E_2 on the strength of thermal energy. As a result, the level E_2 is virtually empty. Therefore, population inversion is attained between the states E_3 and E_2. A chance photon of energy hv = $(E_3 - E_2)$ emitted spontaneously can start a chain of stimulated emissions, bringing the atoms to the lower laser level E_2. From state E_2, the atoms subsequently under go non–radiative transitions to the ground state E_1 and will be once again available for excitation, making it possible for light to be emitted continuously.

(*a*) The lower laser transition level in this scheme is nearly vacant. Therefore, less pump power is sufficient to achieve population inversion.

(*b*) Four level lasers operate in continuous wave (cw) mode.

44.9 THRESHOLD CONDITION FOR LASING

Light, as it travels back and forth in the optical resonator, gets amplified and it suffers various losses. The losses occur due to the scattering and diffraction of light within the active medium and also due to transmission at the output mirror. For the buildup of oscillations to the required level, it is necessary that the amplification between two consecutive reflections of light from rear end mirror balance the losses. The threshold gain can be found by considering the change in intensity of a beam of light undergoing a round trip within the resonator.

Let us assume that the laser medium fills the space between the mirrors M_1 and M_2 (see Fig. 44.12), which have reflectivity r_1 and r_2 respectively. Let the mirrors be separated by a distance L. Further, let the intensity of the light beam be I_0 at M_1. Then, in travelling from mirror M_1 to mirror M_2, the beam intensity increases from I_0 to $I(L)$, which is given by

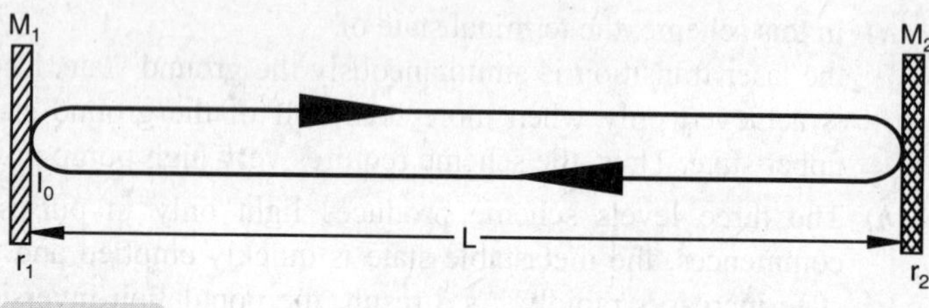

Fig. 44.12: Round trip path of the radiation through the laser cavity

$$I(L) = I_0 e^{(\gamma - \alpha_s)L} \quad \text{...(44.21)}$$

where γ is the gain coefficient and α the loss coefficient of the active medium.

After reflection at M_2, the beam intensity will be $r_2 I_0 e^{(\gamma-\alpha_s)L}$ and after a complete round trip the final intensity will be

$$I(2L) = r_1 r_2 I_0 e^{(\gamma-\alpha_s)2L} \quad \text{...(44.22)}$$

The amplification obtained during the round trip is

$$G = \frac{I(2L)}{I_0} = r_1 r_2 e^{(\gamma-\alpha_s)2L} \quad \text{...(44.23)}$$

The product $r_1 r_2$ represents the losses at the mirrors whereas α_s includes all the distributed losses such as scattering, diffraction and absorption occurring in the medium. The losses are balanced by gain, when $G \geq 1$ or $I(2L) = I_0$. It requires that

$$r_1 r_2 e^{2(\gamma-\alpha_s)L} \geq 1 \quad \text{...(44.24)}$$

or

$$e^{2(\gamma-\alpha_s)L} \geq \frac{1}{r_1 r_2}$$

Taking logarithms on both sides, we get

$$2L(\gamma - \alpha_s) \geq -\ln r_1 r_2$$

$$\gamma - \alpha_s \geq -\frac{1}{2L} \ln r_1 r_2$$

$$\therefore \qquad \gamma \geq \alpha_s - \frac{1}{2L}\ln r_1 r_2 \qquad ...(44.25)$$

or

$$\gamma \geq \alpha_s + \frac{1}{2L}\ln \frac{1}{r_1 r_2} \qquad ...(44.26)$$

Eq. (44.26) is known as the ***condition for lasing***. It shows that the initial gain must exceed the sum of the losses in the cavity. This condition is used to determine the threshold value of pumping energy for lasing action.

γ will be dependent on how hard the laser medium is pumped. As the pump power is slowly increased, a value of γ_{th} called ***threshold value*** is reached and the laser starts oscillating. The threshold value γ_{th} is given by

$$\gamma_{th} = \alpha_s + \frac{1}{2L}\ln \frac{1}{r_1 r_2} \qquad ...(44.27)$$

Eq. (44.27) states the condition when the net gain would be able to counteract the effect of losses in the cavity and is known as the ***threshold condition for lasing***. The value of γ must be atleast γ_{th} for laser oscillations to commence.

44.10 MODES OF THE LASER BEAM

The laser beam radiated from the laser cavity is not random but is determined by the dimensions of the cavity. The dimensions of the cavity determine certain discrete resonant conditions. They are known as *resonant modes*. The light waves oscillating at modes that match the oscillation modes of the cavity are sustained. The laser modes governed by the axial dimensions of the resonant cavity are called the ***longitudinal modes***, and the modes determined by the cross-sectional dimensions of the laser cavity are called ***transverse modes***.

(a) Longitudinal Modes: A light wave which moves inside the laser cavity from right to left is reflected by the left mirror. Reflected light moves to the right and is reflected from the right mirror, and so forth. So, two waves of the same frequency and amplitude moving in opposite directions creates a standing wave.

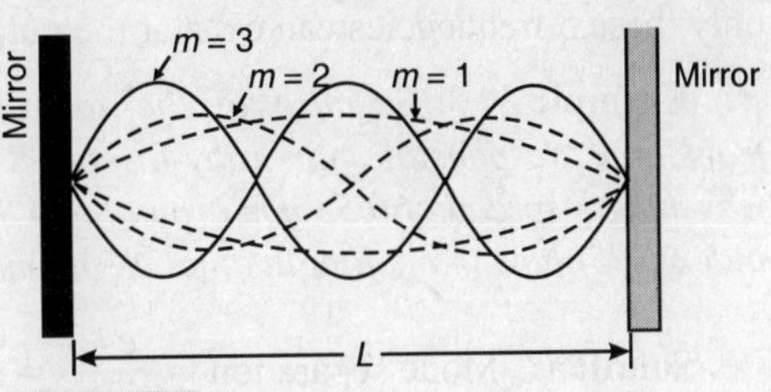

Fig. 44.13: Longitudinal Modes

In order to create a standing wave, the wave must start with the same phase at the mirror (Fig. 44.13). Therefore, the optical path from one mirror to the other and back must be an integer multiplication of the wavelength. Thus,

$$2\mu L = m\lambda_m \qquad (m = 1, 2, 3, ...)$$

or

$$\lambda_m = \frac{2\mu L}{m} \qquad ...(44.28)$$

Light waves are amplified strongly if, and only if, they satisfy the above condition. Only those wavelengths that satisfy Eq. (44.28) can exist inside the cavity. Waves of other wavelengths interfere destructively with each other as they pass back and forth between the mirrors. Thus, they attenuate very quickly. Because of its relatively longer length as compared to the wavelength of light, the resonator may support simultaneously several standing waves.

These standing waves are known as the ***longitudinal modes***.

$$\nu_m = \frac{mc}{2\mu L} \qquad ...(44.29)$$

Thus, frequency difference between two adjacent modes (m and $m \pm 1$) is

$$\Delta\nu = \frac{c}{2\mu L} \qquad ...(44.30)$$

The frequencies given by the Eq. (44.29) are the *allowed frequencies* inside a laser cavity of length L. However, all these allowed frequencies will not be emitted from the laser, since there are other limiting conditions. Only those frequencies (modes) that have amplification above the lasing threshold, to overcome absorption will be emitted out of the laser (see Fig. 44.14)

Fig. 44.14 shows the gain versus frequency curve of a certain active medium. The ***lasing threshold*** and ***possible longitudinal modes of the laser*** are indicated therein. The region under the curve and above the lasing threshold encloses the lasing range. For example, Fig. 44.14 consists of only 5 frequencies out of those allowed inside the cavity and which are above the lasing threshold. So, only these 5 frequencies can exist at the output of this laser.

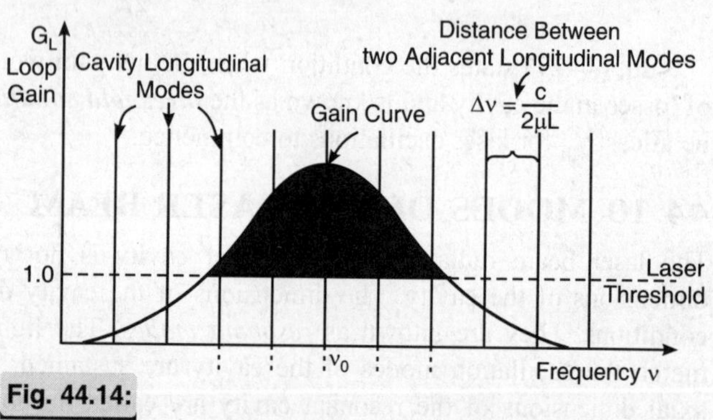

Fig. 44.14: Gain curve of a laser and allowed frequencies.

Example 44.3: *Because of the interaction of chromium ion with ruby lattice, the transitions responsible for ruby-laser-emission are spread over an energy, resulting in wavelength spread of 0.53 mm around 694.3 nm. If the length of ruby rod is 2 cms (refractive index 1.75) how many longitudinal cavity modes would the ruby-laser emission contain?*

Solution: Mode separation $= \frac{c/\mu}{l} = \frac{3\times10^8 / 1.75 \text{ m/s}}{2\times10^{-2}\text{ m}} = 8.6 \times 10^9$ Hz

Frequency spread of laser emission

$$\Delta\nu = (c/\lambda^2)\Delta\lambda = \frac{3\times10^8 (\text{m/s})\times0.53\times10^{-9}(\text{m})}{(694.3\times10^{-9}\text{ m})^2} = 330 \times 10^9 \text{ Hz.}$$

No. of cavity modes = $\Delta\nu$ / mode separation = $330 \times 10^9 / 8.6 \times 10^9 = 38.5$

i.e., = **38 modes**

Example 44.4: *If the half-width of the He-Ne laser operating at wavelength 6328Å is 1500 MHz, what must be the length of the laser cavity to ensure that only one longitudinal mode oscillates?*

Solution:

The length of cavity is given by $L = \frac{mc}{2\Delta\nu}$

$$= \frac{1 \times 3 \times 10^8 \text{ m/s}}{2 \times 1.5 \times 10^9 \text{ / s}} = \mathbf{0.1\ m.}$$

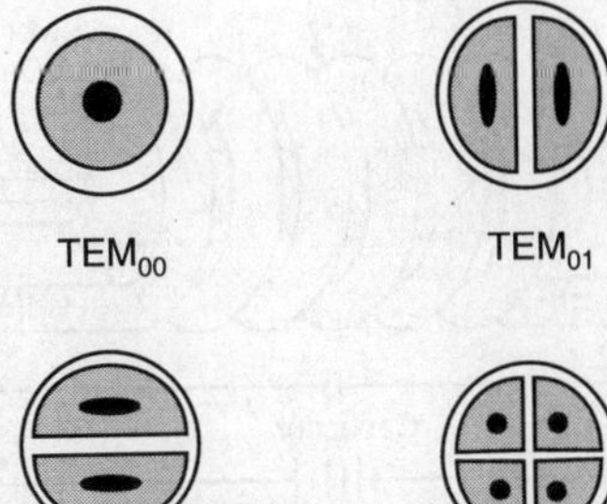

Fig. 44.15: Beam profiles of transverse modes

(b) Transverse Modes: The transverse modes illustrate the intensity distribution across the cross-section of the laser beam. The shape of the optical cavity determines the transverse modes. The allowed modes in an optical cavity are designated as TEM_{mn}, where *T*, *E*, and *M* stand for transverse, electric, and magnetic modes, respectively, and m and n are integers. The simplest transverse mode is TEM_{00}, which has a smooth cross-section profile with a Gaussian peak in the middle, as shown in Fig. 44.15.

Integers *m* and *n* of TEM_{mn} mode stand for minima between the edges of the beam in two perpendicular directions. A TEM_{01} beam has a single minimum dividing the beam into two bright spots. A TEM_{11} beam has two perpendicular minima one in each direction, dividing the beam into four bright spots. The larger the values of *m* and *n*, the more bright spots are contained in the laser beam. Most laser cavities are designed to produce only the TEM_{00} mode.

44.11 TYPES OF LASERS

Lasers are divided into different types basing on different considerations. We divide them here on the basis of the material used. Some of the important types of lasers are

(*i*) Solid-state lasers — Examples: Ruby laser, Nd:YAG laser etc
(*ii*) Gas lasers — Examples: Helium-Neon laser, CO_2 laser etc
(*iii*) Semiconductor diode lasers — Examples: GaAs laser, InP laser etc

Most lasers emit light in the red or IR regions. Lasers work in continuous mode or in a pulsed mode.

44.11.1 Ruby Laser

Ruby laser belongs to the class of solid-state lasers. Ruby is basically Al_2O_3 crystal containing about 0:05% of chromium atoms. Cr^{3+}ions are the actual active centers while aluminum and oxygen atoms are inert. Chromium ions have absorption bands in the blue and green regions.

(a) Construction: Ruby rod is taken in the form of a cylindrical rod of about 4 cm in length and 1 cm in diameter. Its ends are grounded and polished such that the end faces are exactly parallel and are also perpendicular to the axis of the rod. The end faces of the ruby rod are silvered so that they form the optical resonator. The rear face is made totally reflecting while the front face is made partially reflecting.

The laser rod is surrounded by a helical photographic flash lamp filled with xenon (Fig. 44.16a). Whenever activated by the power supply the lamp produces flashes of white light.

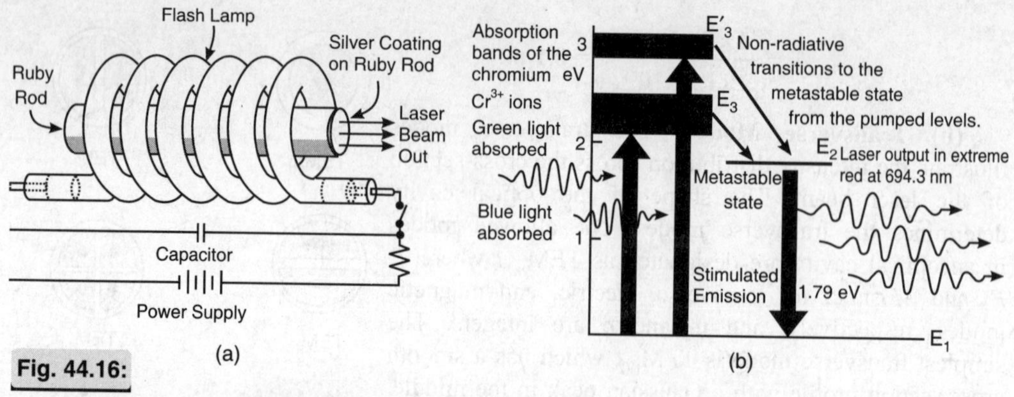

Fig. 44.16: (a) Schematic of a ruby laser (b) Energy levels and transitions in a ruby laser

(b) Working: The energy levels of Cr^{3+} ions in the crystal lattice are shown in Fig. 44.16 (b). The energy level structure of the Cr^{3+} ions is characterized by two absorption bands and a metastable state.

(c) The Pumping Mechanism

(*i*) When the xenon flash lamp is switched on, the discharge generates an intense burst of white light lasting for a few milliseconds.

(*ii*) The Cr^{3+} ions are excited to the energy bands E_3 and E_3' by the green and blue components of white light.

(*iii*) The Cr^{3+} ions undergo non – radiative transitions from these energy levels to level E_2. E_2 is a metastable state.

(*iv*) The metastable state E_2 has a lifetime of approximately 1000 times more than the lifetime of E_3 and E_3' levels. Therefore, Cr^{3+} ions accumulate at E_2 level.

(*v*) The metastable level E_2 is the upper laser level, while E_1 is the ground level and constitutes the lower laser level.

(d) Population Inversion

(*i*) The upper laser level E_2 will be rapidly populated, as the excited Cr^{3+} ions quickly make downward transitions from the upper energy bands.

(*ii*) When more than half of the Cr^{3+} ion population accumulates at E_2 level, the state of population inversion is established between E_2 and E_1 levels.

(e) Lasing action

(*i*) A chance photon is produced when a Cr^{3+} ion makes a spontaneous transition from E_2 level to E_1 level.

(*ii*) This spontaneous photon stimulates another excited ion to make a downward transition.

(*iii*) This stimulated photon and the initial photon trigger many excited ions to emit photons.

(*iv*) Red photons of wavelength 6943 Å travelling along the axis of the ruby rod are repeatedly reflected at the end mirrors and light amplification takes place.

(*v*) On attaining sufficient energy, the laser beam emerges out through the partially reflecting mirror.

(*vi*) The laser emission occurs in the visible region at a wavelength of 6943 Å.

(vii) Once stimulated transitions commence, the metastable state gets depopulated very rapidly and the state of population inversion disappears and lasing action ceases.

(viii) The laser becomes active once again when population inversion state is reestablished.

(ix) Therefore, the output of the laser is not a continuous wave but occurs in the form of pulses of microsecond duration.

(f) Salient Features

(i) Uses three-level pumping scheme

(ii) The active centers are Cr^{3+} ions

(iii) Light from a xenon flash lamp is the pumping agent

(iv) Poor efficiency

(v) Operates in pulsed mode

44.11.2 Nd: YAG Laser

Nd: YAG laser is one of the most popular types of solid state laser. It is a four-level laser. Yttrium aluminium garnet, $Y_3Al_5O_{12}$, commonly called YAG is an optically isotropic crystal. Some of the Y^{3+} ions in the crystal are replaced by neodymium ions, Nd^{3+}. Doping concentrations are typically of the order of 0.725% by weight. The crystal atoms do not participate in the lasing action but serve as a host lattice in which the active centres, namely Nd^{3+} ions reside.

(a) Construction: Fig. 44.17 illustrates a typical design of Nd: YAG laser. The system consists of an elliptically cylindrical reflector housing the laser rod along one of its focus line and a flash lamp along the other focus line. The light leaving one focus of the ellipse will pass through the other focus after reflection from the silvered surface of the reflector. Thus the entire flash lamp radiation gets focused on the laser rod. The YAG crystal rods are typically of 10 cm in length and 12 mm in diameter. The two ends of the laser rod are polished and silvered and constitute the optical resonator.

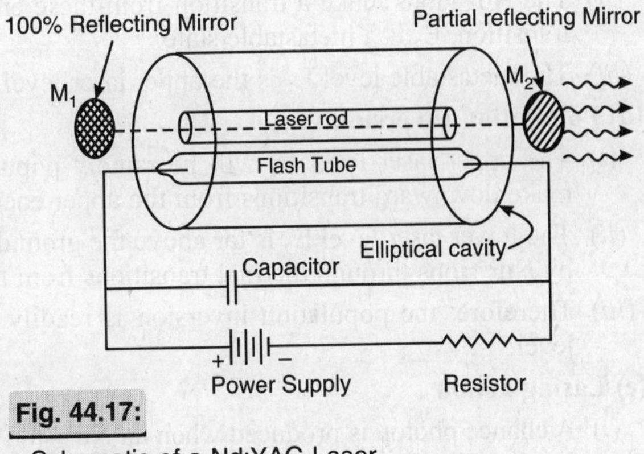

Fig. 44.17: Schematic of a Nd:YAG Laser

(b) Working: A simplified energy level diagram for the neodymium ion in YAG crystal is shown in Fig. 44.18. The energy level structure of the free neodymium atom is preserved to a certain extent because of its relatively low concentration. However, the energy levels are split and the structure is complex.

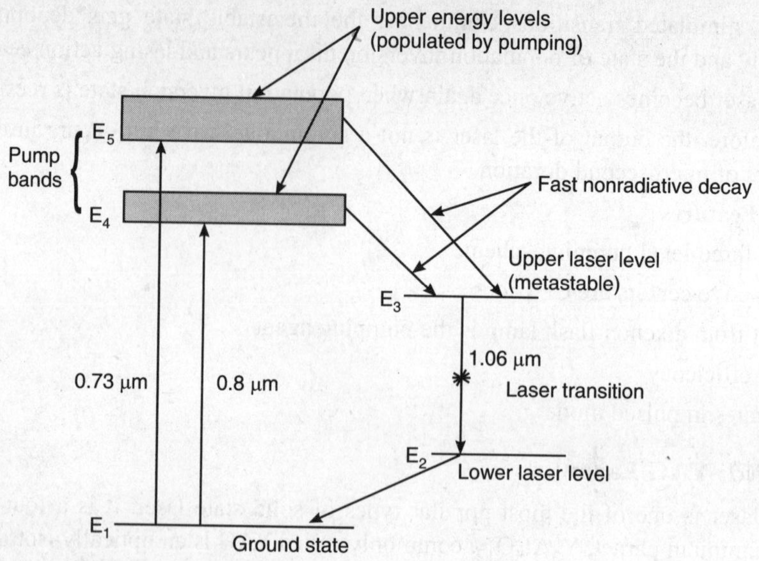

Fig. 44.18: Energy levels and transitions in a Nd : YAG laser

(c) The Pumping Mechanism

(*i*) When the krypton flash lamp is switched on, the Nd^{3+} ions are excited to the upper energy bands E_4 and E_5.

(*ii*) The Nd^{3+} ions make a transition from these energy levels to level E_3 by non-radiative transition. E_3 is a metastable state.

(*iii*) The metastable level E_3 is the upper laser level, while E_2 forms the lower laser level.

(d) Population Inversion

(*i*) The upper laser level E_3 will be rapidly populated, as the excited Nd^{3+} ions quickly make downward transitions from the upper energy bands.

(*ii*) The lower laser level E_2 is far above the ground level and hence it can not be populated by Nd^{3+} ions through thermal transitions from the ground level.

(*iii*) Therefore, the population inversion is readily achieved between the E_3 level and E_2 level.

(e) Lasing action

(*i*) A chance photon is produced when an Nd^{3+} ion makes a spontaneous transition from E_3 level to E_2 level.

(*ii*) This spontaneous photon stimulates another excited atom to make a downward transition.

(*iii*) This stimulated photon and the initial photon trigger many excited atoms to emit photons.

(*iv*) Photons thus generated travel back and forth between the two end mirrors and gain in strength very rapidly.

(*v*) On attaining sufficient energy, the laser beam emerges out through the partially reflecting mirror.

(*vi*) The laser emission occurs in infrared (IR) region at a wavelength of about 10,600 Å (1.06 μm).

(*vii*) The Nd^{3+} ions return to the ground state E_1 from the lower lasing level E_2, on their own through non-radiative transitions.

(f) Salient Features

(*i*) Uses four-level pumping scheme

(*ii*) The active centers are Nd^{3+} ions

(*iii*) Light from a xenon or krypton flash lamp is the pumping agent

(*iv*) Low efficiency (1%) and moderate power output (watts)

(*v*) Operates in CW/pulsed mode

44.11.3 Helium-Neon Laser

Gas lasers are the most widely used lasers. They range from the low power helium-neon laser used in college laboratories to very high power carbon dioxide laser used in industrial applications.These lasers operate with rarefied gases as the active media and are excited by an electric discharge. In gases, the energy levels of atoms involved in the lasing process are narrow and as such require sources with sharp wavelength to excite atoms. Finding an appropriate optical source for pumping poses a problem. Therefore optical pumping is not used in gas lasers. The most common method of exciting gas laser medium is by passing an electric discharge through the gas. Electrons present in the discharge transfer energy to atoms in the laser gas by collisions.

The first gas laser was He-Ne laser, which was invented in 1961 by Ali Javan, William R. Bennett, Jr. and Donald R.Herriott.

(a) Construction: A Helium - Neon laser is illustrated in Fig. 44.19. It is made of a long discharge tube filled with a mixture of helium and neon gases in the ratio 10:1. Neon atoms are the active centers and have energy levels suitable for laser transitions while helium atoms help in exciting neon atoms.

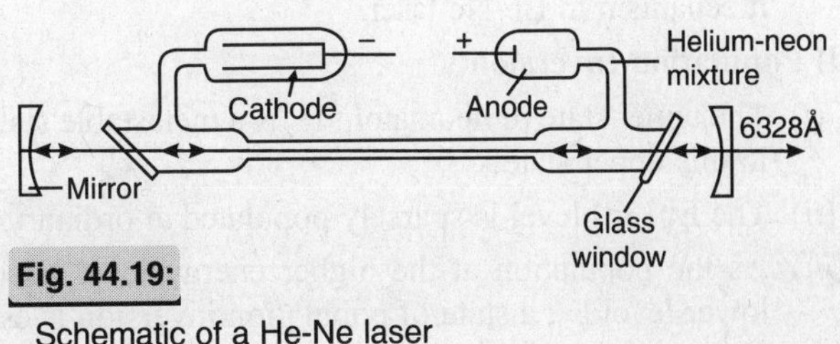

Fig. 44.19: Schematic of a He-Ne laser

Electrodes are provided in the discharge tube to produce discharge in the gas. They are connected to a high voltage power supply. The tube is hermetically sealed by inclined windows arranged at its two ends. On the axis of the tube, two mirrors are arranged externally, which form the Fabry-Perot optical resonator. The distance between the mirrors is adjusted to be $m\lambda/2$ such that the resonator supports standing wave pattern.

(b) Working: The energy levels of helium and neon are shown in Fig. 44.20.

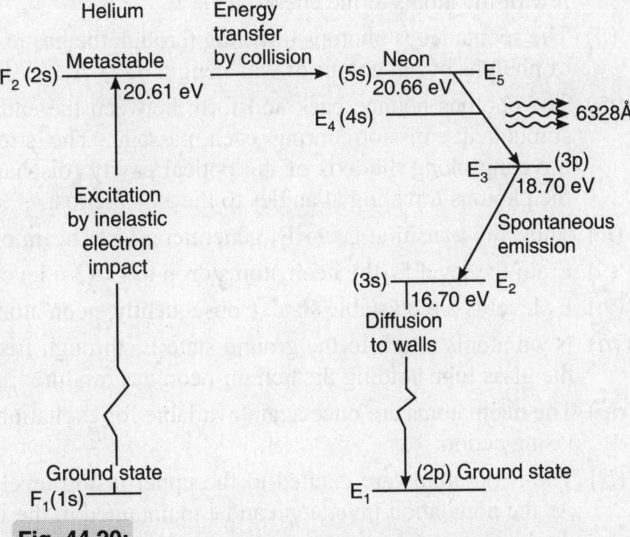

Fig. 44.20: Energy levels of helium and neon and Laser transitions.

(c) Pumping Mechanism

(*i*) When the power is switched on, a high voltage of about 10 kV is applied across the gas mixture. It ionizes the gas.

(*ii*) The electrons and ions produced in the process of discharge are accelerated towards the anode and cathode respectively. They collide with helium and neon atoms on the way.

(*iii*) The energetic electrons excite helium atoms more readily, as they are lighter.

(*iv*) One of the excited levels of helium F_2 (2s) is at 20.61 eV above the ground level. It is a metastable level and the excited helium atom cannot return to the ground level through spontaneous emission.

(*v*) However, the excited helium atom can return to the ground level by transferring its excess energy to a neon atom through collision. Such an energy transfer can take place when the two colliding atoms have identical energy levels. Such an energy transfer is known as **resonant energy transfer.**

(*vi*) The neon energy level E_5 (5s) is at 20.66 eV, which is close to the excited energy level F_2 of helium atom. Therefore, resonant transfer of energy occurs between the excited helium atom and ground level neon atom. The kinetic energy of helium atoms provides the additional 0.05 eV required for excitation of the neon atoms.

(*vii*) Helium atoms drop to the ground state after exciting neon atoms. This is the pumping mechanism in He-Ne laser.

(d) Population Inversion

(*i*) The upper state of neon atom E_5 is a metastable state. Therefore, neon atoms accumulate in this upper state.

(*ii*) The E_3 (3p) level is sparsely populated at ordinary temperatures.

(*iii*) As the population at the higher energy level E_5 is greater than the population at the lower level E_3, a state of population inversion is established between E_5 and E_3 levels.

(e) Lasing Action

(*i*) Random photons of red colour of wavelength 6328 Å are emitted spontaneously by a few of the atoms at the energy level E_5.

(*ii*) The spontaneous photons traveling through the gas mixture prompt stimulated emission of photons of red colour of wavelength 6328 Å.

(*iii*) The photons bounce back and forth between the end mirrors, causing more and more stimulated emission during each passage. The strength of the stimulated photons traveling along the axis of the optical cavity (discharge tube) builds up rapidly while the photons traveling at angles to the axis are lost.

(*iv*) Thus, the transition $E_5 \rightarrow E_3$ generates a laser beam of wavelength 6328 Å.

(*v*) From the level E_3 the neon atoms drop to E_2 (3s) level spontaneously.

(*vi*) E_2 level is a metastable state. Consequently, neon atoms tend to accumulate at E_2 level.

(*vii*) Neon atoms return to the ground state E_1 through frequent collisions with the walls of the glass tube holding the helium-neon gas mixture.

(*viii*) The neon atoms are once again available for excitation to higher state and participate in lasing action.

(*ix*) The neon atoms are excited to the upper lasing level continuously through collisions. As the population inversion can be maintained in the face of continuous laser emission, the laser operates in continuous wave mode.

(f) Role of helium atoms: The role of helium atoms in the laser is to excite neon atoms and to cause population inversion. The probability of energy transfer from helium atoms to neon atoms is more, as there are 10 helium atoms per 1 neon atom in the gas mixture. The probability of reverse transfer of energy from neon to helium atom is negligible.

(g) Necessity of narrow glass tube: During the operation of the laser, it is necessary that the atoms accumulating at the metastable level E_2 are brought to the ground state E_1(2p) quickly; otherwise the number of atoms at the ground state will go on diminishing and the laser ceases to function. The only way of bringing the atoms to the ground state is through collisions. Therefore, to increase the probability of atomic collisions with the tube walls, the discharge tube is made narrow.

(h) Salient Features

(*i*) Uses four-level pumping scheme

(*ii*) The active centers are neon atoms

(*iii*) Electrical discharge is the pumping agent

(*iv*) Low efficiency and low power output

(*v*) Operates in CW mode

44.11.4 Carbon Dioxide Laser

The carbon gas laser is a very useful and efficient laser. It is a four-level molecular laser and operates at 10.6 μm in far IR region.

(a) Construction: The schematic of typical CO_2 laser is shown in Fig. 44.21. It is basically a discharge tube having a bore of cross section of about 1.5 mm^2 and a length of about 260 mm. The discharge tube is filled with a mixture of carbon dioxide, nitrogen and helium gases in 1:4:5 proportions respectively. CO_2 molecules are the active centres. Lasing transitions occur between the vibrational levels of the electronic ground state.

Fig. 44.21: Schematic of a carbon dioxide laser

(b) Energy levels of CO_2 molecule: Fig. 44.22 shows the vibrational modes and rotations of CO_2 molecule.

(*i*) The electron energy levels of an isolated atom are discrete and narrow. However, in case of molecules the energy spectrum is complicated due to many additional features.

(*ii*) Each electron energy level is associated with nearly equally spaced vibrational levels and each vibrational level in turn has a number of rotational levels.

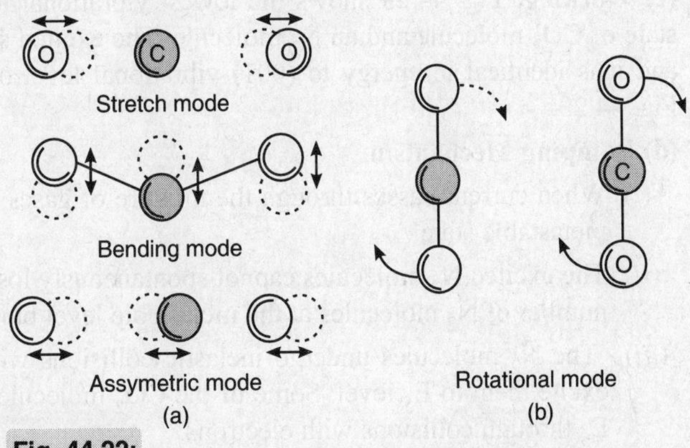

Fig. 44.22: (a) Vibrational modes and (b) Rotational modes of a CO_2 molecule

(*iii*) CO_2 molecule is a linear molecule consisting of a central carbon atom with two oxygen atoms attached one on either side.

(*iv*) It undergoes three independent vibrational oscillations known as the ***vibrational modes*** (Fig. 44.20). These vibrational degrees of freedom are quantized. At any one time, a CO_2 molecule can vibrate in a linear combination of three fundamental modes.

(*v*) The energy states of the molecule are represented by three quantum numbers (m n q).

(*vi*) These numbers stand for the amount of energy related with each mode. For example, the number (020) shows that the molecule in this energy state is in the pure bending mode with two units of energy.

(*vii*) Each vibrational state is related with rotational states corresponding to the rotation of CO_2 molecule about its centre of mass. The separations between vibrational-rotational states are to a large extent lesser on the energy scale compared to the separations between electron energy levels.

The nitrogen molecule N_2 is also characterized by similar vibrational levels.

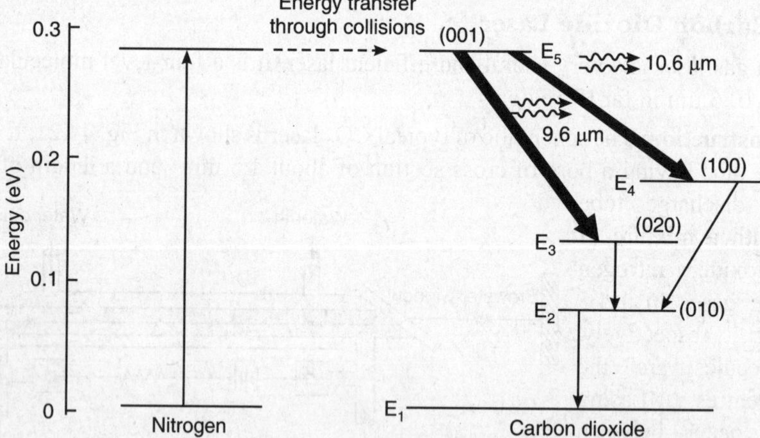

Fig. 44.23: Energy levels of nitrogen and carbon dioxide molecules and transitions between the levels

(c) Working: Fig. 44.23 shows the lowest vibrational levels of the ground electron energy state of CO_2 molecule and an N_2 molecule. The excited state of an N_2 molecule is metastable and it is identical in energy to (001) vibrational level of CO_2 molecule, indicated as E_5 in Fig. 44.23.

(d) Pumping Mechanism

(*i*) When current passes through the mixture of gases, the N_2 molecules get excited to the metastable state.

(*ii*) The excited N_2 molecules cannot spontaneously lose their energy and consequently, the number of N_2 molecules at the metastable level builds up.

(*iii*) The N_2 molecules undergo inelastic collisions with ground state CO_2 molecules and excite them to E_5 level. Some of the CO_2 molecules are also excited to the upper level E_5 through collisions with electrons.

(e) Population Inversion

(*i*) The E_5 level is the upper lasing level while the (020) and (100) states marked as E_3 and E_4 levels act as the lower lasing levels.

(*ii*) As the population of CO_2 molecules builds up at E_5 levels population inversion is achieved between E_5 level and the levels at E_4 and E_3.

(f) Lasing action

(*i*) Random photons are emitted spontaneously by a few of the atoms at the energy level E_5.

(*ii*) The spontaneous photons traveling through the gas mixture prompt stimulated emission of photons.

(*iii*) The photons bounce back and forth between the end mirrors, causing more and more stimulated emission during each passage. The strength of the stimulated photons traveling along the axis of the optical cavity (discharge tube) builds up rapidly while the photons traveling at angles to the axis are lost.

(*iv*) The laser transition between $E_5 \rightarrow E_4$ levels produces far IR radiation at the wavelength 10.6 µm (1, 06,000Å).

(*v*) The lasing transition between $E_5 \rightarrow E_3$ levels produces far IR radiation at 9.6 µm (96,000Å) wavelength.

(*vi*) E_3 and E_4 levels are also metastable states and the CO_2 molecules at these levels fall to the lower level E_2 through inelastic collisions with normal (unexcited) CO_2 molecules.

(*vii*) This process leads to accumulation of population at E_2 level. As the gaseous mixture heats up, the E_2 level, which is close to the ground state, tends to be populated through thermal excitations. Thus, the de-excitation of CO_2 molecules at the lower lasing level poses a problem and inhibits the laser action.

(*viii*) The helium atoms de-excite CO_2 molecules through inelastic collisions and decrease the population density of CO_2 at E_2 level. It also aids cooling the gaseous mixture through heat conduction.

(*ix*) The CO_2 molecules are once again available for excitation to higher state and participate in lasing action.

(*x*) CO_2 molecules are excited to the upper lasing level continuously through collisions. As the population inversion can be maintained in the face of continuous laser emission, the laser operates in continuous wave mode.

Salient Features

(*i*) Uses four-level pumping scheme

(*ii*) The active centers are CO_2 molecules

(*iii*) Electrical discharge is the pumping agent

(*iv*) High efficiency (40%) and high power output (several kilowatts)

(*v*) Operates in CW mode

44.11.5 Semiconductor Diode Laser

A ***semiconductor diode laser*** is a specifically made *p-n* junction diode that emits coherent light under forward bias. R. N. Hall and his coworkers made the first semiconductor laser in 1962. p-n junction lasers are made to emit light almost anywhere in the spectrum from UV to IR.

Diode lasers are remarkably small in size (0.1mm long). They have high efficiency of the order of 40%. Modulating the biasing current easily modulates the laser output. They operate at low powers. In spite of their small size and low power requirement, they produce power outputs equivalent to that of He-Ne lasers. The chief advantage of a diode laser is that it is portable. Because of the rapid advances in semiconductor technology, diode lasers are mass produced for use in optical fibre communications, in CD players, CD-ROM drives, optical reading, and high speed laser printing etc wide variety of applications.

(i) Semiconductor Materials

(*i*) Among the semiconductors, there are two different groups. They are *direct band gap semiconductors and indirect band gap semiconductors*.

(*ii*) Direct band gap semiconductor is the one in which a conduction band electron can recombine directly with a hole in the valence band.

(*iii*) The recombination process leads to emission of light.

(*iv*) Most of the compound semiconductors belong to this group.

(*v*) Direct recombination of conduction band electron with a hole in the valence band is not possible in indirect band gap semiconductors. Silicon and germanium belong to this group. The recombination of an electron and a hole produces heat in these materials.

(*vi*) Direct band gap semiconductors are formed by group III-V elements and group IV-VI elements.

(*vii*) Lasers are made using direct band gap semiconductors. Gallium Arsenide (GaAs) diode is an example of semiconductor diode laser.

(ii) Principle

(*i*) The energy band structure of a semiconductor consists of a valence band and a conduction band separated by an energy gap, E_g.

(*ii*) The conduction band contains electrons and the valence band contains holes and electrons.

(*iii*) When an electron from the conduction band jumps into a hole in the valence band, the excess energy E_g is given out in the form of a photon.

(*iv*) Thus, the electron-hole recombination is the basic mechanism responsible for emission of light.

(*v*) The wavelength of the light is given by the relation $\lambda = hc/E_g$.

(*vi*) Semiconductors having a suitable value of E_g emit light in the optical region.

(iii) Types of semiconductor diode lasers

Broadly there are two types of semiconductor diode lasers. They are known as *homojunction semiconductor lasers and heterojunction semiconductor lasers*.

(a) Homojunction Semiconductor Laser

A simple diode laser which makes use of the same semiconductor material on both sides of the junction is known as a homojunction diode laser.

Example: Gallium arsenide (GaAs) laser.

(b) Heterojunction Semiconductor Laser

A diode laser which makes use of different semiconductor materials on the two sides of the junction is known as a heterojunction diode laser. These are further classified as single heterojunction diode lasers and double heterojunction diode lasers.

Example: A junction laser having GaAs on one side and GaAlAs on the other side

44.11.5.1 Homojunction semiconductor laser

(a) Construction: Fig. 44.24 shows the schematic of a homojunction diode laser. Starting with a heavily doped *n*-type GaAs material, a *p*-region is formed on its top by diffusing zinc atoms into it. A heavily zinc doped layer constitutes the heavily doped p-region. The diode is extremely small in size. Typical diode chips are 500 μm long and about 100 μm wide and thick. The top and bottom faces are metallized and metal contacts are provided to pass current through the diode. The front and rear faces are polished parallel to each other and perpendicular to the plane of the junction. The polished faces constitute the Fabry-Perot resonator. In practice there is no necessity to polish the faces. A pair of parallel planes cleaved at the two ends of the PN junction provides the required reflection to form the cavity. The two remaining sides of the diode are roughened to eliminate lasing action in that direction. The entire structure is packaged in small case which looks like the metal case used for discrete transistors.

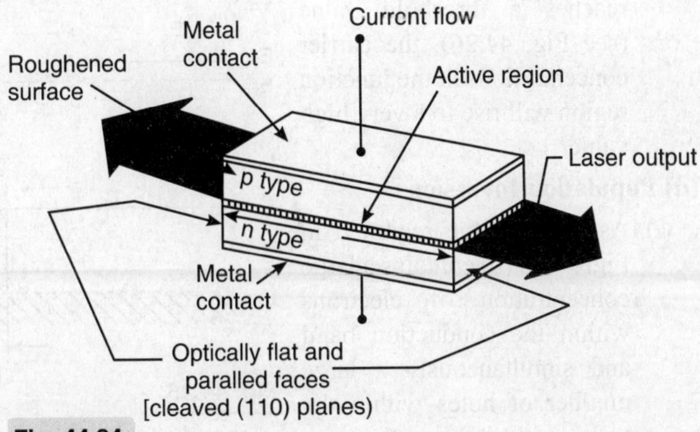

Fig. 44.24: Schematic of homojunction diode laser

(b) Working: The energy band diagram of a heavily doped *p-n* junction is shown in Fig. 44.25.

(*i*) Heavily doped *p*- and *n*- regions are used in making a laser diode.

(*ii*) Because of very high doping on n-side, the donor levels are broadened and extend into the conduction band. The Fermi level also is pushed into the conduction band.

(*iii*) Electrons occupy the portion of the conduction band lying below the Fermi level.

(*iv*) Similarly, on the heavily doped p-side the Fermi level lies within the valence band and holes occupy the portion of the valence band that lies above the Fermi level.

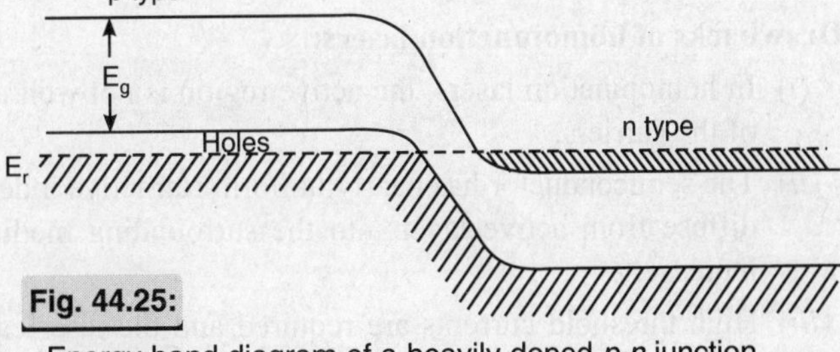

Fig. 44.25: Energy band diagram of a heavily doped *p-n* junction without bias

(*v*) At thermal equilibrium, the Fermi level is uniform across the junction.

(c) The Pumping Mechanism

(*i*) When the junction is forward-biased, electrons and holes are injected into the junction region in high concentrations.

(*ii*) In other words, charge carriers are *pumped* by the dc voltage source.

(*iii*) When the diode current reaches a threshold value (see Fig. 44.26), the carrier concentrations in the junction region will rise to a very high value.

(d) Population Inversion

(*i*) As a result, the region d in Fig. 44.26 (b) contains a large concentration of electrons within the conduction band and simultaneously a large number of holes within the valence band.

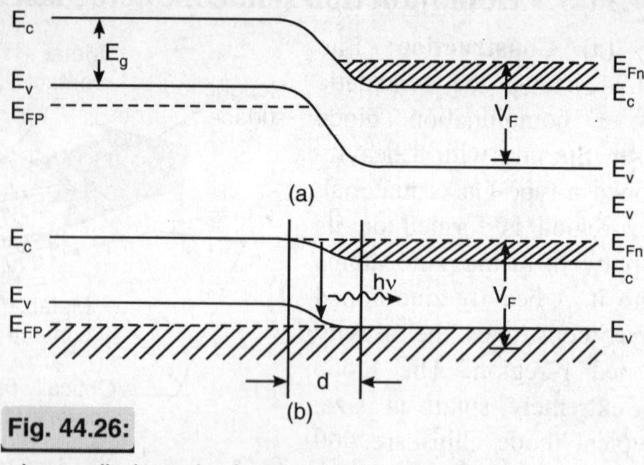

Fig. 44.26: Laser diode under forward bias

(*ii*) Holes represent absence of electrons.

(*iii*) Thus, the upper energy levels in the narrow region are having a high electron population while the lower energy levels in the same region are vacant.

(*iv*) Therefore, the condition of population inversion is attained in the narrow junction region. This narrow zone in which population inversion occurs is called an ***inversion region or active region***.

(e) Lasing action

(*i*) Chance recombination acts of electron and hole pairs lead to emission of spontaneous photons.

(*ii*) The spontaneous photons propagating in the junction plane stimulate the conduction electrons to jump into the vacant states of valence band.

(*iii*) This stimulated electron-hole recombination produces coherent radiation.

(*iv*) GaAs laser emits light at a wavelength of 9000 Å in IR region.

Drawbacks of homojunction lasers:

(*i*) In homojunction lasers, the active region is not well defined due to the diffusion length of the carriers.

(*ii*) The semiconductor has nearly uniform refractive index throughout. Therefore, light can diffuse from active layer into the surrounding medium. As a result the cavity losses increase.

(*iii*) High threshold currents are required and the laser cannot be operated continuously at room temperature.

44.11.5.2 Heterojunction laser

Heterojunction lasers are multilayer-structures designed such that the carriers are confined in a narrow region and population is built up at lower current levels. We study here the structure and working of a double heterojunction (DH) laser.

(a) Construction: In a double heterojunction laser, a GaAs layer is sandwiched between two layers of GaAlAs (Fig. 44.27). The material GaAlAs has a wider energy gap and a lower refractive index than GaAs. Metal contacts on the top and bottom faces provide the passage of current through the diode. The front and rear faces are polished parallel to each other and perpendicular to the plane of the junction and constitutes the Fabry-Perot resonator.

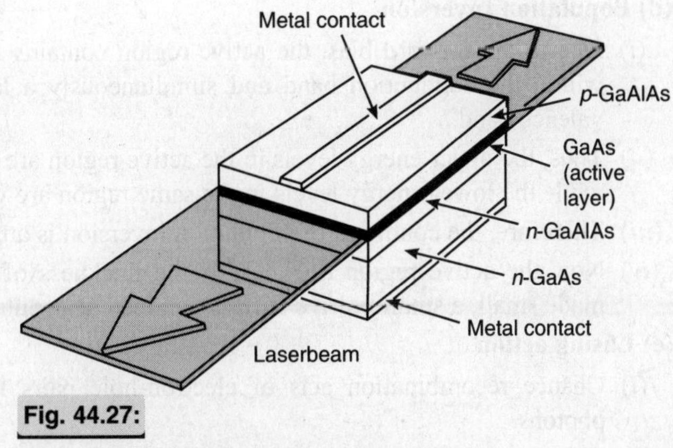

Fig. 44.27: Schematic of heterojunction laser

(b) Working:

(*i*) The basic principle of working of heterojunction diode is similar to that of a homo-junction diode.

(*ii*) P-type narrow band gap GaAs layer at the centre constitutes the active layer in which the lasing occurs.

(*iii*) This layer is flanked by an n-type wide band gap GaAlAs layer on one side and by a *p*-type wide band gap GaAlAs layer on the other side (Fig. 44.28 a).

(*iv*) The refractive indices of GaAlAs layers are smaller than that of GaAs layer (Fig. 44.28c).

(c) Pumping Mechanism

(*i*) When the junction is forward-biased, electrons and holes are injected into the active region in high concentrations.

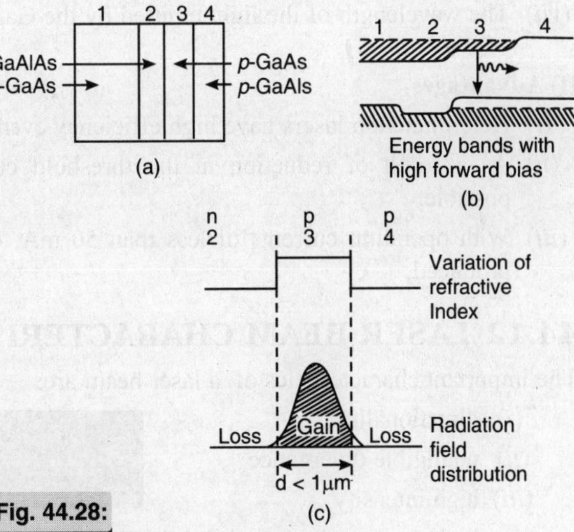

Fig. 44.28: Working of heterojunction laser

(*ii*) When the diode current reaches a threshold value, the carrier concentrations in the active region will rise to a very high value.

(*iii*) The electrons injected from the *n*-type GaAlAs layer into the p-GaAs layer confront energy barrier at the junction where p-type GaAs and *p*-type GaAlAs meet and are reflected back into the active region. Similarly, the holes are reflected by the potential barrier provided by the higher band gap *n*-GaAlAs layer. Thus, the layers in heterostructure confine the charge carriers to the GaAs layer.

(d) Population Inversion

(*i*) Due to the forward bias, the active region contains a large concentration of electrons within the conduction band and simultaneously a large number of holes within the valence band.

(*ii*) Thus, the upper energy levels in the active region are having a high electron population while the lower energy levels in the same region are vacant.

(*iii*) Therefore, the condition of population inversion is attained in the narrow active region.

(*iv*) Now the active region thickness is the thickness of *p*-GaAs layer. If its thickness is made small, a smaller drive current can lead to population inversion.

(e) Lasing action

(*i*) Chance recombination acts of electron-hole pairs lead to spontaneous emission of photons.

(*ii*) The spontaneous photons propagating in the active layer stimulate the conduction electrons to jump into the vacant states of valence band and produce stimulated photons.

(*iii*) The reflection at the GaAs – air interface provides sufficient feed back for laser oscillation.

(*iv*) When the diode losses are off-set by the laser gain, the laser oscillations will start.

(*v*) When the radiation attains appropriate strength, laser beam emerges from the diode.

(*vi*) As the refractive index of GaAs is higher than the refractive index of GaAlAs layers, the light is trapped within the active region and travels in one direction only.

(*vii*) The wavelength of the light emitted by the GaAs layer is 800 nm when its band gap is 1.55 eV.

(f) Advantages

(*i*) Heterojunction lasers have high efficiency even at room temperature.

(*ii*) As a result of reduction in the threshold current density, continuous operation is possible.

(*iii*) With operating currents of less than 50 mA, output powers of about 10 mW can be produced.

44.12 LASER BEAM CHARACTERISTICS

The important characteristics of a laser beam are:

(*i*) directionality

(*ii*) negligible divergence

(*iii*) high intensity

(*iv*) high degree of coherence and

(*v*) high monochromaticity.

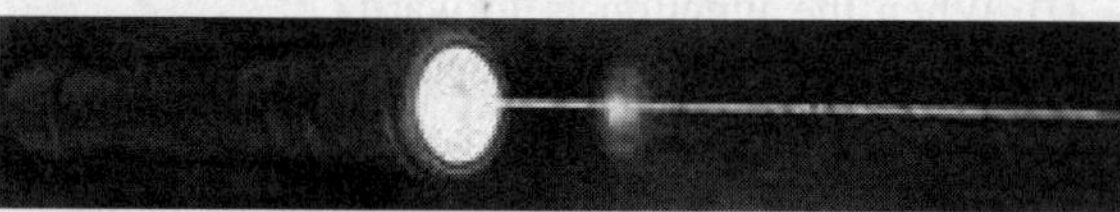

Fig. 44.29: Illustration of Directionality

(*i*) Directionality: The conventional light sources emit light uniformly in all directions. When we need a narrow beam in a specific direction, we obtain it by placing a slit in front of the source of light.

In case of laser, the active material is in a cylindrical resonant cavity. Any light that is travelling in a direction other than parallel to the cavity axis is eliminated and only the light that is travelling parallel to the axis is selected and reinforced. Light propagating along the axial direction emerges from the cavity and becomes the laser beam. Thus, a laser emits light only in one direction.

***(ii)* Divergence:** Light from conventional sources spreads out in the form of spherical wave fronts and hence it is highly divergent.

On the other hand, light from a laser propagates in the form of plane waves. The light beam remains essentially as a bundle of parallel rays. The small divergence that exists is due to the diffraction of the beam at the exit mirror. A typical value of divergence of a He-Ne laser is 10^{-3} radians. It means that the diameter of the laser beam increases by about 1 mm for every meter it travels.

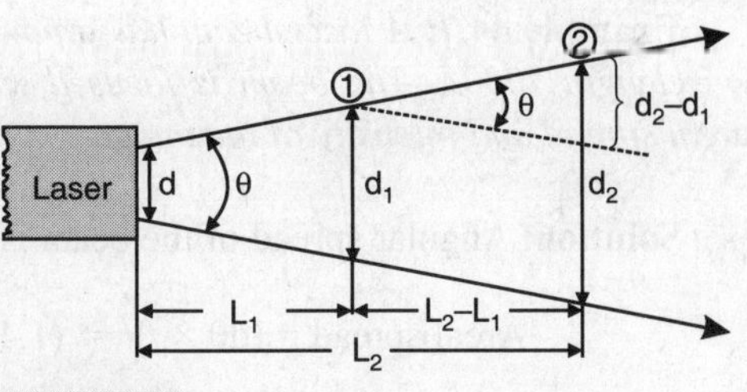

Fig. 44.30: Illustration of Divergence

The extent of divergence can be estimated in a simple way as follows:

If the diameters of spot produced by the laser on a screen which is held at two different distances from the laser are measured, then the angle of divergence is given by

$$\phi = \frac{d_2 - d_1}{l_2 - l_1} \qquad ...(44.31)$$

where d_1 is the spot diameter at the distance l_1 and d_2 is the spot diameter at the distance l_2.

Example 44.5: *Calculate the divergence of light beam issuing out of He-Ne laser, which produces spot diameters of 4 mm and 6 mm at 1m and 2m distances respectively.*

Solution: Beam divergence $\theta = \frac{d_2 - d_1}{2(l_2 - l_1)} = \frac{(6-4)\times 10^{-3}\,\text{m}}{2(2-1)\text{m}} = \mathbf{10^{-3}\ radians}$

***(iii)* Intensity:** The intensity of light from a conventional source decreases rapidly with distance as it spreads out in space. Laser emits light in the form of a narrow beam with its energy concentrated in a small region of space. Therefore, the beam intensity would be tremendously large and stays constant with distance. The intensity of a laser beam is approximately given by

$$I = \left[\frac{10}{\lambda}\right]^2 P \qquad ...(44.32)$$

where P is the power radiated by the laser.

To obtain light of same intensity from a tungsten bulb, it would have to be raised to a temperature of 4.6×10^6 K.

Example 44.6: *A 10 mw laser has a beam diameter of 1.6 mm. What is the intensity of the light assuming that it is uniform across the beam?*

Solution: Intensity of light is given by

$$I = \frac{\text{Power of the laser}}{\text{Area of cross section of the beam}} = \frac{P}{A}$$

$$= \frac{10\times 10^{-3}\,\text{W}}{3.143(0.8\times 10^{-3}\,\text{m})^2} = \mathbf{4.97\ kW/m^2}$$

Example 44.7: *A laser beam has a power of 50mW. It has an aperture of 5×10⁻³m and wavelength 7000Å. The beam is focused with a lens of foal length of 0.2m. Calculate the areal spread and intensity of the image.*

Solution: Angular spread of the beam is $d\theta = \dfrac{\lambda}{d} = \dfrac{7000\times10^{-10}\,\text{m}}{5\times10^{-3}\,\text{m}} = 1.4 \times 10^{-4}$ rad.

$$\text{Areal spread} = (d\theta \times f)^2 = (1.4 \times 10^{-4}\,\text{rad} \times 0.2\text{m})^2 = 4 \times 10^{-9}\,\text{m}^2.$$

$$\text{Intensity} = \frac{\text{Power}}{\text{Area}} = \frac{50\times10^{-3}\,\text{watt}}{4\times10^{-9}\,\text{m}^2} = \mathbf{12.5\ MW/m^2}.$$

Example 44.8: *With He-Ne laser, fringes remained clearly visible when the path difference was increased upto 8m. Deduce the lower limits for (i) Coherence length, (ii) Coherence time (iii) Spectral half width (iv) Q-factor, $\lambda=11.5 \times 10^{-7}$ m.*

Solution: (*i*) We know the phase relation will not exist for distance $> \lambda_o$.

∴ Coherence length $\lambda_o > 8$ m

(*ii*) Coherence time $$T_o = \frac{l_o}{c} > \frac{8}{3\times10^8}s = 2.7\times10^{-8}s$$

(*iii*) Spectral half width $$\Delta\lambda = \frac{\lambda^2}{l_o} = \frac{(11.5\times10^{-5})^2}{8} = 1.6\times10^{-3}\,\text{Å}$$

(*vi*) Quality factor $$Q = \frac{l_o}{\lambda} = \frac{8}{11.5\times10^{-5}} = 7\times10^6$$

Example 44.9: *A laser beam has a power of 50 mW. It has an aperture of 5×10^{-3} m and wavelength 7000Å. A beam is focused with 1 cm of focal length 0.2 m. Calculate the areal spread and intensity of the image.*

Solution: Given $\lambda = 7000$ Å $= 7000 \times 10^{-10}$ m

$d = 5 \times 10^{-3}$ m, focal length of lens $f = 0.2$ m

∴ $$\text{Angular Speed } d\theta = \frac{\lambda}{d} = \frac{7\times10^{-3}}{5\times10^{-3}} = 1.4\times10^{-4}\,\text{radians}$$

$$\text{Areal spread } (d\theta \times f)^2 = (1.4 \times 10^{-4} \times 0.2)^2 = 0.4 \times 10^{-8}\,\text{m}^2$$

$$\text{Intensity} = \frac{\text{power}}{\text{speed}} = \frac{50\times10^{-3}}{0.4\times10^{-8}} = 125\times10^5\,\text{W/m}^2$$

***(iv)* Coherence:** The light that emerges from a conventional light source is a jumble of short wave trains which combine with each other in a random manner. The resultant light is incoherent. Coherence length is one of the parameters used as a measure of coherence. In case of a laser a large number of identical photons are emitted through stimulated emissions and therefore they will be in phase with each other. The resultant light exhibits a high degree of coherence.

The coherence length of light from a sodium lamp, which is a traditional monochromatic source, is of the order of 0.3 mm. On the other hand the coherence length of light emitted by an ordinary helium-neon laser is about 100 m.

(v) Monochromaticity: If light coming from a source has only one frequency (single wavelength) of oscillation, the light is said to be *monochromatic* and the source a *monochromatic source*. Light from traditional monochromatic sources spreads over a wavelength range of 100 Å to 1000 Å. On the other hand, the light from lasers is highly monochromatic and contains a very narrow range of a few angstroms (< 10 Å).

44.13 APPLICATIONS

Lasers find application in almost every field. They are used in mechanical working, industrial electronics, entertainment electronics, communications, information processing, and even in wars to guide missiles to the target. Lasers are used in CD players, laser printers, laser copiers, optical floppy discs, optical memory cards etc. We discuss here a few of the applications in industry.

1. Industrial applications: The large intensity that is possible in the focused output of a laser beam and its directionality makes laser an extremely useful tool for a variety of industrial applications.

(*i*) **Welding:** The laser welding is a contactless process and hence there is no chance of impurities reaching into the joint. As total input power is very small compared to that in conventional welding processes, the work pieces are not spoiled.

The heat-effected zone is relatively small because of rapid cooling. Laser welding can be done even at difficult to reach place. CO_2 lasers are used in welding thin sheets and foils.

(*ii*) **Drilling:** Laser drilling does not require a physical drill bit. Drilling by a laser beam takes place through the intense evaporation of material heated by a sequence of powerful light pulses of short duration of 10^{-4} to 10^{-3}s.

The process is faster and highly reproducible. The drilling operation can be done with extremely high precision and in any direction.

Piercing of holes in a baby bottle nipple and drilling holes in diamond dies are two of the interesting examples of laser drilling.

(*iii*) **Heat treatment for Hardening:** Heat treatment is the process, which is done for some time to harden metals and certain other materials. Heat treatment is common in the tooling and automotive industry. Heat-treating converts the surface layer to a crystalline state that is harder and more resistant to wear. In general CO_2 lasers of about 1 kW output power operating in continuous wave mode are used for heat treatment. As metals are more reflecting at IR frequencies, a heat absorbing coating such as graphite or zinc phosphate is applied on the surface of the work piece to help it absorb laser energy more efficiently. Laser heat treatment requires a low amount of energy input to the work piece. Laser processing is advantageous as it can provide selective treatment of the desirable areas. Heat treatment is used to strengthen cylinder blocks, gears, camshafts etc in the automobile industry. As the method is a non-contact method, stress is not induced in the work-pieces.

(*iv*) **Cutting:** A wide range of materials can be cut by CO_2 lasers. The materials include paper, wood, cloth, glass, quartz, ceramics, steel etc. The advantage of laser cutting is that it is fine and precise. It introduces a minimum mechanical distortion and thermal shock in the material being cut. The process does not introduce any contamination. It is easily automatized and high production rates can be achieved.

(*v*) **Holography:** Holography is a technique for recording and reproducing an image of an object without the use of lenses. A three-dimensional image of the object is seen when the record is illuminated. The permanent record of the interference pattern on the recording medium is called a ***hologram***.

Holography offers a wide variety of fascinating applications in different fields such as *Holographic interferometry, Acoustical holography, Three-dimensional photography, Microscopy, Holographic optical elements* (diffraction gratings).

(*vi*) **Electronics Industry:** Electronics industry uses lasers in the manufacture of electronic components and integrated circuits. Lasers have been used to perforate and divide silicon slices having several hundred circuits. They are also used for the isolation of faulty components in a large integrated circuit by disconnecting the conducting paths by evaporation. Trimming of thick and thin film resistors using lasers is a very common application.

(*vii*) **Measurement of atmospheric pollutants:** Laser is a very useful tool for the measurement of the concentrations of various atmospheric pollutants such as N_2, CO, SO_2 etc gases and particulate matter such as dust, smoke and fly ash. Conventional methods of pollution measurements require that samples of pollutants are to be collected for chemical analysis. Therefore, these methods cannot give real-time data. In contrast, laser methods permit measurements by remotely sensing the composition of atmosphere without the necessity of sample collection or chemical processing.

Another technique uses study of absorption of light beam by pollutants. The existence of specific gases in the atmosphere is detected using absorption spectroscopy techniques. A laser beam is transmitted through polluted sample and the attenuation of intensity of light due to absorption in the sample is detected and recorded. Each chemical absorbs light of characteristic wavelengths and from the absorption spectrum; its existence can be inferred.

A third method uses Raman Effect to detect the pollutants. The Raman Effect involves scattering of light by gas molecules accompanied by a shift in the wavelength of light. Raman shifts are characteristic of each molecular species. Hence, analysis of backscattered laser light reveals the constituents of the gas sample. The ozone concentration high in the atmosphere is determined using this technique.

2. Medical Applications: Lasers are playing nowadays a very important role in diagnostics and surgery. Surgeons use a focused laser beam as a scalpel to perform operations. Laser surgery is a highly sterile process since contact does not occur between the surgical tool and the tissue being cut. Operations are done very fast and the patient does not feel pain.

(*i*) The first big success of lasers in medicine is in the treatment of eyes. Lasers are used for the ***welding of detached retina***. Retina is the light sensitive layer at the back of the eye. Sometimes it becomes torn and gets detached from the back of the eye ball. The damage can spread and lead to total blindness. Conventional open-eye surgery is very risky. Laser treatment is very simple and fast. An argon laser beam is focused on the desired point of the retina. The green beam of the laser is strongly absorbed by the red blood cells of the retina and causes thermal effects which weld the retina back to the eye ball.

(*ii*) ***Cataract*** occurs when the natural lens of the eye becomes cloudy and obstructs vision leading to blindness. In traditional open-eye surgery the natural lens is replaced with a plastic lens. A series of laser pulses of high intensity focused to a small spot on the back of lens produce high electric field sparks and ruptures the membrane.

(*iii*) Another interesting eye surgery is the radial keratomy or refractive surgery. Lasers are used to change the shape of the cornea by making small incisions on it. ArF excimer laser (λ = 193 nm) with a pulse energy of 250 mJ and pulse length of 14 ns is typically used to alter the shape of cornea.

(*iv*) Lasers are employed in destroying kidney stones and gall stones. An optical fibre is used to reache the stone. Laser pulse is launched through the fibre to break the stone into small pieces which can pass through the uterer without pain.
Tunable dye lasers are used in this application.

(*v*) Laser canes and spects assist the blind. The laser cane is equipped with three diode lasers. One laser aims its beam downward, the second upward and the third straight ahead. Distance is sensed by vibration of the handle. The obstacle level is indicated by the pitch of vibration.

QUESTIONS

1. What is LASER? How dose a laser beam differ from a point source of light? Mention any three engineering applications of laser.
2. How will you differentiate laser light from ordinary light? Explain in brief.
3. How is laser light different from an ordinary light?
4. Explain with neat diagram, the processes of absorption of light, spontaneous emission and stimulated emission of light. What are the necessary conditions for their occurrence? Why does spontaneous emission dominate over stimulated emission at normal temperature?
5. With the help of neat sketches, explain the three quantum processes that may occur when light radiation interacts with matter.
6. Explain with neat diagram, the processes of absorption of light, spontaneous emission and stimulated emission of radiation.
7. Explain spontaneous and stimulated emission of radiation.
8. With the help of a well labeled diagram explain the interaction of matter and radiation in the processes of (*i*) absorption (*ii*) spontaneous emission and (*iii*) stimulated emission. Which of these processes is maximized for laser operation?
9. Explain the terms: Spontaneous and stimulated emission. Which of the two emission processes is maximized in LASER operation? How?
10. Explain the terms: stimulated emission, population inversion, pumping and metastable states. Highlight their role in working of a laser.
11. With the help of a well-labeled diagram explain the interaction of matter and radiation in the processes of (*i*) absorption (*ii*) spontaneous emission and (*iii*) stimulated emission. Which of these processes is maximized for laser operation?
12. Explain in laser: (*i*) Why the active media should have preferably broad absorption band?, (*ii*) What is the role of metastable state?
13. Explain the terms: (*i*) Stimulated mission, (*ii*) Population inversion, (*iii*) Metastable state, (*iv*) Optical pumping
14. What are Einstein's coefficients? Explain them.
15. State the necessary condition for stimulated emission and explain the Einstein's A and B coefficients.
16. Derive the relation between Einstein's "A" and "B" coefficients.

17. Explain the conditions for light amplification.
18. What is popular inversion? Explain the necessity of population inversion for lasing.
19. Give the applications of laser.
20. What is meant by pumping? Discuss in brief optical pumping.
21. Explain with sketches the basic principle of operation of lasers.
22. Explain the working of a resonant cavity and its role in laser operation.
23. Distinguish between the following: (*i*) Spontaneous and stimulated emission, (*ii*) Three level and four level lasers.
24. What is resonant cavity? Highlight its importance in the production of laser radiation.
25. Why is the optical resonator required in lasers? Illustrate your answer with neat sketches.
26. Explain the role of end mirrors in a laser.
27. What do you understand by a negative temperature state? How it can be achieved?
28. What is the lifetime of charge carrier in metastable state?
29. What is metastable state? What role do such states play in the operation of laser?
30. What is meant by active material in laser?
31. Explain briefly pumping schemes used in Laser. Explain why two levels laser system is not possible.
32. Explain in brief three and four level pumping scheme. Why four levels scheme is preferred over three levels scheme?
33. Explain three level and four level laser systems. What are the advantages of four level laser system over three level laser system?
34. What do you understand by solid-state laser?
35. Explain the production of lasers by Ruby crystal.
36. Explain the working of ruby laser with the help of neat energy level diagram.
37. Describe the construction and working of Nd:YAG laser.
38. What is the main drawback of ruby laser? Explain the operation of a gas laser with the essential components. How stimulated emission takes place with the exchange of energy between helium and neon atoms?
39. Explain the construction and working of He-Ne laser with the help of energy level diagram.
40. Explain the working of He-Ne laser with the help of a neat energy level diagram.
41. Explain with a neat diagram the principle, construction and working of he-Ne laser. What are its merits and demerits?
42. What is the active material in a He-Ne laser? How population inversion is is achieved in a He-Ne laser?
43. With the help of neat energy level diagram, explain how the population inversion is achieved in He-Ne laser. Explain why increase in diameter of He-Ne tube may reduce lasing efficiency.
44. In He-Ne laser, what is the function of He atoms? Why is it necessary to use a tube of narrow diameters?
45. Describe the construction and working of He-Ne laser.
46. Explain the action of a Helium-neon laser. How is it superior to a Ruby laser?
47. What are laser characteristics? Describe the principle and working of he-Ne laser. Why a narrow discharge tube is used in He-Ne laser?
48. Describe the construction and working of CO_2 laser.
49. Explain the modes of vibrations of CO_2 molecule. Describe the construction and working of CO_2 laser with necessary diagrams.
50. Explain with a neat diagram the principle, construction and working of a semiconductor laser. What are its merits and demerits?
51. What are semiconductor diode lasers? Describe with energy band diagram the construction and working of semiconductor diode laser. Mention the uses of diode lasers.
52. What is the role of length of resonant cavity in supporting different frequency modes?
53. Explain longitudinal modes and transverse modes in a laser beam.

54. List out the properties of Laser.
55. State important characteristics of laser beam. Explain construction and working of any one laser.
56. Explain in brief the following typical characteristics of laser: (i) Coherence, (ii) Intensity.
57. What is the reason on monochromaticity of laser beam?
58. Explain in brief the characteristics of a laser beam.
59. Describe briefly the application of lasers in welding, cutting and drilling. Mention the nature and property of the lasers used.
60. Discuss some of the applications of laser in industry.
61. State a few applications of laser.
62. Explain the applications of laser in medical field.

PROBLEMS

1. A three level laser emits laser light at a wavelength of 5500Å. (*i*) In the absence of optical pumping, what will be the equilibrium ratio of the population of the upper level to that of the lower level? Assume T = 300 K., (*ii*) At what temperature for the conditions of (*i*) above would the ratio be ½?, (*iii*) What conclusion can be drawn on the basis of the results obtained by you in relation to choice of pumping mechanism?
2. A typical He-Ne laser emits radiation of λ = 6328 Å. How many photons per second would be emitted by a one milli-watt He-Ne laser? **[Ans:** 3×10^{15}]
3. Find the relative populations of the two states in a ruby laser that produces a light beam of wavelength 6943 Å at 300 K and 500 K. **[Ans:** 8.75×10^{-19}]
4. Calculate the ratio of spontaneous emission to stimulated emission if the wavelength of the radiation is 5500 Å at 2000K. **[Ans:** 4.6×10^{5}]
5. For the He-Ne laser (λ = 10640 Å), estimate the broadening of the wavelength due to Heisenberg uncertainty principle, assuming that the metastable state has a lifetime of 1 ms. **[Ans:** 1.06×10^{-19} m]
6. If the pulse width of a laser (λ = 10640 Å) is 25 ms and average output power per pulse is 0.8W, how many photons does each pulse contain? **[Ans:** 107×10^{15} m]
7. In an experiment on the divergence of the He-Ne laser beam, the diameters of the beam spot were measured to be 4 mm and 6 mm at distances of 1 m and 2 m respectively. Determine the divergence of the beam. **[Ans:** 1 milliradian]
8. A ruby laser produces a beam of spot diameter 5 mm and divergence of 1 milliradian. It is directed towards the moon. The earth-to-moon distance is 376284 km. Compute the area on the moon that would be illuminated by the laser beam. **[Ans:** 445000 km^2]
9. Estimate the angular spread of a laser beam of wavelength 6930 Å due to diffraction, if the beam emerges through a 3 mm diameter mirror. How large would be the diameter of this beam when it is incident on a satellite 300 km above the earth? **[Ans:** 84.6m]
10. A laser beam λ = 6000 Å on earth is focused by a lens of diameter 2m on to a crater on the moon. The distance of the moon is 4×10^{8} m. Hoe big is the spot on the moon? **[Ans:** 3×10^{-7} rad, 1.4×10^{4} m^2)
11. A ruby laser emits light of wavelength 694.4 nm. If a laser pulse is emitted for 1.2×10^{-11} s and the energy release per pulse is 0.15 J, (*i*) What is the length of the pulse and (*ii*) How many photons are there in each pulse?

45 Non-Linear Optics

Learning Objectives:

By the end of this chapter, you will be able to:

- Understand concept of Wave Propagation and Momentum Conservation.
- Explain Linear medium related concepts.
- Understand principle of Non Linear Polarization.
- Principles of higher-harmonics generation.
- Describe Non-linear Materials and its types.

45.1 INTRODUCTION

Lasers produce coherent radiation at several wavelengths covering a range of wavelengths. But, lasers that can generate light at all wavelengths of interest cannot be fabricated. It was discovered in 1961 that certain non-linear materials could double the frequency of laser light passing through them. It opened up the possibility of converting the frequency of light generated by lasers into another frequency with the help of non-linear materials. Frequencies both higher and lower than the original can be created by nonlinear optics. Thus, nonlinear optics extends the usefulness of lasers. Today, laser pointers are available that generate green light by doubling the infrared output from a tiny diode-pumped neodymium solid-state laser.

Harmonic generation, sum and difference frequency generation and parametric oscillations are some of the important nonlinear processes utilized in laser light frequency transformations. In this chapter we acquaint ourselves with the methods of nonlinear processes.

45.2 WAVE PROPAGATION AND MOMENTUM CONSERVATION

A plane wave propagating through a vacuum in a general direction is represented by the equation

$$E = E_o e^{i(\omega t - k.r)} \qquad ...(45.1)$$

where k is the vector that defines the direction of propagation of the wave.

Propagation vector $\vec{k}$ can be written as

$$\vec{k} = \vec{i}k_x + \vec{j}k_y + \vec{k}k_z \qquad ...(45.2)$$

When a light wave interacts with an isotropic material, it will be affected in the same way whatever may be the direction of the beam with respect to the material. Hence, the refractive index of the material is the same in every direction and the velocity of the beam passing through the material is also the same in every direction. Since the magnitude of **k** is related to the velocity, the components k_x, k_y and k_z would all be the same in case of isotropic crystals. In anisotropic crystals, the light beam experiences different refractive indices in different directions and propagates at different velocities in different directions. Therefore, the components of **k** will have different values in different directions.

If we consider the photon description of light beam, we know that the momentum and wavelength of a photon are related through the following expression.

$$p = \frac{h}{k} = \frac{2\pi h}{2\pi\lambda} = \hbar k$$

Momentum p is a vector quantity and hence we write the above equation in vector form as

$$p = \hbar\,\mathbf{k} \qquad \text{...(45.3)}$$

where **k** is the vector which defines the propagation direction of the wave associated with the photon.

Thus, the significance of **k** in Eq. (45.1) and Eq. (45.3) is the same.

If a beam of light passes from air into a crystal, the momentum of the wave must be conserved at the interface of the two media. It requires that the normal component and parallel component of the momentum must be separately conserved at the boundary. Therefore, the transverse component of the momentum must be continuous across the boundary. There cannot be an abrupt jump or change in the transverse momentum of the wave as the beam enters the medium. In view of Eq. (45.3), it means that the transverse component of k should not change abruptly at the boundary.

45.3 LINEAR MEDIUM

The optical media are basically dielectric materials. They do not allow electric current to pass through when an electric field is applied across them. Instead, they get polarized. We know that the atom consists of a nucleus with shells of electrons surrounding it. Electrons are held in orbit by linear forces, and the radius of the orbit itself is governed by a linear force equation. When light is incident on the atom, it can force the orbits of the valence electrons to stretch in step with the electric field of the incident light. As the nucleus is too heavy to be affected by an electric field, and as electrons in the inner shell are bound tightly to the nucleus, only valence electrons are affected in such an interaction. Valence electrons in the shell move to one side of the atom under the action of the electric field and the entire atom becomes polarized. A positive and a negative end develop on the atom, and an electric dipole with a dipole moment develops in the direction of the applied electric field.

In the presence of an applied electric field, the entire material becomes polarized on a localized level around the intense beam. Dielectrics are polarized, for example, when they are placed between the charged plates of a capacitor. This is a macroscopic charge polarization of the material. The polarization vector, **P**, denotes the extent to which the dielectric is polarized. The electric polarization is parallel to the applied field, **E** and the amount of charge polarization is directly depends on the intensity of the applied electric field according to

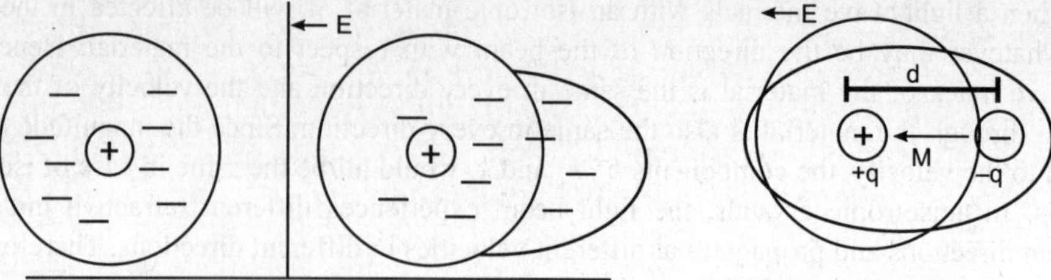

Fig. 45.1:

Effect of applied electric field on an atom

$$\mathbf{P} = \varepsilon_o \chi \mathbf{E} \quad ...(45.4)$$

where ε_o is the permittivity of free space and χ [= $(\varepsilon / \varepsilon_o) - 1$] is a dimensionless constant known as the ***electric susceptibility*** of the medium.

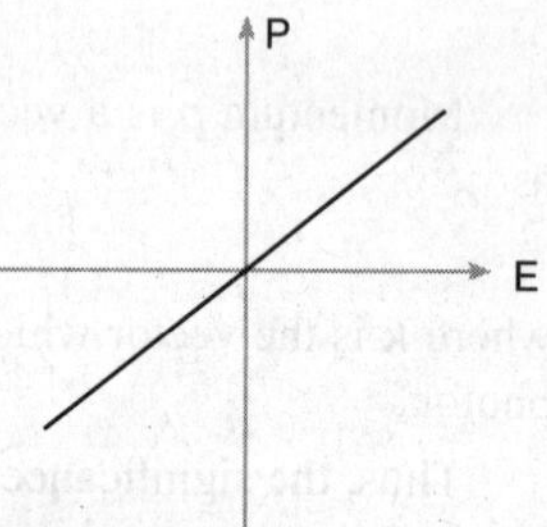

Fig. 45.2:

Linear dielectric medium

The stronger the electric field, the greater will be the polarization and a plot of P versus E is a straight line. Materials, in which such kind of linear relationship holds, are known as ***linear dielectrics***.

Light waves are electromagnetic waves and when they propagate through a dielectric, the electric field of the waves polarizes the dielectric. Hence the optical parameters of the dielectric are closely related to the dielectric polarization. The refractive index of the medium is given by,

$$\mu = \sqrt{\varepsilon} = \sqrt{1 + 4\pi\chi} \quad ...(45.5)$$

As long as the intensity of the light propagating in the dielectric medium is small, the parameters χ and μ are constant quantities and are independent of the intensity of light. Ordinary light sources generate light of field strengths of the order of 10^5 V/m which are very small compared to atomic fields, and therefore cannot affect the optical parameters of the medium.

Eq. (45.4) is a linear relationship. It is valid in case of small applied forces such as those we come across every day, the sunlight and artificial lights. Nonlinear effects are not seen in these cases. The polarization and re-radiation of photons causes a slowing of the velocity of light through the crystal (concept of the refractive index), but no change occurs in the character of the light transitory through the crystal.

45.4 NONLINEAR POLARIZATION

Nonlinear effects are exhibited, when the applied force is very large. When the electric field E in the incident light is large, the parameters χ, ε and μ are not constants but become the functions of E. The directions of **P** and **E** concur in an isotropic medium. We express χ as a power series in the field strength as,

$$\chi(\mathrm{E}) = \chi_1 + \chi_2 \mathrm{E} + \chi_3 \mathrm{E}^2 + ... \quad ...(45.6)$$

and as

$$P = \varepsilon_o \chi(\mathrm{E})\, \mathrm{E}$$

$$\therefore \quad P = \varepsilon_o (\chi_1 \mathrm{E} + \chi_2 \mathrm{E}^2 + \chi_3 \mathrm{E}^3 + ...) \quad ...(45.7)$$

where χ_1 is linear susceptibility and is much greater than the coefficients of the nonlinear terms χ_2, χ_3 and so on.

The nonlinear terms contribute noticeably only at very high-amplitude electric fields.

The second order nonlinear polarization is given by $P_2 = \varepsilon_o \chi_2 E^2$ ...(45.8)

and third order nonlinear polarization by $P_3 = \varepsilon_o \chi_3 E^3$ ...(45.9)

An intense electric field is required for achieving the nonlinear polarization. The electron experiences an internal force of about 10^9 V/cm due to the positive nucleus. Light from ordinary light sources have electric fields of strengths below 100 V/cm. Few sources exist with a high enough electric field to generate nonlinear effects. A focused laser beam can apply an electric field of 10^6 V/cm to 10^7 V/cm on a crystal. Even higher intensities, with even higher electric fields, can be found inside the cavities of most lasers.

Nonlinear polarization leads to nonlinear optical effects. Materials, in which polarization exhibits nonlinear dependence on the field strengths, are called ***nonlinear media***. Fig. 45.3 shows the nonlinear variation of electric polarization with the electric field strength in a nonlinear medium.

In optically isotropic materials, the coefficients of even powers of E in Eq. (45.7) are zero. But in case of anisotropic materials, coefficients of both odd and even powers of E exist. Strictly speaking, any medium becomes *nonlinear* provided the electric field of the incident radiation is very high. With the advent of intense lasers, the nonlinear polarization has assumed importance and made it possible the phenomena of frequency conversion and other non-linear effects.

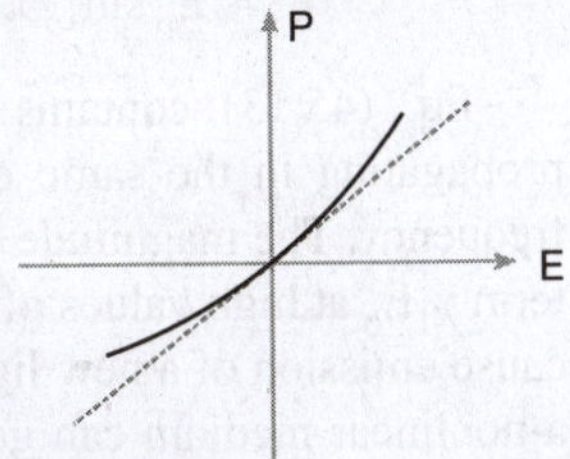

Fig. 45.3: Nonlinear medium

Now if we take into the consideration of electric field oscillations of a light wave, we can write for the electric field of the wave as

$$E = E_0 \sin\omega t$$

The polarization caused by this wave can be expressed as

$$P = \varepsilon_o (\chi_1 E_o \sin \omega t + \chi_2 E_o^2 \sin^2 \omega t + \chi_3 E_o^3 \sin^3 \omega t + \ldots.) \quad \ldots(45.10)$$

Eq. (45.10) can be re-written as

$$P = \varepsilon_o [\chi_1 E_o \sin \omega t + \frac{1}{2}\chi_2 E_o^2 (1 - \cos 2\omega t) + \frac{1}{4}\chi 3 E_o^3 (3 \sin \omega t - \sin 3\omega t) + \ldots] \quad \ldots(45.11)$$

Thus, as the electric field in the incident wave oscillates, the array of dipoles produces an electromagnetic wave, which may be called a ***polarization wave.*** The third term in Eq. (45.11) has a term with order $2v$, that is one component of the polarization in the equation is at twice the fundamental frequency of the original laser beam. It means that as the incident beam falls on the crystal, oscillating electric dipoles absorb light at a frequency v and reradiate it at the same frequency v and also at twice the original frequency $2v$. The total energy of the beam is not changed much, but it is now split between two components. This is the second harmonic of the original beam and the effect is known as *second-harmonic generation* (SHG).

45.5 SECOND HARMONIC GENERATION

In case of anisotropic crystals Eq. (45.7) may be rewritten as,

$$P = \varepsilon_o [\chi_1 E + \chi_1 E^2] \quad \ldots(45.12)$$
$$= P_1 + P_{nl}$$

where P_l and P_{nl} are linear polarization and nonlinear polarization components respectively.

Fig. 45.2 shows the response of the nonlinear medium, in which a symmetrical electric field produces an asymmetrical polarization. Equation (45.11) reveals not two, but three output terms: in addition to the fundamental (v) and second-harmonic ($2v$) term, a constant term $\frac{1}{2}\chi_2 E_0^2$ which represents a DC offset or steady polarization. It also represents an optical rectification analogous to that of a diode rectifying an AC voltage. Indeed, this DC signal (although a minute quantity) can be detected across a nonlinear crystal by placing electrodes on either side of the crystal, and this signal is proportional to the intensity of the light beam passing through the crystal.

Let us consider a plane monochromatic light wave of frequency ν, travelling at the velocity v_1 in a nonlinear medium. The electric field of the wave may be of the form,

$$E = E_o \sin[2\pi v(1 - x/v_1)] + \chi_1 E_o^2 \sin^2[2\pi v(1 - x/v_1)]$$

$$E = E_o \sin[2\pi v(1 - x/v_1) + \frac{1}{2}\chi_1 E_o^2 - \frac{1}{2}\chi_1 E_o^2 \cos[4\pi v(1 - x/v_1)] \quad ...(45.13)$$

Eq. (45.13) contains a term in $4\pi v$ which corresponds to a wave of polarization propagating in the same direction and at the same speed of light wave with twice higher frequency. The magnitude of the term in $4\pi v\{= 2(2\pi v)\}$ approaches the magnitude of the first term $\chi_1 E_o$ at high values of E_o. Secondly, under certain conditions, this polarization wave may cause emission of a new light wave. It means that a strong light of frequency v propagating in a nonlinear medium can generate a new light wave at frequency $2v$. It is a second harmonic and hence the phenomenon is known as *second harmonic generation*. At some initial point the incident wave generates the second harmonic wave and at that point the two waves are coherent. As the incident wave propagates through the crystal, it continues to generate second harmonic waves. All these waves combine constructively only if they satisfy the phase matching condition, Eq. (45.14).

Phase mismatch between the components gives the phenomenon of birefringence. In birefringence the index of refraction depends on the polarization of the individual light component propagating through the crystal. By tilting the crystal such that the index of refraction for both the fundamental and the harmonic component are the same, phase matching can be achieved. When the direction of propagation of light in the crystal is selected such that the indices of refraction for both the fundamental and second harmonic are exactly the same, the crystal is said to be *phase matched*. As the second harmonic component produced in the crystal is in phase with the fundamental component, the constructive interference occurs. This has consequences for efficiency of generation of second harmonic light.

The input wave of frequency v is normally called the ***pump wave***. The ratio of the intensity of the generated second harmonic wave to that of the pump wave is dependent on the ***phase mismatch factor***, Δk, which is given by

$$\Delta k = k(2v) - 2k(v) = 2v\,[\mu(2v) - \mu(v)]$$

In order to obtain an intense second harmonic wave, that is directed like the pump laser beam, one has to achieve the condition $\Delta k = 0$. This implies that

$$\mu(2v) = \mu(v) \quad ...(45.14)$$

Eq. (45.14) is known as the ***phase matching condition***. It is also known as ***index matching condition***.

This condition is satisfied only when the pump and generated waves are of orthogonal polarization. Therefore, the process of second harmonic generation requires a uniaxial double refracting crystal exhibiting high nonlinear susceptibility χ_2. It must be cut in the shape of a rectangular parallelepiped with the axis along the direction of phase matching. The incident light must be plane polarized, with the plane of polarization perpendicular to the plane of principal section of the crystal. When these conditions are satisfied, a wave of frequency ν produces another wave of frequency 2ν, which is the second harmonic. The direction of propagation of the second harmonic coincides with that of the fundamental wave. The plane of polarization of second harmonic will be perpendicular to that of the fundamental wave.

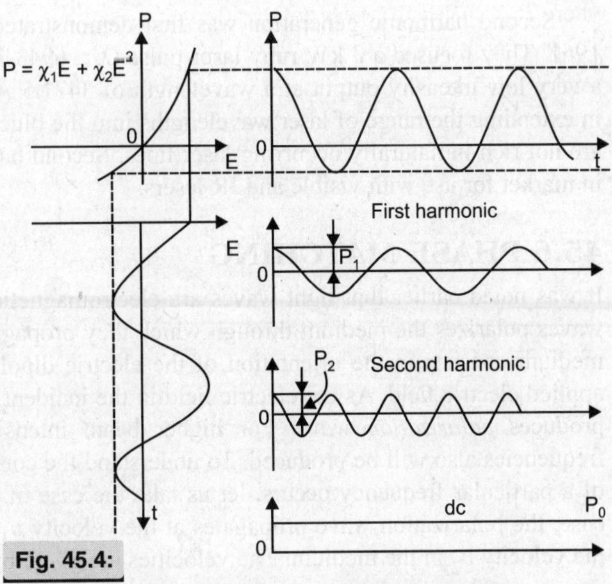

Fig. 45.4: Fourier analysis of the asymmetrical polarization wave into fundamental second harmonic and d.c. components

A commonly used material for second harmonic generation is KDP which is a negative double refracting crystal. In negative crystals, if the fundamental light is a linearly polarized *ordinary wave*, the resulting second harmonic will be an *extraordinary wave*. The reverse is the situation in case of positive crystals. If the fundamental light is a linearly polarized *extraordinary wave*, the resulting second harmonic will be an *ordinary wave*.

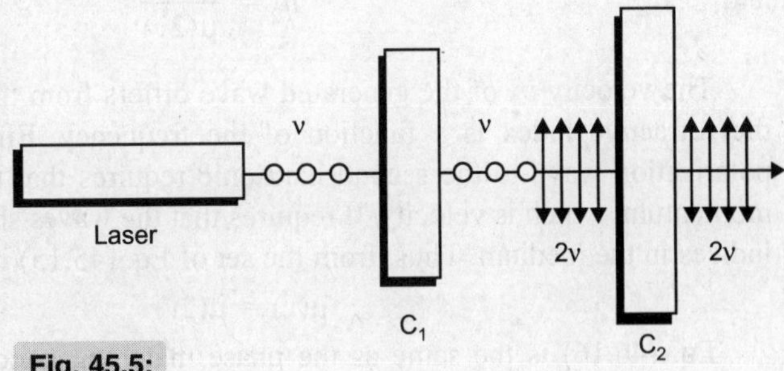

Fig. 45.5: Arrangement of second harmonic generation

The second harmonic generation may also be described from the standpoint of photon concept. According to this picture (see Fig. 45.5), two photons of the fundamental frequency ν, each having energy $h\nu$ combine to produce an energy of $2h\nu$. The energy levels located at $h\nu$ and $2\,h\nu$ are known as virtual levels and shown by dashed lines, as they are not allowed states of the material and as such are not populated. In this process, two photons of frequency ν are destroyed and one photon of frequency 2ν is simultaneously created.

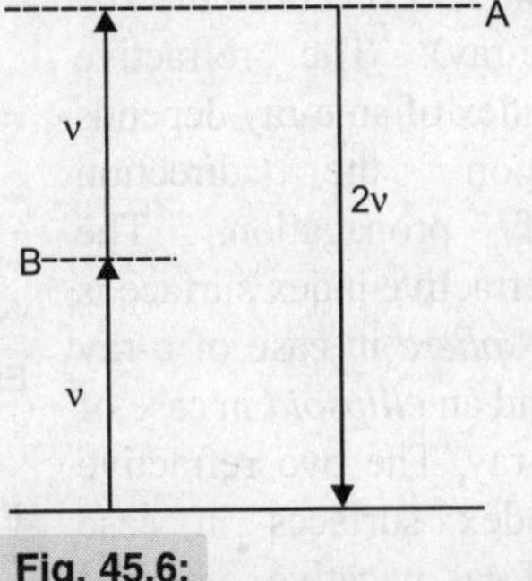

Fig. 45.6: Second harmonic generation

Second harmonic generation was first demonstrated by Franken and his co-workers in 1961. They focused a 3 kW ruby laser pulse (λ = 6943 Å) onto a quartz crystal and obtained a very low intensity output at a wavelength of 3471.5 Å. Second harmonic generation helps in extending the range of laser wavelengths into the blue and UV part of the spectrum, which are not rich in naturally occurring laser lines. Second harmonic generators are now available in market for use with visible and IR lasers.

45.6 PHASE MATCHING

It was noted earlier that light waves are electromagnetic waves and the electric field of the waves polarizes the medium through which they propagate. The polarization response of the medium consists in the orientation of the electric dipoles in the direction of the externally applied electric field. As the electric field in the incident wave oscillates, the array of dipoles produces *polarization wave*. For higher beam intensities, polarization waves of higher frequencies also will be produced. To understand the condition under which the enhancement of a particular frequency occurs, let us take the case of second harmonic generation. In this case, the polarization wave propagates at the velocity v_1, the second harmonic propagates at the velocity v_2 in the medium. The velocities are given by,

$$v_1 = \frac{c}{\mu(\nu)} \quad \text{...(45.15a)}$$

and

$$v_2 = \frac{c}{\mu(2\nu)} \quad \text{...(45.15b)}$$

The velocity v_2 of the generated wave differs from that of the incident wave, v_1 because the refractive index is a function of the frequency. Efficient transfer of energy from the polarization wave to the second harmonic requires that the two waves must be matched in momentum, which is velocity. It requires that the waves should experience identical refractive indices in the medium. Thus, from the set of Eq. (45.15) it follows that,

$$\mu(\nu) = \mu(2\nu) \quad \text{...(45.16)}$$

Eq. (46.16) is the same as the phase matching condition or index-matching condition, Eq. (45.14).

This condition is not satisfied in general cases. It is satisfied *only* in anisotropic crystals known as *uniaxial crystals*. When a light ray is incident on a *uniaxial crystal*, it splits up into two rays, namely, an *ordinary ray* (o-ray) and an *extra ordinary ray* (e-ray). The refractive index of an e-ray depends upon the direction of propagation. The refractive index surface is a *sphere* in case of o-ray and an *ellipsoid* in case of e-ray. The two refractive index surfaces in case of a negative uniaxial crystal are shown in

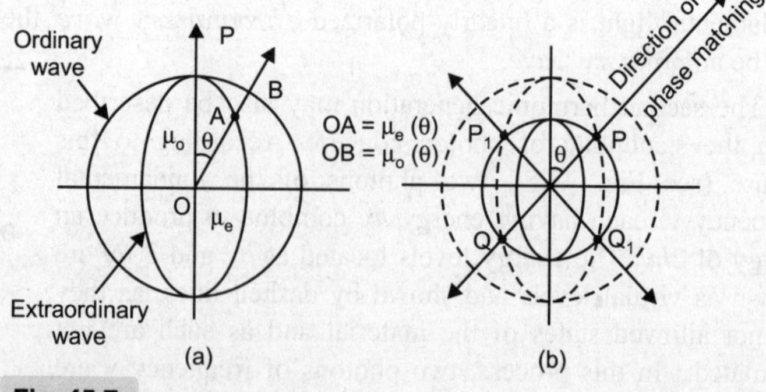

Fig. 45.7: (a) Index ellipsoid and sphere corresponding to a e-ray and o-ray in a negative uniaxial crystal, (b) Index matching in a negative crystal

Fig. 45.7(a). Fig. 45.7(b) shows the refractive index surfaces for the frequency ν and for the frequency 2ν. It is seen that the refractive index surface of the o-ray with frequency ν and that of the e-ray with frequency 2ν intersect; one of the points of intersection is at point *P*. At the point of intersection the velocities υ_1 and υ_2 are equal. It is obvious that the phase-matching condition is satisfied for light waves propagating in the direction *PQ*.

Thus, the direction *PQ* is *the direction of phase- matching*. Along *PQ*,

$$\mu_o(\nu) = \mu_e(2\nu) \qquad ...(45.17)$$

45.6.1 Angle tuning for phase matching

There are two methods for achieving phase matching in double refracting crystals. They are known as *angle tuning* and *temperature tuning*. The refractive index of extraordinary ray μ_e varies with the angle between the direction of propagation and the optic axis. The variation is given by

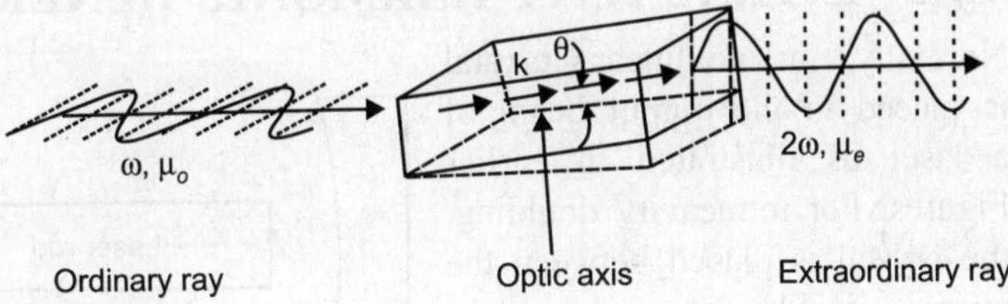

Fig. 45.8: Phase matching using angle tuning

$$\mu_e(\theta) = \left[\frac{\cos^2\theta}{\mu_o^2} + \frac{\sin^2\theta}{\mu_e^2}\right] \qquad ... (45.18)$$

By rotating the crystal in the plane of the optic axis and the incident beam (Fig. 45.8), it is possible select the value of μ_e that will satisfy the phase matching condition Eq. (45.16).

45.7 NONLINEAR MATERIALS

ADP and KDP were two of the first nonlinear materials. They are transparent for wavelengths 200 to 1500 nm but have a small nonlinear coefficient. A more efficient material is KTP, which is commonly used nowadays to double in Nd:YAG and Nd:YVO_4 lasers. Lithium niobate is more useful than ADP or KDP in the infrared. It has a large nonlinear coefficient, making it about 100 times more efficient than KDP as a second-harmonic generator. Table 45.1 shows a few common nonlinear materials and their corresponding coefficients.

Table 45.1: *Nonlinear Optical Coefficients for Various Materials*

Material	Nonlinear Coefficient
Quartz	2.6×10^{-26} to 3.0×10^{-24}
ADP ($NH_4H_2PO_4$)	6.8×10^{-24}
KDP (KH_2PO_4)	3.8×10^{-24} to 4.1×10^{-24}
KTP ($KTiOPO_4$)	4.4×10^{-23} to 1.2×10^{-22}
Lithium Niobate ($LiNbO_3$)	2.3×10^{-23} to 3.9×10^{-22}

45.8 SECOND-HARMONIC CONVERSION EFFICIENCY

Def: Second-harmonic conversion efficiency is the ratio of the power of the second-harmonic output component to the power of the fundamental component. It is directly proportional to the intensity of the electric field, $I = (F_{undamental}/A)$.

For this reason, a number of harmonic generator crystals are kept inside the cavity of the laser where the intensity is highest. Efficiency is also proportional to the square of the length of the second harmonic crystal.

From Eq. (45.11), we find that the relationship between incident power and second-harmonic output power is

$$P_{\text{second harmonic}} = \frac{Kl^2 P_{incident}^2}{A}$$

where K is a proportionality constant, l the length of the crystal, and A the beam area in the crystal.

45.9 INTRACAVITY HARMONIC GENERATION

Normally, a nonlinear crystal is placed in the output beam of a laser as illustrated in earlier Figures. For intracavity doubling, the crystal is placed between the mirrors, inside the resonator as shown in Fig. 45.9.

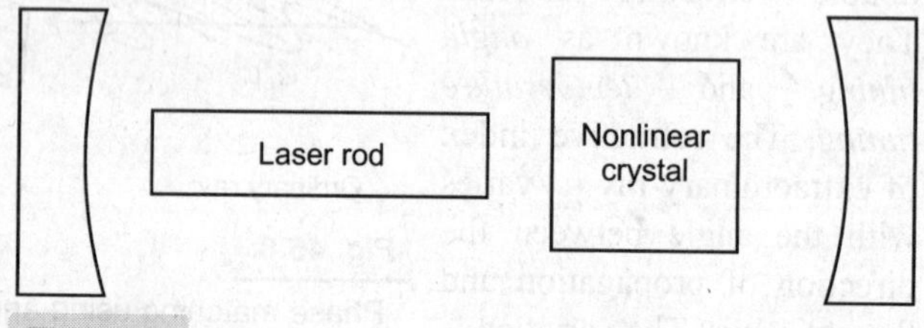

Fig. 45.9: Intracavity doubling with the nonlinear crystal inside the resonator

The efficiency of frequency doubling is improved if the nonlinear crystal is placed inside the resonator. The amount of second-harmonic power generated is proportional to the square of the fundamental power. Pulsed lasers with kilowatts or more peak power can be doubled with an external crystal. CW lasers with milli-watts or more of output are a different class. A 1-W cw laser would produce one one-millionth the second-harmonic power of a pulsed laser with 1 KW of peak power.

If the nonlinear crystal is placed inside the resonator, the effective conversion efficiency would be as high as the conversion efficiency of a pulsed laser. However, there are problems in placing a nonlinear crystal inside a laser. If the crystal introduces even a small loss, it can drastically decrease the circulating power. One percent additional loss can reduce the circulating power of lasers by half, and the advantage of placing the crystal inside the resonator is lost.

It is difficult to fabricate the output mirror of an internally frequency-doubled laser. The mirror must have maximum reflectivity at the fundamental component to keep the circulating power inside the resonator. At the same time, it must transmit the entire second harmonic that falls on it.

The circulating power goes in both directions through the nonlinear crystal. Hence, a second-harmonic beam is generated in both directions. One beam passes out through the output mirror, but the other beam is wasted. Solving this problem involves complex optics and is difficult to implement.

45.10 SUM AND DIFFERENCE FREQUENCY GENERATION

Nonlinear crystals make it also possible to generate beams of sum and difference frequencies when two light waves of different frequencies interact within them. The interaction of waves with frequencies ν_1 and ν_2 generate light waves with frequencies $\nu_1 + \nu_2$ and $\nu_1 - \nu_2$.

Let us consider two coherent waves of frequencies ν_1 and ν_2 propagating simultaneously in a nonlinear medium. Let the waves be represented by $E_1 \sin 2\pi\nu_1 t$ and $E_2 \sin 2\pi\nu_2 t$, respectively. The superposition of the waves leads to,

$$E = E_1 \sin 2\pi\nu_1 t + E_2 \sin 2\pi\nu_2 t \quad ...(45.19)$$

Using Eq. (45.19) in Eq.(45.12), we obtain the following expression for the nonlinear term in the polarization of the medium.

$$\begin{aligned} P_{nl} &= \varepsilon_o \chi_2 E^2 \\ &= \varepsilon_o \chi_2 [E_1^2 \sin 2\pi\nu_1 t + E_2 \sin 2\pi\nu_2 t]^2 \\ &= \varepsilon_o \chi^2 E_1^2 \sin^2 2\pi\nu_1 t + \varepsilon_o \chi_2 E_2^2 \sin^2 2\pi\nu_2 t + 2\varepsilon_o \chi_2 E_1 E_2 \sin(2\pi\nu_1 t)\sin(2\pi\nu_2 t) \\ &= \varepsilon_o \chi_2 E_1^2 (1 - \cos 4\pi\nu_1 t) + \varepsilon_o \chi_2 E_2^2 (1 - \cos 4\pi\nu_2 t) \\ &\qquad + \varepsilon_o \chi_2 E_1 E_2 \{\cos 2\pi(\nu_1 - \nu_2)t - \cos 2\pi(\nu_1 + \nu_2)t\} \end{aligned}$$

or

$$\begin{aligned} P_{nl} &= \varepsilon_o \chi_2 (E_1^2 + E_2^2) - \varepsilon_o \chi_2 E_1^2 \cos 2\pi(2\nu_1)t - \varepsilon_o \chi_2 E_1^2 \cos 2\pi(2\nu_2)t \\ &\qquad + \varepsilon_o \chi_2 E_1 E_2 \{\cos 2\pi(\nu_1 - \nu_2)t - \cos 2\pi(\nu_1 + \nu_2)t\} \quad ...(45.20) \end{aligned}$$

The Eq. (45.20) shows that the interaction of light waves of frequencies ν_1 and propagating in a nonlinear medium generates new light waves of sum (ν_1 + ν_2), difference (ν_1 − ν_2) and double frequencies 2 ν_1 and 2 ν_2. It is to be noted that all the above mentioned frequencies are not generated simultaneously when two waves of frequencies ν_1 and ν_2 propagate through the crystal. If any of these additional frequencies are to be generated, the phase matching must be implemented and it can be implemented only to one of the frequencies at a time.

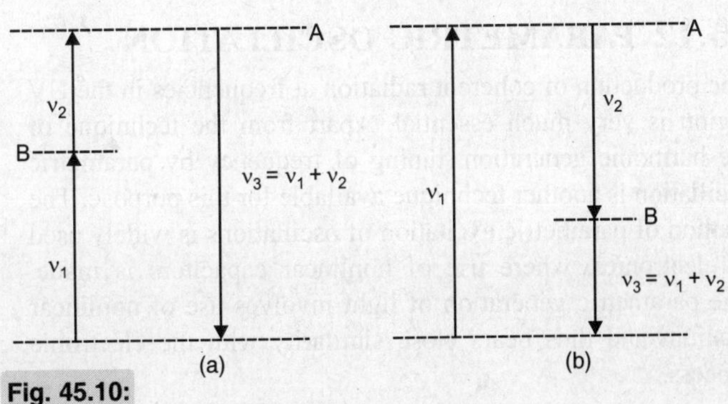

Fig. 45.10: (a) Sum frequency generation, (b) difference frequency generation

The processes of sum and difference frequency generation are depicted in Fig. 45.10. In sum frequency generation the frequencies ν_1 and ν_2 are destroyed and a new photon $\nu_3 = (\nu_1 + \nu_2)$ is created, as shown in Fig. 45.10 (a). In difference frequency generation, the photon of higher frequency ν_1 is destroyed and two new photons of frequencies ν_2 and $\nu_3 = (\nu_1 - \nu_2)$ are created, as shown in Fig. 45.10 (b). Note that ν_2 is already present as one of the input beams; it indicates that ν_2 is amplified in the process.

In sum frequency generation, ν_1 is taken as a weak infrared wave that sums with a strong pump wave ν_2 to generate a high frequency wave at $\nu_3 = (\nu_1 + \nu_2)$. Thus, the light of wavelength of 10.6 μm produced by a CO_2 laser is made to interact with a laser beam of 1.06 μm wavelength produced by Nd: YAG laser to yield a light beam of wavelength of 0.964 μm. Such upconversion is done because it is difficult to detect the radiation at 10 μm while the radiation at 0.964 μm can be easily detected.

45.11 HIGHER-ORDER NONLINEAR EFFECTS

Only a few materials have a nonlinear χ_2 coefficient but all materials have a third-order coefficient. But this coefficient is very small compared to a second-order coefficient and much less efficiency is possible for third-harmonic generation. Calcite, for example, can be used as a third-harmonic generator with efficiencies of less than 10^{-5}.

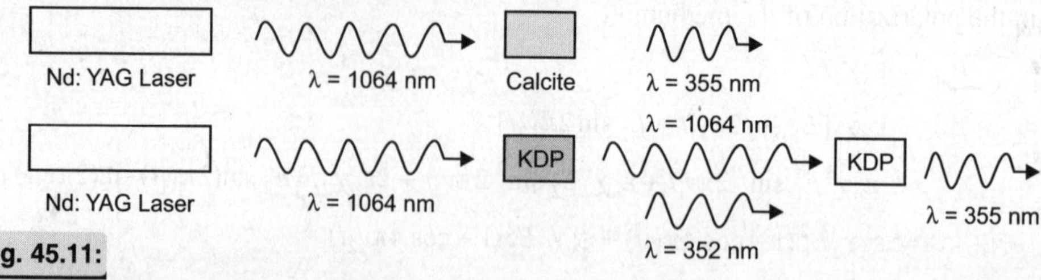

Fig. 45.11: Generation of High-order harmonics

On the other hand, series nonlinear crystals can be used to generate third-harmonic light; the first crystal produces second-harmonic light at 532 nm and the second crystal mixes this light with the original 1064-nm fundamental wavelength. Fig. 45.11 illustrate both methods.

45.12 PARAMETRIC OSCILLATION

The production of coherent radiation at frequencies in the UV region is very much essential. Apart from the technique of the harmonic generation, tuning of frequency by parametric oscillation is another technique available for this purpose. The method of parametric excitation of oscillations is widely used in electronics, where use of nonlinear capacitors is made. The parametric generation of light involves use of nonlinear medium and thus bears close similarity with the electronic process.

Fig. 45.12: Optical parametric oscillation

In sum frequency generation, light at frequencies ν_1 and ν_2 add together to produce the frequency $(\nu_1 + \nu_2)$. On the other hand, if a strong beam at frequency $\nu_1 = (\nu_2 + \nu_3)$ alone is applied to a suitable nonlinear material such as lithium niobate, two beams at lower frequencies ν_2 and ν_3 can be generated. For exciting parametric oscillations, the nonlinear medium is placed between two mirrors, which form a resonant cavity comparable to a laser cavity. The wave of frequency $\nu_1 = (\nu_1 + \nu_3)$ is the *pump wave* and it should be highly intense such that it can induce nonlinear behaviour of the crystal. Secondly the process requires that the phase-matching condition be satisfied. A wave of higher frequency say ν_2, is called the *signal wave* and the lower frequency wave say at ν_3, is called the *idler wave*. Both the *signal and idler* waves are weak and always present in the crystal in the form of noise, which arises due to spontaneous photons. If the pump frequency $(\nu_1 + \nu_3)$ is fixed, then the two frequencies ν_2 and ν_3 are free to spread over a wide range of values. This effect is known as *parametric amplification*. By varying the orientation of the crystal, the values of ν_2 and ν_3 can be varied. If the mirrors of the resonant cavity are reflective at ν_2 and ν_3, and not at the frequency $(\nu_2 + \nu_3)$, resonance of these waves in the cavity enhances the interaction in the nonlinear crystal and builds up the waves.

Thus the system acts as a parametric oscillator. Tuning can be achieved by varying the phase-matching conditions through mechanical or temperature control of the cavity.

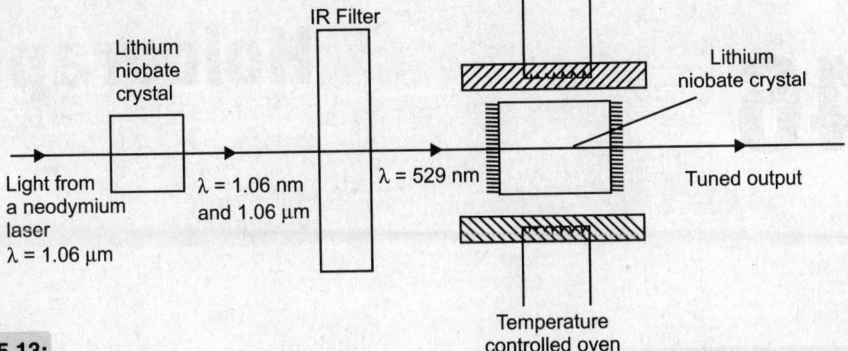

Fig. 45.13: A typical arrangement for observing parametric oscillation in a nonlinear medium

Parametric oscillation was first observed by Giordmaine and Miller in 1965. The schematic of the set up used by them is shown in Fig. 45.13. The output was tuned by changing the temperature of the lithium niobate crystal. A temperature range of about 11° C produced wavelengths in the range 9680 Å to 11540 Å. The conversion efficiency was low as 1%. Currently, conversion efficiency is improved to 40% to 50%.

QUESTIONS

1. What is a nonlinear medium?
2. What is nonlinear polarization and why it assumed importance after the advent of lasers?
3. Explain why phase matching condition must be satisfied to generate coherent radiation at new wavelengths?
4. How is second harmonic generation achieved?
5. What are the different methods of tuning to obtain index-matching?
6. What is meant by parametric oscillation? How a nonlinear medium can be used as a parametric oscillator?
7. Explain what is meant by self-focusing?
8. Explain stimulated Raman scattering? How can it be explained using quantum concept?

46 Holography

Learning Objectives:

By the end of this chapter, you will be able to:

- Understand Principle of holography and its types.
- Learn about Hologram and its properties.
- Classify different types of Holograms.
- Explain different applications of Holograms.

46.1 INTRODUCTION

Images of objects are generally obtained using photographic method. In this method a lens focuses the light reflected from a three-dimensional object onto a photographic film where a two-dimensional image of the object is formed. A negative is first obtained by developing the film and then a positive is obtained through printing. The positive print is a two-dimensional record of light intensity received from the object. It, thus, contains information about the square of the amplitude of the light wave that produced the image but information about the phase of the wave is not recorded and is lost.

In 1948 Dennis Gabor outlined a two-step lens less imaging process. It is radically a new technique of photographing the objects and is known as ***wave front reconstruction***. The technique is also called ***holography***. The word 'holography' is formed by combining parts of two Greek words: 'holos', meaning "whole", and 'graphein' meaning "to write". Thus holography means writing the complete image. Holography is actually a recording of interference pattern formed between two beams of coherent light coming from the same source. In this process both the amplitude and phase components of light wave are recorded on a light sensitive medium such as a photographic plate. The recording is known as a ***hologram***. Holography required an intense coherent light source. Laser was not available when Gabor formulated the idea of holography. Holographic technique became a practical proposition only after the invention of lasers. Leith and Upatnicks prepared laser holograms for the first time. In this chapter we discuss the fundamental concept of holography.

46.2 PRINCIPLE OF HOLOGRAPHY

Holography is a two-step process. First step is the ***recording*** of hologram where the object is transformed into a photographic record and the second step is the ***reconstruction*** in which the hologram is transformed into the image. Unlike in the conventional photography, lens is not required in either of the steps. A hologram is the result of interference occurring between two waves, an object beam which is the light scattered off the object and a coherent background, the reference beam, which the light is reaching the photographic plate directly. In Gabor's original experiments, the reference beam and object beams were *coaxial*. Further advance was made by Leith and Upatnieks, who used the reference, beam at an *offset angle*. That made possible the recording of holograms of three-dimensional objects.

The off-axis arrangement for generating and viewing holograms is described here.

46.2.1 Recording of the hologram

In the off-axis arrangement a broad laser beam is divided into two beams, namely a ***reference beam*** and an ***object beam*** by a beam splitter (Fig. 46.1). The reference beam goes directly to the photographic plate. The second beam of light is directed onto the object to be photographed. Each point of the object scatters the incident light and acts as the source of spherical waves. Part of the light, scattered by the object, travels towards the photographic plate. At the photographic plate the innumerable spherical waves from the object combine with the plane light wave from the reference beam. The sets of light waves are coherent because they are from the same laser. They interfere and form interference fringes on the plane of the photographic plate.

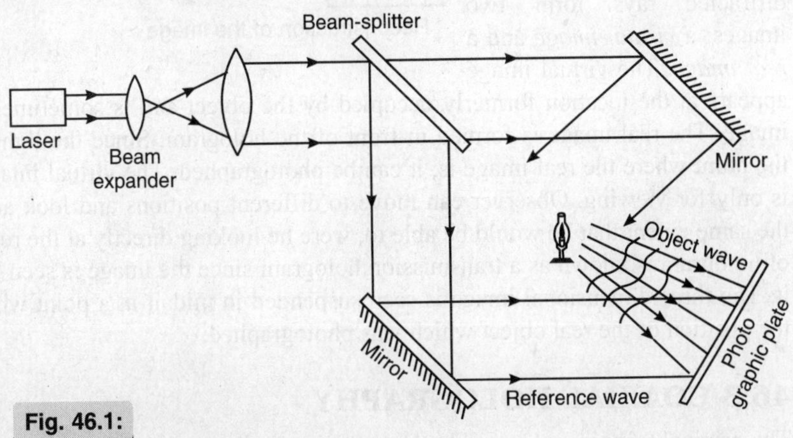

Fig. 46.1: Recording of the hologram

These interference fringes are a series of zone-plate like rings, but these rings are also superimposed, making a complex pattern of lines and swirls. The developed negative of these interference fringe-patterns is a hologram. Thus, the hologram (Fig. 46.2) does not contain a distinct image of the object but carries a record of both the intensity and the relative phase of the light waves at each point.

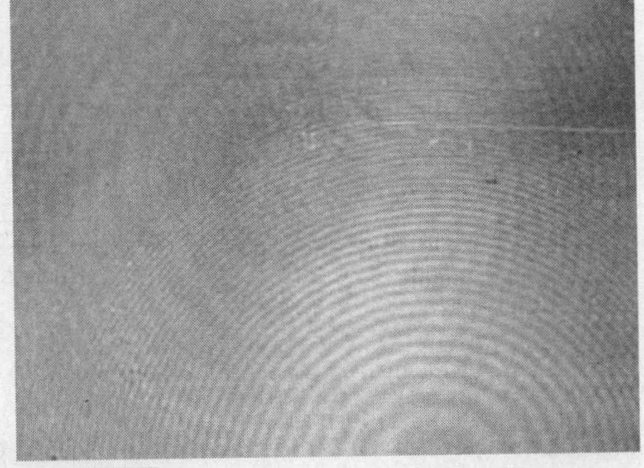

Fig. 46.2: The transmission hologram

46.2.2 Reconstruction of the image

Whenever required, the object can be viewed. For ***reconstruction*** of the image, the hologram is illuminated by a parallel beam of light from the laser (Fig. 46.3). Most of the light passes straight through, but the complex of fine fringes acts as an elaborate diffraction grating. Light is diffracted at a fairly wide angle. The diffracted rays form two images: a *virtual image* and a *real image*. The virtual image appears at the location formerly occupied by the object and is sometimes called as the true image. The real image is formed in front of the hologram. Since the light rays pass through the point where the real image is, it can be photographed. The virtual image of the hologram is only for viewing. Observer can move to different positions and look around the image to the same extent that he would be able to, were he looking directly at the real object. This type of hologram is known as a transmission hologram since the image is seen by looking through it. The three dimensional image is seen suspended in midair at a point which corresponds to the position of the real object which was photographed.

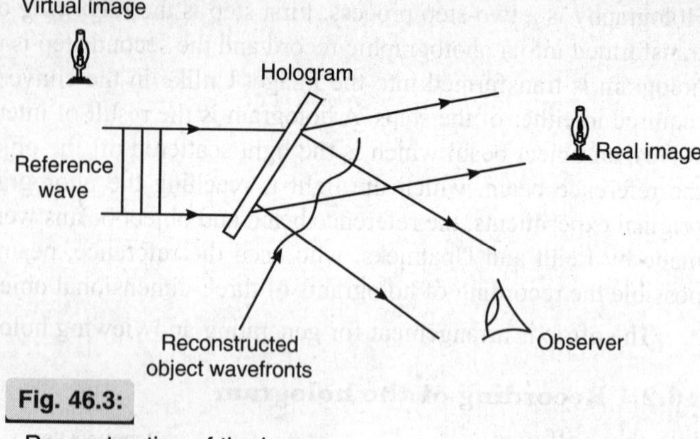

Fig. 46.3: Reconstruction of the image

46.3 COAXIAL HOLOGRAPHY

The original technique adopted by Gabor for recording hologram was a ***coaxial arrangement*** (Fig. 46.4), where he made use of a mercury discharge lamp and collinear object and reference beams. In this arrangement both virtual and real images are on the same axis. The real image is located in front of the virtual image. Thus an observer focusing on one image, always sees it accompanied by the out-of-focus twin image. As such it is inconvenient for viewing or photographing the image.

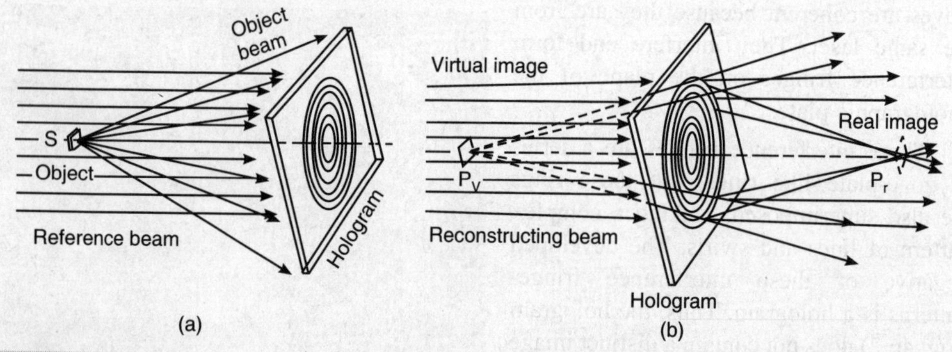

Fig. 46.4: In-line holography (a) Generation (b) Viewing.

46.4 OFF-AXIS HOLOGRAPHY

In 1962, Leith and Upatnicks demonstrated a technique which made it possible to separate the twin images. In this technique a separate coherent reference wave is allowed to fall on the hologram plate during the recording process, at an *offset angle* to the beam from the object. The exposed plate is developed by normal photographic procedures so that the amplitude transmittance of the plate after development is proportional to the exposure.

The advantage of the off-axis configuration is that it generates virtual and real images angularly separated from each other and from the direct beam also.

46.5 THEORY OF HOLOGRAPHY

The general theory of holography is much involved and cumbersome. We illustrate here it by taking the simple example of a point object in a coaxial configuration.

Let the light beam from a coherent source illuminate a point object P (see Fig. 46.5). The beam consists of plane waves. Most of the plane waves reach the photographic plate directly. Part of the light is scattered by the point object and spherical waves are produced. They also reach the photographic plate. The plane waves of the reference beam and spherical waves of the object beam superpose at the plane of the photographic plate and produce interference.

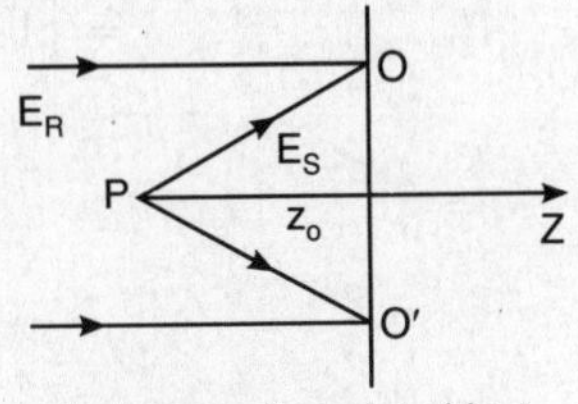

Fig. 46.5: Hologram of a point object

We may write the optical field arriving at point O on the photographic plate as

$$E = E_R + E_S \qquad ...(46.1)$$

where E_R is the field due to the reference beam and E_S is the field scattered from the object.

The scattered field E_S is not simple, both amplitude and phase vary greatly with position. The scattered wave fronts are spherical and concentric around the point of origin. We represent the field of the scattered wave front by

$$E_S = \frac{E_0}{r_0}\exp[i(k r_0 - \omega t)] \qquad ...(46.2)$$

and the field E_R by the plane wave

$$E_R = E_r \exp[i(k z_0 - \omega t)] \qquad ...(46.3)$$

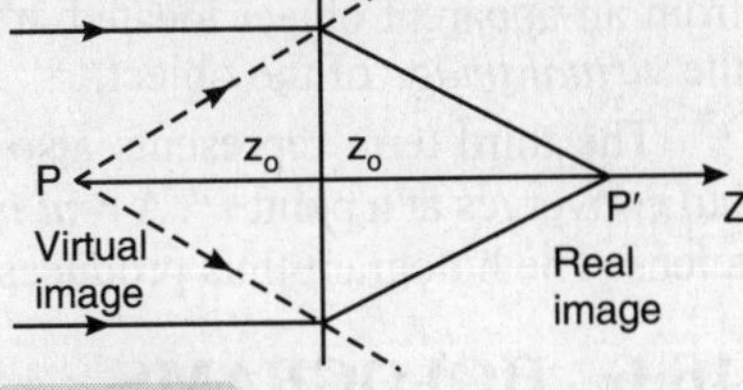

Fig. 46.6: Forming of real and virtual image

where $r_0 = PO$ and z_0 is the distance from P to the plate. The intensity at O is

$$I = |E_R + E_S|^2$$

$$= |E_R|^2 + \frac{|E_S|^2}{r_0^2} + \frac{E_S E_r^*}{r_0}\exp[ik(r_0 - z_0)] + \frac{E_S^* E_r}{r_0}\exp[ik(z_0 - r_0)] \qquad ...(46.4)$$

We combine the last two terms of the above equation and write it as

$$I = |E_R|^2 + \frac{|E_S|^2}{r_0^2} + K\,\frac{\cos[k(r_0 - z_0) + \phi]}{r_0} \qquad ...(46.5)$$

The total intensity I is a function of cosine term and shows a series of maxima and minima. Thus, the interference of the spherical wave E_S and plane wave E_R produces a set of circular interference fringes. If we assume that the plate response is proportional to the intensity I, the power transmission of the plate, T^2 is given by

$$T^2 = 1 - \alpha I \qquad ...(46.6)$$

where α is a constant. Eq. (46.6) can be approximated as

$$T \cong 1 - \frac{1}{2}\alpha I \qquad ...(46.7)$$

When the hologram is illuminated by the reference beam, the field of the transmitted wave may be written as

$$E = TE_2 = \left[1 - \frac{\alpha}{2}I\right] E_r \exp[i(kz_0 - \omega t)]$$

$$= \left[1 - \frac{\alpha}{2}|E_R|^2 - \frac{\alpha}{2}\frac{|E_S|^2}{r_0^2}\right] E_R \exp[i(kz_0 - \omega t)]$$

$$-\frac{\alpha}{2}\frac{E_S^* E_r}{r_0} \exp[ik(r_0 - z_0)] E_r \exp[i(kz_0 - \omega t)]$$

$$-\frac{\alpha}{2}\frac{E_S^* E_r}{r_0} \exp[ik(z_0 - r_0)] E_r \exp[i(kz_0 - \omega t)]$$

$$= \left[1 - \alpha|E_r|^2 - \frac{\alpha}{2}\frac{|E_S|^2}{r_0^2}\right] E_r \exp[i(kz_0 - \omega t)]$$

$$-\frac{\alpha E_S |E_r|^2}{2r_0} \exp[i(kr_0 - \omega t)] - \frac{\alpha}{2}\frac{E_S^* E_r^2}{r_0} \exp[i(2kz_0 - kr_0 - \omega t)] \qquad ...(46.8)$$

The first term in Eq. (46.8) represents the incident plane wave with some attenuation.

The second term represents a spherical wave identical with that emitted by the object except for a constant factor. The wave surface when projected back appears to have come from an apparent object located at the place where the original objected was located. This is the *virtual image* of the object.

The third term represents also a spherical wave, which is identical to the original wave but converges at a point P'. A *real image* is produced at P' which can be photographed without a lens. The hologram thus produces both a real image P and a virtual image P'.

46.6 HOLOGRAMS

Holograms are true three-dimensional images. This is evidenced by the fact that one can move his head while viewing the image and see it in a different perspective. It reveals part of the image which was hidden at another viewing angle.

For example, three images are shown below (Fig. 46.7). They are from the same hologram but are obtained by looking through the hologram at *different angles*. Note that the pawn appears in different perspective in front of the king behind it

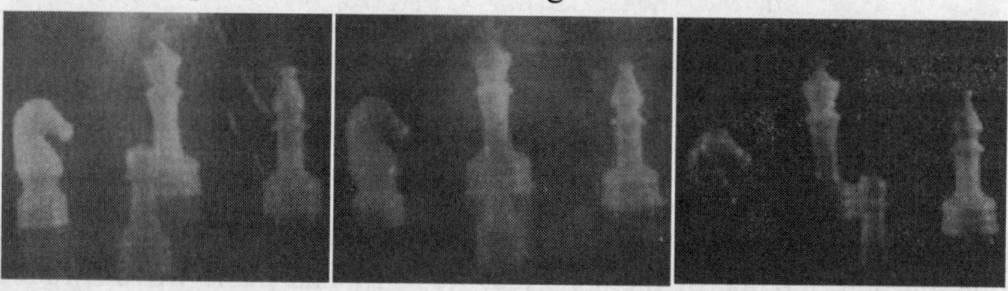

Fig. 46.7:

Viewing the hologram at different angles different perspectives are seen.

Recording a hologram is different from taking a photograph.

(*i*) In hologram, it is necessary that a coherent light source like a laser is used to illuminate the object.

(*ii*) There is a second beam of coherent light which strikes the film on which the hologram is to be recorded. This is called the reference beam. The reference beam and the object beam overlap at the surface of the film and they form an interference pattern.

(*iii*) A high-resolution photographic film is used for recording the fine patterns.

(*iv*) Because the lines that make up the hologram are usually less than a micron, holographic recording is more sensitive to movement and vibrations compared to photographic recording. This is actually a very serious restriction. Therefore, holography technique is restricted to the laboratory where vibration isolation is created.

The fundamental difference between a **hologram** and **an ordinary photograph** is like this.

(*i*) In a photograph the information is stored in an orderly fashion: each point in the object relates to a conjugate point in the image. In a hologram there is no such relationship; light from every object point goes to the entire hologram. This has two main advantages. As the observer moves sideways in viewing the hologram, the image is seen in three dimensions.

(*ii*) If the hologram were shattered or cut into small pieces, each fragment would still reconstruct the whole scene, not just part of the scene.

46.6.1 Orthoscopic and Pseudoscopic Images

A hologram reconstructs two images, one real and the other virtual which are exact replicas of the object. However, the two images differ in appearance to the observer. The virtual image is produced at the same position as the object and has the same appearance of depth and the parallax as the original three dimensional objects. The virtual image appears as if the observer is viewing the original object through a window defined by the size of the hologram. This image is known as ***orthoscopic image***.

The real image is also formed at the same distance from the hologram, but in front of it. In the real image, however the scene depth is inverted. This is due to the fact that the corresponding points on the two images (virtual and real) are located at the same distances from the hologram. The real image is known as ***pseudoscopic image*** and does not give a pleasing sensation as we do not come across objects with inverted depths in normal life.

46.7 IMPORTANT PROPERTIES OF A HOLOGRAM

(*i*) In an ordinary photograph each region contains separate and individual part of the original object. Therefore, destruction of a portion of a photographic image leads to an irreparable loss of information corresponding to the destroyed part. On the other hand, in a hologram each part contains information about the entire object. From even a small pat of the hologram the entire image can be reconstructed if only with a reduced clarity and definition of the image. Therefore, a hologram is a reliable medium for data storage.

(*ii*) It is not useful to record several images on a single photographic film. Such a record cannot give information about any of the individual images. On the other hand, several images can be recorded on a hologram. Therefore the information holding capacity of a hologram is extremely high. While a 6 × 9 mm photograph can hold one printed page, a hologram of the same size can store up to 300 such pages.

(*iii*) On a hologram information is recorded in the form of interference pattern. The type of the pattern obtained depends on the reference beam used to record the hologram. The information can be decoded only by a coherent wave identical to that of the reference wave. The reference wave can be chosen appropriately. Consequently without the knowledge of the shape of the reference wave front the information encoded in the form of interference pattern on the hologram cannot be deciphered.

(*iv*) The reconstruction of the image of the hologram can be done with reference beam of any wavelength if it is coherent and identical to the original reference beam. If the wavelength λ of the reconstructing beam is greater than that λ_o of the reference beam, the reconstructed image will be a magnified image. The magnification will be proportional to the ratio of the two wavelengths.

46.8 CLASSIFICATION OF HOLOGRAMS

Holograms may be classified in a number of different ways depending on their thickness, method of recording, method of reconstruction etc.

(*i*) **Thin Holograms or Plane Holograms:** Holograms may be thin (plane) or thick (volume) (see Fig. 46.8). A hologram may be regarded as *thin* if its emulsion thickness is much less than the fringe spacings. Otherwise it is called a *thick hologram*. Thick hologram is also known as *volume hologram*.

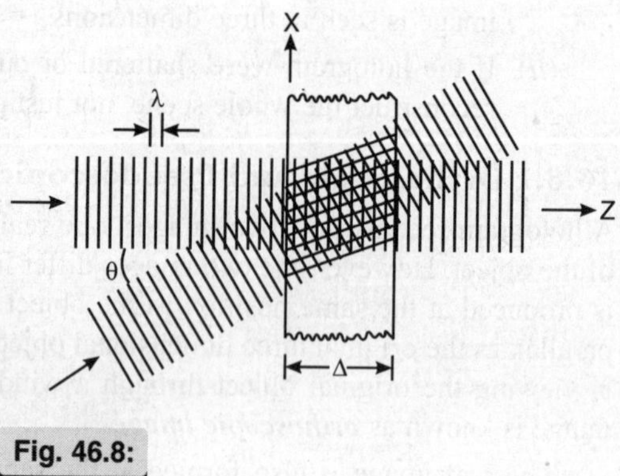

Fig. 46.8: Thick hologram

Thin holograms produce several orders (*i*) zero order which is the directly transmitted reference beam, (*ii*) the first order diffraction producing virtual image, (*iii*) the minus first order diffraction equal in intensity to the first order producing the conjugate image and (*iv*) higher orders of decreasing intensity.

(*ii*) **Volume hologram:** In 1962 Yuri Denisyuk used a process similar to Lippmann colour photography. In this method the object wave is reflected from the object and propagates backward and overlaps the incoming reference wave. The two waves form standing wave pattern. The fringes are recorded by the photo-emulsion throughout its entire thickness to form a ***volume hologram***. The hologram may be regarded as a three-dimensional grating. In the volume holograms there is an interdependence of the wavelength and the scattering angle, because the scattering follows Bragg's law $2d \sin\theta = m\lambda$. Therefore, by successively changing the incident angle or wavelength, a number of holograms can be stored in the medium. Different and mutually incoherent laser beams may be used to produce different component holograms of the object and when they are illuminated, a multicolored image is seen. A consequence of Bragg condition is that the volume hologram reconstructs the virtual image at the original position of the object if the reconstruction beam exactly coincides with the reference beam. However, the conjugate image and higher order diffractions are absent.

(*iii*) **Amplitude and Phase Holograms:** Holograms recorded in photographic emulsions change both the amplitude and the phase of the illuminating wave. The shape of the recorded fringe planes depend on the relative phase of the interfering beams. Consequently, the reconstructed wave is reflected from the hologram according to the density of the silver deposited with the amplitude variation proportional to the amplitude of the object wave. Similarly the phase of the reconstruction wave is modulated in proportional to the phase of the object wave. Thus both amplitude and phase of the object wave are reproduced. An absorption type hologram produces a change in the amplitude of the reconstruction beam. The phase type hologram produces phase changes in the reconstruction beam due to a variation in the refractive index or thickness of the medium. Phase holograms have the advantage over amplitude holograms that no energy is dissipated within the hologram medium and have higher diffraction efficiency.

(*iv*) **Transmission Holograms:** For preparing a transmitting hologram, both the reference beam and object beam are made incident from the same side of the recording medium, as shown in Fig. 46.9 (a). These two beams form interference pattern and the spacing between the maxima is given by

$$\Lambda = \frac{\lambda}{2n\sin\alpha} \quad ...(46.9)$$

where n is the index of refraction of the recording medium and α is the angle shown in Fig. 46.9 (*a*).

In the reconstruction process, the interference fringes act like reflective layers to the incident reference wave. These reflecting layers are perpendicular to the plane of the hologram. The incident reconstruction wave is reflected toward the other side of the hologram. The reconstructed beam appears as if the beam is transmitted through the hologram.

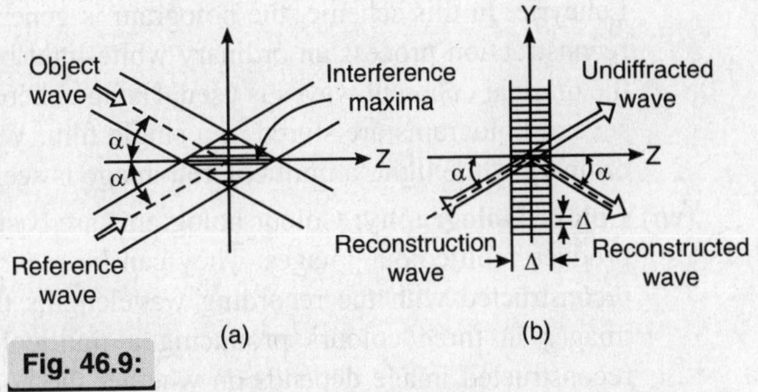

Fig. 46.9: Transmission Hologram (a) Generation (b) viewing.

(*v*) **Reflection Holograms:** The distinction between transmitting and reflection holograms is due to the angle between the reference and object beams. In case of reflection hologram, the object beam is introduced at almost 180° with respect to the reference beam. Further, reflection holograms are often thicker than transmission holograms. There is more physical space for recording interference fringes. The interference pattern consists of vertical strips at a spacing given by

$$\Lambda = \frac{\lambda}{2\,n\cos\alpha} \quad ...(46.10)$$

The reflecting layers formed in the recording medium are parallel to the surface of the hologram. One can think of holograms that are made this way as having multiple **layers** that are only about half a wavelength deep. In the reconstruction process, some of the incident light reflects back toward the light source, and some continues to the next layer, where the process repeats. The light from each layer interferes with the light in the layers above it. This is known as the ***Bragg effect***, and it is a necessary part of the reconstruction of the object beam in reflection holograms. The reconstructed wave is on the same side of the hologram and thus appears as if light is reflected from the hologram.

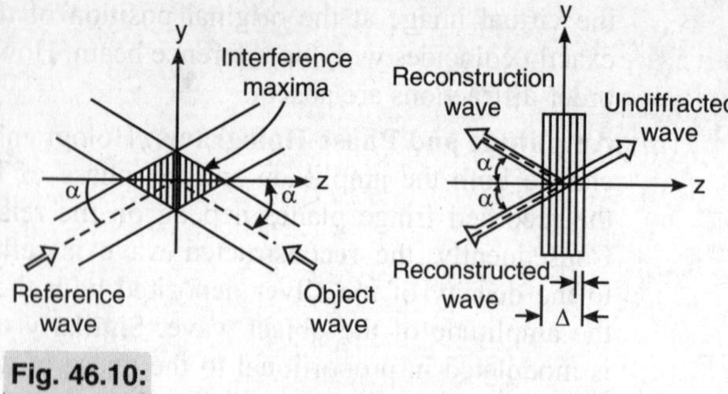

Fig. 46.10: Reflection Holograms (a) Generation (b) viewing.

The Bragg effect can also change the way the hologram reflects light, especially in holograms that one can view in white light. At different viewing angles, the Bragg effect can be different for different wavelengths of light. This means that you might see the hologram as one color from one angle and another color from another angle. The Bragg effect is also one of the reasons why most novelty holograms appear green even though they were created with a red laser.

Reflection holograms are more useful since a laser is not required for reconstruction. They can be viewed in white light.

(*vi*) **White-Light Reflection Holography:** This method is developed by Stroke and Labeyrie. In this scheme, the hologram is generated using coherent light but in the reconstruction process an ordinary white-light beam having a wave front similar to the original coherent waves is used. Using coherent sources at different wavelengths, several holograms are stored in a single film. When the hologram is illuminated by ordinary white light, a multicolored image is seen in reflection.

(*vii*) **Colour Holography:** Colour holograms are basically multiplexed holograms which produce multicolour images. They can be recorded with three wavelengths. When reconstructed with the recording wavelengths the hologram produces overlapping images in three colours producing a multicolour image. The behaviour of the reconstructed image depends on whether the hologram has been recorded in a thin medium or in a thick medium. Colour holograms recorded in a thin recording medium suffer from cross-talk. Volume holograms effectively eliminate cross-talk images utilizing Bragg effect. Both transmission and reflection volume colour holograms can be recorded in thick media. The transmission volume holograms are reconstructed with the laser beams used to record it, while the volume colour reflection holograms are reconstructed with white light due to their inherent wavelength discrimination ability.

(*viii*) **Rainbow Holograms:** Rainbow hologram is a new type of transmission hologram capable of reconstructing a bright, sharp and monochromatic image when viewed in white light. They are made by a double holographic process where an ordinary hologram such as a transmission hologram is used as the object and a second hologram is made through a slit. A horizontal slit limits the vertical perspective of the first image so that there is no vertical parallax in the resultant rainbow hologram.

This slit process removes the coherence requirement on the viewing light so that full advantage can be taken of the image brightness obtained from ordinary room light, while maintaining the three-dimensional character of the image. A true-color hologram image can be observed when the hologram is viewed in the correct plane. If the viewer moves off this plane, different shades of color can still be seen, but the color will be different from that of the original object. The steps involved in making a rainbow hologram are shown in Fig. 46.11.

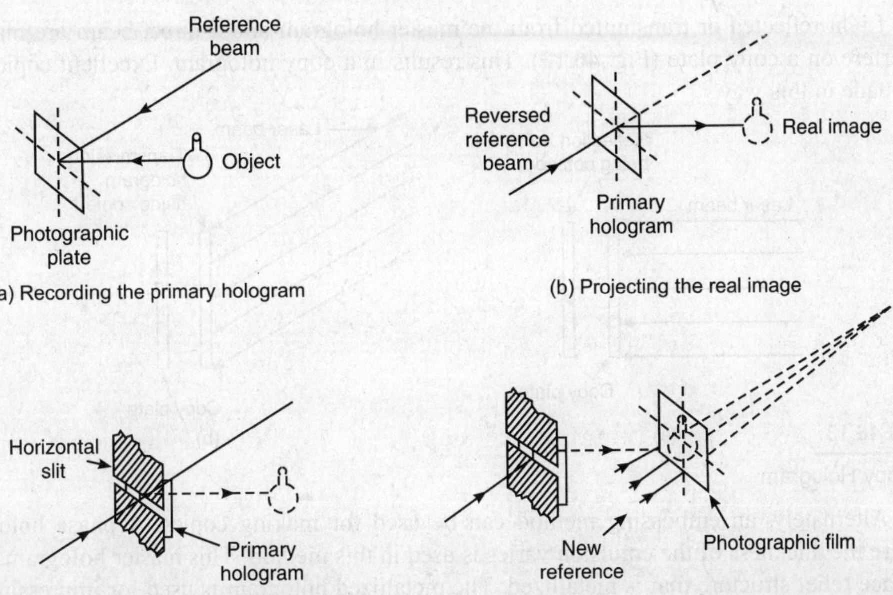

Fig. 46.11: Steps involved in making a rainbow hologram

(*a*) First a conventional transmission hologram of the object is made.

(*b*) The hologram is then illuminated by the conjugate of the original reference beam. It generates a diffracted wave, which is the real image of the object.

(*c*) A horizontal slit is placed over the primary hologram to eliminate vertical parallax.

(*d*) A second and final hologram is recorded which is used for reconstructing the object.

(*ix*) **360° Holograms:** This type of hologram is made on a 360° circular film. A photographic film is mounted on a cylindrical surface surrounding the object to be holographed. A divergent laser beam is made to

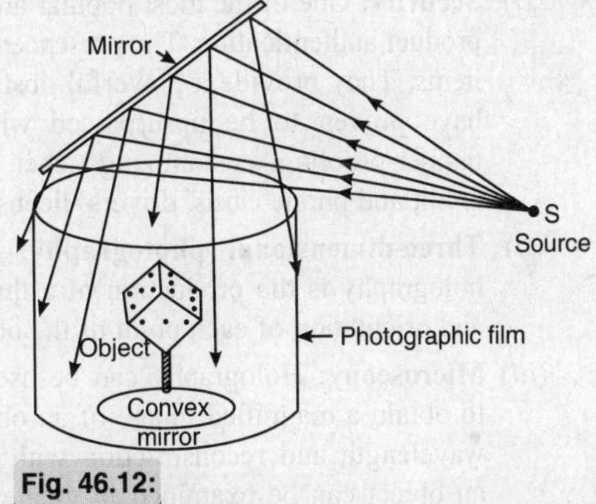

Fig. 46.12: 360° hologram

be incident on the object from the top (Fig. 46.12). A convex mirror at the bottom illuminates the object.

When the cylindrical hologram is illuminated, the virtual image is observed at the centre of the cylinder and can be viewed from all sides.

(*x*) **Copying Holograms:** Many of the commercial products bear a holographic logo and trade mark as a mark of the identity and authenticity of the manufacturer. It requires bulk production of holograms. Large number of copies of the original master hologram can be made using different ways.

Light reflected or transmitted from the master hologram and a direct beam are made to interfere on a copy plate (Fig. 46.13). This results in a copy hologram. Excellent copies can be made in this way.

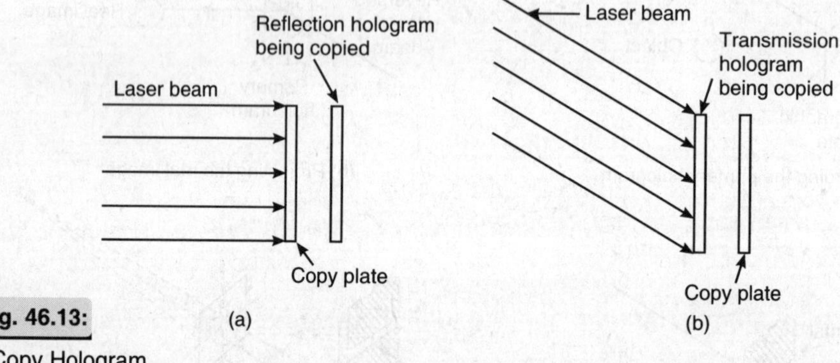

Fig. 46.13: Copy Hologram

Alternately, an embossing method can be used for making copies. A phase hologram where the thickness of the emulsion varies is used in this method. This master hologram has a surface relief structure that is metallized. The metallized hologram is used for impressing the pattern onto thin sheets of plastic.

46.9 APPLICATIONS

Holography can be used for a broad range of applications in different fields. Although holography can solve many problems, it still is a relatively expensive procedure. It is not possible to describe all of them here. Only some typical applications are discussed here.

(*i*) **Security:** One of the most popular areas for the use of holograms is the security and product authentication. The presence of holograms indicates the authenticity of these items. They provide a powerful obstacle to counterfeiting. The security holograms have proven to be unsurpassed when added to documents, anti-counterfeiting, tamper-proofing, customizing ticket protection, identification documents including credit and phone cards, drivers' licenses etc.

(*ii*) **Three-dimensional photography:** One of the most obvious applications of holography is the production of a three-dimensional photograph, with the distance and orientation of each point of the object recorded in the image.

(*iii*) **Microscopy:** Holography can be used in techniques of microscopy. It is possible to obtain a magnified image of an object if recording is done with light of smaller wavelength and reconstruction with light of longer wavelength. Smaller areas in an object can be examined in greater detail. This has great potential in observing micro-objects such as blood cells, amoebas, cancer affected tissues etc.

(*iv*) **Character recognition:** Holography can also be used for character recognition. The complicated wave front from an object is generated from a hologram by the simple wave front of the reference beam. The process is reversible so that reference wave can be generated by the object wave. This principle forms the basis of holographic pattern recognition. This could be used to identify fingerprints etc.

(*v*) **Data storage:** Holograms can also be used for data storage devices and hence are of much use in computer technology. A large amount of information such as 1012 letters/digits can be stored in a cubic cm of a volume hologram. These memories have long lifetime because a small mechanical damage to the portion of a hologram will not erase the stored information.

(*vi*) **Photolithography:** Holography is used in the production of photographic masks used to produce microelectronic circuits.

(*vii*) **Holographic projection** is used to display flight information at the pilot's eye level in an airplane cockpit.

(*viii*) **Holographic interferometry:** One of the most important applications of holography has been in interferometry.

Holographic interferometry *is an optical technique to visualize in dark environment small deformations (200 nm to 100 μm) of objects. It is applied to objects, which are placed in a vibration-free set-up.*

Holographic interferometry is used in vibrational analysis, structural analysis, stress and strain evaluation etc. There are three basic methods of holographic interferometry. They are known as real time, multi-exposure and time-average holography. Nowadays, these techniques are widely used in nondestructive testing.

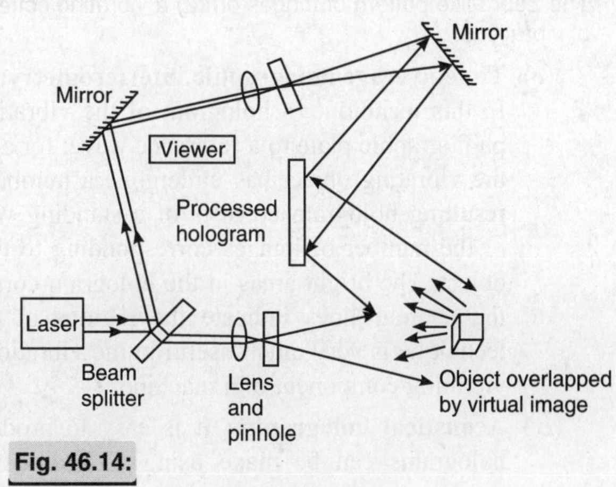

Fig. 46.14: Real time holographic interferometry

(*a*) **Real time holography:** It allows one to observe instantaneously the effects of minute changes in an object as some stress affects it. In this method, first the hologram of the undisturbed objected is obtained. This hologram of the object is superimposed over the object subjected to some small stress (Fig. 46.14). The distortions that appear as a fringe pattern are analysed.

(*b*) **Multi-exposure holography:** It creates a hologram by using two or sometimes more exposures. The first exposure shows the object in an undisturbed state. Subsequent exposures are made on the same image while the object is subjected to some stress. The resulting image depicts the difference between the two states.

The technique where two exposures are made is known as ***double exposure holographic interferometry***. In the technique, a hologram of the undisturbed object is first recorded on the photographic plate with an exposure to a reference wave. Then, before the hologram is removed or developed, the object is stressed and is recorded on the same photographic plate through a second exposure along with the same reference wave. After this double exposure, the hologram is developed. If the hologram is now illuminated by a reconstruction wave, there would emerge from the hologram two object waves - one corresponding to the unstressed object and the other corresponding to the stressed object. These two object waves interfere to produce interference pattern. Thus on viewing through the hologram, the object as well as the surface covered by the interference fringes caused by surface irregularities can be seen. These fringes reveal the distribution of strain in the object.

(a)

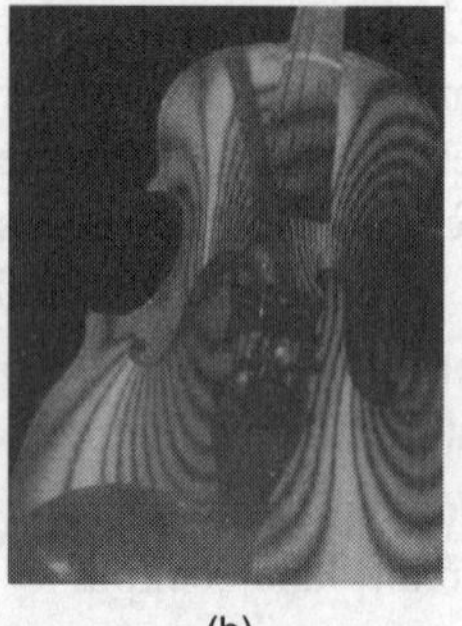

(b)

Fig. 46.15: The Zebra-like pattern of fringes on (a) a vibrating coffee cup (b) a violin shows the nodal pattern of vibrations

(*c*) **Time-average holographic interferometry:** It is used in studying vibrating objects. In this technique, a hologram of the vibrating object is prepared by exposing the photographic plate to a reference wave, for a relatively long period of time such that the vibrating object has undergone a number of oscillations during that time. The resulting hologram consists of a standing wave pattern caused by a superposition of the number of images corresponding to the successive states of vibration of the object. The bright areas in the hologram correspond to undeflected regions whereas the contour lines indicate the regions of constant amplitude of vibration. This technique is very much useful in the vibrational analysis of any vibrating system or vibrating component of a machine.

(*ix*) **Acoustical holography:** It is easy to produce coherent sound waves. Therefore, holograms can be made using ultrasonic waves initially and then visible light can be used for reconstruction of the visual image. Light waves cannot propagate considerable distances in dense liquids and solids whereas sound waves can propagate through them. Therefore, a three dimensional acoustical hologram of an opaque object can be made. By viewing such hologram in visible light the internal structure of the object can be observed. Such techniques will be highly useful in the fields of medicine and technology. In one of the techniques, two submerged coherent sound wave generators emit the reference and the signal waves scattered by an object (Fig. 46.16). On a calm surface of water, these two contributions produce ripples. The ripple

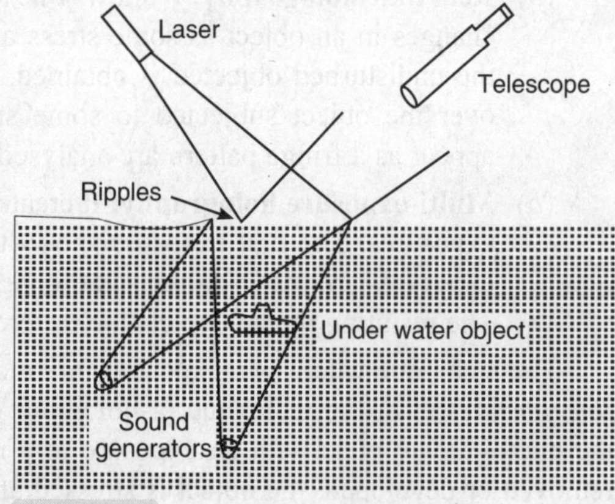

Fig. 46.16: Acoustical holography

pattern is the hologram. The pattern is photographed and reconstructed as and when required. As sound waves can propagate through dense liquids and solids, acoustical holography has an advantage in locating underwater submarines etc and study of internal body organs.

(*x*) **Holographic optical elements:** Traditional optical elements operate on the principle of refraction. A diffractive optical element (DOE) operates on the principle of diffraction. DOEs can be used as a grating, lens or any other type of optical elements. Large optical apertures, lightweight and lower cost are the main features of DOEs. They can offer unique optical properties that are not possible with conventional optical elements. They can be fabricated in a wide range of materials namely, aluminum, silicon, silica, plastics, etc providing the user greater flexibility in selecting the material for a particular application. DOEs can be used to perform more specialized functions like, making the panel instruments of a car visible in the windshield for increased safety.

Some of the important features of DOEs are as follows.

(*a*) Several different optical elements can share the same substrate without interfering with one another. Thus, a single DOE can be used as a lens, beam splitter and spectral filter simultaneously.

(*b*) Diffractive elements are very light, as they are formed in thin films of a few μm thickness only.

(*c*) Because DOEs can generate unique optical functions that are not possible by conventional reflective or refractive optical elements, they provide greater flexibility in system configuration.

(*d*) At least one surface of a conventional glass lens is curved, whereas for a diffractive lens there is no such requirement. A diffractive element can be fabricated on any arbitrary shape of the substrate.

(*e*) They can be made to operate over a narrow wavelength band.

(*f*) The fabrication and replication of DOEs are relatively easy and cheap because no precision shaping of a surface is required.

46.10 MEDICAL APPLICATIONS OF HOLOGRAPHY

Holographic technique is also used in various medical applications like opthalmology, endoscopy, otology, orthopedics and many more. Recent improvements in hologram recording techniques and the availability of tools for the interpretation of holographic interferograms and the success of holographic techniques in imaging through tissues, ophthalmology, dentistry, urology, otology, pathology, and orthopedics shows that holography may emerge as a powerful tool for medical applications. We discuss here some of the applications.

(*i*) **Holographic Endoscope:** Endoscopic holography provides a powerful tool for non-contact, high resolution, 3D imaging and nondestructive measurements inside natural cavities of human body. It combines the features of holography and endoscopy. The holographic endoscopy is of two types. In one of the forms, the hologram is recorded inside the endoscope, while the other form uses an external recording device.

(*a*) Internal Hologram Recording Endoscope: The endoscope accommodates a miniaturized holographic setup inside the instrument and records a reflection hologram. It mainly consists of three parts; a film cartridge, a diaphragm and a single mode optical fibre (core diameter 4 μm) cable. The three parts are assembled in three adjustable stainless steel tubes. The film is placed normal to the endoscope. The holograms are viewed under a powerful microscope allowing for the observation of individual cells. Due to large hologram aperture, the image with a low speckle noise and high lateral resolution is obtained. A lateral resolution of 7 μm has been obtained in the reconstructed image that shows that the technique can be used for cellular structure analysis and may even substitute biopsy in tumour diagnosis. Specific dyes can be used to enhance the contrast of the tissue before recording the holograms as has been used extensively in gynecology and gastrointestinal tract.

(*b*) **External Hologram recording Endoscope:** In the external hologram recording endoscope, a conventional endoscope is used. The system records the hologram outside the endoscope using an external reference beam. An endoscope with extremely small outer diameter can be used. In order to obtain a high signal-to-noise ratio, the holographic endoscope must use gradient-index (GRIN) rod lenses. The speckle noise is reduced by illuminating and imaging the object by the same GRIN lens. An electro-optic crystal can be used as the photographic storage device in the holographic endoscope to provide in-situ recording, reconstruction, and erasure.

Holographic endoscope has been used with success for early recognition of cancerous indurations in the wall of urinary bladder.

(*ii*) **Holography in Ophthalmology:** Recording of a three dimensional image of the eye was one of the earliest applications of holography in the field of ophthalmology. Any retinal detachment or intraocular foreign body can be detected. Holography can also be applied for the measurement of corneal topography and crystalline lens changes and for the study of surface characteristics of both the nerve head and the cornea. Current methods of determining the shape of the central surface miss the central part and its periphery. The major advantage of holographic technique is the ultra high precision (sub-μm range) with which such measurements are possible. The elastic expansion of the cornea can also be measured by holographic interferometry. This information is vital for corneal surgery. The studies made so far show that holography has potential to investigate changes on the cornea, crystalline lens changes, and surface characteristics of both the nerve head and the retina.

(*iii*) **Diffractive bifocal Intraocular Lens:** A very useful application of diffractive optics is in the correction of refractive errors for old persons who have been operated for cataract by the use of a bifocal intraocular lens. Such persons have difficulty in changing the focus of their eyes for near distant and far distant objects. Bifocal lenses are implanted in place of the natural eye lenses. The bifocal lens is a combination of a conventional refractive lens and a diffractive lens, the former focussed to infinity and the later for near distance vision. The efficiency of the diffractive lens is set at 50%, thus both the near and the far foci are accommodated over the whole visual field. The diffractive lens is fabricated on the rear of the conventional lens. When the eyes are focussed for a distant object, a blurred image is superimposed due to the presence of diffractive lens and vice versa, which obviously reduces the image quality. In most cases, the blurred image is discarded by the human visual perception and retinal processing system.

(*iv*) **Holography in Orthopedics:** Holography offers an excellent tool for the contactless study of orthopedic structures, specifically external fixtures to measure strains on fixation pins and rods. Such studies are important in osteo-synthesis with external fixture used for long bone fractures, to prevent dislocations of both fractured ends that are mainly caused by decrease in strength of the fixation pins.

QUESTIONS

1. What is meant by holography? Why is it called wave front reconstruction?
2. Describe a how hologram is generated and image is reconstructed using off-axis configuration?
3. What is a hologram? How does it differ from an ordinary photograph? Describe in short how a hologram is generated and viewed.
4. Describe the recording and reconstruction processes in Holography with the help of suitable diagrams.
5. Write the principle of holography. With neat sketches explain in brief recording of a hologram and reconstruction of images.
6. What is holography? How does it differ from ordinary photographic technique? How do interference and diffraction phenomena related to construction reconstruction of the hologram? Explain in detail with suitable diagram.
7. What is the advantage of off-axis configuration over the coaxial configuration?
8. Explain the principle of holography?
9. Explain some of the important properties of hologram?
10. How are the holograms classified? Explain.
11. Write the differences between holography and photography?
12. Discuss some of the important applications of holography?
13. Explain holographic interferometry.
14. Explain acoustic holography.
15. Describe some of the important medical applications of holography.
16. What are DOEs (diffractive optical elements)? Describe their salient features and applications.

47 Optical Fibres

Learning Objectives:

By the end of this chapter, you will be able to:

- Learn about basic concepts of Optical Fibre.
- Discuss about different aspects related with Optical Fibre –Numerical Aperture, Skip Distance and Modes of Propagation.
- Identify different types of Rays.
- Discuss about different types of Optical Fibre.
- Learn about different types of Losses in Optical Fibre.
- Elaborate about Fibre Optic Communication System.

47.1 INTRODUCTION

In 1870 John Tyndall, a British physicist demonstrated that light can be guided along the curve of a stream of water. Owing to total internal reflections light gets confined to the water stream and the stream appears luminous. A luminous water stream is the precursor of an optical fibre. In the 1950's, the transmission of images through optical fibres was realized in practice. Hopkins and Kapany developed the flexible fibrescope, which was used by the medical world in remote illumination and viewing the interior of human body. It was Kapany who coined the term fibre optics. By 1960, it had been established that light could be guided by a glass fibre. In 1966 Charles Kao and George Hockham proposed the transmission of information over glass fibre, but the fibres available at that time heavily attenuated light propagating through them. In 1970 Corning Glass Works produced low-loss glass fibres.

The invention of solid state lasers in 1970 made optical communications practicable. Commercial communication systems based on optical fibres made their appearance by 1977. Apart from the use as communicational channel, optical fibres are widely used in other areas. Fibro-scopes made of optical fibres are widely used in a variety of forms in medical diagnostics. Sensors for detecting electrical, mechanical, thermal energies are made using optical fibres.

47.2 OPTICAL FIBRE

An optical fibre is a cylindrical wave guide made of transparent dielectric, (glass or clear plastic), which guides light waves along its length by total internal reflection.

It is as thin as human hair, approximately 70 μm or 0.003 inch diameter. Note that a thin strand of a metal is called a *wire* and a thin strand of dielectric materials is called a *fibre*.

Fibre optics is a technology in which signals are converted from electrical into optical signals, transmitted through a thin glass fibre and reconverted into electrical signals.

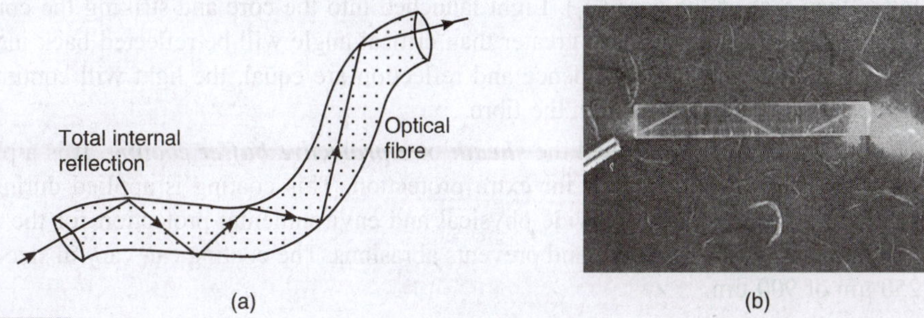

Fig. 47.1: Illustration of a transparent fibre guiding light along its length by total internal reflection

Principle of optical fibre: The *propagation* of light in an optical fibre from one of its ends to the other end is based on the principle of *total internal reflection*. When light enters one end of the fibre, it undergoes successive total internal reflections from sidewalls and travels down the length of the fibre along a zigzag path, as shown in Fig. 47.1 (a). A small fraction of light may escape through sidewalls but a major fraction emerges out from the exit end of the fibre, as shown in Fig. 47.1 (b). Light can travel through fibre even if it is bending (Fig. 47.1c).

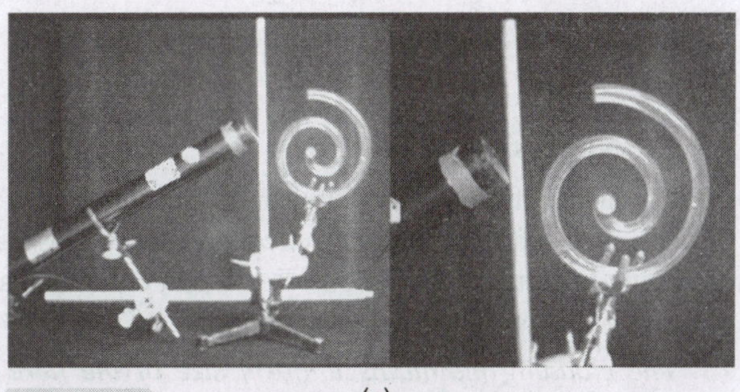

Fig. 47.1: A laser beam repeatedly bounces off the surface of the rod by total internal reflection as it makes its way around the plastic spiral and emerges out at the other end. The whole coil appears to glow red due to the scattering of light by the plastic. However, the path of light is seen clearly in the coil.

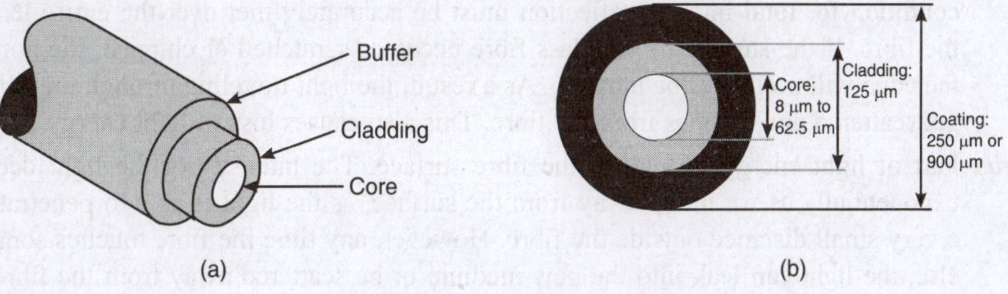

Fig. 47.2: Illustration of Side view and cross sectional view of an optical fibre

Structure: A practical optical fibre is cylindrical in shape (Fig. 47.2a) and has in general three coaxial regions (Fig. 47.2b).

(*i*) The innermost cylindrical region is the light guiding region known as the **core**. In general, the diameter of the core is of the order of 8.5 μm to 62.5 μm.

(*ii*) It is surrounded by a coaxial middle region known as the **cladding**. The diameter of the cladding is of the order of 125 μm. The refractive index of cladding (n_2) is always lower than that of the core (n_1). Light launched into the core and striking the core-to-cladding interface at an angle greater than critical angle will be reflected back into the core. Since the angles of incidence and reflection are equal, the light will continue to rebound and propagate through the fibre.

(*iii*) The outermost region is called the ***sheath*** or a ***protective buffer coating***. It is a plastic coating given to the cladding for extra protection. This coating is applied during the manufacturing process to provide physical and environmental protection for the fiber. The buffer is elastic in nature and prevents abrasions. The coating can vary in size from 250 μm or 900 μm.

To sum up

(*a*) Core is the inner light-carrying member.

(*b*) Cladding is the middle layer, which serves to confine the light to the core.

(*c*) Buffer coating surrounds the cladding, which protects the fibre from physical damage and environmental effects.

47.2.1 Necessity of cladding

The actual fibre is very thin and light entering a bare fibre will travel along the fibre through repeated total internal reflections at the glass-air boundary. However, bare fibres are used only in certain applications. For use in communications and some other applications, the optical fibre is provided with a cladding.

The cladding maintains uniform size of the fibre, protects the walls of the fibre from chipping, and reduces the size of the cone of light that will be trapped in the fibre.

Reasons for the necessity of cladding are:

(*i*) It is necessary that the diameter of an optical fibre remains constant throughout its length and is surrounded by the same medium. Any change in the thickness of the fibre or the medium outside the fibre (when the fibre gets wet due to moisture etc) will cause loss of light energy through the walls of the fibre.

(*ii*) A very large number of reflections occur through the fibre and it is necessary that the condition for total internal reflection must be accurately met over the entire length of the fibre. If the surface of the glass fibre becomes scratched or chipped, the normal to the edge will no longer be uniform. As a result, the light traveling through the fibre will get scattered and escapes from the fibre. This also causes loss of light energy.

(*iii*) Part of light energy penetrates the fibre surface. The intensity of the light decreases exponentially as we move away from the surface, as the light is able to penetrate only a very small distance outside the fibre. However, any time the fibre touches something else, the light can leak into the new medium or be scattered away from the fibre. This effect causes a significant leakage of the light energy out of the fibre. Even a small amount of dust on the surface would cause a fair amount of leakage.

(*iv*) If bare optic fibres are packed closely together in a bundle, light energy traveling through the individual fibres tends to get coupled through the phenomenon of *frustrated total internal reflection*. Cladding of sufficient thickness prevents the leakage of light energy from one fibre to the other.

The fiber is provided with a cladding in order to prevent loss of light energy due to the above reasons.

(*v*) The cladding causes a reduction in the size of the cone of light that can be trapped in the fibre. Light entering the fibre at larger angles will strike the fibre walls at smaller angles (higher modes) and ultimately travel a longer distance. Such higher modes of a light signal will take longer time to reach the end of the fibre than the lower modes. Therefore, a pulse sent through optical fibre spreads out. The spreading would be larger, the larger the cone of acceptance. Such pulse spreading limits the rate of data transmission through the fibre. As fibers with a cladding have smaller cone of acceptance, they carry information at a much higher bit rate than those without a cladding.

Thus, the cladding performs the following important functions:

(*a*) Keeps the size of the fibre constant and reduces loss of light from the core into the surrounding air.

(*b*) Protects the fiber from physical damage and absorbing surface contaminants.

(*c*) Prevents leakage of light energy from the fibre through evanescent waves.

(*d*) Prevents leakage of light energy from the core through frustrated total internal reflection.

(*e*) Reduces the cone of acceptance and increases the rate of transmission of data.

(*f*) A solid cladding, instead of air, also makes it easier to add other protective layers over the fibre.

47.2.2 Optical Fibre System

An optical fibre is used to transmit ***light signals*** over long distances. It is essentially a ***light-transmitting medium***, its role being very much similar to a coaxial cable or wave-guide used in microwave communications. Optical fibre requires a ***light source*** for launching light into the fibre at its input end and a ***photodetector*** to receive light at its output end. As the diameter of the fibre is very small, the light source has to be dimensionally compatible with the fibre core. Light emitting diodes and laser diodes, which are very small in size, serve as the light sources. The electrical input signal is in general of digital form. It is converted into an optical signal by varying the current flowing through the light source. Hence, the intensity of the light emitted by the source is modulated with the input signal and the output will be in the form of light pulses. The light pulses constitute the signal that travels through the optical fibre. At the receiver end, semiconductor photodiodes, which are very small in size, are used for detection of these light pulses. The photodetector converts the optical signal into electrical form. Thus, a basic *optical fibre system* consists of a LED/laser diode, optical fibre cable and a semiconductor photodiode.

47.2.3 Optical Fibre Cable

Optical fibre cables are designed in different ways to serve different applications. More protection is provided to the optical fibre by the "cable" which has the fibres and strength members inside an outer covering called a "jacket".

We study here two typical designs: a single fibre cable and a multifibre cable.

(*i*) **Single Fibre Cable:** Around the fibre a tight buffer jacket of Hytrel is used (Fig. 47.3). The buffer jacket protects the fibre from moisture and abrasion. A strength member is arranged around the buffer jacket in order to provide the necessary toughness and tensile strength. The strength member may be a steel wire, polymer film, nylon yarn or Kevlar yarn. Finally, the fibre cable is covered by a Hytrel outer jacket. Because of this arrangement fibre cable will not get damaged during bending, rolling, stretching or pulling and transport and installation processes. The single fibre cable is used for indoor applications.

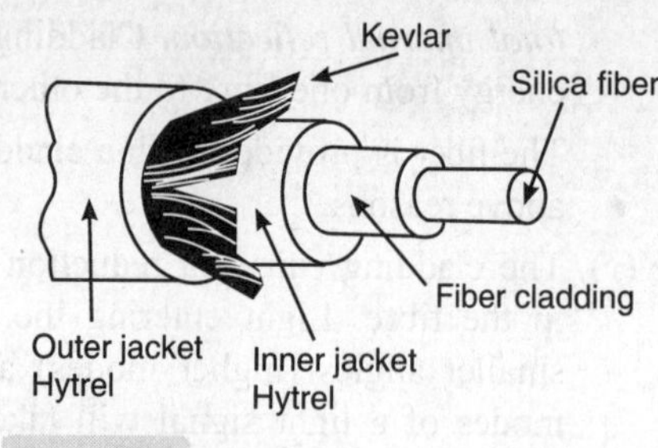

Fig. 47.3: Single fibre cable

(*ii*) **Multifibre Cable:** A multifibre cable consists of a number of fibres in a single jacket. Each fibre carries light independently. The cross-sectional view of a typical telecommunication cable is shown in Fig. 47.4. Usually, optical fibre contains six insulated optical fibre strands along with an insulated steel cable at the center for giving tensile strength. Optical fibre strand comprises a core surrounded by a cladding, which in turn is coated with insulating jacket; all three are independently safeguarded and supported. Electrical transmission, if any required is carried by six insulated copper wires spread in the space between the fibres.

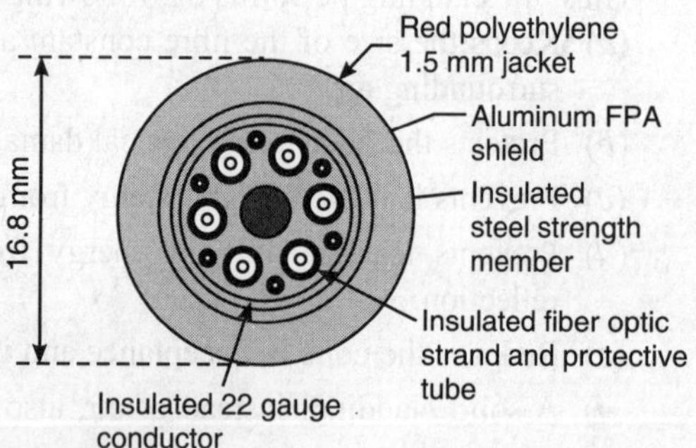

Fig. 47.4: Cross-sectional view of a typical multi fibre cable

47.3 TOTAL INTERNAL REFLECTION

A medium having a lower refractive index is said to be an optically ***rarer medium*** while a medium having a higher refractive index is known as an optically ***denser medium***. When a ray of light passes from a denser medium to a rarer medium, it is bent away from the normal in the rarer medium (see Fig. 47.5a). Snell's law for this case may be written as

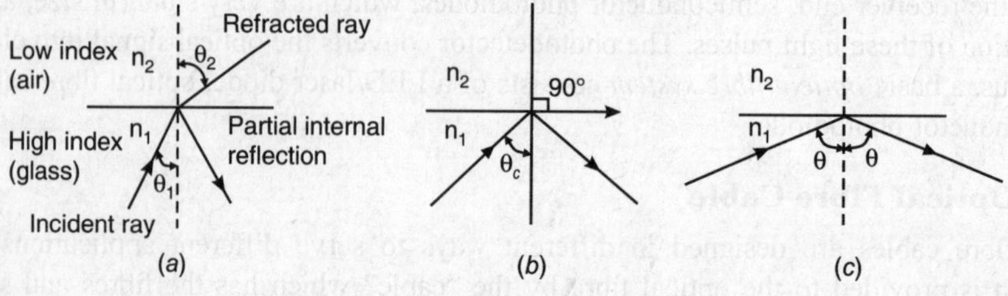

Fig. 47.5: Phenomenon of total internal reflection

$$\sin \theta_2 = \left(\frac{n_1}{n_2}\right) \sin \theta_1 \qquad ...(47.1)$$

where θ_1 is the angle of incidence of light ray in the denser medium and θ_2 is the angle of refraction in the rarer medium. Also $n_1 > n_2$.

When the angle of incidence, θ_1 in the denser medium is increased, the transmission angle, θ_2 increases and the refracted rays bend more and more away from the normal. At some particular angle θ_c the refracted ray glides along the boundary surface so that $\theta_2 = 90°$, as seen in Fig. 47.5 (b). At angles greater than θ_c there are no refracted rays at all. The rays are reflected back into the denser medium as though they encountered a specular reflecting surface (Fig. 47.5 c). Thus,

- If $\theta_1 < \theta_c$, the ray refracts into the rarer medium
- If $\theta_1 = \theta_c$, the ray just grazes the interface of rarer-to-denser media
- If $\theta_1 > \theta_c$, the ray is reflected back into the denser medium.

The phenomenon in which light is totally reflected from a denser-to-rarer medium boundary is known as ***total internal reflection***. The rays that experience total internal reflection obey the laws of reflection. Therefore, the critical angle can be determined from Snell's law.

When $\theta_1 = \theta_c$, $\theta_2 = 90°$.

Therefore, from Eq. (47.1), we get

$$n_1 \sin \theta_c = n_2 \sin 90° = n_2$$

$$\therefore \quad \sin \theta_c = \frac{n_2}{n_1} \qquad ...(47.2)$$

When the rarer medium is air, $n_2 = 1$ and writing $n_1 = n$, we obtain

$$\sin \theta_c = \frac{1}{n} \qquad ...(47.3)$$

47.4 PROPAGATION OF LIGHT THROUGH AN OPTICAL FIBRE

The diameter of an optical fibre is very small and as such we cannot use bigger light sources for launching light beam into it. Light emitting diodes (LEDs) and laser diodes are the optical sources used in fibre optics. Even in case of these small sized sources, a focusing lens has to be used to concentrate the beam on to the fibre core. Light propagates as an electromagnetic wave through an optical fibre. However, light propagation through an optical fibre can as well be understood on the basis of *ray model*. According to the ray model, light rays entering the fibre strike the core-clad interface at different angles. As the refractive index of the cladding is less than that of the core, majority of the rays undergo total internal reflection at the interface and the angle of reflection is equal to the angle of incidence in each case.

Due to the cylindrical symmetry in the fibre structure, the rays reflected from an interface on one side of the fibre axis will suffer total internal reflections at the interface on the opposite side also. Thus, the rays travel forward through the fibre via a

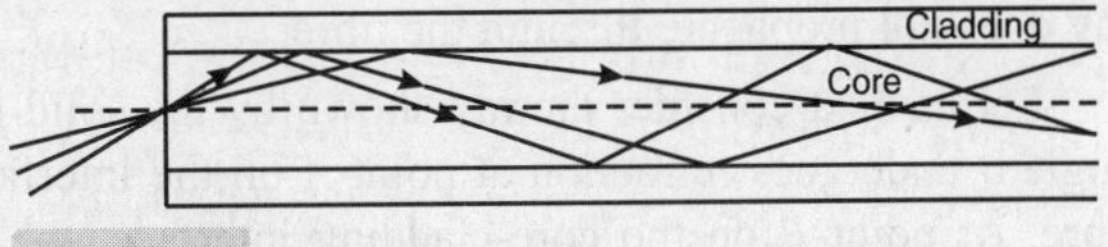

Fig. 47.6: Propagation of light rays through an optical fibre due to total internal reflection.

series of total internal reflections and emerge out from the exit end of the fibre (Fig. 47.6). Since each reflection is a total internal reflection, there is no loss of light energy and light confines itself within the core during the course of propagation. Because of the negligible loss during the total internal reflections, optical fibre can carry the light waves over very long distances. Thus, the optical fibre acts essentially as a wave-guide and is often called a ***light guide*** or ***light pipe***. At the exit end of the fibre, the light is received by a photo-detector.

Total internal reflection at the fibre wall can occur and light propagates down the fibre, only if the following two conditions are satisfied

(*a*) The refractive index of the core material, n_1, must be slightly greater than that of the cladding, n_2.

(*b*) At the core-cladding interface (Fig. 47.7), the angle of incidence ϕ between the ray and the normal to the interface must be greater than the critical angle ϕ_c defined by

$$\sin \phi_c = \frac{n_2}{n_1} \qquad \text{...(47.4)}$$

It is to be noted here that only those rays, that are incident at the core-clad interface at angles greater than the critical angle will propagate through the fibre. Rays that are incident at smaller angles are refracted into the cladding and are lost.

47.4.1 Critical Angle of Propagation

Let us consider a step index optical fibre into which light is launched at one end. The end at which light enters the fibre is called the **launching end.** Fig. 47.7 depicts the conditions at the launching end. In a step-index fibre, the refractive index changes abruptly form the core to the cladding. Now, we consider two rays entering the fibre at two different angles of incidence

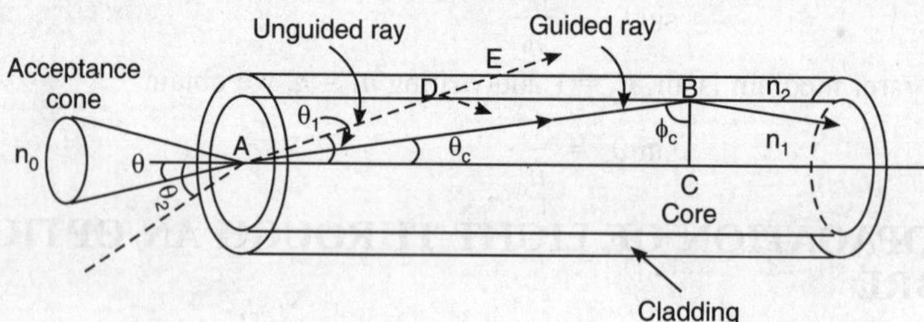

Fig. 47.7: Light rays incident at an angle smaller than critical propagation angle will propagate through the fibre

The ray shown by the broken line is incident at an angle θ_2 with respect to the axis of the fibre. This ray undergoes refraction at point A on the interface between air and the core. The ray refracts into the fibre at an angle θ_1 ($\theta_1 < \theta_2$). The ray reaches the core-cladding interface at point *D*. At point *D*, refraction takes place again and the ray travels in the cladding. Finally, at point E, the ray refracts once again and emerges out of fibre into the air. It means that the ray does not propagate through the fibre.

Let us next consider the ray shown by the solid line in Fig. 47.7. The ray incident at an angle θ undergoes refraction at point *A* on the interface and propagates at an angle θ_c in the fibre. At point *B* on the core-cladding interface, the ray undergoes total internal reflection, since $n_1 > n_2$. Let us assume that the angle of incidence at the core-cladding interface is the *critical angle* ϕ_c, where ϕ_c is given by

$$\phi_c = \sin^{-1}(n_2/n_1) \qquad ...(47.4a)$$

A ray incident with an angle larger than ϕ_c will be confined to the fibre and propagate in the fibre. A ray incident, at the core-cladding boundary, at the critical angle is called a ***critical ray***. The critical ray makes an angle ϕ_c with axis of the fibre. It is obvious that rays with propagation angles larger than θ_c will not propagate in the fibre.

Therefore, the angle θ_c is called the ***critical propagation angle***. From the Δ^{le} *ABC*, it is seen that

$$\frac{AC}{AB} = \sin\phi_c. \quad \text{Also, } \frac{AC}{AB} = \cos\theta_c$$

From the Eq. (47.4 a), $\sin\phi_c = n_2 / n_1$.

$$\cos\theta_c = n_2 / n_1 \qquad ...(47.5)$$

$$\therefore \quad \theta_c = \cos^{-1}(n_2 / n_1) \qquad ...(47.6)$$

Thus, only those rays which are refracted into the cable at angles $\theta_r < \theta_c$ will propagate in the optical fibre.

Example 47.1: *In an optical fibre, the core material has refractive index 1.43 and refractive index of clad material is 1.4. Find the propagation angle.*

Solution: $\cos\theta_c = \dfrac{n_2}{n_1} = \dfrac{1.40}{1.43} = 0.979$

Therefore, propagation angle $\theta_c = \cos^{-1}(0.979) = 11.8°$

Example 47.2: *In an optical fibre, the core material has refractive index 1.6 and refractive index of clad material is 1.3. What is the value of critical angle? Also calculate the value of angle of acceptance cone.*

Solution: Critical angle is given by

$$\sin\phi_c = \frac{n_2}{n_1} = \frac{1.3}{1.6} = 0.8125$$

$$\therefore \quad \phi_c = 54.3°$$

Acceptance angle $\theta_0 = \sin^{-1}\left[\sqrt{n_1^2 - n_2^2}\right] = \sin^{-1}\left[\sqrt{1.6^2 - 1.3^2}\right]$

$$= \sin^{-1}(0.87)$$

$$= \mathbf{60.5°}$$

Angle of acceptance cone = $2\theta_0$ = **121°**

47.4.2 Acceptance Angle

Let us again consider a step index optical fibre into which light is launched at one end, as shown in Fig. 47.8. Let the refractive index of the core be n_1 and the refractive index of the cladding be n_2 ($n_2 < n_1$).

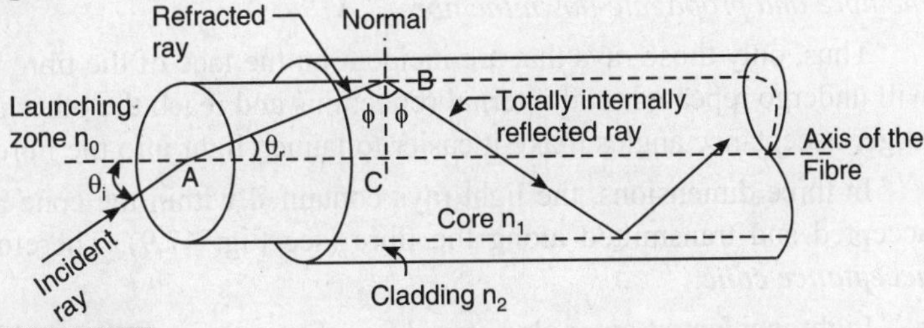

Fig. 47.8: Geometry for the calculation of acceptance angle of the fibre

Let n_0 be the refractive index of the medium from which light is launched into the fibre. Assume that a light ray enters the fibre at an angle θ_i to the axis of the fibre. The ray refracts at an angle θ_r and strikes the core-cladding interface at an angle ϕ. If ϕ is greater than critical angle ϕ_c, the ray undergoes total internal reflection at the interface, since $n_1 > n_2$. As long as the angle ϕ is greater than ϕ_c, the light will stay within the fibre.

Applying Snell's law to the launching face of the fibre, we get

$$\frac{\sin \theta_i}{\sin \theta_r} = \frac{n_1}{n_0} \quad ...(47.7)$$

If θ_i is increased beyond a limit, ϕ will drop below the critical value ϕ_c and the ray escapes from the sidewalls of the fibre. The largest value of θ_i occurs when $\phi = \phi_c$.

From the Δ^{le} ABC, it is seen that

$$\sin \theta_r = \sin (90° - \phi) = \cos \phi \quad ...(47.8)$$

Using Eq. (47.8) into Eq. (47.7), we obtain

$$\sin \theta_i = \frac{n_1}{n_0} \cos \phi$$

When $\phi = \phi_c$,

$$\sin\left[\theta_{i_{\max}}\right] = \frac{n_1}{n_0} \cos \phi_c \quad ...(47.9)$$

But

$$\sin \phi_c = \frac{n_2}{n_1}$$

$\therefore$

$$\cos \phi_c = \frac{\sqrt{n_1^2 - n_2^2}}{n_1} \quad ...(47.10)$$

Substituting the Eq. (47.10) into Eq. (47.9), we get

$$\sin [\theta_i (\max)] = \frac{\sqrt{n_1^2 - n_2^2}}{n_0} \quad ...(47.11)$$

Quite often the incident ray is launched from air medium, for which $n_0 = 1$.

Designating $\theta_i (\max) = \theta_0$, Eq. (47.11) may be simplified to

$$\sin \theta_0 = \sqrt{n_1^2 - n_2^2}$$

$\therefore$

$$\theta_0 = \sin^{-1}\left[\sqrt{n_1^2 - n_2^2}\right] \quad ..(47.12)$$

The angle θ_0 is called the ***acceptance angle*** of the fibre.

Acceptance angle is the maximum angle that a light ray can have relative to the axis of the fibre and propagate down the fibre.

Thus, only those rays that are incident on the face of the fibre making angles less than θ_0 will undergo repeated total internal reflections and reach the other end of the fibre. Obviously, larger acceptance angles make it easier to launch light into the fibre.

In three dimensions, the light rays contained within the cone having a full angle $2\theta_0$ are accepted and transmitted along the fibre (see Fig. 47.9). Therefore, the cone is called the ***acceptance cone***.

Light incident at an angle beyond θ_0 refracts through the cladding and the corresponding optical energy is lost.

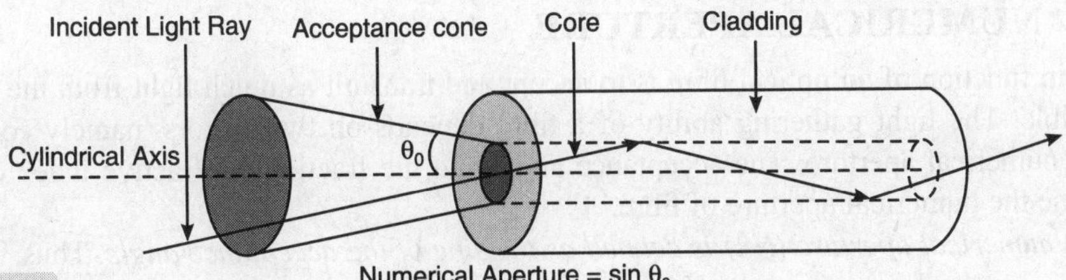

Fig. 47.9: Acceptance cone

Example 47.3: *Calculate the numerical aperture and acceptance angle of an optical fibre from the following data:*

$n_1(core) = 1.55$ and $n_2(cladding) = 1.50$

Solution: $\text{NA} = \sqrt{n_1^2 - n_2^2} = \sqrt{1.55^2 - 1.50^2} = \sqrt{0.153} = \mathbf{0.471.}$

Acceptance angle $\theta_0 = \sin^{-1}\left[\sqrt{n_1^2 - n_2^2}\right] = \sin^{-1}\left[\sqrt{1.55^2 - 1.50^2}\right] = \mathbf{23.02°}$

Example 47.4: *What is the numerical aperture of an optical fibre cable with a clad index of 1.378 and a core index of 1.546?*

Solution:

$$NA = \sqrt{n_1^2 - n_2^2} = \sqrt{1.546^2 - 1.378^2} = \sqrt{0.491} = 0.70$$

Example 47.5: *A fibre cable has an acceptance angle of 30° and a core index of refraction of 1.4. Calculate the refractive index of the cladding.*

Solution:

$$\sin\theta_0 = \sqrt{n_1^2 - n_2^2}$$

$$\therefore \quad \sin^2\theta_0 = n_1^2 - n_2^2$$

$$n_2^2 = n_1^2 - \sin^2\theta_0 = (1.4)^2 - \sin^2 30° = 1.96 - 0.25$$

$$= 1.71$$

$$\therefore \quad n_2 = \mathbf{1.308}$$

Example 47.6: *Calculate the angle of acceptance of a given optical fibre, if the refractive indices of the core and the cladding are 1.563 and 1.498 respectively.*

Solution:

$$\sin\theta_0 = \sqrt{n_1^2 - n_2^2} = \sqrt{(1.563)^2 - (1.498)^2} = 0.4461$$

$$\theta_0 = \sin^{-1}(0.4461) = \mathbf{26.49°}$$

47.5 FRACTIONAL REFRACTIVE INDEX CHANGE

The fractional difference Δ between the refractive indices of the core and the cladding is known as *fractional refractive index change*. It is expressed as

$$\Delta = \frac{n_1 - n_2}{n_1} \qquad ...(47.13)$$

This parameter is always positive because n_1 must be larger than n_2 for the total internal reflection condition. In order to guide light rays effectively through a fibre, $\Delta << 1$. Typically, Δ is of the order of 0.01.

47.6 NUMERICAL APERTURE

The main function of an optical fibre is to accept and transmit as much light from the source as possible. The light gathering ability of a fibre depends on two factors, namely core size and the numerical aperture. The acceptance angle and the fractional refractive index change determine the numerical aperture of fibre.

The numerical aperture (NA) is defined as the sine of the acceptance angle. Thus,

$$NA = \sin \theta_0$$

where θ_0 is the acceptance angle.

But $$\sin \theta_0 = \sqrt{n_1^2 - n_2^2}$$

$$\therefore \quad NA = \sqrt{n_1^2 - n_2^2} \quad \text{...(47.14)}$$

$$n_1^2 - n_2^2 = (n_1 + n_2)(n_1 - n_2) = \left(\frac{n_1 + n_2}{2}\right)\left(\frac{n_1 - n_2}{n_1}\right) 2n_1$$

Approximating $\frac{n_1 + n_2}{2} \approx n_1$, we can express the above relation as $\left(n_1^2 - n_2^2\right) = 2n_1^2 \Delta$.

It gives

$$NA = \sqrt{2n_1^2 \Delta}$$

$$\therefore \quad NA = n_1 \sqrt{2\Delta} \quad (47.15)$$

Numerical aperture determines the light gathering ability of the fibre. It is a measure of the amount of light that can be accepted by a fibre. It is seen from Eq. (47.14) that NA is dependent only on the refractive indices of the core and cladding materials and does not depend on the physical dimensions of the fibre. The value of NA ranges from 0.13 to 0.50. A large NA implies that a fibre will accept large amount of light from the source (Fig. 47.10).

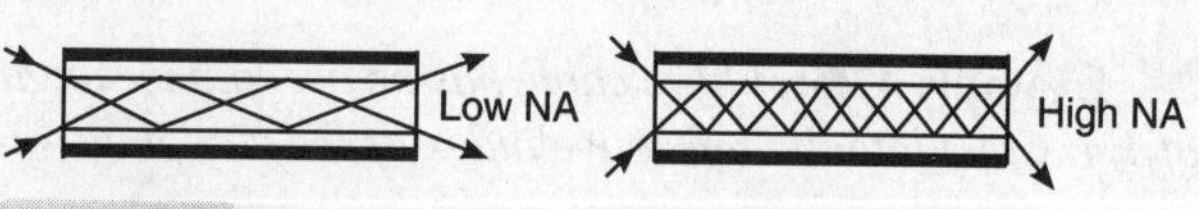

Fig. 47.10: Illustration of the propagation of light through low and high numerical aperture fibres

Example 47.7: *Calculate the fractional index change for a given optical fibre if the refractive indices of the core and the cladding are 1.563 and 1.498 respectively.*

Solution: Fractional index change $\Delta = \frac{n_1 - n_2}{n_1} = \frac{1.563 - 1.498}{1.563} = \frac{0.065}{1.563} = \mathbf{0.0415}$

Example 47.8: *Calculate the refractive indices of the core and the cladding material of a fiber from the following data:*

Numerical aperture (NA) = 0.22 and Δ = 0.012 where Δ is the fractional refractive index change.

Solution: $$NA = n_1 \sqrt{2\Delta}$$

$$0.22 = n_1 \sqrt{2 \times 0.012} = 0.155\, n_1.$$

$$\therefore \quad n_1 = \frac{0.22}{0.155} = \mathbf{1.42}$$

$$\Delta = \frac{n_1 - n_2}{n_1} \quad \therefore \quad \frac{1.42 - n_2}{1.42} = 0.012$$

$$\therefore \quad n_2 = 1.42 - 1.42 \times 0.012 = \mathbf{1.403}$$

Example 47.9: *Find the fractional refractive index and numerical aperture for an optical fibre with refractive indices of core and cladding as 1.5 and 1.49 respectively.*

Solution:
$$\Delta = \frac{n_1 - n_2}{n_1} = \frac{1.5 - 1.49}{1.5} = \mathbf{0.0067}$$

$$NA = n_1\sqrt{2\Delta} = 1.5\sqrt{2 \times 0.0067} = \mathbf{0.174}$$

47.7 SKIP DISTANCE AND NUMBER OF TOTAL INTERNAL REFLECTIONS

We shall now calculate the number of total internal reflections that a light ray undergoes as it travels through an optical fibre of length L. Let a be the radius of the fibre. Let a light ray be incident on one end of the fibre at an angle θ_1 to the axis and refract into the fibre at an angle θ_2, as shown in Fig. 47.11. The ray undergoes the first reflection at B.

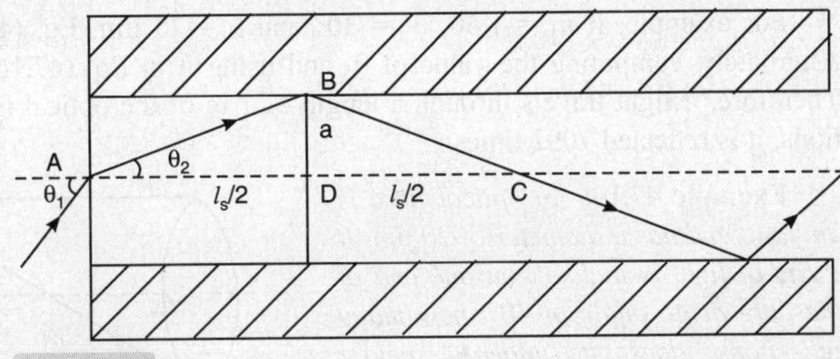

Fig. 47.11: Skip Distance l_s

The distance $AC = l_s$ is known as the ***skip distance*** and represents the distance between two successive reflections of the ray. In the Δ^{le}ABD, $AD = \frac{1}{2}AC = \frac{l_s}{2}$, $BD = a$ and $\angle BAD = \theta_2$.

$$\therefore \quad \tan\theta_2 = \frac{BD}{AD} = \frac{a}{l_s/2}$$

or
$$l_s = \frac{2a}{\tan\theta_2} \qquad \text{...(47.16)}$$

In terms of the incidence angle θ_1, Eq. (47.16) may be rewritten as

$$l_s = 2a\left[\frac{\cos\theta_2}{\sin\theta_2}\right] \quad \text{or} \quad l_s^2 = (2a)^2\left[\frac{1}{\sin^2\theta_2} - 1\right]$$

Using Eq. (47.7) into the above expression, we obtain

$$l_s^2 = (2a)^2\left[\left(\frac{n_1}{n_0 \sin\theta_1}\right)^2 - 1\right]$$

or
$$l_s = 2a\left[\left(\frac{n_1}{n_0\sin\theta_1}\right)^2 - 1\right]^{1/2} \qquad \text{...(47.17)}$$

In case of air, $n_0 = 1$ and $$l_s = 2a\left[\left(\frac{n_1}{\sin\theta_1}\right)^2 - 1\right]^{1/2} \qquad ...(47.17a)$$

The number of total internal reflections in the total fibre length is given by

$$N = \frac{\text{Total length of the cable}}{\text{The } \textit{distance} \text{ travelled during one reflection}} = \frac{L}{l_s}$$

$$\therefore \quad N = \frac{L\tan\theta_2}{2a} \qquad ...(47.18a)$$

Also, $$N = \frac{L}{2a\left[\left(\frac{n_1}{\sin\theta_1}\right)^2 - 1\right]^{1/2}} \qquad ...(47.18b)$$

For example, if $n_1 = 1.50$, $\theta_1 = 30°$ and $a = 25$ μm, Eq. (47.17a) gives $l_s = 141$ μm. Alternately, computing the value of θ_2 and using it in Eq. (47.16), we obtain $l_s = 141$ μm. Therefore, if light travels through a length of 1 m of the optical fibre of the above specifications, it is reflected 7092 times.

Example 47.10: *An optical fibre is 2m long and has a diameter of 20 μm. If a ray of light is incident on one end of the fibre at an angle of 40°, how many reflections does it undergo before emerging from the other end? Refractive index of fibre is 1.3.*

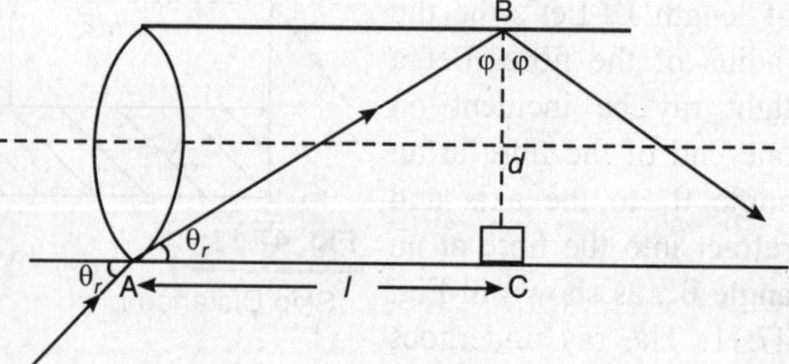

$$\frac{\sin\theta_r}{\sin\theta_i} = \frac{n_1}{n_2}$$

$$\therefore \quad \sin\theta_r = \frac{n_1}{n_2}\sin\theta_i = \frac{(1)(\sin 40°)}{1.3} = 0.4940$$

$$\therefore \quad \theta_r = \mathbf{29.6°}$$

From Fig. *i*, the distance *l* travelled along the fibre during one internal reflection is

$$l = \frac{d}{\tan\theta_r} = \frac{20\,\mu\text{m}}{\tan 29.6°} = \frac{2\times10^{-5}\text{ m}}{0.568} = 3.52\times10^{-5}\text{ m}$$

Number of reflections in the cable is

$$N = \frac{\text{Total length of the cable}}{l} = \frac{2\text{m}}{3.52\times10^{-5}\text{ m}} = \mathbf{56818}$$

47.8 MODES OF PROPAGATION

Light propagates as an electromagnetic wave through an optical fibre and its propagation is governed by Maxwell's equations. Complete understanding of propagation of light waves through optical fibres requires a thorough understanding of solution of these equations in the context of optical fibres. When a plane electromagnetic wave propagates in free space, it travels as a transverse electromagnetic wave. The electric field and magnetic field components associated with the wave are perpendicular to each other and also perpendicular to the direction of propagation. It is known as a TEM wave. When the light ray is guided through an optical fibre, it propagates in different types of modes. Each of these guided modes consists of a variety of electromagnetic field configurations, such as transverse electric (TE), transverse magnetic (TM) and hybrid modes. Hybrid modes are combination of transverse electric and magnetic modes. Fig. 47.6 shows **Propagation** of light rays through an optical fibre due to total internal reflection.

***Modes** can be visualised as the possible number of allowed paths of light in an optical fibre.*

The paths are all zigzag paths excepting the axial direction. Though the rays having propagation angles between $\theta = 0°$ and $\theta = \theta_c$ will be in a position to undergo total internal reflections, all of them will not however propagate along the optical fibre. Only a certain ray directions are allowed. As a zigzag ray gets repeatedly reflected at the walls of the fibre, phase shift occurs. Consequently, the waves travelling along certain zigzag paths will be in phase and undergo constructive interference, while the waves coursing along certain other paths will be out of phase and diminish due to destructive interference.

*The light ray paths along which the waves are in phase inside the fibre are known as **modes**. Each mode is a pattern of electric and magnetic field distributions that is repeated along the fibre at equal intervals.*

Number of modes propagating in a fibre increases when θ_c or Δ increases. Increasing the core refractive index increases the number of propagating modes. On the other hand, increasing the clad refractive index decreases the number of propagating modes. The number of modes that a fibre will support depends on the ratio d/λ, where d is the diameter of the core and λ is the wavelength of the wave being transmitted. The zero order ray travels along the axis is known as the *axial ray*. Note that each mode carries a portion of the light from the input signal.

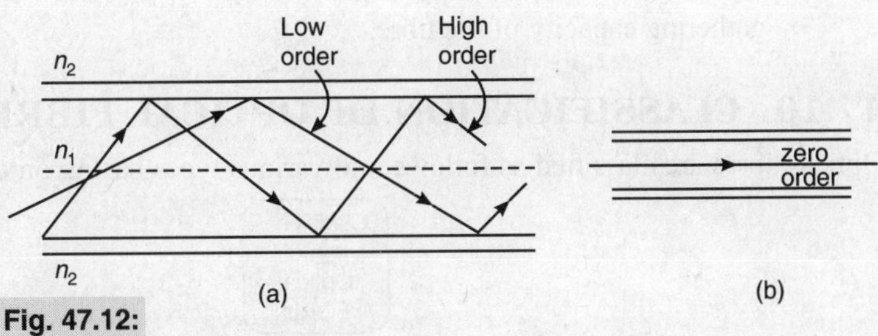

Fig. 47.12: (a) Low and High-order ray paths in a multimode fibre. (b) Axial ray in a single mode fibre

Types of modes: In a fibre of fixed thickness, the modes that propagate at angles close to the critical angle ϕ_C (*i.e.*, critical propagation angle θ_c) are ***higher order modes***, and modes that propagate with angles larger than the critical angle (*i.e.*, lower than the critical propagation angle) are ***lower order modes*** (see Fig. 47.12). In case of lower order modes, the fields are concentrated near the center of the fibre. In case of higher order modes, the fields are distributed more towards the edge of the wave-guide and tend to send light energy into the cladding. This energy is lost ultimately. The higher order modes have to traverse longer paths and hence take larger time than the lower order modes to cover a given length of the fibre. Thus, the higher order modes arrive at the output end of the fiber later than the lower order modes.

47.9 TYPES OF RAYS

The rays that propagate through an optical fibre can be classified into two categories: (*i*) meridional rays and (*ii*) skew rays.

(*i*) **Meridional ray:** A ray that propagates through the fibre undergoing total internal reflection is called meridional ray. It passes through the longitudinal axis of the fibre core (Fig. 47.13 a). The propagation of meridional rays is possible only in the TM or TE modes.

(*ii*) **Skew ray:** The ray that describes angular helical path along the fibre is called a skew ray (see Fig. 47.13 b). These rays do not pass through the axis of the core. These rays are propagated in either hybrid EH or HE modes. Some of these modes produce losses through leakage of radiation. In real situations, the skew rays constitute a substantial portion of the total number of guided rays. They tend to propagate only in the annular region near the outer surface of the core and do not fully utilize the core as the medium. However, they are complementary to the meridional rays and increase the light gathering capacity of the fibre.

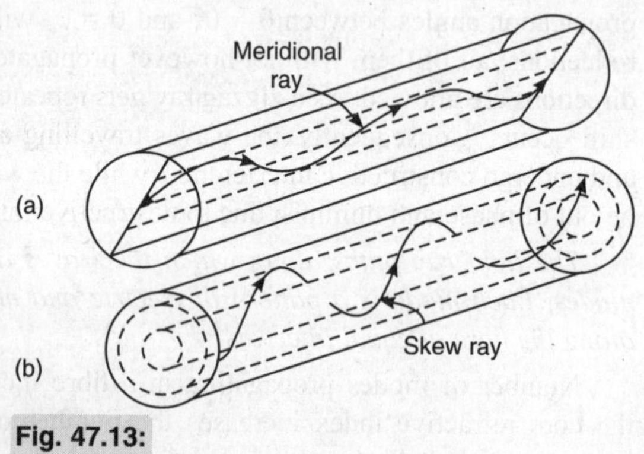

Fig. 47.13: (a) Meridional ray, (b). Skew ray

47.10 CLASSIFICATION OF OPTICAL FIBRES

Optical fibres are classified as follows into various types basing on different parameters.

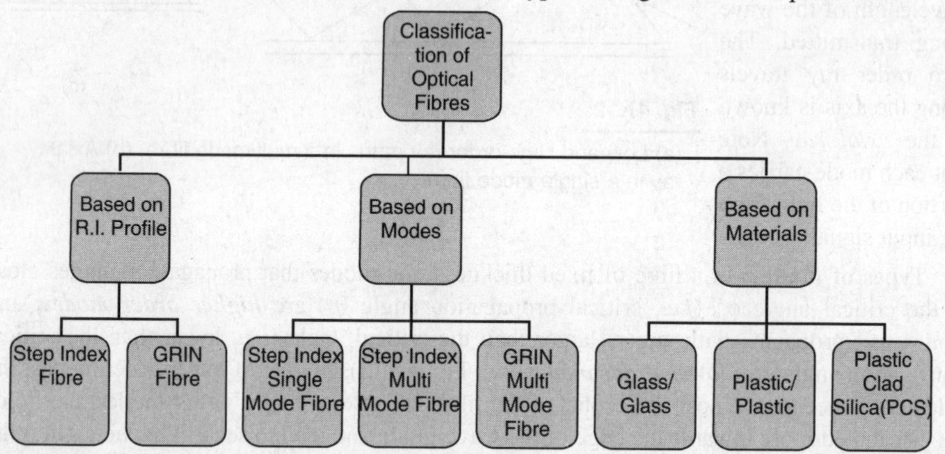

(*i*) Classification basing on refractive index profile: *Refractive index profile* of an optical fibre is a plot of refractive index drawn on one of the axes and the distance from the core axis drawn on the other axis (see Fig. 47.14). Optical fibres are classified into the following two categories on the basis of refractive index profile.

(*a*) Step index fibres and

(*b*) Graded index (GRIN) fibres.

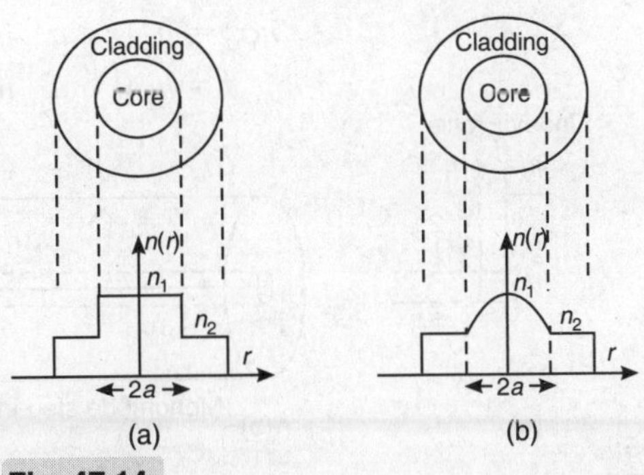

Fig. 47.14: Classification of optical fibres based on R.I. profile (a) Step index fibre (b) GRIN fibre

Step index refers to the fact that the refractive index of the core is constant along the radial direction and abruptly falls to a lower value at the cladding and core boundary (see Fig. 47.14 (a). In case of GRIN fibres, the refractive index of the core is not constant but varies smoothly over the diameter of the core (see Fig. 47.14 b). It has a maximum value at the center and decreases gradually towards the outer edge of the core. At the core-cladding interface the refractive index of the core matches with the refractive index of the cladding. The refractive index of the cladding is constant.

***(ii)* Classification basing on the modes of light propagation:** On the basis of the modes of light propagation, optical fibres are classified into two categories as (a) Single mode fibres (SMF) and (b) Multimode fibres (MMF).

A single mode fibre (SMF) has a smaller core diameter and can support only one mode of propagation.

A multimode fibre (MMF) has a larger core diameter and supports a number of modes.

Thus, on the whole, the optical fibres are classified into three types:

(*a*) Single mode step-index fibre (SMF)

(*b*) Multimode step-index fibre (MMF)

(*c*) Graded index (multimode) fibre (GRIN).

***(iii)* Classification basing on materials:** On the basis of materials used for core and cladding, optical fibres are classified into three categories.

(*a*) Glass/glass fibres (glass core with glass cladding)

(*b*) Plastic/plastic fibres (plastic core with plastic cladding)

(*c*) PCS fibres (polymer clad silica)

47.11 THE THREE TYPES OF FIBRES

We now study the detailed structure and characteristics of the three types of optical fibres.

47.11.1 Single Mode Step Index Fibre

Structure: A single mode step index fibre has a very fine thin core of diameter of 8 μm to 12 μm (see Fig. 47.15 c). Core is generally made of germanium doped silicon and is enclosed by a thick cladding of lower refractive index. The cladding is composed of silica lightly doped with phosphorous oxide. The external diameter of the cladding is of the order of 125 μm. The fibre is surrounded by a non-transparent shielding sheath. The refractive index of the fibre changes abruptly at the core-cladding boundary, as shown in Fig. 47.15 (a).

The variation of the refractive index of a step index fibre as a function of radial distance can be mathematically represented as

$$n(r) = n_1 \, [r < a \quad inside\ core]$$
$$= n_2 \, [r > a \quad in\ cladding] \qquad ...(47.19)$$

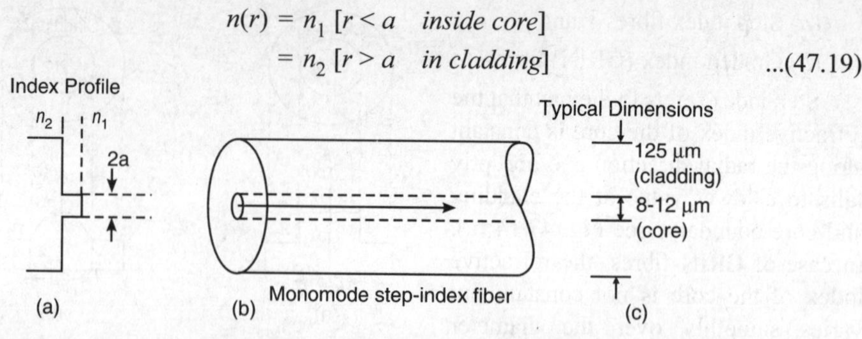

Fig. 47.15: Single mode step index fibre (a) R.I. profile (b) ray paths (c) typical dimensions

Propagation of light in SMF: Light travels in SMF along a single path that is along the axis (Fig. 47.15 b). Obviously, it is the zero-order mode that is supported by a SMF. Both Δ and NA are very small for single mode fibres. It can be achieved by reducing the fibre radius and the relative refractive index change (Δ). The low NA indicates a low acceptance angle and the difficulty in light coupling into the fibre. Expensive laser diodes are required to launch light into the SMF.

47.11.2 Multimode Step Index Fibre

Structure: A multimode step index fibre is very much similar to the single mode step index fibre except that its core is of larger diameter. The core diameter is of the order of 50 to 100 μm, which is very large compared to the wavelength of light. The external diameter of cladding is about 150 to 250 μm (Fig. 47.16 c).

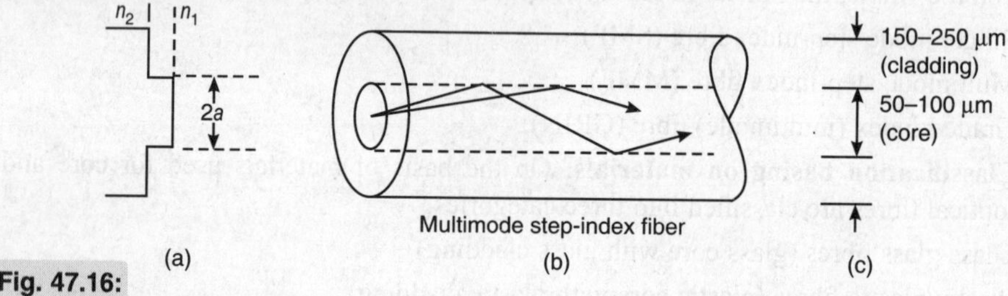

Fig. 47.16: Multimode step index fibre (a) R.I. Profile (b) Ray paths (c) typical dimensions

Propagation of light in MMF: MMF permits a finite number of guided modes. The direction of polarization, alignment of electric and magnetic fields will be dissimilar in rays of diverse modes. Many zigzag paths of propagation are allowed. The length of the path along the axis of the fibre is smaller while the other zigzag paths are longer. Because of this difference, the lower order modes reach the end of the fibre earlier while the high order modes reach after some time delay (Fig. 47.16 b).

47.11.3 Graded Index (GRIN) Fibre

A graded index fibre is a MMF with a core having concentric layers of different refractive indices. The refractive index of the core varies with distance from the fibre axis. It has a high value at the centre and falls of with increasing radial distance from the axis. A typical structure and its index profile are shown in Fig. 47.17 (a). Such a profile causes a periodic focussing of light propagating through the fibre. The size of the graded index fibre is about the same as the step index fibre.

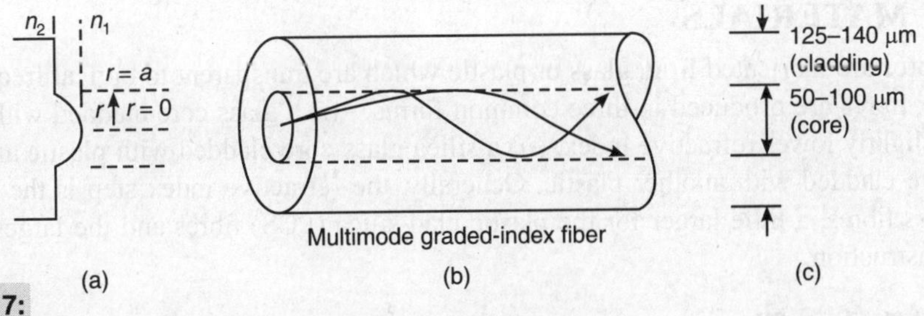

Fig. 47.17:

GRIN fibre (a) R.I. Profile (b) Ray paths (c) typical dimensions

The variation of the refractive index of the core with radius measured from the center is given by

$$n(r) = \begin{cases} n_1\sqrt{1 - \left[2\Delta\left(\dfrac{r}{a}\right)^{\alpha}\right]}, & r < a \text{ inside core} \\ n_2, & r > a \text{ in cladding} \end{cases} \quad ...(47.20)$$

where n_1 is maximum refractive index at the core axis, a the core radius, and α the grading profile index number which varies from 1 to ∞.

when $\alpha = 2$, the index profile is parabolic and is preferred for different applications.

Propagation of light in GRIN fibre: When a light ray travels from a region of larger refractive index to a region of lower refractive index, it is bent away from the normal till the condition of total internal reflection. Then the ray travels back towards the core axis, again being continuously refracted (Fig. 47.18 a). The turning around of the ray may happen even before reaching the core-cladding boundary. Accordingly, continuous refraction is followed by total internal reflection and again continuous refraction towards the axis. In GRIN fibre, rays making larger angles with the axis traverse longer path. However, they travel in a region of lower refractive index at a higher speed of propagation.

Consequently, all rays traveling through the fibre, irrespective of their modes of travel, will have almost the same optical path length and reach the output end of the fibre at the same time (see Fig. 47.18 b). The numerical aperture of a GRIN fibre is given by

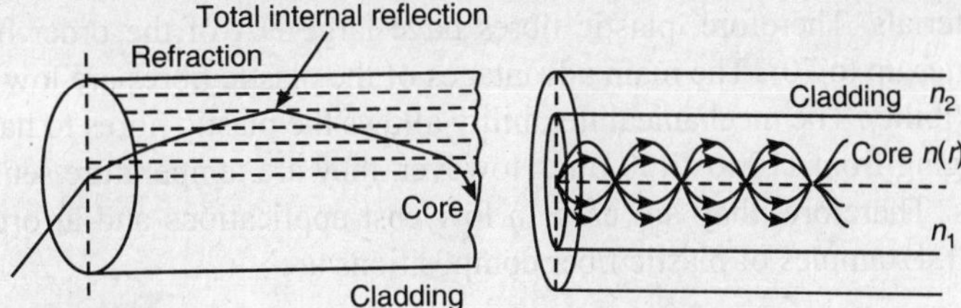

Fig. 47.18:

(a) illustration of refraction at the various high to low index interfaces within graded index fibre, giving an overall curved ray path

(b) Light transmission in a graded index fibre.

$$NA = \sqrt{n^2(r) - n_2^2} \approx n_1(2\Delta)^{\frac{1}{2}}\sqrt{1 - \left(\frac{r}{a}\right)^2} \quad ...(47.21)$$

$$= n_1\sqrt{2\Delta\left[1 - \left(\frac{r}{a}\right)^2\right]}$$

47.12 MATERIALS

Optical fibres are fabricated from glass or plastic which are transparent to optical frequencies. Step index fibres are produced in three common forms – (*i*) a glass core cladded with a glass having a slightly lower refractive index, (*ii*) a silica glass core cladded with plastic and (*iii*) a plastic core cladded with another plastic. Generally, the refractive index step is the smallest for all glass fibres, a little larger for the plastic clad silica (PCS) fibres and the largest for all plastic construction.

47.12.1 All glass fibres

The basic material for fabrication of optical fibres is silica (SiO_2). It has a refractive index of 1.458 at λ = 850 nm. Materials having slightly different refractive index are obtained by doping the basic silica material with small quantities of various oxides. If the basic silica material is doped with germania (GeO_2) or phosphorous pentoxide (P_2O_5), the refractive index of the material increases. Such materials are used as core materials and pure silica is used as cladding material in these cases. When pure silica is doped with boria (B_2O_3) or fluorine its refractive index decreases. These materials are used for cladding when pure silica is used as core material.

Examples of fiber compositions are

(*a*) SiO_2 core – $B_2O_3.SiO_2$ cladding

(*b*) $GeO_2.SiO_2$ core – SiO_2 cladding

The glass optical fibres exhibit very low losses and are used in long distance communications.

47.12.2 All plastic fibres

In these fibres, perspex (PMMA) and polysterene are used for core. Their refractive indices are 1.49 and 1.59 respectively. A fluorocarbon polymer or a silicone resin is used as a cladding material. A high refractive index difference is achieved between the core and the cladding materials. Therefore, plastic fibres have large NA of the order of 0.6 and large acceptance angles up to 70°. The main advantages of the plastic fibres are low cost and higher mechanical flexibility. The mechanical flexibility allows the plastic fibres to have large cores, of diameters ranging from 110 to 1400 μm. However, they are temperature sensitive and exhibit very high loss. Therefore, they are used in low cost applications and at ordinary temperatures (below 80°). Examples of plastic fiber compositions are

(*a*) Polysterene core	n_1 = 1.60	NA = 0.60
- Methyl methacrylate cladding	n_2 = 1.49	
(*b*) Polymethyl methacrylate core	n_1 = 1.49	NA = 0.50
- cladding made of its copolymer	n_2 = 1.40	

47.12.3 PCS fibres

The plastic clad silica (PCS) fibres are composed of silica cores surrounded by a low refractive index transparent polymer as cladding. The core is made from high purity quartz. The cladding is made of a silicone resin having a refractive index of 1.405 or of perfluoronated ethylene propylene (Teflon) having a refractive index of 1.338. Plastic claddings are used for step-index fibres only. The PCS fibres are less expensive but have high losses. Therefore, they are mainly used in short distance applications.

47.13 V-NUMBER

Let us consider a narrow beam of monochromatic light launched on the front end of a step-index fibre, at an angle less than the acceptance angle of the fibre. Let the wavelength of the light be λ_0 and the diameter of the fiber be d. It appears to us from the ray concept that all the rays contained in the beam propagate along the fibre, such that there can be infinite modes of propagation. However, in practice, only a limited number of modes of propagation are possible in an optical fibre. To understand the reason for this behaviour, we have to recall that phase changes occur as the light waves travel forward. The phase shift takes place due to two reasons – (*i*) due to optical path length traversed and (*ii*) due to total internal reflection at the core-cladding interface.

(*i*) When a wave travels a distance l in a medium of refractive index n_1, it undergoes a phase change δ_1 given by

$$\delta_1 = k\, n_1 l = \frac{2\pi l n_1}{\lambda} \qquad ...(47.22)$$

where k is the propagation constant.

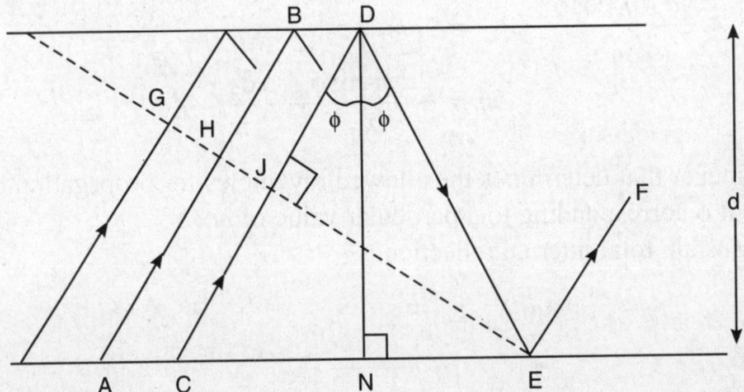

Fig. 47.19: V –number calculations

(*ii*) Whenever, a wave with component normal to the reflecting surface undergoes total internal reflection, the phase shift, δ_2, is given by

$$\delta_2 = 2\tan^{-1}\frac{\sqrt{n_1^2\cos^2\phi - n_2^2}}{n_1\sin\phi} \qquad ...(47.23)$$

In Fig. 47.19, *AB* and *CD* are parallel rays in an incident beam. The line *GJ* is perpendicular to the propagation path of the rays *AB*, *CD* and hence represents a plane wavefront. The points *G* and *J* lying on the same wavefront will be in phase with each other. As the point *E*, which is on the reflected ray *DE*, lies on the wavefront *GJ*, the points *J* and *E* must be in phase with each other. However, moving from the point *J* to *E* along the ray, we find that there occurs a phase shift given by

$$\delta = (JD + DE)\frac{2\pi n_1}{\lambda_0} - 2\delta_2 \qquad ...(47.24)$$

The factor 2 in the Eq. (47.24) takes into account the two total internal reflections at *D* and *E*. In the Δ^{le} *DNE*

$$\frac{DN}{DE} = \cos\phi. \text{ Therefore, } \quad DE = \frac{DN}{\cos\phi} = \frac{d}{\cos\phi}$$

Further, in the Δ^{le} JDE

$$\frac{JD}{DE} = \cos 2\phi. \text{ Therefore, } JD = DE \cos 2\phi$$

$$JD + DE = DE\,(1 + \cos 2\phi) = 2\,DE \cos^2\phi.$$

$$= 2\frac{d}{\cos\phi}\cos^2\phi = 2d\cos\phi$$

Using the above expression into Eq. (47.24), we obtain

$$\delta = \frac{4d\,\pi n_1 \cos\phi}{\lambda_0} - 2\delta_2$$

Now the condition for the wave associated with the ray CD to propagate along the optical fibre is that the phase of the twice reflected wave must be the same as that of the incident wave. That is, the wave must interfere constructively with itself. If this phase condition is not satisfied, the wave would interfere destructively with itself and just die out. It means that the total phase shift must be equal to an integer multiple of 2π radians. Thus,

$$\frac{4d\,\pi n_1 \cos\phi}{\lambda_0} - 2\delta_2 = 2\pi m$$

or

$$m = \frac{2d\,n_1 \cos\phi_m}{\lambda_0} - \frac{\delta_2}{\pi} \qquad \text{...(47.25)}$$

where m is an integer that determines the allowed ray angles for propagation of the wave and ϕ_m is the value of ϕ corresponding to a particular value of m.

In order to sustain total internal reflection,

$$\sin\phi_m \geq \frac{n_2}{n_1}$$

$$\therefore \quad \cos\phi_m \leq \frac{\sqrt{n_1^2 - n_2^2}}{n_1}$$

$$\therefore \quad m \leq \frac{2d\sqrt{n_1^2 - n_2^2}}{\lambda_0} - \frac{\delta_2}{\pi} \qquad \text{...(47.26)}$$

or

$$m \leq \frac{2V}{\pi} - \frac{\delta_2}{\pi} \qquad \text{...(47.27)}$$

where V is given by

$$V = \frac{\pi d}{\lambda_0}\sqrt{n_1^2 - n_2^2} \qquad \text{...(47.28)}$$

V-number is more generally called normalized ***frequency*** of the fibre. Each mode has a definite value of V-number below which the mode is cut off.

Eq. (47.28) can be written as $V = \dfrac{\pi d}{\lambda_0}(NA)$ ...(47.29)

or

$$V = \frac{\pi d}{\lambda_0} n_1 \sqrt{2\Delta} \qquad \text{...(47.30)}$$

The maximum number of modes N_m supported by an SI fibre is given by

$$N_m = \frac{1}{2}V^2 \qquad \text{...(47.31)}$$

Thus, for $V = 10$, N_m is 50. When the normalized frequency V is less than 2.405, the fibre can support only one mode, which propagates along the axial length of the fibre, and the fibre becomes a single mode fibre. It means that for single mode transmission in a MMF, V must be less than 2.405. The wavelength at which the fibre becomes single mode is called ***cutoff wavelength***, λ_c of the fibre.

Using Eq. (47.29), we can write

$$\lambda_c = \frac{\pi d}{2.405}(NA) \qquad \text{...(47.32)}$$

It is seen from the Eq. (47.32) that single mode property can be realized in a multimode fibre by decreasing the core diameter and/or decreasing Δ such that $V < 2.405$.

In case of GRIN fibres, for larger values of V,

$$N_m \cong \frac{V^2}{4} \qquad \text{...(47.33)}$$

Example 47.11: *A step-index fibre is made with a core of refractive index 1.52, a diameter of 29 μm and a fractional difference index of 0.0007. It is operated at a wavelength of 1.3 μm. Find the V-number and the number of modes that the fibre will support.*

Solution: $V = \frac{\pi d}{\lambda_0} n_1 \sqrt{2\Delta} = \frac{3.143 \times 29 \times 10^{-6}\,\text{m}}{1.3 \times 10^{-6}\,\text{m}} \times 1.52\sqrt{2 \times 0.0007} = \mathbf{4.049}$

∴ Number of modes, $N = \frac{1}{2}V^2 = \frac{1}{2}(4.049)^2 = \mathbf{8\ modes}$

47.14 FABRICATION

A number of techniques are available to produce all glass fibres. In one of the methods, known as the double crucible method, fibres are directly produced from the melt.

Double Crucible Technique: The double crucible consists of two concentric platinum crucibles having thin orifices at the bottom. Raw material for the core-glass is placed in the inner crucible and the raw material for cladding is fed to the outer crucible. The double crucible arrangement is mounted vertically (Fig. 47.20) in a furnace. The furnace is maintained at a suitable temperature to take the raw material into molten state. The fibres are drawn through the thin orifices at the bottom of the crucibles. As both the materials are drawn simultaneously, a filament of core glass surrounded by a tube of cladding glass is obtained in the process. The thickness of the fibre is monitored and the fibre is then coated with a polymer. Subsequently, it is passed through a plastic extrusion die to form a plastic sheath over the fibre.

47.15 SPLICING

It is often required to join two optical fibres together to form a continuous optical waveguide. The method and technique for connecting the fibres depends on whether a permanent joint is required or easily disconnected joint is required. The permanent bonding technique is called **splice** technique and easily disconnected joint techniques are called **connectors**. Splicing is analogous to soldering in metal wires. It consists of fusing of two fibre ends and bonding them together in an alignment structure. The generally accepted splicing method is arc fusion splicing, which melts the fiber ends together with an electric arc. For quicker fastening jobs, a "mechanical splice" is used.

(*i*) **Fusion splicing:** Fusion splicing is combining two optical fibres end-to-end using heat so that the light passing through the fibers is not scattered or reflected back by the splice. The two cable ends are fixed firmly inside a splice enclosure and protective coating are removed. The ends are cut with a precision cleaver and are placed into special holders in the splicer. The splicer uses small motors to align the end faces together and emits a small spark between electrodes at the gap to burn off dust and moisture. Then the splicer causes a larger spark that increases the temperature above the melting point of the glass, fusing the ends jointly permanently. The splices offer controlled alignment of fiber optic cables to achieve losses as low as 0.05 dB.

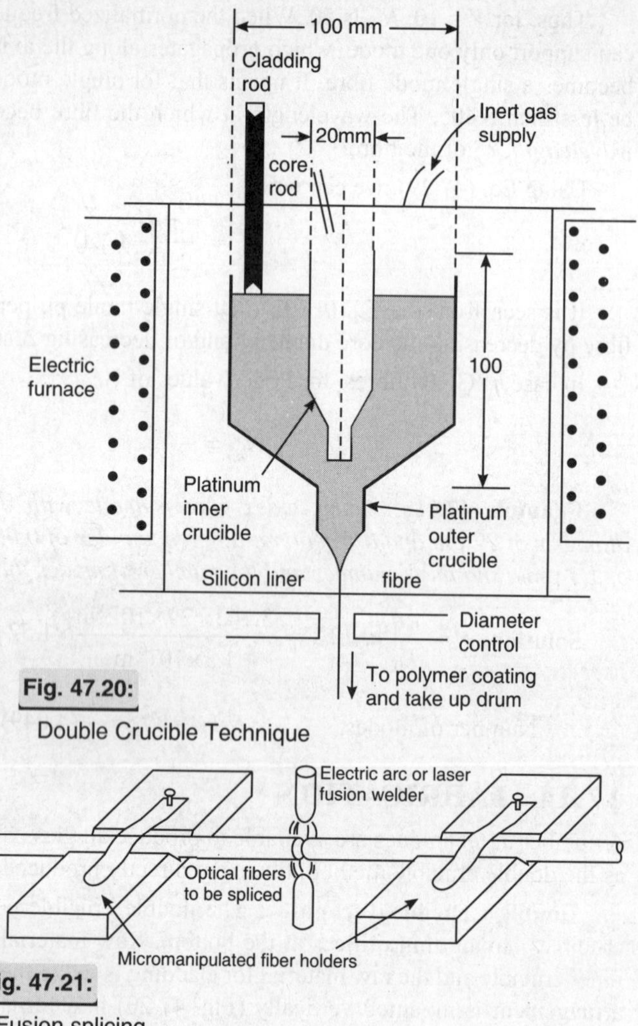

Fig. 47.20: Double Crucible Technique

Fig. 47.21: Fusion splicing

(*ii*) **Mechanical splicing: A mechanical splice** is a junction of two or more optical fibres that are aligned and held in place by a self-contained assembly. The fibers are not permanently joined, just precisely held together so that light can pass from one to another. They are easily applied in the field, require little or no tooling and offer losses of about 0.2 dB.

They are quicker and easier to install but requires stripping, careful cleaning and precision cleaving. The fiber ends are held together by a sleeve

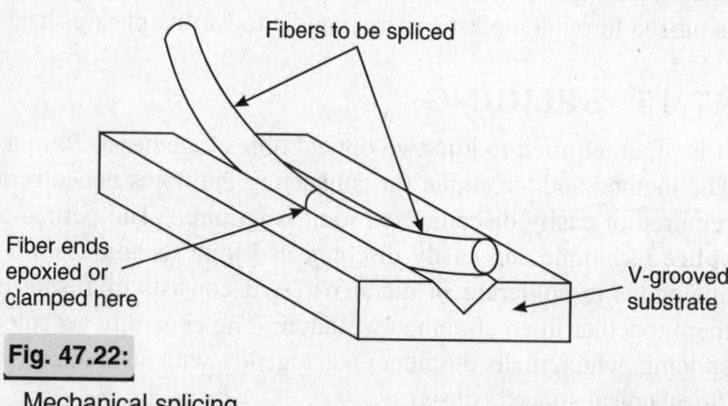

Fig. 47.22: Mechanical splicing

using a clear index-matching gel that improves the transmission of light across the joint. But they have higher optical loss and are less strong. All splicing techniques use an enclosure into which the splice is placed for protection afterward.

(*a*) **V- Groove splice technique:** The V-block is the simplest mechanical splice. The bared fibres to be joined are placed in the groove. Angular alignment is particularly well controlled. The two fibres can slide in the groove until they touch. They are then epoxied permanently into position, so end-separation errors are minimal. If the epoxy is index matched to the fibre, even small gaps can be tolerated with little loss. Lateral misalignment would be negligible in the groove if both fibres had the same core and cladding diameters. A cover plate can be placed over the V-block to protect the splice further.

(*b*) **Elastomer splice technique:** Another splice is essentially a precision sleeve made with elastomeric materials. The elastomer is an elastic material usually made into a cylinder with an opening along its axis. The groove is a little smaller than the fibre but accepts and centers it by expanding slightly when the fibre is inserted. The fibres are inserted from both the ends of the cylinder and touch near its midpoint. The slice can be epoxied for permanent connection. An external splice holder is used for full protection of the splice.

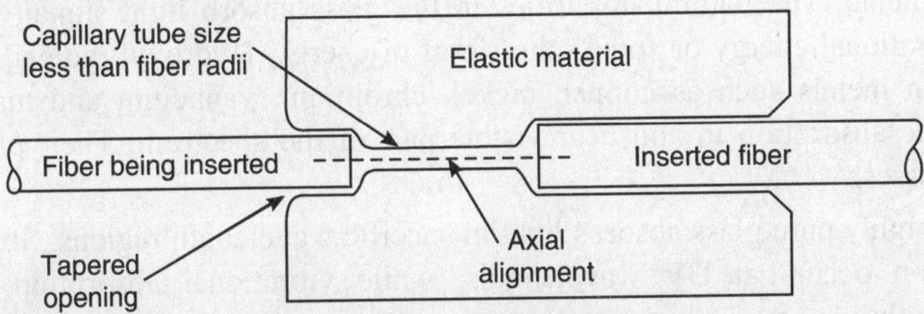

Fig. 47.23:

Illustration of Elastomer splice technique

47.16 LOSSES IN OPTICAL FIBRE

As a light signal propagates through a fibre, it suffers loss of amplitude and change in shape. The loss of amplitude is referred to as *attenuation* and the change in shape as *distortion*.

47.16.1 Attenuation

When an optical signal propagates through a fibre power of the signal decreases exponentially with distance. The loss of optical power as light travels down a fiber is known as **attenuation**.

The ***attenuation of optical signal*** *is defined as the ratio of the optical output power from a fibre of length L to the input optical power.*

If P_i is the optical power launched at the input end of the fibre, then the power P_o at a distance L down the fibre is given by

$$P_o = P_i e^{-\alpha L} \qquad \text{...(47.34)}$$

where α is called the ***fibre attenuation coefficient*** expressed in units of km^{-1}.

Taking logarithms on both the sides of the Eq. (47.34), we obtain

$$\alpha = \frac{1}{L} \ln \frac{P_i}{P_o} \qquad \text{...(47.35)}$$

In units of dB / km, α is defined through the equation

$$\therefore \qquad \alpha_{\text{dB/km}} = \frac{10}{L}\log\frac{P_i}{P_o} \qquad \text{...(47.36)}$$

In case of an ideal fibre, $P_o = P_i$ and the attenuation would be zero.

Different mechanisms of attenuation: There are several loss mechanisms responsible for attenuation in optical fibres. They are broadly divided into two categories: *intrinsic* and *extrinsic* attenuation. Intrinsic attenuation is caused by substances inherently present in the fiber, whereas extrinsic attenuation is caused by external forces such as bending.

(*i*) **Intrinsic Attenuation:** Intrinsic attenuation results from materials inherent to the fiber. It is caused by impurities present in the glass. During manufacturing, there is no way to eliminate all impurities. When a light signal hits an impurity in the fiber, either it is scattered or it is absorbed. Intrinsic attenuation can be further characterized by two components; ***Material absorption and Rayleigh scattering***

(*a*) **Absorption by material:** Imperfection and impurities in the fiber gives material absorption to 3-5% of fiber attenuation. The main general impurity is the hydroxyl molecule (OH-) which remains as a residue and enter the fiber-optic cable material through either a chemical reaction in the manufacturing process or as humidity in the environment. The natural impurities in the glass absorb light signal, and convert it into vibrational energy or some other form of energy. Hydroxyl radical ions (OH) and transition metals such as copper, nickel, chromium, vanadium and manganese have electronic absorption in and near visible part of the spectrum. Their presence causes heavy losses.

Even a highly pure glass absorbs light in specific wavelength regions. Strong electronic absorption occurs at UV wavelengths, while vibrational absorption occurs at IR wavelengths.

Losses due to impurities can be reduced by better manufacturing processes. In improved fibres, metal ions are practically negligible. The largest loss is caused by OH ions. These cannot be sufficiently reduced. The absorption of light either through intrinsic or impurity process forms a transmission loss. The absorption losses are found to be at minimum at around 1.3 μm.

Unlike scattering, absorption can be limited by controlling the amount of impurities during the manufacturing process.

(*b*) **Rayleigh scattering:** Rayleigh scattering accounts for the majority (about 96%) of attenuation in optical fiber. The local microscopic density variations in glass cause local variations in refractive index. These variations, which are inherent in the manufacturing process and cannot be eliminated, act as obstructions and scatter light in all directions (Fig. 47.24). This is known as *Rayleigh scattering*. The Rayleigh scattering loss greatly depends on the wavelength. It varies as $1/\lambda^4$ and becomes important at lower wavelengths. Thus, Rayleigh scattering sets a lower limit, on the wavelengths that can be transmitted by a glass fibre at 0.8 μm, below which the scattering loss is very high.

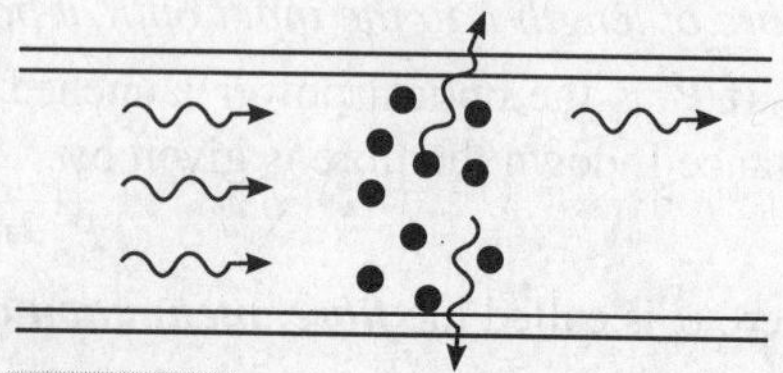

Fig. 47.24: Rayleigh scattering, showing attenuation of an incident stream of photons due to localized variations in refractive index

Any wavelength specifically below 800 nm is impractical for optical communication for the reason that attenuation due to high Rayleigh scattering. Similarly, propagation above 1700 nm is impracticable due to high losses owing to infrared absorption. Fig. 47.25 shows variation of attenuation with wavelength measured for a typical fiber-optic cable.

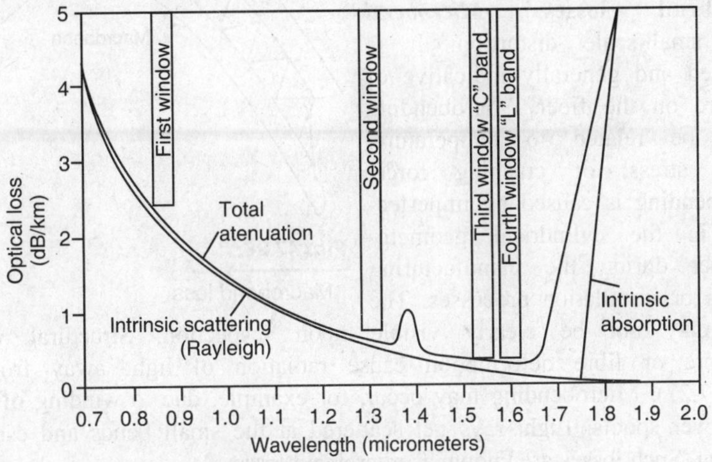

Fig. 47.25: Variation of attenuation with wavelength in a typical fibre-optic cable.

For better performance, the choice of wavelength must be based on minimizing loss and minimizing dispersion. Such windows are selected for communication purposes. It is seen from the attenuation curve that it has a minimum at around a particular band of optical wavelengths. The band of wavelengths at which the attenuation is a minimum is called ***optical window*** or ***transmission window*** or ***low-loss window***. There are three ***principal windows***. These correspond to wavelength regions in which attenuation is low and matched to the capability of a transmitter to generate light efficiently and a receiver to carry out detection.

λ *(nm)*	*Approx. loss (dB/km)*
820–880	2.2
1200–1320	0.6
1550–1610	0.2

From the above data it is seen that the range 1550 to 1610 nm is most preferable. From the point of view of dispersion, the low intramodal dispersion wavelength of about 1300nm is most suitable.

(*ii*) Extrinsic Attenuation or Bending losses: Extrinsic attenuation is caused by two external mechanisms: ***macrobending*** or ***microbending***. Both of them cause a reduction of optical power. If a bend is imposed on an optical fiber, strain is placed on the fiber along the region that is bent. The bending strain affects the refractive index and the critical angle of the light ray in that specific area. As a result, the condition for total internal reflection is no longer satisfied. Hence, light traveling in the core can refract out, and loss occurs.

(*a*) **Macrobend losses:** A ***macrobend*** is a large-scale bend that is visible. When a fibre is bent through a large angle, strain is placed on the fiber along the region that is bent. The bending strain will affect the refractive index and the critical angle of the light ray in that specific area. As a result, light traveling in the core can refract out, and loss occurs. (Fig. 47.26). To prevent macrobends, optical fiber has a *minimum bend* *radius* specification that should not be exceeded. This is a restriction on how much bend a fiber can withstand before experiencing problems in optical performance or mechanical reliability.

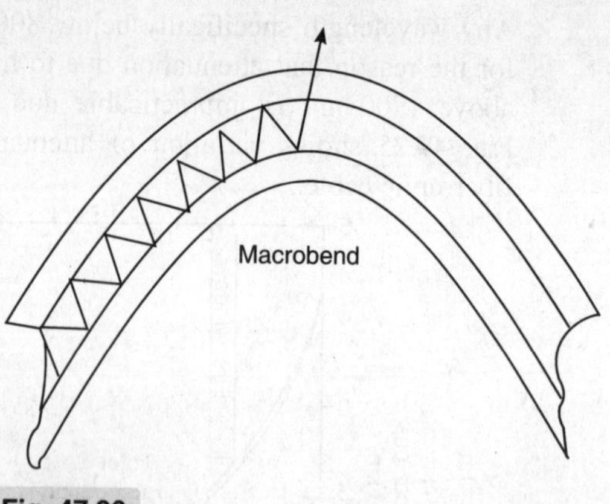

Fig. 47.26: Macrobend loss

(*b*) **Microbend losses:** ***Microbend*** is a small-scale distortion. It is localized and generally indicative of pressure on the fiber. Microbending might be related to temperature, tensile stress, or crushing force. Microbending is caused by imperfections in the cylindrical geometry of fiber during the manufacturing process or installation processes. The bend may not be clearly visible upon inspection. Structural variations in the fibre or fibre deformation cause radiation of light away from the fibre (Fig. 47.27). Microbending may occur, for example, due to winding of optical fibre cable over spools. Light rays get scattered at the small bends and escape into the cladding. Such losses are known as microbend losses.

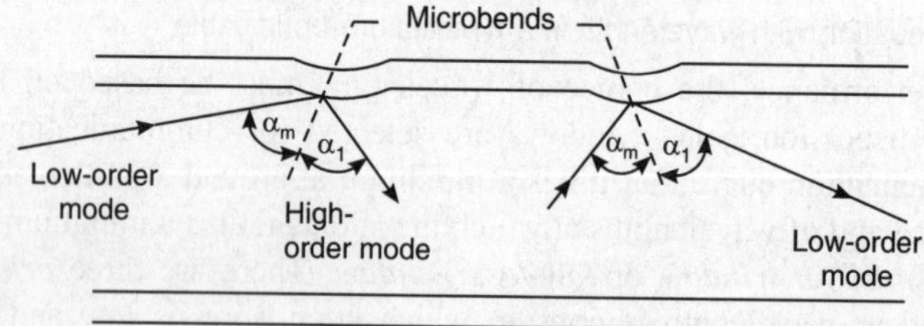

Fig. 47.27: Microbend losses

Example 47.12: *Optical power of 1 mW is launched into an optical fibre of length 100 m. If the power emerging from the other end is 0.3.mW, calculate the fibre attenuation.*

Solution: Attenuation,
$$\alpha = \frac{10}{L}\log\frac{P_i}{P_o} = \frac{10}{0.1\text{ km}}\log\frac{1\text{mW}}{0.3\text{mW}} = \mathbf{52.3\ dB/km}$$

Example 47.13: *What is the attenuation in dB/km, if 15% of the power fed at the launching end of a $\frac{1}{2}$ km fibre is lost during propagation?*

Solution: Attenuation,
$$\alpha = \frac{10}{L}\log\frac{P_i}{P_o} = \frac{10}{0.5\text{ km}}\log\frac{1}{0.15} = \mathbf{16.48\ dB/km}$$

47.16.2 Distortion

In an optical fibre communication system, the information (signal) is coded in the form of discrete pulses of light, which are transmitted through the fibre. The light pulses are of a given width, amplitude and interval. The number of pulses that can be sent per unit time will determine the information capacity of the fibre. More information can be sent by optical cable when distinct pulses can be transmitted in more rapid succession. The pulses travel through the transmitting medium (i.e., optical fibre) and reach the detector at the receiving end. For the information to be retrieved at the detector, it is necessary that the optical pulses are well resolved in time. However, the light pulses broaden and spread into a wider time interval because of the different times taken by different rays propagating through the fibre. This phenomenon is known as ***distortion*** or ***pulse dispersion***.

Hence, even though two pulses may be well resolved at the input end, they may overlap on each other at the output end, as shown in Fig. 47.28. It is obvious that the pulse broadening depends on the length of the travel of the pulses through the fibre. Hence, dispersion is expressed in units of **ns/km** (time/distance).

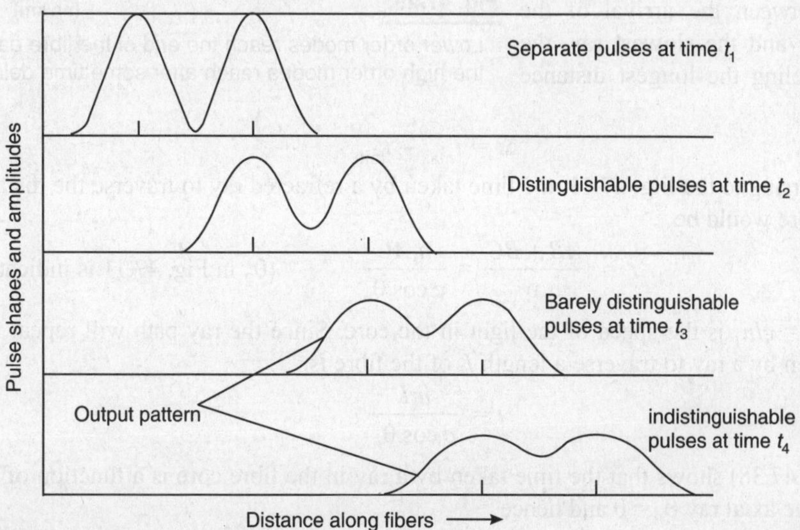

Fig. 47.28:

Distortion of the pulses traveling along a fibre

The following three different dispersion mechanisms determine the distortion of the signal in an optical fibre. They are

(*i*) Intermodal dispersion and

(*ii*) Intramodal dispersion.

Intramodal dispersion is again divided into the following two types.

(*a*) Material dispersion

(*b*) Waveguide dispersion

***(i)* Intermodal Dispersion:** Intermodal dispersion occurs as a result of the differences in the group velocities of the modes. For example, let us consider the propagation of a pulse through a multimode fibre. The power associated with the single pulse gets distributed into the various modes or paths guided by the fibre. The lower order modes (rays reflected at larger angles) travel a greater distance than the higher order modes (lower angle rays). The path length along the axis of the fibre is shorter while the other zigzag paths are longer. Because of this difference, the lower order modes reach the end of the fibre earlier while the high order modes reach after some time delay. As a result, light pulses broaden as they travel down the fibre, causing signal distortion. The output pulses no longer resemble the input pulses (see

Fig. 47.29). This type of distortion is known as ***intermodal*** or simply ***modal dispersion***. This imposes limitation on the separation between successive pulses and thereby reduces the transmission rate and capacity.

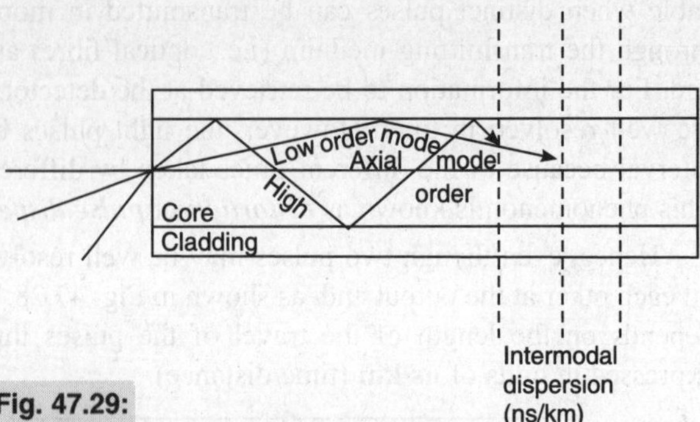

Fig. 47.29: Lower order modes reach the end of the fibre earlier while the high order modes reach after some time delay

Expression for total time delay due to modal dispersion in Step-Index fibre: The total time delay between the arrival of the axial ray and the slowest ray, the one traveling the longest distance is

$$\Delta t = t_{\max} - t_{\min} \quad ...(47.37)$$

Referring to the Fig. 47.11, the time taken by a refracted ray to traverse the distance *ABC* of the fibre would be

$$t' = \frac{AB + BC}{\upsilon} = \frac{n_1 AC}{c \cos \theta_r} \qquad (\theta_2 \text{ in Fig. 47.11 is indicated as } \theta_2)$$

where $\upsilon = c/n_1$ is the speed of the light in the core. Since the ray path will repeat itself, the time taken by a ray to traverse a length *L* of the fibre is

$$t = \frac{n_1 L}{c \cos \theta_r} \quad ...(47.38)$$

Eq. (47.38) shows that the time taken by a ray in the fibre core is a function of the angle θ_r. For the axial ray $\theta_r = 0$ and hence

$$t_{\min} = \frac{n_1 L}{c} \quad ...(47.39)$$

In case of the ray that travels the longest path, $\theta_r = \theta_C$. Therefore,

$$t_{\max} = \frac{n_1 L}{c \cos \theta_C}$$

Using Eq. (47.5) into the above expression, we get

$$t_{\max} = \frac{n_1^2 L}{n_2 c} \quad ...(47.40)$$

Therefore, making use of Eq. (47.40) and Eq. (47.39) into Eq. (47.37), we obtain

$$\Delta t = \frac{n_1 L}{c}\left[\frac{n_1}{n_2} - 1\right] \quad ...(47.41)$$

Using the Eq. (47.13) for fractional refractive index change into Eq. (47.41), we get

$$\Delta t = \frac{n_1 L}{c}\left[\frac{\Delta}{1 - \Delta}\right] \quad ...(47.42)$$

We can also express the Eq. (47.41) in the following form.

$$\Delta t = \frac{n_1 L}{c}\left[\frac{n_1 - n_2}{n_2}\right] = \frac{n_1 L}{c}\left[\frac{n_1 - n_2}{n_2}\right]\left[\frac{n_1 + n_2}{n_1 + n_2}\right]$$

$$= \frac{n_1 L}{c}\left[\frac{(n_1^2 - n_2^2)}{n_2(n_1 + n_2)}\right] = \frac{n_1 L}{c}\frac{(n_1^2 - n_2^2)}{2n_1 n_2} \quad (\text{as } n_1 \approx n_2)$$

or $$\Delta t = \frac{L}{2n_2 c}(NA)^2 \qquad \text{...(47.43)}$$

It is seen from the Eq. (47.43) that the time delay is proportional to the square of the value of NA. Therefore, a large NA fibre allows more modes of propagation of light, which will result in greater modal dispersion. A smaller NA limits the number of modes, hence reduces dispersion. It is further seen that the intermodal dispersion does not depend upon the spectral width of the source. It follows that a light pulse from an ideal monochromatic source would still get broadened.

Example 47.14: *A step-index fibre is with a core of refractive index 1.55 and cladding of refractive index 1.51. Compute the intermodal dispersion per kilometer of length of the fibre and the total dispersion in a 15 km length of the fibre.*

Solution: $$\Delta t = \frac{n_1 L}{c}\left[\frac{n_1}{n_2} - 1\right] = \frac{1.55 \times 10^3 \text{ m}}{3 \times 10^8 \text{ m/s}}\left[\frac{1.55}{1.51} - 1\right] = \mathbf{138\ ns/km.}$$

Total dispersion for 15 km length = $\Delta t \times 15$ km = (138 ns/km) × 15 km = **2.07 μs.**

***(ii)* Intramodal Dispersion:** Intramodal dispersion is the spreading of light pulse within a single mode. The two main causes of Intramodal dispersion are (*a*) material dispersion and (*b*) waveguide dispersion.

(*a*) Material Dispersion: Glass is a dispersive medium. A light pulse is a wave packet, composed of a group of components of different wavelengths. The different wavelength components will propagate at different speeds along the fibre (Fig. 47.30). The short wavelength components travel slower than long wavelength components, eventually causing the light pulse to broaden. This type of distortion is known as **material dispersion**. It is often called the **chromatic dispersion**. Obviously, the spectral width of the source determines the extent of material dispersion.

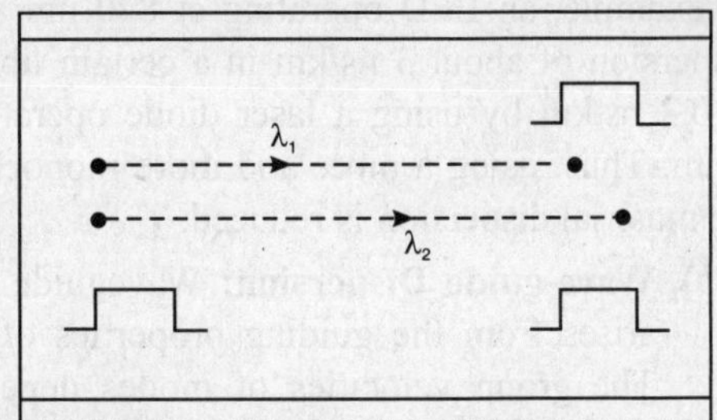

Fig. 47.30: Different wavelength components propagate at different speeds along the fibre

Expression for time delay due to material dispersion: Let us consider a plane wave propagating in fiber core. It is represented by $\Psi \propto \exp(kx - \omega t)$. The wave number k is given by

$$k = \frac{2\pi}{\lambda} = \frac{2\pi}{\lambda_0}\cdot\frac{\lambda_0}{\lambda} = \frac{2\pi}{\lambda_0}\cdot n = \frac{2\pi\nu}{c}\cdot n$$

or $$k = \frac{\omega n}{c} \qquad \text{...(47.44)}$$

and $$\omega = 2\pi\nu = \frac{2\pi}{\lambda}c \qquad \text{...(47.45)}$$

A wave packet of finite spread of wavelengths travels with group velocity υ_g is given by

$$\upsilon_g = \frac{d\omega}{dk}$$

$$\therefore \qquad \frac{1}{\upsilon_g} = \frac{dk}{d\omega} = \frac{d}{d\omega}\left(\frac{\omega n}{c}\right) = \frac{n}{c} + \frac{\omega}{c}\frac{dn}{d\omega} = \frac{1}{c}\left[n + \omega\frac{dn}{d\omega}\right]$$

But $$\frac{dn}{d\omega} = \frac{dn}{d\lambda}\cdot\frac{d\lambda}{d\omega} = -\frac{\lambda^2}{2\pi c}\cdot\frac{dn}{d\lambda}$$

$$\therefore \qquad \frac{1}{\upsilon_g} = \frac{1}{c}\left[n - \frac{\omega\lambda^2}{2\pi c}\frac{dn}{d\lambda}\right] = \frac{1}{c}\left[n - \lambda\frac{dn}{d\lambda}\right]$$

As the signal propagates through the fibre, each spectral component can be assumed to travel independently and to undergo a time delay per unit length in the direction of propagation, which is given by

$$t_{mat} = \frac{L}{\upsilon_g} = \frac{L}{c}\left[n - \lambda\frac{dn}{d\lambda}\right] \qquad ...(47.46)$$

The pulse spread Δt_{mat} for a source of spectral width $\Delta\lambda$ is found by differentiating the Eq. (47.46) with respect to λ and then multiplying by $\Delta\lambda$. Thus,

$$\Delta t_{mat} = \frac{dt_{mat}}{d\lambda}\Delta\lambda = -\frac{L\lambda}{c}\frac{d^2n}{d\lambda^2}\Delta\lambda = D_{mat}(\lambda)L\Delta\lambda \qquad ...(47.47)$$

where $D_{mat}(\lambda)$ is the material dispersion.

$$D_{mat}(\lambda) = -\frac{\lambda}{c}\frac{d^2n}{d\lambda^2} \qquad ...(47.48)$$

From the Eq. (47.48) it is seen that the material dispersion can be reduced either by choosing sources with narrower spectral range or by operating at longer wavelengths. To cite an example, an LED operating at 820 nm and having a spectral width of 38 nm results in dispersion of about 3 ns/km in a certain fibre. In the same fibre, dispersion can be reduced to 0.3 ns/km by using a laser diode operating at 1140 nm and having a spectral width of 3 nm. Thus, using a more and more monochromatic source operating at higher wavelength, the material dispersion is reduced.

(*b*) **Wave-guide Dispersion:** Waveguide dispersion arises from the guiding properties of the fibre. The group velocities of modes depend on the wavelength. Hence, the effective refractive index for any mode varies with wavelength. It is equivalent to the angle between the ray and the fibre axis varying with wavelength which subsequently leads to a variation in the transmission times for the rays and hence dispersion (see Fig. 47.31). Waveguide dispersion is generally small in MMF, but it is important in SMF.

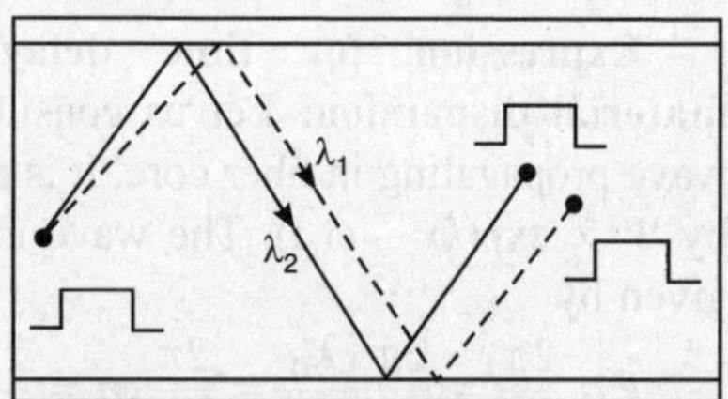

Fig. 47.31: Wave guide dispersion

The intermodal distortion can be reduced if graded index fibre is used. In case of a graded index fibre, the refractive index is larger at the center and it gradually decreases away from the center. A pulse traveling along the axis of the fibre travels along a shorter path but it takes longer time to reach the end of the fibre since it is traveling through a medium of higher refractive index. On the other hand, the pulse traveling away from the axis travels a longer distance but takes lesser time since it is traveling through a medium of lower refractive index. As a result both the pulses reach the end of the fibre simultaneously. Thus, using a GRIN fibre can reduce the problem of intermodal dispersion. Low NA fibres exhibit smaller dispersion. Dispersion may be restricted by a careful selection of low NA fibre and a narrow spectral width fibre.

In a MMF, all three pulse spreading mechanism exist simultaneously. In case of SMF, only material and wave-guide dispersion exist.

Total Dispersion: All the above three dispersions contribute pulse spreading during signal transmission through an optical fibre. The total dispersion introduced by an optical fibre is given by the root mean square value of all the three dispersions. Thus,

$$(\Delta t)_T = \sqrt{(\Delta t)^2_{\text{inter modal}} + (\Delta t)^2_{\text{mat}} + (\Delta t)^2_{\text{wg}}} \qquad \text{...(47.49)}$$

47.17 BANDWIDTH

It is learnt in the above section that various dispersion mechanisms cause broadening of the information signal in time domain. If the pulses spread more, they can interfere with the adjacent pulses resulting in *Inter Symbol Interference* or ISI in short and there can be so much of ISI that it becomes impossible to distinguish between the individual pulses. Therefore, for a given broadening, the pulses have to be separated by a minimum time interval in order to avoid overlapping of the pulses. This would determine the ultimate information-carrying capacity of the system. When the pulse separation is increased, the data transfer rate decreases. Thus, broadening of pulses puts an upper limit on the rate of pulse transmission. To a first approximation, it may be taken that the bandwidth in hertz is equal to the digital bit rate. Thus, $B_T = \frac{1}{\tau} = B$, where τ is the input pulse duration. In other words, the maximum allowable transmission rate is called **bandwidth**. In practice, the fibre bandwidth is expressed in terms of MHz.km, a product of frequency and distance. This is known as **bandwidth-distance** product, which specifies the usable bandwidth over a definite distance. With the increase in distance, different dispersion effects would increase in the optical fibre and as a result the usable bandwidth reduces. The attenuation per kilometer and the bandwidth-kilometer product are the important performance parameters of optical fibres.

47.18 CHARACTERISTICS OF THE FIBRES

(i) Step-index single-mode fibre

(*a*) It has a very small core diameter, typically of about 10 μm.

(*b*) Its numerical aperture is very small.

(*c*) It supports only one mode in which the entire light energy is concentrated.

(*d*) A single mode step index fibre is designed to have a V number between 0 and 2.4.

(*e*) Because of a single mode of propagation, loss due to intermodal dispersion does not exist. With careful choice of material, dimensions, and wavelength, the total dispersion can be made extremely small.

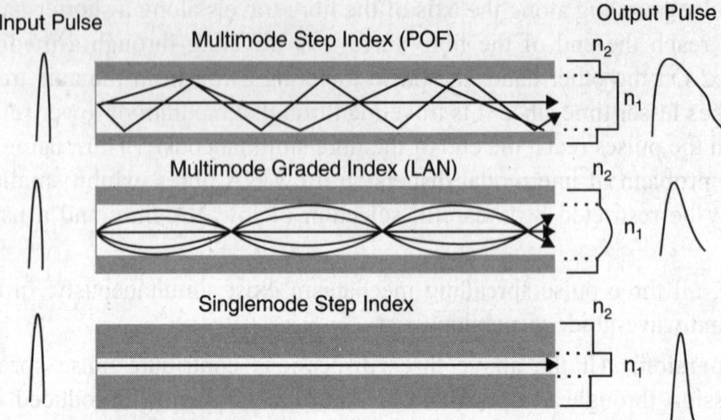

Fig. 47.32:

(*f*) The attenuation is least.

(*g*) The single mode fibres carry higher bandwidth than multimode fiber.

(*h*) It requires a monochromatic and coherent light source. Therefore, laser diodes are used along with single mode fibres.

(*i*) Advantages

- No degradation of signal
- Low dispersion makes the fibre suitable for use with high data rates. Single-mode fiber gives higher transmission rate and up to 50 times more distance than multimode.
- Highly suited for communications.

(*j*) Disadvantages

- Manufacturing and handling of SMF is more difficult.
- The fibre is costlier.
- Launching of light into fibre is difficult.
- Coupling is difficult.

(*k*) Applications

- Used as under water cables

(ii) Step-index multi-mode fibre

(*a*) It has larger core diameter, typically ranging between 50-100 μm.

(*b*) The numerical aperture is larger and it is of the order of 0.3.

(*c*) Larger numerical aperture allows more number of modes, which causes larger dispersion. The dispersion is mostly intermodal.

(*d*) Attenuation is high.

(*e*) Incoherent sources like LEDs can be used as light sources with multimode fibres.

(*f*) Advantages

- The multimode step index fibre is relatively easy to manufacture and is less expensive.
- LED or laser source can be used.
- Launching of light into fibre is easier.
- It is easier to couple multi-mode fibres with other fibres.

(*g*) Disadvantages

- Has smaller bandwidth.
- Due to higher dispersion data rate is lower and transmission is less efficient.
- It is less suitable for long distance communications.

(*h*) Applications

- Used in data links.

(iii) Graded-index multi-mode fibre

(*a*) Core diameter is in the range of 50-100 μm

(*b*) Numerical aperture is smaller than that of step-index multimode fibre.

(*c*) The number of modes in a graded index fibre is about half that in a similar multimode step-index fibre.

(*d*) Has medium attenuation.

(*e*) Intermodal dispersion is zero, but material dispersion is present.

(*f*) Has better bandwidth than multimode step-index fibre.

(*g*) Advantages

(*h*) Either an LED or a laser can be used as the source of light with GRIN fibres.

(*i*) Disadvantages

- The manufacture of graded index fibre is more complex. Hence, it is the most expensive fibre.
- Coupling fibre to the light source is difficult.

(*j*) Applications

- Used in telephone links.

Table 47.1 gives a comparison of the different optical fibres.

Table 47.1: *Comparison of Different Types of Fibres*

S.No.	Feature	SMF	MMF	GRIN
1.	Typical core diametre	10 μm	50 to 100 μm	50 to 100 μm
2.	Δn	Very small	Large	
3.	Numerical Aperture	Small	Large	Smaller than that of MMF
4.	Number of modes	Only one	Many	Many
5.	Attenuation	Least	High	Lower
6.	Dispersion	Zero Intermodal dispersion	Large	Intermodal dispersion is zero. Material dispersion is present.
7.	Bandwidth	>3 GHz-km	<200 MHz-km	200 MHz-km to 3 GHz-km
8.	Advantages	No degradation of signal, High data transfer rate, Highly suitable for communications	Less expensive, LED or laser source can be used, Launching of light is easier, Coupling of fibres is easier.	LED or Laser light source can be used.

S.No.	Feature	SMF	MMF	GRIN
9.	Drawbacks	Costly, Requires a laser source, Coupling is difficult, Launching of light into fibre difficult, Intensity gets reduced.	Degrades signal, less suitable for communications.	
10.	Applications	Under water cables	Data Links	Telephone Lines

47.19 APPLICATIONS

Transmission of light via an optical fibre has a wide variety of applications. We discuss here some of the applications. Broadly, optical fibres have three different applications, apart from other miscellaneous applications.

(*i*) They are used for illumination and short distance transmission of images.

(*ii*) They are used as wave-guides in telecommunications.

(*iii*) They are used in fabricating a new family of sensors.

47.19.1 Illumination and Image Transmission

A large number of fibres whose ends are bound together, ground and polished, form flexible bundles. One of the ends of the bundle acts as an input end while the other acts as an output end. If the relative positions of the fibre terminations at both the ends are not the same, and if no attempt is made to align the fibres in an orderly array, the bundle is said to be an *incoherent bundle*. In such a case, there would not be any correlation in the positions of the fibre terminations at one end of the bundle with that at the other end of the bundle. The primary function of such bundles is simply to conduct light from one region to another. Such ***flexible light carriers*** are relatively easy to make and inexpensive. They are used for illumination purpose.

When the fibres are carefully arranged so that their terminations occupy the same relative positions in both of the bound ends of the bundle, the bundle is said to be coherent. Such a bundle is capable of transmitting undistorted images to a distant place. When one end of such a ***flexible image carrier*** is placed face down flat on an illuminated surface, a point-by-point image of the surface appears at the other end.

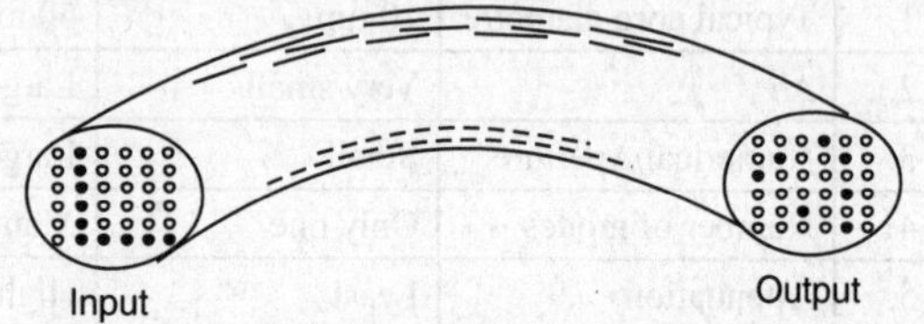

(a) Incoherent bundle-image of letter L scrambled as dark sports

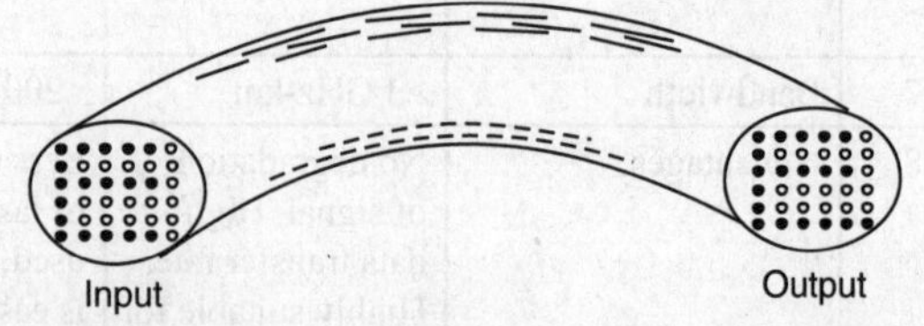

(b) Coherent bundle - image of letter E clearly seen

Fig. 47.33: Fibre Optic bundles

When the fibres are carefully arranged so that their terminations occupy the same relative positions in both of the bound ends of the bundle, the bundle is said to be coherent. Such a bundle is capable of transmitting undistorted images to a distant place. When one end of such a ***flexible image carrier*** is placed face down flat on an illuminated surface, a point-by-point image of the surface appears at the other end.

(a) Endoscopes: The most important application of the coherent bundles is in diagnostic field as an optical endoscope. An endoscope is an optical instrument which facilitates visual inspection of internal parts of a human body. It is also called a fiberoscope. It requires about 10,000 fibres forming a bundle of 1 mm diameter and it can resolve objects with a separation of 70 μm. By allowing direct viewing of what was formerly hidden, a fiberoscope has become a vital diagnostic tool for industry and medicine. The broncho-fiberoscope, gastrointestinal fiberoscope, laparoscope etc are the endoscopes used in medical diagnosis.

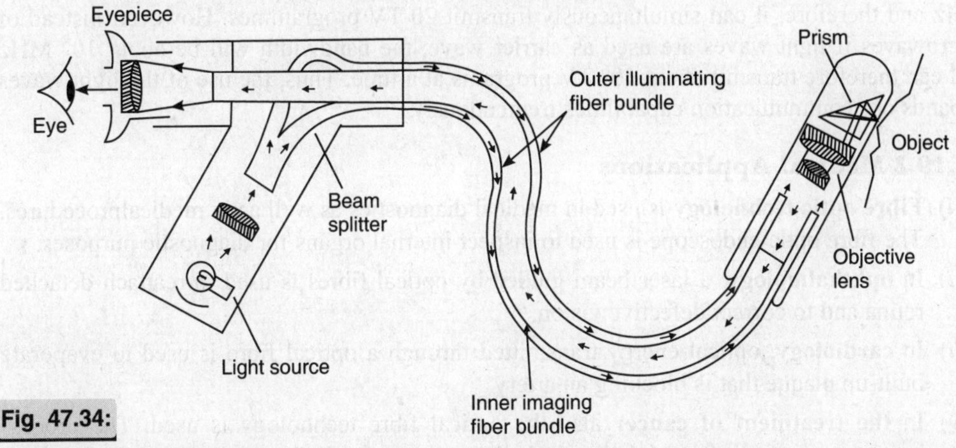

Fig. 47.34: Illustration of Endoscopes

Fig. 47.34 shows the schematic diagram of a flexible endoscope. The endoscopes are designed using low quality, large diameter and short silica fibres. There are two fiber bundles in an endoscope. One of them is used to illuminate the interior of the body and the other is used to collect the reflected light from the illuminated area. A telescope system is added in the internal part of endoscope for obtaining a wider field of view and better image quality. At the object end, there is an assembly of objective lens and prism which are kept in a transparent glass cover and at the viewing end, there is an eye lens. The input end of the endoscope contains a powerful light source. The light rays are focused and coupled to the illuminating fiber bundle. The light rays are finally incident on the surface of the object under study. The light rays reflected from the object surface are received by the objective lens through a prism and are transmitted through the imaging fibre bundle to the viewing end of the scope. Here the eye piece reconstructs the image of the object and one can view the image of the surface of the object. Endoscope pictures can be recorded on a videotape recorder.

47.19.2 Optical Communications

Traditionally, electronic communications were carried out by sending electrical signals through copper cables, coaxial cables or waveguides. In recent years optical fibres are being used, where light signals replace electrical signals. A basic communications system consists of a transmitter, a receiver and an information pathway. Normally, the information to be communicated is a non-electrical message, which is to be converted first into an electrical form. The conversion is done by a transducer. For example, a microphone converts sound waves into currents. Similarly, a video camera converts images into currents. These electrical messages are of low frequency and cannot be transmitted directly. Therefore, they are superposed on a carrier wave of very high frequency. The process of imposing a message signal on a carrier wave is called modulation. Two different techniques of modulation are available. In analog modulation a continuous wave carries the message. In digital modulation message is transmitted in discrete form using binary digits. The message travels along the transmission channel and is received at the receiver. The receiver demodulates the modulated wave and separates out the message and feeds to a transducer such a loud speaker. The bandwidth requirement of the message and the bandwidth of the carrier determine the number of messages that can be simultaneously transmitted on an information channel. For example, a bandwidth of 4 kHz is required for voice transmission while 6 MHz bandwidth is required for TV signal transmission. When signals are transmitted in analog form the carrier should have double the above bandwidth. The normal TV communications has a bandwidth of 250 MHz and therefore, it can simultaneously transmit 20 TV programmes. However, instead of microwaves if light waves are used as carrier wave, the bandwidth will be about 10^8 MHz and can therefore transmit about 10^6 TV programs at a time. Thus, the use of the light waves expands our communication capabilities tremendously.

47.19.3 Medical Applications

(*i*) **Fibre optic technology** is used in medical diagnostics as well as in medicalprocedures. The fibre optic endoscope is used to inspect internal organs for diagnostic purposes.

(*ii*) In **ophthalmology**, a laser beam guided by optical fibres is used to reattach detached retina and to correct defective vision.

(*iii*) In **cardiology**, optical energy transmitted through a optical fibre is used to evaporate built-up plaque that is blocking an artery.

(*iv*) In **the treatment of cancer** also the optical fibre technology is used. The process involves injection of special chemicals that penetrate only the cancerous cells. Infra red energy is transmitted via the fibre illuminates the affected area and is absorbed by the special chemical in the cancerous cells. The heat generated destroys the cancerous cells.

47.19.4 Military Applications

An aircraft, a ship or a tank needs tons of copper wire for wiring of the communication equipment, control mechanisms, instrument panel illumination etc. Use of optical fibre in place of copper reduces weight and further maintains true communication silence to the enemy. For example, a shipboard radar system requires about 250 m of coaxial cable, with a weight of 7 tons and a diameter of 45 cm. These cables can be replaced by optical fibre weighing 20 kg and measuring 2.5 cm in diameter.

Fibre guided missiles are used in recent wars. Sensors placed on the missile, transmit video information through the optical fibre to a ground control and receive commands again.

47.20 FIBRE OPTIC COMMUNICATION SYSTEM

A fibre optic communication system is very much similar to a traditional communications system and has three major components. A ***transmitter*** converts electrical signal to light signals, an ***optical fibre*** transmits the signals and a ***receiver*** captures the signals at the other end of the fibre and converts them to electrical signals.

The block diagram Fig. 47.35 illustrates a typical communications system. The transmitter consists of a light source supported by necessary drive circuits. A transducer converts a non-electrical message into an electrical signal and is fed to a light source. The light source is a miniature source, either a light emitting diode or a semiconductor laser. In either case, light is emitted in the IR range with a wavelength of 850 nm (0.85 μm), 1300 nm (1.3 μm) or 1550 nm (1.55 μm). The light waves are modulated with the signal. Analog modulation is accomplished by changing the intensity of the light beam from the laser diode or LED and digital modulation is accomplished by flashing the laser diode or LED on and off at very fast rate. A pulse of light stands for the number 1 and the absence of light at a specified time stand for zero.

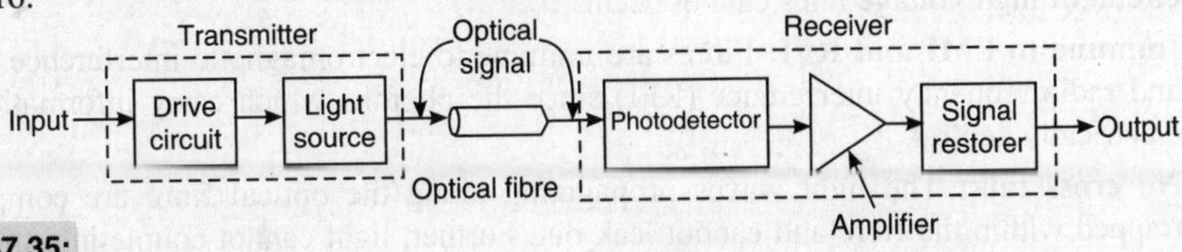

Fig. 47.35:
Illustration of a typical fibre optic communication link

A message can be transmitted by a particular sequence of these 1s or 0s. If the receiver is programmed to recognize such digital patterns, it can reconstruct the original message.

Though the digital modulation requires more complicated equipment such as encoders and decoders and also more bandwidth than analog modulation, it allows greater transmission distance with the same power. This is a great advantage and hence digital modulation has become popular and widely used nowadays. The transmitter feeds the analog or digitally modulated light wave to the transmission channel, namely optical fibre link. The optical signal travelling through the fibre will get attenuated progressively and distorted due to dispersion effects. Therefore, repeaters are to be used at specific intervals to regenerate the signal. At the end of the fibre, an output coupler directs the light from the fibre onto a semiconductor photodiode, which converts the light signals to electrical signals. The photodetector converts the light waves into electrical signals which are then amplified and decoded to obtain the message. The output is fed to a suitable transducer to convert it into an audio or video form.

Applications: Optical fibre communications systems can be broadly classified into two groups: (*i*) local and intermediate range systems where the distances involved are small and (*ii*) long-haul systems where cables span large distances.

(*i*) **Local area networks:** The local area network (LAN) is a computer oriented communication system. LAN operates over short distances of about 1 to 2 km. It is multiuser oriented system. In LAN, a number of computer terminals are interconnected over a common channel allowing each computer to use data and programs from any other. An optical data bus offers a great reduction in cost and increases enormously the information handling capacity.

(*ii*) **Long-haul communication:** One of the most important applications of fibre optic communication is long-haul communication. Long-haul communication systems are used for long distances, 10 km or more. Telephone cables connecting various countries come under this category. A rather sophisticated long-haul network is the NSFNET which links six supercomputer centres throughout U.S.A.

47.21 MERITS OF OPTICAL FIBRES

Optical fibres have many advantageous features that are not found in conducting wires. Some of the important advantages are given here.

(*i*) **Cheaper:** Optical fibres are made from silica (SiO_2) which is one of the most abundant materials on the earth. The overall cost of a fibre optic communication is lower than that of an equivalent cable communication system.

(ii) **Smaller in size, lighter in weight, flexible yet strong:** The cross section of an optical fibre is about a few hundred microns. Hence, the fibres are less bulky. Typically, a RG-19/U coaxial cable weighs about 1100 kg/km whereas a PCS fibre cable weighs 6 kg/km only. Optical fibres are quite flexible and strong.

(iii) **Not hazardous:** As fibres are made of insulating materials, accidents due to short circuit of high voltage lines cannot occur.

(iv) **Immune to EMI and RFI:** Fibres are immune to electromagnetic interference (EMI) and radiofrequency interference (RFI) since the photons which carry information are electrically neutral.

(v) **No cross talk:** The light waves propagating along the optical fibre are completely trapped within the fibre and cannot leak out. Further, light cannot couple into the fibre from sides. In view of these features, possibility of cross talk is minimized when optical fibre is used. Therefore, transmission is more secure and private.

(vi) **Wider bandwidth:** Optical fibres have ability to carry large amounts of information. While a telephone cable composed of 900 pairs of wire can handle 10,000 calls, a 1mm optical fibre can transmit 50,000 calls.

(vii) **Low loss per unit length:** The transmission loss per unit length of an optical fibre is about 4 dB/km. Therefore, longer cable-runs between repeaters are feasible. If copper cables are used, the repeaters are to be spaced at intervals of about 2 km. In case of optical fibres, the interval can be as large as 100 km and above.

Disadvantages: Installation and maintenance of optical fibres require a new set of skills. They require specialized and costly equipment like optical time domain reflectometers etc. All this means heavy investment.

47.22 FIBRE OPTIC SENSORS

Fibre optic sensors are transducers, which generally consist of a light source coupled with an optical fibre and a light detector held at the receiver-end. The fibres used could be either multimode or single mode type. The sensors can be used to measure pressure, temperature, strain, the acoustic field, magnetic field, etc physical parameters. The advantages of these sensors are that they are lighter, occupy lesser volume and are cheaper.

The optical fibre merely carries the light beam in some of the sensors and in others the fibre itself acts as the sensor. We study here a few typical examples of the sensors.

47.22.1 Temperature Sensors

(a) Intensity modulated sensor

Principle: In this type of sensor, temperature is measured by the modulation of intensity of the reflected light from a target, a silicon layer. The operation of the temperature sensor is based on the 1 μm wavelength light-absorption characteristics of silicon as a function of temperature. Depending on the temperature, the amount of light absorbed by the silicon layer varies. The change in intensity of the reflected light is proportional to the change in temperature.

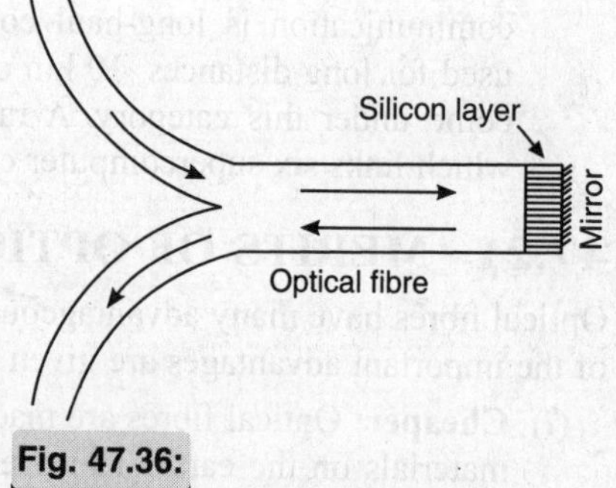

Fig. 47.36: A typical temperature sensor

Construction: Fig.47.36 illustrates a temperature sensor with a multimode fibre. The fibre is coated at one end with a thin silicon layer. The silicon layer is in turn coated with a reflective coating at the back. The silicon layer acts as the sensing element.

Working: The light from a light source is launched into the fibre from one of the ends of one of its branches (see Fig. 47.36). It passes first through the fibre and then through the silicon layer. The mirror coating at the other end of the silicon layer reflects the light back which again travels through the silicon layer. The reflected light emerges out through another branch of multimode fibre and is collected by a photo detector. The amount of the reflected light is converted into voltage by the photodetector. The absorption of light by the silicon layer varies with temperature and the variation modulates the intensity of the light received at the detector. Temperature measurements can be made with a sensitivity of 0.001°C.

(b) Phase modulated sensor

Principle: This temperature sensor is based on phase variation resulting due to the variation of refractive index of the optical fibre under the influence of temperature.

Construction: Fig. 47.37 shows a single mode fibre sensor arranged in what is known as the Mach-Zehnder arrangement. A light source produces light. A beam splitter divides the light into two parts and sends light through the sensing fibre and the reference fibre. Light passing out of the two fibre elements is fed to a detector, which measures the difference in phase of the two light waves. Accurate measurements of the temperature may be obtained from these patterns.

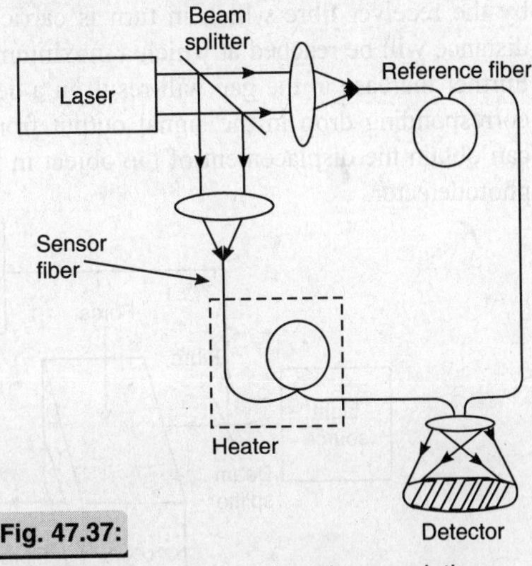

Fig. 47.37: Temperature sensor using phase variations

Working: The light from the source is divided into two parts by the beam splitter. One part is allowed through sensor fibre, and the other part is passed through the reference fibre. Light rays entering the fibres are coherent and have the same phase. Prior to heating, the optical path lengths of the two fibre elements are same and hence both the outputs will be in phase. When the sensor fibre is subjected to heating, the temperature causes a change in the refractive index of the optical fibre. Therefore, the light coming out of the two fibres at the other end will have phase difference due to difference in optical path difference caused by the heating. When the rays are superposed, they interfere and interference pattern will be observed. As temperature increases, the phase difference between the two outputs increases and is observed as a displacement of the fringe pattern. By determining the fringe displacement, we can determine the magnitude of temperature.

47.22.2 Displacement Sensor

Principle: The basic principle employed in displacement sensor consists of using an adjacent pair of fibre optic elements, one to carry light from a remote source to an object whose displacement or motion is to be measured and the other to receive the light reflected from the object and carry it back to a remote photodetector.

Construction: Fig. 47.38 shows the arrangement of a displacement sensor. Two separate optical fibres are positioned adjacent to each other. One of them transmits light coming from a light source. The other fibre receives light reflected from the object under study and passes it on to a photodetector.

Working: Light from the transmitting fibre element is incident on the object under study. The light receiver fibre element is positioned adjacent to the transmitting fibre. If the gap between the object and the fibre elements is zero, the light from the transmit fibre would be directly reflected back into itself and little or no light would go into the receiver fibre. When the object moves away, the gap increases and some of the reflected is captured by the receiver fibre which in turn is carried to the photodetector. As the gap increases, a distance will be reached at which a maximum reflected light is received by the photodetector. Further increase in the gap will result in a decrease in the light at the receiver fibre face and corresponding drop in the signal output from the photodetector. By proper calibration, we can obtain the displacement of the object in terms of the strength of the output signal of the photodetector.

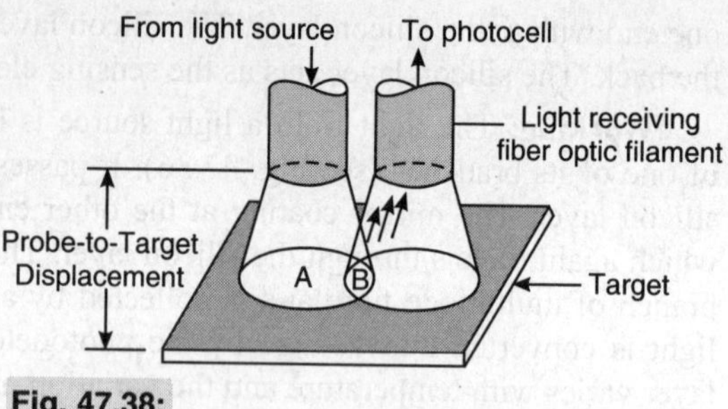

Fig. 47.38: Object moving away from probe, causes increase in reflected light intensity

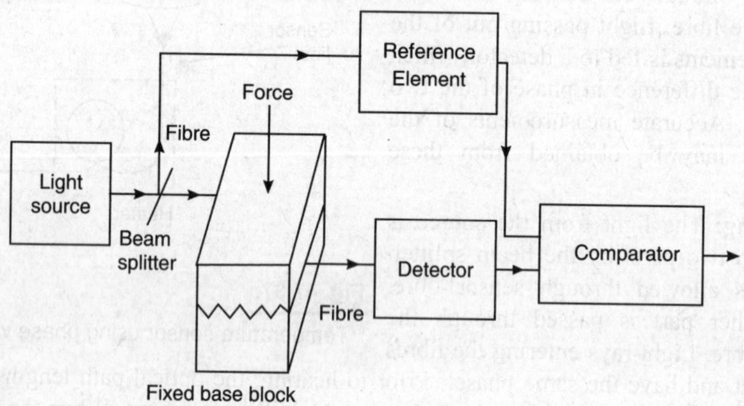

Fig. 47.39: Force sensor using microbend losses

47.22.3 Force Sensor

Principle: This sensor is based on variations of light intensity. When an optical fibre is pressed, a small change occurs in light propagation direction due to microbending of the fibre. As a result, energy from one mode is transferred to another mode through mode coupling. In addition, higher order modes are likely to change into radiation modes. All these effects cause a loss in intensity of the light transmitted through the fibre. Therefore, the change in intensity of the transmitted light is proportional to the force applied on the optical fibre.

Construction: An optical fibre without jacket is held between two corrugated blocks, as shown in Fig. 47.39. Light from a source is divided into two parts by a beam splitter. One part is allowed through the fibre that is held between the blocks, which acts as a sensor element, and the other part is passed through an exactly identical fibre, which acts as a reference element. Photodetectors measure the intensity of transmitted light. A comparator detects the difference between the light intensities.

Working: When a force is applied on the upper corrugated block, the fibre is pressed and microbend losses are introduced in the fibre. The microbendings produce mode coupling such that energy of one mode is transferred to other higher modes. Also, higher modes are converted into leaky modes which reduce the amount of energy transmitted though the fibre. The changes in the light intensity due to these losses are detected by a photodetector and compared with that of the light coming out of the reference element. The change in intensity is related to the force and hence is a measure of the applied force.

47.22.4 Liquid Level Detector

Principle: The liquid level detector described here is based on the principle of total internal reflection.

Construction: A simple liquid level detector is shown in Fig. 47.40. A notch is made at one end of a multimode optical fibre and its other end is chamfered as shown in Fig. 47.40(a). A light source sends light on to the fibre and a photodetector on the other side registers light emerging out from the fibre.

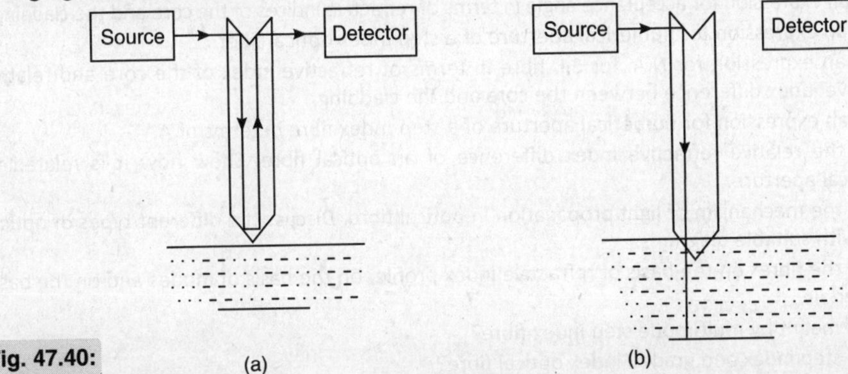

Fig. 47.40: Liquid Level Detecter

Working: The optical fibre is arranged at the desired height in a vessel. The refractive index of the fibre is chosen to be less than that of the liquid whose level is to be detected. Light from the light source is made to be incident on one of the inclined faces of the notch. The light turns through 90° and travels through the fibre. On reaching the chamfered end of the fibre, it gets internally reflected at the fibre-air boundary, if the liquid is below the desired level. Then, it is again turned through 90° at the opposite face, travels back through the fibre to be turned once again through 90° and is detected at the detector (Fig. 47.40a).

When the liquid rises and touches the fibre end, total internal reflection ceases and the light is transmitted into the liquid. Hence, the photodetector does not receive any light (Fig. 47.40b). Thus, an indication of the liquid level is obtained at the detector.

QUESTIONS

1. What is an optical fibre? What is the principle involved in its working?
2. What are optical fibres? Give its applications.
3. Explain core, cladding and sheath.
4. With a neat diagram, explain the structure of an optical fibre.
5. Explain the phenomenon of total internal reflection of light. How is it used in fiber optic communications?
6. Explain the principle of optical fibre as a waveguide for light.

7. What is meant by critical propagation angle of an optical fibre? Obtain an expression for the critical propagation angle.
8. What is meant by critical angle of an optical fibre? Obtain an expression for the critical angle.
9. Describe the propagation of light in optical fibre and obtain expression for numerical aperture.
10. Explain the following terms: (*i*) critical angle, (*ii*) Acceptance cone, (*iii*) Numerical aperture, (*iv*) V-number
11. Deduce an expression for acceptance angle of an optical fibre.
12. Using ray theory, derive the condition for transmission of light within an optical fibre.
13. Explain how light is propagated through an optical fibre and determine the numerical aperture and acceptance angle.
14. Explain with necessary theory, the propagation of light in optical fibres. Derive an expression for numerical aperture.
15. Derive an expression for angle of acceptance of fibre in terms of refractive index of the core and the cladding of an optical fibre. What is meant by acceptance cone?
16. What is meant by acceptance of angle for an optical fibre? Show how it is related to numerical aperture.
17. Derive an expression for acceptance angle in terms of refractive indices of the core and the cladding.
18. Derive an expression for numerical aperture of a step index optical fiber.
19. Derive an expression for N.A. for S.I. fibre in terms of refractive index of the core and relative refractive index difference between the core and the cladding.
20. Derive an expression for numerical aperture of a step-index fibre in terms of Δ.
21. Define the relative refractive index difference of an optical fibre. Show how it is related to numerical aperture.
22. Explain the mechanism of light propagation in optical fibre. Discuss the different types of optical fibres with suitable diagrams.
23. Classify the fibres on the basis of refractive index profile, on the basis of modes and on the basis of materials.
24. What is meant by multimode step index fibre?
25. What is step index and graded index optical fibre?
26. Explain the different types of optical fibre, along with the refractive index profile and mode propagation sketches.
27. Distinguish between single mode and multimode fibres with suitable diagram.
28. What are step-index fibre and graded-index fibre? How light propagates through step index and graded index fibre? Explain how pulse dispersion is minimized in graded index fibre.
29. Describe various mechanisms of attenuation in optical fibres.
30. What is attenuation in an optical fibre? Explain the attenuation mechanisms.
31. List the main components of optical communication system. Describe the basic optical communication system.
32. Explain with neat block diagram the principle of optic fibre communication. Also explain the signal distortion and optical transmission losses in optical fibres.
33. Explain optical communication through block diagram. For long distance communication whether (*i*) mono-mode or multimode and (*ii*) step index or graded index fibre, which are preferable and why?
34. State the main components of optical fibre communication system.
35. Discuss the advantages of optical fibre communication system over the conventional coaxial communication system.
36. Give the block diagram of optic fibre communication system explaining the functions of different blocks. Compare its merits over conventional communication system.
37. What is the principle behind the functioning of an optic fibre? What are the advantages of the optical fibres over coaxial cables?
38. Explain the principle of optic fibre communication. Mention advantages of fibre optic communication.
39. Give the various advantages of optical fibres over conventional cables.

40. Explain the important applications of optical fibre.

41. Discuss any one application of an optical fibre as a sensor.

42. Describe various types of optical fibres. Explain optical fibre sensors.

PROBLEMS

1. An optical fibre has a core material of refractive index of 1.55 and cladding material of refractive index 1.50. The light is launched into the fibre from air. Calculate its numerical aperture.

2. The numerical aperture of an optical fibre is 0.47. If the difference in the refractive indices of the material of its core and the cladding is 0.05, calculate the refractive index of material of the core.

3. An optical fibre has an acceptance angle 26.80°. Calculate its numerical aperture. [**Ans:** 0.4508]

4. Optical fibre refractive indices of core and cladding are 1.53 and 1.42 respectively. Calculate its critical angle. [**Ans:** 68.14°]

5. Consider a fibre having a core of index 1.48, a cladding of index 1.46 and has a core diameter of 30 mm. Show that all rays making an angle less than 9.43° with the axis will propagate through the fibre.

6. A step-index fibre is made with a core of index 1.54, a cladding of index 1.50 and has a core diameter of 50 μm. It is operated at a wavelength of 1.3 mm. Find the V-number and the number of modes that the fibre will support. [**Ans:** 42.15, 888]

7. Using a step index fibre with n_1 = 1.48 and n_2 = 1.46 and the core radius a = 30 μm. Calculate the number of total internal reflections that will occur on its propagation in a length of 1 km fibre.

8. A step-index fibre has a core refractive index of 1.44 and the cladding refractive index of 1.41. Find: (*i*) the numerical aperture, (*ii*) the relative refractive index difference, and the acceptance angle. [**Ans:** 0.292, 0.021, 33.96°]

9. An optical fibre has a numerical aperture of 0.20 and a cladding refractive index of 1.59. Find the acceptance angle for the fibre in water which has a refractive index of 1.33. [**Ans:** 8°47′]

10. Compute the cut-off parameter and the number of modes supported by a fibre which has a core refractive index of 1.54 and the cladding refractive index of 1.50. The radius of the core is 25 μm and operating wavelength is 1300 nm. [**Ans:** 42.15, 888]

11. Find the numerical aperture and acceptance angle of a fibre of core index 1.4 and Δ = 0.02. [**Ans:** 0.28, 32.52°]

12. Compute the total dispersion in 10 km length of a step index fibre, which has a core refractive index of 1.55 and the relative refractive index difference of 0.026. [**Ans:** 138ns]

13. Consider a bare step index fibre having a refractive index of 1.46. The radius of the fibre is 50 μm compute the pulse dispersion per km. [**Ans:** 2238ns]

14. Compute the cut-off parameter and the number of modes supported by a fibre, which has a core refractive index of 1.47 and the cladding refractive index of 1.45. The radius of the core is 50 μm and operating wavelength is 850 nm. [**Ans:** 44.64, 996]

15. A step-index fibre has a normalized frequency V = 26.6 at 1300 nm wavelength. If the core radius is 25 μm, calculate the numerical aperture. [**Ans:** 0.22]

16. Find the core radius necessary for single mode operation at 820 nm of a step index fibre, which has a core refractive index of 1.480 and the cladding refractive index of 1.478. [**Ans:** 4.08 μm]

17. A signal of 100mW is injected into a fibre. The out coming signal from the other end is 40 mW. What is the loss in dB? [**Ans:** 3.98 dB]

18. A communication system uses a 10 km fibre having a fibre loss of 2.5 dB/km. Find the input power if the output power is 1.265 μW. [**Ans:** 400 μW]

19. A fibre length 100m has power input 10 μW and power output 8.8 μW. Find the power loss in dB/km. [**Ans:** 5.55 dB/km]

20. When the mean optical power launched into a 8 km length fibre is 120 μW, the mean optical power at the fibre output is 3 μW. Determine the (*i*) overall signal attenuation in dB through the fibre, (*ii*) signal attenuation per km for the fibre, (*iii*) overall signal attenuation for a 10 km optical link using the same fibre, (*iv*) numerical input/output ratio. [**Ans:** 16 dB, 2.0 dB/km, 20 dB, 100]

48 Modern Engineering Materials

Learning Objectives:

By the end of this chapter, you will be able to:

- Learn about basics of Metallic Glasses and their usage.
- Understand about Shape Memory Alloys and their characteristics.
- Explain basic concepts of Bio-materials and its different types.

48.1 INTRODUCTION

The progress of human civilization is closely linked to the developments in materials. Science and technology have made amazing developments in the design of electronics and machinery using standard materials. Standard materials have been traditionally divided into three basic groups, namely metals, ceramics and polymers. This classification is primarily based on atomic structure and chemical makeup. Later, three other groups of engineering materials are added which are composites, semiconductors and biomaterials. There is a continuous search for materials with improved properties such as good mechanical strength, high stability, large electrical conductivity etc. A number of new materials such as amorphous metals, nonlinear materials, smart materials, etc have been discovered for high-tech applications. These materials have properties that scientists can manipulate.

48.2 METALLIC GLASSES

It has been found in 1970s that some of the metals can be produced in the amorphous state if they are cooled *rapidly* from the liquid state to solid state. Thus, one can produce amorphous metallic solids on cooling from liquid phase to the solid phase at high cooling rates of the order of 10°C/s. The amorphous metallic solids thus formed are known as ***metallic glasses*** or ***met glasses***.

Metallic glasses are amorphous metals with non-crystalline structures that are produced by solidification of liquid alloys.

Metals are malleable, and ductile whereas glass is brittle, transparent and non-magnetic. The metallic glasses share the properties of both metals and glasses. They are strong, ductile, malleable and opaque like a metal, and in addition, they have amorphous structure, brittleness and high corrosion resistance like glass. As they are amorphous solids, they are also known as ***amorphous metals***.

Solids are divided basically into two types; crystals and amorphous solids. Amorphous solids are generally referred to as super cooled liquids, since they do not possess long-range order. The arrangement of atoms within amorphous materials is not periodic and regular. Glass is a familiar example of amorphous solids while metals are examples of crystalline solids. Glass is obtained from liquid glass when cooling rate is high enough to bypass crystallization. Similarly, liquid alloys form metallic glasses on solidification under cooling rates rapid enough to suppress the crystal phase.

Glass transition temperature: The glass transition temperature for metallic alloys is about 20°C to 300°C. The glass-to-liquid transition is reversible and upon heating, metallic glasses exhibit the transition at T_g.

The temperature at which the frozen liquid reaches the glassy state is known as glass ***transition temperature***, T_g.

48.2.1 Types of Metallic Glasses

Metallic glasses are of two types, namely metal-metal glasses and metal – metalloid glasses.

(*i*) **Metal-metal glasses:** Metal-metal glasses are formed through a combination of metals, for example, Cu-Zr, Ni-Nb, Mg-Zn and etc.

(*ii*) **Metal-metalloid glasses:** Metal-metalloid glasses are produced from a combination of metals Fe, Co, Ni and metalloids B, Si, C, P and etc.

The composition ranges are found to be varied for each system. In addition, there are rare-earth transition metal glasses like GdCo, GdFe, GdTbFe which have good magnetic properties. Both the types of glasses are stable at room temperature and above. Some are capable of withstanding temperatures of the order of 773 K without crystallization.

48.2.2 Preparation of Metallic Glass

Metallic glasses are prepared by cooling molten metal to the glassy state at a cooling rate high enough to bypass crystallization. The cooling rate for glass formation varies from material to material. Pure metals are difficult to form in the glassy state because they require very fast rates of cooling. They are produced by atomic deposition technique like sputtering. In general, metallic glasses consist of two or more elements, as they require cooling rates slower than those required for pure metals.

Glass transition temperature *is the temperature at which the transition from liquid to solid occurs. The glass transition temperature for metallic alloys is about 20°C to 300°C.*

There are several techniques available for producing metallic glass.

We discuss here briefly two important techniques.

(*i*) **Melt Spinning:** Melt spinning is one of the common methods used to prepare metallic glass and is used in its large-scale production. The alloy is filled in a refractory tube having a tapered end (Fig. 48.1). The crucible is heated by an r.f. induction heater. The alloy is melted under inert gas atmosphere and when the gas pressure is increased, it is sprayed through the fine nozzle at the end of the refractory tube.

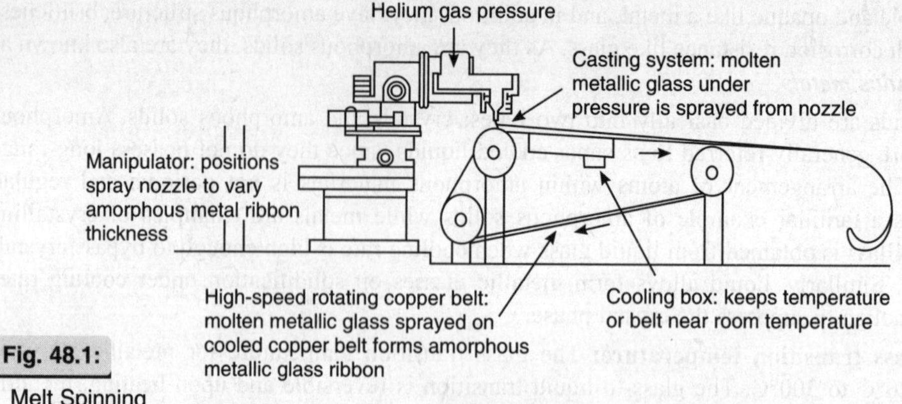

Fig. 48.1: Melt Spinning

The molten metal is directed on to a copper belt, located below the refractory tube and rotating at a very high speed. The molten alloy falling on the cooled copper belt is solidified and forms a thin ribbon of about 0.0025 cm thick and 15 cm wide, which is wound on a spool. The main advantage of this method is that the metallic glass is obtained in the form of continuous ribbon having uniform cross-section and reproducible properties.

(*ii*) **Sputtering:** Sputtering is another common technique, used especially for the formation of amorphous semiconductors and metals. Sputtering is a vacuum deposition process in which atoms are released from a target under the bombardment of positive ions and deposited on a substrate. A sputtering unit consists of a high power diode operating at r.f. frequencies of the order of 14 MHz. The sputtering chamber (Fig. 48.2) is evacuated and a steady flow of ultra high pure sputtering gas is introduced into the chamber. The target (cathode) consists of a base metal partially covered with another metal piece. The substrate is a mirror polished silicon single crystal attached to the anode. When the unit is switched on, the sputtering gas is ionized and the positive ions bombard the surface of the cathode releasing atoms or groups of atoms from it. These atoms fly toward the substrate and form a thin layer of metallic glass on it.

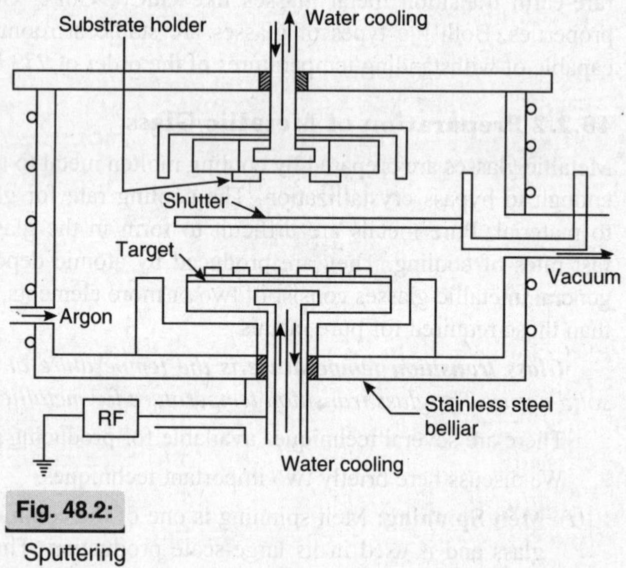

Fig. 48.2: Sputtering

48.2.3 Properties of Metallic Glasses

The metallic glasses exhibit a number of superior properties compared to their crystalline counterparts. Metallic glasses exhibit high mechanical strength, good ductility, magnetic behavior, very high elastic limit, and resistance to wear and corrosion, which set them apart from conventional crystalline materials. As they are non-crystalline in nature, they lack the grain boundaries, dislocation defects and segregated characteristics which are typical of crystalline materials.

(*i*) Thermodynamic properties: Glass transition is a phenomenon that occurs when a liquid undergoes configurationally freezing to become a solid. During this transition, atomic mobility is reduced to a large extent and there is a dramatic increase in viscosity as shown in Fig. 48.3. Hence, the material loses its fluidity.

The heat capacity vanishes at very low temperatures, increases over an intermediate temperature range. Near T_g, an anomalous increase is observed. Fig. 48.4 shows the heat capacity variation with temperature. The thermal conductivity of metallic glass at room temperature is generally of the same order of magnitude as that of crystalline alloys.

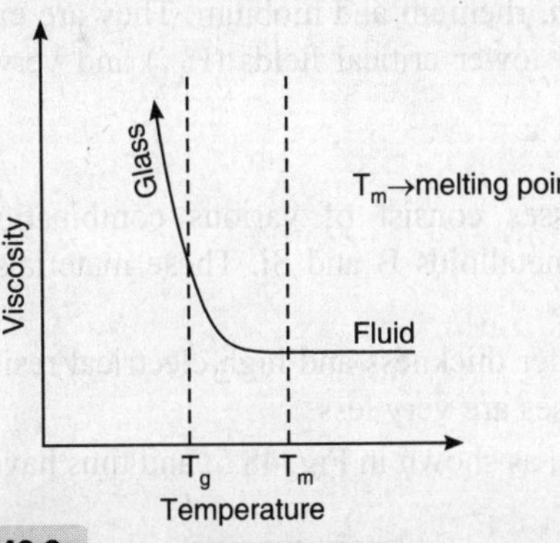

Fig. 48.3: Temperature dependence of viscosity of a supercooled melt

Fig. 48.4: Heat capacity variation as a function of temperature

(*ii*) Structural Properties

(*a*) The metallic glasses do not possess grain boundaries and dislocations.

(*b*) Studies on the structure of state showed that the atomic packing is of *tetrahedral close packing*. These materials do not possess long-range anisotropy.

(*c*) They can be easily made into ribbons.

(*iii*) Mechanical properties

(*a*) Because of the absence of defects and dislocations in their structure, metallic glasses are stronger than metals and alloys.

(*b*) Metallic glasses have better corrosion resistance than the crystalline metal.

(*c*) Metallic glasses are highly ductile.

(*d*) In general, the filaments of metallic glass can be used as low-cost high-performance structural reinforcement elements as they are stronger and stiffer.

(*e*) The elastic constants of metallic glasses are smaller than those observed in the corresponding crystalline materials. By contrast, values of yield strength (σ_y) and hardness (H) are typically much greater than those of crystalline solids. A common measure of the relative strength of solid is the ratio of yield strength to Young's modulus (σ_y/E). For metallic glasses, this ratio is typically 0.02 which approaches the theoretical maximum attainable yield strength. Hence, metallic glasses are among the strongest of known solids. The hardness of metallic glasses corresponds to the yield strength.

(*f*) However, as metallic glasses are very susceptible to brittle fracture, their practical applications are limited.

(*iv*) Electrical properties

(*a*) Electrical resistivity of metallic glasses is high and it does not vary much with temperature.

(*b*) Eddy current losses are very small in metallic glasses due to their high resistivity.

(*c*) The Hall co-efficient of metallic glasses is found to have both positive and negative signs. This indicates that the current through metallic glasses is due to the conduction of electrons and holes as in case of crystalline states.

(*d*) Some metallic glasses exhibit superconductivity; for example metal-metalloid glasses based on molybdenum, ruthenium, rhenium and niobium. They are extreme Type II superconductors with very small lower critical fields (H_{c1}) and very large upper critical fields (H_{c2}).

(*v*) Magnetic properties

(*a*) The amorphous metallic magnetic glasses consist of various combinations of ferromagnetic Fe, Co, and Ni with the metalloids B and Si. These materials have superior soft magnetic properties.

(*b*) Eddy currents are minimized by the smaller thickness and high electrical resistivity of the metallic glasses. Thus, the core losses are very less.

(*c*) They exhibit very narrow hysteresis loops, as shown in Fig. 48.5, and thus have very low hysteresis energy losses.

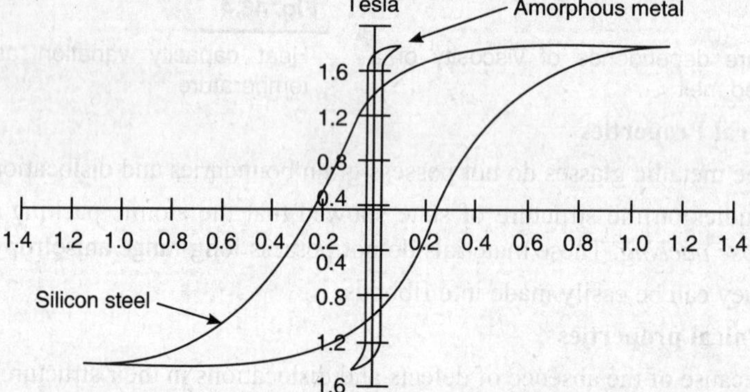

Fig. 48.5: Hysteresis loop

(*d*) These materials do not possess long-range an isotropy and grain boundaries. Therefore, domain walls move with extreme ease in them.

(*e*) These materials possess high maximum permeabilities and hence they can be magnetized and demagnetized very easily.

(*f*) The Curie temperature of metallic glasses is generally slightly lower than for crystalline phase because of the dilution effect of metalloid elements and atomic disorder in metallic glasses.

(*g*) Similarly, the coercivity of ferromagnetic metallic glasses is lower because of random fluctuations in the amorphous phase.

(*vi*) Chemical properties

(*a*) The absence of extended defects like grain boundary or dislocation in the amorphous metals offers a more uniform surface which leads to interesting chemical properties. Due to the formation of protective oxide film chromium containing metallic glasses have excellent corrosion resistance in a variety of chemically hostile environments.

(*b*) Apart from the absence of extended defects, the amorphous metals possess structural fluctuations on the scale of atomic distances. Each local atomic environment is distinct and varies over the surface. Hence, the surface exhibits a distribution of chemically active sites which leads to catalytic properties. Therefore, the amorphous state is more active than the crystalline state. Amorphous iron-nickel-metalloid alloys are used for catalytic synthesis of hydrocarbons by hydrogenation of carbon monoxide.

48.2.4 Applications of Metallic glasses-Soft magnetic materials

The homogeneity, low coercive force and relatively high permeability of metallic glasses are suitable for many magnetic applications. Thin sheets of metallic glasses are used as the core materials for power distribution transformers. Eddy current loss is proportional to the electrical conductivity and inversely proportional to thickness of the sheet. Hence, metallic glasses are used to reduce eddy current losses as they have lower electrical conductivity. The low coercive force reduces the core loss. Hence, the utilization of metallic glasses in transformers reduces the energy loss and thereby increases the efficiency.

Metallic glasses are also increasingly used in inductive components such as reactors, inverter transformers, chokes for magnetic switches, taper recorder heads, magnetic shields etc. Applications in the areas of magnetic sensors and transducers are under development.

The chief advantage of metallic glasses is that they are fabricated using inexpensive metals and they can be fabricated with relative ease in the form of thin tapes.

(*i*) Metallic glasses possess good tensile strength and are superior to common steels. They are used as reinforcing elements in concrete, plastic and rubber.

(*ii*) Metallic glasses possess high ductility, good corrosion resistance and hence used in making springs for different applications.

(*iii*) One of the most important applications of metallic glasses is as inductors in transformers, recording devices, magnetic shielding, motors and others.

(*iv*) A transformer having a core made from metallic glass is found to have eight times less core loss and requires about twenty times less current for excitation, compared to an equal sized commercial transformer having a crystalline silicon steel core.

(*v*) Magnetic recording heads made of metallic glasses are superior in overall performance to those made of ferrites and permalloys, because of their high flux density and high wear resistance.

(*vi*) Metallic glasses offer opportunities for increasing the efficiency of magnets, motors, and transformers used in energy-conversion devices. They would reduce costly losses from power-distribution systems and corrosion damage.

48.3 SHAPE MEMORY ALLOYS

A ***shape memory alloy*** (SMA) is an alloy that memorizes its shape and can be returned to that shape after being deformed on application of heat. When an SMA is below its transformation temperature, it can be deformed into any new shape, which it will retain. Yet, when the material is heated above its transformation temperature it undergoes a change in crystal structure which causes it to return to its original shape. In other words they "remember" their original shape. This effect is known as the ***shape memory effect*** (SME), and the alloys are named ***shape memory alloys*** (SMA). The shape recovery procedure happens due to a reversible solid-solid phase change especially in metal alloys.

Def: Solid-solid phase change is that in which a molecular rearrangement occurs, but the molecules remain closely packed so that the substance remains a solid.

If the SMA encounters any resistance during this transformation, it can generate extremely large forces. Therefore, when the shape memory effect is correctly harnessed, this material becomes a lightweight, solid-state alternative to conventional actuators such as hydraulic, pneumatic, and motor-based systems.

Shape memory alloys *(SMAs) are metallic alloys that when severely deformed at some relatively low temperature, regain their original shape after a thermal (heating/cooling) cycle.*

48.3.1 Two phases

There are two stable solid phases which occur in shape memory alloys. They are

(*i*) The high-temperature phase, called ***austenite***

(*ii*) The low-temperature phase, called ***martensite***.

These two phases have geometrically different crystallographic elementary cells (see Fig. 48.6 b). The martensite can be in one of two forms: **twinned** and **detwinned,** as shown in Fig. 48.6 (a).

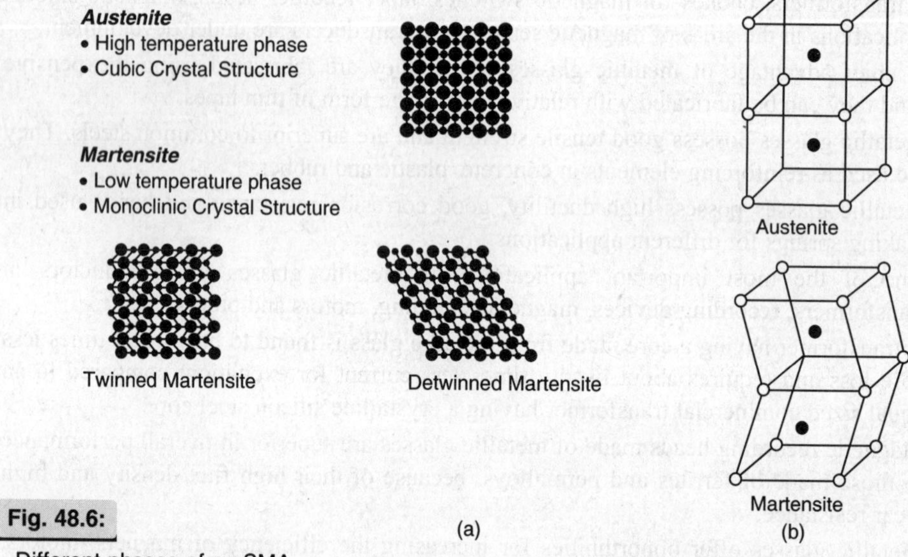

Fig. 48.6: Different phases of an SMA

Martensite is a crystallographically less-ordered phase and therefore is the softer and easily deformed phase of shape memory alloys. This phase exists at lower temperatures. The martensite structure is self-accommodating, and the deformation on transformation to martensite is zero.

Austenite is a more-ordered phase and hence the stronger phase of shape memory alloys. It occurs at higher temperatures. The shape of the austenite structure is cubic. The twinned martensite phase is of the same size and shape as the cubic austenite phase on a macroscopic scale. Change in size or shape becomes visible in shape memory alloys when the twinned martensite is detwinned, that is deformed.

A phase transformation which occurs between these two phases upon heating/cooling is the basis for the unique properties of the SMAs. The phase transformation may occur due to the temperature, or mechanical load applied or due to both temperature and load. The key effects associated with the phase transformation are ***shape memory effect*** and ***pseudoelasticity***.

Let us now study the effect of temperature and mechanical load on the phases of shape memory alloys.

(*i*) **Temperature-induced phase change without mechanical load:** If an SMA is cooled in the absence of applied load the material changes from austenite into twinned martensite. Observable macroscopic shape change does not occur during the transformation. On heating, a material in the martensite phase transforms to austenite by a reverse phase transformation. The above process is shown in Fig. 48.7.

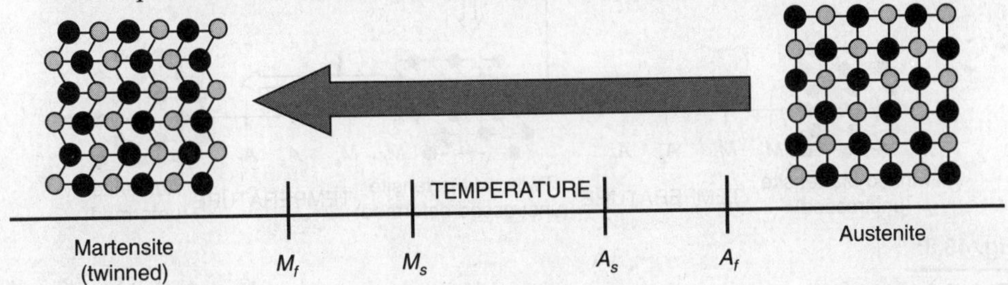

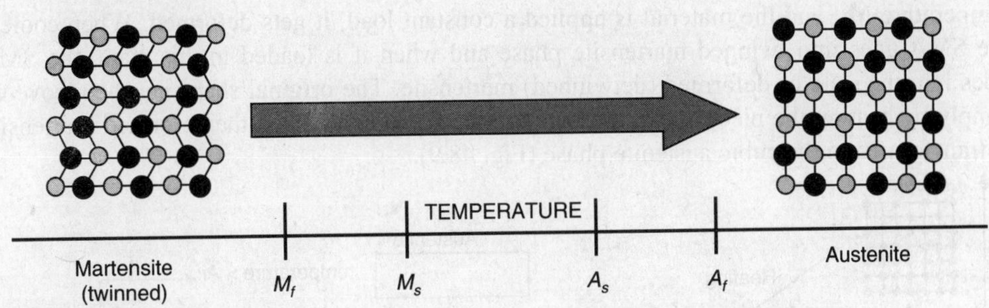

Fig. 48.7:

Temperature-induced phase transformation of an SMA without mechanical loading

Characteristic temperatures: There are four characteristic temperatures characterizing the phase transformation. The four characteristic temperatures are defined in Fig. 48.7.

(*i*) Martensite Start Temperature $\mathbf{M_s}$ is the temperature at which the material starts changing from austenite to martensite.

(*ii*) Martensite Finish Temperature, $\mathbf{M_f}$ is the temperature at which the transformation is complete and the material is fully in the martensitic phase.

(*iii*) Austenite Start Temperature $\mathbf{A_s}$ is the temperature at which austenite phase first appears in the martensite.

(*iv*) Austenite Finish Temperature $\mathbf{A_f}$, is the temperature at which the reverse phase transformation is completed and the material is in the austenitic phase.

Temperature-Induced Transformation with Applied Mechanical Load: At a temperature above $\mathbf{M_s}$, the specimen is entirely in the austenite phase. When the specimen is cooled below $\mathbf{M_f}$, it transforms entirely to the twinned martensite state, but the macroscopic volume of the specimen does not change - a condition known as self-accommodation. By applying small loads the specimen can be easily *detwinned* or *deformed* (Fig. 48.8), and the deformed shape remains after removing the loads.

Now, if the specimen in the detwinned state is heated to above the temperature $\mathbf{A_f}$, the original un-deformed shape may be recovered (Fig. 48.8).

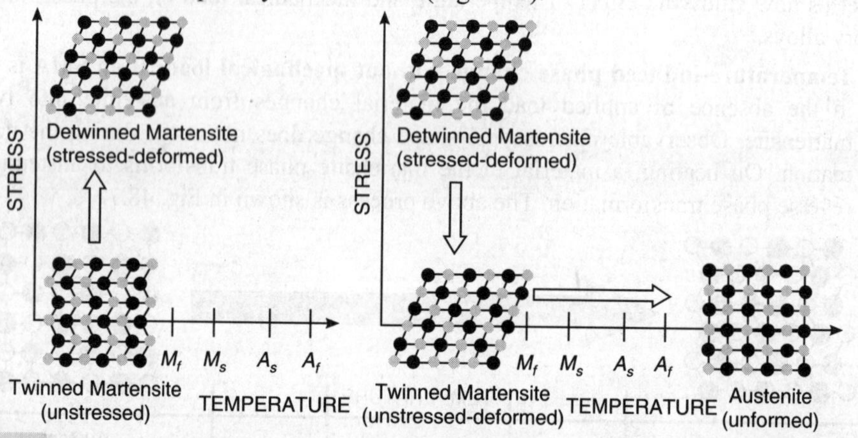

Fig. 48.8: Deforming of martensite phase under the action of applied load

Shape Memory Effect: When a piece of shape memory alloy is cooled below the temperature $\mathbf{M_f}$, and the material is applied a constant load, it gets deformed. When cooled, the SMA goes into twinned martensite phase and when it is loaded in this state, the SMA goes into the state of deformed (detwinned) martensite. The original shape can be recovered simply by heating the piece above the temperature $\mathbf{A_f}$. When heated the deformed martensite is transformed to the cubic austenite phase (Fig. 48.9).

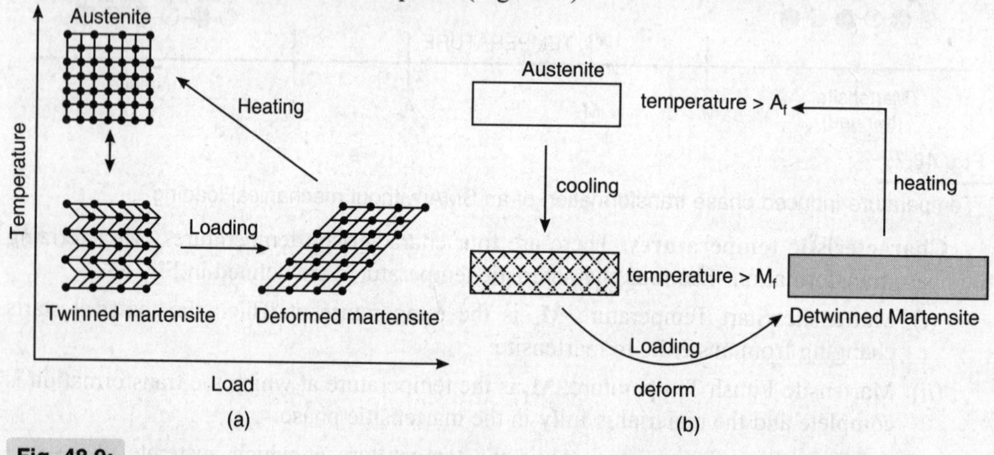

Fig. 48.9: Schematics of Shape Memory Effect

The molecular rearrangement occurs due to the heat transferred to the material.

Note that an SMA undergoes phase changes while remaining a solid. The phase changes occur below its melting point. These phase changes "involve the rearrangement of the position of particles within the crystal structure of the solid". Thus, the alloy can retain its shape without melting. Under the transition temperature, the alloy is in the martensite phase and can be bent into various shapes. To get the initial shape, the metal must be held in position and heated to about 500 °C. The high temperature causes the atoms to arrange themselves into a high symmetry, often cubic arrangement known as the austenite phase.

Load-induced phase change at constant temperature-Pseudo-elasticity: A phase transformation may be induced in a sample purely due to mechanical load and without a change in its temperature. Let the alloy be in the austenite phase. Referring to Fig. 48.9, the load on the shape memory alloy is raised till the austenite changed into martensite. Once the loading is reduced, the martensite begins to change back to austenite. Thus, the material behavior resembles elasticity. This effect is known as *Pseudoelastic Effect* and the phenomenon ***pseudo-elasticity*** or ***super elasticity***. The pseudoelastic stress-strain diagram is shown in Fig. 48.10.

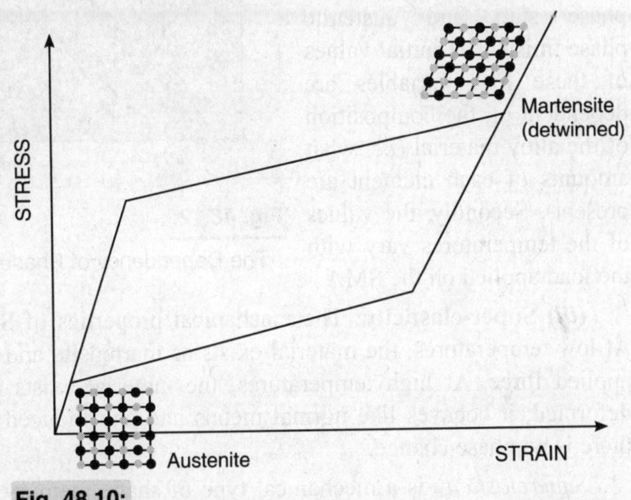

Fig. 48.10: Pseudoelastic stress-strain diagram

48.3.2 Characteristics of SMA

***(i)* Transformation hysteresis:** The transition from the martensite phase to the austenite phase is only dependent on temperature and stress, The transformation of SMA from one solid phase to another solid phase does not occur sharply at a particular temperature, but occurs over a range of temperature, as shown in Fig. 48.11. In this figure, ξ (T) represents the martensite fraction. Most of the transformation occurs over a narrow temperature range. Further, the heating and cooling transformations exhibit hysteresis. That is, the transformation on heating does not overlap on the transformation on cooling (Fig. 48.11). The magnitude of the hysteresis varies with the alloy type and is typically in the range 10-50°C.

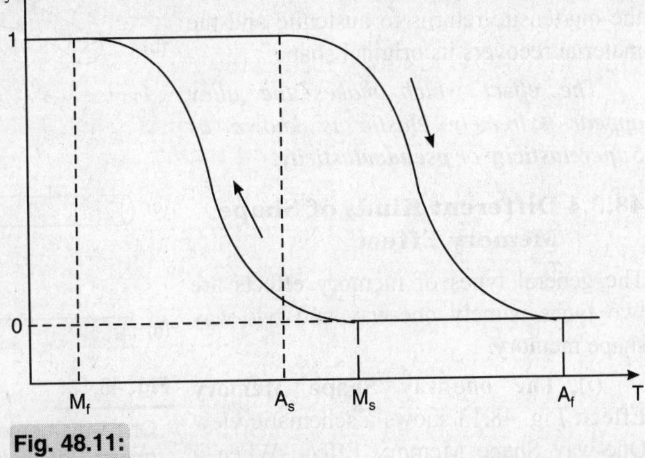

Fig. 48.11: Transformation hysteresis

(ii) Variation of Phase Change Temperature on Loading: Fig. 48.12 shows the characteristic temperatures of a shape memory alloy. $\mathbf{M_s}$ and $\mathbf{M_f}$ are the temperatures of martensite phase start and martensite phase finish respectively; whereas $\mathbf{A_s}$ and $\mathbf{A_f}$ are respectively the temperatures of austenite phase start and austenite phase finish. The initial values of these four variables are dependent on the composition of the alloy material (*i.e.* what amounts of each element are present). Secondly, the values of the temperatures vary with the load applied on the SMA.

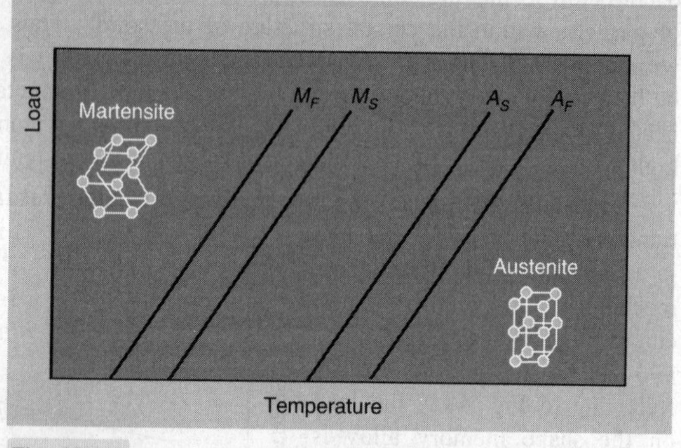

Fig. 48.12: The Dependency of Phase Change Temperature on Loading

(iii) Super-elasticity: The mechanical properties of SMAs depend on the temperature. At low temperatures, the material exists as martensite and is deformed by a relatively small applied force. At high temperatures, the material exists as austenite, which is not easily deformed; it behaves like normal metals and any induced strain is not recoverable because there is no phase change.

Superelasticity is a mechanical type of shape memory as opposed to thermally induced shape memory. In this case, a small force induces considerable deformation but when the force is removed, the material automatically recovers its original shape without the need for heating. Shape memory alloys show a super elastic behaviour if deformed at a temperature which is slightly above their transformation temperatures. The applied stress changes the austenite to martensite and significant deformations happen for a quite small applied stress. When the stress is removed, the martensite returns to austenite and the material recovers its original shape.

The effect which makes the alloy appear extremely elastic is known as ***Superelasticity*** *or* ***pseudoelasticity***.

48.3.4 Different Kinds of Shape Memory Effect

The general types of memory effects are two types namely one-way and two-way shape memory.

(i) The one-way Shape Memory Effect: Fig. 48.13 shows a schematic view One-way Shape Memory Effect. When a shape memory alloy is below $\mathbf{A_s}$ the metal

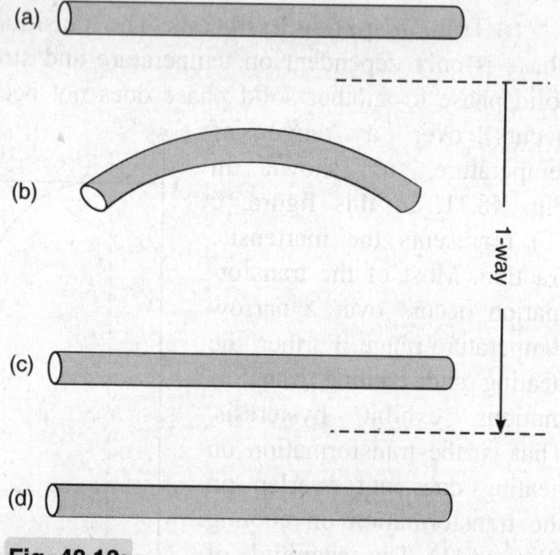

Fig. 48.13: One-way memory effect: (a) Material in martensite phase, (b) material deformed, (c) sampleheated and (d) sample cooled again

can be bent or stretched into a variety of new shapes and will hold that shape until it is heated above the transition temperature. The shape returns back to its original shape on heating. When the metal cools again it will remain in the hot shape, unless distorted again.

When the sample is cooled below the martensite finish temperature, the martensite units produced from austenite take many different orientations for assuming minimum energy configuration. If we heat the material above the austenite start temperature, the martensite units convert totally into austenite and the accumulated deformation is totally recovered. One-way SME is a one time operation and will not be repeated indefinitely. The schematic of a stress-strain-temperature curve is shown in Fig. 48.14.

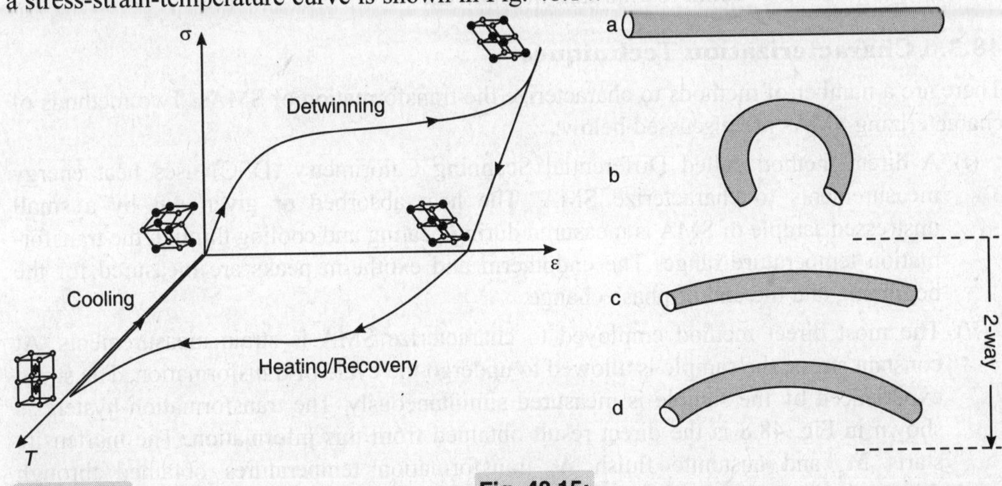

Fig. 48.14: Schematic of a stress-strain-temperature curve showing the shape memory effect

Fig. 48.15: Two-way memory effect: (a) Material in martensite phase, (b) material deformed, (c) sample heated and (d) sample cooled again

***(ii)* The two-way Shape Memory Effect:** In the Two-way Shape Memory Effect, the material remembers two different shapes; one at low temperatures and one at the high temperature. Fig. 48.15 shows a schematic view of the effect. This effect can also be obtained without the application of an external force. In usual conditions, a shape memory alloy *remembers* its high-temperature shape, but upon heating to recover the high-temperature shape, instantly *forgets* the low-temperature shape. But, it can be *trained* to *remember* to leave some reminders of the deformed low-temperature condition in the high-temperature phases.

By thermo-mechanical treatments, micro-stresses in the parent phase can be produced which in turn makes a stress-induced martensitic transformation. On further cooling, a spontaneous and reversible deformation is produced. Just upon heating and cooling cycles, even with no stress applied two-way shape memory effect is achieved. The two-way SME can be repeated indefinitely, Fig. 48.16.

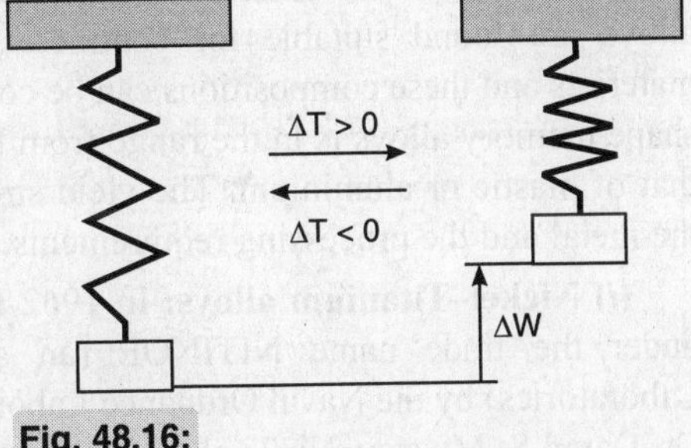

Fig. 48.16: A two-way SME spring. This device is able to produce work (ΔW) by thermal cycles

48.3.5 Manufacture

There are various ways to manufacture shape memory materials. Some of the techniques of producing nickel-titanium alloys are vacuum melting techniques such as electron-beam melting, vacuum arc melting or vacuum induction melting. These are specialist techniques used to keep impurities in the alloy to a minimum and ensure the metals are well mixed. The ingot is then hot-rolled into longer sections and then drawn to turn it into wire. Hot working to this point is done at temperatures between 700° C and 900° C. There is also a process of cold working of Ni-Ti alloys. Carbide and diamond dies are used in the process to produce wires ranging from 0.075 mm to 1.25 mm in diameter.

48.3.6 Characterization Techniques

There are a number of methods to characterize the transformation of SMAs. Two methods of characterizing SMAs are discussed below:

(*i*) A direct method called Differential Scanning Calorimetry (DSC) uses heat energy measurements to characterize SMA. The heat absorbed or given out by a small unstressed sample of SMA is measured during heating and cooling through the transformation temperature range. The endotherm and exotherm peaks are measured for the beginning and the end of phase change.

(*ii*) The most direct method employed to characterize SMA is strain measurements. At constant stress, the sample is allowed to undergo the cycle of transformation. The strain experienced by the sample is measured simultaneously. The transformation hysteresis shown in Fig. 48.8 is the direct result obtained from this information. The martensite starts $\mathbf{M_s}$ and austenite finish $\mathbf{A_f}$ transformation temperatures obtained through this method are slightly higher than those obtained from DSC method. The strain measurement is made by applying stress over the sample. On the other hand, DSC test is made with no stress applied. The increase in the stress is attributed to the increase in transformation temperatures $\mathbf{M_s}$ and $\mathbf{A_f}$. This test is the direct indicative of shape memory effect in any material. Difficulty in specimen preparation and the dependence of results on the way the test is conducted are the disadvantages of this method.

48.3.7 Engineering SMAs

The shape memory effect was first observed in AuCd in 1951 and since then it has been observed in numerous other alloy systems. However, copper base alloys and nickel-titanium alloys are found suitable for commercial exploitation. They are known as engineering materials and these compositions can be contrived to any shape and size. The yield strength of shape memory alloys is in the range from lower than that of conventional steel to higher than that of plastic or aluminium. The yield stress for NiTi can reach 500 MPa. The high cost of the metal and the processing requirements are the limitations of SMAs.

(*i*) Nickel–Titanium alloys: In 1962-63, the nickel-titanium alloys were first developed under the trade name NITINOL (an acronym for Nickel Titanium Naval Ordnance Laboratories) by the Naval Ordnance Laboratory. Their remarkable properties were studied by Dr. David S. Muzzey. Ni–Ti alloy is an extraordinary intermetallic binary compound because it has moderate solubility for excess nickel or titanium. This solubility enables alloying with many other elements thereby getting systems with modified mechanical and transformation properties. Its ductility is comparable to ordinary alloys. It has greater shape memory strain and excellent corrosion resistance compared to copper – base alloys. Also, it tends to be much more thermally stable. Because of the reactivity of titanium, the melting of this alloy must be done in vacuum or in an inert atmosphere. Methods like plasma – arc melting, electron beam melting and vacuum – induction melting are commonly adopted for this purpose. Welding, brazing or soldering of this alloy is generally difficult. The material responds to grinding and shearing. Even punching can be done if thickness is small. The properties of Nitinol are particular to the exact composition of the metal and the way it was processed.

Physical Properties of Nitinol

(*a*) Density: 6.45gms/cc
(*b*) Melting Temperature: 1240-1310° C
(*c*) Resistivity (hi-temp state): 82 μohm-cm
(*d*) Resistivity (low-temp state): 76 μohm-cm
(*e*) Thermal Conductivity: 0.1 W/cm-° C
(*f*) Heat Capacity: 0.077 cal/gm-° C
(*g*) Latent Heat: 5.78 cal/gm; 24.2 J/gm

Mechanical Properties of Nitinol

(*a*) Ultimate Tensile Strength: 754 - 960 MPa
(*b*) Typical Elongation to Fracture: 15.5 percent
(*c*) Typical Yield Strength (hi-temp): 560 MPa
(*d*) Typical Yield Strength (lo-temp): 100 MPa
(*e*) Approximate Elastic Modulus (hi-tem): 75 GPa
(*f*) Approximate Elastic Modulus (lo-temp): 28 GPa
(*g*) Approximate Poisson's Ratio: 0.3

Actuation

(*a*) Energy Conversion Efficiency: 5%
(*b*) Work Output: ~1 Joule/gram
(*c*) Available Transformation Temperatures: –100 to +100° C

***(ii)* Copper–Zinc-Aluminium (CuZnAl) alloys:** CuZnAl was the first copper based SMA containing 15-30 wt% Zn and 3-7 wt% Al. The transformation temperature is a function of the alloy composition and the thermo-mechanical treatments used in their preparation. Though they are the cheapest of the commercial SMAs, their memory properties are modest with a maximum recoverable strain of about 5%. As the ternary alloys have a very large grain size, they are brittle. The addition of < 1% of a grain refiner such as titanium or zirconium resolves the brittleness problem. Copper base alloys have medium corrosion resistance and they are susceptible to stress corrosion cracking. They can be melted in air and having wider range of potential transformation temperatures. Addition of manganese decreases the transformation temperatures of Cu-Zn-Al. It often replaces aluminium giving better ductility. Aging at low temperatures shift the transformation temperatures. The thermal stability of Cu-Zn-Al alloy is generally limited at higher temperatures.

***(iii)* Copper–Aluminium-Nickel (CuAlNi) alloys:** Copper-aluminum-nickel (CuAlNi) typically contain 11-14.5% Al and 3-5% Ni and have transformation temperatures in the range 80-200°C dependant on their composition; particularly sensitive to the aluminum content. The alloy processing is difficult. The final heat treatment has to be tightly controlled to produce an alloy with the desired transformation temperature. The production is more expensive than CuZnAl and cheaper than NiTi. The major advantages of the CuAlNi system are its wide range of transformation temperatures, its stability for applications above 100°C, small hysteresis and relatively low cost.

48.3.8 Applications of Shape Memory Alloys

The shape memory effect and pseudoelasticity of these alloys are being utilized in wide variety of applications in various fields.

For a long time, products containing SMAs are in use, for instance indestructible spectacle frames; these can be bent and twisted to a notable extent and then recover their original shape. The medical and aerospace and marine industries are the largest consumers of shape memory components. Some of their applications are outlined below.

(*i*) Vena-Cava Filters or Blood-clot filters: Vena-cava filter is a device used to trap blood clots. This device inserted as a small cylinder reverts to an umbrella shaped filter to trap small blood clots. Filter prevents blood clots from travelling to parts of the body. The under cooled SMA is collapsed and inserted into the vein and the body heat recovers its functional shape.

(*ii*) Cryofit hydraulic couplings: These couplings are used to join the metal tubings. They are prepared as cylindrical sleeves having diameters slightly less than that of the metal tubings and are fitted in the appropriate position. Upon warming, the martensite change to austenite structures and the coupling attains its functional shape and strongly holds the ends. But the tube does not allow the SMA coupling to recover its manufactured shape fully. The stress created as the coupling attempts to recover its shape is great enough to create a joint. Thus, recovery in shape is constrained and is being exploited to get superior coupling system. Such couplings are used widely in aerospace industry. The military has been using Nitinol couplers in F-14 fighter planes since the late 1960s. These couplers join hydraulic lines tightly and easily.

(*iii*) Super-elastic applications: The pseudoelastic properties of SMAs are being used in many applications such as eye glass frames, guide wire for steering catheters into vessels and arch wires for orthodontic correction. In addition to these because of the biocompatibility, SMAs are used as muscle wires too. Moreover, SMAs are used in surgical tool, coffee pots, space shuttle, thermostats, vascular stents and they even have aeronautical applications.

(*iv*) Stents for veins: The stent is an apparatus used to treat coronary ailment. The chilled stent is brought into position through a probe and expands to its original size when warmed up to body temperature. Stent to opens arteries and increase blood flow as it is used for supporting weak vein walls and broadens the narrow veins. SMA stent replaces similar stainless steel stents that are expanded with a little balloon.

(*v*) Dental and Orthodontic Archwires: These devices work similar to a spring. They apply a continuous and gentle force correcting misaligned teeth, as opposed to the periodic and uncomfortable tightening required by stainless steels.

(*vi*) Other applications in medical field:

(*a*) Tweezers to remove foreign objects through small incisions.

(*b*) Anchors with Nitinol hooks to attach tendons to bone for shoulder surgery and as a guide for catheters through blood vessels.

(*c*) Nitinol eye-glass frames can be bent totally out of shape and return to their parent shape upon warming.

(*vii*) Super-ellatic glasses: These glasses are made from a super-elastic metal alloy, Ni-Ti alloys that can be bent severely without permanent damage.

(*viii*) Actuators and micromanipulators: SMAs are useful as actuators materials that change shape, stiffness, position, natural frequency and other mechanical characteristics in response to temperature or electromagnetic fields.

Nitinol actuators are useful as engine mounts and suspensions (control vibration). The actuators can avert the damage of buildings and bridges.

Nitinol is useful in robotics actuators and micromanipulators to simulate human muscle motion.

48.3.9 Advantages and disadvantages of Shape Memory Alloys

Advantages

(*i*) Bio-compatibility

(*ii*) Diverse Fields of Application

(*iii*) Good Mechanical Properties (strong, corrosion resistant)

Disadvantages

(*i*) Relatively expensive

(*ii*) Have poor fatigue properties

(*iii*) Technology is still in infant stage.

48.4 BIOMATERIALS

A **biomaterial** may be a natural or man-made material that performs, augments, or replaces a natural function. To be specific, it is a non-viable material used in medical devices intended to interact with biological systems.

*A **biomaterial** is a synthetic material used to make devices to replace part of a living system or to function in intimate contact with living tissue.*

Biomaterials are used to make devices that would be in close or direct contact with the body to augment or replace faulty materials. Biomaterials have had a major impact on the contemporary medicine and patient care and in improving the quality of lives of humans. They are used in many of today's medical devices, including, artificial skin, artificial blood vessels, total artificial hearts, pacemakers, dental fillings, wires, plates and pins for bone repair, total artificial joint replacements. The most common classes of materials used as biomedical materials are polymers, metals, and ceramics. These three classes are used singly and in combination to form most of the implantation devices available today.

Biomaterials can be metals, ceramics, polymers, glasses, carbons, and composite materials. Such materials are used as molded or machined parts, coatings, fibers, films, foams and fabrics. They are different from other materials in that they possess a combination of properties, including chemical, mechanical, physical and biological properties that render them suitable for safe, effective and reliable use within a physiological environment. Biomaterials are rarely used on their own but are more commonly integrated into devices or implants. The field of biomaterials uses ideas from medicine, biology, chemistry, materials science and engineering. Engineers play a major role in this field, but they have to work closely with synthetic chemists to optimize materials properties and physicians to ensure that the device is useful in clinical applications.

Some common medical devices comprised of biomaterials are illustrated here:

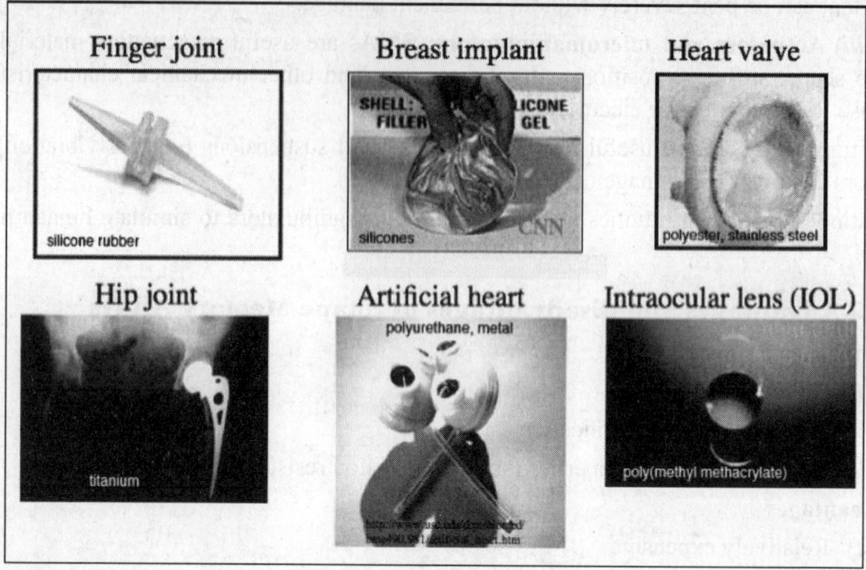

Fig. 48.17:

Illustration of common medical devices comprised of biomaterials

48.4.1 Criteria

The criterion of synthetic and modified natural materials as biomaterials:

Biocompatibility: Biocompatibility is the ability of a material to perform with a correct host reaction such as blood clotting, resistance to bacterial colonization and normal healing in specific applications such as hemodialysis (artificial kidney) membrane, a urinary catheter or hip joint prosthesis.

Toxicity: A biomaterial should not be poisonous.

Mechanical and Performance Requirements: Understanding of the biomaterials and devices require the design principles from physics, chemistry, mechanical engineering, chemical engineering, and materials science. The major categories of understanding are

(*a*) ***Mechanical performance:*** A hip prosthesis must be sturdy and firm. A tendon material must be strong and flexible. A heart valve leaflet must be flexible and tough. A dialysis membrane must be strong and flexible, but not elastomeric. An articular cartilage substitute must be soft and elastomeric.

(*b*) ***Mechanical durability:*** A catheter may only have to perform for 3 days. A bone plate may fulfill its function in 6 months or longer. A leaflet in a heart valve must flex 60 times per minute for at least for 10 or more years. A hip joint must not fail under heavy loads for more than 10 years.

(*c*) ***Bulk physical properties:*** The dialysis membrane has a specified permeability, the articular cup of the hip joint must have high lubricity, and the intraocular lens has clarity and refraction requirements.

48.4.2 Metallic biomaterials

Metallic biomaterials are used specially for load-bearing implants, such as hip and knee prostheses and fracture fixation wires, pins, screws and plates. They are also used as parts of artificial heart valves, as vascular stents and as pacemaker leads.

Metallic biomaterials can be divided into three subgroups:

(*i*) stainless steels

(*ii*) cobalt-based alloys

(*iii*) titanium metals and alloys

These materials have biocompatibility, appropriate mechanical properties, corrosion resistance and reasonable cost. They are very effective in binding the fractured bone, do not corrode and do not release harmful toxins when exposed to body fluids and therefore can be left inside the body for a long period of time.

48.4.3 Ceramic and Glass Biomaterials

The biomaterials are used not as much as metals or polymers. Some ceramics that were used for biomedical applications are alumina, zirconia, bio-glass, tri-calcium phosphate. Generally, ceramics and glasses are used as components of hip implants, dental implants, middle ear implants, and heart valves. They do not experience corrosion and show exceptional strength when loaded in compression. The brittleness and poor tensile (or bending) properties are major drawbacks. Only alumina is used as a bearing surface in joint replacements.

48.4.4 Polymeric Biomaterials

Polymers are the most widely used materials in biomedical applications. They are used for making cardiovascular devices and for replacement and augmentation of various soft tissues. Other applications are vascular grafts, heart valves, artificial hearts, breast implants, contact lenses etc.

Biodegradable polymers are used for sutures, controlled drug delivery, tissue engineering, and facture fixation. Compared with metals and ceramics, polymers have much lower strengths and moduli and generally not suitable for bearing loads (such as body weight) except very high molecular weight polyethylene (used as a bearing surface in hip and knee replacements).

Table 48.1: *Examples of Biomedical Applications of Polymers*

Polymers	Applications
Polyethylene, Polyvinyl chloride, Polyester, Silicone rubber, Polytetrafluroethylene	Cardiovascular implants
Ultra-high-molecular-weight polyethylene, Polymethyl methacrylate	Orthopedic implants
Polylactic acid, Polyglycolic acid	Tissue engineering

48.4.5 Longer-lasting medical implants

Now medical implants such as orthopedic implants and heart valves are made of bio-compatible titanium and stainless steel alloys. These materials are relatively non-porous. Hence, human tissue does not penetrate the implants, thereby reducing their effectiveness. Moreover, these metal alloys wear out quickly requiring recurrent expensive surgeries. Nano-ceramics, nanocrystalline zirconia (zirconium oxide) can be synthesized as aerogels by sol-gel techniques. Nanocrystalline silicon carbide (SiC) is used for artificial heart valves primarily due to its low weight, high strength, and extreme hardness, wear resistance, inertness and corrosion resistance.

48.4.6 Titanium and Titanium Alloys

Titanium and some of its alloys are used for dental and orthopedic applications. These materials are biologically inert biomaterials. Due to their excellent corrosion resistance they remain unchanged when implanted into human bodies. The human body is able to recognize these materials as foreign, and tries to isolate them by encasing them in fibrous tissues.

However, they do not produce any adverse reactions and are tolerated well by the human body. Furthermore, they do not induce allergic reactions such as has been observed on occasion with some stainless steels, which have induced nickel hypersensitivity in surrounding tissues.

The favorable performance of titanium implants is attributed to the chemically very stable oxide film that forms on the metal surface. The oxide film protects the underlying metal from corrosion. The dielectric constant ε of titanium oxide is close to that of water. The result is that the titanium surface does not lead to excessively strong interaction with proteins in the extracellular matrix; rather the surface is in some way 'water-like', interacting gently with the hydrophilic outer surface of the protein molecules.

In applications where high mechanical strength is necessary titanium-based alloys are appropriate than pure titanium. Hydroxyapatite coated on the surface of titanium provides a bioactive surface which plays a part in bone bonding so that bone cements and other mechanical fixation devices are not necessary. Titanium based alloys possess suitable mechanical properties such as strength, bend strength and fatigue resistance to be used in orthopaedics and dental applications. Pure titanium has limited applications such as dental implants.

Ti-6Al-4V alloy has been used for decades as an implant material due to its excellent properties. The material was originally developed as a high temperature aerospace alloy. However, vanadium is a potentially toxic element and as an alternative the Ti-6Al-7Nb alpha-beta alloy, Protasul-100 was developed between 1978 and 1982, and proved to be highly biocompatible.

QUESTIONS

1. Distinguish between an amorphous and a crystalline solid.
2. What are metallic glasses? Why is it easier to form metallic glass from alloys than from pure metals?
3. What is meant by glass transition temperature?
4. Explain briefly two important techniques that are used to produce metallic glass.
5. Describe the salient properties of metallic glasses.
6. Explain the electrical properties of metallic glasses.
7. How does the hysteresis of a metallic glass differ from that of a crystalline material?
8. Why are metallic glasses used in transformers?
9. Discuss some of the applications of metallic glasses.
10. What are shape memory alloys?
11. Explain briefly the main features of shape memory alloys.

12. Explain one-way shape memory effect.
13. Explain two-ways shape memory effect
14. Explain some of the applications of shape memory alloys.
15. Distinguish between an amorphous and a crystalline solid.
16. What are metallic glasses? Why is it easier to form metallic glass from alloys than from pure metals?
17. What is meant by glass transition temperature?
18. Explain briefly two important techniques that are used to produce metallic glass.
19. Describe the salient properties of metallic glasses.
20. Explain the electrical properties of metallic glasses.
21. How does the hysteresis of a metallic glass differ from that of a crystalline material?
22. Why are metallic glasses used in transformers?
23. Discuss some of the applications of metallic glasses.
24. What are shape memory alloys?
25. Explain briefly the main features of shape memory alloys.
26. Explain one-way shape memory effect.
27. Explain two-ways shape memory effect.
28. Explain pseudoelasticity.
29. Explain some of the applications of shape memory alloys.
30. What are biomaterials? State some of their applications.
31. What are the criteria that the biomaterials have to satisfy for being used in devices?
32. Classify biomaterials and discuss their applications.
33. Describe the salient properties of titanium and its alloys that make them superior biomaterials.

49 Nanotechnology

Learning Objectives:

By the end of this chapter, you will be able to:

- Define nanoscale and its significance.
- Understand need of Nanotechnology and its related characteristics.
- List different Production Techniques for Nanotechnology.
- Elaborate properties and different types of Nano-materials.
- Identify application of Nano materials in various fields.

49.1 INTRODUCTION

One of the outstanding features of technological progress in the last century is the continuous miniaturization of technical devices and components. A strong impact for the development arose from the necessity to reduce the geometry of integrated circuits in order to follow the demands for a higher complexity of circuits and devices. It resulted in the development of various micro-technological processes. The development of microelectronics indicate a fundamental approach known as top down approach. The top down approach starts with a large-scale object or pattern and progressively reduces its dimensions. The process is similar to sculpture a statue from a big a rock. Nature adopts the opposite approach in building the materials. Simple atoms and molecules are collected, consolidated and fashioned into a complex structure. This is known as bottom-up approach. Bottom up approach is a widespread process in biology where, for example, enzymes assemble amino acids to construct living tissue that forms the organs of the body.

On 29th December 1959, Physicist and Nobel Laureate, Richard Feynman presented a visionary and prophetic lecture at the annual meeting of American Physical Society, entitled "There is Plenty of Room at the Bottom" where he speculated that the bottom-up approach could some day be adopted by scientists for preparation of materials and fabrication of devices to atomic specifications; he said "the principle of physics, as far as I can see, do not speak against the possibility of manoeuvering things atom by atom." Feynman's sparkling discussion of the problems and promise of miniaturization was the starting point for the use of nanoscale machines to build complex products including other nano-machines. However, it was not until the 1980s that a notable research activity occurred and significant developments resulted.

With emerging new technologies such as scanning microscopes, it is possible to manipulate and move atoms and molecules to form new structures and materials. This gave opportunities to develop mechanical, electrical, magnetic, and other properties in materials that are not otherwise possible. Nanotechnology has been developing along two different lines. One of them is leading to technological developments on the nanometer scale, usually 0.1-100 nm. Much of current interest and enthusiasm is focused on this. The other line anticipates designing and building atom by atom nano-machines and devices which would bring in the next industrial revolution.

49.2 NANOSCALE

The word "nano" is derived from a Greek word meaning extremely small and means a billionth (10^{-9}) part of a unit. A *nanometer* or nm is one thousand millionth of a metre, *i.e.*, $1\ nm = 10^{-9}\ m = 10^{-3}\mu m = 10\ \AA$. One nanometer spans 3 to 5 atoms lined up in a row. For comparison, a single human hair is about 80,000 nm wide and a red blood cell is approximately 7,000 nm wide. Scientists and engineers are nowadays interested in the nanoscale, which may be taken as 100 nm to 0.2 nm approximately. Below this lies the atomic scale 0.1 nm. Therefore, the nano-world is a borderland between the quantum world and the macro world. Table 49.1 shows some examples of size from molecular to macro.

Table 49.1: *Some examples of size from molecular to macro*

Size (nm)	Examples	Terminology
0.1-0.5	Individual chemical bonds	Molecular/atomic
0.5-1.0	Small molecules	Molecular
1-1000	Proteins, DNA, inorganic nanoparticles	Nano
10^3-10^4	Devices on a silicon chip, living cells (bacteria: 1 μm; yeast: 5 μm), human hair (50 μm)	Micro
>10^4	Normal bulk matter	Macro

49.3 THE SIGNIFICANCE OF THE NANOSCALE

Many of the properties of solids depend on the size of the solid. Microscopic details become averaged out in bulk materials. In case of bulk materials, the properties such as density, elastic modulus, resistivity, dielectric constant etc are averaged properties. Many properties of materials change in the nanometer range. As materials become smaller and smaller, eventually a point is reached where the averaging no longer works. Two principal factors cause the properties of nanomaterials to differ significantly from other materials; statistical mechanical effects and quantum mechanical effects. The importance of surface to volume ratio, quantum confinement, dominance of electromagnetic forces and random molecular motion at nanoscale needs to be understood in a better way.

(*i*) **Surface to volume ratio:** Nanomaterials have a relatively larger surface area when compared to the same mass of material produced in a larger form. This can make nanomaterials ***more chemically reactive and affect their strength or electrical properties***.

In nanomaterials, many atoms, more than half, will be near interfaces. Surface properties such as energy levels, electronic structure, and reactivity can be quite different from interior states, and give rise to quite different material properties. Fig. 49.1 illustrates the concept of surface to volume ratio.

(*ii*) **Quantum confinement:** Quantum effects can begin to dominate the behaviour of matter at the nanoscale affecting the optical, electrical and magnetic behaviour of materials. Quantum confinement can be observed once the diameter of a material is of the same magnitude as the de Broglie wavelength of the electron wave function.

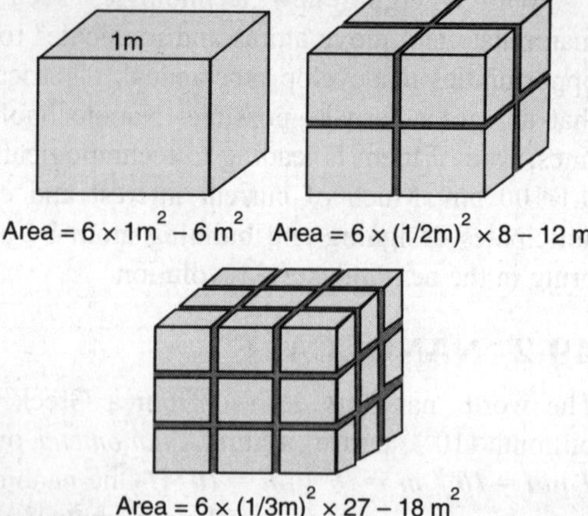

Fig. 49.1: Illustration of surface to volume ratio

Def: Quantum confinement occurs when one or more dimensions of nano-crystals approaches a size called exciton Bohr radius. When the confining dimension decreases to nanoscale, the energy spectrum becomes discrete.

Quantum confinement comes when electrons and holes in a material are confined to a potential well.

(*a*) ***Quantum dot*** is a small sphere where all three dimensions of nanoparticles confines exciton Bohr radius. It is a zero-dimensional potential well.

(*b*) ***Quantum wire*** is a structure where height and both breadth are small and length is long. It is a one-dimensional potential well.

(*c*) ***Quantum well*** is a structure where height is about the Bohr exciton and breadth and length are long. It is a two-dimensional potential well.

(*iii*) **Dominance of electromagnetic forces:** We know four types of forces such as strong nuclear force, weak nuclear force, gravitational force and electric and magnetic force.

(*a*) **Nuclear forces:** Week and strong nuclear forces are the interactions between the particles that compose the nucleus. Strong nuclear force is a huge force and it holds nuclei together. It acts only at nuclear distances, 10^{-5} nm. Weak nuclear force is a small force and it is responsible for nuclear beta decay. It acts only at nuclear distances, 10^{-5} nm. As these forces are significant at extremely short distances than nano scale, they are insignificant.

(*b*) **Gravitational Forces:** It is a long ranged force. Gravitational forces are directly proportional to the masses of the two objects and inversely proportional to the square of the distance between the objects. As the mass of nano scale particles is very small and gravitational forces becomes insignificant.

(*c*) **Electric and Magnetic forces:** These forces are dominant at atomic and nano scales. Normally, electric and magnetic forces are 10^{39} × gravitational forces. Electric and magnetic forces are the forces of attraction and repulsion between objects based on their charge and satisfy inverse square law. Since electric and magnetic forces are not affected by the mass, they can be strong even with nano sized particles. The chemical and physical behaviour depends on magnetic and electrostatic forces which are very significant at the nanoscale.

(*iv*) **Random molecular motion:** All molecules in a substance exhibit random molecular motion due to their kinetic energy. Random motion involves molecules moving around in space, rotating around their bonds and vibrating along their bonds. Random kinctic motion at micro scale is not very influential. But at nano scale these motion can be influential. If the size or one of the dimensions of nanomaterial is smaller than the critical length, the nanomaterial exhibits properties that are different from the corresponding bulk material.

49.4 NANOTECHNOLOGY

The term nanotechnology was first coined in 1974 by Norio Taniguchi of the Tokyo Science University. In 1986, K. Eric Drexler wrote "Engines of Creation" and popularized the term nanotechnology. Originally, nanotechnology means developing the things from the bottom up starting from molecular level. Nowadays it is the study; control of phenomena and materials at length scales below 100 nm. Artisans had taken advantage of nanosized materials since a long time. In the 4th century A.D. Roman glass makers were fabricating glasses containing nanosized metals.

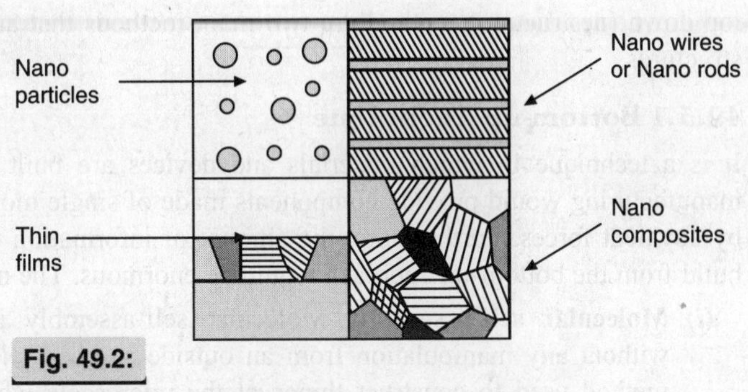

Fig. 49.2: Classification of nanomaterials following Siegel

(*i*) **Nanoscience:** It is the study of the fundamental principles of molecules and structures with at least one dimension roughly between 1 and 100 nanometers.

(*ii*) **Nanotechnology:** It is the design, characterization, production and application of structures, devices and systems by controlling the shape and size at nano scale.

(*iii*) **Nanomaterials:** Siegel classified nano-structured materials (Fig. 49.2) as

(*a*) *Thin films or surface coatings:* Materials that have one dimension in the nanoscale and are extended in the other two dimensions.

(*b*) *Nanowires and nanotubes:* Materials that are nanoscale in two dimensions and extended in one dimension.

(*c*) *Precipitates and colloids:* Materials that are nanoscale in three dimensions.

49.4.1 Why Nanotechnology?

With nanotechnology it is possible to control matter on every important length scale. It enables design of materials. The developments in nanotechnology can affect domains such as:

(*i*) **Materials:** new materials which are harder, more durable and resistant, lighter and less expensive can be produced.

(*ii*) **Electronics:** electronic components will become smaller and smaller, allowing the design of more powerful computers.

(*iii*) **Energy:** a vast increase in the potential of solar energy generation.

(*iv*) **Health and biotechnology:** great expectations are held in the areas of prevention, diagnostics and treatment.

Developments in these domains would impact a broad range of industries such as pharmaceuticals, cosmetics, consumer appliances, communications, security and safety, and space exploration.

49.5 PRODUCTION TECHNIQUES

Production or fabrication or synthesis of nanomaterials and nanoscale devices is not quite easy and simple. As nano-particles are attached to each other via electrostatic forces, they are very hard to handle and move things that are so small. Bottom-up method (construction) and top-down (destruction) method are two main methods that are used to make the nano scale structures.

49.5.1 Bottom-up Technique

It is a technique in which materials and devices are built up atom by atom. Bottom-up manufacturing would provide components made of single molecules which are held together by covalent forces. Furthermore, the amount of information that could be stored in devices build from the bottom-up approach would be enormous. The major methods are:

(i) **Molecular self-assembly:** Molecular self-assembly is the assembly of molecules without any manipulation from an outside source. Self assembly is a manufacturing method used to construct things at the microscale which is comprised of structures with atleast one dimension that is less than 100 μm. Many biological systems use self assembly technique to assemble various molecules and structures. Vast amount of highly ordered ***molecular machinery*** is responsible for survival of living systems. According to ***molecular biology***, the information required for building a living cell is stored in the DNA which is transferred from the DNA to the proteins by the processes called ***transcription*** and ***translation***. The self-assembly involves ***convergent chemical synthesis***.

(ii) **Positional Assembly:** It is based on mechanical engineering functionality of the components such as gears, bearings, motors, and structural members that would enable programmable, ***positional assembly*** to atomic specification. We can call them as molecular robotics.

The molecular manufacturing systems are expected to be designed to use some form of ***convergent assembly***. Convergent assembly means that smaller parts can be assembled into larger parts; larger parts can be assembled into still larger parts, and so forth. As a first step would be to develop nanoscopic machines, called ***assemblers*** that can programmed to manipulate atoms and molecules. Eric Drexler believes that assemblers could first replicate themselves, building other assemblers. Each generation would build another resulting in exponential growth. Assemblers and replicators could work together to automatically construct products, and could eventually replace all traditional labor methods. This possibly will decrease manufacturing costs, thereby making consumer goods plentiful, cheaper and stronger.

(iii) **Sol-Gel Process:** This process is widely employed to prepare oxide materials. It can be understood as

Sol-Gel = Sol (a colloidal solution made of solid particles few hundred nm in diameter, suspended in a liquid phase) + Gel (a solid macromolecule immersed in a solvent)

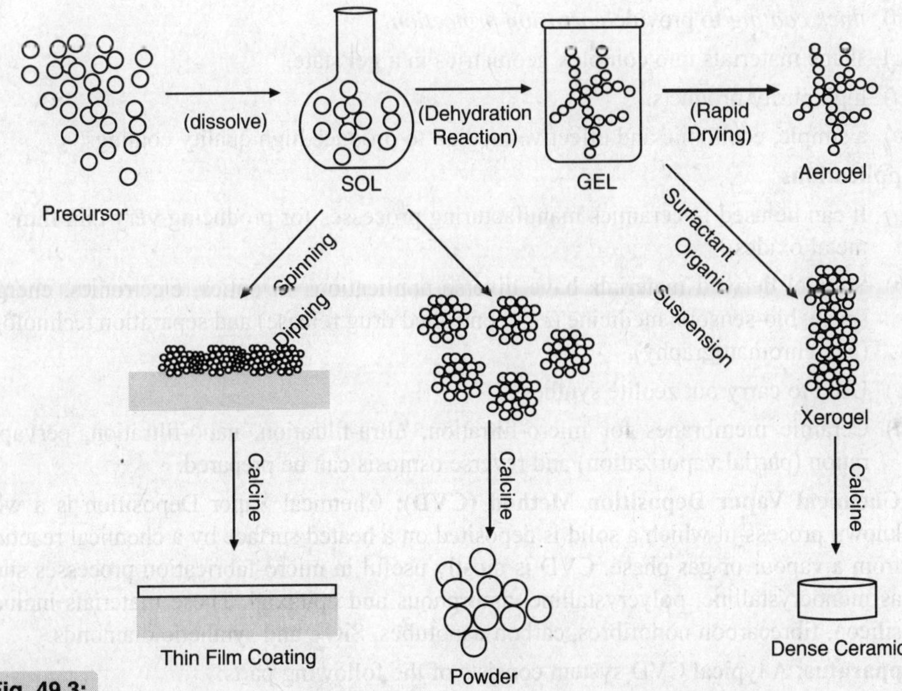

Fig. 49.3:

Schematic representation of sol-gel process of synthesis of nanomaterials

The sol-gel process is summarized in Fig. 49.3. It is discussed with the example of production of SiO_2 and can be understood in the following steps.

Step i - Preparation of stable solution: A stable solution of the alkoxide or solvated metal precursor (raw material) is formed.

Step ii - Formation of Gel: An oxide (M-OH-M) or alcohol bridged (network) gel forms by a polycondensation or polyesterification reaction that leads to increase in the viscosity of the solution.

Step iii - Aging of Gel: The polycondensation reactions continue until the gel becomes a solid mass along with contraction of the gel network and expulsion of solvent from gel pores.

Step iv - Drying of the gel: Water and other volatile liquids are removed from the gel network by drying. If the solvent such as water is extracted under super critical or near super critical conditions, the product is an *aerogel*. If isolated by thermal evaporation the resulting product is termed a *xerogel*.

Step v - Dehydration: By drying, M-OH groups are removed and gel stabilizes. Heating the xerogel at temperatures up to 800°C gel gets stabilized.

Disadvantages:

(*a*) Controlling the growth of the particles is very difficult.

(*b*) Stopping the newly formed particles from agglomeration is difficult.

Advantages: Sol-Gel synthesis is superior of all the available processes as it can produce

(*a*) *thin bond-coating* to provide excellent adhesion between the metallic substrate and the top coat.

(*b*) *thick coating* to provide *corrosion protection*.

(*c*) shape materials into complex geometries in a gel state.

(*d*) high purity products.

(*e*) a simple, economic and effective method to produce high quality coatings.

Applications

(*a*) It can be used in ceramics manufacturing processes for producing very thin films of metal oxides.

(*b*) Sol-Gel derived materials have diverse applications in optics, electronics, energy, space, bio-sensors, medicine (*e.g.* controlled drug release) and separation technology (*e.g.* chromatography).

(*c*) Used to carry out zeolite synthesis.

(*d*) Ceramic membranes for micro-filtration, ultra-filtration, nano-filtration, pervaporation (partial vaporization) and reverse osmosis can be prepared.

(*iv*) Chemical Vapor Deposition Method (CVD): Chemical Vapor Deposition is a well known process in which a solid is deposited on a heated surface by a chemical reaction from a vapour or gas phase. CVD is mostly useful in micro-fabrication processes such as monocrystalline, polycrystalline, amorphous and epitaxial. These materials include silicon, fibrecarbon nonofibres, carbon nanotubes, SiO_2, and synthetic diamonds.

Apparatus: A typical CVD system consists of the following parts:

(*a*) sources and feed lines for gases

(*b*) gas mass flow controllers for sending the gases into the system

(*c*) reaction chamber or reactor

(*d*) system for heating up the wafer on which the film is to be deposited

(*e*) temperature sensors

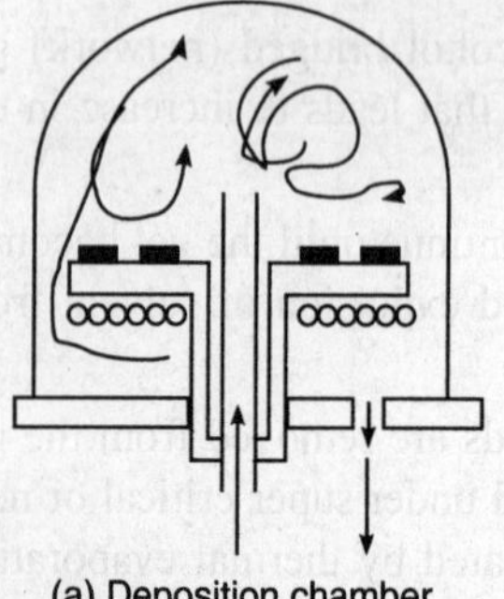

(a) Deposition chamber

(b) CVD process

Fig. 49.4: A typical CVD system

Working Concept: Chemical reaction of gaseous precursor (raw material) on a heated substrate yields a fully dense deposit. Fig. 49.4 shows a typical CVD system with deposition chamber and CVD process. By controlling the temperature, pressure and concentrations, a newly formed molecule sticks to the wafer

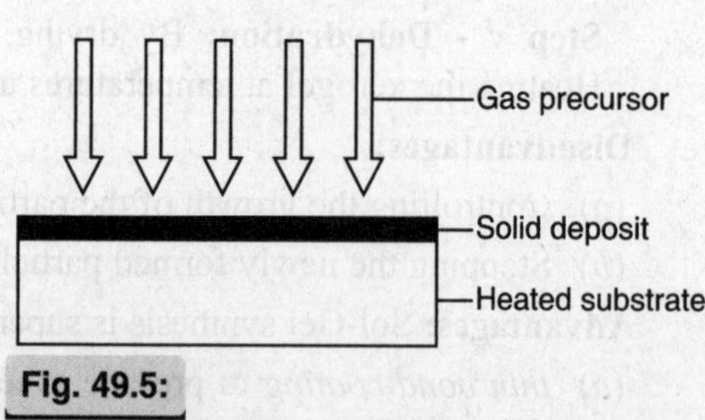

Fig. 49.5: Working Concept of CVD

surface and participates in the formation of a nanolayer. A simplified concept diagram is shown as Fig. 49.5.

Basic CVD Process: It has following steps:

(*a*) Basing on the nanomaterial to be produced, a required reactant gases and diluent inert gases are introduced into the reaction chamber.

(*b*) the reactants get adsorbed on the surface of the substrate; and undergo chemical reactions with the substrate to form the film and

(*c*) The gaseous by-products of the reactions are desorbed and evacuated from the reaction chamber.

Types of chemical vapor deposition: A number of forms of CVD are in wide use. These processes differ in the means by which chemical reactions are initiated (e.g., activation process) and process conditions.

Types basing on temperature:

(*a*) **Hot-wall CVD:** The heating system heats up not only the wafer, but the walls of the reactor also.

(*b*) **Cold-wall:** The heating systems heats up wafer only.

Types based on operating pressure:

(*a*) Atmospheric pressure CVD (APCVD) reactors operate at atmospheric pressure and are therefore the simplest in design.

(*b*) Low-pressure CVD (LPCVD) reactors operate at medium vacuum (30-250 Pa) and higher temperature than APCVD reactors.

Advantages of CVD

(*a*) Useful for coatings or freestanding structures of a wide range of metals and ceramics.

(*b*) Can produce mesh or near-mesh intricate shapes.

(*c*) Gives high purity deposits (>99.995% purity).

(*d*) Conform consistently to outline of substrate surface and internal passages with high length-to-diameter ratios and coats powders also.

(*e*) Has near-theoretical as-deposited density.

(*f*) Has controllable thickness and morphology and can concurrently coat several apparatus.

(*g*) Forms alloys.

(*h*) Creep into fiber performs and foam structures.

Applications

(*a*) Biocompatibility between natural and synthetic materials can be increased by copper capping and anti-corrosive coating.

(*b*) It can be used on large range of industrial components such as aircraft and land gas turbine blades, timing chain pins for the automotive industry, radiant grills for gas cookers and items of chemical plant to resist various attacks by carbon, oxygen and sulphur.

(*c*) Used to provide the photo resist adhesion for semiconductor wafers, silane adhesion for microarrays such as protein, tissue.

(*d*) Coating on *Micro-Electro-Mechanical Systems* (MEMS) to reduces stiction. Stiction is the friction which tends to prevent stationary surfaces from being set in motion.

(*v*) **Physical Vapour Deposition (PVD):** It is a vaporization coating technique where the raw materials (precursors) are in solid form. It has following steps

(*a*) **Evaporation:** Material to be deposited as coating is taken in the form of target and is bombarded by a high energy source (such as a beam of electrons or ions or heating by a filament) in a vacuum chamber. The source material gets heated and starts to boil and then evaporate.

(*b*) **Transport:** In transportation vaporized atoms from the target material move to the substrate to be coated.

(*c*) **Reaction:** The transported atoms react with appropriate gas in transportation stage.

(*d*) **Deposition:** Here coating builds up on the substrate surface. Fig. 49.6 shows a schematic diagram of the principles behind a common PVD method.

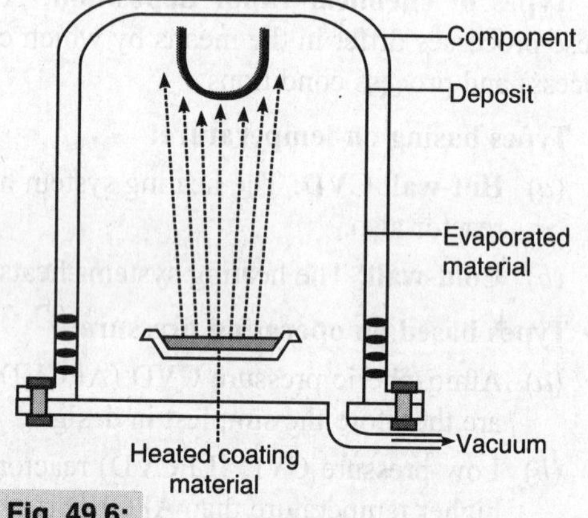

Fig. 49.6: Physical Vapour Deposition system

Advantages

(*a*) Almost any type of inorganic and organic material can be used.

(*b*) The process is more environmental friendly than processes such as electroplating.

Disadvantages

(*a*) Extremely difficult to coat undercuts and similar surface features.

(*b*) High capital cost is required.

(*c*) Operates at high vacuums and temperatures.

(*d*) PVD requires skilled operators.

(*e*) The rate of coating deposition is usually quite slow.

Applications

(*a*) PVD coatings are generally used to improve hardness, wear resistance and oxidation resistance.

(*b*) Coatings are useful in a wide range of applications such asaerospace, automotive, surgical/ medical, dyes and moulds for material processing, cutting tools and fire arms.

(*vi*) **Inert gas condensation:** In this technique, a metallic or inorganic material is vaporized using resistive heating in an atmosphere of 1-50 mbar He or any other inert gas like Ar, Ne, and Kr. The ultrafine particles (100 nm) are formed by collision of evaporated atoms with residual gas molecules. Clusters rapidly condense in the form of small crystallites. The particles are collected by a rotating cylindrical device which is cooled with liquid nitrogen. Subsequently; the nanoparticles are removed from the surface of the cylinder by means of a scraper in the form of a metallic plate (Fig. 49.7). Evaporation is carried in Ta or Mo refractory metal crucibles. This technique is very useful to produce composite materials by mixing two or more evaporation sources. The particle size can be controlled by the evaporation rate and condensation gas pressure.

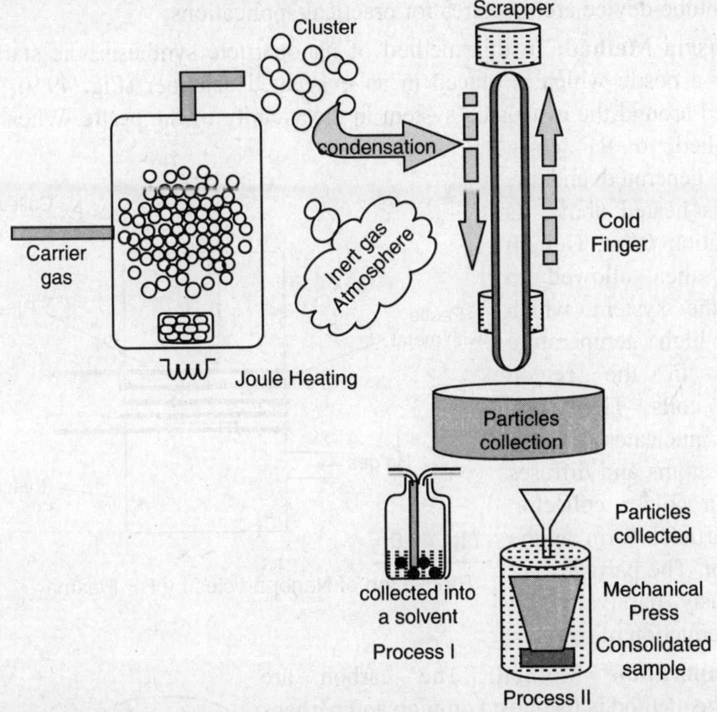

Fig. 49.7:

Schematic of typical set-up for gas condensation synthesis of nanomaterials.

(*vii*) **Laser ablation:** In this method a laser beam is used for producing clusters directly from a solid sample. In dual-pulsed laser technique for producing carbon nano tubes, a quartz tube containing a graphite target with a 50:50 catalyst mixture of Cobalt and Nickel are heated to 1200°C in flowing argon (Fig. 49.8). Carbon is evaporated from the graphite by an intense pulsed laser beam. They move from the high temperature zone to the cold copper collector and condense into nanotubes. The use of two successive laser pulses reduces the amount of carbon deposition as soot. The second laser pulse splits up the larger particles ablated by the first one and supplies them into the growing nanotube structure. Nanotubes of 10-20nm in diameter and 100µm long can be produced. By varying the growth temperature, the catalyst composition, and other process parameters, the average nanotube diameter and size distribution can be controlled.

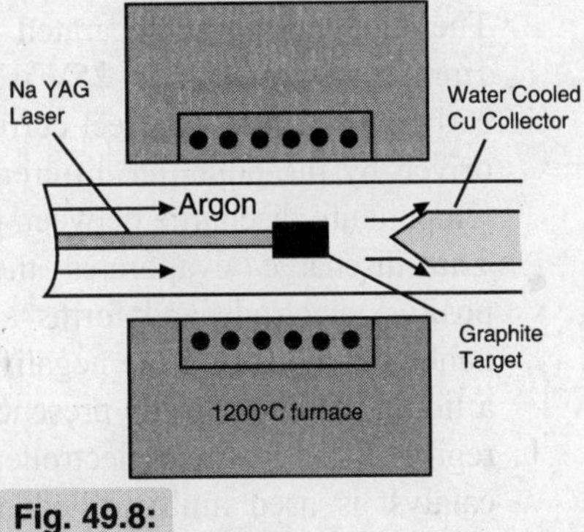

Fig. 49.8:

Schematic drawing of a laser ablation apparatus

Drawbacks:

(*a*) Increasing the production to the industrial level is difficult.

(b) CNTs grow in highly tangled form mixed with unwanted forms of carbon and/or metal species.

(c) CNTs produced are difficult to purify, manipulate and assemble for building nanotube-device architectures for practical applications.

(viii) RF Plasma Method: In this method of nanoparticle synthesis, the starting metal is kept in a pestle which is placed in an evacuated chamber (Fig. 49.9). RF coils are wrapped around the evacuated system in the vicinity of the pestle. When high voltage is applied to RF coils, heat is generated and the metal is heated above its evaporation point. Helium gas is then allowed to enter the system which forms high temperature plasma in the region of the coils. The metal vapour nucleates on the He gas atoms and diffuses up to a colder collector. Nanoparticles form at the collector. The particles are generously passivated by the introduction of oxygen.

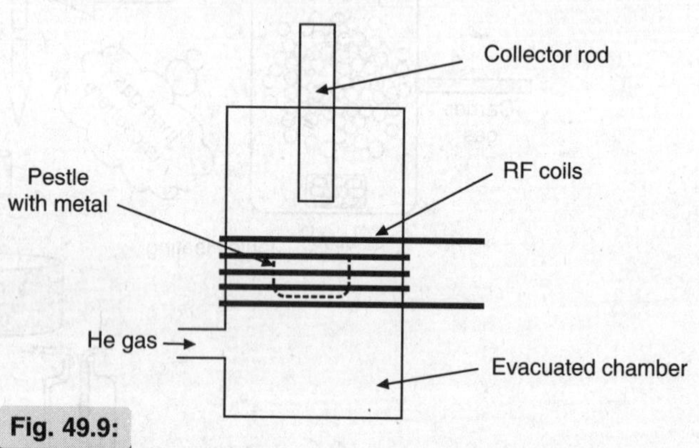

Fig. 49.9: Production of Nanoparticles by RF Plasma

(ix) Arc-evaporation Method: The carbon arc discharge method is the most common and perhaps easiest way to produce Carbon nano tubes CNTs. The method is also called *plasma arcing*. In this method, two carbon rods of 5-20 μm diameters are placed end to end, in an enclosure that is usually filled with inert gas at low pressure (Fig. 49.10). The electrodes are separated by approximately 1mm. A potential of 20-25V is applied across the carbon electrodes. A direct current of 50 to 100 A, driven by the potential difference, creates a high temperature discharge between the two electrodes. The discharge evaporates the surface of the positive electrode and forms a small rod-shaped nanotube deposit on the negative electrode. This is similar to electroplating process in a liquid medium. In the presence of the catalysts such as cobalt, nickel or iron central region of the positive electrode, single-walled nanotubes (SWNT) are produced. If no catalyst is used multi-walled nanotubes (MWNT) are produced. Using this method SWNT of diameters 1-5 nm with a length of 1 μm can be produced.

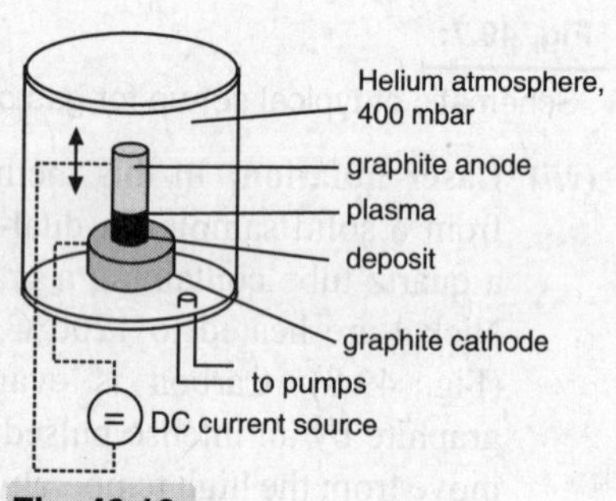

Fig. 49.10: Experimental set-up of an arc discharge apparatus.

Drawback: This technique produces a complex mixture of components, and requires further purification.

(x) Vapor Phase Epitaxy (VPE): Epitaxy technology is quite similar to CVD processes. The technology is primarily used for deposition of silicon. If the substrate is an ordered semiconductor crystal it is likely to build on the substrate with the same crystallographic orientation. Here the substrate acts as a seed for the deposition. When an amorphous/polycrystalline substrate surface is used, the film will also be amorphous or polycrystalline. In VPE number of gases can be introduced in an induction heated reactor where the substrate is heated. The temperature of the substrate typically must be at least 50% of the melting point of the material to be deposited. It is used for producing silicon on insulator (SOI) substrates. Fig. 49.11 shows a typical vapor phase epitaxial reactor.

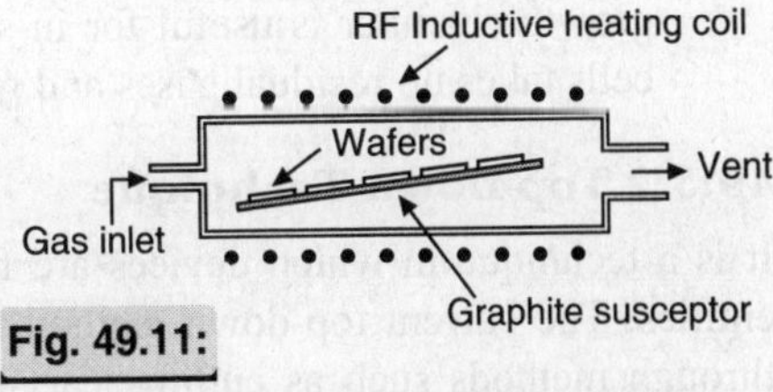

Fig. 49.11: Typical vapor phase epitaxial reactor

(*xi*) **Molecular beam epitaxy:** Epitaxial growth is carried by molecular beam epitaxy method. Here the interaction of one or several molecular or atomic beams happens on a surface of a heated crystalline substrate. The solid sources materials are placed in evaporation cells to give an angular distribution of atoms or molecules in a beam. The substrate is heated to the required temperature and as per the need constantly rotated for the growth homogeneity. MBE provides the controlled epitaxial growth of a new layer by very low rates of impinging atoms. Fig. 49.12 shows a scheme of a typical MBE system.

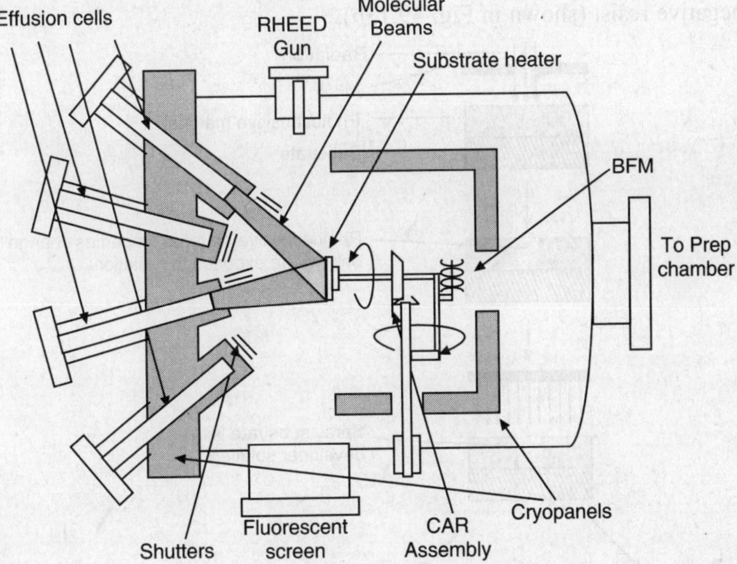

Fig. 49.12: A typical MBE system

The effusion cells, thermal evaporation cells (Knudsen cells) provide the molecular beam for either the bulk constituents or the dopants. Shutter of cells is useful for 'switching' the beam on and off. The substrate is mounted on a heatable and rotatable substrate holder.

The entire arrangement is kept in ultra-high vacuum environment for the formation of a molecular 'beam' and purity and quality of the grown layer. RHEED attached of the growth chamber is useful for in-situ analysis. A cryopanel around the sample and the cells takes up residual gases and gives a clean substrate environment.

49.5.2 Top-Down Technique

It is a technique in which devices are fabricated by removing existing material from larger entities. The current top-down method for manufacturing involves the construction of parts through methods such as cutting, carving and molding. Using these methods it is possible to fabricate a remarkable variety of machinery and electronic devices. However, the sizes at which one can make these devices are severely limited by the ability to cut, carve and mold.

(*i*) **Lithography:** Lithography is the transfer of a pattern to a photosensitive material (typically a photo-resist) by selective exposure to a radiation source such as light. Optical lithography is a technique that generates patterns on the surface which is used in fabricating integrated circuits. If we selectively expose a photosensitive material to radiation, the pattern of the radiation on the material is transferred to the material exposed.

If the resist is placed in a developer solution after selective exposure to the light source, it will etch away one of the two regions (exposed or unexposed). If the exposed material is etched away by the developer and the unexposed region is resilient, the material is considered to be a positive resist (shown in Fig. 49.13a). If the exposed material is resilient to the developer and the unexposed region is etched away, it is considered to be a negative resist (shown in Fig. 49.13b).

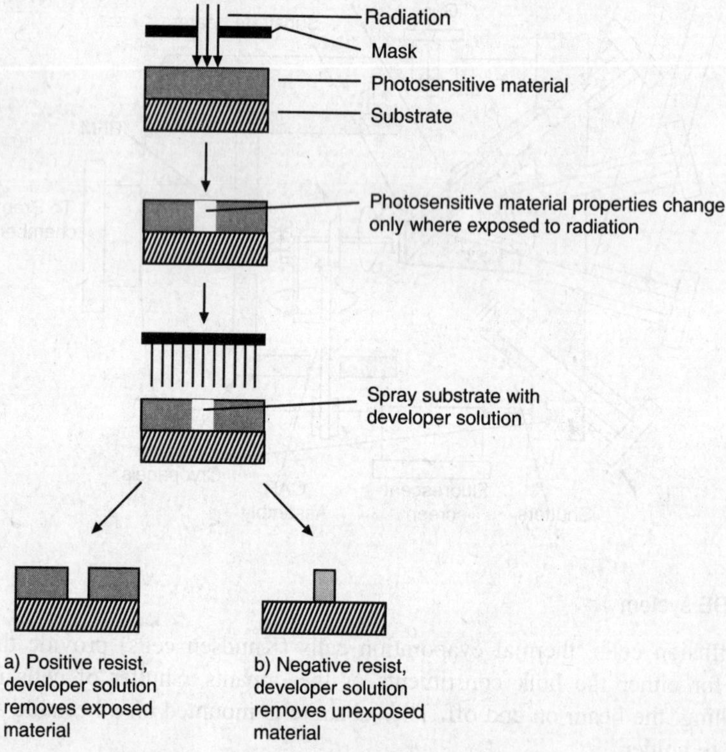

Fig. 49.13:

Lithography - Pattern definition (a) positive resist (b) negative resist

Electron Beam Lithography: Electron beam lithography is used in making nanostructures. A finely focused beam of electrons is used in this method to scan a specific pattern over the surface of a material. It can produce structures of 10 nm resolution. This process is not amenable for large scale production, since this process requires the electron beam to hit the surface of the material point by point in a serial manner.

(ii) Nanoimprint Lithography (NIL): Nanoimprint Lithography is a low-cost, high-production rate technology. Unlike conventional lithography, nanoimprint lithography patterns a resist by physically deforming the resist shape with a mold having a nano-pattern on it. The process is illustrated in Fig. 49.14. Fig. 49.14 (a) shows a mold of a hard material made by electron beam lithography is pressed into a resist, a softer material to make an imprint, Fig. 49.14 (b) shows the mold is then lifted off and 49.14 (c) shows the remaining soft material is removed through etching groves.

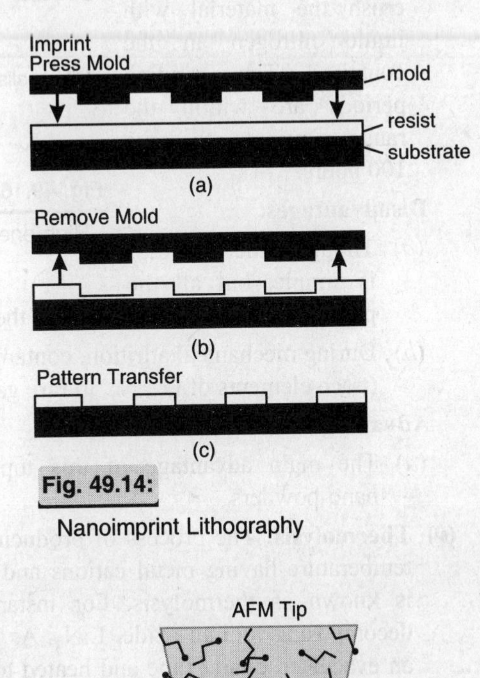

Fig. 49.14: Nanoimprint Lithography

Nanoimprint lithography can produce patterns of 10 nm resolution at low cost because it does not require the use of sophisticated radiation beams for generating patterns.

(iii) Dip Pen Nanolithography (DPN): It is a lithography technique where photoresist is not used. In DPN the tip of an atomic force microscope cantilever acts as a "pen," which is coated with a chemical compound or mixture acting as an "ink," and put in contact with a substrate, the "paper". The atomic force microscope tip transfers molecules to the surface of the substrate via a solvent meniscus (Fig. 49.15). It is a single step process to create nanopatterns on a substrate. This technique allows surface patterning on scales of fewer than 100 nano metres.

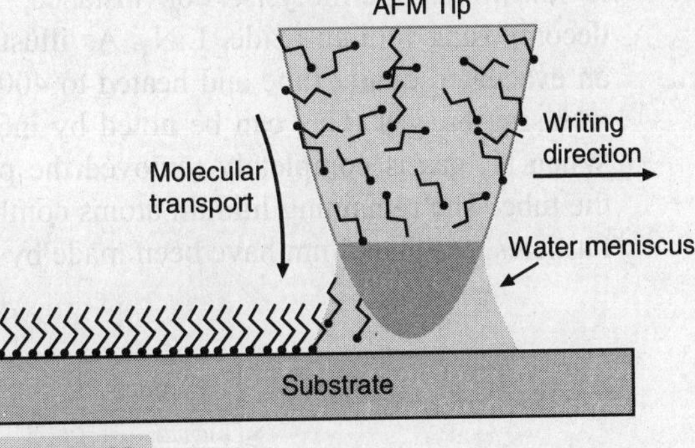

Fig. 49.15: Dip Pen Nanolithography

(iv) High-energy ball milling: High energy ball milling is a top-down approach technique. Coarse grained materials are crushed mechanically in rotating drums by hard steel and tungsten carbide balls. Powder samples are reduced to nanometer range by mechanical deformation produced by ball milling process. Magnetic and catalytic nanoparticles are usually produced by this method. Fig. 49.16 shows an illustration of high-energy ball milling. Here, a container is packed with stainless steel balls of a few millimeters in diameter. The material to be crushed is added as a powder of about 50 μm diameter grain size. Rotating shafts crush the material with liquid nitrogen in the container. The grinding periods are within the range of minutes to some 100 hours.

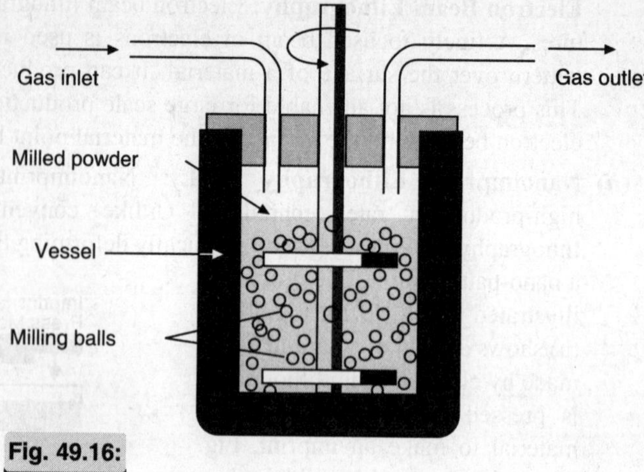

Fig. 49.16: High-energy Ball milling

Disadvantages:

(*a*) Though the process is simple, but all the particles are not broken down to the required particle size.

(*b*) During mechanical attrition, contamination by the milling tools (Fe) and atmosphere (trace elements of O_2, N_2, in rare gases) can be a problem.

Advantages:

(*a*) The main advantage of this top-down approach is high production rates of nano-powders.

(*v*) Thermolysis: The process of producing nanoparticles by decomposing solids at high temperature having metal cations and molecular anions or metal organic compounds is known as thermolysis. For instance, small lithium particles can be made by decomposing lithium azide, LiN_3. As illustrated in Fig. 49.17, the material is kept in an evacuated quartz tube and heated to 400°C. At about 370°C the LiN_3 decomposes, releasing N_2 gas. This can be noted by increased the pressure on the vacuum gauge. When N_2 gas is completely removed the pressure drops back to its original value in the tube. The remaining lithium atoms combine to form small colloidal metal particles. Particles less than 5 nm have been made by this method.

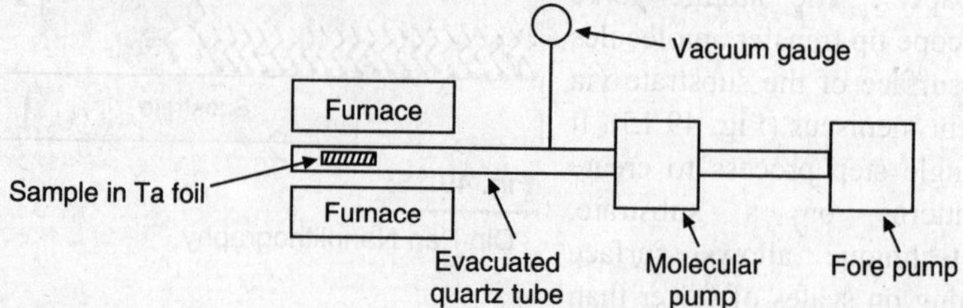

Fig. 49.17: Thermolysis

49.6 PROPERTIES OF NANOMATERIALS

Properties of nanomaterials depend on the sizes of the atom clusters that constitute the material. The subunits can be atoms of a single element, molecules or even combinations of atoms of several elements in subunits with equal stoichiometries (alloys, etc.), for example Na_n, $(SF_6)_n$, $(H_2O)n$, $(Cu_3Au)_n$, $(TiO_2)_n$ and $(ClCH_3C_6H_3CO_2H)n$ etc. The properties of nanoclusters are solely guided by the number of subunits they contain. Condensed matter nanoparticles are generally termed nanoclusters. A nanocluster is a nanometer sized particle made up of equal subunits.

Properties of Metal Nanoclusters

(*i*) **Magic Numbers:** Metal nanoparticles often show a size preference *i.e.* **magic numbers** of metal atoms. When lead metal nanoclusters were formed, magic numbers were observed at 7 and 10 atoms. It means that formation of clusters of 7 and 10 atoms is more likely than other clusters and these clusters are more stable than clusters of other sizes. The existence of magic numbers suggests that clusters can be viewed as super-atoms.

(*ii*) **Jellium Model:** The Jellium model considers a cluster of atoms as a large atom. The positive nuclear charge of each atom is assumed to be smeared out over the entire volume of the cluster while the valence electrons are free to move within this homogeneously distributed, positively charged background. The electron energy levels can be obtained by applying the Schrödinger equation for this system in a manner analogous to that for the hydrogen atom. The results showed the magic numbers correspond to those clusters having a size in which all the energy levels are filled.

(*iii*) **Geometric structure:** Usually the crystal structure of a large nanocluster is the same as the bulk structure of the material. There may be different lattice parameters. For example Cu clusters tend to have an FCC structure but smaller clusters of Cu (e.g. Cu_{55}, Cu_{147} ...) have perfect icosahedral structures. This happens due to a size preference *i.e. magic numbers* of metal atoms.

For example a magic number for aluminium is 13. Instead of a face centered cubic structure, the Al_{13} cluster is an icosahedron (ICOS) of 12 aluminium atoms with an additional aluminium atom in the center (Fig. 49.18 b).

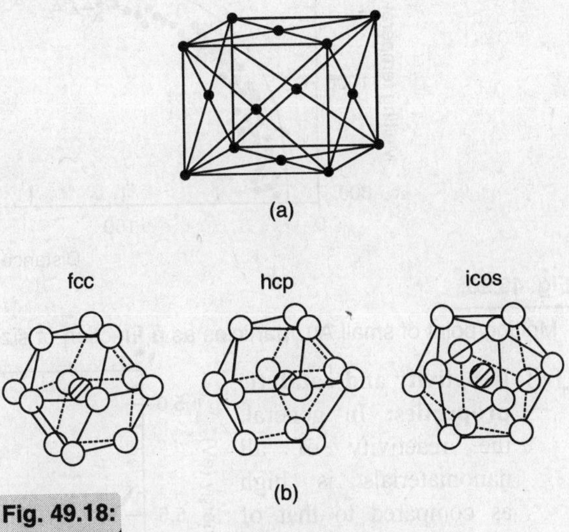

Fig. 49.18:
(a) The unit cell of bulk aluminium
(b) Three possible structures of Al_{13}: a FCC, HCP structure and icosahedral structure.

(*iv*) **Inter-particle spacing:** The effect of decreasing cluster size is to create more surface sites. This changes the surface pressure and result in change in the inter-particle spacing. This effect is illustrated in Fig. 49.19 for the case of Cu_n particles. It is seen that the inter-particle spacing decreases with size due to competition between the long-range electronic forces and the short range core-core repulsion.

(*v*) Melting temperatures: The change in inter-particle spacing and the large surface to volume ratio in particles have a combined effect on material properties. This affects, for example, the thermodynamic properties of which the simplest example is the melting point. Fig. 49.20 shows the melting point of Au_n particles as a function of size. The melting point decreases with size and the rate of decrease increases substantially at very small sizes.

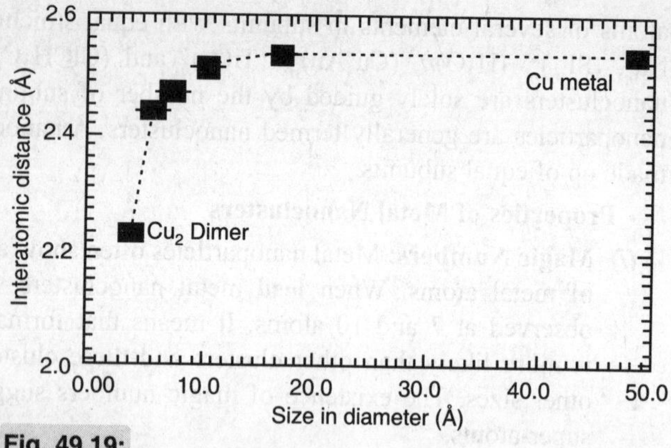

Fig. 49.19: Interatomic distance in Cu_n as a function of siz

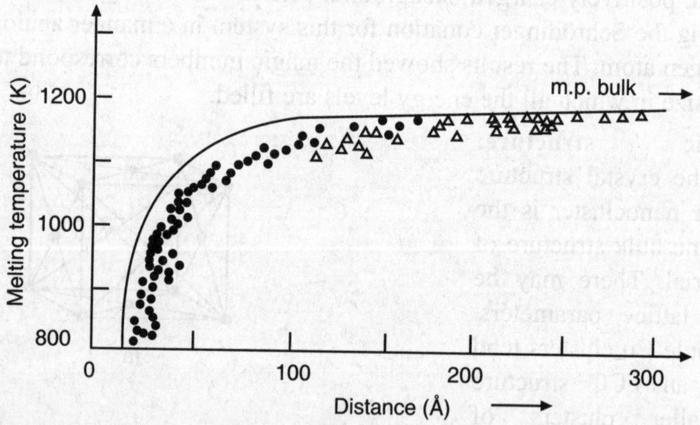

Fig. 49.20: Melting point of small Au_n particles as a function of size

(*vi*) Reactivity and catalytic properties: In general, the reactivity of all nanomaterials is high as compared to that of bulk due to their higher surface to volume. As the size is reduced from the bulk, the delocalized electronic states are transformed to more localized molecular bonds. Due to this change the ionization potential is affected. The ionization potential

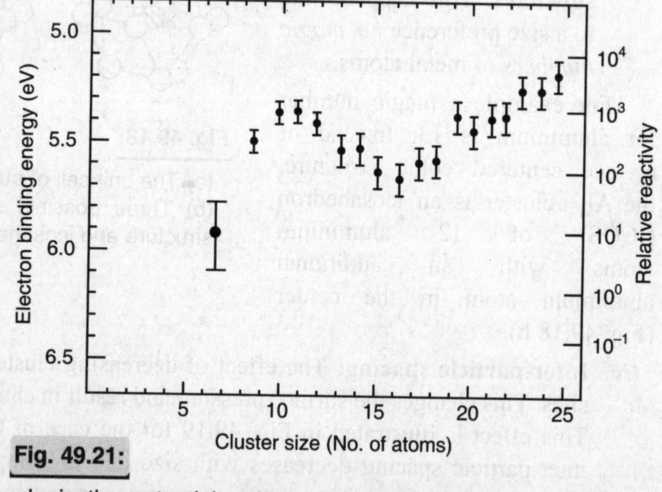

Fig. 49.21: Ionization potential and reactivity of Fe_n clusters.

at small sizes is higher than the bulk work function and show marked fluctuations as a function of size. Fig. 49.21 shows this for the case of Fe_n clusters reacting with hydrogen. The large surface to volume ratio and the variations in geometry and electronic structure have a strong effect on catalytic properties.

(vii) **Electronic Structure:** Atoms and molecules have discrete energy levels. However, in solid materials, these orbitals overlap to form bands. In the case of a metal, one of these bands is partially filled. This makes a metal electrically-conductive. As the bulk shrink in size, the continuous density of energy states in bulk is replaced with a set of discrete energy levels, as illustrated in Fig. 49.22.

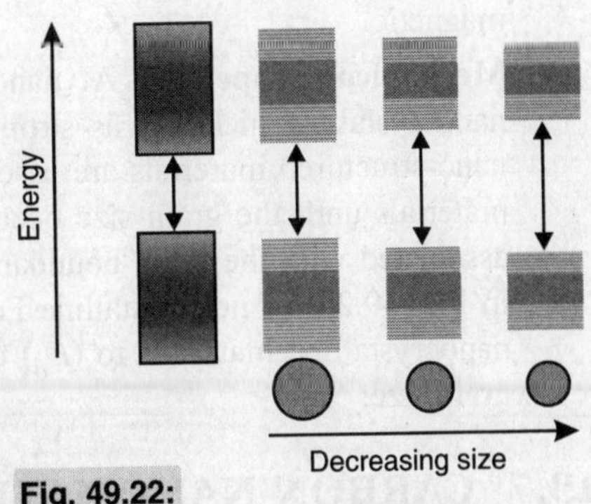

Fig. 49.22: Changes in energy band structure as the size of the cluster decreases.

(viii) **Optical Properties:** The energy level separations in a nanocluster are dependent on the size of the clusters, and material, which therefore affect the energies needed for the transitions of electrons to excited states. Clusters of different sizes will have different absorption spectra but they lie generally in the visible range of the spectrum. Hence, the clusters of different sizes exhibit different colours. For e.g., nanoscale gold particles can be orange, purple, red or greenish depending on the size of the cluster.

(ix) **Magnetic Properties:** Nanoclusters of magnetic materials exhibit totally new class of magnetic properties. As cluster size decreases, materials exhibit **ferromagnetic** behaviour. In some cases, even clusters made up of ***nonmagnetic atoms can have a net magnetic moment***. For example rhenium clusters show a pronounced increase in their magnetic moment when they contain less than 20 atoms. For clusters with less than 15 atoms these moments are fairly large.

Fig. 49.23 shows the plot of the magnetic moment per atom of a cluster on the number of atoms in the cluster. It is seen that the magnetic moment increases as the cluster size decreases. Thus, nanoclusters are more magnetic than the bulk material. At small sizes, the clusters become spontaneously magnetic and the magnetism disappears in clusters

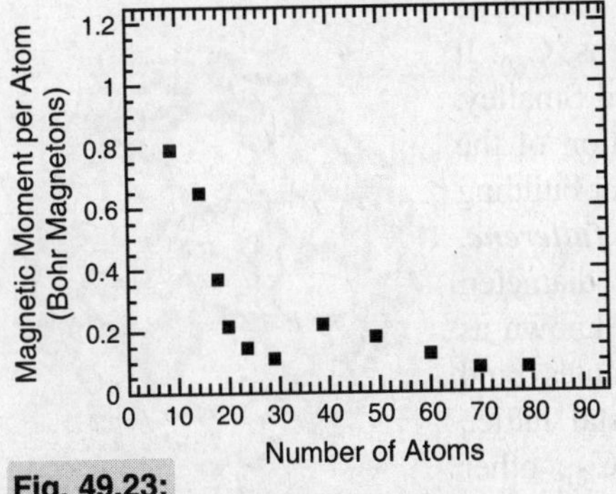

Fig. 49.23: Plot of the magnetic moments versus number of atoms.

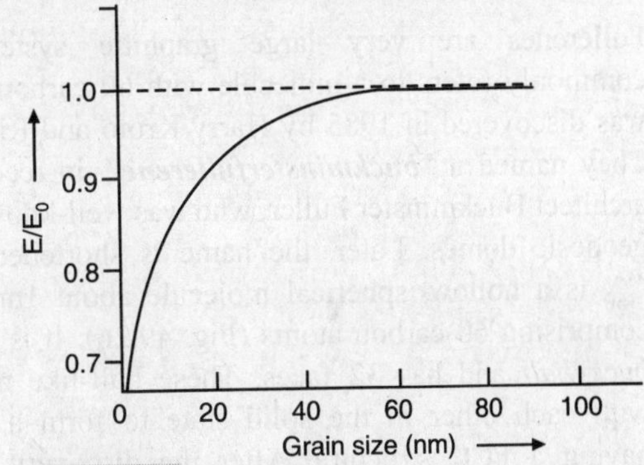

Fig. 49.24: A plot of (E/E_0) versus grain size.

containing more than 80 atoms. Even the clusters of nonmagnetic solids are found to be magnetic.

(*x*) **Mechanical properties:** At nano scale, strength of metals enhances. For instance, nanocrystalline nickel is as strong as hardened steel. The intrinsic elastic moduli of nano-structured materials are essentially the same as those for conventional grain size materials until the grain size becomes very small, e.g. < 5 nm. The number of atoms associated with the grain boundaries and triple junctions become very large as shown in Fig. 49.24 for noncrystalline Fe. Fig. 49.24 shows ratio of Young's modulus (E) of nanocrystalline materials to (E_o) of conventional grain size materials as a function of grain size.

49.7 CARBON NANOMATERIALS

In 1980, we knew of only three forms of carbon, namely diamond, graphite, and amorphous carbon. Diamond and graphite have different physical structures and properties; however their atoms are both arranged in covalently bonded networks. Fullerenes and carbon nanotubes (CNTs) are the two important nanomaterials. The ***fullerenes and carbon nanotubes are a class of allotropes of carbon*** which conceptually are graphene sheets rolled into spheres or tubes respectively. Fig. 49.25 shows the four different types of carbon structures: diamond, fullerene, graphite, and nanotube structure.

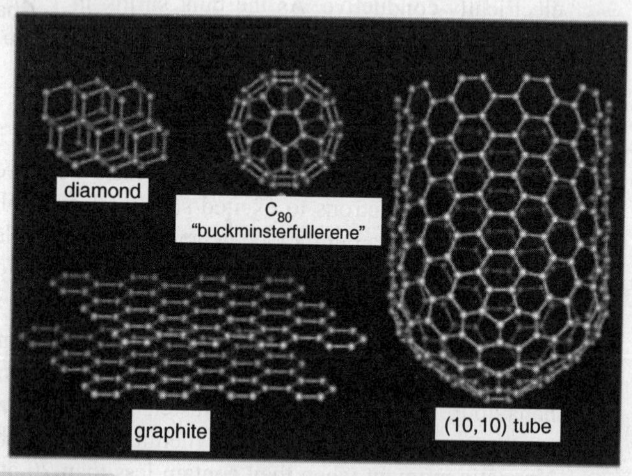

Fig. 49.25: The four different types of carbon structures.

49.8 FULLERENES

Fullerenes are very large graphitic systems. Fullerene commonly refers to a molecule with 60 carbon atoms, C_{60}. It was discovered in 1985 by Harry Kroto and Richard Smalley. They named it "***buckminsterfullerene***", in recognition of the architect Buckminster Fuller, who was well-known for building geodesic domes. Later, the name is shortened to ***fullerene***. C_{60} is a hollow, spherical molecule about 1nm in diameter, comprising 60 carbon atoms (Fig. 49.26). It is also known as *buckyball* and has 32 faces. These ball-like molecules bind with each other in the solid state to form a crystal lattice having a FCC structure. After the discovery of C_{60}, other related molecules (C_{36}, C_{70}, C_{76} and C_{84}) composed of only carbon atoms, but possess different geometric structures, were

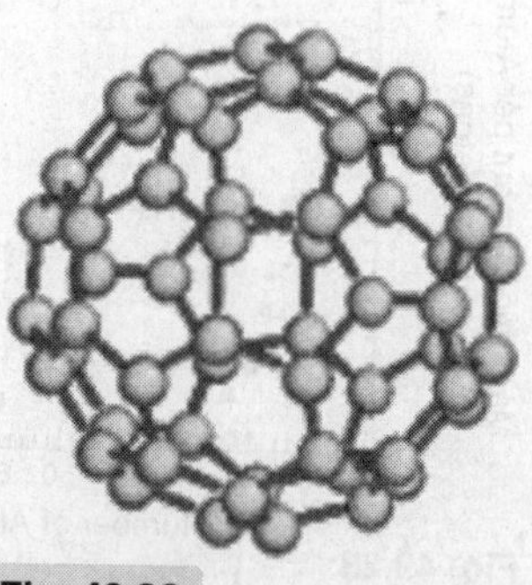

Fig. 49.26: The C_{60} molecule

also discovered. This new class of carbon molecules came to be known as the fullerenes. There are now thirty or more forms of fullerenes. The term fullerene was given to any *closed carbon cage*.

Geometric structure: In C_{60}, 60 carbon atoms are placed at the vertices of a regular truncated icosahedron. Each carbon site is equal to every other site. The average nearest neighbor C-C distance in C_{60} (1.44 Å) is approximately the same to that in graphite (1.42Å). Each carbon atom in C_{60} is trigonally bonded to other carbon atoms, the same as that in graphite, and most of the faces on the regular truncated icosahedron are hexagons. There are 20 hexagonal faces and 12 additional pentagonal faces in each C_{60} molecule with a molecule diameter of 7.10 Å.

Synthesis: Fullerenes are synthesized by arc discharge between graphite electrodes in around 200 torr of He gas. The heat generated at the contact point between the electrodes evaporates carbon to form soot and fullerenes. They condense on the water cooled walls of the reactor. The carbon soot contains up to ~15% fullerenes: C_{60} (~13%) and C_{70} (~2%). The fullerenes are next separated from the soot by use of liquid chromatography. Fig. 49.27 shows the schematic diagram of fullerene soot production chamber.

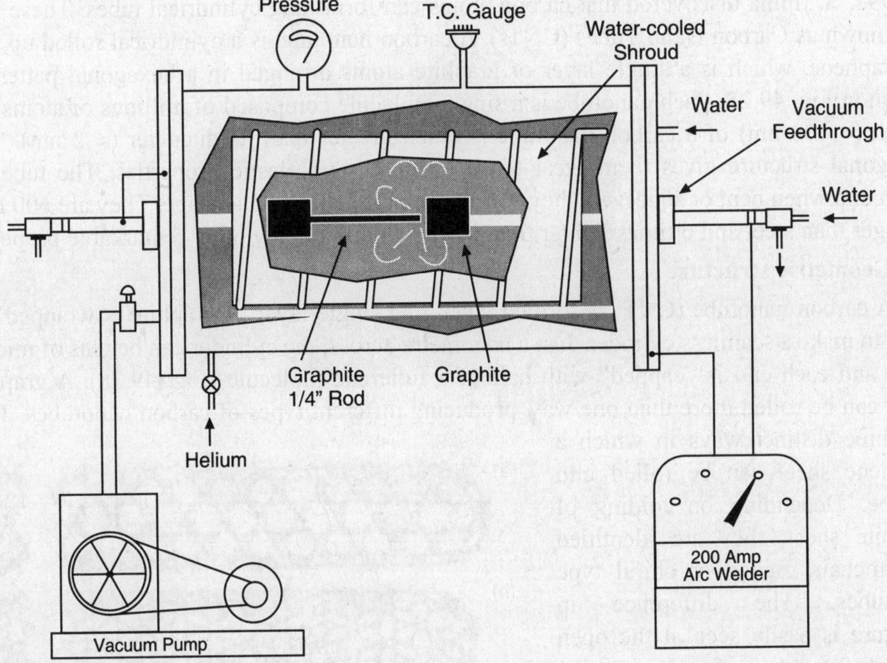

Fig. 49.27:

Synthesis of Fullerenes.

Applications: Some of the potential applications of fullerenes are discussed here.

(*i*) When three electrons are added to C_{60}, ionic solids of the general formula A_3C_{60} are formed; A is any metal in Group I (lithium, sodium, potassium, rubidium, and cesium). These metals display superconductivity at some what low temperatures.

(*ii*) Useful dopant atoms (endohedral) can be placed inside the hollow fullerene ball. C_{60} has the right size to fit into an enzyme, which would kill the virus such as AIDS.

(*iii*) Useful in industrial applications such as optical devices, chemical sensors and chemical separation devices and etc.

(*iv*) Useful in the production of diamonds and carbides as cutting tools or hardening agents.

(*v*) For producing polymers of rugged structure and high reactivity.

(*vi*) Fullerenes have good catalytic properties, experiments shown that fullerenes with alkali metals possess catalytic properties resembling those of platinum.

(*vii*) Fullerenes are a better storage medium for hydrogen than metal hydrides.

(*a*) They are useful in the development of new batteries and even of non-polluting automobiles based on fuel cells.

(*b*) A thin layer of the C_{70} fullerene when deposited on a silicon chip provides a vastly improved template for growing thin films of diamond.

(*c*) Useful for miniature 'ball bearings' to lubricate surfaces, drug delivery vehicles and in electronic circuits.

49.9 CARBON NANOTUBES

In 1991, S. Iijima discovered that carbon atoms can form long cylindrical tubes. These tubes are known as Carbon Nano Tubes (CNTs). A carbon nanotube is a cylindrical rolled up sheet of graphene, which is a single layer of graphite atoms arranged in a hexagonal pattern, as shown in Fig. 49.25. Each nanotube is a single molecule composed of millions of atoms. The length (≈ 100 nm) of a carbon nanotube is much greater than its diameter (≈ 2 nm). Their hexagonal structure gives them great tensile strength and elastic properties. The tubes are tough and when bent or squeezed, they spring back to their original shape. They are 600 times stronger than steel and 6 times lighter than it. They exhibit many other remarkable properties.

Geometric structure

A carbon nanotube (CNT) is formed when one single layer of graphite is wrapped onto itself to make a seamless cylinder. Just a nanometre across, the cylinder can be tens of microns long, and each end is "capped" with half of a fullerene molecule (Fig. 49.28). A graphene sheet can be rolled more than one way, producing different types of carbon nanotubes. There are three distinct ways in which a graphene sheet can be rolled into a tube. Depending on folding of graphite sheet, they are identified as armchair, zigzag or chiral type nanotubes. The difference in structure is easily seen at the open end of the tubes. As in graphite, nanotubes are composed of sp^2 bonding. Nanotubes obviously line up themselves into "ropes" held together by Van der Waals forces. Under high pressure, nanotubes can unite together and sp^2 bonding becomes sp^3. This provides strong, unlimited-length wires through high-pressure nanotube linking.

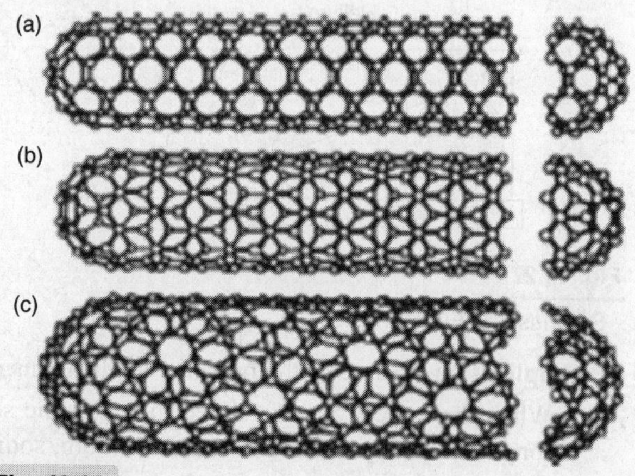

Fig. 49.28: Models of different single wall nanotubes: (a) armchair, (b) zigzag and (c) chiral structure

Types of Carbon Nanotubes

The carbon nanotubes are typically a few nanometres in diameter and several micrometres to centimetres long. Carbon nanotubes are of two types, single-walled nanotubes (SWNT) and multi-walled nanotubes (MWNT).

(*i*) **Single-walled nanotubes (SWNT):** Single-walled nanotube consists of a graphene sheet rolled on leading to a single cylinder (Fig. 49.29 a). Two halves of a Fullerene molecule closes the structure at both ends. Single-wall nanotubes can be thought of as the fundamental cylindrical structure, and these form the building blocks of both multi-wall nanotubes and the ordered arrays of single-wall nanotubes called ropes. The SWNTs have diameters from 1nm to 5 nm and are usually well over 1 mm in length. Single Walled Nanotubes (SWNT) can be considered as long wrapped graphene sheets. The length to diameter ratio of SWNTs is generally about 1000, so that they can be considered as nearly one-dimensional structures.

(*ii*) **Multi-walled nanotubes (MWNT):** A multi-walled carbon nanotube (MWNT) is the arrangement of several coaxial tubes of graphene sheets forming a tube-like structure (Fig. 49.29 b. Each MWNT has from 2 to 50 such tubes. The separation between neighbouring tubes is roughly equal to that between the layers in natural graphite and is of 0.34-0.36nm. MWNTs have inner diameters from 1.5 to 15 nm and outer diameters from 2.5 to 30 nm. MWNTs are usually well over 1 mm in length.

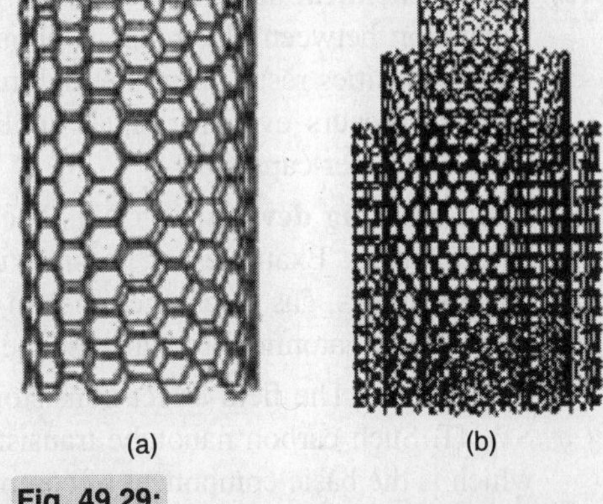

Fig. 49.29:
(a) A single-walled carbon nanotube (SWNT),
(b) Multi-walled carbon nanotubes (MWNT)

Synthesis of Carbon Nanotubes

CNTs are synthesized for the first time in 1991 by Sumio Iijima. There are a number of methods of making CNTs and fullerenes such as arc evaporation, pulsed laser deposition and Chemical Vapour Deposition (CVD) etc. The details of above methods are already discussed in section 49.5.1.

Properties of Carbon Nanotubes

(*i*) **Tensile strength:** Because of sp^2 carbon-carbon bonds, carbon nanotubes can bend like a rubber tube even with a high tensile strength. The tensile strength, (breaking strain) of a carbon nanotube can be up to 63 GPa, around 50× higher than steel. Young's modulus (stiffness of a material- rate of change of stress with applied strain) of the best nanotubes can be as high as 1000 GPa which is approximately 5× higher than steel.

(*ii*) **Thermal conductivity:** Carbon nanotubes conduct heat by vibrations of the covalent bonds between the carbon atoms. The atoms twist around themselves and transmit heat through the material. The thermal conductivity of carbon nanotubes is 10 times that of silver.

(*iii*) **Electrical conductivity:** Some CNT have superior conductivities than that of copper, whereas others act more like semiconductors. The molecular structure gives this difference in properties that result in a different band structure and thus a different band gap.

(*iv*) **Chemical reactivity:** A SWNT consists of two separate regions with different physical and chemical properties. The first is the sidewall of the tube and the second is the end cap of the tube. C-atoms placed in hexagons and pentagons form the end cap structures. The other is symmetric cylinder. A smaller nanotube diameter results in increased reactivity.

Applications of CNTs

(*a*) **Energy storage:** The efficiency of fuel cells is known by the rate of electron transfer at the carbon electrodes, which is the fastest on nanotubes. The small dimensions, smooth surface topology and perfect surface specificity make CNTs as better energy storages.

(*b*) **Hydrogen storage:** Because of their cylindrical and hollow geometry and extremely small diameters carbon nanotubes can store hydrogen in the inner cores through a capillary effect.

(*c*) **Electrochemical super-capacitors:** Super-capacitors have a high capacitance. As the separation between electrodes is about a nanometer for nanotubes in electrodes, very large capacities result from the high nanotube surface area. So, large amount of charge injection occurs even for small applied voltage. This is used for energy storage in nanotube super-capacitors.

(*d*) **Field emitting devices:** In CNTs because their elongated shape have very large field amplification. Examples of possible uses for CNTs as field emitting devices are flat panel displays, gas discharge tubes in telecom networks and electron guns for electron microscopes, atomic force microscope (AFM) tips and microwave amplifiers.

(*e*) **Transistors:** The field-effect transistor can be constructed of only one semiconducting SWNT. Such carbon nanotube transistors can be coupled together as a logical switch, which is the basic component of computers.

(*f*) **Nanoprobes and Sensors:** Because of their flexibility, nanotubes are suitable in scanning probe instruments (MWNT, STM and AFM).

(*g*) The lightness of CNTs gives them great potential in applications such as aerospace.

49.10 NANOWIRES

Nanowires are wires with a very small diameter, sometimes as small as 1 nanometer. Nanowires are 2-D materials. They have an aspect ratio (length-to-width ratio) of 1000 or more. Scientists hope to use them to build tiny transistors for computer chips and other electronic devices. In the last couple of years, carbon nanotubes have overshadowed nanowires.

Types of nanowires:

(*i*) Metallic nanowires (e.g., Ni, Pt, Au)

(*ii*) Semiconducting nanowires (e.g., Si, InP, GaN)

(*iii*) Insulating nanowires (e.g., SiO_2, TiO_2)

(*iv*) Molecular nanowires composed of repeating molecular units either organic (e.g., DNA) or inorganic (e.g., $Li_2Mo_6Se_6$)

Properties of nanowires:

Nanowires have many interesting properties that are not seen in bulk materials. This is because electrons in nanowires are quantum confined laterally and occupy energy levels that are different from the traditional energy bands in bulk materials. The quantum confinement manifests in discrete values of electrical conductance. The discrete values are often referred to as the quantum of conductance and are integer values of $\frac{2e^2}{h} \approx 12.9\ \text{k}\Omega^{-1}$. They are the inverse of the well known resistance unit $\frac{h}{e^2} \approx 25812.8$ ohms which is known as the von Klitzing constant R_K.

Production of nanowires

Nanowires can be suspended, deposited or synthesized.

(*i*) A suspended nanowire can be produced by chemical etching of a bigger wire, or bombarding a bigger wire with high energetic particles.

(*ii*) Vapour-Liquid-Solid (VLS) synthesis is commonly used technique for producing nanowires which uses laser ablated particles or a feed gas such as silane as source material. The source is then exposed to a catalyst. The catalysts used are liquid metal nanoclusters. The source enters these nanoclusters and begins to saturate it. Once super-saturation is reached, the source solidifies and grows outward from the nanocluster. The length of the nanowire can be adjusted by turning off the source.This process often produces crystalline nanowires in case of semiconductor materials.

(*iii*) Inorganic nanowires are synthesized in a single-step vapour phase reaction at elevated temperature.

Applications

(*i*) Nanowires can be used to create active electronic elements such as p-n junctions, logic gates.

(*ii*) Nanowires have potential applications in high-density data storage, ether as magnetic read heads or as storage media, and electronic and optoelectronic nanodevices, for metallic interconnects of quantum devices and nanodevices.

49.11 QUANTUM DOTS

A quantum dot holds discrete number of electrons. Number of molecules together in small space produces the quantum dot effect. Dots have been made ranging from 30 nm to 1 micron in size and holding from zero to hundreds of electrons. Addition or removal of an electron changes the properties of a quantum dot. Adjusting electric fields in the neighbourhood of the dot, for instance by applying a voltage to a nearby metal gate, can change this number. As the quantum dots are fabricated in solids, they have many electrons in them. Almost all of these are tightly bound to atoms in the solid.

In a quantum dot, the energies at which electrons and holes can exist are limited in the particles. As energy is related to wavelength, the optical properties of the particle can be finely tuned depending on its size. So, particles can be made to emit or absorb specific colours of light, merely by controlling their size. By using an external UV light on nano-crystals made from semiconductor materials such as zinc sulphide, cadmium selenide, indium phosphide or lead sulphide, the nano-crystal will absorb the light. As a consequence of the crystal being stimulated by the absorbed light, it will re-emit the light, typically of a certain colour, basing on the size of the quantum dot.

Further, the position of a single electron in a quantum dot might attain several states, so that a quantum dot could represent a byte of data. Alternatively, a quantum dot might be used in more than one computational instruction at a time.

Applications

Recently, quantum dots have found applications in composites, solar cells and fluorescent biological labels. The quantum dots are expected to be used in various other applications such as the following:

(*i*) Quantum computers

(*ii*) Domestic and office lighting applications

(*iii*) Television screens and monitors

49.12 DENDRIMERS

A dendrimer is a tree-like highly branched polymer molecule. Dendrimers are synthesized from monomers with new branches added in discrete steps to form a tree-like architecture. There are many types of dendrimers; the smallest is several nanometres in size.

Applications:

(*i*) Used in conventional applications such as coatings and inks.

(*ii*) Environmental clean-up could be assisted by dendrimers as they can trap metal ions, which could then be filtered out of water with ultra-filtration techniques.

(*iii*) Can act as nanoscale carrier molecules that are useful in drug delivery for cancer applications. They have high drug-carrying capacity that can release a heavy payload without damaging tissue. A single dendrimer can carry a molecule that recognizes cancer cells, a therapeutic agent to kill those cells and a molecule that recognizes the signal of cell death.

49.13 NANOCOMPOSITES

Nanocomposites are the materials with a nanoscale structure. They have improved the macroscopic properties of products. Usually, nanocomposites are clay, polymer or a combination of these materials with nanoparticle building blocks. Carbon nanotubes are very useful in nanocomposites. At present, carbon fibres and bundles of MWCNTs are used in polymers to control or augment conductivity. In some particular nanocomposites, nanoparticles are used as fillers in a matrix; for example, carbon black used as a filler to reinforce car tyres. Because of the stiffness of carbon nanotubes they are good in structural applications. For example, they may be used as reinforcements in high strength, low weight, and high performance composites.

49.14 NANO-LAYERS

Nano-layer is a two-dimensional structure, layer or film on the surface of a solid or liquid that has nanoscale thickness. They are used in electronic device manufacture, chemistry and engineering. In integrated circuit industry many devices rely on atomic level thin films for their operation. Monolayers, one atom or molecule deep, are usually prepared and used in chemistry. Studies are under progress in the control of the composition and smoothness of surfaces and the growth of films. A solid nanoscale layer that separates layers of a much thicker ferromagnetic is useful in the effects of tunnel magneto-resistance and giant magneto-resistance. They are produced by PVD and CVD process.

49.14.1 Nanolayer Devices

(*i*) **Giant Magneto-Resistance (GMR):** In 1988 a new effect called ***giant magneto-resistance*** **(GMR)** was discovered by A. Fert and P. Grünberg in synthetically fabricated magnetic metallic multilayers of nanometer thickness. Magneto-resistance is a phenomenon in which the application of a magnetic field changes the resistance of a material. The magnetoresistance effect occurs in metals only at very high magnetic fields and at low temperatures.

For example, in pure copper at a 4K a field of 10 Wb/m^2 produces a factor of 10 times change in the resistance. Large fields and low temperatures magneto-resistance in metals had limited practical applications. Alternate layers of nanometer thickness of a ferromagnetic material and a non-ferromagnetic metal such as Fe/Cr and Co/Cu deposited on a substrate gives a significant reduction of the electrical resistance of the multilayer. This effect was found to be much larger than other magneto-resistive effects that had ever been observed in metals and was, therefore, called "giant magnetoresistance".

A schematic of a three-layered structure is illustrated in Fig. 49.30 (a). The bottom and top layers are Co layers which are ferromagnetic layers and the middle layer is a Cu layer which is a conductive, nonmagnetic interlayer. In ferro-magnetic materials, the spins from unpaired electrons spontaneously align parallel to give the material a total magnetic moment. In the absence of the magnetic field the magnetizations of the ferromagnetic layers are antiparallel. Copper is normally an excellent conductor, but when it is only a few atoms thick, electron scattering causes its resistance to increase significantly. Resistance to the flow of current in a direction perpendicular to such a multilayer system is high.

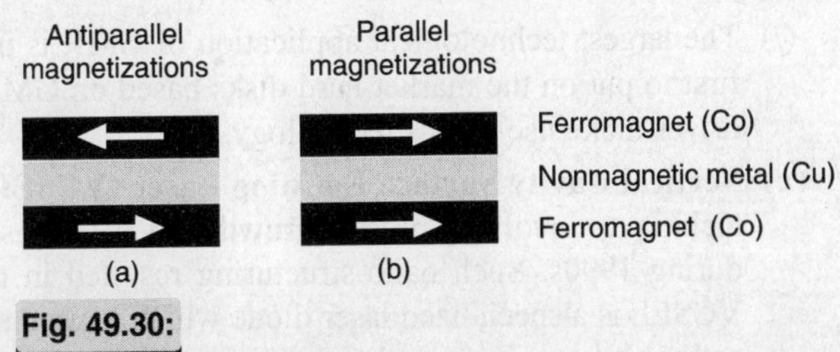

Fig. 49.30: Concept of Giant Magneto-Resistance

When an external magnetic field is applied parallel to the layers, the magnetic moments in successive ferromagnetic layers align in the direction of the magnetic field (see Fig. 49.30 b) and the resistance of the multilayer system decreases, as shown in Fig. 49.31.

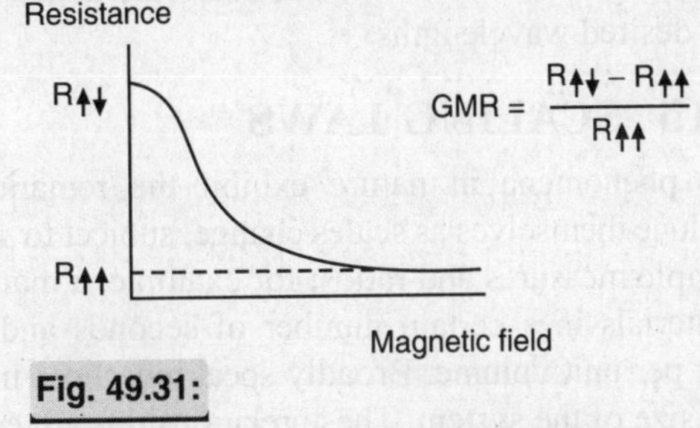

Fig. 49.31: Giant Magneto-Resistance in multilayer systems

The GMR effect occurs because of the dependence of electron mean free path on the orientation of the electron spin with respect to the direction of magnetization. A conduction electron moving through a ferromagnetic material passes through more easily if it encounters electron spins oriented in the same direction as its own spin orientation. Therefore, electrons whose spins are aligned along the direction of magnetization of ferromagnetic layer will travel further in the layer without being scattered.

In a multilayer system, the first magnetic layer polarizes the electron spins. The second layer disperses the spins strongly if its moment is not allied with the moment of the first layer.

If the moment of the second layer is aligned, it allows the spins to pass (Fig. 49.32). So, the resistance varies basing on whether the moments of the magnetic layers are antiparallel (high resistance) or parallel (low resistance).

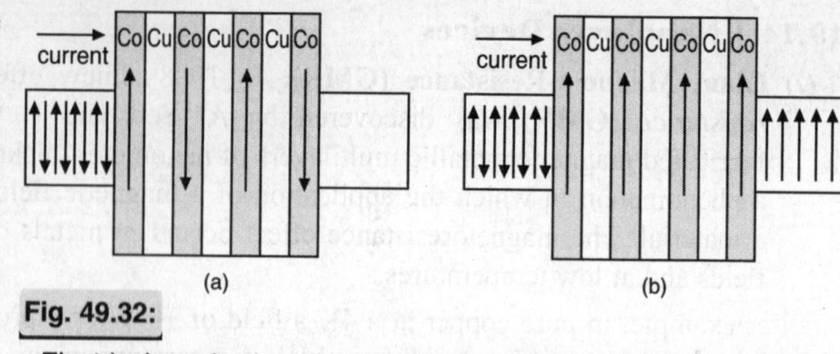

Fig. 49.32: Electrical conduction of multilayered magneto-resistive structure

Best possible layer thicknesses increase magnetic-layer antiparallel coupling, which is necessary to keep the system in the high-resistance state when no field is applied. When an external field rises above the antiparallel coupling, the moments in the magnetic layers align and reduce the resistance. For spin-dependent scattering to be a significant part of the total resistance, the layers must be thinner (to a magnitude of several nanometers) than the mean free path of electrons. In a typical GMR medical sensor the thickness of a conducting layer is about 3 nm.

Applications of GMR

(*i*) The largest technological application of GMR is in the data storage industry. IBM was first to put on the market hard disks based on GMR technology, and nowadays all disk drives make use of this technology.

(*ii*) **Vertical Cavity Surface Emitting Laser (VCSEL):** One-dimensional nanotechnology techniques involving precise growth of very thin semiconductor layers were developed during 1990s. Such nano-structuring resulted in the production of efficient VCSELs. VCSEL is a specialized laser diode which is used as optical source in optical communication. It has nanoscale layers of compound semiconductors grown into their structure – alternating dielectric layers as mirrors and quantum wells. Quantum wells confine the charge carriers in well-defined regions and provide the energy conversion into light at desired wavelengths.

49.15 SCALING LAWS

Many phenomena in nature exhibit the remarkable property of self-similarity and they reproduce themselves as scales change, subject to *scaling laws*. Machines can be characterized by simple measures and ratios; for example, a manufacturing system can handle its own mass of materials in a certain number of seconds and a motor will handle a certain amount of power per unit volume. Broadly speaking, these numbers vary in predictable ways according to the size of the system. These relationships are called ***scaling laws***. Scaling law calculations can show why technology based on nanoscale devices is expected to be extremely powerful.

Def: Scaling laws are extremely simple observations about how physics works at different sizes. They provide a very simple approach to understanding the nanoscale.

Application of Scaling Laws: The scaling laws can compare the relative performance of systems at different scales and the technique works for any systems with the relevant properties-the strength of a steel cable scales the same as a muscle. Any property that can be summarized by a scaling factor can be used in this kind of calculation.

Strength versus weight is proportional to area divided by volume ($1/L$). Though strength and mass are completely different kind of things, but both can affect the performance of systems, and they both scale in predictable ways.

As *Frequency of operation* $\propto$ (1/*L*), if a machine shrinks its frequency of operation will increase proportionally. Similarly, quantities like power and volume can be combined to form useful measures like power density.

As power density $\propto$ (1/*L*) and it varies inversely with machine size.

For example, suppose an engine measuring 10 cm on one of its sides, produces 1,000 watts of power. Then an engine measuring 1 cm on one of its sides should produce 10 watts of power, *i.e.*, 1/100 of the ten-times-larger engine. Then, one thousand (1000) 1-cm engines would take the same volume as one 10-cm engine, but produce 10,000 watts. So according to scaling laws, by building 1,000 times as many parts, and making each part 10 times smaller, we can get 10 times as much power out of the same mass and volume of material. Thus, when the design was shrunk by a factor of ten, power output increased ten-fold.

There is another scaling law regarding functional density $\propto 1/L^3$. If we can build machine parts at nanoscale, then we can pack 10^{18} more parts into the same volume. If the parts can be built using a massively parallel process like chemistry, and if reliability is high and the design is fault-tolerant so that the collection of parts will last for the life of the product, then it would be very much worth doing.

Friction and wear are important factors in mechanical design. Friction is proportional to force. Thus, friction $\propto L^3$. This implies that frictional power is proportional to the total power used, regardless of scale.

Performance of machines improves as machines shrink in size. Scaling laws show what we can expect once we develop the ability to build nanoscale systems: performance vastly higher than we can achieve with today's large-scale machines.

49.16 TOOLS FOR CHARACTERIZATION OF NANOMATERIALS

49.16.1 Transmission Electron Microscope (TEM):

The transmission electron microscope is an extremely useful tool for material characterization. The resolution of TEM is about 0.2 nm, which is about 1000x better than ordinary light microscope. The first TEM was developed in 1933 by Max Knoll and Ernst Rukasa and the first commercial TEM in 1939. In 1986, Ruska was awarded the Nobel Prize in Physics for the development of transmission electron microscopy.

TEM can be used to study the growth of layers and to analyze the quality, shape, size and density of quantum wells, wires and dots. The TEM operates on the same basic principles as the light microscope but uses electrons instead of light. Because the wavelength of electrons is much smaller than that of light, the resolution much better than that from a light microscope.

Principle: A beam of high velocity electrons accelerated under vacuum, focused by condenser lens (electromagnetic bending of electron beam) onto specimen and emergent electron beam is focused by objective lens. Final image forms on a fluorescent screen or camera for the image viewing.

Components: The major components are vacuum system, specimen stage, electron gun, electron lens and computer. A general layout of TEM describing the path of electron beam and a ray diagram for the diffraction mechanism in is shown in Fig. 49.33. The acceleration voltages, 120 to 200 kV determines the velocity, wavelength and hence the resolution of the microscope.

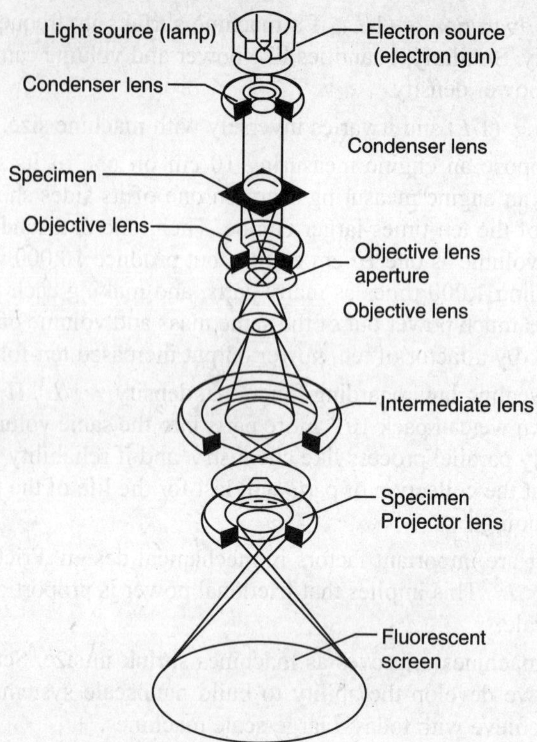

Fig. 49.33:

Path of electron beam in a TEM and a ray diagram

TEM Applications:

(*i*) Transmission Electron Microscope is ideal for life sciences, nanotechnology, medical, biological and material research, forensic analysis, gemology and metallurgy, industry and education.

(*ii*) TEMs provide topographical (accurate representation of the physical features of an area), morphological (structure of things), compositional and crystalline information.

(*iii*) The images allow researchers to view samples on a molecular level, making it possible to analyze structure and texture.

(*iv*) Useful in the study of crystals and metals, but also has industrial applications.

(*v*) TEMs can be used in semiconductor analysis and production and the manufacturing of computer and silicon chips.

(*vi*) Technology companies use TEMs to identify flaws, fractures and damages to micro-sized objects; this data can help fix problems and/or help to make a more durable, efficient product.

Advantages: A Transmission Electron Microscope has number of advantages.

(*i*) TEM offer the most powerful magnification.

(*ii*) TEM have a wide-range of applications and can be utilized in a variety of different scientific, educational and industrial fields.

(*iii*) TEM provides information on element and compound structure.

(*iv*) TEM provides images are of high-quality and detailed in their nature.

(*v*) TEM yields information of surface features, shape, size and structure.

(*vi*) TEM is easy to operate with proper training.

Disadvantages:

(*i*) TEMs are large and very expensive.

(*ii*) Laborious sample preparation and operation and analysis requires special training.

(*iii*) Samples are limited to those that are electron transparent, able to tolerate the vacuum chamber and small enough to fit in the chamber.

(*iv*) Images are black and white.

49.16.2 Scanning Electron Microscope (SEM)

In Scanning electron microscope (SEM) electrons are used to form an image for a resolution better than 1 nanometer. SEM is useful because of its higher magnification, larger depth of field, greater resolution and compositional and crystallographic information it gives.

Principle: In a scanning electron microscope, the specimen is exposed to a narrow electron beam from an electron gun, which rapidly moves over or scans the surface of the specimen. This causes the release of a shower of secondary electrons and other types (backscattered electrons, diffracted backscattered electrons, photons) of radiations from the specimen surface. Due to the very narrow electron beam, SEM micrographs have a large depth of field giving structure of a sample.

Components: Electron Source, Electron Lenses, Sample Stage, Detectors for all signals of interest, Display / Data output devices.

Scanning process: The electron beam, which typically has an energy ranging from 0.2 keV to 40 keV, is focused by one or two condenser lenses to a spot about 0.4 nm to 5 nm in diameter. The beam passes through pairs of deflector so that it scans in a rectangular pattern of image of the sample surface. When the primary electron beam interacts with the sample, the energy exchange happens between the electron beam and the sample. This results in the reflection of high-energy electrons, secondary electrons and EM radiation. Images may be captured a high-resolution cathode ray tube, but in modern machines they are digitized and saved. Fig. 49.34 shows electron specimen interactions and Fig. 49.35 shows the principle of the scanning electron microscope indicating BSE as backscattered electrons, SE as secondary electrons, SC as specimen current, EBIC as electron beam induced current, CRT as cathode ray tube and X as x-rays. Fig. 49.36 shows comparison of principal features of an optical microscope, a TEM and a SEM drawn to emphasize the similarities of overall design.

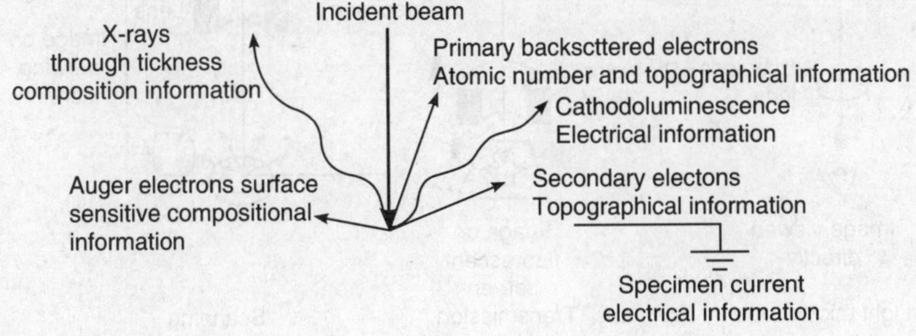

Fig. 49.34:

Electron specimen interactions

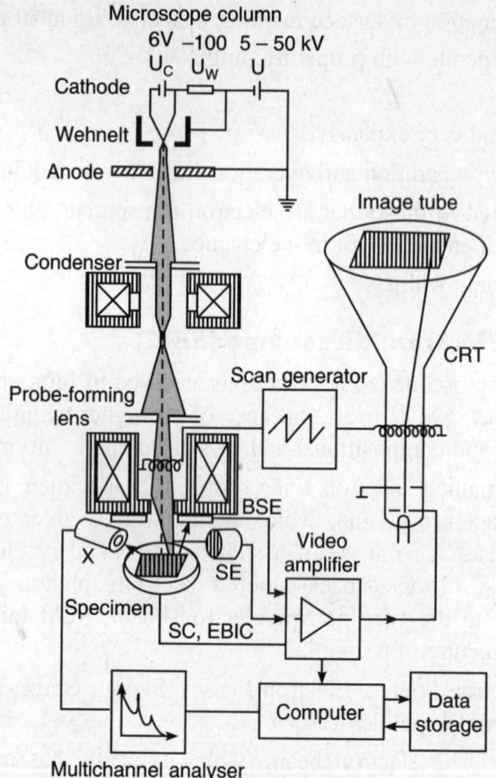

Fig. 49.35:

Principle of SEM

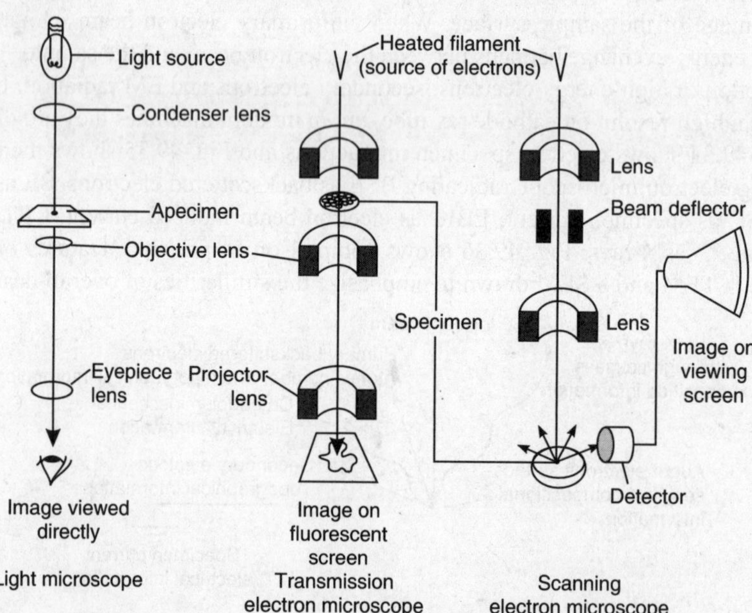

Fig. 49.36:

Comparison of Principal features of an optical microscope, TEM and SEM

49.16.3 Analysis of nano materials by X-Ray Diffraction (XRD)

X-ray diffraction (XRD) is a powerful method for the study of nanomaterials. The wavelength of X-rays is on the atomic scale, so X-ray diffraction (XRD) is a primary tool for probing structure of nano-materials. The intensities measured with XRD can provide quantitative, accurate information on the atomic arrangements at interfaces. X-ray diffractograms of nanomaterials provide a wealth of information from phase composition to crystallite size, from lattice strain to crystallographic orientation.

Experimental Arrangements: In powder or polycrystalline diffraction it is important to have a sample with a smooth plane surface. A typical powder XRD instrumentation consist of four main components such as X-ray source, specimen stage, receiving optics and X-ray detector.

Analysis: XRD analysis is based on Bragg's law and Bragg's angle. When a beam of X- rays is incident on the sample, X-rays are scattered by each atom in the sample. If the scattered beams are in phase, they interfere constructively and give maximum intensity.

Applications: The most traditional use of XRD is:

(*i*) **Identification:** Phase identification, investigation of high/low temperature phases, solid solutions and determinations of unit cell parameters of new materials.

(*ii*) **Texture analysis:** The determination of the preferred orientation of the crystallites in polycrystalline aggregates is referred to as texture analysis.

(*iii*) **Advantages:** It is a powerful and rapid technique for identification of an unknown material. In most cases, it provides an unambiguous phase determination. It requires minimal sample preparation. XRD units are widely available. The Data interpretation is relatively straight forward.

49.16.4 Scanning Tunnelling Electron Microscope (STM)

The scanning tunneling electron microscope (STM) and the atomic force microscope (AFM) are the scanning probes that enabled development of nanotechnology. The instrument was invented in the early 1979 by Gerd Binnig and Heinrich Rohrer, who were awarded the 1986 Nobel Prize in Physics for their work. The scanning tunnelling electron microscope uses electron tunneling to produce images of surfaces down to the scale of individual atoms.

Principle: The concept of tunneling current is the principle of STM. If two conductors are taken close to each other with a finite distance which is equal to the spread of the electronic wave, the electrons from one sample flow into the other. The electrons "tunnel" through the barrier of the order of several atomic spacings into the adjacent sample. The tunneling current is extremely sensitive measure of the distance between two conducting samples.

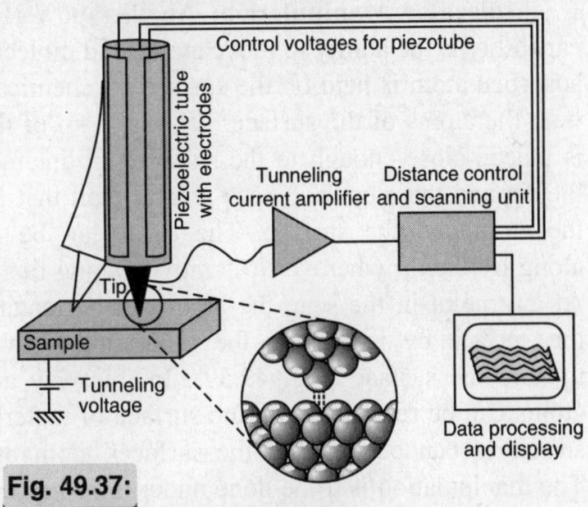

Fig. 49.37: Schematic diagram of STM

Working: The schematic diagram of a scanning tunneling electron microscope is shown in Fig. 49.37. In the scanning tunneling microscope the sample is

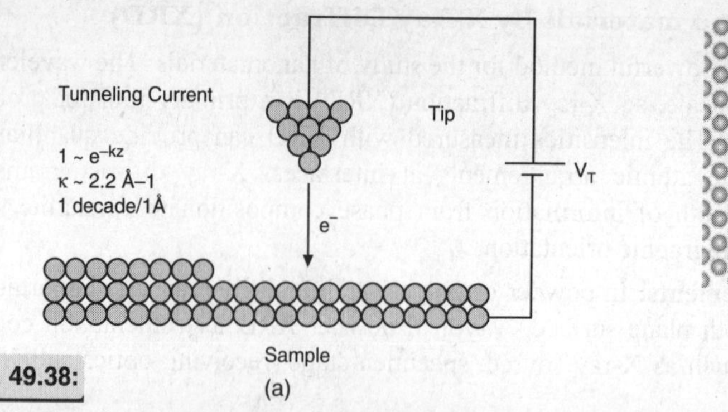

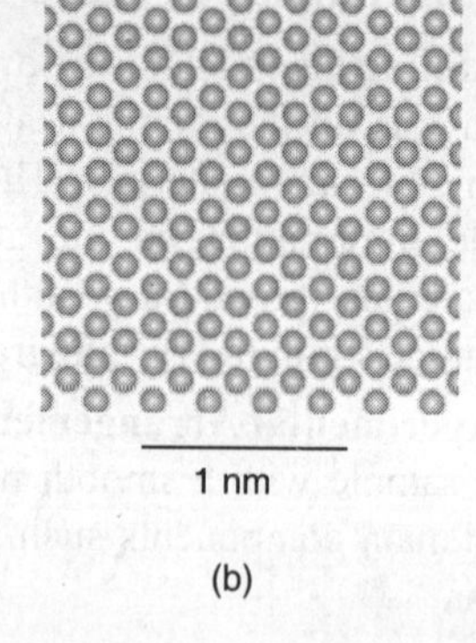

Fig. 49.38:

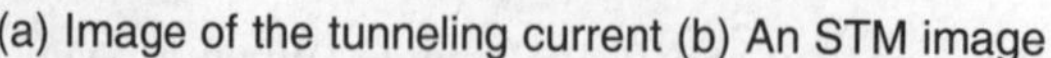

(a) Image of the tunneling current (b) An STM image

scanned by a very fine metallic tip. The tip is mechanically connected to the scanner, an *XYZ* positioning device. The sharp metal needle is brought close to the surface to be imaged. The distance is of the order of a few angstroms. A bias voltage is applied between the sample and the tip. When the needle is at a positive potential with respect to the surface, electrons can tunnel through the across the gap and set up a small "tunneling current" in the needle. The tunneling current is exponentially dependent on the junction width. In typical systems a tip-sample separation of 0.5 nm will produce currents of ~1 nA for biases of 1 V. By scanning the tip along the surface and monitoring the current the surface topography can be understood. The tip motion is controlled using piezoelectric materials with sub-Angstrom resolution. As the sensitivity of the STM is very large the electronic corrugation of surface atoms and the electron distribution around them can be detected.

Fig. 49.38 (a) shows the tunneling tip is scanned over the specimen, producing an image of the tunneling current. An image of the Ag (001) surface taken in this manner is shown in Fig. 49.38 (b). Each bright spot in the image corresponds to a single Ag atom.

Molecular Manipulation: Another powerful STM capability is the ability to move atoms and molecules. An adsorbed atom is held on the surface by chemical bonds with the atoms of the surface. When the tip of the STM is placed close enough to the surface, the interaction of the tip and the atom becomes greater than that between the atom and the surface. The atom can be dragged along by the tip wherever it is moved along the surface. At any point in the scan the atom can be reattached to the surface by increasing the separation between the tip and the surface (Fig. 49.39). In this way adsorbed atoms can be rearranged on the surface of materials and structures can be built on the surfaces atom by atom. The manipulation is to be done under ultra-high-vacuum conditions in order to keep the surface of the material

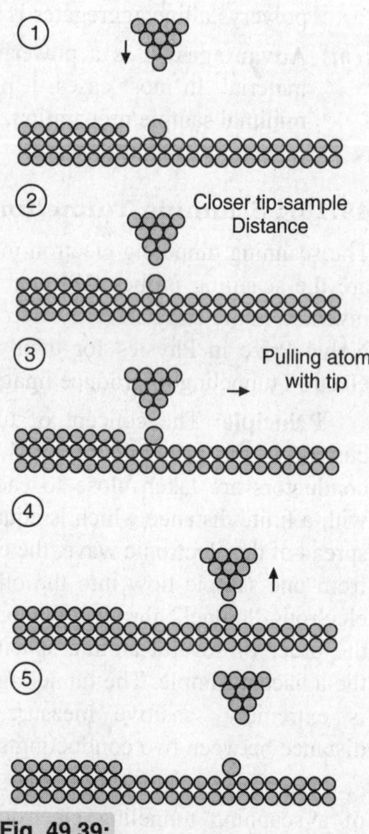

Fig. 49.39:

Schematic diagram of molecular manipulation with the STM

clean. Secondly, the surface of the material has to be held at liquid helium temperature in order to reduce the thermal vibrations which disturb the arrangement of atoms on the surface.

49.16.5 Atomic Force Microscope

The atomic force microscope (AFM) was invented in 1986 by Binnig, Quate and Gerber. The Atomic Force Microscope was developed to overcome a basic drawback with STM - that it can only image conducting or semiconducting surfaces. The AFM, however, has the advantage of imaging almost any type of surface, including polymers, ceramics, composites, glass, and biological samples.

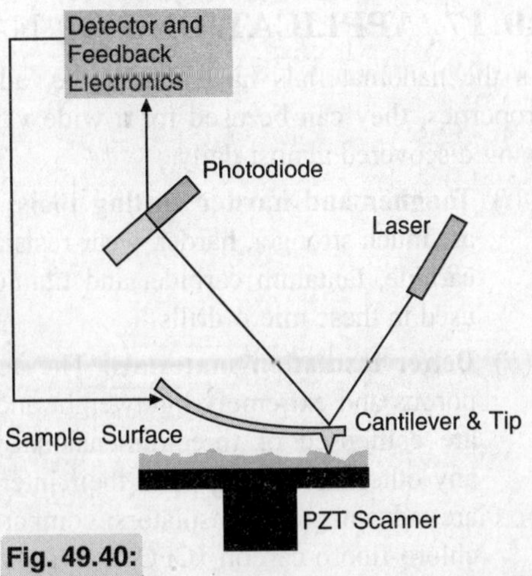

Fig. 49.40: Beam deflection system, using a laser and photodetector to measure the beam position.

Working: The AFM consists of a micro-scale cantilever shaped much like a diving board with a sharp tip (probe) at its end which is used to scan the sample surface (Fig. 49.40). The cantilever is typically made of silicon or silicon nitride. The radius of curvature of the tip is of the order of nanometers. A laser is positioned such that its light strikes at an oblique angle at the very end of the cantilever. When the tip is brought into proximity of a sample surface, the tip is repelled by or attracted to the surface. These forces between the tip and the sample lead to a deflection of the cantilever according to Hooke's law. As the cantilever bends the light from the laser is reflected onto an array of photodiodes. A plot of the laser deflection and tip position on the sample surface gives the resolution of the hills and valleys that constitute the topography of the surface.

Applications:

(*i*) With an interferometer technique, changes in the bending of the cantilever can be measured. Since the cantilever obeys Hooke's Law for small displacements, the interaction force between the tip and the sample can be found.

(*ii*) The AFM can work in contact mode (tip touching the sample), or tapping mode (tip can tap across the surface) much like the cane of a blind person.

Advantages:

(*i*) (As AFM operation does not require a current between the sample surface and the tip). It can move into potential regions inaccessible to STM

(*ii*) It can be used to visualize the surfaces of conductors, insulators, organic materials, biological macromolecules, polymers, ceramics and glasses imaged in different environments, such as liquids, vacuum, and low temperatures.

(*iii*) With modifications to AFM, it can be used as

(*a*) lateral force microscope (LFM)

(*b*) magnetic force microscope (MFM)

(*c*) force modulation microscope (FMM)

(*d*) chemical force microscope (CFM)

49.17 APPLICATIONS OF NANOMATERIALS

As the nanomaterials have distinctive, advantageous chemical, physical, and mechanical properties, they can be used for a wide variety of applications. Many new applications are being discovered almost daily.

(*i*) **Tougher and harder cutting tools:** Cutting tools made of nanocrystalline materials are much stronger, harder, wear-resistant and erosion-resistant and last longer. Tungsten carbide, tantalum carbide, and titanium carbide are nanocrystalline carbides that are used in these micro-drills.

(*ii*) **Better insulation materials:** The aerogels synthesized by the sol-gel technique are porous and extremely lightweight and can withstand 100 times their weight. Aerogels are composed of three-dimensional, nonstop arrangement of particles with air (or any other fluid) trapped at their interstices. So they are used for insulation. Aerogels are advantageous insulators compared to polyurethane foams (PUF), which emit chloro-fluoro-carbon (CFC) gases that damage the UV protecting ozone layer. They are also being used as materials for "smart" windows, which darken when the sun is too bright and they allow more light when the sun is not shining too brightly. Aerogels have also found their application as acoustic impedance matching in ultrasonic distance sensors. Due to their smallest impedance aerogels are useful to boost the efficiency of transducers, which are common components of cameras and of robotic systems that gauge distances by emitting ultrasonic waves.

(*iii*) **Ductile, machinable ceramics:** Ceramics are very hard, brittle, and hard to machine. So, they are difficult to use. However, with a reduction in grain size, these ceramics have increasingly been used. Nanocrystalline zirconia, a hard, brittle ceramic, becomes superplastic. It can be deformed up to 300% of its original length. Nanocrystalline ceramics, such as silicon nitride (Si_3N_4) and silicon carbide (SiC), have been used in such automotive applications as high-strength springs, ball bearings, and valve lifters, because they possess good formability and machinabilty combined with excellent physical, chemical, and mechanical properties. They are also used as components in high-temperature furnaces.

(*iv*) **Low-cost flat-panel electrochromic displays:** Electrochromic devices are like LCDs. Nanocrystalline materials, such as tungstic oxide ($WO_3.xH_2O$) gel, are used in public billboards and sticker boards to convey information. Electrochromic devices display information by changing color when a voltage is applied. When the polarity is reversed, the color is faded. The resolution, brightness, and contrast of these devices greatly depend on the tungstic oxide gel's grain size.

(*v*) **Elimination of pollutants:** Nanomaterials can be used as catalysts to react with harmful and toxic gases as carbon monoxide and nitrogen oxide in automobile catalytic converters and power generation equipment. This can avoid environmental pollution from burning petrol and coal.

(*vi*) **High-power magnets:** The magnets made of nanocrystalline yttrium-samarium-cobalt grains possess very unusual magnetic properties due to their extremely large surface area. They are useful in quieter submarines, automobile alternators, land-based power generators, and motors for ships, ultra-sensitive analytical instruments, and magnetic resonance imaging (MRI) in medical diagnostics.

(*vii*) **High energy density batteries:** Nanocrystalline materials prepared by sol-gel techniques are useful for separator plates in batteries because of their foam-like structure, which can hold considerably more energy. Nanocrystalline nickel and metal hydrides (Ni-MH) batteries requires far less frequent recharging and to last much longer.

(*viii*) **High-sensitivity sensors:** Sensors made of nanocrystalline materials are extremely sensitive to the change in their environment. The nanocrystalline material sensors have applications as smoke detectors, ice detectors on aircraft wings, automobile engine performance sensor, etc.

(*ix*) **Aerospace components with enhanced performance characteristics:** Aerospace components made of nanomaterials are stronger and can operate at higher temperatures; aircrafts can fly faster and more efficiently for the same amount of aviation fuel. Spacecrafts operate at much higher temperatures than aircrafts and at higher speeds. Nanomaterials have the more fatigue life by an average of 200-300%.

(*x*) **Sunscreen:** Many sunscreens contain nanoparticles of zinc oxide or titanium oxide. Smaller particles are less visible; therefore, when the sunscreen is rubbed into skin, it does not give whitish tinge.

(*xi*) **Self-cleaning glass:** Glass with nanoparticles makes glass ***photo-catalytic*** and ***hydrophilic***. In photo-catalytic effect nanoparticles become energized and break down and loosen dirt molecules on the glass when the UV radiation falls on them. Hydrophilic means that when water makes contact with the glass, it spreads across the glass evenly, which helps wash the glass clean. Thus, the glass is a self-cleaning glass.

(*xii*) **Clothing:** Fabrics with a thin layer of zinc oxide nanoparticles, give better protection from UV radiation. Clothes having nanoparticles in the form of little hairs that help repel water and other materials, making the clothing stain-resistant.

(*xiii*) **Scratch-resistant coatings:** Addition of aluminum silicate nanoparticles to scratch-resistant polymer coatings used for cars to eyeglass lenses makes the coatings more effective, increasing resistance to chipping and scratching.

(*xiv*) **Nanomachines and nanodevices:** They are in the early stage of development, and are still in conceptual stages. Nano-sized machines do exist in biological systems and the study of biological machines is expected to provide us insights that will help us in the design of mechanical nanomachines. Computer simulation has been used to evaluate the potential of various nanomachine concepts. One example is the idea of making a gear which is an essential part of any sized nanomachine. Current researchers are developing interactive nanogears fabricated from Fullerene. Synthesis of nanogear systems involves adding a knobby molecule like benzene to the outer walls of a carbon nanotube. If some of the teeth or gear molecules are electronically charged they can be made to rotate in response to the applied stimulus.

QUESTIONS

1. What do you mean by NANO?
2. Define: (*i*) Nano Science, (*ii*) Nano Technology
3. Define nanomaterial? Give classification of nanomaterials?
4. Define top down and bottom up approaches?
5. What are the induced effects due to increase in surface area of nanoparticles?
6. Why [surface area/volume] ratio is very large for nanoparticles compared to bulk materials?

7. Write a short note on self-assembly.
8. Explain the working of scanning electron microscopy (SEM) with a neat sketch.
9. Explain the working of atomic force microscopy (AFM) with a neat sketch.
10. What is the difference between STM &AFM?
11. Explain photolithography (optical lithography) with a neat sketch.
12. Explain Electron beam lithography with a neat sketch.
13. Classify nanomaterials and give examples for them.
14. Write a short note on Nanomaterials.
15. Define nanocomposite and classify nanocomposites.
16. With a neat sketch, explain mechanical milling process for synthesis of nano particles. List advantages and disadvantages also.
17. Describe various techniques of Physical Vapour Deposition.
18. Explain gas condensation process for synthesis of nanopowders with a neat sketch?
19. Explain laser ablation process for synthesis of nanopowders with a neat sketch?
20. Explain SOL-GEL synthesis for producing nanomaterials? Explain with the help of a neat sketch?
21. List any four processes to produce nanopowders?
22. List any four day to day live commercial applications of nanotechnology?
23. What are composites? Explain the classification and advantages of composite materials.
24. What are nanomaterials? Write a note on carbon nanotubes.
25. Define carbon nanotube? What are the types of carbon nanotubes?
26. List the methods for producing carbon nanotubes and explain any one of the method with a neat sketch?
27. Highlight the properties of carbon nanotubes?
28. Give any two excellent properties of carbon nanotubes?
29. Discuss the applications of carbon nanotubes.
30. Discuss the physical properties of carbon nanotubes in relation to their structure.
31. Write a brief note on: (*i*) Nanotechnology, (*ii*) Carbon nanotubes
32. Explain scaling laws.

50 Liquid Crystals

Learning Objectives:

By the end of this chapter, you will be able to:

- Learn basics of Liquid Crystal Phase and shape of Liquid crystal molecule.
- Understand concept of Polymorphism.
- Explain different types of liquid crystals in detail.
- Identify different applications of liquid crystals.

50.1 INTRODUCTION

Usually matter exists in three different forms or states – solid, liquid, and gas. In addition to these well-known basic phases of matter, certain organic compounds exhibit a single or a sequence of phase transitions between the solid and the normal liquid (isotropic) phase. The physical properties like the mechanical, optical and structural properties in these intermediate phases are between those of the crystalline solid and the isotropic liquid phase. Therefore these phases are referred to as *liquid crystalline* phases. Liquid crystals were discovered in 1888 by the Austrian botanist Friedrich Reinitzer when he observed that an ester of cholesterol, namely cholesteryl benzoate ($C_6H_5COOC_{27}H_{45}$) first melted from a solid to a cloudy liquid which changed into a clear liquid at a higher temperature. This intermediate liquid like state was first named by Lehmann in 1889 as "flowing crystals" and then in 1900 as "liquid crystal".

Liquid crystalline properties are displayed by numerous types of systems, such as certain groups of organic molecules, micellar solutions of surfactants, main and side chain polymers and an abundant of biological systems.

50.2 LIQUID CRYSTALS

Crystals possess both positional and orientational order. But in an isotropic liquid, the molecules have neither positional nor orientational order and they are distributed randomly. There is no preferred direction in a liquid. There are some substances which exhibit an additional state in between these two states of matter, which possesses both order and fluidity at the same time. They are known as liquid crystals. As liquid crystals are phases of matter existing between crystalline solid and isotropic liquid phases they are also called "*mesophases*" meaning intermediate phase. Fig. 50.1 shows liquid crystal phase.

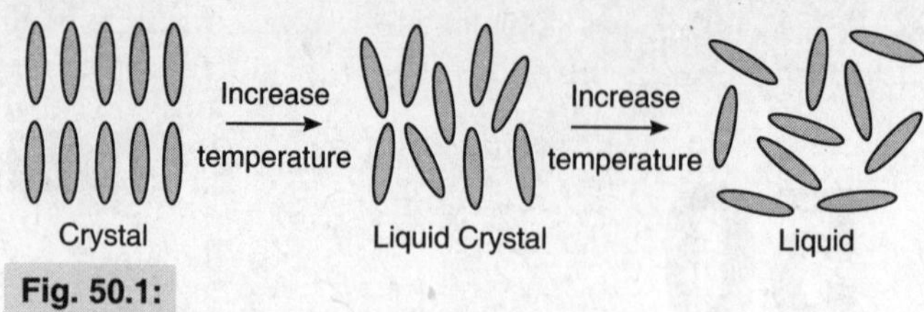

Fig. 50.1:
Liquid Crystal Phase

Liquid crystals are fluids that exhibit molecular order to a certain extent. The molecules in mesophases have liquid-like order at least in one direction and show some degree of orientational order and sometimes some positional order as summarized in Fig. 50.2.

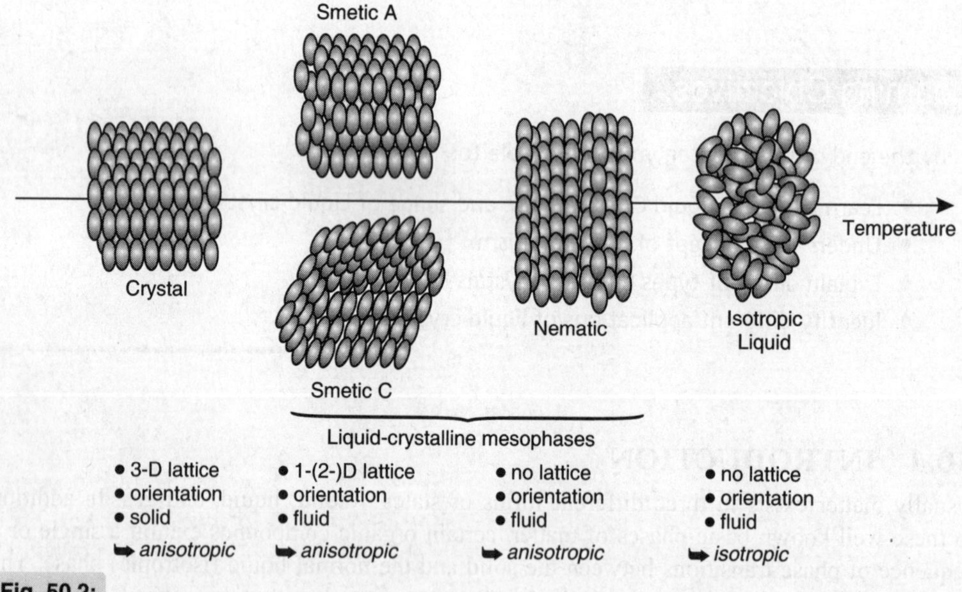

Fig. 50.2:
Illustration of molecular order in different phases

Some important terms related to Liquid Crystal Phase

(i) **Domain:** When a solid melts into liquid crystal at melting point T_m, the positional order is lost while orientational order of the molecular long axes remains to some extent. The molecules cluster locally.

Def: Each cluster in which molecules are aligned in the same direction, *n*, is termed a domain. Molecular orientation is more or less the same in each domain, whereas the domains in the phase orient in different directions.

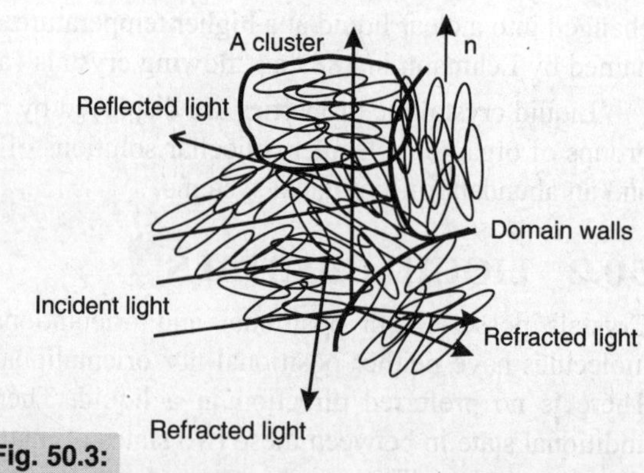

Fig. 50.3:
Domains in the phase orienting in different directions

Fig. 50.3 shows domains in the phase orienting in different directions. The molecular order extends up to the entire domain size which may be of the order of a few microns (μm) but usually does not extend to the macroscopic scale.

(*ii*) **Turbidity:** The most noticeable attribute of a liquid crystal is a characteristic cloudiness or turbidity. Thermal fluctuations cause the molecules to deviate temporally as well as spatially. The spontaneous orientational fluctuations cause light passing through the LC to be scattered. Density fluctuations etc may also cause such fluctuations but are weaker than orientational fluctuations. When the system is heated, the turbidity disappears at the *clearing point* (Fig. 50.4).

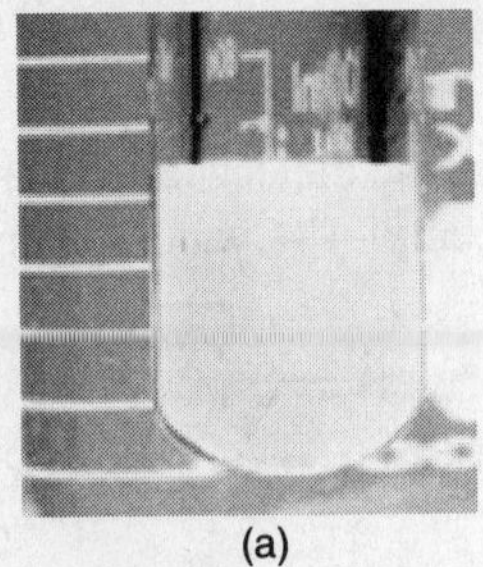
(a)

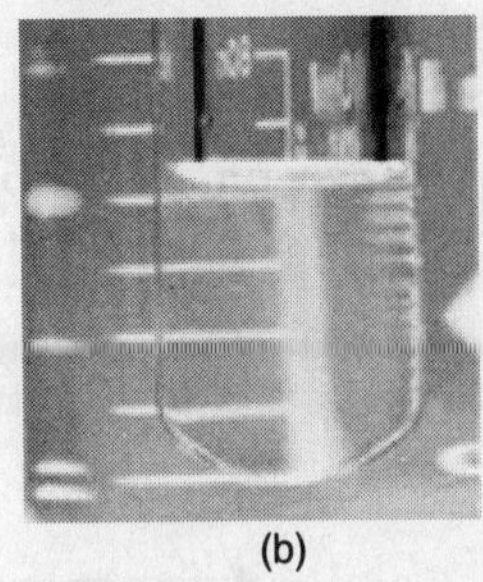
(b)

Fig. 50.4: Turbidity (a) Liquid Crystal state, (b) Isotropic liquid

(*iii*) **Director:** It gives the direction of preferred axis at each point in a sample (Fig. 50.5).

Def: Director is a unit vector **n**, which represents the local direction of alignment of the molecular long axes in the liquid crystal phase.

In most liquid crystals both directions of the vector ***n***, $+n$ and $-n$ are equivalent.

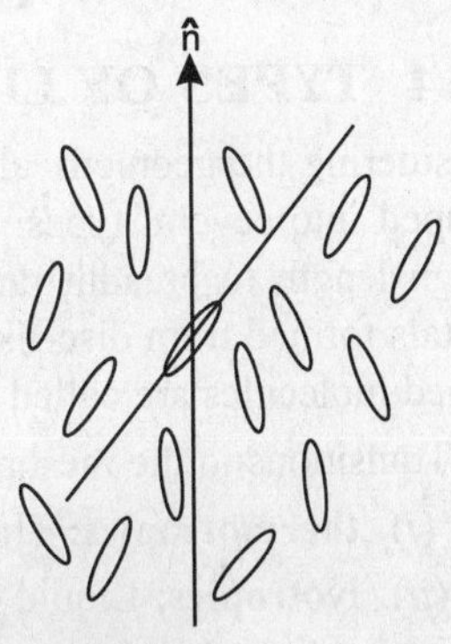

Fig. 50.5: Molecular order in a nematic liquid crystal

(*iv*) **Anisotropy:** The anisotropic nature of liquid crystals is responsible for their unique optical properties.

Def: Anisotropy is the property of substances to show different values in physical properties when measured along axes in different directions.

Types of Order: The liquid crystalline state orders are

(*a*) **Positional order:** Positional order is the extent to which translational symmetry is exhibited.

(*b*) **Orientational order:** Orientational order is a measure of the tendency of the molecules to line up along the director on a long-range basis.

(*c*) **Bond orientational order:** Bond orientational order indicates the line joining the centres of nearby molecule without needing a regular spacing along that line.

50.3 TYPICAL SHAPE OF A LIQUID CRYSTAL MOLECULE

A liquid crystal molecule consists of two or more ring systems connected by a central linkage group, Fig. 50.6 (a). The molecules are generally made from by small organic molecules, with typical lengths in the nanometer range.

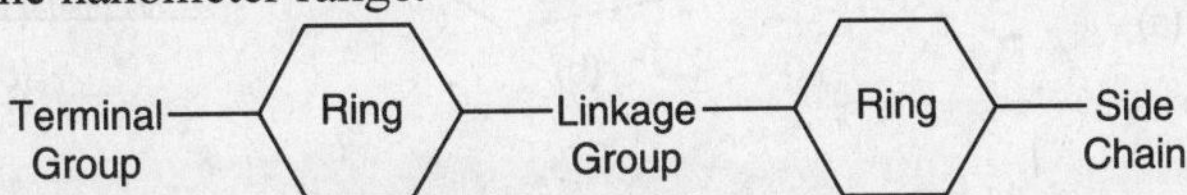

Fig. 50.6 (a): Molecular shape and structure of LC molecules

The rings provide the short range molecular forces required to form the liquid crystal phase. The central linkage group determines the chemical stability of liquid crystals. Compounds having a single bond in the center are the most stable. At one side of the rings a long chain is located and at the other end a terminal group is connected. The long chain strongly influences the elastic constants and the transition temperature of the liquid crystal phases.

The terminal group gives the dielectric constant and its anisotropy. An example of a liquid crystal moleculeis 5CB (4-pentyl-4'-Cyanobiphenyl) is shown in Fig. 50.6 (b).

CH_2 CH_2 CH_2 CH_2 CH_2 —CN

b (Short direction)

a (Long direction)

Fig. 50.6 (b):

Typical shape of a liquid crystal molecule of Length a = 2nm and width b = 0.5nm

50.4 TYPES OF LIQUID CRYSTALS

Considering the geometrical structure of the mesogenic molecules, the liquid crystals can be grouped into several types. The liquid crystals formed from the rod shaped molecules with a high length to breadth ratio are called calamitics or *calamitic liquid crystals*. The liquid crystals formed from disc-like molecules are called as *discotics* and those formed from banana shaped molecules are called *bent-core liquid crystals*.

Transitions to the mesophases occur in two different ways

(*i*) **thermotropics:** The liquid crystals that are obtained by the thermal processes

(*ii*) **lyotropics:** Liquid crystals obtained by the influence of solvents

Amphotropic materials are liquid crystals that form thermotropic as well as lyotropic mesophases. Liquid crystals are also formed from certain macromolecules (e.g. long-chain polymers) and are called as "liquid crystal polymers (LCPs)".

Thermotropic liquid crystals exhibit three types of mesophases (Fig. 50.7) namely the nematic, cholesteric and smectic based on the degree of positional and orientational order of the constituent molecules. There are a number of different smectic phases and these are characterized by the packing formation and the tilt angle with respect to the layer normal. We shall discuss about these at an appropriate stage.

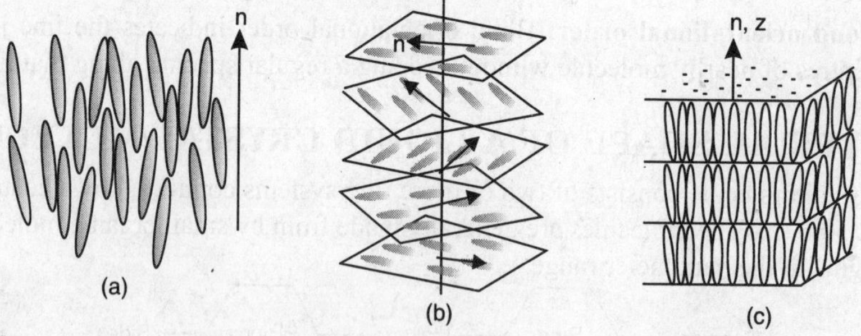

Fig. 50.7:

Mesophases of Thermotropic liquid crystals (a) Nematic, (b) cholesteric and (c) smectic

Thermotropic liquid crystals can be further classified into two types:

(*i*) **Enantiotropic liquid crystals:** Enantiotropic liquid crystals are those substances that can be transformed into the liquid crystal state by either decreasing the temperature of a liquid or by rising of the temperature of a solid. Mesomorphic transition occurs in these substances on heating and these transition reverses in the opposite direction on cooling.

(*ii*) **Monotropic liquid crystals:** In these substances liquid crystal state is obtained from either an increase in the temperature of a solid or by a decrease in the temperature of a liquid only, but not both. There are many compounds which on heating do not exhibit mesophase but on cooling, exhibit a mesophase.

50.5 POLYMORPHISM

Many Liquid crystalline substances form only smectic mesophase or only nematic mesophase. However some form both types of mesophase, smectic followed by nematic. There are substances exhibiting more than one smectic phase. This phenomenon is known as **polymesomorphism**. Above the clearing temperature (T_c) the liquid crystal becomes an isotropic liquid.

Def: Polymesomorphic liquid crystals are those which exhibit two or more liquid crystal phases and the process is known as polymesomorphism.

Crystalline Solid ⇌ Smectic-I ⇌ Smectic-II ⇌ Nematic or Cholesteric ⇌ Isotropic Liquid

50.6 SEQUENCE RULE AND REENTRANT BEHAVIOUR

Considering all structures known in calamitic liquid crystals a hypothetical sequence was derived:

Isotropic (I) - N - SmA - SmC - SmB - SmI - SmF - Crystal B - J - G - E - K - H- Crystal for achiral materials and *I - BP- N*- TGB - SmA - SmC* - SmI* -SmF** for chiral LCs.

Sequence rule is a general rule and according to which smectic phases occur at low temperatures while the nematic phase occurs at higher temperatures. If a variation of temperature or pressure results in two or more phase changes and finally reaches a state that is macroscopically similar to the initial state, the system is said to undergo a reentrant phase transition,

Def: Reentrant phase transition of a system is reaching a state of phase transition which is macroscopically similar to the initial state after the variation of temperature or pressure giving two or more phase changes.

50.7 TYPES OF CALAMITIC LIQUID CRYSTALS

Nematic (N) phase, Cholesteric Phase (Chiral nematic) (N*) and Smectic phases are three types of calamitic liquid crystals.

50.7.1. Nematic (N) phase

The nematic phase is the simplest liquid crystal phase of the calamitic phases. It is the closest to the isotropic liquid state. The word nematic comes from the Greek word '*nema*' which means 'thread' referring to the thread-like texture seen under microscope.The nematic phase is characterized by ***long-range orientational order***. *With long-range order we mean that either the orientation or position, or both, of a molecule is correlated with the orientation or position of molecules a long distance away.*

In the nematic phase, the molecules are, on average, aligned with their long axes parallel to each other. Their orientations are all alike. However, there is no long-range correlation between the centers of mass of the molecules and hence no long range *positional* order. It implies that the molecules have three-dimensional translational symmetry. The molecules appear to be able to rotate about their long axes and also there seems to be no preferential arrangement of the two ends of the molecules if they differ. A simplified picture of the relative arrangement of the molecules in the nematic phase is shown in Fig. 50.7 (a).

All molecules are aligned approximately parallel to each other in the liquid crystal phase. The molecules in a liquid crystal phase maintain unidirectional orientational ordering (Fig. 50.8). The director, *n* specifies the direction of the preferred orientation of liquid crystal molecules as a whole.

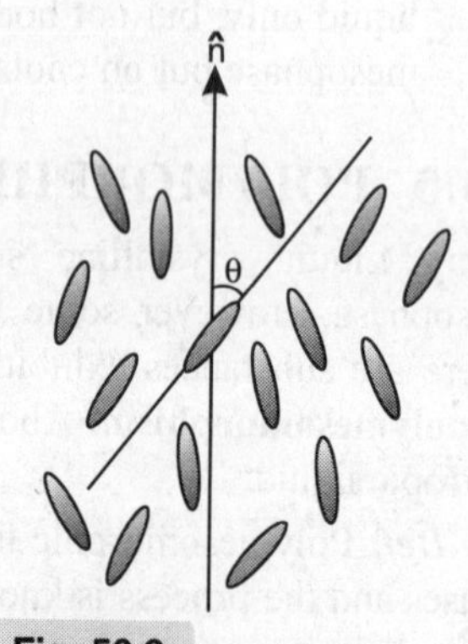

Fig. 50.8: Molecular order in a nematic liquid crystal

The domains in a nematic sample are oriented in different directions and scatter the incident light and appear turbid. Using some techniques, a thin nematic sample may be uniformly aligned such that the domains and the corresponding directors, ***n*** orient in a single direction. As such all the molecules in the sample lie more or less parallel to each other. Such a sample may be called **liquid single crystal**. Such a liquid single crystal has full cylindrical rotational symmetry around the direction and it is optically ***uniaxial***. To describe the degree of order present in a material, the scalar **order parameter** S is commonly used ($0 < S < 1$).

Def: The order parameter is defined as the amount of orientational order a liquid crystal phase exhibits.

Using the 2nd Legendre polynomial

$$S = \langle P_2 (\cos \theta)\rangle = \langle 3/2\cos 2\theta - 1/2\rangle \quad ...(50.1)$$

where θ is the angle between the director and the specified long axis of a molecule (Fig. 50.8).

The brackets in Eq. (50.1) denote an average over all of the molecules in the sample. In an isotropic liquid, the average of the cosine terms is zero, and therefore the order parameter is equal to zero, $S = 0$. For a perfect crystal, the order parameter evaluates to one, $S = 1$. Typical values for the order parameter of a liquid crystal range between 0.3 and 0.9, with the exact value a function of temperature, as a result of kinetic molecular motion. This is illustrated for a nematic liquid crystal material in Fig. 50.9.The tendency of the liquid crystal molecules to point along the director leads to ***anisotropy***. The nematic liquid crystal phase is by far the most important phase for applications.

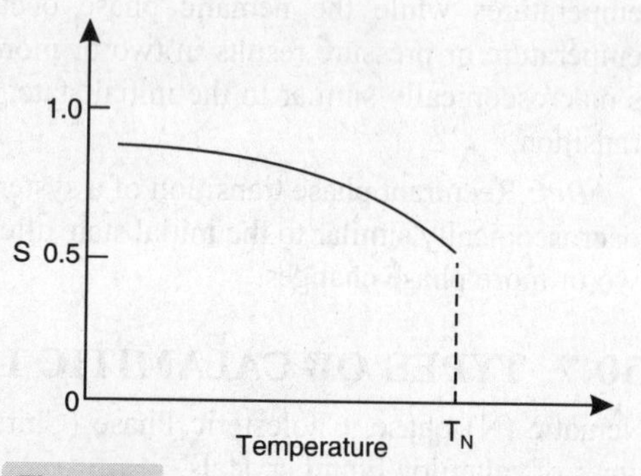

Fig. 50.9: Scalar order parameter *S* versus temperature

50.7.2 Cholesteric Phase (Chiral nematic) (N*)

The cholesteric phase is formed by liquid crystal materials consisting of chiral molecules. Molecules which are not identical to their mirror image are said to be ***chiral***. Generally, chirality indicates handedness. The left and right hands are chiral because one is the mirror image of the other, but they cannot be superimposed directly onto one another. Initially the phase was limited to the materials, which were derivatives of cholesterol. This phase appears in other organic compounds also which consist of chiral nematogenic molecules. Commonly this arises from the asymmetry of the tetrahedral substituted carbon atom.

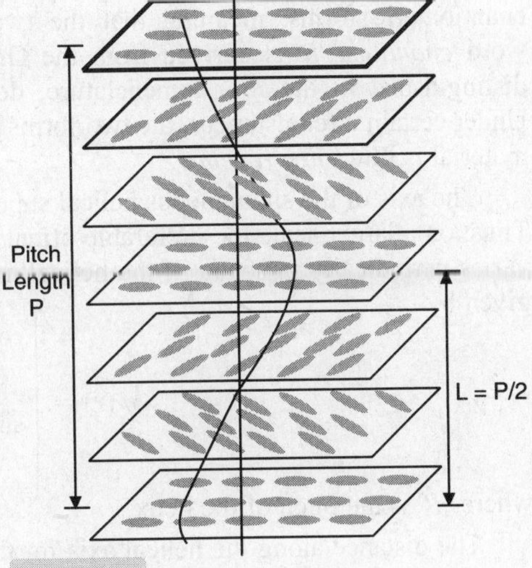

Fig. 50.10: Cholesteric phase-pitch

On a local scale, cholesteric liquid crystal structure is typically similar to nematics. In each layer the cholesteric and nematic ordering are very much similar. Centers of mass lie in the molecular plane chaotically and the **director *n*** also *lies in the layer plane*. Therefore, the phase possesses only orientational order. The thickness of each molecular layer is about 0.5–0.6 nm.

When the molecules align in upper layers, the molecular asymmetry causes the molecules to shift their orientation to rotate by a small angle relatively to the director of the underlying layer. Such shift accumulates progressively and the ends of the molecules and the director tip spirals to form a helix. The ***shift (twist) angle*** of the director in transition from layer to layer be α. In cholesteric phase α is 15 angular minutes on the average. The twist angle α depends on both the temperature and the molecular structure of the cholesteric liquid crystal as well as some other factors. The distance it takes for the director to rotate one full 360° twist in the helix is known as the ***pitch*** which is an important characteristic of the cholesteric phase (Fig. 50.10).

Def: The pitch, P is the distance required for rotating the average direction of the molecules through 360° which is the full turn of the helix.

One pitch length of helix may consist of about 1000 molecular layers. In the molecular layers corresponding to the director twist angles of 0°, π, and 2π the major axes of liquid crystal molecules have the same orientation that makes the layers indistinguishable. It is easy to see the molecular orientation is repeated at a distance equal to the period $L = P/2$. The pitch is much longer than the size of a single molecule. It is to be noted that the structure of the chiral nematic phase repeats itself every half-pitch, since in this phase directors at 0° and ±180° are equivalent. The molecules can move freely and exchange their positions in each molecular layer but their spiral arrangement with the period L remains unchanged.

The cholesteric structure can also be realized with nematics by adding a small concentration of chiral dopant molecules to the nematic phase. Higher concentrations of dopant molecules increases twisting power, and shorten the pitch. The nematic phases formed by chiral molecules are generally called ***chiral nematic phase*** and designated with **N*** and the analogous non-chiral phase with *N*.

A substance may possess either the nematic liquid crystal phase or the chiral nematic liquid crystal phase, but not both. A chiral molecule is a molecule which may exist in two enantiomeric forms, meaning that the two forms are mirror images of each other. The word *enantiomeric* is derived from the Greek, meaning 'opposite' and the two forms are distinguished in the *R/S* nomenclature, denoting right- and left-handedness, respectively. Under certain circumstances, the two forms may coexist in equal quantities, in which case the material is said to be *racemic*.

The axis of the spontaneous helical superstructure is perpendicular to the local director ***n***. Thus, on a larger scale the preferable orientation of the molecules, the cholesteric director, ***n***, varies periodically in space. If the helical axis is oriented along z-direction, the director ***n*** is given by

$$n(r) = \begin{pmatrix} \cos\left(\frac{2\pi}{P}z + \varphi_0\right) \\ \sin\left(\frac{2\pi}{P}z + \varphi_0\right) \\ 0 \end{pmatrix} \qquad \text{...(50.2)}$$

where '*P*' is the pitch of the helix.

The distance along the helical axis for the director to rotate 2π is called the pitch, *P*, and the helical twist is defined as $q_0 = 2\pi/P$.

The magnitude of the pitch of a cholesteric phase can vary considerably, from values as small as $P = 100$ nm to large values of many micrometers. Owing to the head-tail symmetry $+n = -n$, the periodicity *L* of the *N** phase along the helix axis is given by half the pitch (Fig. 50.10).

The spatial period, *L*, is equal to

$$L = \frac{P}{2} = \frac{\pi}{|q_0|} \qquad \text{...(50.3)}$$

The sign of q_0 distinguishes between left and right helices and its magnitude determines the spatial period. A nematic ($q_0 = 0$) can be considered as a cholesteric with infinite pitch.

Apparently, the lead *P* of the spiral is

$$P = \frac{2\pi}{\alpha}d \qquad \text{...(50.4)}$$

where α is the shift (twist) angle of the director in transition from layer to layer; *d* is the thickness of the molecular layer of the cholesteric crystal.

The angular shift of the director as we pass from layer to layer in cholesteric crystals is about 15 angular minutes. The twist angle α depends on temperature as well as on the molecular structure of the cholesteric liquid crystal and on some other factors. Taking $\alpha = 0.5°$ and $d = 0.6$ nm we obtain the lead of the cholesteric spiral as 400 nm.

Helical twist, right handed or left handed depends on the molecular conformation as shown in Fig. 50.11. A right-handed helical super structure is considered to have positive pitch *P* and a left-handed helix a negative pitch.

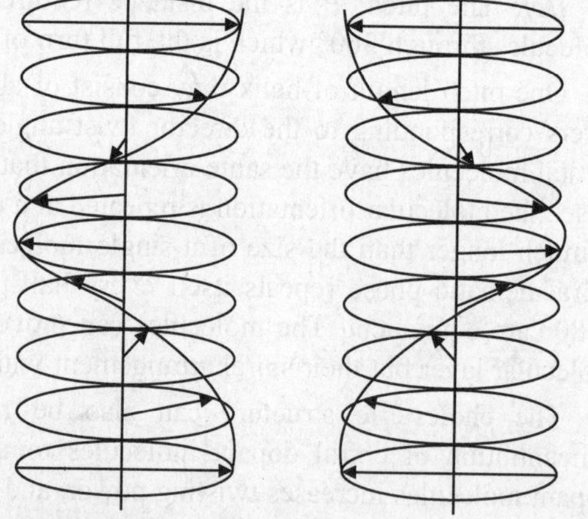

Fig. 50.11: Helical twist may be right handed or left handed

The cholesteric and chiral nematic phases are subclasses of the same family. If the constituent molecules are non-chiral, then the mesophase will be nematic.

If, on the other hand, the constituent molecules are chiral, then the mesophase will be cholesteric. However, chiral nematic phase is the term used to indicate both the phases.

(*i*) **Chiral Dopant:** Chiral liquid crystal can be obtained by introducing chiral center or mixing nematic molecules with chiral dopants. When a chiral dopant is mixed with a nematic liquid crystal, the pitch of the mixture is related to the concentration of the chiral dopant.

The relationship is

$$1/P = (\text{HTP})X_C \qquad \text{...(50.5)}$$

where

HTP is the macroscopic Helical Twisting Power which characterizes the chiral dopant and depends slightly on the nematic host.

X_C is concentration of the chiral dopant.

(*ii*) **Selective Reflection:** The helical arrangement is responsible for the unique optical properties of the phase, such as selective reflection. The cholesteric phase, when illuminated with white light, reflects light of a certain wavelength depending on the helical pitch of the material (Fig. 50.12). This is known as *selective reflection.*

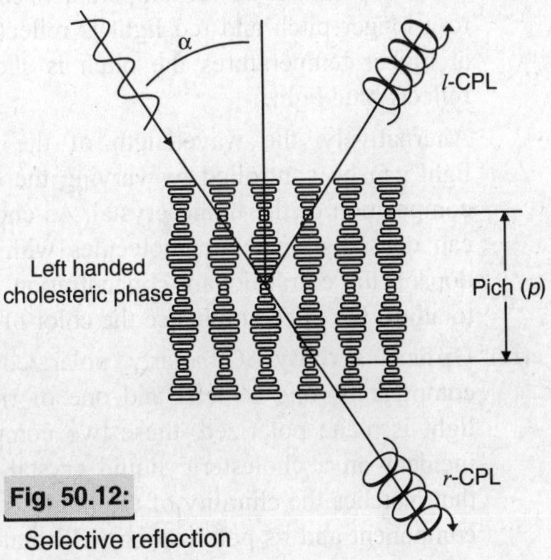

Fig. 50.12: Selective reflection

The natural white light propagating along the phase may be viewed as composed of two components. The electric vector of one of the components rotates clockwise and that of the other component rotates counter-clockwise. They are called right circularly polarized light (*r*-CPL) and left circularly polarized light (*l*-CPL) respectively. The reflected and transmitted light will also become circularly polarized, with their signs depending on the sign of the cholesteric helix. The reflected light of a left-handed helix will be left-handed (*l*-CPL), the transmitted light is right-handed circularly polarized (*r*-CPL) which is transmitted through the medium without any loss. As only circularly polarized light with one handedness is reflected, the intensity of the maximum reflection of unpolarized light from a chiral phase is 50% only. This phenomenon is known as ***selective reflection*** and ***circular dichroism***. Selective reflection gives the chiral phase a brilliant, iridescent color when illuminated with white light.

The maximum selective reflection occurs due to the first-order Bragg diffraction at a wavelength $\lambda = np$ (for normally incident light).

$$\lambda_o = n_{av} P \qquad \text{...(50.6)}$$

where λ_o is the wavelength in air and n_{av} is the average refractive index of the medium.

$$n_{av} = \sqrt{n_e^2 + n_o^2} \qquad \text{...(50.7)}$$

The reflection occurs within a bandwidth

$$\Delta\lambda = \Delta n\, P \quad ...(50.8)$$

where $\Delta n = n_e - n_o$. In reality, the Bragg reflection occurs in quite a large band of wavelengths defined by $Pn_o < \lambda < Pn_e$.

The wave length changes with change in temperature. By increasing the temperature, the angle at which the director changes is made larger, and it tightens the pitch of the helix. Similarly, decreasing the temperature of the molecules increases the pitch length of the chiral nematic liquid crystal. It leads to the change of the wavelength of reflected light according to the temperature.

(*iii*) **Thermochromic Effect:** As the cholesteric phase selectively reflects light of a particular wavelength, the reflected color varies with temperature and is known as ***thermochromic***. Fig. 50.13 shows the dependence of the wavelength with temperature. Lower temperatures correspond to a longer pitch and red light is reflected while at higher temperatures the pitch is shorter and reflects blue light.

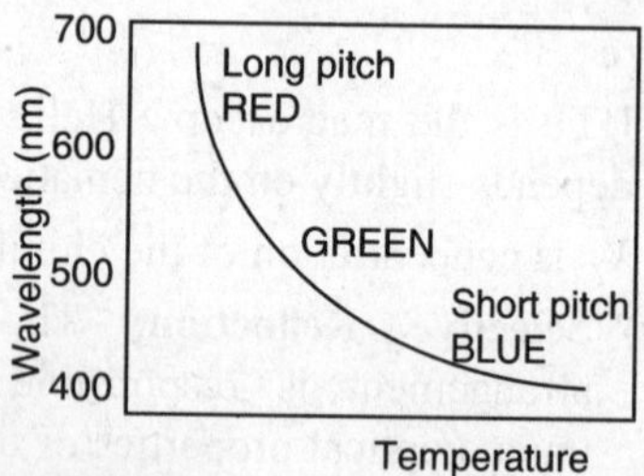

Fig. 50.13: The usual temperature dependence of the colour of a cholesteric liquid crystal

Alternatively, the wavelength of the reflected light can be controlled by varying the chemical composition of the liquid crystal. As cholesterics can consist of nematic molecules with a chiral dopant, the chiral dopant concentration is varied to adjust the pitch and hence the color of the reflected light.

(*iv*) **Optical activity:** Circularly polarized light may be viewed as containing two components, one of left- and one of right-circularly polarized light. If the incident light is plane polarized, these two components are equal. If plane polarized light is incident on a cholesteric liquid crystal, the component of circularly polarized light that matches the chirality of the liquid crystal structure will travel faster than the other component and its polarization will rotate faster with respect to the polarization of the other component. When the two components emerge out of the liquid crystal, their polarizations will rotate at the same initial rate and the resultant light will again be linearly polarized. However, since one of the rotations is in advance of the other, the resultant light will now be linearly polarized along a different angle which depends on thickness of the liquid crystal. This effect is called the *optical activity*. Cholesteric monocrystalline layer of 1 mm thickness rotates the polarization plane by an angle of about 18,000°. Nematic and smectic crystals do not exhibit optical activity.

Def: Optical activity is a measure of the change in polarization angle per unit thickness, and is strongly dependent on wavelength.

50.7.2.1 Chiral Frustrated Phases - Blue Phases

When chiral molecules cannot form helixes they stabilize into ***frustrated phases***. Two examples of frustrated phases are blue phases (BP) and twist grain boundary (TGB) phases. Three distinct blue phases, namely BPI, BPII and BPIII are identified. BPI exhibits body-centered cubic and BPII simple cubic symmetry, whereas BPIII has the same symmetry as the isotropic phase.

Def: The blue phases (BPs) are a set of liquid crystal phases which occur within a small temperature range at the boundary of the helical phase and isotropic liquid phase of highly chiral, thermotropic liquid crystals.

Presently the structures of blue phases are discussed below and Twist Grain Boundary (TGB) Phases in 50.7.3.1.

Blue phases are distinct thermodynamic phases that appear over a narrow temperature range approximately 0.5° to 2° C between chiral nematic and isotropic liquid phases. BP I*, BP II* and BP III*, all separated by first order transitions. Blue phases I and II show cubic symmetry and the misty blue BPIII is called blue fog. BPI has body center cubic symmetry whose elementary cell size changes with temperature. BPII shows simple cubic symmetry and there is no temperature change in elementary cell size. BPIII is almost amorphic. BPI and BPII both have long-range orientational order while the BPIII is isotropic. They are called blue phases because the first discovered phase appears blue in reflected light. However, blue phases are now known to reflect other colors in the visible range.The first observation of the blue phase has been attributed to Reinitzer.

Fundamental properties of BPs: The work of Armitage and Price in 1975 showed that the blue phases are stable, distinct, thermodynamic phases of liquid crystals.

(*i*) The blue phases BP I and BP II have 3D orientational ordering with periods of up to 500nm. Therefore, the blue phase shows no peaks in X-ray diffraction and is not detected by X-rays. Single crystals of BPI floating in the isotropic liquid have been grown (Fig. 50.14) which show very clear faces and the shape of the crystal would indicate a cubic symmetry.

(*ii*) The blue phase containing many domains oriented differently exhibits a mosaic of color. They do not have positional order.

(*iii*) BP I and BP II phases exhibit a non-vanishing elastic shear modulus (about 106 times smaller than that of a conventional solid).

(*iv*) The blue phases display a much higher viscosity than either the helical or isotropic phase. Their viscosity is about 106 times larger than the viscosity of their corresponding helical phases.

(*v*) The blue phases are optically isotropic, *i.e.* birefringence is absent.

The structure of BPs: It is apparent from the orientation of the molecules in the helical phase that the interaction energy of two chiral molecules is minimized when the two molecules orient at a slight angle to one another. In the helical phase, molecules are at slight angles to one another along the pitch axis. However, the local interaction energy of neighboring molecules is less if the orientations of the molecules twist in a second dimension. Such a configuration gives rise to a "double-twist tube". But, the tube becomes strained and distorted as the strain increases. Hence a single large double-twist tube is not possible. As an alternative, the blue phase consists of several of these double-twist tubes, arranged in a lattice. However, the directors of one double-twist tube cannot line up with the directors of neighboring double-twist tubes always. Therefore, it is not possible to extend such a lattice in three dimensions. Hence, BPI and BPII phases may be viewed as lattices of double-twist tubes (Fig. 50.17). Thus, blue phases are special cases where a "double twist" structure fills up large volumes. In double twist structure the director twists simultaneously about two independent directions (Fig. 50.15). The directors rotate along the pitch axis and as well the direction perpendicular to it, and achieve a rotation of 90° across the diameter.

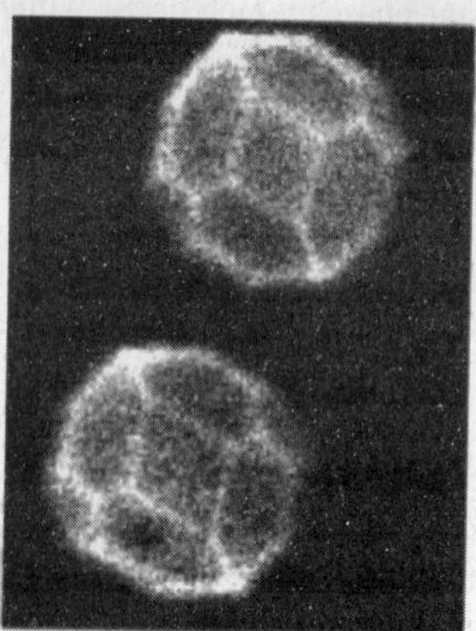

Fig. 50.14:

Micrograph of crystals (0.1-0.2mm) of BPI floating in the isotropic phase.

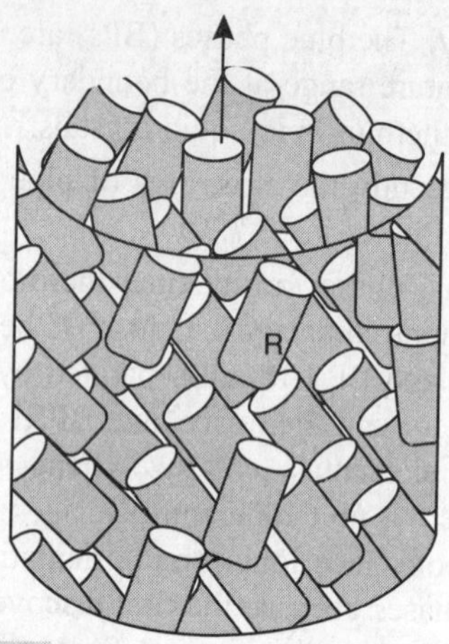

Fig. 50.15:

Double twist structure of the director.

The double twist structure is more stable than the normal helical structure only up to a certain radius of the double twist cylinder. A bigger structure can be composed from the double twist cylinders, however at the points where the cylinders are in contact, the defects occur - disclination, as illustrated in Fig. 50.16.

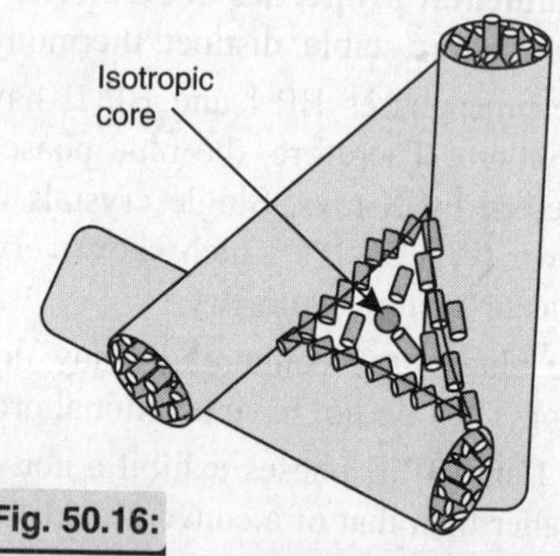

Fig. 50.16:

Structure of BP. Intersection of three double twist cylinders; arrows show disclination

Fig. 50.17 illustrates cubic blue phases BP I and BPII as double-twist cylinders. The double twist cylinders arrange perpendicular to each other to form the three dimensional cubic lattice. Depending on the temperature, the energetic considerations lead to the experimentally observed simple cubic symmetrical structures for BP I and BCC symmetrical structures for BP II.

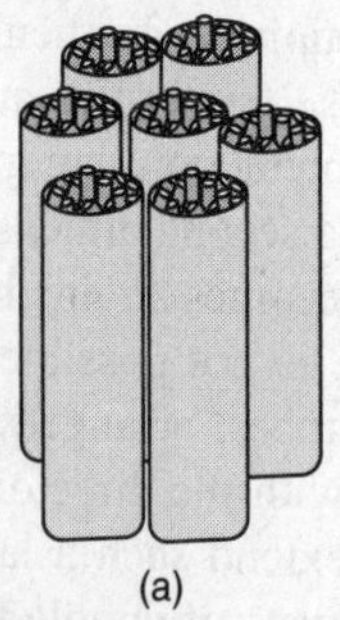

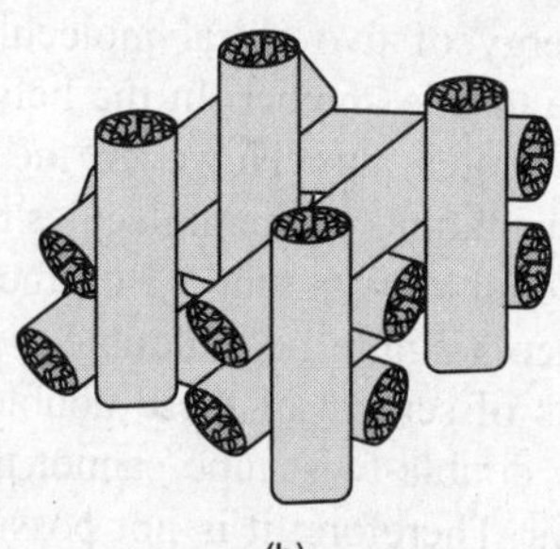

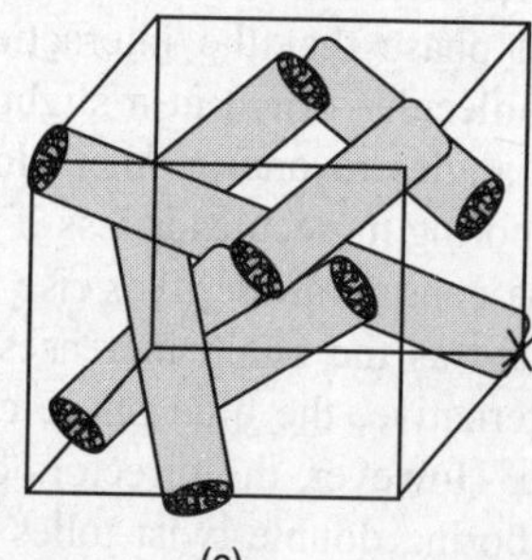

Fig. 50.17:

Arrangement of double twist cylinders as they are believed to exist in BPII

Expanding the stable temperature range of BP through composite BPLC mixtures: The temperature range for BPLC is very narrow (0.5-2°C). As such BPLC cannot be used in commercial devices. Attempts have been made to broaden the temperature range of the BP using reactive LC monomers that can be photo-polymerized. Polymer stabilization of BPs is found to widen the temperature window to more than 60°C which includes the room temperature.

50.7.3. Smectic Phases

Smectic means soap-like appearance and behavior. The smectic phases are layered phases. As in the nematic phase, the rod-like molecules in smectic liquid-crystal phases are orientationally ordered; the molecules tend to align their long axis along the director **n**. In addition to this nematic-like *orientational* order, all smectic phases show a positional order of their molecules along at least one spatial dimension.

Thus, smectic phases exhibit a ***two dimensional order.*** Molecules in this phase show a degree of translational order not present in the nematic. Smectics have stratified structures, with well-defined layer spacing. Motion is restricted to within these layers, and separate layers are observed to flow past each other. The spacing of the molecules within each layer is not uniform.

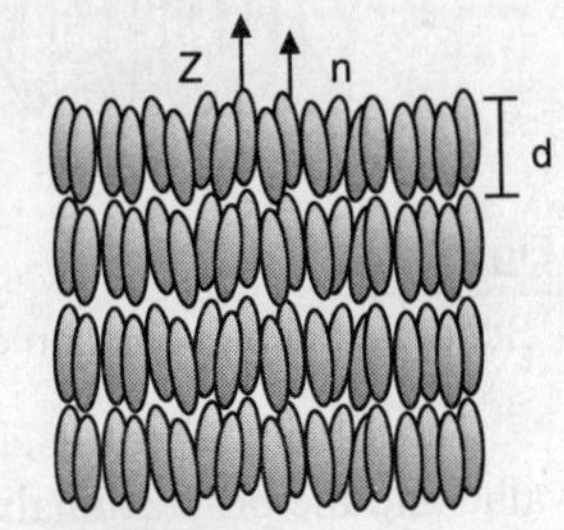

Fig. 50.18: Smectic-A phase

Different smectic phases are characterised by different forms of packing such as SmA, SmB, SmC, etc, ranging from A to K. As many as 12 different variations of smectic phases categorized on the basis of the degree of local positional order and upon whether the molecules are tilted with respect to the layer normal.

(*i*) **Smectic-A phase (SmA)**: The simplest is the Smectic A phase made of non-chiral and non-polar molecules. The phase is characterized by a one-dimensional layered structure or positional ordering, in which each layer is essentially a two-dimensional liquid and the phase may be thought of as a stack of two-dimensional fluid layers. (Fig. 50.18). The smectic A phase is the least ordered smectic phase. The molecules are arranged in diffuse layers. Within the layers, the molecules are parallel and their long axes are on average perpendicular to the layer planes and give a one-dimensional density wave for the centre of molecular mass.

Within the layers, the centers of gravity of the molecules are ordered at random and therefore there is no particular positional order in the layer. The molecular packing within the layers is liquid-like. Thus, smectics A possess the one-dimensional quasi long-range positional order and within the layers molecules show a relatively high mobility. Because of the disorder within the layers, the layers are not well defined. Within the layers, the centers of gravity of the molecules are ordered at random. Likewise, there is no correlation between the lateral positions of the molecules in successive layers. The layer thickness is equal to the molecule length. The director, ***n***, lies along the layer normal, Z-direction, which is perpendicular to the smectic plane.

(*ii*) **Chiral SmA phase:** The chiral SmA* phase is characterized by a diffuse layer structure in which the director ***n*** is oriented perpendicular to the layer plane, and cannot be distinguished mechanically from the achiral SmA phase. The chiral version is structurally the same as that of achiral SmA.

In 1977 Garoff and Meyer have shown that when an electric field E is applied parallel to the layers of a chiral SmA* phase, it induces a molecular tilt θ relative to the layer normal z in a direction orthogonal to the field. This is known as ***electroclinic effect.*** The electroclinic effect (ECE) is an electro-optical effect that consists in a tilt of the optical dielectric tensor of a liquid crystal (LC) material upon application of an external electric field.

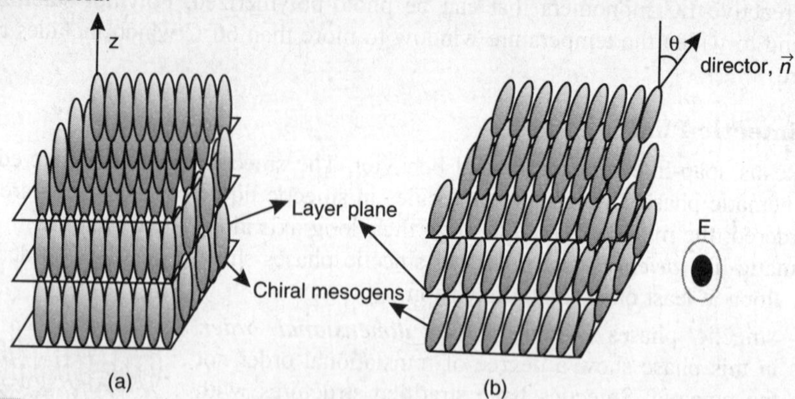

Fig. 50.19: Schematic representation of (a) orthogonal layered geometry, (b) Tilted layered structure

The tilt induced tilt angle θ varies linearly with the applied field E up to a saturation value and the proportionality coefficient, the ***electroclinic coefficient***, is a property of the LC material. Reversal of the field direction leads to reversal of the induced tilt direction.

The tilt angle θ is directly proportional to the electric field

$$\theta = e_c E \quad ...(50.9)$$

where e_c is the electroclinic coefficient, given by

$$e_c = k/(T - T_{AC^*}) \quad ...(50.10)$$

The angle θ can be quite high, often ~10° - 15° close to the SmA*-C* transition.

(*iii*) **Smectic C phase (SmC):** SmC phase is similar to the SmA phase. It is a layered structure and each layer may be thought of as a two-dimensional liquid film. The SmC phase differs from the SmA phase in that the director of each layer is inclined at an angle to the layer normal and this angle being identical for all layers (Fig. 50.20). Furthermore, the tilt angle, θ, which the molecular long axis makes with the layer normal, is a strong function of temperature. To a first approximation, after a second-order transition to the SmC phase from a SmA phase, the tilt angle varies in the following manner with respect to temperature, T, as described by equation

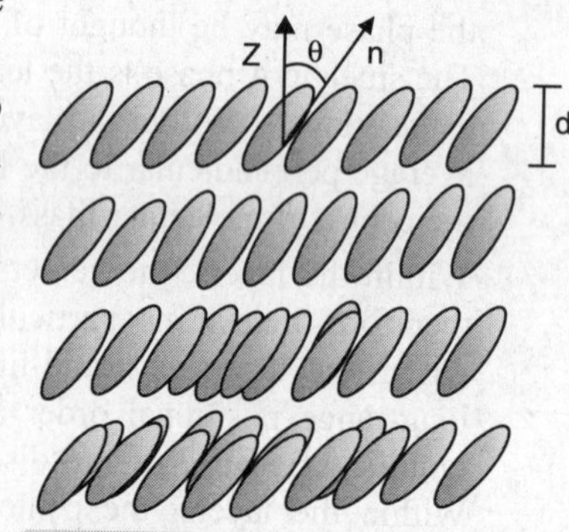

Fig. 50.20: Smectic-C

$$\theta(T) = \theta_0(T_c - T)^{\alpha} \quad ...(50.11)$$

where T_c is the transition temperature, T is the temperature, θ_0 is a constant, and α is an exponent (which has a theoretical value of 0.5).

The angle θ which is an order parameter for this phase can range from zero to as high as 45–50° deep in the SmC phase. The tilt angle in this phase has been shown to vary with temperature and it often increases monotonically with decreasing temperature. The temperature dependence of molecular tilt in the SmC phase is shown in Fig. 50.21.

With increasing temperature SmC having a small slope angle transform into SmA. Angle θ in the SmC depends on the temperature range and can decrease down to zero. But other smectics with the small slope angle β show no temperature dependence. SmC with the great angle β do not transform into SmA and their angle β is independent of the temperature.

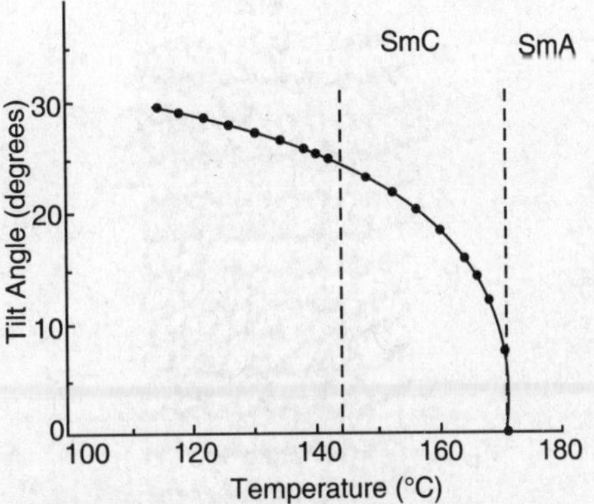

Fig. 50.21: Temperature dependence of molecular tilt in the SmC phase of TBBA determined by high-resolution x-ray diffraction

The transition from SmA to SmC phase at temperature TAC, at a temperature T below is given by

$$d(T) = d(T_{AC}) \cos \theta \qquad ...(50.12)$$

where d is the smectic layer spacing.

A consequence of the tilting of the molecules is that this phase exhibits biaxial optical and physical properties while the SmA phase is generally optically uniaxial. X-ray studies indicate that the layer thickness in this phase is considerably less than the molecular length.

(*iv*) **Chiral Smectic C Phase:** Chiral smectic-C* phase is formed by chiral molecules and exhibits a helical structure which is in some aspects similar to the helical structure in a cholesteric liquid crystal. In the helix each layer is similar to that of the usual SmC phase.

One of the more significant features of the SmC* phase is that it is ***ferroelectric***, with a non-zero spontaneous polarization. Typically the polarization is a very strong function of temperature near the transition (from higher temperature phase) to the SmC* phase and fairly slower farther from the transition. Recall that the SmC* phase consists of tilted molecules and typically the polarization is coupled to the tilt angle which also is zero at the transition to the (un-tilted) SmA phase.

In fact

$$\theta = \theta_0 (T_c - T)^{\alpha} \qquad ...(50.13)$$

$$P = P_0 (T_c - T)^{\beta} \qquad ...(50.14)$$

where θ is the tilt angle, P is the spontaneous polarization, θ_0 and P_0 are constants, T_c is the transition temperature, T the temperature, and α and β are constants, theoretically predicted to be equal to 0.5.

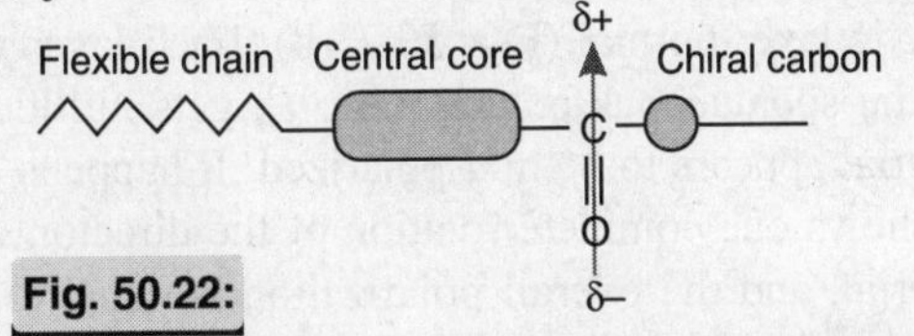

Fig. 50.22: Structure of the liquid crystal molecule DOBANBC

The chirality of the SmC* phase causes a twisting of the molecules in the phase and the molecules arranged in a helical fashion as in Fig. 50.23. In a macroscopic sample, the molecular polarization follows this helix and the total spontaneous polarization would be zero and hence the name *helielectric phase*. If a strong electric field is applied the helix is unwound and there arises a finite spontaneous polarization which can be measured.

In 1974, Meyer predicted theoretically that liquid crystals that form a liquid crystal phase with low symmetry should exhibit ferroelectricity. The structure of the liquid crystal molecule (DOBANBC) is shown in Fig. 50.22 where the director tilt of each layer precesses around a circle upwards through the structure. DOBANBC was found to exhibit *ferroelectricity*

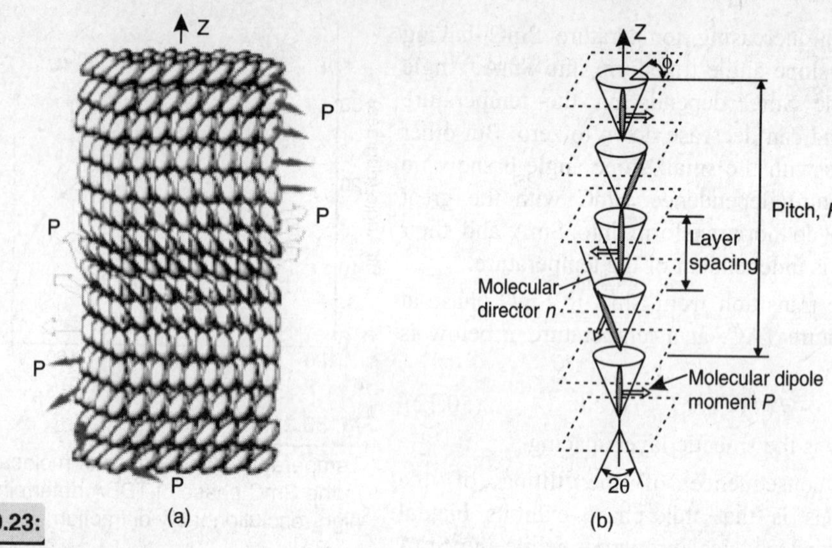

Fig. 50.23: The chiral SmC phase.

(*a*) **Helielectric Phase:** Liquid crystal molecules usually possess a permanent dipole moment μ oriented at some arbitrary angle with respect to the molecular long axis.

As the chiral centre hinders the molecular rotation, the chiral molecules do not rotate freely along their long axes. Consequently, it generates a non-compensated part of the transverse molecular dipole moment μ.

Due to random head-tail alignment of molecules the perpendicular dipole moment component to the layer cancels while an in-layer component of the dipole moment remains, and creates the layer dipole moment *P*. In the macroscopic helical structure the director $\vec{n}$ precesses around the layer normal *Z*. The azimuthal tilt direction φ changes by a small amount when going from layer to layer (Fig. 50.23 b).

As the direction of tilt also rotates gradually from layer to layer, it causes the direction of the polarization to rotate (see the letter **'P'** in Fig. 50.23 a). Apart from that, the director rotates due to chirality of the structure and the direction of dipole moment *P* rotates about the smectic layer normal (Fig. 50.23 b). Each layer within the helix has a polarization direction; but the spontaneous polarization, *P*, gets nullified within one pitch of the helix, *p*, and the material appears to be non-polarized. It happens throughout the bulk of the phase. Fig. 50.23 (b) shows one complete rotation of the director, which corresponds to the pitch length of the material, and the overall polarization is equal to zero. So, the smectic C phase is known as ***helielectric***. The helical pitch P will be in the range from the wavelength of visible light up to arbitrary large values. In most compounds P amounts to a few μm.

(*b*) **Ferroelectricity:** The helix can be unwound by the application of an electric field ***E*** perpendicular to the smectic layer normal. Helical structure is unwound and the polarization vector P_s of each layer is forced to align along ***E*** while the tilt direction aligns in each layer perpendicular to the electric field ***E*** (Fig. 50.24). The director **n** aligns throughout the sample parallel to the plane perpendicular to ***E***. Thus, a macroscopic polar phase is formed. The helical structure can be unwound by surface treatment also. Hence, the smectic *C* phase is called as ***ferroelectric***.

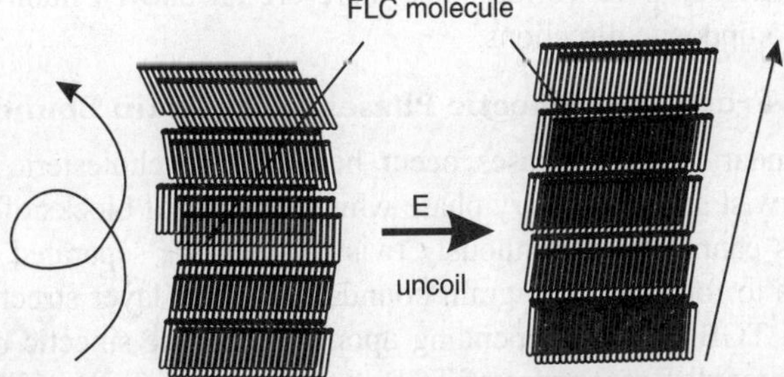

Fig. 50.24:

Tilt direction aligns in each layer perpendicular to the electric field **E**

(*v*) **Antiferrolectric and Ferri- electric Smectic Phases:** Antiferroelectric and ferrielectric smectic phases are similar to ferroelectric smectic phases, although the molecules tilt in an opposite sense in alternating layers as shown in Fig. 50.25 and Fig. 50.26. In consequence, the layer-by-layer polarization points in opposite directions. These two phases are somewhat like the frustrated smectic phases in which local dipoles become antiparallel. In these phases, the layer spacing and the polarization direction are not related in the same manner as in the SmC* phase. These phases, by analogy to the situation in magnets, are called *antiferroelectric* liquid crystals and *ferrielectric* liquid crystals.

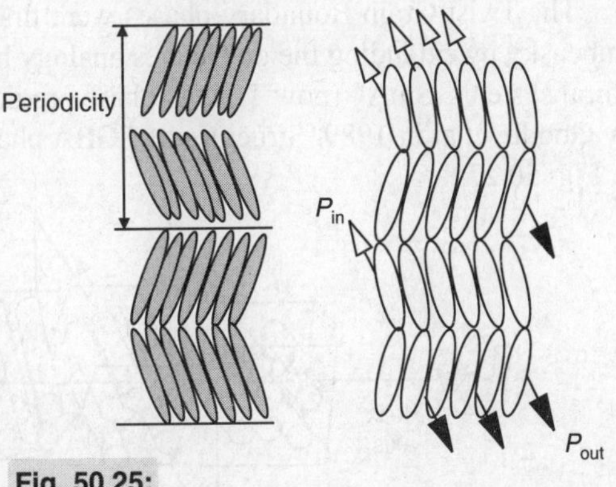

Fig. 50.25:

Antiferroelectric smectic phases

Fig. 50.25 schematically depicts the SmC* phase, the proposed structure of the antiferroelectric phase and Fig. 50.26 shows the simplest ferrielectric phase. All phases possess chirality. The SmC* phase is characterized by the spontaneous polarization of each layer pointing in the same direction.

In the ***antiferroelectric phase*** the average tilt angle and the average spontaneous polarization are both zero since the phase is characterized in adjacent layers by alternating directions of the spontaneous polarization and the molecular tilt. An adequately strong electric field will change the antiferroelectric phase into the ferroelectric phase. Antiferroelectrics repeat their helical structure every 180° of rotation about the layer normal in contrast to 360° for the ferroelectric phase.

The ferrielectric liquid crystal phase is characterized by an uneven numbers of layers stacked with tilt to the left and tilt to the right as illustrated in Fig. 50.26. In this phase local dipoles array antiparallel. The phase exhibits a

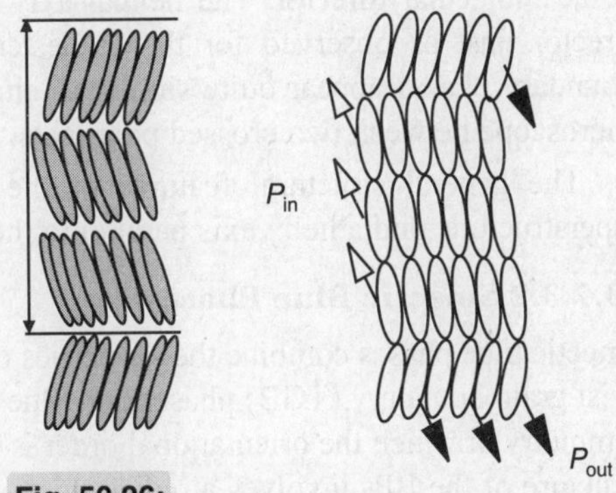

Fig. 50.26:

Simplest Ferrielectric Phase

net but low spontaneous polarization since there are an uneven number of molecules with polarization in two opposite directions.

50.7.3.1 Chiral Frustrated Smectic Phases-Twist Grain Boundary (TGB) Phases

Twist Grain Boundary (TGB) phases occur between the cholesteric and smectic phases. Molecules forms twist grain boundary phase which consists of blocks of smectic layers. Since the smectic layers cannot be continuously twisted, they are separated from one another by screw dislocations to form periodic grain boundaries. Thus, layer structure as well as helical twist is present in TGB phases. Depending upon whether the smectic blocks are of SmA or SmC or Sm C*, three different types of TGB phases such as TGBA, TGBC and TGBC* are observed. Their temperature range of stability is generally very small, of the order of ~1 K.

The Twist Grain Boundary phases were first theoretically predicted in 1988 by Renn and Lubensky by extending the de Gennes analogy between liquid crystals and superconductors to chiral systems. SmA* (now TGBA) phase predicted by Renn and Lubensky was first observed by Goodbyet al in 1989. Structure of TGBA phase predicted by Renn and Lubensky is shown in Fig. 50.27.

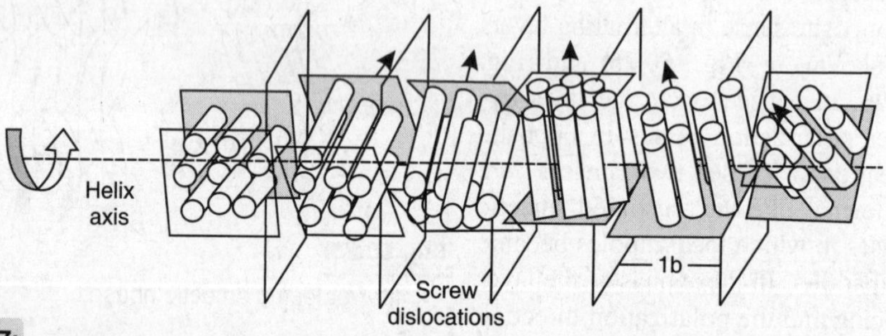

Fig. 50.27:

Schematic image of the TGB phase, arrow indicating director orientation.

The structure of the TGB phase in Fig. 50.27 consists of fluid smectic layers (SmA or SmC), broken up into slabs of a width in the order of lb ~100 nm. The grains are mediated by a regular lattice of screw dislocations with a distance on the order of ld ~10 nm between them. In the slabs, molecules are arranged in layers with their director normal to the smectic layers. Neighbouring slabs (and hence molecular director in the slabs) are twisted with respect to each other by angle $\Delta\chi$, thereby forming helical structure with the helix axis ($\hat{P}$) normal to the molecular director. The helical axis is oriented perpendicular to the local molecular director, just as observed for the cholesteric phase. This is the reason why Twist Grain Boundary phases appear quite similar to chiral nematic phases when viewed in a polarizing microscope between two crossed polarizers.

The general structural features of the TGB phase are a layered structure, a helical superstructure, and a helix axis parallel to the smectic layer plane.

50.7.3.2 Smectic Blue Phases

Smectic blue phases combine the properties of the two types of frustrated mesophases, namely twist grain boundary (TGB) phases and blue phases (BP). Blue phases show an unusual cubic symmetry in which the orientational order is periodic and long range in three dimensions. The structure of the BPs involves a radial twist of the director called a double cylinder. However, this double twisted structure cannot extend perfectly into three-dimensional space. Geometrical models of blue phases therefore consist of cubic networks of double twist cylinders separated by disclination lines (Fig. 50.28).

Since smectic layers cannot be continuously twisted, the TGB phases consist of blocks of pure smectic material (which can be either smectic-A for TGBA or smectic-C for TGBC) separated by parallel, regularly spaced grain boundaries, formed by a periodic array of screw dislocations (Fig. 50.29). Such a dislocation arrangement allows helical twist. Thus, in TGB phases, as in blue phases, the frustration is relieved by the presence of defects. Fig. 50.29 shows geometrical model of a smectic double twist cylinder. The surfaces represent the smectic layers and the black lines wrapping around the cylinder represent the screw dislocations. Smectic order persists between cylinders and gives rise to the continuous smectic ring (region II).

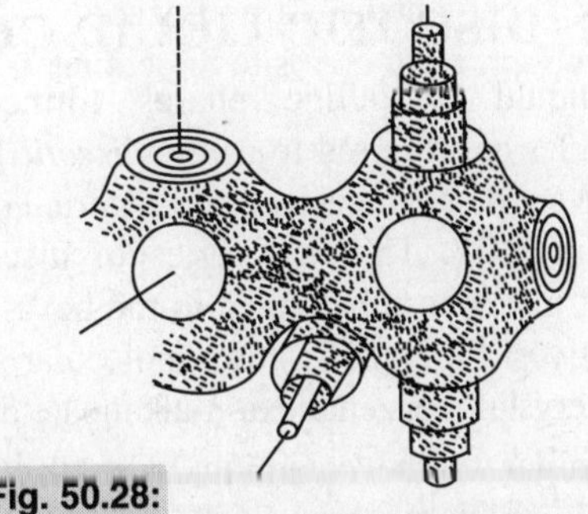

Fig. 50.28:

Geometrical model of blue phases involving double twist cylinders and a network of disclination lines

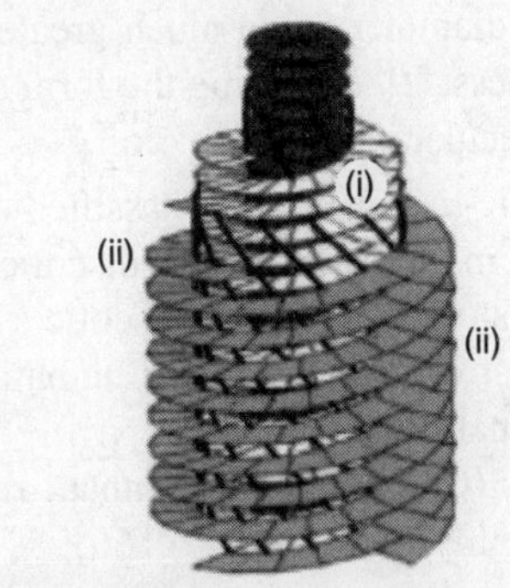

Fig. 50.29:

Periodoc array of screw dislocations

Smectic blue phases (BPSm) appear in a very narrow temperature range with the following phase sequence: TGB–BPSm1–BPSm2–BPSm3–Iso, without any intermediate cholesteric state between the BPSm and TGB phases.

The third smectic blue phase (BPSm3) has, like the classical BP3 phase, an amorphous structure of the same macroscopic symmetry as the isotropic phase. The x-ray scattering experiments show that the BPSm phases exhibit, in contrast to BPs, quasi-long-range smectic order with a typical persistence length of about 60 nm.

A smectic blue phase can be regarded as the extension to 3D of the TGB phase. Indeed, in contrast to the cholesteric and to the blue phases where the twist occurs at the molecular level, the twist occurs at the scale of the smectic slabs for both TGB and smectic blue phases. Thus, smectic blue phases represent an original physical system of condensed matter exhibiting a double frustration: the extension of chirality in three spatial dimensions, and helical twist competing with smectic order.

Smectic Blue Phases show a local structure occurring in a volume of about $(10 \text{ nm})^3$ and a superstructure with a periodicity of some hundred nm or a few pm which are incompatible with respect to each other. The impossibility of matching these two structures by a continuous molecular arrangement with constant degree of order (frustration) leads to a regular lattice of defects.

In principle, both TGB and BP phases can be considered as heterogeneous systems consisting of the respective high-temperature and low-temperature phases. Another similarity between TGB and BP is the occurrence of a "fog phase" (N^*_L and BPIII, respectively). The transition from the regular lattice of defects to the non-ordered fog phase can be considered as an example of lattice melting.

50.8 DISCOTIC LIQUID CRYSTALS

The liquid crystalline phases formed by disk-shaped molecules are referred to as the ***discotic phases*** or ***discotics***. They were first synthesized and reported by S. Chandrasekhar et al in 1977. The first series of discotic compounds to exhibit mesophase belonged to the hexa-substituted benzene derivatives. Fig. 50.30 shows molecular structure of discotic liquid crystal, benzenehexa-n-alkanoate derivatives.

1

$R = C_4H_9$ to C_9H_{19}

Fig. 50.30: Molecular structure of the first series of discotic liquid crystal

Similar to the calamitic LCs, discotic LCs possess a general structure comprising a planar (usually aromatic) central rigid core surrounded by a flexible periphery, represented mostly by pendant chains (usually four, six, or eight), as illustrated in Fig. 50.31. As can be seen, the molecular diameter '*d*' is much greater than the disc thickness, '*t*', imparting the form anisotropy to the molecular structure.

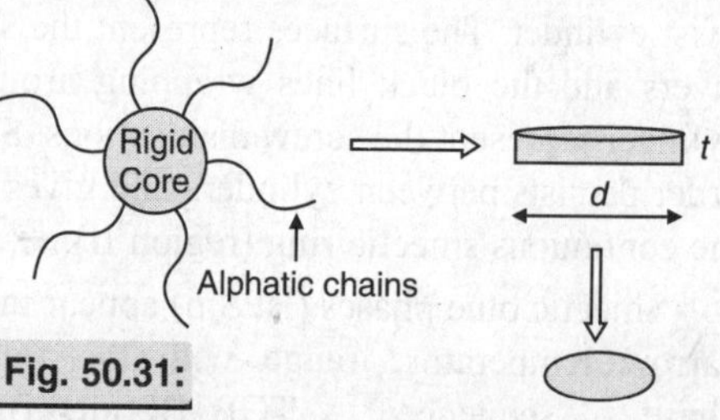

Fig. 50.31: Schematic of the general shape of discotic LCs, where $d >> t$.

Discotic-mesogensare classified primarily into two main categories: discotic nematic mesophases (N_D) and discotic columnar mesophases. Fig. 50.32 shows nematic discotic (N_D), nematic columnar (N_{col}), columnar hexagonal (Col_h) and columnar rectangular (Col_r).

Both types of mesophases are very important for various device applications. Nematic DLCs can be polymerized to form negative birefringence films which have been commercialized as compensation foils to enlarge the viewing angle of commonly used LCDs. They can also be used as active component in LCDs.

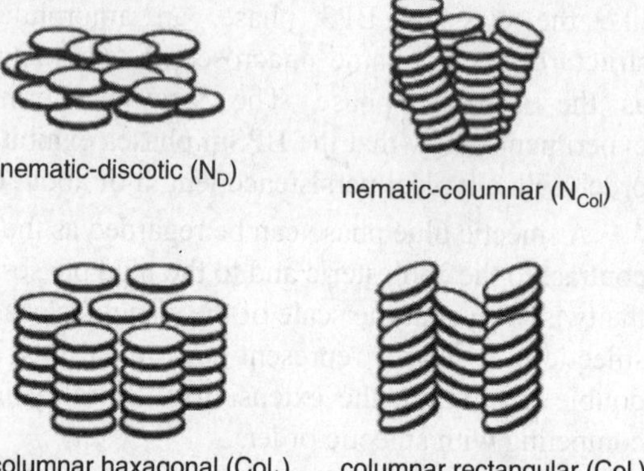

Fig. 50.32: Discotic liquid crystal phases

A majority of DLCs are derived from polycyclic aromatic cores having strong π-π interactions and, therefore, arrange themselves in columns. The separation in between two molecules (core–core separation) in a columnar mesophase is usually of the order of 3.5 Å. Therefore, there is considerable overlap of p-orbitals. As flexible long aliphatic chains surround the core, the separation between two columns (intercolumnar separation) is usually 20–40 Å, depending on the lateral chain length. Therefore, interactions between neighbouring molecules within the same column are much stronger than interactions between neighbouring columns. The columnar mesophase may be described as molecular wires. Due to this unique property, DLCs have been extensively studied for the energy and charge migration and their device applications such as, one-dimensional conductors, photoconductors, light emitting diodes, photovoltaic solar cells, field-effect transistors and gas sensors have been sought.

(*i*) **Nematic discotic phase:** Nematic discotic (N_D) is the least ordered mesophase, where the molecules have only orientational order being aligned on average with the director as illustrated in Fig. 50.33. There is no positional order. The discotic nematic phase is denoted by N_D where the subscript D is to avoid confusion with the normal nematic phase. This phase is characterized by the short axis of the molecule aligning parallel to the director. The N_D phase is diamagnetically and optically negative. This is in contrast with the rod-like nematics which are mostly optically and diamagnetically positive.

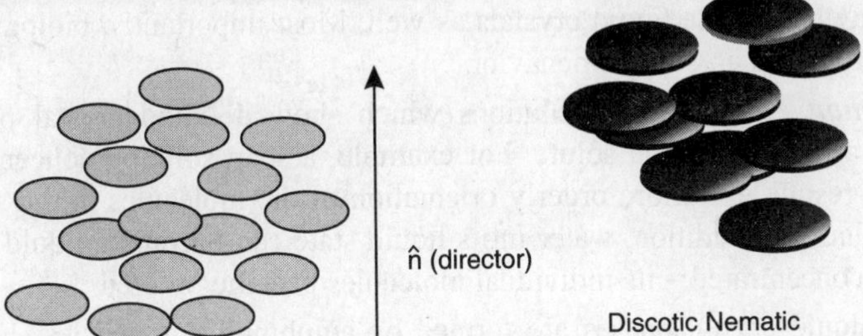

Fig. 50.33:

Schematic of the N_D phase

(*ii*) **Columnar Phases:** Discotic columnar mesophases are obtained when the disc-like molecules stack on top of each other. The simplest columnar phase consists of stacked disks, forming a one-dimensional liquid-like structure. The columnar nematic phase has no positional order but the columns themselves form a two-dimensional lattice. Other columnar phases have both long-range orientational and two-dimensional positional order. Hexagonal, rectangular, and other lattice types have been identified. Tilted phases have also been reported. Fig. 50.33 illustrates the N_D phase and the columnar phases. The columns themselves may be organized into rectangular or hexagonal arrays as in Fig. 50.34.

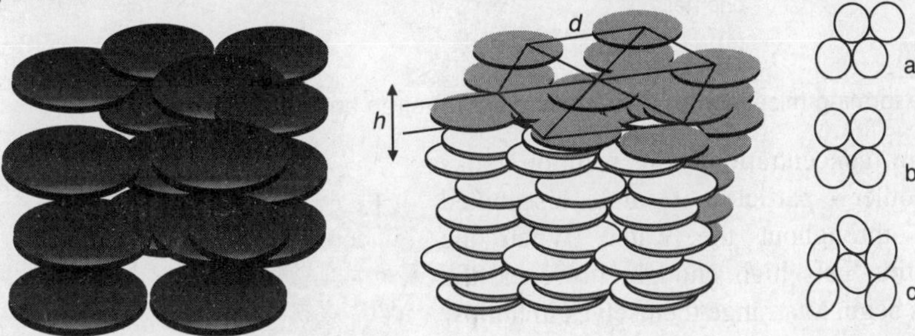

Fig. 50.34:

Columnar phase formed by the disc-shaped molecules (a) Hexagonal, (b) rectangular, and (c) herringbone

The columnar phase is a class of liquid-crystalline phases in which molecules assemble into cylindrical structures to act as mesogens. Nematic columnar liquid crystals have no long-range order and are less organized than other columnar liquid crystals. Other columnar phases with long range order are classified by their two-dimensional lattices: hexagonal, tetragonal, rectangular, and oblique phases. The discotic nematic phase includes nematic liquid crystals composed of flat-shaped discotic molecules without long-range order. In this phase, molecules do not form specific columnar assemblies but only float with their short axes in parallel to the director.

50.9 LYOTROPIC LIQUID CRYSTALS

Lyotropic liquid crystals were actually discovered long before thermotropic liquid crystals were known. In 1850, their texture was noticed in a mixture of myelin and water. However, the significance of liquid crystals was not understood, so most research has been done on thermotropics. Lyotropic liquid crystals are found in countless everyday situations. Soaps and detergents form lyotropic liquid crystals when they combine with water. In the kitchen, cake batters may harbor the liquid crystals as well. Most importantly, biological membranes display lyotropic liquid crystalline behavior.

Lyotropic liquid crystals are solutions which show a liquid crystal phase at certain concentrations of solvent in the solute. For example, a soap solution concentration above a certain amount results in a more orderly orientation of the molecules in the soap, creating a liquid crystal phase. In addition, water in its liquid state can become a liquid crystal when it becomes more concentrated – its individual molecules become ordered.

Lyotropic liquid crystal phases are formed by amphiphilic molecules (Fig. 50.35). The molecules consist of two distinct parts: a polar, often ionic, head and a nonpolar, often hydrocarbon tail. (Not all surfactants, however, form lyotropic liquid crystals.) The head is attracted to water, or hydrophilic, and the tail is repelled by water, or hydrophobic. When dissolved in high enough concentrations, the molecules arrange themselves so that the polar heads are in contact with a polar solvent and/or the nonpolar tails are in contact with a nonpolar solvent.

"Head" H_2O

Na

one "tail"

(a) (b)

Fig. 50.35: (a) Soap forming micelles, (b) Schematic diagram of an amphiphilic molecule

At low concentrations, the solution looks like any other - particles of solute distributed randomly throughout the water. When the concentration gets high enough, however, the molecules begin to arrange themselves in hollow spheres, rods, and disks called micelles. As the concentration increases, the micelles begin to arrange themselves into loose patterns. These patterns are the actual liquid crystal aspects of the molecular behavior. One of the first liquid crystal phases has micelles forming a structure similar to a face-centered or body-centered cubic crystal lattice. The illustration below (Fig. 50.37a) shows a cubic crystal structure. Micelles take the place of individual atoms, ions, or molecules. It should also be noted that the pattern is not as stable or as rigid as that of a solid crystal like graphite or table salt, hence the term liquid crystal.

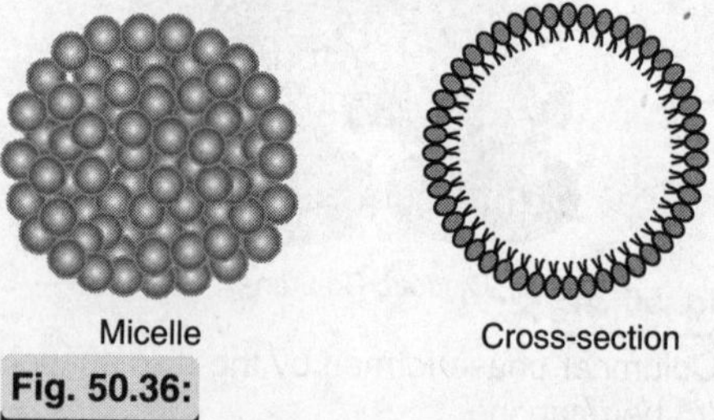

Fig. 50.36: Lyotropic liquid crystalat low concentrations

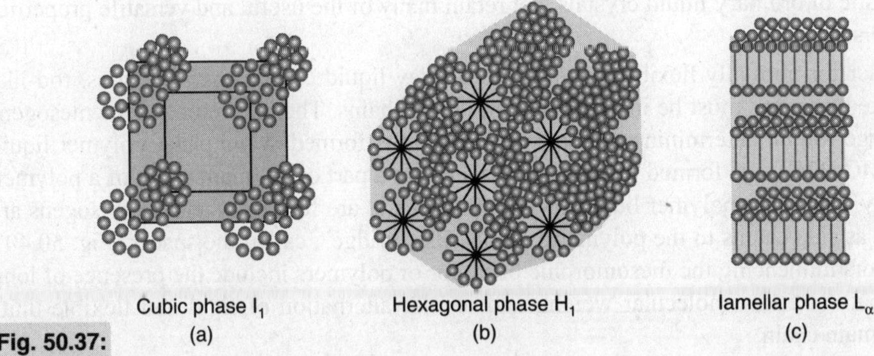

Fig. 50.37: Lyotropic liquid crystal at increased concentrations

Rod-shaped micelles often form into hexagonal arrays made out of six rods grouped around a central one for a total of seven, as illustrated in the Fig. 50.37 (b). In the enlargement of a single rod, notice that the micelle surface is composed of hydrophilic heads. The hydrophobic tails are isolated inside the micelle. Hexagonal liquid crystals generally exist in solutions that are forty to seventy percent amphiphile. The liquid crystals may come apart if too much water or salt is added to the solution, but many varieties can absorb oil by expanding the diameter of the rod-shaped micelle.

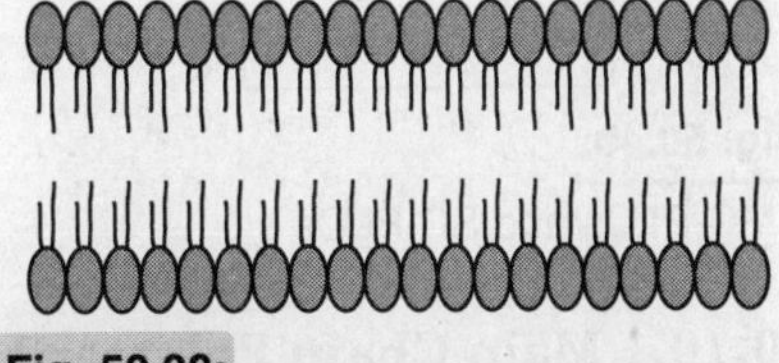

Fig. 50.38: Liquid crystaldouble layer of molecules

At even higher concentrations the molecules move into another liquid crystalline phase - the lyotropic liquid crystal bilayer. This structure (Fig. 50.38) has a double layer of molecules arranged a bit like a sandwich with polar heads taking the place of the bread and nonpolar tails as the filling. This pattern is similar to that of smectic liquid crystals in the thermotropic category. Because the sheet-like layers can slide easily past each other, this phase is less viscous than the hexagonal phase, at least in the direction of the sliding, despite its lower water content. Fig. 50.39 shows the presence of phospholipid in lyotropic liquid crystal arrangement.

Fig. 50.39: Phospholipid in (a) cell membranes, (b) bilayer lyotropic liquid crystals

50.10 POLYMER LIQUID CRYSTALS

Polymers are large molecules. They consist of one or more continually repeating monomer units. Polymers are common materials and we are familiar with many of them such as Polyethelene, Nylon, Terylene, Teflon etc. PLCs are a class of materials that combine the properties of polymers with those of liquid crystals. They exhibit the same mesophase characteristic of ordinary liquid crystals, yet retain many of the useful and versatile properties of polymers.

In order for normally flexible polymers to display liquid crystal characteristics, rod-like or disk-like mesogens must be incorporated into their chains. The placement of the mesogens plays a large role in determining the type of PLC that is formed. Main chain polymer liquid crystals (MC-PLCs) are formed when the mesogens are a part of the main chain of a polymer. Conversely, side chain polymer liquid crystals (SC-PLCs) are formed when the mesogens are connected as side chains to the polymer by a flexible "bridge", called the spacer (Fig. 50.40). Other factors influencing the mesomorphic behavior of polymers include the presence of long flexible spacers, a low molecular weight, and regular alternation of rigid and flexible units along the main chain.

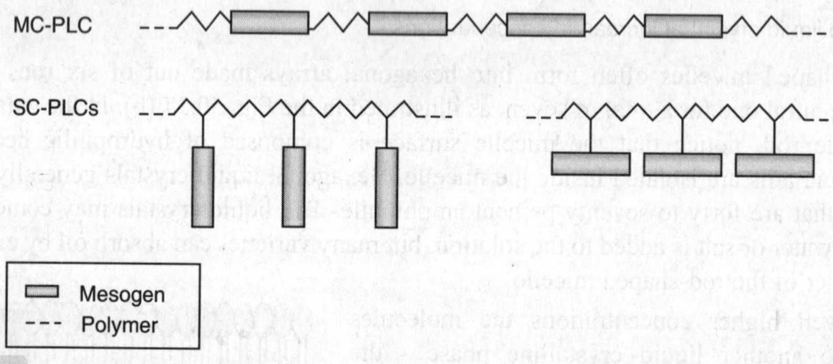

Fig. 50.40: MC-PLCs and SC-PLCs

50.10.1 Main Chain Polymer Liquid Crystals

Main chain polymer liquid crystals are formed when rigid elements are incorporated into the backbone of normally flexible polymers. These stiff regions along the chain allow the polymer to orient in a manner similar to ordinary liquid crystals, and thus display liquid crystal characteristics. There are two distinct groups of MC-PLCs, differentiated by the manner in which the stiff regions are formed.

PHNA poly (hydroxynapthoic acid)

Fig. 50.41: Main Chain Polymer Liquid Crystals-PHNA

The first group of main chain polymer liquid crystals is characterized by stiff, rod-like monomers. These monomers are typically made up of several *aromatic* rings which provide the necessary size. Fig. 50.41 shows an example of this kind of MC-PLC.The second and more prevalent group of main chain polymer liquid crystals is different because it incorporates a *mesogen* directly into the chain. The mesogen acts just like the stiff areas in the first group. Generally, the mesogenic units are made up of two or more aromatic rings which provide the necessary restriction on movement that allow the polymer to display liquid crystal properties. Another characteristic of the mesogen is its *axial ratio*. Experimental results have concluded that these molecules must be at least three times long as they are wide. Otherwise, the molecules are not rod-like enough to display the characteristics of liquid crystals.

This second group is different from the first in that the mesogens are separated or "decoupled" by a flexible bridge called a *spacer*. Decoupling of the mesogens provides for independent movement of the molecules which facilitates proper alignment. Fig. 50.42 shows a main chain polymer liquid crystal. Notice the flexible spacer (methylene groups) and the stiff mesogen (aromatic ring and double bonds).

$$\left[-O-(CH_2)_2-O-\overset{O}{\overset{\|}{C}}-C_6H_4-\overset{O}{\overset{\|}{C}}- \right]_n$$

PET poly (p-phenyleneterephthalate)

Fig. 50.42: Main Chain Polymer Liquid Crystals-PHNA-PET

50.10.2 Side Chain Polymer Liquid Crystals

Side Chain Polymer Liquid Crystal materials are formed when mesogenic units are attached to the polymer as side chains. Side chain polymer liquid crystals have three major structural components; the backbone, the spacer, and the mesogen (Fig. 50.43). The versatility of SC-PLCs arises because these structures can be varied in a number of ways.

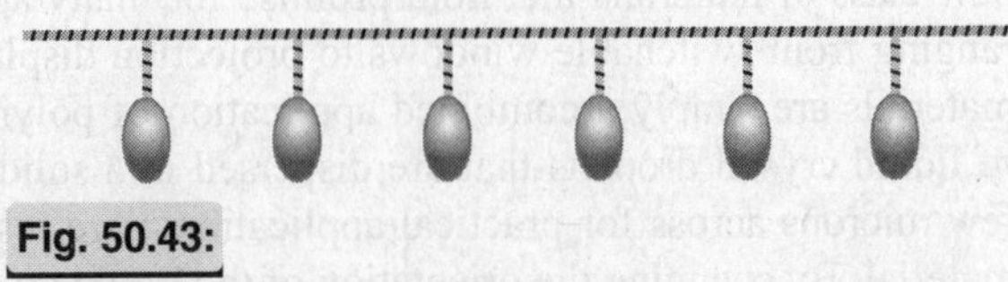

Fig. 50.43: Side Chain Polymer Liquid Crystals

(*i*) **The Backbone:** The backbone of a side chain polymer liquid crystal is the element that the side chains are attached to. The structure of the backbone can be very important in determining if the polymer shows liquid crystal behavior. Polymers with rigid backbones typically have high glass transition temperatures, and thus liquid crystal behavior is often difficult to observe. In order to lower this temperature, the polymer backbone can be made more flexible.

(*ii*) **The Mesogen:** Perhaps the most important part of a side chain polymer liquid crystal is the mesogen. It is the alignment of these groups that causes the liquid crystal behavior. Usually, the mesogen is made up of a rigid core of two or more aromatic rings joined together by a functional group. Fig. 50.44 shows a typical repeating unit in a side chain polymer liquid crystal. Notice the spacer of methylene units and the mesogen of aromatic rings.

(*iii*) **The spacer:** The structure of the spacer is an important determining factor in side chain polymer liquid crystals. Generally, the spacer consists of two to four methylene (CH_2) groups attached together in a line (Fig. 50.44). Accordingly, the spacer length has a profound effect on the temperature and type of phase transitions. Usually, the *glass transition temperature* decreases with increasing spacer length. Short spacers tend to lead to nematic phases, while longer spacers lead to smectic phases. In Fig. 50.44 even though the polymer may be in a tangled conformation, orientation of the mesogens is still possible because of the decoupling action of the spacer.

$$\left[-CH_2-CH- \right]_n \quad O{=}C-O-(CH_2)_{11}-O-C_6H_4-CH{=}N-C_6H_4-CN$$

Fig. 50.44: The Spacer and Mesogen

50.10.3 Phases of Polymer Liquid Crystals

Side chain polymer liquid crystals show a tendency to separate the backbone from the mesogenic side groups. This is achieved in the smectic phase where the mesogenic groups and backbone chains form individual layers. Many side chain polymer liquid crystals form the "double comb" configuration when side chains point away from the backbone in an alternating fashion (Fig. 50.45). The double comb configuration allows the polymers to form layers characteristic of the smectic phase.

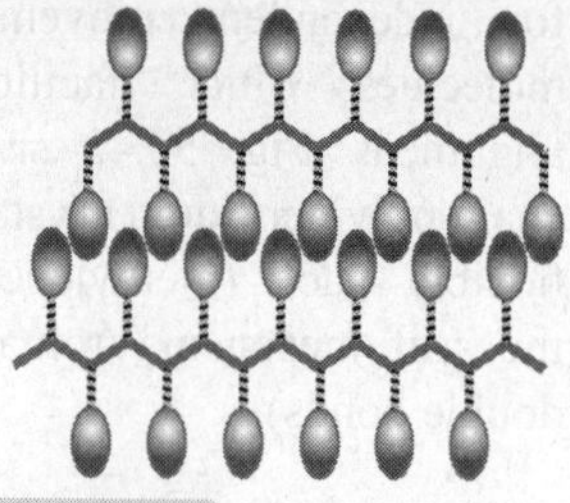

Fig. 50.45: Double comb configuration

50.10.4 Polymer-Dispersed Liquid Crystals

Polymer-dispersed liquid crystals (PDLCs) are a relatively new class of materials that hold promise for many applications ranging from switchable windows to projection displays. These materials are simply a combined application of polymers and liquid crystals. PDLCs consist of liquid crystal droplets that are dispersed in a solid polymer matrix. These tiny droplets (a few microns across for practical applications) are responsible for the unique behavior of the material. By changing the orientation of the liquid crystal molecules with an electric field, it is possible to vary the intensity of transmitted light.

50.11 PHYSICAL PROPERTIES OF LIQUID CRYSTALS

50.11.1 Anisotropy

As a consequence of the anisotropic shape of the constituting molecules, orientational order occurs in the mesophases. As a result of order, most physical properties of liquid crystals are anisotropic. In general, the magnitude of a physical property measured along the average molecular orientation, *i.e.* along the director vector, *n*, is different from that measured orthogonal to the director. Examples are the dielectric permittivity, the magnetic susceptibility, optical birefringence or the heat diffusion. Additionally, there are new physical qualities, such as elastic or frictional torques (rotational viscosity) which do not appear in simple liquids.

50.11.2 Surface Orientation

In the absence of any prior treatment of the substrate, the director of a liquid crystal is free to point in any direction on the surface. However, a specific orientation of LC molecules on the substrate (a glass slide) may be obtained if the substrate is treated before mounting the sample on it.

(*a*) Homogeneous planar orientation of the director is obtained by spreading a thin polymer coating usually a polyimide and rubbing the surface with velvet cloth in a single direction. It is observed that liquid crystal molecules in contact with that surface align with the rubbing direction as shown in Fig. 50.46.

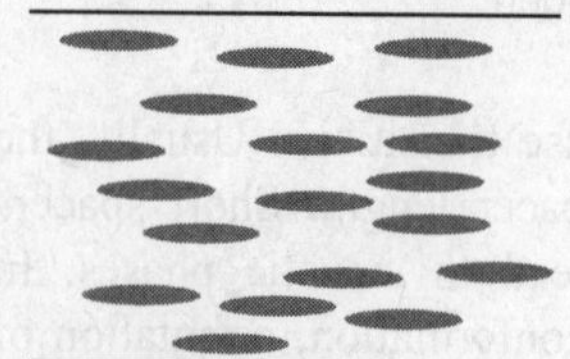

Fig. 50.46: Planar orientation

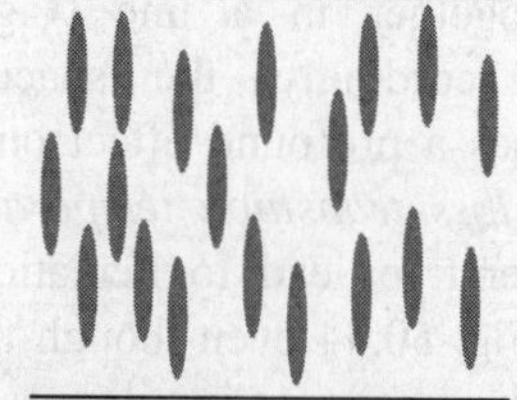

Fig. 50.47: Homeotropic orientation

(*b*) In order to get homeotropic orientation as shown in Fig. 50.47 substrate can be washed with surfactant, lecithin for example.

(*c*) Tilted orientation as shown in Fig. 50.48 can be obtained with oblique evaporation of SiO thin film on the substrate surface.

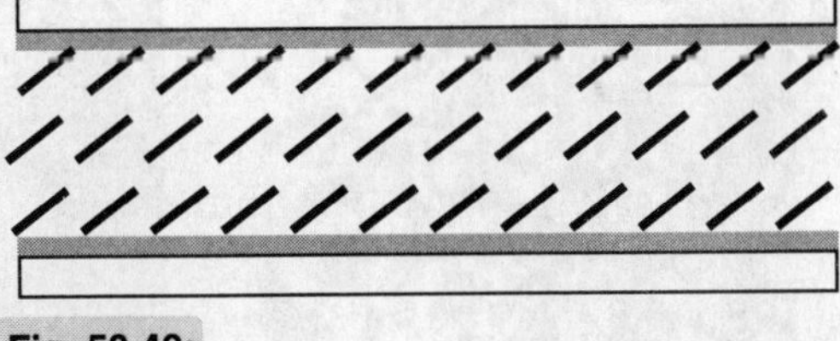

Fig. 50.48: Tilted orientation

50.11.3 Liquid Crystal Textures

The term texture refers to the orientation of liquid crystal molecules in the vicinity of a surface. Each liquid crystal mesophase can form its own characteristic textures, which are useful in identification. In a polarising microscope, the light is polarized by passing it through a polarizing filter. It then passes through the sample, and then through a second polarizing filter called the analyzer (Fig. 50.49). When a liquid crystal material is placed on a microscope slidewith a cover slip and the slide is heated and viewed using a polarizing microscope, textures characteristic of each type of liquid crystal are seen. The textures have different domains within the material. Each domain corresponds with a different way that the liquid crystal molecules are oriented. The existence of different kinds of defects gives rise to the characteristic features in the textures of liquid crystals.

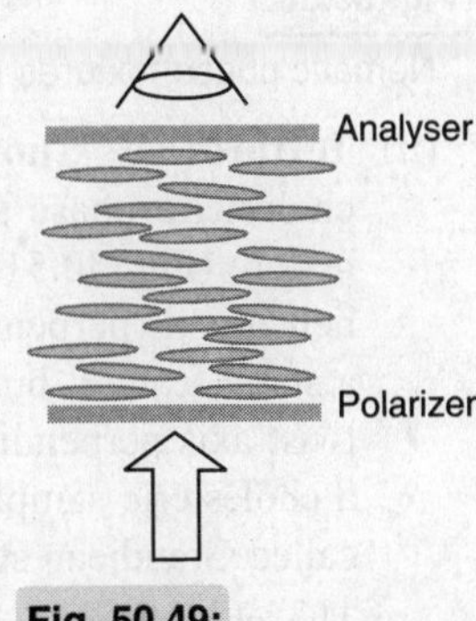

Fig. 50.49: Liquid Crystal Textures

(*i*) **Textures of nematic phase:** Several types of nematic textures are observed – mono-domain texture, Schlieren texture, threaded texture, marbled textures etc. A homogenously aligned nematic phase does not exhibit specific textures. When observed between crossed polarizers and whenever the optic axis (the director) is oriented along one of the polarizer directions, the sample appears black. Maximum intensity is observed when the optic axis is oriented at 45° between polarizer and analyzer. In homeotropic arrangement, surface treatments of the glass slides produces layers in which the director is arranged perpendicular to the slides and the sample appears black without exhibiting specific textures in any orientation of the slides.

Nematics between untreated glass plates, orient with their director parallel to the substrates. If this orientation is not homogeneous, but varies slowly in the plane of the substrate, ***Schlieren textures*** (threaded textures) are observed. Schlieren textures display dark brushes which have irregular curved shapes [Fig. 50.50 (a)]. At certain points these dark brushes meet and these points indicate *disclinations* in the structure.

Nematic ***marbled textures*** are obtained when the glass slide surfaces are not treated for alignment or with thin specimens. They consist of different domains with different molecular orientation (Fig. 50.50 b). On observing under the crossed polarizers they appear as different domains each having a constant coloured interference patterns. Different areas of the sample appear quite uniform but may slightly change in colour, due to a change in birefringence from variation of the director.

Sometimes on cooling a liquid crystal in isotropic melt state, the nematic phase usually begins to separate at the clearing point in the form of *nematic droplets*. The sample forming the nematic droplets has a coexistence of two phases, isotropic and nematic, with the nematic droplets being surrounded by the isotropic phase (Fig. 50.50 c). Fig. 50.50 shows some of the Nematic phase textures.

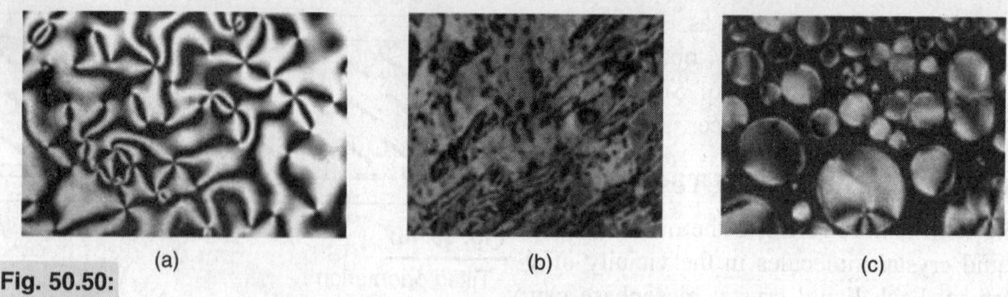

Fig. 50.50:

Nematic phase textures (a) Schlieren, (b) marble and (c) Nematic droplets

(*ii*) **Textures of Cholesteric phase:** The most commonly observed textures of the cholesteric phase spread between two untreated glass substrates are the *oily streaks* texture (Fig. 50.51 a). Here the director is oriented parallel to the glass plates and the helical axis perpendicular to the glass plates. The oily streaks network is generally not a static feature, but may coarsen with time, leaving a uniformly oriented sample with twist axis perpendicular to the substrate plane. *Grandjean textures* are observed when a cholesteric sample is placed in a wedge cell (Fig. 50.51 b). They exhibit dark lines called Grandjean steps, periodically.

The cholesteric fan-like textures are observed for strongly twisted materials, while cholesteric *fan-shaped textures* that exhibit *focal conics* are found for materials with a slightly smaller twist (Fig. 50.51 d). *Polygonal cholesteric textures* are characterized by the formation of a network of polygonal structures (Fig. 50.51 e). Fig. 50.51 shows Cholesteric phase textures.

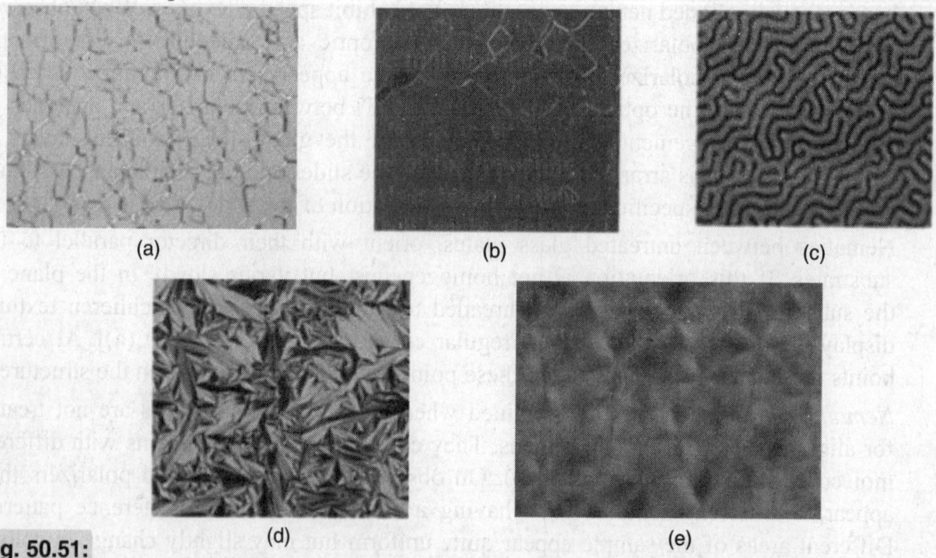

Fig. 50.51:

Cholesterictextures (a) Oily streaks, (b) Grandjean, (c) Fingerprint, (d) Focal conic, (e) Polygonal

(*iii*) **Textures of blue phase:** The appearance of the BPIII phase is hard to detect by polarizing microscopy, as generally only a dark bluish to black, foggy image without any structure is observed. BPII and BPI appears in the form of platelet textures. These platelets exhibit colours between crossed polarizers, but also with one polarizer or no polarizers at all (Fig. 50.52 a). A sample of one of the blue phases will usually have regions in which the defect lattices are oriented in different directions. The different colours are due to the different orientations of the system of double twist cylinders *i.e.* different "crystal planes" are observed, which show Bragg scattering at different wavelengths, because the condition for constructive interference depends on the distance between lattice planes and their orientation with respect to the direction of incident light. Even though they are called Blue Phases, different colours may be observed, depending on the wavelength of the selective reflection (Fig. 50.52).

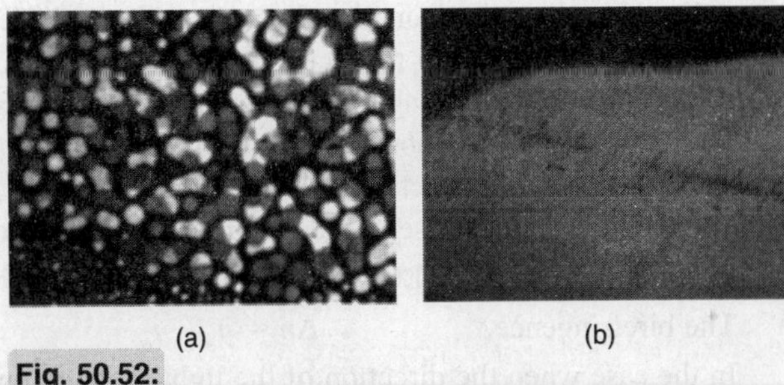

Fig. 50.52: Blue phase" in cholesterics (a) BPI platelets, (b) BPIII, the fog phase

(*iv*) **Textures of smectic phase:** Smectic A and C phases are the most common smectic phases. In homeotropic alignment where the layers are parallel to the glass slides, the molecules are oriented perpendicularly to the planes, and therefore, the light in the polarizing microscope is propagating parallel to the optic axis in the sample. In well oriented sample cells the light is almost completely extinguished and small bright ribbons usually appear only at the edges of the specimen or around air bubbles.

In unaligned samples, the direct transition to a fluid smectic phase is often characterized by the growth of elongated structures, called smectic *bâtonnets*. The most commonly observed natural SmA appearance is that of the *fan-shaped* texture. Its topology consists of focal domains Fig. The fan-shaped or focal conic textures are preferably observed in rather thin sample preparations. The chiral smectic C (SmC*) phase exists as striated fan shaped texture if the smectic texture preceding it exhibits a simple fan shaped texture (Fig. 50.53).

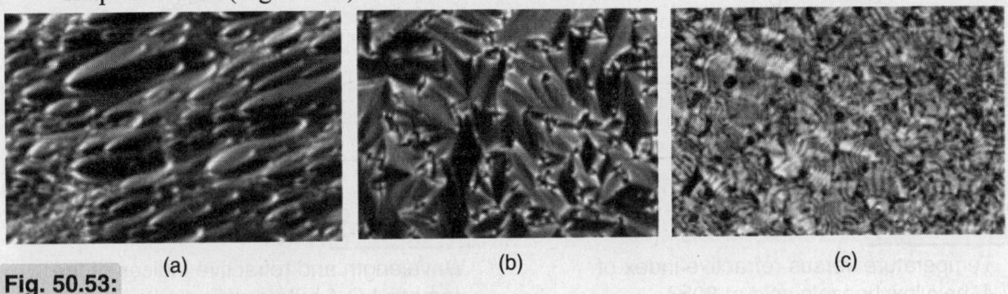

Fig. 50.53: Textures of smectic phase (a) focal conic domains, (b) focal conic fan texture (c) fan texture of smectic C* phase

50.11.4 Optical Anisotropy

Liquid crystals due to their shape and polarization anisotropy are birefringent. They exhibit different properties for light traveling with the electric vector propagating parallel and perpendicular to the director or optic axis. The electric vector of incident plane polarized light entering a liquid crystalline phase is split into two components, ordinary (o) and extraordinary (e) rays. The electric field of the o-ray is always perpendicular to the optic axis, so its refractive index n_o is a constant independent of propagation direction. The electric field of e- ray lies in a plane that contains the optic axis, so its refractive index $n_e(\theta)$ varies with the ray propagation angle with respect to the optic axis. Thus the birefringence of the medium, $\Delta n(\theta)$, depends on the propagation direction and is defined as $\Delta n(\theta) = n_e(\theta) - n_o$. Usually, ne $> n_o$ and, therefore, Δn is positive ($\Delta n > 0$) and varies in the range from values close to zero to about 0.4. Interference between the e-ray and the o-ray that travel through the medium with different velocities gives rise to the coloured appearance of these thin films.

In the case of uniaxial LCs the optic axis coincides with the director *n*.

The birefringence $$\Delta n = n_e - n_o \quad ...(50.15)$$

In the case when the direction of the light propagation is tilted with respect to the optical axis, the refractive index for the ordinary wave is equal to n_o, but the refractive index for the extraordinary wave is equal to some effective value given by

$$n_{eff}^2 = \frac{n_e^2 n_o^2}{n_e^2 \cos^2\theta + n_o^2 \sin^2\theta} \quad ...(50.16)$$

where θ is an angle between the optical axis and the light propagation direction.

Then birefringence $$\Delta n = n_{eff} - n_o \quad ...(50.17)$$

In the case of propagation along the optical axis, the refractive index is n_o for both modes, ordinary and extraordinary. In general, birefringence Δn of LCs decreases as the wavelength of the incident light or the temperature increases, $\Delta n \propto S(T)$.

Fig. 50.54 (a) shows relation between temperature and refractive index with Tc as clearing temperature, n_{iso} as refractive index of isotropic. Fig. 50.4 (b) shows relation between wavelength and refractive indices.

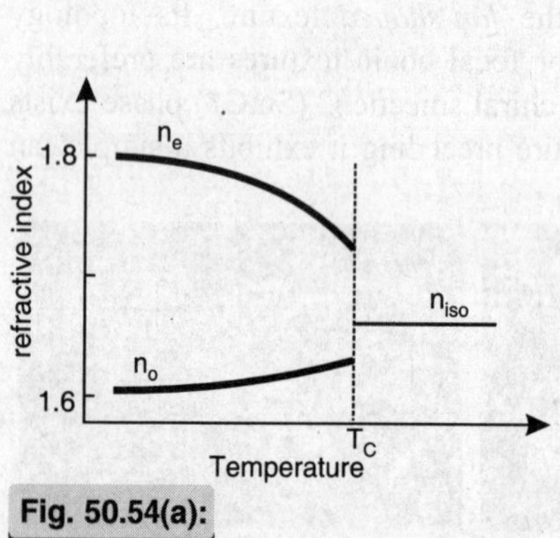

Fig. 50.54(a): Temperature versus refractive index of 4′-hexyloxybenzoic acid at 80°C

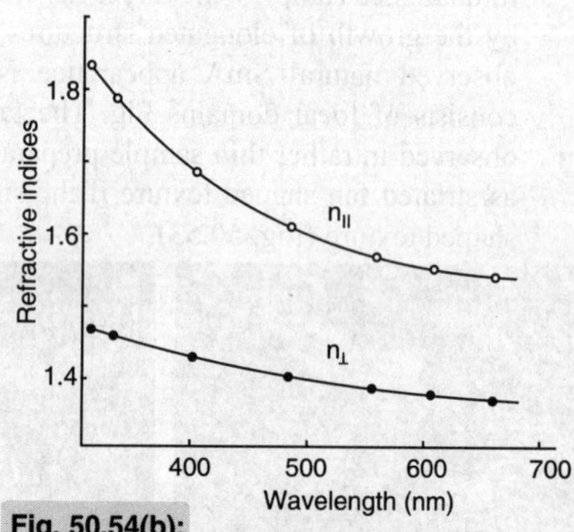

Fig. 50.54(b): Wavelength and refractive indices of the nematic LC 4-butoxyphenyl ester

The unpolarized light incident upon a LC is split into the ordinary and extraordinary waves, which travel with different velocities through the material. They emerge from the LC with some phase difference that depends on the thickness of material *d*

$$\Delta\varphi = 2\pi\, \Delta n\, d/\lambda \quad ...(50.18)$$

where λ is a light wavelength.

Thus, varying the thickness of the LC material produces different phase shifts and the outcoming light polarizations.

50.11.5 Dielectric Anisotropy

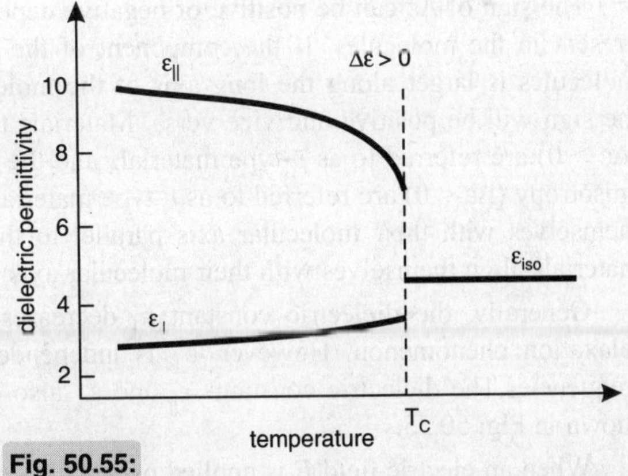

Fig. 50.55: Temperature dependence of dielectric permittivity in liquid crystals with $\Delta\varepsilon > 0$: T_c - clearing temperature; ε_{iso} - permittivity of isotropic liquid

Dielectric properties of liquid crystals are related to the response of liquid crystal molecules to the application of an electric field. Permittivity is a physical quantity that describes how an electric field affects and is affected by a dielectric medium and is determined by the ability of a material to polarize in response to an applied electric field, and thereby to cancel, partially, the field inside the material. In the liquid crystal materials consisting of non-polar molecules, there is only an induced polarization, which consists of two parts: the electronic polarization (which is also present at optical frequencies) and the ionic polarization. In the liquid crystals with polar molecules, there is in addition to the total induced polarization, the orientation polarization, due to the tendency of the permanent dipole moments to orient themselves parallel to the field.

Considering the uniaxial liquid crystal phases in a macroscopic coordinate system *x*, *y*, *z*, with the *z* axis parallel to the director **n**, it is possible to distinguish two principal permittivities, parallel to the director $\varepsilon_{\parallel} = \varepsilon_{zz}$, and perpendicular to the director $\varepsilon_{\perp} = 1/2(\varepsilon_{xx} + \varepsilon_{yy})$. Then the dielectric anisotropy $\Delta\varepsilon = \varepsilon_{\parallel} - \varepsilon_{\perp}$ can take positive and negative values. The graph of temperature dependence of dielectric permittivities for a typical liquid crystal (Fig. 50.55) shows that magnitude of $\Delta\varepsilon$ usually depends on temperature.

$$\Delta\varepsilon \propto S(T)$$

For non-polar liquid crystals the situation is relatively simple but in the case of polar molecules with permanent dipole moments, the orientation polarization must be added, which, by Maier and Meier, leads to an expression for the dielectric anisotropy

$$\Delta\varepsilon \propto \left[\Delta\alpha + \frac{\mu^2 F}{2k_B T}(3\cos^2\beta - 1)\right] \quad \text{...(50.19)}$$

where $\Delta\alpha$ is a polarizability anisotropy

μ is a molecular dipole moment

F is parameter depending on the reaction field factor

k_B is a Boltzmann constant and

T is temperature

Then $\Delta\varepsilon$ is determined by the angle β between the molecule axis and the direction of the off-axis molecular dipole moment μ. The dipole contribution to $\Delta\varepsilon$ is positive for $\beta < 54.7°$ and negative for $\beta > 54.7°$, which may result in negative $\Delta\varepsilon$ depending on the relative magnitude of two contributions. The larger the anisotropy the smaller electric field is needed to make the liquid crystal respond to it.

The anisotropy of the dielectric permittivity

$$\Delta\varepsilon = \varepsilon_{\parallel} - \varepsilon_{\perp} \quad \text{...(50.20)}$$

The sign of $\Delta\varepsilon$ can be positive or negative depending on the permanent dipole moments present in the molecules. If the component of the net dipole moment of the liquid crystal molecules is larger along the long axis of the molecule than that along the transverse axis the sign will be positive and vice versa. Materials that exhibit positive dielectric anisotropy ($\Delta\varepsilon > 0$) are referred to as *p*-type materials and the materials that exhibit negative dielectric anisotropy ($\Delta\varepsilon < 0$) are referred to as *n*-type materials. The *p*-type LC materials tend to align themselves with their molecular axis parallel to the applied electric field, whereas *n*-type materials align themselves with their molecular axis perpendicular to the applied field.

Generally, the dielectric constant $\varepsilon_{\parallel}$ decreases with increasing frequency due to the relaxation phenomenon. However, $\varepsilon_{\perp}$ is independent of frequency over a large range of frequencies.The dielectric constants $\varepsilon_{\parallel}$ and $\varepsilon_{\perp}$ also change as a function of temperature, as shown in Fig. 50.55.

When an electric field E is applied over a dielectric medium such as a liquid crystal, the field will polarize the medium, an electric dipole moment **P** will be induced, proportional to the field strength and to the permittivity of the medium. Since a liquid crystal is dielectrically anisotropic the polarization and the field will be directed differently for any director orientation that is not along **E**.

A dielectric torque

$$\Gamma_{\varepsilon} = \mathbf{P} \times \mathbf{E} \qquad ...(50.21)$$

acts on the director, turning to align the material along or perpendicular to the field.

50.11.6 Diamagnetic anisotropy

Liquid crystals are very sensitive to a magnetic field. They are diamagnetic, like most organic materials. The application of an external magnetic field to a mesophase results in magnetization **M** which is defined in terms of magnetic induction **B** and the magnetic field strength **H** as

$$\mathbf{M} = \mu_o (\mathbf{B} - \mathbf{H}) \qquad ...(50.22)$$

where, μ_o is the permeability of free-space

The diamagnetic properties of uniaxial mesophases can be described by the two susceptibilities and $\chi_{\parallel}$ and $\chi_{\perp}$, which are the resulting susceptibilities when the magnetic field is applied parallel and perpendicular to the molecular long-axis or the director, respectively. The anisotropic diamagnetic susceptibility anisotropy is defined as

$$\Delta\chi = \chi_{\parallel} - \chi_{\perp} \qquad ...(50.23)$$

50.11.7 Electrical conductivity

The electrical conductivity of liquid crystals is mainly due to residual ionic impurities, the nature of which is usually unknown. The conductivity is usually of the order of 10^{-9} $(\Omega\text{cm})^{-1}$. The conductivity of the material mainly depends on temperature and resistance of the material. In liquid crystals the conductivity increases because of residual ionic impurities.

The anisotropic conductivity is

$$\Delta\sigma = \sigma_{\parallel} - \sigma_{\perp} \qquad ...(50.24)$$

where, $\sigma_{\parallel}$ and $\sigma_{\perp}$ are the principal components of conductivity parallel and perpendicular to the director.

Usually for nematics $\Delta\sigma$ is positive as the ions can move more freely along the long axis than perpendicular to it.

In smectics usually $\Delta\sigma$ is negative as it is easier for the ions to flow in the layers than perpendicular to the layers.

50.11.8 Elastic Constants

In undeformed liquid crystals the director **n** points in the same direction throughout the liquid crystal. When an external force exerts and exceeds a certain value, deformation will take place. A deformed liquid crystal is one in which the director changes its direction from point to point. The relative position of the liquid crystal molecules will be changed. They were forced to splay, twist and bend until equilibrium. When the system is in equilibrium, it is in minimum energy status. Generally speaking, the deformations in liquid crystals can be described in terms of three basic types of deformation: *splay, twist* and *bend* (Fig. 50.56). Not only external field can cause these deformations, the alignment of the substrates also can cause these deformations. For example, wedged cells makes the liquid crystal molecules arrange themselves in splay and TN cell makes the molecules arranged in twist.

The first is a "splay," where molecules are spread by external stress, the second is a "twist" where molecules are twisted by an external stress, and the third is a "bend" where molecules are bent by an external stress.

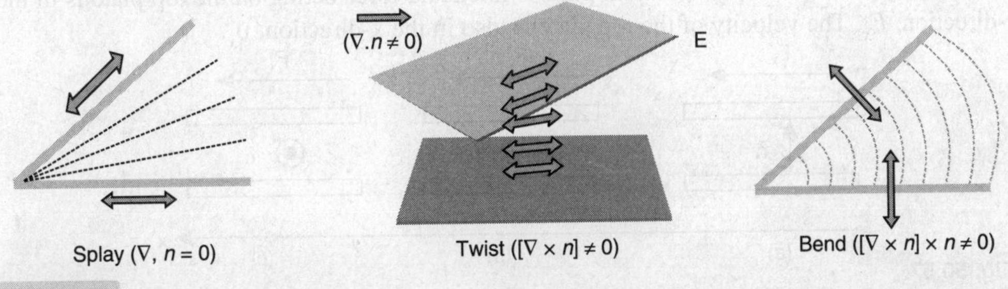

Fig. 50.56:

Representations of elastic deformation for a nematic material

The relationship between the deformation and stress is expressed respectively by splay elasticity, k_{11}, twist elasticity, k_{22} and bend elasticity k_{33}. k_{11}, k_{22} and k_{33} are known as Frank's elastic constants. The elasticity of liquid crystals is of the order of 10^{-6} to 10^{-7} dyne and is much lower than that of ordinary elastic material. This makes alignment modification of liquid crystals by the application of electrical field, magnetic field and external stress much easier. The elastic contributions to the free energy are given through Frank's elastic energy equation for a nematic liquid crystal system

$$F = \frac{1}{2}\left[K_{11}(\Delta.n)^2 + K_{22}(n.\nabla \times n)^2 + K_{33}(n \times \nabla \times n)^2\right] \quad ...(50.25)$$

where k_{11}, k_{22} and k_{33} are comparable to the traditional spring constant in Hooke's Law and depend mainly on the molecular characteristics of a given material.

The elastic constants are molecular parameters and describe the restoring forces on a molecule within a liquid crystalline phase in response to an external force that distorts the medium from its lowest energy configuration. The elastic constants co-determine the spatial and temporal response of the director to the applied external electric and magnetic fields. The equilibrium position is then restored upon removal of the field by elastic forces that originate at the interface between the LC and the orientation layers that cover the device substrates.

More important than their individual values is the ratio of bend/splay (k_{33}/k_{11}), which needs to be low for optimum electro-optical characteristics with typical values in the range 0.6 to 0.8. The values and ratio of the elastic constants are more difficult to obtain through molecular engineering than are the viscosities, and the optical and dielectric anisotropies.

50.11.9 Viscosity

The viscosity of any material is a collective property resulting from the interaction of the molecules with one another. The flow velocity in the liquid crystalline state is also an anisotropic property that depends on the direction of flow of individual molecule with respect to the director at any one point within the medium. Liquid crystals are fluids that show anisotropy in their flow behaviour. Three parameters are required to characterize the nematic state as the constituent molecules are anisotropic. These are η_1, which is perpendicular to the direction of flow, but parallel to the velocity gradient; η_2, which is parallel to the direction of flow, but perpendicular to the velocity gradient and η_3 which is perpendicular to the direction of flow and to the velocity gradient. This can be understood by imagining measuring the viscosity of a liquid crystal by placing it between two flat plates and measuring the force necessary to slide one plate over the other. In (Fig. 50.57), the plates lie in *xy* plane and are separated by a distance *d*. The bottom plate is fixed and force acting on the top plate is in the *x*-direction, F_x. The velocity of the top plate is also in the x-direction, υ_x.

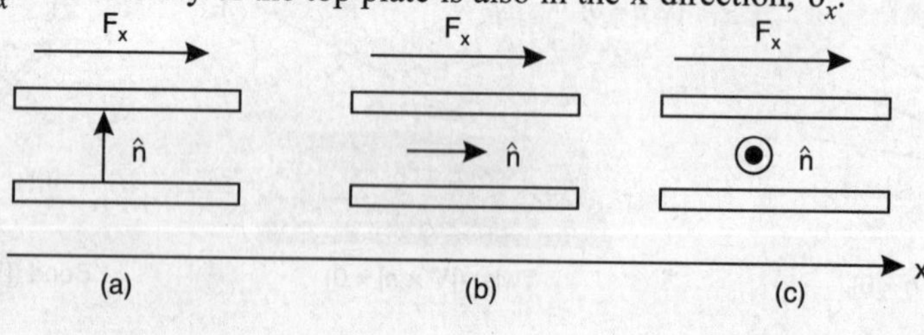

Fig. 50.57: Measuring the three viscosities of nematic liquid crystals

The force necessary to slide the plates over each other is different for each case, being nearly the same cases for (b) and (c) but much higher for case (a). No such effect exists for a liquid placed between the two plates. The resulting three viscosities, called the Miesowics coefficients, are η_1, η_2, and η_3. Individual viscosity coefficients influence the optical response times in an electro-optic display device.

50.11.10 Electric and Magnetic Field Effects

The response of liquid crystal molecules to an electric field is the major characteristic utilized in industrial applications. The ability of the director to align along an external field is caused by the electric nature of the molecules. Permanent electric dipoles result when one end of a molecule has a net positive charge while the other end has a net negative charge. When an external electric field is applied to the liquid crystal, the dipole molecules tend to orient themselves along the direction of the field. In the Fig. 50.58, the arrows **E** represent the electric field vector and the arrows **F** show the electric force on the molecule.

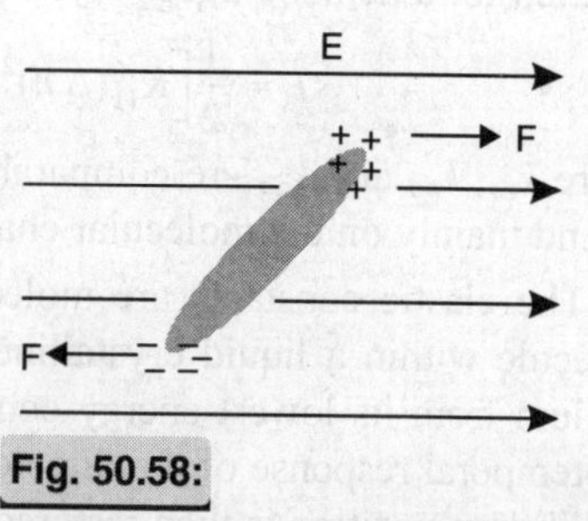

Fig. 50.58: Influence of external electric field

Even if a molecule does not carry a permanent dipole, it can still be influenced by an electric field. In some cases, the field produces slight re-arrangement of electrons and protons in molecules such that an induced electric dipole results. While not as strong as permanent dipoles, orientation with the external field still occurs.

The effects of magnetic fields on liquid crystal molecules are analogous to electric fields. Because magnetic fields are generated by moving electric charges, permanent magnetic dipoles are produced by electrons moving about atoms. When a magnetic field is applied, the molecules will tend to align with or against the field.

Fredrik's transition: Fredericks transition refers to a transition from a uniform director configuration to a deformed director configuration of a homeotropically or planarly aligned nematic liquid crystal (NLC) layer upon application of a magnetic or electric field of sufficiently large magnitude. The behavior of a nematic liquid crystal is determined by two competing torques: (1) the elastic torque within the material that tries to align the NLC molecules along the easy direction as determined by the anchoring boundary conditions; and (2) external-field-induced torque that tends to align the NLC molecules either parallel or orthogonal to the field. A uniform director distribution occurs when only one of the torques is present. When both torqueses exist, there is a competition between them. Once the applied field strength exceeds a threshold value, the uniform NLC orientation will deform and the Fredericks transition (Fig. 50.59) takes place.

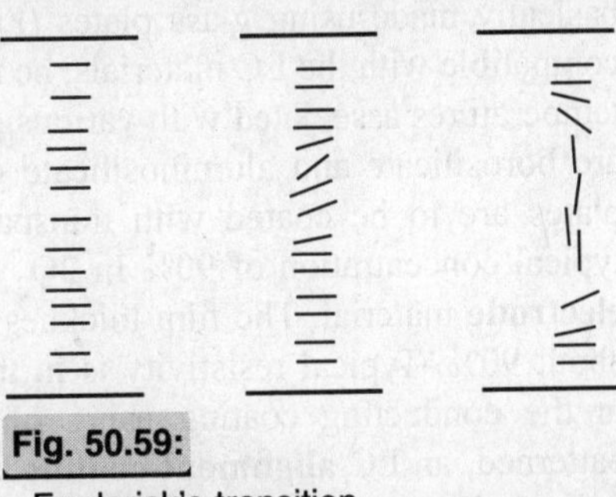

Fig. 50.59: Frederick's transition

The Freedericks transition is simply a transition from a uniform director configuration to a deformed director configuration; *i.e.*, at any point in the LC, the order of the molecules relative to one another remains the same.

The threshold electric field, E_c

$$E_c = (\pi/d)\,(K_{ii}/|\Delta\varepsilon|)^{1/2} \qquad \text{...(50.26)}$$

Thus the threshold voltage V_{th} of the LC electro-optic effect is given by

$$V_{th} = E_c d = \pi\,(K_{ii}/|\Delta\varepsilon|)^{1/2} \qquad \text{...(50.27)}$$

where d is the thickness of the LC and k_{ii} is the appropriate elastic constant.

When the field is perpendicular to the initially homogeneous orientation of the director, $K_{ii} = K_{11}$ or K_{22}.

When the field is parallel to the initially homogeneous orientation, $K_{ii} = K_{33}$.

In the case of a twisted orientation, $K_{ii} = K_{11} + (K_{33} - 2K_{22})/4$

The Fredericks transition plays a fundamental role in the operation of many liquid crystal displays because the director orientation can be controlled easily by an external magnetic or electric field. One striking feature is that the magnitude of the Fredericks threshold voltage is essential to the operation of many liquid-crystal devices. If the Fredericks threshold voltage could be reduced reliably, this would make it possible for the manufacture of very low-power liquid crystal displays and other liquid-crystalline devices.

50.12 APPLICATIONS OF LIQUID CRYSTALS

50.12.1 Liquid Crystal Displays (LCDs)

LCDs are light valves. The principle of operation of liquid crystal display is radically different from all other display devices. LCDs are light modifiers, not light producers. All the other devices are self-illuminating as they produce their own light. The LCD does not make its own light, but operates by modifying light from other sources. Therefore, power consumption is low in the LCD. The external light modified by the LCD may be ambient light or a special light source installed within the device just to supply the LCD some light to modify.

50.12.2 Fabrication of LCDs

There are several types of LCDs utilizing different LC materials and LCD modes. The general display assembly processes and materials are very similar for all these LCDs. An LCD is basically made using glass plates (Fig. 50.60 a). The *glass substrate* should be chemically compatible with the LC materials, be flat, and maintain dimensional stability under processing temperatures associated with various display fabrication steps. Typical display glass materials are borosilicate and aluminosilicate glasses, with a thickness of 0.7 or 1.1 mm. The glass plates are to be coated with transparent conducting film. Indium tin oxide (ITO) with a typical concentration of 90% In $2O_3$ and 10% SnO_2 is used as the **transparent conducting electrode** material. The film thickness is typically 50 to 300 nm and light transmission of is about 90%. Typical resistivity is in the range of 2×10^{-4} W. cm. Electrodes are **patterned** in the conducting coating using **photolithoprocess**. After the transparent electrodes are patterned, an LC **alignment coating** is applied. The most commonly used TN LCD requires a tilted **alignment** to eliminate reverse tilt disclinations. Inorganic films such as an obliquely evaporated SiO, as well as organic materials such as polyimide, can be used as alignment materials. For TN LCD, the alignment process is selected to achieve a pretilt angle of around 2 to 5°.

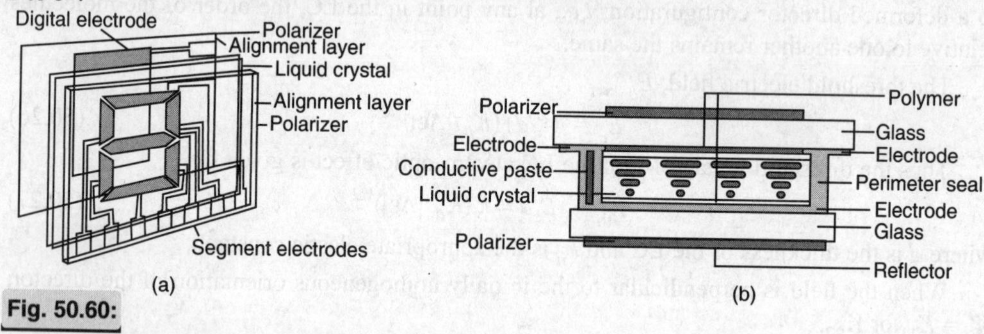

Fig. 50.60: Fabrication of LCDs

Spacing materials are applied in the active area of the display as well as in the peripheral area, where the two glass substrates are **sealed**. Typical spacing materials include glass or plastic fibers (cylinders) or spheres (balls), with a tightly controlled size(diameter) distribution. A sealing adhesive material such as an epoxy is then applied at the seal area with an opening (fill hole) left for LC material injection after the sealing process. After the spacers and seal materials are applied on the first(active) substrate, it is precisely aligned to the second substrate, and the seal is cured. After the empty cell is fabricated, it is filled with the **LC material**. Because the cell thickness (spacing) is small (~5 to 10 μm), it is filled using special techniques. After the cell is completely filled, the fill hole is capped by using an epoxy adhesive that is chemically compatible with the LC material.

The **external components** of an LCD include polarizers, reflectors, and a backlight assembly (Fig. 50.60 b). A polarizer is attached on the front as well as back surfaces of the LCD after the cell assembly is complete. Also, in the most commonly used normally white mode TN LCD, the polarizers are attached with their polarization axis crossed.

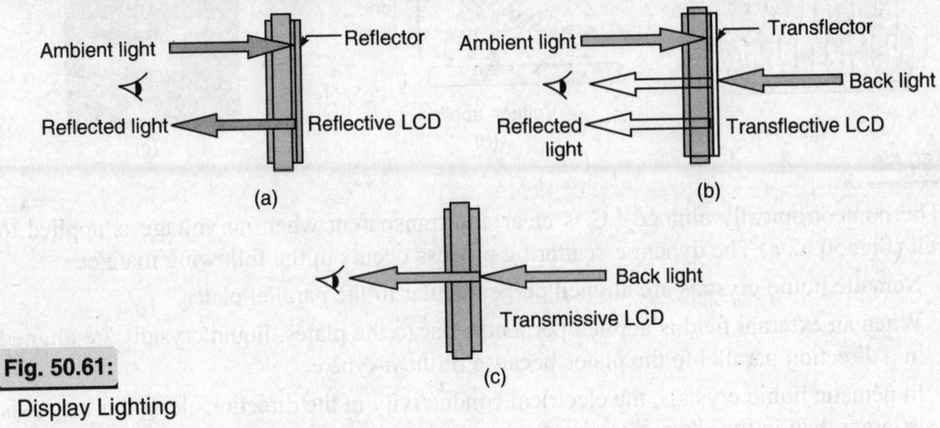

Fig. 50.61: Display Lighting

50.12.3 Modes of Operation of LCDs

LC displays based on many different modes of operation have been developed. Historically, the phase change (PC) effect was discovered first in 1968 by Wysoki et al. The same year, dynamic scattering (DS) mode and guest–host (GH) mode were announced by Heilmeier et al. Then in 1971, the TN mode was reported by Schadt and Helfrich and electrically controlled birefringence (ECB) was reported by Schiekel and Fahrenschon and Hareng et al. The DS effect is based on the an isotropy of the conductivity. Because of higher voltage operation and higher power consumption, the DS mode is not currently used. The TN and STN effects are most widely used among all the LCDs.

Dynamic Scattering mode LCD: The display cell consists of two glass plates each coated with tin oxide (SnO_2) on the inside with transparent electrodes separated by a liquid crystal layer 5 to 50 micrometer thick. The oxide coating on the front sheet is etched to produce a single or multi segment pattern of characters with each segment properly insulated from each other.

LC displays can be realized with DSM effect in the transmissive and reflective mode of operation. The driving voltages required for light scattering are in the range of several ten volts and the non-trivial current is depending on the area of the activated segments. The DMS effect was thus not suited for battery powered electronic devices.

Because the DSM LCDs suffered from relatively high power consumption, limited life, and poor contrast, the search continued for a workable LCD. James Fergason at Kent State invented an improved display based on the twisted nematic field effect in 1969.

A Dynamic Scattering mode cell requires the following ingredients:

(*a*) a liquid crystal with negative dielectric anisotropy (aligns the LC long axis perpendicular to the electric field)

(*b*) homeotropic alignment of the LC (*i.e.* perpendicular to the substrate planes)

(*c*) doping of the LC with a substance that increases the conductivity of the LC to allow a current to flow.

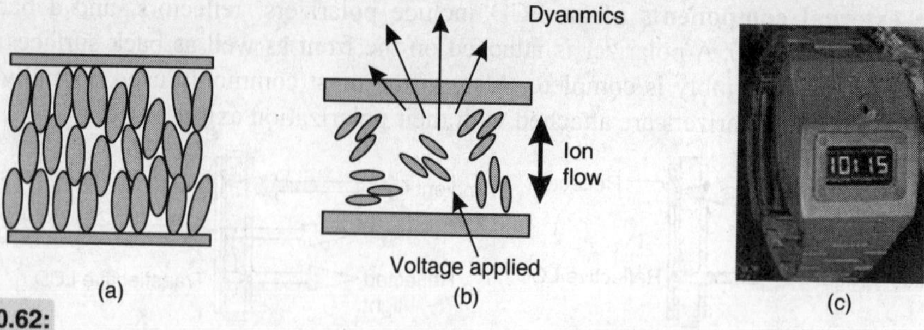

Fig. 50.62:

The homeotropically aligned LC is clear and transparent when no voltage is applied to the cell (Fig. 50.62 a).The dynamic-scattering process occurs in the following manner.

(*i*) Nematic liquid crystals are aligned perpendicular to the parallel plates.

(*ii*) When an external field is applied perpendicular to the plates, liquid crystals are aligned in a direction parallel to the plates because of the *n*-type.

(*iii*) In nematic liquid crystals, the electrical conductivity in the direction along the long axis is larger than in the short-axis directions, which causes charge buildup.

(*iv*) The induced field and external field give rise to a shear torque on the molecules, which causes a circular motion of the molecules. With increasing voltage and current, the electric field is trying to align the long molecular axis of the LC perpendicular to the field while the ion transport through the layer has the tendency to align the LC perpendicular to the substrate plates. As a result, a striped repetitive pattern is generated in the cell, of which the building blocks are named "Williams domains".

(*v*) By further increasing the external field, the molecules go into a mechanically unstable state or turbulence (Fig. 50.62 b). The molecules now randomly scatter the light incident on the molecules and appear milky white. Upon further increase of the voltage the regular pattern is replaced by a totally turbulent state which is strongly scatters light (Fig. 50.62 b). Fig. 50.62 c shows a wrist-watch with a DSM LCD.

50.12.4 Twisted Nematic LCD

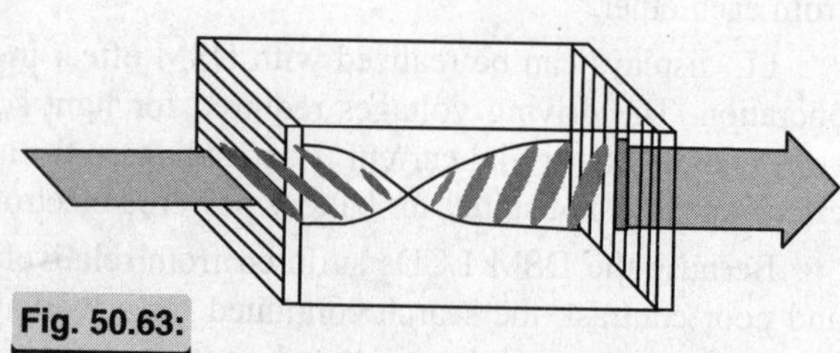

Fig. 50.63: Twisted nematic arrangement in the liquid crystal cell

Twisted Nematic LCD consists of nematic LC material with a positive dielectric anisotropy with a layer thickness of about 5 μm, sandwiched between two transparent electrodes. The surfaces of the transparent electrodes are coated with a polyimide alignment layer and rubbed to orient LC molecules at the substrate surfaces along the rubbing direction with a small (~3°) pretilt angle. The molecules on the two substrates are oriented 90° from each other as shown in Fig. 50.63; i.e., the LC molecular axis rotates (twists) continuously through 90° from the first substrate to the second substrate. It is called a twisted molecular arrangement. Therefore the liquid crystal display device is called a twisted nematic liquid crystal display (TN LCD). The two sides of LCD are attached with crossed polarizers with their polarization direction being parallel to the LC director orientation on that substrate.

Working: Fig. 50.64 is a schematic diagram of a TN LCD. Ambient light enters from the left and is polarized by a polarizing filter. The polarized light passes through the front glass wall of the display, then through a transparent, electrically conductive coating on the glass, and into the liquid crystal. As the linearly polarized light propagates through the liquid crystal, the optical vector is rotated through 90° by the twisted molecular arrangement. At the back of the display, the light passes through another electrically conductive coating, glass plate, and polarizing filter. This rear polarizing filter is placed with its axis at 90 degrees to the front filter. The polarized light passes through this rear filter, because its polarization was also rotated by 90 degrees by the liquid crystal. At the back of the LCD is a mirror that reflects the light back through the cell. The light retraces its path, and an observer sees the display as relatively bright.

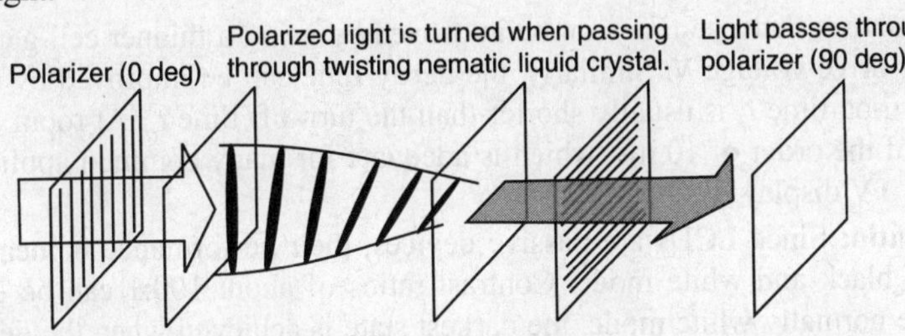

Fig. 50.64: A schematic diagram of TN LCD

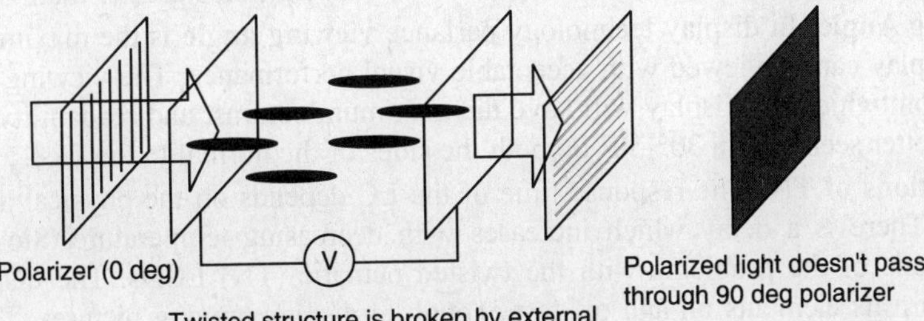

Fig. 50.65: A schematic diagram of TN LCD when an electric potential is applied

When an electric potential is applied (Fig. 50.65) between the two conductive coatings, the resulting electric field affects the twist in molecular arrangement and align the molecules in the liquid crystal along the field direction. When this happens, the plane of polarized light passing through the cell is no longer rotated by 90 degrees, and it cannot pass through the rear filter. Therefore, it is no longer reflected back to an observer, and the area appears dark. By the selective charging of areas of the conductive coating, patterns of dark digits or letters against a bright background are formed. Fig. 50.66 shows voltage-transmission characteristics of a TN cell for NB and NW modes of operation.

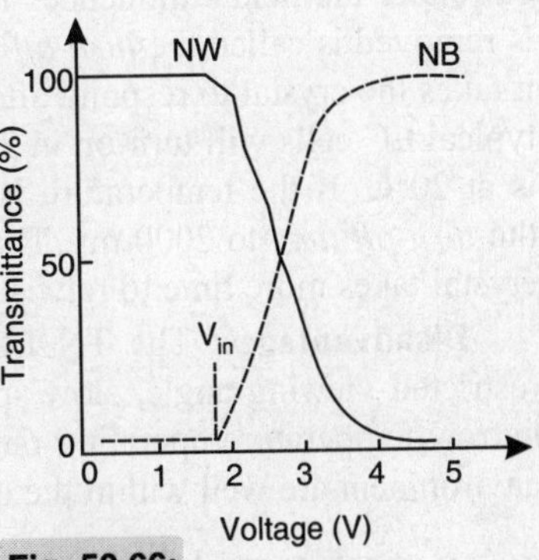

Fig. 50.66: Voltage-transmission characteristics of a TN cell

The threshold voltage of a TN mode LCD is given by

$$V_{th} = \pi.\sqrt{\{[K_{11} + (K_{33} - 2K_{22})/4]/\varepsilon_0.\Delta\varepsilon} \quad ...(50.28)$$

V^{th} depends on the dielectric anisotropy and elastic constants, and is generally in the range of 2 to 3 V, with the maximum operating voltage being in the range of 5 to 8 V. This low voltage driving, coupled with low current due to high resistivity of LC materials, contributes to the very low power consumption (~1 mW/cm^2) of LCDs.

The response times measured by the rise and decay times of the display are

$$\tau_r = \gamma d^2/\Delta\varepsilon\,(V^2 - V_{th}^2)$$

$$\tau_d = \gamma d^2/\,(\Delta\varepsilon.V_{th}^2) \quad ...(50.29)$$

where γ is the rotational viscosity coefficient.

Eq. (50.28) shows that rise time can be improved by using a thinner cell gap d, a higher $\Delta\varepsilon$, and a higher drive voltage V. Similarly, the decay time can be improved by reducing the cell gap. The turn-on time t_r is usually shorter than the turn-off time t_d. At room temperature, these times are of the order of 10 ms, which is adequate for many common applications such as computer and TV displays.

Contrast Ratio: Since LCDs are passive devices, their performance is measured in the contrast ratio of black and white mode. Contrast ratios of about 100:1 can be expected for this mode. In the normally white mode, the darkest state is achieved when the device is fully turned on and the transmitted of light is determined by alignment and efficiency of polarizers. For the normally white mode it is possible to obtain contrast ratios greater than 1000:1.

Viewing Angle: In display technology parlance **viewing angle** is the maximum angle at which a display can be viewed with acceptable visual performance. The viewing angle is the direction from which the display will have the maximum contrast and readability. The visual display is better seen within 30°-50° on both the sides of the normal to the face.

Limitations of TN: The response time of the LC depends on the physical alignment of molecules. There is a delay which increases with decreasing temperature. Slow operating speed was one of the problems with the twisted nematic (TN) LCDs. The display simply could not turn its elements on and off fast enough to display moving pictures. The speed of any particular LC cell operation is greatly dependent on its temperature and drive method. The time it takes the LC cell to respond to this electric field is called the *turn-on time*. Turn-on time is the sum of two factors. One factor is the turn-on delay before the electric field is established around the crystal, and the other is the rise time it takes the crystal to straighten out under the field's influence. The time X takes the crystal to retwist after the electric field is removed is called the *turn-off time*. Turn-off time is the sum of turn-off delay and the time it takes the crystal to respond after the field is gone (called decay time). In terms of real time, typical LC cells will turn-on in 20 ms and turn-off in 100 ms when the LC cell's temperature is at 20°C. If the temperature is lowered to 0°C, the turn-on time increases to 90 ms, and the *turn-off time* to 2000 ms. The longer turn-off time than turn-on time is simply the liquid crystal takes more time to retwist itself after the electric field is removed.

Disadvantages: The TN LCD suffers from several disadvantages, among which are, restricted viewing angle, slow speed only when displaying moving graphics (like TV), and narrow temperature operating range. However, the operating temperatures within the human environment are well within the display's range.

50.12.5 High Twisted Nematic (HTN) LCDs

HTN (High Twisted Nematic) displays are based on a higher molecular twist (usually 110°) than TN (90°) and therefore offer wider viewing angles and improved contrast. In fact, these HTN products offer viewing characteristics close to those of STN technology. As low operating voltage as 2.5V and marginal extra cost over TN means that the products are well suited to hand-held applications.

50.12.6 Super Twisted Nematic (STN) LCDS

Although Twisted Nematic LCDs may be driven in a time multiplexed fashion to increase the amount of information displayed, they are restricted in terms of reduced contrast and limited viewing angle because change in on/off voltage is small. This problem was eliminated with the invention of the supertwisted nematic (STN) LCD in the early 1980s. Super Twisted Nematic LCDs have a twist that is greater than 90° but less than 360°. Currently most STN displays are made with a twist between 180° and 270°. Fig. 50.67(a) illustrates the STN effect. In this type of display, the LC material undergoes a greater than 90° twist from plate to plate; typical values range from 180 to 270°. To sustain a twist angle greater than 90° requires a nematic liquid crystal with an intrinsically twisted structure known as a chiral nematic. Chiral nematics are ordinary nematic liquid crystals doped with a few percent of optically active material. The higher twist angles cause steeper brightness-voltage curves which put the on and off voltages closer together. The steeper thresholds allow multiplex rates greater than 32 to be achieved. The LC in the cell is "supertwisted" that will give it the ability to use a high multiplex rate. As the twist is increased, the LC molecules in the middle of the layer are aligned with the applied electric field by smaller changes in voltage (Fig. 50.67 b). This gives rise to a very steep transmission vs. voltage curve, allowing up to 240-line multiplexing.

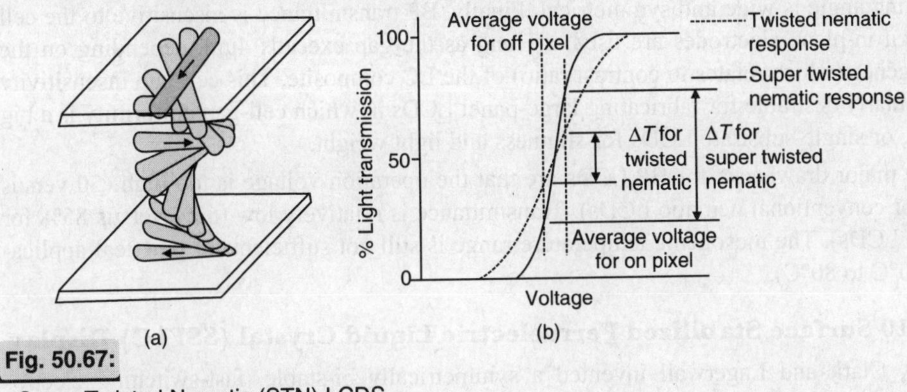

Fig. 50.67: Super Twisted Nematic (STN) LCDS

The polarizers in this case are not mounted parallel to the LC at the surface but rather at some angle. The cell, therefore, does not work on a light "guiding" principle, as in Twisted Nematic LCDs, but instead on a birefringence principle. The position of the polarizers, the cell thickness, and the birefringence of the LC are carefully chosen to result in a particular color in the "off" state. Usually, this is a yellow-green to maximize the contrast ratio.The STN technology comes in two colorations, Green STN and Silver STN. The STN-Green has Dark Violet / Black characters on a Green background. The STN-Silver has Dark Blue / Black characters on a Silver background. The response time of a typical STN display is of the order of 150 ms. These displays typically have a lower contrast ratio and a narrow viewing angle.

50.12.7 Film Compensated Super Twisted Nematic (FSTN) LCDs

The most recent advance has been the introduction of Film compensated Super Twisted Nematic (FSTN) displays. This adds a retardation film to the STN display that compensates for the color added by the birefringence effect. This allows a black and white display to be produced and provides for a higher contrast and wider viewing angle.The FSTN technology comes in a single coloration, Black characters on a White / Gray background.

50.12.8 Double Super Twisted Nematic (DSTN)

DSTN was the first commercial black and white conversion of the STN display. DSTN displays are actually two distinct STN filled glass cells glued together. The first is a LCD display, the second is a glass cell without electrodes or polarizers filled with LC material for use as a compensator which increases contrast and gives the black on white appearance. DSTN provides better contrast than STN and FSTN, and offers automatic contrast compensation with temperature. Its response time is significantly enhanced. DSTN reduces the tendency of a screen to be slightly red, green or blue. Since its polarizer mode is negative, DSTN LCDs need backlighting which is provided by either LED or CCFL. It provides a resolution up to 122×32 dots. DSTN is suitable for use in/with automobiles, gasoline pumps, etc.

50.12.9 Blue Phase Displays

In comparison to conventional, nematic LCDs, polymer-stabilized BP-LCDs primarily exhibit four advantages. First, the switching speed is superior to current technologies, response time is in the sub-millisecond range, which helps to minimize motion image blur. A second advantage of the BP is that it requires no alignment layer which simplifies the manufacturing process and reduces costs. Third, the dark state of a BP-LCD is optically isotropic, which means that its viewing angle is wide and symmetrical. Finally, BP transmittance is insensitive to the cell gap when in-plane electrodes are used, as long as the gap exceeds 4μm, depending on the birefringence (which relates to contrast ratio) of the LC composite. This cell-gap insensitivity is particularly desirable for fabricating large-panel LCDs in which cell-gap uniformity is a big concern, or single-substrate LCDs for slimness and light weight.

The major drawbacks for BP LCDs are that the operation voltage is too high (50 versus $5V_{rms}$ for conventional nematic LCDs). Transmittance is relatively low (65% versus 85% for nematic LCDs). The mesogenic temperature range is still not sufficient for practical applications (40°C to 80°C).

50.12.10 Surface Stabilized Ferroelectric Liquid Crystal (SSFLC) Display

In 1980, Clark and Lagerwall invented a symmetrically bi-stable, fast-switching electro-optical device described as the 'surface-stabilized ferroelectric liquid crystal' (SSFLC) display device. It is required in liquid crystal displays that the liquid crystal director is oriented in a specific single direction. In the SmC* phase, the LC director twists from layer to layer.

The SSFLC cell consists of a liquid-crystalline material sandwiched between parallel glass plates.The plates have been coated with an electrode layer (ITO) and then an alignment material (e.g., a polymer coating such as polyimide, nylon or PVA which has been rubbed in a single direction) to ensure a planar and unidirectional oriented specimen of the liquid crystal molecules due to liquid crystal-surface interactions. The cell gap, d, should be about four times less than the helical pitch P of the liquid crystal (d~P/4). The inner surfaces of the substrates are coated with alignment layers which promote parallel alignment of the liquid crystal on the surface. The helix of the SmC* phase spontaneously unwinds due to elastic interactions with two polyimide-coated glass slides. The molecules in contact with the orientation layer are forced to lie in the plane of the glass plates. By forcing the tilt plane of each SmC* layer to orient in the plane of the glass plates 'bookshelf geometry' (Fig. 50.68 a&b) two director orientations can exist (+θ, – θ), which correspond to two opposite orientations of P_S along the polar axis, as shown in Fig. 50.69.

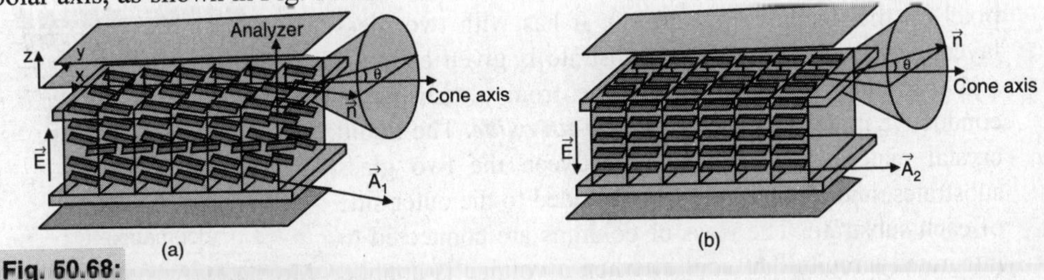

Fig. 50.68:

Surface Stabilized Ferroelectric Liquid Crystal (SSFLC) Display

Ideally, the smectic C layers are perpendicular to the plane of the cell in the so called bookshelf configuration. The polarizers are crossed (90°) with one polarizer parallel to the optic axis of one bistable state. When the molecular optical axis is aligned parallel to the plane of polarization, the light passes through the liquid crystal material and is absorbed by the analyzer and so the cell appears dark. On the other hand, when an external DC field is applied, the direction of spontaneous polarization would be in the opposite direction and forces the long molecular axis to rotate around the smectic C cone (ideally 45°) with respect to the polarizers; the material rotates the plane of incident light through 90° and therefore the cell appears bright. In this way, the light that passes through the SSFLC can be switched ON and OFF by reversing the polarity of the applied field.

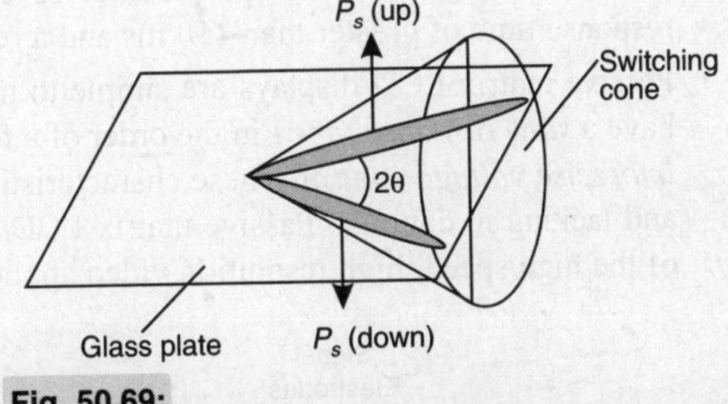

Fig. 50.69:

The two molecular orientations in the SSFLC device

The SSFLC display offers multiple advantages compared to the conventional twisted-nematic liquid crystal displays. The switching time (τ_s) is in the order of microseconds as opposed to milliseconds for twisted nematic LC cells. In addition, the SSFLC cell affords higher viewing contrast, wider viewing angles, bistability (the two surface stabilized-states are maintained by surface interactions in the absence of an electric field), and low power consumption.

50.12.11 Driving Liquid Crystal Displays

Small liquid crystal displays such as used in calculators and other devices have direct driven segments. Voltage can be applied across one segment without interfering with other segments of the display (Fig. 50.70). This is impractical for a large display with a large number of picture elements (**pixels**). A ***flat panel liquid crystal display (LCD)*** is made up of a matrix of pixels filled with liquid crystal material and arrayed in front of a light source. There are mainly two types of flat panel LCD displays - ***passive matrix*** and ***active matrix***.

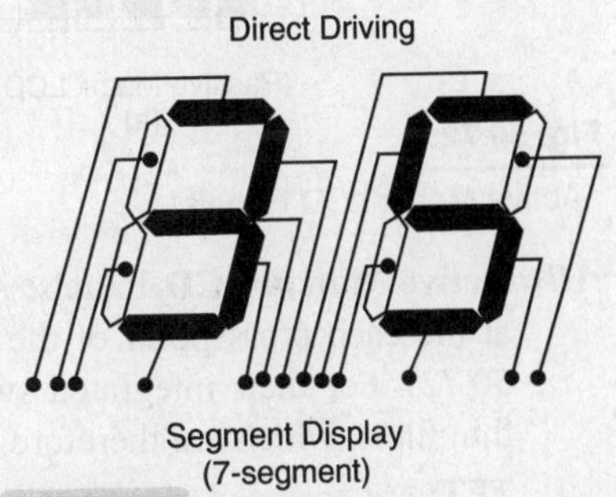

Fig. 50.70:

Direct drive seven segment display

(*i*) **Passive-matrix LCD Panels:** Passive-matrix LCDs use a simple grid to supply the voltage to a particular pixel on the display (Fig. 50.71). It has with two glass layers called ***substrates***. One substrate is given columns and the other is given rows made from a transparent conductive material, usually ***indium-tin oxide***. The liquid crystal material is sandwiched between the two glass substrates, and a polarizing film is added to the outer side of each substrate. The rows or columns are connected to integrated circuits that control when a voltage is applied to a particular column or row. The point of intersection of the row and column represents the designated pixel on the LCD panel to which a voltage is applied to untwist the liquid crystal at that pixel to control the passage of light.

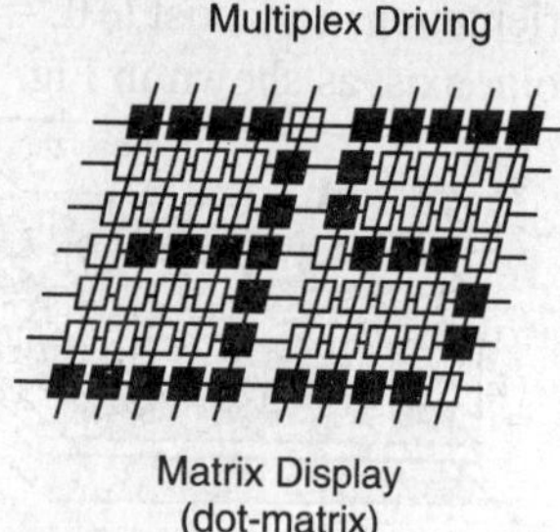

Fig. 50.71: Passive-matrix LCD Panels

In passive-matrix LCDs (PMLCDs) there are no switching devices, and each pixel is addressed for more than one frame time. The effective voltage applied to the LC must average the signal voltage pulses over several frame times, which results in a slow response time of greater than 150 ms and a reduction of the maximum contrast ratio.

Passive matrix LCD displays are simple to manufacture, and therefore cheap, but they have a *slow response time* - in the order of a few hundred milliseconds - and a relatively *imprecise voltage control*. These characteristics render images that are somewhat fuzzy and lacking in contrast. Passive matrix LCD displays are therefore unsuitable for most of the high speed, high resolution video applications.

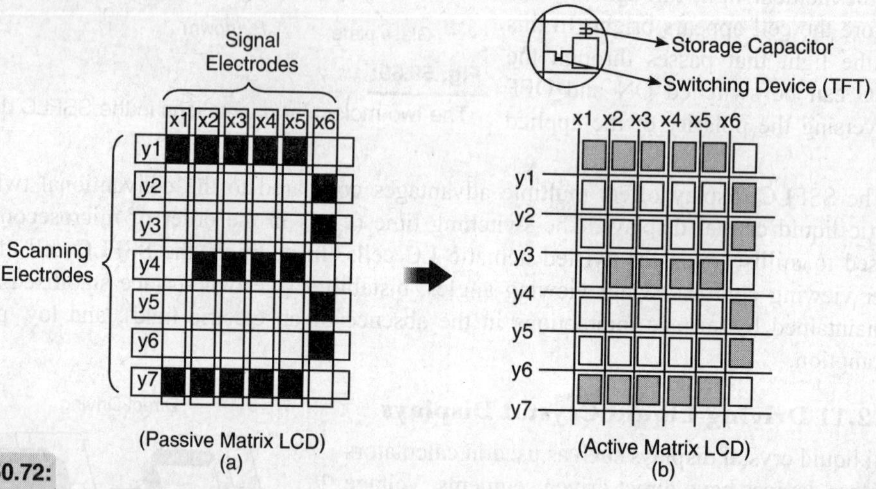

Fig. 50.72: Active Matrix LCD Panels

(*ii*) **Active Matrix LCD Panels:** A switching device and a storage capacitor are integrated at the each cross point of the electrodes in active-matrix LCDs (AMLCDs) (see Fig. 50.72). For their integrated switching devices they use transistors made of deposited thin films, which are therefore called thin-film transistors (TFTs).

TFTs are arranged in a matrix on a glass substrate to control each picture element (or pixel). Switching on one of the TFTs will activate the associated pixel.The use of an active switching device embedded onto the display panel itself to control each picture element helps reduce cross-talk between adjacent pixels while drastically improving the display response.

An active matrix display does not suffer from many of the limitations of the passive display. It can be viewed at an angle of up to 45 degrees and has a contrast of 40:1, meaning that the brightness of an "on" pixel is 40 times greater than an "off" pixel. It does, however, require a more intense back lighting system because the TFT's and the gate and source lines are not very transparent and therefore block a fraction of the light.

50.13 OTHER APPLICATIONS

(*i*) **Liquid Crystal Thermometers:** Chiral nematic (cholesteric) liquid crystals reflect light with a wavelength equal to the pitch. Because the pitch is dependent upon temperature, the color reflected also is dependent upon temperature. Liquid crystals make it possible to accurately gauge temperature just by looking at the color of the thermometer. By mixing different compounds, a device for practically any temperature range can be built.

(*ii*) **Thermal Mapping:** These applications can be extended to medicine and electronics. Special liquid crystal devices can be attached to the skin to show a "map" of temperatures. This is useful because often physical problems, such as tumours, have a different temperature than the surrounding tissue. Liquid crystal temperature sensors can also be used to find bad connections on a circuit board by detecting the characteristic higher temperature.

The "mood ring", a popular novelty a few years ago, uses the unique ability of the chiral nematic liquid crystal.

(*iii*) **Omni-directional Lasing:** Cholesteric liquid crystals are potential candidates for lasing. For these applications, mostly droplets of short pitch cholesteric liquid crystals are used. In short-pitch cholesteric droplets, selective Bragg reflection in the visible wavelength range can happen along the droplet radius. This is attractive as the helix of the cholesteric or other chiral liquid crystal showing visible reflection can act as a resonant cavity for mirror-less lasing if one adds a fluorescent dye with the emission wavelength tuned to the reflection band of the cholesteric. The spherical geometry of the cholesteric droplets leads to omni-directional lasing, *i.e.* the laser light is emitted radially in all three dimensions. At the same time, the ease in tuning the cholesteric pitch, e.g. by temperature changes, allows for tuning of the lasing wavelength, although the range is limited by the requirement to match the emission wavelength of the laser dye.

(*iv*) **Spatial Light Modulators:** Liquid crystal based SLMs offer several advantages including large modulation depth, no moving parts, low power dissipation, potential for large aperture operation, and low cost. In order to modulate the phase of incident light, a nematic liquid crystal modulator is aligned in a planar conformation. Here the liquid crystal director (*i.e.* long axis of the molecules) is oriented parallel to the polarization of the incident light. Upon application of a voltage, the molecules tilt in a direction parallel with the direction of propagation of the optical field. This causes the incident light to encounter a reduced refractive index. The change in refractive index translates directly to a change in the optical path, and consequently a phase shift for the incident light. If enough voltage is applied, the variation in refractive index ranges from the extraordinary index (for no applied voltage) to the ordinary index (maximum tilt of the molecules).

Fabrication of high performance SLMs can be achieved using electro-optic effects in chiral smectic liquid crystal materials. These materials offer electro-optic effects with higher switching speeds and better contrast ratios than nematics.

(*v*) **LCD projector:** An LCD projector works on transmissive technology. They are commonly used in business meetings, presentations and seminars. Inventor Gene Dolgoff developed the first LCD projector in 1984. A powerful light source emits a beam of intense, white light. Beam of white light bounces off a group of mirrors that includes two **dichroic mirrors**, which are coated in a special film that reflects only a specified wavelength of light (Fig. 50.73). So the white light hits the mirrors, and each reflects a beam of colored light on through the projector: one red, one green and one blue. The beams of red, green and blue light each pass through a liquid crystal display composed of thousands of tiny pixels. It comes down to tiny, colorless pixels that either block light or allow it to pass through when triggered by an electric current. All three of the LCD screens in the projector display the same image or moving images, only in gray scale. When the colored light passes through these three screens, they relay three versions of the same scene: one tinted red, one tinted green and one tinted blue. But of course the final image we see is not red, green or blue; it is full color. So inside the LCD projector, the three tinted versions of this scene recombine in a **dichroic prism** (a finely crafted combination of four triangular prisms) to form a single image composed of not three colors but millions of colors. The light of this vibrant, colorful version of the scene then passes through a projector lens and onto the big screen.

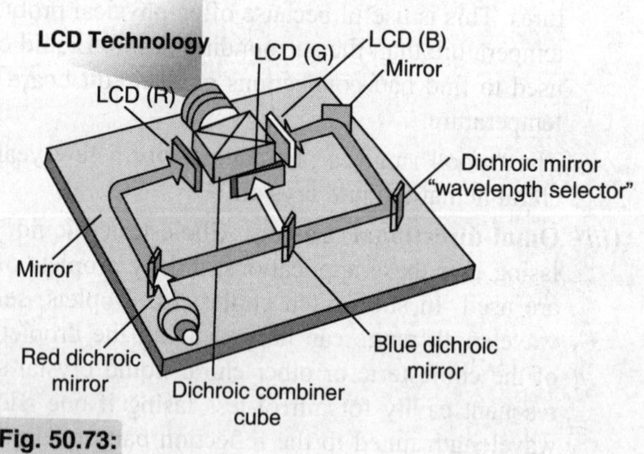

Fig. 50.73: Illustration of LCD projector

Advantages:

(*a*) Higher lumen output can be delivered by an LCD projector at a lower cost compared to a digital light processing (DLP) projector.

(*b*) Provide greater brightness with lower energy consumption.

(*c*) LCD projectors do not suffer from rainbow effects and dithering. Their image sharpness and provides greater magnification.

Disadvantages:

(*a*) Comparatively bulkier than other projectors with lesser portability.

(*b*) The LCD panels have a limited lifetime; they do not have high contrast, and may suffer from black and dead pixels.

(*c*) Maintenance can be higher in the case of LCD projectors when compared with other projectors.

50.14 LIQUID CRYSTAL POLYMERS APPLICATIONS

(*i*) **Pumping residual oil:** Because of their ability to dissolve both polar and nonpolar substances, a mixture of water and an amphiphilic compound can be pumped into a depleted oil well in order to remove much of the residual oil.

(*ii*) **Coating of specific drugs:** A lyotropic liquid crystal can coat a drug to keep it from being destroyed in the digestive tract. The drug can then be taken orally, and after it reaches the proper location in the body, the liquid crystal breaks down and the drug is released.

(*iii*) **Making stable hydrocarbon foam:** Hydrocarbon foams have been difficult to produce in the past because the surface tension of the hydrocarbon is low enough that adsorption to an oil-soluble surfactant would have no significant effect. Without the adsorption, the hydrocarbon simply behaves as a liquid. When lyotropic liquid crystal molecules change from inverse micelles to lamellar sheets, they lower the surface tension enough for foam to form. The hydrocarbon and the surfactant can dissolve in each other, and the surfactant cannot dissolve in water, although water can dissolve in the surfactant and mix into the liquid crystal.

(*iv*) **Many substances are more soluble in lyotropic liquid crystals:** One example is the drug hydrocortizone. It is often taken in topical applications, but its uses have been limited because the highest concentration possible has been only 1%. When the drug was blended into a liquid crystal of lecithin and water, the concentration went up to 4%. In time, liquid crystals may become a primary solvent for topical medications.

(*v*) **LCP high - Strength fibers:** An application of polymer liquid crystal that has been successfully developed for industry is the area of high strength fibers, e.g. Kevlar fibers, which are used to make such things as helmets and bulletproof vests, is just one example of the use of polymer liquid crystal in application calling for strong, lightweight materials. Ordinary polymers have never able to demonstrate the stiffness necessary to compete against traditional materials like steel. It has been observed that polymers with long strength chains are significantly stronger than their tangled counter parts. As these technologies continue to develop an increasing variety of new materials with strong and light-weight properties will become available.

(*vi*) **Polymer - Dispersed Liquid Crystal shutter:** In the *polymer-dispersed* liquid crystal display, very small bubbles of liquid crystal are produced inside a transparent polymer. If field is not applied, these bubbles of liquid crystal will take on many different orientations. A simple diagram illustrating the electro-optical effect of a polymer-dispersed liquid crystal (PDLC) shutter is shown in Fig. 50.74 (a) and (b) for the off and on states, respectively. Suppose the polymer matrix material is optically isotropic and has refractive index n_p. The LC directors within the droplets are determined by the polymer-LC interaction at each droplet boundary. They have no preferred orientation but vary nearly randomly from droplet to droplet in the absence of external field.

The index mismatch ($n_e > n_p$) between the LC droplets and the host polymer results light scattering. Because the cell thickness (~ 10 μm) is much larger than the droplet size (~ 0.5 μm), the incident light will be scattered many times before emerging from the film. Thus, little light is transmitted through the cell. The degree of the off-state scattering depends on the size, birefringence, and concentration of the liquid crystalline droplets, and the film thickness. When the applied field is sufficiently strong, the directors in the droplets are reoriented along the field direction. Its effective refractive index is equal to n_o which is similar

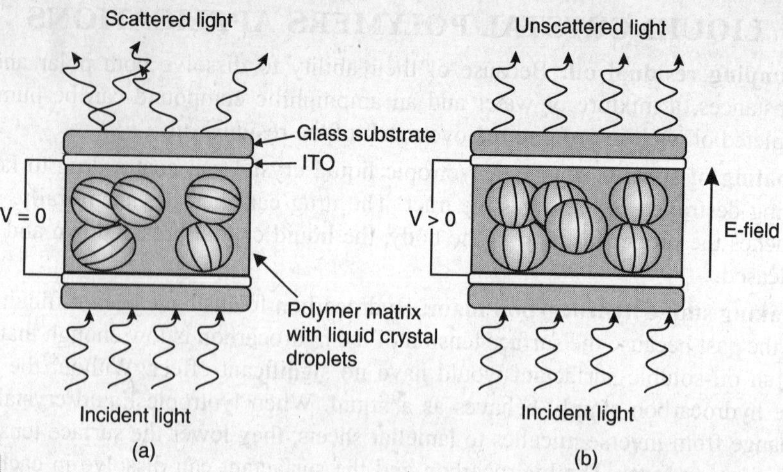

Fig. 50.74: Action of polymer-dispersed LCDs (a) off-state (b) on-state

to the index of the polymer, n_p. Thus, the normally incident light acts as the ordinary ray and transmits through the cell. The PDLC film becomes clear. This mode of operation is called normal mode.

PDLCs have been used for privacy, in exhibits, for safety visors used by pilots, and in heads-up displays.

QUESTIONS

1. What are liquid crystals? How are they different from crystals and liquids?
2. Explain positional and orientational order in liquid crystals.
3. How are the liquid crystals classified?
4. What is a twisted nematic display? How is it made?
5. Explain the construction and working of a segment LCD.
6. What are matrix displays and how are they classified?
7. Explain the construction and working of a passive matrix LCD.
8. Explain the construction and working of an active matrix LCD.
9. Explain the construction and working of a TFT LCD.

Index

D

E